MW01617812

THEORY AND DESIGN OF BRIDGES

THEORY AND DESIGN OF BRIDGES

PETROS P. XANTHAKOS
Consulting Engineer
Washington, DC

A Wiley-Interscience Publication
JOHN WILEY & SONS, INC.
New York Chichester Brisbane Toronto Singapore

This text is printed on acid-free paper.

Library of Congress Cataloging in Publication Data:

Xanthakos, Petros P.
Theory and design of bridges / Petros P. Xanthakos.
p. cm.
Includes bibliographical references and index.
ISBN 0-471-57097-4
1. Bridges—Design and construction. 2. Load factor design.
I. Title.
TG300.X36 1993
624'.25—dc20 92-21520

Printed in the United States of America

10 9 8 7 6 5 4 3 2 1

CONTENTS

PREFACE

Bridge engineering, which began with stone and wooden structures as early as the first century B.C., has undergone a dramatic evolution in terms of analysis and use of materials. This program is exemplified by the extension of the design in the elastic-plastic range, the use of high-strength steel and concrete, and the introduction of a probabilistic approach to supplement, and often replace, the deterministic methodology. As a result, authors always endeavor to keep abreast of the latest developments by expanding the body of a book and by adding new concepts. Thus, this text presents a synthesis of old and new material, and integrates the classical concepts with modern methods. The underlying premise is that traditional approaches can be integrated with modern formulations to produce a merge where both methods are workable alternatives.

The significance of this contrast is that it shows how advanced analytical techniques can be used to solve complex problems and also to explain overall behavior. If advanced techniques are equally used to analyze a bridge structure and simplify its overall design, the resulting validity will underscore the goal of modern practice in producing economies. It follows, therefore, that engineers should have a range of options from which to choose, and this should encompass bridge types, design philosophies, and construction procedures.

Although certain bridge types are unlikely to be design alternatives in the future, they have been included in this text because of their engineering relevance and also because they represent an impressive volume in the present bridge inventory. Examples are certain concrete structures such as through girders, flat slabs and cantilever decks, and two-girder steel bridges.

To bring matters to a focus, and consistent with current practice, the previously separate treatment of steel and concrete bridges has been combined, and this has made it easier to compare and explain the similarities and differences in behavior. On the other hand, certain aspects common to all bridge types are treated in the context of a definite and coherent structural thought.

There is an obvious urgency to improve the clarity and understanding of bridge behavior but without sacrificing the design skills. The text has attempted to accomplish this goal by explaining and making popular design approaches that articulate bridge response under service conditions and at failure. The domain of structural performance is thus described in both the elastic and inelastic range, because bridges and their components are commonly designed, and expected to perform, elastically and plastically.

These considerations provided the incentive to pursue an independent review of bridge theory and behavior, rather than compiling a design manual. Bridge design should be based however on relevant specifications and should demonstrate compliance with applicable standards to ensure credible results. Yet, optimum solutions can be obtained only when the designer understands the assumptions and limitations of analysis. In retrospect, codes are changed continuously but bridge behavior may not, and this difference is important in choosing the contents and the subject matter. In the same context, predicting bridge behavior from physical models has become common practice, and the results confirm the value of testing in verifying structural response.

A main feature is the introduction to the LRFD specifications, resulted from research requested by AASHTO and initiated by the National Cooperative Highway Research Program (NCHRP). This document, still under development draws from completed and recent bridge research signifying new provisions and major areas of change. It addresses load models and load factors, nominal resistance and resistance factors, structural analysis, concrete and steel structures, decks and deck systems, and bridge substructure elements and foundations. The associated methodology constitutes the "limit states" approach which may encompass strength, fatigue and fracture, serviceability, and extreme events. The underlying philosophy moves the design toward a more rational and probability-based procedure supported by reliability theories. Implementation of this approach is expected to result in more uniformly reliable bridges.

Among the many new topics covered in this document is the proposed new live load model, limit states for soils and foundations, and an explicit consideration and coverage of redundancy and ductility.

The book is organized into 14 chapters. Chapter 1 presents a brief history of bridge engineering, esthetics and economic considerations, and draws attention to bridge management. A conclusion drawn from the current bridge inventory relates to the large number of bridges that must be strengthened, rehabilitated, or replaced.

Chapter 2 reviews design methods and loads. Although the text focuses on the classical force and displacement methods amenable to hand calculations, with computer usage these may be supplemented by refined procedures such as finite difference and finite element techniques, folded late methods, finite strip methods, grillage analogy, series or other harmonic methods, and yield line theories. Load models and distribution continue their place as prominent features. Current load models and combinations are examined in conjunction with proposed deviations based on a new load regime of vehicular live load, impact, and braking force.

Chapter 3 deals with the common types of concrete bridges, including those built prior to World War II. Prestressed concrete is expected to continue as a prevalent form of bridge superstructures, and trends in prestressed applications are documented. The range of options is expanded if the design takes into account the criteria for optimum solutions as they relate to allowable compression, ultimate strength, protection against cracking, load ratios, and maximum immediate and long-term deformations. Attention is drawn to the use of strut-and-tie models and to the new approach to shear design. Whereas the use of higher strength concrete should not be prohibited, the implementation of this option should not represent merely a notion for high-performance materials.

Steel bridges are covered in Chapters 4 through 6. The range includes I-beams, plate and box girders, and horizontally curved systems. Ductility, brittle fracture, and redundancy are key topics, whereas emphasis is placed on composite construction because of the associated advantage. Allowable stress design is a standard method, whereas the load factor concept is an alternate method for designing simple and continuous beams and girders of moderate lengths. Inelastic procedures are only applied to continuous bridges. Plate girders may be homogeneous or hybrid, and with stiffened webs or unstiffened webs. Steel box girders are discussed within an extended scope of design that includes short-to-moderate span multibox composite, simple-box composite, and long-span orthotropic deck.

Steel orthotropic plate decks are reviewed in Chapter 7. An appropriate method of analysis is the equivalent-orthotropic slab method where the deck is simulated by a continuous two-dimensional system with different stiffness in the two principal directions. Although in some instances the savings in materials have been offset by increases in fabrication costs, this bridge type is expected to continue to be favored in certain cases.

Segmental concrete bridges are reviewed in Chapter 8 in the context of the 1989 AASHTO guide specifications. This document is comprehensive in nature and articulates the effect of creep, shrinkage, temperature differentials, and shear lag. Longitudinal analysis should include load groups and stresses during erection, and loads and stresses in the final structural system. The text concentrates on the observed performance of segmental bridges and reviews results from recent research here and abroad. Case studies include curved prestressed segmental bridges, and moment redistribution.

Chapter 9 combines trusses, movable bridges, and cable-stayed bridges. The popularity of truss bridges appears to have diminished as other forms, such as steel box girders and cable-stayed structures, appear to dominate the intermediate span range. Trusses, however, exhibit structural characteristics that will continue to attract attention. Chapter 9 also gives a brief review of movable bridges and their basic forms, namely bascule, vertical lift, and horizontal swing. Because cable-stayed bridges are treated extensively in at least two other major publications, in this chapter the discussion is limited to their basic features and design principles. Focus is on the analytical requirements of multispan stayed bridges and on fatigue effects from wind-induced vibrations.

Chapter 10 discussed arch bridges on concrete and steel. Graphical solutions and force diagrams provide the background of arch action and of the line-of-thrust theory. Theoretical principles and formula derivations form a clear concept of elastic behavior and provide the background for a more complex design approach. Of direct interest is the section on buckling and geometry imperfections with the associated stability considerations. The literature has been enriched with recent work here and abroad, particularly for slender steel arches, that addresses in-plane linear and nonlinear stability, in-plane ultimate load, and ultimate strength. Both the current AASHTO and the LRFD specifications address concrete and steel arches under combined flexure and axial load, and relate arch components to compression members and slenderness effects.

The term "special bridges" used in Chapter 11 refers to unusual member geometry, structural configurations and combinations, and support conditions. Prefabricated superstructure elements are combined with innovative substructure designs, and this may prove advantageous for bridges and grade separation structures in urban area.

Various topics relevant to design are discussed in Chapter 12, and include lateral wind bracing, unintended composite action, deflections, settlement, temperature effects, strength and fracture, and fatigue. Since the first fatigue resistance provisions were formulated in 1972, several major fatigue studies have been completed, and more than 1500 test results have been added to the data base. Extensive fatigue tests also been performed in Japan, Britain, Germany, Canada, and by ORE and ICOM. This record has provided a formidable background for reviewing fatigue problems and solutions. A trend is also evident toward recognizing fracture-critical nonredundant steel bridge members and developing guidelines for a better understanding of the design and behavior of nonredundant bridges.

Chapter 13 deals with bridge details such as bearings, links and hangers, field splices, expansion joints, diaphragms and cross frames, and hinge details for column bases. The text also reviews the distortional response of curved concrete box girders and concludes that, unless these deformations are inhibited by the presence of transverse diaphragms, they may induce stresses approaching and exceeding the normal bending stresses. A brief discussion focuses on construction aspects of bridge decks such as the pouring sequence of the deck slab and the requirement of construction joints.

Substructure types and methods of design are reviewed in Chapter 14. The 1992 AASHTO specifications provided a comprehensive review of substructure systems and foundation types, and extended the design criteria accordingly. Both the allowable strength design, and the load factor method are standard procedures for substructure analysis. It appears, however, that service load procedures are either too conservative or somewhat unrealistic particularly for members subjected to combined axial load and flexure, and this has prompted many designers to use the load factor approach in the analysis of reinforced concrete pier columns.

In office practice, the design of bridges utilizes computers and many versatile software packages. Special computer programs have been developed, ranging from simple formula applications to elaborate analyses. With rapidly improving and expanding computer technology, the most precise but complex analytical techniques become routine options. The designer should be cautioned, however, that a computer program is only a tool, and hence the designer should clearly understand the basic assumption of the program and all output should be verified. Thus, any method of analysis that satisfies equilibrium and compatibility and has stress-strain relationships implanted in the process is acceptable. Simple and complex methods are liberally interchanged in the text, but the latter are not explicitly addressed.

In general, the book is oriented toward the needs of practicing engineers, but the material may be reorganized to accommodate one or two courses in bridge design at the undergraduate and graduate levels. Appropriate sections may be selected from Chapters 2 through 10 and 12 through 14. There is ample flexibility to allow the instructor to ensure structural continuity and place emphasis on topics that are consistent with the requirements and goals of the course.

In addition to the references provided in the text, valuable sources of material have been the publications, reports, books, journals, and manuals of the FHWA, AASHTO, TRB, ASCE, AISC, ACI, PCA, AWS, PCI, ASTM, universities, and technological institutes. This record represents an enormous contribution to bridge technology, and is acknowledged with deep appreciation.

My special thanks are extended to my wife, whose unselfish dedication, commitment, and tenacity under the most difficult circumstances made this book possible.

PETROS P. XANTHAKOS

Great Falls, Virginia
May 1993

CHAPTER 1

INTRODUCTION

1-1 BRIEF HISTORY OF BRIDGE ENGINEERING

Probably the first-known documents dealing with construction materials and structure types are the books about architecture by Marcus Vitruvios Pollio in the first century B.C. The fundamentals of statics were developed by the Greeks, and were exemplified in scientific works and engineering applications by Leonardo da Vinci, Cardano, and Galileo. Engineers in the 15th and 16th centuries, seemingly unaware of this record, relied solely on experience and tradition for building bridges and aqueducts. By the end of the 17th century, Leibnitz, Newton, and the Bernoulli brothers were using infinitesimal calculus, and the state of the art was rapidly changing. Lahire (1695) and Belidor (1729) published works about the theoretical analysis of structures, and provided the framework for the field of mechanics of materials that became the focal point of work in France during the 18th century. Notable training centers were established in France and became quite famous. Interestingly, some of the most prominent American bridge engineers were trained there. Among those to be mentioned are C. Ellet, Jr., R. Modjeski, and L. F. G. Bouscarey, who completed his studies in 1873 to become later the chief engineer of the Cincinnati Southern Railroad. The impact of trained engineers on bridge design was felt 1850 onwards (Bresse, 1880).

Kuzmanovic (1977) describes stone and wood as the first bridge building materials. Iron was introduced during the transitional period from wood to steel. According to known record, concrete was used in France as early as 1840 for a bridge 39 ft (12.0 m) long to span across the Garoyne Canal at Grisolcs, but reinforced concrete was not used in bridge work until the beginning of the century. Prestressed concrete was introduced in 1927.

Stone and Wooden Truss Bridges Stone bridges of the arch type were constructed in Rome and other European cities in the Middle Ages. These arches were half-circular, with the flat arches beginning to dominate bridge work during the Renaissance period. This design was markedly improved by Perronet at the end of the 18th century, and was structurally adequate to accommodate the upcoming railroad loads. In terms of analysis and use of materials, stone bridges have not changed much. Lahire (1695) developed the first theoretical treatment by introducing the pressure line concept (see also Chapter 10), used in practical designs in the early 1770s. Coulomb developed the arch theory in model tests where typical failure modes were considered. These results were published by Frezier (1739). Bresse (1880) developed the theory for the bending and stability of curved bars, and applied it to arches thereby taking into account changes in curvature as well as changes in length. Culman (1851) introduced the elastic center method for fixed-end arches, and showed that three redundant parameters can be found by the use of three equations of compatibility.

The first major bridge work using wooden trusses was in the 16th century, when Palladio built triangular trusses to construct bridges with spans up to 100 ft (30.5 m). Palladio also focused on the three basic principles of bridge design: convenience, appearance, and endurance. Several timber bridges were constructed in Western Europe beginning in the 1750s with spans up to 200 ft (61 m). However, during the 19th century significant progress in timber bridges was made in the United States and Russia (Hertwig, 1950). Contributing factors to this choice were scarcely populated areas with large distances, major rivers to cross, and an abundance of suitable timber. Favorable economic considerations included the initial low cost and fast construction. Under these conditions wooden bridges provided the ideal solution. Town (ASCE 1976) developed and patented the lattice bridge in 1820, and this wooden truss also became the prototype of the early nonlattice bridges. In 1840 Howe introduced and patented a truss system that became the standard for many early railroad bridges.

A further development in wooden trusses was the arch type with or without ties. A detailed account of American bridges was provided by Culmann (1851, 1852). In his review, Culmann emphasized the American practice and the many original ideas. On the theoretical side, Culmann proposed new methods for calculating stresses, and these included statically redundant trusses. One of the outstanding wooden trusses was developed by

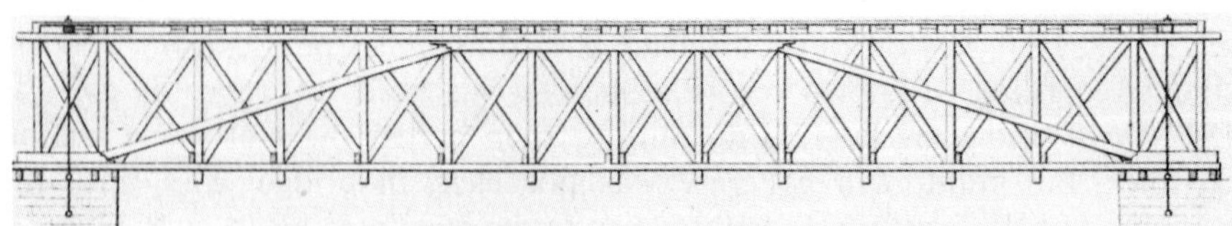

FIGURE 1-1 Truss type developed by Long (1839).

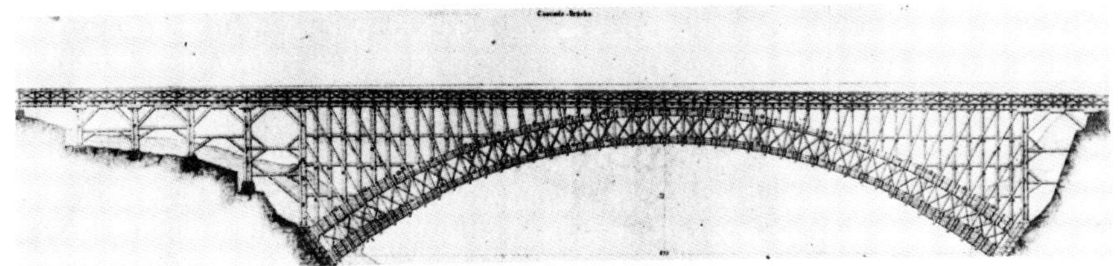

FIGURE 1-2 Cascade Bridge.

Long (1839), and is shown in Figure 1-1. Another notable bridge of this period is the Cascade Bridge of the Erie Railroad spanning a valley 175 ft (533 m) deep and 300 ft (91.4 m) wide, as shown in Figure 1-2.

Iron Bridges The transition from wooden to steel bridges probably did not start until about 1840, but the first recorded use of iron in bridges was the chain bridge built in 1734 across the Oder River in Prussia. The first all-cast-iron bridge was built in 1779 by Darby at Coalbrookdale according to a design prepared by Pritchard (De Mare, 1954). The first truss completely made of iron was built in 1840 in the United States by Trumbull, followed by England in 1845, Russia in 1857, and Germany in 1853. Also in 1840 Whipple built the first iron arch truss bridge across the Erie Canal at Utica. The same engineer built his first railroad bridge, a bowstring truss, in 1853 near Troy, New York, for the Rensselaer and Saratoga Railroad.

The Impetus of Analysis The theory of structures, developed in the 19th century, focused on truss analysis, and the first book on the subject was written by Whipple (1847). Pope write the first book on bridges in 1811. In 1846 Warren introduced his triangular truss, and in 1850 Bload developed a correct method for calculating forces in this truss type while Humber (1857) provided important test results about the forces in the Warren truss. I beams fabricated from plates became popular in England and were used in short-span bridges. Further progress in truss design is attributed to Schwedler (1862), Ritter (1877), and Zimmerman.

In 1866 Culmann explained the principles of cantilever truss bridges, and one year later the first cantilever bridge was built across the Main River at Hassfurt, Germany, with a central span of 425 ft (129.5 m). The first cantilever bridge in the United States was built in 1875 across the Kentucky River. The most impressive railway cantilever bridge in the 19th century was the Firth of Forth Bridge, built between 1883 and 1890, with span magnitudes of 1711 ft (521.5 m).

At about the same time, structural steel was introduced as a prime material in bridge work, although its quality was often poor. Between 1874 and 1883 three major bridges were built of structural steel: (a) the Eads Bridge in St. Louis, (b) the Brooklyn Bridge in New York, and (c) the Glasgow Bridge in Missouri. By 1890 the advantages of steel were generally accepted, and its use was expanded accordingly.

Toward the end of the 19th century, Maxwell contributed further improvements to the analysis of truss bridges, particularly in graphical solutions. He also pointed out that a unique reciprocal correspondence exists between the geometry of the truss and the force diagram, and to each point corresponds one closed force polygon and vice versa. For certain statically indeterminate trusses, Maxwell gave a force method solution based on the equality of external and internal deformation work. In 1872 Cremona published his book about graphical statics, and his solution became known as the Cremona diagram. The force method was also redefined by Mohr, who made significant contributions to the analysis of structures. Mohr pointed out that the differential equation of a force polygon and of an elastic line are mathematically the same. Based on this principle, he developed a graphoanalytical method for finding deflections using corrected moment diagrams as fictitious loadings.

Kuzmanovic (1977) mentions two important box girder bridges, the Conway and Britannia bridges designed by Stephenson. These were tubular structures for the railroads across the Conway and Menai straits. The Conway Bridge consisted of a single span 412 ft (125.6 m) long, but the Britannia Bridge was a continuous structure with spans of 230 (70) 460 (140), 460 (140), and 230 (10) ft (m). These designs were the first important examples of box girders. Further knowledge was gained by the work of Fairbairn, who conducted tests to determine the best and most favorable cross section (rectangular, circular, elliptic) and who also experimented with plate girders and their stiffening requirements.

Clapeyron, who introduced the three-moment equation in 1849, made an analysis of the Britannia box girder in 1857 and determined that the bending stresses were not balanced; the negative moment stresses at the interior supports were 2.5 to 3.0 times larger than the stresses at the midspans, and therefore this bridge needed a more efficient distribution of plate thickness.

New Analytical Methods Structures of a high order of redundancy could not be analyzed with the classical methods of the 19th century. The introduction of reinforced concrete at the beginning of this century in multispan frames imposed new analytical requirements. The importance of joint rotation was already demonstrated by Manderla (1880), and Bendixen (1914) developed relationships between joint moments and angle rotations from which the unknown moments could be obtained (the so-called slope deflection method).

More simplifications in frame analysis were made possible by the work of Calisev (1923), who used successive approximations to reduce the system of equations to one simple expression for each iteration step. This approach was further refined and integrated by Cross (1930) in what is known as the method of moment distribution.

Among the recent developments to be mentioned in the analytical procedures is the extension of design in the elastic–plastic range, better known as the strength or ultimate design method. Plastic analysis was introduced with some practical observations by Tresca (1864), and was formulated by Saint–Venant (1870). The concept of plasticity attracted researchers and engineers after World War I mainly in Germany, with the center of activity shifting to England and the United States after World War II, with Baker, Prager, Van Den Broek (1948), and others. The probabilistic approach is a new design concept that tends to replace the present deterministic methodology.

At the present time concepts having a strong place in bridge practice include composite construction, new structural systems such as prefabricated members, orthotropic plates, segmental construction, curved girders, and cable-stayed bridges. Interest in box sections remains strong and expresses efforts to reduce the flange plate thickness of long-span bridges and to avoid the danger of brittle fracture. The distortion and stability of box sections still remain the main consideration in design.

Suspension Bridges The first suspension bridge in the United States was built in Pennsylvania in 1796. Several bridges were built in the first quarter of the 19th century, and the largest was constructed in Great Britain by Talford, with a center span of 570 ft (173.7 m). The first theoretical treatment of suspension bridges was proposed by Navier (1823). Vicat was the first to observe creep in iron, and he also invented the method of spinning cables at the bridge site. Ellet (1823) was the first engineer in the United States to use wire cables instead of chains. He designed and built the world record span of 1010 ft (307.8 m) over the Ohio River at Wheeling between 1847 and 1849, considered the most beautiful and largest of its kind. Another notable engineer was Roebling, who designed the Brooklyn Bridge.

The main problem of early suspension bridges was a lack of stiffness, and this led to the concept of a stiffening girder suspended by cables as the main supporting element. The stability of suspension bridges was further improved by introducing stabilizing trusses, and the first attempt to calculate the deflections of these trusses was made by Ritter (1877). The best practical design procedures were developed by Melan (1888). Bleich generalized Timoshenko's approach of using trigonometric series by applying this method to various boundary conditions of the stiffening trusses. Problems of dynamic stability were investigated after the Tacoma Bridge collapse, and this work led to significant contributions.

1-2 BRIDGE APPEARANCE AND ESTHETICS

Although in many civil engineering works esthetics has been practiced almost intuitively, particularly in the past, bridge engineers have not ignored or neglected the esthetic discipline. Recent scientific research appears to lead to a rationalized esthetic design methodology (Grimm and Preiser, 1976). Work

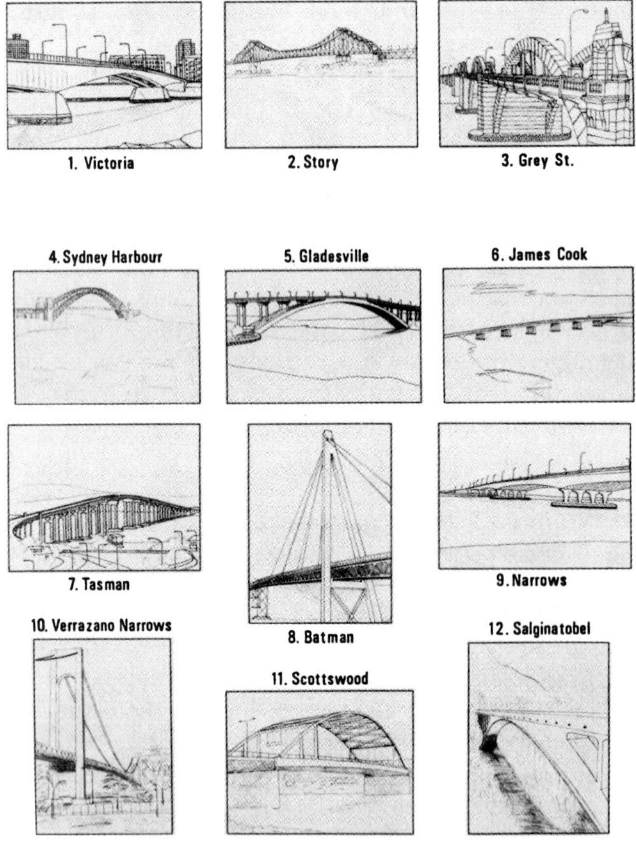

FIGURE 1-3 Sketches of 12 bridges used in a survey (listed in Table 1-1).

has been done on the esthetics of color, light, texture, shape and proportions, and other perceptual modalities, and this direction is both theoretically and empirically oriented.

In the United States as well as in most European countries, esthetic control mechanisms are now integrated into the land-use regulations and design standards. In addition to concern for esthetics at the state level, federal concern also focuses on the effects of the manufactured environment on human well-being. Besides esthetics in environmental planning, other perspectives are directed toward improving quality and appearance in the design process. Good potential for the upgrading of esthetic quality in bridges can be seen in the predesign evaluations of superstructure and substructure types aimed at improving the appearance of the structure.

Public Response to Bridge Appearance Figure 1-3 shows sketches for the 12 bridges listed in Table 1-1. These bridges were rated for appearance in

TABLE 1-1 List of Bridges Used in Survey

Bridge (1)	Location (2)
Surveys 1971, 1978-1, and 1978-2	
1. Victoria	Brisbane (Queensland), Australia
2. Story	Brisbane (Queensland), Australia
3. Grey Street	Brisbane (Queensland), Australia
4. Sydney Harbour	Sydney (New South Wales), Australia
5. Gladesville	Sydney (New South Wales), Australia
6. James Cook	Sydney (New South Wales), Australia
7. Tasman	Hobart (Tasmania), Australia
8. Batman	Launceston (Tasmania), Australia
9. Narrows	Perth (Western Australia), Australia
10. Verrazanno Narrows	New York, U.S.A.
11. Scotswood	Newcastle-on-Tyne, England
12. Salginatobel	Near Schiers, Switzerland

a public survey (O'Conner, Burgess, and Egan, 1980) at two different times in 1971 and 1978. In these surveys, respondents were asked to select a specific rating, 0 for extremely poor appearance to 9 for most pleasing and excellent. This scale appears to have been satisfactory, and was used in its entirety. In interpreting the results, the number giving each rating was counted and used to compute the mean rating and the standard deviation (SD).

The results of public opinion show that the Narrows Bridge (Perth) and the Victoria Bridge are the most popular. In both cases the SDs are small (1.4–1.5). The Verrazano Narrows Bridge also has a high rating. In the 1971 survey, the Story and Grey Street bridges have the lowest ratings, whereas in the 1978 survey the Scotswood Bridge had an even lower rating.

Interestingly, the results demonstrate a definite bias by city, and the highest score is that given by respondents residing in the city in which the bridge is located. The rating also differentiates familiarity with the bridge, age difference, and difference of opinion between men and women. Thus, certain groups have a higher regard for slender simple shapes, whereas others prefer solidity and more complexity of form.

A second survey taken in 1978 was based on photographs. Good agreement between photographs and sketches was obtained for the following bridges: Victoria, Sydney Harbour, Gladesville, Verrazano Narrows, and Salginatobel. The most comparable ratings for the 12 bridges of Table 1-1 appear to be those based on photographs, from the 1978 surveys, and by excluding respondents who claim familiarity with the bridge. After adjusting the ratings to reflect these considerations, the bridges are listed in rank order in Table 1-1.

O'Connor, Burgess, and Egan (1980) conclude that public opinion on bridge appearance can be sampled reliably by proper techniques, and simple sketches may be used to obtain results compatible with those obtained with photographs. There are variations in the ratings according to respondent group, age, and gender.

Bridge Esthetics on Regional Basis Billington (1981) suggests that bridge esthetics characterizes regional trends, and that leading bridge engineers have developed individual styles within well-defined regions. The six men considered in this study are Telford (1757–1834), Roebling (1806–1869), Eiffel (1832–1923), Maillart (1872–1940), Ammann (1879–1965), and Menn (1927–).

Telford is credited for working with metal. One of his metal arches rests on hollow masonry piers above the valley and carries the canal over the river Dee. The original work is still in serviceable condition after approximately 175 years of use. However, Telford is distinguished for his flat-iron lattice arches that constitute his mature style in medium-span bridges. With one exception, these bridges were built in the west of England and Wales in hilly country cut by narrow streams. This was the region where the Industrial Revolution probably began at the time Telford began to practice and experiment with the use of new materials (cast iron) and a new form (the lattice arch).

In 1844 Roebling won a competition to build the first suspension bridge carrying a canal over the Allegheny River. Roebling built his first suspension bridge for a roadway in Pittsburgh over the Monogahela River in 1845, and his next major works were the 815-ft (248.4 m) span Niagara Falls rail and road bridge completed in 1855 and the Cincinnati Bridge completed in 1866. In 1869 Roebling presented his plans for the Brooklyn Bridge, but he died in July of the same year. His eldest son became chief engineer for the bridge. The Brooklyn Bridge, built essentially as designed by Roebling, became a major example of bridge design and was officially made a National Historic Landmark in 1964 (McCullough, 1972). The features that distinguish his style are massive masonry towers, a thin deck supported by vertical suspenders, and cable stays radiating out from the towers. Invariably, his bridges were built in Pennsylvania, New York, and Ohio, and the three major works span large open waterways.

Eiffel was born in 1832, and by 1878 he was regarded as the leading engineer of metal structures in France (Harris, 1975). Major bridges designed by Eiffel include the viaduct over the Creuse river at Busseak in 1864, the four viaducts between Gannat and Commentry between 1867 and 1869, the Maria Pie Viaduct over the Douro River in Portugal in 1877, and the Gărabit Viaduct in the Massif Central, shown in Figure 1-4. This bridge was completed in 1884 as the longest spanning arch in the world, with an opening of 165 m (540 ft) (Eiffel, 1888). The last design in the Massif Central reflects the problems of heavy wind loads that tend to cause overturning effects on high structures. His style is characterized by his solution to widen the towers and the arches in the lateral direction at the supports and near the base, giving

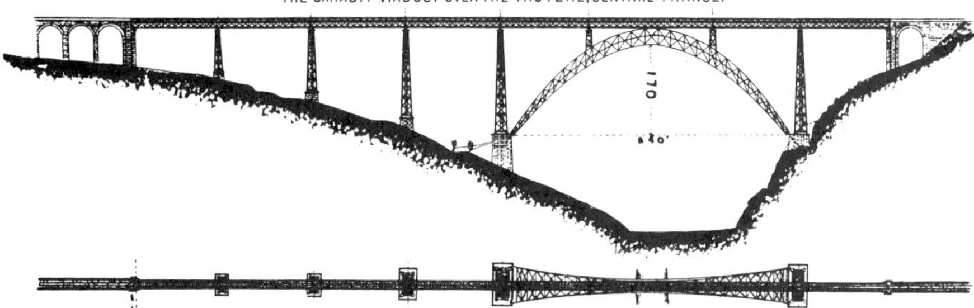

FIGURE 1-4 Garabit Viaduct.

the strong visual impression of stability. This style is exhibited in Figure 1-4 and in the Eiffel Tower in Paris.

Between 1920 and 1940 Maillart produced more than 30 bridge designs of undisputed originality. He is credited for developing two major new bridge forms in reinforced concrete: the hollow box and the deck-stiffened polygonal arch (Billington, 1981). His early hollow boxes were three-hinged arches, but in later designs Maillart also used this form in straight spans. The deck-stiffened type is exhibited in the Schwandbach Bridge built in 1930, and shown in Figure 1-5. These bridges were built in narrow valleys in the hills and mountains of Switzerland, and were located primarily in two regions: the Graubunden and Bern. His style is distinguished by very thin polygonal

FIGURE 1-5 Schwandbach Bridge.

FIGURE 1-6 Felsenau Bridge.

arches or very thin hinged points on the hollow-box forms (Billington, 1981). The concrete is fully exposed, and the bridge shape contrasts with the mountain environment. Interestingly, the works of Maillart were accepted because they were cost-competitive.

Ammann was a distinguished engineer of suspension bridges. His first major design was the George Washington Bridge in New York, developed in 1923. From 1924 until World War II, Ammann was the chief bridge engineer for the Port Authority of New York and New Jersey. Besides the George Washington Bridge with a span of 3500 ft (1067 m), other major works by Ammann include the Bayonne Bridge in 1931 with an arch span of 1652 ft (503.5 m), the Triborough Bridge in 1936 with a span of 1380 ft (420.6 m), the Bronx–Whitestone Bridge in 1939 with a span of 2300 ft (701 m), the Throgs Neck Bridge in 1961 with a span of 1850 ft and (564 m) the Verrazano Narrows Bridge in 1964 with a span of 4260 ft (1298 m). With a few exceptions, Ammann's long-span designs are located in New York. His style

FIGURE 1-7 Ganter Bridge.

in suspension bridges is distinguished by strong vertical towers with a single cross member at the top and high horizontal decks.

Menn is credited for updating the Swiss building code for reinforced and prestressed concrete. As a bridge engineer, he has made a series of major bridge designs, including the Felsenau Bridge built in 1974 (see Figure 1-6) and the Ganter Bridge completed in 1980 (see Figure 1-7). Menn is strongly influenced by the works of Maillart, but his style evolved as increasing labor costs made arch scaffolding and formwork more costly. This is shown in the deck-stiffened arch bridges where an increase in spacing between cross walls and the accompanying reduction in formwork was made possible by prestressing the deck (see also Chapter 10). The Felsenau Bridge shows this trend in two ways. First, the use of slender columns reduces materials and improves the visual impression by suggesting lightness. Second, for the relatively wide bridge, only a single box is used with very long slabs cantilevered laterally to complete the roadway. The horizontal curvature makes the design more complicated. The Ganter Bridge shown in Figure 1-7 has a center span of 174 m (570 ft), and the column support is 130 m (426 ft). The bridge consists of a prestressed cantilever girder in which the main prestressing tendons rise well above the girder at the columns. The tendons are concreted into walls over the central part of the span to give a unique profile as shown in Figure 1-7.

The Value of Critique It is apparent from the foregoing discussion that bridge esthetics is not an isolated concept, but should be examined in the context of structural requirements and budget constraints. In addition, engineers must also consider utilitarian esthetics, which is a composite of physical factors and visual design aspects. With bridges, these include location, alignment, roadway characteristics and details, bordering conditions, vistas and views, and the presence of open space and manufactured complexes.

However, certain common ideas are inherent in the foregoing examples. Tentatively, these are structural lightness and smooth lines, exposed and undistorted appearance, and a composition that considers harmony with the surrounding environment. Environmental fitness does not mean only physical harmony, but implies compatibility with the foundation conditions, the wind forces, the temperature variations, and the many factors influencing bridge design.

Billington (1981) notes that for the six engineers discussed previously the expression of bridge form was self-developed. The style was restricted to a well-defined region of small area and of consistent topology; the political context dictated public debate and often open competition; gradually emerging experience through completed work made the last projects the best works; and a gradually evolving personal style matured to a clear concept.

1-3 ELEMENTS OF GEOMETRIC DESIGN

The geometric design of bridges relates to the location and proportioning of the visible elements of the structure, but it does not include structural design. Geometric design practices by state highway departments and other supervising agencies are not entirely uniform. Considerable variation exists in the laws of each state controlling the size, weight, and distribution of traffic and motor vehicles. Differences in the financing ability also exist and influence the decision to modify the design standards.

Bridge design standards relate to horizontal alignment and profile, clearances, location of substructure elements, roadway cross section, and framing plan. Bridges over waterways are governed mainly by vertical clearance, pier location, and horizontal clearance. Bridges in rural areas follow the geometric configuration of the main highway.

Various types of structures can be used to separate the grades of two intersecting highways, roads, or a highway and a railroad. The type best suited, however, should give drivers the minimum sense of restriction. Whereas drivers take practically no notice of a structure, they react to sudden, erratic changes in speed and direction. A structure that avoids these problems has liberal clearance on the roadways at both levels. Piers, abutments, walls, and so on are suitably offset from the traveled way. The structures should also conform to the natural lines of the roadway approaches in alignment, profile, and cross section. Although this relates to many variables, it does not preclude standardization, particularly in structural elements.

Overpasses For the overpass highway, a suitable structure is the deck type. The supports are underneath and out of sight. The bridge has unlimited clearance vertically, and the clearance laterally is controlled mainly by the location of curbs, parapets, and railings. These are chosen to enhance the concept of safety, yet they should be of a total height and openness to give little feeling of narrowness.

Deck-type bridges require less maintenance and accommodate widening if required. Prestressed deck elements allow longer spans in relation to depth. Spans of highway grade separations are rarely long enough to require through trusses, but even in this case plate girder bridges are preferred over trusses.

Underpass From the standpoint of vehicle operation, the most desirable structure for an underpass roadway will span the entire roadway section from the top of slopes when the road is cut. This solution is seldom practical, and substantial cost savings and vertical depth reductions are possible if substructure elements are provided at the edges of roadway shoulders. On divided highways, center supports should be used where the median is wide enough to provide the necessary clearance. This arrangement usually results in a four-span bridge. Most states, however, favor an unobstructed view by eliminating the shoulder piers, and this policy has resulted in a popular two-span continuous structure.

Examples Typical examples of grade separation structures include (a) single-span bridges with full abutments, an arrangement generally pleasing and offering little sense of restriction; and (b) bridges with open-end spans, in lieu of solid abutments and wing walls, with one, two, or three intermediate piers depending on the width of the median and on horizontal clearance requirements.

Rural Bridges For most states, the classification and development of design criteria for rural primary highways is a function of the appropriate planning agencies. These have the responsibility to develop, modify, revise, and interpret the physical geometries of proposed bridges in order to remain compatible with AASHTO standards and specifications for rural highways.

This policy is reflected in the vertical clearance (highway or railroad over highway). All new structures spanning the interstate and the primary systems must be proportioned to furnish a minimum vertical clearance of 16 ft 3 in. Bridges spanning other systems must provide a minimum vertical clearance of 14 ft 6 in. Through highway trusses must provide a minimum vertical clearance of 17 ft 3 in. Structures over railroads must have a vertical clearance of 23 ft between the top of the rail and the low point of the deck beams.

Most states require a horizontal clearance between the right edge of the through pavement and the adjacent pier or abutment of at least 30 ft. The only exception to this policy is for very high unit cost bridges resulting from either extreme skews or from railroad loading. The minimum horizontal clearance between the left edge of the pavement and the adjacent pier on divided highways with medians greater than or equal to 64 ft must be 30 ft.

For medians less than 64 ft, the pier can be placed at the center of the median. Interpretation of the foregoing considerations may result in a two-span structure with semifull abutments or in a four-span structure with open-end bents based on structural analysis and cost comparison.

Urban Bridges The assignment of geometric design criteria on urban highways is, likewise, similar except that more stringent requirements are warranted to serve the higher urban traffic volumes. However, the prime consideration for service is predicated on the relationship of structure capacity to public benefit. Urban geometrics is thus developed to accommodate higher traffic volumes at lower speeds. Because vertical clearances are not a function of traffic capacity, the general criteria may be the same as for rural bridges, except in highly developed areas where conditions may warrant a clearance less than 16 ft 3 in. and close to 14 ft 6 in.

The foregoing represent some of the warrants for construction at grade separations and highlight their influence on bridge type.

Structure Widths AASHTO has developed criteria for roadway widths for various volumes of traffic, and these recommendations are disseminated in the document "A Policy on Geometric Designs of Highways and Streets" (1990).

1-4 ECONOMIC EVALUATION AND RELEVANT FACTORS

Bridge types can be identified in terms of (a) main constituent materials (concrete, steel); (b) structural system; and (c) interaction with substructure (continuous spans, simple spans, rigid frames). Invariably, the span length for bridges that do not fall into the category of grade separation structures will articulate the bridge type (e.g., steel box girders, cable-stayed bridges, and suspension bridges).

The selection of materials and structural form for the superstructure is a complex procedure because it must take into account all factors affecting design. It is also influenced by the quality and cost of fabrication and construction procedures, foundation conditions and requirements, bridge height, and erection constraints.

The relationship of span to structural type becomes obvious from the analysis of statistical data. Table 1-2 relates span length to superstructure type. The last column, showing the maximum span in service, is probably the most significant indication of structural feasibility and its dependence on span length. As the span range covers 400- to 700-ft (122–213-m) spans, the choice becomes broader in terms of both structure type and materials. The choice of bridge type and materials is further enhanced in the small-span range.

TABLE 1-2 Span Length Range for Various Superstructure Types

Structural Type	Material	Range of Spans (ft)	Maximum Span in Service (ft)
Slab	Concrete	0–40	
Girder	Concrete	40–700	682, Bendorf
	Steel	100–860	856, Sava I
Cable-stayed girder	Concrete	≤ 800	771, Maracaibo
	Steel	300–1100	1050, Knie
Truss	Steel	300–1800	1800, Quebec (rail)
			1576, Greater New Orleans (road)
Arch	Concrete	300–1000	1000, Gladesville
	Steel truss	800–1700	1675, Bayonne
	Steel rib	400–1200	1200, Port Mann
Suspension	Steel	1000–4500	4260, Verrazano

Guidelines and comments on choice of bridge type are given in the following chapters. For conventional grade separations and crossings, the selection of a suitable structure type is usually governed by applicable standards. Certain site conditions require a practical correlation within the bridge type. Among these, we mention the foundation constraints, horizontal curvature, and skew angle. Once one or two suitable types are tentatively identified, they should be subjected to intensive investigation, particularly because cost trends, new materials, and design and construction procedures change continuously and thus influence economy and structure type.

Selection of Span

Typical unit prices for concrete and steel, the two predominant materials in bridge construction, are given in Table 1-3. These values are clearly average and suitable for comparative purposes only. These prices refer to 1969

TABLE 1-3 Unit Costs in Composite Steel and Concrete Bridge and Pile Foundation (1969 Dollars)[a]

Item	Cost
Steel girder, in place	\$0.20/lb
Steel reinforcement, in place	\$0.20/lb
In situ concrete, in place	\$50.00/yd^3
Concrete in piles, in place	\$100.00/yd^3
Prestress	\$0.05/kip-ft

[a]These costs are suitable for comparative purposes only. True unit prices in many parts of the world are higher than these values.

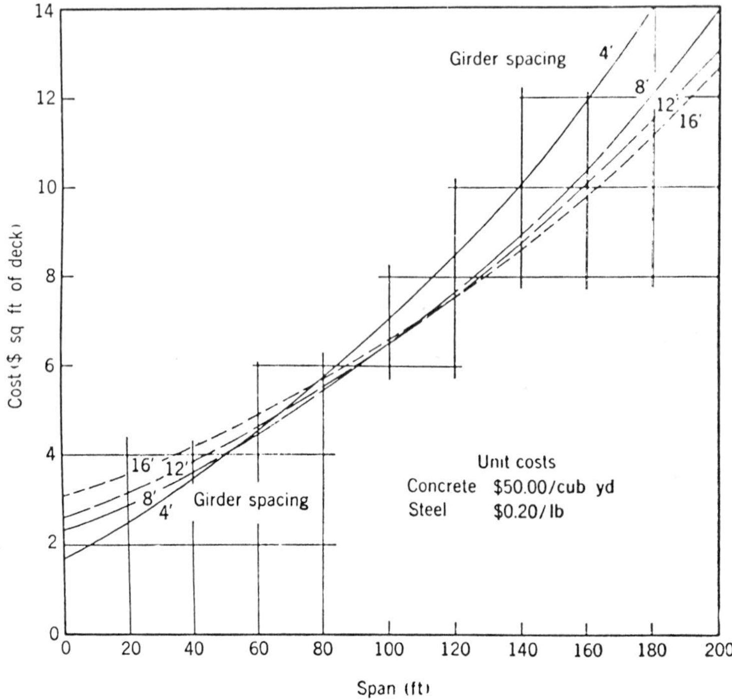

FIGURE 1-8 Theoretical costs (1969 dollars) of steel and concrete composite girder bridge.

dollars, and will be used merely to demonstrate the cost approach in selecting span lengths.

The cost for complete superstructure is given in graphical form in Figure 1-8 for the case of simply supported composite girder bridges. These costs refer to 1969 dollars. The bridges consist of cast-in-place concrete supported on welded plate girders. The lower bound of these curves gives a linear variation of cost and span, expressed by

$$C_U = 1.5 + \frac{L}{20} \tag{1-1}$$

where C_U = superstructure cost, dollars/ ft^2
L = span length, ft

O'Connor (1971) has demonstrated that a computer program can be developed to handle a particular pier with one, two, or three rows of circular prestressed concrete piles, a solid wall, and a cast-in-place concrete pile cap. The top of the pile cap is assumed to be at ground level. Longitudinal forces

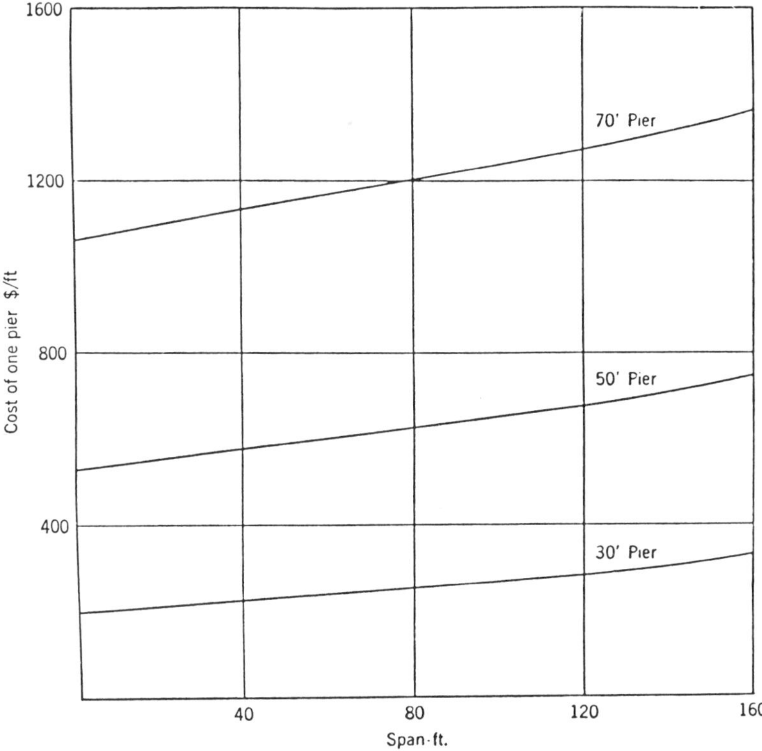

FIGURE 1-9 Foundation costs for bridges with various span lengths (1969 dollars).

from the superstructure are assumed to be balanced by a horizontal force developed as passive resistance against the pile cap. Such a program is intended to obtain minimum cost solutions. The parameters included are related to pile characteristics and pier dimensions. The bridge is assumed to have an infinite width and to consist of an infinite number of successive simply supported spans. All data refer to 1969 dollars.

The costs per foot-width of substructure are given in Figure 1-9 for variable pier heights, namely 30 (3.1), 50 (15.2), and 70 (21.3) ft (m), and evidently the foundation cost increases linearly with span, although at a slow rate. For instance, a 30-ft pier has a cost equal to $200 + 0.7L$ (dollars). The corresponding surcharge to the superstructure cost per unit area of deck is this value divided by L, or

$$C_L = \frac{200}{L} + 0.7 \tag{1-2}$$

Superstructure and substructure costs are given in Figure 1-10. They are presented separately and are also added to obtain the total bridge cost

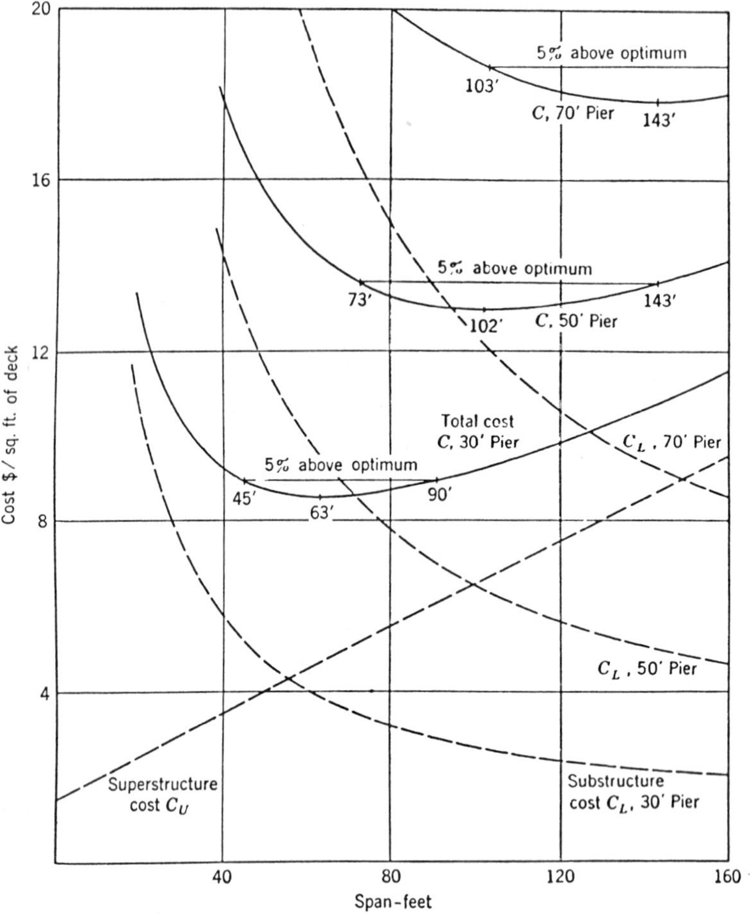

FIGURE 1-10 Superstructure, substructure, and total bridge costs for various spans (1969 dollars).

(O'Connor, 1971). Apparently, for each pier height there is an optimum span which may be computed as follows:

$$C_U = A_1 + A_2 L \qquad C_L = A_3 + A_4/L \tag{1-3}$$

or

$$C = C_U + C_L = A_1 + A_3 + A_2 L + A_4/L \tag{1-4}$$

For the optimum condition,

$$\frac{dC}{dL} = A_2 - \frac{A_4}{L^2} = 0 \tag{1-5}$$

which gives

$$L = L_o = \left(\frac{A_4}{A_2}\right)^{1/2} \tag{1-6}$$

From (1-1), $A_2 = 1/20$, and, from (1-2), $A_4 = 200$ (for a 30-ft pier), or, entering (1-6),

$$L_o = \sqrt{4000} = 63 \text{ ft}$$

which agrees with the value obtained in Figure 1-10.

From (1-6) it follows that the optimum span varies as $\sqrt{A_4/A_2}$. Thus, if the superstructure cost C_U and the substructure cost C_L are doubled or reduced both by one-half, the optimum span remains the same. If the cost of the substructure is doubled, the optimum span increases by $\sqrt{2}$. If we can assume that C_U and C_L have the forms given in (1-3), the parameters A_2 and A_4 may be obtained from exact design of any two spans and then used to predict the optimum span.

The total bridge cost is, however, rather insensitive to nominal deviations from the optimum span. Let us assume, for example, that the optimum cost C_o corresponds to the optimum span L_o. Then

$$L_o = \left(\frac{A_4}{A_2}\right)^{1/2} \quad \text{and} \quad C_o = A_1 + A_3 + A_2 L_o + A_4/L_o$$

Consider now the special case where $A_1 + A_3 = 0$. Then $C_o = 2\sqrt{A_2 A_4}$. Let $L = kL_o$, then $C = A_2 L + A_4/L$ or

$$C = A_2 k L_o + \frac{A_4}{kL_o} = \left(k + \frac{1}{k}\right)\sqrt{A_2 A_4}$$

from which it follows that

$$\frac{C}{C_o} = \frac{C}{2\sqrt{A_2 A_4}} = \frac{1}{2}\left(k + \frac{1}{k}\right) \tag{1-7}$$

Numerically, we can consider the case $C/C_o = 1.05$. Then $k = 0.73$ or 1.37; that is, for spans between 0.73 and 1.37 times the optimum span, the cost exceeds the optimum by no more than 5 percent. These results may be compared by reference to Figure 1-10. This shows a 5 percent cost increase for spans equal to 0.71 and 1.43 times the optimum.

Selection of Bridge Type

The general principles governing the selection of a suitable bridge type (including feasible alternatives) are presented at the beginning of this section, and are articulated in subsequent chapters. Certain factors will be considered here as they relate to the economic evaluation of a proposed bridge.

Bidding Practice The cost of a bridge must remain within the specified budget range. Although very frequently this is not the case, particularly in an unstable economy where material and labor costs fluctuate widely, engineers are nonetheless cautioned to study economic trends to ensure that a proposed project will not be affected by economic cycles.

Invariably, the design process and bidding stage are related, and where contractual bidding practices are expected to vary under economic or legal pressures, they may affect the design methodology accordingly.

Single Bridge Design This is typical for relatively small bridges such as conventional grade separations, small river crossings, and structures over single railroad lines. Design drawings are detailed for every bridge segment, member, and element, and a bill of material is included. Construction specifications become a part of the contract documents, and are supplemented in many instances by special provisions. The contractor (usually the low bidder) executes the project in strict accordance with the bidding documents. If changing conditions are encountered, they are covered and authorized by a change order, and paid for as extras.

This methodology works better in a fairly stable economy, and its apparent shortcomings are articulated in its lack of flexibility to accommodate sudden price changes, proprietary designs, and patented construction methods.

Design and Build This is chosen more often in certain European countries where the intent of the bidding practice is to have the contractors prepare and submit their own design of the project. In this case, the initial plans are more general, and are refined and further detailed after a first choice is made. The contractor is responsible for producing a complete design, plans, and details. The adequacy of this design is confirmed by a proof engineer retained by the owner.

Value Engineering This concept is mandatory among certain U.S. and state agencies, and requires a reevaluation of the complete design by a third party considering also other viable options. A value engineering proposal must indicate a substantial cost savings, and this is intended to preclude minor changes whereby the cost of processing offsets the anticipated savings. Economic benefits resulting from a value engineering proposal generally are shared by the contractor and the agency after allowance is made for design and processing costs.

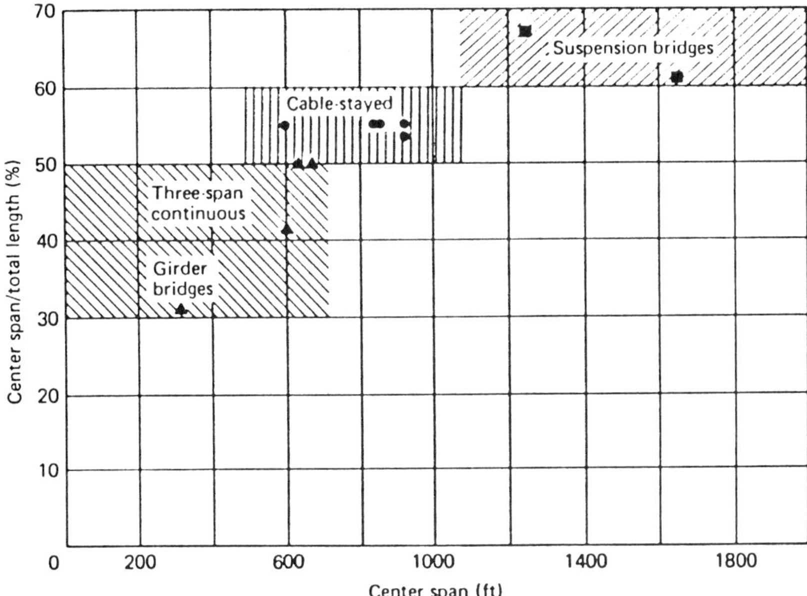

FIGURE 1-11 Bridge type comparison and range of center span, three-span continuous systems.

Alternate Designs This method has been developed with support from the Federal Highway Administration in order to produce systems that incorporate the best features of the single-design and the design-and-build approach. Its objectives are to (a) provide better competition between materials and construction procedures and (b) provide contract flexibility in the context of procedures and expertise.

The associated policy statement (FHWA, 1983) emphasizes the engineering and economic evaluation of acceptable alternate designs, and suggests options for structure components (piling, expansion joints, bearings, prestressing systems, etc.). Approval should be expected to be based on the goals of safety, cost efficiency, and esthetics, in addition to meeting all the structural requirements of the project. When comparative economic estimates are reasonably close, the respective designs should be articulated with complete sets of contract documents for advertising and bidding. The statement concludes that the intent is to take advantage of the evolving state of the art of bridge construction and fluctuating market conditions without compromising the project requirements.

Economic Analysis An example of study on the economical range for three major bridge types is shown in Figure 1-11, comparing the center span lengths to the total bridge lengths for three-span continuous girder bridges, cable-stayed bridges, and suspension bridges. The limits of economical application are 700 ft (213.3 m) for the center span of the girder bridge with a ratio of center span to total bridge length of 30 to 50 percent. The suspension bridge is indicated by economy for center spans of 1000 ft (305 m) and larger, and has a center span–total length ratio of 60 to 70 percent. The cable-stayed bridge covers the intermediate range with center spans of 700 to 1000 ft, and a respective ratio of 50 to 60 percent (Thul, 1966).

Interestingly, as construction experience is extended and design expertise is enhanced, the applicability of bridge types becomes valid within a broader range.

Example of Economic Comparison Podolny and Scalzi (1986) discuss the economic feasibility of the Sitka Harbor Bridge, considering six different bridge types. Candidate systems are summarized in Table 1-4, ranking the various types on a relative basis using the cable-stayed bridge as a base assigned the value of 1.00. The advantages and disadvantages associated with each bridge type are discussed by Gute (1973).

Bridge type I is a plate girder system requiring a main span of 250 ft (76.2 m), skewed to accommodate the fender system along the navigation channel. The piers and fenders would be in 52 ft (15.8 m) of water, rendering the system the most expensive option. Increasing the main span to 450 ft (137 m) would move the piers beyond the deep water but would also increase the superstructure cost because of the longer center span.

Bridge types II and III have spans of 300, 450, and 300 ft. In this range, the continuous plate girder became less competitive because of increased cost. The orthotropic box girder (type III) had a cost only 4 percent higher than the cable-stayed system, but a serious disadvantage was the large depth required at midspan. The design indicate a required superstructure depth at midspan close to 14 ft, compared to 6 ft required for the tied arch and cable-stayed structure. The use of a through truss or cantilever truss was considered but rejected because of higher maintenance requirements and as less appealing esthetically.

TABLE 1-4 Cost Study Data and Bridge Comparison, Sitka Harbor Bridge (From Podolny and Scalzi, 1986)

Type	Description	Cost Ratio (Cable-Stayed Girder = 1.00)
I	Plate girder with fenders	1.15
II	Plate girder continuous	1.13
III	Orthotropic box girder	1.04
IV	Through tied arch	1.04
V	Half-through tied arch	1.06
VI	Cable-stayed box girder	1.00

Types IV, V, and VI require small short piers, and have reduced end spans (150 ft). The two tied arch systems would require high superstructures in the center of the channel, which also serves as the approach path for seaplanes. The cable-stayed bridge, type VI, was selected in this case because it satisfied cost, safety, and functional considerations.

The generally competitive design that has emerged in the last decades has mandated economic feasibility studies as a routine step in the design process. Alternate designs are thus useful, particularly if the concept is compatible with the bidding process so that it does not exclude the application of value engineering.

1-5 BRIDGE APPRAISAL AND INVENTORY

Although the practical concepts of bridge appraisal and structure inventory are not directly relevant to the scope of this book, it is prudent to identify the recording and coding procedures used in the inspection and evaluation of bridges in the national (interstate) and local systems.

The NBIS (National Bridge Inspection System), prepared by the FHWA (1988) as Report FHWA-ED-89-044, has been endorsed by ASSHTO, and is available to states for recording and coding the data elements that comprise the NBIS inventory data base. In this manner, an accurate record can be compiled to accommodate the needs for future action and legislation.

The contents of this document cover almost 120 items that encompass structure characteristics, classification, traffic, maintenance, inspection, repairs, improvements, demolition, and replacement. These coded items are intended to be an integral part of the data base, and can be used to meet federal as well as state requirements.

States are thus encouraged to pursue the development of an inventory record based on specific appraisal and sufficiency rating formulas and complete the tabulation of pertinent elements and information data on individual structures. This will allow the FHWA to effectively monitor and manage a national bridge program.

Among the most relevant items of the coding guide is item 43, main structure type. The inspection and description of the bridge is first recorded on the basis of the predominant materials, such as concrete, steel, prestressed concrete, timber, masonry, iron, and others. The second recording reflects the predominant type of design and/or type of construction. In this appraisal, the bridge is described according to the grouping presented in Table 1-5.

The functional characteristics of type of service that the bridge provides are identified in terms of 10 code items, namely: highway, railroad, pedestrian, highway–railroad, waterway, highway–waterway, railroad–waterway, highway–waterway–railroad, relief for waterway, and "others."

TABLE 1-5 Code Number and Bridge Description (Investigation and Appraisal Guide, FHWA)

Code	Description
01	Slab
02	Stringer/multibeam or girder
03	Girder and floor beam system
04	T beam
05	Box beam or girders—multiple
06	Box beam or girders—single or spread
07	Frame
08	Orthotropic
09	Truss—deck
10	Truss—through
11	Arch—deck
12	Arch—through
13	Suspension
14	Stayed girder
15	Movable—lift
16	Movable—bascule
17	Movable—swing
18	Tunnel
19	Culvert
20[a]	Mixed types
21	Segmental box girder
22	Channel beam
00	Other

[a]Applicable only to approach spans.

1-6 BRIDGE NEEDS

The national bridge inventory program, briefly outlined in Section 1-5, has shown that a large percentage of bridges have reached or are approaching the end of their useful lives. Since the 1940s, design lane widths and design and legal loads have increased. Deicing measures and inadequate maintenance funding have combined with increased traffic to accelerate the deterioration of many bridges.

Procedures for strengthening existing bridges have been proposed by AASHTO through NCHRP Project 12-28 (Klaiber, Dunker, Wipf, and Sanders, 1986). A prime task of this project was to determine what bridge types can be strengthened effectively and economically. In this context strengthening is indicated for bridges in fair to excellent condition, but which require increased load-carrying capacity because of current loads. Strengthening criteria could be defined through a comprehensive bridge management system considering such factors as actual bridge capacity versus required capacity, functional adequacy, and long-term economic planning. Until such a

TABLE 1-6 Number and Percentage of 15 Common Bridge Types (National Bridge Inventory)

National Bridge Inventory Item 43	Main Structure Type	Number of Bridges	Percentage of Bridges
302	Steel stringer	130,892	27.2
702	Timber stringer	58,012	12.0
101	Concrete slab	42,450	8.8
402	Continuous steel stringer	36,488	7.6
310	Steel through truss	31,206	6.5
104	Concrete tee	26,798	5.6
502	Prestressed concrete stringer	26,654	5.5
201	Continuous concrete slab	21,958	4.6
102	Concrete stringer	16,884	3.5
505	Prestressed concrete multiple box	16,727	3.5
303	Steel girder–floor beam	9,224	1.9
204	Continuous concrete tee	7,467	1.6
111	Concrete deck arch	6,245	1.3
501	Prestressed concrete slab	5,561	1.2
504	Prestressed concrete tee	4,687	1.0
	Total	441,253	91.8

comprehensive system is generalized in the entire United States, three basic approaches are proposed to articulate the bridge types suitable for strengthening.

National Bridge Inventory By definition, this is the compilation of structure inventory and appraisal data collected in conjunction with the requirements of the National Bridge Inspection Standards (NBIS), prepared and maintained by each state for all bridges subject to the NBIS. This aggregation contains records on more than 575,000 highway bridges with spans greater than 20 ft (6 m), culverts of bridge length, and tunnels. These records are prepared according to the coding guide discussed in Section 1-5. Items considered most relevant include the construction year, the main structure type, superstructure condition, estimated remaining life, inventory rating, structural condition rating, and type of proposed improvement. These items are further combined to determine bridge life.

Data reliability is enhanced and interpretation errors are avoided by appropriate computer programming that rejects records containing blanks or unauthorized characters.

It appears that masonry through trusses, steel slabs, or other unusual and fictitious bridge types are less than 1 percent of the total. Thus, the 15 most common bridge types shown in Table 1-6 were selected for study. These represent approximately 92 percent of the bridges included in the national bridge inventory, approximately 481,000 structures. The remainder of the 575,000 national bridge inventory records are for tunnels and culverts.

Bridge Strengthening Needs A direct approach to determining bridge needs is by reference to the improvements recommended by the bridge inspector. Often, these are tempered by functional obsolescence, available funding programs, and experience level in rehabilitation and strengthening methods. For the 15 most common types shown in Table 1-6, some form of improvement is recommended for almost one-half of the bridges. Where improvements are recommended, the types of improvement are ranked in Figure 1-12 (Dunker, Klaiber, and Sanders, 1987), and evidently the overwhelming choice is replacement. Figure 1-12 also shows that only 0.9 percent of the bridges were recommended for strengthening. This small percentage may indicate the unavailability of strengthening options, or it may mean that the inspection did not recognize strengthening as a means to extend bridge life.

Bridges recommended for strengthening are ranked by type in Figure 1-13 (Dunker, Klaiber, and Sanders, 1987). Steel stringer bridges account for more than one-half of the total, followed by steel through trusses, steel girder floor beams, timber stringers, and concrete slabs.

A second approach, which is more general because it accounts for almost the entire bridge record on the National Bridge Inventory, is to examine the structural adequacy and safety factor (S1) derived from superstructure and

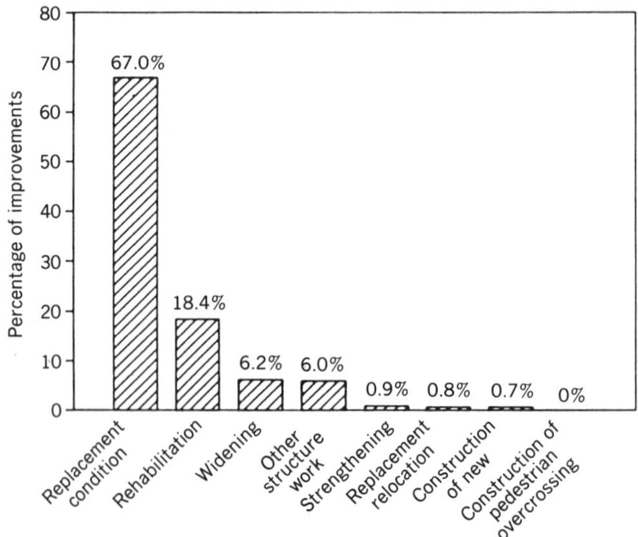

FIGURE 1-12 Bridge improvements recommended by inspection (National Bridge Inventory).

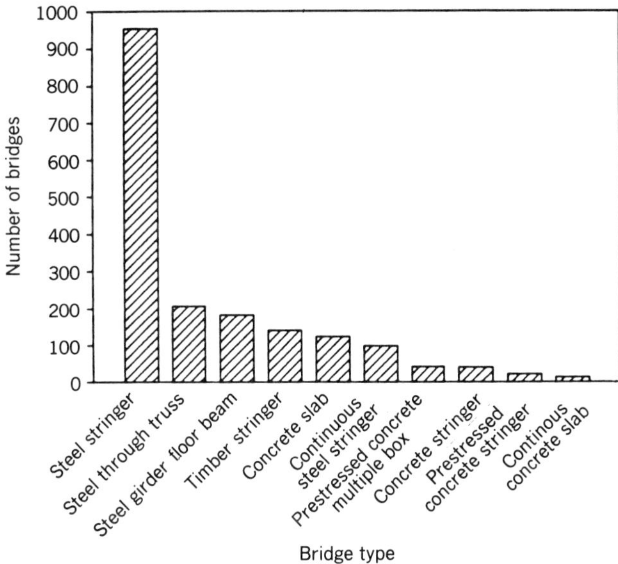

FIGURE 1-13 Recommendations for bridge strengthening according to inspection, ranked by bridge type.

substructure condition rating and inventory, as detailed by the FHWA guide for bridges (see Section 1-5), remaining life, and anticipated retirement. These data are directly obtained or computed from the bridge records. Low structural adequacy and safety usually extrapolate a need for strengthening.

Remaining life is also an indicator of the need for some strengthening. Bridges found by inspection to have a relatively short remaining life may be

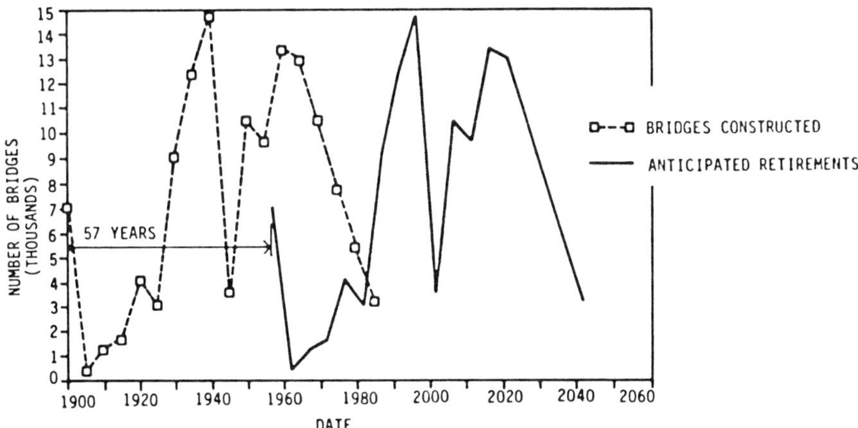

FIGURE 1-14 Number of steel stringer bridges constructed and anticipated retirements in 5-year increments (National Bridge Inventory).

ideal candidates for strengthening. Bridge types with a large number of anticipated retirements in the near future, ranked according to maximum number in any 5-year period, are steel stringers (I beams), timber stringers, steel girder floor beams, and concrete deck arches. For steel stringers, the anticipated retirement process is shown graphically in Figure 1-14. Similar diagrams are available for timber stringer and steel through-truss bridges (Dunker, Klaiber, and Sanders, 1987) and show similar urgency. As of 1985, the number of anticipated retirements is either at a high level (Figure 1-14) and expected to continue or small but with a large projected increase.

These bridge records are very consistent in identifying the steel I-beam, timber stringer, and steel through-truss types as the primary groups for which strengthening is required in order to prolong life. Secondary groups include the concrete slab, concrete tee, concrete girder, steel girder floor beam, and concrete deck arch bridge types.

1-7 PRINCIPLES OF BRIDGE MANAGEMENT

Introduction

The development of a bridge management system (BMS) has followed the bridge inventory and appraisal program, and is intended to provide a model at the network level that can be implemented by small and medium transportation agencies. In addition, the system ensures the effective use of available funds and articulates the influence of various funding levels on the bridge network.

Six major modules have been identified as the minimum requirements for developing and programming the system. These are: the BMS data base module; the network level maintenance, rehabilitation, and replacement selection module; a maintenance module that assigns maintenance programs in a rational and continuing way within the system; the historical data analysis module; a project level interface module; and the reporting module. These components may be adjusted to the needs of the transportation agency, and additional modules can be added and modified as needed.

A second phase of the program completes the concept development for a network level BMS, computerizes system programming, and validates the system and transportation agencies.

Program Initiative The BMS reflects the magnitude of the bridge problem. More than 100,000 bridges are judged to be structurally deficient because of deterioration or distress, and as many are considered functionally obsolete or inadequate for current requirements. This problem appears to be growing in terms of acceleration of the functional obsolescence and in the context of economic support. Thus, the BMS has been developed to adapt appropriate technology, economics, systems engineering, planning techniques, and management optimization of bridge resources, with the ultimate objective to provide the most effective treatment in each case.

Bridge management systems focus on the following tasks.

1. Selecting engineering methods to assess the future and present needs of existing bridges (inventory, inspection, capacity, maintenance, rehabilitation, replacement, and funding).
2. Developing guidelines for determining cost-effective alternatives with and without economic constraints.
3. Identifying priority treatment of the problem area, from posting and preventive maintenance to replacement.
4. Ensuring flexibility to handle a broad range of policy approaches and accommodate future expansion.
5. Selecting methods to ascertain standards of data reliability.

An integrated model must compare the administration and public costs of gradual structural deterioration and functional obsolescence with the costs and benefits of routine maintenance, interim repairs, partial rehabilitation, major reconstruction, and replacement for each structure. This assessment involves a set of activities that are related to the transportation infrastructure and include (a) predicting bridge needs; (b) articulating bridge conditions; (c) allocating funds for construction, replacement, rehabilitation, and maintenance operations; (d) identifying bridge priorities for action and finding cost-effective alternatives; (e) scheduling and performing minor repairs; (f) monitoring and rating bridges; and (g) keeping an appropriate data base of information.

Within the scope of assessment, the BMS must also analyze different funding levels and compare different spending policies; study maintenance, rehabilitation, and replacement actions; analyze project options and different timing alternatives; and predict the consequences of different scenarios.

Benefits In the context of engineering management, the BMS benefits all users at the administrative, executive, and technical level. The associated framework provides the flow of data related to managing the bridge network, and ensures access to data collection, decision making, and technical implementation. In particular, the technical action is based on input and editing of condition data, whereas details are available for project level design, current costs, and effectiveness of a particular response. A further benefit is access to planning and programming data.

Specific BMS Features A life-cycle costing submodule expands the options from one action based on current need to a set of actions to be taken on a bridge over a period of time. Cost and effectiveness (benefit) analyses are therefore based on a cycle of future activities taking into account their subsequent consequences on bridge condition and serviceability.

In implementing this submodule, emphasis should be placed on improving the cost and effectiveness criteria used in program selection because these parameters are based on anticipated resource allocation and consequences over the near future rather than the next needed action. The uniform annual cost method is normally used to produce the cost value. With respect to effectiveness, the model BMS considers the average efficiency value over the analysis period as its default effectiveness measure.

An optimization submodule expands selection on a bridge from one choice to multiple alternatives. Optimization techniques are therefore incorporated into the system to articulate the most beneficial set of actions available across the entire bridge network. This determination is, however, subject to budget constraints and other considerations. Flexibility is thus essential so that the optimization techniques, applicable criteria, and external constraints can be user-defined in order to produce the effective program. An advantage in this case is that the user is not required to determine the best choice on a case-by-case basis. Instead, several possible actions and decision rules are incorporated into the analysis to produce a program upon applying the optimization algorithm.

A model frequently used is a derivative of the incremental cost–benefit analysis method (McFarland, Rollins, and Dheri, 1983). This procedure articulates the optimum alternative using an incremental cost–benefit ratio, and evidently availability of funds is a constraint. In the model BMS, however, effectiveness is used in place of benefit (as expressed in monetary terms), and a minimum acceptable cost-effective threshold is a necessary parameter. In addition, the budget constraint is expanded from one budget amount to several.

For a complete review of the essential elements of a network level bridge management system, see Hudson, Carmichael, Moser, and Hudson (1987).

Current FHWA BMS Approach

A report compiled by the Federal Highway Administration (FHWA, 1989) addresses the long-term needs of bridges at the national level and establishes the underlying philosophy necessary to articulate the priorities for maintenance, rehabilitation, and replacement. Despite a growing program aimed at replacing or rehabilitating more than 10,000 bridges a year, the percentage of bridges that become structurally deficient or functionally obsolete continues to escalate. This fact provides a quick indication of the scope and intent of the current BMS program, and forms the basis for the goals of the Highway Bridge Replacement and Rehabilitation Program (HBRRP).

Although projected expenditures reflect the extent of bridge needs, they are not refined estimates for several reasons: (a) National Bridge Inventory data do not address maintenance or other improvement options except replacement and rehabilitation; (b) the criteria for evaluating the data in prioritizing needs do not distinguish between deficient bridges on different classes of highways; and (c) the numerical sufficiency rating, used as an eligibility criterion, has shortcomings when used to express needs and priorities.

Certain conclusions, however, can be drawn, namely: (a) the combined effect of all bridge improvement programs is not reducing the total backlog of bridge needs or the number of bridges found deficient, and significant relief is not in sight given the budget deficits and other discouraging factors; and (b) although National Bridge Inventory data reveal the magnitude, they do not distinguish among the critical needs or the sequence for meeting these needs.

Analysis of long-term needs shows that bridges are added and removed at about the same rate from the list of structures eligible for federal funding, but this rate of additions and removals should not indicate that the goal is to maintain the status quo. The total cost of replacement, rehabilitation, and maintenance is roughly proportional to the average age of structures in the highway network. With a high percentage of new bridges, the program cost is relatively low; with a high percentage of old bridges, the cost is high. The histogram of Figure 1-15 shows the current age distribution of bridges. Each bar represents the number of bridges in existence, built within a 5-year period. The beginning year of each 5-year period is shown. This inventory indicates a large number of relatively young bridges, starting in the 1950s.

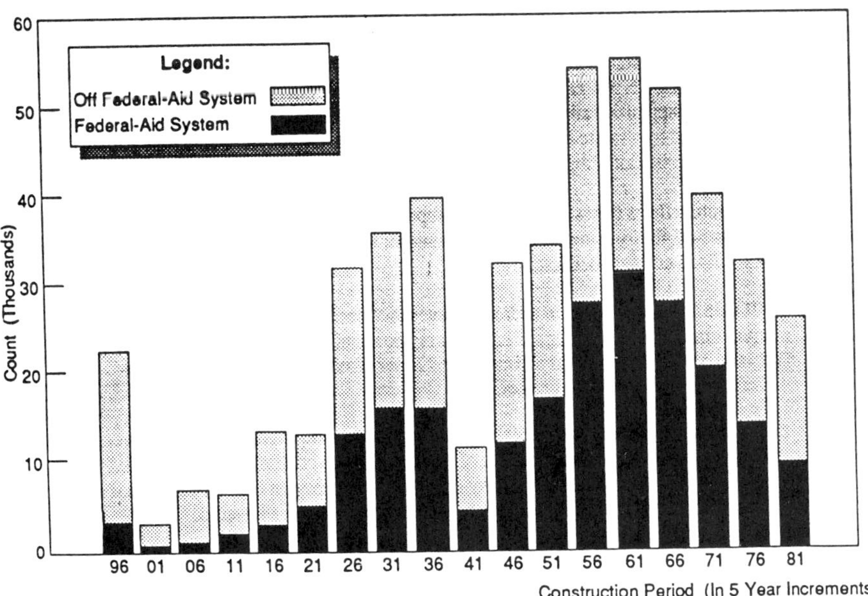

FIGURE 1-15 Number of existing bridges, built during a 5-year period.

The life expectancy of the 575,000 bridges is normally 60 to 70 years based on a realistic estimate.

The criteria used to establish the goal or standard assume a critical function in determining the level of bridge needs. According to this philosophy, the option that costs less over the long run represents the present need. Alternatively, an equally valid definition states that needs can be viewed as simply the most beneficial improvement to each bridge on the state highway system. Net benefits are usually defined as the difference between savings in user costs and agency costs over a given period of time. Needs can also be defined on the basis of minimizing total costs over time. When total costs are minimized, net benefits are maximized.

Contents of BMS Program The FHWA BMS program covers various aspects for developing and introducing a comprehensive bridge management philosophy.

Introduction This is a general analysis of the merits of the system as they relate to bridge decision making. Data organization, presentation of results, and systems analysis precede selection of possible alternatives and strategies.

Needs Defined by Level of Service Goals This program component is a practical approach that uses specified bridge characteristics, traffic volume, and highway functional classification to identify needed bridge improvements. By setting the level of service goals at a minimum point for traffic service, critical needs are determined.

Priority Ranking Formulas This task analyzes and compares empirical formulas developed by states for ranking bridges based on need, and also discusses the limitations of the sufficiency rating as a ranking criterion.

Levels of Service for Maintenance This phase in the program introduces a procedure for selecting maintenance service levels. This is achieved through the use of decision theory and mathematical optimization whereby optimal bridge components are selected for a given budget.

Estimating Service Life This extrapolates the results of studies to bridge service life predictions. The background is provided by statistical analysis of the rate of deterioration in bridge conditions and appraisal ratings, allowing the service life and remaining life to be approximated. These factors, resulting from maintenance and rehabilitation, are critical in bridge program decisions.

Cost-Effective Improvement Strategies Relevant to the decision-making process at project and network levels are the users of life-cycle cost analysis and cost–benefit analysis. These tasks provide an alternative to the various approaches pursued in the program. In particular, a network level approach to priority optimization that considers agency and users costs and benefits is developed and illustrated by case histories.

Needs Prediction Models Procedures for modeling future bridge needs on a systemwide basis relate to the ability to cope with accruing problems depending on the available or necessary resources at the proper time. Some simple modeling techniques developed at the FHWA and state level are presented.

Data Collection and Management This task focuses on topics and issues to be considered when restructuring an existing data base to support a comprehensive BMS. Adequate and reliable data bases require analysis of factors to determine which items are relevant, essential, desirable, or marginal. Among factors that might weigh in the decision to include or exclude data items are: (a) the relative cost and trends in completing collection; (b) data relevance to the analytical process selected in the BMS; (c) time dependency as it may relate to the need to estimate life expectancy or to compare life performance of bridge components, repair methods, and so on; (d) data homogeneity as it applies to the meaning of conditions or appraisal rating, and as it may detract from the usefulness of past data; and (e) accessibility because this determines data usefulness in a systemwide level analysis.

Conclusions It may appear from these remarks that bridge programs are managed using a particular system suited to a unique bridge set, bridge problems, and agency organizational structure. Current management systems have evolved through adaptation and modification.

A comprehensive BMS has a natural relationship to bridge programming and the project selection process that may be already in place. Thus, it can articulate and stengthen current bridge aspects such as inspection, priority ranking, programming, and project implementation on a systemwide basis.

REFERENCES

AASHTO, 1990. "A Policy on Geometric Design of Highways Streets", AASHTO, Washington, D.C., 1088 pp.

ASCE, 1976. "American Wooden Bridges," Committee on History and Heritage, ASCE, Historical Publication No. 4.

Belidor, 1729. *La Science des Ingenieurs*, Paris.

Bendixen, A., 1914. *Die Methode der Alpha-Gleichungen zür Berechnung von Rahmenkonstrukionen*, Berlin.

Billington, D. P., 1981. "Bridge Design and Regional Esthetics," *J. Struct. Div. ASCE*, Vol. 107, No. ST3, March, pp. 473–486.

Bleich, H. H., 1935. *Die Berechnung Verankerten Hangebrücken*, Vienna.

BRASS-Girder, 1986. "Bridge Rating and Analysis Structural System User Manual," July.

Bresse, J. A., 1880. *Cours de Mécanique Appliquée*, Vols. 1 and 3, Paris.

Burgess, M. D., 1978. "The Esthetics of Bridges," Thesis, Univ. Queensland, St. Lucia, Australia.

Calisev, K. A., 1922–1923. *Technicki list*, Nos. 1–2, 1922, and Nos. 7–21, Zabreb, Yugoslavia, 1923.

Cremona, L., 1872. *Le Figure Reciproche nella Statica Grafica*, Milan.

Cross, H., 1930. "Analysis of Continuous Frames by Distributing Fixed-End Moments," *Proc. ASCE*, Vol. 56, May, pp. 919–928.

Crouch, A. G. D., 1974. "Bridge Esthetics—a Sociological Approach," *Civ. Eng. Transactions*, Institution of Engineers, Vol. CE16, No. 2, pp. 138–142, Sydney, Asutralia.

Culmann, K., 1851. "Der Bau der holzernen Brücken in der Vereinigten Staaten von Nordamerika," *Allgemeine Bauzeitung*, Vol. 16, Vienna, pp. 69–129.

Culmann, K., 1852. "Der Bau der eisernen Brucken in England und Amerika," *Allgemeine Bauzeitung*, Vol. 17, Vienna, Austria.

Culmann, K., 1866. *Die Graphische Statik*.

De Mare, E., 1954. *The Bridges of Britain*, B. T. Batsford, London, p. 145.

Dubrova, E., 1972. "On Economic Effectiveness of Application of Precast Reinforced Concrete and Steel for Large Bridges (USSR)," *ABSE Bulletin*, Vol. 28.

Dunker, K. F., F. W. Klaiber, and W. W. Sanders, 1987. "Bridge Strengthening Needs in the United States," *Transp. Research Record* 1118, TRB, National Research Council, Washington, D.C., pp. 1–8.

Egan, M. J., 1978. "The Esthetics of Bridges," Thesis, University Queensland, St. Lucia, Australia.

Eiffel, G., 1888. "Les Grandes Constructions Métalliques," lecture, French Assoc. for Advancement of Science, March, p. 22.

Ellet, C., Jr., 1823. *A Popular Notice on Suspension Bridges*.

FHWA, 1984. "Fifth Annual Report to Congress—Highway Bridge Replacement and Rehabilitation Program," Office of Eng., Bridge Div., Washington, D.C., May.

FHWA, 1986. "Cost Effective Bridge Management Strategies," Vols. 1 and 2, Final Report No. FHWA/RD-86/110, June, Washington, D.C.

FHWA, 1988. "Recording and Coding Guide to the Structure Inventory and Appraisal of the Nation's Bridges," FHWA-ED-89-044, Fed. Hwy. Admin., U.S. Dept. Transp., Washington, D.C.

FHWA, 1989. "Bridge Management Systems," Demonstration Project No. 71, Report No. FHWA-DP-71-01, March, Washington, D.C.

Fitzpatrick, M. W., D. A. Law, and W. C. Dixon, 1980. "The Deterioration of New York State Highway Structures," Report FHWA/NY/SR-80/70, Eng. Research and Development Bureau N.Y. Dept. Transp., Albany, Dec.

Frezier, 1739. *Traité de la coupe des pierres*, Vol. 3, Strasbourg, France.

Grimm, C. T. and W. F. E. Preiser, 1976. "Civil Engineering Esthetics," *J. Struc. Div. ASCE*, Vol. 102, No. ST8, Aug., pp. 1531–1536.

Gute, W. L., 1973. "Design and Construction of the Sitka Harbor Bridge," Meeting Reprint 1957, ASCE National Struct. Eng. Meeting, April 9–13, San Francisco.

Harris, J., 1975. *The Tallest Tower*, Houghton Mifflin, Boston.

Hertwig, A., 1950. *Leben und Schaffen der Reichsbahn-brückenbauer Schwedler, Zimmermann, Labes, Schapter*, Verlag W. Ernst und Sohn, Berlin.

Hudson, S. W., R. F. Carmichael, L. O. Moser, and W. R. Hudson, 1987. "Bridge Management Systems," NCHRP Report 300, Transp. Research Board, National Research Council, Washington, D.C.

Humber, W., 1857. *A practical treatise of cast and wrought iron bridges and girders*, London.

Klaiber, F. W., K. F. Dunker, T. J. Wipf, and W. W. Sanders, 1986. "Methods of Strengthening Existing Highway Bridges," Preliminary Draft Final Report, NCHRP 12-28(4), ISU-ERI-Ames-87049.

Kuzmanovic, B. O., 1977. "History of the Theory of Bridge Structures," *J. Struct. Div. ASCE*, Vol. 103, No. ST5, May, pp. 1095–1111.

Lahire, P., 1695. *Traité de Méchanique, Paris.*

Long, S. H., 1839. "Improved brace bridge specifications of a patent for an improved bridge," granted patent.

Manderla, H., 1880. "Die Berechnung der Sekundarspannungen," *Allgemeine Bauzeitung*, Vol. 45, p. 34.

McClure, R. and S. Marquiss, 1982. "Bridge Prioritization Methods," Pennsylvania Dept. Transp. Report on Bridge Management, April.

McCullough, D., 1972. *The Great Bridge*, Simon & Schuster, New York.

McFarland, W. F., J. B. Rollins, and R. Dheri, 1983. "Documentation for Incremental Benefit-Cost Technique," Texas Transportation Inst., Fed. Hwy. Admin., Dec.

Melan, J., 1888. *Theorie der eisernen Bogenbrücken und der Hangenbrücken*, 2nd. ed., Berlin.

Navier, L. M., 1823. *Rapport et Mémoire sur les ponts suspendues*, Paris.

O'Connor, C., 1971. *Design of Bridge Superstructrues*, Wiley, New York.

O'Connor, C., M. D. Burgess, and M. E. Egan, 1980. "Bridge Appearance," *J. Struct. Div. ASCE*, Vol. 106, No. ST1, Jan., pp. 331–349.

Olsen, J. L., 1971. "The Esthetics of Bridges," Thesis, Univ. Queensland, St. Lucia, Australia.

Podolny, W., Jr., 1973. "Economic Comparison of Stayed Girder Bridges," *Highway Focus*, Vol. 5, No. 2, U.S. Dept. Transp., Fed. Hwy. Admin., Washington, D.C., Aug.

Podolny, W. and J. R. Scalzi, 1986. *Construction and Design of Cable-Stayed Bridges*, 2nd. ed., Wiley, New York.

Pope, T., 1811. *A Treatise on Bridge Architecture.*

Princeton Unviversity, 1978. *The Bridges of Christian Menn*, Princeton Univ. Att Museum, Princeton, N.J.

Ritter, W., 1877. *Zeitung für Bakwesen*.

Saint-Venant, B., 1870. "Mémoire sur l'établissement des équations différentielles des mouvements intérieurs opéres dans les corps solides ductiles," *Comptes Rendus*, Académie de Science, Vol. 70, Paris, p. 473.

Schwedler, J. W., 1862. *Theorie der Brückenbalken-Systeme*, Vol. 1, Z. Bakwesen, Berlin.

Thul, H., 1966. "Cable-Stayed Bridges in Germany," *Proc. Conf. Struct. Steelwork*, Institution of Civ. Engineers, Sept. 26–28, London.

Timoshenko, S. P., 1953. *History of Strength of Materials*, McGraw-Hill, New York.

Tresca, H., 1864. "Mémoire sur l'écoulements des corps solides," *Comptes Rendus*, Académie de Science, Vol. 59, p. 754, Paris.

Van Den Broek, 1948. *Theory of Limit Design*, Wiley, New York.

Whipple, S. C. E., 1847. *A Work on Bridge Building*, Utica, N.Y.

Wilkes, F. M., 1983. *Capital Budgeting Techniques*, 2nd. ed., Wiley, New York.

Zuk, W., 1973. "Methodology for Evaluating the Esthetic Appeal of Bridge Designs," *Hwy. Research Record* 428, pp. 1–4.

CHAPTER 2

DESIGN METHODS AND LOADS

2-1 SPECIFICATIONS AND STANDARDS

The design philosophy followed throughout this book is based on the current "Standard Specifications for Highway Bridges," 15th edition (1992), adopted by the American Association of State Highway and Transportation Officials (AASHTO). The compilation of this document began in 1921 with the organization of the Committee on Bridges and Structures of the American Association of State Highway Officials (AASHO). The specifications and design methodology were gradually developed and expanded, and as the several divisions were considered and approved they were made available for use by state highway departments and organizations engaged in bridge design. A complete specifications document was introduced in 1926, revised in 1928, and printed in 1931.

Following the first standard specifications published in 1931, the association issued revised editions in 1935, 1941, 1944, 1949, 1953, 1957, 1961, 1965, 1969, 1973, 1977, 1983, and 1989. The present 15th edition is the result of constant research and development in steel, concrete, and timber design. It appears, therefore, that in terms of content and scope, these specifications will continue as a developing document, revised from time to time to reflect new knowledge and continuing technical progress. Annual interim specifications are generally added to the document and supplement the material. The intent of AASHTO is to provide a standard or guide for the preparation of state specifications and for reference by practicing engineers.

In this context, the specifications stipulate minimum requirements that are consistent with current practice, but modifications may be indicated to accommodate local or special conditions. Because they apply primarily to the most common and usual bridge types, additional design guidelines may be necessary for unusual or exceptionally long bridges. In conjunction with the AASHTO document, entirely relevant to the theory and design of bridges are the current "Manual of Steel Construction" (AISC), the current "Concrete Code" of the ACI, and current ASTM specifications. In neighboring Canada, important technical documents in bridge design are the "Ontario Highway Bridge Design Code," and the publication "Design of Highway Bridges," National Standard of Canada. Other AASHTO specifications are referred to in the course of the book.

2-2 STRUCTURAL ANALYSIS IN BRIDGES

Structural Forms and Continuity

Continuity in bridges is exhibited essentially in two forms: continuous spans and rigid frames. Invariably, the investigation requires three basic steps in a typical order: selection of suitable methods of analysis, consideration of the physical constants entering the problem, and application of these elements to design. A critical factor is the sensitivity of the analysis to variations in material properties and to other underlying assumptions.

Three sets of elementary physical constants are considered in the geometric relationships. These are the changes of length of members subjected to axial forces, the rotation of elements subjected to bending moments, and the twisting of members under the effect of torque application. These parameters are expressed in appropriate terms that provide deformation constants used to quantify the deformation events. Uncertainties and variations in the moment of inertia can produce considerable deviations from the true results of the analysis, and these are more critical if the structural materials are not truly elastic and do not complete the elastic recovery.

Stiffness and Flexibility A typical deformation–force relationship can be presented in the form $\Delta = f_a F$, where Δ = deformation (axial elongation), F = force, and $f_a = l/EA$ = axial flexibility (l = member length, E = elastic modulus, and A = area). In the inverse form, a force–deformation relationship is derived given by $F = k_a \Delta$, where $k_a = EA/l$ = axial stiffness.

Similar concepts can be developed to describe the flexural deformation problem. Consider, for example, the beam shown in Figure 2-1*a*. For this condition, the moment acting on an element at distance x is $M = M_1(1 - x/l)$, as shown on the moment diagram of Figure 2-1*b*. If the member is prismatic, EI = constant. In this case we can write $\phi = f_b M_1$, where f_b = flexural flexibility = $l/\alpha EI$. The parameter α depends on the variation of M along the member length, which, in turn, depends on the end support conditions.

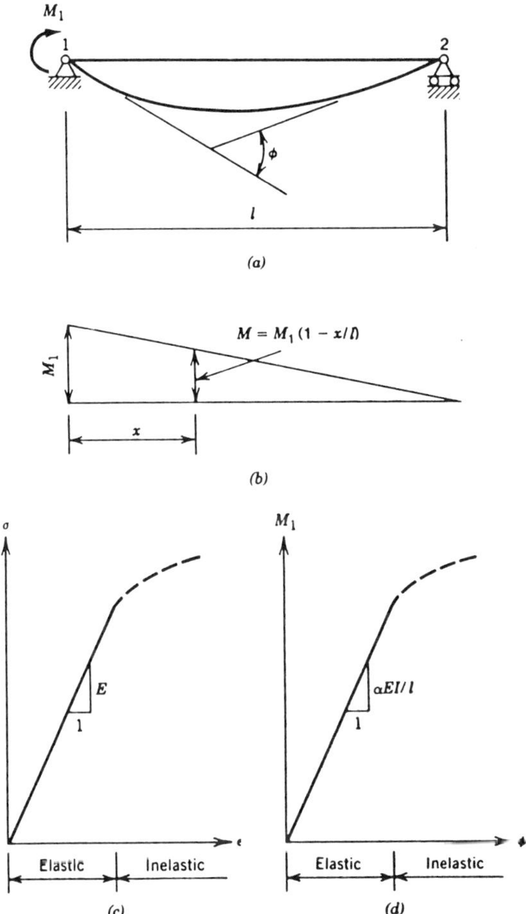

FIGURE 2-1 Flexural force–deformation behavior: (*a*) member subjected to flexure; (*b*) moment diagram; (*c*) stress–strain diagram; (*d*) moment–rotation relationship.

Expressed in the form of a force–deformation relationship, we can now write $M_1 = k_b\phi$, where k_b is a measure of the flexural stiffness, or $k_b = \alpha EI/l$. Because a linear stress–strain diagram was assumed, the resulting moment–rotation diagrams correspond to this linearity as shown in Figures 2-1*c* and *d*. If the members were stressed beyond the elastic range, the resulting moment–rotation curve would reflect this inelastic behavior as shown by the dashed lines.

In problems involving torsion, similar relationships may be developed to correlate the angle of twist with the twisting moment, expressed in terms of a torsional stiffness that is also a function of the modulus of shear rigidity $G = E/2(1 + \mu)$. Other terms used in the book include *true stiffness*, *effective stiffness*, *rotational stiffness*, and so on.

Simple and Continuous Beams A single-span beam on simple supports is unrestrained at the ends, and thus it has no negative moments. If extended to adjacent spans, the same beam develops negative moments at the support according to the degree of restraint and its actual stiffness. The restraint will have its effects, whether the beam is a member of a series in a continuous system or a member of a frame. If the restraint at an interior support can be determined numerically, we can estimate the negative moment induced at the ends of the span in question because the other variable (statical member stiffness) is a function of the I/L value and the restraining effect.

In a continuous system the actual maximum moments can be calculated by loading each span individually and then superimposing the results, for either distributed or concentrated loads. Although this solution is not the most formidable, it can be used to obtain the algebraic summation of the results. Consider, for example, the moment diagrams for the five-span continuous beam shown in Figure 2-2, where, for simplicity, all spans are taken equal. With all spans loaded, the negative moment at the third support has a coefficient of -0.0790 as shown in Figure 2-2g. However, the maximum moment coefficient at this support is -0.1112 and occurs when only the second, third, and fifth spans are loaded as shown in Figure 2-2i.

Because the dead load acts on all spans simultaneously, its sequential application is not necessary. Live loads, either truck or lane, must, however, be analyzed separately, and in a series of continuous beams one span will act to restrain the others.

Examples of Methods of Analysis

Continuous beams may be analyzed by one of the classical methods, or by a combination of matrix techniques and computer applications. Whereas bridge analysis appears to have been challenged to adapt the latter, the value of the classical approach should not be discounted. In this context, a brief review of beam analysis by compatibility and equilibrium methods is considered essential.

The Three-Moment Equation This is particularly useful in determining the internal support moments of a continuous beam. For a beam of m spans, the internal moments at the $(m-1)$ support points are the redundants. For a beam extending over three or more unyielding supports with simple bearings, the equation is expressed in the basic form

$$M_1 l_2 + 2M_2(l_1 + l_2) + M_3 l_2 = -\frac{w_1 l_1^3}{4} - \frac{w_2 l_2^3}{4} \tag{2-1}$$

Equation (2-1) relates the support moments of the first two spans with uniform load w_1 and w_2, respectively. Thus, for the beam of m spans, the

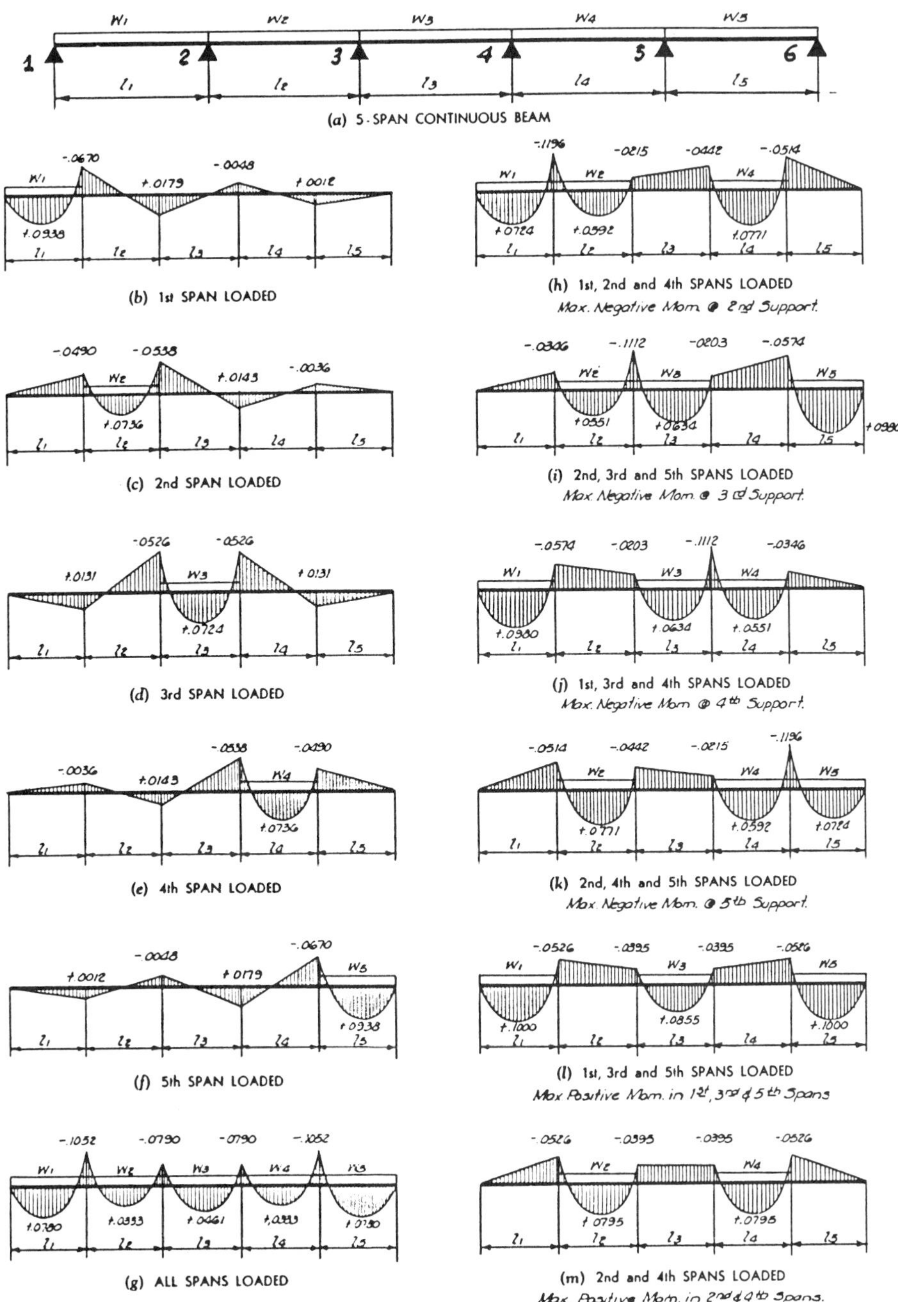

FIGURE 2-2 Bending moment diagrams for a five-span symmetrical continuous beam, simply supported.

equation would be applied at each of the $(m-1)$ interior supports to provide the $(m-1)$ compatibility relationships that are necessary to determine the $(m-1)$ redundant moments.

If concentrated loads are considered, the three-moment equation must be modified accordingly. For fixed-end beams there is an additional redundant moment at each fixed end. Although a special equation form must be formulated in this case, a convenient artifice is to replace the fixed end by an imaginary span of zero length and then apply the three-moment equation at the end support points.

One of the major advantages of using the internal moments as the redundants is that the flexibility matrix used in the solution is a banded matrix. This greatly simplifies the solution of the simultaneous equations. Because the solution of the resulting compatibility equations yields member-end moments, the shear and moment diagrams are easily constructed. It is essential, however, to follow strict adherence to sign convention. For example, positive moments are those that cause compression on the top fibers of the beam, and support displacements are positive when upward.

The Slope Deflection Equation The member force–displacement equations that are needed for the slope deflection method may be developed by referring to Figure 2-3. Essentially, the approach is an equilibrium method that accounts for flexural deformation but ignores axial and shear deformation.

In the undeformed position, the member is along the x axis, and the deformed beam has the configuration shown. The positive axes, together with the positive member-end force components and displacement components, are also shown. For this sign convention, the boundary conditions require

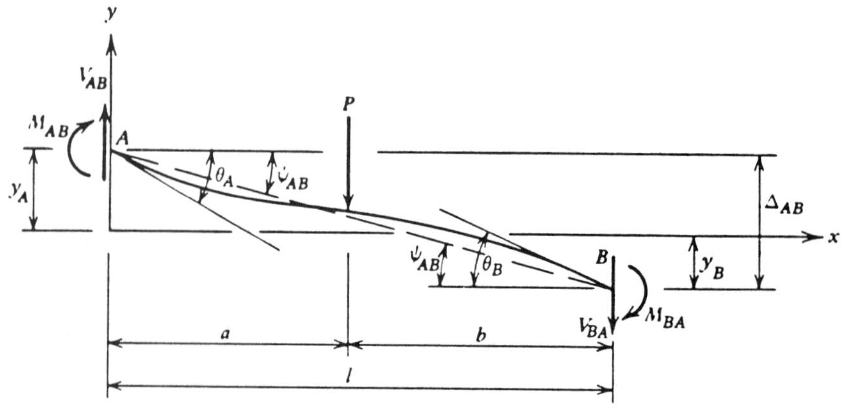

FIGURE 2-3 Deviation of slope deflection equation.

that, at $x = 0$,

$$y(x=0) = y_A \qquad \frac{dy}{dx}(x=0) = -\theta_A \tag{2-2}$$

and, at $x = l$,

$$y(x=l) = -y_B \qquad \frac{dy}{dx}(x=l) = -\theta_B \tag{2-3}$$

As the beam deforms, the moment–curvature relationship requires that

$$M = EI\frac{d^2y}{dx^2} \tag{2-4}$$

The moment as a function of x is given by $M = M_{AB} + V_{AB}x - P(x-a)$, which combined with (2-4) yields

$$\frac{d^2y}{dx^2} = \frac{M_{AB}}{EI} + \frac{V_{AB}x}{EI} - \frac{P}{EI}(x-a) \tag{2-5}$$

Integrating (2-5) and applying the boundary conditions provides simultaneous solutions for M_{AB}, V_{AB}, M_{BA}, and V_{BA}.

Matrix Methods A more generalized approach to the problem of analysis is to establish a complete interactive relationship between member-end forces and the associated member-end displacements. In matrix form, if $\{F\}$ is the vector of member-end forces and $\{\delta\}$ is the vector of member-end displacements, then

$$\{F\} = [k]\{\delta\} \tag{2-6}$$

where $[k]$ is the member stiffness matrix. The details concerning the determination of the stiffness coefficients are covered in appropriate references. Likewise, the force–deformation relationship can take the form

$$\{\delta\}_r = [f]\{F\}_r \tag{2-7}$$

where $\{\delta\}_r$ and $\{F\}_r$ are reduced versions of the displacement and force matrices, respectively, and $[f]$ is the member flexibility matrix.

When equilibrium methods are formulated in a general matrix format, the resulting equations contain the structure stiffness matrix, derived from the synthesis of the individual member stiffness matrices. In this context, the matrix method resulting from a generalization of the equilibrium approach is referred to as the *stiffness method*. Conversely, when compatibility methods are expressed in a general matrix formulation, the governing equations involve the structure flexibility matrix that results from a synthesis of the individual member flexibility matrices. In this case, the matrix method evolving from a generalization of the compatibility formulation is called the *flexibility method*.

Clearly, the foregoing provide only a crude definition of the matrix concept. With this method, the structural properties of a bridge member are taken together, and the resulting synthesis is used to calculate load behavior and structural response. The matrix approach is particularly suited to computer applications that must process a large number of reiterate calculations. In addition, the direct stiffness method is becoming the most common method of solution because of its generality and ease of programming. For a detailed study of these methodologies, see West (1989), Arbabi (1991), and Armenakas (1991).

Useful Guidelines In analyzing continuous beams the following guidelines are useful.

1. The summation of moments at a joint equals zero, or $\Sigma M = 0$.
2. If any loaded span is subjected to restraint, bending moments will be induced at the ends so restrained.
3. To produce maximum negative live load moment at any interior support, load the span on each side of that support and each alternate succeeding span.
4. To obtain maximum positive live load moment in any span, load this span and each alternate succeeding span.
5. For dead load moments, consider all spans simultaneously loaded.
6. A span loaded individually yields negative moments at its supports. The moment developing at the next support is positive, thereafter changing to negative and positive alternately and for each succeeding support or unloaded span.
7. For uniformly distributed load, bending moment curves are parabolic.
8. For concentrated loads, bending moment diagrams are triangular.
9. For unloaded spans of a continuous beams, bending moment diagrams are straight lines crossing the beam axis at the points of contraflexure.

End Restraint

The uniformly loaded beam shown in Figure 2-4*a* is simply supported. Because the ends are not restrained, the moments are zero at these points and positive in the span. Next, the same beam is considered in a continuous system consisting of a beam series. The adjoining spans restrain the beam

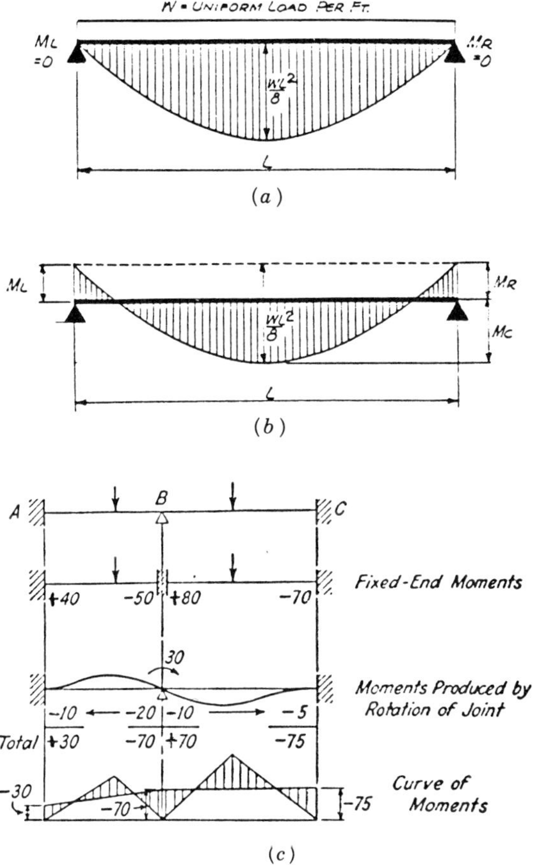

FIGURE 2-4 (*a*) Bending moment diagram for a beam simply supported; (*b*) bending moment diagram for a beam restrained at its ends as in continuous spans; (*c*) application of the method of moment distribution.

against rotation at its ends and introduce negative moments as shown in Figure 2-4*b*. The moment curves in Figures 2-4*a* and *b* are the same, and both have a center ordinate $wL^2/8$. In Figure 2-4*b*, however, the effect of negative moments is shifting of the diagram with a corresponding reduction of the positive moment. We can write

$$M_C = \frac{wL^2}{8} - \frac{M_L + M_R}{2} \tag{2-8}$$

where M_C is the center (positive) moment and M_L and M_R are the negative moments at the left and right support, respectively. For a restraint of any measurable degree introduced at the ends of a beam, a negative moment will result. Once this restraint is quantitatively known, the negative moments at the ends can be determined.

True Stiffness For a member subjected to flexure, the true rotational stiffness at one end involves the actual value as well as the end restraint of the member at the far end. The member may be fully fixed or hinged at its far end, or it may have a restraint between these extreme conditions. Full fixity requires infinite restraint (zero rotation and infinitely stiff implying that $1/EI = 0$), which is practice is difficult to attain. Likewise, ideally free or hinged ends are difficult to provide although they are commonly assumed in bridge analysis. In contemplating fixed- and free-end conditions, we only need to define the extremes of the range. In most practical problems the end conditions lie in the region of partial restraint expressed in terms of relative values, nonetheless susceptible to quantitative analysis.

For a member ideally fixed at the ends, the rotational stiffness is one extreme limit. Next, we consider a hinged or partially restrained member. From the slope deflection principles and referring to Figure 2-3, we note that as the restraint at one end increases from a hinged condition (zero restraint) to fixity (full restraint) the angular rotation at the same end decreases from a maximum to zero.

If the restraining factor at end B is denoted as R_B, we can express the angular rotation at the same end as $(1 - R_B)\theta_B$, noting that this expression satisfies the general requirements. If $R_B = 0$ meaning no restraint, the angular rotation is θ_B, which is the maximum value for end B hinged. If $R_B = 1$ meaning that the end B is fixed, the rotation is zero. A well-known relation expresses the modified stiffness S_m for end A of beam AB in terms of the factor I/L and the end condition at point B. Accordingly, $S_{MA} = (3 + 0)I/L$ if end B is hinged, and $S_{MA} = (3 + 1)I/L$ if end B is fixed. Thus, a hinged member is 25 percent less stiff than a fully restrained member. In other words, a beam fixed at one end only is 75 percent as stiff (offers 75 percent as much resistance to rotation) as a beam fixed at both ends. Likewise, a beam half-fixed at its far end would be $(1 + 0.75)/2 = 0.875$ as stiff as a fully restrained member.

The stiffness of an end span of a continuous beam freely supported at its outer end is $0.75I/L$, and this is also true for supporting columns of a frame hinged at their far ends. Where the end spans of continuous beams can be considered fixed, their stiffness is the respective I/L values.

For a series of continuous members in a beam, the restraint at a near end (considered) of a member is affected only by the modified stiffness of the adjacent end of the adjoining span, or

$$R = \frac{S_M}{S_G} \tag{2-9}$$

where S_M is the modified stiffness of the adjacent end (adjoining span), and

$S_G = I/L$ is the value of the member considered. Because the distribution of moments at a joint (or support) is directly proportional to the stiffness of the restraining members at the joint, it follows that the theory of continuity is based on relative restraints that are functions of the member stiffness. With the restraining stiffness being equal to the restraint of the member considered, for several members meeting at a joint we can write

$$R = \frac{\Sigma S_M}{S_G} \tag{2-10}$$

where ΣS_m = the sum of modified stiffness at the adjacent end of the adjoining spans, and $S_G = I/L$ = the value of the member considered.

Moment Distribution

According to this method, an iterative-type solution is carried out with emphasis on the physical process of alternatively clamping and releasing joints until equilibrium is achieved. Introduced by Professor Hardy Cross in 1929 (Cross and Morgan, 1932), it is widely used to analyze continuous beams and frames, although with the advent of digital computers direct solutions are feasible. The method allows the determination of moments, shears, and reactions for a given set of loads. Its application requires knowledge of the following: (a) the moments developed at the ends of loaded spans when these ends are considered fixed (FEM); (b) the resisting moment developed at a joint on the end sections of members connecting at the joint; and (c) the resisting moment developed at the fixed end of a beam by action of a moment applied at the other end that is not fixed. The slope deflection procedure provides the last two relations.

Let us consider the beam shown in Figure 2-4c where joint B is clamped and therefore temporarily restrained against rotation. Next, we calculate the end moments on the two fixed-end beams for this condition, shown on the sketch as fixed-end moments. Evidently, the joint at B is unbalanced. There is a counterclockwise moment of 50 on the left side and a clockwise moment of 80 on the right, producing an unbalanced moment of 30 clockwise. As the joint is released, it will rotate clockwise until the moments on the two sides balance. The moment of 50 will be increased by some amount, say 20, and the moment of 80 will be decreased by 10, so that $50 + 20 = 80 - 10$ or $50 + x = 80 - (30 - x)$, so that the sum of the changes in moments on the two sides of the joint will be the unbalanced moment of 30. This unbalanced moment at B is therefore distributed between the connecting beams in a specific way, and in proportion to the moment necessary to rotate each beam through a given angle at B. This distribution is a function of the relative member stiffness.

Referring again to Figure 2-4c, we note that when a positive moment rotates beam AB at B, a negative moment is introduced at A as shown by the curvature of the beam. Likewise, when a negative moment rotates beam BC at B, a positive moment is produced at C as shown. It follows therefore that fractions of the distributed moments are carried over to the other ends of the beams with the same sign, the carry-over factor being one-half.

The analysis of continuous beams by moment distribution follows a systematic procedure. The first step is to determine the stiffness I/L for each member throughout the beam. These stiffnesses are then used to determine the distribution factors at each joint that will be released during the moment distribution process, and evidently only relative stiffness quantities are needed to determine the distribution factors. These relative stiffnesses show the relative magnitude of member stiffnesses for members meeting at a joint (or support). The fixed-end moments are then calculated, with all supports assumed fixed against rotation. The moment distribution process of sequential release, balance, and carry-over is followed until each released joint is in equilibrium. The process registers member-end moments according to the slope deflection sign convention.

The moment distribution is not an approximate method of analysis, because if enough cycles are used the procedure will converge to the exact solution. In the usual case, however, the analysis can be concluded when the carry-over moments become small enough to be inconsequential.

Useful references on fixed-end moments, data, and miscellaneous tables are provided by Rogers (1953) and Kleinlogel and Haselbach (1963), who include complete data on multibay frames and beams.

Influence Lines

Figure 2-5*a* shows the moment diagram for a four-span continuous beam with a single load acting at point 14. If the same point load moves along the beam from the left to the right end, for every position the moment produced at point 14 will be different. Thus, for a particular point we can indicate by an ordinate the moment corresponding to every point of application of a moving load, and the resulting diagram is an *influence line* curve. Figure 2-5*b* shows the influence line for point 14 of the same continuous beam; the moments produced at this point are negative if the load P acts in spans 1 and 3, and positive if the load is applied in spans 2 and 4. The largest moment is produced when the load is applied at point 14 itself. Similar influence lines can be drawn for all points along a beam.

Critical Influence Lines for Bending Moments For every span there is one influence line that contains the absolute largest moment in the span, but as a rule this line does not coincide with any influence line of the 10th-point division. For a loaded span, the longest ordinate is under the load itself. In unloaded spans the longest ordinates are at the point where the support moment has the longest ordinate. For 10-division tables, the critical influence lines for the span (positive) moments of the loaded spans are situated as

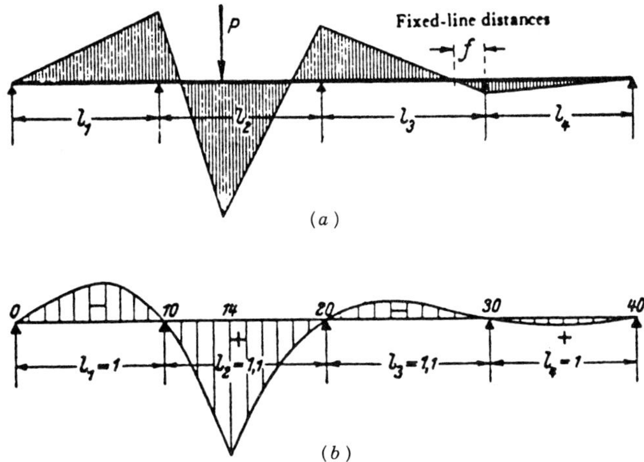

FIGURE 2-5 (*a*) Bending moment diagram for a four-span continuous beam for a single load applied at point 14; (*b*) moment influence line for point 14 of the same beam.

shown in Table 2-1. The largest ordinate of the support moment (negative) influence lines (and also of the critical influence lines of the unloaded spans) occurs for the load position shown in Table 2-2.

Influence Lines for Shears The influence line for the left end support 0 of a four-span continuous beam is shown in Figure 2-6*a* as a unit load moves along the beam. Usually, shear ordinates are given as a fraction of the unit load and are thus independent of the span length. The shear at any point in span l_1 such as point 04 may be obtained merely by adding -1 to the ordinate for point 0 to point 04, that is, by displacing the branch of the curve downward to point 04 as shown in Figure 2-6*b*. Likewise, the influence line for a point just to the left of point 10 is obtained as shown in Figure 2-6*c* merely by displacing the entire curve for span l_1 downward in a parallel translation. The same procedure can be followed to construct the influence

TABLE 2-1 Location of Critical Influence Lines for Positive (Span) Moments[a]

	Two Spans	Three Spans	Four Spans
In span 1	Between points 4 and 5	Between points 4 and 5	Between points 4 and 5
In span 2	Between points 15 and 16	In point 15	Between points 13 and 16
In span 3		Between points 25 and 26	Between points 24 and 27
In span 4			Between points 35 and 36

[a]*Note:* Within the limits 1 : 0, 5 : 1, to 1 : 2 : 2 : 1.

TABLE 2-2 Location of Critical Influence Lines for Negative (Support) Moments[a]

	Two spans	Three spans	Four spans
For support moment M_{I}	In 5, 77 and 14, 23	5, 77 and 24, 33	5, 77 and 34, 23
In the end spans		Between 13 and 14	Between 13 and 14
In the central spans			Between 23 and 24
For support moment M_{II}			5, 77 and 34, 23
In the end spans		5, 77 and 24, 23	Between 16 and 17
In the central spans		Between 16 and 17	Between 23 and 24

[a]*Note:* Within the limits 1 : 0, 5 : 1, to 1 : 2 : 2 : 1.

line for shear at the right side of point 10, producing the curve shown in Figure 2-6*d*. Influence lines for the shear at point 14 in the second span and just to the left of point 20 are shown in Figures 2-6*e* and *f*, respectively.

Müller–Breslau Principle In the foregoing analysis, responses are shown as ordinates of the influence lines for respective response functions. To construct the complete set of influence lines, a separate indeterminate analysis is essential for each position of the unit load. Alternatively, influence line coefficients may be directly obtained from tables if span ratios do not deviate from these tables.

A different approach is suggested by the Müller–Breslau principle. This evolves from a direct application of Betti's law, stemming from a unique application of virtual work concepts. The underlying theory is that the influence line for any response function is given by the deflection curve that results when the restraint corresponding to that response function is removed and a unit displacement is introduced in its place.

The Müller–Breslau principle is used primarily as a qualitative tool for verifying the shape of influence lines, and thus remains as a major feature in the analysis of statically indeterminate structures. Alternatively, analytical procedures may be introduced to calculate the ordinates of the deflected structure that result when the unit displacement is applied. Any of the methods used in determining the deflected configuration of the beam is appropriate, although certain methods have certain inherent advantages. It is essential that the induced deflected shape contain no displacement discontinuities other than the unit displacement corresponding to the response function for which the influence line is constructed.

The application of the principle to continuous beams reduces to the determination of the ordinates of the deflected structure. Because influence lines are normally desired at several locations along the beam, a suitable deflection analysis must be selected to meet these needs. Conveniently, it is possible to combine the conjugate beam method with the tubular procedure

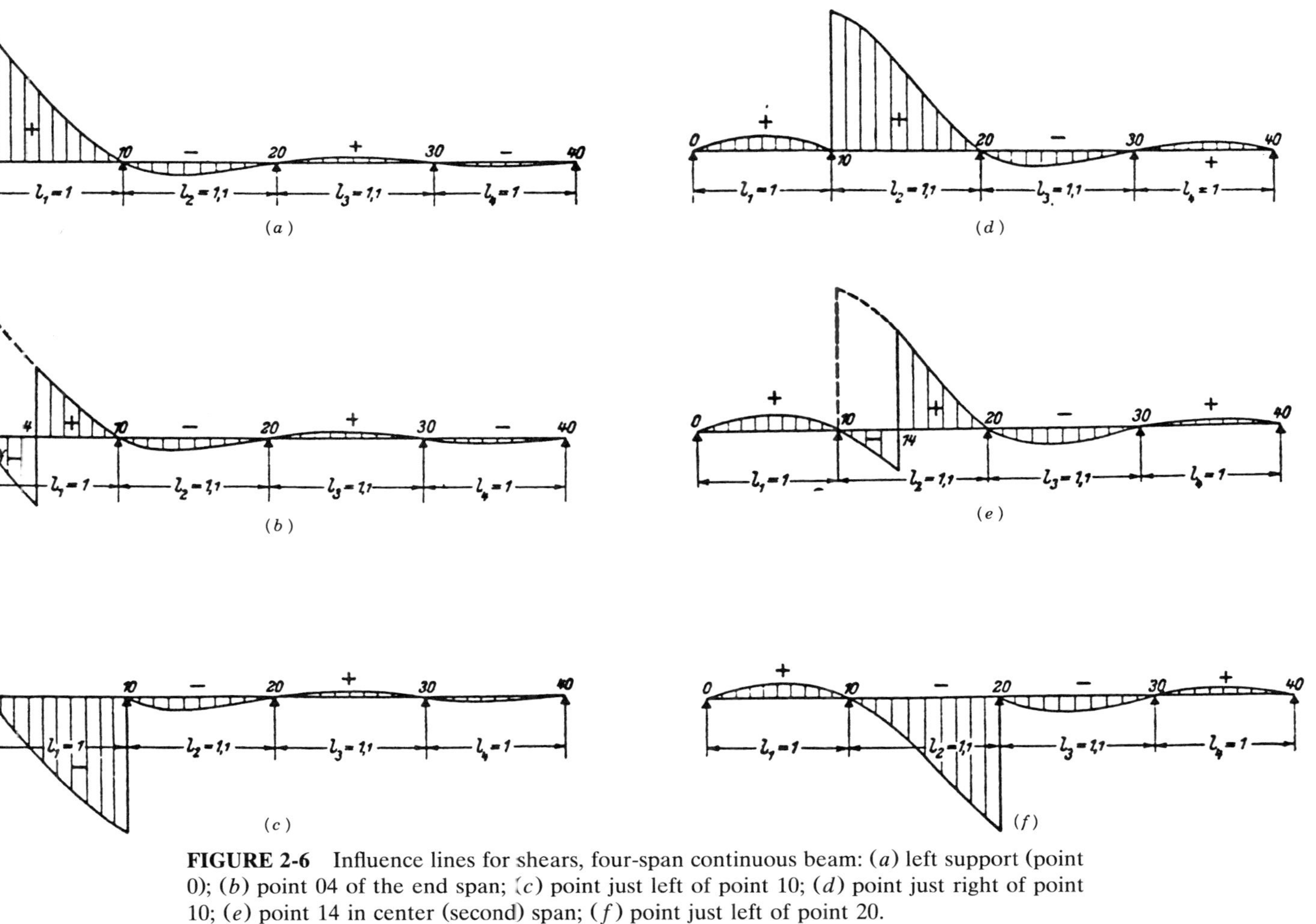

FIGURE 2-6 Influence lines for shears, four-span continuous beam: (*a*) left support (point 0); (*b*) point 04 of the end span; (*c*) point just left of point 10; (*d*) point just right of point 10; (*e*) point 14 in center (second) span; (*f*) point just left of point 20.

suggested by Newmark (1943). Numerical examples of the principle are given by West (1989).

Variable Moment of Inertia

Direct use of moment distribution methods and influence line tables normally implies continuous beams with a constant moment of inertia. Most often, however, continuous beams are deepened near the supports or made stronger for structural optimization. This configuration results in an increased moment of inertia, with a direct effect on moments. Thus, the negative moments at the supports become larger with an accompanying reduction in the positive span moments.

Haunches affect the fixed-end moments, the member stiffness, and the carry-over factors. Of these, the most important is the change in fixed-end moments. If all the beams are haunched so that their proportions are the same, their relative stiffness will not change. Although procedures are available to deal with variable I, the associated work is tedious and time consuming. The solution in this case should be obtained with the use of digital computers.

2-3 DESIGN METHODS: CONCRETE BRIDGES

Both AASHTO specifications and the ACI building code allow two alternate design procedures. In the *service load* design method, or *allowable stress design* (ASD), working or unfactored loads provide the basis for concrete strength assessment. In flexure, the maximum elastically computed stresses cannot exceed allowable or working stresses (usually 0.4–0.5 times the concrete and steel strengths).

The working stress method implies that the ultimate limit state is automatically satisfied if allowable stresses are not exceeded, but depending on the variability of materials and loads this is not always true. Thus, it is often necessary to consider the deflection limit state and the crack-width limit state. Inconsistencies in working stress design are pointed out by MacGregor (1976) and Ellingwood, Galambos, MacGregor, and Cornell (1980). The most serious shortcomings are the inability to quantify the variability of resistances and loads, the approximation of the level of safety, and the inability to consider groups of loads where one increases differently from the others.

The strength design method (or load factor method) is essentially limit states design with emphasis on ultimate limit states, with the serviceability limit states checked after the original design is completed. According to this philosophy, the required strength of a section is the strength that must be developed to resist the factored loads and forces applied to the structure in combinations stipulated in relevant criteria. The "(design) strength" refers to ϕR_n, factored resistance, whereas "required strength" refers to the load effects computed from factored loads $\gamma(\beta_D D + \beta_L(L + I) + \cdots)$.

Probabilistic Analysis of Safety Factors Let R denote resistance and let S denote load effects expressed in terms of a quantity, for example, bending moment. For a given distribution of load effects, the probability of failure can be reduced if resistance is increased. Thus, the term $Y = R - S$ represents the safety margin. By definition, failure will occur if Y is negative. The probability of failure, P_f, expresses the chance that a particular combination of R and S will yield a negative Y.

The function Y has a mean value $\bar{Y}$ and a standard deviation σ_Y. The parameter $\bar{Y}/\sigma_Y$ is called the safety index. If Y follows a standard statistical distribution, and if $\bar{Y}$ and σ_Y are known, the probability of failure is obtained from statistical tables as a function of the type of distribution and the value of $\bar{Y}/\sigma_Y$. For a given value of the safety index, we can estimate the number of failures for x number of structures during their lifetimes.

Because strengths and loads vary independently, it is expedient to have one factor or series of factors to account for variability in resistance, and a second series of factors to account for variability in load effects. These are referred to, respectively, as resistance factors ϕ, and load factors γ and β. The derivation of probabilistic equations for calculating the values of ϕ and γ are based on the assumption that both the strength and the load effects can be represented by log-normal distributions. The coefficient β is computed so as to differentiate the variability of load effects between different load types and groups. For example, for live loads this coefficient is larger than for dead loads due to the greater variability of the former. Engineers involved in strength analyses normally would not compute the values of ϕ, γ, and β because appropriate design codes specify the values to be used. Thus, for Group I loads AASHTO specifies the coefficients β for dead and live load plus impact as 1 and 1.67, respectively, whereas the load factor γ is taken as 1.3. This leads to a simple relationship between factored resistance ϕR_n and load effects (required strength) as follows:

$$\phi R_n \geq 1.3[D + 1.67(L + I)] \tag{2-11}$$

Likewise, the factor ϕ depends on the type of load effects (e.g., flexure, shear, torsion, etc.) and on the specific characteristics of the loaded member (e.g., conventional reinforced concrete, prestressed concrete, factory produced, cast-in-place, etc.). The application of (2-11) is demonstrated in Chapter 3.

2-4 DESIGN METHODS: STEEL BRIDGES

The current AASHTO specifications as well as the AISC manual provide for two alternate design methods. AASHTO considers the service load procedure (allowable or working stress design, ASD) the standard design approach for all structure types. The application of the method implies that structural steel members are proportioned on the basis of design loads and forces, allowable stresses, and functional limitations for the appropriate material under service conditions.

According to the strength design method (load factor design), engineers can choose an alternate procedure to proportion simple and continuous beam and girder bridges of moderate length. This approach considers multiples of the design loads. Furthermore, to ensure serviceability and durability, the design must also focus on control of permanent deformations under overloads, on fatigue characteristics under service loads, and on control of live load deflections under working conditions.

Service live loads are defined as vehicles that may operate on a road system without special permits; examples are the standard truck and lane loadings. For design purposes, service loads include the dead, live, and impact loads described in Section 3 of AASHTO. Overloads are live loads that can be allowed on a bridge on special occasions provided they do not cause permanent damage. For design purposes the maximum overload is limited to $1.67(L + I)$, which is the factored live load effect in (2-11). Moments, shears, and other load applications are computed assuming elastic behavior of the structure, although exceptions are noted. Bridge members are then proportioned so that their computed factored resistance is at least equal to the total effect of the factored loads.

Methods of Static Analysis For the purpose of elastic analysis, Sanders and Elleby (1970) classify steel beam bridge systems according to the manner in which these systems are idealized. The main classifications are (a) orthotropic plate concepts that consider the bridge system an elastic continuum to be treated as an equivalent plate; (b) grid systems idealized as an equivalent grillage of interconnected longitudinal and transverse beams, cross members, and diaphragms; and (c) girder–plate redundant techniques where the interacting forces between the slab and longitudinal girders are treated as the redundants of the system.

Introducing the load factor method in the context of elastic analysis results in a procedure that is consistent with the criterion of preventing the undesirable effects of yielded bridge members. Thus, stresses are determined elastically, and members are proportioned by their strength to carry these stresses. Some authors suggest, however, that the intent of elastic structural behavior throughout the useful cycle is inconsistent with the proportioning of members according to their ultimate strength. Here, the argument is that with stresses determined by elastic distribution, the stresses would have to be plastically redistributed in some structures in order to mobilize the strength of the members so proportioned. As a point of perspective, a limit analysis method should consider plastic redistribution based on a collapse mechanism resulting from this mode of yielding (see also the following sections).

Examples of inelastic methods of analysis are found in composite bridge systems. Kuo and Heins (1973) have, for instance, developed a finite-difference technique to formulate the load deformation equations. Subsequently, the stiffness of the elements in these equations is redefined during the course of loading according to the moment–curvature response of the element.

Plastic methods of analysis of bridge superstructures have been introduced since the mid-1970s. Examples are (a) predictions of ultimate load behavior of composite bridges by reference to yield-line theory, and (b) ultimate load behavior of horizontally curved girders (Yoo and Heins, 1972) that led to practical design techniques.

Extensive work has been carried out in this area, and useful references include Guyon (1946, 1949), Massonnet (1950, 1954, 1965, 1967), Sanders and Munse (1960), Heins and Looney (1966, 1968), Heins and Galambos (1972), Newmark (1938, 1943, 1948), Newmark et al. (1946, 1948), Kuo (1973), Reddy and Hendry (1969), Lash and Nagaraja (1970), and Eyre and Galambos (1973).

Whereas the introduction of load factor design criteria in steel bridge systems has been a noble step forward, the ASCE–AASHTO (1975) Committee also recommended research with emphasis on the following areas.

1. Ultimate strength of a bridge against plastic collapse because of the practical significance of this failure where extreme overloads are involved. However, only bridges in a specific span range are subject to this analysis, because short bridges are controlled by axle loads and larger bridges usually have their strength dictated by dead loads so that a single vehicle would produce a secondary effect only.
2. Effects of heavy moving loads on fatigue and ultimate strength.
3. Shakedown characteristics when subjected to heavy moving loads.
4. Ultimate behavior of bridges subjected to earthquakes.
5. Effects of service life and environmental conditions (corrosion, etc.) on ultimate strength.
6. Interaction between bending and torsion in the inelastic range.
7. Improved design criteria for shear connectors in curved girder bridges and in the negative moment regions of continuous composite bridges.

2-5 BRIDGE LOADS

A noble goal in bridge analysis and design is to understand (a) what loads and forces act on the structure, (b) how they are distributed, and (c) how they should be applied to the various components of the superstructure and the substructure.

The loads reviewed in this section are those stipulated by AASHTO, and those that represent the best estimates and criteria developed by task committees from a joint ASCE–AASHTO effort. In some instances, engineering opinion endorses current practice, whereas in others it appears to follow a radical departure from established codes and standards.

In general, loads must be considered together with allowable stresses or load factors. For instance, a moderate design load and a low allowable stress may be combined to yield a more severe condition than a heavy load with a higher allowable stress.

AASHTO Loads

Section 3 of the AASHTO specifications stipulates the loads and forces to be considered in the design of bridge structures. These are dead load, live load, impact or dynamic effect of live load, wind load, and other forces such as longitudinal forces, centrifugal force, thermal forces, earth pressure, buoyancy, shrinkage stresses, rib shortening, erection stresses, ice and current pressure, and earthquake stresses.

The foregoing loads, groups, and combinations thereof are identified and detailed in the AASHTO document, and need not be repeated here. Besides these conventional loads, AASHTO also recognizes indirect load effects such as friction at expansion bearings and stresses resulting from differential settlement of structure components.

Vehicle Loading for Short-Span Bridges

Considerable effort has been made both in the United States and in Canada to develop a live load model that represents the actual highway loading more realistically than the H or the HS AASHTO models. Proposals for such a design traffic loading have been made by various states and provinces (California, Louisiana, and Ontario), and in some cases it has been adopted by the authority having jurisdiction. However, the consensus of opinion among engineers is that additional documentation should be presented to warrant changes in the standard highway loading, and the AASHTO model thus remains the applicable loading in bridge design (see also proposed LRFD specifications).

Vehicle Loading for Long-Span Bridges

Rational design of long-span bridges often is inhibited by the lack of a design code. The live loading is in this case set by the design consultant. Certain types of bridges are relatively insensitive to errors in estimating these loads, because their design is governed by the dead loads (e.g., heavy truss bridges). Other types, however, are very sensitive to errors or to changes in the live load, and examples include single box girders and cable-stayed bridges. Although most of the world's long-span bridges are found in North America, very little effort has been made to establish representative loads for long-span bridges with few exceptions (Ivy et al., 1953). Such loads would be considered for any bridge or component of a bridge that is outside the range of the code normally appropriate for a shorter bridge under the same conditions.

Relevant Studies The credible load occurring on a short-span bridge is the heaviest truck or trucks that can travel on the bridge deck, but this is not the case for a long-span bridge because this structure will not be entirely covered by the heaviest possible vehicles. This claim was tested in a bridge in Vancouver, British Columbia, Canada (Navin et al., 1976), with a main span of 473 m (1550 ft). Two methods were used to predict long-span bridge loading: a purely analytical solution of probability equations, and a procedure based on the random-scatter capability of a computer to simulate incoming traffic (Buckland, Navin, Zidek, and McBryde, 1980). Basic assumptions included stationary traffic where maximum loading will occur only when the traffic is stationary and bumper-to-bumper. When traffic begins to move, vehicle distance increases and load intensity decreases. For the maximum loading, therefore, the traffic is stationary and allowance is not made for impact.

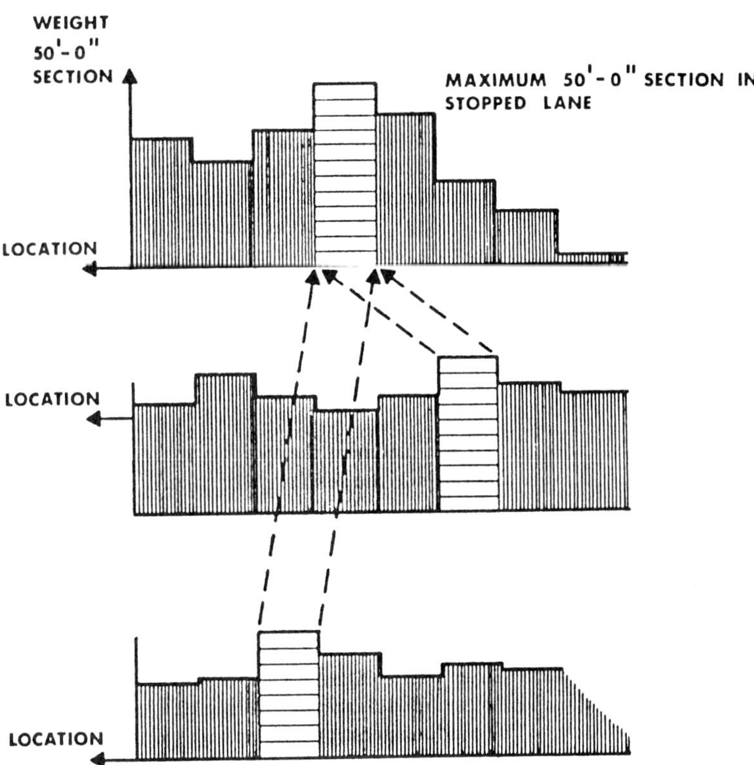

FIGURE 2-7 Line up of segments; computer simulation program. (From Buckland, Navin, Zidek, and McBryde, 1980.)

A computer simulation program written for these studies feeds random traffic onto the bridge, but allows random stoppages in one or more traffic lanes on or off the bridge. It then scans the stopped traffic and searches for the maximum load, moment, or shear at 15, 30, 61, 122, 244, 488, 976, and 1951 m (50, 100, 200, 400, 800, 1600, 3200, and 6400 ft). These sections are referred to as the loaded length L. In order to find the maximum moment and shear, the program considers the loaded length L as a simply supported beam, and calculates the center moment and end shears. Having found the maximum load, moment, and shear in one lane, the program proceeds as follows: (a) it searches for the maxima in other lanes (the traffic in these lanes may be stopped or moving); (b) it moves the traffic in the second, third, and other lanes so that the most heavily loaded 15-m (50-ft) segments line up as shown in Figure 2-7; (c) it then finds the maximum load, moment, and shear on three lanes and six lanes for each of the loaded lengths as well as for the lane previously calculated. The maximum values are printed for each time period of 3 months, and the mean and standard deviations are computed. Maximum loading, moment, and shear are then predicted for any required return period.

For each loaded length L, equivalent parameters are selected so as to give the maximum shear or moment. In simulating a simple beam, the concentrated load P and the uniform load U acting as shown in Figure 2-8 are

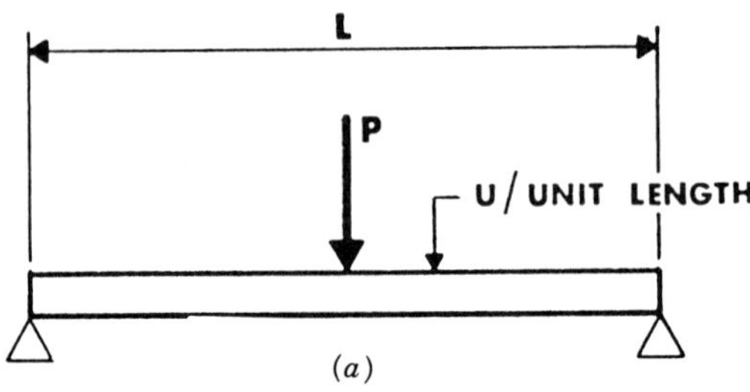

(*a*)

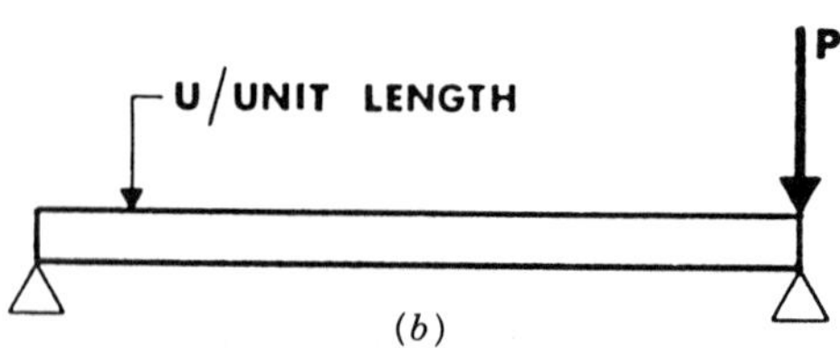

(*b*)

FIGURE 2-8 Program simulation; concentrated and uniform loads for simple spans: (*a*) configuration for maximum moment; (*b*) configuration for maximum shear. (From Buckland, Navin, Zidek, and McBryde, 1980.)

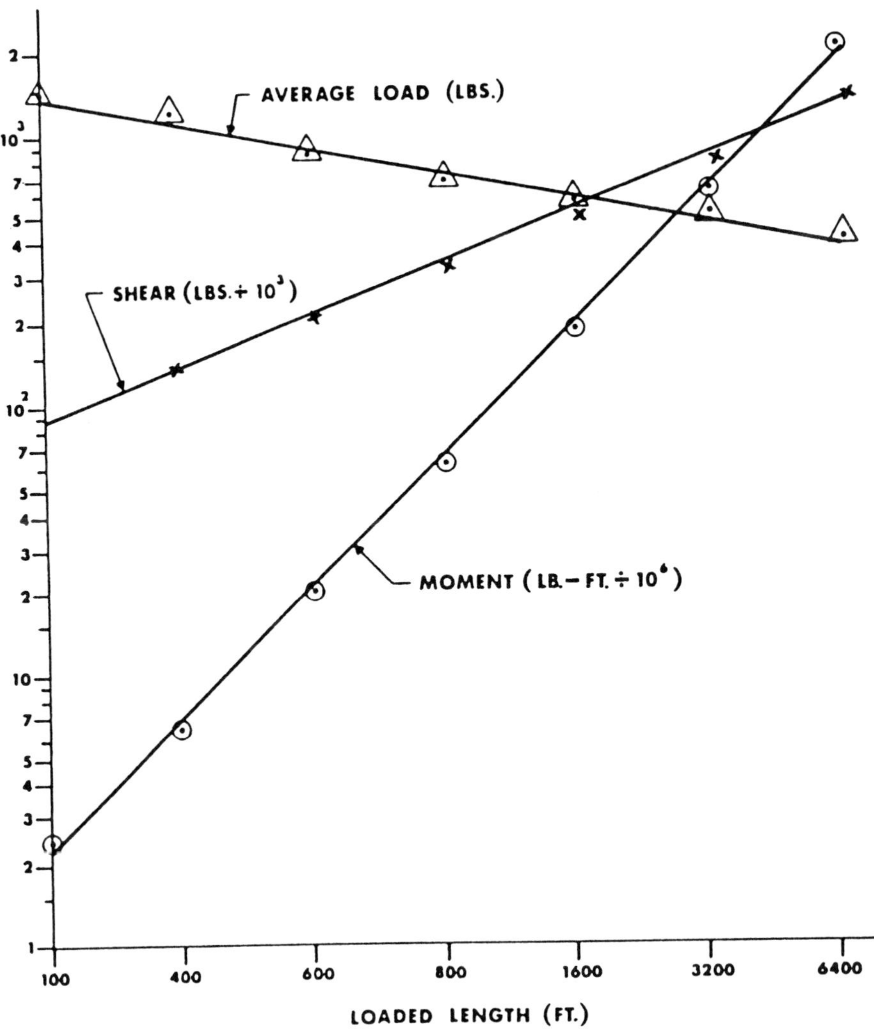

FIGURE 2-9 Maximum load, shear, and moment versus simply supported length for 7.4 percent heavy vehicles. (From Buckland, Navin, Zidek, and McBryde, 1980.)

suitable. Then, if W is the total weight, $W = P + UL$. The maximum moment M and shear S can now be represented as $M = PL/4 + UL^2/8$ and $S = P + UL/2$.

Results for a single lane for a return period of 5 years, with 2000 events per year, are summarized and plotted versus the loaded length. For 7.4 percent heavy vehicles, taken as the base load case, values of shear, moment, and average load arc shown in Figure 2-9 on a logarithmic scale, and evidently the curves are essentially linear. If the parameters P and U are

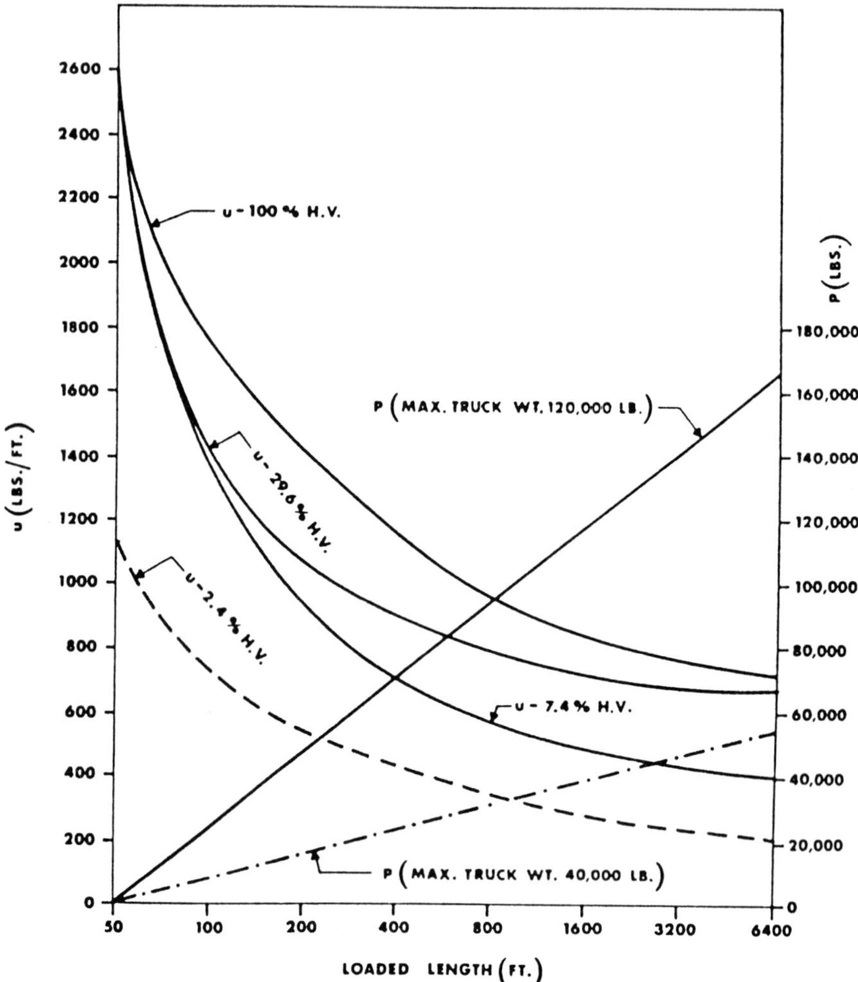

FIGURE 2-10 Parameters P and U for different percentages of heavy vehicles. (From Buckland, Navin, Zidek, and McBryde, 1980.)

computed from the foregoing simple relationships for four load cases, the smooth curves shown in Figure 2-10 are derived.

For the case of 7.4 percent heavy vehicles, the accuracy with which simply supported weights, shears, and moments are computed is shown in Figure 2-11. The average weight is the least accurate but within acceptable deviation. The shear fluctuates 2 percent from 200 to 3200 ft (61–976 m), and the moment is overestimated by 6 percent maximum.

The concentrated load P increases as the loaded length increases, and this can be explained by the hypothetical traffic distribution shown in Figure 2-12. For a short-loaded length, L_1, the loading is almost completely uniform,

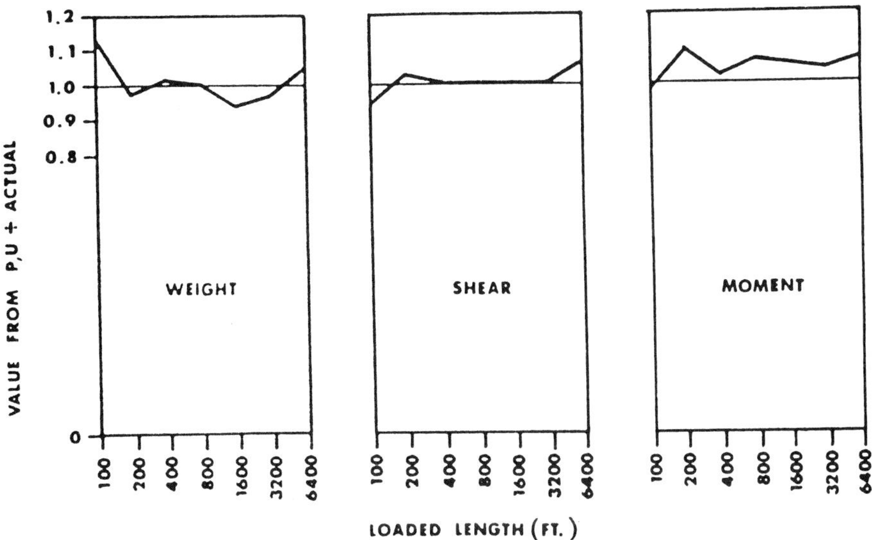

FIGURE 2-11 Accuracy of parameters. (From Buckland, Navin, Zidek, and McBryde, 1980.)

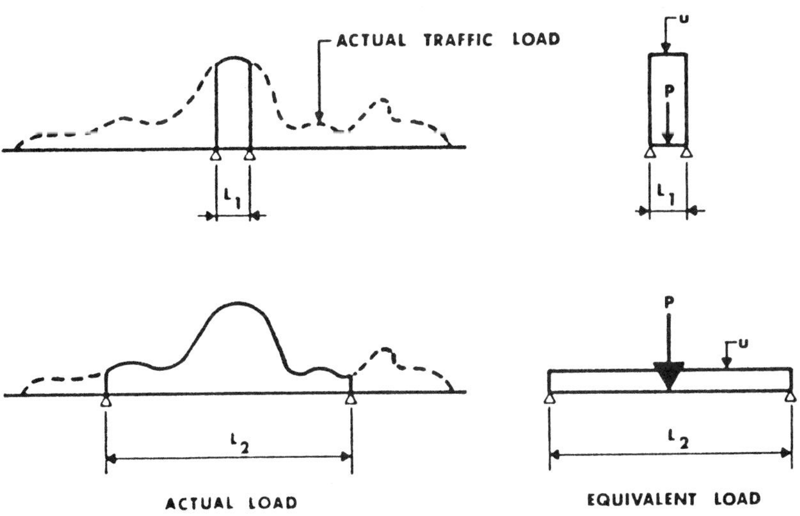

FIGURE 2-12 Equivalent loads for short and long spans. (From Buckland, Navin, Zidek, and McBryde, 1980.)

that is, U is large and P is small. For a larger length L_2, the load approximates a small uniform load U but a large concentrated load P.

Based on the results of this study, the following conclusions are drawn.

1. The maximum loading occurs with traffic stationary.
2. The loading can be represented by a uniform load and a concentrated load for predicting moments and shears.
3. The average load decreases as the loaded length increases.
4. The concentrated load increases with increasing loaded length.
5. For a given loaded length, a single uniform and a single concentrated load can be used to represent maximum moment and maximum shear.
6. The loading is not unduly sensitive to increases in the intensity of truck traffic.
7. Further research is necessary, and more data should be forthcoming to ensure the accuracy of predictions.

Recommended Design Loads The ASCE Committee on Loads and Forces for Bridges (1981) recommends a basic lane load consisting of a uniformly distributed load U and a single concentrated load P, as given in Table 2-3 or as shown in Figure 2-13. If more than one span is loaded, only

TABLE 2-3 Uniform and Concentrated Lane Loads for Design of Long-Span Bridges (From ASCE Task Committee, 1981)

Loaded Length [ft (m)] (1)	Concentrated Load, P [lb (kN)] (2)	Uniform Load, U [lb/ft (kN/m)]		
		7.5% H.V.[a] (3)	30% H.V.[a] (4)	100% H.V.[a] (5)
50 (15.25)	0	2,600 (38)	2,600 (38)	2,600 (38)
100 (30.5)	24,000 (107)	1,400 (20.4)	1,500 (21.9)	1,750 (25.5)
200 (61)	48,000 (214)	940 (13.7)	1,100 (16)	1,425 (20.8)
400 (122)	72,000 (320)	710 (10.4)	950 (13.9)	1,170 (17.1)
800 (244)	96,000 (427)	570 (8.3)	830 (12.1)	960 (14)
1,600 (488)	120,000 (534)	485 (7.1)	740 (10.8)	840 (12.3)
3,200 (975)	144,000 (640)	440 (6.4)	700 (10.2)	770 (11.2)
6,400 (1,950)	168,000 (747)	400 (5.8)	680 (9.9)	720 (10.5)

[a]% H.V. denotes the average percentage of heavy vehicles [buses and trucks greater than 12,000 lb (53 kN)] in the traffic stream.

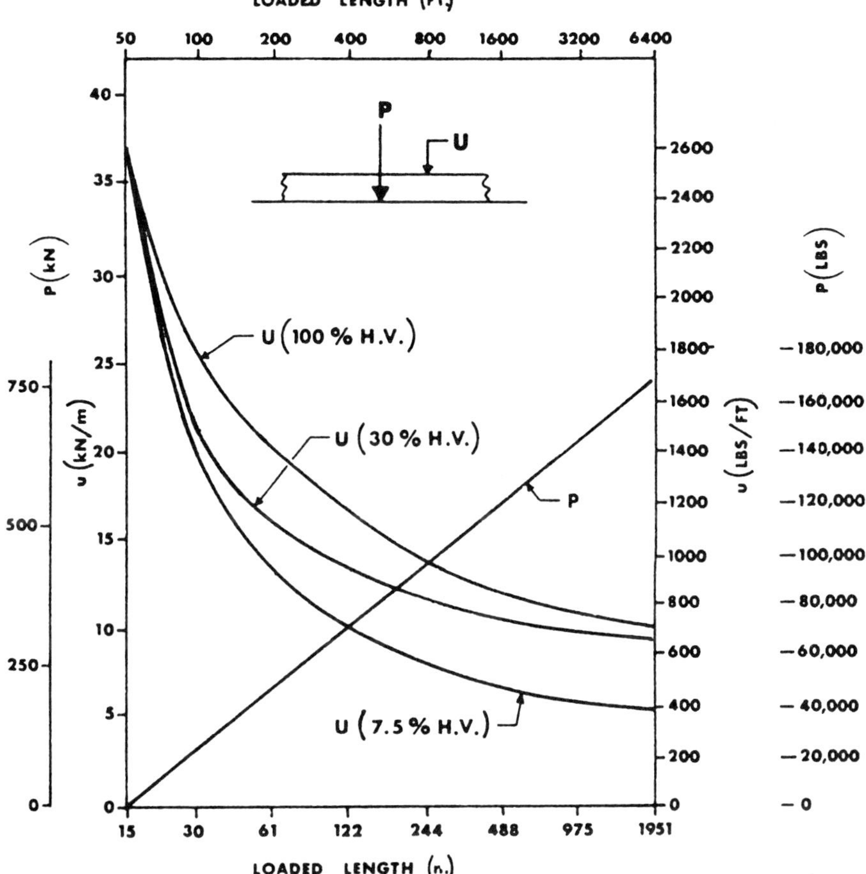

FIGURE 2-13 Parameters P and U (P = concentrated load per lane, U = uniform load per lane, and % H.V. = average percentage of heavy vehicles in traffic flow). (From ASCE Task Committee, 1981.)

one concentrated load P will be used per lane. No allowance will be made for impact.

The loaded length is the length producing the maximum effect from the uniform load U. The concentrated load P, appropriate for the loaded length, will be added and placed at a position yielding the maximum effect, but not outside the length covered by the uniform load. If two or more lengths of bridge are loaded, whether adjacent or not, the loaded length is the sum of the various loaded lengths and the concentrated load is applied once. In general, all loads will be applied in the centers of their respective lane, but a mean position may be chosen at the engineer's discretion.

If more than one lane is loaded, the lane producing the maximum effect has the basic lane load of Table 2-3, and this is marked as lane 1 in

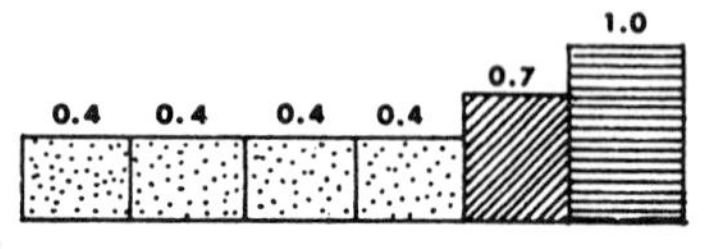

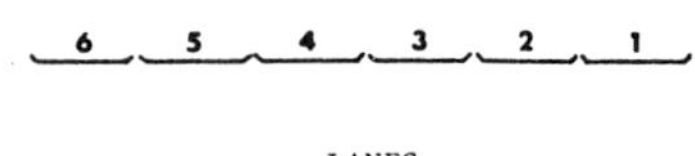

FIGURE 2-14 Multiple-lane distribution. (From ASCE Task Committee, 1981.)

Figure 2-14. All other adjacent loaded lanes are obtained by an appropriate reduction factor as shown. All lanes will have the same loaded length, but some may have zero load if this maximizes the total effect. The concentrated load may not be in the same position for all lanes.

If the maximum effect is obtained by placing the most heavily loaded lane on one side of the bridge for part of the loaded length and on the other side for the remainder (an arrangement producing maximum torque), the loaded length will be taken as the total length of the most heavily loaded lane as shown in Figure 2-15. In this case the most heavily loaded lane cannot occur in two lanes at the same point along the bridge.

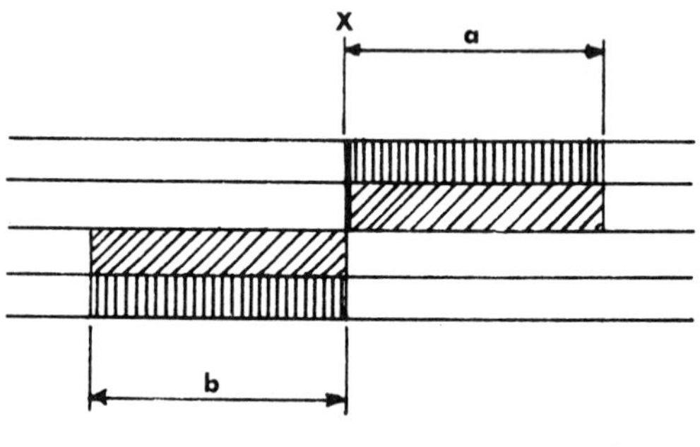

Lane producing maximum effect (basic lane load)

Lane producing next greatest effect (0.7 × basic lane load)

concentrated load (P in outside lane, 0.7 P in next lane)

Loaded length = a + b

FIGURE 2-15 Lane loading to produce maximum torque at point *X*.

The foregoing guidelines are not to be used in the design of deck slabs, floor beams, stringers, or other members for which the standard truck or lane load governs.

Impact and Dynamic Effects

Impact is still estimated according to Article 3.8 of AASHTO, but the tendency is to introduce the more descriptive term *dynamic allowance for traffic loadings*. AASHTO has indicated the need for further studies for both straight and curved girders, and has included an extensive survey of relevant theory for review and consideration.

Among the many factors influencing the dynamic response of bridges is the condition of the bridge approach and the springing of the vehicle. Extremely large impacts can be induced in a bridge due to initial vibration of the moving load if an area of settlement occurs in the roadway in advance of the bridge. For unsprung loads the most important variable is the velocity of the load. If the mass of a load changes, the effect is a change in velocity at which the maximum impact occurs. The change in the maximum impact caused by a change in the mass of the load is insignificant for single-span bridges, but can be appreciable for continuous spans.

Measured impacts for short spans tend to be higher than those predicted by the specifications, but for longer spans observed impacts are nearly the same as those estimated according to the specifications. Impact and dynamic effects are reviewed in greater detail in other sections. It appears, however, that for optimum design the dynamic allowance should be simple and preferably independent of vehicle and bridge variables. Furthermore, it should be applicable to (a) continuous, cantilever, and simple spans; (b) curved and straight bridges; (c) concrete and steel structures; (d) load factor and working stress design; (e) all ranges of design live loads; (f) all load effects such as shear, moment, and torsion; and (g) the entire range of bridge vibration frequencies.

Fatigue

The probability of failure of steel members and their connections under service loads is analyzed in conjunction with a specified range of allowable stresses for fatigue. The expected distribution of truck traffic is represented by a fatigue design truck that is the same as the HS 20 truck. Its gross weight, however, may be selected so that the number of cycles to failure for the fatigue design truck is the same as the total number of cycles to failure for the different trucks in the distribution. If specific information is available on the expected distribution of truck traffic, the gross weight W_F is determined from

$$W_F = \left(\sum \alpha_i W_i^3\right)^{1/3} \tag{2-12}$$

where α_i = the fraction of trucks with weight W_i. If specific information is not available, W_F may be taken as 50 kips. The truck loading for fatigue design should be placed in the lane that gives the greatest stress range for the effect considered while the other lanes are assumed unloaded (see also the applicable sections in Chapter 12).

The number of cycles of maximum stress range to be considered in the design is selected from AASHTO Table 10.3.2A unless traffic and loadometer surveys provide more representative criteria. Fatigue effects and the response of steel members to fatigue are reviewed in more detail in Chapter 12.

Earthquake Effects

Guide specifications for seismic design of bridges were first published by AASHTO in 1983. The current document includes revisions incorporated in 1985, 1987, and 1988. It is considered comprehensive in nature, and introduces several new concepts that mark a significant departure from previous provisions. This document is now standard specifications.

Probably a major turning point in the development of seismic design criteria was the 1971 San Fernando earthquake in California. Until that time, only minor bridge damage was caused directly by vibration effects (Gates, 1976). Previous failures observed worldwide (Iwasaki, Penzien, and Clough, 1972) involved (a) tilting, settlement, and overturning of substructures; (b) displacement at supports and anchor bolt breakage; and (c) settlement of approach fills and wingwall damage. Following the 1971 San Fernando event, significant vibration effects on bridges were observed (Fung et al., 1971), resulting from very large vertical and horizontal ground accelerations possibly exceeding $0.5g$ (Scott, 1973).

The major damage in San Fernando, particularly vibration effects, was concentrated within the narrow region close to and possibly within the causative fault zone. Some bridges withstood the extreme ground shaking with negligible to moderate damage and were able to continue to carry traffic almost without interruption. Where collapse-type failures occurred, deficiencies in details, particularly at connections, were credited with a major role in bridge response.

Current seismic procedures for bridge design (sizing of individual members, connections, and supports) are based on internal forces derived by modifying the results from a linear elastic analysis. The intent is that the columns may yield during an earthquake but that damage to connections and foundations is minimized. For a rational approach, the design criteria must consider site-dependent characteristics and the vibrational characteristics of the structure, and must also incorporate improved details for all bridges subject to seismic activity. In addition, attention has been given to upgrading the earthquake resistance of existing bridges. Earthquake loads and effects are also discussed in the following sections in the context of existing specifications.

Longitudinal (Braking) Forces

There is indication that the longitudinal (braking) force that can be applied by one vehicle is at least equal to the weight of the vehicle (Transportation Research Board, 1975), although it is argued that the design truck may not exert 100 percent friction because of restraints. Because it is unlikely that a large number of vehicles will exert the maximum braking force simultaneously, reduction in load intensity for multiple lanes is justified.

It is worth noting that the 5 percent longitudinal load required by AASHTO is far less than the same type of load specified by other codes. For example, a comparison of longitudinal loadings for one lane 100 ft long gives the following results.

1. AASHTO: 4.1 kips (18.2 kN) for HS 20
2. British code: 100 kips (444.8 kN) for HA
3. Canadian: 101 kips (449.2 kN) for MS 250
4. French: 66 kips (293.6 kN) for type B
5. Ontario: 23.6 kips (105 kN) for OHBD truck
6. ASCE: 57.6 kips (256.2 kN) for HS 20

The foregoing data, however, do not provide an accurate comparison of load effects because some of these loads are applied in combinations of groups, allowing a stress increase.

Friction

AASHTO stipulates that provisions should be made to transfer the forces from the superstructure to the substructure so as to reflect the effect of friction at expansion bearings or shear resistance at elastomeric bearings. The presence of a longitudinal force at expansion bearings often causes debate and may lead to inconsistencies in the design. McDermott (1978) describes the results of model tests carried out to determine the distribution of longitudinal forces to fixed and expansion bearings. The report assesses the experimental variables, analyzes the test data, and presents a theory for longitudinal force distribution. Because of inaccuracies in the calibration system and the variability of friction coefficients, a design procedure could not be experimentally verified (see also the following sections).

Friction coefficients for sliding-type expansion bearings vary considerably. For example, Missouri, New York, and Iowa specify friction forces as 0.14, 0.15, and 0.25 of the dead load reaction, respectively, for steel bearing on steel. Recommendations on appropriate friction coefficients are predicted on

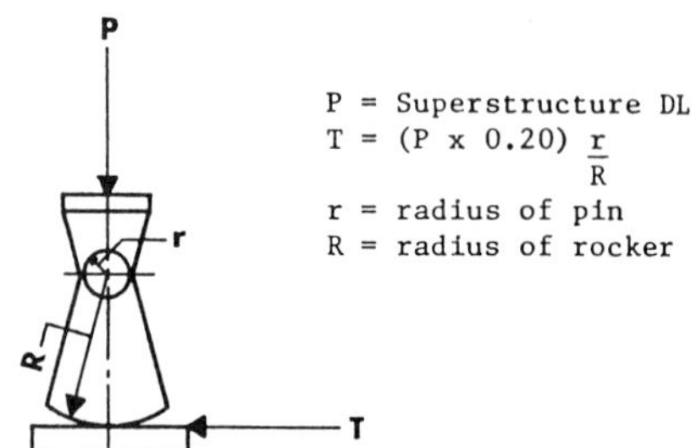

FIGURE 2-16 Forces on rocker bearings.

good maintenance procedures and inspection. Rusted or frozen bearings will develop a higher frictional resistance.

For sliding-type bearings, the longitudinal force due to friction recommended by ASCE (1981) should be based on the following fraction of the dead load reaction:

Steel bearing on steel	0.2
Steel bearing on self-lubricating bronze plate	0.1
Polytetrafluorethylene (PTFE) on same or stainless steel	0.06
Elastomeric bearings	Shear as per AASHTO

For rocker-type bearings, this force should be estimated as 20 percent of the dead load on the pin, and then reduced in proportion to the radii of the pin and rocker as shown in Figure 2-16.

Temperature

Consideration of thermal effects concentrates on the principal uncertainties regarding temperature gradients through the deck–girder system, contribution of thermal factors to overall stresses, effect of restraint on thermal movement induced by friction or shear at the expansion device, and flexibility of substructure frames to expand in relation to their foundation portions. Considerable work has been done in areas related to thermal effects on bridges (see also Chapter 12), but a rational procedure to analyze and quantify these effects is yet to come.

It has been demonstrated that nonlinear temperature gradients can develop during the daily heating-to-cooling cycle, leading to thermal and continuity stresses several times the live-load-induced stresses. This is particularly true for continuous and composite system. Factors impacting on

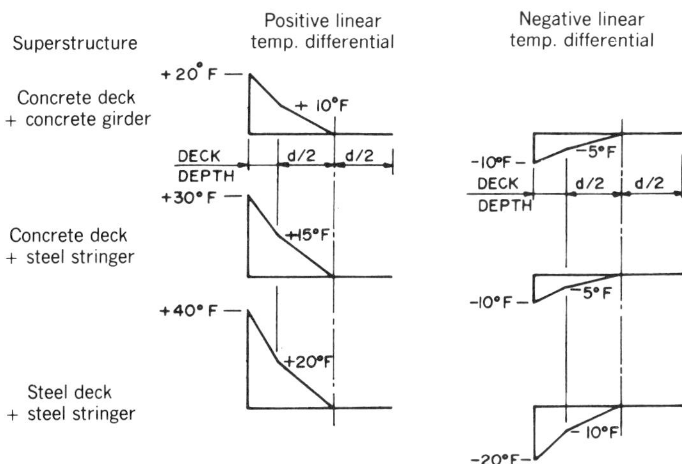

FIGURE 2-17 Temperature ranges and distribution along deck–girder depths. (From ASCE Task Committee, 1981.)

thermal stresses are the air temperature, radiation, wind speed, and thermal characteristics of materials.

The rise and fall in temperature is fixed by the locality in which the bridge is to be constructed, and is computed from an assumed temperature at the time of erection. Although there are no specific recommendations, most designs are based on the assumption that the erection temperature is 65°F, to be adjusted depending on bridge site location.

AASHTO (Article 3.16) provides a general range of temperature changes for metal and concrete structures and for cold and moderate climates. A probable temperature range and distribution along deck–girder depths is shown in Figure 2-17, and is applicable to concrete and steel superstructures and combinations of these materials. Thermal effects are also discussed by AASHTO guide specifications (1989) and in Chapter 12.

Wind Forces

The wind loads specified in Article 3.15 (AASHTO) are for wind velocity of 100 mph. For Group II and V loadings, they may be reduced or increased in the ratio of the square of the design wind velocity to the square of the base wind velocity. Alternatively, the horizontal wind load on the superstructure may be estimated from

$$W_h = \frac{z^{0.2} V_{30}^2 C_h}{600} \tag{2-13}$$

where W_h = horizontal wind load (lb/ft^2) of exposed area

z = height (ft) of the top of the floor system above ground or water level, but not less than 30 ft

V_{30} = fastest mile wind speed (mph) at the 30-ft height with a mean recurrence interval of 100 years

C_h = shape factor for horizontal wind load

The parameter V_{30} is obtained from local wind data. The shape factor C_h is 1.5 or greater for plate or box girder bridges, and 2.3 or greater for truss bridges unless wind-tunnel data indicate a lower value.

Likewise, the vertical wind load on the superstructure may be calculated from

$$W_v = \frac{z^{0.2} V_{30}^2 C_v}{600} \tag{2-14}$$

where W_v is the vertical wind load (lb/ft^2) of plan area, z and V_{30} are as before, and C_v is the shape factor for vertical loads (usually 1.0).

The forces transmitted to the substructure by the superstructure may be calculated for a wind at right angles to the bridge and for a range of wind angles skewed off from this direction. The division into horizontal and vertical components can be as shown in Table 2-4, and the horizontal and vertical loads are those given by (2-13) and (2-14). The percentage of base load to be applied for a given skew angle is the same as that given in AASHTO tables (Article 3.15.2.1). The loads given previously are subject to reduction for appropriate group loadings.

For the standard girder and slab bridges with maximum spans of 125 ft, most designers have found it practical and entirely acceptable to use the simplified wind load analysis described in Article 3.15.2.1.3 of AASHTO.

TABLE 2-4 Skew-Angle Factor for Loads on Superstructure (From ASCE Task Committee, 1981)

	Truss Spans		Girder Spans	
Skew Angle (1)	Transverse Horizontal or Vertical (2)	Longitudinal Horizontal (3)	Transverse Horizontal or Vertical (4)	Longitudinal Horizontal (5)
0	1.00	0.00	1.00	0.00
15	0.93	0.16	0.88	0.12
30	0.87	0.37	0.82	0.24
45	0.63	0.55	0.66	0.32
60	0.33	0.67	0.34	0.38

Long Spans and Flexible Bridges Bridges with moderate-to-medium spans (< 400 ft) are unlikely to sustain damage from wind-induced vibrations because of their low flexibility. Longer bridges are, however, flexible and they may be prone to wind-induced vibrations. Cable-stayed and suspension bridges, in particular, exhibit dynamic response under wind. A common problem of bluff-section bridge decks is vibrations induced by vortex shedding, flutter, and the action of wind gustiness. In addition, similar problems may arise in individual structural members or in freestanding towers during erection. In such cases, a wind-tunnel investigation and a detailed analysis of the dynamic response is indicated (see also Sections 9-14 and 9-21).

Snow and Ice

The effect of snow is usually compensated by a corresponding reduction in live load. In the United States, snow loading may be neglected in areas below 2000 ft east of 105° west longitude, and in areas below 1000 ft west of 105° west longitude. Effects of loadings from avalanches should be considered where appropriate.

Ice Forces generated from floating ice on piers should be determined by considering the site characteristics and the condition of the ice. Ice effects may be (a) dynamic action of ice sheets or floes carried by stream flow, wind, or currents; (b) static action due to thermal movement of continuous ice sheets; (c) static action of ice jams; and (d) vertical action resulting from adhering ice sheets in fluctuating water levels. AASHTO quantifies these effects in Article 3.18.2.

Overturning Forces

AASHTO requires the effect of forces tending to overturn bridges to be calculated for Groups II, III, V, and VI, assuming the wind direction is at a right angle to the longitudinal axis of the structure. In addition, an upward force should be applied at the windward quarter point of the transverse superstructure width. This force is 20 $\mathrm{lb/ft^2}$ of the deck and sidewalk plan for Groups II and V, and 6 $\mathrm{lb/ft^2}$ for Group III and VI combinations.

Comments on Highway Loadings

Live Load The H and HS truck loads, and by analogy the lane loading, have been found to best represent live load effects for bridges in the United States. By extension, certain or all parts of the AASHTO specifications have been introduced or adopted by other countries, particularly where well-planned highway systems are a comparatively recent development.

Examples of the design load requirements in countries other than the United States are summarized in Figure 2-18, and involve both conventional

(a) British HA loading

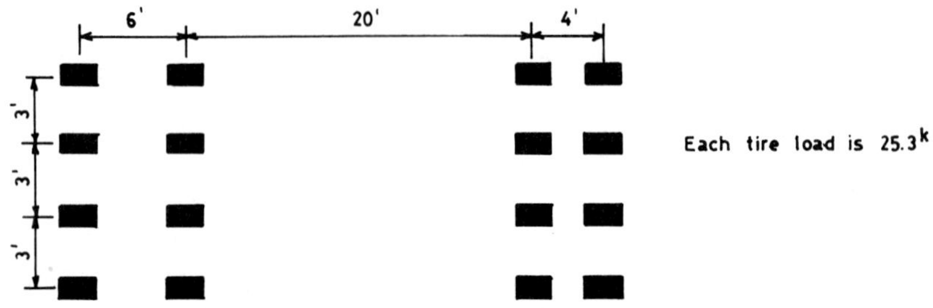

(b) British HB - Abnormal loading

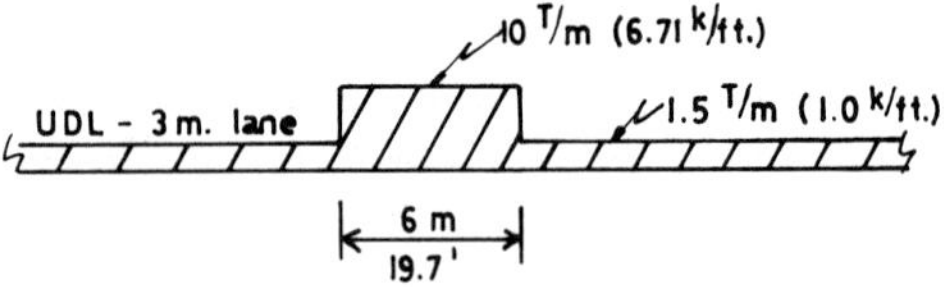

(c) German loading DIN 1078

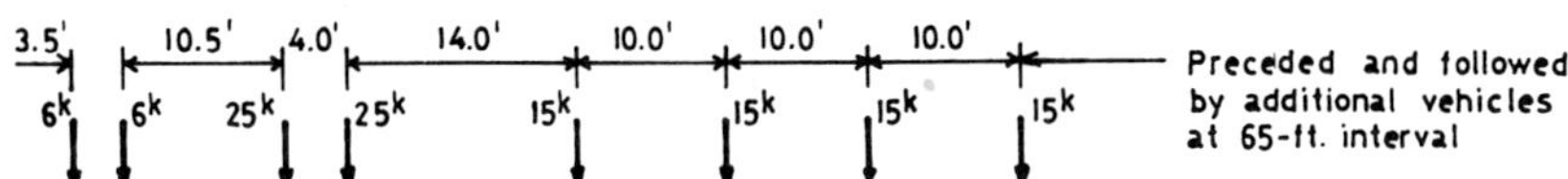

(d) Indian - Class A train of vehicles

Note:

A meaningful comparison would have to include dynamic allowance, multiple lane loading, distribution to structural elements, etc.

FIGURE 2-18 Design highway live loads for various countries.

and military loadings. In certain countries, for example England, the inability of the railroad system to carry exceptionally heavy loads has prompted the adoption of unusually heavy highway loadings.

Whether the AASHTO loads represent the best live load effects is academic. The selection of design vehicles for a highway system requires careful consideration of various factors. Bridges should not be expected to suffer damage as a result of normal traffic or because of an occasionally single overload. On the other hand, bridge obsolescence due to heavy vehicle traffic should be deterred. Thus, the design loads should not exceed the heaviest vehicles expected on the structure during its lifetime. In this respect, the standard AASHTO loadings, combined with overload provisions and the alternate military loading, are sufficient for design purposes and represent a logical exercise of resourceful judgment. Where unusually long spans are involved, the loadings discussed in the foregoing sections should be considered.

Longitudinal Forces Accelerating or braking forces from a moving vehicle are applied in the direction of the bridge axis, and transmitted from the wheels to the deck. The associated effects depend on the acceleration or deceleration, with the maximum force resulting from the sudden application of braking. Given the weight of a vehicle, W (kips), the acceleration of gravity, g (32.2 ft/sec^2), the change in velocity, Δv (ft/sec), and the time interval, Δt (sec), the resulting force is

$$F = \frac{W}{g}\left(\frac{\Delta v}{\Delta t}\right) \tag{2-15}$$

For example, a truck of weight W moving at 60 mph (88 ft/sec) and having its brakes suddenly applied will induce a longitudinal force that is dependent on the time necessary to change the velocity from 60 mph to zero. If this time is taken as 6 sec and the deceleration is assumed uniform, the longitudinal force is

$$F = \frac{W}{32.2} \times \frac{88}{6} = 0.46W$$

If the coefficient of friction between the rubber tires and the dry roadway is assumed to be greater than or equal to 0.5, the entire force can be transmitted to the deck at this level.

Comparatively, the 5 percent provision stipulated in the specifications may result in realistic estimates for long-span bridges, but may underestimate the braking effect on relatively short structures. For example, a 300-ft-long bridge with two lanes in the same direction will have a longitudinal force according to AASHTO of $F = 0.05 \times 2 \times (0.64 \times 300 + 18) = 21.0$ kips. Under the same conditions a 100-ft-long bridge has a force $F = 0.05 \times 2(0.64 \times 100 +$

18) = 8.2 kips. The comparative force from an HS 20 truck suddenly braking may be $F = 0.46 \times 72 = 33$ kips, and for a 300-ft-long bridge the possibility of more vehicles than one truck braking simultaneously should not be excluded.

The variability in the criteria for the longitudinal force is demonstrated by the state specifications. For example, the 1955 Ohio State Highway Department specifications stipulated that the forces due to traction and momentum should be considered as longitudinal forces with a magnitude 10 percent the vertical live load on at least one-half but not more than two-thirds of the lanes, applied at the point of contact between superstructure and substructure.

Reduction in Load Intensity Where maximum stresses are produced in any member by loading two or more lanes simultaneously, the percentage of the resulting live load stresses will be as follows:

Two lanes loaded	100 percent
Three lanes loaded	90 percent
Four or more lanes loaded	75 percent

This reduction reflects the improbability of coincident maximum loading. The reduction in load intensity on transverse members such as floor beams should be determined using the number of traffic lanes across the roadway width that must be loaded to produce maximum stress.

2-6 AASHTO COMBINATIONS OF LOADS

AASHTO groups and combinations of loads are shown in Table 2-5. The groups listed in the appropriate column represent various combinations of loads and forces to which a bridge and its members may be subjected. Each bridge component including its foundation should be proportioned to withstand safely all relevant group combinations that are applicable to the bridge type or site. A typical group loading combination is given by

$$\begin{aligned} \text{Group } (N) = \gamma[&\beta_D \cdot D + \beta_L(L + I) + \beta_C \text{CF} + \beta_E E \\ &+ \beta_B B + \beta_S \text{SF} + \beta_W W + \beta_{\text{WL}} \text{WL} \\ &+ \beta_L \cdot \text{LF} + \beta_R(R + S + T) + \beta_{\text{EQ}} \text{EQ} \\ &+ \beta_{\text{ICE}} \text{ICE}] \end{aligned} \tag{2-16}$$

TABLE 2-5 Group Loadings and Coefficients γ and β (From AASHTO Standard Specifications, 1992)

	Col. No.	1	2	3	3A	4	5	6	7	8	9	10	11	12	13	14	
			β FACTORS														
	GROUP	γ	D	$(L+I)_n$	$(L+I)_p$	CF	E	B	SF	W	WL	LF	R+S+T	EQ	ICE	%	
SERVICE LOAD	I	1.0	1	1	0	1	β_E	1	1	0	0	0	0	0	0	100	
	IA	1.0	1	2	0	0	0	0	0	0	0	0	0	0	0	150	
	IB	1.0	1	0	1	1	β_E	1	1	0	0	0	0	0	0	**	
	II	1.0	1	0	0	0	1	1	1	1	0	0	0	0	0	125	
	III	1.0	1	1	0	1	β_E	1	1	0.3	1	1	0	0	0	125	
	IV	1.0	1	1	0	1	β_E	1	1	0	0	0	1	0	0	125	
	V	1.0	1	0	0	0	1	1	1	1	0	0	1	0	0	140	
	VI	1.0	1	1	0	1	β_E	1	1	0.3	1	1	1	0	0	140	
	VII	1.0	1	0	0	0	1	1	1	0	0	0	0	1	0	133	
	VIII	1.0	1	1	0	1	1	1	1	0	0	0	0	0	1	140	
	IX	1.0	1	0	0	0	1	1	1	1	0	0	0	0	1	150	
	X	1.0	1	1	0	0	β_E	0	0	0	0	0	0	0	0	100	Culvert
LOAD FACTOR DESIGN	I	1.3	β_D	1.67*	0	1.0	β_E	1	1	0	0	0	0	0	0	Not Applicable	
	IA	1.3	β_D	2.20	0	0	0	0	0	0	0	0	0	0	0		
	IB	1.3	β_D	0	1	1.0	β_E	1	1	0	0	0	0	0	0		
	II	1.3	β_D	0	0	0	β_E	1	1	1	0	0	0	0	0		
	III	1.3	β_D	1	0	1	β_E	1	1	0.3	1	1	0	0	0		
	IV	1.3	β_D	1	0	1	β_E	1	1	0	0	0	1	0	0		
	V	1.25	β_D	0	0	0	β_E	1	1	1	0	0	1	0	0		
	VI	1.25	β_D	1	0	1	β_E	1	1	0.3	1	1	1	0	0		
	VII	1.3	β_D	0	0	0	β_E	1	1	0	0	0	0	1	0		
	VIII	1.3	β_D	1	0	1	β_E	1	1	0	0	0	0	0	1		
	IX	1.20	β_D	0	0	0	β_E	1	1	1	0	0	0	0	1		
	X	1.30	1	1.67	0	0	β_E	0	0	0	0	0	0	0	0		Culvert

$(L + I)_n$ - Live load plus impact for AASHTO Highway H or HS loading

$(L + I)_p$ - Live load plus impact consistent with the overload criteria of the operation agency.

* 1.25 may be used for design of outside roadway beam when combination of sidewalk live load as well as traffic live load plus impact governs the design, but the capacity of the section should not be less than required for highway traffic live load only using a beta factor of 1.67. 1.00 may be used for design of deck slab with combination of loads as described in Article 3.24.2.2.

$$** \text{ Percentage} = \frac{\text{Maximum Unit Stress (Operating Rating)}}{\text{Allowable Basic Unit Stress}} \times 100$$

For Service Load Design

% (Column 14) Percentage of Basic Unit Stress

No increase in allowable unit stresses shall be permitted for members or connections carrying wind loads only.

β_E = 1.00 for vertical and lateral loads on all other structures.

For culvert loading specifications, see Article 6.2.

β_E = 1.0 and 0.5 for lateral loads on rigid frames (check both loadings to see which one governs). See Article 3.20.

For Load Factor Design

β_E = 1.3 for lateral earth pressure for retaining walls and rigid frames excluding rigid culverts.

β_E = 0.5 for lateral earth pressure when checking positive moments in rigid frames. This complies with Article 3.20.

β_E = 1.0 for vertical earth pressure

β_D = 0.75 when checking member for minimum axial load and maximum moment or maximum eccentricity ..... For

β_D = 1.0 when checking member for maximum axial load and minimum moment .... Column Design

β_D = 1.0 for flexural and tension members

β_E = 1.0 for Rigid Culverts

β_E = 1.5 for Flexible Culverts

For Group X loading (culverts) the β_E factor shall be applied to vertical and horizontal loads.

where N = group number
γ = load factor, see Table 2-5
β = coefficient, see Table 2-5
D = dead load
L = live load
I = live load impact
E = earth pressure
B = buoyancy
W = wind load on structure
WL = wind load on live load–100 lb/linear foot
LF = longitudinal force from live load
CF = centrifugal force
R = rib shortening
S = shrinkage
T = temperature
EQ = earthquake
SF = stream flow pressure
ICE = ice pressure

For service load design (working stress), the percentage of the basic unit (allowable) stress for the various groups is given in column 14 of Table 2-5.

For load factor design, the γ and β factors given in Table 2-5 are intended basically for designing members by the load factor method. Thus, loads may not be increased by the load factors when designing foundations, piles, and so on. The load factor method may not be applied to the analysis of foundation stability (overturning, sliding, soil overstressing, etc.), except when using the LRFD approach.

If long-span bridges are designed by the load factor method, the γ and β coefficients stipulated for this procedure represent general conditions and may be increased if the expected loads, service conditions, and construction materials are different from those identified in the specifications. Alternatively, structures may be analyzed for an overload selected by the operating agency. Size and configuration of the overload, loading combinations, and load distribution should be consistent with procedures adopted by the agency. This load is applied to Group IB of Table 2-5. For loadings less than H 20, the Group IA loading combination governs and should be used.

2-7 DISTRIBUTION OF LONGITUDINAL FORCES TO BRIDGE BEARINGS

Background and Assumptions

Longitudinal forces at bridge bearings are induced by traction (acceleration and deceleration of the live load), thermal movement along the bridge axis, and wind acting on the structure as well as on the live load. A frictional force may also be considered for rockers, rollers, and sliding-type expansion devices. These forces may be distributed as follows.

1. We may assume that expansion bearings resist only thermal forces, whereas the fixed bearing resists all longitudinal forces associated with wind and traffic loads. This would place expansion piers under Groups IV, V, and VI.

2. Alternatively, we may assume again that expansion bearings resist only thermal forces, but the resulting effect is likely to be released at any expansion pier by sudden live load vibrations. In this context, expansion piers can become inoperable, and the forces released in this manner must now be resisted at the fixed pier.

3. Distribution according to a third approach assumes that expansion bearings will carry thermal forces as well as wind and live loads up to their frictional capacity, the latter computed as the dead load reaction multiplied by a friction coefficient. The expansion bearings are assumed to resist these forces until they become large enough for the bearings to slip. At this point the longitudinal force resisted at the expansion bearing reaches a maximum, and any additional load must be resisted by the fixed pier.

The third approach seems to be followed by most designers, but is subject to the following comments: (a) AASHTO Article 3.2.1 clearly stipulates that provisions must be made to reflect the effect of friction at expansion bearings or shear resistance at elastomeric bearings; (b) treating friction as a resisting rather than as an applied force, it is omitted from group loads and group load application; (c) designing an expansion pier for the full frictional resistance can yield erroneous results unless the actual load allocated at this bearing exceeds the friction (e.g., an erroneous result is an expansion pier made heavier than the fixed pier); and (d) the percentage of stress to be used with friction is not specified and therefore rests with engineering discretion.

4. With massive, stiff abutments at the ends of a structure, many designers prefer to distribute the longitudinal as well as the lateral forces at these locations, and they design the intermediate piers for vertical loads only.

It appears from these remarks that none of the foregoing theories is completely acceptable because of the arbitrary, although convenient, assumptions. The rationalization of a design approach is further complicated by the variability of the friction coefficients. Other factors influencing the distribution of longitudinal forces to bridge bearings include the relative stiffness of substructure elements and their foundations, the overall thermal movement, and the relationship of the total longitudinal force to the maximum resisting force available at expansion bearings. After a slip occurs at a bearing, the remaining devices (fixed bearings and expansion bearings that have not slipped) will resist additional longitudinal effects.

TABLE 2-6 Longitudinal Force Distribution Test (From McDermott, 1978)

Induced Internal Forces	Applied Load	
	Normal Force, N	Horizontal Force, F_L
Strong-Axis Test		
Zero Expansion Contraction	77 to 177 lb	25 to 150 lb
Weak-Axis Test		
Zero Expansion Contraction	57 to 177 lb	50 to 150 lb

Small-Scale Tests

McDermott (1978) has studied the load effects at bridge bearings in small-scale model tests supplemented by a survey of existing two- and three-span bridges. Typically, these bridges had a ratio of design external force to internal resisting force at expansion bearings in the range 0.41 to 1.96. The actual ratio of superstructure-to-substructure stiffness varied from 3 to 248.

For the model tests the study emphasized the effect of the superstructure and substructure stiffness ratio, taken as 1.7 and 20.7 for the strong and weak pier axis, respectively. The induced forces and the range of externally applied loads are given in Table 2-6, and evidently three cases are considered: (a) internal forces zero, (b) maximum expansion, and (c) maximum contraction. The desired range of external forces to maximum resisting forces was obtained by varying the external longitudinal load and the normal load at the expansion bearings. An additional test was carried out with all bearings fixed to determine the load distribution as a function of pier and abutment stiffness.

The model structure is shown in Figures 2-19 and 2-20. The bearing devices consist of steel plates attached to the top of abutments and the fixed pier as shown in the details. The plates are either fixed (no internal force generated) or expansion (allowed to slide). Thermal forces are simulated at the expansion bearings by turning the screw mechanism in detail A until slip occurs. I sections are used in the structural members including the fixed pier where the bending axis is changed to study strong and weak axis effects. The abutments consist of steel pipes, fixed at the base as in the pier.

For the test setup shown in Figure 2-21, a bending moment is induced at the top of the pier corresponding to the force F_L. This moment is determined by the relative stiffness of the beam and column, the maximum resisting force

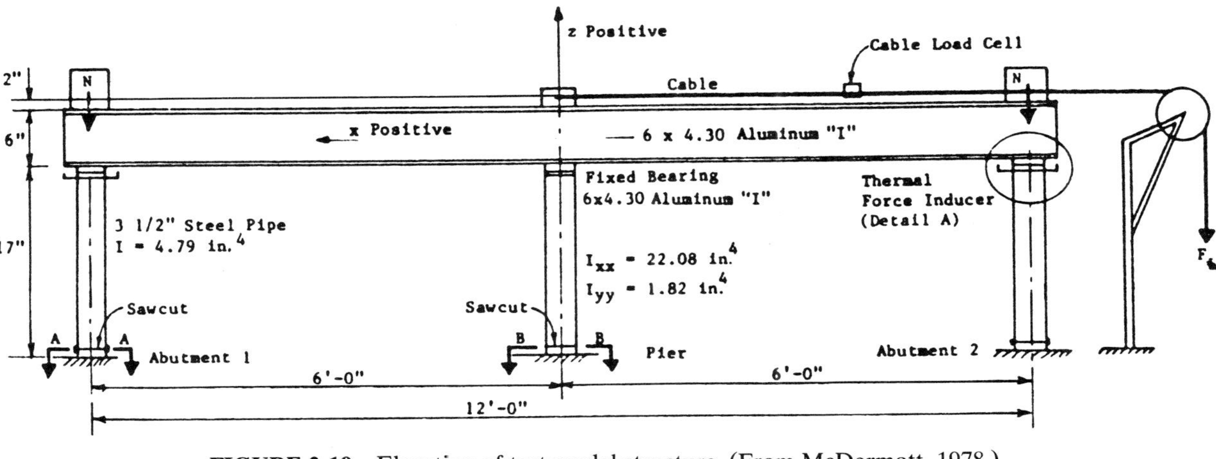

FIGURE 2-19 Elevation of test model structure. (From McDermott, 1978.)

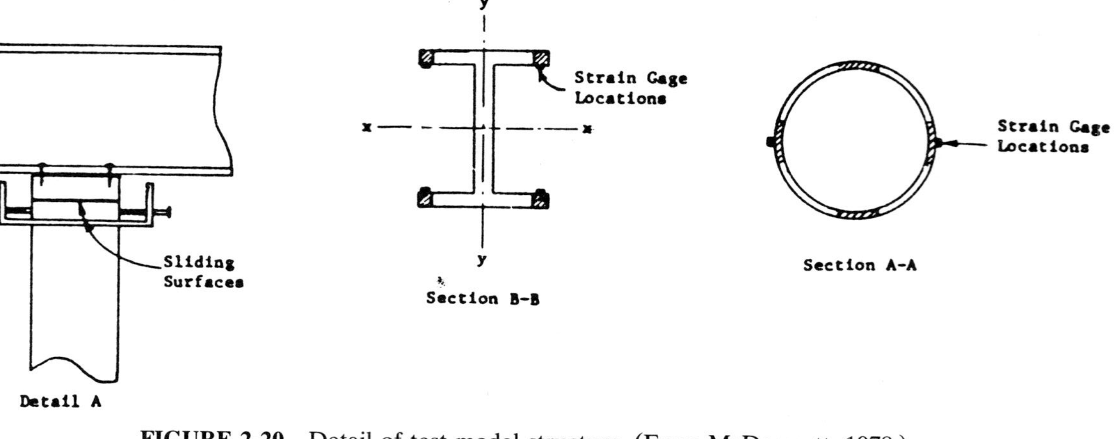

FIGURE 2-20 Detail of test model structure. (From McDermott, 1978.)

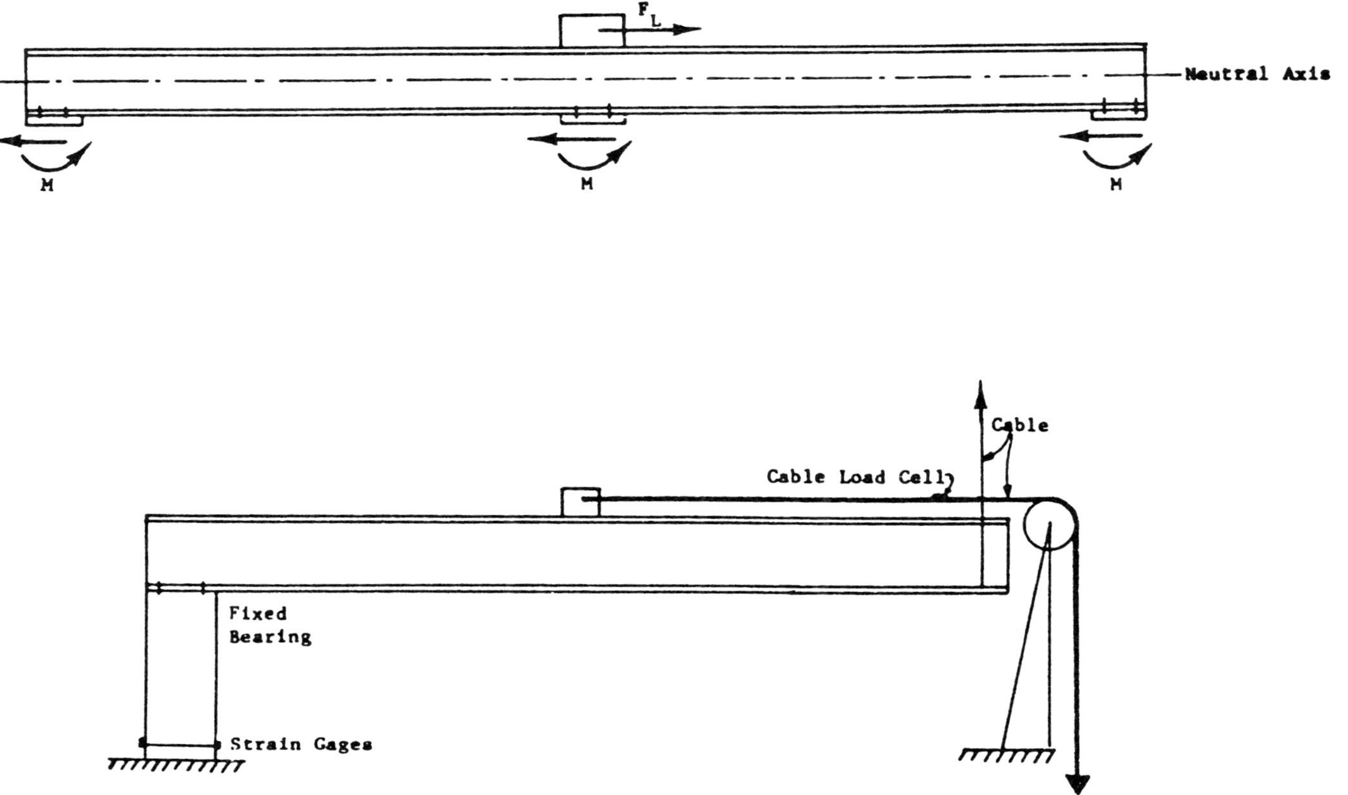

FIGURE 2-21 Simulation of bending moments in test model. (From McDermott, 1978.)

at expansion bearings, and the point of application of force F_L. As a result, the moment at the pier base depends not only on pier height but also on factors affecting the top moment.

Two important features of the test structure are duplicated in the calibration procedure. First, the horizontal load is applied at the same height above the instrumented section as in testing. Second, the stiffness of the horizontal member is included in restraining vertical deflection. Certain calibration errors became apparent, however, in relating strain measurements to the test structure. An equilibrium check between longitudinal loads and internal resisting forces showed that in weak pier axis tests the applied load was consistently less than the sum of the resisting forces, whereas in most strong axis tests it was greater. The difference was as high as 8 percent.

Test Procedure Resistance to horizontal loads was measured at the two abutments and at the center pier. Readings were taken with no loads on the system, after a change in applied force, following the application of vertical loads, horizontal loads, and simulated thermal forces, and after removal of these loads.

In this system a vertical load placed over a bearing simulates dead load beam reaction and induces potential frictional resistance. Simulated thermal forces are introduced to a certain level before a horizontal force is applied. Because of the variability of the friction coefficient, induced forces are not equivalent (linearly proportional) to the normal load. In order to satisfy static equilibrium, some force must be resisted at the fixed pier even if no direct force is applied at that point. After the thermal force is induced, horizontal loads are applied over the center fixed pier 2 in. above the top flange of the beam. Longitudinal force distribution is then tabulated with the internal force and the longitudinal force in place. A partial summary of the longitudinal force distribution is given in Table 2-7.

TABLE 2-7 Partial Summary of Longitudinal Force Distribution (From McDermott, 1978)

Test	Vertical Load N (lb)	Applied Longitudinal Load F_h (lb)	Induced Lateral Force			Longitudinal Force Distribution		
			Abut. 1	Pier	Abut. 2	Abut. 1	Pier	Abut. 2
1A	77	−141.6	0.0	0.0	0.0	3.7	116.6	12.0
5A	127	−49.1	0.0	0.0	0.0	15.1	21.4	11.9
7A	77	−94.1	−24.8	2.7	16.7	9.6	60.5	17.3
$9A_E$	127	−61.5	−46.8	8.3	31.8	−23.3	48.9	32.4
2BB	127	−88.6	0.0	0.0	0.0	42.5	14.9	41.0
$6BB_C$	77	−90.3	16.6	8.0	−24.3	20.6	32.0	42.9
10BB	97	−91.1	0.0	0.0	0.0	44.6	16.1	40.9

TABLE 2-8 Apparent Coefficients of Friction (From McDermott, 1978)

Test	Friction Coefficient when Slip Occurs[a]	
	Abutment 1	Abutment 2
$7A_E$	0.32	0.22
	0.28	0.24
	0.31	0.26
$7A_C$	0.12	0.19
	0.03	0.24
	0.02	0.11
$8A_E$	0.28	0.28
	0.31	0.23
	0.34	0.25
$8A_C$	0.15	0.20
	0.16	0.24
	0.18	0.24
$9A_E$	0.37	0.25
	0.31	0.26
	0.30	0.21
$9A_C$	0.19	0.27
	0.13	0.07
	0.11	0.29
$6BB_E$	0.45	0.55
	0.44	0.56
	0.40	0.61
$6BB_C$	0.22	0.32
	0.39	0.42
	0.30	0.37
$7BB_E$	0.41	0.46
	0.41	0.42
	0.47	0.42
$7BB_C$	0.36	0.37
	0.51	0.44
	0.40	0.33
$8BB_E$	0.33	0.48
	0.44	0.42
	0.36	0.44
$8BB_C$	0.20	0.40
	0.37	0.42
	0.32	0.31
$11BB_E$	0.35	0.41
	0.48	0.68
	0.53	0.67
$11BB_C$	0.34	0.54
	0.35	0.42
	0.33	0.54

[a]Friction coefficient μ = Induced thermal force/N.

Variability of Friction Coefficients Friction coefficients are calculated by dividing the induced (simulated) thermal force by the normal load over the bearing, and are given in Table 2-8.

The wide range of variation is demonstrated by a minimum friction coefficient of 0.02 and a maximum friction coefficient of 0.68, and is not explained by conventional theories of frictional behavior. According to the test format, internal forces were induced until slip occurred, and at this stage the friction force was recovered and assumed to have its maximum value. Although no increase should be expected, it was observed at abutment 1 at compression and at abutment 2 at expansion after applying the longitudinal load. Because the normal load did not increase, this can be explained by a variation in the friction coefficient.

In practice, the performance of bearings can be affected by the presence of dust, humidity, rust, and oxide films, temperature change, surface films, and the extent of contamination. In theory, these factors cannot be singly or jointly considered in quantitative terms, and care was taken to eliminate these effects from the tests. McDermott (1978) suggests that the average friction coefficient for steel on steel in these tests should be close to 0.4, but some discrepancies might have resulted from inaccuracies in measurement.

Theoretical Models

A suggested theoretical model is presented in Figure 2-22, and shows how the frictional forces at expansion bearings affect the distribution of forces. The typical forces used on the test system are the applied forces, longitudinal F_L and normal N, resisted at the fixed and expansion bearings. The maximum resisting force at the expansion bearing is μN, where μ is the coefficient of friction. If μ can be assumed constant, the total frictional resistance of the system is $\mu \Sigma N$. Whenever the force is known at the expansion bearings, the force at the fixed bearing is estimated as the difference $F_L - \mu \Sigma N$. This analysis should not mean, however, that the force

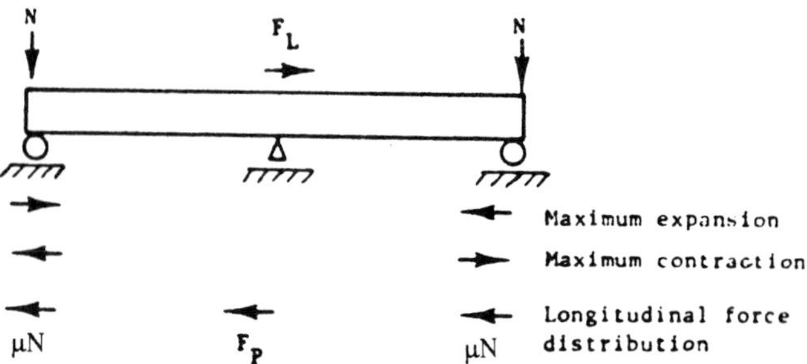

FIGURE 2-22 Magnitude and direction of longitudinal resisting forces on the test structure. (From McDermott, 1978.)

at the fixed bearing is zero until the difference $F_L - \mu\Sigma N$ becomes positive, because superstructure and substructure stiffness affects the distribution of force F_L before expansion bearings reach their maximum. When forces at expansion bearings are not sufficient to cause slip, a predictable percentage of longitudinal force is resisted at the fixed bearing. This is a function of the ratio of pier-to-abutment stiffness, and the force resisted at the pier increases

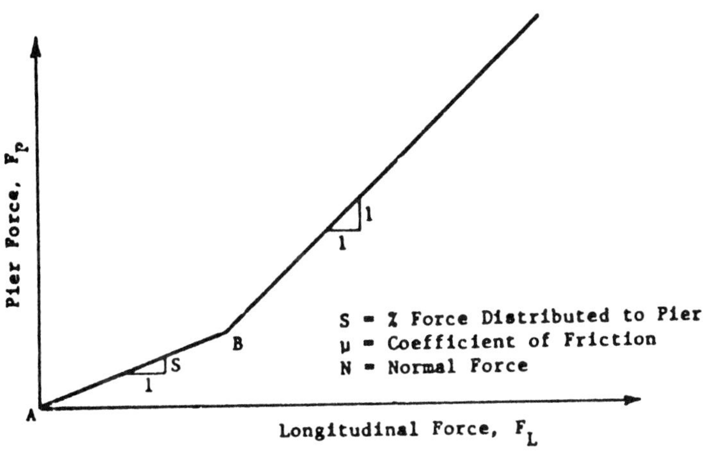

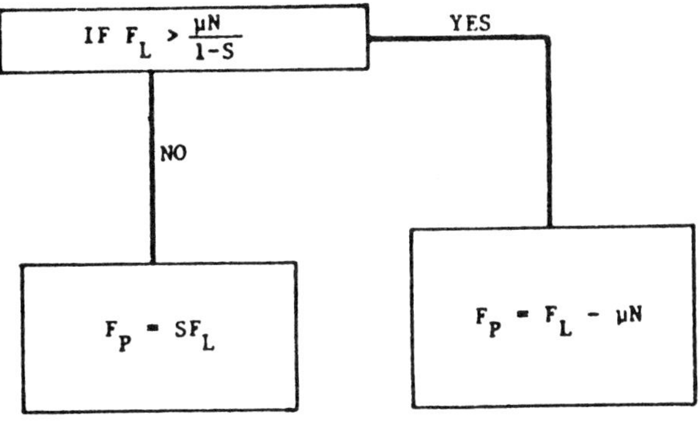

FIGURE 2-23 Theoretical distribution of longitudinal force. (From McDermott, 1978.)

as this ratio increases. If the force reaches a maximum at expansion bearings and slip occurs, additional longitudinal forces are resisted at the fixed pier.

The preceding discussion highlights the behavior of the test bridge, in this case a two-span symmetrical structure. For three or more spans, the fixed pier may not necessarily coincide with the point of zero movement under thermal expansion and contraction. In this case the analysis must differentiate between actual applied loads, such as wind force and traction, and forces manifested as movement, such as temperature changes.

The theoretical distribution for the test problem is shown in Figure 2-23. Between points A and B, the distribution is essentially controlled by the relative stiffness of each substructure element, and a percentage S of the longitudinal force is always resisted at the fixed bearing. The slope of line AB is influenced by forces initially present in the system. If the maximum thermal force is induced (just before slip occurs), the resisting force at the expansion bearing is μN, but its direction depends on temperature change, expansion, or contraction. Because one bearing is in the same direction as the longitudinal resisting forces, it has no resisting capacity under any additional load, which must now be resisted by the fixed and other expansion bearings. These elements will receive the additional load in proportion to their relative stiffness. As the longitudinal force continues to increase, the resisting capacity manifested as friction will be depleted in the direction of load application, and at this point any additional load is resisted by the fixed pier only. This behavioral model does not fully explain the effects of thermal expansion in real structures unless the analysis considers the point of zero movement. Furthermore, live load vibrations may release the thermal forces.

Conclusions

Figure 2-24 shows the percentage of force carried by the fixed pier, and apparently a relatively constant fraction of the applied load is distributed to the fixed pier. A consistent percentage of the longitudinal force is resisted by the pier until slip occurs at the expansion bearings.

Prediction, however, of the distribution beyond point B is inconsistent because of the variability of the friction coefficients, inhibiting conclusions as to when bearing friction forces reach a maximum. Figure 2-25 shows scattered data from the test for normal-force tests for both the weak and strong axis versus the theoretical distribution lines. Two relationships are plotted: (a) maximum initial internal force, where one abutment is effective in resisting longitudinal loads; and (b) zero internal force, where both abutments are effective. The theoretical line is plotted with friction coefficients that represent average experimental data. The variability of the friction coefficients may explain why the data do not coincide with the plot. Because the two lines intersect in a fashion that depends on μ, accurate predictions should not be attempted.

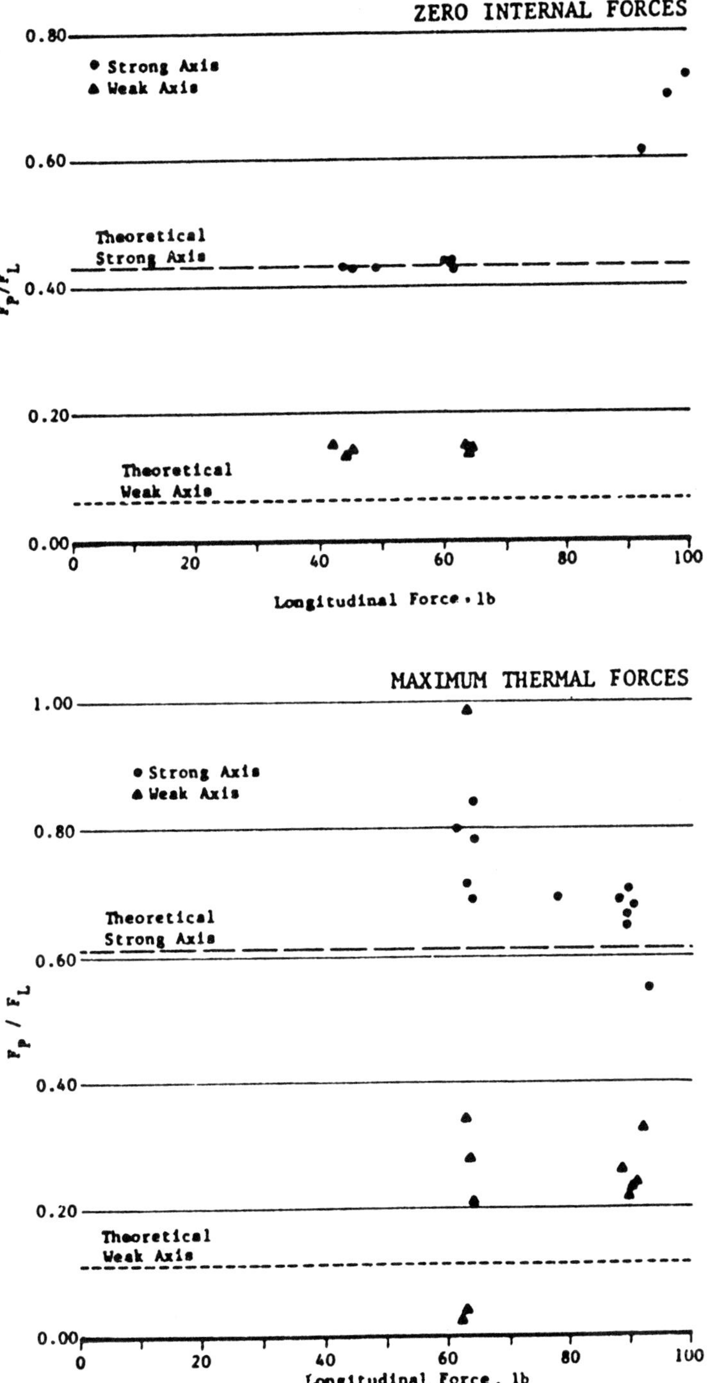

FIGURE 2-24 Force distribution to fixed pier. (From McDermott, 1978.)

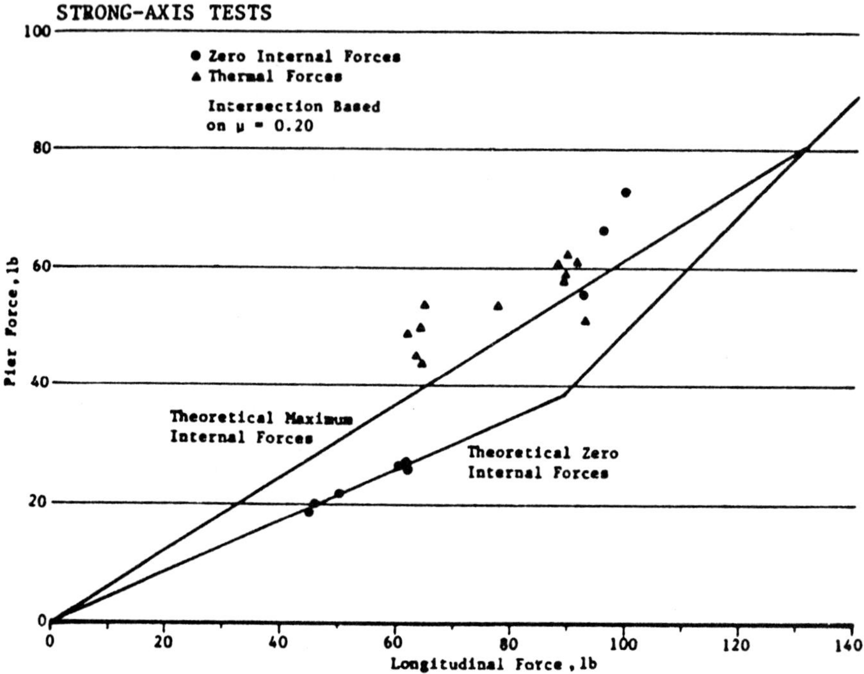

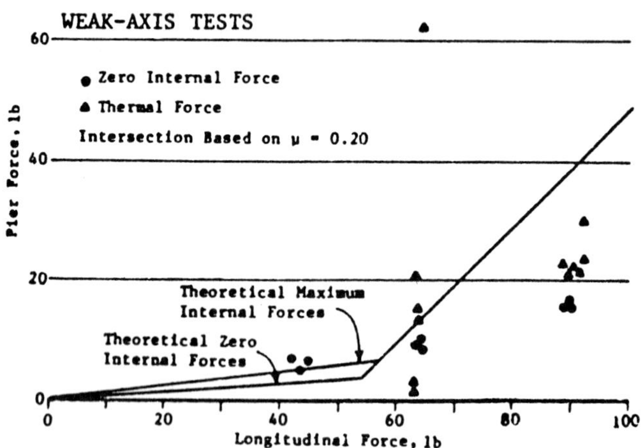

FIGURE 2-25 Fixed pier force; 127-lb normal force. (From McDermott, 1978.)

Design Remarks For design purposes, it is assumed that a bilinear relationship exists between the distribution of force at the fixed pier and the applied longitudinal load, represented by the diagram of Figure 2-23. As the expansion bearings begin to slip, any additional longitudinal load is carried at the fixed bearing.

For real structures, abutments have considerable stiffness compared to the piers. It is very likely that initially a high percentage of force will be resisted at these locations before slip occurs, and then any additional force will be transferred to the fixed pier.

A bilinear relationship between pier force and longitudinal load may not always exist in a real structure. This is possible essentially in bridges of equal spans. For bridges with unequal spans, the dead load reactions will be different, changing the frictional capacity at the expansion bearings. In addition, thermal forces are different between spans, and the associated movement will determine when slip occurs. If slip occurs at one location before the other, more longitudinal load may be resisted by the other expansion bearings and the fixed pier. If slip occurs simultaneously at all expansion locations, any new loads must be resisted by the fixed pier. The possibility of friction force release should be considered when designing the fixed pier.

2-8 APPLICATION OF LOADS: NUMERICAL EXAMPLE

Fixed Pier

The bridge shown in Figures 2-26*a* and *b* is a four-span continuous steel plate girder superstructure with concrete deck. The span lengths are 50 (15.2), 72 (21.9), 72 (21.9), and 51 (15.5), ft (m), giving approximately a total bridge length of 250 ft (76 m). From an analysis of the superstructure, the dead and live load reactions have been determined at each substructure element. The application of loads will be demonstrated for fixed pier 2. The special features of this bridge are reviewed and discussed in other sections (see Chapter 5). A basic feature of this structure is that the longitudinal deck beams frame into transverse box girders at the piers. The associated procedures for shop assembly resulted in the use of numerically controlled equipment to handle fabrication details.

The transverse box girders at the piers are supported by two individual concrete columns on square spread footing transferring the loads to underlying rock. Special bearing devices allow rotation of the superstructure elements in both the longitudinal and transverse direction. The basic design philosophy was to design the fixed pier for all longitudinal forces plus a corresponding portion of the transverse loads. A basic pier plan is shown in Figure 2-26*c*. The live load is HS 20.

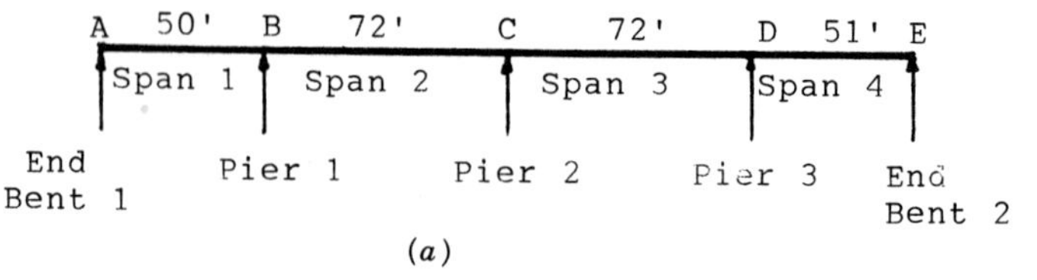

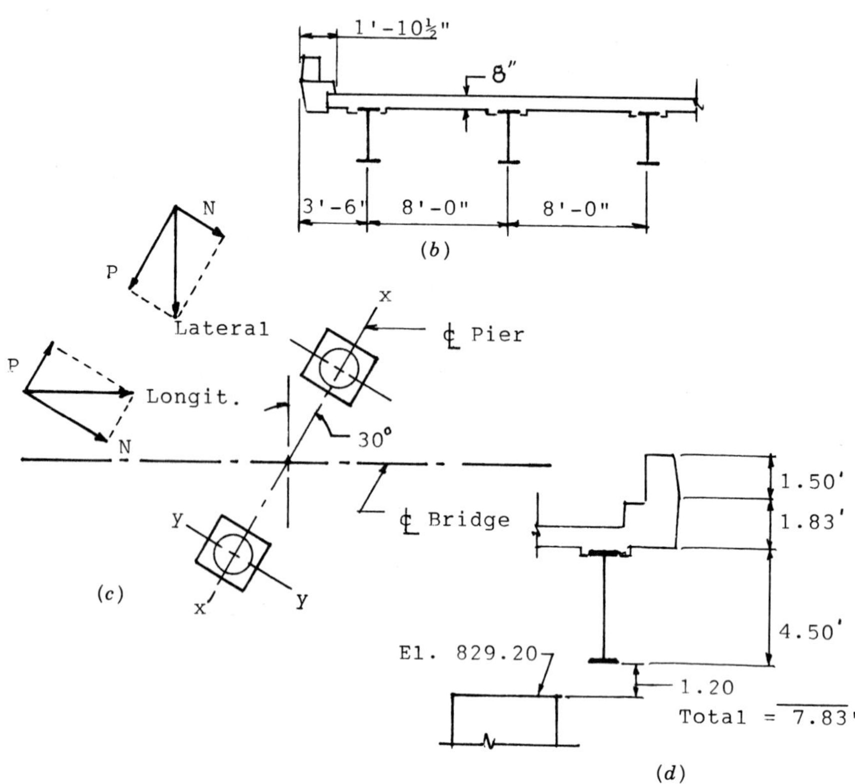

FIGURE 2-26 Part of I-75–I-275 four-level interchange: (*a*) I-75 bridge elevation; (*b*) partial deck cross section; (*c*) fixed pier plan; (*d*) fascia detail.

1. *Dead Load Reactions* The following results were obtained from superstructure analysis:

Interior girders (5) = 110 × 5	=	550 kips
Exterior girders (2) = 99 × 2	=	198 kips
Total DL (superstructure)	=	748 kips
Weight of box girder	=	56 kips
Total DL	=	804 kips or 402 kips/column

2. *Live Load Reactions* Again, from superstructure analysis we obtain directly

$$\text{Total LL} = 154 \times 0.9 = 139 \text{ kips/column}$$

where the coefficient 0.9 is applied to three lanes loaded.

Note that the pier is skewed at 30°, as shown in Figure 2-26*c*. Hence, all loads (longitudinal and transverse) are resolved into two components parallel and normal to the pier axis, P and N, respectively. For convenience, we write $\sin 30° = 0.500$, and $\cos 30° = 0.866$.

3. *Wind on Structure* The exposed superstructure, as seen from an elevation parallel to the bridge, is shown in Figure 2-26*d* in relation to the pier top. According to relevant standards, the bridge is designed for a wind velocity of 84 mph, so that the wind intensity reduction coefficient is $84^2/100^2 = 0.71$. Both transverse and longitudinal winds are applied simultaneously at the elevation of the center of gravity of the exposed area of the superstructure. Because the maximum span lengths are less than 125 ft, Article 3.15.2.1.3 of AASHTO is applicable.

We now estimate the wind forces on the structure as follows:

		N	P
Lateral	$= 75 \times 7.83 \times 0.050 \times 0.71 = 20.8$ kips	10.4	18.1
Longitudinal	$= 250 \times 7.83 \times 0.012 \times 0.71 = 16.7$ kips	± 14.5	∓ 8.4

Keeping the lateral wind direction the same and reversing the longitudinal wind direction, we obtain the following N, P combinations:

(a) $N = 10.4 + 14.5 = 24.9$ kips

$P = 18.1 - 8.4 = 9.7$ kips $\qquad \sqrt{24.9^2 + 9.7^2} = 27$ kips

(b) $N = 14.5 - 10.4 = 4.1$ kips

$P = 18.1 + 8.4 = 26.5$ kips $\qquad \sqrt{26.5^2 + 4.1^2} = 27$ kips

suggesting that for optimum design the footing must have a square shape.

4. *Wind on Live Load* Likewise, the wind on live load is computed using the following simplified method:

	N	P
Lateral $= 75 \times 0.10 = 7.5$ kips	3.8	6.5
Longitudinal $= 250 \times 0.04 = 10.0$ kips	± 8.7	∓ 5.0
Total used	$N = 12.5$ kips	$P = 1.5$ kips

5. *Wind on Pier (One Column)* Assume a column height of 23 ft and a column diameter of 4 ft 6 in.:

	N	P
Lateral $= 23 \times 4.5 \times 0.04 = 4.1$ kips	2.1	3.6

acting at an elevation of 817.7.

6. *Longitudinal Forces (One Lane)*

	N	P
One lane $= 0.05(0.64 \times 250 + 18) = 8.9$ kips	7.7	4.5

acting 6 ft above the top of the slab.

7. *Temperature Forces* Assume that both bearings are fixed in the direction of the pier axis and that the differential expansion temperature is 60°F. If the steel box girder expands and contracts in the direction of the pier, the associated elongation or shortening will induce a corresponding force at the top of the columns. For a distance c. to c. of column of 46 ft, the total expansion per column is $0.5 \times 46 \times 12 \times 60 \times 6.5 \times 10^{-6} = 0.108$ in. Ignoring the reinforcing steel, the moment of inertia of the column is $I = \pi d^4/64 = 417{,}680$ in.^{4}, and the force resulting from this movement is $F = 3EI\Delta/l^3 = 19.3$ kips acting at the top of the column.

Footing Design Other loads to be considered are the weight of the column, 55 kips; the weight of the footing, 45 kips (assumed 10 ft × 10 ft × 3 ft); and the weight of earth on top of the footing, calculated as 30 kips. The bottom of the footing elevation is taken as 803.20, giving a pier height of 26 ft.

Group I $= D + L$ (CF $= E = 0$)

$$
\begin{aligned}
&\text{Dead load} = 402 + 55 + 45 + 30 &&= 532 \text{ kips}\\
&\text{Live load (three lanes)} &&= \underline{139 \text{ kips}}\\
&\qquad\text{Total} &&= 671 \text{ kips/footing}
\end{aligned}
$$

Group II $= D + W$ (CF $= E = B =$ SF $= 0$)

$$
\begin{aligned}
&\text{Wind on structure: From } N = 24.9, && M_{x-x} = 0.5 \times 24.9 \times 31.10 && = 387 \text{ ft-kips}\\
&\qquad\qquad\qquad\quad\;\; \text{From } P = 9.7, && M_{y-y} = 0.5 \times 9.7 \times 31.10 && = 151 \text{ ft-kips}\\
&\text{Wind on column: From } N = 2.1, && M_{x-x} = 2.1 \times 14.5 && = 30 \text{ ft-kips}\\
&\qquad\qquad\qquad\quad\;\; \text{From } P = 3.6, && M_{y-y} = 3.6 \times 14.5 && = 52 \text{ ft-kips}
\end{aligned}
$$

Summary, Group II (one footing): $F = 532$ kips, $M_{x-x} = 417$ ft-kips, and $M_{y-y} = 203$ ft-kips.

Group III $= D + L + \mathrm{LF} + 30\%W + \mathrm{WL}$ $(\mathrm{CF} = E = B = \mathrm{SF} = 0)$

Dead load + Live load: $F = 671$ kips

Longitudinal forces: From $N = 7.7$, $M_{x-x} = 2.7 \times 7.7 \times 0.5 \times 37.8 = 393$ ft-kips

(Three lanes) From $P = 4.5$, $M_{y-y} = 2.7 \times 4.5 \times 0.5 \times 37.8 = 230$ ft-kips

Wind on live load: From $N = 12.5$, $M_{x-x} = 12.5 \times 0.5 \times 37.8 = 236$ ft-kips

From $P = 1.5$, $M_{y-y} = 1.5 \times 0.5 \times 37.8 = 28$ ft-kips

Wind on Structure: For Group III, the wind should not be reduced. Thus, the loads estimated for Group II are converted by multiplying by the ratio $1.0/0.71 = 1.41$ or $N = 24.9 \times 1.41 = 35.1$ kips, and $P = 9.7 \times 1.41 = 13.7$ kips.

For 30% $N = 35.1 \times 0.3 = 10.5$, $M_{x-x} = 10.5 \times 0.5 \times 31.10 = 164$ ft-kips

For 30% $P = 13.7 \times 0.3 = 4.1$, $M_{y-y} = 4.1 \times 0.5 \times 31.10 = 64$ ft-kips

30% Wind on column $M_{x-x} = 9$ ft-kips

$M_{y-y} = 16$ ft-kips

Summary, Group III: $F = 671$ kips, $M_{x-x} = 802$ ft-kips, and $M_{y-y} = 338$ ft-kips.

Group IV = Group I + T: $F = 671$ kips, $M_{y-y} = 19.3 \times 26 = 502$ ft-kips.

We should note that for Groups II and III, an overturning force should be added to the load effects, applied at the windward quarter points of the transverse superstructure width. For a bridge width of 55 ft, this force is $55 \times 0.02 = 1.10$ kips/ft of length, and the force allocated to the fixed pier is $75 \times 1.10 = 82.5$ kips. The overturning moment is, therefore, $M_{\mathrm{OT}} = 82.5 \times (55/4) = 1134$ ft-kips. Because the superstructure is not rigidly connected to the two pier columns, this moment is resolved into two equivalent forces, one downward and one upward, acting on each column. The magnitude of this force is $1134/46 = 25$ kips for group II, and $0.3 \times 25 = 7.5$ kips for group III, and may be neglected.

Soil Pressure For this example, the allowable soil pressure is $q_{\mathrm{all}} = 15$ kips/ft^2.

Group I. Soil pressure $q = 671/100 = 6.71$ kips/ft.

Group III. Compute $e_x = 802/671 = 1.20$ ft, and $e_y = 338/671 = 0.50$ ft. Then

$$q = \frac{\Sigma V}{B^2}\left(1 \pm \frac{6e_x + 6e_y}{B}\right) = \frac{671}{100} \times \left[1 \pm \frac{6(1.20 + 0.50)}{10}\right]$$

$$= 6.71 \times (1 \pm 1.02) = 13.6 \text{ kips/ft}^2 \quad \text{max}$$

$$= -0.13 \text{ kips/ft}^2 \quad \text{min}$$

The small negative value indicates that a very small area is in tension (no contact), but for preliminary design this is acceptable. Whether the allowable soil pressure for this group should be taken as $15 \times 1.25 = 18.75$ kips/ft^2 is a matter of judgment and interpretation. In the context of our design philosophy, the allowable soil pressure for group III should not exceed the basic 15 kips/ft^2.

Expansion Pier

For this structure, bearing devices at expansion piers and abutments are similar to the bearing detail shown in Figure 2-16. In this case the effect of friction is very small and can be disregarded.

1. *Dead Load Reactions* The following results were obtained from superstructure analysis:

$$
\begin{aligned}
&\text{Interior girders (5)} = 112 \times 5 &&= 560 \text{ kips}\\
&\text{Exterior girders (2)} = 101 \times 2 &&= \underline{202} \text{ kips}\\
&\qquad \text{Total DL} &&= 762 \text{ kips}\\
&\text{Weight of girder} &&= \underline{\ 56} \text{ kips}\\
&\qquad \text{Total DL} &&= 818 \text{ kips or 409 kips/column}
\end{aligned}
$$

$$
\begin{aligned}
\text{Weight of substructure: Column} &= 57 \text{ kips}\\
\text{Footing} &= 27 \text{ kips (assume 8.5 ft} \times 8.5 \text{ ft} \times 2.5 \text{ ft)}\\
\text{Earth} &= \underline{\ 27} \text{ kips}\\
\text{Total} &= 111 \text{ kips}\\
\text{Total dead load (one column)} &= 409 + 111 = 520 \text{ kips}
\end{aligned}
$$

2. *Live Load* This is the same as in the fixed pier or 139 kips/column. the overall pier height is 26.5 ft, giving a column height of 24 ft.

3. *Temperature Forces* Both bearings are fixed in the direction of the pier. Hence, expansion or contraction of the steel box girder will induce the same movement in the top of the column, or 0.108 in./column. For a column diameter of 4 ft 6 in. and a height of 24 ft, the calculated force at the top induced by this movement is $F = 17.0$ kips. Then $M_{y-y} = 17 \times 26.5 = 451$ ft-kips.

The soil pressure is now as follows (for footing 8 ft 6 in. square): For Group I, the total load $F = 520 + 139 = 659$ kips and $q = 659/8.5^2 = 659/72.25 = 9.1$ kips/ft^2. For Group IV, we first compute $e_y = 451/659 = 0.68$. Then

$$
\begin{aligned}
q = 9.10 \times \left(1 \pm \frac{6 \times 0.68}{8.5}\right) = 9.10 \times (1 \pm 0.48) &= 13.5 \text{ kips/ft}^2 \quad \text{max}\\
&= 4.7 \text{ kips/ft}^2 \quad \text{min}
\end{aligned}
$$

Because the intent of an engineering solution is to make a reliable analysis of the problem, the foregoing procedure as applied to the design at hand is adequate and yields conservative results in the context of economy.

2-9 IMPACT AND DYNAMIC ANALYSIS

Impact Considerations

From the brief review of impact in Section 2-5, it appears that our present understanding of the subject is not fully articulated. For example, factors leading to dynamic effects include smoothly rolling loads, entry transients, vehicular excitation by deck or approach roughness, shock due to offset joints or debris, and resonance phenomena in longer spans.

Research by Csagoly and Dorton (1977) has resulted in conclusions that prompted many bridge codes to take a more conservative approach to the dynamic allowance for traffic loadings, including the LRFD specifications introduced in Section 2-13. The current Ontario Bridge Code specifies the following: (a) a dynamic load allowance not less than 0.4 for deck slabs and deck systems governed by wheel loads; (b) a factor of 0.35 for floor beams and stringers spanning less than 39 ft (12 m); (c) an impact factor for main load-carrying members that is a function of the calculated first flexural frequency of the members (a range 0.30–0.55 is permitted and a minimum value 0.35 is specified); (d) reduction in dynamic load allowance for permit load required to proceed at reduced speeds; (e) modification factors for reducing impact at ultimate limit states involving more than one truck; and (f) optional use of theoretical and experimental dynamic analysis to establish practical load impact factors.

Rational design methods (Huang and Veletsos, 1960; Wright and Walker, 1971) address the purpose served by a dynamic load allowance and articulate its interaction with other design considerations. Although these methods have been advocated by researchers, most codes still continue to treat dynamic loading as an increment to the static design live load. Among the most compelling concerns of these codes are (a) the fact that measured dynamic increments of strain in bridge structural members have sometimes equaled the live load strains, and (b) the fact that the current AASHTO design load including impact is among the lighter ones on a worldwide basis. An offsetting factor is that measured dynamic strain increments have been superimposed on static live load strains well below the design level.

The ASCE Committee on Loads and Forces on Bridges (ASCE, 1981) suggests consideration of the probabilistic nature of the bridge impact problem, and focuses on two aspects. First, bridges are characterized by uncertainties regarding the traffic conditions and the features of the vehicles including their initial motion. The second aspect relates to the chance of producing the maximum dynamic effect due to a fluctuation of vehicle velocity. This depends on the chance of the maximum amplitude of an increment curve coinciding with the maximum static influence curve. Such probabilistic considerations are incorporated in the Ontario Bridge Code. In addition, the concept of the present impact factor is intended to cause an increase of the vehicle forces acting on the bridge, but without a separate consideration of the bridge inertia force. For design purposes, the dynamic effect may be considered as the result of the bridge inertia force and the applied vehicle forces. Thus, if a secondary member is designed, its own inertia force associated with the bridge motion should be considered.

There is concern that concrete bridge decks may be overstressed by impact-induced forces. Several reasons can be cited: (a) direct contact with the load, (b) possible shock effects because of localized deck or approach irregularities, (c) obstacles left on the roadway, and (d) evidence that large strains may occur in concrete deck slabs due to dynamic wheel loading.

Optional Impact Analysis

In the Guide Specifications for Horizontally Curved Highway Bridges (1980) and in the Interim Specifications (1981, 1982, 1984, 1985, and 1986), AASHTO proposes an impact analysis that may be reviewed along with further investigation. Accordingly, the dynamic effects, including centrifugal forces, produced by the HS truck load may be considered using the impact factors shown in Table 2-9. The design value for a given quantity of live load effect, considering the dynamic factor, is the product of that quantity (obtained from static analysis) and $(1 + I)$ or

$$\text{Dynamic Live Load Value, DLLV} = \text{Static Live Load Value} \times (1 + I) \tag{2-17}$$

where I is the impact factor given in Table 2-9. This factor is valid within the following parameter values:

$50 \text{ ft} \leq L \leq 200 \text{ ft}$ where L = span length

$200 \text{ ft} \leq R_c \leq 1000 \text{ ft}$ where R_c = radius of centerline of bridge

$V \leq 70$ mph

Number of I girders ≤ 6

Number of continuous spans ≤ 2

Weight of vehicle/weight of bridge ≤ 0.6

If these ranges are exceeded, a dynamic analysis is indicated.

Theoretical Background The foregoing impact factors have resulted from a finite-element analysis developed by Shore and Rabizadeh (1974) at the University of Pennsylvania. In this program, the bridge deck was discretized by annular sector thin plate elements, and the curved girders were discretized by thin-walled curved beam elements, both developed on small displacement theory. A general computer program was formulated to analyze the dynamic response of horizontally curved bridges due to moving forces and masses, as well as the free-vibration analysis of the bridge due to specified initial conditions.

TABLE 2-9 Optional Impact Factors for Curved Bridges (Steel I Girder) (From AASHTO, 1980)

Quantity	Impact Factor, I
Reactions and shear forces	0.30
Moments in longitudinal girders	0.25
Torsional moments in longitudinal girders	0.40
Moments in slab	0.20
Bimoments in longitudinal girders	0.25
Forces and moments in diaphragms	0.25
Deflections	0.25

Dynamic Analysis

The foregoing brief remarks highlight the current trend of considering dynamic analysis in certain bridge types. Typically, however, analysis for vehicle- and wind-induced vibrations is not to be considered in bridge design. When a vehicle crossing a bridge is in a nonstatic situation, the bridge is analyzed by statically placing the vehicle at various locations and applying the dynamic load allowance. Flexible bridges, however, and long slender components may have dynamic force effects that exceed the impact formulas, and these cases may require analysis for moving load.

In general, dynamic effects due to moving vehicles may be attributed to two origins: (1) hammering effect manifested as the dynamic response of the wheel assembly to riding surface discontinuities, such as deck joints, cracks, potholes, and delaminations; and (2) dynamic response of the bridge as a whole to passing vehicles (this may be due to long undulations in the roadway pavement such as those caused by settlement of fill, or to resonant excitation as a result of similar frequencies between bridge and vehicle).

Elastic Dynamic Responses When an analysis for vehicle-induced vibrations is required, it is necessary to specify the surface roughness, speed, and dynamic characteristics of the relevant vehicle. According to the LRFD specifications (Section 2-13), impact is derived as a ratio of the extreme dynamic force effect to the static force effect. Design limitations relate to the fact that the deck surface roughness is a major factor in vehicle–bridge interaction, because it is difficult to estimate long-term deck deterioration effects at the design stage.

Wind-induced vibrations and unstable wind–structure interaction may be critical for slender or torsionally flexible structures. These should be analyzed for lateral buckling, excessive thrust, and divergence.

Inelastic Dynamic Responses Examples include a major earthquake or ship collision. The minimum analysis requirements for seismic effects under the LRFD specifications are shown in Table 2-17 (Section 2-13).

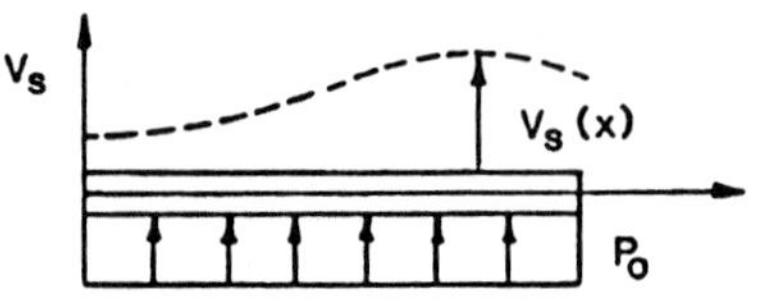

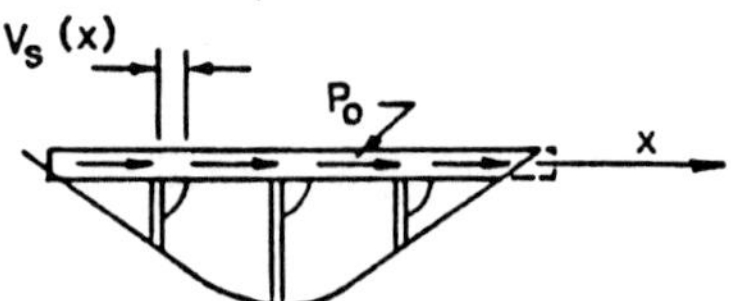

FIGURE 2-27 Bridge deck subjected to assumed transverse and longitudinal loading.

A recommended method of analysis is the single-mode spectral method, which may be used for both transverse and longitudinal earthquake motions. Examples are given by AASHTO (1983) and ATC (1981). This involves the following steps.

Step 1. Calculate the static displacements $V_s(x)$ due to an assumed uniform loading P_0 as shown in Figure 2-27.

Step 2. Calculate factors α, β, and γ from the expressions

$$\alpha = \int V_s(x)\,dx$$

$$\beta = \int W(x)V_s(x)\,dx \tag{2-18}$$

$$\gamma = \int W(x)V_s(x)^2\,dx$$

where P_0 = a uniform load arbitrarily set equal to unity (k/ft), $V_s(x)$ = deformation corresponding to P_0, and $W(x)$ = dead load of the bridge superstructure and tributary substructure.

Step 3. Calculate the period of the bridge using the expression

$$T_m = 2\pi\sqrt{\frac{\gamma}{P_0 g \alpha}} \tag{2-19}$$

where g = acceleration of gravity (ft/sec^2).

Step 4. Calculate the equivalent static earthquake loading $p_e(x)$ from the expression

$$p_e(x) = \frac{\beta C_{\mathrm{sm}}}{\gamma} W(x)V_s(x) \tag{2-20}$$

where C_{sm} = dimensionless elastic seismic response coefficient and $p_e(x)$ = intensity of the equivalent static seismic loading applied to represent the primary mode of vibration (kips/ft).

Step 5. Apply loading $p_e(x)$ to the structure and determine the resulting member force effects.

When coupling occurs in more than one of the three coordinate directions within each mode of vibration, the multimode spectral analysis should be used. A linear dynamic analysis using a three-dimensional model should represent the structure. Rigorous methods of analysis are recommended for critical structures or those that are geometrically complex and close to active faults. Time history methods of analysis are indicated in this case, with provisions for both the modeling of the structure and the selection of the input time histories of ground acceleration.

Spectra of Deck Surface Roughness, Field Investigation

In the dynamic context, impact coefficients, fatigue intensity, and human perception of moving vehicles are relevant considerations. Likewise, road surface roughness along the bridge is one of the causes of vehicle vibrations. Because it is measured only rarely, most studies are based on a convenient roughness coefficient a, usually 0.001 $cm^2/m/c$ calculated using an exponential function of the power spectral density (PSD). However, a dynamic analysis using $a = 0.001$ gives results that are lower than the actual dynamic response of the bridge (Honda, Kajikawa, and Kobori, 1982).

The roughness characteristics of a bridge deck may be defined in the longitudinal and vertical directions, but the former is selected because it is a cause of bridge vibration under moving vehicles. From measurements of road surface roughness on 56 bridges, Honda, Kajikawa, and Kobori (1982) have developed data expressing this concept in terms of the PSD, assumed from a stationary normal probability process with a zero mean value. The PSD is calculated by the maximum entropy method (Healey et al., 1977), also called the *nonlinear estimation method*. It requires a calculating time greater than the fast Fourier transform and the Blacknan–Tukey method, but gives better results in resolving the power and stability of a spectrum (Hino, 1978).

Typical graphs of the PSD are shown in Figures 2-28a through d for simple and continuous girder bridges, a truss bridge, and a Langer bridge (stiffening arch). The full line and dotted line in these graphs represent a measuring position of 0.5 m and 2.0 m from the centerline, respectively, and evidently there are no major differences in the PSDs between the two measuring positions, meaning that the characteristics of load surface roughness are the same. In these examples, the PSD can be approximated by an

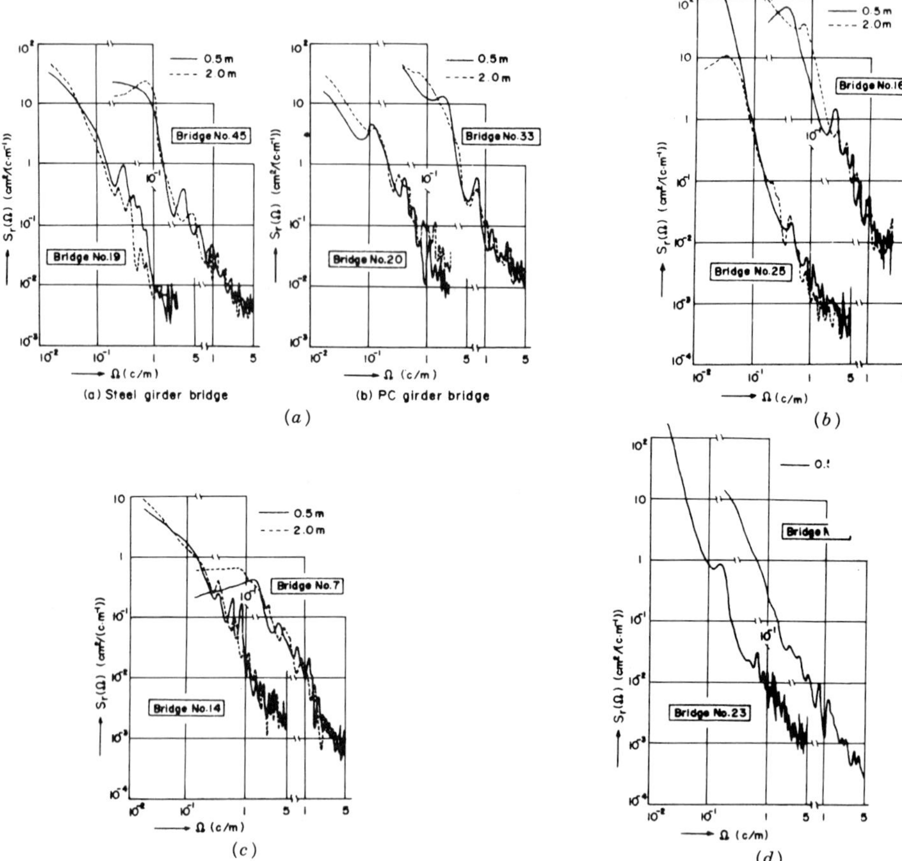

FIGURE 2-28 PSD graphs for various types of bridges: (*a*) simple girder bridge; (*b*) truss bridge; (*c*) Langer girder bridge (stiffening arch); (*d*) three-span continuous girder bridge. (From Honda, Kajikawa, and Kobori, 1982.)

exponential function expressed as $S_r(\omega) = a\omega^{-n}$, where ω = roughness frequency, a is as defined, and n = spectral roughness exponent.

The relationship between n and a is shown in Figure 2-29 for bridge types and pavements. In this nomenclature N is the number of measured lines, AP is asphalt pavement, and CP denotes concrete pavement. The n and a values are scattered in the range 1.3 to 2.5 and 0.001 to 0.06 $cm^2/m/c$. The relationship between n and a is essentially independent, based on the corrective coefficient $p = 0.152$. No significant difference between the n and a values can be observed between bridge types and pavements.

Honda, Kajikawa, and Kobori (1982) also give average values of a and n for various bridge systems and road pavements, shown in Table 2-10. Clearly,

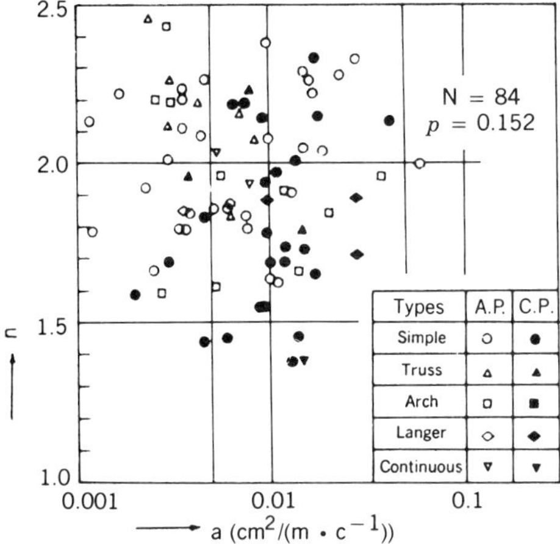

FIGURE 2-29 Diagram of relationship between n and a. (From Honda, Kajikawa, and Kobori, 1982.)

the road surface conditions on concrete pavement are worse than on asphalt, and the slope of the PSD is more gentle because of the high frequency in the roughness of concrete pavement. A comparison of the girder and nongirder system shows that the road surface conditions are essentially similar.

These results highlight the random nature of vibrations induced by deck roughness, and suggest the value of analysis by a random probability process. Caution is, however, necessary in interpreting the data thus obtained because shock effects of vertical offsets in joints or debris accumulation on decks may

TABLE 2-10 Values of $\bar{a}$ and $\bar{n}$ of Each Structural System of Bridges and Pavements of Road Surface (From Honda, Kajikawa, and Kobori, 1982)

(1)	Pavement (2)	Total (3)		$\bar{a}$ $(cm^2/m/c)$ (4)		$\bar{n}$ (5)	
Girder	Asphalt	35	60	0.0080	0.0098	2.04	1.92
	Concrete	25		0.0115		1.79	
Nongirder	Asphalt	18	24	0.0096	0.0116	2.02	1.96
	Concrete	6		0.0135		1.90	
Whole	Asphalt	53	84	0.0088	0.0107	2.03	1.94
	Concrete	31		0.0125		1.85	

make the random roughness excitation irrelevant to local components of the structural system.

2-10 DISTRIBUTION OF LOADS IN CONCRETE BOX–SLAB BRIDGES, MODEL STUDY

Theoretical and experimental studies on the wheel load distribution in slab and concrete box bridges have been carried out by Massonnet (1950, 1954) Reese (1966), and Scordelis (1966). Model studies on existing structures are provided by Litle and Hansen (1963), Massonnet (1953), Nelson et al. (1963), and Newmark and Siess (1954).

Pilot Tests

Box beam bridges with a separate concrete slab have been tested at Lehigh University (Macias-Rendon and Van Horn, 1973). The prototype represents a bridge with 0° skew and a clear span of 62.25 ft. Figure 2-30 shows a typical cross section. The cross section of maximum moment is 3.55 ft from midspan, and corresponds to a test vehicle consisting of a three-axle combination applied to approximate the standard HS 20 truck. A scale of 1/16 yielded a model size of 25 × 50 in., with a typical section shown in Figure 2-31.

The model load was applied with the axis of each vehicle in each of the five loading lanes shown in Figure 2-31. With the load on lane 3, the moment carried by the edge box beams was about 25 percent of the total moment in the model. The corresponding value from field tests on the prototype bridge was close to 21 percent. The difference is explained by (a) the interaction between slab and curb and between parapet and curb (better in the model because of stronger adhesive materials), and (b) the neoprene bearing pads used in the prototype bridge actually restraining stretching of the bottom fibers of the box beams.

The tests were intended to generate data that can be subjected to a proper statistical analysis. The sampling size was kept at four, so that with the presence of two rejectable outliers, two acceptable values still remain and can be averaged. The investigators used the Dixon criterion, applicable when the population mean and the standard deviations are unknown, and the sample at hand is the only source of information (Hoel, 1965; Natrella, 1966). A 50 percent risk of rejecting an observation that really belongs in the group was used throughout the experimental work.

For the main computer program, the following computational assumptions were made: (a) longitudinal slab strains vary linearly over the width of the slab corresponding to each beam; (b) longitudinal reactions are distributed among beams in proportion to individual moment percentages; (c) the mechanical jack and the model vehicle did not impose longitudinal restraint on the model; (d) initial individual slab widths extend to the midspacing of

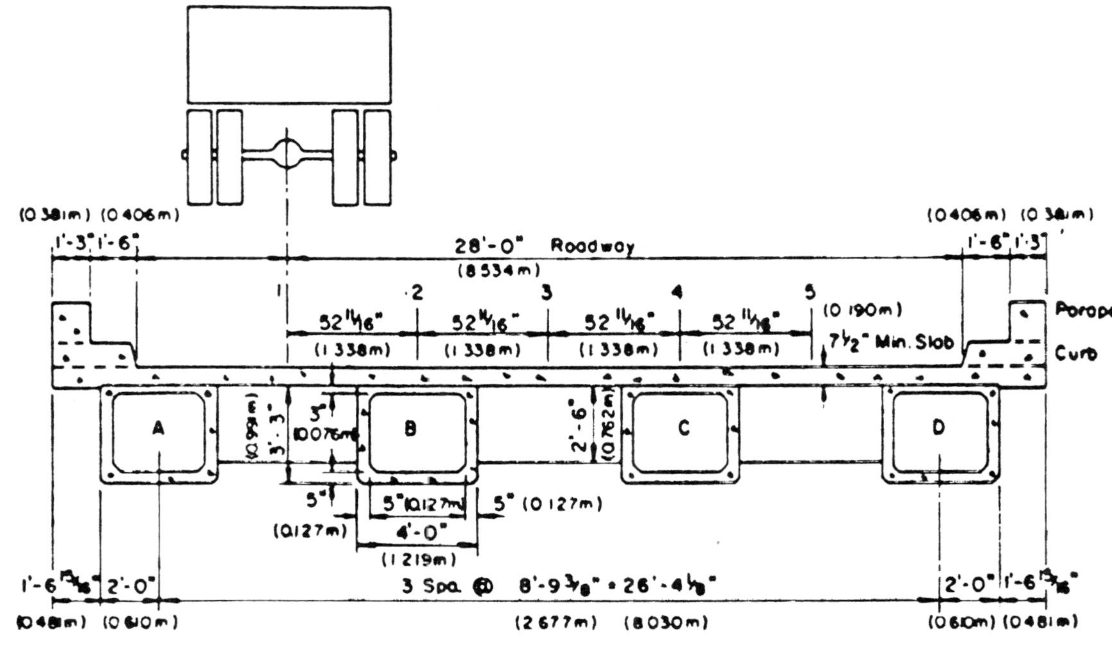

FIGURE 2-30 Cross section of Berwich Bridge prototype model. (From Macias-Rendon and VanHorn, 1973.)

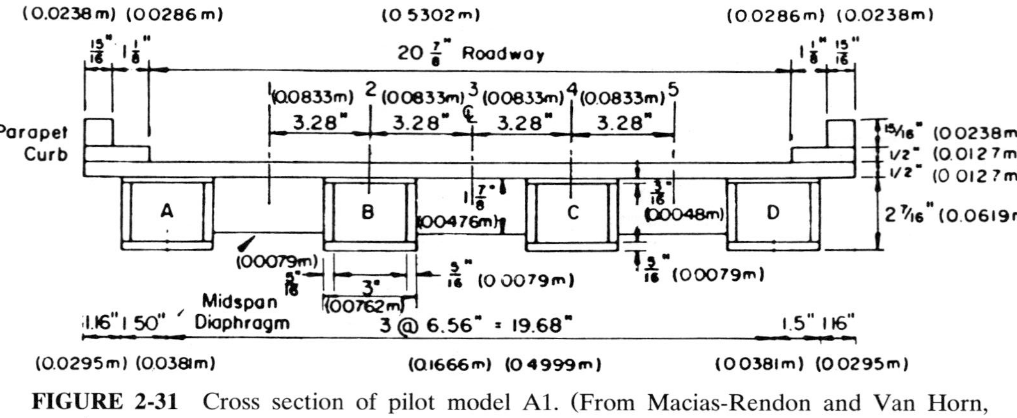

FIGURE 2-31 Cross section of pilot model A1. (From Macias-Rendon and Van Horn, 1973.)

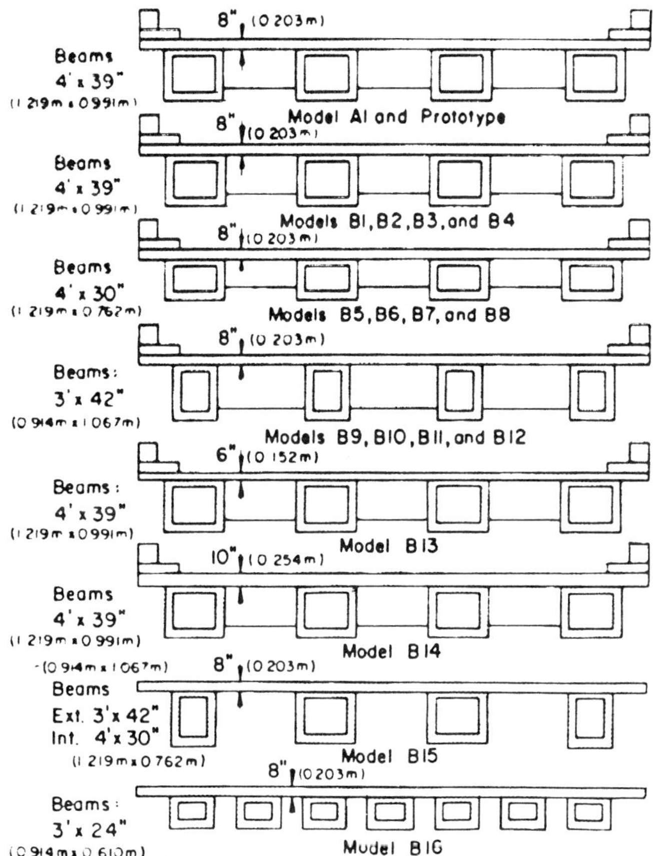

FIGURE 2-32 Basic cross sections tested. (From Macias-Rendon and Van Horn, 1973.)

the box beams; and (e) successive approximations on individual bending moments have a negligible correction after four cycles and two extrapolations.

Results from Model Tests

The first series of the test program, including pilot model A1, is shown schematically in Figure 2-32. All midspan cross sections are shown together with significant dimensions of the prototypes. In addition to model A1, seven basic sections of slab and box beams were used in the first series. In the second series of tests, 10 bolted models were used and included (a) five new sections consisting of 3 × 24 beams (3 ft wide, 24 in. deep) and (b) five new variations of the four-beam bridge with 4 × 39 beams to study the individual effect of parapets and midspan diaphragms.

The bending moments for individual beams are computed from stress blocks of beam elements and individual slab widths, converted from measured longitudinal strains. The contribution of curb and parapet is assigned to the exterior beams. In the first series of models, the range of discrepancies between the average of the sum of the computed beam moments near midspan and the applied moments with the load on the five lanes (based on static analysis) was from −19.8 to +4.4 percent with a mean discrepancy of −8.4 percent. In the second series, the discrepancy was slightly less.

Interpretation of Results For a standard HS 20 truck laterally located at the center of the roadway, the typical effects on the load distribution for the exterior beams of the deck are summarized as follows.

1. In four-beam bridges, the presence of diaphragms at both ends and at midspan increases the bending moment carried by the exterior beams at midspan by 15 percent. This value represents an average increase of 3 percent of the truck load. This is shown in Figure 2-33 for models B3 and B4. The increase in bending moments for the exterior beams ranges from 9 to

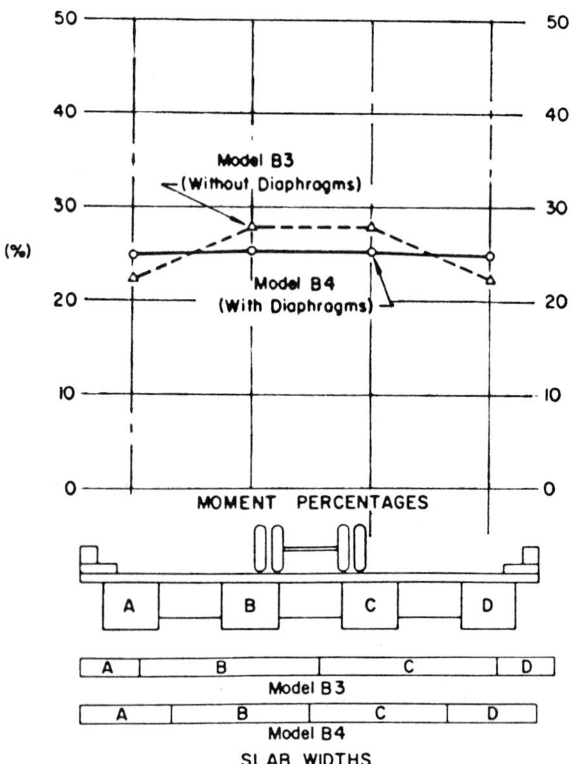

FIGURE 2-33 Models B3 and B4, lane 3. (From Macias-Rendon and Van Horn, 1973.)

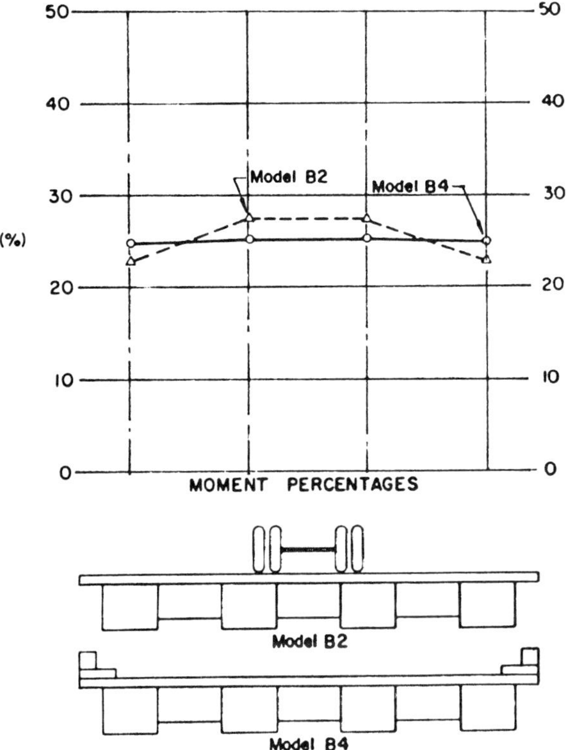

FIGURE 2-34 Models B2 and B4, lane 3. (From Macias-Rendon and Van Horn, 1973.)

23 percent for the various cross sections. In a four-beam bridge with 4 × 39 beams, the midspan diaphragms accounted for 93 percent of the combined effect of all diaphragms.

2. In four-beam bridges, the presence of curbs and parapets (assumed monolithically cast with the slab) results in a 12 percent increase in the bending moment carried by the outside beams near midspan, as illustrated in Figure 2-34. For a four-beam model with 4 × 39 beams, the parapet accounted for about 52 percent of the combined effect of curbs and parapets.

3. In three four-beam models, with 3 × 42, 4 × 39, and 4 × 30 beams, the moments near midspan carried by the outside beams are 24.3, 24.8, and 26.2, respectively, of the total bending moment, as shown in Figure 2-35.

4. For a four-beam bridge with 4 × 39 beams, an increase of slab thickness from 6 to 10 in. increased the bending moment carried by the outside beams by 2.4 percent, as illustrated in Figure 2-36.

5. The moment percentages shown in Figure 2-37 clearly show the uniformity achieved in the lateral load distribution of a seven-beam bridge with 3 × 24 beams.

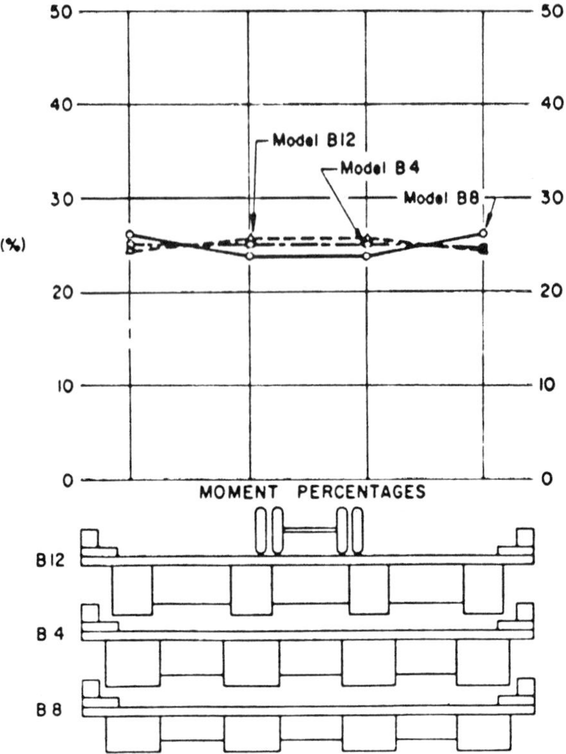

FIGURE 2-35 Models B12, B4, and B8, lane 3. (From Macias-Rendon and Van Horn, 1973).

Macias-Rendon and Van Horn (1973) obtained correlation between field and pilot model tests. The field analysis produced the moments carried by the individual beams giving due consideration to possible restraints from the supports. Taking into account the ratio of 0.9 in the modulus of elasticity between the slabs and the beams in the prototype, the moment carried by the outside beams increased from 21 to 24.3 percent of the total maximum moment in the prototype, in close agreement with the value of 25.5 percent obtained in the field.

These results articulate some of the effects on lateral load distribution produced by a stiff slab, the presence of curbs, parapets, and diaphragms, and variations in slab thickness, beam size, and beam spacing.

2-11 DISTRIBUTION OF LOADS ACCORDING TO AASHTO SPECIFICATIONS

The provisions of Part C, Section 3, are not intended for orthotropic deck bridges. The document also recognizes the complexity of the theoretical

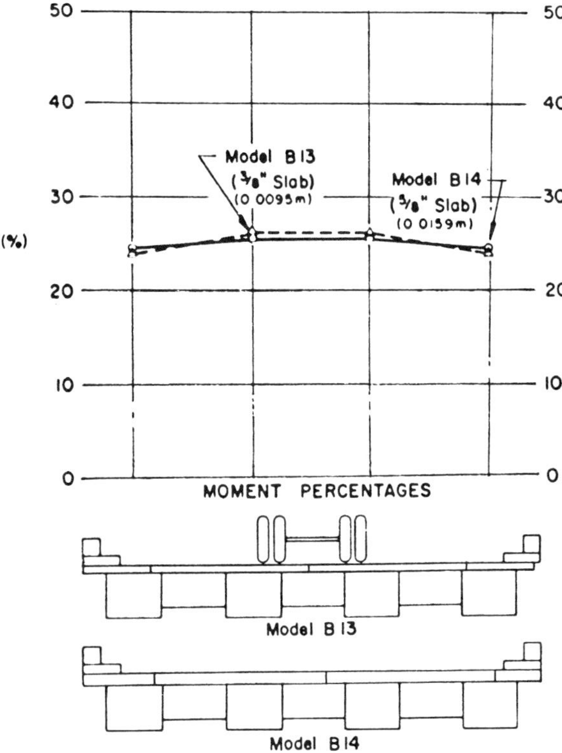

FIGURE 2-36 Models B13 and B14, lane 3. (From Macias-Rendon and Van Horn, 1973).

analysis involved in the distribution of loads, and suggests the application of these empirical methods to the design of normal highway bridges (see also Sections 2-12 and 2-13). The following cases are covered: (a) distribution of loads to stringers, longitudinal beams, and floor beams; (b) distribution of loads to concrete slabs; (c) distribution of wheel loads on timber flooring; (d) distribution of wheel loads on steel grid floors; and (e) distribution of loads for bending moments in spread box girders. A tire contact area is also defined and assumed as a rectangle with an area (in $in.^2$) of $0.01P$ and a length in the direction of the traffic–width of tire ratio of 1/2.5, in which P is the wheel load (in lb).

Stringers, Longitudinal Beams, and Floor Beams AASHTO (Table 3.23.1) gives the fraction of wheel load to be applied to interior stringers and beams. The effect of more than one lane loaded is not differentiated but lumped in the load distribution factor. This distribution is not intended to give an exact account of wheel load carried by an interior beam or stringer, but rather provides a uniform method of analysis for the usual types of

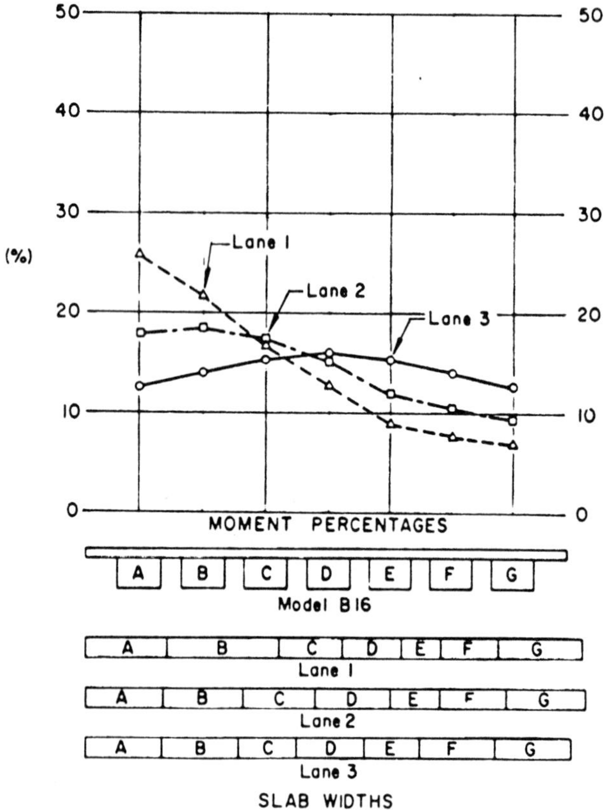

FIGURE 2-37 Model B16, lanes 1, 2, and 3. (From Macias-Rendon and Van Horn, 1973.)

highway bridges. The complexities of the problem have been long recognized, yet the AASHTO procedure has been found to work well.

For outside beams and stringers, the guidelines contain three possible distributions, namely: (a) the analysis may be carried out by applying to the stringer or beam the reaction of the wheel load obtained by assuming the floor acts as a simple span; (b) the exterior beam will not have less carrying capacity than an interior beam; and (c) for a bridge on four or more beams, the fraction of wheel load should not be less than $S/5.5$ if $S \leq 6$ ft, or $S/(4.0 + 0.25S)$ if $S > 6$ ft. Where $S \geq 14$ ft, the load on each interior and exterior stringer is the reaction of the wheel loads, assuming the floor acts as a simple span.

The application of these guidelines is illustrated in Figure 2-38. For the cross section shown in Figure 2-38*a*, the axle load can be positioned as shown. Applying provision (a), we obtain a wheel load distribution of $(8/7 + 2/7) = 1.42$, which exceeds the requirements of provision (b) or (c), in this

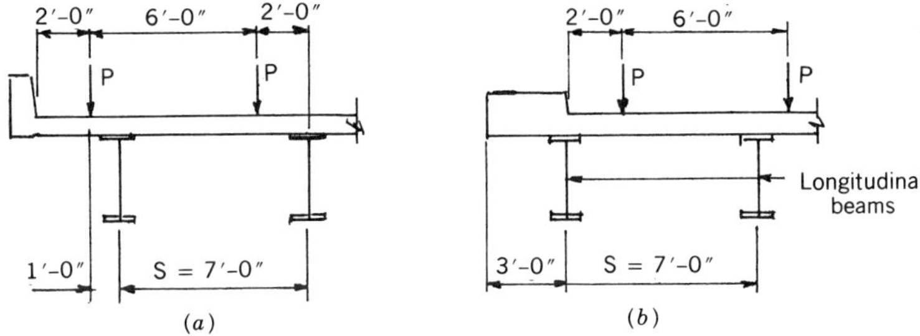

FIGURE 2-38 Distribution of wheel load to outside stringers and beams.

case less than 1.42. Hence, the distribution 1.42 should be used in the design. For the section shown in Figure 2-38*b*, the distribution stipulated by provision (a) yields a wheel load factor of $5/7 = 0.71$, which is grossly inadequate to satisfy the requirements of provision (b). In this case, the distribution should be $S/5.5$ if the intent is to provide the same load-carrying capacity as in the interior beams, or $S/5.75$ if the intent is to satisfy provision (c).

For bridges where longitudinal beams are omitted and the floor is supported directly on floor beams, the load distribution is stipulated in AASHTO Table 3.23.3.1. No transverse distribution can be assumed. The application of this provision is illustrated in Figure 2-39. Because the bridge system consists of a concrete slab supported on transverse beams, the fraction of wheel load applied to each floor beam is $S/6$, until $S > 6$ ft in which case the reaction of live load is computed assuming simple beam action for the slab. Let us consider now wheel load P acting directly above floor beam 1. If $S \leq 6$ ft, the provision suggests that a fraction of P will be distributed to the adjoining floor beams by the stiff slab (stiffness is considered a function of the slab thickness and floor beam spacing). If, however, $S > 6$ ft, the effect of slab stiffness begins to diminish, and the simple beam method yields a live load reaction of at least P. For $S \leq 6$, the live load reaction is reduced linearly with S. The foregoing remarks are valid for the regular truck load shown in Figure 2-39*a*. For the military load shown in Figure 2-39*b* (two 24-kip axle

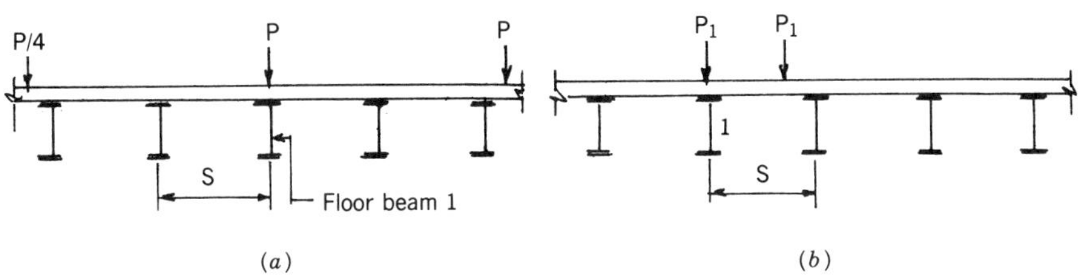

FIGURE 2-39 Distribution of live load in transverse beams: (*a*) partial longitudinal section, standard HS 20 truck; (*b*) military load. Concrete slab on transverse floor beams.

loads spaced 4 ft apart), the application of this provision is subject to interpretation, because it is unlikely that this guideline was prepared with the military load in mind. This author has used the military load as follows: for the load P_1 directly above beam 1 a distribution $S/6$ is used, but for the second load P_1 the distribution assumes simple span theory. In practice, this question may seldom arise, however, because floor beam spacing generally is greater than 6 ft in most two-way systems.

From the foregoing it follows that the wheel load fraction carried by a beam may be determined by one of the following procedures: (a) by applying the general formulas contained in the AASHTO provisions; (b) by placing the load in the most unfavorable position with respect to the member under consideration and computing the corresponding reactions for a set of valid assumptions; and (c) by comparing or combining results from (a) and (b). In most instances, no elaborate procedures are warranted, and the simple method recommended by AASHTO should be used (see also subsequent sections).

Interestingly, one of the early procedures for determining the portion of truck load carried by one beam or girder was based on the assumption that the axle load is distributed uniformly across the traffic lane and at right angles to its centerline, and that the roadway slab is made up of freely supported slabs so that its reaction on the girder may be computed by the rules of simple statics. This produced the simple formula

$$W = \frac{1}{2}\left(\frac{s + s_1}{c}\right)P \tag{2-21}$$

where P = truck load (one or both axles)
W = portion carried by one beam or girder
s_1, s = girder spacing on both sides of the girder under consideration
c = clearance width of the design truck

Likewise, early procedures for determining the portion of truck load to be allocated to a transverse floor beam recognized its dependence on the floor beam spacing. Using the truck load shown in Figure 2-40 as design live load for a floor beam spacing less than 6 ft, the rear axle load ($0.8P$) was distributed by the slab to adjoining floor beams. In this case the loading taken by one floor beam was

$$W = \frac{s}{6.0} \times 0.8P$$

(s is the beam spacing, or the same as in current AASHTO guidelines).

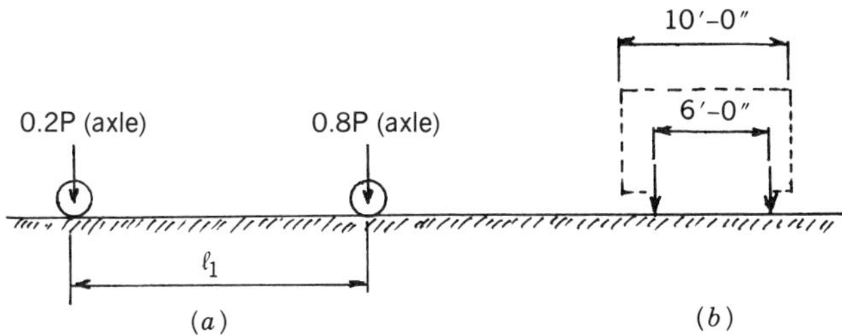

FIGURE 2-40 Early design truck load (similar to current H loading): (*a*) truck elevation; (*b*) truck cross section.

For floor beam spacing s larger than the axle spacing l_1 but smaller than the spacing of trucks l_2,

$$W = \left(0.8 + 0.2\frac{s - l_1}{s}\right)P$$

For floor beam spacing larger than the truck spacing l_2,

$$W = \left[0.8 + 0.2\frac{2s - (l_1 + l_2)}{s}\right]P$$

Load Distribution in Concrete Slabs This topic, covered in Article 3.24 of the specifications, has its theoretical origin in the work of Westergaard (1930), commonly referred to as the Westergaard theory. Other studies that have contributed to the formulation of this procedure include Jensen (1936), Newmark (1939), and University of Illinois Bulletin 346 (1940). The load distribution is covered in Chapter 3, and need not be repeated here.

Multibeam Decks Usually, multibeam decks are constructed with precast reinforced or prestressed concrete beams placed side by side on supports. The interaction between beams is developed by continuous longitudinal shear keys and lateral ties (bolts) that may, or may not, be prestressed (see also the following sections).

In determining the live load bending moment, the fraction of wheel load applied to each beam is computed from S/D, where S is the width of the precast member and D is a factor depending on the number of traffic lanes, a specific relevant span length, and elastic properties such as the moment of inertia, Poisson's ratio, and the Saint-Venant torsion constant.

For the bridge rehabilitation program of the structure carrying the Southwest Highway over a network of railroads and two creeks in Chicago, about 22 spans consisting of multibeam decks (prestressed concrete boxes) were structurally modified. The new work consisted of removing the existing asphalt surface on the top of the beams and replacing it with a 4-in. dense concrete overlay. A mat of small-diameter epoxy-coated bars embedded in the overlay provided the structural continuity between adjacent box beams. The new stiff deck serves to distribute the live load and reduce the fractions carried by each box beam (Xanthakos, 1981). For this bridge the width of precast beams is 3 ft. Using a somewhat different approach based on applicable state standards, the fraction of wheel load carried by one box beam was computed as 0.54, ignoring the effect of the concrete overlay. A more detailed analysis indicated that this could be reduced to 0.40 if the interaction with the concrete overlay was included. With either distribution, however, the box beams were stressed within the range of allowable values.

Recent studies conclude that the girder width should be limited to the range 4 to 10 ft, beyond which a special analytical investigation may be necessary. The same studies also recommend end diaphragms to ensure proper load distribution. In addition, there is a strong indication that interior diaphragms may cause a reduction in the live load distribution to the interior girders and a corresponding increase in the live load distribution to the exterior girders predicted under Article 3.23.4.3.

2-12 DISTRIBUTION OF LOADS BASED ON RESULTS OF NCHRP PROJECT 12-26

New code proposals under review may partially or wholly be incorporated into the AASHTO specifications. Currently in a draft form, the methodology for calculating wheel load distribution factors is based on power curves. Some formulas include stiffness and inertia terms, but in order to predict the most accurate factors an iterative approach must be used. Given the complexity of the theoretical analysis involved in wheel load distribution, the empirical method presented in this draft is recommended for the design of normal highway bridges.

Essentially, the draft addresses in its entirety Part C of Section 3 of the AASHTO specifications, and proposes new procedures for (a) distribution of loads to stringers, longitudinal beams, and floor beams; (b) distribution of loads and design of concrete slabs; (c) distribution of loads on timber flooring; (d) distribution of loads and design of composite wood–concrete members; (e) distribution of steel grid floors (but not applicable to orthotropic bridges); and (f) tire contact area. The following are points of interest.

1. For steel I beams and prestressed concrete girders, the wheel load distribution factor is a function of the beam spacing, the span length, the slab thickness, and a longitudinal stiffness parameter.

2. For exterior beams, the distribution factor is the factor for the interior beams multiplied by a coefficient $e = (7 + d_e)/9.1 \geq 1.0$, where d_e is the

edge distance of the traffic lane (ft), calculated as the distance between the center of the exterior beam and the edge of the exterior lane (face of curb).

3. For skew bridges and where the skew angles of two adjoining supports are close, the bending moments corresponding to a normal continuous or simple beam analysis may be reduced. This reduction is implemented by applying a reduction factor to the normally obtained distribution factor.

4. However, for continuous superstructures, the moments must be increased. Correction factors are applied to the moments obtained from a continuous beam analysis. For steel I beams, prestressed concrete girders, and concrete T beams, the correction factor is 1.10 and 1.05 for negative and positive moment, respectively. No adjustment is necessary for steel box girders and precast concrete beams other than box beams used in multibeam decks.

5. Compatible formulas are given for end shears in simple, continuous, and skew bridges.

6. The distribution of loads and design of concrete slab are still based substantially on the Westergaard theory (1930). However, for Case B (main reinforcement parallel to traffic), the distribution width E is determined from a specific range of bridge length and width in conjunction with the number of traffic lanes. For skew supports, the distribution width may be decreased by an appropriate factor. For continuous concrete bridges, the distribution width E must be reduced by 10 percent, prompting a corresponding increase in the negative moment. No changes are introduced for cantilever slabs.

We should note that the numerical value of stiffness terms (K_8/Lt_s^3) and I/J is close to unity for most common bridges. These factors may be taken as

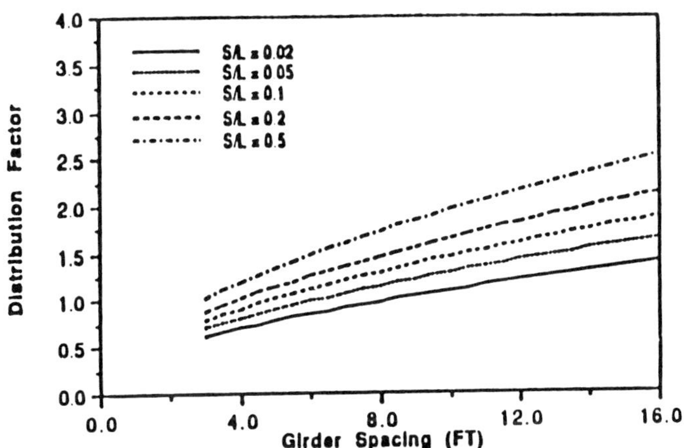

FIGURE 2-41 Live load distribution factor for beam and slab bridges, multiple-lane loading ($K_8/Lt_s^3 = 1.0$).

unity for initial design or where greater accuracy is not warranted. The effect of various bridge parameters is demonstrated and better understood if these formulas are presented graphically. Figures and curves thus obtained may be used in place of formulas in obtaining distribution factors. Such diagrams are plotted in Figure 2-41 for beam and slab bridges, multilane loading, and taking $K_8/Lt_s^3 = 1$. The distribution factor is obtained graphically for various S/L ratios (S is the beam spacing and L is the span length).

Comparison of results obtained from the various distribution procedures is documented and quantified in numerical bridge design examples in the following sections.

2-13 LRFD SPECIFICATIONS (1992)

Throughout this book the reader will come across terms such as *ultimate strength design*, *strength design*, *plastic design*, *load factor design*, and the more recent *load and resistance factor design*. All these constitute the *limit states* approach which may encompass strength, fatigue and fracture, service, and extreme event. The underlying philosophy is to move toward a more rational and probability-based procedure that can provide a formidable substitute for *allowable stress design*, the principal philosophy of structural design in the last 100 years. A review of the current status of the limit states concept and its use in design is presented by Haaijer (1983) and Kennedy (1985).

The proposed LRFD specifications for bridges are the result of research requested by AASHTO and initiated by the National Cooperative Highway Research Program (NCHRP) as Project 12-33. This document draws from completed and recent bridge research, and signifies new provisions and major areas of change. It addresses load models and load factors, structural analysis, concrete and steel structures, decks and deck systems, foundations, and bridge substructure elements. Special requirements for movable bridges are also discussed.

Design Philosophy

Bridges should be designed for the specified limit state. In every case, the structural system including its components, joints, and connections must be designed to reach the design failure mechanism before any other mechanism is developed. Thus, unintended overstrength of a member or component where hinging is predicted should be avoided, because it may result in the formation of a plastic hinge at an undesirable location and with adverse effects.

Service limit states represent restrictions on stress, deformations, and crack width under regular service conditions. The nonstrength status relegates function as the chief objective.

Fatigue and fracture limit states are selected as restrictions and limitations on the stress range under normal service conditions to make the structure safe against the expected stress range variations and cycles.

Strength limit states will ensure the strength and stability of the structure, partially or wholly, necessary to resist the statistically possible load combinations that may act on a bridge during its design life.

Extreme-event limit states are taken to ensure the structural survival of a bridge during a major earthquake or when a vessel, vehicle, or ice flow collides with it.

Commentary The basis of any design approach is to show that (a) if loads, material strengths, stiffness, and so on, are as expected, the bridge will perform satisfactorily within acceptable deformations, and (b) even if both loads and material properties are significantly worse than expected, there will still be an adequate factor of safety against collapse.

Under this premise, if a bridge member of component reaches a point where it no longer satisfies the requirements for which it was designed, then it is said to have reached a limit state. Limit states should be classified as either ultimate limit states in which consideration is given to the worst credible values that the associated variables could reach or serviceability limit states where the most probable values are used. In principle, all limit states should be examined explicitly, but, in practice, it usually becomes apparent that one is more critical than the others and therefore it may not be necessary to investigate all the states to the same extent and detail.

A bridge, its foundation, and the ground on which it rests constitute a combined system of ground and structure, and their implicit interaction must be recognized. An ultimate limit state is reached when (a) failure in the ground occurs without failure in the structure, involving loss of stability or causing substantial rigid body movement; (b) failure in the structure is manifested, involving loss of stability or structural fracture, without failure occurring in the ground; and (c) failure occurs in the structure and ground together.

Serviceability limit states include (a) excessive deformation of the ground leading to excessive (unacceptable) differential settlement, heave, lateral movement, and so on of the structure; (b) excessive deformation (deflection) of the structure; and (c) excessive cracking of the structure.

Basic LRFD Methodology Each component and connection must satisfy a modified version of (2-11). For other than the strength limit state, resistance factors may be taken as 1.0. All limit states are considered of equal importance. Accordingly,

$$n \sum \gamma_i Q_i \leq \phi R_n = R_f \tag{2-11a}$$

where $n = n_D n_R n_i$

γ_i = load factor (statistically based multiplier applied to force effects)

ϕ = resistance factor (statistically based multiplier applied to nominal resistance)

n = load multiplier (factor relating to ductility, redundancy, and operational importance)

n_D = factor relating to ductility

n_R = factor relating to redundancy

n_i = factor relating to operational importance

Q_i = force effect (deformation or stress, i.e., thrust, shear, torque, or moment caused by applied loads, imposed deformations, or volumetric changes)

R_n = nominal resistance (based on permissible stresses, deformations, or specified strength of materials)

R_f = factored resistance = ϕR_n

Ductility Ductility, also discussed in other sections, may be defined as the amount of permanent strain (strain exceeding the proportional limit) up to the point of fracture. Ductility is important because it permits yielding locally due to high stresses and thus causes the stress distribution to change. In this context, ductile behavior provides warning of structural failure by large inelastic deformations. Under cyclic loading large reversed cycles of inelastic deformation dissipate energy and benefit structural response.

Behavior that is ductile in a static context but not during dynamic response should be avoided. Examples include shear and bond failures in concrete members and loss of composite action in flexural members. If, by means of confinement or other measures, a member or connection made of brittle material can sustain inelastic deformations without loss of its load-carrying capacity, the member may be considered ductile. Bridge components designed according to proper specifications would normally provide adequate ductility. For unusual and important structures in high-seismic zones, ductility must be ensured by specifying a minimum ductility factor expressed as

$$\mu = \frac{\Delta_\mu}{\Delta y} \tag{2-22}$$

where Δ_u = maximum plastic (ultimate) deformation

Δ_y = deformation at the elastic limit

The ductility capacity of structural members, joints, and connections may be ascertained by full- or large-scale tests or by analytical models based on actual material behavior. The ductility capacity of a structural system can be determined by integrating local deformations over the entire system.

For the strength limit state for all members, the ductility factor can be taken as follows: $n_D = 1.05$ for nonductile components and connections, and $n_D = 0.95$ for ductile components and connections. For other limit states, n_D can be taken as 1.0.

Redundancy Redundancy is discussed in Section 5-17. Examples of nonredundant bridges are presented in Chapters 3 and 5. Whereas both the standard AASHTO specifications and the LRFD document specify the use of multiple-load path structures, there may be compelling reasons to the contrary. Main tension members whose failure would be expected to cause bridge collapse are designated as fracture-critical, and the associated structural system must be designed as nonredundant.

For the strength limit state, the redundancy factor should be taken as follows: $n_R = 1.05$ for nonredundant members, and $n_R = 0.95$ for redundant members. For other limit states, $n_R = 1.00$.

Operational Importance This concept applies to the strength and extreme-event limit states. The classification is based on social–survival or security–defense requirements. A bridge may be declared to be of operational importance, in which case n_i must be taken as 1.05; otherwise, $n_i = 0.95$.

Load Events and Load Factors

Load Groups Loads and forces are characterized as permanent and transient, summarized in Table 2-11. Load groups, combinations, and load factors are defined for the limit states indicating a wide range of multiple performance levels. A summary of load combinations and load factors is given in Tables 2-12 and 2-13.

Strength I is the basic load combination relating to normal vehicular use of the bridge without wind, and corresponds to the group loading generally applicable to bridge superstructures.

Strength II load combination reflects the use of bridges by permit vehicles without wind. The permit vehicle should not be assumed to be the only vehicle on the structure unless traffic is controlled by an escort vehicle. Applying a distribution factor procedure to a loading that involves a heavy permit load can be very conservative, unless lane-by-lane distribution factors are available. Refined methods may be used to remedy this situation.

Strength III load combination relates bridge exposure to maximum wind velocity which prevents a significant live load from being on the bridge.

Strength IV load combination refers to bridges with very high dead–live load force effect ratios. A standard calibration process has been carried out for a large number of bridges with spans less than 200 ft, and spot checks have been made on bridges with spans up to 600 ft. For the primary

TABLE 2-11 Load Summary and Designations (LRFD Specifications)

Permanent Loads
DC = Dead load of structural components and nonstructural attachments
DD = Downdrag
DW = Dead load of wearing surfaces and utilities
EA = Earth pressure load
EF = Dead load of earth fill
ES = Earth surcharge load
Transient Loads
BR = Vehicular braking force
CE = Vehicular centrifugal force
CR = Creep
CT = Vehicular collision force
CV = Vessel collision force
EQ = Earthquake
FR = Friction
IC = Ice load
IM = Vehicular dynamic load allowance
LL = Vehicular live load
PL = Pedestrian live load
SE = Settlement
SH = Shrinkage
TG = Temperature gradient
TU = Uniform temperature
WA = Water load and stream pressure
WL = Wind on live load
WS = Wind load on structure

components of longer bridges, the ratio of dead–live load force effects is high and could result in resistance factors different from the set found acceptable for small-and medium-span bridges. It is believed that load combination IV will govern when the dead–live load force effect ratio exceeds about 7.0.

Strength V load combination relates to normal vehicular use with wind velocity not exceeding 55 mph. Vehicles are expected to become unstable if wind velocity exceeds 55 mph.

The extreme-event limit state reflects conditions created by seismic events, ice loads, and collision by vessels and vehicles. Under these extreme conditions, the structure is expected to undergo considerable inelastic deformation whereby locked-in force effects due to TU, TG, CR, SH, and SE will be relieved. The 0.50 live load factor is compatible with the low probability of the presence of maximum vehicular live load when extreme events occur.

Service I load combination applies to the normal operational use of the bridge with 55 mph wind. All loads are taken at their nominal values and extreme load conditions are excluded.

TABLE 2-12 Load Combinations and Load Factors (LRFD Specifications)

Load Combination / Limit State	DC DD DW EA EF ES	LL IM CE BR PL	WA	WS	WL	FR	TU TG CR SH SE	Use One of These at a Time: EQ	IC	CT	CV
Strength I	γ_p	1.70	1.00	—	—	1.00	0.50/1.20	—	—	—	—
Strength II	γ_p	1.30	1.00	—	—	1.00	0.50/1.20	—	—	—	—
Strength III	γ_p	—	1.00	1.40	—	1.00	0.50/1.20	—	—	—	—
Strength IV											
EA, EF, ES, DW	γ_p	—	1.00	—	—	1.00	0.50/1.20	—	—	—	—
DC only	1.5										
Strength V	γ_p	1.30	1.00	0.40	0.40	1.00	0.50/1.20	—	—	—	—
Extreme event	γ_p	0.50	1.00	—	—	1.00	—	1.00	1.00	1.00	1.00
Service I	1.00	1.00	1.00	0.30	0.30	1.00	1.00/1.20	—	—	—	—
Service II	1.00	1.30	1.00	—	—	1.00	—	—	—	—	—
Service III	1.00	0.80	1.00	—	—	1.00	—	—	—	—	—
Fatigue—LL, IM, and CE only	—	0.75	—	—	—	—	—	—	—	—	—

Service II load combination corresponds to the overload provisions of the AASHTO standard specifications. Its intent is to prevent premature yielding of steel structures due to vehicular live load, approximately halfway between the service I and Strength I limit states, for which wind effects are insignificant.

Service III load combination relates essentially to prestressed concrete structures with the primary objective of crack control.

The fatigue and fracture load combination relates to gravitational vehicular live load and dynamic response. The load factor reflects a load level that has been found to be representative of the truck population with respect to a large number of return cycles.

TABLE 2-13 Load Factors for Permanent Loads, γ_p (LRFD Specifications)

Type of Load	Load Factor: Maximum	Minimum
DC: Component and attachments	1.25	0.90
DD: Downdrag	1.80	0.45
DW: Wearing surfaces and utilities	1.50	0.85
EA: Earth pressure	1.50	0.90
EF: Earth fill	1.35	0.90
ES: Earth surcharge	1.50	0.75

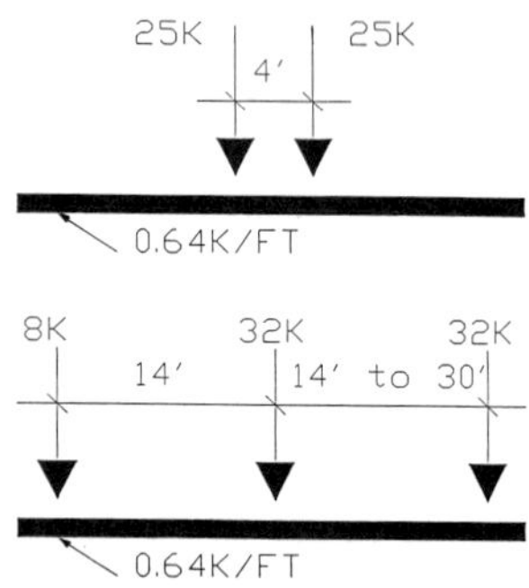

FIGURE 2-42 Live load models, LRFD specifications; axle loads. *Note:* These models consider (*a*) the effect of the design tandem combined with the effect of the design lane load, and (*b*) the effect of the design truck combined with the effect of the design lane load. The effects of an axle sequence and the lane load are superimposed to obtain extreme values. The lane load is not interrupted to provide space for the axle sequence, but interruption is needed only for patch loading patterns to produce extreme force effects.

Uplift, treated as a separate load in the AASHTO standard specifications, becomes a Strength I load combination. When the dead load reaction is positive but the live load causes a negative reaction, the load combination would be 0.9DC + 0.85DW + 1.7(LL + IM). If both reactions were negative, the load combination would be 1.25DC + 1.50DW + 1.7(LL + IM).

Vehicular Live Load The vehicular live load on bridges or incidental structures consists of the following models and combinations thereof: (a) design truck, (b) design tandem, and (c) design lane load. These load models are shown in Figure 2-42. The design truck is the same as the HS 20 truck unit. The design tandem consists of a pair of 25-kip axles spaced 4 ft apart. The design lane load is similar to the AASHTO lane load and consists of a uniformly distributed load in the longitudinal direction of 0.64 kips/ft. The live load is assumed to occupy 10 ft transversely.

The live load model, consisting of either a truck or tandem coincident with a uniformly distributed large load, was developed as a notional representation of shear and moment produced by a group of vehicles routinely permitted on highways by various states under grandfather exclusions to weight limitations. These results are based on a study conducted by the Transportation Research Board (TRB) (Cohen, 1990), and the load model is called "notional" because it is not intended to represent any particular truck.

Comparison between the force effects produced by a single exclusion truck per lane and the new load model shows a fairly close grouping, with the implication that the new load model has general applicability and can have a single load factor.

Application of Design Vehicular Live Loads The effects of an axle sequence and the lane load are superimposed in order to obtain extreme values. This is a significant departure from the standard AASHTO approach where either the truck or the lane load with an additional concentrated load is used for extreme effects. The lane load is not interrupted to provide space for the axle sequences of the design tandem or truck; interruptions are needed only for patch loading patterns to produce extreme force effects.

For negative moment and reaction at interior piers of continuous bridges, the extreme force effect should be determined for the loading combination consisting of 90 percent of the effect of two design trucks spaced a minimum of 50 ft between the load axle of one truck and the rear axle of the other truck, and 14 ft between the two 32-kip axles, combined with 90 percent of the effect of the design lane load, if this gives a larger value than the live load shown in Figure 2-42.

Axles that do not contribute to the extreme force effect should be neglected, and the lane load should be positioned longitudinally for extreme effect.

Fatigue Load For components other than decks, the fatigue load consists of the design truck load but with a constant spacing of 30 ft between the 32-kip axles. The variable spacing of 14 to 30 ft may be used to simplify the design process by reducing the number of load conditions to be considered, but in this case the resultant stress range may be grossly overestimated.

The number of cycles of maximum stress range to be considered in fatigue design is correlated to the $ADTT_{SL}$ (single-lane average daily truck traffic). This frequency is applied to all bridge components. In the absence of more data, the single-lane ADTT is taken as $ADTT_{SL} = p \times ADTT$, where ADTT = average number of trucks per day in one direction over the design life, and p = a number depending on the number of lanes available to trucks.

Dynamic Load Allowance (IM) Impact coefficients reflecting dynamic effects are based on the percentages shown in Table 2-14, and are considerably higher than those stipulated by the current impact formula. This dynamic allowance may be reduced for components other than joints and components of the deck if the design warrants. In Table 2-14 "Components of the Bridge Deck" include the slab or plate that directly supports the wheel loads. "All Other Components" include the girders, beams, bearings (except elastomeric bearings), columns, and above-ground foundations.

TABLE 2-14 Dynamic Load Allowance, Percentage IM (LRFD Specifications)

Component	IM
Deck joints—all limit states	75%
Components of the bridge deck—all limit states	50%
All other components	
Fatigue and fracture limit state	15%
All other limit states	33%

Braking Force (BR) The braking forces are taken as 25 percent of the axle weights of the design truck or tandem per lane placed in all design lanes carrying traffic headed in the same direction. The braking force (longitudinal force LF) is estimated from energy principles relating this force to vehicle weight as

$$b = \frac{v^2}{(2ga)} \tag{2-23}$$

where a = uniform deceleration, v = speed of vehicle, and b = fraction coefficient as a function of vehicle weight. Using a = 400 ft and v = 55 mph gives $b = 0.25$. Only one design truck should be considered, because other vehicles represented by the design lane load are expected to brake out of phase.

The same coefficient of 0.25 is obtained from (2-15) if the time is taken as 11 sec. The apparent shortcomings of (2-15) and (2-23) reflect the condition that a minor variability in the parameters causes a marked variation in the coefficients F and b; hence, these solutions do not represent a rational approach to the braking force problem.

Wind Load (WL and WS) Typically, a bridge structure should be examined separately under wind pressure from two or more different directions in order to ascertain the combination producing the most critical wind effects. The specifications articulate the phenomenon of aeroelastic instability, focusing on bridge types and components that are likely to be wind-sensitive.

A typical effect due to wind is excitation due to vortex shedding. This is the escape of wind-induced vortices behind the member, and tends to excite the component at its fundamental natural frequency in harmonic motion. Flexible bridges, such as cable-stayed, or very long spans of any type may require special studies involving simulation of the local wind environment at the bridge site. Vortex-induced oscillating stresses must be kept below the "infinite life" fatigue stresses. This subject is briefly reviewed in Section 9-21.

Earthquake Effects The specifications establish provisions for bridges to minimize their susceptibility to damage from earthquakes. The provisions apply to slab, girder, box girder, and truss superstructures with spans up to 500 ft.

The acceleration coefficient A to be used in the application of these provisions is determined from appropriate contour maps. Competent technical advice should be sought for sites located close to active faults. With the coefficient A thus selected, the bridge will be assigned to one of the four seismic zones in accordance with Table 2-15. These seismic zones reflect the variation in seismic risk across the country and relate to different requirements for methods of analysis, minimum support lengths, column details, and

TABLE 2-15 Seismic Zones (LRFD Specifications)

Acceleration Coefficient	Seismic Zone
$A \leq 0.09$	1
$0.09 < A \leq 0.19$	2
$0.19 < A \leq 0.29$	3
$0.29 < A$	4

foundation design procedures. The seismic zones in Table 2-15 correspond to the seismic performance categories A, B, C, and D, respectively, articulated by the seismic provisions of AASHTO.

In addition, site effects should be included in the determination of seismic loads, depending on soil conditions. Essentially, four soil profiles are introduced to define the site coefficient to be used in modifying the acceleration coefficient. These profiles represent different subsurface conditions selected on the basis of a statistical study of spectral shapes developed on soils close to seismic activity zones.

Methods of Structural Analysis

In general, the specifications give preference to elastic methods of analysis, but inelastic methods are encouraged for specific cases. Inelastic redistribution of force effects in some types of structures is explicitly outlined.

Certain advantages associated with elastic analysis have been outlined in the foregoing sections. An inherent inconsistency is that the analysis is based on material linearity, but the resistance model may be based on inelastic behavior for the strength limit states. This inconsistency has existed with previous and current load factor design methods. Thus, the load factors developed in the LRFD specifications are based on probabilistic principles combined with analysis using linear material models.

With the advent of rapidly improving computer technology, the classical force and displacement methods (amenable to hand calculations) can be supplemented or replaced by finite-difference and finite-element techniques, folded-plate methods, finite-strip methods, grillage analogy, series or other harmonic methods, and yield line theories.

Structural Material Behavior Materials are considered to behave linearly up to the elastic limit and inelastically thereafter. For normal beam–slab bridges, stiffness characteristics may be based on full participation of uncracked concrete. For concrete and composite members, stiffness should be consistent with the anticipated behavior and may involve cracked or uncracked sections. Extreme-event limit states may be accommodated in either the elastic or the inelastic range.

Inelastic analysis should be applied only to components that contain materials that are truly ductile or can be made to behave in a ductile manner. Only factored loads may be used in the elastic range, and no superposition of force effects may be applied. The order of load application should be consistent with the load sequence on the actual bridge.

Geometry Small-deflection theory applies where the deformation of the structure does not result in a significant change in force effects (e.g., beam-type bridges, trusses, and tied arches).

If the deformation of the structure results in a significant change in force effects, the effects of deformation should be considered. Such structures include suspension bridges, very flexible cable-stayed bridges, arches other than tied arches, and some frames. In such cases, large-deflection theory (or second-order analysis) may be necessary.

With large-deflection analysis, moment magnification is not required. However, deformation effects increase the eccentricity of axial forces, and the result is a loss of stiffness (an apparent softening of the member). The axial compressive stress becomes a significant percentage of the Euler buckling stress, and must be included in the analysis.

Static Analysis The influence of plan geometry is essential. When transverse distortion of a superstructure is small compared to the longitudinal deformation, it does not markedly affect the load distribution; hence, an equivalent beam representation is appropriate. The limit of such an idealized single beam is expressed in terms of the length–width ratio defined as the *plan aspect ratio*. If this ratio exceeds 2.5, the equivalent beam is applied.

Distribution of Loads: Approximate Methods

Deck Slabs For an approximate method of analysis, the deck is subdivided into strips perpendicular to the supporting members. This approach is acceptable for decks other than fully filled and partially filled grids. Each strip is loaded for extreme effect using the dual wheels of the design truck. The strips may be treated as continuous or simply supported beams, as appropriate. The width of equivalent strips is a function of the spacing of supporting components, the deck depth, and the span length.

Beam–Slab Bridges Approximate methods may be used for beam–slab bridges where (a) the width is constant, (b) the number of beams is not less than four, (c) the slab overhang does not exceed 3 ft, and (d) the curvature in plan satisfies certain criteria. Interestingly, results of analysis of continuous beam–slab bridges show that the distribution coefficients for negative moments exceed those obtained for positive moments by about 10 percent. On the other hand, the stresses near an internal bearing are reduced by the

TABLE 2-16 Common Deck Superstructures Covered in the Approximate Method

SUPPORTING COMPONENTS	TYPE OF DECK	TYPICAL CROSS-SECTION
Steel Beam	Cast-in-place concrete slab, precast concrete slab, steel grid orthotropic steel, glued/spiked plank, stressed wood	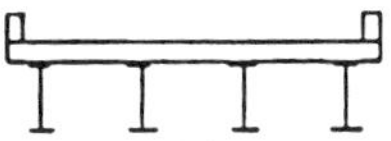 (a)
Closed Steel or Precast Concrete Boxes	Orthotropic steel deck on steel boxes Cast-in-place concrete slab on either steel or concrete boxes	 (b)
Open Steel or Precast Concrete Boxes	Cast-in-place concrete slab	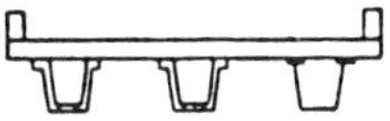 (c)
Cast-in-Place Concrete Multi-cell Box	Monolithic	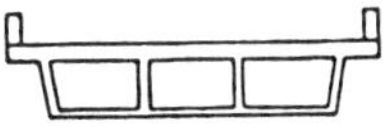 (d)
Cast-in-Place Concrete Tee Beam	Monolithic	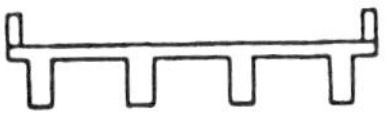 (e)
Precast Solid, Voided or Cellular Concrete Boxes with Shear Keys	Cast-in-place Concrete Overlay	 (f)
Precast Solid, Voided or Cellular Concrete Box is Transverse Post-Tensioning	N/A	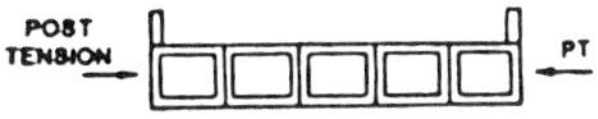 (g)
Precast Concrete Channel Section with Shear Keys	Cast-in-place concrete overlay	 (h)

TABLE 2-16 *(Continued)*

SUPPORTING COMPONENTS	TYPE OF DECK	TYPICAL CROSS-SECTION
Precast Concrete Double Tee Section with Transverse Post-Tensioning	N/A	POST TENSION PT (i)
Precast Concrete Tee Section	N/A	POST TENSION PT (j)
Precast Concrete I or Bulb-Tee Sections	Cast-in-place Concrete	(k)
Wood Beams	Cast-in-place concrete or plank, glued/spiked panels or stressed wood	(l)

fanning of the reaction force by about the same amount, so that the two tend to cancel each other.

The approximate distribution described in this section applies to deck cross sections that are consistent with one of the cross sections shown in Table 2-16.

The live load distribution factor for moment is essentially the same as that stipulated in NCHRP Project 12-26, discussed in Section 2-12, an example of which is shown in Figure 2-41. A factor of 0.5 is applied because these loads are per lane.

Truss and Arch Bridges For trusses analyzed as planar structures, the live load may be obtained from the lever rule (assuming the slab acts as a simple beam). If space analysis is used, either the lever rule or direct loading of the deck may be used. For loads other than the dead weight of members and wind loads theorem, the truss may be analyzed as a pin-connected assembly.

Distribution of Loads: Refined Methods Refined methods relate to the experimentally observed response of bridges. Consideration is given to aspect ratios of elements, positioning and number of nodes, and other features of topology that may influence the results of the analysis. Criteria are provided for deck slabs, beam–slab bridges, cellular and box bridges, truss bridges, arch bridges, cable-stayed bridges, and suspension bridges. Grid analysis appears to be favored (see also Section 2-4) and signifies the intent to recognize the three-dimensional effect of bridge superstructures with a definite interconnection of longitudinal and transverse beams, cross frames, and diaphragms. Live load forces in diaphragms should be calculated by grid and finite-element analyses, and preferably computed by influence surfaces analogous to the forces of the main longitudinal members. For cellular and box beams, flexural as well as torsional effects should be considered in the analysis. For truss bridges, load applied to the deck or floor beams, rather than the truss joints, will yield an analysis that better quantifies out-of-plane action.

For arch bridges, the analysis should consider rib shortening (see also Chapter 10). A three-hinged arch is statistically determinate, so that stresses resulting from temperature changes and rib shortening are essentially eliminated. Arches fixed at the abutments have live load moments and reactions that are smaller than those of hinged arches.

The distribution of force effects in cable-stayed bridges may be determined by spatial structural analysis or by planar analysis. These bridges should be investigated for nonlinear effects resulting from (a) the change in cable sag at all limit states, (b) the deflection of deck superstructure and towers at all limit states, and (c) material nonlinearity at the extreme-event limit states. Cable-stayed bridges should be investigated for the loss of any one cable stay.

Distribution of Negative Moment in Beam – Slab Bridges The redistribution mentioned in Article 10.48.1.3 of the AASHTO standard specifications may be considered due to the inelastic behavior of continuous deck–beam bridges. In this case inelastic behavior due to shear or uncontrolled buckling is not permitted. Redistribution of flexural moments may be considered in the longitudinal direction only. Reduction in the negative moment over the supports should be accompanied by a commensurate increase in the positive moment in the span.

Dynamic Analysis In general, analysis of vehicle- and wind-induced vibrations is not considered in bridge design. When a vehicle crosses a bridge, the structure is analyzed by considering static vehicle locations and applying a dynamic load allowance to account for the dynamic response caused by the movement. Flexible bridges, however, may be excited by live load moment beyond the impact allowance. Flexible continuous bridges may be particularly susceptible to vibrations and thus require analysis for moving live load.

Dynamic models should include relevant aspects of the structure and the excitation. Relevant aspects of the structure are the distribution of mass, the distribution of stiffness, and damping characteristics. Relevant aspects of excitation are the frequency of the forcing action, the duration of application, and the direction of application.

Inelastic dynamic response is enhanced during extreme events where energy is dissipated by (a) elastic and inelastic deformation of a mass colliding with the structure; (b) inelastic deformation of the bridge and its components; (c) permanent displacements of both; and (d) inelastic deformation specially introduced. Energy absorbed by inelastic deformation may be assumed to be concentrated in plastic hinges and yield lines. The location of these sections may be determined by successive approximation to obtain a lower bound solution for the energy absorbed.

For the purpose of establishing deformation limitations, sections of components subjected to inelastic deformation should be demonstrated to have a ductile response or should be made ductile by confinement. The analysis should also determine that shear and bond failure do not precede the formation of a flexural plastic mechanism. Where the geometric integrity of the structure is disrupted by large deformations, the resulting effects should be considered.

Analysis for Earthquake Effects Minimum analysis requirements for seismic effects are stipulated in Table 2-17.

Single-span bridges and bridges in seismic zone 1 do not require seismic analysis, regardless of their importance and geometry. These bridges may, however, require special connections between superstructure and abutments as well as a minimum set width.

Minimum analysis requirements for multispan bridges should be taken from Table 2-17, where the notation is as follows:

* = no seismic analysis required
SM = single-mode elastic method
MM = multimode elastic method
TH = time history method

Essential bridges are those that must be open to emergency vehicles and for defense–security purposes after an earthquake event. A regular bridge

TABLE 2-17 Minimum Analysis Requirements for Seismic Effects (LRFD Specifications)

		Multispan Bridges					
		Other Bridges		Essential Bridges		Critical Bridges	
Seismic Zone	Single Span	Regular	Irregular	Regular	Irregular	Regular	Irregular
1	No seismic design required	*	*	*	*	*	*
2		SM	SM	SM	MM	MM	MM
3		SM	MM	MM	MM	MM	TH
4		SM	MM	MM	MM	TH	TH

has no abrupt change in stiffness or mass along its length. A curved bridge may be considered regular if the subtended angle between abutments is less than 60°. All other bridges are considered irregular.

Analysis by Physical Models Scale model testing may in some instances be indicated to establish and verify structural behavior. Dimensional and material properties as well as boundary conditions should be modeled with sufficient accuracy. For dynamic analysis consideration must be given to inertial scaling, load and excitation, and damping functions. Factored dead loads should be simulated for strength limit states.

Existing bridges may be instrumented, and the results thus obtained may be used for structural assessment in order to establish force effects.

2-14 PARALLEL GIRDER SYSTEMS

A simple interconnected system of parallel girders is the open grid shown in Figure 2-43. For effective transfer of load, the cross girders (or diaphragms) should be structurally continuous through the interior main girders. More commonly, the main structure and the deck beams are integral with a continuous deck slab. For analysis purposes, the slab may be subdivided into (a) areas acting as upper flanges of beams and (b) strips acting as additional transverse or longitudinal elements. Figure 2-44*a* shows three main girders joined only by a slab at the upper flange level. The effective cross section of a single girder is as shown in Figure 2-44*b* and includes part of the slab. The slab may also be divided into equivalent transverse beams to form the open-grid system shown in Figure 2-44*c*.

Evidently, this idealization will result in some degree of approximation if it ignores the capacity available across the transverse cuts to transmit axial forces, horizontal and vertical shears, and bending and twisting moments. Unreasonable discontinuities in the stresses in the slab may result.

Alternatively, a complete girder bridge may be regarded as a stiffened system, articulated by the typical cross sections in Figure 2-45. The case shown in Figure 2-45*b* is a single multicell tubular girder. The cases shown in

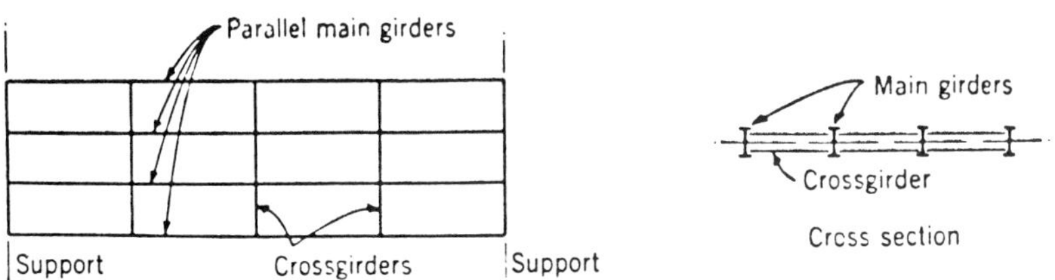

FIGURE 2-43 Open-grid system.

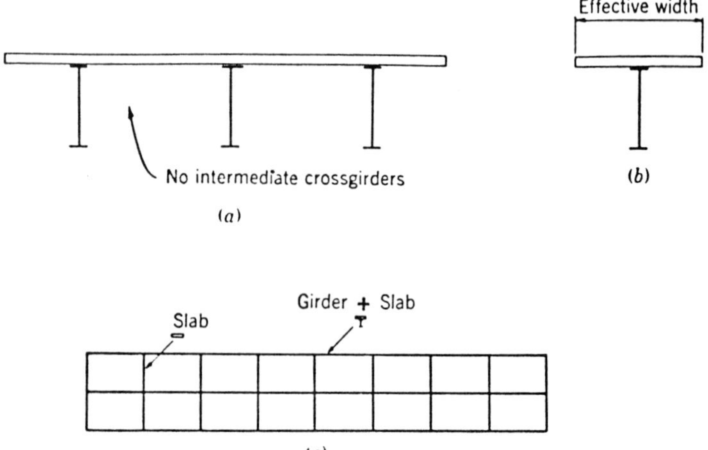

FIGURE 2-44 Idealization of deck slab to form equivalent open grid: (*a*) bridge cross section; (*b*) girder cross section; (*c*) equivalent open grid.

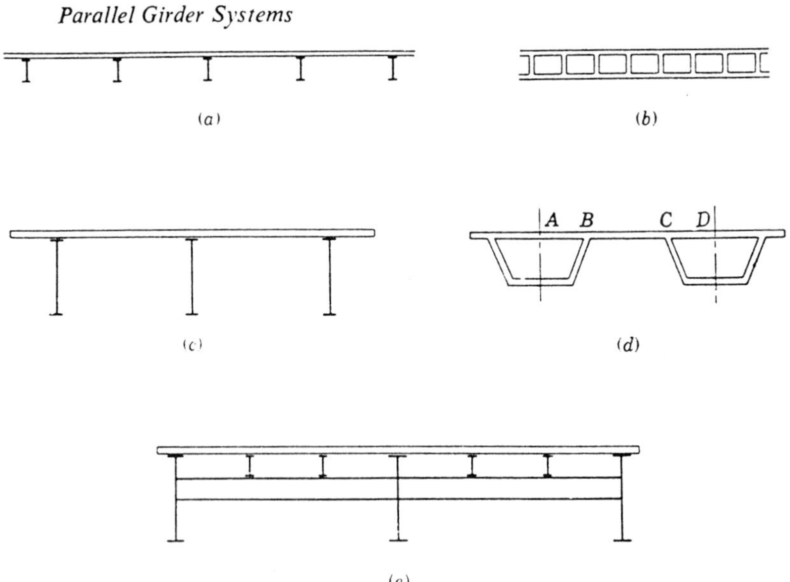

FIGURE 2-45 Typical cross section; plain and stiffened deck systems.

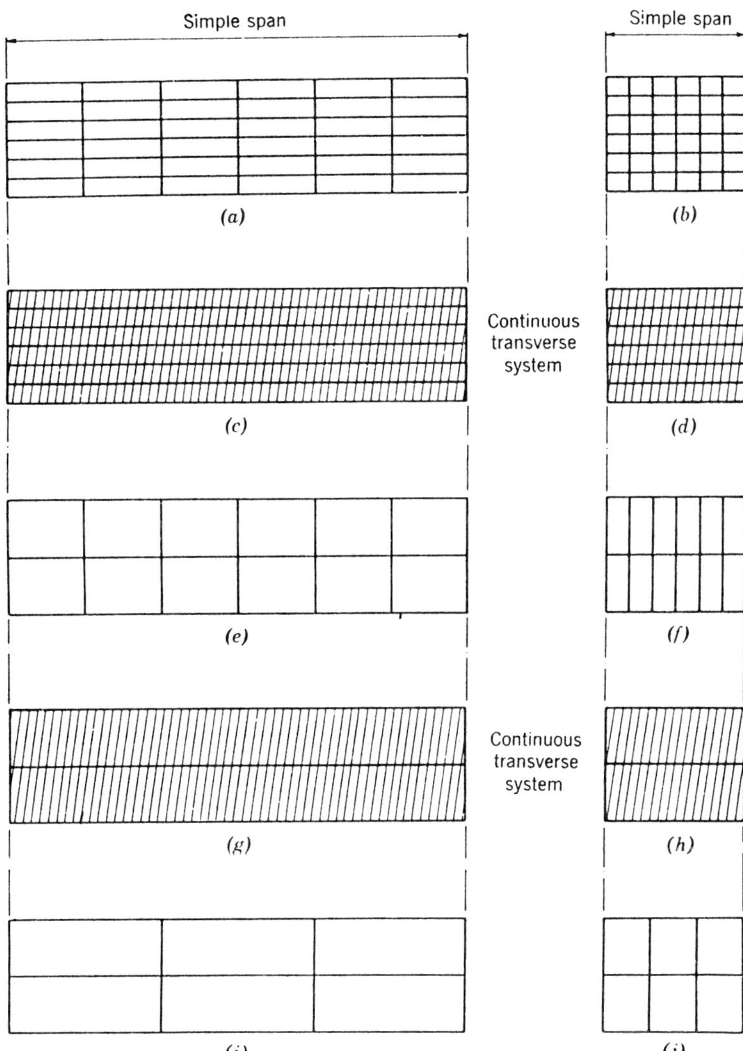

FIGURE 2-46 Typical plan geometries and continuity.

Figures 2-45*a* through *d* may be constructed with or without cross girders or diaphragms. The case shown in Figure 2-45*e* incorporates cross girders as an essential part of a subdivided deck system.

The plan geometry varies within a broad range of configurations. As shown in Figure 2-46, the basic variables are the span–width ratio (also called the aspect ratio), the number of longitudinal girders, and the type of transverse connection. In addition, the structure may be a simple span or continuous. There are three primary structural combinations.

1. Both longitudinal and transverse elements are continuous or closely spaced, and typical cases are shown in Figures 2-46*a* through *d*. In Figures 2-46*c* and *d*, the transverse system is provided only in the slab and is distributed continuously along the length.
2. The system has only a few longitudinal members connected by a dense (continuous or closely spaced) transverse network as shown in Figures 2-46*g* and *h*.
3. Both longitudinal and transverse systems consist of a few widely spaced members as shown in Figures 2-46*i* and *j*.

A certain degree of idealization of the real structure is required with any method of analysis, but as this process is formulated the results of certain analyses may be more accurate and relevant to certain geometries. For example, the structure mentioned previously in combination 1 may be replaced by an equivalent orthotropic slab; likewise, a bridge with a large span–width ratio and a few longitudinal members adequately connected by cross girders and diaphragms may be analyzed as an equivalent single or line member.

The following methods of analysis are possible: (a) open-grid analysis, (b) Guyon–Massonnet–Bares othrotropic plate theory, (c) Wagner theory, (d) folded-plate analysis, and (e) special methods for structures with a continuous transverse system.

A brief review of each method is provided in this section merely to describe the underlying assumptions and the range of applicability.

Open-Grid Analysis An open grid of the type shown in Figure 2-43 may be analyzed using relevant computer programs (Lightfoot and Sawko, 1960; Livesley, 1964; Sawko, 1965). The most common solution is obtained by the displacement method based on member stiffness. There are three unknown displacements at every joint: a vertical displacement v normal to the grid and rotations θ_x and θ_z about axes lying in the grid. The grid is assumed to be a plane and displacement in this plane is ignored (rotation $\theta_y = 0$).

In the total stiffness matrix the joint loads W, M_x, and M_z are expressed in terms of the corresponding displacements using member stiffness. The joint loads are known and appear directly in the analysis. Loads applied to a member between joints produce fixed-end moments and reactions that appear as joint loads. Final moments in the member represent the sum of the initial fixed-end moments and those corresponding to displacements.

The bridge may be simply supported or continuous and have any proportions, but the practical number of unknowns provides a limit to the complexity of the structure that can be solved. Alternatively, the grid may be solved by the force method, forming a base structure by inserting a cut in each transverse member. If all members have zero torsional stiffness, the only inferior redundant forces are vertical and the number of unknowns is reduced considerably. Likewise, if the transverse members are torsionally weak, the moments M_x disappear.

The designer, however, is cautioned that certain difficulties may arise as follows (O'Conner, 1971).

1. A vertical misalignment may exist between the centroidal and shear center axes of adjacent members. This may give rise to stress components not predicted by the grid analysis (Hondros and Marsh, 1960).

2. The members may have a finite width as in the cross section shown in Figure 2-45*d* consisting of two tubular sections connected by a slab. In the grid analysis the slab may be replaced by a set of discrete beams of equivalent stiffness, but the length of these beams is much less than the distance between centerlines of the main members. This difficulty may be overcome by adjusting the member stiffness. For the example shown, the member may be taken as having length *AD* with an infinite stiffness over *AB* and *CD*.

3. The omission of torsional bending may give rise to errors, but it is the best available approximation in a majority of cases.

4. The analysis may lead to stress distributions that are incompatible across the deck slab, although in many bridges the slab area is sufficiently large having a confining effect on stresses.

Orthotropic Plate Theory This analysis was developed by Guyon (1946), expanded by Massonnet (1950) to include torsion, and further extended by Bares (1965). The method is presented in Chapter 7.

Wagner Theory Notable applications of this theory include a single member of open cross section (e.g., torsion of beams with open section) and single or multicell closed tubes.

The basic assumption is that the cross section does not distort, thus implying the presence of a closely spaced system of rigid diaphragms. For a bridge with a high span–width ratio, this assumption is reasonable although the structure may consist of a series of parallel members. For analysis purposes, the complete structure is replaced by an equivalent line at the shear center of the cross section. A load intersecting this line produces bending only, but an eccentric loading will introduce twisting moments.

Subject to certain restrictions, the Wagner method may be used to determine the load distribution in a parallel girder system. It offers the advantage of incorporating the deck plate in the analysis, and may be used for simple- as well as continuous-span bridges.

Folded-Plate Analysis This method has been used for tubular girders but may also be applied to parallel girder systems. It divides any applied load into Fourier components applied at the joints of the cross section. Each component is applied separately and analyzed by folded-plate theory.

Parallel Girder Systems

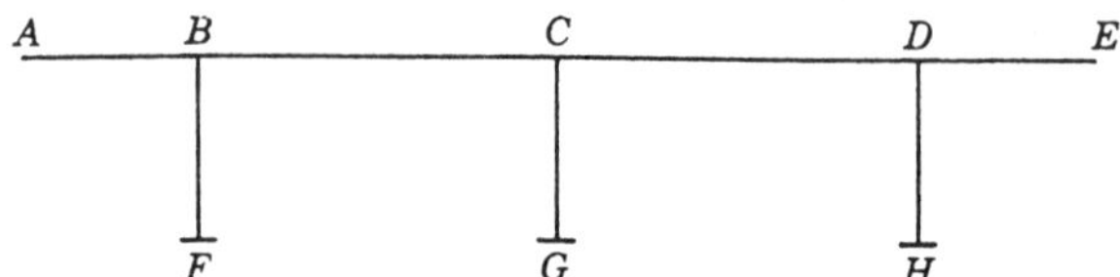

FIGURE 2-47 Cross section of girder bridge without cross girders.

The best results are obtained with simple-span bridges without intermediate cross girders and diaphragms. The method may be made exact by including sufficient Fourier components. Proper allowance is made for all parts by members of the structure, and the effects of shear distortions are included. It may be used for open or closed sections, such as those shown in Figures 2-45*c* and *d*, and for all plan proportions without diaphragms, such as those shown in Figures 2-46*g* and *h*.

For the structure shown in Figure 2-47, the lower flanges may be replaced by concentrated areas whose stiffness against longitudinal deformation may be readily calculated. Including the bending stiffness of the deck slab in the analysis, the relevant displacements consist of rotations θ_z at each of the five deck points A–E, together with displacements u, v, and w at all eight points A–H of the cross section.

Special Methods for Continuous Transverse Systems For longitudinal girders connected either by the deck alone or by the deck in conjunction with many closely spaced cross members, it is reasonable to replace the transverse system by a continuous elastic connection between the main members (Hendry and Jaeger, 1956, 1958). For the cross section shown in Figure 2-48*a*, the tubular members *BCGH* and *DEJK* are linked by the slab *CD* without diaphragms. If the main girders exhibit large torsional stiffness, the only allowable relative movement is as shown in Figure 2-48*b*. For a unit length of slab, it is possible to estimate the force V required for a relative displacement δ equal to unity, based on distortion in both the slab and the girder walls. The slab is equivalent to a spring restraint of this stiffness.

Any load applied to one girder may be regarded as the sum of the components shown in Figure 2-48*c*. The symmetrical component produces equal deflections of the girders and does not complicate the distribution problem. For the antisymmetrical loading, the midpoint of the slab is a point of contraflexure and does not deflect. Thus, each girder may be treated as a beam on elastic foundation.

The analysis may be extended to include twisting of the main members and systems consisting of more than two main members. A comprehensive review of these methods in a simplified form is given by Bakht and Jaeger (1985).

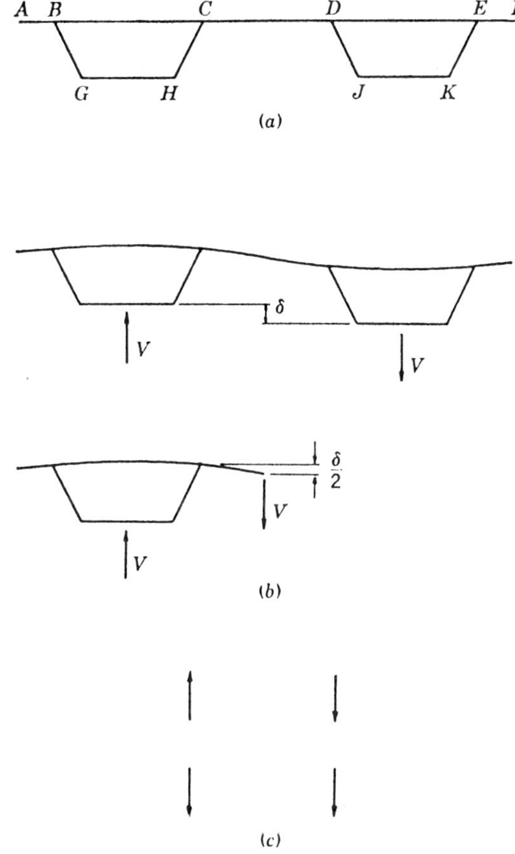

FIGURE 2-48 Deck slab replaced by continuous elastic restraint: (*a*) cross section of bridge without cross girders; (*b*) deformation; (*c*) component loads.

Commentary Results obtained from the use of these methods suggests that the most effective diaphragm or cross frame is one provided at the load itself. Once this interaction is established, the behavior is not very sensitive to the spacing of the remaining cross frames and diaphragms. For moving loads, diaphragm location and spacing give rise to continuous changes in behavior.

The convenience of replacing an arbitrary longitudinal line load by a set of harmonic components should not be underestimated. For a simply supported girder, these components are sinusoidal. For continuous girders, they may be taken as similar to the natural modes of vibration of the girder considered alone (Hendry and Jaeger, 1958). If the load acting on one girder is replaced by a series of harmonics, the share of this load taken by adjoining girders tends to be dominated by the lower order of harmonics, and it is often reasonable to approximate the share by the first harmonic alone. The net load in the loaded girder is then found by subtracting the share on all other girders from the total load.

A useful review of the literature on grillages and stiffened plates is given by Kerfoot and Ostapenko (1967). For a slab on flexible beams, most useful data are provided by Newmark (1938). The case of hollow-slab bridges formed by precast concrete box girders joined by a shear key and transverse ties represents a popular bridge type (Cusens and Pama, 1965; Nasser, 1965; Pool, Arya, Robinson, and Khachaturian, 1965).

The popularity of composite bridges relates to the efficiency of incorporating the floor slab as part of the upper flange of the main and secondary girders. The phenomenon known as shear lag is customarily dealt with by replacing the wide flange by an imaginary uniformly stressed flange having a reduced width, called the *effective width*.

2-15 STRUT-AND-TIE MODEL

These are conceptual models used to proportion reinforcement and concrete sections in regions of concentrated loads and supports, and by extension in areas of geometric discontinuities. Where conventional methods are not adequate because of nonlinear strain distribution, the strut-and-tie model can provide an acceptable solution in approximating load paths and force effects. This model is fairly new, and is included in the LRFD specifications.

Strut-and-tie models should be considered in the design of deep footings and pile caps where the distance between the applied load and the supporting reactions is less than about twice the member thickness.

Conventional section-by-section design assumes that the steel reinforcement depends on the calculated values of V_u, M_u, T_u, and N_u (factored shear force, bending moment, torsion, and axial load), and is not related to the support details or to the manner in which the force effects are applied. The conventional design also assumes that the shear flow remains constant and that the longitudinal strains will vary linearly over the depth of the beam. However, for members such as the beam shown in Figure 2-49, these assumptions are not valid because the shear stresses on a section just to the right of support A will be concentrated near the bottom face. The behavior of a component such as this deep beam can be better predicted if the flow of forces through the complete structure is considered. This means a study of the flow of compressive stresses going from load P to supports A and B together with the required tension force to be developed between these supports.

In cracked reinforced concrete, the load is essentially carried by compressive stresses in the concrete and tensile stresses in the reinforcement. After significant cracking, the principal compressive stress trajectories in the concrete approach straight lines and can be represented by straight compressive struts. Tension ties are used to model the main reinforcement. For the

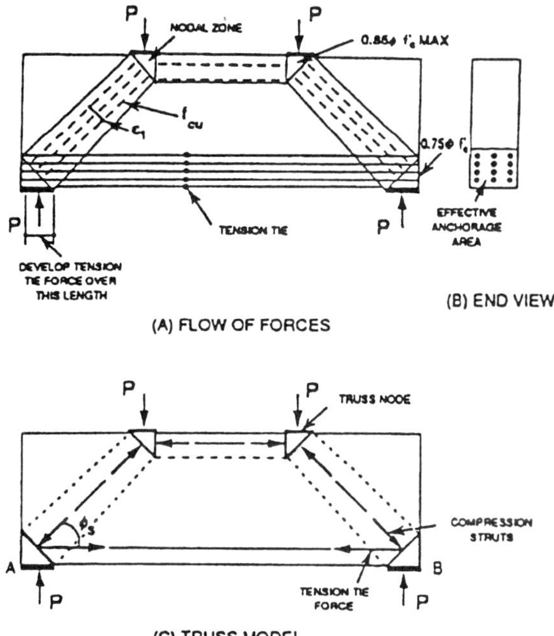

FIGURE 2-49 Strut-and-tie truss model for a deep beam.

strut-and-tie truss model shown in Figure 2-49, compressive struts represent the zones of high unidirectional compressive stress in the concrete. Regions subjected to multidirectional stresses (where the struts and ties meet the joints of the truss) are represented by nodal zones.

The LRFD specifications discussed in Section 2-13 consider strut-and-tie models a viable design option. These specifications include provisions for (a) structural optimization considering the factored resistance of struts and ties; (b) proportioning of compressive struts; (c) proportioning of tension ties; (d) proportioning of node regions; and (e) crack control reinforcement. The last provision is intended to control the width of cracks and to ensure a minimum ductility for the member in order to enhance the redistribution of internal stresses if necessary.

REFERENCES

AASHTO, 1980. "Guide Specifications for Horizontally Curved Highway Bridges," and Interim Spec. 1981, 1982, 1984, 1985, 1986, Washington, D.C.

AASHTO, 1983. "Guide Specifications for Seismic Design of Highway Bridges," and Interim Specifications 1985, 1987–1988, Washington, D.C.

Anger, G., 1956. *Tend Division Influence Lines for Continuous Beams*, Ungar, New York.

Applied Technology Council, 1978. "Tentative Provisions for the Development of Seismic Regulations for Buildings," ATC Report No. 3-06, June, Berkeley.

Applied Technology Council, 1982. "Comparisons of United States and New Zealand Seismic Design Practices for Highway Bridges," ATC-12, Calif.

Arbabi, F., 1991. Structural Analysis and Behavior, McGraw-Hill, New York.

Armenakas, A. E., 1991. Modern Structural Analysis: The Matrix Method Approach, McGraw-Hill, New York.

ASCE, 1981. "Recommended Design Loads for Bridges," Committee on Loads and Forces on Bridges, Committee on Bridges of the Strut. Div., *J. Struct. Div. ASCE*, Vol. 107, No. ST7, July, pp. 1161–1213.

ASCE–AASHTO, 1975. "State-of-the-Art Report on Ultimate Strength Design of I-Beam Bridge Systems," Subcommittee of Joint ASCE–AASHTO Task Committee on Metals of the Struct. Div.

ATC, 1981: "Seismic Design for Highway Bridges," ATC-6, Applied Technology Council, Berkeley, CA.

Bakht, B. and L. G. Jaeger, 1985. "Bridge Analysis Simplified," McGraw-Hill, New York.

Bares, R., 1965. "Complement à la méthode Guyon-Massonet de calcul des ponts à poutres multiples," *Ann. Tran. Publ.*, Belg., 1 (Jan.), pp. 7–69; 2 (April), pp. 145–71.

Basu, S. and M. Chi, 1981. "Analytic Study for Fatigue of Highway Bridge Cables," Report No. FHWA-RD-81-090, FHWA, U.S. Dept. Transp., Washington, D.C.

Basu, S. and M. Chi, 1981. "Design Manual for Bridge Structural Members under Wind-Induced Excitations," Report No. FHWA TS-81-206, FHWA, U.S. Dept. Transp., Washington, D.C.

Beal, D. B., 1982. "Load Capacity of Concrete Bridge Decks," *J. Struct. Div. ASCE*, Vol. 108, No. ST4, April, pp. 814–831.

British Standards Institution, 1978. "Steel, Concrete and Composite Bridges. Part 2: Specification for Loads," B.S. 5400, British Standards Institution, London.

Buckland, P. G., et al., 1978. "Traffic Loading of Long Span Bridges," *Proc. Conf. Bridge Eng.*, Trans. Research Board, *Transp. Research Record* 665, Sept., pp. 146–154.

Buckland, P. G., et al., 1978. "Proposed Vehicle Loads for Long Span Bridges," ASCE Spring Conv., April, Preprint 3148, Pittsburgh, Pa.

Buckland, P. G., F. P. D. Navin, J. V. Zidek, and J. P. McBryde, 1980. "Proposed Vehicle Loading for Long-Span Bridges," *J. Struct. Div. ASCE*, Vol. 106, No. ST4, April, pp. 915–932.

Canadian Standards Association, 1978. "Design of Highway Bridges," National Standards of Canada CAN 3-S6-M78, Rexdale, Canada.

Chapman, H. E., 1973. "Earthquake Resistant Design of Bridges and the New Zealand Ministry of Works Bridge Design Manual," *Proc. 5th World Conf. on Earthquake Eng.*, Vol. 2, Rome, pp. 2242–2251.

Collins, M. P. and D. Mitchell, 1991. *Prestressed Concrete Structures*, Prentice-Hall, Englewood Cliffs, N.J.

Cross, H. and N. D. Morgan, 1932. *Continuous Frames of Reinforced Concrete*, Wiley, New York.

Csagoly, P. F. and R. A. Dorton, 1977. "The Development of the Ontario Bridge Code," Ontario Ministry Transp. and Communications, Canada.

Cusens, A. R. and R. P. Pama, 1965. "Design of Concrete Multibeams Bridge Decks," *Proc. ASCE, J. Struct. Div. ASCE*, Vol. 91, No. ST5, Oct., pp. 255–278.

Dhan, E., C. Massonnet, and J. Seyvert, 1955. "Experimental Research on Multiple Beam Bridges," *Annales des Travaux Publics de Belgique*, No. 2, Brussels.

Dowrick, D. J., 1987. *Earthquake Resistance Design*, Wiley, New York.

Ellingwood, B., T. V. Galambos, J. G. MacGregor, and C. A. Cornell, 1980. "Development of a Probability Based Load Criterion for American National Standard A58," NBS Special Publ. 577, National Bureau of Standards, U.S. Dept. Commerce, Washington, D.C.

Ellingwood, B., J. G. MacGregor, T. V. Galambos, and C. A. Cornell, 1982. "Probability Based Load Criteria: Load Factors and Load Combinations," *J. Struct. Div. ASCE*, Vol. 108, No. ST5, May, pp. 978–996.

Elms, D. G. and G. R. Martin, 1979. "Factors Involved in the Seismic Design of Bridge Abutments," *Proc. Workshop on Earthquake Resistance of Hwy. Bridges*, Applied Technology Council, Berkeley, Jan.

Eyre, D. G. and T. V. Galambos, 1973. "Shakedown of Grids," *J. Struct. Div. ASCE*, Vol. 99, No. ST10, Oct., pp. 2049–2060.

Ferritto, J. M. and J. B. Forest, 1977. "Determination of Seismically Induced Soil Liquefaction Potential at Proposed Bridge Sites," Fed. Hwy. Admin., Office of Research and Development, Washington, D.C.

Frangopol, D. M. and R. Nakib, 1991. "Redundancy in Highway Bridges," *Eng. Journ.*, AISC, Vol. 128, No. 1.

Fung, G., et al., 1971. "Field Investigation of Bridge Damage in the San Fernando Earthquake," Bridge Dept., Div. of Highways, Calif. Dept. Transp., Sacramento.

Gates, J. H., 1979. "Factors Considered in the Development of the California Seismic Design Criteria for Bridges," *Proc. Workshop on Earthquake Resistance of Hwy. Bridges*, Applied Technology Council, Berkeley, Jan.

Gates, J. H., 1976. "California's Seismic Design Criteria for Bridges," *J. Struct. Div. ASCE*, Vol. 102, No. ST13, Dec., pp. 2301–2314.

Guyon, Y., 1949. "Analysis of Slab Bridges," *Ann. Ponts Chaussées*, Vol. 119, No. 29, pp. 555–589.

Guyon, Y., 1946. "Calcul des ponts a poutres multiples solidarisées par des entrefoises," ("Analysis of Wide, Multiple Beam Bridges Stiffened by Cross Beams") *Ann. Ponts Chaussées*, Vol. 166, Nos. 9, 10, Paris, pp. 553–612.

Haaijer, G., 1983. "Limit States Design—A tool for Reducing the Complexity of Steel Structures," paper presented at AISC National Eng. Conf., March.

Healey, A. J., E. Nathman, and C. C. Smith, 1977. "An Analytical and Experimental Study of Automobile Dynamics with Random Roadway Inputs," *Trans. ASME*,

Journ. of Dynamic Systems, Measurement and Control, Series G., Vol. 99, Dec., pp. 284–292.

Heins, C. P. and C. F. Galambos, 1972. "Highway Bridge Field Tests in the U.S., 1948–1970," *Public Roads*, U.S. Dept. Transp., Fed. Hwy. Admin., Feb., pp. 271–291.

Heins, C. P. and C. T. G. Looney, 1966. "An Analytical Study of Eight Different Types of Highway Structures," Civ. Eng. Report No. 12, Univ. Maryland, College Park, Sept.

Heins, C. P. and C. T. G. Looney, 1966. "The Solution of Continuous Orthotropic Plates on Flexible Supports as Applied to Bridge Structures," Civ. Eng. Report No. 10, Univ. of Maryland, College Park, March.

Heins, C. P. and C. T. G. Looney, 1968. "Bridge Analysis Using Orthotropic Plate Theory," *J. Struct. Div. ASCE*, Vol. 93, No. ST2, Feb., pp. 565–592.

Heins, C. P. and C. T. G. Looney, 1969. "Bridge Tests Predicted by Finite Difference Plate Theory," *J. Struct. Div. ASCE*, Vol. 95, No. ST2, Feb., pp. 249–265.

Henderson, W., 1954. "British Highway Bridge Loading," *Proc. Institution Civ. Engineers*, Road Paper No. 44, March, pp. 325–373.

Hendry, A. W. and L. G. Jaeger, 1958: *The Analysis of Grid Frameworks and Related Structures*, Chatto and Windus, London; Prentice-Hall, Englewood Cliffs, N.J., 1959, 308 pp.

Hendry, A. W. and L. G. Jaeger, 1959. *The Analysis of Grid Frameworks and Related Structures*, Prentice-Hall, Englewood Cliffs, N.J.

Ho, M. and Y. Nakamura, 1982. "Aerodynamic Stability of Structures in Wind," IABSE Surveys S-20/82, Int. Assoc. for Bridge and Struct. Eng., ETH-Hongger-berg, Zurich, May, pp. 33–56.

Hino, M., 1978. Spectral Analysis, Asakura Book, Co., Tokyo, Japan, pp. 183–227.

Hoel, P. G., 1965. Introduction to Mathematical Statistics, Wiley, New York.

Honda, H., Y. Kajikawa, and T. Kobori, 1982. "Spectra of Road Surface Roughness on Bridges," *J. Struct. Div. ASCE*, Vol. 108, No. ST9, Sept., pp. 1956–1966.

Hondros, G. and J. G. Marsh, 1960: "Load Distribution in Composite Girder-Slab Systems," *Proc. ASCE, J. Struct. Div.*, 86(ST11), 79–109 (Nov.), paper 2645.

Huang, T. and A. S. Veletsos, 1960. "Dynamic Response of Three-Span Continuous Highway Bridges," Tenth Progress Report on Highway Bridge Impact Investigation, Univ. Illinois, Urbana, Sept.

Imbsen, R. A., R. V. Nutt, and J. Penzien, 1979. "Evaluation of Analytical Procedures used in Bridge Seismic Design Practice," *Proc. Workshop on Earthquake Resistance of Hwy. Bridges*, Applied Technology Council, Berkeley, Calif., Jan.

Ivy, R. J., et al., 1953. "Live Loading for Long Span Highway Bridges," *Trans. ASCE*, Paper No. 3708, June.

Iwasaki, T., J. Penzien, and R. Clough, 1972. "Literature Survey—Seismic Effects on Highway Bridges," Report EERC 72-11, Earthquake Eng. Research Center, Univ. Calif. Berkeley, Nov.

Japan Society of Civil Engineering, 1977. "Earthquake-Resistant Design of Bridges," Bridge and Struct. Committee, Japan Society Civil Engineering, Tokyo.

Jensen, V. P., "Moments in Simple Span Bridge Slabs with Stiffened Edges," *Univ. Illinois Bull. No.* 315, Urbana.

Jensen, V. P., "Solutions for Certain Rectangular Slabs Continuous over Flexible Supports," Univ. Illinois, Bull. No. 303, Urbana.

Kennedy, D. J. L., 1985. "North American Limit States Design," *Proc. 1985 Int. Eng. Symp. Struct. Steel*, Chicago, AISC, May.

Kerfoot, R. P. and A. Ostapenko, 1967. "Grillages under Normal and Axial Loads—Present Status,"Lehigh Univ., Fritz Eng. Lab. Report 323.1, June.

King, Csagoly, P. F., and Fisher, 1975. "Field Testing of the Aquasabon River Bridge," Ontario, personal communication.

Kleinlogel, A. and A. Haselbach, 1963. "Multibeam Frames," Ungar, New York.

Kuo, T. C., 1973. "Elasto-Plastic Behavior of Composite Slab Highway Bridges," Ph.D. Thesis, Univ. Maryland, College Park.

Kuo, T. C. and C. P. Heins, 1973. "Live Load Distribution on Composite Highway Bridges at Ultimate Load," Civ. Eng. Report No. 53, Univ. Maryland, College Park.

Larrabee, R. D. and C. A. Cornell, 1981. "Combinations of Various Loads Processes," *J. Struct. Div. ASCE*, Vol. 107, No. ST1, Jan., pp. 223–239.

Lash, S. D. and R. Nagaraja, 1970. "The Ultimate Load Capacity of Beam and Slab Bridges," Dept. Highways Report No. RR159, Ontario, Canada, May.

Lightfoot, E. and F. Sawko, 1960. "Analysis of Grid Frameworks and Floor Systems by Electronic Computer," *Struct. Engineer*, Vol. 38, No. 3, March, pp. 79–87.

Little, W. A. and R. J. Hansen, 1963. "The Use of Models in Structural Design," *Journ. Boston Society of Civ. Eng.*, Vol. 50, No. 2, April.

Livesley, R. K., 1964. *Matrix Methods of Structural Analysis*, Pergamon Press, London.

MacGregor, J. G., 1976. "Safety and Limit States Design for Reinforced Concrete," *Canadian Journ. Civ. Eng.*, Vol. 3, No. 4, Dec., pp. 484–513.

MacGregor, J. G., 1983. "Load and Resistance Factors for Concrete Design," *ACI J., Proc.*, Vol. 80, No. 4, July–August, pp. 279–287.

MacGregor, J. G., 1988. "Reinforced Concrete—Mechanics and Design," Prentice-Hall, Englewood Cliffs, N.J.

Macias-Rendon, M. A. and D. A. Van Horn, 1973. "Model Study of Beam-Slab Bridge Superstructures," *J. Struct. Div. ASCE*, Vol. 99, No. ST9, Sept., pp. 1805–1821.

Massonnet, C., 1950. "Méthode de calcul des ponts à poutres multiples tenant compte de leur résistance à la torsion," (Analysis Method for Multiple Beam Bridges Taking Account of their Torsional Resistance"), *Publ. Int. Assoc. for Bridge and Struct. Eng.*, No. 10, Zurich, Switzerland, pp. 147–182.

Massonnet, C., 1953. "Study of Structures by Means of Small Scale Models without Use of Microscopes," *Bull. of Center of Studies, Research and Scientific Tests, Civ. Eng.*, Structures, Univ. of Liege, Vol. VI, Belgium.

Massonnet, C., 1954. "Extension to the Method of Analysis of Multiple Beam Bridges," *Ann. des Travaux Publics de Belgique*, Vol. 107, Brussels, pp. 680–748.

Massonnet, C., 1967. "Complete Solutions Describing the Limit of Reinforced Concrete Slabs," *Concrete Research*, Vol. 19, No. 58, pp. 13–32.

Massonnet, C. and M. A. Save, 1965. *Plastic Analysis and Design*, Volume 1, *Beams and Frames*, Blaisdell, New York.

McDermott, R. J., 1978. "Longitudinal Forces on Bridge Bearings," *FHWA Report No. FHWA-NY-78-SR58, National Tech. Information Service, Springfield*, Va.

Murakami, E., et al., 1972. "Actual Traffic Loadings on Highway Bridges and Stress Levels in Bridge Members," Inter. Assoc. for Bridge and Struct. Eng., Paper V1a, Ninth Congress, Amsterdam, May.

Nasser, K. W., 1965. "Design Procedure for Lateral Load Distribution in Multi-Beam Bridges," *J. Prestressed Concrete Inst.*, Vol. 10, No. 4, Aug., 54–68.

Navin, F. P. D., et al., 1976. "Design Traffic Loads for the Lions Gate Bridge," Transp. Research Board, Washington, D.C.

Natrella, M. G., 1966. "Experimental Statistics," National Bureau of Standards Handbook 91, Washington, D.C.

Nelson, H. M., A. Beveridge, and P. D. Arthur, 1963. "Tests on a Model Composite Bridge Girder," *Publications, IABSE*, Vol. 23, Zurich, Switzerland.

Newmark, N. M., 1938. "A Distribution Procedure for the Analysis of Slabs Continuous over Flexible Beams," *Eng. Experiment Station Bull.* No. 304, Univ. Illinois, Urbana.

Newmark, N. M., 1939. "A Distribution Procedure for the Analysis of Slabs Continuous over Flexible Beams," *Univ. of Ill. Bull. No.* 304, Urbana.

Newmark, N. M., 1943. "Numeric Procedures for Computing Deflections, Moments, and Buckling Loads," *Trans. ASCE*, Vol. 108, paper 1161.

Newmark, N. M., 1943. "Numerical Procedures for Computing Deflections, Moments, and Buckling Loads," *Trans. ASCE*, Vol. 108, paper 2202.

Newmark, N. M., 1948. "Design of I-Beam Bridges," *Proc. ASCE*, March, pp. 305–331.

Newmark, N. M., C. P. Siess, and W. M. Packham, 1948. "Studies of Slab and Beam Highway Bridges, Part II, Tests of Simple-Span Skew I-Beam Bridges," *Eng. Experiment Station Bull.* No. 375, Univ. of Illinois, Urbana.

Newmark, N. M., C. P. Siess, and W. M. Packham, 1946. "Studies of Slab and Beam Bridges, Part I, Tests of Simple-Span Right I-Beam Bridges," *Eng. Experiment Station Bull.* No. 363, Univ. of Illinois, Urbana.

Newmark, N. M., and C. P. Siess, 1954. "Research on Highway Bridge Floors," *Univ. of Illinois Bull. No.* 52, Urbana.

O'Conner, C., 1971. *Design of Bridge Superstructures*, Wiley, New York.

Ontario Ministry of Transportation and Communications, 1979. "Ontario Highway Bridge Design Code (OHBDC)," Toronto.

Pool, R. B., A. S. Arya, A. R. Robinson, and N. Khachaturian, 1965. "Analysis of Multibeam Bridges with Beam Elements of Slab and Box Section," *Univ. Illinois, Eng. Experiment Station Bull.*, No. 483, 93 pp.

Reddy, V. M. and A. W. Hendry, 1969. "Ultimate Load Behavior of Composite Steel-Concrete Bridge Deck Structures," *Indian Concrete Journ.*, Bombay, India, May.

Reese, R. T., 1966. "Load Distribution in Highway Bridge Floors: A Summary and Examination of Existing Methods of Analysis and Design and Corresponding Test Results," M.S. Thesis, Brigham Young Univ., Provo, Utah.

Richards, R. and D. G. Elms, 1977. "Seismic Response of Retaining Walls and Bridge Abutments," Report No. 77-10, Univ. Canterbury, Christchurch, New Zealand, June.

Rogers, P., 1953. *Fixed End Moments*, Ungar, New York.

Sanders, W. W. and W. H. Munse, 1960. "The Lateral and Longitudinal Distribution of Loading in Steel Railway Bridges," *Civ. Eng. Studies Structural Research Series*, Report 208, Univ. of Illinois, Urbana.

Sanders, W. W. and H. A. Elleby, 1970. "Distribution of Wheel Loads on Highway Bridges," NCHRP Report No. 83, Hwy. Research Board, Washington, D.C.

Sawko, F., 1965. "Bridge Deck Analysis—Electronic Computer vs. Distribution Methods," *Civ. Eng.*, London, Vol. 60, No. 705, April, pp. 534–538.

Schlaich, J., K. Schafer, and M. Jennewein, 1987. "Towards a Consistent Design of Structural Concrete," *PCI J.*, Vol. 32, No. 3, May–June, pp. 74–151.

Scordelis, A. C., 1966. "Analysis of Simply Supported Box Girder Bridges," Univ. of California, Berkeley.

Scott, R. F., 1973. "The Calculation of Horizontal Accelerations from Seismoscope Records," *Bull. Seismological Soc. America*, Vol. 63, No. 5, Oct.

Seed, H. B. and P. B. Schnabel, 1972. "Accelerations in Rock from Earthquakes in Western United States," Report EERC 72-2, Earthquake Eng. Research Center, Univ. Calif., Berkeley.

Shore, S., and R. Rabizadeh, 1974. "Static and Dynamic Analysis of Horizontally Curved Box Girder Bridges," CURT Report No. T0174, Research Project HR-2(111), Univ. of Pennsylvania, Dec.

Structural Engineers Association of California, 1959. "Recommended Lateral Force Requirements," Seismology Committee, SEAC, San Francisco, Dec.

Structural Engineers Association of California, 1975. "Recommended Lateral Force Requirements and Commentary," 1975 Edition, SEAC, San Francisco.

Transportation Research Board, 1975. "Effects of Studded Tires," National Cooperative Hwy. Research Program (NCHRP), Synthesis of Highway Practice 32, Washington, D.C.

Tseng, W. S. and J. Renzien, 1973. "Seismic Response of Highway Overcrossings," *Proc. Fifth World Conf. Earthquake Eng.*, Vol. 1, Rome, pp. 942–951.

Turkstra, C. J. and H. D. Madsen, 1980. "Load Combinations in Codified Structural Design," *J. Struct. Div. ASCE*, Vol. 106, No. ST12, Dec., pp. 2527–2543.

University of Illinois, 1940. "Highway Slab Bridges with Curbs: Laboratory Tests and Proposed Design Method," *Bull.* No. 346, Urbana.

Veletsos, A. S. and T. Huang, 1970. "Analysis of Dynamic Response of Highway Bridges," *J. Eng. Mechanics Div.*, Vol. 96, No. EM5, Oct., pp. 590–620.

West, H. H., 1989. Analysis of Structures, Wiley, New York.

Westergaard, H. M., 1930. "Computation of Stresses in Bridge Slabs due to Wheel Loads," *Public Roads*, March.

White, D. W. and J. F. Hajjar, 1991. "Application of Second-Order Analysis in LRFD," *Research to Practice*, *Eng. J. AISC*, Vol. 28, No. 4.

Wiegel, R. L., 1970. *Earthquake Engineering*, Prentice-Hall, Englewood Cliffs, N.J.

Wolchuk, R., 1990. "Steel-Plate Deck Bridges," in Gaylord and Gaylord, *Structural Engineering Handbook*, 3rd. ed., McGraw-Hill, New York, pp. 19-1–19-27.

Wood, H. H. and P. C. Jennings, 1971. "Damage to Freeway Structures in the San Fermando Earthquake," *Bull. New Zealand Soc. Earth. Eng.*, Vol. 4, No. 3, Dec., pp. 347–376.

Wright, R. N. and W. H. Walker, 1971. "Criteria for the Deflection of Steel Bridges," *American Iron and Steel Inst. Bull.*, No. 19, New York.

Xanthakos, P. P., 1981. "Comprehensive Investigation and Structural Report, Bridge Carrying Southwest Highway over B&O.C.T. RR, Melina and Stony Creek," In-House Report, Illinois Dept. Transp., 45 p.

Yoo, C. H. and C. P. Heins, 1972. "Plastic Collapse of Horizontally Curved Bridge Girders," *ASCE Journ. Struct. Div.*, Vol. 98, No. ST4, April, pp. 899–914.

CHAPTER 3

REINFORCED CONCRETE BRIDGES

3-1 CHARACTERISTICS OF REINFORCED CONCRETE

Basic Considerations

The use of reinforced concrete for bridge construction goes back to the turn of the century. In 1901 Maillart built a three-hinged box section with solid spandrels over the Rhine River at Tavanas, Switzerland, which became the prototype for similar structures in the next 40 years (Heins and Lawrie, 1984). Since then, this material has become universally accepted for bridge work in a variety of dominant structural forms. The range of applications covers all-concrete structures and noble combinations of concrete decks supported on steel beams or girders. The broad acceptability of reinforced concrete is related to the availability of structural materials such as reinforcing bars and the constituents of concrete: sand, gravel, and cement. The choice is enhanced by the relative simple skills required at the site. Reinforced concrete bridges may be composed of cast-in-place concrete formed and cast in its final location, or they may include elements of precast prestressed concrete produced under factory conditions and subsequently erected at the construction site.

Concrete is strong in compression but weak in tension. As a result, cracks develop when the applied loads or restrained temperature and shrinkage changes introduce tensile stresses exceeding the tensile strength of concrete. Steel reinforcing bars are commonly embedded in concrete to develop the tension forces necessary for moment equilibrium after the concrete has cracked. In addition, concrete members are proportioned for adequate strength and stiffness to prevent unwarranted deformations and excessive deflections. Constructability is also a factor to be considered.

Factors Affecting Choice of Concrete

The choice of reinforced concrete in lieu of other structural forms is based on the following considerations.

Economy and Suitability of Materials Frequently, the choice is dictated by economy, expressed in terms of the total cost including materials, labor, and maintenance. Long-term economy depends on durability and maintenance cost, and concrete inherently requires less maintenance than its steel or timber counterpart. This is particularly true if dense, air-entrained concrete has been used for exposed surfaces, and adequate drainage is provided. However, special precautions are necessary for concrete exposed to salts such as deicing chemicals.

Structural Stability Concrete is placed in a plastic condition providing the opportunity to obtain the desired shape by proper choice of forms and finishing techniques. The associated structural shapes include flat slabs, beams, T sections, hollow sections, and configurations that are stable and satisfy the design. The stiffness and mass of concrete add rigidity to a bridge structure.

Availability of Materials Sand, gravel, cement, and concrete-mixing facilities are usually available for in situ delivery or preparation of fresh concrete, and reinforcing bars are readily transported to most job sites. As a result, the choice is particularly favored in remote areas.

However, certain considerations tend to inhibit the use of reinforced concrete, and these include the following.

Low Tensile Strength This does not deter the use of concrete because steel reinforcement is commonly used to carry the tensile stresses and limit the crack width. Nevertheless, in certain cases these cracks may enhance water penetration together with deicing salts, accelerating the deterioration of concrete.

Forms and Shoring The construction of reinforced concrete bridges usually requires three operations: (a) the erection of forms, (b) stripping and removal of these forms, and (c) propping or shoring of new concrete to support its weight until proper strength is attained.

Relative Strength per Unit Weight or Volume Because the compressive strength of concrete is a fraction (5–10 percent) of the steel strength whereas its unit density is higher than the unit density of steel, a concrete structure needs much larger volume and greater weight for the same loading. As a result, longer spans are built more economically using steel.

Time-Dependent Volume Changes Both steel and concrete undergo about the same thermal expansion and contraction. Steel, however, releases a lesser mass of material to be heated and cooled, and also is a better conductor than concrete; hence, a steel bridge is affected by temperature changes differently than concrete. The latter undergoes drying shrinkage that, if restrained, can cause deflections or cracking. Deflections tend to increase with time due to creep of concrete under sustained loads.

Strength and Design Criteria

Compressive and Tensile Strength The compressive strength of concrete, designated as f'_c, represents the 28-day strength determined from a 6-in. round cylinder 12 in. long. For a maximum usable compressive strain of 0.003 in./in., the specified concrete strength can vary from 3000 to 11,000 psi. Conventional reinforced concrete bridges usually specify 3000 to 4000 psi, prestressed concrete 5000 to 6000 psi, and special structures 6000 to 11,000 psi.

The tensile strength is essential because it affects the extent and frequency of cracking; it is determined by split cylinder tests in which the cylinders are compressed when positioned on their sides.

Steel reinforcement can consist of bars, wire fabric, or wires. However, only deformed steel bars are considered, normally grade 40 or 60 ($f_y = 40{,}000$ or 60,000 psi).

Allowable Stresses: Service Load Design For flexure, the allowable extreme fiber stress in compression for the concrete is $f_c = 0.40f'_c$. For $f'_c = 3500$ psi, $f_c = 1400$ psi. Likewise, the extreme concrete fiber stress in tension is $f_t = 0.21f_r$, where f_r is the modulus of rupture determined as specified by AASHTO. For grade 60 reinforcement, the allowable tensile stress in the steel is $f_s = 24{,}000$ psi.

Elasticity, Creep, and Shrinkage The modulus of elasticity of concrete is a function of its compressive strength. A plot of a typical stress–strain curve yields three tangent lines: an initial modulus (slope at origin), a tangent modulus (slope at stress $0.5f'_c$), and a secant modulus (slope from origin to $0.5f'_c$). According to AASHTO, the modulus of elasticity is $E_c = w^{1.5}33\sqrt{f'_c}$ (in psi), and for normal-weight concrete ($w_c = 145$ pcf), $E_c = 57{,}000\sqrt{f'_c}$. The modulus of elasticity of non-prestressed steel reinforcement is taken as $E_s = 29{,}000{,}000$ psi.

Creep and shrinkage are time-dependent deformations and must be included in the design. The response of concrete to creep can be related to the initial elastic deformation or strain. In this case the creep coefficient is

$$C_t = \frac{\text{Creep strain}}{\text{Initial elastic strain}} \tag{3-1}$$

AASHTO incorporates the effects of creep by multiplying by the factor

$$\left[2 - 1.2\left(\frac{A'_s}{A_s}\right)\right] \geq 0.6 \tag{3-2}$$

where A'_s = area of compressive reinforcement
A_s = area of tension reinforcement

Shrinkage is defined as the volume change in the concrete with time. According to AASHTO, the shrinkage strain is 200×10^{-6} in./in. (shrinkage coefficient of 0.0002).

Thermal Expansion and Contraction Provisions for temperature changes must be made for simple spans exceeding 40 ft. In continuous bridges, the design should consider thermal stresses, or provisions should be made for movement caused by temperature changes. For normal-weight concrete, the thermal coefficient is 0.000006 per degree Fahrenheit (0.0000108 per degree Celsius).

Criteria for Load Factor Design: Standard Specifications (AASHTO)

Essentially, the criteria for load factor design involve factoring the design loads (dead and live) and then comparing the results with the factored nominal strength. This is expressed in the form

$$P_u = \gamma[\beta_D(\text{DL}) + \beta_L(\text{LL} + I)] \leq {}_u\phi R_n \tag{3-3}$$

where γ = load factor
β = coefficient as per AASHTO
DL, LL + I = dead load, live load plus impact, respectively
P_u = required strength of concrete section,
ϕR_n = factored nominal strength

The factors γ, β are specified according to the loading combinations (see also Chapter 2). The factored nominal strength of the section is determined as illustrated in the following sections.

3-2 TYPES OF REINFORCED CONCRETE BRIDGES

In general, a reinforced concrete bridge structure may consist of deck slabs, T beams (deck girders), through and box girders, rigid frames, and flat slab types. Combinations of these with precasting or prestressing produce additional structural forms and enhance bridge versatility. Typical examples of concrete bridge superstructures are shown in Figures 3-1 and 3-2.

A major advantage in the use of concrete is the broad variety of structural shapes and forms. In the selection of the proper type of bridge, however, cost is usually the determining criterion. Occasionally, the selection is complicated by factors such as the ratio of dead to live load, appearance, depth constraints and available headroom, limited construction time, labor costs, and difficulties in formwork because of the support height or because of traffic maintenance requirements during construction. In this case steel bridges may be more cost-effective.

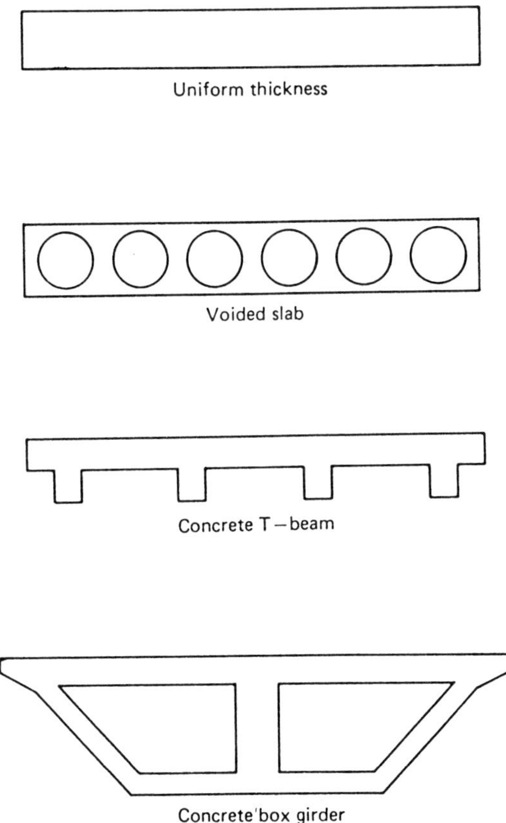

FIGURE 3-1 Typical types of concrete superstructures. (From Heins and Lawrie, 1984.)

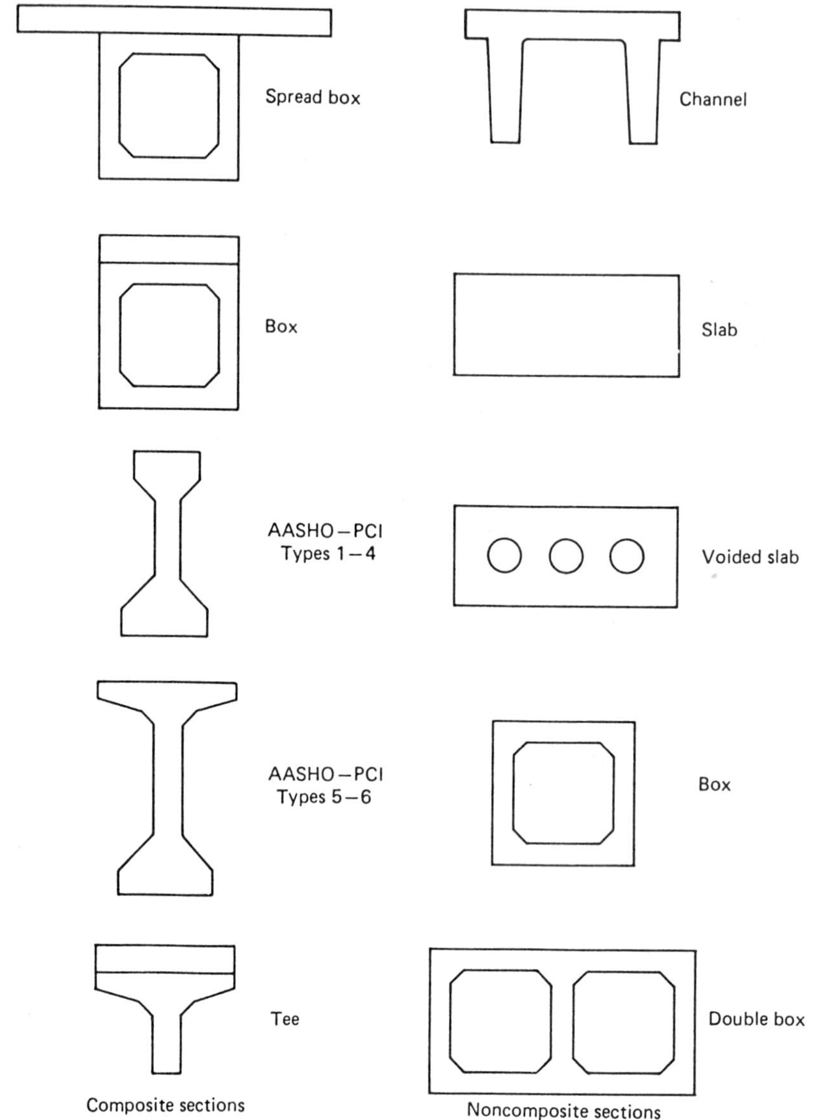

FIGURE 3-2 Typical concrete sections (precast and prestressed) used in bridge superstructures. (From Heins and Lawrie, 1984.)

For single-span bridges, suitable structure types include (a) simply supported deck or through girders; (b) right-angle rigid frames; (c) right-angle frames with concealed cantilevers with or without counterweights; (d) simply supported girders with concealed cantilevers with or without counterweights; and (e) two short concealed spans, each on either side of the opening, provided with a cantilever extending into the opening and supporting a short center span. The simply supported structure in category (a) is statically determinate and simple in design, but has a higher cost. When unyielding foundations are feasible, the rigid frames in types (b) and (c) provide more economical solutions. Girders with cantilevers as in types (d) and (e) should be considered for longer spans where small girder depth must be maintained.

For bridges with several spans, feasible arrangements include (a) simply supported girder spans in a series, (b) a combination of girders with cantilevers and short spans between these cantilevers, (c) continuous girders on independent supports, and (d) multispan rigid frames with rigidly connected elements. This range, therefore, covers statically determinate and indeterminate systems. Ordinarily, the overall bridge cost is higher for simple spans and lower for rigid frames. This suggests that the former should be used at sites where reasonably unyielding foundations are not attainable. Where heavy piers are required as in river crossings, or where the structure is to be supported on existing piers, type (c) with continuous spans is suitable.

A rigid connection between heavy piers and a flexible superstructure is incompatible with structural theory, and should not be attempted. Likewise, the rigid frame in type (d) is feasible where the elasticity of the vertical supports and the superstructure is compatible. In types (b) and (c), the advantages of introducing cantilevers in the end spans should be analyzed because this may reduce the cost of abutments.

3-3 SLAB BRIDGES: SIMPLE SPANS

Slab bridges normally require more concrete and reinforcing steel than girder bridges of the same span, but the formwork is simpler and less expensive; hence, they are more economical when these cost factors balance favorably. In the United States, the cost of formwork is high compared to the cost of materials, and slab bridges have been found economical for spans up to 30 ft (9 m). In Europe, the relatively low cost of formwork favors other types of concrete bridges, and simply supported slabs are seldom used for spans greater than 18 ft. The small overall superstructure depth in slab bridges is a favorable factor at grade separations.

Structural Configurations Various types of slab bridges designed and built before World War II have the cross sections shown in Figure 3-3. In these examples, the slab is bordered along both sides by curbs either forming an integral part or poured separately after the slab had cured and the forms removed. Sidewalks, when used, were supported on thinner slabs and separated from the main roadway slab by longitudinal joints, or the slab was poured as shown in Figure 3-3*b*. In other examples, the sidewalk was cantilevered as shown in Figure 3-3*e*.

Slab bridges designed according to current AASHTO standards must be provided with an edge beam if the main reinforcement is parallel to traffic.

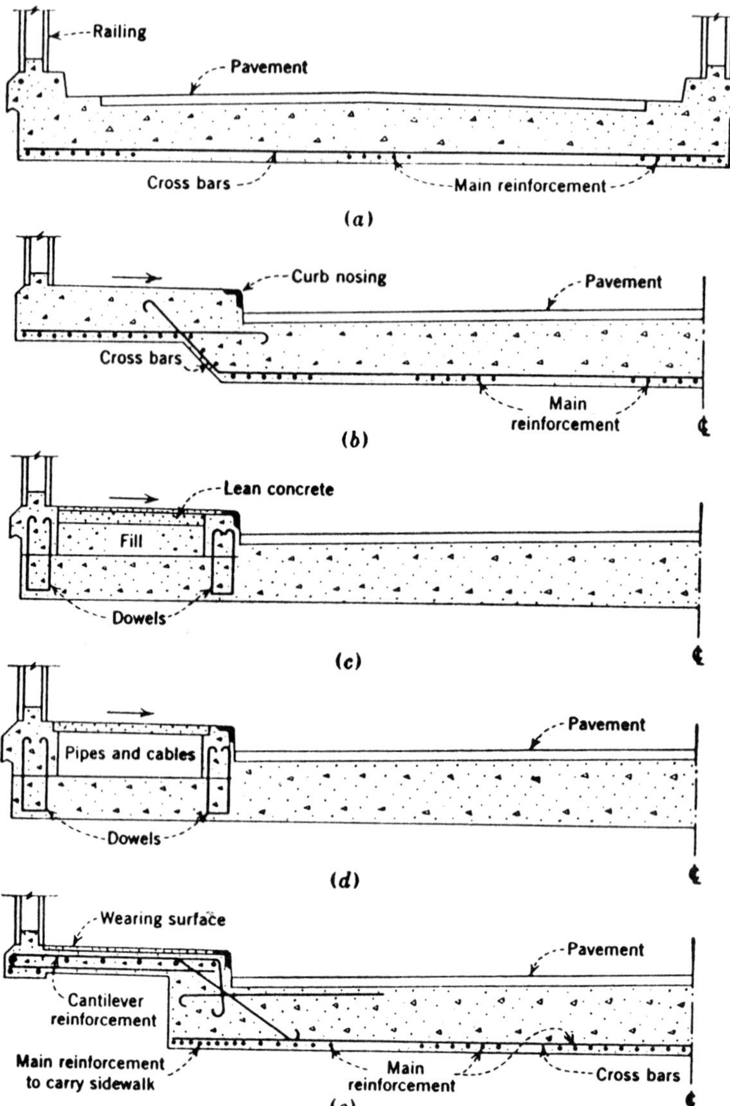

FIGURE 3-3 Typical cross sections of slab bridges designed and built before World War II. (From Taylor, Thomson, and Smulski, 1939.)

This beam may consist of an additionally reinforced slab section, a beam integral with but deeper than the slab, or an integral reinforced section of slab and curb.

Distribution of Concentrated Loads A concentrated (wheel) load placed on a wide slab is distributed laterally over a width of slab appreciably greater

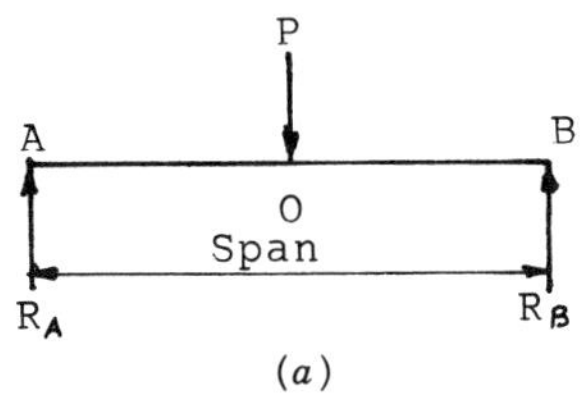

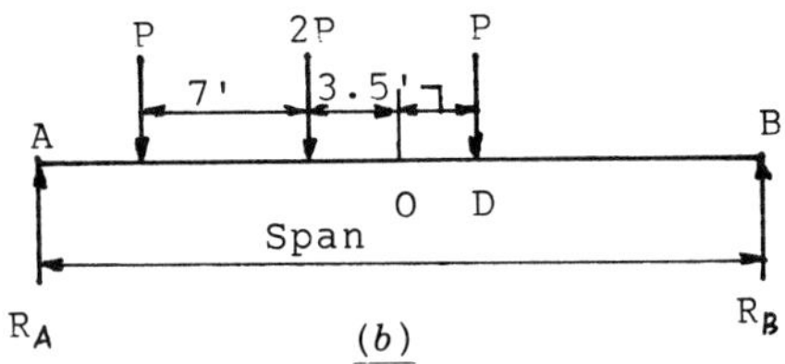

FIGURE 3-4 Concentrated live loads and position for maximum moment in simple span.

than the width of the contact area of the load with the slab (see also Chapter 2). Field tests show that in a slab loaded with a concentrated load the deflection and stresses are greater directly under the load, and decrease gradually, diminishing to zero at some distance from the load. In practice, this condition is approximated by assuming a uniform transverse distribution over a smaller width. According to AASHTO, the distribution width is $E = 4 + 0.06S$ (maximum 7.0 ft), where S is the design span. This value of E is for wheel load, whereas lane loads are distributed over a width of $2E$. Longitudinally reinforced slabs are designed for the appropriate HS loading.

Likewise, edge beams must be provided with main reinforcement parallel to traffic (longitudinal). According to AASHTO, edge beams must resist a live load moment $M_{LL} = 0.10PS$, where P is the wheel load and S is the design span (see also Chapter 2).

Maximum Live Load Moments For a typical HS truck loading, the maximum live load moment in simple spans is produced by the loading position shown in Figure 3-4. The single concentrated load at the center O of the span controls for spans up to 24 ft, and in this case the maximum live load moment is equal to $PS/4$ at the center O. For spans greater than 24 ft, the double load placed as shown in Figure 3-4*b* governs. In this case the bending moment under load P at point D reaches a maximum value when the bisector of the distance between that load and the resultant $2P$ is at the center O of the span. This is valid for spans up to about 35 ft.

The maximum moments, shears, and reactions for simple spans can be found directly by reference to AASHTO tables. These values are for one lane.

Design for Skew Crossings Where the skew angle is relatively small, a slab is normally designed for main reinforcement parallel to traffic, and the design span is the distance along the centerline of the roadway. When the skew angle is large or the roadway is wide in relation to the span, the structure may be designed to span the direct angle distance between supports. The triangular portion of the slab at each side now has no support along its edge, and should therefore be supported by an edge beam or a parapet girder extended above the roadway to obtain sufficient depth.

Example of Slab Design The design of a simply supported slab bridge will be illustrated in a typical example. The clear span is 25 ft, and the live load is HS 20-44. The supports are assumed to be 1 ft wide, so that the design span length S (distance between centers of supports) is 26 ft. The dead load is the weight of the slab plus 25 lb/ft^2 for future wearing surface. If a railing is placed after the slab has cured, its weight may be assumed uniformly distributed across the entire width of the slab.

Strength parameters and allowable stress are as follows:

$$f_c' = 3500 \text{ psi} \qquad f_c = 0.4 \times 3500 = 1400 \text{ psi}$$

$$f_y = 60{,}000 \text{ psi} \qquad f_s = 24{,}000 \text{ psi}$$

$$E_c = 57{,}000\sqrt{3500} = 3{,}370{,}000 \text{ psi} \qquad E_s = 29{,}000{,}000 \text{ psi}$$

$$n = 9$$

Using the working stress method, we compute $k = 0.34$, $j = 0.89$, $K = 211$, and $a = 1.78$.

The distribution of wheel load $E = 4 + 0.06S = 5.56$ ft. The live load moment for one lane is found from AASHTO tables, or $M_{\text{LL}} = 222$ ft-kips for the truck load position shown in Figure 3-4*b*. The impact factor is estimated as $I = 50/(26 + 125) = 33$ percent (use 30 percent). The live load plus impact moment is (per foot-width of slab)

$$M_{\text{LL}+I} = \frac{222}{5.56 \times 2} \times 1.30 = 20.0 \times 1.30 = 26.0 \text{ ft-kips}$$

Next, we assume a slab thickness of 18 in., producing a slab dead weight of $1.50 \times 0.15 = 0.225$ ksf. Including the future wearing surface, the total dead load is $w = 0.25$ ksf, producing a dead load moment of

$$M_{\text{DL}} = 0.25 \times \frac{26^2}{8} = 21.1 \text{ ft-kips}$$

The total moment is

$$M = 26.0 + 21.1 = 47.1 \text{ ft-kips}$$

With the foregoing data, we can now estimate the minimum effective thickness d and the area of steel A_s per foot-width of slab. Thus,

$$\min d = \sqrt{\frac{47.1}{0.211}} = 15 \text{ in.}$$

For a selected slab thickness of 18 in. and using 1 in. cover for bottom

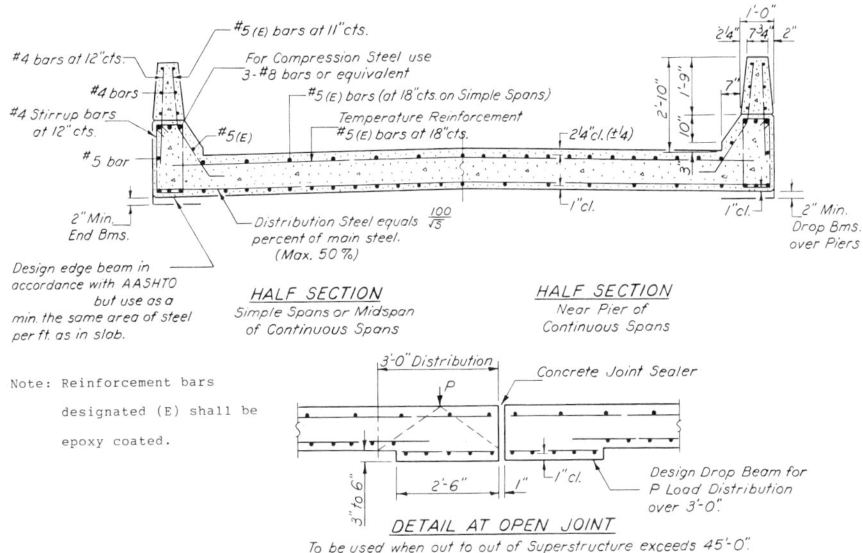

FIGURE 3-5 Typical cross section of a slab bridge: (left half) simple spans or midspan of continuous spans; (right half) near pier of continuous spans.

reinforcement, $d = 16.5$ in., OK. Likewise, we compute

$$A_s = \frac{47.1}{16.5 \times 1.78} = 1.6 \text{ in.}^2/\text{ft}$$

Use #9 at 7.5-in. centers or #8 at 6-in. centers.

Distribution reinforcement is provided as per AASHTO. The required steel is $100/\sqrt{S} = 20$ percent of the main reinforcement or $0.20 \times 1.6 = 0.32$ in./ft, which is provided by #5 bars at 12-in. centers. Temperature reinforcement will be provided in the top of the slab in both the lateral and longitudinal directions. This reinforcement is #5 bars at 18-in. centers.

A typical cross section is shown in the left half of Figure 3-5. The parapet section above the construction joint is usually built after the slab and curbs are cast; hence, it represents a superimposed dead load. This parapet is provided with open joints spaced at 14- to 20-ft centers. When the superstructure width exceeds 45 ft fascia-to-fascia, it is good practice to split the deck using a 1-in. open joint as shown in the detail, and design it for a 3-ft wheel load distribution. Most state standards specify epoxy-coated bars for reinforcement exposed to deicing chemicals and salts.

The edge beam consists of a portion of the slab together with the curb section, and is approximated as a rectangle 1 ft 7 in. wide and 2 ft 6 in. high. Its weight is $1.58 \times 2.50 \times 0.15 = 0.590$ ksf (use 0.6 ksf). This produces a dead load moment of

$$M_{\mathrm{DL}} = 0.6 \times \frac{26^2}{8} = 50.7 \text{ ft-kips}$$

Likewise, we compute the live load plus impact moment as

$$M_{\mathrm{LL}+I} = 0.10 \times 16 \times 26 \times 1.3 = 54.1 \text{ ft-kips}$$

The total moment is

$$M = 50.7 + 54.1 = 104.8 \text{ ft-kips}$$

$$\min d = \sqrt{\frac{104.8}{0.211 \times 1.58}} = 17.7 \text{ in.} \qquad \text{provided } 28.0 \text{ in.,} \qquad \text{OK}$$

The tension reinforcement for the edge beam is

$$A_s = \frac{104.8}{1.78 \times 28} = \frac{104.8}{49.8} = 2.10 \text{ in.}^2 \qquad \text{use three \#8}$$

Although compression steel is not required in this example, the standard of Figure 3-5 shows three #8 bars in the top of the curb.

Shear (diagonal tension) and bond stress in slabs designed for bending moment according to the foregoing method should be considered satisfactory. However, we will check the maximum shear stress according to AASHTO. From tables, the maximum end shear is 46.8 kips for one lane. The shear per foot-width of slab is

$$V_{\mathrm{LL}+I} = \frac{46.8}{11.12} \times 1.3 = 5.50 \text{ kips}$$

Likewise,

$$V_{\mathrm{DL}} = 0.25 \times 13 = \underline{3.25} \text{ kips}$$

$$\text{Total} = 8.75 \text{ kips}$$

The maximum shear stress is

$$v = 8750/(12 \times 16.5) = 44.2 \text{ psi}$$

and the allowable shear is

$$v_{\mathrm{al}} = 0.95\sqrt{f_c'} = 56 \text{ psi} \qquad \text{OK}$$

In the foregoing example, the distribution E is independent of the bridge width and the number of traffic lanes, that is, the design ignores the length–width ratio. The slab will be checked now using a bridge width $W = 27$ ft and a wheel load distribution according to NCHRP Project 12-26.

For main reinforcement parallel to traffic, the distribution width E is $(3.5 + 0.06\sqrt{L_1 W_1})$, where $L_1 = S = 26$ ft and $W_1 = W = 27$ ft. Then $E = 3.5 + 1.6 = 5.10$ ft, compared to $E = 5.56$ ft. The difference, due partly to the change in the fixed coefficient and partly to the span–width ratio effect, is relatively small yet sufficient to warrant a corresponding change in the reinforcement requirements.

3-4 SLAB BRIDGES: CONTINUOUS SPANS

Characteristics

Where reasonably unyielding foundations can be provided, a superstructure consisting of a continuous slab that extends over several supports may be more economical. Continuous concrete bridges in units of three, four, or five spans are adaptable to most stream crossings and grade separations. The substructure usually consists of concrete pile or frame bents. For long spans (probably up to 125 ft), the continuous T girder offers obvious economy, whereas the continuous slab presents advantages for spans less than 35 ft. For spans exceeding the range of solid-web T girders, the continuous hollow-girder bridge is most suitable and economical. Because continuous-girder bridges are ideally proportioned when the interior–end span ratio is between 1.3 and 1.4 (for loadings and unit stresses commonly used in practice), this bridge type is more satisfactory than a series of simple spans especially when the piers can be placed on the stream bank or outside the main channel for stream crossings, and at the sides of the roadway for grade separations.

In general, continuous bridges require single bearings at interior supports, thus reducing the width of the pier cap. The continuity also implies fewer expansion joints. With longer spans, the depth of sections follows closely the bending moment variation, from a minimum at the center to a maximum at the supports. The effect of dead load on the design is reduced accordingly. Reduction in deck depth, particularly at midspan, imparts to the continuous bridge economic and esthetic advantages. The longer interior spans, necessary for structural reasons, are combined with the haunched soffits and improve the appearance of the bridge.

Slab Bridges The choice between a continuous slab and a continuous girder depends on the span lengths as well as on the available clearance. When the headroom is restricted, or where the length and height of the approaches are related to the depth of the bridge structure, a continuous slab results in a lower profile. Concrete bridges have been built with spans up to 70 ft and slab–span ratios of 1/32.

Continuous slabs have the following advantages: (a) nominal superstructure depth, improving the alignment of the approaches; (b) simplified layout of reinforcement in both the top and the bottom, implied by the absence of stirrups; (c) a workable formwork and a smaller area of exposed concrete surface, resulting in a lower cost of surface finish; and (d) better distribution of live loads laterally and longitudinally, resulting in fewer critical sections in the design. Slab bridges have, however, certain disadvantages articulated mainly in the higher cost of materials and the associated greater dead loads in comparison with girder bridges.

Example of Bridge Layout

It is necessary to lay out a highway bridge over a small stream for which the site conditions are as follows.

1. Waterway area required 630 ft.^{2}
2. Depth of flow at high water 9 ft.
3. Stream cross-sectional profile as shown in Figure 3-6.
4. The stream is not subject to severe floods and carries only a small amount of drift.
5. Subsoil conditions permit pile driving.
6. Floodplain extends 120 ft west of centerline of channel.
7. Fill at point 60 ft west of centerline of channel is 7 ft above original ground line, and 3 ft at point 40 ft east of centerline.
8. A crossroad is located 70 ft east of centerline of channel.
9. The required clearance above high water is estimated as 18 in.

These conditions are satisfied by a three-span continuous-slab bridge with uniform slab depth, having spans 33, 43, and 27 ft as shown. The superstructure consists of pile bents, and a 4-in. slope protection wall is provided at abutment bent 1.

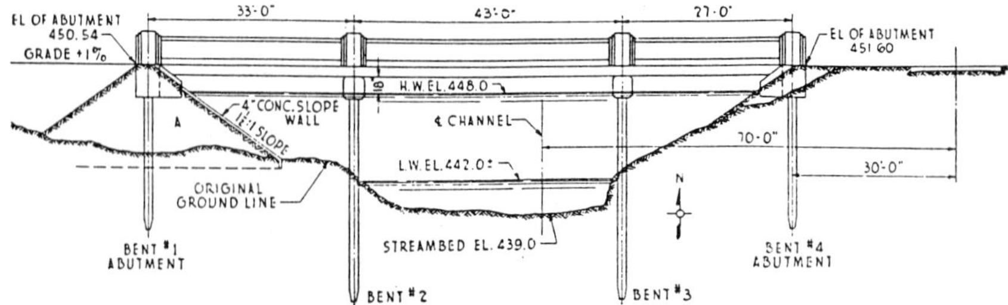

FIGURE 3-6 Three-span unsymmetrical continuous-slab bridge over a small stream; general bridge layout.

This layout accommodates 650 ft^2 of waterway, and the subsoil conditions combined with the absence of floods make concrete pile bents a logical choice for the substructure. Preliminary comparative estimates indicate a 25 percent cost reduction over that of a single span of shorter length with closed abutments on foundation piles. The location of the crossroad fixes the location of bent 4 in order to allow room for turnout. Note that bents 2 and 3 are shifted 3 ft east of the position that would give a symmetrical layout in order to place bent 3 at a compatible location with reference to the stream bank. The deck slab is made a constant depth because no appreciable savings can be realized by introducing a slab of variable depth.

With the foregoing layout, the floodplain is confined to the minimum that satisfies the design requirements, and additional channel excavation is not required. The 4-in. slope protection at open bent 1 will prevent erosion of fill that might occur because of flow restrictions at flood stage. Backfill A should be firmly compacted, and the slope wall placed before the deck is constructed.

Design Procedure for Continuous Slabs

Depth Ratio It is not necessary to define numerically the depth of the slab at the various critical sections; however, the ratio r of increase in depth at the supports to the depth at the center of the spans can be assumed. Actual dimensions can be finalized later by simple calculations.

Where pier and abutment location is not fixed by physical factors and geometric requirements, structural economy and optimization will dictate the ratio of interior to end spans. Thus, for end spans up to 35 ft, this ratio is usually 1.26; for end spans of 35 to 50 ft, the optimum ratio is 1.30. Based on these limits, the parameter r is taken as zero (constant slab depth) for end spans up to 35 ft. For end spans of 35 to 50 ft, r can be assumed to be 0 to 0.4 at the outer end of the end spans, 0.4 at the first interior support, and 0.5 at all other interior supports. In practice, it is economical to use $r = 0$ at the outer end of the end spans regardless of length. Where two or more units of several continuous spans are used, the converging free ends are usually haunched for appearance (see also the example of Section 11-9).

Live Load Support Moments Live load moments at interior supports due to a unit load in any position in the various spans can be obtained with the use of influence lines drawn for the correct distribution and carry-over factors, and in this respect reference to Table 2-2 is useful. For the usual HS truck load, the variable axle spacing V may be chosen within the indicated range (14–30 ft), so that one concentrated load can be placed on each span on either side of the support to produce maximum negative moment.

Live Load Moments in Spans Likewise, maximum live load moments in spans for truck loading are computed from influence lines for the points indicated in Table 2-1. The exact location of maximum combined dead and live load positive moment in the end spans is usually determined by trial, so that live load moments at several sections near the anticipated critical area must be computed and then combined with the dead load moments.

For certain span lengths, the critical position of loading may cause one wheel load of the standard truck to be off the structure while the other is on, or may result in a wheel load on an adjoining span not normally loaded for maximum moment at the particular section under consideration.

Dead Load Moments Dead load moments are commonly computed at the same critical points as live load moments. This may involve moment distribution or direct reference to tables. Where the analysis involves uniform and haunch loads, the fixed-end moments must be adjusted accordingly.

Depth of Slab at Center of Span A trial value for the depth at the center of the span, h_c, can now be assumed. For the stresses and strength specified in Section 3-1, h_c will be about 1/28 to 1/32 of the length of the longest intermediate span, and slightly smaller for haunched slabs. If the slab is haunched, the depth at the supports will be $h_c(1 + r)$. Once the slab depth has been tentatively selected, dead load moments are checked and adjusted if necessary.

Maximum Moment Curves Curves of maximum moments can be drawn in several simple steps. First, we locate closely the points of maximum positive moment in the end spans; then we find the maximum and negative moments at the 0.7 point of the end spans and at the 0.2 and 0.8 or 0.3 and 0.7 points of the intermediate spans. These, in addition to the maximum span and support moments, enable us to draw moment curves that are sufficiently accurate for design purposes by passing parabolas through the points of positive moments and straight lines through the points of negative moments.

Shear Normally, a slab thickness that satisfies the moment requirements should be adequate for shear. If the slab thickness at the supports is not adequate, it should be increased accordingly.

Example of Continuous-Slab Bridge

Figure 3-7 shows a three-span simply supported continuous-slab bridge, with a span ratio of 44/34 = 1.294 (for design purposes we will use 1.3). The design live load is HS 20, and the dead load includes an allowance of 25

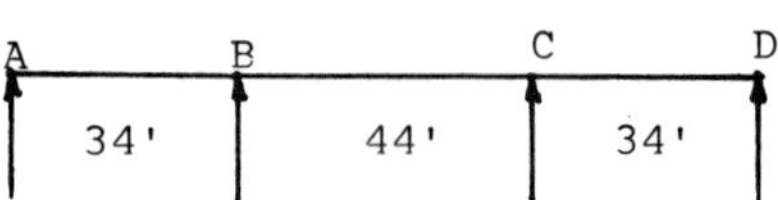

FIGURE 3-7 Three-span continuous-slab bridge; span lengths and simple supports.

lb/ft^2 for future wearing surface. Strength parameters and allowable stresses are as in the example of Section 3-3.

The distribution E of wheel load is as follows:

Spans AB and CD	$E = 4 + 0.06 \times 34 = 6.04$ ft
Span BC	$E = 4 + 0.06 \times 44 = 6.64$ ft

The procedure outlined in the foregoing sections should normally be used, but for simplicity we will limit the calculations to the following: (a) estimation of support moments, (b) estimation of span moments, and (c) shear at supports. The impact coefficient for moments is estimated at 30 percent. Assuming a slab thickness (uniform) of 21 in., the total dead load weight (slab plus wearing surface) is $w = 0.29$ ksf.

Moments in Span AB (End Span) Figure 3-8*a* shows the truck load position that produces maximum positive live load moment in span AB (point 0.4). This usually involves a previous trial whereby the two wheel loads are moved along the span and the moments estimated at critical positions. Note that the actual distance between points 0.4 and 0.8 is $4 \times 3.4 = 13.6$ ft, or roughly the minimum distance V stipulated for wheel loads. Next, the live load moment is obtained readily with the use of influence line coefficients (AISC, 1966). For this span, the wheel load distribution is $16/6.04 = 2.65$ kips per foot-width of slab. The live load plus impact moment is now

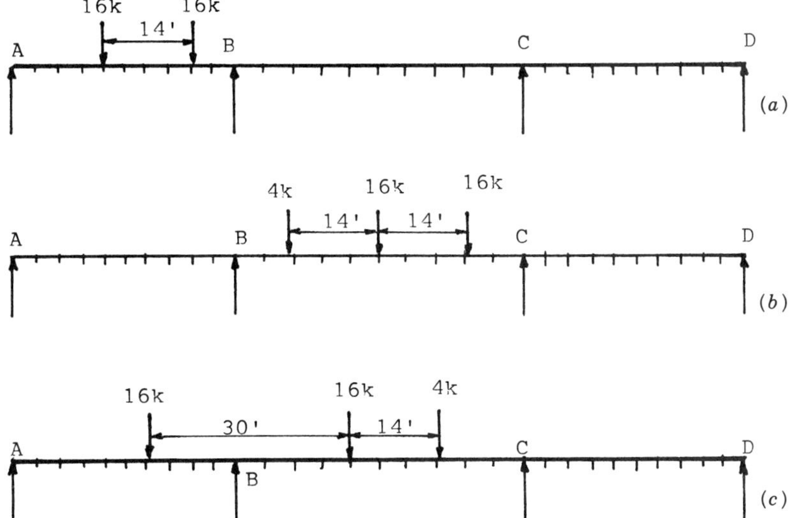

FIGURE 3-8 Position of live load for maximum moments; bridge of Figure 3-7.

$$M_{LL+I} = 2.65 \times 34 \times (0.2082 + 0.0500) \times 1.3 = 30.3 \text{ ft-kips}$$

The dead load moment is computed from appropriate formulas for $N = 1.3$ (ratio of center to end span) as

$$M_{DL} = 0.0664 \times 0.29 \times 34^2 = 22.2 \text{ ft-kips}$$

Note that the maximum dead load moment in span AB occurs at point 0.36, slightly different from point 0.4 used for live load moments. The total maximum positive moment in span AB is now

$$M_{AB} = 30.3 + 22.2 = 52.5 \text{ ft-kips}$$

Moments in Span BC (Center Span) The maximum live load moment is obtained by placing the truck load as shown in Figure 3-8*b*. For this position the maximum moment occurs at point 0.5. The front wheel load is just to the left of point 0.2, and the rear axle is just to the right of point 0.8. The wheel load distribution is as follows: front wheel, $4/6.64 = 0.60$; middle and rear wheels, $16/6.64 = 2.41$ kips per foot-width of slab.

Again, with the use of influence line coefficients, we obtain

$$\begin{aligned} M_{LL+I} &= [0.60 \times 34 \times 0.055 + 2.41 \times 34 \times (0.2176 + 0.0550)] \times 1.30 \\ &= 30.5 \text{ ft-kips} \end{aligned}$$

The dead load moment is

$$M_{DL} = 0.0758 \times 0.29 \times 34^2 = 25.4 \text{ ft-kips}$$

so that the total maximum moment (positive) in span BC is

$$M_{BC} = 30.5 + 25.4 = 55.9 \text{ ft-kips}$$

Moments at Interior Supports B and C Both truck and lane loading will be considered. For truck loading, the position that produces maximum negative moment at interior support B is shown in Figure 3-8*c*, and the front wheel load is just to the right of point 0.6 in span BC. The moment contribution is derived as follows:

$$\begin{aligned} &\text{Load in span } AB, M = 2.65 \times 0.0907 \times 34 && = \quad 8.2 \text{ ft-kips} \\ &\text{Load in span } BC, M = 2.41 \times 0.1154 \times 34 + 0.60 \times 0.065 \times 34 && = \quad \underline{10.8} \text{ ft-kips} \\ &\qquad\qquad\qquad\qquad \text{Total } M_{LL} && = -19.0 \text{ ft-kips} \end{aligned}$$

The lane loading specified by AASHTO will be distributed over a width $2E$, in this case the average value of $(6.04 + 6.64)/2 = 6.34$ ft. Thus, the lane load acting on a foot-width of slab is $18/6.34 \times 2 = 1.42$ kips concentrated load, and $0.640/12.68 = 0.05$ kip/ft uniform load. Maximum negative moment due to lane loading is obtained by placing two concentrated loads, one in each span *AB* and *BC*, and by loading spans *AB* and *BC* with the uniform lane loading. This moment is

$$M_{\text{LL}} = 1.42 \times 34 \times (0.0907 + 0.1154) + 0.05 \times 34^2 \times 0.1522$$
$$= -18.8 \text{ ft-kips}$$

almost the same as the truck load moment. Therefore, the total live load plus impact moment at the support is

$$M_{\text{LL}+I} = 19 \times 1.3 = -24.7 \text{ ft-kips}$$

The dead load moment is

$$M_{\text{DL}} = -0.1355 \times 0.29 \times 34^2 = -45.3 \text{ ft-kips}$$

and the total maximum negative moment at support *B* is

$$M_B = -(24.7 + 45.3) = -70.0 \text{ ft-kips}$$

Slab Thickness With the foregoing data, we can now estimate the minimum effective slab thickness *d*. Thus,

$$\min d = \sqrt{\frac{70}{0.211}} = 18.2 \text{ in.}$$

For the selected slab thickness of 21 in. and using 2 in. cover for the top reinforcement, $d = 18.4$ in., OK.

Reinforcement The required reinforcement for the negative and positive moments is now estimated for the three critical locations.

At the interior supports, the required steel at the top of the slab is

$$A_s = \frac{70}{18.4 \times 1.78} = \frac{70}{32.8} = 2.13 \text{ in.}^2/\text{ft}$$

Use #9 at 5.5 in. $\quad A_s = 2.18 \text{ in.}^2/\text{ft}$

In end spans *AB* and *CD*, the required steel at the bottom of the slab is

$$A_s = \frac{52.5}{19.4 \times 1.78} = \frac{52.5}{34.5} = 1.52 \text{ in.}^2/\text{ft}$$

Use #8 at 6 in. $\quad A_s = 1.58 \text{ in.}^2/\text{ft}$

In center span BC, the required steel at the bottom of the slab is

$$A_s = \frac{55.9}{34.5} = 1.62 \text{ in.}^2/\text{ft}$$

Use #8 at 5.5 in. $\quad A_s = 1.72 \text{ in.}^2/\text{ft}$

Distribution reinforcement in the bottom of the slab transverse to the main reinforcement is intended to ensure the lateral distribution of concentrated live loads. According to AASHTO, the required steel is as follows: For span AB the distribution reinforcement is $100/\sqrt{34} = 17$ percent, or $0.17 \times 1.52 = 0.26$ in.^{2}/ft. For span BC the distribution reinforcement is $100/\sqrt{44} = 15$ percent, or $0.15 \times 1.62 = 0.24$ in.^{2}/ft, which is provided by #5 bars at 15-in. centers.

A typical cross section showing details of the reinforcement is shown in Figure 3-5. The left half represents a section near the midspan, and the right half shows a section near the interior supports.

3-5 DECK GIRDER BRIDGES

Deck girder bridges are divided into three main types according to the interaction between the girder and slab. Thus, we distinguish (a) girder-and-slab systems, where the slab spans transversely between longitudinal girders as in steel beam design, providing a typical T-beam action; (b) girder, floor beam, and one-way slab, where the slab is supported by floor beams spanning two or more longitudinal girders; and (c) girder, floor beam, and two-way slab supported along the four edges. The last type is the most complex in technical terms, but it may yield overall economy.

Girder-and-Slab Bridge (T Beams)

Several deck sections of T-beam superstructures are shown in Figure 3-9, built mainly during the 1920s and 1930s. Because the current intent of applicable specifications and standards is to make the exterior girders of a capacity not less than that of the interior girders, the usual practice is to cantilever the projecting section of the slab so as to equalize the combined loading on all girders. In the simple cross section in Figure 3-9*c*, the exterior girders appearing in elevation view conceal the superstructure and produce a pleasing effect.

T-beam decks consist of a vertical rectangular stem with a wide top flange, usually a transversely reinforced slab forming the riding surface for traffic. The usual range of stem thickness is 14 to 22 in., and is dictated by the required horizontal spacing of the main bottom reinforcement. A T-beam design that accommodates positive moments may not necessarily provide the strength required for negative moment because of some possible loss of

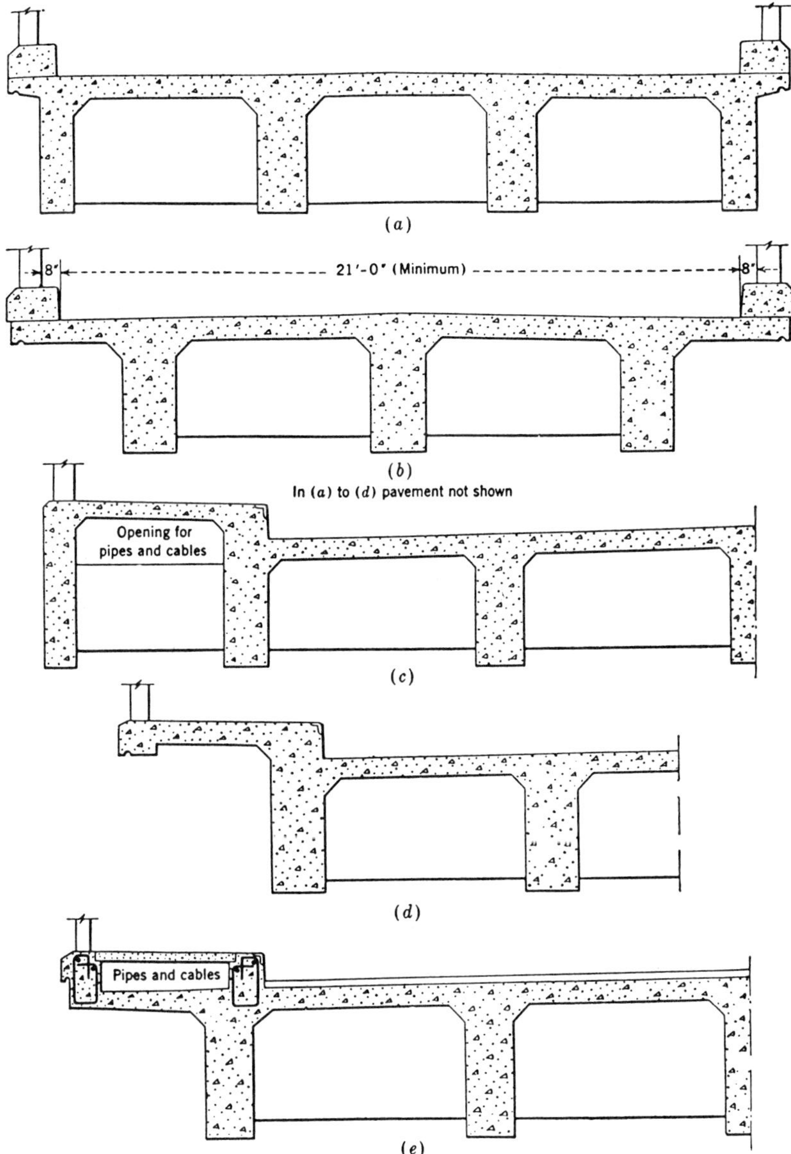

FIGURE 3-9 Typical cross sections of deck girder (T-beam) bridges. (From Taylor, Thomson, and Smulski, 1939.)

strength by the wide compression flange in the cracked tensile zone of the section. This problem is remedied by (a) thickening the stem in the areas of negative moment, (b) providing a partial bottom slab, and (c) providing compression reinforcement.

Based on span lengths, T-beam bridges are the next class of structures beyond the range of longitudinally reinforced slabs, and may be considered for spans 45 to 90 ft long. Where T-beam decks are constructed monolithically with the substructure, the resulting advantage is the considerable bending strength of the entire structure.

Girder Spacing The lateral spacing of longitudinal girders has a marked effect on the cost of the bridge. Where the range of slab thickness is defined by standards and design manuals, the usual approach is to use the maximum girder spacing that accommodates the selected slab thickness. Alternatively, comparative estimates are necessary before a final scheme is selected. For a slab thickness of 7 to 9 in., the girder spacing may vary between 6 and 9 ft. The economical girder spacing yields a minimum cost of formwork and materials in the slab and the girders, and from experience the optimum spacing is 7 to 9 ft. However, where vertical supports for the formwork are difficult and expensive, girder spacing should be increased accordingly.

Cross Beams These are used primarily to stiffen the main girders laterally and to control torsion of the outside girders. In the usual designs, the slab is considered partially restrained along the outside girders, whereas the interior panels approach the condition of full fixity. A further function of the cross beams is to equalize the deflection of a partly loaded deck, but this effect is not always recognized in determining the distribution of live load. Exterior girders are affected by cross beams to a lesser degree.

Cross beams (diaphragms) should be placed at the ends of T girders unless other means are included to resist lateral forces and to maintain section geometry. However, where structural analysis ensures adequate strength, girders of considerable length can be built without cross beams. AASHTO recommends one intermediate diaphragm at the point of maximum positive moment for spans exceeding 40 ft.

Because live load deflection may be based on the assumption that the superstructure flexural members act jointly and are subjected to equal deflections, the intent of the design should be to reduce live load deflection by distributing the loads laterally. This action is enhanced if sufficient reinforcement is provided in the cross beams and particularly where they connect to the main girders. If the diaphragms are designed only as stiffening struts, bottom reinforcement should not be less than 0.4 percent of the effective cross section of the member.

Large Skews For wide crossings with large skew angles, it may be more desirable to place the girders at right angles to the supports as shown in Figure 3-10. In this case, at each side of the crossing there is a triangular section where a girder may rest at one end at the abutment and at the other end on the parapet girder. The latter carries heavy loads, and it may be necessary to increase its depth by extending the member above the roadway deck.

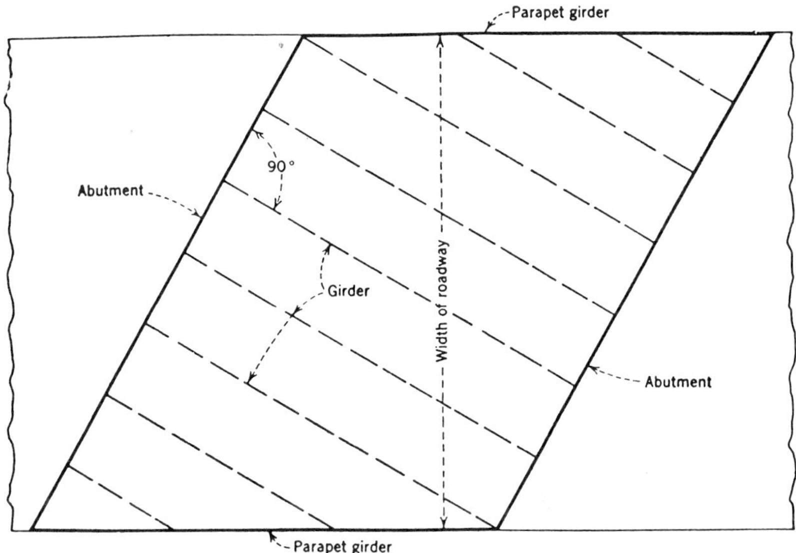

FIGURE 3-10 Arrangement and girder layout in wide skew crossing. (From Taylor, Thomson, and Smulski, 1939.)

Example of T-Beam Bridge Design

A T-beam superstructure will be designed for a two-span continuous bridge with span lengths of 70 ft. The roadway width is 30 ft face-to-face of curb, resulting in the cross section shown in Figure 3-11*a*. Span lengths and support type are shown in Figure 3-11*b*.

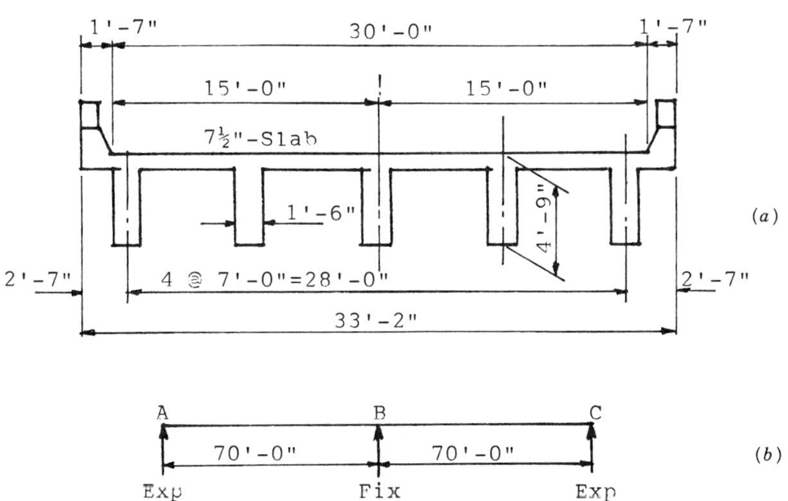

FIGURE 3-11 Two-span continuous T-beam bridge: (*a*) typical deck cross section; (*b*) beam elevation.

According to AASHTO, the minimum superstructure depth is $0.065 \times 70 = 4.55$ ft (use 4 ft 9 in. as shown). The curb and parapet detail is as shown in Figure 3-5. The curb portion is placed monolithically with the slab, whereas the parapet is poured later and its weight is therefore distributed to all the girders. We select a slab 7.5 in. thick, and we include provisions for a 25 lb/ft^2 future wearing surface. The live load is HS 20.

Past construction of reinforced concrete T beams often had 6- to 9-in. fillets at the slab–stem intersection. Modern construction has eliminated their use, and there is no evidence to indicate that problems may have resulted.

The bridge will be designed according to service load procedures. For uniformity, we will use the same strength and stress parameters for slab-and-T-beam design. There are no requirements for overload capacity. Likewise, specified strengths and allowable stresses are as in the previous examples for slab bridges, and the criteria stipulated in Article 8.15.3 of AASHTO for flexure will be applicable.

Design of Slab For main reinforcement perpendicular to traffic, the design span is the clear span, or $S = 5.5$ ft. The live load plus impact moment is

$$M_{LL+I} = \left(\frac{5.5 + 2}{32}\right) \times 16 \times 1.3 \times 0.8 = 3.90 \text{ ft-kips}$$

where the factor 0.8 is applied to the live load moment to reflect the continuity of the slab. The dead load moment includes the weight of the slab $0.625 \times 0.15 = 94$ lb/ft^2, plus 25 lb/ft^2 wearing surface, or dead load $w = 0.119$ kip/ft,

$$M_{DL} = \frac{0.119}{10} \times 5.5^2 = 0.36 \text{ ft-kip}$$

The total moment is

$$M = 3.90 + 0.36 = 4.26 \text{ ft-kips}$$

To ensure adequate stiffness and reinforcement clearance, the recommended minimum slab thickness is $t \geq (S + 10)/30$, or 0.625 ft (7.5 in.), whichever is greater. According to this criterion, the 7.5-in. thickness is exceeded if $S > 8.75$ ft.

The reinforcement in the top of the slab is obtained using 2.25-in. clearance and assuming #5 bars. In this case $d = 7.50 - 2.25 - 0.31 = 4.94$ in., and

$$A_s = \frac{4.26}{1.78 \times 4.94} = 0.48 \text{ in.}^2/\text{ft}$$

Use #5 at 7.5 in. $\quad A_s = 0.50$ in.^{2}/ft

Likewise, the reinforcement in the bottom of the slab is obtained using 1-in. clearance, so that $d = 7.50 - 1.00 - 0.31 = 6.19$ in., and

$$A_s = \frac{4.26}{1.78 \times 6.19} = 0.39 \text{ in.}^2/\text{ft}$$

Use #5 at 9 in. $\quad A_s = 0.41 \text{ in.}^2/\text{ft}$

The amount of distribution reinforcement in the bottom of the slab as a percentage of the main reinforcement is $220/\sqrt{S} = 93$ percent (use maximum 67 percent), or

$A_s = 0.41 \times 0.67 = 0.27 \text{ in.}^2/\text{ft}$ Use #5 at 12 in. $\quad A_s = 0.31 \text{ in.}^2/\text{ft}$

Note that for main reinforcement perpendicular to traffic, the specified amount of distribution reinforcement will be used in the middle half of the slab span, and not less than 50 percent of the specified amount will be used in the outer quarters of the slab span.

Interior Girders The dead loads acting on the interior girders are as follows:

$$\text{Weight of slab} = 7 \times 0.625 \times 0.15 = 0.66 \text{ kip/ft}$$
$$\text{Weight of girder} = 4.125 \times 1.50 \times 0.15 = 0.93 \text{ kip/ft}$$

The superimposed dead load is

$$\text{From parapet} = 2 \times 1.75 \times 0.83 \times 0.15/5 = 0.09 \text{ kip/ft}$$
$$\text{From W.S.} = 30 \times 25/5 = 0.15 \text{ kip/ft}$$
$$\text{Total dead load } w = 1.83 \text{ kips/ft}$$

A diaphragm is placed at midspan as a concentrated load (assume 1.0 ft thick × 5.50 ft × 3.00 ft). The concentrated load at midspan is

$$P_{\text{DL}} = 1.00 \times 5.50 \times 3.00 \times 0.15 = 2.48 \text{ kips}$$

The distribution of wheel loads in longitudinal girders is $S/6.0 = 7/6 = 1.17$. The impact factors are the same for span and support moments, and the impact coefficient is $I = 50/(70 + 125) = 26$ percent.

The fraction of wheel load applied to interior girders is $1.17 \times 4 = 4.68$ kips for the front wheels and $1.17 \times 16 = 18.72$ kips for the rear wheels. The intent of this analysis is to establish the maximum structural requirements by computing maximum negative support and maximum positive span moments.

Dead load moments are computed as follows. At support B,

$$M_{\mathrm{DL}} = -1.83 \times 0.125 \times 70^2 = -1120 \text{ ft-kips}$$

At point 0.375 of AB,

$$M_{\mathrm{DL}} = 1.83 \times 0.0703 \times 70^2 = 630 \text{ ft-kips}$$

Moments from the concentrated dead load P_{DL} are computed as follows. At support B,

$$M_{\mathrm{DL}\,P} = 2.48 \times 0.0938 \times 2 \times 70 = -33 \text{ ft-kips}$$

At point 0.4 of AB,

$$M_{\mathrm{DL}\,P} = 2.48 \times (0.1625 - 0.0375) \times 70 = 22 \text{ ft-kips}$$

The maximum live load span moment is at 0.4 of span AB, and is produced by the truck position with the heavy wheels at points 0.4 and 0.6. This moment is obtained with the use of influence line coefficients as follows:

$$\begin{aligned}
&\text{Span } AB, M_{\mathrm{LL}} = (4.68 \times 0.1008 + 18.72 \times 0.2064 + 18.72 \times 0.1216) \times 70 &&= 463 \text{ ft-kips}\\
&\qquad \text{Impact} = 26\% &&= \underline{120} \text{ ft-kips}\\
&\text{Span } AB, M_{\mathrm{LL}+I} &&= 583 \text{ ft-kips}
\end{aligned}$$

The maximum live load negative moment at support B is produced by lane loading and two concentrated loads at points 0.6 and 1.4:

$$\begin{aligned}
&\text{From uniform load,} && M_B = 0.64 \times 0.125 \times 70^2 &&= 392 \text{ ft-kips}\\
&\text{From concentrated loads,} && M_B = 18 \times 2 \times 0.096 \times 70 &&= 242 \text{ ft-kips}\\
&\text{One lane,} && M_B &&= -634 \text{ ft-kips}
\end{aligned}$$

For the fraction of lane load carried by one girder, the moment is

$$\begin{aligned}
&M_B = -634 \times 0.5 \times 1.17 &&= -371 \text{ ft-kips}\\
&\text{Impact} = 26\% &&= -\ \underline{96} \text{ ft-kips}\\
&\text{Support } B,\ M_{\mathrm{LL}+I} &&= -467 \text{ ft-kips}
\end{aligned}$$

Summary of Moments Interior girders:

$$M_{AB} = 630 + 22 + 583 = 1235 \text{ ft-kips}$$
$$M_B = -(1120 + 33 + 467) = -1620 \text{ ft-kips}$$

Exterior Girders The dead load acting on the exterior girders (weight of the slab) will be taken as the portion from the center of the outside slab to the fascia. Because we have assumed that the curb will be placed monolithically with the slab, its weight will be added to the dead load of the exterior girders.

The following loads are computed:

$$\begin{aligned}
&\text{Weight of slab} = 6.08 \times 0.625 \times 0.15 &&= 0.57 \text{ kip/ft}\\
&\text{Weight of girder} &&= 0.93 \text{ kip/ft}\\
&\text{Weight of curb} = 1.33 \times 1.00 \times 0.15 &&= \underline{0.20} \text{ kip/ft}\\
&\text{Dead load } w &&= 1.70 \text{ kips/ft}\\
&\text{Superimposed dead load} &&= \underline{0.24} \text{ kip/ft}\\
&\text{Total dead load } w &&= 1.94 \text{ kips/ft}
\end{aligned}$$

Likewise, the diaphragm introduces a concentrated load at midspan of $2.48 \times 0.5 = 1.24$ kips.

For flexural analysis, the live load carried by the exterior girders is the reaction of the wheel load, assuming the slab acts as a simple span between girders, and this fraction is $(7.0 + 1.0)/7 - 1.14P$ (note that for most severe effects one wheel load is placed directly over the exterior girder).

By analogy, we compute the dead load moments as follows. At support B,

$$M_{\text{DL}} = -1.94 \times 0.125 \times 70^2 = -1187 \text{ ft-kips}$$

At point 0.375 of AB,

$$M_{\text{DL}} = 1.94 \times 0.0703 \times 70^2 = 668 \text{ ft-kips}$$

The dead load moments from the diaphragm at midspan are one-half the respective moments for the interior girders.

The live load plus impact moments can be estimated directly by considering the load distribution ratio $1.14/1.17 = 0.97$. These moments are

$$\text{Span } AB, \quad M_{\text{LL}+I} = 583 \times 0.97 = 568 \text{ ft-kips}$$
$$\text{Support } B, M_{\text{LL}+I} = -467 \times 0.97 = -453 \text{ ft-kips}$$

Summary of Moments Exterior girders:

$$M_{AB} = 668 + 11 + 568 = 1247 \text{ ft-kips}$$

$$M_B = -(1187 + 17 + 453) = -1657 \text{ ft-kips}$$

Characteristics of T Girders The effective flange width overhanging on each side of the stem should not exceed six times the slab thickness, or one-half the clear distance between stems (AASHTO requirements). From Figure 3-11 we obtain $6 \times 7.5 = 45$ in., so that one-half the clear distance (33 in.) between stems controls.

An exaggerated deflected view of a T girder is shown in Figure 3-12*a*. Because of its continuity, this member develops positive moments at midspan (section *A–A*) and negative moments over the supports (section *B–B*). At midspan the compression zone is shown in Figures 3-12*b* and *d*. It may be rectangular, or the neutral axis may shift down into the stem zone giving a T-shaped compression zone. The support section shown in Figure 3-12*c* and the midspan section shown in Figure 3-12*b* both have a rectangular compression zone, and are therefore analyzed as rectangular beams with the beam width *b* taken as shown. The section shown in Figure 3-12*d* has its compression zone in a T shape, and the normal coefficients *k* and *j* should be taken

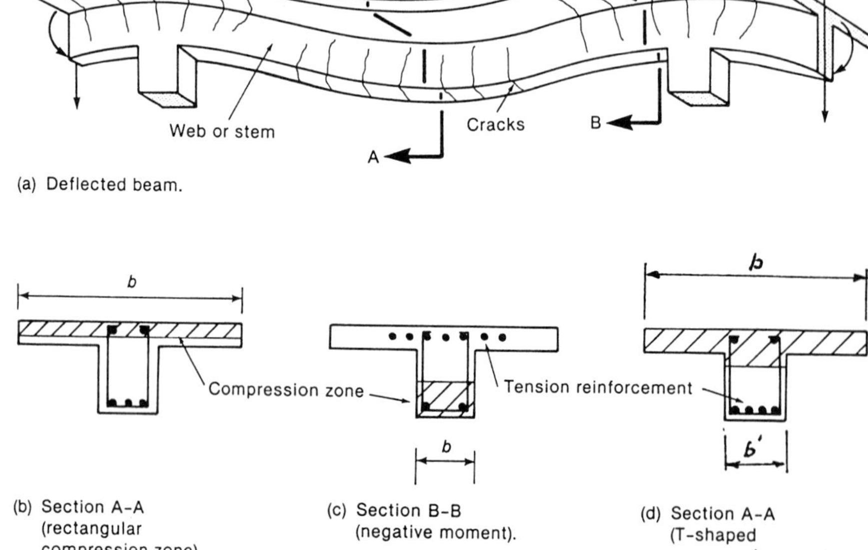

FIGURE 3-12 Positive and negative moment regions in a T beam.

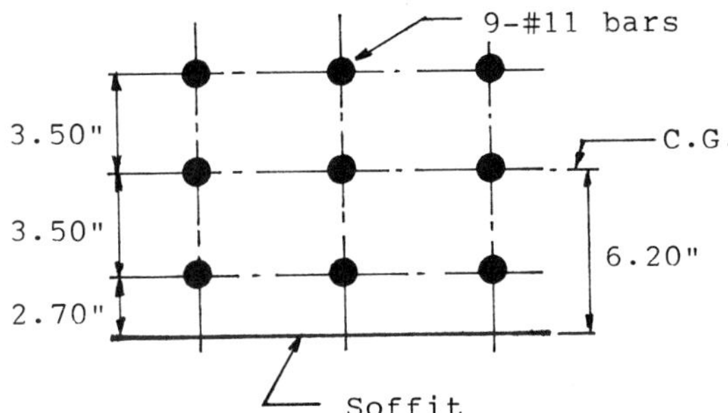

FIGURE 3-13 Arrangement of bars at midspan, bottom of stem, T girders.

from diagrams used solely for T-beam design. In this case both dimensions b and b' (width of flange and stem, respectively) are relevant.

T-Girder Design at Midspan The design moment is 1247 ft-kips. First, assume a rectangular section 1 ft 6 in. × 4 ft 9 in. Next, assume $j = 0.90$ and $d = 4$ ft 3 in. = 51 in.

The approximate A_s required is

$$A_s = \frac{1247}{1.80 \times 51} = 13.6 \text{ in.}^2$$

Try nine #11 bars, $A_s = 14.04$ in.^{2}, arranged as shown in Figure 3-13. For this arrangement, actual $d = 57 - 6.2 = 50.8$ in. By reference to diagrams for T beams, we obtain

$$\frac{t}{d} = \frac{7.5}{50.8} = 0.148 \qquad pn = \frac{14.04 \times 9}{73 \times 50.8} = 0.034$$

$$k = 0.25 \qquad j = 0.93 \text{ (obtained graphically)}$$

We can now compute the actual stresses and compare them with the allowable.

$$f_s = \frac{1247 \times 12}{14.04 \times 0.93 \times 50.8} = 22.75 \text{ ksi} < 24 \text{ ksi}$$

$$f_c = \frac{22.75 \times 0.25}{9(1 - 0.25)} = 843 \text{ psi} < 1400 \text{ psi}$$

T-Girder Design at Interior Supports Next, the negative moment section at B is designed. We will use the negative moment for interior girders of 1620 ft-kips. Assuming #11 bars placed just below the main transverse reinforcement in the slab, the effective depth is

$$d = 57.0 - 3.5 = 53.5 \text{ in.}$$

The approximate required tensile reinforcement is

$$A_s = \frac{1620}{1.78 \times 53.5} = 17 \text{ in.}^2 \qquad \text{Use 11 \#11 bars} \qquad A_s = 17.16 \text{ in.}^2$$

The approximate required depth based on $K = 211$ for balanced design is

$$d = \sqrt{\frac{1620}{0.211 \times 1.5}} = 72 \text{ in.} > 53.5 \text{ in.}$$

The section therefore needs compressive reinforcement. We can now estimate the factor F for $b \times d = 18 \times 53.5$, as $F = 18 \times 53.5^2/12{,}000 = 4.3$. Next, we compute

$$M - KF = 1620 - 211 \times 4.3 = 1620 - 907 = 713 \text{ ft-kips} \qquad \text{(Positive)}$$

which is the residual moment not taken by the concrete, and must be resisted by the compressive reinforcement. For #11 bars and 1.5-in. clearance at the bottom of the girder,

$$d' = 1.5 + 0.5 + 0.70 = 2.70 \text{ in.}$$

Then

$$\frac{d'}{d} = \frac{2.70}{53.5} = 0.050$$

Now we estimate the factor c for 24,000/9/1400 and $d'/d = 0.05$. The factor c is approximately 1.59. Therefore,

$$A'_s = \frac{M - KF}{cd} \qquad \text{or} \qquad A'_s = \frac{713}{1.59 \times 53.5} = 8.4 \text{ in.}^2$$

$$\text{Use six \#11 bars} \qquad A'_s = 9.36 \text{ in.}^2$$

The girder cross section at midspan and at the support with the arrangement of the reinforcement is shown in Figure 3-14.

Side-face reinforcement should be placed in beams according to applicable AASHTO standards.

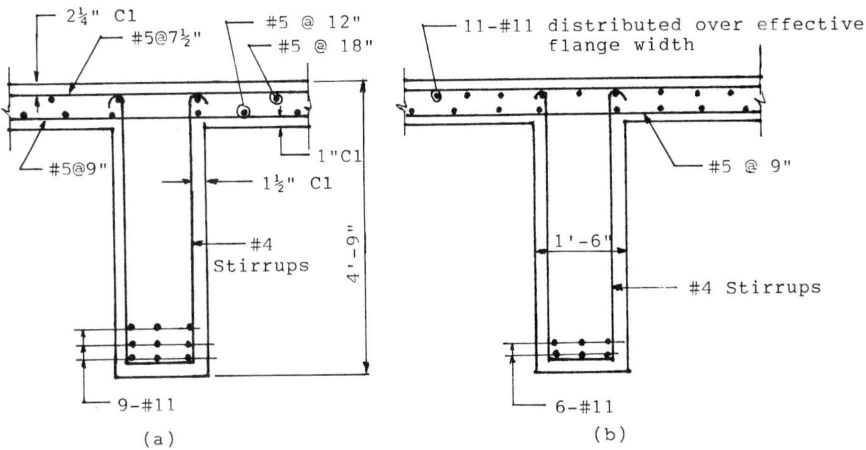

FIGURE 3-14 Typical girder T section showing main reinforcement in flange and stem; bridge of Figure 3-11: (*a*) at midspan; (*b*) at support.

Shears and Reactions: Interior Girders Recall the dead load $w = 1.83$ kips/ft, the concentrated load at midspan is 2.48 kips, and the live load wheel distribution is 1.17.

For a complete girder analysis and design of stirrups, shears should be tabulated at the 10th points. For live load shears, the truck load controls for moderate or short spans. For a two-span continuous bridge with equal spans, the truck load produces maximum shears up to spans of about 105 ft. For greater span lengths, the lane loading controls. For the influence line curve shown in Figure 2-6*b* (point 4 in the end span of a four-span continuous beam), it is obvious that the heaviest wheel should be placed at this point with the other wheels as close as possible toward support 10. The exact point for reversing truck direction as we compute maximum live load shears in the first span depends on the magnitude and sign of the shear due to the different end moments. If tables are used to estimate the live load shears, they should indicate the location of truck reversal. Interestingly, maximum live load reactions at the interior support are governed by the truck load up to spans of about 50 ft. For greater spans, the lane load controls the interior support reaction.

Reactions and shears at support A (Figure 3-11) are as follows:

Dead load	$R_A = 0.375 \times 1.83 \times 70$	$= 48.0$ kips
Concentrated dead load	$R_A = 2.48 \times 0.406$	$= 1.0$ kip
	Total dead load R_A	$= 49.0$ kips

According to AASHTO Article 3.23.1, lateral distribution of the wheel load for shear is obtained by assuming the floor to act as simple span between beams. For loads in other positions of the span, the distribution for shear is determined as for moment. Accordingly, the reactions and shears computed from appropriate tables (AISC, 1966) are modified for the effect of the axle load adjacent to the end support by adding the value

$$P = 16\left(3 - \frac{10}{S} - Q\right)$$

where S is the girder spacing and $Q = 1.17$ (wheel load distribution).

P is estimated directly as

$$16\left(3 - \frac{10}{7} - 1.17\right) = 6.4 \text{ kips}$$

Thus, the live load reaction at A is

$$R_{A(\text{LL}+I)} = (60.2 \times 0.5 \times 1.17 + 6.4) \times 1.26 = 52.4 \text{ kips}$$

The total reaction at A is

$$R_A = 49.0 + 52.4 = 101.4 \text{ kips} \qquad (\text{also } V_A)$$

Likewise, the maximum shear on either side of support B is computed as follows:

Dead load	$V_B = 0.625 \times 1.83 \times 70$	$= 80.00$ kips
Concentrated dead load	$V_B = 2.48 \times 0.594$	$= 1.48$ kips
	Total dead load V_B	$= 81.48$ kips

The live load shear at B is likewise estimated directly from the same tables, and is

$$V_{B(\text{LL}+I)} = (65.5 \times 0.5 \times 1.17 + 6.4) \times 1.26 = 56.6 \text{ kips}$$

The total shear at B is

$$V_B = 56.6 + 81.5 = 138.1 \text{ kips}$$

At support A, the unit shear is computed from $v = V/b_w d$, or $v = 101.4/18 \times 50.8 = 112$ psi. Because the girder is subject to shear and moment only, the allowable shear stress carried by the concrete, v_c, may be taken as $0.95\sqrt{f'_c} = 56.6$ psi (a more detailed calculation is given by AASHTO in Article 8.15.5.2). The shear that must be carried by the stirrups is therefore $v - v_c = 112 - 57 = 55$ psi, and the stirrup spacing for #4 stirrups (two legs) is

$$s = \frac{0.40 \times 24{,}000}{55 \times 18} = 9.7 \text{ in.}^2 \qquad \text{say \#4 at 9 in.}$$

At support B, the unit shear is $v = 138.1/18 \times 53.5 = 144$ psi, and the shear that must be carried by the stirrups is $144 - 57 = 87$ psi. For #4 stirrups (two legs), the spacing is

$$s = \frac{0.40 \times 24{,}000}{87 \times 18} = 6.2 \text{ in.} \qquad \text{Use \#4 at 6 in.}$$

Comments on Two-Span Continuous Bridges In the last 10 to 15 years, two-span continuous structures have been used more frequently, and this trend is expected to continue. Experience gained through construction of a large number of grade separations at both the Interstate and Local Highway System indicates that appreciable advantages may result from two-span rather than four-span structures. These advantages are articulated in bridge esthetics, reduction in the number of fixed and expansion bearings, and the elimination of shoulder piers that present an obstruction to snow removal. The absence of shoulder piers also accommodates additional traffic lanes to meet future traffic demands, and equally important is the increased safety resulting from the elimination of obstructions and the improvement of sight clearance (see also Section 1-3).

Procedure for Multiple-Span T Girders In a continuous-girder bridge, the depth of sections should follow closely the moment requirements, varying from a minimum at the center to a maximum at the supports. In this case the effect of dead load on the design is reduced accordingly, and the variation of section from the center of the spans to the supports responds to the stress requirements.

If the T girders in a continuous unit are haunched, the parameter r (ratio of increase in depth at the supports to the depth at the center of the spans) can be taken as 0.3 at the two intermediate supports of a three-span bridge, 0.5 for the center support of a four-span bridge, and 0.3 at the second and fourth support of the same unit. Interestingly, these relative girder depths are valid for span ratios (interior to end) of 1.3 : 1 to 1.4 : 1, which is the optimum range for continuous T-girder units with end spans greater than 35 ft.

The width b' of the girder stem is dependent on the girder spacing, slab thickness, and length of span, but more importantly on the arrangement of the reinforcing steel. A fair approximation of b' is

$$b' = 0.0025\sqrt{b}\,(L) \tag{3-4}$$

where $b = S$ = girder spacing (in.)
L = length of end span (in.)

According to (3-4), the width b' of the example shown in Figure 3-11 should be $0.0025\sqrt{84} \times 840 = 19$ in., or very close to the selected 18-in. thickness.

Because the depth at the supports is dependent on the girder spacing, span lengths, and width of stem, it can be expressed as a function of these variables (Portland Cement Association, "Continuous Concrete Bridges," Second Edition). For a typical highway loading and normal allowable stresses, the depth at the supports may be estimated as

$$h_s = (1.93)(b')^{-0.11} \cdot b^{0.25} \cdot L^{0.83} \tag{3-5}$$

where b, b', and L are as in (3-4). Solution of the foregoing exponential equation is not necessary if the values of h_s are obtained directly with the help of the graphs of Figure 3-15.

T Girders, Floor Beams, and Slabs

Where a relatively wide girder spacing is indicated and requires a thick slab (considerably thicker than 7.5 in.), the design may not be economical because of larger quantities of slab materials and the heavier dead loads. In this case the choice is improved by the use of transverse floor beams spaced to permit the use of thin slabs (preferably 7.5 in. thick). In this type of deck, the slab is designed as a continuous beam over a number of supports with the main reinforcement parallel to traffic.

The floor beams are usually cast monolithically with the slab, but because their stiffness is considerably smaller than that of the girders, the restraining effect is correspondingly reduced. However, no distinction is made in the selection of the effective (design) span length or the coefficient to be applied to the moments because of continuity. Thus, the distribution of loads and design of slabs supported by floor beams is in accordance with AASHTO (Article 3.24.1), unless more refined methods are used.

Floor Beams These are designed for a combination of dead load, live load, and impact. With the floor supported directly on floor beams, the beams are designed for the most unfavorable truck position between the T girders. For floor beam spacing less than 6 ft, the load may be considered as transferred by the slab to the adjoining floor beams, so that the fraction carried by one floor beam is $S/6$, where S is the floor beam spacing (ft.) If $S > 6$, the load on the beams should be the reaction of the wheel loads, assuming the floor acts as a simple beam.

One-span floor beams, such as in through girders, may be considered as restrained at the ends by the girders, but the degree of restraint will depend on the rigidity of the floor beam and the torsional resistance of the girder. This resistance is increased by neighboring floor beams acting as struts,

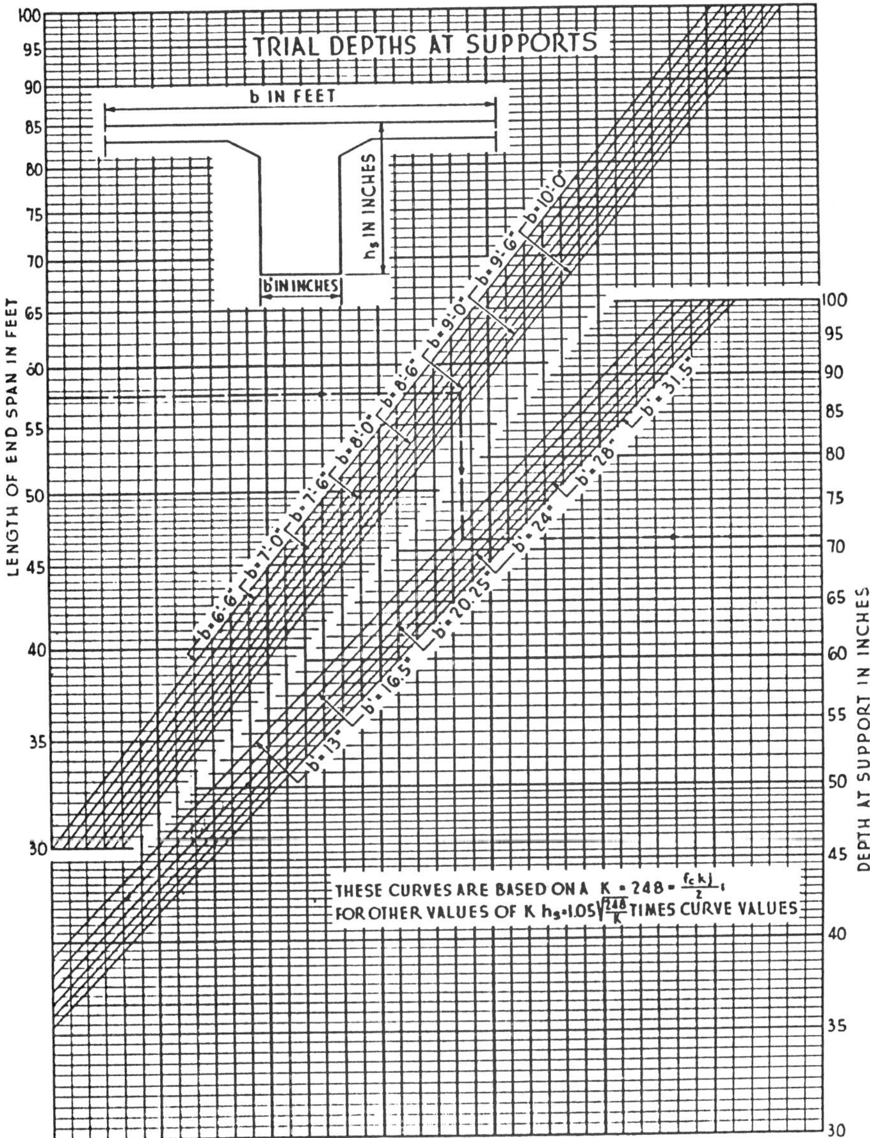

FIGURE 3-15 Trial girder depth at supports; continuous T girders. (From Portland Cement Association, "Continuous Concrete Bridges," Second Edition.)

because usually one floor beam at a time is subjected to maximum live load moment. End restraint in floor beams reduces all positive moments, and this reduction may be assumed to be 20 percent of the static bending moment. The maximum negative moment introduced at the ends of the floor beam because of the restraint may be taken as 50 percent of the static moment. Larger negative moments should be used if it appears advisable or after an exact analysis, but they should not exceed the fixed-end moment. The reinforcement should be doweled into the girders. Floor beams over three or more girders should be treated as continuous units.

T Girders, Floor Beams, and Two-Way Slabs

In this arrangement the floor beam spacing is made appreciably larger than usual and close to the spacing of longitudinal girders. In square crossings the floor beams divide the slab into a number of square or rectangular panels supported on four sides and designed as two-way slabs (see AASHTO Article 3.24.6). The resulting advantages are (a) for the same girder spacing the slab thickness is decidedly smaller, (b) in a two-way design the concentrated loads are distributed in two directions, (c) all reinforcement is effective in resisting flexural stresses, and (d) the decrease in slab thickness implies reduced dead loads.

AASHTO stipulates that the distribution width E for the live load taken by either span should be determined as provided for other types of slabs, and the moments obtained in this manner will be used in designing the center half of the short and long span. The reinforcement in the outer quarters may be reduced by 50 percent.

Shear Strength of Two-Way Slabs: Case Study The shear strength of concrete slabs provided with lateral restraints is markedly higher than predicted by flexural theories (Batchelor and Tissington, 1972; Park, 1964; Tong and Batchelor, 1971). However, current design procedures for two-way slabs are based on elastic theory and ignore the effect of in-plane (compressive membrane) stresses on slab strength, although this effect may result in more efficient use of reinforcement. A study by Batchelor and Tissington (1976) proposes a method for predicting the shear strength of isotropically reinforced two-way bridge slabs.

The theoretical background is presented by ASCE–ACI (1974) Joint Task Committee 426, which also treats the effects of compressive membrane action. Important contributions to this topic are made by Elstner and Hognestad (1956), Tong (1969), and Moe (1961). Punching failures have been studied by Gesund and Dikshit (1971) and Gesund and Kaushik (1970) using the yield line theory. Tong and Batchelor (1970, 1971) have proposed an extension of the yield line theory to account for compressive membrane strength enhancement of partially restrained two-way slabs. Accordingly, the flexural capacity V_{total} can be closely predicted by

$$V_{\text{total}} = V_{\text{flex}} + V_{\text{mem}} \tag{3-6}$$

where V_{flex} = flexural capacity of the slab calculated by the yield line theory
V_{mem} = compressive membrane contribution to slab strength

It is also proposed that, in order to prevent collapse upon first cracking, reinforcement should be provided to satisfy a minimum reinforcement ratio

$$p_{\min} = 0.25\frac{f_r}{f_y} > 0.0025 \tag{3-7}$$

where f_r = modulus of rupture of concrete
f_y = specified yield strength of reinforcement
$p = A_s/bd$ = reinforcement ratio
A_s = area of tension reinforcement per unit width of slab

According to these theories, the strength enhancement is thought to be generated by the membrane moment resulting from the eccentricity of the membrane (in-plane) forces along assumed yield lines. These are said to depend on the lateral stiffness of edge restraint to the slab, on support conditions, and on the magnitude of compressive stresses generated in the slab.

Batchelor and Tissington have used a series of scaled strength models of a hypothetical bridge to investigate these effects. The bridge has an 80-ft span and contains three isotropically reinforced square slab panels in the longitudinal direction, as shown in Figure 3-16. The variables in the study are model scale, boundary conditions, and slab reinforcement percentage.

The tests are divided into the A series and the C series. The former has all three-panel and single-panel specimens supported along the entire length of beams and diaphragms. All specimens in the C series have simple supports at the ends of the longitudinal girders. As shown in Figure 3-16, the single-panel specimens, bounded by opposite beams and diaphragms, represent one panel of the corresponding bridge specimens. Slab reinforcement in the longitudinal (W) direction of the single panels is anchored in both diaphragms, and in all other aspects the reinforcement in these panels is arranged as shown in the detail of Figure 3-16. All panels are isotropically reinforced with 0.33 percent reinforcement.

The following main nondimensional parameters are used in the study: (a) flexural load function $\phi_0 = V_{\text{test}}/V_{\text{flex}}$, (b) total load function $\psi = V_{\text{test}}/V_{\text{total}}$, (c) reinforcement index $\omega = pf_y/f'_c$, and (d) deflection ratio $\Delta/\Delta_{\text{cr}}$. In the foregoing, V_{test} is the observed shear strength of the slab and Δ is the observed net slab deflection.

Load–Deflection Relationship For the C series, typical load–deflection relationships were obtained for an assumed value of $\Delta_{\text{cr}} = d/2$, where Δ_{cr} is the critical value of Δ at which the maximum compressive membrane moment is developed and d is the effective slab depth. The results show that smaller scale models exhibit more ductile behavior than larger ones. Observed values of Δ_{cr} range from $0.45d$ to $0.87d$ and depend on the stiffness of the surrounds.

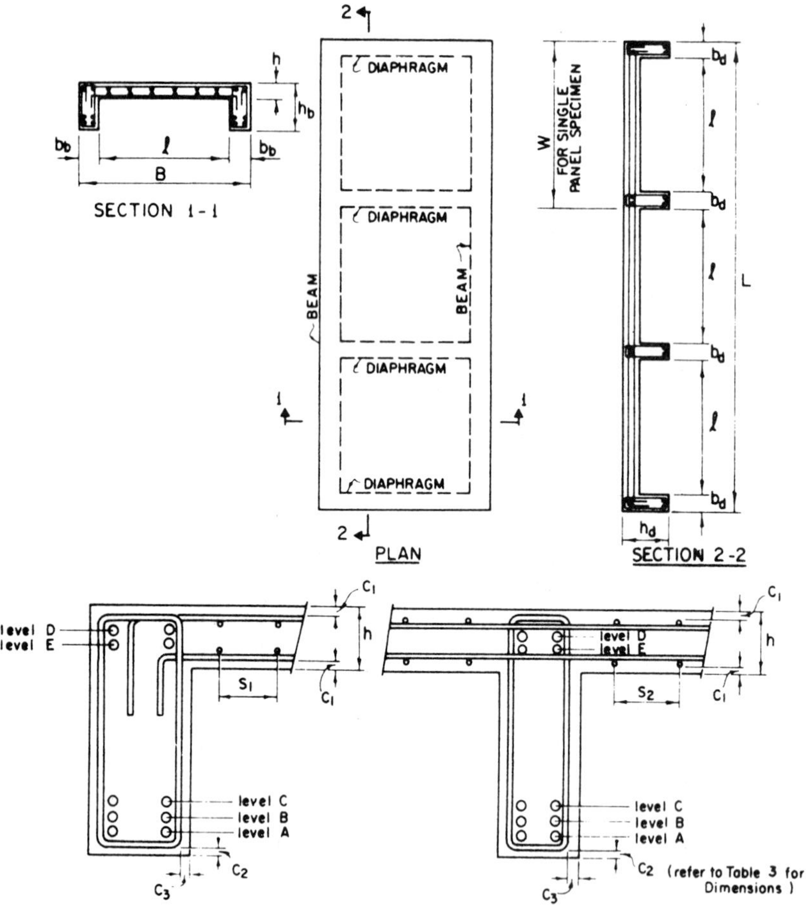

FIGURE 3-16 General layout of bridge specimens and reinforcement details of two-way slabs. (From Batchelor and Tissington, 1976.)

Effect of Model Scale A nonlinear relationship between ϕ_0 and ω was linearized by plotting ϕ_0 versus $(1/\omega)$. The results clearly show that there is no effect of model scale on the shear strength of the slabs. It appears also that the minimum percentage in the panels could be reduced to 0.28 percent.

Comparison of Methods of Analysis The proposed method does not assume one-way slab action for the center panels as does the method originally proposed by Tong (1969) and Tong and Batchelor (1971). The proposed method also assumes that failure in the end panels occurs when the weakest surrounding element (diaphragm) has cracked completely.

Relationship Between V_{test} and V_{total} Plots were obtained for V_{test} versus V_{total} for the slabs of the C series (variable scale, constant nominal ω) and a T series (constant scale, variable ω). The plots are based on results proposed by Tong and Batchelor (1971), and include the relationship recommended for predicting the ultimate shear capacity V_u. The results reveal some apparent scale effects.

The effect of support conditions on shear strength is also apparent. The beam action present in the models of the C series induced compressive in-plane stresses in the slabs and led to enhanced strength. For example, the center panels of the C series failed at loads approximately 1.46 of those of the A series. This enhancement is possible because the beams and diaphragms of the latter series are restrained against vertical displacements.

Relationship Between ψ and ω Plots were obtained for the C and the T series, and load to the following relationships.

For the center panels,

$$V_u = [2.152 - 1.543 \log(100\omega)] V_{total} \qquad R = 0.920 \qquad (3\text{-}8)$$

For the end panels,

$$V_u = [1.606 - 0.861 \log(100\omega)] V_{total} \qquad R = 0.949 \qquad (3\text{-}9)$$

For the single panels,

$$V_u = [1.388 - 0.777 \log(100\omega)] V_{total} \qquad R = 0.956 \qquad (3\text{-}9a)$$

where R is the correlation coefficient.

Conclusions Batchelor and Tissington (1976) present the following conclusions.

1. The effects of scale on the shear strength of slabs are negligible, provided that appropriate nondimensional slab parameters are used. A key parameter is the reinforcement index ω.
2. The important influence of support conditions is documented where the beams and diaphragms are restrained against vertical displacement. These premature failures indicate that slab shear strength can be markedly enhanced by the compressive stresses caused by beam action in the models.
3. Equations (3-8) and (3-9) may be used to predict shear strength of bridges for the slab types investigated.
4. Bridge slabs with minimum isotropic reinforcement according to (3-7) can be expected to perform satisfactorily in terms of cracking and

deflection. For instance, deflections in slab panels were less than 1/600 of the span at working loads based on a load factor of 2.5 for live load. All slabs in the models had a span–thickness ratio of 25.

3-6 BOX GIRDER BRIDGES

Characteristics

Multicell reinforced concrete box girders become practical at about the maximum optimum span length of a T-girder bridge. They are usually considered for spans of 95 to 140 ft. Beyond this range it is probably more economical to select a different type of bridge, such as a posttensioned box girder or a steel girder superstructure, because of the massive increase in volume and materials, rendering the structure relatively inefficient.

Box girder decks are cast-in-place units that can be constructed to follow any desired alignment in plan, so that straight, skew, and curved bridges of various shapes are common in the highway system. Because of the high torsional resistance, a box girder structure is particularly suited to bridges with significant curvature. The high torsional strength also allows the bridge to be designed as a unit without considering individual girders.

Interchange ramp structures typically require sharp curved alignment, which is a main reason for selecting this type of bridge at interchanges on freeways. This is particularly evident in California where about 70 to 80 percent of all bridges (computed on the basis of deck area) are multicell concrete box girders.

This construction facilitates esthetic treatment where both the side view and the underside of the superstructure develop smooth and harmonious lines. Sloping exterior webs are often part of this treatment, and haunching the girder soffit to maximum depths at interior supports is common. Occasionally, the lower box corners are rounded to reduce the effect of massive appearance. Monolithic construction of the substructure and superstructure offers structural advantages and also enhances appearance. Pier caps can be placed within the box, so that the superstructure can be rigidly connected to pier shafts.

AASHTO Article 8.10.2 stipulates that the entire slab width can be assumed effective for compression. For integral bent caps, the effective flange width overhanging each side of the bent cap should not exceed six times the least slab thickness, or 1/10 the span length of the bent cap. For cantilevered bent caps, the span length is taken as twice the length of the cantilever span.

For span lengths of 95 to 140 ft commonly in use, a depth–span ratio of 0.055 is recommended. For haunched structures, the depth–span ratio is about 0.05 in the spans and about 0.08 at the supports.

Design Fundamentals Because box girders have a top and a bottom flange, they should be designed as T beams for both positive and negative moment. The presence of the bottom flange is effectively utilized in negative moment regions to control compression stresses, and alleviates the need for compressive reinforcement. Increasing the bottom slab thickness in areas of negative moment is common, as is the thickening of girder webs adjacent to supports to control shears. Article 8.11 of AASHTO limits the minimum bottom flange thickness to 1/16 of the clear span between girder webs, or 5.5 in. The bottom slab reinforcement should not be less than 0.4 percent of the flange area in the longitudinal direction, or 0.5 percent of the flange area in the transverse direction.

End diaphragms may be omitted where adequate justification is provided by the structural analysis. AASHTO Article 8.12.3 stipulates that straight box girders and curved box girders with an inside radius of 800 ft or more do not require intermediate diaphragms. For an inside radius less than 800 ft, intermediate diaphragms may be required. If diaphragms are used, the recommended maximum spacing is 40 ft (see also subsequent sections).

Structural Response of Concrete Box Girder Bridges: Case Studies

Bridges with 0° Skew Scordelis, Bouwkamp, and Wasti (1973) have investigated the structural response of box girder bridges on a scale model of a typical two-lane bridge in California, with two equal spans of 101.5 ft each. A typical cross section is shown in Figure 3-17, and a plan and elevation are

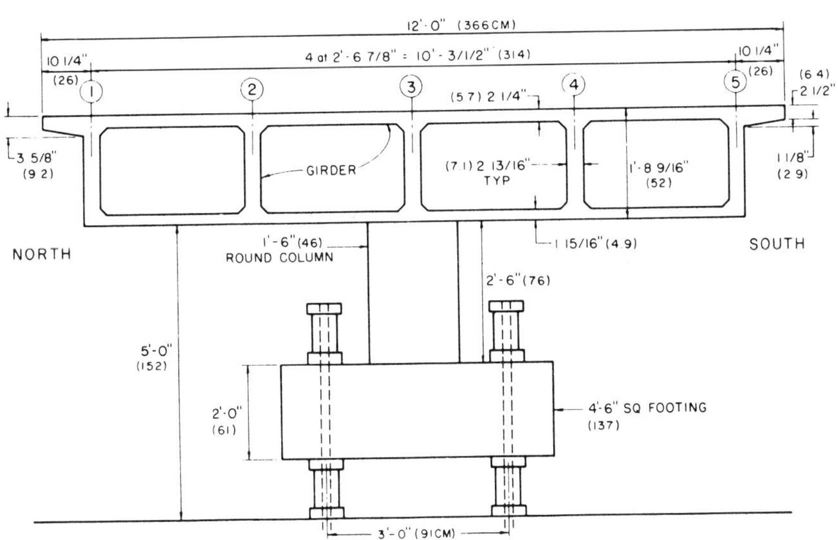

FIGURE 3-17 Typical cross section of box girder bridge model; model scale: 1 : 2.82. (From Scordelis, Bouwkamp, and Wasti, 1973.)

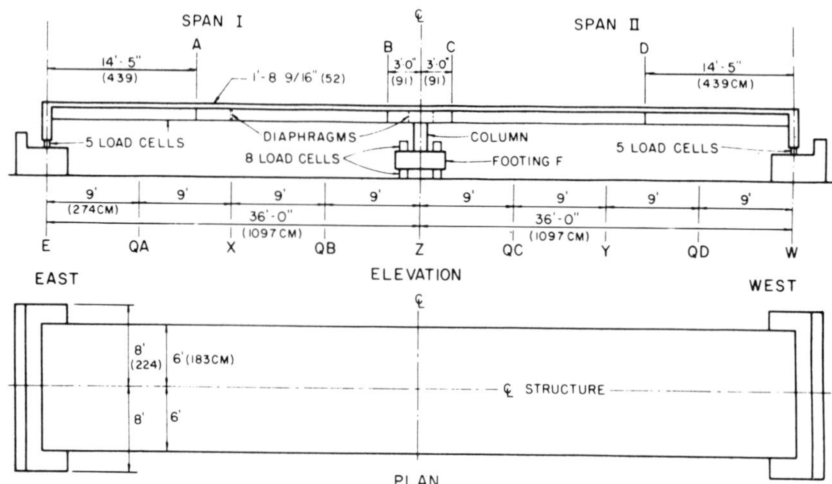

FIGURE 3-18 Plan and elevation of box girder bridge model with transverse locations; model scale: 1 : 2.82. (From Scordelis, Bouwkamp, and Wasti, 1973.)

shown in Figure 3-18. Point loads were chosen to produce stresses of the order of working ranges, 24 to 30 ksi total tensile stress in the reinforcement. The theoretical methods treated the structure as an elastic uncracked homogeneous member, and the distribution of the total internal moment at the instrumented sections A, B, C, and D was predicted by two computer programs. The calculated theoretical deflections show that, as expected, the deflection under load is greater at the midspan section $1Y$ without a diaphragm than at section $1X$ which has a diaphragm.

Response Before Overload Stress Levels A loading phase of the initial application of conditioning loads to produce a steel stress of 30 ksi is taken as representative of the response before overload stress levels. After removal of the conditioning load, point load combinations were used to induce the specified working stress levels.

Good agreement was verified between the theoretical and experimental results for reactions, deflections, and moments, and this comparison suggests the validity of superposition of the theoretical and experimental data obtained in the study.

Response After Overload Stress Levels After an overload sequence of conditioning loads inducing stress levels of 40, 50, and 60 ksi, point load combinations produced working stress levels of the order of 30 ksi. This stage was intended to articulate the structural response after an overload event expected to accentuate the process of cracking, deflection, and stress development.

Summary of Results Results and conclusions from this program are summarized as follows.

1. The relationship between total center or west reactions and applied point load remains essentially unchanged after conditioning overloading up to 60 ksi.

2. For a point load at 1*Y*, after each higher conditioning overload the deflection increases at 1*Y* directly under the load, whereas on the opposite side of the bridge at the same location the deflection decreases. The ratio of experimental to theoretical deflection ranges from 1.25 to 1.60. This suggests that, beyond the 30-ksi working stress conditioning load, an analysis based on the uncracked section predicts the transverse distribution between 1*Y* and 5*Y*, but with the experimental values being 60 percent higher than the theoretical values. After higher conditioning overloads, the theory is no longer valid in predicting the transverse distribution of deflections.

3. The transverse distribution of the total moment at a section (expressed as percentage to each girder) can be predicted by theory at working stress levels for both single point loads and uniform loading across the bridge width. After overload stress levels, agreement between theory and tests decreases for single point loads, yet is still satisfactory for uniform loading. Because actual critical girder design moments are based on several wheel loads acting on the bridge, the theoretical method should adequately predict design moments even after overload events.

Bridges with Skew Scordelis, Bouwkamp, Wasti, and Seible (1982) have also presented analytical and experimental results from a study of a scale, 45° skew, two-span four-cell reinforced concrete box girder bridge model. This is a replica of a typical California prototype structure shown in Figure 3-19, with two spans and a skew center-bent diaphragm. The bridge is assumed to be an elastic, homogeneous, isotropic, and uncracked structure.

Skew box girder geometry is often dictated by the lack of space in congested urban areas and by complex intersections. Current specifications make no distinction for load distribution in straight, curved, and skew box girder bridges by ignoring the effects of curvature and skew angle. California standards specify an increase in shear for the girder ends at the obtuse corners of the span.

The conclusion drawn from this study is that a simple one-dimensional beam model can be used to predict the total reactions, centerline deflections, and longitudinal total moments for preliminary designs. Finite-element programs allow more accurate assessment of the longitudinal and transverse distribution of experimental reactions, strains, and moments. Deflections may be included in the program if magnification factors are used to consider the effect of cracking.

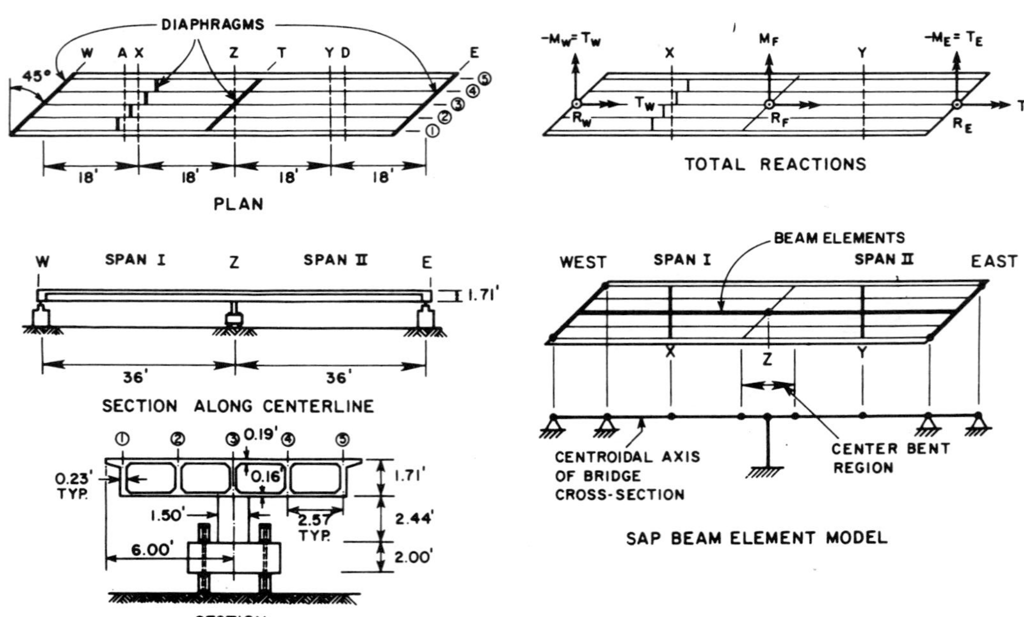

FIGURE 3-19 Skew box girder bridge model (in ft): (*a*) dimensions of skew box girder bridge model; (*b*) one-dimensional model of skew box girder bridge. (From Scordelis, Bouwkamp, Wasti, and Seible, 1982.)

For dead load and conditioning load applications, the effect of skew angle is a reduction in positive moments at midspan and a similar reduction of negative moments over the center bent. For point loads on the acute side of the span, the midspan moments in the skew bridge are larger than in corresponding positions for straight and curved bridges. For the same point loads on the obtuse side of the span, however, the midspan moments in the skew bridge are smaller than in the corresponding positions for straight and curved bridges.

For the skew box girder bridge model subjected to heavy concentrated loads, the distribution of moments transversely to the girders is essentially nonuniform. This challenges the validity of load distribution factors typical for straight bridges. Alternatively, analysis using the CELL program may be more appropriate and closer to the actual structural response.

After all overloads up to the 60-ksi stress level, the structural response of the skew bridge model under point loads yielding working stresses does not change, but larger magnification factors are needed in predicting experimental deflections to account for increased cracking. The mere fact that the bridge model has been temporarily subjected to overloads up to 60 ksi does not modify bridge behavior under working loads, with the exception of the effects of cracking.

Comments on These Investigations Libby (1974) points out that the nonlinear deflections for a 100-kip load analyzed by Scordelis, Bouwkamp, and Wasti (1973) for the bridge of Figure 3-18 show good correlation between the shapes of the actual and theoretical deflections, but this loading is much more severe than the AASHTO bridge design loading. Conversely, the main purpose of the point loads in this case was to obtain experimental results that could be compared with theoretical predictions.

There is general agreement (Libby, 1974; Scordelis et al., 1975) that for the practical design of straight multicell concrete box girder bridges with widths equal to or less than their spans, the flexural stresses due to longitudinal bending can be assumed to be resisted by the box girder section as a unit without transverse rotation of the section. In this context, dividing the box into interior and exterior girders, as required by AASHTO, for the purpose of analysis has no merits and appears to be unrealistic. For bridges that are wider than their spans or where the live load moment (longitudinal) is large compared to the dead load moment, transverse elasticity is admissible and should be considered in the design.

For bridges of unusual curvature, this procedure is still adequate but with some minor modifications. However, for box girder bridges with large skews or for bridges with widths much greater than their spans, a rational and simple design procedure is yet to come. Interestingly, the results obtained for multicell concrete box girder bridges should not arbitrarily be applied to composite structures consisting of a concrete slab on individual thin-walled steel boxes.

Topics Relevant to Box Girders

Irregular and Skew Supports Figure 3-20 shows two examples of irregular internal support arrangements due to geometric conditions at the crossing. If diaphragms are used, they have the same skew configuration.

Kristek (1974) has introduced solutions for box girders with two axes of symmetry and a variable cross section along the longitudinal direction. The analysis is feasible even with a sudden change in the cross-sectional dimensions, with arbitrary support type at the ends, and for irregularly placed internal supports and skew diaphragms. It also provides a solution to the case of right diaphragms that may be deformable and made in plate-bracing or truss-bracing form.

The influence of irregularly placed supports and skew diaphragms is demonstrated in examples of loading producing sole bending at the usual right-supported structures. In this case a sudden gap in the moment diagram has been noted because the skew diaphragm is stressed not only by shear but also by bending in its place, thereby transmitting part of the total bending moment along the length.

These principles can be illustrated in the example of Figure 3-21, showing a box girder cross section supported by two intermediate web supports with a

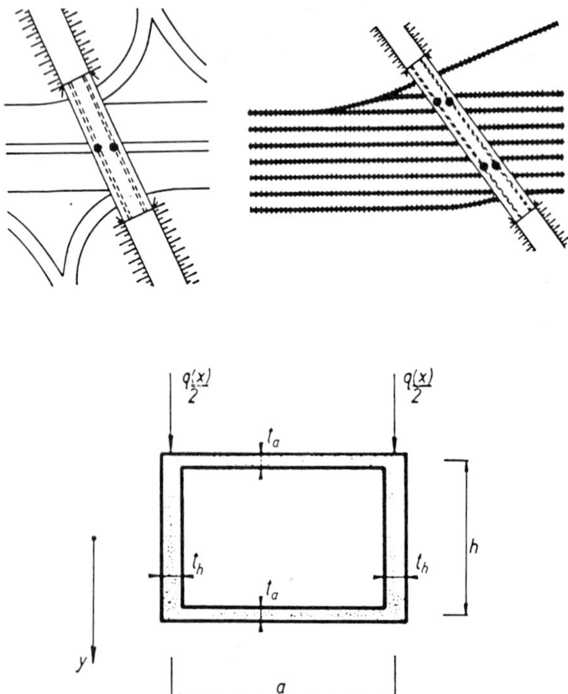

FIGURE 3-20 Examples of complex skew crossing; box girder bridges.

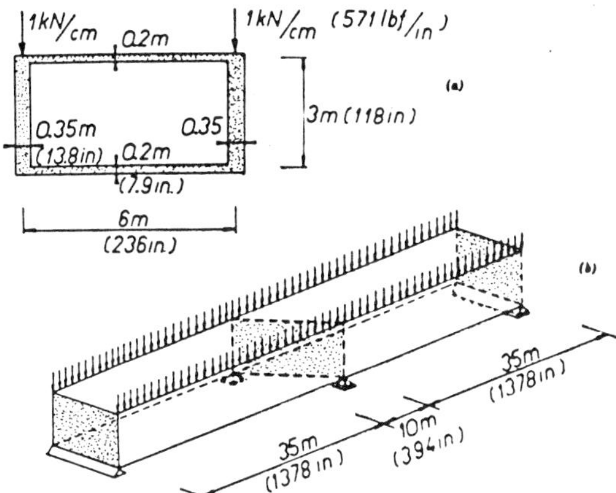

FIGURE 3-21 Solved example; irregular supports at skew diaphragm for box girder bridge. (From Kristek, 1974.)

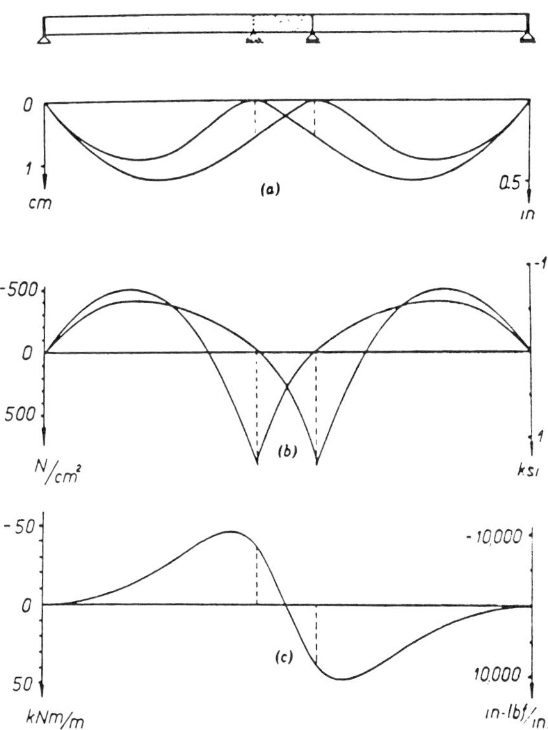

FIGURE 3-22 Deflections, longitudinal normal stresses, and transverse flexural distortion moments of corners of cell of solved example of Figure 3-21. (From Kristek, 1974.)

FIGURE 3-23 Shape of deformed cross section at point of action of support reaction; bridge of Figure 3-21. (From Kristek, 1974.)

skew. At the ends the bridge is simply supported and has right diaphragms. The intermediate support points are also connected with a skew diaphragm of thickness $t = 0.3$ m. The concrete structure has a modulus $E_c = 38{,}500$ MN/m^2 (5890 ksi), and is loaded symmetrically with two uniformly distributed live loads of 100 kN/m each.

The load and the cross section are symmetrical about the vertical axis, yet torsional effects are present. Figure 3-22*a* shows the resulting vertical deflection along the web lines of the box, and Figure 3-22*b* shows the longitudinal normal stresses at the upper corners along these lines. Figure 3-22*c* shows the transverse flexural distortion corner moments.

These results confirm the presence of considerable transverse flexural distortion stresses resulting from irregularly spaced supports. A second point is the discontinuity in the normal stress diagram at a point lying above the internal supports, caused by the additional moment and bimoment factors at the connection of the box and the skew diaphragm. The deformed cross section at the point of action of the support reaction is shown in Figure 3-23. The function of the skew diaphragms is to preserve the vertical position of the supported left web, so that the deflection of the right web at this point is caused by the distortion effect.

In addition to this study, the following are useful references: Kristek (1970, 1970a, 1971), Richmond (1969), and Vlasov (1959).

Box Girder Bridge Diaphragms with Openings The appearance of a box girder bridge is enhanced if the superstructure can be supported on single slender piers as shown in Figure 3-24. The diaphragm at each support is usually provided with an opening for access and utility installation.

Prestressing is often used to reduce the high tensile stresses in the diaphragms, but the location of prestressing cables and the amount of feasible prestressing are often prohibited by the geometry of the system. Normally, the critical zone in the diaphragm is the lower beam above the pier. Because the stress distribution is essentially nonlinear (Sargious and Dilger, 1977), it is usually determined using analytical methods such as finite-element techniques.

Solutions are presented by Sargious, Dilger, and Hawk (1979), and can be used to determine stresses and forces at critical diaphragm locations with openings and under the effect of external loads and prestressing forces. These investigators also provide design examples.

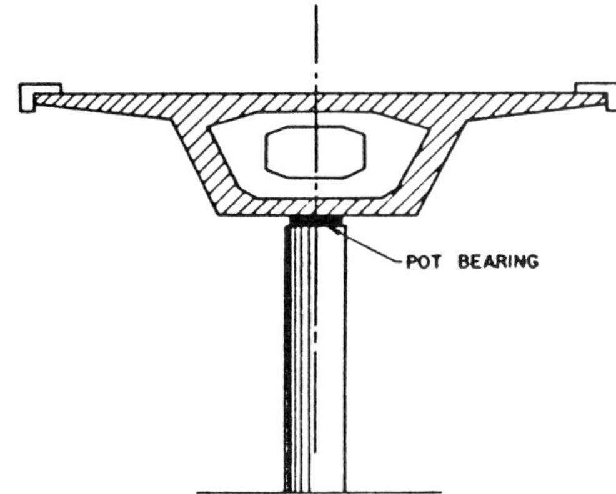

FIGURE 3-24 Box girder bridge supported on single piers.

Example of Box Girder Bridge

A two-span continuous concrete box girder bridge with spans of 100 ft each has a cross section shown in Figure 3-25. The transverse reinforcement in the deck slab is designed as in a conventional T-beam bridge and will not be repeated here. For service loads, the stress levels are as in previous examples, and the design live load is HS 20.

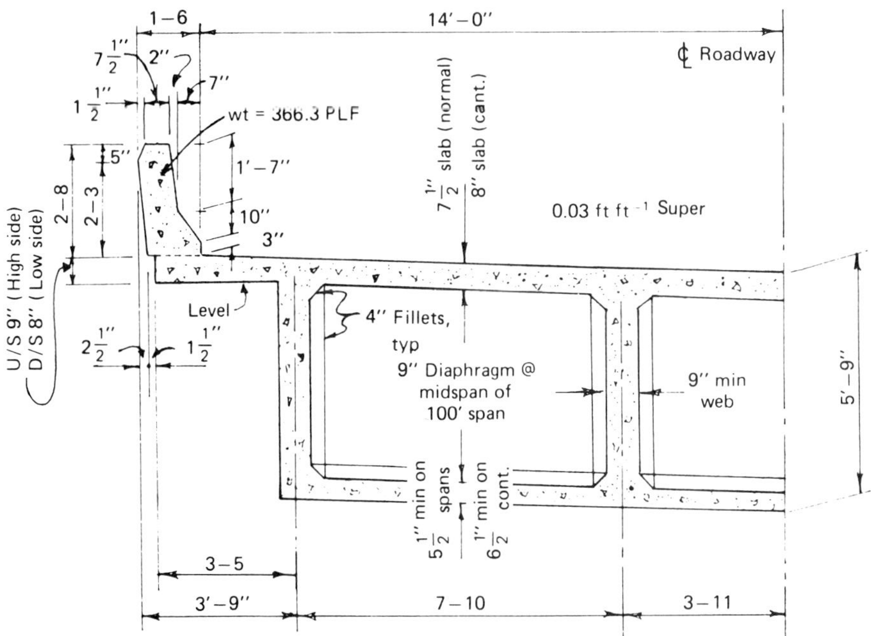

FIGURE 3-25 Half typical section for box girder of design example. (From Heins and Lawrie, 1984.)

The bottom flange thickness and reinforcement requirements are determined according to AASHTO Articles 8.11.2 and 8.17.2.3. Using a nominal girder spacing $S = 7.83 - 0.75 = 7.08$ ft, we compute the bottom slab thickness as $t = (1/16)(7.08) = 0.44$ ft (use 5.5 in.).

The bottom flange reinforcement parallel to the girders is $A_s = 0.004(0.46)(24.25)(144) = 6.4$ in.^{2}. Likewise, the transverse bottom flange reinforcement is $A_s = 0.005(5.50)(12.0) = 0.33$ in.^{2}/ft (use #5 bars at 9-in. alternating top and bottom, $A_s = 0.41$ in.^{2}/ft).

Dead Load The weight of the bottom slab, stems, top slab in boxes, fillets, and forms is calculated as 6.81 kips/ft of deck. The weight of the cantilever slabs, parapets, and future wearing surface is calculated as 2.08 kips/ft of deck. Therefore, the total dead load is $w = 8.89$ kips/linear foot.

The dead load moments are computed as follows. For positive moment in span,

$$M_p = 8.89 \times 0.0703 \times 100^2 = 6250 \text{ ft-kips}$$

For negative moment at support,

$$M_n = -8.89 \times 0.125 \times 100^2 = -11{,}112 \text{ ft-kips}$$

Likewise, the dead load shears are computed as follows. For shear at end support,

$$V_e = 0.375 \times 8.89 \times 100 = 333 \text{ kips}$$

For shear at interior support,

$$V_i = 0.625 \times 8.89 \times 100 = 557 \text{ kips}$$

Live Load For the design of box girders, the distribution factor as a unit for the exterior and interior girders is combined. Thus, the distribution factor is obtained by dividing the out-to-out slab width by seven, or DF $= 30.33/7 = 4.33$ lines of wheels. Using impact $I = 50/(100 + 125) = 22$ percent, the LL + I coefficient per total box girder is computed as $4.33 \times 1.22 = 5.28$ (lines of wheels).

From ASIC tables, the live load plus impact moment is calculated as follows. For positive moment in span,

$$M_p = 1234(1/2)(5.28) = 3258 \text{ ft-kips}$$

For negative moment at support,

$$M_n = -1146(1/2)(5.28) = -3025 \text{ ft-kips}$$

Likewise, the live load plus impact shears are calculated as follows. For shear at end support,

$$V_e = (63.7)(1/2)(5.28) = 168 \text{ kips}$$

For shear at interior support,

$$V_i = (67.8)(1/2)(5.28) = 179 \text{ kips}$$

Moment Design At the interior support, the total moment is $M = -(11{,}112 + 3025) = -14{,}137$ ft-kips per box. The approximate required area of steel reinforcement is computed assuming $j = 0.90$ and $d = \pm 65$ in. Thus,

$$A_s = \frac{14{,}137(12)}{24(0.90)(65)} = 121 \text{ in.}^2 \qquad \text{Try 54 \#14} \qquad A_s = 121.5 \text{ in.}$$

The width of the entire box at the bottom (out-to-out of web) is $b = 291$ in. Also, $d = 69.00 - 2.00 - 0.75 - 0.85 = 65.4$ in. At the support, we select the bottom slab thickness $t = 7$ in. Using these section dimensions and neglecting the box webs (stems), the stresses may be computed as

$$f_s = \frac{M}{A_s jd} \qquad \text{and} \qquad f_c = \frac{f_s k}{n(1-k)}$$

Next, we compute

$$pn = \frac{121.5(9)}{291(65.4)} = 0.056 \qquad \text{and} \qquad \frac{t}{d} = \frac{7}{65.4} = 0.107$$

The factors k and j are computed from diagrams as $k = 0.36$ and $j = 0.95$. Therefore, the stresses in the steel and the concrete are

$$f_s = \frac{14{,}137(12)}{121.5(0.95)(65.40)} = 22{,}470 \text{ psi} \qquad \text{OK}$$

$$f_c = \frac{22{,}470(0.36)}{9(0.64)} = 1404 \text{ psi} \approx 1400 \text{ psi} \qquad \text{OK}$$

The design for the maximum positive moment in the span is completed in a similar manner.

Shear Design The maximum shear at the interior support is $557 + 179 = 736$ kips. We assume that the webs resist the entire shear, and at the supports we select a web thickness of 12 in. Then total $b = 4 \times 12 = 48$ in. The shear stress $v = V/bd = 736/(48 \times 65.40) = 234$ psi and the allowable shear stress $= 0.95\sqrt{3500} = 56$ psi. Note that $v - v_{\text{al}} = 234 - 56 = 178 < 4\sqrt{f'_c}$. Using #5 stirrups with two legs per stirrup, $A_v = 8 \times 0.31 = 2.48$ in.^{2}. Therefore, the spacing is

$$S = \frac{2.48(24{,}000)}{178(48)} = 7 \text{ in.}$$

3-7 THROUGH GIRDER BRIDGES

Through girder bridges, popular mainly in the pre–World War II period, are structures in which the main longitudinal girders extend above the roadway. The vertical clearance below the bridge is therefore determined by the thickness required for the floor. The latter may consist of (a) a solid slab spanning between the main girders; (b) closely spaced floor beams supporting a thin slab; and (c) a composite system of longitudinal beams, floor beams, and slab. As a rule, through girder bridges are less economical than other types of concrete bridges, but are a good choice under limited headroom, or where the width is appreciably smaller than the span.

In highway and road construction, through girder bridges are seldom used because of the need to use intermediate girders which is objectionable in terms of highway standards, and because this type of bridge does not lend itself to widening. On the other hand, in railroad construction multitrack through girder bridges are common and include intermediate girders placed between adjacent tracks. Bridges of this type are usually simply supported spans.

Girder–Slab Interaction Usually, the main girder spacing exceeds 14 ft; hence, the live load on each girder is the reaction of wheel loads, assuming the slab acts as a simple beam. For example, this distribution may be applied to the bridge of Figure 3-26*a* showing a slab spanning between two main girders. This arrangement is economical for narrow bridges.

In computing bending moments for the slab, it may be necessary to establish the actual effect of the restraint induced to the slab as partial fixity where it joins the heavy girders, especially where the construction is monolithic. The restraint level is affected by the torsional rigidity of the girder and the stiffness of the slab, and both these parameters can be expressed analytically. In lieu of an exact solution, a reduction factor of 0.8 may be applied to the positive moment obtained by statical analysis, and 50 percent of the previously determined maximum static moment may be applied at the supports as negative moment.

For relatively narrow and long bridges and where the deck stiffness is appreciable, the bridge may be analyzed as a single unit, assuming that deflections along the same cross section are the same at both girders. If through girders are used in continuous units, the section over the supports becomes an inverted partial T beam where the slab resists the compression.

Main Girders In the simplest form, the width and depth of the girders are made constant, producing plane surfaces that simplify formwork. For the purpose of analysis, each main girder is a rectangular beam because the slab, being in the tensile zone, is not considered affective in resisting stresses.

The ratio of girder depth to span length is greater than in deck girder bridges (T beams), usually ranging between $1/8$ and $1/10$, and the design is

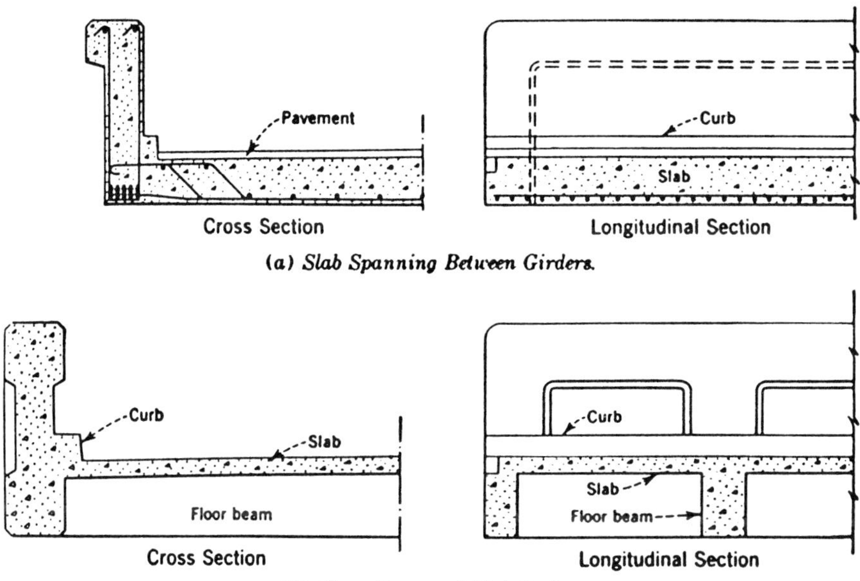

FIGURE 3-26 A typical through girder-and-slab bridge, with the slab spanning between girders, or girders and floor beams. (From Taylor, et al., 1939.)

more economical when compression stresses are resisted entirely by the concrete. However, where the depth and width of the girders are restricted, compressive reinforcement consisting of longitudinal bars is provided.

In bridges designed for two traffic lanes, each girder should be considered as carrying one line of trucks, although the effect of eccentricity may increase this load under a strict interpretation of the specifications. For bridges that accommodate three lanes of traffic or more, the slab may be assumed to act as a simple beam and the truck load distributed accordingly.

Numerical Design Example

A through girder bridge with a floor system consisting of a slab will be designed for the following data: span, 50 ft; clear roadway width between girders, 16 ft; live load, HS 20; strength and stress parameters, as in previous examples (note that the clear roadway width is selected arbitrarily).

Design of Slab For main reinforcement perpendicular to traffic, the bending moment for live load is determined according to Article 3.24.3.1 (AASHTO) which is valid up to spans of 24 ft. Using $S = 16$ ft (clear span), we obtain

$$M_{\mathrm{LL}+I} = \frac{16 + 2}{32} \times 16 \times 1.3 = 11.7 \text{ ft-kips}$$

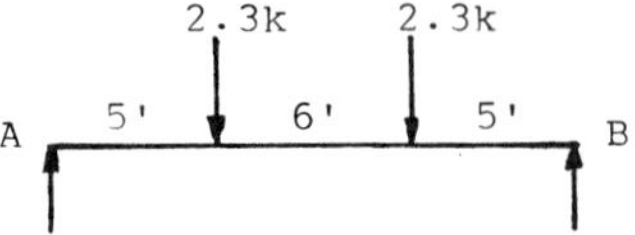

FIGURE 3-27 Approximate truck position for maximum live load moment in slab; reinforcement perpendicular to traffic, through girder bridge.

However, we will check the moment assuming a longitudinal wheel load distribution $4 + 0.06L = 7$ ft, and the truck position shown in Figure 3-27. The fraction of wheel load per foot-width of slab = 16/7 = 2.3 kips. Therefore, the live load moment is

$$\begin{aligned} M_{\mathrm{LL}} &= 2.3 \times 5 = 11.5 \text{ ft-kips} \\ \text{Impact} = 30\% &= \underline{\ 3.5\ } \\ M_{\mathrm{LL}+1} &= 15.0 \text{ ft-kips} \end{aligned}$$

Now, assume a 12-in.-thick slab, weight = 150 lb/ft^2, and a future wearing surface, total dead load $w = 0.175$ kip/ft. Then

$$\begin{aligned} M_{\mathrm{DL}} &= 0.175 \times 0.125 \times 16^2 = 5.6 \text{ ft-kips} \\ \text{Total moment} &= 15.0 + 5.6 = 20.6 \text{ ft-kips} \end{aligned}$$

We may apply a reduction moment factor of 0.8 on account of the end restraint, or $M = 20.6 \times 0.8 = 16.5$ ft-kips. The minimum thickness d is now estimated as

$$\min d = \sqrt{\frac{16.5}{0.211}} = 9 \text{ in.}$$

For a cover of 1 in. and #8 bars, the available d is 10.5 in. We now estimate the bottom reinforcement

$$A_s = \frac{16.5}{10.5 \times 1.78} = 0.89 \text{ in.}^2/\text{ft} \qquad \text{Use \#8 at 10 in.} \qquad A_s = 0.95 \text{ in.}^2/\text{ft}$$

Approximately 50 percent of the static moment, or 10.3 ft-kips, will be assumed to exist as restraint at the junction with the girders. For a cover of 2.75 in., the effective d is $12.00 - 2.75 - 0.50 = 8.75$ in. The area of steel required by the top of slab is now

$$A_s = \frac{10.3}{8.75 \times 1.78} = 0.66 \text{ in.}^2/\text{ft} \qquad \text{Use \#7 at 10 in.} \qquad A_s = 0.72 \text{ in.}^2/\text{ft}$$

Design of Girders First, we obtain the fraction of wheel load by placing one wheel 2 ft from the face of the girder and assuming the slab transfers the load as a simple beam. The wheel load fraction is easily obtained as

$Q = 1.375P$. The impact factor is $50/(50 + 125) = 28.6$ percent. From tables, we obtain the live load moment (one truck) as $M_{\text{LL}} = 628$ ft-kips.

With these data, we now estimate the design live load plus impact moment as

$$M_{\text{LL}+I} = 0.5 \times 628 \times 1.375 \times 1.286 = 556 \text{ ft-kips}$$

Next, we assume a girder 5 ft 3 in. deep and 2 ft wide (ratio = $50/5.25 = 9.5$). The dead load is now as follows:

Weight of slab + Wearing surface	$= 8 \times 0.175$	$= 1.40$ kips/ft
Weight of girder	$= 5.25 \times 2.00 \times 0.15$	$= 1.57$ kips/ft
Total dead load		$= 2.9$ kips/ft

Therefore, the dead load moment is

$$M_{\text{DL}} = 2.97 \times 0.125 \times 50^2 = 927 \text{ ft-kips}$$

The total design moment is now

$$M = 556 + 927 = 1483 \text{ ft-kips}$$

The depth required for balanced design, based on $K = 211$, is

$$d = \sqrt{\frac{1483}{0.211 \times 2}} = 59 \text{ in.}$$

For #11 bars in two layers, the effective d is $63 - 2.70 - 1.75 = 58.5$ in. We now estimate the reinforcement required to resist tension, or

$$A_s = \frac{1483}{1.78 \times 58.5} = 14.3 \text{ in.}^2 \qquad \text{Use 10 \#11} \qquad A_s = 15.6 \text{ in.}^2$$

arranged in two layers, five bars in each layer.

The reinforcement is arranged as shown in Figure 3-28. Distribution reinforcement in the bottom of the slab is $220/\sqrt{S} = 55$ percent, or $0.55 \times 0.95 = 0.52$ in.^{2}/ft, or #6 bars at 10 in. $= 0.53$ in.^{2}/ft. Temperature reinforcement in the top of the slab is provided with #5 bars at 18 in. Three #8 bars are included in the top of the girder. The design should be completed by estimating shears and reactions for stirrup design as in the example of Section 3-5.

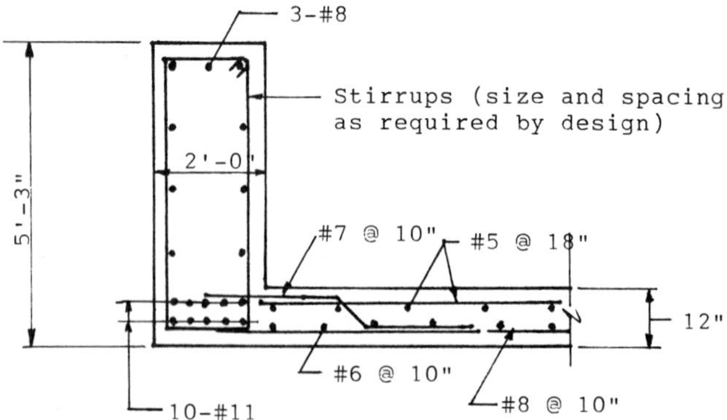

FIGURE 3-28 Through girder bridge; dimensions and reinforcement details.

3-8 TOPICS RELEVANT TO SLAB BRIDGES

Flat Slabs

Flat slab bridges consist of a reinforced concrete slab extending in four directions and supported directly by isolated individual concrete columns without beams or girders. In building construction, flat slabs are widely used because of economy and structural compatibility with heavier live loads. These merits, however, have not been fully demonstrated in bridge work, the main reason being that the proper arrangement of columns is not always feasible under the usual geometric conditions. The practical use of flat slabs is thus limited to certain examples, briefly discussed in this section.

Compared to other bridge types, flat slab design is likely to require the minimum construction thickness, and this is a clear advantage where headroom is limited. With a properly designed flat slab, the cost of the bridge may be 20 to 25 percent less than the cost of other concrete types, and this saving results mainly from favorable formwork costs for both superstructure and columns. Because flat slabs are built monolithically with the supporting columns, expansion bearings are not needed, and because the superstructure is of uniform cross section and reinforced in two directions, temperature and shrinkage changes are more efficiently resisted.

Arrangement of Columns In modern grade separations or long overpasses, the geometry and the usually unrestricted horizontal alignment dictate the location of supports in the longitudinal direction. Transversely, the spacing of columns has likewise limited choices. Examples of column spacing are shown in Figure 3-29. These solutions are typical and must be confirmed for the intended live load (truck or pedestrian). For structures of considerable width, the transverse spacing may depend mainly on economy.

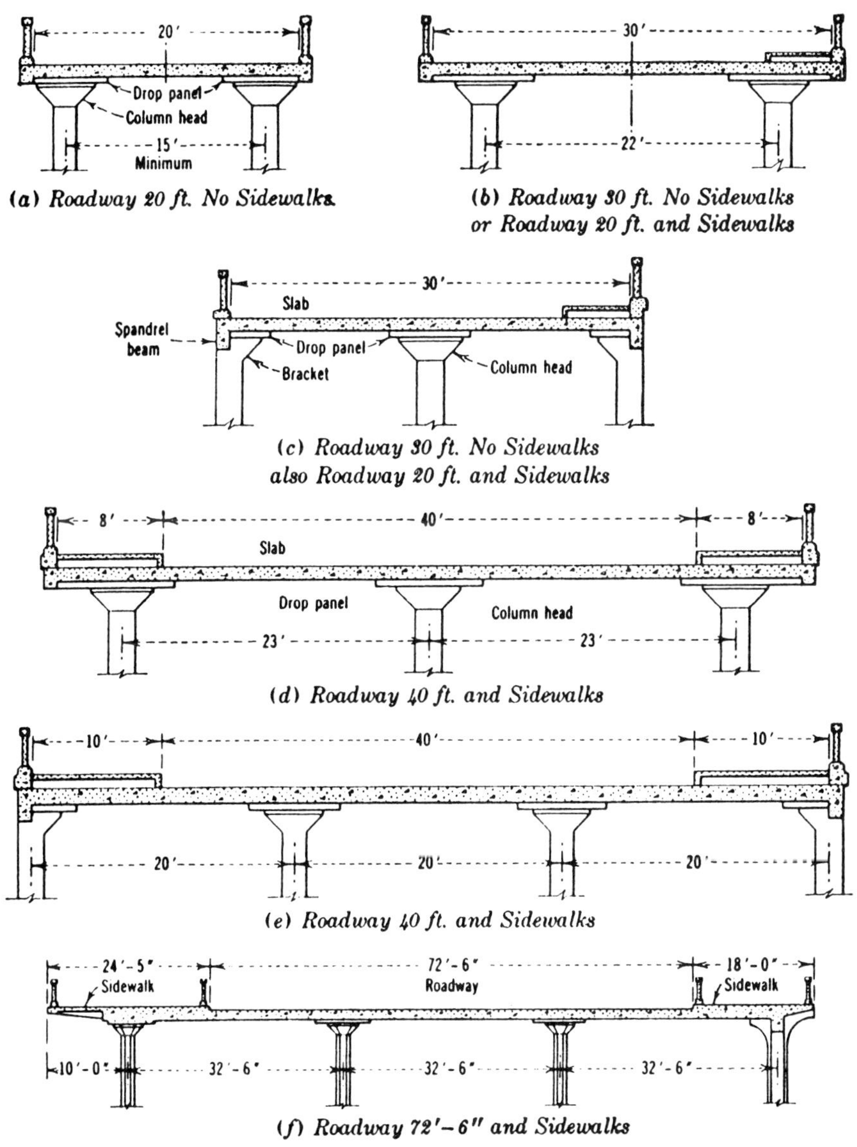

FIGURE 3-29 Typical cross sections of flat slab bridges. (From Taylor et al., 1939.)

Elements of Flat Slab Bridges Structurally, a flat slab bridge consists of (a) a continuous slab, (b) drop panels at the columns, (c) column heads, (d) columns, (e) spandrel beams, and (f) footings and foundation elements.

A typical design unit of a flat slab is the panel bordered on four sides by lines connecting the centers of four supporting columns, usually rectangular or square. Where skew structures are unavoidable, the panel may be rhomboidal or rhombic, and in odd parts of the bridge odd-shaped panels may be necessary. Where the panels are square or nearly square, the design is more economical and results in a constant slab thickness. For preliminary rule-of-thumb estimates, the slab thickness t for a unit extending over several panels may be estimated from the following:

$$t = 0.02l\sqrt{w} + 1.5 \tag{3-10}$$

where t = slab thickness (in.)
l = largest span (ft)
w = a uniformly distributed load (lb/ft^2) representing dead load, live load, and impact

A flat slab is usually more economical if it is provided with drop panels at the columns. For preliminary estimates, the total length or width of a drop panel may be taken as 0.375 of the span length in the same direction. The function of drop panels is to resist the normally greater moments and shears at the columns, so that the total construction thickness at this location is governed by negative bending moments at the column strips or by shear stresses. From experience, the thickness of the drop panel is about one-half of the slab thickness but this is not a fixed criterion.

Most flat-slab bridges in the past had column heads, flaring out at the top in the shape of a truncated cone or pyramid. This arrangement is considered effective in strengthening the slab if the angle of the flaring with the vertical does not exceed 45°. Column heads help to increase the shear resistance of the slab at critical column locations and also reduce the effective span of the slab. For a preliminary analysis, bending moments in the slab may be estimated using a theoretical (effective span) l_e as follows:

$$l_e = l\left(1 - \frac{2}{3}\frac{c}{l}\right) \tag{3-11}$$

where l = actual span, center-to-center of columns
c = effective diameter of round column head, measured where the thickness at the edge, below the slab or the drop panel, is at least 1.5 in.

The strength of flat slab construction depends on the rigidity of the columns, in particular, the exterior columns. In effect, a flat slab bridge is a rigid frame where the slab represents the horizontal members and the columns are the vertical units. Because of this interaction, the design of a flat slab bridge should state clearly the underlying assumptions and related criteria. A typical design procedure involves the following steps.

1. Select the column spacing with the intent to make the slab panels as nearly square as possible. For continuous units, bending moments in the slab are balanced if the outside panels are made smaller.
2. Analyze the need for drop panels and select the diameter of the column head which fixes the theoretical (design) span.
3. Estimate the preliminary dead load and an equivalent uniformly distributed live load including impact.
4. Determine bending moments in the slab at critical panel sections. Use the largest positive moment to obtain the slab thickness, and the largest negative moment at the column to obtain the total thickness of the slab plus the drop panel. Where final dimensions differ from assumed values, adjust the design accordingly.
5. Check shear stresses at the edge of the column head.
6. After all dimensions are finalized, estimate the areas of steel required at various critical sections and decide on the preferred steel bar arrangement (straight bars, bent bars, etc.).
7. Determine the dimensions and reinforcement of the supporting columns.

Analysis of Rectangular Slab Bridges by the Method of Coefficients

A series solution of the isotropic plate equation is available for concentrated loading on slabs. Curves of distribution coefficients for longitudinal and transverse moments have been developed for orthotropic decks since 1969 (Cusens and Pama, 1969). This reference contains figures and diagrams for selecting distribution coefficients for simply supported slab bridges. Design parameters can be extrapolated provided the point of interest and the load location lie at the transverse centerline of the bridge slab. The expression for moments and deflections has been rewritten by Hossain (1975) using a two-part identical cosine series to yield a system of coefficients that is only a fraction of a full influence surface table.

Figures 3-30*a* and *b* show plan, elevation, and typical cross sections of concrete slab bridges relevant to this analysis. Several bridges of this type have been built in Ontario, Canada, and have provided adequate flexural and torsional strength. Nonetheless, the load distribution characteristics depend on the number and location of isolated supports, the width—length ratio, and the actual flexural and torsional rigidity.

Hossain (1975) has modified the Guyon–Massonnet load distribution theory (Guyon, 1946, 1949; Massonnet, 1950, 1954, 1959; Massonnet and Bares, 1968), beginning with the governing differential equation of an isotropic plate

$$D\left(\frac{\partial^4 \omega}{\partial x^4} + 2\frac{\partial^4 \omega}{\partial x^2\, \partial y^2} + \frac{\partial^4 \omega}{\partial y^4}\right) = P(x, y) \qquad (3\text{-}12)$$

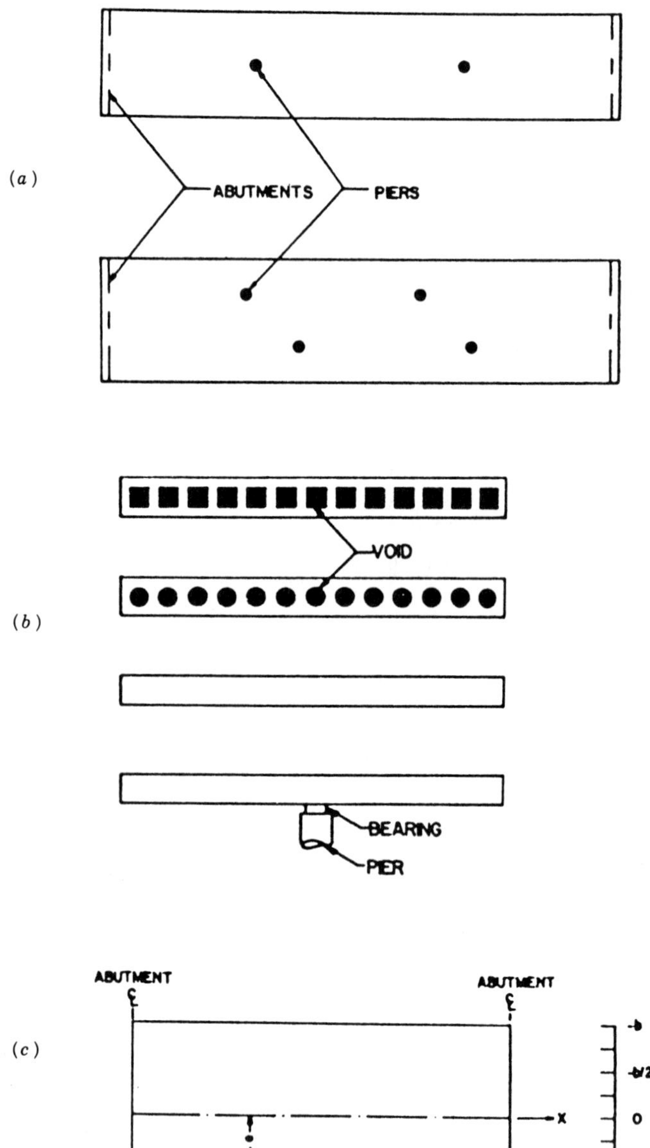

FIGURE 3-30 (*a*), (*b*) Plan, elevation, and cross sections of slab bridges on isolated supports; (*c*) simply supported slab subjected to concentrated load *P*.

where ω = deflection normal to the coordinate axes x and y
$D = (Eh^3)/[12(-\nu^2)]$
E = elastic modulus
ν = Poisson's ratio
P = load
x, y = coordinates
h = slab thickness

For the simply supported plate of length $2a$, shown in Figure 3-30c and subjected to a concentrated load P of eccentricity e, (3-12) is the governing equation (see also Section 7-4).

The solution of (3-12) yields expressions for the deflection, longitudinal moments, transverse moments, and twisting moments (torsion) in terms of relevant parameters. The influence of Poisson's ratio is significant, particularly in transverse moments and torsion (Rowe, 1962).

Useful data and values for the distribution coefficients are tabulated by Massonet and Bares (1968). A similar method of analysis has been developed by Cusens and Pama (1969) for bridge slabs subjected to a concentrated load.

Influence Coefficients Influence coefficients can be developed from the equations giving the deflection, longitudinal moment, transverse moment, and twisting moment (Hossain, 1975). The effect of a concentrated load is then obtained as the product of the influence coefficient and the load.

Coefficients developed in this manner for deflections and transverse moments have the same trigonometric identities as for longitudinal moments. The coefficients derived for twisting moments are related to a parameter that is a function of (a) the coordinate x, (b) the distance d of the concentrated load, and (c) the length $2a$ of the bridge.

Slab Bridges on Isolated Supports Experimental studies relevant to this type of bridge are limited. One of the few reported studies is for the Cumberland Basin Scheme (Best, 1964; Best and West, 1965). This model consisted of three columns and a short cantilever section beyond each end. Because the present method is for slabs simply supported at ends but having intermediate isolated supports, the cantilever effects were superimposed on the longitudinal moment field.

Application of the method of coefficients gave good agreement between the measured and computed stress distribution considering pier reactions, support conditions, and the actual Poisson ratio. This conclusion is valid for the uniformly distributed load and the concentrated live load.

Comments for Slabs on Isolated Supports A complete examination of bridge behavior for slabs on isolated supports is not warranted. However, the following remarks are appropriate.

1. If the width–length ratio $2b/2a$ exceeds 0.05, a simple beam analysis may not be satisfactory, and a more rigorous study should be undertaken, particularly for dead load.

2. The behavior of slabs with internal voids (see also the following sections) is, in principle, similar to the behavior of solid slabs. The torsional parameter α is defined as

$$\alpha = \frac{\gamma_P + \gamma_E}{2\sqrt{p_P p_E}}$$

where γ_P, γ_E = torsional stiffness per unit width and length, respectively
p_P, p_E = flexural stiffness per unit width and length, respectively

Based on Sattler's work (1955, 1956), Hossain (1970) has found that if $\alpha > 0.7$ and $\theta \leq 0.2$, the coefficients for solid slabs give results that have a maximum error of 6 percent in distribution patterns. The factor θ is a flexural parameter given by

$$\theta = \frac{b}{2a}\left(\frac{p_P}{p_E}\right)^{1/4}$$

Cellular and Voided Slab Bridges

Cellular and voided slab bridges are shown in Figures 3-31a and b, respectively. Unlike the typical multibeam deck constructed of precast concrete beams developing interaction by the presence of continuous longitudinal

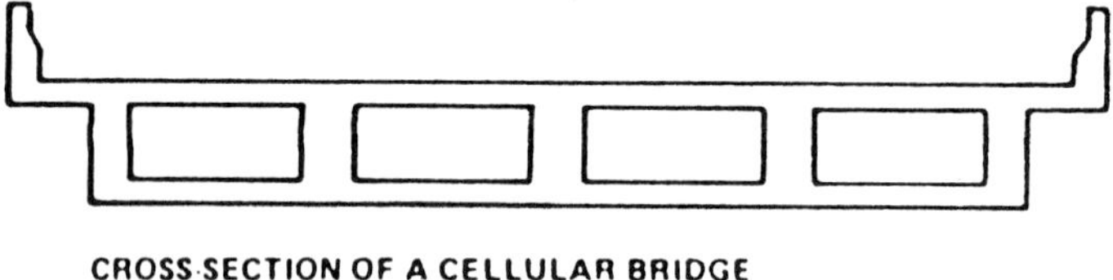

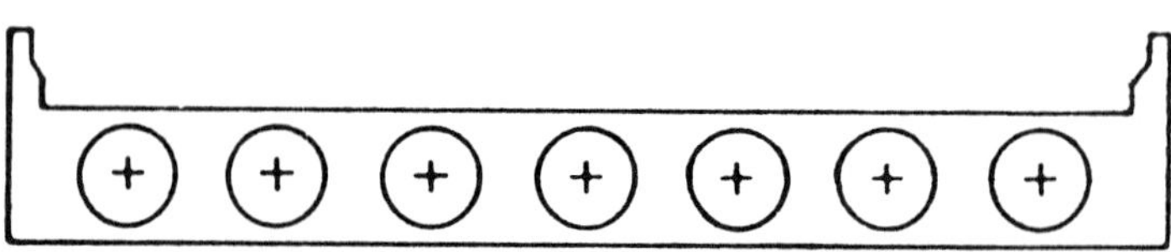

FIGURE 3-31 Typical cross sections of cellular and voided slab bridges.

shear keys and lateral bolts, the slabs in Figure 3-31 are monolithic units. What distinguishes them from conventional slabs is the deformable cross section. Because of this difference, these slabs should be designed considering cell distortion, although this effect is not explicitly articulated. Rigorous methods of analysis have been presented by Bakht et al. (1979). Bakht, Jaeger, and Cheung (1981) have also developed a simplified method of analysis.

If the cells are prevented from distorting in the transverse direction, the bridge can be idealized as a conventional isotropic plate (Crisfield and Twemlow, 1971). By contrast, the absence of transverse diaphragms causes the cells to distort, thereby increasing the transverse flexibility of the structure. The result is a reduced ability to distribute loads transversely, hence a higher concentration of longitudinal moments and shears. Solutions are proposed by Massonet and Gandolfi (1967), Sawko (1968), Hook and Richmond (1970), and Robertson et al. (1970). Conclusions are reflected in AASHTO design criteria for multibeam hollow decks.

When a bridge cross section is subject to distortion, the pattern of transverse distribution of longitudinal moments requires three nondimensional parameters α, θ, and δ for its characterization. These parameters are functions of the longitudinal flexural rigidity, longitudinal torsional rigidity, transverse torsional rigidity, coupling rigidity, transverse shear rigidity, and the physical dimensions of the bridge (Bakht, Jaeger, and Cheung, 1981). The same three parameters may also characterize the distribution of other structural responses such as shears and deflections, but not necessarily in the same pattern.

For cellular and voided slab bridges, one of these three parameters has an almost constant value ($\alpha = 1$). Thus, the conclusion is reached that for cellular structures, the transverse distribution of deflections, longitudinal moments, and shears for vehicle loads is characterized by the system

$$\alpha = 1.0 \qquad \theta = \frac{b}{L} \qquad \text{and} \qquad \delta = \frac{\pi^2 b}{L^2}\left[\frac{D_x}{S_y}\right]^{1/2} \tag{3-13a}$$

where b = half-width of bridge
L = span length
D_x = longitudinal flexural rigidity per unit width
S_y = transverse shear rigidity per unit length

Given the two variables θ and δ, the distribution of loads in cellular bridges could have been predicted using appropriate charts developed for the practical range of the θ–δ space, but such charts do not articulate the extent to which cell distortion causes the structure to respond differently from its counterpart where cell distortion is absent. An alternative method is the use of coefficients in the form of magnifiers which can be used in conjunction with AASHTO criteria and also reflect the influence of cell distortion. Thus, for $\alpha = 1$, $\theta = b/L$, and $\delta \neq 0$, the moment magnifier is

$$\lambda_m = \frac{M_{xm}}{M_{xa}} \tag{3-13b}$$

where M_{xm} = maximum moment with cell distortion
M_{xa} = maximum moment when cell distortion is absent

A similar magnifier λ_s is introduced for longitudinal shears.

This method is vehicle independent, and the effect of axle spacing can be disregarded. The magnifier always has a value greater than unity and depends on the number of lanes in the bridge and the number of lanes that are loaded.

The suggested approach for voided slabs is essentially the same. Bakht et al. (1979) propose specific relationships for estimating the slab parameters, from which the factors θ and δ can be calculated. For most voided slabs, α ranges from 0.85 to 0.95, but this variation is not critical.

3-9 RIGID-FRAME CONCRETE BRIDGES

One-span rigid-frame bridges were introduced in the 1920s, and followed a period of uninterrupted construction until World War II. This type of bridge became popular on a regional basis (e.g., in the states of New York and Connecticut).

In a rigid-frame bridge, the abutments and the deck are cast as a unit, and this solution is favored where solid foundations are easily obtainable. In the United States, most rigid-frame bridges have spans of about 100 ft, but single-span rigid frames have been used for spans up to 150 ft.

Characteristics In structural terms, one-span rigid frames are structures consisting of horizontal members one span long, each rigidly connected with the vertical supporting members. The vertical members at their lower ends must resist horizontal thrusts produced by the frame action (see also the discussion on arches in Chapter 10). If the base of either vertical support is free to move horizontally, the structure is statically determinate.

The most common types of rigid-frame bridges are shown in Figure 3-32. In Figure 3-32*a* the ends of the vertical members are restrained from any horizontal or vertical movement, but are free to turn and rotate, so that they cannot resist or transfer bending moments at this point. This frame is considered hinged at the ends. The frames shown in Figure 3-32*b* have the ends of the vertical members restrained from movement and also against

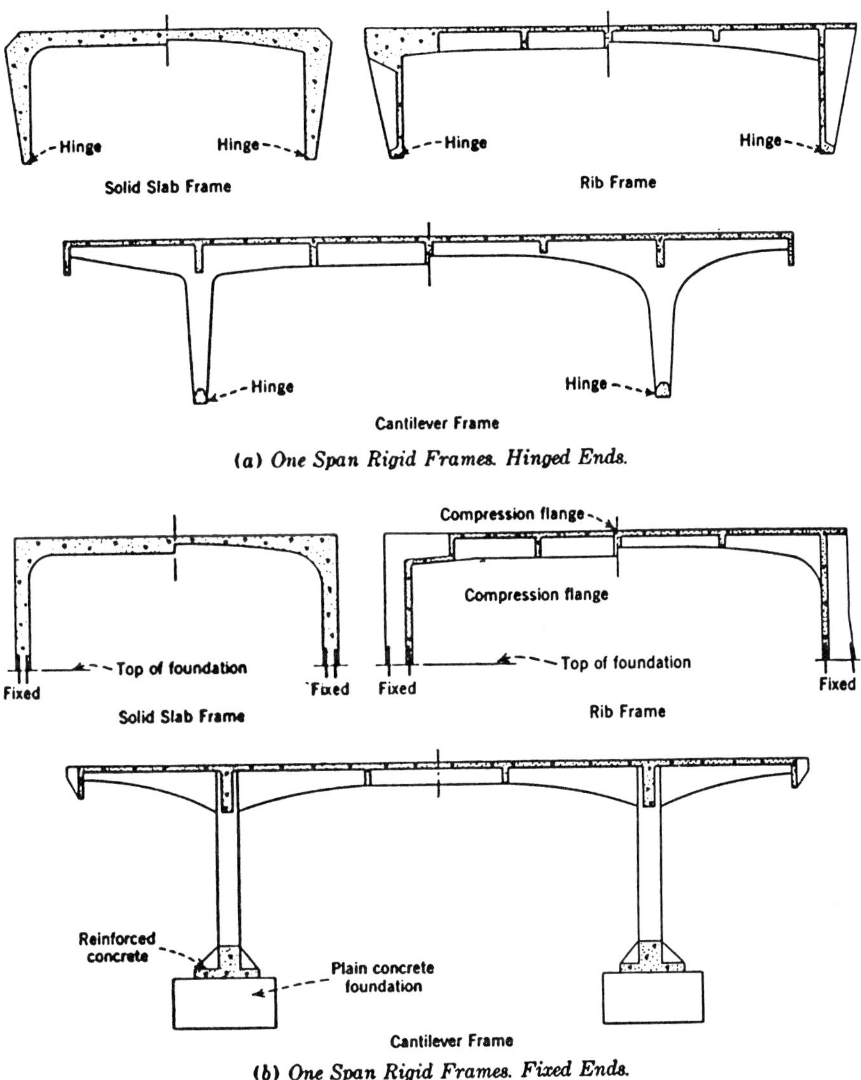

FIGURE 3-32 Typical one-span rigid-frame bridges. (From Taylor, et al., 1939.)

rotation. These members are fixed to the foundations, so that they can resist and transfer bending moments. These frames are called fixed at the ends. A third condition, not shown, is an intermediate response between hinges and full fixity. The ends are now partially restrained against rotation and can therefore resist bending moments in part. For partially restrained ends it is necessary to establish the probable intermediate moment resisted at this location, usually by means of the fixed-point method, which establishes the

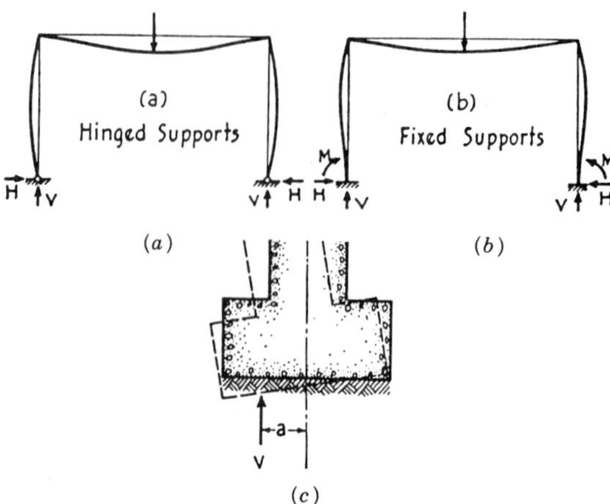

FIGURE 3-33 Rigid-frame bridges: (*a*) hinged supports; (*b*) fixed supports; (*c*) restraint offered by narrow footing.

lower fixed point in the vertical member according to the expected degree of restraint at the ends.

The structural deformation of a frame with hinged ends is shown in Figure 3-33*a*. The legs of this frame rest on footings that are free to rotate, producing the corresponding soil reactions H and V as shown. The frame shown in Figure 3-33*b* has the ends of the legs completely restrained, producing the structural deformation shown and the three reactions H, V, and M. The effect of rotation shown in Figure 3-33*c* is that the resultant of the soil reactions becomes eccentric. If the resultant V is moved a distance a from the theoretical midpoint, the product Va is the moment that counteracts the rotation and restrains the footing. Under the condition shown in Figure 3-33*c*, the soil offers little restraint against rotation of the footing. For this case analyses show that the restraining moment is small compared to the actual moment M required for fixity, and that the stresses in the frame are only slightly affected by the moment Va. From these comments, it follows that for ordinary rigid-frame bridges with comparatively narrow footings it is reasonable to assume hinged conditions.

Selection of Frame Dimensions For a preliminary analysis, it is necessary to establish the approximate frame dimensions. The following procedure is applicable to rigid frames of the type illustrated in Figure 3-34, and involves the following steps.

1. Lay out the top of the deck ABA' according to the roadway requirements.
2. Determine the clear span L from horizontal clearance data.

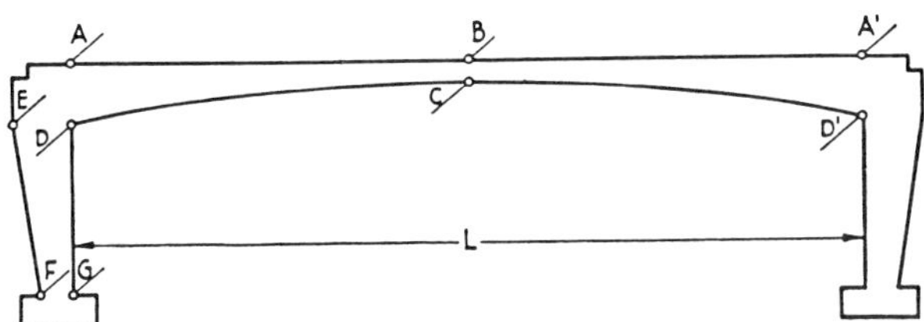

FIGURE 3-34 Layout procedure for a rigid-frame bridge.

3. Establish the dimensions *AD* and *ED* equal to about $L/15$.
4. Establish the dimension *BC* equal to about $L/35$. This value may be reduced to $L/40$ if the frame is founded on essentially firm foundation.
5. Draw the soffit curve *DCD'*.
6. Determine the elevation of *F* and *G* from geometric requirements and foundation conditions.
7. Select *FG* equal to 1 ft 6 in. for 30-ft spans, about 2 ft 6 in. for 60-ft spans, and about 3 ft 6 in. for 90-ft spans.
8. Connect points *E* and *F* with a straight line.

We should note that this layout accommodates heavy highway loading and is compatible with strength–stress parameters typical in highway bridges.

Analysis Deflection of axes and rotation of joints control the distribution of moments, thrusts, and shears. A better conception and a more effective working ability of rigid-frame analysis is provided if we can understand the physical significance of the problem in relation to the mathematical treatment. Usually, this involves moment distribution methods normally covered in computer programs which need not be repeated in this text.

The analysis may involve (a) correction for deck curvature, (b) frame dimensions and determination of axes and coefficients, (c) selection of frame constants, (d) dead and live load analysis, (e) changes in length of deck and horizontal displacement, (f) earth pressure, (g) dissymmetry and sidesway, and (h) estimation of total moments and shears. Typically, the critical sections for determining stresses are at the crown and at the corners. A design example is presented in Section 11-8.

3-10 FUNDAMENTALS OF PRESTRESSED CONCRETE BRIDGES

The first comprehensive guidelines for prestressed concrete bridges were developed by the Bureau of Public Roads in 1954. This document disseminates the criteria for the design, materials, and construction of bridge superstructures consisting of prestressed and posttensioned concrete members, and is now considered the progenitor of the current AASHTO specifications. These specifications, covered in Section 9 of the AASHTO document, cover general requirements and materials, analytical aspects, design methodology, and detailing.

Working Principles of Prestressed Concrete

Lin (1955) sets forth the general principles of prestressed concrete: the prestressing introduces internal stresses of such magnitude and distribution that the stresses resulting from given external loads are counteracted to the desired degree. In reinforced concrete members, the prestress is commonly introduced by tensioning the steel reinforcement. For members subjected to flexure and shear, understanding their behavior is enhanced from three perspectives.

First, we may consider the member as essentially a concrete structure with the tendons supplying the prestress to the concrete. In this respect, the concrete is subjected to two systems of forces, prestress and external load. Tensile stresses due to external action are counteracted by the compressive stresses due to prestress, and the associated cracking is prevented or delayed. As long as cracks are absent, the stresses, strains, and deflections caused by the two systems can be considered separately and superimposed if necessary.

The rectangular simple beam shown in Figure 3-35 is prestressed by a tendon through its centroidal axis. The resulting stresses due to prestress and

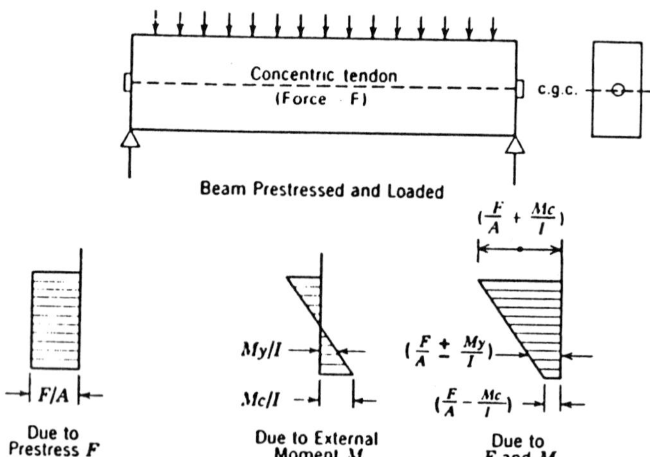

FIGURE 3-35 Stress distribution across a concentrically prestressed concrete section. (From Lin and Burns, 1981.)

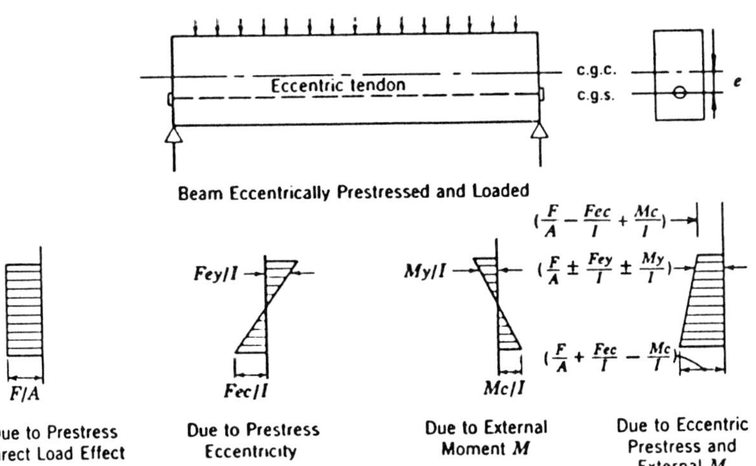

FIGURE 3-36 Stress distribution across an eccentrically prestressed concrete section. (From Lin and Burns, 1981.)

the external moment are as shown, so that the resulting stress distribution is

$$f = \frac{F}{A} \pm \frac{My}{I} \tag{3-14}$$

where F = prestressing force
A = cross-sectional area of the member
M = external moment
y = distance from the centroid to the fiber considered
I = moment of inertia of the section

Now, we apply the prestress eccentrically with respect to the centroidal axis as shown in Figure 3-36. The eccentricity e introduces a moment Fe, resulting in a stress diagram as shown. The stress distribution for the lowermost and uppermost fiber is

$$f = \frac{F}{A} \pm \frac{Fec}{I} \mp \frac{Mc}{I} \tag{3-15}$$

The second perspective considers steel and concrete acting together, with the steel taking the tension and the concrete resisting compression. The two materials form a couple resisting the external moment as shown in Figure 3-37. Because the high-tensile steel must be elongated considerably before its strength is fully developed, if simply buried in the concrete it will cause the latter to crack excessively. Hence, it is necessary to prestretch the steel and

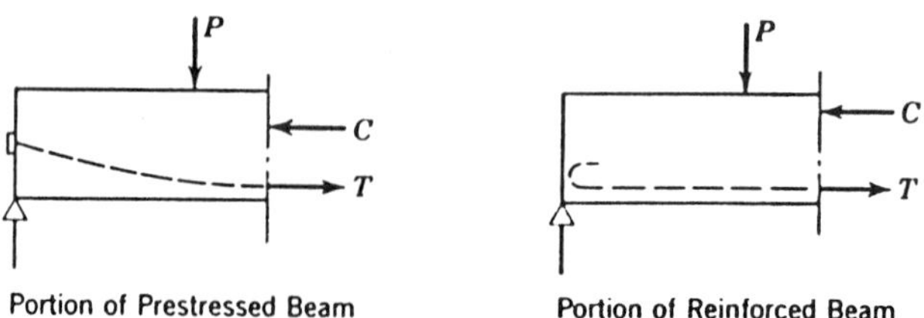

FIGURE 3-37 Internal resisting moment in prestressed and reinforced concrete beams. (From Lin and Burns, 1981.)

anchor it against the concrete, and in this process we produce desirable stresses and strains in both materials.

In this context, prestressed concrete is an extension of conventional reinforced concrete to include steels of higher strength, but the application is still based on the necessity of resisting external moments by an internal couple.

The third point of view is to treat a prestressed concrete beam as a steel member with the characteristics of a suspension bridge where the wires form the load-carrying elements (Lin, 1955). The prestressing tendons are self-anchored against the concrete and stiffened by this material.

Important Characteristics

For bridge structures of nominal span length, two basic configurations are produced: (a) I-beam sections used with a cast-in-place deck and (b) multi-beam deck sections with an integral or separate riding surface. The latter are normally used for shorter spans and where construction time is limited, because of the simplicity of the erection phase. The primary advantage of prestressed concrete is durability, which can be extended with little or no maintenance. The absence of tension cracking should further increase the service life of both concrete and reinforcement.

Under certain circumstances a prestressed concrete beam can support its own weight with no increase in section properties beyond what is required for live load. However, in relatively long span bridges this advantage is diminished because the dead load dominates.

The concept of prestressing has prompted the introduction and use of high-strength steels, and this has improved the control of losses in prestress due to concrete shrinkage and creep. By contrast, in conventional reinforced concrete the allowable steel tensile stress must be limited to avoid excessive cracking of the concrete. Furthermore, precast members have the benefit of better quality control, and although they have a higher cost per unit volume they yield a lower cost per unit of load capacity. This relates to the more efficient utilization of materials because the compression capacity of the complete concrete section is available, instead of only the uncracked portion.

Complete structural units can be factory manufactured, either prestressed or posttensioned, and their feasible size depends mainly on the available clearance during transportation. Often, it is feasible to use precast bridge girders up to spans of 100 to 120 ft. Alternatively, the units may be cast as a series of segments to be joined during erection by the prestress.

Structural Behavior

Current design methodologies are based on strength (load factor design) and on behavior at service conditions (allowable stress design) at all load stages that may be critical during the life of the structure. Stress concentrations due to the prestressing as well as the effects of temperature and shrinkage should be considered. In addition, for monolithic members the following assumptions are made: (a) strains vary linearly over the depth of the member throughout the entire load range (elastic analysis); (b) before cracking, stresses are linearly proportional to strain; and (c) after cracking, tension in the concrete is neglected (AASHTO Article 9.13).

Loss of Prestress This includes (a) elastic shortening of concrete, (b) creep and shrinkage in concrete, (c) creep (relaxation) in steel, (d) loss due to anchorage take-up, (e) loss due to bending of members, and (f) frictional loss.

For pretensioned concrete, elastic shortening is manifested as the prestress is applied and transferred to the concrete. As a result, the prestressing steel shortens with it, and this represents a direct loss of prestress in the steel. With posttensioning, the problem is different; for a single tendon, the concrete shortens as the tendon is jacked against it, but because the force in the steel is measured after the concrete has shortened elastically there is no loss of prestress to be accounted for. If more than one tendon is used and stressed in succession, the prestress is gradually applied and the shortening in the concrete increases accordingly, so that the loss of prestress may differ in each tendon. In addition, certain stressing procedures may alter the elastic shortening losses further.

Because creep and shrinkage may be twice the elastic shortening, they are more critical. Furthermore, the loss due to elastic shortening may be counterbalanced for posttensioned members, whereas the loss due to creep cannot be easily compensated for, except where the steel is not yet bonded to the concrete. Shrinkage, on the other hand, varies widely with the proximity of the concrete to moisture and the time of application of prestress. Certain codes recommend a total shrinkage of 0.0003 for pretensioning, but for transfer at 2 to 3 weeks, a shrinkage coefficient of 0.0002 is considered sufficient.

Creep in steel (relaxation) is a decrease of stress, and a corresponding loss of load in the tendon, with time while the tendon is held under constant strain. This behavior is manifested by the gradual replacement of elastic

TABLE 3-1 Estimate of Prestress Losses (From AASHTO, 1992.)

Type of Prestressing Steel	Total Loss	
	$f_c' = 4000$ psi	$f_c' = 5000$ psi
Pretensioning strand	—	45,000 psi
Posttensioning[a]		
Wire or strand	32,000 psi	33,000 psi
Bars	22,000 psi	23,000 psi

[a]Losses due to friction are excluded. Friction losses should be computed according to Article 9.16.1.

strain by plastic strain causing the subsequent relaxation of elastic stresses (Xanthakos, 1991). AASHTO makes a distinction between the various types of steel. Stress relaxation increases rapidly with temperatures above 20°C, and thus in warm climates it should be adjusted accordingly. Under average conditions a relaxation loss of 1 to 5 percent is not uncommon, and can be approximated at 3 percent as a fair assumption.

Loss due to anchorage take-up may be prevented or avoided by the use of proper stressing procedures and stressing equipment for posttensioned members. Likewise, loss of prestress due to bending of a member depends on the direction of bending and the location of the tendon. Any change in prestress is controlled by the type of prestressing, whether pretensioned or posttensioned. However, if the prestress from the steel on the concrete is considered to be a force applied at the ends, the change in stress is not a change in prestress.

In lieu of a prestress loss breakdown, AASHTO allows an average estimate of the total loss of prestress due to concrete shrinkage, elastic shortening, creep in concrete, and relaxation in the steel, as shown in Table 3-1.

Friction Loss Friction loss in posttensioned steel normally occurs prior to anchoring but should be estimated and checked during stressing operations. Extensive work on this subject indicates that initially there is some friction in the jacking and anchoring system, so that the stress in the tendon is less than the value indicated by the pressure gage. More serious frictional loss occurs, however, between the tendon and the surrounding material, whether concrete or sheathing, and whether lubricated or not. This frictional loss is conveniently considered in two parts, the length effect and the curvature effect. In posttensioned beams friction loss in the steel should preferably be based on experimentally established wobble and curvature coefficients.

Deflections Under working loads prestressed concrete beams should not be expected to crack; hence, their deflections can be predicted with sufficient accuracy. These deflections, however, differ from those of ordinary reinforced beams in the effect of prestress. Controlled deflections due to prestress can be advantageously utilized to give desired cambers and to offset deflections due to exterior loads, but can also cause certain problems. Methods for computing deflections due to prestress are presented in the design examples in the following sections.

3-11 PRECAST PRESTRESSED CONCRETE I-BEAM BRIDGES: SIMPLE AND CONTINUOUS SPANS

Strength and Design Criteria

Allowable Stresses: Service Load Design Ordinarily, the design of precast prestressed concrete members should be based on $f'_c = 5000$ psi. Prestressed I beams may be designed for $f'_c = 6000$ psi if the former cannot produce the required capacity for a specific beam depth, and if it is reasonable to expect that the higher strength will be obtained consistently. For the prestressing steel, the ultimate strength is $f'_y = 270{,}000$ psi.

The allowable stresses in the prestressing steel are as follows.

Pretensioned Members

$0.70f'_s$	for stress relieved strands
$0.75f'_s$	for low relaxation strands

Slight overstressing up to $0.85f'_s$ for short periods of time may be permitted to offset seating losses, provided the stress after seating does not exceed the preceding values.

Posttensioned Members

$= 0.70f'_s$ With provisions for overstressing up to $0.90f^*_y$

Stress at service load after losses

$$= 0.80f^*_y$$

(f^*_y is the yield point stress of prestressing steel as defined in Article 9.15 of AASHTO.)
(Note that service load consists of all loads identified in Article 3.2 of AASHTO, but does not include overload provisions.)

Likewise, the allowable stresses in the concrete are as follows:

Compression

Pretensioned members	$0.60f'_{ci}$
Posttensioned members	$0.55f'_{ci}$

Tension: In precompressed tensile zones, temporary allowable stresses are not specified. In tension areas without bonded reinforcement, the allowable tensile stress in concrete is 200 psi or $3\sqrt{f'_{ci}}$. Where the calculated tensile stress exceeds this value, bonded reinforcement will be provided. The maximum tensile stress should not exceed $7.5\sqrt{f'_{ci}}$, where f'_{ci} is the compressive strength of concrete at the time of initial prestress. Note that the foregoing are temporary stresses before losses due to creep and shrinkage.

The allowable stresses at service load for concrete after losses have occurred are as follows:

Compression	$0.40f'_c$
Tension in the Precompressed Tensile Zone	
(a) For members with bonded reinforcement	$6\sqrt{f'_c}$
except for severe corrosive conditions	$3\sqrt{f'_c}$
(b) For members without bonded reinforcement	0

The modulus of elasticity of the prestressing steel strand, E_s, may be assumed as 28,000,000 psi, unless more specific data are available. The modulus of elasticity E_c is computed as in conventional concrete, except at the time of transfer of stress it can be calculated from $E_{ci} = w^{1.5}33\sqrt{f'_{ci}}$.

Design Theory Prestressed concrete members must meet the strength (load factor) and working stress requirements specified by AASHTO, at all load stages that may be critical during the life of the structure from the time prestressing is first applied. For precast prestressed concrete (PPC) beams, the intent of the normal design procedure is to consider allowable working stresses and to check initial stresses and nominal moment capacity. Most PPC beams utilize pretensioned prestressing where the strands are stressed before the concrete is cast around the strands. When the concrete has cured to sufficient strength, the strands are cut and the prestressing is transferred to the beam. Posttensioning is occasionally used with precast sections, and in this case the strands are installed in ducts after concrete placement and curing, and then jacked, anchored, and grouted.

Standard Beam Sizes Most standards in the United States are based on beam sizes and shapes such as the 54-in. beam shown in Figure 3-38, giving dimensions, strand patterns, reinforcement details, and section properties.

Design Charts and Tables These are very useful and are provided by most manuals for the flexural design of the standard prestressed concrete beams; they apply to simply supported or continuous units. Given the design span length, the highway loading, and the beam spacing, or the design span and the modified external moment, the number of strands and the strand

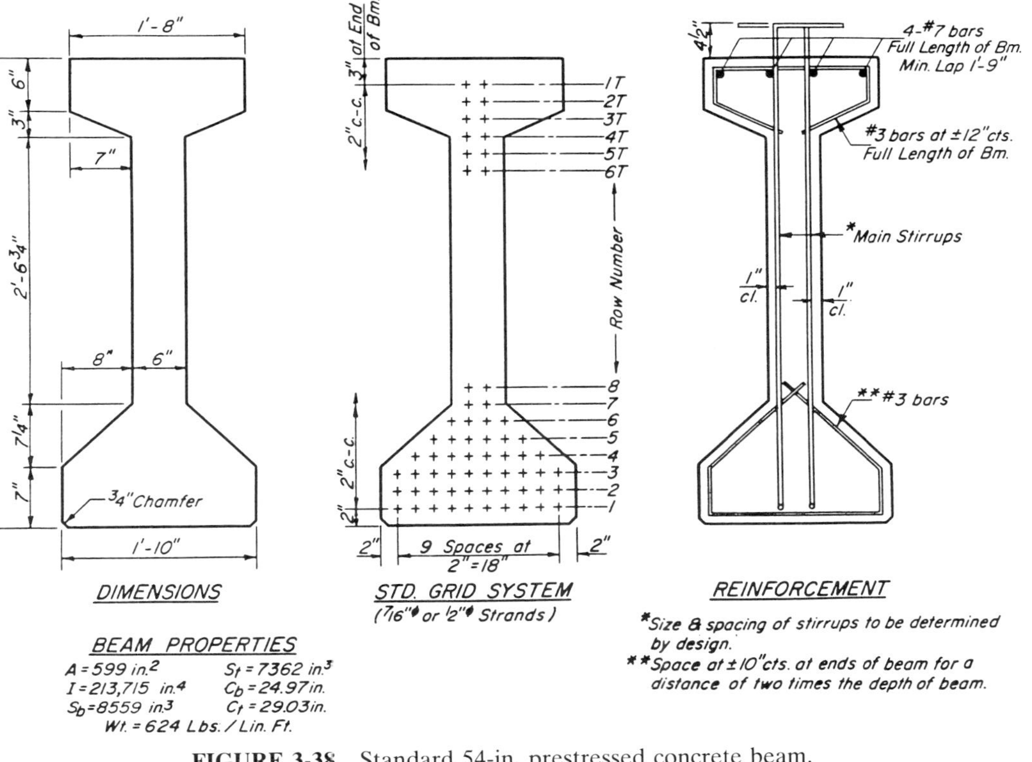

FIGURE 3-38 Standard 54-in. prestressed concrete beam.

pattern can be selected on a preliminary basis. These charts are usually computed for the HS 20 loading and a 7.5-in. slab, and they make allowance for a superimposed dead load. With these preliminary data available, final calculations are expeditiously carried out.

Design Example of PPC Beams

Problem Design a two-span precast prestressed concrete beam bridge for HS 20-44 loading, with equal design spans of 70 ft. The superstructure consists of six 54-in. I beams of the type shown in Figure 3-38 spaced at 7 ft 3 in. centers. The slab is 7.5 in. thick, but the effective thickness is taken as 7 in.

We assume the following.

Precast Concrete

$$f'_c = 5000 \text{ psi}$$
$$f'_{ci} = 4000 \text{ psi}$$
$$f'_s = 270 \text{ ksi}$$
$$f'_{si} = 0.70 f'_s = 189{,}000 \text{ psi}$$

(f'_{si} is the initial stress in prestressing steel before losses).

Cast-in-Place Concrete

$$f''_c = 3500 \text{ psi}$$
$$f_c = 1400 \text{ psi}$$
$$f_y = 60{,}000 \text{ psi}$$
$$f_s = 24{,}000 \text{ psi}$$

Dead Loads: Noncomposite

Weight of beam	= 624 lb/ft
Slab = 150 × 0.625 × 7.25	= 680 lb/ft
Fillet (assume)	= 10 lb/ft
Dead load w_n	= 1.314 kips/ft
Composite dead load	
Wearing surface 25 × 7.25	= 181 lb/ft
Curb, parapet	= 133 lb/ft
Composite dead load w_c	= 0.314 kip/ft

For simplicity, we assume that the interior girders control, so that the exterior girders will be made the same. The slab is designed as in the preceding examples. The PPC beams are designed as simple spans with the beam section carrying the dead load of the slab and girder (noncomposite); the composite beam-and-slab section carries the superimposed (composite) dead load plus live load with impact.

The section properties for the 54-in. beam are

A = 599 in.^{2} = cross-sectional area
I = 213,715 in.^{4} = moment of inertia
S_b = 8559 in.^{3} = noncomposite section modulus for bottom fiber
S_t = 7362 in.^{3} = noncomposite section modulus for top fiber
C_b = 24.97 in. = distance from bottom fiber to neutral axis
C_t = 29.03 in. = distance from top fiber to neutral axis

For the properties of the composite section, we first compute the modular ratio of elasticity:

Non-prestressed
$$E_c = 150^{1.5} \times 33 \times \sqrt{3500} = 3.58 \times 10^6 \text{ psi}$$

Prestressed
$$E_c = 150^{1.5} \times 33 \times \sqrt{5000} = 4.29 \times 10^6 \text{ psi}$$

$$n = \frac{3.58}{4.29} = 0.8$$

Note that the effective flange width is controlled by the beam spacing. This width is then multiplied by the modular ratio n, and the composite beam section is as shown in Figure 3-39.

From the composite beam section, the following parameters are computed $\bar{y}$ = 39.56 in., I_c = 500,015 in.^{4}, S_{bc} = 12,639 in.^{3}, and S_{tc} = 34,627 in.^{3}.

Computation of Moments and Stresses These are computed for noncomposite loads, composite dead loads, and live loads plus impact.

Noncomposite Moments

M_{DL} (Beam) = 0.125 × 0.624 × 70^2 = 382 ft-kips (Positive in span)

M_{DL} (Slab and fillet) = 0.125 × 0.690 × 70^2 = 423 ft-kips

M_{DL} (Total) = 382 + 423 = 805 ft-kips

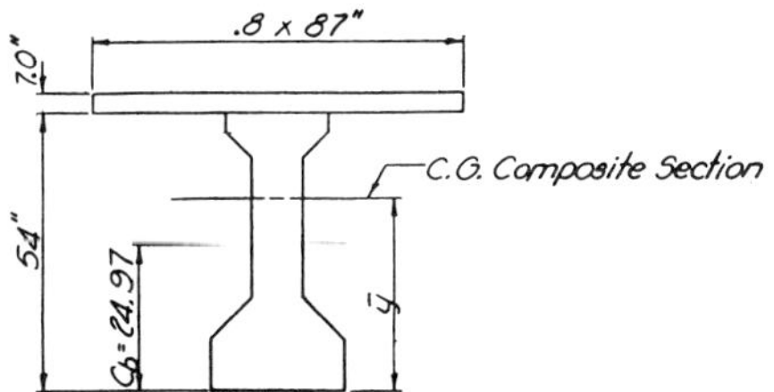

FIGURE 3-39 Composite section of design example.

Composite Moments: (Assume the system is continuous over the support)

M_{DL} (Composite) $= 0.07 \times 0.314 \times 70^2 = 108$ ft-kips (Positive in span)

M_{DL} (Composite) $= -0.125 \times 0.314 \times 70^2 = -192$ ft-kips (Negative at support)

Live Load plus Impact Moments: (Two-span continuous beam)

$$M_{\text{LL}+I} = 0.353 \times 16 \times \frac{7.25}{5.5} \times 1.26 \times 70 = 657 \text{ ft-kips}$$

(Positive in span)

$$M_{\text{LL}+I} = -\left(0.125 \times 0.64 \times \frac{7.25}{11} \times 1.26 \times 70^2 + 0.096 \times 2 \times 18 \times \frac{7.25}{11} \times 1.26 \times 70\right)$$

$$= -(326 + 201) = -527 \text{ ft-kips} \qquad \text{(Negative at support)}$$

(Note that live load plus impact moments are obtained with the use of influence line coefficients.)

Next, we compute service stresses f_b and f_t for the bottom and top fiber of the prestressed beam, respectively. These are as follows:

Beam dead load	$f_b = \dfrac{12 \times 382}{8559} = 536$ psi	$f_t = \dfrac{12 \times 382}{7362} = 623$ psi
Total noncomposite DL	$f_b = \dfrac{12 \times 805}{8559} = 1130$ psi	$f_t = \dfrac{12 \times 805}{7362} = 1310$ psi
Composite DL	$f_b = \dfrac{12 \times 108}{12{,}639} = 103$ psi	$f_b = \dfrac{12 \times 108}{34{,}627} = 38$ psi
Live load plus impact	$f_b = \dfrac{12 \times 657}{12{,}639} = 624$ psi	$f_b = \dfrac{12 \times 657}{34{,}627} = 228$ psi
Total service stresses	$f_b = 1857$ psi (Tension)	$f_t = 1576$ psi (Compression)

Allowable Stresses

Final $f_b = 6\sqrt{f'_c} = 6\sqrt{5000} = 425$ psi (Tension)

Initial $f_b = 0.60 f'_{ci} = 0.60 \times 4000 = 2400$ psi (Compression)

Final $f_b = 0.40 f'_c = 0.4 \times 5000 = 2000$ psi (Compression)

Initial $= 7.5\sqrt{f'_{ci}}$ or $f_t = 475$ psi (Tension)

Limits of Prestress and Strand Pattern The expected loss of prestress may be estimated from Table 3-1 for $f'_c = 5000$ psi. A more conservative approach is to assume 20 percent loss, and estimate the required prestress accordingly.

$$\begin{aligned}
\text{Bottom fiber (final)} = \text{Total stress} &= 1857 \text{ psi} && \text{(Tension)}\\
\text{Allowable} &= \underline{\ 425} \text{ psi} && \text{(Tension)}\\
\text{Required prestress} = 1432/0.8 &= 1790 \text{ psi} && \text{(Compression)}\\
\text{Bottom fiber (initial)} = \text{Beam dead load} &= 536 \text{ psi} && \text{(Tension)}\\
\text{Allowable} &= \underline{2400} \text{ psi} && \text{(Compression)}\\
\text{Allowable prestress} &= 2936 \text{ psi} && \text{(Compression)}\\
\text{Top fiber (final)} = \text{Total stress} &= 1576 \text{ psi} && \text{(Compression)}\\
\text{Allowable} &= \underline{2000} \text{ psi} && \text{(Compression)}\\
\text{Top fiber (initial)} = \text{Beam dead load} &= 623 \text{ psi} && \text{(Compression)}\\
\text{Allowable} &= \underline{\ 475} \text{ psi} && \text{(Tension)}\\
\text{Allowable prestress} &= 1098 \text{ psi} && \text{(Tension)}
\end{aligned}$$

The bottom fiber stress must be between 1790 and 2936 psi under the effect of the prestress force, and the top fiber must have tension less than 1098 psi.

We now select 1/2-in-diameter strands, F_i per strand is $0.7 \times f'_s = 0.7 \times 41.3 = 28.9$ kips, and arrange the strands as follows (Figure 3-38, grid system detail)

$$\begin{aligned}
&\text{Row 1} = 10 \text{ strands}\\
&\text{Row 2} = \underline{\ \ 6} \text{ strands}\\
&\text{Total} \ = 16 \text{ strands} \qquad \text{Initial prestress} = 16 \times 28.9 = 462 \text{ kips}
\end{aligned}$$

Next, we compute

$$e_{\text{℄}} = C_b - \frac{\Sigma N\bar{y}}{N} = 22.22 \text{ in.}$$

Initial Stresses The two critical sections are (a) midspan of beam and (b) end of beam. At midspan of the beam,

$$f_{cit} = \frac{462}{599} - \frac{462 \times 22.22}{7362} + \frac{382 \times 12}{7362} = 0 < 475 \text{ psi} \qquad \text{(Tension)}$$

$$f_{cib} = \frac{462}{599} + \frac{462 \times 22.22}{8559} - \frac{382 \times 12}{8559} = 1437 \text{ psi} < 2400 \text{ psi}$$

(Compression)

At the end of the beam,

$$f_{cit} = \frac{362}{599} - \frac{462 \times 22.22}{7362} = 624 \text{ psi} > 475 \text{ psi} \qquad \text{(Tension)}$$

Therefore, strands must be draped. If the two top strands are draped, $e_e = 16.35$ in. Then

$$f_{cit} = \frac{462}{599} - \frac{462 \times 16.35}{7362} = 255 \text{ psi} < 475 \text{ psi}$$

$$f_{cib} = \frac{462}{599} + \frac{462 \times 16.35}{8559} = 1655 \text{ psi} < 2400 \text{ psi}$$

At this stage, the loss of prestress should be computed according to Article 9.16 of AASHTO. This loss is 20.6 percent (computations not shown).

Service Stresses For a computed loss of prestress of 20.6 percent, the force remaining in the strands (effective prestress) is $0.794 \times 462 = 367$ kips. Hence, the stresses at service loads are

$$f_{cb} = \frac{367}{599} + \frac{367 \times 22.22}{8559} - \frac{805 \times 12}{8559} - \frac{(108 + 657)12}{12{,}639}$$
$$= 289 \text{ psi} < 425 \text{ psi}$$

$$f_{ct} = \frac{367}{599} - \frac{367 \times 22.22}{7362} + \frac{805 \times 12}{7362} + \frac{(108 + 657)12}{34{,}627}$$
$$= 1080 \text{ psi} < 2000 \text{ psi}$$

Design for Ultimate Positive Moment

Theoretical Background The methodology for predicting the ultimate flexural capacity of a prestressed concrete member is essentially the same as for a plain reinforced concrete member, except in the stress–strain relationship between prestressing and intermediate grade steel. For a PPC I beam, the ultimate moment capacity (flexural strength) requires knowledge of the location of the neutral axis. If the neutral axis lies in the flange (provided by the cast-in-place slab), the section is assumed rectangular. If the neutral axis is in the beam, the section is a flanged section. For the usual configuration of PPC I-beam bridges, the section is rectangular.

The underlying principle is that as the ultimate capacity is approached, the steel is yielding. Next, we define the following expressions:

$$p^* = \frac{A_s^*}{bd} \tag{3-16}$$

where p^* = ratio of prestressing steel
A_s^* = total area of prestressing steel
b = width of rectangular member (or effective width of flange)
d = distance from extreme compression fiber to centroid of prestressing steel

and also

$$f_{su}^* = f_s'\left(1 - \frac{0.5p^*f_s'}{f_c''}\right) \tag{3-17}$$

where f_{su}^* = average stress in prestressing steel at ultimate load
f_s' = ultimate strength of prestressing steel
f_c'' = compressive strength of cast-in-place concrete (28 days)

A refinement in Eq. (3-17) was introduced in the 1992 AASHTO specifications. Thus, the numerical factor 0.5 is replaced by the ratio γ^*/β_1. The factor γ^* articulates the type of prestressing steel as follows:

$$\begin{aligned}\gamma^* &= 0.28 \text{ for low-relaxation steel}\\ &= 0.40 \text{ for stress-relieved steel}\\ &= 0.55 \text{ for bars}\end{aligned}$$

The factor β_1 is as stipulated in AASHTO Article 8.16.2.7 ($\beta_1 = 0.85$ for concrete strengths up to and including 4000 lb/in.^{2}).

Likewise, we write

$$kd = \frac{1.4dp^*f_{su}^*}{f''c} \tag{3-18}$$

where kd is the depth of the compression zone. The design flexural capacity (strength) for rectangular sections or flanged sections in which the neutral axis lies in the flange (i.e., $kd <$ flange thickness) is assumed as (AASHTO, Article 9.17.2)

$$\phi M_n = M_u = \phi A_s^* f_{su}^* d\left(1 - \frac{0.6p^*f_{su}^*}{f_c''}\right) \qquad \text{where } \phi = 0.90 \tag{3-19}$$

In order for the intended failure mechanism to be satisfied, AASHTO specifies the maximum percentage of steel. Therefore, for rectangular sections the reinforcement index is limited as follows:

$$p^*\frac{f_{su}^*}{f_c''} \leq 0.36\beta_1 \tag{3-20}$$

If the reinforcement index exceeds the value stipulated in (3-20), the design flexural capacity (for rectangular sections) is estimated from

$$\phi M_n = \phi\left[\left(0.36\beta_1 - 0.08\beta_1^2\right)f_c'' bd^2\right] \qquad (3\text{-}20a)$$

Likewise, there should be a sufficient amount of prestressing steel to develop an ultimate capacity in flexure at the critical section at least 1.2 times the cracking capacity. The cracking stress is $7.5\sqrt{f_c'}$, and for $f_c' = 5000$ psi, the cracking stress is 530 psi.

Numerical Example For the example of PPC beams presented in the foregoing section, the factored moment is

$$M_u = 1.3[D + 1.67(L + I)]$$

or

$$M_u = 1.3[805 + 108 + 1.67(657)] = 2610 \text{ ft-kips}$$

Next, we compute

$$A_s^* = 16 \times 0.153 = 2.45 \text{ in.}^2$$
$$d = 54 + 7.0 - \bar{y} = 54 + 7.0 - 2.75 = 58.25 \text{ in.}$$
$$p^* = 2.45/(87 \times 58.25) = 0.000483$$

Also,

$$f_{su}^* = 270\left(1 - 0.5 \times 0.000483 \times \frac{270}{3.5}\right) = 265 \text{ ksi} \qquad \text{OK}$$

In order to locate the neutral axis, we compute

$$kd = 1.4 \times 58.25 \times 0.000483 \times 265/3.5 = 2.98 \text{ in.} < 7.0 \text{ in.}$$

Hence, the section is rectangular. The design moment capacity of the section is

$$\phi M_n = 2.45 \times 265 \times 58.25\left(1 - 0.6 \times 0.000483 \times \frac{265}{3.5}\right)(0.9)$$
$$= 2770 \text{ ft-kips} > 2610$$

Likewise, the reinforcement index is

$$0.000483 \times 265/3.5 = 0.037 < 0.3$$

In order to check for the minimum prestressing steel, we first compute the cracking moment

$$M_{cr} = \frac{12{,}639}{12}\left(\frac{367}{599} + \frac{367 \times 22.22}{8559} - \frac{805 \times 12}{8559} + 0.530\right) + 805$$

$$= 1824 \text{ ft-kips}$$

Then

$$1.2M_{cr} < M_n \qquad \text{or} \qquad 1.2 \times 1824 = 2189 \text{ ft-kips} < 3080 \qquad \text{OK}$$

Design for Ultimate Negative Moment

Theoretical Background The advantages of continuous prestressed concrete beams were considered as early as the mid-1950s (Lin, 1955; Morice and Lewis, 1955). The philosophy of prestressing was, in fact, extended beyond the original concept of supplying a compressive force to a beam cross section, and included such procedures as the deliberate adjustment of the ratio of midspan to support bending moments in a continuous unit. Because complete structural units may be factory produced, either pretensioned or posttensioned, the feasible size of these units depends on the available clearance during transportation, the proximity of the casting yard, and the maximum practical weight.

There is overall agreement that continuity in reinforced concrete yields economy, and this may be attained to a greater extent in prestressed construction. This is more evident considering the ultimate capacity of continuous beams, although certain general principles hold true within the elastic range. In both forms of analysis, elastic or plastic, a resisting couple exists at each intermediate section of the beam. For both ranges, with one-half of the beam a free body, there are two resisting moments in a continuous beam, but only one in a simple beam. Within the elastic range, the positive and negative moments acting on the beam may not be the same, so that one of these moments will control the design. Because of the variation of moments along the continuous beam, the concrete section and the amount of steel must often be varied accordingly. In this case the peaks of the negative moments are reinforced with non-prestressing steel, and advantage can be taken of the redundant reactions to obtain favorable lines of pressures in the concrete. Designs are then based on ultimate strength but applying the principles of limit design, and this appears to be the intent of current methodologies. Continuity in prestressed bridge girders is discussed in detail in a subsequent section.

Tendon Profiles For a beam acted on by a centroidal prestress force F, the effects of a moment M on the beam may be considered if we introduce an eccentricity e to the center of F, where $e = M/F$. If some form of statistically indeterminate beam is loaded by exterior loads, the resulting bending moment together with a uniform compression could be produced in the unloaded beam by a prestressing tendon with an eccentricity $e = M/F$. The introduction of a prestressing tendon in the shape of a bending moment profile counteracts the effect of external loads. Such a tendon profile is called a *concordant* profile. Developed from external loading conditions, it is not only feasible but also represents a useful solution for practical design purposes.

Whereas these principles are valid in theory, the application of standard highway loadings produces significant reversal of live load moments in most continuous bridges, and the severity of this problem is more evident if the live loads are much heavier than the dead loads, as in short continuous spans of equal length. Because peaks of maximum negative moments may control the number of tendons required for the entire length of the beam, they must be strengthened with the use of deeper sections or by adding prestressed and non-prestressed reinforcement over the portions where they are needed.

Details of Continuity In practice, continuous bridges of PPC beams are built with standard details. A bridge is made continuous over the piers by casting a continuous concrete diaphragm between the ends of adjoining beams. The concrete in the slab is placed not less than 45 min or not more than 90 min after the diaphragm has been poured, mainly to ensure a monolithic construction and to control differential shrinkage. Longitudinal reinforcement is placed in the slab to resist negative moments. With this arrangement, the members are assumed to be fully continuous for live and composite loads, and the moments are computed using a constant moment of inertia.

Procedure for Ultimate Negative Moment: Continuous Spans By encasing the beam ends with cast-in-place concrete, the design is essentially similar to that for ultimate positive moment. The negative moment reinforcement is proportioned by strength design to resist a moment equal to $1.3[D + 1.67(L + I)]$.

AASHTO (Article 9.7.2.3) stipulates that the effect of initial precompression in the beams due to the effective prestress (after losses) may be neglected if the maximum initial compressive stress at the end is less than $0.4f'_c$ (f'_c is the compressive strength of prestressed concrete), and if the reinforcement index p of the negative reinforcement in the deck is less than 0.015, where $p = A_s/bd'$.

The design negative resisting moment is calculated using the compressive strength of the beam concrete, regardless of the strength of the diaphragm concrete. Positive moments that may occur in negative moment regions should be considered in the design. Such positive moments may be caused by creep and shrinkage in the girders and deck slab or by live loads in remote spans. Positive moment reinforcement over piers that is not prestressed is based on an allowable working stress of $0.6f'_y$ but not more than 36 ksi. In the foregoing notation, A_s is the total area of longitudinal intermediate grade steel in the slab, b is the width of the bottom flange, and d' is the distance from the extreme compressive fiber to the centroid of intermediate grade steel.

We can also write

$$kd = 1.4pd'\frac{f'_y}{f'_c} \tag{3-21}$$

where kd = depth of compressive zone
f'_y = yield point of intermediate grade reinforcing steel
f'_c = compressive strength of prestressed concrete

For a rectangular section (neutral axis within the bottom flange of the prestressed beam), the design flexural capacity is

$$\phi M_n = M_u = A_s f'_y d'\left(1 - \frac{0.6pfy'}{f'_c}\right) \qquad (\phi = 1) \tag{3-22}$$

which is similar in form to (3-19).

The minimum amount of non-prestressed longitudinal reinforcement provided in the cast-in-place slab is 0.25 in.^{2} per foot-width of slab. Standard practice is to place #5 bars at 12-in. centers in the top of the slab, providing 0.31 in.^{2}/ft. In continuous spans additional steel is provided over the supports, and consists of #6 bars at 12 in. centers placed between the #5 bars. The #6 bars should extend a distance not less than $0.75D$ on each side of the pier center, where D is the distance to the dead load point of contraflexure. Longitudinal distribution reinforcement in the bottom of the slab contributes to resistance and should be included in computing ultimate moment capacity.

Numerical Example Figure 3-40 shows a cross section for the example analyzed in the foregoing sections. We compute the center of gravity of the reinforcement as $d' = 57.66$ in. Also, $A_s = 7.30$ in.^{2}.

Then

$$p = 7.30/(22 \times 57.66) = 0.00575 < 0.015$$

and

$$0.4f'_c = 0.40 \times 5000 = 2000 \text{ psi} > 0.9 \times 1655 = 1490 \text{ psi}$$

Next, we compute

$$kd = 1.4 \times 0.00575 \times 57.66 \times 60/5 = 5.57 \text{ in.} < 7 \text{ in.}$$

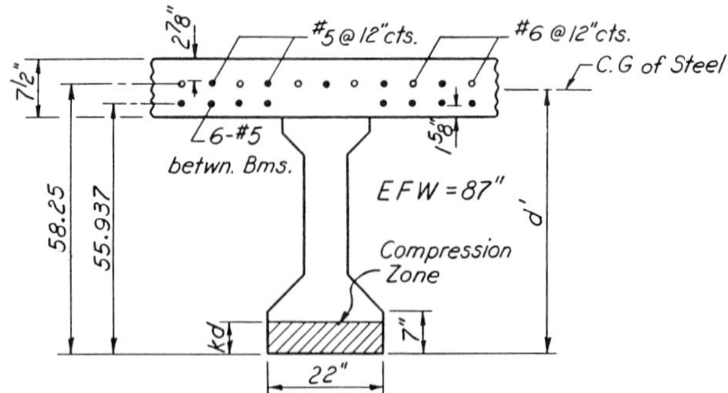

FIGURE 3-40 Cross section of PPC beam and slab over pier, showing reinforcement details and dimensions, for design example.

Hence, this is a rectangular section. The design moment capacity is

$$\phi M_n = 7.30 \times 60 \times 57.66(1 - 0.6 \times 0.00575 \times 60/5) = 2017 \text{ ft-kips}$$

The applied factored moment is

$$M_u = 1.3[192 + 1.67(527)] = 1391 \text{ ft-kips} < 2017$$

The compressive stress over the pier is

$$f_b = \frac{367}{599} + \frac{367 \times 16.35}{8559} + \frac{527 \times 12}{12{,}639} = 1815 \text{ psi} < 0.6f'_c = 3000 \text{ psi}$$

The investigation should be completed by checking the ultimate capacity at the cutoff point $0.75D$ after making allowance for the anchorage length.

Design for Shear

Theoretical Background Web reinforcement in PPC I beams consists of stirrups perpendicular to the axis of the beam. They should extend to a distance d from the extreme compression fiber and should be carried as close to the compression and tension surfaces as the cover permits. Web reinforcement should be anchored at both ends.

Experience shows that the principal tension due to factored loads invariably governs over working loads. It follows therefore that the amount of web reinforcement necessary to develop the required capacity for ultimate loading is determined from the following:

$$A_v = \frac{V_s s}{f'_y d} \tag{3-23}$$

where A_v = area of web reinforcement
s = longitudinal spacing of web reinforcement
V_u = factored shear, with $\phi = 0.90$, or $V_u \leq \phi(V_c + V_s)$
V_c = nominal shear carried by the concrete
V_s = nominal shear strength provided by web reinforcement
d = distance from extreme compressive fiber to centroid of prestressing steel
b' = width of web of prestressed beam
f_y' = yield strength of non-prestressed steel < 60 ksi

and

$$j = 1 - \frac{0.6p^* f_{su}^*}{f_c''}$$

The shear strength V_c provided by the concrete will be the lesser of the values V_{ci} and V_{cw}. The shear strength V_{ci} need not be less than $1.7\sqrt{f_c'}\,b'd$, and d need not be less than $0.8h$, where h is the depth of the beam. The shear strength V_{cw} is computed from the equation

$$V_{cw} = \left(3.5\sqrt{f_c'} + 0.3f_{pc}\right)b'd + V_p \tag{3-24}$$

where f_{pc} = compressive stress in concrete due to effective prestress force only (after allowance for all prestress losses) at extreme fiber of section where tensile stress is caused by externally applied loads
V_p = vertical component of prestress force at section

In addition to the foregoing requirements, the design must also satisfy the following: (a) V_s should not be taken greater than $8\sqrt{f_c'}\,b'd$; (b) the spacing of web reinforcement should not exceed $0.75h$, or 24 in., and when V_s exceeds $4\sqrt{f_c'}\,b'd$ the maximum spacing should be reduced by one-half; and (c) the minimum area of web reinforcement is $A_v = 50b's/f_{sy}$, where b' and s are expressed in inches and f_{sy} is the yield strength of non-prestressed reinforcement (psi).

A prestressed concrete beam without web reinforcement will normally fail in a region of high moment and not necessarily near the end where maximum shear occurs. The usual mode of failure is manifested in inclined shear cracks originating from flexural cracks, and failure is thus a combination of bending and shear stresses.

For simply supported members subjected to moving loads, it is sufficient to investigate the shear within the middle half of the span. The area of web reinforcement required at the quarter points should be used in the exterior quarters of the span.

Numerical Example Given: design span = 75 ft, beam spacing = 7.25 ft, loading = HS 20, and slab thickness = 7.5 in. Also, the following beam data are available (54-in. beam): $f'_c = 5000$ psi, prestress = twenty-two 1/2-in.-diameter strands, $f^*_{su} = 263$ psi, $p^* = 0.00067$, $f''_c = 3500$ psi, $e_c = 21.70$ in. (distance from centroid of prestressing steel to centroid of prestressed beam at midspan of beam), and $e_e = 17.61$ in. (distance from centroid of prestressing steel to centroid of prestressed beam at end of beam).

The dead load is 1.314 kips/ft (noncomposite) and 0.314 kip/ft (composite). The beam properties are shown in Figure 3-38.

The computed dead load and live load plus impact shears at the supports and at the center are as follows. At the supports,

$$V_{\text{DL}} = 61.1 \text{ kips} \qquad V_{\text{LL}+I} = 51.9 \text{ kips}$$

At midspan,

$$V_{\text{DL}} = 0 \qquad V_{\text{LL}+I} = 22.3 \text{ kips}$$

Next, we compute the factored shear at the support and at midspan, using $\phi = 0.9$. At the support,

$$V_u = \frac{1.3}{0.9}(61.1 + 1.67 \times 51.9) = 213.4 \text{ kips}$$

At midspan,

$$V_u = \frac{1.3}{0.9}(0 + 1.67 \times 22.3) = 53.64 \text{ kips}$$

These values are used to draw the maximum ultimate shear diagram shown in Figure 3-41.

The next step in the design requires estimation of $d = e + c_t + t_{\text{effective}}$. At the support,

$$d = 17.61 + 29.03 + 7.0 = 53.64 \text{ in.}$$

At midspan between draping points,

$$d = 21.70 + 29.03 + 7.0 = 57.73 \text{ in.}$$

Note, however, that d need not be less than $0.8h = 0.8 \times 54 = 43.2$ in. The shear resisted by the concrete is computed as

$$V_c = 1.7\sqrt{5000} \times 6 \times 43.2 = 31.1 \text{ kips}$$

As long as we compute d as $0.8h$, the effect of the strand profile (whether straight or draped) is ignored; hence, the shear carried by the concrete at the

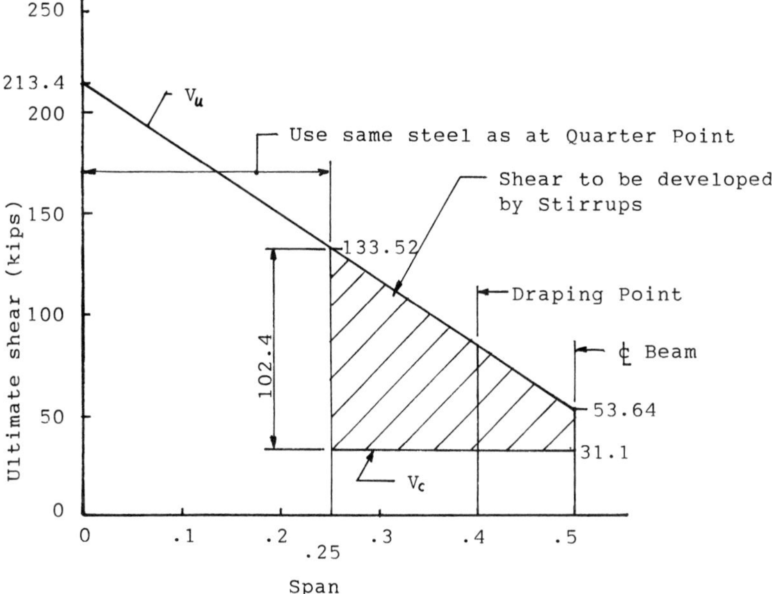

FIGURE 3-41 Maximum ultimate shear diagram for design example.

quarter point is still 31.1 kips, so that

$$V_s = V_u - V_c = 133.5 - 31.1 = 102.4 \text{ kips}$$

Check the value

$$8\sqrt{f'_c}\,b'd = 8\sqrt{5000} \times 53.64 \times 6 = 182 \text{ kips} > 102.4$$

For #5 stirrups with two legs, $A_v = 0.62$ in.^{2}. Therefore, the required spacing is

$$s = \frac{0.62 \times 60 \times 56.2}{102.4} = 20.4 \text{ in.} \quad \text{say 20-in. centers}$$

Continuous Spans In continuous beams the negative moments over the supports reduce the effect of prestressing, and the prestressed beam approaches the condition of a conventionally reinforced concrete beam. The analysis for shear is thus different. The effect of prestressing may be neglected in the negative moment area between the support and the quarter point, and this assumption converts the beam to a conventional member reinforced with tension reinforcement in the slab. In the middle half of the span between quarter points, the shear may be investigated as a simple-span prestressed beam. Shear is discussed further in the following sections.

Development of Horizontal Shear (Composite Action)

Full transfer of the ultimate horizontal shear can be assumed if contact surfaces are left clean and intentionally roughened, where minimum vertical ties are provided and stirrups are fully anchored into all intersecting components, and the beam webs are designed to resist the entire vertical shear.

Alternatively, in lieu of the preceding requirements, ultimate horizontal shear stresses can be computed from the expression

$$v = \frac{V_u Q}{I'b} \tag{3-25}$$

where v = ultimate horizontal shear stress

V_u = maximum shear caused by the factored loads

Q = moment area of cast-in-place slab about the centroid of the composite member

I' = moment of inertia of the composite section

b = width of top flange of prestressed beam

The minimum web reinforcement extended into the cast-in-place slab should not be less than #3 bars at 12-in. centers. The spacing of vertical ties should not be greater than four times the average thickness of the composite flange, and in no case greater than 24 in.

Numerical Example For the horizontal shear design of the example of Figure 3-39, we will assume that Article 9.20.4.4 of AASHTO is satisfied, and that the contact surfaces are clean and roughened. Then the shear capacity at the contact surface is 300 psi.

Next, we compute the maximum (ultimate) shear $V_u = 133.5$ kips. The moment area Q of the slab is $Q = 0.8bt\bar{y}$, where $\bar{y}$ is the distance from the centroid of the composite section to the centroid of the slab, equal to 17.94 in. The factor 0.8 is the modular ratio of the concrete. Then

$$Q = 0.8 \times 87 \times 7 \times 17.94 = 8740 \text{ in.}^3$$

From (3.25) we obtain

$$v = \frac{133.5 \times 8740}{500{,}015 \times 20} = 117 \text{ psi} < 300 \text{ psi}$$

Minimum vertical ties, two #5 bars at 24 in. = 0.31 in.^{2} > 0.22 in.^{2}/ft.

3-12 PRECAST PRESTRESSED DECK BEAM BRIDGES

Most multibeam bridges today consist of pretensioned prestressed concrete sections. Protection of the roadway surface is essential, and many sections

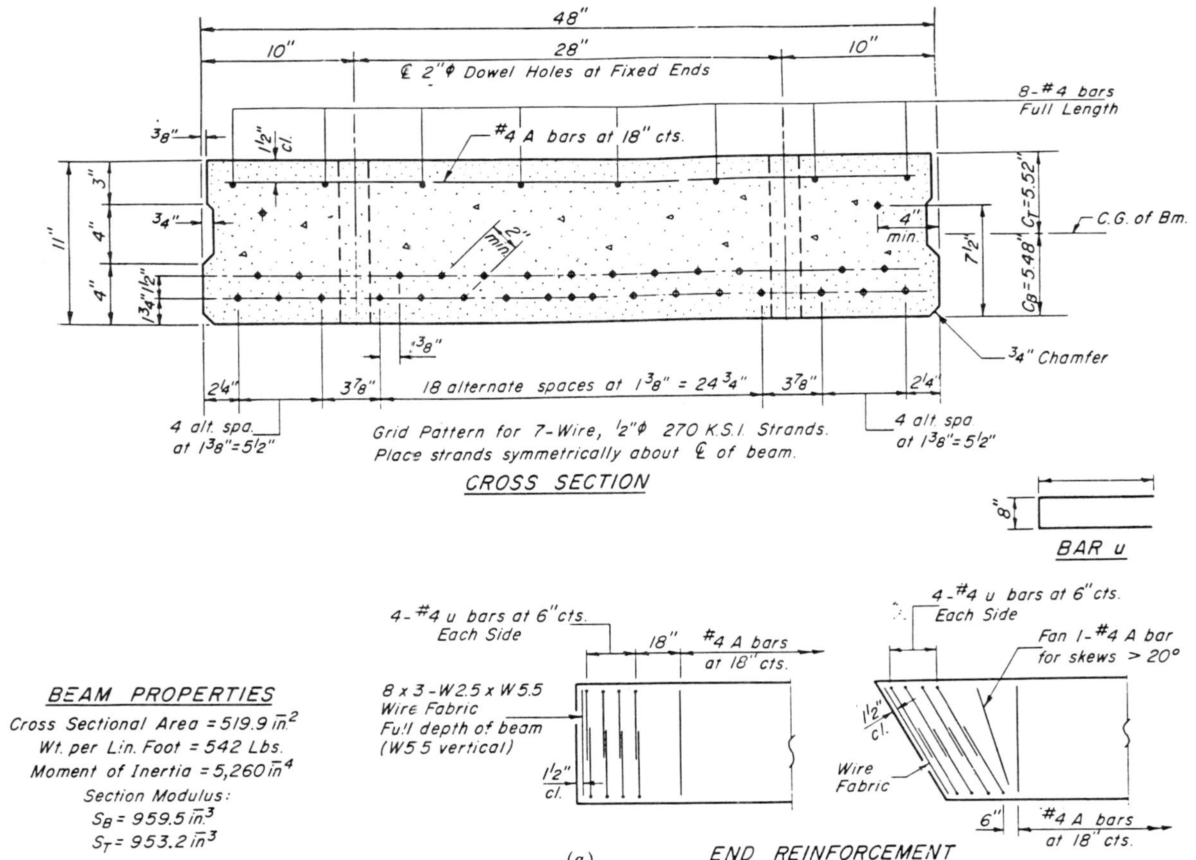

FIGURE 3-42 Typical standard deck beam sections of prestressed concrete.

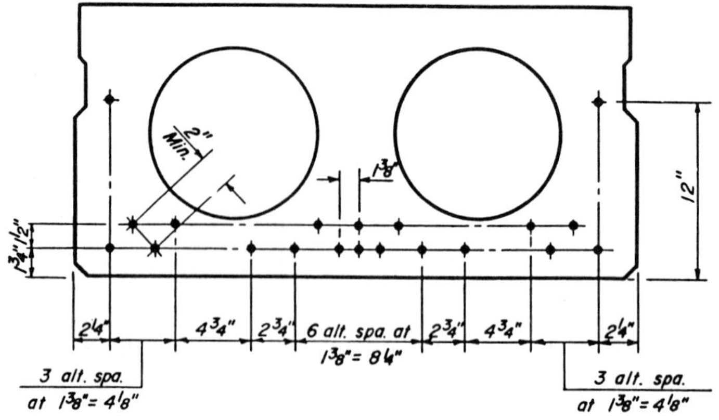

(b)

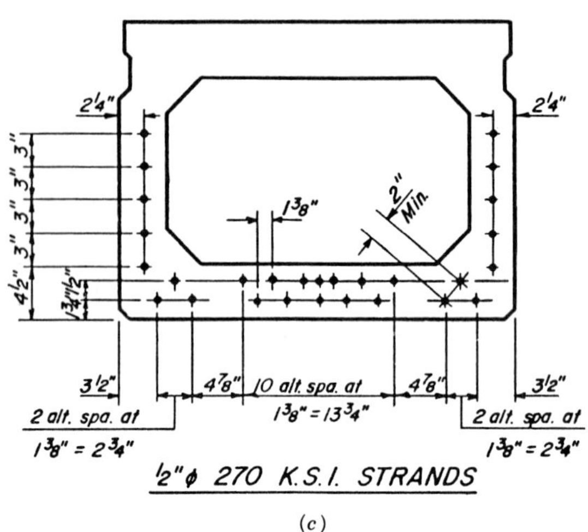

(c)

FIGURE 3-42 *(Continued).*

are manufactured with a thin top deck for riding surface. This, however, if allowed to deteriorate, requires replacement of the entire superstructure. Alternatively, an asphaltic wearing surface is placed as a riding surface and also to even transverse and longitudinal irregularities inherent in construction. There is a good reason to expect that these wearing surfaces can collect and trap moisture mixed with deicing salts at the top of the concrete unless an efficient waterproof membrane system is used as protection.

Standard precast prestressed deck beams are available in depths of 11, 17, 21, 27, and 33 in., and in various widths. Details for representative sections are shown in Figures 3-42*a* through *c*. The beam shown in Figure 3-42*a* has a solid rectangular shape; the beam shown in Figure 3-42*b* has a voided slab shape; and the beam shown in Figure 3-42*c* has a cellular cross section.

Practical span lengths vary within a relatively narrow range, but depend mainly on the section. For preliminary estimates and rule-of-thumb design, the probable span range for each type can be inferred with the help of appropriate charts and diagrams and other design aids.

Design Procedure The underlying principle is that the slab and box units are likely to act as multibeam sections because of the large surface along which interaction is developed as a result of the transverse posttensioning and the presence of continuous longitudinal shear keys. Note that according to AASHTO (Article 3.23.4) the lateral ties may, or may not, be prestressed. In calculating bending moments, no longitudinal distribution of wheel load is assumed. The fraction of wheel load applied to a beam is S/D, where S is the width of the precast member, and D is a factor that depends on the number of traffic lanes, the bridge dimensions (width and length), the moment of inertia, the Saint-Venant torsion constant, and Poisson's ratio for the beams.

3-13 TOPICS RELEVANT TO PRESTRESSED CONCRETE BRIDGES

Deflections and Camber

Deflections due to prestress may be computed by considering the concrete member a free body separated from the tendons, which are replaced by a system of forces acting on the concrete as shown in Figure 3-43. This requires

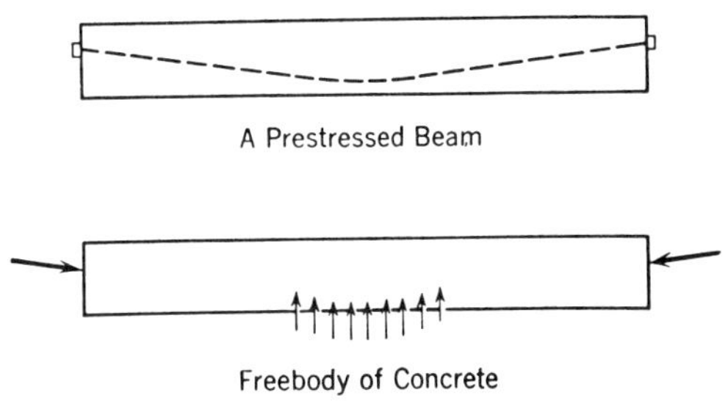

FIGURE 3-43 Method for computing deflection due to prestress. (From Lin and Burns, 1981.)

computation of the proper components of forces at the end anchorages of the tendons. The method is applicable to both simple and continuous beams, but for simplicity certain assumptions must be made (see also AASHTO Article 9.13.3.3).

Acting simultaneously with the prestress is the weight of the beam, which produces deflections depending on the support conditions. These can be computed by the usual elastic theory. The resultant deflections of the beam at transfer are obtained by summing algebraically the deflections due to prestress and those due to the beam weight. Camber is thus the result of the difference between the upward deflection caused by the prestress and the downward deflection caused by the weight of the beam.

Camber will vary with the age of the member because of two factors. Loss of prestress tends to decrease the deflection, whereas creep tends to increase it. Studies of these two parameters show that the camber of a beam (essentially an I shape) at erection is approximately 1.8 times the initial camber based on 100 percent of the prestress force. A correction factor is thus introduced in the elastic analysis to reflect this change.

Deflections due to external loads are similar to those for non-prestressed beams. As long as the concrete has not cracked, the beam is treated as a homogeneous body and the elastic theory is applicable. When cracks begin to occur, the nature of deflection is certain to change. At the very start of the process, the effective section that resists moments is cracked, and as cracking becomes deeper a greater portion of the moment of inertia is lost. However, only that section of the beam subjected to higher moment has cracked while the rest may still remain intact. Thus, the deflection of the beam will increase faster as shown in Figure 3-44. As the live load is removed, the beam will return to a nearly initial position except for come cracks that have already developed.

Estimation of Camber The procedure for computing camber is based on the following assumptions:

1. The prestress producing upward deflection is somewhat between the initial and the final effective value. However, it is sufficiently accurate to use a constant value, that is, the total initial prestressing force. For 1/2-in.-diameter strands, this is $F_i = 28.9$ kips.
2. The component of the prestress along the beam axis is assumed constant, unless the inclination of the tendons becomes excessive. Thus, two procedures are presented, one for straight and one for draped strands.
3. The tendons are treated as a whole instead of individually, and computations are based on the center of gravity.
4. Deflections due to shear are small and can be neglected.

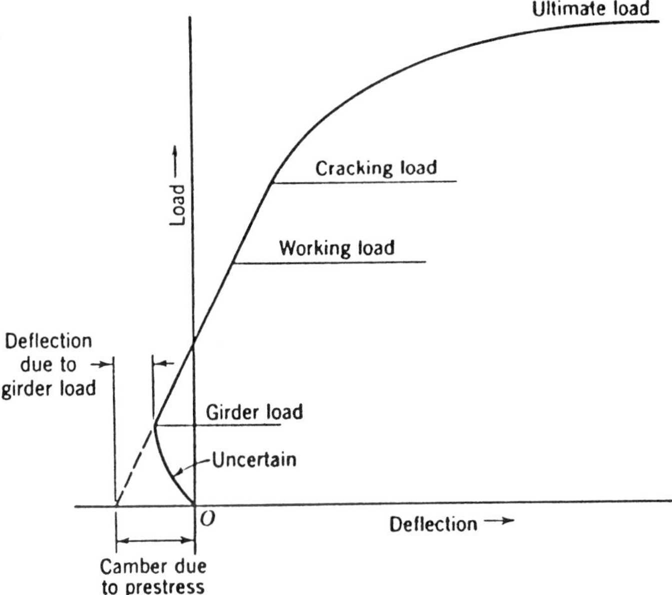

FIGURE 3-44 Load-deflection curve for a prestressed beam. (From Lin and Burns, 1981.)

The deflection for straight strands is estimated from the equation

$$\Delta_c^p = \frac{0.125 F_i e L^2 144}{0.55 E_{ci} I} \tag{3-26}$$

where Δ_c^p = upward deflection at midspan of the beam due to the prestressing force (in.)

L = beam length (ft)

e = eccentricity of prestress (in.)

I = moment of inertia of the prestressed beam (in.^{4})

E_{ci} = flexural modulus of elasticity of the concrete at the time of strand release (psi)

F_i = total initial prestressing force (lb)

The deflection for draped strands (harped at points 0.4 and 0.6) is given by

$$\Delta_c^p = \frac{F_i L^2 144}{0.55 E_{ci} I}(0.0983 e_c + 0.0267 e_e) \tag{3-27}$$

where e_c = eccentricity of the prestressing steel at midspan of the beam (in.)

e_e = eccentricity of the prestressing steel at the end of the beam (in.)

The downward deflection due to the dead load (weight) of the beam is

$$\Delta_c^b = \frac{5w_b L^4 1728}{0.55 \times 384 E_{ci} I} \tag{3-28}$$

where Δ_c^b = downward deflection due to weight of beam (in.)
w_b = weight of beam (lb/ft)
L = beam length (ft)
E_{ci} = flexural modulus of elasticity of the concrete at the time of strand release (psi)
I = moment of inertia of the prestressed beam (in.^{4})

The resulting camber, at the time of erection, from the prestressing force and the weight of the beam is therefore

$$\text{Camber} = \Delta_c^p - \Delta_c^b$$

The practical value of these computations is in adjusting the seat elevations to ensure the specified slab thickness throughout the span. In addition, it is necessary to compute deflections of the beams due to the weight of the slab. These deflections are then used to adjust the grade line (theoretical profile elevations).

Numerical Example For the example of Figure 3-38, we first compute the modulus of elasticity:

$$E_{ci} = 150^{1.5} \times 33 \times \sqrt{4000} = 3.83 \times 10^6 \text{ psi}$$

and

$$F_i = 28{,}900 \times 16 = 462{,}400 \text{ lb}$$

Using (3-27), we obtain $\Delta_c^p = 1.9$ in. upward. Using (3-28), we obtain $\Delta_c^b =$ 0.75 in. downward. Therefore,

$$\text{Camber} = 1.90 - 0.75 = 1.15 \text{ in. upward}$$

Considerations of Optimum Design for Prestressed Concrete Tension Members

Procedures for optimum design of prestressed concrete tension members should consider maximum allowable compression, ultimate strength, protection against cracking and decompression, minimum reinforcement, and maximum immediate and long-term deformations. These design criteria can be

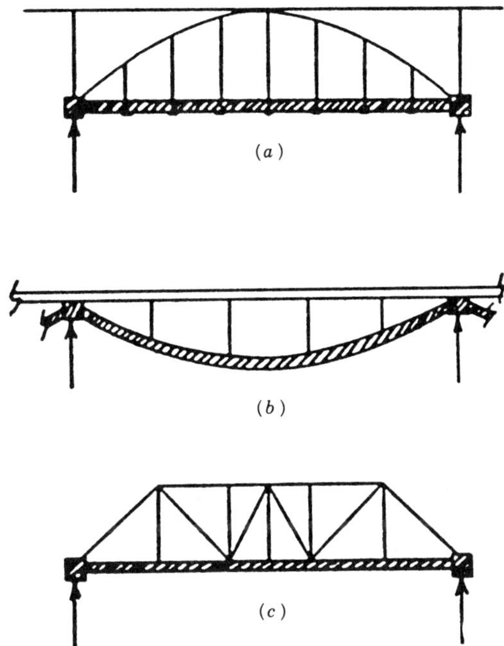

FIGURE 3-45 Structures where tension members are used: (*a*) arch bridge; (*b*) inverted suspension bridge; (*c*) truss bridge.

expressed in terms of two unknown variables, namely, the area of prestressing steel and the cross-sectional area of the concrete. The relationships representing the various criteria may then be assembled in a feasibility domain bound mainly by linear functions. Using design approximations acceptable for practical applications, Naaman (1982) has reduced the analysis to a linear programming problem in which the objective is to minimize the cost and/or the weight of the member.

Besides the conventional beam-type bridge, examples of tension members in bridge construction are shown in Figure 3-45. Morandi (1969) built several bridges featuring prestressed concrete tension elements used in a cable-stayed configuration. Lin and Kulka (1973) designed and built the inverted suspension bridge over Rio Colorado in Costa Rica, where a catenary-shaped prestressed concrete ribbon spanning 146 m (480 ft) was used to support the horizontal deck. Such bridges may represent a comparatively economical solution for crossing deep valleys (Matsushita and Sato, 1979; Lin and Burns, 1981).

In general, prestressed concrete tensile members have not generated a great need for experimental research. Wheen (1979) has explained the influence of the variables on the load–deformation response of prestressed concrete ties, and confirms that in their precracking range these members essentially behave as predicted theoretically. Because prestressed concrete tension members are expected to perform in the linear elastic uncracked range under service loads, their analysis and design are relatively simple (Lin and Burns, 1981; Nilson, 1978). However, current design criteria have been extended to include cracking, maximum elongation, and ultimate resistance.

Service Stresses, Cracking, and Ultimate Load The net area A_n of a concrete section subjected to a prestressing force is

$$A_n = A_g - A_{ps} \tag{3-29}$$

where A_g = gross sectional area of the member
A_{ps} = area of prestressing steel

Based on criteria that establish the margin of safety against cracking under force effects as well as a safety margin against decompression, Naaman

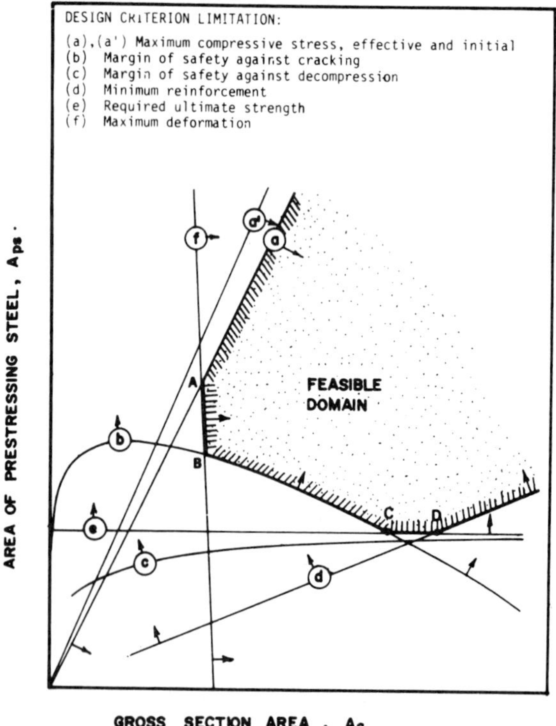

FIGURE 3-46 Geometric representation of design criteria and feasible domain. (From Naaman, 1982.)

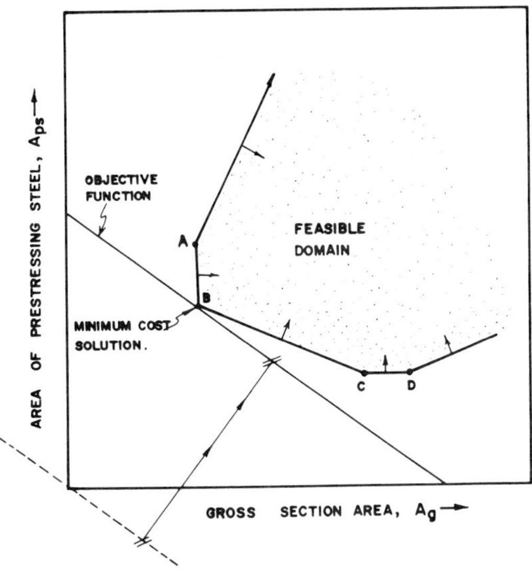

FIGURE 3-47 Graphical solutions for minimum-cost design problem. (From Naaman, 1982.)

(1982) proposes a procedure based on the two unknown variables A_g and A_{ps}. A geometric representation of feasible sets of these parameters is shown in Figure 3-46, and is defined as the feasible domain. A wide range of choices is possible, but for optimum design the selection must yield the lowest cost. Interestingly, most criteria yield straight-line representations, except for safety against cracking and decompression which yield curves b and c.

Naaman (1982) also presents a methodology for minimum-cost design based on unit weights and unit costs for the prestressing steel and the concrete. A general optimization problem is formulated subject to appropriate constraints representing design criteria. Because the objectives and constraints are expressed in terms of linear functions, the problem involves linear optimization and is solved by a linear programming algorithm. An example of a linear solution is given graphically in Figure 3-47. The representative line can be moved graphically in a parallel translation from the zero point to a minimum cost for which only one point of the line belongs to the feasible domain. This point is the vertex B of the graph $ABCD$, and gives the minimum-cost solution.

Continuity in Precast Prestressed Bridge Girders

One of the main uncertainties in the design of continuous prestressed girders is the prediction of the elastic, inelastic, time-dependent, and ultimate positive and negative moments at the cast-in-place connection diaphragms over the piers. This uncertainty is related to the different loading and construction stages, time-dependent effects, and the details used to make connections. Oesterle, Glikin, and Larson (1989) have developed procedures that can ensure more rational designs, based on experimental and analytical programs. Recommendations are given for service moments and strength design.

Design for Service Moments Results of studies of time-dependent restraint moments and service load moments at supports of bridges made continuous indicate that there are no structural advantages in providing positive moment reinforcement at supports. The time-dependent positive restraint moment generally induces a crack in the bottom of the diaphragm concrete, but with the application of live load this crack must close prior to inducing negative moment at the continuity connection. Positive moment reinforcement reduces the crack size and thus increases the apparent live load continuity. However, the positive restraint moment resulting from the presence of reinforcement at the bottom of the support connection also increases the positive midspan resultant moment. Hence, positive reinforcement across the girders at the supports does not benefit flexural behavior.

Positive service moments at midspan consist of simple-span moments due to girder and deck weight, moments acting on the continuous structure induced by superimposed dead load and live load plus impact, and time-dependent restraint moments at the supports. The results show, however, that the time-dependent behavior can influence the continuous behavior to the extent that the effective continuity for live load plus impact can vary from 0 to 100 percent.

Time-Dependent Restraint Moments In analyzing the bridge to determine time-dependent restraint moments, full structural continuity is assumed. This applies with or without positive moment connections at the supports. Superimposed dead load applied shortly after continuity is established should be included in restraint moment analysis. Relevant time factors should reflect the expected construction schedule and should also account for the variability of creep and shrinkage behavior because time-dependent effects can vary significantly (see also AASHTO Article 9.13.3.3).

Midspan Service Moments The continuity moment, used to compute load effects at midspan under service conditions, is defined as the sum of restraint, additional dead (superimposed) load, and live load plus impact moments. If the average of the continuity moments for the two supports is positive, time-dependent effects have reduced the effective structural continuity for live load plus impact to 0 percent. In this case the positive midspan moment should be calculated as the sum of simple-span moments for superimposed dead load and live load plus impact.

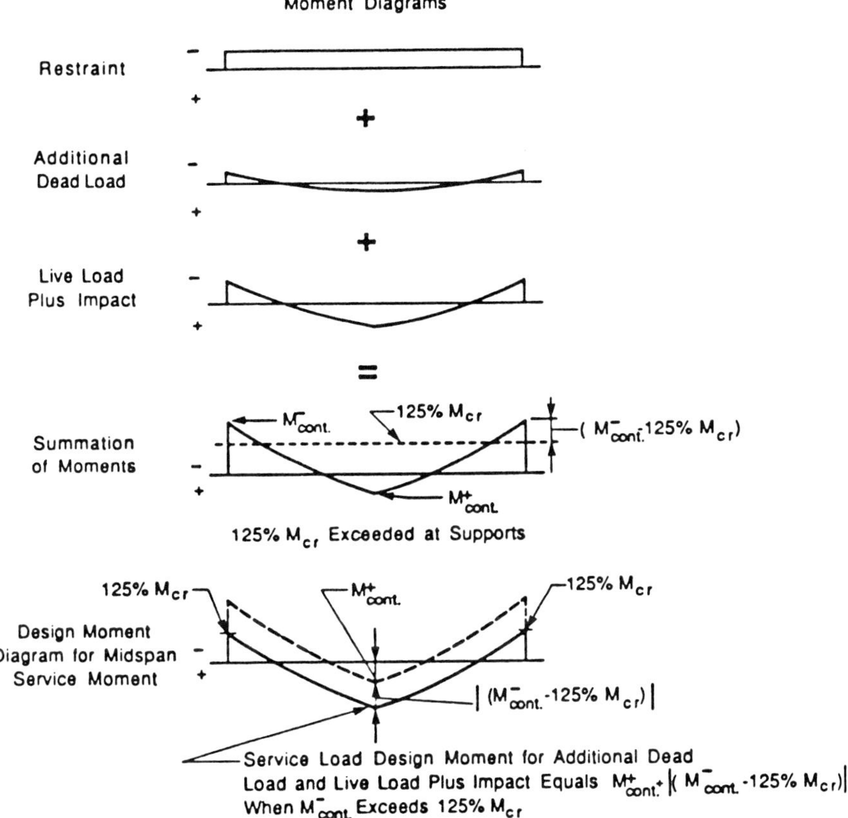

FIGURE 3-48 Determination of service load design moments at midspan.

If the average of the continuity moments for the two supports is negative and does not exceed 125 percent of the negative cracking moment, partial or full structural continuity exists. In this case the positive span moment may be calculated by adding the simple-span and continuous-span moments to the average calculated restraint moment.

The summation of moments (excluding simple-span moments due to the weight of the deck and girder) for a typical symmetrical span is illustrated in Figure 3-48, and is self-explanatory. If no analysis is carried out to determine time-dependent restraint moments and no positive moment reinforcement is provided at the supports, the midspan service moment should be taken as the sum of simple-span superimposed dead load and live load plus impact moments.

Negative Moment at Supports Likewise, the effects of possible negative restraint moment should be included. Results of time dependent analysis indicate that maximum negative moment and maximum potential for cracking in the deck occur approximately 50 days after casting the deck and the diaphragm. Therefore, the restraint moment should be calculated for the same time. For checking fatigue limits, the maximum negative moment at an early age is essentially a transient condition. Although restraint moments can remain negative for the life of the structure, they reach a reduced and relatively constant level in about 2 years.

Service Moments at Supports The service moment at the supports should consist of the sum of peak negative restraint moment, superimposed dead load moment, and live load plus impact moment computed for full structural continuity. Negative service moment stresses in steel and concrete should be calculated using cracked, transformed section properties. Concrete compressive stress should include stress due to prestress force acting on the girder only.

Strength Design Because stresses and strains in the girder from creep and shrinkage are self-limited, the presence or absence of time-dependent restraint moments has no effect on the strength of the structure.

For positive moment at midspan, the factored design moment should include dead load (girder and deck) acting on the simple beam, and superimposed dead load and live load plus impact acting on the fully continuous structure. For negative moment at the supports, the factored design moment consists of the superimposed dead load and live load plus impact using full continuity (see also the foregoing design example).

Construction Sequence Continuity performance appears to be highly relevant to the age of the girder when the deck and diaphragms are cast. Higher negative restraint moments at the support connections are likely when continuity is established at a late girder age. Full continuity for live load can be ensured, depending on the age of the girder at the time the deck and diaphragms are cast and on the creep coefficient for the girder concrete.

Casting the deck prior to casting the diaphragm increases the resultant positive moment at midspan, but inhibits the potential for deck cracking. Casting the diaphragm before casting the deck decreases the resultant midspan positive moments only slightly, but increases the potential for deck cracking. Thus, there is no obvious advantage to sequencing the casting of the deck and diaphragms.

Modifications to AASHTO Specifications Pertinent revisions are recommended for AASHTO Articles 9.7.2 and 9.1.3. These include the definition of effective restraint moments and procedures for calculating positive and negative moments for service load and strength design.

3-14 PRESTRESSED WAFFLE SLAB BRIDGES

One of the early posttensioned waffle slab bridges is reported by Lin et al. (1969). In the past waffle slabs were considered incompatible with the predominantly one-way supporting systems. With its severe skews and irregularly shaped platforms, however, waffle construction has been shown to provide structural benefits (Kennedy and Ghobrial, 1980). In the same context, waffle slabs may be compared to voided slabs prone to cracking. The associated advantages are (a) dead load reduction, (b) more efficient live load distribution, (c) enhanced deck durability, and (d) reduced dead load moments and smaller deflections.

Posttensioned waffle slab bridges can be analyzed by superimposing a bending analysis to an in-plane stress analysis assuming no coupling (Kennedy and El-Sebakhy, 1980). The results of an uncoupled analysis are valid if the associated deflections are small. The context of orthotropic concrete structure in this case means that orthotropy is the result of geometry rather than of material as shown in Figure 3-49. Because of variations in sections the member is actually anisotropic, but if the elastic constants are chosen to represent the bending and twisting behavior, the applicability of orthotropic plate theory is valid.

From these considerations Kennedy and El-Sebakhy (1980) have developed analytical expressions for bending and twisting moments and shears. In-plane stress analysis involves stresses written conveniently as a Fourier series. The final results are obtained by appropriate superpositions, whereby

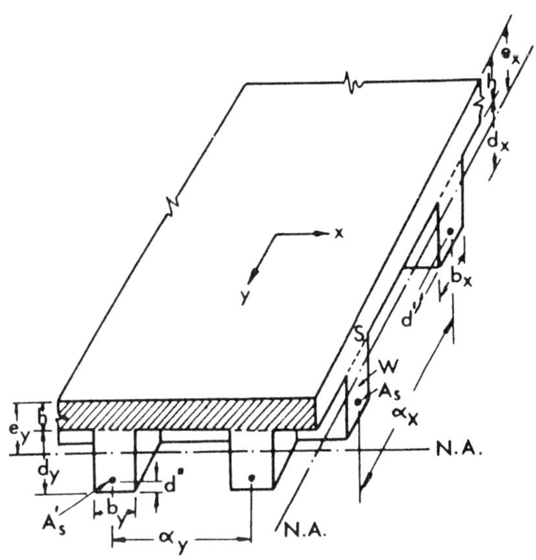

FIGURE 3-49 Geometry of prestressed concrete waffle slab structure with reference axes. (From Kennedy and El-Sebakhy, 1980.)

deflection and moments are found from the bending analysis, and total stresses and strains are derived from the bending and in-plane analysis.

Test Studies Test studies have been carried out by Kennedy and El-Sebakhy (1980) to compare the structural response between a waffle slab (system I) bridge and a uniform prestressed slab (system II) bridge, both having the same volume of concrete and the same amount of steel. The calculated flexural rigidities of system I are almost double those of system II, whereas its torsional rigidity is one-fourth that of system II. Other differences are as follows.

1. The deflection of system I is more favorable than in system II. A comparison of the camber of the two systems articulates this difference. The considerable reduction in the degree of bending for system I reduces the secondary stresses induced under sustained load.
2. For a live load of 1 kip applied at the center, the resulting strains in system I were compressive when system II exhibited some tension in the central section of the bridge.
3. If a triangular stress distribution was assumed with zero tensile stress in the bottom fiber of the center section, system I yielded a live load capacity several times that of system II for both rectangular and skew bridges.

Continuous Bridges Similar studies have been extended to continuous waffle slab units over line piers by super-imposing the in-plane and bending solutions (Gupta and Kennedy, 1978), adjusted to accommodate the appropriate boundary conditions. For slabs supported on isolated columns, a solution cannot be developed unless the column reaction is evaluated first.

Recent Feasibility Studies Kennedy (1987) reports feasibility studies undertaken to investigate the structural efficiency of waffle slab bridges compared with the solid slab and the slab-on-girder (one-way, ribbed slab) type. The following three categories are considered.

1. Category I includes two-span continuous structures, as shown in Figure 3-50*a*, with skew ranging from 0° to 45°. The concrete volume is kept constant, as shown in Figure 3-50*c*.
2. Category II has the same plan as category I, but the bridges are continuous over two isolated supports, as shown in Figure 3-50*b*. Likewise, the concrete volume is kept constant.
3. Category III has a plan configuration as in category II, except that cross sections are chosen to have constant depth, as shown in Figure 3-50*d*.

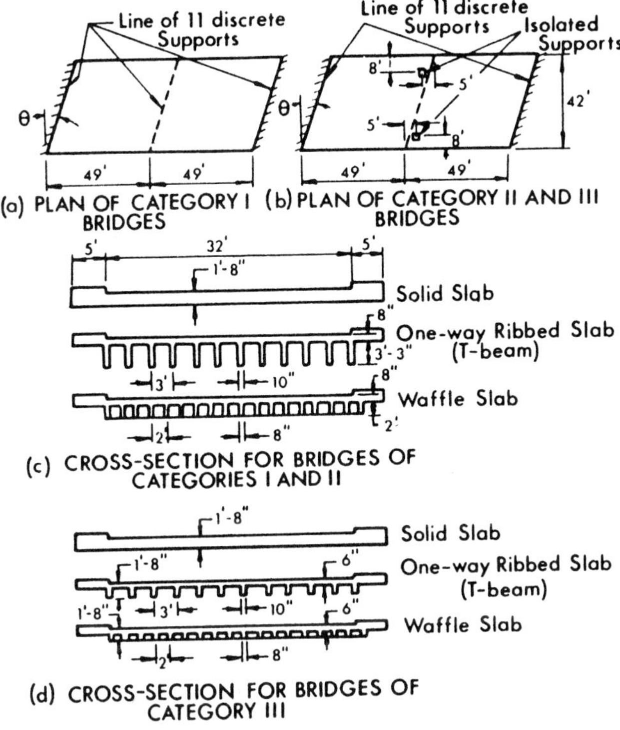

FIGURE 3-50 Details of bridges in feasibility study.

Elastic Analysis The linear elastic response of prestressed waffle bridges requires realistic estimates of the orthotropic rigidities. In addition, the following assumptions are made: (a) the number of ribs is large enough for the real bridge to be replaced by an idealized model with continuous properties and (b) the deck area is magnified by the factor $1/(1 - \mu^2)$ to allow for the influence of the Poisson's ratio μ.

Calculated rigidities derived from these assumptions are affected by creep that also influences the modular ratio n. These influences can be readily accounted for by adjusting E_c to a time-dependent modulus of deformation (Saeed and Kennedy, 1970) and by augmenting a value of n according to accepted behavior.

Ultimate Load Analysis Kennedy (1987) assumes yield line patterns of failure for ultimate load analysis. This approach is based on (a) test results from posttensioned waffle slab bridge models, (b) test results from simply supported waffle slab bridge models, and (c) results from a parametric study of a progressive failure analysis. In general, the predicted collapse load is found to be lower than the actual collapse load because some load-enhancing

TABLE 3-2 Ratio of Estimated Cost of Prestressed Concrete Bridges in Solid Slab or One-Way Ribbed Slab (T-Beam) Construction to That in Waffle Slab Construction

Bridge Category[a]	Type of Construction	Bridge Skew Angle θ			
		0°	15°	30°	45°
I	Solid slab	1.09	1.12	1.11	1.16
	One-way ribbed slab (T beam)	1.62	1.61	1.59	1.68
II	Solid slab	1.18	1.33	1.30	1.29
	One-way ribbed slab (T beam)	2.07	2.20	2.02	2.28
III	Solid slab	1.10	1.20	1.15	1.28
	One-way ribbed slab (T beam)	1.22	1.46	1.45	1.51

[a]See Figure 3-50.

effects are ignored (e.g., strain hardening in the prestressing steel, torsional resistance of the waffle slab, and, to a lesser extent, membrane action).

Results of Studies The economic advantages of the bridge categories of Figure 3-50 are compared in Table 3-2. These cost estimates reflect the volume of concrete in the superstructure, the amount of prestressing steel, and the associated formwork. They do not include, however, the substructure cost, scaffolding to support the formwork, and the bridge approaches. Unit prices are based on 1983 Canadian dollars. The table gives the ratio of the estimated total cost of each bridge in solid slab or one-way ribbed slab construction to the estimated total cost of the corresponding bridge in waffle slab construction.

In addition to these economic advantages, Kennedy (1987) concludes that the presence of transverse ribs in waffle slabs benefits the serviceability limit state by providing better crack control and by reducing local deformations due to heavy wheel loads for better live load distribution.

In general, waffle slab bridges should be much stiffer than similar reinforced concrete superstructures, and with a smaller amount of secondary stresses produced by sustained loads. The potential advantages become greater with an increasing width–span ratio or skew angle.

3-15 PRESTRESSED CONCRETE BOX GIRDERS

General Principles

The recent widespread use of posttensioned concrete box girders in bridge decks is partly attributed to the availability of dependable larger tendons and suitable jacking systems. Greater tendon lengths have become practical with low-friction rigid conduits, and high-strength cast-in-place concrete is used to improve efficiency.

Design Requirements For cast-in-place box girders with normal span and girder spacing, where the slabs are treated as an integral part of the supporting girder, the entire slab width may be assumed to be effective in compression. For members of unusual proportions, AASHTO stipulates methods of analysis that consider shear lag principles in determining stresses in longitudinal bending.

Strength and Deformation of Pretensioned Box Girders Taylor and Warwaruk (1981) have used models to analyze prestressed concrete box girders of arbitrary cross section subjected to bending, torque, and shear. A constitutive matrix is developed for (a) uncracked concrete and (b) conditions upon cracking. Relevant parameters include the postcracking member stiffness and the cracked concrete shear modulus, derived as proposed by Houde and Mirza (1974). In addition to the classical plate theory, finite-element models are developed to simulate diaphragm action for the general diaphragm shape and thickness.

Failure Criteria The tensile and compressive strength of concrete is taken as proposed by Kupfer, Hilsdorf, and Rush (1969) but with some simplifications. Beyond cracking, the shear rigidity across the concrete crack is developed by the aggregate interlock and dowel mechanisms. Failure in a structural context is assumed to have occurred when one of the following conditions arises: (a) a concrete element crushes, (b) a major prestress reinforcement element is stressed beyond its yield strength, and (c) the member becomes unstable as its strength reserve cannot sustain the imposed load increments.

Experimental Programs Taylor and Warwaruk (1981) have supplemented the analytical assessment by an experimental test program that characterizes the complex cross-sectional geometry of concrete box girders and combines bending, torsion, and shear load patterns. The two different box versions used in these tests are the rectangular shape and the trapezoidal shape shown in Figures 3-51 and 3-52, respectively.

Comparison of Analytical and Test Results The analytical results are influenced by the following physical aspects: (a) any errors in computing the initial concrete modulus are magnified in predicting uncracked and cracked beam deformations (this is because the stiffness contribution of reinforcement in an uncracked underreinforced beam is small); (b) in the analytical model, the maximum moment lever arm is defined by finite-element mesh geometry and can give rise to modeling inaccuracy in shallow, thick-flanged box beams; (c) in uncracked box girder walls the surface torsional shear stress is larger than at wall midthickness, and because plane stress concrete elements accommodate only uniform shear flow, the analytical model overestimates the torsional cracking strength; and (d) in some instances, transverse

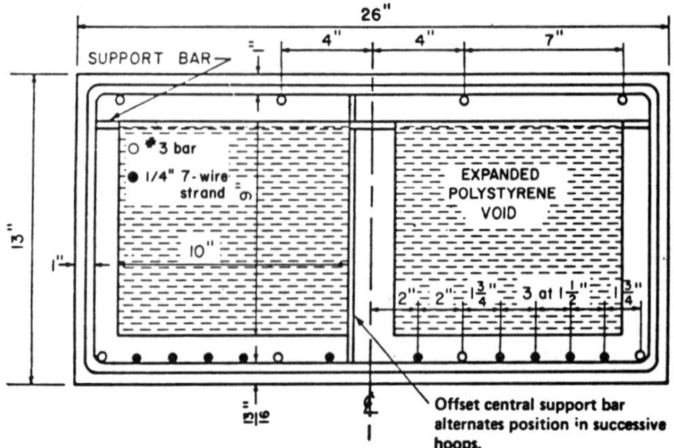

FIGURE 3-51 Rectangular beam cross section. (From Taylor and Warwaruk, 1981.)

reinforcement is not modeled effectively, resulting in a premature shear failure.

In this study, the correspondence between model and experimental stiffness in the elastic region is close (Taylor and Warwaruk, 1981). However, the results indicate that the analytical model consistently underestimates the beam failure loads. Thus, for all the beams tested in the experimental program, bending moment dominated performance. Coupled with the shallow, thick-flanged section geometry of Figures 3-51 and 3-52, the comparatively shorter model bending moment arm at failure results in premature yielding of the tension reinforcement. However, for underreinforced box

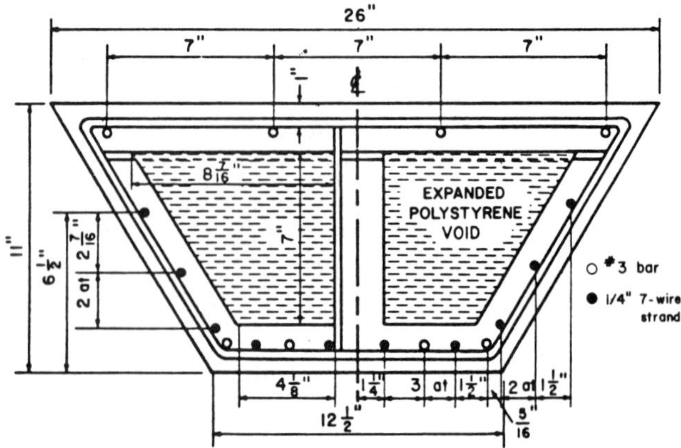

FIGURE 3-52 Trapezoidal beam cross section. (From Taylor and Warwaruk, 1981.)

girders of smaller wall thickness–depth ratios, the discrepancy in predicting strength is diminished.

The analytical model performance compares to current theory in terms of strength characteristics. For pure torsion, model and theoretical ultimate capacity predictions are close. However, strength comparison under pure shear conditions is difficult because analytical loading combinations do not isolate the critical cross section from concentrated load effects or the presence of large moments.

Model Study of Standard Prestressed Bridge

Standard precast concrete trapezoidal girders have been used in typical two-span bridges with spans in the range of 100 to 150 ft. Several states and provinces (e.g., Ontario, Canada) have standardized this concept and developed typical girder configurations. When used in conjunction with a central pier column, the outline of a prototype structure is as shown in Figure 3-53.

The construction sequence is outlined in Figure 3-53*a* and involves a two-span bridge with a total length of 230 ft. For a two-lane roadway, four girders are required, each containing a pair of precast pretensioned troughs placed in series in each span. These units are made continuous with an in situ central pier cap using longitudinal prestressing, and they develop composite action with the deck as in the foregoing examples. The deck is cast simultaneously with the pier cap that is transversely prestressed to allow the use of a single circular column as the central pier. The system has high torsional capacity and a relatively small construction depth (about 1/27 of the span).

Although the analysis of the system indicates a structural adequacy in terms of flexural and shear stresses, some concern remains about the integrity of the interface section at the junction with the pier cap as shown in Figure 3-54. This involves the behavior of the structure in the negative moment region (at the junction of the pier cap with the precast girder) and the fact that reliance is placed entirely on the longitudinal prestress to ensure integrity at the interface section, with the latter located in a zone where high bending moments and shears occur simultaneously.

These problems have been studied by Batchelor, Campbell, McEwey, and Csagoly (1976) under working and ultimate load conditions with emphasis on (a) the ultimate strength and mode of failure of the interface section under conditions producing maximum shear at this location, (b) the amount of slip, if any, that may be expected between the precast girder and the cast-in-place pier cap at working and ultimate loads, and (c) the ultimate load and mode of failure at the section directly over the pier.

Theoretical Background The details of the theoretical model are articulated by Batchelor, Campbell, McEwey, and Csagoly (1976), who also present the criteria correlating the model and the prototype. The parameters M and

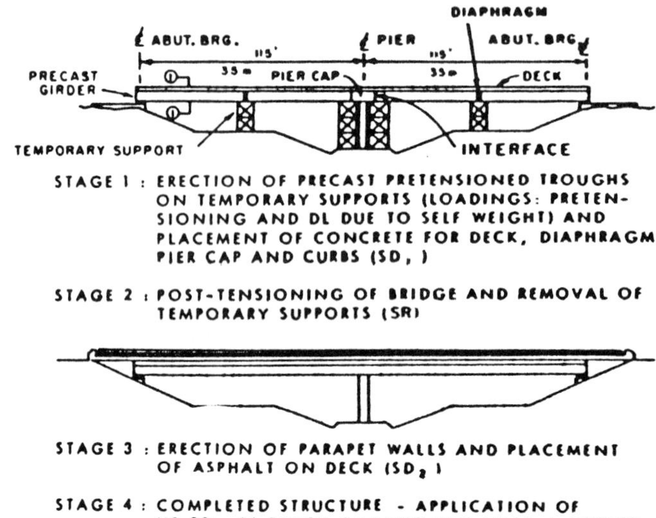

(*a*)

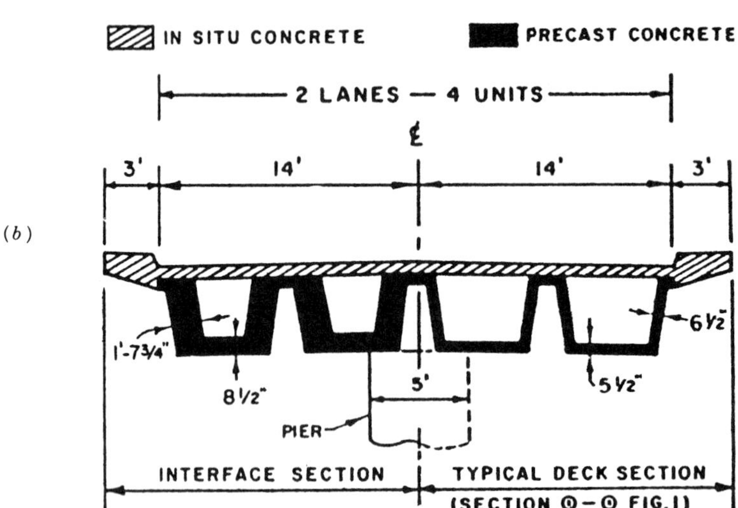

(*b*)

FIGURE 3-53 (*a*) Outline of prototype structure; (*b*) typical cross section. (From Batchelor, Campbell, McEwey, and Csagoly, 1976.)

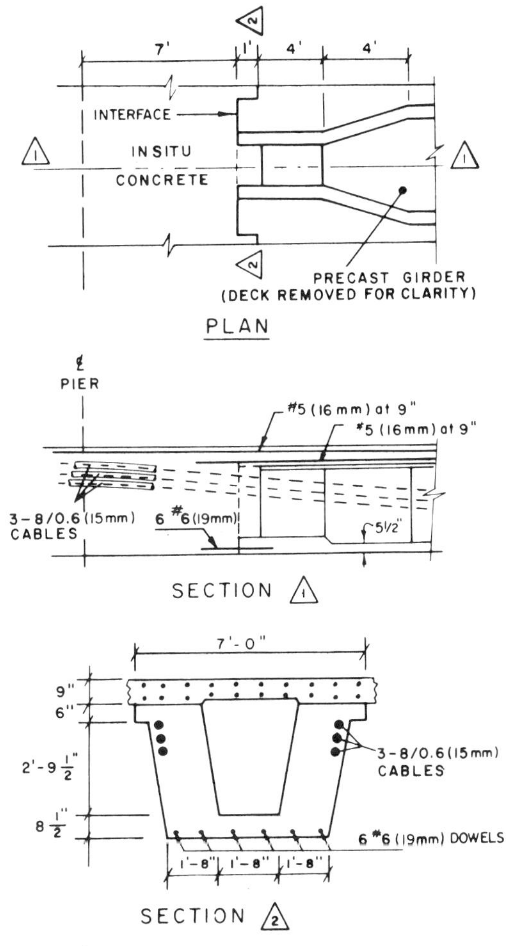

FIGURE 3-54 Details of prototype interface section between precast girders and in situ pier cap. (From Batchelor, Campbell, McEwey, and Csagoly, 1976.)

V (moment and shear, respectively) corresponding to ultimate load conditions were determined according to the 1969 AASHO factors, or $U = 1.5D + 2.5\,(L + I)$, where U denotes the factored effects for ultimate capacity. Secondary moments and shears due to prestress were included using a load factor of 1.

Analysis of Data Using a scale model, Batchelor, Campbell, McEwey, and Csagoly (1976) carried out tests, one in the working load range and four beyond this range. Results from the working load test show that the response of the model is essentially elastic. There was an apparent nonlinearity between stresses and strains resulting from a modeling criterion (M/Vd = constant). Measured concrete strains were small, and cracks were not observed. The conclusion is, therefore, that the capacity of the prototype interface section is adequate at working load, and cracking in the negative moment region is unlikely at this load level.

Results from ultimate load tests indicated that cracking first occurred at a model moment $M_{cr} = 38.3$ ft-kips. From theoretical considerations using simple beam theory and uncracked section properties, the cracking moment was calculated as 38.1 ft-kips. Using limited tensile strength, the cracking moment was calculated as 37.3 ft-kips for an assumed cracking strength of $7.5\sqrt{f'_c}$. Calculations based on $f'_c = 4000$ psi and a cracking strength of $7.5\sqrt{4000}$ gave a cracking moment of 38.8 ft-kips for exact scaled section properties. Using the scale analogy between the model and the prototype, Batchelor, Campbell, McEwey, and Csagoly (1976) concluded that the pier section of the prototype will not crack until it is subjected to a moment exceeding twice the working load moment.

The load–deflection relationship was linear up to first cracking, and the model completed elastic recovery even from loads exceeding those at first cracking. Shear strains measured at the interface did not disclose any slip at this location.

Estimation of the ultimate capacity of the interface sections under combined moment and shear was inferred at flexural–shear failure under ultimate model moments of 107 to 113 ft-kips. Theoretical ultimate moments were computed as 86.6 (AASHO, 1969) and 92.0 ft-kips based on the capability of deformations and the equilibrium of forces. The difference is explained as capacity enhancement due to the confinement of the section within the crushing zone.

3-16 PRINCIPLES OF STRENGTH DESIGN: CONVENTIONAL REINFORCED CONCRETE DECKS

AASHTO Philosophy

The load factor equations simply state that the design strength (nominal strength multiplied by a strength reduction factor) must exceed the required resistance (factored load effects). The proportioning of a member is controlled, however, by various stages of behavior: elastic, cracked, and ultimate. The latter may have to be altered when the limiting stages are the elastic and cracked conditions for working loads, but these proportions should not give a resistance less than the ultimate state. A member is thus satisfactory if its proportions satisfy the various limiting stages under the most credible loading conditions. In this context, member response is better weighted throughout the various stages of behavior, and the structural safety is more uniform.

At the ultimate stage, load effects are resisted by the tensile yield strength of steel and by the compressive strength of concrete. Where high-yield-strength steels are used and flexural behavior is involved, excessive crack widths and deflections may develop and thus require additional controls. Serviceability is therefore considered in strength design and includes limitations with reference to (a) structural fitness at overload conditions and (b) safety against excessive cracking, deflection, vibrations and permanent set, and fatigue of materials.

Ultimate Loads The load magnification coefficients β and overall load factors γ are applied to the working loads, and are taken from Table 2-5 (AASHTO Table 3.22.1A). These factors are derived taking into account the probability of a limiting stage such as accidental damage, possible increase in loads, construction defects, and possible stress redistribution. The load factors are also modified for different loading groups, but Group I normally controls superstructure design.

Typical values of the strength reduction factor ϕ are shown in Table 3-3. These values may be increased linearly from the values for compression members to the values for flexure as the design axial load decreases from its designated value to zero.

Fatigue and Crack Control In investigating stresses at service loads to satisfy fatigue and crack control requirements, the straight-line theory of stress and strain is used together with the assumptions stated in AASHTO Article 8.15.3.

The fatigue strength of reinforced concrete superstructures is discussed in Section 12-14. Failures of reinforced beams subjected to repeated loads may occur in shear or diagonal tension, but the same beams under static load may fail in steel tension.

With relatively high dead-to-live load ratios, the fluctuating stresses at working loads would normally be less than half the compressive and shear strengths of concrete, so that its fatigue strength will not be the governing factor. Where high live-to-dead load ratios are involved, and where repeated

TABLE 3-3 Typical Reduction Factors, ϕ, for Strength Design (AASHTO Specifications)

Load Effect	Value of ϕ
Flexure	0.90
Shear	0.85
Bearing on concrete	0.70
Axial compression with spirals	0.75
Axial compression with ties	0.70

applications are expected, consideration of the fatigue strength of concrete is indicated.

For crack control, AASHTO provides criteria for the distribution of flexural reinforcement if $f_y > 40{,}000$ psi. Crack control is best obtained if the reinforcement bars are well distributed over the effective concrete area, and if this area has the same centroid. For a given area of concrete around a bar, flexural crack width is independent of bar diameter, but crack spacing and width decrease as the thickness of concrete cover decreases. It follows therefore that cracking is inhibited if the volume of concrete around each bar is minimal and the steel tensile stress is low.

Deflections In reinforced concrete decks, deflections depend on the elastic and inelastic properties of the steel and concrete as well as on shrinkage and creep. For members subjected to bending, deflections may control the strength and serviceability criteria. The effects of shrinkage and creep should not be ignored, and thus the calculation of the final camber becomes more uncertain as the span–depth ratio increases. Accordingly, the advantages of strength design are enhanced as the span length increases and high-strength steel is used. With increasing spans, however, the dead load begins to dominate, and creep effects increase accordingly.

LRFD Specifications

The general principles of the LRFD philosophy are briefly discussed in Chapter 2 and include strut-and-tie models (Section 2-15) to proportion members near supports and concentrated loads where conventional methods are not adequate because of nonlinear strain distribution.

Fatigue need not be investigated for concrete slabs with primary reinforcement perpendicular to traffic or for slabs designed according to the stipulated empirical methods. In regions of compressive stress due to permanent loads plus prestress, fatigue must be considered only if this compressive stress is less than twice the maximum tensile load stress resulting from the fatigue load combination as specified in Table 2-12. If fatigue must be considered, the stress range must be determined using the fatigue load combinations specified in Table 2-12. The section properties should be based on either cracked or uncracked sections under dead load plus prestress plus 1.5 times the fatigue load, assuming an allowable tensile stress of $0.95\sqrt{f_c'}$.

Resistance factors For the strength limit states (strength and stability), the factored resistance is the product of the nominal resistance and a resistance factor. For flexure and bearing on concrete, this factor is the same as in Table 3-3. For shear, the factor is 0.90. A resistance factor of 0.80 is stipulated for compression in anchorage zones.

Design Considerations The specifications address the importance of equilibrium and strain compatibility in the analysis, and require the investigation of the effects of imposed deformations due to shrinkage, temperature change, creep, and support movement. Based on experience, the redistribution of load effects as a result of creep and shrinkage is not considered necessary for most common structure types. In addition, a structure as a whole and its components should be proportioned to resist sliding, overturning, uplift, and buckling.

Design for Flexural and Axial Forces The design assumptions are articulated for (a) service and fatigue limit states and (b) strength and extreme-event limit states. For usual designs, the rectangular stress distribution may be used, defined by an equivalent rectangular concrete compressive block of $0.85f'_c$ over a zone bounded by the edges of the cross section and a straight line located parallel to the neutral axis at a distance $a = \beta_1 c$ from the extreme compression fiber. The distance c is measured perpendicular to the neutral axis. The factor β_1 should be taken as 0.85 for concrete strengths less than 4 ksi. For concrete with $f'_c > 4$ ksi, β_1 is reduced at a rate of 0.05 for each 1.0 ksi of strength in excess of 4.0 ksi, but β_1 should not be less than 0.65. This distribution is essentially the same as in AASHTO Article 8.16.27.

The factor β_1 is related to rectangular sections. For flange sections with the neutral axis in the web, β_1 has been found experimentally to be a sufficient approximation. The factored resistance M_r is the product ϕM_n, where M_n is the nominal resistance and ϕ is the resistance factor.

Control of Cracking The provisions for crack control apply to all concrete components in which service loads cause tension in the gross section exceeding $0.22\sqrt{f'_c}$.

Satisfactory crack control is ensured when the steel reinforcement is well distributed over the zone of maximum concrete tension. Thus, several bars at moderate spacing are more effective in controlling cracking than one or two larger bars of the same area. Laboratory work with deformed reinforcing bars confirms that crack width at the service limit state is proportional to steel stress. However, two significant variables reflecting steel detailing are the thickness of concrete cover and the area of concrete in the zone of maximum tension surrounding each individual bar.

Moment Redistribution Where bonded reinforcement is provided at internal supports of continuous beams and the ratio c/d_c does not exceed 0.28, negative moments determined by elastic analysis at strength limit states may be increased or decreased but not more than 6.8 percent. Positive moments should be adjusted to satisfy statics. In this case c is the distance from the extreme compression fiber to the neutral axis, and d_c is the thickness of the concrete cover measured from the extreme tension fiber to the center of the bar.

Deflection and Camber Long-term deflections should be calculated taking into account creep and shrinkage factors that include the effect of aggregate characteristics, humidity, relative member thickness, maturity at time of loading, and length of time under load. Camber calculations should be based on the modulus of elasticity and the maturity of concrete when loads are added or removed.

Compression Members The analysis should consider the effects of axial loads, variable moment of inertia on member stiffness, fixed-end moments, deflections on moments and forces, and the duration of loads. Provisions should be made to transfer all force effects from compression components, adjusted for second-order moment magnification, to adjacent components. If the connection is by a concrete hinge, longitudinal reinforcement should be centralized within the hinge to minimize moment intensity, and should be developed on both sides of the hinge.

Tension Members For members in which the factored loads induce tensile stress throughout the cross section, the axial force shall be assumed to be resisted only by the reinforcement.

Shear and Torsion Regions of members that can be reasonably assumed to have plane sections can be designed for shear and torsion using two basic methods. With the sectional model, the member is checked by comparing the factored shear forces and the factored shear resistance at various sections along its length. This model is appropriate for typical bridge girders, slabs, and other components where conventional beam theory is valid. The theory assumes that the response at a particular section depends only on the calculated force effects and does not consider the mechanism of how these force effects are introduced. Although the strut-and-tie model can be applied to flexural regions, it is more appropriate near discontinuities where the actual flow of forces must be considered in more detail.

The general requirements for torsion are based on the factored torque resistance T_r, taken as $T_r = \phi T_n$, where T_n is the nominal torque resistance and ϕ is the resistance factor. Consideration of torsion is based on a quantitative criterion. If the factored torsional moment is less than 0.25 times the pure torsional cracking load, it is assumed to cause only a small reduction in shear capacity or flexural capacity and hence can be neglected.

3-17 PRINCIPLES OF STRENGTH DESIGN: PRESTRESSED CONCRETE DECKS

General Considerations

The analysis presented in the foregoing sections generally applies to predictions associated with elastic behavior. The moment producing initial cracks in the prestressed beam is thus computed from elastic theory. Cracking starts when the stress in the extreme fiber exceeds the modulus of rupture, taken as $7.5\sqrt{f_c'}$ according to the ACI code. Some investigators question the validity of this analysis although available data show that the elastic theory is sufficiently accurate up to the point of cracking. A further concern is whether the usual bending test for the modulus of rupture can give values of the tensile strength of concrete under prestress effects.

Assuming elastic theory, we rewrite (3-15) as

$$f_r = -\frac{F}{A} - \frac{Fey}{I} + \frac{My}{I} \tag{3-30}$$

which gives the value of f_r initiating cracking. Rearranging the terms, we obtain the cracking moment as

$$M_{\text{cr}} = Fe + \frac{FI}{Ay} + \frac{f_r I}{y} \tag{3-31}$$

where $f_r I/y$ = resisting moment due to the modulus of rupture of concrete
Fe = resisting moment due to the eccentricity of prestress F
FI/Ay = moment due to the direct compression of the prestress

Ultimate Moment The methods proposed for determining the maximum flexural resistance of prestressed sections are either purely empirical or highly theoretical. The empirical methods are basically simple, but limited to the conditions of the test. The theoretical approach is intended mainly for research and is unnecessarily complicated for design purposes. The compromise is a rational approach jointly presented by AASHTO and ACI, consistent with test results but neglecting refinement of an inconsequential nature, so that reasonably correct answers are obtained in a consistent manner.

The design flexural strength is specified for rectangular and flanged sections (AASHTO Article 9.17). The members may be assumed to act as uncracked sections under combined axial and bending effects within the specified service loads. The analysis is articulated for bonded and unbonded members.

Bonded Beams In this case failure is assumed to occur in flexure without shear, bond, or anchorage failure that may decrease strength. The method involves the simple principle of a resisting couple in a prestressed beam, as shown in Figure 3-55. At the ultimate load, the couple is made of two forces T' and C' acting with a lever arm a'. The steel supplies the tensile force T', and the concrete supplies the compressive force C'.

Failure may start either in the steel or in the concrete. A general case is the failure of an underreinforced section where failure begins with excessive steel elongation and ends with the crushing of concrete. Failure of an overreinforced section where the concrete is crushed before the steel is stressed in the plastic range is fairly uncommon. Another unusual mode of

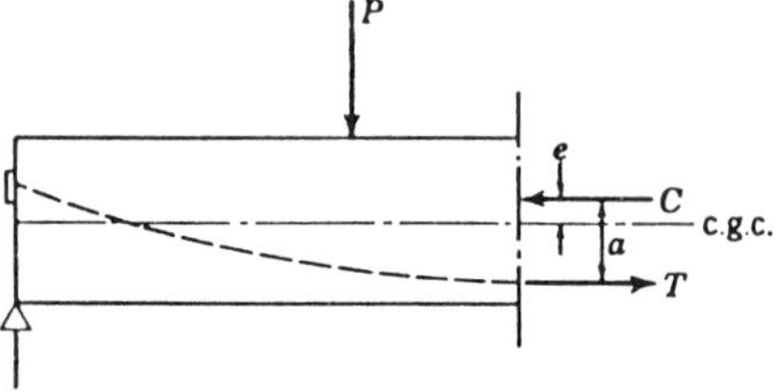

FIGURE 3-55 Internal resisting couple C'–T' with arm a'. (From Lin and Burns, 1981.)

failure is when a highly reinforced section fails by the breaking of the steel immediately following the cracking of concrete (Lin and Burns, 1981).

A sharp line of demarcation between an overreinforced and an underreinforced beam cannot be drawn because the transition from one type to another occurs gradually as the percentage of steel changes. A clear distinction between unbalanced and balanced conditions is academic for all practical purposes because most prestressing steels do not exhibit a definite yield point. Appropriate guidelines are thus introduced by limiting the reinforcement index p^* to ensure that the prestressed steel will be the first material to enter the yield range, and these lead to (3-20).

For underreinforced bonded beams, the steel is presumably stressed to a level that approaches its ultimate strength at the point of failure for the beam in flexure. At this ultimate stage it is sufficiently accurate to assume that the steel is stressed to a value of f_{su}^*. Provided the effective prestress is not less than $0.5f_s'$, the approximate value for f_{su}^* is given by (3-17).

The derivation of an expression for the ultimate resisting moment is based on the following considerations. First, referring to Figure 3-55, the ultimate compressive force equals the ultimate tensile force in the steel, or $C' = A_s f_{su}^*$. Next, it is necessary to locate the center of pressure C'. Among the many plastic theories assuming a stress block that has the shape of a rectangle, trapezoid, or parabola, we choose a simple rectangle. The depth of the compression zone, kd, is now computed from $C' = k_1 f_c' kbd$, where $k_1 f_c'$ is the average compressive stress in concrete at rupture. It follows that the lever arm is $a = d(1 - k/2)$. Although some discrepancies exist in the values of k_1 computed according to elastic theory and those resulting from cube strength tests, variations in these values do not appreciably affect the lever arm a. For design purposes a value of $k_1 = 0.85$ is assumed. In this case for a rectangular section, for the compression area and underreinforced conditions, the ultimate resisting capacity is expressed by (3-19).

Unbonded Beams Although a reliable computation method of the ultimate strength is not available, it is generally agreed that unbonded beams are weaker than bonded members by 10 to 30 percent. Among the reasons for the lower ultimate strength is the appearance of several large cracks in the concrete instead of many small well-distributed cracks.

A general formula for f^*_{su}, the stress in steel at ultimate load, is $f^*_{su} = f_{se} + \Delta f_s$, where f_{se} is the effective prestress in the steel and Δf_s is the additional stress in the steel produced as a result of bending up to the ultimate load. Results from tests (Lin and Burns, 1981) indicate a broad range of values for Δf_s ranging from 10,000 to 80,000 psi. A more conservative approach is followed by AASHTO where Δf_s is taken as 15,000 psi.

Composite Sections Most PPC beam-and-slab bridges are composite, as in the example of Section 3-11. For the usual construction and erection procedure, the stress distribution) for various stages of loading is shown in Figure 3-56 and is explained as follows.

1. (Figure 3-56*a*) Under the initial prestress and the weight of the beam, there is heavy compression in the bottom fibers and possibly some small tension in the top fibers. The resisting couple C–T is formed with a small lever arm.
2. (Figure 3-56*b*) After losses have occurred, the effective prestress and the accompanying stress redistribution lower the compression in the bottom and result in a smaller tension, and possibly some compression, at the top.
3. (Figure 3-56*c*) The addition of the slab causes the stress diagram shown.
4. (Figure 3-56*d*) The effects in Figures 3-56*b* and *c* are now added. The compression at the bottom becomes smaller, and some compression exists at the top. The lever arm for the C–T couple further increases.
5. (Figure 3-56*e*) As the live load is added the moment is resisted by the composite section, and the stresses are as shown.
6. (Figure 3-56*f*) The effects in Figures 3-56*d* and *e* are added to produce the stress diagram shown.

There may be slight tension or some compression in the bottom fibers, whereas the compressive stresses at the top are increased. The resisting

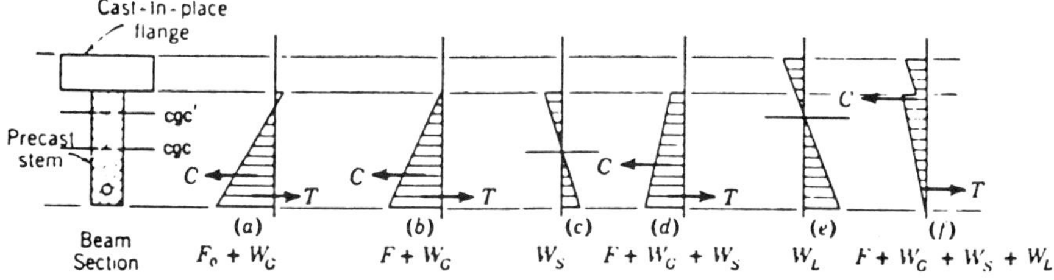

FIGURE 3-56 Stress distribution for a composite section. (From Lin and Burns, 1981.)

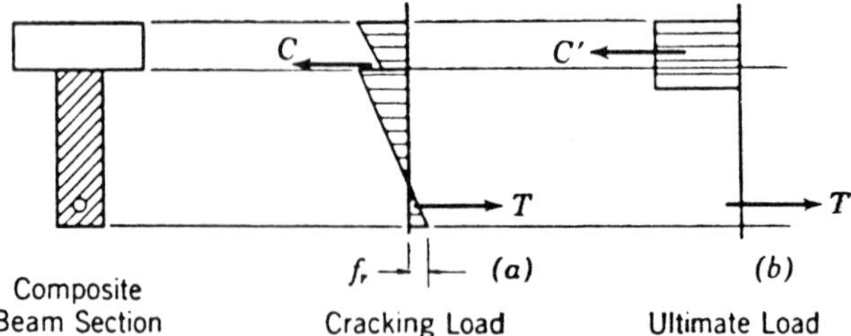

FIGURE 3-57 Stress distributions for cracking and ultimate loads. (From Lin and Burns, 1981.)

couple C–T acts with an appreciable lever arm. Usually, the force is applied in the flange.

The foregoing comments highlight the stress distribution under working load conditions. For overloads, the stresses are distributed as shown in Figure 3-57. When the first cracks occur, the lower fibers reach a tensile stress presumably equal to the modulus of rupture. This pattern is manifested when the live load stresses in Figure 3-56*e* plus an overload dead weight are high enough to cause the stresses shown in Figure 3-57*a*, computed by elastic theory.

Under ultimate conditions, however, the elastic theory is not entirely valid. As a first approximation, the ultimate resisting couple is represented by the system shown in Figure 3-57*b*. As long as failure in bond and shear is controlled and prevented, the ultimate strength of the composite section can be predicted as in a simple prestressed member.

LRFD Specifications

In the current (third) draft of the specifications, the provisions for the design of prestressed and partially prestressed concrete are combined because of the apparent similarity. On the other hand, the concept of partial prestressing results in a unified theory where conventional reinforced and prestressed concrete become special boundary cases. As a design concept, partial prestressing allows one or more (in combination) of the following design solutions: (a) a concrete member with a combination of prestressed and non-prestressed reinforcement designed to resist the same force effects simultaneously, (b) a prestressed concrete member designed to crack in tension under service load, and (c) a prestressed concrete member where the effective prestress in the prestressed reinforcement is kept lower than its maximum allowable value.

Prestressed and partially prestressed concrete members should be designed for both initial and final prestressing forces.

Loss of Prestress Prestress losses in members constructed and prestressed in a single stage include (a) instantaneous loss due to anchorage set, friction, and elastic shortening and (b) time-dependent loss due to creep, shrinkage, and relaxation. Prestress loss in multistage construction and posttensioning should be determined by considering the time elapsed between stages.

Losses for partially prestressed concrete can be estimated in a similar manner. The following comments are useful: (a) instantaneous prestress loss is the same as in a fully prestressed members; (b) the average concrete stress in a partially prestressed member is normally smaller than in the fully prestressed stage and therefore prestress loss due to creep is also smaller (if cracking exists under sustained load, the loss of prestress due to creep can be neglected); (c) if the prestressing steel is tensioned to the same initial level as in a fully prestressed member, the relaxation loss would be the same (however, since creep has a lesser effect, the relaxation loss would be slightly higher); (d) all other factors being equal, prestress loss due to shrinkage should be the same; (e) the presence of considerable non-prestressed reinforcement in partially prestressed concrete results in smaller prestress loss; and (f) because a partially prestressed concrete member can crack under sustained load, the prestress loss in the steel may be balanced by the increase in the steel stress at cracking (this increase is needed to maintain equilibrium and account for the loss of tensile capacity contribution by the concrete section).

Losses for Deflection Calculations For camber and deflection calculations of prestressed nonsegmental members with spans not exceeding 160 ft, made of normal-weight concrete with a strength in excess of 3.5 ksi at the time of prestress, f_{cgp} and f_{cdp} may be computed as the stresses at the center of gravity of the prestressing steel averaged along the length of the member. In this case f_{cgp} is the concrete stress due to the prestressing force at transfer and the self-weight of the member at sections of maximum moment, and f_{cdp} is the concrete stress due to all dead loads at the same section for which f_{cgp} is calculated.

Other Provision Other special concerns associated with partially prestressed members relate to losses under various prestressing ratios, crack control, and fatigue behavior of prestressing steel in a cracked section.

The provisions also address the stability and integrity of tendons against crushing or pulling out of curved girders, the design of posttensioned anchorage zones, and the durability of prestressed concrete members. Criteria for crack width control emphasize the maximum steel stress, and rules are provided to ensure the uniform distribution of flexural tension steel and nominal steel along the side faces between the tension face and the neutral axis.

Curved tendons are known to induce deviation forces that are radial to the tendon in the plane of its curvature. Curved tendons with multiple strands or wires also induce out-of-plane forces perpendicular to the plane of tendon curvature. Without adequate reinforcement, the tendon deviation forces may rip through the concrete cover on the inside of the tendon curve, or unbalanced compressive forces may push off the concrete on the outside of the curve.

New Provisions For Shear

The suggested shear design model takes into account residual tensile stresses in cracked concrete. The model is applied to both prestressed and non-prestressed members and quantifies the effects of longitudinal reinforcement, magnitude of moment, and axial force and member size (Collins, Vecchio, Adebar, and Mitchell 1991).

Tests by Vecchio et al. (1986) of reinforced concrete panels subjected to pure shear have shown that tensile stresses exist even after cracking, and can significantly enhance the ability of concrete to resist shear stresses. The cracked concrete transmits load in a complex manner. In a modified compression field model, this behavior can be represented without considering all the details. Thus, the crack pattern is idealized as a series of parallel cracks forming an angle θ with the longitudinal direction. Only the average stress state and the stress state at a crack are considered. As these two states are statically equivalent, the loss of tensile stress in the concrete at the crack is replaced by higher steel stresses and, after yielding of the reinforcement, by shear stresses on the crack interface. The shear stresses transmitted across the crack depend on the crack width.

The average principal tensile strain ϵ_1 in the cracked concrete is a damage indicator that controls the average tensile stress f_1 in the cracked concrete, the ability of the diagonally cracked concrete to resist compressive stresses f_2, and the shear stress v_{ci} that can be transmitted across a crack.

Approach to Shear Design The modified compression field model (Vecchio et al., 1986) considers the influence of residual tensile stresses, but its application requires certain simplifying assumptions. Referring to Figure 3-58, the shear stresses are assumed uniform over the effective shear area $b_w jd$. The largest longitudinal strain ϵ_x within the effective shear area is used to calculate the principal tensile strain ϵ_1. In principle, ϵ_x can be calculated from a plane section analysis that considers the influence of axial load, moment, and shear. For design purposes, it is approximated as

$$\epsilon_x = \frac{(M_u/jd) + 0.5N_u + 0.5V_u \cot\theta - A_{ps}f_{se}}{E_s A_s + E_p A_{ps}} \qquad (3\text{-}32)$$

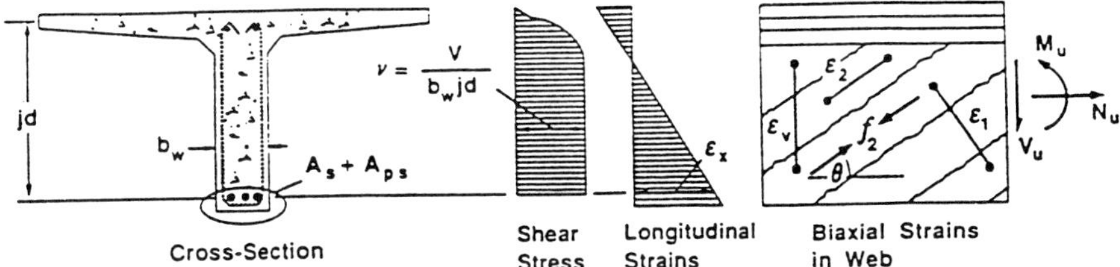

FIGURE 3-58 Beam subjected to shear, moment, and axial loads; modified compression field theory. (From AASHTO, 1992.)

where A_s and A_{ps} are the areas of non-prestressed and prestressed steel, respectively, on the flexural side of the member.

From strain compatibility, the principal tensile strain ϵ_1 is related to ϵ_x and the principal compressive strain ϵ_2 by

$$\epsilon_1 = \epsilon_x + (\epsilon_x - \epsilon_2)\cos^2\theta \tag{3-33}$$

As ϵ_x becomes larger and the inclination θ becomes smaller, the "damage indicator" ϵ_1 becomes larger.

The shear strength V_u can be expressed as

$$\begin{aligned} V_u &= V_c + V_s + V_p \\ &= \beta\sqrt{f_c'}\,b_w jd + \frac{A_u f_y}{s} jd \cot\theta + V_p \end{aligned} \tag{3-34}$$

where V_C = shear strength provided by residual tensile stresses in the cracked concrete

V_s = shear strength provided by tensile stresses in the stirrups

V_p = vertical component of force in the prestressing tendons

The values of θ and β obtained from the modified compression field model are given in Table 3-4 for members with web reinforcement and in Table 3-5 for members without web reinforcement. Analytically, the factor β is based on the expressions

$$\beta = \frac{0.18}{0.3 + \dfrac{24w}{a + 16}} \qquad \text{but} \qquad \beta \le \frac{0.33\cot\theta}{1 + \sqrt{500\epsilon_1}} \tag{3-35}$$

The expressions in (3-35) are based on shear stresses that can be transmitted across diagonal cracks. These are a function of crack width w and the maximum aggregate size a.

TABLE 3-4 Values of β and θ for Members with Web Reinforcement

		Longitudinal Strain $\epsilon_x \times 1000$				
v/f'_c		0	0.5	1.0	1.5	2.0
≤ 0.05	β	0.437	0.251	0.194	0.163	0.144
	θ	28°	34°	38°	41°	43°
0.10	β	0.226	0.193	0.174	0.144	0.116
	θ	22°	30°	36°	38°	38°
0.15	β	0.211	0.189	0.144	0.109	0.087
	θ	25°	32°	34°	34°	34°
0.20	β	0.180	0.174	0.127	0.090	0.093
	θ	27°	33°	34°	34°	37°
0.25	β	0.189	0.156	0.121	0.114	0.110
	θ	30°	34°	36°	39°	42°

The crack width is taken as $\epsilon_1 s_{m\theta}$, where $s_{m\theta}$ is the average spacing of diagonal cracks. The second expression in (3-35) is based on an average residual tensile stress in cracked concrete with a cracking stress of $0.33\sqrt{f'_c}$. In deriving the values shown in Tables 3-4 and 3-5, the crack spacing $s_{m\theta}$ is taken as 300 mm for members with web reinforcement. For members without reinforcement, the spacing of diagonal cracks is assumed as $s_{mx}/\sin\theta$, where s_{mx} is taken as shown in Figure 4-59.

Yielding of the longitudinal reinforcement is prevented if

$$A_s f_y + A_{ps} f_{ps} \geq \frac{M_u}{jd} + 0.5V_u + (V_u - 0.5V_s - V_p)\cot\theta \qquad (3\text{-}36)$$

TABLE 3-5 Values of β and θ for Members Without Web Reinforcement

x		Longitudinal Strain $\epsilon_x \times 1000$				
(mm)		0	0.5	1.0	1.5	2.0
125	β	0.406	0.263	0.214	0.183	0.161
	θ	27°	32°	34°	36°	38°
250	β	0.384	0.235	0.183	0.156	0.138
	θ	30°	37°	41°	43°	45°
500	β	0.359	0.201	0.153	0.127	0.108
	θ	34°	43°	48°	51°	54°
1000	β	0.335	0.163	0.118	0.095	0.080
	θ	37°	51°	56°	60°	63°
2000	β	0.306	0.126	0.084	0.064	0.052
	θ	41°	59°	66°	69°	72°

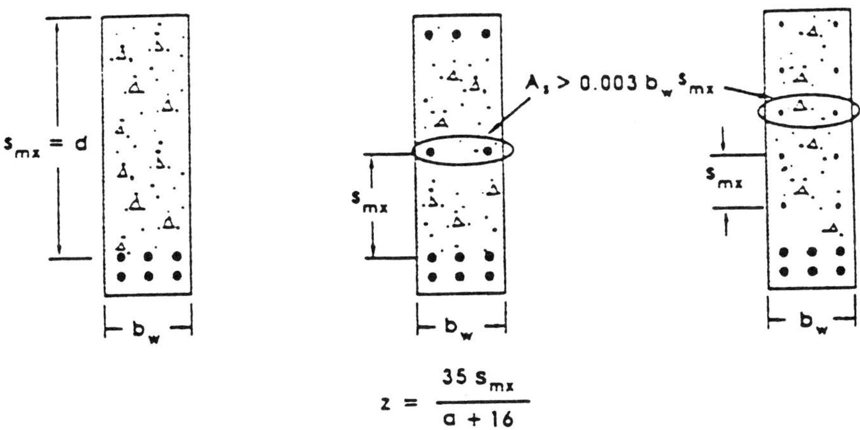

FIGURE 3-59 Crack spacing parameter z. Member without crack control reinforcement. (From AASHTO, 1992.)

Influence of Member Size Vecchio and Collins (1988) have shown that the modified compression field theory predicts shear capacity of members with web reinforcement fairly well (variation coefficients of about 10 percent). The effect of axial tension on the shear capacity of members without web reinforcement is also predicted accurately, with an 11 percent coefficient of variation (Bhide and Collins, 1989).

For members without (web) crack control reinforcement, as member size increases the crack spacing $s_{m\theta}$ also increases; hence, for a given value of strain ϵ_1 the crack width will increase. An increase in crack width produces an accompanying decrease in the shear stress that can be transmitted across the crack and thus reduces the shear capacity of the member. Referring to Table 3-5, members having large amounts of longitudinal reinforcement or prestressed concrete members (with low values of ϵ_x) are less sensitive to member size than members lightly reinforced or members subjected to high moments (high ϵ_x values). For example, if $\epsilon_x = 0$, the shear stress at failure increases by a factor of 1.33 as the size decreases by a factor of 11. Likewise, if $\epsilon_x = 0.002$, the shear stress increases by a factor of 3.10.

Design of Stirrups The number of stirrups required to resist a given shear V_u can be determined from the expression

$$\frac{A_u f_y}{s} jd \geq \left(V_u - \beta\sqrt{f'_c}\, b_w jd - V_p\right) \tan\theta \tag{3-37}$$

where the factors β and θ depend on the longitudinal strain parameter ϵ_x, reflecting the effects of the moment, axial load, prestressing, and longitudinal reinforcement ratios. For members without web reinforcement, β and θ also depend on member size. The sectional design model summarized in this section is indicated for these regions of a member where plane sections can be reasonably assumed to remain plane. If member regions have static or geometric discontinuities, strut-and-tie models are more appropriate (Collins and Mitchell, 1991).

3-18 DESIGN EXAMPLES: STRENGTH DESIGN METHOD (LOAD FACTOR)

Slab in a T-Beam Bridge

The slab of Figure 3-11 will be checked for ultimate strength. The procedure is based on the assumption that sufficient reinforcing steel is provided to satisfy (a) ultimate strength and (b) distribution of flexural reinforcement (crack control). Reference is made to AASHTO Eq. (8-15).

For design moment strength, we write

$$\phi M_n = \phi\left[A_s f_y d\left(1 - 0.6\frac{p f_y}{f'_c}\right)\right] \tag{3-38}$$

where M_n = nominal moment strength

$\phi = 0.9$

A_s = area of tension reinforcement

$p = A_s/bd$ = tension reinforcement index

f_y = specified yield strength of steel

f'_c = specified compressive strength of concrete

We also compute the factored moment as

$$M_u = 1.3(0.36 + 1.67 \times 3.90) = 8.93 \text{ ft-kips} \leq \phi M_n$$

Likewise, we compute

$$\phi M_n = 0.9\left[0.50 \times 60 \times \frac{4.94}{12}\left(1 - \frac{0.6 \times 0.0084 \times 60}{3.5}\right)\right] = 10.15 \text{ ft-kips}$$

or

$$\phi M_n > M_u$$

The criterion of crack control is satisfied if

$$f_s = \frac{z}{(d_c A)^{1/3}} \leq 0.6 f_y \tag{3-39}$$

where A = effective tension area (in.^{2}) of concrete surrounding the flexural tension reinforcement and having the same centroid as that reinforcement, divided by the number of bars

d_c = thickness of concrete cover measured from extreme tension fiber to center of the closest bar (in.)

z = 130 kips/in., assuming that this member is exposed to aggressive conditions and a corrosive environment

Continuous T-Girder Bridge

Figures 3-60*a* and *b* show elevation and cross section, respectively, for a four-span continuous T-girder bridge of constant depth. The slab thickness is 7.5 in., and the span ratio is 70/50 = 1.4. We will assume a stem thickness of 18 in. Strength and stresses are taken as in previous examples. The live load is HS 20.

Design of Slab The design span is the clear distance S = 6 ft 4 in = 6.33 ft. First, we compute the live load plus impact moment

$$M_{\text{LL}+T} = \left(\frac{6.33 + 2}{32}\right) \times 16 \times 1.3 \times 0.8 = 4.33 \text{ ft-kips}$$

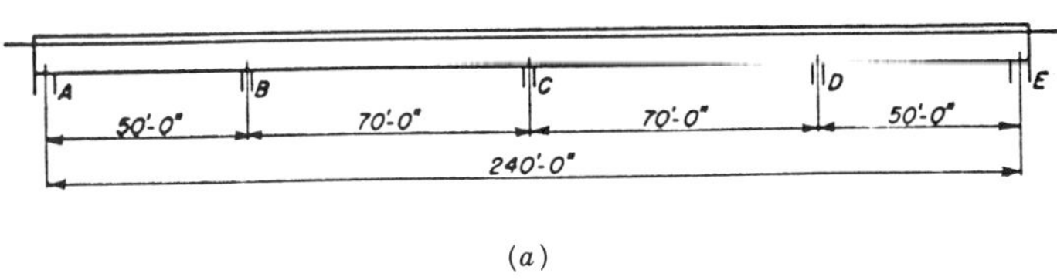

(*a*)

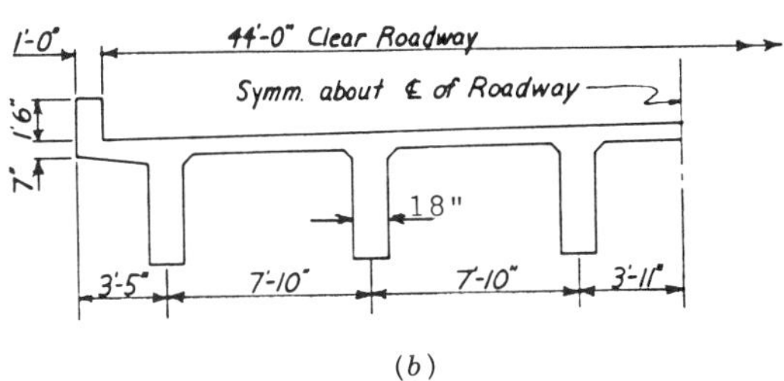

(*b*)

FIGURE 3-60 Four-span continuous T-girder bridge: (*a*) elevation showing dimensions; (*b*) typical cross section.

Recall from previous examples that dead load $w = 0.119$ kip/ft, and

$$M_{DL} = \frac{0.119 \times 6.33^2}{10} = 0.48 \text{ ft-kip}$$

The total moment is $4.33 + 0.48 = 4.81$ ft-kips.

For reinforcement in the top of the slab and #5 bars, $d = 4.94$ in., so that

$$A_s = \frac{4.81}{1.78 \times 4.94} = 0.55 \text{ in.}^2/\text{ft} \qquad \text{Use \#5 at 6.5 in.} \qquad A_s = 0.57 \text{ in.}^2/\text{ft}$$

For reinforcement in the bottom of the slab, $d = 6.19$ in., so that

$$A_s = \frac{4.81}{1.78 \times 6.19} = 0.44 \text{ in}^2/\text{ft} \qquad \text{Use \#5 at 8.5 in.} \qquad A_s = 0.44 \text{ in.}^2/\text{ft}$$

For this example, we will check the ultimate resistance in the bottom reinforcement. We compute

$$M_u = 1.3(0.48 + 1.67 \times 4.33) = 10.02 \text{ ft-kips}$$

Note that $p = 0.44/(12 \times 6.19) = 0.0059$. Next, we compute

$$\phi M_n = 0.9\left[0.44 + 60 \times \frac{6.19}{12}\left(1 - \frac{0.6 \times 0.0059 \times 60}{3.5}\right)\right] = 11.52 > 10.02$$

Interior Girders From Figure 3-11, recall that the minimum superstructure depth is 4.55 ft (use 4 ft 9 in.). Assume that the parapet is placed after the deck. The dead loads are as follows:

Weight of slab	$= 7.83 \times 0.625 \times 0.15$	$= 0.73$ kip/ft
Weight of girder	$= 4.125 \times 1.50 \times 0.15$	$= 0.93$ kip/ft
From parapet	$1.50 \times 1.00 \times 0.15/3$	$= 0.075$ kip/ft
FWS	$44 \times 25/6$	$= 0.183$ kip/ft
Total dead load w		$= 1.93$ kips/ft

A diaphragm will be placed at midspan but its weight may be disregarded.

The distribution of wheel load is $7.83/6 = 1.31$. The impact coefficient is 0.286 for end spans, 0.270 for supports B and D, 0.256 for support C, and 0.256 for spans BC and CD. For the span ratio of 1.4, moments and shears are extrapolated directly from AISC tables (Moments, Shears, and Reactions).

Design Moment		At Working Loads (ft-kip)		At Ultimate Load (ft-kip)
Span AB				
$M_{DL} = 0.063 \times 1.93 \times 50^2$	=	305	$\times$ 1.30 =	397
$M_{LL+I} = \frac{512}{2} \times 1.31 \times 1.286$	=	430	$\times$ 2.17 =	933
Total M	=	735		1330
At Support B				
$M_{DL} = -0.145 \times 1.93 \times 50^2$	=	−700	$\times$ 1.30 =	−910
$M_{LL+I} = \frac{474}{2} \times 1.31 \times 1.27$	=	−394	$\times$ 2.17 =	−855
Total M	=	−1094		−1765
Span BC				
$M_{DL} = 0.0865 \times 1.93 \times 50^2$	=	418	$\times$ 1.30 =	543
$M_{LL+I} = \frac{604}{2} \times 1.31 \times 1.256$	=	497	$\times$ 2.17 =	1078
Total M	=	915		1621
At Support C				
$M_{DL} = -0.173 \times 1.93 \times 50^2$	=	−837	$\times$ 1.30 =	−1088
$M_{LL+I} = \frac{536}{2} \times 1.31 \times 1.256$	=	−440	$\times$ 2.17 =	−955
Total M	=	−1277		−2043

Point C: Assume effective d = 53 in., then at ultimate conditions approximate

$$A_s = \frac{2043 \times 12}{60 \times 0.85 \times 53 \times 0.90} = 10.1 \text{ in.}^2$$

For working stress, and again using d = 53 in., the approximate required tensile reinforcement is

$$A_s = \frac{1217}{1.78 \times 53} = 12.9 \text{ in.}^2$$

The required depth for balanced design is

$$d = \sqrt{\frac{1277}{0.211 \times 1.5}} = 63 \text{ in.}$$

The section therefore needs compression reinforcement. We estimate the factor $F = 4.3$, and then we compute $M - KF = 1217 - 211 \times 4.3 = 310$. Likewise, we compute $d' = 2.7$ in., $d'/d = 0.05$, and $c = 1.59$. Then

$$A'_s = \frac{310}{1.59 \times 53} = 3.7 \text{ in.}^2$$

Now, select top steel 13 #9 bars, $A_s = 13$ in.^{2}, and also bottom steel 4 #9 bars, $A'_s = 4$ in.^{2}. Next, we compute $(A_s - A'_s)/bd = (13 - 4)/18 \times 53 = 0.0094$ and also

$$0.85\beta_1\left(\frac{f'_c d'}{f_y d}\right)\left(\frac{87{,}000}{87{,}000 - f_y}\right) = 0.85 \times 0.85 \times \frac{3500 \times 2.7}{60{,}000 \times 53} \times \frac{87{,}000}{27{,}000}$$
$$= 0.0068$$

Because $0.0094 > 0.0068$, the design moment strength (ultimate) ϕM_n is computed from

$$\phi M_n = \phi\left[(A_s - A'_s)f_y(d - a/2) + A'_s f_y(d - d')\right] \tag{3-40}$$

where

$$a = \frac{(A_s - A'_s)f_y}{0.85 f'_c b}$$

First, we compute $a = (13 - 4)60/0.85 \times 3.5 \times 18 = 10$ in. The design moment strength is now

$$\phi M_n = 0.90\left[(13 - 4)60\left(\frac{53 - 5}{12}\right) + 4 \times 60\left(\frac{53 - 2.7}{12}\right)\right] = 2849 \text{ ft-kips}$$

In addition, the design must check the maximum reinforcement. The ratio of reinforcement p provided should not exceed $0.75p_b$ which would produce balanced strain conditions for the section (the portion p_b balanced by compression reinforcement need not be reduced). Balanced strain conditions exist at a cross section when the tension reinforcement reaches the strain corresponding to its specified yield strength f_y just as the concrete in compression reaches its assumed ultimate strain 0.003.

For a rectangular section, the balanced reinforcement ratio p_b is given by

$$p_b = \left[\frac{0.85\beta_1 f'_c}{f_y}\left(\frac{87{,}000}{87{,}000 + f_y}\right)\right] + p'\left(\frac{f'_s}{f_y}\right) \tag{3-41}$$

where

$$f'_s = 87{,}000\left[1 - \left(\frac{d'}{d}\right)\left(\frac{87{,}000 + f_y}{87{,}000}\right)\right] \leq f_y$$

As a first approximation, we assume $f'_s \approx f_y$, then we compute $p' = 0.0042$. The balanced reinforcement ratio is

$$p_b = \left[\frac{0.85 \times 0.85 \times 3.50}{60}\frac{(87{,}000)}{(147{,}000)}\right] + 0.0042 = 0.028$$

or $0.75p_b = 0.028 \times 0.75 = 0.021$. The actual reinforcement ratio is

$$p = \frac{13}{53 \times 18} = 0.013 < 0.021$$

Span BC: First, we assume $j = 0.90$ and $d = 51$ in. The design moment is 915 ft-kips (working load)

$$\text{Approximate required} \qquad A_s = \frac{915}{1.80 \times 51} = 10.0 \text{ in.}^2$$

Try eight #10 bars, $A_s - 10.16$ in.^{2}, arranged as shown. The moments about the soffit are as follows:

$$2 \times 1.27 \times 9.70 = 24.64$$
$$3 \times 1.27 \times 6.20 = 23.62$$
$$3 \times 1.27 \times 2.70 = 10.29$$

$$10.16 \qquad\qquad 58.55 \qquad \bar{y} = 58.55/10.16 = 5.9 \text{ in.}$$

$$\text{Actual} \qquad d = 51 \text{ in.} \qquad \text{OK}$$

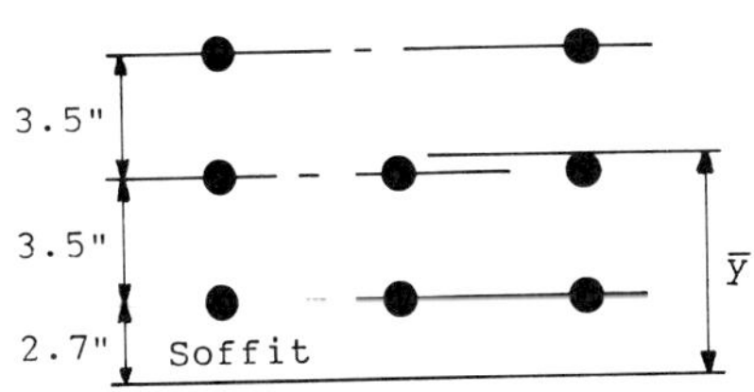

The next step is to compute the depth of the equivalent rectangular stress block a. If the compression flange thickness (in this case the slab thickness) is equal to or greater than the dimension a, the design moment strength ϕM_n may be computed from the same equations as for negative moment, that is, the section is rectangular. This moment is then compared to the factored moment of 1621 ft-kips.

If the compression flange thickness is less than a, the design moment strength may be computed from AASHTO equations (8-19), (8-20), and (8-21). For T-girder and box girder construction, the width of the compression face, b, should be taken as defined in AASHTO Article 8.10.

Load Capacity of Skew RC Box Girders: Case Study

Analytical and experimental results of the ultimate strength behavior of a skew RC box girder bridge have been obtained by Scordelis, Wasti, and Seible (1982). Based on postulated collapse mechanisms, failure loads were imposed by successively increasing the applied load beyond the yield strength.

Ultimate Strength The skew of the substructure elements modifies the distribution pattern of reactions at all locations. Girder moments in the span and interior support are reduced compared to straight bridges. Based on a linear elastic analysis, the distribution of the end support reactions under two sets of five 20-kip loads at locations X and Y shows a definite concentration toward the obtuse corner, as shown in Figure 3-61a. Also shown are the resultant force and moment reactions at the abutments, representing a single reaction with an eccentricity 3.45 ft from the center toward the obtuse corner. In effect, this shortens the span length and results in reduced girder moments.

Figure 3-61b shows the total longitudinal moments for span II, computed from the reactions shown in Figure 3-61a. For the same spans and cross section but with 0° skew, the maximum moments would be 562 ft-kips and −676 ft-kips at midspan and support, respectively. Thus, the introduction of skew supports reduces the moments by 6 to 7 percent.

The bending moment capacity (ultimate) is computed from the actual yield strength of the longitudinal reinforcement, as shown in Figure 3-61c. Also shown is the dead load moment diagram. At section Y, a moment capacity of 1010 ft-kips is attained if a live load of 181 kips is placed at midspan, producing the straight-line diagram shown. The maximum negative moment of −1314 ft-kips at section Z is less than −1372 (yield moment capacity). The calculated total moment at the center-bent region, based on the internal longitudinal forces in the girders, is −1112 ft-kips.

Yield Moments From the data of Figure 3-61, it appears that yield moments first develop at the loaded right midspan sections X and Y, but the bridge skew and bar cutoff locations inhibit the explicit manifestation of the

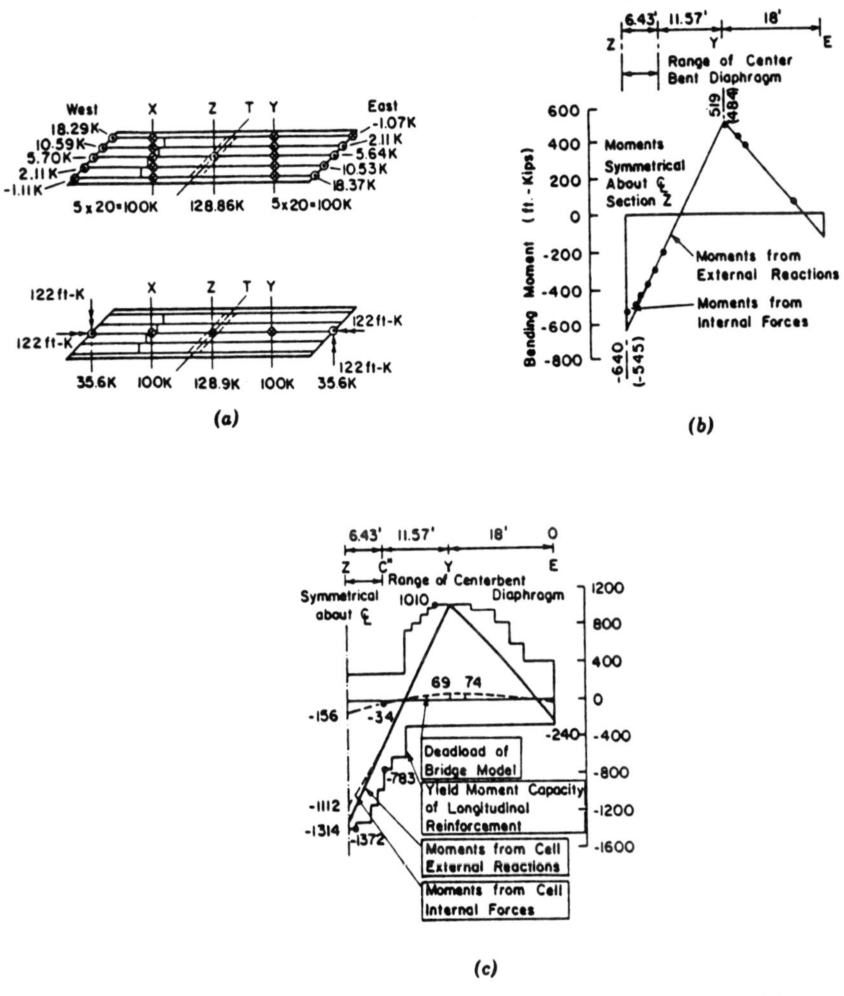

FIGURE 3-61 Skew bridge model reactions and moment capacities: (*a*) typical distribution of support reactions for normalized 100-kip per span conditioning loads; (*b*) total moments in undiaphragmed span due to normalilzed 100-kip per span conditioning loads; (*c*) resisting moment capacity and moments due to dead load and live loads of 181 kips at sections *X* and *Y*. (From Scordelis, Wasti, and Seible, 1982.)

critical regions, namely skew sections *B* and *C* and right sections *Z*, *B″*, and *C″*, shown in Figure 3-62. Sections *B* and *C* are about 2.6 ft on each side of the center support, and their relatively high moment capacity is caused by the considerable longitudinal reinforcement. Accordingly, they are not immediately critical. A yield moment failure along sections *B″* and *C″* is possible as is a failure at location *C* based on a no-twist deflection compatibility of the bridge segments in each span between yield hinges and end supports.

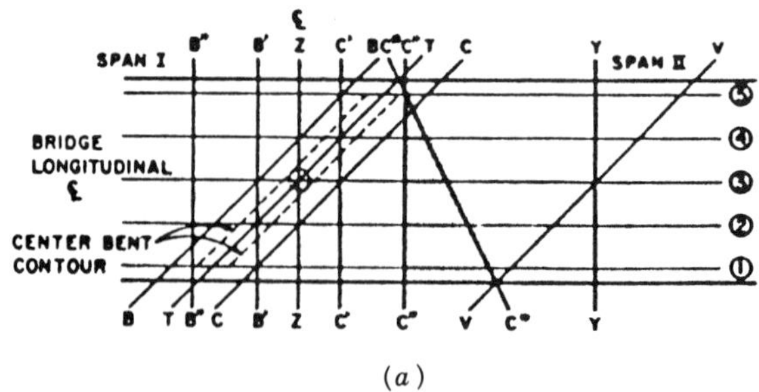

(a)

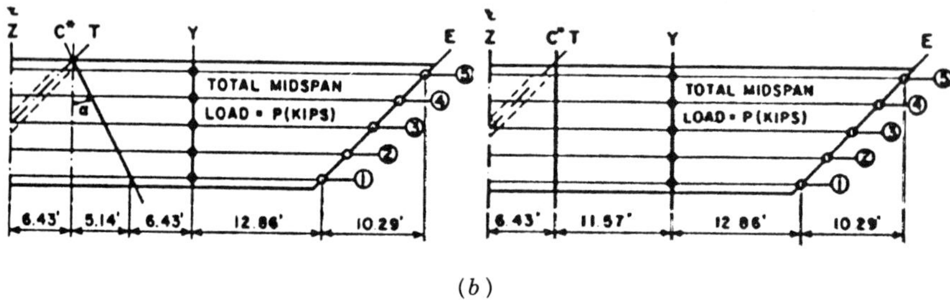

(b)

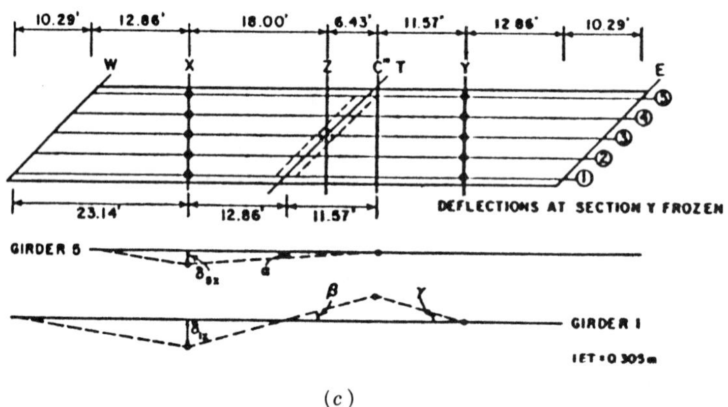

(c)

FIGURE 3-62 Yield hinges and postulated collapse mechanisms: (*a*) possible critical bridge sections for support hinges; (*b*) possible yield hinges for span II; (*c*) collapse mechanism for span I. (From Scordelis, Wasti, and Seible, 1982.)

The basic collapse mechanisms are shown in Figure 3-62*b*. Section C^* is located such that the distances 6.43 and 11.57 between that section and section Y give the same ratio as the distances 12.86 and 23.14 between sections E and Y. Following failure of the skew bridge model in span II (without diaphragm), the crack pattern of the top deck indicated that yield hinges may be located in the cracked triangular area of the top deck bounded by sections C'' and C^*.

Collapse Mechanism: Span II (No Diaphragm) Referring to Figures 3-61*c* and 3-62*b*, we can identify two possible collapse mechanisms. For the Y–C'' pattern, the live load moment capacities are 941 and 749 ft-kips for sections Y and C'', respectively. At collapse, and assuming the bridge segment rotates at an angle θ about the skew end support section E, the vertical deflections at locations $1Y$ and $5Y$ are calculated as

$$\delta_{1Y} = 12.86\theta/\sqrt{2} = 9.1\theta \qquad \delta_{5Y} = 23.14\theta/\sqrt{2} = 16.4\theta$$

The resulting rotations at support hinge C'', adjusted for out-of-plane twisting of the bridge segment between sections Y and C'', are computed as

$$\alpha_{1C} = 0.79\theta \qquad \alpha_{5C} = 1.41\theta$$

Based on the plastic moments 941 and -749 ft-kips, the collapse load per span is calculated as 198 kips.

A similar analysis of the second possible collapse mechanism with hinges formed at sections Y and C^* gives a collapse load of 226 kips, which compares well with the experimental loads of 207 and 206 kips. It appears therefore that the actual collapse mechanism may lie between the two assumed critical conditions.

Collapse Mechanism: Span I (With Diaphragm) This span withstood loads of 240 and 243 kips. During this application the deformations of span II (already collapsed) were maintained in order to adjust the collapse observations about span I. At collapse of span I, a mechanism with hinges at sections X, C'', and Y was noted. Based on this failure pattern, the collapse load was calculated as 245 kips, which is in good agreement with the experimental results.

Shear and Torsion Scordelis, Wasti, and Seible (1982) have also analyzed the collapse loads causing shear stresses to exceed the shear flow capacity. The shear flow for girder 1 due to the combined dead load and midspan load of 100 kips at sections X and Y is calculated and compared with the shear capacity of the girder (stirrups and concrete). The live load capacity derived from combined shear and torsion effects is about 182 kips compared with the experimental collapse load of 207 kips, and this agreement converges further if the shear capacity of the stirrups is adjusted. Yielding of stirrups in exterior girders 1 and 5 does not necessarily cause collapse because redistribution of forces may take place, transferring more load to the interior girders (William and Scordelis, 1970).

3-19 HIGH-STRENGTH CONCRETE IN BRIDGES

The use of high-strength concrete has been considered both in basic research and user-oriented research, where the introduction of new materials is normally associated with new design concepts. Alternatively, the implementation of this option may demonstrate the notion of high-performance materials. Thus, in addition to the high compressive strength, other prominent characteristics may also be considered, such as increased durability, frost–thawing resistance, abrasion resistance, and imperviousness to water and gas.

High-Strength Concrete (HSC) Experimental Bridge

Malier and Pliskin (1990) summarize the requirements of a HSC bridge.

1. It should represent the standard bridges without introducing unusual or exceptional characteristics.
2. It should accommodate construction in urban and industrial sites.
3. It should develop the stipulated strength (60 MPa or 8600 psi) using locally available materials.
4. It should satisfy the design requirements of prestressed concrete.
5. It should provide compatibility with field instrumentation to articulate its performance and confirm the conceptual approach.

Preliminary studies show that the concrete quantity may be reduced by 25 to 30 percent through the use of high-strength concrete in the superstructure, leading to a 20 percent volume reduction in the substructure.

Test Bridge An experimental bridge built across the Yonne River near Joigny, France, developed a minimum strength that exceeded the characteristic strength adjusted for a standard deviation (in this case 3 MPa or 0.43 ksi). The average tensile strength reached 5 MPa (0.7 ksi) on 28-day samples.

The concrete for this bridge was placed in one continuous phase with fresh mix supplied from two concrete plants. In a continuous pouring operation, 1300 yd^3 of concrete were placed in a 24-hr period. The structure was prestressed longitudinally with 13 external tendons allowing a relatively simple and accurate measurement on the time evolution of the tensile forces and deformations. During the time-setting phase, the thermal behavior of the concrete was monitored and checked with a special finite-element program. Creep and shrinkage studies have been planned for the early period of the structure. The intent of this experimental bridge is to check the long-term performance of structures built with a 60-MPa characteristic strength concrete, and assess its durability compared with ordinary concrete bridges.

Thus far, economic comparisons appear to indicate that, considering the higher unit price but the reduced concrete volume, high-strength concrete results in less initial cost. These benefits may be enhanced by the improved durability of the finished structure.

High-Strength Concrete in Prestressed Concrete Box Beams

The use of high-strength concrete in box beams in conjunction with geometry modifications has been studied by Schemmel and Zia (1990). A range of applications was established in terms of span capacity with emphasis on long-span simply supported beams. The results indicate that the maximum attainable span of box beams can be increased further through the use of high-strength concrete combined with a modified internal void shape.

In a parametric study, the flexural analysis outline followed the standard AASHTO specifications, noting however, some basic differences between normal and high-strength concrete related to the modulus of elasticity, the modulus of rupture, and in some cases to the creep of concrete. The equations expressing the modulus of elasticity E_c and the modulus of rupture f_r were as recommended by ACI Committee 363 (American Concrete Institute, 1984), or

$$E_c = 40{,}000(f_c')^{1/2} + 1{,}000{,}000 \text{ psi} \tag{3-42}$$

$$f_r = 11.7(f_c')^{1/2} \text{ psi} \tag{3-43}$$

The study concluded that some uncertainties remain regarding the creep of high-strength concrete and its effect on prestress loss. Thus, it was assumed that the creep strain would remain nearly the same as in normal-strength concrete, and that prestress loss would be expressed as in current AASHTO equations. High-strength concrete was considered in the range of 6000 to 12,000 psi.

Modified Box Beams Typical standard AASHTO box beams are shown in Figure 3-63*a*. The variation of width and depth is as shown, and the shape, size, and location of the internal voids are such that the wall thickness remains unchanged for all members. Modified versions are shown in Figures 3-63*b* and *c*. Type A has a bottom flange larger than the standard beam and also a thinner web. Its cross section permits two rows of prestressing strands across the bottom. Type B has a wider web, allowing two columns of strands and a thicker bottom flange. Both 0.5-in.- and 0.6-in.-diameter seven-wire prestressing strands were used in the beams. Only low-relaxation grade 270 steel was considered.

The results show that in most cases the maximum capacity of a beam increases with increasing concrete strength. Using a 6000 psi concrete as a reference strength, the increase in maximum span ranges from 0 to 25 percent. The data also show that when 0.6-in.-diameter strands are used, there are no advantages in using compressive strength greater than 8000 psi. However, with 0.5-in. strands, the span length increase continued with a concrete strength increase up to 10,000 psi.

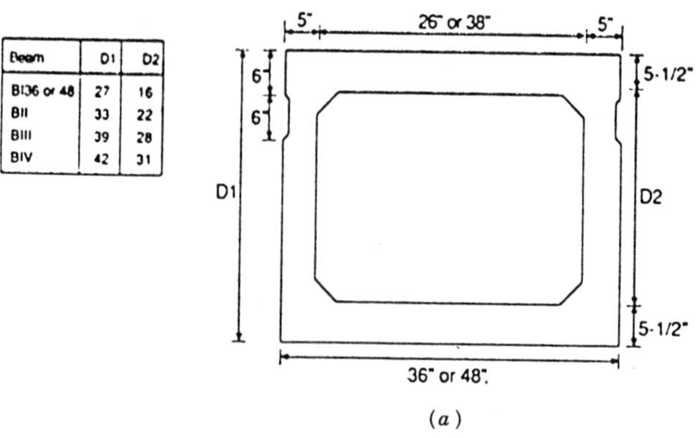

Beam	D1	D2
BI36 or 48	27	16
BII	33	22
BIII	39	28
BIV	42	31

(*a*)

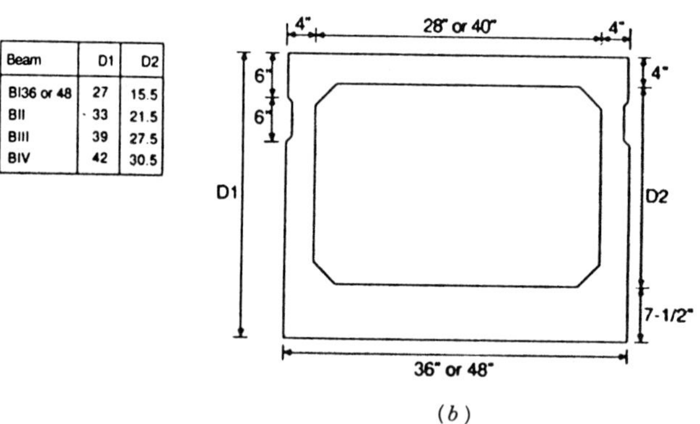

Beam	D1	D2
BI36 or 48	27	15.5
BII	33	21.5
BIII	39	27.5
BIV	42	30.5

(*b*)

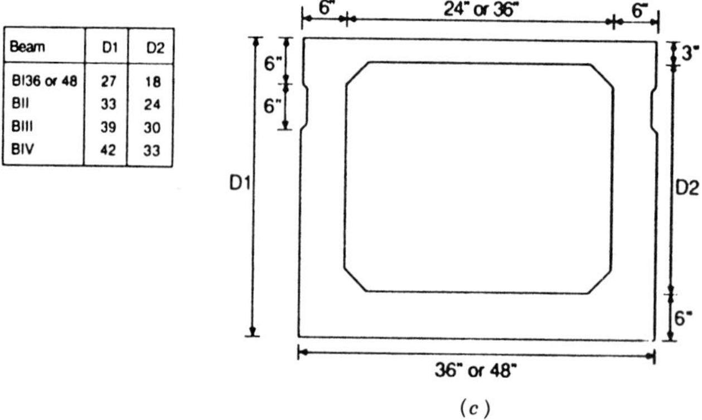

Beam	D1	D2
BI36 or 48	27	18
BII	33	24
BIII	39	30
BIV	42	33

(*c*)

FIGURE 3-63 (*a*) Standard AASHTO bridge box beams; (*b*) modified type A bridge box beams; (*c*) modified type B bridge box beams.

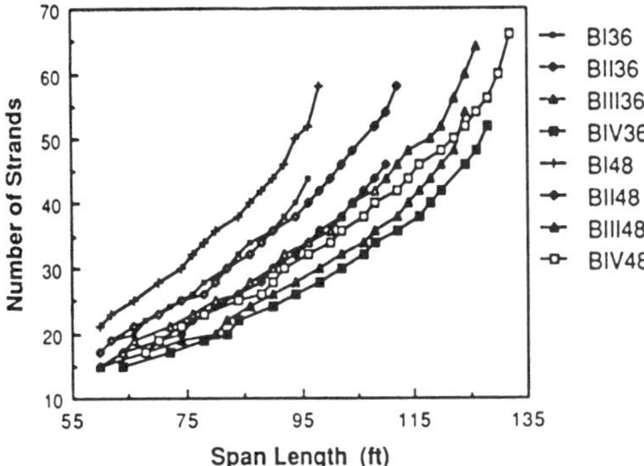

FIGURE 3-64 Span capacity curves for all standard box beams (f'_c is 12,000 psi; strand diameter is 0.5 in.).

The main advantage of using 0.6-in. strands is that fewer strands are required for the same span, although maximum span lengths tend to be shorter with 0.6-in. strands.

Span versus strand curves are plotted for all standard box beams for deck configurations with adjacent beam spacing, and 0.5-in. strands with 12,000 psi concrete, in Figure 3-64. The plots show that the 4-ft sections require more strands per beam compared to the 3-ft sections for the same span. However, the total number of strands required for a bridge are less with the 4-ft sections because fewer members are needed with adjacent box beams. Also, the total volume of concrete is less with the 4-ft beams.

Based on a detailed analysis and economic study considering the modified box beams shown in Figure 3-63, Schemmel and Zia provide the following conclusions and recommendations.

1. For spans up to 95 ft, 3-ft beams at wide spacing are more economical, but for longer spans the 4-ft beams should be selected.
2. The modified type A box beams shown in Figure 3-63*b* may be more suitable than the current standard because they provide a fully optimized structural configuration.
3. Further modification of the standard box beam is warranted. For example, the 27-in.-deep beam is not as cost-effective as the others. The 42-in.-deep beam is one of the most economical for all spans, and a deeper section may even enhance these benefits further.
4. Although studies appear to indicate less shrinkage and creep with high-strength concrete, the associated effects on prestress loss are yet to be fully established and warrant additional research.
5. Live load deflections should be further analyzed as a serviceability requirement but also in the context of structural relevance. It may

appear that the advantages of high-strength concrete might be greatly diminished if the deflection criteria are similar to steel bridges.

6. More demonstration projects are needed to validate the structural concept of high-strength concrete bridges and develop criteria and specifications for either standard or modified sections.

3-20 TRENDS IN PRESTRESSED CONCRETE

Bridges using prestressed concrete were introduced in the United States in the late 1940s, but since then several types of prestressed concrete bridges have become standard construction. Evolving from small simple-span structures, the concept has been applied to major projects utilizing continuity, drop-in spans, pretensioning and posttensioning, long girders, precast beams, cast-in-place girders, and bridge segments.

The standard I beam shown in Figure 3-38 will continue to dominate the prestressed concrete alternatives, particularly where the size of the project justifies a casting yard near the site. Box beams have been used widely in Pennsylvania and throughout the Midwest. Very often they constitute simple spans with the boxes adjacent to one another, joined by grouted shear keys and tie rods and provided with a wearing surface. Usual configurations are shown in Figures 3-42*a*, *b*, and *c*. This bridge system allows rapid construction and has been a forerunner in bridge replacement programs with a minimum traffic disruption. Many states have used spread box beams, by placing them about 4 ft apart and providing a composite deck. Spread box construction has the advantage of less construction depth than I-beam types. Both types of composite box beam construction can accommodate negative moment steel in the deck to achieve continuity. If used as a composite type, the box beam shown in Figure 3-42*c* has a thinner top slab because of the cast-in-place deck incorporated in the design.

Other integral deck-type beams include special double tees, single tees, bulb tees, and quad tees in addition to the box beam or precast solid slab. These types also lend themselves to quick bridge replacement.

Prestressed I-beam bridges can be constructed in spans of 40 to 120 ft, simply supported or with continuity as shown in the foregoing sections. A special design is shown in Figure 3-65, and consists of a center cantilever notched for drop-in girders. This arrangement can accommodate spans exceeding 160 ft without sacrificing the economy of I-beam construction.

The case-in-place posttensioned box girder built on scaffolding has gained wide use in several states, most notably California. Other combinations of precast units with cast-in-place concrete and posttensioning are feasible. Predominant types are segmental I beams and segmental box girders for long spans discussed in Chapter 8.

The viability of prestressed concrete and its choice in bridges should be the result of economy, durability, fast construction and low maintenance, availability, and control of stresses. In structural terms, the last factor is

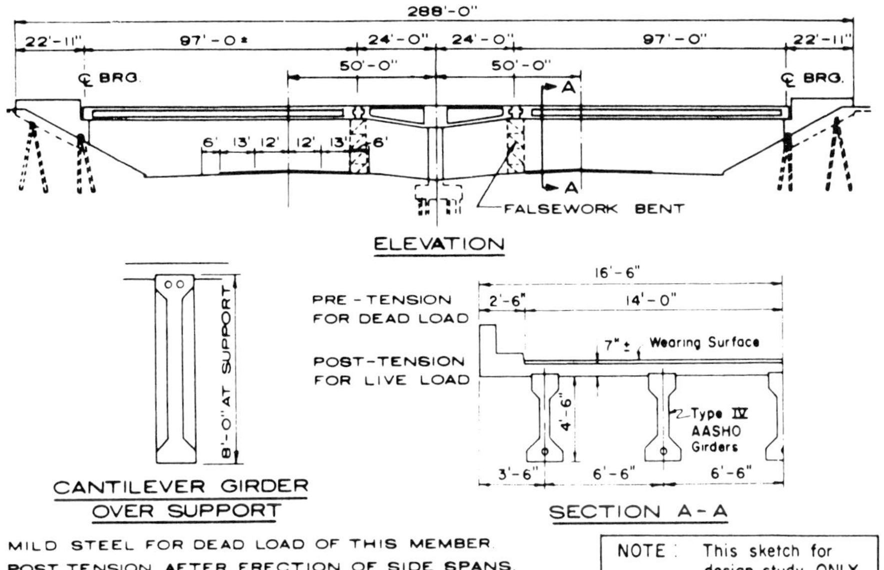

FIGURE 3-65 Center cantilevers notched for drop-in girders of design study example.

probably the most important. The ability to resist stresses both with material and with counterstresses articulates the availability of solutions and may satisfy optimization criteria. On the other hand, plant capacity and availability is essential for the economic use of prestressed members.

The efficient coordination between design and industry is also exemplified in structures with standard details. Examples are (a) elastomeric bearing pads with no embedded steel; (b) elimination of end blocks for posttensioned beams; (c) elimination of protrusions, with all geometry changes accomplished by insetting; (d) stirrups designed to be fabricated in cages that can be placed after the strands are stressed; (e) reasonable tolerance on hold-up and hold-down points to allow economical production; and (f) specifications based on performance.

REFERENCES

AASHO, 1969. "Standard Specifications for Highway Bridges," American Assoc. of State Highway Transp. Officials, Washington, D. C.

AASHTO, 1992. "Standard Specifications for Highway Bridges" and Interim 1990, 1991, Amerian Assoc. of State Highway and Transp. Officials, Washington, D.C.

ACI-ASCE, 1973. "Shear Strength of Reinforced Concrete Members," *J. Struct. Div. ASCE*, Vol. 99, No. ST6, June.

ACI-ASCE, 1974. "The Shear Strength of Reinforced Concrete Members—Slabs", Joint ASCE–ACI Task Committee 426, *J. Struct. Div. ASCE*, Vol. 100, No. ST8, Aug., pp. 1543–1591.

AISC, 1966. *Moments, Shears and Reactions for Continuous Highway Bridges*, American Inst. Steel Construct., N.Y.

Aldridge, W. W. and J. E. Breen, 1970. "Useful Techniques in Direct Modelling in Reinforced Concrete Structures," *Models for Concrete Structures*, ACI Publication No. 24, Paper No. SP-24-5, ACI.

American Concrete Institute. ACI Building Code Requirements for Reinforced Concrete (ACI 318-1983) and Commentary, ACI.

American Concrete Institute, 1984. "State of the Art Report on High-Strength Concrete," *J. American Concrete Inst.*, Committee 363, Vol. 81, No. 4, July–Aug., pp. 364–411.

Aster, H., 1975. "The Analysis of Rectangular Hollow RC Slabs Supported on Four Sides," Ph.D. Thesis, Technological Univ. Stuttgart, Germany.

Aswad, A., 1985. "Rational Deformation Prediction of Prestressed Members," *Deflections of Concrete Structure*, ACI Publ. SP-86, Detroit, pp. 263–274.

Bakht, B., 1979. "Use of Simplified Methods of Analysis Specified in the OHBD Code," Struct. Research Report 79-SPR-8, Research and Development Div., Ontario Ministry of Transp. and Communications, Ontario, Canada.

Bakht, B., Cheung, M. S. and T. S. Asis, 1979: "Application of a Simplified Method of Calculating Longitudinal Moments to the Proposed Ontario Highway Bridge Design Code," *Canadian Journ. of Civil Engineering*, 6(1), pp. 36–50.

Bakht, B., L. G. Jaeger, and M. S. Cheung, 1981. "Cellular and Voided Slab Bridges," *J. Struct. Div. ASCE*, Vol. 107, No. ST9, Sept., pp. 1797–1813.

Bask, A. K. and J. M. Dawson, 1970. "Orthotropic Sandwich Plates," Supplement, *Proc. Institution Cov. Engineers*, London, pp. 87–115.

Batchelor, B. de V., and P. Y. Tong, 1970. "An Investigation of the Ultimate Shear Strength of Two-way Continuous Bridge Slabs Subjected to Concentrated Loads," Report No. RR 167, Ministry of Transp. and Communications, Downsview, Ontario, Canada, Sept.

Batchelor, B. de V., et al., 1974. "An Investigation of a New Type of Prestressed Concrete Trepezoidal Box Girder Bridge," Report on OJT and CRP Project Q56 to the Ministry of Transp. and Communications, Queen's Univ. Ontario, Canada.

Batchelor, B. de V., T. I. Campbell, D. W. McEwey, and P. Csagoly, 1976. "Model Study of New Standard Prestressed Concrete Bridge," *J. Struct. Div. ASCE*, Vol. 102, No. ST9, Sept., pp. 1789–1806.

Batchelor, B. de V. and I. R. Tissington, 1972. "A Study of Partially Restrained Two-Way Bridge Slabs Subjected to Concentrated Loads," Report No. RR 178, Ministry of Transp. and Communications, Ontario, Canada, March.

Batchelor, B. de V. and I. R. Tissington, 1976. "Shear Strength of Two-Way Bridge Slabs," *J. Struct. Div. ASCE*, Vol. 102, No. ST12, Dec., pp. 2315–2331.

Bazant, Z. P., 1972. "Prediction of Creep Effects Using Age-Adjusted Effective Modulus Method," *American Concrete Inst. J.*, Vol. 69, No. 4, April, pp. 212–217.

Bazant, Z. P. and L. Panula, 1980. "Creep and Shrinkage Characterization for Analyzing Prestressed Concrete Structures," *J. Prestressed Concrete Inst.*, May–June, pp. 86–122.

Bazant, Z. P. and F. H. Wittmann, 1982. *Creep and Shrinkage in Concrete Structures*, Wiley, New York.

Beeby, A. W., 1983. "Cracking, Cover and Corrosion of Reinforcement," *Concrete International, Design and Construction*, Vol. 5, No. 2, Feb., pp. 34–40.

Best, B. C., 1964. "Certain Aspects of Model Testing for the Cumberland Basin Scheme," *Proc. Cement and Concrete Associate Meeting on Model Testing*, London, March 17.

Best, B. C. and R. West, 1965. "Tests on a Model of a Continuous Skewed Slab Supported on Columns," *Building Science*, Vol. 1, No. 1.

Bhide, S. B. and M. P. Collins, 1989. "Influence of Axial Tension on the Shear Capacity of Reinforced Concrete Members," *ACI Struct. J.*, Vol. 86, No. 5, Sept.–Oct.

Birkeland, N. SW., 1974."How to Design Prestressed Concrete Beams of Minimum Cross Section," *J. American Concrete Inst.*, Vol. 71, No. 12, Dec., pp. 634–641.

Bouwkamp, J. G., A. C. Scordelis, and S. T. Wastl, 1971. "Structural Behavior of a Two Span Reinforced Concrete Box Girder Bridge Model," Struct. Eng. and Struct. Mechanics Report No. 71-5, Vol. 1, Univ. California, Berkeley, April.

Bouwkamp, J. G., A. C. Scordelis, and S. T. Wastl, 1974. "Ultimate Strength of a Concrete Box Girder Bridge," *J. Struct. Div. ASCE*, Vol. 100, No. ST1, Jan. pp. 31–49.

Burdet, O. L., 1990. "Analysis and Design of Anchorage Zones is Posttensioned Concrete Bridges," Ph.D. Thesis, Univ. Texas, Austin, May.

Bureau of Public Roads, 1954. *Criteria for Prestressed Concrete Bridges*, U.S. Dept. Commerce, Washington, D.C.

Cheung, Y. K., I. P. King, and O. C. Zienkieicz, 1968. "Slab Bridges with Arbitrary Shape and Support Conditions," *Proc. Institution Civ. Engineers.*, Vol. 40, No. 1, London, May, pp. 9–36.

Christiansen, K. P., 1963. "The Effect of Membrane Stresses in the Ultimate Strengthof the Interior Panel of a Reinforced Concrete Slab," *Struct. Engineer*, Vol. 41, No. 8, London, Aug., pp. 261–265.

Crisfield, M. A., and R. P. Twemlow, 1971. "The Equivalent Plate Approach for the Analysis of Cellular Structures," *Civ. Eng. and Public Works Rev.*, March.

Collins, M. P. and D. Mitchell, 1986. "Rational Approach to Shear Design—The 1984 Canadian Code Provisions," *ACI J.*, Vol. 83, No. 6, Nov.–Dec.

Collins, M. P. and D. Mitchell, 1991. *Prestressed Concrete Structures*, Prentice-Hall, Englewood Cliffs, N.J.

Collins, M. P., F. J. Vecchio, P. Adebar, and D. Mitchell, 1991. "A Consistent Shear Design Model," *Proc. Colloquium Struc. Concrete*, Stuttgart, Germany, April 10–12.

Comartin, C. D. and A. C. Scordelis, 1972. "Analysis and Design of Skew Box Girder Bridges," Struct. Eng. and Struct. Mechanics Report No. UC SESM 72-14, Univ. Calif., Berkeley, Dec.

Cusens, A. R., and R. P. Pama, 1969. "Distribution of Concentrated Loads on Orthotropic Bridge Decks," *Struct. Engineer*, Vol. 47, No. 9, Sept.

Cusens, A. R. and R. P. Pama, 1975. *Bridge Deck Analysis*, Wiley, New York.

Dilger, W. H., 1982. "Creep Analysis Using Creep-Transformed Section Properties," *J. Prestressed Concrete Inst.*, Vol. 27, No. 1, pp. 98–117.

Elfgren, L., 1972. "Reinforced Concrete Beams Loaded in Combined Torsion, Bending a Shear-A Study of Ultimate Load-Carrying Capacity," Div. of Concrete Structures, Publ. 71 : 3, Chalmers Univ. of Technology, Göteborg.

El-Sebakhy, I. S. and J. B. Kennedy, 1986. "Progressive Failure Analysis of Orthotropic Bridges," *J. Structures and Computers*, Vol. 24, No. 3, pp. 401–411.

Elstner, R. C., and E. Hognestad, 1956. "Shearing Strength of Reinforced Concrete Slabs," *American Concrete Inst. J.*, Vol. 53, No. 1, July, pp. 29–58.

Ferguson, P. M., J. E. Breen, and J. O. Jirsa, 1982. *Reinforced Concrete Fundamentals*, Wiley, New York.

FHWA, 1969. Strength and Serviceability Criteria, Reinforced Concrete Members, Ultimate Design, Washington, D.C., Oct.

FHWA, 1982. "Optimized Sections for Major Prestressed Concrete Bridge Girders," Report No. FHWA/RD-82/005, U.S. Dept. Transp., Feb..

Freyermuth, C. L., 1969. "Design of Continuous Highway Bridges with Precast, Prestressed Concrete Girders," *J. Prestressed Concrete Inst.*, Vol., No. 2, April.

Gesund, H., and Y. P. Kaushik, 1970. "Yield Line Analysis of Punching Failures in Slabs," *Publications*, Intern. Assoc. for Bridge and Structural Engineering, Zurich, Switzerland, Vol. 30–1, pp. 41–60.

Gesund, H. and O. P. Dikshit, 1971. "Yield Line Analysis of the Punching Problem at Slab/Column Intersections," *Cracking, Deflection and Ultimate Load of Concrete Slab Systems*, Special Publ. 30, American Concrete Inst., Detroit, pp. 177–201.

Glikin, J. D., S. C. Larson, and R. G. Oesterle, 1987. "Computer Analysis of Time Dependent Behavior of Continuous Precast, Prestressed Bridges," Special Publ., ACI.

Gupta, D. S. R. and J. B. Kennedy, 1978. "Continuous Skew Orthotropic Plate Structures, *J. Struct. Div. ASCE*, Vol. 104, No. ST2, Feb., pp. 313–328.

Guyon, Y. 1972. *Prestressed Concrete*, Wiley, New York.

Guyon, Y., 1946. "Calcul des pont larges a poutres multiples solidarisees par des entretoises," *Annales des Ponts et Chaussees*, No. 24, Sept.–Oct., pp. 553–612.

Guyon, Y, 1949. "Calcul des ponts dalles," *Annales des Ponts et Chaussees*, Vol. 119, No. 29, pp. 555–589; and Vol. 119, No. 36, pp. 683–718.

Hanson, N. W., 1960. "Precast-Prestressed Concrete Bridge, 2: Horizontal Shear Connectors," *J. PCA Res. and Dev. Laboratories*, Vol. 2, No. 2, May.

Hatcher, D. S., 1980. "Standard Prestressed Concrete Beam Design," *J. Struct. Div. ASCE*, Vol. 106, No. ST1, Jan. pp. 23–37.

Heins, C. P. and R. A. Lawrie, 1984. *Design of Modern Concrete Highway Bridges*, Wiley, New York.

Hook, I. M. A. and B. Richmond, 1970. "Western Avenue Extension, Precast Concrete Box Beams in Cellular Bridge Decks," *Struct. Engineer*, vol. 48, No. 3, London, March.

Hossain, I., 1970. "Analysis of Rectangular Slab Bridges on Isolated Supports by Orthotropic Plate Theory," Ph.D. Thesis, Univ. Waterloo, Ontario, Canada.

Hossain, I., 1975. "Coefficient to Analyze Rectangular Slab Bridges," *J. Struct. Div. ASCE*, Vol. 101, No.1 St1, Jan. pp. 151–167.

Houde, J. and M. S. Mirza, 1974. "A Finite Element Analysis of Shear Strength of Reinforced Concrete Beams," *Symp. on Shear in Reinforced Concrete*, ACI Special Publ., SP-42, Detroit.

Kabir, A. F. and A. C. Scordelis, 1974. "Computer Program for Curved Bridges on Flexible Bents," Struct. Eng. and Struct. Mechanics Report No. UC SESM 74-10, Univ. Calif., Berkeley, Sept.

Jobes, H. J., 1981. "Applications of High-Strength Concrete for Highway Bridges," Report FHWA/RD-82/097, FHWA, U.S. Dept. Transp., Oct., 228 pp.

Kennedy, J. B., 1987. "Prestressed Waffle Slab Bridges," Transp. Research Record 1118, TRB, National Research Council, Washington, D.C., pp. 29–38.

Kennedy, J. B. and B. Bakht, 1983. "Feasibility of Waffle Slabs for Bridges," *Canadian J. Civ. Eng.*, Vol. 10, No. 4, pp. 627–638.

Kennedy, J. B., and I. S. El-Sebakhy, 1980. "Prestressed Waffle Slab Bridges-Elastic Behavior," *J. Struct. Div. ASCE*, Vol. 106, No. ST12, Dec., pp. 2443–2462.

Kennedy, J. B., and I. S. El-Sebakhy, 1985. "Ultimate Loads of Orthotropic Bridges Continuous Over Isolated Interior Supports," *Canadian J. Civ. Eng.*, Vol. 12, No. 3, pp. 547–558.

Kennedy, J. B. and M. Ghobrial, 1980. "Waffle Slab Construction for Concrete Bridge Decks—Phase I," Report 11-26, Industrial Research Inst. Univ. Windsor, Ontario, Canada.

Kristek, V., 1970. "Tapered Box Girders of Deformable Cross Sections," *J. Struct. Div. ASCE*, Vol. 96, No. ST8, August, pp. 1761–1793.

Kristek, V., 1974. "Irregular Supports and Skew Diaphragms of Box Girders," *J. Struct. Div. ASCE*, Vol. 100, No. ST5, May, pp. 197–931.

Kristek, V., 1970a. "Box Girders of Deformable Cross Section—Some Theory of Elasticity Solutions," *Proc. Institution of Civ. Eng.*, Proc. Paper 7317, Oct., pp. 239–253.

Kristek, V., 1971. closure to "Tapered Box Girders of Deformable Cross Sections," *Journ. Struct. Div., ASCE*, Vol. 97, No. ST7, Proc. Paper 8225, July, pp. 2021–2023.

Kumar, A., 1986. "Continuity in Composite Bridge Construction," *Proc. Second Int. Conf. on Short and Medium Span Bridges*, vol. 1, Ottawa, Canada, August, pp. 93–107.

Kupfer, H., H. Hilsdorf, and H. Rush, 1969. "Behavior of Concrete Under Biaxial Stress," *Proc. ACI*, Vol. 66, Aug., pp. 656–666.

Leming, M. L., 1988. "Properties of High Strength Concrete: Investigation of High Strength Characteristics Using Materials in North Carolina," Final Report, North Carolina Dept. Transp., Raleigh, July.

Leonhardt, F., 1987. "Cracks and Crack Control in Concrete Structures," *IABSE Proc.*, P109/87, pp. 25–44.

Libby, J. R., 1974. "Structural Response of Concrete Box Girder Bridge," *J. Struct. Div. ASCE*, Vol. 100, ST9, Sept., pp. 1962–1963.

Lin, C. S. and A. C. Scordelis, 1971. "Computer Program for Bridges on Flexible Bents," Struct. Eng. and Struct. Mechanics Report No. 71-24, Univ. of Calif., Berkeley, Dec.

Lin, T. Y., 1955. Prestressed Concrete Structures, Wiley, New York.

Lin, T. Y., and N. H. Burns, 1981. *Design of Prestressed Concrete Structures*, Wiley, New York.

Lin, T. Y. and F. Kulka, 1973. "Construction of Rio Colorado Bridge," *J. Prestressed Concrete Inst.*, Vol. 18, No. 6, Nov.-Dec., pp. 92–101.

Lin, T. Y., F. Kulka, and Y. C. Yange, 1969. "Posttensioned Waffle and Multispan Cantilever System with Y-Piers Composing the Hegenberger Overpass," *Concrete Bridge Design*, SP23-27, American Concrete Inst., Detroit, pp. 495–506.

MacGregor, J. G., 1988. *Reinforced Concrete, Mechanics and Design*, Prentice-Hall, Englewood Cliffs, N.J.

Mailer, Y. and L. Pliskin, 1990. "Bridge at Joigny: High-Strength Concrete Experimental Bridge," *Transp.* Research Record 1275, TRB, National Research Council, Washington, D.C., pp. 19–22.

Massonnet, C., 1950. "Contributions au calcul des ponts a poutres multiples," *Annales des Travaus Publics de Belgique*, Vol. 103, No. 3, Brussels, Belgium, pp. 377–422; Vol. 103, No. 5, pp. 749–796; Vol. 103, No. 6, pp. 927–964.

Massonnet, C., and R. Bares, 1968. Analysis of Beam Grids and Orthotropic Plates, Frederick Ungar Publ. Co., N.Y.

Massonnet, C. and A. Gandolfi, 1967. "Some Exceptional Cases in the Theory of Multigrade Bridges," IABSE, Publ.

Matsushita, H. and M. Sato, 1979. "The Hayahi-No-Mine Prestressed Bridge," *J. Prestressed Concrete Inst. J.*, Vol. 24, No. 2, March–April, pp. 90–109.

McCormick, Rankin and Associates Ltd., 1973. "Precast Trapezoidal Beams for Bridges," Report to the Ontario Precast Concrete Manufacturers Assoc., Ontario, Canada.

Morandi, R., 1969. "Some Types of Tied Bridges in Prestressed Concrete," Special Publ. SP-23, First Int. Symp. on Concrete Bridge Design, American Concrete Inst., Detroit, pp. 447–465.

Morice, P. B. and H. E. Lewis, 1955. "Prestressed Continuous Beams and Frames," *ASCE Struct. J.*, Sept., p. 1055.

Morris, D., 1978. "Prestressed Concrete Design by Linear Programming," *J. Struct. Div. ASCE*, Vol. 104, No. ST3, March, pp. 439–452.

Mostafa, T. and P. Zia, 1977. "Development Length of Prestressing Strands," *J. Prestressed Concrete Inst.*, Sept.–Oct.

Naaman, A. E., 1976. "Minimum Cost Versus Minimum Weight of Prestressed Slab,s" *J. Struct. Div. ASCE*, Vol. 102, No. ST7, July, pp. 1493–1505.

Naaman, A. E., 1982. "Optimum Design of Prestressed Concrete Tension Members" *J. Struct. Div. ASCE*, Vol. 108, No. ST8, August, pp. 1722–1738.

Naaman, A. E., "Partially Prestressed Concrete: Review and Recommendations," *PCI J.*, Vol. 30, No. 6, Nov.–Dec., pp. 30–71.

Naaman, A. E., 1982. *Prestressed Concrete Analysis and Design*, McGraw-Hill, New York.

Neville, A. M., W. H. Dilger, and J. J. Brooks, 1983. *Creep in Plain and Structural Concrete*, Construction Press, London, 361 pp.

Nilson, A. H., 1978. *Design of Prestressede Concrete*, Wiley, New York.

Oesterle, R. G., J. D. Glikin, and S. C. Larson, 1989. "Design of Prestressed Bridge Girders Made Continuous," NCHRP Report 322, Transp. Research Board. National Research Council, Washington, D.C.

Oesterle, R. G., J. D. Glikin, and S. C. Larson, 1989. "Design of Precast Prestressed Bridge Girders Made Continuous," Transp. Research Board, National Research Council, Washington, D.C., Nov.

Ontario Ministry of Transportation and Communications, 1979. Ontario Highway Bridge Code, Ontario, Canada.

Park, R., 1964. "The Ultimate Strength and Long Term Loading Behaviour of Uniformly Loaded Two-way Concrete Slabs with Partial Restraint at all Edges," *Magazine of Concrete Research*, London, Vol. 16, No. 48, Sept., pp. 139–152.

Portland Cement Association. "Concrete Members with Variable Moment of Inertia," ST103, PCA.

Portland Cement Association. "*Continuous Concrete Bridges*, 2nd ed.

Portland Cement Association. "Continuous Hollow Girder Concrete Bridges."

Portland Cement Association. "Frame Analysis Applied to Flat Slab Bridges," ST64-2, PCA.

Portland Cement Association, 1982. "Tennessee's Holston River Bridge Jointless for Over 2650 ft," *Bridge Report* No. SR248.01E, Skokie, Ill.

Post-Tensioning Institute. *Post-Tensioning Manual*, Current Ed., Glenview, Ill.

Prestressed Concrete Institute, 1975. "Recommendation for Estimating Prestress Loss," PCI Committee on Prestress Losses, *J. Prestressed Concrete Inst.*, Vol. 20, No. 4, July–August, pp. 44–75.

Prestressed Concrete Institute, 1978. *Precast Segmental Box Girder Bridge Manual*, PCI, Chicago.

Rabbat, B. G., T. Takayanagi, and H. G. Russel, 1982. "Optimized Sections for Major Prestressed Concrete Bridge Girders," Report FHWA/RD-82/005, FHWA, U.S. Dept. Transp. Feb. Richmond, B., 1969. "Trapezoidal Boxes with Continuous Diaphragms", *Proc. Institution Civ. Engineers*, London, Aug., p. 641–650.

Robertson, J. C., Pama, R. P., and A. C. Cusens, 1970. "Transverse Shear Deformation in Multi-Cell Box Beam Bridges," I.A.B.S.E. publication.

Rowe, R. E., 1962: *Concrete Bridge Design*, Wiley, New York.

Sargious, M. A., and W. H. Dilger, 1977. "Prestressed Concrete Deep Beams with Openings," *J. Prestressed Concrete Inst.*, Vol. 22, No. 3, May–June, pp. 64–79.

Sargious, M. A., W. H. Dilger, and H. Hawk, 1979. "Box Girder Diaphragms with Openings," *J. Struct. Div. ASCE*, Vol. 105, No. ST1, Jan., pp. 53–65.

Sargious, M. A., and G. Tadros, 1974. "Stresses and Forces in Deep Beams Prestressed with Straight Cables," *J. Prestressed Concrete Inst.*, Vol. 19, No. 4, July–Aug., pp. 86–99.

Sattler, K., 1955. "Betrachtungen zum Berechnungsverfahren von Guyon-Massonnet fuer freiliegende Traegerroste und Erweiterung des Verfahrens auf beliebige Systeme," *Bauingenieur*, Vol. 30, No. 3.

Sattler, K., 1956. "Betrachtungen ueber Traegerroste mit Steifigkeitsunterschieen zwischen Rand-und Innentraegern," *Bauingenieur*, Vol. 34, Nos. 1 and 2, pp. 1–53.

Sawko, F., 1968. "Recent Developments in The Analysis of Steel Bridges using Electronic Computers," presented at the June BCSA Conf. on Steel Bridges, held at London.

Schemmel, J. J., and P. Zia, 1990. "Use of High Strength Concrete in Prestressed Concrete Box Beams for Highway Bridges," Transp. Research Record 1275, TRB, National Research Council, Washington, D.C., pp. 12–18.

Schlaich, J., K. Schafer, and M. Jennewein, 1987. "Towards a Consistent Design of Structural Concrete," *PCI J.*, Vol. 32, No. 3, May–June, pp. 74–151.

Scordelis, A. C., 1983. *Berkeley Computer Program for the Analysis of Concrete Box Girder Bridges*, Martinus-Nijkoff, The Hague.

Scordelis, A. C., J. G. Bouwkamp, and S. T. Wasti, 1973. "Structural Response of Concrete Box Girder Bridge," *J. Struct. Div. ASCE*, Vol. 99, No. ST10, Oct., pp. 2031–2048.

Scordelis, A. C., J. G. Bouwkamp, S. T. Wasti, and F. Seible, 1982. "Ultimate Strength of Skew RC Box Girder Bridge," *J. Struct. Div. ASCE*, Vol. 108, No. ST1, Jan. pp. 105–121.

Scordelis, A. C., J. G. Bouwkamp, and S. T. Wasti, 1975. "Structural Response of Concrete Box Girder Bridges," Discussion, *ASCE Str. Journ.*, May, pp. 1141–1145.

Scordelis, A. C. and P. K. Larsen, 1977. "Structural Response of Curved RC Box Girder Bridge," *J. Struct. Div. ASCE*, Vol. 103, No. ST8, Aug., pp. 1507–1524.

Scordelis, A. C., S. T. Wasti, and F. Seible, 1982. "Structural Response of Skew RC Box Girder Bridge," *J. Struct. Div. ASCE*, Vol. 108, No. ST1, Jan. pp. 89–104.

Shioya, T., 1989. "Shear Properties of Large Reinforced Concrete Members," Special Report No. 25, Inst. Tech. Shimizu Corp., Tokyo, Feb.

Sinno, R. and H. L. Furr, 1972. "Computer Program for Predicting Prestress Loss and Camber," *J. Prestressed Concrete Inst.*, Vol. 17, No. 5.

Somayaji, S., 1982. "Prestressed Concrete Flexural Member: Design," *J. Struct. Div. ASCE*, Vol. 108, No. ST8, Aug. pp. 1781–1796.

Tadros, M. K., A. Ghali, and W. H. Dilber, 1975. "Time-Dependent Prestress Loss and Deflection in Prestressed Concrete Members," *J. Prestressed Concrete Inst.*, Vol. 20, No. 3, May–June, pp. 86–98.

Taylor, F. W., S. E. Thomson, and E. Smulski, 1939. *Reinforced Concrete Bridges*, Wiley, New York.

Taylor, G. W., and J. Warwaruk, 1981. "Strength and Deformation of Pretensioned Box Girders," *ASCE Str. Journ.*, Vol. 107, No. ST5, May, pp. 775–788.

Timoshenko, S., 1940. *Theory of Plates and Shells*, McGraw-Hill, New York.

Tissington, I. R., 1970. "A Study of Partially Restrained Two-Way Bridge Slabs Subjected to Concentrated Loads," Thesis, Queen's Univ., Kingston, Ontario.

Tong, P. Y., 1969. "An Investigation of the Ultimate Shear Strength of Two-way Continuous Slabs Subjected to Concentrated Loads," thesis presented to Queen's Univ. at Kingston, Ontario, Canada, in partial fulfillment of the requirements for the degree of Doctor of Philosophy.

Tong, P. Y., and Batchelor, B. de V., 1971: "Compressive Membrane Enhancement in Two-way Bridge Slabs," *Special Publ.* 30, Cracking, Deflection and Ultimate Load of Concrete Slab Systems, American Concrete Institute, Detroit, Michigan, pp. 171–286.

University of Illinois, 1976. "Time Dependent Behavior of Noncomposite and Composite Posttensioned Concrete Girder Bridges," *Structural Research Ser. No.* 430, *Civ. Eng. Studies*, Urbana, October.

University of Illinois, 1980. "Summary Report: Field Investigation of Prestressed Reinforced Concrete Highway Bridges," Project IHR-93, Urbana, pp. 1–14.

Vecchio, F. J. and M. P. Collins, 1986. "Modified Compression Field Theory for Reinforced Concrete Elements Subjected to Shear," *ACI J.*, Vol. 83, No. 2, March–April.

Vecchio, F. J. and H. P. Collins, 1988. "Predicting the Response of Reinforced Concrete Beams Subjected to Shear Using Modified Compression Field Theory," *ACI Struct. J.*, Vol. 85, No. 3, May–June.

Virginia Highway and Transportation Council, 1981. "Jointless Bridges," Report No. FHWA/VA-81/48, June.

Vlasov, V. Z., 1959. "Thin-Walled Elastic Beams," PST Catalogue No. 428, U.S. Dept. Commerce, Washington, D.C.

Wallace, M. R., 1976. "Studies of Skew Concrete Box Girder Bridges," *Transportation Research Record* 607—*Bridge Testing, Design, and Evaluation*. Transp. Research Board, National Research Council, Washington, D.C.

Wang, C. K. and C. G. Salmon, 1973. *Reinforced Concrete Design*, Intext, New York.

West, R. and L. A. Clark, 1971. "Bibliography of Experimental Work on Slab and Pseudo-Slab Bridges," Departmental Note DN/7004, Cement and Concrete Assoc., London, June.

Westergaard, H. M., 1930. "Computation of Stresses in Bridge Slabs Due to Wheel Loads," *Public Roads*, March, pp. 1–23.

Wheen, R. J., 1979. "Prestressed Concrete Members in Direct Tension," *Journ. Struct. Div., ASCE*, Vol. 105, No. ST7, July, pp. 1471–1487.

William, K. J. and A. C. Scordelis, 1972. "Cellular Structures of Arbitrary Plan Geometry," *J. Struct. Div. ASCE*, Vol. 108, No. ST7, July, pp. 1377–1394.

Wilson, E. L., 1972. "SOLID SAP—A Static Analysis Program for Three-Dimensional Solid Structures," Struct. Engineering and Struct. Mechanics Report No. 1 UC SESM 71-19, Univ. Calif., Berkeley, Sept., revised Dec. 1972.

Xanthakos, P. P., 1991. *Ground Anchors and Anchored Structures*, Wiley, New York.

CHAPTER 4

STEEL I-BEAM BRIDGES

4-1 CHARACTERISTICS OF I-BEAM BRIDGES

Throughout this book we have classified bridges according to their structural arrangement. Alternatively, these structures may be articulated according to the service performed, such as pedestrian and railroad bridges, or according to the cross section of the superstructure. Thus, a deck bridge is a system that has a floor resting on top of suitable carrying members, such as beams or girders, so that overhead bracing is not required. If the floor is connected to the lower portion of the load-carrying structural system, the structure is identified as a through bridge. A double-deck bridge has vehicular facilities on two different levels, both of which can be through decks, or one can be a through deck and the other an open deck.

Classified according to the structural layout of the principal components, beam, girder, and truss bridges can be the simple-span continuous, or cantilever type. Some engineers prefer the cantilever arrangement to the continuous bridge because it has favorable moments along its length and is not subjected to settlement stresses. The cantilever is also easier to analyze. However, this type requires special hinge connections and is less rigid than a continuous unit. A derivative of the cantilever version is the movable bridge discussed in other sections.

Classified according to the makeup of the main supporting members, an I-beam bridge utilizes I rolled sections as the main support system of the superstructure. When the spans exceed a certain limit length, built-up plate girders are used and the structure is referred to as a plate girder bridge. Plate girders forming an integral unit with a steel deck are called orthotropic bridges. For still longer spans, truss bridges and cable-stayed bridges are more economical. For very long spans, the suspension bridge is probably the only available solution, with high-strength-steel cables carrying the main loads. Suspension bridges are usually stiffened with trusses, and the cables may vary from two to three or more. For all bridge types, the number of beams, girders, box girders, trusses, or cables is at least two and possibly three or more.

I-Beam Superstructures An I-beam floor system consists of the roadway and the supporting rolled beams. Floor systems are usually provided with a concrete slab, 7.5 to 9 in. thick, with reinforcement perpendicular to traffic. The construction may be independent where the concrete slab at the I beams does not develop structural interaction, or composite where the live loads are resisted jointly by the slab–I-beam action. Selection of the appropriate design for a particular set of conditions is governed by several considerations such as span lengths, deflection limitations, construction time, and overall cost.

Simple-span I-beam bridges support the roadway directly on the top flanges of a series of rolled beams placed parallel to the direction of traffic and extending from abutment to abutment. A rolled-beam simple-span highway bridge is likely to be economical for spans up to 60 ft, and for railroad traffic probably up to 50 ft. Composite construction of slab–I-beam decks has been found attractive and economical for spans up to 100 ft. The plate girder scheme begins to become economical for spans greater than 70 ft.

The majority of steel I-beam bridges at both the Interstate and local system have some kind of a continuous beam as the main structural member, although there is a wide variation in the span lengths and ratios, number of spans, and width of superstructure. Where the bridge length is sufficiently long to warrant multiple spans, the selection usually focuses on simple or continuous units. The associated advantages and disadvantages for each scheme are examined and analyzed in detail in the following sections. A fairly simple comparison of the economy made possible through the use of continuous-beam design is tabulated in Table 4-1, showing the weight saving resulting from the continuity and the bearing, expansion device, and field splice requirements. The examples involve continuous units versus a similar number of equal simple spans.

Given the superstructure width, an optimum deck cross section may be developed by considering several beams at close spacing, or fewer beams at a larger spacing. The top reinforcement (main) usually controls the design of the concrete slab because of the greater clearance (normally 2.25 in.) that must be provided at the top. From these considerations it follows that optimum design must balance the number of beams and the slab thickness. Using a slab thickness of 7.5 in. and a current H 20 loading, the maximum spacing of steel beams is approximately 8.25 ft, but at this spacing the main transverse reinforcement requirements may be excessive.

TABLE 4-1 Comparison of Simple and Continuous I-Beam Bridges

	Problem 1–114-ft Crossing		Problem 2–240-ft Crossing		Problem 3–360-ft Crossing	
	Continuous Spans	Two 57-ft Simple Spans	Continuous Spans	Three 80-ft Simple Spans	Continuous Spans	Four 90-ft Simple Spans
Discription	33WF141	36WF170	36WF194 and 36WF230	36WF280	36WF190 and Cover Plates	36WF280
Weight—one line of beams—lb	16,070	19,380	48,940	67,200	68,520	100,800
Weight saving due to continuity	17%		27%		32%	
Required number of shoes	3	4	4	6	5	8
Required number of dams	2	4	2	6	2	8
Required number of field splices	1	0	2	0	3	0

The ratio of the beam depth to the span length should not exceed 1/25, and for composite design the ratio of the overall depth (concrete slab plus steel beam) should not be less than 1/25. For continuous bridges, the span length is considered as the distance between the dead load points of contraflexure. If depths less than those specified are used, the beam sections should be increased so that the maximum live load deflection is not any greater had this ratio not been exceeded.

Members in simple or continuous spans should be designed so that the deflection due to service live load plus impact does not exceed 1/800 of the span, except on bridges in urban areas used partly by pedestrians where this ratio should not exceed 1/1000. AASHTO, however, stipulates that for bridges having cross-bracing or diaphragms sufficient in depth or strength to ensure lateral distribution of loads, the deflection may be computed by loading the entire deck, considering all beams as acting together and having equal deflection. When a beam is part of a composite deck, the service live load may be assumed to act on the composite section, and the moment of inertia of the gross cross-sectional area may be used in computing deflections. We should note that the intent of Article 10.6.4 is not to articulate the effect of diaphragms on the lateral load distribution but rather to impose a uniform methodology in interpreting the live load deflection limitations (see also subsequent sections).

Likewise, end reactions of continuous beams should be checked for possible uplift. Because the floor slab and diaphragm system are assumed sufficiently rigid for the resulting uplift distribution, uplift analysis may be based on all beams acting together and having a uniform reaction under the critical loading. All lanes should be loaded simultaneously and include impact.

Exterior beams should have the same section and capacity as the interior beams, even though analysis indicates that they could be less (see also Chapter 2). If special conditions exist dictating design requirements that exceed the capacity of the interior beams, the design should take into account modification of the fascia portion and details to equalize the loads. Using the New Jersey type of parapet, the usual overhang dimension is 3 to 3.5 ft.

4-2 STRUCTURAL MATERIALS

AASHTO specifications list acceptable steels in Article 10.2.2 and Table 10.2A. Several grades of steel are available for use in bridge construction. Grades meeting AASHTO requirements are as follows.

1. *Structural Steel*, AASHTO designation M 270, Grade 36. This is equivalent to A36 steel. Essentially in the carbon steel family, it has a copper content between 0.40 and 0.60 percent and a maximum manganese content of

1.65 percent. No minimum content is specified for other alloy elements. This widely used steel in rolled sections has a minimum yield strength of 36 ksi and a tensile strength that can vary from 58 to 80 ksi. The minimum quoted elongation is 20 percent in an 8-in. gage length. Plate thickness is up to 4 in. The steel is readily weldable and is usually the most economical grade for short-span bridges. It must be painted for protection from moisture.

2. *High-Strength Low-Alloy Steel*, AASHTO designation M 270, Grade 50. This is equivalent to A572, Grade 50 steel. The steel is not heat-treated to obtain the necessary stress levels. It has a minimum yield strength of 50 ksi and a minimum tensile strength of 65 ksi.

3. *High-Strength Low-Alloy Steel*, AASHTO designation M 270, Grade 50W. This is equivalent to A588 steel. It has a minimum yield strength of 50 ksi and a minimum tensile strength of 70 ksi.

A572 and A588 steels are two relatively new grades that can readily be used in bridge work. These materials constitute a specific class of steels in which enhanced mechanical properties, and in most cases good resistance to atmospheric corrosion, are obtained by incorporating moderate proportions of one or more alloying elements other than carbon. These steels are generally intended for use where weight savings are possible through greater strength and where improved durability is obtained because of other desirable characteristics. The combination of greater strength and low cost imparts to these high-strength steels certain economic advantages, but their effective utilization requires competent judgment and careful analysis.

For higher resistance to atmospheric corrosion, A588 is usually specified because its corrosion resistance is four times that of A36 steel. This steel has been left unpainted in many bridges. In this condition it forms a thin iron oxide film on the surface in the presence of atmospheric moisture. This steel is readily weldable by the electric-resistance, submerged arc, manual arc, and gas metal arc welding processes. Low-hydrogen electrodes are specified by the American Welding Society (see also Chapter 5).

Where section thickness is reduced by the use of high-strength steel, it is essential to rely on greater corrosion resistance in comparison with other types of steel in order to ensure equal service life. Although corrosion resistance may not be expressed quantitatively because of its dependence on variable factors, performance of a steel structure can, however, be compared for various materials under similar conditions. A large number of corrosion tests under variable conditions of exposure have articulated the sensitivity of various steels and permit assessment of the trends in weight loss due to corrosion. Figure 4-1 shows the loss in thickness of specimens exposed on racks in the industrial atmosphere of Kearny, New Jersey, for a period of 12 years. Steel L is structural carbon steel with a low residual copper content. Steel K is a structural copper steel. The remaining curves represent the high-strength low-alloy steels.

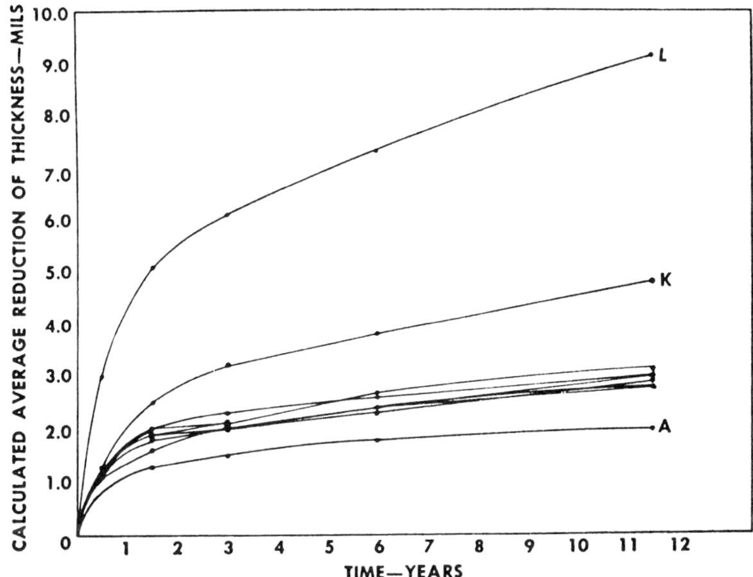

FIGURE 4-1 Time–corrosion curves for steels in industrial atmosphere of Kearny, New Jersey.

4. *Quenched and Tempered Low-Alloy Steel*, AASHTO designation M 270, Grade 70W. This is equivalent to A852 steel. It has a minimum yield strength of 70 ksi and a minimum tensile strength of 90 ksi.

5. *High-Yield-Strength, Quenched and Tempered Alloy Steel*, AASHTO designation M 270, Grades 100/100W. This is equivalent to A514 steel. These steels contain alloying elements exceeding those contained in carbon steel, and are heat-treated to obtain strength and notch toughness. The yield strength is 90 to 100 ksi, and the tensile strength is 100 to 110 ksi in thickness up to 2.5 in., inclusive. This steel has high corrosion resistance.

The yield strength is of prime importance because it is the property that determines working unit stresses. The ratio of yield to tensile strength for structural carbon steel is about 0.60, but for high-strength steels this ratio is in the range of 0.70 to 0.80.

The notch toughness, as measured in a notched-bar impact test, reflects the behavior in actual structures. High-strength steels exhibit superior notch toughness characteristics, whether in terms of energy absorbed in breaking a specimen at room temperature or in terms of the refrigerated temperature to which they preserve their toughness.

It appears that structural steels for bridge construction represent three general groups: (a) carbon steels, (b) high-strength low-alloy steels, and

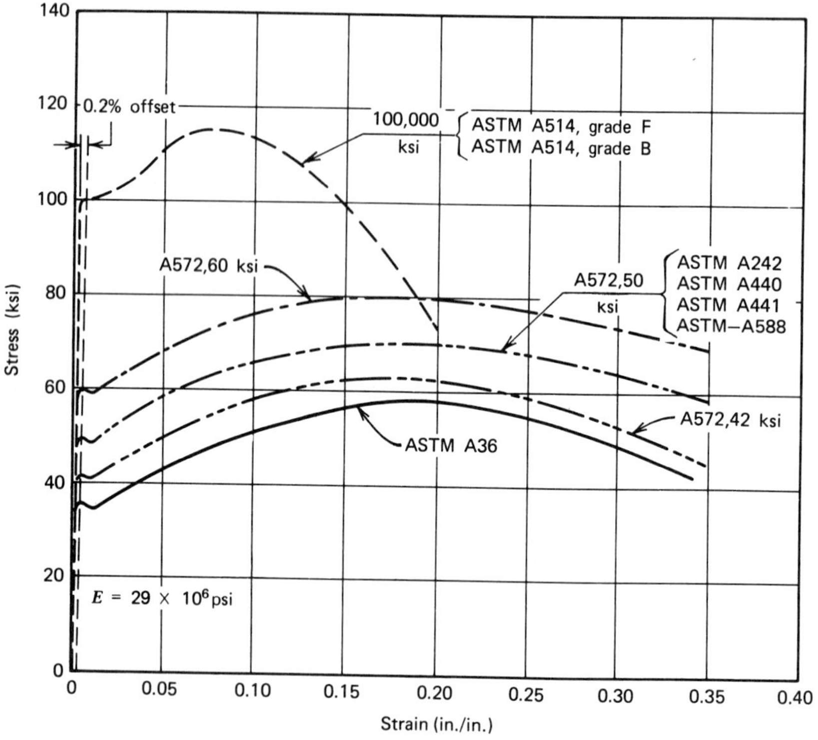

FIGURE 4-2 Stress–strain curves for various steels.

(c) heat-treated alloy steels. Figure 4-2 shows stress–strain curves for steels in each group. In general, the physical properties are determined by the type and content of the alloying elements, the amount of carbon, the cooling rate, and the mechanical process of the steel such as rolling and stressing.

Whereas the general availability of these steels enhances the design possibilities, engineers are cautioned that not all grades can be supplied by all rolling mills, and availability should therefore be checked on a local basis.

Brittle Fracture Considerations Failures of some bridges have been documented as the result of brittle fracture in steel (Scheffey, 1971). This type of failure is enhanced by (a) higher service stresses and lower service temperatures, (b) more complex structural arrangements leading to favorable principal stress patterns and high stress concentration, and (c) wide use of welding (see also Section 12-6).

It is almost certain that solids under a uniform tension acting in all directions are able to resist only certain definite stresses. If the three principal stresses are equal tensile stresses, solid materials break without preceding permanent deformation. Brittle fracture, however, involves materials under uniaxial or multiaxial tensions and occurs suddenly and preferentially without prior permanent deformation under such stresses. If a brittle material should break under any combination of stresses as soon as at least one of the three principal stresses is a tensile stress and reaches a certain maximum value independently of the others, this may be explained by a *maximum stress theory*. Thus, unlike ductile materials, brittle materials do not deform appreciably or permanently before rupture.

The transition from ductile to brittle behavior in steel plates has been studied in static-tension tests showing that it is possible to produce brittle fracture in large flat steel specimens under static tension at normal temperatures. The influence of the rate of loading, time of load, stored energy in the system, and specimen size on fracture stresses is significant. The notch toughness characteristics mentioned in the preceding paragraphs constitute an index sufficient to ensure against brittle behavior, because they define the ability to deform plastically in the presence of a notch.

The following guidelines will help to avoid brittle fracture in bridges (Rolfe, 1972).

1. Flaws should be restricted in the finished steel, and the material behavior should be documented by superior toughness data.
2. Stress concentrations should be avoided as should residual stresses because they can initiate crack propagation.
3. Provisions should be made to lower stress levels if the number of stress cycles is expected to be high because fatigue stresses can increase the size of a flaw.
4. In very cold regions, the possible decrease in crack toughness should be considered.
5. Transition from ductility to brittleness can occur in steel plates under a triaxial state of stress.
6. Structural elements subjected to high-impact stresses have an enhanced tendency to brittle behavior because the loading and timing rate has a favorable influence on this phenomenon.

In Article 10.3.3 AASHTO defines the Charpy V-notch impact requirements for main load-carrying members subjected to tensile stresses. These impact requirements vary depending on the type of steel, type of construction, whether welded or mechanically fastened, and the average minimum service temperature (see also Barsom, 1975; AASHTO specifications for Fracture Critical Non-Redundant Steel Bridges, 1978 and Interim).

Uncoated Weathering Steels

Experience with the performance of weathering steel in highway bridges shows that the majority of these bridges are in good condition although there have been localized areas of corrosion on many structures and some have experienced excessive attack by deicing salts. The general conclusion is that weathering steel can provide a satisfactory service life with limited maintenance if the structural details are designed so as to prevent accelerated attack. In this context vulnerable areas should be painted and contamination with chlorides should be inhibited.

Fatigue Consideration Fatigue tests have been performed on 8-year weathered AASHTO M270 Grade 50W (ASTM A709 Grade 50W) transverse stiffeners under constant loading in air and aqueous environments (Albrecht and Sidami, 1987). The possibility of fatigue damage to uncoated weathering steel members as a result of corrosion has been considered by AASHTO task forces and by the FHWA (1989). Although the results from these investigations are not explicitly stated, they have led to a lower stress range (AASHTO Table 10.3.1A) for unpainted weathering steel, A709, all grades, when used as base metal (fatigue Category A).

Guidelines on Uses of Weathering Steels The principal factors that determine the ultimate performance of weathering steel bridges are location and design. Environmental differences may exist from one location to another with respect to amount and type of atmospheric pollution, extent of rainfall, variations in humidity, temperature and the prevailing winds, and the amount of airborne salinity in marine environments. Because of these factors the corrosion performance of weathering steel can differ in degree from location to location.

Location Consideration The proposed site for a structure should be evaluated for local environmental effects before uncoated weathering steel is selected for application. This should include a site inspection, study of available meteorological data, and appropriate tests. The latter should emphasize atmospheric exposure tests with carbon steel and weathering steel panels to determine the level of atmospheric contaminants such as chlorides and sulfur oxides. Environmental locations to be avoided are (a) those exposed to highly corrosive chemical and industrial fumes; (b) those subject to high rainfall and humidity or where there is constant wetness; (c) depressed roadways that create tunnel-like conditions; (d) those that are low level water crossings that can lead to highly humid or frequently wet conditions; (e) those subject to salt spray or significant salt-laden fogs at coastal areas; and (f) those where the steel may be continuously submerged in water, buried in soil, or covered by vegetation.

Design Details Proper attention to the design details can contribute markedly to the satisfactory long-term performance of weathering steel bridges. Items to be considered include expansion joints, leakage and drainage, painting below joints, integral abutments, connections, handling and storage of weathering steels, and the welding process.

At present, an FHWA moratorium prohibits the use of electroslag (ESW) and electrogas (EGW) process for welding bridge members subjected to tensile or reversing stresses, and efforts are continuously made to improve the welding characteristics of weathering steels.

Excellent reviews of this subject are provided by the related references listed at the end of this chapter.

4-3 DESIGN REQUIREMENTS

The basic design method of flexural members involves the use of an elastic section modulus except when utilizing compact sections under strength design. Allowable stress design is the standard design method for all structure types, whereas the load factor concept is an alternate method for designing simple and continuous beams and girders of moderate lengths (see also Section 2-4 and subsequent sections).

Diaphragms or cross frames are provided at each support and in all bays, spaced at intervals not to exceed 25 ft. Diaphragms for rolled beams should be at least one-third and preferably one-half the beam depth. Cross frames should be as deep as practicable, and intermediate cross frames should be the cross or vee type. End diaphragms are usually proportioned to transmit the lateral forces to the bearings.

All fixed ends must be firmly anchored to the beam seat. Bridges less than 50 ft long do not require provisions for deflection. Spans 50 ft or greater must be provided with a bearing type allowing beam rotation, such as a hinge, curved bearing plate, elastomeric pads, or a suitable pin.

Bridges less than 50 ft long may be designed to slide on metal surfaces, such as plates, and provisions for deflection are not necessary. Spans 50 ft or greater should be free to expand or contract by means of rollers, rockers, or sliding plates. Alternatively, elastomeric bridge bearings may be used to transmit loads and accommodate movement between the deck and supporting substructure.

4-4 SLAB – I-BEAM BRIDGE WITH SIMPLE SPAN

A simple-span I-beam bridge (noncomposite) is the simplest structural system in the context of analysis and design, but is seldom economical. For steel I-beam bridges longer than about 40 ft, composite design usually provides an economical solution. For spans less than 40 ft, all-concrete bridge decks offer competitive schemes. Therefore, in this section we will consider an I-beam bridge merely to demonstrate the design methodology. We will also include

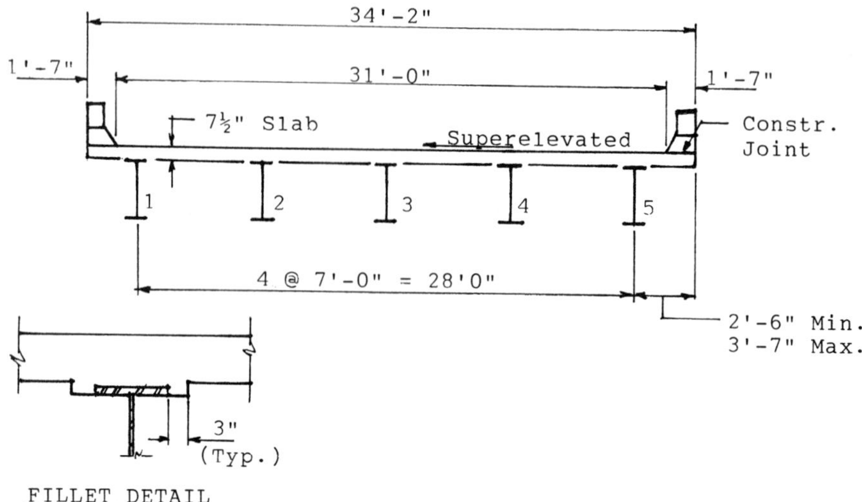

FIGURE 4-3 Typical cross section, I-beam bridge, 60 ft long.

examples of slab analysis (main and cantilever) and compare the structural requirements between interior and exterior beams.

The bridge in this example is 60 ft long center-to-center of bearings and has a roadway width of 31 ft (see also Figure 4-3). A New Jersey type of parapet is incorporated giving a dimension O. to O. of deck of 34 ft 2 in. We choose five beams spaced at 7-ft centers. Although the deck is on a horizontal curvature, the steel framing consists of straight beams. This scheme results in overhang dimensions as shown. For the exterior fascia beam, the slab overhang is 2 ft 6 in. at the supports and 3 ft 7 in. maximum near the center of the span. All beams are embedded in the concrete by means of the fillet detail shown in Figure 4-3. The top flange of the beams is therefore laterally supported against buckling. The design live load is HS 20.

Design of Slab Using $f_c' = 3500$ psi, $f_c = 0.4 \times 3500 = 1400$ psi, $f_y = 60{,}000$ psi, and $n = 9$, we compute $k = 0.34$, $j = 0.89$, $K = 211$, and $a = 1.78$. For main reinforcement perpendicular to traffic and assuming a flange width of 12 in., the effective (design) span S is 6 ft 6 in. = 6.5 ft. We use a minimum slab thickness of 7.5 in., and include provisions for a 25-lb/ft^2 future wearing surface. The live load plus impact moment is

$$M_{LL+I} = \frac{(6.5 + 2)}{32} \times 16 \times 1.3 \times 0.8 = 4.42 \text{ ft-kips}$$

where the factor 0.8 is applied with regard to the continuity of the slab. The dead load is the weight of the slab $0.625 \times 0.15 = 94$ lb/ft^2, plus 25 lb/ft^2

wearing surface, or dead load $w = 0.119$ kip/ft. We now compute

$$M_{\text{DL}} = \frac{0.119}{10} \times 6.5^2 = 0.50 \text{ ft-kip}$$

The total moment is

$$M = 4.42 + 0.50 = 4.92 \text{ ft-kips}$$

Thus,

$$\min d = \sqrt{4.92/0.211} = 4.8 \text{ in.}$$

The reinforcement in the top of the slab is obtained using 2.25-in. clearance and #5 bars. This gives $d = 7.50 - 2.25 - 0.31 = 4.94$ in. > 4.8 in. The required reinforcement is

$$A_s = \frac{4.92}{1.78 \times 4.94} = 0.56 \text{ in.}^2/\text{ft}$$

Use #5 at 6.5 in. $\quad A_s = 0.57$ in.^{2}/ft

Likewise, the reinforcement in the bottom of the slab is calculated using 1-in. clearance, giving $d = 6.19$ in., and

$$A_s = \frac{4.92}{1.7 \times 6.19} = 0.45 \text{ in.}^2/\text{ft}$$

Use #5 at 8 in. $\quad A_s = 0.47$ in.^{2}/ft

The required distribution reinforcement in the bottom of the slab is $220/\sqrt{S} = 86$ percent (use maximum 67 percent, or $A_s = 0.47 \times 0.67 = 0.31$ in.^{2}/ft. Use #5 bars at 12 in., $A_s = 0.31$ in.^{2}/ft. This distribution reinforcement is used in the middle half of the slab span, and not less than 50 percent of the specified amount is used in the outer quarters. Longitudinal reinforcement in the top of the slab is #5 bars at 12 in.

In the past, the practice was to use straight bars, both top and bottom, alternating with truss bars bent at the quarter panel points to accommodate the bending moment profile where the moment is essentially positive in the central panel between beams and negative above and adjacent to the beams. Most states have discontinued this practice, and main bars are now placed straight throughout the deck panels in both the top and the bottom.

Because the bridge has a variable overhang dimension, we will check the steel reinforcement requirements in the cantilever slab. In this case the centerline of the wheel load is placed 1 ft from the face of the curb. Using the maximum overhang of 3.58 ft, this gives $x = 1$ ft, where x is the distance from the wheel load to the point of support (center of exterior beam). For reinforcement perpendicular to traffic, the distribution is $E = 0.8x + 3.75 = 4.55$ ft. The live load plus impact moment is now

$$M_{LL+I} = \frac{16}{4.55} \times 1.0 \times 1.3 = 4.55 \text{ ft-kips}$$

The dead load moment includes the slab and the curb–parapet section. This is computed as follows:

$$\begin{aligned} &\text{Slab,} && 3.58 \times 0.75 \times 0.15 \times 1.79 = 0.72 \text{ ft-kip} \\ &\text{Curb,} && 1.57 \times 1.00 \times 0.15 \times 2.66 = 0.61 \text{ ft-kip} \\ &\text{Parapet,} && 1.75 \times 0.75 \times 0.15 \times 3.07 = \underline{0.61} \text{ ft-kip} \\ &&& \text{Total } M_{DL} = 1.94 \text{ ft-kips} \end{aligned}$$

The total moment is

$$M = 4.55 + 1.94 = 6.49 \text{ ft-kips}$$
$$\min d = \sqrt{6.49/0.211} = 5.55 \text{ in.}$$

requiring a total slab thickness of $5.55 + 2.25 + 0.31 = 8.11$ in., which can be provided by lowering the exterior beams by about 1/2 in. below the theoretical beam elevation that accommodates the 7.5-in. slab and also by extending the fillet on the inside of the beam. Using $d = 5.50$ in., we calculate

$$A_s = \frac{6.49}{1.78 \times 5.5} = 0.66 \text{ in.}^2/\text{ft} \qquad \text{Use \#5 at 5.5 in.} \qquad A_s = 0.68 \text{ in.}^2/\text{ft}$$

The foregoing analysis shows the sensitivity of A_s to the structural requirements of the slab as soon as the live load moment begins to act on the cantilever section, and establishes an upper limit in the overhang slab dimension for the curb–parapet type shown in Figure 4-3.

Design of Interior Beams For the service load design method (allowable stress), the working steel stresses are summarized in AASHTO Table 10.32.1A. For structural carbon steel (equivalent to A36), the design (allowable) stress is 20 ksi. This applies to axial tension in the beams and compression in the top flange if the latter is supported laterally its full length by embedment in concrete. Note that this is $0.55Fy$, where Fy is the minimum yield strength, or 36 ksi. The beam-span-to-beam-depth ratio limitation implies a minimum beam depth of $60/25 = 2.4$ ft, or 30 in.

The dead loads acting on the interior beams are as follows:

$$\text{Weight of slab} = 7 \times 0.625 \times 0.15 = 0.66 \text{ kip/ft}$$

$$\text{Weight of beam etc., say} = 0.15 \text{ kip/ft} \qquad \text{(Assumed)}$$

The superimposed dead load is

$$\text{From curb} = 0.24/2.5 = 0.10 \text{ kip/ft}$$

$$\text{From parapet} = 0.20/2.5 = 0.08 \text{ kip/ft}$$

$$\text{From W.S.} = 31 \times 25/5 = \underline{0.16} \text{ kip/ft}$$

$$\text{Total dead load} = 1.15 \text{ kips/ft}$$

The dead load moment is

$$M_{\text{DL}} = \frac{1.15 \times 60^2}{8} = 518 \text{ ft-kips}$$

The fraction of wheel load applied to the interior beams is $S/5.5 = 7/5.5 = 1.27$, and the impact factor is $50/(60 + 125) = 27$ percent. The live load moments may be computed by positioning the truck as shown in Figure 3-4, or by direct reference to Appendix A of AASHTO. For a span of 60 ft, the moment for one lane is 806 ft-kips (without impact). Therefore, the live load plus impact moment per beam is

$$M_{\text{LL}+I} = 806 \times 0.5 \times 1.27 \times 1.27 = 649 \text{ ft-kips}$$

The total design moment is $518 + 649 = 1167$ ft-kips. This will require a section modulus of $1167 \times 12/20 = 700$ in.^{3}. Note that none of the lightweight beams in the 33- or 36-in. range can provide the required Section modulus. Thus, we select W36 $\times$ 230 with a section modulus of 837 in.^{3}. The actual beam weight plus diaphragms plus fillets is closer to 0.25 kip/ft. The final dead load moment is $1.25 \times 60^2 \times 0.125 = 563$ ft-kips, giving a total moment of $563 + 649 = 1212$ ft-kips. The adjusted section modulus is now $1212 \times 0.6 = 727$ in.$^3 < 837$ in.^{3}.

Design of Exterior Beams Beam 5 in the cross section of Figure 4-3 has a maximum slab overhang of 3.57 ft at the center of the span, and 2.50 ft minimum at the ends. Therefore, it controls the design. The dead load from the weight of the slab is as follows:

$$\text{Weight of slab near center} = (3.58 + 3.50) \times 0.625 \times 0.15 = 0.66 \text{ kip/ft}$$

$$\text{Weight of slab near ends} = (2.50 + 3.50) \times 0.625 \times 0.15 = 0.56 \text{ kip/ft}$$

$$\text{Weight of beam etc.} \qquad \text{Say} = 0.26 \text{ kip/ft}$$

$$\text{Superimposed dead load weight} = 0.34 \text{ kip/ft}$$

The distribution of the wheel load is calculated for $S > 6$ ft, and is $S/(4.0 + 0.25 \times 7) = 1.22$. This procedure applies to a span with a concrete floor supported by four or more beams. Noting that near the center of the span a wheel load can be placed directly above the exterior beam, we may compute the wheel load distribution, assuming simple beam action of the slab. In this case the distribution is $(1 + 1/7) = 1.14$ and does not control (see also Section 2-11 and Figure 2-38).

By comparison the dead and live loads acting on the exterior beam are somewhat less than those applied to the interior beams. Hence, the exterior beams are the same as the interior.

End Shear (For Interior Beams) The dead load shear is $1.25 \times 30 = 37.5$ kips. The live load shear is calculated according to AASHTO Article 3.23.1, and for maximum value the truck load is placed as shown in Figure 4-4. For the wheel load acting above the end support, the lateral distribution is obtained assuming the slab acts as a simple beam, or the wheel load fraction is $(1 + 1/7) = 1.14$. For the intermediate positions, the distribution is as for moment, or wheel fraction $= 1.27$. From Figure 4-4 we calculate the live load plus impact shear as

$$V_{LL+I} = \left(16 \times 1.14 + 16 \times \frac{46}{60} \times 1.27 + 4 \times \frac{32}{60} \times 1.27\right) \times 1.3 = 47.5 \text{ kips}$$

The total end shear is $V = 37.5 + 47.5 = 85.0$ kips

$$\text{Average web shear} = \frac{85.0}{35.9 \times 0.76} = 3.11 \text{ ksi} \qquad \text{OK.}$$

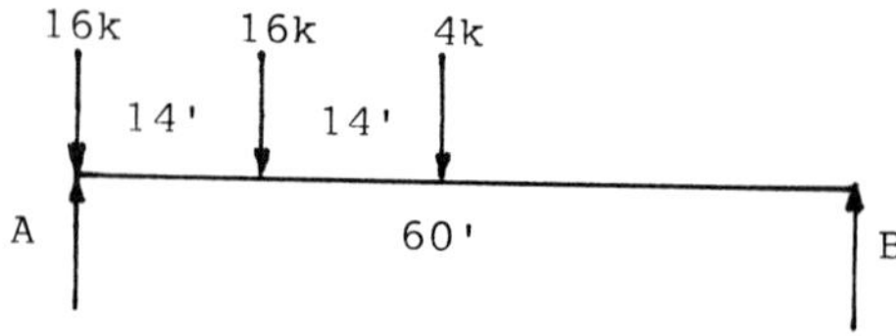

FIGURE 4-4 Position of truck load for maximum end shear.

Live Load Deflection The live load deflection is calculated for two lanes loaded simultaneously, assuming all five beams act as one unit. Note that although the roadway width is 31 ft, strict interpretation of design traffic lanes does not warrant the use of lane fractions. Because the deflection caused by the front wheel load is very small and may be neglected, as a first approximation we compute the live load deflection by considering the rear axle loads concentrated at the center, using the resultant load. Then the deflection is computed from the simple formula $\Delta = PL^3/48\ EI$, or

$$\Delta = \frac{128 \times 60^3 \times 12^3}{48 \times 29 \times 10^3 \times 5 \times 15{,}000} = 0.46 \text{ in.}, \qquad \text{or } 0.60 \text{ in. with impact}$$

The allowable deflection is $60 \times 12/800 = 0.90$ in., or almost 1.5 times the computed deflection; hence, a more detailed analysis of live load deflection is not warranted.

In the foregoing example, the design method is based on working stress theory. Ultimate strength design for concrete slabs is reviewed in Chapter 3, and for beams in subsequent sections of this chapter.

Design According to Proposed Distribution The proposed Table 3.23.1 of the AASHTO specifications (Imbsen, 1991) gives formulas for computing the wheel load fraction (g) for both front and rear axles. This fraction is the same for concrete floors on steel I beams, prestressed concrete girders, and concrete T beams (see also Section 2-12).

First, we estimate the stiffness term (Kg/Lt_s^3), where K_g is the longitudinal stiffness parameter. We calculate

$$K_g = n\left(I + Ae_g^2\right) = 9(15{,}000 + 68 \times 22^2) = 431{,}200$$

and

$$Lt_s^3 = 60 \times 12 \times 7.5^3 = 303{,}840$$

Then the stiffness term is $431{,}200/303{,}840 = 1.4$.

For the interior beams, we can now use Figure 2-41 which gives the distribution factor for multilane loading and a stiffness term of 1.0, noting that $S/L = 7/60 = 0.117$. From the graph we obtain directly $g = 1.23$, which is very close to the distribution factor used (1.27), and the difference does not warrant redesigning the beams.

For the exterior beams, the wheel load distribution factor is given by Table 3.23.2 of the proposed AASHTO specifications. This factor is $g = eg_{\text{interior}}$, where $e = (7 + d_e)/9.1$ and d_e is the edge distance of the traffic lanes, or the distance between the center of the outside beam and the edge of the exterior lane (face of curb). From Figure 4 3, $d_e - 2$ ft (approximate) at the center of the bridge. Next, we calculate $e = (7 + d_e)/9.1 = 0.99$ but not less than 1.0. Because the wheel load distribution is now $g = 1.23$, the exterior beams are the same as the interior.

4-5 COMPOSITE STEEL BEAM – CONCRETE CONSTRUCTION

Composite steel–concrete construction is used extensively in buildings and bridges in a variety of forms and configurations. Besides composite beams and girders with a homogeneous cross section, other noble composite forms are concrete-encased steel beams, composite slabs, steel–concrete composite columns, and composite plates and shells. Early references for research on composite beams are provided by Viest (1960), Stevens (1965) on encased columns, and Gardner and Jacobson (1967) on concrete-filled steel tubes.

A considerable economy may be attained in a beam–slab system if the latter participates with the beam in resisting subsequent loads such as live and superimposed loads. The combined action is manifested if horizontal shear is resisted at the concrete–steel interface. Once the friction between the two materials is exceeded and slippage is about to occur, mechanical devices of sufficient shear capacity develop the common interaction, and both the steel and concrete act as an integral unit.

Composite Beams and Concrete Slab

The elastic behavior of composite beams has been well understood since the later 1950s. Elastic design methods provided the basis of the 1957 AASHTO specifications. An important step was the 1960 tentative recommendations for building design by the Joint ASCE–ACI Committee on Composite Construction that included certain modifications to reflect ultimate strength concepts.

Ultimate Strength In this context the design of shear connectors takes into account a limit in the slip between the slab and the beam. However, initial tests by Slutter and Driscoll (1965) of pushout specimens and composite beams showed that these limits were unnecessary because composite beams develop the full flexural capacity of the cross section provided the sum of the ultimate strengths of individual connectors between the point of zero and the maximum moment is at least equal to the horizontal shear. Tests summarized by the ASCE–ACI report show that the ultimate strength of a composite beam is essentially independent of the history of loading; hence, a beam built with temporary supports and a beam built without temporary supports have the same ultimate strength, but temporary supports during construction influence the magnitude of deflection.

The development of the theory for the ultimate strength of beams with inadequate shear connectors established the lower range of shear connector requirements and, although of limited practical usefulness, provided a rational basis for treating the ultimate design concept. The introduction of a purely ultimate strength procedure for composite beams in buildings in Great Britain represents the second most significant step in understanding ultimate strength behavior during the 1960s.

In independent studies by Barnard (1965), the effect of the shape of the stress block on the ultimate flexural capacity of composite beams was further analyzed. Daniels and Fisher (1967) reported tests of composite beams with simulated moving loads. Lew (1970) investigated the effects of shear connector spacing on the ultimate strength of beams, and Reddy and Hendry (1970) summarized the results of studies of simply supported composite beams carried out in Britain. The shear strength of composite compact sections was studied by Johnson and Willmington (1972). Further studies of the effects of local buckling on the ultimate strength of continuous composite beams were reported by Climenhaga and Johnson (1972).

Inelastic Methods of Analysis Study of the elastic–plastic behavior and load-carrying capacity of simple-span composite beams is possible by inelastic methods of analysis (Dai and Siess, 1963; Baldwin et al., 1965; Yam and Chapman, 1968). Both the beams and the slab are assumed to deflect equally at every point along the span. A linear strain distribution is assumed over the depth of the composite beam, and strain discontinuity is accepted at the beam–slab interface. The concrete is assumed to develop no tensile strength and to produce a trapezoidal stress–strain relationship in compression. For the steel the stress–strain relationship is taken as elastic–perfectly plastic, or elastic–strain hardening. The load–slip dependence for the connectors is either a smooth parabolic curve or idealized as three straight lines.

Work on continuous composite beams was reported by Wu, Slutter, and Fisher (1971), and included the effects of shrinkage and prestressing of the concrete slab.

Torsion Consideration The behavior of composite beams under torsion is relevant to the analysis of curved beams and box girder bridges. Elastic methods have been proposed for homogeneous beams with either open or closed section. The elastic analytical techniques developed for composite beams are mere extensions of the procedures whereby pure Saint-Venant torsional properties and stresses are modified according to the shear modulus ratios, and warping torsional properties and stresses are modified by modulus of elasticity ratios. Heins and Kuo (1972a) give tables listing the pertinent torsional properties of composite bridge members. Heins and Kuo (1972b) have also studied the torsional response of composite T-beam models, and the results show that these analytical procedures predict the elastic response adequately. The ultimate torsional capacity is determined by the diagonal tension strength of the concrete slab. Tests on simple-span curved composite T beams are reported by Colville (1973). Under combined bending and torsion, axial deformation of the studs may permit the slab to rotate less than in steel beams. This investigator also presents a simplified method for designing stud connectors in curved composite members.

Effective Width of Slab Both American and British codes and specifications provide guidelines for calculating the effective flange width in bridges, but the derivation of these rules is not explained. The use of one-fourth of the beam span as one limitation of the effective width may be related to the theory of elasticity (Timoshenko and Goodier, 1951). This rule is also supported by Beschkine (1937–1938). The provision relating effective slab width to the stem width plus a multiple of slab thickness has probably appeared in earlier specifications regarding protection against crippling or buckling in steel compression members (see also subsequent sections).

The effective flange width has been studied by Mackey and Wong (1961), Lee (1962), and Reddy and Hendry (1970). The rules and formulas have been reviewed, and new procedures are proposed for simple-span composite beams. Studies on the effective slab width in negative moment regions (slab in tension) of continuous composite beams with continuous slab reinforcement have been carried out by Garcia and Daniels (1971), and led to the conclusion that the effective slab width may be taken constant throughout the beam length.

Reinforcement Several early reported failures of composite T beams were caused prematurely by longitudinal splitting of the concrete slab. Adekola (1959) reports studies showing that transverse tensile stresses were developed in the slab and must be resisted by transverse reinforcement. Roderick et al. (1967) continued the analytical and experimental investigation of various modes of failure (pullout of studs, shearing of studs, and longitudinal splitting) and studied the requirements of transverse reinforcement by testing scaled models of T beams.

The effect of transverse bending on the reinforcement has also been studied by Johnson, Van Daley, and Kemp (1967), leading to proposed design methods of transverse reinforcement necessary to ensure adequate longitudinal ultimate shear strength in the positive and negative moment regions.

The principal studies of longitudinal reinforcement requirements in negative moment regions of continuous composite beams have been carried out by Daniels (1972), and Johnson (1970) in the United States. Heavy longitudinal reinforcement in the slab appears to increase the vertical shear strength and shear stiffness of a composite beam. This made it practical to design composite beams assuming that a portion of the vertical shear is carried by the concrete slab. Other studies in conjunction with static and fatigue tests indicated that the total area of continuous longitudinal reinforcement in the negative moment regions of bridge members should equal or exceed 1 percent of the total cross-sectional area of the slab. This is necessary to prevent premature fatigue failure in the top layer of longitudinal steel (see also Section 12-14).

Deflection, Creep, Shrinkage, and Thermal Effects Long-term static tests on composite beams demonstrated considerable increases in deflection due to creep and shrinkage, and prompted the recommendation that time-dependent deflection be taken the same as instantaneous deflection (McGarraugh and Baldwin, 1971; Janss, 1972; Hasse, 1969). In Great Britain, Menzies (1968) determined that, following construction, the effects of temperature changes obscured the small effects caused by shrinkage and creep. In the United States, Roll (1971) analyzed stresses and deflections caused by differential shrinkage and creep in both shored and unshored beams, and obtained good correlation between measured and theoretical deflections.

The effects of creep, shrinkage, and temperature on a two-span continuous composite test beam have been studied by Ciolina (1971), from which a coefficient of shrinkage of 3.1×10^{-4} was obtained, which is somewhat different from the coefficient of 4×10^{-4} recommended by the French code. A mathematical model for analyzing continuous composite beams has been proposed by Wu, Slutter, and Fisher (1971), and includes the effects of shrinkage, prestressing, and loading.

Continuous Beams Results from investigations in the United States and Great Britain lead to the conclusion that continuous beams with adequately anchored longitudinal reinforcement placed continuously over the negative moment regions can be analyzed by simple plastic theory. Plastic hinges with moment–rotation capacity can be developed under combined negative moment and shear if adequate transverse reinforcement is provided and the design addresses the problem of compression flange buckling and web buckling of interior supports (Johnson, Van Daley, and Kemp, 1967).

The possibility of transverse cracks in the slab in negative moment areas complicates the analysis of continuous composite beams. Under the convenient assumption that concrete resists no tension (fully cracked section), it is possible to compute stress resultants and deflections under gravity loads. However, research (Daniels and Fisher, 1967) suggests that a cracked concrete slab in negative moment areas continues to participate with the longitudinal reinforcement, but the extent of this participation is reduced as the loading increases. Certain significant variables affecting this participation are thus isolated, and this includes the relative crack widths and crack patterns for continuous composite bridge beams. These conclusions form the basis for calculating the range of stress in the reinforcement and in the tension flange for fatigue considerations.

The practice in the past was to omit shear connectors in the negative moment region of continuous composite beams. Usually, this is defined as the portion between field splices, normally placed at the points of dead load contraflexure. The main intent was to avoid the deleterious effect of welded shear connectors on the fatigue strength of tension flanges. However, if the reinforcement in this region is continuous across the positive moment area, overstressing and premature fatigue failure of the anchorage connectors can occur well before the service (design) life has been consumed. To avoid this, additional connectors must be placed near the contraflexure points to develop the tension force in the longitudinal direction (initially recommended in the 1969 AASHO specifications).

Types and Strength of Shear Connectors

The three usual types of shear connectors include (a) shear studs, (b) steel channels, and (c) steel rods bent to form a spiral. Studs are the most common shear connector, and resemble a bolt without threads but with a round head. Studs are attached to the compression flange by an automatic welding process, in which the studs are inserted into a gun and held against the flange of a steel beam. A pull of the trigger closes the circuit by supplying the necessary voltage to melt the weld material at the end of the stem. The time necessary to weld the stud is approximately 1 sec. Alternatively, a small section of steel channel with its long axis transverse to the beam is placed across the top flange and properly welded. A circular spiral laid along the top flange is welded to it at the contact point. Shear connector details are discussed in the following sections.

The *useful capacity* of a shear connector is defined as the point where the load–residual slip relationship becomes nonlinear. According to early guidelines, if this point was not readily discernible, the useful capacity was taken as the load corresponding to a residual slip of 0.003 in. These considerations were derived primarily from static pushout and beam tests, and the associated conservative approach did not require attention to fatigue problems. In 1960, however, a modified version of the AASHO formula was proposed by the Joint ASCE–ACI Committee, and provided the first step in establishing more liberal shear connector criteria.

Static Strength of Studs The strength of shear connectors obtained from pushout tests is generally lower than the strength obtained from beam tests (Slutter and Driscoll, 1965). The same investigators have also determined that the magnitude of slip at the beam–slab interface does not significantly affect the development of the ultimate moment resisting capacity provided the total strength of the connection is adequate to resist the ultimate compressive force in the concrete. These results formed the basis for the 1961 AISC specifications on ultimate strength design, with an assumed safety factor of 2.5. Studies at Imperial College (Chapman, 1964; Chapman and Balakrishman, 1964) led to the same conclusions. Thus, both American and British standards permit uniform spacing of connectors, consistent with the results of beam tests demonstrating that plastic redistribution of forces occurs before failure.

The need to specify a minimum tensile strength for stud connectors was demonstrated by Hawkins (1970) in static tests of full-scale T beams and pushout specimens. At working load and at ultimate load, loss of composite action was clearly greater for hot-forged studs than for cold-headed studs, because of the lower tensile strength of the former. Goble (1967) has shown that even with different yield strengths, comparable ultimate shear loads are obtained from tests of cold-headed studs of the same tensile strength.

Studies at Lehigh University resulted in the development of a new ultimate strength relationship for stud connectors, applicable to both normal-weight and lightweight concrete (Slutter and Fisher, 1971). This interaction was valid for stud shear connector capacity for concrete strengths in the range 2.7 to 5.1 ksi and for densities of 90 to 148 pcf. The same studies have also shown that the strength of shear connectors increases with increasing stud height, but beyond a height equal to four diameters the results are moderate. Other tests have shown that for heights less than four diameters, stud pullout and shear capacity are reduced. Based on pushout tests, Menzies (1971) has recommended lower static and fatigue strengths of studs, in normal concrete than those specified in Code of Practice 117 (British Standards Institute) together with a concrete strength that can be extended to 7500 psi. Most of the work described in this paragraph has provided the basis for the current AASHTO standards on shear studs.

Fatigue Strength The fatigue strength of shear connectors has been investigated in pushout and small-scale beam tests. Other tests involved full-sized beams. From these, a procedure was developed whereby the fatigue design is based on the shear stress range, that is, the difference between the maximum and minimum stresses. In the United States, the research has focused mainly on stud connectors, but in Great Britain, fatigue studies included studs, channels, and bar connectors.

Fatigue tests in Japan with direct tension specimens and beams subjected to negative bending confirm a substantial reduction of the tension flange fatigue strength in the presence of shear connectors, and this reduction tends to be smaller in bending than in tension tests. These conclusions agree with results of tests on steel plates with studs completed in the United States and Germany.

Haunches Pushout tests with haunched, unreinforced slabs led to a procedure for predicting the strength of connectors based on a shear failure–plane theory. In other pushout tests, the strength of stud connectors in vertical-sided haunches 2 and 4-in. deep was compared with the strength of studs in unhaunched slabs for a wide range of concrete strengths. The strength of studs in haunches with vertical sides is reduced because of less containment for the triaxially stressed concrete around the stud connector. A limit on the slope of haunch sides of one vertical to three horizontal is included in Code of Practice 117 (BSI).

Deep haunches in composite beams have been studied in pushout and beam tests. Failure by bursting of the concrete around the stud connectors shows the importance of sufficient side cover and reinforcement in deep haunches. Johnson (1972) has proposed an ultimate strength method for designing the reinforcement in deep haunched composite beams in the new British bridge code.

Effect of Beam Flange Thickness Tests of stud connectors welded to light-gage box beams carried out by Vergun and Shah (1968) show that most of the specimens failed by tearing of the base metal and rotation of the studs. The ultimate strength of connectors ranged from 62 to 73 percent of the strength calculated from AISC formulas. In other investigations conducted by Goble (1968) on the behavior of 1/2- 5/8-, and 3/4-in.-diameter stud connectors welded to thin beam flanges, the conclusion was reached that the ratio of stud diameter to flange thickness should not exceed about 2.7. This is necessary to prevent the reduction of connector strength caused by pullout from the beam flange. These tests involved A36 steel.

High-Strength Bolts The Yatsumichi railroad bridge in Japan was constructed in 1969 with precast slabs embedded in epoxy mortar and connected to the steel girders with high-strength bolts. Static and fatigue tests show that these beams are stronger under repeated loading.

High-strength friction bolts were first used as shear connectors for the Ems pedestrian bridge in 1960. Dallam and Harpster (1968) studied the strength of these bolts in pushout and beam tests. Other tests were made by Holtz and Kulak (1972) and Marshall et al. (1971). Although a first slip could be safely calculated using a coefficient of friction of 0.45 for precast slabs on steel, a higher coefficient of friction was considered appropriate for cast-in-place slabs.

Summary of AASHTO Guidelines

Article 10.38 of AASHTO deals with composite beams and girders. The composite moment of inertia is the basis for computing stresses in the steel and the concrete, but the analysis should be consistent with predetermined material properties. Where dead loads act on the composite section, the effect of creep must be considered.

In simple spans and in the positive moment regions of continuous spans, the composite section should be proportioned so that the neutral axis lies below the top of the steel beam. In the negative moment regions of continuous spans, only the slab reinforcement can be considered to be part of the composite action with the steel beams. If the main beams or girders are not supported by intermediate falsework, they must be investigated for structural stability during the curing time of the concrete.

Maximum compressive and tensile stresses in beams and girders that are not provided with temporary supports during the concrete placement are the result of stresses calculated from dead loads acting on the noncomposite steel beam alone and superimposed loads (dead and live) acting on the composite section.

A continuous composite bridge may be built with shear connectors placed only in the positive moment regions, or throughout the length of the bridge. The positive moment regions may be designed with composite sections as in simple spans. Shear connectors should be provided in negative regions where the reinforcement embedded in the concrete is considered part of the composite section. If the latter is not a precondition, shear connectors in negative moment regions are not necessary, but additional shear connectors should be placed near the points of dead load contraflexure. In the negative moment regions of continuous spans, the minimum longitudinal reinforcement should not be less than 1 percent of the cross-sectional area of the slab.

The provisions regarding live load plus impact deflections are also applicable to composite beams and girders. When the girders are not provided with falsework or other effective supports during the placement of the concrete slab, the deflection due to the weight of the slab and other permanent dead loads added before the concrete has attained 75 percent of its required 28-day strength should be computed on the basis of the noncomposite section.

Yield and Plastic Moment: LRFD Specifications

The yield moment M_y of a composite section is the sum of the moments applied separately to the steel and to the short-term and long-term composite sections to cause first yielding in either steel flange when any web yielding is disregarded. M_y depends on the ratio of the moments applied to the steel and composite sections, and is needed for the strength limit state for the following types of composite design: compact positive bending sections in continuous spans; negative bending sections designed by alternate procedures; homogeneous sections with stiffened webs subjected to combined moment and shear exceeding specified limits; and noncompact sections used at the last plastic hinge location in inelastic designs.

The plastic moment M_p of the composite section is the first moment of plastic forces about the plastic neutral axis. Plastic forces in steel portions are calculated using the yield stress of structural steel and reinforcing steel. Plastic forces in the concrete (compression zone) are based on a rectangular stress block. The position of the plastic neutral axis is determined by the equilibrium condition that a net axial force does not exist.

4-6 TRENDS IN COMPOSITE BRIDGES

Composite bridges have become standard practice both in the United States and abroad. Although standards and specifications are available, few if any include provisions for bridges with precast and prestressed concrete decks.

With beam or girder bridges, both simple and continuous, composite steel–concrete construction has been extended to box girders, arches, rigid frames, cable-stayed, and suspension bridges. This discussion focuses, however, on bridge decks, with emphasis on simple and continuous steel bridges.

A cracked slab is subject to the adverse effects of water and deicing salts, and this condition is particularly serious when the wearing surface constitutes an integral part of the concrete slab. Alternatively, an asphalt surface may be used as protection, but watertightness cannot be assumed even with a protective layer between the asphalt and the concrete slab. Several state codes stipulate an increased concrete cover on the top reinforcement to provide for cracking and deck wear, but this is inconsistent with the practice of placing the reinforcement near the surface of the slab to control cracking.

In the United States, most composite configurations are in simple spans or in the positive moment regions of continuous spans. Prestressed bridge decks are seldom employed for composite bridge construction. Because prestressing adds considerably to the cost of continuous composite bridges, the present trend is to avoid this design unless depth restrictions control, where slab cracking must be prevented or where continuity must be provided for a slab composed of prestressed precast concrete elements (see also Section 3-11).

Prestressed Cast-in-Place Bridge Decks The prestressing concept in bridges has resulted in two philosophies, both evolved from the prestressing of decks in continuous bridges to control slab cracking in negative moment regions. One solution is to precompress the slab to control cracking under the combined effects of live load, dead load, and related influences such as creep, shrinkage, and differential temperature. The slab in this case is analyzed as a homogeneous uncracked section, and the treatment constitutes total prestressing although small tensile stresses may be allowed. On the other hand, partial prestressing is an alternate solution whereby total prestressing is introduced to handle dead loads and long-term effects while the slab is considered uncracked. However, under the influence of live load, some cracking in the concrete must be accepted. Limitations are placed on the crack opening by stipulating lower values for tensile stresses in reinforcement steel.

Transverse precompression of the slab by posttensioning cables is introduced to control longitudinal cracking and creep deformations of the deck under dead load. Either complete or partial prestressing may be used. However, transverse posttensioning is costly and has been used occasionally for wide bridge decks on two main girders.

Longitudinal prestressing methods may be used, separately or combined, to prevent or limit tensile stresses in the concrete slab in negative moment regions. One method, used in Europe, involves the concept of vertical adjustment of the intermediate supports. These are lowered after the concrete has cured and a composite section has been obtained. Bending moments are thus introduced in the composite structure, and the concrete is precompressed in the negative moment region. Some doubts exist about the effectiveness of countering shrinkage because the benefits are balanced by creep in the slab.

Tensile stresses of the concrete in the negative moment regions may also be reduced by a proper pouring sequence, in conjunction with the use of either continuous or traveling formwork. The deck in the positive moment areas is cast first, followed by the portions in the negative moment regions. After the concrete has cured rendering the composite action effective, only the additional loads, such as live and superimposed dead loads, will influence the composite structure.

At present, the use of longitudinal prestressing appears to be limited because of the initial cost and the difficulty of controlling prestress loss. A further problem is cracking at the location of the anchorage of the posttensioning cables. Continuous bridges are more susceptible to adverse effects in view of support settlement.

Precast Elements for Bridge Decks In many instances, the cost of continuous formwork warrants consideration of alternatives based on the concept of prefabricated elements. Prefabricated deck sections of unit length and, where feasible, of full deck width may offer noble solutions. This form of bridge construction is discussed in other sections.

Design Considerations for Prestressed and Precast Composite Bridges In most cases, specifications and design philosophies for prestressed and precast composite structures are not available, and engineers must therefore resort to codes for prestressed concrete and structural steel. Experience indicates that, in assessing the structural safety of the prestressed deck, the ultimate moment capacity of the section should form the basis, because prestressing and long-term effects such as creep and relaxation do not change the ultimate moment capacity. Where sufficient plastic hinge rotation cannot be predicted because of local instabilities, safety against this factor should be checked. In addition, the fatigue strength must be evaluated, and because the stress range is a controlling factor the design should address the live load effects. The influence of studs, stiffeners, and other attachments warrants ample evaluation in the negative moment regions. This detail is more important in high-strength-steel bridges where no increase in fatigue strength should be expected for the higher grade steel at welded details and connections.

4-7 COMPOSITE I-BEAM BRIDGE: SIMPLE SPAN

Economic Considerations

If an I-beam rolled section is used along with a nominal concrete slab 7.5 in. thick in composite design, the centroidal axis of the composite section is normally just below the steel compression flange. The incorporation of the concrete slab in the beam properties increases the composite moment of inertia, but also increases the distance to the extreme tension fiber and thus the section modulus for the tension steel increases only moderately. For a more efficient composite section, the steel bottom flange must be larger than the steel compression flange, thus shifting the centroidal axis of the composite section closer to the centroidal axis of the steel beam. This can be achieved in two ways: (a) by selecting a welded plate girder where wide choice exists in specifying the size and thickness of the top and bottom flange and (b) by welding a cover plate (partial length) to the tension flange of a rolled steel beam.

Experience shows that maximum economy is attained with the lightest and shallowest beam combined with a large cover plate. The thickness of the cover plate is limited to two times the thickness of the flange to which the cover plate is attached. The cover plate can be wider or narrower than the flange. The end detail of a narrower cover plate is simplified because transverse welds at the ends are not required. This design is also preferable in terms of fatigue strength of the steel rolled section at this location. In either case, the difference in width between the steel flange and the cover plate should allow sufficient space on each side to lay a fillet weld.

Partial-length cover plates should extend beyond the theoretical end by a terminal distance, or to a section where the stress range in the beam flange is equal to the allowable fatigue stress range for base metal adjacent to or connected by fillet welds, whichever is greater. The theoretical end of the cover plate is the section where the stress in the flange alone does not exceed the allowable service load stress exclusive of fatigue considerations.

Economy Graphs For rule-of-thumb or preliminary estimates, direct reference to appropriate graphs is useful. Two examples of economy graphs are shown in Figures 4-5 and 4-6. The criteria for developing these charts are based on an HS 20 truck load and a wheel load distribution of $S/5.5$. Cover plate size (area) and length for the bottom flange can be found from companion graphs. Total cover plate lengths include extensions beyond the theoretical length to the point where the allowable fatigue stress in the beam flange equals the allowable fatigue stress adjacent to fillet welds for 500,000 maximum load applications, or a terminal distance 1.5 times the beam flange width. Impact is included, and unshored construction is assumed. The charts show the old nomenclature for rolled beams and are taken from AISC (1969).

The left portion of the charts depicts a variable ordinate scale that takes into account the span and superimposed dead load. The dead load (resisted by the beam only) is the weight of the steel beam plus cover plate and the concrete structural slab. An integral wearing surface, a future wearing surface, curbs and parapets, or any other loads are treated as superimposed dead loads.

The charts give the lightest rolled section of each beam depth series less than 36 in. Beam spacing can affect overall bridge economy. Normally,

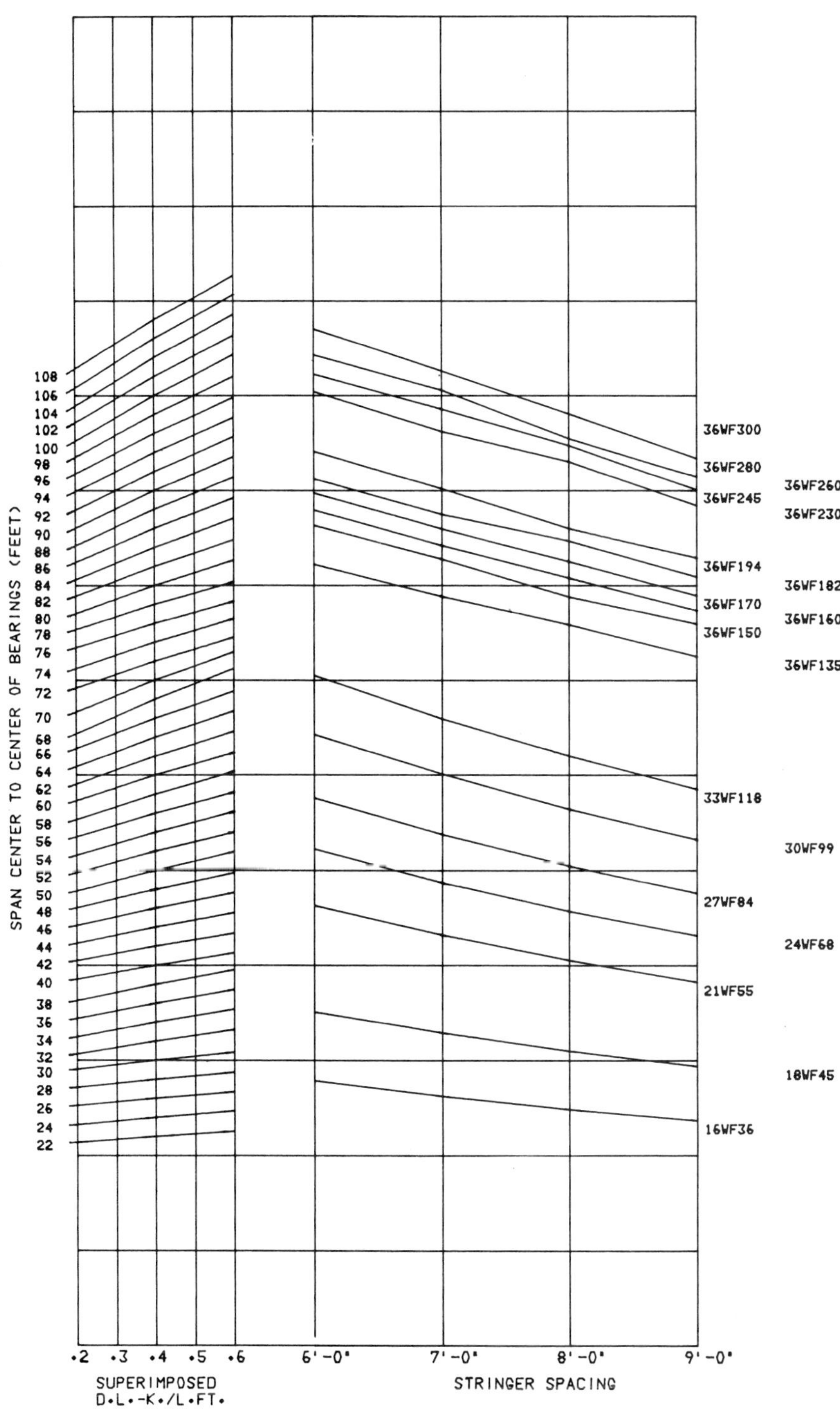

FIGURE 4-5 Economy graph of preliminary composite beam selection; steel yield strength 36 ksi, 7-in. structural slab thickness. (From AISC, 1969.)

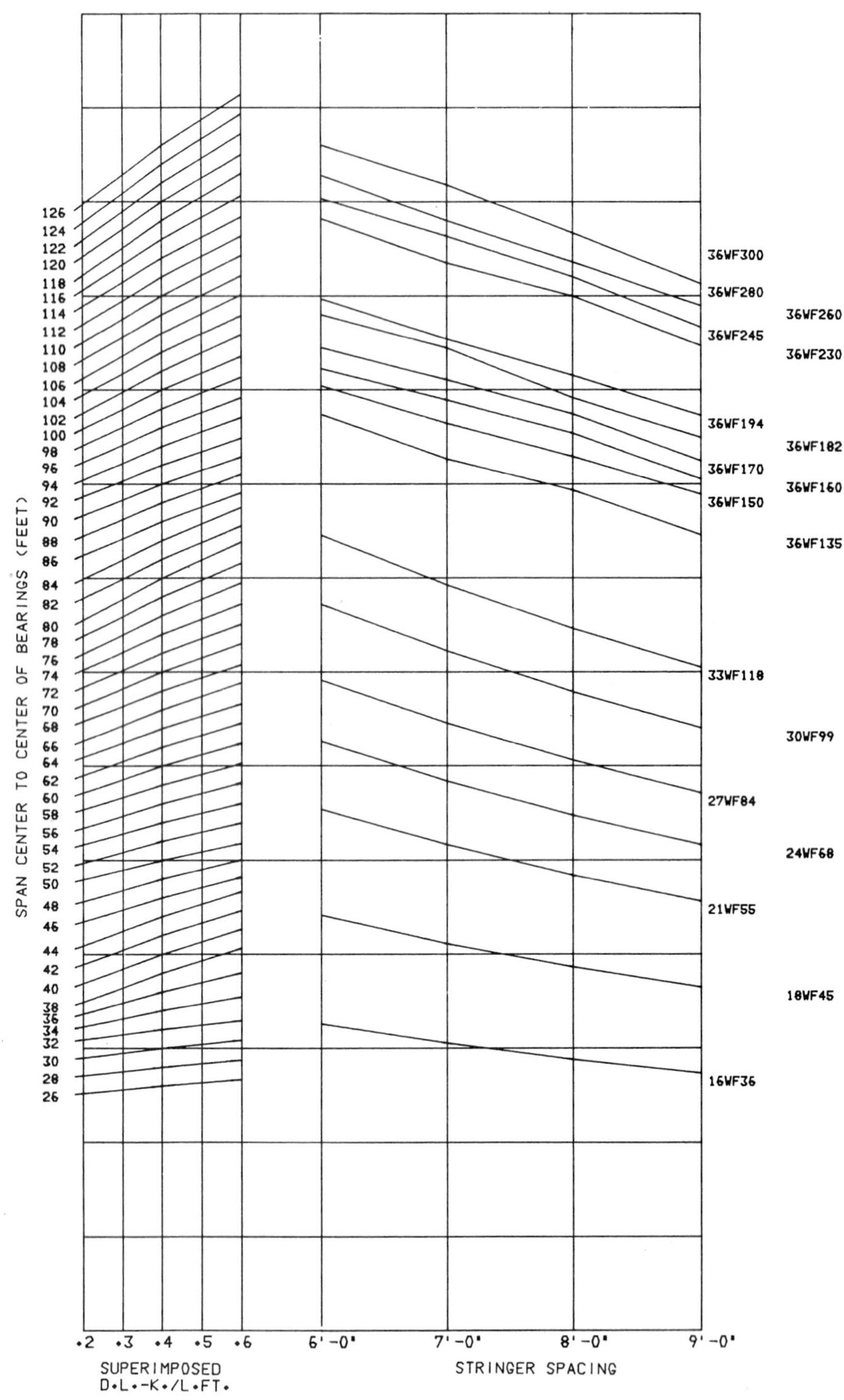

FIGURE 4-6 Economy graph for preliminary composite beam section; steel yield strength 50 ksi, 7-in. structural slab thickness. (From AISC, 1969.)

economy is maximized by using the least number of beams consistent with maximum allowable overhang of the slab beyond the exterior beams. Unless relevant standards dictate maximum beam spacing in conjunction with optimum slab thickness, fewer beams in a cross section may result in minimum structural steel weight, together with fewer bearings and diaphragms.

Example The following data are given for a composite bridge:

Span, center-to-center = 55 ft
Beam spacing = 7 ft 6 in.
Superimposed dead load = 0.45 kip/ft
Structural slab thickness = 7 in.
Material = AASHTO M 270, Grade 36

From the economy chart of Figure 4-5, we find the intersection of the curve for a 55-ft span with the vertical line corresponding to a superimposed dead load of 0.45 kip/ft. Then we proceed horizontally to the right until intersecting the vertical line corresponding to a beam spacing of 7 ft 6 in. Thus, we determine that W30 × 99 (new designation) will probably yield the most economical design.

Next, we refer to the companion graph for W30 × 99 shown in Figure 4-7, and we intersect the curve for the 55-ft span with the vertical line representing the superimposed dead load of 0.45 kip/ft. From this point, we move horizontally to intersect the beam spacing of 7 ft 6 in. Moving downward to the right and parallel to the curves, we intersect the cover plate area axis and find that the cover plate area must be 10.75 in. Moving to the left and parallel to the dashed curves, we intersect the cover plate length line and determine that the total length is 40 ft.

The preliminary (initial) selection is therefore: beam, W30 × 99; cover plate, 9 × 1.25, 40 ft long. Graphs for the entire range of slab thicknesses and rolled beams can be found in the AISC publication Simple Span Steel Bridges (composite beam design charts).

In order to check the efficiency and economy of composite design, we will compare the foregoing design with a simple noncomposite beam. The weight of the slab is 7.5 × 0.583 × 0.15 = 0.66 kip/ft. Assuming a beam weight of 0.15 kip/ft, the total dead load is 0.66 + 0.15 + 0.45 = 1.26 kips/ft, giving a dead load moment

$$M_{\mathrm{DL}} = 0.125 \times 1.26 \times 55^2 = 476 \text{ ft-kips}$$

Using a live load distribution factor of 7.5/5.5 = 1.36 and an impact factor of 50/(155 + 125) = 28 percent, the live load plus impact moment is

$$M_{\mathrm{LL}+I} = 0.5 \times 717 \times 1.36 \times 1.28 = 623 \text{ ft-kips}$$

Therefore, the total design moment is 476 + 623 = 1100 ft-kips. This will require a section modulus of 1100 × 0.6 = 660 in.^{3}, which is provided by a

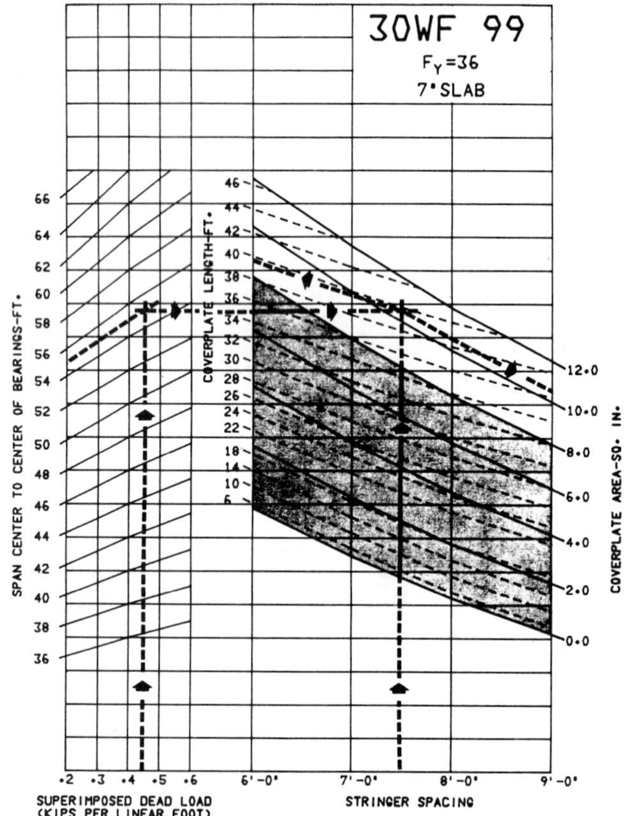

FIGURE 4-7 Economy graph for preliminary cover plate selection; W30 × 99 beam, 7-in. slab, simple spans. (From AISC, 1969.)

W36 × 194 beam. Economic comparison involves the cost of the beam, cover plate, welding, and shear connectors for the composite design versus the cost of a single beam for the noncomposite scheme.

Design Example (Without Shoring)

For a simple composite I-beam bridge, the following are given:

Span = 70 ft, center-to-center of bearings

Structural slab = 7.5 in.

Loading = HS 20; number of beams = 5; spacing = 7 ft 6 in.

Steel-AASHTO M270, grade 36

Future wearing surface = 25 psf

Haunch (fillet) = 1 in. over beam

Curb and parapet are as shown in Figure 4-3 and are to be treated as superimposed dead load

The dead load for noncomposite action consists of the following:

$$\text{Slab} = 7.50 \times 0.625 \times 0.15 = 0.70 \text{ kip/ft}$$

$$\text{Beam etc. (assume)} = 0.20 \text{ kip/ft}$$

$$\text{Haunch} = \underline{0.025} \text{ kip/ft}$$

$$\text{Dead load } w = 0.925 \text{ kip/ft}$$

The superimposed dead load consists of the following:

$$\text{Curb, parapet} = 0.18 \text{ kip/ft}$$

$$\text{FWS} = 7.50 \times 0.025 = \underline{0.19} \text{ kip/ft} \quad \text{(approximately)}$$

$$\text{Superimposed dead load } w = 0.37 \text{ kip/ft}$$

The dead load moment is

$$M_{\text{DL}} = 0.125 \times 0.925 \times 70^2 = 566 \text{ ft-kips}$$

The superimposed dead load moment is

$$M_{\text{SDL}} = 0.125 \times 0.37 \times 70^2 = 226 \text{ ft-kips}$$

Wheel load per beam is from $S/5.5 = 7.5/5.5 = 1.36$. From Figure 2-41, with $S/L = 7.50/70 = 0.11$, $g = 1.27$ (obtained graphically); use a wheel load distribution of 1.3. Likewise, we compute the impact factor $I = 50/(125 + 70) = 0.256$. The live load moment (one lane) is obtained directly from AASHTO tables as 986 ft-kips. The live load plus impact moment per beam is

$$M_{\text{LL}+1} = 0.5 \times 986 \times 1.3 \times 1.26 = 807 \text{ ft-kips}$$

The Effective width of the flange in a composite beam is as follows:

$$\text{One fourth of span length} = 70/4 \quad 17.5 \text{ ft}$$

$$\text{Distance center-to-center of beams} = 7.5 \text{ ft}$$

$$\text{12 times slab thickness} = 12 \times 7.5/12 = 7.5 \text{ ft} \qquad \text{Use 7.5 ft}$$

The transformed width of the flange is $7.5 \times 12/9 = 10$ in. ($n = 9$) for $f'_c = 3500$ psi. Also, $7.5 \times 12/27 = 3.3$ in. ($n = 27$).

By reference to the AISC graphs, we select W36 × 135 and a bottom cover plate 10 in. × 1.25 in. Next, we compute the properties of the composite section according to the details shown in Figure 4-8. Axis x–x is the neutral axis of the beam. Note that the weight of the beam plus the weight of the cover plate is approximated as 0.18 kip/ft, which is close to the assumed weight.

The composite beam properties are as follows:

Beam and Cover Plate	A	x	Ax	$Ax^2 + I_0$
WF	39.8	—	—	7820
PL	12.5	−18.40	−248	4233
	52.3		−248	12,053
$x_0 = 248/52.3 = 4.74$ in.				$-248 \times 4.74 = -1175$
				$I_{comp} = 10{,}878$ in.^{4}
$17.78 + 4.74 = 22.52$	$10{,}878/22.52 = 483$			S, top/steel
$19.03 - 4.74 = 14.29$	$10{,}878/14.29 = 761$			S, bottom/steel

With Slab, $n = 9$	A	x	Ax	$Ax^2 + I_0$
WF + PL	52.3		−248	12,053
Slab = 10 × 7.5 =	75.0	22.53	1690	38,452
	127.3		1442	50,505
$x_0 = 1442/127.3 = 11.33$				$-1442 \times 11.33 = -16{,}338$
				$I_{comp} = 34{,}167$ in.^{4}
$17.78 + 1.25 + 11.33 = 30.36$	$34{,}167/30.36 = 1125$			S, bottom/steel
$17.78 - 11.33 = 6.45$	$34{,}167/6.45 = 5297$			S, top/steel
$6.45 + 8.50 = 14.95$	$34{,}167/14.95 = 2285$			S, top/concrete

With Slab, $n = 27$	A	x	Ax	$Ax^2 + I_0$
WF + PL	52.3		−248	12,053
Slab = 3.3 × 7.5 =	24.8	22.53	559	12,705
	77.1		311	24,758
$x_0 = 311/77.1 = 4.03$				$-311 \times 4.03 = -1253$
				$I_{comp} = 23{,}505$ in.^{4}
$17.78 + 1.25 + 4.03 = 23.06$	$23{,}505/23.06 = 1019$			S, bottom/steel
$17.78 - 4.03 = 13.75$	$23{,}505/13.75 = 1709$			S, top/steel
$13.75 + 8.50 = 22.25$	$23{,}505/22.25 = 1056$			S, top/concrete

Steel stress in the bottom flange (tension)

$$f_s = \frac{566 \times 12}{761} + \frac{226 \times 12}{1019} + \frac{807 \times 12}{1125} = 8.93 + 2.66 + 8.61 = 20.20 \text{ ksi}$$

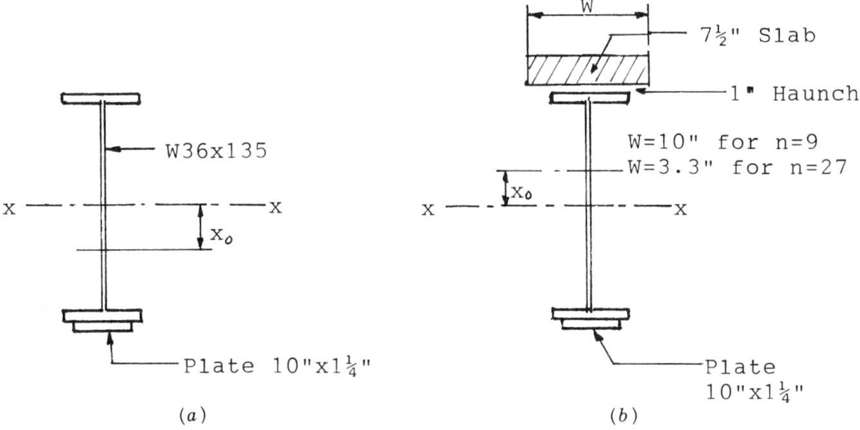

FIGURE 4-8 Typical cross section for composite beam: (*a*) steel beam and cover plate; (*b*) steel beam, cover plate, and slab.

Steel stress in the top flange (compression)

$$f_s = \frac{566 + 12}{483} + \frac{226 \times 12}{1709} + \frac{807 \times 12}{5297} = 14.06 + 1.59 + 1.83 = 17.48 \text{ ksi}$$

Stress in top of the concrete (compression)

$$f_c = \frac{226 \times 12}{1056 \times 27} + \frac{807 \times 12}{2285 \times 9} = 238 + 470 = 718 \text{ psi} < 1400 \text{ psi}$$

The 1 percent overstressing for the bottom steel is acceptable and the design is therefore satisfactory.

Length of Cover Plate Our practice has been to detail the end of the cover plate as shown in Figure 4-9. The terminal distance is 1.5 times the cover plate width because the cover plate is welded across its ends. The cover plate is tapered as shown, and has an end width of 3 in.

The required length can be determined by calculating the moments at several locations inward from the bearings. From our initial reference to design charts, we have estimated an approximate cover plate length of 48 ft. This suggests that moments may be computed for distances 5, 10, and 15 ft from the bearings.

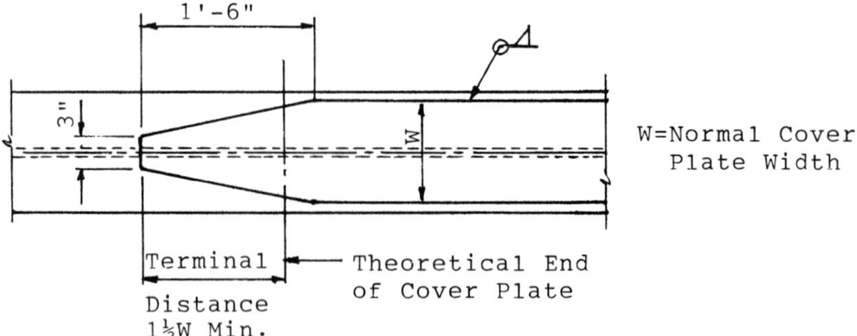

FIGURE 4-9 Cover plate end detail.

The composite properties for the beam and slab (without cover plate) are as follows:

$n + 9$	A	x	Ax	$Ax^2 + I_0$
WF	39.8	—	—	7820
Slab	75.0	22.53	1690	38,452
	114.8		1690	46,272
$x_0 = 1690/114.8 = 14.72$			$-1690 \times 14.72 =$	$-24,877$
				$I_{\text{COMP}} = 21,395$ in.^{4}
$17.78 + 14.73 = 32.51$	$21,395/32.51 = 658$			S, bottom/steel
$3.05 + 8.50 = 11.55$	$21,395/11.55 = 1852$			S, top/concrete

$n = 27$	A	x	Ax	$Ax^2 + I_0$
WF	39.8	—	—	7820
Slab	24.8	22.53	559	12,705
	64.6		559	20,525
$x_0 = 559/64.6 = 8.65$			$-559 \times 8.65 =$	4835
				$I_{\text{comp}} = 15,690$ in.^{4}
$17.78 + 8.65 = 26.43$	$15,690/26.43 = 594,$			S, bottom/steel

At any distance x along the beam, measured from the end bearing, the dead load moment is

$$35wx - \frac{wx^2}{2}$$

where w is the uniform load (dead or superimposed). The live load influence

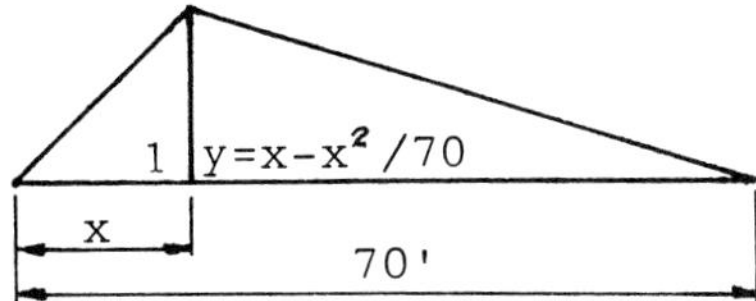

FIGURE 4-10 Bending moment influence diagram; simple span 70 ft long, unit load, point at distance x from left support.

diagram is shown in Figure 4-10. At $x = 5$ ft, $y = 4.67$, and

$$M_{LL} = 16\left(4.67 + 4.67\frac{51}{65}\right) + 4\left(4.67\frac{37}{65}\right) = 133 + 11 = 144 \text{ ft-kips}$$

At $x = 10$ ft, $y = 8.57$, and

$$M_{LL} = 16\left(8.57 + 8.57\frac{46}{60}\right) + 4\left(8.57\frac{32}{60}\right) = 243 + 18 = 261 \text{ ft-kips}$$

At $x = 15$ ft, $y = 11.79$, and

$$M_{LL} = 16\left(11.79 + 11.79\frac{41}{55}\right) + 4\left(11.79\frac{27}{55}\right) = 330 + 23 = 353 \text{ ft-kips}$$

All live load moments must be multiplied by the distribution factor and the impact coefficient to obtain the live load plus impact moment per beam. At the three locations, dead load, superimposed dead load, and live load plus impact moments are tabulated as follows (DL $w = 0.925$, SDL $w = 0.37$)

x	M_{DL}	M_{SDL}	M_{LL+I}
5	150	60	235
10	278	111	425
15	381	152	575

From the foregoing data, we now compute the bending stresses (tension) in the beam without the cover plate. These are tabulated as follows:

x	Stress (DL)	Stress (SDL)	Stress (LL + I)	Total
5	4.1	1.2	4.3	9.6
10	7.6	2.2	7.8	17.6
15	10.4	3.1	10.5	24.0

The theoretical cutoff point is between 10 and 15 ft from the bearing, and straight-line interpolation provides sufficient accuracy. Then

$$x = 10 + \frac{2.4}{6.4}(5) = 11.9 \text{ ft}$$

The cover plate length (theoretical) is therefore

$$70 - 2 \times 11.9 = 46.2 \qquad \text{say } 46.5 \text{ ft}$$

The terminal distance is $1.5 \times 10 = 15$ in. $= 1.25$ ft, giving a total cover plate length of $46.5 + 2.5 = 49$ ft.

Because of repetitive loading considerations, the design must check the allowable fatigue stress at the end of the cover plate. For a base metal of partial length welded cover plate narrower than the flange having square or tapered ends, with or without welds across the ends and a flange thickness of 0.79 in. $\leq$ 0.8 in., the stress category is E. For 500,000 cycles and stress category E, the specifications allow a range of stress $F_{sr} = 13$ ksi. In this case the stress range is simply the live load plus the impact stress. At the end of the cover plate (10.5 ft from the support), this stress is about 8 ksi, and clearly less than the allowable stress range.

Shear Connectors We choose to distribute the live load for a truck axle above the end bearing assuming simple beam action and placing two trucks in 10-ft lanes. For the end axle, the live load distribution is

$$1 + \frac{1.5}{7.5} + \frac{3.5}{7.5} = 1.67 \qquad \text{and} \qquad 1.3 \text{ for other axles}$$

Note that according to the new (*proposed*) shear distribution formulas, the live load end shear is obtained from AASHTO (NCHRP Project 12-26) Table 3.23.6 as the algebraic sum $0.4 + 7.5/6.0 - (7.5/25)^2 = 1.56$. Therefore, using 1.67 as the distribution coefficient is more conservative for the axle load at the end.

The shear at the end of the beam is as follows:

$$V_{\text{DL}} = 0.93 \times 35 = 32.6 \text{ kips}$$

$$V_{\text{SDL}} = 0.37 \times 35 = 13.0 \text{ kips}$$

$$V_{\text{LL}+I} = \left[1.67 + 16 + 1.3\left(16 \times \frac{56}{70} + 4 \times \frac{42}{70}\right)\right] \times 1.26 = 58.6 \text{ kips}$$

which is also the range of shear at the end of the beam.

The range of shear at 7 ft from the end of the beam (Figure 4-11)

$$\text{Positive} \qquad V_{\text{LL}+I} = 1.30\left(16 \times \frac{63}{70} + 16 \times \frac{49}{70} + 4 \times \frac{36}{70}\right) \times 1.27$$

$$= 46.0 \text{ kips}$$

$$\text{Negative} \qquad V_{\text{LL}+I} = 1.30\left(16 \times \frac{63}{70} - 16\right) \times 1.27 = -2.6 \text{ kips}$$

FIGURE 4-11 Live load arrangement for shear at 7 ft from support: (*a*) positive; (*b*) negative.

obtained by reversing the truck direction as shown in Figure 4-11*b*, giving a range of shear of 46.0 − (−2.6) = 48.6 kips.

The range of shear at 14 ft from end of the beam (Figure 4-12)

$$\text{Positive} \qquad V_{\mathrm{LL}+I} = 1.30\left(16 \times \frac{56}{70} + 16 \times \frac{42}{70} + 4 \times \frac{28}{70}\right) \times 1.28$$

$$= 39.9 \text{ kips}$$

$$\text{Negative} \qquad V_{\mathrm{LL}+I} = 1.30\left(16 \times \frac{56}{70} - 16\right) \times 1.28 = -5.3 \text{ kips}$$

obtained by reversing the truck direction as shown in Figure 4-12*b*, giving a range of shear of 39.9 − (−5.3) = 45.2 kips.

The range of shear at midspan of beam (Figure 4-13)

$$\text{Positive} \qquad V_{\mathrm{LL}+I} = 1.30\left(16 \times \frac{1}{2} + 16 \times \frac{21}{70} + 4 \times \frac{7}{70}\right) \times 1.30$$

$$= 22.3 \text{ kips}$$

$$\text{Negative} \qquad V_{\mathrm{LL}+I} = 1.30\left(4 \times \frac{63}{70} + 16 \times \frac{49}{70} + 16 \times \frac{1}{2} - 36\right) \times 1.30$$

$$= -22.4 \text{ kips}$$

obtained by reversing the truck direction as shown in Figure 4-13*b*, giving a range of shear of 22.3 − (−22.4) = 44.7 kips.

Design of Stud Shear Connectors for Fatigue: For 5-in. × 7/8-in. shear studs, $H/d > 4$. From Article 10.38.5.1 (AASHTO), the allowable range is $Z_r = \alpha d^2$, where $\alpha = 10{,}600$ for 500,000 cycles, or $Z_r = 10{,}600(7/8)^2 = 8200$

FIGURE 4-12 Live load arrangement for shear at 14 ft from support: (*a*) positive; (*b*) negative.

FIGURE 4-13 Live load arrangement for shear at midspan: (*a*) positive; (*b*) negative.

lb. The range of shear is computed by the formula

$$S_r = \frac{V_r Q}{I} \tag{4-1}$$

where S_r, V_r, Q, and I are defined in AASHTO Article 10.38.5.1.1. At the end of the beam (without the cover plate), the distance from the neutral axis of the composite section to the centroid of the transformed concrete area is $11.55 - 3.75 = 7.8$ in. We compute $Q = 75 \times 7.8 = 585$ in.^{3} and recall that $I = 21{,}395$ in.^{4}.

With the cover plate present, $A = 75(14.95 - 3.75) = 11.2 \times 75 = 840$ in.^{3} and $I = 34{,}167$ in.^{4}. The shear per inch must satisfy the equation

$$nZ_r = S_r \times (\text{Spacing}) \tag{4-2}$$

where n is the number of studs per row, from which we obtain

$$\text{Spacing} = \frac{nZ_r}{S_r} \tag{4-3}$$

Using the foregoing data, we tabulate the stud spacing as follows ($n = 2$):

Distance from Support (End) (ft)	V_r (kip)	I (in.^{4})	Q (in.^{3})	S_r (kip/in.)	Spacing (in.)
0	58.6	21,395	585	1.60	10.25
7	48.6	21,395	585	1.33	12.30
14	45.2	34,167	840	1.11	14.9
℄	44.7	34,167	840	1.10	15.0

Using two studs per section, the spacing is as shown in Figure 4-14*a*. Alternatively, we may choose equal spacing throughout the beam length. From the preceding table,

$$\text{Average } V_r = (58.6 + 44.7)/2 = 51.7 \text{ kips}$$

$$\text{Average } I/Q = (21{,}395/585 + 34{,}167/840)1/2 = 38.6 \text{ in.}$$

$$\text{Spacing} = 16.4 \times 38.6/51.7 = 12.2 \text{ in.} \qquad \text{say 12 in.}$$

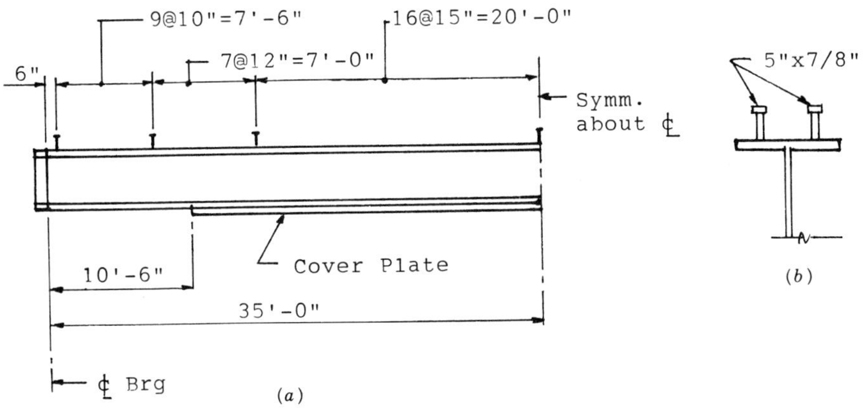

FIGURE 4-14 Beam elevation and stud spacing details.

Check Shear Connectors for Ultimate Strength: The number of connectors required for fatigue must be checked to ensure that the ultimate strength provisions are satisfied (AASHTO Article 10.38.5.1.2). The number of shear connectors must equal or exceed the number given by the following:

$$N_1 = \frac{P}{\varphi S_u} \tag{4-4}$$

where N_1 = number of connectors between points of maximum positive moment and adjacent end supports

S_u = ultimate strength of shear connectors, or $0.4d^2\sqrt{f_c'E_c}$

φ = reduction factor of 0.85

P = force in the slab, defined hereafter as P_1 or P_2, where

$P_1 = A_sF_y$, $P_2 = 0.85f_c'bt_s$, where

A_s = total area of steel section including cover plate

F_y = specified minimum yield point of steel used

f_c' = compressive strength of concrete at 28 days

b = effective flange width of composite section

t_s = thickness of concrete slab

We calculate

$$P_1 = 52.3 \times 36 = 1883 \text{ kips}$$

$$P_2 = 0.85 \times 3.5 \times 90 \times 7.50 = 2009 \text{ kips} \qquad \text{Use } P = 1883 \text{ kips}$$

Next, we compute $0.4d^2\sqrt{f_c'E_c} = 33{,}500$ lb, and $N_1 = 1883/0.85 \times 33.5 = 66$

studs. From Figure 4-14, the number of studs between the support and centerline is 66, or the number required for ultimate strength.

Weld Design of Cover Plate to Beam The weld connecting the cover plate to the flange in its terminal distance should be continuous and of sufficient size to develop a total stress not less than the stress in the cover plate computed at its theoretical end. Welds connecting the cover plate to the beam flange will be continuous and according to AASHTO Article 10.23.2. The shear per unit length in the fillet weld is $f_v = VQ/I$.

The theoretical cutoff point of the cover plate is 12 ft from the end support, and the fillet weld of this location is designed for both total stress and fatigue considerations. For total stress, the incremental values of f_v for dead, superimposed, and live load plus impact are calculated at 12 ft from the support. For dead load,

$$V_{\text{DL}} = (35 - 12) \times 0.93 = 21.4 \text{ kips}$$

$$Q = 12.5 \times (17.78 - 4.74 + 0.62) = 171 \text{ in.}^3$$

$$I = 10{,}878 \text{ in.}^4 \qquad \text{(Steel section only)}$$

$$f_v = \frac{21.4 \times 171}{10{,}878} = 0.34 \text{ kip/in.}$$

For superimposed dead load,

$$V_{\text{SDL}} = 23 \times 0.37 = 8.5 \text{ kips}$$

$$Q = 12.5 \times (17.78 + 4.03 + 0.62) = 281 \text{ in.}^3$$

$$I = 23{,}505 \text{ in.}^4$$

$$f_v = \frac{8.5 \times 281}{23{,}505} = 0.10 \text{ kip/in.}$$

For live load plus impact, we compute

$$V_{\text{LL}+I} = 1.30\left(16 \times \frac{58}{70} + 16 \times \frac{44}{70} + 4 \times \frac{30}{70}\right) \times 1.27 = 41.4 \text{ kips}$$

$$Q = 12.5 \times (17.78 + 11.33 + 0.62) = 29.73 \times 12.5 = 372 \text{ in.}^3$$

$$I = 34{,}167 \text{ in.}^4$$

$$f_v = \frac{41.4 \times 372}{34{,}167} = 0.45 \text{ kip/in.}$$

$$\text{Total } f_v = 0.34 + 0.10 + 0.45 = 0.89 \text{ kip/in.}$$

The minimum size of the fillet weld for a maximum plate thickness of 1.25 in. is 5/16 in. The allowable shear stress from AASHTO Article 10.32.2 is

$$F_v = 0.27F_u \tag{4-5}$$

where F_v = allowable basic shear stress
F_u = tensile strength of the electrode classification but not greater than the tensile strength of the connected part, in this case 58 ksi

From (4-5) we calculate $F_v = 0.27 \times 58 = 15.7$ ksi. With a shear on the throat of two 5/16-in. fillet welds, 0.89 kip/in., the unit stress is

$$f_v = \frac{0.89}{2(5/16)(0.707)} = 2.03 \text{ ksi} < 15.7 \text{ ksi}$$

The welds must now be checked for fatigue stresses. The range of shear at 12 ft from the support is approximately 45 kips. We now calculate

$$f_v = \frac{45 \times 372}{34{,}167} = 0.49 \text{ kip/in.} \qquad \text{OK}$$

The allowable fatigue stress for category B (longitudinal fillet weld) and 500,000 cycles is 29 ksi (AASHTO Table 10.3.1A, and the 5/16-in. fillet weld is therefore more than adequate to accommodate this stress. The development length of the cover plate based on weld strength is calculated as follows. The strength of the cover plate is $12.5 \times 20 = 250$ kips. The strength of two 5/16-in. fillet welds is

$$2 \times \frac{5}{16} \times 0.707 \times 15.7 = 6.9 \text{ kips/in.}$$

The length of weld necessary to keep the total stress in the cover plate less than 20 ksi is $L = 250/6.9 = 36$ in., which is more than satisfied by one-half of the theoretical cover plate length.

Exterior Beams For a roadway width (face-to-face of curb) of 30 ft and a framing plan of five beams at 7.5-ft spacing, the fascia cross section is as shown in Figure 4-15. The exterior beams must be designed for the same live load, the same superimposed dead load, and a dead load corresponding to a slab strip 5 ft 4 in. wide as shown in Figure 4-15, which is also the effective width of the slab in the composite section. According to the intent of the design, a mandatory horizontal construction joint will be provided between the slab and the curb, and between the curb and the parapet. Vertical open joints in the curb–parapet section should be provided to the top of the slab and at suitable intervals to inhibit composite action in this section.

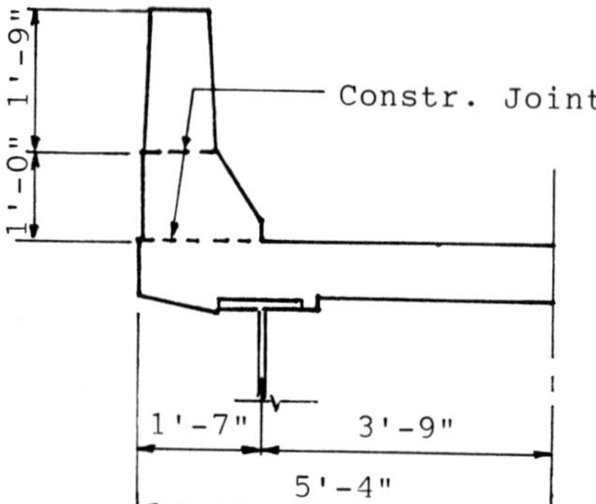

FIGURE 4-15 Fascia section of bridge for design example.

Design Example (With Shoring)

The foregoing example of a composite bridge will be analyzed assuming that shoring is temporarily used to support the dead weight of the slab until the concrete has attained its 7-day strength. The loads are rearranged as follows:

$$\text{Dead load (noncomposite)} = 0.20 \text{ kip/ft} \qquad \text{(beam, plate, etc.)}$$

$$\text{Superimposed dead load} = 0.73 + 0.37 = 1.10 \text{ kips/ft}$$

The live load is the same as in the foregoing example. The moments are now as follows:

$$M_{\text{DL}} = 0.20 \times 70^2 \times 0.125 = 123 \text{ ft-kips}$$

$$M_{\text{SDL}} = 1.10 \times 70^2 \times 0.125 = 674 \text{ ft-kips}$$

$$M_{\text{LL}+I} = \qquad 807 \text{ ft-kips}$$

Because we use the same beam and cover plate, the intent of the analysis is to demonstrate the reduction in the stress (tensile) made possible by the use of temporary shoring. Steel stresses in the bottom flange are now

$$f_s = \frac{123 \times 12}{761} + \frac{674 \times 12}{1019} + \frac{807 \times 12}{1125} = 1.93 + 7.94 + 8.61 = 18.48 \text{ ksi}$$

This reduction may justify redesigning the bridge using a lighter beam or cover plate, and in this case the saving in the weight of structural steel must be compared to the cost of temporary shoring.

Influence of Plastic Flow

For temporary or instantaneous loads, the assumption that the concrete in the composite sections behaves elastically is reasonable and fairly accurate. However, the performance of composite bridges demonstrates that plastic flow may, under certain conditions, cause overstressing in the concrete. The following assumptions are thus made: (a) only permanent loads causing compressive stresses in the concrete induce plastic flow, whereas temporary loads such as moving live loads have inconsequential effects; (b) the amount of plastic flow varies with the magnitude of the permanent compressive concrete stress, and low compressive stresses produce little plastic flow; and (c) the age of the concrete when the permanent load is applied does not influence the total amount of plastic flow, although it affects the length of the flow period, and the greater this age the longer the flow period.

Plastic flow in a composite beam induces tensile stresses in the concrete, compressive stresses in the top flange, and small tensile stresses in the bottom flange of the beam. The neutral axis of the stress curve is close to the bottom flange. The dead load producing the plastic flow is assumed to be carried by the composite section with the usual value of n (9 or 10) with the usual stress curve. When these two curves interact and are added, the resulting stress curve has its neutral axis between the neutral axes of the component curves. Because plastic flow stresses are difficult to analyze, the position of the resulting neutral axis is computed using a lower value of E_c. Thus, the combined effect of dead load and plastic flow stresses in the concrete and steel is approximated by using a higher value of n for loads producing plastic flow. This behavior and its underlying assumptions has prompted AASHTO to recommend multiplying the value of n by 3, which gives higher stresses and shears.

Possible Composite Sections

Preferably, the neutral axis of the composite section should be located below the top flange of the steel beam, because in this position the entire concrete section is in compression. Figures 4-16*a* and *b* show, however, two cases where the neutral axis is in the concrete section. For the neutral axis in the slab, as shown in Figure 4-16*a*, the properties of the composite section are estimated from the following:

$$kd = \frac{nA_s}{b}\left[-1 + \sqrt{1 + \frac{2bd}{nA_s}}\right] \tag{4-6}$$

$$I = \frac{b(kd)^3}{3n} + I_s + A_s(d - kd)^2 \tag{4-7}$$

For the neutral axis in the concrete haunch as shown in Figure 4-16*b*, the

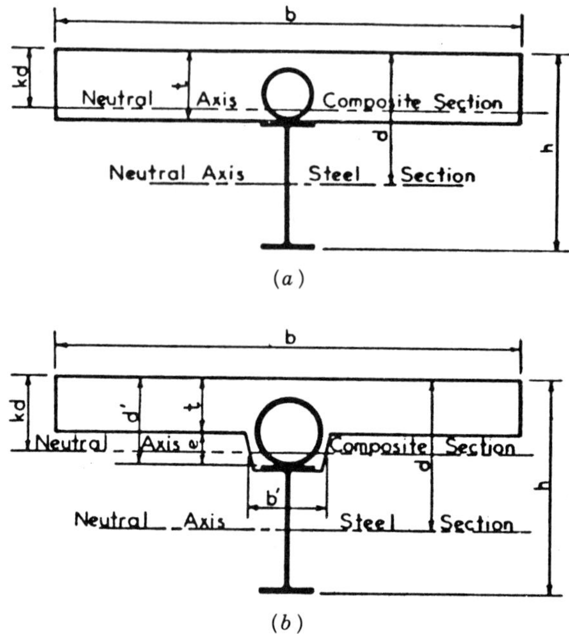

FIGURE 4-16 Position of neutral axis in composite beams: (*a*) beam and slab without haunch; (*b*) beam and slab with haunch.

properties of the composite section are estimated from the following:

$$kd = \frac{nA_s + (b - b')t}{b'}\left[-1 + \sqrt{1 + \frac{b'[2nA_s d + (b - b')t^2]}{[nA_s + (b - b')t]^2}}\right] \tag{4-8}$$

$$I = \frac{b(kd)^3 - (b - b')(kd - t)^3}{3n} + I_s + A_s(d - kd)^2 \tag{4-9}$$

where all terms correspond to the notation of Figure 4-16 and also to the following:

A_s = cross-sectional area of steel member
n = modular ratio
d = distance from top of slab to center of gravity of steel section
kd = distance from top of slab to neutral axis of composite section
I_s = moment of inertia of the steel section about its neutral axis
I = moment of inertia of composite section about its neutral axis

If the composite section has an irregular concrete area, as in Figure 4-17, a trial design may be made by substituting for the irregular section a slab of

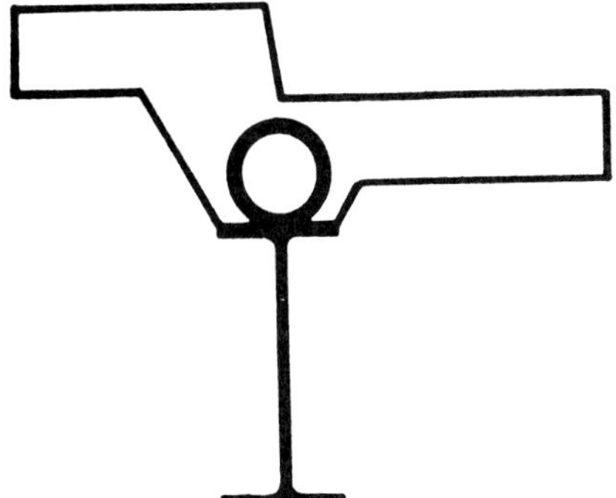

FIGURE 4-17 Irregular composite section.

uniform thickness with an area and width equal to the original concrete area. The slab is located so that its center of gravity is at the same distance above the steel flange as the original irregular concrete area. After a trial design is completed, the actual composite section should be checked for stresses and deflection.

The condition shown in Figure 4-17 may arise where a curb is placed monolithically with the slab and is analyzed as part of the composite section of the exterior beam, probably shifting the neutral axis into the concrete section and above the steel top. The composite design in this case becomes more complex as soon as the composite action of the adjacent interior beam begins to dominate and the neutral axis shifts below the steel top.

In composite beams with temporary intermediate supports (shoring) during placement of the concrete slab and curing of the concrete, the entire dead load is carried by the composite section causing permanent compressive stresses in the concrete, thereby inducing plastic flow. In this case plastic flow has a greater effect and may be considered in various ways. By using $3n$ as the modular ratio, the design will show greater bottom flange (tensile) stresses. Because plastic flow produces additional compression on the top flange, the stresses in this flange will be increased when the neutral axis lies in the steel. Because concrete stresses are in this case reduced by plastic flow, it may be advisable to determine concrete stresses by neglecting plastic flow.

Deflection of Composite Beams

Because composite construction is stiffer than the steel beam alone, live load deflection seldom controls the design under the current AASHTO requirements. The moment of inertia of the gross cross-sectional area is used to calculate deflections, and all beams are considered as having equal deflection for the number of trucks that the bridge deck can accommodate.

For the 70-ft-long bridge analyzed in the preceding example, we can compute the live load deflection as in the example of Section 4-4 by considering the resultant of the rear axles acting at midspan and by using the moment of inertia of the composite section as a constant. The deflection is

$$\Delta = \frac{128 \times 70^3 \times 12^3}{48 \times 29 \times 10^3 \times 5 \times 34{,}167} = 0.32 \text{ in.}$$

or with impact 0.40 in., which is below the allowable 1.0 in.

4-8 NONLINEAR ANALYSIS OF COMPOSITE BEAMS

Analysis of composite beams that includes the effects of interface slip, yielding of the steel, residual stresses, and nonlinearity in the concrete is presented by Ansourian and Roderick (1978). The method also permits predictions of behavior throughout the loading range up to the occurrence of collapse.

The standard procedure presented in Section 4-7 assumes an elastic structure and ignores interface slip. Allen and Severn (1961) have proposed rigorous solutions for the simple case of a slab covering a series of parallel beams simply supported and subjected to symmetrical loading. In this approach, the slab is idealized as a thin plate subjected to in-plate displacements u and v and to transverse (bending) displacement w. The displacements, membrane stresses, and moments in the slab are obtained from solutions of biharmonic equations, but, in practice, these are unlikely to be attempted because of difficult boundary conditions that are not adequately expressed analytically.

Alternatively, a composite beam-and-slab system can be represented in various ways by finite elements (Ansourian, 1975), where the slab is idealized either as a solid brick-type element or as a thin plate at the midsurface, whereas the beam is represented by conventional beam elements and the eccentricity of its axis relative to the slab midsurface is enforced with rigid links. These links are not actual physical members, but their function is simulated numerically by making the displacements along the beam axis depend on the displacements of the slab nodes located vertically.

Results obtained for a floor system consisting of a concrete slab rigidly connected to the top flange of a series of parallel beams of I section encastered at the ends are shown in Figure 4-18. The distribution of membrane stress is clearly different from the distribution assumed in simple T-beam theory. The stress is a maximum at the beam and falls away toward the center of the slab. The distribution peaks at the maximum negative moment section, and is essentially unaffected by the type of loading, uniform or concentrated along the beam.

For the single beam shown in Figure 4-19, both experimental results and finite-element analysis show an error in midspan deflection less than 3 percent. At the center section, the distribution of stress in the steel beam is

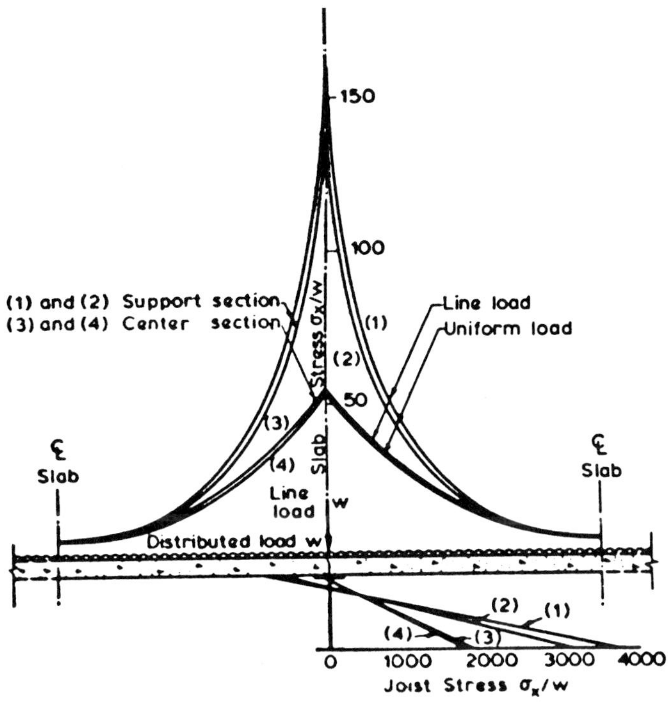

FIGURE 4-18 Effect of loading on distribution of longitudinal stresses. (From Ansourian and Roderick, 1978.)

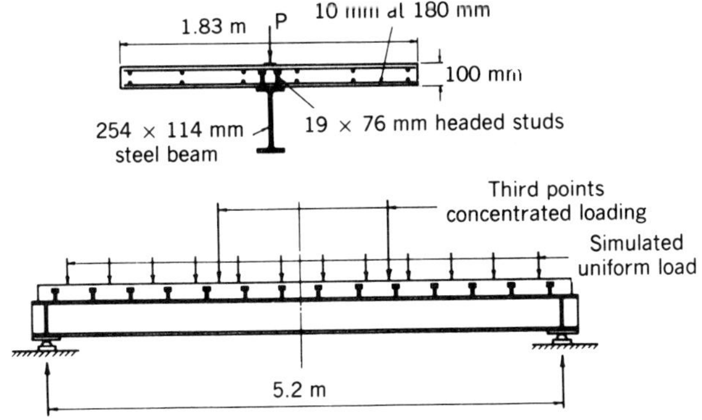

FIGURE 4-19 Typical test beam. (From Ansourian and Roderick, 1978.)

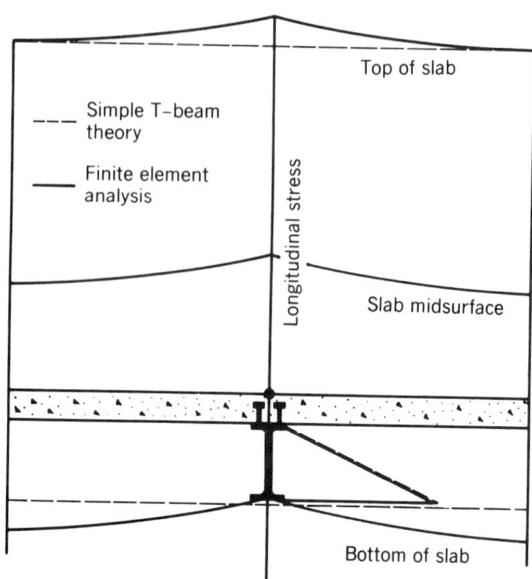

FIGURE 4-20 Stresses at midspan for isolated beam. (From Ansourian and Roderick, 1978.)

practically the same, calculated either by finite-element or T-beam theory, and the difference in peak slab stress is only 7 percent, as shown in Figure 4-20. Along the full length of the beam, the error in bottom flange stress is markedly small. From these results we can conclude that very little error is introduced when simple T-beam theory is applied to an isolated composite beam.

When the load–slip relationship of the shear connection is linear, the interface slip may be determined as proposed by Siess, Viest, and Newmark, (1952). The greatest loss of composite action occurs at concentrated loads or at sections of maximum moment. In practice, beams having an adequate shear connection may experience the primary effect of slip in the form of larger deflections, of the order of 10 percent, simultaneously with a 10 percent reduction in the force acting on the connectors.

Nonlinear Analysis For a composite beam loaded to failure, two principal factors affect its behavior: yielding of the steel and the nonlinear response of the concrete. Strain hardening may be significant with a deep bending moment gradient, which rarely happens in simply supported beams (see also Section 4-5). Thus, refined analyses focus on the effect of slip and residual stress (Ansourian and Roderick, 1978).

The shape of the load–slip curve for connectors is obtained from pushout tests, and an example is shown in Figure 4-21. In a composite T beam, the resisting moment is relatively insensitive to the stress–strain diagram of the concrete (Barnard and Johnson, 1965), and the usual bilinear relationship is applicable with a maximum concrete stress of $0.85f_c'$.

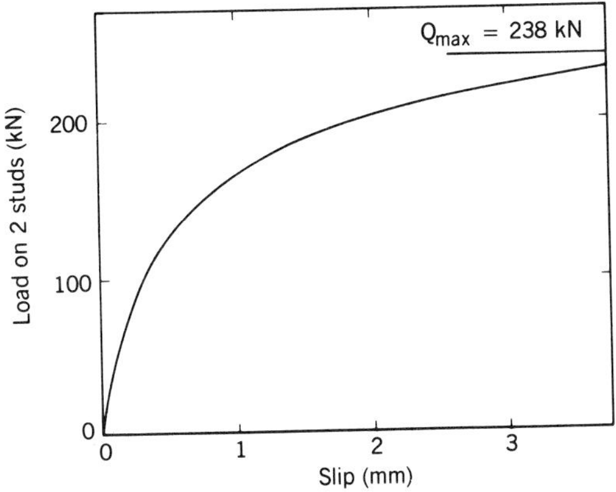

FIGURE 4-21 Typical pushout characteristic for 19-mm stud. (From Ansourian and Roderick, 1978.)

Figure 4-22 shows a section bent to a curvature p and stressed well to the plastic range. At the steel–concrete interface there is a strain discontinuity ϵ_{sp}, caused by the flexibility of the connector. The concrete is nonlinear to a depth of $n_1 d$, and the steel yields to a depth of $n_2 d$. The force components are derived by subtracting, from the force for the wholly elastic case, correction terms taking account of the nonlinearity. For example, the compressive force in the slab is $F_c - F_3$, where F_c is the elastic component and F_3 is the correction term. Thus,

$$F_c = \frac{b}{2} E_c p (k_1 d)^2 \qquad F_3 = \frac{b}{2} E_c p (n_1 d)^2 \tag{4-10}$$

where E_c is the elastic modulus of concrete. Horizontal equilibrium of the cross section requires

$$F_c - F_3 = F_s - F_1 - F_2 \tag{4-11}$$

where F_s is the elastic force in the joint and F_1 and F_2 are the correction terms for yielding in the lower flange and web, respectively. From these,

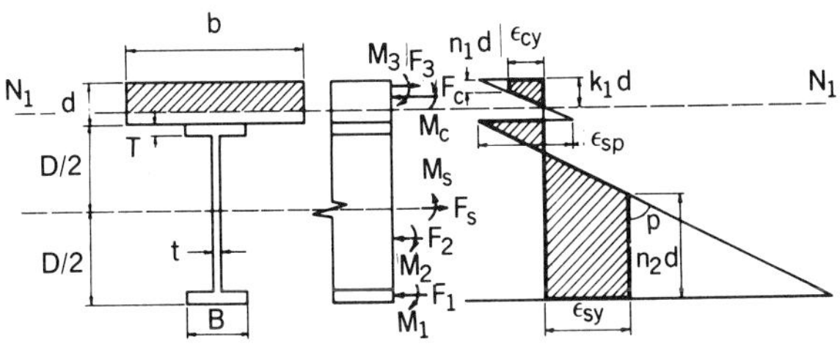

FIGURE 4-22 Distribution of stress. (From Ansourian and Roderick, 1978.)

expressions can be derived for the depth of yielding in the beam and the nonlinear depth in the slab (Ansourian and Roderick, 1978).

Two more expressions are required before $k_1 d$ can be obtained explicitly. The first expresses equilibrium of the actions on the cross section with the externally applied moment, and the second ensures compatibility of the slip strain with the connector displacements.

Using the foregoing procedure and, where appropriate, similar expressions for cases of extensive yielding, the beam problem may be solved iteratively at any level of applied load. The curvature and the extent of yielding at every section are adjusted until equilibrium is attained with the external moment. Further iteration is necessary to ensure that the final slip strain is compatible with the connector forces. The convergence of the iterative procedure is rapid at low loads but becomes very slow as the loads increase, at which stage the moment–curvature and force–slip relationships become markedly nonlinear (Ansourian and Roderick, 1978). The use of numerical techniques can speed convergence.

Effect of Residual Stresses Comparison of deflections obtained experimentally and theoretically shows that above a certain load the observed deflection increases more rapidly than predicted by the assumed nonlinearity of concrete and by the effect of slip. These larger deflections are attributed to the effect of residual stresses in the beam. Whereas residual stresses depend on section geometry, method of manufacture, rate of cooling, and extent of cold working, in rolled beams of medium size residual stresses are distributed parabolically in the flanges and the web as shown in Figure 4-23*a*. For simplicity, this distribution is replaced by the diagram shown in Figure 4-23*b*, with a uniform tensile stress in the flanges and a corresponding compressive stress in the web.

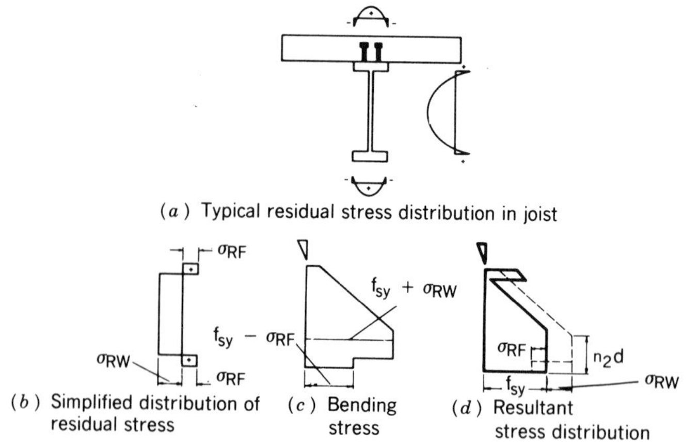

FIGURE 4-23 Residual stresses. (From Ansourian and Roderick, 1978.)

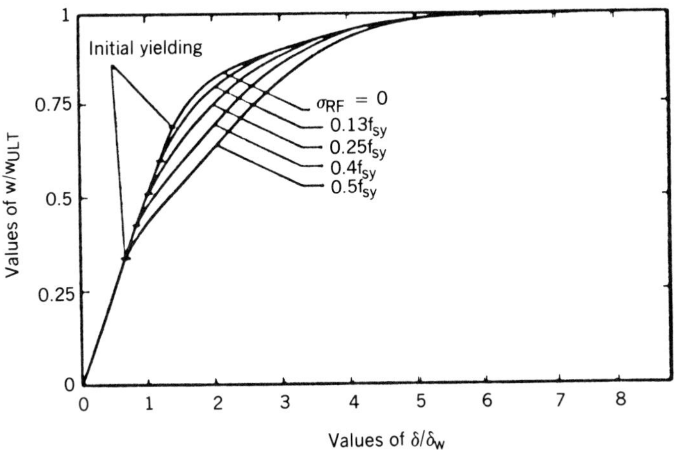

FIGURE 4-24 Effect of residual stress on midspan deflection. (From Ansourian and Roderick, 1978.)

Before any load is applied the beam is in equilibrium, and at any section the residual stresses must satisfy the equation

$$2\sigma_{RF}A_F + \sigma_{RW}A_W = 0 \tag{4-12}$$

where A_F and A_W are the cross-sectional areas of the flange and web, respectively. If a certain level of residual stress is assumed in the flanges, the stress in the web is therefore

$$\sigma_{RW} = -\frac{2A_F}{A_W}\sigma_{RF} \tag{4-13}$$

This means that the effect of residual stresses is to reduce the moment at which yielding develops in the flange and to raise the moment for web yielding. The curvature developed under these conditions is calculated by replacing the yield stress in the flange by an effective yield stress $f_{sy} - \sigma_{RF}$ and replacing the yield stress in the web by $f_{sy} + \sigma_{RW}$. Then the stresses balancing the applied moment are as shown in Figure 4-23*c* when yielding extends to the web.

The effect of varying the residual stress on the midspan deflection for the beam of Figure 4-19 is shown in Figure 4-24 for σ_{RF} values ranging from zero to half the yield stress. As the residual stress is increased, initial yielding is initiated at a lower applied load, but after it has progressed through the thickness of the bottom flange, the beam begins to respond elastically. At this stage the stiffness of the beam is less than the initial stiffness because, although the web is elastic, the bottom flange is plastic. After yielding extends through the depth of the joist, all traces of residual stresses disappear.

4-9 TOPICS RELEVANT TO COMPOSITE BEAMS

Plastic Rotation

Full-scale beam tests by Ansourian (1982) have provided data on the sagging rotation capacity of composite beams with ductility parameters in the range 0.65 to 3.0. When secondary failures associated mainly with the shear connection are avoided through proper design, the rotation capacity is controlled principally by the extent of strain hardening developed in the steel beams when the slab crushes. This occurrence is represented by a single ductility parameter x, expressed as a function of the section geometry and of the mechanical properties of the steel and the concrete. The value $x = 1.0$ corresponds to a beam that has the neutral axis at collapse at a level for which the strain at the start of strain hardening, e_{sh}, is attained in the lower flange at the same time as the crushing strain of the concrete, e_u, is reached at the top of the slab.

In these tests the beams were loaded with a single concentrated load at the center. For the beam with the most severe brittle characteristics ($x = 0.65$), the maximum steel strain at collapse was 0.013, or $0.8e_{sh}$. Conversely, in the most ductile beam this strain was over 0.035, or $2.7e_{sh}$. The midspan deflections at unloading were $2.4\delta_e$ and $8.8\delta_e$ (elastic component of deflection) for the brittle and ductile beam, respectively. The plastic rotation ratio θ_{pr} (ratio of ultimate to elastic rotation) ranged from 1.5 to 5.8.

These data provided the basis for introducing minimum values for the ultimate deflection ratio, δ_{ur}, and the plastic rotation ratio, θ_{pr}, in the range $x = 1.0$ to 3.5. With a minimum value of $x = 1.4$, it was concluded that sufficient rotation is available to develop the plastic design collapse load, calculated conventionally, for the worst combination of span and loading.

Temperature Distribution

The temperature distribution in composite bridges has been studied by Emanuel and Hulsey (1978) using finite-element models (see also Section 12-5). These effects were investigated for air temperature extremes likely to occur at the prototype (Columbia, Missouri) during the life of the structures, consisting of a concrete deck on steel beams.

The results confirm that the temperature distribution depends on the ambient air temperature, the short-wave radiation absorbed by the bridge deck, the heat transfer due to long-wave thermal radiation, the film coefficient, and the thermal characteristics of the materials. The presence of a thin asphalt layer can produce a higher thermal gradient within the bridge cross section. A nonlinear temperature distribution produces thermal stresses within isotropic homogeneous material.

For a composite steel beam bridge, the magnitude of stress depends on the temperature distribution, the internal induced forces generated by differences in the coefficients of thermal expansion, and the boundary support conditions. A steady-state thermal condition never exists in a bridge structure. The convection constant (defined as the cooling produced by the wind) markedly affects the temperature distribution; on a still, hot day the temperature differential within the deck may reach 39°F (22°C), whereas on a cold winter day this differential can be 31°F (17°C).

4-10 COMPOSITE BRIDGES WITH PRECAST DECKS

Construction Procedure One form of construction, developed in Switzerland, is the stage deck jacking method; it involves the following main steps: (a) cast a given length of concrete deck (20–25 m) in the bridge axis near the abutment or at midspan, depending on constructional convenience; (b) jack this section onto the flanges of the steel beams (by liberating the casting bed); and (c) prepare the casting bed for concreting a second deck length (Beguin, 1978).

The finished deck is thus produced in a series of segments in incremental casting, and when in place each slab section is connected to the steel beams by shear attachments. The associated advantages are as follows: (a) the same formwork is used repeatedly, (b) concrete pouring is done at a fixed location, and (c) the entire deck is made structurally continuous by overlapping the reinforcing bars at the ends of segments. The method is feasible with I beams, twin-girder systems, and open-box sections.

During jacking the freshly hardened deck sections slide on special strips, usually built on standard I steel profiles. The formwork may be held in place by ground anchors when the casting is done outside the bridge, or it may be suspended from the main beams for interior casting. When the concrete sections are in their final position, the shear connectors are welded to the flanges at predetermined openings, and these holes are filled with concrete.

The movement of deck segments normally requires the use of hydraulic jacks. A padding evens out the contact between the concrete and the head of the ram. For bridges longer than 70 m, sliding shoes accommodate movement, but better sliding details can be developed from a consideration of the friction forces. A sliding surface having a central groove for lubrication can reduce friction markedly. Deck steering and path control are achieved by the combined action of the main and auxiliary jacks in conjunction with visual checking.

Design Considerations Factors to be considered in the design are related to (a) the physical process of incremental launching and (b) the effect of the launching forces on the placed deck and steel structure. In the final position, the loads and composite action are as in conventional construction, and because the shear connectors are installed after the deck segments are in place, the concrete weight is resisted by the steel beams alone.

Factors Governing Launching The jacking forces are applied parallel to the plane of the deck, and must overcome the frictional resistance. For a body moving at low speeds, this reaction is neither typical nor continuous, but is manifested in a series of "sticks" and "slips." The stick–slip phenomenon is articulated by Bowden and Leben (1939), and further data on this process are provided by Morgan, Muskat, and Reed (1941). Experience with stage launching shows that deck segments are subjected to stick–slip effects depending on (a) the mass of the driven deck, (b) the segment length; (c) the mass of the steel structure; (d) the elasticity of the links; and (e) the jacking velocity (Beguin, 1978).

Steering a concrete deck strip along a prescribed path (sometimes several hundred feet) may cause the strip to veer off its course. In this case the deck must have its path corrected with the help of a transverse force.

Structural Steel Support The absence of concrete fillets into which the top flange is normally embedded is not a structural deterrent, because this flange is fully supported laterally by the shear studs.

The simple beam shown in Figure 4-25*a* is assumed to have a uniformly distributed friction load of intensity r. The beam has a cross section with the center of gravity at midheight h. If the beam is acted upon or held longitudinally by a tie fixed at point C, the eccentricity is small, and the only force acting at distance x is an axial force $N = rx$.

We now consider the n-span continuous beam shown in the top of Figure 4-25*b*, where n consecutive spans are uniformly loaded along their entire length by distributed friction forces of intensity r. The beam is held at the level of the top flange at point F. The internal forces may be determined according to the following two cases.

1. The continuous beam is loaded as in the original beam but is held at midheight point C as shown in the middle of Figure 4-25*b*, so that the only internal force in the beam is the axial tension $N = rh$.

2. The continuous beam is not subjected to friction but to a terminal couple $M = Rh$ applied at the end as shown in the bottom of Figure 4-25*b*. For equal spans and the same bending stiffness, the bending moment generated in each section by the terminal couple M may be obtained from a second-order linear difference equation. For the n equal continuous span, the bending moment Mj over the support j ($j = 1, 2, \ldots, n$) is

$$Mj = -M(-1)^{n+j}\frac{\sinh j\gamma}{\sinh n\gamma} \tag{4-14}$$

where $\cosh \gamma = 2$.

Using appropriate trigonometric functions, an expression for Mj may also be written in terms of continued fractions (Beguin, 1978). Based on this analysis, the rotation of sections over supports can be derived and used to

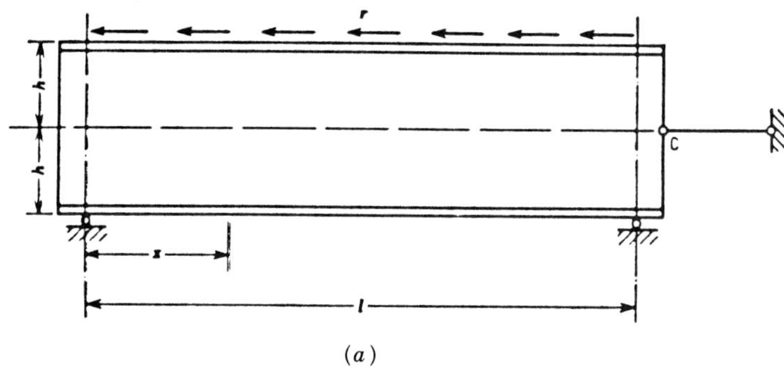

(a)

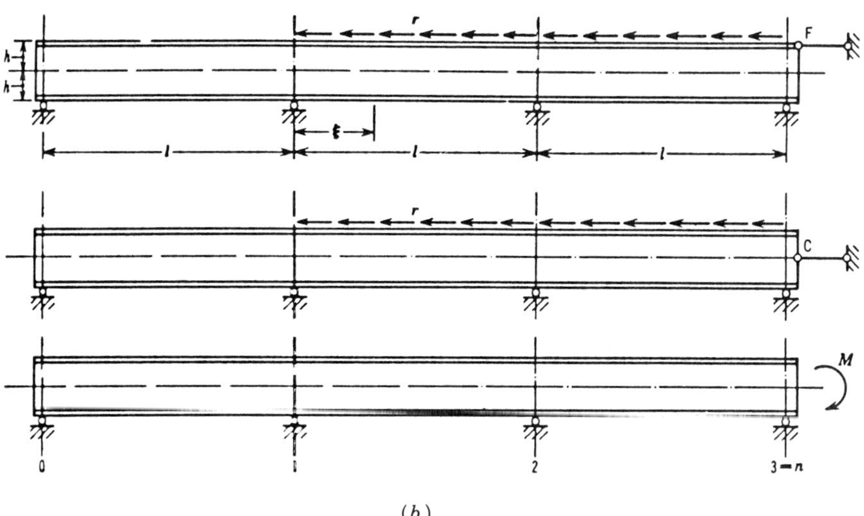

(b)

FIGURE 4-25 (a) Simple span under uniform friction loading; (b) continuous beam under uniform friction loading. (From Beguin, 1978.)

handle more complex loadings. If the jacking force is symmetrical with respect to the bridge axis, the friction forces are essentially distributed between the two beams. Asymmetrical driving of the deck can cause bimoments and warping of the steel sections. A case of asymmetrical jacking is solved by Beguin (1978) in a two-step analysis.

Effect of Deck Eccentricity Over Supports The effect of deck slip on its elastic support depends mainly on the width of the sliding shoes. Any resulting eccentricity should preferably be less than half this width in order to limit transverse bending of the top flange. With larger eccentricities this bending may become critical.

The following stresses are induced near the shoe: (a) a longitudinal stress in the top flange and web due to normal loads, (b) a vertical compressive stress, (c) a transverse bending stress, and (d) a shear stress caused by shear and torque. An eccentric shoe also causes a slight twist of the flange locally, and this can lead to yielding over a limited area.

Because the shoes may transfer concentrated loads at any point along the beam, the use of web stiffeners is not practical. In this case buckling theories must be considered and interpreted with caution (Carskaddan, 1969). In addition, it appears that the introduction of additional web stiffeners may not be justified given the temporary nature of this loading. Invariably, the shoes are left in place, and the resulting gap between the flange and the concrete must therefore be filled with cement grout.

Further Improvements It is essential to detect deck eccentricity during launching before it becomes detrimental and to overcome the undesirable features of the stick–slip motion. An eccentric position of the shoe over the beam web can be avoided if the shoe is provided with a lateral guide. Because sliding panels must bear on two adjacent beams for stability, bridges suitable for incremental launching should preferably have a framing system of 2, 4, 6, and so on beams.

Transverse prestressing may be applied in conjunction with conventional reinforcement to compensate for the effects of transverse bending moments. Longitudinal prestressing has seldom been used. The presence of friction of the slab along the top flange in cast-in-place decks inhibits the action of prestressing. With incremental launching this adherence is not a problem, however, and longitudinal posttensioning may be beneficial.

With the stage deck jacking method, shear connection is achieved in the last stage of constructions. At this time most of the shrinkage in the concrete has occurred, whereas the jacking action imparts to the concrete a definite compression. By combining the driving jacks and the prestressing tendons, the undesirable friction between the steel and deck may be further reduced.

Where the deck section size and weight can accommodate the usual crane and hoisting capabilities available at the site, in situ precast members should be considered. For example, a slab bridge with an overall deck width of 32 to 33 ft can be constructed in prefabricated deck sections of full width and 10 to 15 ft long. These sections have an approximate weight of 45 kips and are within the hoisting range of most cranes. Prefabricated decks are discussed in subsequent sections. The main advantages relate to the elimination of the jacking stage and the associated costs.

4-11 CONTINUOUS-BEAM BRIDGES

Multiple-span bridges are required where a single-span structure becomes too long for an economical solution. The bridge system in this case may consist of a series of simple spans or the design may be continuous over the piers.

Simple spans require less engineering effort; they do not require field splices, making possible faster field erection; differential support settlements do not have to be considered in the beam design; and expansion devices must accommodate single-span movement only.

On the other hand, continuous spans allow a reduction of material or longer spans and fewer piers for the same steel section; they result in less deflection and live load vibration effects; they require a single bearing at the supports and fewer expansion joints; and they enhance improvements in appearance through variation in span length and beam depth.

Where the span ratio (center to end) is structurally favorable (1.25–1.35), the advantages of continuous bridges become obvious, and comparison of alternate designs will reflect a corresponding cost difference between simple and continuous beams. The routine availability of efficient deep foundation systems practically precludes differential support settlement and ensures moments and shears as predicted by the design. For relatively short spans, continuity makes little cost difference, and concrete-type bridges should be considered.

A continuous bridge, whether concrete or steel, usually implies a beam system of a variable moment of inertia. For short spans, the continuous steel beam is a wide-flange rolled section with welded cover plates in the region of maximum negative moments or with heavier sections between field splices. Continuity means two or more spans, but five continuous spans are a structural upper limit in terms of analysis and functional characteristics. Two-span continuous beams have a slight economy over simple bridges. The usual three- and four-span continuous bridge has the center spans one fifth to one-third longer than the end spans.

Methods of Analysis Continuous beams are analyzed by the moment of inertia method. The general principles of structural continuity are discussed in Section 2-2. Variations in the moment of inertia from the assumed model can produce considerable deviations from the true results of the analysis, and are thus more critical where the structural materials are not truly elastic and they do not complete elastic recovery. With current methods of analysis, this problem is remedied through the use of computer programs that consider these variations and also provide parametric solutions. However, a complete understanding of concepts such as stiffness and restraint, moment distribution, and influence lines for moments and shears is essential and will help designers in solving continuous-beam problems where hand calculations become necessary. The applicability of these concepts is illustrated in Chapters 2 and 3.

For continuous superstructure, the provisions of NCHRP Project 12-26 (Article 3.23.2.5) stipulate that longitudinal bending moments must be increased by a correction factor c applied to the moments obtained from a continuous-frame analysis. Values of correction factors for different floor and beam types are listed in Table 3.23.4 (NCHRP Project 12-26). For steel I beams, the correction factor is 1.05 for positive moments, and 1.10 for negative moments.

Design Example: Two-Span Continuous Beam (Noncomposite)

In this example, we consider a bridge with two equal spans, 70 ft each. Other design parameters and data are the same as in the composite bridge analyzed in Section 4-7. The live load distribution is 1.3, and a constant moment of inertia is assumed. We recall that the total dead load plus superimposed dead load is $0.93 + 0.37 = 1.30$ kips/ft.

Moments and shears are obtained directly from AISC tables (Moments, Shears, and Reactions for Continuous Highway Bridges) for $N = 1.0$ (span ratio) and a total beam length of 140 ft. A beam elevation is shown in Figure 4-26.

Dead load moments are computed as follows:

$$\text{Positive DL moment in span } AB = 1.30 \times 70^2 \times 0.0703 = 448 \text{ ft-kips}$$

$$\text{Negative DL moment at support} = -1.30 \times 70^2 \times 0.125 = -796 \text{ ft-kips}$$

Likewise, live load plus impact moments are calculated as follows (include the correction factor c):

$$\text{Positive in span } AB = 792 \times 0.5 \times 1.30 \times 1.256 \times 1.05 = 678 \text{ ft-kips}$$

$$\text{Negative over support} = -634 \times 0.5 \times 1.30 \times 1.256 \times 1.10 = -569 \text{ ft-kips}$$

Note that truck load governs the positive moment in the span, and lane loading determines the negative live load moment over the support.

The total moments are

$$\text{Positive} \quad M_T = 448 + 678 = 1126 \text{ ft-kips}$$

$$\text{SM} = 1126 \times 0.6 = 676 \text{ in.}^3$$

$$\text{Negative} \quad M_T = -(796 + 569) = -1365 \text{ ft-kips}$$

$$\text{SM} = 1365 \times 0.6 = 819 \text{ in.}^3$$

(Assume full lateral support)

The choice involves the following:

Scheme A: Positive moment W36 × 194, SM = 665 in.^{3}
Negative moment W36 × 230, SM = 837 in.^{2}

Scheme B: Positive moment W36 × 194
Negative moment W36 × 194 plus cover plates over support

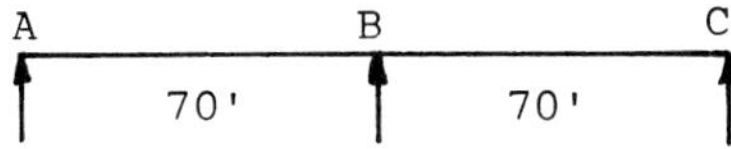

FIGURE 4-26 Beam elevation; two-span continuous steel beam bridge.

In scheme A, the heavier beam section is provided between field splices placed at the points of dead load contraflexure. For an exact solution, dead and live load moments must be recalculated taking into account the variable moment of inertia. An increase in this parameter in the vicinity of interior support produces an increase in the maximum negative moments and interior reactions. For a uniformly loaded beam continuous in two spans, cover plates placed along one-fourth of the span length each side of the interior support and producing a 50 percent increase in the moment of inertia will probably cause an increase in the negative moment of 5 percent. This is not a linear function, and the increase in negative moment rapidly becomes accelerated for greater increases in the moment of inertia. In the positive moment area an overstressing of $676/665 = 1.6$ percent is acceptable.

For scheme B, we select cover plates, top and bottom, of 11 in. × 1/2 in. Note that the minimum cover plate width is 3/8 in., or $w/24 = 11/24 = 0.46$ in.

The additional moment of inertia of the two plates is

$$
\begin{aligned}
I &= 11 \times 0.5 \times 2 \times 18.5^2 = 3762 \text{ in.}^4 \\
I \text{ of beam} &= \underline{12{,}100 \text{ in.}^4} \\
\text{Total } I &= 15{,}862 \text{ in.}^4
\end{aligned}
$$

which gives a section modulus of $15{,}862/18.75 = 846$ in.^{3}. The increase in the section modulus is $846/665 = 27$ percent, and the cover plate is thus acceptable.

Length of Cover Plates for Scheme B The cover plate is detailed as shown in Figure 4-9, and the terminal distance is again 1.5 times the cover plate width. In order to determine the theoretical cutoff point, we compute the moment at 7 and 14 ft from the interior support (10th points).

The maximum dead load shear just at one side of the interior support is $0.625 \times 1.30 \times 70 = 57$ kips:

$$\text{Dead load moment at 7 ft} = -796 + 57 \times 7 - 1.30 \times \frac{7^2}{2} = -429 \text{ ft-kips}$$

$$\text{Dead load moment at 14 ft} = -796 + 57 \times 14 - 1.30 \times \frac{14^2}{2} = -125 \text{ ft-kips}$$

The maximum live load moment at 7 ft from the interior support requires two distinct steps because it is produced by the lane loading. First, we load both spans with the uniform lane loading and compute the moment as in the case of dead load. Then we place two concentrated loads at two critical points on either span to produce maximum moment at point 0.1 of span 2.

From the uniform load of 0.64 kip/ft, the corresponding shear is $0.625 \times 0.64 \times 70 = 28$ kips, and the moment at B is $0.125 \times 0.64 \times 70^2 = -392$ ft-kips:

$$\text{Moment at 7 ft} = -392 + 28 \times 7 - 0.64 \times \frac{7^2}{2} = -212 \text{ ft-kips}$$

From the two concentrated loads, each 18 kips, the moment is

$$-(0.0864 + 0.0356) \times 18 \times 70 = -154 \text{ ft-kips}$$

or total

$$M_{\text{LL}} = -(212 + 154) = -366 \text{ ft-kips} \qquad \text{(One lane)}.$$

The moment per beam (plus impact) is computed as

$$M_{\text{LL}+I} = -366 \times 0.50 \times 1.30 \times 1.26 = -300 \text{ ft-kips}$$

At 7 ft from the support, the maximum moment is $-(429 + 300) = -729$ ft-kips and requires a section modulus of $729 \times 0.6 = 437$ in.^{3}. Therefore, the theoretical cutoff point for the cover plate is less than 7 ft from the support and can be determined by linear interpolation, or

$$\frac{x}{7} = \frac{154}{382} = 2.8 \text{ ft} \qquad \text{say 3 ft}$$

The terminal distance is $1.5w = 1.5 \times 11 = 17$ in. = 1.5 ft, giving a total cover plate length of $4.5 \times 2 = 9$ ft.

In order to compare schemes A and B, we locate the point of dead load contraflexure. This is determined from the expression $-796 + 57x - 1.30(x^2/2) = 0$, which gives $x = 18.5$ ft.

Shears and Reactions The maximum reaction at the interior support is obtained for the lane loading, and is calculated directly from AISC tables:

$$\text{Dead load reaction at } B = 1.25 \times 1.30 \times 70 = 114 \text{ kips}$$

$$\text{Live load reaction at } B = 82 \times 0.5 \times 1.30 \times 1.20 = 64 \text{ kips}$$

The total reaction at B is $114 + 64 = 178$ kips, producing an average web shear stress of $178/(36.5 \times 0.77) = 6.4$ ksi, OK. The reaction at B controls the maximum web shear.

Whereas this example demonstrates the design methodology of continuous beams, it is very unlikely that the noncomposite scheme will be the final choice. Thus, the foregoing analysis normally should be completed by considering a simple-span composite design, such as the one reviewed in Section 4-7, or a two-span continuous composite beam.

Design Example: Optimum Four-Span Continuous Beam

The objective of this example is to select span lengths and locate intermediate piers for a four-span continuous-beam bridge 360 ft long so as to require one size of rolled beam (with cover plates at interior supports). The deck is a concrete slab, and the beam spacing is 6 ft 6 in. For the purpose of this analysis, we assume a dead load of 0.8 kip/ft per beam, and a live load consisting of the HS 15 truck.

From $S/5.5$ we obtain a distribution coefficient of $6.5/5.5 = 1.18$. From Figure 2-41, using a beam spacing of 6.5 ft and assuming $S/L = 0.1$, we estimate the distribution factor close to the same value. The bridge supports are identified as A, B, C, D, and E. The live load moments per beam are the tabular values found in the AISC tables multiplied by (0.75×0.59). Using N values (ratio of center to end span) 1, 1.1, 1.2, and 1.3, we calculate the dead load, live load, and impact moments, and the results are tabulated in Table 4-2. These moments are plotted versus the corresponding values for spans

TABLE 4-2 Summary of Moments (Positive); Four-Span Continuous Beam for Various Values of N (Span Ratio)

	Span AB			
N	1.0	1.1	1.2	1.3
$M_D =$	500	438	381	331
$M_L =$	474	451	429	410
$M_I =$	110	107	104	101
	1084	996	914	842
	Span BC			
$M_D =$	236	278	317	356
$M_L =$	385	404	421	436
$M_I =$	90	92	94	96
	711	774	832	888

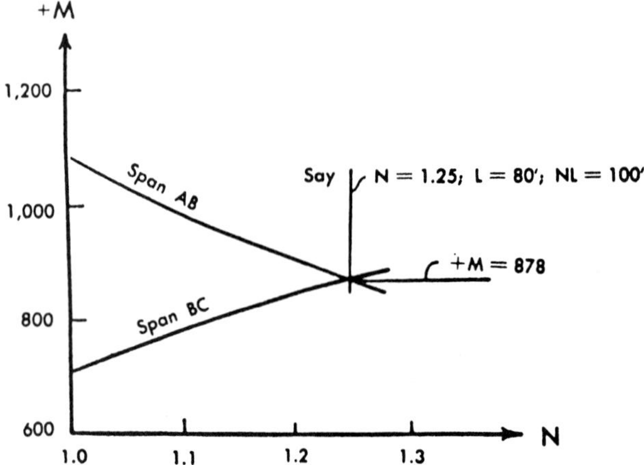

FIGURE 4-27 Graphical presentation of moment versus span ratio N, for a four-span continuous beam; center and end spans (AB is end span, BC is center span).

AB and BC (the end and center span, respectively) to produce the graphs of Figure 4-27. Note that all moments are positive span moments.

Where the two graphs intersect, they define the optimum span ratio N, in this case $N = 1.25$. Using a total beam length of 360 ft and $N = 1.25$, we compute the span lengths as 80, 100, 100, and 80 ft. The positive moment corresponding to $N = 1.25$ may be recalculated or found directly from the graphs of Figure 4-27. This moment is 878 ft-kips, and for $F_s = 20$ ksi the required section modulus is $878 \times 0.6 = 527$ in.^{3}. This is provided by a W36 × 160 with a section modulus of 542 in.^{3}. For $N = 1.25$, the moments at support B (first interior pier) are

$$M_{\text{DL}} = 0.5(0.1226 + 0.1328) \times 0.8 \times 80^2 = \quad 654 \text{ ft-kips}$$

$$M_{\text{LL}+I} = 0.5(906 + 911) \times 0.443 \times 1.233 = \quad \underline{496} \text{ ft-kips}$$

$$\text{Total } M \qquad = -1150 \text{ ft-kips}$$

This is increased by 5 percent, or $M = 1150 \times 1.05 = -1210$ ft-kips.

$$\left.\begin{array}{r}\text{Required SM} = 1210 \times 0.6 = 726 \text{ in.}^3, \text{W36} \times 160 \\ \text{Two plates 10 in.} \times 5/8 \text{ in.}\end{array}\right\} \quad \text{SM} = 749 \text{ in.}^3$$

For $N = 1.25$, the moments at support C (interior center pier) are

$$M_{DL} = 0.5(0.1187 + 0.1448) \times 0.8 \times 80^2 = 675 \text{ ft-kips}$$

$$M_{LL+I} = 0.5(949 + 1001) \times 0.433 \times 1.22 = 528 \text{ ft-kips}$$

$$\text{Total } M = -1203 \text{ ft-kips}$$

$$\text{and multiplied by 1.05,} \quad M = -1265 \text{ ft-kips}$$

$$\left.\begin{array}{r}\text{Required SM} = 1265 \times 0.6 = 759 \text{ in.}^3, \text{W36} \times 160 \\ \text{Two Plates 10 in.} \times 3/4 \text{ in.}\end{array}\right\} \quad \text{SM} = 790 \text{ in.}^3$$

This example demonstrates the procedure in determining the bridge proportioning, sometimes referred to as the *optimum* span ratio. In general, this may vary from 1.20 to 1.35, depending on the total bridge length and the ratio of live to dead load moments.

4-12 CONTINUOUS COMPOSITE BEAM BRIDGES

Design Considerations

Continuous composite beam bridges can result in further economy and shallower construction. This is possible because the limiting values of the depth–span ratios are 1/25 and 1/30 (concrete slab plus beam and steel beam alone, respectively), and the span length can be taken as the distance between dead load points of contraflexure. Although composite construction can be used for dead and live loads with temporary supports, this scheme is not usually recommended.

Near midspan, live load stresses and stresses resulting from superimposed dead loads are computed as in the example of Section 4-7 for simple spans, using two values of n. The steel section for positive moment may be designed with or without a cover plate on the bottom flange of the beam.

For continuous beams, negative dead and live load moments at the supports usually are greater than positive moments near midspan. Experience shows, however, that the additional flange area can be obtained by attaching cover plates to the beam, top and bottom, so that the same rolled beam may be used throughout particularly if the section at the positive moment does not require a bottom cover plate. Alternatively, over the supports a satisfactory combination of plates and reinforcing bars (longitudinal) can be developed by using a smaller top cover plate (or by omitting this plate) than bottom plate and then adding longitudinal bars in the slab. The top cover plate may be omitted for short spans, but is required for longer spans. If welded plate girders are used, the required variation in flange area

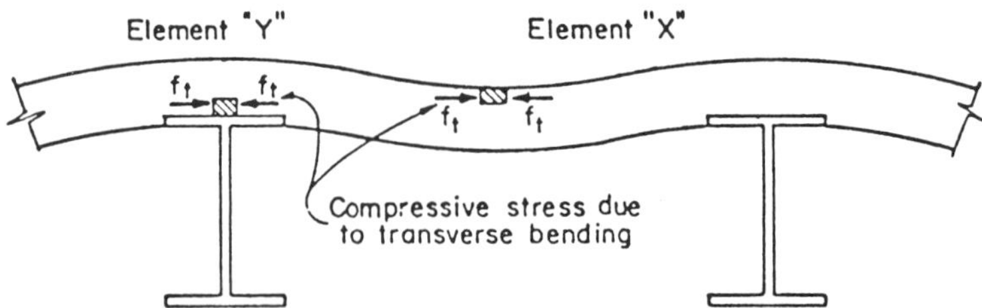

FIGURE 4-28 Longitudinal and transverse compressive stresses in slab because of composite action.

is obtained by selecting flange plates of proper thickness. In negative moment regions, only the slab reinforcement can be considered to act compositely with the steel beams in calculating resisting moments. However, on the tension side of the neutral axis, it may be considered as contributing to the moment of inertia for deflection calculations and for determining stiffness factors in continuous-beam analysis.

From the foregoing examples, we can see that composite action produces concrete stresses in the longitudinal direction, but, in practice, these are seldom combined with the usual stresses caused by transverse bending in the slab. As a matter of interest, this combined effect is illustrated in Figure 4-28. Consider, for example, the slab element X located at the center of the slab span near the center of the beam span. Theoretically, the maximum stress in the concrete is obtained by combining f_g (the compressive stress in the composite section), f_w (the compressive stress due to local longitudinal bending of the slab), and f_t (the compressive stress due to transverse bending of the slab). This combination yields the maximum principal stresses.

In practice, however, this investigation is seldom necessary because it is unlikely that the position of the wheel loads can produce maximum values of these stresses simultaneously. Interestingly, this condition is also manifested in concrete T-beam decks, but is commonly ignored as long as both f_g and f_t do not exceed the allowable working stress.

For the slab element Y, the usual practice is to ignore the effect of f_g, but where conditions warrant the design should account for the combined stress caused by the shear connectors and the transverse bending. When composite action is used for live load only, the magnitude of f_g is close to 600 psi and very seldom exceeds 700 psi, so that it is not necessary to check the principal stress. If composite action resists live load and initial dead load, f_g may become critical and the principal stresses should be investigated.

Design Example (Without Shoring)

The bridge shown in Figures 2-26*a* and *b* carries I-75, at the four-level interchange with I-275. The superstructure consists of an 8-in. concrete slab on rolled beams composite in the positive moment areas. The wheel load fraction is based on $S/5.5 = 1.455$ (for interior beams). From the graphs of Figure 2-41 and using $S/L = 0.1$, the distribution factor is 1.25. Hence, the initial design is more conservative. The design criteria are summarized as follows.

The design stresses are $f_s = 20$ ksi, $f_c = 1.2$ ksi, and $n = 8$ and $n = 24$ for live load and superimposed dead load, respectively. The effective slab thickness is reduced by 1 in. Approximately 10 percent of the concrete dead load is added to allow for the weight of forms in computing steel dead load stress. For dead load the composite flange is assumed laterally unsupported.

From Figure 2-26*b* the loads are computed as follows:

Dead load, slab $= 8 \times 0.667 \times 0.15$	$= 0.800$ kip
Fillets	$= 0.020$ kip
Beam, etc.	$= \underline{0.180}$ kip
Dead load w	$= 1.000$ kip

and adjusted for forms, dead load $w = 1.10$ kips/ft.

Superimposed dead load, W.S. $= 51.25 \times 25/7$	$= 0.180$ kip
Curb and parapet	$= 0.170$ kip
Railing	$= \underline{0.010}$ kip
Superimposed dead load w	$= 0.36$ kip

For computer analysis, the beam elevation is shown in Figure 4-29. This indicates the range of composite and noncomposite design, the field splices, and the variation in the moment of inertia of the steel only (noncomposite). The composite properties are computed as follows:

$n = 8$	$I_{comp} = 23{,}630$ in.^{4}
Section modulus	752 in.^{3}, bottom steel
	5148 in.^{3}, top steel
	1954 in.^{3}, top of concrete
$n = 24$	$I_{comp} = 17{,}538$ in.^{4}
Section modulus	687 in.^{3}, bottom steel
	1675 in.^{3}, top steel
	976 in.^{3}, top of concrete

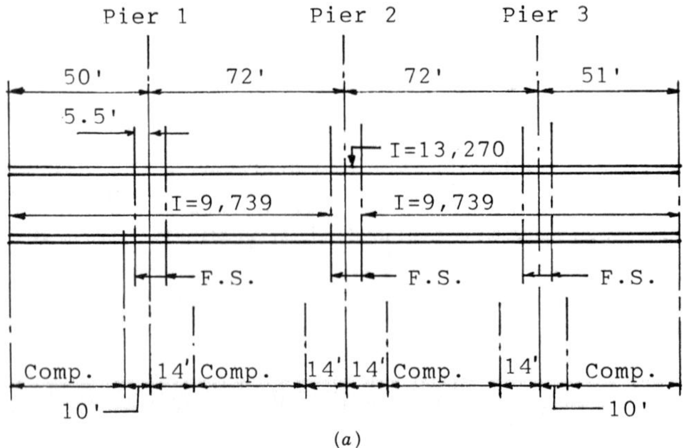

(a)

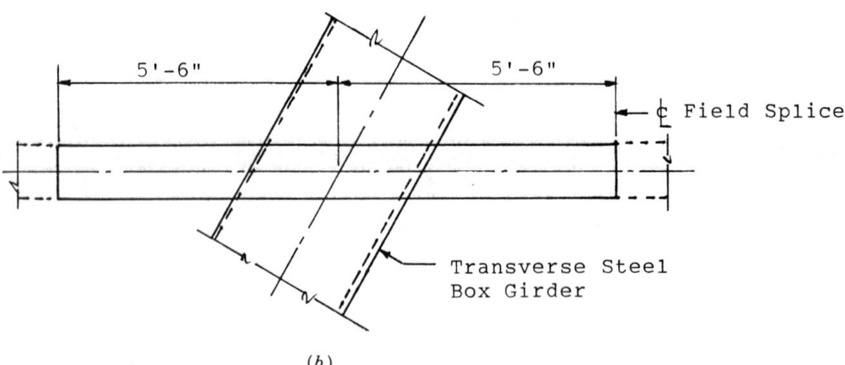

(b)

FIGURE 4-29 (*a*) Beam elevation showing data relevant to moment analysis; (*b*) transverse box girder and beam plan at interior supports.

Note that the computer performs three moment analyses: (a) a run that is based on the I of the steel section alone, giving dead load moments; (b) a run based on the composite I for $n = 8$ and the steel beam in noncomposite sections, giving live load moments; and (c) a run based on the composite I for $n = 24$ and the steel beam in noncomposite sections, giving superimposed dead load moments. However, the analyses show that the effect of using $n = 24$ on the stresses due to superimposed dead load is negligible, and a single analysis could have been made for both the live load and the superimposed dead load using $n = 8$.

Selection of Sections At pier 1, the total moment is -858 ft-kips

$$\text{Required Section modules} = 858 \times 0.6 = 515 \text{ in.}^3$$

Use W36 × 160 Section modules = 541 in.^{3}

At pier 2, the total moment is −1113 ft-kips

$$\text{Required Section modules} = 1113 \times 0.6 = 667 \text{ in.}^3$$

$$\text{Use W36} \times 194 \qquad \text{Section modules} = 664 \text{ in.}^3$$

At pier 3, the total moment is −865 ft-kips

$$\text{Required Section modules} = 865 \times 0.6 = 519 \text{ in.}^3$$

$$\text{Use W36} \times 160 \qquad \text{Section modules} = 541 \text{ in.}^3$$

The dead load stresses at the interior supports (compressive bottom flange laterally unsupported) are

$$\text{For W36} \times 160 \qquad f_x = 407.8 \times 12/541 = 9.05 \text{ ksi}$$

$$\text{For W36} \times 194 \qquad f_s = 534 \times 12/664 = 9.65 \text{ ksi}$$

Note that the selection of a heavier rolled-beam section at the interior supports is compatible with the steel box girder supporting the longitudinal beams as shown in Figure 4-29*b*. The location of field splices is determined by clearance requirements. The beam sections over the supports are fabricated with the transverse box girders, shipped, and erected as one piece. A maximum lateral dimension of 10 ft was stipulated as shown to accommodate shipping and erection.

Stresses in the composite sections are computed as follows:

End spans:

$$\text{Bottom flange} \qquad f_s = \frac{169.8 \times 12}{541} + \frac{62.6 \times 12}{687} + \frac{510.8 \times 12}{752} = 13.01 \text{ ksi}$$

$$\text{Top flange} \qquad f_s = \frac{169.8 \times 12}{541} + \frac{62.6 \times 12}{1625} + \frac{510.8 \times 12}{5148} = 5.41 \text{ ksi}$$

$$\text{Top of concrete} \qquad f_c = \frac{62.6 \times 12}{976 \times 24} + \frac{510.8 \times 12}{1954 \times 8} = 424 \text{ psi}$$

Center spans:

$$\text{Bottom flange} \qquad f_s = \frac{243.9 \times 12}{541} + \frac{98.1 \times 12}{687} + \frac{645.4 \times 12}{752} = 17.43 \text{ ksi}$$

$$\text{Top of concrete} \qquad f_c = \frac{98.1 \times 12}{976 \times 24} + \frac{645.4 \times 12}{1954 \times 8} = 545 \; psi$$

Because the rolled section changes at field splices adjacent to pier 2, the stresses at these locations should be checked for the lighter beam. All splices are located at 5.5 ft from the centerline of the pier. From the computer results, we obtain $M_{19} = -594$ ft-kips and $M_{21} = -594$ ft-kips, at 7.2 ft from the centerline of the pier. Also recall that $M_{20} = -1113$ ft-kips. By straight-line interpolation, we calculate the following:

Moment at 5.5 ft from pier = -716 ft-kips

Moment at 3.0 ft from pier = -897 ft-kips (SM = $897 \times 0.6 = 538$ in.^{3})

Hence, the rolled beam W36 × 160 becomes sufficient at 3 ft from the centerline of the pier, and the location of the field splices is thus acceptable.

Likewise, the maximum and minimum moments at field splice locations are always negative, and therefore fatigue strength considerations do not control.

The design of shear connectors was demonstrated in the example of Section 4-7, and will not be repeated here.

4-13 PRESTRESSED DECKS IN CONTINUOUS COMPOSITE BRIDGES: CASE STUDY

The inevitable cracking (transverse) of the concrete deck in the region of intermediate supports tends to inhibit the composite advantages from being fully realized. Thus, most engineers confine composite design to simple spans or to the sagging regions of continuous spans. Transverse cracking reduces the stiffness of the bridge and increases deflections. In this context, the effect of the interaction between prestressing a portion of the slab and shear connectors on the elastic response and crack control of continuous composite bridges becomes quite relevant to the optimum design.

The theoretical analysis of this problem is presented by Kennedy and Grace (1982), based on orthotropic plate theory and reliable estimates of the various rigidities. These results are verified by tests on scale models of a two-span continuous bridge of composite construction.

Summary of Analysis Grace (1981) has demonstrated that the effect of the composite action between the concrete slab and steel beams can be included in the computation of moments if the bridge deck is treated as an orthotropic plate with eccentric stiffeners, that is, a plate stiffened on one side by longitudinal and transverse members. In this case orthotropy is assumed to be the result of geometry. In addition to the basic assumptions (Huber, 1923; Kennedy and Gupta, 1976), the following are admissible in the context of composite construction: (a) the rigidities of both longitudinal and transverse members are uniformly distributed throughout the deck; (b) the bridge is represented by an idealized substitute orthotropic plate of uniform thickness; (c) the neutral plane in each of the two orthogonal directions coincides with the center of gravity of the total section in the corresponding direction; and (d) the area of the flange plate is magnified by a factor of

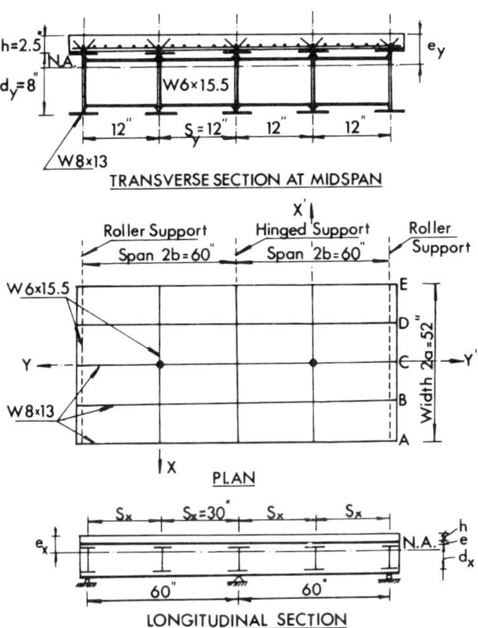

FIGURE 4-30 Geometry of two-span continuous composite bridge structure. (From Kennedy and Grace, 1982.)

$1/(1 - \mu^2)$ to allow for the influence of Poisson's ratio μ. Past experience shows that the preceding assumptions are valid in practice (Clifton et al., 1963; Fisher, Daniels, and Slutter, 1972; Iwamoto, 1962; Kennedy and Bali, 1979).

Results of tests on the two models shown in Figure 4-30 indicate that there is complete interaction between the concrete deck and the longitudinal beams. With rigid connections between the transverse diaphragms and the longitudinal beams, the interaction between the concrete deck and the diaphragms in enhanced. From these considerations, expressions are derived for the orthotropic rigidities, D_x, D_y, D_1, and D_2, of a composite bridge with rigidly connected diaphragms and an uncracked concrete deck (Kennedy and Grace, 1982).

Likewise, the transverse and longitudinal torsional rigidities per unit width, D_{xy} and D_{yx}, can be expressed in terms of the shear modulus $E_c/2(1 + \mu)$ and the torsional constants $I_{xy} = I_{yx} = \frac{1}{2}kh^3$, where k is a factor dependent on the cross-sectional dimensions of the concrete deck (Rowe, 1962).

Experimental Study The two models shown in Figure 4-30 are two-equal-span continuous bridges. They are identical, except that model II was prestressed longitudinally in the region of intermediate support where model

I was conventional nonprestressed. The model scale is 1/8 in. plan and 1/3 vertically. High early-strength cement was used to provide 7-day concrete strengths of 5 ksi and 8 ksi for models I and II, respectively. Further details are provided by Kennedy and Grace (1982).

The concrete deck of model II consists of two segments; the first is a prestressed deck, 33-in. long (model scale), centered about the intermediate support and reinforced transversely at the bottom; the second segment is a concrete slab reinforced at the bottom by steel mesh. The prestressing wires are extended some distance on either side of the bearing plates to form a bond between the prestressed deck portion and the two concrete deck portions cast subsequently.

Below the cracking load condition, each bridge was loaded by a single concentrated load at three positions: (a) midspan of an exterior beam, (b) midspan of the first interior beam, and (c) midspan of the middle beam. Both models were tested to near collapse using two equal concentrated loads in each span applied through a heavy distribution beam to the midspan of the middle beam. The ends of both bridges were tied down to resist possible uplift caused by eccentric loading.

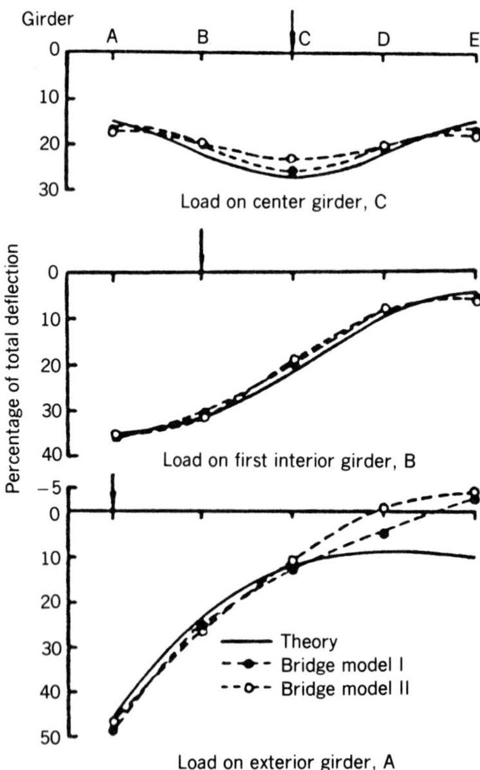

FIGURE 4-31 Comparison between experiment and theory for deflection distribution at midspan of bridge models I and II. (From Kennedy and Grace, 1982.)

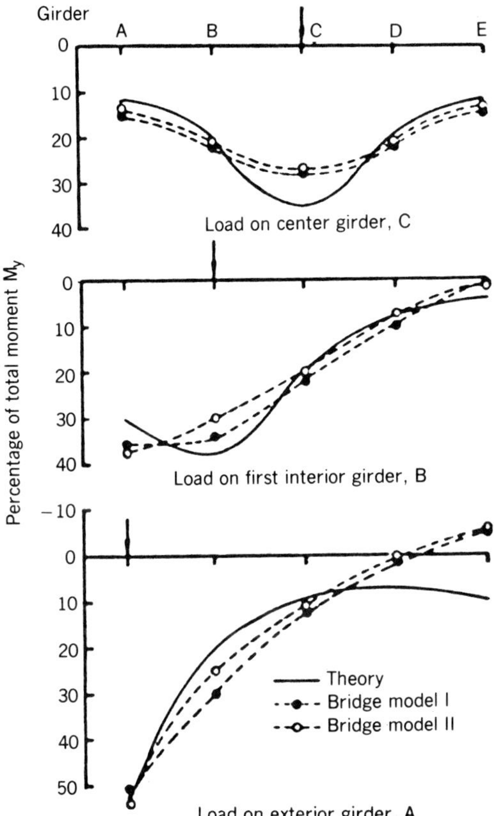

FIGURE 4-32 Comparison between experiment and theory for longitudinal moment M_y; distribution at midspan of bridge models I and II. (From Kennedy and Grace, 1982.)

Figure 4-31 shows a comparison of the deflections in the form of transverse distribution of midspan for all three load positions. The deflection of each beam is expressed as a percentage of the total deflection of the five beams at midspan, and evidently close agreement exists between the theory and the test results. The transverse distribution of the longitudinal moment, M_y, at midspan is shown in Figure 4-32 for the three positions of the applied single load. Likewise, theoretical predictions and test results compare favorably. Results of the distribution for the longitudinal moment, M_y, at the support are shown in Figure 4-33. In the latter case, two relevant theories are shown, designated as I and II, for models I and II, respectively. Theory II accounts for (a) the applied moment due to prestressing and changes in this moment due to an increase in the prestressing force caused by the applied load, (b) the secondary moment resulting from prestressing a continuous structure, and (c) the primary moment due to external load.

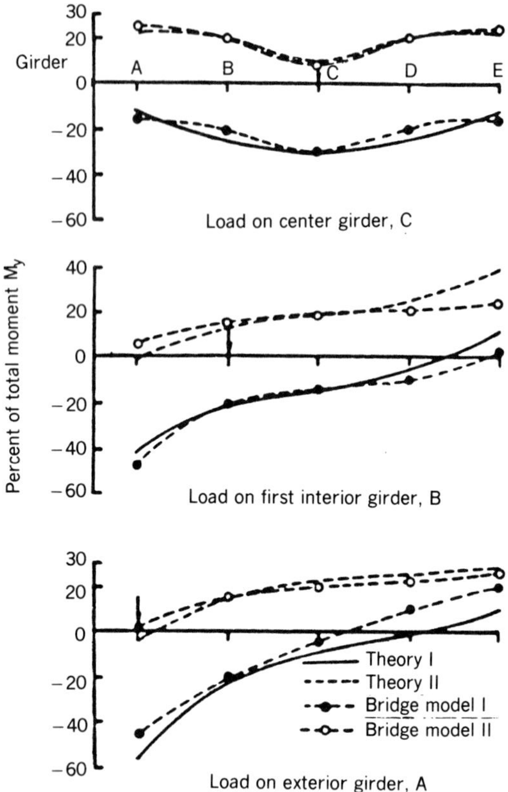

FIGURE 4-33 Comparison between experiment and theory for longitudinal moment M_y; distribution at intermediate support of bridge models I and II. (From Kennedy and Grace, 1982.)

The difference in bridge behavior between models I and II due to prestressing is evident by reference to Figure 4-33. A final comparison is made in Figure 4-34, showing the transverse distribution of the transverse moment, M_x. A discrepancy between the test results and the theoretical predictions for beam E, due to an eccentric load applied at beam A, is probably caused by the fact that the middle support is not tied down. Thus, when the load was applied to beam A, an upward deflection and lift-off of beam E resulted.

Cracking This condition was induced in both models by a two-span loading system. For model I, the first cracks developed near the intermediate support at a load of 40 kips, and measured 0.009 mm in width, distributed randomly. Hairline cracks were detected in the longitudinal direction. At a load of 80 kips, the transverse cracks widened to 0.1 mm, and severe cracking occurred at a load of 180 kips when some cracks reached a width of 0.3 mm.

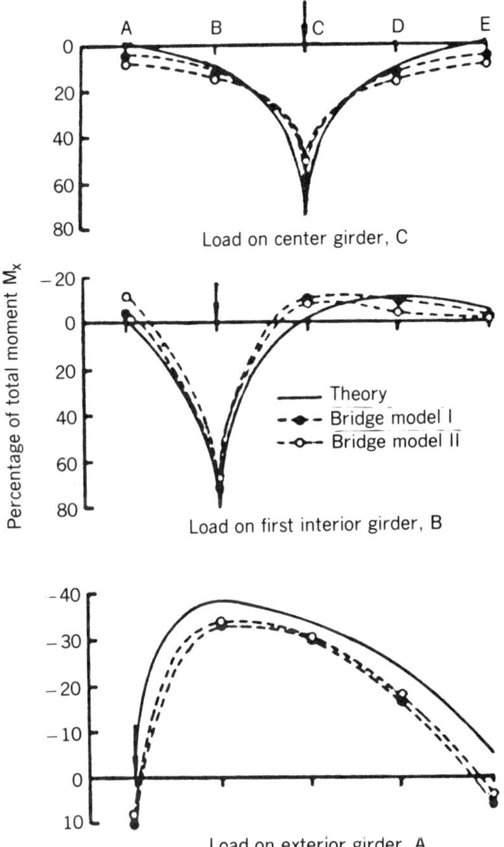

FIGURE 4-34 Comparison between experiment and theory for transverse moment M_x; distribution at midspan of bridge models I and II. (From Kennedy and Grace, 1982.)

At a load of 80 kips, no cracks were detected for the prestressed model II. At a load of 180 kips, this model developed transverse cracks 0.15 mm wide. For the same model the separation of the two joints (between prestressed and non-prestressed sections) was monitored throughout the load applications. These joints were located at the points of contraflexure and received no special treatment. Thus, with a load of 50 kips acting at midspan of beam C, the joint separated by 0.005 mm near the top but with no detectable separation at middepth of the deck.

With the two-span loading system the joint separation increased to 0.01 mm at a load intensity of 100 kips, but at 180 kips this separation decreased to 0.005 mm when transverse cracking at the intermediate support reduced the section stiffness and shifted the contraflexure point closer to the support. Hence, this theoretical treatment does not reflect such joint separation in the

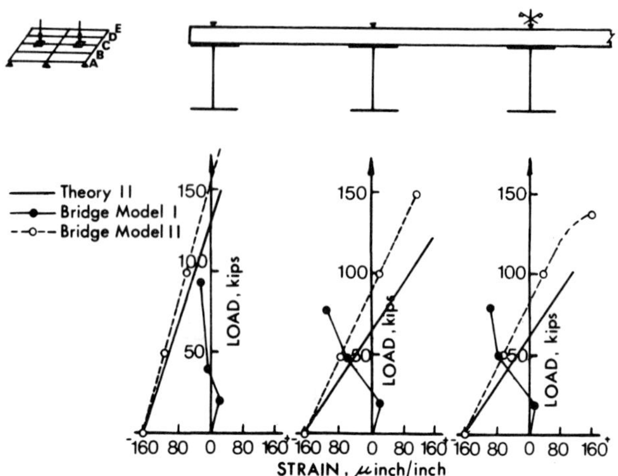

FIGURE 4-35 Comparison between bridge models I and II for longitudinal strain at top of concrete deck at intermediate support. (From Kennedy and Grace, 1982.)

deck, and the loss of stiffness due to this separation was estimated by the method of column analogy.

Comparison of Behavior At a load of 50 kips, model I cracked transversely at the intermediate support, whereas model II remained crackless. This cracking reduced the bridge stiffness for model I, reflected in the comparison of midspan deflections. Comparison of the results indicates that model I deflected about 15 percent more than model II, and this difference could increase with increasing applied load due to the nonlinearity of cracked concrete. The same observation applies to the longitudinal strains at midspan, based on the same results. We should note that transverse cracking at the intermediate support has no marked influence on the transverse distribution of deflections or on the longitudinal and transverse moments at midspan for the continuous structure.

The relationship between applied load and longitudinal strain at the top of the concrete deck at the intermediate support with beam C subjected to two-span loading is shown in Figure 4-35. For model I, the strains are tensile and become compressive only as a result of gage instability following severe cracking. For model II, the strains remain compressive, compatible with the prestressing until the load reaches a very high intensity.

The two bridge models were subjected to maximum loads of 280 and 240 kips, respectively. These intensities caused severe cracking in model I but moderate cracking in model II. From plastic analysis, the predicted failure load in either bridge model would be about 320 kips. Kennedy and Grace (1982) suggest that in the analysis of a prototype bridge the design of the deck in the transverse direction should be considered for wheel loads applied between beams, and the results superimposed on those obtained from orthotropic plate analysis presented in this study.

Commentary A convergence series solution using the concept of equivalent orthotropy can be used to analyze continuous composite bridges under static load. The concept of prestressing involves realistic estimates of the various rigidities, but the interaction of the prestressed deck (in the negative moment region) with the composite response is explicit.

Similar studies have emphasized crack control in the hogging moment region, and Fisher, Daniels, and Slutter (1972) have recommended increasing the percentage of longitudinal reinforcing steel in the slab over the hogging moment region to at least 1 percent (see also Section 4-5). This provision complies with current AASHTO requirements.

The use of simple plastic theory for continuous composite beams must ensure that the first hinges to form can maintain their strength while developing sufficient rotation for the plastic moment to be reached at the other hinge locations (Johnson, 1975). For certain combinations of spans, loadings, and cross sections, the rotation requirements may be severe. Parametric studies and tests tend to confirm the conclusion that slenderness limits for steel sections where buckling can occur must be more restrictive than those for all-steel sections if plastic design is to be used, and for an internal span carrying a heavy point load, premature flexural failure under the load can occur if the adjacent spans are less stiff than the loaded span.

In these cases plastic design methods must be used with caution. Protection against poor use of these procedures must rely on limit state design requirements such as the stipulation that the flexural stress in the structural steel and reinforcement due to loads at the serviceability limit state should not exceed a certain limit. However, this limit is likely to be exceeded at the supports of continuous beams designed by plastic theory, for example, in the bottom flange for unshored construction and in the slab reinforcement of shored construction. The question then arises as to whether or not safe predictions can be made of the deflections and crack widths on an elastic basis. These difficulties tend to make plastic theory less attractive for composite structures (see also other sections).

4-14 STRENGTH DESIGN METHOD: LOAD FACTOR DESIGN (CURRENT AASHTO TERMINOLOGY)

The introduction to the strength design method presented in Sections 2-4 and 2-13 highlights current trends in the design of steel bridges. For a combined dead and live load, the basic expression of load factor design is

$$\phi R_n \geq \gamma_R[\gamma_D D + \gamma_L(L + I)] \tag{4-15}$$

where ϕ = resistance (strength) reduction factor

R_n = nominal (ultimate) resistance (strength) for moments, shear, force, and so on

γ_R = load factor reflecting the uncertainties of structural analysis
γ_D, γ_L = load coefficients applied to dead and live load, respectively

For the usual nomenclature, the factored nominal strength ϕR_n is the design strength of a member, whereas the second term gives the required strength. By rearranging (4-15), we obtain

$$R_n \geq \frac{\gamma_R}{\phi}[\gamma_D D + \gamma_L(L + I)] \qquad (4\text{-}16)$$

In the AASHTO specifications (group I), the load factors and coefficients are as follows: $\gamma_R = 1.3$, $\gamma_D = 1.0$, and $\gamma_L = 1.67$, giving a required strength equal to $1.3[D + 1.67(L + I)]$. This in effect means that the factor of safety for dead load is 1.3, whereas for live load this factor is 2.17.

The left side of (4-15) denotes the resistance or strength that can be developed in the component or system, and the right term represents the loads expected to be carried. The γ factors are overload coefficients applied to dead and live loads to obtain the sum of the factored loads. A bridge must have adequate stiffness and toughness to permit proper functioning during its service life, but the design must also provide for some reserve strength to accommodate the possibility of overload. In addition, a provision must be included to recognize the possibility of understrength.

The concept of the LRFD methodology was developed in the 1970s and adapted by the AISC in 1986 (Galambos, 1972; Pinkham and Hansell, 1978). The adoption of probabilistic methods to steel design is explained by Galambos (1981) and by Ravindra and Galambos (1978). The general format of the LRFD criteria is given by (4-15) and means that the strength provided in the design must be at least equal to the factored loads. The detailed application of the LRFD approach is discussed in the proposed LRFD specifications (see also Section 5-15).

Load Distribution The intent of current standards is to apply a wheel load distribution factor to a single beam, and compute the maximum dead load, live load and impact moments, shears or forces to be sustained by the stress-carrying member. The response of the member, however, stimulates its interaction with the bridge system.

In addition to the topics discussed in Section 2-13, new distribution factors have been proposed by Sanders and Elleby (1970) as well as by other investigators. Kuo and Heins (1973) have developed the following expression for the wheel load distribution factor at ultimate load

$$(\mathrm{DF})_p = 3.45 - 1.809\gamma + 0.315\gamma^2 \qquad (4\text{-}17)$$

where $(\mathrm{DF})_p$ is the ultimate load or plastic distribution factor and γ is the ratio of the number of girders or beams to the number of lanes. The range of applicability of (4-17) covers straight simple spans of composite I beams in bridges with 4 to 9 beams or girders, 6- to 9-ft beam spacing, and 2 to 4 lanes.

A study by Kuo (1973) shows that, unless fatigue or deflection problems arise, additional material savings can be realized using (4-17). Furthermore, work on inelastic effective slab width by Fan and Heins (1974) demonstrates that the standard evaluation of effective slab width of composite beams needs modification for load factor design (see also subsequent sections).

Comparison of Methods: Allowable Stress Versus LRFD For allowable stress design (working load), (4-15) may be reformulated as follows:

$$\frac{\phi R_n}{\gamma_R} \geq \sum Q_i \tag{4-18}$$

where ΣQ_i is the summation of load effects (dead and live plus impact). In this philosophy all loads are assumed to have the same average variability. The entire variability of loads and strength is placed on the strength side of the equation. In terms of working stress, the same requirement is expressed as

$$f_s < \frac{F_y}{\mathrm{FS}} \qquad \text{or} \qquad f_s < \frac{F_{\mathrm{cr}}}{\mathrm{FS}} \tag{4-19}$$

where f_s = maximum stress calculated from elastic theory for the working loads

F_y = specified minimum yield stress

F_{cr} = critical buckling stress

Initial yielding or elastic buckling are not, however, the two applicable limit states, and other relevant conditions involve connections, postbuckling strength, different factors of safety (FS) for different situations, and so on.

The rationale of (4-18) is derived from the strength capable of being achieved if the structure is overloaded. Where the section is ductile and buckling does not occur, strains greater than the "first yielding" strain can exist on the section. Such ductile inelastic behavior may accommodate higher loads to be carried than possible if the bridge had remained entirely elastic. In this case the allowable stress must be adjusted upward. When the strength is controlled by buckling or some other behavior so that the stress does not reach yield condition, the allowable stress must be adjusted downward. Serviceability requirements such as deflection limits require routine investigations for either the allowable or the load factor method. It may appear, therefore, that the allowable stress method is a version of ultimate strength design based on elastic behavior, but conceptually this approach is difficult to understand.

In the load factor approach, the various types of working loads are multiplied by load factors indicating the variability and different levels of uncertainty with which these loads are determined. For example, the dead load stress can be predicted with greater accuracy than the maximum live load plus impact stress. On short-span bridges the live load plus impact stresses are much greater than dead load stresses, but for long-span bridges the opposite is true. The new approach can thus yield greater economy for long-span bridges, whereas in relatively short bridges the results tend rather to articulate smaller stresses.

The rationality of LRFD as outlined by most investigators suggests that as a design philosophy it may relegate working stress design to a second option, although many bridges will continue to be designed by allowable stress methods. The same working stress approach will be needed to evaluate bridges of the past. The load factor design method can yield advantages associated with weight saving in structural steel. However, the method makes design in all materials more compatible. The variability of loads is actually unrelated to the type of material used, and this is a distinct advantage in future specifications. The LRFD procedure results in a bridge system having elements of a uniform live load capacity, and this means a better balance of the proportioned structure. Safer structures may thus result because the method enhances awareness of structural behavior. Changes in overload factors and resistance factors are easier to make than changes to allowable (working) stresses. For a complete treatise of the LRFD approach, reference is made to Smith (1991), Salmon and Johnson (1990), and more recently to Modjeski and Masters (1992).

4-15 DESIGN EXAMPLES: SIMPLE BEAMS (LOAD FACTOR APPROACH)

Simple-Span I-Beam Bridge (Noncomposite)

The example of Section 4-7 will be analyzed using the load factor approach. The bridge is a simple I-beam structure, 70 ft long and noncomposite.

Recall that the total dead load and superimposed dead load is 0.93 + 0.37 = 1.30 kips/ft. The wheel load distribution factor is taken as 1.30, and the impact factor is 1.256.

The dead load moment is $1.30 \times 70^2 \times 0.125 = 796$ ft-kips. The live load plus impact moment is obtained directly from the example of Section 4-7 as 807 ft-kips. The total factored moment is therefore

$$M_u = 1.3(796 + 1.67 \times 807) = 2786 \text{ ft-kips}$$

Assuming that a rolled I beam meets the requirements of compact sections, the required plastic section modulus, Z, can be obtained from the expression (see also Section 4-17)

$$M_u = F_y Z \qquad \text{or} \qquad Z = \frac{M_u}{F_y} \tag{4-20}$$

which is AASHTO Eq. (10.91). For $F_y = 36$ ksi, we calculate

$$Z = \frac{2786 \times 12}{36} = 929 \text{ in.}^3$$

From the AISC manual, we select W36 × 230, $Z = 943$ in.^{3}.

Using the allowable stress method, the total moment is 796 + 807 = 1603 ft-kips. The required section modulus is 1603 × 0.6 = 962 in.^{3}. A W36 × 260 beam provides a section modulus of 952^3 in., and the 1 percent overstressing is acceptable. Note, however, that the total dead load is computed assuming that the correct allowance is made for the weight of the beam so that recalculation of the dead load moment may be necessary. Although the dead load and live load plus impact moments are nearly the same, a weight saving of 12 percent is possible if the design follows the load factor method.

Compact Section Requirements The projecting compression flange element must satisfy the following (AASHTO Article 10.48.1).

(a) Projecting compression-flange elements

$$\frac{b'}{t} \leq \frac{2055}{\sqrt{F_y}} \tag{4-21}$$

For W36 × 230, $b'/t = 7.85/1.26 = 6.23$, and $2055/\sqrt{36{,}000} = 10.82$ OK.

(b) Web thickness

$$\frac{D}{t_w} \leq \frac{19{,}230}{\sqrt{F_y}} \tag{4-22}$$

For W36 × 230, $D/t_w = 33.36/0.761 = 43.8$, and $19{,}230/\sqrt{36{,}000} = 101$ OK.

In the foregoing,

b' = width of the projecting flange element
t = flange thickness
D = clear distance (web height) between flanges
t_w = web thickness

The beam qualifies therefore as a compact section.

End Shear According to AASHTO Article 10.48.8, for beams with unstiffened webs the shear capacity is limited to the plastic or buckling shear force as follows.

$$V_u = CV_p \tag{4-23}$$

$$V_p = 0.58F_y Dt_w \tag{4-24}$$

where

$$C = 1$$

[The constant C is equal to the buckling shear stress divided by the shear yield stress, and because $(D/t_w) < (6000\sqrt{k})/\sqrt{F_y}$ where $k = 5$, C is taken as 1.]

From (4-23), we obtain

$$V_u = V_p = 0.58 \times 36 \times 33.36 \times 0.761 = 529 \text{ kips}$$

By simple examination of the loads, it is apparent that V_u does not control.

Deflections The deflected shape of a bridge due to service conditions can be economically controlled by precambering the steel beams. Precambering is achieved by erecting the beams with built-in deformations so that the bridge deflects from its theoretical no-load shape when the maximum service dead load is placed on the structure. For uniformly simply supported beams, the maximum deflection is

$$\Delta_{\max} = \frac{5wL^4}{384EI} \tag{4-25}$$

which upon substituting for $M = 0.125wL^2$, $f_s = Mc/I$, and $c = d/2$, gives

$$\Delta_{\max} = \frac{10f_s L^2}{48Ed} \tag{4-26}$$

Live load deflection should be checked according to the AASHTO provisions as in allowable stress design. Continuous lateral support (unbraced length $L_b = 0$) is assumed for the top flange (compression) if this member is encased in the concrete slab and provided the slab is prevented from translation in the direction perpendicular to the web of the beam. The unbraced length should be checked for noncompact sections.

Overload capacity is also investigated according to AASHTO Article 10.57.1. The moment caused by $D + 1.67(L + I)$ should not exceed $0.8F_yS$. If beams are designed for group 1A loading, the moment caused by $D + 2.2(L + I)$ should not exceed $0.8F_yS$, where S is the elastic section modulus. However, $D + 1.67(L + I) > 0.8F_yS$, and heavier section must be selected (W36 × 245).

Simple Span I-Beam Bridge (Composite)

Nominal (Ultimate) Moment Strength of Fully Composite Section According to LRFD 13.2 The nominal moment strength M_n of a composite section with the slab in compression (positive moment) depends on the yield stress F_y and section properties of the steel beam, the concrete slab strength f_c', and the strength of the shear connectors providing the interface shear transfer between the slab and beam. This strength concept was first applied to design practice as recommended by the ASCE–ACI (1960) Joint Committee on Composite Construction, and further modified by Slutter and Driscoll (1965). Viest (1974 chaired a state-of-the-art review, and Pinuham and Hansell (1979) treated the subject in the context of the LRFD concept.

The nominal moment strength M_n is considered in two cases, depending on web slenderness, as follows:

1. For $D/t_w \leq 640\sqrt{F_{yf}}$, M_n based on plastic stress distribution of the composite sections, and $\phi = 0.85$, where F_{yf} is the yield stress in the flange (in ksi).
2. For $D/t_w > 640\sqrt{F_{yf}}$, M_n based on elastic stress superposition, and $\phi = 0.90$.

Considering the nominal strength M_n in plastic stress distribution, two different cases are possible depending on whether the plastic neutral axis (PNA) is in the flange (concrete slab) or in the beam. The concrete is assumed to develop only compressive forces because its tensile strength is negligible at the strains corresponding to maximum strength.

Plastic Neutral Axis (PNA) in the Slab By reference to Figure 4-36 and assuming the Whitney rectangular stress distribution (Wang and Salmon, 1985) of a uniform stress of $0.85f_c'$ acting over a depth a, the compressive

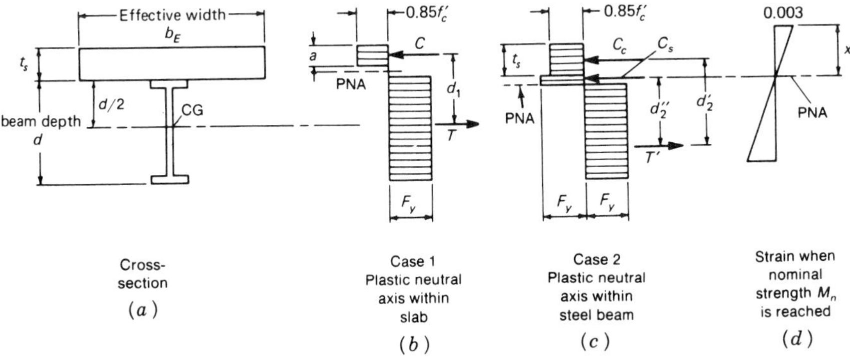

FIGURE 4-36 Plastic stress distribution at nominal moment strength M_n (PNA is the plastic neutral axis).

force in the slab is

$$C = 0.85 f'_c a b_E \tag{4-27}$$

The tensile force T in the beam is the yield stress multiplied by the beam area, or

$$T = A_s F_y \tag{4-28}$$

Setting $C = T$ gives

$$a = \frac{A_s F_y}{0.85 f'_c b_E} \tag{4-29}$$

The nominal moment strength M_n is obtained from Figure 4-36b as

$$M_n = Cd_1 \qquad \text{or} \qquad Td_1 \tag{4-30}$$

With the slab capable of developing a compressive force at least equal to the full yield strength of the steel beam, the PNA lies in the slab, and this condition is common for most fully composite bridges. In this case the nominal moment strength can be expressed in terms of the steel force as

$$M_n = A_s F_y \left(\frac{d}{2} + t_s - \frac{a}{2} \right) \tag{4-31}$$

The usual procedure is to assume that $a \leq t_s$ and calculate its value from (4-29). If this assumption is verified, the maximum moment strength is computed from (4-31). We should note that the nominal strength is independent of whether or not the system is shored during construction. Although service load stresses are different, the nominal strength at plastic stress distribution is the same for shored and unshored composite beams.

Numerical Example For the example of Section 4-7, we assume that a is in the slab. Then we compute

$$A_s F_y = 52.3 \times 36 = 1882$$

and

$$0.85 f'_c b_E = 0.85 \times 3.5 \times 90 = 268$$

The value of a is $1882/268 = 7$ in. < 7.5 in.

The moment arm must be adjusted because the beam has a bottom cover plate. From previous data, we calculate this arm as $22.5 + 8.5 - 3.5 = 27.5$ in. The nominal moment strength is therefore

$$M_n = 1882 \times 27.5/12 = 4310 \text{ ft-kips}$$

Applying the strength reduction factor of 0.85, the design moment strength is $4310 \times 0.85 = 3663$ ft-kips > 2786 ft-kips (factored moment).

Plastic Neutral Axis (PNA) in the Steel Beam If the depth a exceeds the slab thickness, the stress distribution is as shown in Figure 4-36c. The compressive force in the slab is $C_c = 0.85f_c'b_Et_s$. In addition, there is a compressive force in the steel beam for the portion above the PNA, shown in Figure 4-36c as C_s. The tensile force T' is less than A_sF_y, and for equilibrium we have

$$T' = C_c + C_s \tag{4-32}$$

also

$$T' = A_sF_y - C_s \tag{4-33}$$

from which we obtain

$$C_s = \frac{A_sF_y - C_c}{2} = \frac{A_sF_y - 0.85f_c'b_Et_s}{2} \tag{4-34}$$

In terms of the compressive forces C_s and C_c, the nominal moment strength M_n for this case is

$$M_n = C_cd_2' + C_sd''_2 \tag{4-35}$$

where the moment arms d_2' and d_2'' are clearly shown in Figure 4-36c.

The steel beam must be capable of accommodating plastic strains in both tension and compression for the nominal strength condition. With the PNA located lower in the steel section, more local buckling may influence the steel behavior. This possibility has dictated the D/t_w limits mentioned at the beginning of this section. Local buckling in the flange is also of concern, but the presence of shear connectors and the encasement of the flange in the concrete make the steel section exempt from compact flange criteria.

AASHTO Procedure for Computing Maximum Strength The proportioning of composite beams to satisfy load factor design criteria involves the following steps.

Step 1. Check whether the steel section satisfies the compactness requirements of Article 10.50.1.1.2. The factor D in (4-22) may be replaced by $2D_{cp}$ for simple spans, where D_{cp} is the distance from the compression flange to the neutral axis in plastic bending. The compression depth of the composite section including the slab should not exceed $(D + t_s)/7.5$, where D is the beam depth and t_s is the slab thickness.

Step 2. If the top flange is in compression, check the ratio of the projecting flange plate width to plate thickness b'/t. This ratio should not exceed $2200/\sqrt{1.3f_{dn}}$ where f_{dn} is the top flange compressive stress due to noncomposite dead load. Note that this criterion does not relate to the full plastic condition, but it is an elastic stress requirement. Referring to the example of Section 4-7, $f_{dn} = 14.06$ ksi from which we compute $2200\sqrt{1.3 \times 14{,}060} = 16.3$. For the rolled beam W36 $\times$ 135, $b'/t = 5.67/0.74 = 7.2 < 16.3$.

Step 3. Check the maximum shear due to noncomposite dead load with a load factor of $\gamma = 1.3$. This should not exceed the shear buckling capacity of the web expressed by (4-23). If this criterion is satisfied, the moment of first yield should be computed considering the application of the dead and live loads to the steel and composite sections.

Step 4. For compact sections, compute the maximum strength as the resultant moment of the fully plastic stress distribution acting on the section. This involves computation of the compressive force in the slab, compressive force in the steel (top flange), and tension force in the steel beam, according to a procedure similar to the foregoing analysis. In computing the compressive force in the slab, the designer may include the yield strength AF_y of the steel reinforcement that lies in the compression zone of the slab.

4-16 DESIGN EXAMPLES: CONTINUOUS BEAMS (LOAD FACTOR APPROACH)

In the design of continuous beams with compact sections complying with (4-21) and (4-22), AASHTO stipulates that negative moments over supports at overload and maximum load conditions determined by elastic analysis may be reduced by a maximum of 10 percent. This reduction must be accompanied by an increase in maximum positive moment in adjacent spans equal to the average decrease of the negative moments at the adjacent supports. Comparable provisions are included in the LRFD document (see also Section 2-13).

Current steel design approaches have brought together concepts relating to both load and resistance factor design (LRFD) and allowable stress design (ASD). In addition, plastic analysis is introduced for application to continuous beams, but this procedure may be applicable with "compact sections" having adequate lateral bracing. Plastic analysis is allowed under Chapter N of the ASD specification (AISC, 1989) and may be applicable under the LRFD method. However, plastic analysis is not required to be used, whereas elastic analysis is always permitted under AISC codes. The reduction permitted by AASHTO in the negative moment areas partially reflects the effects of plastic behavior.

Two-Span Continuous I-Beam Bridge (Noncomposite)

The two-span continuous beam bridge analyzed in Section 4-11 is considered under the load factor approach. We recall that the bridge has two equal spans, each 70 ft long. Dead and live load plus impact moments are as follows:

Positive	M_{DL} in span $= 448$ ft-kips
Negative	M_{DL} at support $= -796$ ft-kips
Positive	$M_{\text{LL}+I}$ in span $= 678$ ft-kips
Negative	$M_{\text{LL}+I}$ at support $= -569$ ft-kips

The maximum load (factored) moments are calculated as follows:

Positive in span	$M_u = 1.3(448 + 1.67 \times 678) + 114 = 2167$ ft-kips
Negative at support	$M_u = 1.3(796 + 1.67 \times 569) \times 0.90 = 2043$ ft-kips

Assuming a compact section, the required plastic modulus Z is

$$Z = \frac{2167 \times 12}{36} = 722 \text{ in.}^3$$

A W36 $\times$ 194 rolled beam provides a plastic section modulus of $Z = 768$ in.^{3}, so that the design is identical to the beam selection stipulated by the allowable stress method reviewed in Section 4-11 for the same bridge. Note, however, that the same section is sufficient over the negative moment region, and the net saving is thus the elimination of the cover plates at the interior supports.

Because the intent of AASHTO Article 10.48.1.3 is permissible rather than obligatory, we may choose not to decrease the negative moment. In this case, the design (maximum) moments are -2270 ft-kips at the support and 2054 ft-kips in the span. Assuming again a compact section, the design requires a plastic modulus

$$Z = \frac{2054 \times 12}{36} = 685 \text{ in.}^3$$

which is provided by a W36 $\times$ 182 ($Z = 718$ in.^{3}). Cover plates must be included, however, in the negative moment region. Both rolled section choices will be checked for compact section requirements. For W36 $\times$ 194,

$$b'/t = 5.68/1.26 = 4.51 < 10.82$$
$$D/t_w = 33.96/0.77 = 44 < 101$$

By inspection W36 $\times$ 182 also qualifies as a compact section.

Braced Noncompact Sections Compact sections are symmetrical I-shaped beams with high resistance to local buckling. These members are able to form plastic hinges with an inelastic rotation capacity almost three times the elastic rotation corresponding to the plastic moment.

Braced noncompact sections of rolled beams meeting certain requirements stipulated in Article 10.48.2 of the AASHTO specifications may be designed for a maximum moment $M_u = F_y S$, where S is the elastic section modulus. A limit, however, exists on the spacing of lateral bracing for the compression flange, L_b.

Examples of lateral bracing for the compression flange of a rolled WF beam are shown in Figure 4-37. In all cases the top flange is assumed to be the compression flange. Continuous lateral support is assumed for the section shown in Figure 4-37*a*. For the usual shear stud spacing, continuous lateral

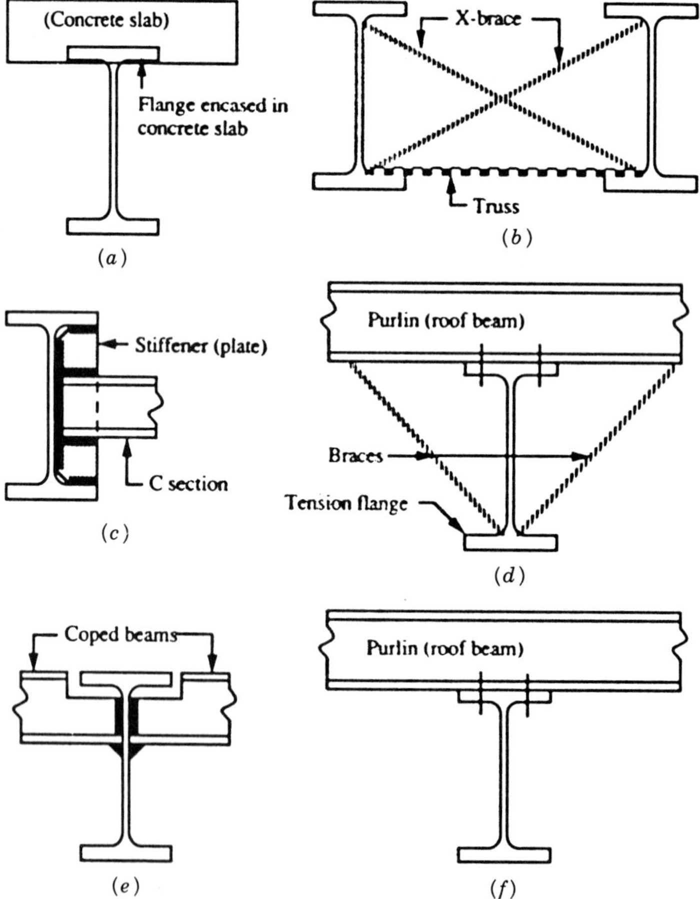

FIGURE 4-37 Examples of lateral support for WF rolled section beams. (From Smith, 1991.)

support ($L_b = 0$) is provided for the top beam flange. The braces shown in Figure 4-37*b* are more suitable for plate girders. The cross beams or diaphragms shown in Figures 4-37*c* and *e* are acceptable for load factor analysis. In this case the spacing is calculated according to AASHTO Article 10.48.2.1.

4-17 BASIC PRINCIPLES OF PLASTIC ANALYSIS

Behavior of Laterally Stable Beams Figure 4-38 shows the stress distribution of a typical rolled-beam section subjected to increasing bending moment. In the service load range (working stress), the section is elastic as in Figure 4-38*a*, and the elastic condition continues until the extreme fiber stress reaches the yield strength F_y, Figure 4-38*b*. When the strain ϵ reaches ϵ_y (beginning of plastic region), further strain increase causes no increase in stress. The elastic–plastic stress–strain behavior is an acceptable idealization for structural steels with $F_y \leq 65$ ksi.

For the condition shown in Figure 4-38*b*, where $f = F_y$, the nominal moment strength M_n, referred to as the yield moment, is computed as

$$M_n = S_x F_y = M_y \tag{4-36}$$

where S_x is the section modulus for elastic conditions. When the stage shown in Figure 4-38*d* is reached, every fiber has a strain equal to or greater than $\epsilon_y = F_y/E_s$, and the beam is stressed in the plastic range. The nominal moment strength M_n is the plastic moment, and is computed from

$$M_n = F_y Z = M_p \tag{4-37}$$

which is the same expression as (4-20). The parameter $Z = \int y\,dA$ is called the plastic modulus.

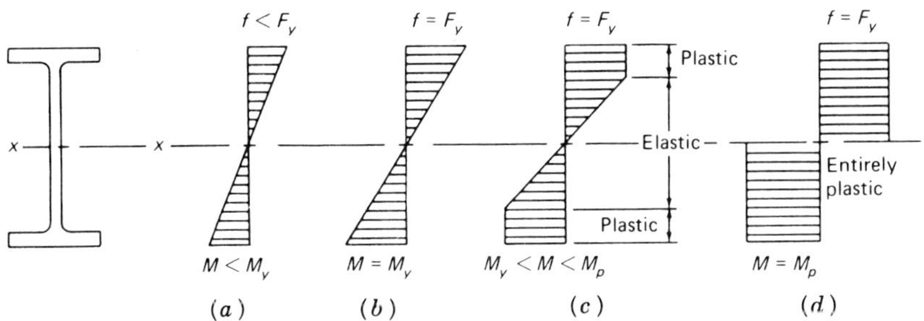

FIGURE 4-38 Stress distribution diagrams at different stages of loading, rolled I-beam section.

The ratio M_p/M_y depends on the cross-sectional shape and is not dependent on the material properties. Obviously,

$$\frac{M_p}{M_y} = \frac{Z}{S} = \xi \tag{4-38}$$

where the ratio ξ is called the shape factor. For most rolled I-beam sections, the shape factor ranges from about 1.09 to about 1.18 (with reference to the strong axis x–x).

Design procedures recognize the beam behavior discussed in this section. Thus, AASHTO articulates this behavioral pattern in the load factor design methods for compact and noncompact sections. For compact sections, the stage shown in Figure 4-38*d* is allowed to be reached, and the maximum moment M_u is simply the plastic moment $M_p = F_y Z$. For noncompact sections, the full plastic stage simply is not allowed, and the maximum moment strength M_u is the yield moment $M_y = F_y S_x$ (see also Section 4-14). The same philosophy is applied to composite beams; for compact sections acting compositely with the slab, the maximum moment is computed on the basis of a fully plastic stress distribution, whereas noncompact sections are designed for the moment at first yield. However, the maximum moment strength computed in this manner is not multiplied by a strength reduction factor.

Plastic Strength of Continuous Beams If the concentrated load shown in Figure 4-39 is increased toward the value W_n that causes the collapse condition, the portion of the beam where the moment does not exceed M_y behaves elastically (meaning straight line θ). In the region near the plastic moment M_p, an inelastic length jL exists producing a large midspan curvature. The beam has two rigid parts (straight lines) connected by a hinge at B known as a plastic hinge, and having a concentrated angle change or hinge rotation θ_u.

A plastic hinge is a zone of yielding due to flexure. The plastic hinge length is thus the beam region length where the moment exceeds the yield moment. Its actual length depends on the shape of the cross section and the loading configuration. Referring to Figure 4-39, the relationship between the bending moments and the beam segments AB and AB' is established as

$$\frac{AB}{M_p} = \frac{AB'}{M_y} \quad \text{or} \quad \frac{L/2}{M_p} = \frac{L/2 - jL/2}{M_y}$$

which yields

$$j = 1 - \frac{M_y}{M_p} = 1 - \frac{1}{\xi} \tag{4-39}$$

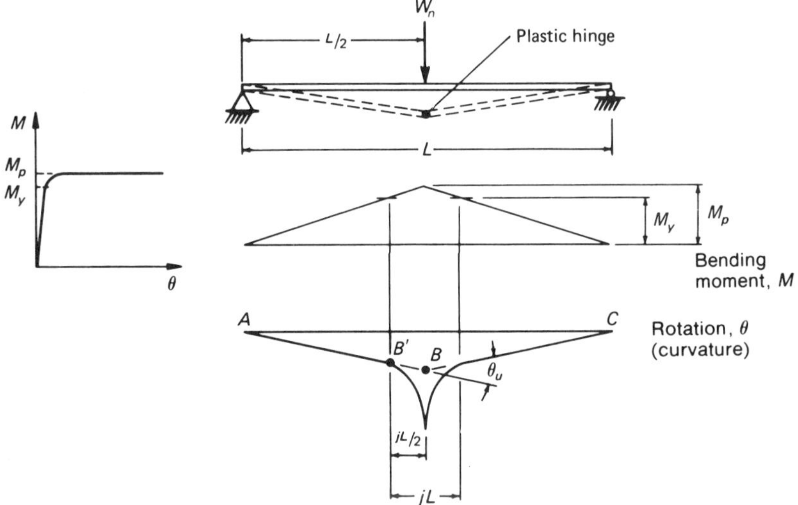

FIGURE 4-39 Moment–curvature relationship for a plastic hinge.

When the load W_n has been reached, the beam has no more resistance to bending at the maximum moment point until strain hardening occurs. Relying on strain hardening is neither easy nor necessary, and it is conservative to ignore the associated extra strength.

In a continuous beam, the points of maximum moment occur at the interior supports and at the point of zero shear. In the context of strength and plastic analysis, the interior span of a continuous beam behaves the same as a beam with fixed ends. Thus, researchers choose to refer to the plastic strength of statically indeterminate beams.

The first plastic hinge forms at the supports, and then redistribution of moment occurs until a plastic hinge forms at the point of zero shear. Each beam segment between plastic hinges can move without any increase in load. A system of such segments manifests a *plastic collapse mechanism*, which is an unstable structure until strain hardening occurs. The deflected shape of the PCM (plastic collapse mechanism) is a series of straight beam segments between plastic hinges. When the PCM forms, each beam segment is bent due to the plastic moment, but further motion of the system does not involve additional bending of individual segments, so that the PCM behavior is identical to a linkage of straight bars connecting real hinges.

The collapse mechanism can form in any span of a continuous beam. For a span subjected to a uniform load, there is only one possible beam mechanism (three possible plastic hinges in the interior spans, and two possible plastic hinges in the end spans), because simple ends are treated as real hinges. For a beam subject to n concentrated loads at n locations between the member ends, there are n possible beam mechanisms and $n + 2$ possible plastic hinge locations (interior spans), but only $n + 1$ possible plastic hinge locations if one member is a real hinge (end spans).

Equilibrium Method of Analysis At the collapse condition with the plastic limit load W_n acting, the requirements of equilibrium still apply, but a collapse condition is reached when the load W_n is large enough to cause the plastic moment M_p to occur at one location. When a sufficient number of plastic hinges has been developed to allow instantaneous hinge rotations without producing increased resistance, the plastic mechanism is said to have occurred. The objective of the method is to find an equilibrium moment diagram where $M_u \leq M_p$ at a sufficient number of locations to produce a PCM, where M_u is the maximum moment due to factored loads and M_p is the required plastic hinge strength. The steps involved in the procedure are as follows.

1. Select the appropriate redundants. For a continuous beam these are the moments at the supports.
2. Draw the moment diagram for the factored loads.
3. Draw the moment diagrams for the redundants.
4. Assume that a plastic hinge forms at sufficient locations to produce a CPM.
5. Solve the moment equilibrium equation for M_p.
6. Check the design to confirm that $M_u \leq M_p$.

4-18 DESIGN EXAMPLE: CONTINUOUS COMPOSITE I-BEAM BRIDGE (LOAD FACTOR APPROACH)

Section 4-12 presented an example of a four-span continuous composite I-beam bridge of moderate span lengths based on allowable stress procedures. In this section we will consider the same bridge using the strength design method.

For composite beams, the maximum strength in the negative moment regions is computed as in continuous noncomposite beams and in accordance with AASHTO Articles 10-48 and 10-49. The concrete is assumed to be inactive in carrying tensile stresses. Where the longitudinal slab reinforcement is continuous over interior supports, it may be considered to act compositely with the steel beam. For reference, the beam elevation is shown in Figure 2-26.

Negative Moments For the purpose of comparison with the allowable stress design of Section 4-12, we choose not to apply the adjustment factors mentioned in AASHTO Article 10.48.1.3.

At pier 2, the total moment is -1113 ft-kips, where $M_{DL} = -690$ ft-kips and $M_{LL+I} = -423$ ft-kips. This gives a maximum factored moment $M_u = 1.3 \times (690 + 1.67 \times 423) = -1815$ ft-kips. Assuming a compact section, the required plastic modulus is

$$Z = \frac{1815 \times 12}{36} = 605 \text{ in.}^3 \qquad \text{W36} \times 160 \qquad Z = 625 \text{ in.}^3$$

At pier 3, $M_{DL} = -524$ ft-kips and $M_{LL+I} = -342$ ft-kips, giving a maximum factored moment $M_u = 1.3 \times (524 + 1.67 \times 342) = -1423$ ft-kips. The required plastic section modulus is

$$Z = \frac{1423 \times 12}{36} = 474 \text{ in.}^3 \qquad \text{W36} \times 135 \qquad Z = 510 \text{ in.}^3$$

These sections must be checked for compactness requirements. For W36 × 135, $b'/t = 5.70/0.79 = 7.21 < 10.82$, and $D/t_w = 33.97/0.60 = 57 < 101$.

Positive Moments Consider a W36 × 160 in the composite sections, and assume that a is in the concrete slab. The force in the steel section is $A_s F_y = 47.1 \times 36 = 1696$ kips. Using $b_E = 84$ in. and $f_c' = 3000$ psi, we calculate $0.85 f_c' b_E = 0.85 \times 3 \times 90 = 230$. The value of a is $1696/230 = 7.37$ in., which is within the 7.5-in. structural thickness. The nominal moment strength is

$$1696 \times (18.00 + 7.50 - 3.70) = 1696 \times \frac{21.8}{12} = 3080 \text{ ft-kips}$$

and for a strength reduction factor of 0.85, the design moment strength is $3080 \times 0.85 = 2618$ ft-kips.

For span 2, the moments are $M_{DL} = 342$ ft-kips and $M_{LL+I} = 645$ ft-kips, giving a factored moment of $1.3(342 + 1.67 \times 645) = 1845$ ft-kips < 2618.

Try a W36 × 135, $A_s F_y = 39.8 \times 36 = 1433$, giving $a = 1433/230 = 6.23$ in. The nominal moment strength is

$$1433 \times (17.78 + 7.50 - 3.12) = 1433 \times \frac{22.16}{12} = 2636 \text{ ft-kips}$$

providing a design moment strength of 2240 ft-kips.

4-19 TOPICS RELEVANT TO LOAD FACTOR DESIGN OF CONTINUOUS COMPOSITE BRIDGES: CASE STUDIES

Distribution Factor at Ultimate Load

The load factor procedure articulated in the foregoing sections actually constitutes a pseudoplastic design method. Its application requires assessment of induced elastic forces and then factoring these forces to obtain a quantity assumed to represent full plastification of the member under consideration. One of the major deficiencies on which most researchers agree relates to the use of elastic distribution factors to evaluate the induced plastic moments. Kuo and Heins (1973) have proposed a distribution factor at ultimate load for simply supported beams, expressed by (4-17), which relates only to transverse moment distribution. Most investigators agree, however, that the real advantages of plastic design are demonstrated in continuous bridges, but in this case the associated transverse and longitudinal distribution is not explicitly considered.

From the foregoing it follows that the load-carrying capacity of continuous composite beams is controlled by two structural elements: (a) the concrete slab and (b) the longitudinal steel beams. Heins and Kurzweil (1976) examine the strěngth of the slab using the yield line theory (Johansen, 1954). Subsequent work on this subject has been done by Das, Malairongs, and Smith (1972) and Li and Blackwell (1969). The ultimate strength of the beams is determined using simple plastic beam theory involving the plastic hinge collapse mechanism. The bridge format is shown in Figure 4-40.

Local Collapse Mode The slab panels of the generalized bridge shown in Figure 4-40*a* can be resolved into two types according to the boundary conditions: (a) interior panel between interior supports and longitudinal beams, giving four continuous plate supports at the edges, and (b) exterior panel located between the interior and end bridge span supports, giving three continuous plate supports and one simply supported edge. The analysis of the local collapse mode of the two slab panels is conducted in three stages involving uniform dead load, concentrated live loads, and combined loads. The various yield load patterns considered are shown in Figures 4-40*c* and *d*, with the particular parameters for the live load case shown in Figure 4-40*e*. Assuming the yield patterns shown in Figure 4-40, Heins and Kurzeil (1976) have developed expressions for the slab panels relating moments, plate parameters, and loads.

For the beams, the general collapse mode analysis deals with the effects that the dead and live loads have on the failure of the slab and steel beams acting compositely. Plastic hinges are assumed to occur at the bridge supports dividing the bridge into spans generalized as interior and exterior units. This is consistent with our previous analysis where interior supports develop

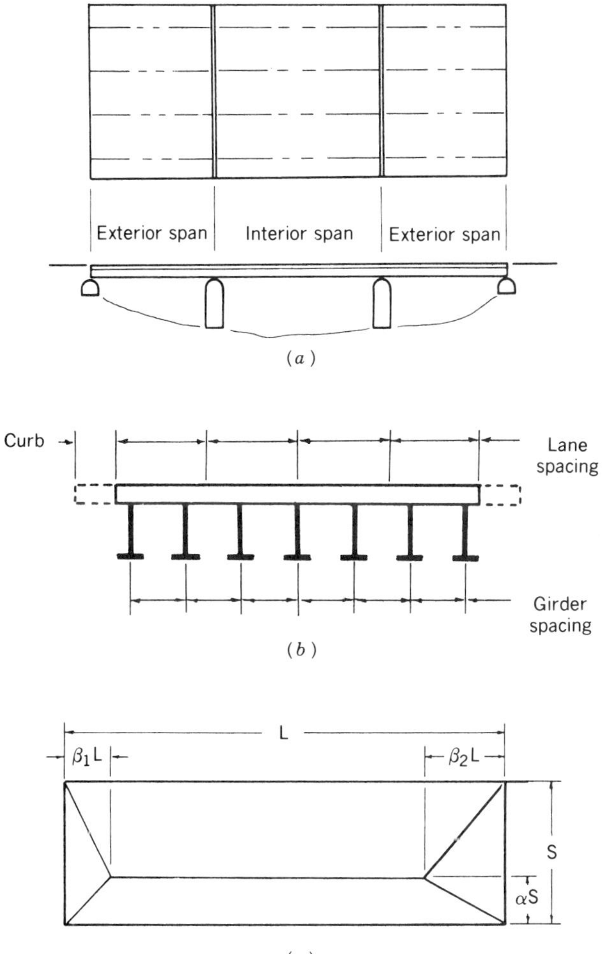

FIGURE 4-40 Bridge prototype used in study: (*a*) general arrangement showing plan and elevation; (*b*) bridge section; (*c*) assumed local failure pattern under uniform loading; (*d*) local failure mode cases, live load; (*e*) local failure pattern geometry, live load. (From Heins and Kurzweil, 1976.)

plastic hinges whereas end supports are treated as real hinges. Lane loads are HS 20-44 trucks traveling in the same direction. The steel beams are assumed to act as discrete elements of the structure and not as uniformly distributed elements with respect to the slab. Because it is conjecture to predict the behavior of the composite section intersected by an oblique yield line, the failure pattern has yield lines at right angles to the bridge as shown in Figure 4-41. The four cases considered for the failure pattern are shown in

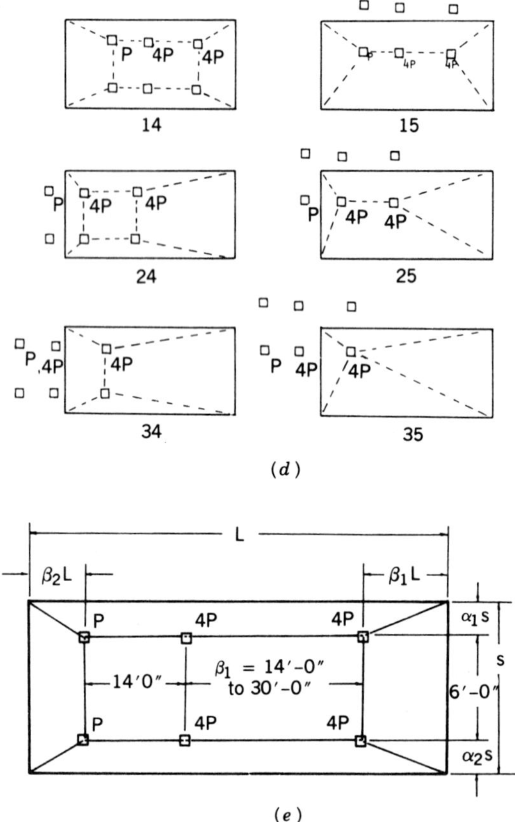

FIGURE 4-40 *(Continued)*

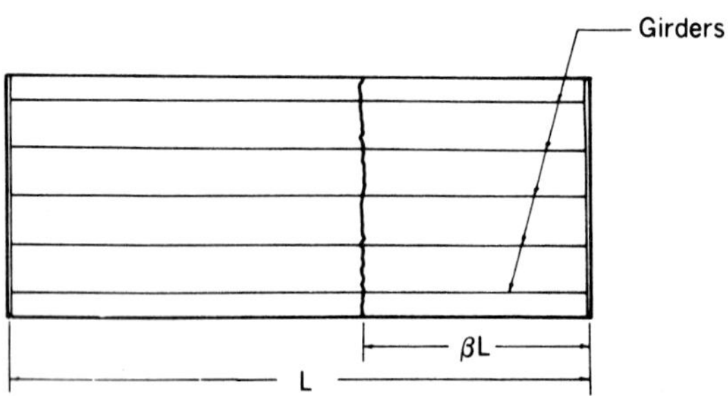

FIGURE 4-41 Generalized failure mode showing yield line on bridge. (From Heins and Kurzweil, 1976.)

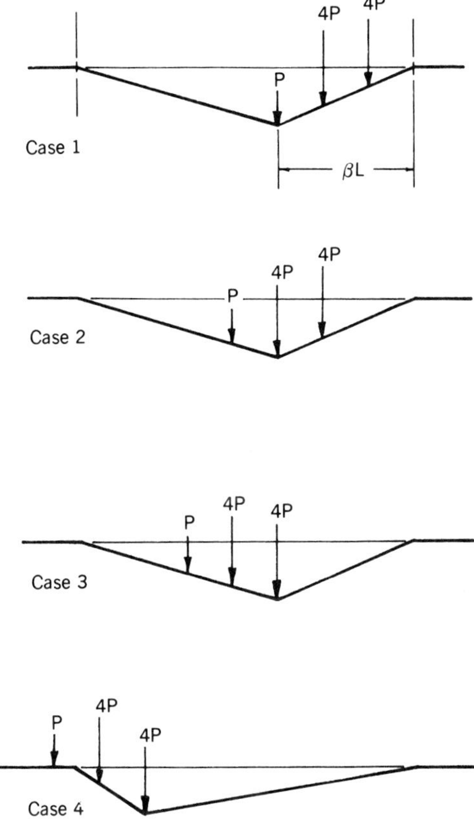

FIGURE 4-42 General failure mode cases for various truck load positions. (From Heins and Kurzweil, 1976.)

Figure 4-42. Each case corresponds to a yield line developing under one of the truck axles, and case 2 results in the worst design condition yielding the maximum ultimate moment.

In this analysis the ultimate moment of the bridge is assumed to consist of the sum of the composite sections across the bridge width. Tests by Burdette and Goodpasture (1971) indicate that as each girder or beam is stressed to its ultimate capacity the total bridge moment capacity approaches the value obtained considering the bridge as a single wide beam.

Heins and Kurzweil (1976) have developed generalized expressions for the plastic limit loads (concentrated live loads and uniform dead load) by energy methods (principles of virtual work). In this analysis a parameter p is introduced to distinguish between negative and positive moment capacities, and expresses the ratio of negative and positive ultimate moment of the composite section. The ultimate moment capacity of the bridge acting as one unit is

$$M_{cs} = \frac{jM}{n} \tag{4-40}$$

where M_{cs} = ultimate moment required for composite section
M = average lane moment obtained from energy methods
j = number of lanes
n = number of girders

Parametric Study Using the AASHTO load factors for dead and live loads, the ultimate slab moments were calculated for each of the slab panel parameter sets. The resulting moments are functions of various slab parameters as shown in Figures 4-43*a*, *b*, and *c*. The parameters μ, γ_1, and γ_2 are slab properties (see Heins and Kurzweil, 1976). Figure 4-43*a* shows the variation of the ultimate slab moment with beam length and spacing, and Figure 4-43*b* shows the variation of the ultimate slab moment with beam length and factored dead load. Figure 4-43*c* shows the variation of the ultimate slab moment with γ_2, the ratio of the negative to the positive transverse ultimate slab moment. These data show that the general shape of the curves is not influenced by the slab parameters, but there is a trend to generate a family of similar curves.

In establishing the general beam collapse mechanism, two considerations are recognized: (a) under plastic analysis, the longitudinal moment distribution effect must be taken into account, and (b) because at general failure the bridge acts as a single unit, the ultimate moment is evaluated as an average over the total bridge width for each traffic lane. The average ultimate lane moments for the interior and exterior bridge spans subjected to combined loading of uniform dead load and concentrated live wheel loads are given by Heins and Kurzweil (1976), based on a lane width of 12 ft. The results show the variation of the average ultimate lane moment as a function of the span length for the interior and exterior spans, and the two curves diverge as the span length increases.

Design Data For the purpose of developing design aids for the ultimate strength analysis of a specific bridge geometry, Heins and Kurzweil (1976) derived design equations for the ultimate slab moment and for the average ultimate lane moment. This work is based on empirical expressions using a least-squares fit of baseline data, adjusted by correction factors. For the beams, the curve-fitting program involves data from the average ultimate lane moment as a function of span length, factored dead load, and ratio of

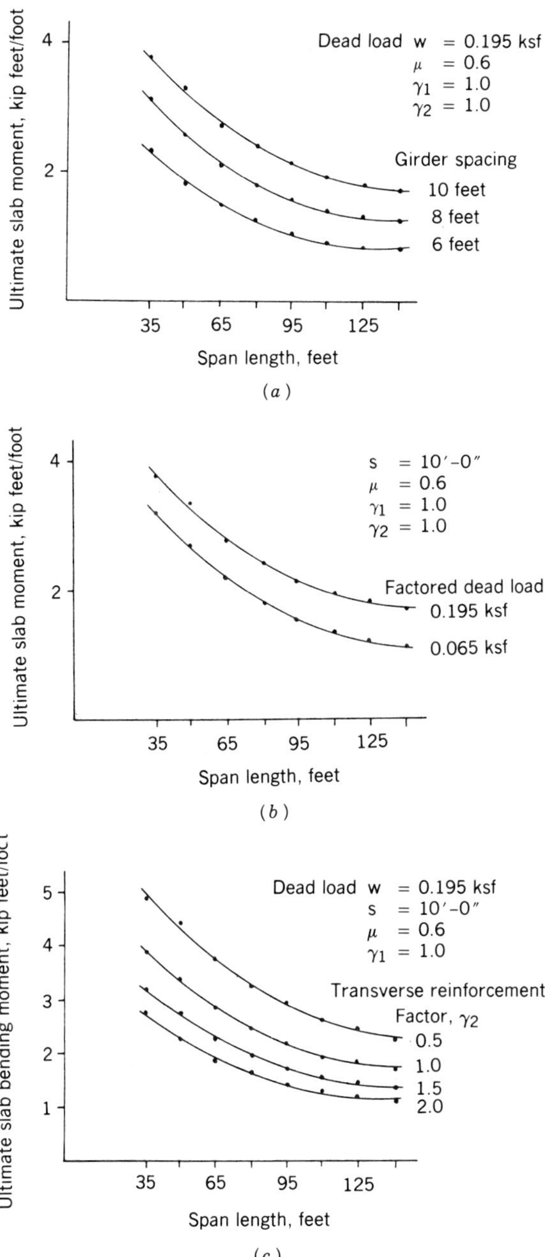

FIGURE 4-43 (*a*) Variation of ultimate slab moment with girder spacing; (*b*) variation of ultimate slab moment with dead load; (*c*) variation of ultimate slab moment with transverse reinforcement ratio. (From Heins and Kurzweil, 1976.)

negative to positive ultimate moment. These design equations are modified to give the ultimate composite section moment. The ultimate composite section moments are as follows:

$$M_{c_{\text{int}}} = \frac{2j}{(1+p)n}(0.765wL^2 + 24.81L - 366) \tag{4-41}$$

$$M_{c_{\text{ext}}} = \frac{2j}{(1+p)n}(1.029wL^2 + 34.77L - 495) \tag{4-42}$$

where M_c = ultimate composite section moment (interior or exterior span)
p = ratio of negative to positive ultimate composite section moment
j = number of lanes
n = number of composite beams
L = span length (ft)
w = factored dead load (ksf)

A design comparison of an example with span lengths of 40 and 50 ft in a parametric study leads to the conclusion that reinforced concrete slab design may be based on service behavior rather than ultimate strength. The final composite beam design based on both ultimate and service overload criteria shows that significant savings of material can be realized by the load factor approach.

Effective Composite Beam Width at Ultimate Load

The effective width of the concrete slab at ultimate load has been investigated by Heins and Fan (1976), considering the stress redistribution after plastification of parts of the structure. Using plate theory, the effective width is defined as a function of the longitudinal stress at the top surface of the plate, expressed as

$$2be = \frac{2\int_0^{b/2} (\bar{\sigma}_y)\, dx}{(\bar{\sigma}_y)_{\max}} \tag{4-43}$$

where $2be$ = effective width
b = beam spacing
$\bar{\sigma}_y$ = longitudinal stress at top surface of the slab
$(\bar{\sigma}_y)_{\max}$ = maximum $\bar{\sigma}_y$ occurring at the junction of the slab and beam

Other investigators (Adekola, 1968) define the effective width as

$$2be = \frac{2\int_0^{b/2} \sigma_y \, dx}{(\sigma_y)_{\max}} \tag{4-44}$$

where σ_y = longitudinal stress in the middle surface
$(\sigma_y)_{\max}$ = peak longitudinal stress in the middle surface at the girder–slab intersection

Another expression is

$$2be = \frac{F_g}{(\sigma_y)_{\max}} \tag{4-45}$$

where F_g is the axial force in the girder (beam).

Equations (4-44) and (4-45) were used in conjunction with plate theory to evaluate the effective width of various composite bridges shown in Table 4-3. According to Bethlehem Steel Corporation (1967), these represent examples of the most economical simple-span composite highway bridges. The live load is HS 20. The analysis assumes f'_c = 3 ksi, A36 steel, and a Poisson's ratio of μ = 0.1.

The results obtained from these calculations are shown in Table 4-4. These effective widths are computed at midspan of the beams. At ultimate load, the effective widths computed from (4-45) are much larger than those obtained by (4-44), but for the interior section the results are reversed.

TABLE 4-3 Composite A36 Steel I-Beam Bridges (From Heins and Fan, 1976)

Span Length (ft) (1)	Girder Spacing (ft) (2)	Number of Girders (3)	Distance Between Edge Girder (ft) (4)	WF Sections (5)	Cover Plate Section (in.) (6)	Cover Plate Length (7)
40	7.0	4	21.0	W24 × 68	13 × 1/2	30
	8.0	4	24.0	W24 × 68	13 × 5/8	32
	9.0	4	27.0	W24 × 68	13 × 3/4	32
60	7.0	4	21.0	W30 × 99	14-1/2 × 1	46
	8.0	4	24.0	W33 × 118	15-1/2 × 3/4	42
	9.0	4	27.0	W33 × 118	15-1/2 × 7/8	44
80	7.0	4	21.0	W36 × 160	14 × 1-1/4	54
	8.0	4	24.0	W36 × 170	14 × 1-1/2	56
	9.0	4	27.0	W36 × 182	14 × 1-1/2	56
100	7.0	4	21.0	W36 × 280	14-1/2 × 1-1/2	60
	8.0	4	24.0	W36 × 300	14-1/2 × 1-1/2	60

Note: Thickness of concrete deck is 8.5 in.; 1 in. = 25.4 mm; 1 ft = 0.305 m.

TABLE 4-4 Effective Widths of Composite Bridges

Span Length (ft)	Girder Spacing (ft)	Aspect Ratio, b/L	Effective Width by (4-45) (in.)				Effective Width by (4-44) (in.)				AASHTO	
			Edge Girders		Interior Girders		Edge Girders		Interior Girders			
			Elastic	Plastic	Elastic	Plastic	Elastic	Plastic	Elastic	Plastic	Edge	Interior
(1)	(2)	(3)	(4)	(5)	(6)	(7)	(8)	(9)	(10)	(11)	(12)	(13)
40	7.0	0.175	43.86	46.74	71.30	63.67	38.28	34.12	76.10	79.06	40	84.0
	8.0	0.200	43.44	45.86	85.86	78.48	44.27	35.52	84.11	89.43	40	96
	9.0	0.225	42.19	44.14	99.31	94.23	50.50	36.33	91.39	98.39	40	102
60	7.0	0.117	49.87	58.16	71.57	61.26	40.41	39.35	80.68	81.68	42	84.0
	8.0	0.133	48.34	62.35	89.22	72.29	46.43	44.38	90.58	91.77	48	96
	9.0	0.150	45.35	57.23	107.02	90.08	52.62	42.92	99.96	102.45	51	102
80	7.0	0.0875	51.06	61.94	72.50	60.59	41.20	40.70	82.13	82.36	42	84.0
	8.0	0.100	50.62	63.25	89.87	74.94	47.23	43.09	92.88	93.67	48	96
	9.0	0.1125	49.15	67.02	107.51	88.84	53.35	47.87	103.29	104.39	51	102
100	7.0	0.070	52.08	60.03	72.45	62.72	41.53	39.87	82.82	82.98	42	84.0
	8.0	0.080	53.03	66.00	88.86	74.77	47.56	44.50	94.02	94.23	48	96

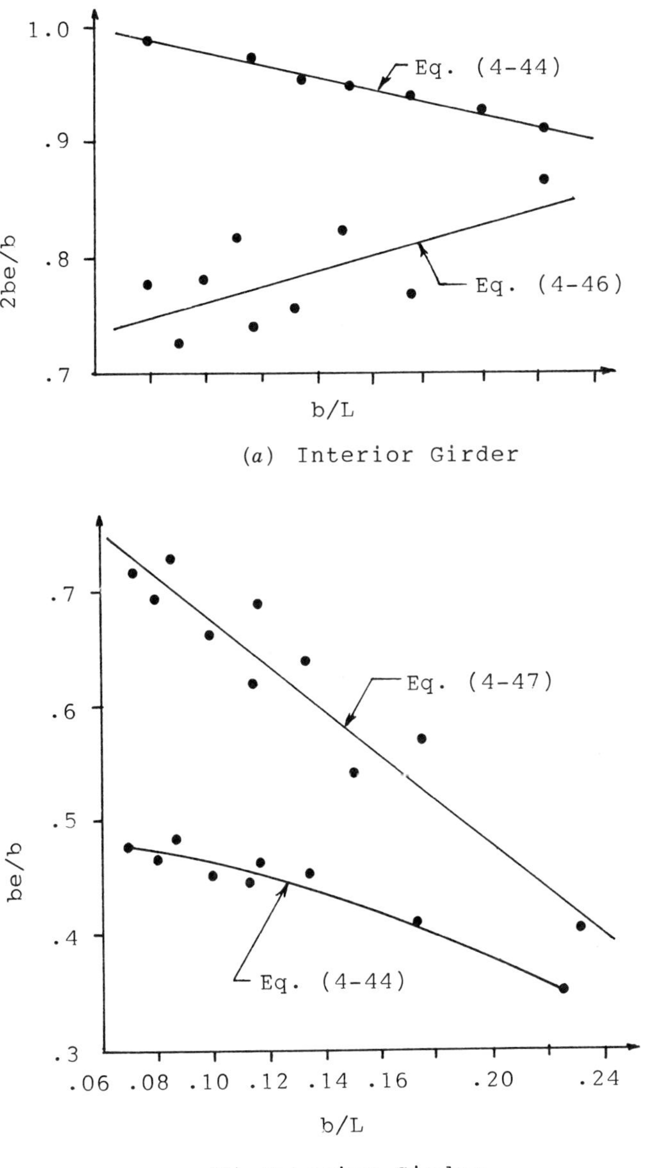

FIGURE 4-44 Effective width at ultimate load. (From Heins and Fan, 1976.)

The ratio of computed effective width at ultimate load to the beam spacing is plotted versus the ratio of beam spacing to span length in Figure 4-44 for the bridges of this study. Based on (4-44), the effective width varies from 0.409 to 0.737 of the girder spacing for the edge section and 0.873 to 0.722 for the interior section. As the ratio b/L increases, the value be/b decreases for the exterior sections but increases for the interior sections, which is explained by the fact that the greater the ratio b/L the more difficult it is to distribute the load to the edge sections. There are two extremes for which we can establish the effective width: (a) when b/L approaches zero, the effective width is the same for interior and exterior beams and equal to the total slab width divided by the number of beams; and (b) when b/L is very large, the effective width of a multibeam composite system is approximately equal to the single isolated T beam that has a very large flange width.

A linear regression analysis on the data of Figure 4-44 results in a linear set of equations. Articulating the scatter that is probably due to the effect of the location of loadings, Heins and Fan (1976) derive the following:

For the interior girders,

$$\frac{2be}{b} = 0.617\frac{b}{L} + 0.702 \tag{4-46}$$

in which $2be \geq$ slab width/number of beams; the slab width $= b \times$ (number of beams $-$ 1).

For the exterior beams,

$$\frac{be}{b} = -1.98\frac{b}{L} + 0.873 \tag{4-47}$$

in which $be \leq$ slab width/number of beams; the slab width $= b \times$ (number of beams $-$ 1), where b is the beam spacing.

Equations (4-46) and (4-47) are derived from the study of a series of composite bridges consisting of four-beam systems, with a beam spacing of 7.75 ft and two-lane truck loading. However, they may be extended to bridges with three and four lanes provided the beam spacing does not exceed 9 ft.

Further Case Studies

Ultimate Load Capacity Considerations The applicability of an elastic distribution factor to evaluate the ultimate moment capacity has generally been found unrealistic because it does not reflect the ultimate load response. A study by Heins and Kuo (1975) addresses this topic and presents results for simple-span composite bridges. The technique used for the elastic–plastic response considers the M–ϕ curve of the bridge elements and introduces a corresponding modification of the element properties.

In general, the stiffness of a section between the elastic and fully plastic state may be expressed as

$$EI = \frac{M/M_y}{\phi/\phi_y}(EI)_e \tag{4-48}$$

where M = moment at a given load
M_y = moment at first yield of the section
ϕ = curvature at a given load
ϕ_y = curvature at first yield of the section
$(EI)_e$ = elastic stiffness of section

In order to make use of (4-48), the M/M_y versus ϕ/ϕ_y function must be known. From a study of composite beam bridges, Heins and Kuo (1973) have developed data that resulted in a subcomputer program, and can readily be used to determine the M–ϕ curve (see also Section 12-3). If a member is subjected to any loading condition, the curvature ϕ is first evaluated from member deflections and is then normalized giving ϕ/ϕ_y. The intercept of ϕ/ϕ_y on the M/M_y curve gives the moment M required to cause ϕ. At this state of loading, the bending stiffness is computed from (4-48), and the structure is analyzed using the new stiffness. The resulting system deflections are compared to the previous deflections. When the difference is less than 0.001 in., the structure is in equilibrium and the next loading may be applied.

Beginning with the AASHTO distribution factor DF $= S/K$, where S is the beam spacing and K is a constant (normally 5.5), the program evaluates the bending moment of each beam at different loading stages. After one beam has yielded, it cannot resist the same increment of loading due to a reduction in it stiffness. In order to achieve equilibrium of the system, the additional beams must absorb the difference. This phenomenon of load redistribution is associated with plastic response.

For simple-span composite bridges, Heins and Kuo (1975) have obtained the distribution factor for ultimate loading by plotting the $(\text{DF})_p$ values as a function of γ, where γ is the number of girders per number of lanes. A mean regression line through these data gives the appropriate equation. Adding two standard deviations to the mean value to contain 5 percent of the data gives the final design criterion expressed by (4-17), with the factor K approaching 7.0.

LFD Criteria for Composite I Beams Heins (1980) has proposed effective slab widths and load distribution factors under factored loads for ultimate plastic conditions of composite steel I-beam bridges. Essentially, these criteria reflect the effects of plastification and are based on analytical and experimental programs, some of which are presented in the foregoing sections. Thus, for load factor design, the following are proposed.

Lateral Distribution for Bending Moments These may be determined by applying to the beam the fraction of a wheel load (both front and rear) calculated from the expression $DF = 2N_L/N_G$, where N_L is the number of traffic lanes and N_G is the number of beams or girders. This is applicable for $2 < N_L < 4$ and a beam spacing of 6 to 9 ft.

Composite Beams These should be proportioned to satisfy the criteria expressed by (4-46) and (4-47).

4-20 ALTERNATE LOAD FACTOR DESIGN: BRACED COMPACT SECTIONS (AASHTO, 1991)

The 1991 edition of the AASHTO "Guide Specifications for Alternate Load Factor Design of Steel Bridges Using Braced Compact Sections" contains revisions of the corresponding articles of the 1989 and 1992 AASHTO specifications and gives procedures for calculating the resistance of continuous beams by the mechanism method. The overload provisions are modified to reflect the effect of local yielding, and the use of prismatic members is encouraged to improve the overall structural performance.

Rather than reducing the negative moments over supports with an accompanying increase in maximum positive moments at overload and maximum load (determined by elastic analysis), AASHTO Article 10.48.1.3 is revised to allow the calculation of the load-carrying capacity from a plastic mechanism analysis (see also Section 4-17). The intent of the 1991 AASHTO document is therefore to expand the load factor design by allowing plastic analysis and shakedown to be used in determining the strength and inelastic behavior of beams using braced compact sections. This procedure is subject to the following limitations and modifications.

1. The procedure is limited to steel with a yield stress not exceeding 50 ksi.
2. When a composite section reaches the plastic moment in positive bending, no further rotation is permitted.
3. For sections in negative bending that are composite with the deck reinforcement and for noncomposite sections in positive and negative bending, an effective plastic moment should be determined if plastic rotations are required for mechanism formation.
4. Bearing stiffeners should be provided at support locations where plastic hinges occur.

One of the goals of the 1991 guide specifications is to eliminate details such as cover plates to rolled beams and multiple splices in the flanges of welded beams by using prismatic sections.

Symmetrical Beams and Girders Compact sections are members that can reach the plastic moment with limited rotation as a plastic hinge (see also Figure 4-39). If significant plastic rotation is required, an effective plastic moment may be determined according to AASHTO Article 10.50A (1991 specifications).

Composite Beams and Girders The web slenderness requirements for a compact section expressed by (4-22) apply to symmetrical sections. At the interior supports of composite continuous bridges where negative bending occurs, the longitudinal reinforcement in the slab is commonly assumed to act compositely with the beams, shifting the neutral axis toward the slab. Yielding shifts the neutral axis further upward and toward the slab. The depth of web in compression thus increases and has an important effect on the rotation capacity of unsymmetrical sections. Therefore, for unsymmetrical sections where the distance from the neutral axis to the compression flange exceeds $D/2$, (4-22) must be modified by replacing the parameter D by $2D_{cp}$, where D_{cp} is the distance to the compression flange from the neutral axis at the plastic moment.

Mechanism Strength This topic, treated in AASHTO Article 10.50A, addresses the concept of an effective plastic moment to be used in strength analysis. The procedure is applicable when the plastic rotation requirements exceed three times the elastic rotation corresponding to the plastic moment.

The effective yield strengths of the compression flange and web, F_{yfe} and F_{ywe}, are derived from the plastic design slenderness requirements of the AISC specifications (eighth edition). Sections that satisfy these criteria are termed ultracompact. These sections have sufficient inelastic rotation capacity at rotating hinges to allow the plastic mechanism analysis using the plastic moment M_p at the rotating hinges.

If the projecting compression flange–slenderness ratio b'/t for the section at a rotating hinge exceeds the limit $1565/\sqrt{F_{yf}}$ and/or the effective web slenderness ratio D/t_w at the plastic moment for that section exceeds the limit $12{,}377/\sqrt{F_{yt}}$ (ultracompact sections), then the corresponding effective yield strength will be less than F_{yf} (yield strength of the compression flange), and the effective plastic moment M_{pe} given in Article 10.50A will be less than M_p. If both ratios are below these limits, M_{pe} will equal M_p. The web is assumed to have a depth equal to 95 percent of the total depth of the section.

F_{ywe} is limited to F_{yf}, and the web reduction factor R_w (used to compute M_{pe}) is normalized with respect to F_{yf} because plastic web buckling is governed by the flange strain. F_{yf} is also limited to 50,000 psi or less because research on plastic design has been limited to materials with $F_y \leq 50{,}000$ psi.

Permanent Deformations (Overload) Controlled local yielding and concomitant inelastic rotation are allowed at interior support sections at overload that result in the formation of favorable residual stress patterns (autostress) and positive automoments that remain in the structure after removal of the live load. The automoments and autostress ensure that the structure will shake down under repeated loadings not exceeding the highest overload.

The automoments cause elastic moment redistribution by effectively reducing the peak elastic negative moments at interior supports and slightly increasing the elastic maximum span moments. For more than two continuous spans, the automoment formed at one pier causes additional elastic moments at other piers that affect the automoments at those piers by shifting the beam line. An iterative computational procedure is required to account for this carry-over effect to determine the final automoment distribution in the beam.

Design Example A complete design example is included in the AASHTO 1991 guide specifications to illustrate the autostress procedures. The bridge of this example is a two-span continuous structure (100–100 ft) composite for positive and negative moment. The section in the positive moment region consists of the steel beam and the slab. In the negative moment region, the section consists of the steel beam and the longitudinal reinforcing steel in the slab. The beam spacing is 8 ft 4 in. and accommodates a slab thickness of 7 in. The material strength $F_y = 50{,}000$ psi, $f'_c = 4000$ psi, modular ratio $n = 8$, and slab reinforcement $F_y = 60{,}000$ psi. The live load is HS 20-44. The analysis considers compact section requirements, plastic properties, design for maximum moment and shear, uplift, lateral bracing, bearing stiffeners, provisions for overload, crack control, and design for service loads.

4-21 PRESTRESSED STEEL BRIDGES

The concept of prestressing in structural beams is extended to steel beams, plain or composite. It may be applied to the design of new steel bridges, or used in conjunction with the rehabilitation and strengthening of existing bridges by introducing a suitable prestressing system. The prestressing of a steel beam or girder can be achieved by the use of high-strength-steel cables anchored at the bottom flange (ASCE–AASHO, 1968; Finn, 1964; Knee, 1966; Magnel, 1950; Samuley, 1955). The purpose of prestressing is not to overcome tensile deficiencies of the material (as in the case of concrete structures), but to create initial stresses to counteract the stresses caused by external loads.

Prestressed steel bridges are treated by Troitsky (1990) in a broad context of bridge types and uses. Hence, the review of this section is limited to certain bridge forms merely to demonstrate and document the technical feasibility of prestressing.

Prestressed Composite Steel Beams The analysis of composite steel beams, prestressed with high-strength cables before the concrete slab is cast, was first presented by Hoadley (1963), who made comparisons between the moment capacity of conventional and prestressed composite steel beams. A more accurate method was introduced by Reagan and Krahl (1967) showing that for working stress the use of high-strength cables for prestressing is not as effective as the addition of a cover plate for conventional composite design. The reason for this is that prestressing in the former case does not appreciably increase the moment of inertia of the cross section because of the small tendon area compared to the area of a cover plate.

Anand and Talesstchi (1973) present a design methodology for simply supported prestressed composite steel beams where the prestressing is achieved by welding a high-strength-steel cover plate on the flange of an initially stressed rolled beam that is released before casting the concrete slab. Published results documenting this procedure are provided by ENR (1961).

Stresses Due to Prestressing For a simply supported rolled steel beam, the mechanism of prestressing is shown in Figure 4-45. In stage I, jacking forces P are applied at the one-third points of the span as shown. The allowable maximum bending moment due to these forces is

$$Mj = F_b S_b \tag{4-48}$$

where $S_b = I_b/(d/2)$ and F_b is the allowable bending stress. At this stage, the stresses in the extreme fibers are $\pm F_b$ (+ for tension and − for compression). In stage II, a high-strength-steel cover plate is welded to the

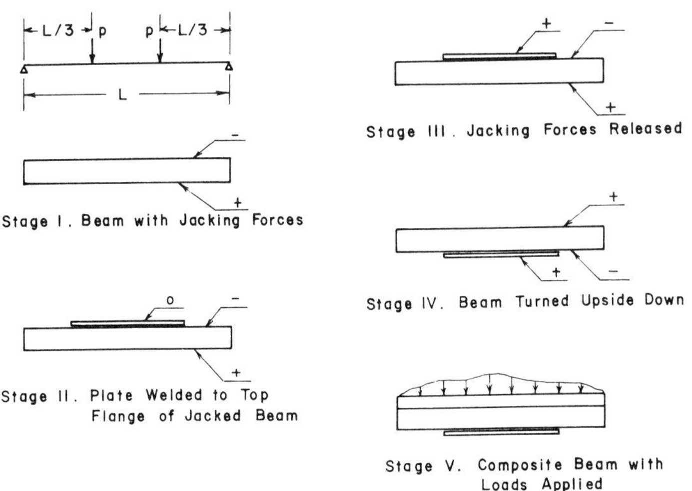

FIGURE 4-45 Prestressing mechanism and stages. (From Anand and Talesstchi, 1973.)

top flange of the beam (stress in the cover plate = 0). The jacking forces are released in stage III. Accordingly, the stresses in the extreme fibers change by Mj/S_{np} and Mj/S_{ap}, where S_{np} and S_{ap} are the section moduli of the extreme fibers of the beam at the cover plate and away from the cover plate, respectively. At this stage, the cover plate is stressed by $+Mj/S_p$, where S_p is the section modulus of the extreme fiber of the cover plate.

The stresses remaining in the beam and cover plate are given by

$$f_{np} = -F_b + \frac{Mj}{S_{np}} = -F_b + \frac{F_b S_b}{S_{np}} \tag{4-49}$$

$$f_{ap} = F_b - \frac{Mj}{S_{ap}} = F_b - \frac{F_b S_b}{S_{ap}} \tag{4-50}$$

and

$$f_p = \frac{Mj}{S_p} - \frac{F_b S_b}{S_p} \tag{4-51}$$

where f_{np}, f_{ap}, and f_p are stresses in the extreme fibers of the beam at the cover plate and away from the plate, respectively.

In stage IV, the beam is turned upside down. In stage V, the prestressed steel beam is cast with the concrete slab and ready to be subjected to the bridge loads. These superimposed loads (dead and live) are carried by the composite steel beam action. The resulting stresses represent the algebraic sum of stresses due to prestressing and to the bridge loads. Because the stresses in the beam due to the bridge loads have opposite signs, the result is an increase in the elastic moment capacity of the beam. Formulas and approximate equations based on a ratio t_p/d varying between 0.02 and 0.10 (t_p is the cover plate thickness and d is the beam depth) are given by Anand and Talesstchi (1973).

Design Examples In one example, a composite beam (36-ft span) with a 4-in. concrete slab is designed using temporary shores. The two solutions are (a) a prestressed beam using A36 steel for the rolled beam and A572, Grade 60 for the cover plate, and (b) a non-prestressed steel beam (A36) for both the beam and the cover plate. A comparison shows a 13 percent reduction in weight (in favor of the prestressed beam), but a 3 percent reduction in the cost of the section.

In a second example, the span length, slab thickness, loads, and material properties are kept the same as in the first example, but the composite beam is designed without temporary shores. The reduction is now 9 percent by weight and 3 percent for the total cost.

In a third example, a composite beam 50 ft long with a 5-in. slab is designed using temporary shores. The cover plate material is chosen with

A514 steel with an F_y of 100 ksi. The use of prestressing results in a reduction of 20 percent in weight and 12 percent in the cost of the steel section. For the unshored construction, the corresponding reductions are 18 and 6 percent, respectively.

This author has used prestressing in rolled steel beams with a cover plate and composite design for single-span bridges. In the 70-ft-span range, the resulting weight reduction was 14 percent, with a corresponding cost saving of 7 percent.

REFERENCES

AASHTO, 1978. "Guide Specifications for Fracture Critical Non-Redundant Steel Bridge Members." (Also Interim)

AASHTO, 1981. "Standard Specifications for Welding of Structural Steel Highway Bridges."

AASHTO, 1988. "Commentary on Bridge Welding Code 01.5-88."

AASHTO, 1991. "Guide Specifications for Alternate Load Factor Design Procedures for Steel Beam Bridges Using Braced Compact Sections."

Adekola, A. O., 1959. "Interaction Between Steel Beams and a Concrete Floor Slab," thesis submitted to the Imperial College of Science and Technology, Univ. of London, in partial fulfillment of the requirements for the degree of Doctor of Philosophy.

Adekola, A. O., 1968. "Effective Widths of Composite Beams of Steel and Concrete," *Struct. Engineer*, Sept.

AISC, 1969. *Simple Span Steel Bridges*, American Inst. Steel Construction, Chicago.

AISC, 1986. *Load and Resistance Factor Design*, American Inst. Steel Construction, Chicago.

AISI, 1987. "Suggested Autostress Procedures for Load Factor Design of Steel Beam Bridges," *Bull.* No. 29, American Iron and Steel Inst., April.

AISC, 1989. "Specifications of Structural Steel Buildings—Allowable Stress Design and Plastic Design," American Inst. of Steel Construction, Chicago.

Albrecht, P., and A. H. Naeemi, 1984. "Performance of Weathering Steel in Bridges," NCHRP Report 272, National Cooperative Highway Research Program, Transportation Research Board, Washington, D.C., July.

Albrecht, P., and M. Sidani, 1987. "Fatigue Strength of 8-Year Weathered Stiffeners in Air and Salt Water," NCHRP Project 10-22, National Cooperative Highway Research Program, Transportation Research Board, Washington, D.C., Oct.

Albrecht, P., Coburn, S. K., Wattar, F. M., Tinklenberg, G. L., and W. P. Gallagher, 1989. "Guidelines for the Use of Weathering Steel in Bridges," NCHRP Report 314, National Cooperative Highway Research Program, Transportation Research Board, Washington, D.C., June.

Allen, D. N. deG. and R. T. Severn, 1961. "Composite Action Between Beams and Slabs under Transverse Load," *Struct. Engineer*, Vol. 39, London, May, pp. 149–154.

American Association of State Highway and Transportation Officials, 1990. "Recommendations of Weathering Steel Fatigue Study Group," AASHTO Technical Committee T-14, Feb.

Anand, S. C. and A. Talesstchi, 1973. "Prestressed Composite Steel Beam Design," *J. Struct. Div. ASCE*, Vol. 99, No. ST3, March, pp. 301–319.

ANSI/AASHTO/AWS, 1988. Bridge Welding Code D1.5-88.

Ansourian, P., 1975. "An Application of the Method of Finite Elements to the Analysis of Composite Floor Systems," *Proc. Institution Civ. Engineers*, Part 2, Vol. 59, London, Dec., pp. 699–726.

Ansourian, P., 1982. "Plastic Rotation of Composite Beams," *J. Struct. Div. ASCE*, Vol. 108, No. ST3, March, pp. 643–659.

Ansourian, P. and J. W. Roderick, 1978. "Analysis of Composite Beams," *J. Struct. Div. ASCE*, Vol. 104, No. ST10, Oct., pp. 1631–1645.

ASCE, 1971. Commentary on Plastic Design in Steel, Joint Committee, Welding Research Council and ASCE, Manual and Reports on Practice No. 41.

ASCE, 1974. "Composite Steel-Concrete Construction," Report of Task Committee on Composite Construction, *J. Struct. Div. ASCE*, Vol. 100, No. ST5, May, pp. 1085–1139.

ASCE–AASHO, 1968. "Development of Use of Prestressed Steel Flexural Members," Report by Subcommittee of the Joint ASCE–AASHO Committee on Steel Flexural Members, *J. Struct. Div. ASCE*, Vol. 94, NO. ST9, Sept., pp. 2033–2060.

ASCE–AASHTO, 1975. "State-of-the-Art Report on Ultimate Strength of I-Beam Bridge Systems," Subcommittee on Ultimate Strength, Joint ASCE–AASHTO Task Committee on Metals, Struct. Div., *J. Struct. Div. ASCE*, Vol. 101, No. ST5, May, pp. 1085–1096.

ASTM, 1988a. "Specification for Structural Steel (A36-88c)," American Society for Testing and Materials, Philadelphia.

ASTM, 1988b. "Specification for High-Strength Low Alloy Steels (A572-88c)," American Soc. for Testing and Materials, Philadelphia.

ASTM, 1988c. "Specification for High-Strength Low-Alloy Structural Steel (A588-88a)," American Soc. for Testing and Materials, Philadelphia.

Aulthouse, F. D., 1989. "Economics of Weathering Steel in Highway Structures," FHWA Publ. FHWA-TS-89-016, Federal Highway Administration Forum on Weathering Steels for Highway Bridges, Alexandria, Virginia, June.

AWS, 1987. *Welding Handbook*, Vol. 1, *Welding Technology*, American Welding Soc., Miami.

AWS, 1988. "Structural Welding Code—Steel," 11th ed., American Welding Soc., Miami.

Baldwin, J. W., Henry, J. R., and G. M. Sweeney, 1965. "Study of Composite Bridge Stringers, Phase II," Univ. of Missouri, May.

Barnard, P. R., 1965. "Series of Tests on Simply Supported Composite Beams," *J. Amer. Concrete Inst.*, Vol. 62, No. 4, April, pp. 443–456.

Barnard, P. R. and R. P. Johnson, 1965. "Ultimate Strength of Composite Beams," *Proc. Institution Civ. Engineers*, Vol. 32, London, Oct., pp. 161–179.

Barsom, J. M., 1975. "The Development of AASHTO Fracture-Toughness Requirements for Bridge Steels," American Iron and Steel Inst., Washington, D.C.

Beedle, L. S., 1958. *Plastic Design of Steel Frames*, Wiley, New York.

Beguin, G. H., 1978. "Composite Bridge Decking by Stage-Deck Jacking," *J. Struct. Div. ASCE*, Vol. 104, No. ST1, Jan., pp. 171–189.

Beschkine, L., 1937–38. "Determination of the Effective Width of the Compression Flange of T-Beams," (in French), *Publications*, Intern. Assoc. for Bridge and Structural Engineering, Fifth Volume, p. 65.

Bethlehem Steel Corporation, 1976. *Economics of Simple Span Highway Bridges*, Bethlehem Steel Corp., Pennsylvania.

Bowden, F. P. and L. Leben, 1939. "The Nature of Sliding and Analysis of Friction," *Proc. Royal Soc.*, Series A, Vol. 169, London, pp. 371–390.

Burdette, E. G. and D. W. Goodpasture, 1971. "Full Scale Bridge Testing—An Evaluation of Bridge Design Criteria," Univ. Tennessee, Knoxville, Dec.

Carskaddan, P. S., 1969. "Bending of Deep Girders with A514 Steel Flanges," *J. Struct. Div., ASCE*, vol. 95, No. ST10, Proc. Paper 6839, Oct., pp. 2219–2242.

Carskaddan, P. S., G. Haaijer, and M. A. Grubb, 1982. "Computing the Effective Plastic Moment," *AISC Eng. J.*, First Quarter.

Chapman, J. C., 1964. "Composite Construction in Steel and Concrete—The Behavior of Composite Beams," *Struct. Engineer*, Vol. 42, No. 4, April, pp. 115–125.

Chapman, J. C., and S. Balakrishnan, 1964. "Experiments on Composite Beams," *Struct. Engineer*, Vol. 42, No. 11, Nov., pp. 369–383.

Ciolina, F., 1971. "Composite Steel-Concrete Bridge Structures," (in French), Annales de L'Institute Technique du Batiment et des Travaux Publics, July-Aug., p. 37.

Climenhaga, J. J., and R. P. Johnson, 1972: "Local Buckling in Continuous Composite Beams," *Struct. Engineer*, Vol. 50, No. 9, Sept., pp 367–374.

Colville, J., 1973. "Tests of Curved Steel-Concrete Composite Beams," *J. Struct. Div., ASCE*, Vol. 99, No. ST7, Proc. Paper 9867, July, pp. 1555–1570.

Dai, P. K. H. and C. P. Siess, 1963. "Analytical Study of Composite Beams with Inelastic Shear Connection," Univ. Ill. Struct. Research Ser. 267, June, 113 pp.

Daniels, J. H., 1972. "Recent Research on Composite Beams for Bridges and Buildings," *Civ. Eng. Transactions*, The Inst. of Eng., Australia, Vol. CE14, No. 2, Oct., pp. 228–233.

Daniels, J. H., and J. W. Fisher, 1967. "Static Behavior of Composite Beams with Variable Load Position," Fritz Eng. Lab. Report 324.3, Lehigh Univ., March, 52 p.

Daniels, J. H., and J. W. Fisher, 1967. "Static Behavior of Continuous Composite Beams," Lehigh Univ., Fritz Eng. Laboratory Report 324.2, March, 79 p.

Daniels, J. H. and J. W. Fisher, 1968. "Fatigue Behavior of Continuous Composite Beams," Hwy. Research Record No. 253, Hwy. Research Board, pp. 1–20.

Das, M., Malairongs, K., and R. B. L. Smith, 1972. "Yield-Line Analysis of Reinforced Concrete Slabs Using a Digital Computer," Pub. SP-33, American Concrete Institute.

Das, S. C. and J. W. Baldwin, 1967. "Shear Connectors in Haunched Composite Beams," *Eng. Experiment Station Bull.*, Univ. Missouri-Columbia, August.

Davidson, J. H. and J. Longworth, 1969. "Composite Beams in Negative Bending," Report No. 20, Dept. Civ. Eng., Univ. Alberta, Edmonton, Canada, May.

Emanuel, J. H. and J. L. Hulsey, 1978. "Temperature Distribution in Composite Bridges," *J. Struct. Div. ASCE*, Vol. 104, No. ST1, Jan., pp. 65–78.

Fan, H. M. and C. P. Heins, 1974. "Effective Width of Composite Bridges at Ultimate Load," Civ. Eng. Report No. 57, Univ. Maryland, College Park, June.

Federal Highway Administration, 1989. "Uncoated Weathering Steel in Structures," FHWA Tech. Advisory (T5140.22), Washington, D.C., Oct.

Finn, E. V., 1964. "The Use of Prestressed Steel in Elevated Roadway," *Struct. Engineer*, Vol. 42, No. 1, Jan.

Fisher, J. W., J. H. Daniels, and R. G. Slutter, 1972. "Continuous Composite Beams for Bridges," Prel. Report, Int. Assoc. Bridge and Struct. Eng., Amsterdam, May, pp. 113–123.

Flint, A. R. and L. S. Edwards, 1970. "Limit State Design of Highway Bridges," *Struct. Engineer*, Vol. 48, No. 3, March, pp. 93–108.

Galambos, T. V., 1972. "Load Factor Design of Steel Buildings," *AISC Eng. J.*, American Inst. Steel Construction 9, 3 (July), 108–113.

Galambos, T. V., 1981. "Load and Resistance Factor Design," *AISC Eng. J.*, American Inst. Steel Construction, Vol. 18, No. 3, pp. 74–82.

Galambos, T. V. and M. K. Ravindra, 1978. "Proposed Criteria for Load and Resistance Factor Design," *AISC Eng. J.*, American Inst. Steel Construction 15 1 (First Quarter), 8–17.

Garcia, I. and J. H. Daniels, 1971. "Negative Moment Behavior of Composite Beams," Fritz Eng. Labor. Report No. 359.4, Lehigh Univ., Bethlehem, Pa.

Gardner, N. J. and E. R. Jacobson, 1967. "Structural Behavior of Concrete Filled Steel Tubes," ACI Vol. 64, No. 7, July, pp. 404–413 (and *J. Suppl.* No. 2, Title No. 64-38).

Goble, G. G., 1967. "The Influence of Stud Yield Strength on the Strength of Composite Specimens," Test Report, Case-Western Reserve Univ.

Goble, G. G., 1968. "Shear Strength of Thin Flange Composite Specimens," *AISC Eng. J.*, American Inst. Steel Construction, Vol. 5, No. 2, April, pp. 62–65.

Godfrey, G. G., 1988. "The Use of Weathering Steels in Composite Highway Bridges," *Intern. Symp. on Steel Bridges*, London, Feb.

Grace, N. F. F., 1981. "Effect of Prestressing the Deck in Continuous Bridge of Composite Construction," thesis presented to the Univ. of Windsor, at Windsor, Ontario, Canada, in partial fulfillment of the requirements for the degree of Master of Science.

Guyon, Y., 1972. "Composite Steel-Prestressed Concrete Construction," *Annales de l'ITBTP*, No. 298, Oct., pp. 131–156.

Hansell, W. C. and I. M. Viest, 1971. "Load Factor Design for Steel Highway Bridges," *AISC Eng. J.*, American Inst. Steel Construction, Vol. 8, No. 4, Oct., pp. 113–123.

Hansell, et al., 1980.

Hasse, G., 1969. "Effect of Creep on Statically Indeterminate Composite Beams and Frames of Variable Cross Section," (in German), *Verein Deutscher Ingenieure, Berichte*, Part 4, No. 16, Nov., 111 p.

Hawkins, N. M., 1970. "The Strength of Stud Shear Connectors," Research Report No. R141, Univ. Sydney, May.

Heins, C. P., 1980. "LRF Criteria for Composite Steel I-Beam Bridges," *J. Struct. Div. ASCE*, Vol. 106, No. ST11, Nov., pp. 2297–2312.

Heins, C. P. and H. M. Fan, 1976. "Effective Composite Beam Width at Ultimate Load," *J. Struct. Div. ASCE*, Vol. 102, No. ST11, Nov., pp. 2163–2179.

Heins, C. P. and R. O. Holliday, 1974. "Ultimate Load Response of a Composite Multi-Girder Bridge Model," Civ. Eng. Report No. 54, Univ. Maryland, College Park, March.

Heins, C. P., and J. T. C. Kuo, 1972a. "Torsional Properties of Composite Girders," *AISC Eng. J.*, American Inst. Steel Construction, Vol. 9, No. 2, April, pp. 79–85.

Heins, C. P., and J. T. C. Kuo, 1972b. "Composite Beams in Torsion," *J. Struct. Div. ASCE*, Vol. 98, No. ST5, Proc. Paper 8921, May, pp. 1105–1117.

Heins, C. P. and T. C. Kuo, 1973. "Live Load Distribution on Composite Highway Bridges at Ultimate Load," Civ. Eng. Report No. 53, Univ. Maryland, College Park, April.

Heins, C. P. and T. C. Kuo, 1975. "Ultimate Live Load Distribution Factor for Bridges," *J. Struct. Div. ASCE*, Vol. 101, No. ST7, July, pp. 1481–1496.

Heins and Kuo, 1976.

Heins, C. P. and A. D. Kurzweil, 1976. "Load Factor Design of Continuous Span Bridges," *J. Struct. Div. ASCE*, Vol. 102, No. ST6, June, pp. 1213–1228.

Hoadley, P. G., 1963. "Behavior of Prestressed Composite Steel Beams," *J. Struct. Div. ASCE*, Vol. 89, No. ST3, June, pp. 21–34.

Holtz, N. M., and G. L. Kulak, 1972. "High-Strength Bolts Used as Shear Connectors," Test Report, Dept. of Civ. Eng., Univ. of Alberta, Edmonton, Canada, Jan.

Huber, H. T., 1923. "Die Theorie Der Kreuzweise Bewehrtren Eisebetonplatten Nebst Anwendungen Auf Mehrere Bautechnisch Wichtige Aufgaben uber Rechteckige Platten," *Der Bauingenieur*, Vol. 4, pp. 354–360.

Hulsey, J. L., 1976. "Thermal Effects on Composite-Girder Bridge Structures," Ph.D. Thesis, Univ. Missouri, Rolla.

Imbsen and Associates, 1991. "Distribution of Loads in Bridges," NCHRP Project 12-26, Transp. Research Board, National Research Council, Washington, D.C.

Iwamoto, K., 1962. "On the Continuous Composite Girder," Bulletin 339, Hwy. Research Board, Tokyo.

Janss, 1972.

Johansen, K. W., 1954. "Pladeformier," *Formelsamling Polyteknish Forening*, 2nd Ed., Copenhagen.

Janss, J., 1972. "Research on Composite Structures Carried Out by the CRIF at the Univ. of Liege," (in French), *Preliminary Report*, Intern. Assoc. for Bridge and Structural Engineering, 9th Congress, Amsterdam, pp. 125–132.

Johnson, R. P., 1970. "Longitudinal Shear Strength of Composite Beams," *ACI J.* Vol. 67, No. 6, June, pp. 464–466.

Johnson, R. P., 1972. "Design of Composite Beams With Deep Haunches," *Proc. Institution Civ. Engineers*, Vol. 51, Jan., pp. 83–90.

Johnson, R. P., 1975. "Composite Steel-Concrete Construction, Discussion, *ASCE Struct. J.*, May, p. 1156.

Johnson, R. P., K. Van Daley, and A. R. Kemp, 1967. "Ultimate Strength of Continuous Composite Beams," *Proc. Conf. Struct.* Steelwork, London, pp. 27–35.

Johnson, R. P. and R. T. Willmington, 1972. "Vertical Shear Strength of Compact Composite Beams," *Proc. Institution Civ. Engineers*, Suppl. Vol., Jan., pp. 1–16.

Kennedy, J. B. and S. K. Bali, 1979. "Rigidities of Concrete Waffle-Type Slab Structures," *Can. J. Civ. Engineers*, Vol. 6, No. 1.

Kennedy, J. B. and D. S. R. Gupta, 1976. "Bending of Skew Orthotropic Plate Structures," *J. Struct. Div. ASCE*, Vol. 102, No. ST8, Proc. Paper 12302, Aug., pp. 1559–1574.

Kennedy, J. B., and N. F. Grace, 1982. "Prestressed Decks in Continuous Composite Bridges," *ASCE Struct. J.*, Nov., pp. 2384–2410.

King, D. C., R. G. Slutter, and G. C. Driscoll, 1965. "Fatigue Strength of $\frac{1}{2}$-inch Diameter Stud Shear Connectors," Hwy. *Research Record* 103, Hwy. Research Board, pp. 78–106.

Knee, O. W., 1966. "The Prestressing of Steel Girders," *Struc. Engineer*, Vol. 44, No. 10, Oct.

Kuo, T. C., 1973. "Elasto-Plastic Behavior of Composite Slab Girder Highway Bridges," Ph.D. Thesis, Univ. Maryland, College Park.

Kuo, T. C. and C. P. Heins, 1973. "Live Load Distribution on Composite Highway Bridges at Ultimate Load," Civ. Eng. Report No. 53, Univ. Maryland, College Park, April.

Lay, M. G. and P. D. Smith, 1965. "Role of Strain Hardening in Plastic Design," *J. Struct. Div. ASCE*, Vol. 91, No. ST3, June, pp. 25–43.

Lee, J. A. N., 1962. "Effective Width of Tee Beams," *Struct. Engineer*, Vol. 40, No. 1, Jan., pp. 21–27.

Lew, H. S., 1970. "Effect of Shear Connector Spacing on the Ultimate Strength of Steel-Concrete Composite Beams," National Bureau of Standards Report No. 10 246, U.S. Dept. Commerce, Washington, D.C., Aug.

Li, S. and R. V. Blackwell, 1972. "Design Yield Moment of Continuous Concrete Deck Slabs Under Vehicular Loading," ACI, Publ. SP-33.

Lowe, P. A. and A. R. Flint, 1971. "Prediction of Collapse Loadings for Composite Highway Bridges," *Proc. Institution of Civ. Engineers*, London, April.

Mackey, S. and F. K. C. Wong, 1961. "Effective Width of Composite Tee-Beam Flange," *Struct. Engineer*, Vol. 39, No. 9, Sept., pp. 277–285.

Magnel, G., 1950. "Prestressed Steel Structures," *Struct. Engineer*, Vol. 28, No. 11, Nov.

Magnel, 1969.

Malairungs and Smith, 1972.

Marshall et al., 1971.

Marshall, W. T., Nelson, H. M., and H. K. Banerjee, 1971. "An Experimental Study of the Use of High-Strength Friction-Grip Bolts as Shear Connectors in Composite Beams," *Struct. Engineer*, vol. 49, No. 4, April, pp. 171–178.

McGarraugh, J. B., and J. W. Baldwin, 1971. "Lightweight Concrete-on-Steel Composite Beams," *AISC Eng. J.*, American Inst. Steel Construction, Vol. 8, No. 3, July, pp. 90–98.

Menzies, J. B., 1968. "Structural Behavior of the Moat Street Flyover, Coventry," *Civ. Eng. and Public Works Rev.*, vol. 63, Sept., pp. 967–971.

Menzies, J. B., 1971. "CP 117 and Shear Connectors in Steel-Concrete Composite Beams Made with Normal-Density or Lightweight Concrete," *Struct. Engineer*, Vol. 49, No. 3, March, pp. 137–154.

Modjeski and Masters, 1991. "LRFD Bridge Design Code," NCHRP Project 12-33, Transp. Research Board, National Research Council, Washington, D.C.

Morgan, F., M. Muskat, and D. W. Reed, 1971. "Studies in Lubrication-X Friction Phenomena and the Stick-Slip Process," *J. Appl. Phys.*, Vol. 12, Oct., pp. 743–752.

O'Connor, C., 1971. *Design of Bridge Superstructures*, Wiley, New York.

Pinkham, C. W., and W. C. Hansell, 1978. "An Introduction to Load and Resistance Factor Design for Steel Buildings," *AISC Eng. J.*, American Institute of Steel Construction, 15 1 (First Quarter), 2–7.

Plum, D. R., 1972. "Strength of Studs in Composite Construction," *Proc. Institution Civ. Engineers*, Vol. 51, London, Feb. pp. 319–335.

"Prestressing Steel Stringers Reduce Bridge Weight by 25%, 1961. *Engineering News-Record*, Vol. 167, No. 16, Oct. 19, pp. 32–33.

Priestley, M. J. N., 1976. "Linear Heat-Flow Analysis of Concrete Bridge Decks," Research Report 76-3, Dept. Civ. Eng., Univ. Canterbury, New Zealand, Feb.

Ravindra, M. K. and T. V. Galambos, 1978. "Load and Resistance Factor Design for Steel," *J. Struct. Div., ASCE*, Vol. 104, No. ST9, Sept., 1337–1353.

Reagan, R. S., and N. W. Krahl, 1967. "Behavior of Prestressed Composite Beams," *J. Struct. Div. ASCE*, Vol. 93, No. ST6, Proc. Paper 5663, Dec., pp. 87–108.

Reddy, V. M. and A. W. Hendry, 1970. "Ultimate Strength of a Composite Beam Allowing for Strain Hardening," *Indian Concrete J.*, Vol. 44, No. 9, Sept., pp. 388–396.

Roderick, J. W., Hawkins, N. M., and L. C. Lim, 1967. "The Behavior of Composite Steel and Lightweight Concrete Beams," *Civ. Eng. Transactions*, The Inst. of Engineers, Australia, Vol. CE9, No. 2, Oct., pp. 265–275.

Rolfe, S. T., 1972. "Designing to Prevent Brittle Fracture in Bridges," *Proc. Specialty Conf. Safety and Rehabilitation of Metal Structures*, ASCE, pp. 175–216.

Roll, F., 1971. "Effects of Differential Shrinkage and Creep on a Composite Steel-Concrete Structure," ACI Special Publ. SP-27-8.

Rotter, J. M. and P. Ansourian, 1979. "Cross-Section Behavior and Ductility in Composite Beams," *Proc. Institution Civ. Engineers*, Part 2, Vol. 67, London, June, pp. 453–474.

Rowe, R. E., 1962. Concrete Bridge Design, C. R. Books, Ltd., London.

Salmon, C. G. and J. E. Johnson, 1990. *Steel Structures—Design and Behavior*, Harper Collins, New York

Samuley, F. J., 1955. "Structural Prestressing," *Struc. Engineer*, Vol. 33, No. 2, Feb.

Sanders, W. W. and H. A. Elleby, 1970. "Distribution of Wheel Loads on Highway Bridges," NCHRP Report No. 83, Hwy. Research Board, Washington, D.C.

Scheffey, C. V., 1971. "Point Pleasant Bridge Collapse," *Civ. Engineer* July, p. 41.

Schmitt, R. J., 1989. "Development and Application of Weathering Steels," FHWA Publ. FHWA-TS-89-016, Federal Highway Administration Forum on Weathering Steels for Highway Bridges, Alexandria, Virginia, June.

Schmitt, R. J., 1989. "Selection of Weathering Steel for A Given Location," FHWA Publ. FHWA-TS-89-016, Federal Highway Administration Forum on Weathering Steels for Highway Structures, Alexandria, Virginia, June.

Slutter, R. G., and G. C. Driscoll, 1965. "Flexural Strength of Steel-Concrete Composite Beams," *J. Struct. Div. ASCE*, Vol. 91, No. ST2, Proc. Paper 4294, April, pp. 71–99.

Slutter, R. G., and J. W. Fisher, 1971. "The Strength of Stud Shear Connectors in Lightweight and Normal-Weight Concrete," *AISC Eng. J.*, American Instit. Steel Construction, Vol. 8, No. 2, April, pp. 55–64.

Smith, J. C., 1991. *Structural Steel Design—LRFD Approach*, Wiley, New York.

Stevens, R. F., 1965. "Encased Stanchions," *Struct. Engineer*, Vol. 43, No. 2, Feb., pp. 59–66.

"Suggested Guidelines for Improving Performance of Weathering Steel Bridges," *Bethlehem Steel Technical Bulletin*, Jan., 1988.

Thurlimann, B., 1956. "Simple Plastic Theory," *Proc. AISC National Eng. Conf.*, pp. 13–18.

Troitsky, M. S., 1990. *Prestressed Steel Bridges*, Van Nostrand Reinhold. Florence, Kentucky.

United States Steel Corporation, 1982. "Finite-Element Modeling Techniques for Analysis of Live Load Distribution in a Stringer Bridge," *Research Bull.*, April.

Viest, I. M., 1960. "Review of Research on Composite Steel—Concrete Beams," *J. Struct. Div. ASCE*, Vol. 96, ST6, June, pp. 1–21.

Viest, I. M. (Chairman), 1974. "Committee Steel-Concrete Construction," Report, Task Committee of the Committee on Metals of the Struct. Div., *J. Struct. Div. ASCE*, Vol. 100, No. ST5, May, pp. 1085–1139.

Vincent, G. S., 1969. "Tentative Criteria for Load Factor Design of Steel Highway Bridges," *Bull.* No. 15, American Iron and Steel Inst., March.

Wang, C. K. and C. G. Salmon, 1985. *Reinforced Concrete Design*, Harper and Row, New York.

Wood, W. E., and J. H. Devletian, 1987. "Improved Fracture Toughness and Fatigue Characteristics of Electroslag Weldments," Report No. FHWA/RD-87/026, Federal Highway Administration, Washington, D. C., Aug.

Wu, Y. C. and R. G. Slutter, 1971. "Continuous Composite Beams Under Fatigue Loading," Fritz Eng. Lab. Report No. 359.2, Lehigh Univ., Bethlehem, Pa.

Wu, Y. C., R. G. Slutter, and J. W. Fisher, 1971. "Analysis of Continuous Composite Beams," Fritz Eng. Lab. Report No. 359.5, Lehigh Univ., Bethlehem, Pa.

Yam, L. C. P. and J. C. Chapman, 1968. "Inelastic Behavior of Simply Supported Composite Beams of Steel and Concrete," *Proc. Institution Civ. Engineers*, Vol. 41, Dec., London, pp. 651–683.

Zuk, W., 1965. "Thermal Behavior of Composite Bridges," Hwy. Research Record No. 76, Hwy. Research Board, pp. 231–253.

Zuraski, P. D. and J. E. Johnson, 1985. "Research on the Remaining Life in Steel Bridges," *Proc. ASCE Specialty Conf. Probabilistic Mechanics and Struct. Reliability*, Berkeley, Calif., Jan. 11–13, 1984.

CHAPTER 5

WELDED PLATE GIRDER AND BOX GIRDER BRIDGES

5-1 INTRODUCTION

The simplest form of a plate girder is the built-up beam consisting of two flange plates welded to a web to form an I section. The choice of flange and web plates leads to a more efficient arrangement and utilization of material than is possible with rolled beams. Box girders are also popular and offer distinct advantages in many cases.

Better understanding of plate girder behavior, higher-strength steels, and improved welding techniques have combined to make plate girders economical in many instances once considered ideal for trusses. In general, simple spans 70 to 150 ft long have traditionally been the domain of plate girders. For bridges, continuous spans frequently using haunches are typically selected for spans larger than 90 ft. Several three-span continuous plate girders have been built in the United States with center spans exceeding 400 ft, and structures with longer spans are likely to be feasible in the future. The largest plate girder bridge in the world is a three-span continuous structure over the Save River at Belgrade, Yugoslavia, with spans of 246, 856, and 246 ft. This bridge consists of a double box girder with depth varying from 14 ft 9 in. at midspan to 31 ft 6 in. at the pier.

The principal difference between the design of a rolled beam and a plate girder is articulated in the greater freedom and flexibility in proportioning the cross section by choosing the flange and web plates. The larger depth usually selected for a plate girder often results in relatively thin webs making web-buckling problems more relevant to the design. Box girders are made of two flange plates and two web plates imparting to them high torsional strength and rigidity.

5-2 FLOOR SYSTEMS

Reinforced Concrete Concrete slabs supported on longitudinal stringers and transverse floor beams are typical floor systems. A concrete slab may also be directly supported on top of the main longitudinal girders. In the latter case shear attachments may be provided to create composite action.

Steel Grid Steel grid floors may offer the following advantages: (a) reduced dead weight reflected in the required size of stringers, floor beams, and girders; (b) snow and rain do not remain on the grid floor, eliminating the need for a roadway crown and drainage facilities; and (c) the grid floor can be installed conveniently and quickly.

Steel Plate Steel plates welded to the bridge structure and properly stiffened have been used as bridge floors. Attaching a comparatively thin plate to the top flange of longitudinal members produces a built-up section with increased strength and stiffness. This system is often called *battleneck flooring*.

Typical Floor Systems Figure 5-1*a* shows a design utilizing a steel grid floor resting on longitudinal stringers and on the main girders. The floor beams (transverse) are set lower so that the stringers are flush with the top of

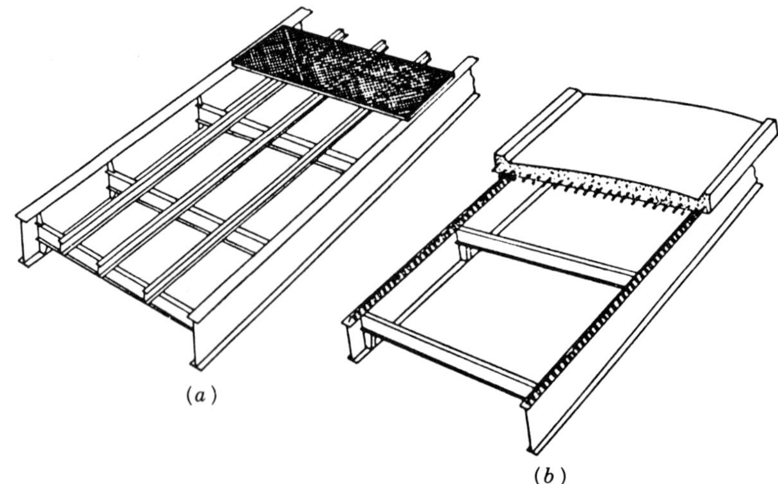

FIGURE 5-1 Two-girder bridge systems: (*a*) steel grid floor; (*b*) concrete slab and transverse floor beams.

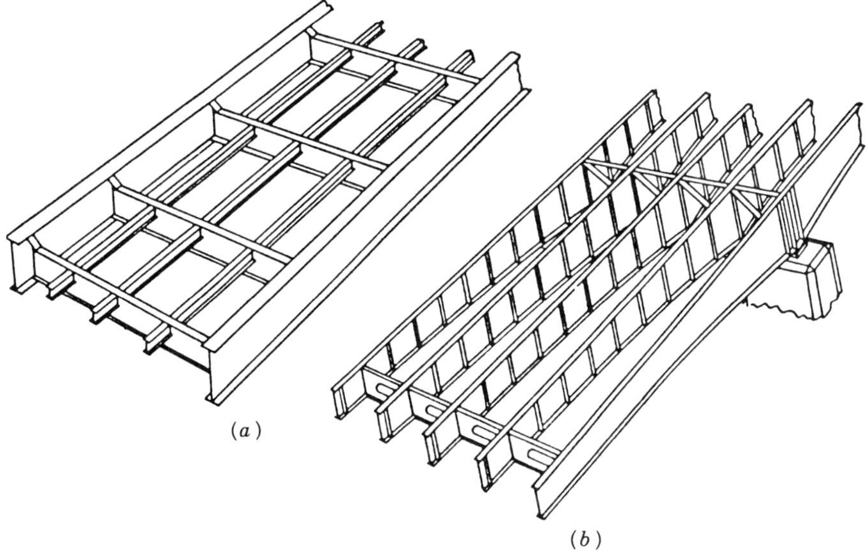

FIGURE 5-2 Plate girder bridge systems: (*a*) two girders with floor beams and stringers; (*b*) multiple-girder bridge with haunches over interior supports.

the girders. Brackets are shop-welded to the girders to receive the floor beams. The floor system shown in Figure 5-1*b* consists of two main longitudinal girders, transverse floor beams, and a concrete slab attached to the girders by means of shear connectors.

For the design shown in Figure 5-2*a*, the bridge consists again of a two-girder system, but the top portion of the girders helps to form the curb. This means that the floor beams are set lower in order to bring the bridge concrete slab below the top flange of the girders. This produces a compact structure and efficient design but requires strict attention to the geometry of the structure, cross-section configuration, and vertical profile. A very popular design is the multigirder bridge with concrete deck shown in Figure 5-2*b*. The example shown is a continuous structure with the girders haunched over the interior supports. For improved appearance the outside girders are shown with intermediate stiffeners placed on the inboard side only.

The advantages of box girder construction are demonstrated by (a) a flat surface available for attaching other members; (b) a less critical corrosion problem because of the flat surface and better inside protection; (c) a considerable increase in torsional stiffness offered by the closed section; and (d) improved lateral stability. Examples of box girder designs are shown in Figure 5-3. In Figure 5-3*a* the bridge consists of two box girders with floor beams cantilevered for optimum results. The floor beams are flush with the box girders to accommodate a concrete slab supported on four sides. Along

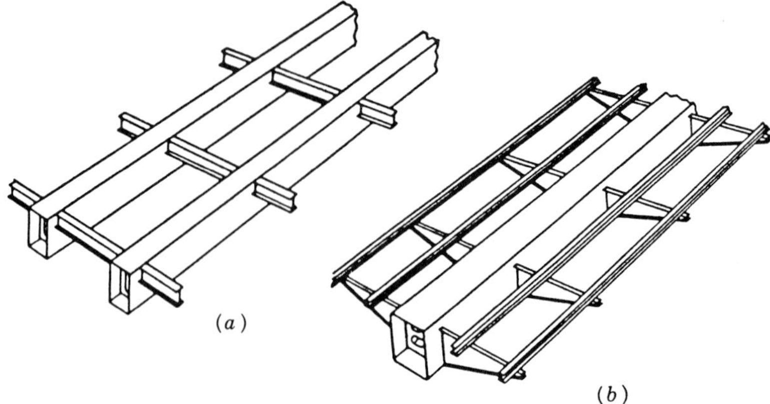

FIGURE 5-3 Box girder bridges: (*a*) two-girder bridge with floor beams; (*b*) single box girder bridge with brackets and longitudinal stringers.

the main girders the system is provided with shear connectors for composite action.

The design shown in Figure 5-3*b* utilizes a single box girder with floor beams extended as brackets to support the bridge floor through longitudinal stringers. A concrete slab provides the bridge deck. An alternative design may consist of a large-diameter fabricated pipe section.

Torsional Resistance Considering the advantages of bridge floors that have an inherent lateral stability and torsional resistance, plate girders can be designed to articulate these benefits.

A member subjected to a torsional moment develops shear stresses. One set of stresses occurs at right angles to the axis of the member and the other set occurs along its length. As shown in Figure 5-4, the shear stress *b* acts normal to the axis of the member and causes it to twist. A flat section or open section provides very little resistance to twisting effects. The cross member is subjected to the shear stress *a* and, likewise, twists. If a diagonal member is placed in the structure as shown, both shear stresses *a* and *b* will act on it. However, the components of these stresses act at right angles to the diagonal element and cancel each other so that no twisting is applied to the member. The stresses combine to introduce tension and compression in line with the members, thus subjecting the diagonal member to bending that can be effectively resisted.

An example of a bridge floor system designed to accommodate twisting effects is shown in Figure 5-5. This floor makes use of diagonal members framed to produce a grid-type structure resistant to twisting and lateral movement (Kavanagh, 1970).

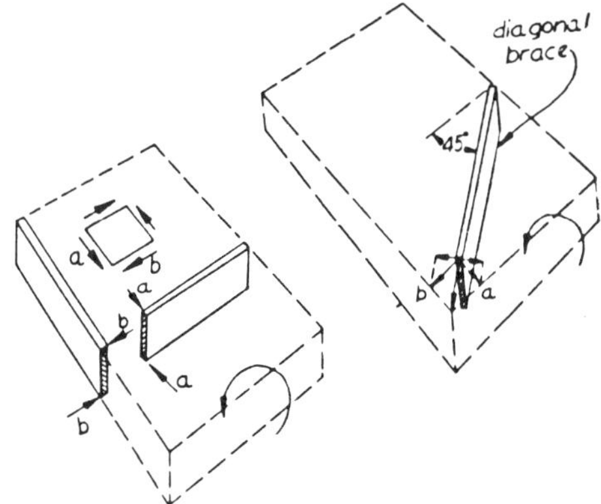

FIGURE 5-4 Torsional effects on bridge members.

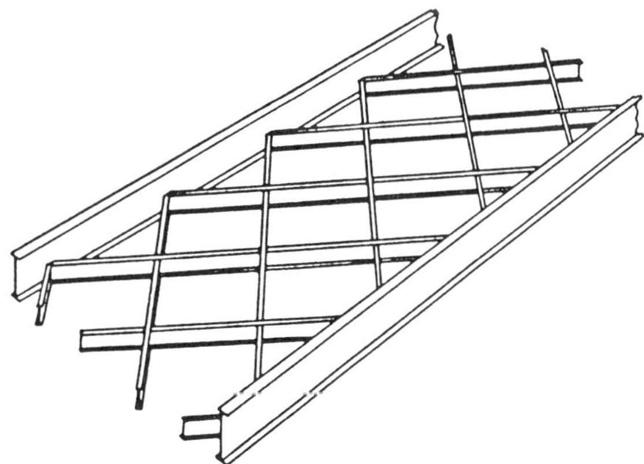

FIGURE 5-5 Bridge floor provided with diagonal bracing to resist torsional effects.

5-3 CORRELATION BETWEEN FLANGE AND WEB AREA: FLANGE AREA FORMULA

The depth of a plate girder and the girder spacing in a bridge depend on many factors such as headroom or vertical clearance, length–depth ratio, deflection limitations, and bridge appearance. Even when the choice depends only on economic considerations, the analysis will indicate that a broad range of combinations may be available (see also Section 5-6).

The tendency toward the use of wide, thin flanges is justified where it is possible to adjust the flange area by varying the width rather than the thickness provided buckling is controlled. In choosing the web, the design may consider a thick web with moderate stiffening requirements or a slender web with considerable stiffening. The correct choice is related to the contrast between material and fabrication costs, but may also be influenced by appearance. An unobstructed smooth web face is suggested for esthetic reasons. On the other hand and despite the presence of stiffeners, a thin web may be expected to produce savings in material cost and weight. A girder with a slender web is efficiently stiffened by a series of longitudinal stiffeners, spanning between cross frames. In this manner, it is possible to avoid transverse fillet welds to the main material especially in the region of tensile stresses, which improves the performance of the structure in relation to fatigue and brittle fracture (see also subsequent sections).

Because the design moments and shears vary along the girder length, it is customary to vary the girder depth accordingly, usually by introducing a curve or straight-line change (haunches). An alternative is to use a stronger steel in the region of higher moments and retain the simplicity of constant depth. This design is referred to as a multisteel or hybrid girder, and requires an optimization analysis in selecting the best combination of steels.

In proportioning the cross section, the usual procedure is to determine a web that has the necessary capacity in shear and to select two flanges that together with the web provide the required section modulus. Because it is more convenient to proportion by area than by moment of inertia, a procedure must be available for a tentative determination.

Let A_f denote the area of one flange, h_g the distance between the centers of gravity of the two flanges, h_w and t the height and thickness of the web plate, respectively, and h_o the overall girder depth. Neglecting the moment of inertia of each flange about its own axis, we can write

$$I = 2A_f\left(\frac{h_g}{2}\right)^2 + \frac{th_w^3}{12} \tag{5-1}$$

where I is the moment of inertia of the section. Because h_w and h_g are approximately equal and setting $th_w = A_w$ (area of web), we obtain

$$I = \frac{h_g^2}{2}\left(A_f + \frac{A_w}{6}\right) \tag{5-2}$$

Considering also the relation $I/c = M/f$ gives

$$A_f = \frac{M}{fh_g^2h_o} - \frac{A_w}{6} \tag{5-3}$$

Noting that fh_g/h_o is the flexural stress at the center of gravity of the flange, denoted by f_g, yields

$$A_f = \frac{M}{h_g f_g} - \frac{A_w}{6} \tag{5-4}$$

which correlates A_f and A_w in terms of relevant parameters.

5-4 BRIDGE PLATE GIRDERS WITH VARIABLE DEPTH

Characteristics of Haunched Girders The sloping bottom flange of a parabolic haunch has a vertical component of its compressive force that reduces the shear stress τ_{sy} in the girder web in this region. In addition, the concave compression flange produces a radial compressive stress σ_y in the web depending on the radius of curvature of the flange, as shown in Figure 5-6*b*.

In contrast, the fish belly haunch shown in Figure 5-6*a* provides no appreciable reduction in shear in the critical portion of the web near the supports because the slope of the bottom flange is small in this area. The convex compressive flange produces a radial tensile stress σ_y in the web which is greater than the radial compressive stress in the parabolic haunch. This is because of the sharper curvature of the fish belly haunch.

Referring to the Huber–Mises formula

$$\sigma_{\text{cr}} = \sqrt{\sigma_x^2 - \sigma_x\sigma_y + \sigma_y^2 + 3\tau_{xy}^2} \tag{5-5}$$

we can note that both these factors will result in a lower value of the yield criterion σ_{cr} for the parabolic haunch. This result compared with the yield strength of the steel would indicate a higher factor of safety.

Haunched girders should not result in an unreasonable cost increase for longer spans. The web is usually trimmed by flange cutting so that a gradual curve adds little to the cost. In most instances the curved flange plates are pulled into place against the curved web.

Modified Shear The horizontal force F_h in the sloping flange is equal to M/d, where d is the vertical distance between the two flanges. This force may also be found by multiplying the flange area by the bending stress in the flange using the section modulus of the girder. Knowing F_h, both the actual force in the flange (F_x) and the vertical component (F_v) are easily computed

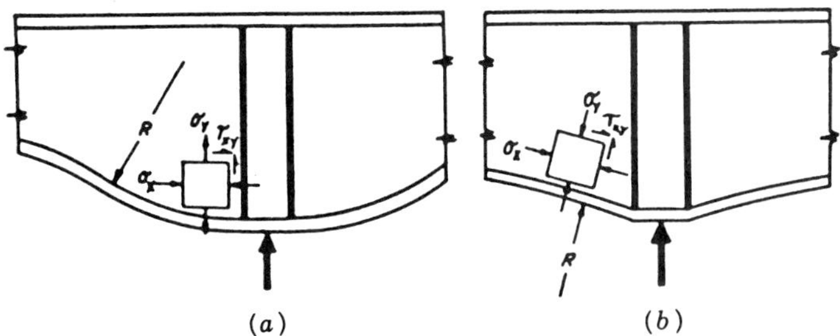

FIGURE 5-6 Stress components in girder haunches: (*a*) fish belly haunch; (*b*) parabolic haunch.

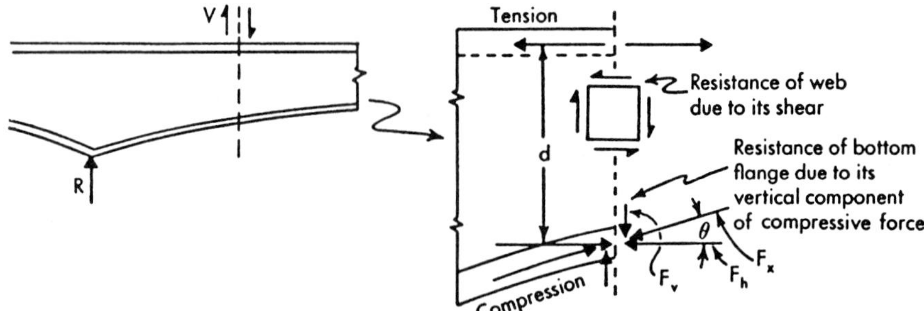

FIGURE 5-7 Modified shear and stress conditions, continuous parabolic haunched girder.

as

$$F_x = \frac{F_h}{\cos\theta} = \frac{M}{d\cos\theta} \qquad \text{and} \qquad F_v = F_h \tan\theta = \frac{M}{d}\tan\theta \qquad (5\text{-}6)$$

where θ is the angle between the tangent to the curve and the horizontal as shown in Figure 5-7.

"Modified" shear is the resulting shear force in the web after the vertical component F_v of the flange force is subtracted or added, depending on its direction.

For the continuous parabolic haunched girder shown in Figure 5-7, the external shear is

$$V = A_w\tau_w + \frac{M}{d}\tan\theta$$

and the modified shear is

$$V' = A_w\tau_w = V - \frac{M}{d}\tan\theta$$

because in this case the vertical component is subtracted from the web shear.

For the continuous fish belly haunched girder shown in Figure 5-8, the external shear is $V = A_w\tau$. In this case the flange force has no vertical component; hence, there is no reduction of shear in the web.

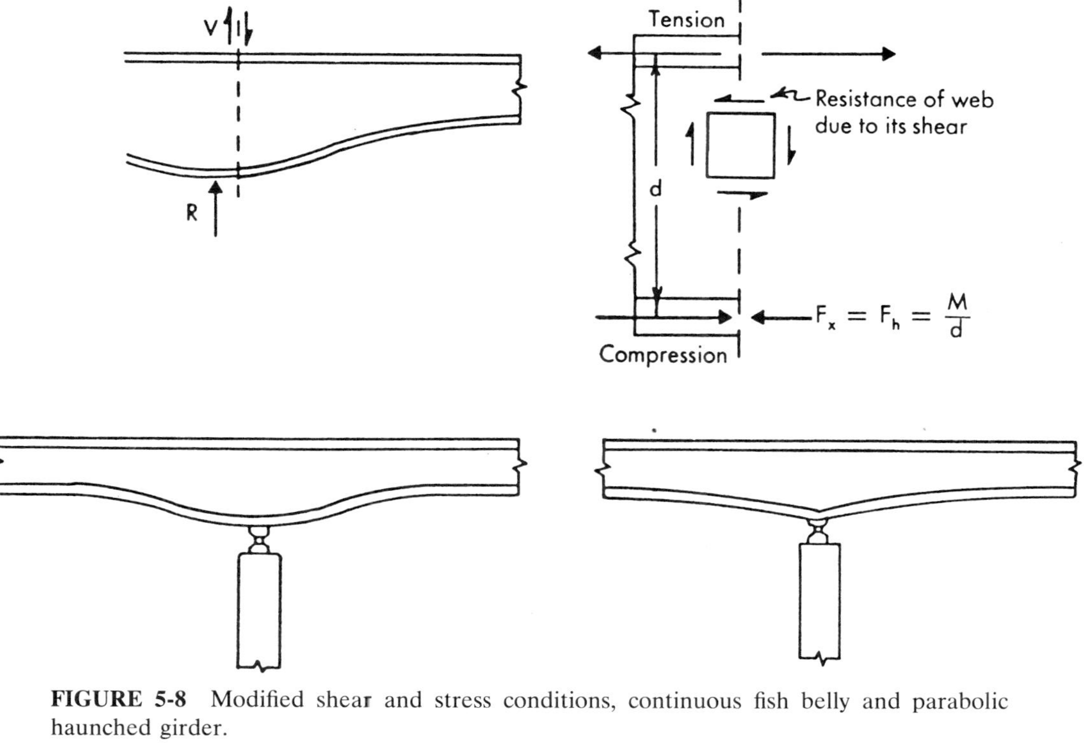

FIGURE 5-8 Modified shear and stress conditions, continuous fish belly and parabolic haunched girder.

5-5 BASIC PRINCIPLES OF WELDING

The weldability of a steel measures the result of producing a crack-free and sound structural joint or connection. Some of the readily available structural steels are more suited to welding than others, and are discussed in Chapter 4. Usually, welding procedures are based on a steel's chemistry instead of the published maximum alloy content. When a mill produces a run of structural steel, it maintains a complete record of its chemical content on the shapes made from a particular ingot. If concern exists about a particular chemistry, the designer may request a mill test report.

Allowable Stresses For service load design, AASHTO gives a detailed summary of allowable stresses for steels in Table 10.32.1A. The yield point and ultimate strength of the weld metal should be equal to or greater than the minimum specified value of the base metal. Allowable stresses on the effective area of the weld metal are as follows.

Butt Welds The same as the base metal joined, except for metals of different yields where the lower-yield material governs.

Fillet Welds The allowable stress is

$$F_v = 0.27F_u \tag{5-7}$$

where F_v is the allowable basic shear stress and F_u is the tensile strength of the electrode classification but not greater than the tensile strength of the connected part.

Plug Welds The allowable basic shear stress is $F_v = 12{,}400$ psi for resistance to shear stresses only.

Basic Types of Joints

Butt Joints A butt joint is used mainly to join the ends of flat plates of the same or nearly the same thickness. The principal advantage is in the elimination of the eccentricity developed in single lap joints. The principal disadvantage is the preparation (beveling, grounding flat) necessary for the members to be connected. Because only little adjustment is possible, the pieces must be carefully aligned, detailed, and fabricated. As a result, most butt joints are made in the shop where the welding process can be better controlled.

Lap Joints A lap joint is the most common type. It offers the advantage of easy fitting, because pieces intended to be joined do not require the preparation typically necessary with other types of joints. Lap joints utilize fillet

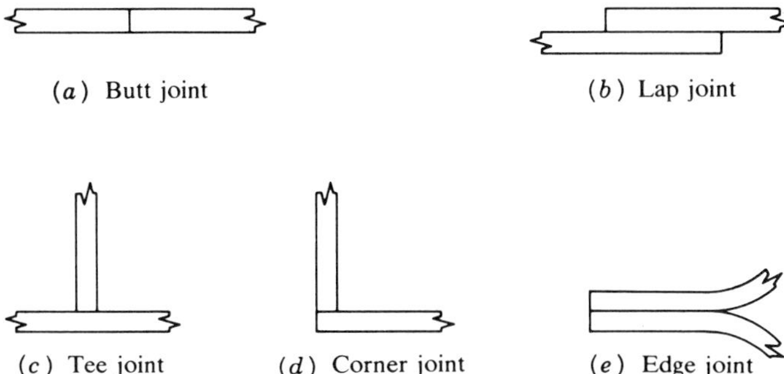

FIGURE 5-9 Basic types of welded joints.

welds and are therefore equally suited to shop or field welding. A further advantage of the lap joint is the ease in which plates of different thickness can be joined.

Tee Joints This type of joint is used to fabricate built-up sections such as I shapes, plate girders, bearing stiffeners, and brackets. I tis especially useful in that it permits sections to be built up of flat plates that can be joined by either fillet or groove welds.

Corner Joints These are used mainly to form built-up rectangular box sections and box beams with high torsional resistance.

Edge Joints These are not necessarily structural connections, but are most frequently used to keep two or more plates in their correct alignment.

Examples of the basic joint types are shown in Figure 5-9.

Strength of Butt Welds

Irregularities in the weld affect the stress distribution according to the stress paths shown in Figure 5-10. Stress paths are defined as lines indicating the direction of the principal stresses, whereas the spacing between these stress paths shows the magnitude of the stress. With widely spaced lines the stress is low; conversely, closely spaced lines indicate that failure is likely to start there. If the mechanical properties of the weld metal are different from the base metal, the stress distribution is not uniform.

Stress concentrations will occur at sharp reentrant corners of butt welds connecting bars of different cross section, thereby reducing the strength of the connection considerably. To minimize stress concentration, a gradual transition from one section to the other must be provided. Tests on butt-welded specimens of variable section but with adequate transition show that the tensile strength is almost the same as that of constant-section specimens.

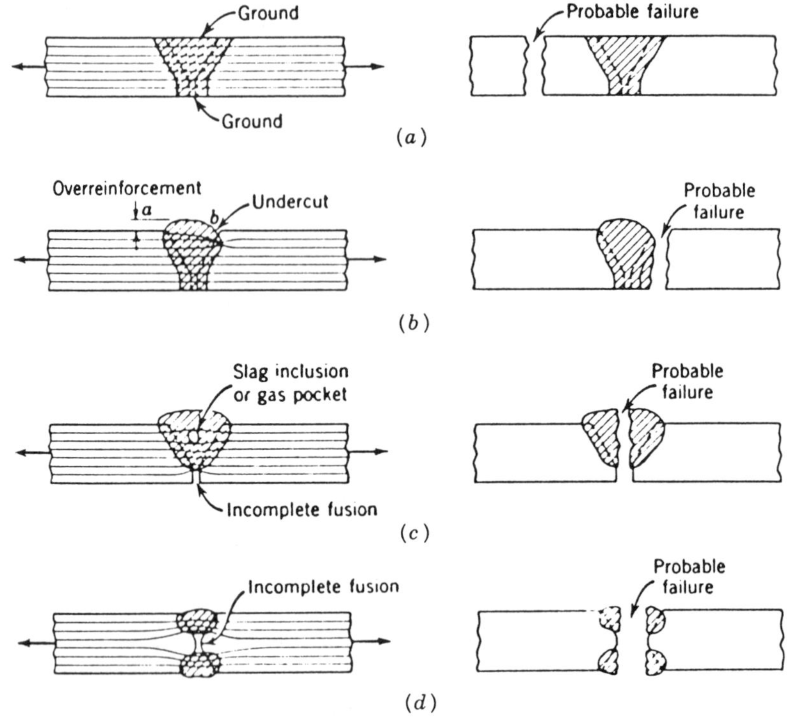

FIGURE 5-10 Stress paths and locations of failure; plates joined by butt welds.

Tension results of butt-welded specimens indicate that their average tensile strength is nearly the same as the average tensile strength of the base metal, but strength variation is greater for the welded than for the unwelded specimen because of imperfections and nonhomogeneity.

Imperfections occur in some types of butt welds at such a frequency that the best estimate of strength is obtained as the product of the ultimate strength of the base metal and an "effective" weld thickness.

Strength of Fillet Welds

Unlike the full continuity attained with butt welds, fillet welds introduce eccentricities of force transmission and discontinuity of shape in the connected parts. Consequently, the actual stress distribution is extremely complex. Although a rigorous solution for stress distribution in fillet welds cannot be obtained, useful results are available from approximate solutions based on the theory of elasticity or experimental measurements on physical models.

In actual structures, plates, bars, and other connected members are never rigid, and some allowance must be made for their deformations. In the elastic region, deformation is proportional to the stress, but any difference in deformation between two connected members results in a variation of shearing strain in the weld. When the connected members or the weld or both are stressed beyond the elastic limit, behaving more plastically than elastically, the shear in the weld becomes more uniform. For side fillets under static loads at or near failure, the plates and the welds are stressed beyond the elastic limit, and the assumption of uniform stress distribution throughout the weld length is closer to reality than at low or intermediate loads.

Stress distribution in end fillet welds is more complex because of the local deformations within the weld and adjacent plate material. Within the elastic limits the strains at the end of the fillet welds are considerably greater than at the center. Furthermore, tests show that long side fillet welds have smaller ultimate unit strength than short ones, but this difference is small because of stress redistribution in the plastic range. The ultimate nominal shearing strength of fillet welds is based on tests of specimens subjected to pure tension. For end fillet welds, the nominal shearing strength is not less than 0.875 times the minimum tensile strength of the base metal, and for side welds it is not less than two-thirds of the same.

Design Considerations: Elastic and Plastic Range

Elastic Butt welds are preferable to fillet welds on the basis of strength, particularly fatigue. Butt welds, however, result in higher residual stresses, require more preparatory work, and impose strict limitations on length tolerance.

Two other factors to be considered in the choice of weld size are the minimum weld size necessary to prevent quick cooling resulting in weld brittleness, and the maximum weld size as determined by practical restrictions in obtaining the proper shape. Whereas the minimum practical weld size is 3/16 in., the most economical size is 5/16 in., which is generally the maximum weld manually obtainable in one pass. The cost–strength ratio increases rapidly with an increase in size.

In general, it is more economical to use a small-size continuous weld rather than a larger discontinuous weld, with both deposited in one pass. In bridge design intermittent welds should not be used to transmit calculated stresses.

Plastic For structures proportioned on the basis of plastic theory, all connections should be capable of resisting moments, shears, and axial loads acting as a result of the applied ultimate loads. Therefore, the welds must be proportioned to resist the forces produced at ultimate loads using unit stresses that have been increased accordingly.

For butt weld stresses the assumption is made that the weld metal is capable of developing the tensile yield stress on the base metal on the minimum throat area.

For fillet welds it is commonly assumed that the weld material can develop at least the shearing yield stress of the base metal on the minimum throat area. A safe design value is obtained by multiplying the elastic design allowable stress value for the weld by the ratio F_y/F_w, where F_y is the yield strength and F_w is the allowable tensile stress of the base material.

AASHTO stipulates that the yield point and ultimate strength of the weld metal in groove and fillet welds should be equal to or greater than that of the base metal when detailing fillet welds for quenched and tempered steels. However, the welding procedure and weld metal should be selected to ensure sound welds.

Acceptance criteria for welds and welding processes are reviewed in some detail in subsequent sections. For current requirements refer to the ANSI–AASHTO–AWS D1.5 bridge welding code, and subsequent AASHTO Interim Specifications.

5-6 TOPICS RELEVANT TO OPTIMUM PROPORTIONING OF PLATE GIRDERS

Continuous Girders

Considerable analytical work has been done in the area of optimization of plate girder design (Annamalai, Lewis, and Goldberg, 1972; Azad, 1978; Chong, 1976; Holf and Heithecker, 1969; Matlock, 1962; Schilling, 1974). The usual approach is to relate the section characteristics to a given set of bending moment and shear. An example of this methodology is presented in Section 5-3, and correlates flange and web area in terms of applied moment and allowable steel stress. The web may be assumed either as a continuous variable or as a variable conforming to the required plate thickness.

A procedure for the optimum design of a continuous plate girder based on elastic analysis is presented by Azad (1980). In this example, the web depth is assumed constant, and for simplicity the loading on the girder is restricted to a uniformly distributed load. The number of flange cutoff points in each span and the minimum length of each flange plate are decided from practical considerations.

For a continuous beam with n spans, the bending moment distribution for each span is known once the support moments M_i are determined. Each span may be divided into segments based on the number of plate cutoff points and the shape of the moment curve. A typical arrangement is shown in Figure 5-11 where segments of arbitrary length are identified for an exterior and an interior span.

Optimization Criteria Either the total cost or the total weight of the girder may be considered as an objective function, although the weight is chosen by most investigators as the relevant parameter. In this case and assuming a

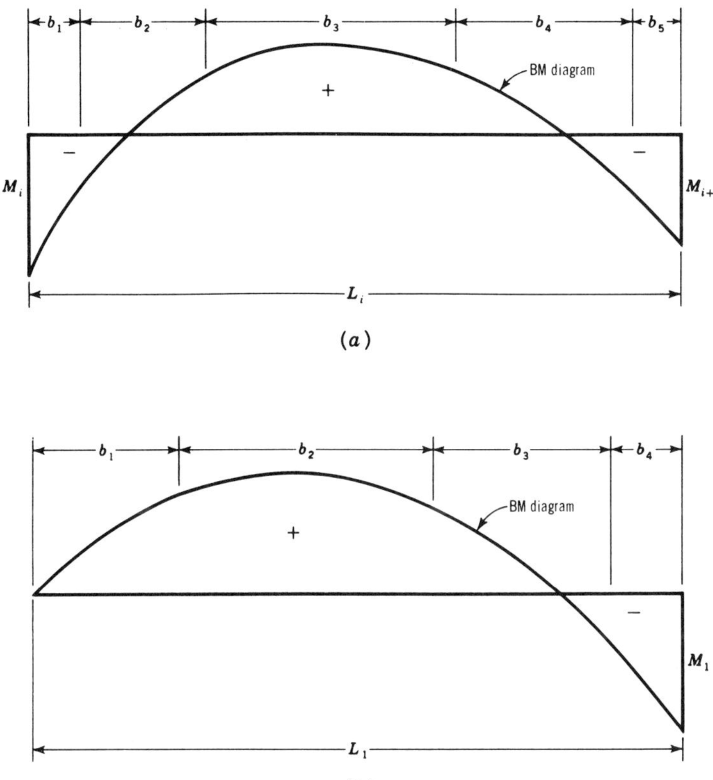

FIGURE 5-11 Division of interior and exterior span into segments: (*a*) interior span divided into five segments; (*b*) exterior span divided into four segments.

girder with constant web height h and thickness t, but with variable flange plates, the weight W for a span L is

$$W = p\left(htL + \sum_{i}^{n} 2A_{fi}b_i\right) \tag{5-8}$$

where A_{fi} is the area of one flange, b_i is the length of the appropriate segment, and n is the number of segments in the span. Noting that the section modulus is

$$S_x = A_f h + \frac{h^2 t}{6} = \frac{M}{F_s}$$

and combining this with (5-8) gives

$$W = p\left[htL + \sum_{i}^{n}\left(\frac{2M_ib_i}{F_sh} - \frac{htb_i}{3}\right)\right] \tag{5-9}$$

Because F_s is constant, (5-9) shows that for a girder with a chosen web the weight W is minimum when the absolute value of ΣM_ib_i approaches a minimum. If this criterion is applied to all spans of a continuous unit, it can be shown that the total ΣM_ib_i for all spans must be a minimum to yield the minimum girder weight.

Usually, a two-stage iterative procedure is required to minimize ΣM_ib_i. The first step is to determine the optimum locations of the cutoff points based on the number chosen. The resulting flange plate lengths are independent of the optimum web height, determined later by considering the design moment and shear.

In the first stage of the analysis, the support moments are determined by iteration for assumed lengths b_i. In the second stage, the values of moments are kept unchanged and the segment lengths are varied to determine the summed product Mb, that is, a satisfactory convergence is obtained. Values of b corresponding to the minimum ΣM_ib_i are obtained by a search process using the principles of dynamic programming (Taha, 1976). These values of b are then fed back into the first stage to recalculate the moments and this is followed by the second-stage computation to evaluate b again. This procedure is repeated until M values are compatible with the values of b. A generalized computer program is developed to obtain the values of the design moments and segment lengths. Azad (1980) shows the important steps of the main program in a flowchart.

Selection of Web A web choice is suggested based on the procedure outlined by Azad (1978). Using a modified section modulus $\bar{S}$ for the beam as a whole in place of that required for the maximum moment, we write

$$\bar{S} = Z/F_s\bar{L} \tag{5-10}$$

where Z is the summation of absolute values of ΣM_ib_i for all spans of the girder and $\bar{L}$ is the summed length of all spans. From the generalized form of (5-9) and setting $dW/dh = 0$, one obtains the web depth h for a chosen thickness t as

$$h = \sqrt{\frac{3\bar{S}}{t}} \tag{5-11}$$

provided, however, the depth–thickness ratio satisfies the criteria stipulated by AASHTO in Article 10.34.3. For direct reference Figure 5-12 plots the

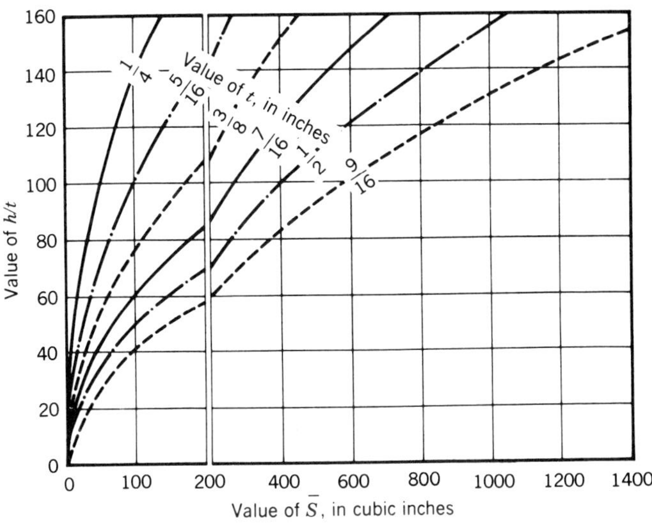

FIGURE 5-12 Optimum h/t for given $\bar{S}$ and t. (From Azad, 1978.)

ratio h/t versus the modified section modulus $\bar{S}$. Note that the upper limit of the ratio is 160 which satisfies AASHTO requirements. Note also that if the calculated compressive stress in the flange equals the allowable compressive stress, the web thickness (with or without stiffeners) should be increased according to AASHTO Article 10.34.3.1.2.

In the foregoing analysis the number and size of intermediate vertical stiffeners are not considered parameters, so that the results are valid for either option. A design example of this method is given by Azad (1980).

Optimum Girder Depth

Approximate expressions for obtaining the optimum girder depth are developed by Bresler, Lin, and Scalzi (1969), based on the required moment resistance and essentially without regard to other related elements. These are different from (5-4) in that they give the weight of the girder as a function of the depth. They include, however, various constants associated with the girder type (riveted or welded), the magnitude and shape of the bending moment diagram, the presence of stiffeners, gross or net cross-sectional area, and so on.

Three cases are distinguished: (a) when the shear is heavy and controls the design of the web; (b) when the web thickness is constant, such as the required minimum thickness; and (c) when the web plate is relatively thin, but must be increased in certain sections to satisfy minimum buckling requirements.

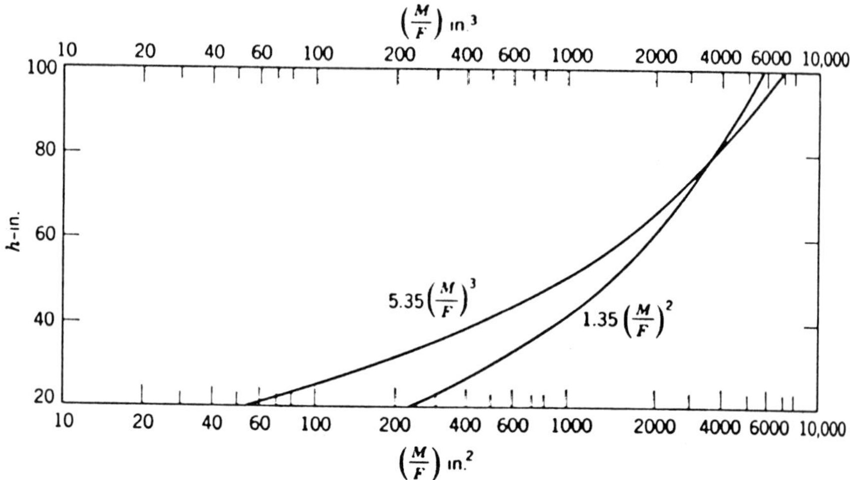

FIGURE 5-13 Graphs for obtaining optimum girder depth, typical girders. (From Bresler, Lin, and Scalzi, 1968.)

For a typical girder, the web height h is expressed as a function of the ratio $M/F_s t$ or the ratio M/F_s where M is the bending moment, F_s is the allowable steel stress at the extreme fiber, and t is the web thickness. The two expressions for the depth h are plotted in Figure 5-13. The graphs may be used for preliminary web selection noting that there is only a slight difference between the optimum depths of riveted and welded girders.

Additional Requirements In addition to minimum cost or weight considerations, a plate girder must also be selected to accommodate the following functions and service requirements: (a) vertical stiffness to satisfy applicable deflection limitations; (b) lateral stiffness to prevent lateral–torsional buckling of the compression flange; and (c) stiffness to improve buckling or postbuckling of the web plate. The design must also check the assumption that the minimum cost is equivalent to the minimum weight.

5-7 LONGITUDINALLY STIFFENED PLATE GIRDERS

For plate girders provided with longitudinal stiffeners, the web plate thickness is one-half of the same plate girder without longitudinal stiffeners. It appears therefore that longitudinal stiffeners, as shown in Figure 5-14, can increase the bending and shear strength of a plate girder, although they are not as effective as transverse stiffeners. Also called stability stiffeners, they are frequently specified in highway bridges for esthetic reasons. The

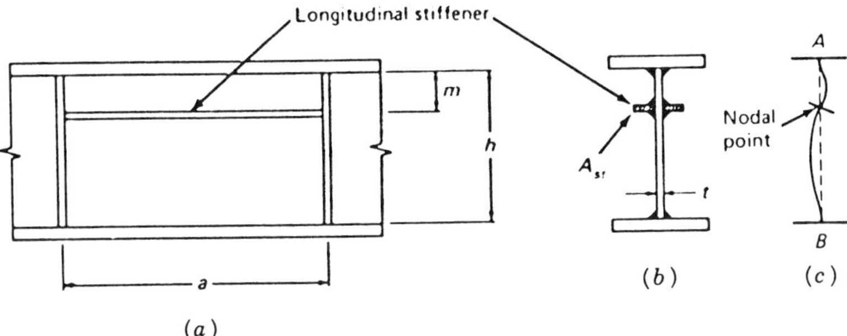

FIGURE 5-14 Effect of longitudinal stiffener on plate girder web stability: (*a*) side view; (*b*) cross section; (*c*) buckled shape.

ASCE–AASHTO Task Committee (1978) provides a full review of the theory and design of longitudinally stiffened plate girders.

The primary function of longitudinal stiffeners is to control lateral web deflections, as shown in Figure 5-14*c*, and hence the bend-buckling strength. Their design involves two requirements: (a) a moment of inertia to ensure adequate stiffness to manifest a nodal line along the stiffener plate and (b) an area that can share the axial compression while acting integrally with the web.

Strength of Plate Girder in Pure Bending

This characteristic is controlled by the instability of the compression flange or by the yielding of one or both of the girder flanges. Previous work on this topic includes Basler and Thurlimann (1963), Hoglund (1973), and Fujii (1968a, 1968b). The strength of unsymmetrical plate girders has been studied by Chern and Ostapenko (1970). The effect of longitudinal stiffeners on the bending strength has been investigated by Rockey and Leggett (1962) and Cooper (1965). The effect of longitudinal stiffeners on the fatigue strength has been articulated in tests conducted by Yen and Mueller (1966).

Girders with Transverse but Without Longitudinal Stiffeners In one of the models, the bending strength is determined from the compression flange and a portion of the web as a column subjected to lateral, torsional, and vertical buckling (Basler and Thurlimann, 1963). This concept has been refined by the strength models presented by Fujii (1968a, 1968b) and Hoglund (1973). Chern and Ostapenko (1970) have presented a model that includes unsymmetrical and hybrid girders. The same physical behavior is assumed, but the ultimate bending moment is expressed as the sum of the moment prior to and after flexural buckling of the web, written as

$$M_u = M_b + M_t \tag{5-12}$$

where M_b is the moment causing flexural buckling of the web and M_t is the postbuckling moment. In (5-12), the prebuckling moment is calculated from the elastic buckling stress associated with flexural buckling of the web. The postbuckling moment is the additional moment that would cause instability of the compression flange area after flexural buckling occurred.

Girders with Transverse and Longitudinal Stiffeners The additional strength due to longitudinal stiffeners is attributed to the control of the lateral deflection of the web as shown in Figure 5-14. This increases the flexural stress carried by the web and also improves the bending resistance of the flange because of greater web restraint. Rockey and Leggett (1962) have determined that the optimum location for a longitudinal stiffener is 0.22 times the web depth from the compression flange, assuming that the web is fixed at the flanges and simply supported at the transverse stiffeners. The same location is 0.20 times the web depth if the web is assumed simply supported at all four edges.

The presence of longitudinal stiffeners increases the maximum allowable slenderness ratio for the web, required to develop the yield stress moment capacity of the girder without buckling, from 170 to about 400 for mild steel.

The increase in bending strength of a plate girder with a longitudinal stiffener can be predicted as suggested by Cooper (1965). This procedure is verified by tests. The range of increased strength is from 14 to 26 percent, and is due principally to the increased resistance to out-of-plane buckling of the web. This behavior inhibits premature vertical buckling of the compression flange. In addition, the presence of longitudinal stiffeners improves the fatigue resistance.

Shear Strength of Plate Girders

Girders with Transverse but Without Longitudinal Stiffeners The ultimate shear capacity of a plate girder web is the sum of two parts, or

$$V_u = V_{\mathrm{cr}} + V_p \tag{5-13}$$

where V_{cr} is the load corresponding to the elastic buckling of the web panel, also referred to as the beam action strength, and V_p is the postbuckling strength due to the action of a diagonal tension field that forms after the web panel buckles (Basler, 1961; Skaloud, 1971). Because of this action, a plate girder can develop an ultimate shear strength that is several times the buckling strength.

All of the theoretical models proposed for predicting the shear strength include the beam action contribution. For the stiffened web panel shown in Figure 5-15 and subjected to a uniform shear, the stresses will be resisted by the "beam action" until the panel buckles.

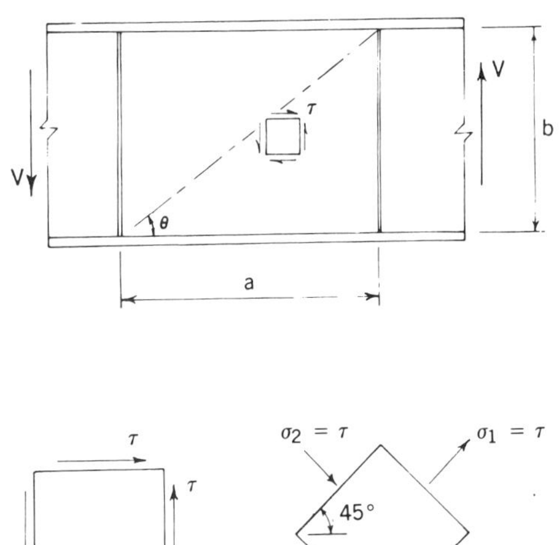

FIGURE 5-15 Beam action shear stress in plate girder web.

According to an approach by Rockey and Skaloud (1972), the postbuckling strength of a symmetrical girder is the result of the diagonal tension field stress that is influenced by the stiffness of the flanges. Other researchers have developed strength models that also recognize the effect of the flanges on the postbuckling shear strength. Fujii (1968a, 1968b) assumes internal plastic hinges in the flanges in midpanel positions. Komatsu (1971), Calladine (1973), and Hoglund (1973) introduce a failure mechanism similar to that proposed by Rockey and Skaloud (1972) with interior plastic hinges in the flanges.

Girders with Transverse and Longitudinal Stiffeners The effect of longitudinal stiffeners is to increase the beam action shear strength because the web panel size is reduced. With one longitudinal stiffener, the optimum position or shear effect is at the middepth making the subpanels of equal size and shear resistance. With the stiffener at a different location, the larger panel will buckle first. The beam action shear strength is

$$V_{\text{cr}} = V_{\text{cr}_1} + V_{\text{cr}_2} + \cdots \tag{5-14}$$

where V_{cr_1} denotes the critical shear strength of subpanel 1, and so on.

The postbuckling shear strength of longitudinally stiffened girders is evaluated in two ways. Cooper (1965) assumes that each subpanel develops its own tension field after buckling. Porter, Rockey, and Evans (1975) assume that only one tension field is developed between the flanges and transverse stiffeners even if longitudinal stiffeners are present.

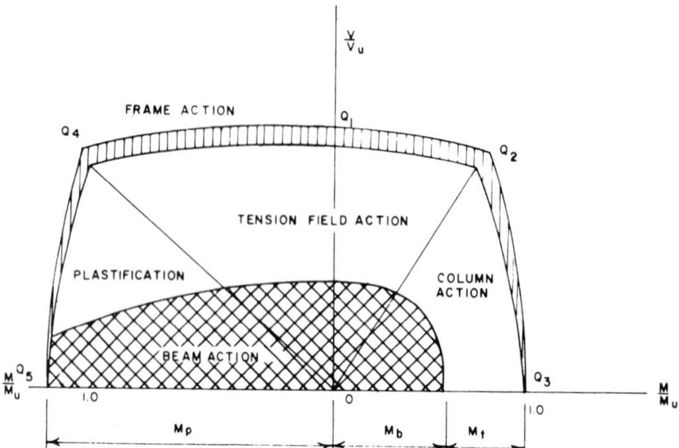

FIGURE 5-16 Interaction curve of general nonsymmetric longitudinally stiffened girder.

Rockey and Leggett (1962) have suggested a model after observing from tests that an overall tension field develops in the web. The same approach is used to obtain the tension field shear resistance of a longitudinally stiffened girder as for an unstiffened girder. However, in computing the tension field stress associated with failure, the critical beam action shear corresponding to buckling of the largest subpanel is used in the yield criterion.

Combined Bending and Shear

Girders with Transverse but Without Longitudinal Stiffeners One approach (Chern and Ostapenko, 1969) assumes that the ultimate capacity is dictated by web failure, instability of the compression flange, or yielding of the tension flange. The complete interaction behavior is shown schematically in Figure 5-16. Curve $Q_2Q_1Q_4$ represents failure of the web, curve Q_2Q_3 represents buckling of the compression flange, and curve Q_4Q_5 represents yielding of the tension flange. The ultimate shear under combined stress is developed based on the region that controls failure.

Rockey, Evans, and Porter (1974) postulate a model that includes three additional factors not included in the pure bending or pure shear models. These factors are: (a) the reduction in the shear buckling stress of the web due to the presence of bending stresses; (b) the influence of in-plane bending stresses on the value of the diagonal tension field stress at failure; and (c) the reduction in the magnitude of the plastic modulus of the flanges resulting from the axial compressive and tensile strength.

After a panel has buckled, the tension field shear is computed by using a postbuckling shear force modified to include the effect of combined bending and shear. The tension field stress is modified to include both bending and shear stresses in the yield criterion, and an interaction equation is used to

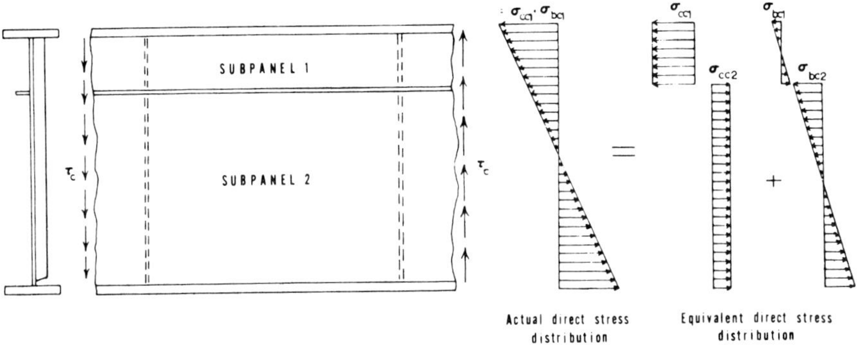

FIGURE 5-17 Stress distribution in panels of plate girder under shear and bending.

calculate the flange plastic moments when axial stress resulting from bending acts on the flanges.

Transversely and Longitudinally Stiffened Girders Most investigators treat the shear strength under combined shear and bending as in the case of pure shear. However, modifications are necessary for the buckling stresses, tension field stresses, and plastic moments in the flange to account for the combined stress effect. Other investigators include the contribution from the longitudinal stiffener in computing the frame action strength, and modify the plastic moment capacity of the stiffener to account for the combined stress effects.

Computation of the buckling stress in the subpanels involves an interaction approach applied to each subpanel stress condition as shown in Figure 5-17. The state of stress in a subpanel is a shear stress τ_c, a pure bending stress σ_{bc}, and an axial stress σ_{cc}. The buckling condition is

$$\frac{\sigma_{cc}}{\sigma_{crc}} + \left(\frac{\sigma_{bc}}{\sigma_{cp}}\right)^2 + \left(\frac{\tau_c}{\tau_{\mathrm{cr}}}\right)^2 = 1.0 \qquad (5\text{-}15)$$

where σ_{crc} = beam action combined stress

σ_{cp} = web critical buckling stress in pure bending

τ_{cr} = critical shear stress

The postbuckling strength of the panel is calculated as in the case without longitudinal stiffeners.

Experimental Data

A considerable record is available from tests on transversely stiffened girders with and without longitudinal stiffeners (Johnston, 1976). It appears that the foregoing theories for predicting the strength of plate girders in bending show good agreement with the experimental evidence, and there is good correlation between theory and experimental results.

Cooper's theory gives conservative estimates of the shear strength consistent with the assumptions. The Chern–Ostapenko and Rockey theories supplement the experimental data and are in good agreement with these tests.

Design Approach

Current bridge design practice is to locate stiffeners one-fifth of the web depth from the compression flange to force a linear bending stress distribution in the girder cross section by controlling flexural buckling. Stiffeners are often used at interior supports for high shear. In this case the most effective location is between one-fifth and one-half of the girder depth, but no guidelines are available for selecting the optimum location. Most frequently, longitudinal stiffeners are placed on one side of the web for economy.

The stiffeners must be stiff enough to force a node in a buckled web, and must also be capable of supporting axial compression resulting from bending and from the diagonal tension field in regions of high shear. The AASHTO design formulas in Article 10.34.5 are based on the requirement that the stiffener should provide a nodal point in the buckled web by resisting local and overall stiffener buckling. These formulas are derived, however, from linear buckling analysis; hence, they ensure resisting capacity up to web buckling. In order to develop the ultimate strength of girders and longitudinally stiffened webs, the stiffness of the stiffener should be increased beyond the elastic web-buckling value.

The primary reason for using longitudinal stiffeners is economy associated with the use of a thinner web, but unless the girder depth exceeds the 6- to 7-ft range, this economy is not conclusive. Because the use of stiffeners increases fabrication costs, many designers tend to avoid them. These stiffeners do not necessarily have to be continuous, but they must be cut where they intersect transverse intermediate stiffeners if they lie on the same side of the web. This means that they must be cut into short lengths and then inserted between the transverse stiffeners, resulting in increased welding time and production cost.

5-8 DESIGN EXAMPLE: OPTIMUM BRIDGE PROPORTIONING

Summary

This example presents a summary and sample calculations from the structural investigation and study of the bridge carrying Route 1A, known as the Salem–Beverly Connector, in Massachusetts. In the final configuration, this structure is divided into three main sections (Xanthakos, 1970).

1. *Off-Shore Section* Selected bridge spans have an average length of 165 ft, 12-ft clearance to the mean water level, and poor foundation conditions necessitating expensive substructure. Piers are normal to the roadway.

2. *Land Section* The structure has average span lengths of 110 ft and is elevated to provide clearance at local street and railroad crossings. The piers are placed normal to the roadway except at crossing where some severe skews require complex framing.

3. *Beverly Harbor Section* Average span lengths are 190 ft, and the structure is elevated to provide 45-ft clearance to the high water level. The average water depth is 20 ft. The piers are set normal to the roadway, and the foundations are extended to rock. The total bridge length is approximately 5500 ft.

Superstructure It was the intent of the design to provide equal span lengths in continuous units with cantilevers as shown in Figures 5-18*a* and *b*. This concept was preferred by the client because it yields orderly and pleasing proportions, but was also found to be structurally feasible and optimum for the spans involved and more particularly adaptable to the two-girder system. An economic range of the center–end span ratio was established between 1.15 and 1.20. Setting $x = 20$ and $L = 130$ in Figure 5-18 gives an end span of $130 - 20 = 110$ ft, and a span ratio of $130/110 = 1.18$. Left of pier 1, the cantilever action introduces a negative moment at the support and balances the moments in span 1–2.

A typical slab thickness of 8 in. was assumed in all schemes, and therefore the design utilized the maximum possible beam or girder spacing compatible with 8-in. slab thickness.

The structural investigation covers four schemes. A fifth option, the two-girder systems with cantilevered floor beams, was rejected for esthetic reasons.

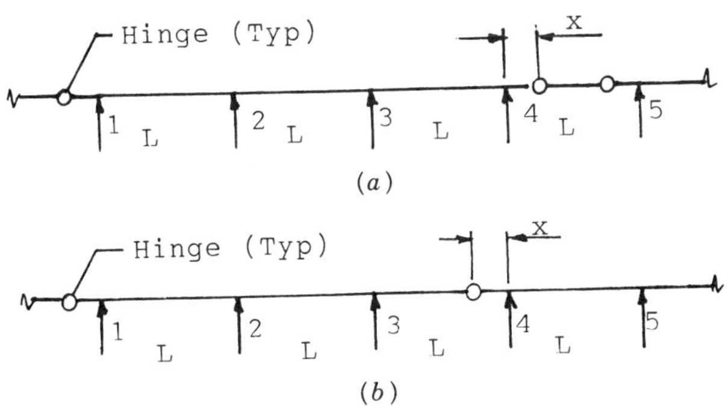

FIGURE 5-18 Continuity configurations and span lengths, Route 1A, Salem–Beverly Connector.

The main bridge has two separate roadways, each carrying four traffic lanes. Along the approach section, each roadway has three traffic lanes.

Scheme 1: Two-Girder System with Floor Beams and Stringers This is essentially similar to the type shown in Figure 5-1*a* except the steel grid is replaced by a concrete slab. Plate girders were investigated for span lengths ranging from 110 to 210 ft with 20-ft intervals. The floor beam spacing considered 16-, 20-, and 24-ft intervals, with the 16-ft spacing found to be the most economical. Because floor beam cantilevers or outside brackets were considered objectionable due to their appearance, the main girders were spaced to give a maximum slab overhang not exceeding 3 ft. This layout requires exceptionally heavy floor beams for bridges with wide decks. Hence, the advantages of this scheme must be established with reference to the roadway width. It appears to be an economical choice for spans as short as 120 ft and for normal roadway widths (in the region of 24 ft), and may be economical for span lengths exceeding 150 ft regardless of the roadway width.

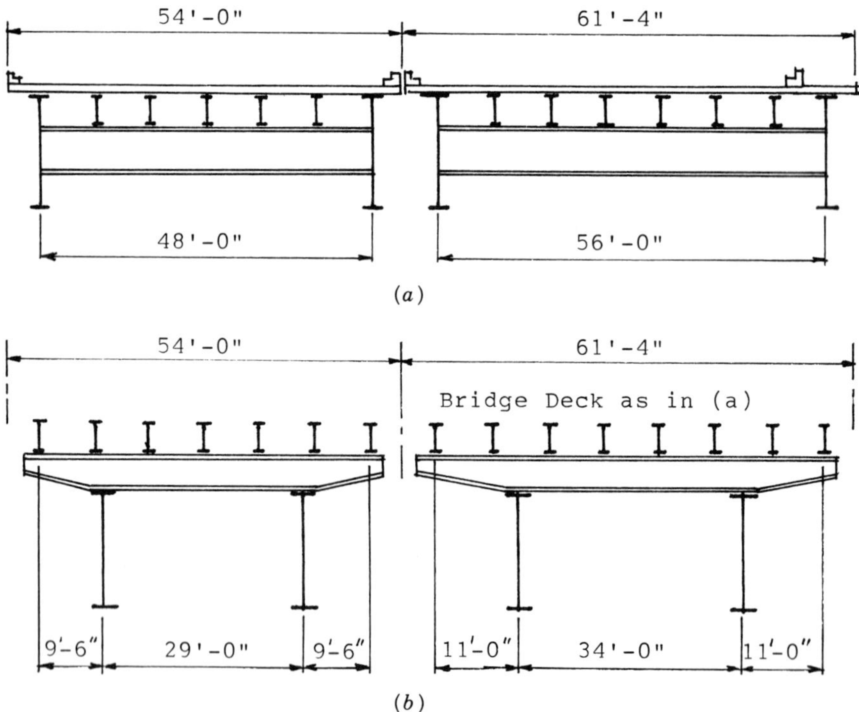

FIGURE 5-19 Typical deck section for main bridge, scheme 1: two-girder system with floor beams and stringers, Salem–Beverly Connector: (*a*) conventional floor beams; (*b*) cantilevered floor beams.

Because the logical substructure choice is single columns for each girder, reduction in pier length is not involved in variable roadway width. The main advantage over the multigirder system is the actual live load distribution.

For spans exceeding 125 ft, the weight of the bottom bracing was calculated based on actual designs, and a percentage was applied to the girder weight for miscellaneous steel (stiffeners, connection plates, etc.). For spans less than 125 ft, a 10 percent increase was applied to the girder weight for miscellaneous steel. Haunches over the interior supports were considered for spans 190 ft and longer. This scheme is shown in Figure 5-19.

Scheme 2: Multiple-Girder System Typical deck sections for the main bridge and the approach roadway are shown in Figure 5-20. Welded plate girders were investigated for composite and noncomposite construction, with the former yielding definite weight saving. This advantage is maximized in the span range of 90 to 140 ft. For example, for 130-ft spans in the main bridge (15 girders), composite design yields a weight saving of 450 lb per linear foot of deck, but this is somewhat reduced by the cost of the shear connectors. For spans exceeding 170 ft, the cost saving is only 5 percent, offset by the cost of the shear devices. Hence, beyond this span range the effectiveness of composite construction is questionable. The explanation relates to the smaller live load moments in comparison to dead weight moments for larger spans. For example, for 90-ft spans the positive live load moment is more than twice

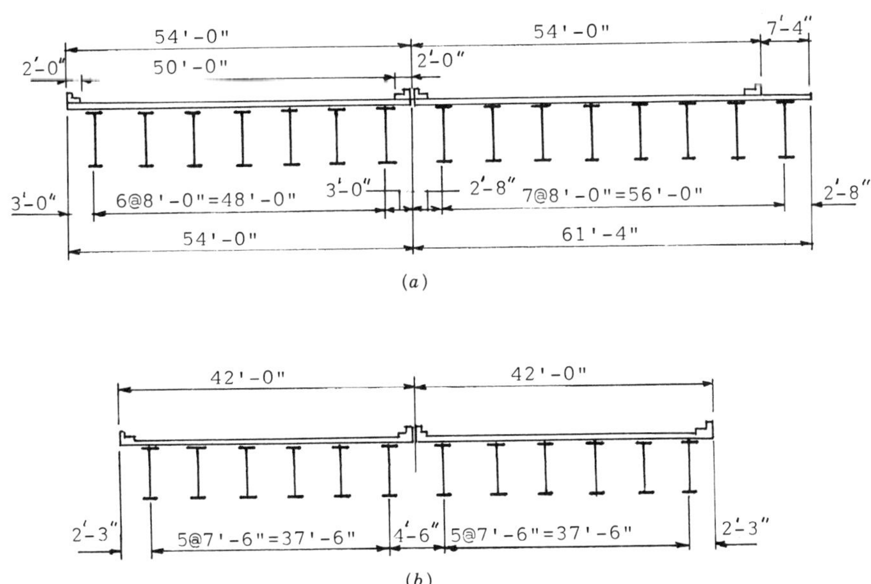

FIGURE 5-20 Typical deck section, scheme 2: multiple-girder system, Salem–Beverly Connector: (*a*) main bridge; (*b*) approach roadway.

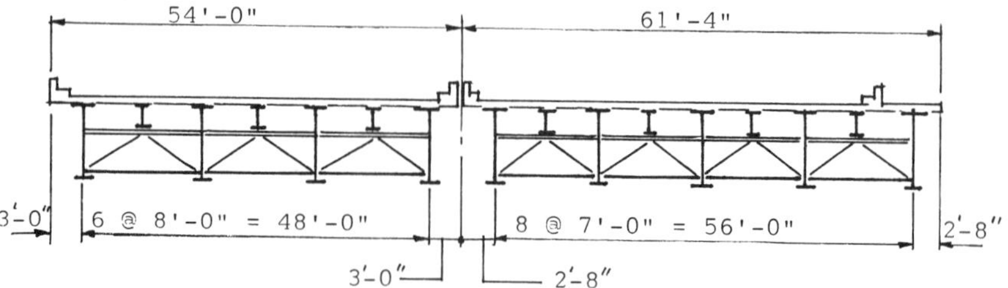

FIGURE 5-21 Typical deck section, scheme 3: combined girders and stringers, Salem–Beverly Connector.

the dead load moment, but for 210-ft spans the positive live load moment is about 0.8 the dead load moment based on HS 20 loading.

Scheme 3: Girder and Stringer System This scheme, shown in Figure 5-21, is a derivative of schemes 1 and 2. The main advantage of this system is the utilization of the lateral cross frames as main supporting members at no considerable increase in their weight. Further economy is achieved if the exterior girders are designed for the actual dead and live load rather than using the design requirements of the interior girders. In order to make this system compatible with the two-girder system, especially in the longer span range, the cross frames and the bottom bracing were designed and their weight calculated accordingly. In addition, a nominal percentage was applied to the girder weight for miscellaneous steel.

This scheme is particularly suitable where an even number of girder–stringer panels can be arranged in multiples of 8-ft spacing, thus allowing maximum use of the 8-in. slab. Other schemes are possible in connection with fascia dimensions greater than 3 ft. The weight of this scheme is similar to scheme 1 for spans of 170 ft and greater. Scheme 3 yields, however, less steel weight for spans less than 170 ft.

Scheme 4: Box Girders These were chosen to maintain compatibility with the multigirder system and the 8-in. slab. A typical cross section for this alternative is shown in Figure 5-22. The top-flange spacing is 7 or 8 ft. The principal advantage in cost analysis is the improved live load distribution ratio in conjunction with the high torsional stiffness of the member. Assuming lane widths of 12 ft and assuming the number of box girders is the same as the number of lanes,the live load resisted by each box girder is approximately one truck (two wheel loads) according to AASHTO Article 10.39.2. Each girder in the multigirder system was designed for 1.45 wheels or 0.73 truck, and because two girders in the multigirder system correspond to one box girder, the potential reduction in the design live load is $1.00/1.45 = 0.69$. Hence, the box girders may be designed for 69 percent of the live load applied to the multigirder system.

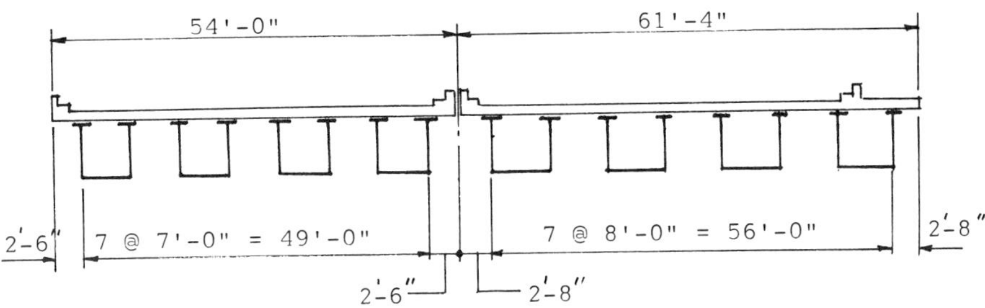

FIGURE 5-22 Typical deck section, scheme 4: multiple box girders, Salem–Beverly Connector.

This advantage is maximized in the 110- to 130-ft span range where the live load moments approach the highest percentage of the total moment, and is diminished in spans greater than 170 ft. For rectangular box girders, a beam spacing of 7 or 8 ft, and a minimum plate thickness of 3/8 in., the design is optimized in spans greater than 130 ft. Below this range, selecting a bottom flange plate 84 in. by 3/8 in. or 96 in. by 3/8 in. is a mere compliance with minimum plate thickness requirements.

Comments on Superstructure Schemes Schemes 1 and 3, the preferred schemes, essentially reflect the concept of a two-way structural system, where supporting members are combined in the longitudinal and transverse direction. The two schemes are compatible throughout the span range investigated for this bridge. A significant parameter influencing cost, and therefore choice, is the width of the deck slab because this factor determines the floor beam requirements. Consider, for example, the bridge system shown in Figure 5-23. The structure shown in Figure 5-23*a* is a relatively narrow deck (probably two lanes), and the two-girder system with floor beams and stringers may be the design choice. As the deck width increases, this solution becomes uneconomical, and the scheme shown in Figure 5-23*b* becomes the choice. As the deck width continues to increase, the scheme shown in Figure 5-23*b* approaches cost limitations, and the scheme shown in Figure 5-23*c* prevails. This articulation may be valid regardless of the span length, because the weight of stringers, floor beams, and cross frames is essentially independent of this factor. Hence, the range over which each system will be favored is not generalized with reference to the span lengths, and the most pertinent parameter is essentially the roadway width.

Substructure Because the objective of this investigation was to select suitable superstructure schemes and proportion the span lengths in relation to the substructure and foundation costs, the substructure elements were studied within the same analytical framework.

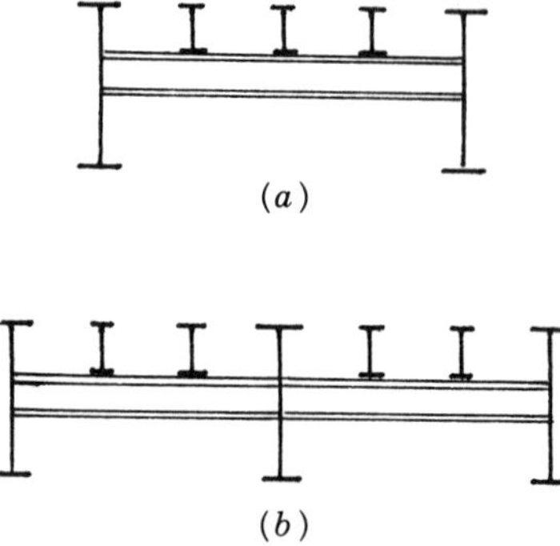

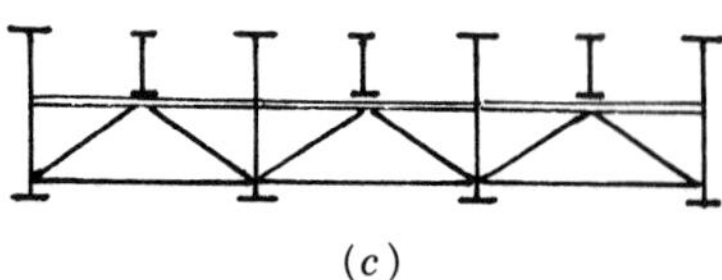

FIGURE 5-23 Typical bridge systems: (*a*) two-girder bridge with floor beams and stringers; (*b*) three-girder bridge with floor beams and stringers; (*c*) four-girder system with cross frames and stringers.

For scheme 1 the piers consist of individual columns without pier caps. For the remaining schemes the pier type consists of two single hammerheads. Because the rock profile is rather erratic, the average foundation conditions were approximated.

Piers for scheme 1 require less concrete above the water level, but more concrete below this level because they are longer. Piers for the other schemes are more massive above the water but the 12-ft overhangs in the hammerheads reduce the pier length below the water level. The difference in the pier costs between schemes 1 and 3 is about 1.5 percent and, therefore, inconsequential.

Sample Calculations (Allowable Stress Design)

Concrete Data The concrete is class D concrete with $f_c = 1200$ psi and $n = 10$. The stipulated 8-in.-thick concrete slab is sufficient for 9-ft girder spacing and HS 20 live load (the assumed flange width is 16 in.).

Steel Data Two grades of structural steel are specified, A588 and A441. A588 (current AASHTO M 270 Grade 50) is used in load-carrying main members because it is readily weldable by various processes.

The live load distribution is based on conventional AASHTO methods. For example, in parallel beams and stringers the distribution factor is $S/5.5$ and independent of the span length. For the 8-ft girder spacing shown in Figure 5-20a, the wheel load fraction is $8/5.5 = 1.45$. Using $L = 110$ and 210 ft (the span range investigated for most schemes), we compute S/L as 0.073 and 0.038, respectively. According to Figure 2-41, the distribution

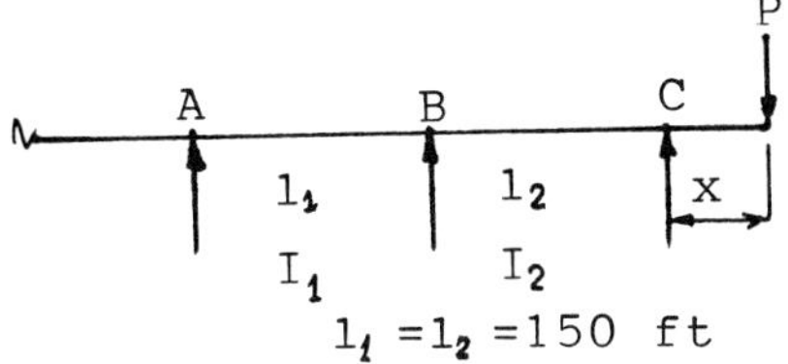

FIGURE 5-24 Live load deflection in cantilevered portion of girder.

factor varies from 1.10 to 1.25. The upper value corresponds to the 110-ft span, and the lower value to the 210-ft span. It is conceivable that these values may change the upper optimum limits of each scheme, but not appreciably.

Length of Cantilever For a roadway width of 50 ft and using four traffic lanes and seven girders, the truck load per girder is $4 \times 0.75/7 = 0.43$ truck. If the suspended span is assumed to have a length of 115 ft, the resulting live load reaction is 66 kips (truck, one lane, simple span) giving a live load plus impact reaction of $0.43 \times 66 \times 1.3 = 37$ kips/girder.

The live load deflection of the cantilever is calculated referring to Figure 5-24. Selecting $x = 18$ ft and assuming I (of cantilever) $= 50{,}000$ in.^{4}, the deflection is the algebraic summation of two deflections Δ_1 and Δ_2 due to the cantilever action and beam continuity at support B, respectively. Next, we calculate

$$\Delta_1 = \frac{37{,}000 \times 18^2 \times 12^2 \times 168 \times 12}{3 \times 29 \times 10^6 \times 60{,}000} = 0.67 \text{ in.}$$

In order to determine the deflection Δ_2 due to the beam continuity at support B, we calculate $M_B = M(2k+2)$, where $k = I_2 l_1 / I_1 l_2$. Assume $I_1 = I_2 = 100{,}000$ and $l_1 = l_2$. Then $k = 1$ and $M_B = M/4$. The deflection is now computed as

$$\Delta_2 = \frac{M_B l_2}{6EI} = \frac{37{,}000 \times 18^2 \times 12^2 \times 150 \times 12}{4 \times 6 \times 29 \times 10^6 \times 100{,}000} = -0.04 \text{ in.}$$

The net deflection is $0.67 - 0.04 = 0.63$ in. The allowable deflection is $L/300 = 0.72$ in.

Multiple-Girder System A typical girder unit is shown in Figure 5-18*a*, and consists of a three-span continuous section with two cantilevers and a suspended simple span. The length unit is therefore $4L$. Relevant data for the entire series of 90 to 210 ft are summarized in Table 5-1 for the two roadways (15 girders) of the bridge section and for composite and noncomposite design.

TABLE 5-1 Summary of Girder Data, Multiple-Girder Scheme

		Composite				Noncomposite			
Span Length (ft)	One Unit Length (ft)	Girder Depth Used	Girder Weight /Linear Foot	×1.10	Weight/Steel Linear Foot Deck	Girder Depth Used	Girder Weight /Linear Foot	×1.10	Weight/Steel Linear Foot Deck
90	360	48 in.× 3/8 in.	146#	161#	2415#	48 in.× 3/8 in.	164#	180#	2700#
110	440	54 in.× 3/8 in.	172#	189#	2835#	54 in.× 3/8 in.	195#	215#	3225#
130	520	60 in.× 3/8 in.	203#	223#	3345#	50 in.× 3/8 in.	231#	254#	3810#
150	600	72 in.× 7/16 in.	249#	274#	4110#	72 in.× 7/16 in.	271#	298#	4470#
170	680	78 in.× 1/2 in.	293#	322#	4830#	78 in.× 1/2 in.	313#	344#	5160#
190	760	84 in.× 9/16 in.	352#	387#	5805#	84 in.× 9/16 in.	377#	414#	6210#
210	840	84 in.× 9/16 in.	397#	437#	6555#	84 in.× 9/16 in.	426#	468#	7020#

Two loadways = 15 girders
10 percent for miscellaneous steel

Two-Girder System The floor beam spacing and size of the stringers are interrelated. Stringer size, floor beam size, and floor beam spacing are determined from a simultaneous analysis. Stringers are usually made continuous over several spans subject to shipping and erection limitations. A feasible stringer length may accommodate four bays. The ends are usually attached to the floor beams introducing a partial restraint between hinged and fixed conditions.

Consider the three-span continuous unit shown in Figure 5-25, where the end supports are assumed simple for convenience. The dead load moments are

$$\text{Span } AB \qquad M_{\mathrm{DL}} = 0.08wL^2$$

$$\text{Support } B \qquad M_{\mathrm{DL}} = -0.10wL^2$$

The dead load is as follows:

$$\begin{aligned} \text{Slab} = 8.00 \times 0.667 \times 0.15 &= 0.800 \\ \text{Fillets} &= 0.040 \\ \text{Beam (assumed)} &= \underline{0.060} \\ \text{Total } w &= 0.900 \text{ kip/ft} \end{aligned}$$

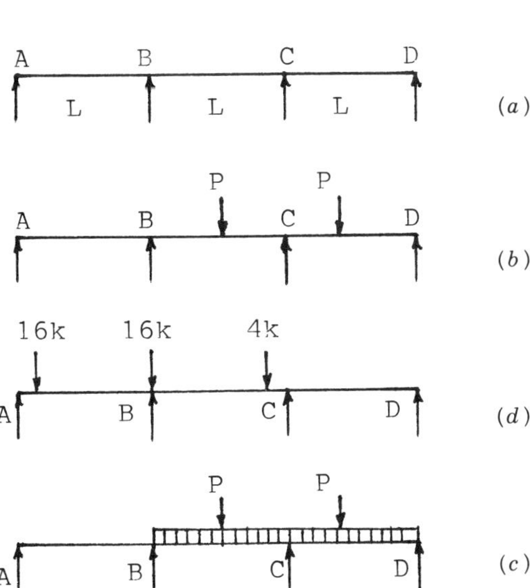

FIGURE 5-25 (*a*) Stringer elevation; (*b*) position of truck load for maximum negative moment; (*c*) position of lane loading for maximum negative moment; (*d*) position of truck load for maximum reaction.

Using $L = 20$ ft, we compute

$$\text{Span } AB \qquad M_{\text{DL}} = 28.8 \text{ ft-kips}$$
$$\text{Support } B \qquad M_{\text{DL}} = -36.0 \text{ ft-kips}$$

The negative live load moment at the interior support may be caused by the truck or lane loading placed as shown in Figures 5-25*b* and *c*, respectively. for $S/L = 8/20 = 0.4$, the distribution factor from Figure 2-41 is found graphically as 1.6. Using, however, the value 1.45, we obtain the following wheel loads: $1.45 \times 16 = 23.2$ kips (truck), and $1.45 \times 18 \times 0.5 = 13.1$ kips, $0.64 \times 0.5 \times 1.45 = 0.46$ kip/ft (lane).

From Figure 5-25*b*,

$$M_{\text{LL}} = -23.2 \times 20 \times (0.102 + 0.080) = -84.4 \text{ ft-kips} \qquad \text{(Truck)}$$

and from Figure 5-25*c*,

$$M_{\text{LL}} = -13.1 \times 20 \times 0.182 - 0.1167 \times 0.46 \times 20^2$$
$$= -69.2 \text{ ft-kips} \qquad \text{(Lane)}$$

The maximum negative live load moment plus impact, at support B, is $M_{\text{LL}+I} = -109.7$ ft-kips. The total moment at support B is $-(36.0 + 109.7) = -145.7$ ft-kips. For A36 steel, the required section modulus is $145.7 \times 0.6 = 87.4$ in.^{3}. Use W18 × 50, with a section modulus of 89.1 in.^{3}.

Using $L = 16$ ft, we compute

$$M_{\text{DL}} = -23.0 \text{ ft-kips} \qquad \text{(Support } B\text{)}$$

and

$$M_{\text{LL}+I} = -23.2 \times 16 \times (0.102 + 0.080) \times 1.3 = -87.8 \text{ ft-kips}$$

The total moment at support B is -110.8 ft-kips. The required section modulus is $110.8 \times 0.6 = 66.5$ in.^{3}. Use W18 × 40, with a section modulus of 68.4 in.^{3}.

Floor Beams Note that the floor beams are designed after the stringer configuration and the number of spans in a continuous unit are analyzed. Both the dead and live load reactions at floor beams A and B (Figure 5-25) are unequal, but for expedience all floor beams are made the same. Note also that the floor beams should not be designed for maximum live load reaction of stringers occurring simultaneously, because this condition is unattainable. The floor beam of interior supports B and C (Figure 5-25) controls the design.

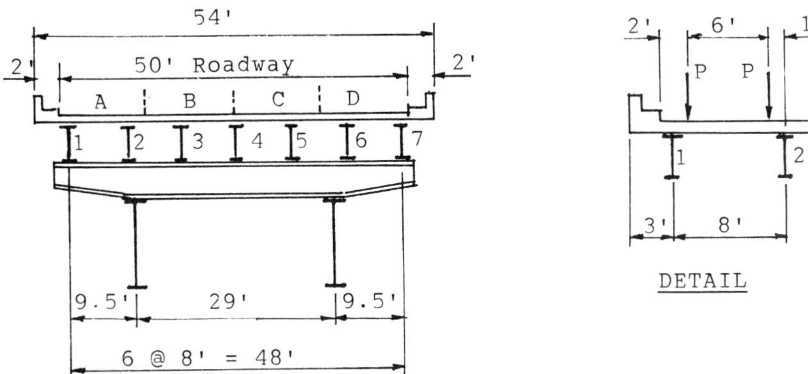

FIGURE 5-26 Typical floor beam elevation, cantilever scheme of two-girder system, roadway without sidewalk.

Using $L = 20$ ft, the dead load reaction at B is $0.9 \times 20 \times 1.1 = 19.8$ kips (interior girders). The dead load reaction for exterior stringers is 18.2 kips. The maximum live load reaction is obtained by placing the truck as shown in Figure 5-25*d*, and its direction is determined from a consideration of the influence line coefficients. The reaction at B is $P(0.46 + 1.00) + (P/4)(0.32) = 1.54P$, which is the live load contribution from all three axles.

A typical floor beam elevation for the two-girder system is shown in Figure 5-26, and articulates the cantilever effect in equalizing negative and positive moments. For maximum cantilever moment we load only one lane as shown in detail. Assuming simple beam action of the slab, the resulting reactions for stringers 1 and 2 are $R_1 = R_2 - P$. It follows therefore that the reaction at support B is $1.54 \times 16 = 24.6$ kips for both stringers 1 and 2, giving a cantilever moment (including impact) of

$$M_{LL+I} = -(24.6 \times 9.5 + 24.6 \times 1.5) \times 1.3 = -352 \text{ ft-kips}$$

Likewise, we compute the dead load moment as

$$M_{DL} = 18.2 \times 9.5 + 19.8 \times 1.5 = -203 \text{ ft-kips}$$

giving a total (negative) cantilever moment of $M = -555$ ft-kips.

The design must also check the positive moment in the span. From symmetry and referring to Figure 5-27*a*, the dead load reactions are $R_1 = R_2 = 67.7$ kips. At the center (beam line 4), the dead load moment is

$$M_{DL} = 67.7 \times 14.5 - 18.2 \times 24 - 19.8 \times 16 - 19.8 \times 8 = 70 \text{ ft-kips}$$

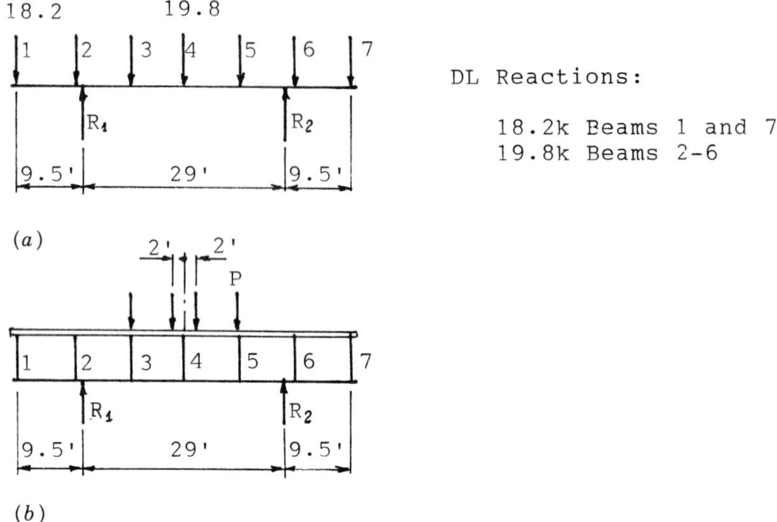

FIGURE 5-27 Loads for positive moment in span: (*a*) dead load; (*b*) live load.

Assuming simple beam action for the slab and referring to Figure 5-27*b* where the two center lanes are loaded as shown, we calculate the reaction for stringers 3, 4, and 5:

$$S_3 = S_5 = 16 + 16 \times 2/8 = 20 \text{ kips} \qquad S_4 = 24 \text{ kips}$$

Along the floor beam line these reactions are multiplied by 1.54, giving

$$S_3 = S_5 = 30.8 \text{ kips} \qquad \text{and} \qquad S_4 = 37 \text{ kips}$$

From symmetry, $R_1 = R_2 = 49.3$ kips. The positive live load plus impact moment in the span is therefore

$$M_{\text{LL}+I} = (49.3 \times 14.5 - 30.8 \times 8) \times 1.3 = 610 \text{ ft-kips}$$

giving a total positive moment of $M = 680$ ft-kips. The required section modulus is $680 \times 0.6 = 408$ in.^{3}. Use W36 × 135, with a section modulus of 440 in.^{3}.

For an infinite number of spans, each three-span continuous stringer unit is 60 ft long and has seven stringers and three floor beams. The steel weight is therefore

$$7 \times 60 \times 50 + 3 \times 50 \times 135 = 21{,}000 + 20{,}250 = 41{,}250 \text{ lb}$$

This gives a steel weight per linear foot of deck of 41,250/60 = 687 lb.

The floor beam spacing is optimized by repeating the foregoing procedure using L = 16 ft, 24 ft, and so on, and then calculating the steel weight per unit length of deck. For this example, the analysis indicated the 16-ft spacing as the most economical in terms of net steel weight.

Steel Weight and Cost Comparison: All Schemes The gross weight of steel per linear foot of deck for the two roadways is shown in Figure 5-28. The multiple-girder system consists of 12 girders, because this summary is for

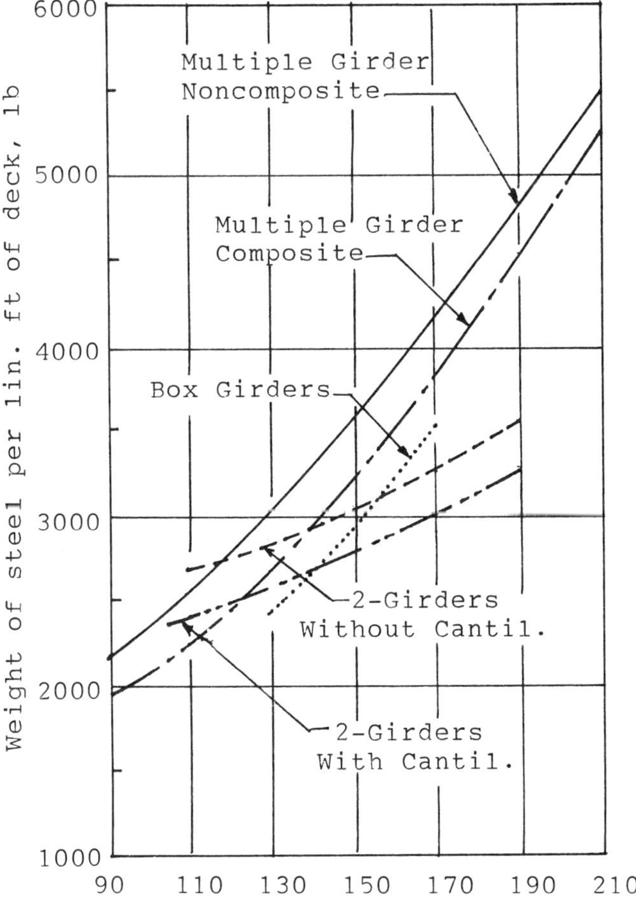

FIGURE 5-28 Gross weight of steel per linear foot of deck, two roadways; multiple-girder system: 12 girders; two-girder system: 4 girders; box girder system: 6 girders. Approach section.

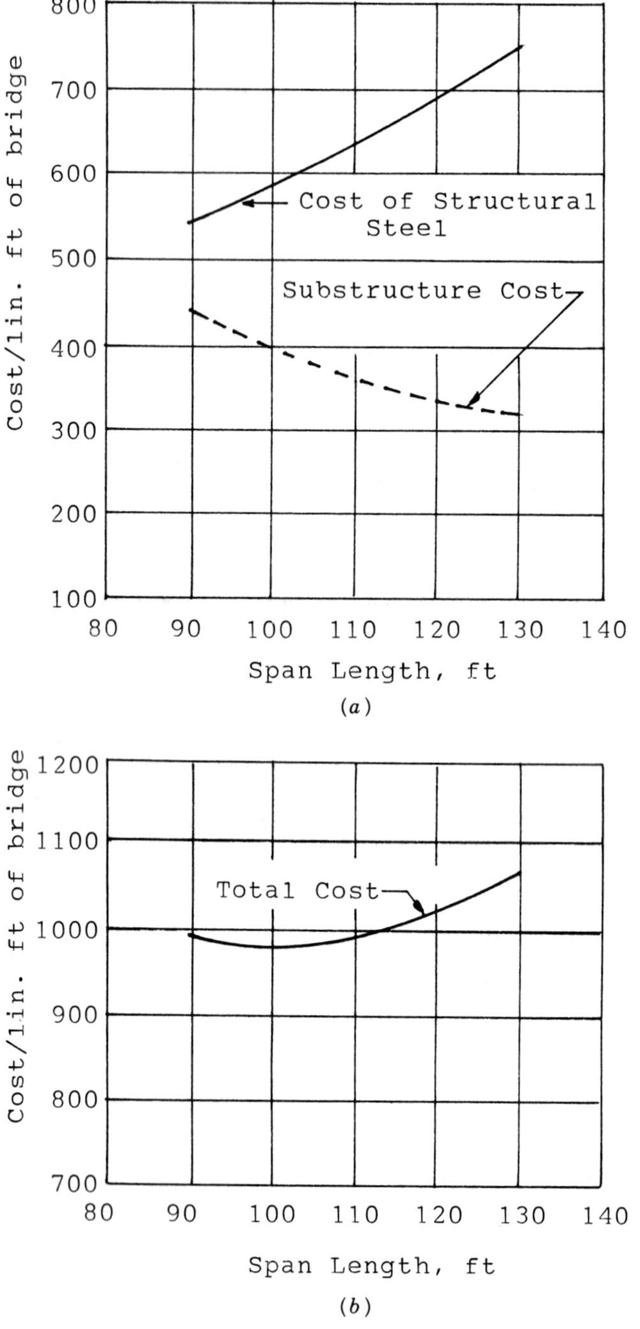

FIGURE 5-29 Bridge cost per linear foot (1970 dollars), approach bridge section on land, multiple-girder system: (*a*) superstructure and substructure cost graphs; (*b*) total cost graph.

the approach section shown in Figure 5-20*b*. Likewise, the two-girder system has four girders, and the box girder system involves six open-top girders. For comparison, the graphs include a two-girder system with and without cantilever floor beams. The graphs supplement the conclusions summarized in the initial review.

A cost summary relating a bridge unit length (ft) to the span length in terms of the cost of structural steel and the substructure cost is shown in Figure 5-29*a* for the multiple-girder system. This summary is for the bridge approach section on land and therefore reflects the particular characteristics of the bridge at this location (deck height, pier height, foundation conditions, etc.). The composite total cost is shown in Figure 5-29*b*, and evidently the 90- to 110-ft spans range in the optimum for these conditions and merit further investigation. Similar comparisons were made for all three bridge sections.

General Comments

The reasonable accuracy and simplicity of results support the validity of the final conclusions regarding superstructure type and optimum span length, although further details for the final structure are not available. The division of the project into the off-shore section, the land section, and the harbor section is consistent with the design considerations in terms of site conditions, clearance restrictions, erection procedures, and method of foundation construction. On the other hand, sufficient data were obtained in order to derive the relative bridge cost, and these involved a comprehensive analysis of superstructure and substructure types. Other comments are as follows.

1. A concrete deck 8-in. thick was assumed throughout the study. However, comparative bridge studies are not necessarily limited to a fixed concrete deck parameter and may include variations in slab thickness. Examples where thick slabs are selected will be found in current bridge designs.

2. The state authority recommends the use of hinges in long waterway crossings as the preferred design. However, the presence of hinges in continuous spans is functionally desirable because it eliminates double bearings at expansion piers.

3. The initial design was based on A36 steel, because cost comparison with higher-strength steels favored this choice. Using 1970 dollars, the unit prices for the three grades of steel are as follows: A36, 28 cents/pound; A441, 34 cents/pound; and A588, 45 cents/pound. A cost-effectiveness comparison is made as follows:

A441 versus A36: strength ratio is $23/20 = 1.15$; cost ratio is $34/28 = 1.21$.
A588 versus A36: strength ratio is $27/20 = 1.35$; cost ratio is $45/28 = 1.61$.

Although the use of A36 steel was in this case the most effective, A441 and

A588 steels were recommended because of their improved resistance to atmospheric corrosion and extended durability. Based on the section modulus requirements and steel unit prices, the following analogy is also indicative of materials efficiency:

A36 steel	SM = $0.6M$ (Moment)	Cost = $0.6 \times 28 = 16.8$
A441 steel	SM = $0.52M$	Cost = $0.52 \times 34 = 17.7$
A588 steel	SM = $0.44M$	Cost = $0.44 \times 45 = 19.8$

The lowest cost-efficiency factor (16.8) indicates the maximum materials efficiency, and is for A36 steel.

4. In comparing superstructure schemes, unit steel prices for each scheme were adjusted to reflect the simplicity or complexity of fabrication and erection, the amount of welding, cutting and assembling, the presence of haunches, and other design features.

5-9 DESIGN EXAMPLE: TWO-GIRDER SYSTEM WITH CONCRETE SLAB

This example involves the 341-ft-long bridge presented in Figures 5-30 and 5-31, and consisting of three spans arranged as shown. The regional preference is for hinges, producing a center suspended span and two end spans cantilevered over the piers. In this scheme the abutments are fixed so that the piers must accommodate the expansion.

The following design data are specified. Concrete slab, $f'_c = 3500$ psi, $f_c = 1400$ psi, $f_s = 24{,}000$ psi, $n = 9$, from which we compute $k = 0.34$, $j = 0.89$, $K = 211$, and $a = 1.78$; structural steel, A36 Grade, $f_s = 20{,}000$ psi; live load, HS 20; weight of railing, 100 lb/ft.

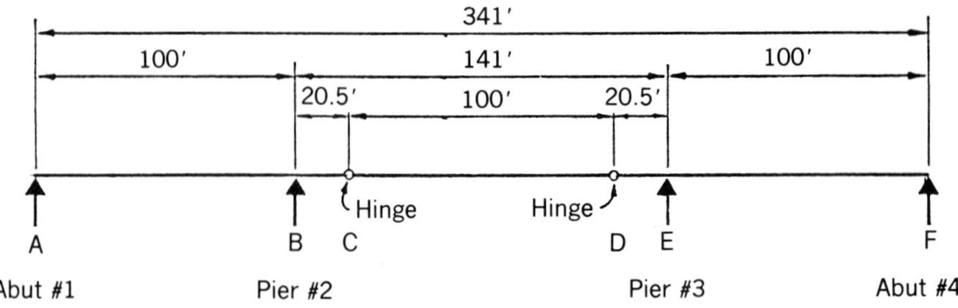

FIGURE 5-30 Typical girder elevation; three-span plate girder bridge.

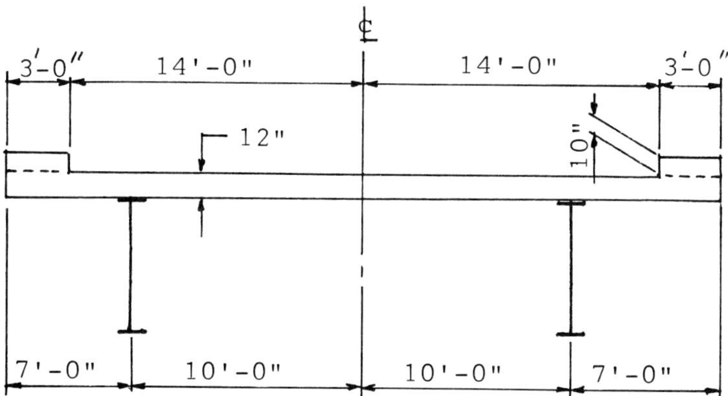

FIGURE 5-31 Typical deck section for bridge of Figure 5-30.

Design of Slab Assuming a 12-in. slab, the dead load moments are computed for the cantilever portion and the midspan section.

For the cantilever moment,

$$
\begin{aligned}
\text{Slab} &= 7.00 \times 0.15 \times 3.5 &&= 3.70 \text{ ft-kips} \\
\text{Curb} &= 0.83 \times 3.00 \times 0.15 \times 5.5 &&= 2.06 \text{ ft-kips} \\
\text{Railing} &= 0.1 \times 6 &&= \underline{0.60} \text{ ft-kips} \\
&\text{DL cantilever moment} &&= -6.36 \text{ ft-kips}
\end{aligned}
$$

For the span moment, compute the total weight of the slab

$$34 \times 1 \times 0.15 + 6 \times 0.83 \times 0.15 + 0.200 = 6.05 \text{ kips} = 3.03 \text{ kips/girder}$$

$$
\begin{aligned}
\text{Moment} &= 3.03 \times 10.0 - 2.55 \times 8.5 - 0.38 \times 15.50 - 0.1 \times 16.0 \\
&= 1.12 \text{ ft/kips}
\end{aligned}
$$

The cantilever live load moment is obtained by placing the wheel 1 ft from the curb, giving $x = 3$ ft = distance from wheel load to point of support. The distribution is $E = 0.8 \times 3 + 3.75 = 6.15$ ft, giving a wheel load reaction of $16/6.15 = 2.60$ kips/ft.

$$\text{Cantilever} \qquad M_{LL+I} = 2.60 \times 3.00 \times 1.30 = -10.14 \text{ ft-kips}$$

The positive live load moment in the span is computed next,

$$\text{Positive} \qquad M_{LL+I} = \frac{20 + 2}{32} \times 16 \times 1.30 = 14.30 \text{ ft-kips}$$

The total moments are now

$$\text{Cantilever} \qquad M = -16.50 \text{ ft-kips}$$
$$\text{Positive Span} \qquad M = 15.42 \text{ ft-kips}$$

The two moments are approximately equal, meaning that the cantilever–span ratio is satisfactory.

Required min $d = \sqrt{16.50/0.211} = 9.0$ in. Assuming #7 bars and 2.25-in. clearance, $d = 9.3$ in.

$$A_s = \frac{16.50}{1.78 \times 9.3} = 0.99 \text{ in.}^2/\text{ft} \qquad \text{Use \#7 at 7 in.} = 1.03$$

The design of the slab should be completed by estimating the bottom steel and then developing moment envelopes in both the negative and positive areas to determine cutoff points for the reinforcement. Distribution reinforcement is calculated as in previous examples.

Design of Girders This phase involves three steps: (a) obtain maximum values of DL + LL + I at controlling locations, (b) construct envelope curves for the total moments, and (c) select girder sections to suit the envelope curves.

Assuming a girder weight of 0.4 kip/ft (including bracing), the total dead load w per girder is 3.4 kips/ft. Referring to Figure 5-30, the dead load moments and shears are computed as follows:

$$M_{BC} = -(3.4 \times 20.5^2 \times 0.5) - 3.4 \times 50 \times 20.5 = -4200 \text{ ft-kips}$$
$$M_{CD} = 0.125 \times 3.4 \times 100^2 = 4250 \text{ ft-kips}$$

Likewise, we calculate the shears at supports A and B and hinge C:

$$V_A = 3.40 \times 50 - 4200/100 = 170.0 - 42.0 = 128.0 \text{ kips}$$
$$V_B \text{ left} = 170.0 + 42.0 = 212.0 \text{ kips}$$
$$V_B \text{ right} = 3.4 \times 141/2 = 242.0 \text{ kips}$$
$$V_C \text{ right} = 100 \times 3.4 \times 0.5 = 170.0 \text{ kips}$$
$$\text{Reaction at } B = 212 + 242 = 454.0 \text{ kips}$$

The dead load moment and shear diagrams are shown in Figure 5-32.

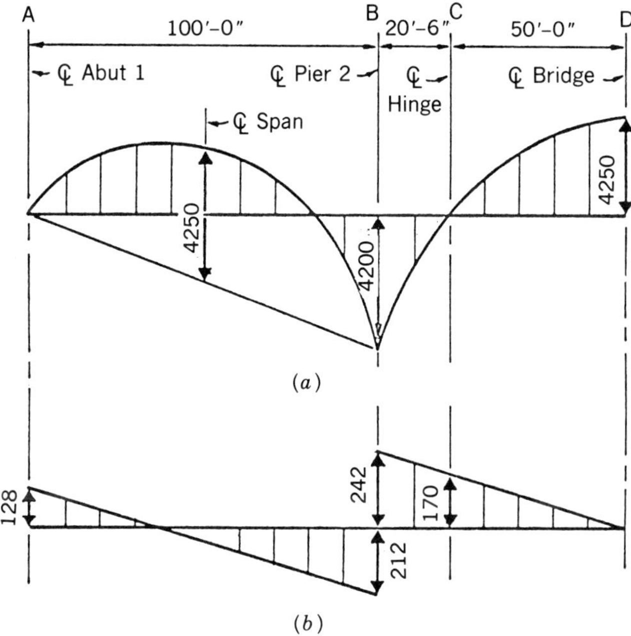

FIGURE 5-32 Dead load moment and shear diagrams: (*a*) dead load moments (ft-kip); (*b*) shear (kip).

Live Load The interpretation of the AASHTO provisions regarding traffic lanes is as follows: (a) for a 28-ft bridge roadway width, there will be two traffic lanes defined by the centerline of the deck; and (b) the loads may be placed within their individual traffic lanes so as to produce the maximum stress under consideration. The loads are placed therefore as shown in Figure 5-33. Taking moments about the left girder line, we obtain

$$\text{Fraction of lane per girder} = (5.0 + 19.0)/20 = 1.20 \text{ lanes}$$

The impact factor is $50/(125 + 100) = 22.2$ percent, giving

$$\text{Total LL} + I \text{ per girder} = 1.22 \times 1.20 = 1.47 \text{ lanes}$$

For the maximum negative live load moment at B, we consider both lane and truck loads. For the lane load, a concentrated load of 18 kips is placed at C, and a lane load of 0.64 kip/ft is placed in span BD. The moment is

$$\begin{aligned} M_{BC} &= 18 \times 20.5 &&= 369 \text{ ft-kips} \\ &+0.64 \times 20.5^2 \times 0.50 &&= 134 \text{ ft-kips} \\ &+0.64 \times 50 \times 20.50 &&= 656 \text{ ft-kips} \\ &\text{Total } M_{BC} &&= -1159 \text{ ft-kips (one lane)} \end{aligned}$$

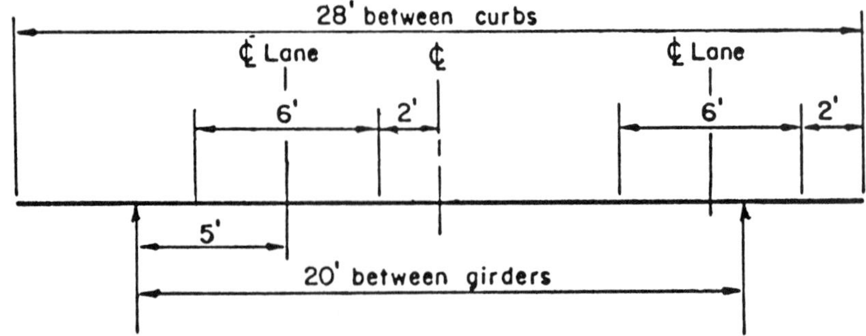

FIGURE 5-33 Position of lane loads for maximum live load reaction at girders.

For the truck load, the simple beam reaction for the truck load and a 100-ft span is 65.3 kips, giving

$$M_{BC} = 65.3 \times 20.5 = -1339 \text{ ft-kips} \qquad \text{(Controls)}$$

$$\text{Total } M_{BC} \text{ (live load + Impact)} = 1339 \times 1.47 = -1970 \text{ ft-kips}$$

The maximum live load plus impact moment in span *AB* or *CD* is

$$M_{AB} = M_{CD} = 1524 \times 1.47 = 2240 \text{ ft-kips}$$

obtained directly from AASHTO tables. Live load shears are obtained by considering both truck and lane loadings. Live load plus impact envelope curves from moments and shears are shown in Figure 5-34.

Selection of Sections At center span *CD*, $M = 2240 + 4250 = 6490$ ft-kips. We choose a 96-in.-deep web stiffened longitudinally. The minimum web thickness is $96/330 = 0.29$ in.; use a 5/16-in.-thick web. Select also plates 16 in. by 2.25 in. top and bottom flange.

$$I_{\text{web}} = 0.312 \times 96^3/12 = 23{,}000 \text{ in.}^4$$

$$I_{\text{plate}} = 72 \times 49.125^2 = \underline{173{,}740} \text{ in.}^4$$

$$\text{Total } I = 196{,}740 \text{ in.}^4$$

$$\text{SM} = 196{,}740/50.25 = 3915 \text{ in.}^3 \qquad f_s = 6490 \times 12/3915 = 19.89 \text{ ksi}$$

At support *B*, $M = -(1970 + 4200) = -6170$ ft-kips. Use again a web 96 in. by 5/16 in. and a flange plate 16 in. by 2.25 in.

$$f_s = 6170 \times 12/3915 = 19.0 \text{ ksi}$$

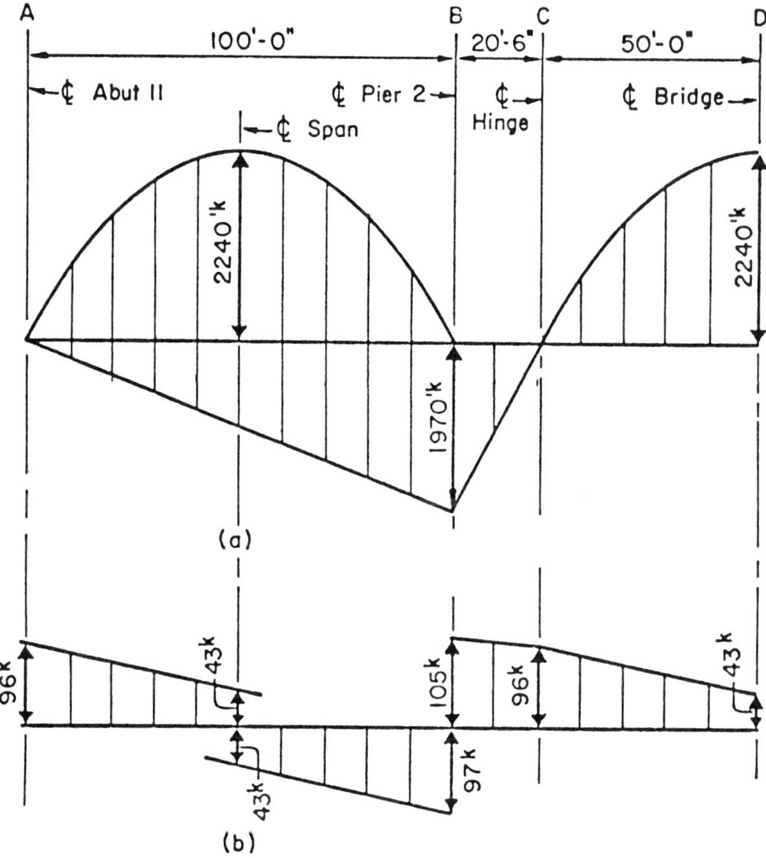

FIGURE 5-34 Moment and shear envelopes for live load plus impact: (*a*) moment diagrams; (*b*) shear diagrams.

For partially unsupported girders, the allowable stress is (1989 AASHTO)

$$F_b = 20{,}000 - 7.5\left(\frac{l}{b}\right)^2 \tag{5-16}$$

where l is the length (in.) of unsupported flange between lateral connections, knee braces, or other points of support, and b is the flange width. Using $l = 10$ ft $= 120$ in., and $b = 16$ in., $F_b = 20{,}000 - 400 = 19{,}600$ psi, OK. Note however, that (5-16) has been modified in the 1992 AASHTO specifications (AASHTO Table 10.32.1A). The allowable stress F_b at support B should be check using the new criteria.

Near the center of span AB, $M = 2240 + 2310 = 4550$ ft-kips. The required section modulus is $4550 \times 0.6 = 2730$ in.^{3}. Select a web 96 in. by 5/16 in. and plates 16 in. by 1.5 in.

$$I_{\text{web}} \qquad\qquad\qquad = 23{,}000 \text{ in.}^4$$
$$I_{\text{plate}} = 48 \times 48.75^2 = \underline{114{,}096} \text{ in.}^4$$
$$\text{Total } I \qquad = 137{,}096 \text{ in.}^4 \qquad \text{SM} = 2770 \text{ in.}^3 \qquad \text{OK}$$

Location of Flange Splices The points of plate thickness change may be determined either by plotting the maximum moment envelope (DL + LL + I) and then superimposing the capacity of preselected flange plates, or by computing the maximum total moment at various points and selecting suitable flange plates. Some previous experience combined with total selections gives satisfactory results. Most designers stipulate a minimum difference in plate thickness not less than 1/2 in., and a minimum plate length of 10 ft, as the cost of the additional butt weld tends to offset the material savings.

For this example, the recommended flange splices should correspond to three flange plates, 2 1/4, 1 1/2, and 3/4 in. thick. Thus, flange plates should be changed in 3/4-in. increments.

Bearing Stiffeners As a sample calculation, we will determine the size of the bearing plate for pier 2 (support B). We recall the following:

$$\text{DL reaction} = 454 \text{ kips}$$
$$\text{LL} + I \text{ reaction} = \underline{163} \text{ kips}$$
$$\text{Total reaction} = 617 \text{ kips}$$

According to AASHTO (Article 10.34.6), bearing stiffeners are designed as columns. We assume three pairs of stiffeners at 6-in. centers and an allowable column stress of 15,500 psi.

$$\text{Required area of column section} = \frac{617}{15.5} = 39.8 \text{ in.}^2$$

$$\text{Area of web acting as column section} = \left(18 \times \frac{5}{16} + 12\right)\frac{5}{16} = 5.5 \text{ in.}^2$$

$$\text{Required stiffener area} = 39.8 - 5.5 = 34.3 \text{ in.}^2$$

The bearing stiffeners must extend as nearly as practical to the outer edge of the flange plates. We select six plates 7-in. wide.

$$\text{Required thickness} = 34.3/6 \times 7 = 0.82 \text{ in.} \qquad \text{Select } 7/8 \text{ in.} \times 7 \text{ in.}$$

$$\text{Minimum thickness by specifications} = \left(\frac{7}{12}\right)\sqrt{\frac{36{,}000}{33{,}000}} = 0.61 \text{ in.}$$

The total column area of the stiffeners is

Six plates 7 in. × 7/8 in. area	= 36.7 in.^{2}
Web plate 17 1/2 in. × 5/16 in. area	= 5.4 in.^{2}
Total area	= 42.1 in.^{2}

Compute

$$I = \frac{3 \times 0.875 \times 14.31^3}{12} = 641 \text{ in.}^4 \qquad r = \sqrt{\frac{641}{42.1}} = 3.9$$

$L/r = 96/3.9 = 24.6$, use $K = 1$, $f_s = 16{,}980 - 0.53 \times 24.6^2 = 16{,}660$ psi.

$$\text{Actual stress} = 617/42.1 = 14{,}655 \text{ psi} < 16{,}660 \qquad \text{OK}$$

Assuming stiffeners are coped 1 in. to clear the flange–web weld, the bearing area is $(7 - 1) \times 6 \times 0.875 = 31.5$ in.^{2}.

$$\text{Actual stress in bearing} = 617/31.5 = 19{,}600 \text{ psi} < 29{,}000 \text{ psi allowed}$$

The bearing stiffeners are arranged as shown in Figure 5-35.

Intermediate Stiffeners The maximum shear occurs to the right of support B (pier 2), and has a magnitude of $(242 + 105) = 347$ kips. The average shear stress in the web is $347/96 \times 0.312 = 11.58$ ksi $< F_y/3 = 12$ ksi. However, the second criterion related to web thickness is not satisfied because the web thickness is less than $D/150$; hence, transverse intermediate stiffeners must be provided. The stiffener spacing must meet the analytical stipulations of AASHTO Article 10.34.4.2.

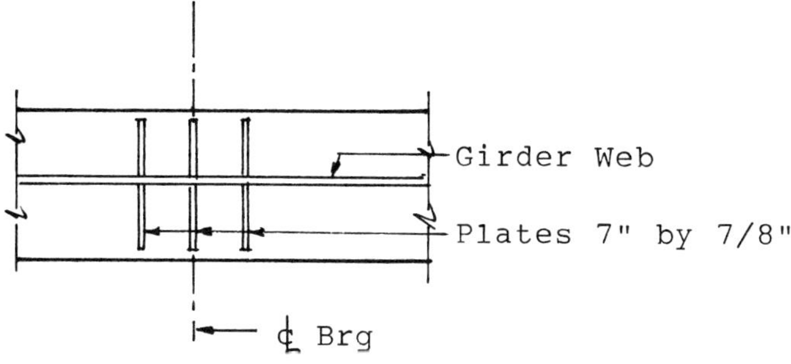

FIGURE 5-35 Detail of bearing stiffeners.

For this example, however, the spacing d_o was calculated using previous criteria, namely,

$$d_o = \frac{11{,}000t}{\sqrt{f_w}} \tag{5-17}$$

or

$$\min d_o = \frac{11{,}000 \times 0.312}{\sqrt{11{,}580}} = 32 \text{ in.},$$

to the right of support B.

Note, however, that the girder panel to the right of support B is subjected to simultaneous action of shear and bending moment where the shear stress exceeds $0.6F_v$. In this case the bending stress must be limited to $(0.754 - 0.34f_v/F_v)F_y$ (AASHTO Article 10.34.4.4), where the value of F_v is obtained in connection with the spacing d_o, but not to exceed 12,000 psi $(F_y/3)$. The value of f_v is computed for the live load producing the maximum moment at the section under consideration. The maximum live load moment at B is produced by the truck load placed as shown in Figure 5-36. The simple beam reaction at C is also the shear at B, or the live load plus impact shear at B is $65.3 \times 1.47 = 96.0$ kips. The total shear is $242 + 96 = 338$ kips, giving $f_v = 338/29.95 = 11.28$ ksi and $f_v/F_v = 11.28/12.00 = 0.94$. The allowable bending stress is now

$$F_s = (0.754 - 0.34 \times 0.94)F_y = 15{,}800 \text{ psi}$$

The required section modulus is $6220 \times 12/15.8 = 4665$ in.^{3}. The first selection of plates 16 in. by 2.25 in. gives a section modulus of 3915 in.^{3}, which is not adequate, and the design must therefore be revised. Select a web 96 in. by 5/16 in. and plates 16 in. by 2.75 in.

$$\begin{aligned} I_{\text{web}} &\qquad\qquad\qquad = 23{,}000 \text{ in.}^4 \\ I_{\text{plate}} &= 88 \times 49.375^2 = \underline{214{,}500 \text{ in.}^4} \\ I &\qquad\qquad\qquad = 237{,}500 \text{ in.}^4 \end{aligned}$$

The section modulus provided is $237{,}500/50.75 = 4680$ in.^{3}, OK.

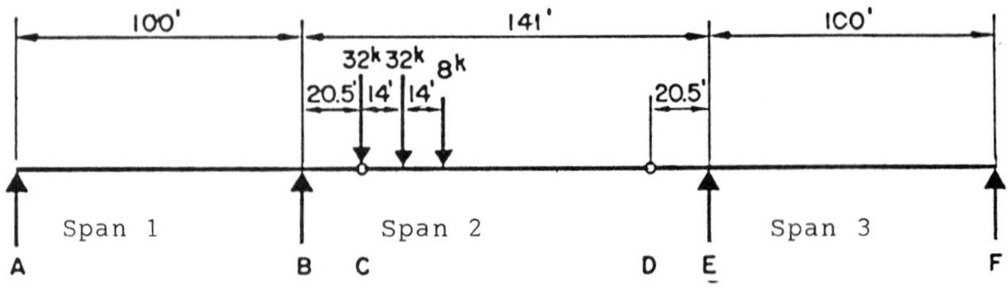

FIGURE 5-36 Position of truck load for maximum cantilever moment at support B.

Longitudinal Stiffeners These are located at a distance $D/5$ from the inner surface of the compression flange. They are proportioned so that

$$I = Dt_w^3\left(2.4\frac{d_o^2}{D^2} - 0.13\right) \tag{5-18}$$

where I = minimum stiffener moment of inertia about the edge in contact with the web

D = unsupported distance between flanges (in.)

t_w = web plate thickness (in.)

d_o = actual distance between transverse stiffeners

The thickness of the stiffener should not be less than $b'\sqrt{f_b}/2250$, where b' is the width of the stiffener and f_b is the calculated compressive stress in the flange. Using d_o = 60 in. and referring to (5-18), we compute

$$I_{\text{min}} = 96 \times (0.312)^2 + \left(2.4 \times \left(\frac{60}{96}\right)^2 - 0.13\right) = 2.37 \text{ in.}^4$$

Select a stiffener 4 in. by 5/16 in.

$$\text{Actual } I = (1/3)(5/16)4^3 = 6.67 \text{ in.}^3 \qquad \text{Satisfactory}$$

$$\text{Minimum } t = 4\sqrt{20{,}000}/2250 = 0.25 \text{ in.} = 1/4 \text{ in.} \qquad \text{OK}$$

Arrangement of Stiffeners In spans 1 and 3, for simplicity and esthetics the longitudinal stiffener for the upper flange is made continuous from the bearing stiffeners between the pier and abutment. The same stiffener is made continuous in the suspended span. The stiffener for the bottom flange is extended 40 ft left of the pier and to the hinge right of the pier. To avoid cutting and interference between stiffeners, the longitudinal stiffeners are placed on the exterior of the girder and the transverse stiffeners are placed on the inside face. The longitudinal stiffeners are shown in Figure 5-37.

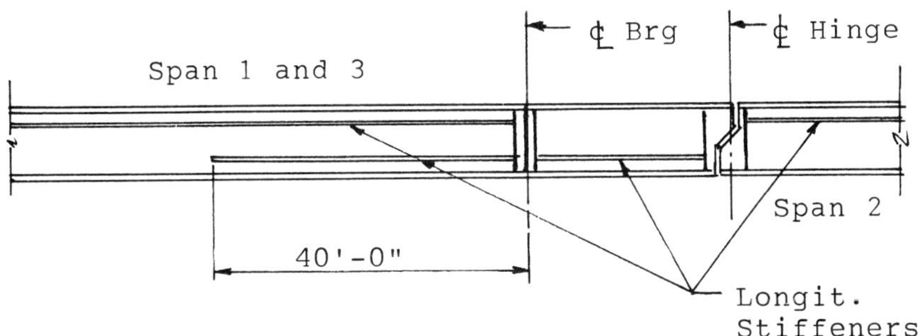

FIGURE 5-37 Arrangement of longitudinal stiffeners.

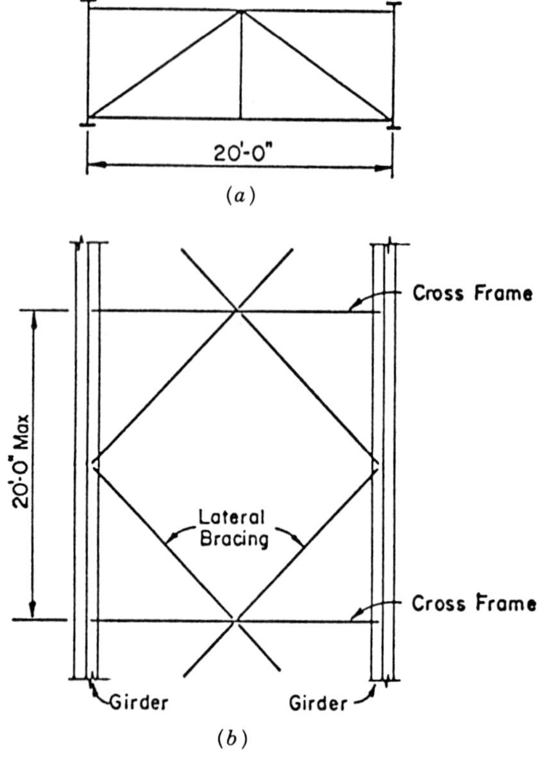

FIGURE 5-38 Suggested girder bracing: (*a*) typical cross frame; (*b*) lateral (bottom) bracing.

Girder Bracing The suggested bracing system is shown in Figure 5-38. The typical cross frame shown in Figure 5-38*a* consists of two types: (a) end cross frames at piers and abutments designed to transmit lateral wind forces to the girder bearings and (b) intermediate cross frames designed to satisfy the L/r requirements.

Bottom lateral bracing has been found to be quite effective in providing stability during and after construction, and will be used for all spans. Because the spacing of cross frames must be compatible with the transverse stiffener spacing, and also uniform with a span, we consider two possible alternatives: (a) cross frames at 25 ft and (b) cross frames at 20 ft with stiffeners at 5 ft. The 20-ft cross frame spacing is selected however, because it will result in a diagonal lateral bracing close to 45°.

End Cross Frame For simplicity, we consider a total superstructure depth exposed to wind (including the railing) of 12 ft. The total wind force is $0.05 \times 12 = 0.6$ kip/ft. Assuming the same condition of continuity as for the

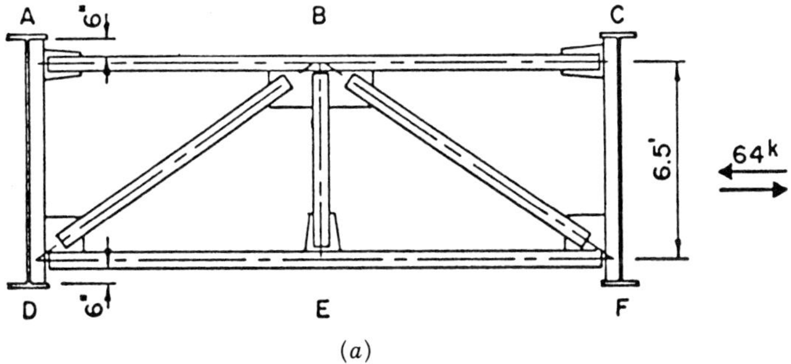

(a)

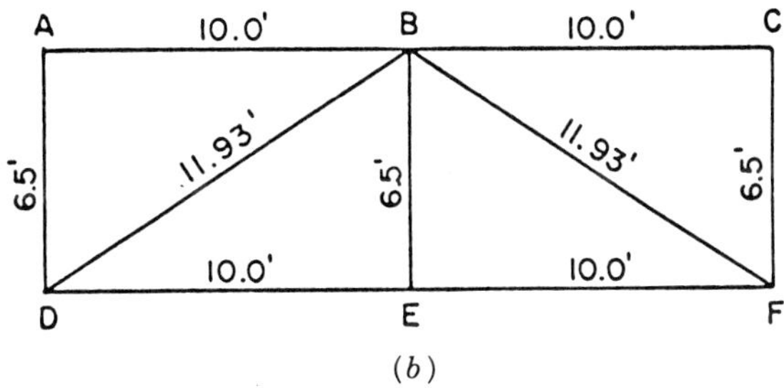

(b)

FIGURE 5-39 (*a*) Lateral wind loading and end cross frame detailing; (*b*) frame dimensions.

vertical loads, the wind reaction at pier 2 (support B) is

$$R = \frac{0.60}{3.4} \times 454 = 80 \text{ kips}$$

This reaction is factored for 125 percent allowable stress, or $R = 80/1.25 = 64$ kips, and is assumed to act as a horizontal shear force across the frame. The loading and frame are shown in Figures 5-39*a* and *b*.

Assume a horizontal section across the frame. Members DB and BF are equally effective in tension and compression. The horizontal component in DB is $64/2 = 32$ kips (compression), giving

$$\text{Force } DB = 32 \times 11.03/10 = 38 \text{ kips} \qquad \text{(Compression)}$$

Likewise, we obtain

$$\text{Force } BF = 38 \text{ kips} \qquad \text{(Tension)}$$

TABLE 5-2 Summary of Forces, End Cross Frame at Pier 2, Caused by Lateral Wind

Member	Wind from Right	Wind from Left
AB	32 kips tension	32 kips compression
BC	32 kips compression	32 kips tension
DB	38 kips compression	38 kips tension
BF	38 kips tension	38 kips compression
DE	0	0
EF	0	0
BE	0	0

Summing up the horizontal forces at B, we obtain

$$\text{Force } AB = 32 \text{ kips} \qquad \text{(Tension)}$$

and

$$\text{Force } BC = 32 \text{ kips} \qquad \text{(Compression)}$$

At points D and F, the horizontal force component in DB and BF will be resisted by the bearings so that the stress in DE and EF is zero. By inspection, the stress in BE is also zero. The maximum and minimum stresses for either wind direction are summarized in Table 5-2.

The minimum section size (tee, angle, or WF) is selected to satisfy the slenderness ratio requirements (usually $L/r = 140$), basic allowable unit stress, or applicable state and local standards.

Lateral Bracing The members of the lateral bracing (shown in Figure 5-38) are designed to satisfy the L/r requirements. The system suggested for this example keeps the unsupported length of the laterals to a minimum.

If large deformation stresses could be induced into the laterals from the lower flange, struts should be placed at the panel points laterally to take the transverse component of this stress and thus support the flange laterally. In this example, the deformation stresses (due to live load plus impact) are small and may be neglected. With high-strength steels, however, these stresses are considerable and should be investigated.

Stresses in Fillet Welds The flange plate-to-web weld connection (fillet welds) should be checked at critical locations based on fatigue considerations. Such critical locations are manifested at supports A, B, and C. By inspection, the remainder of the fillet welds are not critical, and their size is determined by plate thickness requirements.

An example of fatigue stress analysis at points of stress reversal is included in subsequent sections. Fatigue is treated in detail in Chapter 12.

5-10 STEEL BOX GIRDERS: GENERAL PRINCIPLES

Box girders are particularly effective in resisting bending stresses because of the wide bottom flanges. The closed shape of the section provides considerable rigidity in resisting torsional effects on the bridge. In addition, the box is not as susceptible to corrosion as ordinary steel members because half of the steel surface is contained within the section. A box girder bridge is considered an esthetically pleasing structure. All these advantages have received broad attention from designers.

Steel box girder bridges can generally be classified into the three categories shown in Figure 5-40: (a) short-to-moderate span multibox composite types, suitable for the normal-grade separation structure; (b) longer-span composite bridges, in the range of 200 to 400 ft; and (c) long span orthotropic deck structures.

Current AASHTO specifications (Article 10.39) pertain to the design of simple and continuous bridges of moderate length supported by two or more single-cell composite box girders; hence, these provisions apply to the type shown in Figure 5-40*a*, and are based on the work carried out by Mattock and Fountain (1967).

Relevant Specifications (FHWA-TS-80-205)

Studies and reviews of the research in this field have been carried out by the ASCE–AASHTO Subcommittee on Box Girder Bridges (ASCE–AASHTO, 1967, 1971) and by the ASCE–AASHTO Subcommittee on Ultimate Strength of Box Girders (ASCE-AASHTO, 1974). The new specifications extend the scope of steel box girder design to include the entire range shown in Figure 5-40 (Wolchuk, 1980). These criteria are applicable to all types of box girders with parallel flanges (except for horizontally curved bridges). Because current AASHTO provisions have been found practical and adequate for the design

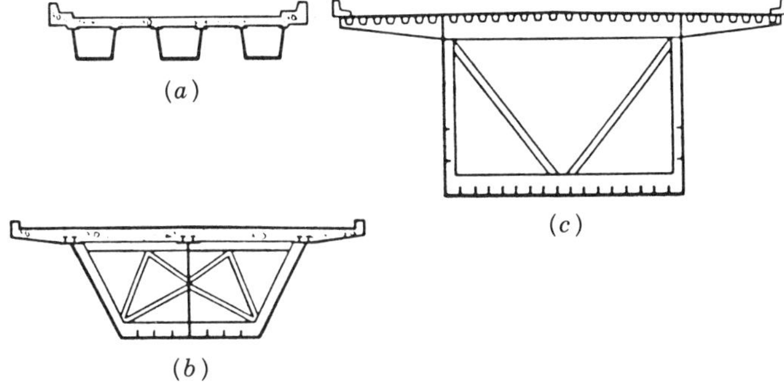

FIGURE 5-40 Basic types of steel box girder bridges: (*a*) short-to-moderate-span multibox composite; (*b*) longer-span composite; (*c*) long-span orthotropic deck. (From Wolchuk, 1980.)

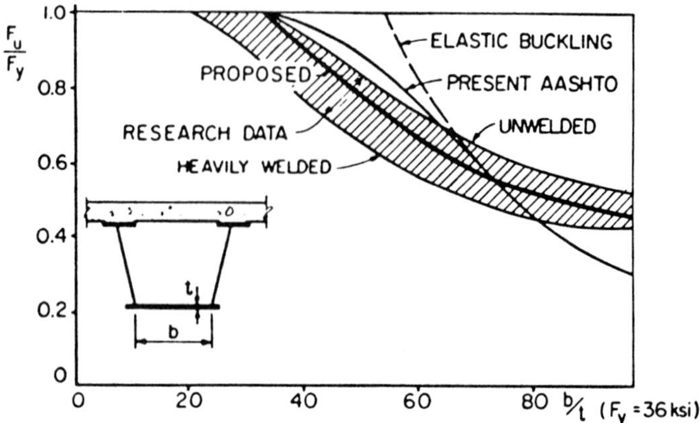

FIGURE 5-41 Strength of unstiffened flange panels. (From Wolchuk, 1980.)

of moderate-span multibox composite structures, however, the proposed specifications may be considered optional for such structures.

Flanges Current AASHTO provisions on the strength of unstiffened flanges are based on elastic buckling for slender plate panels (b/t ratio) and on a transition curve derived by analogy from the column behavior but unrelated to plate strength tests (see Figure 5-41). Studies of plate behavior have shown that the strength of plates in compression is affected by geometric imperfections and residual stresses. The strength of stocky plates is smaller than given by the transition curve, but the strength of very slender plates is greater than predicted by elastic buckling theory. The actual strength range is represented by the shaded area in Figure 5-41, with the upper and lower limits corresponding to unwelded and heavily welded panels, respectively. Residual stresses in unstiffened flanges are relatively low; hence, a strength curve approaching the upper limit is considered appropriate. Interestingly, the proposed strength curve lies below the present AASHTO curve in the practical range of flange plate slenderness.

Similar considerations are applied to compression flanges stiffened longitudinally and supported at suitable intervals by transverse stiffeners. In this case the ultimate strength is also affected by the out-of-straightness of the longitudinal stiffeners, as shown in Figure 5-42. The out-of-straightness Δ may be positive, with stiffeners bent toward the box interior, or negative, with stiffeners bent the opposite way. Considering these parameters, the strength of the flange can be determined by a second-order numerical method (Little, 1976), which, although quite complex, may be applied by means of simple rules.

The strength of a stiffened flange in compression is found with the use of an interaction diagram considering two parameters: (a) the L/r ratio of one

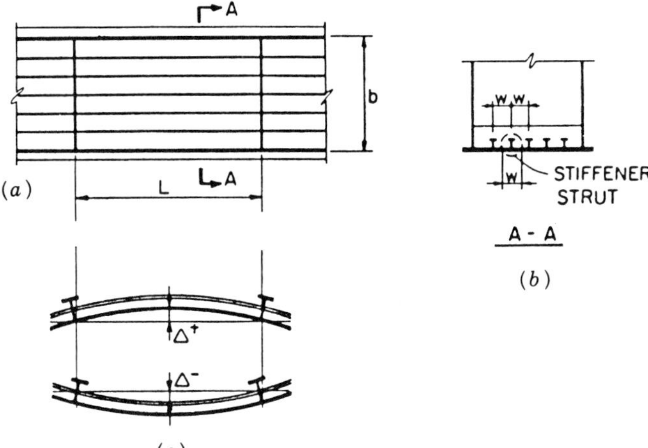

FIGURE 5-42 Steel box girders with stiffened flange: (a) plan of bottom flange; (b) section A–A; (c) stiffener out-of-straightness. (From Wolchuk, 1980.)

stiffener strut between transverse stiffeners and (b) the w/t ratio of a plate panel between the longitudinal stiffeners, where w is the longitudinal stiffener spacing and t is the plate thickness. This interaction is shown in Figure 5-43. With these two parameters computed, the critical stress is read directly from the diagram as shown in the example.

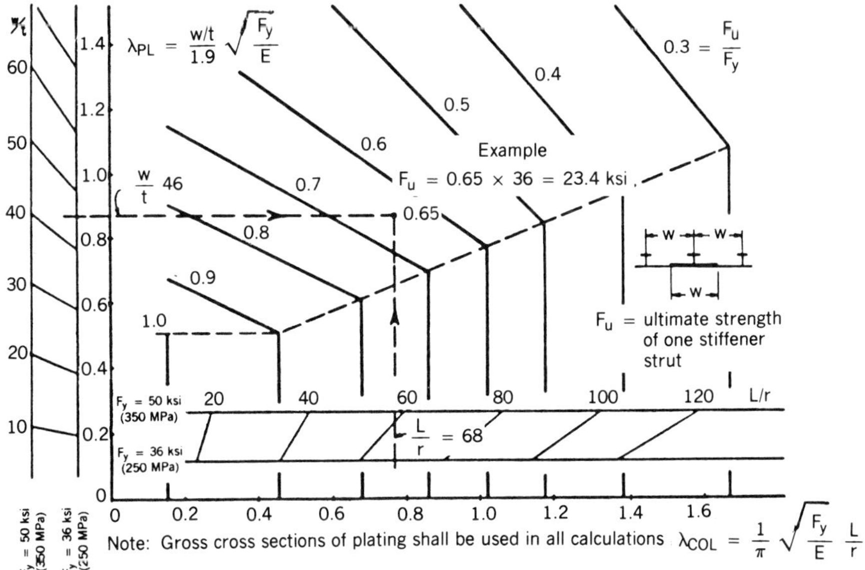

FIGURE 5-43 Strength of stiffened flange in compression. (From Wolchuk, 1980.)

Webs The present AASHTO provisions for web design address the associated problems from the empirical relationship between the moment and shear capacity of an entire beam section. This approach may not be entirely suitable for the webs of large box girders where a consistent analysis of the effects of coincident flexure and shear is necessary. The proposed approach addresses web strength independently of the flange properties. The calculations reflect the two contributions to the ultimate web strength: (a) the beam shear strength (elastic buckling) and (b) the tension field strength (postbuckling). The critical elastic shear buckling stress is computed for the combined action of shear and flexural stresses by appropriate interaction equations (Wolchuk, 1980). The beam shear strength V_B is obtained by multiplying the critical buckling stress by the web area.

The tension field strength V_T of the panel is applicable to the webs of plate girders where the flange rigidity is always adequate to ensure sufficient anchorage of tension field forces, and these forces do not control the stability of the flange. Because this is not always the case with box girders, tension field design must be introduced with caution and by utilizing only the lower limit of the tension field strength corresponding to the assumptions. The tension field strength, or the post-buckling strength, is the vertical component of the tension field force in the panel. The total web strength is the sum $V_B + V_T$.

Web stiffeners are articulated in the three basic types shown in Figure 5-44 as unstiffened webs, transversely stiffened webs, and transversely and longitudinally stiffened webs. The assumed web design strength depends on the adequacy of the web stiffeners. Two basic criteria are prescribed for stiffener adequacy. For rigidity, the stiffeners must provide supports to the web subpanels, that is, they must remain straight and undeflected during web buckling and postbuckling and up to the ultimate design capacity. This implies a minimum required moment of inertia (Wolchuk, 1980) as shown in

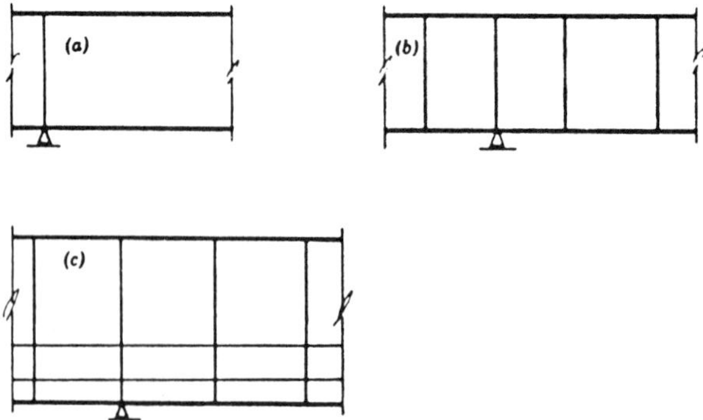

FIGURE 5-44 Types of webs: (*a*) unstiffened webs; (*b*) transversely stiffened webs; (*c*) transversely and longitudinally stiffened webs. (From Wolchuk, 1980.)

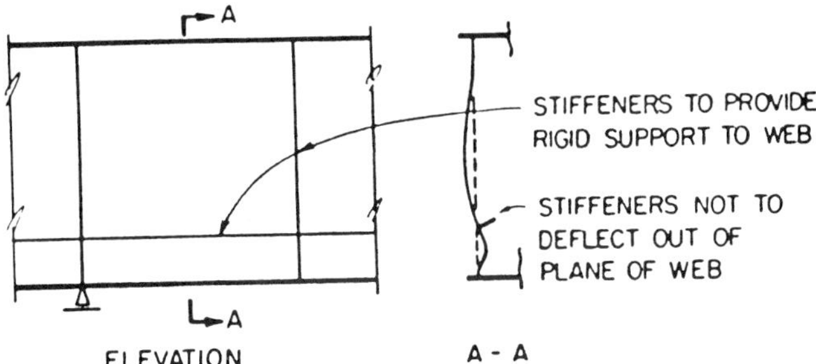

FIGURE 5-45 Web stiffeners: rigidity requirements. (From Wolchuk, 1980.)

Figure 5-45, whereby the rigidity coefficient γ^* is multiplied by an empirical factor m.

The second strength requirement is shown in Figure 5-46. The transverse stiffener is subject to direct load from the deck P and to a vertical force P_{VT} from tension field action. The longitudinal stiffener is subject to a compression force P_H due to the flexural stress in the web. These axial forces must be considered, and the vertical and horizontal stiffeners designed as eccentrically loaded struts.

Other Provisions These address the effective width of the flanges and the determination of the stress distribution across the flange due to shear lag. Also given are rules for preventing local torsional instability of stiffeners in

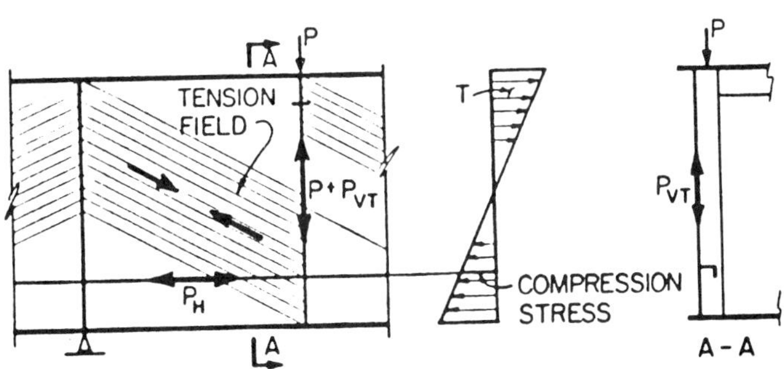

FIGURE 5-46 Web stiffeners: strength requirements. (From Wolchuk, 1980.)

compression, design criteria for tension flanges, intermediate cross frames, transverse flange stiffeners, end diaphragms, and miscellaneous details.

Comments and additional considerations regarding this topic can be found in the references listed at the end of this chapter, and in subsequent sections.

Optimum Span Range

Designers experienced with steel box girder bridges suggest that this bridge type may not be competitive for spans shorter than 500 ft with a concrete deck, and possibly 750 ft with a steel orthotropic deck (Seim and Thoman, 1981). A prestressed concrete bridge in this span range requires a box girder configuration. In this context engineers often choose to compare the cost of steel box girders with concrete box girders because of their similarity in appearance. However, if the comparison is made on the bridge type considering only the economic span range, the steel plate girder or steel truss may be more competitive with the concrete box.

The proposed "Design Specifications for Steel Box Girders" were utilized as the basis of the design criteria for the webs and flanges for a composite steel box girder bridge with five continuous spans of 310, 400, 450, 400, and 310 ft. The trapezoidal box section consists of three webs with maximum and minimum depths of 21.4 and 10.4 ft, respectively. Refinements in the design approach were introduced, however, during the analysis. For example, a simplification was made by ignoring the tension field strength which markedly reduced the design complexity without greatly increasing the cost.

Proportioning of Box Girders

Results of studies on the optimum design of steel box girder bridges are reported by Heins and Hua (1980), and have led to the development of a series of equations for rapid evaluation of plate geometry.

Parametric Study The bridge configurations considered in this study are (a) single-span, (b) two-span, and (c) three-span composite structures, all meeting the AASHTO criteria. The specific geometry includes the following variables.

1. *Span Length L* *Simple-span* 50, 100, *and* 150 *ft*; *two-span*, $L_1 = 50$, 100, *and* 150 *ft*, $L_2 = NL_1$, *where* $N = 1.0$, 1.2, 1.4, *and* 1.6; *and three-span*, $L_1 = L_3 = 50$, 100, *and* 150 *ft*; $L_2 = NL_1$, *where* $N = 1.0$, 1.2, 1.4, *and* 1.6.
2. *Web Depth* *Ratio* $d/L = 1/25$.
3. *Top Flange* *Ratio* $b/t \leq 23$ (*positive region*).
4. *Bottom Flange Width* 80, 100, *and* 120 *in*.
5. *Bottom Flange Stiffeners* $ST7.5 \times 25$ (*negative moment region*).

6. *Composite Concrete Slab* 8.5 *in. thick.*
7. *Steel Grade* $A36$, $F_y = 36$ *ksi.*
8. *Modular Ratio* $n = 9$, $3n = 27$, $f'_c = 4000$ *psi.*
9. *Unit Weight* *Steel*, 490 *pcf*; *concrete*, 150 *pcf.*
10. *Relevant Parameters* *These include parapet, wearing surface, miscellaneous concrete (fillet, haunch), and miscellaneous steel* (12 *percent added for bracing etc.).*

The procedure for determining the correct plate geometry for the various bridges is based on the following steps.

1. Fix span length.
2. Select web depth $d = 12L/25$.
3. Select bottom flange width $w = 80$ in.
4. Select web thickness.
5. Select top flange width $b \leq 23t$.
6. Determine dead load moments.
7. Determine location of cross-sectional changes.
8. Revise sections, and compute dead load, live load, and resulting stresses, and revise according to specifications.
9. Set bottom flange width $w = 100$ in., and repeat procedure.

Design Equations The induced force F in the flange may be expressed as follows

$$F = A_b F_b C_1 + C_2 + \cdots \tag{5-19}$$

where A_b is the bottom flange area (in.^{2}), F_b is the bending stress (kg/in.^{2}), and C_1, C_2 are coefficients. The maximum induced moment M is

$$M = Fd \tag{5-20}$$

where d is the girder depth (in.). From (5-19) and (5-20) we obtain

$$A_b = \frac{CM}{F_b d} + K \tag{5-21}$$

Once the terms C, M, F_b, d, and K are calculated, the area of the bottom flange can be determined. This area can be related to the top flange by

$$A_t = RA_b \tag{5-22}$$

Single-Span Bridge Heins and Hua (1980) have derived the following design equations.

For the bottom flange,

$$A_b = 1.9\left(\frac{M}{dF_b}\right) + B(x) \tag{5-23}$$

and

$$B(x) = -0.0013L^2 + 0.096L - 9.8 \tag{5-24}$$

where M is the maximum induced moment (kg-in.) due to DL, SDL, and LL, d is the web depth (in.), F_b is the allowable (design) stress (ksi), and L is the span length (ft).

For the top flange area, (5-22) can be used, where $R = (84W + 3050 + 315L - L^2)/50{,}000$, W is the bottom flange width (in.), and L is the span length (ft).

Two-Span Bridge For the positive moment region,

$$A_b = \left(\frac{M}{dF_b}\right)C + B(L, N) \tag{5-25}$$

where
$$C = 9 \times 10^{-5}L^2 - 0.03L + 4$$
$$B(L, N) = E(L)(N) + F(L)$$
$$E(L) = 9 \times 10^{-4}L^2 + 0.4L - 49$$
$$F(L) = -3 \times 10^{-3}L^2 + 0.5L - 5$$

$$R = A_t/A_b$$

where
$$R = S(L)(\text{TL})D(L)$$
$$S(L) = -8.8 \times 10^{-5}L + 0.02$$
$$D(L) = 1.5 \times 10^{-4}L^2 - 0.03L + 0.006$$
$$\text{TL} = \text{total length (ft)}$$

For the negative moment region,

$$A_b = 3.4\left(\frac{M}{dF_b}\right) + B(L, N) \tag{5-26}$$

$$
\begin{aligned}
\text{where} \quad B(L, N) &= E(L)N^3 + F(L)N^2 + G(L)N + H(L) \\
E(L) &= 0.09L^2 - 16L + 1095 \\
F(L) &= -0.3L^2 + 52L - 3798 \\
G(L) &= 0.3L^2 + 56L + 4326 \\
H(L) &= -13L^2 + 19L - 1609
\end{aligned} \tag{5-27}
$$

$$R = A_t/A_b = 1.0 \tag{5-28}$$

where L is the first span length and N = span 2/span 1.

Three-Span Bridge Using the design data and moments, the design equations are as follows.

For the positive moment regions,

$$A_b = \left(\frac{M}{dF_b}\right)C + B(L, N) \tag{5-29}$$

where

$$C = -9.9 \times 10^{-3}L + 3.3 \qquad B(L, N) = E(L)(N) + F(L)$$

$$E(L) = 0.3L - 51 \tag{5-30}$$

$$F(L) = -0.36L + 33 \qquad R = A_t/A_b = 7 \times 10^{-3}\text{TL} - 0.02L \tag{5-31}$$

where TL is the total span length (ft) and L is the first span length (ft).

For the negative moment regions,

$$A_b = 3.2\left(\frac{M}{dF_b}\right) + B(L, N) \tag{5-32}$$

$$
\begin{aligned}
\text{where} \quad B(L, N) &= E(L)N^3 + F(L)N^2 + G(L)N + H(L) \\
E(L) &= -0.05L^2 + 4.4L - 486 \\
F(L) &= 0.22L^2 - 21L + 2086 \\
G(L) &= -0.29L^2 + 31L - 2943 \\
H(L) &= 0.13L^2 - 15L + 1369
\end{aligned} \tag{5-33}
$$

$$R_t/R_b = 1.1 \tag{5-34}$$

Nominal Shear Strength (Including Tension Field Action)

The work of Wolchuk (1980) is based on the ability of a plate girder to behave in a manner similar to a truss (Basler, 1961). As shown in Figure 5-47, the tension forces are carried by the membrane action of the web (tension

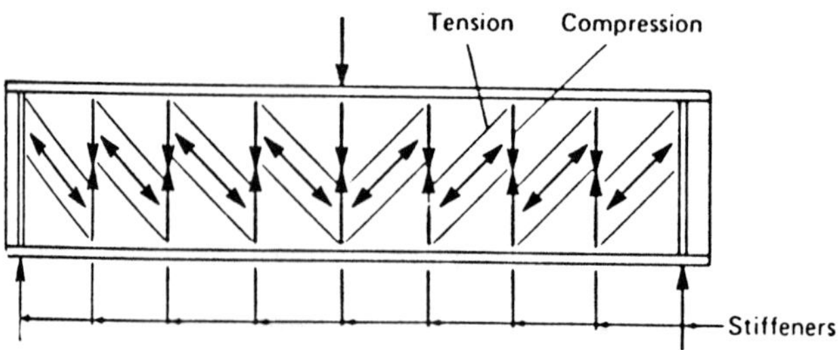

FIGURE 5-47 Tension field action.

field action), whereas the compression forces are carried by the transverse stiffeners. The work of Basler led to a theory that agrees with tests and provides criteria to ensure truss action. The consideration of this action increases the shear strength based on elastic buckling to a condition corresponding to the shear yield. The elastic and postbuckling strength is shown in Figure 5-48. Without truss action the shear is represented by line *ABCD*, but with the inclusion of tension field action the shear approaches line *ABE*.

The shear strength V_T arising from the tension field action develops a band of tensile forces occurring after the web has buckled under diagonal compression (principal stresses in ordinary beam theory). Equilibrium is maintained by transferring the force to the vertical stiffeners. As the girder load increases, the angle of the tension field changes to accommodate the greatest carrying capacity (see also Section 5-16).

Investigation of girder webs based on membrane stresses should include the optimum direction of tension field action, shear strength arising from this condition, failure criteria, force in the vertical stiffeners, and nominal shear strength from the elastic buckling and postbuckling contributions. Appropriate references on this subject for further study are given at the end of this chapter, and include Salmon and Johnson (1990).

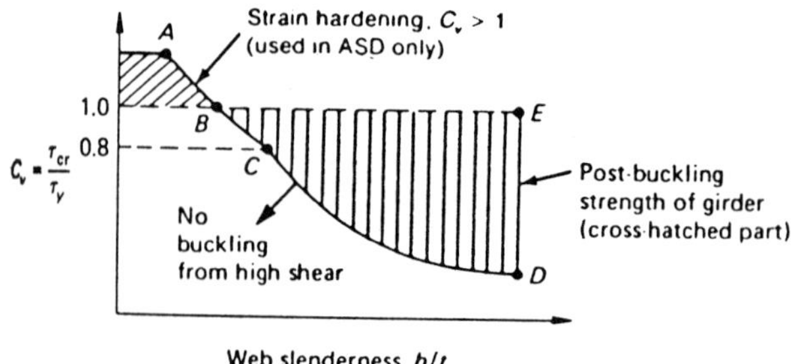

FIGURE 5-48 Shear capacity available, considering postbuckling strength.

5-11 DESIGN EXAMPLE: STEEL BOX GIRDER BRIDGE

This example is presented to demonstrate the applicability of current AASHTO criteria to steel box girder design. The span lengths are not necessarily within the optimum span range for this type of bridge, and site conditions normally would warrant consideration of alternate designs in both steel and concrete configurations. Typical bridge elevation and cross section are shown in Figure 5-49.

The following parameters and data are given:

Span lengths = 120 ft
Slab = 7.5 in. (composite)
Spacing to boxes = 7 ft 10 in.
Roadway width = 44 ft (face-to-face of parapets)
Steel grade = A572 at interior supports, A36 elsewhere
Haunches = 1 in. by 3 in. over beam flanges
Future wearing surface = 25 psf
Concrete $f_c' = 3000$ psi
Loading = HS 20
Truck loading = 5×10^5 cycles

Dead Load (noncomposite):

$$\begin{aligned}
&\text{Slab} = 0.583 \times 15.67 \times 0.15 &&= 1.47 \text{ kips/ft} \\
&\text{Haunch} &&= 0.01 \\
&\text{Steel girder (estimated)} &&= \underline{0.42} \\
&\qquad \text{Total DL } w &&= 1.90 \text{ kips/ft}
\end{aligned}$$

Superimposed dead load (composite):

$$\begin{aligned}
&\text{Parapet} = 1.0 \times 1.50 \times 0.150 \times 2/3 &&= 0.15 \text{ kip/ft} \\
&\text{Railing} &&= 0.01 \\
&\text{Wearing surface} = 0.025 \times 15.67 &&= \underline{0.36} \\
&\qquad \text{Total SDL } w &&= 0.52 \text{ kip/ft}
\end{aligned}$$

In order to make the analysis feasible with conventional calculators without reference to computer programs, we compute moments and shears assuming a uniform moment of inertia. In this case we choose to use AISC tables and influence line coefficients.

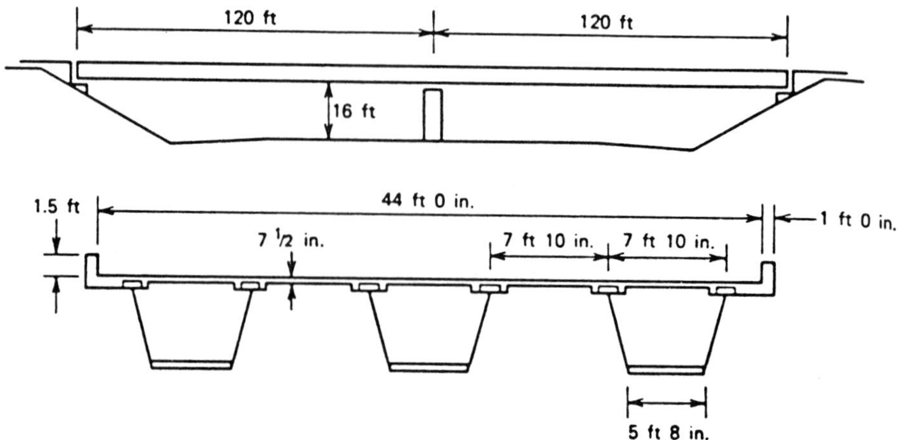

FIGURE 5-49 Elevation and cross section of two-span continuous steel box girder bridge.

Dead Load Moments These are calculated at 0.20 points as follows:

At 0.2	M_{DL}	$= 0.055 \times 1.9 \times 120^2$	$= 1504$ ft-kips
0.4		$= 0.070 \times 1.9 \times 120^2$	$= 1915$
0.6		$= 0.045 \times 1.9 \times 120^2$	$= 1231$
0.8		$= -0.020 \times 1.9 \times 120^2$	$= -547$
Support B		$= -0.125 \times 1.9 \times 120^2$	$= -3420$

Reaction and shears are computed as $R_A = V_A = 1.90(120/2) - 3420/120 = 85.5$ kips, $V_B = 114.0 + 28.5 = 142.5$ kips, $R_B = 285$ kips.

Superimposed Dead Load Moments Following the same procedure, we compute moments and shears due to superimposed dead load $w = 0.52$ kip/ft.

At 0.2	M_{SPL}	$= 0.055 \times 0.52 \times 120^2$	$= 412$ ft-kips
0.4		$= 0.070 \times 0.52 \times 120^2$	$= 524$
0.6		$= 0.045 \times 0.52 \times 120^2$	$= 337$
0.8		$= -0.020 \times 0.52 \times 120^2$	$= -150$
Support		$= -0.125 \times 0.52 \times 120^2$	$= -936$

Reaction and shears are computed as $R_A = V_A = 0.52 \times 60 - 936/120 = 31.2 - 7.8 = 23.4$ kips, $V_B = 31.2 + 7.8 = 39.0$ kips, $R_B = 78$ kips.

Live Load According to AASHTO Article 10.39.2, we compute $N_w = 44/12 = 3.67$ (use 4); $R = 4/3 = 1.33$; and $W_L = 0.1 + 1.7 \times 1.33 + 0.85/4 = 2.58$ wheels or 1.29 lanes, live load distributed to each box girder.

$$\text{Impact} = 50/(125 + 120) = 0.204$$

Maximum positive and negative moments are found directly from AISC tables.

Support B $\quad M_{LL+I} = -1567 \times 1.29 \times 1.204 = -2434$ ft-kips

Span AB $\quad M_{LL+I} = 1530 \times 1.29 \times 1.204 = 2375$ ft-kips

(occurring about 50 ft from the left support A, or approximated at the 0.40 point). Live load plus impact moments are also computed at all 0.20 points and are summarized as follows:

At $0.2M_{LL+I} = 1789$ ft-kips $\qquad$ At $0.9M_{LL+I} = 359$ ft-kips
$0.6M_{LL+I} = 2132$ ft-kips
$0.8M_{LL+I} = 1082$ ft-kips (positive) $\qquad M_{LL+I} = -974$ ft-kips (negative)

These results are plotted in Figure 5-50, showing moment envelopes for both positive and negative moments.

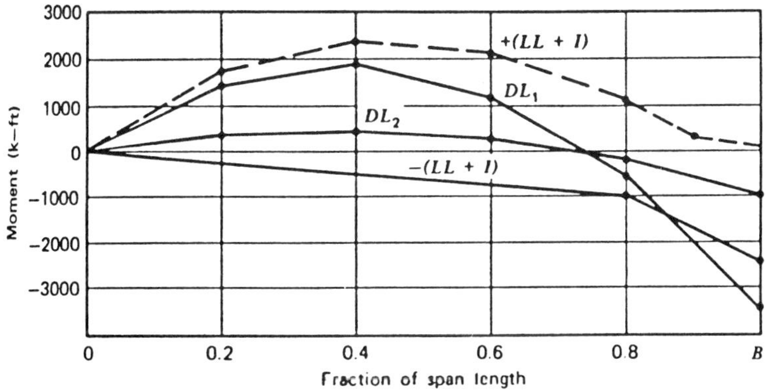

FIGURE 5-50 Moment envelope.

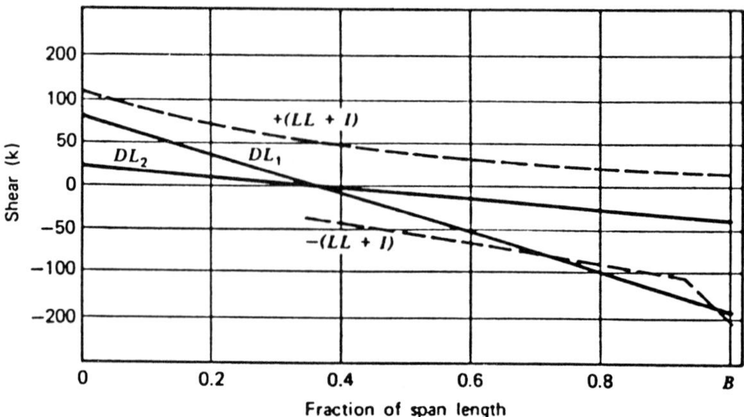

FIGURE 5-51 Shear envelope, bridge of Figure 5-49.

Live load plus impact shears are likewise computed at the 0.20 points, and are summarized as follows.

At left support	$R_A = V_A = 101$ kips (Truck load)
At 0.2	$V_{LL+I} = 74$
0.4	$= 49$
0.6	$= 26$
For span BA, at 0.4	$V_{LL+I} = 43$ kips
Point 0.6	$= 68$
Point 0.8	$= 90$
At B	$V_{LL+I} = 106$ kips
For lane loading	$R_A = V_A = 92$ kips
At B	$R_B = 179$ kips

A shear envelope for dead and live loads is shown in Figure 5-51.

Section Properties After preliminary trials, the sections are shown in Figure 5-52 for three locations along the girders, corresponding to the girder elevation shown in Figure 5-53.

The section properties are calculated following the procedure used in previous examples. These properties are as follows.

1. Section at pier ($0.9L$ to B):

$I = 110{,}759$ in.^{4} $\quad S_T = 3608$ in.^{3} $\quad S_B = 3929$ in.^{3} $\quad \bar{y} = 30.70$ in.

2. Section at $0.7L$ to $0.9L$:

$I = 87{,}310$ in.^{4} $\quad S_T = 2813$ in.^{3} $\quad S_B = 3184$ in.^{3} $\quad \bar{y} = 31.04$ in.

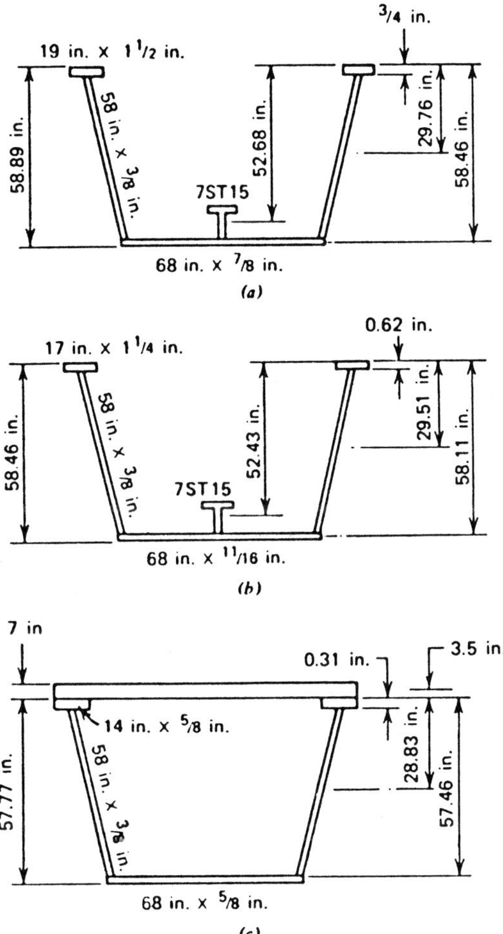

FIGURE 5-52 Typical box girder sections, bridge of Figure 5-49; (*a*) section at point 0.9*L* to *B*; (*b*) section at point 0.7*L* to 0.9*L*; and (*c*) section at point *A* to 0.7*L*.

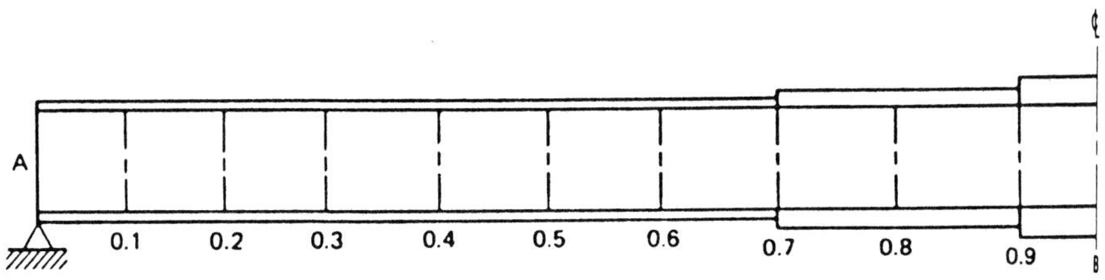

FIGURE 5-53 Typical box girder elevation, bridge of Figure 5-49.

3. Section from A to $0.7L$:

$$I = 55{,}649 \text{ in.}^4 \qquad S_T = 1555 \text{ in.}^3 \qquad S_B = 2531 \text{ in.}^3 \qquad \bar{y} = 35.78 \text{ in.}$$

   The dimension $\bar{y}$ is the distance from the centroid (neutral axis) to the top of the top flange.

4. Composite section (from support A to approximately point $0.7L$). The effective flange width is the least of the following dimensions:

$$\begin{aligned} \text{One-quarter of span length} &= 120 \times 0.25 = 30 \text{ ft} \\ \text{Center-to-center of girders} &= 15.67 \text{ ft} \\ \text{Twelve times slab thickness} &= 7 \times 7/12 = 7 \text{ ft, or } 2 \times 7 = 14 \text{ ft, use} \end{aligned}$$

For $n = 30$, the transformed flange width is $14 \times 12/30 = 5.6$ in. The properties of the composite section are

$$I = 99{,}677 \text{ in.}^4 \qquad \text{Steel} \quad S_T = 3987 \text{ in.}^3 \qquad S_B = 3042 \text{ in.}^3$$
$$\text{Concrete} \quad S_T = 3115 \text{ in.}^3 \qquad \bar{y} = 28.5 \text{ in.}$$

where the dimension $\bar{y}$ is the distance from the centroidal axis to the center of gravity of the concrete slab.

For $n = 10$, the transformed flange width is $14 \times 12/10 = 16.8$ in. The properties of the composite section are

$$I = 141{,}067 \text{ in.}^4 \qquad \text{Steel} \quad S_T = 9468 \text{ in.}^3 \qquad S_B = 3291 \text{ in.}^3$$
$$\text{Concrete} \quad S_T = 6441 \text{ in.}^3 \qquad \bar{y} = 18.4 \text{ in.}$$

Unstiffened Compression Flange Using the basic allowable stress $0.55F_y$, the flange requirements are

$$\frac{b}{t} = \frac{6140}{\sqrt{F_y}}$$

For $F_y = 50$ ksi, $b/t = 27.5$; for $F_y = 36$ ksi, $b/t = 32.4$. Actual b/t is 13.6 and 22.4, both less than 32.4.

Bending Stresses Referring to the moments computed in the foregoing sections, the maximum stresses may be determined at various sections.

Support B The allowable stress for A572 steel is 27 ksi. The total moment is -6790 ft-kips.

$$f_{\text{top}} = \frac{6790 \times 12}{3608} = 22.6 \text{ ksi} \qquad f_{\text{bot}} = \frac{6790 \times 12}{3929} = 20.7 \text{ ksi}$$

Fatigue at Support B This roadway has an ADT < 2500, which corresponds to a truck loading of 500,000 stress cycles. Pertinent structural categories are butt-welded flanges and fillet-welded webs to flanges (category B), stiffeners to webs (category C), and shear studs to flanges (category C). Obviously, category C governs, allowing a live load stress range $F_{sr} = 21$ ksi. We recall that $M_{B(LL+I)} = 2434$, which is the actual moment range. Hence,

$$F_{LL+I} = \frac{2434 \times 12}{3608} = 8.1 \text{ ksi} < 21 \qquad \text{OK}$$

Allowable Compressive Stress at Support B (Bottom Flange) First, we check the ratio

$$\frac{w}{t} = \frac{6140}{\sqrt{50{,}000}} = 27.5$$

Actual $w/t = 68/0.875 = 77.7 > 27.5$. We also note that $w/t > 45$; hence, a longitudinal stiffener is needed (AASHTO Article 10.39.4.2.5).

As shown in Figure 5-52*a*, a structural 7 ST 15 tee is provided at midwidth giving $w = 34$ in. The required stiffness is

$$I_s = \phi t^3 w$$

where $\phi = 0.125k^3$ and k is a buckling coefficient not to exceed 4. Assuming $k = 3$ and using $w = 34$ in. and $t = 0.875$ in., we compute

$$I_s = 0.125 \times 27 \times 0.875^3 \times 34 = 76.9 \text{ in.}^4$$

For the 7 ST 15 tee, the stiffness about the base is

$$I_{\text{base}} = 19.0 + 4.42(5.35)^2 = 145.5 \text{ in.}^4 > 76.9 \qquad \text{OK}$$

With the compression flange stiffened longitudinally and for a basic allowable stress $0.55F_y$, the ratio should not exceed the value given by

$$\frac{w}{t} = \frac{3070\sqrt{k}}{\sqrt{F_y}} = \frac{3070\sqrt{3}}{\sqrt{50{,}000}} = 23.8 < 34.0$$

Therefore, the allowable stress is governed by formula (10-77), Article 10.39.4.3.3, AASHTO. Substituting the values of F_y, k, w, and t in this formula, we compute the allowable $F_b = 23.7$ ksi. The actual compressive stress in the bottom flange is 20.7 ksi, and therefore the flange is adequate.

Stresses at Point 0.9 (A36 Steel) The maximum and minimum moments at this point are

$$M_{\max} = -(2000 + 500 + 1700) = -4200 \text{ ft-kips}$$

$$M_{\min} = -(2000 + 500 - 359) = -2141 \text{ ft-kips}$$

$$f_{\text{top}} = \frac{4200 \times 12}{2813} = 17.9 \text{ ksi} \qquad f_{\text{bot}} = \frac{4200 \times 12}{3184} = 15.8 \text{ ksi}$$

Allowable = 20 ksi, OK. For fatigue, the actual moment range is 4200 − 2141 = 2059 ft-kips, or

$$F_{\text{sr}} = \frac{2059 \times 12}{2813} = 8.78 \text{ ksi} < 21 \qquad \text{OK}$$

Stresses at Point 0.4L (Composite Section, A36 Steel) We recall M_{DL} = 1915 ft-kips, M_{SDL} = 524 ft-kips, and $M_{\text{LL}+I}$ = 2375 ft-kips. Stress in the steel

$$f_s = \frac{1915 \times 12}{2531} + \frac{524 \times 12}{3042} + \frac{2375 \times 12}{3291} = 9.1 + 2.1 + 8.7 = 19.9 \text{ ksi}$$

Stress in the concrete

$$f_c = \frac{524 \times 12}{3115 \times 30} + \frac{2375 \times 12}{6441 \times 10} = 0.067 + 0.442 = 0.51 \text{ ksi} < 1.2 \text{ ksi}$$

Web Stresses and Stiffeners Criteria on web–depth ratio limitations are:

Steel section $d/L \geq 1/25$

Composite section: Concrete and steel $d'/L \geq 1/25$

Steel $d/L \geq 1/25$

By reference to Figure 5-52, these ratios are satisfied.

For web plates not stiffened longitudinally, the web thickness must satisfy the criterion

$$t \geq \frac{D\sqrt{f_b}}{23{,}000}$$

but in no case should $t \leq D/170$, where D is the web depth and f_b is the calculated compression stress in the flange. The maximum compression stress

TABLE 5-3 Summary of Moments and Shears for Bridge of Figure 5-49

	Combined Moment					
Location	DL_1	DL_2	LL + I (+)	LL + I (−)	$M+$	$M-$
0.2	1504	400	1789	—	3693	—
0.4	1915	509	2375	—	4827	—
0.6	1231	327	2132	—	3690	—
0.8	−547	−145	1082	−974	1082	1666
B	−3420	−909	—	−2419	—	6748

	Shear (kip)			
	DL_1	DL_2	LL + I	Reaction (kip)
A	85.5	23.4	101	209.9
B	143	39	106	288
$0.8L$	100	25	90	215

is 20.7 ksi at support B, where $D = 56.5$ in. Therefore,

$$t_w = 56.6 \times \frac{\sqrt{20{,}700}}{23{,}000} = 0.353 \text{ in.}$$

The 3/8-in. web is satisfactory without longitudinal stiffeners for all sections.

The requirements for intermediate transverse stiffeners should be investigated according to AASHTO Article 10.34.4. For this example, however, the webs are stiffened at intervals not to exceed $d \leq (11{,}000)t\sqrt{f_v}$, where t is the thickness of the web plate, or 0.375 in., and f_v is the average shear stress at the section considered.

The design web shear V_w is computed from $V_w = V_v/\cos\theta$, where V_v is the vertical shear and θ is the angle of web inclination to vertical. The total vertical shear is summarized in Table 5-3 together with the total moments. The total web shear V_w, web stress f_v, and required stiffener are computed and tabulated in Table 5-4.

TABLE 5-4 Shear and Web Stiffener Data

Section	V_v	$\cos\theta$	$V_w = V_v/\cos\theta$	$A_t = 2A_w$	$f_v = (V_w/A_t)$	$\sqrt{f_u}$	d_{max} (in.)
End							
Support A	209.9	0.973	215.7	$2 \times 58 \times \frac{3}{8} = 43.5$	4.96	70.4	58.6
$0.8L$	215.0	0.973	221.0	$2 \times 58 \times \frac{3}{8} = 43.5$	5.10	71.4	57.8
Center							
Support B	288.0	0.973	296.0	$2 \times 58 \times \frac{3}{8} = 43.5$	6.8	82.5	50.0

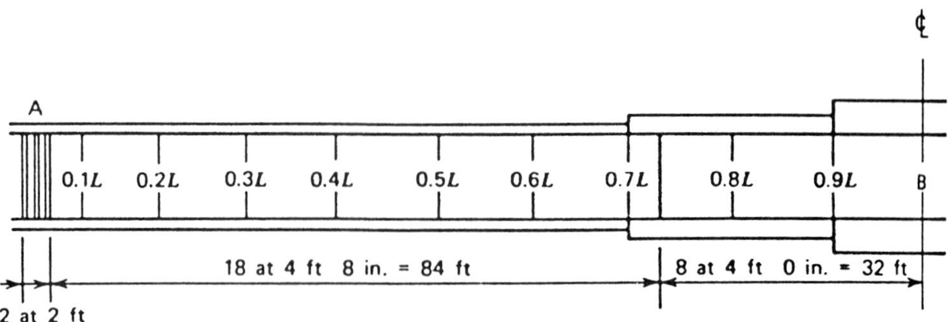

FIGURE 5-54 Girder stiffener spacing.

Referring to Figure 5-54, the following sequence of stiffener spacing is derived, beginning at support A: 2 spaces at 2 ft 0 in., 18 spaces at 4 ft 8 in., and 8 spaces at 4 ft 0 in., for a total of 120 ft.

A stiffener plate 4 in. by 5/16 in. satisfies the moment of inertia requirements stipulated in AASHTO Article 10.34.4.7.

Bearing stiffeners are provided according to AASHTO Article 10.34.6.1 and designed as columns. This procedure, however, has been demonstrated in previous examples, and will not be repeated here.

5-12 HYBRID GIRDERS

A hybrid girder is a member that has the main plates fabricated from more than one type of steel such as the design example of Section 5-11. AASHTO considers this concept in the design of girders utilizing a lower-strength steel in the web than in one or both flanges. This articulation applies to composite and noncomposite plate girders and composite box girders. At any cross sections where the bending stress in either flange exceeds 55 percent of the minimum specified yield strength of the web steel, the compression flange area should not be less than the tension flange area. In this case the top flange area includes the transformed area of any portion of the slab or longitudinal steel bars acting compositely with the steel girder.

In most plate girder bridges, A36 steel is adequate for the shear stresses in the web, whereas the use of higher-strength steel in the flanges will result in a more reasonable plate size provided the strength ratio compares favorably to the cost ratio. However, for very deep girders the extreme fiber stress in bending is only slightly smaller than the stress in the flange plate. In this case, if the actual bending stress in the flange is near the allowable value (e.g., 27 ksi), the extreme fiber stress in the web will most likely exceed the allowable 20 ksi.

Allowable Stresses Recognizing these conditions, AASHTO specifies a bending stress for the web that may exceed the allowable stress for the web steel provided the actual stress in the flange steel does not exceed the allowable stress for the flange steel reduced by a factor R given by

$$R = 1 - \frac{\beta\psi(1 - \alpha)^2(3 - \psi + \psi\alpha)}{6 + \beta\psi(3 - \psi)} \tag{5.35}$$

where α = minimum specified yield strength of the web divided by the minimum specified yield strength of the tension flange

β = area of the web divided by the area of the tension flange

ψ = distance from the outer edge of the tension flange to the neutral axis (transformed section for composite girders) divided by the depth of the steel section

Numerical Example Consider a plate girder designed to resist a bending moment of 5800 ft-kips. We specify A588 steel for the flanges (allowable bending stress 27 ksi) and A36 steel for the web (allowable stress 20 ksi).

We select flange plates 20 in. by 5/8 in. and a web plate 72 in. by 7/16 in. The moment of inertia of the section is

$$I = 13{,}610 + 32.5 \times 2710 = 101{,}685 \text{ in.}^4$$

$$f_b = \frac{5800 \times 12 \times 37.625}{101{,}685} = 25.7 \text{ ksi}$$

We compute

$$\alpha = 36/50 = 0.72 \qquad \beta = 31.5/32.5 = 0.97 \qquad \psi = 0.5$$

The value of R is calculated as

$$R = 1 - \frac{0.97(0.72)(1 - 0.72)^2(3 - 0.5 - 0.5 \times 0.72)}{6 + 0.97(0.5)(3 - 0.5)} = 0.984$$

The allowable bending stress for A588 is $27 \times 0.984 = 26.6$ ksi > 25.7, OK.

This analysis shows that the flange plate is reduced in thickness from 2.25 in. (A36 steel) to 1.625 in. (A588 steel). If this reduction is indicated by cost savings, it must be compared with the unit price for both steel grades.

5-13 TORSIONAL LOADING

This condition is manifested by the application of a force that tends to cause the member to twist about its structural axis. Torsion is usually referred to in terms of torsional moment or torque T, expressed as the product of the externally applied force and the moment or force arm. The principal deflection caused by torsion is measured by the angle of twist or by the vertical movement of one corner of the frame.

Structural rolled or built-up sections are very efficient in resisting torsion. This resistance is greatly improved with the use of torsionally rigid sections. Three basic procedures are available for the best utilization of steel where torsional loads are present.

1. Use closed sections (boxes) where possible.
2. Introduce stiffeners and diagonal bracing.
3. Make rigid end connections.

The solid or tubular round closed section is ideal for torsional loading because the shear stresses are uniform around the circumference of the member. A second choice is a closed square or rectangular tubular section. If the latter consists of plates with uniform thickness as shown in Figure 5-55*a*, the torsional resistance factor R is given by

$$R = \frac{2b^2d^2t}{b + d} \tag{5-36}$$

If the section consists of plates with different thickness as shown in Figure 5-55*b*, the torsional resistance factor is

$$R = \frac{2b^2d^2}{\dfrac{b}{t_b} + \dfrac{d}{t_d}} \tag{5-37}$$

Members having a box section, when butt-welded directly to a primary member, have the fully rigid end connections required for high torsional resistance.

Applications Figure 5-56 shows a typical girder elevation for ramp C, which is part of the four-level I-75/I-275 interchange in Kentucky (Xanthakos, 1971). The bent at pier 3 consists of a closed steel box section supported on two individual columns, marked as supports A and B in Figure 5-57*a*, showing the box girder elevation and the position of the longitudinal girders. A typical box girder cross section is shown in Figure 5-57*b*.

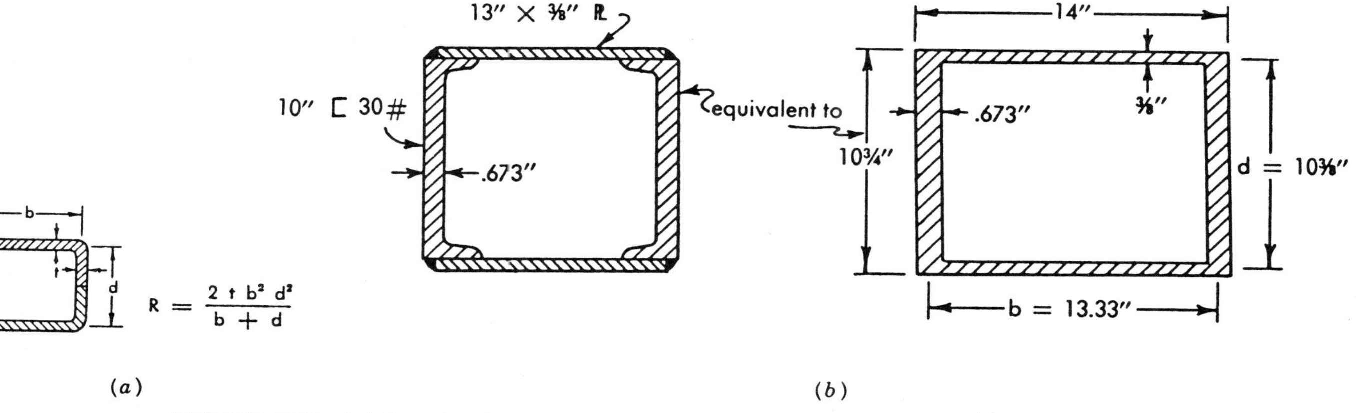

FIGURE 5-55 (*a*) Rectangular section with plates of uniform thickness; (*b*) rectangular box with plates of different thickness.

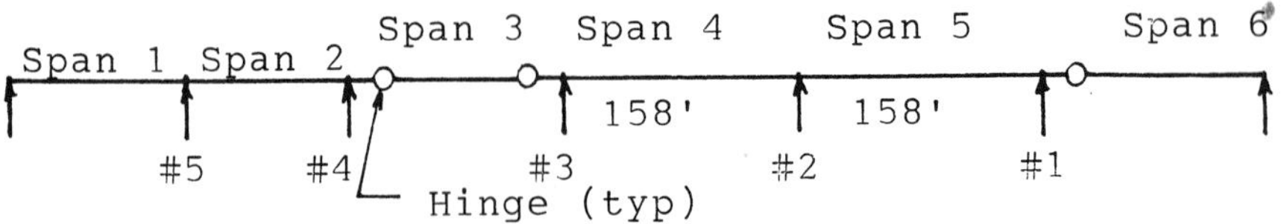

FIGURE 5-56 Typical girder elevation, ramp *C*, I-75/I-275, four-level interchange, Kentucky.

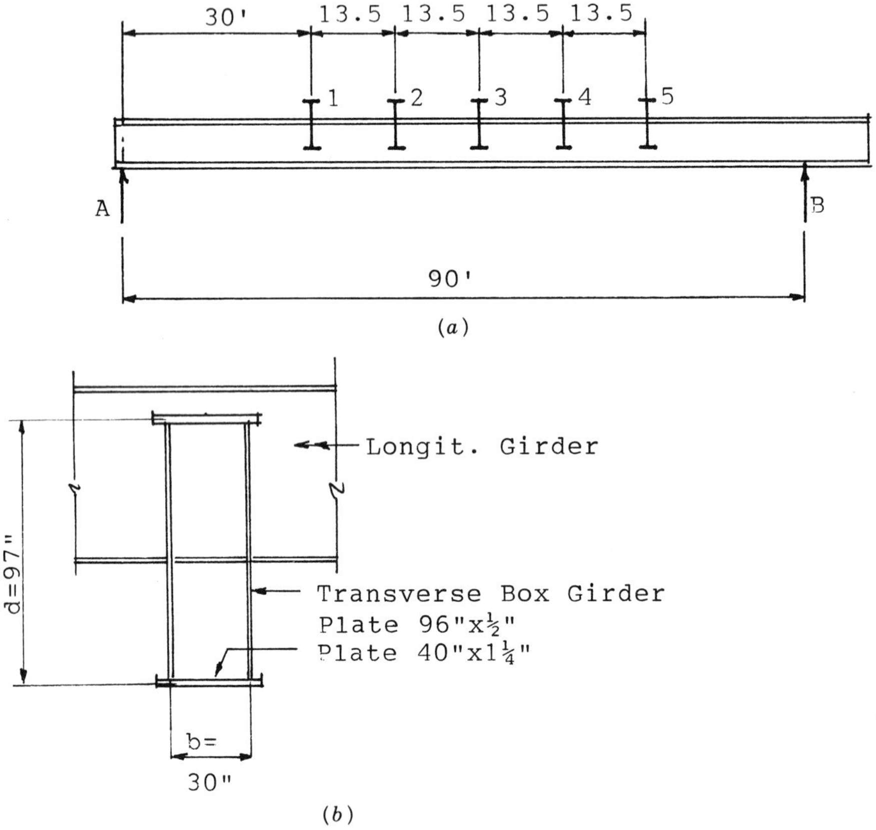

FIGURE 5-57 (*a*) Box girder elevation at pier 3, ramp C; (*b*) box girder section.

The following assumptions are made regarding the support conditions.

1. The longitudinal girders are continuously welded to the bent box girder so that the end connections are rigid. The ends of the girders are therefore restrained against rotation and must be designed for this condition.

2. The ends of the box girder are fixed against rotation in either direction by appropriate bearings.
3. The loading from either span may be applied sequentially, resulting in a differential condition of structural response.

From Figure 5-57*b* the torsional resistance factor is

$$R = \frac{2 \times 30^2 \times 97^2}{(30/1.25) + (96/0.5)} = 78{,}408 \text{ in.}^4$$

The torque in the central portion of the box girder is related to the end moment of the longitudinal connecting girders. Each longitudinal girder carries a uniform dead load $w = 1.5$ kips/ft $= 125$ lb/in., and has an average moment of inertia $I = 50{,}000$ in.^{4}. Referring to Figure 5-56 and considering beam 3-2 simply supported and without end restraints, the end moment M is zero, and the end rotation is given by

$$\theta_e = \frac{wL^3}{34EI} = \frac{125 \times 158^3 \times 12^3}{24 \times 30 \times 10^6 \times 50{,}000} = 0.024 \text{ radian}$$

For the same beam restrained against rotation at the end ($\theta_e = 0$), the end moment is

$$M_e = \frac{wL^2}{12} = \frac{1.5 \times 158^2}{12} = 3120 \text{ ft-kips}$$

Next, we determine the torque that must be applied to each beam location at the central section of the supporting box girder to cause it to rotate by the same amount as the end rotation of the supported beams (assuming simple ends), or by 0.024 radian. Referring to Figures 5-57*a* and 5-58, to torque T_i that must be applied at point i to cause rotation θ_i is found from the expression

$$\theta_i = \frac{T_i ab}{LE_s R} \quad \text{or} \quad T_i = \frac{\theta_i LE_s R}{ab} \tag{5-38}$$

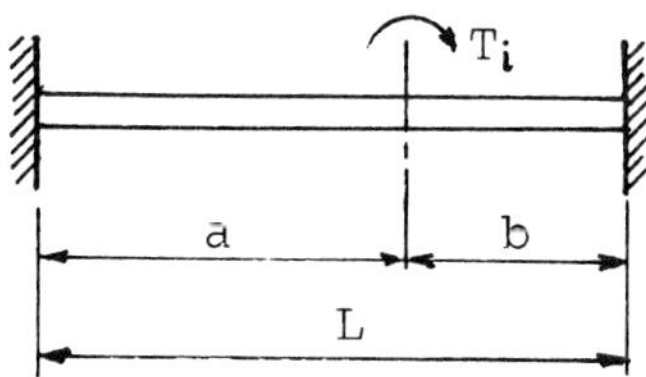

FIGURE 5-58 Rotation and torque for box girder, restrained at the ends against rotation.

$$\text{For beam 1, } T_1 = \frac{(90 \times 12) \times (12 \times 10^6)(78{,}408)(0.0241)}{(30 \times 12)(60 \times 12)} = 7800 \text{ ft-kips}$$

$$\text{For beam 2, } T_2 \qquad = 6970 \text{ ft-kips}$$

$$\text{For beam 3, } T_3 \qquad = 7450 \text{ ft-kips}$$

$$\text{For beam 4, } T_4 \qquad = 10{,}260 \text{ ft-kips}$$

$$\text{For beam 5, } T_5 \qquad = 27{,}900 \text{ ft-kips}$$

(Note that $E_s = 12 \times 10^6$ psi.)

A moment–rotation chart shows the relationship in Figure 5-59. The straight lines through the origin ($M = \theta = 0$) represent the applied torque T and angular rotation θ at the central section of the supporting box girder. A

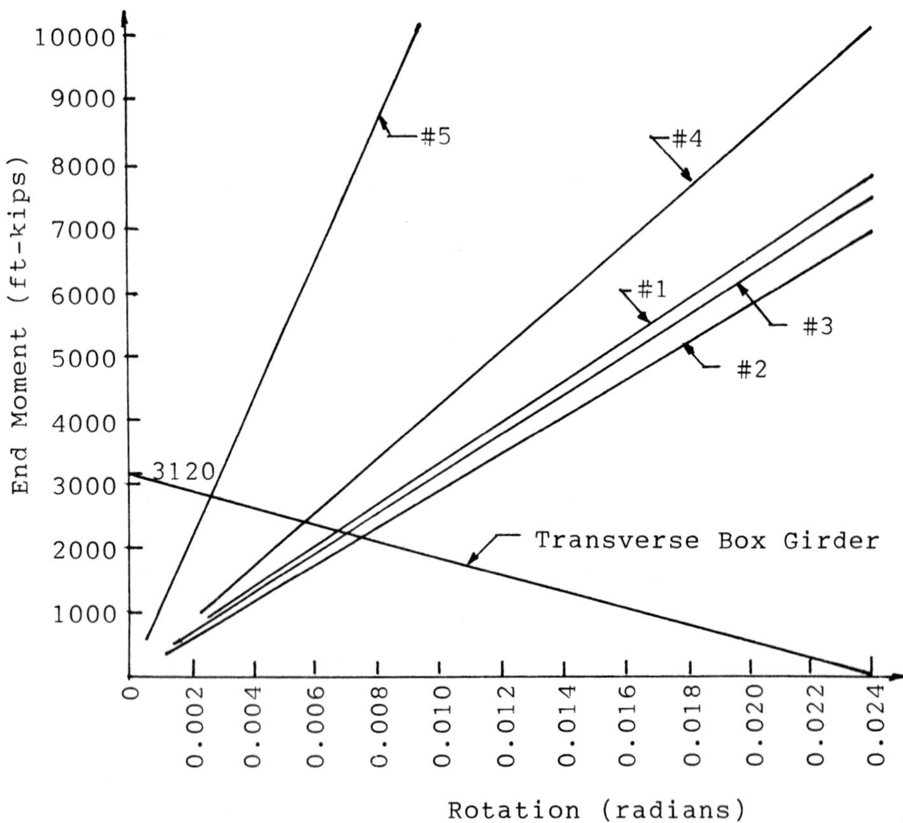

FIGURE 5-59 Moment–rotation chart, box girder of Figure 5-57.

straight line in the opposite direction represents the end moment M_e and rotation θ_e of the supported beams. The intersecting points define the actual end moment or torque T and the corresponding rotation at the point under consideration. From a graphical solution, we determine

$$T_1 = 2300 \qquad T_2 = 2200 \qquad T_3 = 2250 \qquad T_4 = 2450 \qquad T_5 = 2850 \text{ (ft-kips)}$$

The end torque at support B, assumed torsionally fixed, for the box girder is

$$T_B = 2850 \times \frac{84}{90} + 2450 \times \frac{70.5}{90} + 2250 \times \frac{57}{90}$$

$$+ 2200 \times \frac{43.5}{90} + 2300 \times \frac{30}{90} = 7830 \text{ ft-kips}$$

Torsional Shear Stresses in Box Girder The shear stress in the flange is

$$\tau_b = \frac{T}{2[A]t_b} = \frac{7830 \times 12 \times 1000}{2 \times 97 \times 30 \times 1.25} = 12{,}900 \text{ psi}$$

where $[A] = bd$ (Figure 5-57*b*).

Likewise, the shear stress in the web is calculated as

$$\tau_d = \frac{7830 \times 12 \times 1000}{2 \times 97 \times 30 \times 0.50} = 32{,}250 \text{ psi} > \text{allowable}$$

Because the design is not satisfactory, three possible solutions are considered: (a) provide a roller-type bearing at supports A and B of the box girder to allow rotation in the direction of the beam; (b) arrange the concrete placement on either side of the bent so as to avoid differential loading; and (c) increase the web thickness and web spacing. For example, if the web is 1 in. thick and the spacing is 4 ft, the shear stress is $32{,}250 \times 0.5 \times 2.5/4 = 10{,}060$ psi, OK. However, a spherical bearing that accommodates rotation in both directions satisfies the functional requirements of the superstructure and is the preferred choice.

Torsional deflection in plate girders is reviewed in more detail in subsequent sections dealing with curved bridges.

5-14 FATIGUE CONSIDERATIONS

The analysis of the probability of fatigue of steel members or connections under service loads and the allowable range of stress for fatigue are covered in AASHTO Article 10.3. Fatigue is covered in detail in Sections 12-7 through 12-13. Certain considerations, however, relate to the scope of this chapter and are reviewed here in general terms.

Nature of Fatigue Loading Increased vehicular traffic intensity and extensive use of weldment have made fatigue an integral part of the design considerations of steel bridges. When a structural member is subjected to constant-amplitude cyclic stress, it will fail after a number of cycles of stress applications. In reality, the fatigue strength of materials is random. The failure is progressive over a period of time and begins with a plastic movement within a localized region. Although the average unit stresses across a cross section may be well below the yield strength, a nonuniform stress distribution may cause the yield point to be exceeded within a small area and lead to plastic movements, eventually manifesting a minute crack. The localized plastic movement further aggravates the nonuniform stress distribution, and further plastic movement causes the crack to progress (see also Section 12-8).

The fatigue load may be represented by a stress varying from a maximum value $\sigma_{\max}$ to a minimum $\sigma_{\min}$, and these define the range of stress. Tension and compression stresses have opposite signs, and therefore they are added to obtain the stress range. The cycle is represented by the ratio

$$K = \frac{\sigma_{\min}}{\sigma_{\max}} \tag{5-39}$$

One approach to the problem is to let the variable stress σ_v be the ordinate and let the steady or mean stress σ_m (shown in Figure 5-60) be the abscissa. Referring to Figure 5-61, when the mean stress σ_m is zero, the variable stress σ_v becomes the value for a complete reversal of stress σ_r. This value would be determined by tests, and becomes point b in the diagram. When the stress σ_v becomes zero, the maximum resulting mean stress σ_m is equal to the ultimate stress σ_u for a steady load; this becomes point a.

The following stresses are distinguished in Figure 5.61:

σ_r = fatigue strength for a complete reversal at stress
σ_v = variable stress superimposed upon steady stress
σ_u = ultimate (or yield) strength under steady load
σ_m = mean (average) stress .

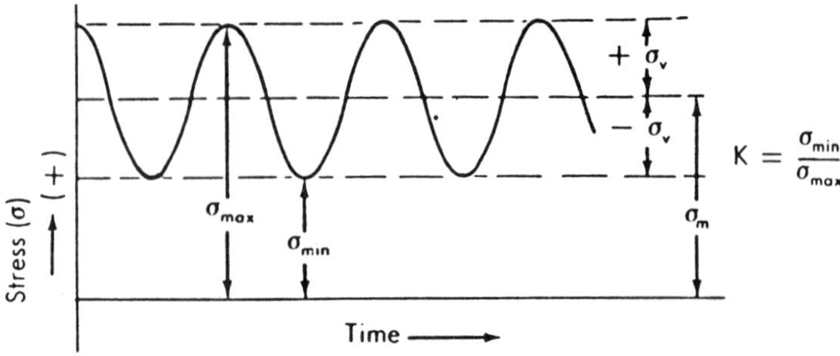

FIGURE 5-60 Typical fatigue load pattern.

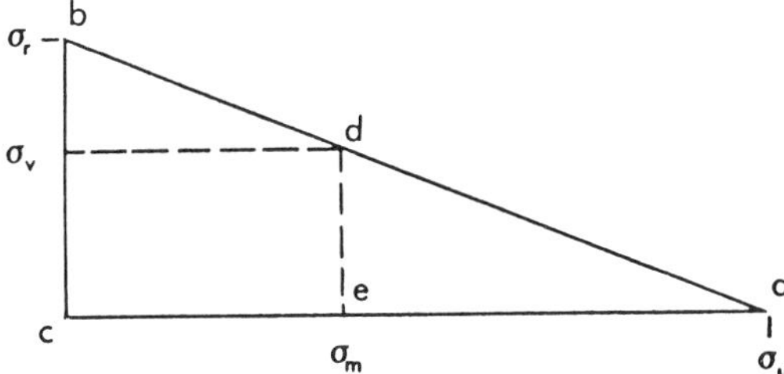

FIGURE 5-61 Stress reversal and ultimate strength under steady load.

The line connecting points a and b indicates the relationship between the variable stress σ_v and the mean stress σ_m for any type of fatigue cycle and for a given fatigue life N. The straight-line approach yields conservative values, and almost all of the test data will lie just outside of this line. From Figure 5-61, we obtain

$$\frac{\sigma_v}{\sigma_r} + \frac{\sigma_m}{\sigma_u} = 1 \tag{5-40}$$

Fatigue Strength Fisher et al. (1974) have carried out extensive tests regarding the effects of weldments on the fatigue strength of steel beams that cover most of the types commonly used in highway bridges. Under the passage of vehicles, sections of structural members of bridge superstructures are not necessarily subjected to constant-amplitude cyclic stresses, and the number of stress cycles experienced by the sections in their lifetime are not known with certainty. However, for design purposes the specifications assume constant-amplitude cyclic stresses as in Figure 5-60, and assign several values to the number of stress cycles for which the bridge must be designed, depending on the type of road (see also Section 12-9).

Associated with a given value of stress amplitude, a probability function of fatigue life may be determined as shown in Figure 5-62. For any specific value S_1, the associated probability function provides a measure of the likelihood of failure. Given the value S_1, the probability that the member may fail after N_1 cycles is the shaded area under the probability function. The derivation on the S–N diagram is improved if a point is marked off so that the shaded area to its left is equal to a prescribed value, for example, 0.05. The curve joining the points is refereed to as the P–S–N curve. If the number of cycles N is stated and the probability of fatigue failure must be limited (e.g., to 0.05), the allowable stress S_o from which the member section may be selected is determined from the P–S–N curve.

The intent of the fatigue provisions is to assign the vehicle loading and the number of stress cycles, in conjunction with the member category. Allowable stresses are then determined from the P–S–N curves. In this manner, the random characteristics of fatigue strength are explicitly retained, but those of vehicular traffic are only implicitly incorporated (see also Section 12-9).

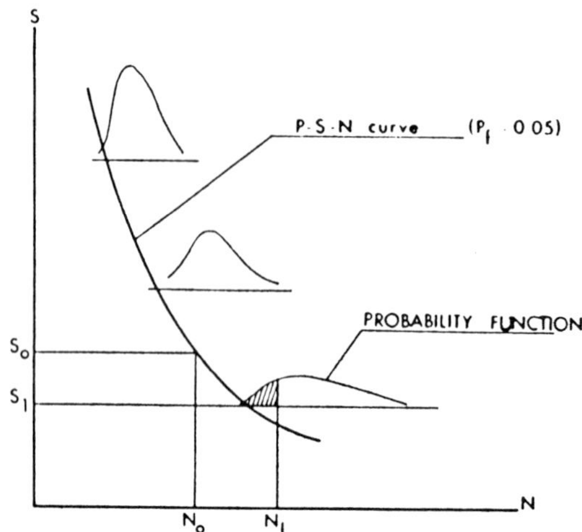

FIGURE 5-62 Fatigue life and scatter in S–N relationship.

Materials fatigue is thus a complex phenomenon, and there exists no universally accepted theory even for simple structural members under deterministic loading conditions. The fatigue behavior of structures subjected to random dynamic loads still constitutes a formidable problem because of its sensitivity to materials characteristics, structural geometry, loading history, and physical conditions.

Topics such as the relative severity of the fatigue problem, combined fatigue stresses, influence of welded connections and joint design, and further research needed are discussed in some detail in Chapter 12.

Infinite Fatigue Life The proposed LRFD specifications (Modjeski and Masters, 1992) articulate an infinite fatigue life if the single-lane ADTT equals or exceeds 2000 trucks per day. The fatigue resistance is taken as

$$F_n = (\Delta F)\text{threshold}$$

where F_n is the nominal infinite life fatigue resistance and (ΔF)threshold is the constant-amplitude fatigue limit. The latter is termed the allowable fatigue stress range for over 2,000,000 cycles on a redundant load path structure in the last column of AASHTO Table 10.3.1A.

Inasmuch as fatigue is the initiation and/or propagation of cracks due to a repeated variation of normal stress with a tensile component, it can be categorized as either load-induced or distortion-induced fatigue. The latter is caused by secondary compatibility stresses not quantified in the typical analysis and design of a bridge.

5-15 LRFD APPROACH: BASIC PRINCIPLES

Resistance Factors

The AASHTO standard specifications stipulate a reduction factor ϕ as follows: (a) $\phi = 0.85$ to be applied to the ultimate strength of shear connectors and (b) a factor ϕ is applied to the nominal strength R for A325 and A490 high-strength bolts subjected to applied axial tension or shear.

The LRFD specifications extend the factor ϕ to cover a wide range of design cases as follows:

For flexure	$\phi_f = 1.00$
For shear	$\phi_v = 1.00$
For axial compression, steel only	$\phi_c = 0.85$
For axial compression, composite	$\phi_c = 0.95$
For tension, fracture in net section	$\phi_u = 0.95$
For tension, yielding in cross section	$\phi_y = 0.90$
For bearing	$\phi_b = 1.00$
For A325 and A490 bolts in tension	$\phi_t = 1.00$
For A325 and A490 bolts in shear	$\phi_s = 0.55$
For shear connectors	$\phi = 0.85$

Resistance factors ϕ are also assigned to welds and weld metal.

Bending of Straight I Sections

This discussion applies to bending of rolled or built-up straight steel sections, composite or noncomposite, stiffened or unstiffened, homogeneous or hybrid, prismatic or nonprismatic, and symmetrical about the vertical axis in the plane of the web. These are analyzed in conjunction with decks of reinforced concrete, prestressed concrete, orthotropic steel, and timber flooring. Both the flanges and the web can have different F_y values. Nonprismatic members include haunched sections, variable in depth or with constant depth, with variations in the flange size.

Composite hybrid sections consisting of a 50-ksi bottom flange and a 36-ksi web and top flange are permitted, because these are assumed to be particularly efficient in positive moment regions. Likewise, sections with a higher-strength steel in the web are permitted to accommodate high shears, but they are not considered hybrid girders by definition. Particular emphasis is placed on the strength limit state and on the control of permanent deflections, as well as on the fatigue requirements for the webs.

Negative Flexure Slab Reinforcement This provision applies to continuous beams and girders, with or without shear connectors in negative moment regions. The logitudinal reinforcement should not be less than 1 percent of the cross-sectional area of the slab, and should have a minimum yield strength not less than 60 ksi. In this strength range, the steel bars are expected to remain elastic even if inelastic redistribution of negative moments occurs. The intent is to enhance elastic recovery after the overload is removed, and this should tend to close the slab cracks (AASTO, 1991; Haaijer, Carskaddan, and Grubb, 1987).

The use of shear connectors is encouraged in the negative moment regions. If they are omitted in this section of the bridge, the longitudinal reinforcement should be extended into the positive bending region.

Procedures for Strength Limit State Elastic procedures should be used for analyzing bending moments (calculating load effects) in simple spans, because the moments in simple-span members are not affected by yielding. For compact sections, the calculated elastic moments are compared with the plastic moment capacity of the sections. We should note that the current (1992) AASHTO specifications recognize inelastic behavior by utilizing the plastic moment capacity of compact sections and by permitting an arbitrary 10 percent redistribution of peak negative moments at both overload and maximum load.

For continuous spans, the strength limit state may be checked by either elastic or inelastic procedures. However, inelastic procedures should be used only for composite or noncomposite members with constant depth and a specified minimum yield strength not exceeding 50 ksi.

Elastic Procedures The design factored moment M_u,, due to the factored loads, computed at each cross section by elastic procedures (assuming materials linearity) should not exceed the corresponding factored bending resistance ϕM_n, or

$$M_u \leq \phi M_n = M_r \tag{5-41}$$

where the nominal bending resistance M_n is specified for composite and noncomposite sections. A moment redistribution is permitted as in Article 10.48.1.3 of the current AASHTO specifications. The application of this provision is illustrated in Figure 5-63 for a two-span continuous beam. A 10 percent reduction of the negative moment due to the factored loads is applied to the interior support. A corresponding increase is applied to the positive factored moment in the span. At the point of the maximum possible moment (assumed at 0.4 pt), the increase in the positive span moment is $0.04M$ as shown.

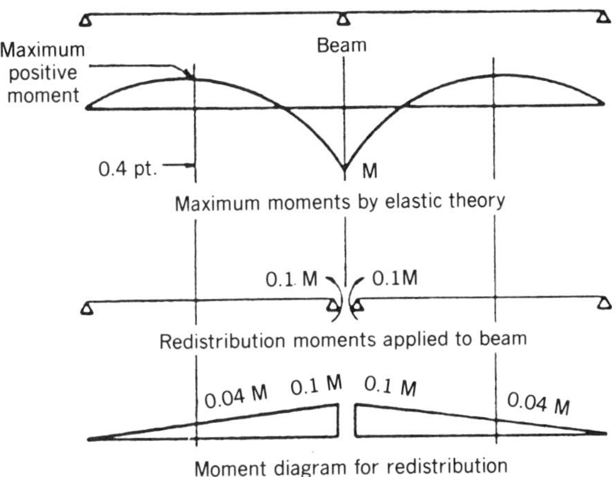

FIGURE 5-63 Moment redistribution at interior supports for compact sections, partially associated with plastic design of continuous beams.

Likewise, the design shear V_u, due to the factored loads, computed at each relevant cross section using elastic analysis should not exceed the corresponding factored shear resistance ϕV_n, or

$$V_u \leq \phi V_n = V_r \tag{5-42}$$

where V_n is the nominal shear strength.

Bending Stresses The extreme fiber stress f_f in both flanges for either positive or negative bending moment should not exceed the following:

Composite sections	$f_f \leq 0.95 R_h F_{yf}$	(5-43*a*)
Noncomposite sections	$f_f \leq 0.80 R_h F_{yf}$	(5-43*b*)

where R_n and F_{yf} are the flange stress reduction factor and yield stress, respectively. The factor R_h reflects the nonlinear variation of stresses due to local yielding of lower-strength webs under bending stresses, and does not apply if a suitable longitudinal stiffener is provided. The factor R_h is unity for nonhybrid selections and for hybrid sections where the stress in both flanges does not exceed the yield stress of the web.

Fatigue Requirements for Webs Certain provisions are introduced to control the flexing of webs due to bending or shear under repeated live load. They involve a check of the maximum web-buckling stresses rather than of the stress ranges caused by cyclic loading, and they do not apply to sections with longitudinally stiffened webs.

Composite Sections If adequate shear connectors are provided throughout the length of the bridge, sections in both positive and negative bending are considered composite. When calculating stresses from bending moments in negative bending regions, the composite section consists of the steel girder plus the longitudinal reinforcement.

The sequence of loading is reflected in calculating the stresses caused by loads applied separately to the steel alone, to the short-term composite section, and to the long-term composite section.

Both the yield moment and the plastic moment resistance are relevant. The former, denoted as M_y, is needed only for the strength limit state check of certain types of composite sections. The plastic moment resistance M_p is defined as the resultant moment of the fully plastic stress distribution acting on the composite section (see also Figure 4-36), and the position of the plastic neutral axis (PNA) is determined ignoring tension in the concrete.

Compact sections are proportioned to reach their full plastic moment capacity. In continuous spans, however, yielding above the moment M_y can cause a redistribution of moments. The attainment of a plastic mechanism is presented by limiting the moment capacity of positive bending sections to a value less than or equal to M_p. Furthermore, the shape factor M_p/M_y for composite sections in positive bending can reach a value of 1.5, implying that considerable yielding must occur to get from M_y to M_p with a corresponding reduction in the effective stiffness. In continuous spans, the reduction in stiffness can shift moments from positive to negative moment regions, that is, actual negative moments may be higher than predicted by elastic analysis.

Noncompact sections can reach the specified minimum yield stress in at least one flange. These sections must satisfy, however, certain web slenderness, compression flange slenderness, and compression flange bracing requirements.

Provisions regarding lateral–torsional buckling are introduced for members exhibiting a variation in flange size and/or section depth between brace points, as well as for members with a constant cross section between brace points. The behavior of a compression flange under the effect of lateral buckling between brace points is analogous to that of a column, but the torsional stiffness of the bending member enhances the buckling resistance of the compression flange. The effect of variation in the compressive force along the length between brace points is considered by using a factor C_b that approaches a minimum value of 1 when the force, and corresponding moment, is constant over the entire length.

Noncomposite Sections The yield moment strength M_y of a noncomposite section is the moment required to cause first yielding in either flange when any web yielding in hybrid sections is disregarded. The plastic moment resistance M_p is the resultant moment of the full plastic stress distribution acting on the section (see also Figure 4-38).

Noncomposite members meeting the compact section requirements are allowed to reach their full plastic moment capacity, that is, the nominal bending resistance is $M_n = M_p$. The limitations for moments above M_y specified for compact composite positive bending sections in continuous spans do not apply in this case because the shape factor M_p/M_y is much lower for noncomposite sections. Therefore, the shift in moment from the positive to the negative moment region is small and can be neglected.

Noncompact sections may reach the yield moment subject to certain restrictions. For noncompact sections in both positive and negative bending, the nominal bending resistance of each flange is limited by

$$F_n = R_b R_h F_{yf} \tag{5-44}$$

where R_b and R_h are flange stress reduction factors and F_{yf} is the yield stress capacity.

For compression flanges that do not satisfy the bracing requirements, the lateral–torsional buckling provisions for composite sections apply.

Shear The factored shear resistance is given by (5-42). Expressions for the nominal shear resistance V_n are provided for (a) sections without stiffeners, (b) sections with transverse stiffeners only, and (c) sections with both transverse and longitudinal stiffeners.

Concurrent moment affects the specified shear strength only if tension field action is utilized in carrying the shear (see also Section 5-10). Separate interaction equations are given to define this effect for compact sections and noncompact sections to reflect the design philosophy. Compact selections are designed in terms of moments, whereas noncompact sections are designed in terms of stresses.

Shear Connectors Simple-span composite bridges should have shear connectors throughout their length. Continuous-span composite bridges should preferably be provided with shear connectors along the entire bridge length.

Inelastic Procedures Inelastic methods of analysis may be applied to constant-depth continuous beams, either composite or noncomposite, that have a specified yield strength not exceeding 50 ksi and that satisfy certain requirements discussed in this section. The general principles of inelastic (plastic) analysis are reviewed in Section 4-17 including collapse mechanisms and equilibrium requirements. Basically, the proposed procedures are similar to the alternate load factor design (AASHTO, 1991) adapted as guide specifications.

Unlike the elastic methods where the required cross section at any location is explicitly defined by the moment envelope, inelastic procedures permit different combinations of cross sections to be used in both the positive and negative bending regions. Suitable inelastic analytical methods include (a) the mechanism method (ASCE, 1971) and (b) the unified autostress method (Schilling, 1989, 1991).

Computer programs are generally required to apply these methods efficiently for continuous beams and girders with more than two spans. The two methods are applicable to both compact and noncompact sections provided the plastic rotation characteristics are known. Because, however, these characteristics are not yet adequately established for the full range of noncompact section geometries, inelastic procedures in this case are not permitted for either the strength or service limit state.

The ultimate load-carrying capacity of a continuous beam is said to have been reached when enough plastic hinges have developed to form a mechanism. With the exception of the last hinge, all others must sustain more plastic rotation after reaching the plastic moment capacity. Because it is conceivable that the additional amount may exceed that provided by compact sections satisfying the web slenderness, compression flange slenderness and compression flange requirements are specified for composite and noncomposite bridges. Hence, certain restrictive conditions are introduced to ensure that the sections can sustain sufficient plastic rotations up to incipient mechanism formation.

Mechanism Method Bending at sections required to sustain plastic rotations must satisfy the expression

$$M_n = M_{pe} \tag{5-45}$$

where M_n is the nominal moment resistance at a plastic hinge required to sustain plastic rotations necessary to form a mechanism and M_{pe} is the nominal effective plastic moment resistance.

For bending at other sections, the nominal bending resistance must satisfy the expressions

$$M_n = M_p \qquad \text{(Compact section)} \tag{5-46a}$$

$$M_n = R_b R_h M_y \qquad \text{(Noncompact section)} \tag{5-46b}$$

where R_b, R_h, and M_p are as given previously and M_y is the yield moment.

The nominal effective plastic moment resistance is the resultant moment of the fully plastic stress distribution based on specific effective yield stresses. The introduction of an effective plastic moment (usually smaller than the full plastic moment) is intended to enable certain flange and web sections with a slenderness ratio too high to be designed by the mechanism method. The ALFD guide specifications give an empirical procedure for calculating the effective plastic moment for compact sections (Grubb and Carskaddan, 1981; Haaijer, Carskaddan, and Grubb, 1987). In this procedure the effective plastic moment is calculated by applying effective yield stresses to the flanges and web of the section, depending on the specified flange and web slenderness ratios. When these ratios are below limiting values, the effective yield stress may be taken as the actual yield stress; otherwise, the effective yield stresses are below the actual yield stresses, and the effective plastic moment is below the actual plastic moment.

Unified Autostress Method This procedure is described in detail in the ALFD guide specifications and in Schilling (1991). It may be used to proportion flanges and webs of girders designed by inelastic methods. Factored plastic rotation curves are used at all locations were yielding is assumed to occur. Each factored plastic rotation curve is obtained from the nominal plastic rotation curve by multiplying the ordinates by appropriate reduction factors ϕ_f.

Nominal plastic rotation curves are shown in Figure 5-64 for negative bending sections. These curves should not be applied, however, to plastic rotations greater than 0.008 radian. They are straight-line approximations of two plastic rotation curves (ALFD, 1991), and cover the loading portion of the plastic rotation curve that is needed in the permanent deflection service limit state check. The curves are independent of the geometric proportions of the sections, except as they may affect the maximum moment capacity.

The unified autostress method may also be used for calculating the actual moment redistribution resulting from yielding in negative bending in continuous beams.

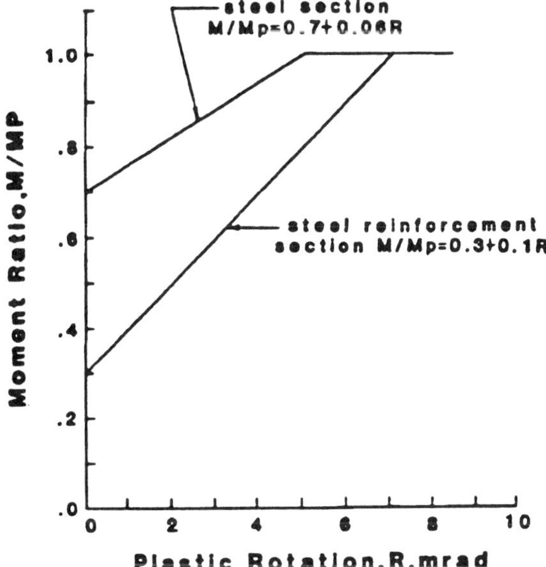

FIGURE 5-64 Nominal plastic rotation curves.

Bending of Straight Box Girders

This discussion applies to straight steel multiple or single composite box sections, stiffened or unstiffened, homogeneous or hybrid, with constant depth or haunched, and symmetrical about the vertical axis in the plane of the web. The design requirements include (a) the strength limit state, (b) fatigue consideration for webs, (c) constructibility provisions, and (d) other applicable limit states. Data and procedures regarding the design of long-span steel box girder bridges are included in the document "Proposed Specifications for Steel Box Girder Bridges" (Wolchuk, 1980), discussed in Section 5-10, and also in Wolchuk (1990).

Multiple box sections must be proportioned as shown in Figure 5-65. In addition, when nonparallel box sections are used, the distance center-to-center of adjacent flanges at supports should neither be greater than 135 percent nor less than 65 percent of the distance center-to-center of the flanges of each adjacent box.

A single box section may have its top either open or closed with a steel plate, but the member should be positioned to have the center of gravity of the dead load as close to the shear center of the box as is practical. If the girder is closed with a single top flange plate, the plate should be checked for buckling prior to curing of the concrete deck. Shear connectors should be placed across the top deck to ensure plate support against buckling.

Analysis Analysis of single box sections may consist of a single-line model, but both torsional and flexural effects must be investigated. The single box should not be considered torsionally rigid unless adequate internal bracing is provided to maintain the cross section.

Live loads at extremes of the deck can induce critical torsional loads, whereas the bending moments are nominal.

Strength Limit State The factored resistance F_r in terms of stress is likewise expressed as $F_r = \phi_f F_n$, where ϕ_f is the resistance factor for flexure and F_n is the nominal bending resistance.

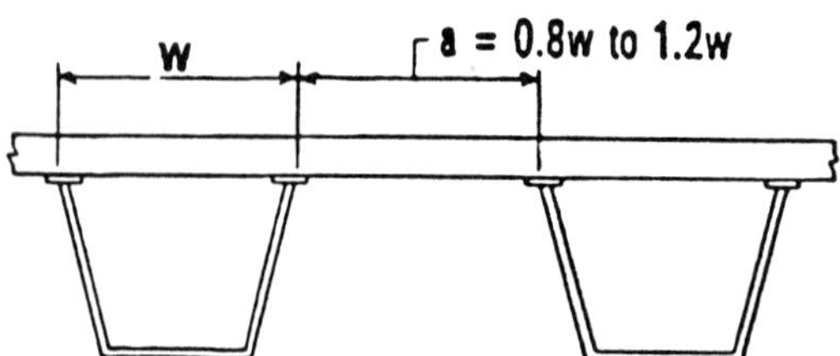

FIGURE 5-65 Proportioning of multiple box sections, parallel boxes.

Positive Bending Sections The nominal bending resistance for each flange of multiple box sections and the top flange of single box sections should be determined as in I plate girders, noncompact sections. The tensile strength of the bottom flange of single box sections is affected by torsional shear stresses. The von Mises yield criterion is applied to the effect of shear stress. The combined result of torsional shear and normal flexure is difficult to determine, and the worst case of either may be superimposed to obtain a conservative estimate. Thus, the nominal bending resistance of the bottom flange of single box sections is given by

$$F_n = R_b R_h F_{yf} \sqrt{1 - 3\frac{f_v}{F_{yf}}} \qquad (5\text{-}47)$$

where F_{yf} is the specified minimum yield stress of the bottom flange (ksi), f_v is the maximum torsional shear stress in the flange due to the factored loads (ksi), and R_b, R_h are flange stress reduction factors as given previously.

For simple spans, the bottom flanges of multiple and single box sections are fully effective in resisting bending if the width does not exceed one-fifth of the span. If this ratio is exceeded, only a width of one-fifth of the span should be considered effective in resisting bending. For continuous spans, this criterion is applied to the distance between points of dead load contraflexure.

The foregoing criteria are the result of analysis of actual box girder designs using a series of folded-plate equations (Goldberg and Leve, 1957). The bridges included in this study had span–flange width ratios ranging from 5.65 to 35.3. The effective flange width was estimated as 0.89 to 0.99 times the total flange width. On this basis, it appears reasonable to assume the flange plate to be fully effective provided its width does not exceed one-fifth of the span.

Negative Bending Sections For multiple and single box sections, the nominal bending resistance is determined for two cases: (a) for compression flanges with longitudinal stiffeners and (b) for compression flanges without longitudinal stiffeners. There are no specific checks on compression flange slenderness or compression flange bracing at negative bending sections.

The provisions for compression flanges with longitudinal stiffeners only are based on the theory of elastic stability (Timoshenko and Gere, 1961). A general discussion of the problem of reduction of critical buckling stresses due to the presence of torsional shear is given by Johnston (1966).

Shear For multiple box sections, one-half of the distribution factor for moment should be used in calculating the live load vertical shear in each box section web. For inclined webs the depth of the web plate should be measured on the slope. Each web should be designed for shear V_{ui} due to the factored loads taken as $V_{ui} = V_u/\cos\theta$, where θ is the angle of inclination of the web plate to the vertical.

For single box sections, the absolute value of the maximum flexural and torsional shears should be added.

Stiffeners The specifications contain provisions for (a) web stiffeners (transverse intermediate stiffeners and longitudinal stiffeners) and (b) compression flange stiffeners (longitudinal and transverse). When longitudinal compression flange stiffeners are used, preference should be given to at least one transverse stiffener placed on the compression flange near the point of dead load contraflexure. If the design includes both longitudinal and transverse stiffeners, stresses in the bottom flange should be evaluated using appropriate methods of analysis (SSRC, 1988).

Constructibility Box sections should be checked for strength and stability during construction, including sequential deck placement (see also Section 5-16). Possible eccentric loads during construction should be considered. These may include uneven placement of concrete and construction equipment. The individual box section geometry should be maintained during each construction phase. The need to provide temporary or permanent interior diaphragms or cross frames, exterior diaphragms or cross frames, top lateral bracing, or other support means should be investigated to control deformations during fabrication, erection, and placement of the concrete deck.

Connections of overhang deck forms to the top flanges may cause lateral bending of the top flanges, and these effects should be considered. Top flanges of through-type box sections should be designed as braced only at points where internal cross frames or top lateral bracing are attached for loads applied prior to curing of the concrete deck.

Welded Connections

The factored resistance of a welded connection is governed by the resistance of either the base metal or the deposited weld metal. The nominal resistance of fillet welds is determined from the effective throat area, whereas the strength of the connected parts is governed by their respective thickness.

Complete-Penetration Groove-Welded Connections The factored resistance R_r of complete-penetration groove-welded connections subjected to tension or compression normal to the effective area or parallel to the axis of the weld is taken as

$$R_r = \phi_{cp} F_y \qquad (5\text{-}48a)$$

where F_y is the specified minimum yield stress of the connected material and

ϕ_{cp} is the resistance factor for tension/compression in complete-penetration welds, the same as in base metal.

The factored resistance of complete-penetration groove-welded connections subjected to shear on the effective area should be taken as the lesser of the two values given by the following:

$$R_r = 0.6\phi_{cp}F_y$$
$$R_r = 0.6\phi_{e1}F_{exx} \qquad (5\text{-}48b)$$

where F_{exx} is the classification strength of the weld metal (ksi) and ϕ_{e1} is the resistance factor for the weld electrode, or 0.80.

In groove welds the maximum forces are usually tension or compression. Tests show that groove welds of the same thickness as the connected parts have adequate strength to develop the factored resistance of the connected parts.

Partial-Penetration Groove-Welded Connections The factored resistance of partial-penetration groove-welded connections subjected to tension or compression parallel to the axis of the weld or compression normal to the effective area should be taken as in (5-48*a*).

The factored resistance of partial-penetration groove-welded connections subjected to tension normal to the effective area should be the lesser of the following values:

$$R_r = \phi_{cp}F_y$$
$$R_r = 0.6\phi_{e1}F_{exx} \qquad (5\text{-}49a)$$

where ϕ_{e1} is the resistance factor for the weld electrode, or 0.80.

Likewise, the factored resistance of partial-penetration groove-welded connections subjected to shear parallel to the axis of the weld is the lesser of the factored nominal resistance of the connected material or the factored resistance of the weld metal given by

$$R_r = 0.6\phi_{e2}F_{exx} \qquad (5\text{-}49b)$$

where ϕ_{e2} is the resistance factor for the weld electrode, or 0.75.

Fillet-Welded Connections Fillet welds subjected to shear have a factored resistance that depends on the strength of the weld metal and the direction of the applied load (this may be parallel or transverse to the weld). In either case the weld fails in shear, but the plane of rupture is different. Because shear yielding is not critical in welds, the factored shear resistance is based on the shear strength of the weld metal multiplied by a suitable resistance factor. Hence, the factored resistance is taken as the lesser of either the factored resistance of the connected material or as given by

$$R_r = 0.6\phi_w F_{exx} \tag{5-50}$$

were $\phi_w = 0.75$.

If fillet welds are subjected to eccentric loads producing a combination of shear and bending stresses, they must be proportioned on the basis of a direct vector addition of the shear forces on the weld.

Splices

Provisions are included for connection elements in tension and shear, and apply to splice plates, gusset plates, angles, brackets, and other connection devices.

The factored resistance R_r in tension is taken as the least of the three values given by

For yielding $\quad R_r = \phi_y P_{ny}$

For fracture $\quad R_r = \phi_u P_{nu} \quad$ (5-51*a*)

For block shear rupture $\quad R_r$ is as specified for this condition

where P_{ny} = nominal tensile resistance for yielding
P_{nu} = nominal tensile resistance for fracture
ϕ_y = resistance factor for yielding in tension, or 0.90
ϕ_u = resistance factor for fracture in tension, or 0.95

The factored resistance R_r in shear is taken as

$$R_r = 0.58\phi_v A_g F_y \tag{5-51b}$$

where A_g = gross area of the connection element
F_y = specified minimum yield stress of the connection element
ϕ_v = resistance factor for shear

Bolted Splices Web splices of bending members should be designed for the portion of the factored design moment resisted by the web, and for the moment due to the eccentricity of the shear produced by the factored loads. The splices should also be checked for a shear equal to the shear due to the factored loads multiplied by the ratio of the design moment and the moment due to the factored loads of the splice.

For bolt groups subjected to eccentric shear, a usual approach is to treat the group as an elastic cross section subjected to direct shear and torsion. A vector analysis is carried out assuming that there is no friction and the plates are rigid and the bolts are elastic. Recent experimental work has led to the concept of plastic analysis assuming that the eccentrically loaded fastener group rotates about an instantaneous center of rotation and the deformation of each fastener is proportional to its distance from the center of rotation (Kulak, Fisher, and Struik, 1987; Polyzois and Frank, 1986).

At the strength limit state, the bending stress in the splice plates computed on the net section should not exceed the specified minimum yield stress of the splice plates. The shear due to the factored loads should not exceed the factored shear resistance.

Flange splice plates should be designed for the portion of the bending moment distributed to these members, and should not include the portion of the moment resisted by the web. Preferably, splices should be located at points where there is an excess of section.

5-16 STRENGTH OF STEEL BOX GIRDER BRIDGES

The usual distinction in steel box girder bridges articulates (a) large, long-span composite box girders or box girders with steel orthotropic deck systems and (b) smaller shorter-span bridges with a concrete deck acting compositely with the steel girder. Most of the steel box girder bridges built in the United States are of the second type with average spans of 200 ft (see also Section 5-10).

Construction Failures of Long-Span Box Girders Since 1969 several major and minor failures of box girder bridges have occurred. The four most important construction failures are described in the following sections.

Milford Haven Road Bridge This bridge is located in Wales, United Kingdom, and failed in June 1970 (Merrison Committee, 1973). It consists of a welded box girder superstructure with a main span of 700 ft, claimed to be Europe's longest bridge without cable support. The trapezoidal box is 20 ft deep, 66 ft wide at the top, and 22 ft wide at the bottom.

For the first span on each side of the bridge, the box sections were erected on falsework and then welded together. On the next span the box sections were cantilevered using a mobile crane moving along the deck as construction proceeded. The final section of the 252-ft side span had almost reached the next high pier when the span buckled on top of the pier, and the unsupported cantilever end collapsed.

The proximate cause of the failure was buckling of the load-bearing diaphragm over the pier. Three unforeseen hazards were cited: (a) the diaphragm could have been as much as 3/4 in. out of flat, making it susceptible to local buckling; (b) the bearing on the pier was out of line with the neutral axis of the diaphragm and could impose bending moment; and (c) some bolts intended to be loose in oversized holes were tight, and under load movement they could have torn the longitudinal stiffeners from the bottom flange and made it unstable in compression. In addition, subsequent calculations indicated that the diaphragm would be prone to failure when the reactions at the pier supporting the cantilevered span exceeded 900 long tons. At the time of failure the reaction was 963 long tons.

Fourth Danube Bridge, Vienna, Austria The design of this bridge incorporated three different types of construction: (a) prestressed concrete box girders with a cantilevered upper slab, (b) orthotropic steel box girders, and (c) prestressed concrete girders with a prestressed deck slab. The steel superstructure had twin boxes in three spans: 393 ft, 688 ft, and 269 ft. The bridge was erected by cantilevering from each bank. Whole prefabricated boxes could not be erected because of the long cantilever and the small average effective depth (16.4 ft). Two inner web plates and the deck plate between them were erected initially, then the outer web plates were mounted and the box completed. Due to asymmetry the closing point was 397 ft into the center span from the longer end span.

At the closing point the web plates were connected to be followed by the bottom and deck plates scheduled to be mounted the following day. By that evening both twin box girders had buckled at two points: at the center of the 393-ft left span, and about 197 ft from the right bank pier in the center span (Sattler, 1970; Heckel, 1971). The resulting shortening of the superstructure damaged the abutment bearings and the support at the pier. The closing point, having only four web plates with small flanges, had to withstand the full dead weight moment of the steel loads, and the web plates were stressed beyond the elastic limit. As a result, the small upper flanges were twisted and the upper sections of the webs plastically buckled. Three causes for the buckling were articulated.

1. The theoretical buckling stress curve used as a basis for the safety check was 7 percent higher than the actual buckling stresses in the elastic–plastic transitional range, and the bottom plates where the damage occurred were within this range of maximum deviation.

2. Deformations caused by welding seams in the bottom plate and its ribs and variations in plate thickness represented deviation from the assumed ideal straight plane of the panel.

3. Differential temperature in the deck–steel box system, particularly a temperature drop during the night, was stated as the principal cause. The roadway deck was heated above air temperature under sunshine, but cooling after sunset caused additional compression stresses in the bottom plates.

Two more representative examples of construction failures are the West Gate Bridge in Melbourne, Australia (Rixon, 1971), and the Rhine River Bridge at Koblenz (Engineering News Record, 1972). A report by the Technical University of Karlsruhe indicated that the Rhine River Bridge failed because of inadequate stiffening across the transverse weld seam in the lower flange. Without direct support, the flange plate at this point had a theoretical safety factor of 1.79, but in the presence of the smallest wave in the flange plate this factor was reduced to 1.0. The conclusion was reached that the linear theory used to calculate the critical buckling stress on the steel plates was inadequate.

These failures brought out several fundamental problems associated with the ultimate limit state of box girder bridges such as (a) the margin of safety during construction as it relates to theoretical aspects of failure probabilities, (b) the inadequacy of the linear buckling theory for sections that are wide in comparison to their length, and (c) understanding of the ultimate capacity of partially completed steel box girders. Subsequent research in the United Kingdom, Australia, Belgium, Germany, and the United States addressed these and other problems and led to the current design and construction procedures.

Current Practice Both the standard AASHTO specifications and the proposed LRFD document recognize that the design must address two stages: the requirements during construction and the requirements during service (based either on the working stress or load factor approach).

In addition to strength and rigidity, stability is a more critical consideration before deck curing because lateral–torsional buckling can occur in unsupported top flanges in compression. Whereas this problem is most serious for horizontally curved bridges, top flange bracing is usually provided even in straight girders for handling and erection. Most designers assume Saint-Venant torsion for box girders closed by lateral bracing.

If shoring is not used, it may be best to specify pouring the deck full width for multibox bridges. If partial-width pours with longitudinal construction joints are made, the girders supporting the first pour deflect below the level of the adjacent unloaded girder. Because this prevents proper deck forming and keying of the construction joint for the adjacent pour, the design should make provisions to include external diaphragms between boxes stiff enough to equalize deflections as the deck is poured.

Analysis Straight multiple box bridges are generally analyzed by methods used for straight beams. A transverse load distribution is assumed and determined using the design specifications distribution factor. Thereafter, the analysis proceeds using conventional statics or classical continuous beam theory. Finite-element methods may be applied to bridge systems where transverse differential deflection is anticipated, that is, for girders with lengths that are not great compared to the transverse dimensions, for complex components such as diaphragms and cross frames, or for complicated framing schemes. Unless, however, these programs are production oriented, they may not have the economic capability as a sole basis for design. Quite relevant to the analysis is the generation of envelopes for live load effects, and it is therefore essential to use computer programs with the capability of generating influence lines to obtain design values.

Ongoing Research Section 5-10 and the foregoing discussion highlight some of the principles and problems associated with the design of steel box girder bridges. Following the collapse and failure of the steel box bridges documented in this section, research programs were extended to form the basis for new design rules and to identify areas where further research is needed. Dowling (1974) presents results on ongoing programs in Great Britain, which are summarized in the following discussion.

Box girders and plate girders have three main differences: (a) unlike plate girders, a box girder has wide thin flange plates, stiffened by an orthogonal grid of stiffeners necessitating consideration of shear lag and out-of-plane buckling; (b) differences in flange configuration and behavior articulate web response accordingly, and thus tension field theories developed in the context of plate girder webs may not be readily extrapolated to box girder webs (see also Section 5-10); and (c) in plate girder reactions are transmitted to the supports by bearing stiffeners acting as struts, whereas in box girders these reactions must be transmitted through plate diaphragms subjected to a complex system of stresses.

In addition, there are differences between the stability of perfect and imperfect plated structures, reflected in the fact that classical elastic theory is not satisfactory in the presence of initial geometric imperfections and locked-in residual stresses.

It is generally agreed that the limit state philosophy should be adapted as a basis for formulating design procedures. The three principal limit states are (a) the limit state at collapse, (b) the limit state of unserviceability, and (c) the limit state of fatigue (Dowling, 1974).

The application of this approach to box girders is illustrated in Figure 5-66. For the compact I sections shown in Figure 5-66*a*, the stress–strain behavior is expressed by the stress–strain characteristics shown in Figure 5-66*b*. Collapse is not reached when first yield occurs at a moment M_y, but when the external moment reaches the full plastic moment capacity of the section M_p, where $M_p \approx 1.15M_y$, as shown in Figure 5.66*c*.

The behavior of the box girder shown in Figure 5-66*d*, subjected to both moment and shear, is fairly complex. Because plane sections do not remain plane, simple beam theory is not adequate to predict flange stress distribution σ_f. If the web is assumed stocky and only the flange is slender, first yield will occur in the outer panels of the flange where direct and shear stresses are maximum due to shear lag. Because this plate panel and the stiffener bounding may be slender, the applied stresses are magnified at the plate surface due to nonlinear elastic out-of-plane buckling. This results in one of the forms of stress–strain behavior shown in Figure 5-66*e*. Two assumptions may now be made: (a) the load-carrying capacity of the box is exhausted

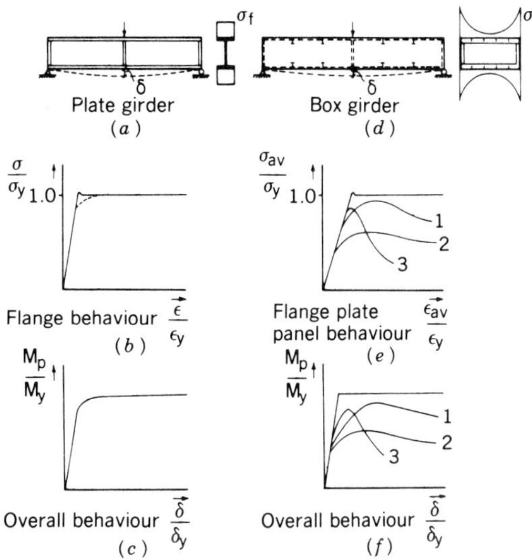

FIGURE 5-66 Behavior of plate girder and box girder. (From Dowling, 1974.)

when yielding of the outer panel occurs and (b) the flange has the capacity to redistribute the stresses across its width so as to equalize the available capacity in the lower-stressed central areas.

In reality, the load-carrying capacity depends on whether (a) the outer panels continue to carry their share of the load as redistribution takes place as shown in Figure 5.66*e*, curve 2, or (b) the outer panels rapidly shed load as shown in curve 3. It appears therefore that in order to understand the behavior of the box girder, we must reliably predict the behavior of each component up to and beyond collapse, including the effect of boundary conditions.

Shear Lag Effects A parametric study by Moffat and Dowling (1972) provided the basis for developing shear lag rules. Among the parameters considered are flange orthotropy, ratio of flange to web area, cross-sectional shape, aspect ratio of flanges, position of point loads, combination of loads, and support conditions.

Restrained Warping Effects With the exception of circular sections and square thin-walled sections, warping of the cross section occurs when the sections twist. If this warping is resisted, restrained warping stresses are induced in the section. A simple formula may be used to predict peak warping stress in a box that will rarely exceed 1.5 tons/in.^{2} (Benscoter, 1954; Gent and Shebini, 1972).

Cross-Sectional Distortion Effects Eccentric loading deflects and rotates the box, and also distorts the cross-sectional shape between internal diaphragms. Although this distortion is small in terms of displacement, it may give rise to significant transverse bending and longitudinal warping stresses. Work by Billington (1974), Crisfield (1973), and Billington, Ghavami, and Dowling (1972) has established an approach for predicting these effects.

Global Analysis A satisfactory analysis may be achieved using grillage methods where the "effective breadth" concept is used to account for shear lag, and an "effective torsional rigidity" accounts for the cross-sectional distortion of the box cells.

Imperfections Two types of imperfections are generally distinguished, residual stresses and geometric imperfections, as shown in Figure 5-67.

Residual stresses are also produced by the differential cooling effects caused by welding. Although considerable data are available (Dibley and Manoharan, 1973; Guile and Dowling, 1972), it is only possible to reliably predict the average midplane compressive stresses induced in the plate of a longitudinally stiffened panel.

Results of studies on geometric imperfections suggest that with longitudinal stiffeners in many cases the initial bow is in the opposite direction. With torsionally weak stiffeners, the initial shape must be considered in relation to the torsional buckling mode, as shown in Figure 5-67*f*. Transverse stiffener distortions are as shown in Figure 5-67*d*.

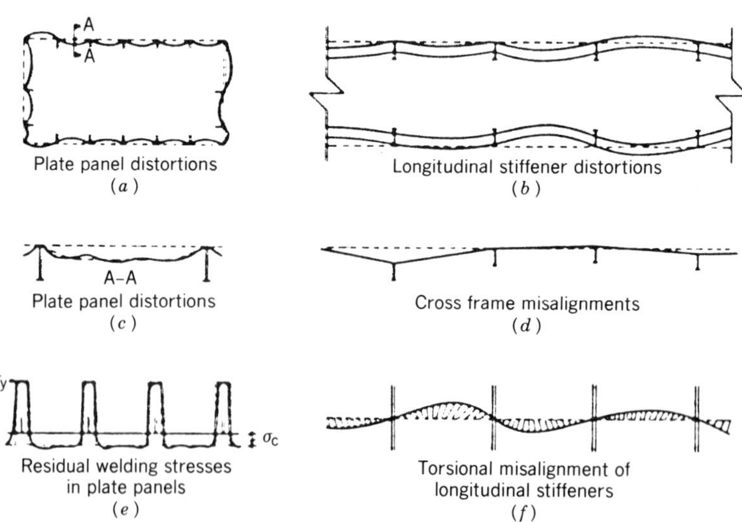

FIGURE 5-67 Typical imperfections, present in steel box girders. (From Dowling, 1974.)

If webs are less closely and more regularly stiffened than flanges, both the magnitude and direction of the initial bow in the stiffeners and plates are more difficult to predict (Figure 5-67*a*). However, both webs and diaphragms tend to be less imperfection sensitive than compression flanges.

Compression Flanges The load end shortening curves of uniaxially loaded rectangular plate panels may be predicted by an elastoplastic large-deflection analysis (Dwight and Moxham, 1969). The effect of residual stresses and the initial lack of flatness can be allowed for in the analysis, although it is difficult to measure the geometric imperfections. Residual stresses have the predominant weakening effect on practical plates in compression. Analyses also show that geometric imperfections can significantly reduce plate strength. Cylindrical out-of-plane deformations have little weakening effect. Hence, particular attention was placed on the effects of transverse butt welds between plates and cross frames (Dorman and Dwight, 1973). Results of tests show that fillet welds do not reduce plate strength, and there is no significant weakening in overall strength. However, stepping at a butt weld, particularly where a thinner plate is connected, reduces strength.

The approach adapted in the design rules states that for uniaxial compression the ultimate stress is

$$\sigma_{\text{ult}} = \sigma_y \left(1 - \frac{\sigma_y}{8a\sigma_{\text{cr}}} \right) < a\sigma_y \tag{5-52}$$

where σ_{ult} = collapse stress
σ_y = yield stress
σ_{cr} = critical buckling stress
a = applied stress to cause first surface yield/yield stress

The factor a is obtained from large-deflection elastic analysis and varies with b/t, assumed initial imperfections, and the degree of end restraint present.

For plates under more complex edge loading than uniaxial compression, the design criterion relates the critical buckling stress and factor a to the equivalent stress in an ideally flat plate as defined by the von Mises formula, expressed by (5-5).

For stiffened panels, the modes of failure are articulated by referring to Figure 5-68. The stiffened panel shown in Figure 5-68*a* may fail by (a) local plate buckling between stiffeners or of the stiffener outstand at the end where maximum compression is applied, (b) buckling of the longitudinally stiffened panel between transversals initiated either by plate or stiffener failure, and (c) overall buckling of the stiffened panel involving both longitudinal and transverse stiffeners. Examples of these modes of failure are given by Dowling et al. (1973), Horne and Narayanan (1974), and Dorman and Dwight (1973).

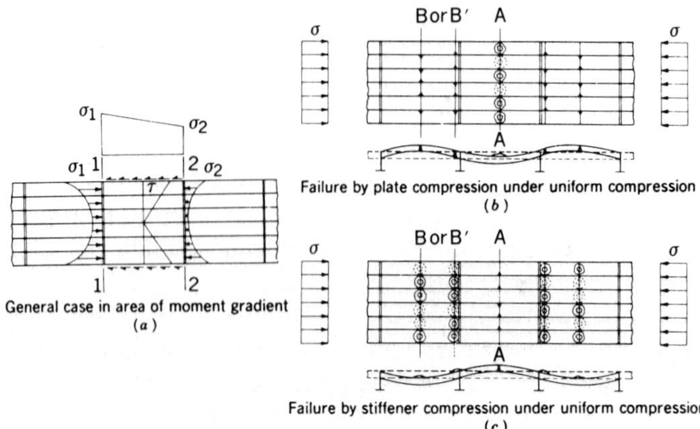

FIGURE 5-68 Modes of failure of compression flanges in box girders. (From Dowling, 1974.)

The possible mechanisms of failure of stiffened panels with open stiffeners of stocky cross section are shown in Figure 5-68 for panels loaded by uniform compression at each end. The initial deflections in each bay will be different so that collapse will be concentrated in one bay only. If failure occurs by compression of the plate (Figure 5-68*b*), bucklets will form at the center of the collapsing panel where the plate stresses are magnified by buckling of the stiffeners between transversals (Dowling, 1974). In the alternative mode shown in Figure 5-68*c*, collapse of the panel is initiated as the stiffener tips reach yielding stage in compression midway between transversals leading to a plastic tripping mechanism.

The design rules for stiffened flanges are based on the work of Culver and Dym (1972) and Dowling and Virdi (1974), where the longitudinally stiffener plate is treated as a series of columns with T sections using an appropriate column formula (Dowling, 1974).

Webs Figure 5-69 shows girders with thin stiffened flanges. In a series of tests by Rockey, Evans, and Porter (1973) involving girders and boxes with vertical stiffeners only, the results indicated that the full tension field capacity of the web plate panels as predicted by theory could not always be achieved. Full web response in some instances was prevented by local failure of the flange by transverse bending as shown in Figure 5-69, section *A–A*. A similar mode of failure was observed to occur in model box tests, as shown in section *B–B* of Figure 5-69. In the latter case, however, the effect was even more acute as the flange plate was less effectively supported than in the plate girder specimen. The main difficulty therefore in applying the tension field theory is to predict the "effective boundary member" resisting tension field forces in the web plate panels.

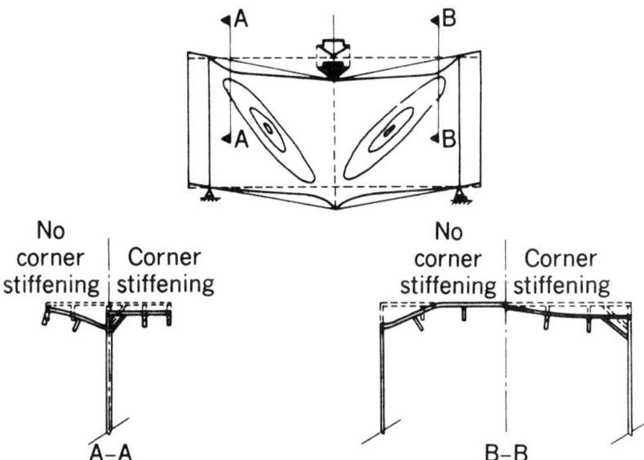

FIGURE 5-69 Failure of web plate panels in girders and boxes with thin flange. (From Dowling, 1974.)

Results of tests on box girders with longitudinally stiffened webs (one stiffener at the neutral axis and another at the compression zone at one-fifth the web depth) showed that the ultimate strength of the web corresponded to first collapse of a web panel. Because of the differences in behavior articulated between plate and box girders, the British design rules were modified to specifically identify those cases where physical and structural restraint is available for the web panels to allow the full tension field capacity to develop.

The design rules for stiffeners are based on theoretical parametric studies (Richmond, 1972). Two phenomena, however, influence the design: (a) the destabilizing effect of in-plane bending and shear stresses and (b) the tension field forces in restrained web panels.

Support Diaphragms After the diaphragm failure that precipitated the collapse of the Milford Haven Road Bridge, this topic has received detailed analytical and experimental consideration.

Unstiffened Diaphragms Referring to Figure 5-70, an unstiffened diaphragm supported on a single bearing produces the stress system shown. For convenience, the bearing may be classified as soft or hard. The former deforms with the diaphragm and produces a uniform vertical stress distribution, whereas the latter does not deform, thus resulting in nonuniform bearing. Three stresses are identified: (a) shear stresses τ along the web–diaphragm boundary caused by flexure of the box girder between supports (their distribution is not easily predicted by simple beam theory); (b) reactive vertical direct stresses σ_V, generally nonuniform and tapering to zero at the top edge; and (c) horizontal direct stresses σ_H produced by the diaphragm bending in its own plane as a deep beam (the transverse compressive component of these stresses acts in the lower part of the diaphragm and contributes to the destabilizing effects of other stress systems).

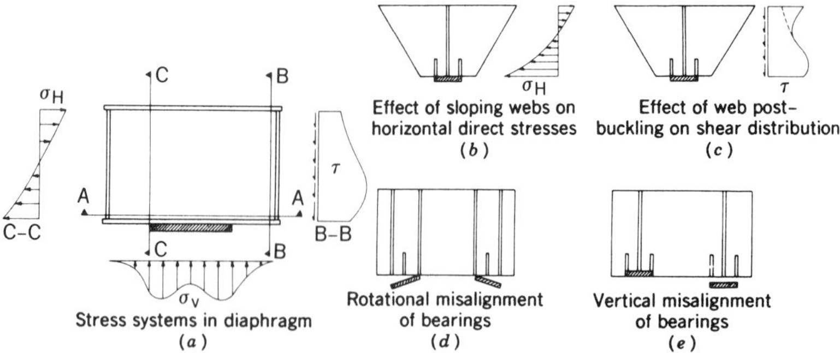

FIGURE 5-70 Load-bearing diaphragm behavior. (From Dowling, 1974.)

The design rules incorporate procedures for calculating the three basic stress systems and for limiting the maximum equivalent in-plane stress of the diaphragm in terms of the yield stress and the calculated elastic critical buckling load (Mair and Linn, 1972; Dean and Disney, 1974).

Stiffened Diaphragms Based on a series of tests by Dowling et al. (1973), stiffened diaphragms were investigated for sloping webs, tension field action, openings, bearing misalignment, and orthogonal stiffening. The resulting procedures produced very conservative designs as they treat the elements in isolation and ignore all membrane and flexural restraints, and assume no redistribution of load between elements.

Current research in Great Britain focuses on the three main problems, namely: (a) the reduction in strength in compression flanges due to imperfections, (b) the difficulty in applying available tension field theories to box girder design, and (c) the stress systems affecting the design of load-bearing diaphragms.

Theoretical Aspects of Strength An excellent review of relevant topics is given by O'Connor (1971), and includes analysis of tubular girders, folded-plate analysis, and numerical examples.

5-17 REDUNDANCY DESIGN OF TWO-GIRDER BRIDGE SYSTEMS

AASHTO defines redundant load path structures as systems with multiload paths such that a single fracture in a member does not lead to collapse. Unlike this response, the consequences of fracture of nonredundant load path structures are assumed to be so severe that collapse is typically assumed to follow.

An alternate definition of redundancy is proposed by Daniels, Kim, and Wilson (1989) following the investigation of the after-fracture behavior of two-girder steel bridges. Accordingly, redundant load path structures may be new, existing, or rehabilitated steel bridges where at least one alternate load path exists and prevents collapse after fracture of a main load-carrying member. The redundancy rating is used to assess the after-fracture condition of the bridge. If this rating falls below the level necessary to maintain safety and serviceability of the bridge, the remedy is to retrofit the existing alternate load path or introduce a new one.

Redundant Bracing System Design Daniels, Kim, and Wilson (1989) give general guidelines for designing an alternate path to provide structural redundancy. These criteria are also applicable to existing bridges. An example is a bridge consisting of a two-girder system with K cross frames and bottom bracing. The as-built structural members and connections can be retrofitted to provide the redundant alternate load path. A second example is a two-girder system with floor beams and longitudinal stringers but with only bottom laterals. In this case, an alternate load path must be provided independently of the bracing system.

Fracture Mechanism Fracture and fatigue failure are discussed in other sections. Invariably, we can assume that fracture can occur anywhere along a plate girder, but fatigue stresses may cause fracture at (a) details with low fatigue strength, (b) zones of high tension range, (c) details prone to displacement-induced fatigue, and (d) locations exhibiting defects because of physical deterioration (Fisher et al., 1974; Shanafelt and Horn, 1984; Klaiber et al., 1987). Potential fracture locations may be identified in conjunction with the detection of fatigue cracking.

Behavior Before Fracture The usual design criterion is that the two girders support all the vertical loads and are therefore the only load paths available for transmitting these loads to the bearings. This assumption is reasonable for zero-skew bridges with symmetrical cross section and symmetrically loaded about the longitudinal axis. For skew bridges, for unsymmetrical cross sections and for unsymmetrical loading, the girders deflect unequally, stressing the bracing numbers accordingly. Torsion and rotation about the longitudinal axis of the bridge induce additional stresses. However, in the simple design model and in terms of load capacity, a solution is obtained by ignoring the bracing members.

It can be argued that this approach is valid for static loads but not for dynamic or cyclic loads, even if static loads are adjusted for impact factors, because fatigue cracking is a function of the real live load stress range at a detail. For main members such as the two girders, the real stress range may be greater or smaller than the calculated stress range. Furthermore, stress ranges are not calculated for details relevant to bracing members because these members are not included in the load model. It is not uncommon to detect strains in bracing members, signifying displacement-induced stresses

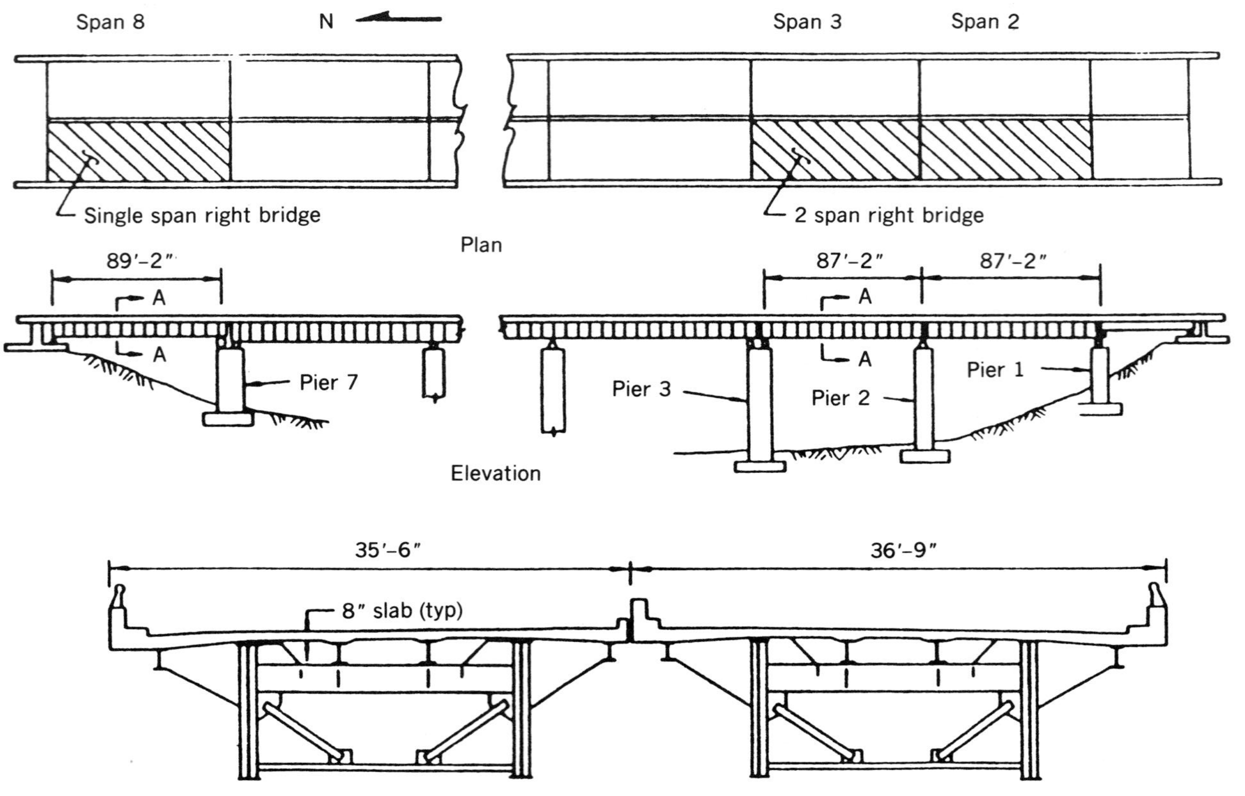

FIGURE 5-71 Plan, elevation, and cross-section of the study bridge. (From Daniels, Kim, and Wilson, 1989.)

approaching the yield level and setting up the possibility of fracture of the main girder.

Behavior After Fracture Studies of after-fracture redundancy of welded two-girder systems are reported by Daniels, Wilson, and Chen (1987), and include the bridge shown in Figure 5-71. Spans 2 and 3 of the southbound lanes represent a two-span continuous unit, and span 8 is the simple unit. This bridge was designed in 1961 for HS 20 loading.

For this example, welded and bolted connections are assumed to develop the member strength. The deck is composite in the positive moment regions throughout the elastic–plastic range. The bottom lateral bracing has an X shape and is connected to the girder bottom flanges. K-type cross frames are

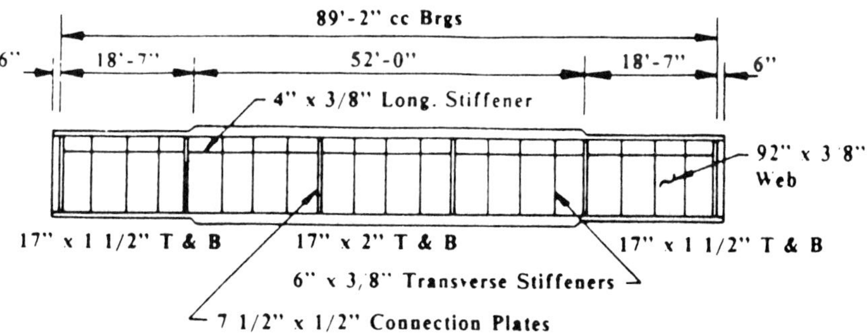

(*a*) Span 8 - Simple Span

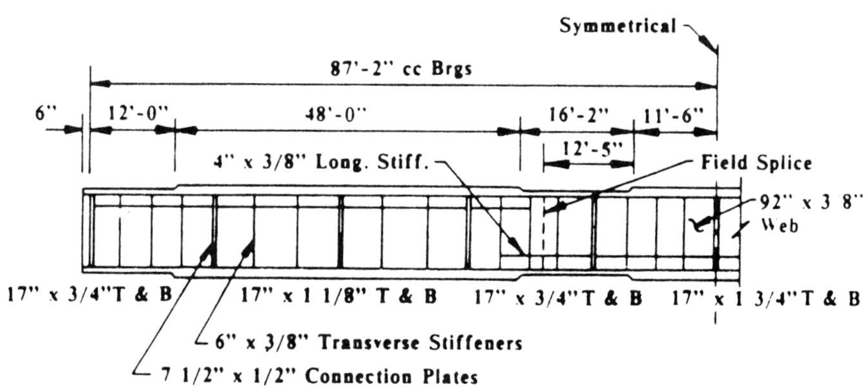

(*b*) Spans 2 & 3 - Continuous

FIGURE 5-72 Simple- and continuous-span bridge girders.

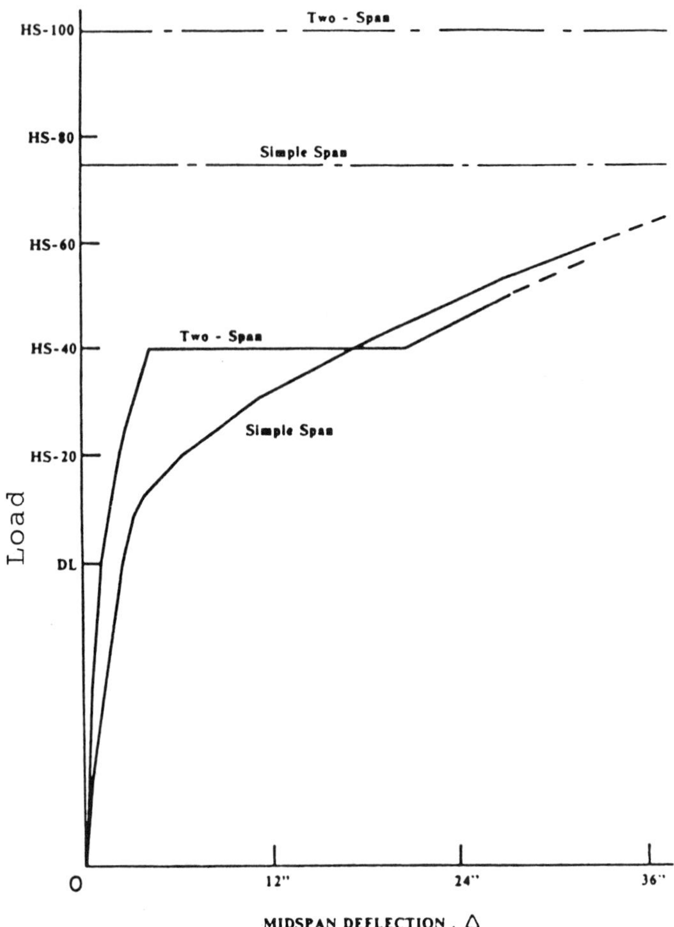

FIGURE 5-73 Comparison of load versus midspan deflection for simple-span and two-span continuous bridges.

spaced at 17- to 18-ft intervals. Typical girder elevations are shown in Figure 5-72.

A near full-depth fracture scheme is postulated at midspan of one of the girders for the simple span with the top assumed to remain intact. For the two-span continuous unit, the same fracture is assumed at midspan of one of the girders in the span containing the field splice. Incremental loading is applied until the maximum capacity is reached or large deflections occur.

The resulting deflection for the two bridge sections is shown in Figure 5-73. Damage first occurs as buckling and yielding of the bottom horizontals and some K bracings. At service dead load plus two lanes of HS 20 live load with impact, more damage occurs corresponding to 6-in. deflection, and includes yielding of the diaphragm connection plates, transverse stiffeners, and buckling of two bottom lateral diagonals. The maximum bridge capacity is in excess of HS 60, but the deflection is now more than 30 in. or too large for safe vehicular traffic.

For the two-span unit, the deflection at HS 40 load suddenly increases by about 16 in. because of compression flange buckling of the fractured girder. The midspan fracture requires the girder to cantilever from the interior support to midspan, resulting in higher negative bending moments. This increased compression stress leads to lateral flange buckling, transforming the girder into a simple fractured span. The maximum capacity of the two-span unit is HS 50.

These results show that for equal deflections the two-span unit has smaller live load capacity, hence less after-fracture redundancy. This may be explained by the smaller cross-sectional (average) area, referring to Figure 5.72. When the continuous unit is converted into a simple span after fracture, it has a smaller section capacity.

Potential Alternate Load Paths The redundancy of the simple-span bridge depends mainly on the elastic–plastic strength of the bracing members. Considerable inelastic behavior is necessary to achieve after-fracture redundancy, suggesting a suitable configuration and pattern for the bracing system.

An alternate load path can provide redundancy if one of the girders is fractured and can perform the "fail-safe" function. Survivability of the alternate load path depends on its initial status (primary or secondary structural system) prior to fracture. Members and connections of the primary system (deck, stringers, floor beams, and girders) are stressed by applied loads, but members of the secondary system (bracing, diaphragms, and top and bottom laterals) are stressed as a result of deflection of the primary system. If the alternate load path is within the primary structural system prior to girder fracture, it may sustain the same fatigue damage. An example of a nonredundant load path is a system of tension flanges welded or bolted to the existing girder. Under live load reversal these flanges will be subjected to the same stress range as the primary system so that detail connections are fatigue critical.

Alternatively, a redundant bracing system of top and bottom laterals and diaphragms has a secondary function prior to fracture because this system is essentially subjected to deflections and axial deformations. It may, however, be prone to fatigue due to alternating stresses, and its efficiency is thus maximized if details and connections are designed accordingly. In this context, after girder fracture the bracing system and its connections become part of the primary system and act as an alternate load path.

Redundant Bracing System Requirements Top and bottom lateral bracing should include (a) X-type configurations along the full span placed just above or below the flanges, (b) identical diagonal members for the full length of the bottom and top lateral bracing, (c) diagonal members designed for dead and live load for the worst case of girder fracture, (d) connections designed and detailed to minimize displacement-induced fatigue stresses, and (e) sufficient flexural stiffness to resist vibration stresses.

Full-depth interior and end diaphragms should be present, connected to all four flanges at the same locations of top and bottom lateral bracing connections. The cross-bracing should resist the applied loads mainly in axial tension or compression, and cross frames should be designed to resist the loads by bending. Connections of load-carrying members should develop member strength.

The emerging concept is a three-dimensional, pseudo space truss with the two girders as the vertical sides. The diaphragms connect the four chords to distribute torsion and provide transverse rigidity. The structure can support vertical loads even if fracture occurs in the span. This principle is illustrated in Figure 5-74 where a prismatic space truss has a missing panel. Side *ADHE* represents the fractured girder, and the fracture is modeled by removing chord *GF* and the diagonals *BG* and *CF*. Side *JKML* is the uncracked girder. The top and bottom lateral bracings are represented by lines *JKDA* and *LMHE*, respectively. Two end and two interior cross-bracing diaphragms are provided.

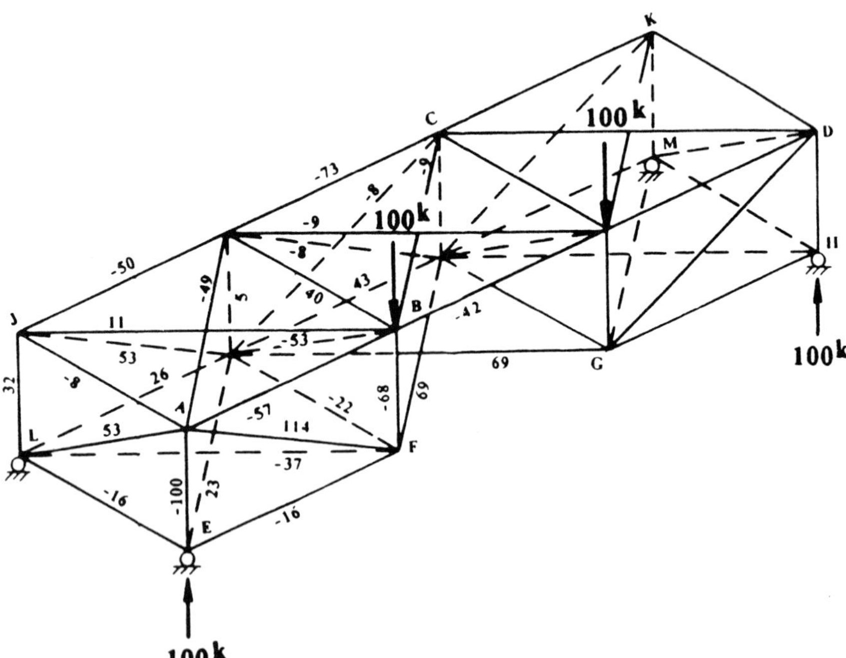

FIGURE 5-74 Load-carrying space truss with missing side panel. (From Daniels, Kim, and Wilson, 1989.)

The truss is analyzed for two 100-kip vertical loads applied as shown, assuming pinned ends for members and equal cross-sectional area. Members EF and EL are 4 units long, and JL is 3 units high. The structure is stable and highly indeterminate. The axial forces are shown only in two bays due to symmetry, and positive values indicate tension. Corresponding diagonal and lateral members of the top and bottom lateral bracing trusses have opposite signs, and because of symmetry cross-bracing diagonals carry equal forces. Likewise, reactions at E and H are equal to the applied loads, and thus no reactions appear at points L and M.

Design Principles A detailed analysis for the development of redundant bracing systems and the associated requirements is given by Daniels, Kim, and Wilson (1989). In general, redundancy design may be carried out using the allowable stress method. It may follow simplified but conservative criteria or it may have finite-element modeling as a basis.

Alternatively, redundancy design may be implemented using the load factor approach. Load factors corresponding to either the operating or the redundancy level may be used with either the simplified method or where finite-element modeling can be shown to be advantageous (e.g., when more complex analyses are involved).

Structural redundancy was a major topic at the 1984 Fourth Specialty Conference on Probabilistic Mechanics and Structure Reliability of the ASCE (ASCE, 1984). Redundancy in bridges is discussed in detail in the State-of-the-Art Report of the Joint ASCE–AASHTO Committee on Flexural Members (ASCE–AASHTO, 1985). The reader will also find useful references on bridge redundancy at the end of this chapter.

REFERENCES

AASHTO, 1991. "Guide Specifications for Alternate Load Factor Design Procedures for Steel Beam Bridges Using Braced Compact Sections."

Anderson, K. E., and K. P. Chong, 1986. "Least Cost Computer-Aided Design of Steel Girders," *Eng. J. AISC*, Vol. 23, No. 4 (4th Quarter) pp. 151–156.

Annamalai, N., A. D. M. Lewis, and J. E. Goldberg, 1972. "Cost Optimization of Welded Plate Girders," *J. Struct. Div. ASCE*, Vol. 98, No. ST10, Oct., pp. 2235–2246.

ASCE, 1971. *Plastic Design in Steel*, 2nd ed.

ASCE–AASHTO, 1974. "Steel Box Girder Bridges—Ultimate Strength Considerations," Subcommittee on Ultimate Strength of Box Girders, Joint ASCE–AASHTO Task Committee on Flexural Members, *J. Struct. Div. ASCE*, Vol. 100, No. ST12, Dec., pp. 2433–2448.

ASCE–AASHTO, 1967. "Trends in the Design of Steel Box Girder Bridges," Subcommittee on Box Girders, Joint ASCE–AASHTO Task Committee on Flexural Members, *J. Struct. Div. ASCE*, Vol. 93, No. ST3, June, pp. 165–180.

ASCE–AASHTO, 1971. "Program Report on Steel Box Girder Bridges," Subcommittee on Box Girders," Joint ASCE–AASHTO Task Committee on Flexural Members, *J. Struct. Div. ASCE*, Vol. 97, No. ST4, April, pp. 1175–1186.

ASCE–AASHTO, 1978. "Theory and Design of Longitudinally Stiffened Plate Girders," Report of Task Committee, Joint ASCE–AASHTO Committee on Flexural Members, *J. Struct. Div. ASCE*, Vol. 104, No. ST4, April, pp. 697–716.

ASCE–AASHTO, 1985. "State-of-the Art Report on Redundant Bridge Systems" Committee on Flexural Members, *J. Struct. Div. ASCE*, Vol. 111, No. ST12, Dec., pp. 2517–2531.

Azad, A. K., 1978. "Economic Design of Homogeneous I-Beams," *J. Struct. Div. ASCE*, Vol. 104, No. ST4, April, pp. 637–648.

Azad, 1979.

Azad, A. K., 1980. "Continuous Steel I Girders. Optimum Proportioning," *J. Struct. Div. ASCE*, Vol. 106, No. ST7, July, pp. 1543–1556.

Basler, K., 1961. "Strength of Plate Girders in Shear," *J. Struct. Div. ASCE*, Vol. 87, No. ST7, Oct., pp. 151–180.

Basler, K., 1961. "Strength of Plate Girders in Shear," *Trans. ASCE*, Vol. 128, Part II, pp. 693–719.

Basler, K., et al., 1960. "Web Buckling Tests on Welded Plate Girders," *Welding Research Council Bull.* Ser. No. 64, Sept.

Basler, K., and B. Thurlimann, 1963. "Strength of Plate Girders in Bending," *Trans. ASCE*, Vol. 128, Part II, pp. 655–682.

Benscoter, B. V., 1954. "A Theory of Torsion Bending for Multi-Cell Beams," *J. Appl. Mech.*, Vol. 21, March.

Billington, C. J., 1974. "The Theoretical and Experimental Elastic Behavior of Box Girder Bridges," Ph.D. Thesis, Univ. London, London.

Billington, C. J., K. Ghavami, and P. J. Dowling, 1972. "Parametric Study of Cross-Sectional Distortion due to Eccentric Loading," CESLIC Report BG16, Imperial College, London, Sept.

Blodgett, O. W., 1966. *Design of Welded Structures*, James F. Lincoln Arc Welding Foundation, Cleveland, Ohio.

Bresler, B., T. Y. Lin, and J. Scalzi, 1968. *Design of Steel Structures*, Wiley, New York.

Bresler, B., Lin, T. Y., and J. B. Scalzi, 1969. Design of Steel Structures, Wiley, New York.

Calladine, C. R., 1973. "A Plastic Theory of Collapse of Plate Girder Under Combined Shearing Force and Bending Moment," *Struct. Engineer*, London, Vol. 51, No. 4, April, pp. 147–154.

Calladine, 1976.

Carskaddan, P. S., 1980. "Autostress Design of Highway Bridges, Phase 3: Interior-Support-Model Test," Research Laboratory Report, U.S. Steel Corp., Monroeville, Pa., Feb.

Chatterjee, S. and P. J. Dowling, 1976. "The Design of Box Girder Compression Flanges," *Steel Plated Structures, An Intern. Symp.*, Crosby Lockwood Staples, London.

Chern, C. and A. Ostapenko, 1969. "Ultimate Strength of Plate Girders Under Shear," Fritz Eng. Lab. Report No. 328.7, Lehigh Univ., Bethlehem, Pa., Aug.

Chong, K. P., 1976. "Optimization of Unstiffened Hybrid Beams," *J. Struct. Div. ASCE*, Vol. 102, No. ST2, Feb., pp. 401–409.

Cooper, P. B., 1965. "Bending and Shear Strength of Longitudinally Stiffened Plate Girders," Fritz Eng. Lab. Report No. 304.6, Lehigh Univ., Bethlehem, Pa., Sept.

Cooper, P. B., 1967. "Strength of Longitudinally Stiffened Plate Girders," *J. Struct. Div. ASCE*, Vol. 93, No. ST2, April, pp. 419–451.

Crisfield, M. A., 1973. "Large-Deflection Elasto-Plastic Buckling Analysis of Plates Using Finite Elements," TRRL LR593, Transp. and Road Research Lab., Crowthorne, England.

Csagoly, P. F. and L. G. Jaegar, 1979. "Multi-Load-Path Structures for Highway Bridges," Transp. Research Record 711, TRB, National Research Council, Washington, D.C. pp. 34–39.

Culver, C. G., 1972. "Instability of Horizontally Curved Members—Design Recommendations for Curved Highway Bridges," Carnegie-Mellon Univ., Pittsburgh, Pennsylvania Dept. Transp., June.

Culver, 1974.

Culver, C. and J. Mozer, 1971. "Horizontally Curved Highway Bridges—Stability of Curved Box Girders," Report No. B3, Carnegie-Mellon Univ., Pittsburgh, Dec.

Culver, C., and C. Dym, 1972. "Elastic Buckling of Stiffened Plates," *J. Struct. Div. ASCE*, Vol. 98, No. ST11, Proc. Paper 9374, Nov., pp. 2641–2645.

Daniels, J. H., W. Kim, and J. L. Wilson, 1989. "Recommended Guidelines for Redundancy Design and Rating of Two-Girder Steel Bridges," NCHRP Report 319, TRB, National Research Council, Washington, D.C.

Daniels, J. H., J. L. Wilson, and S. S. Chen, 1987. "Redundancy of Simple Span and Two-Span Welded Steel Two-Girder Bridges," Final Report, Pennsylvania Dept. Transp. Research Project 84-20.

Dean, J. E., and P. L. Disney, 1974. "The Measurement of Residual Stresses and Initial Imperfections on the Cleddau Bridge," Construction Industry Research and Information Assoc. Project RP/183 Report.

Dibley, G. E., and A. Manoharan, 1973. "Experimental Behavior of a Two Span Continuous Box Girder," presented at the Feb. 13–14, Inst. Civ. Engineers Intern. Conf. on Steel Box Girder Bridges, London, Feb.

Dowling, P. J., 1974. "Strength of Steel Box Girder Bridges," *Proc. ASCE–EIC–RTAC Joint Transp. Eng.* Meeting, Montreal.

Dowling, P. J., et al., 1973. "Experimental and Predicted Collapse Behavior of Rectangular Stiffened Steel Box Girder," Intern. Conf. on Steel Box Girder Bridges, Inst. Civ. Engineers, London, Feb. 13–14.

Dowling, P. J., J. A. Loe, and J. A. Dean, 1973. "The Behavior up to Collapse of Load Bearing Diaphragms in Rectangular and Trapezoidal Stiffened Steel Box Girders, *Proc. Intern. Conf. Steel Box Girder Bridges*, Institution of Civ. Engineers, London.

Dowling, P. J. and K. S. Virdi, 1974. "Outline of a Design Method for Stiffened Compression Flanges," CESLIC Report BG36, Imperial College, London.

Dubas, P., 1976. "Plated Structures with Closed-Section Stiffeners," *Steel Plated Structures, An Intern. Symp.*, Crosby Lockwood Staples, London.

Elgaaly, M., 1983. "Web Design Under Compressive Edge Loads," *Eng. J. AISC*, Vol. 20, No. 4, 4th Quarter, pp. 153–171.

Engineering News Record, 1970. "Cantilevered Box Girder Bridge Collapses During Construction," June 11.

Engineering News Record, 1970. "Australian Box Girder and Pier Collapse," Oct. 22.

Engineering News Record, 1972. "Rib Gaps Blamed for Bridge Failure," Nov. 23.

Fisher, J. W., et al., 1974. "Fatigue Strength of Steel Beams with Welded Stiffeners and Attachments," NCHRP Report 147, TRB, National Research Council, Washington, D.C.

Fujii, T., 1968. "On Improved Theory for Basler's Theory," *Proc. Eighth Congress, Intern. Assoc. for Bridge and Struct. Eng.*, N.Y., Sept., pp. 477–487.

Fujii, T., 1968. "On Ultimate Strength of Plate Girders," *Japan Shipbuilding and Marine Eng., Tokyo, May.*

Galambos, T. V., 1988. *Guide to Stability Design Criteria for Metal Structures*, 4th Ed., Wiley, New York.

Garson, R., 1972. "Highway Bridge Fatigue Life Prediction," Ph.D. Thesis, Case Western Reserve Univ., Cleveland.

Gent, A. R. and V. K. Shebini, 1972. "Parametric Study Report on Torsional Warping," CESLIC Report, Imperial College, London, June.

Goldberg, J. E. and H. L. Leve, 1957. "Theory of Prismatic Folded Plate Structures," *IABSE Publications*, Vol. 16.

Gorman, M. R., 1984. "Structural Redundancy," *Proc. Fourth ASCE Specialty Conf. on Probabilistic Mech. and Struct. Reliability*, Berkeley, Calif. pp. 45–49.

Grubb, M. A., 1989. "Autostress Design Using Compact Welded Beams," *Eng. J. AISC*, Vol. 26, No. 2, 4th Quarter.

Grubb, M. A., and P. S. Carskaddan, 1981. "Autostress Design of Highway Bridges, Phase 3: Moment-Rotation Requirements," Research Lab. Report, U.S. Steel Corp., Monroeville, Pa., July.

Guile and Dowling, 1972.

Haaijer, G., P. S. Carskaddan, and M. A. Grubb, 1987. "Suggested Autostress Procedures for Load Factor Design of Steel Beam Bridges," *Bull.* No. 29, AISC, April.

Haaijer, G., C. G. Schilling, and P. S. Carskaddan, 1979. "Bridge Design Procedures Based on Performance Requirements," Transp. Research Record 711, TRB, National Research Council, Washington, D.C., pp. 30–33.

Heckel, R., 1971. "The Fourth Danube Bridge in Vienna, Damage and Repair," *Proc. Conf. Developments in Bridge Design and Construction*, Univ. College, Cardiff, Wales, March–April.

Heins, C. P., 1975. *Bending and Torsional Design of Structural Members*, D. C. Heath, Lexington, Mass.

Heins, C. P., and C. K. Hou, 1980. "Bridge Redundancy: Effects of Bracing," *J. Struct. Div. ASCE*, Vol. 106, No. ST6, June, pp. 1364–1367.

Heins, C. P., and L. J. Hua, 1980. "Proportioning of Box Girder Bridges," *J. Struct. Div. ASCE*, Vol. 106, No. ST11, Nov., pp. 2345–2349.

Heins, C. P. and H. Kato, 1981. "Load Redistribution of Cracked Girders," *J. Struct. Div. ASCE*, Vol. 108, No. ST8, Aug., pp. 1909–1915.

Heins, C. P.and J. C. Oleinik, 1976. "Curved Box Beam Bridge Analysis," *Computers and Structures J.*, Vol. 6.

Hoglund, T., 1973. "Design of Thin-Plate Girders in Shear and Bending," Royal Inst. Technology, Bull. No. 94, Stockholm.

Holf, E. C., and G. C. Heithecker, 1969. "Minimum Weight Proportions for Steel Girders," *J. Struct. Div. ASCE*, Vol. 95, ST10, Proc. Paper 6838, Oct., pp. 2205—2217.

Hua, L. J., 1979. "Optimization of Box Girder Bridges," Ph.D. Thesis, Univ. Maryland, College Park.

Johnston, B. G., 1966. *Design Criteria for Metal Compression Members*, 2nd ed., Wiley, New York.

Johnston, B. G., 1976. *Guide to Stability Design Criteria for Metal Structures*, 3rd Ed., Wiley, New York.

Kavanagh, T. C., 1970. Design Concepts for the Rio Bianco River Bridge, Personal Communication.

Klaiber, F. W., et al., 1987. "Methods of Strengthening Existing Highway Bridges," NCHRP Report 293, TRB, National Research Council, Washington, D.C.

Komatsu, S., 1971. "Ultimate Strength of Stiffened Plate Girders Subjected to Shear," *Proc. Colloquium on Design of Plate and Box Girders for Ultimate Strength*, Intern. Assoc. for Bridge and Struct. Engineers, London, pp. 49–65.

Kulak, G. L., Fisher, J. W., and J. H. A. Struik, 1987. "Guide to Design Criteria for Bolted and Riveted Joints," 2nd ed., Wiley, New York.

Little, G. H., 1976. "Stiffened Steel Compression Panels—Theoretical Failure Analysis," *Struct. Engineer*, London, Vol. 54, No. 12, Dec., pp. 489–500.

Loveall, C. L., 1986. "Advances in Bridge Design and Construction," *Proc. AISC National Eng. Conf.*, Nashville, Tenn., June.

Lyse, I. and H. J. Godfrey, 1935. "Investigation of Web Buckling in Steel Beams," *Trans. ASCE*, Vol. 100, pp. 675–706.

Mair, R. I. and P. T. K., Linn, 1972. "Stress Analysis of Rectangular Load-Bearing Diaphragms," Report Dept. Environment, London, Nov.

Massonnet, C. and R. Maquoi, 1971. "Discussion on a Report By P. Dubas," Seminar of IABSE, London, pp. 381–389.

Mattock, A. H., and R. S. Fountain, 1967. "Criteria for Design of Steel-Concrete Composite Box Girder Highway Bridges," U.S. Steel Corp., August.

Mattock, R. B., 1962. "I-Section Efficiency," Tech. Reference, Stran-Steel Corp., Houston, Tex., April.

Merrison Committee, 1973. "Inquiry into the Basis of Design and Method of Erection of Steel Box Girder Bridges—Interim Report," Dept. Environment, St. Christopher House, London.

Modjeski and Masters, 1992. "Development of Comprehensive Bridge Specifications and Commentary: LRFD Approach," 3rd draft, NCHRP, TRB, National Res. Board, Washington, D.C.

Moffatt, K. R. and P. J. Dowling, 1972. "Parametric Study on the Shear Lag Phenomenon in Steel Box Girder Bridges," CESLIC Report BG 17, Imperial College, London, Sept.

O'Connor, C., 1971. *Design of Bridge Superstructures*, Wiley, New York.

Ostapenko, A. and C. Chern, 1970. "Strength of Longitudinally Stiffened Plate Girders Under Combined Loads," Fritz Eng. Lab. Report No. 378.10, Lehigh Univ., Bethlehem, Pa., Dec.

Parr, D. H. and S. P. Maggard, 1972. "Ultimate Design of Hollow Thin-Walled Box Girders," *J. Struct. Div. ASCE*, Vol. 98, No. ST7, July, pp. 1427–1442.

Petzold, E. H., 1974. "Behavior and Design of Large Steel Box Girder Bridge," Thesis, Washington Univ., St. Louis, Mo.

Polyzois, D., and K. H. Frank, 1976. "Effect of Overspan and Incomplete Masing of Faning Surfaces on the Slip Resistance of Bolted Connections," *AISC Eng. J.*, 2nd Quarter.

Polyzois, D., and K. H. Frank, 1986. "Effect of Overspray and Incomplete Masking of Faying Surfaces on the Slip Resistance of Bolted Connections," *AISC Eng. J.*, 2nd Quarter.

Porter, D. M., K. C. Rockey, and H. R. Evans, 1975. "The Collapse Behavior of Plate Girders Loaded in Shear," *Struct. Engineer*, London, Vol. 53, No. 8, Aug., pp. 313–325.

"Report of the Royal Commission into the Failure of the West Gate Bridge," C. H. Rixon, Govt. Printer, Melbourne, Australia, 1971.

Richmond, B., 1972. "Report on Parametric Study on Web Panels," Report for the Dept. Environment, Maunsell and Partners, London, July.

Roberts, T. M. and C. K. Chong, 1981. "Collapse of Plate Girders Under Edge Loading," *J. Struct. Div. ASCE*, Vol. 107, No. ST8, Aug., pp. 1503–1509.

Rockey, K. C., H. R. Evans, and D. M. Porter, 1973. "Ultimate Load Capacity of Stiffened Webs Subjected to Shear and Bending," *Proc. Intern. Conf. on Steel Box Girder Bridges*, Institution Civ. Engineers, London.

Rockey, K. C., H. R. Evans, and D. M. Porter, 1974. "The Ultimate Strength Behavior of Longitudinally Stiffened Reinforced Plate Girders," *Symp. on Nonlinear Techniques and Behavior in Struct. Analysis*, Transport and Road Research Lab., Crowthorne, England, Dec.

Rockey, K. C. and D. M. A. Leggett, 1962. "The Buckling of a Plate Girder Web Under Pure Bending When Reinforced by a Single Longitudinal Stiffener," *Proc. Institution Civ. Engineers*, London, Vol. 21, Jan., p. 161.

Rockey, K. C. and M. Skaloud, 1972. "The Ultimate Load Behavior of Plate Girders Loaded in Shear," *Struct. Engineer*, London, Vol. 50, No. 11, Jan., pp. 29–48.

Salmon, C. G. and J. E. Johnson, 1990. *Steel Structures: Design and Behavior*, HarperCollins, New York.

Sattler, K., 1970. "Discussion of the Causes of Failure of the New Danube Bridge at Vienna," *Tiefbau*, Vol. 12, Guetersloh, Germany, pp. 948–950.

Schilling, C. S., 1974. "Optimum Properties of I-Shaped Beams," *J. Struct. Div. ASCE*, Vol. 100, No. ST12, Proc. Paper 11007, Dec. pp. 2385–2401.

Schilling, C. G., 1989. "A Unified Autostress Method," Report on Project 51, AISI, Nov.

Schilling, C. G., 1991. "Unified Autostress Method," *Eng. J., AISC*, Vol. 28, No. 4, Fourth Quarter.

Seim, C., and S. Thoman, 1981. "Proposed Specifications for Steel Box Girder Bridges," Discussion, *J. Struct. ASCE*, Dec., pp. 2457–2458.

Shanafelt, G. O. and W. B. Horn, 1984. "Guidelines for Evaluation and Repair of Damaged Steel Bridge Members," NCHRP Report 271, TRB, National Research Council, Washington, D.C.

Skaloud, M., 1971. "Ultimate Load and Failure Mechanism of Thin Webs in Shear," *Proc. Colloq. Design of Plate and Box Girders for Ultimate Strength*, Intern. Assoc. for Bridge and Struct. Engineering, London.

SSRC Task Group 20, 1988. "A Specification for the Design of Steel Composite Columns," *AISC Eng. J.*, 4th Quarter.

Stephenson, H. K., 1957. "Highway Bridge Live Load Based on Laws of Chance," *J. Struct. Div. ASCE*, Vol. 83, No. ST4, July, pp. 1–23.

Sweeney, R. A. P., 1979. "Importance of Redundancy in Bridge-Fracture Control," *Transp. Research Record* 711, TRB, National Research Council, Washington, D.C. pp. 23–29.

Taha, H. A., 1976. Operations Research, The Macmillan Co., N.Y., pp. 208–238.

Timoshenko, S. P. and J. M. Gere, 1961. *Theory of Elastic Stability*, McGraw-Hill, New York.

Tupula Yamba, F., 1980. "Buckling Behavior of Steel Box Girder Bridges," Reports No. 5, 6, and 7, Experimental Investigation, Dept. Civ. Eng., Concordia Univ., Montreal, Canada, June.

United States Steel Group, 1978. "Steel/Concrete Composite Box Girder Bridges: A Construction Manual," *AISC Marketing*, Pittsburgh, Pa., Dec.

Vachajitpan, P., and K. C. Rockey, 1978. "Design Method for Optimum Unstiffened Girders," *J. Struct. Div. ASCE*, Vol. 104, No. ST1, pp. 141–155.

Wolchuk, R., 1980. "Proposed Specifications for Steel Box Girder Bridges," also known as Report No. FHWA-TS-80-205, *J. Struct. Div. ASCE*, Vol. 106, No. ST12, Dec., pp. 2463–2474.

Wolchuk, R., 1981. "Design Rules for Steel Box Girder Bridges," Proc., Intern. Assoc, for Bridge and Structural Engineering, p-41/81, Zurich, May.

Wolchuk, R., 1982. "Proposed Specifications for Steel Box Girder Bridges," Discussion, *J. Struct. Div. ASCE*, Vol. 108, No. ST8, Aug., pp. 1933–1936.

Wolchuk, R., 1990. "Box Girders," Section 19, Steel Plate Deck Bridges, Str. Eng. Handbook, McGraw-Hill, New York, 3rd ed.

Wright, R. N., S. R. Abdel-Samad, and A. R. Robinson, 1968. "BEF Analogy for Analysis of Box Girders," *J. Struct. Div. ASCE*, Vol. 94, No. ST7, July.

Xanthakos, P. P., 1970. "Salem-Beverly Connector, Route 1A," Struct. Investigation and Report, In-House Design.

Xanthakos, P. P., 1971. "I-75, I-275 4-Level Interchange, Kentucky," In-House Report. Chicago.

Yen, B. T., and J. A. Mueller, 1966. "Fatigue Tests of Large-Size Welded Plate Girders," *Welding Research Council Bull.* No. 118, Nov.

CHAPTER 6

STEEL BRIDGES CURVED IN PLAN

6-1 GENERAL CHARACTERISTICS

Bridge superstructures with horizontal curvature generally have higher costs than comparable structures on straight alignment due to increased design, fabrication, and construction costs. In most instances, however, the extra cost is nominal and offset by the associated functional improvement. In the past, curved bridges had the deck formed to follow the roadway curvature, but were supported on straight beams and girders with changing direction to accommodate the deck alignment. Since the early 1960s, curved spans and framing systems have become standard features of highway interchanges and urban expressways.

A curved deck may still be placed on a series of straight beams or girders if the curvature is not very steep and the maximum slab overhang resulting from this arrangement is compatible with the practical slab thickness. Roadway curvature with small radius is common in access ramps and elevated roadways where the plan alignment is restricted by site conditions. In such cases clearance requirements and structural optimization may indicate a curved framing system that limits the cross-sectional variation and may also be economically competitive. The appearance of a curved framing system is more pleasing compared to straight girders placed on a chord configuration.

Structural Requirements Invariably, vertical loads acting on a curved bridge cause primary twisting moments, requiring an appropriate torsional strength for the structure. This factor limits the range of feasible solutions to the following structural systems; (a) curved slabs, (b) curved parallel I plate girders properly stiffened for torsional rigidity, (c) tubular (box) girders, and (d) arches with a slab rib. Other feasible solutions include trusses with a closed upper and lower lateral system, but these may be fairly complicated in geometry, fabrication, and erection. Most bridges fall into categories (b) and (c), and these are the subject of this chapter.

Curved steel plate girders may have the flanges cut to the correct curved shape, or they may be fabricated in straight elements jointed to shape. Tubular curved girders are difficult to fabricate and erect but have the advantage of extra torsional stiffness. A curved alignment places additional considerations on the support requirements. Furthermore, the provision of two or more vertical reactions at a cross section will cause restraint against torsional rotation and extend the degree of indeterminacy of the structure. As a result, torsional moments will be developed tending to reduce the vertical reaction at the bearing on one side of the bridge and eventually causing the structure to lift off these bearings. In addition, temperature changes must be considered.

6-2 SIMPLIFIED METHOD OF ANALYSIS

The rigorous analysis presented in the following sections requires calculation of distortion coefficients and the solution of simultaneous equations and is therefore possible with the use of computers. The simplified method presented in this section is of particular value because of its sufficient accuracy and because it helps the engineer understand curved beam behavior.

Analytical Procedure Consider the two-span continuous curved girder shown in Figure 6-1. The first step of the simplified method is to isolate each curved girder and straighten it to its full developed length. The external loads are then applied to the developed girder, and the moments are computed as in straight continuous beams. The moment diagram thus obtained is called the *primary moment diagram*.

The second step is to construct a similar diagram having as its ordinates the ordinates of the primary moment diagram divided by the horizontal radius of the curved girder. This is referred to as the M/R diagram, and its purpose is explained with the help of Figure 6-1. For the portion of the flange (straight) shown in Figure 6-1*a* and ignoring flexural stresses carried by the web, we can state that the internal force F at any point along the flange is equal to the moment M divided by the depth d of the beam, or

$$F = M/d \tag{6-1}$$

If the flange is curved with a radius R, radial components of the internal forces F are developed as shown in Figure 6-1*b*, designated as a distribution force q. The magnitude of q is obtained from the equilibrium condition of a

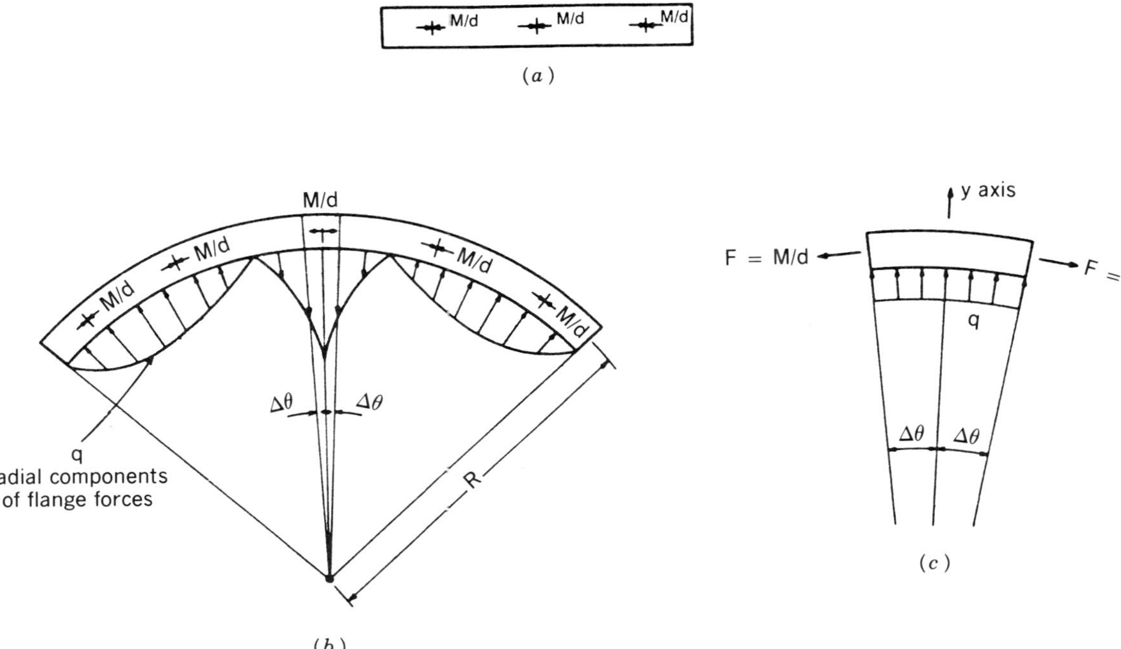

FIGURE 6-1 Girder plan and forces in plan due to curvature: (*a*) portion of straight flange, plan view; (*b*) curved flange, plan view; (*c*) small segment of curved flange. (From Scalzi, 1963.)

very small segment of the girder as shown in Figure 6-1c, where the direction of q is reversed. We will note that q and F vary along the girder length, but for a very small segment they may be considered constant. In the y direction, equilibrium requires

$$2qR\,\Delta\theta = 2F \sin \Delta\theta$$

But for small angles, $\Delta\theta = \sin \Delta\theta$, giving

$$q = F/R \tag{6-2}$$

Combining (6-1) and (6-2) yields

$$qd = M/R \tag{6-3}$$

The M/d internal forces developed at the top and bottom flanges are equal but in opposite directions. Therefore, their radial components q are also equal in magnitude but opposite in direction, representing a couple or torque equal to qd. In effect, the M/R diagram is a torque diagram per unit length acting on the girder due to curvature.

This M/R diagram is now applied as a distributed load acting laterally on the developed length of the girder, which is now considered supported at each point of torsional restraint, that is, at each diaphragm or floor beam. Because the M/R loading is really torque per foot, the support reactions at the floor beams are the concentrated resisting torques developed by the floor beams to restrain twisting of the curved girder. The girder is continuous over the support at each floor beam, but this continuity can be ignored and the reactions at the floor beams due to the loading can be determined by simple beam theory. With these reactions computed, a shear diagram can be constructed representing the internal torque diagram of the curved girder. After the concentrated torques at both ends of the floor beams are known, the end shears at each floor beam are computed from statics.

These shears are then applied as vertical concentrated loads at each floor beam location, and moments are likewise computed for the developed girder. These moment diagrams are referred to as the *secondary moment diagrams*. This procedure constitutes a convergence process whereby the M/R values are applied until the necessary convergence is attained. Any effect, however, beyond the secondary moment diagram is negligible, and the first torque diagram is sufficient for girder design.

A final solution is obtained by considering the following relationships.

1. The final moment in the girder at any point is the sum of the developed girder moment under external loads and the moment due to floor beam end shears.
2. The end shears at each floor beam are directly related to its end moment by conventional statics.
3. The end moments in the floor beams divided by the depth of the girder give the radial forces applied to the girder at the floor beam points.

From these relationships final equations can be written with the intermediate floor beam end shears as the unknowns. With the end shears known, the torque and moments in the girder are easily computed. Using matrix algebra, influence lines for end shears, moments, and torques can be conveniently determined. The method is extended to multispan, multigirder systems in Section 6-8.

Design Example Figure 6-2 shows the plan, cross section, and girder elevation for a two-span continuous curved girder system. This bridge will be

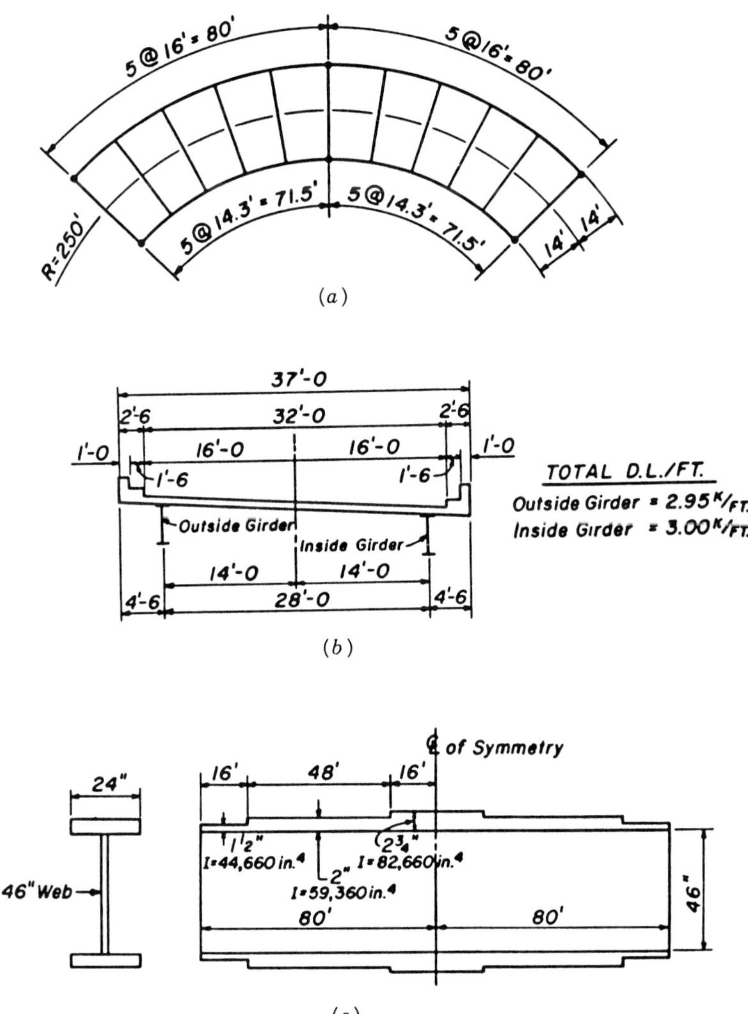

FIGURE 6-2 Two-span curved girder bridge: (*a*) curved girder framing plan and floor beams; (*b*) typical deck section; (*c*) elevation and section of outside girder. (From Scalzi, 1963.)

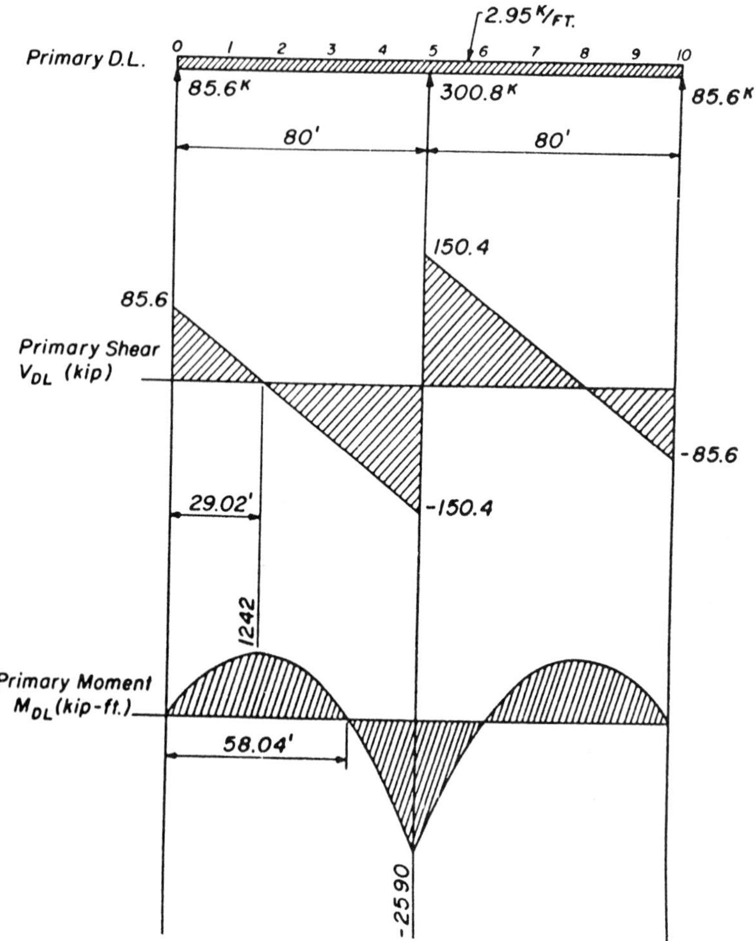

FIGURE 6-3 Primary dead load, primary shear, and primary dead load moment diagram for the bridge of Figure 6-2. (From Scalzi, 1963.)

analyzed for dead load effects. The total dead load is as follows: outside girder, $w = 2.95$ kips/ft; inside girder, $w = 3.00$ kips/ft.

First, we isolate the outside girder and develop it to its full length. The uniform dead load is then applied to the developed two-span continuous girder, moments and reactions are computed, and the primary moment and shear diagrams are drawn as shown in Figure 6-3.

The next step involves calculation of the M/R ordinates, from which the M/R diagram is drawn. This diagram is applied as laterally distributed load to the developed girder length which is assumed to be a series of simple spans supported by the floor beams. Simple beam reactions for the M/R

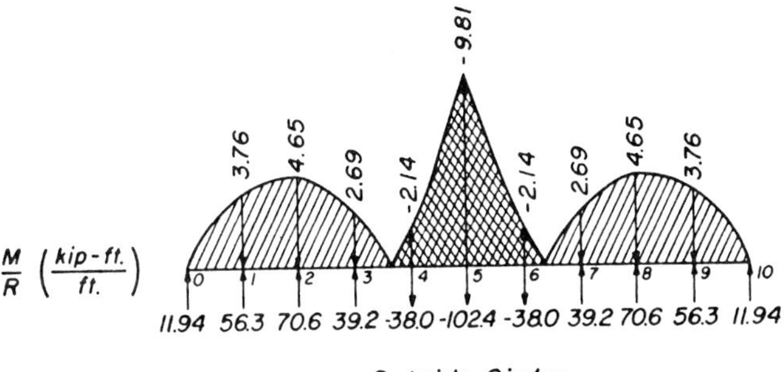

		Outside Girder		Inside Girder
$R_0 = R_{10} = \frac{16}{24}\left[(7\times0)+(6\times3.76)-4.65\right]^{\circ}$ 2nd degree curve	=	11.94^{KI}	$\times 0.813^{*}$	= 9.71^{KI}
$R_1 = R_9 = \frac{16}{12}\left[0+(10\times3.76)+4.65\right]^{\circ}$	=	56.3^{KI}		= 45.8^{KI}
$R_2 = R_8 = \frac{16}{12}\left[3.76+(10\times4.65)+2.69\right]$	=	70.6^{KI}		= 57.4^{KI}
$R_3 = R_7 = \frac{16}{12}\left[4.65+(10\times2.69)-2.14\right]$	=	39.2^{KI}		= 31.9^{KI}
$R_4 = R_6 = \frac{16}{12}\left[2.69-(10\times2.14)-9.81\right]$	=	-38.0^{KI}		= -30.9^{KI}
$R_5 = \frac{16}{12}\left[-2.14-(10\times9.81)-2.14\right]$	=	-102.4^{KI}	$\times 0.813$	= -83.3^{KI}

° See Page 155, Bibliography (1)

* For Inside Girder Reactions, use the following relationship:

$$\frac{\text{I.G. Reaction}}{\text{O.G. Reaction}} = \left(\frac{\text{I.G. Radius}}{\text{O.G. Radius}}\right)^2 \times \frac{\text{I.G.D.L.}}{\text{O.G.D.L.}} = \left(\frac{236}{264}\right)^2 \times \frac{3.00}{2.95} = 0.813$$

FIGURE 6-4 Torque loading diagram and simple torque beam reaction due to dead load for the bridge of Figure 6-2. (From Scalzi, 1963.)

loading are computed at each floor beam and represent the torque acting at these points. These data are shown in Figure 6-4.

From the torque values shown in Figure 6-4, we compute the end shears resulting from the torsional effects. These computations are shown in Figure 6-5 together with the corresponding secondary load concentrated at floor beam points, the secondary shear, and the secondary moment diagram.

The stresses in the girder due to torque are computed, noting that the torsion in this member is induced by the radial forces q in the flanges. Therefore, analyzing the stresses in the girder due to torque is the same as analyzing the stresses in the flange due to the q forces. Considering the top or bottom flange as a horizontal beam, the applied load q is resisted by concentrated reactions from the floor beams. These reactions are calculated merely by dividing the end moments of the floor beams by the depth of the

$$V = \frac{T + T'}{L}$$

$T_0 = 11.94^{K'}$ $\qquad T_0' = 9.71^{K'}$

$L = 28'$

$V_0 \qquad V_0'$

$$V_0 = V_{10} = (11.94 + 9.71) \div 28 = 0.77^K$$
$$V_1 = V_9 = (56.3 + 45.8) \div 28 = 3.65^K$$
$$V_2 = V_8 = (70.6 + 57.4) \div 28 = 4.57^K$$
$$V_3 = V_7 = (39.2 + 31.9) \div 28 = 2.54^K$$
$$V_4 = V_6 = (-38.0 - 30.9) \div 28 = -2.46^K$$
$$V_5 = (-102.4 - 83.3) \div 28 = -6.63^K$$

(*a*)

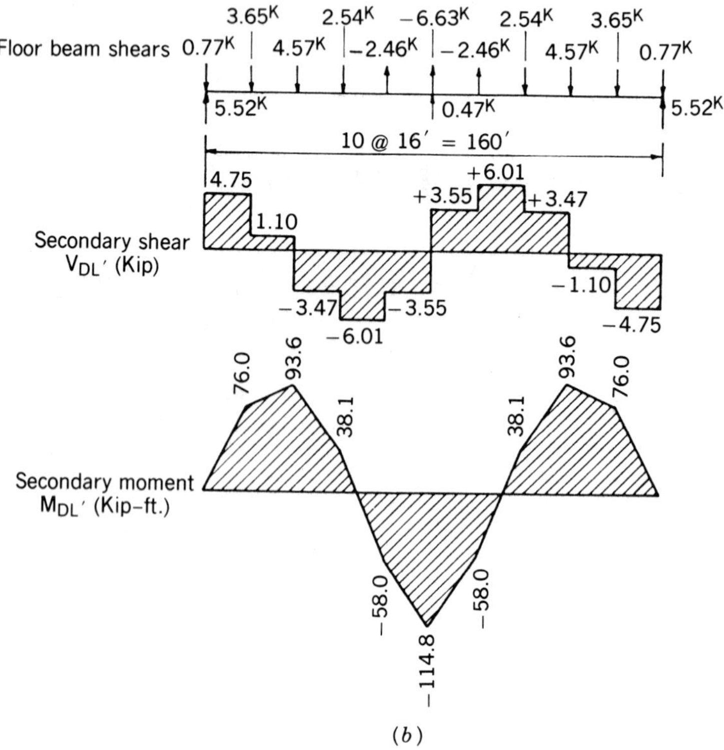

(*b*)

FIGURE 6-5 (*a*) Secondary dead load on outside girder and secondary shear diagram; (*b*) secondary moment diagram; for the bridge of Figure 6-2. (From Scalzi, 1963.)

girder (see also Figure 6-4). In this manner we obtain the loading diagram q on the flange as well as the reactions from the floor beams necessary to maintain equilibrium. This operation is illustrated in Figure 6-6, where, for simplicity, the girder is treated as simply supported between floor beams.

Likewise, the maximum flexural stress in the girder can be computed at critical locations from the foregoing data. For example, the dead load

Point of investigation – over center support
(assume 2 span continuous beam)

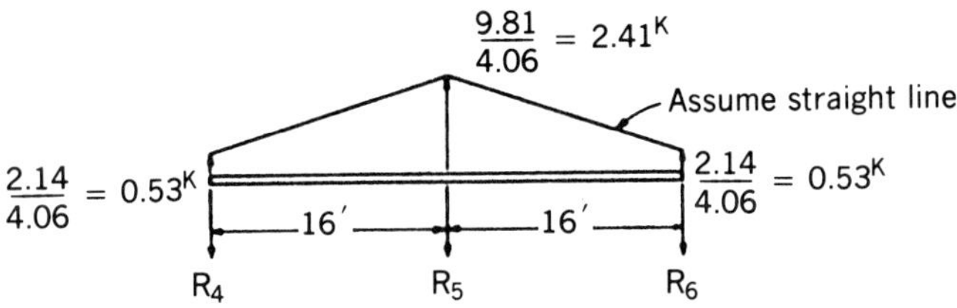

Lateral moment at center support:

Taking advantage of symmetry:

$$0.53 \times 16^2 \times \frac{1}{8} = 17.0^{K'}$$

$$(2.41 - 0.53) \times 16^2 \times \frac{1}{15} = 32.1^{K'}$$

$$\text{Total } M_{Lat.} = 49.1^{K'}$$

Lateral bending stress in flange at center support:

$$f = \frac{M_{Lat.}}{Z_{flange}} = \frac{49.1 \times 12}{1/6 \times 2.75 \times 24^2} = 2\ 23 \text{ k.s.i.}$$

FIGURE 6-6 Stress in outside girder due to dead load torque. (From Scalzi, 1963.)

stresses due to bending at the center support are

$$\text{Primary bending} \qquad f_s = \frac{Mc}{I} = \frac{2590 \times 12 \times 25.75}{82{,}660} = 9.68 \text{ ksi}$$

$$\text{Secondary bending} \qquad f_s = \frac{114.8 \times 12 \times 25.75}{82{,}660} = 0.43 \text{ ksi}$$

$$\text{Lateral bending} \qquad f_s = \underline{2.23} \text{ ksi}$$

$$\text{Total } f_s = 12.34 \text{ ksi}$$

The same procedure must be repeated for live load. For each critical location the position of live load causing maximum effects is determined from influence line coefficients. When the load is positioned along the girder for maximum moment, the analysis is carried out by repeating the same steps.

Types of Curved Girder Framing Two types of curved girder framing are distinguished, according to the action of the system in resisting torsion: closed framing and open framing.

Closed framing is a system of curved girders tied together by diaphragms or floor beams and also horizontal lateral bracing at the girder flange levels. This interaction forms a closed box section that can be treated and analyzed as a single box girder in resisting torsion. The box may be single or multiple depending on the number of individual curved girders that comprise the box. The internal shear flow in the box girder must be treated accordingly.

Open framing consists of curved girders tied together by diaphragms or floor beams but without horizontal lateral bracing. In this case torsion must be resisted by individual curved members, interacting, however, through the effect of diaphragms or floor beams.

Considering the two systems, closed-framing systems should encounter no special problems because they are treated as single box girders. Special attention must be given, however, to open-framing systems where the interaction of several curved girders due to external loads must be analyzed. This structural type is also more sensitive to torsional loads, suggesting the expedience of a careful analysis before a safe and economical design is available.

6-3 EFFECTS OF SKEW ANGLE AND ELEVATED SUPPORTS

A curved girder bridge with skewed supports is shown in Figure 6-7. The left support intersects the imaginary radial support at an angle α, measured clockwise from the radial direction as shown. The skew angle at the right support is $\theta - \alpha$, where θ is the total subtended angle defined by L_c/R_c, in which L_c is the arc length along the centerline of the bridge and R_c is the radius to the bridge centerline.

A survey of curved bridges (Brockenbrough, 1973) suggests the following range of characteristics: number of lanes, 2 to 4; α, 0° to 30°; lane width, 10 ft; girder spacing, 8 ft. These parameters were used by Funkhouser and

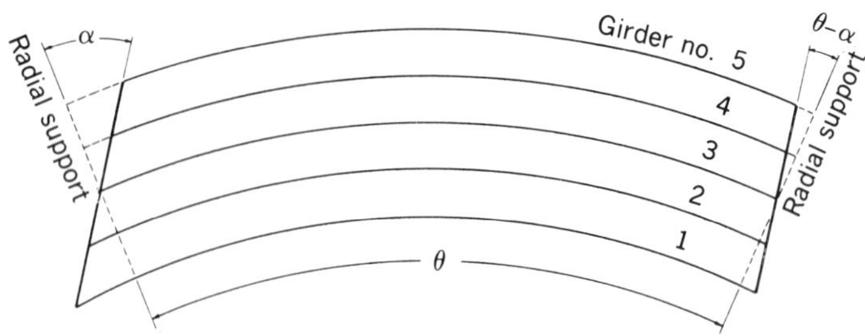

FIGURE 6-7 Typical skewed curved bridge. (From Funkhouser and Heins, 1974.)

Heins (1974) in conjunction with 50-, 75-, and 100-ft spans to study the effect of skew angle. Bridge curvature was represented by 150-, 300-, 450-, 600-, and 1200-ft radii.

The effect of supports that do not lie at the same elevation was examined from the same set of parameters, but with the addition of an elevation differential $\Delta h = 2$, 4, and 6 ft. For the elevated study the skew angle α was taken as zero, thus giving radially supported bridges. The elevation differences produce slopes or grades ranging from 2 to 12 percent. HS 20 truck loads were positioned within the span to produce maximum moment. Only one truck per lane was included, placed along the center of the respective lane.

Bridges with Skewed Supports Partial results of the study are shown graphically in Figures 6-8*a* through *d* for the ratio of the maximum functions (shear, moment, torsion, and deflection, respectively) produced in a curved bridge with skew angle α to the maximum functions produced in a radially supported bridge of the same radius and span length versus the subtended angle for various span lengths. The term "torsion" represents the Saint-Venant torsion, which is not a serious limitation for composite girders where the warping moment is generally negligible.

For any of these functions, the ratio K_{skew} is given by

$$K_{\text{skew}} = \frac{F_\alpha}{F_r} \times 100 \tag{6-4}$$

where F_α are the maximum functions (shear, moment, torsion, deflection) of a skew-supported bridge and F_r are the maximum functions for a radially supported bridge. These curves can be used as follows.

1. Calculate the maximum F_r functions using analysis of radially supported bridges (Heins and Siminou, 1970).
2. Using data relevant to the bridge (radius and span length), determine the subtended angle $\theta = L_c/R_c$. For the given skew angles α and θ, determine the factor K_{skew} referring to the graphs.
3. Determine the maximum functions for the skewed bridge from (6-4).

Elevated Bridges For three-lane sloped curved bridges, K factors are tabulated in Table 6-1. The values given are ratios of the maximum functions in the elevated bridge to the maximum functions in the radially supported bridge. The maximum functions considered in the elevated bridge are identified in Figure 6-9.

Columns 1 to 3 in Table 6-1 give the general geometric data that describe a bridge (radius, span, and slope). Column 4 is the ratio of the maximum deflection Δ in the elevated bridge to the maximum deflection Δ^* in the

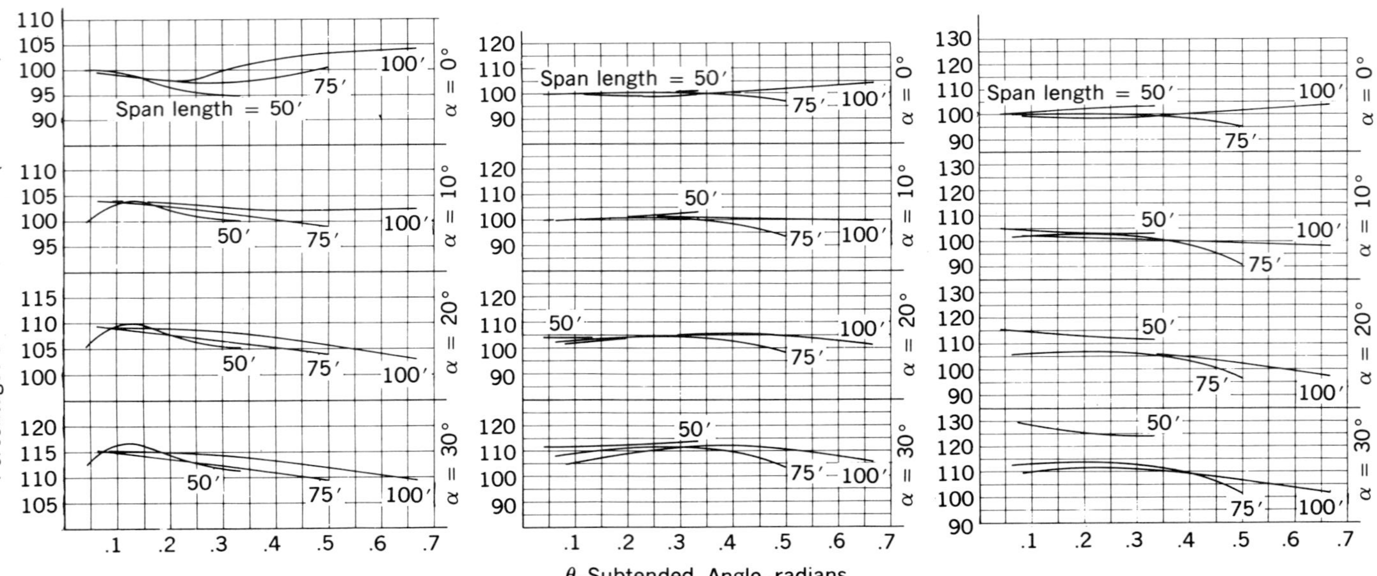

(*a*)

FIGURE 6-8 Three-lane curved bridges with skewed supports: (*a*) shear ratios; (*b*) moment ratios; (*c*) torsion ratios; (*d*) deflection ratios. (From Funkhouser and Heins, 1974.)

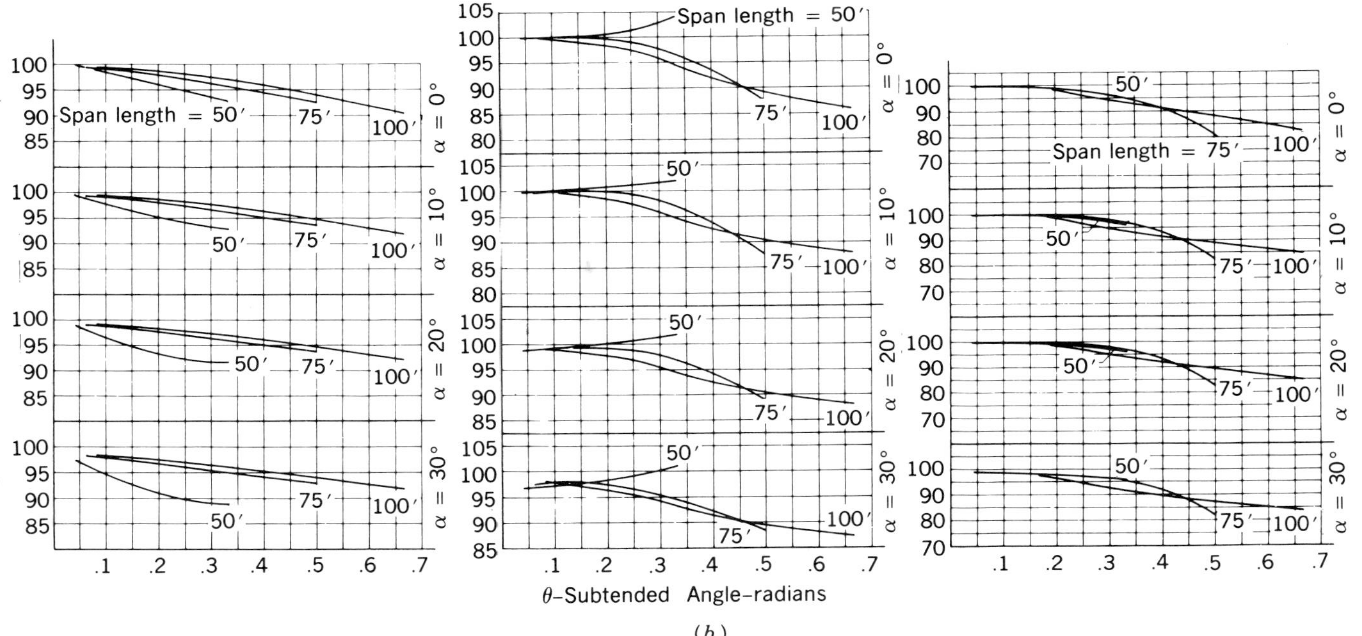

(b)

FIGURE 6-8 *(Continued)*

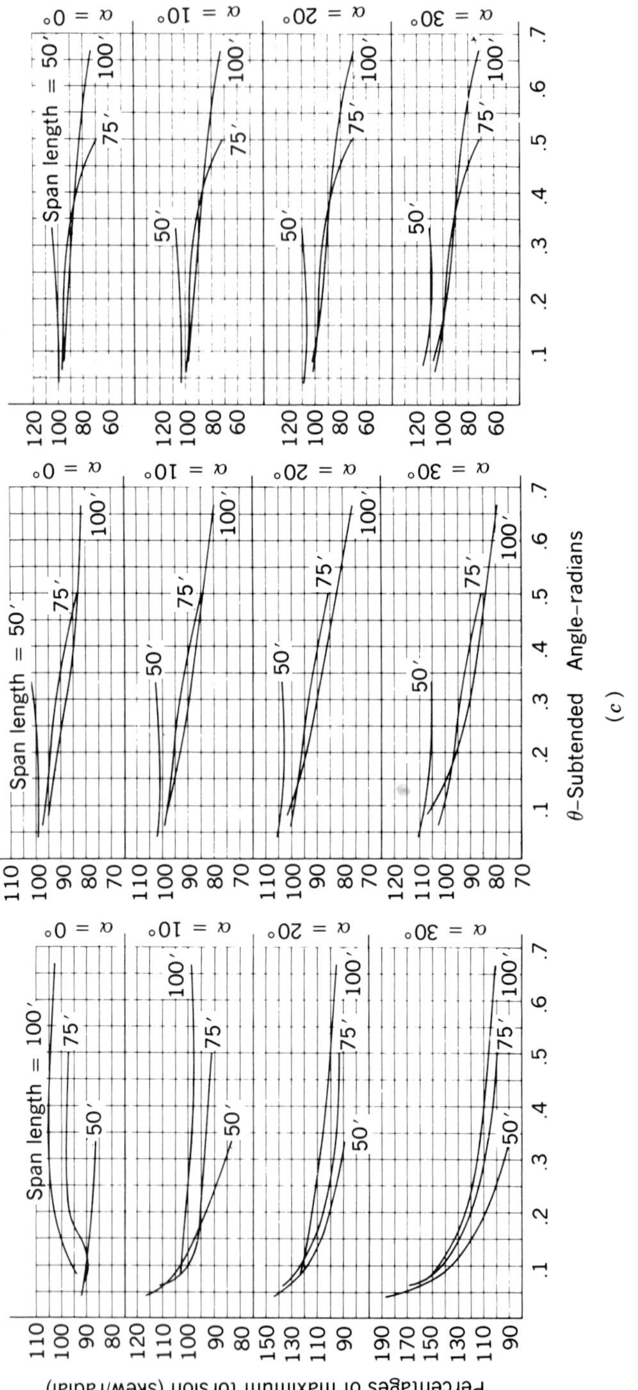

FIGURE 6-8 *(Continued)*

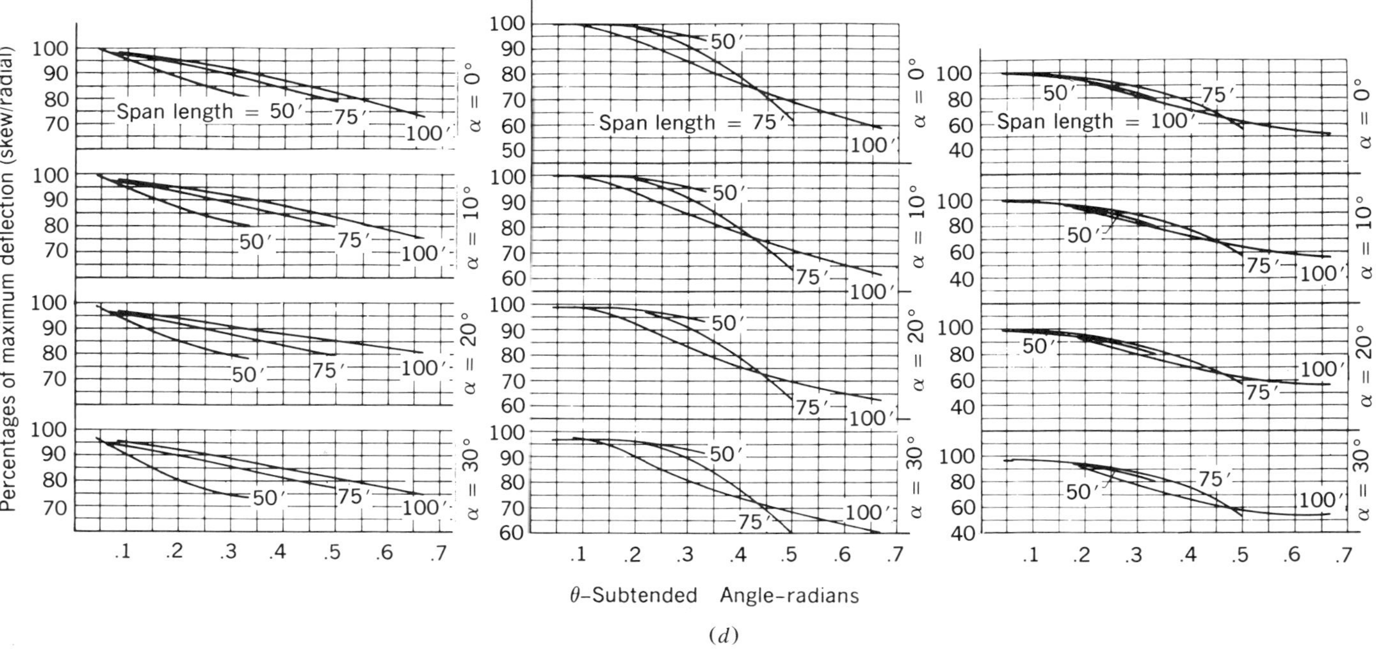

(*d*)

FIGURE 6-8 *(Continued)*

TABLE 6-1 ***K* Factors for Three-Lane Sloped Curved Girder Bridges (From Funkhouser and Heins, 1974)**

Radius (ft) (1)	Span (ft) (2)	Slope (as a Percentage) (3)	Δ/Δ^* (4)	δ/Δ^* (5)	T/T^* (6)	M/M^* (7)	m/M^* (8)	V/V^* (9)	v/V^* (10)	$(P/A)/(M/S)^*$ (11)
150	50	4	0.998	0.054	0.991	0.997	0.030	0.996	0.174	0.021
		8	0.993	0.076	0.966	0.991	0.053	0.986	0.301	0.037
		12	0.988	0.112	0.936	0.985	0.066	0.972	0.367	0.048
	75	2.67	0.996	0.027	0.986	0.999	0.022	0.998	0.207	0.024
		5.33	0.986	0.052	0.947	0.995	0.042	0.993	0.389	0.046
		8	0.971	0.077	0.894	0.989	0.060	0.985	0.531	0.063
	100	2	0.994	0.020	0.989	0.997	0.022	0.996	0.219	0.028
		4	0.979	0.039	0.961	0.989	0.044	0.985	0.424	0.054
		6	0.956	0.057	0.919	0.977	0.063	0.968	0.601	0.076
300	50	4	0.998	0.040	0.993	0.998	0.015	0.998	0.086	0.010
		8	0.995	0.079	0.973	0.994	0.026	0.991	0.150	0.019
		12	0.990	0.117	0.945	0.986	0.033	0.980	0.184	0.024
	75	2.67	1.000	0.027	0.992	0.999	0.011	0.998	0.116	0.012
		5.33	0.999	0.053	0.970	0.997	0.021	0.994	0.220	0.024
		8	0.999	0.079	0.936	0.994	0.028	0.988	0.301	0.033
	100	2	0.999	0.020	0.996	1.000	0.010	1.000	0.119	0.014
		4	0.999	0.040	0.985	0.999	0.020	0.999	0.232	0.027
		6	0.998	0.060	0.967	0.999	0.029	0.998	0.333	0.038
450	50	4	0.999	0.040	0.993	0.999	0.010	0.998	0.058	0.007
		8	0.997	0.079	0.974	0.994	0.018	0.992	0.100	0.014
		12	0.993	0.117	0.946	0.988	0.022	0.982	0.123	0.019
	75	2.67	0.998	0.026	0.992	0.998	0.007	0.998	0.079	0.008
		5.33	0.994	0.053	0.968	0.993	0.014	0.994	0.150	0.016
		8	0.988	0.078	0.932	0.986	0.020	0.986	0.205	0.022
	100	2	1.001	0.020	0.995	0.999	0.007	1.000	0.096	0.009
		4	1.003	0.041	0.980	0.998	0.013	1.002	0.185	0.018
		6	1.006	0.061	0.957	0.996	0.019	1.005	0.265	0.026
1200	50	4	0.999	0.039	0.993	0.999	0.004	0.998	0.022	0.006
		8	0.997	0.079	0.973	0.995	0.007	0.993	0.037	0.011
		12	0.995	0.118	0.945	0.989	0.010	0.983	0.046	0.016
	75	2.67	0.999	0.026	0.991	0.998	0.004	0.998	0.041	0.004
		5.33	0.995	0.052	0.965	0.994	0.007	0.993	0.078	0.007
		8	0.989	0.078	0.925	0.986	0.010	0.986	0.108	0.010
	100	2	0.999	0.020	0.992	0.999	0.004	0.999	0.054	0.004
		4	0.997	0.040	0.971	0.996	0.007	0.997	0.105	0.007
		6	0.993	0.059	0.937	0.992	0.010	0.993	0.149	0.011

Note: 1 ft = 0.305 m.

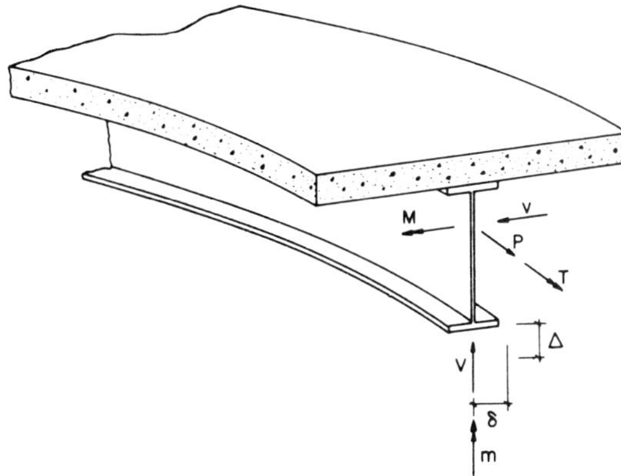

FIGURE 6-9 Forces and deformation notation in curved elevated bridges. (From Funkhouser and Heins, 1974.)

horizontally–radially supported bridge. Column 5 gives the ratio of the maximum lateral deflection δ in the elevated bridge to the maximum vertical deflection Δ^* in the horizontally–radially supported bridge. Column 6 is the ratio of torsion T produced in the elevated bridge to T^* in the horizontally–radially supported bridge. Columns 7 and 8 give bending moment ratios, where M is the major axis and m is the minor axis bending induced in the elevated bridge, and M^* is the bending about the major axis in the horizontally–radially supported bridge. Likewise, shear ratios are given in columns 9 and 10, where V is the major axis shear, v is the minor axis shear in the elevated bridge, and V^* is the major axis shear in the horizontally–radially supported bridge. Column 11 is a ratio of the maximum axial stress produced in the elevated bridge to the maximum bending stress computed in the horizontally–radially supported bridge.

6-4 SPECIFICATIONS AND DESIGN RECOMMENDATIONS: I GIRDERS

The design of steel I-girder bridges is governed by the AASHTO Guide Specifications for Horizontally Curved Bridges (1980, and Interim up to 1990). These guidelines are explicit and detailed. This section therefore focuses on the theoretical and empirical basis of the AASHTO design methodology. The current AASHTO document has evolved from the design recommendations compiled by a task committee jointly formed by ASCE–AASHTO (1977).

Transverse Intermediate Stiffeners Article 10.34.4 of the standard AASHTO specifications considers the added postbuckling shear from tension field action. However, there has not been sufficient evidence that this postbuckling strength will develop in curved girders; hence, the following procedure is suggested.

To determine if intermediate stiffeners are required, reference is made to Article 10.34.4.1 of the AASHTO specifications. If stiffeners are required, their spacing is determined using a formula that is simply equivalent to the shear buckling strength of a flat web panel supported at the flanges and stiffener boundaries. The maximum intermediate stiffener spacing is limited to the web depth D.

Methods of Analysis Among the methods of analysis suggested for I-section curved girder bridges are seven programs that provide some or all of the information needed to proportion the individual members (United States Steel, 1963; Bell and Heins, 1969; Lavelle, 1973a, 1973b, 1973c; Lavelle, Greig, and Wemmer, 1971; Brennan and Mandel, 1973; Shore, 1973). The development of design aids has been enhanced by several analytical methods that accurately represent the response of the structural system under dead load (Murphy and Heins, 1973; Dabrowski, 1968) and live load (Bell and Heins, 1970; Heins, 1967). Useful equations for initial aid to design are provided by Chu and Pinjarkar (1971). These allow estimation of probable induced girder forces and moments, in addition to a preliminary determination of girder properties and diaphragm spacing prior to trial-and-error analysis.

Live Load Distribution As mentioned previously, the curved bridge may constitute an open system (approaching a plane grid) or a closed system (approaching a space frame). Several cases of live load distribution are distinguished.

Case 1 The bridge is composite (simple or continuous) and has bottom lateral wind bracing in at least two bays. In this case the resulting maximum live load (bending and warping) bottom flange stress in the outside girder, f_{be}, computed using plane grid analysis is adjusted by the following:

$$f_{be} = f_{ube}(\mathrm{DF})_b \tag{6-5}$$

where f_{ube} is the maximum live load flange stress in the outside exterior girder based on grid analysis and $(\mathrm{DF})_b$ is a distribution factor bottom flange.

Case 2 The bridge has curved girders and bottom bracing is not considered. The live load distribution is again based on the interaction of all bridge elements. The concept of a distribution factor requires a relationship between the response of the loads in a system to those developed in a single isolated girder subjected to a set of wheel loads. For composite design, the distribution factor DF is the ratio (curved system function)/(curved single-girder function). Expressions of the distribution factor are given as functions of the girder spacing S, the radius R, a factor $\overline{N} = R/100$, and the span length L. These distribution factors are applied to the wheel load, and then introduced in the analysis of the individual girder. Two distinct distribution factors are developed: the first applies to the vertical normal bending moment, and the second applies to the bimoment (lateral flange bending).

Amplification Factor This relates single curved girder response to that of an equivalent straight girder. The amplification or modification factor K_i is thus the ratio (curved single girder function)/(single straight girder). Applicable expressions are given by

$$K_{\text{moment}} = \frac{0.15}{\overline{N}}\left(\frac{L}{R}\right) + 1 \tag{6-6}$$

$$K_{\text{bimoment}} = \frac{35\overline{N}(L/R) - 15}{0.108L - 1.68}\left(\frac{L}{R}\right) \tag{6-7}$$

where $\overline{N}$, L, and R are as given previously.

Reduction Factor Although not present in the current AASHTO specifications, this factor allows a reduction to be applied to the simple-span data to give preliminary load effects in continuous spans. Brennan (1970) has evaluated the maximum load effects in a two- or three-span curved bridge of four, six, and eight girders under various critical conditions. A study of the results produced the data shown in Table 6-2. These values of the reduction factor are relative to the number of spans and independent of the number of girders. The results are valid for equal span lengths in the continuous unit, with maximum span length not to exceed 100 ft. For a given two- or three-span continuous unit, preliminary force data are obtained merely by multiplying the function in the single-span by the coefficient shown.

TABLE 6-2 *K* Reduction Factor for Maximum Function in Two- and Three-Span Bridges

Number of Spans (1)	$K_{\text{bending moment}}$ (2)	$K_{\text{deflection}}$ (3)	K_{rotation} (4)	$K_{\text{Saint-Venant}}$ (5)	$K_{\text{warping torsion}}$ (6)	K_{bimoment} (7)	K_{shear} (8)
Two span	0.75	0.70	0.70	0.75	0.45	0.35	1.00
Three span	0.65	0.60	0.70	0.75	0.40	0.35	0.90

Fatigue The difference between the stress conditions in a curved and straight girder is the contribution of torsional stresses resulting in a complex pattern, particularly for the curved plate. A rigorous stress analysis is suggested for fatigue considerations and effects.

Allowable Flange Normal Stress The present formulas are intended to prevent local buckling prior to reaching a specific stress level. Although studies indicate that curvature has little effect on local buckling for most practical curved girders (Culver and Nasir, 1971), the stress gradient across the flange is significant.

If the flange width–thickness requirements for straight girders are applied to curved girders, allowable design stresses are derived by applying a factor of safety to the stress at the flange tip (Culver, 1972a, 1972b; Mozer and Culver, 1970). An obvious inefficiency in this case is that the major portion of the flange is not utilized effectively. Alternatively, allowable stresses based on a factor of safety applied to the full plastic moment under combined bending and torsion could be used (Mozer, Ohlson, and Culver, 1971; Mozer, Cook, and Culver, 1972).

Current allowable compression flange stresses have been developed by Culver (1972a, 1972b) for two cases. The first is applicable to noncompact flanges and provides a factor of safety against initial yield at the flange tip. The second is applicable to compact girders and provides the same factor of safety against full plastification under combined bending and nonuniform torsion (Culver and McManus, 1971). In a series of comprehensive investigations by these two researchers, it was established that the lateral buckling strength of a curved girder segment between the points of lateral support is dependent on the direction and magnitude of the warping bimoment at the bracing points. The lateral buckling strength is enhanced if the bimoment tends to bend the compression flange toward the center of curvature.

Allowable Web Shear Stress This is the same as the allowable shear stress specified in Article 10.32 of the standard AASHTO specifications. Tests by Culver and Mozer (1970) show that the fully plastic shear strength of curved girder webs can be reached if the web plate is adequately proportioned or stiffened.

Curved Composite Girders Shear stresses are induced by the combined effect of bending and torsion. According to present design concepts, both elastic and ultimate shear must be assessed. Results of analysis, however, show that the elastic torsional forces developed in this case (Colville, 1972) are negligible and thus the straight girder fatigue theory may be used.

Warping effects have been examined by Heins and Bonakdarpour (1971), and show that the maximum change in the warping torsion T_w per unit length is much less than 2 kip-in./in. Assuming this maximum value, the shear flow is estimated as 0.049 kip/in. Therefore, both warping and pure torsional shear flows may safely be neglected in most practical problems, and the design of shear connectors for curved girders is the same as in straight members. The range of horizontal shear S_r is therefore computed from the formula

$$S_r = V_r Q/I \tag{6-8}$$

where V_r is the range of shear due to live load and impact taken as the difference in the minimum and maximum shear envelope excluding dead load, Q is the statical moment about the neutral axis of the composite section of the transformed compressive concrete area, and I is the moment of inertia of the transformed composite section.

In the ultimate strength state, the forces due to bending stresses depend on the location of the neutral axis of the composite beam. The force to be developed by the shear connectors at ultimate load is $P = 0.85 f'_c A_c$, which is the same as in previous analyses. An additional factor to be considered, however, is the eccentricity of the longitudinal force due to bending illustrated in Figure 6-10. The force P at the section of maximum moment is not linear with the shear connectors, and because of this eccentricity the load on each connector must account for the torsional effect.

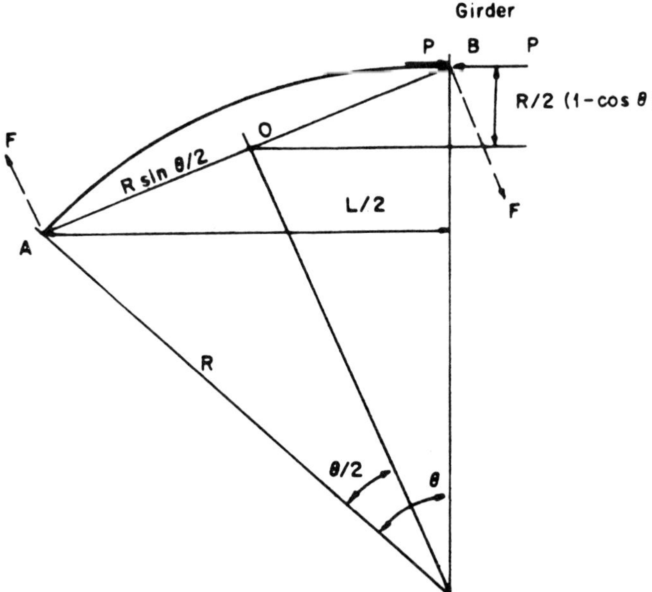

FIGURE 6-10 Eccentricity of P.

The following assumptions are made.

1. Shear connectors are spaced uniformly along line AOB, but located along curved axis AB.
2. The force in each connector due to the eccentricity of P with respect to the centroid (point O) is proportional to the distance of the connector from the same point.

The resultant force P_c in the extreme connector is given by

$$P_c = \sqrt{\overline{P}^2 + F^2 + 2\overline{P}F \sin\frac{\theta}{2}} \tag{6-9}$$

where $\overline{P} = P/N$ (N is the number of connectors between points of maximum positive moment and adjacent end supports or dead load points of contraflexure, or between points of maximum negative moment and adjacent dead load points of contraflexure); and

$$F = \frac{P(1 - \cos\theta)}{N_s 4K \sin\theta/2}$$

where

$$K = 0.166\left(\frac{N}{N_s} - 1\right) + 0.375$$

and N_s is the number of shear connectors placed at each section.

Curved Hybrid Girders As shown by Mozer and Culver (1970) and Mozer, Ohlson, and Culver (1971), curved plate girders with webs having lower yield strength than the flanges behave as homogeneous girders up to yielding of the flanges. The design in this case can be based on initial yielding of the flanges. To account for the lower web yield strength, the allowable flange stresses in the specifications (Article 1.20) are multiplied by a reduction factor, equation (a). This factor is the ratio of the flange yield moment for the hybrid section to the yield moment for a homogeneous section of the flange steel. Reduction factors based on this analysis are plotted in the graphs of Figure 6-11 as a function of the area ratio $\beta = A_w/A_f$.

Because the stress in the compression flange of composite steel girders is low, the tension will govern the design in positive moment regions, and the criteria are the same as for noncomposite girders. In the negative moment region of a continuous composite girder, the steel section essentially behaves

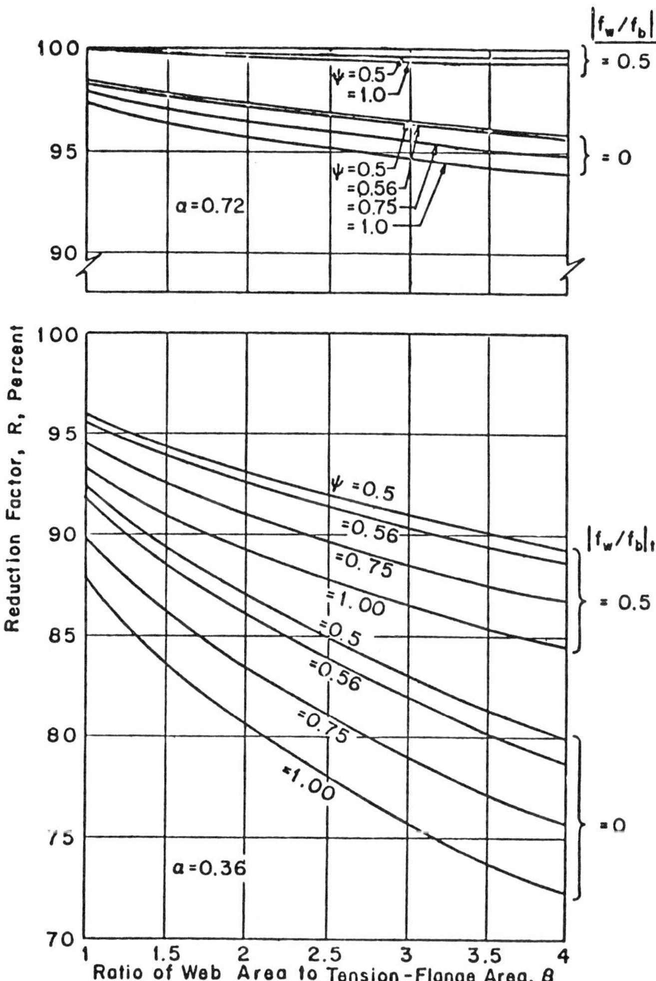

FIGURE 6-11 Hybrid girder reduction factors, noncomposite girders and positive moment region of composite girders.

as noncomposite, and therefore the reduction factor given previously applies. For warping–bending stress ratios in the compression flange equal to or greater than the limit of equation (f), the compression flange will yield a lower moment than the web, and no reduction is necessary to reflect hybrid conditions. For intermediate values of this ratio, the total stresses at the compression flange tip will control yield according to equation (e). Reduction factors in this case are plotted in the graphs of Figure 6-12 as a function of the area ratio A_w/A_f.

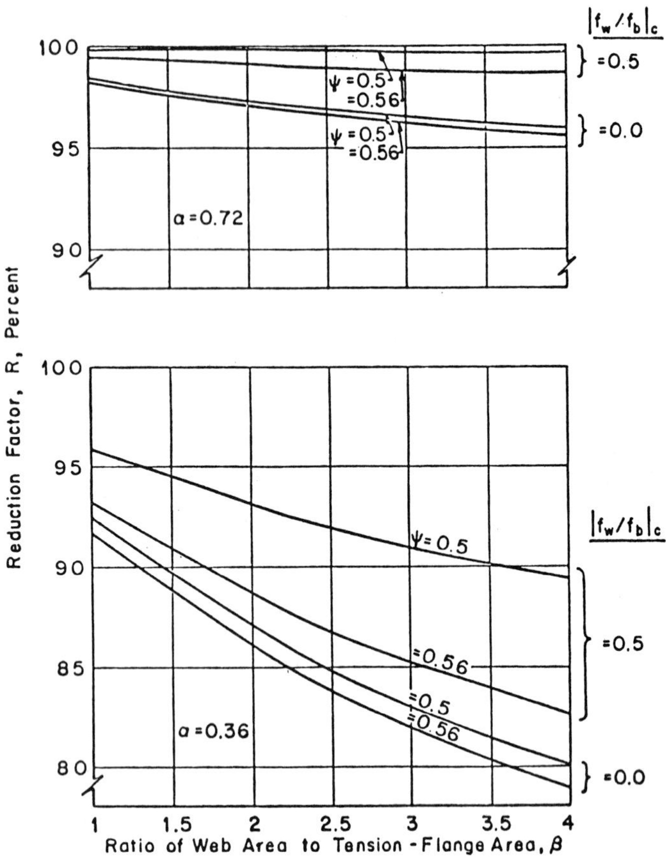

FIGURE 6-12 Hybrid girder reduction factors, negative moment region of composite girders.

6-5 SPECIFICATIONS AND DESIGN RECOMMENDATIONS: STEEL BOX GIRDERS

General Principles

Specifications and guidelines for steel box girder curved bridges are incorporated in the same document as I girders, discussed in Section 6-4. Points of direct interest are the following.

1. If the webs of the steel section are connected by a common steel flange plate across the top, the steel section will be analyzed as a closed member under the wet concrete curing.
2. If the top flange consists of steel lip plates welded to individual web plates, the steel section will be analyzed as an open section under the curing concrete and associated construction loads.

3. Individual box girders not connected by a common flange plate will be provided with intermediate diaphragms or cross frames to distribute the torsional forces unless the analysis indicates that these are not required.
4. The effect of warping stresses due to nonuniform torsion and geometry deformation will be included in the design, particularly for fatigue, unless a rational analysis concludes that these effects are inconsequential.

Methods of Analysis Among the available analytical procedures recommended for curved steel box girder analysis is a program developed as part of the CURT project (FH-11-7389) initiated in 1969. This program involves finite-element techniques and calculates stresses and deformations in curved box girders due to bending torsion and geometry distortion. The size of the program limits its use to the final analysis stage in the design cycle. Accordingly, approximate methods are given for preliminary design and section selection.

Live Load Distribution Based on the work of Heins (1972), Ho and Reilly (1971), and Johnston and Mattock (1967), distribution factor expressions have been developed for curved box girders. The live load bending moment and torsional moment for each box girder is determined by applying a wheel distribution factor DF as follows (wheel load front and rear):

$$\text{Bending Moment} \qquad \mathrm{DF} = \frac{S}{-2.6 + 0.6(GS/N)} \tag{6-10}$$

$$\text{Torsional Moment} \qquad \mathrm{DF} = \frac{S}{-15.3 + 1.5(GS/N)} \tag{6-11}$$

where G is the number of girders (2, 3, 4, or 5), S is the girder spacing centerline-to-centerline of boxes (ft) (10 ft $\leq S \leq$ 16 ft), and N is the number of lanes (2, 3, 4, or 5). GS/N should not be less than 12 or greater than 16.

Beam Forces The specifications include the effect of curvature by determining the ratio of the force function in the curved box to the same function in the equivalent straight box. Thus, the functions obtained from straight box analysis are multiplied by the following factors to give the same functions for the curved box:

$$\text{Bending moment} \qquad 1.0 + (A/R) + (B/R^2)$$

$$\text{Torsional moment} \qquad C + (D/R^2)$$

where $A = 4.5 + 0.028L$, $B = 0.2L^2 - 11L + 300$, $C = 0.2 - 0.0005L$, $D = 36L - 1000$, L is the span length (ft), and R is the radius (ft).

The foregoing expressions consider the box beams acting as a system, but the effect of moments due to local distortion at the cross section is not considered.

Dead Load Effects For boxes that do not have a full-width top plate connected to the webs, the design is based on the principles of open section for wet concrete and other construction loads. In order to prevent lateral buckling of the lip plates under these loads, the bending stresses are limited to the specified values for curved I girders. The warping stresses in the lip plates required in the allowable stress analysis may be determined as suggested by United States Steel (1967).

Torsion For bridges consisting of separate box girders, consideration should be given to the use of intermediate diaphragms between individual boxes in order to cause the bridge to act as one unit under torsional loading. A suggested method to determine load effects in the boxes and in the diaphragms is as proposed by Heins (1972), Lavelle, Greig, and Wemmer (1971), and Brennan (1970).

Warping Effects Warping and shear forces caused by nonuniform torsion are generally small in box girders, although there is evidence that deformation of the cross section induces additional warping effects as well as transverse bending of significant magnitude. The presence of cross frames and intermediate diaphragms reduces these stresses. Studies by Kristek (1970) show, however, that warping stresses may still be high at intermediate diaphragm spacing, producing optimal reduction in transverse bending, and these should be considered in box design in conjunction with geometry distortion. Distortional effects have been studied by Wright (1969) and Wright, Abnel-Samad, and Robinson (1968) for straight box girders, and this analysis has been extended to curved boxes by Dabrowski (1968), Shore and Wilson (1973), and Chu and Pinjarkar (1971).

Fatigue should receive equal attention because fatigue strength may be under- or overestimated unless a three-dimensional analysis is carried out.

Specifications and design recommendations for steel I and box girders are outlined in subsequent sections.

Design Procedures

The CURT project mentioned in the preceding section represents the largest concentrated research effort for investigating the response of steel box girder bridges. The program was sponsored by the Federal Highway Administration between 1969 and 1976. A consortium of university research teams (CURT) participated in the project methods that have evolved in the past years for analyzing and designing curved steel box bridges and that cover a broad range of static and dynamic approaches, all based on inherent assumptions

TABLE 6-3 Curved Box Girder Methods of Analysis

Method (1)	Solution (2)	Reference (3)
Stiffness	Static analysis	
	Planar grid	58
	Space frame	11–15
	Finite element	9, 22, 33, 86, 87, 91
	Finite strip	86
Differential equation	Finite difference	40, 41, 53, 54, 55, 74,
	Closed form	105
Approximate	Design aids	54, 75, 76, 99, 101
Differential equation	Dynamic analysis	33, 49–52, 70, 81, 89, 90
Finite element		

Note: Reference numbers in Column 3 are as shown in the *Journal of the Structural Division*, ASCE, Vol. 104, No. ST 11, Nov., 1978 pp. 1734–1739.

and with certain limitations and suitability formats. Table 6-3 gives a summary of methods and solutions with appropriate references.

Static Analysis In this procedure, vibration and moving effects of loads are not considered. As can be seen from Table 6-3, most solutions are based on the stiffness method, which in turn is based on a solution of simultaneous equations developed from compatibility and equilibrium relationships.

Planar Grid This method treats the curved girder bridge on a planar frame loaded normally to the frame (Lavelle and Bolck, 1965). Several improvements to the original version have been made (Lavelle, Greig, and Wemmer, 1971). For example, power series expansions are used for angle functions to eliminate problems of computational instability, and advantage is taken of the banded symmetrical characteristics of the structure stiffness matrix in the solution process. More recent refinements (Lavelle, 1976) make the method suitable to closed-section box girders and truss-type diaphragms. The algorithm assumes that the member is prismatic, the cross section does not change shape, and bracing is provided at the top flange to make the section closed. The slab is included in computing torsional properties. Uniform torsion theory is assumed, utilizing the box girders as a single curved line. The method is not recommended for structures of significant skew because the actual width of the box is considerable.

Space Frame This method idealizes a curved box bridge as a three-dimensional space frame (Brennan and Mandel, 1974), and applies to boxes with open or closed sections, slabs, truss-type diaphragms, and lateral bracing. Curved members are approximated, however, as chord segments (Hildebrand,

1956), and the solutions of the equations are by the Gauss–Jordan procedure. The method does not incorporate warping stiffness.

Finite Elements Several approaches have been developed. One version by Meyer and Scordelis (1971) presents two variations to the solution of curved box girders. The first variation applies to box girders of constant depth, and is programmed to analyze cellular structures. The second variation is for solutions of curved box girder bridges with an arbitrary three-dimensional frame, characterized by plane stress elements. A special feature is that a rotational degree of freedom is added to the commonly used two translational degrees of freedom per node.

Chu and Pinjarkar (1971) have presented a variation of the finite-element approach for box girders, where the top and bottom slabs are idealized as horizontal sector plates, and the walls of the boxes represented as vertical cylindrical shell elements. Both membrane and bending action for the plate and shell elements are considered, and equations for stiffness of sector plate elements are included. Shell element stiffness is based on a solution by Hoff. The method, however, must be extended to continuous units with intermediate diaphragms to be practical in curved box girder design.

Shore and Lansberry (1972) have developed a fully compatible annular segment finite element for curved bridges as part of the CURT program. The approach applies to thin plates subjected to membrane and bending stresses, using the stiffness method of analysis. Six types of elements are chosen: (a) cylindrical shell elements used to idealize webs of curved girders; (b) annular elements used to represent the flanges or the roadway deck; (c) quadrilateral plate elements representing solid diaphragms; (d) straight beam elements that simulate longitudinal girders or cross beams; (e) triangular elements assumed to idealize skewed sections of the roadway slab; and (f) circular curved I sections used as curved longitudinal girders. Classical thin-plate theory is used to express element stiffness and stress resultant matrices. The element stiffness and stress matrices are derived from a homogeneous, linear elastic, cylindrically orthotropic material. Only linear portions of the strain–displacement relationship are used, and this limits the analysis to small displacements. The form of displacement expansion for all three displacement components (radial, tangential, and transverse) makes the element fully compatible.

This option is useful in the analysis of complex, built-up structures such as horizontally curved bridges. Two criteria must be satisfied by the assumed displacement expansion to ensure convergence to the correct solution. First, the displacement expansion can represent all states of constant strain, including the rigid-body displacement states. Second, the displacement expansion satisfies the conditions of compatibility along the element boundary as well as in the element. For bending problems, this implies that the displacement expansion and its first partial derivatives are continuous within the element and across interelement boundaries. For membrane problems, the displacement expansion must be continuous within the element and across interelement boundaries. The element exhibits a high degree of accuracy in calculating deflections and stress resultants.

A stiffness method at initial stress using finite elements is presented by Bazant and Nimeiri (1974). This method approximates long-span box girder bridges as thin-walled beams of closed section that undergo longitudinal warping and transverse distortion, and uses a discrete subdivision in the longitudinal direction rather than the transverse direction. It appears that the subdivision into beam elements was more suitable because in narrow and shallow cross sections the transverse distributions can be expressed by fewer parameters than the longitudinal distributions. The curved beams are approximated by straight finite elements, and the requirements for convergence and accuracy are satisfied. Full continuity is obtained by using a procedure allowing for transverse shear deformations, which does not require the cross section to be normal to the beam axis. A serious restriction of the method is that it is limited to a single box girder.

Finite Strip Scordelis (1974) has presented an analytical solution using a finite strip method and a harmonic analysis approach. The plate element extends longitudinally over the entire span and transversely between designated points on the cross section. The applied loads are considered in a harmonic analysis.

Differential Equation Solutions A closed-form solution has been presented by Konishi and Komatsu (1962, 1963). Differential equations are developed and solved for certain boundary conditions to give moments, twist, angle of rotation, deflection, Saint-Venant torsion, secondary shear, stress resultants, deformations, and influence surfaces.

Finite-difference solutions are presented by Komatsu, Nakai, and Nakanishi (1971) for the skewed effects on curved box girders. The analysis is based on simple torsion theory, and utilizes a transfer matrix for static behavior.

Analytical methods have also been developed by Heins and Oleinik (1976), Oleinik (1974), and Yoo et al. (1975) for single- and multiple-span box systems, based on the Vlasov differential equations. The solution automatically applies to variable stiffness and continuity. Plane bending, warping torsion, and distortional stresses are considered.

Dynamic Analysis In general, the dynamic analysis of curved bridges is a complex problem to which research has responded with caution. The current impact and fatigue criteria, in effect, account for live load dynamic behavior.

Differential Equations Solutions have been presented by Komatsu and Nakai (1966, 1970) and Nakai and Kotoguchi (1975) for the free and forced dynamic response of curved box girder bridges. Practical formulas are given

TABLE 6-4 Curved Girder Computer Programs

Name and Reference (1)	Origination (2)	Computer System (3)	Brief Description (4)
CBRIDG (33)	McGill University, Montreal, Quebec, Canada	IBM 360/75	Three-dimensional finite-element model for static and dynamic analysis of prismatic cellular curved box girders with vertical web element (multiple cells).
CELL (103)	University of California, Berkeley, CA	CDC 6400	Finite-element analysis of cellular structures of constant depth with horizontal plate and vertical web elements (multiple cells).
CUR BOX 1 (75)	University of Maryland, College Park, MD	UNIVAC 1108	Finite-difference solution to analyze composite or noncomposite single-span curved box girders (single cell).
CUR BOX 2 (41)	University of Maryland, College Park, MD	UNIVAC 1108	Finite-difference solution for design of continuous two-span or three-span box girders. Moment shear, torsion, and bimoment envelopes developed (single cell).
CUGAR 3 (59)	University of Rhode Island, Kingston, RI	IBM 360—Also Burroughs and UNIVAC	Planar-grid solution based on uniform-torsion theory idealizing the box girder as a single curved line simplified input–output (multiple cells).
DYNCRB/BG (89,90)	University of Pennsylvania, Philadelphia, PA	IBM 370/165	Finite-element method of analysis to compute frequencies and mode shapes. Effects of moving loads on box girder bridge dynamic response (multiple cells).
FINPLA 2 (68)	University of California, Berkeley, CA	CDC 6400	Finite-element analysis of general nonprismatic cellular structures of varying width and depth (multiple cells).
STACRB (91)	University of Pennsylvania, Philadelphia, PA	IBM 370/165	Finite-element method to analyze open or closed, straight or curved girder bridges (multiple cells).
SU3D (11,16)	Syracuse University, Syracuse, NY	IBM	Space-frame analysis of composite or noncomposite, single- or multiple-span continuous curved I-girder and box girder bridges via the stiffness method (multiple cells).

for displacements and frequencies, and the work is extended to cover forced vibrations including moving loads.

Finite Elements Fann (1973) has presented a finite-element procedure for static-vibration and free-vibration analysis. The model is part of a proposed idealization scheme for prismatic cellular bridges with straight or curved alignment.

Rabizadeh and Shore (1975) have developed a method for dynamic analysis of box girder bridges with constant radius. The underlying assumptions are: (a) the webs are orthogonal to the deck and bottom flanges; (b) superelevation of the deck is neglected; (c) the supports have a radial direction; (d) the deck is assumed to be homogeneous, cylindrically orthotropic, and linearly elastic; (e) other members of the bridge are homogeneous, isotropic, and linearly elastic; and (f) the effect of damping is neglected. The weight ratio of the design vehicle to the superstructure weight is taken as less than 0.3, and therefore the mass of the vehicle is neglected. This method of analysis was developed as part of the CURT project, and has been incorporated into a computer program.

Table 6-4 gives a summary of curved girder computer programs. Because this table is not completely current, engineers should inquire with the appropriate sources.

6-6 EXAMPLES OF CURVED STEEL BOX GIRDER BRIDGES: A SURVEY

A 1978 survey of curved steel box girder bridges identified a total of 82 structures in the United States, Canada, Europe, and Japan (ASCE–AASHTO, 1978). These bridges fall into several subcategories of curved boxes. Most of the early structures are true steel boxes with all four sides consisting of steel. Later types are through boxes open at the top when fabricated. In this case the top of the box is provided with the cast-in-place concrete deck. Open boxes required top bracing to prevent buckling during handling. The double or multiple box bridge with floor beams or diaphragms between girders is suitable to longer spans where greater structure depth is required. The floor beams and diaphragms are usually connected to the boxes with full moment connections.

In this survey, 10 examples were orthotropic deck structures. Seven bridges were simple spans, 22 had two spans, and the remainder consisted of three or more spans. Table 6-5 gives the location of the bridges included in the survey, and identifies certain structural characteristics. The only nonhighway bridges in the list are the two structures on the Washington Metro.

TABLE 6-5 Summary of Curved Steel Box Girder Bridges (1978 Survey)

United States (1)	Number of Structures (2)	Foreign (3)	Number of Structures (4)
Alaska	1	Austria	1
Colorado	7	Belgium	2 (1 steel deck)
Connecticut	1	Germany	8 (7 steel deck)
Dallas/Fort Worth Airport	1	Great Britain	13
Florida	4	Italy	1
Louisiana	1	Japan	6 (2 steel deck)
Maryland	1	Scotland	1
Massachusetts	1	Sweden	3
New York	2	Switzerland	3
Oregon	2		
Pennsylvania	15		
Virginia	5		
Washington Metro	2		
West Virginia	1		

Note: All bridges had concrete decks, except as noted; all except two are highway bridges.

Size and Depth of Box Curved box bridges cover a broad range of spans and radii. Few bridges, however, have average spans greater than 350 ft, which seems to be a self-imposed limit related to economic considerations.

The size of the box is a function of span, roadway width, and radius. From actual data, an enclosed area of box per foot of bridge has been determined. The average span length, radius, (bridge width)/(box depth), and external bracing are significant variables. Using a least-squares fit of the data gives the following relationship:

$$\begin{aligned} A_b &= 1.077 \quad (L^{1.14}) + 650.6(R^{-0.15}) - 104.8(W/D) - 88.26(K) \\ t &= (10.95) \qquad\qquad (4.34) \qquad\qquad (1.74) \qquad\qquad (2.59) \end{aligned} \tag{6-12}$$

where A_b = enclosed area of box per foot-width (in.^{2}/ft)

L = span (ft)

R = radius (ft)

W = bridge width (ft)

D = box depth (in.)

K = 1 without external bracing, and 2 with external bracing

t = distribution statistic, or test values for significance using the Student distribution

The equation explains 70 percent of the variation, and has a maximum error of 61 percent and an average error of 20 percent. It indicates past trends from actual practice, and should not be used for actual design. In the examples listed in Table 6-5, the span–depth ratio L/D varies from 12 to 40, and the ratio of box depth to bottom flange width ranges from 0.2 to 3.0 with most values in the range 0.2 to 1.0.

The bottom flange has a width–thickness ratio ranging from 10 to 300, but this wide variation does not relate to any single parameter. Likewise, the span–box width ratio varies with no discernible pattern. As the box becomes deeper and narrower, the plate varies in thickness for the particular span and radius. Specific limits are imposed on the slenderness ratio for buckling and shear lag.

The bridges have a concrete slab thickness that varies in a wide range, from 6.5 to 12.5 in. This is probably because of a variation in the framing methods. In some structures, open sections are used as floor stringers framed into transverse floor beams supported on the primary boxes.

Supports Support details for curved boxes must be designed to minimize horizontal restraint at the bearings. Inasmuch as boxes are much stiffer in the lateral direction than plate girders, temperature changes will induce lateral forces at the restraining points. Exempting friction, the ideal horizontal support system must be statically determinate and should not develop thermal forces. In a single continuous box, a statically determinate horizontal support is designed by fixing each of three supports in one direction only and freeing all other supports in all directions. One way is to fix one interior support in the tangent direction, the two end supports in the radial direction, and provide complete freedom at all other supports. If such a system is part of a multiple-girder system where the girders interact with the deck and diaphragms, all other bearings in the remaining girders should be free in all directions to satisfy the statical determinacy of the bridge.

In any event, friction will upset the degree of statical determinacy, particularly with large reactions from dead loads, because this factor may cause bearings to respond as fixed whereas theoretically they are free to move. If this fixity is induced to the piers, it will cause them to deflect with corresponding force and moment effects. Most bearings will behave as fixed until friction is overcome and slip can occur.

Another important detail is seating the box on its bearings associated with the difficulty of fabricating the box girder to exact geometry. For a given span, four-point support (bearing shoes at all four corners) of a stiff box girder will require special care to avoid poor fit and rocking.

Almost, all four basic types of bearings have been used in steel box girder bridges. The extent of this use is as follows: steel, 21 percent; steel–self-lubricating bronze, 12 percent; elastomeric pad, 24 percent, elastomeric pot, 38 percent; and others, 5 percent. The shoe type is correlated with geometric parameters in a rather consistent pattern. For example, all-steel bearings

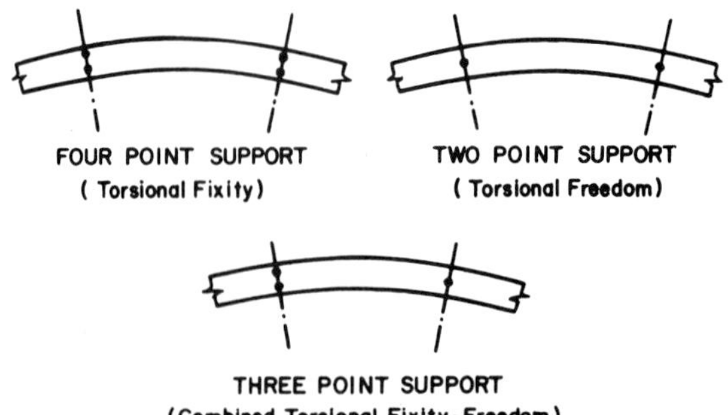

FIGURE 6-13 Bridge support options, curved steel box girders.

were not used in more steeply curved structures (specifically with central angles greater than 8° and radii less than 1000 ft); pot bearings were used over the full range of parameters; elastomeric pads were not used in bridges with more than three continuous spans; and steel shoes were not used for bridges with less than three continuous spans. These results may indicate a trend to establish criteria for support systems.

Two different framing schemes are used to transmit torques into the substructure. One way is to design the bearings so that the boxes are torsionally fixed at some or all supports, merely by providing two shoes inducing a couple at each bearing line under each girder. For a single-box structure, this degree of torsional rigidity is the absolute requirement, especially with severe curvature, but may also be used in multiple-girder bridges.

The alternative is to allow torsional freedom at the bearings but connect the girders together by diaphragms at the supports. This is applicable to multiple-girder bridges. Torsional freedom is ensured by a single bearing that can rotate radially. In effect, this arrangement converts torque at the supports into vertical reactions at the bearings rather than into couples. Various bridge support options are shown in Figure 6-13.

Results from the survey confirm that engineers make provisions for longitudinal rotation in all cases unless the boxes are integrally framed into the piers. In the majority of examples, longitudinal translational restraint at interior supports was provided as previously outlined. As in straight girders, the longitudinal fixing was located at one of the interior supports to minimize the expansion length. The problem of excessive restraint in the curved box girder system was recognized in all instances, and all the supports were not fixed transversely as would normally be the case with straight girders.

6-7 ANALYTICAL PROCEDURE: SINGLE SPAN

Dead Load Effects

Bending Moment As outlined in the foregoing sections, the analysis of curved systems is highly indeterminate, and generally requires specialized

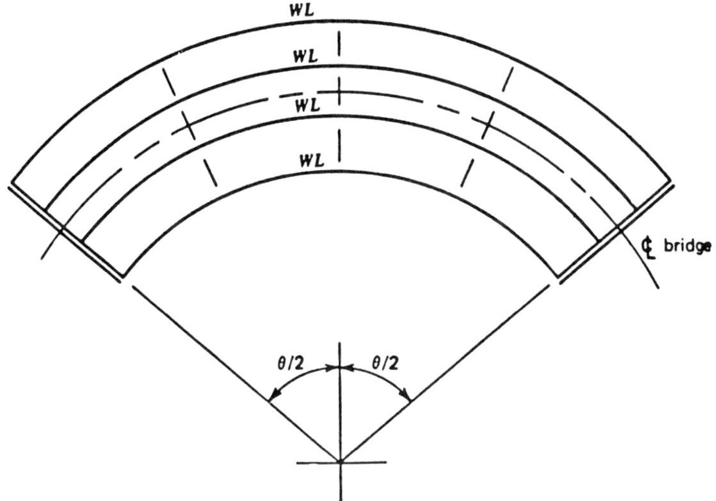

FIGURE 6-14 Bridge layout; curved girder dead load system. (From Heins and Firmage, 1979.)

computer techniques. Invariably, however, these programs require girder stiffness data so that a preliminary evaluation of the structural requirements is essential. The simplified procedure suggested by Nettleton (1976) may be used for estimating load effects on curved girder bridges.

The bridge structure shown in Figure 6-14 is subjected to a dead load W. The resulting total overturning moment may be estimated by considering the effective moment to be developed about the chord of each girder of the system as shown in Figure 6-15.

The average moment arm Y_{av} is related to the moment arm e by

$$Y_{av} = e\left(1 - \frac{1 - \cos\theta/4}{1 - \cos\theta/2}\right) \tag{6-13}$$

For the range of θ usually applicable to curved bridges, (6-13) yields

$$Y_{av} = \tfrac{3}{4}e \tag{6-14}$$

which is within less than 1 percent for any value of θ. The overturning moment resulting from this configuration is

$$M_G = \tfrac{3}{4}eWL \tag{6-15a}$$

and for N girders the total overturning moment effect is therefore

$$M_T = N\tfrac{3}{4}eWL \tag{6-15b}$$

Referring to the cross section of the bridge shown in Figure 6-16, we can

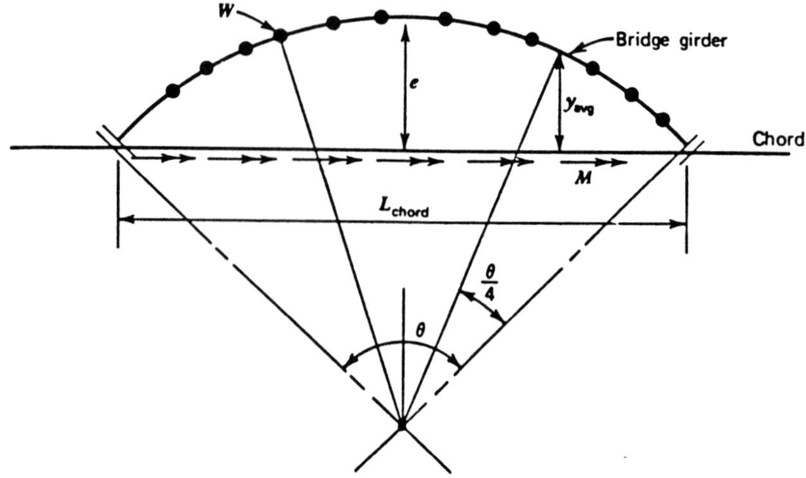

FIGURE 6-15 Isolated curved girder and chord series, dead load. (From Heins and Firmage, 1979.)

compute the effect of the total torque M_T on the system. If this system rotates rigidly about the bridge centerline, the induced forces resulting from M_T are

$$M_T = \frac{3S}{2}G_1 + \frac{S}{2}G_2 + \frac{S}{2}G_3 + \frac{3S}{2}G_4 + \cdots + XG_N \qquad (6\text{-}16)$$

If a linear distribution of forces is assumed as in the rivet problem, the

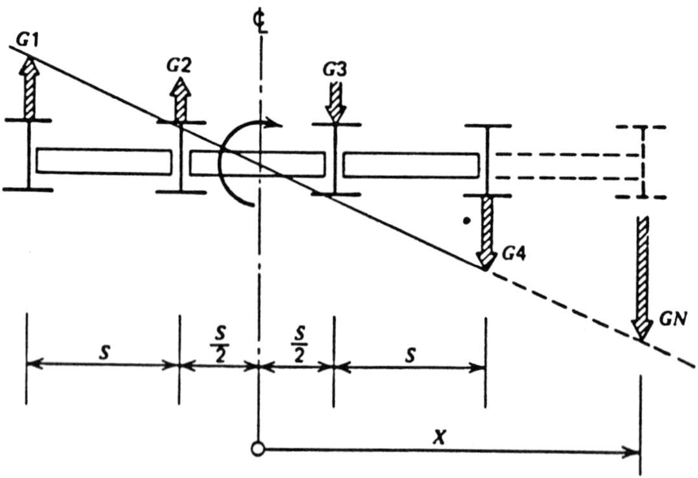

FIGURE 6-16 Bridge cross section showing torque effects. (From Heins and Firmage, 1979.)

following analogy is derived:

$$\frac{G_1}{3S/2} = \frac{G_2}{S/2} = \frac{G_3}{S/2} = \frac{G_4}{3S/2} = \frac{G_N}{X} \tag{6-17}$$

The forces relative to G_N and X are now

$$G_N = \frac{G_N X}{X} \qquad G_1 = \frac{3S/2}{X} G_N \qquad G_2 = \frac{S/2}{X} G_N \ldots$$

If these values are substituted into (6-16), the result is

$$M_T = \left(\frac{3S}{2}\right)\left(\frac{3S/2}{X}\right)G_N + \left(\frac{S}{2}\right)\frac{S/2}{X}G_N + \cdots + \frac{X^2}{X}G_N$$

$$M_T = \left(\frac{3S}{2}\right)^2 \frac{G_N}{X} + \left(\frac{S}{2}\right)^2 \frac{G_N}{X} + \cdots + X^2 \frac{G_N}{X}$$

$$M_T = \frac{G_N}{X}\left[\left(\frac{3S}{2}\right)^2 + \left(\frac{S}{2}\right)^2 + \cdots + X^2\right]$$

In general, therefore, the following relationship is valid:

$$M_T = \frac{G_N}{X} \sum X^2 \tag{6-18}$$

which is rearranged to give the forces on each girder as follows:

$$G_N = \frac{M_T X}{\Sigma X^2} \qquad G_1 = \frac{M_T(3S/2)}{\Sigma X^2} \cdots \tag{6-19}$$

where ΣX^2 is the sum of the squares of the distance to each girder. Using (6-15*b*), the girder force G_N is

$$G_N = \frac{M_T X}{\Sigma X^2} = \left[N\left(\frac{3}{4}\right)eWL\right]\frac{X}{\Sigma X^2} \tag{6-20}$$

which represents the additional vertical induced load to girder *N*. This corresponds to the secondary load discussed in Section 6-2, and is added to the primary effective dead load *WL*. The girder moment is now computed as

$$M_G = \frac{(WL + G_N)L}{8} \tag{6-21}$$

The bending stress due to dead load is calculated from the conventional expression $\sigma_b = M_G/S$, where S is the section modulus.

Warping Moment Stresses developed as a result of warping or lateral bending of the flanges may manifest the distortion shown in Figure 6-17. These lateral bending or warping stresses can be approximated by considering the effects of the bending stresses on the curved flanges, for example, between the two transverse diaphragms shown in Figure 6-18.

These resultant stresses give forces F as follows:

$$F = M_G/h \tag{6-22}$$

where h refers to the notation of Figure 6-19. Examination of the equilibrium conditions for this flange section shows that, in order for the vertical forces F_v to be balanced by external loading, an internal moment must be present. This is referred to as the warping moment (see also Section 6-2). Considering

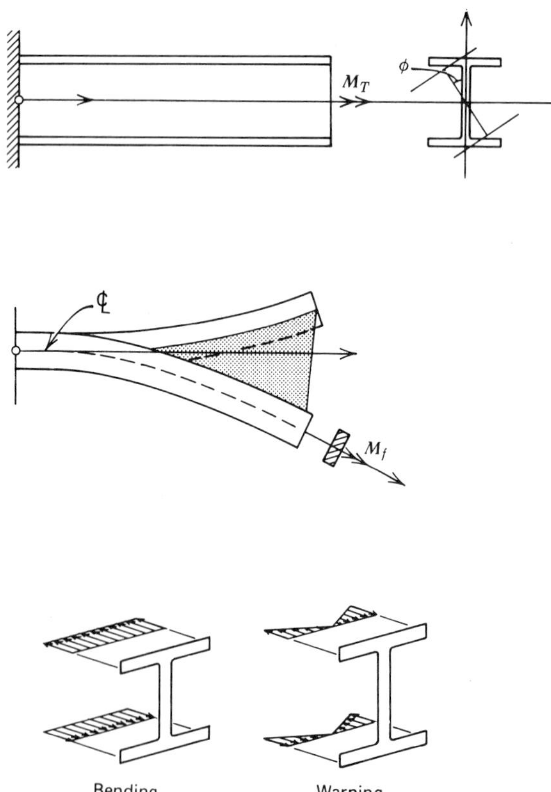

FIGURE 6-17 Lateral bending or warping of WF-shape section. (From Heins and Firmage, 1979.)

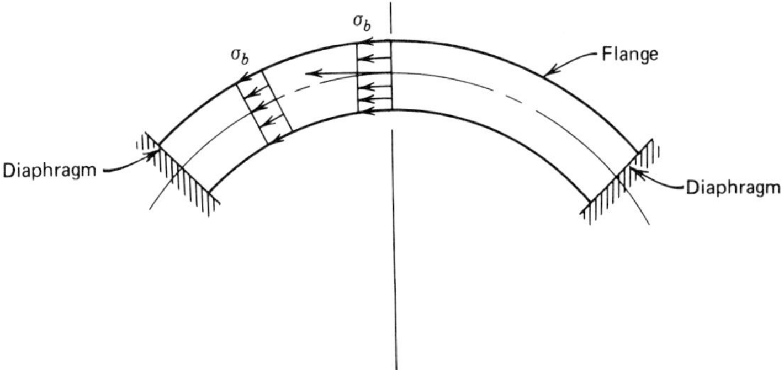

FIGURE 6-18 Bending stresses in curved flange. (From Heins and Firmage, 1979.)

a load q applied to the flange as shown, we can write

$$F_v = 0 \qquad \text{or} \qquad 2F \sin \Delta = Sq$$

Alternatively, we rewrite (6-2) as follows (see also Eq. 6-3):

$$q = \frac{M_G}{h} \frac{1}{R} \tag{6-23}$$

where all the symbols correspond to the notation of Figure 6-19. Referring now to Figure 6-20, the induced moment is calculated from

$$M_{f(\text{end})} = \frac{q(kL)^2}{12} \tag{6-24}$$

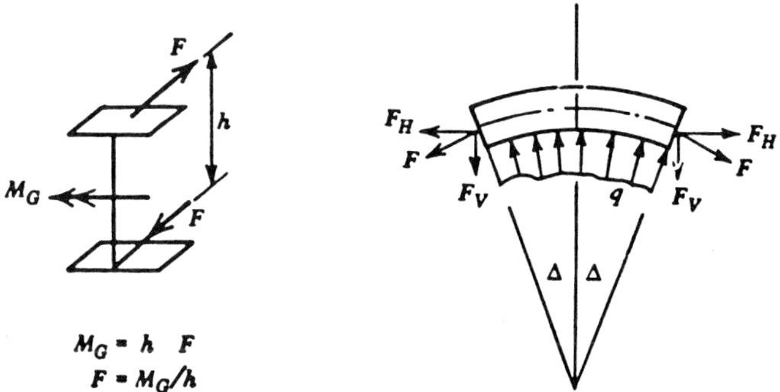

FIGURE 6-19 Flange equilibrium in curved girder. (From Heins and Firmage, 1979.)

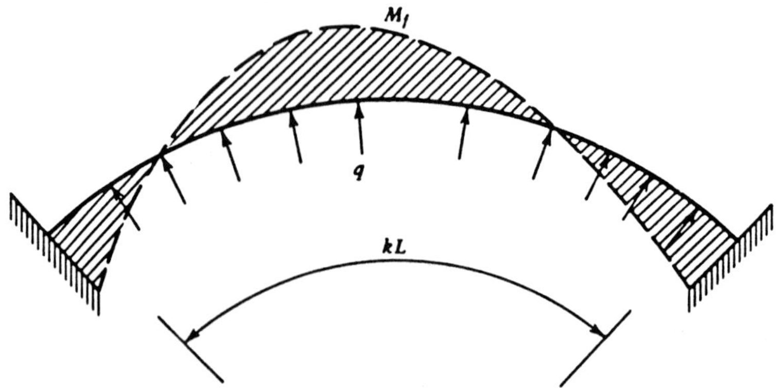

FIGURE 6-20 Bending moments in flange, assumed fixed at diaphragm locations. (From Heins and Firmage, 1979.)

where kL is the diaphragm spacing, M_G is a constant, and the flange is taken as fixed at the diaphragm. From (6-23) and (6-24) we obtain the resulting warping moment

$$M_w = \frac{M_G(kL)^2}{12hR} \tag{6-25}$$

If the flange width and thickness are b and t, respectively, the section modulus in the vertical axis is $S = b^2t/6$, and the stress resulting from the warping effect is $\sigma_w = M_w/S$, which is added to σ_b.

Distortional Effects If the elements of the girder system are subjected to variations in shears, the system will be distorted, thus inducing normal stresses. Properly spaced diaphragms reduce this effect (Dabrowski, 1972; Heins and Martin, 1977). The induced distortional normal stress due to dead load is approximated by

$$\frac{(\sigma_d)}{\sigma_b + \sigma_w} = \frac{A(X)^3 + B(X)^2 + C(X) + D}{100} \tag{6-26}$$

where σ_d is the distortion stress, σ_b and σ_w are as given previously, X is the diaphragm spacing/span length, and A, B, C, D are constants given in Table 6-6. If the coefficients shown in this table do not correspond to the design parameters, interpolation may be used.

TABLE 6-6 Dead Load Distortion Coefficients

Girder Length (ft)	Radius (ft)	A	B	C	D
50	250	0	19.75	4.722	0.3027
	500	0	44.38	−2.619	0.9935
	750	0	61.38	−8.986	1.667
75	100	0	73.11	−13.67	2.175
	250	0	−15.71	18.20	−1.196
	500	0	−1.732	17.13	−1.221
	750	0	8.400	15.32	−1.129
	1000	0	15.72	13.70	−1.025
100	250	−14.93	4.022	10.89	−0.5629
	500	−24.60	28.10	7.729	−0.4398
	750	−9.880	32.48	6.910	−0.4023
	1000	3.445	33.40	6.564	−0.3824
125	250	−181.2	70.24	0.9687	−0.06552
	500	−208.6	101.3	−2.562	0.05658
	750	−203.7	111.0	−3.913	0.1090
	1000	−187.4	111.8	−4.163	0.1200
150	250	−254.1	108.2	−3.548	0.1287
	500	−258.5	131.4	−6.327	0.2254
	750	−233.5	132.8	−6.709	0.2411
	1000	−217.3	133.2	−6.918	0.2508

Live Load Effects

Bending Moment The standard truck load will be considered in this analysis. The load application is assumed to introduce two effects: (a) torsional moments about the chords of the respective loaded curved girder, referred to as local loading; and (b) torsional moments about the centerline of the entire bridge caused by the combined truck load or center of gravity of these loads, designated as system loading.

Local Loading For simplicity, we assume the truck load developed along the curved girder as shown in Figure 6-21. As in dead load effects, the induced moment has the form

$$M = yP \tag{6-27}$$

where y is the moment arm to the respective axle loads. For the truck position shown in Figure 6-21, $y = e$ for the center axle load, and $y = (e - c)$ for the other two axle loads. The c distance is computed assuming a curve

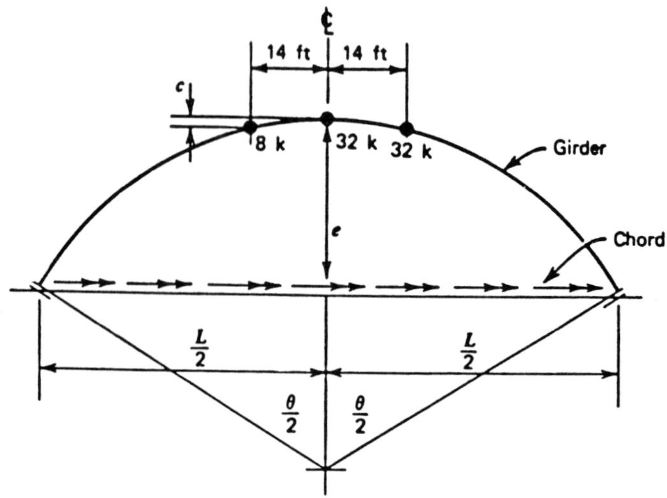

FIGURE 6-21 Plan geometry for isolated girder and position of truck load. (From Heins and Firmage, 1979.)

form $y = AX^2 + B$. From the plan geometry, we derive $y = e[1 - (X/0.5L)^2]$. For $X = 14$ ft, the y value is $e[1 - (14/0.5L)^2]$, and the offset is $c = e - y$, or $c = (14/0.5L)^2 e$, or $c = \overline{K}e$ and $c/e = \overline{K}$. Values of $\overline{K}$ are computed for $L = 140$, 100, and 70 ft, and are 0.040, 0.078, and 0.160, respectively.

The total bending moment about the chord is now

$$M_G = 8(e - c) + 32(e - c) + 32e$$

or

$$M_G = 72e\left[1 - \frac{40}{72}\overline{K}\right] \tag{6-28}$$

Values of the factor inside the bracket are computed for $L = 140$, 100, and 70 ft, and are 0.98, 0.96, and 0.91, respectively; hence, an average value 0.95 is used. Therefore,

$$M_G = 72e(0.95) \tag{6-29}$$

For several trucks on the bridge and assuming that each is concentrated on a given girder, the total moment is

$$M_T = (0.95e)72N_T \tag{6-30}$$

where N_T is the number of trucks.

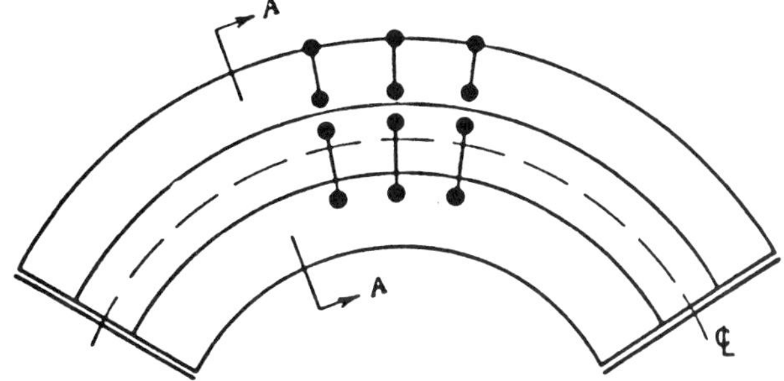

FIGURE 6-22 Plan and position of truck loads, system effect. (From Heins and Firmage, 1979.)

System Loading The position of trucks in plan and cross section is shown in Figures 6-22 and 6-23, respectively.

From the cross section of Figure 6-23, we can write

$$M_T = \overline{X}x(2)(72) \tag{6-31}$$

and for N_T trucks

$$M_T = \overline{X}N_T 72 \tag{6-32}$$

Combined Loading The combined effect is obtained by summing the M_T values from (6-30) and (6-31) which gives

$$M_T = 72N_T(0.95e + \overline{X}) \tag{6-33}$$

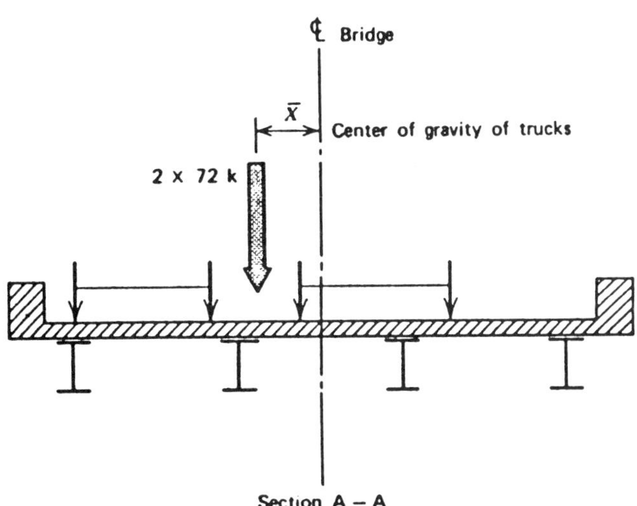

FIGURE 6-23 Cross section and position of truck loads, system effect. (From Heins and Firmage, 1979.)

where N_T, e, and $\overline{X}$ are as previously defined. We should note that (6-33) is similar to (6-15*a*). The forces due to the live load torque are derived as before as

$$G_N = 72N_T(0.95e + \overline{X})\frac{X}{\Sigma X^2} \tag{6-34}$$

This additional force effect is added to the primary line of the truck wheels to obtain the combined effect.

Bending Stress The normal bending stress is $\sigma_b = M_B/S$, where

$$M_B = M_P(1 + CE) \qquad CE = 2N_T(0.95e + \overline{X})\frac{X}{\Sigma X^2} \tag{6-35}$$

and M_P is the primary truck load moment (obtained for straight girders).

Warping Stress The warping stress σ_w is obtained from the appropriate section modulus and the warping moment $M_w = M(kL)^2/12hR$, where $M = q$.

Distortional Stress As in dead load effects, the distortional stress is calculated from an expression similar to (6-26) as follows:

$$\frac{\sigma_d}{\sigma_b + \sigma_w} = \frac{A(X)^3 + B(X)^2 + C(X) + D}{100} \tag{6-36}$$

where σ_d, σ_b, σ_w, and X are as given previously, and the coefficients A, B, C, D are constants given in Table 6-7.

Design Example

A bridge curved in plan is shown in Figures 6-24 and 6-25. The framing system consists of four curved composite girders with the end supports placed radially as shown.

The following information is given: span, 100 ft center-to-center bearings outside girder; structural slab, 7.5 in.; girder spacing, 8 ft; outside radius, 600 ft; FWS, 25 psf; live load, HS 20; steel grade, A36; diaphragm spacing, 20 ft; bridge constructed without shoring.

Dead Load Computed noncomposite dead load $W = 1.006$ kips/ft. Computed superimposed (composite) dead load $W = 0.294$ kip/ft. The dead load moments are calculated for the outside girder. If the chord and arc lengths are assumed essentially equal, we can calculate the dead load moment from

TABLE 6-7 Live Load Distortion Coefficients

Girder Length (ft)	Radius (ft)	*A*	*B*	*C*	*D*
50	250	0	−7.115	12.88	4.155
	500	0	6.160	14.17	4.573
	750	0	16.75	14.30	4.482
	1000	0	26.65	12.53	4.632
75	250	0	−78.09	40.65	−0.3318
	500	0	−90.80	54.60	−1.585
	750	0	−88.91	58.57	−2.004
	1000	0	−84.77	59.83	−2.168
100	250	662.8	−418.8	94.61	−3.479
	500	663.6	−407.8	100.1	−3.944
	750	699.2	−411.3	102.9	−4.164
	1000	710.8	−405.2	102.8	−4.181
125	250	−2067.0	761.1	−79.58	4.989
	500	−2709.0	1027.0	−107.4	6.038
	750	−2807.0	1085.0	−113.5	6.269
	1000	−3072.0	1193.0	−125.6	6.727
150	250	4783.0	−1949.0	265.5	−9.159
	500	5209.0	−2077.0	283.4	−9.873
	750	5474.0	−2159.0	293.8	−10.28
	1000	5632.0	−2207.0	299.8	−10.51

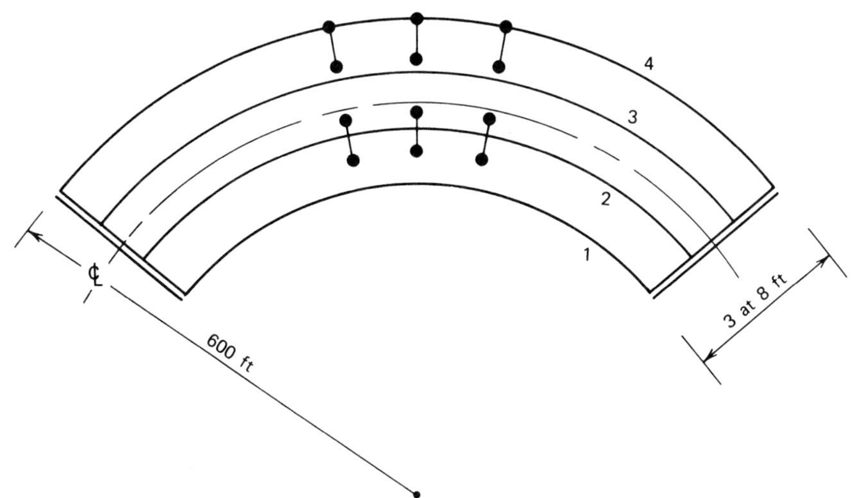

FIGURE 6-24 Single span curved steel girder bridge. (From Heins and Firmage, 1979.)

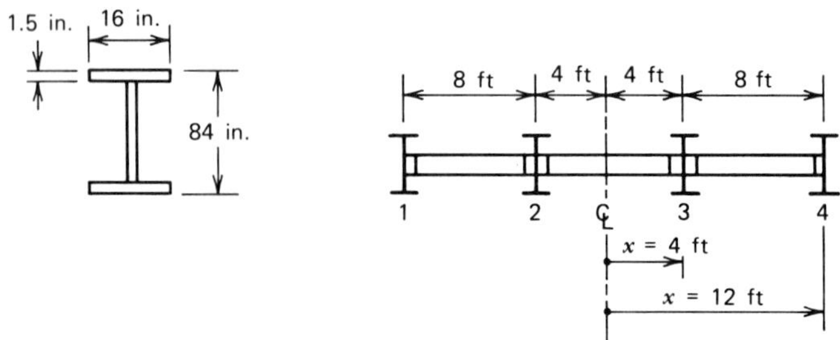

FIGURE 6-25 Cross sections showing the dead load system. (From Heins and Firmage, 1979.)

(6-20) and (6-21), where G_N is the total force on the girder due to curvature. Also $N = 4.0$, $e = 2.0$ ft, $W = 1.01$ kips/ft, and $L = 100$ ft.

The term ΣX^2 is the sum of the squares of the distance to each girder, and $\Sigma X^2 = 2(4^2 + 12^2) = 320\ \text{ft}^2$.

The total force G_N on the outside girder, where $X = 12$ ft, is

$$G_4 = 4.0(3/4)(2)(1.01)(12/320)L = 0.227\ \text{kip/ft} \times L$$

The total load per foot of outside girder is therefore

$$N_T = (W + G_4) = (1.01 + 0.227) = 1.237\ \text{kips/ft}$$

The moment from the noncomposite dead load is easily computed as

$$M_{G4} = \frac{W_T L^2}{8} = \frac{1.237 \times 100^2}{8} = 1546\ \text{ft-kips}$$

Following the same procedure for the superimposed dead load, we compute the additional load on the outside girder due to curvature as

$$G_4 = 0.227 \times \frac{0.294}{1.01} = 0.065\ \text{kip/ft}$$

giving a total $W_T = 0.311$ kip/ft and a moment $M_{G4} = 0.125 \times 0.311 \times 100^2 = 388$ ft-kips.

Live Load With the truck positioned as shown in Figure 6-26, the additional load caused by girder curvature is calculated from (6-34). Given $N_T = 2$ (number of trucks), $e = 2.0$ ft, $\overline{X} = 3.0$ ft (center of gravity of all trucks to center of bridge), $x = 12$ ft, $\Sigma X^2 = 320\ \text{ft}^2$. With these values, (6-34) gives

$$G_4 = 72 \times 2(0.95 \times 2 + 3) \times 12/320 = 24.46\ \text{kips}$$

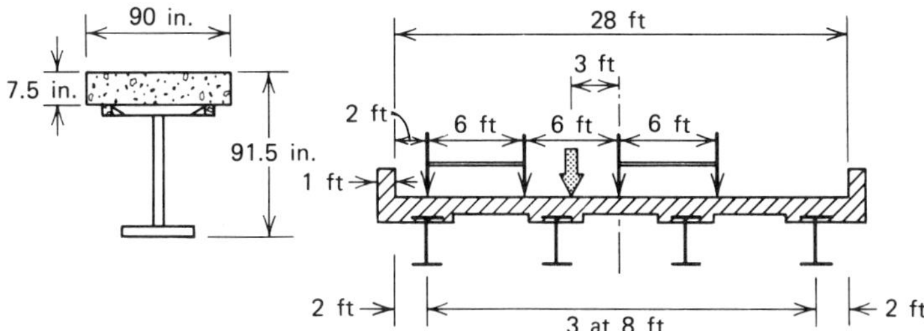

FIGURE 6-26 Cross section showing the live load system. (From Heins and Firmage, 1979.)

Alternatively, we can directly apply (6-35), where M_P is the AASHTO moment. Using an impact factor of 0.222, we calculate

$$M_{\text{AASHTO}} = 762.1 \times 1.222 = 931 \text{ ft-kips}$$

Likewise, we compute the term CE as

$$CE = 2 \times 2(0.95 \times 2 + 3) \times 12/320 = 0.74$$

Therefore,

$$M_{\text{LL}+I} = 931 \times 1.74 = 1620 \text{ ft-kips}$$

The final design moments are

$$M_{\text{DL}} = 1546 \text{ ft-kips}$$
$$M_{\text{SDL}} = 388 \text{ ft-kips}$$
$$M_{\text{LL}+I} = 1620 \text{ ft-kips}$$

Section Selection The following equation may be used to calculate the bottom and top flange area of steel (Hacker, 1957):

$$A_{Sb} = \frac{12}{F_b}\left(\frac{M_{\text{DL}}}{d_{\text{cg}}} + \frac{M_{\text{SDL}} + M_{\text{LL}+I}}{d_{\text{cg}} + t}\right)$$

$$A_{St} = A_{Sb}\left(\frac{50}{190 - L}\right)$$

where A_{Sb} = bottom flange area (in.^{2})
A_{St} = top flange area (in.^{2})
t = thickness of concrete slab (in.)
d_{cg} = distance between centers of gravity of steel flanges (in.)
F_b = allowable flange stress (ksi)

As a first selection, we assume $d_{cg} = 84$ in. Then

$$A_{Sb} = \frac{12}{20}\left(\frac{1546}{84} + \frac{2008}{91.5}\right) = 24.1 \text{ in.}^2$$

Select a flange 16×1.5, $A = 24$ in.^{2}, and a web $1/2$ in. thick. The steel section properties are $I_T = 103{,}832$ in.^{4} and $S_T = S_B = 2472$ in.^{3}.

Composite Section n = 30 Using an effective flange width for the composite section of $12 \times 7.5/12 = 7.5$ ft, the composite section properties are computed as $I_T = 141{,}488$ in.^{4}, $S_T = 3515$ in.^{3}, and $S_B = 2761$ in.^{3}.

Composite Section n = 10 The composite section properties are calculated as $I_T = 184{,}302$ in.^{4} and $S_B = 2985$ in.^{3}

Stresses in Bottom Flange The noncomposite dead load induces a stress at the bottom flange

$$\sigma_{\text{DL}} = \frac{1546 \times 12}{2472} = 7.51 \text{ ksi}$$

From superimposed load,

$$\sigma_{\text{DL}} = \frac{388 \times 12}{2761} = 1.68 \text{ ksi}$$

$$\text{Total } \sigma_{\text{DL}} = 9.19 \text{ ksi}$$

The warping moment M_w is computed from (6-25), where $M_G = 1546$, or $M_w = 12.5$ ft-kips. The dead load warping stress is

$$\sigma_w = \frac{6M_w}{b_t^2 t_f} = \frac{6 \times 12.5}{16^2 \times 1.5} \qquad \text{or} \qquad \sigma_w = 2.3 \text{ ksi}$$

The dead load distortion is computed from (6-26), where σ_d is the distortion stress. Using $X = 20/100 = 0.2$, we obtain $A = -18.71$, $B = 29.95$, $C = 7.40$, and $D = 0.425$; substituting these into (6-36) gives $\sigma_d = 0.23$ ksi.

Thus, the total dead load stress is

$$\sigma_T = \sigma_b + \sigma_w + \sigma_d = 9.19 + 2.30 + 0.23 = 11.72 \text{ ksi}$$

The live load plus impact stress for the bottom flange is

$$\sigma_{\text{LL}+I} = 1620 \times 12/2985 = 6.52 \text{ ksi}$$

and the corresponding warping stress is found from

$$M_w = \frac{1620 \times 12 \times 20^2}{12 \times 82.5 \times 600} = 13.1 \text{ ft-kips}$$

The warping stress is

$$\sigma_w = \frac{6 \times 13.1 \times 12}{16^2 \times 1.5} = 2.5 \text{ ksi}$$

As in dead load, the distortion stress is computed using (6-36), where $X = 0.2$, $A = 649.36$, $B = -409.2$, $C = 101.2$, $D = 4.03$. From these coefficients, $\sigma_d = 0.45$ ksi, giving a total stress of

$$\sigma_T = 6.52 + 2.5 + 0.45 = 9.47 \text{ ksi}$$

The total stress, dead plus live load, is therefore

$$\sigma_T = 11.72 + 9.47 = 21.2 \text{ ksi}$$

or 6 percent higher than the allowable, say OK.

We should note that the foregoing procedure is for simple, single-span bridges. If the curved girder bridge is continuous, the method must be modified introducing a "modified" span length between supports in considering torsional effects.

6-8 EXTENDED METHOD OF ANALYSIS

The method presented in Section 6-2 has been modified and further simplified for multigirder systems (United States Steel, 1967). Although the procedure is still approximate, it gives satisfactory agreement with other more exact and more complicated methods. This approach is now known as the *V*-load method (United States Steel, 1984) because a considerable part of the torsional load is approximated by sets of vertical shears referred to as *V* loads. The method was initially developed for noncomposite open-frame systems, with bridge piers on radial alignment. Most curved bridges, however, have concrete slabs acting compositely with the curved girders and bottom bracing, inducing closed-frame conditions. In addition, in many instances bridge piers and abutments are placed on a skew angle. The method of analysis presented in this section extends the *V*-load procedure to composite open-frame bridges with any general support configuration. The results are compared to those obtained from finite-element analysis (United States Steel, 1984).

Theoretical Background

From the discussion of Section 6-2, we conclude that full-depth cross frames, floor beams, or diaphragms provide the primary resistance to the torsional loads caused by the bridge curvature. In this interaction, the cross frames inhibit large warping stresses by restricting the lateral bending of the girder flanges. They are therefore primary load-carrying members.

The approximate calculation of the internal torsional load caused by curvature is illustrated in Figure 6-1. The distribution of the radial force $q = M/dR$ follows the shape of the vertical moment diagram shown in Figure 6-1*b*. These radial forces cause lateral bending of the girder flanges, and this effect is what we refer to as warping stresses. Note that the radial component of the top flange is directed outward where the flange is in compression (positive bending) and inward where the flange is in tension (negative bending).

Articulation of V Loads Figure 6-27*a* shows a segment of the curved top flange of the outside girder in compression. Note that the symbols are somewhat different from Figure 6-1. Also note that the lateral forces H_1 and H_2 are simple beam reactions of the unit q times the cross-frame spacing d shown in Figure 6-27*a*. Simple moment equilibrium requires

$$VD = (H_1 + H_2)h \tag{6-37}$$

or, in terms of moments and geometric parameters,

$$V = \frac{M_1 + M_2}{(RD)/d} \tag{6-38}$$

where M_1, M_2 are final moments in curved girders 1 and 2, respectively, expressed as

$$\begin{aligned} M_1 &= M_{1P} + M_{1V} \\ M_2 &= M_{2P} + M_{2V} \end{aligned} \tag{6-39}$$

where M_{1P}, M_{2P} = primary bending moments
M_{1V}, M_{2V} = secondary shear moments (V loads)

The parameters R, D, and d are as shown in Figure 6-27. Setting $(RD)/d = K$, (6-38) becomes

$$V = (M_1 + M_2)/K \tag{6-40}$$

The V-load moments are assumed to be proportional to the respective girder

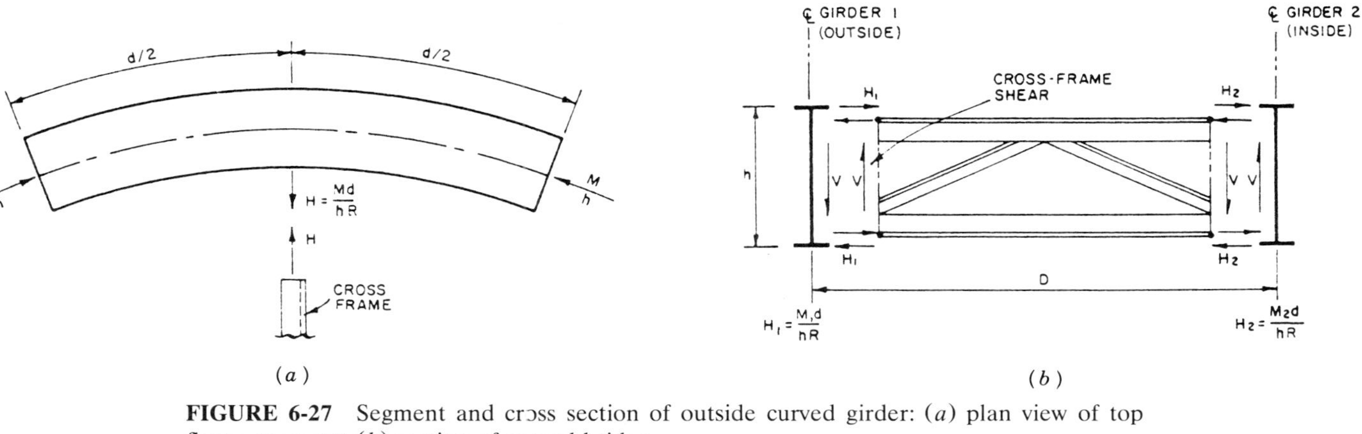

FIGURE 6-27 Segment and cross section of outside curved girder: (*a*) plan view of top flange segment; (*b*) section of curved bridge.

depths, or

$$M_{2V} = -M_{1V}(L_2/L_1) \tag{6-41}$$

From (6-39) and (6-41), we obtain

$$M_1 + M_2 = M_{1P} + M_{2P} + \left[M_{1V}\left(1 - \frac{L_2}{L_1}\right)\right] \tag{6-42}$$

Note, however, that M_{1V} is generally small compared to the primary moments, and the term $(1 - L_2/L_1)$ is also small; hence, the term in the brackets is neglected. Equation (6-42) now becomes

$$M_1 + M_2 = M_{1P} + M_{2P} \tag{6-43}$$

From (6-40) and (6-43), we obtain

$$V = (M_{1P} + M_{2P})/K \tag{6-44}$$

which is essentially the same as the relationship $V = (T + T')/L$ shown in Figure 6-5.

V Loads for Multiple-Girder Systems Consider a three-girder system. In this case it is reasonable to assume that the cross-frame shears between the outside and center girder and the shears between the center and inside girder are equal. This means that the V loads on the straight outside and inside girders are equal and opposite, and therefore there are no V loads on the straight center girder because its cross-frame shears cancel.

For curved systems with more than three girders, this interaction is no longer true, and the distribution of cross-frame shears across the section must consider the relative stiffness of the adjacent girders. The following assumptions are made: (a) girders in the same section have the same stiffness, (b) girder shears across the section are self-equilibrating, and (c) the loading on girders is increased outside the longitudinal centerline of the system, and decreased inside the longitudinal centerline.

An idealized behavior based on these assumptions considers the shear on a girder to be proportional to its distance to the bridge centerline, implying a linear shear distribution across the section as in the procedure discussed in Section 6-7. In this process a curved bridge cross section rotates as a rigid body under torsion induced by bridge curvature, thus shifting the load toward the outside girder. This behavior is enhanced if a rigid concrete slab is present. For bridges with large girder stiffness variation, however, the girder shear distribution may be nonlinear, and this procedure will not give a valid expression of the system behavior.

Consider the four-girder system shown in Figure 6-28*a* where the girders are equally spaced. The section shown is in the positive moment region and is

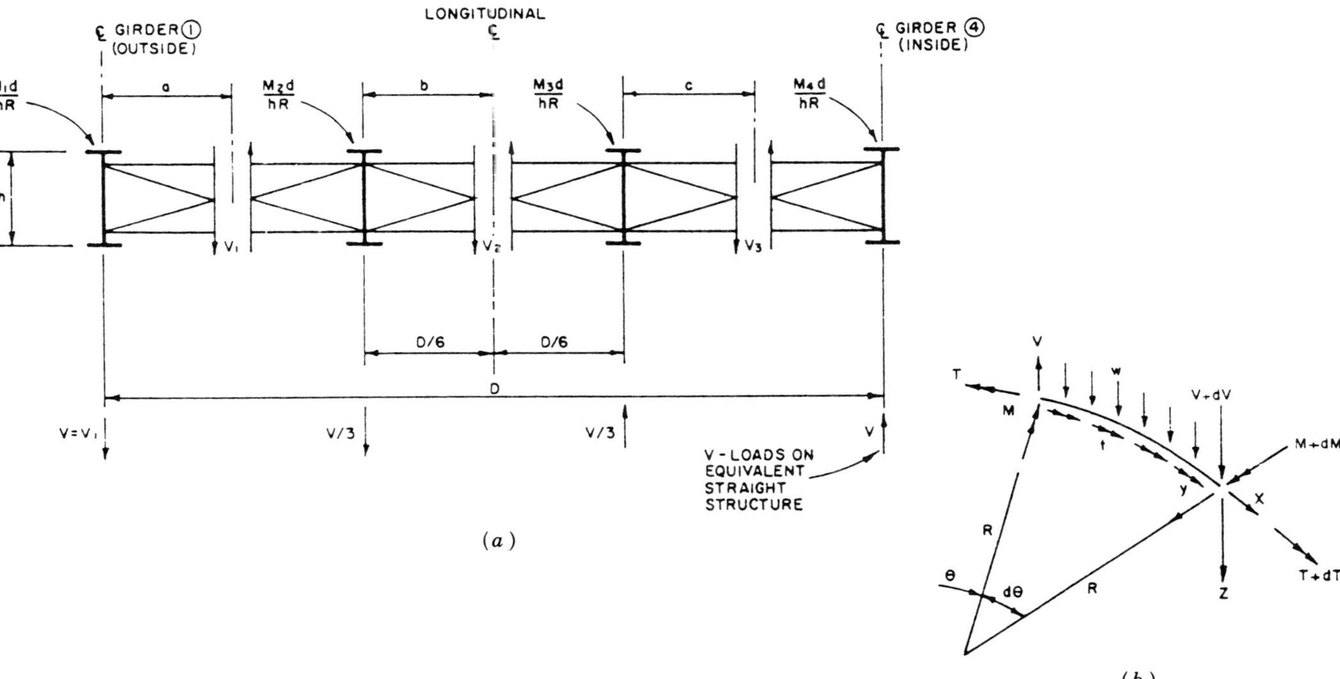

FIGURE 6-28 Four-girder curved system: (*a*) cross section; (*b*) equilibrium of curved girder segment.

therefore subjected to a net counterclockwise torque induced from internal radial forces H in the girder flanges (similar to the H forces shown in Figure 6-27*b*). For torque equilibrium, the secondary internal cross-frame shears V_1, V_2, and V_3 are developed at inflection points assumed at distances a, b, and c as shown. The shears on the inside and outside girders are assumed equal, or $V_1 = V_3 = V$. Because the girder shear is assumed proportional to its distance to the bridge centerline, it follows that the magnitude of the shears on the two interior girders is equal to $(1/3)V$. It also means that $V_2 = V_1 + (1/3)V = (4/3)V$. (The factor 1/3 is the proportionality factor based on bridge cross-sectional geometry.) The V loads to be applied to the straight girders as secondary shears are shown in the following discussion.

Moment equilibrium yields the following equations:

$$V_1 a = H_1 h = M_1 d/R \tag{6-45a}$$

$$V_1[(D/3) - a] + V_2 b = M_2 d/R \tag{6-45b}$$

$$V_2[(D/3) - b] + V_3 c = M_3 d/R \tag{6-45c}$$

$$V_3[(D/3) - c] = M_4 d/R \tag{6-45d}$$

where M_1, M_2, M_3, and M_4 are the final moments in the respective girders. Noting that $V_1 = V_3 = V$ and $V_2 = (4/3)V$ gives

$$V = \frac{M_1 + M_2 + M_3 + M_4}{(10RD)/(9d)} \tag{6-46}$$

Likewise,

$$M_1 + M_2 + M_3 + M_4 = M_{1P} + M_{2P} + M_{3P} + M_{4P} \tag{6-47}$$

which yields

$$V = \frac{M_{1P} + M_{2P} + M_{3P} + M_{4P}}{(10RD)/(9d)} \tag{6-48}$$

The last equations may be expressed in a more general form as

$$V = \frac{\Sigma M_P}{CK} \tag{6-49}$$

where ΣM_P is the summation of primary moments in each girder at a particular cross frame and C is a coefficient depending on the number of girders in the system, and $K = RD/d$, where R, D, and d are for the outside girder. Values of the coefficient C are given in Table 6-8 assuming girders are equally spaced.

TABLE 6-8 Values of Coefficient C as a Function of the Number of Girders

Number of Girders in System	Coefficient, C
2	1
3	1
4	10/9
5	5/4
6	7/5
7	14/9
8	12/7
9	15/8
10	165/81

Equation (6-49) gives the V loads on the outside and inside girders of the equivalent straight structure at the cross frames for multigirder systems. Proper proportionality factors are applied to these values to obtain the V loads acting on other (interior) straight girders of the section.

Effects of Torsion Referring to Figure 6-28b, the three equations expressing equilibrium of the segment are

$$\sum F_z = 0 \qquad \frac{dV}{dx} + w = 0 \tag{6-50a}$$

$$\sum M_y = 0 \qquad \frac{dM}{dx} + \frac{T}{R} - V = 0 \tag{6-50b}$$

$$\sum M_x = 0 \qquad \frac{dT}{dx} - \frac{M}{R} + t = 0 \tag{6-50c}$$

where w is the externally applied uniformly distributed load, t is the externally applied uniformly distributed torque, and V, M, and T are the internal shear, bending moment, and torque, respectively, acting at a given cross section. Because M and T appear in both (6-50b) and (6-50c), they are coupled and cannot be solved directly. Where the central angle θ is small, as is the case with most curved bridges, the T/R term in (6-50b) may be dropped. If T/R is nearly equal to zero, the bending moment caused by the applied loads may be initially computed as in the straight girder. Because t is usually zero, the approximate torsional moment per unit length dT/dx is equal to M/R, from (6-50c), where M is approximately equal to the primary bending moment, that is, (6-3) is derived.

For bridges with sharp curvature, the T/R term is significant and the bending moments are influenced by the torsional moments. Differentiating

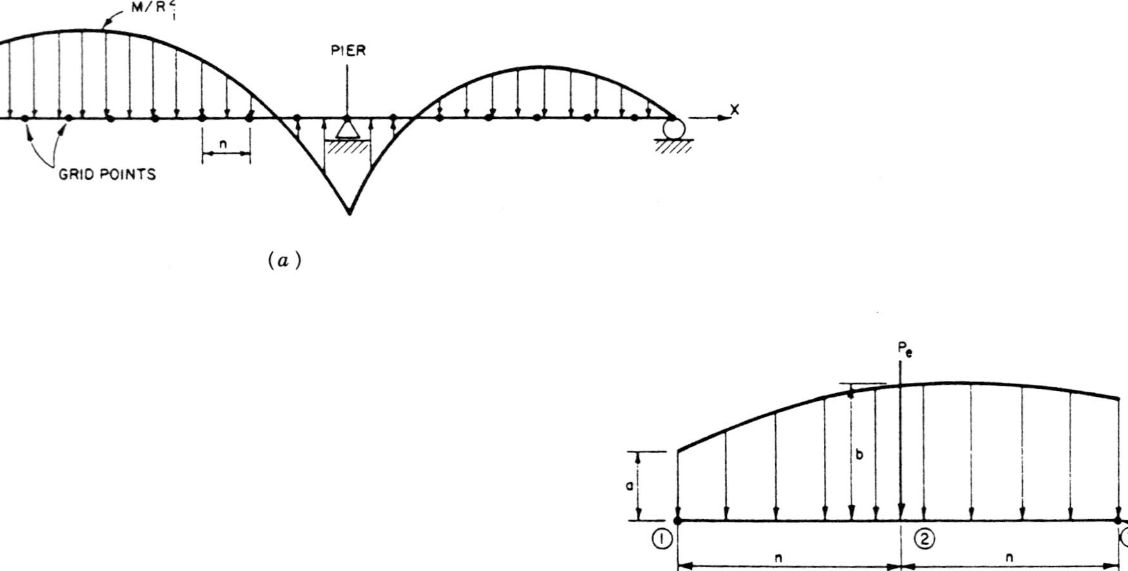

FIGURE 6-29 Loading diagrams; (*a*) M/R_1^2 loading, straight girders; (*b*) equivalent concentrated loads.

(6-50b) and substituting R_i (girder radius) for R gives

$$\frac{d^2M}{dx^2} + \frac{1}{R_i}\left(\frac{dT}{dx}\right) - \frac{dV}{dx} = 0 \tag{6-51}$$

Substituting (6-50a) and (6-50c) into (6-51) gives

$$\frac{d^2M}{dx^2} + \left(\frac{M}{R_i^2} - \frac{t}{R_i}\right) + w = 0 \tag{6-52}$$

In (6-52) the terms d^2M/dx^2 and w represent distributed loads; hence, the quantity $(M/R_i^2 - t/R_i)$ must also be a distributed load. This load may be applied to each girder as a torque load as shown in Figure 6-29a (again assuming that $t = 0$), which is essentially similar to the diagrams of Figure 6-1. In this case M is the primary moment M_p. Moments from the M/R_i^2 loads could then be added to the primary moments from the external loads to obtain more exact moments for calculating the V loads. This M/R_i^2 moment correction is not, however, statically correct; hence, it is only an indicator to assess the accuracy of the V-load assumptions.

Finite-Element Analysis

Results from finite-element analysis will be presented for three curved bridge schemes representing typical highway structure concepts. The layouts of the three bridges are similar except for support configurations. The three bridges are as follows.

Scheme A: radial supports

Scheme B: parallel skew supports

Scheme C: two parallel skew supports and one radial support

Plan views of the three schemes are shown in Figure 6-30. All three bridges are noncomposite. The framing systems are two-span continuous, nonprismatic, four-girder structures with unequal spans and compound radii. All schemes are initially open-framed (no horizontal lateral bracing).

As a typical reference, the left span is span 1, and the right span is span 2. Note that the radius changes as shown. Cross-frame spacing is selected in order to minimize warping effects. Note also that schemes B and C have intersecting cross frames at the interior pier. The skew angle is approximately 41°.

Typical cross sections and frame details are shown in Figures 6-31a and b, respectively. Other relevant data include the design stresses f'_c, the modular ratio, and the longitudinal area of steel A_s. The centrifugal force resulting from the deck superelevation is ignored in the analysis, and the effective

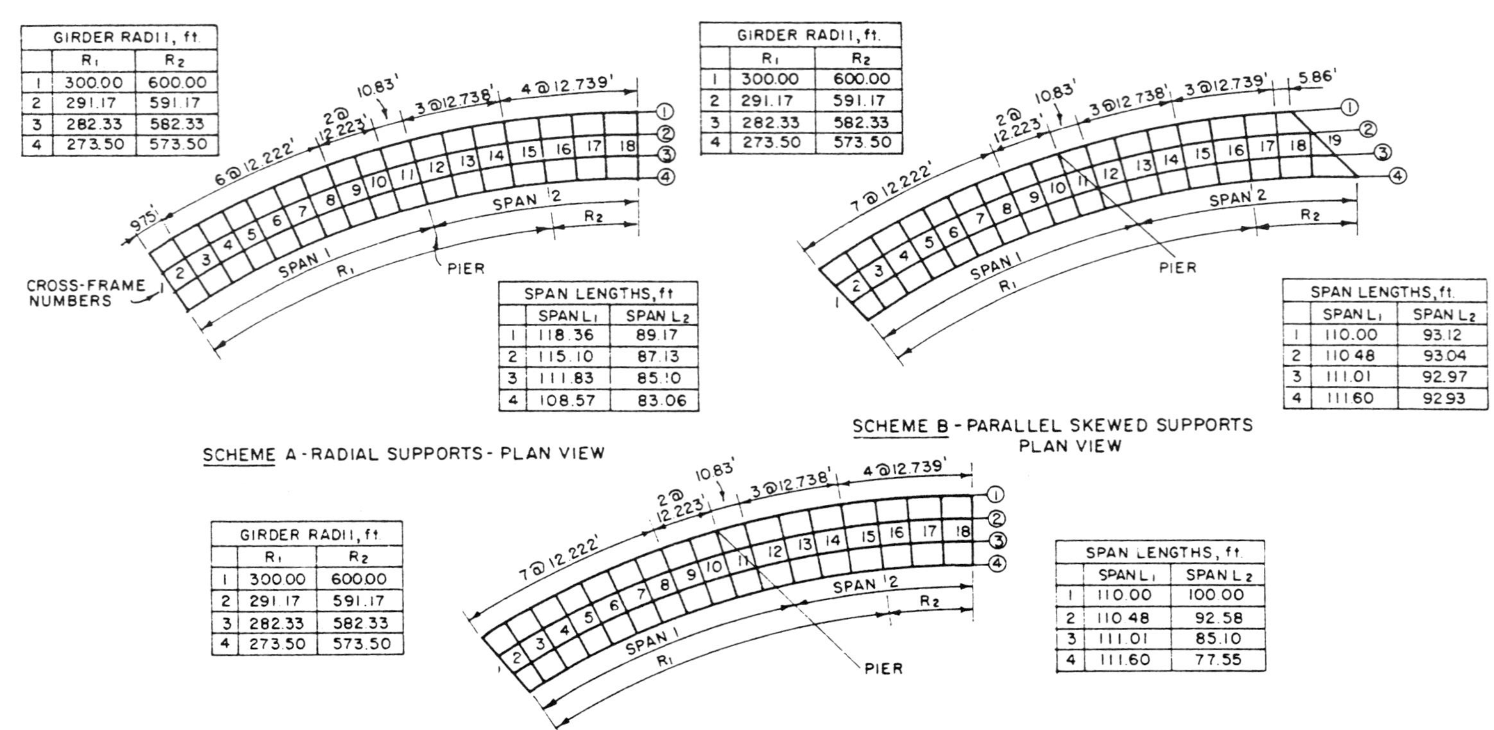

FIGURE 6-30 Plan views and framing of three two-span continuous curved bridges.

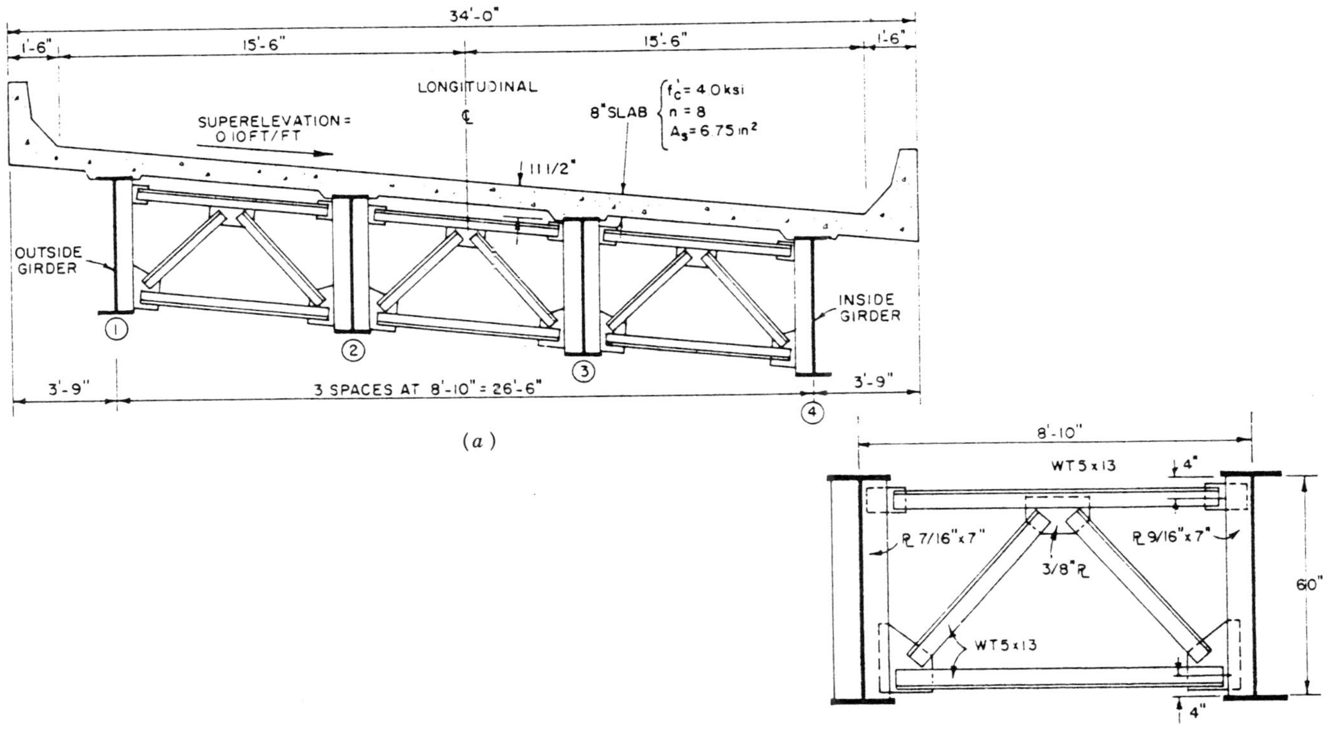

FIGURE 6-31 (*a*) Typical deck cross section, all schemes; (*b*) typical cross-frame details.

structural slab thickness is taken as 7.5 in. The typical cross frame has a K configuration and consists of structural tees (WT5 × 13). The centroids of the horizontal tees are offset 4 in. from the top and bottom of the web. In all schemes the girders have 60-in.-deep webs.

In the context of finite-element analysis, a preliminary girder design for all schemes was based on a straight-girder study adjusted for the curvature effects, using $F_y = 50$ ksi, and the load factor method of the standard AASHTO specifications. The girders were checked for fatigue under 500,000 cycles of truck load and 100,000 cycles of lane load. All webs were transversely stiffened.

Finite-Element Modeling A detailed description of the finite-element models is given by United States Steel (1984). Noncomposite models were generated for all three schemes. Composite models, including concrete slab elements, were generated for schemes A and C. The program used in this case provides an extra degree of freedom to account for warping by allowing the user to introduce the warping stiffness of the girder in the input.

Live Load Analysis Article 1.4 of the AASHTO specifications stipulates that if a rational analysis considers the system as a plane grid and not a space frame and bottom lateral bracing is contemplated, the live load stresses are estimated according to Section E of the same article, that is, (6-5) is applicable, which also gives the distribution factor $(DF)_b$. A distinction is made in the derivation of $(DF)_b$ for bridges where all bays have bottom bracing and those where bracing is in every other bay. This methodology involves a rational analysis that considers the system as a whole and develops influence surfaces for specified girder locations.

With the V-load method, however, the girders are analyzed as isolated straight girders; hence, the loads are distributed laterally using the same distribution factors. Three of the live load moments are investigated: negative moment over the pier and maximum positive moments in each span. In this procedure it is a fair approximation to assume that the corresponding three loading positions are the same for the curved and the straight bridge models.

Figure 6-32a illustrates the calculation of the longitudinal equivalent grid point wheel loads for girder 4 of scheme C. The middle wheel P_2 is placed at the approximate location $0.4L$ for maximum moment in span 1. Figure 6-32b shows the calculation of the lateral equivalent grid point wheel loads for girder 4 of scheme C. The cross section shown is at grid point line A adjacent to the front wheel P_1. According to the computer program, concentrated wheel loads must be applied at the grid points for the plate elements. Because the grid point spacing does not necessarily correspond to the wheel and axle spacing, equivalent grid point wheel loads must be computed in both the longitudinal and lateral directions.

Applying the AASHTO stipulations of lane width and spacing with two lanes loaded, the equivalent longitudinal wheel loads are shifted radially

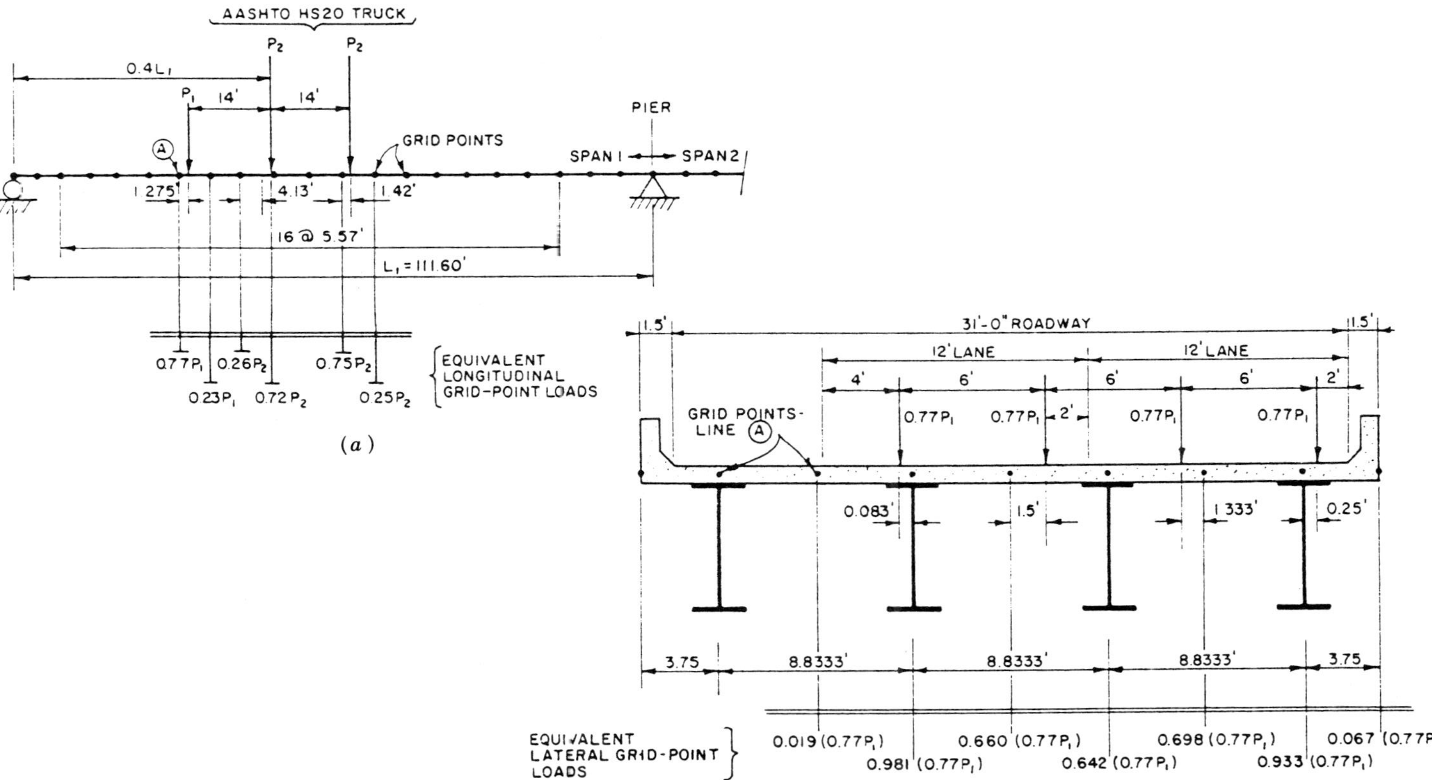

FIGURE 6-32 (*a*) Longitudinal equivalent grid point loads; (*b*) equivalent lateral grid point loads.

inward in their design lanes to produce the maximum wheel load fraction on girder 4. The equivalent lateral grid point loads are then obtained by summing up the statically equivalent wheel load reactions. These grid point loads are applied to the finite-element model.

V-Load Live Load Analysis The same methodology is used to predict the effect of the V loads resulting from the primary load analysis. However, equivalent lateral grid point loads are not used because the models represent isolated girders. A separate lateral distribution factor is required to compute the live load moments due to V loads. For a given position of the trucks, the V loads act concurrently on all the girders; hence, the AASHTO distribution factors applied alternately to each girder would result in excess V load on the structure.

For the V-load distribution across the section, the sum of the V loads must equal the number of wheels on the structure, and this leads to a simple wheel load lateral distribution factor as follows:

$$(\mathrm{DF})_{V\ \mathrm{load}} = 2N_L/N_G \tag{6-53}$$

where N_L is the number of lanes loaded and N_G is the number of girders in the section. For this example, the V-load distribution factor is 1.0, and applies to all girders.

Bending Moments for Open-Frame Systems

Dead Load A sample calculation of the V loads for DL1 (dead load noncomposite) of cross frame 6 in span 1 of scheme A is presented in Table 6-9. The cross-frame number corresponds to Figure 6-30. The following comments are appropriate: (a) the primary moments are determined by applying the dead load to noncomposite straight girders and (b) the V loads are calculated using (6-49) where a minus sign indicates a downward force. Also shown at the bottom of the table are the calculated V loads in the straight girders at other selected cross frames.

The calculated V loads across the section are plotted in Figure 6-33, which also shows the girder shears in the corresponding computer model. There is satisfactory agreement, suggesting that the assumption of linear girder shear distribution across the section is valid. The deviation observed at the pier (cross frame 11) is not critical because the V loads at this location go directly into the supports and have no influence on the moments.

Similar results have also been obtained for schemes B and C, and show deviations of the order of 3 to 7 percent. The assumption of essentially the same vertical girder stiffness is relevant to the V-load principle because it implies the linear proportionality of shears to the distance from the bridge centerline. In this example the girders have nearly the same vertical stiffness.

TABLE 6-9 Calculation and Values of *V* Loads for DL1 (Noncomposite)

Scheme A—radial supports
Noncomposite = 1.210 kip/ft—exterior girders 1 and 4
Dead load—DL1 = 1.227 kip/ft—interior girders 2 and 3

Cross frame 6—span 1

C = 10/9 = 1.1111 Proportionality factor = 1/3

$$K = \frac{RD}{d} = \frac{(300)(26.5)}{12.222} = 650.5 \text{ ft}$$

M_{1P} = primary moment in girder 1 = +1162.6 kip-ft
M_{2P} = primary moment in girder 2 = +1118.7 kip-ft
M_{3P} = primary moment in girder 3 = +1052.2 kip-ft
M_{4P} = primary moment in girder 4 = + 994.2 kip-ft

$$\sum M_P = +4327.7 \text{ kip-ft}$$

$$V \text{ load on girder } 1 = \frac{(4327.7)}{(1.1111)(650.5)} = -6.0 \text{ kips}$$

V load on girder 2 = 1/3(−6.0) = −2.0 kips

V load on girder 3 = −1/3(−6.0) = +2.0 kips

V load on girder 4 = −(−6.0) = +6.0 kips

Cross Frame Number	*V* Loads (kips)			
	Girder 1	Girder 2	Girder 3	Girder 4
1	0.0	0.0	0.0	0.0
6	−6.0	−2.0	+2.0	+6.0
7	−4.5	−1.5	+1.5	+4.5
Pier—11	+9.0	+3.0	−3.0	−9.0
14	−0.5	−0.2	+0.2	+0.5
15	−0.9	−0.3	+0.3	+0.9
18	0.0	0.0	0.0	0.0

Composite Dead Load The uniform superimposed dead load (DL2) on all composite girders for all three schemes is 0.41 kip/ft. A comparison between the finite-element model and the V-load analysis shows a maximum deviation of about 7 percent in the results.

M/R² Moment Correction The M/R^2 correction mentioned in the foregoing section was included in this example for the dead load analysis. Considering the actual radii of the outside girder (600 and 300 ft), the conclusion emerges that the factor M/R^2 is almost negligible.

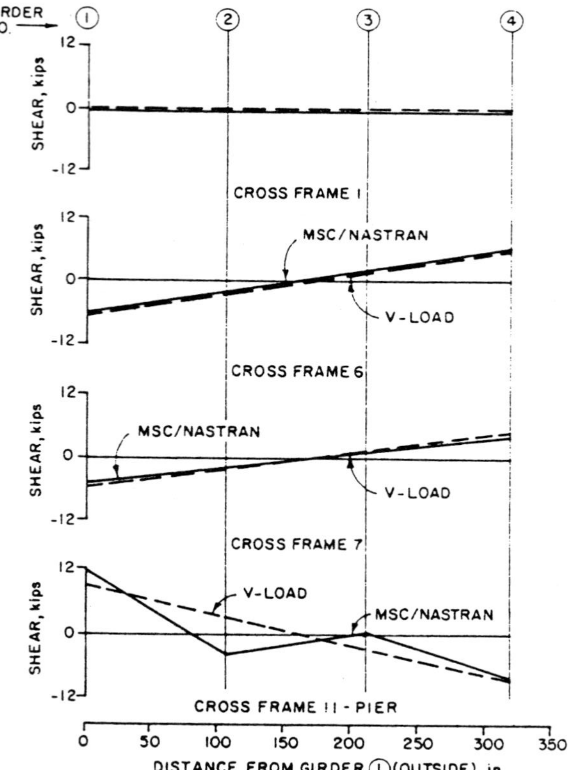

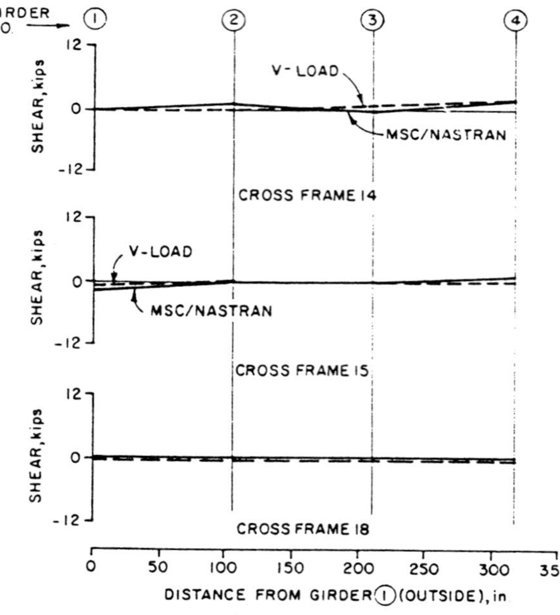

FIGURE 6-33 Comparison of shear, *V*-load method versus finite-element model; scheme A, dead load (noncomposite).

Live Load Live load moments for the *V*-load method and the finite-element method are compared in Table 6-10 for the four girders of scheme A. All moments are factored and include impact according to AASHTO. It appears that the *V*-load analysis for the interior girders gives results that vary up to 42 percent, which is conservative compared to the finite-element moments. The *V*-load analysis of moments for the exterior girders gives results that are within 12 percent of the moments calculated by the finite-element program. Note that the distribution factors used to determine the original primary live load moments in the developed straight girders for each of the three loading positions are as follows: 1.42 for exterior girders 1 and 4, and 1.61 for interior girders 2 and 3.

In an ancillary study, the *V*-load analysis of live load effects was repeated using different live load distribution factors to compute the primary moments. These factors were simply calculated considering the equilibrium of a straight bridge cross section undergoing a rigid-body rotation, assuming that the cross frames are stiff enough to make the bridge respond as a rigid body and that all girders have the same vertical stiffness. The resulting distribution factors were 1.61 for the exterior and 1.20 for the interior girders.

The percentage errors between finite-element analysis and *V*-load analysis were reduced to well within 10 percent. Interestingly, for scheme C, an error in negative moment at the pier is introduced by the approximate loading positions of the truck in the span, which were different for the computer program and the *V*-load analysis (due to the skew support).

Bending Moments for Closed-Frame Systems

A closed-frame mechanism was postulated with the addition of bottom lateral bracing (making the girders behave like quasi-box girders), causing a reduction in the lateral live load distribution factor. In order to check this assumption, lateral bracing elements were added to the curved noncomposite and composite scheme A models in the plane of the bottom flange elements. Structural tees WT6 × 32.5 ($A = 9.54$ in.^{2}) were chosen to satisfy the slenderness ratio $L/r < 140$. Bracing was provided in alternate bays (outside bays only), except in the composite model where bracing was provided in every other bay and in all bays.

For dead load there is a definite redistribution of load in all girders with the bracing present. In general, the dead load moments decrease, although some moments increase slightly. Live load moments decrease markedly in almost all girders. For combined dead and live load, there is little difference between bracing all the bays and bracing only the outside bays.

It appears from these results that a bridge with bottom lateral bracing responds to loads differently, and the *V*-load assumptions are not entirely valid. More particularly, the composite bridge section behaves like a multicellular box girder resisting the applied torque mainly through internal Saint-Venant shear flow.

TABLE 6-10 Comparison of Live Load Plus Impact Moment for the *V*-Load Method and Finite-Element Analysis (Scheme A)

Live load—two AASHTO HS 20 trucks—1.3*[5/3($L + I$)] L = AASHTO HS 20 truck wheel load
Scheme A—radial supports I = AASHTO impact factor

Location	Primary Moment (kip-ft) (1)	Girder 1 (Outside MSC/NASTRAN Moment (kip-ft) (2)	*V*-Load-Analysis Moment (kip-ft) (3)	Percentage Error $100\frac{(3)-(2)}{(2)}$
Positive moment in span 1 ($0.42L_1$)	+2681.1 ($0.35L_1$)	+3648.0 ($0.35L_1$)	+3562.6 ($0.35L_1$)	−2.3
Negative moment—pier	−1032.9 ($0.35L_1$)	−1525.4 ($0.35L_1$)	−1335.3 ($0.35L_1$)	−12.5
Positive moment in span 2 ($0.54L_2$)	+2035.5 ($0.55L_2$)	+2541.6 ($0.55L_2$)	+2400.0 ($0.55L_2$)	−5.6

Location	Primary Moment, (kip-ft) (1)	Girder 2 MSC/NASTRAN Moment (kip-ft) (2)	*V*-Load-Analysis Moment (kip-ft) (3)	Percentage Error $100\frac{(3)-(2)}{(2)}$
Positive moment in span 1 ($0.42L_1$)	+2973.7 ($0.35L_1$)	+2342.2 ($0.35L_1$)	+3267.8 ($0.35L_1$)	+39.5
Negative moment—pier	−1330.6 ($0.54L_1$)	−1155.6 ($0.54L_1$)	−1438.3 ($0.54L_1$)	+24.5
Positive moment in span 2 ($0.53L_2$)	+2222.0 ($0.55L_2$)	+1787.3 ($0.55L_2$)	+2367.5 ($0.55L_2$)	+32.5

Girder 3

Location	Primary Moment, (kip-ft) (1)	MSC/NASTRAN Moment (kip-ft) (2)	V-Load-Analysis Moment (kip-ft) (3)	Percentage Error $100 \frac{(3) - (2)}{(2)}$
Positive moment in span 1 ($0.37L_1$)	+2906.9 ($0.35L_1$)	+1895.2 ($0.35L_1$)	+2627.4 ($0.35L_1$)	+38.6
Negative moment—pier	−1358.6 ($0.54L_1$)	−875.3 ($0.54L_1$)	−1246.3 ($0.54L_1$)	+42.4
Positive moment in span 2 ($0.53L_2$)	+2137.3 ($0.55L_2$)	+1528.2 ($0.55L_2$)	+2076.0 ($0.55L_2$)	+35.8

Girder 4 (Inside)

Location	Primary Moment, (kip-ft) (1)	MSC/NASTRAN Moment (kip-ft) (2)	V-Load-Analysis Moment (kip-ft) (3)	Percentage Error $100 \frac{(3) - (2)}{(2)}$
Positive moment in span 1 ($0.37L_1$)	+2525.4 ($0.35L_1$)	+1914.5 ($0.35L_1$)	+1711.2 ($0.35L_1$)	−10.6
Negative moment—pier	−1213.8 ($0.54L_1$)	−944.9 ($0.54L_1$)	−866.6 ($0.54L_1$)	−8.3
Positive moment in span 2 ($0.53L_2$)	+1833.0 ($0.55L_2$)	+1680.7 ($0.55L_2$)	+1572.6 ($0.55L_2$)	−6.4

Torsional Stresses

Internal torsional loads are assumed to be resisted by shears developed in the diaphragms or cross frames, whereas any remaining torque must be resisted internally by the girders. An open section, such as an I girder, provides two forms of torsional resistance: Saint-Venant stiffness and warping torsional stiffness. These induce additional bending and shear stresses on the section that must be taken into account. Warping stresses are particularly critical to I girders.

Pure Saint-Venant torsion is illustrated in Figure 6-34*a*, where the section is an unrestrained I girder subjected to equal and opposite torques at the ends. The result is twisting by an angle ϕ, and because the section is unrestrained it will tend to warp out-of-plane as shown. The twisting also causes a shear flow f_{SV} set up as shown. The Saint-Venant shear stress is given by

$$\tau_{SV} = \frac{tT_{SV}}{K_{SV}} \tag{6-54}$$

where t is the thickness of the flange or web and K_{SV} is the Saint-Venant torsional stiffness.

Warping torsion, referred to in the foregoing sections, occurs in pure form when the out-of-plane warping of the section is completely restrained. Warping-torsion stresses are caused by restraint of warping, partly due to end support conditions or by variation of torque along the span as in a curved bridge. The derivation of warping stresses is illustrated in Figure 6-34*b*. The girder section is twisted by an angle ϕ, resulting in a lateral flange deflection u as shown. The resisting torque T_w and the self-equilibrating flange shear forces V_{fw} are related by

$$T_w = V_{fw}h \tag{6-55}$$

Assuming that the stresses occur only in the flanges and have a parabolic distribution, the maximum warping shear stress at the junction of the flange and web is

$$\tau_w = \frac{3}{2}\frac{V_{fw}}{A_f} \tag{6-56}$$

where A_f is the cross-sectional flange area. This warping shear stress is generally small compared to the average shear stress due to bending and may be neglected. The maximum warping normal stress at the flange tip is given by

$$\sigma_w = \frac{M_{fw}}{S_f} \tag{6-57}$$

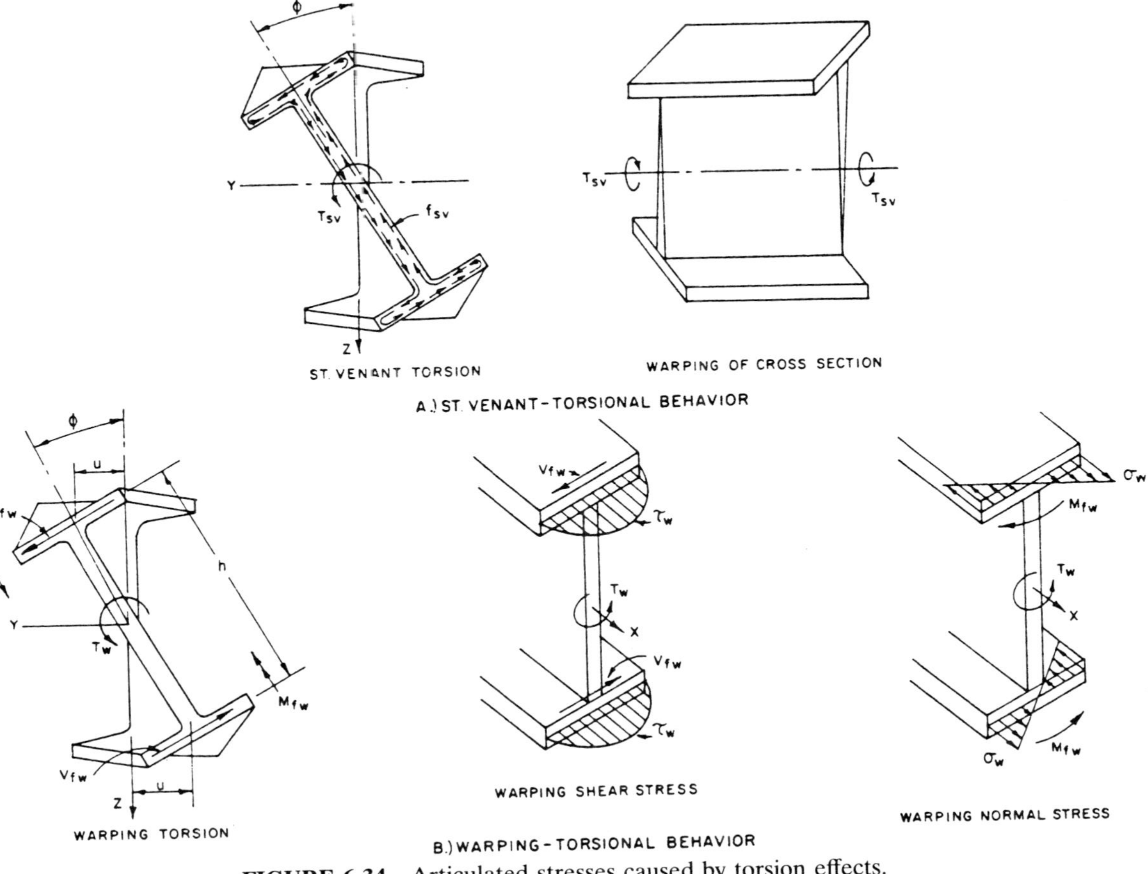

FIGURE 6-34 Articulated stresses caused by torsion effects.

where S_f is the section modulus of the rectangular flange about its strong axis. As shown in the foregoing sections, this stress is significant and may reach the range of 5 to 10 ksi.

Lateral Flange Warping Moments and Shears As already shown, the presence of cross frames reduces lateral bending and the warping stresses in the flanges. As discussed previously, a conservative assumption is that the lateral distribution force on each flange has a constant value $M/(hR)$, where M is the total vertical bending moment in the girder at each cross-frame location. Because the cross frames are assumed to act as rigid supports for the flanges, an approximate flange warping moment can be computed considering the fixed-end moment in a straight beam, or

$$M_{fw} = \frac{Md^2}{12hR} \tag{6-58}$$

where M is the total vertical bending moment (primary plus V load) in the girder at the cross frame. Likewise, the flange warping shear at each cross frame may be approximated by

$$V_{fw} = \frac{\Delta M_{fw}}{d} \tag{6-59}$$

where ΔM_{fw} is the difference in flange warping moments at adjacent cross frames.

Step-by-Step Procedure

The foregoing method of analysis may be implemented according to the following steps (United States Steel, 1984).

1. Determine the primary moments due to dead loads (initial and superimposed) and live loads in the isolated developed straight girder, using either the AASHTO or a modified distribution factor. For bridges with very sharp curvature, the effect of torsion on the primary moments should be included.
2. Compute the V loads in the outside and inside girders from the summation of the primary moments using (6-49). The corresponding V loads in the interior girders are determined by factoring these V loads by the proportionality factor. Note that this factor is 1.0 for the outside and inside girders. The V loads increase the loads on girders outside the longitudinal centerline and decrease the loads on girders inside this centerline. To compute the live load V loads, first factor the primary moments by the ratio of the V-load lateral distribution factor for the applied truck or lane loads, and then use these reduced primary moments M_p' in (6-49). The V-load wheel load lateral distribution factor is given by (6-53).

3. Compute the V-load moments by applying these loads to the respective developed straight girders at each cross-frame location.

4. Add the V-load moments to the primary moments from step 1 to obtain the final moments in the curved girders.

5. Compute the approximate flange warping moments at each diaphragm location using (6-58). Superimpose the maximum warping normal stresses on the longitudinal bending stresses in the flanges.

6. Determine shears, deflections, and reactions in the curved girders using the same procedure as in bending moments.

For live load analysis, these steps must be repated for truck or lane load at selected positions along each developed straight girder. From these steps, moment shear and flange warping moments can be developed. The V loads for a particular girder for truck or lane load at a specific position are computed by placing the truck or lane load at the same position. Because different loads and distribution factors may be required to calculate live load shears, reactions, and deflections, different sets of V loads may be needed.

In actual bridge design, the foregoing analysis must be repeated several times through the design cycle as changes are made, and the process may thus become too tedious for hand calculations. In most cases therefore the use of computers may be mandatory.

REFERENCES

AASHTO, 1980. "Guide Specifications for Horizontally Curved Highway Bridges," and Interim 1981, 1982, 1984, 1985, 1986, and 1990, Washington, D.C.

ASCE–AASHTO, 1977. "Curved Girder Bridge Design Recommendations," Task Committee of the Joint ASCE–AASHTO Committee on Flexural Members.

ASCE–AASHTO, 1978. "Curved Steel Box Girder Bridges: A Survey," Task Committee of the Joint ASCE–AASHTO Committee on Flexural Members.

Bazant, Z. P. and M. E. Nimeiri, 1974. "Stiffness Method for Curved Box Girders at Initial Stress," *J. Struct. Div. ASCE*, Vol. 100, No. ST10, Oct., pp. 2071–2090.

Bell, L. C. and C. P. Heins, 1968. "The Solution of Curved Bridge Systems Using the Slope-Deflection Fourier Series Method," Civ. Eng. Dept., Report No. 19, Univ. Maryland, College Park, June.

Bell, L. C. and C. P. Heins, 1969. "Curved Girder Computer Manual," Civ. Eng. Dept., Report No. 30, Univ. Maryland, College Park, Sept.

Bell. L. C. and C. P. Heins, 1970. "Analysis of Curved Girder Bridges," *J. Struct. Div. ASCE*, Vol. 96, No. ST8, Aug., pp. 1657–1673.

Brennan, P., 1970. "Analysis for Stress and Deformation of a Horizontally Curved Girder Bridge Through a Geometric Structural Model," Syracuse Univ. Report for N.Y. Dept. *Transp.*, Albany, N.Y.

Brennan, P. and J. A. Mandel, 1973. "Appendix B: Three-Dimensional Analysis of Horizontally Curved Bridges," CURT Final Report, Syracuse Univ., N.Y.

Brennan, P. J., and J. A. Mandel, 1974. "Users Manual: Program for Three-Dimensional Analysis of Horizontally Curved Bridges," Research Report, Syracuse Univ., CURT Program, Dec.

Brockenbrough, R. L., 1973. "Survey of Curved Girder Bridges," *J. Civ. Eng. ASCE*, Vol. 43, No. 2, Feb., pp. 54–56.

Carskadan, P. S., 1980. "Autostress Design of Highway Bridges; Phase 3, Interior Support Model Test (AISI Project 188)," Research Report 97-H-045(019-5), Feb.

Chu, K. and S. G. Pinjarkar, 1971. "Analysis of Horizontally Curved Box Girder Bridges," *J. Struct. Div. ASCE*, Vol. 97, No. ST10, Oct., pp. 2481–2501.

Colville, J., 1972. "Shear Connector Studies on Curved Girders," C. E. Report No. 45, Univ. Maryland, College Park, Feb.

Culver, C., 1972. "Design Recommendations for Curved Highway Bridges," Carnegie-Mellon Univ. Report, Pennsylvania Dept. Transp., June.

Culver, C., 1972. "Flange Proportions for Curved Plate Girders," Hwy. Research Board, Washington, D.C.

Culver, C. and P. McManus, 1971. "Instability of Horizontally Curved Members—Lateral Buckling of Curved Plate Girders," Carnegie-Mellon Univ. Report, Pennsylvania Dept. Transp., Sept.

Culver, C. and J. Mozer, 1970. "Horizontally Curved Highway Bridges—Stability of Curved Box Girders (Report No. B1)," Carnegie-Mellon Univ. Report, Research Project HPR-2(111), Sept.

Culver, C., and G. Nasir, 1971. "Inelastic Flange Buckling of Curved Plate Girders," *J. Struct. Div. ASCE*, Vol. 97, No. ST4, April, pp. 1239–1256.

Dabrowski, R., 1968. *Gekzummte dunnwandige Träger—Theorie und Berechuung*, Springer-Verlag, New York.

Dabrowski, R., 1972. "Thin-Walled Girders," Cement and Concrete Assoc., London.

Fann, A. R. M., 1973. "Static and Free Vibration Analysis of Curved Box Girders," *Struct. Dynamics Ser.* No. 73-2, McGill Univ., Montreal, March.

Funkhouser, D. W. and C. P. Heins, 1972. "Skew and Elevated Support Effects on Curved Bridges," Civ. Eng. Dept. Report No. 46, Univ. Maryland, College Park, Feb.

Funkhouser, D. W. and C. P. Heins, 1974. "Skew and Elevated Support Effects on Curved Bridges," *J. Struct. Div. ASCE*, Vol. 100, No. ST7, July, pp. 1379–1396.

Hacker, J. C., 1957. "A Simplified Design of Composite Bridge Structures," *J. Struct. Div. ASCE*, Vol. 93, No. ST11, Nov.

Heins, C. P., 1967. "The Presentation of the Slope-Deflection Method for the Analysis of Curved Orthotropic Highway Bridges," C. E. Report No. 15, Univ. Maryland, College Park, June.

Heins, C. P., 1972. "Design Data for Curved Bridges," C. E. Report No. 47, Univ. Maryland, College Park, March.

Heins, C. P., and B. Bonakdarpour, 1971. "Behavior of Curved Bridge Models," *Public Roads J.*, Vol. 36 No. 11, Dec.

Heins, C. P., and D. A. Firmage, 1979. *Design of Modern Steel Highway Bridges*, Wiley, New York.

Heins, C. P. and J. O. Jin, 1982. "Load Distribution of Braced Curved I-Girder Bridges," C. E. Report, Univ. of Maryland, College Park, April.

Heins, C. P., and J. Siminou, 1970. "Preliminary Design of Curved Bridges," *American Inst. Steel Construction Eng. J.*, Vol. 7, No. 2, April.

Heins, C. P. and T. L. Martin, 1977. "Distortional Effects of Curved I-Girder Bridges," C. E. Report No. 65, Univ. Maryland, College Park, Feb.

Heins, C. P. and J. C. Oleinik, 1976. "Curved Box Beam Bridge Analysis," *J. Computers and Structures*, Vol. 6, Pergamon Press, London, June, pp. 65–73.

Hildebrand, F. B., 1956. *Introduction to Numerical Analysis*, McGraw-Hill, New York.

Ho, Y. S. and R. J. Reilly, 1971. "Analysis and Design of Curved Composite Box Girder Bridges," C. E. Report No. 41, Univ. Maryland, College Park, June.

Johnston, S. B. and A. H. Mattock, 1967. "Lateral Distribution of Load in Composite Box Girder Bridges," Hwy. Research Record, No. 167, Hwy. Research Board, Washington, D.C., Jan.

Kolbrunner, C. F. and K. Basler, 1969. *Torsion in Structures, An Engineering Approach*, Springer-Verlag, New York.

Komatsu, S. and H. Nakai, 1966. "Study on Free Vibration of Curved Girder Bridges," *Trans. Japanese Soc. Civ. Eng.*, No. 136, Dec., pp. 35–60.

Komatsu, S. and H. Nakai, 1970. "Application of the Analogue Computer to the Analysis of Dynamic Response of Curved Girder Bridges," *Trans. Japanese Soc. Civ. Eng.*, No. 2, Part I, pp. 143–149.

Komatsu, S., and H. Nakai, 1970. "Fundamental Study of Forced Vibration of Curved Girder Bridges," *Trans. Japanese Soc. Civ. Eng.*, No. 2, Part I. pp. 37–42.

Komatsu, S., H. Nakai, and M. Nakanishi, 1971. "Statical Analysis of Horizontally Curved Skew Box Girder Bridges," *Trans. Japanese Soc. Civ. Eng.*, No. 3, Part 2, pp. 134–135.

Konishi, I. and S. Komatsu, 1962. "On Fundamental Theory of Thin Walled Curved Girder," *Trans. Japanese Soc. Civ. Eng.*, No. 87, Nov., pp. 35–48.

Konishi, I. and S. Komatsu, 1963. "Three Dimensional Analysis of Simply Supported Curved Girder Bridges," *Trans. Japanese Soc. Civ. Eng.*, No. 90, Feb., pp. 11–28.

Kristek, V., 1970. "Tapered Box Girders of Deformable Cross Section," *J. Struct. Div. ASCE*, Vol. 96, No. ST8, Aug., pp. 1761–1793.

Lavelle, F. H., 1973. "The CUGAR 2 Algorithm," CURT Tech. Report No. 5(L), Univ. Rhode Island, Kingston, April.

Lavelle, F. H., 1973. "The CUGAR 3 Algorithm," CURT Tech. Report No. 7(L), Univ. Rhode Island, Kingston, May.

Lavelle, F. H., 1973. "Appendix L: CUGAR 2 and CUGAR 3," CURT Final Report, Part 3, Univ. Rhode Island, Kingston, Aug.

Lavelle, F. H., 1976. "The CUGAR Algorithm," CURT Program Report No. 56, Univ. Rhode Island, Kingston, Jan.

Lavelle, F. H. and J. S. Bolck, 1965. "Program to Analyze Curved Girder Bridges," *Eng. Bull.* No. 8, Univ. Rhode Island, Kingston.

Lavelle, F. H., R. A. Greig, and H. R. Wemmer, 1971. "CUGAR 1—A Program to Analyze Curved Girder Bridges," *Eng. Bull.* No. 14, Sept., Vol. 1, User's Information; Vol. 2, System Information.

McCormick, C. W. "MSC–ASTRAN Users Manual," McNeal-Schwendler Corp., Los Angeles.

Meyer, C., and A. C. Scordelis, 1971. "Analysis of Curved Folded Plate Structures," *J. Struct. Div. ASCE*, Vol. 97, No. ST10, Oct. pp. 2459–2580.

Mozer, J., J. Cook, and C. Culver, 1972. "Horizontally Curved Highway Bridges—Stability of Curved Plate Girders (Report No. P3)," Carnegie-Mellon Univ., CURT Program, Pittsburgh, Nov.

Mozer, J. and C. Culver, 1970. "Horizontally Curved Highway Bridges—Stability of Curved Plate Girders (Report No. P1)," Carnegie-Mellon Univ., CURT Program, Pittsburgh, Sept.

Mozer, J., R. Ohlson, and C. Culver, 1971. "Horizontally Curved Highway Bridges—Stability of Curved Plate Girders (Report No. 2)," Carnegie-Mellon Univ., CURT Program, Pittsburgh, Sept.

Murphy, E. L. and C. P. Heins, 1973. "Dead Load Analysis of Single Span Curved Bridges," C. E. Report No. 52, Univ. Maryland, College Park, June.

Nakai, H., and H. Kotoguchi, 1975. "Dynamic Response of Horizontally Curved Girder Bridges Under Random Traffic Flows," *Proc. Japanese Soc. Civ. Eng.*, No. 244, pp. 117–128.

Nakai, H., and C. Hong Yoo, 1988. Analysis of Design of Curved Steel Bridges, McGraw-Hill, New York.

Nettleton, D., 1976. "Analysis of Plate Girder Curved Bridges," Fed. Hwy. Admin., In-House Report, Washington, D.C.

Oleinik, J. C., 1974. "Design Criteria for Distortional Analysis of Curved Steel Box Beams,' Thesis, Univ. Maryland, College Park.

Rabizadeh, R. O. and S. Shore, 1975. "Dynamic Analysis of Horizontally Curved Box Girder Bridges," *J. Struct. Div. ASCE*, Vol. 101, No. ST9, Sept., pp. 1899–1912.

Richardson, Gordon and Associates 1976. "Curved-Girder Workshop," Fed. Hwy. *Admin.*, Washington, D.C.

Scalzi, J. B. 1963. "Analysis of Horizontally Curved Steel Bridge Girders," U.S. Steel, Pittsburgh.

Schilling, C. G., 1982. "Lateral Distribution Factors for Fatigue Design," *ASCE Struct. J.*, Sept.

Scordelis, A. C., 1974. "Analytical Solutions for Box Girder Bridges," *Development in Bridge Design and Construction*, Crosby, Lockwood and Son, Univ. College, Cardiff, Wales, April.

Shore, S., 1973. "Static Analysis of Curved Bridges (STACRB)," CURT Final Report, Part 3: Computer Program User's Manual, Univ. Pennsylvania, Philadelphia, Aug.

Shore, S. and C. R. Lansberry, 1972. "A Fully Compatible Annular Segment Finite Element," CURT Program, Report T0272, Univ. Pennsylvania, Philadelphia, April.

Shore, S. and J. L. Wilson, 1973. "Users Manual for the Static Analysis of Curved Bridges (STACRB)," CURT Report No. T0173, Research Project HPR-2(111), Univ. Pennsylvania, June.

Spates, K. R. and C. P. Heins, 1968. "The Analysis of Single Curved Girders with Various Loadings and Boundary Conditions," Univ. Maryland, College Park.

Tung, D. H. H. and R. S. Fountain, 1970. "Approximate Torsional Analysis of Curved Box Girders by the M/R Method," *AISC Eng. J.*, Vol. 7, No. 3, July.

United States Steel Corporation, 1963. "Analysis of Horizontally Curved Steel Bridge Girders," U.S. Steel Struct. Report ADUCO 91063, Pittsburgh.

United States Steel Corporation, 1967. "Horizontally Curved Bridges," Hwy. Structures Design Handbook, Vol. I, Pittsburgh.

United States Steel Corporation, 1982. "Finite Element Modeling Techniques for Analysis of Live Load Distribution in a Stringer Bridge," U.S. Steel Research Bull., Pittsburgh, April.

United States Steel Corporation, 1982. "Live-Load Lateral Distribution for the Sunshine Skyway Approach Spans," U.S. Steel Research Bull., Pittsburgh, April.

United States Steel Corporation, 1984. "V Load—A Computer Program for the Analysis of Horizontally Curved Steel Bridge Girders," U.S. Steel Engineers and Consultants, Pittsburgh.

Vincent, G. S., 1969. "Tentative Criteria for Load Factor Design of Steel Highway Bridges," *Bull.* No. 15, American Iron and Steel Inst., March, N.Y.

Williamson, D. M., 1974. "Analysis for Cross-Sectional Deformation of Curved Box Girders with Internal Diaphragms," MS Thesis, Carnegie-Mellon Univ., Pittsburgh.

Wright, R. N., 1969. "Design of Box Girders of Deformable Cross Section," Hwy. Research Record, No. 295, HRB, Washington, D.C.

Wright, R. N., S. R. Abnel-Samad, and A. R. Robinson, 1968. "BEF Analogy for Analysis of Box Girders," *J. Struct. Div. ASCE*, Vol. 94, No. ST7, July, pp. 1719–1743.

Yoo, C. H., et al., 1975. "An Analysis of a Continuous Curved Box Girder Bridge," Hwy. Research Record, No. 579, HRB, Washington, D.C.

CHAPTER 7

ORTHOTROPIC DECK BRIDGES

7-1 DEVELOPMENT OF THE STEEL PLATE DECK

As a concept, the orthotropic bridge theory is the result of efforts to obtain optimum structural performance of steel materials. As a practical application, orthotropic bridges became feasible through improved welding techniques and progress in methods of structural analysis.

Unlike the reinforced concrete slab used in conventional bridge construction, an orthotropic bridge employs a stiffened steel plate to support the live loads. Orthotropy is derived from the expression *orthogonal anisotropic*, meaning material properties exhibiting differences at right angles. Three factors have contributed to the articulation of the orthotropic system: (a) long experience with the so-called battledeck floor, (b) the structural representation of gridwork and cellular systems, and (c) the development of refined methods of analysis.

Battledeck Floor This system, shown in Figure 7-1, was proposed in the early 1940s by the AISC following a series of practical tests (Lyse and Madsen, 1938; AISC, 1938). Rolled I beams, used as stringers, are spaced 2 ft apart and simply supported between floor beams. Continuous steel plates 3/16 to 1/4 in. thick are placed with their edges along the stringer centerline.

Extensive tests at Lehigh University in the late 1930s showed that the deck plating had much higher strength reserve than predicted by conventional bending theory, and this resulted in semiempirical formulas recommending a 40 percent increase in allowable stress.

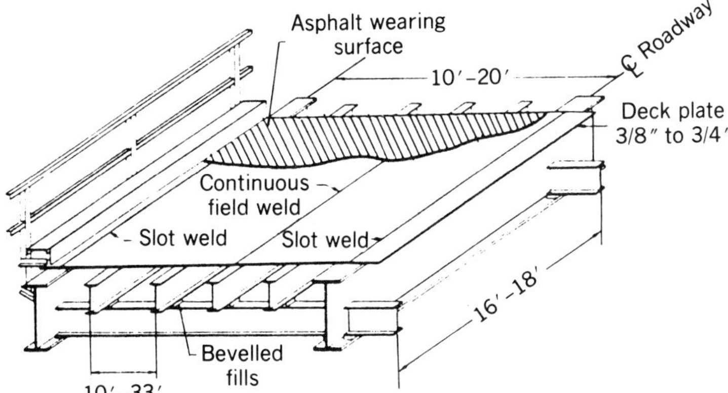

FIGURE 7-1 Detail of battledeck floor. (From Troitsky, 1967.)

Battledeck floors have been used in relatively few bridges in the United States. The system was considered for the Triborough Bridge and computed as a separate T section in conjunction with the floor elements. However, the deck plate should have been assumed to act with the floor beam as a single component and incorporated into the design of the main bridge structure.

Gridworks Grid systems have been mentioned in the foregoing sections, and examples are shown in Figure 7-2. Unlike the conventional bridge design where the deck slab, stringers, floor beams, and main girders are considered to act independently and analyzed separately, gridwork systems are assumed to perform as one integral unit interconnected as shown in Figures 7-2*a* and *b*. Although the interconnected bridge girder system differs considerably in constructional form from the open-grid frameworks, slabs, and stiffened plates, all these systems are interrelated and may be analyzed by the same theoretical approach. In reality, a bridge deck formed by the intersecting system of stringers and cross beams with a deck plating rigidly connected on top is no longer a true grid. However, by intersecting imaginary cuts in the deck plate between stringers and cross beams and selecting appropriate section properties, an approximate equivalent grid system is derived as shown in Figure 7-2*c*.

With respect to static behavior, gridwork systems are considered in the following context: (a) if the composing beams do not possess any resistance to torsion, the grid acts as if the beams were only connected at their joints with hinges transferring only axial stresses, and is called a hinged gridwork; and (b) if the beams develop resistance to torsion or they can transfer moments in two directions through stiff joints, the framework is called a stiff gridwork.

Cellular System An example of cellular construction is shown in Figure 7-3, and is characterized by closely spaced gridwork. This, used together with a plating, is developed into a steel deck of cellular form. The deck consists of

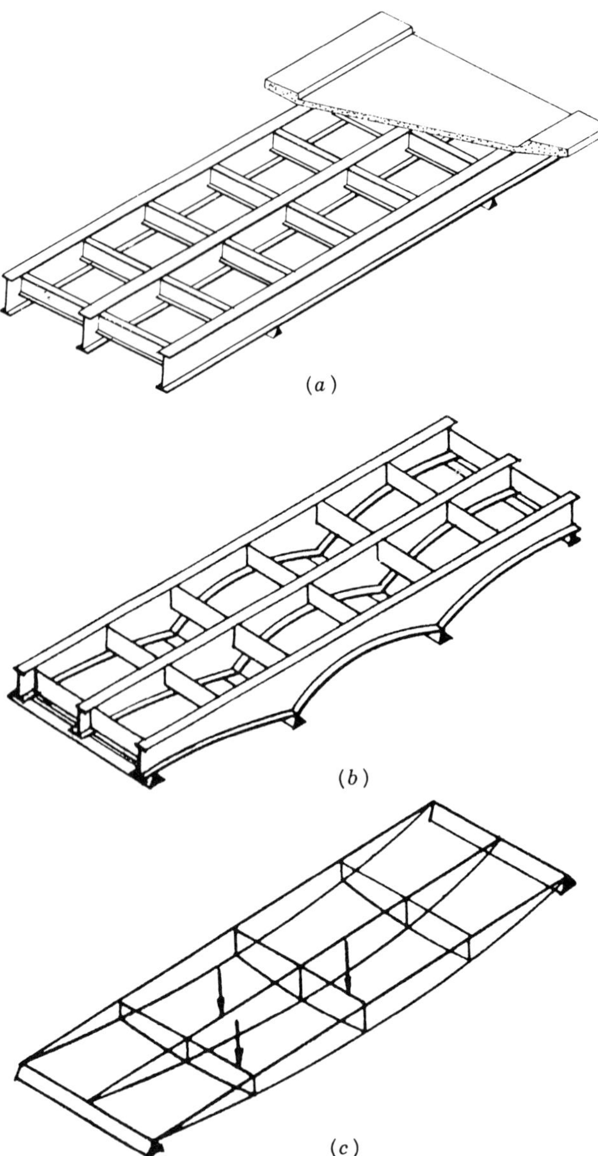

FIGURE 7-2 (*a*) Gridwork-type bridge; (*b*) continuous grid system; (*c*) equivalent grid system. (From Troitsky, 1967.)

a thin steel plate stiffened in two directions by a relatively shallow gridwork of welded ribs spaced on 1- to 2-ft centers.

The considerable amount of manual welding prompted changes in this system, and specifically increased stiffener spacing resulting in larger panels. Several bridges of this type were erected in Germany before 1936. A study of more recent cellular bridges (Heins and Looney, 1968) shows the following

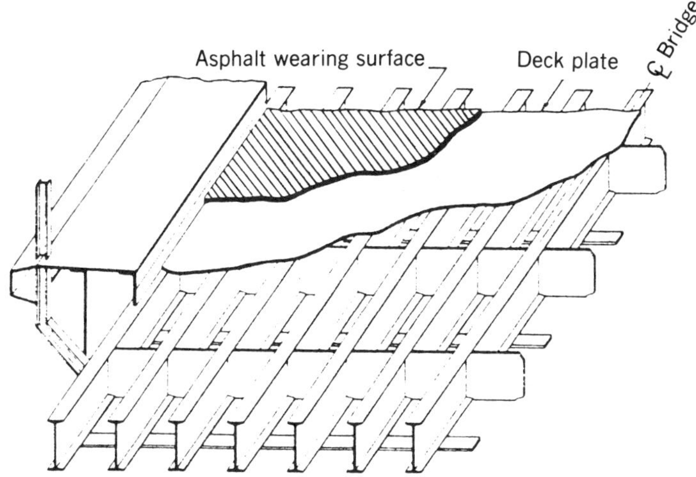

FIGURE 7-3 Cellular-type bridge. (From Troitsky, 1967.)

trend:

1. *Open Cellular Decks*
   Size of rib, 3/8 × 8 in. to 1 × 12 in.
   Rib spacing, 12 to 16 in.
   Floor beam spacing, 4 to 7 ft
2. *Closed Cellular Decks*
   Cell spacing, 24 to 28 in.
   Size of cell, 5/16 × 12 × 12 in.
   Floor beam spacing, 4 to 15 ft

Examples of open- and closed-rib construction are shown in Figures 7-4*a* and *b*, respectively, whereby the ribs are used to stiffen the deck in one direction, creating a larger inertia per unit of width at right angles to the stiffeners than on a section parallel to the stiffeners.

7-2 THE ORTHOTROPIC SYSTEM

Construction Details

The rib types shown in Figure 7-4 are repeated in amplified form in Figure 7-5, characterized as torsionally soft or open type or as torsionally stiff or closed type. Invariably, the ribs reinforce a continuous steel plate with a thickness of 3/8 to 1 in., depending on the rib spacing, loading requirements, and allowable local deflections. As a rule, the ribs are arranged in the longitudinal direction.

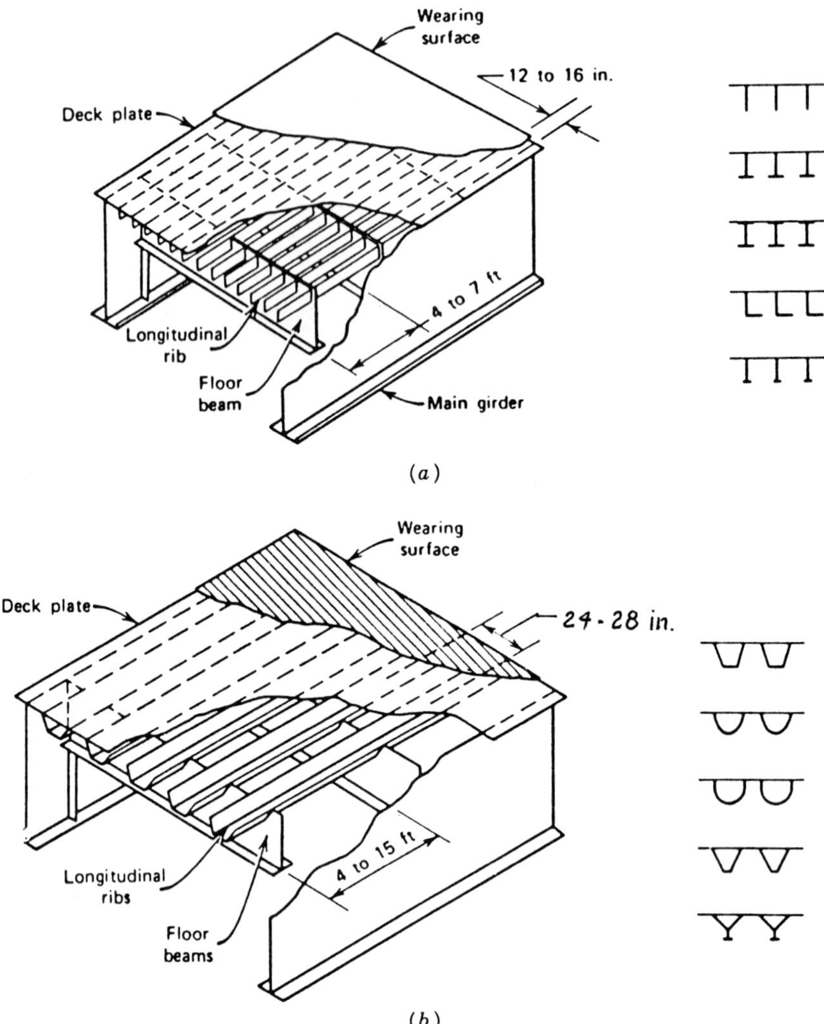

FIGURE 7-4 Cellular decks: (*a*) open ribs; (*b*) closed ribs. (From Heins and Firmage, 1979.)

Open ribs are usually made from flat bars, bulb shapes, inverted T sections, angles, and channels. Because they participate in the girder action, the resulting stresses are high; hence, the ribs are made continuous through slots in the web of the floor beams.

Closed ribs are available in trapezoidal, semicircular, triangular, and combined shapes, with the most common type being the trapezoidal. In this form, the ribs provide considerable torsional rigidity and contribute to the transverse load distribution. Likewise, the closed ribs continue through slots in the web of the floor beams.

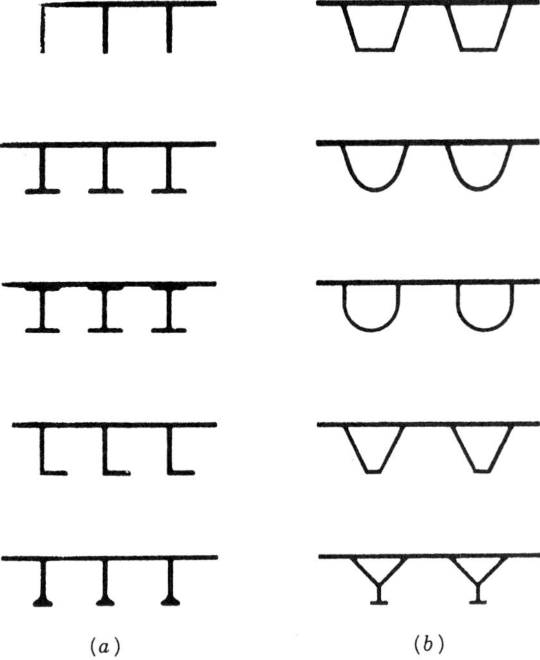

FIGURE 7-5 Rib types: (*a*) torsionally soft or open type; (*b*) torsionally stiff or closed type. (From Troitsky, 1967.)

Floor beams are spaced from 4 to 5 ft, but in exceptional cases this spacing may be increased. As a rule of thumb, the floor beam spacing should be increased if the main girder spacing is large. In addition, the dimensions of the floor beams are also governed by erection requirements and allowable deflection.

Comparison of Deck Types

Open Ribs Open ribs are relatively simple to fabricate, and their dimensions can easily be varied to accommodate the different parts of the deck. The field splicing of ribs is also relatively simple, and the bottom of the open-rib deck allows easy access for inspection and maintenance. Disadvantages of the open-rib deck are the relatively small wheel load distribution in the transverse direction and the closer beam spacing. As a result, the steel deck with open ribs is heavier per square foot. In addition, the amount of welding required to fabricate an open-rib deck is almost twice as much as for the closed-rib deck.

Closed Ribs These systems have enhanced load distribution capacity because of the greater flexural and torsional rigidity in the transverse direction.

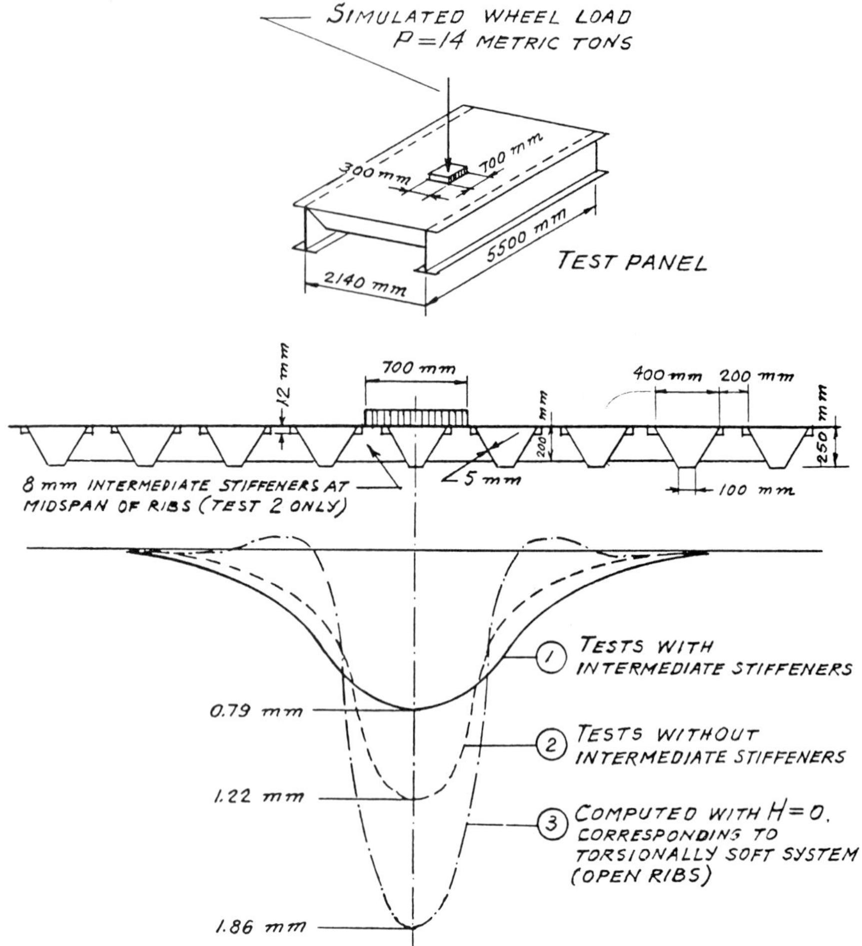

FIGURE 7-6 Results of tests comparing the deflection effects of orthotropic decks with closed and open ribs (H is the effective torsional rigidity). (From Troitsky, 1967.)

The load-distributing effect of the longitudinal closed boxes is demonstrated in Figure 7-6, where this capacity is compared with open ribs. Under the concentrated load of 14 tons (metric), the superiority of the boxes is obvious. In the same test, the behavior of the deck plates for different positions of the concentrated load was also studied, and the results show that the orthotropic deck with closed ribs provides the most efficient design.

The greater rigidity of the closed sections contributes to better load distribution with a corresponding stress reduction in the deck plating. Better control of distortional effects, fewer welds, and somewhat less steel are further advantages. The reduction in the bridge weight is associated with wider floor beam spacing normally possible with closed ribs. Wider spacing also means less welding.

Inherent disadvantages include fabrication difficulties and more complicated field splices of the ribs. Closed ribs also require fabrication and erection precision to ensure proper fit at the splices.

Connections The deck plate, the ribs, and the floor beams must be connected to provide an integral unit. Because the stresses in the welds connecting the deck plate with the longitudinal ribs and the floor beams are usually low, a nominal amount of weld is required at these joints. However, a proper welding sequence must be carefully followed to minimize residual stresses. The amount of weld affords the opportunity to subassemble the sections for automatic downhand welding and efficient fabrication methods. Because a large number of identical deck sections are required, they may be set up in a jig and automatically submerged-arc-welded with nominal time and cost.

Field splices may be either bolted or welded. With bolted connections, it is easier to fit the members because deformations do not develop in the field, and the work is independent of weather conditions. An obvious disadvantage of welded splices is the protruding splice plate and bolt heads leaving a thin wearing surface over these elements. Another disadvantage of a bolted splice is the loss of material in the net cross section. Where the deck plating is fabricated together with the upper parts of the main girders, the connection between the girders and the deck is made as in a longitudinal girder splice.

Welded field splices do not present the foregoing disadvantages, but the work requires utmost precision. In addition, transverse and longitudinal shrinkage of field welds tends to deform the plating.

Experience indicates that erecting orthotropic deck bridges with combined types of splices involving bolting and welding makes it almost impossible to obtain the required fit of the deck elements. The main advantage of all-welded field connections is jointless construction without additional splice plates or protruding elements.

Wearing Surface

With an orthotropic deck system, a light and yet traffic-resistant roadway covering is an essential part of the design. The wearing surface of a steel plate deck must satisfy the following requirements.

1. The covering should be lightweight to reduce the dead weight, but of sufficient thickness to cover irregularities and protrusions.
2. The surfacing should be skid resistant to increase friction and minimize ice hazards.
3. Stability and durability are essential and must be maintained over the expected temperature range under traction and braking forces.

4. The wearing surface must provide permanent corrosion protection, be resistant and impervious to water and deicing chemicals, inhibit crack development, and maintain good bond to the top of the steel.

In general, the preferred wearing surface is a bituminous mix such as asphalt concrete and asphalt mastic. In order to achieve good bond, an appropriate insulating material is first applied to the steel plate, which also serves as a protective layer against corrosion. Bituminous-mix wearing surfaces usually consist of a high-quality hot mix of a low-penetration asphalt and a dense graded aggregate, with a thickness in the range of 1.5 to 2.5 in., placed in two courses.

Tests confirm that the wearing surface contributes to the distribution of concentrated loads such as truck wheels and increases somewhat the rigidity of the steel plate.

Corrosion Protection

Protection against corrosion focuses primarily on areas inaccessible after erection. These are (a) the inside of closed or box ribs and (b) the surface of the deck plate. According to foreign practice, there is no reason for concern about corrosion on the inside surface of box ribs, and no provisions for protection are necessary other than making the ribs airtight during fabrication.

The reduction of the wall thickness of closed ribs to 3/16 in. may sometimes be justified considering their good resistance to corrosion when welded airtight and where the smooth outside surface can be easily maintained. When closed ribs are field-spliced with high-strength bolts, hand holes are required and airtightness is ensured by welding diaphragm plates inside the rib section on either side of the splice.

The protection of the plate surface, however, is more specialized because the deck surface acts compositely with the steel plate and this action induces shear stresses at the interface. A bituminous-mix wearing surface placed directly on the steel deck does not provide satisfactory protection of the steel deck, and a suitable insulating and bonding layer must be applied to the deck. For materials and method of application, reference is made to appropriate specifications and standards.

Main Girders

In the past the main girders of orthotropic bridges were bolted, but welding has been used in most recent structures. Single web girders are usually shop-welded and fabricated with the bottom flange and nominal top flange. The deck plating is connected in the field.

The shop fabrication of box girders is accomplished through a sequence of subassemblies. Before subassembly, the parts of the girder are prepared separately. Deck panels forming the top flanges are prepared by welding ribs and floor beams, whereas bottom flange plates are usually reinforced with longitudinal and transverse stiffeners.

The deck panels are placed in the upside-down position, and transverse stiffening frames are set in place. Then the web plates are installed and tack-welded to the deck panels. After the bottom plate is set in place in an upside-down position and tack-welded, complete welding follows usually starting at the center stiffening frame and proceeding toward the ends.

7-3 RELEVANT SPECIFICATIONS

The standard AASHTO specifications include provisions for orthotropic deck superstructures in Article 10.41. Appropriate methods of analysis include the equivalent–orthotropic slab method or the equivalent grid method. The equivalent stiffness properties should be selected to simulate the actual deck. A method of elastic analysis such as the thin-walled-beam procedure that accounts for torsional distortion effects is suggested for the design of main girders in box girder orthotropic bridges. This design should be checked for lane or truck loading arrangements producing maximum distortional effects.

AASHTO Article 10.41 also covers wheel load contact area, effective width of the deck plate, allowable stresses, diaphragms, stiffness requirements, closed ribs, and wearing surface.

Besides the requirements of a wearing surface discussed in the foregoing sections, this material should also provide: (a) sufficient ductility to accommodate expansion and contraction of the deck plate without cracking or debonding and (b) sufficient fatigue strength to withstand flexural cracking due to deck plate deflections.

Commentary on Specifications

Design Requirements The equivalent–orthotropic slab method, used in the AISC manual (Wolchuk, 1963), simulates the deck by a continuous two-dimensional slab with different stiffness in the longitudinal and transverse directions. The equivalent grid method simulates the deck by a grid of one-dimensional beams (Bouwkamp, 1967; Bouwkamp and Powell, 1967; Erzurumlu and Toprac, 1970; Heins and Looney, 1966, 1968). In both methods the stiffness properties of the equivalent structure must be chosen to adequately approximate the behavior of the actual deck.

For the thin-walled-beam method of analysis of box girders (Vlasov, 1967), also referred to as ordinary folded-plate theory (De Fries-Suene and Scordelis, 1964), the following assumptions are made: (a) the membrane (in plane) stresses produced in each plate element by longitudinal bending can be calculated by elementary beam theory applied to each element and (b) transverse plate bending stresses can be calculated assuming the plate elements act as one-way slabs spanning longitudinal joints. The thin-walled-beam method accounts for the effects of torsional distortions of the cross-sectional shape, which are significant in box girders (Galambos, 1971); hence, the method is satisfactory in designing box girder orthotropic systems. In this case arrangements producing maximum distortional effects, such as a checkerboard pattern, should be investigated.

The wheel load distribution stipulated in Article 10.41.2 (AASHTO) is based on many investigations, showing that a 45° spread of pressure through the wearing surface is reasonable (Farago and Chan, 1960; Henderson, 1954; Teller and Buchanan, 1937; Troitsky, 1967).

Effective Width of Deck Plate Both the AISC (Wolchuk, 1963) and Lincoln (Troitsky, 1967) manuals give approximate methods of design charts, and these are adequate for design purposes. Theoretical calculations indicate that for uniform loading, 94 percent of a slab or plate joining T beams is effective as the top flange of the beams if the ratio of the span to the distance between beams is about 5, or if the ratio of the span to the distance from a beam to an unsupported edge of the slab or plate is 10. If these ratios are smaller, a smaller percentage of the plate is effective.

Allowable Stresses Three interrelated but articulated structural systems induce stresses in the deck. These are (a) the deck plate as it carries the wheel loads to the ribs and cross beams, (b) the ribs and cross beams both utilizing the deck plate as the top flange act as gridwork or as an orthotropic slab to carry the loads to the main girders, and (c) the main girders utilize the deck system (including ribs) as the top flange and carry the loads to reaction points.

The wheel loads are supported on the deck by a combination of bending and membrane action. Where the cross-beam spacing is at least three times the spacing of the longitudinal ribs, most of the load is carried transversely by the ribs. Bending usually predominates at working loads, and the stresses may be calculated by elementary plate-bending theory (Glockner, Verna, and de Paiva, 1971). At higher loads, significant membrane stresses are present, and the deck can now carry loads far exceeding those predicted by plate-bending theory and also exceeding actual loads. Hence, it is not necessary to check the static strength of the deck plate, nor is it necessary to consider the local deck plate stresses in designing the main girders, longitudinal ribs, or cross beams.

The AISC manual concludes, based on experimental results, that with open-rib deck configurations fatigue would not be a critical design factor. Because, however, these results are difficult to extrapolate to other configurations (closed ribs), AASHTO specifies an allowable stress of 30,000 psi based on fatigue considerations. This is approximately 85 percent of the fatigue limit for base metal, and is the same for all steels.

The actual deck plate stresses are generally smaller than calculated because (a) the impact is less than 30 percent, (b) the calculation methods are approximate but conservative (Dowling, 1966), and (c) the participation of the wearing surface causes a reduction of deck plate stresses.

The stresses in longitudinal ribs caused by a combination of rib and girder bending may exceed the basic allowable stress by 25 percent because of the limited structural significance and low probability of occurrence of local peak stresses in the ribs. Ribs, beams, and girders at intersecting points may develop stress concentrations adversely affecting the fatigue characteristics if they occur in regions of tensile stresses. With a large variety of joint details available, special care is suggested in designing these details, and references that give data on fatigue tests should be considered (Davis and Toprac, 1966; Erzurumlu and Toprac, 1970; Hansch and Mueller, 1961; Klöppel and Roos, 1960).

Both analytical and experimental studies show that wheel loads cause significant transverse bending stresses in the webs of closed longitudinal ribs, and these stresses change sign as the lateral position of the wheel load is changed. Peak values usually occur at the connection between the rib web and deck plate. Accordingly, the fatigue strength of this connection should be investigated (Seim and Feriverda, 1972).

Thickness of Plate Elements The width–thickness ratios based on local buckling are appropriate for orthotropic systems. Local buckling has been suggested as a possible cause of failure during erection of certain box girder bridges (see also previous sections), and therefore adequate consideration should be given to local buckling due to erection overloads.

Maximum Slenderness of Longitudinal Ribs Deck overall buckling resulting from compressive stresses may be prevented if the system exhibits sufficient stiffness. The slenderness formula is derived from pinned-end column theory, but greater slenderness is permitted if justified by analysis.

Diaphragms The location, stiffness, and strength of diaphragms and cross frames should be determined from girder analysis, particularly with boxes. The most severe loading condition for diaphragms normally occurs during erection, and is intensified with the cantilever erection method.

Stiffness Requirements The suggested deflection limitation $L/500$ is a compromise between the ratio permitted for other bridge types and the ratio allowed by other codes (e.g., the German code allows $L/225$). Limiting the deflection of the deck plate to $L/300$ is justified considering the intent to prevent deterioration of the wearing surface.

The following empirical formula has been suggested to determine the deck plate thickness not withstanding the stresses in the plate:

$$t \geq 0.007\sqrt[3]{p}\,(a) \tag{7-1}$$

where t = plate thickness (in.)
a = spacing of web ribs (in.)
p = design tire pressure (designed wheel load divided by contact area) (psi)

Other more precise procedures for calculating deck plate deflections are given by Bouwkamp (1967) and Dowling (1966).

Wearing Surface The exact physical properties necessary to design a satisfactory wearing surface are yet to be fully established.

Closed Ribs The intent of sealing requirements is to inhibit corrosion. Inspection of hollow sealed bridge members up to 50 years old has shown the absence of significant internal corrosion. Therefore, closed ribs sealed against the entrance of moisture are permitted.

Proposed LRFD Specifications

Orthotropic Plate Models In the context of analysis, orthotropic plate modeling may be assumed to have the stiffness of its components uniformly distributed along the cross section of the deck. If the torsional stiffness of the deck is contributed by a solid plate of uniform thickness, the torsional stiffness may be calculated as in an isotropic plate model. Alternatively, the torsional stiffness may be established by physical testing, three-dimensional analysis, or other verified approximations. The method should reflect the fact that the accuracy of orthotropic plate analysis is sharply reduced for systems consisting of a small number of components under concentrated loads. By definition, an orthotropic system is a plate having significantly different structural properties in the two principal directions.

Analysis Force effects in orthotropic decks may be determined by elastic methods such as equivalent grillage, finite strip, or finite element. Alternatively, an approximate method of analysis may be followed.

In the approximate method of analysis, the effective width of the deck plate acting compositely with one longitudinal rib can be determined from Table 7-1, where s is the rib span. The effective width of the deck acting as a flange of a cross beam or of a main longitudinal supporting component may either be calculated by analysis or be taken as shown in Figure 7-7.

The effective spans shown as L_1 and L_2 in Figure 7-7 are taken as the actual span for simple spans and as the distance between points of dead load contraflexure for continuous spans. For more details on the development of this figure, reference is made to Moffat and Dowling (1976) and Wolchuk (1990).

In the analysis the deck plate should be assumed to act as a common flange of the ribs, the floor beams, and the main longitudinal components of

TABLE 7-1 Effective Width of Deck Plate Acting with a Rib (LRFD Specifications)

Calculation of		a_o	$a_o + e_o$
Rib section properties		$a_o = a$	$a_o + e_o = a + e$
Flexural effects due to dead load		$a_o = 1.1a$	$a_o + e_o = 1.1\ (a + e)$
Flexural effects due to wheel loads	s = 8.0 FT	$a_o = 1.3a$	$a_o + e_o = 1.3\ (a + e)$
	8 < s < 14	linear interpolation between s = 8.0 and 14.0 FT	
	s > 14.0 FT	$a_o = 1.6a$	$a_o + e_o = 1.6\ (a + e)$

the bridge. The provisions address, in particular, the wearing surface, considered an integral part of the system. Its contribution to the stiffness of the members may be considered if structural and bonding properties are demonstrated over the temperature range −20 to +20°F. A wearing surface acting compositely with the steel deck plate can reduce stresses and deformations. Its effectiveness depends on the thickness, elastic modulus, and bond performance. Thus, the long-term composite action between the deck plate and wearing surface should be documented by static and cyclic load tests.

Decks with Open Ribs For rib spans not exceeding 15 ft, the load on one rib due to wheel loads may be calculated as the reaction of a transversely continuous deck plate on rigid ribs. For rib spans exceeding 15 ft, the effect of rib flexibility on the lateral wheel load distribution may be considered by elastic analysis. For rib spans less than 10 ft, the flexibility of the cross beams must be considered, and the rib may be analyzed as a continuous beam supported on the cross beams.

Decks with Closed Ribs The specifications recommend the Pelican–Esslinger method for analyzing decks with closed ribs (see also the following

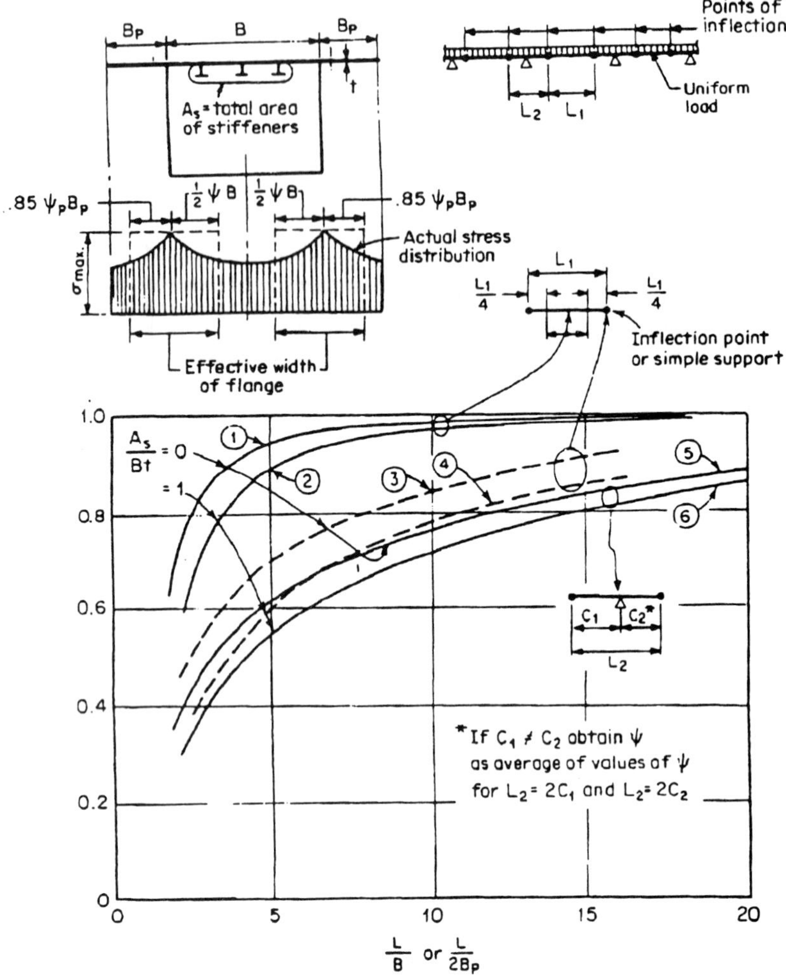

FIGURE 7-7 Effective width of deck (LRFD specifications).

sections). For spans shorter than 20 ft, the load effects on a closed rib may be calculated from wheel loads placed on one rib only, disregarding the effects of adjacent wheels.

Design As mentioned previously, the orthotropic deck is a composite system and therefore develops composite overall stresses that may be added to stresses generated in the deck locally. The loads considered in the design of each element should not be mixed freely but articulated, and even the governing positions of the same load for local and overall effects should be different. The bridge must therefore be analyzed separately for all load regimes, the one that governs should be applied.

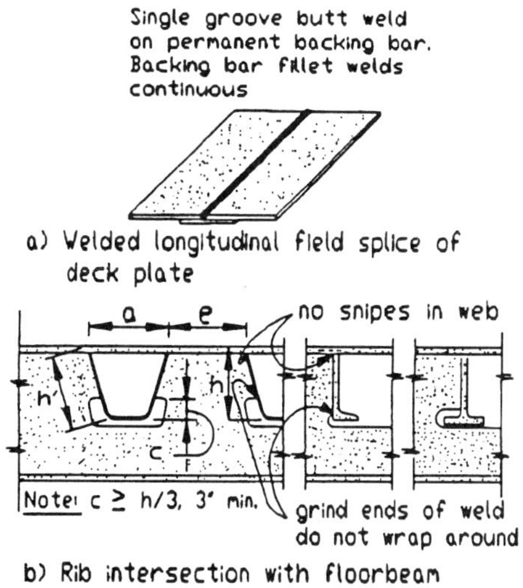

Single groove butt weld on permanent backing bar. Backing bar fillet welds continuous

a) Welded longitudinal field splice of deck plate

no snipes in web

Note: c ≥ h/3, 3" min.

grind ends of weld do not wrap around

b) Rib intersection with floorbeam

FIGURE 7-8 Detailing requirements for orthotropic decks (LRFD specifications).

Limit States At service limit states, the radius of curvature of a rib should not be less than 1/1000 of the span length. In addition, the maximum relative live load deflection between the web of a main girder and the rib adjacent to it, or between two adjacent ribs, should not exceed 0.10 in. The relative deflection criterion is intended to prevent debonding and fracture of the wearing surface caused by excessive flexural distortion of the deck.

As a result of tests indicating a large degree of redundancy and load redistribution between first yield and deck failure, the combination of local and global effects may be reduced by 25 percent at strength limit states. However, the deck, as a component of the overall system, may be exposed to in-plane compression; consequently, the effects of compressive instability should be investigated.

Detailing Requirements The specifications articulate plate thickness and welding requirements. Deck and rib splices should be either high-strength-bolted or welded, as shown in Figure 7-8. Ribs should be run continuously through cutouts in the webs of cross beams as shown.

Combined Bending

Longitudinal The factored moment resistance of girders, longitudinal ribs, and deck plates must be such that

$$\frac{M_{fg}}{M_{rg}} + \frac{M_{fr}}{M_{rr}} + \frac{M_{fp}}{M_{rp}} < 1.33 \tag{7-2}$$

where M_{fg} = factored applied moment of girder
M_{rg} = factored moment resistance of girder
M_{fr} = factored applied moment of longitudinal rib
M_{rr} = factored moment resistance of longitudinal rib
M_{fp} = factored applied longitudinal moment of deck plate as a result of the plate carrying wheel loads to adjacent beams
M_{rp} = factored moment resistance of deck plate in carrying wheel loads to adjacent beams

For deck configurations where the transverse beam spacing is at least three times the longitudinal rib web spacing, the third term in (7-2) may be omitted.

Transverse The factored moment resistance of the transverse beams and deck should satisfy the following:

$$\frac{M_{tb}}{M_{rb}} + \frac{M_{tp}}{M_{rp}} \leq 1.0 \tag{7-3}$$

where M_{tb} = factored applied moment of transverse beam
M_{rb} = factored moment resistance of transverse beam
M_{tp} = factored applied transverse moment of deck plate as a result of the plate carrying wheel loads to adjacent longitudinal ribs
M_{rp} = factored moment resistance of deck plate in carrying wheel loads to adjacent ribs

Likewise, for deck configurations where the spacing of transverse beams is at least three times the spacing of longitudinal rib webs, the second term in (7-3) may be omitted.

Compressive Forces in the Deck Longitudinal ribs, including the effective width of the deck plate, should be designed as individual columns simply supported at cross beams, unless a rigorous analysis shows that overall buckling of the deck will not occur as a result of compressive forces induced by bending of the girders.

Diaphragms Diaphragms or cross frames must be provided at supports with sufficient stiffness to transmit lateral forces to the bearings and also to resist transverse rotation, displacement, and distortion. Intermediate diaphragms and cross frames must be provided where indicated by the analysis of the girders, with sufficient stiffness to resist transverse distortions.

7-4 DECK PLATE: THEORY AND BEHAVIOR

Basic Considerations

In this section we consider the deck system as an independent structural element denoted as system II. The stress in the deck as a part of the main carrying members (system I) is not considered. The bridge deck system, consisting of a deck plate, longitudinal ribs, and transverse floor beams using the deck plate as their common flange, is highly indeterminate.

In the equivalent grid method, the deck plate is assumed slit between the longitudinal ribs, which are treated as individual beams between panel points of the grid system, with the deck plate strips acting as upper flanges. The effect of the deck plate rigidity perpendicular to the ribs is disregarded and must be considered separately.

In the equivalent plate theory (orthotropic slab method), it is assumed that the rigidities of both the floor beams and the longitudinal ribs are uniformly distributed throughout the deck in the direction perpendicular to the respective member. Thus, the actual discontinuous structure of the steel plate deck is represented as an idealized substitute orthotropic slab.

Of the two approaches, the orthotropic plate concept is relatively simple and offers practical advantages. A deck with closed ribs is treated as a continuous orthotropic plate. A simplification of the orthotropic plate equation allows decks with open ribs to be treated as continuous beams.

Properties of Ideal Orthotropic Plate In the foregoing sections we have defined an orthogonal anisotropic plate as a plate having different elastic properties in two principal directions, designated as x and y in Figure 7-9. Because the plate has a constant thickness and its material is continuous, the different elastic properties in the two principal directions must be due to different moduli of elasticity, or $E_x \neq E_y$, and different Poisson ratios, $\nu_x \neq \nu_y$, of the material. Thus, the orthotropic plate theory assumes the plate to be an anisotropic material.

Except for the condition of anisotropy, deformations and stresses of an orthotropic plate are calculated by ordinary elastic theory based on the same assumptions of an isotropic plate. These assumptions are as follows.

1. The material is homogeneous and continuous between the outer surfaces of the plate (this precludes the existence of voids, slits, or other geometric irregularities within the plate).
2. The plate thickness is uniform and small compared with other dimensions, so that stresses normal to the middle plane of symmetry of the plate (x, y) and the effect of shear stresses on plate deformations may be disregarded.
3. The loaded plate deforms elastically, and straight lines normal to the middle surface of the plate remain straight and normal to the deformed middle surface of the loaded plate.

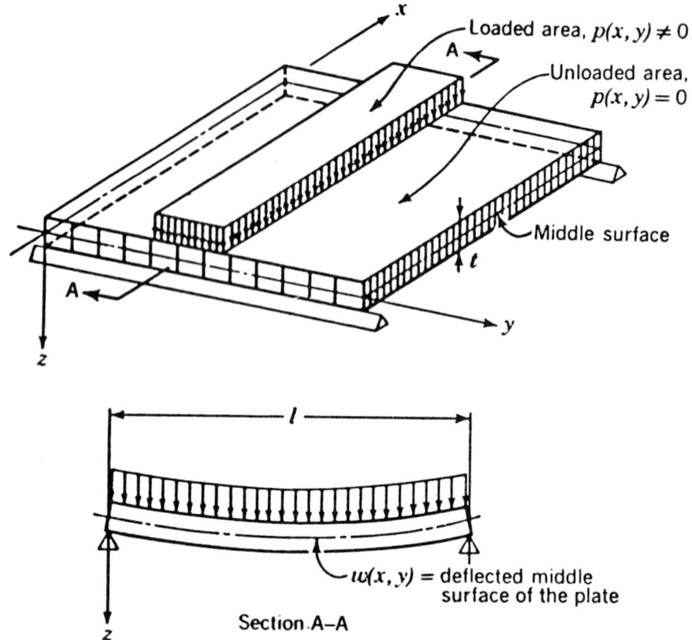

FIGURE 7-9 Basic designation of orthotropic plate as an anisotropic system.

4. The deflections of the loaded plate are small compared with the plate thickness, a condition that precludes the occurrence of membrane stresses.

Rigidity Coefficients We define the elastic properties of an orthotropic plate in terms of three rigidity coefficients: (a) the flexural rigidity in the x direction, D_x; (b) the flexural rigidity in the y direction, D_y; and (c) the effective torsional rigidity, H.

The parameters D_x and D_y are usually expressed in units of kip-in.^{2}/in., and they characterize the resistance to flexure of a plate strip of unit width and thickness t in the x and y directions. They are defined as follows:

$$D_x = \frac{E_x t^3}{12(1 - \nu_x \nu_y)} \qquad D_y = \frac{E_y t^3}{12(1 - \nu_x \nu_y)} \tag{7-4}$$

The effective torsional resistance H determines the resistance of a plate element to twisting, and is expressed as

$$2H = 4C + \nu_y D_x + \nu_x D_y \tag{7-5}$$

The factor $2C$ is the torsional rigidity coefficient, defined as the reciprocal of the angle of twist of a plate element with $dx = dy = 1$ due to the action of

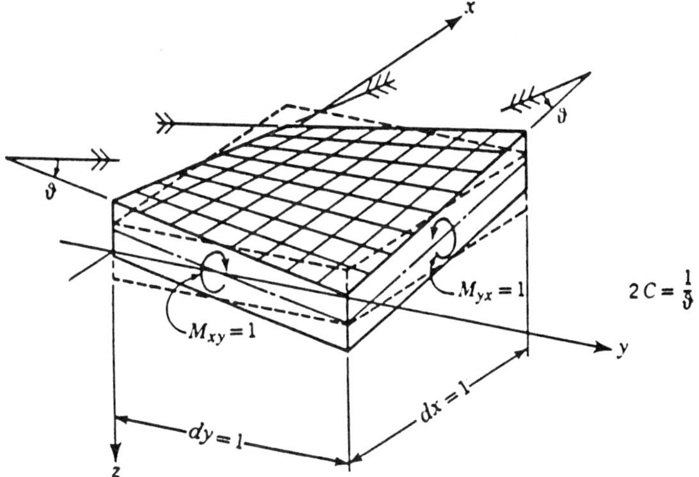

FIGURE 7-10 Plate element subjected to twisting.

twisting moments $M_{xy} = M_{yx} = 1$, or

$$2C = 1/\theta \tag{7-6}$$

The plate element deformation under the action of M_{xy} and M_{yx} is shown in Figure 7-10.

Comparison with Isotropic Plate An isotropic plate may be regarded as a special case of an orthotropic plate. With $E_x - E_y$ and $\nu_x = \nu_y$, (7-4) becomes

$$D_x = D_y = \frac{Et^3}{12(1 - \nu^2)} \tag{7-7}$$

which is the usual expression for plate rigidity.

The torsional rigidity coefficient for an isotropic plate is given by

$$2C = \frac{Et^3}{12(1 + \nu)} \tag{7-8}$$

Combining (7-5), (7-7), and (7-8), we obtain the expression for H as

$$2H = \frac{Et^3}{6(1 - \nu^2)} = 2D \qquad \text{or} \qquad H = D \tag{7-9}$$

so that for an isotropic plate the effective torsional rigidity is the same as the flexural rigidity.

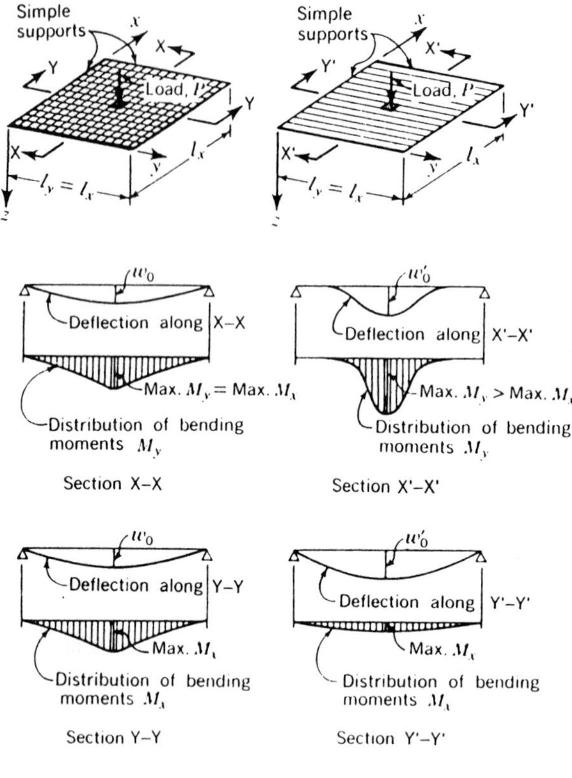

(a) ISOTROPIC PLATE, $D_x = D_y$ (b) ORTHOTROPIC PLATE, $D_x < D_y$

FIGURE 7-11 Comparison of deflections and bending moments in a square isotropic and a square orthotropic plate.

We can compare an isotropic plate (a steel plate or a concrete slab of constant thickness) with an orthotropic plate in the context of structural behavior by referring to Figure 7-11. Both plates are square, supported on four sides, and loaded with a load P, distributed over a small area in the center of the plate. In Figure 7-11*a*, because of symmetry and because $D_x = D_y$, the load is equally carried in both directions, and the corresponding deflections and moments in the x and y directions are identical.

For the orthotropic plate shown in Figure 7-11*b*, $D_x < D_y$. Accordingly, the deflections along sections $X'–X'$ and $Y'–Y'$ are dissimilar, except for the common ordinate under the load, ω'_0. The load is carried predominantly in the direction of the greater rigidity y, as can be seen from the larger bending moments M_y acting in the stiffer direction. Furthermore, the maximum values of the bending moments M_y in the orthotropic plate are greater than the corresponding values of the moments in the isotropic plate. In the orthotropic plate the moments do not extend over the entire span l_x, but are confined to a plate strip near the load.

It follows that a square orthotropic plate with $D_y > D_x$ may be compared with an elongated isotropic plate with $l_x > l_y$. A load application at the center of such a plate will be carried mainly in the y direction, and if the ratio l_x/l_y is large, the load effects will not extend to the edges $x = 0$ and $x = l_x$, as in the case of the orthotropic plate.

In the extreme case, the rigidity values D_x and H of an orthotropic plate may be very small compared with D_y. In this case M_y in the plate due to an applied load may be assumed to act only within the strip corresponding to the width of the load. Accordingly, the plate is modeled as a series of beams lying side by side.

Deck Plate Behavior

The analysis of an anisotropic plate was first presented by Gehring and Boussinesq. A complete solution was first presented by Huber (1914). The differential equation giving the relationship between deflection and the loading of an orthotropic plate, often referred to as Huber's equation, is

$$D_x \frac{\partial^4 \omega}{\partial x^4} + 2H \frac{\partial^4 \omega}{\partial x^2 \partial y^2} + D_y \frac{\partial^4 \omega}{\partial y^4} = p(x, y) \qquad \text{(7-10)}$$

where ω is the deflection of the middle surface of the plate at a point (x, y) as shown in Figure 7-9. The parameters D_x, D_y, and H are the rigidity coefficients, and $p(x, y)$ is the loading intensity at any point expressed as a function of the coordinates x and y. Equation (7-10) is a nonhomogeneous differential equation because the function $p(x, y)$ does not contain the deflection ω. A function $\omega(x, y)$ that satisfies (7-10) represents a solution of the differential equation.

In areas of the plate where no direct vertical load is applied (as shown in Figure 7-9), the load function $p(x, y)$ is zero, and the deflection ω of the unloaded portion of the plate is obtained as

$$D_x \frac{\partial^4 \omega}{\partial x^4} + 2H \frac{\partial^4 \omega}{\partial x^2 \partial y^2} + D_y \frac{\partial^4 \omega}{\partial y^4} = 0 \qquad \text{(7-11)}$$

which is homogeneous because the unknown function ω appears in all terms. The homogeneous equation generally represents a plate loaded by bending moments and live loads applied only along the plate edges.

The methods used for solving (7-10) and (7-11) of an orthotropic plate are essentially the same as in the isotropic plate, except that the integration constants and the arguments of the functions must include the characteristic rigidities of the plate, D_x, D_y, and H. Suggested procedures are given by the AISC manual (Wolchuk, 1963).

Application of Huber's Theory to Deck Analysis

A flat plate stiffened by ribs in one or two directions may be treated as an orthotropic plate provided the rib spacing is sufficiently smaller than the span

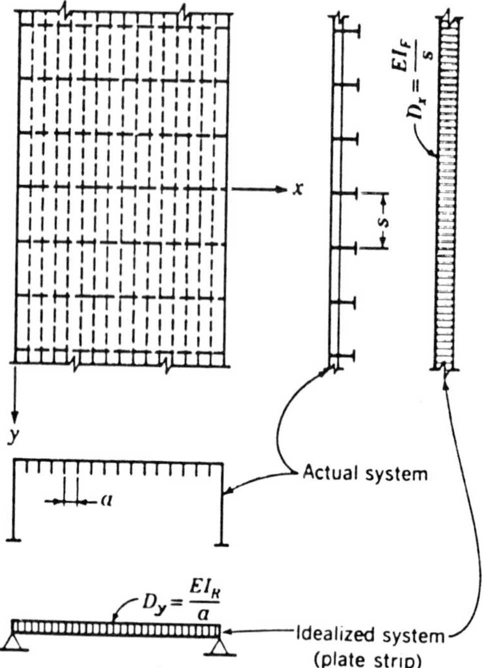

FIGURE 7-12 Bridge deck treated as an orthotropic plate strip.

to ensure full participation of the flat plate in the flexural action of the ribs. A deck system having a steel plate, longitudinal ribs, and transverse floor beams as components may be represented in two ways.

Model A The rigidities of both longitudinal ribs and transverse floor beams are assumed to be continuously distributed throughout the deck. A substituted idealized deck system is thus obtained as an orthotropic plate strip supported on the main girders as shown in Figure 7-12. This approach is justified for closely spaced floor beams.

For the cross section in the x direction, the substitution involves continuous instead of separate elements, or instead of the bending stiffness of each rib, we introduce the distributed stiffness I_R so that $D_y = EI_R/a$. Likewise, in the y direction $D_x = EI_F/s$.

Model B The rigidity of the longitudinal ribs is only assumed to be uniformly distributed, whereas the transverse floor beams are treated as individual members. In this case the deck system is analyzed as a continuous orthotropic plate supported on the main girders and the floor beams.

In both cases bending moments are calculated for the substitute orthotropic plate, and are used to obtain bending moments and stresses in the individual members.

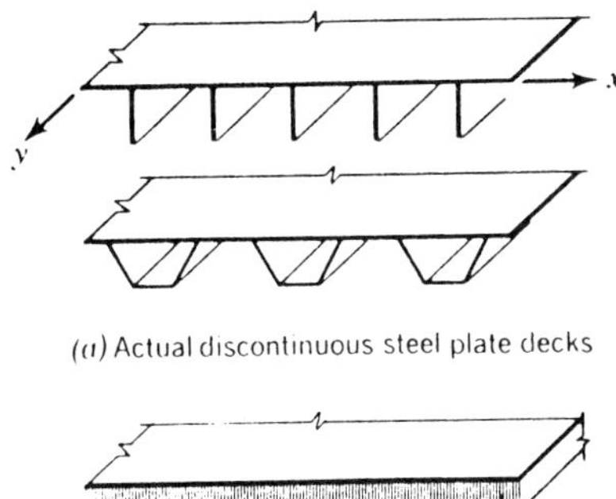

(a) Actual discontinuous steel plate decks

dx

(b) Idealized system – plate with slits

FIGURE 7-13 Actual steel decks and their idealized representation as an orthotropic plate.

Validity of Assumptions The typical orthotropic bridge deck of a steel plate stiffened by a series of closely spaced longitudinal ribs and floor beams is different from conventional bridge decks. Structurally, it is neither a gridwork nor an orthotropic plate. The most important properties of the actual system at variance with the assumptions of the general orthotropic plate theory are (a) the discontinuity and (b) the asymmetry of the steel plate deck. These two variations are demonstrated in Figure 7-13. The two different rigidities in the perpendicular directions result from different geometries rather than from different elastic properties. Because the stiffening ribs are placed on one side of the deck plate, the result is asymmetrical. Due to the eccentric arrangement of the stiffeners, the neutral surfaces of the deck for bending in the x and y directions do not coincide.

Effect of Discontinuity For model A, only the floor beam moments are obtained from orthotropic plate analysis. Additional steps are needed to calculate moments and stresses in the longitudinal ribs under wheel loads placed between panel points as shown in Figure 7-14. For the approach of model B, the effects of system discontinuity on stresses are articulated as follows.

1. With open ribs and with D_x and H much smaller than D_y, the effective width (obtained in the idealized system) is hardly any larger than the width of the applied load. For a loading width equal to the rib spacing, the effective width of the idealized system may be smaller than the width over which the live load effect extends in the actual plate (see also Figure 7-15). In this case the moment in the directly loaded rib calculated from an orthotropic plate analysis would be larger than the actual moment, as can be seen from Figures 7-15a and c.

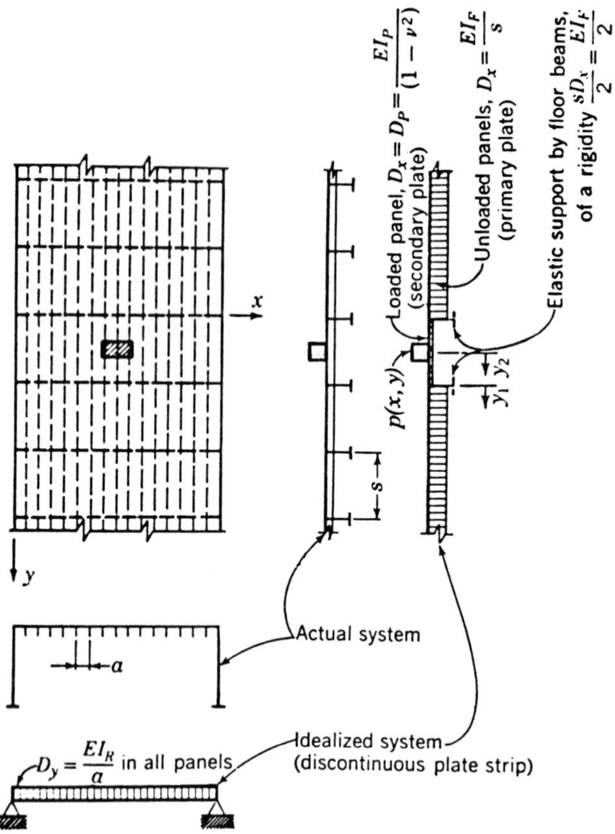

FIGURE 7-14 Bridge deck treated as an orthotropic plate strip, with consideration of the discontinuity in the loaded panel.

2. In some instances involving unusual loading cases, conditions may be reversed and the calculated moment from the substitute system may be smaller than the actual moment. Thus, representing a deck with open ribs may not always be possible to the extent desired.

3. In a deck with closed ribs, having a better load-distributing capacity than the open system in the transverse direction, the load distribution obtained in the substitute model is in good agreement with the actual deck. A certain inaccuracy in the rib moment and stresses may result when the moment per unit width, M_y (computed in the substitute plate at the center of the loaded rib), is multiplied by $(a + e)$, as shown in Figure 7-16, to give the total moment in the rib. The total rib moment thus obtained is larger than the actual average moment over the width of the rib, as shown in Figure 7-16a. With constant loading width, this discrepancy tends to decrease as the relative transverse system rigidity is increased.

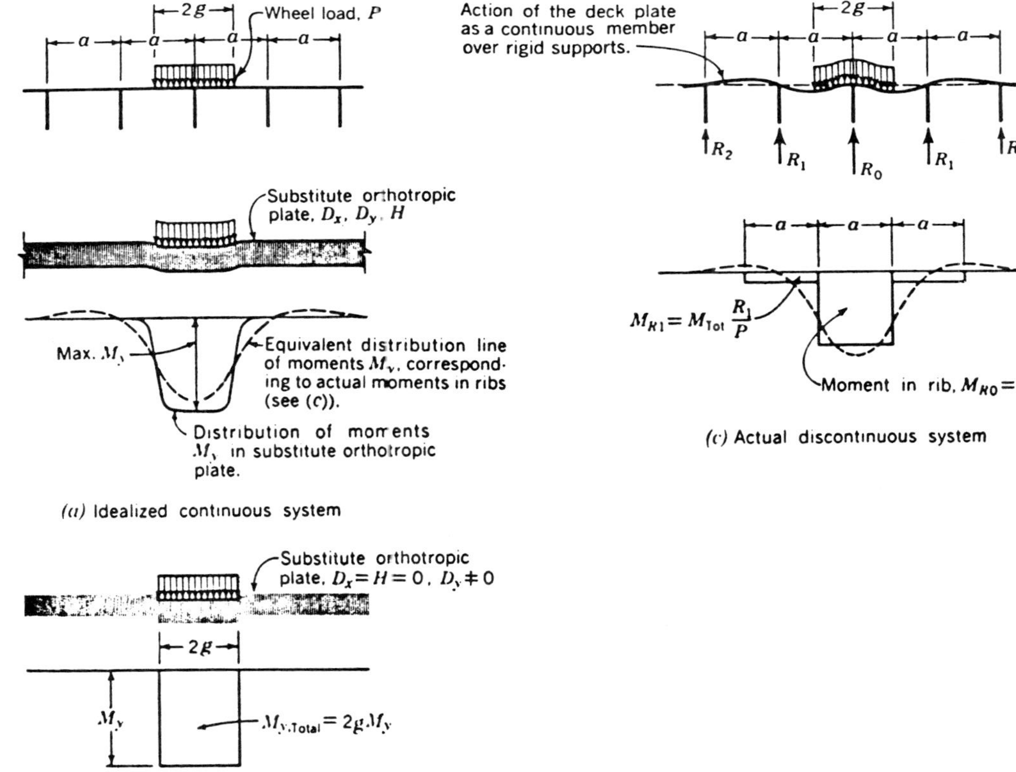

FIGURE 7-15 Determination of the bending moments in the open ribs.

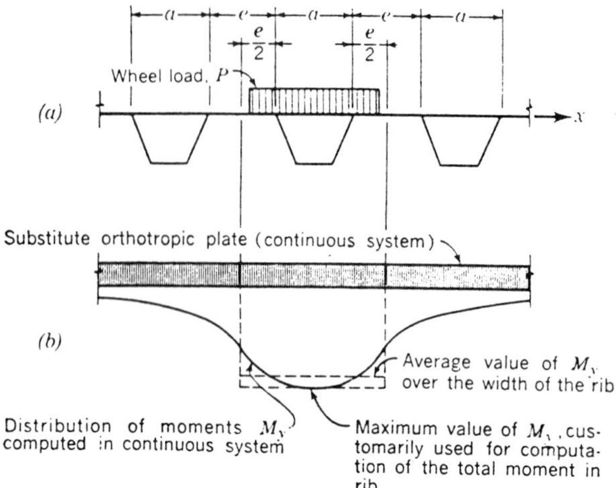

FIGURE 7-16 Determination of the bending moments in the closed ribs.

7-5 DESIGN APPROACHES

Basic Concepts

With the development of the orthotropic deck system, various approaches were suggested to determine the load distribution and internal forces. Huber's equations were first applied to stress computation by Cornelius (1951).

In a subsequent publication, Cornelius (1952) gave general expressions for the integration constants and for the various coefficients appearing in equations for deflections, moments, and shears of an orthotropic plate using $\nu_x = \nu_y = 0$ and $H < \sqrt{D_x D_y}$, the latter assumption corresponding to the structural properties of steel plates with open ribs.

Numerical evaluations of the expressions for the bending moments for certain specific cases are given by Homberg (1951), Homberg and Weinmeister (1956), Olsen and Reinitzhuber (1959), and Benner (1939).

Current methods for analyzing stresses in orthotropic deck systems (based on Huber's theory) include the following four main groups.

1. Ideal gridwork system
2. Deck as uniform medium orthotropic plate strips
3. Application of influence surfaces
4. Semiempirical methods

Ideal Gridwork System

The basis of this method is the orthortropic plate theory whereby the deck is considered as statically equivalent to open grids. The method is simple but approximate, and was proposed by Guyon (1946) and Massonet (1950), who extended the approach to include the effects of torsion.

Essentially, the objective of the analysis is to determine how a concentrated load or system of loads is distributed among longitudinal beams at the bridge considering the various degrees of transverse stiffness and torsional resistance. The actual bridge deck structure, a system of discrete interconnected longitudinal and transverse elements, is replaced by an elastically equivalent slab system whose structural properties in the two orthogonal directions are uniformly distributed along their length. The parameters necessary to define the structural properties are as follows.

Effective Width of Bridge This is defined in terms of the number of longitudinal beams and the beam spacing. A solid slab may be considered as a series of rectangular beams placed side by side, from which it follows that the equivalent width is equal to the actual width.

Elastic Equivalence The elastic equivalence between a general interconnected beam system and an orthotropic plate indicates that both are governed by Huber's equation (7-10).

Flexural Parameter This has a fundamental effect on the load-distributing characteristics of the deck structure. In a grillage without torsion, the flexural parameter alone determines the distribution of an applied load throughout the system.

Torsional Parameter For a bridge deck without torsional stiffness, this parameter is zero. For a solid slab (the structure with maximum torsional stiffness), the value of this parameter is unity. All other structural forms between these two extremes have torsional parameter values between zero and unity. Thus, for an infinite plate freely supported at four sides and for sinus-like distributed loading, Guyon gave solutions disregarding torsional stiffness, whereas Massonet used torsional stiffness approaching unity.

Distribution Coefficient This parameter was developed as the relationship between the actual deflection of a longitudinal bridge section under a loading system and the deflection of the bridge with the same loading system distributed uniformly across the bridge. Values of this coefficient may be extrapolated from diagrams prepared by Guyon and Massonet for the assumed torsional stiffness.

Longitudinal Bending Moments The distribution coefficients introduced for deflections also apply to longitudinal moment. Thus,

$$M_L = K'M_{\text{av}} \tag{7-12}$$

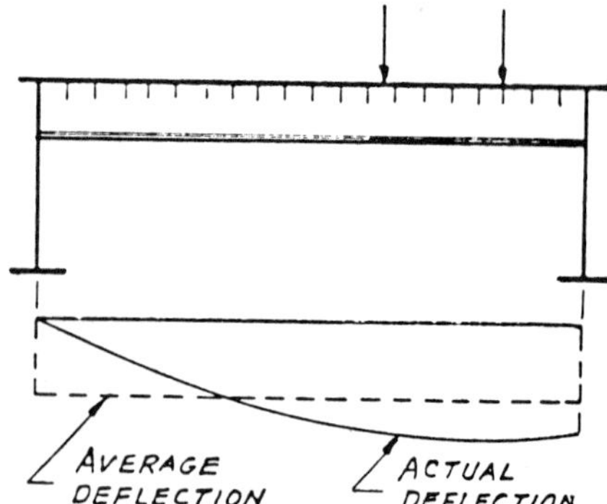

FIGURE 7-17 Actual and average transverse deflections. (From Troitsky, 1967.)

where K' is the distribution coefficient for the actual value of flexural stiffness for the bridge deck under consideration. The "mean" effects are found by the uniform distribution of each individual load at the transverse section where it is applied. The next step is the determination of the moment profile of a beam or slab carrying a longitudinal distribution of point loads.

Transverse Bending Moments The effect of a concentrated load is the unequal deflections across a transverse section and the associated distortion as shown in Figure 7-17. This distortion induces transverse bending stresses, according to a transverse bending moment given by an infinite series.

Orthotropic Plate Strips

Cornelus (1952) treats the actual orthotropic plate as equivalent to a continuum, and thus the spacing between floor beams is not taken into account. Expressions for deflections, slopes, bending moments, torsional moment, shear forces, and reactions are obtained from a general solution for each case of loading represented by a Fourier series. The plate is usually treated as a rectangular plate or infinitely long plate strips simply supported along the main girders, the latter regarded as rigid.

The solution for the idealized plate with assumed continuous properties adequately describes the elastic behavior as a whole, but it still disregards local deformations and stresses occurring as a result of actual discontinuity under a wheel load placed between panel points (see also the foregoing sections). For example, stresses in longitudinal ribs under loads placed between the floor beams cannot be determined by considering the deck as an orthotropic plate strip with floor beam rigidities uniformly distributed. Thus, the original method was jointly refined by Mader (1957).

Under the new concept, the actual orthotropic deck is considered as equivalent to one continuum, that is, the rigidity of the floor beams is assumed to be continuously distributed only in areas outside of the loaded panel, whereas the panel under load is treated as a secondary orthotropic

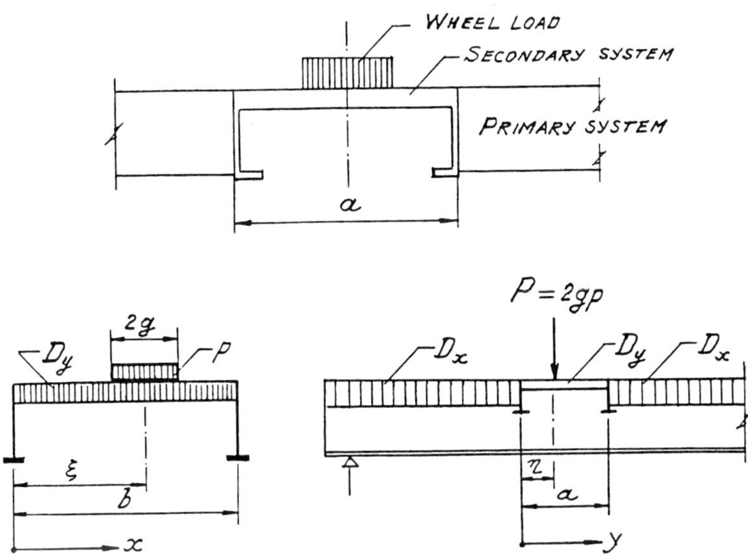

FIGURE 7-18 Primary and secondary systems for the orthotropic plate according to Mader (1957).

plate elastically supported by the two adjoining floor beams and continuous with the primary orthotropic plate of the remaining deck as shown in Figure 7-14.

Constants are given for bending moments and deflections distributed along an edge of a plate, and allow calculations in orthotropic plates with fixed edges, continuous plates over three or more supports, and plates supported on two or three edges. The original Cornelius method does not distinguish between span and support moments in ribs because of the assumption of uniformly distributed floor beam stiffness. The modified approach articulates the orthotropic plate into a primary and secondary continuum. In the latter, these ribs and plate form another system, spanning between elastic supports at floor beams, as shown in Figure 7-18, where

$$D_y = \frac{EI_y}{a} \qquad D_x = \frac{Et^3}{12(1-\nu^2)} \qquad H = \sqrt{D_x D_y} \tag{7-13}$$

Eight loading systems analyzed by Mader (1957) are shown in Figure 7-19. It appears that it is usually sufficient to evaluate the maximum support moments of the one rib above one cross beam and the span moment between two cross beams. After the necessary forces are evaluated, the shear forces, slope angle, and deflections are calculated as in conventional statics.

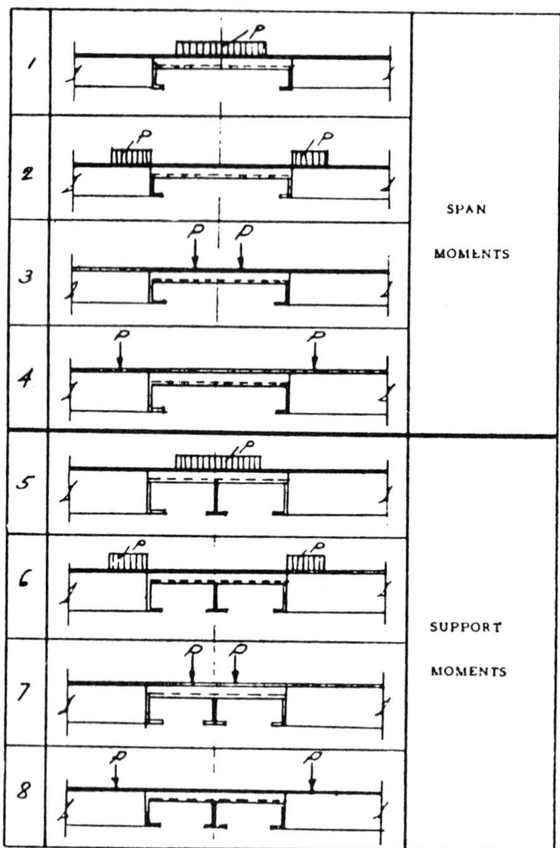

FIGURE 7-19 Loading systems analyzed by Mader (1957).

Application of Influence Surfaces

Influence surfaces developed for orthotropic plates have been proposed by Hawranek and Steinhardt (1958), Homberg and Weinmeister (1956), Krug and Stein (1961), and Thürlimann (1960). The moments in a steel plate deck can be obtained by using influence surfaces in two ways.

1. The longitudinal and transverse bending moments in a deck system consisting of the steel plate, longitudinal ribs, and transverse floor beams are calculated from influence surfaces developed for moments in an orthotropic plate strip. The local bending moments in the longitudinal ribs acting continuously in rigid supports are then superimposed on the overall moments to obtain final moments in the ribs.

2. The longitudinal bending moments are calculated from an influence surface for an orthotropic plate panel, consisting of the deck plate and the longitudinal ribs only and supported on the main girders and floor beams.

The effects of deck continuity and floor beam elasticity and final bending moments in the floor beams are then computed as a second step.

In both cases the influence surfaces supply only part of the solution; the other part of the solution will be computed by other methods. Procedures to be used in conjunction with influence surfaces include: (a) a gridwork theory developed by Homberg and Weinmeister (1956), which considers the orthotropic plate as a member consisting of an infinite number of closely placed longitudinal and cross beams; (b) the method of transformed coordinates proposed by Girkmann (1956), which transforms the load coordinates such that it is possible to use reference tables based on the condition $D_x = D_y$ for any ratio D_x/D_y; (c) the method proposed by Olsen and Reinitzhuber (1959), analyzing orthotropic plates supported on two edges with different spans and loading by using influence surfaces; (d) the method introduced by Fisher (1952) based on the application of the sinus-affinitat between bending moments and deflections; and (e) the method proposed by Krug and Stein (1961), which evaluates the influence surfaces for orthotropic plates in order to obtain the most favorable analyzing conditions for the required accuracy.

For the theoretical expression of these concepts, reference is made to the AISC manual (Wolchuk, 1963), the Lincoln Foundation manual (Troitsky, 1967), and the direct references mentioned.

Simplified Methods

For several orthotropic deck bridges, the system has been designed by one of the foregoing approaches and analytical procedures. Then a model or full-scale test was performed on the designed plating to determine its actual capacity.

A simplified, yet sufficiently accurate, method has been developed by Pelikan and Esslinger (1957), based on an abbreviated rather than full form of Huber's equation, and obtained by eliminating parameters of least importance. The deck plate with longitudinal ribs is treated as a continuous orthotropic plate supported on rigid main girders and elastic floor beams. The design, completed in two steps, may be further simplified by charts and tables prepared for a specific loading. The method is suggested by the standard AASHTO and proposed LRFD specifications and is outlined in subsequent sections. Design charts based on AASHTO loads are also included.

7-6 PELIKAN–ESSLINGER METHOD

General Considerations and Method Outline

In addition to the methods summarized in the foregoing sections, solutions of orthotropic deck plates have been obtained by finite-difference (Heins and Looney, 1968, 1969) and finite-element techniques (Powell and Ogden, 1969;

Cusens and Pama, 1975). Invariably, these solutions require tedious and lengthy computations, feasible only with the use of computers.

The approach presented in this section provides a practical and relatively simple procedure for designing orthotropic steel bridges. The main simplified assumption is that the orthotropic plate is continuous and is supported rigidly by the main girders and elastically by the floor beams. Other assumptions are that bending moments in a steel plate treated as an orthotropic plate depend on (a) the loading, (b) the floor beam spacing, (c) the main girder spacing, and (d) the magnitude and ratio of the three characteristic rigidities of the substitute orthotropic plate representing the actual system.

Design Steps For expediency, the bridge deck is divided into three distinct systems whose individual actions are added to obtain the final bridge response.

System I. The local behavior of the deck plate spanning the distance between supporting ribs.

System II. The response of the deck plate, ribs, and transverse floor beams forming the bridge deck.

System III. The action of the main longitudinal girder, interacting with the deck plate and longitudinal ribs, assumed to be a large beam spanning between supports.

Whether the plate deck has open or closed ribs, two computation steps are usually necessary. In the first step the maximum values of the bending moments in the longitudinal ribs and in the floor beams are computed assuming that the floor beams are infinitely rigid. In the second step the effects of the elastic flexibility of the floor beams are determined, and the bending moments obtained in the first steps are adjusted.

In general, system I may not be included in the consideration of the total stress. The behavior of this system is based on the flexural theory of the first order, assuming small deflections and no membrane stresses, whereas the actual load-carrying capacity of the plate is much higher and cannot be predicted by first-order theory. In this context, the general response of systems II and III is shown in Figure 7-20.

As a practical matter, some of the factors influencing the design have little importance and may be eliminated from the computations.

Decks with Closed Ribs With the assumption $D_x = 0$, the differential equation of the orthotropic plate is reduced to

$$2H\frac{\partial^4 \omega}{\partial x^2 \partial y^2} + D_y \frac{\partial^4 \omega}{\partial y^4} = p(x, y) \tag{7-14a}$$

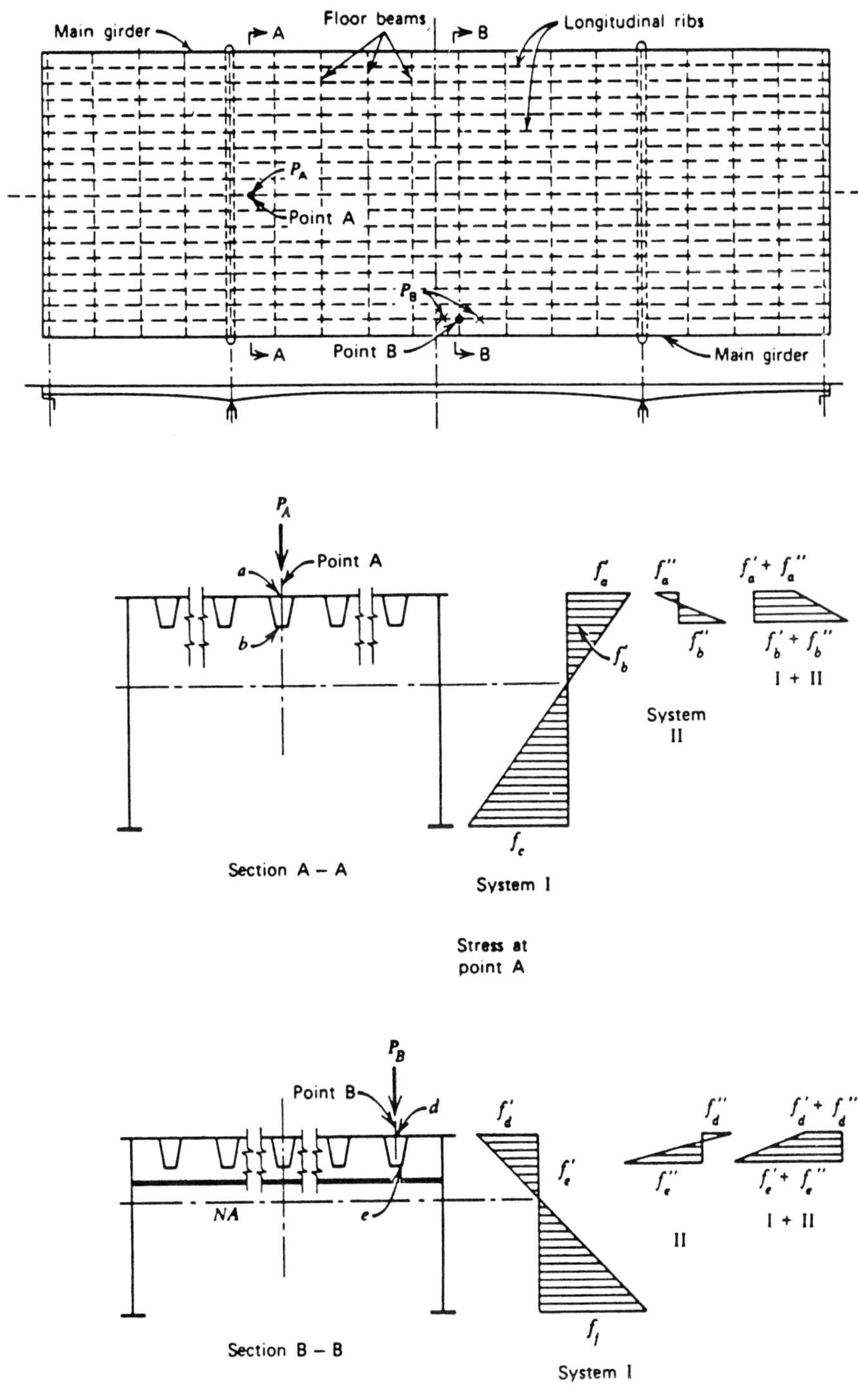

FIGURE 7-20 Two-system response of orthotropic deck bridge according to the Pelikan–Esslinger method. (From Heins and Firmage, 1979.)

The solution of the abbreviated equation (7-14) and the resulting formulas for bending moments are considerably simpler than the original equation. Furthermore, the moments computed assuming $D_x = 0$ are greater than those determined from the full equation; hence, the results obtained by the simplified method are on the safe side.

Decks with Open Ribs The assumptions $D_x = 0$ and $H = 0$ accommodate a bridge deck with open ribs of usual proportions, that is, a relatively thin plate and comparatively short rib spans. Such a system has very small load-distributing capacity in the transverse direction. Thus, we can write

$$D_y \frac{\partial^4 \omega}{\partial y^4} = p(x, y) \tag{7-14b}$$

which defines an idealized structural system representing the actual steel plate deck with open ribs. This idealized system may be visualized as a series of infinitely narrow plate strips placed side by side and running continuously in the y direction as shown in Figure 7-15b. In this system the effective width is the width $2g$ of the applied loading.

If the deck plate is unusually thick, or if the span of the ribs exceeds 6 ft, the effect of plate rigidity appears to dominate, and a correction applied to the original simplified computational is indicated.

Rigidity Conditions

For a deck with closed ribs on rigid supports, the necessary parameters are D_y and H. The transverse flexural rigidity D_x does not enter into the moment analysis, but is used in determining H.

Effective Plate Width The effective width of a plate acting with one rib may be smaller, equal to, or larger than the rib spacing. In general, for design purposes it may be obtained by referring to Table 7-1. The theoretical background of the effective plate width is discussed in this section.

An infinitely wide plate strip of span s_1 stiffened by a longitudinal rib and subjected to flexure is shown in Figure 7-21. The plate strip is axially loaded by shear forces V introduced at the junction with the rib. Because the rib and the plate act together, they have the same unit elongation or contraction at each point along the junction. The effective plate width a_0 is defined as the width of a plate strip that has the same contraction as the actual plate at the junction with the rib.

If the plate is stiffened by one rib only, the effective width is approximately one-third of the rib span s_1, and is independent of the loading as shown in Figure 7-22a. Additional ribs spaced far apart have no influence on the effective width as shown in Figure 7-22b. If the ribs are closely spaced but

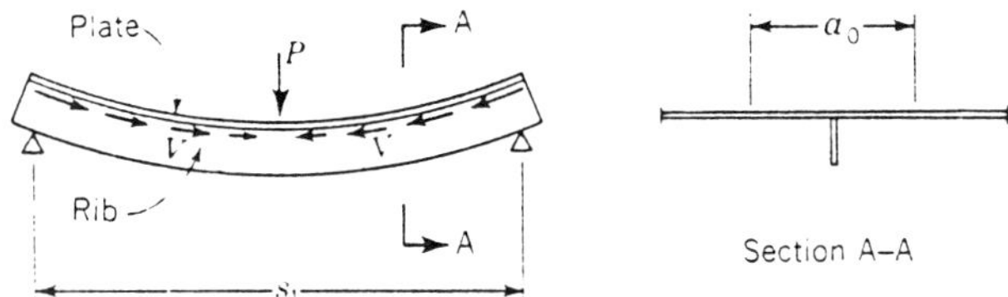

FIGURE 7-21 Composite action of a plate and rib subjected to flexure.

only one is loaded, the effective width is still $s_1/3$ as shown in Figure 7-22*c*. If, however, the closely spaced ribs are all uniformly loaded, the effective widths approach the actual rib spacing as shown in Figure 7-22*d*.

Effective Span The effective span of the rib, s_1 used in the computation of the effective width of a plate, is defined as the average length of the positive moment area of the rib as shown in Figure 7-23. In the design of ribs continuous over rigid supports, the average value of the effective span is

$$s_1 = 0.7s \tag{7-15}$$

In computing the effects of the floor beam flexibility, the effective span of the ribs is always large, or $s_1' = \infty$.

Effective Width of Plate for Equal Loading of Ribs For a plate stiffened by uniformly spaced ribs, the effective width a_0 is easiest to determine in the case of equal sinusoidal live loads applied to all ribs, as shown in Figure 7-24. The effective width ratios obtained from this case are given in Table 7-2. In practical terms, we may assume that the effective width is essentially the same for other load types equally applied to all ribs. Note that for flexible floor beams, $s_1' = \infty$; hence, $a/s_1 = 0$ and $a_0/a = 1.10$ or $a_0 = 1.10a$. This

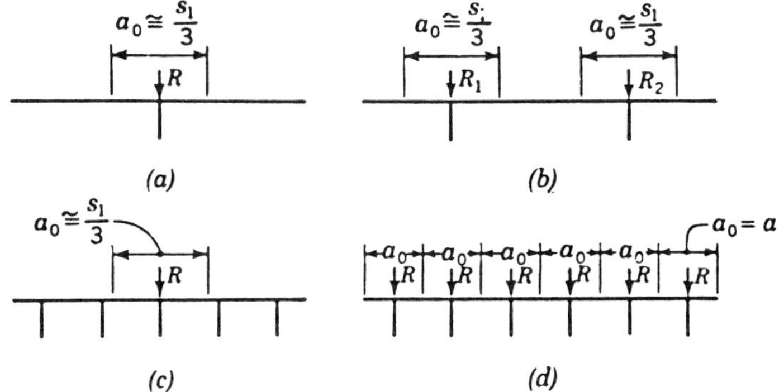

FIGURE 7-22 Effective width of plate with ribs and load.

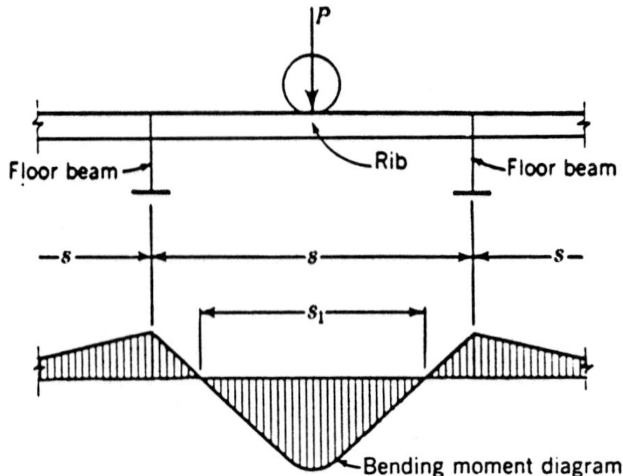

FIGURE 7-23 Effective span of ribs.

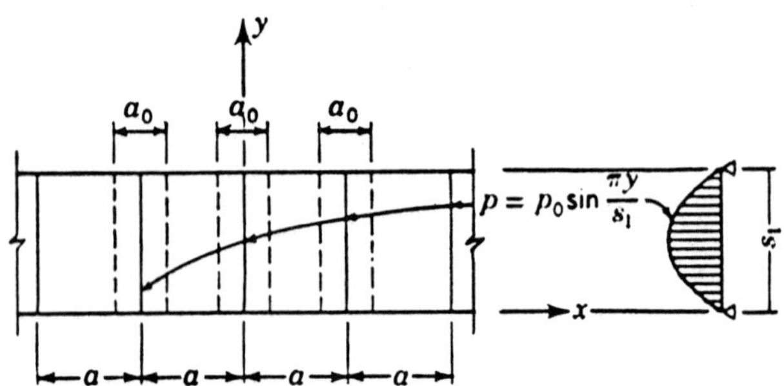

FIGURE 7-24 Effective width of plate for equal loading of all ribs.

TABLE 7-2 Effective Width of Plate Ratios for All Ribs Equally Loaded

a/s_1	a_0/a	a_0/s_1
0.0	1.099	0.000
0.2	1.005	0.201
0.4	0.808	0.323
0.6	0.620	0.372
0.8	0.480	0.384
1.0	0.383	0.383
∞	0.000	0.363

a is the rib spacing, a_0 is the effective width of the plate acting with one rib, and s_1 is the effective span.

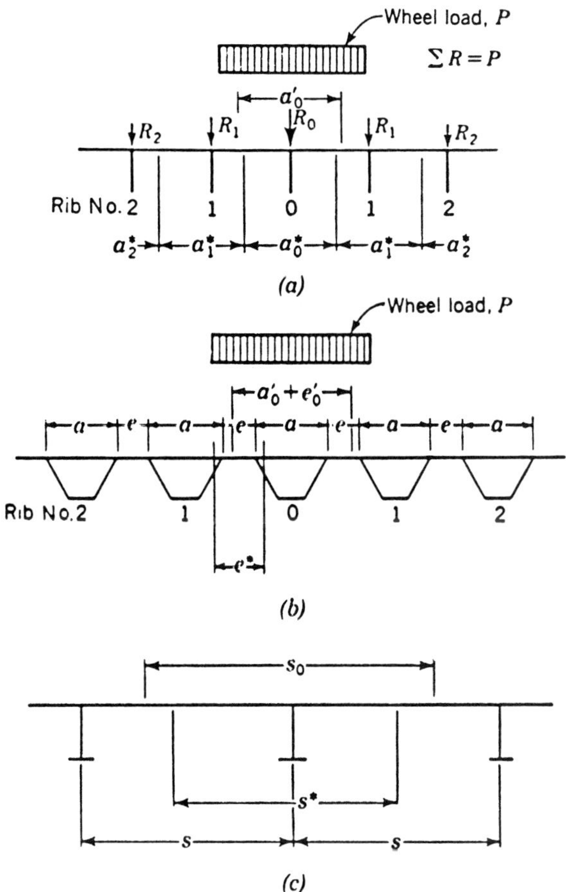

FIGURE 7-25 Effective width for unequal loading of the ribs.

would be the effective width for computing flexural effects due to dead load (uniform), and is the value given in Table 7-1.

Effective Width of Plate for Unequal Loading of Ribs As shown in Figure 7-25, loads on individual ribs due to a wheel effect are not equal. The rib directly under the wheel receives a load R_0, the adjoining rib a load R_1, and so on; hence, the effective width ratios given in Table 7-2 do not apply in this case.

These ratios, however, are approximately correct if an "effective spacing" a^* is used instead of the actual rib spacing a. The value of a^* can be computed only if the final values of loads or moments on the ribs are known. For example, the idealized effective spacing is derived from the condition that the effective distances of the ribs are proportional to the load effects on individual ribs. Referring to Figure 7-25, we can write

$$a_0^* = \left(\frac{2M_0}{M_0 + M_1}\right)a \qquad a_1^* = \left(\frac{M_1}{M_0 + M_1} + \frac{M_1}{M_1 + M_2}\right)a \qquad \text{etc.} \quad (7\text{-}16)$$

With the values of a^* determined, the corresponding effective plate widths a_0' may be found as before. This solution is approximated in the effective width values shown in Table 7-1 for live load.

Section Properties of Ribs and Floor Beams These may be computed by reference to applicable specifications such as Table 7-1, or may be based on a more detailed analysis. Because they depend on the effective plate width, the procedure should articulate the two conditions of loading, that is, equal and unequal loading on ribs.

Open Ribs For continuous ribs on rigid floor beams, stresses due to wheel loads cause unequal loads on individual ribs; hence, the effective spacing may be computed from (7-16).

For a deck with open ribs, the plate is assumed to act continuously and distribute the load on rigid supports, with the rib reaction determined accordingly. In considering the effects of floor beam elasticity, all ribs may be assumed equally loaded. In either case the effective rib spacing can be taken as the actual rib spacing, and the criterion of Table 7-1 is thus derived with $a_0 = a$.

Closed Ribs The top of the ribs is designated as a, and the width of the plate between ribs as e, in Figure 7-25 and Table 7-1. Because these are generally different, their contribution to the effective plate width is determined separately.

If the ribs are equally loaded, their effective spacing equals the actual spacing, and the effective width is computed from

$$a_0 + e_0 = \frac{a_0}{a'}a + \frac{e_0}{e'}e \qquad (7\text{-}17)$$

or $a_0 + e_0 = a + e$, as shown in Table 7-1.

The wheel loads on individual ribs on rigid supports are not equal (step 1). However, in computing the flexural rigidity D_y of the substitute orthotropic plate, necessary for determining the bending moment in closed ribs, the effective width as computed from Table 7-1 and (7-17) is used, because it represents a fair representative effective width of the plate acting with the ribs affected by the wheel load. In this case the flexural rigidity coefficient is computed from

$$D_y = \frac{EI_R}{a + e} \tag{7-18}$$

where I_R is the moment of inertia of one rib based on $a_0 + e_0$.

Floor Beams In computing dead load stresses and the rigidity ratio of the floor beam to the rib, the values of the moment of inertia and section modulus of the floor beam include the effective width of the deck plate, s_0, as shown in Figure 7-25*c*. This is computed from

$$s_0 = s\frac{s_0}{s} \tag{7-19}$$

where s is the actual floor beam spacing as shown in Figure 7-25*c*, and s_0/s is a coefficient obtained using the actual floor beam span l as the effective floor beam span. If the floor beams are restrained or continuous at the ends, the effective span must be adjusted.

In computing live load stresses in floor beams, the section modulus must be based on the effective width s_0' related to the unequal loading, or $s_0' = s^*(s_0/s^*)$, with

$$s^* = \frac{2F_0}{F_0 + F_1}s \tag{7-20}$$

where F_0 is the load on the floor beam under consideration and F_1 is the load on the adjoining floor beam.

For practical purposes it is sufficient to use the same section modulus considering all floor beams equally loaded.

Torsional Rigidity For a deck with open (torsionally soft) ribs, the torsional rigidity H may be disregarded. For a deck with closed (torsionally stiff) ribs, H may be considerable. The general expression of the effective torsional rigidity is given by (7-5). Assuming $\nu_x = \nu_y = 0$, this reduces to

$$H = 2C \tag{7-20a}$$

The value of $2C$, represented by the twisting angle θ, results from two equal contributions in the two perpendicular directions as shown in Figure 7-10. With closed ribs, the torsional rigidity is supplied essentially in one direction by rib resistance to the twisting moment M_{yx} acting in a plane perpendicular to the ribs, whereas resistance to the twisting moment M_{xy} is negligible. Therefore, in a deck system represented by an idealized orthotropic plate, the effective torsional rigidity is assumed to be equal to one-half of the torsional rigidity of the ribs per unit width.

Torsional Deformations In computing the effective torsional rigidity H, we commonly assume that the deck consists of the hollow ribs and the deck plate only, as shown in Figure 7-25*b*, and is not additionally stiffened between floor beams.

Certain flexural deformations and stresses in the deck plate are also imparted to the side walls and the bottom of the ribs, because the two systems are rigidly connected by welding. These flexural stresses in the direction transverse to the ribs are generally small and disregarded. They are, however, important in determining the effective torsional rigidity. Thus, this parameter in the transverse direction, $D_x = D_P$, enters into the analysis indirectly through a reduced effective torsional rigidity value.

Computation of Effective Torsional Rigidity This is defined as the torsional rigidity of an ideal system, free from secondary flexural deformations, in which the work of deformations due to torsion alone is equal to the work of deformations due to torsion and secondary flexure of the actual ribs. Thus, with deformations being inversely proportional to rigidity, the torsional rigidity of the ideal system, or the effective torsional rigidity of the ribs, is smaller than the rigidity value obtained without considering secondary flexural deformations. A reduction factor μ is thus applied to a theoretical rigidity value to obtain the effective torsional rigidity of the substitute model.

Effective torsional rigidity of an orthotropic plate representing a steel plate stiffened by closed ribs is

$$H = \frac{1}{2}\left(\frac{\mu GK}{a + e}\right) \tag{7-20b}$$

where H = effective torsional rigidity (kip-in.^{2}/in.)
G = elastic shear modulus of steel (kip/in.^{2})
K = section property characterizing torsional resistance (in.^{4})
a, e = dimension as shown in Table 7-1 and Figure 7-25*b* (in.)
μ = reduction coefficient (dimensionless)

The rib section property K is given by

$$K = \frac{4A^2}{(u/t_R) + (a/t_P)} \tag{7-20c}$$

where A = area enclosed by one closed rib (in.^{2})
u = developed width of one rib plate (in.)
t_R = rib thickness (in.)
t_P = deck plate thickness (in.)

Formulas for the reduction coefficient μ are given in the AISC manual (Wolchuk, 1963).

7-7 SOLUTIONS FOR STEEL PLATE DECK ON RIGID SUPPORTS

Deck with Open Ribs

The assumption $D_x = 0$ and $H = 0$, applicable to a deck with open ribs, yields

$$\frac{\partial^4 \omega}{\partial y^4} = \frac{P}{D_y} \tag{7-21}$$

which is the same as the deflection equations of a beam. Thus, the analysis of decks with open ribs reduces to a linear problem, and the ribs may be designed by continuous beam formulas.

Influence Lines Bending moments and reactions are computed from influence lines of a continuous beam. We recall that the influence line for the bending moment at any point of a continuous beam is defined as the deflection line of the beam due to a unit rotation at that point. Similarly, the influence line for the reaction at a support is the deflection line due to a unit deflection at the support where the reaction is sought. Typical influence lines are shown in Figure 7-26 (see also Section 2-2).

By referring to the notation of Figure 7-26, the bending moments follow the general expression

$$M = \frac{\eta}{s} sP \tag{7-22}$$

where η/s is the influence line ordinate for a unit span, expressed as a function of the ratio y/s. The bending moment acting on one rib, M_R, is given by

$$M_R = M_{\text{total}}\left(\frac{R}{P}\right) \tag{7-22a}$$

where M_{total} is computed from (7-22) and R/P is the ratio of the total load on the rib to the total wheel load (see also the following sections).

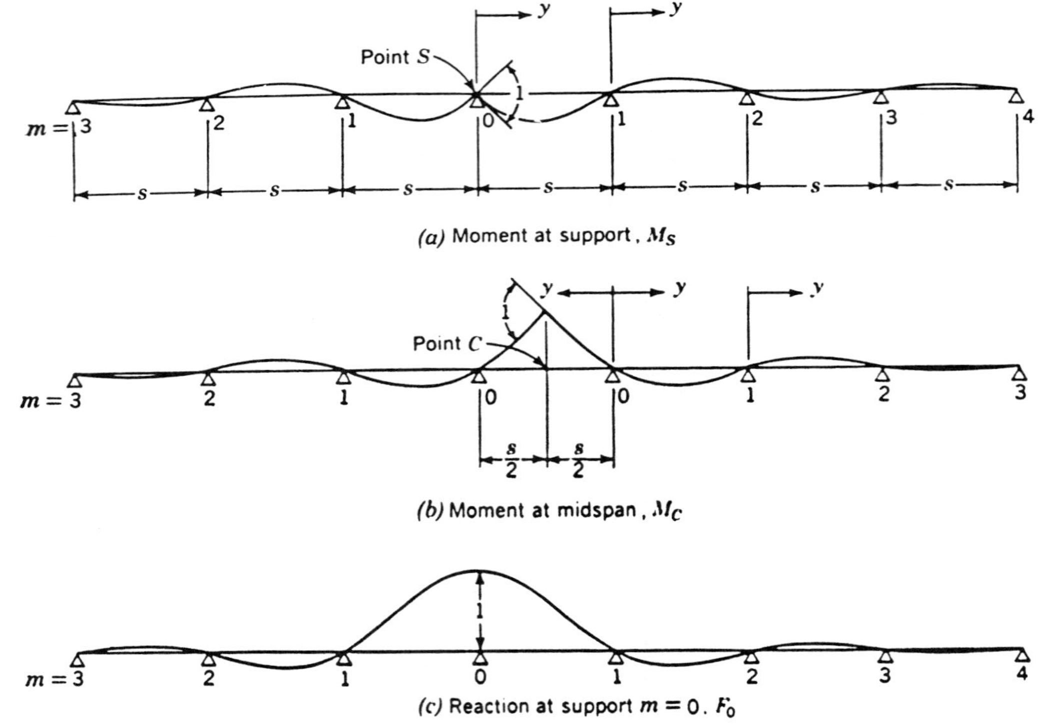

FIGURE 7-26 Influence lines for continuous beam on rigid supports.

Bending Moment at Supports For concentrated load, the equation of the influence line in panel 0–1 for the moment M_s at the support $m = 0$, Figure 7-26a, is

$$\left(\frac{M_s}{sP}\right)_{0-1} = -0.5\frac{y}{s} + 0.866\left(\frac{y}{s}\right)^2 - 0.366\left(\frac{y}{s}\right)^3 \tag{7-23}$$

For panels 1–2, 2–3, 3–4, ..., (m)–$(m+1)$, see Figure 7-27a, the support moment M_s at support m is given by

$$\left(\frac{M_s}{sP}\right)_m = \left[-0.5\frac{y}{s} + 0.866\left(\frac{y}{s}\right)^2 - 0.366\left(\frac{y}{s}\right)^3\right](-0.268)^m \tag{7-23a}$$

The maximum ordinate of the influence line occurs at $y/s = 0.38$, and at this location

$$\frac{M_s}{sP} = -0.085 \tag{7-24}$$

The support moment M_s due to a concentrated load P is computed directly from (7-22) through (7-24).

For a distributed load, Figure 7-27c, of length $2c$, the bending moment M_s needs to be computed only for the load located in panel 0–1. The maximum

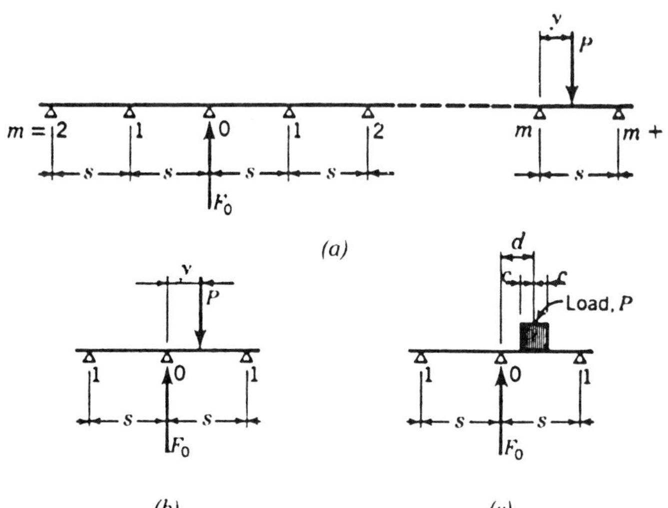

FIGURE 7-27 Loads used in computation of the moment M_s and reaction F_0 at support $m = 0$ of a continuous rib.

value is obtained at $d = 0.38s$, and is computed from

$$\left(\frac{M_s}{sP}\right)_{0.38} = -0.085 + 0.149\left(\frac{c}{s}\right)^2 \tag{7-25}$$

Bending Moment at Midspan For a concentrated load at any point, three cases are articulated.

1. For the bending moment M_c at midspan of panel 0–0, Figure 7-28a, and for a load in the same panel

$$\left(\frac{M_c}{sP}\right)_{0-0} = 0.183\frac{y}{s} + 0.317\left(\frac{y}{s}\right)^2 \tag{7-26}$$

where $2y < s$.

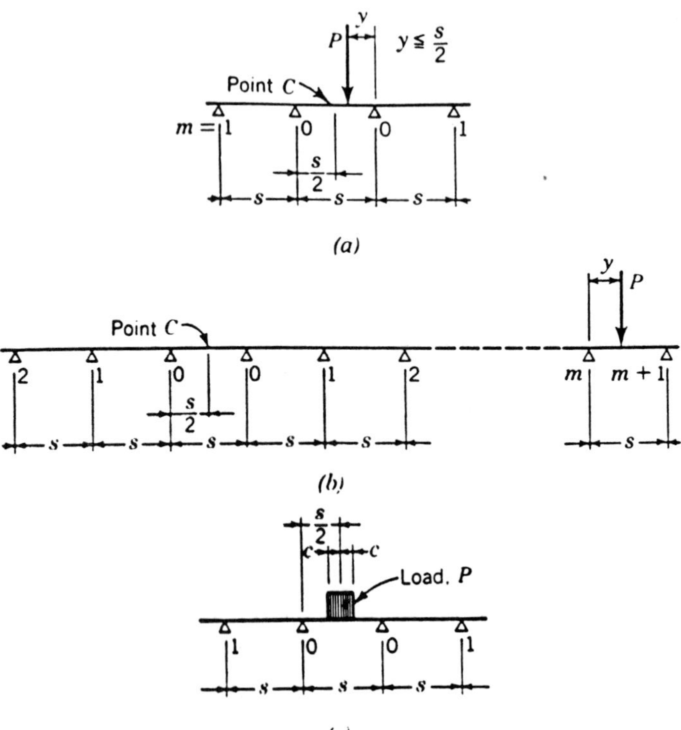

FIGURE 7-28 Loads used in the computation of the moment M_c at the midspan of panel 0–0 of a continuous rib.

2. The maximum span moment is for the load at midpoint of panel 0–0 as shown in Figure 7-28*b*, or $y = 0.5s$. At this location

$$\left(\frac{M_c}{sP}\right)_{0.5} = 0.171 \tag{7-27}$$

3. For the distant panels 0–1, 1–2, 2–3, and (m)–$(m+1)$, Figure 7-28*b*, the equation of the influence line is

$$\left(\frac{M_c}{sP}\right)_m = \left[-0.183\left(\frac{y}{s}\right) + 0.317\left(\frac{y}{s}\right)^2 - 0.134\left(\frac{y}{s}\right)^3\right](-0.268)^m \tag{7-28}$$

where m is the smaller of the two support numbers enclosing the panel under consideration.

For a distributed load, as shown in Figure 7-28*c*, located at midspan, the equation of the bending moment is

$$\frac{M_c}{sP} = 0.171 - 0.250\frac{c}{s} + 0.106\left(\frac{c}{s}\right)^2 \tag{7-29}$$

Reactions These forces must be computed at floor beam and rib locations.

Floor Beam Reactions on Ribs Usually, for computing floor beam reactions F on the ribs, it is sufficient to use concentrated loads P. Reactions are needed for the following cases.

1. The reaction F_0 at support $m = 0$ for the load in panel 0–1, Figure 7-27*b*, is

$$\frac{F_0}{P} = 1 - 2.196\left(\frac{y}{s}\right)^2 + 1.196\left(\frac{y}{s}\right)^3 \tag{7-30}$$

2. For a load at any panel (m)–$(m+1)$, as shown in Figure 7-27*a*, the equation of the influence line for F_0 is

$$\frac{F_0}{P} = \left[-0.804\frac{y}{s} + 1.392\left(\frac{y}{s}\right)^2 - 0.588\left(\frac{y}{s}\right)^3\right](-0.268)^{m-1} \tag{7-31}$$

where m is the smaller of the two support numbers enclosing the panel.

3. Reactions $F_1, F_2, \ldots, F_m$ at the distant floor beams due to a load at a specific position are calculated from (7-30) or (7-31).

Rib Reactions on Deck Plate By referring to Figure 7-15c, the reactions $R_0, R_1, \ldots, R_m$ are computed assuming the steel deck plate acts as a continuous isotropic member on rigid supports. Expressions are available for a distributed load P with a width smaller than the rib spacing. If the actual width of the wheel load exceeds the rib spacing, the rib reactions must be obtained by superposition. The following cases are considered.

1. Load in panel 0–1 in any position as shown in Figure 7-29*a*

$$R_0 = P\left[1 - 2.196\left(\frac{d}{a}\right)^2 + 1.196\left(\frac{d}{a}\right)^3 + \left(\frac{f}{a}\right)^2\left(-0.183 + 0.299\frac{d}{a}\right)\right] \tag{7-32}$$

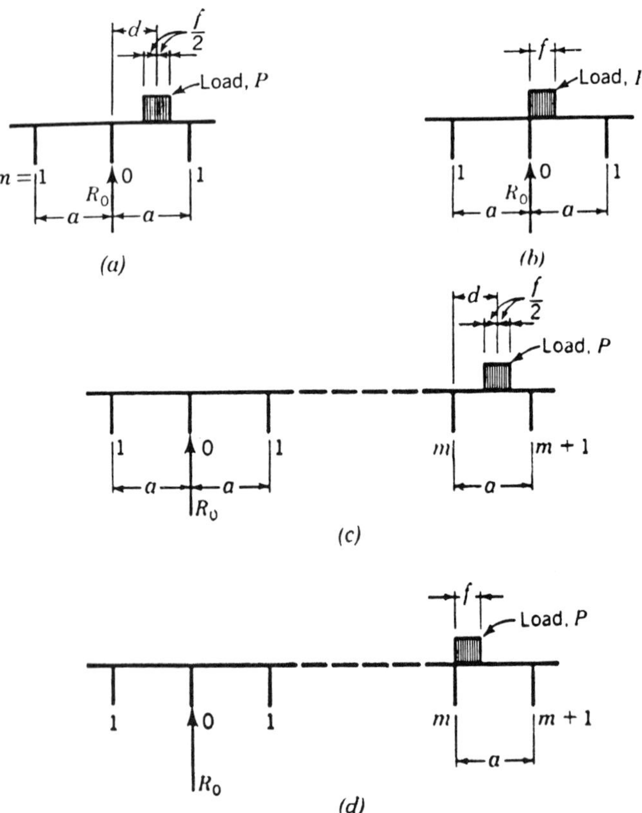

FIGURE 7-29 Loads used in the computation of the reaction of a rib.

2. Load in panel 0–1 at support 0 as shown in Figure 7-29*b*

$$R_0 = P\left[1 - 0.732\left(\frac{f}{a}\right)^2 + 0.299\left(\frac{f}{a}\right)^3\right] \qquad (7\text{-}33)$$

Influence Line Ordinates Influence line coefficients for bending moments at supports and midspan and for reactions at supports of a continuous beam on rigid supports are given in Table 7-3, and conform to the mathematical expressions of the foregoing sections.

For the purpose of expediting design computations for more complex problems, reference is also suggested to various charts contained in the AISC and Lincoln Foundation manuals (Wolchuk, 1963; Troitsky, 1967).

Effects of Deck Plate Rigidity In the foregoing analysis, the deck plate is assumed to act continuously, supported on rigid ribs. With longer rib spans or a thick deck plate, plate rigidity and rib flexibility on load distribution may have to be considered. This case may be treated as a continuous member on elastic supports, relieving the rib directly under the load and distributing the load to adjoining ribs, as shown in Figure 7-30.

Thus, if a more accurate analysis is necessary, the moment relief $\Delta' M_R$ in a rib due to rib flexibility may be computed and subtracted from the value of the moment computed for rigid ribs. The solution of this problem is found in several references such as the AISC manual (Wolchuk, 1963). For rule-of-thumb estimates, the moment reduction $\Delta' M_R$ due to rib flexibility will be less than 3 percent of the moment computed without flexibility if the relative rigidity (ratio of deck plate and rib rigidity) γ' does not exceed the following values:

With $a = 12$ in. $\quad \gamma' \leq 0.006$

With $a = 16$ in. $\quad \gamma' \leq 0.004$

Deck with Closed Ribs

Assuming $D_x = 0$, the differential equation of the orthotropic plate reduces to (7-14*a*). In deriving expressions for influence surfaces necessary to obtain bending moments in the ribs, only the homogeneous equation for the unloaded plate is needed. The equation corresponding to this condition is

$$D_y \frac{\partial^4 \omega}{\partial y^4} + 2H \frac{\partial^4 \omega}{\partial x^2 \partial y^2} = 0 \qquad (7\text{-}34)$$

TABLE 7-3 Influence Ordinates for a Beam on an Infinite Number of Rigid Supports

Point						0.5′	0.4′	0.3′	0.2′	0.1′	0.0
η_C/s-Moment at midspan						+0.1708	+0.1239	+0.0834	+0.0493	+0.0215	0
Point	0.0	0.1	0.2	0.3	0.4	0.5	0.6	0.7	0.8	0.9	1.00
η_C/s-Moment at midspan	0	−0.0153	−0.0250	−0.0300	−0.0311	−0.0290	−0.0246	−0.0187	−0.0121	−0.0056	0
η_S/s-Moment at support	0	−0.0471	−0.0683	−0.0819	−0.0849	−0.0792	−0.0673	−0.0512	−0.0331	−0.0514	0
ϑ-Reaction	+1	+0.9792	+0.9217	+0.8346	+0.7252	+0.6005	+0.4678	+0.3342	+0.2069	+0.0931	0
Point	1.0	1.1	1.2	1.3	1.4	1.5	1.6	1.7	1.8	1.9	2.0
η_C/s-Moment at midspan	0	+0.0041	+0.0067	+0.0080	+0.0083	+0.0078	+0.0066	+0.0050	+0.0033	+0.0015	0
η_S/s-Moment at support	0	+0.0112	+0.0183	+0.0220	+0.0227	+0.0212	+0.0180	+0.0137	+0.0089	+0.0041	0
ϑ-Reaction	0	−0.0671	−0.1098	−0.1317	−0.1364	−0.1274	−0.1082	−0.0823	−0.0553	−0.0247	0
Point	2.0	2.1	2.2	2.3	2.4	2.5	2.6	2.7	2.8	2.9	3.0
η_C/s-Moment at midspan	0	−0.0011	−0.0018	−0.0022	−0.0022	−0.0021	−0.0018	−0.0013	−0.0009	−0.0004	0
η_S/s-Moment at support	0	−0.0030	−0.0049	−0.0059	−0.0061	−0.0057	−0.0048	−0.0037	−0.0024	−0.0011	0
ϑ-Reaction	0	+0.0180	+0.0294	+0.0353	+0.0366	+0.0341	+0.0290	+0.0221	+0.0143	+0.0066	0

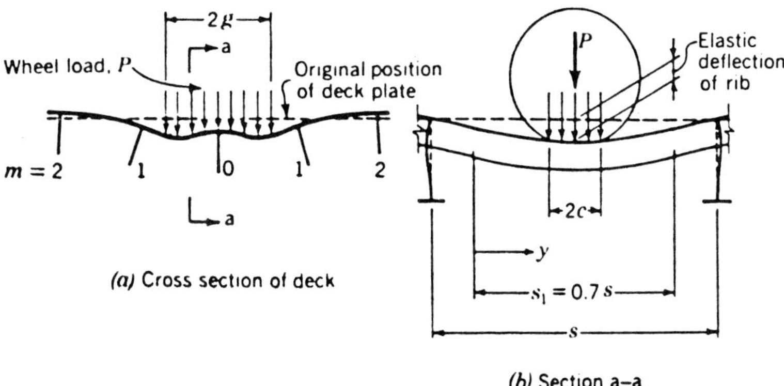

FIGURE 7-30 Longitudinal ribs acting as elastic supports of the deck plate.

The general solution of (7-34) may be obtained as an infinite series

$$\omega = \sum_{n=1}^{\infty} \omega_n \tag{7-35}$$

The actual deck loading is expressed through the Fourier analysis as a summation of sinusoidal component loads

$$Q_x = \sum_{n=1}^{\infty} Q_n \sin \frac{n\pi x}{b} \tag{7-36}$$

where Q_x is the load per unit width at location x, extending over the width b of the deck. The factor Q_n is represented by a Fourier series

$$Q_n = \frac{2Q_0}{l} \int_{x_1}^{x_2} \sin \frac{n\pi x}{l} dx \tag{7-36a}$$

where Q_0 is the actual load on the bridge deck per unit width, equal to $P/2g$, and l is the span length as shown in Figure 7-31 equal to the width of the bridge deck.

It can be shown that the bending moment in the y direction at any location x, of a chosen line y = constant, of a continuous plate due to a sinusoidal load in any position y is (see also Figure 7-32)

$$M_n = Q_n \sin \frac{n\pi x}{b} \eta_n \tag{7-37}$$

where the influence ordinate η_n is a function of the load position y and is

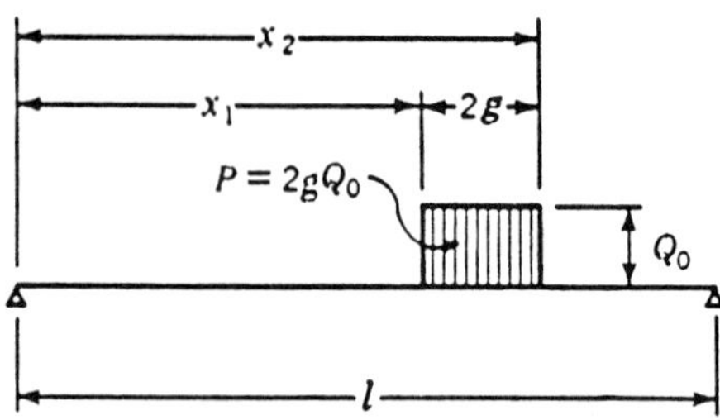

FIGURE 7-31 Load distribution for Fourier series.

not a function of x. The expression for η_n may be called the influence line for the moment in the plate.

The moment M_y per unit width of plate at any location x is given by

$$M = Q_0 s \sum_{n=1}^{\infty} \frac{Q_{nx}}{Q_0} \frac{\eta_n}{s} \tag{7-38}$$

and the moment acting on one rib of the actual steel plate deck is usually obtained by multiplying the moment per unit width, expressed by (7-38), by the width of the rib $(a + e)$, or

$$M_R = M(a + e) \tag{7-39}$$

However, a more accurate value may be obtained by integrating the moment distribution curve over the width of the rib. For design purposes, the bending moments are required at the supports (moment M_s in Figure 7-32) and at midspan between supports (moment M_c in Figure 7-33).

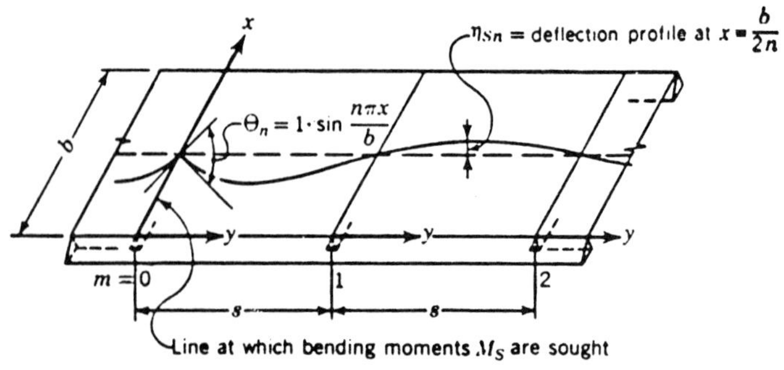

FIGURE 7-32 Bending deflection of deck due to actual loads, support moment.

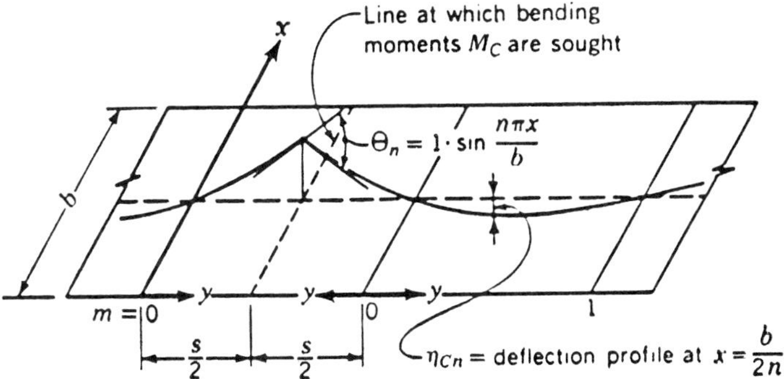

FIGURE 7-33 Deflection of deck due to actual loads, midspan moment.

Bending Moment at Supports The influence line η_{sn} for a moment at support $m = 0$ is obtained as the profile at the location $x = b/2n$ of the deflection surface caused by a unit rotation $1 \cdot [\sin(n\pi x/b)]$ at the support $m = 0$, as shown in Figure 7-32. The moment at support 0, $M_{0n} = M_{0n} \sin(n\pi x/b)$, necessary to produce a unit rotation, is

$$M_{0n} = \frac{\kappa}{(1 - \kappa^2)} \frac{D_y \alpha^2 s}{\alpha^*} \sin \frac{n\pi x}{b} \tag{7-40}$$

and the corresponding moment at support 1 is

$$M_{1n} = \kappa M_{0n} \tag{7-41}$$

With these moments applied at the edges of the plate panel 0–1, the four boundary conditions are

$$y = 0 \qquad \omega = 0 \qquad M = M_{0n}$$
$$y = s \qquad \omega = 0 \qquad M = M_{1n}$$

from which the integration constants in the equation of the deflection line, η_{sn}, are computed. The dimensionless influence ordinate η_{sn}/s is given as

$$\frac{\eta_{sn}}{s} = \frac{M_0^*}{s} \kappa^m \left(C_1 \sinh \alpha y + C_2 \operatorname{Cosh} \alpha y + C_3 \frac{y}{s} + C_4 \right) \tag{7-42}$$

where

$$C_1 = \frac{-\kappa + \cosh \alpha s}{\sinh \alpha s} \qquad C_2 = -1, \qquad C_3 = \kappa - 1 \qquad C_4 = +1 \tag{7-43}$$

and

$$\frac{M_0^*}{s} = \frac{1}{\alpha^*} \frac{\kappa}{(1 - \kappa^2)} \tag{7-44}$$

The parameters α, α^*, and κ are defined as follows:

$$\alpha = \frac{n\pi}{b}\sqrt{\frac{2H}{D_y}} \qquad \alpha^* = 1 - \frac{\alpha s}{\sinh \alpha s} \qquad \kappa = -k + \sqrt{k^2 - 1} \tag{7-45}$$

The bending moment M_s at the support per unit width of the plate, at the location x, is computed by (7-38) by substituting $\eta_n = \eta_{sn}$ from (7-42). It can be seen that the process is tedious, and shortcuts applicable to the determination of M_s are available.

Bending Moment at Midspan By referring to Figure 7-33, the influence line η_{cn} for the moment M_c at midspan of panel 0–0 is obtained as a profile at the location $x = b/2n$ of the deflection surface caused by a unit rotation $1 \cdot [\sin(n\pi x)/b]$, applied at the midspan as shown. The boundary conditions at the edges $y = 0$ and $y = s/2$ are

$$\begin{array}{lll} y = 0 & \omega = 0 & M = M_{0n} \\ y = s/2 & \omega = \frac{1}{2}\sin(n\pi x/b) & V^* = 0 \end{array}$$

where V^* is the substitute shear representing the combined shear and the twisting moment that have to be equal to zero at $y = s/2$. From these boundary conditions, the four integration constants in the expression for the deflection line, η_{cn}, in panel 0–0 are computed.

For a load in panel 0–0 as shown in Figure 7-28a, the dimensionless influence ordinate η_{cn}/s may be expressed as

$$\frac{\eta_{cn}}{s} = \frac{\sinh \alpha y}{2\alpha s \cosh(\alpha s/2)} + \frac{M_0^*}{s}(C_1 \sinh \alpha y + C_2 \cosh \alpha y + C_3 \alpha y + C_4 \tag{7-46}$$

with $y < s/2$ and where

$$C_1 = \tanh(\alpha s/2) \qquad C_2 = -1 \qquad C_3 = 0 \qquad C_4 = +1 \tag{7-47}$$

and

$$\frac{M_0^*}{s} = \frac{\kappa}{\alpha^*(1 - \kappa)} \frac{1}{2\cosh(\alpha s/2)} \tag{7-48}$$

and the parameters α, α^*, and κ are as defined by (7-45).

The bending moment M_c at midspan per unit width of the plate is likewise computed from (7-38) by substituting $\eta_n = \eta_{cn}$ from (7-46).

Shortcut Formulas for Numerical Solution Several references suggested for orthotropic solutions, such as the AISC and the Lincoln Foundation manuals, provide simplified formulas and charts for designing decks with open or closed ribs and AASHTO loads. Usually, the bending moments are represented as functions of two parameters, the span s and the rigidity ratio H/D_y of the deck system.

The spans of closed ribs in most existing bridges do not exceed 8 ft. Because comparative studies indicate that longer rib spans are economically feasible, the charts are extended to include spans $s = 25$ ft.

The rigidity ratio H/D_y of the deck without additional stiffening diaphragms between floor beams usually ranges from very small values to about 0.15, with the usual range 0.04 to 0.08. Charts are available, however, for $H/D_y = 0.6$ to demonstrate the possible effect of large torsional rigidity (this may be achieved by transverse stiffening members) on the load distribution. With H/D_y ratios beyond 0.2, the corresponding decrease in bending moments is insignificant.

7-8 EFFECTS OF FLEXIBLE FLOOR BEAMS

Basic Considerations

The influence of floor beam flexibility may be considered assuming the deck has complete continuity. In this case the flexural rigidity of the stiffened plate is uniformly distributed across the width of the bridge.

In most instances the stiffness of the floor beams inhibit large-deflection curvatures when subjected to loading. As a result, the torsional and bending moments originating in the transverse direction produce small additional stresses in the orthotropic deck and can be disregarded. In this context, the orthotropic deck plate can be separated into infinitely small longitudinal strips as continuous torsionless beams supported by the elastic floor beams.

The longitudinal flexural rigidities of these strips become quite relevant when the variations in deflection of interacting floor beams are significant. This can occur particularly where the floor beam spans are large, irrespective of the small curvature.

Flexibility of Floor Beam In the first stage of the Pelikan–Esslinger analysis, the bending moments of a steel plate deck are computed assuming rigid or unyielding floor beams. In the actual bridge, however, the deck will deflect under load, and any differential deflection of the adjoining floor beams will cause a redistribution of the bending moments. A load placed directly on a floor beam will cause deflection which, distributed to adjoining beams not directly loaded, will cause some relief in the loaded beam.

The deflection of the floor beams will also have an effect on the ribs. This effect is an increase in positive moments under the load in the midspan of the ribs, and a decrease in negative moments where the ribs are supported by the floor beams. These effects are maximized at the midspan of the bridge deck between the main girders, where the floor beam deflections are maximum, and almost negligible near the main girders, where the floor beam deflection is small.

The Condition of Elastic Supports For a continuous beam on elastic supports, the reaction $\overline{F}_m$ at any support m is proportional to the vertical movement δ_m at the support, or

$$\overline{F}_m = \delta_m k_m \tag{7-49}$$

where k_m is the elastic spring constant of the support expressing the resistance of the support to vertical movement. Bending moments and reactions in a loaded continuous beam depend on a dimensionless parameter γ expressing the ratio of beam stiffness to support rigidity.

Typical influence lines $\overline{\eta}$ and $\overline{\theta}$ of a continuous beam on elastic supports, with equal spans and uniform beam rigidity and uniform elastic constant at all supports, are shown and compared with similar lines η and θ of a beam on rigid supports in Figure 7-34. The influence ordinates for a continuous beam on elastic supports are likewise computed by conventional analytical techniques.

In estimating the effects of floor beam flexibility, the influence line ordinates $\overline{\eta}_s$, $\overline{\eta}_c$, and $\overline{\theta}_0$ for the bending moments at support, midspan, and support reaction, respectively, are required only at the supports. Charts are available in the AISC manual giving graphical solutions for various values of the rigidity coefficient γ.

The Elastic Support Theory in the Analysis of Steel Plate Decks Continuous over Floor Beams of Uniform Rigidity

The parallel strips discussed in the preceding section may be treated as continuous beams on elastic supports if the reactions of any support satisfy (7-49). This is possible only if the loading is distributed over the deck width in such a manner that the resulting floor beam deflections at each point are proportional to the plate strip reactions at the same points.

For floor beams simply supported at the main girders, this condition is satisfied by a sinusoidal loading extending over the entire deck width, causing sinusoidal reactions at any floor beam, which in turn cause sinusoidal deflections of the floor beams. Therefore, in considering floor beam flexibility, the actual load on the deck is represented by a Fourier series consisting of sinusoidal component loads.

For design purposes, it is generally sufficient to include only the first sinusoidal component load $Q_{1x} = Q_1 \sin \pi x/l$ as shown in Figure 7-35. In computing the effects of floor beam flexibility, each plate strip of unit width at location x is loaded by the corresponding loading at the same location, Q_{1x}, and treated as a beam on elastic supports.

Formulas for the dimensionless loading coefficients, Q_{1x}/Q_0, are given in appropriate manuals (AISC or Lincoln Foundation) for several loading cases:

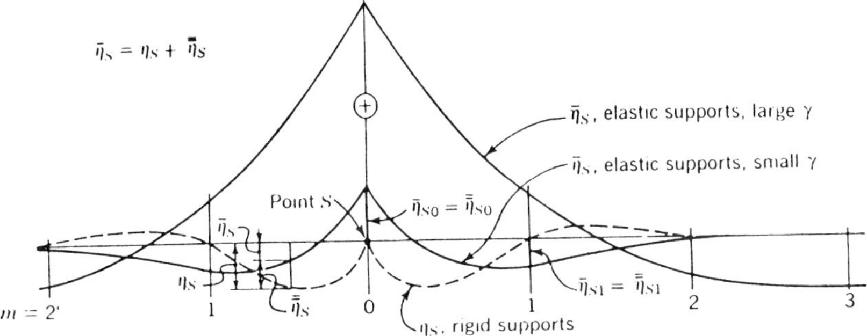

(a) Influence lines η_S and $\bar{\eta}_S$ for the bending moment at support 0 (point S)

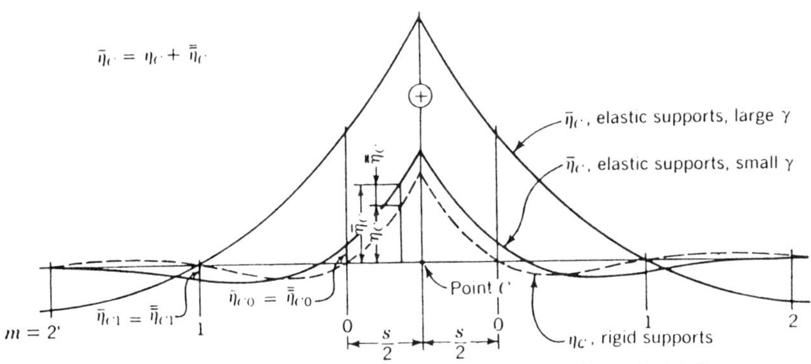

(b) Influence lines η_C and $\bar{\eta}_C$ for the bending moment at midspan (point C)

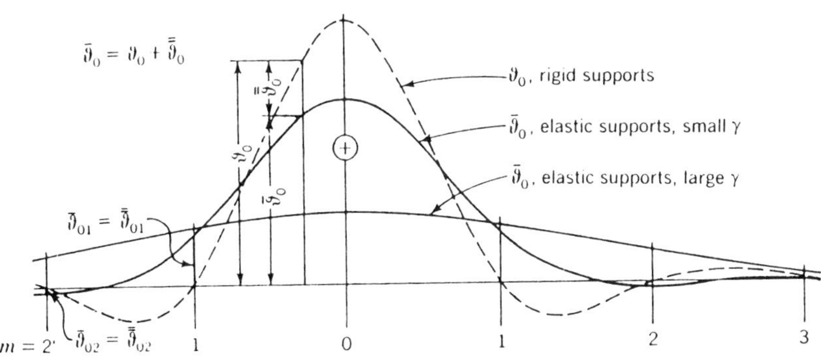

(c) Influence lines ϑ_0 and $\bar{\vartheta}_0$ for the reaction at support 0

FIGURE 7-34 Typical influence lines for continuous beams on elastic and on rigid supports.

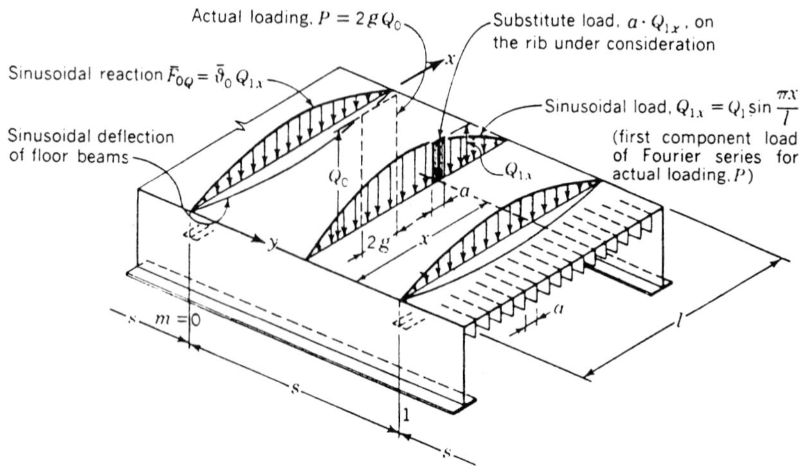

FIGURE 7-35 Sinusoidal loading on the deck, causing sinusoidal reactions and deflections of the floor beams.

(a) one wheel load at the center of the plate strip considered, (b) one axle symmetrically placed, (c) axles symmetrically placed, (d) one axle in any position, (e) two axles in any position, and (f) many axles in any position.

Relative Rigidity Coefficient, γ For the system shown in Figure 7-35, the floor beams act as elastic supports for the individual plate strips. For a sinusoidal variation of reactions and deflections, the spring constant k as defined by (7-49) is

$$k = \bar{F}/\delta \tag{7-49a}$$

and has the same value at all points along the floor beam.

It can be shown that the value of the spring constant is

$$k_n = \frac{n^4 \pi^d EI_F}{l^4} \tag{7-50}$$

where EI_F is the floor beam rigidity and l is the floor beam span.

The relative rigidity coefficient γ_n is the ratio of the stiffness of the deck strip to the spring constant of the support k_n. The expressions for γ_n are as follows:

Deck with open ribs $$\gamma_n = \frac{EI_R/s^3 a}{k_n} \tag{7-51}$$

Deck with closed ribs $$\gamma_n = \frac{EI_R/s^3(a + e)}{k_n} \tag{7-51a}$$

It follows that the relative rigidity coefficient γ of a deck system is not constant for all loading cases, but depends on the number n of the sinusoidal component loading applied. Using only the first component load ($n = 1$), the relative rigidity coefficients are

Deck with open ribs
$$\gamma = \frac{l^4 I_R}{a s^3 \pi^4 I_F} \tag{7-52}$$

Deck with closed ribs
$$\gamma = \frac{l^4 I_R}{(a + e) s^3 \pi^4 I_F} \tag{7-52a}$$

In the foregoing equations, I_F and I_R denote the moment of inertia of the floor beam and the rib, respectively. If the beam has a variable depth, an equivalent uniform moment of inertia may be used.

Bending Moment Corrections due to Floor Beam Flexibility From Figures 7-34a and b, we can see that the bending moment at any point consists of two contributions. The first is caused by loads P acting on a system with unyielding supports, and corresponds to the influence ordinate η. The second is due to the vertical movement of the supports and corresponds to the difference $\bar{\bar{\eta}} = \bar{\eta} - \eta$, where $\bar{\eta}$ is the influence ordinate of the beam on elastic supports. Therefore, the influence ordinate $\bar{\bar{\eta}}$ expresses the effect of support elasticity only.

Moment Increase in Ribs The additional moment ΔM in the ribs at any point i of the beam, due to a vertical movement of the supports under a load or a group of loads P, is given by

$$\Delta M = P\bar{\bar{\eta}} = \sum_{m=0}^{\infty} F_m \bar{\eta}_{im} \tag{7-53}$$

where F_m is the reaction at support m of a continuous beam on rigid supports due to the same loads P and $\bar{\eta}_{im}$ is the influence ordinate at support m for the bending moment at point i under consideration, of a continuous beam on elastic supports.

Application of this formula to a bridge deck supported on flexible beams implies that the actual loads P are replaced by their first sinusoidal component loads Q_{1x} as shown in Figure 5-35. In this case the additional bending moment ΔM_R of the ribs due to floor beam deflection is given by the following:

Deck with open ribs
$$\Delta M_R = A_0 s a \frac{Q_{1x}}{Q_0} \sum \frac{F_m}{P} \frac{\bar{\eta}_{im}}{s} \tag{7-54}$$

Deck with closed ribs
$$\Delta M_R - Q_0 (a + e) \frac{Q_{1x}}{Q_0} \sum \frac{F_m}{P} \frac{\bar{\eta}_{im}}{s} \tag{7-54a}$$

where $Q_0 = P/2g$ = wheel load per inch of deck width (Figure 7-31)
Q_{1x} = value at the rib under consideration (location x) of the first sinusoidal component load of the actual load P

and other parameters are as given previously.

The additional moment ΔM_R usually has a positive value, meaning an increase in midspan moment M_c and a decrease in support moment M_s. Where several traffic lanes may be loaded simultaneously, the wheel loads and truck positions may be different in the lane where the moment is computed and in the adjoining lanes. In such cases the total moment change in the ribs is the sum of the effects of the loads in the lanes involved, subject to a load reduction coefficient for multiple lane loading.

Moment Relief ΔM_F in Floor Beams From the influence lines of Figure 7-34*c*, it follows that a load P acting near a floor beam on a deck supported by flexible floor beams induces a reaction that is smaller than in a similar deck with rigid beams (unyielding supports).

Likewise, if only the first term of the series is considered, the expression for ΔM_F is

$$\Delta M_F = \left[\frac{F_0}{P} - \frac{\bar{F}_0}{P}\right] Q_1 \left(\frac{l}{\pi}\right)^2 \sin\frac{\pi x}{l} \tag{7-55}$$

where the ratios F_0/P and $\bar{F}_0/P$ are the influence ordinates at the locations of the individual loads for the reaction at support $m = 0$ for the continuous beam on rigid and elastic supports, respectively.

For a load or group of loads P acting at any position on the deck, the ratio $\bar{F}_0/P$ is computed from

$$\frac{\bar{F}_0}{P} = \sum_{m=0}^{\infty} \frac{F_m}{P} \bar{\theta}_{0m} \tag{7-56}$$

where $\bar{\theta}_{0m}$ is the influence ordinate at support m for the reaction at support $m = 0$ of a beam on elastic supports.

Shear Corrections due to Floor Beam Flexibility By analogy, floor beam flexibility will affect shear in both the ribs and the floor beams.

Shears in Ribs Because in general the rib reactions in the elastic system are smaller, shear corrections ΔV_R may be computed from the same considerations. In most cases, however, the shear stresses in the ribs are low and unlikely to control the design; hence, adjustments in shear values will not be necessary.

Shears in Floor Beams The expression for the shear relief ΔV_F in the floor beams can be found as for the moment relief ΔM_F. Thus, the final design shear in the floor beam is

$$\bar{V}_F = V_F - \Delta V_F \tag{7-57}$$

The shear V_F for the rigid floor beam due to loads P on the deck is obtained from the influence line for shear at the point under consideration with the values of deck reactions F_0, or

$$V_F = \sum F_0 \theta_v \tag{7-58}$$

where θ_v is the influence ordinate for shear in the rigid beam.

In developing the influence line for the final shear $\bar{V}_F$ in the elastic floor beam, the loads P are expressed in terms of their reactions F_0 on the floor beams as

$$P = \frac{F_0}{(F_p/P)} \tag{7-59}$$

It can also be shown that the shear relief ΔV_{FP} due to a concentrated load P at location d at the floor beam is

$$\Delta V_{FP} = F_0 \frac{2}{\pi} \left[1 - \frac{\Sigma (F_m/P) \bar{\theta}_{0m}}{F_0/P} \right] \cos \frac{\pi d}{l} \sin \frac{\pi x}{l} \tag{7-60}$$

These values are entered into (7-57) to obtain $\bar{V}_F$.

The influence ordinates for shear in an elastic floor beam, given by the foregoing equations, are shown graphically in Figure 7-36, and are self-explanatory.

Use of Charts Various charts are available for determining the effects of beam flexibility and for AASHTO loads. They may be found in the AISC and Lincoln Foundation manuals with appropriate comments explaining their use.

Miscellaneous design aids are also contained in the same references to help with the analysis of bridge decks with floor beams of nonuniform rigidity, and for bridge decks with floor beams elastically restrained at the main girders or continuous over more than two girders.

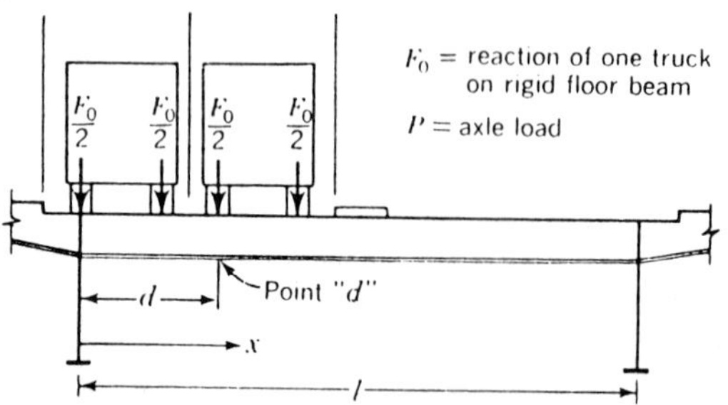

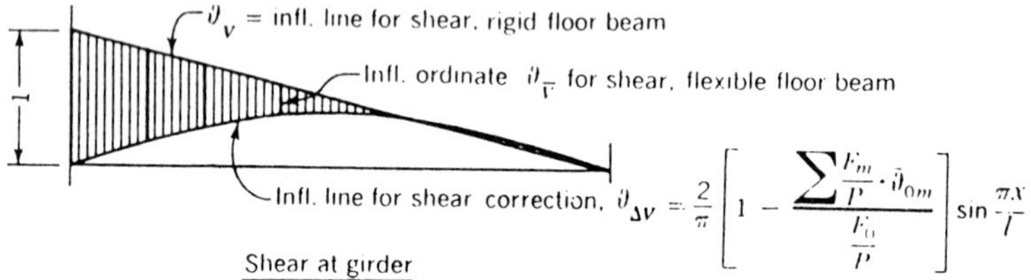

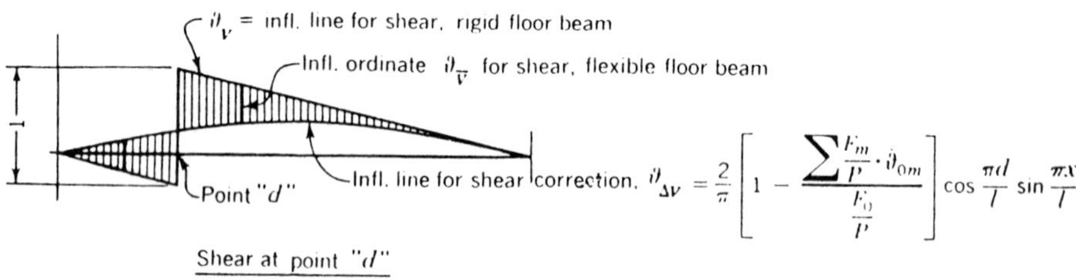

FIGURE 7-36 Influence lines for shears in a flexible floor beam.

7-9 DESIGN CONSIDERATIONS OF THE PLATE DECK

The deck of an orthotropic steel deck bridge, acting locally to transmit the wheel loads to the longitudinal ribs, behaves as a continuous isotropic plate in the range of design loads. When the loads, however, exceed the design range, membrane stresses occur in the plate, and the associated stress increments are no longer proportional to the applied loads.

Tests show that the ultimate capacity of the deck plate, near which the plate responds as a membrane, is quite large and of the order of 15 to 20 times the ultimate load computed from ordinary flexural theory. In this respect, the bridge deck plate possesses sufficient structural capacity to withstand overload provisions.

Because the relationship between applied loads and stresses in the higher load range is nonlinear, the local stress in the plate under design loads cannot be considered a measure of actual deck strength. From these remarks, it follows that computation of local stresses in the deck plate is not always warranted.

The fatigue strength of a deck plate under the effect of dynamic loads or stress reversal has been found to be considerable. For design purposes, therefore, fatigue of low-alloy steel deck plates of the usual proportions and subjected to standard wheel loads is not a critical factor in the design, and should be checked only in extreme cases. The minimum deck plate thickness may thus be based on deflection limitations, or may be dictated by longevity characteristics related to deterioration, wear, and rust.

Structural Behavior of Plate as an Independent Element

Deck Plate Subjected to Small Loads Tests on models and actual deck plates subjected to small loads of the order of standard truck wheel loads confirm an isotropic plate behavior. These results, however, are only approximate because in measuring local stresses it is not always possible to isolate the effects of plate action as a flange of the ribs, floor beams, and main girders.

Tests on decks with open ribs are reported by Klöppel (1951, 1958), Klöppel and Roos (1960), and Radojkovic (1958), showing that the stresses in the deck plate over the rib drop sharply because of the strengthening of the plate by the weld and the rib, so that the maximum stress occurs at the toe of the weld. Stress measurements in decks with closed ribs are reported by Pelikan and Esslinger (1957), and show that under usual working loads the measured stresses in the deck plate and the ribs in the transverse direction are in good agreement with the results obtained by analytical methods.

Analytical Stress Determination Under Small Loads A rigorous analysis of the deck plate under wheel loads is complicated for the following reasons: (a) the plate support conditions are not "knife edges" because they have a definite width and restrain rotation; (b) the exact magnitude and distribution of stresses over the supports, where the deck is stiffened by welds, is difficult to determine analytically; (c) small membrane stresses may occur even at design loads, and thus an analysis based on first-order theory is not representative; (d) the ribs supporting the deck plate are not completely rigid, and this flexibility tends to increase plate stresses between ribs and decrease plate stresses over the ribs; and (e) in actual bridge decks a composite action between the steel plate and wearing surface tends to reduce considerably stresses and deflections.

Deck Plate on Open Ribs For the wheel load P distributed over an area $2g \times 2c$ as shown in Figure 7-37, the bending moments and stresses will occur predominantly in the x direction, and may be computed by continuous beam formulas. In reality, the plate extends in the y direction beyond the width of the applied load. Thus, the critical maximum moment in the x direction per unit width will be smaller in the full plate. The ratio of moment M_P of the full plate to the corresponding moment M_B in the continuous

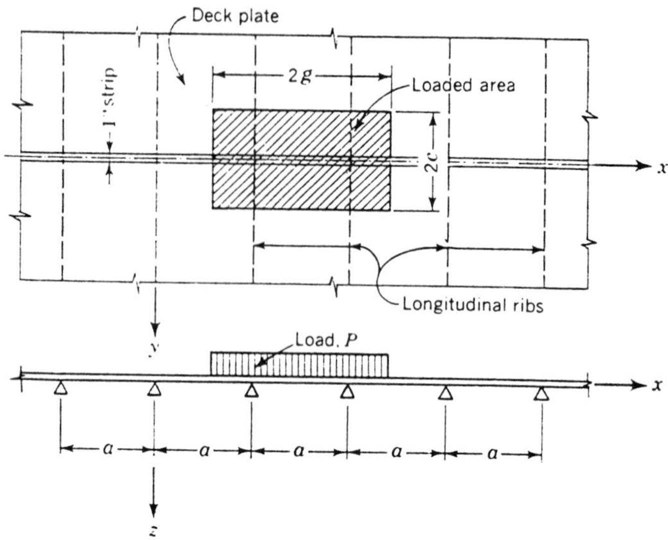

FIGURE 7.37 Deck plate supported on open ribs treated as a continuous plate on rigid knife-edge supports.

beam per unit width (usually 1 in.) is defined as the plate factor ψ, or

$$\psi = M_P/M_B \tag{7-61}$$

The plate factor ψ (< 1) is independent of the plate thickness and varies with the ratio of the loading width to the plate span, $2c/a$, the load distribution in the x direction, and the location of the bending moment. Values of ψ for moments at midspan and over the supports have been determined by Pelikan and Esslinger (1957) and are shown in Figure 7-38.

The effects of elastic flexibility in the ribs are small and may be disregarded in most cases. If deflections must be computed, the 1-in.-wide plate strip can be first treated as a simple beam, and then they analyzed by applying to it the moments computed for the continuous plate.

Deck Plate on Closed Ribs In this case the stresses consist of two components: (a) stresses due to direct loading of the deck plate supported on rigid supports, as shown in Figure 7-39a; and (b) stresses due to load transfer from the directly loaded rib to the adjoining ribs in the actual deck acting as an orthotropic plate, as shown in Figure 7-40a.

For stresses due to direct loading, the system is first simplified by replacing the rigid connections with knife-edge supports assuming a uniform spacing a for these supports, as shown in Figure 7-39b. Using a symmetrical loading is a further simplification. The structure is first treated as a 1-in.-wide frame,

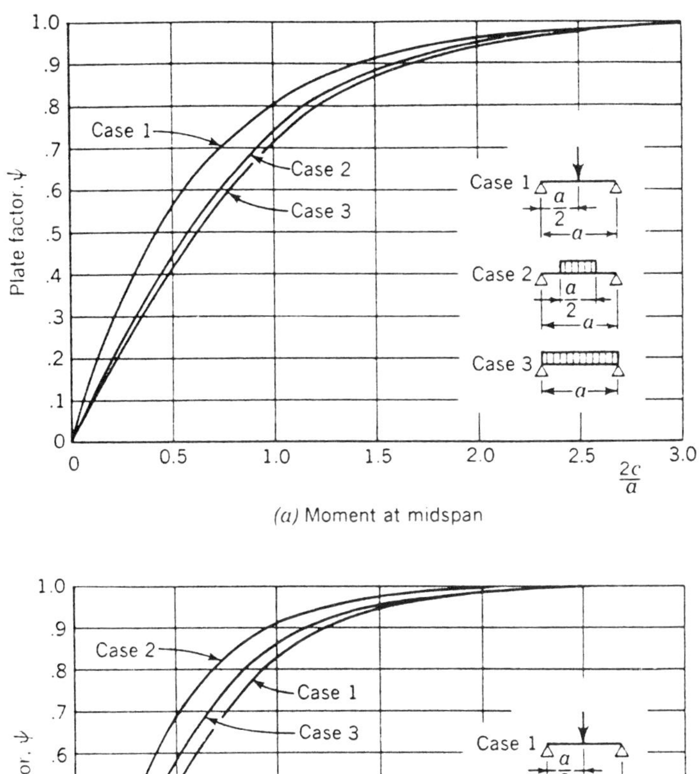

(a) Moment at midspan

(b) Moment at support

Notes: 2c = length of load in direction of supports.
Plate moment = ψ·moment in a beam.

FIGURE 7.38 Plate factors ψ for moments in continuous isotropic plate. (From Pelikan and Esslinger, 1957.)

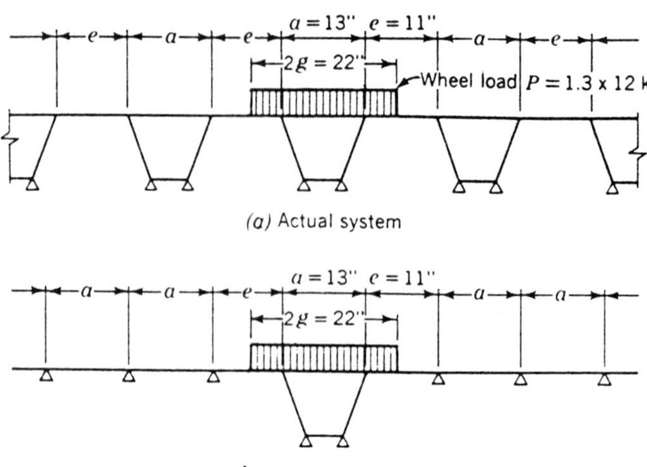

(a) Actual system

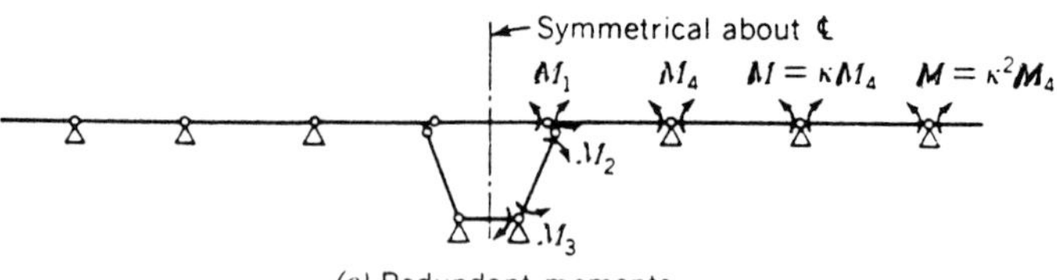

(b) Simplified system

Symmetrical about ℄
M1 M4 M = κM4 M = κ²M4
M2
M3

(c) Redundant moments

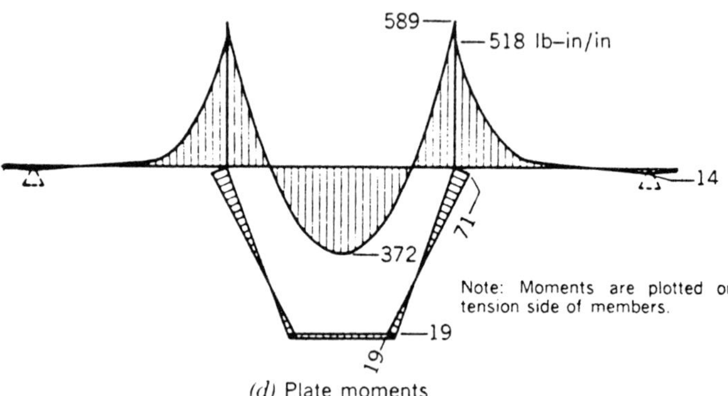

(d) Plate moments

FIGURE 7.39 Computation of the bending moments in a deck plate supported by closed ribs. Moments due to direct loading.

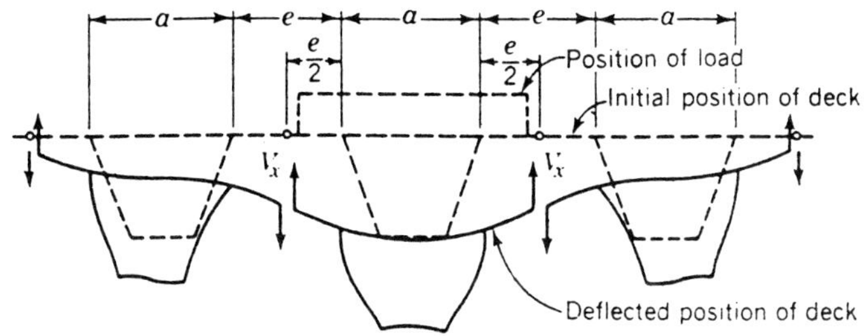

(a) Deformations due to transverse shear transfer

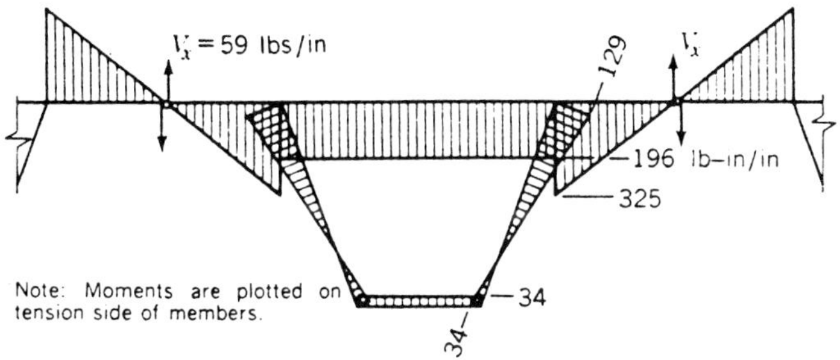

(b) Bending moments due to $V_x' = 59$ lb/in

FIGURE 7.40 Computation of the bending moments in a deck plate supported by closed ribs. Moments due to transverse shear transfer.

and the moments in the directly loaded rib are obtained from frame analysis applying appropriate plate factors. For the loading case shown, the number of redundants is four, as shown in Figure 7-39*c*. These redundant moments in the unloaded portions of the plate may be expressed as a function of M_4 and the carry-over coefficient κ as shown. A typical moment diagram, with the moment values modified by the appropriate plate factors, is shown in Figure 7-39*d*.

For stresses due to transverse shear transfer, the associated shear force V_x is transmitted by the deck plate acting in flexure, as shown in Figure 7-40*a*, from the directly loaded rib. Its magnitude may be computed considering the deck plate and the ribs as an idealized orthotropic plate with continuously distributed elastic properties. Shears computed in the idealized system, however, are valid only at the midpoint between the ribs because the actual deck is discontinuous, with the smallest element having a width $a + e$. Thus, the shear force V_x is computed from the general expression of shear in an orthotropic plate, and is given by

$$V_x = \frac{8Q_0 s}{b}\frac{H}{D_y}\sum\frac{M_{cn}}{Qs}\sin\frac{n\pi g}{b}\sin\frac{n\pi}{2}\cos\frac{n\pi x}{b} \tag{7-62}$$

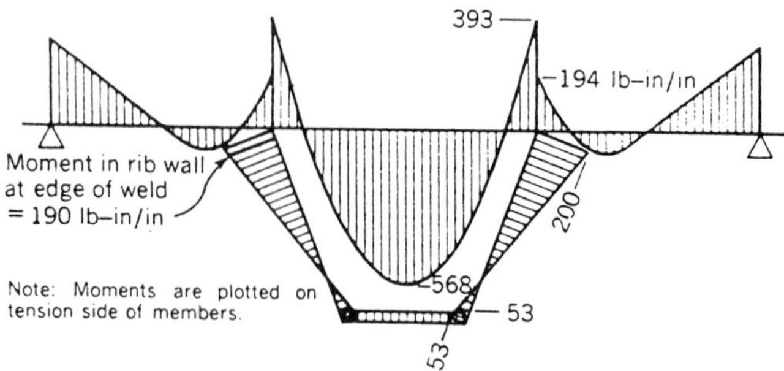

FIGURE 7.41 Computation of the bending moments in a deck plate supported by closed ribs. Total bending moments in deck plate in x direction, obtained by superposition of the effects of direct loading and transverse shear transfer (Figures 7-39 and 7-40).

with the same notation as before. This expression is used to compute the shear V_x for the rib directly loaded, and at location $x = (b/2) + (a + e)/2$. With the shear V_x calculated, the bending moments in the deck plate and the rib walls are determined as shown in Figure 7-40b.

The total local moments and stresses in the deck plate are now obtained by superimposing the two effects: stresses due to direct loading and stresses from transverse shear transfer, as shown in Figure 7-41. For plate stresses over the ribs, the moments may be reduced to the values at the face of the weld as in the case of open ribs.

Deck Plate Subjected to Large Loads "Large loads" are those under which stresses in the deck plate computed according to plate theory of the first order exceed the normally allowable limit. Tests on steel deck plates show that under increasing loads the elastic behavior extends beyond the predictions of ordinary flexural theory, and this is evidenced by the absence of permanent deformations upon unloading (Klöppel, 1951, 1958; Lyse and Madsen, 1938). As the loads and deflections increase, membrane stresses develop in the plate and cause horizontal reactions at supports and internal compressive stresses in the portions of the plate adjoining the loaded area. At higher loads, plastic hinges form over the supports, and the flexural stresses are replaced almost entirely by membrane stresses. In spite of the plastic conditions, the deflections are still limited because of the restraining effect of the adjoining unloaded portions of the plate. In the final loading stage, strain hardening effects appear, and thus the full tensile strength of the material is reached before rupture occurs at ultimate load.

Theoretical and experimental investigations show that stresses and deflections of a steel plate in combined flexure and tension are influenced by axial prestress and initial dishing of the plate.

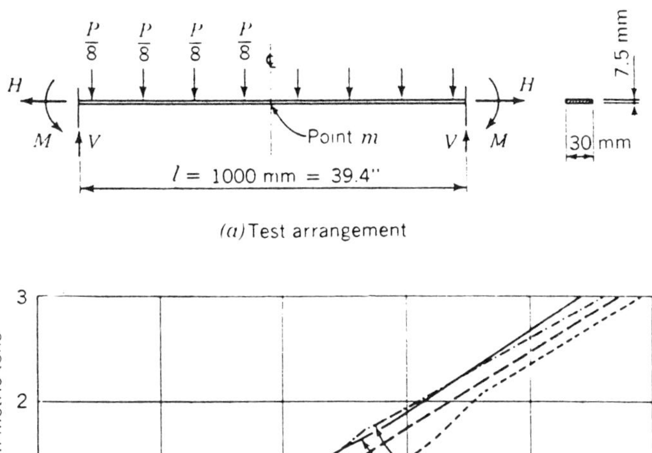

(b) Load-deflection diagram

FIGURE 7.42 Effect of axial prestressing on deflections of a slender bar. (From Klöppel, 1958.)

The effect of a tensile prestressing force H on the deflection of a loaded slender flat bar is shown in Figure 7-42, and evidently the prestressing markedly reduces deflections under vertical loads. The presence of prestressing in a steel plate deck may be the result of transverse contraction of the fillet weld connecting the plate and the ribs.

Initial dishing of the deck plate between the ribs tends to reduce the stresses and deflections. Even a small initial deflection of less than 1/300 of the span can markedly decrease the deflection under an applied load. An initial dishing of the order of 5 percent of the span stiffens the system considerably. Unintended dishing of the deck plate may occur following warpage of the deck system due to weld shrinkage.

Fatigue Strength of Deck Plate Results of early fatigue tests on steel deck plates are shown in Table 7-4 (Klöppel and Roos, 1960). Comparison of the test loads with actual wheel loads may be made using the ratio of the upper values of the unit pressure in the tests and the pressure under a 12-kip wheel, plus impact. However, using a loaded area 22 × 12 in. and all other

TABLE 7-4 Fatigue Tests on Steel Deck Plate

Test No.	Unit Pressure of Applied Pulsating Loading (psi) From	To	No. of Load Cycles	Remarks
1	18	195	2.2×10^6	Test discontinued with no crack apparent
2	18	311	6.75×10^6	Test made after completion of test 1 on the same specimen. Visible crack at 6.75×10^6 cycles
3a	80	240	5×10^6	Test discontinued with no crack apparent
3b	80	300	5×10^6	Test continued on the same specimen. No crack apparent
3c	107	461	2.2×10^6	Test continued on the same specimen. Crack along longitudinal rib at 2.2×10^6 cycles
4	226	338	6.88×10^6	Visible crack at 6.75×10^6 cycles. Failure at 6.88×10^6 cycles

The gross contact area is 21.6×8 in.

conditions being equal, the maximum stress in the deck would be 20 percent higher.

For test 2 the maximum principal stress in the top fiber of the plate above the toe of the weld (where the crack occurred) was measured greater than 45 ksi, for a steel yield strength of 47.5 ksi, suggesting that the fatigue strength in this case was approximately equal to F_y.

Factors Affecting the Fatigue Strength of a Deck Plate The high fatigue strength documented in these tests may be partly explained by the fact that where the crack occurred, the weld is on the compressive region of the plate. Thus, local stress concentrations at the toe of the weld would not affect the maximum tensile stress at the top fiber of the plate. A second factor contributing to the high fatigue strength is the rapid decrease of stress in the direction of depth of the combined plate–rib section. The view that fatigue design requirements are not critical in deck plates is further supported by the following: (a) with rib spacing and wheel dimensions of the common order, the stresses in the deck plate depend directly on the contact pressure under load and therefore the design wheel load is far from causing conditions leading to fatigue failure; (b) stress fluctuations in the normal (practical) range do not appear to influence fatigue strength; (c) deck plate deflections are nominal within the stipulated plate thickness range, and below the values that may affect the resistance to fatigue cracks; (d) the actual stresses in the steel deck plate are very likely to be smaller than computed; and (e) if a fatigue crack in the plate deck develops, it is likely to originate in the direction of the ribs, which is the longitudinal direction.

Fatigue is discussed further in Chapter 12.

Use of Empirical Formulas

For estimating deck plate thickness based on specified deflection, an approximate formula may be used as proposed by Klöppel (1958). According to this formula, the deflection of the plate is expressed as

$$\omega_m = \frac{5}{6}\frac{1}{384}\frac{pa^4}{EI} \tag{7-63}$$

where ω_m is the maximum deflection of the deck plate under a wheel load, p is the wheel load unit pressure, a is the rib spacing, and $I = t_P^3/12$, with t_P the plate thickness.

If the deflection is limited to 1/300 of the span length, in this case the open-rib spacing, the condition $\omega_m \leq a/300$ yields

$$t_P \geq a\sqrt[3]{\frac{125}{16}\left(\frac{p}{E}\right)} \tag{7-64}$$

which for $E = 29 \times 10^6$ psi results in $t_P \geq 0.007a\sqrt[3]{p}$, that is, (7-1) is derived. The unit pressures p under standard AASHTO wheel loads including 30 percent impact are 59 and 67 psi for the 12- and 16-kip wheel, respectively. Deck plate thickness obtained by (7-1) for these wheel pressures is shown in Figure 7-43 as a function of the rib spacing a. Also shown for comparison is the thickness obtained from deflections computed assuming a continuous deck plate with the actual wheel dimensions (see also the design example in the following section).

We should note that the condition of a constant deflection ratio of the deck corresponds, with a constant uniform unit load, to an approximately constant value of stress in the plate, for any deck plate span. For example, in a deck with open ribs, the maximum theoretical stress in the deck plate whose thickness is computed from (7-1) will be 32 to 35 ksi. Computed stresses in actual designs are usually lower because plate thickness is rounded to the next-higher standard size.

Ultimate Capacity of Plate Tests show that a flat steel bar fixed at both ends and with a uniformly distributed load as shown in Figure 7-42a acts at failure as a cable, with the load carried by axial stresses only, sustained by the support reactions. For a bar of span a and cross-sectional area A acting in

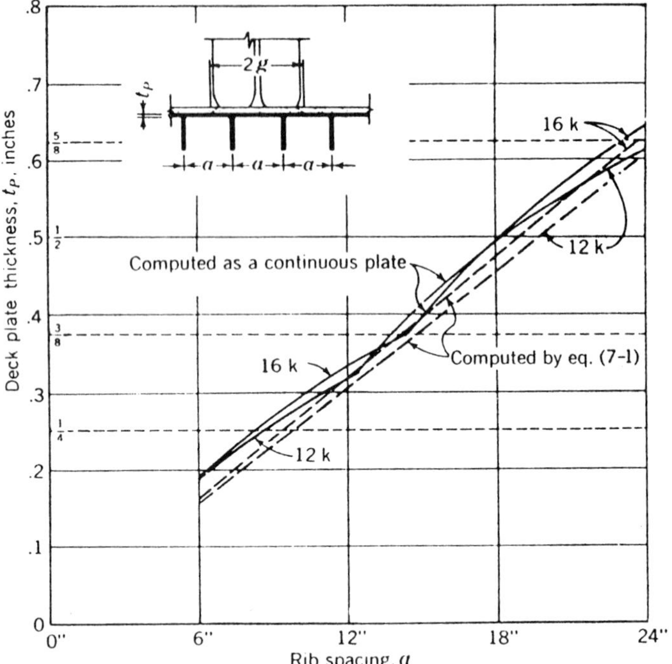

Note: AASHO wheel loads and dimensions are used as follows:
12 k wheel: P = 12 x 1.3 = 15.6 k, loaded area = 12" x 22"
16 k wheel: P = 16 x 1.3 = 20.8 k, loaded area = 12" x 26"

FIGURE 7.43 Deck plate thickness required for a deflection $w_m = (1/300) \times$ plate span. (*Note*: Using the AASHTO distribution and assuming a 2-in. wearing surface, the distribution area would be $24 \times 12 = 288$ in.^{2} and $28 \times 12 = 336$ in.^{2} for the 12-kip and 16-kip wheel load, respectively, giving somewhat smaller unit pressures.)

this manner, the following is true

$$H = \left(pa^2\right)/8\omega_m \tag{7-65}$$

where H is the horizontal reaction at support, p is the loading per unit length, and ω_m is the maximum deflection at midspan. The unit elongation in this case is

$$\epsilon = \frac{1}{24}\frac{p^2a^2}{H^2} \tag{7-66}$$

The ultimate value of the horizontal reaction H_u is expressed as

$$H_u = F_u A \tag{7-67}$$

Combining (7-66) and (7-67), we obtain

$$\epsilon_u = \frac{1}{24}\left(\frac{pa}{F_u A}\right)^2 \tag{7-68}$$

Hence, the ultimate load (per unit length) on the bar is derived as

$$p_u = \frac{4.9F_u A}{a}\sqrt{\epsilon_u} \tag{7-69}$$

The value of ϵ_u may be taken as the ultimate unit elongation at maximum load, obtained from ultimate load tests, usually 14 percent.

For a plate with a uniform load extending only over a portion of the plate width, (7-69) gives overly conservative results. Hence, an empirical correction coefficient $k = 1.25$ is proposed (Klöppel, 1958) so that (7-69) is adjusted as follows:

$$p_u = 1.25\frac{4.9F_u A}{a}\sqrt{\epsilon_u} \tag{7-69a}$$

where, for a unit strip of plate, $A = t =$ plate thickness.

Design Example

Deck with Open Ribs For an orthotropic steel bridge, the deck plate is 3/8 in. thick. The ribs are 8 in. by 1/2 in. thick, and are spaced at 12-in. centers. The plate is loaded by a 12-kip wheel load, with loaded area dimensions $2g \times 2c = 23 \times 11$ in. The unit pressure p including 30 percent impact is 60 psi. The plate is treated as a continuous beam.

Beam Moments Referring to Figure 7-44a, the moments for a 1-in.-wide beam loaded as shown are calculated by integration of the equations of the influence lines.

- For moment at support 0: using (7-23),

$$M_0 = 2\int_{x=0}^{''}\left[-0.5\left(\frac{x}{a}\right) + 0.866\left(\frac{x}{a}\right)^2 - 0.366\left(\frac{x}{a}\right)^3\right] pa\,dx = -890 \text{ lb-in.}$$

- For moment at support 1: in a similar manner using (7-23) and (7-23a), the moment M_1 at support 1 is computed as $M_1 = -317$ lb-in.

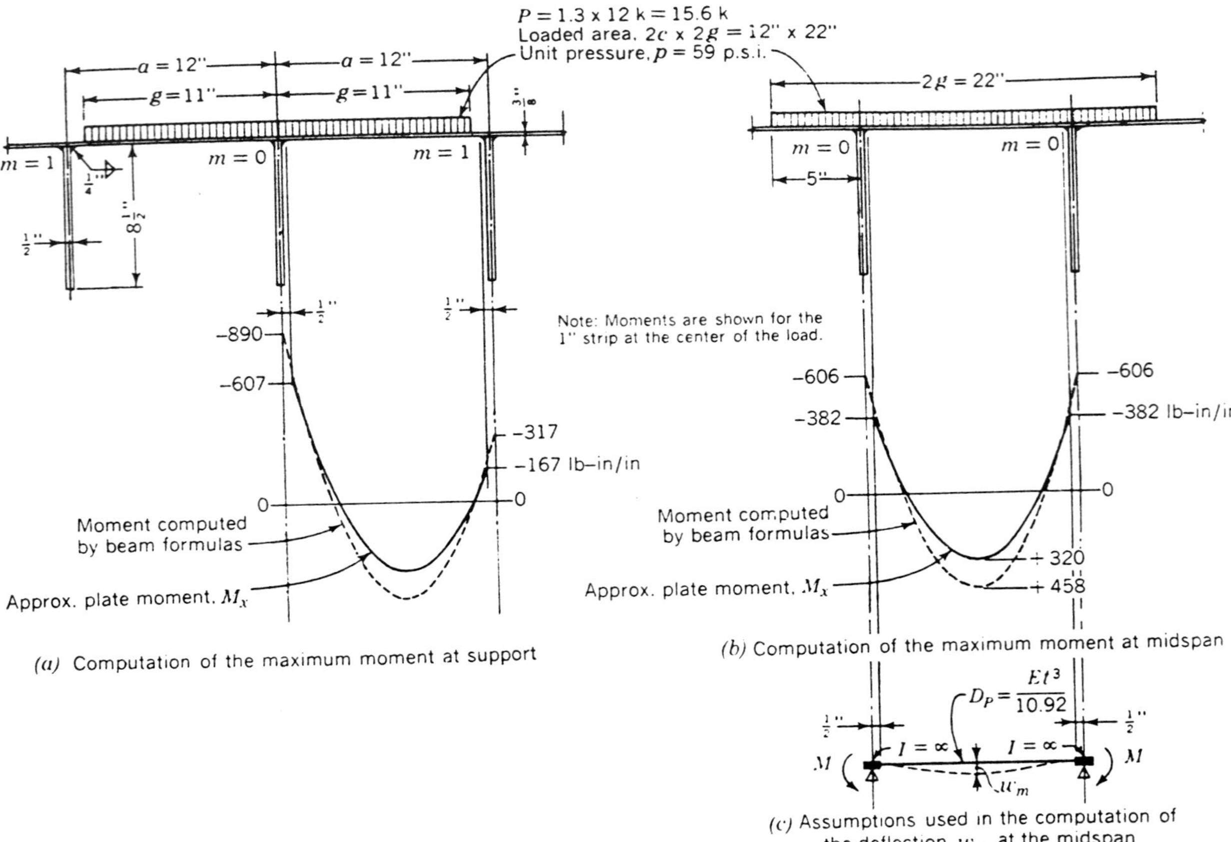

FIGURE 7.44 Computation of the bending moments in a deck plate supported by open ribs.

- For moment at support 0 at the toe of weld: the value of the shear V_0 at support 0 is obtained as

$$V_0 = R_0 + \frac{-M_0 + M_1}{a} = \frac{(60)(11)(6.5)}{12} + \frac{890 - 317}{12} = 399 \text{ lb}$$

The bending moment at the toe of the weld at a distance 1/2 in. from the rib centerline is

$$M_T = -890 + (399)(0.5) - (60)(0.5)^2/2 = -698 \text{ lb-in.}$$

- For moment at midspan: for the loading position shown in Figure 7-44*b*, this moment is computed by integration of (7-26) as $M_c = 458$ lb-in.

Plate Moments and Stresses For bending moments at support, the plate factor ψ from Figure 7-38*b*, case 3, is obtained for $2c/a = 12/12 = 1.0$ as $\psi_s = 0.87$. The maximum moment at the toe of the weld is therefore $M_T = -0.87(698) = -607$ lb-in./in. The section modulus of the 3/8-in. plate is $S = (0.375)^2/6 = 0.0234$ in.^{3}/in., and the maximum stress is therefore

$$f_{max} = 607/0.0234 = 25.94 \text{ ksi}$$

For the bending moment at midspan, the plate factor is obtained from Figure 7-38*a*, case 3, as $\psi_c = 0.70$. The adjusted maximum moment at midspan is $M_c = 0.70(458) = 320$ lb-in./in., and the resulting stress is

$$f = 320/0.0234 = 13.68 \text{ ksi}$$

Maximum Deflection This is computed approximately by applying the computed moment to a simple beam as shown in Figure 7-44*b*. The plate rigidity used in the computation is $D_P = Et^3/10.92$, and the rigidity of the beam at the ribs is assumed to be infinitely large, as shown in Figure 7-44*c*, to consider the plate stiffness at the supports. The computed deflection is $\omega_c = 0.024$ in.

Effects of Rib Flexibility The relative rigidity coefficient γ' is computed as 0.0041. Hence, the error induced by disregarding the effects of rib flexibility on the bending moments of the deck plate is less than 3 percent.

Deck with Closed Ribs An orthotropic plate deck with closed ribs has the configuration shown in Figure 7-45. The loading consists of one 12-kip wheel with a unit pressure of 59 psi as shown in Figure 7-39*a*.

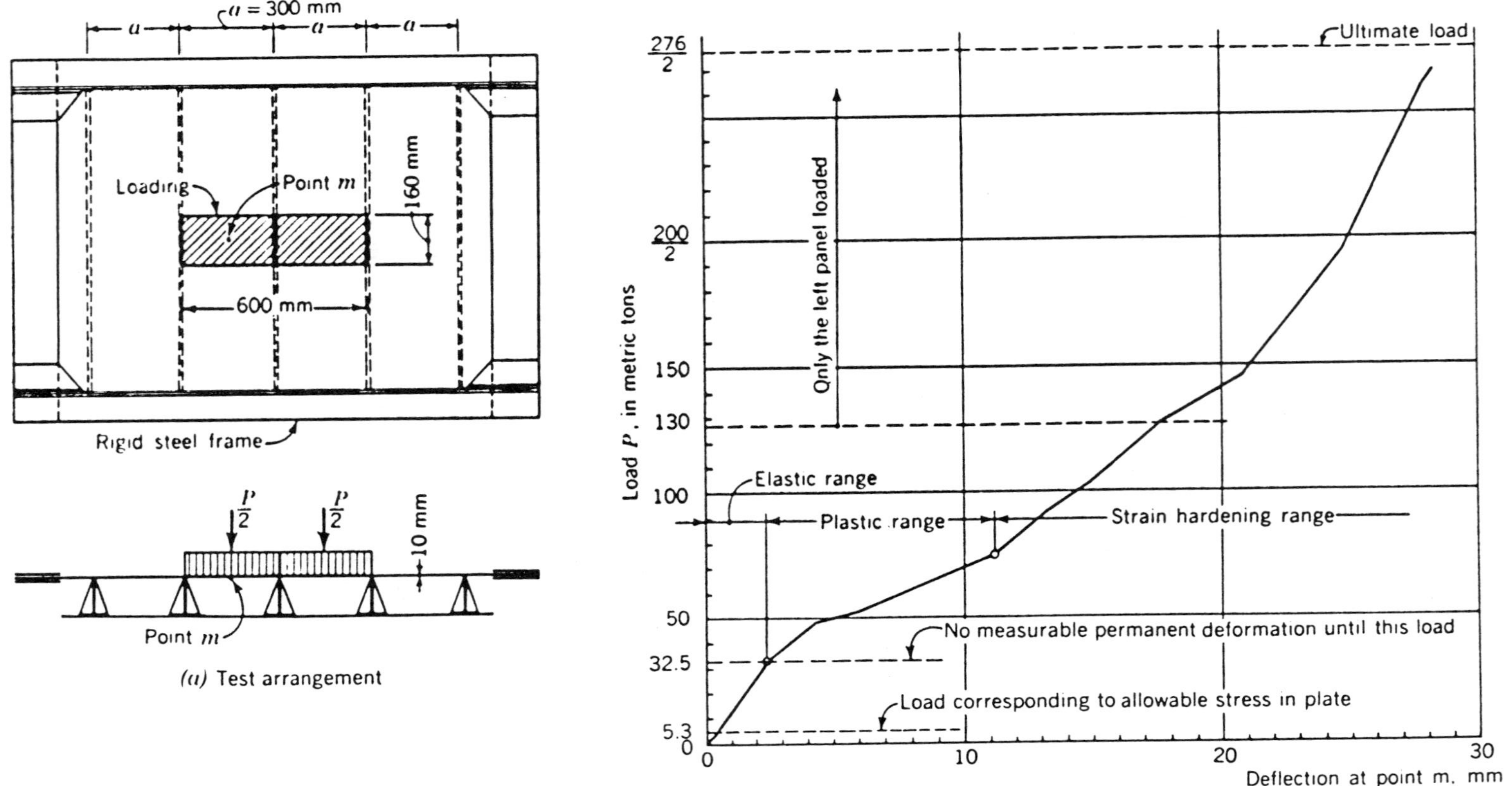

FIGURE 7.45 Ultimate load test on a steel deck plate.

Effects of Direct Loading For this analysis the ribs are considered rigid as shown in Figure 7-39*a*, and the simplified system corresponding to this condition is shown in Figure 7-39*b*. The redundant moments shown in Figure 7-39*c* are obtained from a statically indeterminate analysis for a 1-in. strip, and are as follows:

$M_1 = -701$ lb-in. $\quad M_2 = -85$ lb-in. $\quad M_3 = 23$ lb-in. $\quad M_4 = 17$ lb-in.

The moment at midspan of the deck plate panel under load is calculated as $M_c = 547$ lb-in. The plate moments are obtained by introducing the plate factors based on the most heavily loaded span $a = 13$ in. In this case $2c/a = 12/13 = 0.92$. From Figure 7-38, case 3, the plate factors are extrapolated as:

$$\psi_s = 0.84 \qquad \text{for moment at support}$$

and

$$\psi_c = 0.68 \qquad \text{for moment at midspan}$$

The plate moments are computed as follows:

$$M_1 = -(701)(0.84) = -589 \text{ lb-in.}$$
$$M_2 = -(85)(0.84) = -71 \text{ lb-in.}$$
$$M_3 = (23)(0.84) = 19 \text{ lb-in.}$$
$$M_4 = (17)(0.84) = 14 \text{ lb-in.}$$
$$M_5 = (547)(0.68) = 372 \text{ lb-in.}$$

These moments are plotted in Figure 7-39*d*.

Effects of Transverse Shear Transfer The resulting shear from the transverse load transfer at the point midway between the loaded rib and the adjacent rib, as shown in Figure 7-40*a*, is calculated from (7-62) using values of M_{cn}/Qs given in AISC tables.

The value of

$$\sum_{n=1}^{35} \frac{M_{cn}}{Qs} \sin\frac{n\pi g}{b} \sin\frac{n\pi}{2} \cos\frac{n\pi x}{b}$$

at location $x = (b/2) + (a + e)/2 = 154/2 + (13 + 11)/2 = 89$ in. is computed as -0.174. These parameters are entered into (7-63) and give

$$V_x = \frac{(8)(0.709)(180)}{154}(0.051)(0.174) = -0.059 \text{ kip/in.}$$

The moments in the deck plate and in the rib walls induced by a shear of 59 lb/in. on both sides, as shown in Figure 7-40*b*, are calculated as follows:

$$M_1 = 196 \text{ lb-in.} \qquad M_2 = -129 \text{ lb-in.} \qquad M_3 = 34 \text{ lb-in.}$$

The application of shape factors to these moments is not required because the shear V_x is determined from an analysis of the deck acting as a plate.

Total Moments and Stresses Superposition of the effects from the two previous strips gives total moments, as shown in Figure 7-41.

The maximum stress in the 3/8-in.-thick deck plate occurs at midspan and is

$$f = M/s = 568/(0.375^2/6) = \pm 24.20 \text{ ksi}$$

The stress in the side wall of the ribs at the junction with the plate is

$$f = 190/(0.250^2/6) = \pm 18.20 \text{ ksi}$$

Likewise, the stress in the bottom of the rib is

$$f = 53/(0.250^2/6) = \pm 5.09 \text{ ksi}$$

The stress increments in the y direction caused by stresses in the plate in the x direction are based on Poisson's ratio of 0.3.

In the deck plate $\quad \Delta f_y = \pm(0.3)(24.2) = \pm 7.25$ ksi

In the bottom rib $\quad \Delta f_y = \pm(0.3)(5.09) = \pm 1.53$ ksi

Ultimate Load Test on Deck Plate

Results of a full-scale loading test (Klöppel, 1958) are shown in Figure 7-45 for a 3/8-in.-thick deck plate supported on ribs spaced at 12-in. centers. After the total load reached 122 tons (metric), only the left loading (shaded) area continued to be loaded until the ultimate load was reached. The deck plate steel corresponds to A7 carbon structural steel.

From the load–deflection diagram, it appears that the plate behaves elastically until the load exceeds 32.5 tons (corresponding to 3.25 times the German standard design wheel load). The plastic range is extended from 32.5 to 75 tons, and is indicated by a flatter curve. For a load exceeding 75 tons, the deflection increments become smaller, again indicating strain hardening effects. The ultimate capacity of the plate is reached at the load $P/2 = 276/2 = 138$ tons.

If the same system is analyzed as a continuous isotropic plate, a theoretical allowable load of 5.3 tons is obtained if an allowable stress of 1600

kg/cm^2 (22.7 ksi) is assumed at the toe of the weld. The test load of 32.5 tons (resulting in no permanent deformations) would by the same analysis correspond to a stress of 136 ksi, which far exceeds the ultimate strength. Thus, in this case a first-order plate theory in the analysis of the deck plate is clearly inappropriate and misleading for loads exceeding the usual limits. Furthermore, the stress under working loads cannot be used to infer ultimate static strength.

7-10 FLOOR BEAM FORCES AND DEFORMATIONS

Development of computer-based techniques (Heins and Looney, 1968) that evaluate the response of continuous orthotropic plates on flexible supports has resulted in analytical procedures that predict live load effects on transverse floor beams (Perry and Heins, 1972). The entire bridge is considered as an interacting flexible system, and equations are available for estimating the elastic and rigid support effects in a single step.

Considering the effects of various parameters on the loaded system, the following expression is proposed:

$$\log\left(\frac{f}{f^*}\right) = -0.11N - 0.86(1 - \log S) + (-0.0485N + 0.046)\log\left(\frac{D_y}{EI}\right) \tag{7-70}$$

where $f/f^* = (\delta/\delta^*), (M/M^*), (V/V^*)$
δ = girder system deflection
M = girder system moment
V = girder system shear
δ^* = simple-beam deflection
M^* = simple-beam moment
V^* = simple-beam shear
D_y = primary plate bending stiffness/width
EI = floor beam stiffness

The girder system values (f) represent the response of the floor beams while interacting as a component of the orthotropic system. The (f^*) values are the functions as computed when the floor beams are isolated as simple beams and are subjected to a set of simple HS 20 axles, as determined by the number of lanes (similar to the $S/5.5$ distribution factor).

As an approximate method, the application of (7-70) has been found to give satisfactory results, and compares well with the rather extensive procedures necessary to account for deck and floor beam flexibility.

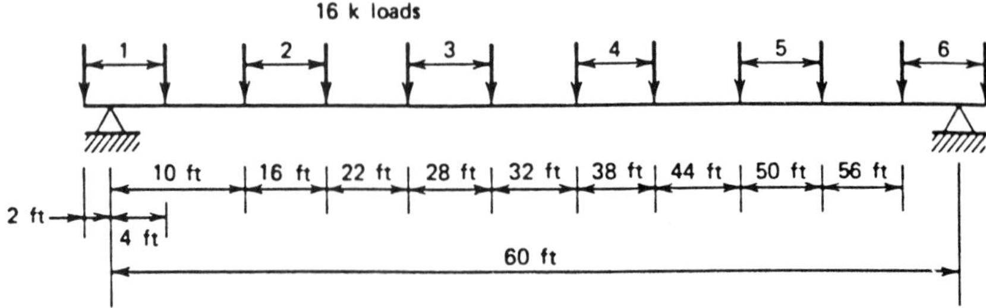

FIGURE 7.46 Truck loading for approximate analysis. (From Heins and Firmage, 1979.)

Numerical Example As an example, we consider the position of trucks across the bridge width shown in Figure 7-46.

In this example, the span can accommodate six lanes, but in this case the wheel loads overhanging the supports cause cantilever moments that reduce the positive moments produced by the interior loads. Thus, for preliminary design the floor beam is loaded with five lanes. Applying (7-70) to the moments gives

$$\log_{10}\left(\frac{M}{M^*}\right) = -0.11N - 0.86(1 - \log S) + (-0.0485N + 0.046)\log_{10}\left(\frac{D_y}{EI}\right)$$

where $N = 5$ lanes, $D_y = 2.76 \times 10^4$ kip-ft^2/ft, $S = 12$ ft, and $EI = 4.73 \times 10^6$ kip-ft^2.

Substituting these values into the preceding equation gives

$$\frac{M}{M^*} = 0.906$$

The induced simple beam moment caused by the loads shown in Figure 7-46 is computed as $M^* = 1192$ ft-kips. With these data, the maximum bending moment is now computed as

$$M = \frac{M}{M^*} \times M^* = 0.906 \times 1192 = 1080 \text{ ft-kips}$$

7-11 ORTHOTROPIC PLATE BEHAVIOR UNDER ALTERNATE LOADS: CASE STUDY

In the preceding sections we made comments on the high ratio of the ultimate plastic load to the limit elastic load, normally observed with steel orthotropic plates. However, the behavior under repeated loads stressing the

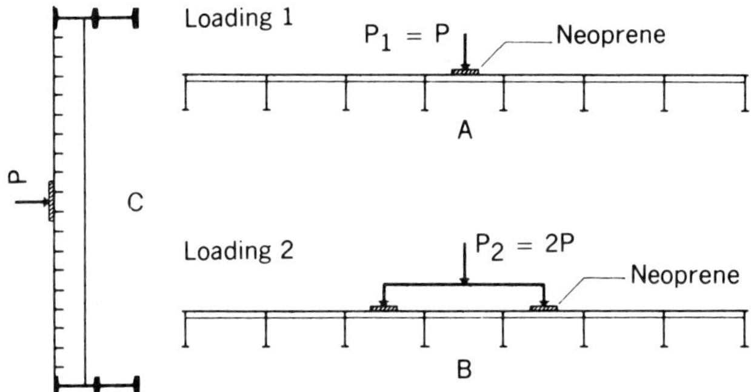

FIGURE 7.47 Loading conditions for test model. (From Ceradini, Gavarini, and Petrangeli, 1975.)

plate beyond the elastic range is yet to be fully established. As a matter of expediency, this behavior should not be identified with high-cycle fatigue, generally conditioned by technological details such as welding (Seim and Feriverda, 1972).

Tests have been conducted by Ceradini, Gavarini, and Petrangeli (1975) on a model previously studied in the elastic range. This model is a system with open ribs on cross beams supported by a two-girder structure. The overall stiffness imparts to the plate sufficient longitudinal rigidity. The model represents an element of bridge deck on a scale of 1 : 5.

Typical sections of the model with the loading conditions are shown in Figure 7-47. The static loading program involved a loading sequence $P_1, O_1, P_2, O_2, P_1, \ldots,$ where P_1 and P_2 are as shown, and O_1 and O_2 indicate the state of the plate after unloading.

The first cycles were obtained with $P = 1.1$ tons (1 ton = 2 kips), chosen to keep stresses in the elastic range. At a subsequent load level of $P = 1.65$ tons, the first yielding appeared. With continuing load increments, the plastic range was defined, when a steady situation was reached after a sufficient number of cycles. This steady condition is defined as the repetition of the deflection of a preceding cycle to within a close range. The applied load levels are as follows: 1.10, 1.65, 2.20, 2.75, 3.30, 4.40, 5.50, 6.60, 7.70, 8.80, 9.90, 11.0, 12.10, and 13.75 tons.

Test Results It appears that the behavior of the model is essentially elastic for $P = 1.10$ tons. The deflection of the central directly loaded ribs is typical of a beam on elastic supports. The first indication of yielding is at $P = 1.65$ tons, although between 1.65 and 4.40 tons the center ribs yielded. At 4.40 tons the central transverse beams and ribs began to yield. At 7 tons a transverse beam showed signs of yielding. At 9.9 tons yielding of ribs and transverse beams occurred, and at 11 tons the yielding was extended to more ribs. At 13.75 tons the entire plate was yielding, and plastic flow was essentially constant under load 2, signaling the possibility of collapse.

The behavior of the plate model is also documented from further results related to (a) the evolution of residual displacements after the number of cycles required at each loading level to obtain stabilization and (b) the structural strain hardening partially attributed to material behavior caused essentially by the evolution of plasticity in the structure as well as by relevant geometric variations of the model. From a study of the residual deflections of the central ribs after shakedown (O_1 and O_2) and by observing the noncoincidence of these residual configurations in the steady structure, it was possible to detect the occurrence of alternating plasticity.

Ceradini, Gavarini, and Petrangeli (1975) provide a qualitative presentation of moment–plastic rotation diagrams in two sections of the model. These clearly indicate the shakedown of the structure as a whole, whereas the hysteresis cycle shows alternating plasticity in the loaded ribs.

Interpretation A comparison of the total vertical displacement (theoretical and observed) for the two locations and up to $P = 4.4$ tons shows the variation of displacements with every cycle for the associated load positions. The relationship of plastic rotations versus moments is obtained theoretically for hinges developed at the two points. Thus, the occurrence of the two stages is clearly demonstrated. For $P < 2.75$ tons, shakedown occurs; the steady situation is characterized by the coincidence of residual displacements O_1 and O_2. For $P = 3.3$ tons, stabilization was again reached but with associated alternating plasticity in one of the sections, and for $P = 4.4$ tons, the hysteresis cycle appeared in the other section. The continuous beam model concept is no longer valid when the load exceeds the level $P = 4.4$ tons, because the effects of plasticity in the transverse beams begin to dominate. At this load level, the total carrying capacity is essential, and loading condition 1 becomes irrelevant.

Load Level $>$ 4.4 Tons to Collapse Ceradini, Gavarini, and Petrangeli (1975) consider a model of a rigid perfectly plastic grid with longitudinal ribs and transverse beams connected with a plane membrane (the deck plate). As the load continues to increase, the grid elements become plastic forming plastic hinges, and local mechanisms take place in the system. Eventually, the displacements of these mechanisms become large, activating a membrane load-carrying capacity. This capacity transfers part of the load to other grid elements, not as yet yielded, until new more extended mechanisms are manifested, and so on. Although this is a crude behavioral approximation, the suggested response involves a deck plate with three distinct functions: a membrane, upper chord of longitudinal beams, and upper chord of transverse beams.

Conclusions Because of the simplicity of the testing methods, the results give essentially qualitative indications of plastic strength, yet they confirm the effects of alternating loads on orthotropic steel plates. Thus, the very high load-carrying capacity of this structure (about eight times the elastic limit) under static load is confirmed, but the behavior under alternating loads is far less favorable. An important factor in this case is the phenomenon of alternating plasticity in the directly loaded ribs at load level, slightly above the elastic limit (about 1.7 to 2 times).

The rigid plastic grid model, connected with a membrane that begins to work when large plastic displacements occur, allows interpretation of the global behavior of the plate for larger loads and up to collapse.

7-12 THEORETICAL ASPECTS OF ORTHOTROPIC SKEW STEEL DECKS: CASE STUDY

Plastic Analysis

Results from the analysis of isotropic and orthotropic rectangular plates have been reported by Ang and Lopez (1968), Bhaumic and Hanley (1967), Lin and Ho (1968), Kagan (1964), and Wegmuller (1974). Solutions for predicting the elastoplastic structural response of eccentrically stiffened skew steel plates are presented by Kennedy and Chowdhury (1977).

These solutions are derived by finite-difference techniques and are based on the following assumptions: (a) large-deflection behavior of plates is introduced; (b) the material of the plate is elastic–perfectly plastic; (c) no reversal of stress such as unloading occurs in the plastic region; (d) no instability will occur before reaching the collapse load; and (e) the structure is free of residual stresses, or the presence of these stresses does not influence the ultimate strength of systems whose load capacity is not limited by instability.

Analytical Expression Beginning with two second-order equations, the associated functions are transformed into oblique coordinates and are then expressed in finite-difference forms in terms of the lateral deflection w (Chowdhury, 1975) and Airy stress function F (Lekhnitsky, 1968). By applying the appropriate finite-difference equations at the various nodal points of the skew network on the equivalent orthotropic plate, two groups of simultaneous equations are generated and expressed in matrix form.

A method of successive approximations together with an iterative scheme is used for solving these equations based on the von Mises yield criterion applied to the stiffening rib and steel deck plate. There are approximately nine steps in the iterative process leading to the elastic and elastoplastic solutions. The entire process is programmed for a suitable computer.

Experimental Studies Kennedy and Chowdhury (1977) have confirmed the foregoing analytical approach from tests on rectangular and skew orthotropic plate models of various geometries and elastic properties conforming to ASTM standards. All specimens exhibited a very close elastic–perfectly plastic material behavior up to a strain of 1.5 percent beyond which strain hardening took place. The models were first tested in the elastic stress domain under uniformly distributed and concentrated lateral load applied symmetrically and asymmetrically with the following boundary conditions: (a) simple supports on all edges and (b) simple supports on two opposite edges and free along the other two edges. Subsequently, the models were tested in the elastoplastic stress range, under uniformly distributed load with all edges simply supported.

Results These studies compare the analytical and test results for a special rectangular model subjected to uniformly distributed load and with all edges simply supported. In the lower bound (first yield in the y direction), the theoretical and experimental loads are 6.25 and 6.52 psi, respectively, a difference of about 5 percent, with corresponding deflections of 0.158 and 0.156 in. The predicted and experimental load–deflection characteristics from the beginning to final collapse load agree fairly well. Noting that at the terminal stage of full yielding of the stiffening rib the ratio of center deflection to the shorter span is approximately $1/20$, we can consider this deflection as an indication of imminent collapse.

Comparison of a second model (45° skew), stiffened eccentrically by unidirectional ribs, shows likewise good agreement, which is consistent particularly in the lower bound values. The yielding sequence of the stiffening ribs for the second model at the various nodal points and under uniformly distributed load shows that first yielding occurs at the center in the longitudinal direction and propagates to the center of the adjoining stiffening ribs, thereafter continuing in the same sequence to adjacent locations. Yielding occurs in the region of the obtuse corner before it occurs in the region of the acute corner. This should be expected as a result of the intensity and sense of the membrane stresses.

At the collapse load of 18.75 psi, compressive membrane stresses are present along the transverse direction near the skew supports, whereas tensile membrane stresses exist along the longitudinal direction. As the corners are approached, both stresses in these directions become compressive near the acute corner and tensile near the obtuse corner.

Effect of Skew These effects are examined by comparing model 1 with skew model 3 (eccentrically stiffened in both directions). In order to study the variation in the load-carrying capacity, the rigidity ratio and skew aspect ratio of model 3, with a 45° skew angle, are chosen to be comparable to those for model 1 with a rectangular plan. Comparison of the two models shows that the load-carrying capacity of the skew model is 25 percent at yield initiation

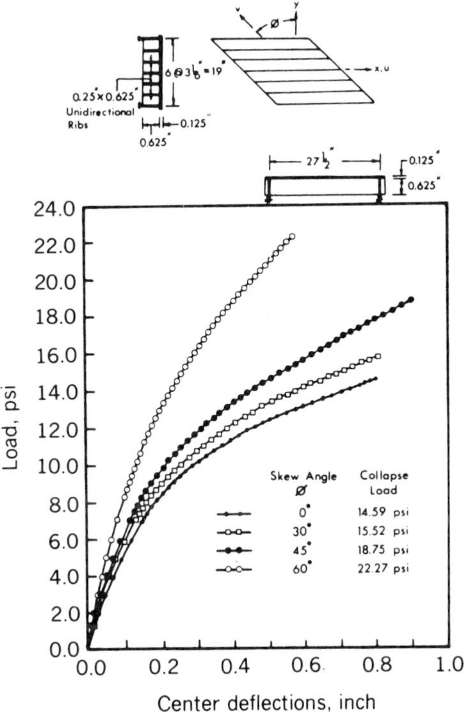

FIGURE 7.48 Variation of load–deflection with skew for uniform load. (From Kennedy and Chowdhury, 1977.)

and 18 percent at collapse, higher than the corresponding load capacities of the rectangular model.

Using skew angles of 0°, 30°, 45°, and 60° and the remaining geometry as in model 2, theoretical solutions were obtained to analyze the influence of skew on the load-carrying capacity of eccentrically stiffened steel plate structures. Results are compared in Figure 7-48. A direct observation is that the collapse load ˙ncreases with skew, with a 50 percent increase in this magnitude at a skew of 60° compared to the corresponding value of the rectangular model. In addition, the corresponding deflection markedly decreases with increasing skew.

Load–deflection data for a concentrated load at the center are shown in Figure 7-49, and similar trends are noticed. For example, the collapse center load for the 60° model is about 70 percent higher than in the corresponding rectangular structure.

The influence of skew on the effective bending and torsional rigidities is articulated by Chowdhury (1975). Once yield is initiated at the section of maximum stress intensity, considerable redistribution of stresses occurs with continuing load increments. Comparison of the effective bending rigidities at

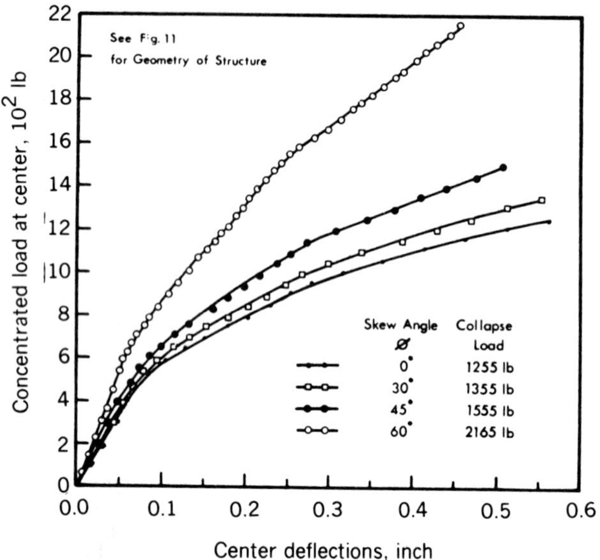

FIGURE 7.49 Variation of load–deflection with skew for concentrated load at center. (From Kennedy and Chowdhury, 1977.)

the center of a rectangular structure and a 60° skew structure shows that the latter exhibits a much longer range of stress redistribution after first yield initiation. It is this characteristic that contributes to the increase of the overload capacity of eccentrically stiffened skew plates with increasing skew.

Conclusions The following conclusions are appropriate.

1. The concept of equivalent orthotropy is valid in the elastoplastic state when the structural rigidities are properly modified.
2. The collapse, or ultimate load, of such structures increases considerably with increasing skew.
3. Membrane stresses become considerable in the elastoplastic state and exhibit tensile and compressive configurations in the vicinity of the obtuse and acute corners, respectively.

Postbuckling Considerations

The postbuckling behavior of rectangular plates subjected to edge forces has been studied analytically (Walker, 1969) and more recently by finite-element techniques (Colville, Becker, and Furlong, 1973; Gallagher, 1972; Shye and Colville, 1979; Yang, 1971, 1972). Solutions of the postbuckling behavior of simply supported orthotropic skew plate structures subjected to combined in-plane biaxial compression and edge shear are presented by Kennedy and Prabhakara (1980). The analysis is formulated using a perturbation technique and double series representing the transverse deflection and force functions. Numerical results applicable to the design of geometrically as well as materially orthotropic skew plate structures are obtained to explain postbuckling behavior.

Using the foregoing techniques, the nonlinear coupled partial differential equations are reduced to a system of uncoupled linear partial differential equations solved by the double Fourier series method. The solution converges rapidly with respect to the number of terms in the Fourier series as well as with respect to the order of approximation for the range of plate deflections considered.

Kennedy and Prabhakara (1980) draw the following conclusions:

1. The convergence of the postbuckling results becomes less rapid (a) when the maximum deflection in the plate is more than twice the plate thickness, (b) for skew angles greater than 45°, and (c) for an elastic moduli ratio greater than 10. A high-order approximation and a large number of terms in the series are required to arrive at reasonably accurate results.

2. The edge postbuckling load for a fixed maximum plate deflection increases with increasing elastic moduli ratio and with increasing skew angle, but decreases with increasing plate aspect ratio.

3. At small postbuckling loads, the bending moment at the center of the plate increases with skew angle, but the trend is reversed with large loads. In the latter case, the small bending moments in plates with small skews are followed by large deflections.

4. At the center of the plate, the membrane force transverse to the edge compression is tensile and increases with the edge compressive load. The membrane force at the plate center and in the direction of the edge load increases at a rate slower than at the edge load after the plate has buckled due to tensile membrane forces developed as a result of bending of the plate.

7-13 COMBINED SYSTEM OF PLATE AND GIRDER

Development of Orthotropic Plate Equation

The direct analysis of orthotropic plates on flexible supports requires the use of a computer program. A plate is assumed to have a specified grid or nodes as shown in Figure 7-50. The plate is subjected to a load q and deforms an amount w_0, w_a, w_b, w_e, and w_r at the respective nodes 0, a, b, e, and r as shown. The deflected surface $w(x, y)$ may be assumed to be represented by a parabola

$$\overline{w} = Ax^2 + Bx + C + Dy + Ey^2 \tag{7-71}$$

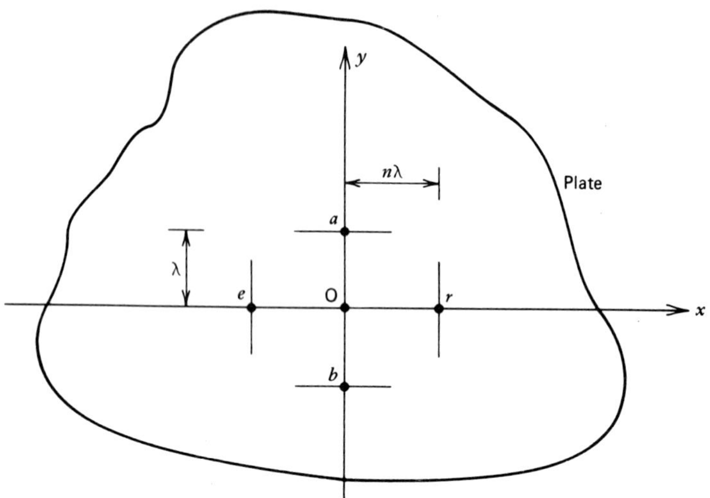

FIGURE 7.50 Ideal orthotropic plate with a specific grid or nodes; plate difference mesh. (From Heins and Firmage, 1979.)

such that $\overline{w}$ is identical to w at the five given points, e, r, 0, a, and b. For the conditions at these points to exist, the following equations must be satisfied:

At $x = y = 0$, $\quad \overline{w} = w_0 \qquad \overline{w} = w_0 = C$ (a)

At $x = -n\lambda \quad y = 0 \quad \overline{w} = w_e$

$$\overline{w} = w_e = A\overline{n\lambda}^2 - Bn\lambda + C + D(0) + E(0)$$

$$w_e = A\overline{n\lambda}^2 - Bn\lambda + C \tag{b}$$

At $x = +n\lambda \quad y = 0 \quad \overline{w} = w_r$

$$w_r = A\overline{n\lambda}^2 + Bn\lambda + C \tag{c}$$

At $x = 0 \quad y = -\lambda \quad \overline{w} = w_b$

$$w_b = C - D\lambda + E\lambda^2 \tag{d}$$

At $x = 0 \quad y = \lambda \quad \overline{w} = w_a$

$$w_a = C + D\lambda + E\lambda^2 \tag{e}$$

Solving for the constants A, B, C, D, and E gives

$$
\begin{aligned}
A &= \frac{1}{2n\lambda^2}(w_r - 2w_0 + w_e) \\
B &= \frac{1}{2n\lambda}(w_r - w_e) \\
C &= w_0 \\
D &= \frac{1}{2\lambda}(w_a - w_b) \\
E &= \frac{1}{2\lambda^2}(w_a - 2w_0 + w_b)
\end{aligned}
\tag{f}
$$

and substituting (f) into (7-71) provides the general expression

$$
\overline{w} = \frac{1}{2\lambda^2}\left[(w_r - 2w_0 + w_e)\frac{x^2}{n^2} + (w_r - w_e)\frac{x\lambda}{n} + w_0 2\lambda^2 + (w_a - w_b)y\lambda + (w_a - 2w_0 + w_b)y^2\right] \tag{7-72}
$$

Equation (7-72) is used in part to evaluate the biharmonic expression given by (7-10) where the second term is replaced by notation q and written as

$$
D_x\frac{\partial^4 w}{\partial x^4} + 2H\frac{\partial^4 w}{\partial x^2\,\partial y^2} + D_y\frac{\partial^4 w}{dy^4} = q \tag{7-73}
$$

Next, the mesh pattern is extended to include more points as shown in Figure 7-51 because all terms would vanish if $\partial^4 w/\partial x^4$ and so on were taken off (7-72). Taking the partial derivatives of (7-72) gives

$$
\begin{aligned}
\frac{\partial^2 \overline{w}}{\partial x^2} &= \frac{1}{n\lambda^2}(w_r - 2w_0 + w_e) \\
\frac{\partial^2 \overline{w}}{\partial y^2} &= \frac{1}{\lambda^2}(w_a - 2w_0 + w_b)
\end{aligned}
$$

The fourth-order differential can be written as

$$
\frac{\partial^4 \overline{w}}{\partial x^4} = \frac{\partial^2}{\partial x^2}\left(\frac{\partial^2 w}{\partial x^2}\right) = \frac{1}{n\lambda^2}\left[\left.\frac{\partial^2 \overline{w}}{\partial x^2}\right|_r - 2\left.\frac{\partial^2 \overline{w}}{\partial x^2}\right|_0 + \left.\frac{\partial^2 \overline{w}}{\partial x^2}\right|_l\right] \tag{g}
$$

where expansion is made with respect to nodes r, 0, and l. These expansions are

$$\left.\frac{\partial^2 \overline{w}}{\partial x^2}\right|_r = \frac{1}{\overline{n\lambda}^2}(w_{rr} - 2w_r + w_0)$$

$$\left.\frac{\partial^2 \overline{w}}{\partial x^2}\right|_0 = \frac{1}{\overline{n\lambda}^2}(w_r - 2w_0 + w_l)$$

$$\left.\frac{\partial^2 \overline{w}}{\partial x^2}\right|_l = \frac{1}{\overline{n\lambda}^2}(w_0 - 2w_l + w_{ll})$$

According to the mesh pattern shown in Figure 7-51 and substituting these expressions into (g) gives

$$\frac{\partial^4 \overline{w}}{\partial x^4} = \frac{1}{\overline{n\lambda}^4}(w_{rr} - 4w_r + 6w_0 - 4w_l + w_{ll}) \tag{h}$$

$$\frac{\partial^4 \overline{w}}{\partial y^4} = \frac{1}{\lambda^4}(w_{aa} - 4w_a + 6w_0 - 4w_b + w_{bb}) \tag{i}$$

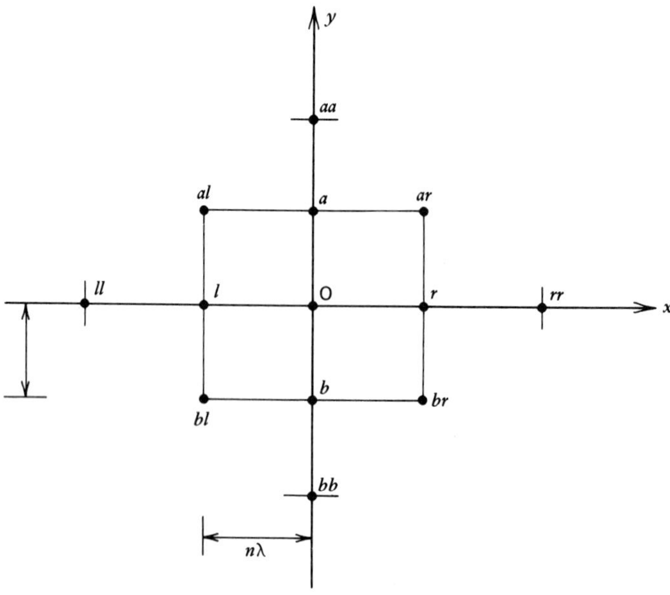

FIGURE 7.51 General difference mesh. (From Heins and Firmage, 1979.)

The mixed partial derivative $(\partial^4 w/\partial x^2 \partial y^2)$ is found by expanding $(\partial^2 w/\partial y^2)$ about nodes r, 0, and l according to the expression

$$\frac{\partial^4 \overline{w}}{\partial x^2 \partial y^2} = \frac{\partial^2}{\partial y^2}\left(\frac{\partial^2 w}{\partial x^2}\right) = \frac{\partial^2}{\partial y^2}\left(\frac{w_r - 2w_0 + w_l}{n^2\lambda^2}\right)$$
$$= \frac{1}{n^2\lambda^2}\left[\left.\frac{\partial^2 w}{\partial y^2}\right|_r - 2\left.\frac{\partial^2 w}{\partial y^2}\right|_0 + \left.\frac{\partial^2 w}{\partial y^2}\right|_l\right]$$

We note, however, that

$$\left.\frac{\partial^2 w}{\partial y^2}\right|_r = \frac{w_{ar} - 2w_r + w_{br}}{\lambda^2}$$
$$\left.\frac{\partial^2 w}{\partial y^2}\right|_0 = \frac{w_a - 2w_0 + w_b}{\lambda^2}$$
$$\left.\frac{\partial^2 w}{\partial y^2}\right|_l = \frac{w_{al} - 2w_l + w_{bl}}{\lambda^2}$$

Hence,

$$\frac{\partial^4 w}{\partial x^2 \partial y^2} = \frac{1}{n^2\lambda^4}\left[w_{ar} - 2w_r + w_{br} - 2w_a + 4w_0 - 2w_b + w_{al} - 2w_l + w_{bl}\right] \tag{j}$$

Setting

$$\beta = \frac{H}{D_y} \qquad \alpha = \frac{H}{\sqrt{D_x D_y}}$$

and substituting (h), (i), and (j) into (7-73) gives the final general orthotropic plate equation in difference form as

$$\left[w_0\left(6n^2 + 8n^2\beta + 6\left(\frac{\beta}{\alpha}\right)^2\right) + 4(w_r + w_l)\left(\left(\frac{\beta}{\alpha}\right)^2 + n^2\beta\right)\right.$$
$$-4(w_a + w_b)(n^2\beta + n^4) + (w_{rr} + w_{ll})\left(\frac{\beta}{\alpha}\right)^2 + (w_{aa} + w_{bb})n^4$$
$$\left. + (w_{ar} + w_{br} + w_{al} + w_{bl}) \cdot 2n^2\beta\right] = \frac{qn^4\lambda^4}{D_y} \tag{7-74}$$

Likewise, moments and reactions can also be expressed in difference form as follows:

$$M_x = -\frac{D_x}{n\lambda^2}(w_l - 2w_0 + w_r) \tag{7-75}$$

$$M_y = -\frac{D_y}{\lambda^2}(w_a - 2w_0 + w_b) \tag{7-76}$$

$$R_x = -D_x\left[\frac{1}{2n\lambda^3}(w_{rr} - 2w_r + 2w_l - w_{ll}) + \frac{2\xi}{2n\lambda^3}(-2_{al} + 2w_l - w_{bl} + w_{ar} - 2w_r + w_{br})\right] \tag{7-77}$$

$$R_y = -D_y\left[\frac{1}{2\lambda^3}(w_{aa} - 2w_a + 2w_b - w_{bb}) + \frac{2\beta}{2n^2\lambda^3}(w_{al} - 2w_a + w_{ar} - w_{bl} + 2w_b - w_{br})\right]. \tag{7-78}$$

where

$$\xi = \frac{H}{D_x}$$

Orthotropic Plate with Interacting Girders

Heins and Firmage (1979) have developed the solution of an orthotropic plate interacting with girders based on the general finite-difference equations of the foregoing section.

The orthotropic plate shown in Figure 7-52 is supported on interacting girders. The girder stiffness is EI_x and EI_y as shown, and the plate has stiffness D_x, D_y, and H as before. Equilibrium of the intersecting plate and girders requires that

$$\sum F = 0 \qquad q_T = q_{PL} + q_{Bx} + q_{By} \tag{7-79}$$

where q_T = externally applied load
q_{PL} = load resisted by plate
q_{Bx} = load resisted by the beam in the x direction
q_{By} = load resisted by the beam in the y direction

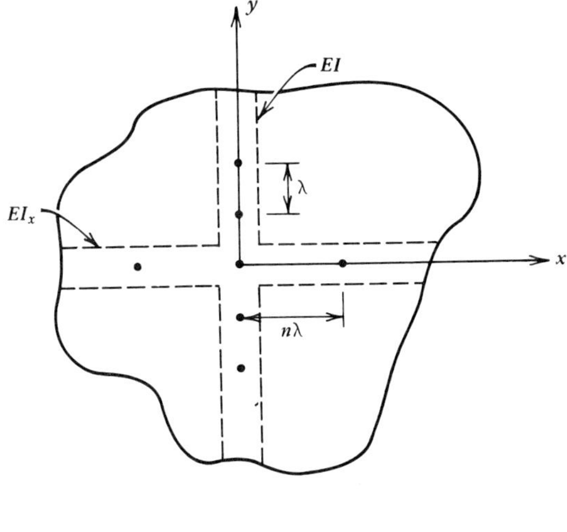

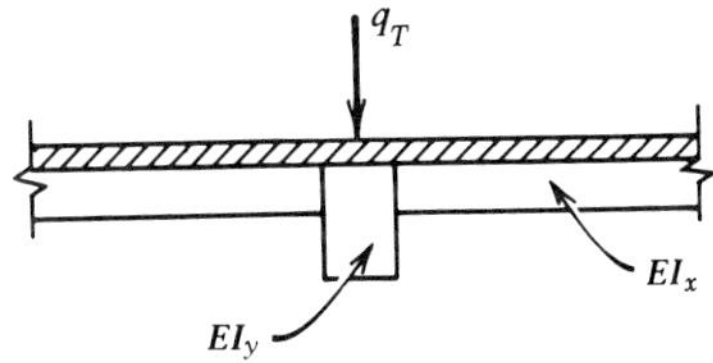

FIGURE 7.52 Presentation of plate interaction with girders. (From Heins and Firmage, 1979.)

From beam theory, we can write $d^4w/dy^4 = P_y/EI_y$, or in difference form

$$\frac{P_y\lambda^4}{EI_y} = (w_{aa} - 4w_a + 6w_0 + 4w_b + w_{bb}) \tag{7-80}$$

Similarly, $d^4w/dx^4 = P_x/EI_x$ and

$$\frac{P_x\overline{n\lambda}^4}{EI_x} = (w_{ll} - 4w_1 + 6w_0 - 4w_r + w_{rr}) \tag{7-81}$$

The forces P_y and P_x given by (7-80) and (7-81) are per unit length of beam. The force equilibrium in the plate and girder, expressed by (7-79), has the parameters shown as force per unit area. Thus, setting

$$q_{Bx} = \frac{P_x}{\lambda} \qquad \text{and} \qquad q_{By} = \frac{P_y}{n\lambda}$$

gives

$$\frac{P_{Bx}}{\lambda} + \frac{P_{By}}{n\lambda} + q_{PL} = q_T, \tag{7-82}$$

Making appropriate substitutions into (7-82) yields

$$Jw_{Bx}[\ ] + Kw_{By}[\ ] + w_{PL}[\ ] = \frac{qn^4 l^4}{D_y}$$

where $J = EI_x/D_y\lambda$, $K = (EI_y/\lambda D_y)n^3$, and [ ] are the mesh point parameters.

Expansion of the last equation gives the general orthotropic form

$$w_0\left[6n^4 + 8n^2\beta + 6\left(\frac{\beta}{\alpha}\right)^2 + 6K + 6J\right] - 4(w_r + w_l)\left[\left(\frac{\beta}{\alpha}\right)^2 + n^2\beta - 4J\right]$$
$$- 4(w_a + w_b)(n^2\beta + n^4 - 4K) + (w_{rr} + w_{ll})\left[\left(\frac{\beta}{\alpha}\right)^2 + J\right]$$
$$+ (w_{aa} + w_{bb})\left(n^4 + K + (w_{ar} + w_{br} + w_{al} + w_{bl})2n^2\beta = \frac{qn^4\lambda^4}{D_y}\right) \tag{7-83}$$

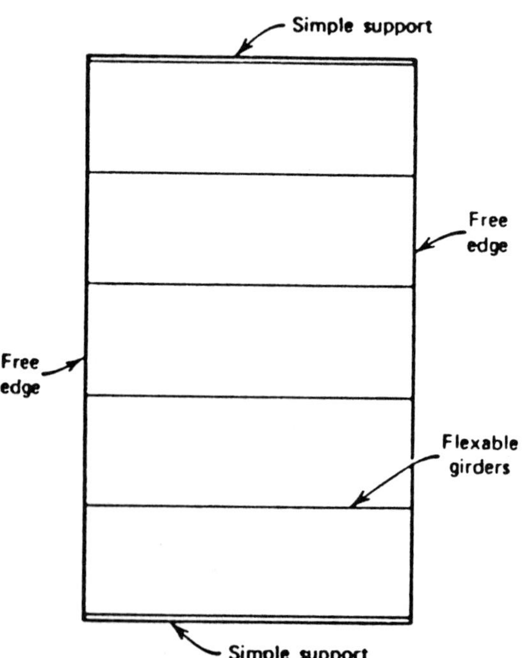

FIGURE 7.53 Bridge boundaries. (From Heins and Firmage, 1979.)

Heins and Firmage (1979) also describe the general orthotropic plate equation in mesh form. If $K = J = 0$ the general equation (7-83) reduces to an equation representing an orthotropic plate without girders. If the plate is isotropic ($\alpha = \beta = 1$) and girder interaction is absent ($K = J = 0$) and for a square mesh ($n = 1$), (7-83) is modified accordingly.

For the usual orthotropic deck bridge, the structure consists of two main longitudinal girders, transverse floor beams, and a continuous orthotropic deck. Such a structure has two free edges and two simple supports as shown in Figure 7-53. Considering the boundary conditions for this case, the general difference plate expression, (7-74), is modified, resulting in mesh patterns for load adjacent to the simple support, adjacent to the free edge, load on the free edge, adjacent to the simple support, and load on the free edge, adjacent to the simple support.

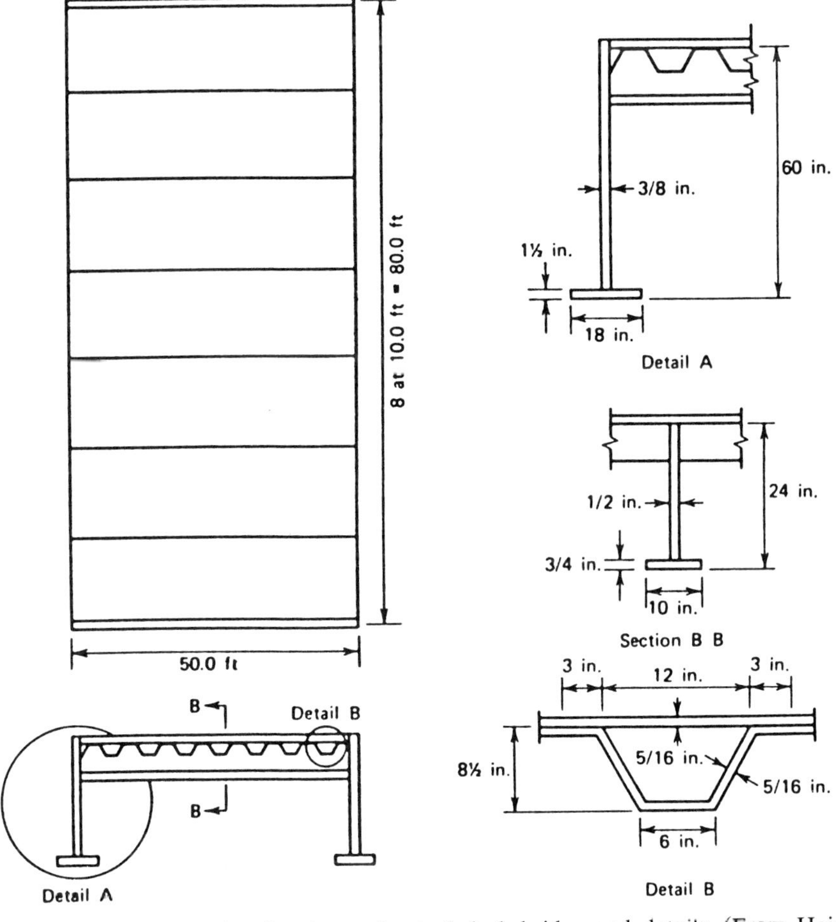

FIGURE 7.54 Example of orthotropic steel deck bridge and details. (From Heins and Firmage, 1979.)

Each pattern is written at the respective point on the structure after the mesh spacing configuration is defined.

The solution of the resulting set of equations provides information at each of the node points. Using the deflections thus computed, the induced moments are determined as

$$M_{x,y} = -\frac{EI}{(\text{spacing})^2}\left(w_a^l - 2w_0 + w_b^r\right)$$

As an example, for the typical bridge (orthotropic plate) shown in Figure 7-54, Heins and Firmage (1979) have developed the mesh pattern shown in Figure 7-55.

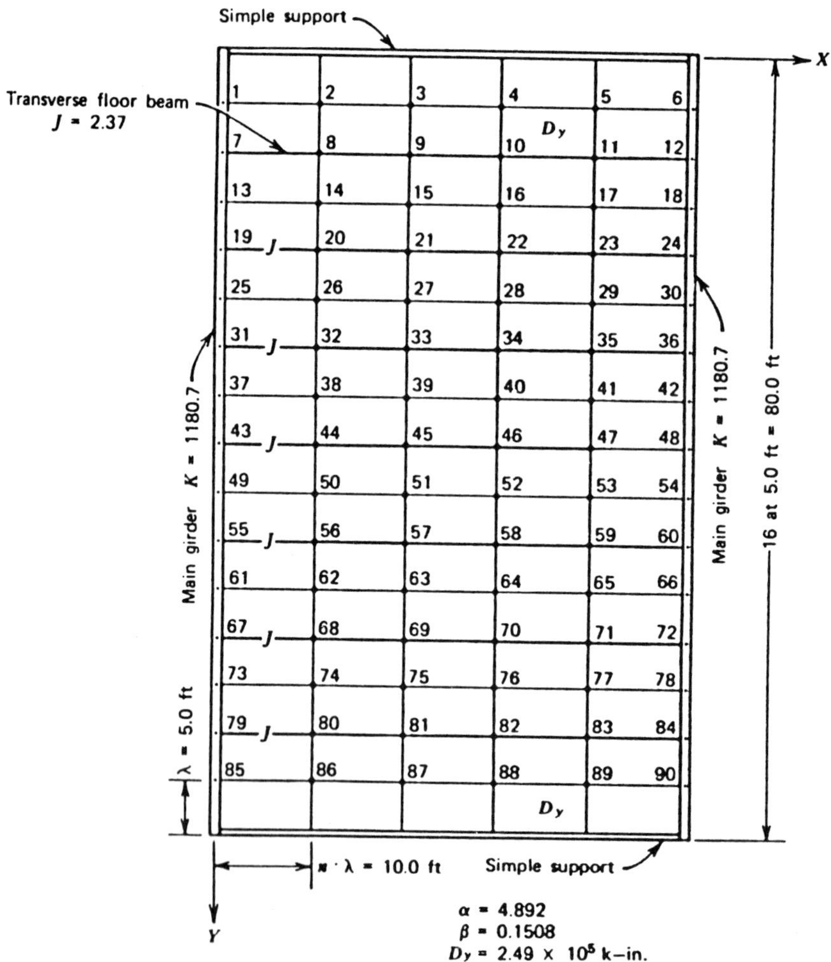

FIGURE 7.55 Suggested mesh pattern for example of Figure 7.54.

7-14 BENDING AND CONTINUITY: SKEW ORTHOTROPIC PLATE STRUCTURES

Orthotropic forms are also found in contemporary concrete bridges as orthogonally or nonorthogonally reinforced slabs, or as slabs and plates stiffened with an orthogonal or skew grid of beams and girders. These structures provide the floor system in bridges, and can be adequately analyzed by the equivalent orthotropic plate theory (Huffington, 1956). Other methods used to provide solutions include finite element (Cheung, King, and Zienkiewicz, 1968; Herrmann, 1967; Monforton, 1972), finite difference (Javor, 1967; Naruoka and Ohmura, 1959), and basic functions (Sawko, 1960). Series solutions have also been introduced (Iyengar et al., 1967). In many instances the results have been limited because of approximations, the magnitude of the skew angle, specific boundary conditions, and the type of load considered.

This section examines solutions to the problem of bent skew orthotropic plate structures developed by Kennedy and Gupta (1976) and Gupta and Kennedy (1978), considering practical skew angles, boundary conditions, and distributed and concentrated loads. The present review is therefore a supplement to Section 7-12.

Bending Considerations: Case Study

The plate structure shown in Figure 7-56 is in reality anisotropic because of variations in sections. By choosing, however, the elastic constants of the equivalent structure to resemble the twisting and bending characteristics of an orthogonally stiffened plate system, it becomes valid to analyze the original structure by orthotropic plate theory (Hoppmann, 1955).

Analytical Solutions The basic differential equation governing the lateral deflection is transformed to skew coordinates. In order to generalize the solution, results are obtained for plate structures with the following boundary conditions: (a) all edges are simply supported; (b) all edges are clamped (restrained); and (c) two opposite edges are simply supported, and the other two elastically supported or free.

Three particular cases are considered in the choice of complementary function, corresponding to the following practical classifications: (a) $H^2 > D_y D_x$ (torsionally stiff and flexurally soft plate), (b) $H^2 = D_x D_y$, and (c) $H^2 < D_x D_y$ (torsionally soft and flexurally stiff plate).

The complete solution to the basic differential equation is presented in the form

$$w = w_c + w_c' + w_p \tag{7-84}$$

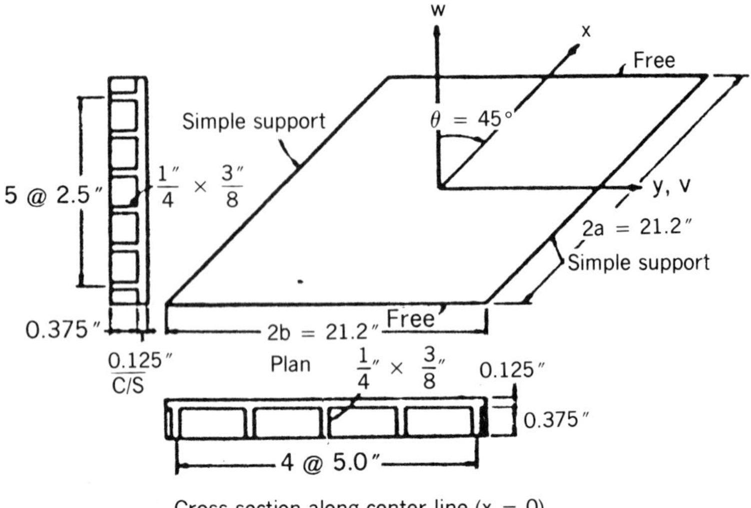

FIGURE 7.56 Example and geometry of skew orthotropic plate structures. (From Kennedy and Gupta, 1976.)

where w is the lateral displacement of the orthotropic plate and w_c, w_c', and w_p are the complementary and particular solution parts, respectively.

The constants, implicit in (7-84), are found from the boundary conditions. The quadrant symmetry is considered by dividing any loading into polar symmetric and antisymmetric components. For a symmetric loading condition, all odd terms in (7-84) vanish, and for the antisymmetric state the even terms vanish. Thus, the number of constants is reduced to one-half for either case. For N number of harmonics in (7-84), a matrix equation of dimension $(8N + 4)$ is generated from the boundary conditions along two adjacent edges. The coefficient of the solution matrix equation can be determined as suggested by Gupta (1974), and the solution of this matrix equation for each of the two loading states yields the arbitrary constants. In this manner, the deflection function w and the bending and twisting moments over the entire domain of the skew orthotropic plate structure are determined.

Parametric Study The influence of skew, aspect ratio, torsional rigidity H, and flexural rigidity ratio D_y/D_x on the bending behavior of skew orthotropic plate structures has been analyzed by Kennedy et al. (1976) in a parametric study. The results obtained from this study also refer to the moments M_x and M_y in bridge slab structures with two opposite edges simply supported and the other two free. Three types of loading are considered: uniform load, concentrated load at the center, and concentrated load very close to the center of one edge.

Effects of Skew These effects are considerable on the moments due to a uniformly distributed load. The moments tend to decrease with increasing skew partly because of a decrease in the effective span. Moments due to a concentrated load, however, are less affected. In general, they show a slight decrease with increasing skew for low values of the ratio D_y/D_x, but they increase with skew at very high D_y/D_x ratios.

Effects of Aspect Ratio b / a In this ratio, $2a$ is the skew width of the orthotropic plate, and $2b$ is the skew length of the orthotropic plate.

The moment M_y caused by a uniformly distributed load and the same moment induced by a concentrated load applied close to the center of an edge increase with increasing aspect ratio. The moment M_x decreases for the former loading and increases for the latter. For a concentrated load applied at the center of a skew orthotropic plate, both moments decrease with increasing aspect ratio.

Effect of Torsional Parameter H For all the three types of loading considered, the moment M_y increases with decreasing H, whereas the moment M_x remains nearly unchanged for a concentrated load at the center of the plate and for a load near the center of a free edge, as well as for a uniformly distributed load when $(b/a) > 1$. However, M_x increases significantly for a uniformly distributed load with decreasing H when $(b/a) \leq 1$. Torsional rigidity reduces deflections and moments and improves the lateral load distribution.

Effects of Rigidity Ratio D_y / D_x For plate structures with small aspect ratios and subjected to a uniformly distributed load, the moment M_x increases with skew when $D_y/D_x = 1$, and appears to attain a stable maximum value at about 45° skew. This variation is not observed with the two concentrated loads. For $D_y/D_x > 1$, the variation in M_x becomes monotonic. In all cases, the moment M_y increases appreciably with an increase in the ratio D_y/D_x.

Effects of Edge Stiffening Edge stiffening does not appear to change the flexural behavior with changing skew. However, the influence of the aspect ratio changes in the presence of edge stiffening. For example, for a concentrated load applied at the center of an edge and for a uniformly distributed load, the moments M_y and M_x decrease with increasing aspect ratio. Edge stiffening does not alter the effects of H and D_y/D_x, and has no influence on the moments at the center of the skew structure for a uniformly distributed and concentrated center load. For a concentrated load near the edge, stiffening decreases the edge moment M_y and increases the edge moment M_x.

Design Example This example demonstrates the use of the foregoing results for preliminary design. A 45° skew prestressed concrete bridge has a

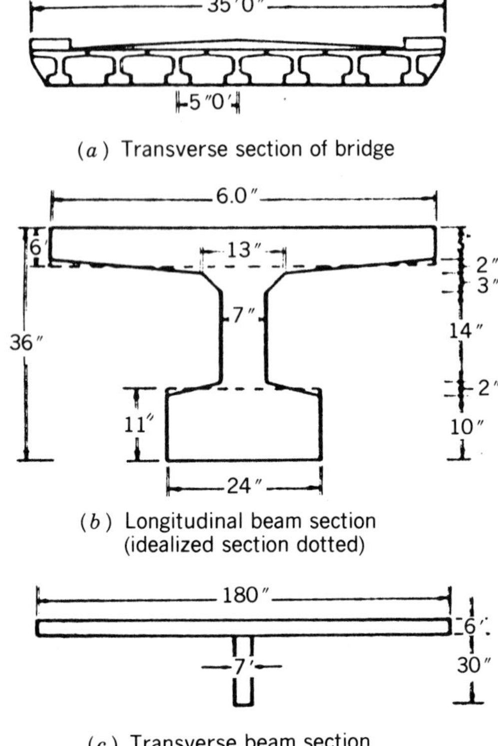

(*a*) Transverse section of bridge

(*b*) Longitudinal beam section (idealized section dotted)

(*c*) Transverse beam section

FIGURE 7.57 Typical bridge section for design example. (From Kennedy and Gupta, 1976.)

longitudinal span of 75 ft and a transverse width of 35 ft. The bridge will be designed for 0.3-ksf (14.37-kN/m^2) dead load and a wheel load equivalent to 50 tonnes (112 kips) at the center. The concrete strength is $f_c' = 5$ ksi (34.5 MN/m^2). Figures 7-57*a*, *b*, and *c* show the transverse deck section, longitudinal beam section, and transverse beam section, respectively. The longitudinal beams are spaced at 5-ft centers, and the transverse beams at 15-ft centers.

The calculated section properties for the longitudinal beams are as follows:

$$\text{Moment of inertia } I = 123{,}400 \text{ in.}^4$$
$$\text{Bottom fiber section modulus } S = 5790 \text{ in.}^3$$
$$D_y = 123{,}400(E/60) = 2057E \text{ in.}^4/\text{inch-width}$$

Likewise, the corresponding properties for the transverse beam are

$$\text{Moment of inertia } I = 75{,}970 \text{ in.}^4$$
$$\text{Bottom fiber section modulus } S = 2525 \text{ in.}^3$$
$$D_x = 75{,}970(E/180) = 422E \text{ in.}^4/\text{inch-width}$$

The calculation of the torsional rigidity of the longitudinal beam section is based on the idealized section shown by the dashed line in Figure 7-57*b*. Thus,

$$
\begin{aligned}
D_{xy} &= G\left[(6)^3(60/6) + (0.255)(7)^3(19) + (0.236)(11)^3(24)\right]\Big/60 \\
&= 82.4E \text{ in.}^4/\text{inch-width}
\end{aligned}
$$

using $G = 0.44E$ and assuming Poisson's ratio for concrete of 0.15. Similarly,

$$D_{yx} = 22.7E \text{ in.}^4/\text{inch-width}$$

Other properties relevant to the design include

$$D_1 = D_2 = (0.15)(6)^3 E/12\left[1 - (0.15)^2\right] = 2.8E \text{ in.}^4/\text{inch-width}$$

Therefore,

$$H = \left[82.4 + 22.7 + (2)(2.8)\right]E/2 = 55.4E \text{ in.}^4/\text{inch-width}$$

Next, we compute the ratio $H/\sqrt{D_x D_y} = 0.06$. The aspect ratio $b/a = 75 \sin 45°/35 = 1.5$.

The relative moment coefficients may be obtained by linear interpolation. For uniform loading, interpolation between values obtained for $b/a = 1$, $b/a = 2.0$, $H = 0.5\sqrt{D_x D_y}$, and $H = 0$, gives a coefficient for

$$
\begin{aligned}
M_y = \{&[37.5 - (0.06)(37.5 - 29.5)/0.5] \\
&+ [44.2 - (0.06)(44.2 - 33.5)/0.5]\}\big/2 = 39.8
\end{aligned}
$$

Similarly, the coefficient for M_x is calculated as 4.4.

For wheel loads equivalent to 50 tons metric (112 kips) at the center, the following moment coefficients are obtained (data not shown): 0.44 for M_y and 0.14 for M_x. Accordingly, the total moments M_y and M_x at the center are

$$
\begin{aligned}
M_y &= \left[(39.8)(300)(75)(35)/(400 \cos 45°)\right] \\
&\quad + \left[(0.44)(50)(2240)/1000\right] = 160 \text{ in.-kips/in.} \\
M_x &= \left[(4.4)(300)(75)(35)/(400 \cos 45°)\right] \\
&\quad + \left[(0.14)(50)(2240)/1000\right] = 28 \text{ in.-kips/in.}
\end{aligned}
$$

With an allowable compressive stress in the concrete of $0.45f'_c = 2.25$ ksi (for applicable code), the required section modulus for the longitudinal beam is $(160)(60)/2.25 = 4280 \text{ in.}^3 < 5790$, OK.

The required section modulus for the transverse beam is (28)(180)/2.25 = 2245 in.^{3} < 2525, OK. Some excess capacity in both the longitudinal and transverse direction may be desirable to accommodate secondary stresses.

Continuous Skew Orthotropic Plate Structures: Case Study

Gupta and Kennedy (1978) have extended the solution to the bending problem to include continuous skew orthotropic plate structures under lateral loads. Relevant factors to be examined include the practical skew angles, distributed and concentrated loads, flexural rigidity ratios, aspect ratios, torsional rigidity, and span ratios.

A typical continuous skew orthotropic plate structure is shown in Figure 7-58. Such a structure may be reinforced, prestressed, or stiffened by a suitable gridwork. Likewise, the application of the orthotropic plate theory is justified. In order to generalize the solution, results are obtained for two- and three-span continuous plate structures.

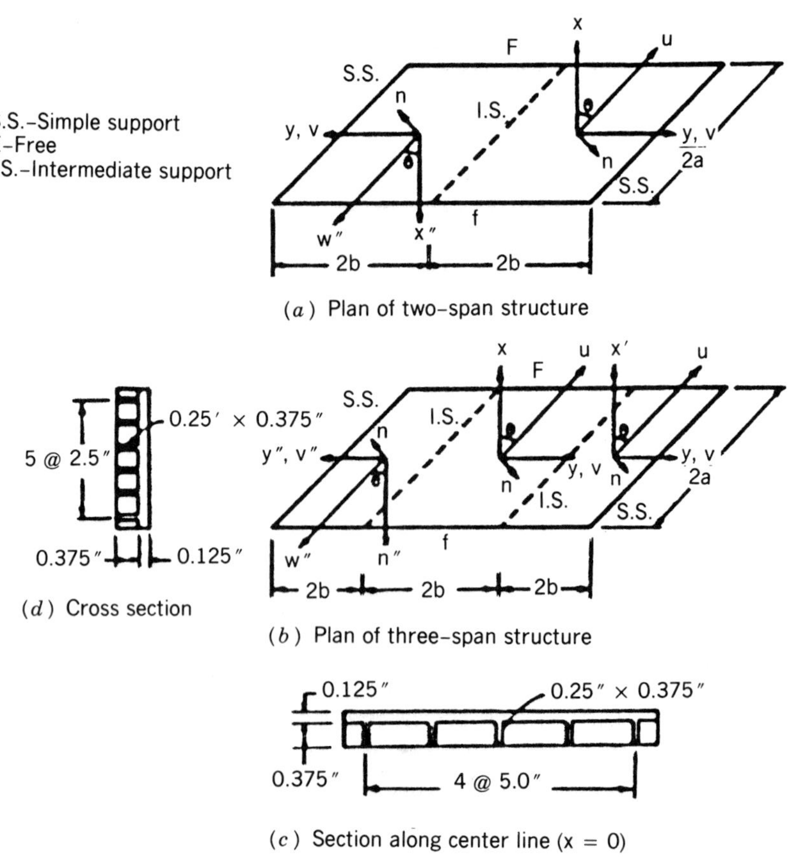

FIGURE 7.58 Typical continuous skew orthotropic plate deck.

Boundary Conditions As in the previous section D_{xy} and D_{yx} are the transverse and longitudinal torsional rigidities per unit width, respectively. Also, D_1 and D_2 are the coupling rigidities per unit width arising from the Poisson ratio effect. The solution given by the basic equation is applicable to the torsionally soft and flexurally stiff plate classification, that is, $H^2 \leq D_x D_y$.

Two-Span Structure A plan of the two-span structure is shown in Figure 7-58*a*. The two spans are assumed to be identical in geometry and stiffness characteristics. The quadrant symmetry is advantageously considered by dividing any loading into polar symmetric and antisymmetric components with respect to the center of the intermediate support. The inclusion of the boundary condition yields eight equations for either the symmetric or the antisymmetric loading component.

Three-Span Structure The two end spans may be assumed to have the same geometry and stiffness properties, offering the advantage of the quadrant geometric symmetry as in the two-span structure as shown in Figure 7-58*b*. There are 12 boundary conditions that must be satisfied.

Structural conformity with these geometric requirements allows the determination of the arbitrary constants in the basic orthotropic equation. For N number of harmonics, a matrix solution equation of dimension $(16N + 8)$ for the two-span structure and $(24N + 12)$ for the three-span plate structure is generated from the corresponding boundary conditions. The solution of the matrix equation for the two loading components yields the deflection function w and therefore the moment and shear for the design.

Using the concept of an equivalent orthotropic plate theory, continuous skew orthotropic plates are analyzed using a Fourier series solution. In addition, these investigators confirm the solution and its convergence in experimental programs for both uniformly distributed and concentrated loads.

Parametric Study A parametric study by Gupta and Kennedy (1978) shows the effect on the deflection and flexural behavior of factors such as skew angle θ, aspect ratio b/a, flexural rigidity ratio D_y/D_x, torsional rigidity parameter $H/\sqrt{D_x D_y}$, and span ratio b'/b. These results are articulated in appropriate diagrams, and may be used directly to generate preliminary designs for the structure types considered in this study. The range covers single-span, two-span continuous, and three-span continuous units with b'/b ratios of 1.0, 0.8, and 0.6, respectively. The results are summarized in Table 7-5.

Bending Moments As expected, the largest moment occurs in single spans, and for 0° skew the orientation of the maximum principal moment at the span center coincides with the longitudinal axis. For small skews, the twisting

TABLE 7-5 Load Effects and Classifications (From Gupta and Kennedy, 1978)

Case	Identification of Results
1	Moment (deflection) at center of single-span slab under uniform load, or under concentrated load at center of single-span slab.
2	Moment at center of one span of a two-span slab when a concentrated load is at center of one span or moment at the center of one span for uniform load,[a] or moment at center of intermediate support when one span is loaded by a concentrated load or when the two-span slab is uniformly loaded.[a]
3	Moments under a concentrated load at center of central span or at center of central span for uniform load,[a] or at center of support for uniform load,[a] or concentrated load at center of central span in three-span slabs, with span ratios $b'/b =$ 1.0, 0.8, and 0.6, respectively.
4	Moments at center of end span either due to a concentrated load at center of end spans or for uniform load,[a] or moment at center of support for uniform load,[a] in three-span slabs, with span ratios $b'/b = 1.0, 0.8$, and 0.6, respectively.

[a]The entire structure is uniformly loaded.

moment M_{xy} at the center is practically negligible, and the principal moments are M_x and M_y. With increasing skew, M_{xy} also increases and causes the maximum principal moment to rise as high as twice the largest value of M_x or M_y, although it is less than the maximum moment at the center for 0° skew.

The maximum principal moment along the edge, equivalent to M_y, occurs at the center of the free edge for 0° skew, and this position shifts toward the obtuse corner with increasing skew. For uniformly loaded plate structures, this maximum moment exceeds the maximum principal moment at the center by about 5 percent for skews less than or equal to 20°, and by about 15 percent for larger skew angles. The maximum moment under a concentrated load along the free edge is typically smaller than the maximum moment under a concentrated load at the center.

For uniformly loaded plates, the ratio of the maximum moment along an interior support to the moment at the center of the same support varies from 1 to 1.06 for 0° skew, 1 to 1.2 for 30° skew, 1 to 1.25 for 45° skew, and 1 to 1.4 for 60° skew. These results are for three-span continuous units with $b/a = b'/b = 1$, and may vary somewhat for other ratios.

For a concentrated load at the center of the structure, the maximum negative moment occurs at the center of the interior support for 0° skew. This position, however, shifts toward the obtuse corner as the skew angle increases. The ratio of the maximum moment along the interior support to the moment at the center of this support varies from 1 to 1.4 for 30° skew, 1 to 1.7 for 45° skew, and 1.4 to 3.0 for 60° skew.

Deflections Variations in deflections are also included in the diagrams. For a uniformly distributed load, the deflection at the center increases rapidly with an increase in skew, and this trend is not consistent for a concentrated load. The effect of edge stiffening is to reduce deflection. Maximum deflections along the free edge are typically greater than at the center for either uniformly distributed or concentrated load located very close to the free edge, and this applies to two- and three-span structures.

The results obtained for deflections, essentially for single spans, may be extended to two- and three-span continuous plates as follows: (a) for two-span continuous skew plates, the maximum deflection may be estimated as 40 and 60 percent of that in single-span slabs for uniformly distributed and concentrated loads, respectively; and (b) for three-span continuous skew plates under a uniformly distributed load, the maximum deflection may be estimated as 50 and 100 percent of that in single-span slabs with 0° and 60° skews, respectively (for a concentrated load the maximum deflection may be taken the same as in the single-span structure).

Design Example As in the previous section, the use of the diagrams (Gupta and Kennedy, 1978) will be demonstrated by a numerical example. A 45° skew two-span continuous prestressed orthotropic concrete bridge has spans 90 ft long and 42.25 ft wide. Details are shown in Figure 7-59 and include a deck section, longitudinal beam section, and transverse beam section. The bridge will be designed for 0.3-ksf dead load and a wheel load equivalent to 40 tonnes (88 kips) at the center. The concrete strength is $f'_c = 5$ ksi. The longitudinal beams are spaced at 5-ft centers, and transverse beams are located at 15-ft intervals.

The calculated section properties for the longitudinal beam are as follows:

$$\text{Moment of inertia } I = 123{,}400 \text{ in.}^4$$
$$\text{Bottom fiber section modulus } S = 5790 \text{ in.}^3$$
$$D_y/E = 123{,}400/60 = 2057 \text{ in.}^4/\text{inch-width}$$

For the transverse beam, the corresponding properties are

$$\text{Moment of inertia } I = 75{,}970 \text{ in.}^4$$
$$\text{Bottom fiber section modulus } S = 2525 \text{ in.}^3$$
$$D_y/E = 75{,}970/180 = 422 \text{ in.}^4/\text{inch-width}$$

The calculation of the torsional rigidity of the longitudinal beam section is again based on the idealized section outlined by the dashed line in Figure

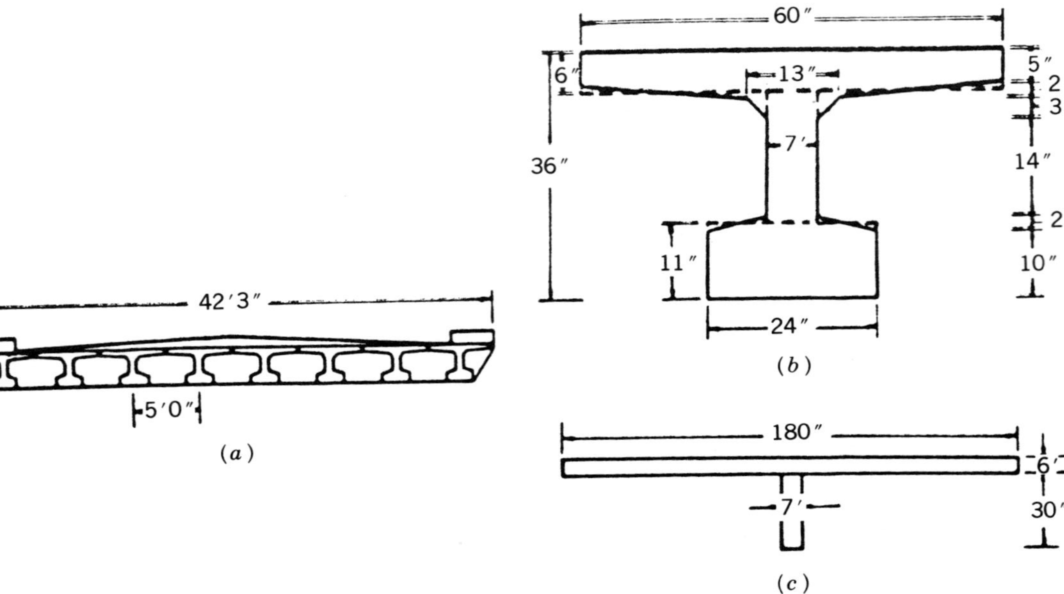

FIGURE 7.59 Bridge details of design example: (*a*) transverse deck section; (*b*) longitudinal beam section; (*c*) transverse beam section.

7-59*b*. According to Jackson (1968), we compute $D_{xy}/E = 181$ in.^{4}/inch-width using $G = 0.44E$ and Poisson's ratio for concrete of 0.15. Likewise, $D_{yx}/E = 179$ in.^{4}/inch-width.

The coupling rigidities D_1 and D_2 arising from Poisson's ratio effect are calculated with reference to the two neutral axes, resulting in $D_1/E = 128$ in.^{4}/inch-width and $D_2/E = 10$ in.^{4}/inch-width. Therefore,

$$H/E = (181 + 179 + 128 + 10)/2 = 249 \text{ in.}^4/\text{inch-width}$$

From the foregoing, we compute the ratio $H/\sqrt{D_x D_y} = 0.27$. The aspect ratio $b/a = 90 \sin 45°/42.25 = 1.5$, and $D_y/D_x = 4.87$.

As in the preceding example, the moment coefficients are obtained by linear interpolation. For uniform loading, interpolating between the values given in the diagrams (not shown) for $b/a = 1$ and 2 and $H/\sqrt{D_x D_y} = 0.5$ and 0.0, we estimate the coefficient for

$$M_y = [19.6 - (0.27)(19.6 - 15.6)/0.5 + 22.4$$
$$-(0.27)(22.4 - 10.4)/0.5]/2 = 19.1$$

In a similar manner, the coefficient for M_x is 2.4.

For wheel loads equivalent to 40 tonnes (88 kips) at the center, the following moment coefficients are obtained from Gupta and Kennedy (1978): 47.1 for M_y and 16.9 for M_x. The total moments at the center are therefore

$$M_y = (19.1)(0.3)(45)^2/100 + (47.1)(40)(2240)/1000/100$$
$$= 158 \text{ in.-kips/in.}$$

and

$$M_x = 30 \text{ in.-kips/in.}$$

The coefficients for the negative moment at the center of the pier support are -33.1 for uniform loading and -14.6 for the wheel loads, respectively, yielding a moment of -214 in.-kips/in. For the 45° skew structure, the negative moment is increased by 15 percent to -246 in.-kips/in. as a fair approximate estimate of the maximum moment along the support.

With an allowable compressive stress in the concrete of $f_c = 0.45 f_c' = 2.25$ ksi (according to applicable specifications), the required section modulus for the longitudinal beam at the center is $(158)(60)/2.25 = 4224$ in.$^3 < 5790$, OK. Likewise, the required section modulus for the transverse beam is computed as 2400 in.$^3 < 2525$, OK.

In order to resist the negative moment at the support, the slab thickness is increased to 27 in. at the pier, giving a section modulus in excess of 120 in.^{3}, which satisfies the minimum section modulus requirement $246/2.25 = 110$

in.^{3}. Further calculations show that this thickness can be linearly reduced to the original slab thickness at a distance approximately one-seventh of the span length.

7-15 DESIGN EXAMPLE

A two-girder, three-span continuous bridge with spans of 360, 444, and 360 ft has the typical cross section shown in Figure 7-60. The orthotropic steel plate deck has open ribs supported on floor beams located at 12-ft intervals. Other data are as follows: rib spacing, 18 in.; wearing surface, 2 in.-thick asphalt; materials, A588, A572, and A36 steel; and live load, AASHTO HS 20. The procedure is as proposed by Hall (1971).

Deck Plate The minimum deck plate thickness is $t_p = 0.007(a\sqrt[3]{p})$, where a is the rib spacing and p is the wheel pressure for a 12-kip wheel load. Referring to AASHTO Article 10.41.2, we obtain the following:

$$\text{Width } 2g \text{ perpendicular to traffic} = 20 + 2 \times 2 = 24 \text{ in.}$$

$$\text{Length } 2c \text{ in direction of traffic} = 8 + 2 \times 2 = 12 \text{ in.}$$

giving a contact area $A = 24 \times 12 = 288$ in.^{2} and $p = 1.3 \times 12{,}000/288 = 54$ psi (including impact).

For $a = 18$ in. and $p = 54$ psi, we obtain $t_p = 0.007 \times (18)\sqrt[3]{54} = 0.44$ in., use a 1/2-in.-thick plate.

The local plate stress may be computed by analyzing the plate as a continuous beam on rigid supports as shown in Figure 7-61. From previous analysis, the shaded area under the curve is $A = 0.4506 \times 2 = 0.901$, giving a support moment $M_s = AwL = 0.901 \times 54 \times 18 = -876$ lb-in. The section modulus of the plate is $S = t^2b/6 = 1/24$, and the induced local stress is

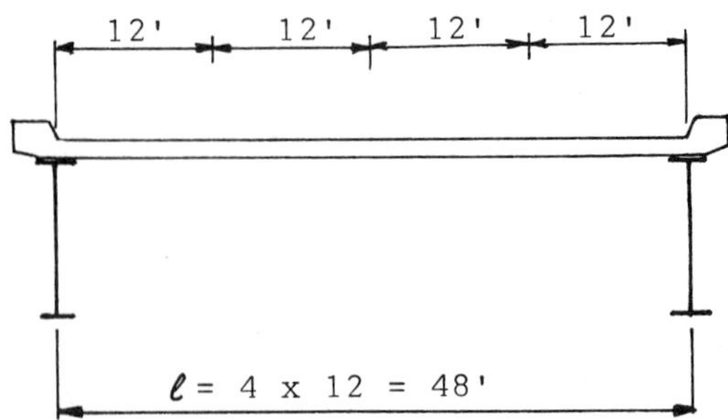

FIGURE 7.60 Typical bridge section for the design example.

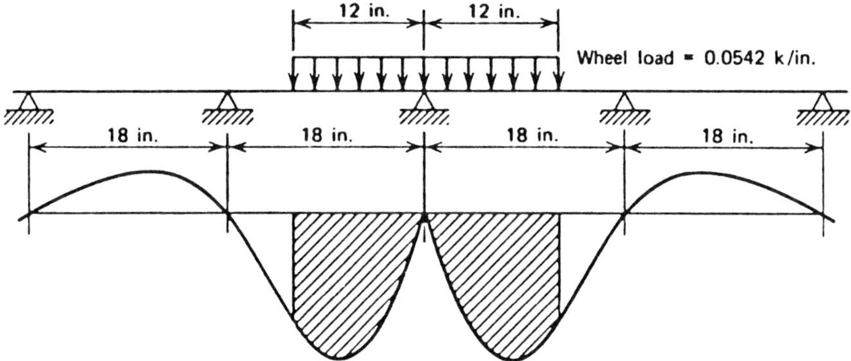

FIGURE 7.61 Uniformly distributed wheel load for computing local stress in deck plate; continuous beam on rigid supports.

calculated as

$$f_T = M/S = 876/(1000/24) = 21.02 \text{ ksi} \qquad \text{(Transverse)}$$

Therefore, the longitudinal stress is ($\mu = 0.3$)

$$f_L = 0.3 \times 21.02 = 6.30 \text{ ksi}$$

Rigid Support For a floor beam spacing of 12 ft (144 in.), the rib length is $l_1 = 0.7l = 0.7 \times 144 = 100.8$ in. The effective width a_0 is $2g/a = 24/18 = 1.33$ (where $2g = 20 + 4$). From the graphs of Figure 7-62, for $B/a = 1.33$ and $a^*/a = 1.74$, we calculate $a^* = 1.74 \times 18 = 31.3$ in. (ideal rib spacing). Also

$$\beta = \pi a^*/l_1 = 3.14 \times 31.3/100.8 = 0.976$$

From the graphs of Figure 7-63, for $\beta = 0.976$ and $a_0/a^* = 0.90$, the effective plate width is

$$a_0 = 0.90a^* = 0.90 \times 31.3 = 28.2 \text{ in.}$$

Rib Geometry For the loading shown in Figure 7-64, the induced forces R_0 and R_0' (acting on the ribs) are calculated as follows (assuming simple beam action between rib supports)

$$R_0 \times 18 = (0.5 \times 12)(6 + 6) \qquad \text{or} \qquad R_0 = R_0' = 6 \times 12/18 = 4 \text{ kips}$$

The total reaction per rib is $R = 8$ kips, giving $M = PL/4 = 8 \times 1.3 \times 12/4 = 31.2$ ft-kips per rib. Assuming that the allowable live load stress is 11 ksi, a preliminary section modulus is $S = 31.2 \times 12/11 = 34$ in.^{3}.

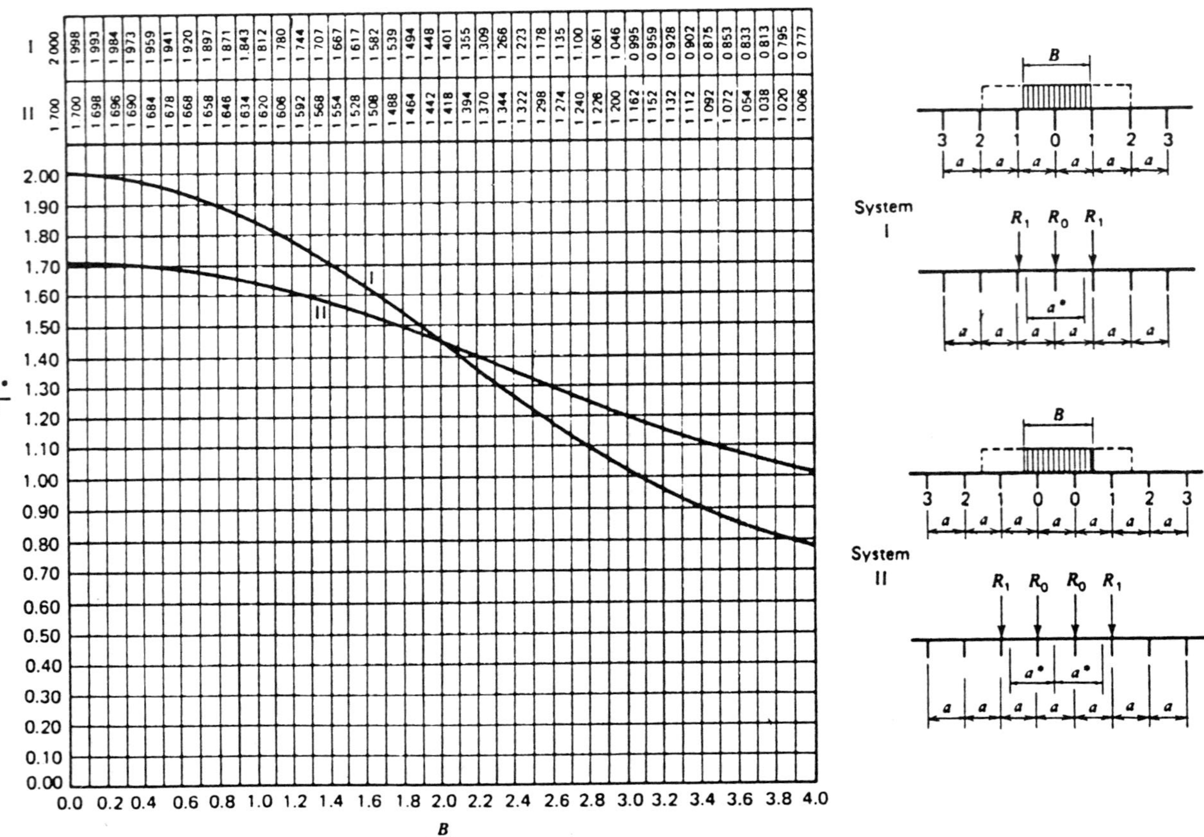

I	II
2 000	1 700
1 998	1 700
1 993	1 698
1 984	1 696
1 973	1 690
1 959	1 684
1 941	1 678
1 920	1 668
1 897	1 658
1 871	1 646
1.843	1 634
1 812	1 620
1 780	1 606
1 744	1 592
1 707	1 568
1 667	1 554
1 617	1 528
1 582	1 508
1 539	1 488
1 494	1 464
1 448	1 442
1 401	1 418
1 355	1 394
1 309	1 370
1 266	1 344
1 223	1 322
1 178	1 298
1 135	1 274
1.100	1 240
1 061	1 228
1 046	1 200
0 995	1 162
0 959	1 152
0 928	1 132
0 902	1 112
0 875	1 092
0 853	1 072
0 833	1 054
0 813	1 038
0 795	1 020
0 777	1 006

FIGURE 7.62 Load on rib under wheel and ideal spacing of flexible ribs. (From Heins and Firmage, 1979.)

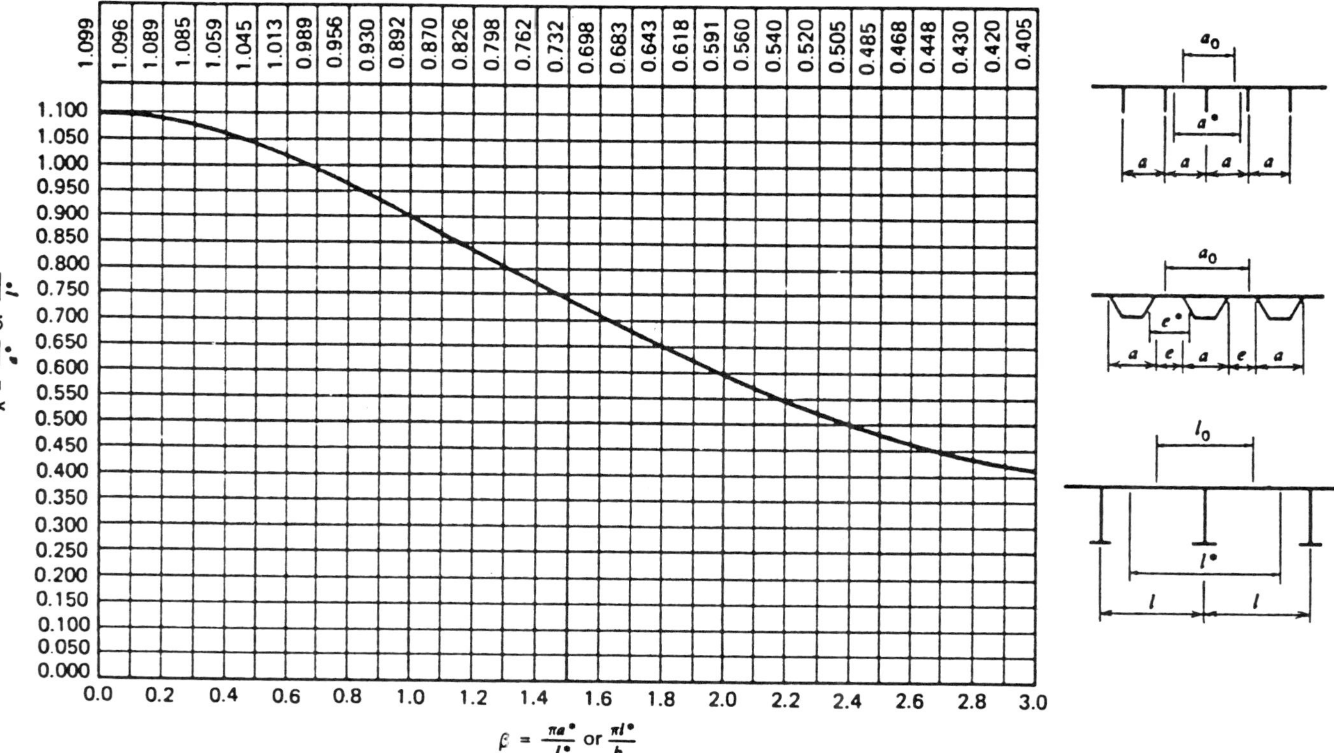

FIGURE 7.63 Effective width of orthotropic deck plate, a and l_0 = effective width of the deck plate; a^* and l^* = ideal span of T ribs, box ribs, or floor beams.

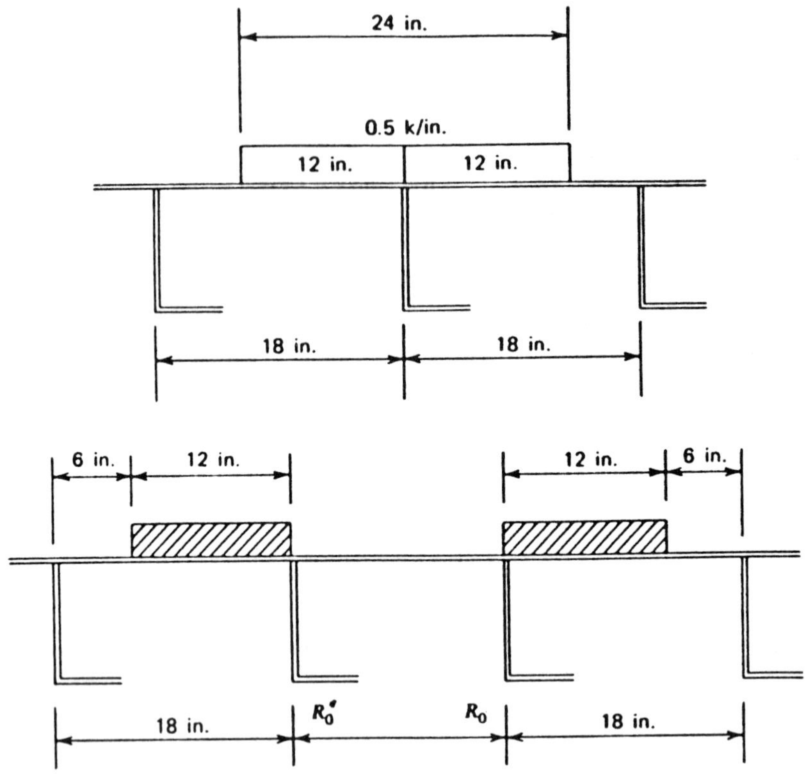

FIGURE 7.64 Suggested loading configuration for preliminary rib selection, simple beam reactions. (From Heins and Firmage, 1979.)

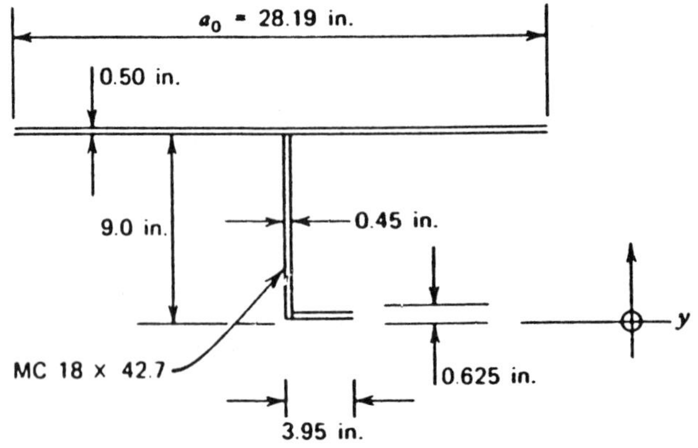

FIGURE 7.65 Rib cross section and geometry. (From Heins and Firmage, 1979.)

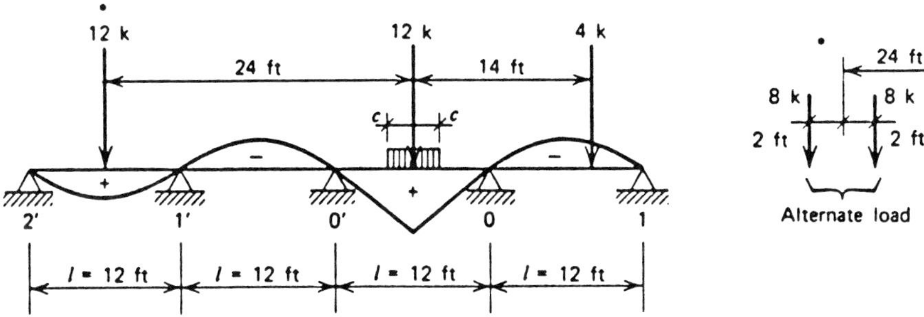

FIGURE 7.66 Truck load position for evaluating moment in midspan of rib.

A channel section MC 18 × 42.7 cut in half as shown in Figure 7-65 has the following section properties:

$$A_T = 20.3 \text{ in.}^4 \qquad I = 219.6 \text{ in.}^4 \qquad S_B = 29.92 \text{ in.}^3 \qquad S_T = 101.7 \text{ in.}^3$$

Therefore, the analysis will continue using the section of Figure 7-65 as the final rib section.

Live Load Moment M_c in Rib The position of truck midspan moment is shown in Figure 7-66. Note that the center wheel is distributed along a length $2c$, whereas the end wheels are applied as concentrated loads.

However, before this moment can be computed, we must determine the live loads on the ribs. Referring to Figure 7-67, for $B/a = 1.33$, $R_0/P = 0.76$, we calculate the following:

For alternate load, wheel = 8 kips, and

$$R_0 = 0.76 \times 8 \times 1.30 = 7.9 \text{ kips}$$

For truck load, wheel = 12 kips, and

$$R_0 = 0.76 \times 12 \times 1.30 = 11.85 \text{ kips}$$

The moment at midspan of the rib deck for the distributed load in span 0′–0 as shown in Figure 7-66 is given by (7-29), or

$$M_{c(0-0')} = 2R_0cs\left[0.171 - 0.250\frac{c}{s} + 0.106\left(\frac{c}{s}\right)^2\right]$$

$$= 2(11.85)(0.5)(12)\left[0.171 - 0.25(0.5/12) + 0.106(0.5/12)^2\right]$$

or

$$M_{c(0-0')} = 22.84 \text{ ft-kips}$$

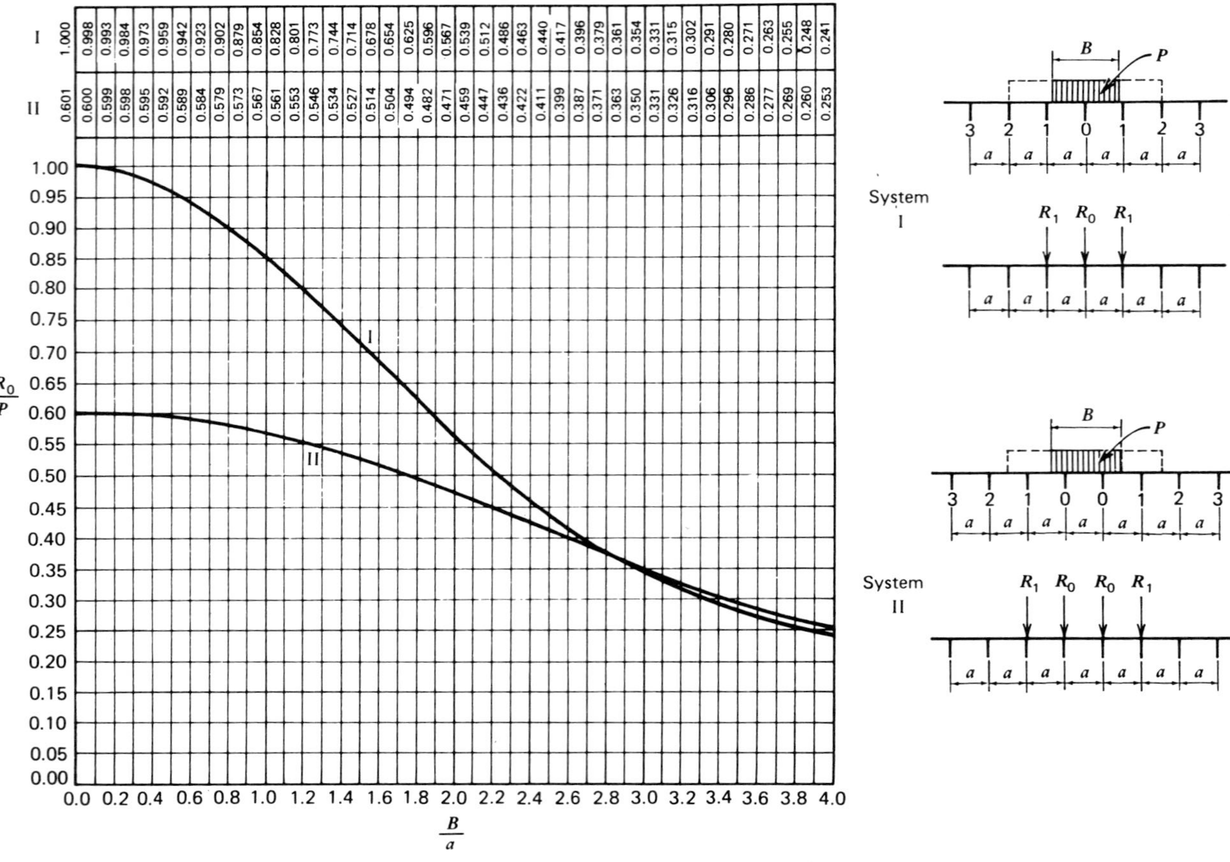

I	1.000	0.998	0.993	0.984	0.973	0.959	0.942	0.923	0.902	0.879	0.854	0.828	0.801	0.773	0.744	0.714	0.678	0.654	0.625	0.596	0.567	0.539	0.512	0.486	0.463	0.440	0.417	0.396	0.379	0.361	0.354	0.331	0.315	0.302	0.291	0.280	0.271	0.263	0.255	0.248	0.241
II	0.601	0.600	0.599	0.598	0.595	0.592	0.589	0.584	0.579	0.573	0.567	0.561	0.553	0.546	0.534	0.527	0.514	0.504	0.494	0.482	0.471	0.459	0.447	0.436	0.422	0.411	0.399	0.387	0.371	0.363	0.350	0.331	0.326	0.316	0.306	0.296	0.286	0.277	0.269	0.260	0.253

FIGURE 7.67 Open-rib loading distribution.

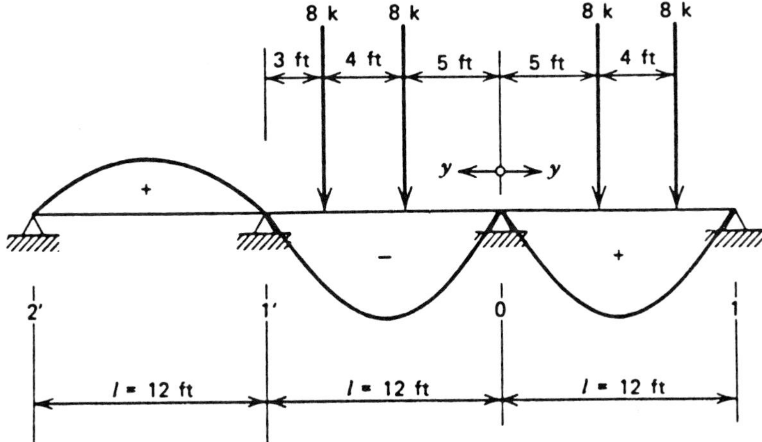

FIGURE 7.68 Truck load position for main rib moment at support M_s.

The contribution of the two end wheel loads positioned in spans 2′–1′ and 0′–1′ is estimated by (7-28), where $y = 8.0$ ft. For the front wheel, $R_0 = 1.3 \times 4.0 \times 0.76 = 3.95$ kips, and for the rear wheel, $R_0 = 11.85$ kips. These moments are computed as -0.95 and 1.10 ft-kips for the front and rear wheel, respectively. Therefore, the total midspan moment is

$$M_c = 22.84 - 0.95 + 1.10 = 23.0 \text{ ft-kips} \qquad \text{with the front wheel}$$

$$M_c = 22.84 + 1.10 = 23.9 \text{ ft-kips} \qquad \text{without the front wheel}$$

Live Load Support Moment M_s in Rib Likewise, the support moment M_s in the rib can be calculated from (7-23*a*) for the truck load positioned as shown in Figure 7-68.

For the loads in spans 0–1 and 0–1′, the term $(-0.268)^m$ is set equal to 1. Applying the previous equation for spans 0–1 and 0–1′ gives, *for an 8-kip load and $y = 5$ ft*,

$$y/s = 0.42 \qquad (y/s)^2 = 0.174 \qquad (y/s)^3 = 0.073 \qquad (s = l)$$

Therefore,

$$\begin{aligned} M_s &= (7.9 \times 12)[-0.5(0.42) + 0.866(0.174) - 0.366(0.073)] \\ &= -7.9 \times 12 \times (0.083) \end{aligned}$$

For an 8-kip load and $y = 9$ ft,

$$y/s = 0.75 \qquad (y/s)^2 = 0.563 \qquad (y/s)^3 = 0.422$$

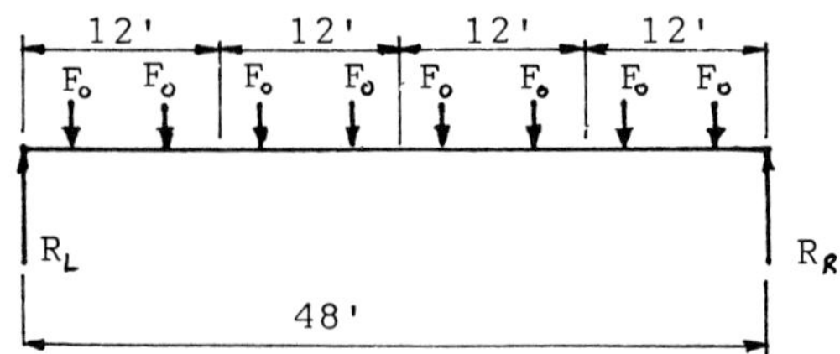

FIGURE 7.69 Truck positions for four lanes loaded; maximum live load moment in floor beams.

Therefore,

$$M_s = (7.9 \times 12)[-0.5(0.75) + 0.866(0.563) - 0.366(0.422)]$$
$$= -7.9 \times 12 \times (0.043)$$
$$\text{Total moment } M_s = -23.9 \text{ ft-kips}$$

Live Load Moment in Floor Beams For all lanes loaded, the trucks are positioned in their respective lanes and induce reactions F_0 as shown in Figure 7-69. Reactions are computed considering the main longitudinal girders as the end supports. From symmetry, we obtain $R_R = R_L = 4F_0$.

$$M_A = [24 \times 4 - (2 + 8 + 14 + 20)]F_0 = 52F_0 \text{ ft-kips}$$

where F_0 is one wheel reaction R_0 with impact. Reduction in load intensity (75 percent) for multiple lane loading gives

$$M_A = 52 \times 0.75 = 39F_0 \text{ ft-kips}$$

For three lanes loaded (extreme right lane unloaded), we can write

$$48R_L = (44 + 38 + 32 + 26 + 22 + 16)F_0$$

or

$$R_L = (178/48)F_0 = 3.7F_0$$
$$M_A = [22 \times 3.7 - (18 + 12 + 16)]F_0 = 35.4F_0 \text{ ft-kips}$$

Reduction in load intensity (90 percent) for multiple lane loads gives

$$M_A = 35.4 \times 0.9 = 31.9F_0 \text{ ft-kips}$$

or four lanes loaded control.

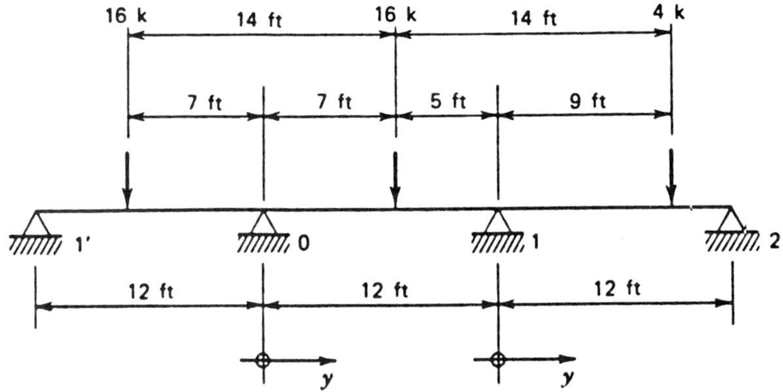

FIGURE 7.70 Longitudinal position of truck for maximum floor beam reaction.

The total wheel load reaction on the floor beam is computed from the application of influence line reaction coefficients, positioning the truck longitudinally as shown in Figure 7-70, which is the position giving maximum reaction at support 0.

The reactions are calculated from (7-31). Thus, for span 0–1 and $y = 7$ ft, $y/s = 0.58$, $(y/s)^2 = 0.34$, $(y/s)^3 = 0.197$, we obtain

$$R_0 = [1 - 2.196(0.34) + 1.196(0.197)] = 0.489 \qquad \text{(Also for span 0–1')}$$

Likewise, for span 1–2, using (7-31) and $y = 9$ ft, $y/s = 0.75$, $(y/s)^2 = 0.56$, $(y/s)^3 = 0.42$, $m = 1$, we calculate

$$R_0 = [-0.804(0.75) + 1.392(0.56) - 0.588(0.42)] = 0.065$$

Therefore, the total reaction at support 0 including impact is computed as

$$F_0 = 1.27[2 \times 16 \times 0.489 + 4 \times 0.065] = 20.23 \text{ kips}$$

The maximum moment induced in the floor beam is

$$M_A = 39F_0 = 39 \times 20.23 = 789 \text{ ft-kips}$$

Dead Load Moment in Ribs Using the geometric configuration shown in Figure 7-65, the dead load resisted by one rib is

2-in. asphalt	= 30 lb
1/2-in. deck plate	= 31 lb
Rib MC 18 × 42.7 (half-section)	= 21 lb
Total dead load w for 18-in. span	= 82 lb/ft

Using continuous moment coefficients, the moments are computed as follows:

At midspan $M_L = wl^2/24 = 82(12)^2/24(1000) = 0.49$ ft-kips

At support $M_L = wl^2/12 = -0.98$ ft-kips

Approximate Floor Beam Section For the computed live load moment of 789 ft-kips and using $f_b = 27$ ksi, the required section modulus for the floor beam is $S = 789 \times 12/27 = 351$ in.^{3}. For first trial, we select a web 42 in. × 7/16 in. and a bottom flange plate 10 in. × 9/16 in.

Dead Load Moment in Floor Beams The dead load acting on one floor beam is as follows:

2-in. asphalt	= 240 lb/ft
Ribs	= 171
Deck plate	= 245
Web and flange plate	= 82
Total dead load w for 12-ft spacing	= 738 lb/ft

The dead load moment in the floor beam is therefore

$$M = 0.125wl^2 = 738(48)^2/8 \times 1000 = 212 \text{ ft-kips}$$

Influence of Elastic Supports As mentioned in the foregoing sections yielding of the floor beams has considerable effect on the effective span of the ribbed plate, which in turn influences the effective plate width. When the floor beams are flexible, the effective span is greater than the floor-beam spacing. Therefore, for a complete solution to the design example, the influence of elastic supports should be quantified.

7-16 TYPES OF ORTHOTROPIC PLATE BRIDGES

Most applications of steel plate orthotropic decks are in conjunction with plate girders. However, this predominance does not exclude the use of orthotropic plates with other structural bridge systems, and noble applications are found in cable-stayed, suspension, arch, truss, and movable bridges.

Plate Girder Bridges Application of the orthotropic deck in plate girders makes it possible to reduce the dead weight of the superstructure and, as a result, to increase the spans beyond the conventional limits. With wider application of the weldable high-strength steels, it is possible to design welded girders with considerable cross-sectional area in the flanges.

The web thickness of the girder may be kept small, usually 9/16 in., by appropriate stiffening of the web with horizontal stiffeners, and this has become standard practice even with very deep girders. Existing bridges have demonstrated that the orthotropic deck can be successfully combined with various plate girder arrangements. These may include a two-girder system, multiple-girder bridges, and two pairs of main girders.

Box girders have been used in conjunction with orthotropic decks. Wide box girders have the top and bottom flanges stiffened by longitudinal ribs. The resulting advantage is the considerable torsional stiffness of the box girder, allowing a slender size and pleasing appearance. The orthotropic bridge may have a box with two webs, a box of multiple-web construction, or a system of inclined webs.

Cable-Stayed Bridges Cable-stayed bridges occupy an intermediate position between unstiffened girders and suspension bridges. Although structural-steel box girders incorporating an orthotropic deck have been used in the past, more recent applications suggest that the system is not always economically competitive with segmental box girders. The result is a trend toward structural-steel composite deck construction with two cable-stayed planes.

Suspension Bridges The stability of a suspension bridge largely depends on the dead weight of the superstructure and the rigidity of the stiffening girders or trusses. Although the use of a plate deck results in a reduced dead weight, this effect is more than offset by the accompanying increase in flexural and torsional rigidity.

The use of orthotropic steel plate decks in suspension bridges is documented in structures with plate girders or trusses as stiffening members. If air slots are required in the deck near the stiffening trusses for increased stability, they may be provided by introducing a horizontal truss between the top chord of the stiffening truss and the steel plate deck.

Arch Bridges In a stiffened arch system, the steel plate deck acts together with the girders and contributes to the flexural rigidity. In a stiffened tied arch bridge, such as the one shown in Figure 10-4*k*, the plate deck also participates in resisting the tensile stresses of the girders acting as ties.

Truss and Movable Bridges In truss-type bridges, the orthotropic deck participates in resisting the chord stresses. An example is the seven-span continuous truss bridge over the river Fulda in Bergshausen, Germany, built in 1962. The deck for this bridge is part of the upper chord. The two trusses are spaced 19 ft 8 in. apart. The orthotropic plate has triangular box section longitudinal ribs with an extended T section. The top chord transmits the web member forces into the deck and provides restraint against buckling. The Ems Canal highway bridge in Duisburg, Germany, has the orthotropic deck in the lower chord system.

The use of the orthotropic plate in movable bridges produces dead weight reduction. In addition, it adds considerable rigidity to the bridge and allows a shallower depth in the superstructure. A solid roadway surface is a further advantage.

REFERENCES

American Institute of Steel Construction, 1938. "The Battledeck Floor for Highway Bridges," Committee on Technical Research.

Ang, A. H. S. and L. A. Lopez, 1968. "Discrete Model Analysis Elastic-Plastic Plates," *J. Eng. Mech. Div. ASCE*, Vol. 94, No. EM1, Feb., pp. 271–293.

ASCE, 1974. "Commentary on Orthotropic Deck Bridge Specification," ASCE Task Committee on Orthotropic Plate Bridges, Committee on Metals, *J. Struct. Div. ASCE*, Vol. 100, No. ST1, Jan. pp. 109–128.

Bares, R. and C. Massonet, 1968. *Analysis of Beam Grids and Orthotropic Plates by the Guyon–Massonet–Bares Method*, Frederick Unger, New York.

Bhaumic, A. K. and J. T. Hanley, 1967. "Elasto-Plastic Plate Analysis by Finite Differences, *J. Struct. Div. ASCE*, Vol. 93, No. ST5, Proc. Paper 5511, Oct., pp. 279–294.

Blodgett, O. W., 1963. "Application of Welding to the Design of Orthotropic Bridges," Lincoln Electric Company Pamphlet No. 1302.158, Feb.

Blodgett, O. W. 1963. "Orthotropic Bridge Decks," Lincoln Electric Company Pamphlet, No. 11-2, p. 4.

Bouwkamp, J. G. and G. H. Powell, 1967. "Structural Behavior of an Orthotropic Steel Deck Bridge," Structures and Materials Research Report No. 67-27, Univ. Calif., Berkeley, Aug.

Ceradini, G., C. Gavarini, and M. P. Petrangeli, 1973. "Experimental Study of Steel Orthotropic Plate Behavior under Alternate Loads," Technical Report Instituto di Scienza delle Costruzloni, Facolta di Ingegnerie, Universita di Roma, Rome, July.

Ceradini, G., C. Gavarini, and M. P. Petrangeli, 1975. "Steel Orthotropic Plates under Alternate Loads," *J. Struct. Div. ASCE*, Vol. 101, No. ST10, pp. 2015–2026.

Cheung, Y., I. King, and O. Zienkiewicz, 1968. "Slab Bridges with Arbitrary Shape and Support Conditions: A General Method of Analysis Based on Finite Elements," *J. Institution Civ. Engineers*, London, Vol. 40, May.

Chowdhury, P. K., 1975. "Elasto-Plastic Analysis of Orthotropic Skew Plates," Ph.D. Thesis, Univ. Windsor, Canada.

Colville, J., E. B. Becker, and R. W. Furlong, 1973. "Large Displacement Analysis of Thin Plates," *J. Struct. Div. ASCE*, Vol. 99, No. ST3, March, pp. 349–364.

Cornelius, W., 1951. *Orthotropic Plate, Ole Neke Köln-Mulheimer Brücke*, Herausgegeben von der Stadt Kölin, p. 85.

Cornelius, W., 1952. "The Theory of Orthogonal Anisotropic Plate," *Stahlbau*, Vol. 21, pp. 21, 43, 60.

Croci, C. and M. P. Petrangeli, 1968. "Indagina Teorico—Sperimentale su un Modello di Impalcato da Ponte in Acciaio a Plastra Ortotropa," *Costruzioni Metalllche*, No. 6.

Cusens, D. R. and R. P. Pama, 1975. *Bridge Deck Analysis*, Wiley, New York.

Davis, H. L. and A. A. Toprac, 1966. "Fatigue Testing of Ribbed Orthotropic Bridge Elements," Research Report No. 77-1, Center for Hwy. Research, Univ. Texas, Austin, June.

De Fries-Skene, A. and A. C. Scordelis, 1964. "Direct Stiffness Solution for Folded Plates," *J. Struct. Div. ASCE*, Vol. 90, No. ST4, August, pp. 15–47.

DIN 1073, 1969. *Steel Highway Bridges, Design Basis and Commentary*, Deutscher Normenausschuss, Bamberg, Germany, Sept.

Dowling, P. J., 1966. "Research on Orthotropic Steel Plate Bridge Decks," British Construction Steelwork Assoc., Conf. on Struct. Steelwork, London.

Erzuzumlu, H. and A. A. Toprac, 1970. "The Deck Elements of the Ammi System," Technical Report P550-12, Structures Fatigue Research Lab., Univ. Texas, Austin, March.

Farago, B. and W. Chan, 1960. The Analysis of Steel Decks with Special Reference to the Highway Bridges at Amara and Kut in Iraq. Proc. of the Institution of Civ. Eng., London, Vol. 16, May, p. 1.

Fischer, G., 1952. Die Berechnung der Stahlfahrbahntafel der Bürgermeister-Smidt-Brücke in Brehmen. *Der Stahlbau*, Vol. 21, pp. 213, 237.

Fung, C., 1965. "An Experimental Approach to Analysis of Continuous 45 Skew Slabs," Thesis, Univ. Toronto, Canada.

Gallagher, R. H., 1972. "Geometrically Nonlinear Finite Element Analysis," *Proc. Specialty Conf. on Finite Element Method in Civ. Eng.*, McGill Univ., Montreal, pp. 1–33.

Gangarao, H. V. S., P. R. Raju, and N. R. Koppula, 1992. "Behavior of Concrete Filled Steel Grid Decks," TRR 92-0818, Transportation Research Board, Washington, D.C.

Girkmann, K. 1956. *Flachentragwerke* Vierte Auflage, Springer, Wien.

Glockner, P. G., K. K. Verna, and H. A. R. de Paiva, 1971. "Study of the Behavior of a Three-Panel Orthotropic Steel Plate Deck," *Trans. Eng. Inst. Canada*, Vol. 14, No. A-3, March.

Goldstein, A., E. Lightfoot, and F. Sawko, 1961. "The Analysis of a Three Span Continuous Grillage Having Varying Section Properties," *Struc. Engineer*, Vol. 39, Aug., pp. 245–254.

Gupta, D. S. R., 1974. "Orthotropic Continuous Skew Plates," Ph.D. Thesis, Univ. Windsor, Canada.

Gupta and Kennedy, 1978.

Guyon, Y., 1946. "Calcul de Ponts Larges à Poutres Multiples Solidarisées par des Entretoises," *Annales Ponts Chaussées*, Vol. 116, p. 553.

Hall, D. H., 1971. "Orthotropic Bridges-PD 2032," Bethlehem Steel Co., Bethlehem, Pa., March.

Hansch, H. and G. Mueller, 1961. *Fatigue Tests on Hollow Rib Connections*, Schweisstechnik, Berlin, May.

Hawranek, A. and O. Steinhardt, 1958. *Theorie und Berechnung der Stahlbrücken*, Springer-Verlag, Berlin.

Heins, C. P. and D. A. Firmage, 1979. *Design of Modern Steel Highway Bridges*, Wiley, New York.

Heins, C. P. and C. T. G. Looney, 1966. "The Solution of Continuous Orthotropic Plates on Flexible Supports as Applied to Bridge Structures," Civ. Eng. Report No. 10, Univ. Maryland, College Park, March.

Heins, C. P. and C. T. G. Looney, 1968. "Bridge Analysis Using Orthotropic Plate Theory," *J. Struct. Div. Am. Soc. Civ. Eng.*, Vol. 93, No. ST2, Feb.

Heins, C. P. and C. T. G. Looney, 1969. "Bridge Tests Predicted by Finite Difference Plate Theory," *Struct. Div. Am. Soc. Civ. Eng.*, Vol. 95, No. ST2, Feb.

Henderson, W., 1954. "British Highway Bridge Loading," *Proc. Institution Civ. Engineers*, London, Part II, June.

Hendry, A. W. and L. G. Jaeger, 1958. *The Analysis of Grid Frameworks and Related Structures*, Chatto and Windus, London.

Herrmann, L. R., 1967. "Finite Element Bending Analysis for Plates," *J. Eng. Mechanics Div. ASCE*, Vol. 93, No. EM5, Proc. Paper 5822, Oct., pp. 13–26.

Hetenyi, M., 1938. *A Method of Calculating Grillage Beams*, S. Timoshenko 60th Anniversary Volume, MacMillan, New York.

Homberg, H., 1951. *Kreuzwerke*, Springer-Verlag, Berlin.

Homberg, H. and J. Weinmeister, 1956. *Einflussflachen für Kreuzwerke*, Springer-Verlag, Berlin.

Hoppmann, W. H., 1955. "Bending of Orthogonally Stiffened Plates," *J. Appl. Mech. ASME*, Vol. 77, June, pp. 267–271.

Hoppmann, W. H., N. J. Huggington, and L. S. Magness, 1956. "A Study of Orthogonally Stiffened Plates," *J. Appl. Mech., Trans. ASME*, Vol. 78, pp. 343–350.

Huber, M. T., 1914. "Die Grundlagen einer Rationellen Berechnung der Kreuzweise Bewehrten Eisenbetonplätten," Zeitschrift des Osterreichischen Ingenieur—u. Architekten—*Vereines*, Vol. 66, No. 30.

Huber, M. T., 1923. "Die Theorie der Kreuzweise bewehrten Eisenbetonplätten," *Bauingenieur*, Vol. 4, pp. 354, 392.

Huffington, N. J., 1955. "Theoretical Determination of Rigidity Properties of Orthogonally Stiffened Plates," *J. Appl. Mech.*, Paper No. S5-A-12.

Iyengar, K. T., R. Sundara, and R. S. Srinivasan, 1967. "Analysis of Orthotropic Skew Plates," *Proc. 1st Australian Conf. on the Mechanics of Struct. Matrices*, Sydney, Australia, pp. 168–187a.

Javor, T., 1967. Skew Slab and Grid Work Bridges, Slovenske Vydavatel'stvo Technickej Literatury, Bratislava, Czechoslovakia.

Kagan, H. A., 1964. "Large Deflection Elasto-Plastic Behavior of Rectangular Two-Way Reinforced Plates," Ph.D. Thesis, New York Univ., New York.

Kennedy, J. B. and P. K. Chowdhury, 1977. "Plastic Analysis of Orthotropic Skew Metallic Decks," *J. Struct. Div. ASCE*, Vol. 103, No. ST3, March, pp. 533–546.

Kennedy, J. B. and D. S. R. Gupta, 1976. "Bending of Skew Orthotropic Plate Structures," *J. Struct. Div. ASCE*, Vol. 102, No. ST8, Aug., pp. 1559–1574.

Kennedy, J. B. and M. K. Prabhakara, 1980. "Postbuckling of Orthotropic Skew Plate Structures," *J. Struct. Div. ASCE*, Vol. 106, No. ST7, pp. 1497–1513.

Klöppel, K., 1951. *Zur Orthotropey Platte aus Stahl, Die Neke Köln-Mulheimer Brücke*, Herausgegeben von der Stadt Köln, p. 78.

Klöppel, K., 1958. "Über zulassige Spannungen im Stahlbau," Stahlbau-Tagung, Baden-Baden, Stahlbau Verlag, Köln.

Klöppel, K. and E. Roos, 1960. "Statische Versuche und Dauerversuche zur Frage der Bemessung von Flachblechen in orthotropen Plätten," *Der Stahlbau*, Vol. 29, p. 361.

Krug, S. and P. Stein, 1961. *Influence Surfaces of Orthogonal Anisotropic Plates*, Springer-Verlag, Berlin.

Kuo, S. S., 1965. "The Computer Analysis of Orthotropic Steel Plate Superstructure for Highway Bridges," Univ. New Hampshire Report, Bureau Public Roads No. CPR 11-9736, Vol. I, II.

Lechnitsky, S. G., 1957. *Anisotropic Plates*, 2nd ed., Moscow.

Lechnitsky, S. G., 1963. *Theory of Elasticity of an Anisotropic Body*, Holden-Day, San Francisco.

Lightfoot, E. and F. Sawko, 1960. "The Analysis of Grid Frameworks and Floor Systems by Electronic Computer," *Struc. Eng.*, Vol. 38.

Lin, T. H. and E. Ho, 1968. "Elasto-Plastic Bending of a Rectangular Plate," *J. Mech. Div. ASCE*, Vol. 94, No. EM1, Feb., pp 199–210.

Loveys, P. C., 1963. "Orthotropic Steel Plate Deck Bridges," Canadian Inst. Steel Construction, Nov.

Lyse, I. and I. E. Madsen, 1938. "Structural Behavior of Battledeck Floor Systems," *Proc. ASCE*, Jan.

Mader, F. W., 1957. "Berechnung Orthotroper Plätten unter Flachenlasten, Randmomenten und Rand durchblegungen," *Stahlbau*, Vol. 26, p. 131.

Mader, F. W. 1957. "Die Berucksichtigung der Diskontinuitat bei der Berechnung orthotroper Plätten," *Der Stahlbau*, Vol. 26, p. 283.

Massonnet, Ch., 1950. "Contrictuion au calcul des ponts a poutres multiples," Annales des Travaux Publics de Belgique, Juin, Octobre, Decembre.

Massonnet, Ch., 1950. "Methode de calcul des ponts a poutres multiples tenant compte de leur resistance a la torsion," Publications Intern. Assoc. for Bridge and Structural Engineering, Vol. 10.

Miller, A. B., 1929. "The Effective Width of a Plate Supported by a Beam," *Selected Eng. Papers*, Institution Civ. Engineers, London, No. 83.

Moffat, K. R. and P. J. Dowling, 1976. "Shear Lag in Steel Box Girder Bridges," *Struct. Engineer*, London, Oct., pp. 438–447.

Monforton, G. R., 1972. "Some Orthotropic Skew Plate Finite Element Results," *J. Struct. Div. ASCE*, Vol. 98, No. ST4, April, pp. 955–960.

Morice, P. B. and G. Little, 1954. "Load Distribution in Prestressed Concrete Bridge Systems," *Struct. Engineers*, Vol. 32, p. 83.

Naruoka, M. and H. Ohmura, 1959. "On the Analysis of Skew Girder by the Theory of Orthotropic Parallelogram Plates," *Publ. of the Intern. Assoc. for Bridge and Structural Eng.*, Vol. 19, pp. 231–256.

Olsen, H. and E. Reinitzhuber, 1959. "Die Zweiseitig Gelagerta Platte," Band I, Wilhelm Ernst und Sohn, Berlin.

Pelikan, W. and M. Esslinger, 1957. "Die Stahlfahzbahn-Berechnung und Konstruck-tion," *M.A.N. Forchungshaft*, No. 7.

Perry, P. G. and C. P. Heins, 1972. "Rapid Design of Orthotropic Bridge Floor Beams," *J. Struct. Div. ASCE*, Vol. 98, No. ST11, November.

Powell, G. H. and D. W. Ogden, 1969. "Analysis of Orthotropic Steel Plate Bridge Decks," *J. Struct. Div. ASCE*, Vol. 95, ST5, May.

Radojkovic, M., 1958. "Die Kurpfalzbrücke über den Neckar in Mannheim," *Stahlbau*, Vol. 27, pp. 29, 70.

Renner, W. A., 1939. "The Bending of Orthogonally Non-Isotropic Slabs," Thesis, Univ. Illinois, Urbana.

Saint-Venant, B., 1964. *Mém. savants étrangers*, Vol. 14, or Navier *Résume des lécons sur l'application de la mécanique*, 3rd ed., Paris.

Sawko, F., 1960. "Analysis of Grid Framework and Related Structures," thesis presented to the Univ. of Leeds, Leeds, United Kingdom in partial fulfillment of the requirements for the degree of Master of Science.

Seim, C. and R. Feriverda, 1972. "Fatigue Study of Orthotropic Bridge Deck Welds," Materials and Research Dept. Report CA-HWY-MR666473(1)72-22, Calif. Div. Highways, April.

Shields, E. J., 1966. "Poplar Street Bridge—Design and Fabrication," *Civ. Eng.*, Feb., pp. 52–55.

Shye, K. Y. and J. Colville, 1979. "Postbuckling Finite Element Analysis of Flat Plates," *J. Struct. Div. ASCE*, Vol. 105, No. ST2, Feb., pp. 297–311.

Teller, L. A. and J. A. Buchanan, 1937. "Determination of Variation in Unit Pressure over the Contact Area of Tires," *Public Roads*, Vol. 18, No. 10, Dec., pp. 195–198.

Thürlimann, B., 1960. "Tables of Influence Values for Moments in Orthotropic Plates," Fritz Lab. Report No. 264.6, Lehigh Univ., Bethlehem, PA.

Troitsky, M. S., 1962. "Orthotropic Design in Modern Bridge Engineering," Paper No. 10, Eng. Inst. Canada, Annual General Meeting, June.

Troitsky, M. S., 1967. "Orthotropic Bridges Theory and Design," The James F. Lincoln Arc Welding Foundation, Cleveland, Ohio.

Van Douwen, A. A., 1968. "Dutch Research on Orthotropic Decks," British Construction Steelwork Assoc. Conf. on Steel Bridges, London.

Wah-Thein and R. Sherman, 1969. "Study of Behavior of Grillages in Elastic and Plastic Ranges," Research Report, Naval Ships Systems Command, Southwestern Rhode Island, Dec.

Walker, A. C., 1969. "Post-Buckling Behaviour of Simply Supported Square Plates," *Aeronautical Quarterly*, Vol. 20, p. 203.

Wegmuller, A. W., 1974. "Full Range Analysis of Eccentrically Stiffened Plates, *J. Struct. Div. ASCE*, Vol. 100, No. ST1, Jan., pp. 143–159.

Wilkins, E. B. and R. M. Hearst, 1965. "Wearing Surfaces for Orthotropic Plate Bridge Decks," British Columbia Dept. Highways Report, Dec.

Williams, D. G., 1969. "Analysis of a Doubly Plate Grillage under In-Plane and Normal Loading," Ph.D. Thesis, Univ. London.

Wolchuk, R., 1959. "Orthotropic Plate Design for Steel Bridges," *Civ. Eng.*, Feb.

Wolchuk, R., 1962. *Orthotropic Steel Plate Deck Bridges*, American Inst. Steel Construction, N.Y.

Wolchuk, R., 1990. "Steel Plate Deck Bridges," in Gaylord and Gaylord, *Struct. Engineer Handbook* 3rd. ed., McGraw-Hill, New York, pp. 19-1 to 19-27.

Wolchuk, R., 1991. "Lessons from Weld Cracks in Orthotropic Decks on Three European Bridges," *J. Struct. Eng. ASCE*, Jan.

Wolchuk, R. and A. Ostapenko, 1992. "Secondary Stresses in Closed Orthotropic Deck Ribs at Floorbeams," *J. Struct. Eng. ASCE*, Feb.

Yang, T. Y., 1971. "A Finite Element Procedure for Large Deflection Analysis of Plates with Initial Deflections," *American Inst. Aeronautics and Astronautics J.*, Vol. 9, No. 8, Aug., pp. 1468–1473.

Yang, T. Y., 1972. "Finite Displacement Plate Flexure by the Use of Metrix Incremental Approach," *Intern. J. Numerical Methods in Eng.*, Vol. 4, No. 3, pp. 415–432.

Yusuff, S., 1952. "Large Deflection Theory for Orthotropic Rectangular Plates Subjected to Edge Compression," *J. Appl. Mech., Trans. ASME*, Vol. 19, Series E, p. 446.

CHAPTER 8

SEGMENTAL CONCRETE BRIDGES

A complete treatment of segmental concrete bridge construction is provided by Podolny and Muller (1982), and deals with all aspects of design and construction methods for all types of bridges as a ready reference source. However, with the 1989 AASHTO specifications for design and construction of segmental concrete bridges, methods of analysis, design, and detailing are articulated. The same document includes construction specifications relevant to materials, installation, geometry control, and tolerances.

The 1989 AASHTO guidelines are comprehensive in nature and represent several new concepts that may be considered significant departures from previous design and construction provisions. They are intended to reflect the observed performance of bridges of this type as well as results from recent research here and abroad. This document was initially prepared as NCHRP Project 20-7/32 carried out by the Post-Tensioning Institute.

This chapter reviews segmental concrete bridges in the context of the latest AASHTO specifications.

8-1 BASIC CONSTRUCTION METHODS

Segmental bridges contemplated under AASHTO specifications include those erected by the following methods.

1. Balanced cantilever
2. Span-by-span with truss or falsework

3. Span-by-span lifting
4. Incremental launching
5. Progressive placement

The choice between cast-in-place and precast construction depends mainly on local conditions and size of project, time allowed for construction, restrictions on access, environmental constraints, and equipment available at the site. The size and weight of segments are limited by transportation and erection capability, and segments exceeding 250 tons are seldom economical. Cast-in-place construction is not affected by these limitations, although the weight and cost of travelers are related to the weight of the heaviest segment.

Balanced Cantilever

This procedure simply cantilevers segments from a pier in a balanced fashion on each side until midspan is reached and a closure is made with a previous half-span cantilever from the preceding pier. The same erection process is repeated until the structure is completed. Unless symmetrical segments are simultaneously erected, the pier will be out of balance by one segment, and its design must accommodate the moment thus induced. Where feasible, temporary bracing may be provided.

For cast-in-place construction, the movable formwork is supported from the previously erected segment or from a movable form traveler while the new segment is being formed, cast, and stressed.

A successful application of balanced cantilever construction is in conjunction with cable-stayed bridges.

Span-by-Span Construction

The balanced cantilever process is suited primarily to long spans where construction activity for the superstructure can be planned at deck level and without the use of extensive falsework. Long viaduct structures with relatively shorter spans are better processed using the span-by-span method.

In this process, the superstructure is constructed in one direction span-by-span using a form traveler, with construction joints or hinges placed at the points of contraflexure. The form carrier in effect provides a factory-type operations transplanted to the job site. It may be supported on the piers, from the edge of the previously completed construction, at the joint location, or from the pier.

Initially, the span-by-span method was intended to accommodate cast-in-place construction. The same principle has been applied in conjunction with precast segments that may be assembled on a steel truss to make a complete span. Prestressing tendons ensure the assembly of the various segments in one span while maintaining full continuity with the preceding span, as shown in Figure 8-1.

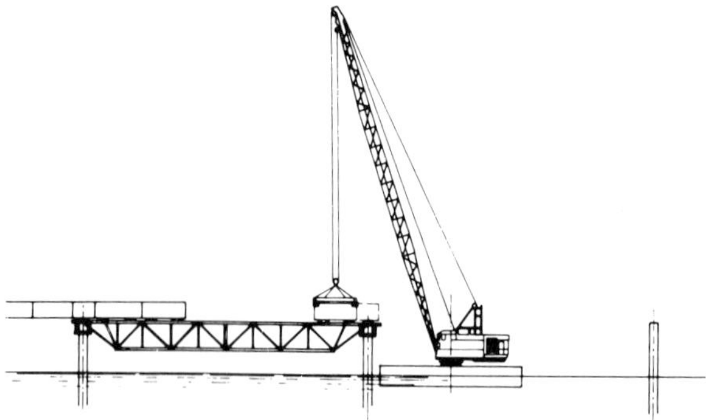

FIGURE 8-1 Span-by-span assembly of precast segments with truss. (From Podolny and Muller, 1982.)

Incremental Launching

This process, also referred to as push-out construction, is shown schematically in Figure 8-2. Segments of the bridge structure are cast in lengths of 30 to 100 ft in stationary forms located behind the abutments. Each individual unit is cast directly against the previous unit, and after it attains sufficient strength it is posttensioned to the previous unit. The assembly is moved foreward, and the same step is repeated for the succeeding unit. Normally, a work cycle of one week is required to cast and launch one segment irrespective of its length.

In order to allow the superstructure to move forward, special low-friction sliding bearings are employed at the piers, provided with suitable lateral guides. The main objective is to judge and overcome the frictional resistance of the superstructure under its own weight at all stages of launching and in all sections.

Progressive Placement

In principle, progressive placement is similar to the span-by-span process, because the construction starts again at one end of the structure and proceeds continuously to the other end. Its origin, however, its derived from the cantilever concept. Precast segments are placed in successive cantilevers on the same side of the same pier rather than by balanced cantilevers on each side of the pier. The method appears practicable and economical in the span range of 100 to 300 ft.

Because of the appreciable bending moment resulting from the cantilever length, a movable temporary stay arrangement is often necessary to limit the cantilever stresses. The erection procedure is shown schematically in Figure 8-3. Precast segments are transported along the completed portion to the end of the deck, and positioned by a swivel crane. About one-third of the span from the pier may be erected by the free cantilever. For the balance, the

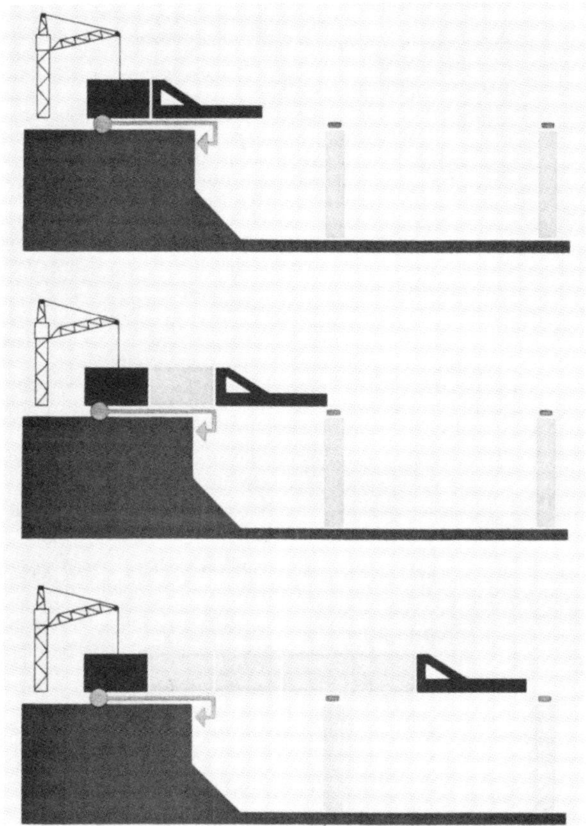

FIGURE 8-2 Incremental launching sequence. (From Podolny and Muller, 1982.)

segments are held in place by temporary ties and stays passing over the temporary tower and anchored in the completed deck as shown.

Applicability

Segmental construction can extend the practical span range of concrete bridges. Transporting and handling constraints limit prestressed I-girders to spans of 130 to 150 ft. Beyond this range, posttensioned cast-in-place box girders on falsework are probably the only feasible concrete alternative. In many instances, however, falsework is neither practical nor feasible, and examples are crossings over deep ravines or over large navigable waterways. Under these conditions, segmental construction is applicable within a wide span range, normally 150 to 800 ft or even longer. With cable-stayed structures the span range can exceed 1300 ft utilizing current materials technol-

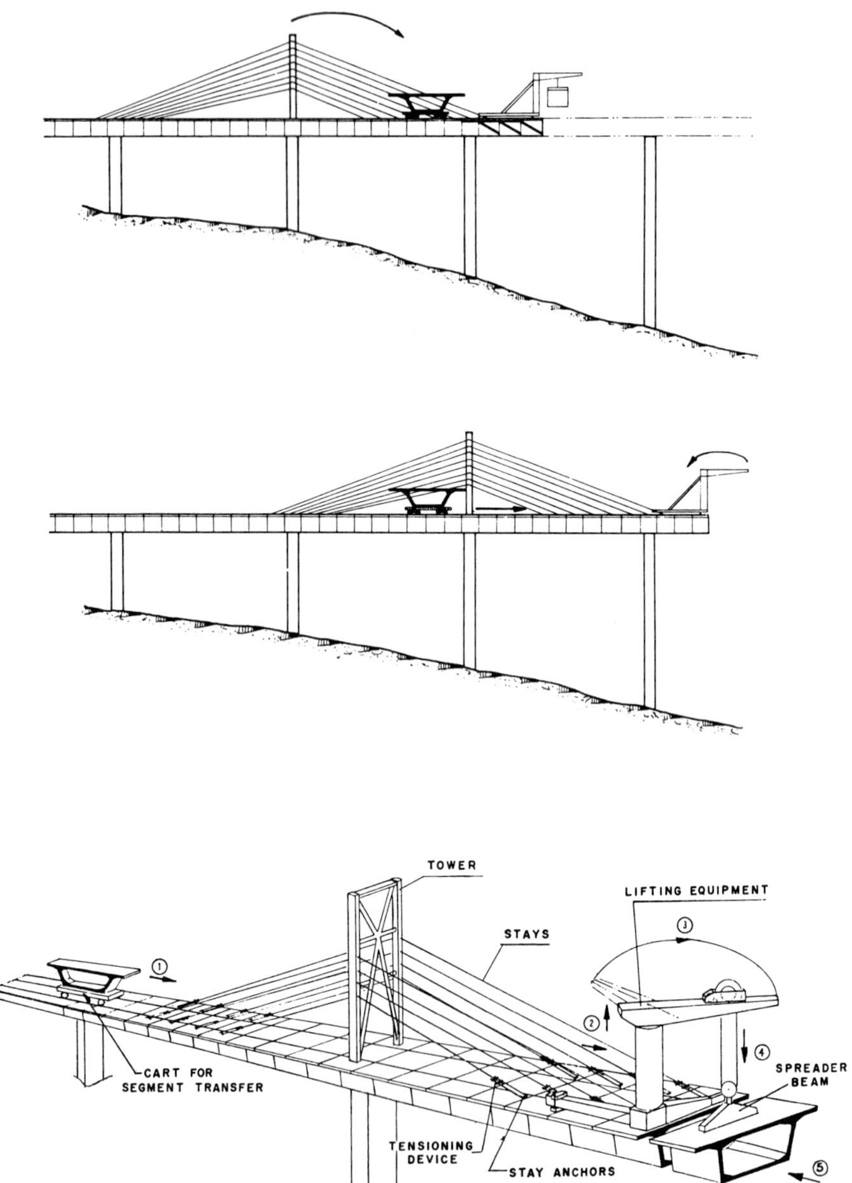

FIGURE 8-3 Progressive placement procedure. (From Podolny and Muller, 1982.)

TABLE 8-1 Range of Application of Bridge Type by Span Lengths[a] (From Podolny and Muller, 1982)

Span	Bridge Types
0–150 ft	I-type pretensioned girder
100–300 ft	Cast-in-place posttensioned box girder
100–300 ft	Precast balanced cantilever segmental, constant depth
250–600 ft	Precast balanced cantilever segmental, variable depth
200–1000 ft	Cast-in-place cantilever segmental
800–1500 ft	Cable stay with balanced cantilever segmental

[a]Current data indicate that segmental construction is practiced for spans up to 400 ft for precast segments, and up to 600 ft for cast-in-place decks.

ogy. Table 8-1 gives the range of application of construction types according to span.

Advantages The main advantages of segmental construction may be summarized as follows.

1. The falsework requirements may be eliminated where erection can be accomplished from completed portions of the bridge. This aspect is particularly important in high-level crossings, where it is necessary to minimize interference with the bridge site, and when heavy traffic must be maintained under the bridge during construction.
2. The structure geometry may be designed to accommodate any horizontal or vertical curvature or roadway crown.
3. The effects of concrete shrinkage and creep are better controlled during erection and in the completed structure because segments will normally cure to full strength before erection.
4. Except for temperature and weather limitations related to mixing and placing epoxy, the construction is relatively insensitive to weather.
5. The durability of the bridge deck is enhanced by precompression of the concrete and better crack control, and also by the use of high-quality concrete produced under conditions ensuring high quality.

Disadvantages Before final acceptance of the method for a given site, the following disadvantages must be considered (applicable to precast segmental construction).

1. A high degree of accuracy and geometry control is mandatory during fabrication and erection of the segments.
2. Extremely variable temperature and weather conditions can have an adverse effect on the mixing and placing of epoxy joint material.

3. The absence of steel reinforcement across the joint and limited tensile stresses may constitute a structural disadvantage.

8-2 SEGMENT DIMENSIONS AND DETAILS

A typical box girder cross section is shown in Figure 8-4. For design purposes, the principal segment dimensions are the top slab width W, the construction depth D, the width of the bottom slab B and by analogy the cantilever length C, the web spacing s, and the segment length L.

Overall Cross-Sectional Dimensions AASHTO stipulates that the overall dimensions should not be less than required to limit the live load plus impact deflection to 1/1000 of the span, computed using the gross section moment of inertial and the secant modulus of elasticity. All traffic lanes must be fully loaded with reduction in load intensity as specified, and the live load uniformly distributed to all flexural members. Interestingly, the deflection limit of 1/1000 of the span was selected arbitrarily, although segmental concrete bridges with normal dimensions readily satisfy this criterion.

The segment width W is usually the width of the bridge deck. Weight limitations, or when the bridge width exceeds about 40 ft, may dictate multiple segments with transverse connections of the top slabs. Single boxes with multiple webs have been used for widths up to 70 ft. For intermediate widths single box sections may be combined with integral floor beams under the roadway slab or boxed cantilevers.

The girder depth D depends mainly on the span, but in order to provide satisfactory deflection behavior the following guidelines may be considered.

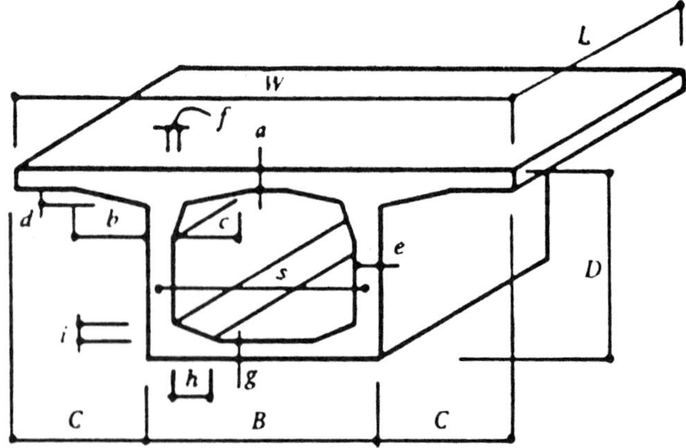

FIGURE 8-4 Typical box girder cross section and principal dimensions. (From Heins and Lawrie, 1984.)

TABLE 8-2 Depth–Span Ratios for Incrementally Launched Girders of Constant Depth

L = 100 ft	= 1/15	$< d_0/L$	< 1/12
L = 200 ft	= 1/13.5	$< d_0/L$	< 1/11.5
L = 300 ft	= 1/12	$< d_0/L$	< 1/11

For constant-depth girders, $1/15 > D/L_s > 1/30$ with an optimum of 1/18 to 1/20, where L_s is the span length between supports. For incrementally launched girders, the girder depth should preferably be calculated according to Table 8-2.

For variable-depth girders with straight haunches, the depth–length ratio at the pier should be $1/16 > D/L_s > 1/20$ with an optimum of 1/18. The same ratio at the center of the span should be $1/22 > D/L_s > 1/28$ with an optimum of 1/24.

For variable-depth girders with circular or parabolic haunches, the depth–length ratio at the pier should be $1/16 > D/L_s > 1/20$ with an optimum of 1/18. At the center of span this ratio should comply with $1/30 > D/L_s > 1/50$.

A single-cell box is indicated when $D/B \geq 1/6$, and a two-cell box should be selected when $D/B < 1/6$. A single box may be used beyond the foregoing range if a more rigorous analysis demonstrates its structural adequacy, but may require longitudinal edge beams at the tip of the cantilever to distribute the live loads. An analysis for shear lag is also indicated in this case. Transverse load distribution is not increased substantially by the use of three or more cells.

The segment length L has a marked effect on the overall economy and production. Because most of this cost is fixed per unit, maximum economy is achieved by using the smallest number of segments consistent with transportation and erection requirements.

The spacing of webs s is related to essentially relevant structural criteria, and, in principle, any web spacing may be considered subject to an appropriate structural analysis. As a general guideline, the cantilever length of the top flange should preferably not exceed 0.45 of the interior span of the top flange.

Detail Segment Dimensions The dimensions of the top slab, bottom slab, web, and haunches largely depend on structural considerations as well as on practical factors related to the production and handling of the segments. AASHTO stipulates a top and bottom flange thickness not less than the following: (a) 1/30 the clear span between webs or haunches (a smaller dimension will require transverse ribs at a spacing equal to the clear span between webs or haunches), or (b) 9-in.-thick top flange where transverse posttensioning is used. Transverse posttensioning is used where the clear span between webs or haunches exceeds 15 ft. Irrespective of these criteria, the selection of the top slab thickness *a* shown in Figure 8-4 must take into account (a) transverse bending moments, (b) compression zone requirements for longitudinal bending moments for spans exceeding 350 ft, (c) local bending stresses induced by wheel loads acting directly over epoxy joints, and (d) local anchorage bearing and splitting stresses for transverse posttensioning (when used). The top slab thickness must also accommodate transverse and longitudinal mild steel reinforcement, longitudinal tendons, and the required concrete cover.

Haunches *b*, *c*, and *d* in Figure 8-4 are selected to accommodate the transverse bending moments and to provide the space for the anchorages of the longitudinal posttensioning tendons. Usually, at least two layers of longitudinal tendons are required and result in a concrete depth of 14 in. A depth of 10 in. may be sufficient for bar tendons.

The minimum web thickness specified by AASHTO is as follows: (a) 8 in. for webs without longitudinal or vertical posttensioning tendons, (b) 12 in. for webs with only longitudinal or vertical posttensioning tendons, and (c) 15 in. for webs with both longitudinal tendons. As an example, the anchorage hardware of 12-strand tendons is likely to require a web thickness of 14 in., and this is also necessary to accommodate the bursting and splitting force from anchorages for 12-strand tendons. As a stiff element, the web in the box section provides considerable moment restraint to the top slab. The associated transverse moments at the web–slab junction are therefore high. An increased concrete thickness helps to reduce the reinforcement required.

The structural significance of the bottom slab is its contribution to section properties. However, with small thickness (5 to 6 in.), it is very difficult to prevent cracking due to the combined effect of the dead load carried between webs and the longitudinal shear between the web and bottom flange. In negative moment areas the bottom slab thickness is controlled by high compressive stresses, and slab thickening near the piers is invariably necessary to keep compressive stresses within allowable limits.

8-3 SPAN ARRANGEMENT AND DETAILS

Span Lengths With the balanced cantilever method, the segments are placed symmetrically about a pier. For a typical three-span structure, the end spans must preferably be reduced to 65 percent of the main span in order to shorten the deck portion adjoining the abutment as shown in Figure 8-5*a*, because this cannot be erected in a balanced cantilever.

Where span lengths must vary, a compromise solution is to introduce an intermediate span whose length is the average of the two flanking spans as shown in Figure 8-5*b*. This makes the cantilever concept compatible with the span on either side of the intermediate span.

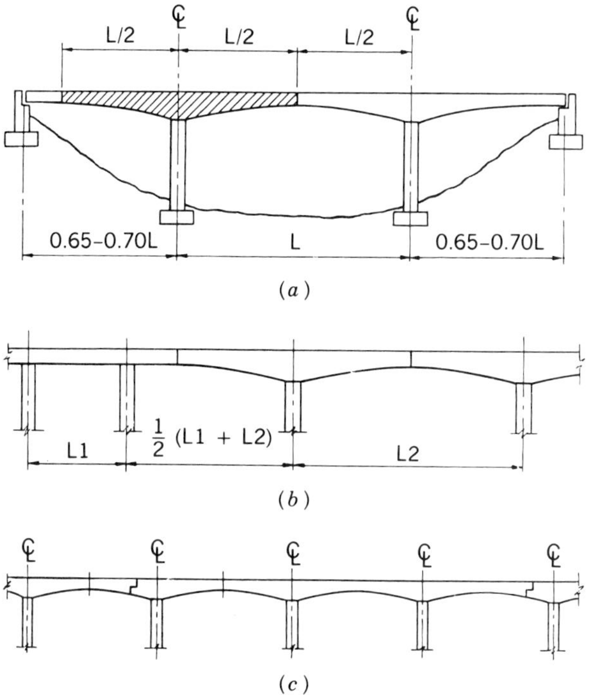

FIGURE 8-5 Balanced cantilever construction showing effects on span lengths and location of expansion joints. (From Podolny and Muller, 1982.)

Individual cantilever sections are made continuous by tendons inserted in positive moment areas upon closure. This alleviates the need for permanent hinges at midspan. Continuous decks without joints have been constructed up to lengths of 2000 ft and have proved satisfactory in terms of functional aspects. Where expansion joints are unavoidable, as in long viaducts, they should preferably be placed near points of contraflexure as shown in Figure 8-5*c* to avoid slope changes that may occur at other locations.

Deck Expansion Expansion joints must be designed for the full range of movement resulting from creep, shrinkage, and temperature effects. They should be located at piers, abutments, or within one segment length of points of contraflexure. Allowance for opening movements of modular compression sealed expansion joint devices must be based on 1.3 times the total anticipated movement. To account for the larger amount of opening movement, expansion devices are set precompressed to the extent possible.

There are no specific criteria as to the maximum continuous length beyond which intermediate expansion joints must be provided. Obviously, given the site conditions and type of bridge, the design and location of expansion joints must take into account the applicable temperature range, pier flexibility, ability of expansion bearings to reduce horizontal loads, quality of deck joints, and potential maintenance problems.

Structural Continuity Data on the relative behavior of a continuous segmental bridge and one with hinges at midspan show that there is no significant difference between the two types. However, when the effect of concrete creep is considered, the two structures respond differently. The hinged spans have a vertical deflection several times that of the continuous bridge. Likewise, live load deflections of the continuous structure may be several times more rigid than in the hinged unit. From these considerations, most designers tend to agree that the use of full continuity and the elimination of hinges benefits bridge performance.

8-4 METHODS OF ANALYSIS

Transverse Analysis The transverse analysis of box girder segments for flexure considers the box as a rigid frame. Wheel loads must be positioned for maximum moments, and elastic methods are applied to determine the effective longitudinal distribution of wheel load at each location. A probable increase in web shear and other effects resulting from eccentric loading should be considered.

AASHTO recommends influence surfaces (Homberg, 1968; Homberg and Rogers, 1965; Pucher, 1969) or other elastic analysis procedures to evaluate live load effects in the top flange. For beam-type segmental bridge decks, the analysis may be according to Article 3.24.3 of the standard AASHTO specifications. Other effects to be included result from elastic and creep shortening due to prestressing, shrinkage, and secondary moments.

These design parameters must be defined in the context of time and with regard to a specific construction schedule and method of erection. Computer programs must normally be used to articulate the time-dependent response of segmentally erected prestressed concrete bridges during construction and under service. Suggested programs by Ketchum (1986), Shushkewich (1986), and Danon and Gamble (1977) are available.

The time-related effects of creep and shrinkage may be evaluated as suggested by ACI Committee 209 (1982). A procedure based on graphical values for the creep and shrinkage parameters is presented by the CEB–FIB model code (1978). Comparison of results from the application of these two references is presented in the appendix of the AASHTO specifications and by Ketchum (1986).

More recent studies (Brant and Vadhanavikkit, 1987) suggest that ACI Report 209 may underestimate creep and shrinkage strains for the large-scale specimens used in segmental bridges. Creep predictions based on the ACI 209 study were consistently 65 percent of the experimental results. Thus, modification of the ACI 209 equations may be indicated to reflect the size and thickness of the segments.

Longitudinal Analysis AASHTO requirements focus on a specific construction method and construction schedule, as well as on the time-related effects of creep, shrinkage, and prestress loss. Secondary moments due to prestressing must be included in stress analysis under working loads. In determining ultimate strength moments or shears, the effects of prestressing with a load factor of 1.0 should be added to the moments and shears due to factored or ultimate dead and live loads.

Erection analysis must consider construction load combinations, stresses, and stability aspects. The associated methodology is outlined in AASHTO Article 8.4.

Likewise, the final structural system must be investigated for redistribution of erection stage moments resulting from the effects of creep and shrinkage, as well as from changes in the statical system including the closure of joints. Thermal effects and thermal gradients in the final structural system are considered according to TRB–NCHRP Report 276 (see also Section 12-5).

Comparative results of analysis of a segmental concrete bridge with creep coefficients ϕ_c of 1, 2, and 3 are given in the appendix of the AASHTO specifications using both the ACI 209 and CEB–FIP models. The final stresses are basically the same for creep coefficients of 1, 2, and 3 using the ACI 209 creep criteria. The analysis with the CEB–FIP creep model exhibits more variation in final stresses, but the associated range is still small considering the large variation in creep coefficients. Although the selection of a creep model has a broader impact on final stresses than that of creep coefficients, the full range of stresses is unlikely to have a practical significance with reference to the performance of the structure. AASHTO considers satisfactory an analysis using a single value of creep coefficient. However, further research is warranted to rectify the differences in the ACI 209 and CEB–FIP creep models. Notwithstanding the accurate determination of creep values, it is suggested that these values have only a minor impact on prestress losses, deflections, and the axial shortening of the structure.

8-5 DESIGN CONSIDERATIONS

General Requirements

Creep and Shrinkage For permanent loads, the response of segmental bridges after closure can be approximated using an effective modulus of elasticity E_{eff} calculated as $E_{\text{eff}} = E_{cm}/\phi_c$, where ϕ_c is the creep coefficient and E_{cm} is the 28-day secant modulus of elasticity of the concrete computed from $E_{cm} = 57{,}000\sqrt{f'_c}$ (f'_c is in psi).

Load Factors An additional load combination is (service load design)

$$(\text{DL} + \text{SDL} + \text{EL}) = \beta_E E + B + SF + R + S + (\text{DT}) \qquad (8\text{-}1)$$

where DL = dead load, structure only
SDL = superimposed dead load
EL = erection load (final stage)
R = creep effects, to be considered in conjunction with any rib-shortening effect
DT = thermal differential

and other terms and designations are as per standard AASHTO.

The basic strength reduction factors ϕ_f and ϕ_v for flexure and shear, respectively, are differentiated for fully bonded tendons and for unbonded or partially bonded tendons, and for two types of joints. Cast-in-place concrete joints and wet concrete or epoxy joints between precast units are designated as type A joints. Dry joints between precast units are considered type B joints. Invariable, type B joints have a smaller strength reduction factor.

The values of ϕ_f and ϕ_v are based on relatively limited test results (Hoang and Pasquignon, 1985; Rabbat and Sowlat, 1987; Koseki and Breen, 1983), and are considered interim provisions as more data are forthcoming.

The proposed ϕ_v values for shear of type A joints in bonded tendons are based on results of ultimate shear tests reported by Kashima and Breen (1975). Comparative shear tests of epoxy and dry joints are shown in Figure 8-6, and indicate that the epoxy joints develop the full strength of monolithically cast specimens. Dry joints develop less strength and show appreciable tendency to slip along the joint. For this reason, type B joints are assigned a lower strength factor.

When the strength reduction factors ϕ are used, the strength provided in the design must be at least equal to the following load factor combinations:

$$\text{For amximum forces and moments} = 1.1(\text{DL} + \text{DIFF}) + 1.3\text{CE} + 2A$$
$$\text{For minimum forces and moments} = \text{DL} + \text{CE} + 2A \qquad (8\text{-}2)$$

where DL = dead load, structure only
DIFF = differential (unbalanced) dead load from one cantilever
CE = weight of specialized construction equipment
A = static segment weight

Construction Load Combinations AASHTO stipulates the erection loads during construction, load combinations, and the associated stresses. The latter are limited to less than the modulus of rupture, that is, construction load stresses should not generate any cracking.

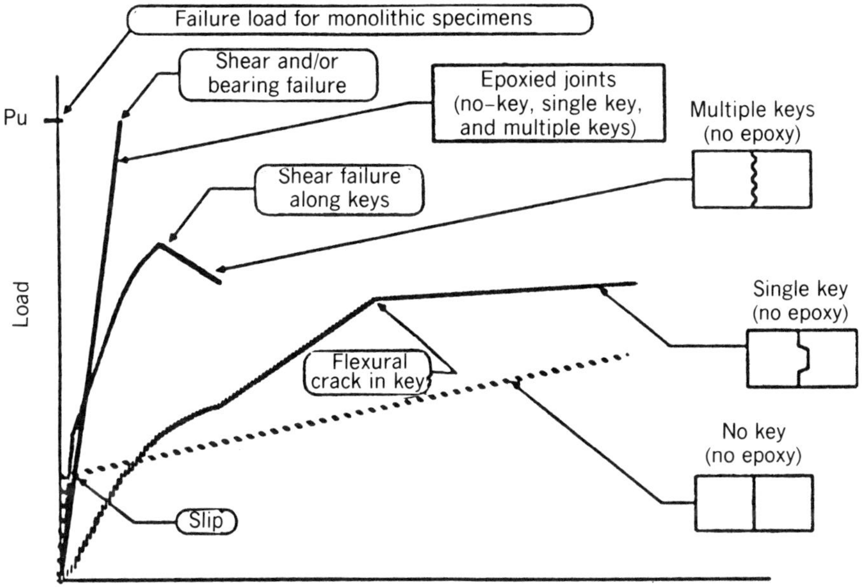

FIGURE 8-6 Relative joint displacement and shear key behavior. (From Koseki and Breen, 1983.)

Besides the conventional dead loads, the stability of the structure must be checked for (a) distributed construction live load, covering miscellaneous item of plant, machinery, and other equipment in addition to the major specialized erection equipment; (b) specialized construction equipment such as a launching gantry, segment delivery trucks, and loads applied to the structure during lifting of segments; (c) impact from equipment; (d) longitudinal forces arising from construction equipment; (e) segment unbalance, applied primarily to balanced cantilever construction; (f) wind uplift on cantilever; (g) accidental release or application of a precast segment load or other sudden impact from a static load, and (h) creep, shrinkage, and thermal effects.

Stresses are checked under service load combinations (see AASHTO Table 8-1), but if more unfavorable conditions should occur for a particular construction system, they must be considered in the analysis. Load factor design for construction conditions is essentially limited to (8-2).

Allowable Stresses The allowable stresses stipulated in AASHTO Article 9.0 cover prestressing steel, prestressed concrete, and anchorage stresses. Allowable stress levels in prestressing steel are according to ACI 318 (1989). However, the maximum tendon jacking force is reduced from $0.85f_s'$ to $0.80f_s'$

for low-relaxation strands (see also Xanthakos, 1991). This change is intended to protect against wire breaking at higher stressing levels, and also reflects the fact that jacking stresses at $0.85f_s'$ are incompatible with the maximum stress of $0.74f_s'$ immediately after prestress transfer.

Attention is focused on concrete splitting forces in anchorage systems. With NCHRP Project 10-29 under way to develop more comprehensive recommendations for proportioning reinforcement for anchorage splitting stresses, the current specifications are interim guidelines. At present the design of anchorage zone reinforcement is based on the work of Stone et al. (1981), Stone and Breen (1981a, 1981b), and PCI (1975).

Bursting or splitting forces occur in front of individual anchors inside the local zone. General zone bursting forces exist beyond the individual tendon local zones, and have a magnitude dependent primarily on the overall concrete dimensions as well as on the stress level, direction, and eccentricity of the total prestressing force. For design purposes, it may be assumed that any local zone reinforcement does not contribute to the strength of the general zone.

Prestress Loss Time-related procedures for calculating prestress loss are suggested by Ketchum (1986), Danon and Gamble (1977), ACI-209 (1982), CEB–FIP (1978), Leonhardt et al. (1965), Dilger (1982), Huang (1982), Tadros et al. (1975), and Tadros, Ghali, and Dilger (1977).

The loss of prestress due to friction and wobble within a duct is calculated from

$$T_0 = T_x e(\mu\alpha + \kappa l) \tag{8-3}$$

where T_0 = tendon stress at jacking end
T_x = tendon stress at x feet from jacking end
μ = friction coefficient (per radian)
α = total angular deviation from jacking end to point x (radians)
κ = friction wobble coefficient (per foot)
l = tendon length from jack end to point x

For a general development of friction loss theory for bridges with inclined webs and for horizontally curved bridges, reference is made to Podolny (1986).

Steel relaxation over a time interval t_1 to t may be estimated for stress-relieved steel and low-relaxation steel. As is the case with elastic shortening and concrete creep, steel relaxation is a time-dependent function influenced by strains induced by stressing as well as by time-related losses due to creep and shrinkage strains. Accordingly, an exact evaluation of steel relaxation requires a computer. Approximate values obtained from the recommended formulas are acceptable for preliminary designs.

Flexural Strength, Shear, and Torsion

Flexural Strength In general, the flexural strength of segmental concrete bridges is based on Articles 9.17, 9.18.1, and 9.19 of the standard AASHTO specifications, excluding Article 9.18.2.1. The minimum reinforcement provided in Article 9.18.2.1 was developed to avoid brittle failure in glossy underreinforced simple-span precast, prestressed sections. Application to segmental concrete bridges implies requirements or more bonded reinforcement for bridges with more conservative design stress levels. Minimum steel reinforcement requirements are covered in the allowable stress and load factor provisions of these specifications.

The value of average stress in prestressing at ultimate load f_{su}^* for unbonded members, normally computed from equation (9.18) in AASHTO Article 9.17.4.1 (standard specifications), may require additional research. For example, the German DIN specification allow a stress increase of only 6 ksi for unbonded cantilever tendons, and no stress increase for fully continuous unbonded tendons.

Shear and Torsion The design for shear and torsion of prestressed concrete segmental bridges is based on ultimate load conditions because little information is available with regard to actual stress distribution at working or service load levels. Useful comments on the general design requirements are summarized as follows.

1. Regions with beam-type action satisfy the Bernoulli hypothesis of linear strain profiles. Discontinuity regions, where linear strain profiles cannot be assumed, usually exist for some distance from a concentrated load or point of geometric discontinuity. Moving wheel loads are not considered as large concentrated loads. For the use of strut-and-tie models, reference is made to Schlaich, Schafer, and Jennewain (1987). We should note that a structure can consist of both beam-type and discontinuity regions.

2. Where dissimilar materials are involved, for shear transfer adequate reinforcement should be provided perpendicular to the vertical planes of web–slab interfaces to transfer flange longitudinal forces at ultimate conditions. Several methods may be utilized for shear transfer design, and among those the strut-and-tie method is useful in proportioning transverse reinforcement to assist the transfer of horizontal shear between elements.

3. The shear effect of moving loads may be considered in terms of maximum factored shear envelopes, making use of these values in calculating the factored ultimate live load shear on the section. Likewise, prestressing is an applied load whose magnitude and direction are carefully controlled. Components of the prestress force can increase or decrease the shear on a cross section. In cantilevered segmental construction the prestress vertical component can reverse the applied shear direction near the supports.

When members are subjected to combined shear and torsion, the design should consider the shear forces resulting from the combined shear flows. A limit of 100 psi is placed on the value of $\sqrt{f_c'}$ in connection with any shear analysis and design where the factor $\sqrt{f_c'}$ appears. This limitation on the effective diagonal tension and interlock components of shear strength contributed by the concrete reflects current ACI criteria.

Transverse reinforcement must be provided in elements (except slabs and footings) where $V_u < 0.5\phi V_c$, where V_u is the factored shear force at the section and V_c is the nominal shear strength provided by the concrete. In this case V_c may be computed from

$$V_c = 2K\sqrt{f_c'}\, b_w d \tag{8-4}$$

where b_w is the minimum web width and K may not exceed 1.0 at any section. The foregoing expression is a simplified way to determine V_c, thus eliminating the need to check V_{ci} and V_{cw} (nominal shear strength provided by concrete when diagonal cracking results from combined shear and moment and nominal shear strength provided by concrete when diagonal cracking results from excessive principal tensile stress in web, respectively). This expression has been checked against a wide range of test data, and is considered conservative yet simple.

Transverse reinforcement may consist of (a) stirrups perpendicular to the axis of the member, or placed at 45° to intercept potential cracks; (b) welded wire fabric; (c) longitudinal bars bent to provide an angle of 30°; (d) well-anchored prestressed tendons; (e) combinations of stirrups, tendons, and bent longitudinal bars; and (f) spirals.

Torsion reinforcement may consist of (1) longitudinal bars or tendons, (b) closed stirrups or closed ties perpendicular to the axis of the member, (c) a closed cage of welded wire fabric, and (d) spirals.

A traditional approach for shear and torsion design for plane-section-type regions is suggested in lieu of the more complex application of strut-and-tie models for moving loads. The conditions under which the simplified version is acceptable are specified in Article 12.3.1.

The design of cross sections subject to shear must be based on $V_u \leq \phi V_n$, where V_n is the nominal shear strength. Any unfavorable effects of prestressing V_p should be considered, and for design purposes V_n may be computed as

$$V_n = V_c + V_s \tag{8-5}$$

where V_c is computed from (8-4). The factor V_s is the nominal shear contribution of the truss model with concrete diagonals at 45°, and is computed from

$$V_s = A_v f_{sy} d/s \tag{8-6}$$

where A_v = area of transverse shear reinforcement within a distance s
f_{sy} = specified yield strength of non-prestressed reinforcement
s = bar (stirrup) spacing

The longitudinal and transverse reinforcement for torsion must be determined from $T_u \leq T_n$, were T_u is the factored torsion in the section and T_n is the nominal torsion resistance.

Strut-and-Tie Truss Models AASHTO stipulates that the design of any region for shear and torsion may be based on an analysis of the internal load paths for all forces acting on the member or regions. The internal load paths should be idealized using appropriate strut-and-tie or space truss models consisting of (a) concrete and compressive reinforcement compression chords; (b) inclined concrete compressive struts; (c) longitudinal reinforcement tension members or ties; (d) transverse reinforcement tension members or ties; and (e) node regions at all points of chords, struts, and ties. These criteria integrate the recommendations of Schlaich, Schafer, and Jennewain (1987) and Marti (1985). The proposed stress limits on struts and nodes stipulated in AASHTO Article 12.4 may be subject to further refinements.

Figures 8-7 and 8-8 illustrate analysis using truss and strut-and-tie models, respectively (Schlaich, Schafer, and Jennewain, 1987).

Fatigue Consideration

For bonded non-prestressed reinforcement, fatigue stress limits are according to Article 8.16.8.3 of the standard AASHTO specifications. For concrete deck slabs with main reinforcement perpendicular to traffic, fatigue stress limits need not be considered.

Prestressed reinforcement is not subject to fatigue considerations because the section should remain uncracked for bridges designed under the allowable stresses of the specifications. In this case fatigue of prestressed reinforcement should not occur under the related small stress range (see also Section 12-14).

Fretting fatigue caused by rubbing between the duct and strands should not occur in uncracked sections.

General and Local Anchorage Zones

A local anchorage zone is the region immediately surrounding each anchorage device, and may be taken as a cylinder or prism with transverse dimensions approximately equal to the sum of the projected size of the bearing plate plus the specified minimum side or edge cover.

The general zone is the region in front of the anchor extending along the tendon axis for a distance equal to the overall depth of the member, also

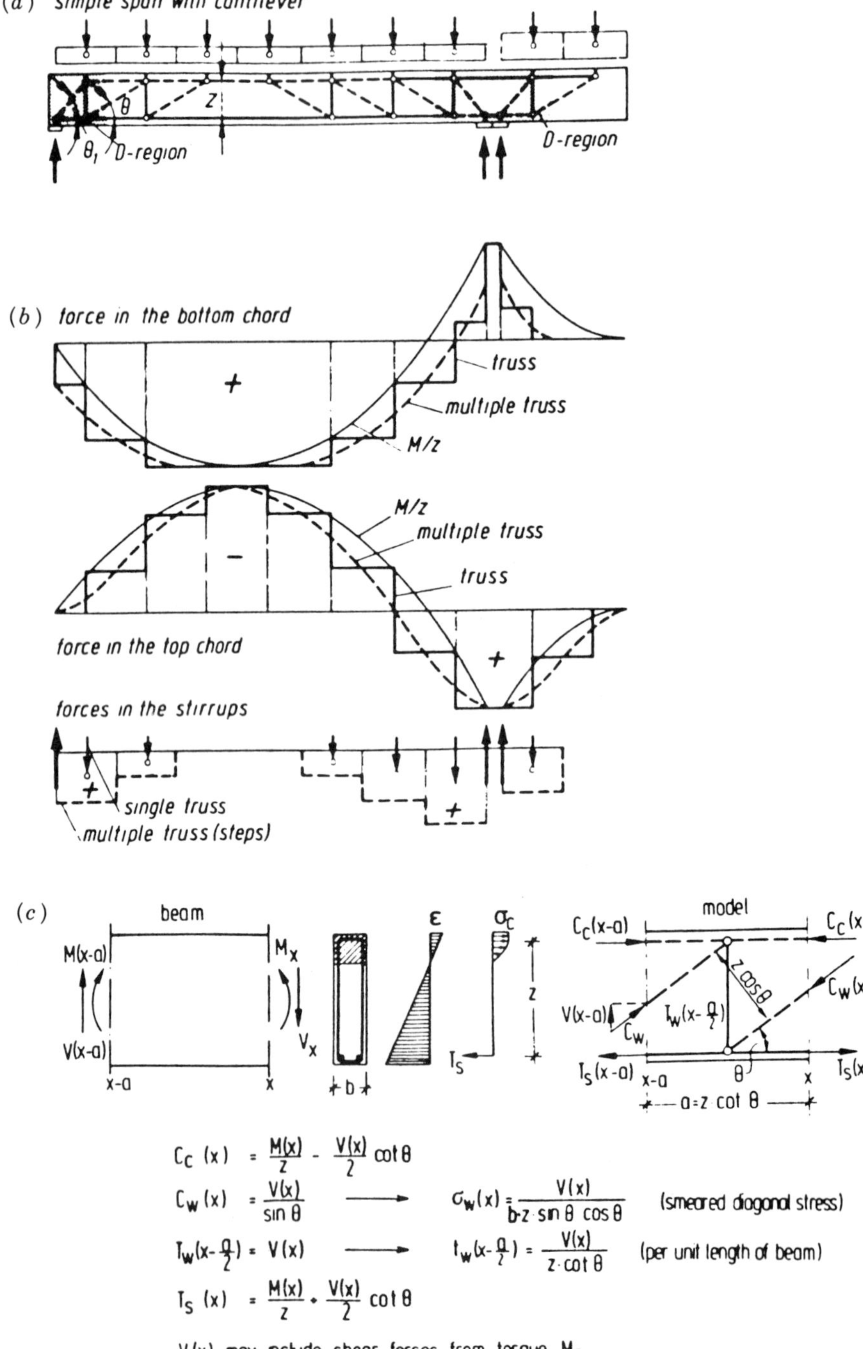

FIGURE 8-7 Truss model of a beam with cantilever: (*a*) models; (*b*) distribution of inner forces; (*c*) magnitude of inner forces derived from equilibrium of a beam element. (From Schlaich, Schafer, and Jennewain, 1987.)

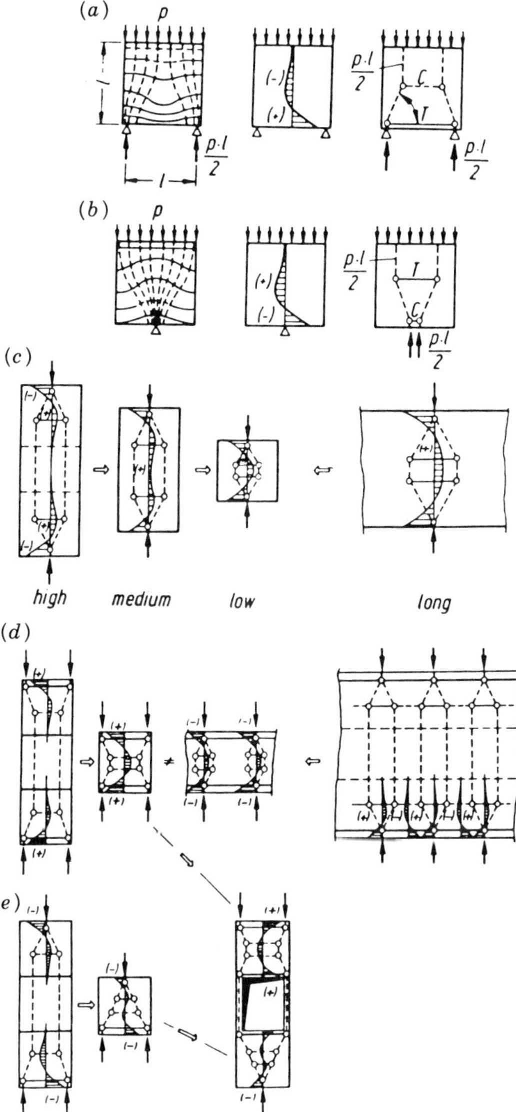

FIGURE 8-8 The two most frequent and most useful strut-and-tie models, (*a*) and (*b*); some of their variations, (*c*) through (*e*). (From Schlaich, Schafer, and Jennewain, 1987.)

considered the height of the general zone. For the case of intermediate anchorages that are not at the end of a member, the general zone extends along the projection of the tendon axis for about the same distance before the anchor.

The design and specifications of any supplementary reinforcement required in the general zone (in addition to the required local zone reinforcement) conservatively assume that the presence of any local zone reinforcement does not contribute to the strength of the general zone.

The general distribution of forces and the amount of reinforcement necessary in the general anchorage zone to counteract the bursting forces of the anchorages may be determined by the strut-and-tie model (Schlaich, Schafer, and Jennewain, 1987). This approach gives a good approximation of the reinforcement quantity and distribution necessary to counteract the tensile forces set up in the general anchorage zone directly in advance of the anchorages as shown in Figure 8-8*b* as well as in the outer regions of the general anchorage zones with eccentrically located anchorages.

In lieu of the foregoing analysis, the total bursting force F_{burst} for an individual anchorage may be taken as

$$F_{\text{burst}} = 0.30\,(1 - d_a/d_{sp})P_j \tag{8-7}$$

where d_a = depth of anchor plate

d_{sp} = total depth of symmetric concrete prism above and below the anchor plate (also assumed to be the length of the anchorage zone)

P_j = tendon jacking force

The center of the bursting force is located approximately 3/8 of the depth of the section in front of the anchorage zone. Tendon inclination, tendon curvature, and the blockout to achieve tendon inclination at the face of the anchorage tend to increase the bursting stresses.

LRFD Specifications

These are based essentially on the 1989 AASHTO specifications for segmental concrete bridges outlined in the foregoing sections. The span length of bridges to which these provisions apply ranges approximately to 800 ft. Bridges supported by stay cables are not specifically covered in this section.

Bridge designs should allow the contractor some latitude in choosing construction methods. To ensure compatibility of the design features and details with proposed construction methods, the contractor should be required to prepare working drawings and design calculations based on the choice of method.

Analysis of the Final Structural System Results of analysis of a segmental concrete superstructure with creep coefficients of 1, 2, and 3 show that final stresses were essentially unchanged using the ACI 209 creep provisions. Because the creep coefficient is usually determined with reasonable accuracy under these requirements, analysis using a single creep coefficient value is considered satisfactory. The use of low and high values in analysis is generally unnecessary. Further research is needed, however, to reconcile the differences in the ACI 209 and CEB–FIP creep models.

In early specification drafts, the resistance factors at the joints (type A or B) were the same as in AASHTO Article 8.3.6. In the current (third draft), a provision is included that requires joints in girders made continuous by unbonded posttensioning steel to be investigated for the simultaneous effects of axial force, moment, and shear that may occur. Shear is assumed to be transmitted through the contact area only.

Construction Load Combinations and Allowable Stresses Stresses due to construction load combinations are essentially the same as in Table 8.1 (1989 edition). The stresses in this table (at service limit states) limit construction load stresses to less than the modulus of rupture for structures with internal tendons and type A joints. The construction load should not therefore generate any cracking.

The distribution and application of the individual erection loads appropriate to a construction phase should be selected to give the most unfavorable effects. The construction load compressive stress in concrete should not exceed $0.50 f_c'$, where f_c' is the compressive strength at the time of the load application.

Construction Load Combination of Strength Limit States The factored resistance determined using the appropriate resistance factors should not be less than the combination of factored loads expressed by (8-2), where, however, the term $2A$ is replaced by $(A + AI)$, where AI is the dynamic response due to accidental release of application of a precast segment load or other sudden application of an otherwise static load to be added to the dead load, taken as 100 percent of load A.

Allowable Stresses at Service Limit States Longitudinal tensile stresses in transverse joints should not exceed the following:

1. Type A joints with minimum bonded auxiliary reinforcement through the joints sufficient to carry the calculated tensile force at a stress of $0.5\, f_y$ with internal tendons: $0.095\sqrt{f_{ci}'}$, tension.
2. Type A joints without the minimum bonded auxiliary reinforcement through the joints with internal tendons: $0.095\sqrt{f_{ci}'}$, except that no tension should be allowed under Service I load.

3. Type B joints with external tendons: 0.200 ksi, compression.
4. Tension caused by transverse bending in the precompressed tensile zone: $0.095\sqrt{f'_{ci}}$, maximum tension, where f'_{ci} is the specified compressive strength of concrete at the time of initial loading or prestressing.
5. Tension in other areas without bonded non-prestressed reinforcement: not allowed.

Temperature, Creep, and Shrinkage Transverse analysis for the effects of differential temperature outside and inside box sections may be necessary for relatively shallow bridges with thick webs and for a possible temperature differential of $\pm 10°F$.

The creep coefficient (the ratio of the strain that exists at certain days after casting to the elastic strain caused when a load is applied) may be determined by a comprehensive test program. Creep strains and prestress losses occurring after closure of the structure cause a redistribution of the force effects, and must be considered.

The specifications articulate the friction and wobble coefficients of (8-3). Tests and experience with segmental concrete bridges have often indicated higher friction and wobble losses due to the movement of ducts during concrete placement and misalignment at segment joints. As a consequence, in situ friction tests are recommended for major projects as a basis for modifying friction and wobble loss values (see also AASHTO Table 10.2). Where gross duct misalignment problems arise, no reasonable values for friction and wobble coefficients can be recommended. Additional ducts are suggested to compensate for high friction and wobble losses or for provisional posttensioning tendons, as well as other contingencies.

Special Provisions Long-term redistribution of creep- and shrinkage-induced stresses, as well as erection stresses due to changes in the support system, should be considered in the design. Because neither the creep and shrinkage characteristics of the concrete nor the construction schedule can be explicitly predicted at the time of the design, and in lieu of specific tests to determine the statistical variation of the mix design, a variation of ± 30 percent is recommended. Variation of the actual schedule from the assumed will also affect redistribution due to creep and shrinkage, necessitating a corresponding allowance.

Joints in precast segmental bridges should be either cast-in-place closures or match cast. Precast segmental bridges using internal posttensioning tendons and bridges located in areas subject to freezing temperatures and deicing chemicals should be provided with type A joints. Other precast segmental bridges may employ type B joints. The prestressing system for type A joints should provide a minimum compressive stress of 30 psi and an average stress of 40 psi across the joint until the epoxy has cured.

Joints for cast-in-place segments should be specified as either intentionally roughened or keyed. Whereas vertical joints normally must be keyed, a properly roughened joint is essential to ensure better bond between the segments and improve shear strength.

Incrementally launched girders can be subjected to moment reversal during the erection process. Temporary piers or a launching nose may be used to reduce launching stresses. Likewise, the force effects due to permissible construction tolerances should be superimposed upon those resulting from gravity loads. These tolerances are as follows.

- In the longitudinal direction between two adjacent bearings, 0.2 in.
- In the transverse direction between two adjacent bearings, 0.1 in.
- Between the fabrication area and the launching equipment in the longitudinal and transverse directions, 0.1 in.
- Lateral deviation at the outside of the webs, 0.1 in.

The horizontal force acting on the lateral guides of the launching bearings should not be less than 1/100 of the vertical support reaction.

8-6 CURVED PRESTRESSED SEGMENTAL BRIDGES: CASE STUDY

Van Zyl (1978) has developed a method for the analysis of curved segmentally erected prestressed concrete box girder bridges. This method is described by Van Zyl and Scordelis (1979), illustrated by numerical examples to demonstrate the capabilities of the computer program and the accuracy of results thus obtained. Deflections and stresses can be computed at any construction stage or service life. Alignment changes during construction can be included, and the effect of stage prestressing determined with sufficient accuracy.

The method applies to either precast segments or cast-in-place construction. The bridge cross section consists of a single box section with vertical or inclined webs with cantilever flanges. Time-dependent material changes, such as creep and shrinkage or humidity and temperature changes, are accounted for. Prestressing in the longitudinal direction is included in the analysis, and tendons may be located in either the webs or flanges and have a linear or parabolic profile. Stressing operations common in segmental construction, such as stress and release operations, slip in tendons, and restressing and removal of tendons, are included. The structural configuration and erection sequence may be specified according to the design. The program can also analyze addition of segments, prestressing, changing of support boundary conditions, application or removal of construction loads, and prescribed displacements. The completed structure can be investigated for various loads and time steps.

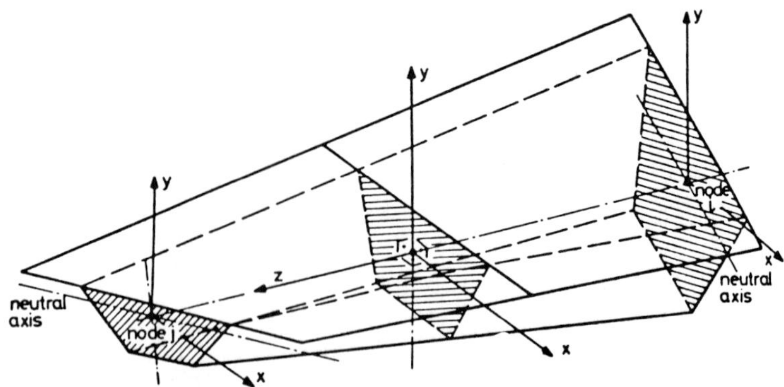

FIGURE 8-9 Typical skew-ended finite element. (From Van Zyl and Scordelis, 1979.)

Stiffness Method

Skew-Ended Finite Element A skew-ended finite element with eight degrees of freedom at each side of its two end nodes is shown in Figure 8-9 (Bazant and El Nimeiri, 1974). The model has three translational and three rotational degrees of freedom, and also a transverse distortional and a longitudinal warping degree of freedom at each end. The three displacement components of a general point of the cross section are axial displacement u, tangential displacement v, and normal displacement w, with positive directions as shown in Figure 8-10a.

These displacements are expressed in terms of the basic displacement distributions or shape functions as follows:

$$u = \sum_{k=1}^{4} U_k(z)\phi_k(s) \qquad v = \sum_{k=1}^{4} V_k(z)\psi_k(s) \qquad w = \sum_{k=1}^{4} V_k(z)X_k(s) \tag{8-8}$$

The shape functions are articulated in two groups. The first group includes V_k and U_k, and describes the longitudinal variation of the displacement patterns. The second group includes $\phi_k(s)$, $\psi_k(s)$, and $X_k(s)$, and describes the transverse variation of the displacement patterns.

Stresses and Strains The nodes are numbered from one end to the other, producing a chain-type structure. In this case the stiffness matrix is banded with a half band width of 16 and a length of eight times the number of nodes. The diaphragm stiffnesses are added to the terms corresponding to the node where the disphragm is located. The last step is to add the support spring stiffnesses to the appropriate diagonal terms.

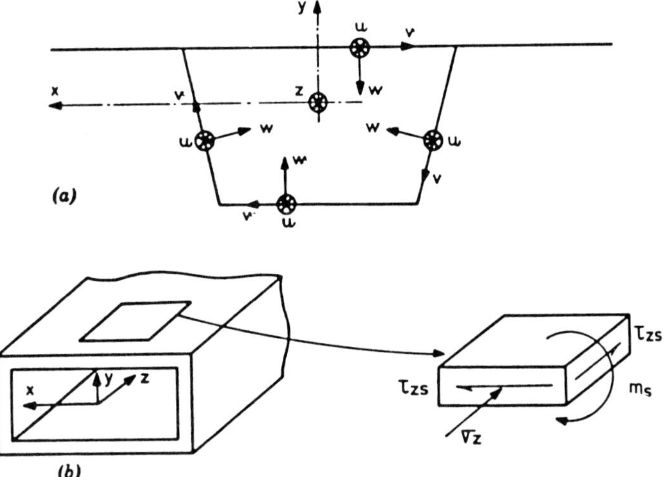

FIGURE 8-10 Positive: (*a*) displacement components; (*b*) stress components. (From Van Zyl and Scordelis, 1979.)

Any point load on an element is identified in terms of its axial component p_z, its normal component p_n, and the tangential component p_s. Positive loads are properly identified. The consistent nodal loads are obtained by multiplying the three components by the appropriate shape functions.

Time-Dependent Behavior

Creep Creep is modeled using a step-by-step method developed by Zienkewicz and Watson (1966). The stress history record is kept by retaining only the running sums of certain expressions. Small time steps are used during which the stresses are assumed to remain constant. At the end of each interval, the incompatibility resulting from the creep strain change during this interval is corrected by an elastic solution. Loads are applied at the end of the step when the instantaneous stress can be added to the stress change required to restore compatibility.

This method has been extended by Kabir (1976) to include the temperature shift function $\phi(T)$. The creep compliance is given in the form of a Dirichlet series in which all parameters can be evaluated using experimental data or the recommendations of ACI Committee 209 (1982).

Shrinkage Shrinkage is defined as material deformation with time under no external loads. The shrinkage strain is determined from experimental curves or as recommended by the ACI criteria.

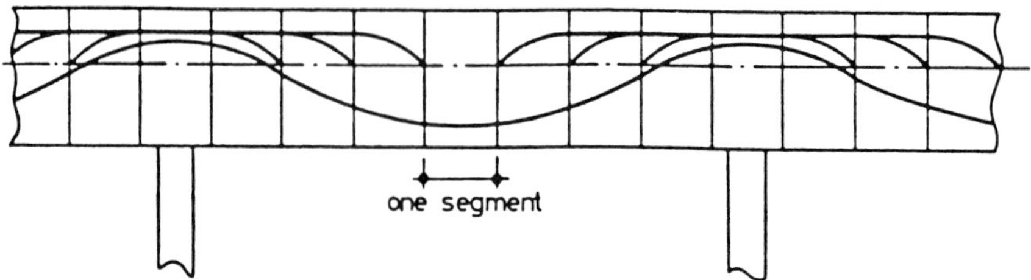

FIGURE 8-11 Typical tendon layout used in segmental bridges. (From Van Zyl and Scordelis, 1979.)

Constitutive Relationship The corresponding constitutive relationship is expressed as

$$\sigma_n = E_n D_0 (e_n - e_n'') \tag{8-9}$$

where the term D_0 is a function of Poisson's ratio ν and the plate thickness δ. Also E_n is the current elastic modulus, e_n is the strain vector, e_n'' is the elastic strain vector (consisting of a combination of creep, shrinkage, and temperature strains), and σ_n is the stress vector.

Prestressing Analysis

A typical tendon layout often used in segmental construction is shown in Figure 8-11. The tendons may have a linear or parabolic profile between inflection points.

The analysis considers prestress losses due to anchorage slip, friction, steel relaxation, and deformation of concrete. It is assumed, however, that the change in the tendon force is the same at both anchorages. The tendon force is assumed to be constant over the element length, and the forces imposed by the tendon on the structure are applied at a cross section halfway between the element ends at a point called the tendon reference point.

Local Force Coefficients With the prestressing applied in stages requiring analyses at intermediate time steps, and load imposed by a tendon on the structure is likewise recomputed. To simplify the computational approach of the equivalent load, a set of local force coefficients is introduced providing a total of nine coefficients at each tendon reference point.

The first two are tendon force coefficients f_t^A and f_t^B, from which the tendon force at the reference point is determined. These account for loss due to friction, and are due to unit tendon forces applied at anchor A or B. The next three coefficients p_a^e, p_t^e, and p_n^e are due to a unit tendon force at the reference point, and reflect essentially the effect of tendon and structure geometry on the equivalent load. The last four coefficients p_a^A, p_t^A or p_a^B, p_t^B

are related to a unit tendon force at the reference point, and compensate for the load introduced by friction between the tendon and duct along the element length.

The notation P_a^A and P_t^A represents the axial and tangential components of the sum of the friction losses over the length of an element produced by a unit tendon force, respectively. At any level of analysis, the three load components p_a, p_n, and p_t can be computed from the tendon force P_i at the point and the local force coefficients corresponding to the controlling anchor. For example, if the anchor at A controls, $P_i = P_A f_t^A$, where P_A is the prestress force in anchor A, and then

$$p_a = \left(p_a^A + p_a^e\right)P_i, \qquad p_n = p_n^e P_i \quad \text{and} \quad p_t = \left(p_t^A + p_t^e\right)P_i$$

Similar expressions are developed if the anchor at B controls.

The local force coefficients p_a and p_t, representing the friction loss over the length of an element, are calculated considering the change in the local force coefficients f_t^A and f_t^B as shown in Figure 8-12. It appears that two sets of coefficients may result: p_a^A and p_t^A used when the tendon force at the point is controlled by anchor A, and p_a^B and p_t^B used for control by anchor B.

Effect of Concrete Deformation Concrete deformations can cause the tendons to extend or shorten, both resulting in a change in the prestressing load prompting further deformation in the concrete. This problem reduces to the computation of a change in anchor force that will produce a certain change in tendon length. Computation of the prestressing load producing a given tendon extension is simplified by introducing the average extension factor A_{ef}, defined as the prestressing force applied at both anchorages of a tendon that will produce a unit extension of the tendon and includes the effect of friction.

For a unit tendon force applied at both anchorage ends, the tendon force at a given reference point i is f_t^A or f_t^B with a corresponding elongation

$$\Delta L_i = \sum_{i=1}^{n} l_i \Big/ (E_s A_s) f_t^c$$

where the superscript c refers to the controlling anchor. The average extension factor is

$$A_{\text{ef}} = E_s A_s \sum_{i=1}^{n} \left(f_t^c l_i\right)^{-1} \tag{8-10}$$

where L is the length of the tendon, ΔL_i is the tendon elongation over length l_i of element i, E_s is the modulus of elasticity of steel, and n is the

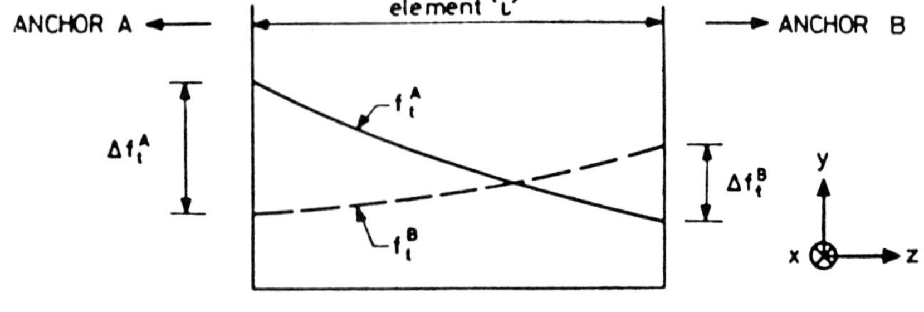

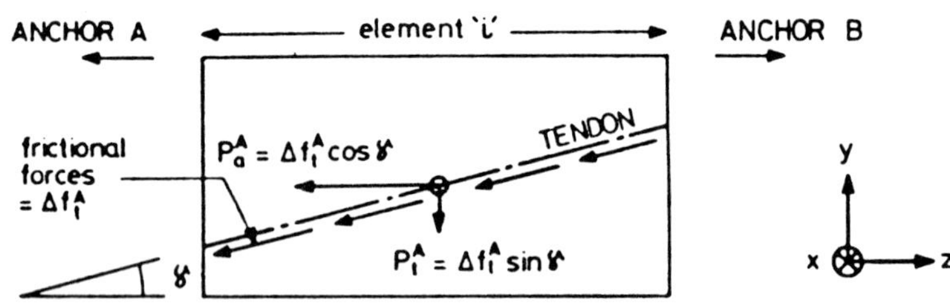

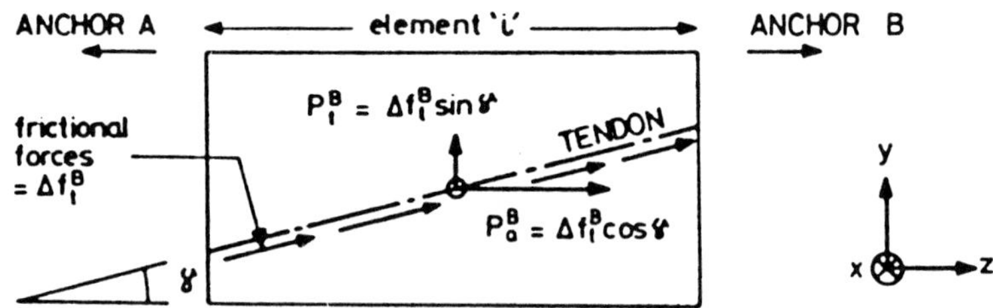

FIGURE 8-12 Local tendon force coefficients. (From Van Zyl and Scordelis, 1979.)

number of elements spanned by the tendon. The factor A_{ef} corresponds to a stiffness factor because it represents the equal anchor forces required at the ends of the tendon to cause a unit total elongation of the tendon subjected also to frictional resistance.

The increase ΔP of the anchor force at the two ends that will induce an extension ΔL of the tendon is computed from $\Delta P = \Delta L A_{ef}$. The parameter ΔP is an estimate of prestress loss caused by deformation of the concrete forcing the tendon to extend or shorten by an amount ΔL. A distinction is made between bonded and unbonded tendons.

Procedure for Segmental Analysis

The stiffness method used in this example divides the structure into discrete elements in the longitudinal direction and thus provides more flexibility in the input format. The initial input includes the complete structure geometry, prestressing geometry and pattern, material data, and construction load data

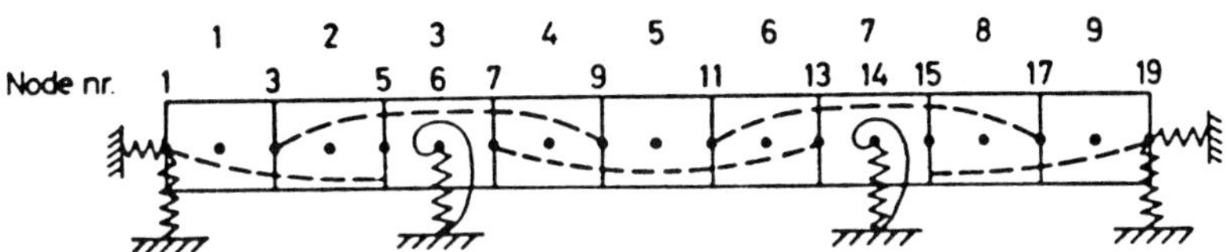

(*a*) SET UP ALL DATA DESCRIBING GEOMETRY, ALL PRESTRESSING GEOMETRY, MATERIAL TYPE DATA AND CONSTRUCTION LOAD TYPE DATA.

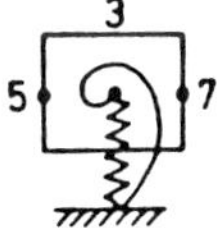

(*b*) SPECIFY THE TIME, THE FIRST SEGMENT TO BE ERECTED, AND THE CONSTRUCTION LOAD. COMPUTE STRESSES, STRAINS AS IN (d.)

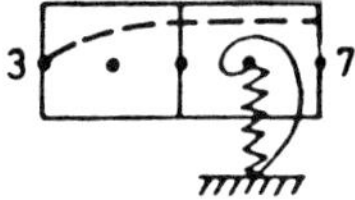

(*c*) SPECIFY THE TIME, NEXT SEGMENT, CONSTRUCTION LOAD AND PRESTRESSING.

(*d*) ANALYZE THE CURRENT STRUCTURE AND COMPUTE ALL THE REQUIRED STRESSES AND STRAINS.

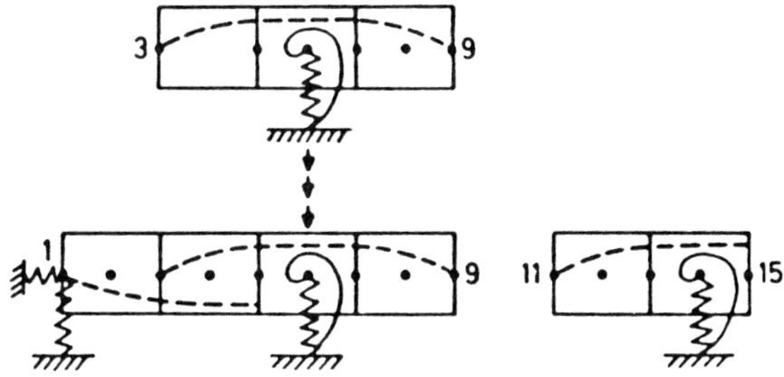

(*e*) REPEAT STEPS c, d FOR ALL SUBSEQUENT SEGMENTS AND TENDONS UNTIL THE STRUCTURE IS COMPLETED.

FIGURE 8-13 Segmental analysis procedure. (From Van Zyl and Scordelis, 1979.)

that will be used in the analysis. Nodes to be supported at any time during construction or after completion of the bridge are specified as supported nodes. Nodes not used initially are assigned zero spring stiffness, and those removed during construction or after completion are given zero stiffness at the appropriate stage. The initial data input also includes a set of time intervals during which temperature and humidity are assumed to be constant. Construction stages are then articulated one by one by identifying which segment is to be erected, the associated construction loads, the tendon stressing procedure, and the time when it takes place.

The entire procedure is shown diagrammatically in Figure 8-13. In practice, the cantilevers from nodes 6 and 14 may be constructed simultaneously, but in the analysis it is more convenient to treat sections of the bridge not yet successively interconnected. This requires only the correct erection time of each segment. The analysis proceeds for the set of interconnected nodes to which the last segment was added, and all unnecessary operations on other parts of the bridge are omitted. Data resulting from one analysis step are used as initial data for the next step, with the exception of initial displacements of a new node which must be computed separately.

The analysis of a structure with cast-in-place segments requires only specification of the appropriate erection time, which is then equal to the time of casting of a segment, and application of the prestressing at a later stage. The procedure may also be modified to handle a cast-in-place bridge built in one operation by using a single segment stretching from the first to the last node.

The time between two successive construction stages is a time interval. A complete printout of the geometry, loads, displacements, stresses, and/or strains is available at the beginning and end of an interval which can be subdivided into any number of time steps.

Computer Program A computer program was developed as part of this analysis, and, as is the case with structures of this complexity, a large-capacity computer is needed to process the program (Van Zyl and Scordelis, 1979).

Example 1

Hood (1976) has presented results from an experimental and analytical investigation of a curved box girder bridge model. The model was made from epoxy resin with a modulus of elasticity of 947 ksi and Poisson's ratio of 0.321. A typical cross section of the actual model is shown in Figure 8-14. The bridge is a two-span continuous unit curved in plan as shown in Figure 8-15*a*. Forty elements have been used in the present study.

The box girder bridge model was subjected to prestress loads according to the tendon profile shown in Figure 8-15*b*. The prestressing consisted of a high tensile 0.064-in.-diameter silver-plated piano wire with an ultimate tensile strength of 300 ksi placed in each web.

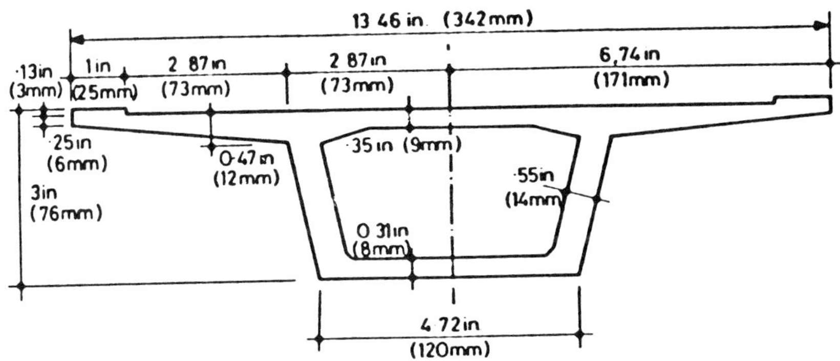

FIGURE 8-14 Typical cross section of model bridge. (From Hood, 1976.)

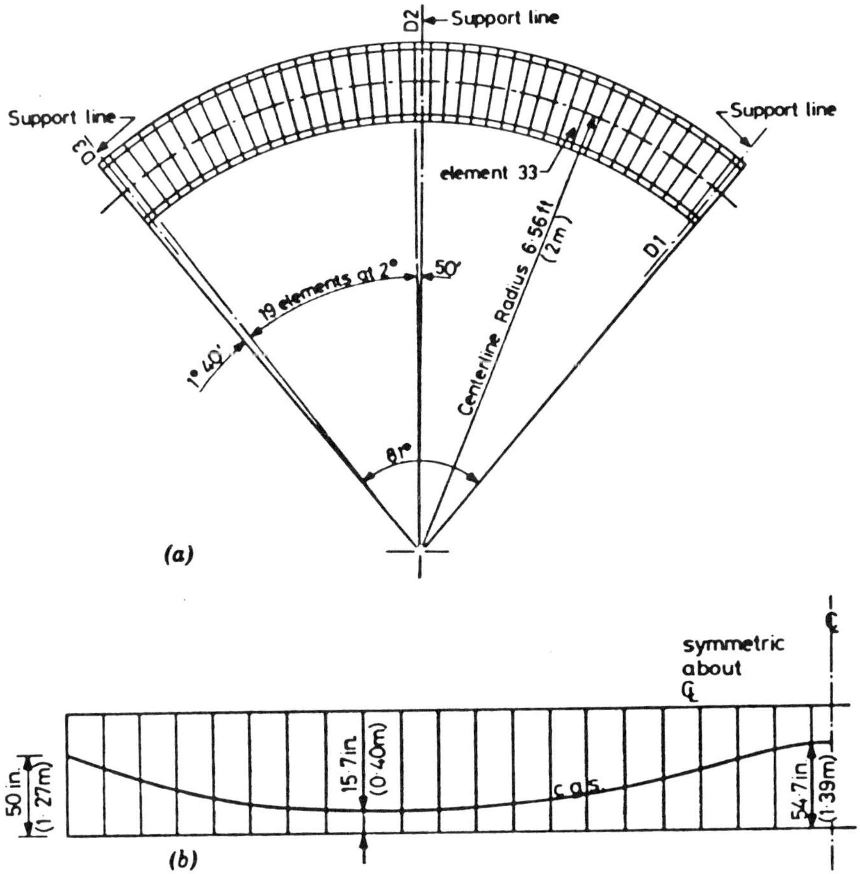

FIGURE 8-15 Box girder bridge of Example 1: (*a*) plan; (*b*) prestressing layout. (From Hood, 1976.)

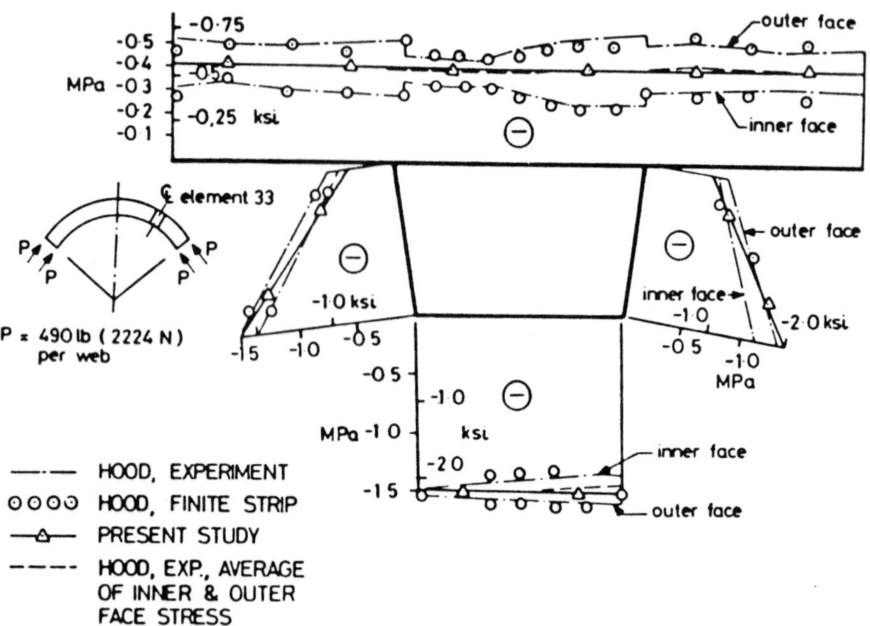

FIGURE 8-16 Tangential stress at center of element 33 due to prestressing.

The tangential stress at midspan due to prestressing is plotted in Figure 8-16, as obtained by Hood (1976) and as computed by the present method. Results obtained by Hood (1976) covered both surfaces of each plate, but in this analysis longitudinal axial stresses (tangential) are computed only at midthickness. A comparison is thus made with the average of the values obtained by Hood on the two surfaces. The computer program used in this analysis gives fairly accurate predictions of tangential stresses. Van Zyl and Scordelis (1979) made a check of the statical moment at the cross section that shows a variation of 2.3 percent between the internal moments and the externally applied moment.

Example 2

One of the early examples of segmental concrete bridges (1973) is the Corpus Christi Intercoastal Canal Bridge with spans of 100, 200, and 100 ft. The bridge width consists of two segments, each with a length of 10 ft and inclined webs. The erection sequence involved a balanced cantilever construction and is shown in Figure 8-17. A modified segment cross section used in the study is shown in Figure 8-18.

For simplicity, only one of the two identical boxes was analyzed, and, because the bridge is symmetrical about its center point, the investigation was reduced to one-quarter of the total bridge. The segment division

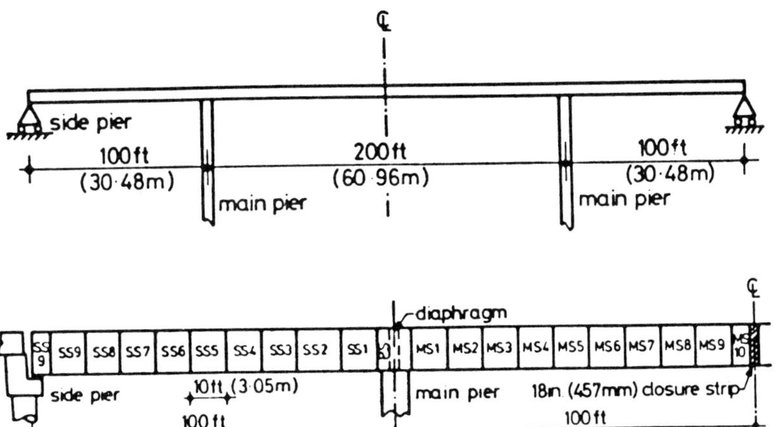

FIGURE 8-17 Erection sequence of Corpus Christi Bridge.

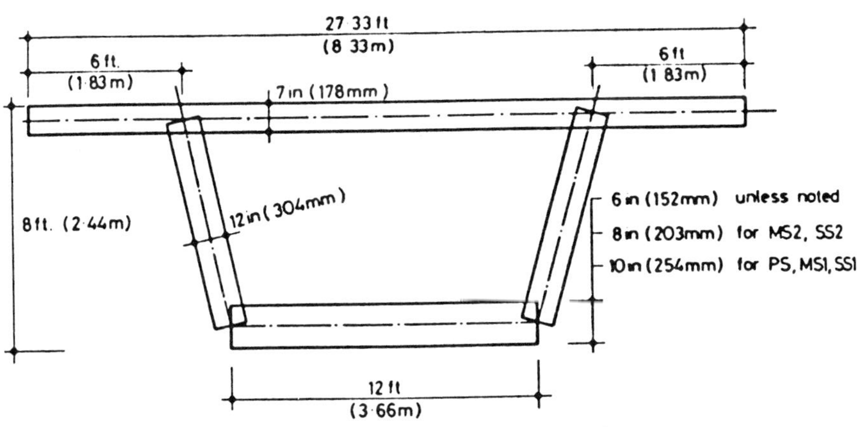

FIGURE 8-18 Cross section of Corpus Christi Bridge used in study. (From Van Zyl and Scordelis, 1979.)

shown in Figure 8-17 is the same as in the prototype. Likewise, the prestressing layout and pattern is the same as in the prototype, except that all tendons in the study are placed in the webs. Nodal points were were taken to coincide with segment ends except where tendon geometry resulted in additional nodes at midlength of a segment.

Segments were assumed to be subjected to a 30-day curing period prior to erection, and were placed in pairs with a 5-day interval between erection of two pairs. Time-dependent material properties were obtained following the ACI criteria for a 28-day compressive strength of 8 ksi, an ambient relative humidity of 40 percent, and a constant temperature of

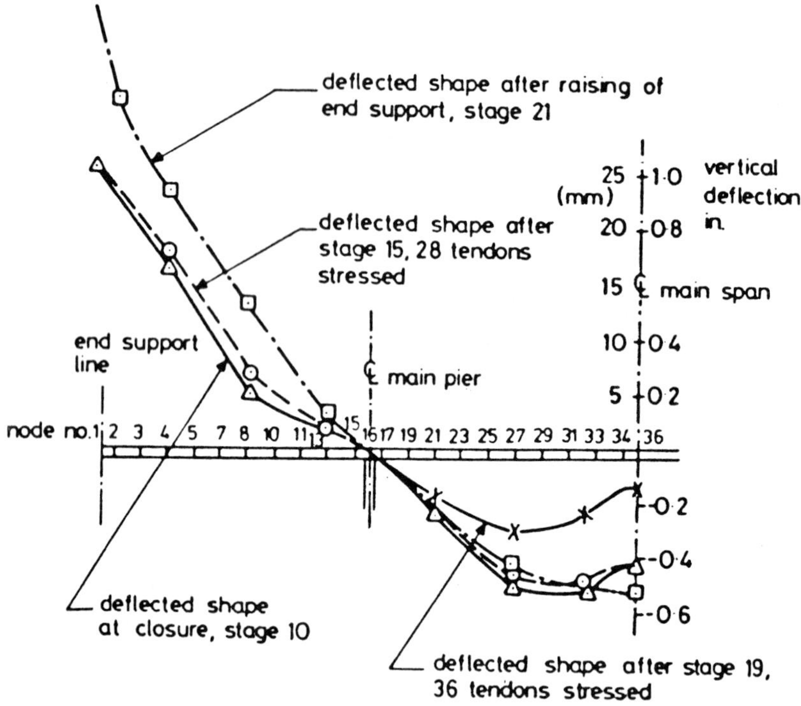

FIGURE 8-19 Deflections of Corpus Christi Bridge during construction as predicted by present method. (From Van Zyl and Scordelis, 1979.)

68°F, although actual field temperatures were much higher. Tendons were stressed to 75 percent of the ultimate tensile strength. Support spring stiffness was taken as recommended by Brown, Burns, and Breen (1974).

Prior to the construction of the prototype, a study was undertaken at the University of Texas at Austin, and included the development of a computer program and the construction and testing of a bridge model (Brown, Burns, and Breen, 1974; Kashima and Breen, 1975). The results obtained from the present study were compared with those obtained in the predesign investigation. Correlation between the deflections obtained by all three methods is satisfactory. The maximum difference at any stage is about 0.3 in., or nearly negligible for a 60-ft cantilever subjected to actual loads.

Deflections are plotted in Figure 8-19 and cover the period until after stressing of the closure tendons. Although data are not available for comparison, the curves show the change in the deflected shape as the structure is made continuous. The strain at the top of segment MS1, shown in Figure 8-17, and the associated variation during construction were compared as predicted in the three studies, and agreement was again satisfactory.

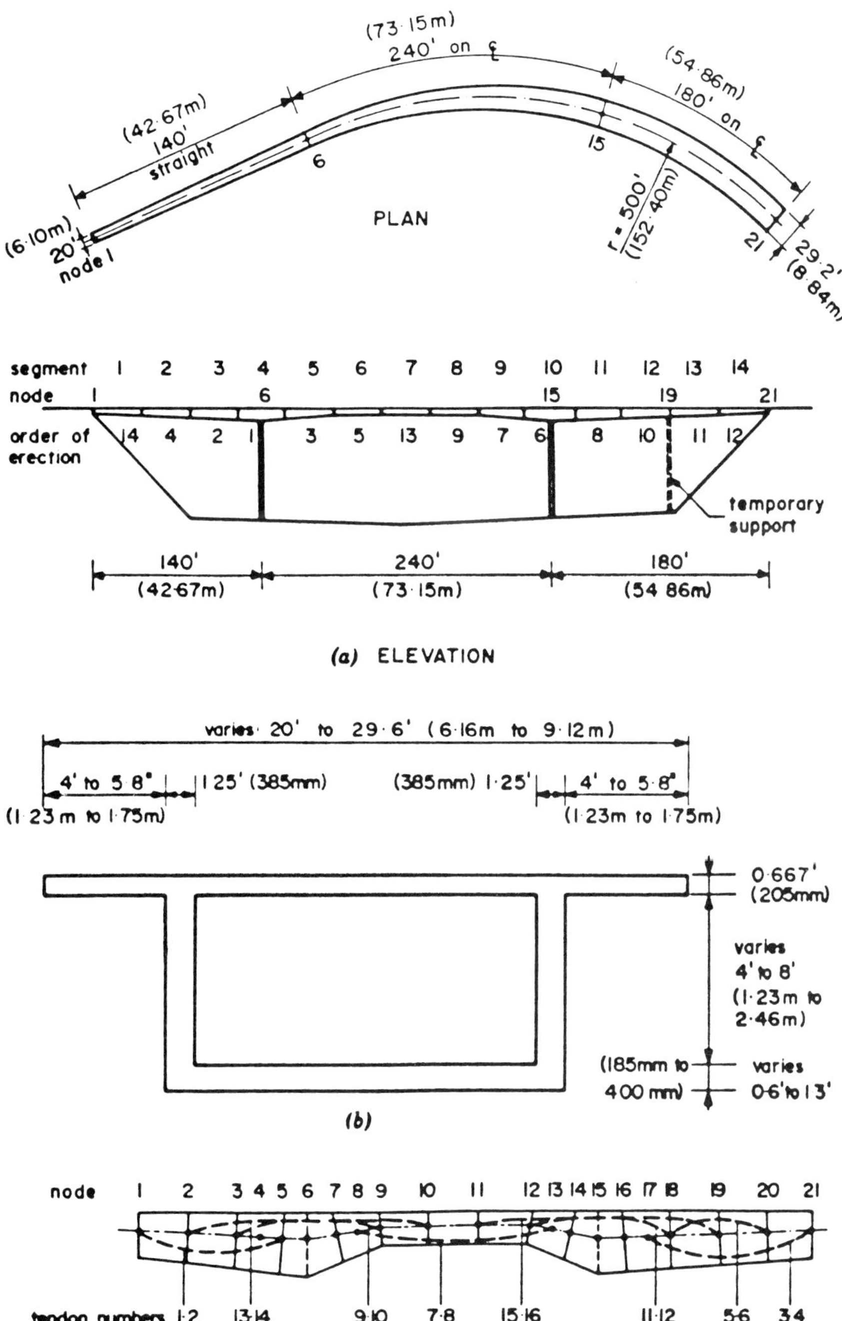

FIGURE 8-20 Design Example 3: (*a*) plan and elevation; (*b*) cross section; (*c*) tendon layout. (From Van Zyl and Scordelis, 1979.)

Example 3

A bridge example analyzed by the present method is shown in Figure 8-20. This structure has longer segments and fewer tendons than normally specified in bridge design, selected to reduce computer effort and to simplify the results of the analysis.

TABLE 8-3 Erection Sequence of Design Example 3 (From Van Zyl and Scordelis, 1979)

Stage (1)	Time (Days) (2)	Number of Segment Erected (3)	Tendons Stressed (4)	Construction Loads (5)	Supports Changed (6)	Given Displacements (7)
1	30	4				
2	35	3				
3	35	5	9, 10			
4	40	2				
5	40	6	13, 14			
6	91[a]					0.0017 rad rotation about z axis at node 6[b]
7	50	10				
8	55	11				
9	55	9	11, 12			
10	60	12				
11	60	8	15, 16	add—10 kips (−44.5 kN) vertical load at node 11[c]		
12	70	13	5, 6	remove load at node 11		
13	75	14	stress 3, 4; remove 5, 6		remove vertical support at node 19	
14	91[c]					
15	91[d]	7	7, 8			
16	91	1	1, 2			
17	150					

[a]The erection of the cantilevers supported on the left pier is complete, but the analysis has to be continued until just prior to the time of closure, which is at 91 days.

[b]The rotation can only be determined by carrying out an analysis up to stage 14.

[c]The analysis of the cantilevers supported on the right pier is carried out to the time just prior to the time of closure.

[d]Midspan closure is affected by placement of segment 7.

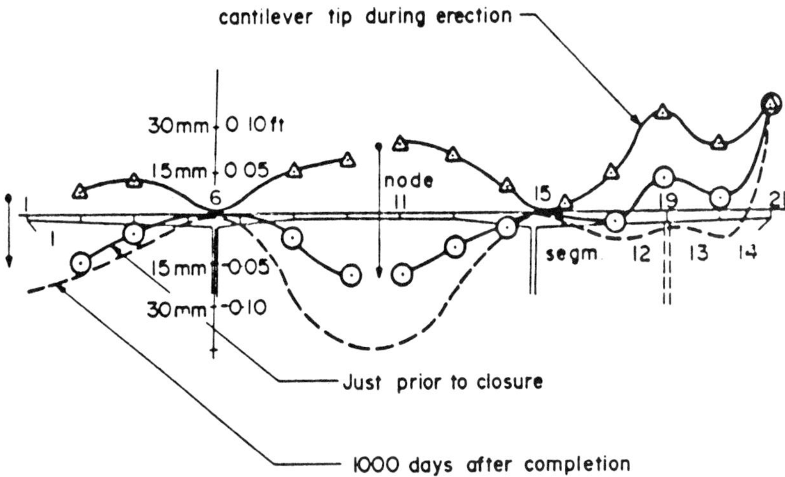

FIGURE 8-21 Deflections of bridge in Example 3 during construction. (From Van Zyl and Scordelis, 1979.)

The bridge has a straight section continued by a curved configuration as shown in the plan. The widths of the box section and cantilever flanges vary linearly along the entire length. Likewise, the depth of the box section and the bottom flange thickness are varied, although compliance with current AASHTO standards is not necessarily demonstrated. The erection sequence is given in Table 8-3, and involves 17 stages. Material properties are assumed to be the same throughout the construction process, but the segments are considered to be cast at different times as in actual practice.

As the construction progresses, both the temperature and ambient relative humidity are changed. At construction stage 14 the temporary support under node 19 is removed, and at stage 15 the temporary tendons 5 and 6 are removed. Before connecting the two cantilevers by erecting segment 7, the bridge is rotated at node 6 to match the two cantilever tips.

The deflections at the cantilever tip are plotted in Figure 8-21 together with the deflections of the continuous structure during stressing of the closure tendons. The graphs articulate the effects of stressing the positive tendons 1, 2, 3, 4, 7, and 8 on the deflection of the structure. The rotation of the supports in stage 6 can be specified only after the structure has been analyzed up to stage 14. A separate analysis was therefore carried out up to stage 14 to determine the rotation. We should also note the deflection of node 11 during the erection of segments 12, 13, and 14. The completed structure was loaded on day 150 with a distributed load of 0.15 ksf over the full width of the top flange and between nodes 8 and 13 along the length.

The final deflected shape in Figure 8-21 is obtained with no alignment changes other than the rotation at node 6. In order to obtain a bridge with a level profile after construction, the data shown in Figure 8-21 can be used to introduce alignment changes between segments as constructions progresses. In this example, the maximum downward deflection of the main span is only 2 in. or 1/2000 of the span, and profile adjustments may not be necessary.

8-7 MOMENT REDISTRIBUTION: CASE STUDY

This example articulates the influence of certain factors on the extent of moment redistribution, prestressing tendon force losses, and internal stress redistribution in a segmental concrete bridge built using the balanced cantilever method (Ketchum, 1987). The same example appears in its entirety in the appendix of the AASHTO specifications. The analysis presented in this section summarizes the results in an abbreviated form.

Four main factors were considered in the investigation: (a) age of the concrete at the time of erection, (b) ultimate creep in the concrete, (c) relocation in the prestressing steel, and (d) creep model used for the concrete.

A special computer program SFRAME developed by Ketchum (1987) is used in the analysis. The program articulates segmental construction and incorporates automated procedures and prestressing options as well as time-dependent behavior, thus allowing the analysis of a broad variety of complex segmental structures. It processes a detailed time-dependent elastic analysis for most prestressed, posttensioned segmental concrete bridges. A complete description of the program is provided by Ketchum (1986).

Description of Bridge Structure This example involves the 250-ft interior span of a straight, constant-depth, multiple-span continuous single-cell box girder bridge. The main cell is 20 ft wide with 10-ft cantilevers giving an overall deck width of 40 ft. The structure is 11 ft deep. The bridge is cantilevered from each pier using ten 12-ft-long segments. Continuity is introduced with a 10-ft closure segment at midspan. Prestressing tendons are stressed both during the cantilever construction and after closure. Ketchum (1987) presents tendon geometry and data for the cantilever tendons and the distribution of continuity tendons. A complete description of the bridge, the design methodology, the loadings, and the analytical aspects of the computer model (including the node and element layout) is also given by Ketchum (1987).

The bridge girder is analyzed for two different construction schedules, two different creep models for the concrete, two different relaxation factors for the prestressing steel, and several different ultimate creep factors for the concrete, giving a total of 10 distinct cases of analysis. In all cases the bridge is investigated for the total design load, including segment dead loads applied during segmental erection, prestressing forces, and superimposed dead load introduced after closure. The 10 cases are as follows.

Case 1: Cast-in-place concrete
Ultimate creep factor for concrete $C = 3$
Relaxation factor for prestressing steel $R = 10$
ACI model for creep for concrete

Case 2: Cast-in-place concrete
Ultimate creep factor for concrete $C = 3$
Relaxation factor for prestressing steel $R = 10$
CEB–FIP model for creep of concrete

Case 3: Precast concrete
Ultimate creep factor for concrete $C = 2$
Relaxation factor for prestressing steel $R = 45$
ACI model for creep of concrete

Case 4: Precast concrete
Ultimate creep factor for concrete $C = 2$
Relaxation factor for prestressing steel $R = 45$
CEB–FIP model for creep of concrete

Case 5: Precast concrete
Ultimate creep factor for concrete $C = 1$
Relaxation factor for prestressing steel $R = 45$
ACI model for creep of concrete

Case 6: Precast concrete
Ultimate creep factor for concrete $C = 2$
Relaxation factor for prestressing steel $R = 45$
ACI model for creep of concrete

Case 7: Precast concrete
Ultimate creep factor for concrete $C = 3$
Relaxation factor for prestressing steel $R = 45$
ACI model for creep of concrete

Case 8: Precast concrete
Ultimate creep factor for concrete $C = 1$
Relaxation factor for prestressing steel $R = 45$
CEB–FIP model for creep of concrete

Case 9: Precast concrete
Ultimate creep factor for concrete $C = 2$
Relaxation factor for prestressing steel $R = 45$
CEB–FIP model for creep of concrete

Case 10: Precast concrete
Ultimate creep factor for concrete $C = 3$
Relaxation factor for prestressing steel $R = 45$
CEB–FIP model for creep of concrete

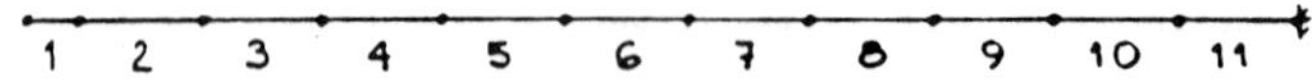

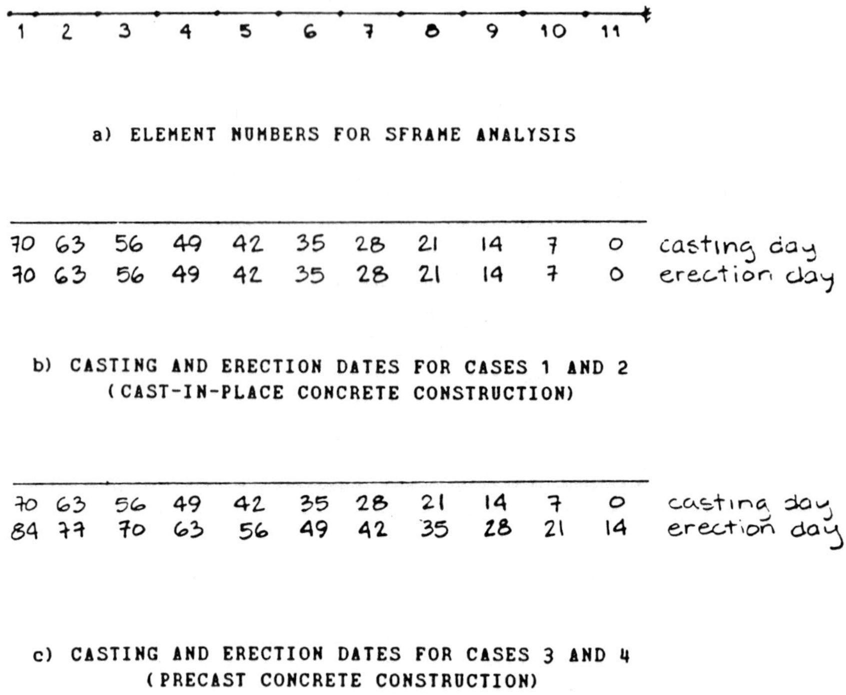

a) ELEMENT NUMBERS FOR SFRAME ANALYSIS

70 63 56 49 42 35 28 21 14 7 0 casting day
70 63 56 49 42 35 28 21 14 7 0 erection day

b) CASTING AND ERECTION DATES FOR CASES 1 AND 2 (CAST-IN-PLACE CONCRETE CONSTRUCTION)

70 63 56 49 42 35 28 21 14 7 0 casting day
84 77 70 63 56 49 42 35 28 21 14 erection day

c) CASTING AND ERECTION DATES FOR CASES 3 AND 4 (PRECAST CONCRETE CONSTRUCTION)

FIGURE 8-22 Segment casting and erection dates. (From Ketchum, 1987.)

The creep factor C always refers to the ultimate creep factor for a load applied when the concrete is 7 days old. The program automatically accounts for the reduction in this factor for loads applied when the concrete ages.

Cases 3 and 4 are intended to illustrate the combined effects of the age of the concrete at initial loading, the ultimate creep factor for the concrete, and the relaxation factor for the prestressing steel on the structural behavior of the girder. Cases 5 through 10 illustrate the effect of the ultimate creep factor and the time-dependent creep model on the structural response of the bridge.

Cast-in-place or precast segments are modeled by adjusting the casting date relative to the erection date. Cast-in-place segments are cast on the same day they are erected. Precast segments are cast 14 days before they are erected. Segment casting and erection dates for the 10 segments are summarized in Figure 8-22. Traveling formwork is included for cast-in-place models but not for precast segments. The formwork is removed from the structure just after installation of the closure segment.

Cases 1 through 4 are analyzed both for the actual segmental erection sequence and for a one-step erection sequence with the age of the segments equal to the average segment age at the time of closure of the midspan joint. Cases 5 through 10 are analyzed for the actual segmental erection sequence only. In all cases, the bridge is analyzed for the assumed erection sequence and for 27 years (10,000) days following construction. The results presented in the illustrations are for observation times at the end of construction and at the end of the 27-year period.

Material properties for the time-dependent analysis are as follows:

Concrete strength	$f'_c = 5000$ psi
Ultimate shrinkage strain	$\epsilon = 0.0008$
Unit weight	$w = 155$ pcf

The time-dependent development of creep, shrinkage, and aging is based on standard ACI and CEB–FIP criteria. The prestressing steel has elastic modulus $E = 28{,}000{,}000$ psi, $f'_s = 270$ ksi, and relaxation coefficient $C = 10$ or 45. All prestressing tendons are stressed to $0.7f'_s$ or 190 ksi, and the initial forces in the tendon segments are found based on a curvature friction coefficient of 0.25/radian and a wobble friction coefficient of 0.0004/ft.

Results of Analysis: Cases 1 and 4 For cases 1 and 4, the structural behavior predicted by segmental construction analysis is compared with the same behavior predicted by a single-step construction analysis typical in conventional present-day analysis. A summary of the moment redistribution predicted by both analyses is shown in Figures 8-23 and 8-24 for cases 1 and 4, respectively. Each illustration includes two distinct plots labeled (a) and (b). The plots labeled (a) illustrate the redistribution of moments in the concrete girder, determined by integrating the stresses in the concrete over the cross-sectional area of the girder. The plots labeled (b) show the redistribution of statical moments, defined as the moments in equilibrium with the total external load (in this case dead load) and the reactions. Primary prestressing is not included in the statical moments, but the secondary moments are included.

Each plot shows three different moment diagrams. One curve represents the moments for the one-step construction sequence at an observation time of 27 years (this is 100 percent moment distribution, with prestressing loss based on the same construction sequence). The second curve represents the moments for the segmental bridge at an observation time of 27 years, and the third curve is also for the segmental construction sequence but at an observation time at the end of construction.

A summary of prestress losses predicted by the analyses for cases 1 and 4 is shown in Table 8-4 and 8-5 for the segmental and one-step construction sequence, respectively, at an observation time of 27 years. The upper sections of these tables show long-term losses, and these do not include instantaneous losses due to friction, taking place when the tendon is stressed. The lower

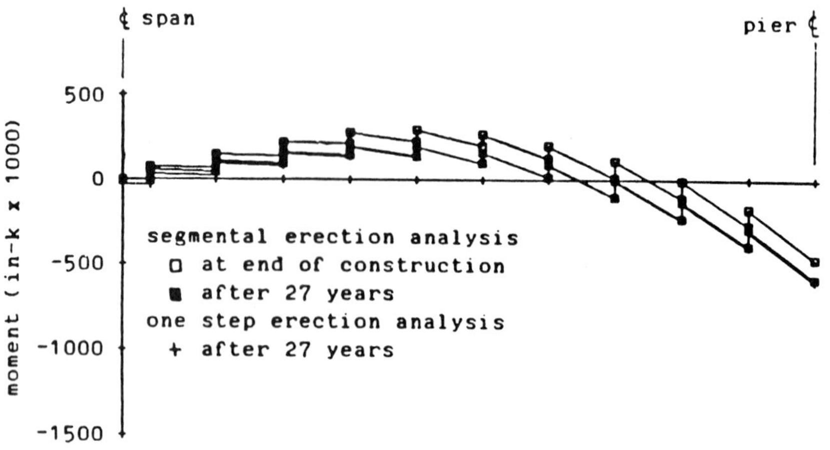

a) REDISTRIBUTION OF MOMENTS IN CONCRETE UNDER TOTAL LOAD

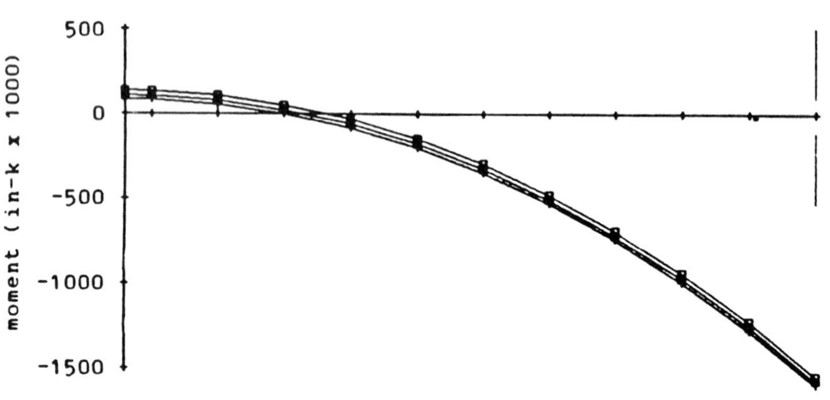

b) REDISTRIBUTION OF STATICAL MOMENTS UNDER TOTAL LOAD

FIGURE 8-23 Results for case 1—cast-in-place concrete, $C = 3$, $R = 10$, ACI model. (From Ketchum, 1987.)

sections of the tables show total losses including instantaneous losses due to friction. In all cases, the difference between the values shown in the two sections represents instantaneous losses due to friction. The losses are tabulated as a percentage loss of the original (initial jacking) force for all tendons passing through a given frame element. Losses in individual tendons are not tabulated.

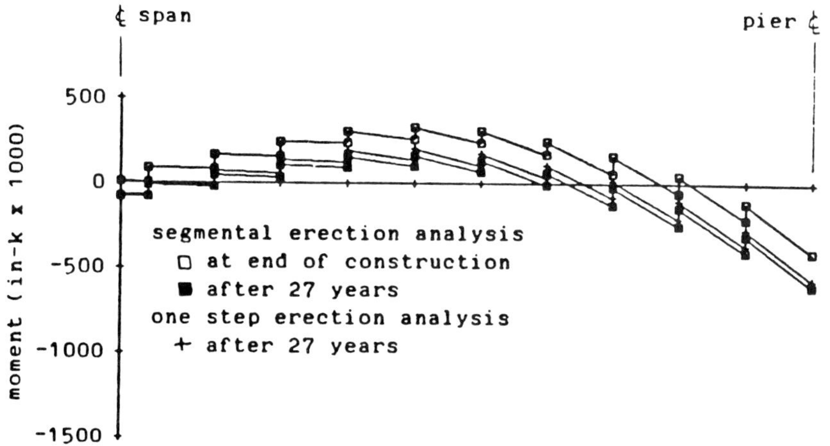

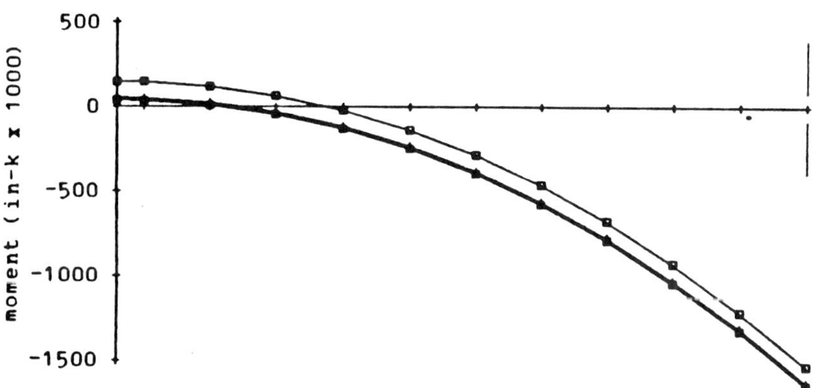

FIGURE 8-24 Results for case 4—precast concrete, $C = 2$, $R = 45$, CEB–FIP model. (From Ketchum, 1987.)

The prestress loss for cast-in-place concrete is markedly higher than for precast segments, probably because of the lower elastic modulus and higher creep potential for the fresh concrete at the time stressing is applied. This difference is significantly higher for the CEB–FIP model than for the ACI model, and this articulates the fundamental features in the two models.

The losses predicted for the one-step construction sequence invariably are much lower than predicted for the segmental construction. This is considered a basic shortcoming in the present analysis approach, which thus may not be the best procedure for predicting prestress losses. For a detailed investigation

TABLE 8-4 Losses of Force in Prestressing Tendons for Cases 1 and 4 at Observation Time 27 Years for Segmental Construction Analysis (Values in Percent for All Tendons in Each Frame Element)

Long-Term Tendon Force Losses (Percent)			Percentage Loss in Tendon Force	
Element Number	Tendon Count	Initial Force	Case 1	Case 4
1	4	1909.	18.41	13.99
2	5	4100.	19.42	13.79
3	6	6266.	21.36	14.89
4	7	8418.	22.59	15.85
5	7	10158.	23.32	16.39
6	7	11909.	23.35	16.50
7	7	13665.	22.78	16.26
8	7	15413.	21.71	15.71
9	8	17599.	20.51	14.85
10	9	19682.	18.96	13.78
11	10	21472.	16.89	12.65
Total Tendon Force Losses (Percent)			**Percentage Loss in Tendon Force**	
Element Number	Tendon Count	Jacking Force	Case 1	Case 4
1	4	2093.	25.57	21.54
2	5	4419.	25.23	19.99
3	6	6744.	26.93	20.92
4	7	9070.	28.16	21.91
5	7	10872.	28.36	21.86
6	7	12675.	27.98	21.54
7	7	14477.	27.11	20.96
8	7	16279.	25.87	20.20
9	8	18605.	24.81	19.47
10	9	20930.	23.79	18.92
11	10	23256.	23.26	19.35

Note: Case 1 = cast-in-place concrete, $C = 3$, $R = 10$, ACI creep model.
Case 4 = precast concrete, $C = 2$, $R = 45$, CEB–FIP creep model.

of the distribution of predicted prestress tendon force losses over the length of each individual tendon, reference is made to Ketchum (1986).

Results of Analysis: Cases 6 and 9 These cases compare the structural behavior predicted by segmental construction analysis for the ACI and the CEB–FIP creep models. Although this behavior is sensitive to the value of the creep factor, this dependence is not illustrated because in both cases the value of the creep factor used is 2.

A summary of moment redistribution is plotted as in Figure 8-23 and 8-24 for the ACI model and the CEB–FIP model, respectively. In each plot two

TABLE 8-5 Losses of Force in Prestressing Tendons for Cases 1 and 4 at Observation Time 27 Years for One-Step Construction Analysis (Values in Percent for All Tendons in Each Frame Element)

Long-Term Tendon Force Loss (Percent)			Percentage Loss in Tendon Force	
Element Number	Tendon Count	Initial Force	Case 1	Case 4
1	4	1909.	13.13	9.14
2	5	4100.	14.22	9.14
3	6	6266.	15.78	10.04
4	7	8418.	17.33	11.08
5	7	10158.	18.57	11.90
6	7	11909.	19.21	12.35
7	7	13665.	19.30	12.46
8	7	15413.	18.90	12.25
9	8	17599.	18.41	12.02
10	9	19682.	17.51	11.56
11	10	21472.	16.02	10.80
Total Tendon Force Losses (Percent)			**Percent Loss in Tendon Force**	
Element Number	Tendon Count	Jacking Force	Case 1	Case 4
1	4	2093.	20.74	17.10
2	5	4419.	20.38	15.68
3	6	6744.	21.73	16.40
4	7	9070.	23.28	17.48
5	7	10872.	23.91	17.67
6	7	12675.	24.10	17.64
7	7	14477.	23.84	17.37
8	7	16279.	23.22	16.92
9	8	18605.	22.83	16.79
10	9	20930.	22.43	16.83
11	10	23256.	22.47	17.65

Note: Case 1 = cast-in-place concrete, $C = 3$, $R = 10$, ACI creep model.
Case 4 = precast concrete, $C = 2$, $R = 45$, CEB–FIP creep model.

different moment diagrams are shown, one representing the moments for an observation time of 27 years, and the other representing the moments for an observation time at the end of construction. Although the effect of the creep factor is not included, a larger creep factor should result in a greater degree of moment distribution as would normally be expected. The extent of moment distribution is greater for the CEB–FIP model than for the ACI model. Whether these differences warrant several investigations for any bridge structure would depend on the corresponding changes in internal girder stresses.

TABLE 8-6 Losses of Force in Prestressing Tendons for Cases 5, 6, 7, 8, 9, and 10 (Values in Percent for all Tendons in Each Frame Element) (From Ketchum, 1987)

Long-Term Tendon Force Losses (Percent)			ACI Creep Model			CEB–FIP Creep Model		
Element Number	Tendon Count	Initial Force	$C = 1$	$C = 2$	$C = 3$	$C = 1$	$C = 2$	$C = 3$
1	4	1909.	11.71	12.31	12.79	12.40	13.99	15.55
2	5	4101.	12.25	13.21	14.13	12.48	13.79	15.06
3	6	6267.	12.85	14.38	15.89	13.06	14.89	16.69
4	7	8418.	13.36	15.42	17.42	13.55	15.85	18.11
5	7	10157.	13.67	16.03	18.30	13.82	16.39	18.90
6	7	11910.	13.62	16.09	18.45	13.81	16.50	19.16
7	7	13663.	13.27	15.68	17.97	13.52	16.26	18.91
8	7	15412.	12.63	14.86	16.95	13.02	15.71	18.29
9	8	17597.	11.94	13.98	15.88	12.33	14.85	17.25
10	9	19681.	11.11	12.87	14.51	11.51	13.78	15.93
11	10	21472.	10.08	11.47	12.75	10.62	12.65	14.56
Total Tendon Force Losses (Percent)			**ACI Creep Model**			**CEB–FIP Creep Model**		
Element Number	Tendon Count	Initial Force	$C = 1$	$C = 2$	$C = 3$	$C = 1$	$C = 2$	$C = 3$
1	4	2093.	19.46	20.00	20.43	20.09	21.54	22.95
2	5	4419.	18.57	19.45	20.30	18.78	19.99	21.17
3	6	6744.	19.02	20.44	21.85	19.22	20.92	22.58
4	7	9070.	19.59	21.51	23.35	19.77	21.91	23.99
5	7	10872.	19.45	21.54	23.66	19.47	21.86	24.21
6	7	12675.	18.83	21.16	23.37	19.01	21.54	24.04
7	7	14477.	18.14	20.42	22.56	18.39	20.96	23.46
8	7	16279.	17.28	19.39	21.37	17.65	20.20	22.64
9	8	18605.	16.71	18.64	20.44	17.08	19.47	21.73
10	9	20930.	16.41	18.07	19.60	16.77	18.92	20.94
11	10	23256.	16.98	18.26	19.44	17.48	19.35	21.12

Note: Case 5 = precast concrete, $C = 1$, $R = 45$, ACI creep model.
Case 6 = precast concrete, $C = 2$, $R = 45$, ACI creep model.
Case 7 = precast concrete, $C = 3$, $R = 45$, ACI creep model.
Case 8 = precast concrete, $C = 1$, $R = 45$, CEB–FIP creep model.
Case 9 = precast concrete, $C = 2$, $R = 45$, CEB–FIP creep model.
Case 10 = precast concrete, $C = 3$, $R = 45$, CEB–FIP creep model.

The prestress losses predicted by the analysis are summarized in Table 8-6 for all cases 5 through 10, for both the ACI and CEB–FIP models, and ultimate creep factors 1, 2, and 3. Likewise, the upper section shows long-term losses only, and the lower section shows total losses including friction. These losses are tabulated as a percentage loss of the original (initial or jacking) force for all tendons passing through a given frame element. The same trends are documented as in Tables 8-4 and 8-5. For both models an increase in creep factor results in a corresponding increase in prestress losses.

TABLE 8-7 Stresses in Concrete at Extreme Fibers for Cases 5, 6, 7, 8, 9, and 10 (From Ketchum, 1987)

Case	Location in Span	Top Stress	Bottom Stress	Notes
5	Center	−148.73	−83.48	$C = 1$, ACI model
	Pier	−467.79	−1850.10	
6	Center	−133.89	−108.34	$C = 2$, ACI model
	Pier	−414.21	−1874.60	
7	Center	−132.31	−109.29	$C = 3$, ACI model
	Pier	−375.87	−1884.50	
8	Center	−65.86	−233.00	$C = 1$, CEB–FIP model
	Pier	−375.61	−1941.10	
9	Center	−.73	−346.30	$C = 2$, CEB–FIP model
	Pier	−256.42	−2023.30	
10	Center	29.84	−396.22	$C = 3$, CEB–FIP model
	Pier	−172.39	−2068.40	

Note: Stresses in psi at top and bottom of section, at midspan and pier.

The predicted stresses at the extreme fiber of the girder for cases 5 through 10 are tabulated in Table 8-7. For a 50 percent change in the creep factor (from $C = 2$ to $C = 1$ or $C = 3$), the stress changes are large enough to warrant several analyses for the same structure, although they have no significant effect on the conformance of the bridge to code requirements inasmuch as the creep factor is usually predicted with sufficient accuracy.

REFERENCES

AASHTO, 1989. "Guide Specifications for Design and Construction of Segmental Concrete Bridges."

ACI, 1982. "Prediction of Creep, Shrinkage and Temperature Effects in Concrete Structures," ACI Committee 209, Report 209 R-82, American Concrete Inst.

ACI, 1985. "Corrosion of Metals in Concrete," ACI Comm. 222, Detroit.

ACI, 1989. "Building Code Requirements for Reinforced Concrete (ACI 318-89)," ACI Committee 318, Detroit.

Ballinger, C. A., W. Podolny, and M. J. Abrahams, 1977. "A Report on the Design and Construction of Segmental Prestressed Concrete Bridges in Western Europe," FHWA, U.S. Dept. Transp., Washington, D.C., Report No. FHWA-RD-78-44.

Bazant, Z. P., 1971. "Numerically Stable Algorithm, with Increasing Time Steps for Integral-Type Aging Creep," Paper H2/3, First Inter. Conf. on Struct. Mech. in Reactor Technology, Vol. 3, Berlin, Sept.

Bazant, Z. P. and M. El Nimeiri, 1973. "Stiffness Method for Curved Box Girders at Initial Stress," *J. Struct. Div. ASCE*, Vol. 100, No. ST10, Oct., pp. 2071–2089.

Breen, J. F., R. L. Cooper and T. M. Gallaway, 1975. "Minimized Construction Problems in Segmentally Precast Box Girder Bridges," Center for Hwy. Research, Univ. Texas, Austin.

Brown, R. C., N. H. Burns, and J. E. Breen, 1974. "Computer Analysis of Segmentally Erected Precast Prestressed Box Girder Bridges," Center for Hwy Research, Univ. Texas, Austin.

Bryant, A. A. and V. Chayatit, 1987. "Creep, Shrinkage-Size, and Age at Loading Effects," *ACI Materials J.*, March–April.

CEB–FIP Model Code for Concrete Structures, 1978. Comite Euro-Inter. de Beton (CEB), available from: Lewis Brooks, 2 Blagdon Rd., New Malden, Surrey, KT3 4AD, England.

Concrete Reinforcing Steel Institute, 1980. "Manual of Standard Practice," CRSI, Chicago.

Danon, J. R. and W. L. Gamble, 1977. "Time-Dependent Deformations and Losses in Concrete Bridges Built by the Cantilever Method," *Struct. Research Ser. No.* 437, Univ. Illinois, Urbana, Jan.

Dilger, W. H., 1982. "Creep Analysis of Prestressed Concrete Structures Using Creek-Transformed Section Properties," *PCI J.*, Jan.–Feb., Vol. 27, No. 1.

DIN 1075, 1981. "*Concrete Bridges*: *Dimensions and Construction*, Deutsche Norm (German Standards), April.

Dokken, R. A, 1975. "CALTRANS Experience in Segmental Bridge Design," *Bridge Notes*, Div. of Structures, Dept. Transp. Calif., Vol. XVII, No. 1, March.

European Concrete Committee, 1970. "International Recommendations for the Design and Construction of Concrete Structures, Principles, and Recommendations," June.

Finsterwalder, U., 1967. "New Developments in Prestressing Methods and Concrete Bridge Construction," *Dywidag-Berichte*, 4-1967, Munich, Sept.

Finsterwalder, U., 1969. "Free-Cantilever Construction of Prestressed Concrete Bridges and Mushroom-Shaped Bridges," *ACI Publ.* SP-23, Paper SP 23-26, ACI, Detroit.

Grant, A., 1975. "Incremental Launching of Concrete Structures," *ACI J.*, Vol. 72, No. 8, Aug.

Heins, C. P. and R. A. Lawrie, 1984. *Design of Modern Concrete Highway Bridges*, Wiley, New York.

Hoang, L. H., and M. Pasquignon, 1985. "Essais de Flexion sur des Poutres en Beton Preconstraintes par des Cables Exterieurs," Vols. 1 and 2, Contrast SETRA-CEBTP, Dossiers de Recherche 910017, Service d'Etude des Structures, Saint Remy Les Chevreuse, Nov.

Homberg, H., 1968. "Fahrbahnplatten mit Verandlicher Dicke," Springer-Verlag, New York.

Homberg, H., and Ropers, W., 1965. "Fahrbahnplatten mit verandlicher Dicke," Springer-Verlag, N.Y.

Hood, J. A., 1976. "An Investigation of a Curved Single Cell Box-girder Bridge," Report No. 138, School of Engineering, Univ., Auckland for the National Roads Board, Auckland, New Zealand, Oct.

Huang, T., 1982. "A New Procedure for Estimates of Prestress Losses," FHWA–Pennsylvania Dept. Transp. Research Project 80-23, Lehigh Univ., Bethlehem, Pa. May.

Kabir, A. F., 1976. "Nonlinear Analysis of Reinforced Concrete Panels, Slabs, and Shells for Time-Dependent Effects," SESM Report No. 76-6, Univ. Calif., Berkeley.

Kashima, S. and J. E. Breen, 1975. "Construction and Load Tests of a Segmental Precast Box Girder Bridge Model," Research Report 121-5, Center for Hwy. Research, Univ. Texas, Austin, Feb.

Ketchum, M. A., 1986. "Redistribution of Stresses in Segmentally Erected Prestressed Concrete Bridges," Report No. UCB/SESM-86/07, Dept. Civ. Eng., Univ. Calif., Berkeley, May.

Ketchum, M. A., 1987. "An Analysis of Moment Redistribution in a Segmentally Erected Posttensioned Concrete Bridge," Post Tensioning Inst., Chicago, July.

Koseki, K. and J. E. Breen, 1983. "Exploratory Study of Shear Strength of Joints for Precast Segmental Bridges," Research Report 248-1, Center for Transp. Research, Univ. Texas, Austin, Sept.

Lacey, G. C. and J. E. Breen, 1969. "Long Span Prestressed Concrete Bridges of Segmental Construction, State of the Art," Research Report 121-1, Center for Hwy. Research, Univ. Texas, Austin, May.

Lacey, G. C. and J. E. Breen, 1975. "The Design and Optimization of Segmentally Precast Concrete Bridges," Research Report 121-3, Center for Hwy Research, Univ. Texas, Austin, Aug.

Leonhardt, F., Kolbe, G., and J. Peter, 1965: "Temperature Differences Dangerous to Prestressed Concrete Bridges," *Beton and Stahlbetonban*, No. 7, pp. 157–163.

Marti, P., 1985. "Basic Tools of Reinforced Concrete Beam Design," *ACI J.*, Vol. 82, No. 1, Jan.–Feb.

Ontario Highway Bridge Design Code, 1983. Ontario Ministry Transp. and Communications, Toronto, Canada.

PCI, 1975. "Recommended Practice for Segmental Construction in Prestressed Concrete," Report PCI Committee on Segmental Construction, *J. Prestressed Concrete Inst.*, Vol. 20, No. 2, March-April.

PCI, 1976. "Prestressed Concrete Segmental Bridges on FA 412 over the Kishwaukee River," *Bridge Bull.* No. 1, PCI, Chicago.

Podolny, W., 1986. "Evaluation of Transverse Flange Forces Induced by Laterally Inclined Longitudinal Posttensioning in Box Girder Bridges," *J. Prestressed Concrete Inst.*, Vol. 31, No. 1, Jan.–Feb.

Podolny, W. and J. M. Muller, 1982. *Construction and Design of Prestressed Concrete Segmental Bridges*, Wiley, New York.

Portland Cement Association, 1974. "John F. Kennedy Memorial Causeway, Corpus Christi, Texas," Bridge Report SR 162.01 E, Skokie, Ill.

Portland Cement Association, 1977. "Napa River Bridge, California," Bridge Report SR 194-01 E, Skokie, Ill.

Posten, R. W., R. L. Carrasquillo, and J. E. Breen, 1987. "Durability of Posttensioned Bridge Decks," *ACI Materials J.*, July–August.

Post-Tensioning Institute, 1985. "Post-Tensioning Manual," 4th ed., "Guide Specification for Post-Tensioning Materials," and "Recommended Practice for Grouting of Post-Tensioned Prestressed Concrete," PTI, Phoenix.

Post-Tensioning Institute, 1988. "Design and Construction Specifications for Segmental Concrete Bridges," NCHRP Report No. 20-7/32, PTI.

Pucher, A., 1969. *Influence Surfaces of Elastic Plates*, 4th rev. ed., Springer-Verlag, New York.

Rabbat, B. G. and K. Sowlat, 1987. "Testing of Segmental Concrete Girders with External Tendons," *PCI J.*, Vol. 32, No. 2, March–April.

Schlaich, J., Schafer, K., and M. Jennewain, 1987. "Towards a Consistent Design of Reinforced Concrete Structures," *PCI Journ.*, May–June, Vol. 32, No. 3.

Shushkewich, K. M., 1986. "Time-Dependent Analysis of Segmental Bridges," *J. Computers and Structures* (Great Britain), Vol. 23, No. 1.

Stone, W. C., Paes-Filha, W., and J. E. Breen, 1981. "Behavior of Post-Tensioned Girder Anchorage Zones," Research Report 208-2, Center for Transportation Research, The Univ. of Texas at Austin, April.

Stone, W. C. and J. E. Breen, 1981. "Analysis of Posttensioned Girder Anchorage Zones," Research Report 208-1, Center for Transp. Research, Univ. Texas, Austin, June.

Stone, W. C. and J. E. Breen, 1981. "Design of Posttensioned Girder Anchorage Zones," Research Report 208-3F, Center for Transp. Research, Univ. Texas, Austin, June.

Tadros, M. K., Ghali, A., and W. Dilger, 1975. "Time-Dependent Prestress Loss and Deflection in Prestressed Concrete Members," *PCI J.*, Vol. 20, No. 3, May–June.

Tadros, M. K., Ghali, A., and W. Dilger, 1977. "Effect of Non-Prestressed Steel on Prestress Loss and Deflection," *PCI J.*, Vol. 22, No. 2, March–April.

Van Zyl, S. F., 1978. "Analysis of Curved Segmentally Erected Prestressed Concrete Box Girder Bridges," SESM Report No. 78-2, Univ. Calif., Berkeley.

Van Zyl, S. F. and A. C. Scordelis, 1979. "Analysis of Curved, Prestressed, Segmental Bridges," *J. Struct. Div. ASCE*, Vol. 105, No. ST11, pp. 2399–2417.

Xanthakos, P. P., 1991. *Ground Anchors and Anchored Structures*, Wiley, New York.

Zienkiewicz, O. C. and M. Watson, 1966. "Some Creep Effects in Stress Analysis with Particular Reference to Concrete Pressure Vessels," *Nuclear Engineering and Design*, No. 4.

CHAPTER 9

TRUSSES, MOVABLE BRIDGES, AND CABLE-STAYED BRIDGES

Part 1: Trusses

9-1 TRUSS TYPES AND CHARACTERISTICS

A truss is essentially a triangulated assembly of straight members. A planar truss may be regarded as a deep girder, where the girder flanges are replaced by the truss chords and the web plate is formed by an open system of web members. It may be used to replace a girder in several cases: as a simply supported or continuous straight girder; in the deck of a stiffened suspension bridge or cable-stayed girder bridge; or as an arch. A planar truss can support loads only in its own plane, and a three-dimensional truss is necessary to support a general system of loads.

In a typical (conventional) truss, the centroidal axes of all members are straight and concurrent at the nodes or points. Because the truss proper is loaded only at the nodes, applied loads are resisted primarily by axial forces induced in the truss members. Bending moments are generally small and have a minor effect on the axial force components interacting for equilibrium. At best, all member bending moments should be close to zero, a condition that may be achieved by inserting frictionless pins at the nodes. Unlike this ideal system, in practice most members are rigidly connected at the joints. However, the resulting restraint induces small moments, further controlled by the truss proportions and fabrication and assembly techniques. At worst, these moments may enhance brittle fracture or fatigue failure by inducing secondary stresses, thus increasing the primary direct stresses.

Truss Components A truss bridge of conventional design consists of the following parts: (a) a deck slab or similar structural system, (b) longitudinal stringers directly supporting the deck slab, (c) cross beams at panel points accepting the load from the longitudinal stringers, (d) the two main truss systems, (e) lateral bracing provided in the planes of the upper and lower chords, (f) end sway frames receiving the horizontal transverse forces from the lateral bracing and transferring these forces to the piers, and (g) additional intermediate sway frames distributing the transverse loads to the lateral system and keeping the system stable during erection. These components are identified in Figure 9-1, showing the skeleton of a through truss highway bridge.

For through trusses, a system of top laterals is always provided, although the live load is applied at the bottom plane. These laterals provide rigidity, stabilize the compression chord, and carry the main part of the wind forces to the bridge portals. They consist of struts and diagonals as shown. The portals of a through bridge cannot be cross-braced and are designed as rigid frames to transmit the load from the top lateral system to the bridge supports.

Sway bracing provided at the panel points ensures the torsional rigidity of the structure. However, the portion of wind load on the top chord carried

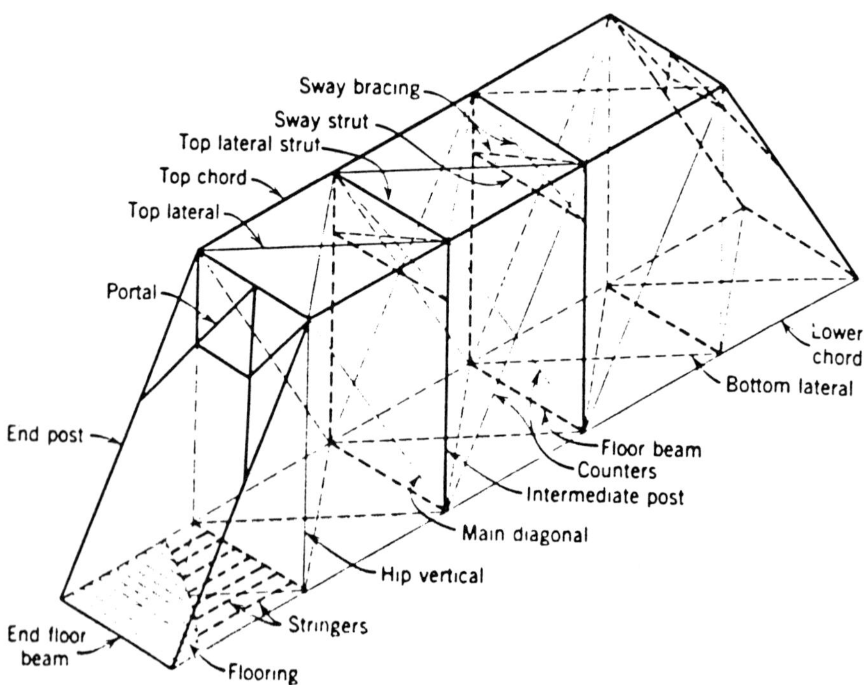

FIGURE 9-1 Typical frame of a through truss highway bridge.

through the sway bracing cannot be analyzed easily and is often designed arbitrarily with the intent of adding rigidity rather than articulating strength.

Because the allowable stresses are generally increased under group loadings including wind, the design of chords in vertical trusses may not be affected by lateral force analysis, especially for wide and shallow bridges. For narrow and high trusses or bridges of unusual proportions, wind forces may govern the design, and requirements of both lateral and torsional rigidity are likely to dictate the design requirements.

Truss Forms The most common bridge truss types are shown in Figure 9-2. In the Warren truss shown in Figure 9-2*a*, the chords carry the bending moment and the diagonals carry the shear. The vertical members carry only panel loads and can be economically designed. However, the secondary stresses are relatively high.

The Pratt truss is by definition an assembly with tension diagonals. For normal loads this definition corresponds to the ordinary arrangement shown in Figure 9-2*b*. An advantage in this case is that the verticals (shorter web members) are in compression rather than the longer diagonals, but this is

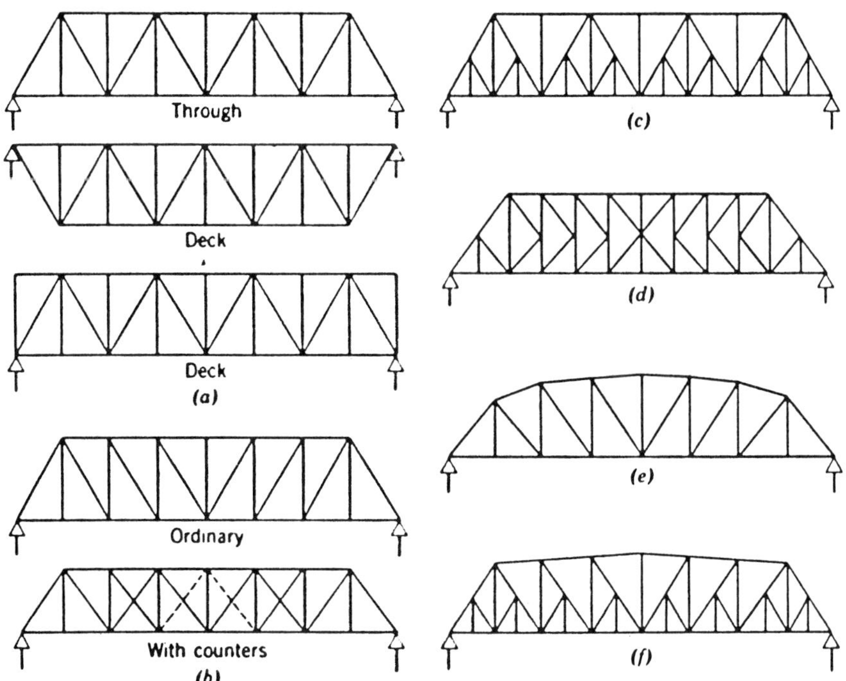

FIGURE 9-2 Typical bridge trusses: (*a*) Warren trusses; (*b*) Pratt trusses; (*c*) subdivided Warren truss; (*d*) K truss; (*e*) curved-chord Pratt truss; (*f*) Pettit truss.

partly offset by the fact that the central compression chord is more heavily loaded than the central tension chord.

The economic height–span ratio is about one-sixth to one-eighth, according to the truss type, loading, span length, and other parameters. The optimum inclination of the diagonals is about 45°, and slight variations from these proportions are not likely to influence overall economy. With increasing span lengths, the economical height also increases. Thus, both the Warren and Pratt trusses will result in long panel lengths if the diagonal inclination remains about 45°. An alternative is to subdivide these trusses as shown in Figure 9-2*c*. However, the subdivided trusses develop high secondary stresses, and K trusses may be preferred as shown in Figure 9-2*d*. K trusses keep the desirable inclination, accommodate the required truss depth, and also limit the stringer span.

As shown in Figures 9-2*e* and *f*, truss chords may be placed on a curved alignment to carry part of the shear and to reduce the stresses in the diagonals. This arrangement results in a slight increase in the fabrication cost which, for medium and long spans, is offset by materials savings.

Special Characteristics A truss bridge thus has two major structural advantages: (a) the primary member forces are axial loads and (b) the open-web system accommodates a greater overall depth than in an equivalent solid-web girder. The increased depth imparts more rigidity to the bridge and results in reduced deflection. These advantages must be compared with increased fabrication and maintenance costs.

In the past, the conventional truss bridge has been found to be economical for medium spans. In general, it has been used for the span range intermediate between the plate girder and the stiffened suspension bridge. Recent advances in design and construction techniques for both steel and concrete girders have tended to increase the economical spans. In addition, the cable-stayed bridge has become a strong competitor to the steel truss for the intermediate span range. It appears now that all these factors have combined to make the truss bridge less favorable in modern designs.

In the past, economical solutions have been documented for highway spans in the range of 500 to 1500 ft. One of the longest truss bridges in service is the 1576-ft main span of the Greater New Orleans Bridge completed in 1958. This is exceeded by the 1800-ft-span Quebec Bridge and the 1700-ft-span Firth of Forth Bridge, both railway bridges.

The truss has practically become the standard stiffening structure for the conventional suspension bridge because of its favorable aerodynamic response, which is the result of the inherent torsional and vertical stiffness. This relative stiffness of a truss bridge is also an erection advantage. The structure may be assembled sequentially by adding members with the use of lifting equipment of conventional capacity. Alternatively, the number of field connections can be reduced by fabricating and erecting the trusses bay-by-bay.

For structural optimization, it is essential to achieve compatibility between the deck and the main structure. The deck may be included to act compositely with the truss chords in resisting axial loads. Alternatively, it may be isolated from the chords by introducing a series of deck expansion joints.

Compared with other solutions, the construction depth of a truss bridge is considerable if the deck is at the upper chord level, but quite nominal if the traffic moves through the truss with the deck at the lower chord level. For grade separations, a through truss bridge is thus a major advantage in overall alignment. In some structures it is expedient to combine both arrangements to provide a through truss over the main span with a small construction depth and approaches with the deck at the upper chord level.

9-2 TOPICS RELEVANT TO TRUSS DESIGN

Truss Connections

Truss members are joined by gusset plates at the joints where the members meet. Connections can be made by riveting, bolting, or welding. The thickness of gusset plates is determined by several factors. A minimum plate thickness is necessary to develop the full strength of the bolts, depending on a single or double shear action. Experience shows that a usual gusset plate thickness is 3/8 to 1/2 in. for light trusses, and 5/8 to 7/8 in. for heavier trusses.

Stresses in gusset plates may be checked using beam theory. Thus, for a direct load P and bending moment M, the nominal stress is $f = (P/A) \pm (Mc/I)$. If shear is also present at the critical section, the resulting shear stresses should be considered, and yielding should be based on the maximum principal stresses.

The foregoing method gives results that are approximate only. Computation of actual stresses is rather complex because it must take into account load concentration, warping of plate sections, and local yielding. Compressive stresses along free edges of gusset plates may cause local buckling unless stiffeners are used.

Stress Transfer Stress transmission through the gusset at a truss joint may be achieved in two ways. If the chord member is continuous through the gusset, the main portion of the stress is transmitted directly within the chord, and only the difference of the chord stresses is carried through the gusset. This arrangement is often used to relieve the guesset plate of any excessive load. If chord splices are necessary, they can be made outside the joint in the lesser-stressed member.

If the chord members are spliced at a joint, the gusset at this location will be subjected to heavy stress because it transmits the entire amount of the chord stresses. For compression chords bearing against each other at the joints, the bearing surfaces are milled for full contact and direct stress transfer, although a portion of the stress must still be transmitted through the gusset as stipulated by the applicable specifications. Gusset plates in tension splices, however, require a complete design of load transfer with the associated additional material. Hence, direct splices for tension chords are normally limited to small spans. For long cantilever spans with the top chord curved over the supports, pin connections may be advantageous.

Welded Connections Component parts of individual truss members may be connected by welds. Welded trusses are basically similar to riveted or bolted trusses, except that the choice of member sections may be dictated by the connection details. Rolled and built-up sections may be welded together, and examples are T sections, H sections, closed boxes, and pipes.

However, application of welded trusses in heavy bridge construction has been rather limited. A major problem can arise from the existence of local moment and torque at welded connections that may produce high stress concentrations and, under repeated cycles, enhance the potential of fatigue.

Secondary Stresses

Triangular trusses are usually designed assuming that the members carry direct stresses only. This is true when transverse loads are not present and moment transfer does not occur at the joints. Direct axial stresses thus computed are termed primary stresses, whereas bending stresses are referred to as secondary stresses. The latter may be produced in truss members by the following conditions: (a) eccentricities in member connections (this condition may arise if the centroids of the sections at a joint do not intersect at one point and moments are thus generated); (b) torsional moments, introduced by members not acting in the plane of truss, such as floor beams; (c) transverse loads on a member such as the weight of the member itself (considered when they are applicable); and (d) truss distortion and rigidity of joints resulting in bending of the members.

Secondary stresses are often intended to denote only those stresses produced by truss distortion and the joint rigidity (see also subsequent sections). Their magnitude varies considerably, and for usual truss forms with members of high slenderness ratios, they range from 5 to 25 percent of the primary stresses. For subdivided trusses and Warren trusses with verticals, certain members may have secondary stresses as high as 40 to 100 percent of the primary stresses. However, the significance of the secondary stresses is not necessarily as critical as in primary stresses. High secondary stresses exist only in some members and in extreme fibers at the ends of these members. Even if these localized stresses reach the yield point, they are unlikely to cause collapse unless they are repeated and associated with fatigue.

Secondary stresses resulting from truss distortions can be limited by reducing the width of members in the plane of bending relative to their length, by choosing truss types with inherent low secondary stresses, and by avoiding the use of vertical hangers and subdivision. Alternatively, pin connections may be used, but friction in the pin holes may not always permit free rotation of the connecting members.

AASHTO stipulates that in designing and detailing truss members, secondary stresses should be as small as practicable. Secondary stresses due to distortion or floor beam deflection need not be considered in members where the width, measured parallel to the plane of distortion, is less than one-tenth of the length. If the secondary stress exceeds 4000 psi for tension members and 3000 psi for compression members, the excess should be treated as primary stress. Stresses due to flexural dead load effects should be included as additional secondary stresses.

Modern numerical methods are used in conjunction with computers to solve the complex problems of analyzing trusses for secondary stresses. Methods of analysis based on classical procedures are presented in the following sections.

Stress Repetition and Reversal

Truss members subjected to stress reversal, as well as their connections, must be checked for fatigue. Stress fluctuations, even without reversal, may cause severe fatigue effects on members or connections at lower stress levels than those at which some form of failure would occur under static load. These effects would be primarily manifested as stress concentrations introduced by constructional details.

Fatigue effects may be avoided or allowed for according to two basic approaches. The first is to reduce the allowable working stress, and the second is to increase the computed forces. Each procedure is based on a method of compensation depending on the range of stress fluctuation, the number of repeated cycles, the quality of the steel, and the type of connection. Both AASHTO and AREA specifications do not consider fatigue unless there is stress reversal, because basic allowable stresses are low enough to accommodate these conditions. For connections and base metal subject to stress reversal, relevant fatigue provisions should apply.

A common practice for dealing with stress reversals in the web members of a truss is to use counter members. The actual stress distribution among the counters is a statically indeterminate problem, but, for simplicity, the members are assumed to act one at a time, taking tension only. This assumption is nearly correct when the members are slender so that they buckle under compression and resist only a small fraction of the stress. The exact amount of compressive force in the counters can thus be computed by considering the buckling strength of the member.

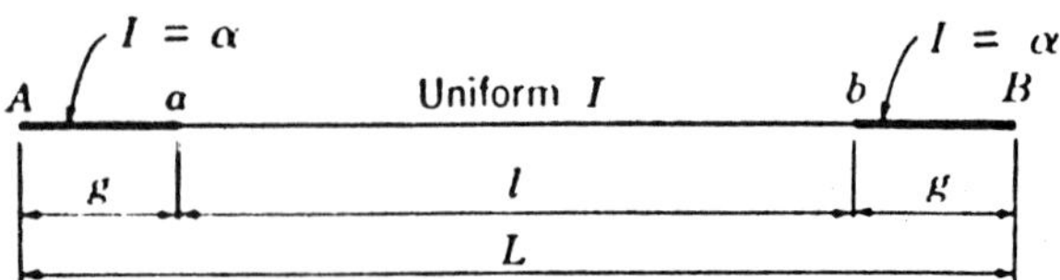

FIGURE 9-3 Typical truss member. (From O'Connor, 1965.)

Counters may be arranged at their intersections in three different ways: by letting them pass by each other without connection between them, by bolting or welding them back-to-back on a gusset plate, and by splicing one of them at the intersecting gusset.

9-3 ANALYSIS OF TRUSSES

Axial Forces The axial forces in a pin-jointed truss can be found directly by applying the principles of structural analysis and frames. Reference is thus made to any book on the subject and to the references included at the end of this chapter.

Secondary Stresses and Buckling With most truss joints rigidly connected, axial strains in the members can cause changes in the nodal geometry with resulting curvatures and moments. These have been referred to as secondary moments and secondary stresses.

Consider the typical truss member shown in Figure 9-3. uniform over much of its length but with ends of infinite rigidity chosen to simulate gusset plates or similar end connections. The equivalent length of these ends is g. Under an axial force P, the change in length is Pl/AE. The axial stiffness is the force P per unit change in length and is equal to AE/l, or, for practical purposes, AE/L, where $L = l + 2g$. The bending rigidity varies with the axial load, reduced by a compressive force and increased by a tensile force.

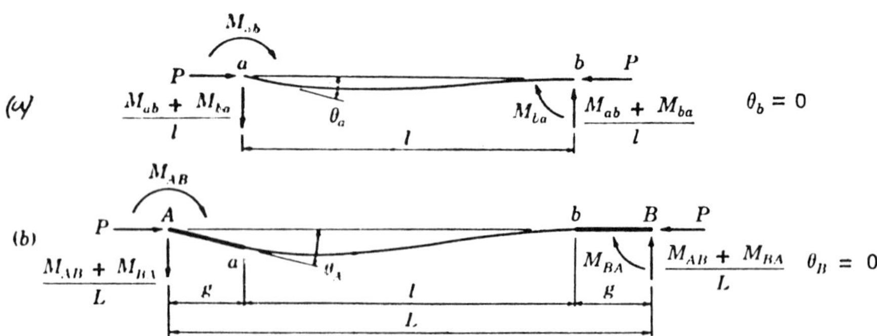

FIGURE 9-4 Deformation of truss member associated with bending. (From O'Connor, 1965.)

For the uniform member ab of length l, shown in Figure 9-4a, fixed at b with a rotation θ_a at a, the end moments M_{ab} and M_{ba} can be written as

$$M_{ab} = s\frac{EI}{l}\theta_a \qquad M_{ba} = sc\frac{EI}{l}\theta_a \tag{9-1}$$

where s is the stiffness coefficient and c is the carry-over factor ($c = M_{ba}/M_{ab}$).

For a compressive load P and the Euler load $P_E = \pi^2EI/l^2$, we can write $\alpha = (\pi/2)(P/P_E)^{1/2}$. Similarly, for a tensile load T, we can write $\beta = (\pi/2)(T/P_E)^{1/2}$, where P_E is the Euler load in absolute magnitude. It can be

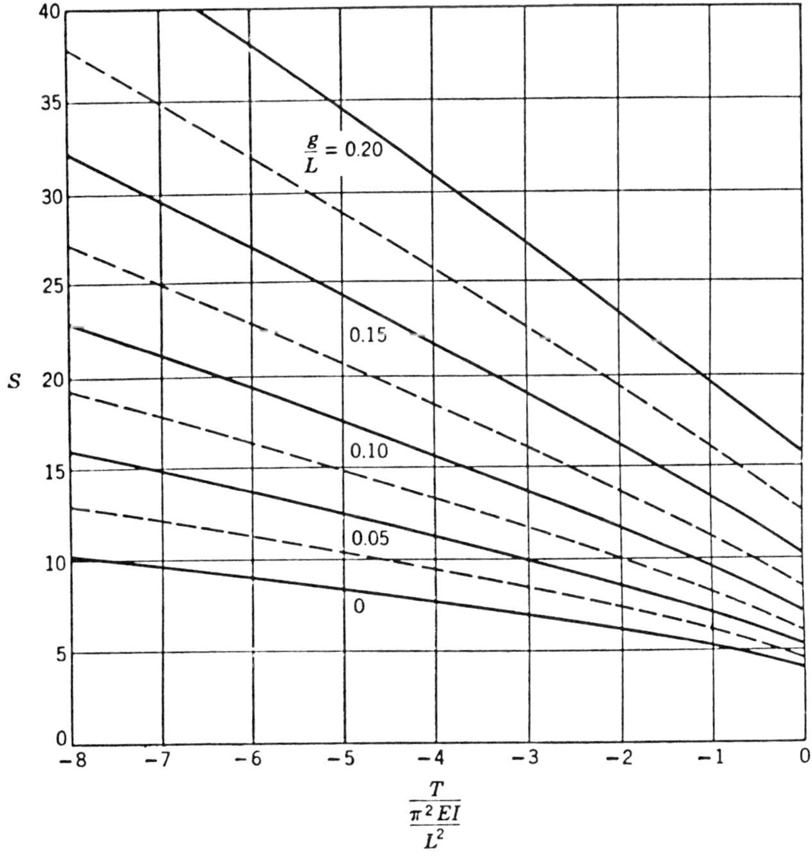

FIGURE 9-5 Stiffness coefficients for tension bars with rigid gusset plates. (From O'Connor, 1965.)

shown that

$$s = \frac{\beta(1 - 2\beta \coth 2\beta)}{\tanh \beta - \beta} \tag{9-2}$$

$$c = \frac{2\beta - \sinh 2\beta}{\sinh 2\beta - 2\beta \cosh 2\beta} \tag{9-3}$$

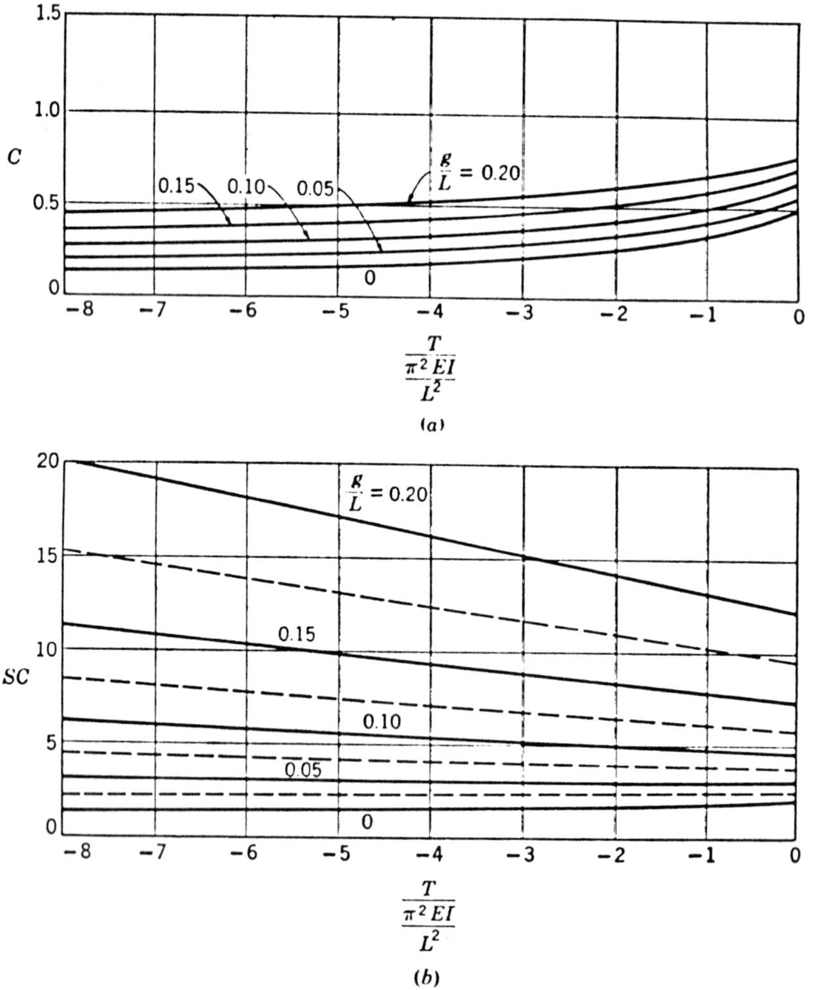

FIGURE 9-6 Carry-over factors and moments for tension bars with rigid gusset plates: (*a*) carry-over factors; (*b*) carried-over moments. (From O'Connor, 1965.)

Considering the compression member AB with rigid supports at each end, as shown in Figure 9-4b, the end moments M_{AB} and M_{BA} are

$$M_{AB} = \frac{EI}{l}\theta_A\left[s\left(1 + \frac{g}{l} + \frac{2g^2}{l^2}\right) + sc\frac{2g}{l}\left(1 + \frac{g}{l}\right) - \frac{Pgl}{E1}\left(1 - \frac{g}{l}\right)\right] \quad (9\text{-}4)$$

$$M_{BA} = \frac{EI}{l}\theta_A\left[s\frac{g}{l}\left(1 + \frac{2g}{l}\right) + sc\left(1 + \frac{2g}{l} + \frac{2g^2}{l^2}\right) - \frac{Pg^2}{EI}\right] \quad (9\text{-}5)$$

where for a tension member P is replaced by $-T$ and β is replaced by α.

Alternatively, these moments may be expressed in the form

$$M_{AB} = s\frac{EI}{L}\theta_A \qquad M_{BA} = sc\frac{EI}{L}\theta_A \quad (9\text{-}6)$$

where s is the stiffness coefficient and c is the carry-over factor for the

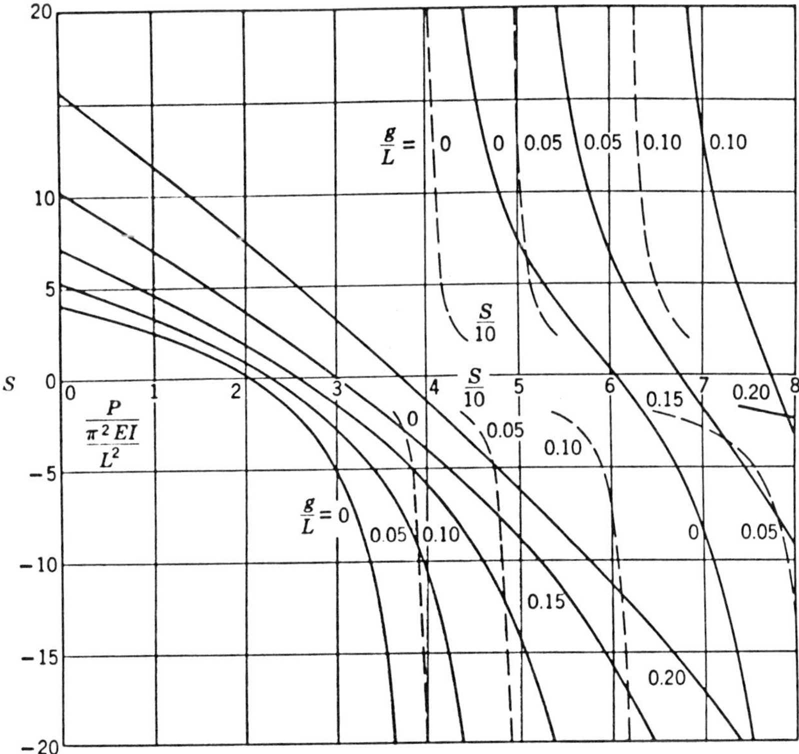

FIGURE 9-7 Stiffness coefficients for compressed bars with rigid gusset plates. (From O'Connor, 1965.)

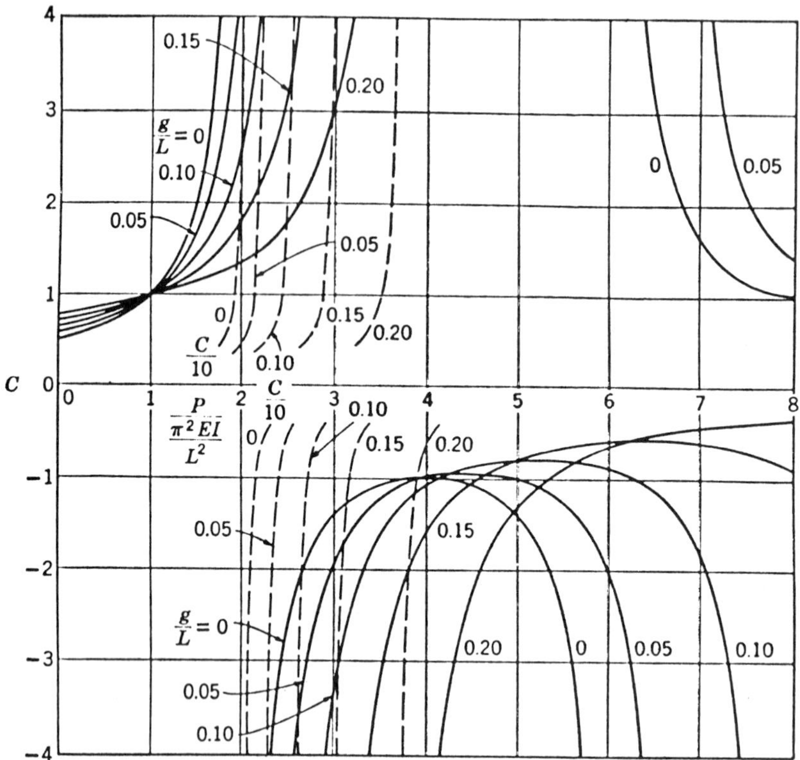

FIGURE 9-8 Carry-over factors for compressed bars with rigid gusset plates. (From O'Connor, 1965.)

complete member based on EI/L, with L being the total length as shown in Figure 9-4*b*.

The functions s, c, and sc for varying values of the axial load are plotted in Figures 9-5 through 9-9. All curves have a finite slope at zero axial force, suggesting that the bending rigidity varies at a finite rate for even small loads.

Direct solutions may be obtained from (9-2) through (9-6), but the analysis is greatly facilitated by suitable computer programming for all member axial forces and end moments (Jennings, 1968). A displacement-type analysis may be programmed as follows (O'Connor, 1970).

1. Calculate the axial forces by a pin-jointed truss analysis.
2. Compute member stiffness and carried-over moments.
3. Incorporate these data into a complete stiffness matrix for the truss.

The format will generally have three displacements per node, two components of linear displacement, and a rotation. Some of these are defined by the known support conditions.

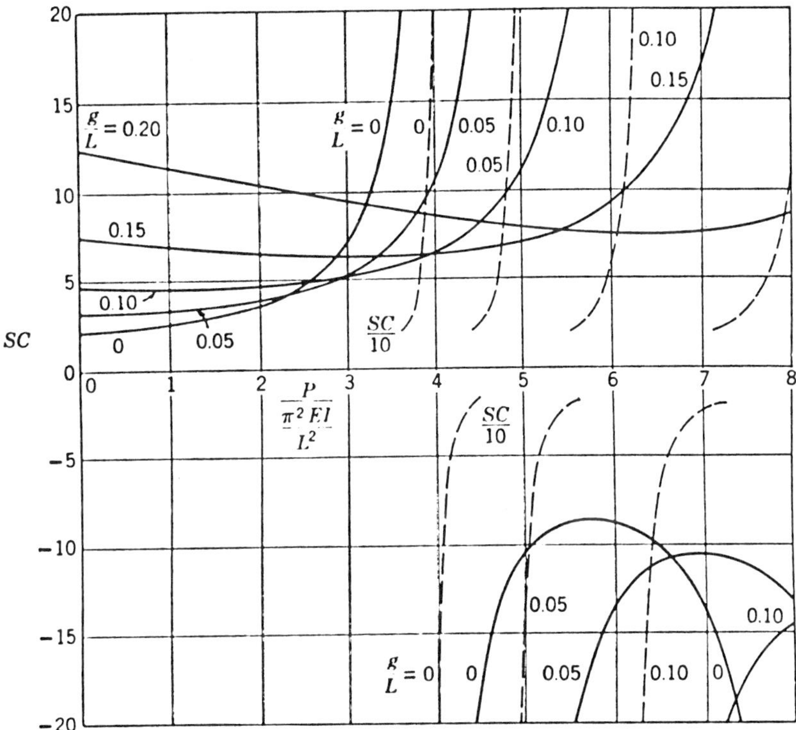

FIGURE 9-9 Carried-over moments for compressed bars with rigid gusset plates.

4. Compute the load matrix and multiply it by the inverse of the stiffness matrix to obtain all displacements.
5. From the stiffnesses, obtain axial forces and moments in the members.

Chu (1959) has analyzed the simple truss with rigid joints shown in Figure 9-10. For the given geometry, the results are sufficiently accurate.The ratio of the secondary bending stress to the primary axial stress is inversely proportional to the ratio L/γ, where L is the member length and γ is the extreme fiber distance in bending. This ratio is plotted in Figure 9-11 for $L/\gamma = 20$, corresponding to a symmetrical member with a length–depth ratio of 10.

At low loads the maximum secondary stress is 0.34 times the primary axial stress, and this may occur in practical bridges with truss members of this length–depth ratio. The graphs show that the secondary stresses vary nonlinearly with applied load, and at $P = 1.93\pi^2 EE/L^2$ the secondary stresses become very large. This load is the truss *buckling load*, and represents the theoretical elastic load for overall buckling of the complete truss.

Chu (1959) summarizes three approximations.

1. For trusses of usual proportions, the effect of secondary moments on the member axial forces is small. Secondary moments may be analyzed

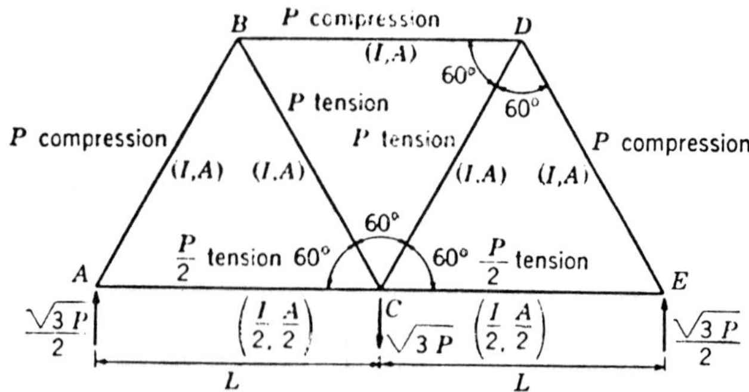

FIGURE 9-10 Simple truss with rigid joints. (From Chu, 1959.)

assuming that the member axial forces are the same as in the equivalent pin-jointed truss. With the axial deformations of all members known, linear displacements of all nodes can be determined. Thus, the solution is reduced to computing member end moments and joint rotations, and is sufficiently accurate as long as the axial member stiffness is large compared to the sidesway bending stiffness. The solution may not be accurate if the truss contains long members of small cross-sectional area mixed with short members with high secondary moments of area.

2. The reduced solution may be formulated as a displacement analysis in the joint rotations. Examples have appeared in most published analyses for truss buckling loads, and a typical reference is Hoff's convergence criterion (Hoff, 1956) based on the moment distribution method by Cross but using stiffness and carry-over factors modified for axial loads.

3. The calculation of secondary moments and stresses may be further simplified by neglecting the influence of axial loads on member bending stiffness (Norris and Wilbur, 1960). The resulting accuracy of the ratio of secondary to primary stresses is sufficient at zero load (left-hand side of Figure 9-11), but the method does not give a correct indication of the variation of secondary stresses with load as the latter increases.

Truss Buckling The truss buckling load determined as shown in the preceding section makes no provision for inelastic behavior, and also ignores truss imperfections manifested as initial curvature, unintended end eccentricities, and residual stresses.

As an example, consider a pin-ended steel member with a slenderness ratio of 60. The theoretical Euler stress is 82 ksi. If the material yields at 36 ksi, an elastic solution is valid up to about 0.45 times the Euler load. For the truss shown in Figure 9-10, the critical load for truss buckling corresponds to 1.93 times the Euler load.

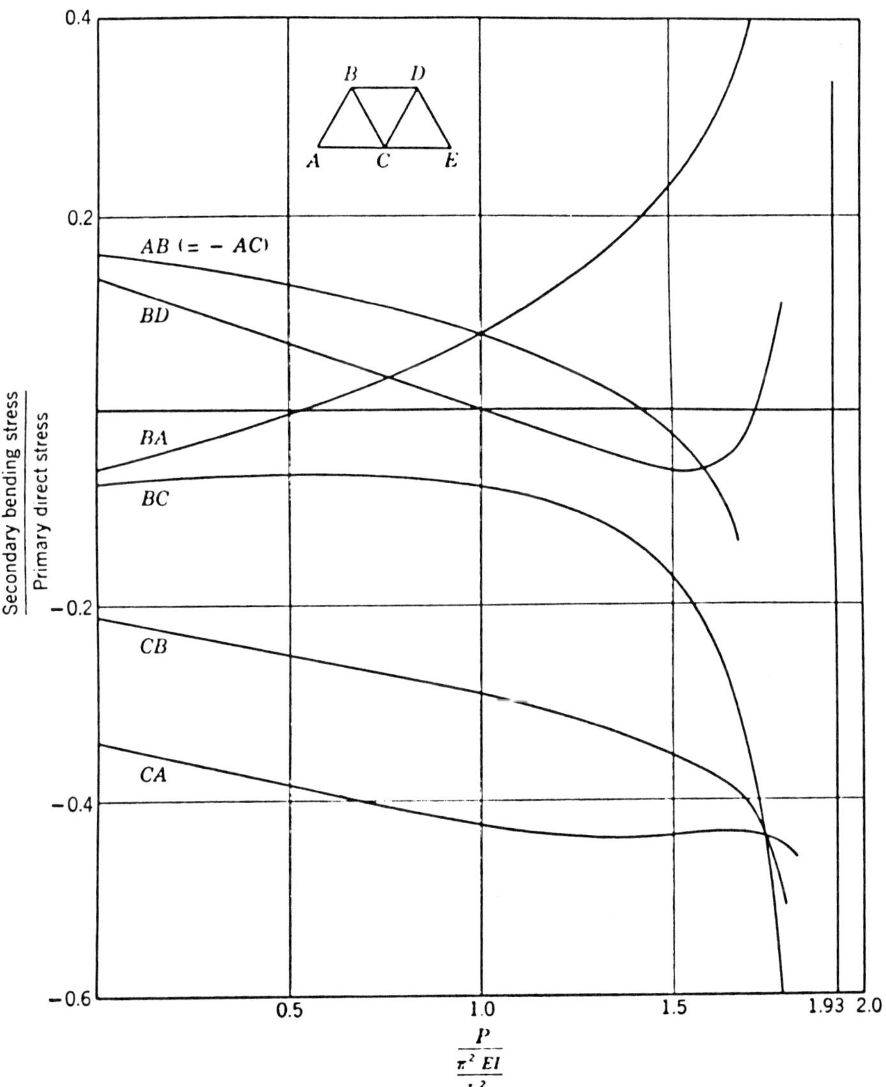

FIGURE 9-11 Variation and dependence of secondary bending stresses on applied load. (From Chu, 1959.)

O'Connor (1965) proposes two possible procedures for inelastic analysis. The first assumes that the effective lengths of all members are the same as those corresponding to the lowest elastic truss buckling load, and the design is carried out treating members as isolated columns. The second assumes an equivalent unstressed, imperfect truss shape, intended to allow for all imperfections, and an elastic analysis is used to compute the load for first yield assuming that this is the failure load.

Relevance of Secondary Stresses At working loads, secondary stresses may contribute to fatigue failure and enhance the possibility of brittle fracture. In a major truss, control of secondary stress levels may be provided by cambering. All truss joints may be laid out with true geometry angles, but the member lengths are fabricated as adjusted for the elongation at full load. Thus, at no load the truss position will lie above the nominal geometry assuming downward loads. Any bending moments and stresses induced in the members during assembly at no load will be reduced to almost zero as the full load is applied. Another alternative is to adjust the lengths of the chords only, but this will not reduce secondary stress levels completely.

Stresses in Gusset Plates Gusset plates must have ample thickness to resist shear, direct stress, and flexure acting on the weakest or critical section of maximum stress. For normal truss joints, gusset plates placed on either side of the primary truss members are mandatory.

In welded construction, transverse fillet welds can be avoided by butt-welding the gusset to the side plates of the truss members. The gusset is loaded through its ends, and the profile near the ends must be designed so that excessive stress concentrations are avoided. With bolted or riveted connections, the gusset loads are distributed across the interior of the gusset so that the edge profile becomes less critical. In this case a gusset plate bounded by a series of straight lines is suitable.

Planar gusset plates of any shape can be analyzed by finite-element methods, but solutions can be obtained by photoelastic techniques. Summaries of available solutions for elastic stress concentrations are given by Heywood (1952), Lipson et al. (1950), Neuber and Hahn (1966), and Peterson (1953).

Deck–Truss Interaction The total change in length at the chord level of a truss may be considerable. For example, a stress of 10 ksi over a half-length of 100 ft results in a movement of 0.4 in. For a deck that must transmit bending moments and shears only, the longitudinal strain at the centroidal axis is zero. If the same deck is located at the chord level, the chord–deck system is incompatible. Two solutions are feasible.

First, the deck may be isolated from the bottom bridge system. For example, it may be supported on longitudinal stringers, which in turn are supported by transverse girders between the truss verticals. The independent deck response is ensured by sliding joints at one stringer end and an open joint at the other.

Alternatively, the deck may be attached to the chords by proper details. The resulting system expands under load as one unit, and incompatible strains are avoided. The longitudinal capacity of the deck cannot be ignored in modern bridge design. The resulting interaction articulates an orthotropic steel deck made integral with the chord system. If the deck is placed at the upper chord level, the resulting arrangement eliminates the need for an upper lateral system.

9-4 DESIGN APPROACH OF TRUSSES: ALLOWABLE STRESS METHOD

General Considerations

Initial important decisions in the design of a truss bridge must address (a) truss type, (b) truss depth, and (c) length of deck panel. Waling (1953) has reported results for simply supported, through-type, double-track railroad bridges, which may also be extended to highway bridges. The two basic truss types considered are Warren and Pratt trusses, with spans of 168 to 300 ft and with 6 to 10 panels. Based on least-weight analysis, the depth d and length L of parallel chord trusses are related as follows:

$$\text{Pratt truss} = d = L(0.381 - 0.018n) \tag{9-7}$$

$$\text{Warren truss} = d = L(0.362 - 0.016n) \tag{9-8}$$

where n is the number of panels.

The optimum d/L ratio appears to be almost independent of the span over the range considered. Interestingly, for six panels the Pratt and Warren designs give nearly equal weights, but for 10 panels the Pratt truss is approximately 10 percent heavier than its Warren counterpart. This analysis is meaningful for initial estimates and is based on an allowance for details (37.5 percent of the base truss members), plus 12.5 percent for bracing.

The depths suggested by (9-7) and (9-8) exceed those normally used; and for $n = 10$ both formulas give $L/d = 5$, which is greater than the $L/d = 10$ (maximum) permitted by AASHTO. It may appear that the cost of a truss is relatively insensitive to the depth for a wide range, and is influenced more significantly by decisions related to detailing, shop joints, slenderness ratios, member cross section, and so on. A highway truss bridge may have an economical L/d ratio ranging from 7 to 10 and even higher provided deflections are within allowable limits. Likewise, the selection of the truss type should have only a small effect on the total cost, and should be determined based on detailing considerations and appearance.

The length of the deck panels should be selected for optimum deck design. Optimum panel lengths are 20 to 40 ft and even higher for large structures. In smaller bridges, an upper bound on the panel length may be dictated by the slope of the diagonals, usually forming an angle with the horizontal less than 45°. As suggested, the axial force capacity of the deck is considerable, and, when possible, its participation in chord forces is advantageous.

Other points of interest in the general design are the following:

1. In general, deck trusses have a smaller depth than through trusses.
2. Larger panels reduce the number of joints, but the weight of the floor increases. These effects tend to offset each other within limits.
3. To keep reasonable floor weights and diagonal angles, subdivided or K trusses should be considered for highway spans greater than 320 ft. (Warren trusses without verticals have been built economically, however, in the span range of 700 to 900 ft.)
4. For parallel chord trusses, the Warren type may be cheaper than the Pratt or Pettit type.
5. Welded members are often more economical than bolted members.

Design Requirements

Dead Loads Floor loads vary considerably with truss type, combination of materials, loading, and design stresses. The actual weights of trusses and bracing may be taken from a preliminary design, by reference to similar structures, or from formulas based on empirical data. An example of an empirical formula for bolted truss weights is given in Table 9-1. The truss weight w (lb/linear foot) is expressed as a function of the total load p (kip/linear foot) and span L. If used directly, these formulas may be applied to preliminary design, and, if properly adjusted, they may be used for final design. They are based on normal-grade steels.

Alternatively, the weight of a truss may be estimated by determining the weight of certain parts or components. If the weight percentage represented by the part under consideration is known, the total weight is readily approximated. For example, it has been estimated that the average proportions between weights of different components of a truss bridge are as follows:

Bottom chords	20%
Top chords	25%
Web members	25%
Bracing	10%
Connections	20%
Total	100%

TABLE 9-1 TRUSS WEIGHT FOR VARIABLE SPAN RANGES

Type of Truss	Span (ft)	Load Range p (kip/ft)	Weight of Truss (kip/ft)
Through Pratt with parallel chords	100–200	3–18	$0.180 + \dfrac{(L - 50)p}{1480}$
Through Pratt with polygonal upper chords	200–300	3–18	$0.180 + \dfrac{(L - 70)(p + 0.3)}{1370}$
Deck Pratt with parallel chords	100–200	3–13	$0.180 + \dfrac{(L - 30)p}{1590}$
	200–300	3–13	$0.180 + \dfrac{(L - 80)p}{1130}$

This procedure is often convenient and reasonably accurate, but can yield erroneous results in certain instances. For example, a truss with a high depth–span ratio will have proportionally more weight in the web, whereas a relatively shallow truss will have more weight in the chords; hence, the foregoing percentages must be adjusted.

Other Loads These include live load, impact, wind, traction, seismic forces, and other loads stipulated and controlled by the design specifications.

Truss Members Truss analysis is generally extended to include the following.

Primary Members These consist of the upper and lower chords, diagonals, and verticals that resist dead and live loads.

Secondary Members These include all bracing and members required for truss stability. A truss is stable if it can resist all lateral, longitudinal, and overturning forces. Both primary and secondary members are designed by conventional statics or by the use of influence lines.

Counters These are used primarily to prevent stress reversal in a member because of variable or fluctuating loads.

Redundant Members These represent a statically indeterminate system and must be analyzed by deflection methods or by methods of least work and virtual work or other energy methods.

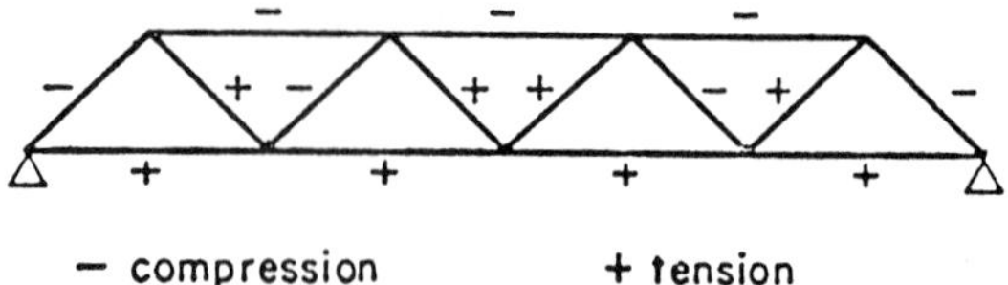

FIGURE 9-12 Typical sign convention for axial forces induced in truss members by loads acting on the structure.

Stresses in Members Primary stresses are computed from the analysis of loads and forces acting on the structure, assuming all joints to be pin-connected. The resulting forces in the truss members act axially as shown in Figure 9-12 under the sign convention indicated.

Secondary stresses were discussed in some detail in the preceding sections together with their significance and methods of analysis. If connections are rigid (bolted or welded), the angles between the members cannot change and the members must therefore bend to accommodate the angular distortion. Secondary stresses are also induced by eccentric joints, beam action of the members, and inadequate or frozen bearings.

Truss Deflection Truss deflection occurring as loads are placed causes unsightly appearance and is usually rectified by introducing an appropriate camber placed in the following ways: (a) by lengthening the top chord as shown in Figure 9-13(*a*) (this is done sometimes arbitrarily by specifying 3/16 in. for each 10 ft of length); (b) by varying the lengths of the diagonals as shown in figure 9-13*b*; (c) by shortening the verticals as shown in figure 9-13*c*; and (d) by a combination of the preceding methods.

In longer spans the main loads (dead load and live load) should be carefully evaluated in terms of the resulting deformations, and the effects on each individual member should be obtained. From these data the truss deflection may be determined by either the principles of work or by constructing a displacement diagram.

Truss deflections can also result from length changes of members because of axial strains, temperature, fabrication and construction tolerance, effects of pin connections, and slippage of bolted connections. These effects are

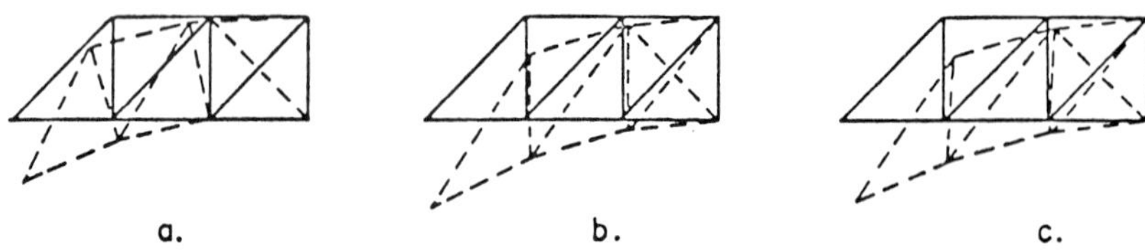

FIGURE 9-13 Camber introduced to compensate for deflections.

superimposed to obtain the total deflection either using algebraic solutions or by graphical methods.

Lateral Trusses These are treated as the main vertical trusses. Forces to be considered include wind, seismic loads, and centrifugal forces. The upper chords of the main truss act as the chords of the upper lateral truss. Lateral bracing is usually provided, although the deck slab may be designed to act as a stiffening member. Bracing of the bridge during construction and before the deck slab is in place should be provided. The lower lateral truss is similar to the upper, and its main chords are the lower chords of the main truss. In deck-type structures the total forces are much less than in the upper lateral truss.

Wind loads are treated essentially as live loads and may be assumed to act only on portions of the structure that produce maximum stresses in the member considered.

In general, if the minimum size member is selected to satisfy the slenderness ratio, the calculated stresses are most likely to be less than the allowable stresses. Because of the long unbraced lengths of these members, it is often advantageous to consider the cross bracing acting in tension only and to neglect the resistance to compression.

The upper lateral truss is also effective in providing support to the compression chord, although the supporting force is not always considered in the design of the bracing.

Overturning Effects Horizontal forces, such as wind applied normal to the structure, do not act at the same horizontal plane as the main truss supports, and thus they produce a torsion or overturning effect. These forces will be transferred from their point of application to the supports through those members that are most stiff, irrespective of the intent of the design. Thus, for a given structure, the analysis should reflect the actual behavior, and the design method should be selected accordingly.

Overturning effects may be dealt with by making the deck continuous and horizontally stiff so that lateral forces are carried to the ends where they are transferred to the main truss supports by means of the end sway bracing.

Alternatively, the deck can be made discontinuous with expansion joints at each panel. In this case the horizontal forces applied to the upper portion of the bridge are transferred to the lower chords by means of bracing at each panel. The analysis must consider relative mass and member stiffness as well as the nature of the forces, which may not be transferred to the extent intended by the design.

Sway Bracing This bracing is required, usually at each panel, to resist the horizontal forces and to prevent the structure from collapsing as shown in Figure 9-14*a*. Because of the unsupported lengths of the bracing, it may be

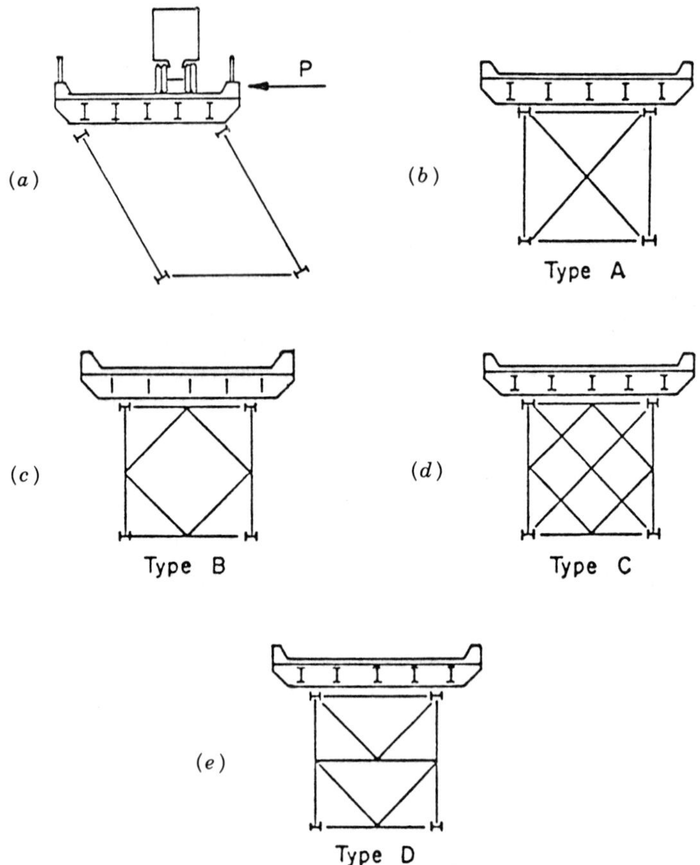

FIGURE 9-14 Sway bracing systems for deck trusses.

more economical to neglect any compression action and design the bracing as acting in tension only.

In some trusses it is possible to transfer the force P shown in Figure 9-14*a* to the lower lateral bracing by means of the sway bracing at each panel, and then through the lower lateral truss to the supports. In other designs a stiff upper lateral truss or a continuous rigid deck carries all the lateral forces at the top portion of the structure to the ends of the bridge where these forces are transferred to the supports by means of a portal or other suitable bracing scheme. The sway bracing is also effective in distributing loads to the main trusses, but this function is not usually considered in its design.

The type A bracing shown in Figure 9-14*b* is the most common. Its disadvantages relate to the complexity of the connections at panel points and the long unsupported lengths in one direction. Type B bracing shown in Figure 9-14*c* has relatively short unsupported members, and in order to be effective it must be induced into some of the horizontal and vertical members by rigid connections normally separated from the truss panel joints.

Type C bracing shown in Figure 9-14*d* should be avoided where possible. The members are supported at short intervals in one direction but are relatively long in the other. The larger number of bracing components means more detailing and a longer assembly process as well as more painting and maintenance. The type D bracing shown in Figure 9-14*e* is often used in trusses that are relatively deep and close together.

Through truss spans should have sway bracing at least 5 ft deep at each intermediate panel point, or at each panel if the design requires. Top lateral struts should be at least as deep as the top chord.

Selection of Members

Truss members for highway bridges generally consist of (a) H sections, either rolled or built up; (b) channel sections; (c) single box sections, usually made with side channels, beams, angles, or plates; and (d) double box sections, similarly made with side channels, beams, angles, or plates. An advantage of wide flange shapes is their immediate availability and reference to direct design data and section properties. One disadvantage is the difference in back-to-back dimensions for the same nominal size members with different weights, making it necessary to provide numerous thicknesses of fills at splices and gusset plates where different members are joined.

Steel plates welded together to make truss members can be arranged to form H shapes, box sections, multiple box sections, and any other arrangements necessary to give the required rigidity and cross-sectional area. Hybrid steel construction may be indicated where appreciable load-carrying capacities result from design and make the use of steels with different strengths a viable alternative. Thicker plates or higher-strength steels may also be used at the ends of members where holes are required for bolted connections. Thus, a member can be designed to its maximum capacity throughout its entire length rather than designing the entire member based on a net section required only at the end.

Compression Members AASHTO limits the slenderness ratio KL/r to 120 for main members and 140 for secondary members. Interestingly, it is often necessary to select a section with cross-sectional area considerably more than is required for stress considerations in order to satisfy the slenderness requirements. In such cases it is possible to disregard any portion of the member that will increase the radius of gyration sufficiently to satisfy the slenderness ratio provided the same area is also disregarded in computing stresses.

As an example we consider a member with a design length of 26 ft (bolted ends) that must carry a compressive force of 120 kips. For $KL/r = 120$ and $K = 1$, the required $r = 26 \times 12/120 = 2.6$ in. Consider a W12 $\times$ 53, min r

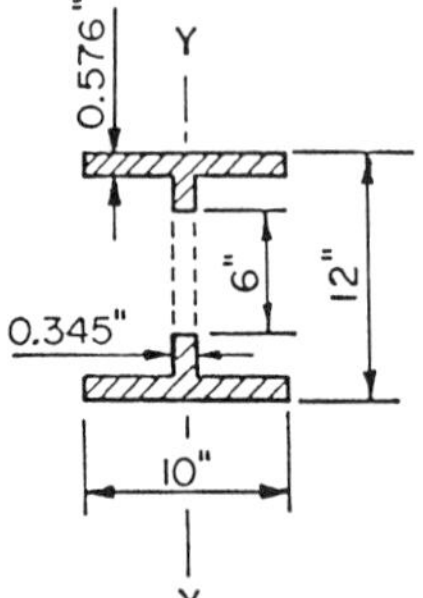

	A	Iyy
12 WF 53	15.59	96.1
Neglected Section	6 x 0.345 = 2.07	$\frac{1}{12} \times 6 \times 0.345^3 = 0$
Net Section	13.52	96.1

FIGURE 9-15 Cross section and properties of design example.

= 2.48. This gives a slenderness ratio of $26 \times 12/2.48 = 126$, or greater than the allowable maximum.

However, we consider the same section but with the center 6-in. portion of the web removed as shown in Figure 9-15, where the new configuration is dimensionally detailed and the new section properties are included. Using the data from the adjusted section configuration, we compute $r = \sqrt{I/A} = \sqrt{96.1/13.52} = 2.66$ in., or $L/r = 28 \times 12/2.66 = 117 < 120$, OK. The allowable $f_s = 16{,}980 - 0.53(KL/r)^2 = 16{,}980 - 0.53 \times 117^2 = 9725$ psi, and the maximum allowable compression load $= 13.52 \times 9725 = 131.4$ kips, OK.

Tension Members For tension members, AASHTO stipulates that, with the exception of rods, eyebars, cables, and plates, the ratio of the unbraced length to the radius of gyration should not exceed 200 for main members, 240 for bracing members, and 140 for main members subjected to stress reversal.

Smaller trusses may be designed using wide flange sections of a single series for all main members. These can be arranged to fit inside gusset plates, and the splices can be made near the panel points by adding splice plates outside the gusset plates and on both sides of the web.

General Guidelines The following guidelines may be useful in selecting main truss members.

1. Proportion the truss to keep secondary stresses to a minimum.
2. Arrange the members to fit together at panel points with a minimum of fills.
3. Study the use of members whose thickness is small enough so that holes can be punched rather than drilled.
4. Arrange parts so that erection is facilitated and maintenance can be provided with the least difficulty.

5. Use shapes that can be readily spliced.
6. Proportion main members so that their gravity axes are as nearly as practicable in the center of the section.

9-5 DESIGN APPROACH OF TRUSSES: LOAD FACTOR METHOD

AASHTO Guide Specifications (1985 and Interim)

The guide specifications for strength design (load factor) of truss bridges apply essentially to truss spans exceeding 500 ft.

Truss Members Article 10.16.2 still applies, except that the 10,000-lb allowance for adjustable counters is treated as dead load and factored accordingly.

Secondary Stresses The provisions of Article 10.16.3 still apply, but the 3000 and 4000-psi allowances are converted into factored allowances and computed as follows:

$$3000 \sum (P_i)(\mathrm{LF}_i) \qquad \text{or} \qquad 4000 \sum (P_i)(\mathrm{LF}_i)$$

where P_i is the percentage of member service load attributed to dead load, live load, or wind load, and LF_i is the appropriate load factor for each type of load.

Member Capacity For tension members, the capacity is evaluated from an interaction equation based on two simplifying assumptions. The first is that the shape of the interaction equation is a straight line connecting points ($P = P_y$, $M = 0$) and ($P = 0$, $M = M_p$). This assumption is conservative for wide flange shapes bent about their major axis and for rectangular shapes. All shapes stipulated in the provisions can be considered in this range. The second assumption is that the plastic shape factor for the net section is the same as for gross effective sections.

The interaction equations for compression members are basically the same as AASHTO equations (10-155) and (10-156) for combined axial load and bending. Box-shaped truss members have an inherent high lateral–torsional stiffness so that the reduction in bending strength because of the lack of lateral support is minimal; the reduction in moment capacity calculated by the applicable equations is small and compatible with these characteristics.

For H-shaped members subjected to bending about their major axis, the recommended bending strength is based on equations used to develop the allowable bending stresses in AASHTO Table 10.32.1A. The more rigorous solution required for truss members is dictated by the longer unbraced lengths.

Thickness of Metal in Compression Members The basic expression for the critical elastic buckling stress is

$$\sigma_{cr} = \frac{K\pi^2 E}{12(1 - \mu^2)(b/t)^2} \tag{9-9}$$

Setting $E = 29{,}000{,}000$ psi and $\mu = 0.3$ and solving for b/t gives

$$\frac{b}{t} = \frac{5120\sqrt{K}}{\sqrt{\sigma_{cr}}} \tag{9-10}$$

The foregoing expression is modified to reflect the observed behavior of plates, indicating that residual stresses and out-of-flatness reduce the strength of plates with intermediate slenderness below the value indicated by simple elastic stability analysis. Using a reduction factor of 0.6, (9-10) reduces to

$$\frac{b}{t} = \frac{3072\sqrt{K}}{\sqrt{\sigma_{cr}}} \tag{9-11}$$

For a simply supported plate (on two edges), the minimum K is 4. Introducing a factor of safety such that the working stress (actual computed compression stress) is $f_a = 0.55\sigma_{cr}/1.25$, (9-11) gives

$$\frac{b}{t} = \frac{4000}{\sqrt{f_a}} \tag{9-12}$$

which is the expression used by AASHTO in Article 10.35.2.5.

For load factor design, the maximum compressive stress is $f_a = 0.85\sigma_{cr}$, which leads to

$$\frac{b}{t} = \frac{5700}{\sqrt{f_a}} \tag{9-13}$$

Eyebar Pins The maximum bearing stress on eyebar pins is $1.35F_y$. This value is based on the OHBDC (1979) utilizing a value of 1.5ϕ, where ϕ is 0.9 for steel. The problem of pin capacity in combined bending and shear is approached with an interaction curve expressed by

$$\frac{6M}{D^3F_y} + \left(\frac{2.2V}{D^2F_y}\right)^3 \le 0.95 \tag{9-14}$$

where D is the pin diameter, M is the factored moment, and V is the factored shear.

Three approximations were considered to derive a suitable interaction curve. Among these are the interaction curve proposed by Drucker (1956) and a similar curve proposed by Hodge (1957), based on the von Mises criterion. However, neither of these studies contains experimental verification of the suggested procedure for eyebar design. Hence, a more conservative interaction curve was introduced as the basic interaction equation, where the limit is reduced to 0.95 (as opposed to the limiting value of 1.0 proposed by Drucker and Hodge). The design equation (9-14) was obtained by substituting the appropriate expressions for plastic shear and moment capacities in terms of the pin diameter.

Gusset Plates Gusset plates are designed for shear, bending, and axial load using the method-of-section approach as in service load design. The maximum stress from combined factored bending and axial loads is the yield stress based on the actual area without reference to plastification concepts.

The maximum shear stress on a section is $F_y/\sqrt{3}$ for uniform shear and $0.74F_y/\sqrt{3}$ for bending shear computed as shear force divided by area. As in allowable stress design, the edge of a gusset plate must be stiffened if its unsupported length exceeds the value given by $11{,}000\sqrt{F_y}$.

LRFD Specifications

Truss Members If the shape of the truss permits, compression chords should be made continuous. If web members are subject to stress reversal, their end connections should not be pinned. If counters are used, they should preferably be rigid.

Secondary Stresses As in the AASHTO standard specifications, secondary stresses include the effects of the flexural dead load moment of the member, truss distortion, and floor beam deflection. The load factors, however, are lumped into a single parameter. Thus, if the secondary stress from factored loads exceeds 6.0 ksi for tension and 4.5 ksi for compression, the excess must be treated as primary stress.

Other Provisions These cover diaphragms, camber, portal and sway bracing, gusset plates, and half-through trusses. A mandatory investigation for sway bracing must include equilibrium, compatibility, and stability analyses for all applicable limit states.

For through truss spans, the portal bracing should be designed to take the full reaction of the top chord lateral system, and the end posts should be designed to transfer this reaction to the truss bearings.

For deck truss spans, the end sway bracing should be proportioned to carry the entire upper lateral load to the supports through the end posts of the truss. Because full-depth sway bracing is easily accommodated in deck trusses, its use is encouraged.

Gusset plates should be used, except for pinned connections. They should be designed for shear, bending, and axial load determined by method-of-section procedures. The maximum stress from these combined factored load effects should not exceed the factored resistance ϕF based on the gross area. However, plastic shape factors, or other parameters implying plastification of the cross section, should not be used.

In half-through trusses, the vertical members and floor beams should be proportioned to resist a lateral force not less then 300 lb/l in ft applied at the top chord panel points, considered as a permanent load for the Strength I limit state. A buckling analysis is recommended for the top chord, considered as a column with elastic lateral supports at the panel points.

Refined Methods of Analysis A refined plane frame or space frame analysis must include: (a) composite action with the deck; (b) continuity among the truss components, where it exists; (c) force effects due to the self weight of components, change in geometry due to deformation, and axial offset at the panel points; and (d) in-plane and out-of-plane buckling. For the upper chords, out-of-plane buckling must be investigated in pony trusses. If the truss derives its stability from transverse frames of which the floor beams are a part, the deformation of the floor beams due to vehicular loads must be considered.

9-6 DESIGN EXAMPLES OF TRUSS BRIDGES

Cantilever Truss Bridge

The structure shown in Figure 9-16 is the Second Carquinez Bridge in California, and incorporates high-strength steels with a combination of welding and bolting.

Design of a Compression Member A lower chord member adjacent to the main tower will be designed. The forces acting on this member have been computed as follows:

Dead load	=	−1676 kips
Live load	=	− 645
Impact	=	− 37
Total dead + live + impact	=	−2358 kips
30-psf wind load on bridge	=	± 996 kips
Windload on live load	=	± 255 kips
Total	=	−3609 kips

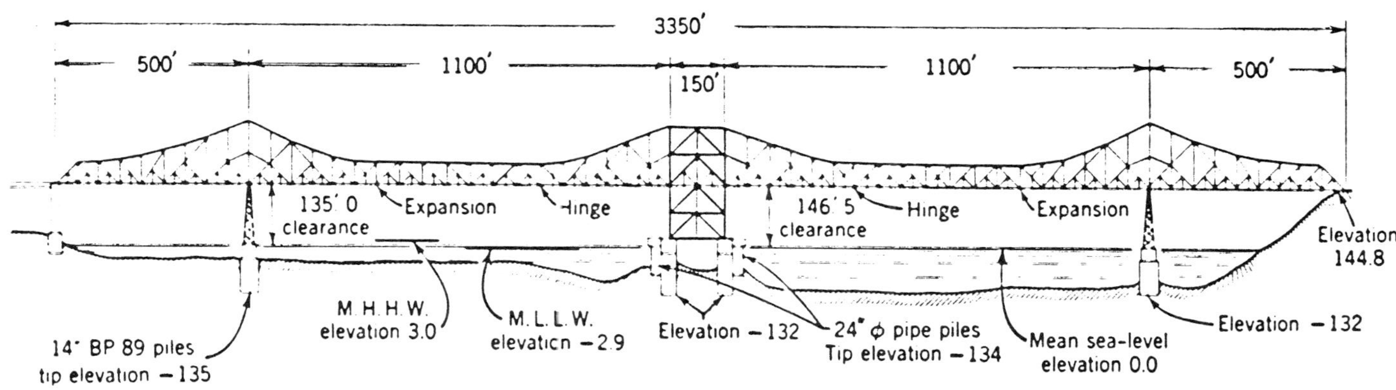

FIGURE 9-16 Elevation view and dimensions of Second Carquinez Bridge.

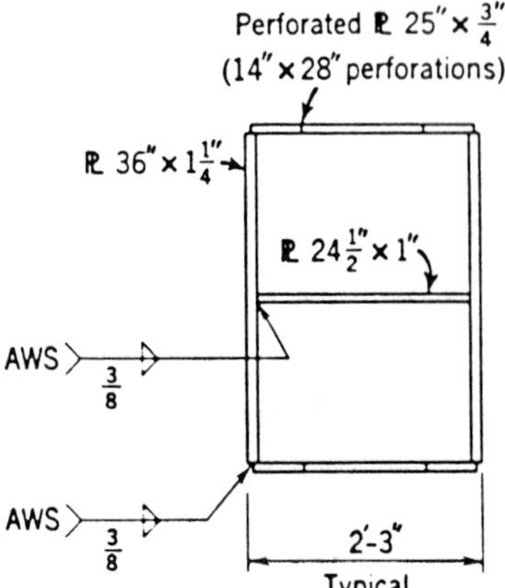

FIGURE 9-17 Section of a compression member.

The increase of 25 percent in the allowable stress when considering wind load is considered by multiplying the total force by the factor $1/1.25 = 0.80$, or design load $= 3609 \times 0.8 = 2887$ kips. Secondary stresses from truss deflection and flexural effects have been computed and found insignificant. The length of the member is 27.78 ft.

The selected section is a double box as shown in Figure 9-17. The steel is grade 50.

The section properties are computed as follows:

Two plates $36 \times 1\frac{1}{4}$ in.	$= 36 \times 2.5 = 90.0$ in.^{2}
One plate $24\frac{1}{2} \times 1$ in.	$= 24.5 \times 1 = 24.5$
Two perforated plates $25 \times \frac{3}{4}$ in. with 14 × 28-in. perforations	$= 11 \times 1.5 = 16.5$
Total net area A	$= 131.0$ in.^{2}

Next, the moments of inertia are calculated as

$$I_x = 8.25 \times 678 + 2.50 \times 3888 = 15{,}290 \text{ in.}^4$$

$$I_y = 45.0 \times 331.5 + 1.0 \times 1227 + 1.5(1302 - 228.7) = 17{,}760 \text{ in.}^4$$

Minimum r is controlled by I_x, or $r = \sqrt{I_x/A} = 10.8$ in., giving a slenderness ratio of $L/r = 27.78 \times 12/10.8 = 30.8$.

For Grade 50 steel, the allowable stress is $f_a = 23{,}580 - 1.03(KL/r)^2 = 22{,}600$ psi, the not to exceed $0.44F_y = 22{,}000$ psi. Hence, the maximum allowable force in the member is $22 \times 131 = 2882$ kips ≈ 2887, OK.

Check for plate thickness using AASHTO Article 10.35.2. For Grade 50 steel and a compressive stress of $0.44F_y$, the 36-in. plates and the 24.5-in. center plate are governed by Article 10.35.2.8 with $b/t \leq 34$, and evidently

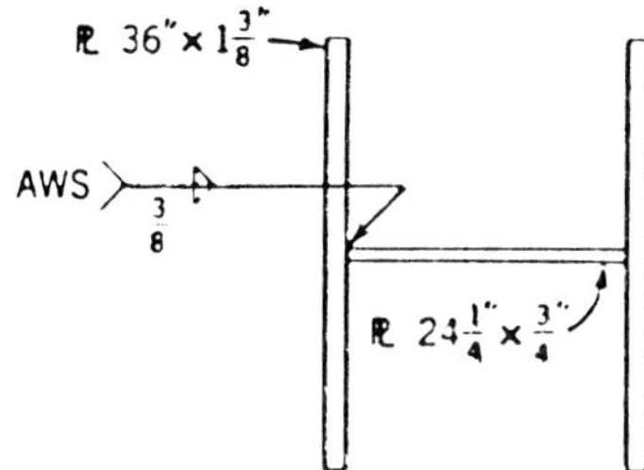

FIGURE 9-18 Section of a tension member.

both plates satisfy this criterion. The perforated plates are governed by Article 10.35.2.10, with $b/t = 41$, or $25/0.75 = 33 < 41$, OK.

Design of a Tension Member An upper chord of the anchor span adjacent to the main tower will be designed. The computed forces acting on this member are

$$\begin{aligned} \text{Dead load} &= 3352 \text{ kips} \\ \text{Live load} &= 665 \\ \text{Impact} &= \underline{39} \\ \text{Total dead + live + impact} &= 4056 \text{ kips} \end{aligned}$$

A 30-psf wind load produces a stress of only 290 kips in this member and therefore will not affect the design. The selected section is shown in Figure 9-18 and consists of Grade 50W steel with the following section properties:

$$\begin{aligned} \text{Two plates } 36 \times 1\tfrac{3}{8} \text{ in.} &= 99.0 \text{ in.}^2 \\ \text{One plate } 24\tfrac{1}{4} \times \tfrac{3}{4} \text{ in.} &= \underline{18.2} \\ \text{Total gross} = \text{net} &= 117.2 \text{ in.}^2 \end{aligned}$$

The minimum moment of inertia is about the horizontal axis and is approximately $I_x = 2.75 \times 36^3/12 = 10{,}700$ in.^{4}, giving minimum $r = \sqrt{10{,}700/117.2} = 9.56$ in. The slenderness ratio is therefore ($L = 120.75$ ft) $L/r = (120.75 \times 12)/(2 \times 9.56) = 75.7$, OK.

The stress in the member is $f_a = 4056/117.2 = 34.6$ ksi < 35, OK.

In order to compensate for the loss of section at the connections, the end of this member is built up by welding a heavier flange plate. The end section consists of the following:

$$\begin{aligned} \text{Two plates } 36 \times 1\tfrac{7}{8} \text{ in.} &= 135.0 \text{ in.}^2 \\ \text{One plate } 23\tfrac{1}{2} \times \tfrac{3}{4} &= \underline{17.6} \\ \text{Gross area} &= 152.6 \text{ in.}^2 \\ \text{Hole deduction } 8 \times 1\tfrac{1}{8} \times 1\tfrac{7}{8} \times 2 &= \underline{33.7 \text{ in.}} \\ \text{Net area} &= 118.9 \text{ in.}^2 \end{aligned}$$

The stresses are now checked for both the gross and the net area. For the gross area,

$$f_a = 4056/152.6 = 26.6 \text{ kis} \qquad (\text{allowable 27 ksi, OK})$$

For the net area,

$$f_a = 4056/118.9 = 34.1 \text{ kis} \qquad (\text{allowable 35 ksi, OK})$$

Design of Gusset Plate

As mentioned in Section 9-3, stresses in gusset plates may be computed by finite-element techniques, or other methods given in the references. In this example a gusset plate is designed according to a method proposed by Whitmore (1962). Although many designers check stresses in gusset plates using conventional beam theory (see also Section 9-2), this investigator found that the stresses in gusset plates do not always follow the distribution expressed by $(P/A) \pm (Mc/I)$. In fact, stresses near plate edges are minimum. For gusset plate design, it is therefore sufficiently accurate to determine maximum direct stresses and maximum shear stresses.

For the diagonal member acting on the gusset plate shown in Figure 9-19 assuming 1/2-in. gusset plates, the number of bolts required is $524/8.36 = 63$, use 34 on each side.

The effective length of the plate is assumed to be the base of a trapezoid with sides inclined at 30° with the axis of the member. At the critical section a–a, the maximum direct stress on the gusset plate is $524/(2 \times 33.2 \times 0.5) = 15.8$ ksi, OK.

For the vertical member, the number of bolts required is $392/8.36 = 47$, use 24 on each side. At the critical section b–b, the maximum stress in the gusset plate is $392/(2 \times 21.2 \times 0.5) = 18.5$ ksi, OK. The critical section for shear is b'–b', giving a maximum shear stress in the gusset plate

$$\frac{3}{2} \times \frac{348}{2 \times 0.5 \times 45} = 11.6 \text{ ksi} \qquad \text{OK}$$

Design of Splice

In smaller trusses the splices of top and bottom chords are ordinarily made at panel points where the gusset plates serve as splice plates. If the size of a member changes, fill plates are necessary between the main members and splice plates, and these may require extra bolts. AASHTO stipulates that splices in truss chords and columns should be located as near the panel points as practicable and usually on the side where the smaller stress occurs. Plates, angles, or other splice elements should be arranged with proper provisions for axial and bending stresses.

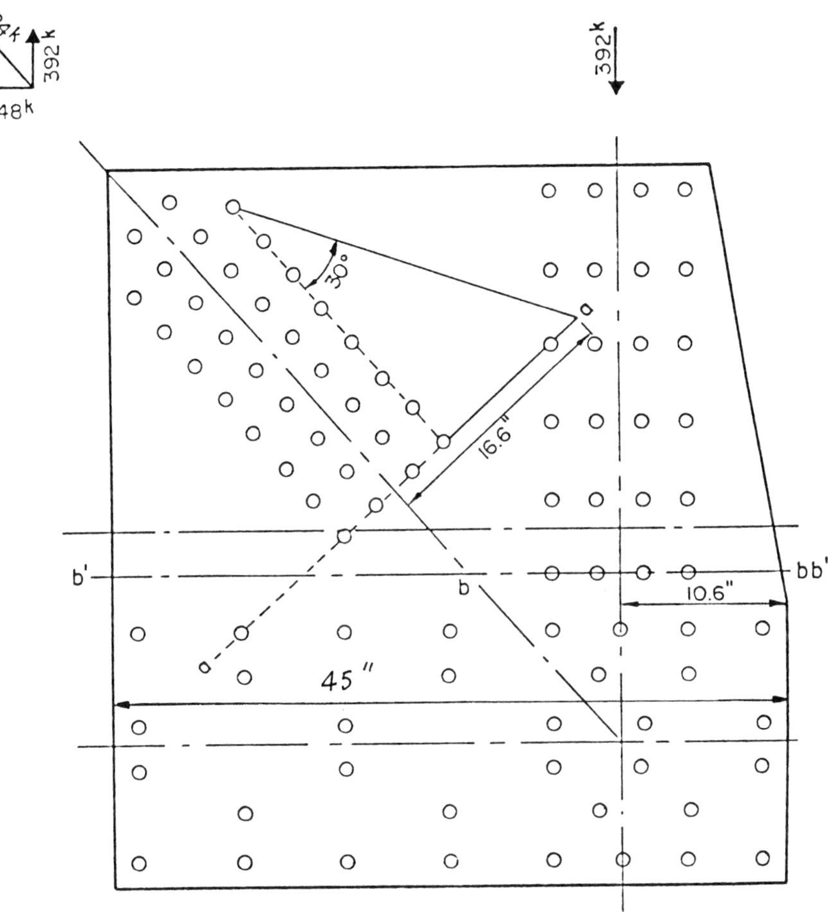

FIGURE 9-19 Gusset plate details of design example. Bolts are 3/4 in. diameter, ASTM A325.

Compression chords should have their ends in close contact at bolted splices. Such members should be detailed and fabricated to have the ends milled for full contact bearing at the splices, and held in place by means of splice plates and high-strength bolts proportioned for not less than 50 percent of the lower allowable design stress of the sections spliced.

For tension members and splice material, the gross section should be used unless the net section area is less than 85 percent of the corresponding gross area, in which case the amount removed in excess of 15 percent should be deducted from the gross area. The maximum allowable tensile stress on the net section should not exceed $0.50F_u$ for service load design or $1.0F_u$ for load factor design, where F_u is the minimum tensile strength of steel. For M270 grade 100/100W steels, the design tensile stress on the net section is $0.46F_u$ for service load design.

There are several approaches for distributing load effects in splices consisting of several members. The design of tension and compression splices is basically similar. The principal differences are that in compression splices the rivets or bolts are assumed to fill the hole completely so that the gross area of the member is considered in resisting stress, and in butt splices some portion of the stress may be assumed to be carried at the contact interface. In tension splices the stresses are transferred from one member to another by rivets or bolts.

In determining the flow of stress at a splice, one of the following three assumptions is made: (a) forces in splice plates are inversely proportional to their distances from the member being spliced; (b) forces in splice plates are proportioned inversely as the moment of their areas about the center of the member being spliced; and (c) forces in each member of a section through a splice are proportioned to their areas.

Example For the splice connection of the tension member shown in Figure 9-20, determine the number of bolts. The design load is 1160 kips, and the net area is 66.8 in.^{2}.

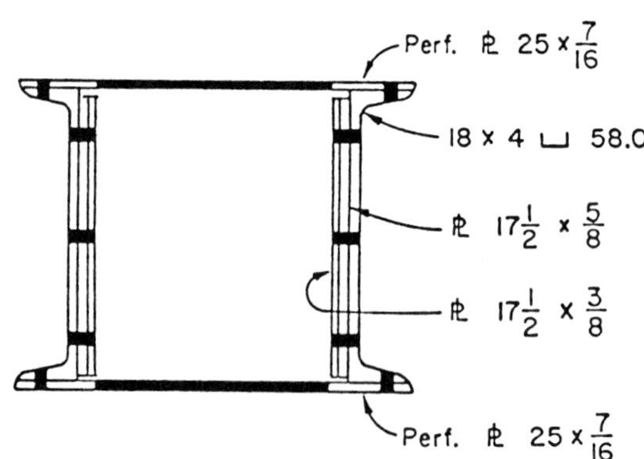

Section	*Net A*
18 × 4 ⊔ 58.0	13.67
⅊ 17½ × ¾	10.88
⅊ 17½ × ⅝	9.06
⅊ 17½ × ⅜	5.43
perf. ⅊ 25 × 7⁄16	5.25
⅊ 15 × ½	6.00

FIGURE 9-20 Section at tension splice of design example.

Using A36 steel with $F_u = 58$ ksi, the strength of the member is $29 \times 66.8 = 1937$ kips. The splice design load is therefore $(1160 + 1937)/2 = 1548$ kips, or $0.75 \times 1603 = 1203$ kips. For convenience, only one-half of the member is considered, or $1548/2 = 774$ kips.

Considering the stiffness of the splice, we assume that the force or portion of the 774 kips distributed to each member is proportioned to the area of the member. Because the top and bottom perforated plates are bolted to the flange of the channels, they are considered as acting together with the channels.

Compression and Tension Member: Load Factor Design

Compression Member The compression member shown in Figure 9-17 will be designed using the load factor approach.

$$\text{Total force } T = D + L + I = -2358 \text{ kips}$$

$$0.77T = 1816 \text{ kips} \qquad \text{or} \qquad D < 0.77T$$

Therefore,

$$\begin{aligned}\text{Group I} &= 1.3[D + (5/3)(L + I)] \\ &= 1.3 \times (1676 + 1.67 \times 682) = -2815 \text{ kips}\end{aligned}$$

Likewise,

$$\text{Group II} = 1.46(D + W) = 1.26 \times (1676 + 996) = -3901 \text{ kips}$$

The factored loads for Group II control the design. The maximum strength of the member is $P_u = 0.85 A_s F_{cr}$, where A_s is the gross effective area, equal to 131 in.^{2}, and F_{cr} is the critical buckling stress.

Using $K = 1$, $L_c/r = 30.8$, and $F_y = 50$ ksi, we compute both terms in AASHTO equation (10-152), and confirm that this is satisfied. Therefore, the critical buckling stress is determined from

$$\begin{aligned}F_{cr} &= F_y\left[1 - \frac{F_y}{4\pi^2 E}\left(\frac{KL_c}{r}\right)^2\right] \\ &= 50\left[1 - \frac{50}{4 \times 3.14^2 \times 29{,}000}(30.8)^2\right] = 47.9 \text{ ksi}\end{aligned}$$

Hence, the maximum strength of the member is

$$P_u = 0.85 \times 131 \times 47.9 = 5317 \text{ kips} \qquad \text{OK}$$

Tension Member Likewise, we compute $0.77T = 3123$ kips, or $D > 0.77T$; therefore, Group I $= 1.5(D + L + I) = 1.5 \times 4056 = 6084$ kips. For the section shown in Figure 9-18, the net area is 117.2 in.^{2}. For Grade 50 steel, the maximum strength is $117.2 \times 50 = 5860$ kips < 6084. Therefore, this section should be revised. Similarly, the section at the end of the member provides a net area of 118.9 in.^{2}, giving a maximum member strength of $118.9 \times 50 = 5945$ kips < 6084, or the section must be increased.

Gusset Plate Subjected to Shear, Direct Force and Bending Moment

A conventional truss for a highway bridge is shown in Figure 9-21*a*, where all members and joints are identified and designated. The truss joint at L_2 is typical of bottom chord connections.

The gusset plate is shop-bolted to member L_2L_3. Open holes through the gusset plate, vertical U_2L_2, and the bottom chord must be provided for the floor beam connections. Diaphragms are incorporated in the truss at the ends of the floor beams. They serve to carry one-half of the floor beam load to the outside plane of the truss and to maintain a right section through the bottom chord. The bottom flange of the floor beam should be flush with the bottom of the chord to permit the lateral connection plates to attach to both the chord and the floor beam.

For the gusset plate at joint L_2, the critical section is along the top line of bolts through the bottom chord. Figure 9-21*b* shows the forces acting on the section of this line. The live load is placed in the position shown in Figure 9-22 for maximum force in U_1L_2. With the live load in this position, the force in U_1L_2 is +312,700 lb, and in U_2L_2 the force is −123,000 lb. The vertical component of the force in U_1L_2 is +240,200 lb. This vertical component is equal to the force in U_2L_2 combined with the floor beam reaction of 117,200 lb under the condition of loading used. One-half of the beam reaction is transferred to the gusset plates below the section under consideration, and the other half of the beam reaction is transferred above that section. The horizontal component of 200,100 lb is a shear force at the section and represents the increase of bottom chord force induced by member U_1L_2.

This system of forces is shown in Figure 9-21*b*, and obviously the section under analysis is subjected to a combination of shear, direct force, and bending moment. The latter depends on the location of the neutral axis of the acting section. The gusset plate resisting the forces consists of two $47 \times 1/2$-in. plates with holes at about 6-in. centers. The net area of the tension side of the plates is equivalent to a plate without holes but with thickness $(6 - 7/8)/6 = 0.85$ in. Because holes are not deducted from the compression area, the center of gravity is found by reference to Figure 9-21*c* as follows:

$$\frac{0.85x^2}{2} = \frac{(47 - x)^2}{2}$$

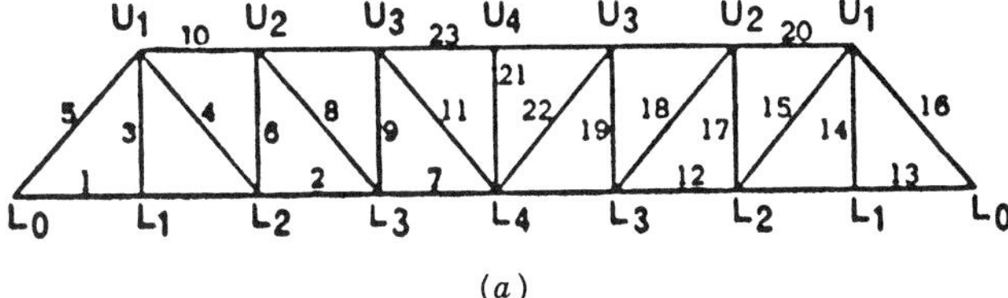

(*a*)

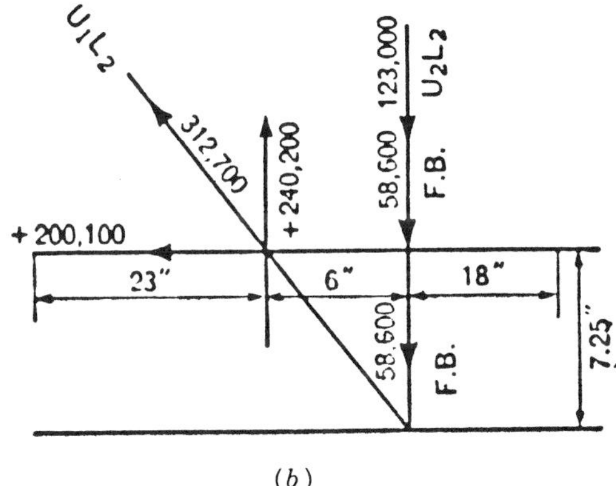

(*b*)

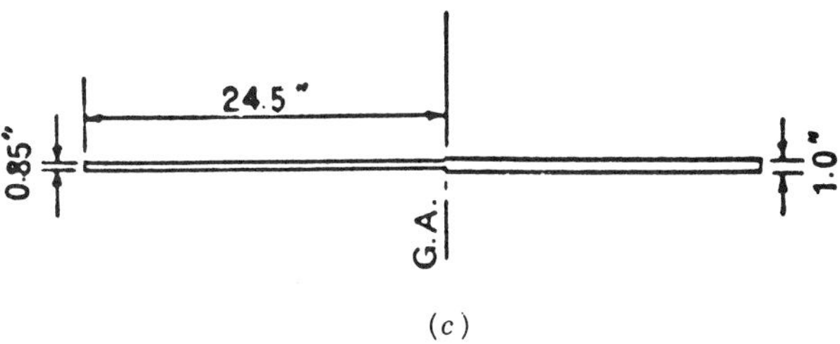

(*c*)

FIGURE 9-21 Truss of design example: (*a*) truss elevation; (*b*) forces acting at joint L_2; (*c*) acting section of gusset plate.

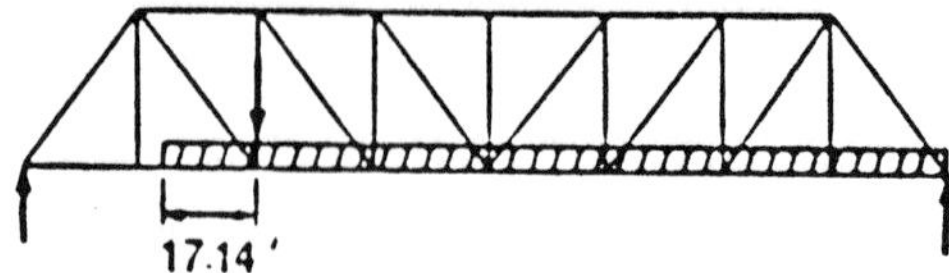

FIGURE 9-22 Live load position for maximum force in member U_1L_2 of truss of Figure 9-21.

which gives $x = 24.4$ in. from the edge of the plate on the tension side. The acting section is therefore as shown, and the area of the section is $A = (0.85 \times 24.4) + (1.0 \times 22.6) = 43.3$ in.^{2}. The moment of inertia of the section is

$$I = (0.85 \times 24.4^3)/3 + (1.0 \times 22.6^3)/3 = 7960 \text{ in.}^4$$

The section is subjected to a direct force of 58,600 lb and a clockwise moment of $(240{,}100 \times 1.4 + 181{,}600 \times 4.6) = 1{,}171{,}500$ in.-lb about the center of gravity of the acting section. The maximum tensile stress in the section

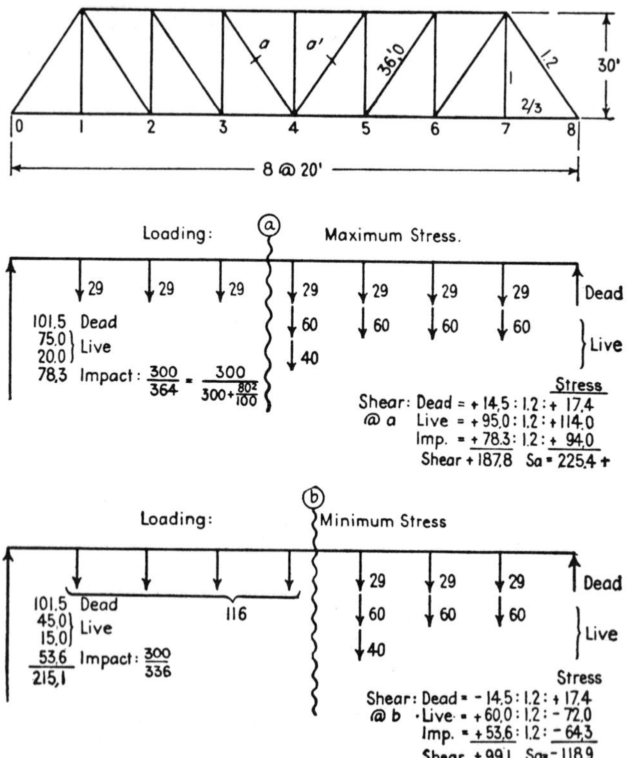

FIGURE 9-23 Truss elevation and stresses in diagonal of design example.

is therefore

$$f_a = \frac{P}{A} + \frac{M_c}{I} = \frac{58{,}600}{43.3} + \frac{1{,}171{,}500 \times 24.4}{7960} = 4950 \text{ psi} \qquad \text{OK}$$

Stress Reversal and Counters

Certain truss web members may be subjected to stress reversal because certain positions of live load can induce a stress of opposite character to the dead load stress and, sometimes, of greater magnitude. For example, early Pratt trusses are made with pin joints and have tension diagonals made with eyebars with enlarged heads bored to receive the pins. Such members buckle under a small load and are quite incapable of resisting compression. In these cases it becomes necessary to provide a second tension diagonal (called a counter) in any panel where the live load might cause a reversal or near approach to a reversal of the usual dead load stress.

AASHTO stipulates that web members subject to stress reversal should not have pinned ends; counters, if used, should be rigid. Adjustable counters should have open turnbuckles. AASHTO also stipulates that in the design of these members an allowance of 10 ksi should be made for initial stress. Only one set of diagonals in any panel should be adjustable.

It appears, however, that unless every counter in a steel truss is adjusted so that it has no initial stress and carries no stress when the main diagonal is in action, a degree of indeterminacy will be introduced. In this context, the use of counters is considered objectionable, and making the main diagonal stiff and capable of carrying both tension and compression may be more economical.

In the following example, proper adjustment is assumed so that only one diagonal in any one panel is in action at a time. For the truss shown in Figure 9-23, the maximum and minimum stresses in diagonal *a* will be computed. The member is stiff and thus can carry both tension and compression. The dead load is 29 kips per panel per truss. The live load is 3 kips/linear foot of deck, with a concentrated load of 40 kips.

The minimum stress in member *a* (compression) occurs with negative shear in panel 3–4, and the maximum value is induced from loading panel

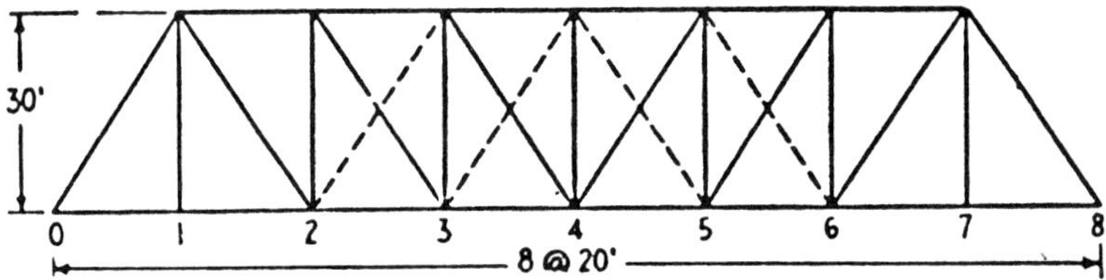

FIGURE 9-24 Location of counters in center panels.

TABLE 9-2 SUMMARY OF CHORD STRESSES FOR THE TRUSS OF FIGURE 9-24

Chord Stresses

Bar	Dead Stress	Live Stress: Uniform	Live Stress: (L)	Live Stress: Excess	Live Stress: Total	Impact 54%	Total $D + L + I$
L_0L_1	+ 67.7	+140.0	L_1	35 × 20 ÷ 30 = + 23.3	+163.3	+ 88.2	+319.2
L_1L_2	+ 67.7	+140.0	L_1	do +23.3	+163.3	+ 88.2	+319.2
L_2L_3	+116.0	+240.0	L_2	30 × 40 ÷ 30 = + 40.0	+280.0	+151.1	+547.1
L_3L_4	+145.0	+300.0	L_3	25 × 60 ÷ 30 = + 50.0	+350.0	+188.9	+683.9
U_1U_2	−116.0	−240.0	L_2	−40.0	−280.0	−151.1	−547.1
U_2U_3	−145.0	−300.0	L_3	−50.0	−350.0	−188.9	−683.9
U_3U_4	−154.7	−320.0	L_4	20 × 80 ÷ 30 = − 53.3	−373.3	−201.5	−729.5

Ratio: Live (uniform Load) to dead = 60/29.

Web Stresses

Bar	Dead Stress	Live Load: Max. + Shear = Vertical Component	Live Load: Excess	Live Load: Total V.C.	Live Load: Stress	Impact: %	Impact: Stress	Total $D + L + I$
U_1L_0	−121.8	60(1 + 2 + 3 + 4 + 5 + 6 + 7)/8 = 210.0	$40 \times \frac{7}{8}$ = − 35.0	−245.0	−294.0	54.0	−158.8	−574.6
U_1L_2	+ 86.9	50(1 + 2 + 3 + 4 + 5 + 6)/8 = 157.5	$40 \times \frac{6}{8}$ = + 30.0	+187.5	− + 225.0	67.6	+152.1	+464.0
U_2L_3	+ 52.1	60(1 + 2 + 3 + 4 + 5)/8 = 112.5	$40 \times \frac{5}{8}$ = + 25.0	+137.5	+165.0	75.0	+123.8	+340.9
U_3L_4	+ 17.4	60(1 + 2 + 3 + 4)/8 = 75.0	$40 \times \frac{4}{8}$ = + 20.0	+ 95.0	+114.0	82.5	+ 94.0	+225.4
U_4L_5	− 17.4	60(1 + 2 + 3)/8 = 45.0	$40 \times \frac{3}{8}$ = + 15.0	+ 60.0	+ 72.0	89.2	+ 64.3	+118.9
U_5L_6	− 52.1	60(1 + 2)/8 = 22.5	$40 \times \frac{2}{8}$ = + 10.0	+ 32.5	+ 39.0	95.0	+ 37.0	+ 23.9
U_1L_1	+ 19.3	60.0	+40.0		+100.0	95.0	+ 95.0	+214.3
U_2L_2	− 53.2	60(1 + 2 + 3 + 4 + 5)/8 = 112.5	$40 \times \frac{5}{8}$ = − 25.0		−137.5	75.0	−103.0	−293.7
U_3L_3	− 24.2	60(1 + 2 + 3 + 4)/8 = 75.0	$40 \times \frac{4}{8}$ = − 20.0		− 95.0	82.5	− 78.4	−197.6
U_4L_4	− 9.7		0		0		0	− 9.7
(U_4L_4)	+ 4.8	60(1 + 2 + 3)/8 = 45.0	$40 \times \frac{3}{8}$ = − 15.0		− 60.0	89.2	− 53.5	−108.7

points L_1, L_2, and L_3 with live load, placing the excess at L_3. Alternatively, we may consider the corresponding member in the other half of the truss, bar a', and compute the stress through it when the shear in panel 4–5 has its maximum positive value. The latter approach is recommended because it facilitates the tabulation of results and because it is usually easier to deal with positive shear.

For the same example, the dead load is articulated as follows: top chord panel points, 9.7 kips; bottom chord panel points, 19.3 kips. Counters are placed in the center panels as shown in Figure 9-24. The chord and web stresses are tabulated in Table 9-2. The letters U and L denote the upper and lower panel points.

Unless the analysis is consistent, the calculation for counters may not be explicit. The first counter is U_4L_5 where the minimum and maximum stresses are -17.4 and $+118.9$ kips, respectively. Assuming this counter to be in action (and, accordingly, the main diagonal in the same panel out of action), the counter carries the dead load shear in the panel, and this stress is compression as far as this element is concerned. If this is not clearly understood, we may refer to the stresses for member a' for section ⓑ shown in Figure 9-23, the only change being the reversal of the three signs for the bar stresses.

Member U_4L_4 will carry its maximum live load stress under the loading that produces maximum live stress in the counter U_4L_5.

9-7 CANTILEVER TRUSSES

Typical examples of cantilever bridges are structures consisting of two shore spans, each with an anchor arm and a projecting cantilever arm, and a suspended span that is essentially a simple bridge supported by the two cantilevers. A radically different arrangement is the Queensboro Bridge in New York, where the suspended spans are omitted and the cantilever arms are joined directly.

The negative vertical reactions at the ends of the anchor arms are usually provided by long eyebars engaging pins on the trusses, and attached by pins to beams anchored deeply in the masonry of abutments or piers. The advantages of the bridge with two cantilever arms and a central suspended span shown in Figure 9-25 are the same as in plate girder and box girder bridges with a similar configuration. The reduction of load effects possible in continuous structures is retained, and the bridge is made statically determinate with respect to reactions because each hinge supplies an equation $M = 0$.

However, it is not always economical to carry the longitudinal forces from braking and traction to a single support. The arrangement shown in Figure 9-25 offers a good solution, with the hanger or rocker at the left end of the

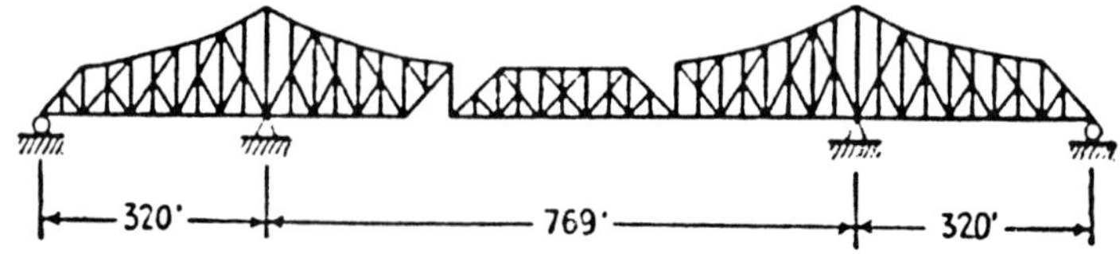

FIGURE 9-25 Typical two-arm cantilever bridge with a central suspended span.

suspended span supplying besides $M = 0$ also the condition $H = 0$, meaning the sum of horizontal forces on either side of the hanger must be zero.

Another method of providing static determinacy with a truss is by omitting the diagonals as shown in Figure 9-26. In this case there is no shear in panels 6–7 and 19–20 without movement of the frame. Accordingly, the absence of diagonals in these two panels gives two equations $S_{6-7} = 0$ and $S_{19-20} = 0$ (or $M_6 = M_7$ and $M_{19} = M_{20}$).

The same concept is also used in draw spans. Instead of no diagonals at all, light bars are used offering little resistance to distortions and are thus assumed to be nonexistent.

Example For the bridge shown in Figure 9-26, it is required to draw influence lines for the vertical reactions at supports 6 and 7 with load at panel point 10.

First, we consider the suspended span as a free body. Obviously, no reactions are developed on this free body, and if a load is carried on this span it is brought to either suspended end. Each shore structure with its anchor and cantilever arm is therefore independent and carries its loads without interacting with the other.

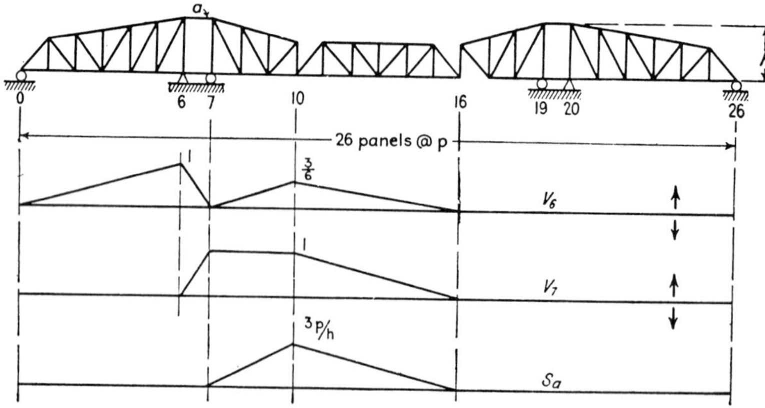

FIGURE 9-26 Cantilever bridge without diagonals.

For a load at point 10, there is no stress in the hanger at 16, and the shear in panel 19–20 is zero. It is also evident that the reactions at points 20 and 26 are zero. Considering the left shore structure, $V_7 = 1\uparrow$ $(S_{6-7} = 0)$. To balance the clockwise couple $1 \times 3p$ (where p is the panel length), there must be a couple at supports 0 and 6 with arm $6p$, or $V_0 = 1/2\downarrow$ and $V_6 = 1/2\uparrow$. The rest of the influence line may be drawn without further computations. The influence line for V_7 is obtained similarly.

For the same example it is required to draw the influence line for bar a (top member of panel 6–7). The free body chosen for the solution is the cantilever arm supported by the reaction at point 7 and the two horizontal bar stresses. The influence line is shown in Figure 9-26 as S_a. In similar problems it is essential to remember that the load moves along a floor system and brings loads to the truss only at panel points.

Part 2: Movable Bridges

9-8 MOVABLE BRIDGES: STRUCTURAL CONSIDERATIONS

When the topography of a bridge site makes it desirable to have the roadway close to the surface of the body of water crossed by the bridge, the vertical underclearance requirements of the navigation passing beneath the bridge may dictate a movable bridge. A movable bridge is a structure that may be moved to permit the passage of navigation. The most important types of movable bridges are (a) bascule bridges, (b) vertical lift bridges, and (c) horizontal swing bridges. The type to be used depends largely on the horizontal and vertical clearance requirements. Whether a low-level movable bridge or a high-level fixed bridge should be used at a given site can usually be determined by an economic analysis.

The design, fabrication, and erection requirements of movable bridges are covered in the AASHTO Standard Specifications for Movable Highway Bridges (1988 Edition) and in Appendix A of the proposed LRFD specifications.

General Requirements

Bascule Bridges A bascule bridge may prove economical where horizontal navigation requirements do not necessitate too long a span where a high vertical clearance is required. Chicago is one of the big cities with a large number of bascule bridges because these conditions typically exist. An early type of bascule bridge is shown in Figure 9-27. Motive power drives a pinion at D that engages the rack, thus opening or closing the span. The required motive power is reduced by the action of counterweight C.

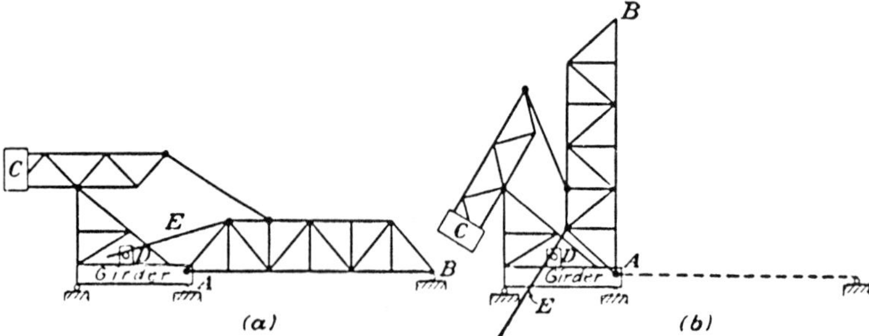

FIGURE 9-27 An early type of bascule bridge.

The dead load stresses in a bascule span change as the bridge is opened or lowered, and it is possible that the dead load stresses in certain members during this operation may exceed the total stresses with the bridge down and subjected to traffic. Let F_H be the stress due to dead load in a member with the span in a horizontal position, and let F_V be the dead load stress in the same member with the span vertical, that is, after a 90° rotation. Both these values are readily computed from the usual methods of structural analysis. With the bridge partly opened and the bottom chord making an angle α with the horizontal as shown in Figure 9-28, each dead panel load may be resolved into two components, one perpendicular and the other parallel to the bottom chord. The components perpendicular to the bottom chord will cause stresses equal to $F_H \cos \alpha$, and the components parallel to the bottom chord will cause stresses equal to $F_V \sin \alpha$. Hence, for any angle α the total dead load stress is given by

$$F_D = F_V \sin \alpha + F_H \cos \alpha \tag{9-15}$$

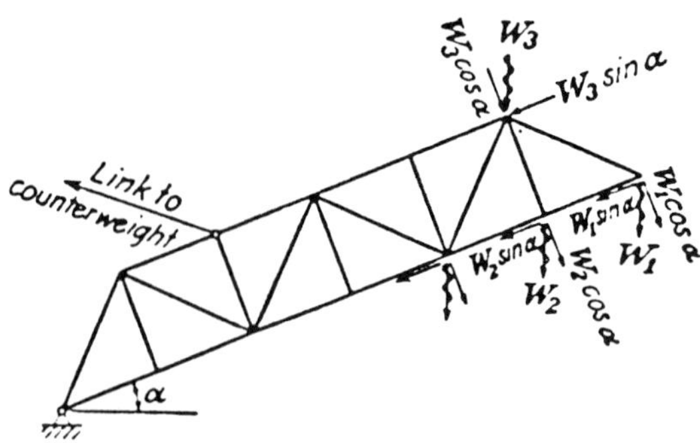

FIGURE 9-28 Semiopen position of a bascule bridge.

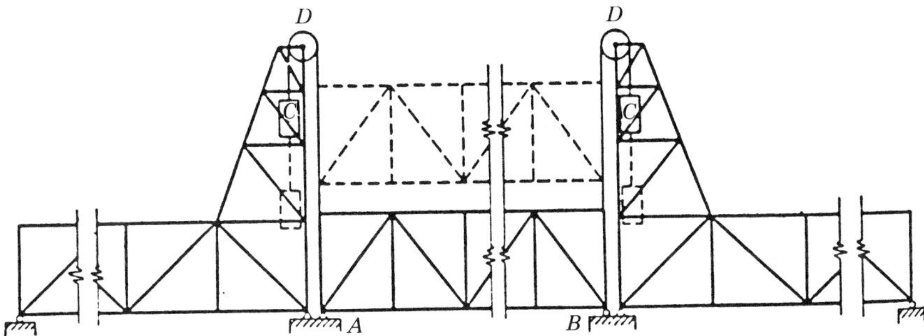

FIGURE 9-29 Typical configuration of a vertical lift bridge.

The value of F_D becomes maximum if

$$\frac{dF_D}{d\alpha} = F_V \cos \alpha - F_H \sin \alpha = 0$$

where $\tan \alpha = F_V/F_H$. Substituting this value into (9-15) gives

$$\max F_D = \sqrt{F_V^2 + F_H^2} \tag{9-16}$$

With the bridge closed, the dead load reaction at the free end is zero because the counterweight holds the dead load in equilibrium, but live loads produce reactions at each end in the same manner as for an end-supported (cantilever) span

Vertical Lift Bridges When the horizontal clearance requirements exceed the vertical clearance requirements for navigation, a vertical lift bridge is likely to be more economical. A typical configuration of a vertical lift bridge is shown in Figure 9-29. The central span is raised or lowered vertically by cables running over sheaves at point D supported at the lower tops. The motive power required to produce this motion is reduced by the counterweights C. These are usually designed to balance the entire dead load weight of the movable span, and therefore the dead load reactions are taken by the cables and transferred to the supports of the fixed spans. However, live loads on the movable span produce reactions on the piers at points A and B.

Horizontal Swing Bridges These structures accommodate unlimited vertical clearance, but the center pier constitutes an obstruction to traffic. Horizontal swing bridges generally are of two types: (a) the center-bearing type shown in Figure 9-30a, or (b) the rim-bearing type shown in Figure 9-30b. In each case the bridge is opened by horizontal rotation about a vertical line. When the bridge is open, the two spans cantilever from the

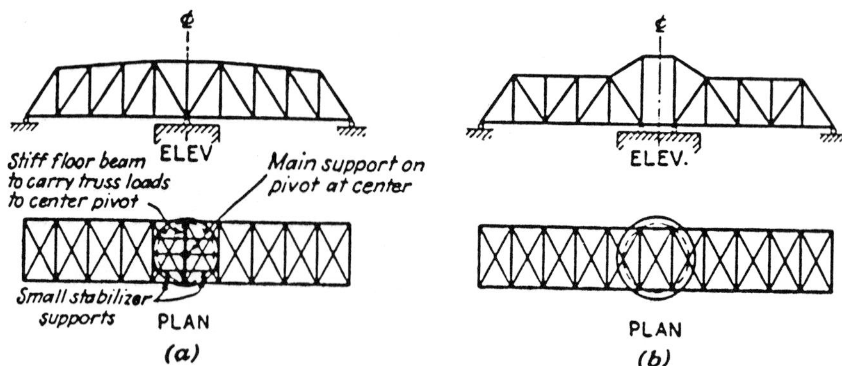

FIGURE 9-30 Typical configurations of horizontal swing bridges: (*a*) center-bearing type; (*b*) rim-bearing type.

center pier and are statically determinate. When the bridge is closed, the truss is a two-span continuous structure and hence statically indeterminate.

When a swing bridge is closed, the dead load reactions induced at the outer ends of the structure depend on the design. If these ends merely touch their supports, any live load on one span may cause uplift at the far end of the other span. This condition is usually avoided by lifting the end supports a slight amount when the bridge is closed. The desired dead load reactions are computed so that they will exceed the maximum live load reaction tending to cause uplift.

Bridge Loads

Live Load The live load on movable bridges is as specified in the current standard AASHTO specifications. However, in computing live load stresses, the live load should be applied either continuously or in separate parts and in such a manner as to produce maximum effects.

Impact As a function of the span length, live load impact is computed as stipulated in the standard specifications.

Dead load stresses in structural parts in which the stress varies with span movement and position (e.g., in bascule bridges) or in parts that move or support moving parts (e.g., in swing-span trusses, vertical lift trusses and towers, and supports for bascule trunions) should be increased 20 percent as a provision for impact or vibratory effects. This impact allowance should not be combined with live load stresses.

In structural parts, stresses induced by machinery or motive power should be increased 100 percent to reflect impact effects. In addition, end floor beams of moving spans should be proportioned for full live load plus twice the normal impact.

Fatigue Structural members and their connections subjected to stress repetition and reversal should be checked for fatigue requirements under the same criteria as in conventional highway bridges. Likewise, secondary stresses associated with stress reversal or those occurring in trusses should be calculated and provided for in the design.

Wind Load The stability of movable bridges should be analyzed assuming the wind loads act transversely, longitudinally, or diagonally at an angle of 45° with the bridge tangent. Exposed areas for transverse wind are as in the standard AASHTO specifications. Exposed areas for longitudinal wind should be taken as one-half of those for transverse wind, except for bascule spans when open when wind loads can also act on the floor. For diagonal wind, the equivalent simultaneous transverse and longitudinal wind loads should be taken as 70 percent of the values for wind acting transversely and longitudinally.

AASHTO specifies the following wind loads and stresses.

1. *Movable Span Closed* The structure is designed as a fixed bridge according to the standard specifications.

2. *Movable Span Open* (a) When the movable span is normally left in the closed position, AASHTO specifies 30 lb/ft^2 on the structure combined with dead load, and 20 percent additional dead load as impact. The allowable stress can be increased by 25 percent.

(b) When the span is open, AASHTO specifies 50 lb/ft^2 on the structure combined with dead load, at 1.33 times the basic allowable stress. For swing bridges, the design provisions will include 50 lb/ft^2 on one arm and 35 lb/ft^2 on the other arm, applied simultaneously.

For open deck grid floor bridges, the area exposed to wind normal to the floor should be taken as 85 percent of the area of a quadrilateral using the actual width and whose length is the span length. For bridges with solid floors or sidewalks, the actual exposed floor surface should be used.

Seismic Loads These loads are articulated in the AASHTO specifications for seismic design. Movable bridges should be designed to resist these loads with the movable span in the open or closed position. However, when the movable span is in one position more than 90 percent of the time, only one-half of the seismic load may be used for design of the other position.

9-9 MOVABLE BRIDGES: SPECIAL FEATURES

Counterweights These are intended to balance the movable spans and their attachments in any position, except that there should be a small positive dead load reaction at the supports when the bridge is seated. Auxiliary counterweights are usually provided for vertical lift spans with vertical movement exceeding 80 ft. The design must also provide for unbalanced conditions of the machinery and power equipment. A swing span of unequal lengths or unbalanced machinery should be balanced by counterweights.

Counterweights are usually made of concrete supported by a steel frame. Concrete counterweights not enclosed in steel boxes must be adequately reinforced. The weight of concrete must be ascertained by means of test blocks cast at least 30 days before concreting begins.

Aligning and Locking Movable bridges are provided with suitable mechanisms to surface and align the bridge and roadway and to fasten them securely in position so that horizontal and vertical displacement is prevented under traffic. End lifting devices are used for swing bridges, and span locks for bascule types. Span locks are also provided in swing and vertical lift bridges normally left in open position.

Traffic Gates Two gates are generally provided at each approach roadway to a movable span bridge. The first acts as a warning device and the second functions as a physical barrier.

Houses for Machinery and Operations These facilities must satisfy the design criteria of bridges, but must also conform to applicable building codes.

Traffic Lights and Bells Traffic signals are mandatory in movable bridges, except those that are manually operated. Warning bells may also be provided to supplement the signal lights.

Audible navigation signals and navigation lights are also standard features, and must conform to the regulations of the U.S. Coast Guard.

Stairways, Walks, and Elevators Metal stairways, platforms, and walks with railings must be provided for safety and access to the operator's house, machinery, trunnions, counterweights, lights, bridge seats, and all points requiring servicing and maintenance. Elevator cars should be fully enclosed, with solid sides and roof, and power-operated with appropriate automatic controls. All these facilities must meet the requirements of local building codes.

9-10 SPECIAL STRUCTURAL DESIGN PROVISIONS FOR MOVABLE BRIDGES

These are covered in detail in the AASHTO Standard Specifications for Movable Highway Bridges (1988 Edition) and in Appendix A of the proposed LRFD specifications. The following summary is provided to articulate certain structural features.

Swing Spans

Loading Conditions and Combinations For swing spans continuous on three or four supports, the stresses in the main girders or trusses should be checked for the following loading conditions.

Case I: *Dead Load* Bridge open, or closed with end wedges not driven.

Case II: *Dead Load* Bridge closed with its end lifted to give positive end reaction, equal to the reaction due to the temperature plus 1.5 times the maximum negative reaction of the live load plus impact, or the force required to lift the span 1 in., whichever is greater.

Case III: *Live Load plus Impact* Bridge closed, with one arm loaded and treated as a simple span, but with end wedges not driven.

Case IV: *Live Load plus Impact* Bridge closed and treated as a continuous structure.

The maximum and minimum stresses should be calculated using the following loading combinations.

- Case I alone, plus 20 percent
- Case I with Case III
- Case II with Case IV

In general, stress computations should show the stresses in the different members for each case of loading, including the combinations that give the maximum and minimum values.

Piers and Foundations Any settlement, movement, or displacement of the supporting piers may render the bridge inoperable. Hence, special design precautions must be taken to prevent and exclude these effects. The foundation should be capable of resisting torsional and horizontal forces produced by movement of the span.

Other Provisions Special structural requirements are also included for the pivot fenders, center bearing, rim bearing, and combined bearing.

Rim girders should be provided with stiffeners, fit close against both flanges. The rim girder flange angles should not be smaller than $6 \times 4 \times 3/4$ in., and for welded construction the flange plates should be at least 1 in. thick. The design should also provide for an end reaction due to a temperature differential between the top and bottom chord of 20° F for truss spans and 15° F for girder spans.

Bascule Spans

The preferred types of bascule spans are the trunnion and rolling lift configurations. Single-leaf bascules should have the forward ends supported. Double-leaf bascules should have the forward ends provided with effective locking mechanisms designed to transmit shear and produce the same deflection under unbalanced live loads.

Loading Conditions and Combinations Stresses in the main and counterweight trusses or girders must be calculated for the following conditions.

Case I: Dead Load Bridge open in any position.

Case II: Dead Load Bridge closed.

Case III: Dead Load Bridge closed with counterweights independently supported.

Case IV: Live Load plus Impact Bridge closed with live load acting.

The maximum and minimum stresses should be calculated for the following loading combinations.

- Case I alone, plus 20 percent
- Case II with Case IV
- Case III with Case IV

In proportioning bridge members, an allowable stress 125 percent of the basic may be used for the Case III with Case IV combination. Members subjected to stress reversal under this combination (involving live load plus impact) should be proportioned without consideration of fatigue.

Segmental and Track Girders The allowable load (lb/linear inch) of live bearing between treads of segments having a diameter greater than or equal to 120 in. should not exceed the value f_b given by

$$f_b = (12{,}000 + 80D)\left[\frac{F_y - 13{,}000}{20{,}000}\right] \tag{9-17}$$

where D is the diameter of the segment (in.). The portions of the segmental and track girders that are in contact when the bridge is closed should be designed for the sum of dead load, live load, and impact stresses. Under this loading, the allowable live loading given by (9-17) may be increased by 50 percent. Segmental and track girders should be reinforced with stiffeners and diaphragms.

Vertical Lift Spans

Loading Conditions and Combinations Stresses in trusses or girders should be computed for the following conditions.

Case I: *Dead Load* Bridge open.

Case II: *Dead Load* Bridge closed.

Case III: *Dead Load* Bridge closed and counterweight independently supported.

Case IV Bridge closed with live load acting.

In computing maximum stresses, the following combinations of these cases should be considered.

- Case I alone, plus 20 percent
- Case II with Case IV
- Case III with Case IV

In proportioning members, AASHTO allows a unit stress 25 percent greater than the basic allowable, to be applied to the combination of Case III with Case IV. Members subjected to stress reversal under this combination or with any other combination should be designed for maximum tensile and compressive stresses without consideration of fatigue.

Towers and lift span, together with all lifting members except counterweight ropes, should be designed for dead load plus 20 percent.

Wind Load Wind loads on the vertical lift span should be considered as in a simple fixed span.

With the span in the closed position, the towers should be designed to resist a wind load of 50 psf, acting in any direction, with a 25 percent increase in the allowable stress. With the span in the open position, the towers should be designed for the wind loads specified in Section 9-8.

Towers and Tower Spans For towers supported partly or wholly by an adjacent fixed span, any live load without impact that can be placed on the supporting span beyond the warning gates should be included in the design of the supporting span.

Lateral bracing of towers should be determined using 25 percent of the total compression in the columns in addition to the specified wind loads.

Anchorage for Cantilever Floor For live load placed on any portion of the floor outside a main truss or girder, suitable provisions should be included for anchorage to resist the resulting negative reaction of the opposite truss or girder.

9-11 EXAMPLE OF SWING BRIDGE

The double-leaf concrete swing bridge shown in Figure 9-31 is built across the Duwammish River in Seattle, Washington. It is one of two designs prepared for this crossing: a posttensioned segmental concrete box girder

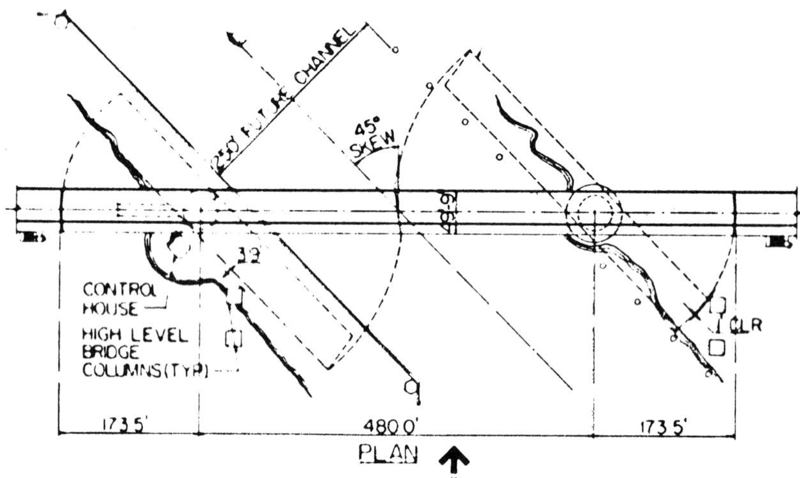

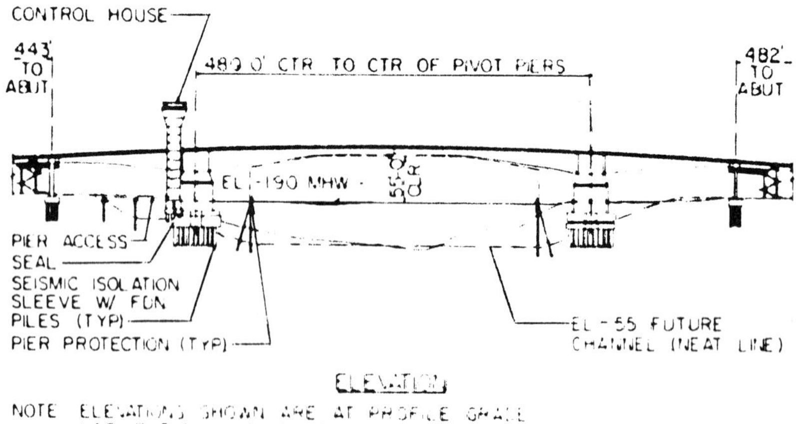

FIGURE 9-31 West Seattle Swing Bridge; plan and elevation. (From Kane, Mahoney, and Clark, 1990.)

(the scheme selected) and a steel box girder with a precast prestressed concrete deck made composite after erection.

Seismicity in the area is moderately active (SPC Zone C with a seismic acceleration coefficient $A = 0.25$). The seismicity studies and seismic design of this structure are discussed by Clark (1984). The new concepts in machinery and movable bridge technology are reviewed by Kane, Mahoney, and Clark (1990).

A primary concern in the design of concrete box girder bridges is control of long-term deformations. In this example, provisions for control of both long- and short-term deformations included posttensioning beyond what is required for stress control, some unbonded tendons, additional future tendons, adjustment of the approach span elevation at the tail span joint, vertical adjustment of each leaf, and simultaneous construction of the two movable leaves.

Unwanted long-term deformations may be excluded if the principle of load balancing is incorporated in the design of longitudinal posttensioning. Thus, the amount of prestress provided is the amount necessary to ensure 100 percent load balancing for the final dead load conditions. In this case about 30 percent more posttensioning was required over the amount that satisfied service load stress conditions. The deck of the box girder is posttensioned transversely, and vertical posttensioning is included in the webs.

Static equilibrium of each leaf about the pivot pier is maintained by providing thicker component section elements in the tail span and by adding ballast concrete so that at least 40 ft of the tail span is solid. Some precast ballast blocks are provided to adjust static equilibrium as necessary.

The pier structure has a diameter of 42 ft, carries the superstructure loads to the foundation, and houses the drive machinery, emergency generators, and part of the control system. The walls of the structure are 32 in. thick and heavily reinforced.

The foundation consists of 48-in.-diameter steel pipe sleeves around each foundation pile. The purpose of the sleeves is to control the elevation at which the foundation piles begin to receive lateral support from the surrounding soil. The annular space is excavated to a set elevation. The pier is located in the slope area of the channel excavation. Thus, without the sleeves, the depth from the footing to the slope surface would vary from zero to 15 ft. Under these conditions, the lateral stiffness of the piles would vary, and significant torsional response to seismic excitation would result from the eccentricity between the center of mass and the center of stiffness. The sleeves serve to eliminate variation in lateral stiffness and the associated eccentricity. The sleeves also support the tremie seal which is separated from the footing.

Machinery Twin hydraulic slewing cylinders rotate each movable leaf from the closed position to the open position for the passage of navigation. Friction is minimized because the structure is supported on the hydraulic fluid of the lift–turn cylinder. The principal source of friction is therefore the pivot shaft in vertical alignment. Hydraulic buffers are designed to stop the leaf from moving at a full rotation speed of 0.57° per second in 0.44° of travel. Open position buffers are located on the roof of the pier house and contact stops on the inner surface of the transition element cone. Tail span buffers are located on the approach span piers and contact the tail span.

Access to all items, parts, and machinery is provided by adequate openings placed in optimum locations to allow inspection and maintenance.

9-12 EXAMPLE OF BASCULE BRIDGE

The Columbus Drive Bascule Bridge in Chicago is the latest of a series of movable bridges built in downtown Chicago across the Chicago River and Sanitary and Ship Canal. Completed in the mid-1970s, it is a double-leaf, trunnion-type bascule span with appreciable fixed bridges at the approaches. The structure consists of welded plate girders, one of the earliest plate girder systems to be used in bascule bridges, and was designed using finite-element techniques (Xanthakos, 1973).

Design Criteria

The structural design is according to applicable AASHTO standards and specifications, with the following modifications and additions (city of Chicago specifications).

Loading The basic design live load is HS 20, except that for lane loading the equivalent uniform unit load is increased from 640 to 1000 lb/ft. The floor system is also checked for the alternate load of two 24-kip axles placed 4 ft apart.

The sidewalk live load is assumed to be 100 psf for the movable part and 150 psf for the fixed part.

Application of Live Load Members of main trusses or girders that require load application to two panels or less of the floor system are designed for truck loads. If the load application is extended to more than two panels of the floor system, the lane loading governs. In addition, the total panel loads are multiplied by a factor of $(60)/(n + 50)$, where n is the number of panels loaded to produce the stress in the member under consideration. Because this is a double-leaf bascule bridge, the load may be placed on both leaves if necessary to produce maximum stress in certain members. In this case the number of panels loaded in both leaves is n.

Impact Roadway stringers and beams affected by loads at expansion joints and breaks in the roadway are proportioned for an impact increment of 75 percent for concentrated loads at the joint and an impact for the remaining loads as specified in the AASHTO formula. The corresponding increment for roadway floor beams and girders is 50 percent for the concentrated loads at the joint.

Castings or weldments at expansion joints and breaks are proportioned for 100 percent impact for the concentrated load at the joint. The brackets, stringers, and deck on movable sidewalks are designed for 15 percent impact. Sidewalks in the fixed bridge are not designed for impact.

Longitudinal Forces These are as per AASHTO, except the magnitude of the longitudinal force is 10 percent of the live load.

Lateral Forces Wind is applied as provided by AASHTO. A transverse force of 240 kips is applied on the bottom chord in the closed position at the second panel point forward from the river pier bearing, and this load may be used in lieu of the wind loads if it produces maximum stresses.

Shear Lock Stresses Shear locks are intended to force one leaf to deflect an equal amount as the other leaf is loaded. The stresses induced at the center lock under unequal loading are computed from the following:

$$F_v = \frac{P}{4}(2 - 3K + K^3) \tag{9-18}$$

where F_v is the shear at the center due to the load P transferred from one leaf to the other, P is the load at any point along the leaf, and K is the ratio of the distance of the point of application of the load P, measured from the center of the span, to the length of the cantilever arm.

Stress Conditions The stresses in main or counterweight trusses or girders are calculated for the following conditions.

Condition 1 Bridge closed.

Condition 2 Bridge open in any position that will produce maximum stress in any member considered.

The following loading cases are considered.

- Case I, condition 1. Dead load only with reactions at the trunnion.
- Case II, condition 1. Live load with reactions at the live load support on the river pier and at the anchor column, with no allowance for the dead load relieved from the trunnion.

- Case III, condition 1. Dead load plus live load with reactions at the live load support on the river pier and at the anchor column with the dead load relieved from the trunnion. Members and parts affected by this condition may be proportioned for a unit stress 25 percent higher.
- Case IV, condition 1. Live load with reactions at the trunnion and at the anchor column for loads placed at the first two panel points of each truss, which is sufficient to cause contact at the live load support. To these stresses is added the stress due to the balance of the live load on the bridge computed as specified in Case II.
- Case V, condition 1. Shear lock load with reactions at the live load support on the river pier and at the anchor column.
- Case VI, condition 2. Dead load only with reactions at the trunnion.
- Case VII, condition 2. Wind load with reactions at the trunnion and the rack.

For determining maximum stresses, the foregoing cases are used in the following combinations.

- Case I alone, plus 20 percent vibration
- Case I plus Case II
- Case I plus Case II plus Case V
- Case III
- Case I plus Case IV
- Case VI plus 20 percent vibration plus Case VII

In considering live load stresses, the live loads are applied either continuously or in detached parts, and in a manner to produce maximum stress.

Design Assumptions: Trunnion Girders As an example, the following additional considerations were introduced in the design of trunnion girders.

1. The center of gravity, dead load of leaf, center of gravity of counterweight, and centerline of trunnion are linear. Otherwise, when the bridge is raised, the dead load resultant force will not pass through the centerline of the trunnion. Without live load, the center of gravity of the resultant load comes through the trunnion, and the entire load is resisted by the trunnion girder.

2. With live load on the leaf, the center of gravity of the combined live load and dead load moves away from the trunnion (see the following diagram). The live load causes upward reactions at the live load bearing and downward reactions at the trunnion, reducing the total load on the trunnion.

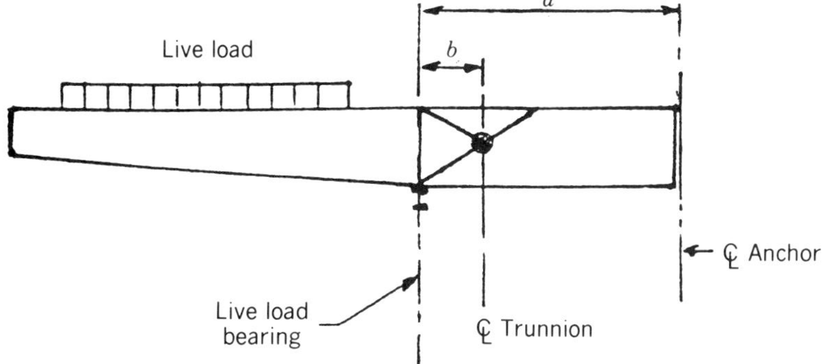

3. When the live load increases and becomes large enough to balance the entire structure over the forward support, the load on the trunnion may approach zero. As more live load is added, the anchor column becomes active (see the preceding diagram), whereas the forward support (live load bearing) takes the entire dead and live load plus a load equal to the negative reaction on the anchor. As the live load center of gravity moves away from the trunnion, the trunnion girder may deflect upward. The rise in the anchor bracket equals the trunnion upward deflection multiplied by the ratio a/b.

4. It follows therefore that two cases of deflections in the counterweight arm must be analyzed: (a) when the leaf is supported on the trunnion and (b) when the leaf is supported on the live load bearing.

5. In determining the actual location of the center of gravity of the entire leaf, allowance is made for variations in the center of gravity of the counterweight mass by 5 to 6 percent of its vertical dimension.

6. At the bottom of the anchor column, a spread base is provided to assist in engaging the desired amount of concrete as anchorage against uplift.

7. If dead load stresses vary as a result of change in position during operation, they are increased by 20 percent to provide for the dynamic effect of the movement, but these increased stresses are not combined with live load stresses.

Part 3: Cable-Stayed Bridges

9-13 GENERAL FEATURES OF CABLE-STAYED BRIDGES

An extensive treatment of cable-stayed bridges is given by Podolny and Scalzi (1986) and Troitsky (1977) and includes the following economic evaluation; concrete, steel, and composite superstructures; vehicular and pedestrian bridges; analysis and structural behavior of cables; analysis and design considerations; and erection and fabrication. These books offer a comprehensive assessment of the subject and are highly recommended to practicing engineers and graduate students at any level of academic and professional activity. A complete bibliography on the subject provides valuable references up to the year 1985.

An extensive bibliography and data on cable-stayed bridges are also provided by the ASCE Committee on Long-Span Steel Bridges (1988) at the date of publication. In the remaining part of this chapter, we will discuss the basic features and design principles of cable-stayed bridges.

Origin and Development

The engineering text *Theory and Practice of Modern Framed Structures* by Johnson, Bryan, and Turneaure (1907) shows a picture of a cable-stayed bridge but gives no details of design or analysis. The first modern cable-stayed bridge was designed and built in Sweden, completed in 1955. This bridge has modest spans of 74, 183, and 74 m. The success of this structure prompted the design and construction of the North Bridge in Dusseldorf, Germany, in 1958, with a central span of 260 m. The clear span requirements on the Rhine River favor this bridge type as the most logical choice. Many cable-stayed bridges have been built over famous waterways and other rivers in Europe, Canada, South America, Japan, and more recently in the United States.

Recent applications in the United States are summarized by Podolny and Scalzi (1986) and include approximately 18 cable-stayed bridges completed, under construction, or in design as of 1986. Notable examples include the Sitka Harbor Bridge in Alaska, the Pasco–Kennewick Intercity Bridge in the state of Washington, the Luling Bridge in Louisiana, and the Quincy Bridge in Illinois. It appears that cable-stayed bridges are suited for clear spans in the 400- to 2000-ft range, and can have girders of steel or prestressed concrete. A favorable aspect of the construction is that the bridge can be erected without falsework in the main span, and this is a considerable advantage over deep canyons or waterways.

Distinctive Components

Cable Arrangement The four basic longitudinal cable configurations are shown in Figure 9-32. The radiating type shown in Figure 9-32*a* is a converging system whereby the cables intersect or meet at a common point at the top of the tower. The harp configuration has the cables parallel and spaced along the girder and the pylon as shown in Figure 9-32*b*. The fan arrangement shown in Figure 9-32*c* combines the radiating and harp types. In the star configuration shown in Figure 9-32*d*, the cables are placed along the pylon and converge at a common point on the girder. Within the basic

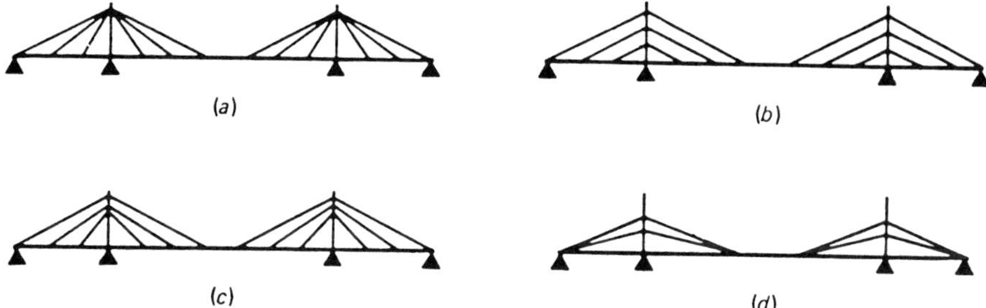

FIGURE 9-32 Longitudinal cable configuration: (*a*) radiating; (*b*) harp; (*c*) fan; (*d*) star. (From Podolny and Scalzi, 1986.)

systems variations are available to fit any conceivable design and esthetic requirement.

In the transverse direction two basic arrangements are generally available: those where the cables are arranged in either two planes or those where the cables are arranged in a single plane. The basic arrangements may have variations represented by four main types.

In type 1 the single-plane configuration has one vertical plane of cable, normally on the longitudinal centerline of the structure. A feasible variation is type 2 where the vertical plane of the cables is positioned laterally from the longitudinal centerline of the bridge. In type 3 the double-plane arrangement has two planes of cables located outside the roadway, either vertical or sloping and intersecting on the bridge center. Type 4 consists of a V-shaped double-plane system, suggested as a means to reduce the tower height without changing the height ratios, avoiding the stay concentration on the tower top and eliminating any lateral sway of the girder deck.

Towers These, often referred to as pylons, may have the simple form of a single cantilever for a single-plane cable arrangement, or may consist of two cantilever members for a double-plane cable structure. Towers may be hinged or fixed at the base, and can also take the form of a portal frame. Base fixity induces considerable bending moments in the pylon, but this disadvantage is offset by the increased rigidity of the structure as a unit. In addition, a fixed base is practical for erection purposes, and may cost less than a heavy pinned bearing.

Alternatively, the tower can have the shape of a transverse A frame fixed or pinned to the pier shown in Figure 9-33*a* in a modified form. Other solutions are the diamond form shown in Figure 9-33*b* and the delta-type tower shown in Figure 9-33*c*.

When several large spans are contemplated, two A frames or two or more portal frames may be placed in the plane of the cables and joined transversely at the top by a portal beam.

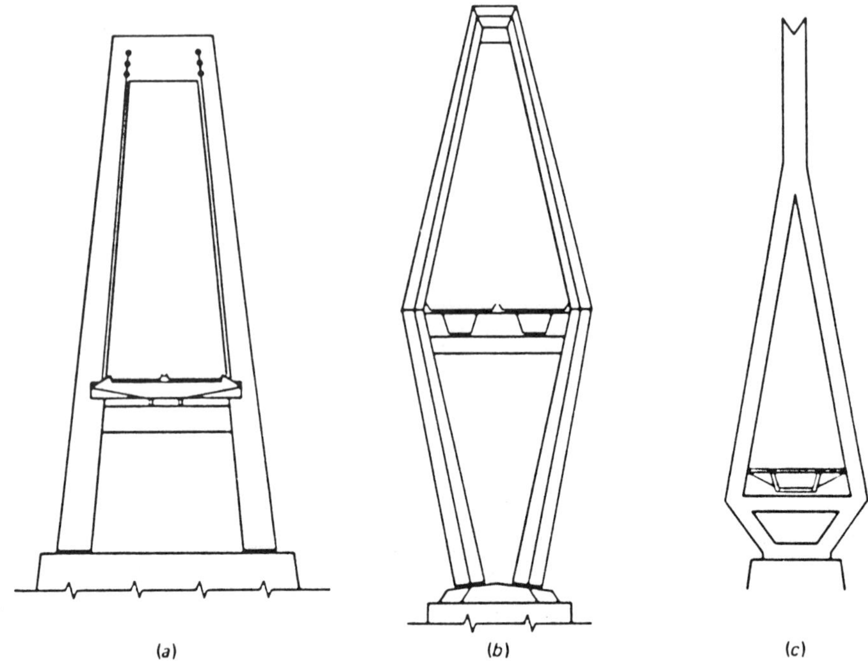

FIGURE 9-33 Alternate tower configuration: (*a*) transverse A frame; (*b*) diagonal shape; (*c*) modified diamond or delta. (From Podolny and Scalzi, 1986.)

Bridge Deck Types The basic structural systems are the stiffened truss and the solid web. Because trusses are seldom selected because of higher fabrication and maintenance costs, the solid-web girder appears to predominate the superstructure types. Solid-web girders incorporated in bridge decks are essentially similar to the structural systems discussed in the foregoing chapters.

Bridge decks in cable-stayed construction are either steel or concrete, and in some cases composite forms have been used. Cast-in-place concrete box girders have been used in certain instances. Precast, prestressed I sections have been placed as "drop-in" sections at the center span between the extremities of the outside cables emanating from the towers. There are also examples of prestressed segmental construction used in conjunction with cable-stayed bridges.

For steel bridges, the box girder with an orthotropic deck provides an efficient design because it provides increased torsional stiffness to resist unsymmetrical live loads and wind forces. The ratio of girder depth to main span length is affected by the allowable live load deflection, but, in general, a quoted economic depth–span ratio is in the range of 1/60 to 1/90.

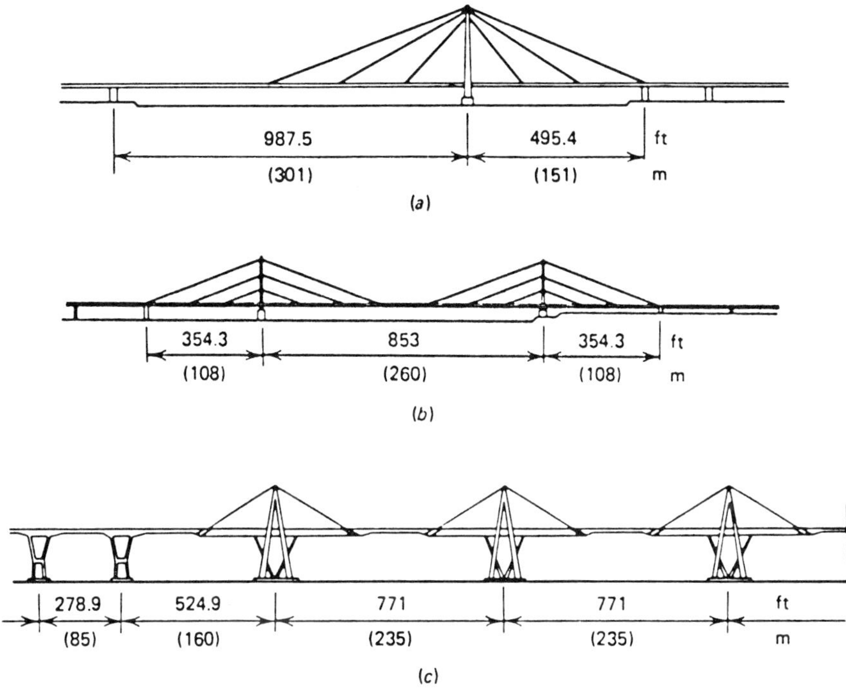

FIGURE 9-34 Span arrangements: (*a*) two-span asymmetrical, Severin Bridge at Cologne, Germany; (*b*) three-span symmetrical, North Bridge at Dusseldorf, Germany; (*c*) multispan, Maracaibo, Venezuela. (From Podolny and Scalzi, 1986.)

Span Arrangement There are three basic span types: (a) two spans, symmetrical or asymmetrical; (b) three spans; and (c) multiple spans. Examples of these types are shown in Figure 9-34. Data on two-span asymmetrical bridges show that the longer span is about 0.60 to 0.70 of the total length. When the back stays of the shorter span are concentrated into a single back stay anchored to an abutment, the ratio of the long span to the total length approaches 0.80. In three-span bridges the ratio of the center span to the total length is about 0.55. Multiple spans normally are of equal length, except flanking spans connecting to approach spans or abutments.

9-14 LOAD EFFECTS IN CABLE-STAYED BRIDGES

Live Load

Design loads, truck or lane load combinations, and application of loads must be consistent with AASHTO specifications (standard or proposed LRFD).

These loadings may be modified to suit local conditions and span lengths beyond the intent of the specifications. Thus, for spans in excess of 500 ft, the recommended live load should be calculated from Table 2-3 and as discussed in Section 2-5. Inasmuch as the maximum traffic load is likely to occur with traffic stationary, engineering discretion may dictate whether impact should be omitted.

Design Approach Studies are under way on the application of load factor design to cable-stayed bridges, but conclusive recommendations may not be forthcoming for some time. Until then, the load factor design approach should be used only in proportioning individual elements and members such as floor beams, stringers, and so on.

Wind Effects

The flexibility of cable-supported bridges exposes the structure to potentially damaging dynamic motion under wind effects. Wind failures are documented by Podolny and Scalzi (1986), and have led to the current wind analysis requirements. Thus, with more state-of-the-art advances forthcoming in predicting aerodynamic bridge deck behavior, wind tunnel tests are recommended on elastically suspended bridge section models to determine the wind velocity causing instability. It appears, however, that cable-stayed bridges with a main span less than 800 ft long, with corresponding flexural frequencies of 0.5 to 0.8 cps and a ratio of torsional, N, to flexural, N_h, frequency of $N/N_h > 2$, undergo a single-degree vortex shedding phenomenon. Larger-span bridges are likely to undergo single-degree or coupled flutter instability. Fatigue effects associated with wind-induced vibrations are discussed in Section 9-21.

Loading Conditions

Cable-stayed bridges are proportioned for the following loads and forces: (1) dead load; (2) live load; (3) dynamic effects on live load, where applicable; (4) wind loads; (5) erection loads; and (6) other loads that may exist, such as longitudinal forces, centrifugal forces, thermal forces, shrinkage, seismic effects, and so forth.

Design Loads Applicable to Cables A standard approach regarding the conditions, loading, and a factor of safety to be used in proportioning the cables is yet to come. At present, the allowable stress is one-third of the effective design breaking strength of the strand where fatigue is not a factor. The design factor is predicated on the assumption of elastic behavior of the structure and the cable. The range of stress in the cable is computed to be less than the prestretched elastic limit. With parallel wire strands utilizing special high-amplitude fatigue sockets, the allowable stress may be increased to 45 percent of the guaranteed ultimate strength. Acceptance of this stress level should be based on performance tests of the strand and the anchorage assembly.

In some instances where fatigue is likely to occur, the recommended design stress is reduced to one-fifth of the ultimate breaking strength of the cable.

Deflections The recommended static deflection of bridge decks under live load should not exceed 1/500 of the span length between piers, unless static and dynamic analysis show that serviceability criteria are met. Although the 1/500 value is in excess of the AASHTO allowable, the consensus of opinion is that more severe requirements would result in overconservative designs.

9-15 CABLE CHARACTERISTICS

Modulus of Elasticity In this structural application, cables are tension members with negligible flexural and torsional rigidity. A cable may consist of structural ropes, strands, locked coil strands, or parallel wire strands. The minimum modulus of elasticity is as follows:

Strand	1/2 to 2 9/16 in. diameter	E = 24,000,000 psi
Strand	2 5/8 in. diameter and larger	E = 23,000,000 psi
Rope	3/8 to 4 in. diameter	E = 20,000,000 psi

The foregoing data are minimum values stated by ASTM specifications to be used for specific sizes and coatings, namely for prestretched strand and rope for class A coating of zinc on the wires.

More recent parallel wire stays consist of ASTM A421, type BA, 1/4-in.-diameter wire with a minimum tensile strength of 240 ksi. The usual configuration is the hexagon, because it provides the most compact grouping and ensures a geometry in which an equal length of individual wires can be maintained for uniform stressing. The minimum modulus of parallel wire strand ranges from 27,500,000 to 28,500,000 psi. The higher modulus is the result of the parallel configuration that allows the group to approach the elastic characteristics of individual wires.

Cable stays consisting of parallel 0.6-in.-diameter low-relaxation ASTM A416 seven-wire prestressing strand are currently favored in U.S. markets. This material is available with a minimum ultimate tensile strength of 270 ksi. The strands have a relatively high breaking strength. Other cable systems used in cable-stayed bridges consist of ASTM A722 type II bars, and a recent example is the Penang Bridge in Malaysia reportedly completed in 1985. The bars have a minimum ultimate strength of 150 ksi, and are available in the usual diameter range of 5/8 to 1 3/8 in.

Interestingly, few results have been disseminated on static tests of structural strand and rope stressed above the prestretching load in the inelastic range.

Equivalent Modulus of Elasticity In cable analysis, the cable force is taken to act along the inclined chord although the cable sags slightly under its own weight. This flexibility is combined with changes in length to induce inelastic response. Several methods have been suggested to account for these inelastic features, and the most popular is the concept of the equivalent modulus of elasticity. In this approach, a straight-line chord member replaces the actual curved chord. A mathematically equivalent modulus for the substitute member is introduced so that its elongation is compatible with that of the member in the curved form. This equivalent cable modulus is

$$E_{\mathrm{eq}} = \frac{E}{1 + \left[\dfrac{(wL)^2 AE}{12T^3}\right]} \tag{9-19}$$

where E is the cable material modulus, L is the horizontal projected length of the cable, w is the weight of the cable per unit length, A is the cross-sectional area of the cable, and T is the cable tension (Ernst, 1965). The function expressed by (9-19) is plotted in Figure 9-35 for $E = 24{,}180$ ksi and $w = 3.035 \times 10^{-4}$ kips/in.^{3}. The cable tension will change, however, as a result of movement of the terminal points of the cable as the structure is loaded; hence, the equivalent cable modulus will also change because of changing geometry. Equation (9-19) is therefore modified to give the equivalent cable modulus E_{eq} at the end of any load increment as

$$E_{\mathrm{eq}} = \frac{E}{1 + \left[\dfrac{(wL)^2 (T_i + T_f) AE}{24T_i^2 T_f^2}\right]} \tag{9-20}$$

where the subscripts i and f denote the initial and final values of the tension, respectively, during the load increment.

Prestretching This is applied as a predetermined tension force to a finished strand or rope to remove the cable looseness inherent in the manufacturing process. The looseness, or constructional stretch, varies with cable size, number of wires in the cable, type of cable in a wire rope, and possibly with the conditions of the stranding equipment. The prestretch load normally should not exceed 55 percent of the rated breaking strength of the cable. While the member is in the prestretching equipment and under a

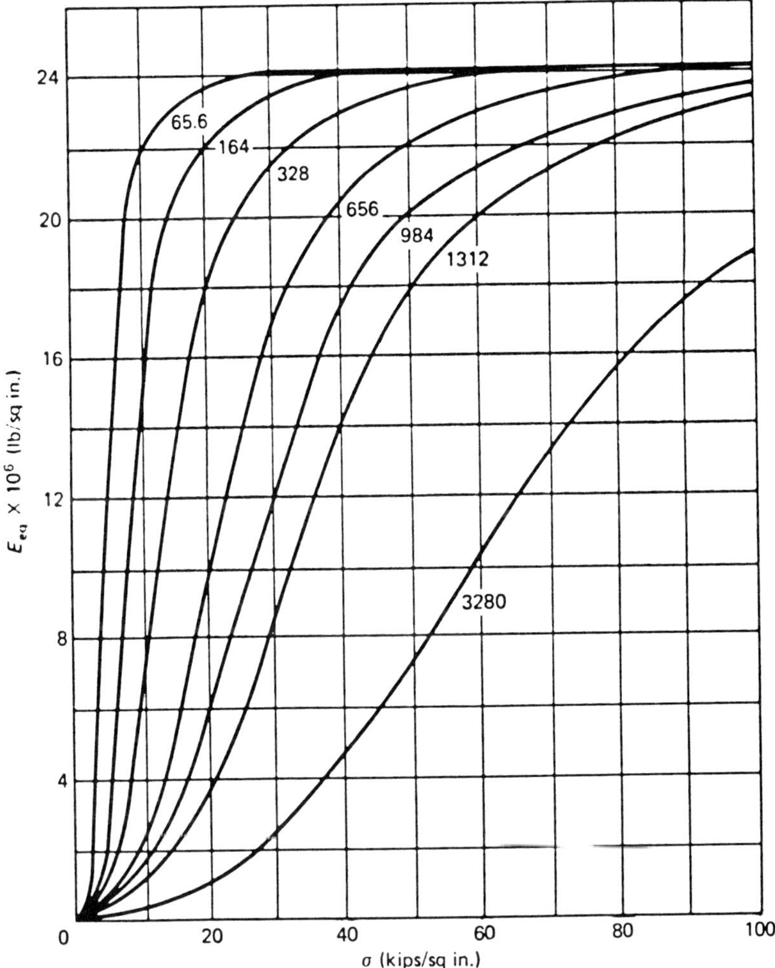

FIGURE 9-35 Equivalent modulus of elasticity. (From Ernst, 1965.)

known load, the exact length of the prestretched rope or strand can be measured.

Static Behavior As longitudinal tension is applied to a spirally wound rope, transverse forces are developed between the individual wires. These tend to reduce the ultimate capacity below the sum of the capacities of the individual wires for two reasons: (a) a combined stress system develops at the contact areas and (b) plastic flow at the contact area causes nicking of the wires with a subsequent reduction in their cross-sectional area (Bert and Stein, 1962; Leissa, 1959; Starkey and Cress, 1959).

If the rope is bent over a saddle,two additional effects are manifested: (a) bending stresses are induced and (b) transverse radial forces are applied to the rope. For a single wire of diameter d bent to a diameter D (at its centerline), the bending stress is given by

$$f = Ed/D \tag{9-21}$$

For a spirally wound rope over a saddle of moderate length, it appears possible that shear movements can occur between adjacent wires, and that the bending stress approaches the value given by (9-21) with d equal to the diameter of the wire. At the other extreme, for a parallel strand rope bent a number of times around a drum with no slip between the component wires, the bending stress approaches an upper limit given by (9-21) with d equal to the rope diameter. In the former case the severity of the bent is measured by the ratio of the wire to the saddle diameter, and in the latter by the ratio of the rope to the saddle diameter.

Fatigue A comprehensive series of tests have been performed at Lehigh University (Fisher and Viest, 1961; Warner and Julsbos, 1966) on a large number of strand specimens, 3/8 and 7/16 in. diameters, seven-wire prestressing strands. Although there are differences in the manufacturing process, these results give a general indication of the behavior of structural strand. The fatigue life is defined as the number of cycles before the first wire breaks in the seven-wire system. However, in structural strand, from 19 to 277 wires, a break in one wire would not be significant. Based on these results, it appears that the range of loading has a more pronounced effect on fatigue life than the maximum, minimum, or mean stresses. Fatigue life is increased in the lower stress ranges although the actual stresses may be higher.

The fatigue characteristics of cables have also been studied by Reemsnyder (1972). This analysis includes tests reported by Fish (1968) for United States Steel Corporation and tests performed by Bethlehem Steel Corporation. Tentative conclusions for structural strand under cyclic tension loads are postulated as follows.

1. Strand diameter, wire size, and wire grade have little or no effect on fatigue life.
2. The fatigue limit (stress at which fatigue failure will not occur) should be expressed in terms of the load range as a percentage of the breaking strength, and not as a function of the mean or maximum load.
3. For a given load range, expressed as a percentage of the breaking strength, fatigue life increases with an increasing number of wires in the strand, but the strand diameter is not relevant.

4. Zinc-cast end fittings, normally used with structural strand, lower fatigue life.
5. The wear caused by the rubbing of wires within the strand affects fatigue life because it decreases slightly fatigue strength.
6. The test data, although limited, indicate a fatigue limit, expressed in terms of the load range, of 15 to 30 percent of the breaking strength of the strand, with the actual value depending largely on construction characteristics.
7. Indication of imminent fatigue failure in strand wires is not directly obvious. With careful measurements, however, a rapid increase in strand temperature can be noted at about 90 percent of the fatigue life under constant cyclic loading.

In a series of tests by Fleming (1974), it was proposed to develop a fatigue life prediction equation in the following form:

$$\log N = A - B(P_r) - C(P_m) - D(P_r)(P_m) \tag{9-22}$$

based on a regression analysis of fatigue data. In this case, P_r and P_m are the load range and mean load, respectively, and N is the number of cycles to failure. The load range and mean load should be expressed as a percentage of the actual breaking strength of the strand.

Foreign Data on Fatigue Table 9-3 shows typical dead and live load stresses in cable-stayed bridges, and evidently the geometry at the structure is such that the stress range is considerable.

Fatigue test results have been reported by Daniel and Schumann (1967), Hess (1960), Hitchen (1957), and Kenyon (1962). These results indicate that fatigue strength is affected by (a) contact between wires in straight rope, (b) additional transverse stresses from externally transverse loads, and (c) bending stresses due to rope curvature of saddles or other details. The locked coil cable has a markedly improved contact surface between the outer wires, and should thus have an improved fatigue performance.

Test results from Havemann (1962) and Hess (1960) are summarized in Table 9-4, and lead to the following conclusions.

TABLE 9-3 Typical Dead and Live Load Stresses

Bridge	Main Span (ft)	DL/ Ultimate	(DL + LL)/ Ultimate	Range/ Ultimate	Range (ksi)
Julicherstrasse	324	0.257	0.408	0.151	29.5
Severin	991	0.201	0.386	0.185	36.9
		0.255	0.368	0.113	23.0

TABLE 9-4 Fatigue Lives of German Locked Coil Wire Ropes (From Havemann, 1962: Hess, 1960)

						Longitudinal Stress (ksi)			Transverse Force			Cycles		Number of Wires Broken			
Bridge	Identification	Length (in.)	Diam. (in.)	Net Area (in.^{2})	Total no. of Wires	Lower	Upper	Range	Length (in.)	Force (kips)	Force per Unit Length	Total $\times 10^6$	To Fracture of first Wire	Grip *A*	Grip *B*	Between Grips	Reduction in area (%)
Rodenkirchen	1	59.1	2.13	3.07	116	60.6	81.8	21.2	15.8	220	13.9	2.063	0.46				
Severin	2	59.1	3.07	6.46	199	41.8	64.6	22.8	15.8	176	11.1	2.002		15(15)	0	82(58)	38
	3	68.6	2.87	5.65	155	42.4	64.6	22.2	15.8	163.8	10.4	2.005		2(2)	25(25)	39(30)	24
	4		1.26	1.042	58	23.9	46.6	22.7	3.25	33.5	10.3	10.93		1	6	2	
	5		1.26	1.042	58	23.9	46.6	22.7	3.25	35.2	10.8	11.04		1	2	3	
	6		1.26	1.042	58	18.2	46.6	28.4	3.25	31.7	9.7	4.55		0	9	26	
	7		1.26	1.042	58	18.2	46.6	28.4	3.25	35.2	10.8	3.79		1	0	22	
	8		1.26	1.042	58	18.2	46.6	28.4	3.25	44.0	13.5	2.26		0	6	31	
North Elbe	9	70.2	2.83	5.59	185	51.2	72.7	21.5	—	—	—	2.210	0.438	22	10	6	22
	10	70.2	2.83	5.59	185	51.2	72.7	21.5	15.8	167.2	10.6	2.089	0.610	—	14	21	20
	11	218.5	2.83	5.59	185	51.2	72.7	21.5	15.8	167.2	10.6	2.035	1.072	1	1	3	3

1. Fatigue behavior is influenced by testing arrangements. Ropes 10 and 11, both from the North Elbe Bridge, have the same nominal stresses but different lives. The longer rope (rope 11) is probably a better approximation of the service conditions.

2. Clamping forces reduce fatigue life. This is shown in results 6, 7, and 8. The inconsistency in results 9 and 10 may be explained by differences in the test arrangement.

3. Fatigue life is markedly influenced by the stress range (tests 7 and 8), and this conclusion is common in fatigue. Thus, the mean stress has little significance, and for this reason the bending stress of a saddle is probably less important than the transverse force.

4. A transverse force of 10 kips/in. was used for many of the tests. Results 2, 3, 4, and 10 appear to indicate that this force is more damaging to the larger ropes, although the results may have been influenced by other factors.

5. Results 1, 9, 10, and 11 indicate that fracture propagation is slow. Thus, if special equipment is available to detect fractured wires during regular inspection of cables in service, fatigue problems may be assessed before collapse is imminent.

9-16 ANALYTICAL CONCEPTS OF CABLE-STAYED BRIDGES

In a cable-stayed bridge the girders are supported at several locations, namely, abutments and piers, usually considered as fixed and nonyielding supports, and at cable points with the cables emanating from the towers. The latter are yielding supports as the cables change length under load and because the towers are also flexible and can move. The structure can therefore be modeled as a continuous beam on both rigid and flexible supports.

Experience shows that the effect of deflections in producing nonlinear behavior in cable-braced girder bridges is relatively small. Felge (1966) reports that this effect on girder bending moments is of the order of 6 to 12 percent. A nonlinear analysis may be completed in the following sequence: (a) carry out a linear analysis using nominal geometry and obtain the deflections, (b) use these deflections to calculate the revised geometry, (c) carry out a second linear analysis using the revised geometry.

Mixed Method of Analysis: Single Plane

According to this method (Smith 1967, 1968; Podolny and Scalzi, 1986), the general behavior is determined by superimposing the action of tower rotation, stay elongation, and tower shortening.This interaction evolves into a mixed-force displacement analysis where modifications are made to account for each action. Additional modifications are introduced for the effects of bending of the towers, base fixity of the towers, shortening of the girder, twist or torsion of the girders, and so on. The compatibility equations are thus modified and the equilibrium conditions formulated to give a single mixed matrix for the total structure.

This method is essentially linear, because it assumes that deflections are proportional to the load at all portions of the structure and for the structure as a whole. The response of the structure is predicted by the general technique of consistent deformations. The positions or locations of the cables serve as restraints or supports along the main girders. These restraints act as springs and are dependent on the induced forces and geometry.

A generalized approach includes the various influences from the cables and the towers. The following phases are considered and superimposed on these effects: (a) rigid supports at cable locations, (b) elastic cable effects at supports, (c) tower shortening, (d) tower rotation, and (e) combined effects.

Rigid Supports of Tower Locations The application of the technique is exemplified in the example shown in Figure 9-36. The double-cable system shown in Figure 9-36*a* has a load P as shown. The tower is pinned at point 2, and the cables are extended to points 1 and 3 which are also pinned to the girder. In Figure 9-36*b* the cables are replaced by supports, and the resulting system is indeterminate to the third degree. Removing the redundant reactions R_1, R_2, and R_3 as shown in Figure 9-36*c* yields a determinate system. However, the associated deformations Δ_1, Δ_2, and Δ_3 caused by removing the reactions R_1, R_2, and R_3 must be zero in the real structure. Essentially, the problem requires determining the values of R_1, R_2, and R_3 to maintain compatibility, or $\Delta_n = 0$. These reactions are determined by considering the effect of individual unit reactions on the system. This is shown in Figure 9-36*d* where $R_1 = 1$ induces deflections δ_{11}, δ_{21}, and δ_{31}. Likewise, $R_2 = 1$ and $R_3 = 1$ are applied separately at points 2 and 3, respectively, as shown in Figures 9-36*e* and *f*. The summation of the combined deformations gives

$$R_1\delta_{11} + R_2\delta_{12} + R_3\delta_{13} = \Delta_1 \tag{9-23a}$$

$$R_1\delta_{21} + R_2\delta_{22} + R_3\delta_{23} = \Delta_2 \tag{9-23b}$$

$$R_1\delta_{31} + R_2\delta_{32} + R_3\delta_{33} = \Delta_3 \tag{9-23c}$$

Effect of Elastic Cables at Supports The same double-cable system is shown in Figure 9-37*a*. The effect of the cables on the displacement at supports 1 and 3 is considered, assuming that the tower does not rotate or shorten. Under these conditions, the cable will elongate and will accommodate a displacement Δ_{V1} as shown in Figure 9-37*b*. Assuming a small angle

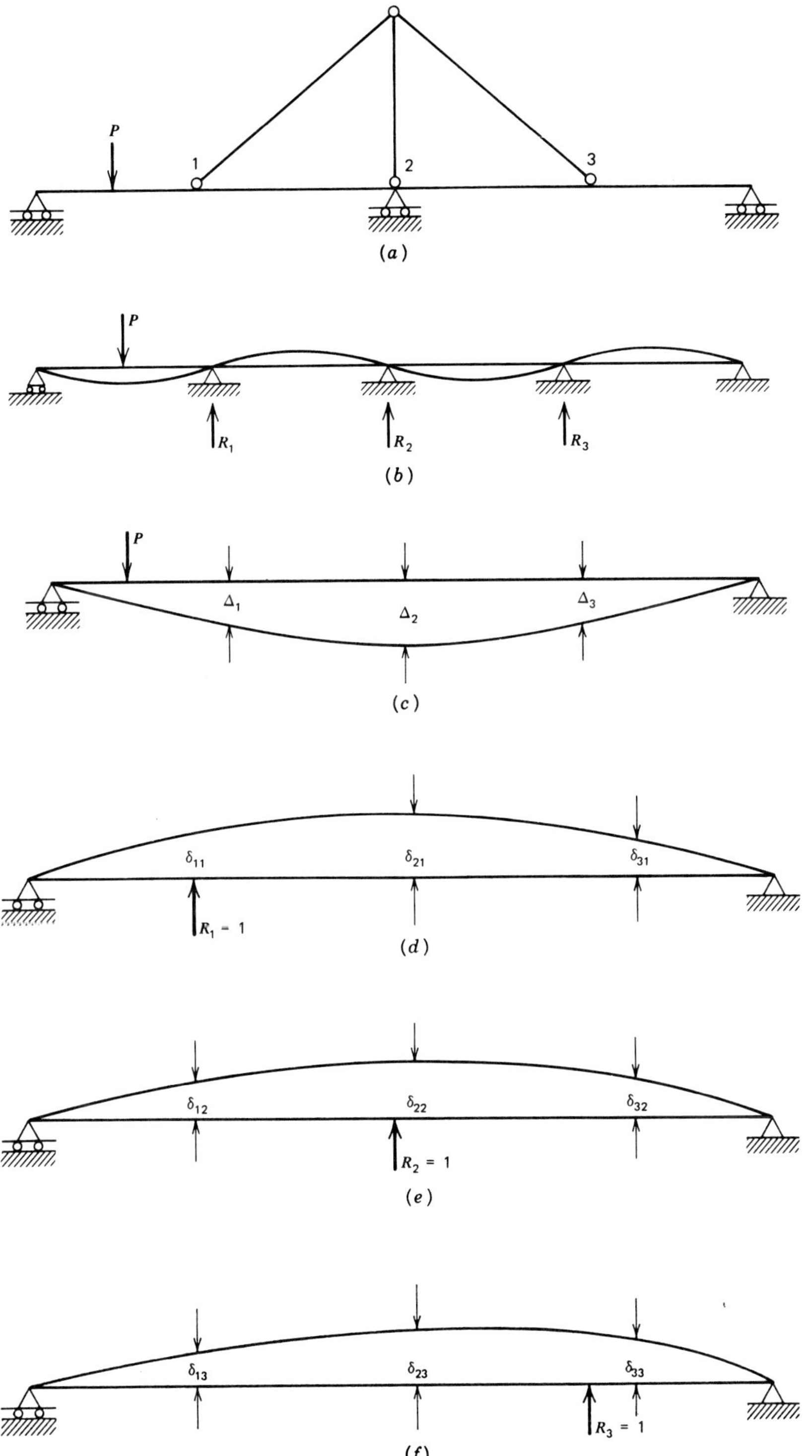

FIGURE 9-36 Example of double-cable systems: (*a*) original double-cable bridge; (*b*) equivalent continuous span; (*c*) simple span, *P* loading; (*d*) simple span, unit loading R_1; (*e*) simple span, unit loading R_2; (*f*) simple span, unit loading K_3.

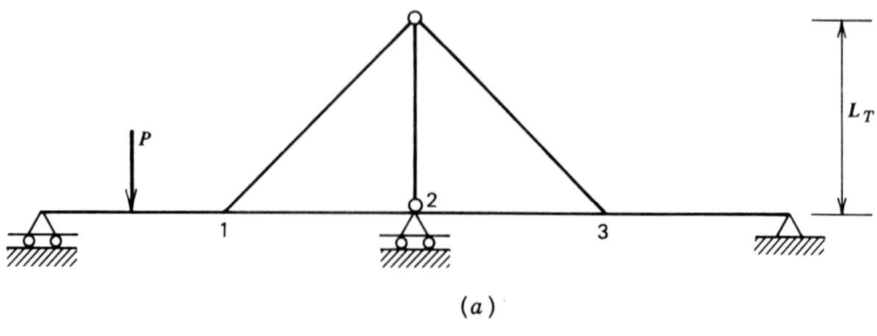

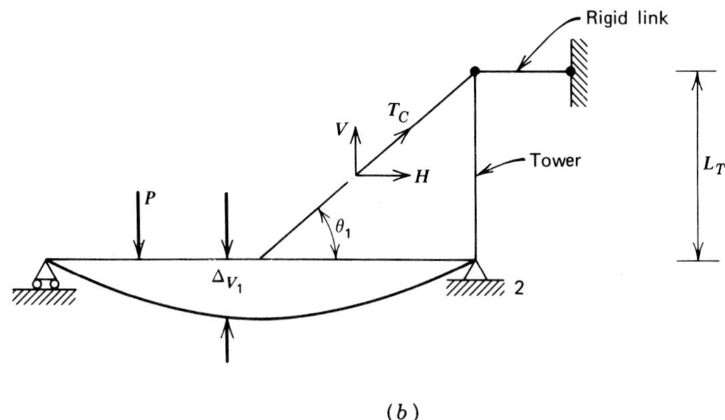

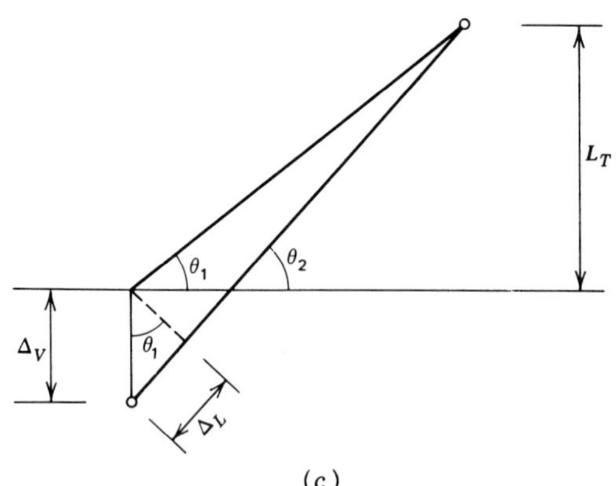

FIGURE 9-37 Double-cable bridge: (*a*) original system; (*b*) equivalent cable forces; (*c*) cable elongation.

change so that $\theta_1 = \theta_2$ still holds as shown in Figure 9-37*c*, we can write

$$\Delta_L = T_C L_C / A_C E_C \tag{9-24}$$

but $\sin\theta = \Delta_L/\Delta_V$, or $\Delta_V = \Delta_L/\sin\theta$, which gives

$$\Delta_V = T_C L_C / A_C E_C \sin\theta \tag{9-25}$$

For a force in the cable, T_C, the vertical component is $V = T_C \sin\theta$, and is equivalent to the reaction R_1. Equation (9-25) is rewritten as

$$\Delta_{V_1} = R_1 L_C / A_C E_C \sin^2\theta \tag{9-26}$$

This displacement coefficient times the redundant force at support point 1 is subtracted from the resultant displacement Δ_1 at support 1. Similar conditions are considered for the cable at support 3, so that (9-23*a*) through (9-23*c*) are modified as follows:

$$R_1\delta_{11} + R_2\delta_{12} + R_3\delta_{13} = \Delta_1 - \Delta v_1 \tag{9-27a}$$

$$R_1\delta_{21} + R_2\delta_{22} + R_3\delta_{23} = \Delta_2 \tag{9-27b}$$

$$R_1\delta_{31} + R_2\delta_{32} + R_3\delta_{33} = \Delta_3 - \Delta v_3 \tag{9-27c}$$

Tower Shortening The vertical component of the cable force causes the tower to shorten as shown in Figure 9-38. If these components are R_1 and R_3

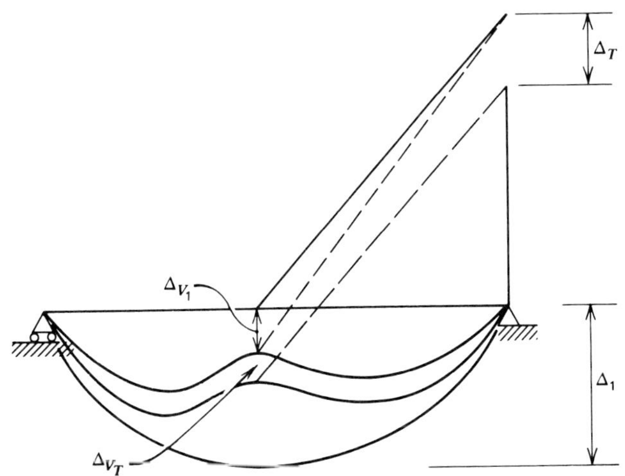

FIGURE 9-38 Tower shortening and effects on bridge girder.

for the cables connected at nodes 1 and 3, respectively, the total shortening is

$$\Delta_{V_T} = (R_1 + R_3)\frac{L_T}{A_T E_T} \tag{9-28}$$

where A_T and E_T denote the cross-sectional area and elastic modulus, respectively, for the tower.

This effect is expressed in the following equations:

$$R_1\delta_{11} + R_2\delta_{12} + R_3\delta_{13} = \Delta_1 - \Delta v_1 - \Delta v_T \tag{9-29a}$$

$$R_1\delta_{21} + R_2\delta_{22} + R_3\delta_{23} = \Delta_2 \tag{9-29b}$$

$$R_1\delta_{31} + R_2\delta_{32} + R_3\delta_{33} = \Delta_3 - \Delta v_3 - \Delta v_T \tag{9-29c}$$

Tower Rotation For a loading that is not applied symmetrically about the tower, the latter rotates as shown in Figure 9-39 causing a vertical displacement Δ_ϕ in the girder expressed as

$$\Delta_\phi = \phi h_1 \tag{9-30}$$

This effect is added to (9-29) to give

$$R_1\delta_{11} + R_2\delta_{12} + R_3\delta_{13} = \Delta_1 - \Delta v_1 - \Delta v_T - \phi h_1 \tag{9-31a}$$

$$R_1\delta_{21} + R_2\delta_{22} + R_3\delta_{33} = \Delta_2 \tag{9-31b}$$

$$R_1\delta_{31} + R_2\delta_{32} + R_3\delta_{33} = \Delta_3 - \Delta v_3 - \Delta v_T + \phi h_3 \tag{9-31c}$$

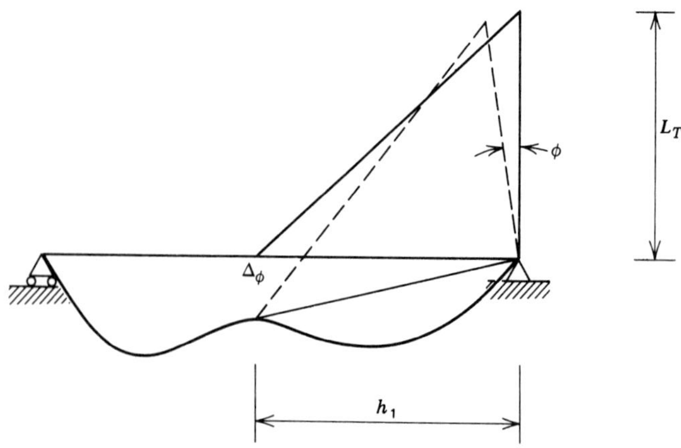

FIGURE 9-39 Tower rotation and effects on bridge girder.

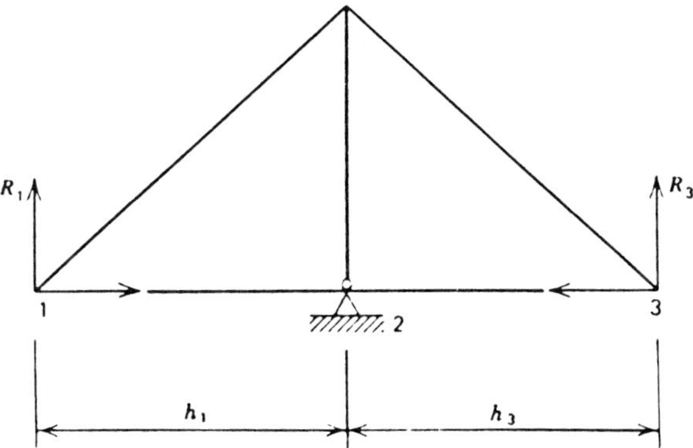

FIGURE 9-40 Cable reactions and equilibrium of forces.

Combined Effect Referring to Figure 9-40 and summing moments, we obtain

$$R_1 h_1 = R_3 h_3 \tag{9-32}$$

The foregoing data are combined in rewriting (9-31) as follows:

$$\begin{aligned} R_1\delta_{11} + R_2\delta_{12} + R_3\delta_{13} + \delta c_1 R_1 + \delta_T(R_1 + R_3) + \phi h_1 &= \Delta_1 \\ R_1\delta_{21} + R_2\delta_{22} + R_3\delta_{23} &= \Delta_2 \\ R_1\delta_{31} + R_2\delta_{32} + R_3\delta_{33} + \delta c_3 R_3 + \delta_T(R_1 + R_3) - \phi h_3 &= \Delta_3 \\ R_1 h_1 - R_3 h_3 0 &= 0 \end{aligned} \tag{9-33}$$

Collecting terms and expressing the foregoing relationships in matrix form gives

$$\begin{bmatrix} (\delta_{11} + \delta c_1 + \delta_T) & \delta_{12} & (\delta_{13} + \delta_T) & h_1 \\ \delta_{21} & \delta_{22} & \delta_{23} & 0 \\ (\delta_{13} + \delta_T) & \delta_{32} & (\delta_{33} + \delta c_3 + \delta_T) & -h_3 \\ h_1 & 0 & -h_3 & 0 \end{bmatrix} \begin{bmatrix} R_1 \\ R_2 \\ R_3 \\ \phi \end{bmatrix} = \begin{bmatrix} \Delta_1 \\ \Delta_2 \\ \Delta_3 \\ 0 \end{bmatrix} \tag{9-34}$$

which is readily solved to obtain the redundant reactions R_1, R_2, R_3, and ϕ as long as all unit displacement terms are known. The column matrix Δ values are also known because they represent the node point displacements for any loads with no restraints present. For a multiple-cable bridge, a matrix similar to (9-34) is developed for each cable (Lazar, 1972; Tang, 1971; Smith, 1967). If the tower is fixed at the base, one additional redundant reaction is present and must be included.

A more detail analysis is presented in several papers and texts (Lazar, Troitsky, and Douglass, 1972; Podolny and Scalzi, 1986), and formalized computer programs are available for more refined solutions. The finite-element approach has also been used.

A design aspect is the control of dead load moments and axial forces by posttensioning of the cables after the girder is erected. This is feasible with one or two cables. With several cables present, however, as one cable is tensioned, it changes the tensioning level in others. Thus, all cables can be pretensioned to a desired level as the structure is built. As the cables are stressed during erection, the girder accepts compressive forces; these can be calculated from statics knowing the prestress.

The value of E_C in (9-25) may be substituted by the equivalent modulus of elasticity discussed in the foregoing sections.

Mixed Method of Analysis: Double Plane

The main difference between single-plane and double-plane cable-stayed bridges is in the structural response. In the single-plane configuration, only vertical reactions of the girder are resisted by the cables, whereas torsional effects are transmitted to the piers through the stiff deck. With a double-cable system, the cables restrain both axial and torsional forces. In this case a three-dimensional analysis is mandatory and must take into account the deflections of the two cable planes in conjunction with the bending and rotation of the girder.

The associated analysis is basically similar to the single-plane structure. Thus, all interior rigid and elastic supports of the girder are released, and a set of deflection compatibility equations is developed for each node and for the stability of the tower. In addition, the double-plane structure requires the inclusion of torque effects, whereby the torque at each interior support is released and additional compatibility equations are written. For the analytical solution, see Podolny and Scalzi (1986) and the references at the end of this chapter.

Nonlinearity

Nonlinear behavior, mentioned briefly at the beginning of this section, may be exhibited in three parts of the bridge, namely the girder, tower, and cables. Nonlinearity effects are present in the girder and the tower when subjected to compressive loads and moments simultaneously. Their extent is largely determined by the relative magnitude of the compressive load compared with the critical Euler value and by the magnitude of the deflection caused by bending. As mentioned previously, these effects are normally small, but for slender girders and pylons they should be analytically ascertained for extreme loading conditions.

Nonlinearity in the cables occurs as the load increases and the cable sag decreases producing an increase in cable chord length with an associated elongation of the cable. From a practical standpoint, the nonlinear effects for the cable may be ignored if an equivalent E is used. When all the details of the design are articulated and expressed in final mathematical form, a nonlinear analysis may be carried out using a preselected overload.

Nonlinear analysis is suggested by the ASCE Task Committee on Cable-Suspended Structures (1977) to study and determine the effects arising from cable sag and girder–pylon deformations caused by combined moments and axial forces.

Typical Influence Lines

Homberg (1955) has presented a series of influence lines for girder moments and cable forces for several geometric configurations shown in Figure 9-41. Note that the notation is somewhat different from that used in Figure 9-32. Thus, in this example the configuration fan is actually the radiating type shown in Figure 9-32.

Influence lines for the geometries shown in Figure 9-41 are presented by Podolny and Scalzi (1986). In their derivation, axial strains in the girder and the tower are disregarded, and the assumption is made that the tower connections are such that tower bending moments are not present.

The area of one cable is expressed by the dimensionless parameter $E_C A_C L_T^2 / E_G I_G$, where E_C is the elastic modulus for the cable, A_C is the cross-sectional area of the cable (constant for all cables), E_G is the elastic modulus for the girder, I_G is the second moment of area of the girder cross section, and L_T is the overall length. Values are tabulated for $E_C A_C L_T^2 / E_G I_G = 250$, 1000, 4000, and 16,000. Practical values are in the range of 2000 to 10,000.

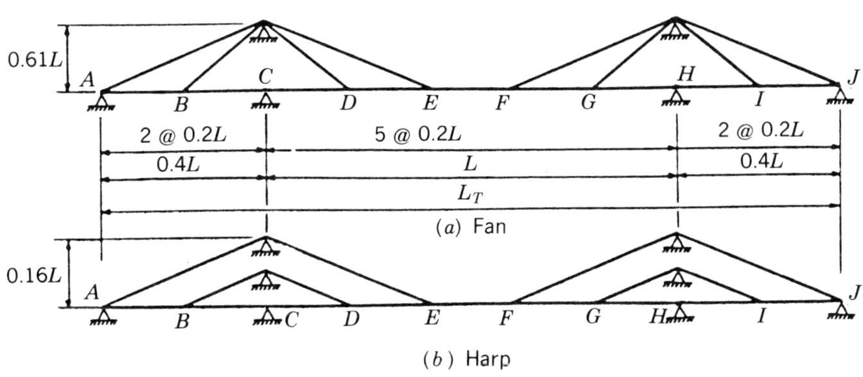

FIGURE 9-41 Geometries used in influence line analysis.

Methods of Analysis A suggested method for obtaining influence lines (not necessarily the simplest with regard to computation time) is as follows.

1. Form a statically determinate base structure by removing the supports C and H and cutting the cables at B, D, E, F, G, and I for the fan design, and at D, E, F, and G for the harp geometry. The vertical components of the forces at these points are represented by the redundants. There are eight such forces for the fan and six for the harp.

2. Compute the deflections corresponding to each of the unit redundants. The flexibility matrix contains terms from the girder deflections plus terms from the cable extensions. Follow the steps discussed in the preceding sections.

3. The influence line for the bending moment is identical to the deflected shape corresponding to a unit angular rotation across an element at the point considered.

4. Compute the initial deflections of the base structure. Insert these values as the constant terms in the system of equations for the redundant force, the coefficients of the unknowns being the flexibility matrix computed previously. Solve for these forces.

5. Apply these forces to the girder alone and compute its deflected shape. Add this to the initial deflected shape to obtain the required influence line.

6. The influence line for the vertical component of a cable force corresponds to a unit relative vertical displacement across the cut in the cable at that point. This unit displacement is introduced in a set of simultaneous equations in the redundant forces, applied to the girder alone. The resulting deflected shape is the influence line for the redundant. The influence line for the cable force is obtained by multiplying each ordinate by $1/\sin\theta$, where θ is the cable slope from the horizontal.

Comparison of Arrangements The influence lines obtained from this procedure may be used for preliminary designs. Recall, however, that the notation conforms to Figure 9-41 (the arrangement denoted as a fan is the same as the one shown as radiating in Figure 9-32). In addition, the following comparison is useful.

1. The harp arrangement has larger bending moment ordinates and smaller cable force ordinates than the fan. This results from the less direct load transfer to the abutments and towers. Thus, the fan may generally be preferred. However, the difference between the two forms is not articulated.

2. For the harp, the influence lines for the bending moments at the cable connections nearest to the towers are only little affected by cable stiffness,

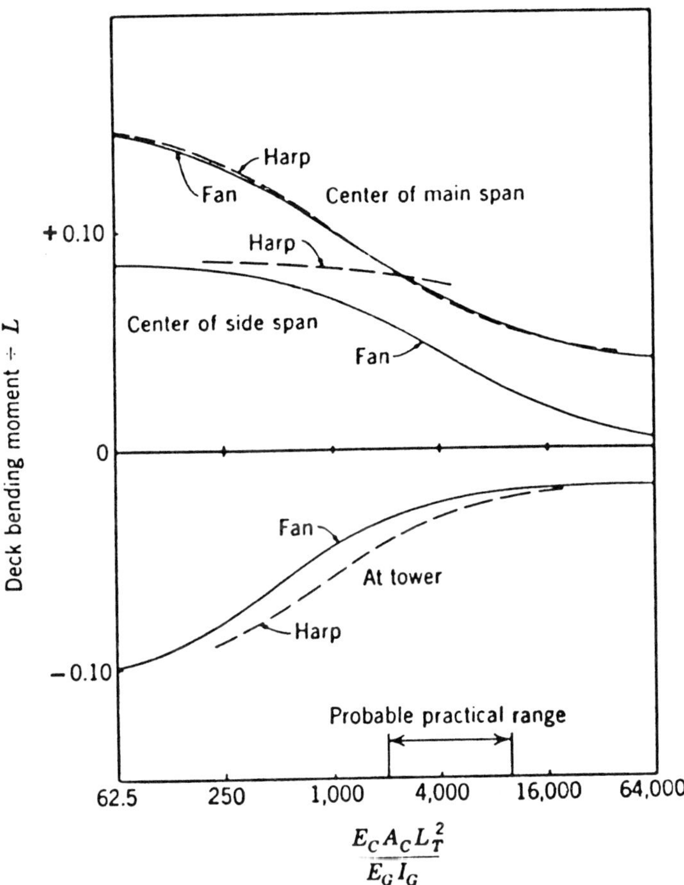

FIGURE 9-42 Variation of deck bending moment with cable stiffness (unit moving load).

that is, these bending moments are hardly benefited by the introduction of cables at these points. With this exception, however, influence lines are markedly affected by the parameter $E_C A_C L_T^2/E_G I_G$.

3. The graphs of Figure 9-42 are actually plots of (a) the maximum positive bending moment at the center of the main span, (b) the maximum positive bending moment at the center of the side span, and (c) the maximum negative bending moment at the tower, for a moving load. Note that the cable stiffness has a small effect on (b) for the harp but a large effect on the other bending moments for either configuration. This influence extends well beyond the normal working range. Note also that the analysis would be grossly inaccurate if cable elongation were ignored.

4. The results of Figure 9-42 also show that for the harp the bending moment at the center of the side spans may become the absolute maximum bending moment.

5. Interestingly, the maximum bending moment at the center of the main span is practically the same for the fan and the harp, and the two curves almost coincide.

6. The plots presented in Figure 9-43 are curves of maximum moment versus location on the girder for a moving unit load. In general, the negative bending moments are smaller than the positive moments. This suggests the benefit of prestressing the main span upwards. For the harp, the graphs also indicate the relatively high bending moments in the side spans near the towers.

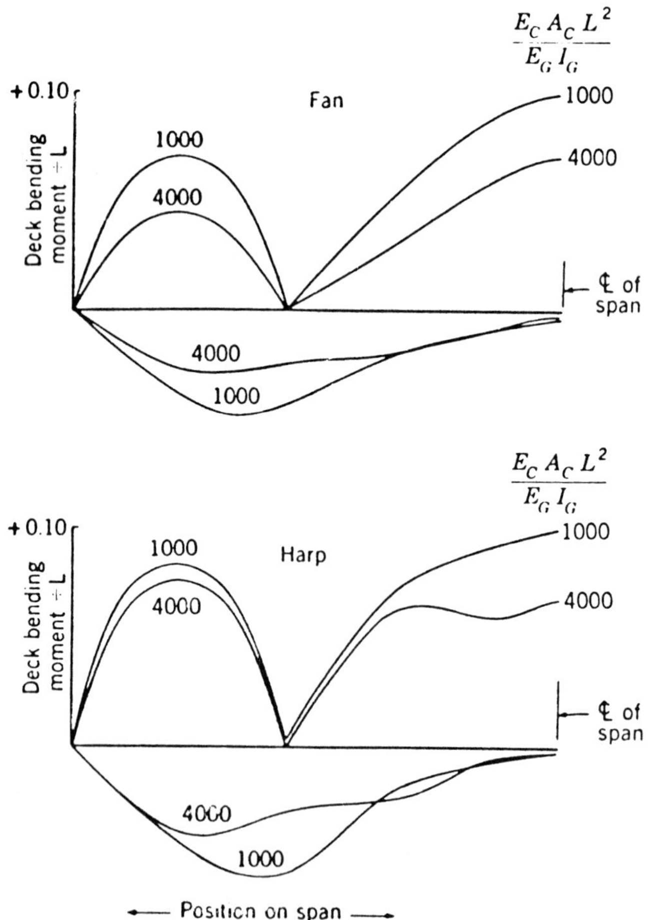

FIGURE 9-43 Curves of maximum deck moment (unit moving load).

7. The influence lines for cable force have negative ordinates for both the fan and the harp. Resultant compressive forces are avoided, however, by moderate dead load tensions. In this respect the fan and the harp behave similarly.

8. A better distribution of design bending moments may be possible for the fan with larger side spans and a smaller gap between the central cables, for example, spans of $0.44L$, L, and $0.44L$ and a central gap of $0.11L$.

9. In these comparisons, equal cable areas are assumed for the harp and the fan arrangements. However, the harp would be expected to have smaller cable forces and therefore smaller sections of cables. As a result, the design bending moments for the harp should be greater than those used in this analysis.

9-17 DESIGN REQUIREMENTS OF CABLE-STAYED BRIDGES

The objectives of the design focus on three main parts of the bridge: (a) the cross section (bridge deck system and location of towers and cables in the cross section); (b) the cable system (type, spacing of girder–cable connections, and cable slopes); and (c) the towers (structural forms, connections to cables, and connections to deck and piers.

Cross Section A single system of towers and cables occupying a vertical plane at the centerline of the deck is generally favored, but is only feasible if the roadway can be provided with a wide median strip. A single-line support results, however, in large torsional moments in the deck, necessitating the use of a deck with at least one torsional element.

For bridges without a median strip or with a large width—span ratio cables may be placed on both sides of the deck. Their plane may be vertical or inclined. The anchorages of these cables may be placed in a tubular girder at each side of the deck.

In either the single- or double-cable plane, the cable forces induce large compressive forces at the deck level, and provisions must be made to resist them, probably by designing the deck as part of the girder system. The use of cables reduces the girder size considerably compared to a simple similar span.

Cable Systems It appears desirable to have a cable system that provides a direct path to the external reactions. Such a system is the fan or the externally anchored harp.

The structural advantages of multiple-cable systems have been demonstrated in practice (Felge, 1966; Thul, 1966). The underlying philosophy of this arrangement is not so much the provision of additional ropes, but rather the distribution of ropes along the span, instead of concentrating the reactions in one or two large cables. The use of smaller cables facilitates their anchoring to the deck. In this case, however, it is difficult to use the fan arrangement with many separate ropes radiating at different angles from a common intersection point of the tower.

The minimum cable slope fixes the tower height. Statistically, minimum slopes (expressed by the tangent of the angle with the horizontal) range from 0.30 to 0.50 with a mean of 0.40. In some countries, for example Germany, there is a tendency toward the use of larger slopes.

Towers Several tower configurations are shown in Figure 9-33. Other terms used to describe tower geometries include (a) single column or pylon, (b) pair of separate columns or double pylon, (c) A frame, and (d) portal frame.

The behavior of the column is related to the details of its connections to the cables and pier. These details should be designed to minimize bending moments in the towers. Alternatively, a fixed base, either to the deck or to the pier, may be indicated to avoid heavy pinned bearings.

9-18 MULTISPAN STAYED BRIDGES

Most cable-stayed bridges have been constructed with two or three spans, because in most navigable waterways only one main span is needed. In these structures the horizontal displacement of the pylon top is controlled by a fixed cable, that is, an upper side span cable anchored to the stiffening girder at the end support.

In stayed bridges with several equal main spans, the inner cable systems do not contain inclined fixed cables, and thus an efficient horizontal fixing of the pylon top is not feasible. With one-sided loading with reference to the inner pylons, the result is rotation of the inner cable systems, inducing considerable vertical deflection of the stiffening girder as shown in Figure 9-44.

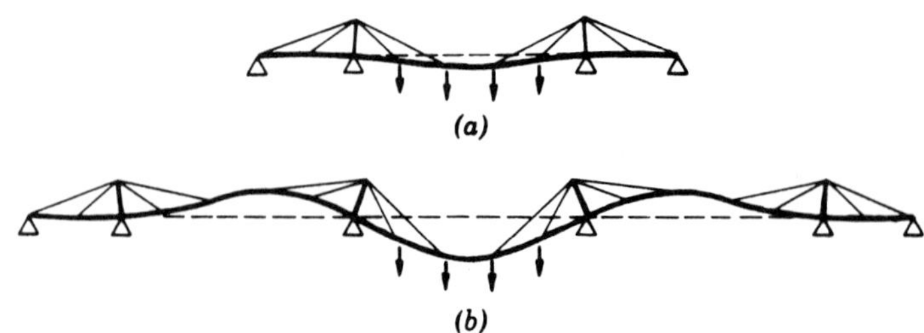

FIGURE 9-44 Deflection of stayed girder bridge with: (*a*) one main span; (*b*) three main spans.

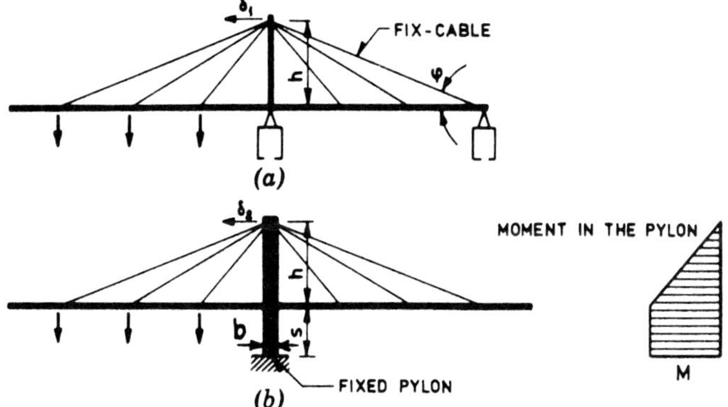

FIGURE 9-45 Fan-shaped cable systems with one-sided loading: (*a*) cable system with fixed-cable; (*b*) cable system connected to fixed pylon.

Several solutions have been studied and analyzed for structural requirements and their associated costs (Gimsing, 1976). A brief review is presented in this section.

Column Pylons Fixed to Pier A pylon fixed to the pier has much less horizontal displacement, but requires large cross-sectional dimensions to limit displacement to the value obtained with a fixed-cable system. Two equal-fan systems with one-sided loading are shown in Figure 9-45. The upper system has a vertical support at the right and therefore contains a fixed cable, whereas the lower system is connected to a moment-rigid pylon assumed to have a constant section.

The horizontal movement of the pylon top is expressed as

$$\delta_1 = \frac{h}{\sin \phi \cos \phi} \frac{\sigma_{fc}}{E_c} \tag{9-35a}$$

$$\delta_2 = \frac{2h^2 + 6sh + 3s^2}{3b} \frac{\sigma_{pb}}{E_p} \tag{9-35b}$$

where σ_{fc} is the tensile stress in the fixed cable, E_c is the modulus of elasticity of the cable, σ_{pb} is the bending stress in the lower part of the pylon, and E_p is the modulus of elasticity of the pylon. Other parameters are as shown in Figure 9-45.

Equal stiffness under one-sided load requires $\delta_1 = \delta_2$. Selecting the following practical values $s = h/2$, $\phi = 20°$, $E_c\sigma_{pb}/E_p\sigma_{fc} = 0.5$, the condition $\delta_1 = \delta_2$ yields $b = 0.31h$. This value of b is in practice unattainable. For example, a bridge with 1000-ft main spans is likely to have pylon heights $h = 170$ ft. For a moment-rigid pylon, this bridge would require a width

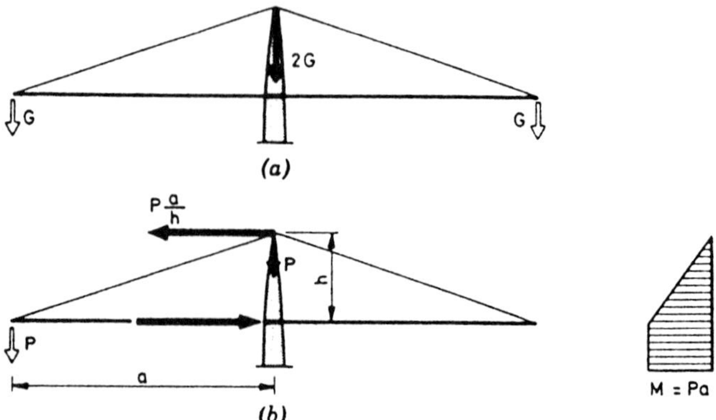

FIGURE 9-46 Forces acting on fixed pylon: (*a*) symmetric load; (*b*) one-sided load.

$b = 50$ ft to give the same stiffness effect available by the fixed cable. By comparison, the same 170-ft pylon in a two-span or three-span bridge usually has a width of 7 to 10 ft.

Triangular Pylon Structures An A-shaped frame can be used to achieve sufficient horizontal stiffness in a fixed pylon, with reasonable member dimensions. Figure 9-46*a* shows the vertical force induced in the pylon top by a symmetrical load. A one-sided load induces a vertical and horizontal force at the pylon top and a horizontal force at the stiffening girder. The horizontal force is Pa/h giving a bending moment of Pa at the girder level. Based on these moments, the pylon configurations shown in Figure 9-47 appear to be good solutions. In this case the moments acting on the pylon structure are converted into axial forces in individual members. At the same time, flexural deformations of the vertical members provide a certain horizontal flexibility of the tower part, and this allows the application of continuous stiffening girders as a practical concept.

The two isometric projections shown in Figure 9-48 are pylons with the tops longitudinally fixed. In the structure on the left, the upper part consists

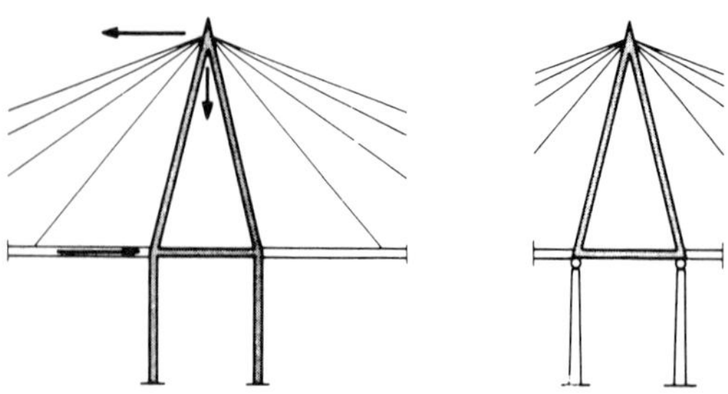

FIGURE 9-47 Triangular pylong structures.

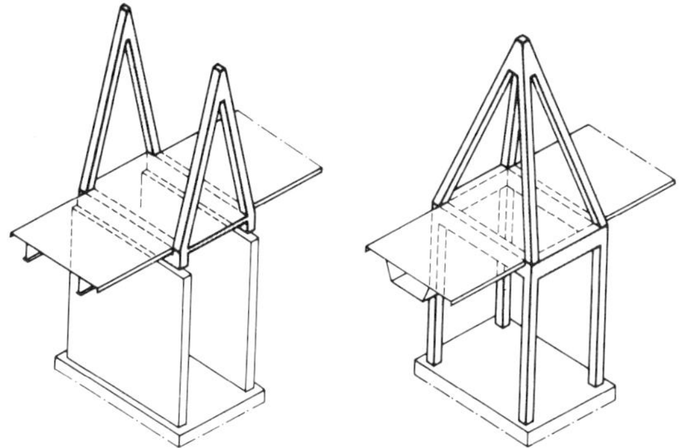

FIGURE 9-48 Isometric projections of triangular pylon structures.

of two freestanding triangular pylons and the lower part, consists of two wall-shaped pier shafts. Heavy transverse girders placed between the pylon legs under the bridge deck ensure the lateral stability. The pylon structure shown on the right of Figure 9-48 has a pyramidal upper part, whereas four legs comprise the lower part. In this manner the pylon top is fixed in all directions. Horizontal legs are placed at the bottom of the pyramid in both the longitudinal and transverse directions.

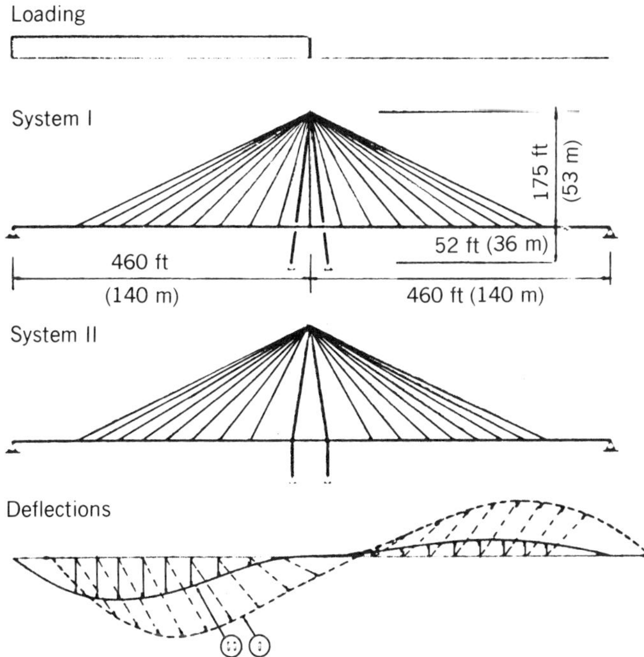

FIGURE 9-49 Deflections of two different stayed girder bridges under one-sided loading.

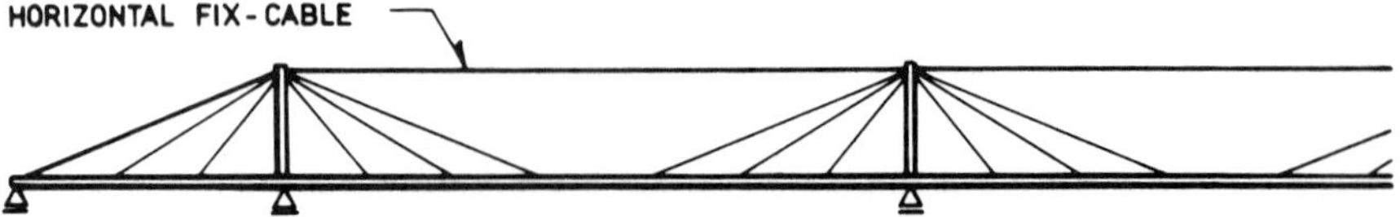

FIGURE 9-50 Stayed girder bridge with horizontal fixed cable between pylon tops.

The importance of a connection transmitting horizontal forces from the stiffening girder to the pylon structure is demonstrated in the two structural systems shown in Figure 9-49. System I has no direct connection between the stiffening girder and the pylon, whereas in system II such a connection is provided and allows the transmission of horizontal forces. In both systems the ends of the stiffening girder are supported on longitudinally movable bearings.

Under a symmetrical load such as dead load, both systems have identical deflections. For one-sided load, however, their behavior is markedly different. Thus, in system I the stiffening girder moves longitudinally toward the unloaded span, and this reduces the action of the stays. The result is increased vertical deflections. In system II the longitudinally fixed stiffening girder inhibits changes in geometry, and this reduces the vertical deflections almost by one-half. The bending moments in the girder and the variation of cable forces in this system are structurally more favorable.

Horizontal Fixed Cable Horizontal movement of intermediate pylons in multispan bridges may also be reduced with the use of a horizontal fixed cable between pylon tops as shown in Figure 9-50. With this connection activated, the pylons are designed as vertical columns with a small flexural stiffness or even a hinged connection at the piers, because no primary moments are to be transmitted.

The free length of the horizontal fixed cable is almost twice the length of the longest cables in the system. The result may be an unfavorable effect on the axial stiffness of the fixed cable because of the larger sag. The reduction of stiffness due to the sag variation may be expressed in terms of the equivalent modulus of elasticity. From analytical considerations, Gimsing (1976) concludes that for a cable length of 1500 ft with E_{eq} = 25,000 ksi, the sag effect will reduce the axial stiffness of a horizontal fixed cable by about 15 percent, which is within the design limits.

Gimsing (1976) has also considered the stresses resulting from temperature variations if a horizontal fixed cable is made continuous between ends. For a typical cable-stayed bridge, stresses induced in this manner are of the order of 5 ksi and therefore nominal.

Deflection of Multispan Bridges The stiffness properties of multispan bridges have been investigated in a parametric study, and the results are shown in Figure 9-51. The deflections of bridges with three 360-m (1180-ft)

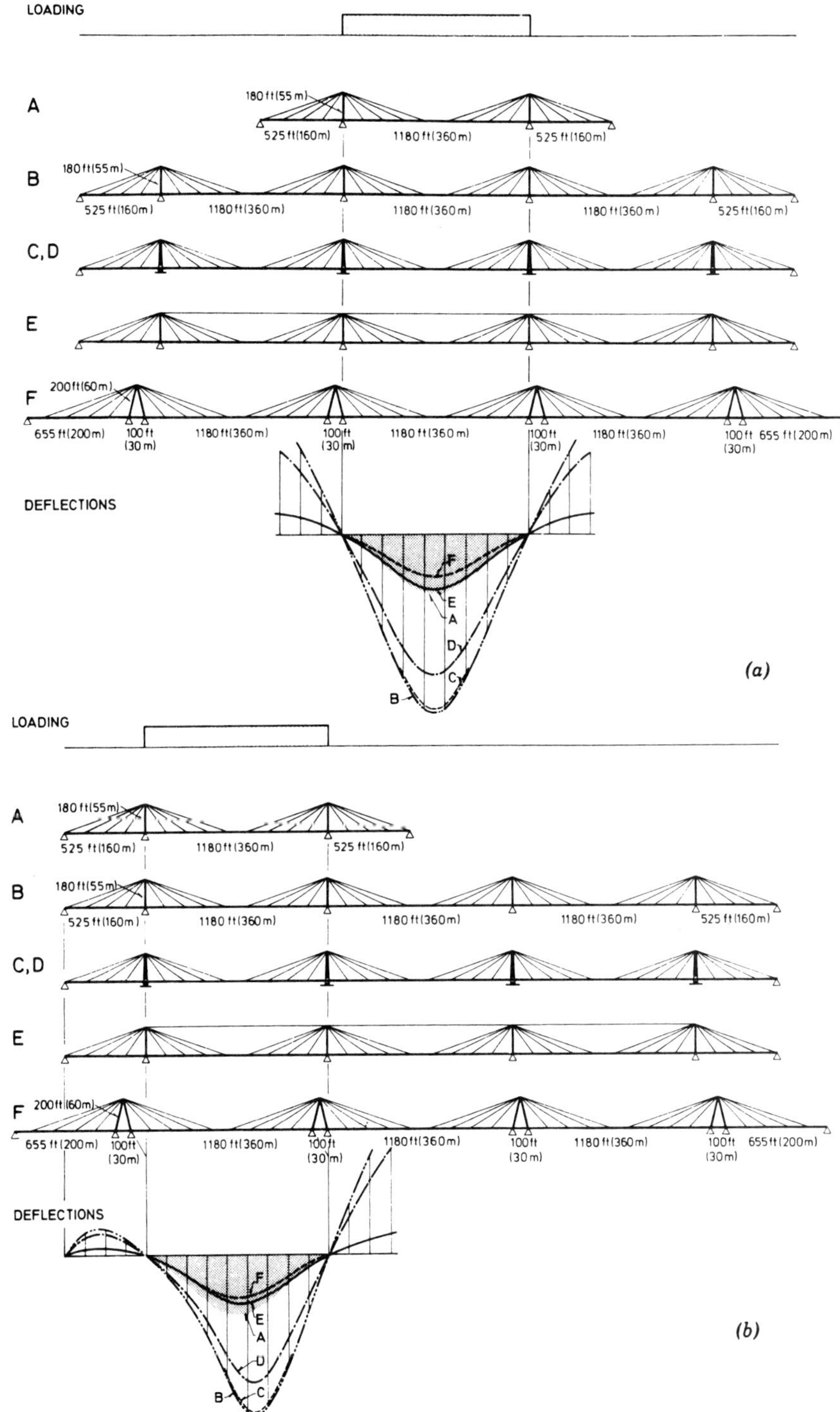

FIGURE 9-51 Relative deflections of stayed girder bridges with different structural systems and loading in: (*a*) central main span only; (*b*) one outer main span.

TABLE 9-5 Cross-Sectional Area and Moment of Inertia for Stiffening Girder and Pylons of Systems A–F (Figure 9-51)

	Stiffening Girder		Pylon	
System (1)	Area, A in ft^2 (m^2) (2)	Moment of Inertia, I_g, in ft^4 (m^4) (3)	Area, A_g in ft^2 (m^2) (4)	Moment of Inertia, I_p, in ft^4 (m^4) (5)
A	10.8 (1.0)	278 (2.4)	4.3 (0.4)	29 (0.25)
B	10.8 (1.0)	278 (2.4)	4.3 (0.4)	29 (0.25)
C	10.8 (1.0)	278 (2.4)	4.3 (0.4)	29 (0.25)
D	10.8 (1.0)	278 (2.4)	6.5 (0.6)	290 (2.50)
E	10.8 (1.0)	278 (2.4)	4.3 (0.4)	29 (0.25)
F	10.8 (1.0)	278 (2.4)	7.5 (0.7)	52 (0.45)

main spans are compared to those of a bridge with only one 360-m main span. The section properties of the stiffening girder and the pylons are given in Table 9-5.

System A is a conventional three-span bridge with fan (radiating) cables and hinge connections between the superstructure and substructure. The structure is proportional for live load according to European standards. The steel yield strength is 50 ksi for the stiffening girder and the pylons, and 220 ksi for the cable wires.

System B is a five-span continuous bridge, corresponding closely to system A in terms of the size of the members and the types of connections between the superstructure and substructure.

Systems C and D have fixed connections between the pylon and the substructure. The pylons are the conventional column type. System C has pylon section properties as in systems A and B, but in system D the pylon has a moment of inertia 10 times larger.

System E incorporates the horizontal fixed cable between pylon tops, and is otherwise similar to system B.

System F has four triangular pylon structures supported on double bearings.

The deflections of the systems under a uniformly applied load are shown at the bottom of Figures 9-51*a* and *b* for the load acting in the span indicated. The deflections in the conventional three-span bridge (system A) are essentially close to the allowable for this structure type. The following conclusions are drawn.

1. Providing base fixity in pylons with common flexural stiffness does not reduce deflections to the extent desired (this is evident by comparing systems B and C).
2. Likewise, base fixity combined with increased stiffness of the pylons by a factor of 10 does not reduce deflections sufficiently (this is evident by comparing systems B and D).
3. The horizontal fixed cable between the pylon tops reduces the deflections of the multispan bridges essentially to acceptable values (system E).
4. Triangular pylon structures supported on double bearings are likewise effective in reducing deflections to allowable values (system F).

Referring to Figure 9-51, the smallest deflection is for system F, but the deflection curve is based on the assumption that the supports under the pylons are completely rigid in the vertical direction. This condition, however, deviates in real structures where the pylons are connected to pier shafts because axial deformations tend to increase the deflections of the stiffened girder.

System comparison is completed by considering the difference in quantities and materials required for each scheme. Relevant data are shown in Table 9-6 per unit length of the systems investigated. The last column in this table gives the relative equivalent quantity of structural steel for the different systems with system A being the reference system (relative quantity = 1.00). In addition, in comparing feasible solutions the erection requirements must also be considered because they influence the total construction cost.

TABLE 9-6 Relative Quantities per Unit Length for Systems in Figure 9-51

System (1)	Structural Steel (Stiffening Girder and Pylons) (2)	Cable Steel (Stays) (3)	Equivalent Relative Quantities of Structural Steel (4)
A	1.00	1.00	1.00
B	0.99	0.77	0.94
C	0.99	0.77	0.94
D	1.02	0.77	0.97
E	0.99	1.25	1.043
F	1.13	0.70	1.046

9-19 BUCKLING OF CABLE-STAYED BRIDGES

Buckling considerations are relevant to cable-stayed bridges because both the girder and towers may be subjected to large compressive forces. Overall buckling of this type may be analyzed by various methods. Tang (1976) presents an analysis where the buckling load is calculated using energy methods.

Analytical Procedure

The accuracy of the critical buckling load P_c depends on the accuracy of the assumed mode of buckling. For a complex structure such as a cable-stayed bridge, a reasonably good approximate mode shape is difficult to obtain. Unacceptable deviations may be avoided by grouping together several possible curves to form curves for the approximation by linear combination

$$W = A_1W_1 + A_2W_2 + \cdots = \sum A_iW_i \tag{9-36}$$

where the terms W_i denote deflections of the girder and towers, and the coefficients A_i can be determined according to the Ritz method (Tang, 1972). A characteristic equation for buckling is obtained as

$$C - \lambda B = 0 \tag{9-37}$$

where $\lambda = P_c$ is the eigenvalue. The elements in matrix C are

$$C_{i,j} = \int EIW_j''W_i''\,ds + \sum(\epsilon_i\epsilon_j E_C A_C I_C) \tag{9-38}$$

and the elements in B are

$$b_{ij} = \int \xi W_i'W_j'\,ds \tag{9-39}$$

where the variable $\xi = P_s/P_0$ expresses the distribution of the axial forces in the girder and towers. Each curve W_i must satisfy all geometric boundary conditions of the bridge, which is achieved by using actual deflection curves of the structure under various loading conditions. About 10 to 20 curves are sufficient for a reasonably good approximation.

Critical Load Because the axial force P_s varies along the girder and towers, a precise definition of the critical load is not possible. The value of P_c is obtained as an eigenvalue depending on the variable ξ. If the latter changes, that is, the axial forces in different portions of the bridge change in relation to each other, the value of P_c will also change. Thus, the critical load is not a single parameter but a group of loads in a fixed relationship, increasing or decreasing in the same proportions.

Tang (1976) proposes to predict the factor of safety against buckling for a given group of loadings than to estimate the buckling load. The factor of safety ν is the ratio

$$\nu = P_{cr}/P_{exist} \tag{9-40}$$

where P_{cr} and P_{exist} are the critical and existing loads, respectively.

Combining the foregoing equations gives

$$V = \frac{\int EIW''^2\,ds + \sum \epsilon_C^2 E_C A_C l_C}{P_0 \int \xi W'^2\,ds} \tag{9-41}$$

Design Examples

Four cable-stayed bridges are shown in Figure 9-52 through 9-55, with relevant data summarized in Table 9-7. About 12 to 18 deflection curves were used to approximate the buckling mode shape. The first and second modes correspond to the lowest and second-lowest critical loads, and are shown in the figures. For simplicity of calculations, all bridge girders and towers are assumed to have constant cross sections. In the first three examples, we assume that the cable forces are modified to eliminate bending moments in the towers, which eliminates the cambering of the towers.

The bridge in Figure 9-52 has a fan-type (radiating) cable configuration and is completely symmetrical except for the end supports (abutments). This unsymmetrical boundary condition will normally exist because rollers must be provided, except one hinged bearing, to accommodate length changes due to temperature and also to resist longitudinal forces. This causes the buckling mode to deviate somewhat from the symmetrical and asymmetrical shapes.

Figure 9-52*d* shows the first and second modes of the same girder under the same proportions of axial loads but without cables. The critical load has a very low value as expected and is only about 15 percent of the cable-stayed girder. For the harp configuration shown in Figure 9-53, the asymmetrical shape of the first mode is probably caused by the higher stiffness of the upper cables that are back-anchored at the side supports. This effect is more evident in the shape of the second mode.

The bridges in Figures 9-52 and 9-53 have the same tower height, girder length, cable spacing, section properties, and loadings that yield the cable forces. The only difference is the upper cable support and slope of the lower cables. Evidently, the fan type has a higher factor of safety.

The bridge shown in Figure 9-54 has a main span of 1500 ft, which is a very large span for a single-tower cable-stayed bridge. Consequently, all back-staying cables are anchored to the piers to stiffen the structure and to prevent excessive live load deflection of the main span.

TABLE 9-7 Summary of Data of Design Examples in Figures 9-52 through 9-55

Figure 9-52	Upper cables	$E = 3{,}600{,}000$ ksf, $A = 1.0$ ft^2
	Lower cables	$E = 3{,}600{,}000$ ksf, $A = 0.68$ ft^2
	Girder	$E = 4{,}320{,}000$ ksf, $I = 250$ ft^4, $A = 12$ ft^2
	Towers	$E = 4{,}320{,}000$ ksf, $I = 180$ ft^4, $A = 15$ ft^2
	Axial forces	
	Upper cables	$I = 13{,}400$ kips
	Lower cables	$I = 4950$ kips
	Girder	$CE = HJ = -15{,}500$ kips,
		$BC = EF = GH = JK = -12{,}000$ kips
	Towers	$P = -19{,}000$ kips
Figure 9-53	Section properties	Same as Figure 9-52
	Axial forces	
	Upper cables	$T = 13{,}400$ kips
	Lower cables	$T = 7800$ kips
	Girder	$CE = HJ = -19{,}000$ kips
		$BC = EF = GH = JK = -12{,}000$ kips, others $= 0$
	Towers	$MN = QR = -12{,}000$ kips, $NP = RS = -19{,}000$ kips
Figure 9-54	Cables	$E = 3{,}400{,}000$ ksf, $A = 1.60$ ft^2
	Girder	$E = 4{,}320{,}000$ ksf, $A = 60$ ft^2, $I = 1200$ ft^4
	Towers	LM, MN: $A = 30$ ft^2, $I = 60$ ft^4,
		NP, PQ: $A = 60$ ft^2, $I = 120$ ft^4
	Axial forces	
	Top cables	$I = 27{,}000$ kips
	Other cables	$I = 24{,}200$ kips
	Girder	$AB = 0.0$, $BC = -25{,}100$ kips, $CD = -47{,}600$ kips
		$DE = -70{,}100$ kips, $EF = -92{,}600$ kips
		$FG = -92{,}600$ kips, $GH = -70{,}100$ kips
		$HI = -47{,}600$ kips, $IJ = -25{,}100$ kips, $JK = 0$
	Tower	$LM = -30{,}120$ kips, $MN = -57{,}120$ kips
		$NP = -84{,}120$ kips, $PQ = -111{,}120$ kips
Figure 9-55	Cables	$E = 4{,}320{,}000$ ksf, $A = 0.0694$ ft^2
	Girder	$E = 617{,}000$ ksf, $A = 125$ ft^2, $I = 43.79$ ft^4
	Tower	$E = 617{,}000$ ksf, $A = 130$ ft^2, $I = 2000$ ft^4

Axial forces
Cable forces and axial forces in girder and towers are based on an uniform load of 24,375 kips per linear foot along the entire bridge girder.

Top cables	$I = 1210$ kips
Bottom cables	$I = 850$ kips
Tower	$P = -23{,}800$ kips
Girder	$P = -15{,}042$ kips

Conversion factors: 1 ft = 0.305 m, 1 kip = 453 kg, 1 ksf = 47.9 kN/m^2.

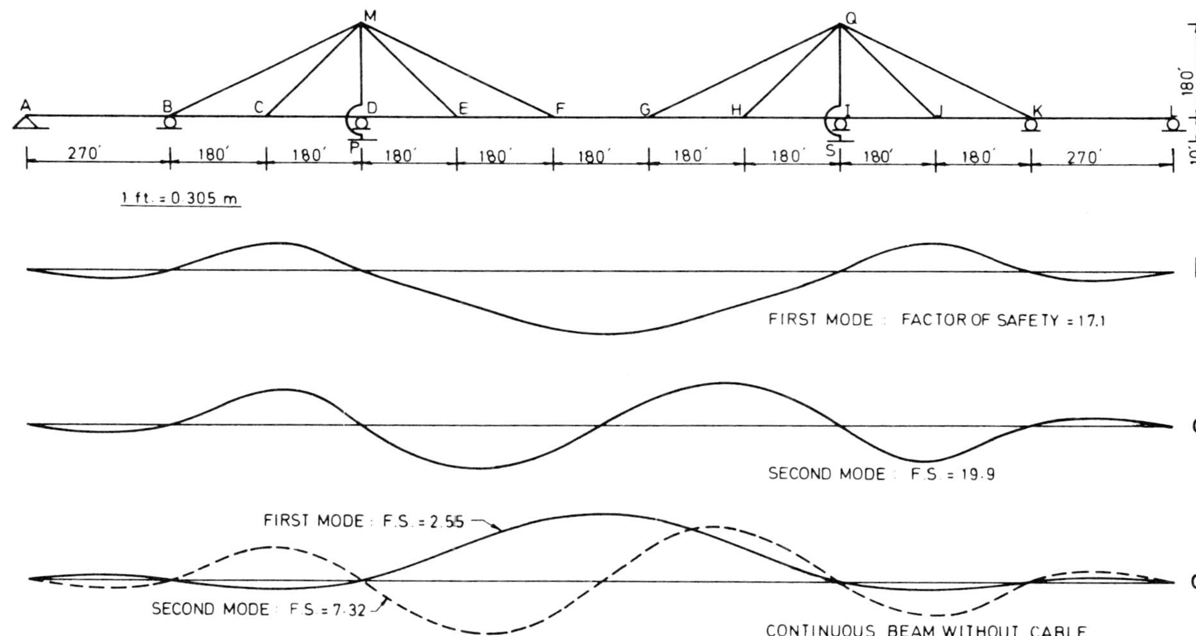

FIGURE 9-52 Cable-stayed bridge for design example.

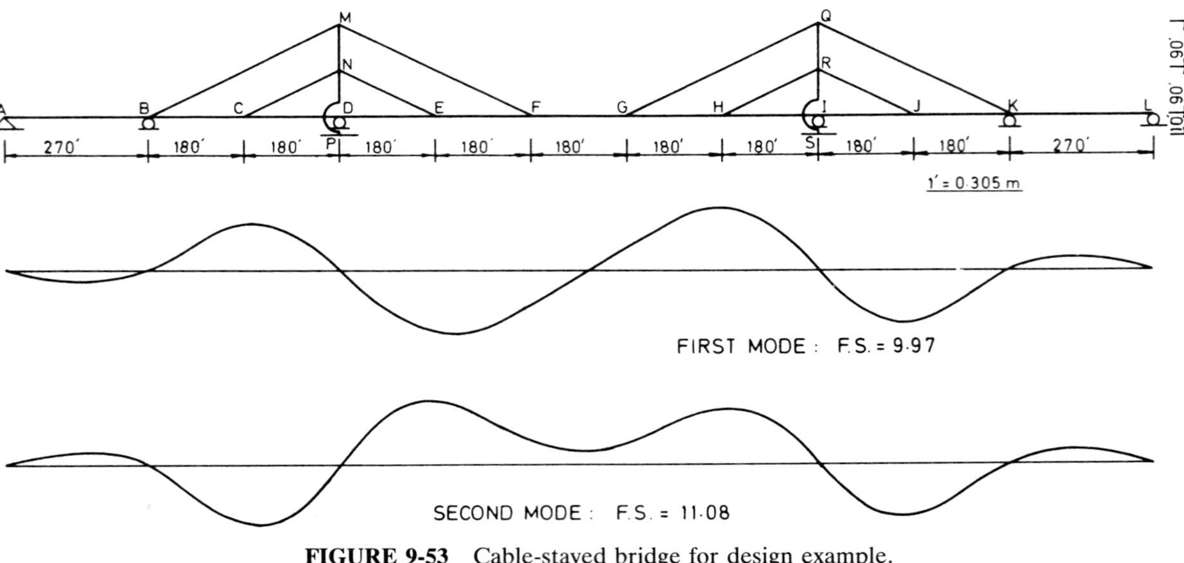

FIGURE 9-53 Cable-stayed bridge for design example.

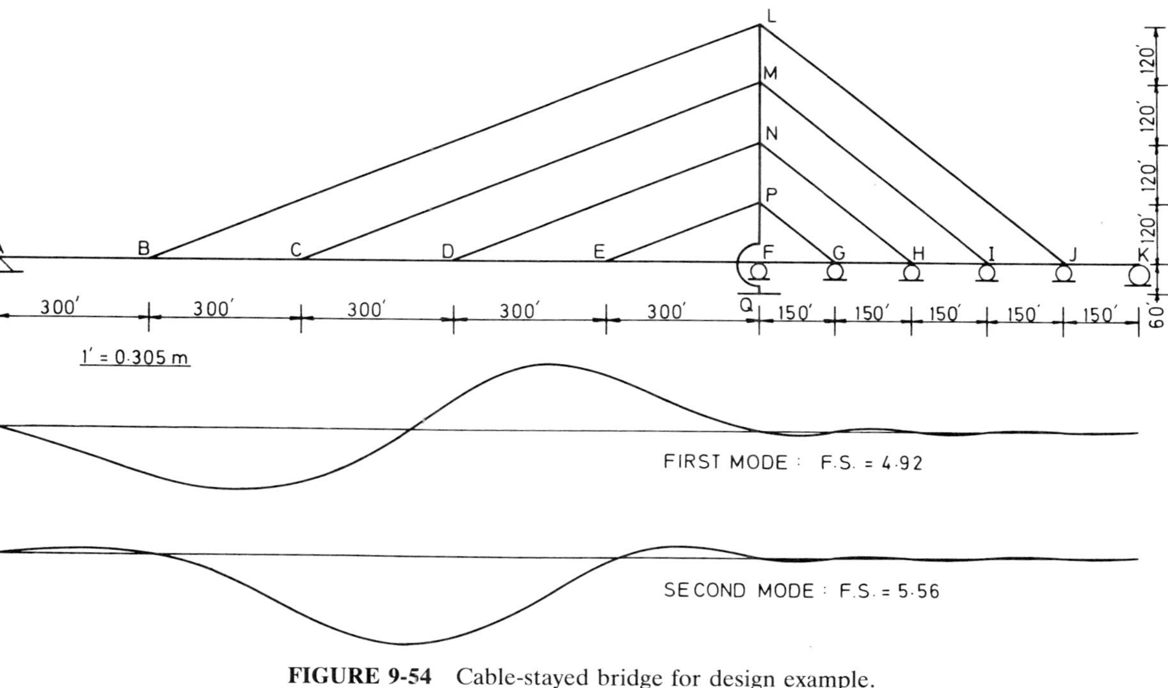

FIGURE 9-54 Cable-stayed bridge for design example.

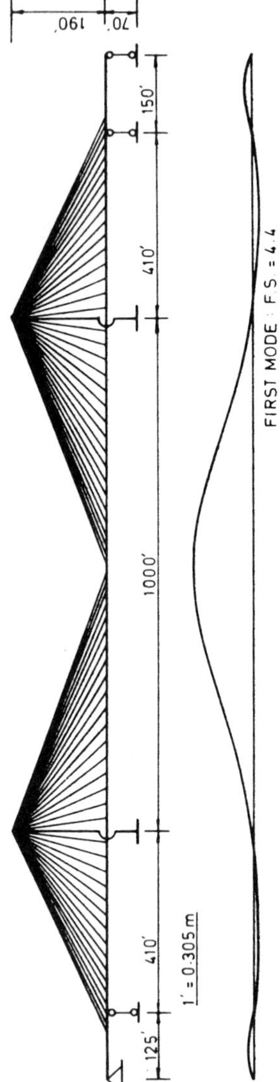

FIGURE 9-55 Cable-stayed bridge for design example.

The bridge shown in Figure 9-55 is a preliminary scheme of a prestressed concrete cable-stayed structure. The small cable spacing accommodates the free cantilever type, widely used for long-span prestressed concrete bridges. The spacing of the cable is varied to permit the use of the same cross section for all cables. The cable forces are not modified, and therefore the forces and moments in the bridge are exactly those caused by the dead and live loads.

A significant characteristic of the bridge of Figure 9-55 is the extreme slenderness of the bridge girder. Because this is a concrete bridge, the actual stiffness EI of the girder section is only 2.3 percent of the stiffness of the bridge girders in Figures 9-52 and 9-53. Yet, this structure has a factor of safety against buckling greater than four.

Nonlinear Cable Behavior In the preceding analysis, the beam–column effect of the girders and towers is considered, but the nonlinear behavior of the cables is not included. At the instant the bridge begins to buckle, however, the displacements are infinitesimally small so that the cables may still be assumed to respond linearly. Because the axial forces in the girders and towers are internal forces induced by external loading, they increase only if the external loading increases. Because the cables become stiffer under increased loading, the assumption of linear cable behavior is conservative.

Approximate Method: Design Example

For preliminary studies, the buckling load of cable-stayed bridges may be determined using approximate methods based on the similarity between the bridge girder and a beam–column on elastic foundation. The bridge shown in Figure 9-55 has very stiff towers compared to the girder, so that the latter approaches a beam on elastic foundation. Although the cable stiffness varies along the bridge, the axial forces vary in a similar pattern, and the two effects tend to compensate each other.

The cable properties at the tower are: $l_C = 189.3$ ft, $E_C A_C = 299{,}808$ kips, and the cable spacing is 35.08 ft. The elastic constant is $\beta = E_C A_C/(\Delta x\, l_C) = 45.15$ ksf. Assuming that the girder is very long and has nonyielding supports, it buckling load based on a beam on elastic foundation is

$$P_c = 2\sqrt{EI\beta} = 2 \times \sqrt{617{,}000 \times 43.8 \times 45.15} = 69{,}852 \text{ kips}$$

The calculated buckling load is $P_c = 4.45 \times 15{,}042 = 66{,}956$ kips, or 4 percent lower than the approximate. The foregoing approximate results are not as satisfactory, however, if the towers are flexible. The reactions of the different cables will, in this case, have a mutual effect on each other, so that the elastic constant β cannot have a reliable estimate.

Alternatively, we may articulate the complex bridge behavior if we consider the displacements of the buckled bridge as shown in Figure 9-56, giving

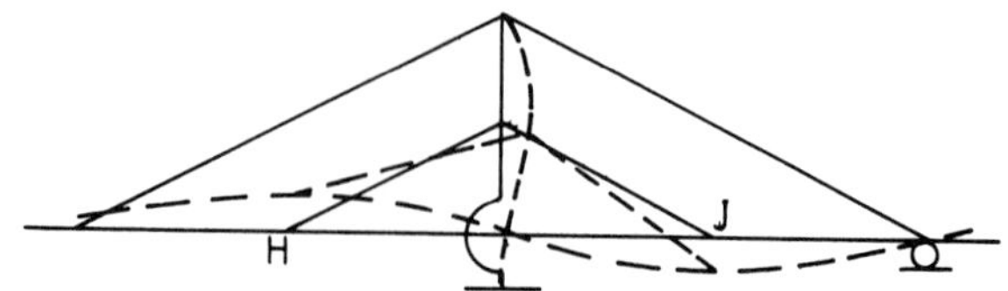

FIGURE 9-56 Buckled shape of bridge shown in Figure 9-53.

the buckling mode of the bridge of Figure 9-53. With the tower relatively flexible, this buckling mode can occur even if the cables are infinitely stiff. At the instant of buckling of the side span, the side span will bend up, whereas the main span will bend down, or vice versa, at the same time, so that the two lower cables will amplify the effect of each other.

From an analysis, the deflection of joint H is computed as follows: $\delta_H = 0.000162$ ft for 1 kip at joint J acting upward, and $\delta_H = 0.000435$ ft for 1 kip at joint H acting downward, so that $\delta_H = 0.000597$ ft for both loads.

The elastic constant of the cable is therefore computed as

$$\beta = \frac{1}{(\Delta x)\delta_H} = \frac{1}{180 \times 0.000597} = 9.3 \text{ kips/ft}^2$$

and the buckling load is

$$P_c = 2\sqrt{4{,}320{,}000 \times 250 \times 9.3} = 200{,}500 \text{ kips}$$

This is 6 percent higher than the calculated buckling load at the tower

$$P_c = 9.97 \times 19{,}000 = 189{,}000 \text{ kips}$$

The same calculation for the bridge of Figure 9-52 yields

$$\delta_H = 0.0000922 + 0.000284 = 0.00036\ l/\text{kip}$$

$$\beta = 1/(180 \times 0.00036) = 15.18 \text{ kip/ft}^2$$

$$P_c = 2\sqrt{4{,}320{,}000 \times 250 \times 15.18} = 256{,}073 \text{ kips}$$

which is about 3.6 percent lower than the calculated value of $17.12 \times 15{,}500 = 265{,}360$ kips at the tower.

9-20 CABLE-STAYED BRIDGES: DESIGN EXAMPLE

Bridges with Two 1970-ft Navigation Spans

For the Great Belt Bridge in Denmark, a preliminary investigation was carried out for the following options: (a) a bridge carrying only road (vehicular) traffic, (b) a bridge carrying only railway traffic, and (c) both rail and road traffic on the same structure. For the main spans across the international navigation channel, two designs were studied and detailed: the first having three truss spans each approximately 1200 ft long, and the second consisting of two cable-stayed spans each 1970 ft long. The study indicated that for the combined road and railway structure, the cable-stayed bridge was compatible with the site conditions, whereas truss spans appeared to be uneconomical. A suspension bridge was considered too flexible for railway traffic (Gimsing, 1976).

A cross section of the stiffening girder is shown in Figure 9-57. The total width of the bridge deck is 147 ft with a double-rail track placed at the center as a depressed section. Between the roadway and railway sections, 8-ft-wide strips accommodate the cables and the pylon legs. The girder system is a multicellular box with an orthotropic deck and a 49-ft-wide horizontal bottom plate at the center connected to inclined bottom plates at the overhangs. Transverse girders and stiffeners (secondary diaphragms) are arranged at 13-ft intervals, and solid (primary) diaphragms are inserted on both sides of each cable anchor point.

The study determined that the application of a horizontal fixed cable in the main spans would not be advantageous in a bridge carrying heavy railway traffic. As an alternative, the two navigation spans could be separated by smaller spans, and this arrangement provided the possibility of several schemes for fixing the tops of the inner pylons.

Several cable systems analyzed in the study are shown in Figure 9-58. The system shown in Figure 9-58*a* has two individual stayed girders placed side by side. Each stayed bridge has a main span 1970 ft long and two side spans 690 ft long. A criterion for determining the ratio of the main span to the side span is that the fixed cable should always be in tension. The application of two stayed girder bridges separated by anchor piers with double shafts would give the advantage of applying four identical cable systems supported by vertical pylons subjected mainly to normal forces. In this case, however, three piers should be founded in the central part of the belt where maximum water depths are found.

For the system shown in Figure 9-58*b*, the central anchor pier is eliminated by placing the two central pylons close enough so that the fixed cable from one pylon can be connected to the stiffening girder at the other pylon and vice versa. In this scheme, however, certain structural problems will exist at the points where the cables intersect. In addition, the normal force in the stiffening girder between the two central pylons is markedly increased because this portion of the girder forms an integral part of both left and right cable-stayed bridge.

The intersecting cables may be avoided as shown in Figure 9-58*c* where the two central pylons are made triangular. Unbalanced horizontal forces are excluded from the dead load by connecting the two central pylons by a horizontal cable, which also distributes the horizontal forces from the live load on the two pylons.

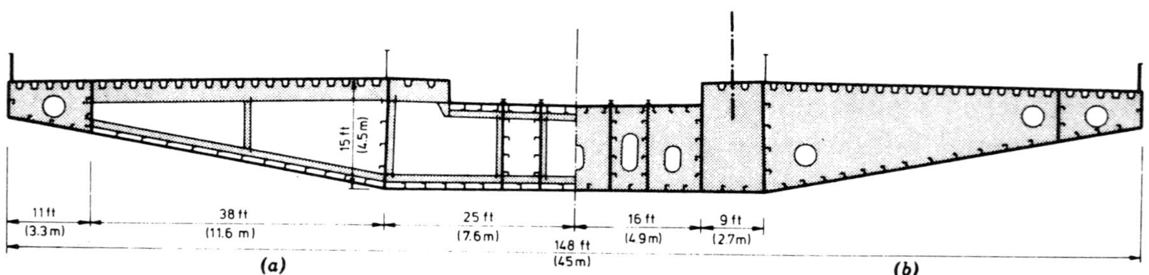

FIGURE 9-57 Cross section of proposed stayed girder bridge carrying both road and railway traffic (Great Belt Bridge in Denmark): (*a*) standard cross section with normal transverse girders and secondary diaphragms; (*b*) cross section at cable anchor points with primary diaphragms.

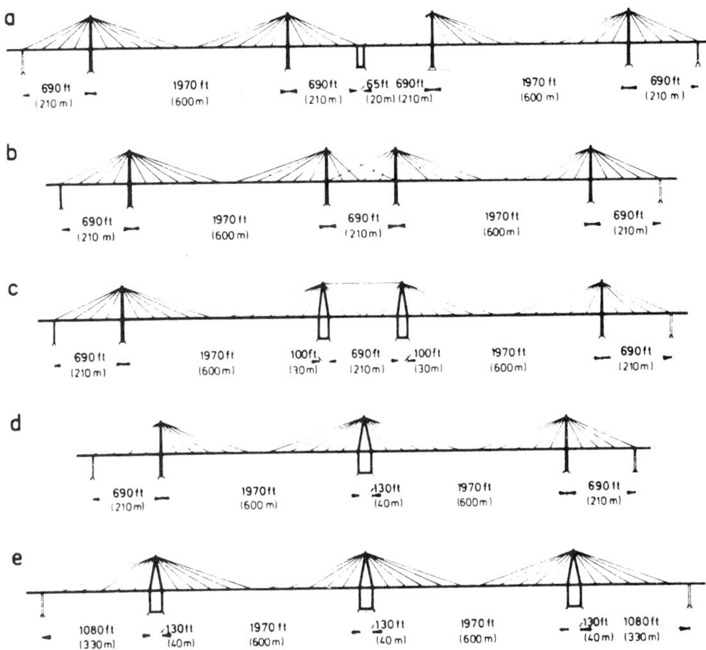

FIGURE 9-58 Cable systems investigated for bridge with two 1970-ft- (600-m) wide navigation spans.

In Figure 9-58*d* the two triangular pylons are replaced by one tower between the two main spans. In this case only one pier must be placed in deep water, but it must be much stronger than the piers in Figures 9-58*a* and *b*. In Figure 9-58*c* the bridge has two different types of pylons, that is, two vertical outer pylons and two central triangular pylons.

A consistent use of the triangular pylon structure is demonstrated in Figure 9-58*e* where the bridge has three identical and symmetrical cable systems. Because the outer pylons also have a triangular shape, fixed cables are superfluous, allowing an increase in the side span length of 690 to 1080 ft.

Based on economic and appearance considerations, Figures 9-58*b*, *c*, and *d* were excluded. Figures 9-58*a* and (*e*) were found feasible, and Figure 9-58*a* was also found to be more compatible with the state of the art and current practice in the field of cable-stayed bridges. This scheme was then compared to the three-span truss.

Erection Procedures Triangular pylons such as those shown in Figure 9-51 system F, offer erection advantages whereby a stable structural system is maintained throughout the erection period. Once the triangular pylons are in place, double cantilevering of the stiffening girder can be carried out as shown in Figure 9-59.

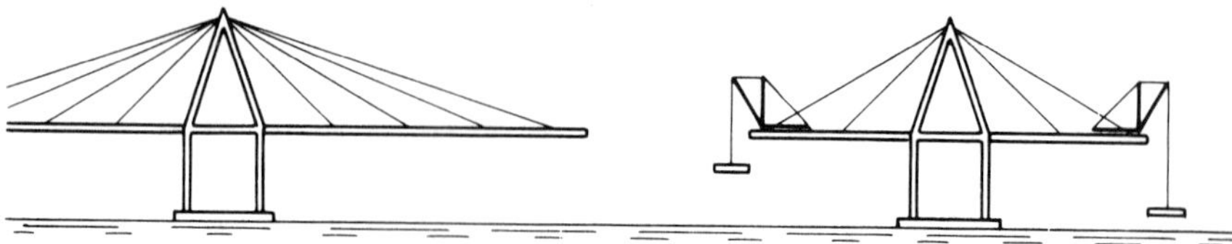

FIGURE 9-59 Erection of stayed girder bridge with triangular pylon structures by double cantilevering.

The application of a horizontal fixed cable as shown in Figure 9-51, system E, does not offer similar erection advantages because the fixed cable requires the presence of the stiffening girder to become active.

9-21 FATIGUE EFFECTS FROM WIND-INDUCED VIBRATIONS

Wind load effects and the associated fatigue behavior of cables has been investigated by Basu and Chi (1981) using procedures of linear elastic fracture mechanics. The same study also includes (a) the formulation of deflections and bending stresses under wind-induced and vortex-shedding vibrations and (b) the formulation of fatigue behavior of bridge cables in terms of fatigue crack initiation and propagation in constituent wires.

Nature of Aeolian Vibration The wind-induced vibration of flexible structural members such as wires and cables can be critical in long slender elastic structures. Near resonance conditions, these can develop flow-induced oscillations by extracting energy from the flow around them. The oscillations, combined with the flow, give rise to a fluid–structure interaction resulting in nonlinear response, and covered by four general phenomena: (a) vortex-induced oscillation, (b) flutter, (c) galloping, and (d) buffeting. These responses may be equally important for a given structural member and flow conditions, but for stay cables the vortex-induced oscillation is the most relevant.

Vortex-Induced Excitation of Stay Cables The mechanism of vortex shedding from a stay cable can be illustrated considering the overall flow pattern around a circular cylinder with increasing Reynolds numbers as shown in Figure 9-60. The Reynolds number Re (a dimensionless parameter characterizing the flow regime) is a function of the flow velocity, the diameter or the depth of the body in flow, and the kinematic viscosity of the fluid. The higher the Reynolds number the higher is the flow velocity, and correspondingly the more turbulent is the flow.

Beyond a Reynolds number of about 40, the vortex starts shedding as shown because of wake instability, and at value of 90 the detached shear

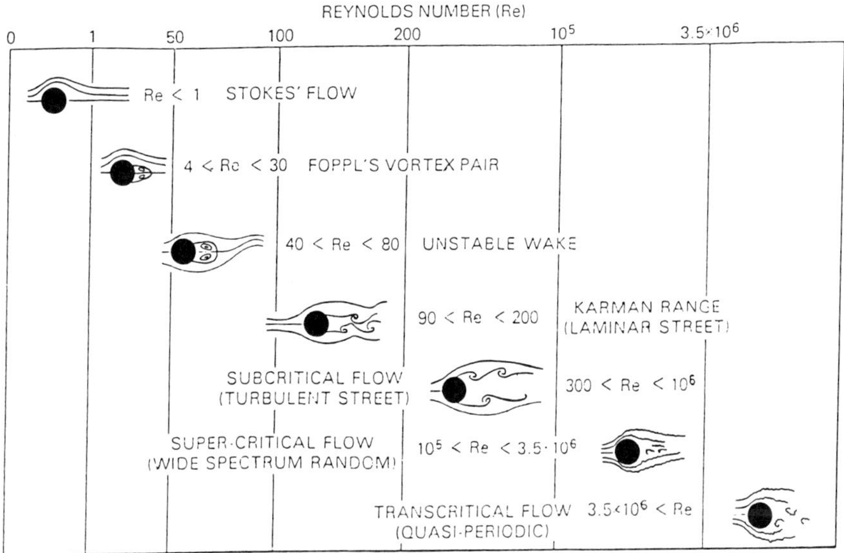

FIGURE 9-60 Overall flow patterns around a circular cylinder for different Reynolds number Re.

layer starts to fold up and forms concentrated vortices. At Re values in the range of 2×10^5 to 3×10^6 the boundary layer undergoes transition and the wake is disorganized. When Re exceeds 3×10^6, a boundary layer becomes fully turbulent. In the supercritical flow there is no well-organized vortex street, and the energy in the wake is diffused into a wide spectrum of frequencies.

Results and Predictions on Fatigue Based on the foregoing behavior, Basu and Chi (1981) predicted the dynamic response and fatigue patterns analytically, and applied these formulations in determining the deflection and bending stresses in typical cables of the Pasco–Kennewick Bridge (in the state of Washington). From these results, these investigators conclude that the deflection is nearly constant for all wind speeds although this may be contrary to the expectation that higher wind velocity will result in higher excursion. Two considerations may provide an explanation of this behavior: (a) the resonance mode contributes primarily to the amplitude of the excursion of deflection (whereas the net contribution of all other modes is insignificant), and this is true even though modal superposition is considered in the numerical computation of the deflection; and (b) the analytical model of vortex excitation considers the wind force harmonic with its magnitude varying quadratically with wind velocity. The former indicates that the nondimensional deflection decreases with the mode number, whereas the latter indicates that the deflection is linearly dependent on the critical wind velocity that increases with the mode number. Thus, the maximum velocity, being a function of both, yields a nearly constant value for all modes.

Conversely, the bending stresses are higher for higher modes, but because of the inherently low moment of inertia of the cable away from the end anchorage, the maximum bending stress is relatively low.

From an analysis of fatigue initiation in a wire, Basu and Chi (1981) determined the endurance limit and the fatigue initiation life consistent with fracture mechanics methodology. The endurance limit for the steel wire used in two bridges (Pasco–Kennewick and Luling) was estimated at 160 ksi. The wire in these cables has an ultimate tensile strength of 240 ksi and a maximum working stress of 108 ksi (excluding cyclic bending stress). This means that an unnotched, dislocation-free single wire can sustain a bending stress up to 52 ksi without exceeding its endurance limit. For wind velocity as high as 52 mph, maximum bending stresses in a cable may be of the order of 2.6 ksi. Even if the cable is 20 times less stiff at the ends, no fatigue cracks will be initiated. Thus, using criteria from classical mechanics, fatigue cracks in a wire do not initiate until 10^7 cycles of load applications.

It may appear that under wind-induced vibration the fatigue life of a high-strength-steel cable is largely dominated by its initiation life. Fatigue crack propagation curves demonstrate that, even at lower ranges of stress fluctuations, the crack initiation life is at least 10 times larger than the crack propagation life, implying that the former constitutes more than 90 percent of the total fatigue life.

Using a simplistic approach to correlate the fatigue life of a wire to a cable, cable failure may be imminent when the effective stress in the unbroken wires in the cable exceeds the ultimate tensile strength. For Pasco–Kennewick group I cable (283 wires, 0.25 in. diameter), this means that approximately 20 percent of the wires can undergo fatigue failure before cable replacement becomes necessary, but this should not be construed to indicate that the fatigue life is 20 times that of a constituent wire. The fatigue life of a cable should be lower than that of a wire, but a rational relationship between the two is yet to be developed. Thus, additional work will be needed, both analytical and experimental, for a complete understanding of the subject.

9-22 LRFD SPECIFICATIONS

These specifications include provisions for refined methods of analysis. The distribution of force effects may be determined by spatial analysis or by planar analysis. The choice should be justified by a consideration of the tower geometry, the number of planes of stays, and the torsional stiffners of the deck superstructure. The nonlinear effects discussed in Section 9-16 may result from (a) a change in cable sag at all limit states, (b) deflection of the deck superstructure and towers at all limit states, and (c) material nonlinearity at the extreme limit state.

Cable sag may be investigated as suggested by ASCE (1991), using an equivalent number modeled as a chord with a modified modulus of elasticity. The modified modulus is as follows.

For instantaneous stiffners,

$$E_{\text{mod}} = E\left[1 + \frac{EAW^2 \cos\alpha}{12H^3}\right]^{-1} \tag{9-42}$$

For changing cable loads,

$$E_{\text{mod}} = E\left[1 + \frac{(H_1 + H_2)EAW^2 \cos\alpha}{24H_1^2 H_2^2}\right]^{-1} \tag{9-43}$$

where E = Young's modulus
W = total weight of cable
A = cross-sectional area of cable
α = angle between cable and horizontal
H, H_1, H_2 = horizontal component of cable force

REFERENCES

AASHTO, 1985. "Guide Specifications for Strength Design of Truss Bridges" (Load Factor Design).

AASHTO, 1988. "Standard Specifications for Movable Highway Bridges.

ASCE, 1977. "Tentative Recommendations for Cable-Stayed Bridge Structures," ASCE Task Committee on Cable-Suspended Structures, *J. Struct. Div., ASCE* Vol. 103, No. ST5, May.

ASCE, 1988. "Bibliography and Data on Cable-Stayed Bridges," ASCE Committee on Long-Span Steel Bridges, *J. Struct. Div., ASCE*, Vol. 103, No. ST10, October.

ASCE, 1991: "Guidelines for Design of Cable-Stayed Bridges," ASCE Committee on Cable-Suspended Bridges, ASCE, New York.

Barnoff, R. M. and W. G. Mooney, 1960. "Effect of Floor Systems on Pony Truss Bridges," *Proc. ASCE, J. Struct. Div.* 86, ST4, April, pp. 25–47.

Baron, F. and S. Y. Lien, 1973. "Analytical Studies of a Cable Stayed Girder Bridge," *J. Computers and Structures*, Vol. 3, Pergamon Press, N.Y.

Basu, S., and M. Chi, 1981: "Analytical Study for Fatigue of Highway Bridge Cables," FHWA Report RD-81/090, Washington, July.

Bert, C. W., and R. A. Stein, 1962: "Stress analysis of Wire Rope in Tension and Torsion," *Wire and Wire Products*, 37(5), 621-4(May); 37(6), 769-70, 772, 816 (June).

Birkenmaier, 1980. "Fatigue Resistance Tendons for Cable-Stayed Construction," *IABSE Proc.* P-30/80, IABSE Periodical, May.

Bolton, A., 1955. "A Quick Approximation to the Critical Load of Rigidity Joined Trusses," *Struct. Engineer*, Vol. 33, No. 3, March, pp. 90–95.

Bolton, A., 1965. "Critical Load of Trusses with Redundant Members," *Struct. Engineer*, Vol. 43, No. 4, April, pp. 128–132.

Bouwkamp, J. C., 1964. "Concept of Tubular Joint Design," *Proc. ASCE, J. Struct. Div.* 90, ST2, April, pp. 77–101.

Chu, K. H., 1959. "Secondary Moments, End Rotations, Inflection Points, and Elastic Buckling Loads of Truss Members," Intern. Assoc. of Bridge and Struct. Eng., Publ. 19, pp. 17–46.

Clark, J. H., 1978. "Relation of Erection Planning to Design," *Cable-Stayed Bridges, Struct. Eng. Ser.* No. 4, June, Bridge Div., FHWA, Washington.

Clark, J. H., 1984. "Foundation Design: West Seattle Bridge," *Trans. Research Record* 982, TRB, National Research Council, Washington.

Daniel, H., and H. Schumann, 1967. "Die Bundesautobahnbrucke uber den Rhein bei Leverkusen," *Stahlbau*, 36(8), 225–236, Aug.

Drucker, D. C. "The Effect of Shear on the Plastic Bending of Beams," *J. Appl. Mech. ASME*, Dec.

Ernst, J. H., 1965. "Der E-Modul von Seilen unter Berücksichtigung des Durchhänges" *Bauingenieur*, Vol. 40, No. 2, Feb., pp. 52–55.

Felge, A., 1966. "Evaluation of German Cable-Stayed Bridges," *Acier-Stahl-Steel*, Vol. 31, No. 12, pp. 523–532, Dec.

Firmage, D. A. and R. J. Goodwin, 1979. "The Economics of Various Parameters of Cable-Stayed Bridges," ASCE Convention and Exposition, Paper 3721, Atlanta, Oct. 23–25.

Fisher, G., 1960. "The Severin Bridge at Cologne," *Acier-Stahl-Steel*, 25(3), 97–107 (March).

Fisher, G., and I. M. Viest, 1961. "The Severin Bridge at Cologne," Acier-Stahl-Steel, vol. 25, No. 3, March.

Fisher, J. W. and J. H. A. Struik, 1974. *Guide to Design Criteria for Bolted and Riveted Joints*, Wiley, New York.

Fleming, J. F., 1974. Latest Development of Cable-Stayed Bridges, Eng. Journ., AISCE, Third Quarter.

Fleming, J. F. and E. A. Egeseli, 1978. "Dynamic Response of Cable-Stayed Bridge Structures," *Cable-Stayed Bridges, Struct. Eng. Ser.* No. 4, June, Bridge Div. FHWA, Washington.

Galambos, T. V. *Guide to Stability Design Criteria for Metal Structures*, Wiley, New York.

Gimsing, J. N. 1966. "Anchored and Partially Anchored Stayed Bridges," *Proc. Intern. Symp. on Suspension Bridges*, Laboratories Nacionel de Engenharia Civil, Lisbon.

Gimsing, N. J., 1973. "Pont sur le Grand-Belt (Danemark)," *Travaux*, No. 463.

Gimsing, N. J., 1976. "Multispan Stayed Girder Bridges," *J. Struct. Div. ASCE*, Vol. 102, No. ST10, Oct., pp. 1989–2004.

Godden, W. G. "Seismic Model Studies of the Ruck-A-Chucky Bridge," Univ. California, Berkeley, July.

Godden, W. G. and M. Aslam, 1978. "Dynamic Model Studies of Ruck-A-Chucky Bridge," *J. Struct. Div., ASCE*, Vol. 104, No. ST12, Dec., pp. 1827–1844.

Goschy, B., 1961. "Dynamics of Cable-Stayed Pipe Bridges," *Acier–Stahl–Steel*, No. 6, June.

Havemann, H. K., 1962. "Die Seilverspannung der Autobahnbrucke uber die Norderelbe-Bericht uber Versuche zur Dauerfestigkert der Drahtseibe," *Stahlbau*, 31(8), 225–232 (Aug.).

Hess, H., 1960. "Die Severinsbrucke Koln," *Stahlbau*, 29(8), 225–261 (Aug.).

Heywood, R. B., 1952. *Designing by Photoelasticity*, Chapman and Hall, London.

Hirai, A. and T. Okubo, 1966. "On the Design Criteria Against Wind Effects for Proposed Honshu-Shikoku Bridges," Paper No. 10, Intern. Symp. on Suspension Bridges, Lisbon, Nov.

Hodge, P. G., 1957. "Interaction Curves for Shear and Bending of Plastic Beams," *J. Appl. Mech. ASME*, Sept. Hoff, N.J., 1956. *The Analysis of Structures*, Wiley, New York.

Hoff, N. J., 1956. *The Analysis of Structures*, Wiley, New York.

Homberg, H., 1955. "Influence Lines for Cable-Stayed Bridges," *Stahlbau*, Vol. 2, No. 2, Feb., pp. 40–44.

Hope, B. B. and J. A. N. Lee, 1964. "Tests on Laboratory Bridge-3," Queen's Univ.–Ontario Joint Hwy. Research Program, Report 23, Oct., 73 pp.

Horne, M. R., 1960. "Elastic Lateral Stability of Trusses," *Struct. Engineer*, vol. 38, No. 5, May, pp. 147–155.

Jennings, A., 1968. "Frame Analysis Including Change of Geometry," *Proc. ASCE, J. Struct. Div.*, vol. 94, ST3, March, pp. 627–644.

Kajita, T. and Y. K. Cheung, 1973. "Finite Element Analysis of Cable-Stayed Bridges," Publ. 33-II, Intern. Assoc. for Bridge and Struct. Eng.

Kane, T. A., T. F. Mahoney, and J. H. Clark, 1990. "West Seattle Swing Bridge," *Trans. Research Record* 1275, TRB, National Research Council, pp. 62-66, Washington, D.C.

Khalil, M. S., W. H. Dilger, and A. Ghali, 1983. "Time Dependent Analysis of PC Cable-Stayed Bridges," *J. Struct. Div. ASCE*, Vol. No. ST8, Aug.

Lazar, B. E., 1972. "Stiffness Analysis of Cable-Stayed Girder Bridges," *J. Struct. Div. ASCE*, Vol. 98, No. ST7, July.

Lazar, B. E., M. S. Troitsky, and M. M. Douglass, 1972. "Load Balancing Analysis of Cable-Stayed Bridges," *J. Struct. Div. ASCE*, Vol. 98, No. ST8, Aug.

Leissa, A. W., 1959. "Contact Stresses in Wire Ropes," *Wire and Wire Products*, 34(3), 307–14, 372 (March).

Leonhardt, F., 1974. "Latest Developments of Cable-Stayed Bridges for Long Spans," *Saetryk af Bygingsstatiske Meddelelser*, Vol. 45, No. 4, Denmark, Danmarko Teckniske Hojskole.

Leonhardt, F. and W. Zellner, 1970. "Cable-Stayed Bridge—Report on Latest Developments," *Proc. Canadian Struct. Eng. Conf.*, Ontario.

Lin, T. Y., Y. C. Yang, H. K. Lu, and C. M. Redfield, 1978. "Design of Ruck-A-Chucky Bridge," *Cable-Stayed Bridges, Struct. Eng. Series* No. 4, June, Bridge Div. FHWA, Washington.

Lipson, C., Noll, G. C., and L. S. Clock, 1950: Stress and Strength of Manufactured Parts, McGraw-Hill, N.Y., 259 pp.

Lipson, S. L. and K. M. Agrawal, 1974. "Weight Optimization of Plane Trusses," *J. Struct. Div. ASCE*, Vol. 100, No. ST5, May, pp. 865–880.

Lipson, S. L. and L. B. Gwin, 1977. "The Complex Method Applied to Optimal Truss Configuration," *J. Computers and Structures*, Vol. 7, pp. 461–468.

Lipson, S. L., and L. B. Gwin, 1977. "Discrete Sizing of Trusses for Optimal Geometry," *J. Struct. Div. ASCE*, Vol. 103, No. ST5, May, pp. 1031–1046.

Lipson, C., Noll, G. C., and L. S. Clock, 1950. Stress and Strength of Manufactured Parts, McGraw-Hill, New York, 259 pp.

Lu, Z. A., 1978. "Dynamic Analysis of Cable Hung Ruck-A-Chucky Bridge," ASCE National Spring Conv., Preprint 3180, Pittsburgh, April 24–28.

McCormac, J. C., 1975. *Structural Analysis*, Intext Educational Publishers, New York.

Modjeski and Masters, 1960. Greater New Orleans Bridge over Mississippi River, Final Report to Mississippi River Bridge Authority, 2 vols.

Neuber, and H. G. Hahn, 1966. "Stress Concentration in Scientific Research and Engineering," *Appl. Mech. Rev.* 19(3), 187–256, March.

Norris, C. H., and J. B. Wilbur, 1960. *Elementary Structural Analysis*, 2nd ed., McGraw-Hill, New York, 650 pp.

O'Connor, C., 1965. "Buckling in Steel Structures-2. The Use of a Characteristic Imperfect Shape in the Design of Determinate Plane Trusses Against Buckling in Their Plane," Univ. Queensland, *Dept. Civ. Eng., Bull.*, 6, July.

Ontario Highway Bridge Design Code, 1979. Ontario Ministry of Transportation and Communication.

O'Connor, C., 1971. Design of Bridge Superstructures, Wiley, New York.

Okauchi, I., A. Yabe, and K. Ando, 1967. "Studies on the Characteristics of a Cable-Stayed Bridge," *Bull. Faculty of Science and Eng.*, Chuo Univ., Vol. 10.

Peterson, R. E., 1953. Stress Concentration Design Factors, Wiley, New York, 155 pp.

Podolny, W., 1971. "Static Analysis of Cable-Stayed Bridges," Ph.D. Thesis, Univ. Pittsburgh.

Podolny, W., 1974. "Cable Connections in Stayed Girder Bridges," *Eng. Journ. AISC*, Fourth Quarter, Vol. 11, No. 4.

Podolny, W., 1974. "Design Considerations in Cable-Stayed Bridges," ASCE Specialty Conf. on Metal Bridges, Nov. 12–13, St. Louis, Mo.

Podolny, W., 1978. "Concrete Cable-Stayed Bridges," *Transp. Research Record* 665, *Bridge Eng.*, Vol. 2, TRB Conf., St. Louis, Mo., Sept.

Podolny, W., 1981. "The Evolution of Concrete Slab Cable-Stayed Bridges," *Concrete Intern. ACI*, Vol. 3, No. 8, Aug.

Podolny, W. and J. F. Fleming, 1972. "Historical Development of Cable-Stayed Bridges," *J. Struct. Div. ASCE*, Vol. 98, No. ST9, Sept., pp. 2079–2095.

Podolny, W. and J. B. Scalzi, 1986. *Construction and Design of Cable-Stayed Bridges*, Wiley, New York.

Reesnyder, M., 1972. Design of Cable-Stayed Girder Bridges, Der Bauingenieur, Aug. and Sept.

Scalzi, J. B., 1971. "Cable-Suspended Roof Construction State-of-the-Art," Task Committee on Special Structures, *J. Struct. Div. ASCE*, Vol. 97, No. ST6, June, pp. 1715–1761.

Simpson, C. V. J., 1970. "Modern Long Span Steel Bridge Construction in Western Europe," *Proc. Institution Civ. Engineers*, London.

Smith, B. S., 1967. "The Single Plane Cable-Stayed Girder Bridges: A Method of Analysis Suitable for Computer Use," *Institution Civ. Engineers*, London, Paper No. 7011, Nov.

Smith, B. S., 1968. "A Linear Method for Analysis for Double-Plane Cable-Stayed Girder Bridges," *Proc. Inst. of Civ. Eng.*, Jan.

Smith, C. V., 1991. "Guidelines for Design of Cable-Stayed Bridges," ASCE Committee on Cable-Stayed Bridges, ASCE, New York.

Starkey, W. L. and H. A. Cress, 1959. "An Analysis of Critical Stress and Mode of Failure of a Wire Rope," *Trans. ASME, Ser. B*, Vol. 81, No. 4, Nov., pp. 307–316.

Steinman, D. B., 1946. "Design of Bridges Against Wind," *Civ. Eng.*, Oct., Nov., Dec.

Steinman, D. B., 1953. *A Practical Treatise on Suspension Bridges*, 2nd ed., Wiley, New York.

Stevens, L. K. and L. C. Schmidt, 1965. "Elastic Critical Loads at Laterally Braced Trusses," Inst. Eng. Australia, *Civ. Eng. Trans.* CE7(2) Oct., pp. 81–91.

Tang, M. C., 1971. "Analysis of Cable-Stayed Girder Bridges," *J. Struct. Div. ASCE*, Vol. 97, No. ST5, May.

Tang, M. C., 1972. "Design of Cable-Stayed Bridges," *J. Struct. Div. ASCE*, Vol. 98, No. ST8, Aug.

Tang, M. C., 1976. "Buckling of Cable-Stayed Girder Bridges," *J. Struct. Div. ASCE*, Vol. 102, No. ST9, Sept.

Thul, H., 1966. "Staehlerne Strassenbruecken in der Bundesrepublik," *Bauingenieur*, 41(5), 169–89 (May).

Troitsky, M. S., 1977. "Cable-Stayed Bridges, Theory and Design," Crosby Lockwood Staples, London.

Troitsky, M. S. and B. E. Lazar, 1971. "Modern Analysis and Design of Cable-Stayed Bridges," *Proc. Institution Civ. Engineers*, London, March.

Waling, J. L., 1953. "Least Weight Proportions of Bridge Trusses," *Eng. Experiment Station Bull. No.* 417, Sept., 56 pp.

Wardlaw,1971. "Some Approaches for Improving the Aerodynamic Stability of Bridge Road Decks," *Proc. Third Intern. Conf. on Wind Effects on Structures*, Tokyo.

Wardlaw, R. L., 1976. "A Review of the Aerodynamics of Bridge Road Decks and the Role of Wind Tunnel Investigation," FHWA, U.S. Dept. Transp., Report No. FHWA-RD-73-76, Washington, D.C.

Warner, S., and K. Julshos, 1966. "The Evolution of German-Stayed Bridges; An Overall Survey," *Acier-Stahl-Steel*, Vol. 31, No. 17, Dec.

West, H. H., 1989. *Analysis of Structures*, Wiley, New York.

Whitmore, R. F., 1962. "Gusset Plate Analysis in Trusses," *Bull. No.* 16, Eng. Experiment Station, Univ. Tennessee.

Williams, D. and W. G. Godden, 1975. "Seismic Response of Long Multiple-Span Highway Bridges," Fifty-fourth Ann. Meeting, Transp. Research Board, National Research Council, Washington, D.C.

Xanthakos, P. P., 1973. "Columbus Drive Bascule Bridge, Chicago," *Struct. Design*, In: House Report.

Zienkiewicz, O. C. and G. S. Holister, 1965. *Stress Analysis*, Wiley, New York.

CHAPTER 10

ELASTIC ARCH BRIDGES

10-1 TYPES OF ARCH BRIDGES

Definition of Arch

The term arch has in several instances been loosely applied, both in American and European engineering literature. As a structural unit, an arch is defined as a member shaped and supported in such a manner that intermediate transverse loads are transmitted to the supports primarily by axial compressive thrusts in the arch. In addition, the member must be sustained by supports capable of developing lateral as well as normal reaction components. For a given loading, the arch shape must be chosen so as to avoid the introduction of bending moments. For typical downward loads, this shape will be concave downwards, and in this case the arch becomes an exact inverse of a suspension bridge cable.

The simple beam shown in Figure 10-1*a* has a fixed bearing at R_1 and a roller-type bearing at R_2. Under the action of any load system, the support at R_2 moves freely as shown and only normal reactions are developed there. Likewise, R_1 will be normal unless inclined loads are applied. The member shown in Figure 10-1*b* has a curved form but statically it is a beam for the same reasons. For the member shown in Figure 10-1*c*, the roller bearing at R_2 has been replaced by a fixed shoe. Under load this structure deflects in a downward direction and shortens axially, thereby exerting lateral thrusts against the supports. The two reactions R_1 and R_2 are now inclined and can be replaced by the horizontal and vertical components as shown. According to our definition, this structure is an arch, a two-hinged arch type.

For the structure shown in Figure 10-1*d*, the supports are fixed against rotation and translation. Lateral thrusts are again developed, but their point

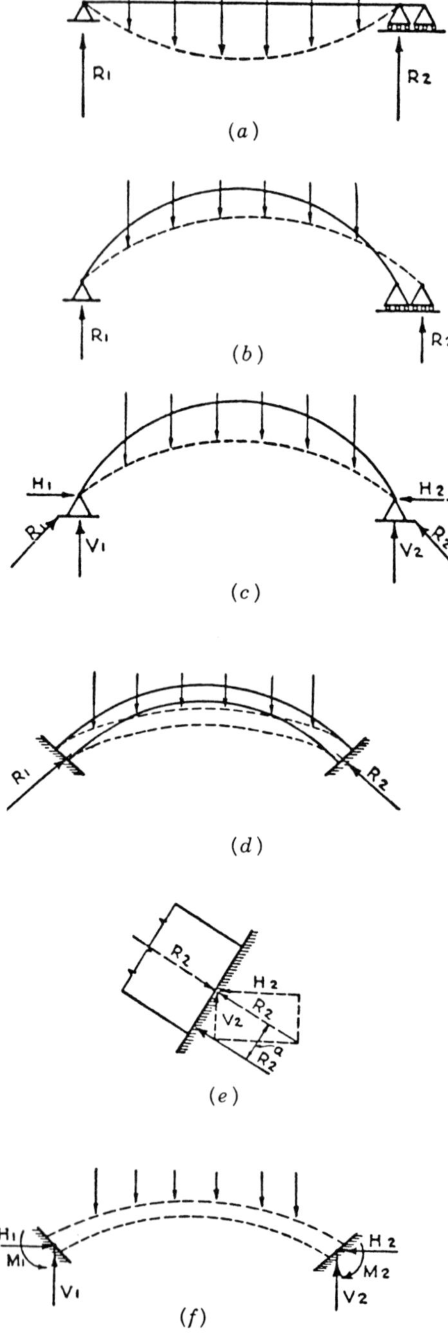

FIGURE 10-1 Various configurations of structural support; beams, hinged arch, and fixed arch.

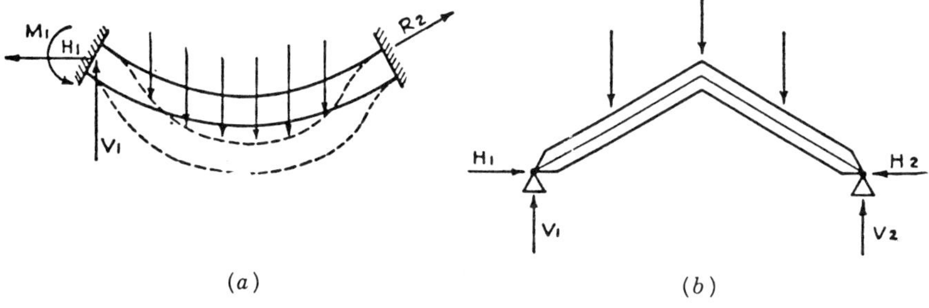

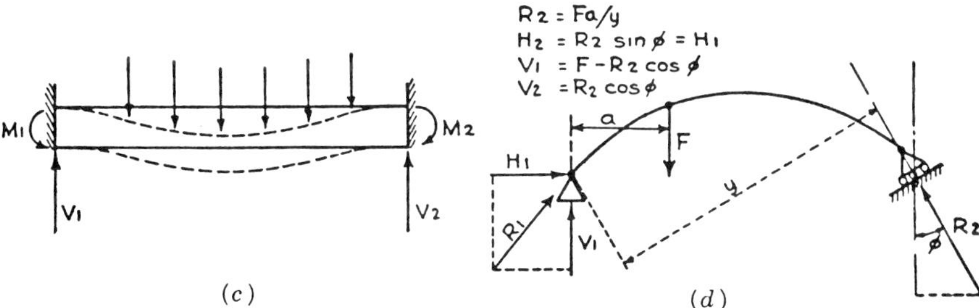

FIGURE 10-2 Statically different structural shapes.

of application is not necessarily at the neutral axis of the rib. We can introduce two equal and opposite forces R_2 at the right support as shown in Figure 10-1*e*, so that the initial reaction R_2 can be resolved into three components, namely, H_2, V_2, and $M_2 = R_2 a$. By analogy, the left reaction is represented by H_1, V_1, and M_1. The static form is now shown in Figure 10-1*f*, and is a fixed or hingeless arch.

The structure shown in Figure 10-2*a* deflects under load and elongates axially, exerting a pull against its supports. By definition, it is a fixed-end suspension system. As the shape changes from Figure 10-1*d* to that of Figure 10-2*a*, the lateral thrusts or horizontal reaction components merely change direction. This means that there is an intermediate point at which these components pass through a zero value as shown in Figure 10-2*c*. In the latter case the reaction components constitute a shear and bending moment at each support, and the structural system is a beam with fixed ends.

Although arches usually have a curved form, by definition this is not a requirement. Figure 10-2*b* is an example of a two-hinged arch with a polygonal axis. The structural unit shown in Figure 10-2*d* appears to be an arch, but both supports are not capable of developing lateral as well as normal reaction components as required by the arch definition. The reaction R_2 will always be normal to the inclined rolled bed, and the only three unknown reaction components (R_2, H_1, and V_1) are estimated as shown from a statically determinate system. By definition this structure is a simple beam although the induced stresses constitute bending moments as well as axial thrusts. The criteria for a credible arch system are therefore explicit and require (a) the supports to be capable of sustaining lateral thrusts and (b) the member to have an axial shape such that these thrusts are indeed developed under load.

Arch Types

Arches during the Roman and Renaissance periods were characterized by stone masonry configurations, described as the *voussoir* or arch block type and clearly distinguished from modern *monolithic* or elastic structures. The latter are analyzed using elastic theory whereby the entire structure is assumed to be a monolithic elastic unit where the true crown thrust is determined using equations that take into consideration the deflection or elastic distortion of the material under load.

The elastic theory is a universally accepted method for arch analysis; hence, we will use the term *elastic arches*. The introduction of structural steel and reinforced concrete in arch bridge construction removed the uncertainty regarding temperature and shrinkage effects and enabled longer spans.

Arch Grouping Arches are grouped according to the following characteristics: (a) materials of construction; (b) structural arrangement of the arch proper; (c) stress distribution in the arch proper; (d) method for supporting the superstructure; and (e) shape or curve of the arch. A distinctive terminology originating from the classic masonry arch is still applicable, and relevant terms are shown in Figure 10-3 which is self-explanatory.

With regard to stress distribution, an arch may be fixed (hingeless) or hinged. The latter type involves three hinge arrangements: the single-hinged typed, the two-hinged type, and the three-hinged type. The fixed type is the most common in concrete. For steel construction both fixed and hinged

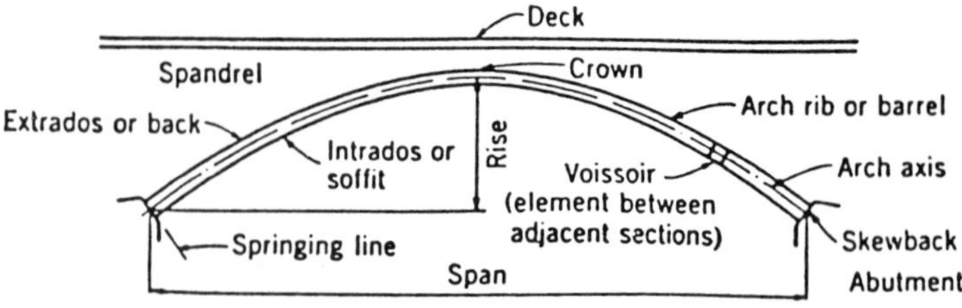

FIGURE 10-3 Arch bridge terminology.

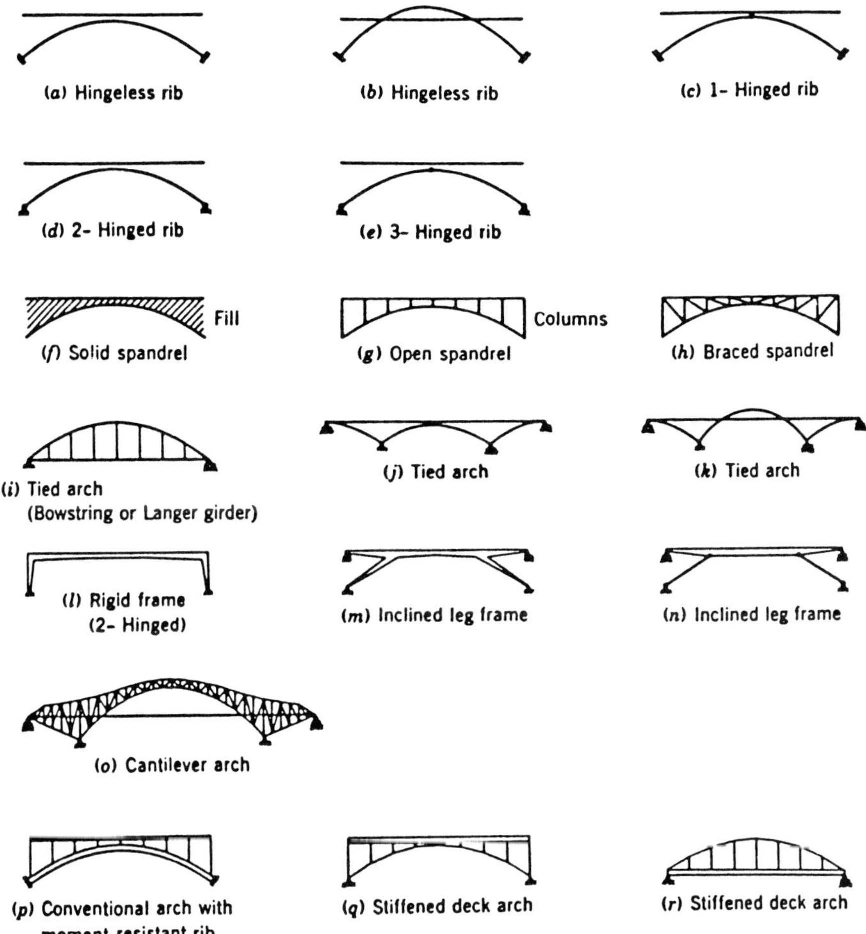

FIGURE 10-4 Types of arch bridges.

arches are common. The fixed arch, the two-hinged arch, and the single-hinged arch are subjected to stresses due to temperature change, settlement or displacement of supports, or because of compression or shrinkage of the materials. The three-hinged arch is statically determinate and free from these stresses. Fixed and hinged types are shown in Figures 10-4*a* through *e*.

The method of supporting the superstructure gives rise to distinctive arch types. If the space between the deck and the arch is filled with solid materials as shown in Figure 10-4*f*, the structure is called a *solid* or *filled spandrel arch*. An *open spandrel arch* has columns or hangers that transmit loads from the deck to the arch as shown in Figure 10-4*g*. If diagonals are added as shown in Figure 10-4*h*, the resulting combinations of arch rib, deck, verticals, and diagonals forms a truss referred to as a *braced spandrel arch*.

The structural arrangement of the arch proper can yield the configurations shown in Figures 10-4*i* through *k*, known as *tied arches*. In tied arches the horizontal reactions to the arch rib are provided by a tie at deck level. The simple case (Figure 10-4*i*) is known as a *bowstring* or *Langer girder* (Balog, 1956; Chandrengsu and Sparkes, 1954; Wittfoht et al., 1963; Zezety, 1962). Alternatively, an arch rib can be replaced by a truss of various forms (Scott and Roberts, 1958). One distinctive structure is the cantilever arch shown in Figure 10-4*o*.

The shape or curve of the arch articulates the structural performance, but a polygonal axis may be designed if necessary. A modern arch bridge may be elliptical, parabolic, semicircular, segmental, a compound circular curve, or a polygonal shape. In present design practice the shape of the axis may be governed by the magnitude and distribution of the superimposed loads. In some instances construction is simplified if the conventional curved arch is replaced by a combination of straight members. Examples are the rigid frame shown in Figure 10-4*l* and the frames shown in Figures 10-4*m* and *n* consisting of horizontal beams at deck level supported by inclined legs as shown. The continuously curved rib is statically ideal where the load is distributed continuously along its length. In the open spandrel arch the dead load of the rib is distributed, but dead and live loads from the deck are applied to the arch as a series of point loads. In this case a conventional arch with a moment-resisting rib may be designed as shown in Figure 10-4*p*. For a practical solution nonzero design bending moments normally result from moving loads and, alternatively, they may be resisted by the deck. The latter structure is a *stiffened deck arch* as shown in Figures 10-4*q* and *r*, and is essentially similar to the stiffened suspension bridge.

10-2 EXAMPLES AND MAIN FEATURES

From the foregoing it follows that the arch form is intended to reduce bending moments in the superstructure with a corresponding reduction in material compared to an equivalent straight, simply supported girder or truss. This efficiency results from the activation of horizontal reactions in the arch rib. As external forces, they can be sustained if suitable ground is available such as an arch foundation on dry rock slopes. Conversely, the conventional curved arch rib is likely to have a higher fabrication and erection cost. The erection problem, in particular, relates to the type of structure, with the cantilever and tied arch being the simplest and the most difficult to construct, respectively. The stability of the latter is markedly improved when the horizontal reactions become available upon completion of the deck.

Arches respond predominantly in compression. For example, the open spandrel arch with the rib below the deck has as main parts a deck, spandrel columns, and arch rib. Of these, the last two are compression members.

However, the stiffened deck design results in a slender arch rib that may have additional structural requirements associated with buckling.

Most of the analysis of elastic arches presented in this chapter applies equally to steel and concrete construction. Nonetheless, the classic arch form shown in Figure 10-4*a* tends to favor the use of concrete as construction material. This arrangement is advantageous where the arch rib can be shaped to resist the dead load without bending moments. This load is referred to as the *form load*. If it is the major fraction of the total load, the live load becomes essentially a minor disturbance applied to a compressed member. Under these conditions, the conventional first-order elastic theory is not adequate because it produces results on the unsafe side, and some form of deflection theory must be introduced in the analysis (O'Connor, 1971). Invariably, the effects of initial imperfections in the arch shape become significant.

Representative Examples

Structural Arrangement In concrete construction, the arch proper usually consists either of a series of independent solid ribs or of a single solid barrel. These arrangements are shown in Figure 10-5 together with other possible configurations.

For the type shown in Figure 10-5*b*, the load is carried to the supports by two or more independent solid ribs generally connected together with cross struts or lateral bracing. A suitable variation is the single solid-rib section shown in Figure 10-5*c*, usually referred to as the *barrel*, which generally is less economical in materials but more rigid under load. In some instances a hollow box, rib, or barrel is used, and sometimes the solid rib is replaced by a webbed or latticed concrete girder (Figures 10-5*d* and *e*), respectively. In the foregoing types the arch rib or barrel is the principal structural unit supporting the bridge deck as shown in Figure 10-5*a*. For barrel arches, the column and girder construction may be replaced by the spandrel walls shown in Figure 10-5*f*.

The braced spandrel shown in Figure 10-4*h*, although frequently used in steel arches, is not common in concrete construction.

Typical arrangements in steel arches are shown in Figure 10-6. For rib arches, plate girder sections, rolled I beams, H beams, or girder beams, and the hollow or box girder section may be used. The braced spandrel type in steel arches requires details and fabrication similar to ordinary trusses.

Examples The Rainbow Bridge shown in Figure 10-7 spans the Niagara River close to the falls, and connects the United States and Canada. It is a hingeless steel rib with a span of 950 ft and rise of 150 ft, giving a rise–span ratio of 0.158.

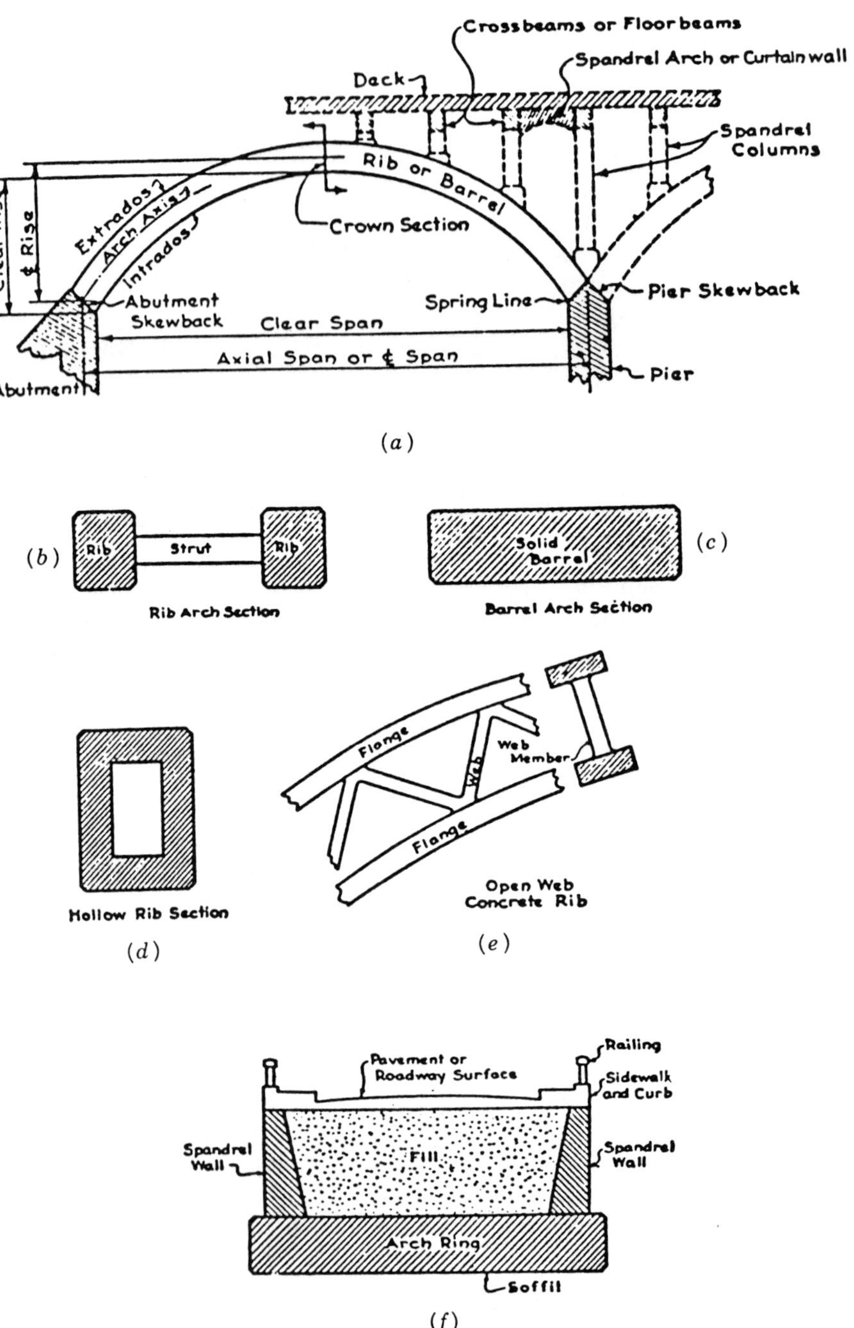

FIGURE 10-5 Structural arrangements in concrete arches.

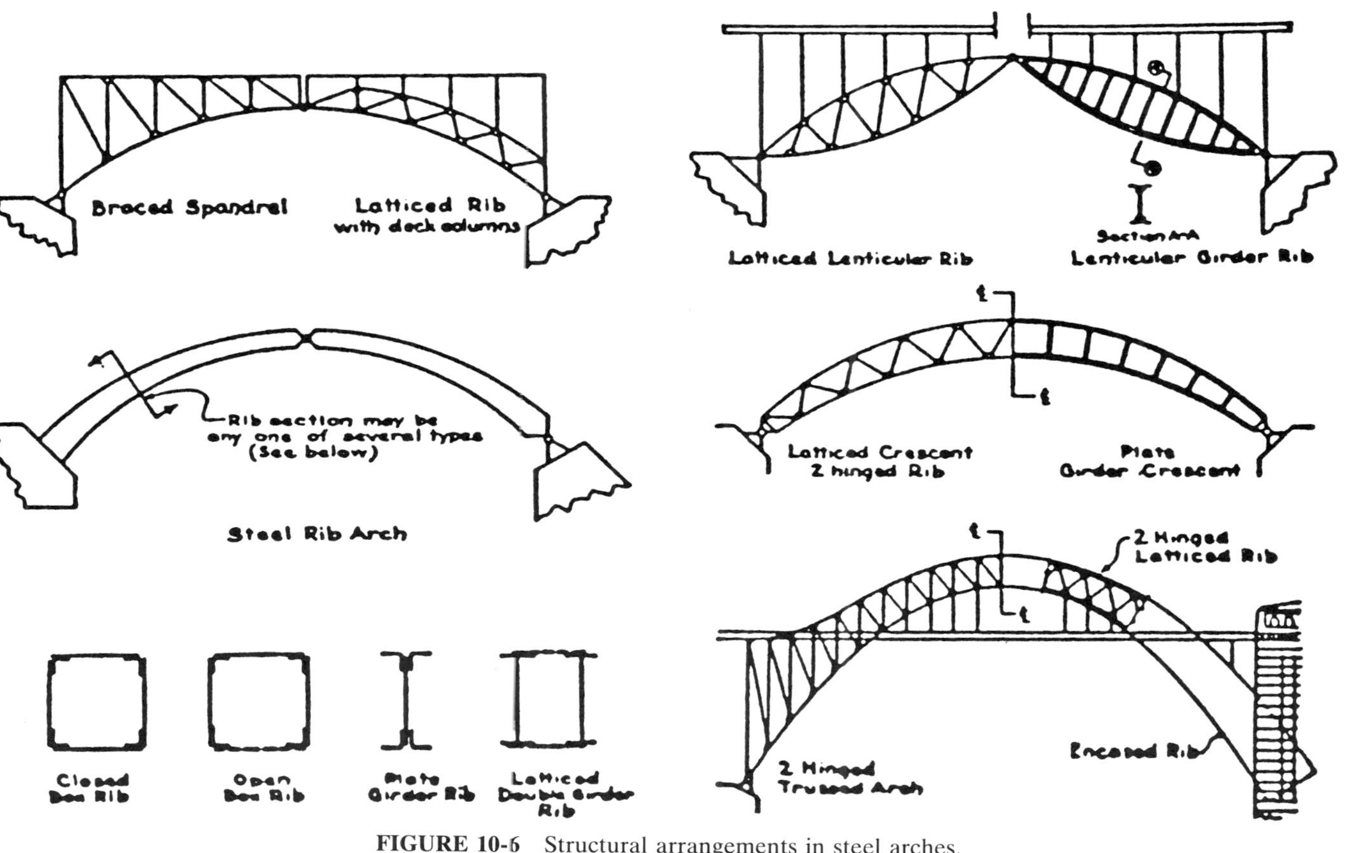

FIGURE 10-6 Structural arrangements in steel arches.

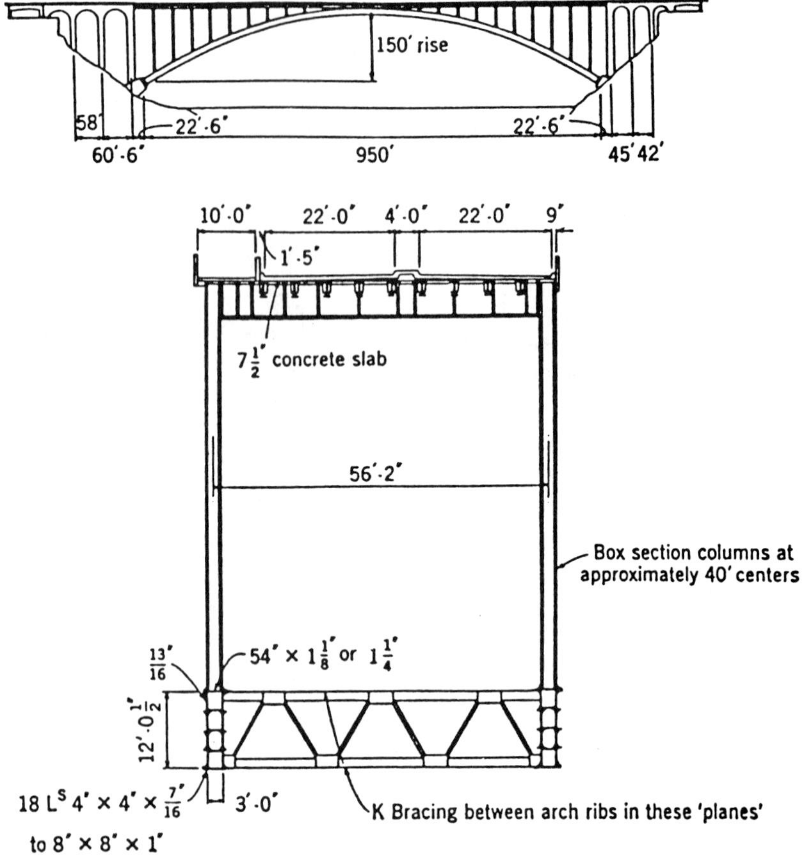

FIGURE 10-7 Rainbow Bridge.

The arch rib consists of twin steel riveted boxes 3 ft wide and 12 ft deep, with an additional horizontal plate at middepth and stay plates connecting longitudinal angle stiffeners at the quarter points of the depth. The bridge has a span–depth ratio of 79. Lateral bracing is also provided in the deck and following the top and bottom faces of the arch rib. Truss diaphragms connect the two arch ribs at the spandrel column locations (Hardesty et al., 1945).

Figure 10-8 shows the bridge at Fehmarnsund, Germany (Bouchet, 1964; Stein and Wild, 1965). This structure connects the German mainland and the island of Fehmarn in the Baltic Sea. The main crossing has a span of 816 ft, and carries a 36-ft roadway and a single-track rail. This concept is an example of the Neilsen system developed in Scandinavia, and consists of a bowstring girder with inclined hangers. The two arch ribs are, likewise, inclined inward and are connected near the crown as shown in the cross section. Large-diameter tubular struts connect the ribs at lower levels. The

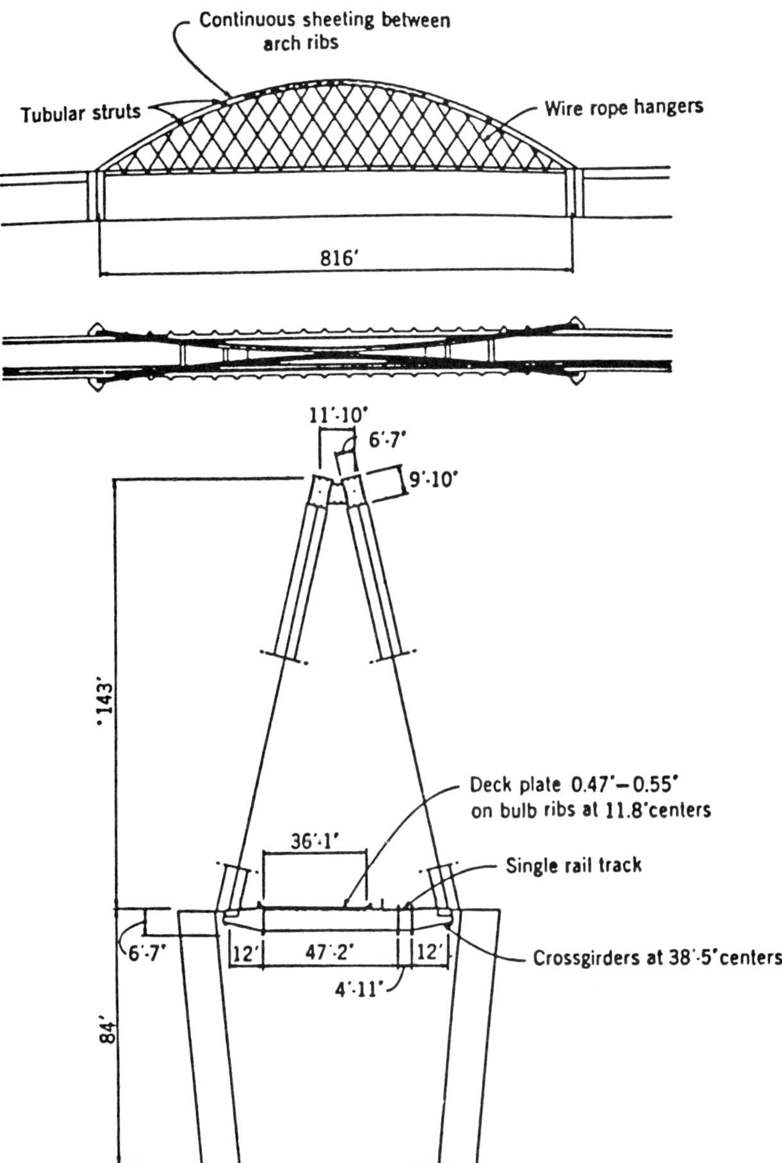

FIGURE 10-8 Bridge at Fehmarnsund, Germany.

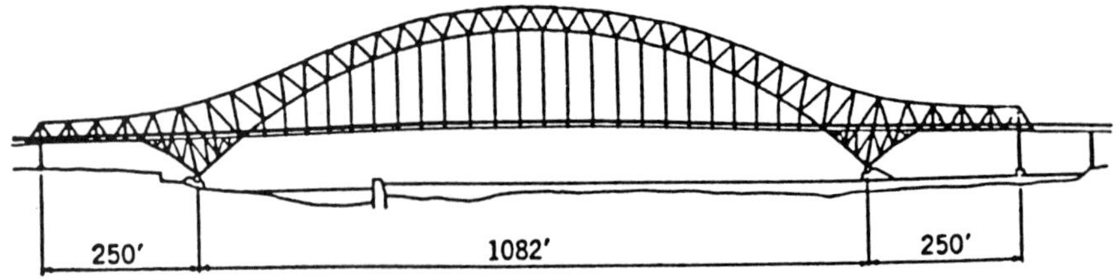

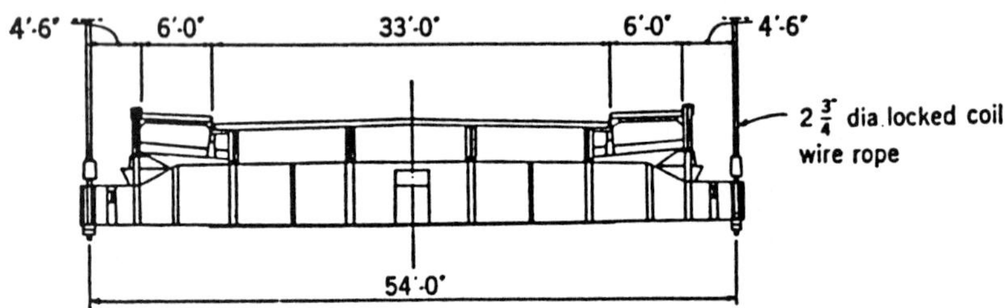

FIGURE 10-9 Runcorn–Widnes Bridge, England.

orthotropic steel deck is supported on three longitudinal stringers between cross girders hung by wire ropes from the arches. The rise–span ratio is 0.176, and the span–depth ratio is 83.

The Runcorn–Widnes Bridge (England) shown in Figure 10-9 is a representative example of a cantilever truss arch. Many bridges of a similar design have been built in the United States; notable examples are the Newark Bay Bridge, turnpike bridges over the Delaware and Susquehannah rivers, and bridges over the Cape Cod and other canals. The present example crosses the mouth of the Massey River near Liverpool and also the Manchester ship canal. The main span is 1082 ft long, and the side–main span ratio is smaller than is customary. The bridge has a concrete deck supported on stringers. These span between cross girders supported by wire ropes from the arch above. Lateral bracing systems are inserted in the deck and at the levels of the upper and lower chords of the arch. There is no lateral bracing between the upper chords of the side spans. The ends of the bridge are held down by two pairs of vertical links.

The Glemstal Bridge near Stuttgart, Germany, shown in Figure 10-10, is a prestressed concrete structure with a main span of 374 ft. The arch configuration approximates an inclined leg frame bridge of the type shown in Figure 10-4m, but the legs are curved and merged into the deck in a manner that allows the structure to retain its normal arch profile. The legs are divided, and the halves slightly splayed as shown. Interestingly, the arch supports the deck where they meet near the crown, and this connection forms a rigid joint. The bridge span between the crown point and the first pier is considerable,

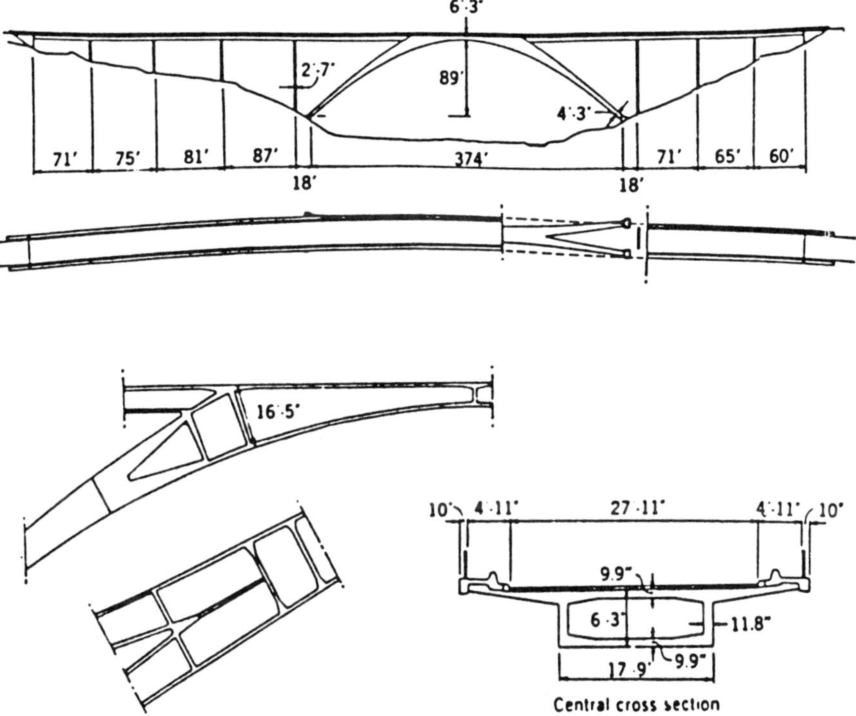

FIGURE 10-10 Glemstal Bridge, near Stuttgart, Germany.

but the associated bending stresses are effectively resisted by the prestressing of the deck.

Technical Comments Collectively, arch bridges are considered the most successful of all bridge types in esthetic terms, and the curved shape is invariably pleasing. The visual advantage is somewhat reduced for the bowstring girder or where the arch rises through the deck, but even in such cases the arch can be made particularly attractive.

The inclined leg frame is suitable for conventional grade separations, for example, overpasses intersecting four-lane divided highways. The bridge shown in Figure 10-11 has five longitudinal frames, each frame consisting of a beam simply supported at the ends on roller bearings. The beam is also rigidly connected to inclined legs at two points for extra support and rigidity, fixed at their ends to the abutments but free to rotate. The structure is thus a two-pinned portal with inclined legs and with horizontal projecting propped cantilevers forming open spandrels.

The bridge frames are rigidly connected by diaphragms and stiffeners that enable all five beams to provide a composite action. The roadway consists of

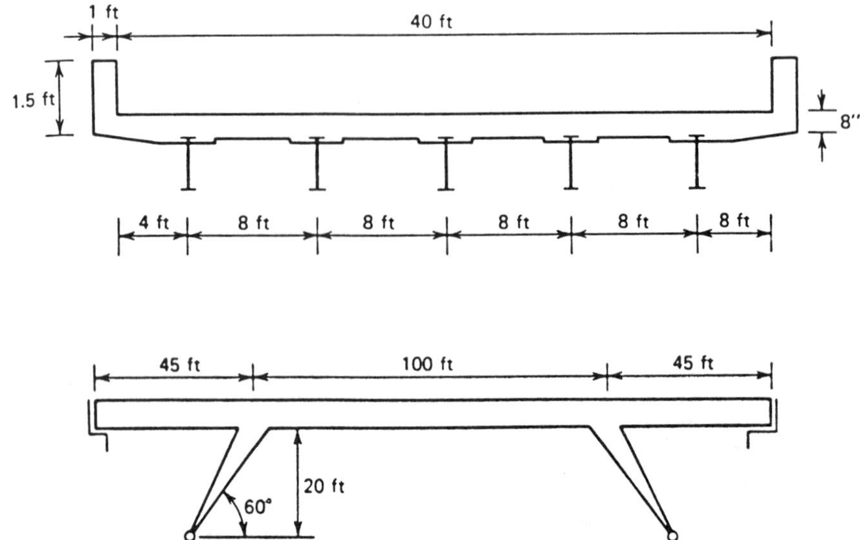

FIGURE 10-11 Welded bridge with inclined leg frame suitable for conventional grade separations.

a concrete slab. The following considerations influence the form of this structure (Clark, 1950).

1. The loads are brought directly to the abutments without the use of transverse or longitudinal secondary load-carrying members.
2. The introduction of the two inclined legs reduces the simply supported beam bending moments at the center of the span. For most and certainly for the worst conditions of loading, the end beam reactions are downward, and the props together with the anchorage act as prestressing agents thereby reducing the stresses at the center of the beam.
3. As a result, the weight of the structure is reduced in the middle section of the span and concentrates more near the supports, further reducing dead load moments.
4. The inclined steel flange plate of the main beam acts as part of the deck and as shuttering for the concrete, thus making the deck unit more economical.
5. The design enables the fabrication of large sections in the shop and reduces field welding to a minimum. At the site, only three structural elements require welding.
6. The bridge may be readily erected without the need of auxiliary, temporary supporting units.

7. The structure exposes relatively small surface areas to transverse or wind forces. All steel parts are readily accessible for painting and can be rendered waterproof and protected against corrosion.
8. The design and construction is relatively simple and economically competitive with other grade separation structures.

10-3 ERECTION PROCEDURES

Arch bridges are constructed by one of the following procedures: (a) use of falsework; (b) progressive cantilevering from the abutments; and (c) special techniques identified and detailed in the design stage. Erection may often be complex and add markedly to the total cost of the structure.

Erection on Falsework

The Fehmarnsund Bridge shown in Figure 10-8 is a typical example of an arch structure erected on falsework. As tied arches, bridges of this type have the common difficulty that the arch supports have no substantial horizontal capacity. Thus, the design had to rely on the substantial approach spans to provide sufficient reaction capacity to resist the tieback system used in the erection of the arch ribs near the abutments. In addition, a central falsework tower has built to provide access and support for the crown of the arches. The arch ribs were erected first, followed by the deck.

The five-span Vila Franca Bridge in Portugal (Upstone and Cardno, 1954) was erected span by span on a special erection truss as shown in Figure 10-12*a*. The truss was moved into position on pontoons floating in the river, and spanned between piers. The reuse prompted by the multiple-span construction rendered this erection procedure economically attractive.

The Kaiserlei Bridge between Frankfurt and Offenbach in Germany (Hartwig, 1965) was supported conventionally from falsework in the river as shown in Figure 10-12*b*, and erected by a floating crane. The deck was erected first, followed by the arch ribs.

Cantilever Erection

Cantilever arch bridges are usually erected by constructing first the side spans on falsework and then cantilevering into the main span. The simplified procedure represents a major advantage of this arch form. For other arches, tiebacks are usually necessary. As an example, the Randbow Bridge shown in Figure 10-7 was erected using the procedure shown in Figure 10-12*c*. The derrick at deck level was provided to accommodate erection and supply materials. The erection of the arch ribs was continued beyond the reach of the first derrick by providing an additional traveler moving on the arch rib

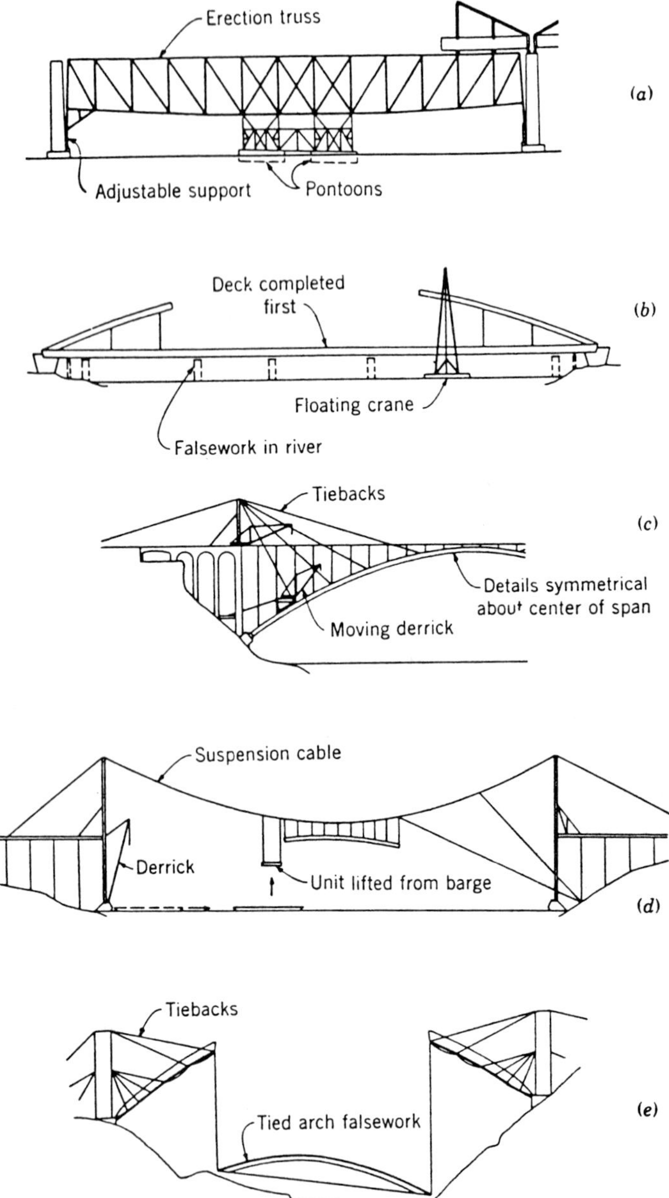

FIGURE 10-12 Erection procedures for arch bridges: (*a*) Marshal Carmona Bridge at Vila Franca de Xira; (*b*) Kaiserlei Bridge; (*c*) Rainbow Bridge; (*d*) bridge over Askerofjord; (*e*) bridge near Caracas.

itself. The moving derrick and the tieback system can be clearly seen in the illustration.

Hazelet and Wood (1961) describe a two-span double-deck, tied arch bridge with a truss rib, built across the Ohio River. In this case one span was erected on falsework and thereafter provided a countersupport for the tieback system used to erect the second span. This procedure satisfied the navigation requirements in the river and partly dictated the design.

Special Techniques

From the infinite variety of special erection procedures, we present two representative examples.

The bridge over the Askerofjord in Sweden (Hartwig and Hafke, 1961) was erected with the use of a pair of suspension cables as shown in Figure 10-12*d*. Arch units were lifted from the bank and placed on barges floated into position, and then lifted with the aid of suspension cables as shown. Erection began at the crown and continued symmetrically to the abutments.

An arch bridge near Caracas, Venezuela (*Engineering News Record*, 1952; Freyssinet, Muller, and Shama, 1953) has concrete ribs cast in situ on a special falsework shown in Figure 10-12*e*. The falsework trusses for the end quarters of the span were hung using tiebacks from the abutment piers. The central falsework truss was assembled in the valley floor and then lifted into position from the cantilever sections.

Arch Rib Closure

Most arch ribs are erected simultaneously from either end starting at both abutments with the intent to meet at the crown. A geometrically correct closure is therefore achieved if the fabrication and erection alignment comply with the requirements of the geometric design. On rib closure, the dead load must be transferred from the temporary supports to the ribs themselves. This is usually attained by jacking the ribs apart at the crown. This implies that provisions must be included to accommodate the jacking devices and make adjustments for possible construction errors. Details are typically worked out in the design stage and incorporated in the construction plans.

10-4 FUNDAMENTAL THEORY OF ARCHES

The First Principle of Arch Action (Line of Pressure)

For a given span and rise, the correct axial curve for an arch depends on the loads alone. This can be illustrated in Figure 10-13. Let us assume that it is necessary to support the loads F_a and F_b by thrust only. With no bending in any member, the load F_a is supported entirely by thrusts 1 and 2, as shown

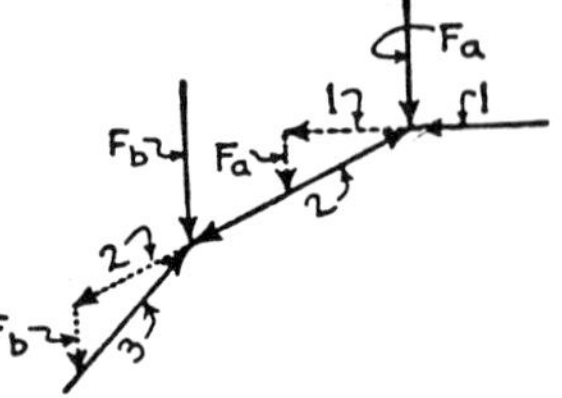

FIGURE 10-13 Forced diagram for line-of-thrust theory.

by the closed force triangle F_{a-2-1}. Likewise, the force F_b must be in equilibrium with thrusts 2 and 3, and so on for the remaining loads and arch elements. A change in any load (for instance, F_b) would require a change in the direction or magnitude of thrust 2 or 3. Any change in thrust 2 would necessitate a change in the amount or direction of thrust 1 to maintain equilibrium with F_a, and so on. The thrust line (or line of pressure) for concentrated loads is therefore a polygon with vertices at the load points, and may be considered so for a uniform loading as well (in the latter case the load spacing and side lengths of the polygon would be infinitesimal, and the number of loads and sides infinite).

The General Equilibrium Polygon From the foregoing it is evident that to support any given load system by thrust only, the force polygon that provides equilibrium at the load points must be determined. The line of thrust, or line of pressure, is defined as the line whose tangent at any point is the line of action of the resultant of all forces acting on one side of that point. We can extend the geometry of Figure 10-13 by considering the funicular or link polygon for the general system of forces shown in Figure 10-14. In this case the reactions are given, including the horizontal components $H_A = H_B = H$, and the external force polygon $0, 1, 2, \ldots, 9, 0$ is drawn accordingly. The line of action of the reaction at A is parallel to 02. Its resultant with F_1 passes

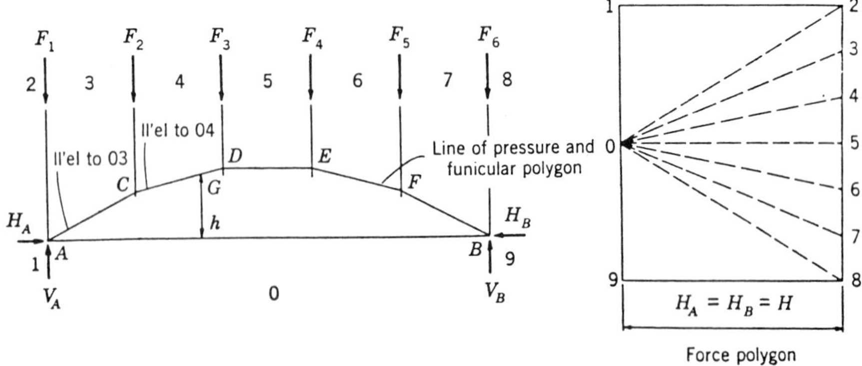

FIGURE 10-14 Funicular polygon for arch with vertical loads.

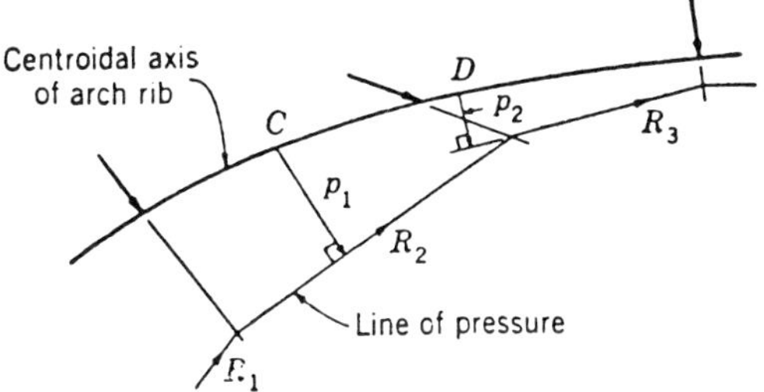

FIGURE 10-15 Bending moments from the line of pressure.

along AC, parallel to 03. Likewise, $\overline{R}_A + \overline{F}_1 + \overline{F}_2$ passes along CD, parallel to 04, and so on. An arch whose centroidal axis coincides with the line of pressure drawn for the correct reactions has zero design bending moments, and the funicular polygon therefore represents the ideal shape for the arch.

The location of any point on the line of pressure can also be found analytically. Consider, for example, point G on sector CD and at a height h above line AB. All forces to the left of G have zero moments about D, or $M_D = 0 = M_{DV} - Hh$, where M_{DV} is the bending moment at D due to the vertical loads and the vertical reaction components. From this it follows that

$$h = M_{DV}/H, \tag{10-1}$$

which shows that the rise in the line of pressure is inversely proportional to the horizontal thrust.

For a curved arch rib, the bending moment is related to the displacement of its centroid from the line of pressure. For example, for the arch shown in Figure 10-15 the bending moment at C is $R_2 p_1$, where R_2 is the resultant of all forces to the left of C and p_1 is perpendicular to R_2. Likewise, the bending moment at D is $R_3 p_2$.

If all interior loads are vertical as in Figure 10-14, the horizontal component of the resultant force at any point is the same and equal to the horizontal component H at the ends. In this case the bending moment at any point on the arch rib is Hv, where v is the vertical intercept between the line of pressure and the centroidal axis of the arch shown in Figure 10-16. This relationship is known as Eddy's theorem.

Invariably, for the line of pressure to be used in obtaining bending moments it is necessary to know the loads and reactions as well as the deflected geometry of the arch rib. If this deflection can be assumed to be negligible, the deflected geometry does not change appreciably from the original, but in certain cases this assumption lacks a technical basis.

Equilibrium Polygon for the Three-Hinged Arch In aligning the arch axis with reference to the thrust polygon (line of pressure), it is prudent to make the two coincide as nearly as possible. A polygonal arch thus derived is

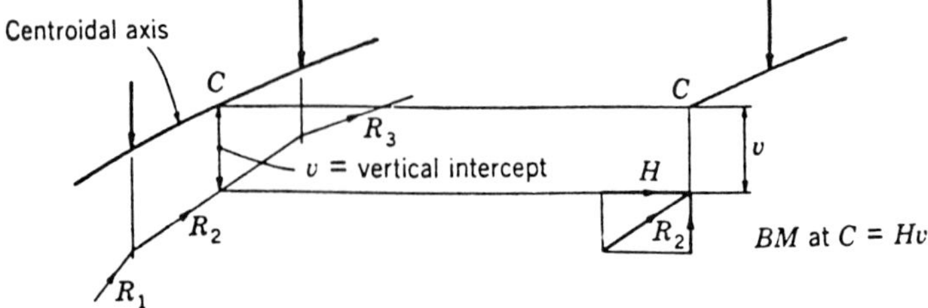

FIGURE 10-16 Bending moments for vertical loads.

therefore the most efficient type for concentrated loads such as shown in Figure 10-14. Except for elastic rib-shortening effects, such an arch would not have dead load moment at any point because there is no angular distortion. Because, however, the usual practice is to build the arch on a curve to avoid an unsightly appearance, some moment will be developed as shown in Figures 10-15 and 10-16. For statically indeterminate types (less than three hinges), this moment will cause the thrust line to shift toward the compression side of the rib as shown. In this respect the difference between the statically determinate and statically indeterminate types is illustrated in the three-hinged arch.

For the three-hinged arch shown in Figure 10-17, the thrust line (line of pressure) must pass through the three hinge centers because these must be points of zero moment. The four reaction components are found from the normal three equations of equilibrium plus a fourth condition requiring the bending moment at the central hinge to be zero. Stated otherwise, for a given load system only one polygon of thrust can be drawn through three points, and this polygon will remain fixed in space regardless of the rib alignment between hinges in relation to it. For the particular case of loads shown in Figure 10-17 (a fixed dead load and a moving live load), the two lines of

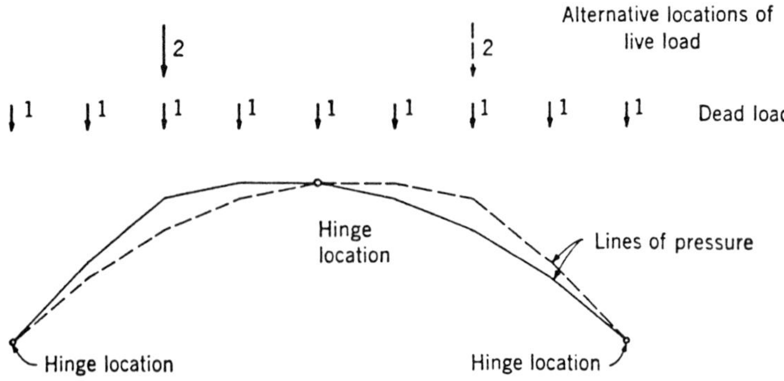

FIGURE 10-17 Lines of pressure for three-hinged arch.

pressure given correspond to two locations of the moving live load. From a comparison of the two lines of pressure, we conclude that the rib axis may be raised or lowered for a particular adjustment of movement that will result in maximum economy.

Fixed Arch Figure 10-18 shows a fixed arch supporting three loads. The theoretical thrust line (based on zero moment at the crown and skewbacks) is shown solid, and the axis is assumed to pass through the polygon vertices. Upon application of the load, negative moments are developed in the rib between load points because of its own action as a curved strut. This distortion clearly produces positive moments at each load point and also at the two skewbacks, although these points lie on the theoretical thrust line. The actual thrust line is shifted toward the extrados at the load points (shown dotted). If an axial curve is laid out circumscribing the dead load polygon, the polygon under load will take a position intermediate between its inscribed and circumscribed positions. By analogy, the same line of reasoning would be valid if the axis were first laid out inscribing the polygon. It appears therefore that an axial curve drawn midway between curves that inscribe and circumscribe the dead load pressure polygon will involve less shifting of this polygon under load and should be preferred to other types. An axis drawn in this manner is termed the *median axis* of the polygon.

The points through which the median axis will pass may be determined if we assume that a flat circular arc is approximated by a parabola. Referring to Figure 10-19, we can write

$$D = KC^2 \tag{10-2}$$

and because for flat curves $B = C/2$ (very nearly), we obtain

$$d = KB^2 = KC^2/4 = D/4 \tag{10-3}$$

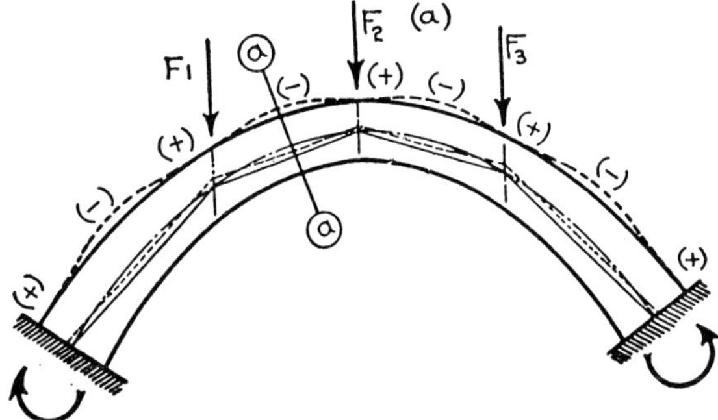

FIGURE 10-18 Lines of pressure for a fixed arch.

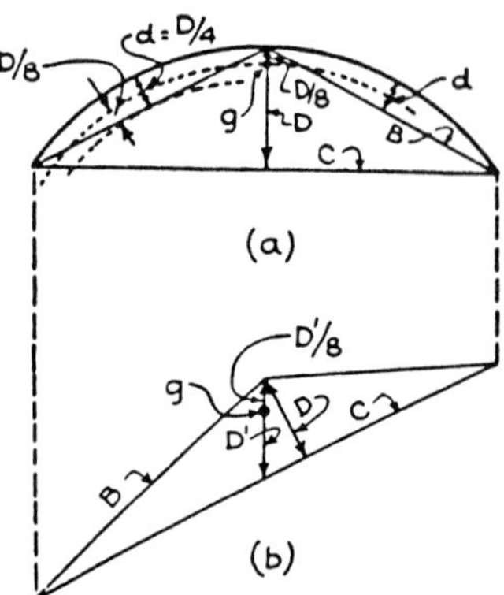

FIGURE 10-19 Median axis location.

The median axis will therefore pass through a point g at a distance D/g below the vertex. If the chord C is inclined as in Figure 10-19*b*, the vertical distance from the vertex to point g will be D'/g, where D' is likewise measured vertically.

Summary The foregoing concepts are useful and generally valid and articulate the behavior of an arch bridge. There are other conditions and aspects that will cause stress and must also be addressed, for example, arch buckling, the amplification of initial imperfections, plastic flow, support displacements, and so on.

For the preliminary work necessary before the first elastic analysis of an arch bridge can be carried out, the following should be included: (a) a superstructure design sufficiently exact to enable the dead load to be estimated; (b) a preliminary assumption with reference to the approximate size and shape of the rib or frame cross section at typical pints along the axis, as a basis for calculating the additional dead load and the structural terms; (c) selection of live load and impact specifications; (d) selection of an appropriate elastic modulus for the material to be used; (e) determination of the shape of the arch axis; and (f) a tentative temperature specification.

Structural Redundancy for Arch Bridges

The three fundamental equations of static equilibrium are

$$\sum H = 0 \qquad \sum V = 0 \qquad \sum M = 0 \tag{10-4}$$

As long as the number of external reaction components does not exceed three, these equations are sufficient and the structure is statically determinate. If the number of reaction components exceeds three, statics alone does not provide a solution and we must resort to deflection theory for structural redundancy.

The elastic theory in this case becomes useful and relevant because for indeterminate structures we must consider elastic distortions in order to develop additional equations to match the number of unknowns. Thus, in addition to the static equations in (10-4), we also introduce elastic equations, and any structure thus becomes completely susceptible to analysis by means of the elastic theory.

Figure 10-20*a* shows a simple curved beam. The support at V_1 is on rollers, implying that H_1 is zero. The unknown reaction components are three, and the three static equations provide a complete solution. In Figure 10-20*b* the roller is converted into a hinged link that still permits free movement of the upper beam about the lower hinge as a center of rotation. The reaction R_1 has a known direction and passes through the lower hinge pin. This reaction together with the components H_2 and V_2 yields three unknowns, and the structure is still statically determinate.

In Figure 10-20*c* the upper hinge is merely moved further. Structurally, the system is not altered, and the new shape known as the *three-hinged arch* is fully determinate from statics although there are four unknown components. In this case the crown hinge provides a fourth condition or equation, because the moment at this point must always be zero. The left-hand reaction (the resultant of H_1 and V_1) for any load to the right of the crown hinge has a known direction because it must pass through said hinge point. Thus, for Figure 10-20*c* the following are derived:

$$R_1 p = Fa \qquad V_1 = R_1 \cos\theta \qquad H_1 = R_1 \sin\theta \qquad H_2 = H_1 \qquad V_2 = F - V_1 \tag{10-5}$$

For the two-hinged arch shown in Figure 10-20*d*, the unknown reaction components are four. Hence, this is the first structural type for which we must resort to the theory of elasticity. Interestingly, any reaction component statically indeterminate is also unnecessary to stability; hence, such reaction components are termed *redundant* or unnecessary forces. In the present case we can remove H_1 merely by placing the left support on a roller nest as shown in Figure 10-20*e*. The structure continues to be stable as a curved simple beam, but the stresses in the rib are modified. For any given load system and with H_1 removed, the structure moves freely horizontally by a displacement Δ_x. It follows now that the original value of H_1 (the redundant force removed) is just sufficient to counteract the displacement Δ_x and to hold the structure in a fixed position horizontally. If we remove H_1 and analyze the rib as a statically determinate structure for a given load system and an unknown horizontal force X applied at the left support, we can formulate an expression for Δ_x in terms of the known loads and the unknown force X. Setting now $\Delta_x = 0$ and solving, we obtain the value of X necessary to hold the structure in its original fixed position at point a. This is the value of the desired reaction H_1.

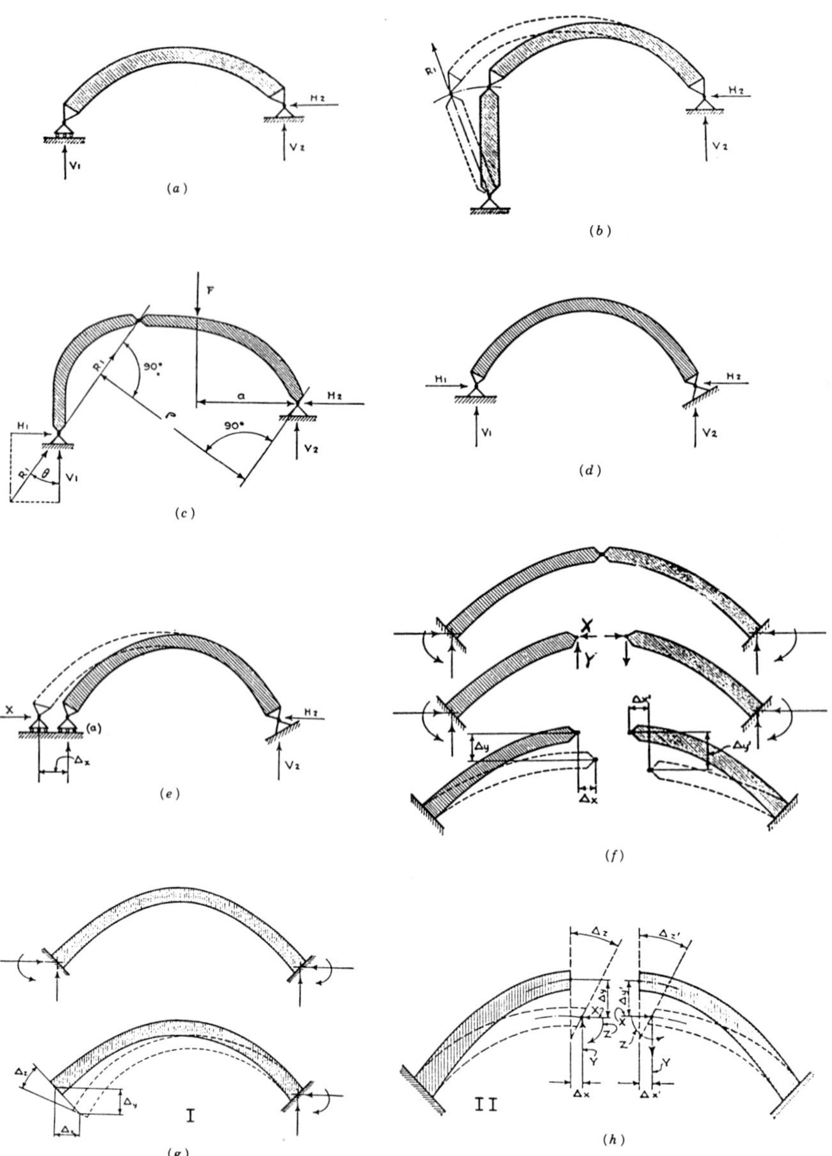

FIGURE 10-20 Structural redundancy of arches.

For the single-hinged arch shown in Figure 10-20f, it may first appear that we have six unknowns (three at each skewback), but by taking each half of the arch as a free body the number is reduced to five. Three equations are provided by statics, and two more are derived from the fact that the crown hinge must have the same relative movement with respect to either skewback. In this context

$$\Delta_x = \Delta_{x'} \qquad \Delta_y = \Delta_{y'} \tag{10-6}$$

The solution is now possible by removing the two redundant crown reactions so as to develop a residual structure or two simple cantilever spans. Each cantilever span is in equilibrium under the action of an external load system, the three reaction components at the skewback, and the two unknown crown forces X and Y introduced to replace the effect of the redundant reactions at this point. The three static equations are sufficient for determining the skewback reactions. The two elastic relationships in (10-6) suffice for computing the unknowns X and Y.

Figure 10-20g shows the fully fixed or hingeless arch type. This case clearly involves six unknowns of which three are statically determinate. For a complete solution three elastic equations are necessary. For scheme I we remove the left support and develop the following relationships to satisfy the fact that the support is fixed and unyielding:

$$\Delta_x = 0 \qquad \Delta_y = 0 \qquad \Delta_2 = 0 \tag{10-7}$$

Alternatively, we may select scheme II shown in Figure 10-20h and develop the following three equations:

$$\Delta_x = \Delta_{x'} \qquad \Delta_y = \Delta_{y'} \qquad \Delta_z = \Delta_{z'} \tag{10-8}$$

Summary and Conclusions The entire procedure of structural redundancy can be summarized as follows.

1. From statics we obtain three equations of equilibrium.
2. For the three-hinged arch the unknown reaction components are four, but the crown hinge provides the fourth condition of equilibrium, so that this structure is statically fully determinate.
3. For all other arch types the unknowns exceed the equations of statics, and for a solution we must resort to elastic theory.
4. Unknown reaction components are divided into two categories: (a) components necessary to maintain structural stability and (b) redun-

dant components necessary to provide a special and fixed position, but which add nothing to stability.

5. Because the redundant reactions are irrelevant to equilibrium, we may remove them and replace them with unknown forces. These can be treated as external loads except that their value is not known. The statically determinate frame resulting from this scenario is called the *residual frame* (or base structure).
6. Using the residual frame as a base, we can write relevant deflection or displacement expressions that involve the given load system and the unknown reactions.
7. We can write additional equations by setting these deflections or displacements to zero or to a certain value. From these we can evaluate the unknown forces of the redundant reactions and proceed with the complete analysis.

In conclusion, the so-called elastic theory can be reduced to the following steps: (a) removal of the redundant forces developing a statically determinate residual frame; (b) replacing the removed redundants by an equivalent number of unknown external forces; and (c) determination of these unknown forces using suitable deflection equations.

Internal Redundancy In the foregoing discussion we have assumed the redundant conditions to be external (reaction components). This procedure serves expediency, and for ribbed or solid webbed arches, it is the only method of analysis. For framed structures, however, we may select certain internal members as redundant conditions.

For the fixed framed arch shown in Figure 10-21, the members a–b, c–d, and e–f may be removed to convert the structure into a statically determinate residual frame, namely, a three-hinged arch.

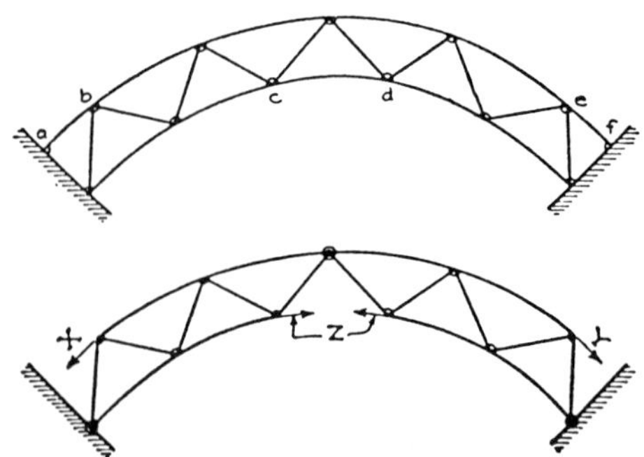

FIGURE 10-21 Internal redundancy for a framed arch.

Multispan Arches on Elastic Piers

The elastic analysis introduced for single-span arches can be expanded to include an arch system of any number of spans and where the intermediate supports are to undergo an elastic yielding under loads.

We can consider the two-span structure shown in Figure 10-22. The end skewbacks are supported by rigid and unyielding abutments. The center pier is also rigid but slender so as to yield or distort under load. In this case we have three sets of unknown reaction components at three locations, for a total of nine, namely,

$$H_1, V_1, \text{ and } M_1$$
$$H_2, V_2, \text{ and } M_2$$
$$H_3, V_3, \text{ and } M_3$$

Three equations are written from statics, and the other six can be obtained as elastic equations for deflections, noting that the movement of point a must be exactly the same when (a) regarded as the right-hand terminal point of arch 1, (b) regarded as the left-hand terminal point of arch 2, and (c) regarded as the upper terminal point of the pier.

Next, we divide the complete elastic system into its three components by means of sections a–a'–a'' and write the following equations:

From statics = three equations

Also
$$\begin{matrix} \Delta_x = \Delta_{x''} & \Delta_y = \Delta_{y''} & \Delta_z = \Delta_{z''} \\ \Delta_x = \Delta_{x'} & \Delta_y = \Delta_{y'} & \Delta_z = \Delta_{z'} \end{matrix} \qquad (10\text{-}9)$$

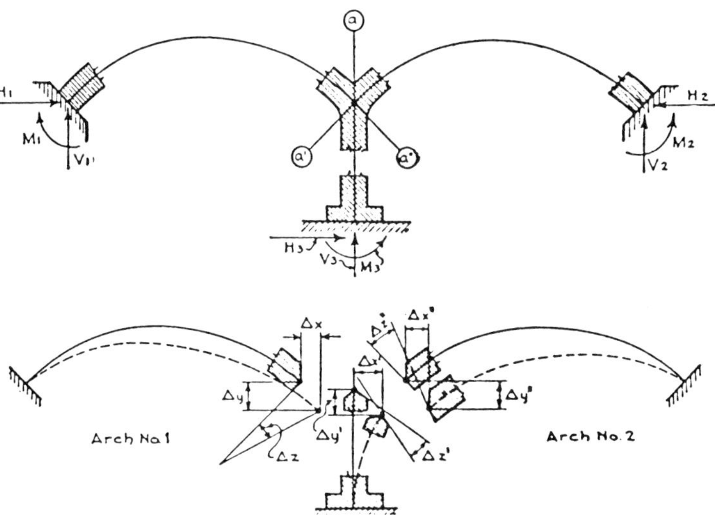

FIGURE 10-22 Two-span arch structure; elastic equilibrium.

for a total number of possible independent equations equal to nine. This number is the same as the number of unknowns; hence, a complete solution is possible.

For a three-span structure there are twelve unknown reaction components, and a complete solution is likewise obtained by writing nine elastic equations.

10-5 FIRST-ORDER ARCH ANALYSIS

First-order arch analysis essentially requires the application of small-deflection theory. In this method the effects of deformations upon force effects in the structure are neglected. Conversely, second-order analysis is based on large-deflection theory, whereby the effects of deformation of the structure upon force effects are taken into account.

A single arch rib with the deck removed is, at the most, indeterminate to the third degree. The hingeless arch considered in the foregoing section and shown in Figure 10-20g will be analyzed now for the general case shown in Figure 10-23. The residual frame chosen in this case (also called the base structure) has a pin at one end (right) and a roller at the other (left). The redundant reaction components are as shown.

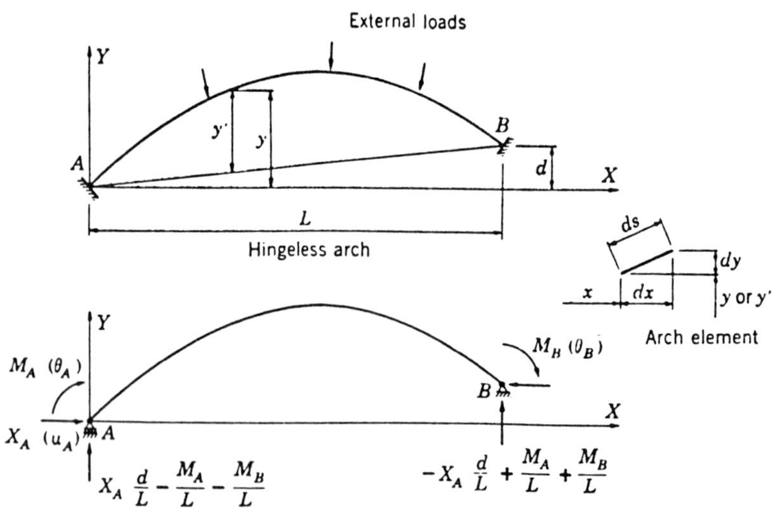

FIGURE 10-23 First-order analysis of a hingeless arch.

At any point (x, y) on the arch, the bending moment M and thrust T are given by

$$M = M_S + M_I \tag{10-10}$$

$$= M_S + M_A\left(1 - \frac{x}{L}\right) - M_B\frac{x}{L} - X_A y' \tag{10-10a}$$

$$T = T_S + T_I \tag{10-11}$$

$$= T_S - \frac{M_A}{L}\left(\frac{dy}{ds}\right) - \frac{M_B}{L}\left(\frac{dy}{ds}\right) + X_A\left\{\left(\frac{dx}{ds}\right) + \frac{d}{L}\left(\frac{dy}{ds}\right)\right\} \tag{10-11a}$$

where M_S and T_S are the moment and thrust, respectively, due to the external loads on the determinate base structure, and M_I and T_I are the moment and thrust, respectively, due to the indeterminate reactions (redundant reaction components on the residual frame). In our notation M is positive if it causes tension in the lower fibers, and T is positive if it is compressive.

Three elastic equations in M_A, M_B, and X_A can be written from the conditions that the rotations θ_A and θ_B and displacement u_A are zero.

Ignoring shear strains, the redundant reaction components are

$$\theta_A = \int \frac{M(\partial M/\partial M_A)}{EI}\,ds + \int \frac{T(\partial T/\partial M_A)}{EA}\,ds \tag{10-12}$$

$$\theta_B = \int \frac{M(\partial M/\partial M_B)}{EI}\,ds + \int \frac{T(\partial T/\partial M_B)}{EA}\,ds \tag{10-13}$$

$$u_A = \int \frac{M(\partial M/\partial X_A)}{EI}\,ds + \int \frac{T(\partial T/\partial X_A)}{EA}\,ds \tag{10-14}$$

$$\frac{\partial M}{\partial M_A} = \left(1 - \frac{x}{L}\right) \qquad \frac{\partial M}{\partial M_B} = -\frac{x}{L} \qquad \frac{\partial M}{\partial X_A} = -y' \tag{10-15}$$

$$\frac{\partial T}{\partial M_A} = -\frac{1}{L}\frac{dy}{ds} \qquad \frac{\partial T}{\partial M_B} = -\frac{1}{L}\frac{dy}{ds} \qquad \frac{\partial T}{\partial X_A} = \frac{dx}{ds} - \frac{d}{L}\frac{dy}{ds} \tag{10-16}$$

Equations (10-12) through (10-14) can be written as

$$\theta_A = \Delta_1 + M_A\delta_{11} + M_B\delta_{12} + X_A\delta_{13} = 0 \tag{10-17}$$

$$\theta_B = \Delta_2 + M_A\delta_{12} + M_B\delta_{22} + X_A\delta_{23} = 0 \tag{10-18}$$

$$u_A = \Delta_3 + M_A\delta_{13} + M_B\delta_{23} + X_A\delta_{33} = 0 \tag{10-19}$$

where

$$\Delta_1 = \int \frac{M_S}{EI}\left(1 - \frac{x}{L}\right)\left(\frac{ds}{dx}\right) dx - \frac{1}{L}\int \frac{T_S}{EA}\left(\frac{dy}{dx}\right) dx \tag{10-20a}$$

$$\Delta_2 = -\int \frac{M_S}{EI}\left(\frac{x}{L}\right)\left(\frac{ds}{dx}\right) dx + \frac{1}{L}\int \frac{T_S}{EA}\left(\frac{dy}{dx}\right) dx \tag{10-20b}$$

$$\Delta_3 = -\int \frac{M_S}{EI}(y')\left(\frac{ds}{dx}\right) dx + \int \frac{T_S}{EA} dx - \frac{d}{L}\int \frac{T_S}{EA}\left(\frac{dy}{dx}\right) dx \tag{10-20c}$$

$$\delta_{11} = \int \frac{1}{EI}\left(1 - \frac{x}{L}\right)^2\left(\frac{ds}{dx}\right) ds + \frac{1}{L^2}\int \frac{1}{EA}\left(\frac{dy}{dx}\right)^2\left(\frac{dx}{ds}\right) dx \tag{10-20d}$$

$$\delta_{12} = -\int \frac{1}{EI}\left(1 - \frac{x}{L}\right)\left(\frac{x}{L}\right)\left(\frac{ds}{dx}\right) dx + \frac{1}{L^2}\int \frac{1}{EA}\left(\frac{dy}{dx}\right)^2\left(\frac{dx}{ds}\right) dx \tag{10-20e}$$

$$\delta_{13} = -\int \frac{1}{EI}\left(1 - \frac{x}{L}\right)y'\left(\frac{ds}{dx}\right) dx - \frac{1}{L}\int \frac{1}{EA}\left(\frac{dy}{dx}\right)\left(\frac{dx}{ds}\right) dx + \frac{d}{L^2}\int \frac{1}{EA}\left(\frac{dy}{dx}\right)^2\left(\frac{dx}{ds}\right) dx \tag{10-20f}$$

$$\delta_{22} = \int \frac{1}{EI}\left(\frac{x}{L}\right)^2\left(\frac{ds}{dx}\right) dx + \frac{1}{L^2}\int \frac{1}{EA}\left(\frac{dy}{dx}\right)^2\left(\frac{dx}{ds}\right) dx \tag{10-20g}$$

$$\delta_{23} = \int \frac{1}{EI}\left(\frac{x}{L}\right)y'\left(\frac{ds}{dx}\right) dx - \frac{1}{L}\int \frac{1}{EA}\left(\frac{dy}{dx}\right)\left(\frac{dx}{ds}\right) dx + \frac{d}{L^2}\int \frac{1}{EA}\left(\frac{dy}{dx}\right)^2\left(\frac{dx}{ds}\right) dx \tag{10-20h}$$

$$\delta_{33} = \int \frac{1}{EI}(y')^2\left(\frac{ds}{dx}\right) dx + \int \frac{1}{EA}\left(\frac{dx}{ds}\right) dx - \frac{2d}{L}\int \frac{1}{EA}\left(\frac{dy}{dx}\right)\left(\frac{dx}{ds}\right) dx + \frac{d^2}{L^2}\int \frac{1}{EA}\left(\frac{dy}{dx}\right)^2\left(\frac{dx}{ds}\right) dx \tag{10-20i}$$

The foregoing expressions include 14 different integrals, five of which relate to external loads. They may be evaluated by Simpson's rule or other methods of numerical integration (Institute of Civil Engineers, 1956). Useful references also include books on the theory of elasticity such as Lowe (1944). The following functions may be tabulated:

$$x, y, \frac{dy}{dx}, \frac{dx}{ds}, \frac{ds}{dx}, EI, EA, M_s, T_s$$

The derivative dx/ds is the reciprocal of ds/dx, and

$$\frac{ds}{dx} = \left[1 + \left(\frac{dy}{dx}\right)^2\right]^{1/2}$$

If the supports are at the same level, the terms in d/L disappear. Likewise, if axial strains are negligible, the associated terms involving EA may be assumed to be zero.

For a two-hinged arch, the only redundant is X_A and can be obtained from (10-19), which becomes

$$u_A = \Delta_3 + X_A\delta_{33} = 0 \tag{10-21a}$$

and

$$X_A = -\frac{\Delta_3}{\delta_{33}} \tag{10-21b}$$

10-6 MISCELLANEOUS TOPICS

Elastic Center

The evaluation of redundant force components and the associated compatibility conditions are invariably based on known deflections of the structure. For the analysis presented in Section 10-5, these conditions correlate bending moments and thrusts with structure geometry, and for the particular case shown in Figure 10-23, this implies that $\theta_A = \theta_B = u_A = 0$, corresponding to the redundant forces M_A, M_B, and X_A. Alternatively, we may choose scheme II of Figure 10-20h.

For example, consider the complete structure cut as shown in Figure 10-24. The moments at this section as well as the thrusts must be such that the relative deflections θ_0, u_0, and v_0 must be zero. The corresponding forces M_0, X_0, and Y_0 produce moments and thrusts as follows:

$$M = M_0 - X_0 y_0 + Y_0 x_0 \tag{10-22}$$

$$T = X_0 \frac{dx_0}{ds} + Y_0 \frac{dy_0}{ds} \tag{10-23}$$

and also

$$\frac{\partial M}{\partial M_0} = 1 \qquad \frac{\partial M}{\partial X_0} = -y_0 \qquad \frac{\partial M}{\partial Y_0} = x_0 \tag{10-24}$$

$$\frac{\partial T}{\partial M_0} = 0 \qquad \frac{\partial T}{\partial X_0} = \frac{dx_0}{ds} \qquad \frac{\partial T}{\partial Y_0} = \frac{dy_0}{ds} \tag{10-25}$$

From the foregoing analysis M_I (moment due to indeterminate reactions) is a linear function of the coordinates x and y and may be expressed as a linear function of any coordinate system x_0, y_0. We can set $M_I = B_1 + B_2 x_0 + B_3 y_0$. However, T_I cannot be expressed in this form. Accordingly, we

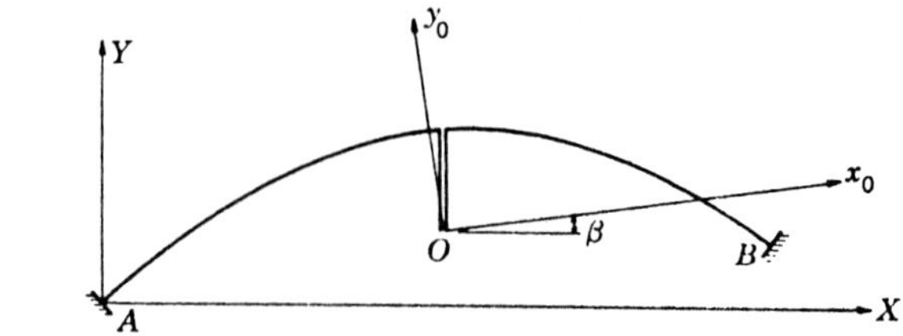

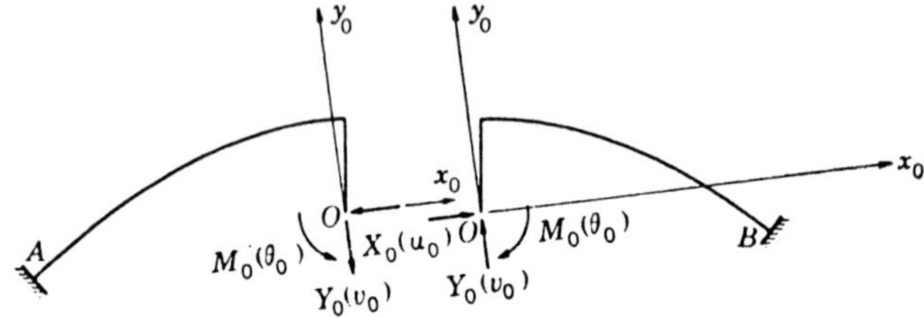

FIGURE 10-24 Alternate cut structure for a hingeless arch.

can write

$$M = M_S + B_1 + B_2 x_0 + B_3 y_0 \tag{10-26}$$

$$T = T_S + T_1 \tag{10-27}$$

The new compatibility conditions are

$$\theta_0 = 0 = \int \frac{M}{EI} \frac{\partial M}{\partial M_0}\, ds + \int \frac{T}{EA} \frac{\partial T}{\partial M_0}\, ds$$

$$= \int \frac{M_S}{EI}\, ds + B_1 \int \frac{1}{EI}\, ds + B_2 \int \frac{x_0}{EI}\, ds + B_3 \int \frac{y_0}{EI}\, ds \tag{10-28}$$

$$u_0 = 0 = \int \frac{M}{EI} \frac{\partial M}{\partial X_0}\, ds + \int \frac{T}{EA} \frac{\partial T}{\partial X_0}\, ds$$

$$= -\int \frac{M_S}{EI} y_0\, ds - B_1 \int \frac{y_0}{EI}\, ds - B_2 \int \frac{x_0 y_0}{EI}\, ds + B_3 \int \frac{y_0^2}{EI}\, ds$$

$$+ \int \frac{T_S}{EA} \left(\frac{dx_0}{ds} \right) ds + \int \frac{T_I}{EA} \left(\frac{dx_0}{ds} \right) ds \tag{10-29}$$

$$v_0 = 0 = \int \frac{M}{EI} \frac{\partial M}{\partial Y_0}\, ds + \int \frac{T}{EA} \frac{\partial T}{\partial Y_0}\, ds$$

$$= \int \frac{M_S}{EI} x_0\, ds + B_1 \int \frac{x_0}{EI}\, ds + B_2 \int \frac{x_0^2}{EI}\, ds + B_3 \int \frac{x_0 y_0}{EI}\, ds$$

$$+ \int \frac{T_S}{EA} \left(\frac{dy_0}{ds} \right) ds + \int \frac{T_I}{EA} \left(\frac{dy_0}{ds} \right) ds \tag{10-30}$$

If axial strains are ignored, (10-28) through (10-30) are simplified by choosing the origin O and the axes x_0, y_0, so that

$$\int \frac{x_0}{EI}\,ds = \int \frac{y_0}{EI}\,ds = \int \frac{x_0 y_0}{EI}\,ds = 0 \tag{10-31}$$

The origin O is defined as the elastic center of the arch rib and is also the centroid of the analogous column. The axes x_0 and y_0 are the principal axes (Carpenter, 1960). We can now obtain

$$B_1 = -\frac{\int (M_S/EI)\,ds}{\int (1/EI)\,ds} \tag{10-32}$$

$$B_2 = -\frac{\int (M_S/EI)\,x_0\,ds}{\int \left(x_0^2/EI\right)ds} \tag{10-33}$$

$$B_3 = -\frac{\int (M_S/EI)\,y_0\,ds}{\int \left(y_0^2/EI\right)ds} \tag{10-34}$$

The indeterminate moments, and by extension the reactions, can be obtained by the use of (10-26). Alternatively, they can be estimated directly by the method of column analogy (Carpenter, 1960). If axial strains cannot be ignored, the equations are correlated through the term T_I and the elastic center approach is no longer advantageous.

Graphical Solution In the foregoing analysis we have dealt with both angular and linear displacements. Because in many cases the total displacement is the sum of several component displacements not necessarily all in the same direction, it is essential to follow a consistent procedure for keeping track of the direction of each force, moment, lever arm, and displacement.

Alternatively, the elastic center may be determined graphically noting that the center of gravity of the elastic load system can be found from any two arbitrary axes drawn through any point on the plane. Figure 10-25 shows a graphical method for this determination. These diagrams are ordinary statical solutions and are self-explanatory.

Temperature Effects

For a uniform change in temperature within the structure geometry, the arch rib changes length axially only. There is no angular distortion because the top fibers expand or contract exactly the same amount as the bottom fibers. In this case we only need to consider axial strains ($wt\,ds$) due to temperature, where w is the coefficient of linear expansion and t is the temperature rise. For the geometry of the hingeless arch of Figure 10-23, the resulting

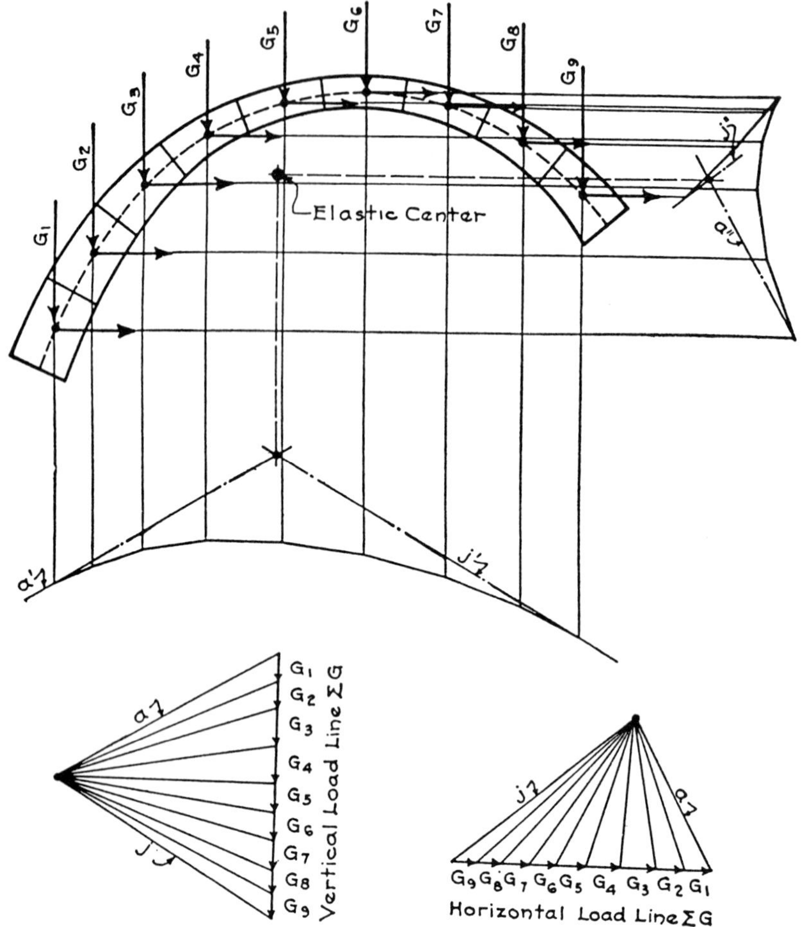

FIGURE 10-25 Graphical method for determining the elastic center.

movement is as shown in Figure 10-26, for a change in length of $wt(L^2 + d^2)^{1/2}$. We can write

$$u_A = -wtL\left(1 + \frac{d^2}{L^2}\right) \tag{10-35}$$

The movement normal to AB will produce rotations

$$\theta_A = \theta_B = \frac{(wt/L)d(L^2 + d^2)^{1/2}}{(L^2 + d^2)^{1/2}} = wt\frac{d}{L} \tag{10-36}$$

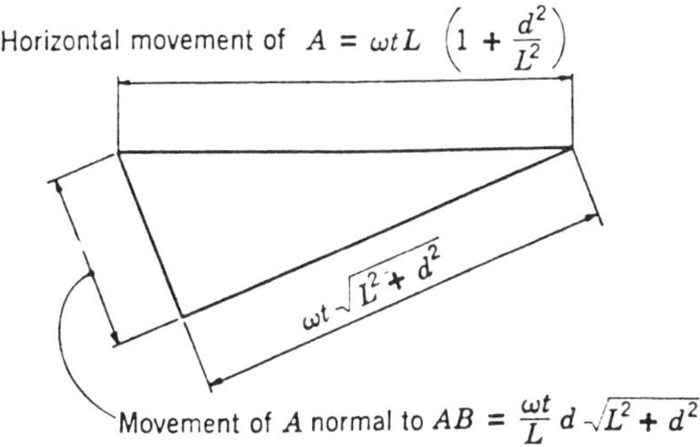

FIGURE 10-26 Temperature movement in original base structure; arch of Figure 10-23.

These angular distortions are entered into (10-17) through (10-19) to obtain relationships in the redundant force components and hence to obtain the resulting arch reactions and bending moments.

With a variable temperature change within the arch geometry, the upper fibers of the rib are brought to a temperature level different from that of the lower fibers. In this case angular displacement results. The bending strains are represented by the expression

$$\phi_{t'} = \frac{wt'\,ds}{h} \qquad (10\text{-}37)$$

where t' represents the difference in temperature between the upper and lower fibers of the section and h is the total depth of the section.

Elastic Rib Shortening due to Loads

The effect of elastic rib shortening under any maximum load combination may be quantified from the fact that for ordinary arches the compressive stress throughout the rib is very close to evenly distributed. Let F be the average compressive fiber stress (lb/in.^{2}) under dead load. Under the action of this fiber stress, each lamina of the rib will shorten axially an amount given by the expression $\lambda = F\,ds/E$. Conversely, a drop t_1 degrees in temperature will shorten each lamina by the amount $\lambda_t = ct_1\,ds$. Setting $\lambda = \lambda_t$ and solving for t_1 yields $t_1 = F/Ec$, where c is the coefficient of thermal expansion. The effect of the elastic rib shortening due to dead load axial stress is therefore equal to that caused by a drop in temperature equal to t_1 degrees.

Likewise, the equivalent temperature effect may be evaluated for any other loading condition and stresses merely by prorating directly from the temperature stresses already obtained. Theoretically, this procedure is only an approximation because it involves an average value of the fiber stress, which varies along the length of the rib. Unless the analysis relates to research work or it involves structures of such magnitude as to warrant extreme accuracy, the procedure is recommended and considered sufficiently accurate.

Support Displacements

In the analysis thus far the supports have been assumed as fixed and unyielding (except for the elastic pier supports in multispan arches), so that the work of the reactions was zero. Where the supports are expected to yield under load, a portion of the external energy will be used up in displacing them, so that the amount of energy absorbed by the frame will be reduced by the same amount. It is possible to evaluate the effect of support displacements by introducing between the true structure and its support an "ideal" section whose unit distortion is such as to completely reproduce the effect of the support proper. This ideal section is assumed in turn to rest upon a rigid support, so that the original method of analysis can be carried out. Support movements are thus handled in the context of this analysis, either if they are known in absolute terms or if they are functions of the reactive forces.

In manual calculations we commonly allow for axial strains in the arch rib in an approximate fashion. The original analysis tends to ignore axial strains and may take advantage of the simplification introduced by the use of the elastic center. From this analysis the arch thrust may be computed at every point. The resulting strains produce movements in the base structure of Figure 10-23 expressed by (10-12) through (10-14), with $M = 0$. Approximate solutions for the effect of axial strains are obtained by using (10-17) through (10-19).

Example of Yielding Foundations For the example shown in Figure 10-27*a*, the left abutment is not truly fixed at the skewback section *a–a*, but distorts under load to some position such as section *b–b*. The base of the abutment rests, however, on solid rock so that the analysis can be completed without significant error if we assume the rock extends clear around to the rock base. This means that we consider section *e–e* in Figure 10-27*b* as the skewback section.

For the condition shown at the right support, the solution is not simple. However, if we can estimate the displacement that the pile footing will undergo under load, we may replace this footing by an "ideal" or "substitute" section of the same basic material as the arch rib proper and having dimensions that will make it deflect or distort under load exactly the same amount as the original footing. Figure 10-27*b* shows a condition where this substitution has been made. Thus, the "equivalent" frame has the skewback sections at sections *e–e* on the left support and *f–f* on the right. This structure may be analyzed as a fixed arch.

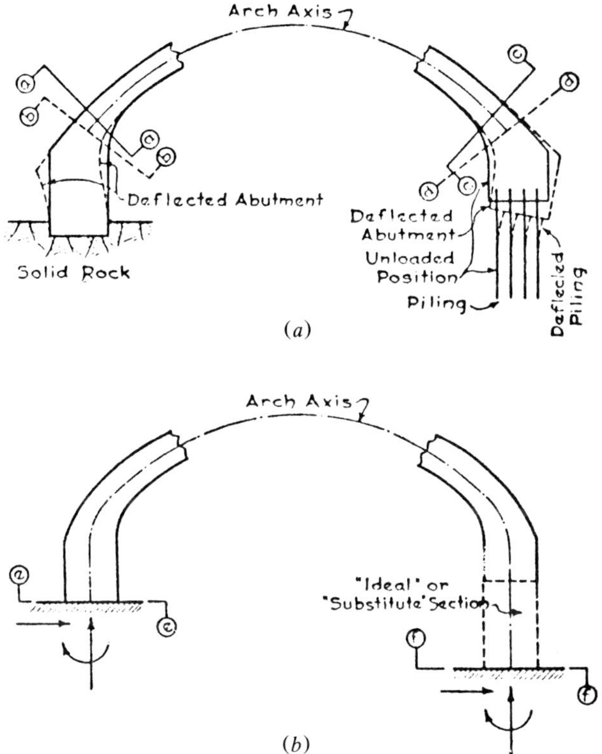

FIGURE 10-27 Equivalent frame for an arch on yielding foundations.

Angular Distortions

Figure 10-28 shows a truss arch. Instead of the linear displacement at point *a*, we must estimate the angular rotation of member *a–b*. For this, we can replace the auxiliary unit load at point *a* with an auxiliary moment couple, and repeat the same procedure. We can write that the angular rotation is

$$\phi = \frac{\Delta_a + \Delta_b}{L_{ab}} \tag{10-38}$$

and

$$\Delta_a = \sum \frac{Ss_a L}{AE} \qquad \Delta_b = \sum \frac{Ss_b L}{AE} \tag{10-39}$$

so that

$$\frac{\Delta_a + \Delta_b}{L_{ab}} = \frac{\Sigma S\left(\dfrac{s_a + s_b}{L_{ab}}\right)L}{AE} \tag{10-40}$$

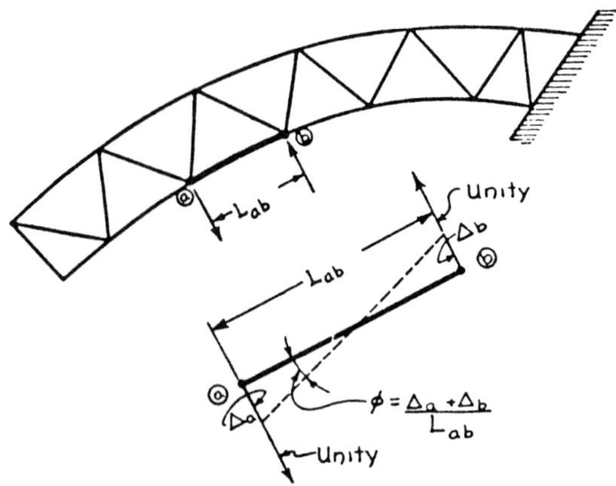

FIGURE 10-28 Effect of angular distortions.

But $(s_a + s_b)/L_{ab} = s_\phi$ (the stress obtained in each member by loading the member a–b with a unit moment couple). From this, it follows that

$$\phi = \sum \frac{Ss_\phi L}{AE} \tag{10-41}$$

10-7 FUNDAMENTALS OF DEFLECTION THEORY

Deflections from Castigliano's Theorem

For framed structures or linkage systems, to find the deflection Δ_a at any point a, measured in any given direction and for any given load system, the following steps are necessary: (a) compute the stress S in each member of the frame due to the given load system and (b) apply a unit auxiliary load at point a, acting in the direction along which it is necessary to calculate the deflection, and compute the resulting stress s_a. The summation $\Sigma(Ss_aL/AE)$ for the entire frame is the deflection desired.

The unit load must be applied in the direction of the desired displacement, and attention must be given to the sign convention. If for any individual member the direction of stress is the same for s_a and S, the product is positive and vice versa.

Deflection Formulas for Ribs and Beams In general, expressions for deflections are obtained as in framed structures, that is, by partial differentiation of the internal work expression.

The deflection resulting from bending strains is

$$\Delta_a = \sum \frac{Mm_a\, ds}{EI} \tag{10-42}$$

where m_a is the moment produced by auxiliary unit loading.

Likewise, the deflection due to axial strains is

$$\Delta_a = \sum \frac{Nn_a\, ds}{AE} \tag{10-43}$$

where n_a is the axial stress produced by the auxiliary unit loading.

Other Deflection Theories and Solutions

Deflections are also determined by finite-difference methods (Hirsh and Popov, 1955), by structural mechanics methods (Asplund, 1966), by interaction techniques, and more recently by finite-element methods. Deflection theories are also formulated by Jennings (1968).

Model Tests O'Connor (1971) reports results on the behavior of a model, parabolic two-hinged steel arch. The model had a span of 60 in., a rise of 9.54 in., and a cross section 1 in. wide by 0.25 in. deep. It was tested for various values of uniform dead load and a superimposed live load uniformly distributed over half the span starting at one support. Figure 10-29 plots the maximum positive moment versus the live load.

This model was intended to simulate a preliminary design of a two-hinged arch for the Rainbow Bridge shown in Figure 10-7. The prototype parameters are as follows:

DL = 8.6 kips/ft at the crown
I = second moment of area of the cross section for bending in its plane
= 119 ft^4 (for two-hinged design)
L = 950 ft
LL = 1.5 kips/ft

The equivalent model loads would be as follows:

$$\text{DL} = (8600/12) \times 0.00361 = 2.59 \text{ lb/in.}$$

$$\text{LL} = 0.174 \times \text{DL} = 0.45 \text{ lb/in.}$$

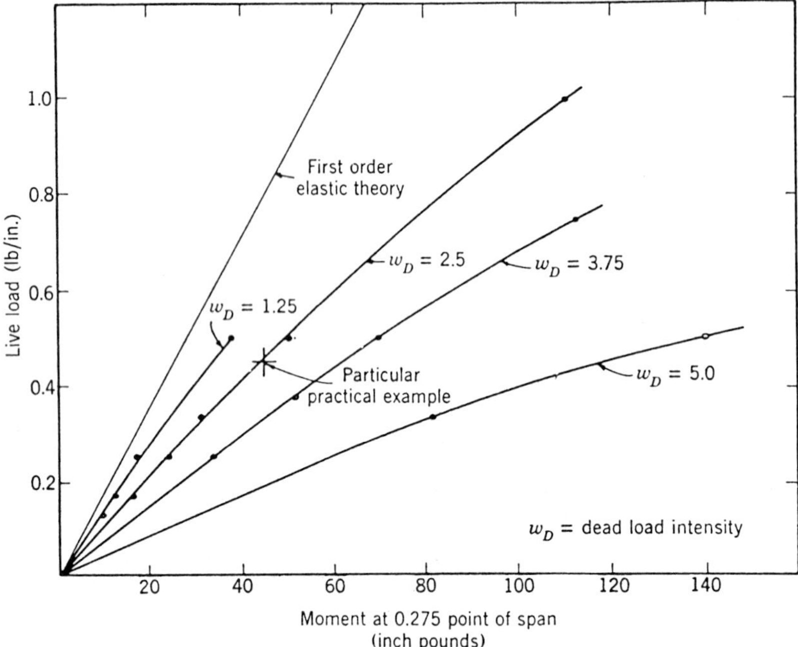

FIGURE 10-29 Graphical presentation of model behavior for a two-hinged parabolic arch.

The final design of the Rainbow Bridge was hingeless and with reduced I. However, from the model study we draw the following conclusions.

1. Although the graphs are nonlinear, the deviation from linear behavior is noncritical over the practical range.
2. The starting slopes of the curves differ considerably from those predicted by first-order elastic theory.
3. The starting slopes increase with the concurrent dead load intensity.
4. A straight line at the starting slope represents fairly the behavior in the practical range.
5. For the equivalent mode loads, the maximum moment is about 1.75 times the value computed by first-order theory.

In effect, an arch of this stiffness and type of loading behaves as a compressed member subjected to a small disturbance, and is similar to a compressed column with a small lateral load. The dead load induces the axial compressive force in the column, and the live load represents the lateral disturbance. The lateral deflection of the column is directly proportional to the disturbing force, provided the axial force remains constant and the transverse deflections do not become excessive. Likewise, the arch responds linearly to live load, and for this behavior to become nonlinear the deflections must be large; this usually does not occur in a practical arch.

First-Order Deflection Analysis It appears from the foregoing that a useful step in arch analysis is to obtain a good estimate of the initial behavior. For the arch shown in Figure 10-30, we will consider a first-order deflection analysis using matrix terminology. The frame consists of a series of straight elements; a typical element ③ connects joints 3 and 4.

The distortions e of an element and the corresponding forces F applied to the element are shown in Figures 10-30b and c, respectively. These obey the following relationship:

$$F = Se \tag{10-44}$$

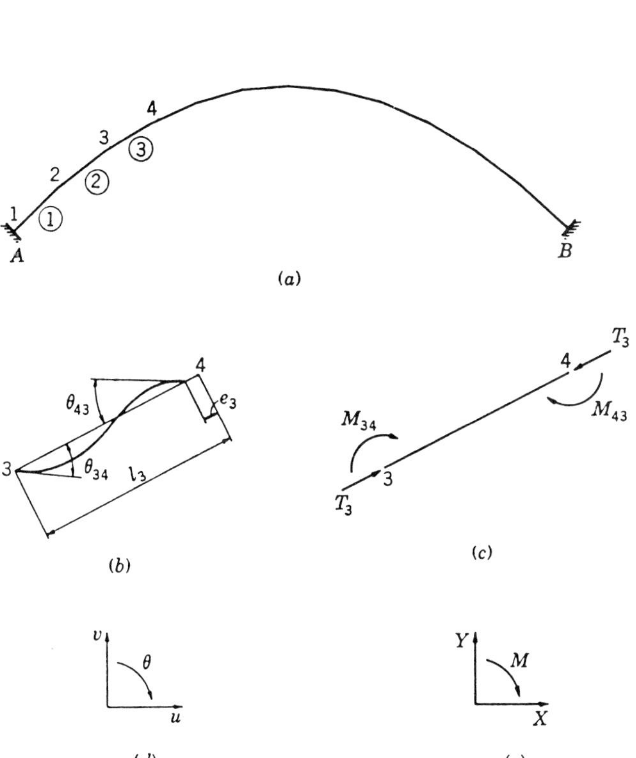

FIGURE 10-30 Matrix arch analysis: (a) structure; (b) element distortions; (c) element forces; (d) joint movements; (e) joint loads.

where S is the element stiffness matrix. We can also write

$$\begin{bmatrix} M_{34} \\ M_{43} \\ T_3 \end{bmatrix} = \begin{bmatrix} \frac{4EI}{l} & \frac{2EI}{l} & 0 \\ \frac{2EI}{l} & \frac{4EI}{l} & 0 \\ 0 & 0 & \frac{EA}{l} \end{bmatrix} \times \begin{bmatrix} \theta_{34} \\ \theta_{43} \\ l_3 \end{bmatrix} \tag{10-45}$$

Possible joint loads P are shown in Figure 10-30c and are connected to the element forces by the relationship

$$P = AF \tag{10-46}$$

where A is the statics matrix commonly computed from statics. Typical terms in A are shown in Figure 10-31 and must be accumulated for each member in the arch frame.

The joint displacements X corresponding to P are shown in Figure 10-30d. Applying the contragredient principle (Asplund, 1966; Hall and Woodhead, 1961), we can write

$$e = A^T X \tag{10-47}$$

so that

$$\begin{aligned} F &= Se = SA^T X \\ P &= AF = ASA^T X \end{aligned} \tag{10-48}$$

ASA^T is the complete structure stiffness matrix, and its inverse $(ASA^T)^{-1}$ is the flexibility matrix. If P are known and X unknown,

$$X = (ASA^T)^{-1} P \tag{10-49}$$

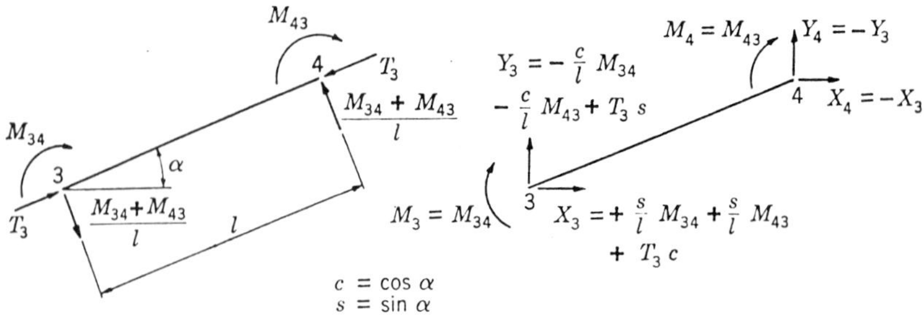

FIGURE 10-31 Element forces and joint loads.

The element forces can therefore be determined from

$$F = SA^TX \tag{10-50}$$

An arch under an initial dead load may be shaped and formed so that this dead load produces thrust forces only, whereas bending moments are absent. However, dead load moments may exist, but in either case we assume that dead load geometry and all forces associated with dead load application are known.

For an improved live load analysis, we can assume that the live load intensity is as small as desired, and its distribution is known. The relationship $F = Se$ in (10-44) remains unchanged, and the element distortions are connected to the joint displacements by the dead load geometry, that is, $e = A^TX$ is valid.

Although the live load may be as small as desired and the corresponding deflections also small, the dead load thrusts are large and the resulting eccentricity (thrust times small deflection) may be appreciable. Thus, an essential correction is necessary because the relationship $P = AF$ (where A is based on dead load geometry) is inadequate.

Let F still denote the true element forces, shown in Figure 10-32 at the deflected geometry. We can replace them by a statically equivalent set at the dead load geometry, and for this it is sufficent to use the thrust as the dead

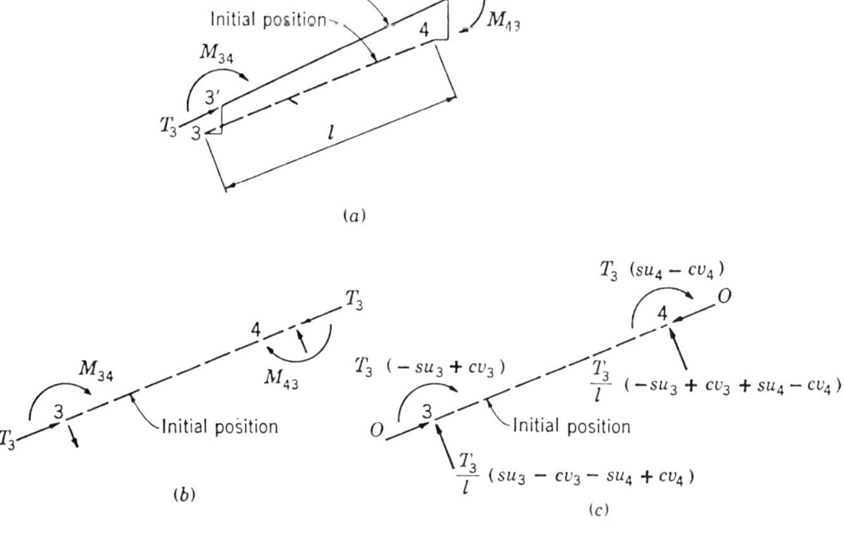

FIGURE 10-32 Corrections to element forces from dead load thrust: (*a*) forces on deflected element; (*b*) element forces at initial geometry; (*c*) additional element forces.

load thrust, noting that the live load may be made sufficiently small. The revised element force system consists of that used in the first-order theory, Figure 10-32b, and a correction system that depends on the joint deflection and the dead load thrust.

The correction system includes moments applied to the element and balancing shears. The moments can be used within the matrix A to determine equivalent joint forces. Alternatively, the moments and shears can be replaced by a statically equivalent system of joint forces, designated as P' and defined as CX. The contributions to C from a typical member are as follows:

$$\begin{bmatrix} X_3 \\ Y_3 \\ M_{34} \\ X_4 \\ Y_4 \\ M_{43} \end{bmatrix} = \begin{bmatrix} -\dfrac{Ts^2}{l} & \dfrac{Tsc}{l} & 0 & \dfrac{Ts^2}{l} & -\dfrac{Tsc}{l} & 0 \\ +\dfrac{Tsc}{l} & -\dfrac{Tc^2}{l} & 0 & -\dfrac{Tsc}{l} & +\dfrac{Tc^2}{l} & 0 \\ (-Ts & +Tc & 0) & 0 & 0 & 0 \\ +\dfrac{Ts^2}{l} & -\dfrac{Tsc}{l} & 0 & -\dfrac{Ts^2}{l} & +\dfrac{Tsc}{l} & 0 \\ -\dfrac{Tsc}{l} & +\dfrac{Tc^2}{l} & 0 & +\dfrac{Tsc}{l} & -\dfrac{Tc^2}{l} & 0 \\ 0 & 0 & 0 & \{+Ts & -Tc & 0\} \end{bmatrix} \times \begin{bmatrix} u_3 \\ v_3 \\ \theta_3 \\ u_4 \\ v_4 \\ \theta_4 \end{bmatrix} \tag{10-51}$$

Terms in parentheses ( ) should be included at the left end of the arch only, and terms in brackets { } should be included at the right end only.

For a vertical dead load we can write $T = H/c$, where H is the constant horizontal component of the thrust and c is the cosine of the angle between the arch centerline and the horizontal. If the horizontal component Δx of the element length l is constant across the span, we can write $T/l = H/\Delta x$.

The total joint loads now become

$$\begin{aligned} P &= AF + P' \\ &= ASA^TX + CX = (ASA^T + C)X \end{aligned} \tag{10-52}$$

or

$$X = (ASA^T + C)^{-1}P$$

so that

$$F = SA^TX \tag{10-53}$$

Usually, the dead load conditions are known. If the arch form differs considerably from the correct shape for the dead load, the foregoing analysis may be used for successive increments of dead load, at each step using the previous deflected geometry and arch thrust (O'Connor, 1971).

A first-order solution may also be introduced in iteration analysis to obtain the linearized second-order result, but with large thrusts the convergence process is slow. A solution for a two-pinned arch similar to the Rainbow Bridge required six trials before reasonable results were obtained (O'Connor, 1971).

10-8 BUCKLING AND GEOMETRY IMPERFECTIONS: STABILITY CONSIDERATIONS

General Stability of Arches

In-Plane Buckling Ideally, arches may experience small deflections. Using the model discussed in Section 10-7, we consider a partial case where all live loads are zero. The displacement X must satisfy the equation

$$(ASA^T + C)X = 0 \qquad (10\text{-}54a)$$

The matrix C can be expressed by the form $w_D D$, where w_D is a scalar quantity representing a typical dead load intensity. We can therefore write

$$(ASA^T + w_D D)X = 0 \qquad (10\text{-}54b)$$

where the critical load w_D is an eigenvalue of this system, and the corresponding eigenvector expresses the deflected shape. Equations (10-54) have zero constant terms. They can yield nonzero values for X only if the determinant of their coefficients is zero. In this case the critical load is defined by the condition

$$|ASA^T + w_D D| = 0 \qquad (10\text{-}54c)$$

which can be solved by trial and error to determine the lowest critical value of w_D.

This approach can be used to develop a differential equation applied to the deflected shape of a buckled arch (Lind, 1962), but this solution is only possible in certain cases, for example, for a circular arch with a uniform radially distributed load (Timoshenko and Gere, 1961). However, numerical results have been tabulated for certain practical cases and are given in the form

$$w_{cr} = \gamma \frac{EI}{L^2} \qquad (10\text{-}55)$$

where γ is a calculated parameter.

For a parabolic arch of span L and rise R and with a uniform dead load, constructed so that the dead load moments are zero, the horizontal thrust component is

$$H = \frac{wL^2}{8R} \tag{10-56}$$

from which we obtain

$$H_{\text{cr}} = \frac{w_{\text{cr}} L^2}{8R} = C_1 \frac{EI}{L^2} \quad \text{where} \quad C_1 = \frac{\gamma}{8(R/L)} \tag{10-57}$$

Values of C_1 are plotted versus R/L for parabolic arches with 0, 1, 2, or 3 hinges in Figure 10-33. The graphs are for constant I and for I approaching sec α, where α is the inclination of the arch. In the latter case the central I should be used in (10-57). Invariably, the critical load is markedly affected by the number of hinges, and is the greatest for the hingeless arch.

Out-of-Plane Buckling A freestanding arch rib may buckle normal to its plane (Waestlund, 1960). Among the factors influencing the critical load are (a) the degree of end fixity of the rib, the two simplest cases being when the rib is completely fixed or completely pinned at the ends; (b) a fixed-end condition prompting the arch to twist as it deflects laterally, with the critical load depending on the ratio of the transverse second moment of area of the arch rib to its torsion constant; and (c) the deck of the arch and its stiffness against horizontal movement.

Other relevant factors are (a) the horizontal stiffness of the deck; (b) the height of the deck with respect to the springing line of the arch; and (c) the bending stiffness of hangers and columns not pinned at both ends.

Studies on the behavior of the unbraced tied arch have been made by Godden (1954) and Godden and Thomson (1959), and constants were tabulated for the critical loads. Similar procedures can be used to analyze other arch forms. For arches below the deck, a convenient bracing can be provided between the ribs. This benefit is lost if the arch rises above the deck because the bracing now may interface with traffic and must be interrupted or omitted. Buckling in this case is most likely to be serious. The Fehmarnsund Bridge in Figure 10-8 is an example of this category. The use of deck to stiffen arch bridges is discussed in more detail in the following sections.

Imperfections in Initial Geometry As discussed in the foregoing sections, any initial shape deviating from the correct shape for the form load will give rise to bending moments. This problem is commonly caused by small accidental imperfections in arch alignment. The corresponding joint displacements X' of the unstressed arch may be assessed as follows.

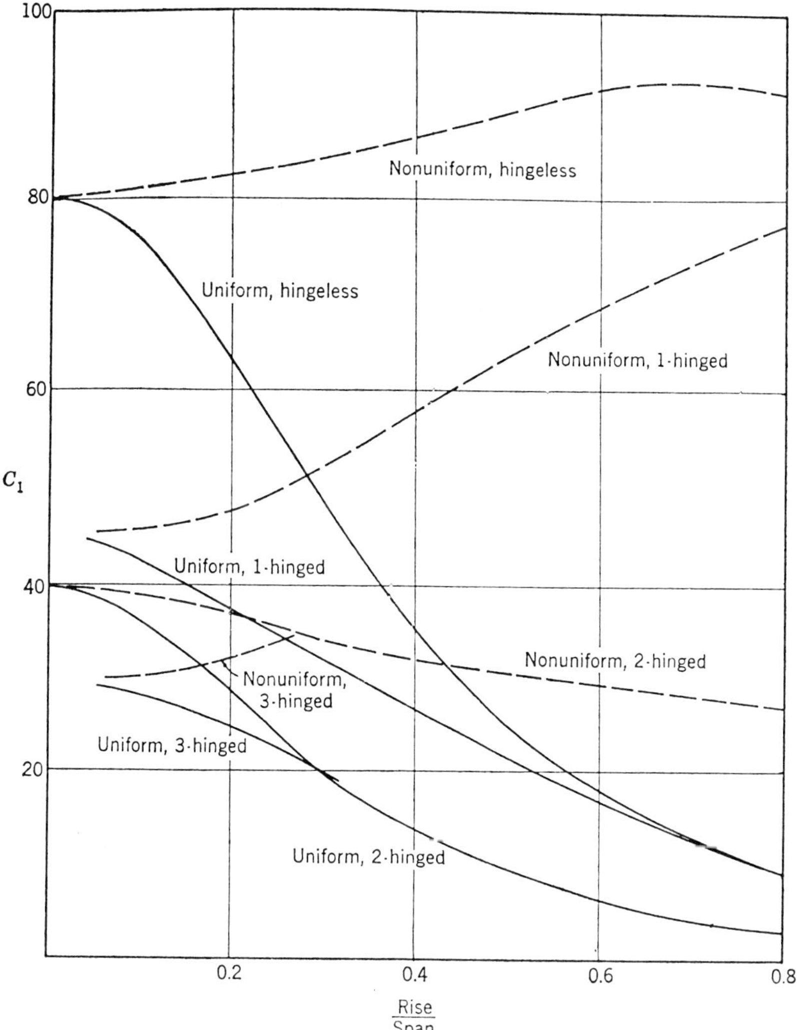

FIGURE 10-33 Coefficients for in-plane buckling of parabolic arch; $H_{cr} = C_1(EI/L^2)$.

From (10-52) we obtain

$$P = 0 = ASA^T X + CX' \tag{10-58}$$

so that

$$X = -(ASA^T)^{-1}(CX') \tag{10-59a}$$

$$F = SA^T X \tag{10-59b}$$

A particular case is when the initial imperfections have the same shape as a characteristic buckled shape for the arch (see also previous sections). Let δ_i be an initial displacement amplified to a final value δ_f by the dead load w_D. From the theory of elasticity (Timoshenko and Gere, 1961),

$$\delta_f = \delta_i \frac{1}{1 - \beta} \tag{10-60}$$

where $\beta = w_D/w_{cr}$. From (10-60) we obtain the additional deflection

$$\delta_f - \delta_i = \delta_i \frac{\beta}{1 - \beta} \tag{10-61a}$$

Bending moments are estimated from the final geometry or from deformations corresponding to the initial deflections.

A live load distribution producing a deflected shape similar to a characteristic shape gives rise to a similar condition. Let m_i denote the initial bending moment at a point due to live load but without dead load. This will be amplified by dead load to

$$m_f = m_i \frac{1}{1 - \beta} \tag{10-61b}$$

For other live load distributions, however, this result is approximate only and provides an overestimation of the amplification factor.

The final bending moments have a linear dependence on the live load because $m_f \propto m_i$ for a given β. The live load response, however, is amplified from the value m_i, predicted by first-order theory, by a factor $1/(1 - \beta)$, and this factor depends on the dead load. For example, if the dead load is less than 5 percent of its critical value, the difference between the first-order analysis and the amplified result will be small and probably less than 5 percent.

Stability Considerations of Steel Arches

In-Plane Linear Stability Many investigators consider arch analysis on the basis of arch response to load and the in-plane mode of failure. Arch characteristics are articulated according to arch type as follows (Galambos, 1988): (a) slender arches of solid-web rolled or built-up sections, subjected primarily to axial forces; (b) slender arches subjected to considerable bending effects and deformations because of unsymmetrical loading; (c) stocky arches, often of truss form, where failure of the chords or flanges may be caused by a combination of axial loads and moments; and (d) arches composed of arch ribs and deck stiffening girders.

In-plane linear stability and its associated problems are discussed by Austin (1971) and Timoshenko and Gere (1961). The critical values of distributed load and the horizontal reaction of fixed two-hinged and three-hinged symmetric arches are summarized for: (a) parabolic shapes subjected to vertical uniformly distributed load; (b) catenary arches under uniform vertical load along the arch axis; and (c) circular arches under hydrostatic loading.

The elastic buckling of nonuniform parabolic arches has been studied for variations in the moment of inertia $I = I_c \sec \phi$ and $I = I_c \sec^3 \phi$, where I_c is the moment of inertia at the crown and ϕ is the angle between the tangent to the arch axis and the horizontal. Likewise, stiffened arches may be analyzed by approximating the critical load and critical horizontal reaction using an equivalent I equal to the sum of the arch and girder moments of inertia (Galambos, 1988).

In-Plane Nonlinear Stability Conditions of nonlinear elastic stability arise where actual loadings produce both axial compression and bending moment on a general cross section of the arch rib with a corresponding change in shape of the arch rib before buckling occurs. Even though the material deforms elastically, the problem is essentially nonlinear.

For symmetrical loading on symmetrical arches, data are provided by Austin and Ross (1976), and include the case of a single concentrated load at midspan. Axial strains are neglected because their influence is small for slender arches.

The critical conditions for antisymmetrical buckling are often expressed in terms of horizontal reactions, and this is particularly meaningful for flatter arches. It appears that the critical horizontal reactions are insensitive to the arch shape and to a lesser degree to the loading, but they vary primarily with the rise–span ratio. This is particularly true for fixed arches.

Several studies have been made for unsymmetrical loading, emphasizing the most important practical cases, such as a uniformly distributed dead load (on a horizontal projection), plus a uniformly distributed live load extending a variable distance from one abutment (these studies are for parabolic arches). Investigations are reported by Deutsch (1940), Kuranishi and Lu (1972), and Harries (1970). A conclusion was that for elastic buckling the total dead plus live load intensity $w = p + q$ at buckling was roughly equal to the buckling value for uniform load over the entire span for the case where half-span live load was used. Other studies have been made by Chang (1973) and Harrison (1982). It should be noted, however, that high live loads acting over partial spans can induce large movements and deformations. The elastic limit load is reached only after very large displacements cause complete distortion of the arch (Harrison, 1982; Yabuki and Vinnakota, 1984). When inelastic behavior is considered, the limit loads for these cases are much less than the corresponding elastic buckling values.

In-Plane Ultimate Load Limit analysis of stocky arches has been presented by Onat and Prager (1953) and Galli and Franciosi (1955). The collapse load of arches has also been reported by Stevens (1957) and Coronforth and Childs (1967). The collapse loads were obtained by the upper and lower bound methods assuming rigid plastic behavior. The method is based on the localized plastic hinge concept and does not consider the longitudinal spread of yielding zones or the effect of deflections on moments. Hence, the studies are valid for very stocky cross sections subjected to predominant flexural moments.

Slender Arches in Pure Compression Slender two-hinged and fixed arches under pure compression buckle by sidesway with a node at the crown, much like a column. The buckling strength in this case is expressed in terms of the axial thrust P at the quarter point of the arch. The elastic critical value is $P_{cr} = \pi^2 EI/(KS)^2$, where S is the length of the curved centroidal axis of the arch rib from the support to the crown and K is the effective length factor.

Komatsu and Shinke (1977) have presented inelastic ultimate strength studies for a two-hinged parabolic arch subjected to uniform load. Residual stresses and geometric imperfections were considered with a broad range of other parameters and rise–span ratios of 0.10, 0.15, and 0.20. The conclusion was that the ultimate thrust value at the quarter point of the arch was accurately predicted with the usual column curves adjusting for yield points, effective lengths, and other parameters.

Slender Arches Under Symmetrical Load The combined effect of axial compression and bending moment prompts the arch to act much like a beam–column. Shinke, Zui, and Namita (1975) have determined that the most relevant practical loading for low bridge arches is the unsymmetrical load discussed in the foregoing section with $S = 0.50$.

Studies of uniform arches subjected to uniform dead load over the entire span and half-span live load ($S = 0.50$) have been reported by Harries (1970) and Kuranishi and Lu (1972), who also considered residual stresses and strain hardening. These investigators concluded that when the effect of yielding is considered, the strength of an arch can be markedly reduced under unsymmetrical loading. The buckling load is less when the live load covers only one-half the span, and the larger the live–dead load ratio the smaller the buckling load becomes. Kuranishi (1973) also reports a second-order analysis for the sandwich cross section that approximates I and box cross sections. The results show that the inelastic ultimate load lies between the load at initial yielding and about 93 percent of that value for the practical range of load ratios.

More recent studies are reported by Kuranishi and Yabuki (1979), Yabuki and Vinnakota (1984), and Shinke, Zui, and Nakagawa (1980). These involve two-hinged parabolic steel arches and arches with stiffening girders. Studies of the effects of an unsymmetrical distributed load supplemented by a single concentrated load at the quarter point of the span (conforming to Japanese design standards) are summarized by Yabuki and Vinnakota (1984). It appears that the limit load depends on factors such as the rise–span ratio, live–dead load ratio, slenderness ratio, yield stress, type of cross section, and residual stresses. For a comprehensive treatment of these effects, see Yabuki and Vinnakota (1984).

Design for In-Plane Stability Arches of considerable span usually have nonuniform cross section and consist of segments with different section properties. The structural system may also be complex, such as the open spandrel type where the arch rib is rigidly connected to the roadway girder. In these cases reference to uniform freestanding arches has limited merit and may be irrelevant. A general approach is to factor the loads, use a second-order elastic analysis for the entire system, and keep the design load below some reference resistance value. Many investigators recommend this procedure with a maximum stress less than 90 percent of the yield stress or with a resistance factor $\phi \leq 0.9$.

Design procedures based on ultimate inelastic strength have been proposed by Kuranishi (1973), Komatsu and Shinke (1977), and Kuranishi and Yabuki (1984b). An interaction-type formula is proposed, similar to beam–column formulas, for the two-hinged parabolic arches under unsymmetrical loading. Practical formulas are also presented for the planar ultimate load of two-hinged and fixed parabolic steel arches as a function of the normal thrust computed at the quarter point of the arch rib by first-order analysis. The ultimate load capacity of arch ribs with variable sections may be evaluated using the mean values of section properties and/or yield stress levels. Practical formulas are also given for the in-plane ultimate strength of parabolic two-hinged steel arch ribs and steel arches with a stiffening girder, expressed in terms of moments and axial thrusts obtained from first-order elastic analysis.

Out-of-Plane Stability The main factors influencing out-of-plane stability are discussed in the foregoing sections, and it appears that the multiplicity of parameters inhibits attempts to formulate a simple but generally applicable theory for determining out-of-plane buckling loads.

Out-of-Plane Buckling of Circular Arches Equal couples applied transversely at the ends can cause buckling. If the ends are fixed against translation but are free to rotate, the critical moment can be calculated (Timoshenko and Gere, 1961). Likewise, two collinear forces applied to the ends can cause lateral buckling (Klöppel and Protta, 1961; Ojalvo, Demuts, and Tokarz, 1969). Forces directed away from each other (pull loads) will cause antisymmetric buckling about the center of the member, and forces directed toward each other will cause a symmetric buckling. The buckling load may be calculated and expressed in a form proposed by Klöppel and Protta (1961).

If uniformly distributed radial forces are directed inward along the centroidal axis and the ends are free to rotate with respect to the principal axes but unable to rotate with respect to the third axis, the buckling load may be calculated as proposed by Timoshenko and Gere (1961). For an arch fixed against rotation at its ends, the uniformly distributed load at buckling may be expressed in the form proposed by Demuts (1969).

Out-of-Plane Buckling of Parabolic Arches The uniformly distributed buckling load may again be expressed as in circular arches, and the effect of the warping torsional rigidity may be similar in principle to the one observed in a circular arch. More data are provided by Stussi (1943), Tokarz (1971), and Kee (1961).

For parabolic arches with tilting loads, it has been demonstrated that the buckling load may be increased if the loads are applied to the arch by a system of vertical hangers connected to a laterally stiff deck at the elevation of the chord. If the loads are applied by means of columns (spandrels) connected to a laterally stiff deck above, the lateral deformation of the arch will be antisymmetric about the crown when the arch is connected to the roadway at this point or if the columns near the crown are very short. Both tied and spandrel-loaded parabolic arches have been studied by Ostlund (1954), Godden (1954), and Shukla and Ojalvo (1971).

Ultimate Strength Under Uniformly Distributed Vertical Loads Residual stresses resulting from welding and initial out-of-plane deflections have a marked effect on the strength of steel arches limited by lateral instability. Analytical results from typical theoretical models show that (a) the residual stresses may reduce the strength by 10 to 20 percent and (b) initial out-of-plane deflections may reduce the strength of a perfectly plane arch by as much as 15 percent (Sakimoto and Komatsu, 1977b; Komatsu and Sakimoto, 1977).

Extensive parametric studies (Sakimoto and Komatsu, 1982) have produced a simple approximate method for calculating the strength of braced or unbraced steel arches that fail by lateral instability. From an analogy between an arch and a column, an equivalent slenderness function λ_a is introduced and related to the ultimate strength of through-type steel arches of box cross section (Sakimoto and Komatsu, 1983a). This function is

$$\lambda_a = \frac{1}{\pi}\sqrt{\frac{\sigma_y}{E}}\,\frac{l}{r_x}K_e K_l K_\beta \tag{10-62a}$$

where σ_y = lowest yield stress among those of different steel grades

l/r_x = slenderness ratio

K_e, K_l, K_β = effective length factors

The coefficient K_e relates to the rotational fixity of the arch rib at the ends with respect to the centroidal axis. The value of K_e is 0.5 for a clamped condition, and 1.0 for a hinged condition. The coefficient K_l relates to the direction of loads, and is 0.65 for the tilting hanger case and 1.0 for nontilting hangers (vertical loads). The coefficient K_β relates to the lateral restraint supplied by the bracing system, and is defined as $K_\beta = 1 - \beta + (2r_x\beta/K_e b)$. The term β is the ratio of the length of the braced portion to the total length. For arches without bracing $\beta = 0$, and $K_\beta = 1$ for an isolated single arch.

The ultimate unit strength σ_u is calculated for the column with the equivalent slenderness function, and is taken as an approximation of the ultimate stress required to buckle the actual arch. The term σ_u for arches is defined as the tangential thrust at the support, N_u, divided by the area A, or $\sigma_u = N_u/A$. The thrust is calculated from a linear theory.

The stiffening effect (constraint to lateral movement) provided by a bridge deck has been studied by Sakimoto and Komatsu (1983b) and Sakimoto and Yamao (1983). The conclusion is that the ultimate strength of through-arch bridges with rigidly connected hangers, such as the tied arch shown in Figure 10-4*i*, increases markedly because the lateral bending stiffness of the hangers constrains lateral arch movement. Similar effects should be expected for deck-type arches, such as the structure shown in Figure 10-4*g* if the spandrels have rigid connections.

When overall lateral buckling is suppressed, only the buckling of rib sections between bracing points must be considered. The slenderness condition of the local instability of an arch segment preceding the overall lateral instability may be determined as suggested by Sakimoto and Komatsu (1983a), expressed as

$$\sqrt{\sigma_y/E}\,(l_p/r_x)_{\max} \geq \sqrt{\sigma_y/E}\,(l/r_x)K_e K_l K_\beta \tag{10-62b}$$

where $(l_p/r_x)_{\max}$ is the slenderness ratio of the most slender arch segment with unbraced length l_p between points of bracing, and σ_y in the left term of (10-62*b*) is the yield stress of the segment with maximum slenderness.

Ultimate Strength Under Vertical and Lateral Uniform Loads Studies of the spatial elastic–plastic behavior and the effects on the ultimate load capacity of through-type braced arches are reported by Sakimoto and Komatsu (1977a) and Sakimoto, Yamao, and Komatsu (1979). Computer analysis under combined vertical and horizontal uniform loads provided a simple approximation method for calculating the ultimate lateral strength of arches partially braced over the central portion. Based on analogy between a laterally loaded arch and a beam–column, the interaction formula is

$$\frac{N}{N_u} + \frac{M}{M_y[1 - (N/N_e)]} \leq 1.0 \tag{10-62c}$$

where N = tangential end thrust computed by linear theory for uniformly distributed design loads

N_u = tangential end thrust of inelastic lateral buckling of the arch under uniformly distributed vertical loads, usually $N_u = A\sigma_u$

M = lateral end moment of individual arch rib subjected to uniformly distributed horizontal loads, excluding contribution of axial load interacting with deflections

M_y = yield moment of the arch rib at the ends

N_e = Euler buckling load computed for a hinged column with length and cross section identical to those of the arch rib at the unbraced end part

The parameter M is approximated by the value computed for the planar system obtained by straightening out the arch horizontally [a simple approximation formula is also given by Sakimoto and Komatsu (1979)].

The effect of lateral horizontal forces on the in-plane strength of arches has also been studied by Kuranishi (1961b), Yabuki and Kuranishi (1973), and Kuranishi and Yabuki (1977). The results suggest that the in-plane strength is not significantly affected by the lateral loads usually associated with actual structures. It follows therefore that through-type steel arches can be designed with a lateral bracing system of sufficient out-of-plane stiffness so that rib design will be determined by in-plane loads.

10-9 LIMIT ANALYSIS OF REINFORCED CONCRETE ARCH BRIDGES; A CASE STUDY

General Considerations

The behavior of reinforced concrete arches in the inelastic range and under ultimate loading conditions has not received full attention, probably because of the inherent computational difficulties. In theoretical investigations carried out by Galli and Franciosi (1955) and Sorgente (1957), the bridge structure geometries were modeled in a manner to minimize the number of redundancies. Collapse loads were analyzed by the upper and lower bound method (Greenberg and Prager, 1952), assuming an elastic–perfectly plastic behavior prior to structural collapse. The limit analysis of rigid plastic arches has been improved by the use of linear programming techniques that eliminate the trial-and-error procedures. Further refinements in elastic–plastic arch theories have enabled the prediction of collapse loads by direct limit analysis (Abdel-Rohman, 1976), by the more traditional methods (Cohn and Abdel-Rohman, 1976), and by extending the computational capabilities in evaluating deformations and overall behavior under shakedown loadings (Ronca and Cohn, 1979).

Ronca (1977) has attempted to expand the previous work on reinforced arch bridges by focusing on the following aspects: (a) extension of the procedure to any arch type (simple, continuous, girder-connected, etc.); (b) complete automization using a general matrix and mathematical programming formulations; (c) refinements in the analytical model by discretizing the structure into finite straight elements and piecewise linearized yield surfaces; and (d) consideration of the limited ductility of reinforced concrete sections.

Assumptions and Criteria

In a limit analysis, Ronca and Cohn (1979) have investigated a bridge consisting of a thin arch and a stiffening girder on a number of discrete supports. The applied service load is considered to vary proportionally up to the collapse of the structure (ultimate load), and produces reversible deformations. The geometry of the structure is fully defined, whereby the arch bridge is considered an assembly of one-dimensional finite elements, and both structure and loads are assumed to act in the same plane.

Both steel and concrete are assumed as elastic–perfectly plastic materials in uniaxial stresses, but a limiting (ultimate) strain is specified for the concrete. A linearized yield interaction curve (M–P) with its associated flow law (normality rule) is adopted for the reinforced concrete sections as shown in Figure 10-34a. The actual structure is discretized into finite prismatic elastic elements linked by rigid perfectly plastic end sections. Plastic deformations are lumped at the nodes (Ronca and Cohn, 1979).

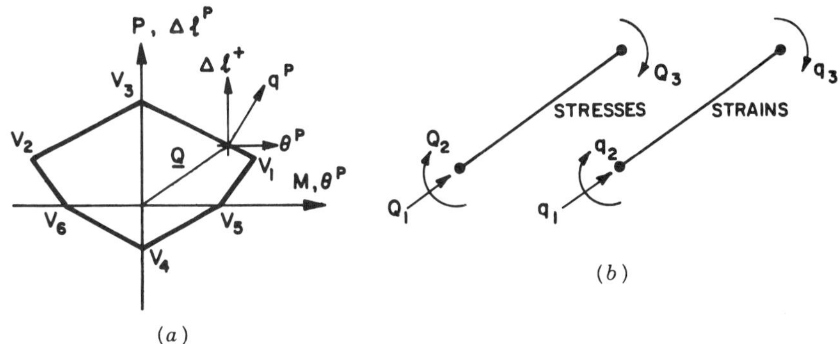

FIGURE 10-34 (*a*) Linearized interaction M–P relationship; (*b*) generalized stresses and strains for typical structural element.

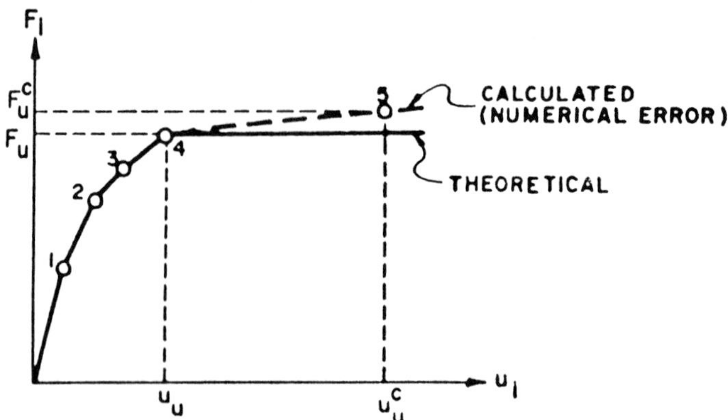

FIGURE 10-35 Typical load–displacement history of structure.

Equilibrium and Compatibility With the foregoing assumptions, the conditions of equilibrium and compatibility can be expressed by a set of linear equations and inequalities. If Q is the vector of generalized stress components in natural variables, as shown in figure 10-34b, and Q^s and Q^{el} are the vector of the self-equilibrated stress component due to plastic deformations and the vector of the elastic stress component in equilibrium with the plastic collapse load sF, respectively, where s is the load factor and F is the applied service load, then

$$Q = sQ^{el} + Q^s \tag{10-63}$$

which is added to the equilibrium matrix in terms of the natural variables. Likewise, Ronca and Cohn (1979) have developed the equations of compatibility, elasticity, and yield conditions as well as the flow rule. Problem formulations and solutions are based on the foregoing criteria and include limit load analysis, deformation analysis at collapse, and historical analysis.

Interestingly, the limit load analysis (historical) in the plastic multiplier variables has the disadvantage shown in Figure 10-35, where a small numerical error in the calculated modified stiffness of the structure can cause gross errors in assessing displacements, although the limit load F_u^c may be a very satisfactory approximation of the true limit load F_u'.

Computer Program

The program layout has a flow diagram consisting essentially of two parts: the elastic program and the LPRB routine. Program capability covers four types of problems: (a) limit load analysis, (b) limit deformation analysis, (c) shakedown analysis, and (d) holonomic historical analysis.

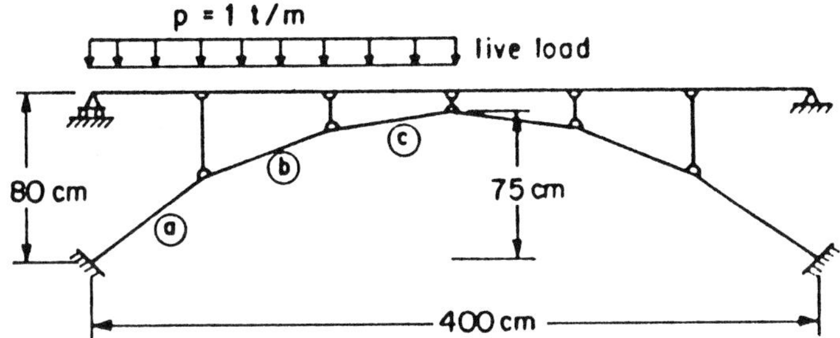

FIGURE 10-36 Arch bridge model example. (From Ronca and Cohn, 1979.)

As an example we consider the reinforced concrete arch bridge model with the geometry and loading shown in Figure 10-36. The shape of the arch is the funicular diagram of its dead weight. The model has a relatively thin rectangular section, 4.5 cm by 50 cm at the crown and 4.5 cm by 105 cm at the supports; a stiffening girder of constant cross section, 15 cm by 50 cm; and a set of five supporting walls, 6 cm by 50 cm, connected to the arch and the girder by idealized pins. The standard concrete strength is 250 kg/cm^2, and the yield limit of the reinforcing steel is 6800 kg/cm^2. The arch geometry is discretized into six elements and is constant between nodes.

The stress distribution under dead load and service live load is shown in Figure 10-37. Section 3 located in the middle of beam element 2 is the first to become plastic. The relatively small elastic moments in the arch justify the assumption on the flexural arch stiffness. The collapse load is calculated from four various behavioral models as follows: (a) negligible flexural rigidity ($M = 0$, $P \neq 0$); (b) negligible axial force ($M \neq 0$, $P = 0$); (c) yield planes allowing the M–P interaction ($M \neq 0$, $P \neq 0$), M_u, for the beam neglecting concrete strength; and (d) as in model (c) but including the concrete strength of the beam.

The collapse stress distribution is illustrated in the diagrams of Figure 10-38 for the four behavioral models. The last model shown in Figure 10-38*d* appears to be the most realistic because it gives a good estimate ($s_s = 4.02$) of the experimental collapse load factor $s_s = 3.84$ for an error of +5 percent.

It appears that the collapse modes are better predicted by models (c) and (d), because these allow for the M–P interaction in the arch. The resulting theoretical collapse mechanism shown in Figure 10-39 is in satisfactory agreement with the actual observed experimental collapse. Because the antisymmetrically loaded structure is eight times statically indeterminate and has eight generalized plastic hinges at collapse, its failure corresponds to a noncomplete antisymmetric mechanism as should be expected. The calculated rotation rates (relative rigid-body rotations) for the plastic hinges of the mechanism of Figure 10-39 are given in Table 10-1.

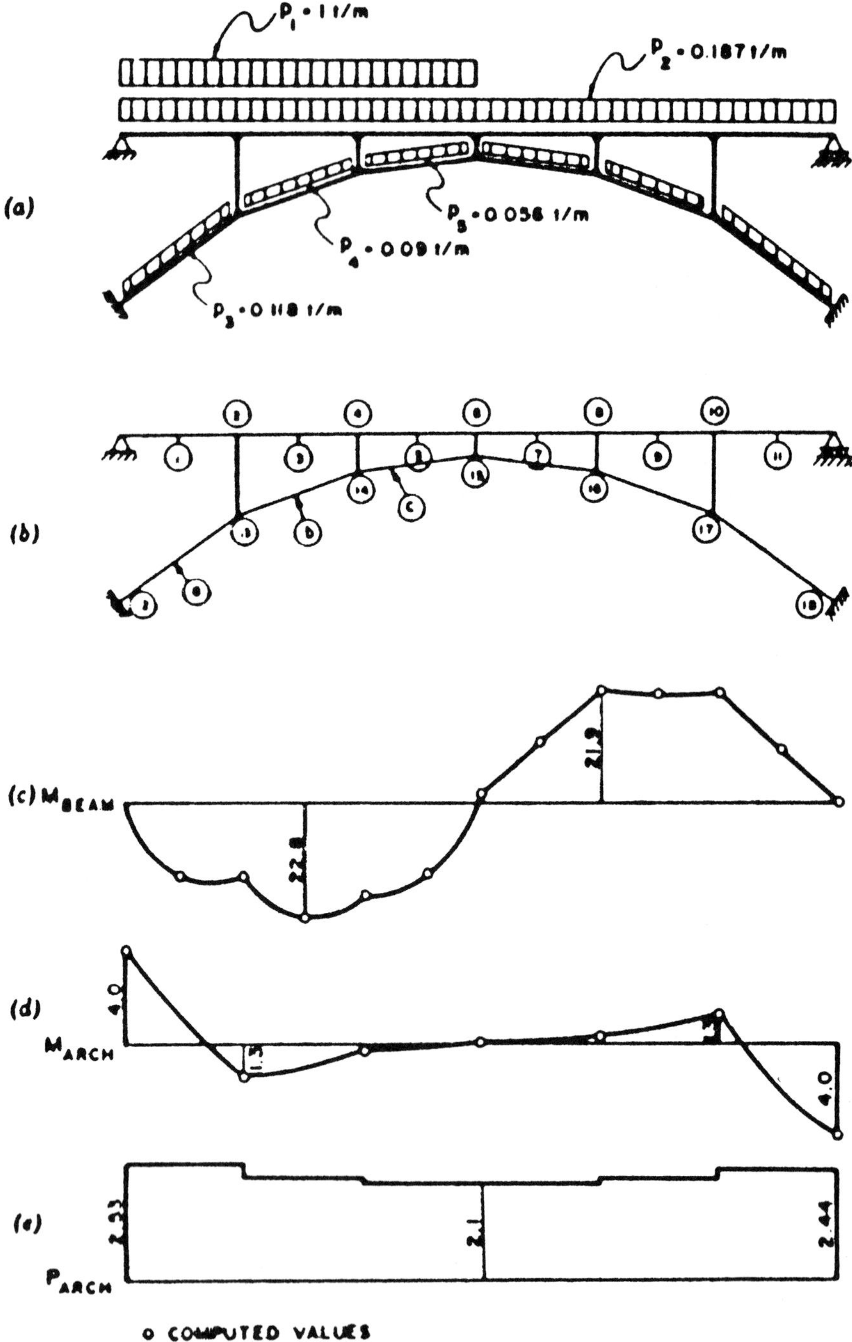

FIGURE 10-37 Stress distribution under dead and service live loads: (*a*) loading conditions; (*b*) identification of critical sections and arch elements; (*c*) beam moments; (*d*) arch moments; (*e*) axial forces in arch.

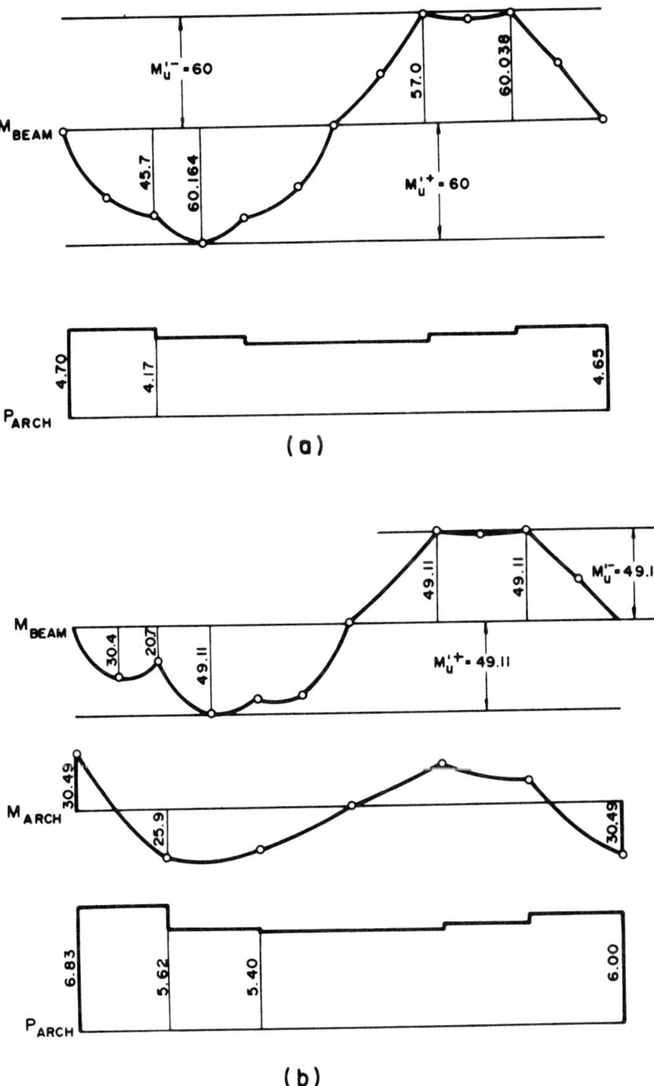

FIGURE 10-38 Collapse stress distributions for: (*a*) $M = 0$, $P \neq 0$ ($s_s = 2.38$); (*b*) $M \neq 0$, $P = 0$ ($s_s = 3.43$); (*c*) $M \neq 0$, $P \neq 0$, concrete strength neglected ($s_s = 3.61$); (*d*) $M \neq 0$, $P \neq 0$, concrete strength considered ($s_s = 4.02$). (From Ronca and Cohn, 1979.)

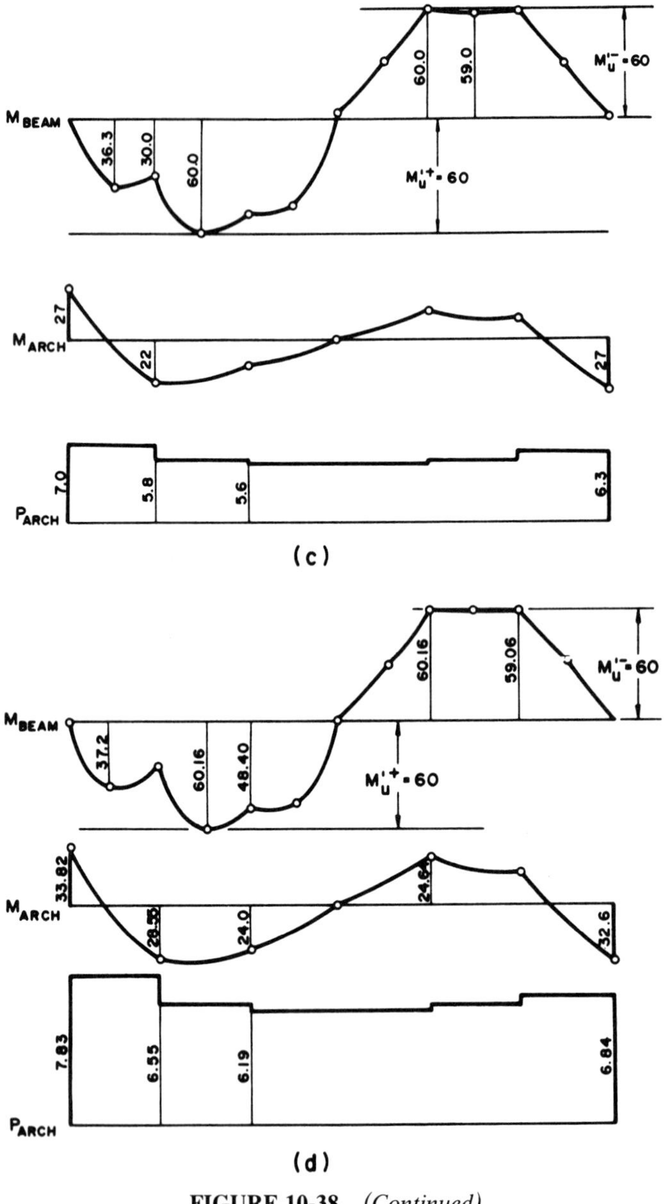

FIGURE 10-38 *(Continued)*

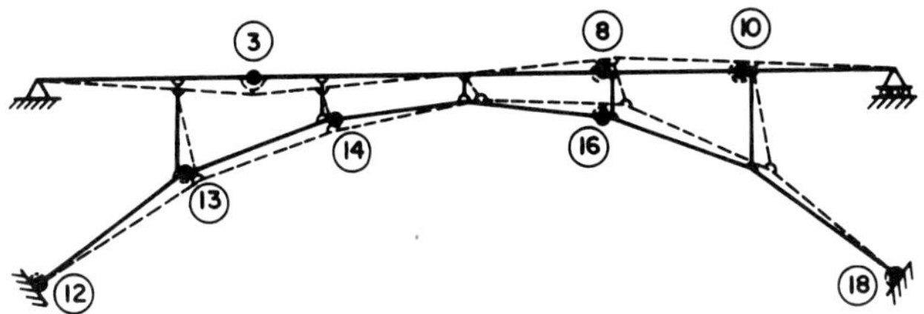

FIGURE 10-39 Theoretical collapse mechanism (calculated $s_s = 4.02$; experimental $s_s = 3.84$). (From Ronca and Cohn, 1979.)

TABLE 10-1 Rotation Rates and Plastic Rotations at Incipient Collapse (From Ronca and Cohn, 1979)

Plastic Hinge (1)	Rotation Rate (2)	Plastic Rotation, θ (radians $\times 10^{-3}$) (3)
3	1.194	136.9
8	1.050	98.4
10	0.003	27.0
12	0.560	55.4
13	0.600	21.6
14	0.600	—
16	1.062	35.87
18	0.405	38.6

10-10 DESIGN PRINCIPLES

The arch types shown in Figure 10-4 demonstrate the need to diversify the design philosophy. On the other hand, for consistency and unified procedure the subject matter of arch design should be restricted to a review of those topics and problems that are basic to arch analysis and of frequent occurrence.

Skewed Arches Because the function of any bridge is to satisfy the conditions of grade, alignment, and clearance, it often becomes necessary to consider an arch structure that is skewed in plan.

If the arch is of the rib type, the skewed structure should not present new or unusual problems because each individual rib may be analyzed as a right arch. Where conditions dictate the use of a barrel arch in concrete, the application of elastic theory becomes complicated unless the barrel is laminated or divided into individual right rib elements using longitudinal joints.

A comprehensive treatment of skewed barrel arches requires finite-element analyses, and is often based on a restatement and extension of the Rathbun (1924) theory.

Specifications For concrete arches, the combined flexure and axial load in the arch ring must be according to AASHTO Articles 8.16.4 and 8.16.5 (strength design method). Slenderness effects in the vertical plane of an arch ring, other than tied arches with suspended roadways, may be evaluated taking the unsupported length l_u as one-half the length of the arch ring, and the radius of gyration r taken about an axis perpendicular to the plane of the arch at the quarter point of the arch span. The strength reduction factor ϕ may be taken as 0.85 (see also Section 10-21).

In arch ribs and barrels the longitudinal reinforcement should provide a ratio of reinforcement area to gross concrete area of at least 0.01, divided equally between the intrados and extrados. In arch barrels upper and lower levels of transverse reinforcement should be provided and designed for transverse bending due to loads from columns and spandrel walls as well as shrinkage and temperature effects.

For solid steel rib arches and service load design, AASHTO criteria are formulated in Article 10.37. If the analysis neglects arch rib deflection, the live load plus impact moment is increased by an amplification factor.

Construction Problems Relevant to Design Construction details that merit attention in the design stage include, among others: (a) articulation (hinges and expansion joints); (b) arrangement of reinforcing metal; (c) drainage problems in the deck, waterstops, and so on; (d) alignment, superelevation, and grade line problems; and (e) bridge appearance and architectural treatment.

With regard to the foundation elements, relevant design topics are (a) the vertical and lateral stability of the rock; (b) protection against erosion; (c) the vertical and lateral stability of the piling, when used; (d) problems related to battered-pile footings; (e) anchorage of the rib reinforcement at the abutments.

Provisions for Horizontal Thrust An important decision in the design stage is the choice between abutments that are capable of providing the necessary horizontal reaction components and the use of a tied arch. This decision depends on cost comparison, clearance requirements for the structure, and associated construction problems.

Factors favoring the use of the elastic arch form with spandrel columns are, in their sequence of importance: (a) strong foundation materials to withstand compression at or near the surface, (b) a high roadway alignment, and (c) site conditions allowing encroachment of the arch rib on the space between abutments. If the site is favorable, this choice should be more economical because the ground strength is utilized to replace structural materials.

The bowstring type is essentially a choice competing with the simply supported truss. The cantilever arch and the tied arch of the Port Mann design are closer approximations to the classic arch forms.

Construction Materials The predominance of compressive forces means that the use of concrete as the major structural material is justified. An arch rib that is prestressed by the thrust may consist of precast segments, but for longer spans the dead weight of concrete becomes an objectionable feature.

A tied structure may take advantage of the tension capacity of an orthotropic steel deck, and in this case the use of steel may be extended to the entire structure. For a bridge that is not self-tied, the major benefit in using a steel deck is dead weight reduction, but this is important for longer spans.

Rib, Barrel, or Truss Section Figures 10-5*b* through *e* show structural configurations, namely, the rib, barrel, and truss sections. The depth of a truss normally provides an increased moment-resisting capacity than is available in a rib section. As a result, the truss should be more economical in material and lighter in dead weight; the weight advantage is a decisive factor in longer spans. In addition, a truss type is also easier to erect. Material cost saving may be offset by higher fabrication costs. In some designs large unbraced member lengths may be penalized by lower allowable stresses. A truss arch should also be expected to have higher maintenance cost.

Rise–Span Ratio The rise–span ratio of a classical arch is dictated by the shape of the funicular polygon. This ratio is also influenced by clearance requirements above and below the structure.

In a bowstring girder traffic clearance requirements are in conflict with the bracing between the ribs and thus influence the possible rise. In this context smaller-span structures may have a larger rise than would be necessary to accommodate an upper bracing system sufficiently long to ensure stability. If the choice is to omit the bracing system completely, the rise–span ratio is selected to optimize structural economy.

Increasing the rise reduces the thrust in the arch rib but increases its length, although to a lesser extent. If the abutment skewback elevation is fixed and the arch must accommodate a deck profile above it, an increase in rise reduces the length of the spandrel columns, as shown in Figure 10-5*a*. Conversely, an increased rise in a bowstring girder increases the cost of the hanger system.

The rise–span ratios for the bridges shown in the foregoing sections are 0.157 for the Rainbow Bridge, 0.175 for the Fehmarnsund Bridge, and 0.238 for the Glemstal Bridge. The high rise for the latter was dictated by site geometry. A typical rise–span ratio is 0.16.

Bracing Requirements In practice, arches are braced against lateral movement either continuously or at specific intervals. An example of continuous bracing is an arch rib continuously connected to a curved roof. Twin arch ribs with lateral bracing often constitute the main structural element of the bridge.

For steel arches, the following properties of the bracing system are relevant to the suppression of lateral buckling: (a) the location and spacing of transverse bars, (b) the distance between arch ribs, (c) the flexural stiffness of the bars in the transverse direction, and (d) the flexural stiffness of the bars in the vertical plane.

Increasing the transverse stiffness has a beneficial effect in suppressing lateral buckling. When pairs of arch ribs are closely braced with stiff transverse bars, the critical load of one arch rib of a set of braced arches can be increased by as much as 250 percent over that of the isolated identical single arch. For actual bridges, however, this increase is theoretical and limited by yielding of the materials.

A simple approximate method has been suggested for calculating the lateral buckling load of braced arches, and involves a planar system obtained by straightening out the arch along a horizontal plane. The compressive force necessary to buckle the longitudinal ribs in the planar system may be computed as in columns with battens, and is a fair approximation of the force required to buckle the actual arch rib. The method is accurate for arches of a torsionally stiff (closed) section, but should be used with caution for arches with an open section.

Twin arches braced with transverse members offer a good choice in terms of esthetics. However, transverse members are not more effective than a bracing system of diagonal members. If the lateral bracing consists of either diagonals, diagonals combined with transverse bars, or a K system, out-of-plane buckling tends to be less of a problem except for through bridges where the bracing cannot be provided throughout.

The stiffness of diagonal bracing members associated with lateral stability has been studied by Sakimoto and Komatsu (1977a, 1977b) and Kuranishi and Yabuki (1981) in relation to the ultimate strength of steel arches under combined vertical and horizontal uniform loads.

Arch Profile The optimum profile of the arch rib follows the correct shape for the form load. If a significant part of the dead load is applied through spandrel columns or hangers, the arch axis should consist of discontinuous segments. The arch axis may also have a shape that deviates from the visible profile. This is achieved in a hollow rib by varying the thickness of the upper and lower flanges. Another solution is a segmental profile. At the outside corners, a fillet of variable size is provided and forms continuous curved boundaries to the exposed face. Fabrication procedures for a steel bridge may dictate a segmental profile.

Effect of Variations in Elastic Modulus Whereas for steel arches the stress–strain function is fairly constant, for concrete this relationship is dependent on several variables such as mix, age, gradation, and quality of aggregates. For concrete design the elastic modulus E_c is related to the parameters w and f'_c, but, in practice, E_c may also be influenced by the stress condition imposed upon the material.

However, any variation in elastic modulus for the particular range of stress likely to be encountered in actual arch response during service may not be sufficient to introduce errors of any consequence. On the other hand, if the behavior of reinforced concrete arches is considered in the inelastic range, the use of mandatory deflection theories will not be entirely applicable. In this case the analysis may be improved by resorting to a general matrix and mathematical programming formulations as discussed in Section 10-9.

Temperature and Shrinkage Effects In steel arches temperature effects are manifested when the arch is closed at a temperature other than the mean. In concrete arch ribs temperature variation occurs as a result of the internal action of the cement during the process of setting, producing heat that raises the internal temperature. In many instances this rise in temperature is sufficient to lift the arch from its falsework, resulting in a premature decentering, but it will return to its original position (bearing on the falsework). The relevant state is thus the average temperature existing at the instant the arch rib starts to act as a load-carrying element. It is difficult to make a completely accurate prediction of this temperature, and for important structures simultaneous measurements of deflections and temperature readings at interior rib locations are essential.

Initial or early shrinkage is distinguished from long-term creep. The coefficient of shrinkage, defined in other sections, represents the effects of mix consistency, aggregate grading, curing conditions, and so forth.

Apart from these considerations, nonlinear temperature gradients can also develop during the average daily cycle, leading to considerable temperature stresses. Temperature monitoring of steel arches in a natural setting with air temperatures of 90 and 95°F in the shade and in the sun, respectively, shows that the steel may reach an average temperature of 115°F, whereas the parts directly exposed to the sun are heated at 130°F and those in the shade at about 105°F, resulting in a temperature differential of 25°F.

End Conditions In a conventional bowstring girder, the ends of the arch rib are rigidly attached to the stiffening girder, which restrains rotation in the plane of the arch (Louis, Guiaux, and Mas, 1962). If out-of-plane buckling of the arch is a potential problem, the arch ends should be fixed against bending rotations also in other directions (Godden, 1954) (see also Section 10-8).

If an arch is slender and externally anchored, fixed ends are desirable to increase in-plane buckling loads and also to minimize the amplification of initial imperfections and live load moments. The choice of end conditions should thus have a technical basis, and the following comments are useful in this regard (a) for an arch of short span, in-plane buckling may not be a serious problem, and design moments for hinged ends are not significantly higher than with fixed ends; (b) if foundations are faulty, a certain degree of articulation will help reduce the sensitivity of the structure to support movements, but if these movements are predictable a good solution is to construct a hingeless arch and analyze it for the effects of support movement.

Spacing of Spandrel Columns In terms of arch rib cost, the effect of spandrel column spacing is almost negligible so that the correct spacing depends on the cost of the deck and the columns. The cost of a spandrel column is less than the cost of a pier of the same height and carrying the same load, because it lacks a footing. Because the cost of the deck decreases with decreasing spandrel column spacing, the latter should vary with column height, being least at the center of the bridge.

Column Slenderness The ribs and spandrel columns of arch bridges are members that may have slender proportions (see also the beginning of this section). For example, the Gladesville Bridge has spandrel columns 2 ft thick and lengths up to 100 ft, giving a maximum length–thickness ratio of about 50. The subject of column slenderness has received considerable attention in research and in literature and will not be treated here. In general, a slender compressive member is defined as a member that has a significant reduction in its axial load capacity due to moments resulting from lateral deflection. According to the ACI code, a "significant reduction" is arbitrarily taken as anything greater than about 5 percent.

For slender members in arch bridges (length–thickness of pin-ended members greater than about 30), the strength depends primarily on the elastic modulus of the concrete. Pertinent factors are not only the concrete strength, but also the aggregate–cement ratio as well as the rate of loading. Hognestad et al. (1955) suggest that for long-term loading E_c is about one-third of the value available for short-term loading. For slenderness ratios exceeding 30, the ultimate strength under an eccentric load may be predicted by the tangent modulus theory.

Because eccentric loading can reduce the ultimate strength significantly, it merits a complete analysis. For example, for a column $3 \times 2 \times 66$ in. ($b \times d \times L$), tests show short-term strengths varying from $0.504F_c'bd$ with an eccentricity of $0.0012L$ to $0.108F_c'bd$ with an eccentricity of $0.0227L$. This column was prestressed to 100 psi, and the value of F_c' was 4.17 ksi.

AASHTO (Article 8.16.5) specifies that the analysis of slenderness effects in compression members should include the influence of axial loads and variable moment of inertia on member stiffness and fixed-end moments, the effect of deflection on moments and forces, and the effect of duration of the loads. An approximate design procedure involves the concept of moment magnifier. The parameters relevant to the analysis are (a) the length of the member (unsupported length l_u); (b) the radius of gyration, r; (c) the effective length factor K, usually taken as 1.0 unless an analysis shows that a lower value is admissible; (d) consideration of slenderness effects by defining the criteria under which these effects may be ignored; and (e) a moment magnifier equation. These AASHTO requirements are essentially the guidelines stipulated in ACI, Section 10.11. Thus, compression members are designed for the factored axial load P_u and a magnified factored moment M_c defined as

$$M_c = \delta_b M_{2b} + \delta_s M_{2s} \tag{10-64}$$

where δ_b δ_s = appropriate moment magnification factors

M_{2b} = value of larger end moment on compression member due to gravity loads that result in no appreciable sidesway, calculated by conventional elastic frame analysis

M_{2s} = value of larger end moment on compression member due to lateral loads or gravity loads that result in appreciable sidesway, defined by a deflection $\Delta > l_u/1500$ and calculated by conventional elastic analysis

The moment magnification method is also discussed in Section 14-5.

Design Approach

Arch ribs may be designed for what we term the *natural stress condition*. Alternatively, the design may proceed by considering methods whereby a portion of these stresses is artificially controlled or modified during erection, so that at the end the final resultant stresses add up to a minimum.

As the centering under an arch rib is released, stresses are progressively induced due to various causes. These are listed not in probable combinations as the conventional load groups, but according to the sequence of occurrence. Thus, we distinguish two main groups as follows.

Group A

1. Dead load proper
2. Elastic rib shortening (due to dead load)
3. Temperature variations from initial to mean
4. Initial and early shrinkage
5. Initial inelastic support displacement
6. Early creep

Group B

7. Live load and impact
8. Seasonal temperature effects
9. Long-term creep
10. Other loads

For group A, conditions 4 and 6 are applicable to concrete arches only. Conditions 2, 3, 4, 5, and 6 result from deformation of the rib or the supports. Group B loads are self-explanatory, and reflect exterior forces and effects. For a load type of a particular magnitude, stresses in group B are unchanged and constant, but group A stresses can be controlled, modified, and occasionally eliminated by means of articulation or arch manipulation during erection. The foregoing provide the basic principle that is the underlying reason for artificial or external stress control. The two main methods for accomplishing this goal are (a) the placement of temporary or construction hinges at the crown and skewbacks, to be keyed in or fixed at a later construction stage; and (b) the so-called Freyssinet method or procedure of rib compensation and adjustment, consisting of the introduction of arbitrary compensating or adjusting stresses at one or more points in the rib using a system of hydraulic jacks or other similar devices. These methods are discussed in some detail in the following sections.

Arch Rib or Barrel Section The unit fiber stresses are computed directly from the arch rib moments and thrusts by means of the following simple relationship:

$$f_s = \frac{F}{A} \pm \frac{Mc}{I}$$

Because this unit fiber stress is a function of both thrusts and moments, it precludes the development of fiber stress influence lines directly from other thrust or moment influence coefficients. However, we can take the center of moments not at the neutral axis but at two points on the section known as the intradosal and extradosal *kern points*, respectively. The fiber stress now becomes a direct linear function of these *kern point moments*. The kern points are located above and below the neutral axis of the rib at a distance r^2/c, where r is the radius of gyration and c is the distance to the extreme fiber.

10-11 ANALYSIS OF FIXED ARCHES

Section 10-5 presents a complete solution for a hingeless arch based on the following generalized procedure (first order): (a) removal of the redundant reactions developing a statically determinate *residual frame*, (b) replacement of the removed reactions by an equal number of unknown external forces, and (c) determination of these new unknown redundant forces by means of suitable deflection equations. Sections 10-6 through 10-8 supplement the analytical procedure.

A complete analysis is carried out in a series of steps, illustrated by extracts from a typical example. The arch chosen for this purpose is the left end span of a group of three spans having an outline and general dimensions as shown in Figure 10-40. The points of fixity are assumed at sections a–a and b–b.

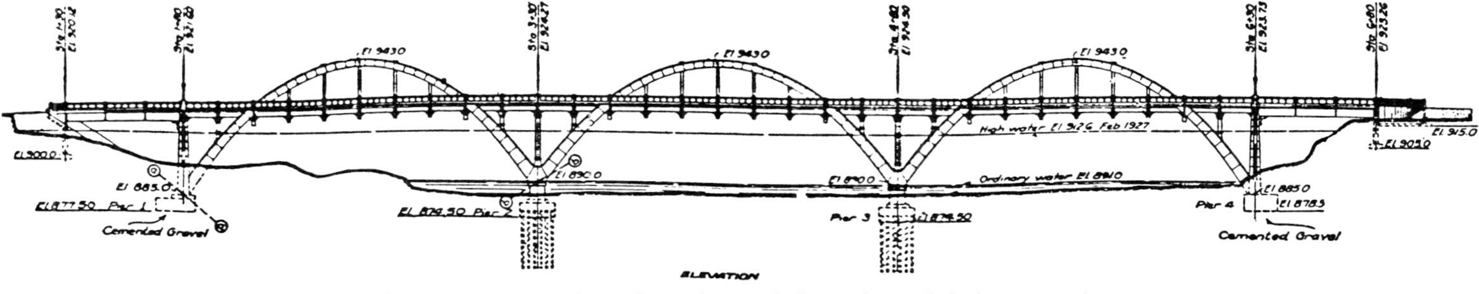

FIGURE 10-40 Arch configuration and dimensions of design example.

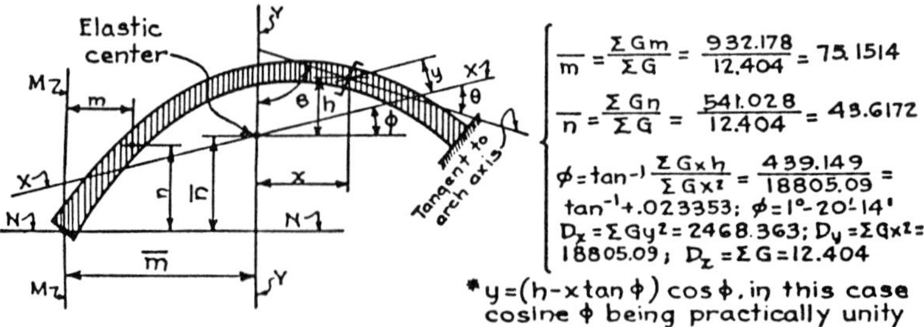

FIGURE 10-41 Elastic center for arch of Figure 10-40.

Step 1. Determine the approximate size and shape of the arch rib from a preliminary investigation and first-order analysis

Step 2. After the arch rib is tentatively selected, divide this member into a convenient number of linear increments or lamina of length ds, and for each increment determine the elastic weight $G = ds/EI$, where I is the moment of inertia at midpoint of the ds division.

Step 3. Determine the elastic center (see also Section 10-6) or center of gravity of the elastic loads ΣG. This may be done analytically by (10-32) through (10-34) or graphically as demonstrated in Figure 10-25.

Step 4. Construct two coordinate axes through the elastic center. The first may be assumed to extend in any direction, preferably vertical for convenience, say axis y–y in Figure 10-41. The other axis x–x must be so directioned as to make the product of inertia ΣGxy of the elastic load system referring to said axes vanish. The direction of the x–x axis is determined by the angle ϕ such that

$$-\cos\phi\left(\sum Gxh - \sum Gx^2 \tan\phi\right) = 0 \tag{10-65}$$

where all terms and parameters are as shown in Figure 10-41.

Step 5. With the conjugate redundant axes located, compute for each increment ds the ordinates x and y. From these calculate the terms Gx, Gy, Gx^2, and Gy^2, and from these values determine the summations ΣGx, ΣGy, ΣGx^2, and ΣGy^2. Also for each increment calculate the terms θ, $\cos\theta$, β, $\cos\beta$, dx, dy, and A (the last term is required if axial strains are not ignored). The effect of axial strains is very small and may be dropped, leaving the following relationships:

$$\begin{aligned} \delta_{xx} &= D_x = \sum Gy^2 \\ \delta_{yy} &= D_y = \sum Gx^2 \\ \delta_{zz} &= D_z = \sum G \end{aligned} \tag{10-66}$$

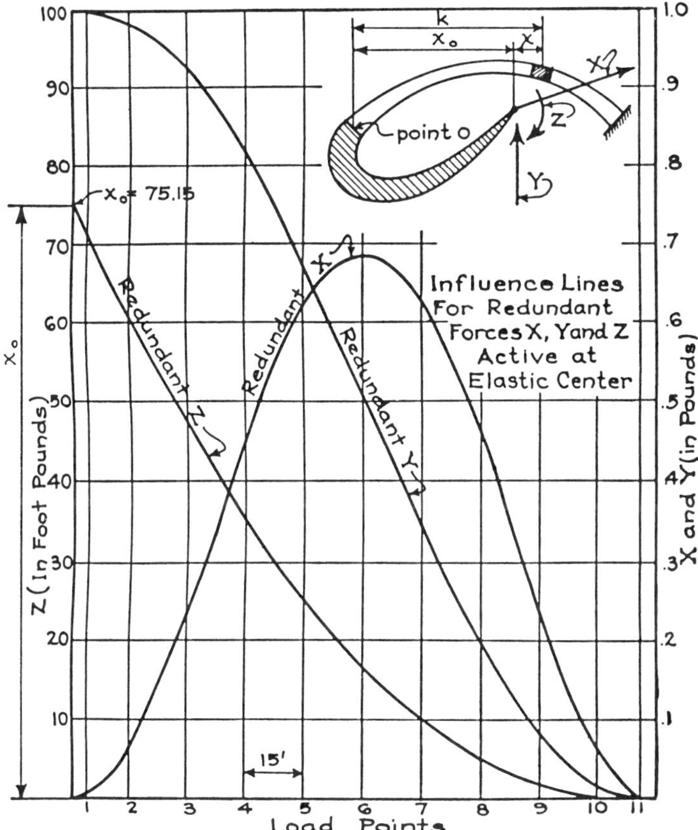

FIGURE 10-42 Influence lines for redundant forces X, Y, and Z, active at elastic center; arch of Figure 10-40.

Note that the coordinates m, n, x, y, and h may be calculated. If the arch axis has a definite mathematical curve, the calculation procedure is entirely feasible and supersedes graphical solutions.

Step 6. With the constants D_x, D_y, and D_z determined, the influence line ordinates can be determined analytically (or graphically) for the three redundant forces X, Y, and Z at the elastic center. These influence diagrams are shown in Figure 10-42.

Step 7. Having determined the influence lines for X, Y, and Z, next calculate the influence lines for the bending moments, thrust, and shear at each critical section. Because the thrust line for ordinary arches lies along a direction nearly normal to each section, it follows that the shear is very small and does not materially affect the fiber stresses.

TABLE 10-2 Influence Line Data for Moment and Thrust; Points 0 and 6 of Figure 10-43

Arch Point	General Data	Load Point	X	-Xy	Y	+Yx	Z	Mo Unit Load x(a)	Total M	X · cosθ	Y · cosβ	Unit Load x Cosβ	N Total
	θ = 56° 40′	1	+.0048	+.201	+.999	-75.076	+71.176	None for this point	-3.699	+.003	+.847	None for this point	+.850
#0	cos θ = +.54951	2											
		3											
	β = 32°	4											
	cos β = +.84805												
	y = -41.862	10											
	x = -75.151	11	+.0045	+.188	+.0012	-.090	+.024		+.122	.002	.001		+.003
	θ = 0°-23′	1	Same as above	-.069	Same as above	-.151	Same as above	-71.0	-.044	+.005	+.030	-.030	+.005
#6	cos θ = +unity (appr)	2						-60.0				.030	
		3						-45.0				.030	
	β = 88°-17′	4						-30.0				.030	
	cos β = +.02996	5						-15.0				.030	
												.030 / 0	+.668 / +.698
	y = +14.387	10											
	x = -.151	11		-.065		0			-.041	.004	0		+.004

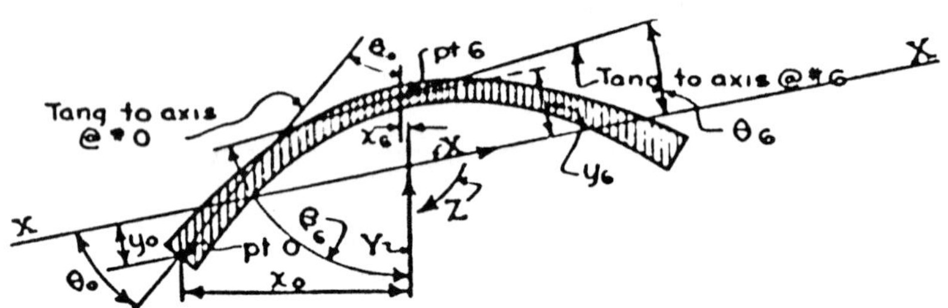

FIGURE 10-43 Arch rib geometry; points 0 and 6 of Table 10-2.

The thrust and bending moment influence line ordinates may be calculated from (10-22) and (10-23). Table 10-2 and Figure 10-43 show data for two critical points, namely points 0 and 6, and their derivation is self-explanatory. Likewise, using the procedure demonstrated in Table 10-2, influence lines are obtained for several points, and are plotted in Figures 10-44 and 10-45. These are derived from redundant influence lines obtained by analytical methods. Figure 10-45 shows the sudden increase in crown thrust as the load passes point 6 due to the slope of the axis at the crown for this unsymmetrical arch.

Step 8. From the moment and thrust influence lines, proceed now to calculate stresses due to primary gravity loads (dead, live, and impact). Table 10-3 compiles dead load moments and thrusts, whereas Table 10-4 gives sample calculations for wheel load stresses plus impact for the truck load

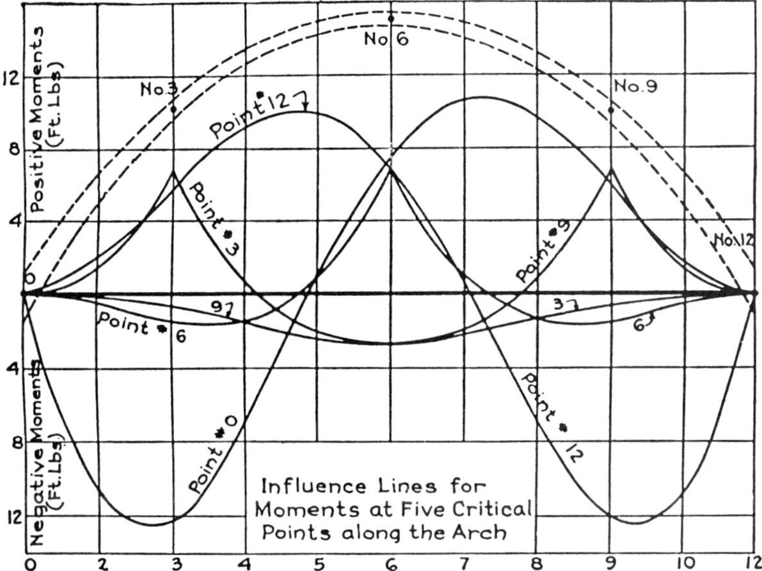

FIGURE 10-44 Influence lines for moments at five critical points along the arch rib.

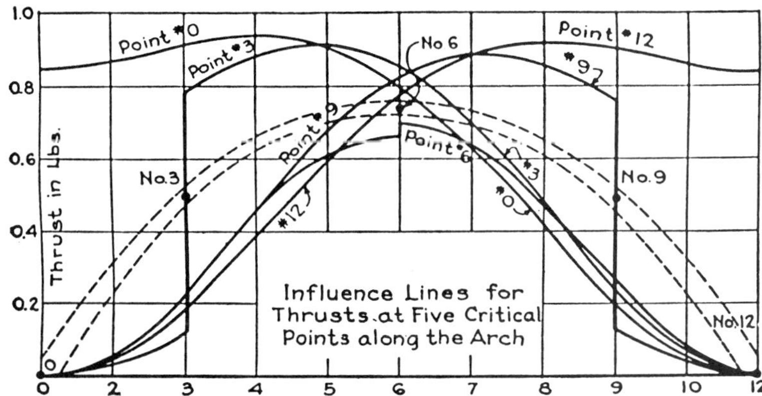

FIGURE 10-45 Influence lines for thrusts at five critical points along the arch rib.

shown in Figure 10-46. These calculations are self-explanatory. Interestingly, the live load for this example consists of a single 15-ton truck for each traffic lane followed or preceded or both followed and preceded by a continuous line or 11.25-ton trucks. The bridge is designed for three traffic lanes as shown.

Step 9. The foregoing steps complete the analysis for primary gravity loads. The next step is the calculation of moments and thrusts due to elastic rib

TABLE 10-3 Summary of Dead Load Stresses

	Load Point	1	2	3	4	5	6	7	8	9	10	11
	Dead Load (1000# units)	55.2	141.2	129.2	109.2	118.7	105.1	118.7	109.2	125.6	139.6	46.3
Stress Pt 0	Mom Coef.	-3.699	-10.784	-12.354	-6.865	+.940	+7.454	+10.622	+9.755	+5.967	+1.716	+.122
Stress Pt 0	Moment	-204.2	-1522.7	-1596.1	-749.7	+111.6	+783.4	+1260.8	+1065.2	+749.5	+239.6	+5.6
Stress Pt 0	Total D.L. Moment = +143000 Ft Lbs											
Stress Pt 0	Thrust Coef.	+.850	.868	.912	.938	.907	.802	.632	.417	.204	.050	.003
Stress Pt 0	Thrust	46.9	122.6	117.8	102.4	107.7	84.3	75.0	45.5	25.5	7.0	.1
	Total Dead Load Thrust = 734800# Eccentricity of DL Thrust $e = \frac{143000}{734800} = .195'$											
Stress Pt 6	Moment Coef	-.044	-.524	-1.535	-1.502	+1.035	+6.836	+1.098	-1.439	-1.539	-.531	-.041
Stress Pt 6	Moment	-2.4	-74.0	-198.3	-164.0	+122.9	+718.5	+130.3	-157.1	-193.3	-74.1	-1.9
Stress Pt 6	Total D.L. Moment = +106.6											
Stress Pt 6	Thrust Coef	.005	.062	.235	.448	.611	.698	.634	.465	.246	.065	.004
Stress Pt 6	Thrust	.3	8.8	30.4	48.9	72.5	73.4	75.3	50.8	30.9	9.1	.2
	Total Dead Load Thrust = 400,600# Eccentricity of D.L. Thrust $e = \frac{106,600}{400,600} = 0.261'$											

TABLE 10-4 Summary of Live Load and Impact Stresses

Stress Point		Stress		15 ton Truck F 9000	15 ton Truck R 36000	11¼ ton Truck F 6750	11¼ ton Truck R 27000	Loaded Length	Impact %	Total Stress Moment	Total Stress Thrust
#0	Pos. Mom	Mom. (+)	Point*	104	90.	148	134				
			Coeff	9.9	10.62		1.9	92	23%	643,000	
			Net M	89,100	382,000		51,300				
		Thrust "N"	Coeff.	.42	.63		.04	92	23%		34,000
			Net N	3800	22700		1100				
	Neg. Mom	Mom. (-)	Point	39	25						
			Coeff.	9.7	12.6			58	28	693000	
			Net M	87300	454,000						
		Thrust "N"	Coeff.	.92	.90			58	28		52100
			Net N	8300	32400						
#6	Pos. Mom	Mom. (+)	Point*	89	75			39	31		
			Coeff	1.3	6.84						
			Net M	11,700	246000					338000	
		Thrust "N"	Coeff	.63	.70			39	31		40500
			Net N	5,700	25,200						
	Neg. Mom	Mom. (-)	Point	44	30	132	118	111	21		
			Coeff.	1.6	1.54	.70	1.7				
			Net M	14400	55400	4700	45900			-146000	
		Thrust "N"	Coeff.	.45	.24	.10	.28	111	21		
			Net N	4000	8,600	6800	7600				32,700

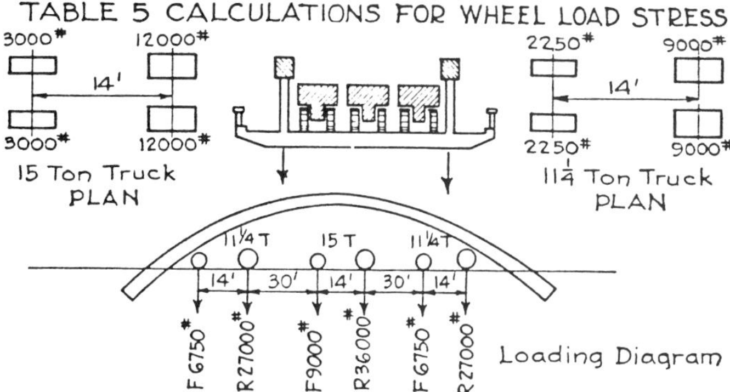

FIGURE 10-46 Truck load for arch rib bridge.

shortening (see also Section 10-6). Shrinkage effects may also be calculated in the same way as temperature stresses. The coefficient of shrinkage for concrete arches varies with several parameters, but for design purposes it can be taken as 0.0002. The equivalent of a drop in temperature is found by dividing the coefficient of shrinkage by the coefficient of temperature. Most of the shrinkage occurs during the first few days subsequent to the placement of concrete, and through the use of preventive means such as suitable keyways in the rib or temporary construction hinges most shrinkage effects can be reduced to about one-third of the potential value (see also subsequent sections).

Table 10-5 summarizes the calculations for the moments and thrusts of the five critical arch points under investigation for a uniform change in temperature of 40°F from the mean. Table 10-6 shows the corresponding moments and thrusts resulting from elastic rib shortening due to dead load and from shrinkage.

A stress summary for the five critical arch points is presented in Table 10-7, and at this stage the analysis is completed except for the effect of

TABLE 10-5 Stresses Caused by Uniform Temperature Changes

Stress Point			Moment					Thrust		
	$\cos\theta$	$\cos\theta$	y	$-X_t y$	x	$+Y_t x$	M	$X_t \cos\theta$	$Y_t \cos\theta$	T
0	+.5495	+ 8480	-41.86	+368.0	-75.15	-2.8	+365,200	+4800	Negligible	+4800
3	+.7698	+.6561	-4.92	+43.3	-45.15	-1.7	+ 41,600	+6800		+6800
6	1.000	+.0300	+14.39	-126.7	- .15	-.1	-126,800	+8800		+8800
9	+.7623	- 6293	-4.28	+37.6	+44.85	+1.7	+ 39,300	+ 6700		+6700
12	+.5250	- 8387	-40.36	+355.0	+74.85	+2.8	+357,800	+4600		+4600

The above stresses are for a 40° Rise, for a 40° drop signs are reversed

TABLE 10-6 Stresses Caused by Elastic Rib Shortening and Shrinkage

Stress Point	Temp Rise 40°		Temp Drop 40°		Rib Shortening(11°)		Shrinkage(20°)	
	Moment	Thrust	Moment	Thrust	Moment	Thrust	Moment	Thrust
0	+365200	+4800	-365200	-4800	-100500	-1320	-182600	-2400
3	+41600	+6800	-41600	-6800	-11400	-1870	-20800	-3400
6	-126800	+8800	+126800	-8800	+34900	-2420	+63400	-4400
9	+39300	+6700	-39300	-6700	-10800	-1840	-19650	-3350
12	+357800	+4600	-357800	-4600	-98400	-1262	-178900	-2300

TABLE 10-7 Summary of Stresses

	ARCH POINT NO. 0				ARCH POINT NO. 3			
	Max. Positive Mom		Max. Neg. Mom.		Max. Positive Mom		Max. Neg. Mom.	
Load Condition	Moment	Thrust	Moment	Thrust	Moment	Thrust	Moment	Thrust
Dead Load	+143000	+734800	+143000	+734800	+124200	+577200	+124200	+577200
Live Load								
Roadway (Inc. Imp)	+643000	+34000	-693000	+52100	+321000	+36900	-169000	+60400
Sidewalk L.L.	+319000	+20200	-269000	+27600	+76300	+15400	-126000	+28600
Temp. Rise (40°)	+365200	+4800	—	—	+41600	+6800	—	—
Temp Drop (40°)	—	—	-365200	-4800	—	—	-41600	-6800
Elastic Rib Shortening	—	—	-100500	-1320	—	—	-11400	-1870
Shrinkage	—	—	-182600	-2400	—	—	-20800	-3400
TOTAL	+1470,200	+793,800	-1467,300	+806,000	+563,100	+636,300	-244,600	+654,130

	ARCH POINT NO 6				ARCH POINT NO. 12			
	Max. Positive Mom		Max. Neg. Mom.		Max. Positive Mom		Max. Neg. Mom.	
Load Condition	Moment	Thrust	Moment	Thrust	Moment	Thrust	Moment	Thrust
Dead Load	+106600	+400600	+106600	+400600	+108500	+712900	+108500	+712900
Live Load								
Roadway (Inc Imp)	+338000	+40500	-146000	+32700	+638000	+38400	-666000	+51100
Sidewalk L.L.	+80200	+14300	-59500	+12800	+292000	+21600	-264000	+28600
Temp. Rise (40°)	—	—	-126800	+8800	+357800	+4600	—	—
Temp Drop (40°)	+126800	-8800	—	—	—	—	-357800	-4600
Elastic Rib Shortening	+34900	-2420	—	—	—	—	-98400	-1262
Shrinkage	+63400	-4400	—	—	—	—	-178900	-2300
TOTAL	+749,900	+439,780	-223,700	+454,900	+1396,300	+777,500	-1456,600	-784438

nonuniform temperature variations. If it can be demonstrated that this effect is relatively small, and can be disregarded.

Step 10. This involves the calculation of the unit fiber stress in the concrete and steel for each point investigated, applying the moment and thrust data of Table 10-7. This procedure is a typical analysis of a rectangular section subjected to combined axial and flexural stresses.

Summary and Comments The five critical sections investigated in the foregoing example demonstrate the design philosophy of any fixed rib arch. In most practical applications, and particularly where computerized techniques are used, the investigation should be extended to intermediate sections and should cover the tenth points along the rib. Table 10-7 presents a liberal compilation of group loading summary and may or may not be in full compliance with the AASHTO grouping and percentage of allowable stresses.

For this example the thrusts that are used are not necessarily the maximum thrusts but the thrusts that occur for the loading producing the moments with which they are combined. The maximum value of the live load thrust occurs for a position corresponding to a fully loaded rib. Because moment stresses generally control, the maximum thrust is seldom used as this is not a critical moment loading. The live load is therefore placed for maximum moment and the corresponding thrust used in connection with this moment.

The dead load is constant and it serves to relieve certain maximum moments while adding to others. Temperature effects are reversible, so that the corresponding moments are added or subtracted. The temperature thrust, however, only adds to the load thrust for a temperature rise, whereas a drop in temperature causes tension or negative thrust. All load thrusts are positive. Shrinkage and rib shortening act in the same manner as a temperature drop and with the same sign.

10-12 ANALYSIS OF HINGED ARCH BRIDGES

Hinged arch bridges have one or more hinges, or points of articulation, in the rib. The introduction of a hinge at any point provides a plane of zero bending, and the structure is free to rotate under load. Hinges represent a definite location of a thrust line at one point and thus add "certainty" to the structural arrangement. Secondary effects, such as temperature stresses, shrinkage, elastic rib shortening, creep, and support displacement, tend to decrease and even diminish as hinges are added, particularly for a three-hinged arch.

On the other hand, a hinged structure is more flexible under load (an obvious disadvantage), and in masonry construction hinge details are difficult to construct and maintain. Masonry arches are generally the fixed or hingeless type except for those locations where support displacements are difficult to control. Metal arches have been constructed with hinges more frequently than without them.

Either type, fixed or hinged, has comparative merits that must be judged and evaluated on an individual basis. For the purpose of analysis, in this section we will consider three-hinged arch spans. In this case once the external reactions are determined, the problem is one of pure statics. The presence of three hinges removes the structural redundancy, rendering the external reaction components completely determinate from statics. In addition, the structure is free to rotate about the central hinge, eliminating stress conditions that may result from temperature, shrinkage, or rib-shortening effects.

Determination of Load Reactions The hinge reaction for an arch of this type may be computed by any of the following three methods: (a) analytically, using the three equations of static equilibrium; (b) graphically, by means of the equilibrium polygon; and (c) from influence lines.

Figure 10-47 illustrates the general analytical procedure for an unsymmetrical span with the crown hinge placed off the horizontal center. For a load F_g at any point g, the reactions are computed from the basic equations of static equilibrium and are as shown. For symmetrical spans with the central hinge placed at the exact crown, the computations are simplified as shown. L denotes the horizontal span.

A graphical solution is obtained by noting that moments must be zero at any of the three hinge points. Hence, the reaction lines for any load condition must pass through these hinge points as shown in Figure 10-48. For

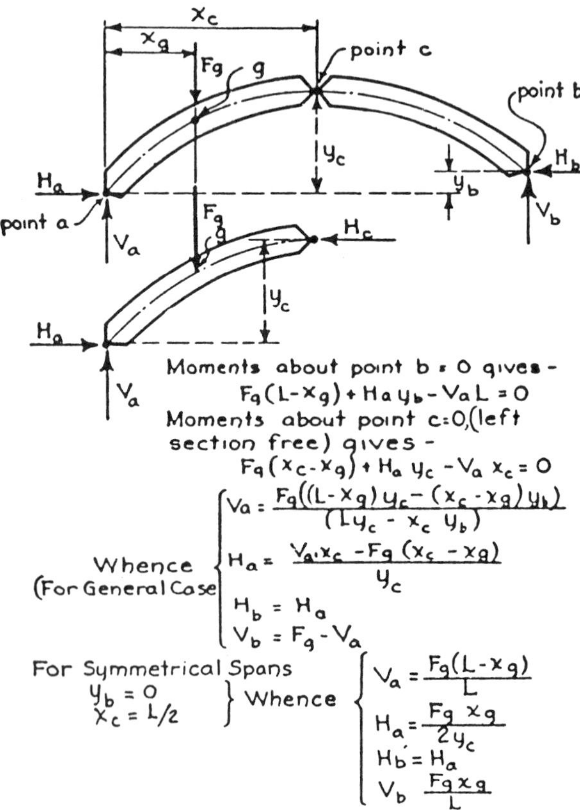

FIGURE 10-47 Configuration and analytical procedure for a three-hinged arch.

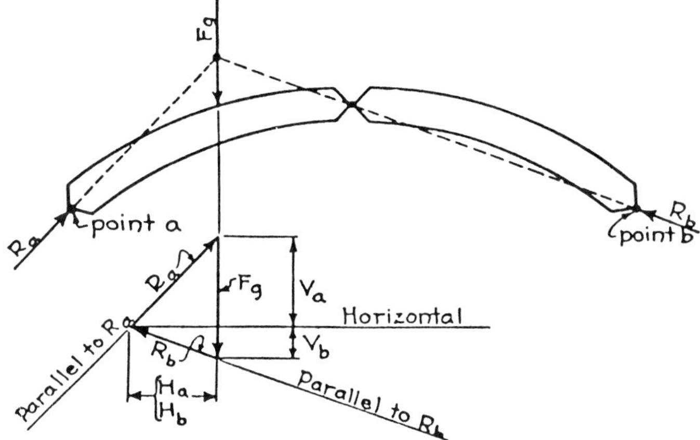

FIGURE 10-48 Equilibrium of forces; three-hinged arch.

a group of loads the graphical approach is extended to the construction of an equilibrium polygon as shown in Figure 10-49. Thereafter, the internal stresses are readily estimated. For the rib arch of Figure 10-49, because segment b' of the equilibrium polygon represents the total thrust T_3 against cross section 3–3 (the resultant of R_a and F_1), it follows that

$$J_3 = T_3 \sin \phi \qquad N_3 = T_3 \cos \phi \qquad M_3 = T_3 r = N_3 c \qquad (10\text{-}67)$$

Influence Lines This procedure is applicable to reactions as well as to the determination of direct stresses. From the equations developed in Figure 10-47, it is apparent that the vertical skewback reactions for symmetrical spans are exactly the same as for a simple beam of the same span. This observation can be used in developing influence lines for stresses in framed arches.

Consider, for example, the spandrel framed arch shown in Figure 10-50. The stress in chord U_1–U_2 is equal to the moment about point L_2 divided by the lever arm p. This moment is also the algebraic sum of the moments M_v and M_h, caused by the vertical loads and the horizontal thrust, respectively.

The influence line for M_v is obtained by laying off a horizontal line and a vertical ordinate kg/L through panel point L_2, and then completing the triangle as shown. The influence line for $M_h(= Hd)$ is a triangle whose vertex lies on the crown vertical with an ordinate $Ld/4y$. Superimposing these two areas yields the influence area for $M = M_v + M_h$, the moment at point L_2. This area is also the influence area for the chord stress in member U_1–U_2 divided by the constant p. From inspection the moment M_v produces compression in U_1–U_2 and the moment M_h causes tension, with the signs of the resultant moment as indicated.

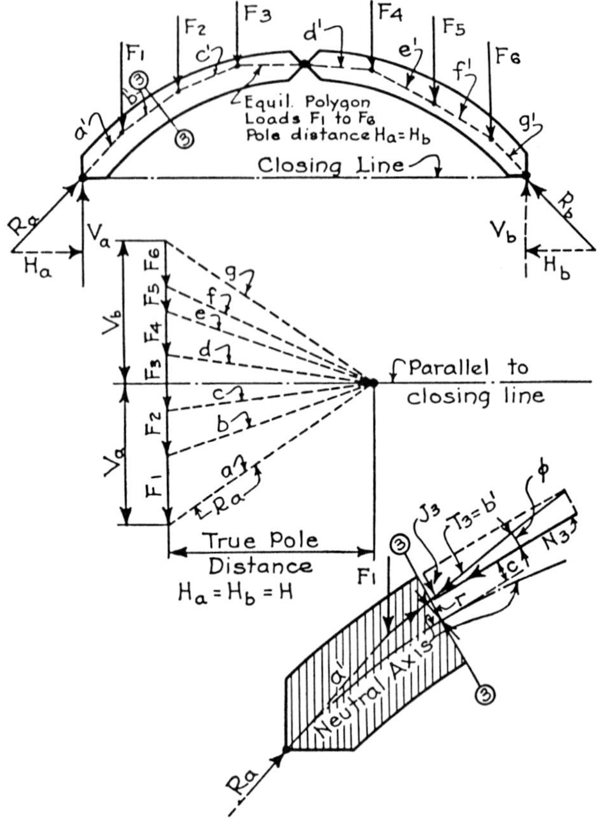

FIGURE 10-49 Equilibrium polygon for a group of loads; three-hinged arch.

The effect of thrust H on the stress in member U_1–U_2 is given by the expression

$$S_h = Hy'/p' \tag{10-68}$$

where y' is the ordinate of the center of moments (point c) for this member (obtained by producing the two chords in the panel to an intersection) and p' is the perpendicular lever arm drawn on the web member from said moment center. The line for S_h is therefore constructed with a central vertical ordinate equal to $Hy'/p' = Ly'/4yp'$. The stress S_v in the same member U_1–L_2 is obtained from the following expressions:

$$S_v = V_a x_c/p' \qquad \text{for loads to the right of } U_1$$
$$S_v = V_b(L - x_c)/p' \qquad \text{for loads to the left of } U_1 \tag{10-69}$$

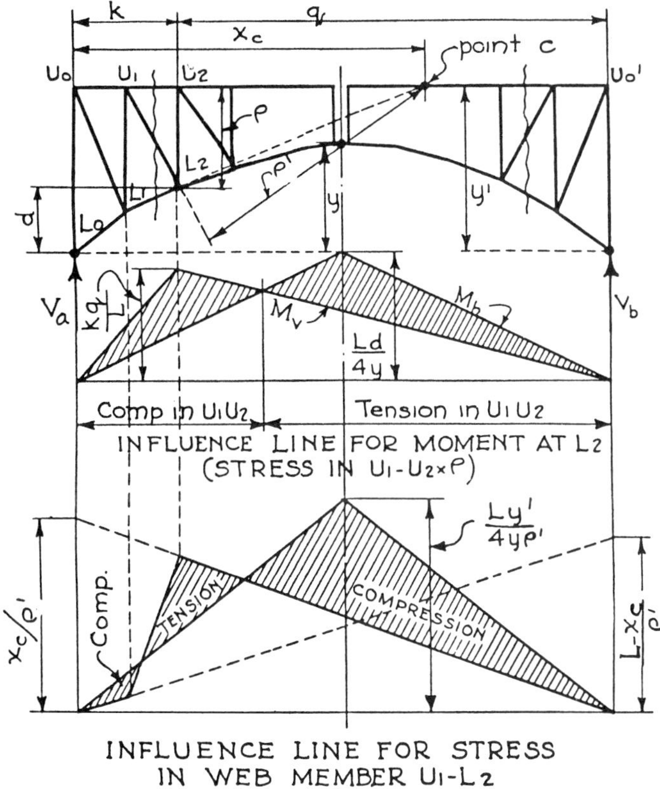

FIGURE 10-50 Configuration and influence lines for a spandrel, three-hinged framed arch.

Because $V_a = 1$ for a unit load at U_0 and zero for a load at U_0' and vice versa for V_b, the influence line for S_v is obtained by constructing the two reaction triangles with ordinates as shown. Superimposing the cross-hatched areas gives the influence area for the stress in member U_1–L_2.

Load Divide Lines Because the reaction lines always pass through the three hinge points, it is possible to determine the limits of live load that will produce stress of a given direction in any given member. For the arch shown in Figure 10-51, we will determine the load that causes maximum compression in the upper chord member U_3–U_4. First, we construct the hinge lines. If a line is drawn through the left skewback hinge and point L_4 (the center of moments for member U_3–U_4), this line will intersect the right-hand hinge line at point *a*. A load at this point will obviously induce no stress in chord member U_3–U_4 because it produces a reaction passing through the center of moments for this member. Any load to the right of point *a* (e.g., point *b*)

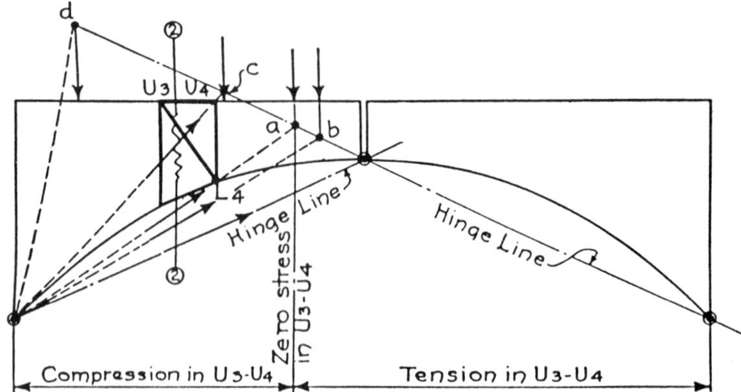

FIGURE 10-51 Load divide lines for maximum load effects; three-hinged arches.

induces a reaction passing through section 2–2 below point L_4. This reaction obviously produces tension in the top chord as shown in Figure 10-52a where the left-hand section of the structure is shown as a free body in equilibrium. As the load passes from point a to the left and is applied, for example, at point c, the reaction line passes above point L_4 and the chord stress changes from tension to compression. As the load continues to move to the left and passes beyond point U_3, the chord stress is still compression as can be seen from Figure 10-52b where the right-hand segment is now shown as a free body.

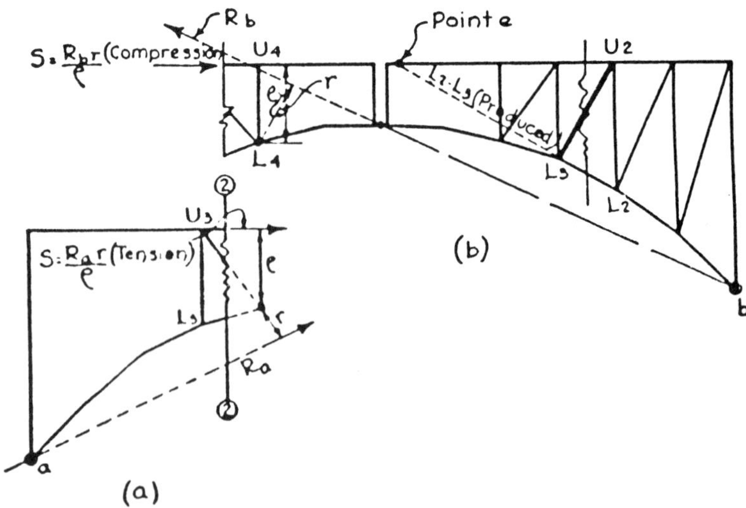

FIGURE 10-52 Free-body diagrams for load divide lines; three-hinged arches.

The load divide point in any other chord member is likewise determined by similar analysis, and the same procedure is used for web members.

Arch Influence Lines from Deflection Diagrams The reciprocal relationship between deflection polygons and influence lines provides the basis for constructing these lines for stresses, reactions, shears, moments, or any other load function in a structural frame or rib. This results from the principles of internal work and is self-evident. These principles are as follows.

1. The influence line for the stress in any member of a structural frame is identical to the deflection polygon resulting from a unit linear distortion in the member.
2. The influence line for the moment about any point in a structural frame or rib is identical to the deflection polygon resulting from a unit angular distortion $\Delta\beta$ at the given point.
3. The influence line for a reaction in any structural frame or rib is identical to the deflection diagram resulting from a unit displacement at the given support.

It is inconceivable that this operation is any easier or more rapid than the conventional method. The concept, however, gives an added meaning to the analysis. It also offers the advantage of constructing influence lines for ordinary arches without rewriting expressions of equilibrium or solving formulas. The fundamental characteristics of the method also accommodate solutions for new and complex structural arrangements.

Illustrative Example Figure 10-53 shows a three-hinged braced spandrel arch. First, we consider the development of the influence line for point q in the bottom chords. The procedure is as follows.

1. Construct line a'–q'–c'', making the angle θ at point q' (θ is unity radians).
2. Locate the "load divide" point ω for the moment at point q by intersecting lines aq and bc. Point ω' is located by drawing a perpendicular through point ω until intersecting line $q'c''$. Because this is a point of zero moment at q, the deflection at this point due to a unit angle change at q must also be zero.
3. Locate point c' by the line $a'\omega''c'$ drawn through point ω''.
4. Complete triangle $c''b'c'$ (b' may be assumed at any point vertically below b).
5. The diagram $a'q'\omega''c''b'c'a'$ is the desired influence line.

Influence diagrams for bending moments at any other point can be developed in a similar manner. Figure 10-53b shows this diagram for point p located in the top chord.

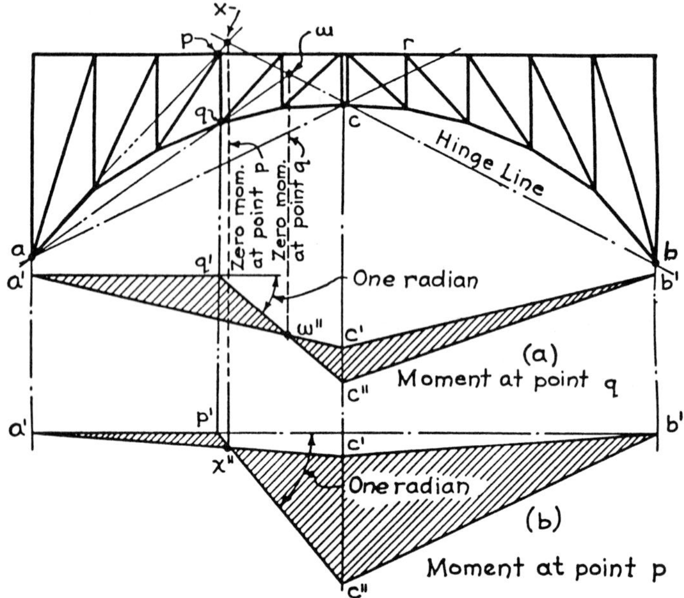

FIGURE 10-53 Influence line for a three-hinged braced spandrel arch determined from deflection polygons.

Figure 10-54 shows the application of the method in determining the horizontal thrust $H_a = H_b$. Point b is given a horizontal displacement $b'b''$. Under the action of this displacement, point b' is first assumed to rotate about the center hinge c' to point b'' and then back about point a' to its original level because the support at b does not yield vertically. The rotation about the center hinge causes point b'' to move upward to point b'''. The

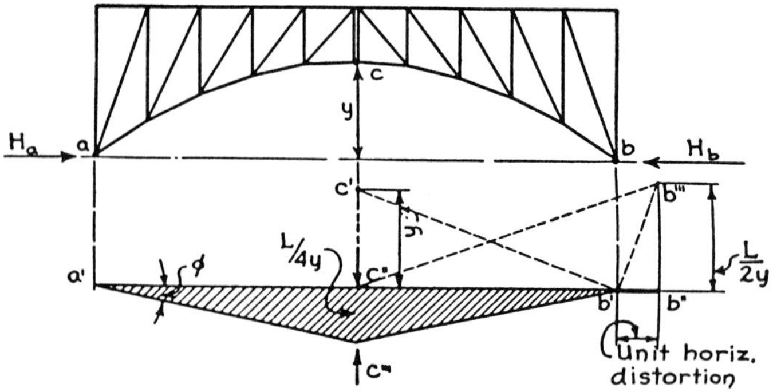

FIGURE 10-54 Influence line for horizontal thrust drawn from deflection polygons.

following relationship exists:

$$b'' - b''' = \frac{L}{2y} \tag{10-70}$$

The rotation of point b''' about point a' back to its original level is through an angle $\phi = \tan^{-1}(1/2y)$, which clearly causes the crown hinge to deflect a distance

$$c'' - c''' = \frac{L}{2} \tan \phi = \frac{L}{4y} \tag{10-71}$$

The diagram $a'c''b'c'''a'$ is therefore the desired influence line (with $c'c''' = L/4y$.

10-13 MULTISPAN ARCHES ON ELASTIC PIERS

In the foregoing examples the effect of elastic distortion of the supports under load has been neglected. However, if a series of arch spans are supported on piers that are relatively high and slender, the elastic distortion becomes an important factor because it can modify the load stresses markedly. In general, the effect of an elastic yielding of the supports is to decrease the horizontal thrusts, to increase the crown moments, and to redistribute stress to adjacent arch spans and intermediate piers.

Interestingly, notwithstanding the fact that the quantitative determination of stresses in arch ribs induced by pier elasticity is comparatively recent, the actual utilization of arch span groups is by no means new.

Basic Principles Figure 10-55*a* shows an ordinary fixed rib arch of single span. If at either end the rib were to be split up into two separate chords or elements as shown in Figure 10-55*b*, the same method of analysis would still apply. Taking this transformation one step further, the fundamental philosophy of the analysis would be applicable if each of the two component segments were extended to form an intermediate pier and an additional span as shown in Figure 10-55*c*. In this transformation the original right-hand voussoir (from section a–a to the skewback) has been replaced by a curved composite voussoir as shown in Figure 10-55*d*.

It is now apparent that if the elastic displacement of the pier and span combination (cross-hatched in Figure 10-55*d*) under a given load F is the same as in the arch block from which it was developed, the elasticity of the central segment a–a to b–b is not changed but remains identical in both cases. It appears therefore that a suitable method of analysis is this development in reverse. Thus, the method consists essentially of the determination of the elastic displacement of each pier and rib combination (beginning at either

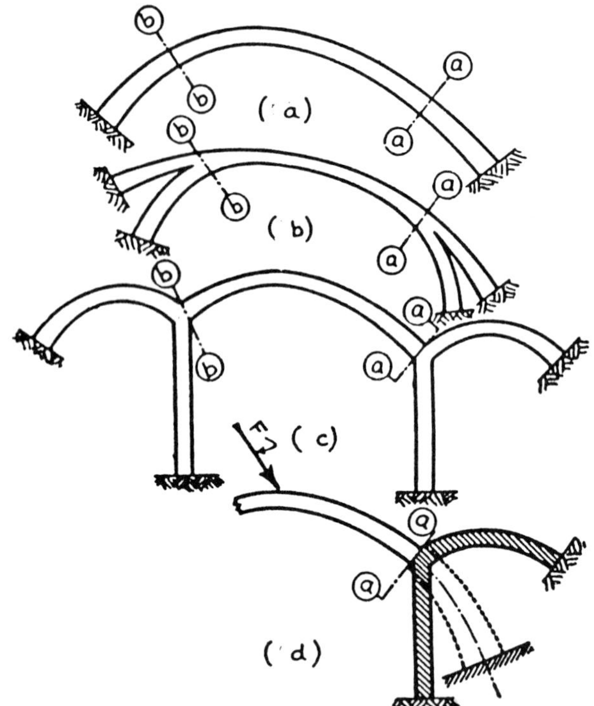

FIGURE 10-55 Conversion of a single fixed rib arch into a multispan arch.

end) and the replacement of such combinations by an ideal or substitute voussoir section having an equivalent elastic displacement.

The complete analysis must be extended one step further, inasmuch as the foregoing procedure is sufficient for the determination of stresses in the central rib by a load acting on that particular rib. In addition, it is necessary to determine what portion of load on any one of the spans is transferred to adjacent spans and what portion is carried down through the intermediate piers.

Elastic Arch Group over Three Spans In general, the method of analysis is rather tedious, but with shortcuts and approximations it is quite possible to pursue a complete set of calculations from which to plot the necessary stress influence lines. Alternatively, a computerized method of analysis will save considerable time and offer the benefit of parametric solutions.

Consider the three-span group shown in Figure 10-56, where the end spans are assumed unsymmetrical. If arch 3 and pier 2 are assumed as an independent elastic system, as shown in Figure 10-56*b*, the first step is to compute the elastic displacement of any point O' in the plane rigidly attached to point b and induced by the following loads: (a) a unit moment

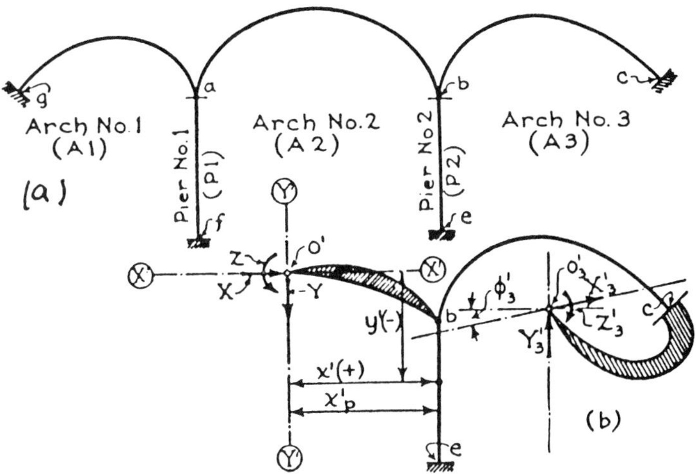

FIGURE 10-56 (*a*) Three-span unsymmetrical group of arches; (*b*) independent elastic system for arch 3 and pier 2.

couple (Z = unity), (b) a unit horizontal force (X = unity), and (c) a unit vertical force (Y = unity).

Each of these displacements may be resolved into three components, namely, horizontal, vertical, and angular. From Figure 10-56*b* and for coordinate dimensions x' and y' measured with point O' as origin, we can write $m_x = y'$, $m_y = x'$, and m_z = unity.

For a unit moment couple (Z = unity) applied at O', the following are true:

$$\begin{aligned} \delta_{zx} &= \sum_e^b M_z m_x G = \sum_e^b M_z G y' \\ \delta_{zy} &= \sum_e^b M_z m_y G = \sum_e^b M_z G x' \\ \delta_{zz} &= \sum_e^b M_z m_z G = \sum_e^b M_z G \end{aligned} \tag{10-72}$$

For a unit horizontal force (X = unity) applied at O',

$$\begin{aligned} \delta_{xx} &= \sum_e^b M_x m_x G = \sum_e^b M_x G y' \\ \delta_{xy} &= \sum_e^b M_x m_y G = \sum_e^b M_x G x' \\ \delta_{xz} &= \sum_e^b M_x m_z G = \sum_e^b M_x G \end{aligned} \tag{10-73}$$

For a unit vertical force (Y = unity) applied at O',

$$\delta_{yx} = \sum_e^b M_y m_x G = \sum_e^b M_y G y'$$

$$\delta_{yy} = \sum_e^b M_y m_y G = \sum_e^b M_y G x' \tag{10-74}$$

$$\delta_{yz} = \sum_e^b M_y m_z G = \sum_e^b M_y G$$

In the foregoing expressions the terms M_z, M_x, and M_y represent moments at sections of the pier *be* regarded as an element of the arch *ebc*, whereas the unit moments m_x, m_y, and m_z are corresponding moments on the section regarded as a cantilever. Thus, for any given point, m_z = unity whereas $M_z = [X'_{z3}y_3 - Y'_{z3}x_3 - Z'_{z3} + 1.0]$, where X'_{z3}, Y'_{z3}, and so forth are the redundant forces induced at point O'_3 (the elastic center of the system $P2$–$A3$) by a unit moment couple Z = unity applied at point O' in the second span. Equations (10-72) through (10-74) represent therefore the displacement components of any point O' in the plane rigidly attached to the elastic system $A3$–$P2$ at point b for the three unit load components applied at point O'.

Ideal or Substitute Single Voussoir The foregoing elastic system is replaced by the section shown in Figure 10-57. The various displacement components of the same point O' corresponding to (10-72) are now written as

$$\delta_{zx} = \sum_d^b M_z m_x G' = \sum_d^b G' y' = \bar{y}' W_1$$

$$\delta_{zy} = \sum_d^b M_z m_y G' = \sum_d^b G' x' = \bar{x}' W_1 \tag{10-75}$$

$$\delta_{zz} = \sum_d^b M_z m_z G' = \sum_d^b G' = W_1$$

where G' represents the elastic weight of each individual laminar or *ds* division in the substitute section and x', y' its coordinates with respect to O' as the origin. The factor W_1 represents the total elastic weight of the substitute section so that

$$W_1 = \sum_d^b G' \qquad \bar{y}' = \frac{\sum_d^b G' y'}{\sum_d^b G'} \qquad \bar{x}' = \frac{\sum_d^b G' x'}{\sum_d^b G'}$$

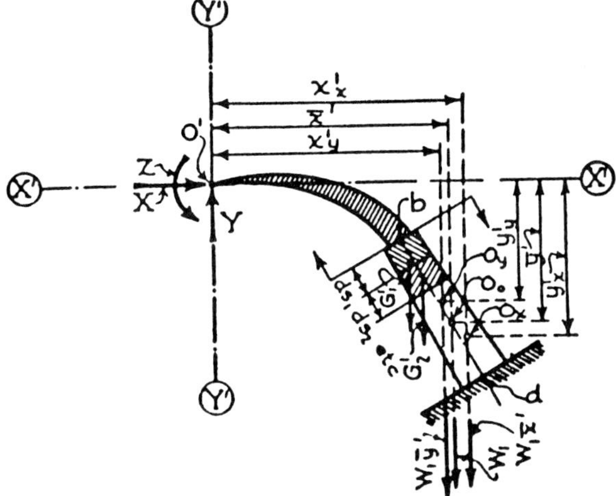

FIGURE 10-57 Ideal or substitute single voussoir.

The total elastic weight W_1 is applied at point O_0 (Fig. 10-57). The sum of the individual elastic moments $G'x'$ and $G'y'$ considered as loads may be replaced by the total values $W_1\bar{x}'$ and $W_1\bar{y}'$, but the point of application of these resultants will not coincide with point O_0 and will be at some other points O_x and O_y.

Likewise, for the determination of the balance of the elastic distortion components, we can write

$$\delta_{xx} = \sum_d^b m_x^2 G' = \sum_d^b G'y'^2 = \bar{y}'y'_y W_1$$

$$\delta_{xy} = \sum_d^b m_x m_y G' = \sum_d^b G'x'y' = \bar{y}'x'_y W_1 = \bar{x}'y'_x W_1 \qquad (10\text{-}76)$$

$$\delta_{xz} = \sum_d^b m_x m_z G' = \sum_d^b G'y' = \bar{y}'W_1$$

$$\delta_{yx} = \sum_d^b m_y m_x G' = \sum_d^b G'x'y' = \bar{y}'x'_y W_1 = \bar{x}'y'_x W_1$$

$$\delta_{yy} = \sum_d^b m_y^2 G' = \sum_d^b G'x'^2 = \bar{x}'x'_x W_1 \qquad (10\text{-}77)$$

$$\delta_{yz} = \sum_d^b m_y m_z G' = \sum_d^b G'x' = \bar{x}'W_1$$

Equating these distortion components and solving, we obtain

$$W_1 = \sum_e^b M_z G \qquad \bar{x}' = \frac{\sum_e^b M_z G x'}{W_1} \qquad \bar{y}' = \frac{\sum_e^b M_z G y'}{W_1} \tag{10-78}$$

which give the elastic weight of the ideal section and the coordinates of its elastic center O_0 referred to the arbitrarily selected axes X'–X' and Y'–Y'. We can note that the effect of the elastic system $A3$–$P2$ is reproduced by an equivalent pier extending from point b (Figure 10-56) to point e, and as each original segment having an elastic weight G has been replaced by one having an elastic weight $G' = M_z G$.

Thus far, the procedure is entirely independent of the location assumed for point O'. We may now proceed to write more equations in order to satisfy the conditions of the problem. Thus, the following are obtained:

$$\sum_e^b M_x G = \bar{y}' W_1 \tag{10-78a}$$

$$\sum_e^b M_x G x' = \bar{y}' x'_y W_1 = \bar{x}' y'_x W_1 \tag{10-78b}$$

$$\sum_e^b M_x G y' = \bar{y}' y'_y W_1 \tag{10-78c}$$

$$\sum_e^b M_y G x' = \bar{x}' x'_x W_1 \tag{10-78d}$$

which provide the basis for determining $\bar{y}'$, x'_y, y'_x, y'_y, and x'_x.

It is obvious that the coordinates of the point of application of the elastic moments ($W_1\bar{x}'$ and $W_1\bar{y}'$) referred to any set of axes such as X'–X' and Y'–Y' are dependent on the position of these axes of reference. These coordinates may be determined for any given set of axes by computing the corresponding moments M_x and M_y for the original elastic system between points b and e, and solving the above equations. In general, the elastic moments do not become relevant until after the axes of reference are chosen.

Example Figure 10-58 combines the ideal voussoir W_1 with arch 2 of Figure 10-56. Because the value and coordinates of the elastic center of the "ideal voussoir" are basically independent of the assumed axes of reference (X'–X' and Y'–Y'), these values may be computed and the elastic weight W_1 added to the system as shown. The true elastic center for the new combined system $P1$–$A2$–$P2$–$A3$ (or $fabW_1d$) may now be readily located and the proper conjugate redundant axes (X'_2–X'_2 and Y'_2–Y'_2) determined as shown.

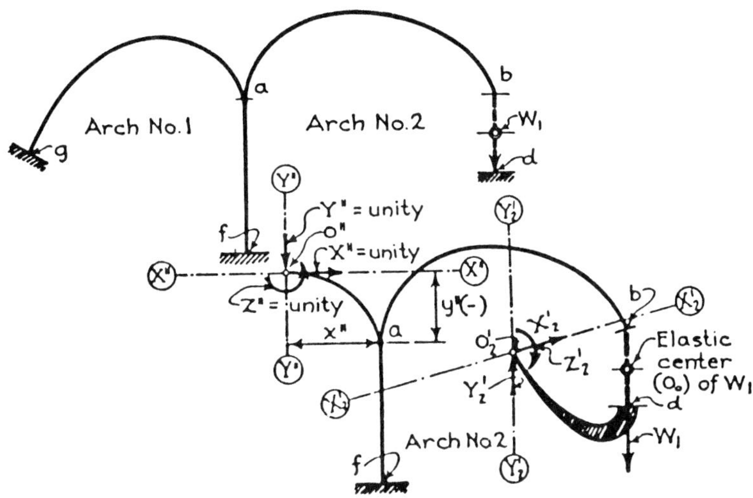

FIGURE 10-58 Ideal voussoir with arch 2 of Figure 10-56.

With these data available, the terms M'_{x2} and M'_{y2} for the pier section b–e are now evaluated and the coordinates x'_x, x'_y, y'_x, and y'_y (referred to the origin O'_2) readily located.

10-14 ARCHES WITH RIGIDLY CONNECTED DECKS

Basic Types

Throughout this discussion the general assumption is that the loading is vertical, because the function of an arch bridge is to support the gravity loads that constitute the major portion of its burden. Where the characteristics of the loading members transmitting the vertical loads to the arch proper are such as to produce secondary forces at the loading points, an arch analysis based on the assumption of purely vertical loads will not indicate the true stress conditions of the structure. This is the case for arches having superstructures rigidly connected to the supporting rib or barrel.

Through Arches This type, shown in Figures 10-59*a* and *b*, includes the through or half-through structure. The load is suspended from the rib by hangers, designed mainly from a consideration of tension so that the slenderness ratio does not impose restrictions. The hangers have thus a small cross section and are flexible to a large degree. This flexibility limits the amount of lateral force or moment that can be transmitted from the deck to the arch.

For this type of construction the assumption of vertical loading is substantially correct, and any forces other than vertical that may be developed at

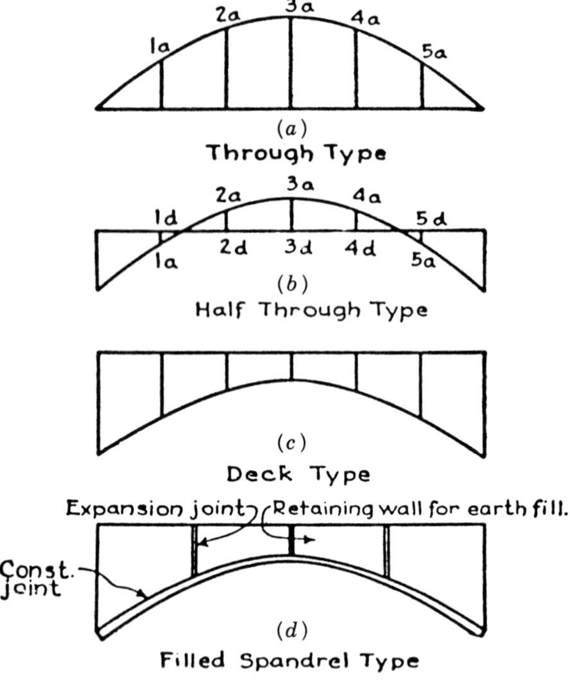

FIGURE 10-59 Deck-type arch bridges.

arch points $2a$, $3a$, and $4a$ are secondary in magnitude. However, if tension hangers such as member 1 are very short, high stresses are likely to develop if connections are rigid, and therefore they should be analyzed as columns.

Open Spandrel Deck Arch For open spandrel deck arches, Figure 10-59*c*, and for the deck portion of the half-through type, Figure 10-59*b*, the secondary stresses are more important than in the through arches because the loading members are in compression. For concrete structures the working stresses are lower, necessitating a larger section. In addition, the columns are subject to slenderness restrictions discussed in the foregoing sections.

Filled Spandrel Arch This type is shown in Figure 10-59*d*. One of the most serious disadvantages of a filled spandrel arch is the indeterminate interaction between the retaining walls and the barrel. This effect continues to exist even with the use of expansion or construction joints that are often depended on for articulation.

In general, the calculated load or deformation stresses are markedly changed by the rigidity of the wall, the weight of which is transferred from point to point of the arch as the radius of curvature varies under changing conditions. In a complete analysis, therefore, the calculated stresses are expected to vary widely from the actual stresses determined from precise measurements.

Methodology for Exact Solution

Consider the deck spandrel arch bridge shown in Figure 10-60. If any deflection such as Δx is produced by forces applied to the arch at the skewback, the general equation between stresses and deflection is

$$\Delta x = -\sum MGy \tag{10-79a}$$

in which the moment M at any rib point b is

$$M = -Xy + Yx + Z + M_0 + M_c \tag{10-79b}$$

This is the same as the usual expression of moment with the added term M_c representing the unknown moments introduced by the rigid superstructure connection at the column footings. These column forces are interrelated with the arch and superstructure movements, and a complete solution would in this case involve the following unknowns:

Unknowns	Number of Unknowns
Horizontal deflection of the deck	1
Horizontal deflection of each arch point assumed fixed (except point c)	6
Angular deflection of all points (except c)	13
Vertical movement of panel points (except c) the same for deck and arch	6
Two stresses (moments and shears) in each of 19 members	38
Total	64

The following equations can be written:

$\sum H = 0$	for the deck	1
$\sum H = 0$	for each arch point	6
$\sum M = 0$	for all points	13
$\sum V = 0$	for each column (except c)	6
Two stress deflection equations (moment and shear) for each of the 19 members		38
Total		64

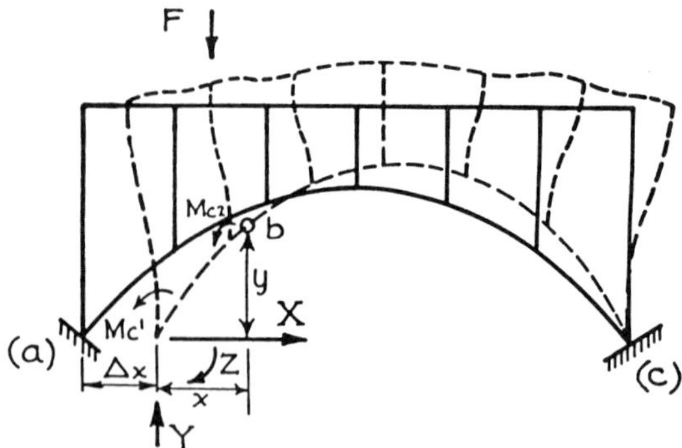

FIGURE 10-60 A deck spandrel arch bridge; initial and deflected position.

This solution is possible but not practical unless a suitable computer program is available. Alternatives to the frame of Figure 10-60 can be formulated by one of the following three courses.

1. Resort to a mechanical analysis. This is one of the most practical methods, but has the inherent disadvantage that the internal stresses must be obtained by cutting the model and inserting a clamp at the point under investigation.
2. Break the continuity of the superstructure by means of hinges or sliding joints. Such devices will not eliminate the secondary stresses, although they may reduce them to an inconsequential value. This is a customary method, but whenever it is applied it is almost certain that the deformation stresses in some members will have values much higher than the designer realizes.
3. Use an approximate analytical solution that will yield secondary stresses sufficiently accurate to enable the designer to determine at what point articulations are necessary and to calculate the stresses that would result if connections were made rigid.

Articulation of Superstructure

This solution provides protection to the superstructure itself from injurious strains that would otherwise be present. With reference to the load effects, the superstructure functions in a manner that benefits the arch proper. Any moment tending to flex the rib is resisted not only by the arch but by the connecting columns, with a subsequent reduction in arch stresses. This reduction is accomplished, however, at the expense of the deck where certain stresses are induced through the rigid connections with the arch (see also Section 10-8).

Because in addition to load stresses there are deformation stresses due to temperature changes, elastic shortening, shrinkage, and creep, more stresses will be induced if these effects occur differentially.

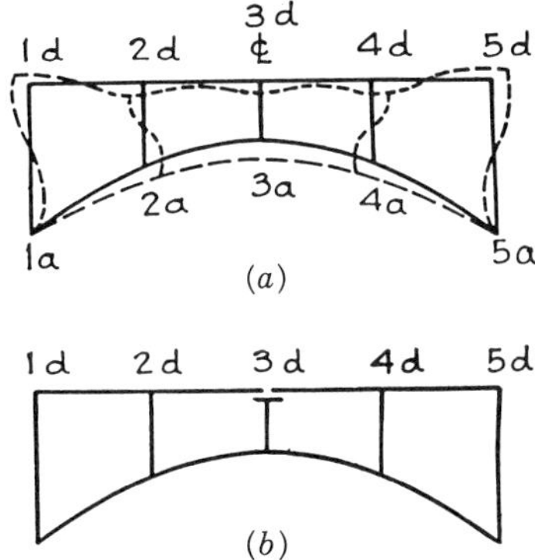

FIGURE 10-61 Articulation of superstructure for a symmetrical arch.

For the symmetrical arch shown in Figure 10-61, any shortening of the arch ring by increasing its radius of curvature would cause the columns to lean inward. If the deck is subjected to a temperature increase relative to the arch rib, it would expand in both directions with the center as the probable point of zero movement. The resulting distortion is shown by the dashed lines. Column 3 will undergo no bending, but the other columns will undergo flexure. The resulting bending moments will vary directly as their horizontal deflections and moments of inertia and inversely as the square of their lengths. Column 1 is long and thus more flexible and may have only little overstress even with rigid top and bottom connections. Column 2 has only one-half the lateral deflection of column 1, but if the length is also half the resulting moment is double.

From the foregoing it appears that if the combined stress due to dead load, live load, thrust, and deformation moment is excessive, two possible solutions may be considered. The first is often followed and involves the introduction of expansion joints in the deck as shown in Figure 10-61*b*. Point 3*d* is thus changed from a fixed to a movable joint. Column 2, being shorter and stiffer than column 1, will now lie very near the point of fixity in the deck under expansion and thus will be relieved considerably from deformation stresses. The freedom of the deck to expand at point 3 will also lessen temperature movement of point 1*d*, with a subsequent reduction in the stresses for column 1.

The second solution will involve a hinge inserted at one end of column 2, or if necessary at both ends. This would reduce all bending stresses in the column to residual values representing friction in the hinges.

Mechanical Analysis of Superstructure (Model Studies)

This method bears the same relationship to purely analytical investigation, but is based on a laboratory procedure that provides a visible check of the analytical results. It is an experimental method of stress analysis based on the arch theory contributed by Clapeyron, Castigliano, Maxwell, Hooke, and others.

Best results are obtained by means of a small-scale replica or model of the structure under consideration combined with principles derived from the fundamental theory of elasticity. Because these principles deal with the structure as an elastic body, it is essential for the model to represent the structure not only in its physical proportions but also in its elastic properties. In this context the elasticity (expressed as the capacity for elastic distortion) of the various members and parts must bear the same relationship to each other in the model and in the structure. If this is accomplished, the absolute elasticity, or quantitative relationship between stress and strain, need not be considered because reactions and stresses are obtained from the deflections of the model. This involves measuring the deflections of the member in question and then reducing these strains to stresses by means of elastic equations.

Consider, for example, the arch of Figure 10-61 and assume that we desire to know the temperature stresses produced in column 2. From the model we find the horizontal and angular deflections at each end of the column corresponding to the assumed temperature change in the length of the span. From these data the moment and shear may be calculated.

Basic Theory of Mechanical Analysis Consider the elastic arch with fixed footings shown in Figure 10-62*a*. It is required to find the reaction produced at point a along line X–X by a load F_b at b.

With the arch free of any load, we assume point a to be removed from its anchorage and moved to a distance Δ_{xx} in the direction X–X only to a new location a'. This is accomplished by applying (a) a distorting force X_0, (b) a linear restraining force Y_0 that prevents movement out of line X–X, and (c) a restraining moment Z_0 that prevents rotation. Because by definition no Y or Z movement takes place, no work is performed by these restraining forces so that the external work is

$$W_E = \tfrac{1}{2}X_0\Delta_{xx} = W_I'$$

Now, with the rib in its deflected position, we add the load F_b and the resulting reactions X_{ba}, Y_{ba}, and Z_{ba}. Because these reactions perform no work, the second increment of work is simply

$$W_I'' = \tfrac{1}{2}F_b\Delta_{bb}$$

where Δ_{bb} represents the displacement of force F_b along its line of action. The total internal work is now

$$W_1 = W_I' + W_I'' = \tfrac{1}{2}X_0\Delta_{xx} + \tfrac{1}{2}F_b\Delta_{bb} \tag{10-80}$$

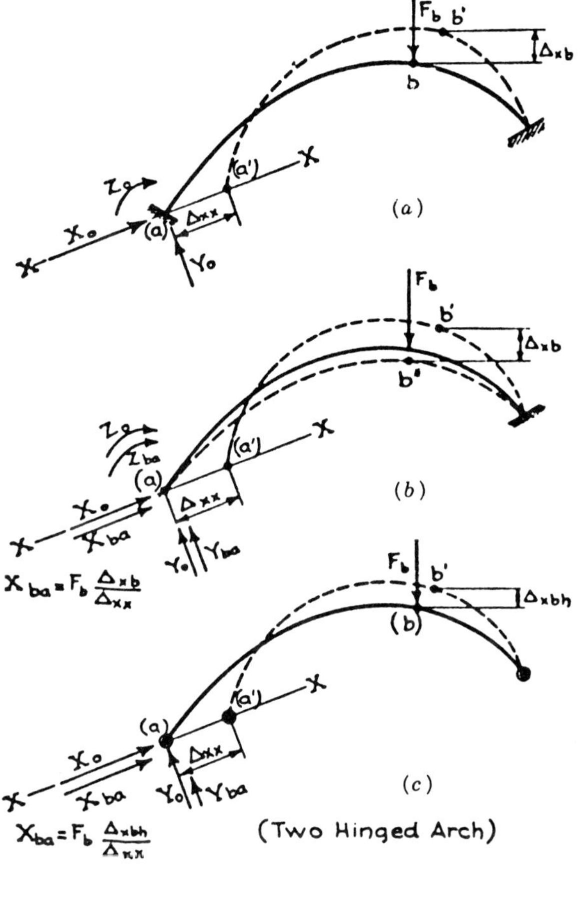

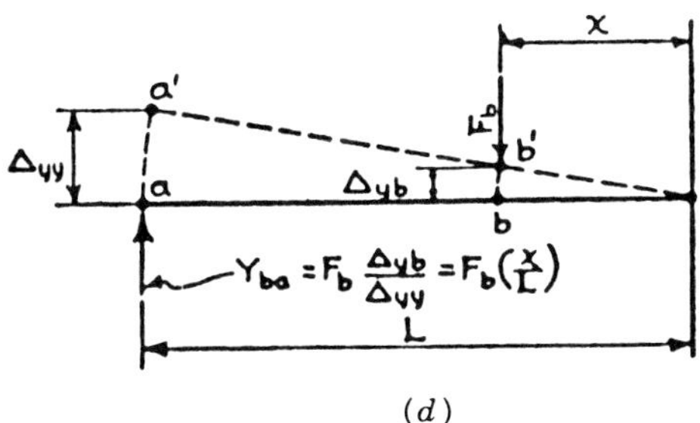

(*d*)

FIGURE 10-62 Model frame for mechanical arch analysis.

The same deflected position may be attained with the same expenditure of energy if we apply X_0, X_{ba}, and F_b simultaneously, in which case

$$W_1 = \tfrac{1}{2}X_0\Delta_{xx} + \tfrac{1}{2}X_{ba}\Delta_{xx} + \tfrac{1}{2}F_b\Delta_{bb} - \tfrac{1}{2}F_b\Delta_{xb} \tag{10-81}$$

where Δ_{xb} is the deflection at b due to the movement Δ_{xx} at a. From (10-80) and (10-81), we obtain

$$X_{ba} = F_b\frac{\Delta_{xb}}{\Delta_{xx}} \tag{10-82a}$$

which states that "the load is to the reaction as any arbitrary movement of the reaction point in the direction of the reaction is to the resulting movement of the load point in the direction of the load," or "any two related forces are inversely proportional to the coincident displacement of their points of application in the direction of the forces."

The latter statement is illustrated in Figure 10-62d showing that the basic relationships between forces and deflections are the same for statically indeterminate as for statically determinate arrangements. The essential difference is that in the latter case the deflections are readily obtained from the dimensions of the structures.

Likewise, for a Y movement at the reaction point a restrained in both X and Z movements, the Y reaction is obtained as

$$Y_{ba} = F_b\frac{\Delta_{yb}}{\Delta_{yy}} \tag{10-82b}$$

Because the expression for any linear reaction is the ratio between two linear deflections, the unit of measurement used in the analysis is not relevant as the same unit is used for both deflections.

For the determination of angular reactions (moments), the procedure is essentially the same, although it warrants an appropriate explanation. Referring to Figure 10-63a, we express the moment Z as a couple Fd, or $F = Z/d$. If this moment is applied so that point a moves to a' and b moves to b', the work done is

$$F\Delta_a + F\Delta_b = Z\left(\frac{\Delta_a + \Delta_b}{d}\right) = Z\Delta_{zz} \tag{10-83}$$

where Δ_{zz} is the angle of rotation (radians). Using this reaction in the work equation for the arch shown in Figure 10-63b, we obtain

$$Z_{ba} = F_b\frac{\Delta_{zb}}{\Delta_{zz}} \tag{10-84}$$

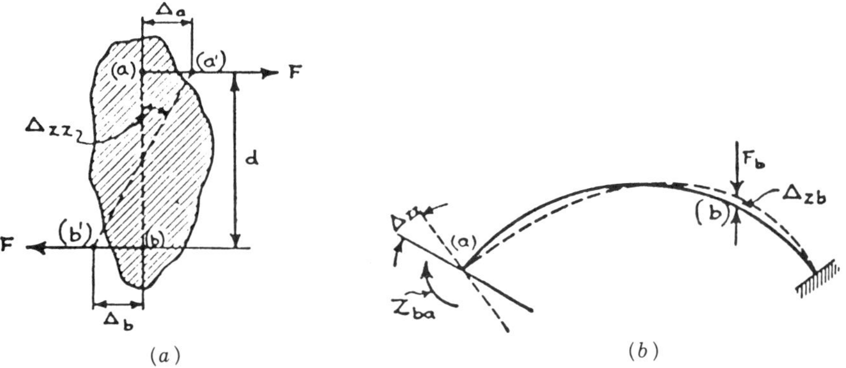

FIGURE 10-63 Moment and frame for mechanical analysis.

Approximate Method of Analysis

For an approximate solution, certain simplifying assumptions are necessary. The problem is approached by considering the columns as links connecting the arch rib and the floor system. The stresses in these links resulting from a relative movement between the arch and deck depend on (a) the angular and horizontal movement of the columns at the bottom where they are attached to the arch; (b) their stiffness (length, elastic modulus, and moment of inertia); and (c) the angular and horizontal movement of the columns at the deck.

10-15 EXAMPLE OF DECK-STIFFENED ARCH BRIDGE

Figure 10-64*a* shows the half-elevation and dimensions for the Landquart Bridge built by Maillart in 1930 (Billington, 1973). This is a deep arch (i.e., it has a small span–rise ratio). The bridge carries a railway loading; hence, a deep arch is important to reduce the horizontal thrust at the abutments and the associated large axial compressions. The horizontal thrust H may be approximated by the simple expression $H = cPL/f$, where P is the concentrated load (locomotive axle weight, for example), L is the span length, f is the rise, and c is a coefficient depending on the location of P on the span.

The line of pressure for dead loads plus one-half the assumed uniformly distributed live load coincides with the axial centroid of the arch. Because the arch rib has a polygonal shape, the elevations at the junctions of the transverse walls are easily varied to give the proper arch axis. The joints lie close to and above the points of a parabola. By including one-half the live load (where the total live load is 6 tons/m), the design allows for some deviation of the arch axis from the line of pressure corresponding to dead load only (the total dead load is 15.4 tons/m), but the resulting bending moments are not significant.

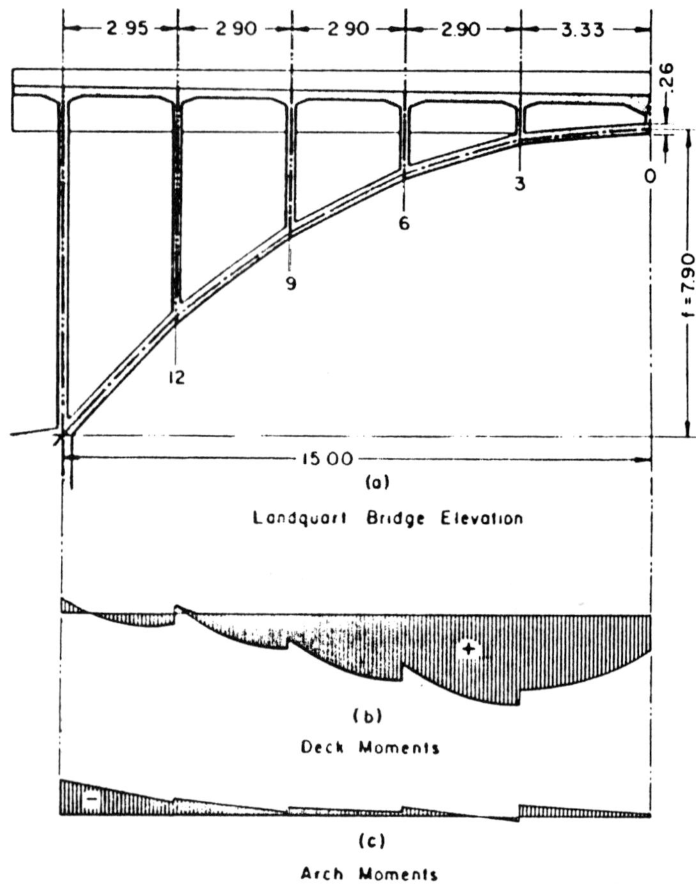

FIGURE 10-64 Half-elevation and bending moment diagrams for the Landquart Bridge. (From Billington, 1973.)

The bending moments in the deck resulting from the vertical displacement of the deck supports as the arch deflects are shown in Figure 10-64*b*, and are considerable. These moments are computed from a standard STRUDL II program which considers the continuity between the arch, spandrels, and deck. A comparison of the moment diagrams shown in Figures 10-64*b* and *c* shows that the deck absorbs most of the flexural action. For example, for a uniform dead load (actual) of the deck of 1.76 tons/m (1.2 kips/ft) and for the largest deck span of 3.33 m (10.8 ft), the continuous-beam moment is $1.76 \times 3.33^2/10 = 1.94$ m-tons (13.9 ft-kips). The actual moment at joint 3 is, however, 21 m-tons (152 ft-kips), or more than 10 times the continuous-beam moment, an increase caused mainly by the displacement of the spandrel supports.

This particular response suggests the basis for the design of deck-stiffened arch bridges. The intent is to dimension the deck girder for large moments by making it stiff, but proportion the arch rib for relatively small moments by keeping it flexible. For this bridge the following assumptions were also made: (a) the arch axis is taken as the line of pressure for the dead load plus one-half the traffic load; (b) the arch rib carries the loads imposed upon it (dead weight plus live load assumed uniformly distributed over the full deck) through axial forces only except for wind and centrifugal forces resisted as lateral bending moments; (c) bending moments in the vertical plane caused by partial live loads, temperature changes, and bracing forces are carried by the stiffening deck girder alone; and (d) these girder moments may be computed from influence lines for a two-hinged parabolic arch.

Assumptions (b) and (c) imply that arch rib bending in the vertical plane is neglected, and all vertical plane bending is carried in the stiffening girder. Assumption (d) means that the vertical plane moments are computed as if the arch is a two-hinged structure, although it is clearly fixed by the thickening that increases continuously from crown to springing.

Although the last three assumptions are dubious, Billington (1973) demonstrates their logic with the help of Figure 10-65, where the stiffened arch is idealized by neglecting the bending stiffness of the spandrels and the live load is taken as uniformly distributed over half the span. Neglecting the very small axial shortening of the walls, the arch and the deck girder must deflect the same at the spandrel locations. As the arch sinks under load in the left half, it will allow the girder to drop as shown but both the arch and girder will rise on the opposite (right) side.

For the two-hinged parabolic arch under a half load w_A, the moment diagram is represented by two antisymmetrical parabolas with maximum $M_A = w_A L^2/64$ at each quarter point. The stiffening girder must deflect similarly and have a maximum moment $M_G = w_G L^2/64$, where w_G is the part of the live load carried by the girder. For a total live load $w = w_G + w_A$, the total moment is $M = M_G + M_A$. Girder deflection can be computed if we consider the girder to be a simply supported beam with vertical displacement $D_G = CM_G L^2/EI_G$, where C is a constant. Arch deflections are computed if the arch is assumed as a parabolic curve with variable moment of inertia such that $I_s \cos \theta_s = I_A$, where I_A is the crown value and θ_s is the slope angle of the arch axis at any point. The actual variation of the bridge does not deviate markedly from this assumption. Noting that $D_A = CM_A L^2/EI_A$ and assuming that $D_A = D_G$, we obtain $M_G/I_G = M_A/I_A$. Then $M_A = M_G r_I$, where $r_I = I_A/I_G$, or, in terms of the total moment, $M_A = M(1 + r_I)$.

For the Landquart Bridge, $I_G = 0.257$ m^4 (29.8 ft^4), $I_A = 0.0055$ m^4 (0.638 f^4), and $w = 6$ tons/m (4.03 kips/ft), so that $M = 6 \times 30^2/64 = 84.5$ m-tons (614 ft-kips). From these data the quarter-span moment is 84.5 m-tons, and with stiffening the arch carries only $M_A = 84.5/(1 + 46.3) = 1.8$ m-tons (13.1 ft-kips), with the girder carrying 82.7 m-tons (600 ft-kips). Billington (1973) shows that a detailed computer analysis reveals the presence of

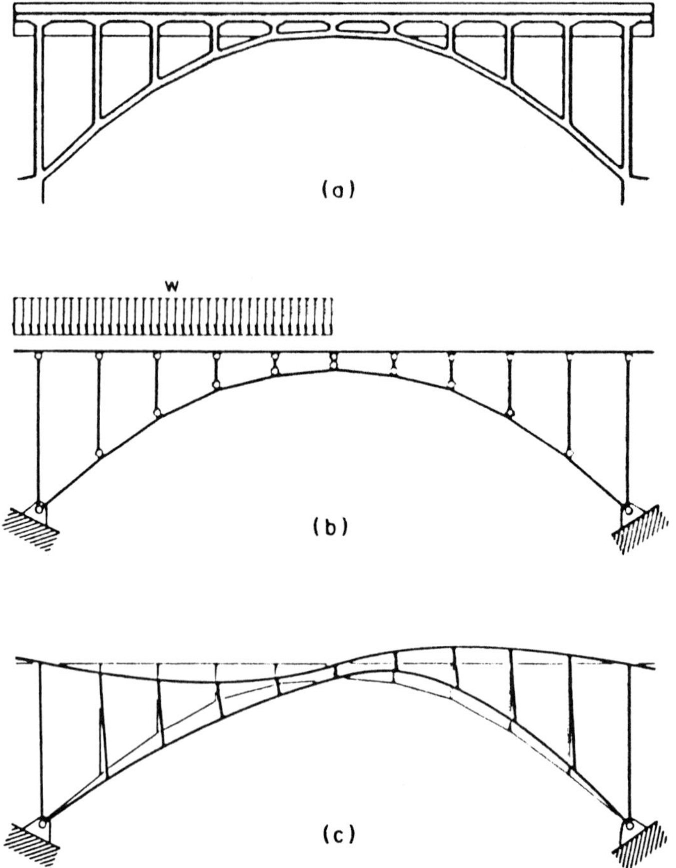

FIGURE 10-65 Idealized deck-stiffened arch behavior; Landquart Bridge. (From Billington, 1973.)

relatively large local moments at the fixed arch springing, contrary to the two-hinged assumption. When, however, these moments are combined with the axial thrust, the result is a cross section entirely under compression. Interestingly, the designer recognized the presence of these moments at these locations by increasing the depth of the arch section near the fixed springings. The result is a structurally adequate arch and a saving in the cost of the bridge by eliminating reinforced concrete hinges (Maillart, 1930, 1931).

Alternate Model Unlike the idealized structure of Figure 10-65 comprised of a solid deck and a solid arch, Kavanagh (1974) considers an alternate model where only the deck is solid, whereas the arch is an articulated bar chain. The difference is fundamental and was early recognized by Müller-Breslau (1922, 1925) as well as by Ostenfeld (1925). These bridge specialists were concerned with structural systems such as suspended bar chains stiffened by girder or truss roadways. In the more popular steel version, it is the girder that clearly stiffens the eyebar chain, much as does the analogous stiffening truss of an eyebar suspension bridge (Kavanagh, 1974).

For these structures the bending moments are divided between the girder and the arch according to the ratio of their moments of inertia as shown in the foregoing analysis. Because, by virtue of its slenderness, the arch is relatively flexible compared with the tie girder, bending moments in the arch are very small and the use of hinges in the arch may be dispensed with.

10-16 ARTIFICIAL OR EXTERNAL STRESS CONTROL

The design approach briefly reviewed in Section 10-10 suggests the possibility of marked economies through the use of stress controls, namely, the placement of temporary or construction hinges and the procedure introduced by Freyssinet whereby arbitrary compensation is introduced by means of suitable techniques.

The Method of Construction Hinges

By employing three temporary or construction hinges, one at each skewback and one at some midsection, the stresses due to conditions 2, 3, 4, 5, and 6 mentioned in group A (Section 10-10) may be entirely eliminated, because the arch bridge is temporarily converted into a statically determinate structure. Stresses caused by condition 1 are modified although not necessarily reduced along the length of the rib. Dead load stresses are, however, reduced at the crown and skewback sections, and this redistribution is advantageous because these locations are frequently the critical or limiting sections of the rib.

Various types of hinges are recommended for this purpose. Figure 10-66 shows the general arrangement of the reinforcing steel in the hinge developed by Mesnager. The bent bars are placed across the central axis of the rib at the center of the gap, and the shear transfer is accommodated by continuous stirrups. The width of the gap is usually made equal to five times the bar diameter. The thrust intensity per unit width that can be carried by this hinge varies according to the bar size.

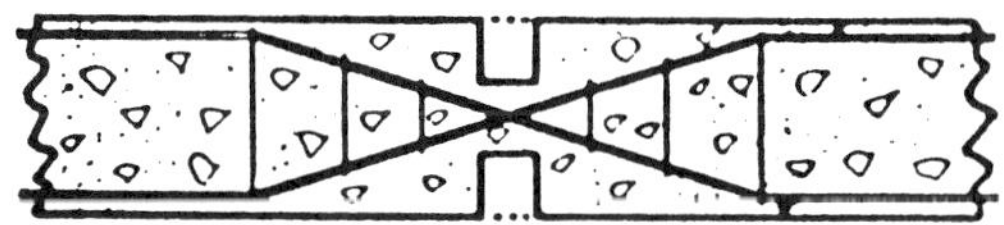

FIGURE 10-66 Type of hinge developed by Mesnager.

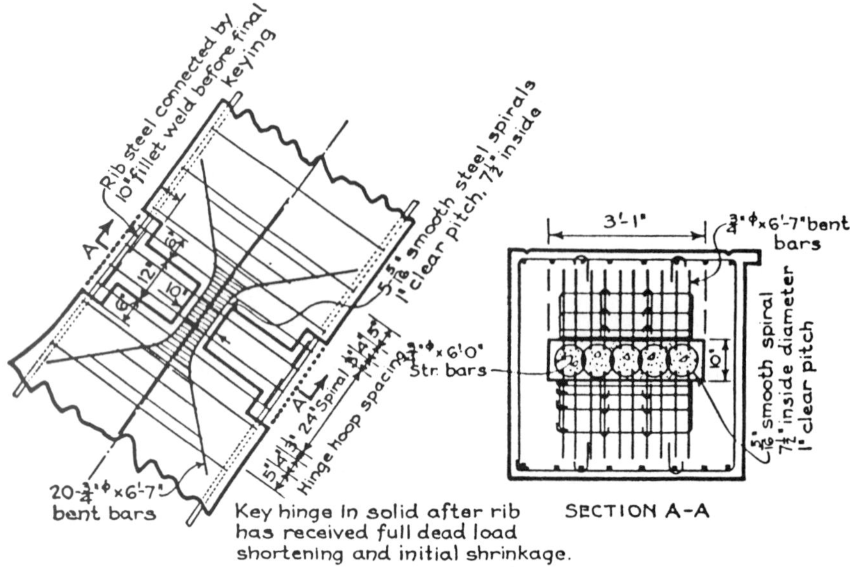

FIGURE 10-67 Details of Considere hinge at skewback; Rogue River Bridge, Grants Pass, Oregon.

Details of a construction hinge of the so-called Considere type are shown in Figure 10-67, and a photographic view of the same hinge is shown in Figure 10-68. The underlying principle with this type of hinge is that hooped concrete under high compressive stress is somewhat plastic, but this hinge is more rigid than the type shown in Figure 10-66. Some concern is expressed about a spiraled hinge on the grounds that the longitudinal tension reinforce-

FIGURE 10-68 Two views of hinge reinforcement; Rogue River Bridge.

ment and the spiraled core form a rigid system resisting rotation. Observations show, however, that the shortening of the spiraled core is so slight that very little tension, if any, is introduced into the longitudinal steel reinforcement.

Measurements of the crown settlement upon striking of the centering showed that the actual vertical movement was 1/16 less than computed, and very little angular restraint of the rib at the hinge points resulted from the hinge as detailed and constructed.

The two types of hinges shown in Figure 10-66 through 10-68 have relative merits but their suitability is probably an open question. For short arch spans or for barrel arches where there is ample room for distribution of the hinge bars, a type similar to Figure 10-66 is simple and more flexible. For relatively large dead load crown thrusts, the practical difficulty of placing the reinforcement bars may lead to the selection of a hinge type similar to Figure 10-67. If the working stress can be kept reasonably low (5000–6000 psi), there is very little loss in flexibility, and the latter type may be employed for any ordinary concrete arch.

Alternatively, it is possible to achieve a certain degree of stress adjustment by using one or two temporary construction hinges rather than three, and in this case the structure would act as a single-hinged or two-hinged arch for conditions 1 to 6, group A loads (Section 10-10). This arrangement would serve to modify and generally reduce the stresses resulting from conditions 2 to 6, although it would not eliminate them entirely. In other instances the keying of the temporary hinges is done prior to the placement of a portion of the dead load, so that stresses due to conditions 1 and 2 are partially reduced under group B. Optimum design is achieved if all dead load, except handrail, curb, and perhaps a portion of the floor slab, is placed before final keying.

If the keying is done at a temperature other than the mean temperature, the stresses under condition 8, group B, will obviously be increased. Likewise, long-term creep (condition 9) is amplified by an early keying of the structure. In summary, stress control through articulation (construction hinges) is feasible and advantageous for (a) arches with a low rise–span ratio for which the deformation stresses are high; (b) arches on yielding foundations where a considerable movement of the skewback is expected on the striking of the center; and (c) through or half-through arches with sidewalks in which (by virtue of space limitations) the rib section is narrow and deep as compared with the rib configuration in the deck type, this extra depth resulting in the accentuation of deformation stresses.

The Freyssinet Method

This method of stress control is also referred to as the method of arch compensation and adjustment, and was developed by Freyssinet. It consists of completely severing the arch and then, by means of hydraulic jacks, introducing at the joint thus formed a calculated stress or strain of a

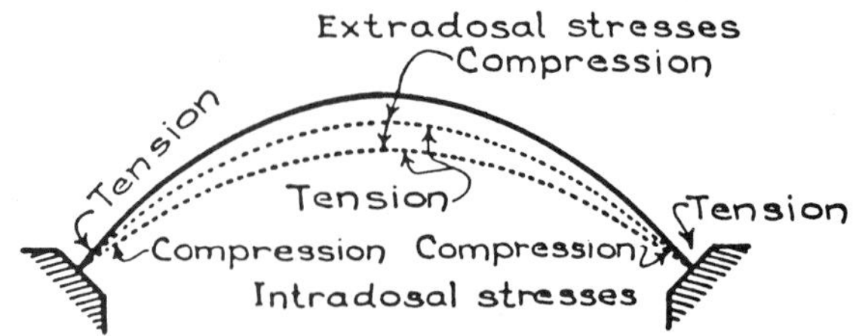

FIGURE 10-69 Arch deformation resulting from ordinary construction.

magnitude and direction sufficient to neutralize or reduce certain elements of stress that would be unavoidable if the arch were built in the usual manner. Considering the difficulties inherent in conventional arch construction, this process constitutes a distinct and important alternative.

Consider, for example, the fixed arch shown in Figure 10-69 constructed of any suitable material. If this structure is closed while resting on its centering, it will be ready to act as an arch but theoretically will be under no stress. If the falsework is now removed, the loading will be transferred to the rib and will produce an axial thrust as discussed in the foregoing sections. This thrust will cause an elastic shortening (see also Section 10-6), forcing the span to assume a shorter curve as shown by the dotted curve in the Figure 10-69. If the temperature decreases or if the material shrinks with age, this condition will be further accentuated and the arch will assume a still lower position. This change in curvature of the rib will set up tensile and compressive stresses called *deformation* stresses. These contribute nothing to the load-carrying capacity of the structure, but are inevitable as a result of the ordinary construction method.

Conversely, if after being swung, the arch is lengthened by an amount equal to its shortening, the axial curve will rise to its original position, the bending at the crown and skewbacks will disappear, and the loads will theoretically be supported by arch thrust only, which is the condition initially desired. Thus, if before striking the falsework a lengthening equal to the anticipated shortening is introduced in the arch, the latter will compress axially without bending. In this process the loads will be transferred to the rib, and the span will be decentered without settlement or flexure.

Besides this arch compensation, the feasibility of controlling the internal stresses may be extended to any point in the arch and may encompass other stress conditions. For example, at the crown the total positive moment usually exceeds the total negative moment. If, under this condition, an artificial negative moment were to be introduced at the crown, the unit stresses of extrados and intrados could be equalized with a corresponding reduction in the structural requirements. The artificial introduction of stress in a rib for this purpose is termed *adjustment* to distinguish it from the concept of *compensation*.

Other advantages associated with this procedure include early decentering, lighter decentering requirements, and reuse of decentering materials. Because the rib may be decentered without flexure, the jacking can be started while the concrete is still fresh and the falsework is thus released and used at another location. It is also practicable to reopen the jacking point and compensate for any delayed shrinkage or creep that might have occurred after the bridge was first jacked and put into service.

Calculation of Arch Compensations The calculation of the horizontal compensation for rib shortening is equivalent to an appropriate drop in temperature as discussed in Section 10-6. In general, we can write

$$\Delta_2 = \sum \frac{N\,dx}{AE} \tag{10-85}$$

where Δ_2 refers to the total shortening due to condition 2 and N represents the total thrust due to the dead load. Likewise, the horizontal shortening due to stress condition 3 is

$$\Delta_3 = \omega t L \tag{10-86}$$

where ω is the coefficient of linear thermal expansion and t is the difference between the mean temperature and that existing at the time of jacking.

The shortening due to stress conditions 4, initial and early shrinkage, and 6, early creep, is calculated as

$$\begin{aligned} \Delta_4 &= QL \\ \Delta_6 &= Q'L \end{aligned} \tag{10-87}$$

where Q and Q' are the shrinkage and creep coefficients, respectively.

The effect of initial inelastic support displacement, or condition 5, can be quantified by measuring the horizontal displacement of the skewbacks at the time of jacking, which is then added to the horizontal compensation. Angular support movement may also be measured at the time of decentering and its effect on the arch calculated by means of elastic theory.

The process of neutralizing the foregoing effects by arch compensation is essentially an artificial spreading or lengthening of the arch axis to counteract certain compressive strains, thereby maintaining the arch in a fixed position.

Calculation of Arch Adjustment The distinction between arch compensation and adjustment is purely arbitrary and serves essentially expediency. As a matter of fact, the compensation for all shortening strains as well as the equalizing of unbalanced stresses may be combined and included in the adjusting forces introduced at the jacking point.

At the commencement of jacking, stresses and strains of unknown magnitude are likely to exit in the rib because of preliminary expansion or shrinkage of the material, the effect of arch reinforcement, the development of cantilever and simple beam action in the arch due to settlement of centering, support movement, and other factors that are not measurable but nonetheless present. If at this stage a certain strain (lengthening) calculated as outlined is introduced at the crown, based on the assumption of zero stress in the rib, the resulting state of elastic equilibrium after jacking will be only approximate. If, however, the arch is first calculated as if fixed and the total stresses including all deformation stresses are computed for this fixed arch at the critical points, an appropriate adjusting moment and thrust may be found by trial that will benefit most points and injure the least. In this procedure the resultant rib stresses are obtained by analyzing the half rib as a statically determinate cantilever in which the total stresses at any point g (Figure 10-70), regardless of the unknown initial state, must be given by

$$M_g = N_j y + J_j x + M_j - \sum Fa \qquad (10\text{-}88a)$$

$$N_g = N_j \cos\theta - \left(J_j - \sum F\right)\sin\theta \qquad (10\text{-}88b)$$

The arch adjustment is effected by introducing the proper jacking forces M_j, N_j, and J_j at the various sections.

The success of the adjustment process is dependent on the following factors.

1. Accuracy of Measurement of the Jacking Force. Using special gage readings to 10 psi, for example, reduces the error at the skewback to negligible values.

2. Angular Accuracy of the Jack Emplacement. This is important if stresses are to be computed from (10-88). Angular error may arise from inaccuracy in construction of the jack emplacements or from rotation of the

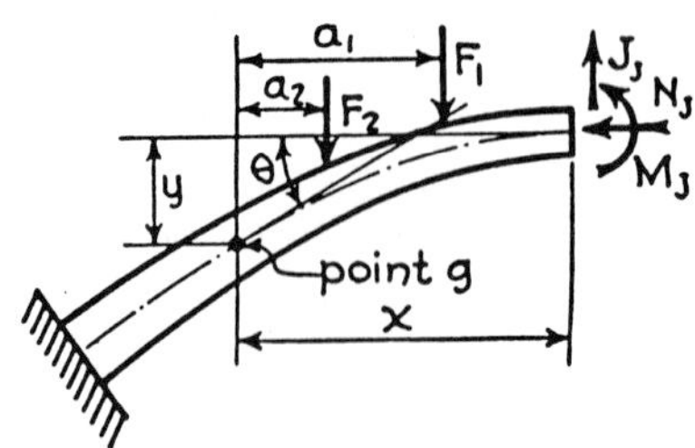

FIGURE 10-70 Equivalent cantilever used to determine resultnig rib stresses.

face during jacking. If this occurs the force N_j will be inclined with a corresponding error.

3. Accurate calculation of Dead Load. A calculation based on actual dimensions and unit weights is essential and allows a reliable check between jacking stresses and deflection.

4. Calculation or Elimination of Shear at the Jacking Point. From Figure 10-70 and (10-88), it is apparent that the shear J_j at the jacking point affects the moment along the entire arch and therefore must be considered. Shear may be developed by friction between the jacks and their emplacements if the tendency for movement in the J direction is unequal on the two sides of the gap. It becomes therefore necessary either to reduce the shear by properly locating the jacking point or to calculate its value.

Summary of Procedure for Single Symmetrical Spans Initially, the arch is constructed upon centering but the structure is completely severed at the crown while suitable emplacements are provided for hydraulic jacks. As soon as decentering is permitted, the jacks are set and the structure lifted from its supports a sufficient distance to accommodate the release of centers. The arch rib can be held in this position using temporary shims until the superstructure is in place or until a compatible construction stage determined from the final keying. At this point the jacks are set again in the emplacements and the structure adjusted to its final position or elastic state desired.

The final adjustment may be implemented in the following two ways.

1. By providing a total horizontal strain or spread to compensate for the horizontal rib shortening caused by stress conditions 1 to 6 listed in Section 10-10. In this case the total compensation is apparently the horizontal crown spread at the time of the initial decentering plus the crown spread effected in the final adjustment.
2. By providing a certain definite crown moment and thrust. This is possible with the use of jacks placed above and below the neutral axis of the rib.

When the desired adjustment is completed, the structure is permanently fixed by filling in the gap resulting from the jacking procedure with concrete. When a connection must be provided that will carry tension as well as compression, the reinforcing system across the gap is welded or otherwise connected for continuity.

We should note that the method of construction hinges eliminates stresses due to conditions 2 to 6 and substitutes for condition 1 a definite stress arrangement. The method of adjustments and compensations substitutes for stress conditions 1 to 6, inclusive, a certain statically determinate stress system that may be varied and controlled within wide limits.

Multiple-Span Arches on Elastic Piers The application of the Freyssinet method to multispan arches will be demonstrated by considering the case of a two-span symmetrical arch group shown in Figure 10-71. This structure may be adjusted by the use of a set of jacks at each crown simultaneously or sequentially. Simultaneous adjustment is preferred because of its simplicity and reliability, and would normally be used where the necessary equipment is available. Because it does not present new problems, we will focus on sequential jacking or on the procedure that must be followed if only one set of jacks is available. With the latter method we must determine the mutual effect of the spans on each other. Obviously, when jacking stresses are first introduced at crown C_1, the increase in these stresses by the subsequent jacking at crown C_2 must be taken into account.

Referring to Figure 10-71, let us assume that it is required to add to each crown the moment and thrust increments ΔM and ΔN, respectively. The next step is to determine the preliminary values of moment and thrust to be produced at C_1 so that, when the jacks are moved to C_2 and the stresses equalized, the moments and thrusts at both crowns will have been increased by the desired amounts ΔM and ΔN.

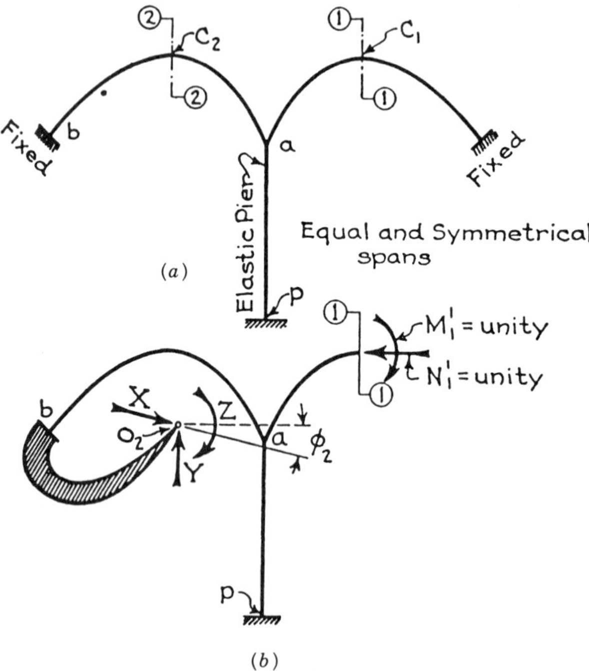

FIGURE 10-71 Application of the Freyssinet method to a two-span symmetrical arch structure.

The initial stresses are assumed equal at both crowns (zero or any other value), and the only acting moments and thrusts are the added values or increments. With equal increments at C_1 and C_2, the final stresses will also be equal. It follows then that both the initial and final shears at both crowns must be zero, and that any crown shear developing because of the preliminary jacking at C_1 must be neutralized by the subsequent jacking at C_2. An additional feature of the two-span case is the possibility of displacement of the skewback at the center support. However, the vertical displacement of this skewback during successive jacking is negligible, and if the crown forces are finally brought to equilibrium, the horizontal and angular forces at this skewback will be in balance.

For the purpose of this analysis, stresses produced by the jacks at their point of emplacement are termed "primary" stresses and denoted by M', N', and J'. These stresses would increase the moment, thrust, and shear at other points producing stresses that are termed "secondary" and designated by M'', N'', and J''.

If primary increments are produced in both spans at C_1 and C_2, either simultaneously or sequentially, both primary and secondary stress components will exist in each span, and the sum of the primary and secondary increments in either span will be equal to the total increment in that span. Thus, for span 1 we may write

$$\text{Total increment} \qquad \Delta M_1 = M_1' + M_1''$$

$$\text{Total increment} \qquad \Delta N_1 = N_1' + N_1''$$

and also the same for span 2. If an equal moment increment ΔM were applied to both spans simultaneously, the crown shears would be zero, but we would have

$$\Delta M = M_1' + M_1'' = M_2' + M_2''$$

as shown in the diagram of Figure 10-72. From symmetry, equal primary components are introduced in each span, and equal secondary stresses will appear in the other span so that

$$\begin{aligned} M_1' &= M_2' = M' \\ M_1'' &= M_2'' = M'' \end{aligned} \tag{10-89}$$

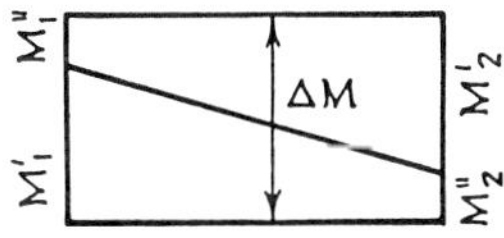

FIGURE 10-72 Relationship between equal moment increments in spans 1 and 2.

Likewise, we can write for thrusts

$$N_1' = N_2' = N'$$
$$N_1'' = N_2'' = N'' \qquad (10\text{-}90)$$

With the foregoing relationships established, we consider the fixed span bC_2ap in Figure 10-71*a* and we will proceed to determine the secondary stresses at C_2 produced by a unit moment and a unit thrust at point C_1. This is accomplished by removing the right half of span 1 as shown in Figure 10-71*b*, inserting the unit crown forces M_1' and N_1', and analyzing the elastic system bC_2ap. For this, we must determine the elastic center O_2 by conjugate angle ϕ_2 and the three redundant forces X, Y, and Z active at O_2 and due to the action of M_1' and N_1' acting at crown C_1.

The crown stresses thus determined are m_m, n_m, and j_m, produced at C_2 by M_1' = unity, and m_n, n_n, and j_n, produced by N_1' = unity. We can now write the equations for total moment and thrust increments as follows:

$$\Delta M = M' + M'm_m + N'm_n$$
$$\Delta N = N' + M'n_m + N'n_n \qquad (10\text{-}91)$$

where all parameters are known except the primary moment and thrust M' and N' that must be determined for the initial jacking at C_1.

Example of Jacking Detail One of the first bridges in the United States to employ the Freyssinet system of decentering and adjustment is the Rogue River Bridge in Oregon. This structure consists of a group of seven symmetrical arch spans, each span 230 ft long with a rise of 47 ft. Each rib was severed at the crown and designed with side brackets to act as emplacements for the hydraulic jacks. Figure 10-73 shows the general details of the emplacements consisting of structural steel H beams fitted with suitable anchorage lugs and a milled plate at the outer end to receive the jack reaction and transmit it to the interior of the rib.

Most Freyssinet arches built in Europe have not utilized structural steel emplacements, but have been designed to transmit the jack reaction by direct bearing on the rib concrete at the point of contact. Local stress concentration is remedied by means of heavy reinforcement mats.

Trends in U.S. Practice

The most important early arch document is probably the 1934 Final Report of the Special ASCE Committee on Concrete and Reinforced Concrete Arches (Morris et al., 1935). This document culminates years of work and presents results from detailed tests on both concrete and plastic models. The studies focus on single-span arch ribs with and without decks, three-span

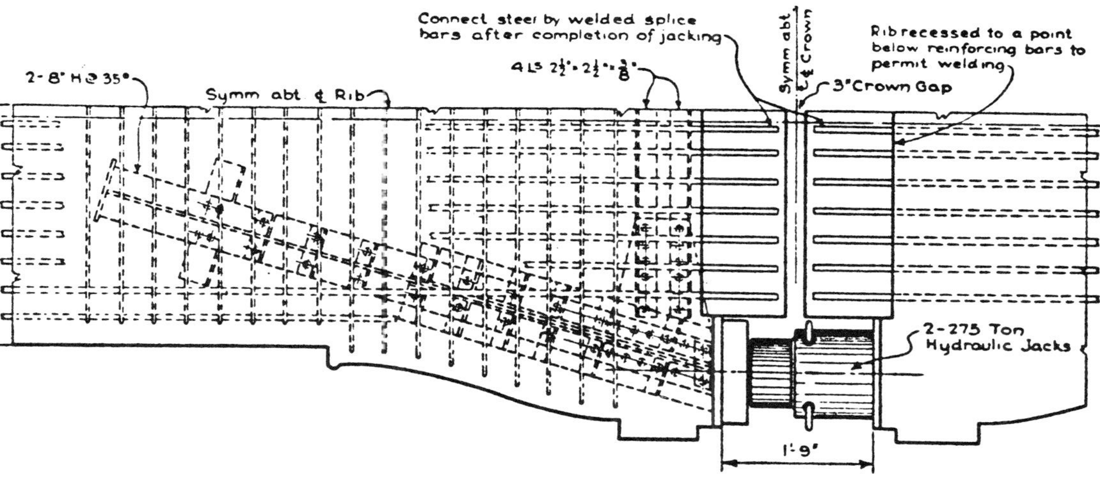

FIGURE 10-73 Partial plan view of rib at crown showing jack emplacement; Rogue River Bridge, Oregon.

arch ribs on slender piers with and without decks, and skew-barrel arches. The effects of climate, support movement, and creep were also analyzed from records on bridges that were in service at the time the report was prepared.

The study emphasizes arch ribs and arches with open spandrels. One of the important conclusions is a method for analyzing single-span arch bridges with an open spandrel and deck, suggested for the design of structures where deck participation is important. The methodology includes the following steps.

1. Determine the dead load stresses analytically assuming that the dead load is carried solely by the rib, unrestrained by the deck.
2. Determine the live load stresses experimentally using elastic models, considering the structure as a whole. The underlying principle is the effect of deck participation on the stresses in the deck and columns rather than on the stresses in the rib.
3. Determine the shrinkage and temperature stresses experimentally using elastic models of the structure as a whole.

Other sections of the study referred to the influence of cutting expansion joints in the deck over the arch for the purpose of reducing the bending moments caused by volume changes due to shrinkage, temperature, and creep. Interestingly, it was concluded that such joints are not recommended and that the deck participation reduces live load moments at the springings.

The analysis for deck participation is admittedly complicated and time consuming, but in this study the deck participation is presented as a problem in analysis rather than a potential design methodology. There is no mention that by making the deck girder stiff the arch can be made much thinner. Indeed, step 2 may well imply that the rib stress will not be markedly influenced by the participation of the deck, an approach clearly in contrast with the examples provided by Maillart and Freyssinet (Billington, 1973).

The significance of this contrast is that it shows a tendency to use advanced analytical techniques to solve complex performance details without emphasizing design insight to explain and understand the overall behavior. The elastic models of the 1930s and the computer programs of today lead to a more accurate assessment of details and also provide a broader explanation of arch behavior. The inherent danger remains, nonetheless, that advanced techniques are often overused in analysis and underused in the simplification of the overall design.

Recent trends in local practice emphasize an approach to the technical meaning of arch concepts and underscore the validity of some of these concepts in demonstrating the inherent economies. Thus, overanalysis and the implied acceptance of automatic computer output should supplement insight into design and should lead to conclusions about the appropriateness of the design itself.

FIGURE 10-74 Edgewood Road arch bridge over I-80.

An example confirming this trend is the Edgewood Road arch over Interstate-80 on the New Jersey approach to the George Washington Bridge, shown in Figure 10-74 (Kuesel, 1974). The concept of the thin, ribbon slab arch stiffened with a relatively deep deck structure is obvious. A parallel concept in steel arches is the bowstring or Langer girder shown in Figure 10-4*j*. This is a tied arch with a thin flexible rib and a deep, stiff tie girder. The concept was not introduced in the United States until 1040, and notable examples include the St. George's arch over the Chesapeake and Delaware Canal in Maryland. With this concept economy is attained by taking the bending in the tension member rather than in the compression elements. A second generation of three-span tied steel arches with the tie at midheight of the arch, but exhibiting the same concept of light rib and stiff deck, includes the Port Mann Bridge near Vancouver, British Columbia, and the Fremont Bridge at Portland, Oregon.

An outstanding contribution to the theory and design of elastic arch bridges is McCullough and Thayer (1931), still a most valuable reference for those involved in the design of arch bridges.

10-17 THE PRESTRESSED BOWSTRING ARCH

The bowstring arch shown in Figure 10-4*j* is a combination of an arch rib and a tie girder. These members are fastened together at the supports and elsewhere connected through equally spaced vertical suspension rods. Any load on the tie girder is resisted jointly by both members, thus making the structure highly indeterminate. Kaldjian (1961) presents methods for determining the redundant forces including the effects of prestressing. A general expression for solving prestressed bowstring arches with extensible suspension rods is derived from the strain energy of the structure. A simplified membrane analogy method is also introduced, and an experimental investigation is summarized.

The Strain Energy Method

Figure 10-75 shows a bowstring arch with equally spaced vertical suspension rods. Artificial cuts or hinges are imagined to be inserted at suitable places to make it statically determinate as shown in Figure 10-75*b*. Moments and shears are then applied to both sides of these hinges to restore the arch to its initial condition. We may express the total strain energy U in terms of these moments and forces.

Let the terms $X_1, X_2, \ldots, X_n$ denote the unknown forces and moments as shown in Figure 10-75*b*. Let M and N denote, respectively, the moment and the axial force on the cut structure produced by the externally applied loads P, and M' and N' the bending moment and axial force, respectively, caused by the unknown parameters X_1, X_2, and X_n. The total strain energy U can now be expressed as a function of all these forces, and also in terms of the lengthy L, the modulus of elasticity E, the moment of inertia I, and the cross-sectional area A.

The relative deflection or rotation at the ith cut is

$$\Delta X_i = \frac{\partial U}{\partial X_i} \tag{10-92}$$

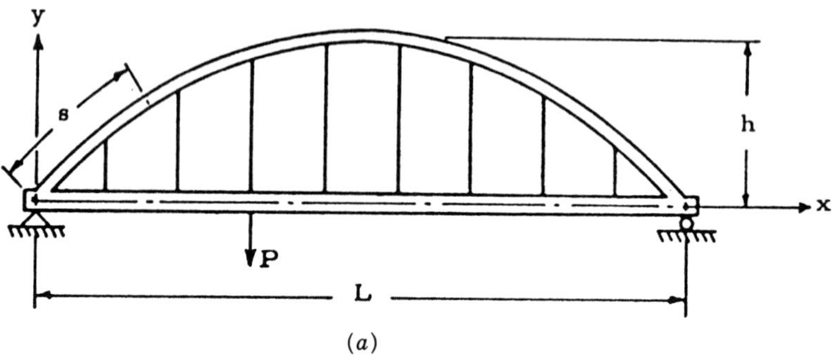

(*a*)

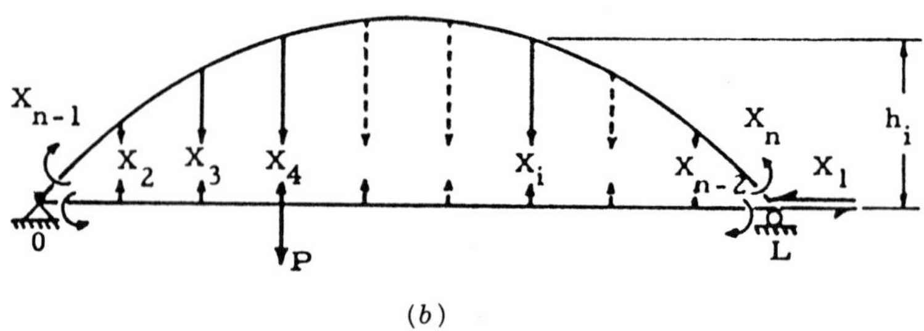

(*b*)

FIGURE 10-75 (*a*) Bowstring arch with equally spaced vertical suspension rods; (*b*) the same arch with artificial cuts and hinges (From Kaldjian, 1961.)

Using an appropriate strain energy equation for U and correlating with (10-92), Kaldjian (1961) derives the following generalized relationship for ΔX_i:

$$\Delta X_i = \sum_0^L \frac{(M_a + M_a')\dfrac{\partial M_a'}{\partial X_i}}{E_a I_a} \Delta s + \sum_0^L \frac{(N_a + N_a')\dfrac{\partial N_a'}{\partial X_i}}{E_a I_a} \Delta s$$
$$+ \sum_0^L \frac{(M_g + M_g')\dfrac{\partial M_g'}{\partial X_i}}{E_g I_g} \Delta x + \sum_0^L \frac{(N_g + N_g')\dfrac{\partial N_g'}{\partial X_i}}{E_g I_g} \Delta x + \sum_{i=2}^{(n-2)} \frac{X_i h_i}{E_r A_r} \quad (10\text{-}93)$$

where the subscripts a, g, and r represent, respectively, the arch rib, the tie girder, and the suspension rod. In (10-93), Δs denotes the length along the centerline of the arch, and Δx refers to the length along the girder or horizontal projection of ΔX. Evidently, n independent equations are provided by (10-93), or one equation per cut, so that the number of equations equals the number of unknowns in the system. These can be solved simultaneously to obtain the values of (X_i)s. Once the unknown forces and moments are determined, we can obtain the actual bending moments, axial forces, and shears at any section of the structure.

Solution of Hinged-End Arch with Six Suspension Rods Such an arch is shown in Figure 10-76 and is parabolic in form. For a rise–span ratio of 1/4, the equation of the centerline of the arch with its origin at midpoint of the tie girder is

$$y = -\frac{x^2}{L} + \frac{L}{4} \quad (10\text{-}94)$$

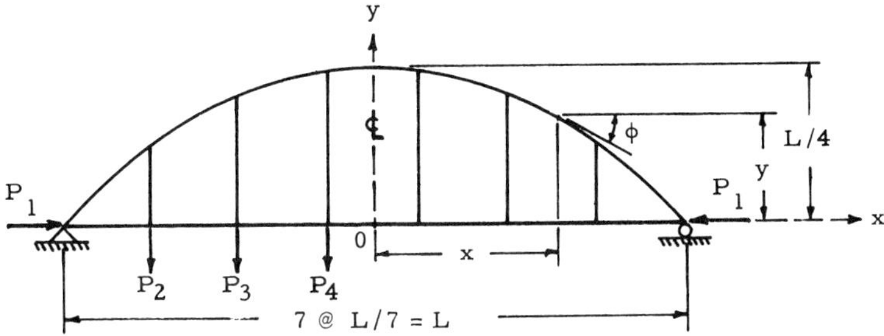

FIGURE 10-76 A bowstring arch with six suspension rods. (From Kaldjian, 1961.)

The cross-sectional area of the arch A_a and the length along the centerline Δs_a are assumed to vary, respectively, as follows:

$$A_a = A_c \sec \phi \qquad \Delta s_a = \Delta s_c \sec \phi = \Delta x \sec \phi \tag{10-95}$$

where ϕ is the angle of the tangent with the horizontal as shown and the subscript c denotes the crown. Assuming the width of the arch is constant throughout, the moment of inertia is $I_a = I_c \sec^3 \phi$.

In order to evaluate (10-93), three tables have been prepared, namely Tables 10-8, 10-9, and 10-10, showing all the forces acting on the structure, as well as the bending moments successively produced in the arch and in the tie girder by all the redundants $X_1, X_2, \ldots, X_n$ and the external forces P_1, P_2, P_3, and P_4.

We can now write

$$m_g = \frac{A_g E_g}{A_c E_a} \qquad m_r = \frac{A_r E_r}{A_c E_a} \qquad m_I = \frac{I_g E_g}{I_c E_a} \tag{10-96}$$

and also note that $(L/r_c)^{-2}$ is identical to $I_c/(L^2 A_c)$ in which r_c is the radius of gyration of the arch rib at the crown, and that $\Delta X_2, \ldots, \Delta X_n$ are

TABLE 10-8 Bending Moments and Axial Forces with Partial Derivations in Arch

Load	$(M_a + M_a')$ & $\frac{\partial M_a'}{\partial X_1}$	Mult. Factor for $(M_a + M_a')$	Mult. Factor for $\frac{\partial M_a'}{\partial X_1}$	N_a' & $\frac{\partial N_a'}{\partial X_1}$	Mult. Factor for N_a'	Mult. Factor for $\frac{\partial N_a'}{\partial X_1}$
X_1 (horizontal, both ends)	3.25, 8.25, 11.25, 12.25, 11.25, 8.25, 3.25	$X_1 \cdot \frac{L}{49}$	$\frac{L}{49}$	-.75927, -.86820, -.96150, 0, -.96150, -.86820, -.75927	X_1	1
X_2	3.0, 5.5, 4.5, 3.5, 2.5, 1.5, 0.5	$X_2 \cdot \frac{L}{49}$	$\frac{L}{49}$	.55780, -.07089, -.03926, 0, .03926, .07089, .09297	X_2	1
X_3	2.5, 7.5, 9.0, 7.0, 5.0, 3.0, 1.0	$X_3 \cdot \frac{L}{49}$	$\frac{L}{49}$	.46484, .35445, -.07851, 0, .07851, .14178, .18593	X_3	1
X_4	2.0, 6.0, 10.0, 10.5, 7.5, 4.5, 1.5	$X_4 \cdot \frac{L}{49}$	$\frac{L}{49}$	.37187, .28356, .15703, 0, .11777, .21267, .27890	X_4	1
X_5	1.5, 4.5, 7.5, 10.5, 10.0, 6.0, 2.0	$X_5 \cdot \frac{L}{49}$	$\frac{L}{49}$	.27890, .21267, .11777, 0, .15703, .28356, .37187	X_5	1
X_6	1.0, 3.0, 5.0, 7.0, 9.0, 7.5, 2.5	$X_6 \cdot \frac{L}{49}$	$\frac{L}{49}$	.18593, .14178, .07851, 0, -.07851, .35445, .46484	X_6	1
X_7	0.5, 1.5, 2.5, 3.5, 4.5, 5.5, 3.0	$X_7 \cdot \frac{L}{49}$	$\frac{L}{49}$	.09297, .07089, .03926, 0, -.03926, -.07089, .55780	X_7	1
7 @ L/7 = L	7 @ L/7 = L			7 @ L/7 = L		

TABLE 10-9 Bending Moments and Axial Forces with Partial Derivatives in Tie Girder

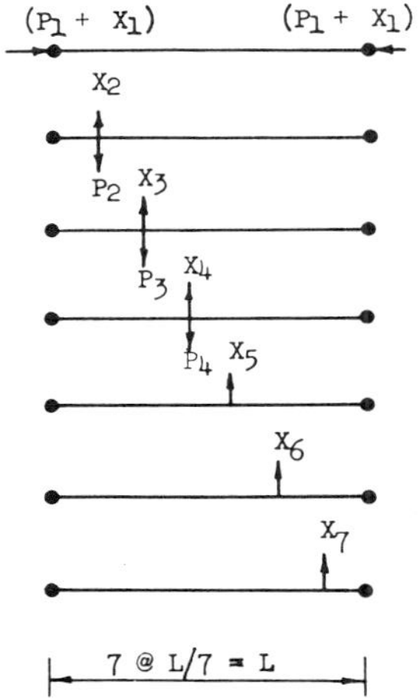

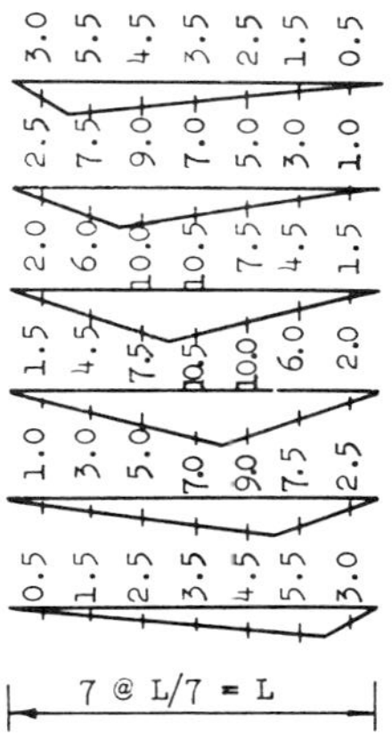

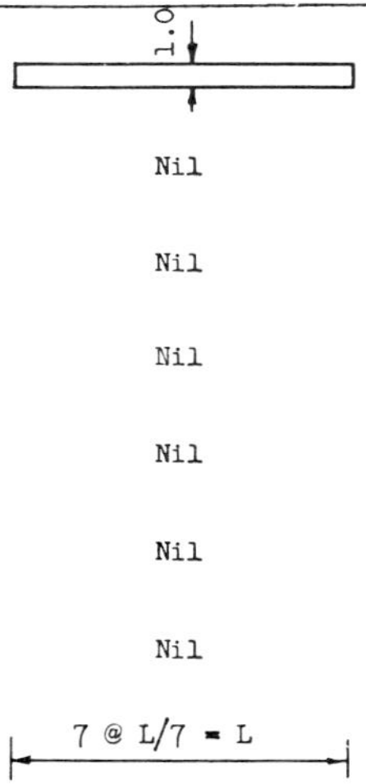

Load	$(M_g + M_g')$ & $\frac{\partial M_g'}{\partial X_i}$	Mult. Factor for $(M_g + M_g')$	Mult. Factor for $\frac{\partial M_g'}{\partial X_i}$	$(N_g + N_g')$ & $\frac{\partial N_g'}{\partial X_i}$	Mult. Factor for $(N_g + N_g')$	Mult. Factor for $\frac{\partial N_g'}{\partial X_i}$
$(P_1 + X_1)$ $(P_1 + X_1)$	Nil	0	0	1.0	$P_1 + X_1$	1
X_2, P_2	3.0 5.5 4.5 3.5 2.5 1.5 0.5	$(P_2 - X_2) \cdot \frac{L}{49}$	$-\frac{L}{49}$	Nil	0	0
X_3, P_3	2.5 7.5 9.0 7.0 5.0 3.0 1.0	$(P_3 - X_3) \cdot \frac{L}{49}$	$-\frac{L}{49}$	Nil	0	0
X_4, P_4	2.0 6.0 10.0 10.5 7.5 4.5 1.5	$(P_4 - X_4) \cdot \frac{L}{49}$	$-\frac{L}{49}$	Nil	0	0
X_5	1.5 4.5 7.5 10.5 10.0 6.0 2.0	$-X_5 \cdot \frac{L}{49}$	$-\frac{L}{49}$	Nil	0	0
X_6	1.0 3.0 5.0 7.0 9.0 7.5 2.5	$-X_6 \cdot \frac{L}{49}$	$-\frac{L}{49}$	Nil	0	0
X_7	0.5 1.5 2.5 3.5 4.5 5.5 3.0	$-X_7 \cdot \frac{L}{49}$	$-\frac{L}{49}$	Nil	0	0
7 @ L/7 = L	7 @ L/7 = L			7 @ L/7 = L		

TABLE 10-10 Axial Forces with Partial Derivatives in Suspension Rods

Load	X_i & $\frac{\partial X_i}{\partial X_i}$	Mult. Factor for X_i	Mult. Factor for $\frac{\partial X_i}{\partial X_i}$
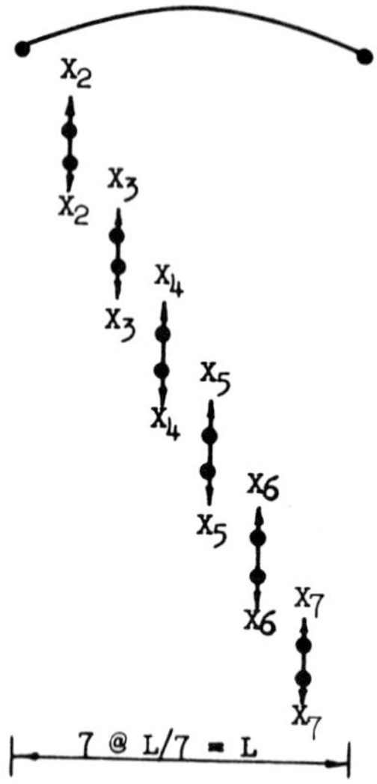 X_2 / X_2	(1) 0 0 0 0 0	$-X_2$	-1
X_3 / X_3	0 (1) 0 0 0 0	$-X_3$	-1
X_4 / X_4	0 0 (1) 0 0 0	$-X_4$	-1
X_5 / X_5	0 0 0 (1) 0 0	$-X_5$	-1
X_6 / X_6	0 0 0 0 (1) 0	$-X_6$	-1
X_7 / X_7	0 0 0 0 0 (1)	$-X_7$	-1
7 @ L/7 = L	7 @ L/7 = L		

equal to zero. From the foregoing we obtain a set of seven equations which are solved to obtain values of $X_1, X_2, \ldots, X_n$:

$$\sum_{n=1}^{7} a_{mn} x_n + c_m = 0 \qquad m = 1, 2, \ldots, 7 \tag{10-97}$$

where $a_{mn} = a_{nm}$ (Kaldjian, 1961). The numerical values of $X_1, X_2, \ldots, X_7$ depend on the external load L/r_c, m_g, m_r, and m_l. Once these parameters are known, the values can be computed.

If an additional prestressing P_1 of the tie girder is desired, a specified initial (vertically restrained) horizontal gap Δ is provided between the ends of the arch rib and the tie girder. The associated redundancy includes a horizontal force X_1 just large enough to close this gap effectively, but because a relative motion exists between the ends of the arch and the tie girder, a different set of equations must be obtained. This is achieved by satisfying the new geometric boundary conditions and by setting the external forces P_1, P_2, P_3, and P_4 equal to zero.

For a bowstring arch with a gap, (10-97) becomes

$$\sum_{n=1}^{7} b_{mn} x_n + d_m = 0 \qquad m = 1, 2, \ldots, 7 \tag{10-98}$$

where $b_{mn} = b_{nm}$.

In order to complete a preliminary design for the arch of Figure 10-76, 50 bending moment and axial force diagrams are prepared from (10-97) and (10-98) for various parameters using computer facilities. Examples of these diagrams are given by Kaldjian (1961).

10-18 EXAMPLE OF NUMERICAL BENDING ANALYSIS

In general, numerical procedures have been used in the elastic analysis of arch bridges by large-deflection, second-order, and classical theories. In particular, arches of high-strength steel are often sufficiently flexible to justify a design based on deflection theory. Austin, Ross, Tawfik, and Volz (1982) have proposed procedures essentially similar to the Newmark analysis of beams and columns, and consisting of simple and accurate repeated numerical integrations. Moments, axial loads, and displacements are evaluated at division points spaced along the rib centroidal axis using elastic analysis. Rib shortening and temperature effects are accurately treated but shearing deformations are ignored. The method is particularly valuable for complex problems involving ribs of arbitrary profile, variable cross section, and any loading.

The example presented in this section provides three levels of sophistication. Initially, a large-deflection analysis approximates the exact solution. The equilibrium relationships are formulated with respect to the deformed arch axis, and large-deflection geometry relates internal deformations to arch displacements. Second, the basic expressions are arranged to allow a simpler deflection theory analysis (second order) as for the large-deflection theory, but the geometric relationships are linearized by making the usual small-deflection approximations. The second deflection theory analysis is useful in the design of slender arches. Finally, the basic expressions are manipulated to obtain a conventional linear analysis (classical analysis). In the latter case displacements of the arch axis are ignored in formulating the equilibrium expressions, and small-deflection geometry is used. This simplified procedure is used to linearly relate stresses and displacements to applied loads, and is useful for arches with negligible buckling tendencies (see also subsequent sections and Section 10-8).

Arch Notation Figure 10-77 shows the unloaded arch axis divided into n segments. Division or node points are numbered successively starting with 1

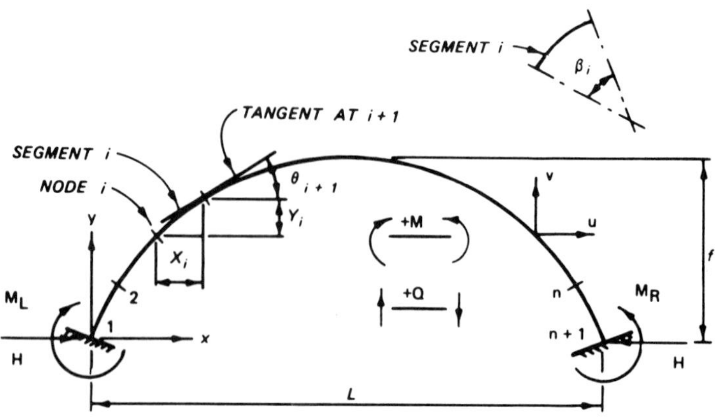

FIGURE 10-77 Arch notation.

at the left support. The segments are likewise numbered starting with 1 at the left end, and the coordinate system is established with the origin at the left support and the x axis horizontal. The shape of the arch axis is defined by the coordinates of nodal points (x_i, y_i), or by the horizontal and vertical projection of each segment as follows:

$$\text{Horizontal projection} \qquad X_i = x_{i+1} - x_i \qquad (10\text{-}99a)$$

$$\text{Vertical projection} \qquad Y_i = y_{i+1} - y_i \qquad (10\text{-}99b)$$

and also by the angle of the tangent at each point with the horizontal, θ_i. The horizontal and vertical displacement of the ith nodal point is denoted by u_i and v_i, respectively, and is positive when in the positive direction of the x and y axes. The moment M is considered positive when it causes compression in the outer fibers, and the axial force P is positive when compressive.

Computation of Moments and Axial Forces

Distributed loads are replaced by substitute statically equivalent forces, concentrated at the node points, and segment couples. These include vertical rib loads distributed uniformly along the arch axis for each segment, vertical deck loads distributed uniformly on a horizontal projection, and concentrated (live) vertical loads acting at node points.

Rib Load Over each segment the rib load q is assumed uniform, but it may vary from segment to segment. If we denote by $\bar{q}_i$ the corresponding equivalent concentrated load at the ith interior node and by $(c_q)_i$ the equivalent couple acting on the ith segment, we can write

$$\bar{q}_i = \frac{h_0}{2}(q_{i-1} + q_i) \tag{10-100a}$$

$$(c_q)_i = -\frac{1}{12}(q_i h_0 Y_i \beta_i) \tag{10-100b}$$

where q_i is the rib load for the ith segment, h_0 is the arc length of the unloaded arch axis of each segment, and β is the effective subtending angle of the segment.

Deck Load Austin, Ross, Tawfik, and Volz (1982) make several assumptions on how the vertical load from the deck acts on the arch axis after the system has experienced large displacements. The vertical deck load p is replaced by a load distributed along the arch axis having an intensity $p \cos \theta_0$, where θ_0 is the angle of inclination of the unloaded arch rib, and p is assumed to act regardless of the magnitude of the displacements.

Likewise, the deck load is considered uniformly distributed along the initial horizontal projection $(X_0)_i$, although it may vary from one segment to another. If p_i denotes the deck load for the ith segment, we can write

$$\bar{p}_i = \frac{1}{2}[p_{i-1}(X_0)_{i-1} + p_i(X_0)_i] \tag{10-101a}$$

$$(c_p)_i = \frac{p_i}{12}[X_i(Y_0)_i(\beta_0)_i - (X_0)_i Y_i \beta_i] \tag{10-101b}$$

where $\bar{p}_i$ is the equivalent concentrated vertical load at node point i and $(c_p)_i$ is the equivalent couple on segment i, positive when clockwise. The segment couple vanishes when the loaded arch is undeflected, when $X_i = (X_0)_i$, $Y_i = (Y_0)_i$, and $\beta_i = (\beta_0)_i$.

Average Vertical Shears Because all loads are represented by equivalent concentrated nodal loads and segment couples, the corresponding vertical shears are constant between these points, and these values are average vertical shears.

In general, we can write

$$Q_i = Q_{i+1} + \bar{p}_{i+1} + \bar{q}_{i+1} + W_{i+1} \tag{10-102}$$

where Q_i is the average shear in the ith segment and W_{i+1} is the concentrated vertical load.

Bending Moments The moment increment from the ith node point to the $(i + 1)$th node point, $(\mathrm{MI})_i$, is

$$(\mathrm{MI})_i = M_{i+1} - M_i = Q_i X_i - HY_i + (c_q)_i + (c_p)_i \tag{10-103a}$$

where H is the horizontal reaction of the arch. Bending moments at the node points are obtained by adding successively from left to right the moment increments, starting with the moment of the left support, or

$$M_{i+1} = M_i + (\mathrm{MI})_i \qquad (10\text{-}103b)$$

Axial Forces The axial forces to the left and right sides of the ith node point, P_{i-} and P_{i+}, respectively, are given by

$$P_{i-} = \{Q_{i-1} - \tfrac{1}{2}[q_{i-1}h_0 + p_{i-1}(X_0)_{i-1}]\} \sin\theta_i + H\cos\theta_i \qquad (10\text{-}104a)$$

$$P_{i+} = \{Q_i + \tfrac{1}{2}[q_i h_0 + p_i(X_0)_i]\} \sin\theta_i + H\cos\theta_i \qquad (10\text{-}104b)$$

Normally, these two values should be the same except where a concentrated vertical load W_i acts and where θ_i is not zero.

Large-Deflection Geometry

Deformations are assumed to be induced only by bending moments, axial forces, and temperature changes. Other load effects, including shear deformations, are neglected; hence, this methodology applies to steel structures. The total curvature and axial strain are related by the expressions

$$\frac{d\theta}{ds} = -\frac{1}{R_0} + a \qquad (10\text{-}105a)$$

and

$$e = \frac{(PL^2)/EI}{(L/r)^2} - a\,\Delta T \qquad (10\text{-}105b)$$

where s = distance along the unloaded arch axis from the left support to a generic point

R_0 = radius of curvature of the arch axis before loading

E = elastic modulus

I = moment of inertia of the cross section

e = axial compressive strain

A = cross-sectional area

r = radius of gyration

a = coefficient of linear expansion = M/EI

ΔT = temperature change

L = span length

Theoretical Background The shape of the arch is defined by the segment projections X_i and Y_i and the inclination angle ϕ_i at the nodes, as shown in Figure 10-78. Alternatively, the shape can be defined by the length l_i and the slope angle θ_i of the hypothetical segment chords connecting a pair of adjacent nodes, plus the angle θ_i. The calculation of the deflected arch shape due to changes in curvature and strains essentially is a problem of geometry involving rotations and changes in length of the segment chords.

The deformations cause a chord length to change from $(l_0)_i$ to l_i and the chord angle to change from $(\phi_0)_i$ to ϕ_i, where the subscript 0 refers to the unloaded state, or $\phi_i = (\phi_0)_i + \Delta\phi_i$ where $\Delta\phi_i$ is the rotation of the chord from the $(\phi_0)_i$ position. From the foregoing we can obtain the expressions for the segment projections as follows:

$$X_i = \frac{l_i}{(l_0)_i}[(X_0)_i \cos \Delta\phi_i - (Y_0)_i \sin \Delta\phi_i] \tag{10-106a}$$

$$Y_i = \frac{l_i}{(l_0)_i}[(X_0)_i \sin \Delta\phi_i + (Y_0)_i \cos \Delta\phi_i] \tag{10-106b}$$

Next, we consider the chord rotations $\Delta\phi$ and the rotations of the tangents to the elastic axis at node points $\Delta\theta$. In the numerical procedure, appropriate parameters are determined from the changes in curvature and axial strain, from which values of $\Delta\phi$ are computed. After the $\Delta\phi$ values are obtained, the values of $\Delta\theta$ are computed accordingly (Austin, Ross, Tawfik, and Volz, 1982). The procedure is the same as suggested by Newmark (1943) for computing deflections, moments, and buckling loads in straight beams, and has been found to give sufficiently accurate results for curved members

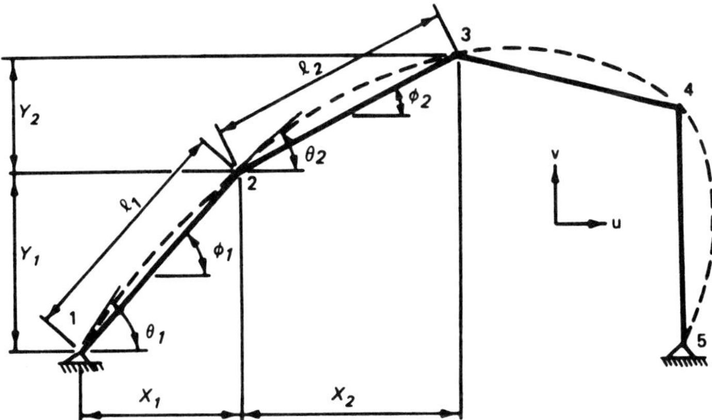

FIGURE 10-78 Definition of arch shape by chords. (From Austin, Ross, Tawfik, and Volz, 1982.)

as well. The $a = M/EI$ function is assumed to vary as a smooth curve over the region involved.

The calculation of the effective (subtending) angle of a segment, β_i, requires two approximations. First, the segment is considered a circular arc as in Figure 10-77 regardless of its actual shape because this is a fair assumption given the very small length of the short segment. Second, the chord length, arc length, and subtending angle are related by an expression approximated as follows:

$$l_i = h_i\left(1 - \frac{\beta_i^2}{24}\right) \tag{10-107}$$

This relationship provides the basis for estimating changes in β from the unloaded arch value caused by bending moments (Austin, Ross, Tawfik, and Volz, 1982). Likewise, the initial arc length h_0 of a typical segment is changed by the axial strain as follows: $h_i = h_0 + \Delta h_i$, where the segment arc length change Δh_i is given by

$$\Delta h_i = -\int_{x_i}^{x_{i+1}} e\,ds \tag{10-108}$$

Discontinuities will be produced in e by a sudden change in the rib cross-sectional area for a concentrated load that is not normal to the rib axis. Discontinuities are produced in de/ds by a discontinuity in the value of a distributed nonnormal load.

Small-Deflection Geometry

Let us assume that moments and axial loads are known at every node point and that it is necessary to find the displacements resulting from a given system defined by the segment projections X_i and Y_i and the angles θ_i and β_i.

The changes ΔX_i and ΔY_i caused by a change in chord length, Δl_i, can be estimated from the geometry of the structure, namely, $X_i = l_i \cos\phi_i$, $Y_i = l_i \sin\phi_i$.

The chord rotations $\Delta\phi_i$ as well as rotations of the tangent at node points, $\Delta\theta_i$, are related to appropriate parameters as in the large-deflection theory (Austin, Ross, Tawfik, and Volz, 1982), and suitable expressions are derived from the boundary conditions. The increase in chord length, Δl_i, is caused by a change in the segment angle $\Delta\beta_i$ and a change in the segment length Δh_i. The following expression is derived:

$$\left(\frac{\Delta l}{l}\right)_i = \frac{h_i}{l_i}\beta_i\frac{\Delta\beta_i}{12} + \frac{\Delta h_i}{h_i} \tag{10-109}$$

where Δh_i is given by (10-108).

Analysis by Large-Deflection Theory

The numerical procedure for large-deflection analysis is presented for a two-hinged parabolic arch of constant cross-sectional area subjected to a uniform rib load and an equal uniform deck load over the left half of the span, shown in Table 10-11. In the appropriate notation, f is the unloaded rise of the arch and n is the number of equal divisions in the rib length.

For a given geometry and loading, rib-shortening effects are manifested by the slenderness ratio L/r, and no rib shortening corresponds to $L/r = \infty$. The example of Table 10-11 has an L/r ratio of 100, representing a moderately slender member. The primary structure is an arch with the left end hinged and the right end on rollers, yielding a horizontal reaction as the redundant force.

For simplicity in the extent of analysis, the arch is divided into four equal segments only (1–2, 2–3, 3–4, and 4–5), and values centered about a vertical line apply to that node. Segment values are tabulated between vertical lines, a sequence allowing the use of a computer program. The dimensions of the numbers in each line are given in the last column.

The geometry of the unloaded arch is presented in Table 10-11a, and the overall solution is the last cycle of a series of iterations. All calculations begin by assuming values of X_i^*, Y_i^*, θ_i^*, β_i^*, and $(H-1)$ from the previous cycle, in which the asterisk denotes assumed values and $(H-1)$ is the assumed horizontal reactions for part 1. The analysis for each cycle consists of two parts; part 1 calculations are shown in Table 10-11c. Lines 10 to 17 refer to the calculation of bending moments and axial thrusts resulting from the interaction of loads and reactions with the assumed deflection pattern given in lines 6 to 9. From this, it follows that we take $X_i = X_i^*$, $Y_i = Y_i^*$, $\beta_i = \beta_i^*$, $\theta_i = \theta_i^*$, and $x_i = x_i^*$.

The balance of Table 10-11 shows the calculations for the deflected configuration of the primary structure corresponding to part 1 bending moments, line 16, and thrust, line 17. The chord rotations must be such that the sum of vertical projections vanishes, or $\Sigma_{i=1}^{n}(Y-1)_i = 0$. Because the correct chord rotation is not known beforehand for any segment, a trial-and-correct technique is used.

After the correct chord rotations have been determined, the panel projections and nodal slope angles for part 1 are computed according to lines 23 to 25. The effect of a small change in horizontal reaction, denoted as $(H-2)$ on part 1 values, is determined in part 2, Table 10-11d. The changes in moment and axial thrust (lines 26–28) result from $(H-2)$ interacting with the assumed deflected configuration, and cause small changes from the displaced configuration found in part 1. Small-deflection geometry is used to calculate these changes. Therefore, the data in lines 31, 33, 34, and 35 imply $h_i = (h-1)_i$, $l_i = (l-1)_i$, $X_i = (X-1)_i$, $Y_i = (Y-1)_i$, $\beta_i = (\beta - 1)_i$, and $x_i = (x-1)_i$. Finally, the horizontal projection increments are summed to give the change in span produced by $(H-2)$. The horizontal reaction

TABLE 10-11 Large-Deflection Analysis of a Two-Hinged Parabolic Arch

UNIFORM RIB LOAD AND EQUAL DECK LOAD ON LEFT HALF OF SPAN

$\frac{f}{L} = 0.15,\ \frac{L}{r} = 100,\ p = q = 17\ EI/L^3,\ n = 4$

Line No. and Quantity	Node 1		Node 2		Node 3		Node 4		Node 5	Common Factors
(a) Undeformed Geometry, $h_o = 0.26428L$										
1. X_o		0.23989		0.26011		0.26011		0.23989		L
2. Y^o		0.10940		0.04060		-0.04060		-0.10940		L
3. θ^o	0.54042		0.30255		0		-0.30255		-0.54042	-
4. ℓ^o		0.26366		0.26326		0.26326		0.26366		L
5. β^o_o		0.2373		0.3044		0.3044		0.2373		-
(b) Assumed Deflected Shape										
6. X^*		0.25268		0.25594		0.26145		0.22993		L
7. Y^*		0.07498		0.06248		-0.01189		-0.12557		L
8. θ^*	0.33551		0.27310		0.13756		-0.28180		-0.66856	-
9. β^*		0.0618		0.1374		0.4212		0.3862		-
(c) Part 1 of Solution; Assume $(H\text{-}1) = 23.000EI/L^2$										
10. $\bar{q}$			4.4928		4.4928		4.4928			EI/L^2
11. C_q		-0.00173		-0.00321		0.00188		0.01816		EI/L
12. $\bar{p}$			4.2500		2.2109		0			EI/L^2
13. C_p		0.00772		0.00132		0		0		EI/L
14. (Q-1)		10.8366		2.0938		-4.6099		-9.1027		EI/L^2
15. (MI-1)		1.0196		-0.9030		-0.9299		0.8133		EI/L
16. (M-1)	0		1.0196		0.1166		-0.8133		0	EI/L
17. (P-1)	26.697		23.915		22.459		23.999		25.083	EI/L^2
18. (h-1)		0.26361		0.26367		0.26367		0.26363		L
19. (β-1)		0.0602		0.1328		0.4154		0.3832		-
20. (ℓ-1)		0.26357		0.26348		0.26177		0.26202		L
21. (α-1)			0.22712		0.03022		-0.17655			-
22. (Δϕ-1)		-0.14550		0.08162		0.11184		-0.06471		-
23. (X-1)		0.25313		0.25615		0.26153		0.23087		L
24. (Y-1)		0.07344		0.06172		-0.01125		-0.12391		L
25. (θ-1)	0.32884		0.26804		0.13710		-0.27642		-0.66015	-
$\Sigma_i(X\text{-}1)_i = 1.00168\ L$										
(d) Part 2; Effects of (H-2) on Part 1										
26. (MI-2)		-0.07498		-0.06248		0.01189		0.12557		(H-2)L
27. (M-2)	0		-0.07498		-0.13746		-0.12557		0	(H-2)L
28. (P-2)	0.94424		0.96294		0.99055		0.96056		0.78472	(H-2)
29. (Δh-2)		-0.000025		-0.000026		-0.000026		-0.000023		$(H\text{-}2)L^3/EI$
30. (Δβ-2)		0.01018		0.02903		0.03683		0.01910		$(H\text{-}2)L^2/EI$
31. $(\frac{\Delta\ell}{\ell}\text{-}2)$		-0.000146		-0.000420		-0.001383		-0.000701		do
32. (α-2)			-0.019540		-0.034690		-0.030682			do
33. (Δϕ-2)		0.038661		0.019121		-0.015569		-0.046251		do
34. (ΔX-2)		-0.002876		-0.001288		-0.000537		-0.005893		$(H\text{-}2)L^3/EI$
35. (ΔY-2)		0.009775		0.004872		-0.004056		-0.010591		do
36. (Δθ-2)	0.04210		0.03192		0.00289		-0.03394		-0.05303	$(H\text{-}2)L^2/EI$
$\Sigma_i(\Delta X\text{-}2)_i = -0.010594\ \frac{(H\text{-}2)L^3}{EI}$; $(\Delta H\text{-}2) = 0.158\ EI/L^2$										
(e) Computed Deflected Shape, $H = 23.158\ EI/L^2$										
37. X'		0.25267		0.25595		0.26144		0.22994		L
38. Y'		0.07499		0.06249		-0.01189		-0.12559		L
39. θ'	0.33551		0.27310		0.13756		-0.28180		-0.66856	-
40. β'		0.0618		0.1374		0.4212		0.3862		-
(f) Other Quantities of Interest										
41. M	0		1.0077		0.0948		-0.8332		0	EI/L
42. P	26.847		24.068		22.616		24.151		25.207	EI/L^2
43. u	0		0.01278		0.00862		0.00995		0	L
44. v	0		-0.03441		-0.01252		0.01619		0	L

TABLE 10-12 Convergence of Numerical Solutions for Parabolic Arch Problem (From Austin, Ross, Tawfik, and Volz, 1982)

Quantity of Interest (1)	Number of Divisions in Arch 4 (2)	8 (3)	12 (4)	16 (5)	24 (6)	48 (7)	Common Factors (8)
			(a) Large-Deflection Solution				
H	23.157	23.057	23.052	23.051	23.051	23.050	EI/L^2
$P(S/4)$	24.066	23.972	23.968	23.967	23.966	23.966	EI/L^2
$M(S/4)$	1.0077	0.9584	0.9560	0.9556	0.9554	0.9554	EI/L
$u(S/4)$	1.2792	1.2107	1.2072	1.2066	1.2064	1.2064	$L/100$
$v(S/4)$	3.4427	3.2184	3.2075	3.2057	3.2050	3.2049	$-L/100$
			(b) Classical Solutions				
H	1.26660	1.26679	1.26680	1.26680	1.26680	1.26680	qL
$P(S/4)$	1.32807	1.32826	1.32826	1.32827	1.32827	1.32827	qL
$M(S/4)$	0.17135	0.17117	0.17116	0.17116	0.17116	0.17115	$qL^2/10$
$u(S/4)$	0.23448	0.23453	0.23454	0.23455	0.23455	0.23455	$qL^44/1000EI$
$v(S/4)$	0.56693	0.56585	0.56580	0.56579	0.56579	0.56579	$-qL^4/1000EI$

increment $(H - 2)$ is therefore evaluated from the condition that the span length is constant, from which it follows that $(H - 2) = 0.158EI/L^2$.

Table 10-11*e* evaluates the computed deflected configuration denoted by primes and the correct value of H by adding the small increments of part 2 to the primary values of part 1. The increments should be small enough so that the approximations of the small-deflection theory do not affect the final results. Because the computed deflected configuration of Table 10-11*e* agrees closely with the assumed configuration of Table 10-11*b*, this iteration is acceptable as the assumed and computed values satisfy the convergence criteria. These are met when the applied load is less than the buckling value.

Table 10-11*f* gives more data of interest. The bending moments and axial loads at each node are obtained by adding the components from parts 1 and 2. The displacements are obtained from the following expressions:

$$u_j = \sum_{i=1}^{j-1} (X' - X_0)_i \qquad v_j = \sum_{i=1}^{j-1} (Y' - Y_0)_i \tag{10-110}$$

Convergence and Accuracy Table 10-12*a* shows results of the same example solved with 4, 8, 12, 16, 24, and 48 segments. The data include the horizontal reaction, axial thrust, bending moment, and horizontal and vertical displacement at the quarter point of the arch rib ($s = S/4$), where S is the total length of the arch rib. For the four-division solution, the values shown in Tables 10-12*a* and 10-11 are slightly different because several significant parameters were used in the computer program. It may appear that 12 divisions are sufficient for the practical purposes of design, and 16 divisions are adequate for research purposes.

Second-Order Analysis

Table 10-13 presents data from a second-order analysis of a fixed circular arch subjected to a uniform rib load of $40EI/L^3$ and a concentrated load of $8EI/L^2$ at the crown. Because the loads, moments, and displacements are symmetrical about the midspan, only the solution for the left half is shown. There are two redundant reactions, the horizontal reaction, and two equal fixed end moments. The analysis therefore requires three parts.

The solution required an iterative procedure, with the last cycle shown. The assumed deflected configuration given in Table 10-13*b* was obtained from the previous iteration. Part 1 of the solution is summarized in Table 10-13*c*. Part 2 of the solution, shown in Table 10-13*d*, gives the effects of a change in horizontal reaction, and part 3, shown in Table 10-13*e*, summarizes the effect of an increment in the end moments. Equal end moment loading (part 3) induces the condition of pure moment in the arch whereby shear and axial loads vanish.

The calculations in all three components are essentially similar, and moments and thrusts are computed as in large-deflection analysis (Austin, Ross, Tawfik, and Volz, 1982).

Classical Analysis

The classical analysis shown in Table 10-14 is for the same problem summarized in the large-deflection analysis of Table 10-11. Invariably, in applying the appropriate equations, the geometry of the undeformed (unloaded) arch is used. The last line in part 1, Table 10-14*b*, gives the increase in span of the primary structure caused by the loads. Likewise, the last line in part 2, Table 10-14*c*, gives the change in span caused by the horizontal reaction only. The total moments M_i, chord strains, $(\Delta l/l)_i$, and chord rotations, $(\Delta\phi)_i$, are estimated in Table 10-14*d* by superimposing the corresponding values in parts 1 and 2. Then ΔX and ΔY are computed from these total quantities, and the displacements are evaluated by summing ΔX and ΔY.

Convergence and Accuracy of Classical Analysis Tables 10-12*b* and 10-15 indicate the convergence and accuracy of the numerical procedure (Austin, Ross, Tawfik, and Volz, 1982). Results of the unsymmetrically loaded two-hinged parabolic arch (classical analysis) solved in Table 10-14 are also summarized in Table 10-12*b* for 4, 8, 12, 16, 24, and 48 divisions. Table 10-12*b* shows that ample convergence is attained with only eight divisions. The numerical solutions of the classical theory are thus very accurate with only a few divisions, although this theory does not yield

TABLE 10-13 Second-Order Analysis of Fixed Circular Arch (From Austin, Ross, Tawfik, and Volz, 1982)

UNIFORM RIB LOAD AND CONCENTRATED LOAD AT CROWN

$\frac{f}{L} = 0.25;\ \frac{L}{r} = 200;\ q = 40\ EI/L^3,\ W = 8EI/L^2\ \ n = 8$

Line No. and Quantity	Node 1	(1–2)	Node 2	(2–3)	Node 3	(3–4)	Node 4	(4–5)	Node 5 (℄)	Common Factors
(a) Undeformed Geometry										
1. X_o		0.09953		0.12096		0.13591		0.14360		L
2. Y_o		0.10485		0.07917		0.04926		0.01672		L
3. θ_o	0.92730		0.69547		0.46365		0.23182		0	-
$h_o = 0.14489\ L;\ \ell_o = 0.14457L;\ \beta_o = 0.2302$										
(b) Assumed Deflected Shape										
4. X*		0.09701		0.12064		0.13810		0.14425		L
5. Y*		0.10715		0.07916		0.04265		0.01146		L
6. θ*	0.92730		0.72086		0.43677		0.17607		0	-
7. β*		0.2048		0.2825		0.2591		0.1764		-
(c) Part 1; Assume $(H\text{-}1) = 30\ EI/L^2$ and $(M_L\text{-}1) = 0.7000\ EI/L$										
8. $W+\bar{q}$			5.7956		5.7956		5.7956		13.7956	EI/L^2
9. C_q		-0.01060		-0.01080		-0.00534		-0.00098		EI/L
10. (Q-1)		24.2846		18.4890		12.6934		6.8978		EI/L^2
11. (MI-1)		-0.8693		-0.1551		0.4681		0.6502		EI/L
12. (M-1)	0.7000		-0.1693		-0.3244		0.1437		0.7939	EI/L
13. (P-1)	39.75		36.65		33.78		31.25		30.00	EI/L^2
14. (Δh-1)		-.000138		-.000127		-.000118		-.000110		L
15. (Δβ-1)		-.02982		.04384		.01795		-.06573		-
16. $(\frac{\Delta\ell}{\ell}\text{-}1)$		-.00038		-.00172		-.00116		.00050		-
17. ($\bar{\alpha}$-1)	0.02541		-0.01591		-0.03948		0.02302		0.08534	-
18. (Δϕ-1)		-.01030		-.02621		-.06569		-.04267		-
$(u\text{-}1)_5 = \Sigma(\Delta X\text{-}1) = 0.006773\ L$; $(\Delta\theta\text{-}1)_1 = (\Delta\phi\text{-}1)_1 - (\bar{\alpha}\text{-}1)_1 = -0.03571$ radians										
(d) Part 2; Effect of (H-2)										
19. (MI-2)		-.10715		-.07916		-.04265		-.01146		(H-2)L
20. (M-2)	0		-0.10715		-0.18631		-0.22896		-0.24042	do
21. (P-2)	0.6000		0.7512		0.9061		.9845		1.0000	(H-2)
22. (Δh 2)		-.000245		-.000301		-.000345		-.000361		$(H\text{-}2)L^3/100EI$
23. (Δβ-2)		0.810		2.165		3.049		3.433		$(H\text{-}2)L^2/100EI$
24. $(\frac{\Delta\ell}{\ell}\text{-}2)$		-0.01726		-0.04370		-0.06100		-0.06849		do
25. (α-2)	-0.2756		-1.5187		-2.6554		-3.2797		-3.4558	do
26. (Δϕ-2)		9.1817		7.6630		5.0076		1.7279		do
$(u\text{-}2)_5 = \Sigma(\Delta X\text{-}2) = -0.0187008\ (H\text{-}2)L^3/EI$; $(\Delta\theta\text{-}2)_1 = (\Delta\phi\text{-}2)_1 - (\bar{\alpha}\text{-}2)_1 = 0.094573\ (H\text{-}2)L^2/EI$										
(e) Part 3; Effect of $(M_L\text{-}3)$										
27. (M-3)	1.0		1.0		1.0		1.0		1.0	$(M_L\text{-}3)$
28. (Δβ-3)		-0.14489		-0.14489		-0.14489		-0.14489		$(M_L\text{-}3)L/EI$
29. $(\frac{\Delta\ell}{\ell}\text{-}3)$		0.00279		0.00279		0.00279		0.00279		do
30. ($\bar{\alpha}$-3)	0.07245		0.14489		0.14489		0.14489		0.14489	do
31. (Δϕ-3)		-.50712		-.36223		-.21734		-.07245		do
$(u\text{-}3)_5 = \Sigma(\Delta X\text{-}3) = 0.095162\ (M_L\text{-}3)L^2/EI$; $(\Delta\theta\text{-}3)_1 = (\Delta\phi\text{-}3)_1 - (\bar{\alpha}\text{-}3)_1 = -0.57956\ (M_L\text{-}3)L/EI$										
(f) Computed Deflected Shape; $(H\text{-}2) = 0.28672\ EI/L^2$ and $(M_L\text{-}3) = -0.014829\ EI/L$										
32. M	0.6852		-0.2149		-0.3926		0.0632		0.7101	EI/L
33. $(\frac{\Delta\ell}{\ell})$		-.00047		-.00189		-.00138		0.00026		-
34. Δϕ		0.02355		0.00113		-.04811		-.03664		-
35. X'		0.09701		0.12064		0.13809		0.14425		L
36. Y'		0.10714		0.07916		0.04265		0.01146		L
37. θ'	0.92730		0.72083		0.43680		0.17613		0	-
38. β'		0.2049		0.2824		0.2590		0.1765		-
39. u	0		-0.00252		-0.00283		-0.00065		0	L
40. v	0		0.00229		0.00228		-0.00432		-0.00958	L

TABLE 10-14 Classical Analysis of Two-Hinged Parabolic Arch (From Austin, Ross, Tawfik, and Volz, 1982)

UNIFORM RIB LOAD AND EQUAL DECK LOAD ON LEFT HALF OF SPAN;

$\frac{f}{L} = 0.15, \frac{L}{r} = 100, n = 4$

Line No. and Quantity	Node 1		Node 2		Node 3		Node 4		Node 5	Common Factors
(a) Undeformed Geometry, $h_o = 0.26428\,L$										
1. X_o		0.23989		0.26011		0.26011		0.23989		L
2. Y_o		0.10940		0.04060		-0.04060		-0.10940		L
3. θ_o	0.54042		0.30255		0		-0.30255		-0.54042	
4. ℓ_o		0.26366		0.26326		0.26326		0.26366		L
5. β_o		0.2373		0.3044		0.3044		0.2373		
(b) Part 1; Assume (H-1) = 0										
6. $\bar{q}$			0.26428		0.26428		0.26428			qL
7. C_q		-0.000572		-0.000272		0.000272		0.000572		qL^2
8. $\bar{p}$			0.25000		0.13006		0			qL
9. (Q-1)		0.65148		0.13720		-0.25714		-0.52142		qL
10. (MI-1)		0.155712		0.035415		-0.066613		-0.124511		qL^2
11. (M-1)	0		0.155712		0.191126		0.124512		0	qL^2
12. (P-1)	0.46488		0.11900		0		0.11599		0.33625	qL
13. (Δh-1)		-0.00722		-0.00106		-0.00116		-0.00575		$qL^4/1000\,EI$
14. (Δβ-1)		-23.225		-48.279		-43.469		-17.728		$qL^3/1000\,EI$
15. $(\frac{\Delta\ell}{\ell}-1)$		0.4330		1.2254		1.1025		0.3296		do
16. $(\bar{\sigma}-1)$			38.502		48.264		31.631			do
17. (Δφ-1)		-61.002		-22.500		25.764		57.395		do
$\Sigma(\Delta X-1)_I = 0.0157006\ qL^4/EI$										
(c) Part 2; Effect of H										
17. (MI-2)		-0.10940		-0.04060		0.04060		0.10940		(H-2)L
18. (M-2)	0		-0.10940		-0.15000		-0.10940		0	(H-2)L
19. (P-2)	0.85749		0.95458		1.00000		0.95458		0.85749	(H-2)
20. (Δh-2)		-0.02406		-0.02598		-0.02598		-0.02406		$(H-2)L^3/1000EI$
21. (Δβ-2)		15.971		35.929		35.929		15.971		$(H-2)L^2/1000EI$
22. $(\frac{\Delta\ell}{\ell}-2)$		-0.4076		-1.0132		-1.0132		-0.4076		do
23. $(\bar{\sigma}-2)$			-27.397		-37.854		-27.397			do
24. (Δφ-2)		46.324		18.927		-18.927		-46.324		do
$\Sigma(\Delta X-2)_I = -0.0123952\ \frac{(H-2)L^3}{EI}$; (H-2) = 1.26667 qL										
(d) Complete Solution; H = 1.2667 qL										
25. M	0		0.01714		0.00113		-0.01406		0	qL^2
26. Δℓ/ℓ		-0.0833		-0.0580		-0.1809		-0.1867		$qL^3/1000\,EI$
27. Δφ		-2.325		1.474		1.790		-1.282		do
28. ΔX		0.2344		-0.0749		0.0256		-0.1851		$qL^4/1000\,EI$
29. u	0		0.2344		0.1595		0.1851		0	do
30. ΔY		-0.5669		0.3811		0.4729		-0.2871		do
31. v	0		-0.5669		-0.1858		0.2871		0	do

sufficient accuracy for most slender arch problems. This is demonstrated from Table 10-12*b*, with $p = q = 17EI/L^3$, which may be compared with the corresponding large-deflection solution of the same problem in Table 10-12*a*. For example, the moment and vertical displacement at the quarter point determined by the classical theory are only 30 percent of the correct values found by the large-deflection theory.

TABLE 10-15 Convergence and Accuracy of Ordinary Linear Solutions for Typical Circular Arches with Symmetrical Loads (From Austin, Ross, Tawfik, and Volz, 1982)

Quantity of Interest (1)	Number of Divisions 6 (2)	12 (3)	18 (4)	24 (5)	Exact Value (6)	Common Factors (7)
(*a*) Hinged Arch with Uniform Rib Load, $f/L = 0.50$; $L/r = 100$						
H	0.25016	0.24982	0.24980	0.24980	0.24980	qL
M_c[a]	0.17645	0.17789	0.17797	0.17798	0.17799	$qL^2/10$
v_c	0.87378	0.88114	0.88132	0.88134	0.88135	$-qL^4/1000EI$
(*b*) Fixed Arch with Concentrated Crown Load; $f/L = 0.15$, $L/r = 200$						
H	1.54163	1.54008	1.53997	1.53996	1.53995	W
M_c	0.50708	0.50778	0.50783	0.50784	0.50784	$WL/10$
M_e	0.31953	0.31790	0.31779	0.31777	0.31776	$WL/10$
v_c	0.45806	0.46206	0.46229	0.46233	0.46235	$-WL^3/1000EI$
(*c*) Fixed Arch with Concentrated Crown Load; $f/L = 0.15$, $L/r = 25$						
H	0.85170	0.85144	0.85142	0.85141	0.85141	W
M_c	0.85583	0.85604	0.85605	0.85605	0.85605	$WL/10$
M_e	0.36662	0.36680	0.36682	0.36683	0.36683	$-WL/10$
v_c	0.29789	0.29801	0.29802	0.29802	0.29802	$-WL^3/100EI$

[a] M_c = crown moment, M_e = fixed end moment, v_c = crown vertical displacement.

Three symmetrically loaded circular arches are summarized in Table 10-15, and exact classical theory solutions are available for these problems. A two-hinged, deep slender arch with uniform rib load is considered in Table 10-15*a*; a shallow, slender fixed arch with concentrated load at the crown is considered in Table 10-15*b*; and a shallow, extremely stocky, fixed arch with concentrated load at the crown is considered in Table 10-15*c*. Results are given for 6, 12, 18, and 24 divisions, and exact values are compared. It appears that all examples converge rapidly to the correct answers for this theory. Table 10-15 also gives an index of accuracy of the equilibrium and geometric relationships used for all three theories and for a wide range of arch properties and loadings.

10-19 DEFORMATION AND ELASTIC BUCKLING BEHAVIOR: CASE STUDIES

Section 10-8 discusses cases of in-plane buckling and stability using specific arch models. Stability considerations of steel arches are also discussed in length, including a review of recent studies.

In the general case, buckling is not only a mathematical concept, because, besides the determination of the buckling load, we must also consider the importance of buckling behavior as it relates to the deflected shape of the arch and the associated buckling modes as a valuable design parameter. In this section we will consider a case study of elastic buckling under symmetrical loading (a special deformation case of arches in the elastic region) and the applicability of curved beam theory to arch buckling analysis.

Elastic Buckling Under Symmetrical Loading

Extensive studies (Timoshenko and Gere, 1961) show that an arch loaded so that the funicular curve coincides with the centroidal axis of the rib will experience very small displacements before buckling. Because, in practice, this is not necessarily the case, arch bridges commonly will experience substantial moments and displacements before buckling.

Austin and Ross (1976) describe the general deformational behavior of symmetrical arches under symmetrical loading in the simple pattern of Figure 10-79. For small loads, the arch deflects symmetrically with a nonlinear load–deflection curve as shown in Figure 10-79*a*. If an antisymmetrical mode does not become dominant, the arch will become unstable when the tangent to the curve becomes horizontal (point A), and it will buckle in a symmetrical pattern. This symmetrical buckling is a snap-through phenomenon. The arch will buckle in an antisymmetrical configuration if the critical load for this condition is less than the critical load for symmetrical bending, as shown in Figure 10-79*b* where a sidesway mode develops at point B. Antisymmetrical buckling is a bifurcation occurrence, very different from the snap-through behavior, whereby the load drops beyond the bifurcation point (B). In some instances the load may increase after bifurcation, in which case the maximum load is attained after large lateral movement.

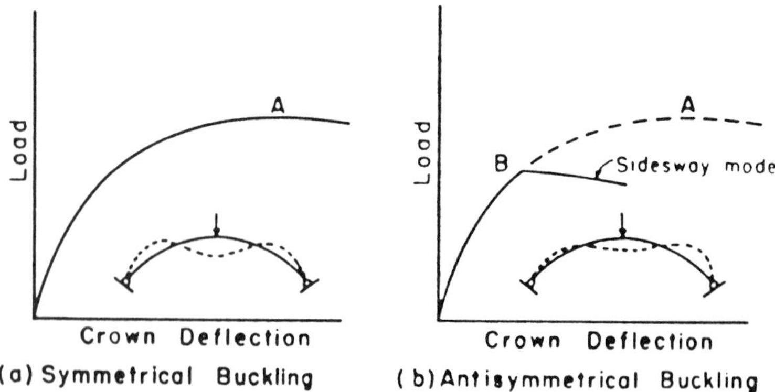

FIGURE 10-79 Buckling behavior with symmetrical loading. (From Austin and Ross, 1976.)

Numerical analyses carried out by Austin and Ross (1976) focus on in-plane elastic buckling in arches that experience prebuckling displacements, and include two-hinged and fixed parabolic and circular arches of constant cross section, subjected to either a concentrated load at the crown or a vertical loading uniformly distributed along the arch rib. The results represent accurate solutions of the exact theory of the inextensional elastica. Rib shortening and shearing deformations are ignored because they are expected to be small for slender arches that buckle elastically. Useful data on the same problem are also given by DaDeppo and Schmidt (1969a, 1969b, 1971, 1972, 1974), who have computed the lowest critical loads and crown deflections for circular arches. These investigators compare the solutions obtained by the exact and classical theories. The postbuckling behavior of arches is also described by Huddleston (1968, 1971), who presents a general large-deflection numerical analysis that includes axial strains (see also previous sections).

Method of Analysis The solutions of the exact theory obtained by Austin and Ross (1976) are based on a procedure described by Ross (1973) and Austin (1971). In many respects this procedure is very similar to Newmark's method for beams and columns, discussed in Section 10-18. A linear eigenvalue analysis is made (see also previous sections) to determine the load that will support a symmetric or antisymmetric perturbation from the equilibrium state defined by the deflected arch profile and the reaction–load ratios established in the bending analysis. This load is called a pseudocritical load; if it is equal to the initially assumed load (corresponding to the equilibrium state tested), this load is the true critical load. The associated iteration procedure works well for the antisymmetrical buckling solutions but is impractical and tedious for symmetrical buckling problems, and is not feasible for describing postbuckling behavior.

Moments, displacements, and so on are computed at division points equally spaced along the arch axis, specifically for 24 divisions in the complete arch. Critical loads and the corresponding reactions and maximum moments are also computed by an approximate theory (classical). A classical theory solution is obtained from a linear eigenvalue analysis of the undeflected arch with the reaction–load ratios computed from ordinary linear bending analysis (see also previous sections). A numerical procedure for the classical buckling theory is also presented by Lind (1966). Solutions of the exact and classical theories are identical when the arch does not deflect under load.

Results and Interpretation Critical values of the concentrated load Q, uniformly distributed load q, and the corresponding values at incipient buckling of the horizontal reaction force H, fixed end moment M_e, height of crown, and maximum moment in the arch M_m are summarized in Tables 10-16 through 10-19, for two-hinged and fixed arches, respectively. All results

TABLE 10-16 Buckling Coefficients for Two-Hinged Arches with Vertical Concentrated Load at Crown

θ		Exact Theory				Classical Theory		
(°)	f_i/L	Q_cL^2/EI	H_cL^2/EI	f_c/f_i	$M_{mc}L/EI$	Q_oL^2/EI	H_oL^2/EI	$M_{m0}L/EI$
(1)	(2)	(3)	(4)	(5)	(6)	(7)	(8)	(9)
(*a*) Parabolic Arches—Antisymmetrical Modes								
—	0.10	17.72	37.2	0.782	1.52	18.83	36.7	1.036
—	0.15	24.9	34.5	0.793	2.13	25.9	33.6	1.437
—	0.20	30.4	31.0	0.810	2.57	30.8	29.9	1.726
—	0.25	33.6	27.0	0.830	2.81	33.6	26.0	1.901
—	0.30	34.8	22.8	0.853	2.88	34.5	22.2	1.977
—	0.35	34.3	18.87	0.876	2.80	34.1	18.69	1.976
—	0.40	32.7	15.44	0.897	2.63	32.7	15.64	1.921
—	0.50	28.0	10.27	0.929	2.22	28.7	10.89	1.725
(*b*) Parabolic Arches—Symmetrical Modes								
—	0.10	20.9	51.3	0.53	2.5	41.9	81.8	2.31
—	0.15	30.0	49.2	0.53	3.6	57.8	74.9	3.20
—	0.20	37.9	46.5	0.52	4.6	68.8	66.8	3.85
—	0.25	44.2	44.1	0.49	5.7	75.1	58.1	4.25
—	0.30	49.0	40.4	0.49	6.4	77.3	49.7	4.43
—	0.35	52.3	37.0	0.47	7.0	76.6	42.0	4.44
—	0.40	54.4	33.9	0.44	7.6	73.8	35.3	4.33
—	0.50	55.3	28.0	0.39	8.4	65.3	24.8	3.93
(*c*) Circular Arches—Antisymmetrical Modes								
50	0.1109	19.32	36.8	0.761	1.73	20.7	36.1	1.173
70	0.1577	25.7	34.1	0.750	2.38	27.2	33.0	1.595
90	0.2071	30.6	30.7	0.738	2.97	31.9	29.0	1.963
106.26	0.2500	33.3	27.2	0.731	3.35	34.2	25.3	2.21
120	0.2887	34.3	23.8	0.729	3.56	34.8	22.0	2.36
140	0.3501	33.4	18.32	0.740	3.61	33.7	16.97	2.47
160	0.4196	29.6	12.65	0.767	3.32	30.0	12.09	2.43
180	0.5000	23.5	7.71	0.805	2.76	24.5	7.79	2.22
(*d*) Circular Arches—Symmetrical Modes								
50	0.1109	22.3	49.5	0.53	2.7	45.9	80.1	2.60
70	0.1577	29.6	46.4	0.51	3.7	60.3	73.1	3.54
90	0.2071	35.5	42.2	0.49	4.6	71.0	64.6	4.37
106.26	0.2500	39.0	38.5	0.47	5.2	76.7	56.9	4.96
120	0.2887	41.1	35.4	0.44	5.8	79.3	50.1	5.37
140	0.3501	42.6	30.3	0.40	6.4	79.4	40.0	5.84
160	0.4196	42.4	25.2	0.35	6.9	75.6	30.5	6.13
180	0.5000	40.6	20.6	0.29	7.2	68.5	21.8	6.22

are expressed in dimensionless form. The notation is as follows:

L = span
E = elastic modulus
I = moment of inertia of cross section
f_i = initial rise of span
θ = angle of opening of the circular arch

The critical load given for antisymmetrical modes is the bifurcation point load. The subscripts c and o indicate values pertaining to the exact and

TABLE 10-17 Buckling Coefficients for Two-Hinged Arches with Vertical Load Uniformly Distributed Along Arch Axis

θ		Exact Theory				Classical Theory		
(°)	f_i/L	$q_c l^3/EI$	$H_c L^2/EI$	f_c/f_i	$M_{mc} L/EI$	$q_o L^3/EI$	$H_o L^2/EI$	$M_{mo} L/EI$
(1)	(2)	(3)	(4)	(5)	(6)	(7)	(8)	(9)
				(*a*) Parabolic Arches—Antisymmetrical Modes				
—	0.10	28.6	36.3	1.002	0.0145	28.6	36.3	+0.0085
—	0.15	38.2	32.9	1.004	0.0422	38.2	32.9	+0.0246
—	0.20	43.4	28.8	1.006	0.0816	43.4	28.7	+0.0473
—	0.25	44.9	24.5	1.008	0.1253	44.9	24.4	+0.0719
—	0.30	43.5	20.5	1.009	0.1664	43.5	20.4	+0.0946
—	0.35	40.4	16.96	1.009	0.1994	40.3	16.75	+0.1133
—	0.40	36.5	13.98	1.009	0.222	36.3	13.70	−0.1267
—	0.50	28.5	9.51	1.008	0.241	28.1	9.16	−0.1470
				(*b*) Circular Arches—Antisymmetrical Modes				
50	0.1109	31.2	35.6	0.994	−0.0412	31.2	35.6	−0.0242
70	0.1577	39.5	32.0	0.989	−0.1081	39.6	32.0	−0.0638
90	0.2071	44.0	27.4	0.981	−0.214	44.1	27.5	−0.1277
106.26	0.2500	44.5	23.2	0.974	−0.323	44.6	23.2	−0.1960
120	0.2887	42.8	19.34	0.968	−0.421	43.0	19.51	−0.261
140	0.3501	37.1	13.75	0.959	−0.549	37.5	14.07	−0.357
160	0.4196	28.9	8.72	0.953	−0.626	29.5	9.14	−0.436
180	0.5000	20.0	4.78	0.950	−0.633	20.7	5.19	−0.475
				(*c*) Circular Arches—Symmetrical Modes				
50	0.1109	63.0	74.6	0.88	0.69	70.3	80.1	−0.0544
70	0.1577	77.6	66.8	0.81	1.31	90.2	73.0	−0.1457
90	0.2071	85.6	57.9	0.75	1.95	103.0	64.2	−0.298
106.26	0.2500	87.7	50.5	0.69	2.5	107.6	56.0	−0.472
120	0.2887	86.7	44.3	0.64	3.0	107.4	48.7	−0.653
140	0.3501	81.7	35.1	0.57	−3.7	100.9	37.9	−0.963
160	0.4196	73.8	27.1	0.45	−4.6	88.4	27.4	−1.307
180	0.5000	64.4	19.07	0.34	−5.4	71.9	18.04	−1.648

classical theory, respectively. For the classical theory, buckling is assumed to occur from the undeflected profile; hence, the height of the crown at buckling is not tabulated for this case.

Data relevant to buckling under the lowest critical load are given for all cases. When the critical load for antisymmetrical buckling controls, data relevant to symmetrical buckling are also included because, in certain practical cases, the sidesway buckling mode may be prevented whereas the symmetrical mode is unrestrained (e.g., in deck arch bridges).

The tabulated maximum prebuckling moments are maximum model values, and their location is as follows: (a) at the crown for all arches with concentrated load at the crown; (b) at the fixed end for all fixed arches under uniform load; (c) at the third node from the hinge when negative, and at the crown when positive, for circular hinged arches under distributed loading; and (d) at the third (if $f_i/L \leq 0.20$) or fourth node from the hinged end for parabolic arches under distributed loading. The loadings and range of the

TABLE 10-18 Buckling Coefficients for Fixed Arches with Vertical Concentrated Load at Crown

		Exact Theory					Classical Theory			
θ (°) (1)	f_i/L (2)	Q_cL^2/EI (3)	H_cL^2/EI (4)	$M_{ec}L/EI$ (5)	f_c/f_i (6)	$M_{mc}L/EI$ (7)	Q_oL^2/EI (8)	H_oL^2/EI (9)	$M_{eo}L/EI$ (10)	$M_{mo}L/EI$ (11)
(*a*) Parabolic Arches—Antisymmetrical Modes										
—	0.50	64.5	36.2	3.85	0.735	6.66	64.9	29.0	1.631	3.37
(*b*) Parabolic Arches—Symmetrical Modes										
—	0.10	23.1	70.4	1.33	0.68	2.3	53.9	125.9	1.651	2.55
—	0.15	33.4	68.0	1.93	0.68	3.4	76.1	117.9	2.28	3.62
—	0.20	42.6	64.9	2.5	0.67	4.4	93.5	108.0	2.72	4.50
—	0.25	50.2	61.4	3.0	0.67	5.3	105.6	97.0	2.99	5.14
—	0.30	56.2	57.8	3.4	0.66	6.1	112.7	85.8	3.11	5.55
—	0.35	60.7	53.1	3.7	0.66	6.6	115.5	74.9	3.11	5.77
—	0.40	63.8	49.4	4.1	0.65	7.2	114.9	64.8	3.02	5.81
—	0.50	66.1	41.4	4.6	0.64	7.9	107.3	47.9	2.70	5.56
(*c*) Circular Arches—Symmetrical Modes										
50	0.1109	24.5	67.9	1.46	0.67	2.5	58.1	122.7	1.892	2.82
70	0.1577	32.4	63.6	1.99	0.67	3.4	76.6	113.5	2.59	3.84
90	0.2071	38.6	58.4	2.5	0.65	4.2	90.7	102.2	3.24	4.76
106.26	0.2500	42.3	53.8	2.9	0.63	4.9	98.5	91.8	3.72	5.41
120	0.2887	44.3	49.8	3.2	0.61	5.4	102.4	82.5	4.08	5.88
140	0.3501	45.4	42.7	3.5	0.60	5.9	103.7	68.6	4.53	6.43
160	0.4196	44.5	35.9	3.7	0.57	6.3	100.0	55.0	4.87	6.80
180	0.5000	41.8	29.6	3.9	0.53	6.6	92.0	42.3	5.09	6.97

TABLE 10-19 Buckling Coefficients for Fixed Arches with Vertical Load Uniformly Distributed Along Arch Axis

θ		Exact Theory				Classical Theory		
(°)	f_i/L	q_cL^3/EI	H_cL^2/EI	$M_{ec}{}^aL/EI$	f_c/f_i	q_oL^3/EI	H_oL^2/EI	$M_{eo}{}^aL/EI$
(1)	(2)	(3)	(4)	(5)	(6)	(7)	(8)	(9)
			(*a*) Parabolic Arches—Antisymmetrical Modes					
—	0.10	60.4	76.1	−0.0382	1.003	60.2	76.2	−0.0221
—	0.15	83.4	70.8	−0.1155	1.006	83.0	70.9	−0.0652
—	0.20	99.3	64.3	−0.234	1.010	98.6	64.4	−0.1305
—	0.25	107.7	57.0	−0.379	1.013	106.9	57.1	−0.208
—	0.30	109.6	49.5	−0.527	1.015	108.8	49.7	−0.286
—	0.35	106.4	42.5	−0.660	1.017	105.7	42.6	−0.354
—	0.40	100.0	36.2	−0.766	1.107	99.4	36.2	−0.407
—	0.50	83.0	25.9	−0.890	1.017	82.2	25.6	−0.464
			(*b*) Circular Arches—Antisymmetrical Modes					
50	0.1109	64.8	75.2	0.1086	0.992	63.5	75.1	0.0627
70	0.1577	83.4	70.0	0.294	0.983	84.6	69.8	0.1689
90	0.2071	95.5	63.3	0.608	0.969	97.6	62.9	0.348
106.26	0.2500	99.9	57.0	0.966	0.954	102.9	56.3	0.554
120	0.2887	99.8	51.2	1.333	0.938	103.3	50.3	0.768
140	0.3501	93.8	42.1	1.932	0.910	97.4	40.7	1.130
160	0.4196	81.9	32.6	2.52	0.880	85.0	31.0	1.515
180	0.5000	66.0	23.4	2.96	0.854	67.9	21.8	1.859
			(*c*) Circular Arches—Symmetrical Modes					
50	0.1109	90.9	110.6	0.49	0.94	106.8	123.0	0.1027
70	0.1577	110.0	98.5	0.95	0.91	138.0	113.9	0.276
90	0.2071	119.4	85.7	1.52	0.88	159.0	102.5	0.567
106.26	0.2500	120.3	75.2	2.04	0.85	167.8	92.0	0.904
120	0.2887	117.2	66.7	2.5	0.82	169.3	82.3	1.259
140	0.3501	107.5	54.2	3.2	0.78	162.4	67.8	1.883
160	0.4196	93.5	43.1	4.0	0.73	146.1	53.2	2.60
180	0.5000	77.2	32.5	4.6	0.67	123.2	39.5	3.37

[a] End moments are the maximum moments in the arch.

rise–span ratios considered in the analysis are considered extreme cases and should cover most practical configurations.

The following comments are appropriate.

1. The lowest critical load corresponds to the antisymmetrical (sidesway) mode, with the exception of fixed circular and parabolic arches with concentrated load at the crown that buckle symmetrically. We should note, however, that the critical load for symmetrical (snap-through) buckling is not much

greater than the critical load for sidesway for the flatter hinged arches with concentrated load at the crown.

2. From the results of Table 10-19, it appears that the critical load and horizontal reaction calculated by the classical theory are in close agreement with the corresponding data of the exact theory for antisymmetrical modes, but in fair disagreement for symmetrical modes. In other words, the classical theory solves bifurcation-type buckling problems better than snap-through buckling phenomena, and performs best when the displacements at the instant of buckling are small (as in the case of antisymmetrical buckling under uniform load). On the other hand, the classical theory still gives good approximations of the antisymmetrical buckling loads for hinged arches with concentrated crown load for which the prebuckling displacements are large (Tables 10-16*a* and *c*). This appears to indicate that prebuckling displacements may have only a small effect on the critical load.

3. We may conclude that the classical theory gives easily obtained, accurate estimates of the critical loads and horizontal reactions for arches with nonuniform cross section, any symmetrical loading, any profile, and so on, for antisymmetrical buckling. However, the bifurcation-type of antisymmetrical buckling occurs only with symmetrical loading. Unsymmetrical loading is of great practical importance and leads to a load–deflection curve similar to that shown in Figure 10-79*a*; the load simply reaches the maximum (critical) value following large displacements. Hence, the elastic theory is not entirely useful in analyzing the buckling load for this case.

4. The classical theory does not articulate bending moments to the extent desired, not even in cases where the displacements at buckling are small, such as parabolic arches under distributed load. This is because in the classical theory the effect of displacement on moments is neglected.

5. Expressing the critical conditions for antisymmetrical buckling in terms of horizontal reactions is a matter of convenience for flatter arches. Figure 10-80 shows graphs of the critical horizontal reaction coefficient versus the rise–span ratio for the antisymmetrical buckling solutions in Tables 10-16, 10-17, and 10-19, and curves for parabolic arches with load distributed uniformly over a horizontal projection (deck load) plus catenary arches with load uniformly distributed along the arch axis. In the latter two cases, the funicular curves of the loading coincide with the arch axes so that prebuckling moments and displacements vanish. The critical loads and horizontal reactions for these cases are given in Table 10-20. From Figure 10-80 it appears that the critical horizontal reactions are clearly insensitive to the arch shape and, to a lesser extent, to the loading. They seem to vary mainly with the rise–span ratio, and this is particularly true for fixed arches. The rise–span ratios vary from 0.10 to 0.30, which is the most practical range. These results suggest that similar curves can be used to make estimates of the critical horizontal reaction for other symmetrical loadings.

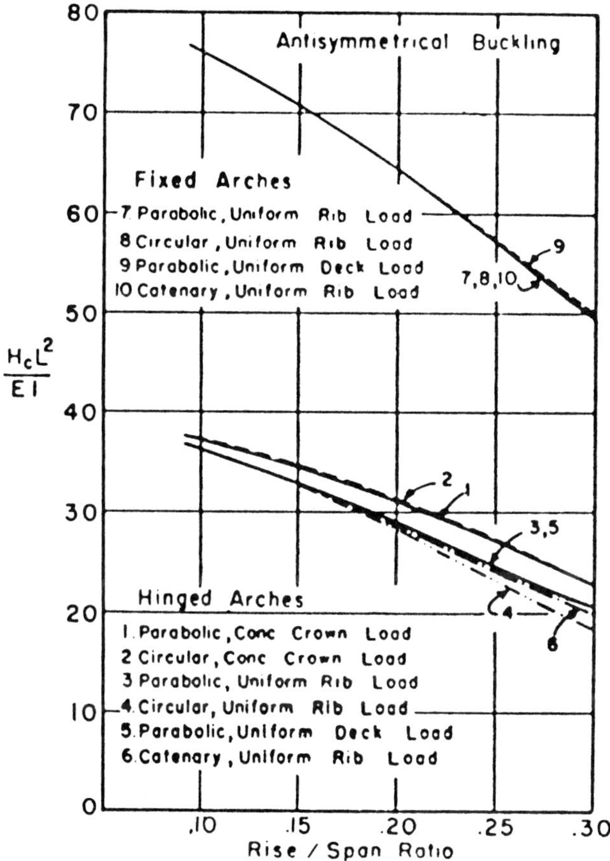

FIGURE 10-80 Variation of critical horizontal reaction with rise–span ratio. (From Austin and Ross, 1976.)

TABLE 10-20 Buckling Coefficients for Parabolic and Catenary Arches Acting in Pure Axial Compression[a]

	Parabolic Profile				Catenary Profile			
	Hinged Ends		Fixed Ends		Hinged Ends		Fixed Ends	
f_i/L (1)	$p_c L^3/EI$ (2)	$H_c L^2/EI$ (3)	$p_c L^3/EI$ (4)	$H_c L^2/EI$ (5)	$q_c L^3/EI$ (6)	$H_c L^2/EI$ (7)	$q_c L^3/EI$ (8)	$H_c L^2/EI$ (9)
0.10	29.07	36.34	60.93	76.16	28.68	36.32	60.13	76.15
0.15	39.47	32.89	85.12	70.93	38.25	32.79	82.66	70.86
0.20	46.10	28.81	103.11	64.45	43.52	28.54	97.96	64.25
0.25	49.19	24.59	114.58	57.29	44.83	24.08	105.92	56.90
0.30	49.46	20.61	120.05	50.02	43.16	19.83	107.43	49.36
0.35	47.80	17.07	120.60	43.07	39.65	16.05	103.96	42.10
0.40	45.01	14.07	117.53	36.73	35.29	12.87	97.18	35.43
0.50	38.20	9.55	105.31	26.33	26.51	8.20	79.30	24.53

[a]All coefficients are for antisymmetrical modes.

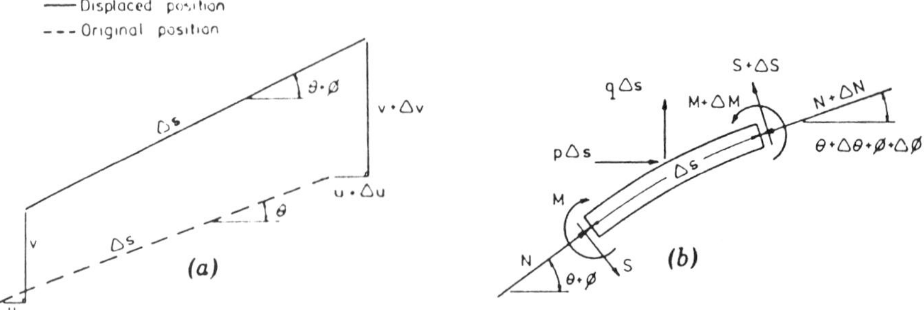

FIGURE 10-81 Arch notation and rotation function ϕ. (From Van der Woude and Cousins, 1979.)

Arch Deformation and Buckling in the Elastic Region: A Mathematical Model

Energy methods, using Lagrangian multipliers to handle geometric restrictions on a Fourier series approximation for the rotation function ϕ, may provide an efficient and simple solution for in-plane deformations of arches. Van der Woude and Cousins (1979) have used this technique to develop a model that includes second-order effects leading to in-plane buckling analysis. Infinite series solutions obtained for a uniform circular arch are shown to be exact and thus establish the correctness level of the approach. Convergence checks indicate that as few as four (and perhaps two) terms in the series approach allow results that fall within standard accuracy. Combined with the linearly elastic model, the energy method may provide a useful tool for studying the nonlinear elastic-in-plane response of arches under general loading conditions.

Uniform Circular Arch The physical model is a plane, freestanding, slender arch loaded in its plane. The loading is assumed to be funicular, rib shortening is neglected, and the material is taken as linearly elastic.

Van der Woude and Cousins (1979) explain the concept of buckling of the arch by introducing an infintesimal disturbance. Let us consider the arch in a slightly deflected state represented by the rotation function ϕ (Figure 10-81). The arch is unstable if this deflected state can be maintained by the loading (funicular) without disturbing forces. The equilibrium condition for the deflected state is established if the normal force N is properly handled, and the result expanded to first-order terms in rotation ϕ and its derivatives (Van der Woude and Cousins, 1978). One case for which an analytical solution is possible is the circular arch under radial load, and, although this case has limited practical applications, the forms of its solution lead to certain general conclusions. These investigators also note that regardless of how the solution is obtained, it depends on the manner in which the load is

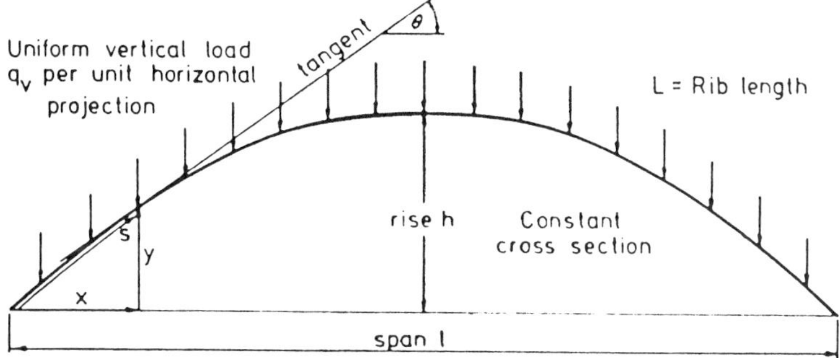

FIGURE 10-82 Parabolic arch under uniform vertical load. (From Van der Woude and Cousins, 1979.)

applied to the buckled arch. In this context the loading is classified as configuration-independent and configuration-dependent (Oran and Reagan, 1969). Configuration dependent loading requires extensions to the basic method, which may be uncertain. In practice, configuration-independent loading occurs more frequently, and an example is the gravity load.

The application of the energy method to a uniform circular arch yields two basic Fourier series for the buckling loads for the antisymmetric and symmetric modes, respectively. These involve a buckling load parameter $\beta^2 = q_r R^3/EI$, where q_r is the radial load per unit rib length, R is the radius of the circular arch under radial load q_r, and EI is the bending stiffness. Numerical results are obtained by summing the series for various values of β^2 and graphing the sums to locate the roots.

Uniform Parabolic Arch Figure 10-82 shows a parabolic arch. The funicular loading for this structure is a vertical load uniformly distributed per unit horizontal projection as shown. The potential energy is expressed in terms of the horizontal span l, and as in the previous example the load q_v (vertical load per unit horizontal projection) is assumed to remain constant as the arch buckles.

For hinged-support conditions, the Fourier series consists of cosine functions. The sets of simultaneous equations in the Fourier coefficients are singular for certain discrete values of load. In order to avoid possible numerical instability in these regions of load, the method of eigenvectors is used. As before, buckling loads are determined by graphic functions against load to determine their roots. In each case a solution was obtained using an appropriate number of terms in the series representation for the rotation function ϕ. Van der Woude and Cousins (1979) present graphically the variation of buckling loads with rise–span ratio h/l. Both modes are represented. These data are supplemented by tables giving the relative values of the first five Fourier coefficients for the basic modes (antisymmetric) for rise–span ratios (h/l) of 0.10 to 0.40.

For fixed supports a Fourier sine series is likewise used, and the analysis is essentially similar to the case of pinned (hinged) supports.

Curved Beam Theory for Arch Buckling Analysis

A nonlinear elastic finite element for a beam initially curved in one plane but deformable in a three-dimensional space is used by Wen and Lange (1981) to calculate the bifurcation buckling loads of arches. The curved beam element can be used to calculate in-plane or out-of-plane buckling loads in arches of different shapes. Geometric nonlinearities are considered by including the effect of rotations on the longitudinal strains. The arch element is assumed to have a constant cross section with two axes of symmetry.

Appropriate stiffness matrices used for linear equilibrium and nonlinear equilibrium analyses were used in this study for incremental equilibrium analysis. Following the presentation by Mallet and Marcal (1968), the stiffness matrices were used to formulate eigenproblems for buckling loads by setting the structural incremental stiffness to zero and assuming that the displacement increases linearly with the applied loads.

The area of classical buckling analysis of curved structures has also been investigated by Ojalvo and Newman (1968), Ojalvo, Demuts, and Tokarz (1969), Shukla and Ojalvo (1971), Tokarz and Sandhu (1972), Walker (1969), Dawe (1971), Sabir and Lock (1973), Mak and Kao (1973), and Ebner and Ucciferro (1972). Invariably, these studies confine the analysis to behavior in a plane.

Formulation of Eigenvalue Problems The method of analysis by Wen and Lange (1981) leads to the following expression for obtaining the critical load:

$$([K] + [N1] + [N2])_{(\overline{Q})}\{\Delta Q\} = \{0\} \qquad (10\text{-}111)$$

where $\{\overline{Q}\}$ = the reference equilibrium position (displacement vector)
$[K]$ = structural linear stiffness matrix
$[N1]$ = first-order structural incremental stiffness matrix
$[N2]$ = second-order structural incremental stiffness matrix
Δ = incremental operator
$\{P\}$ = vector of applied loads
$\{\ \}$ = column vector
$[\ ]$ = rectangular matrix

For a given load vector $\{\overline{P}\}$ and displacement vector $\{\overline{Q}\}$ on the fundamental path, if a nontrivial solution of $\{\Delta Q\}$ is obtained from (10-111), the load $\{\overline{P}\}$

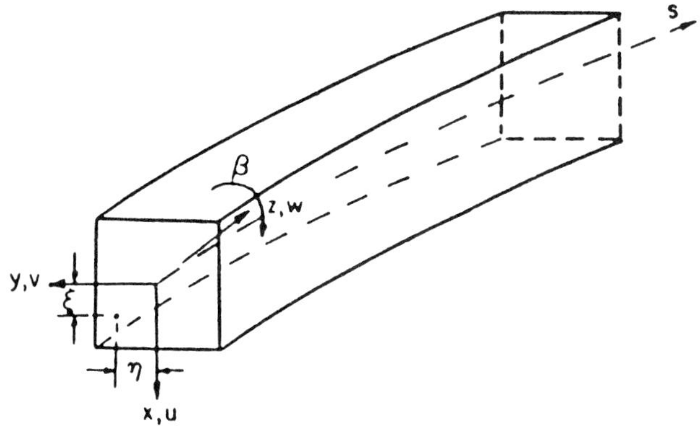

FIGURE 10-83 Curved beam element. (From Wen and Lange, 1981.)

would be the exact buckling load (bifurcation or snap-through load). However, an approximate solution of the buckling load is obtained from (10-111), assuming that the displacement of the arch increases linearly with the applied load until buckling occurs (Mallett and Marcal, 1968). Furthermore, if it can be assumed that at buckling the displacements are sufficiently small, the analysis reduces to a linear eigenvalue problem (Wen and Lange, 1981).

Numerical Results In order to consider the significance of matrix $[N2]$ in the tabulation of buckling loads, Wen and Lange (1981) have solved three types of problems: (a) in plane buckling of a circular arch with an opening angle $2\alpha = 90°$ under a uniformly distributed radial load, (b) out-of-plane buckling of the same arch type under the same loading, and (c) out-of-plane buckling of a parabolic arch subjected to a uniformly distributed vertical load. All three arches are hinged at the ends. At the supports, the nonessential degrees of freedom dw/ds and $d\beta/ds$ (β is the twist of the cross section about the z axis, w is the displacement along the z axis, s is the length along the rib axis) were not restrained. For the out-of-plane cases, the rotation degrees of freedom about the x and z axes at the supports were restrained.

The geometry of a beam (arch) element curved in one plane is shown in Figure 10-83. The centroidal axis curves in the x–z plane with radius R. The x, y, and z axes form a right-hand coordinate system with corresponding displacements u, v, and w.

Table 10-21 shows the values and ratios of the buckling load as computed from the linear eigenproblem to that from the quadratic eigenproblem. In both cases the number of elements was 12. In all cases the ratio is very close to unity, meaning that for these problem types the inclusion of matrix $[N2]$ in the eigenvalue problem would not significantly change the results from the case in which it is neglected.

TABLE 10-21 Linear Versus Quadratic Eigenproblem Solutions

Type of Arch[a] (1)	Type of Load (2)	A (3)	Ixx (4)	Iyy (5)	Buckling Load: Linear (6)	Buckling Load: Quadratic (7)	Buckling Load: Ratio (8)
Circular in-plane	Radial	0.1875	0.9765×10^{-3}	0.8789×10^{-2}	51.51	51.64	0.99
Circular out-of-plane	Radial	1.00	0.5208×10^{-2}	1.333	27.44	27.44	1.00
Parabolic out-of-plane	Vertical	1.50	1.7578×10^{-2}	2.00	189.61	189.67	0.99

[a]Circular arches: $R = 30$, $2\alpha = \pi/2$, $E = 10^7$. Parabolic arch: $H = 9.6$, $L = 48$, $E = 29 \times 10^6$.

Effect of Rotations The first-order incremental stiffness considers three forms of deformation: (a) axial strain, (b) rotation about the y axis, and (c) rotation about the x axis. We should note that in the classical stability analysis the effect of prebuckling rotations is neglected.

In this study solutions were obtained using both the first-order element incremental stiffness matrix, $[n1]$, and the same matrix with rotation terms deleted, $[n1^*]$. The differences between buckling loads obtained by the former and the latter increased with the amount of bending (or prebuckling rotational deformation) in the structure. Thus, for problems involving uniform distributed load that induces little or no bending, results obtained from either solution are only slightly different (within 1–2 percent). However, for a concentrated load at the crown of a semicircular arch, the buckling load determined from $[n1]$ would be about one-third of that using $[n1^*]$. Interestingly, for the concentrated load problem, the buckling load obtained from $[n1^*]$ agrees very closely with experimental results (Langhaar, Boresi, and Carver, 1954) and the bifurcation load computed from nonlinear equilibrium analysis (DaDeppo and Schmidt, 1969b). Other results obtained with the use of $[n1^*]$ are presented in the following discussion.

Convergence of Solutions Four problems studied to determine the convergence of the finite-element model are defined in Figure 10-84. The buckling mode observed for problems 1, 2, and 3 is in-plane and antisymmetrical. For problem 4, it is out-of-plane and symmetrical. The associated buckling loads are listed in Table 10-22 and evidently they converge rather rapidly. Results obtained using only two elements to represent the structure differ by about 25 percent from the converged value.

Table 10-22 also gives a set of reference values for comparison. Although they differ only by small amounts from the finite-element results, this discrepancy may be explained. The buckling loads in the reference values are based

Problem	Structure-Load Systems Considered
1	Circular arch subjected to uniform pressure
2	Circular arch subjected to concentrated load
3	Circular arch subjected to uniform load on horizontal projection
4	Parabolic arch subjected to uniform load on horizontal projection

Problem 1: Circular arch subjected to uniform pressure

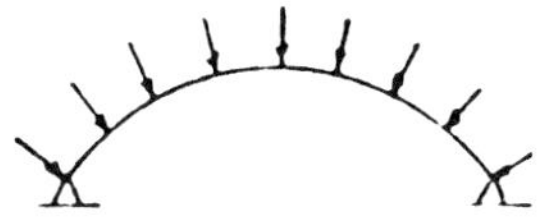

R = 30; 2α = 90°
E = 10^7; A = 0.1875
I = 0.8789×10^{-2}

Problem 2: Circular arch subjected to concentrated load

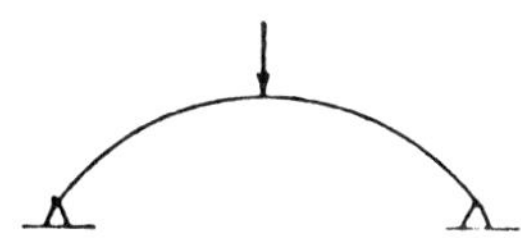

R = 10; 2α = 180°
E = 1.04×10^7; A = 0.0775
I = 6.21×10^{-6}

Problem 3: Circular arch subjected to uniform load on horizontal projection

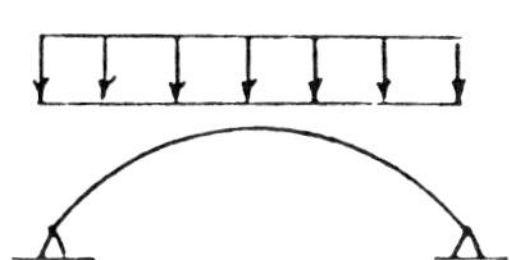

R = 318.2; 2α = 90°
E = 4.176×10^6
A = 2.7
I = 2.03125

Problem 4: Parabolic arch subjected to uniform load on horizontal projection

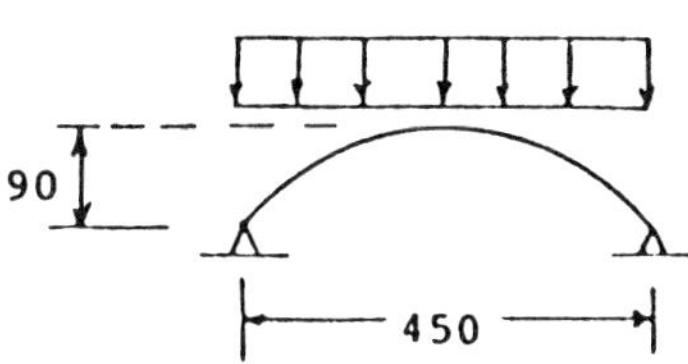

E = 4.176×10^6;
G = 1.606×10^6
A = 2.7; Ixx = 4.45
Iyy = 32.5
K_T = 5.785

FIGURE 10-84 Convergence studies. (From Wen and Lange, 1981.)

on the assumption that the arch axis is inextensible, but this assumption is not made in the finite-element method.

Effect of Cross-Sectional Area This effect on the buckling load p_{cr} is demonstrated in Table 10-23. The parameter A/A_o (cross-sectional area/cross-sectional area used in Table 10-22) is varied over a range of four

TABLE 10-22 Convergence Studies: Buckling Load

	Problem			
Number of Elements (1)	1 (2)	2 (3)	3 (4)	4 (5)
2	64.24	5.08	5.19	9.40
4	52.34	4.52	5.47	7.51
6	51.81	4.40	4.38	7.39
8	51.71	4.37	4.06	7.37
12	51.68	4.37	3.94	7.36
16	51.67	4.37	3.93	7.36
24	51.67	4.37	3.94	7.36
Reference value	48.83	4.17	4.11	7.89
From reference	(21)	(12)	(3)	(22)
Ratio	1.06	1.05	0.96	0.93

orders of magnitude for problems 1 and 4 of Table 10-22, A_o being the value of A used. A total of 12 elements were used for the arch in all cases.

It appears that the effect is generally inconsequential, except for very small values of the ratio A/A_o, or very small values of the slenderness index $\overline{L}/r$. For very small values of A, p_{cr} actually decreases with an increase in A. There is a reason for this, related to the support or boundary conditions.

Consider, for example, the full curve line in Figure 10-85, representing a uniformly compressed arch in equilibrium with the end thrust that has a horizontal component H_0. The presence of hinges requires both ends to move horizontally a distance d, and for this a horizontal force H_1 must be applied, giving a total horizontal force $H = H_0 - H_1$. Let h_A and h_i denote the horizontal stiffness due to area A and moment of inertia I, respectively. Because h_A varies with A for this study, h_I is constant, and d is inversely

TABLE 10-23 Effect of Cross-Sectional Area

	Problem 1			Problem 4		
A/A_o[a] (1)	P_{cr} (2)	H (3)	$\overline{L}/r$[b] (4)	P_{cr} (5)	H (6)	$\overline{L}/r$[c] (7)
10^{-2}	56.3	19.7	10.9	9.15	187	19.3
10^{-1}	52.1	22.3	34.4	7.52	249	60.9
10^{0}	51.8	22.6	108.8	7.36	256	192.5
10^{1}	51.6	22.6	344.1	7.33	257	608.6
10^{2}	51.8	22.6	1088.2	7.21	255	1924.6

[a] A_o = cross-sectional area used in Table 10-22.
[b] $\overline{L}$ = half of curved length of arch.
[c] r = radius of gyration.

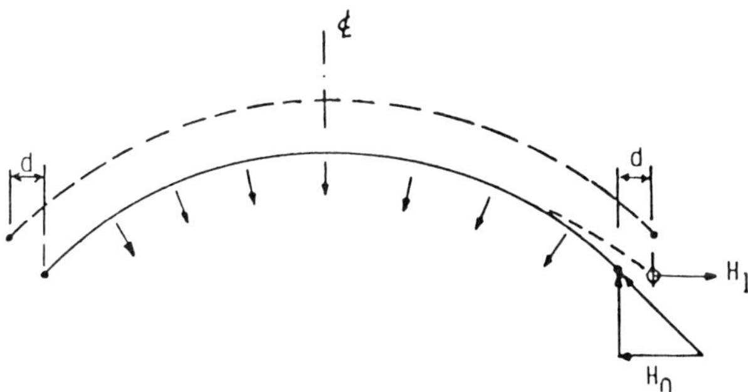

FIGURE 10-85 Effect of axial flexibility. (From Wen and Lange, 1981.)

proportional to A, it follows that

$$H_1 = h_A d + h_1 d = c_1 A(c_2/A) + h_1(c_2/A) = c_3 + c_4 A$$

where all c factors are constant. When A is large, H_1 is basically independent of A. When A is small, H_1 increases with a decrease in A, but an increase in H_1 decreases the total horizontal reaction H, thereby lessening the tendency to buckle with an increase in the buckling load. Values of the horizontal reaction H are shown in Table 10-23.

Summary and Comments Unlike the classical theory based on undeformed geometry where prebuckling deformation is neglected, the present finite-element analysis considers the prebuckling history of the arch, and this deformation is taken linearly proportional to the load. The results appear to indicate that where little prebuckling bending is involved, the buckling loads calculated in this manner are quite accurate. For systems involving substantial prebuckling bending, the procedure will produce uncertain results.

For problems within the domain of the classical theory of elastic stability, the inclusion of rotation terms in $[n1]$ is essentially irrelevant because these rotations are negligible. If the rotations are large, the assumption of a linear relationship between displacement and load may render this method invalid because the actual behavior is nonlinear.

A nonshallow arch with a concentrated load at the crown has a bifurcation load lower than the limit load. In this case the prebuckling deformation is smaller at the former load than at the latter load. At the former load the buckling mode is antisymmetric, but at the latter it is symmetric. For the antisymmetric bifurcation buckling load, the effect of rotation terms in $[n1]$ was sufficient to invalidate the results, but the same prebuckling history may not be large enough to invalidate the classical theory of elastic stability. Interestingly, the classical theory in this case does yield accurate results (see also the foregoing analysis by Austin and Ross, 1976) as does the present procedure of using $[n1^*]$.

For the symmetric buckling mode, the limit load obtained by the classical theory produces errors probably because of significant buckling deformations. For such problems the buckling loads are better obtained by nonlinear equilibrium analysis for a series of load increments testing also for possible instability along the solution path. Both $[n1]$ and $[n2]$ should be used in this case and all terms kept to a minimum unless otherwise indicated.

10-20 DESIGN OF ARCHES BY THE COMPLEX METHOD

Lipson and Haque (1980) present results of optimal arch design using an automated design routine. The free variables of the design space include both the geometric (arch shape) and member-sizing variables (cross-sectional dimensions). The design routine allows multiple-load cases, and provides the optimal arch shape under realistic restrictions on allowable stresses and practical design loads.

The relevant variables are shown in Figure 10-86 for an arch constructed from M straight segments. The variables x_j and y_j denote the coordinates of a typical joint j. A typical member m has a cross section consisting of a symmetrical box whose dimensions are as shown in section A–A. Only symmetrical arches with constant member depth and width along the entire are length are considered.

Optimization Procedure The constrained, nonlinear optimization problem associated with the optimal design is considered in a modified version of the complex method proposed by Box (1965), applied in structural optimization cases (Fu and Levey, 1973; Lipson and Agrawal, 1974; Lipson and Gwin, 1977; Lipson and Russell, 1971).

The optimization process is divided into two levels of analysis. In the first level, the y coordinates, the depth d, and the width b of members are found using the modified version of the complex method. In the second level, the structure is designed for member size (the dimensions t_m and w_m are selected) using the fully stressed optimality criterion. During this level, the skeletal geometry of the structure is regarded as fixed.

The complex method of Box was originally introduced as a mathematical programing scheme to optimize a nonlinear function of variables. It consists of generating a number of feasible points in the design space and then improving these points by successively reflecting the worst of the points through the centroid of the remaining points. The reflecting procedure is repeated until a termination criterion is reached.

Lipson and Haque (1980) have modified the method as follows.

1. An initial point in the design space is chosen. If the point is not feasible, it is made feasible by adjusting one or more coordinates of the design vector.

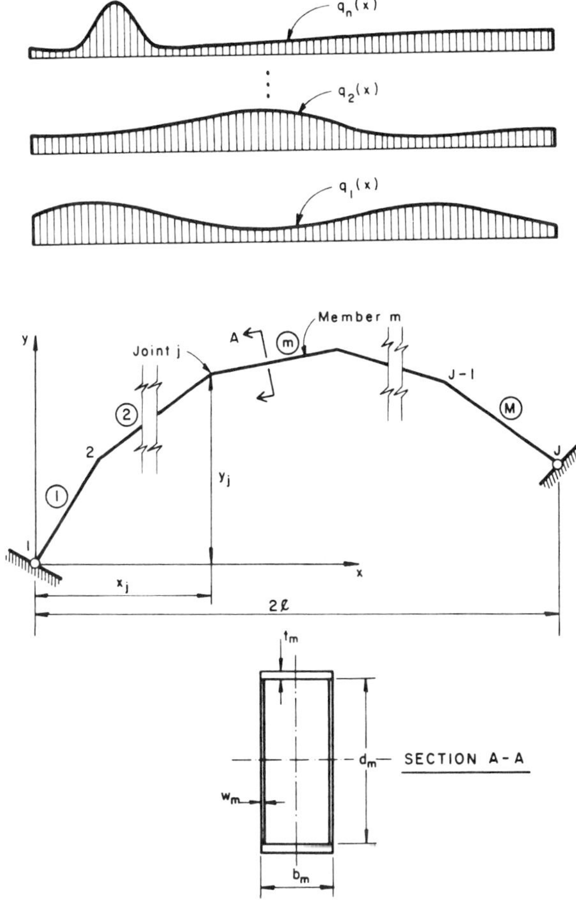

FIGURE 10-86 Variables and definition sketch of arches.

2. As many as $(k - 1)$ points are generated randomly in the interval 0–1, and the procedure ensures that all these points satisfy the explicit constraints. If v_{ip} is the ith coordinate in the pth design vector, then

$$z_{ip} = z_i^L + e_{ip}\left(z_i^U - z_i^L\right) \qquad (i = 1, 2, \ldots, N \qquad p = 1, 2, \ldots, k) \tag{10-112}$$

3. The point in the complex with the worst value of the objective function is identified and reflected through the centroid of all points in the complex excluding the worst point. The reflected distance is set equal to the distance between the worst point and the centroid, multiplied by a factor a. If the reflected point has a function value worse than the worst point, it is moved halfway to the centroid. If the new point shows no improvement, the centroid is used. If the centroid shows no improvement, a new point is generated as in step 2.

4. The procedure in step 3 is repeated until a present termination criterion is reached. This is the case if either the best and worst function values are within β units in γ successive iterations; or if the standard deviation of all function values is less than a value δ_1; or if the mean radius of all the points from the centroid is less than a value δ_2; or if the number of iterations reaches a number δ_3. All the parameters β, γ, δ_1, δ_2, and δ_3 are specified constants.

The design space may be nonlinear, nonconvex, and irregular, and may contain discontinuities. The complex method appears to operate in all such spaces and does not require calculation of partial derivatives.

Check Problem The reliability and performance of this computational approach in optimizing the design of an arch is checked from an optimal solution obtained analytically. The arch is subjected to a uniform load and has a parabolic configuration and uniform cross section. Results obtained analytically are shown in Figure 10-87*a* for the optimum parabolic arch. The exact solution for the optimal shape is presented in Figure 10-87*b*.

The nondimensional arc length, c/l, and the nondimensional axial thrust, P/wl, at the support are plotted versus the nondimensional arch rise, a/l. As the height of the arch approaches zero ($a/l \rightarrow 0$), the arc length approaches the span of the arch, whereas the axial thrust P approaches infinity. For relatively high arches ($a/l \rightarrow \infty$), the axial thrust approaches half of the total uniform load and the arc length goes to infinity. Because the maximum axial thrust in a parabolic arch occurs at the supports, the minimum uniform cross section of the arch to sustain the external load is

$$A = P/\sigma \tag{10-113}$$

where A is the cross-sectional area and σ is the allowable stress. The minimum volume V of the arch with a uniform cross section may be expressed as

$$V = (Pc)/\sigma \tag{10-114}$$

Dividing both sides of (10-114) by wl^2/σ gives

$$\frac{V}{(wl^2)/\sigma} = \left(\frac{P}{wl}\right)\left(\frac{\sigma}{l}\right) \tag{10-115}$$

where all notation conforms to Figure 10-87, and w is the rate of uniformly distributed load (force/length). The quantities in parentheses represent the nondimensional axial thrust and the nondimensional arc length, whereas the left side is a dimensionless volume factor. The variation of the volume factor of parabolic arches of uniform cross section is shown by the solid curve in Figure 10-87*b*. The minimum weight of the parabolic arch occurs at a rise

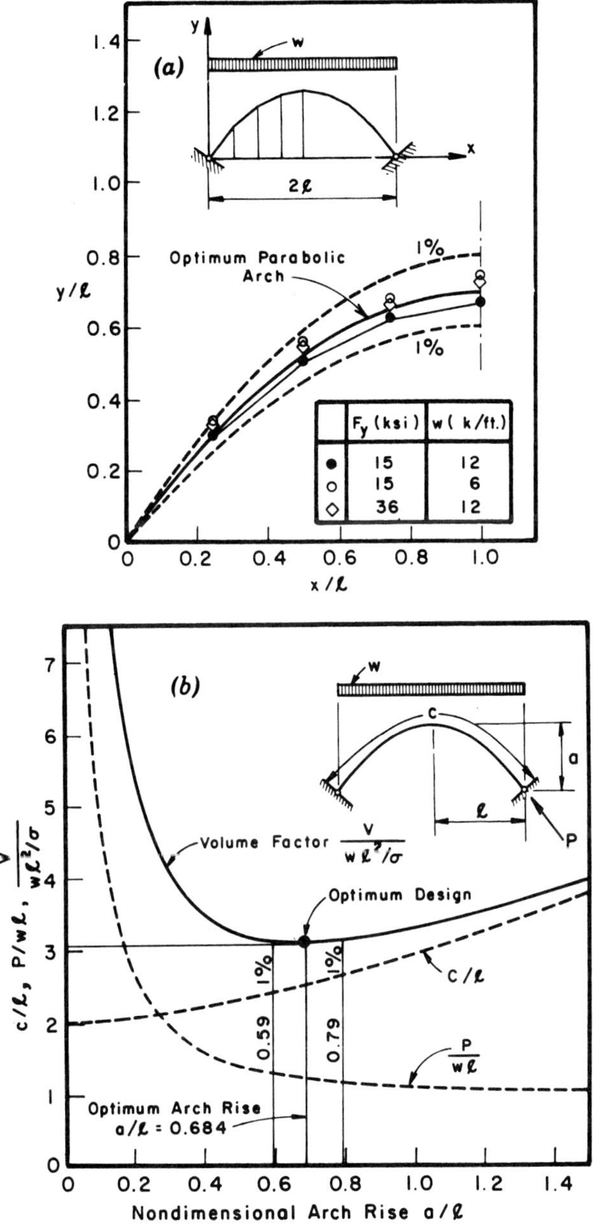

FIGURE 10-87 Comparison of computed optimal arch shapes with optimum parabolic arch (1 ksi = 6.89 MPa, 1 kip/ft = 14.58 kN/m).

$a/l = 0.684$, and the minimum volume factor is 3.113. For this simple case, the volume factor is insensitive to changes in the arch rise near the optimum solution. For example, changing the rise of the arch from the optimum rise of 0.684 to $a/l = 0.79$ raises the optimal weight of the structure by only 1 percent.

For the check problem, the optimal arch shapes were also numerically obtained using the present automated design routine. In this study the arch is represented by eight straight members having their cross-sectional dimensions linked in order to obtain a constant cross section throughout the arc length. The results are shown in Figure 10-87*a*. The curves labeled as 1 percent represent the parabolic arches whose optimal weights are 1 percent heavier than the global minimum. It appears that any parabolic arch shape falling in the shaded area whose cross section is optimally designed will be at most 1 percent heavier than the global minimum corresponding to a parabolic arch with $a/l = 0.684$. It follows therefore that even with an eight-straight-member approximation to the smooth arch curve, the present procedure gives optimal arch shapes within 1 percent of the global minimum. Lipson and Haque (1980) applied these conclusions to a parametric study where optimal designs were analyzed using eight-segment, piecewise straight arches under varied load conditions and span lengths.

Results of Parametric Study In a parametric study of optimal arch design, symmetrical, two-hinged arches with piecewise straight segments were investigated. Material properties are: $E = 29{,}000$ ksi and $F_y = 36$ ksi. The results are discussed and summarized by Lipson and Haque (1980). The following conclusions are appropriate, based on this study.

1. Under uniformly distributed load, the optimal shape is close to a parabolic curve with an arch rise 0.342 times the span length. A slight deviation from this rise (within 10 percent) does not materially increase the optimal weight of the structure.

2. Under two load cases consisting of (a) a uniformly distributed load and (b) an asymmetrical load where half the arch is uniformly loaded at three times the rate at which the other half is loaded, the optimal arch rise is much less than 0.342 times the span length. However, the optimal weight does not significantly increase in the range of the optimal arch rise (within 18 percent). Even for the asymmetrical load case, optimal shapes may be reasonably approximated by parabolic curves, within 5 percent.

3. As the loading deviates further from the uniform load pattern, thereby producing higher bending moments in the arch, the optimal arches become shallower, whereas the arch heights at the 1/4 and 3/4 span lengths tend to increase above the parabolic curve passing through the crown and arch supports.

10-21 RELEVANT SPECIFICATIONS

Standard AASHTO Specifications Under combined flexure and axial load, the strength of a concrete arch ring must conform to the provisions applicable to compression members and slenderness effects (strength design method, Articles 8.16.4 and 8.16.5). In AASHTO equation (8-41), C_m should be taken as 1, and ϕ should be taken as 0.85. The shape of the arch rings should conform as nearly as practicable to the equilibrium polygon for full dead load. Values of the effective length factor K are given for specific ranges of rise–span ratios for the three basic support conditions, that is, the three-hinged arch, two-hinged arch, and fixed arch.

For service load design, solid-rib steel arches are discussed in Article 10-37, providing stipulations for moment amplification and allowable stresses, buckling in the vertical plane, effects of lateral slenderness, and web and flange plate requirements.

Likewise, for strength design method, solid-rib steel arches are discussed in AASHTO Article 10-55, with provisions for moment amplification and allowable stresses and web and flange plate requirements.

Proposed LRFD Specifications As part of the structural analysis and evaluation, consideration should be given to the reduction or elimination of the effects of rib shortening through camber or special construction techniques. For arches with longer spans, large-deflection analysis should be considered in lieu of moment magnification procedures. The underlying principle is that any single-step correction factor should not be expected to accurately model deflection effects over a wide range of stiffness.

Where the distribution of stresses between the top and bottom chords of truss arches is dependent on the manner of erection, the association procedure should be explicitly indicated.

Concrete Arches Stability under long-term loads with a reduced modulus of elasticity may govern. In this state the bending moment in the arch ribs would be small, the elastic modulus would be the long-term tangent modulus, and the relevant moment of inertia would be the transformed section inertia.

In lieu of a rigorous analysis, the effective length for buckling may be estimated as the product of the arch half-span length and appropriate length factors given in the AASHTO specifications. When the approximate second-order correction from moment is applied, the short-term secant modulus may be calculated based on a strength of $0.40f'_c$. Other requirements are as follows.

- Arch ribs should be reinforced as compression members. The minimum 1.0 percent reinforcement should be evenly distributed about the section of the rib, and confinement reinforcement should be included as in columns.

- Unfilled spandrel sections greater than 25 ft in height should be braced with counterforts or diaphragms.
- Spandrel walls should have expansion joints, with appropriate temperature reinforcement according to the joint spacing. The spandrel wall should be jointed at the springline.

Steel Arches Web stability requirements must satisfy conditions as follows: (a) webs with no longitudinal stiffener and (b) webs with one longitudinal stiffener. Likewise, flange stability requirements establish limits to the width–thickness ratio of the flanges.

Slenderness Effects The Rainbow Bridge discussed in Section 10-2 and shown in Figure 10-7 incorporates slenderness features that prompted analyses allowing for the effect of arch deflections in modifying design moments. These were checked by a program of model tests.

Load Factors A conservative estimate of collapse load may be obtained assuming that it occurs at first yield. A suggested procedure for estimating the load factor on live load is as follows.

1. Select a desired load factor for dead load. Apply this factor to dead loads for all subsequent analyses.
2. Make an assessment of the worst unstressed imperfect shape. Use the methods presented in Section 10-8 to obtain the corresponding ultimate dead load stresses f_{DU}.
3. Apply the design live load, producing live load stresses f_L.
4. At the worst section, consider the yield stress $\sigma_y = f_{DU} + Lf_L$, where L is the load factor on live load. Then

$$L = (\sigma_y - f_{DU})/f_L \tag{10-116}$$

Plastic collapse of steel arches is discussed by Lowe (1961) and Stevens (1957). The collapse of concrete arches is yet to be treated comprehensively. The behavior of slender columns is considered in Section 10-10 as well as in other sections.

REFERENCES

Abdel-Rohman, M., 1976. "Contribution to the Analysis of Elastic-Plastic Arches," Thesis, Univ. Waterloo, Ontario, Canada.

Architectural Forum, 1954. "Rio Bianco Bridge," April, pp. 166–167.

Asplund, S. O., 1961. "Deflection Theory of Arches," *J. Struct. Div. ASCE*, Vol. 87, No. ST7, Oct., pp. 125–149.

Asplund, S. O., 1966. *Structural Mechanics: Classical and Matrix Methods*, Prentice-Hall, Englewood Cliffs, N.J.

Austin, W. J., 1971. "In-Plane Bending and Buckling of Arches," *J. Struct. Div. ASCE*, Vol. 97, No. ST5, May, pp. 1575–1592.

Austin, W. J. and Ross, T. J., 1976. "Elastic Buckling of Arches under Symmetrical Loading," *J. Struct. Div. ASCE*, Vol. 102, No. ST5, May, pp. 1085–1095.

Austin, W. J., T. J. Ross, A. S. Tawfik, and R. O. Volz, 1982. "Numerical Bending Analysis of Arches," *J. Struct. Div. ASCE*, Vol. 108, No. ST4, April, pp. 849–868.

Balog, L., 1956: "Truss-tied Arch Bridge has a Long History," *Civ. Eng.* (N.Y.), 26(12), 69, 70 (Dec.)

Baxter, J. W., A. F. Gee, and H. B. James, 1965. "Gladesville Bridge," *Proc. Institution Civ. Engineers*, London, No. 30, March, pp. 489–530.

Billington, D. O., 1973. "Deck Stiffened Arch Bridges of Robert Maillart," *J. Struct. Div. ASCE*, Vol. 99, No. ST7, July, pp. 1527–1539.

Bouchet, A., 1964. "Le Pont de Fehmarsund, élement de la nouvelle liaison Hambourg–Copenhague," *Technique des Travaux*, Vol. 40, Nos. 1–2, Jan.–Feb., pp. 56–64.

Box, M. J., 1965, "A New Method of Constrained Optimization and Comparison with Other Methods," *Computer J.*, Vol. 8, April, pp. 42–52.

Carpenter, S. T., 1960. *Structural Mechanics*, Wiley, New York, 538 pp.

Chandrangsu, S., and S. R. Sparkes, 1954. "A Study of the Bowstring Arch Having Extensible Suspension Rods and Different Ratios of Tie-Beam to Arch-Rib Stiffness," *Proc. Institution Civ. Engineers*, London, Vol. 3, No. 2, Aug., pp. 515–563.

Chang, C. K., 1973. "Effect of Loaded Length on the Buckling Strength of Slender Arches," Thesis, Rice Univ., Houston.

Clark, J. G., 1950. *Welded Deck Highway Bridges*, J. F. Lincoln Arc Welding Foundation, Cleveland.

Cohn, M. Z. and M. Abdel-Rohman, 1976. "Analysis Up to Collapse of Elastic-Plastic Arches," *J. Computers and Structures*, Vol. 6, pp. 511–517.

Coronforth, R. C. and S. B. Childs, 1967. "Computer Analysis of Two-Hinged Circular Arches," *J. Struct. Div. ASCE*, Vol. 93, No. ST2, Feb., pp. 319–338.

DaPeppo, B. A. and R. Schmidt, 1969a. "Nonlinear Analysis of Buckling and Postbuckling Behavior of Circular Arches," *J. Appl. Mech. and Phys.*, Vol. 20, No. 6, pp. 847–857.

DaDeppo, D. A. and R. Schmidt, 1969b, "Sidesway Buckling of Deep Circular Arches Under a Concentrated Load," *J. Appl. Mech. ASME*, Vol. 36, No. 2, June, pp. 325–327.

DaDeppo, D. A. and R. Schmidt, 1971. "Stability of Heavy Circular Arches with Hinged Ends," *J. American Inst. Aeronautics and Astronautics*, Vol. 9, No. 6, June, pp. 1200–1201.

DaDeppo, D. A., and R. Schmidt, 1974. "Stability of Two-Hinged Circular Arches with Independent Loading Parameters," *American Inst. of Aeronautics and Astronautics J.*, Vol. 12, No. 3, March, pp. 385–386.

Dawe, D. J., 1971. "A Finite-Deflection Analysis of Shallow Arches by the Discrete Element Method," *Intern. J. Numerical Methods in Eng.*, Vol. 3, pp. 529–552.

Demuts, E., 1969. "Lateral Buckling of Circular Arches Subjected to Uniform Gravity Type Loading," Dissertation, Ohio State Univ., Columbus.

Dinnik, A. N., 1955. *Buckling and Torsion*, Moscow.

Ebner, A. M. and J. J. Ucciferro, 1972. "A Theoretical and Numerical Comparison of Elastic Nonlinear Finite Element Methods," *J. Computers and Structures*, Vol. 2, pp. 1043–1061.

Freyssinet, E., 1966. "Eugene Freyssinet by Himself," *A Half Century of French Prestressing Technology*, Travaux, Paris, p. 10.

Freyssinet, E., J. Muller, and R. Shama, 1953. "Largest Concrete Spans of the Americas," *Civ. Eng.*, March, Vol. 23, No. 3, pp. 41–55.

Fu, K. C. and G. E. Levey, 1973. "Discrete Frame Optimization by Complex-Simplex Procedure," *J. Computers and Structures*, Vol. 9, pp. 207–217.

Fukasawa, Y., 1963. "Buckling of Circular Arches by Lateral Flexure and Torsion under Axial Thrust," *Trans. Japan Soc. Civ. Eng.*, No. 96, pp. 29–47.

Galli, A. and G. Franciosi, 1955. "Limit Analysis of Thin Arch Bridges with Stiffening Girders," *G. Genio Civ.*, Vol. 93, No. 11.

Galambos, T. V., 1988. *Guide to Stability Design Criteria for Metal Structures*, Wiley, New York.

Garrelts, J. M., 1943. "St. Georges Tied Arch Span," *Trans. ASCE*, Vol. 69, No. 108, Paper No. 2188, pp. 543–554.

Giammatteo, M. and G. C. Gavarini, 1972. "Limit Analysis of Bridges with Arch-Girder Interaction," *Giornale Genio Civile*, Rome, Vol. 11a, Nos. 1–2–3.

Godden, W. G., 1954. "The Lateral Buckling of Tied Arches," *Proc. Institution Civ. Engineers*, London, Vol. 3, No. 2, Aug., pp. 496–514.

Godden, W. G. and J. C. Thomson, 1959. "Experimental Study of Model Tied-Arch Bridge," *Proc. Institution Civ. Engineers*, London, Vol. 14, Dec., pp. 383–394.

Greenberg, H. J. and W. Prager, 1952. "On Limit Design of Beams and Frames," *Trans. ASCE*, Vol. 117, Paper No. 2501, pp. 447–458.

Hall, A. S. and R. W. Woodhead, 1961. *Frame Analysis*, Wiley, New York, 247 p.

Hardesty, S., et al., 1945. "Rainbow Arch Bridge Over Niagara Gorge—A Symposium," *Trans. ASCE*, 110, 2–178.

Harrison, H. B., 1982. "In-Plane Stability of Parabolic Arches," *J. Struct. Div. ASCE*, Vol. 109, No. ST1, Jan., pp. 195–205.

Harrison, H. B., 1982. "In-Plane Stability of Parabolic Arches," *J. Struct. Div. ASCE*, Vol. 108, No. ST1, Jan., pp. 195–205.

Hartwig, H. J., 1965. "Die Kaiserleibrücke," *Stahlbau*, Vol. 34, No. 4, April, pp. 97–110.

Hazelet, C. P. and R. H. Wood, 1961. "Six-Line Tied-Arch Bridge Across Ohio River," *Civ. Eng.*, N.Y., Vol. 31, No. 11, Nov., pp. 43–47.

Hirsch, E. G. and E. P. Popov, 1955. "Analysis of Arches by Finite Differences," *Proc. ASCE*, Vol. 18, No. 829, Nov., 18 pp.

Hognestad, E., N. W. Hanson, and D. McHenry, 1955. "Concrete Stress Distribution in Ultimate Strength Design," *Proc. A.C.I.*, 52(4), 455–480 (Dec.)

Huddleston, J. V., 1968. "Finite Deflections and Snap-Through of High Circular Arches," *J. Appl. Mech. ASME*, Vol. 35, No. 4, Dec., pp. 763–769.

Huddleston, J. V., 1971. "Nonlinear Analysis of Steep, Compressible Arches of Any Shape," *J. Appl. Mech. ASME*, Vol. 38, No. 4, Dec., pp. 942–946.

Jennings, A., 1968. "Frame Analysis Including Change of Geometry," *J. Struct. Div. ASCE*, Vol. 94, No. ST3, March, pp. 627–644.

Kaldjian, M. J., 1961. "Prestressed Bowstring Arch," *J. Struct. Div. ASCE*, Vol. 87, No. ST7, Oct., pp. 1–29.

Kee, C. F., 1961. "Lateral Inelastic Buckling of Tied Arches," *J. Struct. Div. ASCE*, Vol. 87, No. ST1, pp. 23–39.

Kavanagh, T. C., 1972. "Cable-Supported Bridges," *Struct. Steel Designers' Handbook*, F. S. Merritt, ed., McGraw-Hill, New York, pp. 14–71.

Kavanagh, T. C., 1974. "Deck Stiffened Arch Bridges by Robert Maillart," *ASCE J. of Str. Div.*, Vol. 100, No. ST4, April, pp. 829–832.

Klöppel, K. and W. Protta, 1961. "A Contribution to the Buckling Problem of Circular Curved Bars," *Stahlbau*, No. 30, pp. 1–15.

Komatsu, S. and T. Sakimoto, 1977. "Ultimate Load Carrying Capacity of Steel Arches," *J. Struct. Div. ASCE*, Vol. 103, No. ST12, Dec., pp. 2323–2336.

Komatsu, S. and T. Shinke, 1977. "Practical Formulas for In-Plane Load Carrying Capacity of Arches," *Proc. Japan Soc. Civ. Eng.*, No. 267, pp. 39–51.

Kuesel, T. R., 1974. "Deck Stiffened Arch Bridges by Robert Maillart," *ASCE, J. Str. Div.*, Vol. 100, No. ST4, April, pp. 832–833.

Kuranishi, S., 1961a. "The Torsional Buckling Strength of Solid Rib Arch Bridge," *Trans. Japan Soc. Civ. Eng.*, No. 75, pp. 59 67.

Kuranishi, S., 1961b. "Analysis of Arch Bridge Under Certain Lateral Forces," *Trans. Japan Soc. Civ. Eng.*, No. 73, pp. 1–6.

Kuranishi, S., 1973. "Allowable Stress for Two-Hinged Steel Arch," *Proc. Japan Soc. Civ. Eng.*, No. 213, pp. 71–75.

Kuranishi, S. and T. Yabuki, 1977. "In-Plane Strength of Arch Bridges Subjected to Vertical and Lateral Loads," Second Intern. Colloq. Stab. Steel Struct., Prelim. Report, Liège, France, April, pp. 551–556.

Kuranishi, S. and T. Yabuki, 1979. "Some Numerical Estimations of Ultimate In-Plane Strength of Two-Hinged Steel Arches," *Proc. Japan Soc. Civ. Eng.*, No. 287, pp. 155–158.

Kuranishi, S. and T. Yabuki, 1981. "Required Out-of-Plane Rigidities of Steel Arch Bridges with Two Main Ribs Subjected to Vertical and Lateral Loads," Tech. Report, Tohoku Univ., Sendai, Vol. 46, No. 1.

Kuranishi, S. and T. Yabuki, 1984a. "Ultimate Strength Design Criteria for Two-Hinged Steel Arch Structures," *Proc. Japan Soc. Civ. Eng.*, No. 3005/I-2.

Kuranishi, S. and T. Yabuki, 1984b. "Lateral Load Effect on Arch Bridge Design," *J. Struct. Div. ASCE*, Vol. 110, No. 9, pp. 2263–2274.

Langhaar, H. L., A. P. Boresi, and D. R. Carver, 1954. "Energy Theory of Buckling of Circular Elastic Rings and Arches," *Proc. Second U.S. National Congress Appl. Mech.*, p. 437–443.

Lind, N. C., 1962. "Elastic Buckling of Symmetrical Arches," Eng. Experiment Station, Tech. Report 3, Univ. Illinois, Urbana, 33 pp.

Lind, N. C., 1966. "Numerical Buckling Analysis of Arches and Rings," *Struct. Engineer*, Vol. 44, No. 7, July, pp. 245–248.

Lipson, S. L. and A. D. Russell, 1971. "Cost Optimization of Structural Roof System," *J. Struct. Div. ASCE*, Vol. 97, No. ST8, Proc. Paper 8930, Aug., pp. 2057–2071.

Lipson, S. L. and Agrawal, K. M., 1974. "Weight Optimization of Plane Trusses," *J. Struct. Div. ASCE*, Vol. 100, No. ST5, Proc. Paper 10521, May, pp. 865–880.

Lipson, S. L. and L. B. Gwin, 1977. "The Complex Method Applied to Optimal Truss Configuration," *Computers and Structures*, Vol. 7, pp. 461–468.

Lipson, S. L., and L. B. Gwin, 1977. "Discrete Sizing of Trusses for Optimal Geometry," *J. Struct. Div. ASCE*, Vol. 103, No. ST5, Proc. Paper 12936, May, pp. 1031–1046.

Lipson, S. L. and M. I. Haque, 1980. "Optimal Design of Arches Using the Complex Method," *J. Struct. Div. ASCE*, Vol. 106, No. ST12, Dec., pp. 2509–2525.

Louis, H., P. Guiaux, and E. Mas, 1962. "Experimental Investigation of End Stiffening of Bowstring or Vierendeel Type Bridge Girder," *Acier–Stahl–Steel*, Vol. 27, No. 3, March, pp. 119–125.

Lowe, P. G., 1961. "Effect of Axial Thrust on Carrying Capacity of Mild Steel Arch Rib," Inst. Eng. Australia, *Civ. Eng. Trans.*, CE3(2), Sept., pp. 95–101.

Maillart, R., 1930. "Note on Arch Bridges in Switzerland," *Proc. First Intern. Cong. for Concrete and Reinforced Concrete*, Liège, Belgium, Sept.

Maillart, R., 1931. "Leichte Eisenbeton-Brucken in der Schweiz, (Light Reinforced-Concrete Bridges in Switzerland), *Der Bauingenieur*, Berlin, Vol. 10.

Mallett, R. H., and P. V. Marcal, 1968: "Finite Element Analyses of Nonlinear Structures," *J. Struct. Div. ASCE*, Vol. 94, No. ST9, Sept. pp. 2081–2105.

Mak, C. K. and D. Kao, 1973. "Finite Element Analysis of Buckling and Post-Buckling Behavior of Arches with Geometric Imperfections," *J. Computers and Structures*, Vol. 3, pp. 149–161.

Mallett, R. H. and P. V. Marcal, 1968. "Finite Element Analyses of Nonlinear Structures," *J. Struct. Div. ASCE*, Vol. 94, No. ST9, Sept., pp. 2081–2105.

McCullough, C. B. and E. S. Thayer, 1931. *Elastic Arch Bridges*, Wiley, New York.

Miklogsky, H. A. and O. J. Sotillo, 1957, "Design of Flexible Steel Arches by Interaction Diagrams," *J. Struct. Div. ASCE*, Vol. 83, ST2, March, 35 pp.

Morris, C. T., et al., 1935. "Final Report of the Special ASCE Committee on Concrete and Reinforced Concrete Arches," *Trans. ASCE*, Vol. 100, Paper No. 1922, pp. 1427–1531.

Müller-Breslau, H., 1922. *Die Graphische Statik der Baukonstruktionen*, 5th ed., Vol. II, Part 1, Alfred Kroner Verlag, Stuttgart, Germany, pp. 274–294.

Müller-Breslau, H., 1925. *Die Graphische Statik der Baukonstruktionen*, 2nd ed., Vol. II, Part 2, Alfred Kroner Verlag, Leipzig, Germany, pp. 257–259.

Namita, Y., 1968. "Second Order Theory of Curved Bars and Its Use in the Buckling Problem of Arches," *Trans. Japan Soc. Civ. Eng.*, No. 155, pp. 32–41.

Newmark, N. M., 1943. "Numerical Procedures for Computing Deflections, Moments, and Buckling Loads," *Trans. ASCE*, Vol. 108, Paper No. 2202, pp. 1161–1234.

O'Connor, C., 1971. *Design of Bridge Superstructures*, Wiley, New York.

Ojalvo, M., E. Demuts, and F. J. Tokarz, 1969. "Out-of-Plane Buckling of Curved Members," *J. Struct. Div. ASCE*, Vol. 95, No. ST10, pp. 2305–2316.

Ojalvo, I. U., and M. Newman, 1968. "Buckling of Naturally Curved and Twisted Beams," *J. Eng. Mech. Div. ASCE*, Vol. 94, No. EM5, Oct., pp. 1067–1087.

Onat, E. T. and W. Prager, 1953, "Limit Analysis of Arches," *J. Mech. Phys. Solids*, Vol. 1, pp. 77–89.

Oran, C. and R. S. Reagan, 1969. "Buckling of Uniformly Compressed Circular Arches," *J. Eng. Mech. Div. ASCE*, Vol. 95, No. EM4, Aug., pp. 879–895.

Ostenfeld, A., 1925. *Teknisk Statik*, 3rd ed., Vol. II, Jul. Gjellumps Boghanel, Copenhagen, Denmark, p. 225.

Ostlund, L., 1954. "Lateral Stability of Bridge Arches Braced with Transverse Bars," *Trans. Royal Inst. Tech.*, Stockholm, No. 84.

Ronca, P. and M. Z. Cohn, 1979. "Limit Analysis of Reinforced Concrete Arch Bridges," *J. Struct. Div. ASCE*, Vol. 105, No. ST2, Feb., pp. 313–326.

Ronca, P., 1977. "Limit Analysis of Reinforced Concrete Arch Bridges," Thesis, Univ. Waterloo, Ontario, Canada.

Ross, T. J., 1973. "Numerical Large Deflection Bending and Buckling Analysis of Arches," Thesis, Rice Univ., Houston.

Sabir, A. B. and A. C. Lock, 1973. "Large Deflection, Geometrically Nonlinear Finite Element Analysis of Circular Arches," *Intern. Journ. Mech. Sci.*, Vol. 15, pp. 37–47.

Sakimoto, T. and S. Komatsu, 1977a. "A Possibility of Total Breakdown of Bridge Arches due to Buckling of Lateral Bracing," Second Intern. Colloq. Stab. Steel Struct., Final Report, Liège, April, pp. 299–301.

Sakimoto, T. and S. Komatsu, 1977b. "Ultimate Load Carrying Capacity of Steel Arches with Initial Imperfections," Second Intern. Colloq. Stab. Steel Struct., Prelim. report, Liège, April, pp. 545–550.

Sakimoto, T. and S. Komatsu, 1979. "Ultimate Strength of Steel Arches under Lateral Loads," *Proc. Japan Soc. Civ. Eng.*, No. 292, pp. 83–94.

Sakimoto, T. and S. Komatsu, 1982. "Ultimate Strength of Arches with Bracing Systems," *J. Struct. Div. ASCE*, Vol. 108, No. ST5, pp. 1064–1076.

Sakimoto, T. and S. Komatsu, 1983a. "Ultimate Strength Formula for Steel Arches," *J. Struct. Div. ASCE*, Vol. 109, No. 3, pp. 613–627.

Sakimoto, T. and S. Komatsu, 1983b. "Ultimate Strength Formula for Central Arch Girder Bridges, *Proc. Japan Soc. Civ. Eng.*, No. 333, pp. 183–186.

Sakiomot, T. and Y. Yamao, 1983. "Ultimate strength of Deck Type Steel Arch Bridges," Third Intern. Colloq. Stab. Metal Struct., Prelim. Report, Paris, Nov.

Sakimoto, T., T. Yamao, and S. Komatsu, 1979. "Experimental Study on the Ultimate Strength of Steel Arches," *Proc. Japan Soc. Civ. Eng.*, No. 286, pp. 139–149.

Schmidt, R. and D. A. DaDeppo, 1971. "A Survey of Literature on Large Deflections of Nonshallow Arches, Bibliography of Finite Deflections of Straight and Curved Beams, Rings, and Shallow Arches," *J. Industrial Mathematics*, Industrial Mathematics Soc., Vol. 21, Part 2, pp. 91–114.

Schmidt, R. and D. A. DaDeppo, 1972. "Buckling of Clamped Circular Arches Subjected to a Point Load," *J. of Applied Mathematics and Physics*, Vol. 23, pp. 146–148.

Scott, P. A. and G. Roberts, 1958. "The Volte Bridge," *Proc. Institution of Civ. Engineers*, London, No. 9, p. 395, April.

Shinke, T., H. Zui, and T. Nakagawa, 1980. "In-Plane Load Carrying Capacity of Two-Hinged Arches with a Stiffening Girder," *Trans. Japan Soc. Civ. Eng.*, No. 301, pp. 47–59.

Shinke, T., H. Zui, and Y. Namita, 1975. "Analysis of In-Plane Elasto-Plastic Buckling and Load Carrying Capacity of Arches," *Proc. Japan Soc. Civ. Eng.*, No. 244, pp. 57–69.

Shukla, S. N. and M. Ojalvo, 1971. "Lateral Buckling of Parabolic Arches with Tilting Loads," *J. Struct. Div. ASCE*, Vol. 97, No. ST6, pp. 1763–1773.

Sorgente, V., 1957. "Experimental Results on a Stiffened Arch Bridge," *Giornale Genio Civile*, Rome, Vol. 95, Dec.

"Spectacular Venezuelan Concrete Arch Bridge," *Eng. News-Record*, 149(11), 28–32 (Sept. 11, 1952).

Stein, P. and H. Wild, 1965. "Das Bogentragwerk der Fehmarnsundbruecke," *Stahlbau*, Vol. 34, No. 6, June, pp. 171–186.

Stevens, L. K., 1957. "Carrying Capacity of Mild Steel Arches," *Proc. Institution Civ. Engineers*, London, No. 6, March, pp. 511–693.

Stüssi, F., 1943. "Lateral Buckling and Vibration of Arches," *Proc. IABSE Publ.*, Vol. 7, pp. 327–343.

Tadjbakhsh, I. and M. Farshad, 1973. "On Conservatively Loaded Funicular Arches and Their Optimal Design," *Proc. IUTAM Symp. Struct. Optimization*, Warsaw, pp. 215–218.

Timoshenko, S. P. and J. M. Gere, 1961. *Theory of Elastic Stability*, McGraw-Hill, New York.

Tokarz, F. J., 1971. "Experimental Study of Lateral Buckling of Arches," *J. Struct. Div. ASCE.*, Vol. 97, No. ST2, Feb., pp. 545–559.

Tokarz, F. J. and R. S. Sandhu, 1972. "Lateral-Torsional Buckling of Parabolic Arches," *J. Struct. Div. ASCE*, Vol. 98, No. ST5, May, pp. 1161–1179.

Upstone, T. J. and W.N. Cardno, 1954. "The Design and Construction of the Superstructure of the Marshal Carmona Bridge at Vila Franca de Xira, Portugal," *Proc. Inst. Civ. Eng.* (London) 3 (pt. 3, No. 3), 695–737 (Dec.).

Van der Woude, F. and B. F. Cousins, 1978. "Deformation of Arches: Linear Elastic Behavior," Research Report CM-78/3, Univ. of Tasmania, Hobart, Australia.

Van der Woude, F. and B. F. Cousins, 1979. "Deformation of Arches: Elastic Buckling Behavior," *J. Struct. Div. ASCE*, Vol. 105, No. ST12, Dec., pp. 2677–2694.

Waestlund, G., 1960. "Stability Problems of Compressed Steel Members and Arch Bridges," *J. Struct. Div. ASCE*, Vol. 86, No. ST6, June, pp. 47–71.

Walker, A. C., 1969. "A Nonlinear Finite Element Analysis of Shallow Arches," *Intern. J. Solids and Structures*, Vol. 5, pp. 97–107.

Wen, R. K. and J. Lange, 1981. "Curved Beam Element of Arch Buckling Analysis," *J. Struct. Div. ASCE*, Vol. 107, No. ST11, Nov., pp. 2053–2069.

Wittfoht, H., Reisse, H., and H. G. Borck, 1963. "Die neue Fuhrparkbrucke in Hagen," *Beton-u Stahlbetonbau*, 58(6), 129–37 (June).

Yabuki, T. and S. Kuranishi, 1973. "Out-of-Plane Behavior of Circular Arches under Side Loadings," *Proc. Japan Soc. Civ. Eng.*, No. 214, pp. 71–82.

Yabuki, T. and S. Vinnakota, 1984. "Stability of Steel Arch Bridges, A State-of-the-Art Report," *Solid Mech. Arch.*, Vol. 9, Issue 2, Noordhoff Intern. Publ., Leyden, Netherlands.

Yabuki, T., S. Vinnakota, and S. Kuranishi, 1983. "Lateral Load Effect on Load Carrying Capacity of Steel Arch Bridge Structures," *J. Struct. Div. ASCE*, Vol. 109, No. ST10, pp. 2434–2449.

Zezely, B., 1962. "Prestressed Concrete Road Bridge in Yugoslavia," *Concrete Construction Eng.*, 57(9), Sept., pp. 335–344.

CHAPTER 11

SPECIAL BRIDGE TYPES

11-1 TRAFFIC UNDERPASSES IN URBAN AREAS

Traffic underpasses and depressed motorways normally have a uniform cross-sectional configuration, and their grade is usually lowered to provide minimum clearance at intersections and at grade separations. Most intersections, however, are subjected to high traffic volume and may be closed to traffic only during off-peak periods. The use of structural elements that can reduce design effort and minimize delays and public inconvenience is highly favored. Identical, mass-produced prefabricated members can be quickly assembled, reduce forming and labor costs, and minimize lane closure times.

Early problems associated with the use of prefabricated sections have been largely eliminated, and structures assembled in this fashion should be expected to provide long service with minimal maintenance. A problem still facing engineers is the high cost of prefabricated elements in areas where they are not readily available or where construction demand is not sufficient to justify mass production, but even at a higher cost their use may be justified considering their effect on reducing traffic disruption.

The evolution that has taken place in the development of prefabricated members is evidenced in the wide variety of precasting and prestressing of sections such as slab spans, channels, and I beams. Their acceptance and use in bridge construction has been enhanced by improvements in quality control and efficiency at the plant site. Efforts to expand this concept have resulted in the development of substructure panels, usually precast at the site, that are gaining wide acceptance.

Potential Advantages It appears from these brief remarks that in the construction of traffic underpasses and grade separation structures in congested urban areas, the two major drawbacks are (a) the associated high costs and (b) the serious disruption to surface activities. These problems may be remedied and overcome by special designs and construction techniques. Among options prominently mentioned is the use of prefabricated structural members and the design of the permanent structure to act also as the ground support system during construction.

Meaningful cost analysis for general applicability is inhibited by the many variables involved. Reducing the surface disruption, however, should have a direct economic impact, especially on the commercial sector of the area, although the cost of business disruption and public inconvenience is not evaluated quantitatively.

Bridge structures for traffic underpasses and depressed roadways, primarily in urban areas, are synthesized based on the concept of prefabricated elements. Feasible solutions are presented for superstructure and substructure components as well as connections. Interestingly, as the highway program is extended to include maintenance, rehabilitation and replacement, prefabricated members can offer viable options.

11-2 PREFABRICATED ELEMENTS FOR BRIDGE SUPERSTRUCTURES

Prefabricated Concrete Sections

A review of the current literature and standards indicates a wide range of shapes and configurations. Most of the elements currently utilized in bridges are listed in VHTRC–NCHRP Reports 222 (1980) and 243 (1981). These reports conclude that the grouping should focus primarily on the six elements and systems believed to be most promising or most frequently used. These are: (a) precast concrete slab spans, (b) precast box beams, (c) prestressed I beams, (d) precast deck panels, (e) permanent bridge deck forms, and (f) parapet and rail systems. Because other systems appear to have potential in practice, the current review should not be considered complete.

Precast Concrete Slab Spans Precast concrete slab spans are shown in Figure 3-42. They may be fabricated in various lengths and widths to accommodate the range of spans and roadway widths. Solid slabs such as the one shown in Figure 3-42*a* are frequently used for spans up to 30 ft, but more structurally efficient pretensioned or posttensioned voided slabs are commonly used for longer spans. These sections are easy to transport and erect. Special attention should be given to the connection details because premature cracking in the wearing surface and early failures of the system are attributed to keyway and weld–plate failures (Martin and Osborn, 1983). Examples of precast slab spans can be found in many states, but are more frequently used in Illinois, Indiana, Kentucky, New York, North Carolina, Oregon, Virginia, and Alberta (Canada) (Sprinkel, 1985).

Precast Box Beams These sections are shown in Figure 3-42*c*, and they can be used in two configurations as shown in Figure 11-1. Their range covers 50 to 100 ft, and, except for the longer spans, the boxes are easy to transport and erect. Box beams placed side by side are connected in the same way as precast slabs. A wearing surface or concrete overlay is usually provided. Box beams that are spaced apart are tied together with diaphragms, and a cast-in-place slab is added.

Bridges incorporating precast boxes may be found in several states, but are more frequently found in California, Illinois, Indiana, Kentucky, New York, North Dakota, Ohio, Pennsylvania, Texas, Virginia, and West Virginia.

Prestressed I Beams Bridges utilizing precast, pretensioned I beams are discussed in Chapter 3, and typical sections are shown in Figure 3-38. A factor favoring their use is the availability of forms for precasting, whereas the standard cross sections simplify the design and produce cost savings (Panak, 1982). These beams are usually used for spans up to 100 ft, in conjunction with a cast-in-place concrete deck.

Examples of prestressed concrete I-beam bridges can be found in California, Colorado, Georgia, Illinois, Kentucky, Minnesota, Montana, North Dakota, Pennsylvania, Texas, Virginia, and Washington. Modified versions of the AASHTO girder are continuously developed to provide more economical cross sections.

Precast Deck Panels These represent a recent innovation, particularly when they are used in conjunction with steel beams as shown in Figure 11-2. A version of this bridge type is the prestressed deck discussed in Section 4-13. Shear transfer between transverse panels is usually achieved by means of grouted keys or a cast-in-place concrete joint (Biswas, Biffland, Schofield, and Gregory, 1974; TRB, 1976). Transverse panels may be posttensioned in the direction of traffic to improve shear transfer. Composite action between the deck panels and steel beams is developed by shear connectors preattached to the top flange after the voids left in the panels are filled with epoxy mortar.

Deck panels eliminate most on-site formwork and concreting typically required for a steel beam bridge. More precast concrete manufacturers have the facilities and necessary capability to fabricate slab panels. Examples of slab panel applications are the Fremont Street Bridge near Pittsburgh (Smyers, 1984) and the Wilson Memorial Bridge on I-495 in Washington, DC (Lutz and Scalia, 1984).

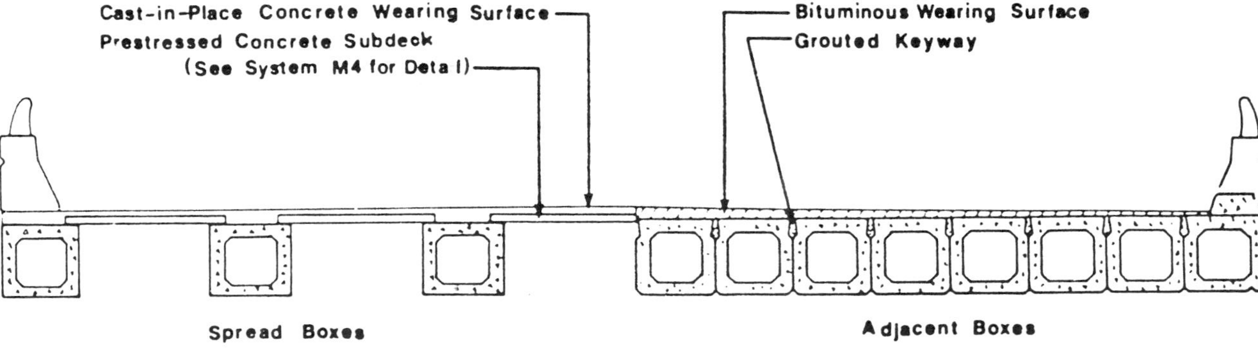

FIGURE 11-1 Precast box beams used for bridge decks.

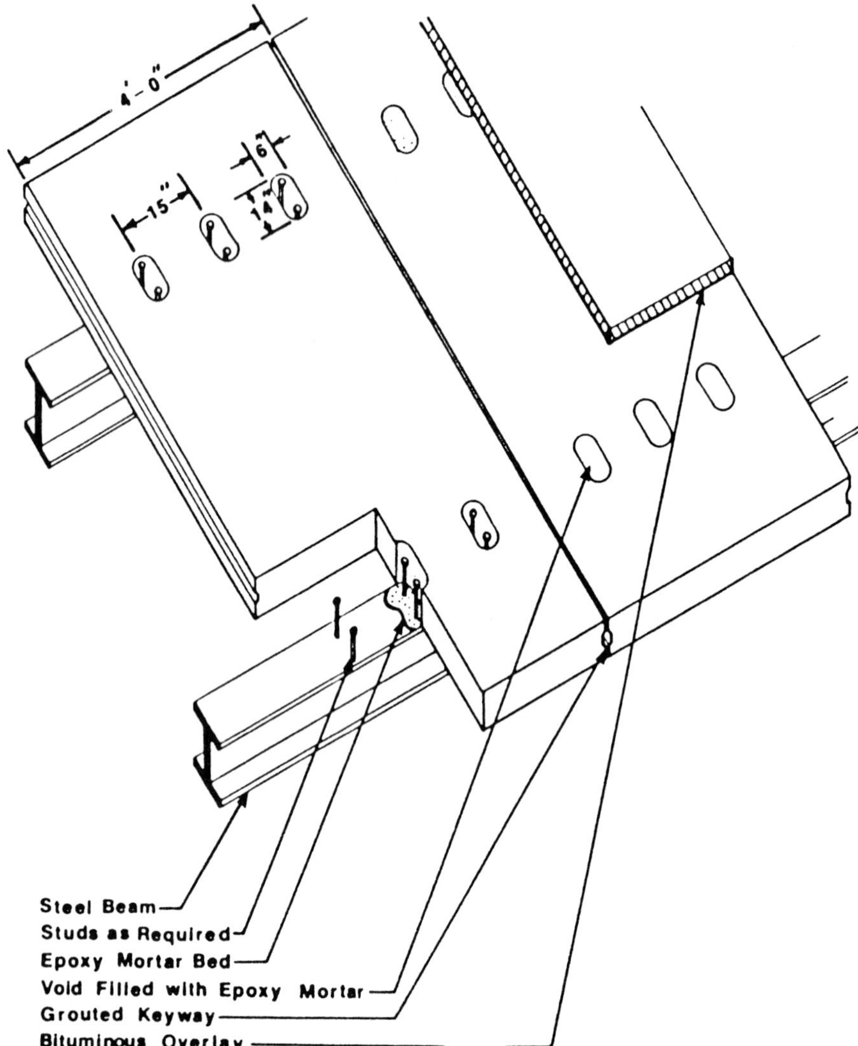

FIGURE 11-2 Precast concrete deck panels on steel I beams.

Permanent Bridge Deck Forms In recent years, steel stay-in-place forms and prestressed concrete subdeck panels have become popular because they eliminate form removal and the associated high cost. Most prestressed concrete manufacturers can fabricate subdeck panels, and the steel forms are provided by most steel fabricators. Details of permanent deck forms are given by NCHRP–TRB (1985).

Precast Parapet Precast parapets have recently been used in some states. The inherent constant shape and uniform section is ideally suited to prefabrication and mass duplication, factors that make precasting economical. Standard precast parapets are typically 8 to 12 ft long and are fabricated upside down to help eliminate honeycombing. The sections must be anchored to the bridge deck to satisfy structural requirements.

Prefabricated Steel Elements

With the exception of steel beams and plate girders, prefabricated steel deck elements are limited in use. Steel members for bridges are prefabricated into the largest subassemblies that can be reasonably shipped and handled. Several innovative concepts involve the use of prefabricated steel or aluminum elements in bridges that are relocatable (Zuk, 1980). As Zuk (1980) aptly concludes, the military is at the forefront of technology for relocatable bridges.

Prefabricated elements of steel are grid systems and orthotropic steel plates. These elements are light and easy to install and therefore suitable for rapid deck installation or replacement. They are, however, relatively expensive and their use must be justified on the basis of reduced dead load and rapid installation. Open steel grid decks have a low skid resistance that can be further improved by filling the grids with concrete mortar.

An example of a steel grid deck on steel beams is shown in Figure 11-3. The system consists of steel stringers and modular open steel grids that may or may not be filled with concrete. A variety of sizes and grid styles are available to accommodate the particular requirements, whereas the panels are selected to fit the size and dimensions of the bridge. Because the grid panels are relatively light and modular, they enhance rapid deck construction or replacement. Where vertical clearance is critical, the grids can provide a shallow deck system.

11-3 PREFABRICATED ELEMENTS FOR BRIDGE SUBSTRUCTURES

Prefabricated Steel Elements

The substructure often requires 60 to 70 percent of the time necessary to complete a bridge (Sprinkel, 1978). Significant time savings can thus be achieved by reducing substructure construction time through the use of prefabricated elements and innovative techniques. In many instances construction schedules can be telescoped by prefabricating and stockpiling the structural members.

Steel elements are widely used in bridge substructures, foundations, and ground support systems to various degrees of prefabrication. Standard rolled sections are commonly used as soldier beams for temporary ground supports or as struts for internal bracing. This construction has minimum prefabrication requirements. The relative simplicity of making field connections and splices by welding is often cited as a principal advantage. Steel sheet piling is

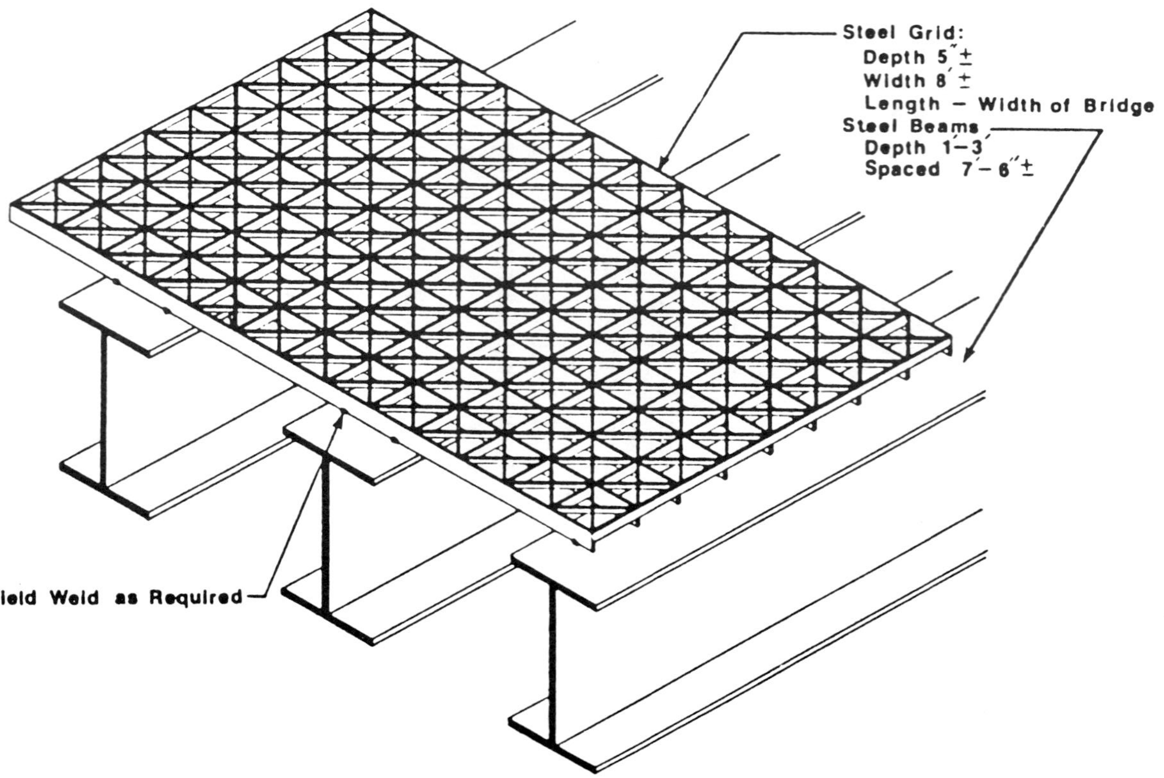

FIGURE 11-3 Steel grid deck on steel beams.

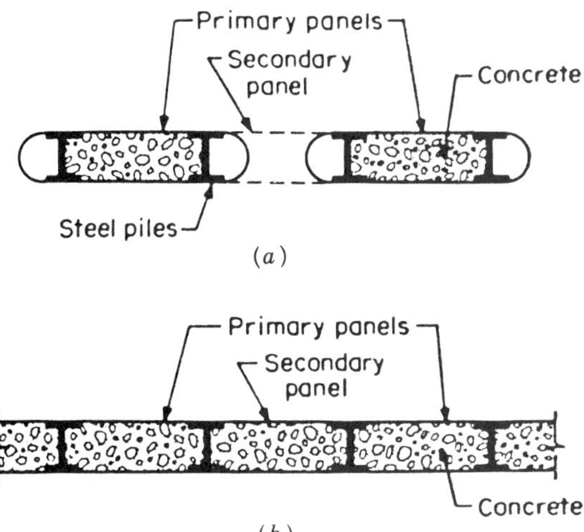

FIGURE 11-4 Typical composite wall: (*a*) outline of excavated panels; (*b*) finished wall.

often used for excavations, and likewise no prefabrication beyond the initial rolling and cutting-to-length is necessary.

Steel soldier piles are combined with reinforced concrete to provide load-bearing elements and ground support systems as shown in Figure 11-4. Where site and ground conditions permit, the installation proceeds as indicated. The primary panels are excavated first, usually with one equipment pass. The steel I beams and the reinforcing cage are assembled and prefabricated as a single unit at the site and inserted in the panel prior to concrete placement.

Prefabricated Concrete Sections

Figures 11-5 and 11-6 illustrate the use of precast concrete abutments and wing walls and prefabricated piling, piers, and caps. Prominent features of the precast abutment shown in Figure 11-5 are the modular panels, easily prefabricated in various shapes to accommodate the abutment configuration and roadway width. The panels are set on cast-in-place concrete pads, temporarily supported, and then connected with weld plates and cast-in-place concrete footing. Other systems and configurations based on modular precast units may be used to obtain a comparable abutment.

The pile substructure schemes shown in Figure 11-6 consist of prestressed concrete or steel H piles with a concrete or steel cap. The piles are driven to the required depth and cut to the required height. Then they are capped with cast-in-place or precast concrete. For abutments, the piling may be backed

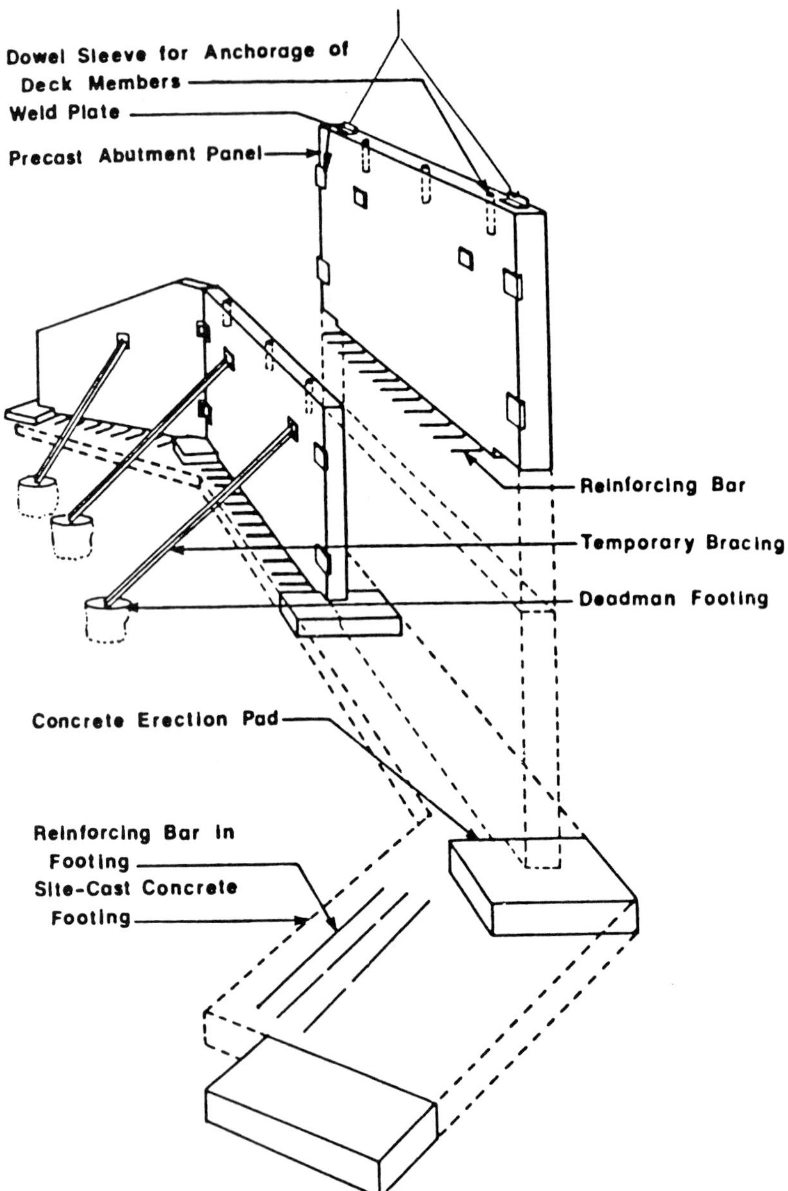

FIGURE 11-5 Precast abutment and wing walls.

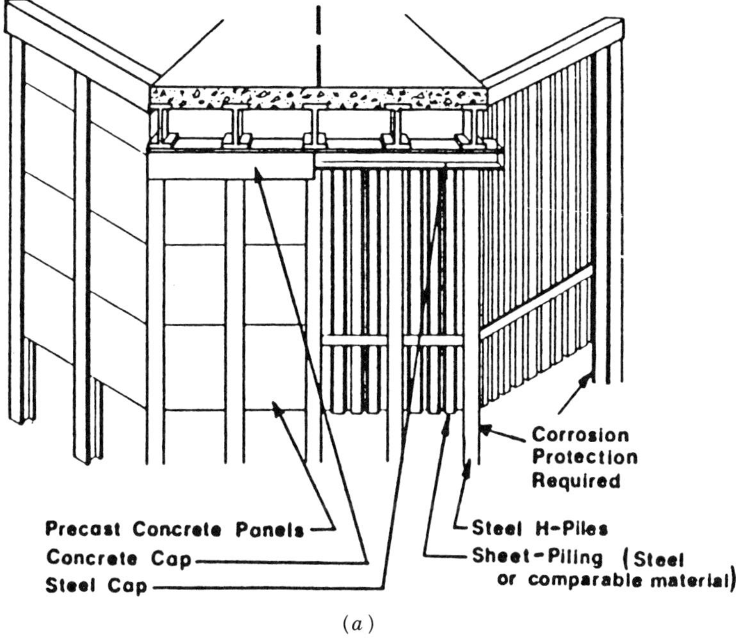

(*a*)

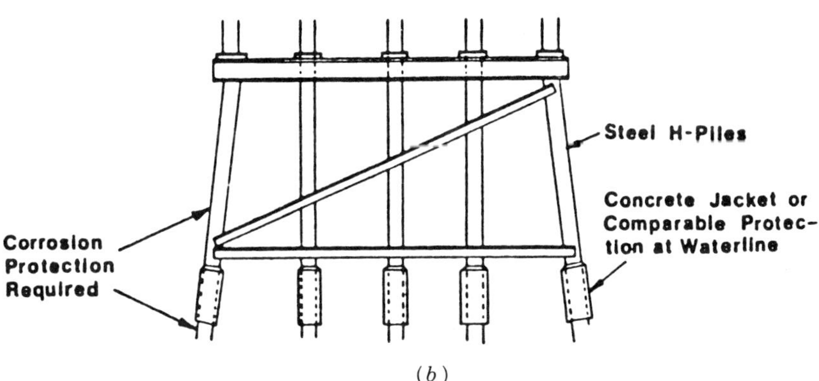

(*b*)

FIGURE 11-6 (*a*) Pile substructure and abutment details; (*b*) pile substructure and pier details.

with a precast concrete plank, a steel bridge plank, or a suitable cribing to retain the soil.

Practical concepts for the use of prefabricated substructure elements are summarized by GangaRao (1978), but with limited applicability because of the many physical differences between bridge sites. The use of standardized elements is further inhibited by variable soil characteristics, bedrock location, and depth at which acceptable bearing is available.

Almost all concepts for substructure applications require the use of either portland cement grout, mortar, concrete, or posttensioning to tie the elements together.

11-4 SLURRY WALLS IN BRIDGE CONSTRUCTION

Cast-in-Place Walls

In urban areas, the four-lane roadway with safety walks is typical for handling vehicular traffic. It may require from 50 to 60 ft face-to-face of walls for an intersection at right angles. For the usual requirements of vertical clearance, the depth of excavation from street level is 19 to 20 ft, which is within the structural range of vertical cantilever systems.

In the covered portion, the roof slab provides the bridge deck and also serves as the uppermost bracing for the walls; it is therefore placed before excavation so that most earth moving is done under cover. In the open (approach) section, the walls may undergo excessive movement if left as free cantilevers. This movement is prevented by bracing the walls, usually using ground anchors near the top and the base slab near the bottom. The initial bracing is combined with temporary strutting until the base slab is in place to reduce the wall embedment. The superstructure deck may consist of the prefabricated elements reviewed in the previous sections.

Because the walls are exposed to public view, face treatment becomes mandatory and includes precast panels, brick, or a separate concrete facing. A typical bridge structure for a traffic underpass is shown in Figure 11-7.

If the width of the underpass exceeds the optimum range of single-span bridges, it may be advantageous to use a center wall. A suitable center wall is the strip panel diaphragm capped with a continuous beam to distribute the superstructure loads (Xanthakos, 1979).

With the abutment walls restrained at the top and bottom, both superstructure ends are fixed at the supports. For spans exceeding 40 ft, provisions should be made to accommodate end rotation of the deck elements.

Prefabricated Walls

Prefabricated diaphragm walls are available in a wide variety of panel types and configurations, but a practical limit in their use is imposed by the weight of the individual units. For the usual crane and hoisting capabilities available at the site, the practical maximum size for a single panel is about 40 ft, which can easily accommodate the usual traffic underpass.

Panel configuration and installation details can be standardized to suit site conditions. Figure 11-8 shows a wall made from typical panels installed so that they slot together. The interlocking is possible with the tongue–groove fitting and ensures high accuracy, resulting in a smooth structure. The type

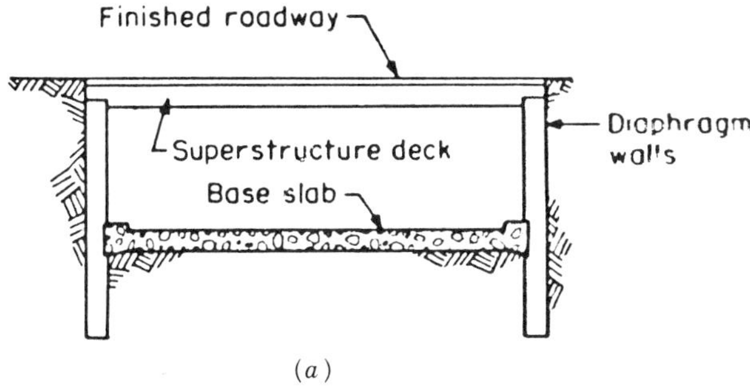

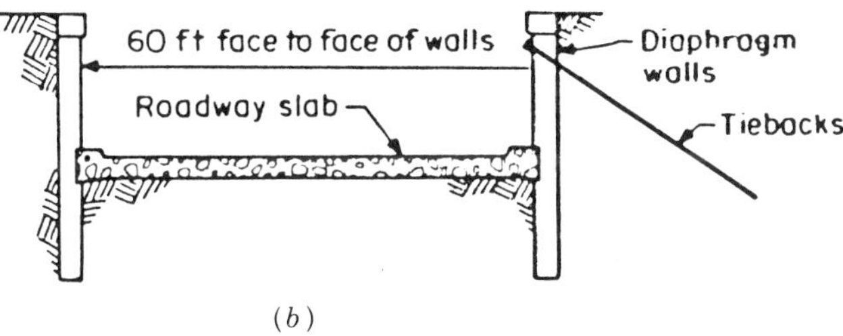

FIGURE 11-7 Diaphragm walls for traffic underpass: (*a*) covered section; (*b*) uncovered section for approach roadway.

shown in Figure 11-9 consists of alternate beam-and-slab sections and is equivalent to the conventional soldier pile wall. The beams are usually made twice as thick as the slabs; a standard thickness is 24 in. for the beams, and 12 in. for the slab panels.

Other configurations and types are feasible and can be developed to fit the job conditions and thus utilize the advantages inherent in precasting. The general appearance of the exposed wall is satisfactory, and the quality control available with precasting ensures the specified concrete strength. On the other hand, these favorable considerations must be balanced against the site conditions. The operation requires careful planning and strict adherence to schedule. Other relevant factors are the minimum job size necessary to offset certain fixed costs inherent in precasting, adverse ground conditions for this type of work, and the proximity of the nearest casting yard. Installation and construction procedures are discussed by Xanthakos (1979).

Alternatively, the beam-and-slab panels shown in Figure 11-9 may be constructed with different lengths. Because the beam sections are relatively narrower than the side panels, they can be made longer without exceeding the maximum practical weight. These members are extended deeper from the bottom of excavation and resist active and passive pressures below this level. The slab sections require only nominal embedment merely to receive the

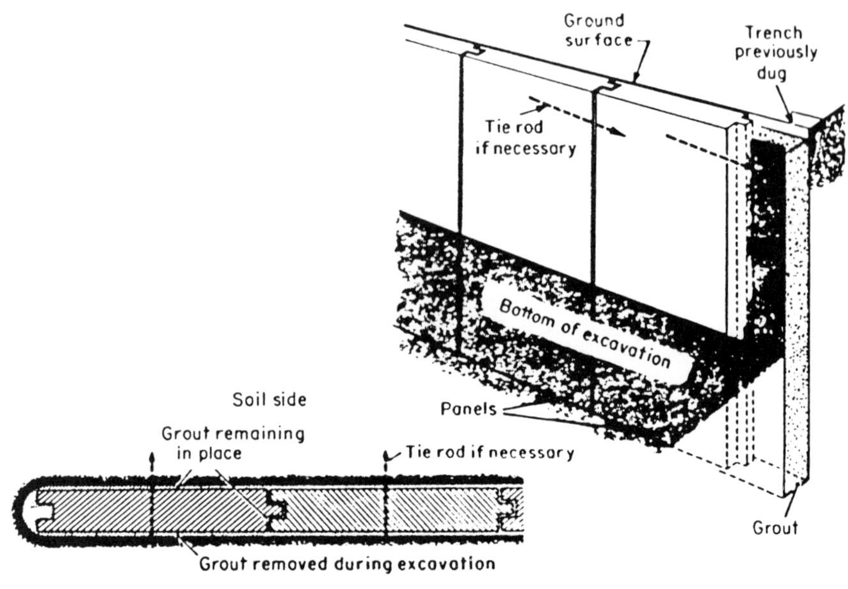

FIGURE 11-8 Prefabricated wall with identical panels.

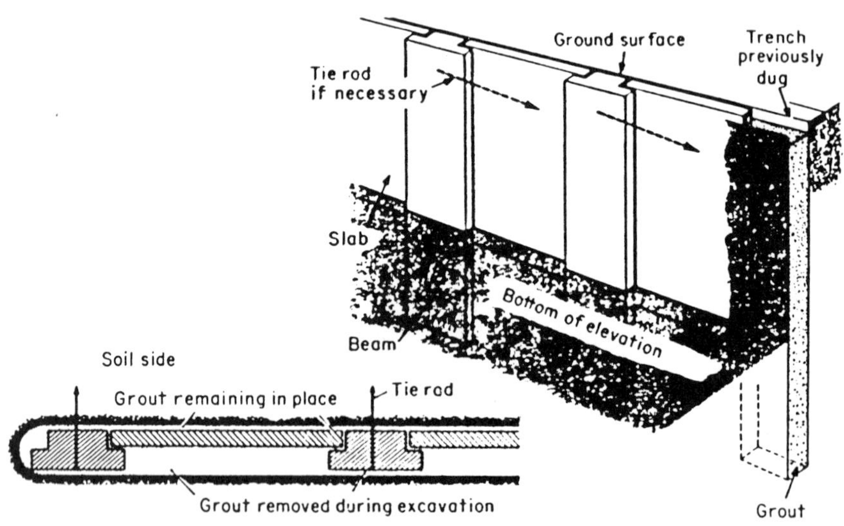

FIGURE 11-9 Prefabricated wall with beam-and-slab panels.

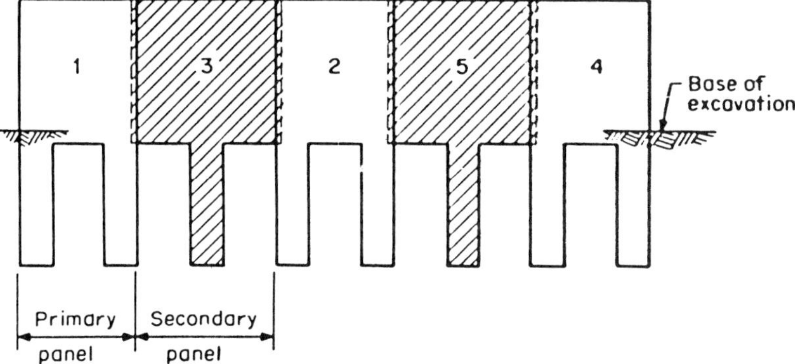

FIGURE 11-10 Installation of a prefabricated wall consisting of beam-and-slab sections: one primary panel consists of two beam sections and one slab section; one secondary panel consists of two slab sections and one beam section.

bottom slab. Thus, in elevation, the wall appears as a series of T shapes or as a wall on stilts. The installation follows the sequence shown in Figure 11-10. Primary panels comprising two beam sections and one slab are inserted in alternate trenches 1, 2, 4, and so on. The secondary panels 3, 5,... are installed between the primary sections and consist of two slabs and one beam. This arrangement involves three units for each panel, which is the optimum daily schedule for one excavating machine and one crane. The installation continues by alternating primary and secondary panels so that the grout in any panel becomes self-standing as the next panel is excavated. When traffic lane closure is restrained, the panel sequence can be modified, and construction restrictions are relaxed if a displacement grout is used (Xanthakos, 1979).

11-5 POSTTENSIONED DIAPHRAGM WALLS

At present, the posttensioning of diaphragm walls is highly specialized. In this application conventional prestressing strands and ducts are arranged in the panel so that they remain unaffected by submerging in bentonite slurry. With the tremie pipes properly located, the presence of ducts should not hinder the flow of fresh concrete. A relevant feature of this system is its behavior while in ground confinement. Results from field tests show that the posttensioning of the strands at nominal eccentricity will not necessarily cause distortion of the panel in the ground. In fact, it may be impossible to introduce tension at the extreme concrete fiber even if the strands are overstressed. This behavior, discussed in the following section, suggests that the soil mass surrounding a posttensioned wall provides an ideal stressing bed of infinite stiffness (Xanthakos, 1979).

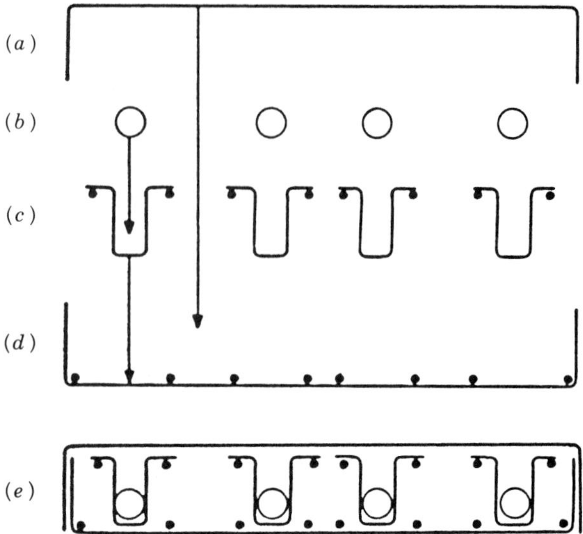

FIGURE 11-11 Method of detailing and assembling a reinforcing cage for posttensioned cast-in-place diaphragm walls: (*a*) upper perimeter bar; (*b*) cable ducts; (*c*) cable brackets with vertical reinforcement; (*d*) lower perimeter bar with vertical reinforcement; (*e*) cage in the assembled position. (*From Gysi, Linder, and Leoni, 1975.*)

Posttensioning Assembly The reinforcing cage is usually provided to serve as an attachment for holding the casing ducts that contain the strands and also to resist handling stresses. Once in place, the cage no longer has a static function except that of preventing the distortion and displacement of the ducts under the effect of differential pressure from the rising fresh concrete. A standard assembly is detailed in Figure 11-11. The cage is spot-welded for increased stiffness, yet it may still be deformable because of its small weight and large size.

The most common procedure for posttensioning is jacking. For example, BBRV units are fixed and firmly held within the cage. Jacks are then used to pull the steel against the hardened concrete, but certain problems can arise with the anchoring of the heavily stressed strands in the base of the wall. Although anchor blocks and similar devices in direct bearing are usually dependable, they should be avoided in this case; looping the bars in a U shape should be used instead. The stress concentration resulting from the small loop radius is resisted by special slot tension bars placed in this area.

Design Considerations

Loads The diaphragm walls supporting the bridge deck of Figure 11-7 must resist the following loads and forces: (a) dead load of superstructure deck; (b) live load from deck traffic; (c) lateral earth pressures corresponding to a particular state of stress (active, passive, at rest); and (d) forces associated with braking or temperature expansion and contraction of the superstructure transmitted to the wall through the fixed bearing. Other loads such as longitudinal wind may reach the walls through the bearings. In addition, for a posttensioned wall, the applied prestress constitutes a permanent state of load activity.

The stressing force and the eccentricity of the stressing steel are determined by the stresses that must be resisted in the final structure. Thus, the general principles of prestressing are still valid, and the objective is to introduce a compressive stress in the concrete before final load application which will compensate tensile stresses.

Elastic Wall Shortening For the application to be effective, the prestress should not dissipate or otherwise be absorbed. The prestressing will normally cause the wall to shorten elastically, and, because it is confined in the ground, the soil in contact may impede the process of elastic shortening and retard or stop it completely. However, comparison of an average soil modulus with that of concrete shows that the latter is 150 to 300 times as stiff; hence, the only impediment will be provided by shear resistance along the face (friction or adhesion) acting opposite to the direction of prestress.

For routine jobs under average conditions, the response of the wall to prestressing can be inferred merely by estimating the probable elastic shortening. For example, a wall 60 ft deep and 24 in. wide acted upon by a prestressing force of 200 kips per foot of length is likely to undergo an elastic shortening of about 0.15 in. (0.4 cm) based on $E_c = 3 \times 10^6$ psi. The corresponding relative displacement at the wall–soil interface will be insufficient to mobilize shear stresses, and therefore the entire prestress will be transferred to the concrete.

Effect of Soil Stiffness An essential feature of the posttensioned wall is that the stressing is carried out while the wall is still fully embedded in the ground, prior to any excavation. In this context the ground constitutes a stressing bed. This results in a particular interaction between the concrete and the soil at stress transfer.

The arrangement and position of the strands and the amount of prestress are chosen to counteract the stresses resulting from the external loads acting on the wall at the end of excavation. Little or no attention is given to the boundary conditions of partial excavation or prior to any excavation. If the tendons are laid out to supply the most desirable and most efficient system of prestress for the final loads, there may be considerable initial tensile stress as in the case for unrestrained posttensioned elements. However, the restraint offered by the soil limits wall deformation and suggests a much different and more favorable response. Tests reported by Braun (1972) show that the introduction of prestress could not cause tension in the concrete, even if the force was applied outside the middle third and the strands were overstressed.

The absence of full flexural deformation of the embedded wall during the eccentric application of the jacking force indicates a considerable intensity of soil response, so that passive resistance is developed (partly) without the movement predicted in soil mechanics taking place. Initially, the amount of this restraint depends on the soil type and its deformability, with the upper limit of the resulting counterthrust approaching the passive resistance. Better results should be expected if the posttensioned wall is in relatively stiff or dense formations. However, the fluidity and flowability of the heavier fresh concrete causes a net gain in the initial stress at rest in most soils, but more particularly in soft or loose materials, and thus accounts for the observed high soil response. This clamping effect should therefore be expected in most soils, although to a varying degree.

Theoretically, a posttensioned wall panel may be treated as a continuous beam supported by springs having a stiffness equivalent to the modulus of subgrade reaction for the soil. Using this approach, Gysi, Linder, and Leoni (1975) derived a solution shown in Figure 11-12, and later confirmed in large-scale tests and in situ measurements. The diaphragm wall is embedded

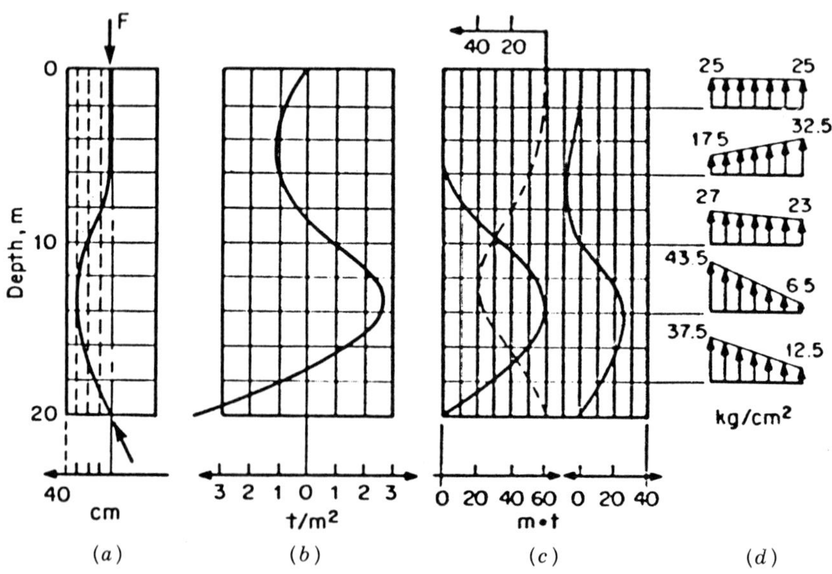

FIGURE 11-12 Prestress, earth pressure, bending moments, and stress in a posttensioned diaphragm wall embedded in silty moraine: (*a*) diagram showing the eccentricity of prestress; (*b*) earth pressure caused by the prestress; (*c*) bending moments caused by the prestress, the earth pressure, and resultant bending moments; (*d*) compressive stresses in the wall at various depths. (*From Gysi, Linder, and Leoni, 1975.*)

in silty moraine having an elastic modulus $E_s = 500\ \text{kg/cm}^2$ (about 6900 psi). The concrete modulus is $E_c = 200{,}000\ \text{kg/cm}^2$ (2.75×10^6 psi), or 400 times the stiffness of the soil. The wall is 20 m (66 ft) deep and 80 cm (32 in.) thick.

Figure 11-12*a* shows the arrangement of prestress. The axis of the tendon is a parabolic curve below 6 m and reaches a maximum eccentricity of 30 cm, or well outside the middle third. Figure 11-12*b* shows the soil pressure against the wall mobilized upon the application of prestress, and Figure 11-12*c* shows the vertical bending moments due to posttensioning, the moments due to earth pressure, and the resultant bending moments. Figure 11-12*d* shows the stress in the concrete section due to the resultant bending moment at various depths, and in spite of the large eccentricities the concrete is in compression.

Without the restraining effect of the soil, the wall would have been stressed in tension in the zone of large eccentricities. For a maximum prestress moment of 60 m-tons corresponding to the largest eccentricity, the resulting tensile stress would have been 11 kg/cm^2 (155 psi).

Loss of Prestress The scarcity of observed or published data on posttensioned diaphragm walls inhibits the development of a comprehensive procedure for estimating prestress losses. In general, the following factors must be considered.

1. *Shrinkage of Concrete* Prestress loss due to shrinkage varies first with the proximity of the concrete to water during hardening and then with the time of application of prestress. The favorable curing conditions and the moist environment keep the initial shrinkage low. However, because the posttensioning is applied much later and the concrete is subjected to a dry atmosphere during excavation, further shrinkage should be expected with a corresponding loss of prestress.

2. *Creep of Concrete* This is a time-dependent deformation resulting from the presence of stress. Two phases are distinguished: (a) before soil excavation, that is, at stress transfer in the fully embedded condition; and (b) after soil excavation, that is, at load transfer in the final stress condition. Creep can be more significant if the prestress in the steel is low and the compression in the concrete is high, but the latter condition is avoided with the relatively large cross-sectional area of the wall. For an average prestress in the concrete of 1000 psi (about 70 kg/cm^2), the loss of prestress due to creep may be 6 to 7 percent.

3. *Steel Relaxation* This process involves a decrease of stress and a corresponding loss of load in the strands with time while the tendon is held under constant strain. It occurs predominantly during the first few hours of loading.

4. *Friction Loss* This occurs within the prestressing cables and depends mainly on the type of cable, the cable curvature, and the roughness at the cable–concrete interface.

5. *Elastic Deformation* This can occur with looped cables if the stressing is applied sequentially. Loss of prestress due to local yielding of weak or contaminated concrete in the loop zone occurs as soon as the jacking force is introduced, and therefore it does not represent a final loss.

Although it is not prudent to generalize the extent of prestress loss, it is conceivable that for the usual curing conditions, type of concrete and steel, amount and time of prestress application, the loss will not exceed 15 percent. More details are given by Xanthakos (1993).

Walls Suitable for Posttensioning

Two main wall types are suitable for prestress: (a) vertical cantilevers fully restrained at the base and (b) walls restrained at the bottom (either by sufficient embedment or by unyielding support) and braced at or near the top. These types are shown in Figure 11-13 with the corresponding lateral earth stress diagrams. In this context walls in traffic underpasses are ideally suited to posttensioning because of the actual support conditions. At the bridge the walls are braced at the top by the superstructure, and adjacent to the bridge by ground anchors (Xanthakos, 1991). Further away from the bridge the walls revert to a simple cantilever. All three conditions are susceptible to posttensioning.

The pattern of the lateral earth stress distribution developed by the application of prestress may be assumed to be essentially similar to the pattern with the excavation completed. This pattern is shown in Figure 11-13 next to the walls, and with top bracing it takes into account a possible preloading of the lateral support. Accordingly, only the magnitude and at

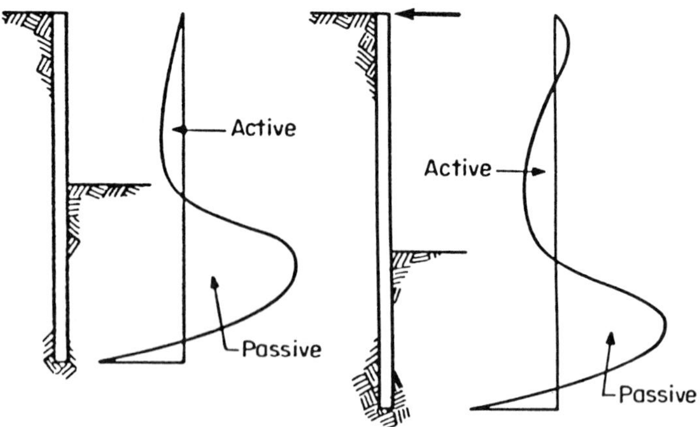

FIGURE 11-13 Walls suitable for posttensioning, and patterns of lateral earth stress distribution.

some point the sign of the stress will change as the excavation is completed, gradually leading to the final active–passive earth stress distribution. The two states for which the walls must be analyzed are the boundary conditions at the initial prestress with the walls fully embedded and the final stage with the excavation completed. Intermediate phases with partial excavation will seldom govern and need not be considered.

The concept of concentric tendons for multibraced or multianchored walls deserves consideration, but this wall type is quite unlikely in conventional traffic underpasses.

11-6 EXAMPLES OF TRAFFIC UNDERPASSES

Posttensioned Walls The first large-scale use of posttensioned walls in permanent structures was at an underpass in Manchester, England. This underpass is 3600 ft long and has posttensioned diaphragm walls over 70 percent of its length. The system includes free cantilever walls, bridge abutments, and an intermediate pier. The design was influenced by two main factors: (a) the location of the project within a built-up area; and (b) the traffic flow on existing loads, which had to be maintained with minimum disruption.

The free cantilever walls have a retained height of 10 to 29 ft. This range is accommodated by three different wall panels 24, 33, and 40 in. thick. Because the design relies on wall embedment below excavation level for lateral stability, the penetration is 68 ft. The panels are 18 ft long with semicircular ends and a nominal gap of 6 in. between them to allow for a perforated vertical drainage pipe. The prestressing elements consist of U-shaped sheathed tendons stressed to 104 to 500 tons (metric) according to the design requirements and in relation to the height of the cantilever and the depth of penetration. A typical wall panel is shown in Figure 11-14 and includes an elevation and section (Fuchsberger, 1980).

The cables in the panels have a standard loop configuration as the bottom anchorage are configured in a loop because this can ensure better results than single anchor heads at the end of each cable. After casting the capping beam and a suitable curing period, the cables were stressed and the ducts were grouted. The panels were cast on the ground, cured, and then lifted by cranes for insertion in the trench.

A plan of a bridge section of this project is shown in Figure 11-15. The Bank Lane Bridge is one of three bridges along the depressed motorway. The abutments consist of T-shaped diaphragm wall panels, posttensioned in the stem by two pairs of U-shaped tendons with a stressing force up to 200 tons (metric) each. The depth of these sections varies from 52 to 72 ft according to the soil conditions and the design load. When the construction of the T panels was completed and the prestress applied, a capping beam was cast on top of the wall to carry the superstructure system. The bridge deck consists of

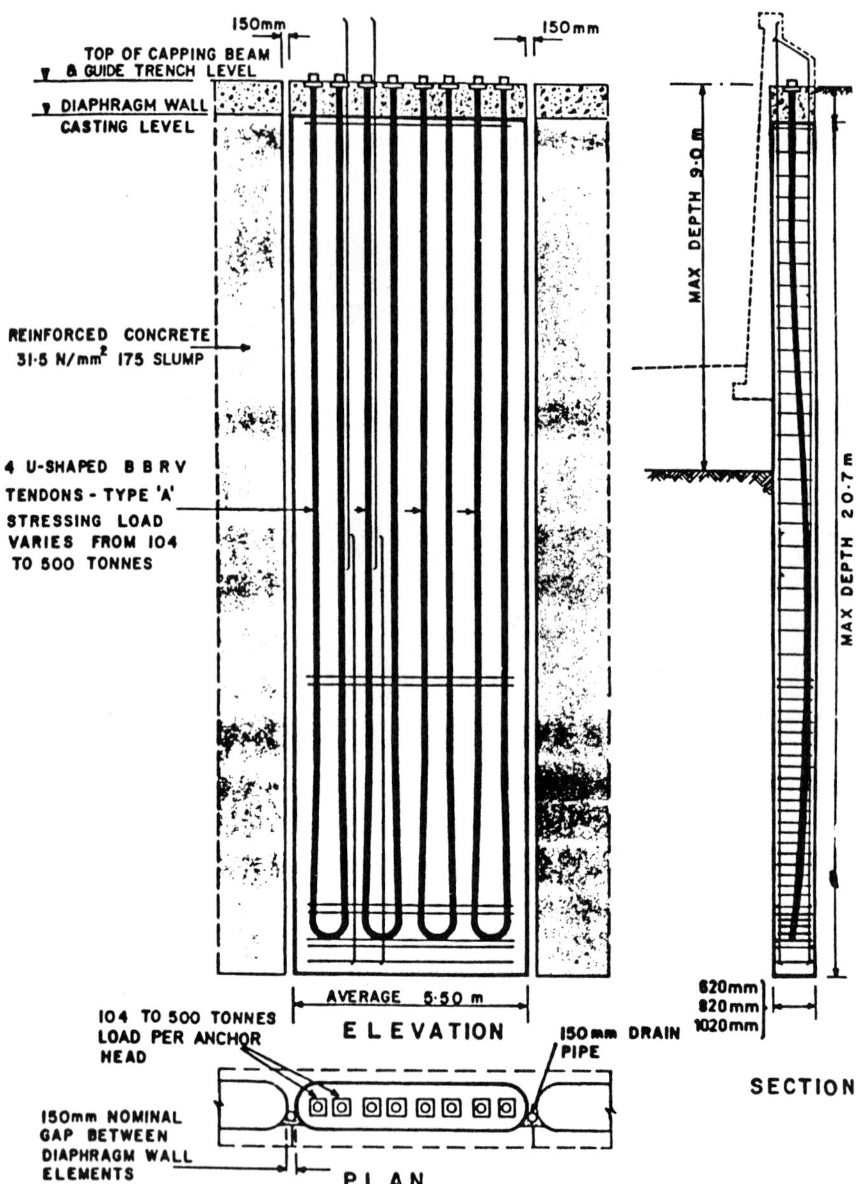

FIGURE 11-14 Typical posttensioned wall element, Manchester underpass. (From Fuchsberger, 1980.)

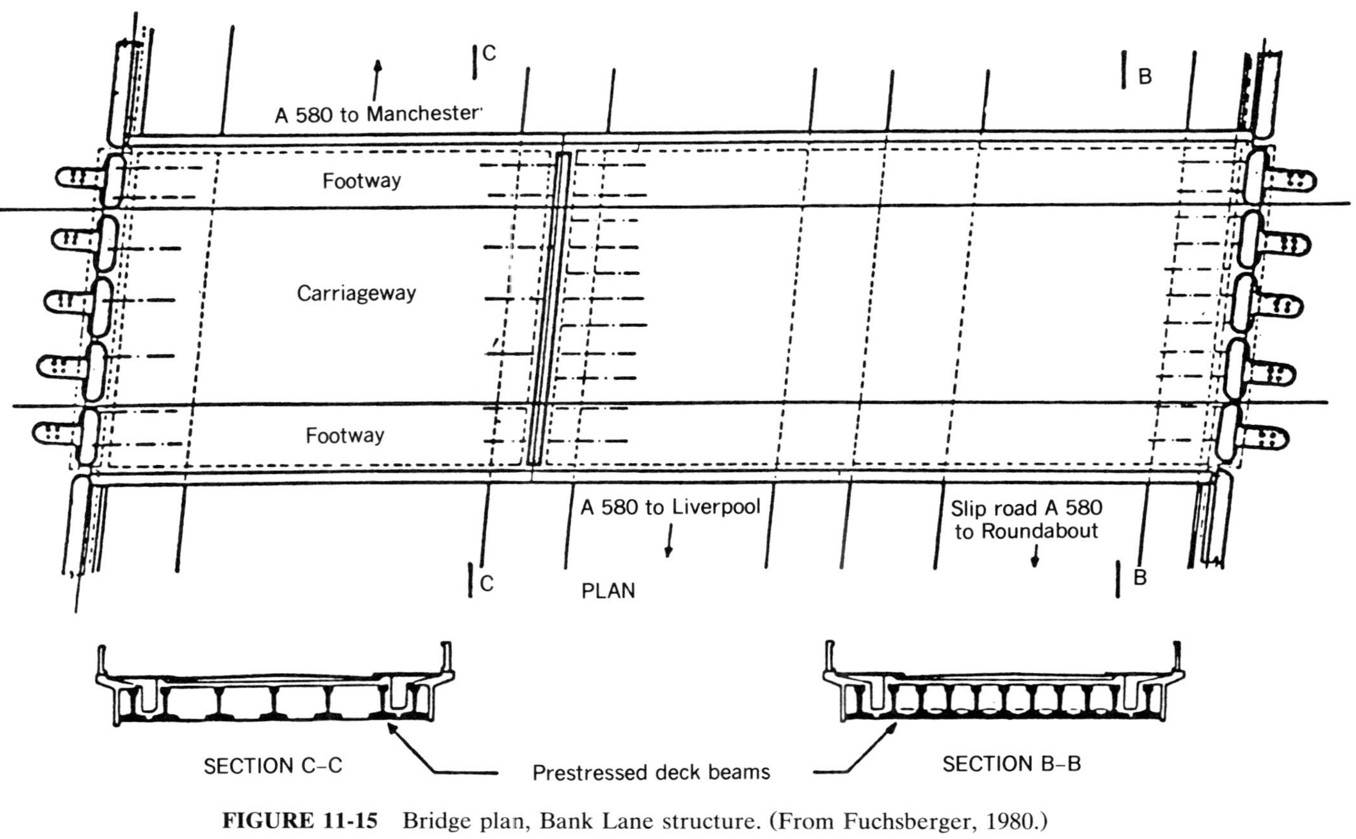

FIGURE 11-15 Bridge plan, Bank Lane structure. (From Fuchsberger, 1980.)

precast prestressed I beams and a cast-in-place concrete slab. The bridges of the three intersections were completed and open to traffic before the bulk excavation for the depressed roadway was commenced. The bridge shown in Figure 11-15 has an intermediate pier built as a diaphragm wall from ground level. After excavation a suitable face treatment was applied.

Prior to posttensioning, a test section was designated to verify the design assumptions and to check the performance of the wall at various stages. Relevant topics included the influence of the soil during stressing, the assumed modulus of subgrade reaction, and the creep occurring between stress transfer and load application. The results show that the entire prestress was transferred to the concrete, and the soil offered no impediment to the process. With the exception of minor tensile stresses in the bottom anchoring zone, only compressive stresses were measured at various cross sections, following a more even trapezoidal distribution than calculated. This leads to the conclusion that the actual soil stiffness as a confining medium was greater than assumed.

Measurements after stress transfer and before excavation gave a good indication of stress loss due to creep. This is an important factor in timing the bulk excavation because not more than about one-half of the total creep should be allowed to occur prior to general earth moving.

Precast Diaphragm Walls A traffic underpass built with precast nonprestressed diaphragm wall panels is shown in Figure 11-16 (Leonard, 1974). The depressed roadway carries Route A13 in the central Paris district. The exterior walls consist of continuous precast panels, but the center pier wall has strip sections. When the walls were in place, a posttensioned deck slab was installed, and earth moving was completed under cover. The bottom part of the trench is filled with plain concrete, because its main function is to transfer the loads at a level where sufficient bearing is available. The deeper wall in the center pier is required because of heavier loads.

11-7 BRIDGE STRUCTURES UNDER ROADWAY EMBANKMENT

Design Considerations

In this category are pedestrian undercrossings and examples of rigid frames such as railroad overheads. The common characteristics are usually articulated in relatively short spans, a live load distribution through earth fills, and end supporting walls that are subjected to considerable lateral earth pressures.

For grade separations at railroad crossings, rigid frames are generally competitive for spans exceeding 30 ft. Structure continuity is desirable to redistribute and balance the bending moments from vertical and lateral loads. For spans less than 30 ft, simple spans with simple joints are economical and usually have both ends fixed against movement in the bridge direc-

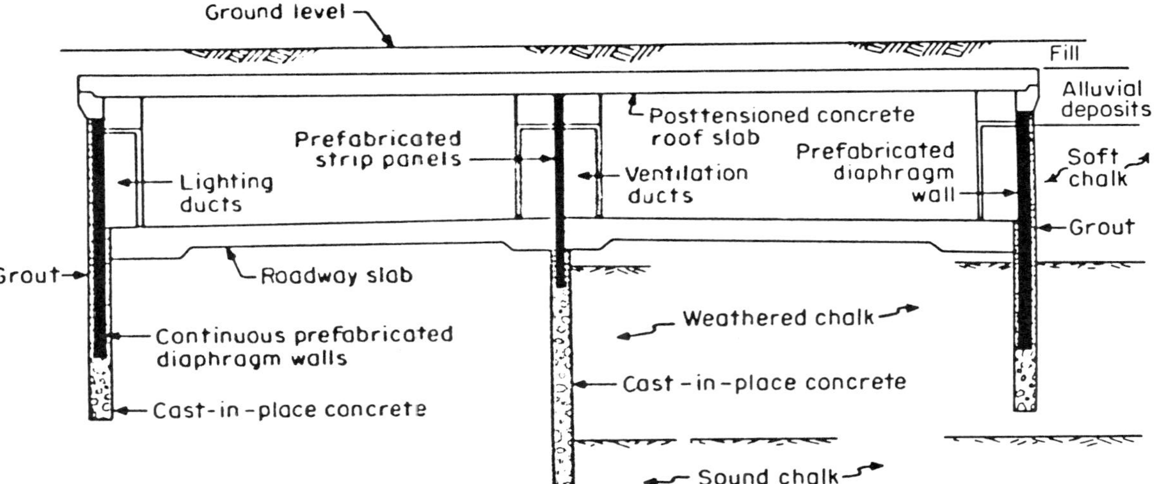

FIGURE 11-16 Cross section of a covered motorway built with prefabricated diaphragm walls. (From Leonard, 1974.)

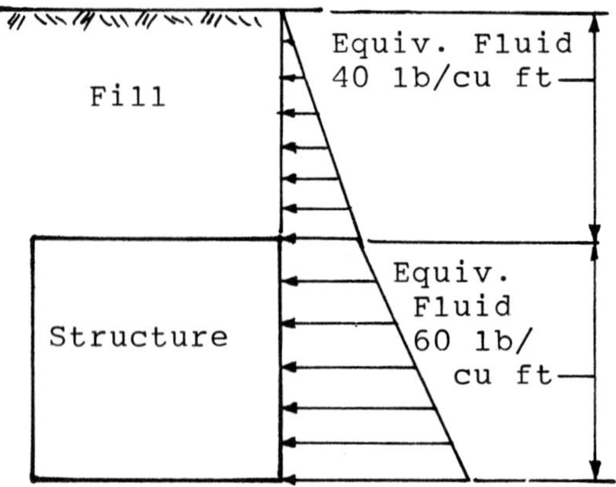

FIGURE 11-17 Equivalent fluid pressure.

tion. Pedestrian undercrossings should typically be investigated for continuous- or simple-span frames with end or intermediate supports.

Dead Loads and Earth Pressure The weight of soil above a bridge structure is commonly taken as 120 lb/ft^3, and this includes compacted sand, earth, gravel, or ballast.

For complicated structures and foundation conditions, the lateral earth pressure should be established from a consideration of soil mechanics principles and the soil–structure interaction. For the usual case, however, an equivalent fluid pressure is recommended as shown in Figure 11-17, consisting of 40 lb/ft^3 for the depth of the fill above the structure and 60 lb/ft^3 for the height of the structure. Interestingly, these lateral pressures may be reduced at the discretion of the engineer if, combined with the vertical loads, they do not yield maximum effects in the slab.

Live Load When the structure has traffic moving on its top deck, the wheel loads are distributed as in ordinary bridges, and the same distribution is assumed when earth fills on structure decks are less than 2 ft deep. For fills more than 2 ft deep, concentrated loads are uniformly distributed over squares with sides equal to 1.75 times the depth of the fill. When such areas overlap and several concentrations converge, the total load is distributed over the area defined by the outside limits of the individual areas, but the total width of distribution should not exceed the widths of the supporting slab. In single spans, the effect of live load may be neglected if the fill is more than 8 ft deep and exceeds the span length. For multiple spans, the live load may be neglected if the depth of fill exceeds the distance between the faces of end supports. However, if live load must be considered and the calculated live load plus impact moment based on distribution through earth fills exceeds the moment computed according to AASHTO Article 3.24, the latter moment must be used.

For fills exceeding 2 ft, live load distribution reinforcement is not required.

Impact This effect decreases quickly with increasing fill. If this cover exceeds 3 ft, impact need not be included. For fill depths of 1 to 3 ft, impact is proportioned from 30 to 10 percent, respectively.

Other Loads Temperature is seldom a problem in this structure type. Long spans should be checked for shrinkage, and provisions should be made in the walls and deck to control shrinkage cracking, usually by means of construction joints. Deep fills generate tension forces along tunnel-type structures when the fills settle, and tension continuity must usually be provided in the floors or footings to keep the joints from pulling apart.

Design Example: Pedestrian Undercrossing

We will design a typical pedestrian undercrossing under a moderate fill merely to illustrate the application of loads and design techniques. The fill exceeds 8 ft; hence, live load will not be considered. However, an arching effect is assumed in the soil mass, which, in effect, reduces the dead load intensity (weight of fill) as follows:

Weight of earth (fill 0–8 ft)	120 lb/ft^3
Weight of earth (fill 8–16 ft)	Varies directly from 120 to 84 lb/ft^3
Weight of earth (fill > 16 ft)	84 lb/ft^3

The weight of the fill is therefore $8 \times 0.12 + 4 \times 0.11 = 1.4$ kips/ft. Assuming a top slab thickness of 12 in., the total dead load weight is $w = 1.55$ kips/ft.

The lateral earth stresses are computed at the centerline of the top and bottom slabs, that is, at depths of 12.5 and 21.5 ft from the top of the fill. These stresses are shown in Figure 11-18, together with the configuration and dimensions of the structure.

For allowable stress design, the design stresses are $f_c = 1200$ psi and $f_s = 20{,}000$ psi, $n = 10$. Span lengths for both the slab and walls are taken center-to-center of the supports (simple-span analysis).

Slab The maximum moment is computed as

$$M = 0.125 \times 1.55 \times 9^2 = 15.7 \text{ ft-kips}$$

$$\text{Required } d = \sqrt{\frac{15.7}{0.197}} = 9 \text{ in.}$$

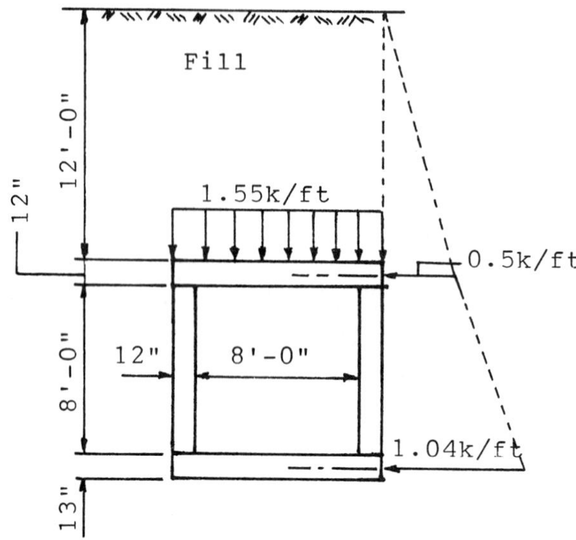

FIGURE 11-18 Cross section, dimensions, loads, and lateral earth pressure; pedestrian undercrossing. The wall thickness in the final design is 10 in.

Assuming #7 bars and 1.5-in. clearance, $d = 10$ in., OK.

$$\text{Required } A_s = \frac{15.7}{1.44 \times 10} = 1.09 \text{ in.}^2/\text{ft} \qquad \text{Use \#7 at 6.5 in.} = 1.11 \qquad \text{OK}$$

Walls The moment due to the assumed lateral pressure diagram shown in Figure 11-18 is

$$M = 0.125 \times 0.5 \times 9^2 + 0.13 \times 9 \times 0.27 \times 9 = 6.1 + 2.9 = 9.0 \text{ ft-kips}$$

The vertical reaction from the weight of the fill is $4.5 \times 1.55 = 7.0$ kips. For the final design, we select a wall thickness of 10 in. Then $d = 8$ in., $d' = 2$ in., and $d'' = 3$ in. Next, we compute

$$e = \frac{9 \times 12}{7} + 3 = 18.4 \qquad E = \frac{18.4}{12} = 1.53$$

Also, $F = 8^2/1000 = 0.064$. To check the compressive reinforcement requirements, we compute

$$NE = 7.0 \times 1.53 = 10.7$$

$$KF = 197 \times 0.064 = \underline{18.6} \qquad (12.6)$$

$$NE - KF = -1.9 \qquad \text{(Negative)}$$

Hence, compressive reinforcement is not required. The area of steel (tension) is

$$A_s = \frac{9.0}{1.44 \times 8} = 0.78 \text{ in.}^2/\text{ft} \qquad \text{Use \#7 at 9 in.} = 0.80 \text{ in.}^2$$

The design should be completed by analyzing the structure as a rigid frame. Although the resulting economy, if any, may not be conclusive, this scheme is a viable alternative because the structure has favorable span and height dimensions, and the applied loads (vertical and lateral) are compatible with the principles of rigid frames. An example is presented in the following section.

11-8 RIGID-FRAME BRIDGE: DESIGN EXAMPLE

As mentioned in Chapter 3, single-span rigid frames are useful for bridges over tracks, streams and canals, and so forth. A rigid-frame structure will be analyzed for traffic moving on the deck surface. Configurations and dimensions are shown in Figure 11-19. The grade separation involves a railroad track that requires a minimum vertical clearance of 23 ft.

We choose to provide tapered legs to reduce somewhat the quantity of concrete and also to favor more moment distribution at the top corners. The legs are assumed hinged at the bottom. The distribution factors are obtained from design aids, and are as follows:

Deck	DF = 0.66
Legs	DF = 0.34

The dead load from the weight of slab is $0.15 \times 1.75 = 0.26$ kip/ft.

First, we calculate the following moments:

Dead load, simple beam:

$$M_{\text{sb}} = \frac{0.26 \times 29.75^2}{8} = 29.1 \text{ ft-kips}$$

Dead Load, fixed end moment:

$$M_{\text{FE}} = 0.26 \times \frac{29.75^2}{12} = 18.5 \text{ ft-kips}$$

The dead load moment distribution is shown in Figure 11-20 with the resulting dead load moment diagrams, and is repeated as baseline in the final (total) moment diagram. We should note that the slight batter of the wall is ignored in the analysis.

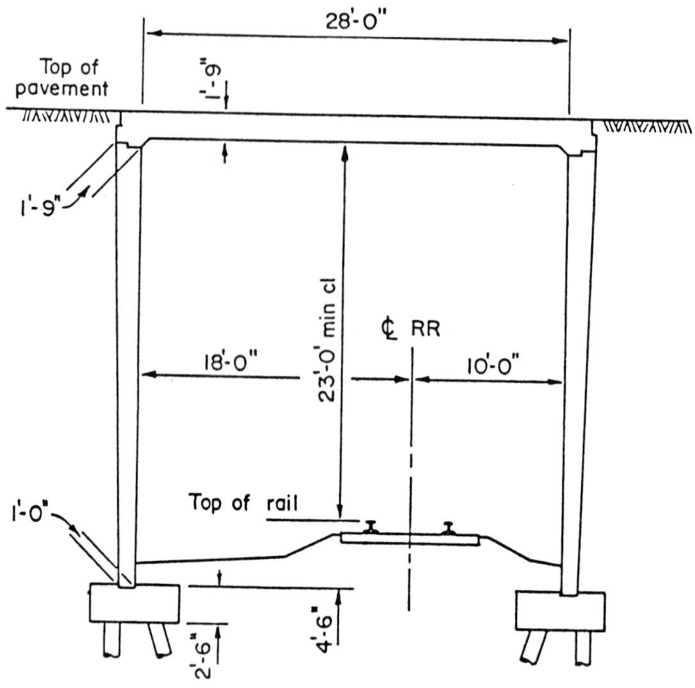

FIGURE 11-19 Typical cross section, rigid frame; structure carrying highway traffic over a rail track.

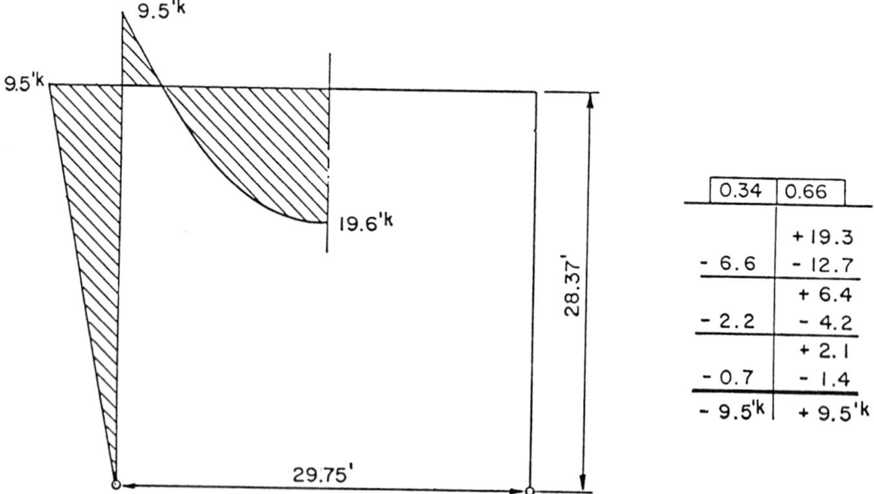

0.34	0.66
	+ 19.3
- 6.6	- 12.7
	+ 6.4
- 2.2	- 4.2
	+ 2.1
- 0.7	- 1.4
- 9.5'k	+ 9.5'k

FIGURE 11-20 Dead load diagram, weight of slab; rigid frame of Figure 11-19.

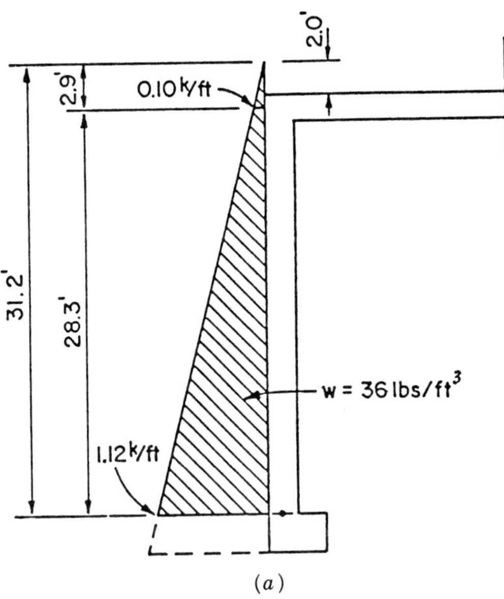

(a)

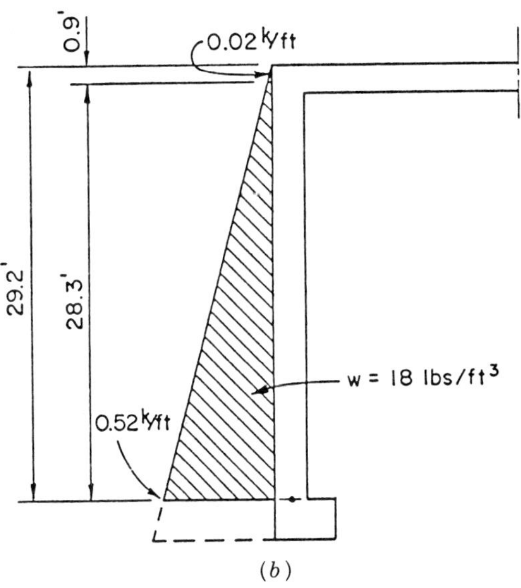

(b)

FIGURE 11-21 Lateral earth pressure diagrams; structure of Figure 11-19: (*a*) full equivalent fluid pressure with 2-ft live load surcharge; (*b*) half fluid pressure.

Minimum I / Maximum I b	Stiffness Factor k	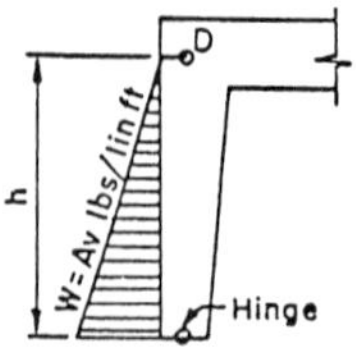 Coefficient of Wh^2	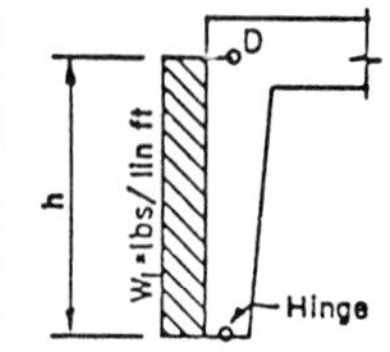 Coefficient of W_1h^2
1.00	3.00	0.117	0.125
0.60	4.41	0.129	0.135
0.30	7.48	0.144	0.148
0.20	10.17	0.153	0.157
0.15	12.66	0.160	0.163
0.12	15.06	0.165	0.167
0.10	17.29	0.169	0.171
0.08	20.59	0.175	0.176
0.06	25.74	0.182	0.182
0.05	29.68	0.187	0.185
0.04	35.1	0.194	0.189
0.03	44.4	0.200	0.194
0.02	62.1	0.213	0.204
		BACKFILL LOADING	SURCHARGE LOADING

FIGURE 11-22 Coefficients for fixed end moment at point D (wall–slab junction); far end hinged.

Lateral Earth Pressures The lateral earth pressure diagram is shown in Figure 11-21*a*, and is based on 36 lb/ft^3 equivalent fluid pressure with a 2-ft live load surcharge. The computation of fixed end moments for triangular loading for tapered beams with hinged-end conditions is complex. Thus, useful data are tabulated in Figure 11-22. The stiffness factors are based on 3 for a wall with a hinged end, and the coefficients shown are for the fixed end moment after the hinge has been released.

From Figure 11-22 and using $b = \min I/\max I = 0.187$, we calculate

$$\begin{aligned} \text{FEM} &= 0.155\, wh^2 + 0.150\, w_1h^2 \\ &= 0.155 \times 0.51 \times 28.3^2 + 0.150 \times 0.10 \times 28.3^2 \\ &= 75.9 \text{ ft-kips} \end{aligned}$$

$$\begin{aligned} \text{SBM} &= 0.128 \times 0.51 \times 28.3^2 + 0.125 \times 0.10 \times 28.3^2 \\ &= 62.2 \text{ ft-kips} \end{aligned}$$

where FEM is the fixed end moment and SBM is the simple beam moment. The moment distribution from this lateral load is shown in Figure 11-23 with the resulting moment diagram, to be superimposed on the total moment diagram.

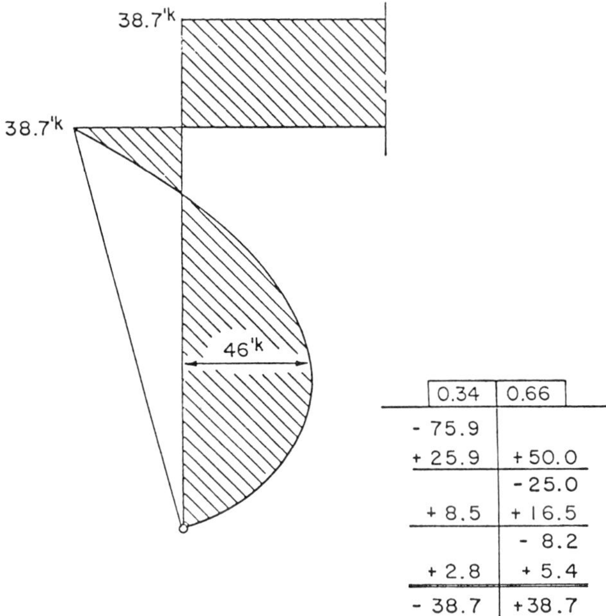

FIGURE 11-23 Moment diagram due to lateral earth pressure; structure of Figure 11-19.

Next, we consider an earth pressure condition that can conceivably exist and reduce the span positive moment. This is shown in Figure 11-21*b* and consists of 18 lb fluid pressure without the 2 ft live load surcharge. The resulting fixed end moment at the top of the wall is 33.5 ft-kips, and will distribute to 17.1 ft-kips. This moment is plotted in the final moment diagram.

Live Load The alternate live load is used, consisting of two axles spaced 4 ft apart. The distribution is $E = 4 + 0.06S = 5.8$ ft. Assuming the bridge width to be 30 ft, the distribution according to the proposed AASHTO NCHRP Project 12-26 is $E = 3.5 + 0.06\sqrt{28.75 \times 30} = 5.3$ (almost the same). Based on $E = 5.8$ ft, the alternate load of two axles is positioned at two locations to produce maximum positive and negative moments, respectively.

For maximum positive moment, the two axles are as shown in Figure 11-24. The simple and fixed end moments are

$$\text{SBM} = 313.6 \times [1.30/(2 \times 5.8)] = 35.3 \text{ ft-kips}$$

$$\text{Left FEM} = 164.7 \times 0.112 = 18.5 \text{ ft-kips}$$

$$\text{Right FEM} = 187.3 \times 0.112 = 21.0 \text{ ft-kips}$$

0.34	0.66	0.66	0.34	
	+ 18.5	- 21.0		
- 6.3	- 12.2	+ 13.8	+ 7.2	
	+ 6.9	- 6.1		
- 2.3	- 4.6	+ 4.0	+ 2.1	
	+ 2.0	- 2.3		
- 0.7	- 1.3	+ 1.5	+ 0.8	
- 9.3	+ 9.3	- 10.1	+ 10.1	
- 0.4	+ 0.4	+ 0.4	- 0.4	Sidesway
- 9.7	+ 9.7	- 9.7	+ 9.7	

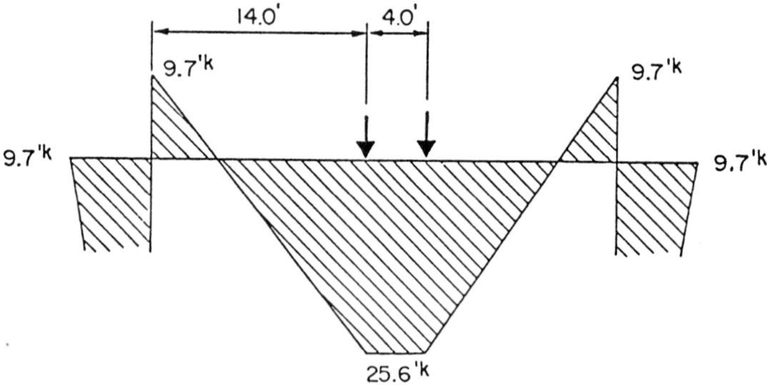

FIGURE 11-24 Axle load position for maximum positive moment, moment distribution, and moment diagram.

The resulting moment distribution is shown in Figure 11-24, together with the moment diagram corresponding to the designated axle position.

Likewise, for maximum negative moment, the two axles are positioned as shown in Figure 11-25. The corresponding simple and fixed end moments are

$$\text{SBM} = 259.8 \times 0.112 = 29.1 \text{ ft-kips}$$

$$\text{Left FEM} = 207.0 \times 0.112 = 23.2 \text{ ft-kips}$$

$$\text{Right FEM} = 109.8 \times 0.112 = 12.3 \text{ ft-kips}$$

The resulting moment distribution and diagram are shown in Figure 11-25 for the two axles placed as shown. Note that when the sidesway correction is applied, this negative moment is less than that obtained from the load diagram of Figure 11-24. The diagram of Figure 11-25, however, provides two relevant points on the moment envelope. These are the extension of the positive moment region close to the left support and the extension of the negative moment region to the left of the right support.

0.34	0.66	0.66	0.34	
	+ 23.2	- 12.3		
- 7.9	- 15.3	+ 8.1	+ 4.2	
	+ 4.0	- 7.7		
- 1.4	- 2.6	+ 5.1	+ 2.6	
	+ 2.6	- 1.3		
- 0.9	- 1.7	+ 0.9	+ 0.4	
- 10.2	+ 10.2	- 7.2	+ 7.2	
+ 1.5	- 1.5	- 1.5	+ 1.5	Sidesway
- 8.7	+ 8.7	- 8.7	+ 8.7	

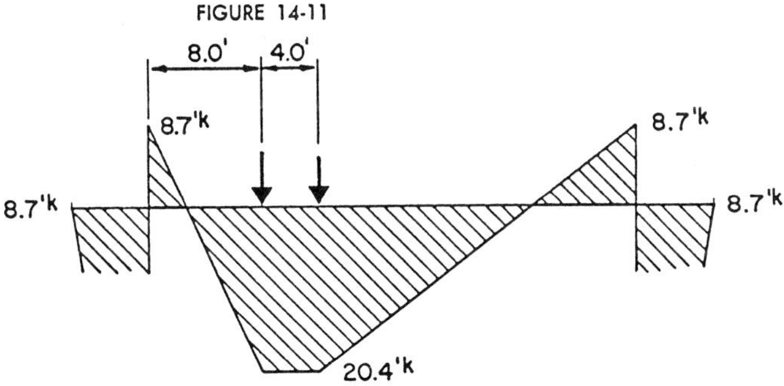

FIGURE 11-25 Axle load position for maximum negative moment, moment distribution, and moment diagram.

Other Loads Temperature stresses may be checked assuming a temperature differential between the top of the walls and the base of 40°F. Given the wall height and the relatively short deck, temperature stresses will not control for the allowable 125 percent stress.

Moment Envelope Moment envelope curves from the preceding calculations are combined in Figure 11-26. As was demonstrated in the example of Section 11-7, the axial loading on the walls from the weight of the slab is small, and the walls may be designed essentially for bending only.

The reinforcement is calculated at four critical sections as follows:

Wall, positive

$$A_s = \frac{42}{1.44 \times 14.5} = 2.01 \text{ in.}^2 \qquad \text{Use \#9 at 6 in.} = 2.00$$

Wall, dead load

$$A_s = \frac{8}{1.44 \times 17.5} = 0.32 \text{ in.}^2 \qquad \text{Use \#5 at 12 in.} = 0.31$$

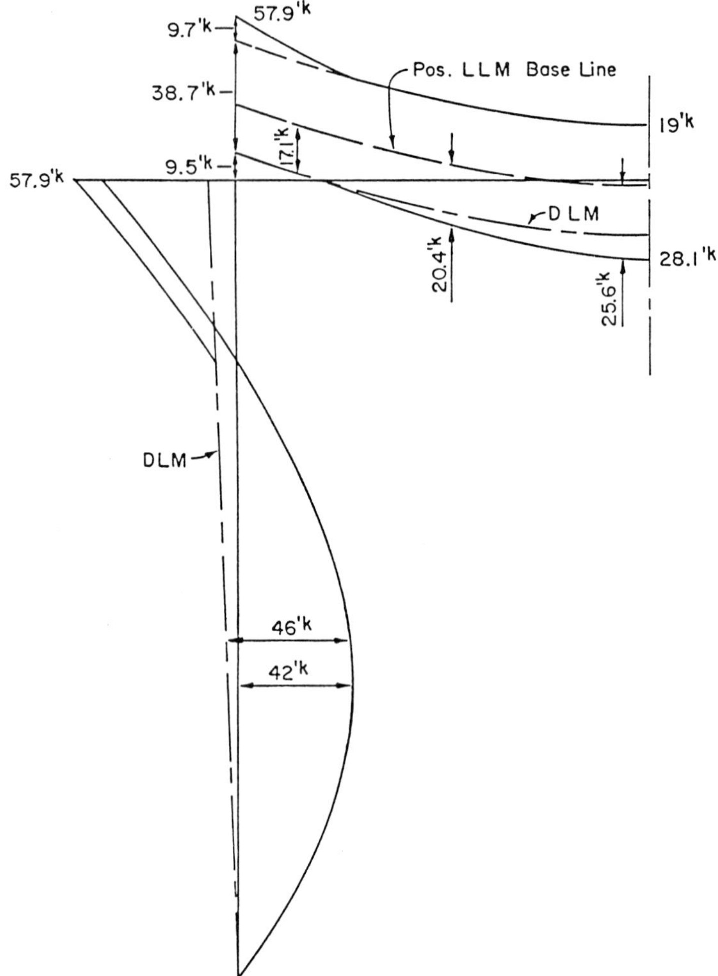

FIGURE 11-26 Moment envelope diagrams.

Heel, negative

$$A_s = \frac{57.9}{1.44 \times 18} = 2.23 \text{ in.}^2 \qquad \text{Use \#10 at 6 in.} = 2.54$$

Slab, positive

$$A_s = \frac{28.1}{1.44 \times 19} = 1.02 \text{ in.}^2 \qquad \text{Use \#7 at 7 in.} = 1.03$$

Footing Block The validity of this analysis requires that the footing block will be prevented from outward (horizontal) movement, because this will change the negative moments at the top corners markedly. If the footings cannot support the horizontal thrust by transferring it effectively to the ground as passive resistance, they should be strutted. Another solution is battered piles as shown in Figure 11-19 to support the resulting shears. Relatively low structures with long spans may require one row of piles to be battered inward.

11-9 SHORT-SPAN BRIDGES WITH HAUNCHES: DESIGN EXAMPLE

The types of slab bridges shown in Figure 3-1 are generally economical for spans in the 30-ft range if simple units are considered, and up to 35-ft spans (end) if continuous units are contemplated. For structures provided with slopes and berms as opposed to full abutments, a shallow superstructure may, however, result in a slightly longer end span and thus increase the total bridge length. This will be demonstrated by the following design example.

We consider a grade separation planned with 2:1 slopes in the end spans and open abutments with berms. Obviously, the berm elevation is determined from the total superstructure depth. A shallower superstructure means a higher berm elevation and therefore somewhat longer end spans. On the other hand, a deeper superstructure means a lower berm elevation which may result in shorter end spans, but will require greater grade difference for the vertical clearance which may again increase the total bridge length. These details are analyzed in the geometric design.

The bridge in this example is 110 to 115 ft long. We will investigate four feasible alternatives, all based on the concept of a three-span continuous unit. These schemes are as follows.

Scheme A This consists of a continuous-slab bridge with parabolic haunches at the interior supports. The assumed slab thickness is 1 ft 5 in. at the end and center spans, and 2 ft 1 in. over the interior supports. The resulting span lengths are 35.0, 43.5, and 35.0 ft, for a total bridge length of 113.5 ft.

Scheme B This is a continuous-slab bridge with constant depth, 1 ft 9 in. The span lengths are basically the same as in scheme A.

Scheme C This is a T-beam concrete bridge with a 7-in. slab and parabolic haunches over the interior supports. The total structure depth is 27.5 in. in the spans and 48.5 in. over the supports. The resulting spans are 33, 44, and 33 ft, for a total bridge length of 110 ft.

Scheme D As in scheme C, this is a T-beam bridge with a 7-in. slab, but with constant depth. The girder is 2 ft 10 in. deep, giving a total structure depth of 3 ft 5 in. The resulting unit has spans 32.5, 40.5, and 32.5 ft, for a total bridge length of 105.5 ft.

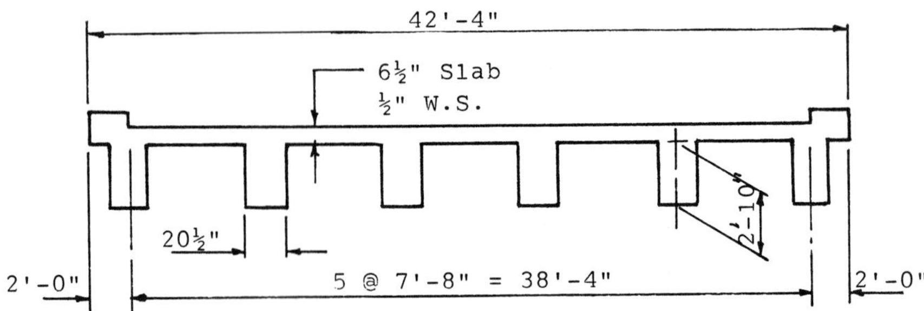

FIGURE 11-27 Typical deck cross section for scheme D.

For all schemes the design live load is HS 20 and alternate. The design stresses are $f_s = 20{,}000$ psi (reinforcing steel) and $f_c = 1200$ or 1400 psi. Allowance is made for a 35-lb/ft^2 future wearing surface, and the effective thickness of the structural slab is taken as 6.5 in. A typical deck cross section is shown in Figure 11-27 for scheme D.

Scheme A The parabolic haunch is shown in Figure 11-28. We note that for $x = 20.75$ ft, $y = 8$ in. $= 0.667$ ft; hence, $K = 8/(20.75^2 \times 12^2) = 1/7750$. At any point along the slab, the depth is determined from the relationship $x^2 = 7750y$, where x is the distance from the origin as shown (in.) and y is measured from the tangent to the curve. For example, at the 05 point (midpoint) of span 1, $x = 4.25$ ft giving $y = 0.34$ in.

Having computed the slab depth at each tenth point for the end and center spans, we calculate the moment of inertia at each of these points and obtain the influence lines based on the exact variation of this parameter.

Dead Load Moments These are computed at three critical points, namely 04, 10, and 15, for the exact weight of the slab. They are as follows:

At supports	$M_{10} = -48.97$ ft-kips
At point 04	$M_{04} = 18.41$ ft-kips
At point 15	$M_{15} = 13.62$ ft-kips

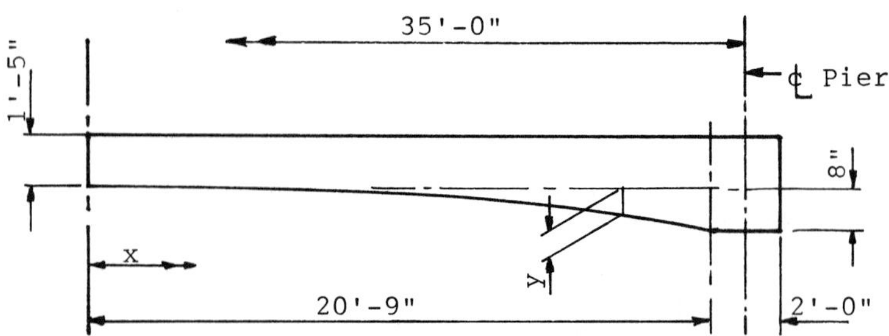

FIGURE 11-28 Detail of parabolic haunch, slab of scheme A.

Live Load Moments Live load distribution is (standard AASHTO formula)

End spans $E = 4 + 0.06 \times 35 = 6.10$ ft
Center span $E = 4 + 0.06 \times 43.5 = 6.60$ ft

According to the proposed AASHTO formula, the distribution is

End spans $E = 3.5 + 0.06\sqrt{35 \times 42.3} = 5.80$ ft
Center span $E = 3.5 + 0.06\sqrt{43.5 \times 42.3} = 6.10$ ft Use 6 ft.

The live load distribution per foot-width of slab is therefore

Truck $16/6 = 2.66$ kips $4/6 = 0.67$ kip
Lane $18/12 = 1.5$ kips (Moment) $26/12 = 2.16$ kips (Shear)
$0.64/12 = 0.053$ kip/ft (Uniform)
Alternate $12/6 = 2$ kips (Spaced 4 ft apart)

Live load moments are likewise computed at points 04, 10, and 15 for truck, lane, and alternate loading, based on influence line coefficients of the exact section. For expediency, we omit the calculations and we summarize the moments as follows:

At supports $M_{10} = -25.05$ ft-kips (Lane load)

or -32.57 ft-kips with impact.

At point 04 $M_{04} = 23.84$ ft-kips (Alternate)

or 30.99 ft-kips with impact.

At point 15 $M_{15} = 22.57$ ft-kips (Truck)

or 29.27 ft-kips with impact.

Total Moments

$$M_{10} = -81.54 \text{ ft-kips}$$
$$M_{04} = 49.40 \text{ ft-kips}$$
$$M_{15} = 42.89 \text{ ft-kips}$$

Reinforcement At supports, the required d is $\sqrt{81.54/0.197} = 20.4$ in. ($f_c = 1200$ psi); the actual $d = 25 - 2.75 = 22.75$ in., OK.

$$A_s = \frac{81.54}{1.44 \times 22.75} = 2.48 \text{ in.}^2/\text{ft} \qquad \text{Use \#10 at 6 in.} = 2.54$$

At point 04, the required d is $\sqrt{49.40/0.197} = 15.8$ in.; the actual $d = 17.00 - 1.50 = 15.5$ in., OK.

$$A_s = \frac{49.40}{1.44 \times 15.5} = 2.20 \text{ in.}^2/\text{ft} \qquad \text{Use \#9 at 5.5 in.} = 2.18$$

At point 15, the required d is $\sqrt{42.89/0.197} = 14.8$ in.

$$A_s = \frac{42.89}{1.44 \times 15.5} = 1.91 \text{ in.}^2/\text{ft} \qquad \text{Use \#9 at 6.5 in.} = 1.95$$

Summary of Quantities The basic quantities (concrete and steel) are computed for this scheme after distribution reinforcement is calculated. They are

$$\text{Concrete} = 285 \text{ yd}^3 \qquad \text{Reinforcement bars} = 58{,}000 \text{ lb}$$

Scheme B Because the objective is to minimize the slab thickness as it is kept constant throughout the bridge, we choose $f_c = 1400$ psi. Then the required d is $\sqrt{81.54/0.249} = 18$ in., or $t = 21$ in., OK.

The moments in this scheme will now be checked using a uniform slab thickness $t = 21$ in., and constant moment of inertia. The span ratio is $43.5/35 = 1.24$.

Dead Load Moments The dead load per foot-width of slab is $1.75 \times 0.15 + 0.035 = 0.03$ kip/ft. The dead load moments are as follows:

$$\begin{aligned}
&\text{At supports} && M_{10} = -0.128 \times 0.3 \times 35^2 = -47.1 \text{ ft-kips}\\
&\text{At point 04} && M_{04} = 0.069 \times 0.3 \times 35^2 = 25.3 \text{ ft-kips}\\
&\text{At point 15} && M_{15} = 0.067 \times 0.3 \times 35^2 = 24.7 \text{ ft-kips}
\end{aligned}$$

Live Load Moments These are computed from influence line coefficients for the actual span ratio, using the live load distribution as in scheme A.

$$\begin{aligned}
&\text{At supports} && M_{10} = -26.00 \text{ ft-kips} && \text{(Includes impact)}\\
&\text{At point 04} && M_{04} = 33.30 \text{ ft-kips} && \text{(Includes impact)}\\
&\text{At point 15} && M_{15} = \text{need not be computed}
\end{aligned}$$

Total Moments

$$\begin{aligned}
M_{10} &= -73.1 \text{ ft-kips}\\
M_{04} &= 58.6 \text{ ft-kips}
\end{aligned}$$

We may consider concrete with $f_c = 1400$ or 1200 psi. For $f_c = 1400$ psi, the required d is $\sqrt{73.1/0.249} = 17.25$ in.; the actual $d = 18.25$, OK. Using

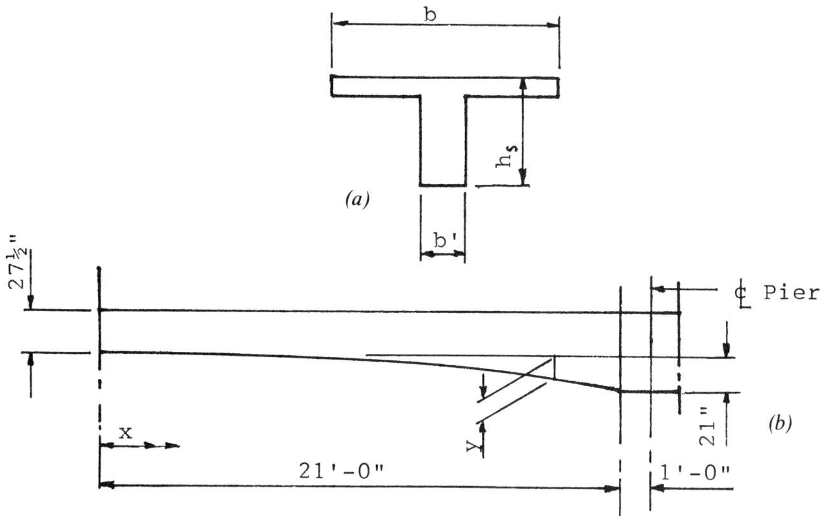

FIGURE 11-29 T-beam cross section and parabolic haunch; scheme C.

$t = 21$ in., we compute the reinforcement requirements as follows:
At supports,

$$A_s = \frac{73.1}{1.46 \times 1.25} = 2.75 \text{ in.}^2/\text{ft} \qquad \text{Use \#10 at 5.5 in.} = 2.77$$

At point 04,

$$A_s = \frac{58.6}{1.46 \times 18.75} = 2.14 \text{ in.}^2/\text{ft} \qquad \text{Use \#9 at 5.5 in.} = 2.18$$

Summary of Quantities The quantities for scheme B are

$$\text{Concrete} = 312 \text{ yd}^3 \qquad \text{Reinforcement bars} = 58{,}500 \text{ lb}$$

Scheme C Referring to Figure 11-29. $b = 7$ ft 8 in. $= 7.67$ ft. We also select $b' = 15.5$ in. and $h_s = 48.5$ and 27.5 in. at the supports and in the span, respectively. The effective slab thickness is taken as 6.5 in., giving a net beam depth of 42 and 21 in., respectively. The span lengths are 33, 44, and 33 ft, and the length of the parabolic curve is 21 ft, as shown in Figure 11-29*b*.

From Figure 11-29*b*, for $x = 21$ ft, $y = 21$ in., giving $K = 1/3024$. The exact moment of inertia is computed at each tenth point for the actual y dimension, and influence line coefficients are calculated using a computer program.

Dead Load Moments For the selected three critical points, the dead load moments are

At supports	$M_{10} = -249$ ft-kips
At point 04	$M_{04} = 67$ ft-kips
At point 15	$M_{15} = 63$ ft-kips

Live Load Moments These are computed using the standard $S/6.0$ distribution factor. Live load plus impact moments are as follows:

At supports	$M_{10} = -317$ ft-kips	(Lane)
At point 04	$M_{04} = 259$ ft-kips	(Alternate)
At point 15	$M_{15} = 191$ ft-kips	(Truck)

Total Moments

$$M_{10} = -566 \text{ ft-kips}$$

$$M_{04} = 325 \text{ ft-kips}$$

$$M_{15} = 311 \text{ ft-kips}$$

Using $f_c = 1200$ psi, the required d is $\sqrt{5.66/1.29 \times 0.197} = 47$ in.; the actual $d = 46.5$ in., OK. The required reinforcement is as follows:

At supports,

$$A_s = \frac{566}{1.44 \times 46.5} = 8.5 \text{ in.}^2 \qquad \text{Use 7 \#10} = 8.9 \text{ in.}^2$$

In the spans,

$$A_s = \frac{326}{1.44 \times 25.5} = 8.9 \text{ in.}^2 \qquad \text{Use 7 \#10} = 8.9 \text{ in.}^2$$

Summary of Quantities The reinforcement steel is approximately estimated as 65,650 lb. The total concrete quantity is 202 yd^3, and includes 19 yd^3 for diaphragms at interior and end supports.

Scheme D A typical cross section for this scheme is shown in Figure 11-27. The spans are 32.5, 40.5, and 32.5 ft, giving a span ratio of 1.25. The live load impact is 30 percent. The total structure depth is 3 ft 4.5 in., and the girder width is 20.5 in.

Dead Load Moments As in the other schemes, we select three critical points. The dead load moments are

At supports	$M_{10} = -220$ ft-kips
At point 04	$M_{04} = 118$ ft-kips
At point 15	$M_{15} = 114$ ft-kips

Live Load Moments These are calculated for a constant moment of inertia and for the actual span ratio. These moments are (including impact)

At supports	$M_{10} = -220$ ft-kips
At point 04	$M_{04} = 279$ ft-kips
At point 15	$M_{15} = 281$ ft-kips

Total Moments

$$M_{10} = -440 \text{ ft-kips}$$
$$M_{04} = 397 \text{ ft-kips}$$
$$M_{15} = 395 \text{ ft-kips}$$

Using $f_c = 1200$ psi, the required d is $\sqrt{440/1.71 \times 0.197} = 36$ in. For a total structure depth of 40.5 in., the actual $d = 37.5$ in., OK. The required reinforcement is as follows:

At supports,

$$A_s = \frac{440}{1.44 \times 37} = 8.4 \text{ in.}^2 \qquad \text{Use 7 \#10} = 8.9 \text{ in.}^2$$

At points 04 and 15,

$$A_s = \frac{397}{1.44 \times 37} = 7.6 \text{ in.}^2 \qquad \text{Use 6 \#10} = 7.6 \text{ in.}^2$$

Summary of Quantities The concrete quantity is 218 yd^3, and includes 16 yd^3 for intermediate and end diaphragms. The quantity of reinforcement bars is approximately estimated as 70,900 lb.

Final Summary: All Schemes A summary of the quantities and costs for each scheme is given in Table 11-1, based on unit prices prevailing at the time this bridge was planned for construction. The higher concrete unit prices for schemes C and D reflect the higher cost of formwork for T beams. A slightly higher unit concrete price for scheme A compared to the same cost for scheme B reflects the curvature of the bottom forms. The cost data shown

TABLE 11-1 Summary of Quantities and Cost: All Schemes

Item	Scheme A	Scheme B	Scheme C	Scheme D
Concrete (yd^3)	285	312	202	218
Reinforcement (lb)	58,000.	58,500.	65,650.	70,900.
Concrete unit price	$130.00	$128.00	$164.00	$164.00
Concrete cost	$37,050.	$39,936.	$33,128.	$35,752.
Reinforcement unit price	$0.24	$0.24	$0.24	$0.24
Reinforcement cost	$13,920.	$14,040.	$15,756.	$17,016.
Total cost	$50,970.	$53,976.	$48,884.	$52,768.

in the table are not current, and for a similar study carried out today they should be adjusted accordingly.

This preliminary analysis provided the basis for structure type selection. Although scheme C is the least expensive, the sponsoring agency chose scheme A for its simplicity, transverse continuity in geometry, and pleasing appearance.

11-10 CANTILEVER BRIDGES

Cantilever bridges, mentioned briefly in Section 3-2, are structures essentially of one span but the longitudinal girders are provided with cantilevers at one or both ends. The ends of the cantilevers may be free or may carry adjoining simply supported spans. Steel plate girder structures with cantilevers are discussed in Chapter 5. These structures are simply supported and therefore statically determinate (moments and shears are calculated by basic rules of statics). Differential foundation settlement does not influence redistribution of load effects.

Bridges with cantilever portions are usually cheaper than simply supported spans but more expensive than continuous girders and rigid frames. Their usefulness is demonstrated where it is necessary to reduce the superstructure depth but without resorting to statically indeterminate designs. Cantilever girders can be used advantageously where sufficiently unyielding foundations, necessary for indeterminate structures, are not easily obtainable. In the past, they have been used for spans up to 220 ft and with ratios of depth to center span as small as 1:35.

In terms of design and construction considerations, we may divide cantilever bridges into two basic groups: (a) bridges with free cantilevers and (b) bridges where the cantilevers receive and support short simple spans.

Bridges with Free Cantilevers

Structures in this category may have the following arrangements: (a) one-span girders with two cantilevers, forming a large center opening and a small overhang at each end; (b) bridges with extensions beyond the end supports to form counterweighed cantilevers; and (c) combinations of any number of spans, each provided with one or two cantilevers.

Single-Span Girders with Two Cantilevers These are statistically determinate. Loads placed on the cantilevers also induce bending moments and shears in the main span. However, loads placed on the main span have no effect on bending moments and shears in the cantilevers.

The following notation is used:

l_1 = length of cantilever
l = length of main span
P_1, P_2, P_3 = concentrated loads on cantilever
a_1, a_2, a_3 = distances from supports of loads P_1, P_2, P_3
M_c = bending moment at supports due to loads on cantilever
V_c = external shear at support due to load on cantilever
M_{cx} = moment in cantilever at point x from support
V_{cx} = shear in cantilever at point x from support

Bending Moments For one-sided loading, moments vary from zero to a maximum at the support in the loaded cantilever, and then to zero at the other support. When both cantilevers are loaded, the moments are M_{c1} at one support and M_{c2} at the other, and vary as a straight line in the span from M_{c1} to M_{c2}. If $M_{c1} - M_{c2}$, the moment in the main span is constant and equal to M_{c1}. For concentrated loads in the cantilevers, $P_1, \ldots, P_s$, the resulting shear and moment diagrams are shown in Figure 11-30*a*.

Moments due to Dead Load For dead load, both cantilevers and the main span are assumed to be loaded simultaneously, although the unit load on the cantilevers may be heavier. The diagrams for shears and moments are shown in Figure 11-30*b*, where M_{c1} and M_{c2} indicate the support moments. Taking line *ab* as a baseline, the diagram in the main span is drawn as shown. The variation in the shear diagram reflects not only variable dead load on the span and cantilevers, but also dead load reactions at the cantilever ends from intermediate spans supported at these points.

Moments due to Live Load For moving live load, the positive moment in the main span is the same as for simply supported bridges. This effect should be considered without loading the cantilevers. The largest negative moment in the main span is produced when both cantilevers are fully loaded, but without placing any live load in the main span.

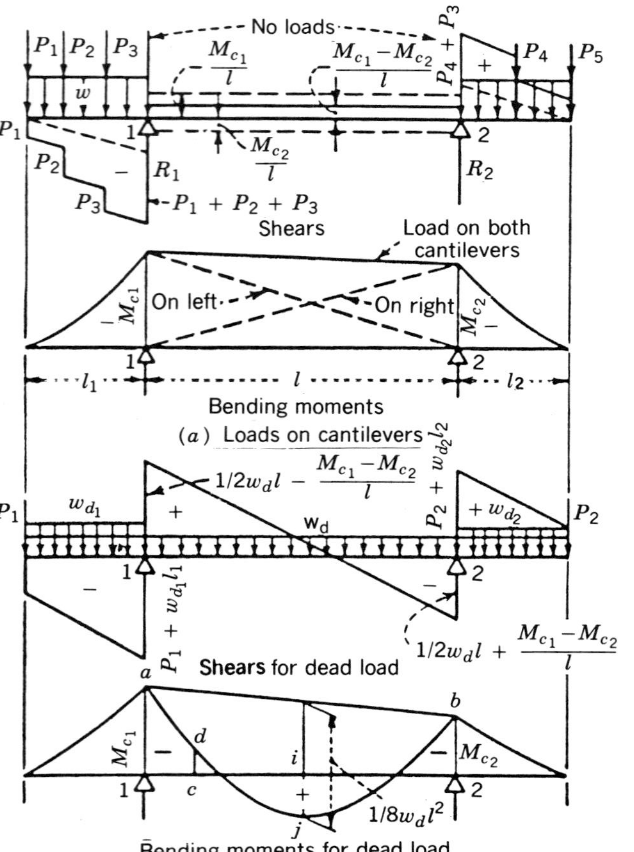

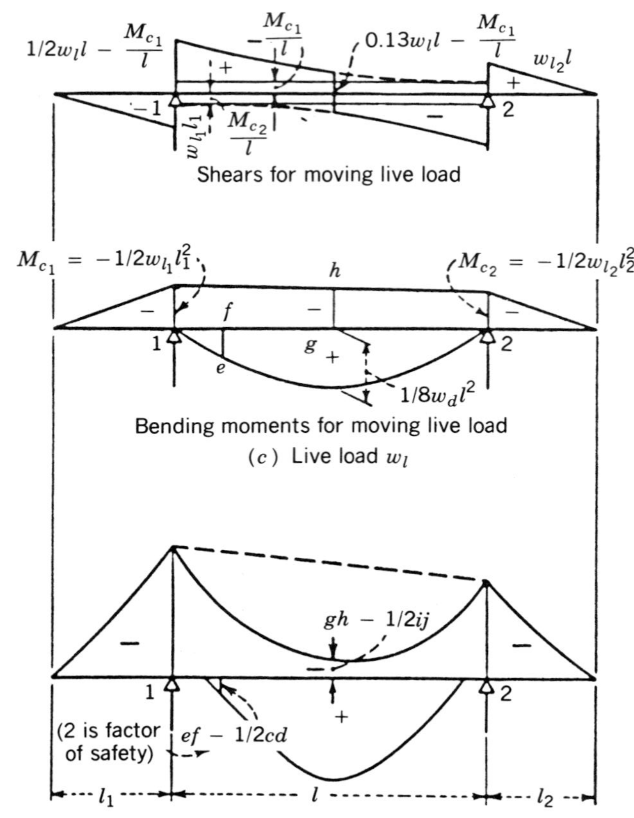

FIGURE 11-30 Moment and shear diagrams for a single span with two cantilevers.

Maximum shear in the main span is produced when one cantilever is loaded, in addition to the maximum and most unfavorable position of load in the main span. Moment and shear diagrams are shown in Figure 11-30*c*.

Combined Bending Moments The combined bending moment diagrams are shown in Figure 11-30*d*, and are self-explanatory.

Single-Span Girder with One Cantilever Bending moment diagrams for a girder with one cantilever are shown in Figure 11-31. In all instances the moment at the right support is zero.

This scheme raises the possibility of uplift at the right (free) end of the girder. Unless this end is anchored, the dead load on the main span must yield a reaction large enough to balance the dead load of the cantilever plus the cantilever live load multiplied by the factor of safety. However, in computing uplift, the bridge may be assumed to act as one unit provided it is sufficiently stiffened by cross beams.

Examples and Applications A single-span with two cantilevers may be used for a bridge with a large center opening and two small side openings. In this configuration abutments are provided at each end to retain the embankment, but they are not end supports for the cantilever. An open expansion joint is therefore provided at these locations.

In some instances additional cost savings may be realized by omitting the abutments and replacing them with apron walls suspended from the cantilever end. The apron walls represent a dead load carried by the cantilever and should be designed to withstand the lateral earth pressure. Although bridges of this type represent past bridge practice, they may well become bridges of the future because of their many advantages.

Bridges of one opening with concealed cantilevers are indicated for large single openings where, because of small available headroom, it is necessary to reduce the superstructure depth. The introduction of negative moment at the ends can contribute markedly to this objective. When an appreciable positive moment reduction is necessary, the cantilevers may be provided with counterweights.

The purpose of the counterweighed cantilevers is to introduce negative dead load moments in the main span, thereby reducing the positive moment. In this design the depth of the load-carrying girders is markedly reduced compared to a simple span with unrestrained ends. As a first step the design is completed in the main span using a floor system yielding minimum dead load. The clearance requirements will dictate the superstructure depth, provided the stipulated span–depth ratio is satisfied. If the controlling vertical clearance is near the midspan, the girders may be haunched. Once the dimensions of the structural system have been determined and the

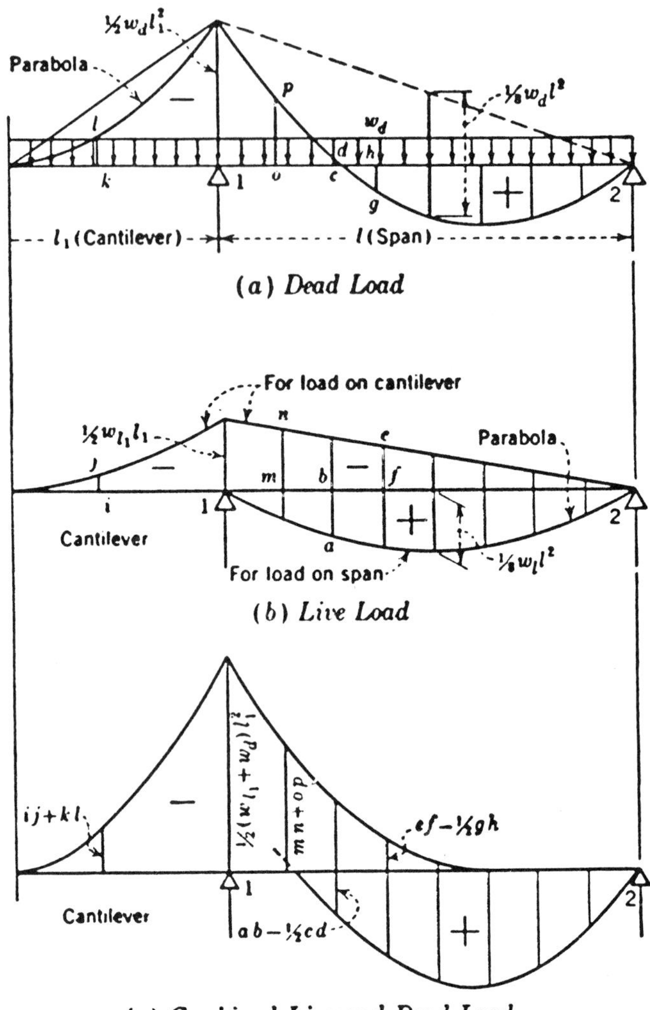

FIGURE 11-31 Moment diagrams for a single span with one cantilever.

desired reduction in the positive moment is established, the cantilever effect is quantified accordingly.

In order to provide a statically determinate system, the bridge must have an expansion bearing at one end (preferably the cantilever), and also vertical expansion joints at both ends. Where the cantilever box also serves as an apron wall, the possible effects must be considered in the design because the freedom of movement may be impaired to some extent both vertically and horizontally due to the interaction with the retained soil.

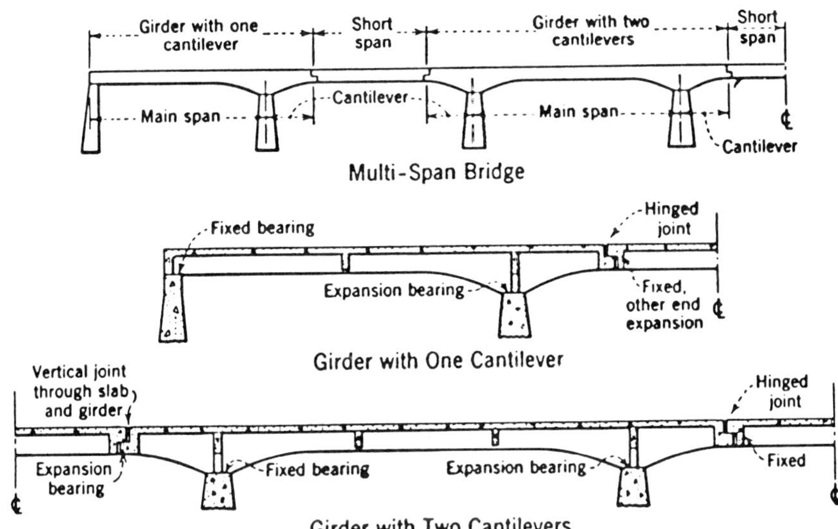

FIGURE 11-32 Combination of girders with cantilevers and short suspended spans.

Cantilevers Supporting Short Simple Spans

This group includes structures consisting of lines of one-span girders, with one or two cantilevers, that carry simply supported spans suspended at the ends. Examples are illustrated in Figure 11-32, and their use is indicated for crossings of three or more openings in locations where it is not practical to select statically indeterminate designs. In effect, the continuous design is changed into simple statically determinate units by providing hinges in alternate spans and at the points of dead load contraflexure. The resulting structure is a series of spans with cantilevers and suspended units.

Because in a girder carrying moving loads the location of the points of contraflexure varies with the loading scheme, it is not possible to reproduce in the substitute structure the same stress conditions that would exist in a continuous unit of the same span lengths. It is possible, however, through a proper cantilever selection to balance the positive moments to an acceptable degree and thus reduce the girder depth at midspan.

Advantages and Disadvantages The advantages of cantilever designs are compatible with those of continuous bridges, namely: (a) material reduction for the load-carrying girders, (b) single bearings at the supports and a resulting reduction in the pier width, (c) concentrically applied loads at the piers, and (d) fewer expansion bearings.

Inherent disadvantages include the design and execution of the details and the arrangement of reinforcement, both requiring previous experience and more technical skills. Compared to continuous bridges, the cantilever design

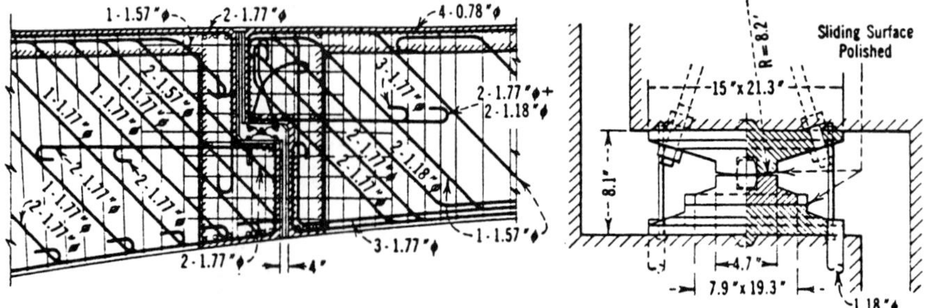

FIGURE 11-33 Expansion bearing for suspended span, bridge across Danube at Dillingen.

offers the advantage of statical determinacy which alleviates the problems of unequal foundation settlement. Thus, cantilever designs should be used only where good foundations cannot be ascertained.

Simple-Span Supports at Cantilever Ends Short girders are usually supported on brackets formed at the ends of cantilevers as shown in the detail of Figure 11-33. Note that the depth of the girder at both the cantilever end and at the end of the suspended section is about the same to provide an uninterrupted line at the bottom. In order to satisfy the shear requirements, the width of the two brackets at each hinge is made larger than the normal girder width, and, in addition, the construction is stiffened with cross beams. Reinforcement in the brackets consists of several bent bars extending into the bracket diagonally as shown to accommodate the shear flow.

Provisions for Expansion and Contraction The bearing details at the hinges should be efficient but inexpensive. At one end of the suspended short span, fixed bearings are normally used allowing, however, rotational girder movement in the vertical plane. Expansion bearings that accommodate rotation should be provided at the other end.

11-11 INTEGRAL BRIDGES

Integral bridges are structures without deck joints. In cold climates such as the northern United States, Canada, and northern Europe, an integrated form of bridge construction has become an attractive option in response to joint-related damage caused by the use of deicing chemicals and the restrained growth of rigid movement.

In multiple spans where intermediate joints are eliminated, the assumption is commonly made that the resulting continuity will induce secondary stresses in the superstructure. These stresses are caused by thermal expansion and contraction, moisture changes and gradients, substructure settlements, and posttensioning. If open joints are eliminated at abutments, secondary stresses will be developed because of restraint introduced by abutment foundations and backfill opposing cyclic movement of the superstructure. The choice of this construction is thus justified if we can demonstrate that for short-to-medium-span bridges of moderate length, more damage and distress are caused by the presence of deck joints than by exposing the superstructure to secondary stresses. Burke (1990) recommends accepting these stresses as part of the design requirements in order to achieve simpler and probably less expensive bridges with enhanced integrity and extended durability.

Continuity in Superstructures This trend, originating in the 1930s (Cross, 1932), has received impetus in current design philosophies. The state of Ohio, for example, has provided the technical background for the movement toward the use of fully integrated continuous construction. Currently, Tennessee appears to be leading the way in long-span continuous bridges. A representative project is the Long Island Bridge at Kingsport, constructed in 1980 as a system of 29 continuous spans without an intermediate joint. This bridge is 2700 ft long. Deck joints and movable bearings have been provided only at the two abutments.

Integral Bridges Currently, 11 states construct continuous bridges with integral abutments for lengths up to 300 ft. Tennessee and Missouri specify this type of construction even for longer bridges. More specifically, Missouri reports examples of steel and concrete bridges with integral decks 500 and 600 ft long, respectively. Tennessee reports examples of bridges 400 and 800 ft long for similar types. At least 20 other state transportation departments have adopted integral construction for continuous bridges.

Structural Distress Although the trend to consider and use integral construction is thus documented, engineers should be cautioned that certain sections and components of these bridges are subjected to very high stresses that are not easily quantified. A recent survey (Emanual et al., 1983) articulates engineering concerns for the potentially higher stresses that might be present in longer bridges. For example, an abutment supported on a single row of piles may be flexible enough to accommodate longitudinal thermal cycling of the superstructure and dynamic end rotations under vehicular traffic. Nonetheless, the piles are routinely subjected to axial and bending stresses that may approach or exceed yield stresses (Wolde-Tinsae, Greimann, and Yange, 1982; Jorgenson, 1983).

Gamble (1984) recommends considering restraint stresses for cast-in-place construction, and cites examples of cracking in continuous concrete frame bridges. In these cases the concrete stress was below the specified yield

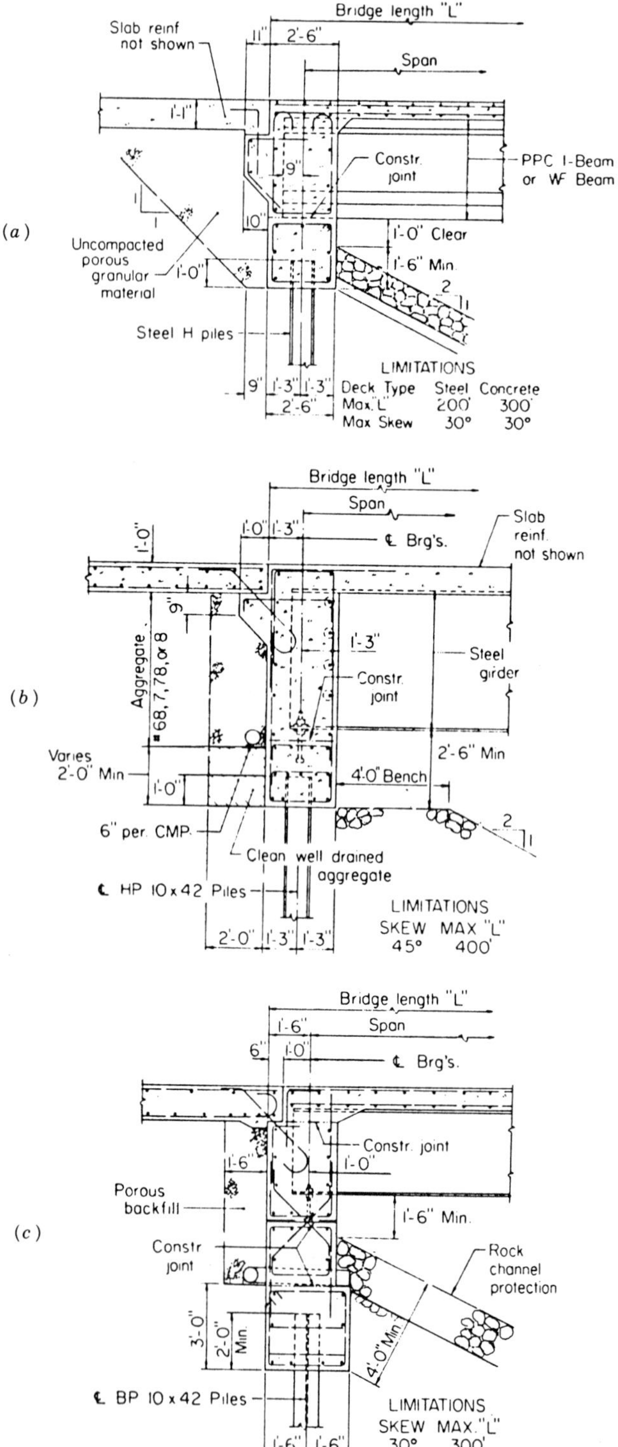

FIGURE 11-34 Abutment details in integral construction: (*a*) Illinois; (*b*) Tennessee; (*c*) Ohio.

strength, but the shear reinforcement did not meet the strength requirements and structural failure was attributed to stiffness and resistance to shrinkage or contraction of the bridge deck.

Where precast concrete or prefabricated members can replace small cast-in-place bridges (see also Section 11-3), the problems associated with initial shrinkage may be eliminated. Consideration, however, should be given to subsequent thermal expansion and contraction.

Integral Bridge Details Abutment–superstructure details are shown in Figures 11-34*a* through *c* for Illinois, Tennessee, and Ohio, respectively. The integral joint connection is between the pile bent, the bridge deck, and the approach slab. In all three details there is a horizontal construction joint between the pile bent and the end concrete diaphragm. For the Illinois detail this diaphragm is cast integrally with the concrete deck, but for the Tennessee and Ohio standard there is another construction joint between the diaphragm and the deck.

The passive pressure developed behind the pile bent as the bridge expands may be controlled in various ways, namely: (a) by limiting the bridge length, skew, and the vertical abutment embedment into the embankment; (b) by using select granular material and uncompacted backfill, as shown in Figure 11-34*a*, and also by providing approach slabs resting on the abutment thus avoiding surcharge loads; (c) by using embankment trenches to shorten wing walls; and (d) by using semiintegral abutments to eliminate passive pressure below bridge seats.

The abutment system becomes more compatible with longitudinal movement if the restraints at this location are inhibited. Among feasible solutions tried by engineers are (a) limiting the foundations of integral bridges to a single row of slender vertical piles, (b) limiting the pile types, (c) orienting the weak pile axis of H piles normal to the direction of movement, (d) using prebored holes filled with fine granular material, (e) providing an abutment hinge to control pile flexure, and (f) using semiintegral abutments for longer bridges as shown in Figure 11-35 to inhibit foundation restraint to longitudinal movement.

Integral Conversion and Retrofitting Burke (1990) documents the current trend toward integral construction and gives examples of conversion of bridges from multiple simple spans to continuous spans. Wasserman (1987) describes and illustrates several such conversions for Tennessee. In addition, the FHWA (1980) has issued a technical advisory on this subject recommending a study of the bridge layout and the joints to determine whether such modifications are feasible and warranted in conjunction with integral designs and unrestrained abutments.

An integral conversion detail at the piers is shown in Figure 11-36. In this example the slab portion of the deck is made continuous by inserting reinforcement bars at the top to resist negative bending. The simply sup-

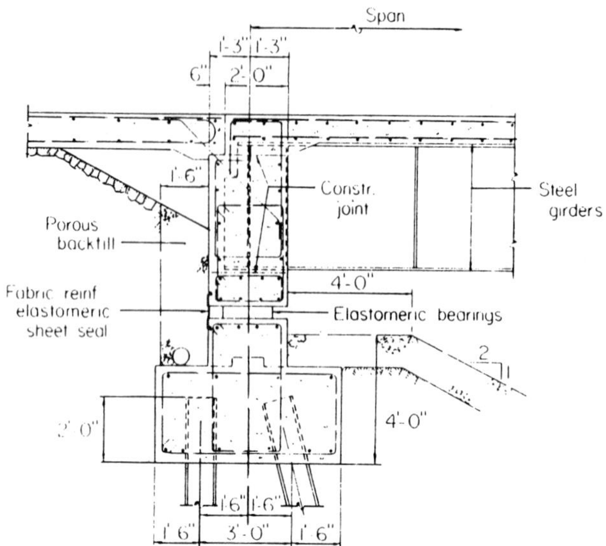

FIGURE 11-35 Semiintegral abutment details used in Ohio.

ported beams remain simple. This conversion is feasible if one or both of the adjacent bearings supporting the beams can accommodate horizontal movement, because this provision will inhibit horizontal forces imposed by beam rotation and slab continuity.

A second detail of pier conversion is shown in Figure 11-37. For deck slabs with bituminous overlay, a membrane may be used for the section over the pier that is removed and replaced. The absence of flexural resistance in this example means that the deck slab may crack as it is exposed to longitudinal

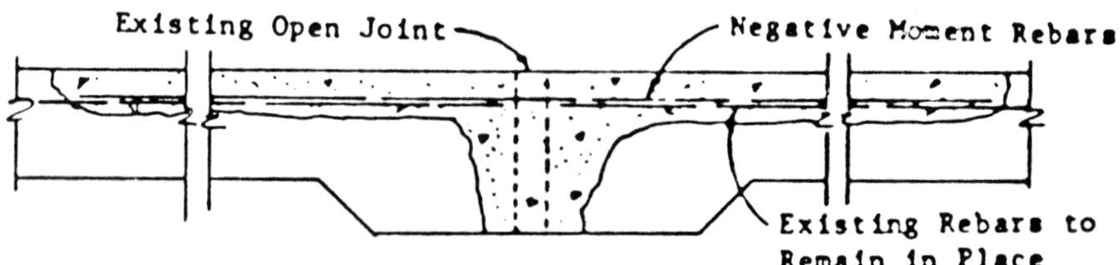

Remove concrete as necessary to eliminate existing armoring, and add negative moment steel at the level of existing top-deck steel sufficient to resist transverse cracking. Generally reconstruct with regular concrete to original grade.

FIGURE 11-36 Conversion of expansion pier into integral construction, Texas.

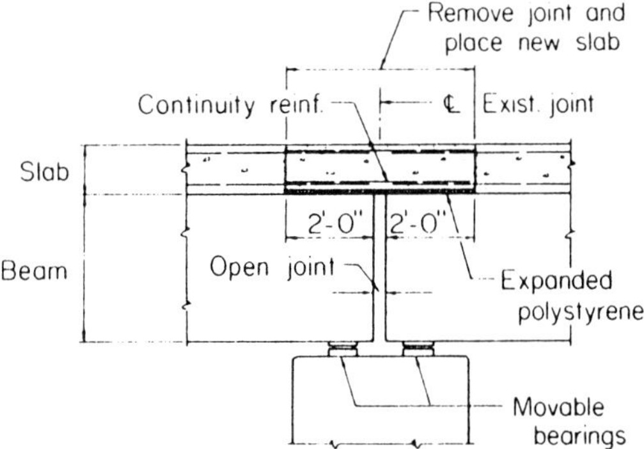

FIGURE 11-37 Integral conversion at pier without deck continuity, Utah.

flexure from beam rotation. Nonetheless, this cracking is preferred for short- and medium-span bridges compared to the adverse effects of an open joint.

Examples of abutment conversion to integral or semiintegral configurations are found in several states (e.g., Ohio). In certain cases, instead of replacing back walls and joints, bearings, and bridge seats, it has been found more economical and functional to reconstruct these details to the pattern of integral bridges.

11-12 OTHER TYPES OF BRIDGES

Double-Deck Roadways These are built in urban areas and along corridors where the right-of-way is limited, but they may also be found across waterways and rivers (e.g., suspension bridges in New York).

Double-deck bridges must satisfy esthetic and utilitarian conditions, and usually require unique designs to meet the associated criteria. Connections to approach roadways and existing streets must be provided with minimum interference with the urban planning, and the design must be carried out with minimum vertical clearance between levels. In addition, the structure must occupy a minimum of space, and have a structural system that can function with minimum depth. Double decks should also have an attractive appearance with the bents blending into the overall structural system.

Delta-Leg Bridges These may be all-concrete or all-steel structures. The rigid delta legs at the piers are usually selected based on a comparative economic study and because they enhance the appearance of the structure. Extensive computer use is usually required in girder analysis and design because the delta legs are built integral with the superstructure.

Trestles These are often described as short decks supported on pile bents (see also Section 14-3). For the usual deck configuration of precast slabs, the spacing of bents is 20 to 30 ft, depending also on the foundation conditions and required pile length.

The piles comprising a trestle bent are connected at the top by a concrete cap, cast in place after the piles are driven and cut off to proper grade. The piles are normally allowed to extend into the cap with suitable reinforcement placed to stiffen the connection.

With a deck slab, the cap may be designed as a continuous beam supported by the piles and loaded by the deck reactions. With a concrete deck and beams, the piles may be placed directly under each beam if this satisfies the economic requirements.

In long trestles or where unusual foundation conditions exist, the bridge is strengthened longitudinally by double bents, usually at every fifth support. Double bents increase the resistance to longitudinal forces and improve stability in soft or loose soil. Alternatively, pier bents may be used in lieu of pile bents, and examples are trestles in deep water or where it is necessary to provide a greater stiffness than can be obtained with pile bents. The piers usually rest on two or three rows of piles.

Special-Purpose Bridges This category includes pedestrian bridges, bridges to carry commercial aircraft loads, relocatable bridges, bridges to carry power and other utility lanes, and bridges often built over expressways and toll roads to provide service facilities. The design criteria and loads are considerably different from highway loadings, and are beyond the scope of this discussion.

Preassembled Steel Space Frame Bridges Preassembled space frames have been used in long span support structures for highway signs. A sectionalized bridge consisting of aluminum alloy has been developed by the U.S. Marine Corps to carry a range of loads over varying spans. Segmental construction has demonstrated the feasibility of off-site construction, including transporting and erecting large sub-assemblies. Considering the advantages of reduced weight and convenience of handling, preassembled steel space frame bridges should be investigated where numerically controlled steel fabrication methods can further reduce shop labor.

The comparative study presented in this section can serve as an example of cost comparison with conventional designs. Characteristics to be considered in alternate solutions should include: (a) structural safety in component use; (b) efficient fabrication; (c) minimum number of component parts; (d) minimum erection time; (e) efficient transportability; and (f) minimal material and labor cost combined with nominal maintenance. The study compares a preassembled steel space frame bridge and three conventional highway grade separation bridges (Sprinkel and Morris, 1976).

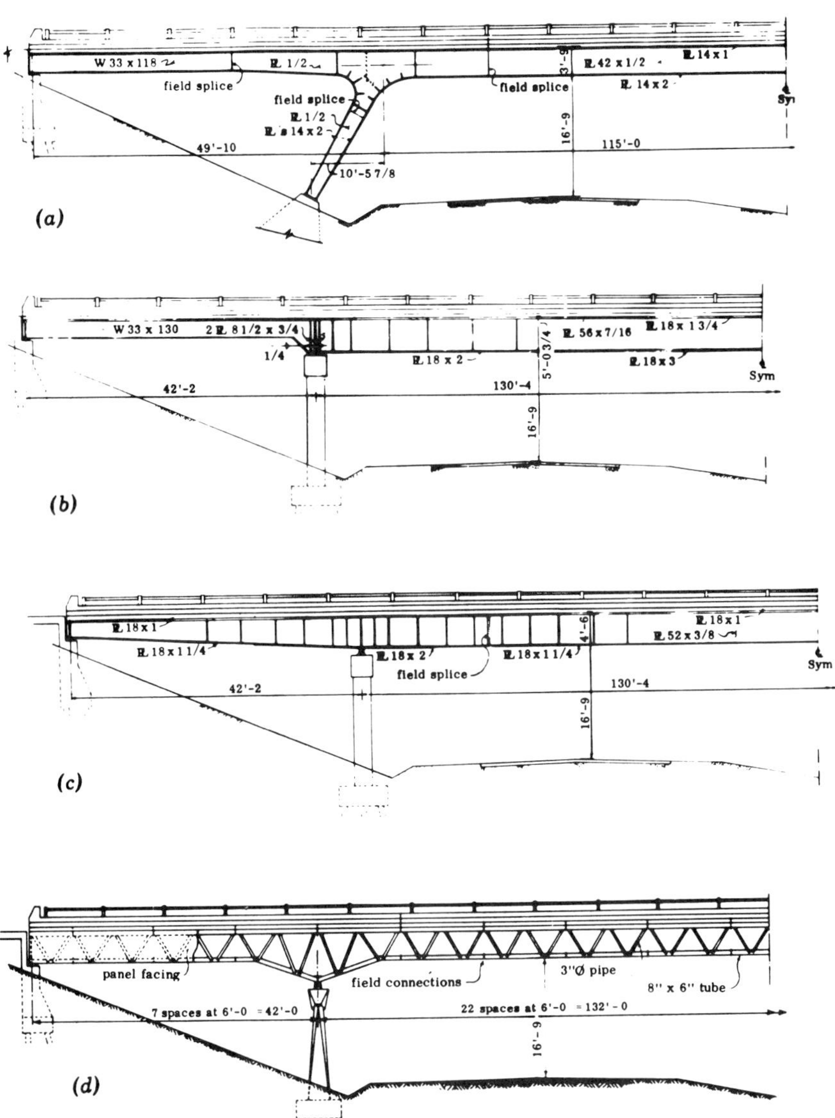

FIGURE 11-38 Alternate plans: (a) plan A: rigid frame; (b) plan B: simple span; (c) plan C: continuous span; (d) plan D: space frame. (From Sprinkel and Morris, 1976.)

The schemes selected for comparison are shown in Figure 11-38. The designs of the conventional bridges shown in parts (*a*), (*b*) and (*c*) represent a rigid frame, three simple spans, and a three-span continuous steel bridge. All three designs have composite concrete decks with shear studs and hybrid steel selection. Approach slabs and the additional height of superstructure as it affects the profile of the approach roadway are not considered. The scheme

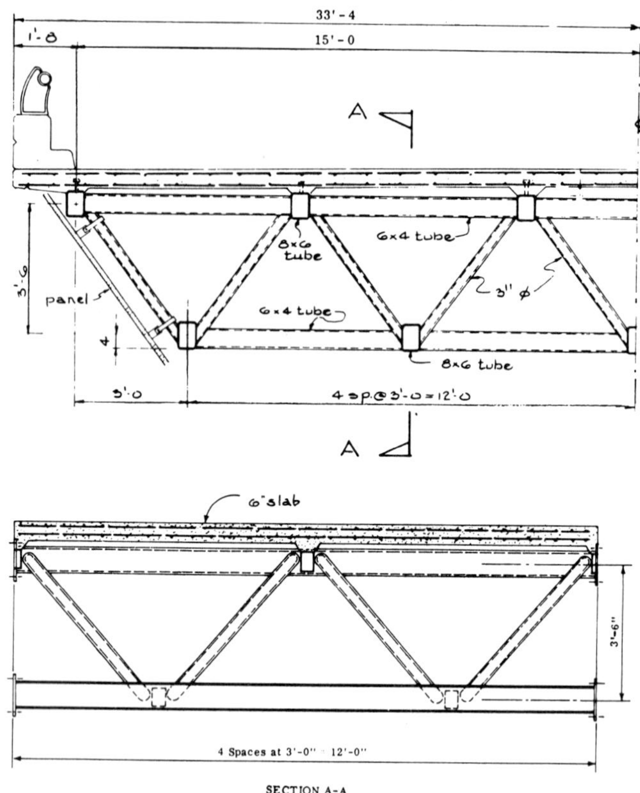

FIGURE 11-39 Space frame modular unit. (From Sprinkel and Morris, 1976.)

shown in Figure 11-38(*d*) is a steel space frame with precast concrete deck erected in transverse segments. A precast pier assembly with a posttensioned cap beam is an optional detail.

The basic modular space frame unit is shown in Figure 11-39. The unit is 30 ft wide (transverse dimension) and 12 ft long, with a 6-in. thick composite concrete deck, cast at an off-site assembly plant. Pipe sections comprise the chords, and rectangular tubes are chosen for the longitudinal and transverse positions to facilitate fabrication and compactness of the joints.

A summary of quantities for all four plans is given in Table 11-2. Additional details for bearing assemblies, splices, and welds were developed for cost estimates. For the structural steel, the material cost per unit weight is considerably higher for tube and pipe sections than for plates and rolled sections. Other relative differences in costs between the four designs reflect fabricator's shop layout, local labor supply, and familiarity with this type of work.

TABLE 11-2 Summary of Quantities for Plans, A, B, C, and D (from Sprinkel and Morris, 1976)

Item (1)	A (2)	B (3)	C (4)	D (5)
A3 concrete (yd^3)	110	140	139	130
A4 concrete (yd^3)	228	228	228	184
Excavation (yd^3)	700	700	700	700
Reinforcement (lb)	63,200	71,000	71,000	71,000
Structural steel (lb)	218,100	239,300	208,700	137,600
Railing (ft)	426	426	426	426
Steel piling (ft)	2,480	2,240	2,240	2,080
Concrete slope (ft^2)	400	400	400	400
Precast piers	0	0	0	0

A range of total costs was established by combining the most economical fabrication estimate with the most economical construction estimate. The same combinations were used to determine the least and most expensive cost of each structure. Based on these data, the total cost for plan (D) is 12% higher than the cost for the conventional plans, and the lowest estimate for this plan is 3% more than the lowest estimate for the conventional plans. It may appear, therefore, that the preassembled space frame bridge can be competitive because of reduced on-site construction time and where quality control is an important consideration. Design areas that may require further study are secondary stresses, structural vibration, and fatigue.

REFERENCES

Anderson, A. R., 1972. "Systems Concepts for Precast and Prestressed Concrete Bridge Construction," HRB Special Report 132, pp. 9–21.

Biswas, M., J. S. Biffland, R. E. Schofield, and A. E. Gregory, 1974. "Bridge Replacements with Precast Concrete Panels," Special Report 148, TRB, National Research Council, Washington, D.C. pp. 136–148.

Braun, W. M., 1972. "Posttensioned Diaphragm Walls in Italy," *Ground Eng.*, March.

Burke, M., 1990. "Integral Bridges," Transp. Research Record 1275, TRB, National Research Council, Washington, D.C., pp. 53–61.

Cross, H., 1932. *Continuous Frames of Reinforced Concrete*, Wiley, New York.

Emanual, J. H., et al., 1983. "Current Design Practice for Bridge Superstructures Connected to Flexible Substructures," Univ. Missouri, Rolla.

FHWA, 1968. "Standard Plans for Highway Bridge, Vol. 1, Concrete Superstructures," Washington, D.C. Aug.

FHWA, 1980. "Bridge Deck Joint Rehabilitation (Retrofit)," *Tech. Advisory* T5140.16, FHWA, U.S. Dept. Transp., Washington, D.C.

Fuchsberger, M., 1980. "The Posttensioned Diaphragm Wall," *Proc. 1980 Symp. Slurry Walls for Underground Transp. Facilities*, FHWA, U.S. Dept. Transp., Aug. 30–31, Cambridge, Mass.

Fuchsberger, M. and H. J. Gysi, 1978. "Posttensioned Diaphragm Wall at Izlams O' Th' Height—Manchester," *Tech. Paper*, Eighth Intern. Congress F.1.8, London, May.

Gamble, W. L., 1984. Bridge Evaluation Yields Valuable Lesson. *Concrete International*, June, pp. 68–74.

GangaRao, H. V. S., 1978. "Conceptual Substructure Systems for Short-Span Bridges," *Transp. Eng. Journ. ASCE*, Jan.

Greimann, A. M., D. M. Wolde-Tinsae, and P. S. Yang, 1983. "Skew Bridges with Integral Abutments," *Transp. Research Record* 903, TRB, National Research Council, Washington, D.C.

Gysi, H. J., et al., 1977. "Behavior of a Prestressed Diaphragm Wall," *Proc. Ninth Intern. Conf. Soil Mech. Found. Eng.*, Vol. 2,3–17, Tokyo.

Gysi, H. J., A. Linder, and R. Leoni, 1975. "Prestressed Diaphragm Walls," *Proc. 1975 European Conf. Soil Mech. Found. Eng.*, Vienna.

Jorgenson, J. L., 1983. "Behavior of Abutment Piles in an Integral Abutment Bridge," Transp. Research Record 903, TRB, National Research Council, Washington, D.C.

Leonard, M., 1974. "Precast Diaphragm Walls Used for the A13 Motorway, Paris," *Proc. Diaphragm Walls Anchorages*, Institution Civ. Engineers, London.

Lutz, J. G. and D. J. Scalia, 1984. "Deck Widening and Replacement of Woodrow Wilson Memorial Bridge," *J. PCI J.*, Vol. 29, No. 3, May–June, pp. 74–93.

Martin. L. D. and A. E. N. Osborn, 1983. "Connections for Modular Precast Concrete Bridge Decks," FHWA/RD-82/106, Aug., pp. 2, 33–58.

Panak, J. J., 1982. "Economical Precast Concrete Bridges," Texas Dept. Highways and Public Transp., Research Report 226-1F, March, pp. 1, 12–18, 27–32.

PCI, 1975. *Short Span Bridges*, Chicago.

Smyers, W. L., 1984. "Rehabilitation of the Fremont Street Bridge," *PCI J.*, Vol. 29, No. 5, Sept.–Oct., pp. 34–51.

Sprinkel, M. M., 1978. "Systems Construction Techniques for Short Span Concrete Bridges," Transp. Research Record 665, Vol. 2, TRB, National Research Council, Washington, D.C., pp. 226–227.

Sprinkel, M. M., 1985. "Prefabricated Bridge Elements and Systems," NCHRP 119, TRB, National Research Council, Washington, D.C., Aug.

Sprinkel, M. M., and S. Morris, 1976. "Preassembled Steel Space Frame Bridges," Virginia Highway and Transportation Research Council, Oct.

Transportation Research Board, 1976. "State of the Art—Permanent Bridge Deck Forms," TRB Circular 181, Sept.

Virginia Highway and Transportation Research Council, 1980. "Bridges on Secondary Highways and Local Roads Rehabilitation and Replacement," NCHRP 222, TRB, National Research Council, Washington, D.C., May, pp. 74–111.

Virginia Highway and Transportation Research Council, 1981. "Rehabilitation and Replacement of Bridges on Secondary Highways and Local Roads," NCHRP 243, TRB, National Research Council, Washington, D.C., Dec., pp. 36–46.

Virginia Prestressed Concrete Association, 1973. "New Approaches in Prestressed, Precast Concrete for Bridge Superstructure Construction in Virginia," Aug., pp. 4, 6, 10, 14, 17, 19, 25.

Wasserman, E., 1987. "Jointless Bridges," *Eng. J.*, Vol. 24, No. 3.

Wolde-Tinsae, A. M., and J. E. Klinger, 1987. "Integral Abutment Bridge Design and Construction," Report FHWA/MD-87/04, Maryland Dept. of Transportation, Annapolis.

Wolde-Tinsae, A. M., L. F. Greimann, and P. S. Yange, 1982. "Nonlinear Pile Behavior in Integral Abutment Bridges," Iowa State Univ., Ames.

Xanthakos, P. P., 1979. *Slurry Walls*, McGraw-Hill, New York.

Xanthakos, P. P., 1991. *Ground Anchors and Anchored Structures*, Wiley, New York.

Xanthakos, P. P., 1993. *Slurry Walls as Structural Systems*, McGraw-Hill, New York.

Zuk, W., 1980. "Kinetic Bridges," Virginia Hwy. Transp. Research Council, VHTRC 81-R6, Charlottesville, July, pp. 2–33.

CHAPTER 12

TOPICS RELEVANT TO DESIGN

12-1 LATERAL WIND BRACING IN PLATE GIRDER BRIDGES: DESIGN EXAMPLE

In the example of Section 5-9 involving a two-girder bridge with spans of 100, 141, and 100 ft, wind load bracing was provided assuming that one-half of the lateral wind is resisted along the plane of the bottom flange.

Until 1979, plate girder bridges with spans of 125 ft or longer were required to have bottom lateral wind bracing. Subsequent revisions stipulated a more rational approach in determining the need for bottom bracing. Current procedures recommend pressures on the windward and leeward sides to be applied simultaneously in the assumed wind direction. Typically, a bridge structure should be examined separately under wind pressure from various directions to determine the combined windward and leeward pattern producing the most critical effect.

Allowable wind-induced stresses in the bottom flange when the top flange in the system is continuously supported (e.g., embedded in the concrete slab) are specified by AASHTO for two cases: when bottom bracing is provided and when this bracing is omitted. The proposed LRFD specifications make similar distinctions for girders attached to concrete decks or other decks of comparable rigidity. An illustrative example of these provisions will demonstrate their application (United States Steel, 1983).

A continuous plate girder bridge with span lengths of 273, 350, and 273 ft has the typical cross section shown in Figure 12-1. Stresses caused by wind loading will be considered only in the exterior girders. The sum of the flange stresses caused by Group II loading (wind and longitudinal dead load

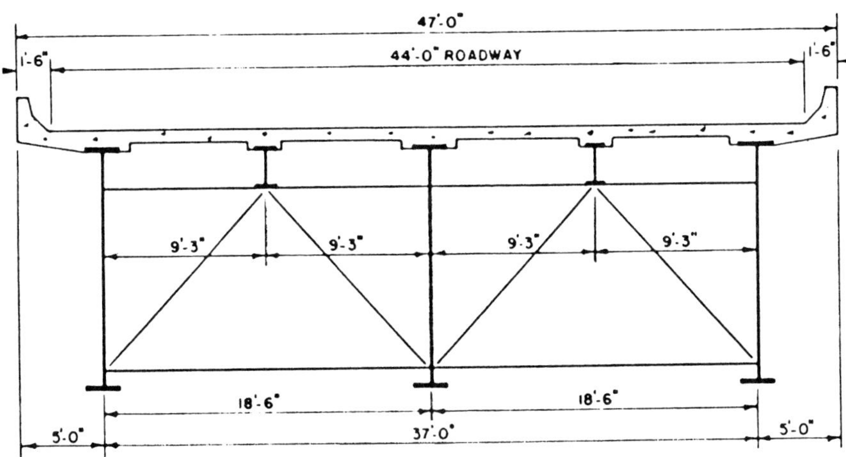

FIGURE 12-1 Typical cross section of bridge.

bending) and Group III loading (wind plus dead and live load) will be checked for tension and compression limit stresses. For tension the limit stress is F_y, and for compression it is the lower between yielding and buckling.

Assuming noncompact sections and a load factor design, we can rewrite AASHTO equation (10.98) as

$$F_{ye} = \left(2200\frac{t}{b'}\right)^2 \leq F_y \tag{12-1}$$

where F_{ye} is the effective compressive limit stress (assumed smaller than the lateral torsional buckling stress). If the actual computed stress exceeds F_{ye}, the design may be rectified by (a) reducing the diaphragm or cross-frame spacing in the overstressed region, (b) increasing the flange width, and (c) adding bottom lateral bracing in the exterior bays of the overstressed span.

Grouping wind, dead, and live loads according to AASHTO Table 3.22.1A, we obtain

$$\text{Group II} = 1.3(D + W)$$
$$\text{Group III} = 1.3(D + L + I + 0.3W)$$

Note that wind on the live load for Group III is assumed to be zero for the bottom flange because this load is resisted at the deck level and carried directly by the concrete slab. For a member that does not meet AASHTO equation (10.100) for noncompact sections, an unbraced condition is assumed whereby the maximum strength is computed from AASHTO equation

(10.102). When the ratio of the smaller moment to the larger moment at the ends of the braced length L_b is less than 0.7, M_u may be increased 20 percent but may not exceed $F_y S$. By rewriting AASHTO equation (10.102), we derive an expression giving the critical buckling stress

$$F_{cr} = \frac{M_u}{S} = F_y\left[1 - \frac{3F_y}{4\pi^2 E}\left(\frac{L_b}{b'}\right)^2\right] \tag{12-2}$$

where b' is the width of the projecting compression flange of the girder in the stepped-down region, L_b is the cross-frame spacing, and S is the section modulus in the stepped-down region. The limit stress is therefore computed from (12-1) or (12-2), whichever yields a lower value.

It should be noted, however, that the 1992 AASHTO edition has modified Equation (10-102). The maximum strength M_u is computed from

$$M_u = M_r R_b$$

where R_b is a factor less than or equal to unity, and M_r is the yield moment defined in terms of relevant parameters. Hence, the present analysis should include the effects of the new criterion if the modified AASHTO Equation (10-102) controls.

The total superstructure depth (not shown in Figure 12-1) is 200 in., and half of the wind load acting on this area is applied in the plane of the bottom flange, or wind $W = 8.3 \times 0.05 = 0.42$ kip/ft of lateral load.

Section 34 ft from Left Support Referring to Figure 12-2 which shows the elevation view of the exterior girder and the cross section of the exterior girder, it is obvious that the section to be analyzed first is the lower flange transition 34 ft from the left support. Relevant data at this location are $L = 273$ ft; $L_b = 24.8$ ft; $t_f = 0.75$ in.; $b_f = 18$ in.; $f_{DL} = 20.0$ ksi; $f_{LL+I} =$ 7.94 ksi (from design).

From AASHTO Article 10.20.2, the maximum induced stress in the bottom flange when the top flange is continuously supported is

$$F_w = RF_{cb} \tag{12-3}$$

where

$$R = (0.2272L - 11)S_d^{-2/3} \tag{12-4}$$

and

$$F_{cb} = \frac{72M_{cb}}{t_f b_f^2} \qquad M_{cb} = 0.08WS_d^2 \tag{12-5}$$

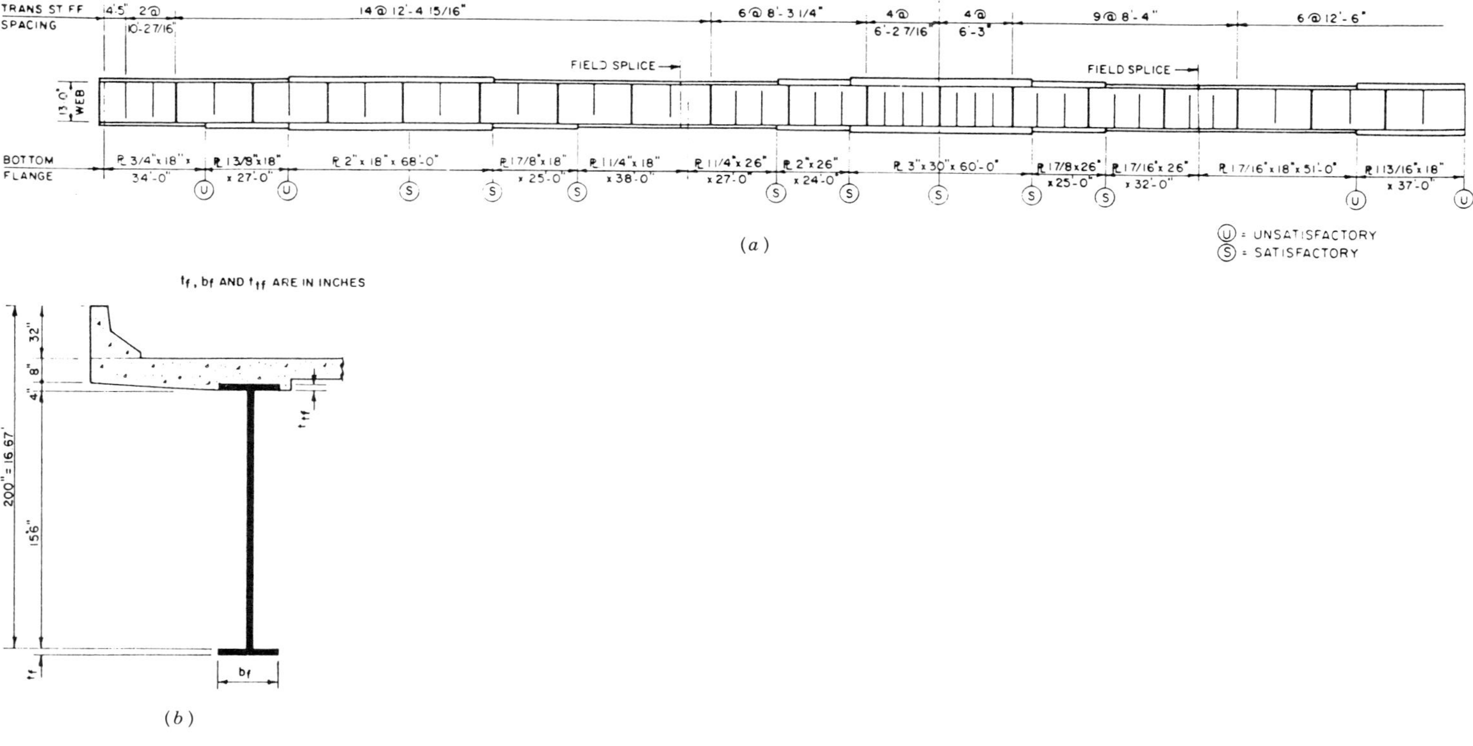

FIGURE 12-2 (*a*) Exterior girder elevation; (*b*) exterior girder and fascia cross section.

where W is the wind load along the exterior flange, $S_d = L_b$ is the diaphragm spacing, and other parameters are as shown. We now compute the following:

$$R = [(0.2272)(273) - 11]24.8^{-2/3} = 6.00$$

$$M_{cb} = (0.08)(0.42)(24.8)^2 = 20.6 \text{ ft-kips}$$

$$F_{cb} = \frac{(72)(20.6)}{(0.75)(18)^2} = 6.10 \text{ ksi}$$

$$F_w = 6 \times 6.10 = 36.6 \text{ ksi}$$

The total factored stresses are now computed as follows ($F_y = 50$ ksi):

$$\text{Group II} = 1.3(D + W) = 1.3 \times (20.00 + 36.6) = 73.6 \text{ ksi} > 50 \text{ ksi}$$

$$\text{Group III} = 1.3(20.0 + 7.94 + 0.3 \times 36.6) = 50.6 \text{ ksi} > 50 \text{ ksi}$$

At this location the bottom flange is in tension; hence, the limit stress is F_y, which is exceeded.

Section 129 ft from Left Support At this location the design parameters are $t_f = 1.875$ in.; $b_f = 18$ in.; $f_{\text{DL}} = 18.6$ ksi; $f_{\text{LL}+I} = 10.85$ ksi (from design). The factors S_d, L, and R are the same as given previously. Likewise, $M_{cb} = 20.6$ ft-kips. We now compute

$$F_{cb} = \frac{(72)(20.6)}{(1.875)(18)^2} = 2.46 \text{ ksi}$$

$$F_w = 6.00 \times 2.46 = 14.75 \text{ ksi}$$

The total factored stresses are

$$\text{Group II} = 1.3(18.60 + 14.75) = 43.35 \text{ ksi} < 50 \text{ ksi} \qquad \text{OK}$$

$$\text{Group III} = 1.3(18.60 + 10.85 + 0.3 \times 14.75) = 44.04 \text{ ksi} < 50 \text{ ksi} \qquad \text{OK}$$

Section 219 ft from Left Support This section is in compression, with $f_{\text{DL}} = -15.90$ ksi and $f_{\text{LL}+I} = -10.00$ ksi (from design). Other relevant parameters are $t_f = 1.25$ in.; $b_f = 26$ in.; and W, L, S_d, and M_{cb} as given previously. Next, we compute

$$F_{cb} = \frac{(72)(20.6)}{(1.25)(26)^2} = 1.76 \text{ ksi}$$

$$F_w = 6 \times 1.76 = 10.55 \text{ ksi}$$

The effective compressive limit stress is now computed from (12-1) as

$$F_{ye} = \left[2200\left(\frac{1.25}{13}\right)\right]^2 = 44.75 \text{ ksi} \qquad \text{(Controls)}$$

Note that F_{cr} computed from (12-2) is 46.56 ksi, and does not control. The limiting value of S_d is

$$\frac{(20{,}000)(26)(1.25)}{(12)(50)(156 + 1.25 + 0.812)} = 6.85 \text{ ft} < 24.8$$

The total factored stresses are

$$\text{Group II} = 1.3(15.90 + 10.55) = 34.39 \text{ ksi} < 44.75 \text{ ksi}$$
$$\text{Group III} = 1.3(15.90 + 10.00 + 0.3 \times 10.55) = 37.78 \text{ ksi} < 44.75 \text{ ksi}$$

Results of the analysis for the entire section are indicated in the girder elevation shown in Figure 12-2*a*, and are obtained using a similar procedure. We should note that at support 2, F_{cr} determined from (12-2) is the controlling limit stress. The requirements applied to Group II and III are satisfied at all sections except for five locations: two flange transition sections adjacent to the end bearing and three sections near the middle of the interior span (United States Steel, 1983).

Possible solutions to lower wind stresses at these locations include (a) increasing the width of the bottom flange (keeping the flange area the same but increasing the width reduces F_W, whereas other stresses remain about the same); and (b) reducing the cross-frame spacing in areas where over-stressing occurs (wind stresses are inversely proportional to the square of the cross-frame spacing S_d, and are thus markedly decreased by nominal adjustments in this dimension).

Introducing lateral bracing is a third solution. In this case the parameter R in AASHTO Article 10.20.2.1 is determined from (10-7) and results in lower wind stresses. In the context of economy, this solution should be considered where the first two options are not feasible. In the foregoing example, changing the configuration of the bottom flange is the most economical solution considering the relatively heavy cross frames (similar to the example of Section 5-8). With multigirder bridges, these cross frames should be lighter, and the corresponding solution of decreasing their spacing should be investigated.

12-2 UNINTENDED COMPOSITE ACTION IN BRIDGES

Analysis of available data on unintended composite action in conventional steel beam-and-slab bridges shows that the presence of a natural or a chemical bond is the prime factor determining the ability of the system to act compositely. Interestingly, there is good evidence confirming composite action in bridges that were not designed to act compositely.

Test Results Suetoh, Burdette, Goodpasture, and Deatherage (1990) have presented a summary of results of tests on bridges designed and built without provisions for composite action but where composite action was documented.

A bridge tested to failure (Burdette and Goodpasture, 1971) consisted of a three-span continuous steel beam with concrete slab, noncomposite. The load at failure was compared with the ultimate load predicted according to current procedures considering the entire section as a wide beam.

The load–deflection diagram for this structure is shown in Figure 12-3. The computed load–deflection curves for the four 27-in. steel rolled beams were first developed assuming noncomposite behavior. However, deflection measurements from the actual behavior show considerable composite action at load levels approaching the yielding load of the steel in the noncomposite bridge. The bond (adhesion or friction) developed at the steel–concrete interface appears to be sufficient to justify composite action. The average shear stress at the steel–concrete interface was approximately 230 psi at a load of 500 kips. The load–deflection diagram prepared assuming composite action in the elastic range matches the observed performance consistently and up to a load representing the capacity of the noncomposite bridge.

Full-scale laboratory and field tests have been performed by Kissane (1985) to determine the resistance to elastic buckling of a steel beam

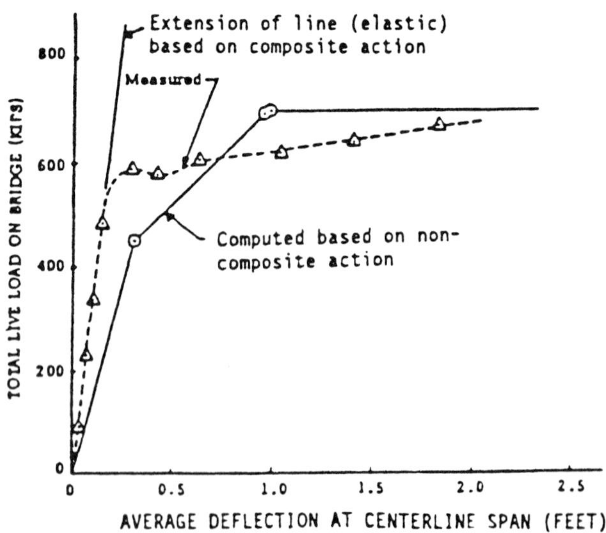

FIGURE 12-3 Load–deflection diagram, measured and computed, bridge tested to failure. (From Burdette and Goodpasture, 1971.)

supporting a noncomposite concrete deck. In the laboratory test the top flange of the beam was sanded to remove mill scale and irregularities before the concrete slab was poured, and the base metal surface was treated with an oil layer to inhibit bond development at the interface. An indicator of composite behavior was provided by the ratio of compression to tension flange strain, showing that a composite action existed and increased at higher loads. Quantitatively, this action represented approximately a 10 percent increase in the bottom flange section modulus above the noncomposite value. Because mechanical and adhesion bond was prevented, the partial composite behavior indicates friction between the steel and the concrete, in spite of the smooth interface.

Diagnostic load tests by Bakht and Csagoly (1980), undertaken to determine the cause of failure of connecting angles in the girder–column system of the Parley Bridge, also articulated the degree of composite action between floor beams and the concrete deck in the absence of shear connectors. Almost 87 percent of the deck slab area was supported by girders and trusses through the floor beams. In quantitative terms, the test showed that the composite action at the floor beam–deck interface varied and could not be expressed.

AASHO (1962) has disseminated results of a study that included 18 beam–slab bridges of a simple-span configuration, 50 ft long. About ten of these bridges have wide flange steel beams without cover plates, and two structures in this group were built compositely with shear connectors. In the noncomposite bridges the top surface of the steel beams were coated with a mixture of graphite and linseed oil to inhibit the formation of bond. In the two composite bridges the steel beams were provided with channels. For design purposes, two bridges were assumed to be fully composite, six bridges were assigned 10 percent composite action to reflect the effect of friction, and the remaining two structures were assumed to be noncomposite.

All bridges were tested with cyclic repeated loads, and the actual location of the neutral axis was in each case determined from strain measurements on the top and bottom of the beams at midspan. Theoretical calculations of the neutral axis were based on composite action for the two fully composite bridges, and on simple behavior for the other steel bridges. Table 12-1 gives the distance of the experimental location for all 30-mph runs within a 10-month observation period.

The results presented in Table 12-1 show a very small difference between the theoretical and measured neutral axis. In effect, this means that the bridges with mechanical shear connectors were fully composite, whereas the others had no composite action. The moment–deflection diagrams also showed a much stiffer section for the composite beams. The report concluded that there is no composite action in bridges without shear connectors.

A structural evaluation combined with live load tests for the Irvine Creek Bridge (Patel, 1984) showed that graphs for theoretical moments fall between the graphs of the moments calculated using the fully composite and noncom-

TABLE 12-1 Difference Between Experimental and Theoretical Location of Neutral Axis

Bridge	Location of Neutral Axis (in.)
1A, NC	0.0
1B, NC	+0.2[a]
2A, NC	+0.1
2B, C	+0.2
3A, NC	+0.2
3B, C	−0.1
4A, NC	+0.2
4B, NC	−0.1
9A, NC	+0.1
9B, NC	0.0

NC, noncomposite; C, composite.
[a]Plus sign indicates position above theoretical.

posite properties. This is interpreted to mean that the stringer sections were acting partially composite with the concrete slab.

Individual Beam-Slab Systems Unintended composite action of beam–slab systems has been observed in tests reported by Viest et al. (1952), Viest (1960), and Bryson and Mathey (1962). These tests show that as long as a bond is present, it provides an effective shear connection. While the bond is unbroken, there is no slip between the slab and the beam. In certain tests this bond withstood 1.7 times the design live load, but the bond broke after the same load was applied 11 times. However, these researchers concluded that if the static load was increased instead of being repeated, at loads approaching the ultimate large-deformation bond stresses at the interface would signal bond failure. In addition, shrinkage and warping of the slab as well as dynamic loading may destroy the bond even at working loads. Thus, a bond is a good shear connector while it exists, but it is unreliable.

12-3 DEFLECTION AND CAMBER

Deflection Limitations

In this context deflection is as computed by the methodology specified in AASHTO Article 10.6. Steel beams or girders with simple or continuous spans are designed so that the live load plus impact deflection does not exceed 1/800 of the span, except for bridges in urban areas used partly by pedestrians where this deflection should not exceed 1/1000 of the span. The deflection of cantilever arms due to live load plus impact is limited to 1/300 of the cantilever arm, except where pedestrian use is involved where this ratio is 1/375. Usually, the moment of inertia of the gross cross-sectional area is used in computing deflections of beams and girders.

These criteria first appeared in the early 1930s when reports on vibration of steel girder bridges indicated a statistical attempt to correlate this dynamic effect with the properties of the structure. Accordingly, the Bureau of Public Roads in 1936 issued specifications limiting the computed live load plus impact deflection to 1/800 of the span for simple and continuous bridges, and in 1938 these specifications were extended to cover cantilever spans.

In the context of bridge design, the primary purpose of limiting depth–span ratios has been to reduce live load deflections. The ratios adopted, however, should show the influence of economic depth under changing specifications for load and unit stresses. For example, the continuing trend toward composite construction for steel beam and girder bridges articulates the question with regard to the continued effectiveness of limitations on computed live load deflection. The difference in economic depth is recognized in the depth–span ratios of noncomposite and composite steel girders, controlled by the specified limits 1/25 and 1/30, respectively. Certain noncomposite bridges may, however, exhibit composite action, although not subject to quantification, and in this case the actual live load deflection is much less than computed by neglecting composite action. On the other hand, bridges provided with shear connectors have a live load deflection very close to the computed live load deflection. If the assumption can be made that composite action exists under service conditions for bridges designed for composite as well as for noncomposite behavior, then the former may have an actual live load deflection 10 to 25 percent greater than in the noncomposite design. In this case a deflection limitation of 1/1000 of the span should be extrapolated to the composite bridge to provide approximately the same stiffness provided by the present limit of 1/800 for noncomposite bridges.

Interestingly, if special considerations exist and must be taken into account when exceeding these limitations, reference is made to AISI (American Iron and Steel Institute) Bulletin 19, Criteria for the Deflection of Steel Bridges.

The limitations on deflection expressed in terms of span appear to have originated because of objectionable vibrations in existing bridges, and since then they have been reexamined on a statistical basis involving comparison of computed deflections. The effect of depth on the deflection of two beams designed for the same unit stress validates a limiting depth–span ratio, particularly as this criterion can be applied directly without trial designs. However, by specifying a minimum depth, control is extended only to the total deflection; hence, this limit is not useful in bridges with high live load–dead load ratios, that is, relatively small bridges. A more relevant and probably more effective criterion is therefore to articulate deflection in terms of structural damage.

The proposed LRFD specifications recognize the results of field tests on composite continuous spans (Baldwin, Salame, and Duffield, 1978; Roeder and Eltvik, 1985) showing that considerable composite action exists in negative moment regions even when shear connectors are not provided in these locations. This finding prompts a convenient stipulation whereby the stiffness of the full composite section may be used over the entire bridge. For the control of permanent deflection, this document mandates a Service II load combination, which is a service limit state relating to a vehicular overload event that may occur several times during the life of the bridge. The combination includes, properly factored, permanent dead load, live load plus impact, and friction. This limit state check is intended to prevent objectionable permanent deflections because of expected severe overload. It essentially affects serviceability and corresponds to the overload provision in the standard AASHTO specifications.

Nonlinear Analysis for Load–Deflection Relationship

Zhou and Nowak (1987) present procedures for the flexural and torsional analysis of simply supported bridges modeled as orthotropic plates. Bridge response is predicted by finite-difference methods.

Section Analysis A force–deformation relationship considers moment–curvature (M–ϕ) and torque–twist interactions. The M–ϕ curve is determined using a computer program developed by Tantani (1986). In this approach the section is idealized as a set of uniform layers as shown in Figure 12-4. Strain is gradually increased in increments, and at each strain level the corresponding moment is calculated using nonlinear stress–strain relationships such as steel and concrete.

A closed-form expression is derived to represent M–ϕ curves for the slab and the composite girder section as follows:

$$\phi = M/EI_e + C_1(M/M_y)^{C_2} \tag{12-6}$$

where EI_e is the elastic bending rigidity, M is the applied moment, M_y is the yielding moment, and C_1, C_2 are constants determined from the following:

$$\phi_y = M_y/EI_e + C_1(M_u/M_y)^{C_2}$$

$$\phi_u = M_u/EI_e + C_1(M_u/M_y)^{C_2}$$

which give

$$C_1 = \phi_y - M_y/EI_e \tag{12-7}$$

$$C_2 = \ln[(\phi_u EI_e - M_u)/(\phi_y EI_e - M_y)]/\ln(M_u/M_y) \tag{12-8}$$

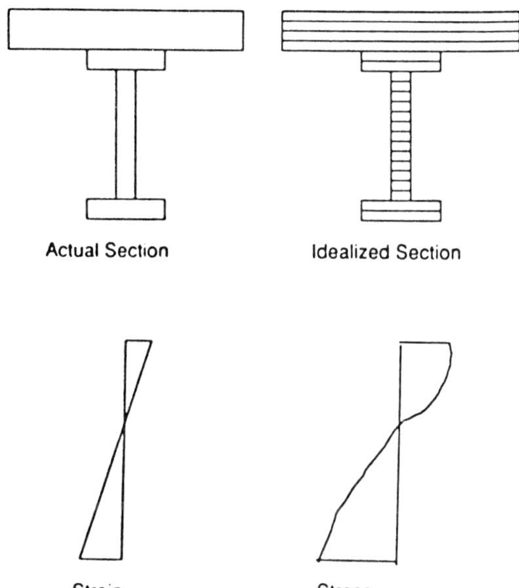

FIGURE 12-4 Idealization of cross section as set of uniform layers. (From Tantani, 1986.)

where M_u is the ultimate moment, ϕ_y is the curvature at yielding, and ϕ_u is the curvature at ultimate strength (see also Section 4-19).

For composite girders values of C_2 range from 20 to 24, and values of C_1 range from 25 to 28 $\times 10^{-5}$ ft^{-1}. For slabs C_2 ranges from about 22 to 26, and C_1 is between 27 and 30 $\times 10^{-5}$ ft^{-1}.

A typical composite girder has the M–ϕ curve shown in Figure 12-5 derived from (12-6), and for comparison the curve points calculated from Tantani (1986) are also shown.

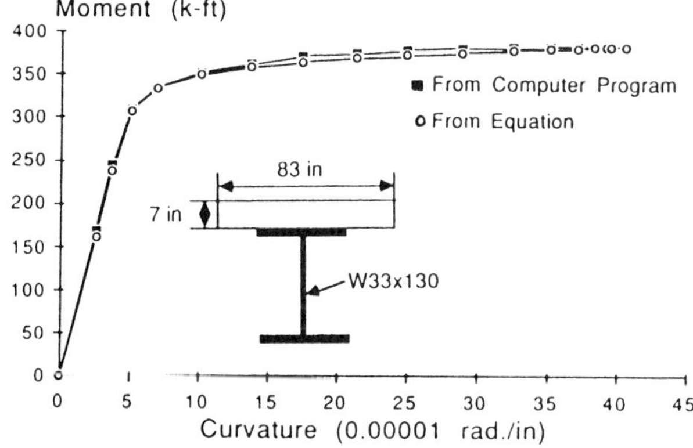

FIGURE 12-5 Curve for a typical girder composite section.

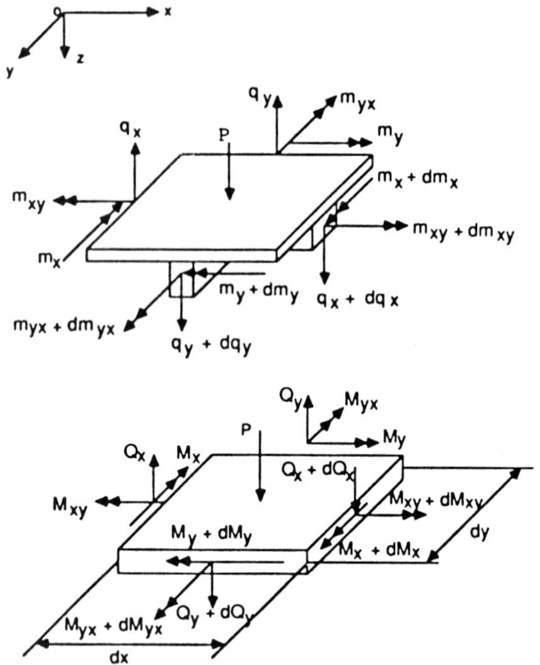

FIGURE 12-6 Orthotropic plate element.

The orthotropic plate modeling is as proposed by Heins and Yoo (1970) and Heins and Kuo (1975). Segments of the longitudinal and transverse members and slab are shown in Figure 12-6 together with the corresponding plate elements. In this presentation m_x, m_{xy}, and q_x denote the bending moment, torsional moment, and shear, respectively, exerted on the transverse member and slab; m_y, m_{yx}, and q_y denote the bending moment, torsional moment, and shear, respectively, acting on the longitudinal member and slab. A differential equation is developed to express the equilibrium of the plate element in terms of the following factors:

M_x = bending moment per unit width of the section in the x direction
M_y = bending moment per unit width of the section in the y direction
M_{xy} = twisting moment per unit width of the section in the x direction
M_{yx} = twisting moment per unit width of the section in the y direction
$P_{(x,y)}$ = applied load per unit area

The total load per plate element P_r is divided into three components acting on the girder, diaphragm, and slab. The compatibility conditions in the force and displacement equations between plate elements are expressed by a system of differential equations. This system is presented in a matrix form as

$$[K]\{W\} = \{P\} \tag{12-9}$$

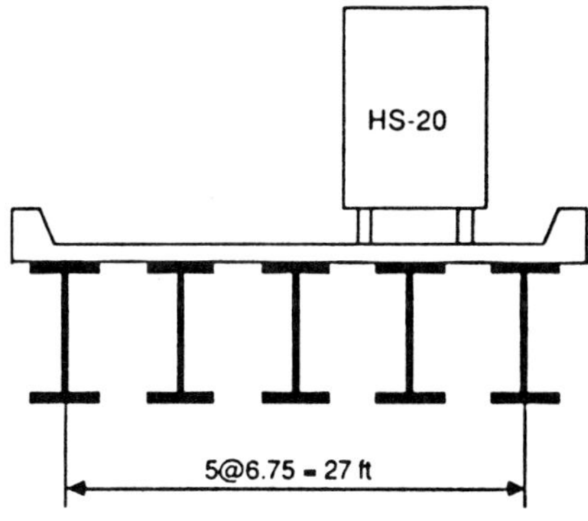

FIGURE 12-7 Cross section of bridge for numerical example, composite girders. (From Zhou and Nowak, 1987.)

where $[K]$ = stiffness matrix
$\{W\}$ = vector of grid point deflections
$\{P\}$ = vector of loads

From this system the unknown grid point deflections are calculated. A system of differential equations is also developed, from which unknown moments and shears are derived.

Numerical Example This procedure is demonstrated for a composite steel girder bridge designed according to AASHTO criteria. The span length is 60 ft, and a typical deck cross section is shown in Figure 12-7. The slab thickness is 7.5 in., and the supporting members consist of W33 × 130 beams. The live load is HS 20 and is positioned as shown. The curve for the resulting live load of total truck weight versus deflection is presented in Figure 12-8.

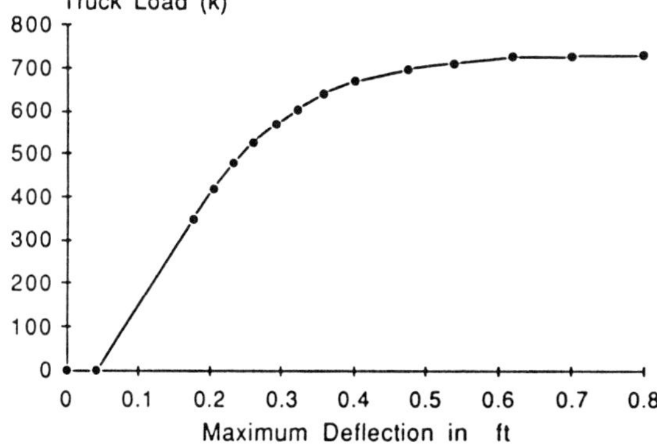

FIGURE 12-8 Load–deflection curve for the composite girder bridge, cross section of Figure 12-7. (From Zhou and Nowak, 1987.)

Camber

Camber is introduced in plate girders to compensate for dead load deflections and vertical curvature required for profile grade. Usually, however, most standards specify the dead load deflection that must be exceeded for camber to be required. A typical criterion is a dead load deflection exceeding 1.5 in. The purpose of camber is to minimize fillet heights and adjust the girder elevation to the geometric profile for improved appearance. When camber is designated on the plans, an allowance is made for fabrication tolerance, usually 1/4 in. In most cases steel mill practice will provide data on the cambering of beams and girders to produce a predetermined design. Because camber is measured in the mill, it will not necessarily be present in the same amount in the section of beam or girder as received due to the release of stress induced during cambering operations. In general, 75 percent of the specified camber is likely to remain. For rolled sections camber will approximate a simple regular curve nearly the full length or between two points specified, and in this case it is designated by the ordinate at the midlength of the portion to be curved.

For steel grid floors provisions for camber are according to AASHTO Article 10.14. Rigid steel units that do not readily follow the required camber should be cambered in the shop.

Example A two-span continuous plate girder bridge with span lengths of 100 ft has composite design in the positive moment regions. The dead load carried by each girder (noncomposite) is 0.90 kip/ft. The composite dead load per girder is 0.165 kip/ft. The appropriate moments of inertia are used in computing dead load deflections for noncomposite and composite loads. Camber ordinates are determined at the quarter and midpoints of the positive moment region, at the inflection point, and at the midpoint of the

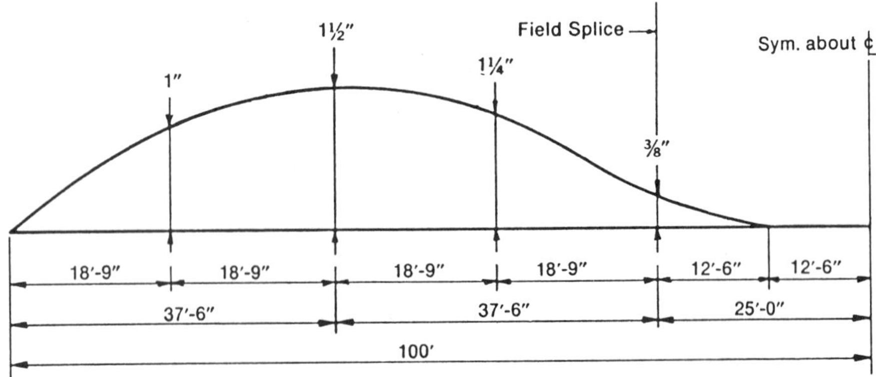

FIGURE 12-9 Camber diagram for a two-span continuous steel plate girder bridge, composite in positive moment regions.

negative moment region. The resulting camber diagram is shown in Figure 12-9.

12-4 FORCES INDUCED BY SETTLEMENT

Analytical Concepts

Although most bridge foundations are designed to preclude differential settlement, deformations due to unstable support conditions may sometimes occur even under the most controllable ground effects. In terms of creep in the concrete, it is conceivable that the extent of distress caused by foundation settlement will also depend on the time rate of settlement, and some structures will thus be able to adjust to settlement if this rate is slow. Criteria for predicting the settlement-induced forces in continuous concrete bridges are therefore useful, particularly if they correlate these effects with creep, stress relaxation, and the time rate of settlement of substructure elements.

Bishara and Jang (1980) introduce a methodology whereby a reinforced concrete flexural member is idealized as a viscoelastic beam. Referring to Figure 12-10 which shows an n-span continuous beam, we can write for elastic analysis

$$[M] = [k][\theta] + [q] \tag{12-10}$$

where $[M]$ = member force matrix

$[k]$ = element stiffness matrix, which is a diagonal matrix with individual element stiffness matrices as its constituents

$[\theta]$ = element deformation matrix

$[q]$ = total effect matrix with fixed end moments

In this expression $[M]$ and $[q]$ are constants, and E_c is replaced by the relaxation modulus $Y(t)$, so that $[\theta]$ is time dependent. The foregoing yield

$$Y(t) = E_c/[1 + \alpha(t)] \tag{12-11}$$

where $\alpha(t)$ is the long time deflection multiplier. Long time deflections are caused principally by the effects of shrinkage and creep. Shrinkage causes stresses similar to those due to temperature changes, although these stresses

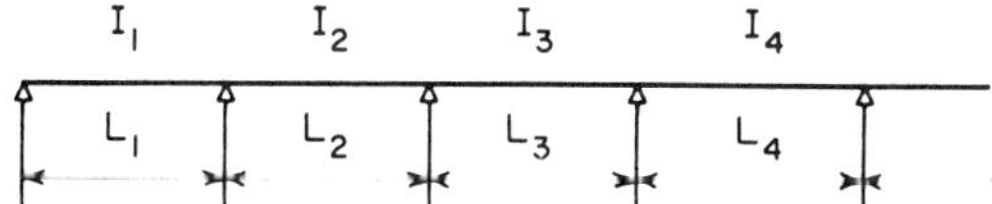

FIGURE 12-10 Continuous beam with n spans.

are not related to the effects of differential settlement. Hence, the latter is excluded from this analysis, and the dominating factor in $\alpha(t)$ is creep.

Creep deflection Δ_{cp} is determined from

$$\Delta_{cp} = K_r C_t \Delta_i \tag{12-12}$$

where Δ_i is the initial elastic deflection and C_t is the creep coefficient, or the ratio of the creep strain ϵ_c to the initial strain ϵ_i, or $C_t = \epsilon_c / \epsilon_i$. Both ϵ_c and C_t are time dependent. The factor K_r is a correction coefficient and is constant for a given problem. If A_s and A'_s express the area of tension and compression reinforcement, respectively, the values of K_r (Branson et al., 1970) are suggested as follows:

$$\begin{aligned} &\text{When } A'_s = 0 && K_r = 0.85 \\ &\text{When } A'_s = 0.5A_s && K_r = 0.60 \\ &\text{When } A'_s = A_s && K_r = 0.40 \end{aligned} \tag{12-13}$$

Linear interpolation is acceptable, as is the average critical positive and negative section values for continued beams. From the foregoing equations we derive $\alpha(t) = K_r C_t$, and

$$Y(t) = \frac{E_c}{1 + K_r C_t} \tag{12-14}$$

The creep coefficient C_t may be evaluated from

$$C_t = K_c \left(\frac{t^{0.6}}{10 + t^{0.6}} \right) C_u \tag{12-15}$$

where t is the time from the date the concrete was placed (days), C_u is the concrete weight factor, and K_c is the overall correction factor reflecting ambient relative humidity, age when loaded, minimum thickness of member, slump, percentage of fines, and air content. The factor C_u ranges from 2.0 to 2.75. In practice, C_u is 2.75 for normal-weight concrete, 2.30 for lightweight concrete, and 2.0 for sand-light concrete.

The variable t in (12-14) relates to t_0, the rate of falsework removal, whereas the variable t in (12-15) is relative to $t = 0$.

The relaxation modulus is now expressed as

$$Y(t) = E_c \left[\frac{10 + (t - t_0)^{0.6}}{10 + \omega_0 (t - t_0)^{0.6}} \right] \tag{12-16}$$

where

$$\omega_0 = 1 + K_r K_c C_u. \tag{12-17}$$

The limit value of $Y(t)$ is found as

$$\lim_{t \to \infty} Y(t) = E_c/\omega_0 \tag{12-18}$$

Because the correction factor for age when loaded is the only time-dependent factor, ω_0 is further expressed as

$$\omega_0 = 1 + K_r C_c C_u p t_0^{-\mu} \tag{12-19}$$

where C_c is termed the ambient condition factor, and the constants p and μ are as follows:

For moist-cured concrete	$p = 1.25$	$\mu = 0.118$
For steam-cured concrete	$p = 1.13$	$\mu = 0.095$

For nonprestressed cast-in-place concrete bridges, the concrete is moist-cured; hence,

$$\omega_0 = 1 + 1.25 K_r C_c C_u t_0^{-0.118} \tag{12-20}$$

Settlement Rate and Induced Forces

Predictions on the rate, type, and amount of settlement are usually made through a soil–structure interaction analysis considering the type of foundation and soil characteristics. Such an analysis is beyond the scope of this book.

Settlement-induced forces are articulated in continuous nonprestressed concrete bridges such as slab, T-girder, and box girder cast-in-place structures. For these bridge types either the interior or the exterior girders of the superstructure may be considered continuous beams on simple supports because the relative stiffness of the column bent is small and should not affect the analysis. Referring to Figure 12-10, if the beam undergoes a set of immediate foundation settlements $\Delta_1, \Delta_2, \ldots, \Delta_{n+1}$ under supports 1 through $n + 1$, we can write

$$[n] = [k][\theta] + [R] \tag{12-21}$$

where $[R]$ is a function depending on Δ_i, I_i, and E_c (Bishara and Jang, 1980).

If a set of settlements imposed at $t = 0$ are held constant, the displacements of all points and all strains are the same as in the corresponding elastic beam (Flügge, 1975). The stresses are derived from those of the elastic problem by multiplying by $Y(t)/E$, where E is the modulus of elasticity of the viscoelastic material. We can now write

$$[M]^v = \frac{Y(t)}{E_c}[M]^e \tag{12-22}$$

where $[M]^e$ is the elastic solution of the forces induced on the joints of the structure by a set of foundation settlements, and $[M]^v$ is the corresponding viscoelastic solution of settlement-induced forces, varying with time.

From elastic analysis we obtain

$$M_i^e = \sum_{j=1}^{n+1} 6E_c\phi_j\Delta_j \tag{12-23}$$

where ϕ_j are constants relating to the stiffness of the superstructure and Δ_j are foundation settlements. Thus, for the general case,

$$M_{ij}^e = 6E_c\phi_i\Delta_i \tag{12-24}$$

and substituting (12-24) into (12-22), we obtain

$$M_{ij}^e = 6Y(t)\phi_j\Delta_j \tag{12-25}$$

where M_{ij}^e is the elastic bending moment at the ith support due to settlement at the jth support.

Combining (12-25) and (12-16) gives

$$M_{ij}^v = 6E_c\phi_j\Delta_j\left[\frac{10 + (t - t_u)^{0.6}}{10 + \omega_u(t - t_u)^{0.6}}\right] \tag{12-26}$$

where

$$\omega_u = 1 + 1.25C_cK_rC_ut_u^{-0.118} \tag{12-27}$$

and t_k is an intermediate time between t_0 and t.

The viscoelastic solution of settlement-induced forces may be derived by a numerical approach based on the concept of a hereditary integral and assuming that I_{eff} does not change with time (Bishara and Jang, 1980).

Example For the two-span continuous reinforced concrete girder bridge shown in Figure 12-11a, the following data are given: average $I_{\text{eff}} = 789{,}610$ in.^{4}; concrete weight = normal and moist cured; $f_c' = 3000$ psi; $f_y = 60{,}000$ psi; date of falsework removal = 28 days after casting; $C_c = 1.0$; $K_r = 0.80$.

The elastic solution for Δ_1, Δ_2, and Δ_3 gives 2.0, 1.5, and 0.5 in., respectively, and these values are summarized in Table 12-2. The force

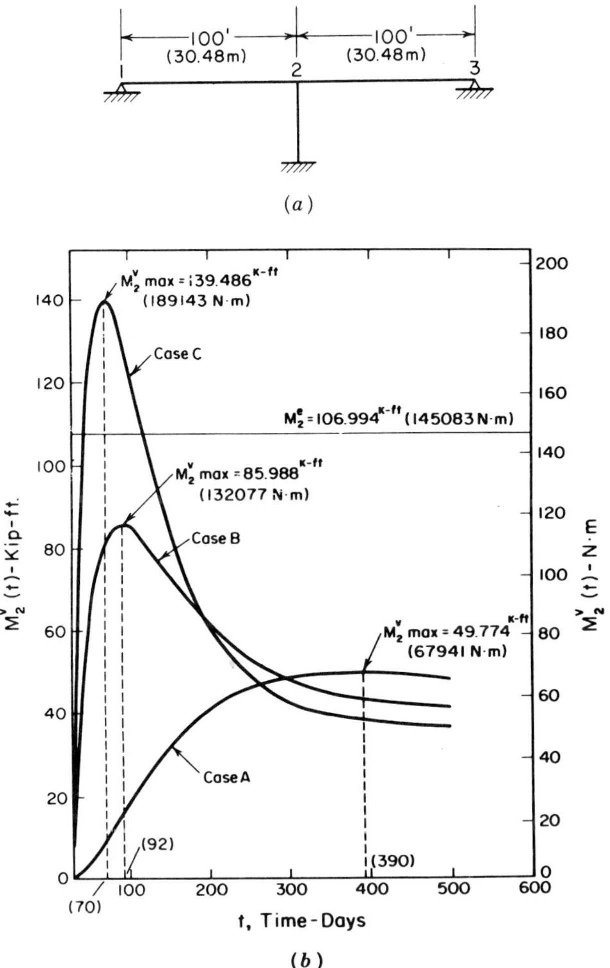

FIGURE 12-11 (*a*) Two-span reinforced concrete girder bridge; (*b*) variation of $M_2^v(t)$ with time, for bridge of Figure 12-11.

induction factors are estimated from a numerical analysis based on the foregoing theory. The elastic moment is also shown.

Figure 12-11*b* shows the variation of M_2^v (viscoelastic moment at pier 2) with time. Evidently, the rate of settlement of pier 2 influences markedly the determination of $M_2^v(t)$. Case A, where the rate of settlement is slower than cases B and C, results in a smaller value for M_2^v maximum (about 47 percent of M_2^e) at a longer time (390 days) after loading. This is explained by the slower rate of settlement for pier 2, allowing the settlement of abutments 1 and 3 to result in larger negative moments to counteract the positive moment

TABLE 12-2 Elastic Solutions for the Example of Figure 12-11 (From Bishara and Tang, 1980)

Δ_f, in in. (mm)	M_2^e, in kip-ft (N-m)
2.0, 0, 0	−427.979
(51, 0, 0)	(−580,339.5)
0, 1.5, 0	641.969
(0, 38, 0)	(870,509.9)
0, 0, 0.5	−106.996
(0, 0, 13)	(−145,086.6)
2.0, 1.5, 0.5	106.994
(51, 38, 13)	(145,086.6)

induced by the settlement of pier 2, in addition to the longer time of stress relaxation in the concrete of the superstructure.

Case B has a rate of settlement comparatively faster than case A and shows a higher value of M_2^v maximum (about 80 percent of M_2^e), occurring 92 days after loading. Case C has a rate of settlement much faster than both cases A and B, and also compared to abutments 1 and 2. The maximum M_2^v in this case is close to $1.3M_2^e$.

Tolerable Movements

In addition to the analytical investigations, the consequences of support movement may be inferred through studies of bridge behavior and field case histories. Useful data are given by Moulton, GangaRao, and Halvorsen (1985), who present criteria for tolerable movements of highway bridges. This study also shows that explicit consideration of the tolerance of bridges to withstand settlement and horizontal movement is desirable and indicated with every bridge design.

Components of Bridge Settlement Figure 12-12 shows the basic components of bridge settlement: uniform displacement, tilt (or rotation), and nonuniform settlement. Uniform settlement is rare and does not adversely affect superstructure stresses, although it may cause problems at the juncture between the approach slabs and bridge deck. Uniform tilt or rotation is more likely for bridges with very stiff superstructures and can develop essentially with single-span bridges. The effects are likewise confined to the bridge ends. Nonuniform or differential settlement results in deformations for bridges with two or more continuous spans. The two different types shown in Figure 12-12 exemplify a regular pattern (settlement increases toward the center from the ends) or an irregular pattern (settlement appears as erratic along the bridge). Nonuniform settlement can affect the utility of the bridge and cause critical stress changes, as discussed in the foregoing sections.

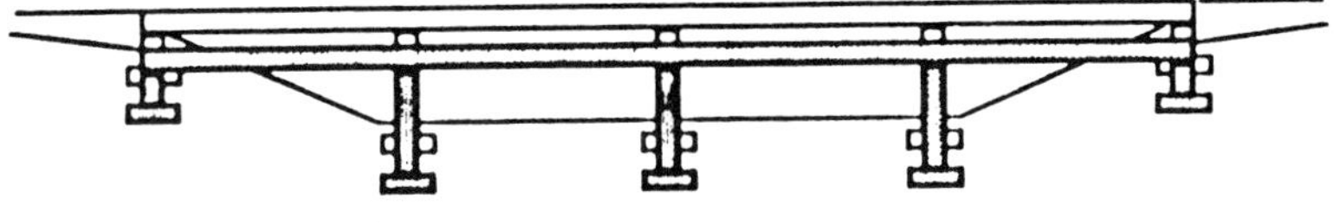

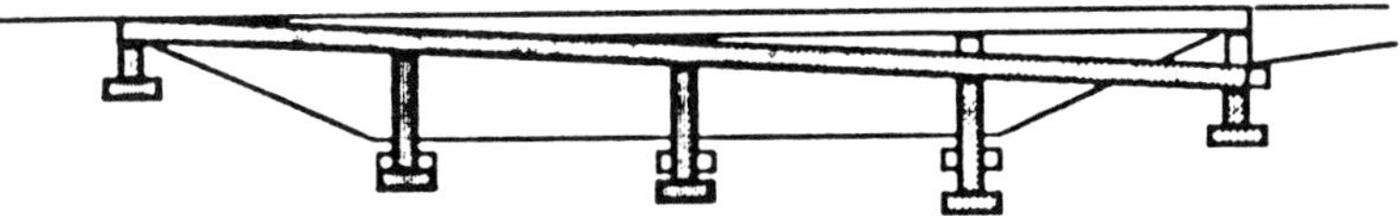

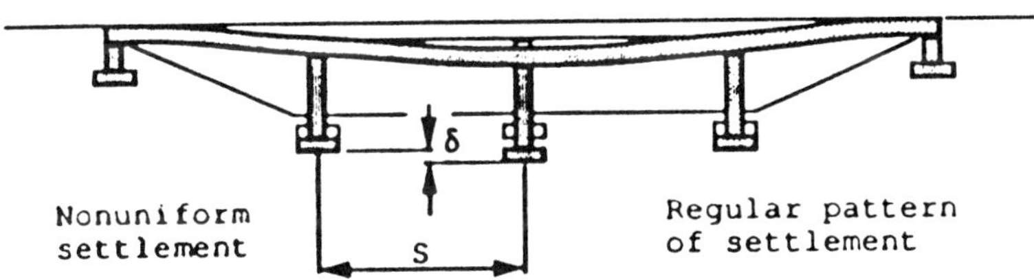

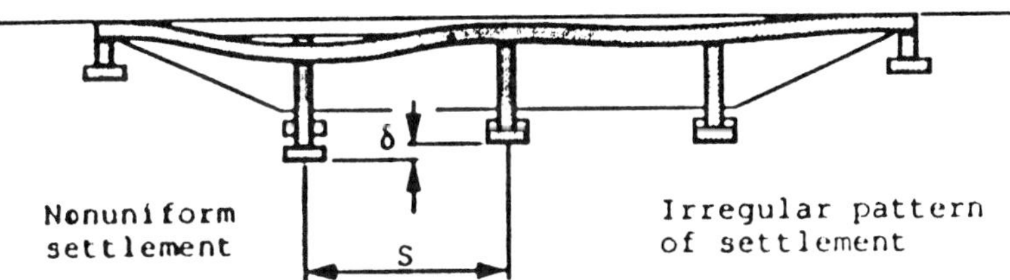

FIGURE 12-12 Components of settlement and angular distortion in bridges.

Criteria for Tolerable Settlements These may be expressed in terms of the angular distortion defined as $A = \delta/S$, where δ and S are as shown in Figure 12-12. Moulton et al. (1985) have found that angular distortion provides a good basis for establishing tolerable movement magnitudes for bridges, and they have provided data from 56 simple-span bridges and 119 continuous-span bridges. About 96 percent of the continuous bridges withstood angular distortion as large as 0.004 without suffering damage considered to be intolerable. Thus, the limit of tolerable angular distortion recommended by these investigators is 0.004 for continuous bridges. For simple spans the criteria were based on 56 cases giving a recommended value of 0.005 as the limit of tolerable angular distortion, corresponding to a 98 percent rate of acceptability among the bridges studied.

Criteria for Tolerable Horizontal Movements Data presented by Moulton, GangaRao, and Halvorsen (1985) show that horizontal movements tend to be more damaging when they are accompanied by settlements than when they are present alone. A horizontal movement less than 1 in. was almost always assessed as being tolerable, whereas movement exceeding 2 in. was likely to be intolerable. It may appear that the value of 1.5 in. suggested by these investigators is a reasonable criterion. This may oversimplify a complex problem but gives a practical guideline for preliminary design.

Suggested Simplified Methods of Analysis A simple method of analysis emerges from the recommendations of Moulton, GangaRao, and Halvorsen (1985) and Yokel (1990). In principle, the shear forces and bending moments caused by settlements are added to the primary force effects to obtain the combined flexural and shear stresses that are then compared with the capacity of the members to determine if the results are tolerable.

Conventional structural analysis often tends to overestimate the effects of settlement. Consider, for example, a two-span continuous bridge with a reinforced concrete deck and assume that the center support settles by an amount δ relative to the end supports (Duncan and Tan, 1991). The bending stress induced in a simple rectangular reinforced concrete bridge deck in this case is

$$f_{\max} = \frac{\delta}{s}\left(\frac{3E}{2}\frac{t}{s}\right)$$

where δ = differential settlement

E = modulus of elasticity

t = deck thickness

s = span length

Cracking is prevented if the maximum bending stress does not exceed the tensile strength f_t of the concrete. The ratio δ/s corresponding to $f_{\max} = f_t$ is

$$\frac{\delta}{s} = \frac{2f_t s}{3Et}$$

Using $f_c' = 4000$ psi, $f_y = 60{,}000$ psi, $E = 3.6 \times 10^6$ psi, and $s/t = 24$, we compute that $\delta/s = 0.0021$. This is, however, one-half the angular distortion considered tolerable based on the survey of damage to bridges reported by Moulton, GangaRao, and Halvorsen (1985), and results in overconservative design.

12-5 TEMPERATURE EFFECTS IN BRIDGES

Basic Principles

Thermal effects are manifested by short-term (daily) as well as by lengthy (seasonal) temperature changes. Several factors influence the longitudinal deformation of a bridge, but these effects are more severe where thermally induced movement is restrained, that is, in integral bridges. Creep and shrinkage contribute to structural deformation, particularly in the early life of the structure.

Ekberg and Emanuel (1967) have provided a survey that articulates the problems associated with the thermal movement of bridges. Thus, there is a tendency to consider these effects more frequently in steel than in concrete bridges, although the general attitude is toward a rational method of design that will account for thermal effects.

In general, certain considerations should be emphasized: (a) thermal stresses can cause damage in many bridges if they are ignored in the design; (b) adequate provisions can be made, similar to those for dead and live loads; (c) temperature criteria can address equally safety and structural damage; and (d) an alternative to providing for free movement is to design the bridge to resist thermally induced stresses.

Studies of temperature effects in composite bridges are presented in Section 4-9, confirming a nonlinear temperature distribution that results in thermal stresses and strains. Seasonal temperature changes cause larger movement than daily changes, and the results may be amplified because either stress or strain, or both stress and strain effects, can be present (Maher, 1970). Thermal strain occurs without stress if the bridge is completely free to move, and thermal stress is caused without strain if the bridge is completely restrained from movement. Stress and strain are usually combined because most bridges have partial restraint against movement. Strain resulting from creep, shrinkage, humidity, moisture swelling, and variations in vehicular load complicates the analysis. Expansion devices may not always perform as expected (i.e., frozen rockers), and the associated effects on expansion and contraction must therefore be considered.

Air Temperature The two basic temperature cycles are (Berwanger, 1920): (a) the daily cycle usually beginning with a low temperature just before sunrise and reaching peak temperature in midafternoon, and (b) the yearly temperature cycle resulting from changes in position and distance of the earth relative to the sun. Attempts to establish the range of temperature and movement for which bridges should be designed have produced the current guidelines listed in the AASHTO specifications, Article 3.16. The proposed LRFD specifications are more explicit in defining a moderate climate and introduce the concept of setting temperature, which is the actual air temperature averaged over the 24 hours immediately preceding the setting process. The setting temperature normally would not be a design parameter, except in large complex structures. A clarification may be noted on the plans, however, explaining how the setting temperature should be determined and the bearings adjusted accordingly.

Bridge Temperature The bottom elements of a bridge ordinarily have the same air temperature, but the upper elements (including the exterior beams) have a temperature that depends on the amount of solar radiation, the wind, and precipitation. The result is a temperature lag or differential that may reverse its sign and vary its magnitude, introducing a variety of temperature distributions.

Maher (1970) assumes a linear temperature distribution in the top slab of a box section for continuous prestressed concrete bridges built with hollow concrete box sections, based on actual observations. The sides and bottom of the box sections were assumed to be at a constant temperature. Studies have been completed by Liu and Zuk (1963) on four types of prestressed concrete members that included (a) prestressed beams with straight tendons, (b) prestressed beams with draped tendons, (c) prestressed beams with straight tendons and composite with the concrete slab, and (d) prestressed beams with draped tendons and composite with the slab. Four different temperature distributions were used for each type or member in the examples. The shear and moment at the interface were determined from three simultaneous equations.

In Britain, Menzies (1968) has related the air temperature to the temperature at middepth of the concrete slab with the following equation:

$$\frac{0.156 d\theta_c}{dt} + \theta_c = \theta_a \tag{12-28}$$

where t is the time (h), θ_c is the concrete temperature (°C), and θ_a is the air temperature (°C).

Barber (1957) has developed and tested an expression relating weather factors to maximum pavement temperature. The coefficients may be varied or modified to allow this expression to be applied to different climates and regions. Zuk (1965a, 1965b) has introduced coefficients for use in Barber's equation for predicting maximum bridge surface temperatures. For a normal concrete deck in the mid-Atlantic states, the maximum surface temperature is

$$T_m = T_a + 0.18L + 0.667(0.50T_r + 0.054L) \tag{12-29}$$

where T_m = maximum surface temperature (°F)
T_a = average daily temperature
T_r = daily range in air temperature
L = solar radiation received on a horizontal surface (gram-calories/cm^2/day).

The maximum surface temperature of a bitumen-covered deck would be

$$T_m = T_a + 0.027L + 0.65(0.50T_r + 0.082L) \qquad (12\text{-}30)$$

where all parameters are the same as in (12-29).

An approximate equation for the maximum differential (temperature lag) between the top and bottom temperatures of a composite steel and concrete bridge is proposed by Zuk (1965a, 1965b) as

$$\Delta T_m = T_m - T_a - \lambda T_r \qquad (12\text{-}31)$$

where λ is a factor indicating the phase lag between the maximum surface and the maximum ambient temperature. For the mid-Atlantic states, lag factors 1/4 of 1/2 were found appropriate for the summer and winter, respectively.

Heat Exchange and Thermal Effects

Heat energy exchange occurs according to three mechanisms: (a) radiation from the sun and reradiation between the surrounding environment and the structure, (b) convection of heat between the surface of the structure and the surrounding environment, and (c) conduction of heat between the surface of the structure and the environment. The mechanism in (a) is the most prevailing mode.

The intensity of solar radiation reaching the earth's surface is dependent on latitude and has an annual variation as shown in Figure 12-13. Radiation penetrating the atmosphere and reaching the surface of a bridge deck is diverted in two ways: it may be reflected, or it may penetrate the surface whereupon it is absorbed and converted to heat. The extent to which this diversion occurs depends on the surface type, color, texture, and so forth. Concrete has a medium absorption capacity; it absorbs only a certain amount of wavelength and reflects the remainder. Emerson (1977) concludes that the influence of the deck surfacing should also articulate the shape of the cross section, whereas Priestley and Buckle (1979) have found that among white surfacing, black surfacing, and no surfacing on a box girder bridge, the maximum surface temperature occurred in the unsurfaced deck. The white surfacing resulted in the lowest temperature.

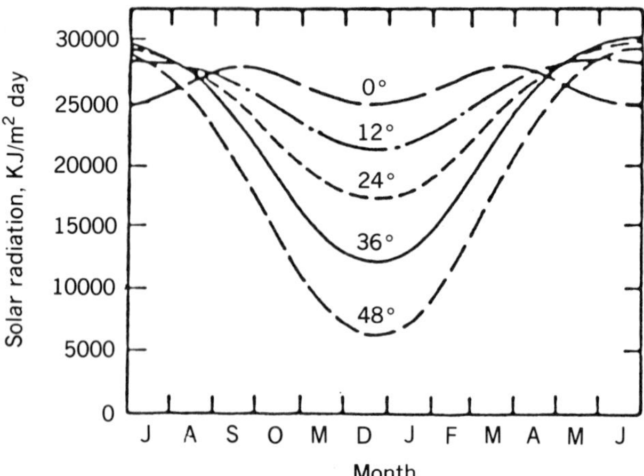

FIGURE 12-13 Variation in solar radiation for various latitudes.

Relevant Temperature Range Results of tests on concrete samples show a variation of 22 percent above and 64 percent below the higher value of the thermal coefficient. Most codes including AASHTO stipulate an average thermal coefficient of 0.000006°F, which is the coefficient commonly used to calculate thermal expansion and contraction of bridges.

Emerson (1979) introduces the concept of "effective" temperature of a bridge. This is the temperature that governs the longitudinal movement of the bridge deck and can be derived by considering both the product of the areas between isotherms and their mean temperature divided by the total area of cross section of the deck. Emerson (1977) and Black, Moss, and Emerson (1976) have correlated the extreme values of the effective bridge temperature with shade temperatures.

Analytical Assessment of Temperature Distribution The temperature gradients in a concrete superstructure will result in rotational distortions producing stresses. These gradients are governed by heat flow through the body and are therefore a function of the density, specific heat, and thermal conductivity of the concrete.

The equations introduced by Zuk (1965a, 1965b) appear to correlate well between measured and computed temperature differential values, and field tests further confirm their accuracy in determining the nonlinear temperature distribution through a concrete deck. Temperature distribution in the steel beams was either uniform or varied moderately. Further studies on this subject are documented by Imbsen et al. (1985), and appropriate references are included at the end of this chapter. Among the analytical procedures to be mentioned here is the finite-element program developed by Will, Johnson, and Matlock (1977) providing two solutions: one for transient heat conduction analysis; the other for the static thermal stress analysis of bridge structures. The program gives the thermally induced movements and stresses and can handle skewed bridges.

Thermal Stress Analysis For a given temperature gradient, a thermal stress analysis may be carried out based on one-dimensional beam theory. A common assumption is that the material has linear stress–strain and temperature–strain characteristics. Thermal stresses are therefore considered independently of stresses or strains imposed by other loading conditions, but the results thus obtained are superimposed. Under temperature effects, initially plane sections are assumed to remain plane, and whereas temperature varies only with depth, it is constant at all points of equal depth. The analysis considers longitudinal and transverse thermal effects independently, but the results are likewise superimposed.

Longitudinal Effects For expedience, the bridge is first analyzed as a statically determinate structure by removing sufficient internal redundancies. Stresses due to the nonlinearity of the temperature profile are also calculated. Appropriate forces and moments are applied to remove inadmissible deformations at the points of released redundancy, and the compatibility of the structure is thus established.

Self-Equilibrating Stresses If an arbitrary unrestrained cross section is subjected to a temperature field which is not plane, its fibers will tend to deform as shown in Figure 12-14*a*. This strain distribution violates, however, our initial assumptions, and because only a portion of the temperature field is responsible for the section's deformation, the final strain profile must be linear as shown in Figure 12-14*b*.

The composite diagram of Figure 12-15 illustrates the strain difference resulting in the self-equilibrating stress. Beginning with the strains resulting from the free expansion of the section fibers and the strains in the resultant plane section, we consider the differences between them, and this results in the self-equilibrating stresses. This step consists of stresses in the artificially restrained structure and stresses resulting from the axial loads and bending moments required to remove the artificial restraints.

For the bridge member shown in Figure 12-16, the temperature varies only in the vertical direction. At the ends of the member, full restraint is provided by the bending moment M and the axial force P. For a nonlinear temperature-induced strain, the longitudinal stresses are

$$\sigma_t(Y) = E\alpha T(Y) \tag{12-32}$$

where $\sigma_t(Y)$ = longitudinal stress at a fiber at distance Y from the center of gravity of the cross section

α = coefficient of thermal expansion

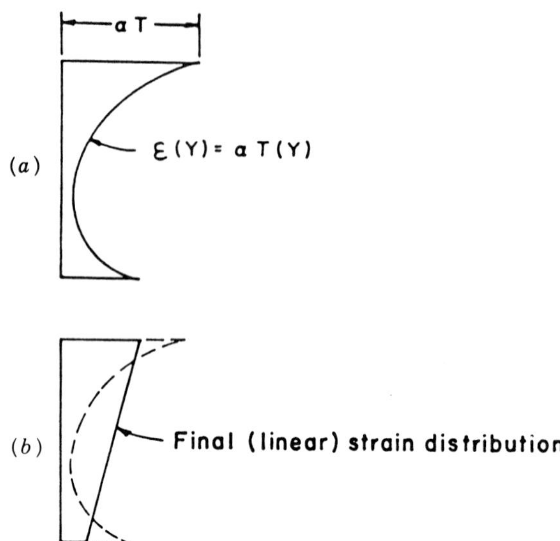

FIGURE 12-14 (*a*) Temperature-induced strain distribution assuming that the section's fibers have no influence on each other; (*b*) final temperature-induced strain distribution.

$T(Y)$ = temperature at depth Y

The stress associated with the axial force P acting on the cross-sectional area A is

$$\sigma_p(Y) = P/A$$

whereas the longitudinal stress associated with the moment M is

$$\sigma_m(Y) = MY/I$$

For a long thin member without end restraints, the longitudinal self-equilibrating stress is derived if we apply the restraining axial force and the

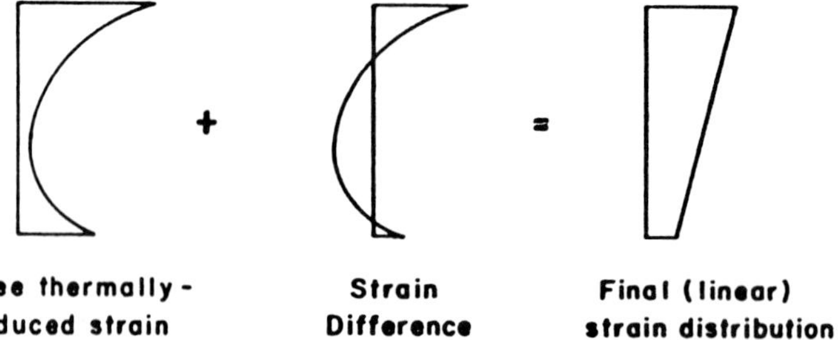

FIGURE 12-15 Strain difference that articulates the results and yields the self-equilibrating stress condition.

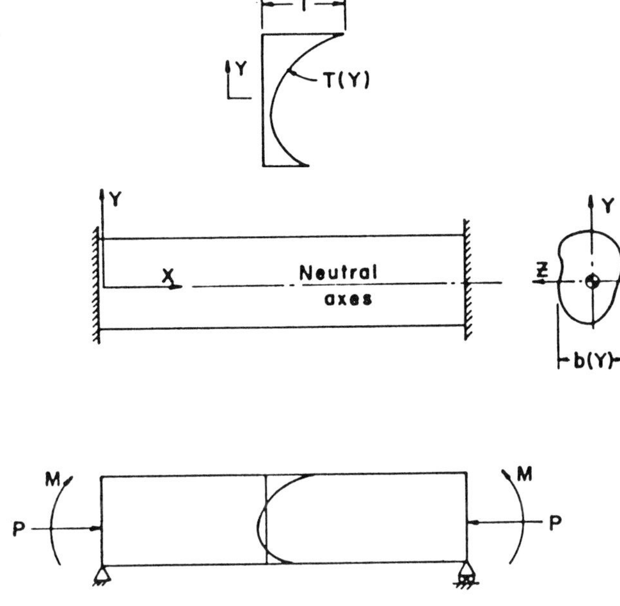

FIGURE 12-16 Member restrained at the ends, with arbitrary cross section and vertical temperature distribution.

restraining end moment (negative) to the stress distribution given by (12-32), or

$$\sigma(Y) = E\alpha T(Y) - (P/A) - (MY/I) \tag{12-33}$$

This process and the resultant are shown graphically in Figure 12-17. We should note that if the temperature variation is linear, self-equilibrating stresses will not exist.

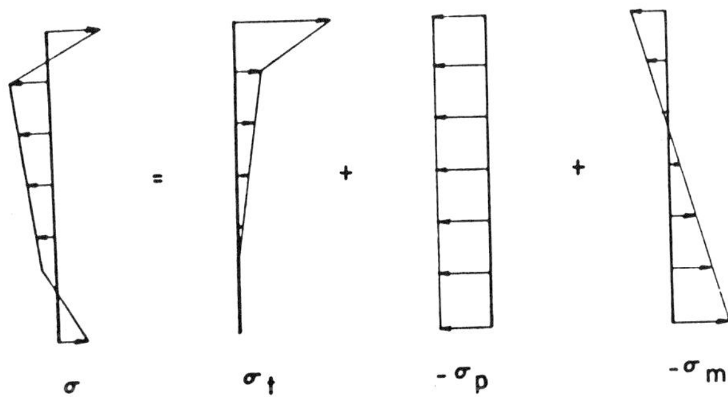

FIGURE 12-17 Distribution and resultant of self-equilibrating stresses.

Transverse Effects For complex sections, such as box girders, transverse induced stresses are analyzed separately and then superimposed on the total effect. Under a vertical temperature gradient, the deck slab in a box girder is subjected to greater temperature variation than the soffit. If the slab is relatively thin (10 in. or less), a linear temperature gradient may be assumed (PTI, 1978; Priestley and Buckle, 1979).

Priestley and Buckle (1979) analyze the transverse response by removing the constraints on the deck slab and allowing it to deform freely. In this case the unrestrained thermal deformation of the heated deck slab involves two components, as shown in Figure 12-18. These are (a) the average transverse increase in length $\Delta L = [L\alpha(T_1 + T_2)]/2$ and (b) a curvature $\psi = [\alpha(T_1 - T_2)]/h$, where T_1 and T_2 are temperature increases for the top and bottom surfaces, respectively, of the deck slab relative to the soffit, L is the distance between web centerlines, and h is the deck slab thickness.

If the system is unrestrained, primary thermal stresses will not be induced because the temperature gradient is linear and the final strain profile is the same as for the unrestrained thermal strain. Secondary thermal stresses will, however, be induced if slab elongation and curvature are restrained. The elongation can be modeled as an initial lack-of-fit case by considering the equilibrium of the section with the initial free elongation of the deck slab. Curvature effects are calculated as moments necessary to restrain rotation.

Interaction of Longitudinal and Transverse Effects The assumed uncoupling in these two processes may result in the approximation of actual effects, and the actual thermally induced stresses may be higher (Priestley and Buckle, 1979). An approximate allowance for the interaction results from Poisson's ratio μ and is obtained by adding μ times the bending component of the longitudinal stress to the transverse stress and vice versa. If the deck is subject to cracking in the transverse direction, Poisson's ratio is not applicable and this adjustment should be ignored.

Examples and Case Studies

Hammersmith Flyover Bridge This structure is a precast, prestressed continuous four-lane viaduct, 2043 ft long between abutments. It has 16 spans, most of which are 140 ft long. The main structural element is a 26-ft-wide continuous hollow-spine beam supported on central columns. At the base, each column rests on a pair of roller bearings, allowing a 10-in. movement range. The entire longitudinal expansion of the superstructure is accommodated at one expansion joint. The design considered 14-in. movement for temperature and humidity, and 3.3 in. for creep and shrinkage, for a temperature range of 60°F.

Movement measurements at one of the pier columns showed a linear relationship with the interior air temperature (inside the box section). From a

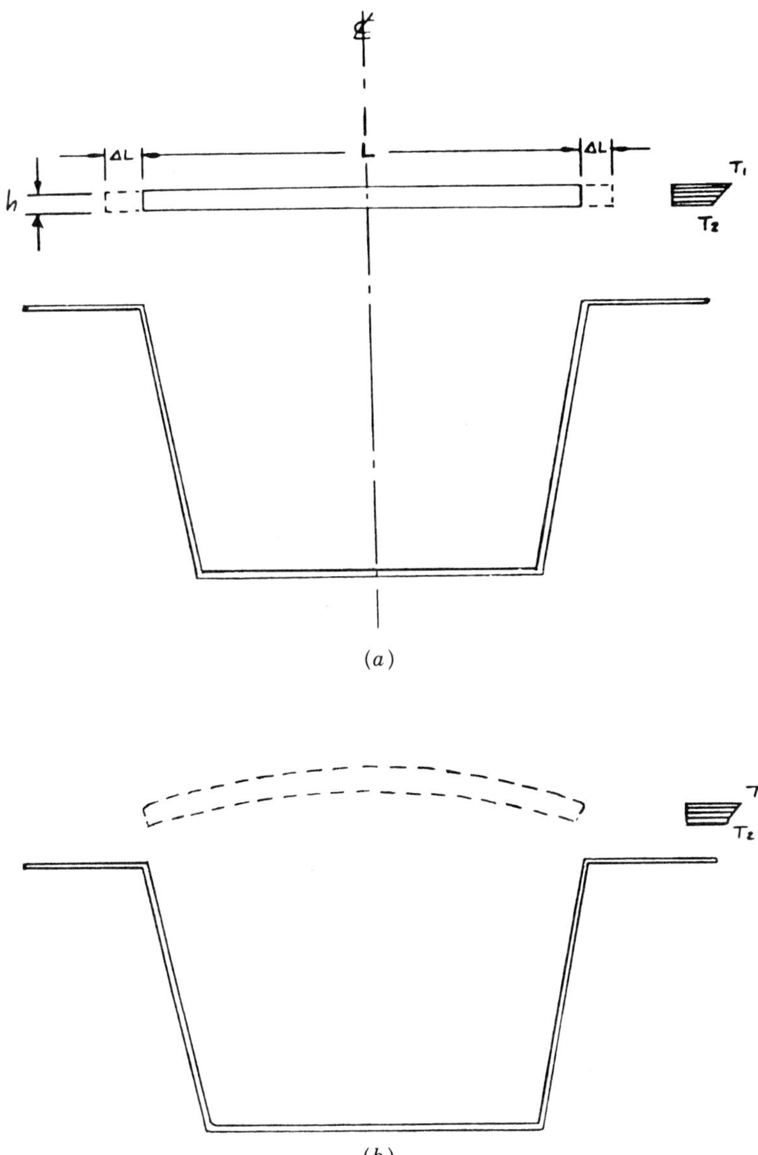

FIGURE 12-18 (*a*) Thermally induced deck slab elongation; (*b*) thermally induced deck slab curvature and end rotation.

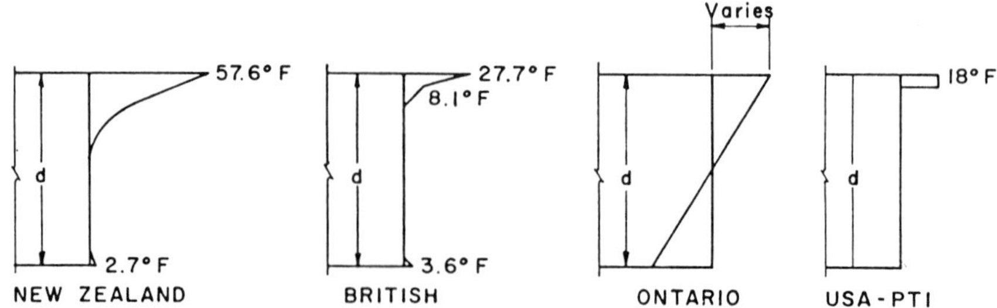

FIGURE 12-19 Typical thermal gradients representing current practices for different countries.

thermal movement of 8.2 in., the coefficient of thermal expansion was calculated as 6.7×10^{-6}, or very close to the assumed design value.

Examples of Temperature Effects The four thermal gradients shown in Figure 12-19 represent a broad range of practice and thermal design criteria. The gradient identified as USA–PTI articulates criteria currently recommended in local practice.

Colorado River Bridge This structure is 560 ft long and consists of five equal spans. Pier heights are about 42 ft. The deck is 49 ft wide and consists of a 7-in.-thick slab supported on seven precast, prestressed I girders provided with end diaphragms. The piers are 2.5 ft thick, fixed at the bottom and pinned at the top, and relatively flexible longitudinally. Driven steel piles provide the foundation. Expansion bearings are provided at the abutments.

The longitudinal variation of the top and bottom stresses was obtained by computing the stresses at the supports and then connecting the plotted points with straight lines. Higher tensile stresses in the bottom fiber were measured at two piers and were attributed to the T shape of the superstructure and the resulting higher location of the neutral axis. Using USA criteria, typical maximum stresses in the top fibers were 250 psi, whereas typical maximum stresses in the bottom fibers reached −300 psi, both caused by positive temperature gradients. Stresses caused by axial forces resulting from pier–wall shear restraints were less than 10 psi and therefore negligible.

West Silver Eagle Road Bridge This structure is 745 ft long and consists of six spans ranging in length from 102 to 135 ft. The superstructure is 52 ft wide, and is a multiple-cell cast-in-place posttensioned box girder without intermediate expansion joints, cast monolithically with the pier columns. The resulting bridge is a continuous frame, with the two pier columns fixed at the top and pinned at the bottom. Pier heights range from 25 to 35 ft, and have drilled subpiers as foundation elements. Expansion bridges are provided at the abutments to accommodate longitudinal movement.

Likewise, the top and bottom fiber stresses were computed along a typical girder centerline. These stresses are less than in the corresponding locations for the previous example, and a probable explanation is the additional resistance provided by the bottom slab.

The effect of monolithic construction is evident at the piers, particularly at locations away from the point of zero movement. Under the New Zealand criterion, the bottom fiber stress at one location changes on either side from −445 to −325 psi, as a result of the rigid connection between superstructure and column. The top fiber stress is likewise changed to a similar extent, but the center piers are less affected because they are close to the point of zero movement.

The bottom slab has a relatively modest flare, and this affects the bottom fiber stresses accordingly. Using the New Zealand gradient, at the right side of the second pier the bottom fiber stress changes from −375 psi at the beginning of the flare to −250 psi at the fascia line. The top fiber stresses are, however, only slightly influenced by the flare configuration.

The variation of the stress diagram with depth for the four gradients is shown in Figure 12-20.

Theoretical Data

In another theoretical study, Zuk (1961) has developed an elastic procedure for determining stresses and strains caused by various linear thermal gradients in composite steel and concrete bridges. Both axial and lateral stresses can be determined by the following.

For a slab,

$$f_{xs(y)} = \frac{F}{2ap} + 3y_s \frac{(Fa - Q)}{2a^3 p} \tag{12-34}$$

$$f_{zs(y)} = m f_{xs(y)} - c_s E_s T(y) \tag{12-35}$$

where $2a$ = slab thickness (in.)

c = coefficient of thermal expansion

E = modulus of elasticity (psi)

F = interface shear (lb)

f_{xs} = longitudinal slab stress (psi)

f_{zs} = transverse slab stress (psi)

m = Poisson's ratio

p = slab width

Q = interface couple (in.-lb)

T = temperature change

y_s = distance measured from midpoint of slab (in.)

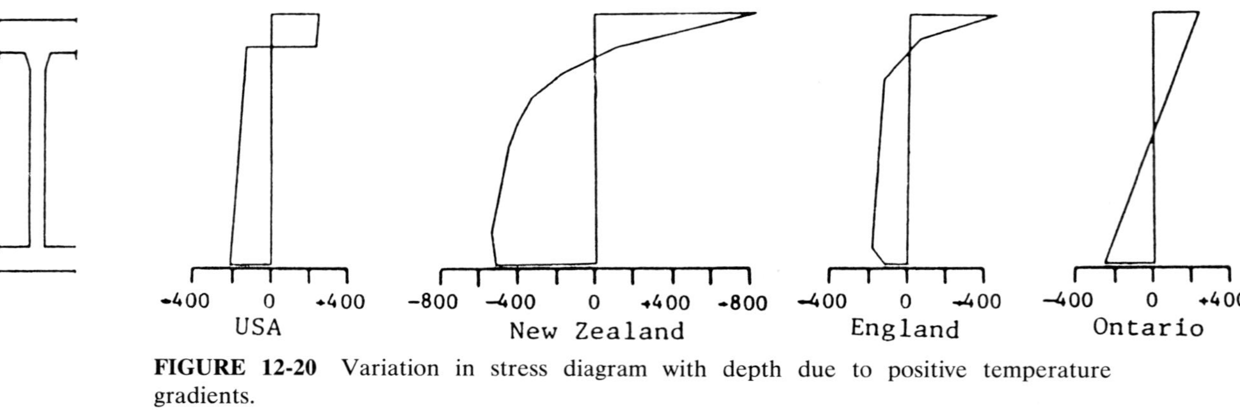

FIGURE 12-20 Variation in stress diagram with depth due to positive temperature gradients.

For a steel beam,

$$f_{xb(y)} = \frac{-F}{A} + y_b \frac{(-Fd_1 - Q)}{I} \tag{12-36}$$

where A = area of beam (in.^{2})
d_1 = distance from centroid to the top interface
f_{xb} = longitudinal steel stress (psi)
I = moment of inertia about centroidal axis of beam (in.^{4})
y_b = distance measured from midpoint of slab (in.)

These theoretical values may be modified to include the effect of creep, slip, and local plastic yielding.

Wah and Kirksey (1969) have used model and field tests together with theoretical investigations to determine the thermal effects on beam–slab bridges. One set of equations was applied to the thermal expansion, and another to normal flexural effects. The thermal loads were allowed to vary through the beam and the slab as well as on the surface. The generalized equations may be used for any temperature variation, and a computer program has been developed to handle the solution of the equations.

Creep effects were found to have a marked influence on the behavior of a heated slab. The slab may arch upward when the heat is first applied because the top is heated more quickly than the bottom. As time progresses the upward deflection decreases and may reverse. The creep deflections tend to offset thermal deflections when the top of the slab is warmer than the bottom.

Thermal stresses have been investigated by Hunt and Cooke (1975). The underlying principle focused on problem formulation and boundary conditions to develop appropriate equations, and a numerical solution was derived using finite-difference approximations. The model involved a box girder concrete bridge. The conclusions drawn from this study are as follows: (a) unsteady temperature distribution can be calculated ignoring horizontal temperature gradients, (b) more work is necessary to determine typical values of heat conduction coefficients for bridges under field conditions, (c) the three-dimensional thermal stress distribution in a prismatic bridge can be estimated as a two-dimensional plane strain problem in thermoelasticity, and (d) uncoupling longitudinal and transverse stresses as suggested by Priestley (1972) leads to accurate and convenient ways for estimating longitudinal normal stresses in regions where transverse stresses are relatively small.

A finite-element analysis is reported by Rahman and George (1979) for computing thermal stresses and displacements in continuous skew slab–girder bridges. A layered girder–slab model was developed to take into account the nonlinear temperature distribution and the material properties. Results from a two-span continuous bridge show that thermal stresses and deflections are considerably influenced by the skew. Uplift tendency is more pronounced over intermediate piers in skew bridges than in rectangular structures, and this tendency is enhanced with increasing skew. The same investigators have checked the applicability of the finite-element program in model tests (Rahman and George, 1980). In this case the model was a similar skew bridge with an intermediate stiffening beam. The analysis provided information on normal and shear stresses, in-plane and out-of-plane displacements, transverse and longitudinal slopes, and twist resulting from nonlinear thermal gradients. The support conditions can have a marked effect on thermal stresses and are considered in the program by specifying displacement restraints. The conclusion is that thermal stresses due to thermal gradients alone are generally low. However, if dead and live loads are also considered, they warrant ample investigation and study.

In their review of thermal stresses, Reynolds and Emanuel (1974) focus on the need for a semiempirical approach in dealing with thermal effects. Because it is difficult to isolate stresses thermally induced from those caused by creep, shrinkage, plastic yielding, and interface slip, caution is warranted in interpreting the validity of theoretical predictions and the results of tests to confirm these predictions. In addition, extrapolating data to other bridges is difficult because of the many variables involved.

From the foregoing it may appear that the task of predicting the actual temperature distribution in bridges is extremely complex. If a uniform distribution is assumed, it may occur only under special conditions such as screening the bridge deck from solar effects due to clouds. The use of a mean temperature does not consider the bending stresses developed when the warmer parts of the bridge expand more than the cooler sections. Assessing the stress level using a nonlinear thermal distribution gives results closer to the actual conditions. A linear temperature distribution can be assumed for concrete box sections through the top slab. The sides and bottom of the box should be expected to have approximately constant temperature.

Suggested Guidelines for Thermal Effects

TRB–NCHRP Report 276 (1985) gives design guidelines for thermal effects in concrete bridge superstructures. These effects result from time-dependent fluctuations in the effective bridge temperature or from temperature differentials within the bridge superstructure. Invariably, all concrete bridges should be designed for the former, and prestressed (posttensioned) bridges should be checked for the effects of the latter.

The values of the anticipated minimum and maximum effective bridge temperature depend on the type of construction and on the air temperature at the site. Appendix A of TRB–NCHRP Report 276 gives relevant data and articulates their applicability.

Differential Temperatures Provisions should be made for stresses and movements resulting from differential temperatures within prestressed concrete bridge superstructures when a long-established history of satisfactory performance for similar bridge superstructure is not available. Guidelines and technical data articulate the variation in the vertical temperature differentials, and apply to superstructures with depths greater than 2 ft. In determining the temperature variations within box girder bridges, the relevant temperatures are increased by 5°F.

Data are also available for negative temperature gradients, whereby temperature differentials result from a rapid cooling of exposed concrete surfaces.

AASHTO Specifications Reference to TRB–NCHRP Report 276 is stipulated by AASHTO for segmental concrete bridges, although it is recognized that additional field research is warranted to verify the temperature gradient range. Although the general conclusion is that consideration of thermal gradients in design of concrete box girder bridges has been demonstrated, additional clarification may lead to desired simplicity.

Transverse analysis for the effects of differential temperature outside and inside box girder sections is not generally mandatory, but may be indicated for relatively shallow bridges with thick webs. In this case the specifications recommend a ±10°F differential temperature. Thermal effects in concrete bridge superstructures for conventional highway bridges are treated in the 1989 AASHTO guide specifications. These contain criteria for thermal analysis and provide design guidelines for thermal effects.

12-6 BASIC CONCEPTS OF STRENGTH AND FRACTURE

Theoretical Background

Ductile and Brittle Behavior A bridge member may react to stresses in three ways: (a) deform elastically, (b) deform plastically, or (c) break. The amount of elastic or plastic deformation (strain) that occurs before fracture depends mainly on the magnitude of the applied loads, how they are combined, cyclic and repetitive nature, and residual stresses. Ductile and brittle behavior can thus change depending fundamentally on the state of stress, and a member that is quite brittle under one system of loading may be ductile under another. An example is the enhanced ductility of marble in compression under hydrostatic pressure, and the response of metals showing that ductility is increased by increasing the average pressure under which the deformation is produced (Maier, 1935; Nadai, 1931, 1937, 1950).

In bending, the outside fibers have the same state of stress as in simple tension, and on the inside the state of stress is as in simple compression.

When a bar specimen is subjected to bending, it elongates in the tension direction and contracts in all lateral directions an amount proportional to its width in each direction (Poisson effect). In bending, the fibers on the outside are free to contract laterally without restraint, and those on the inside are free to expand laterally, but only for narrow specimens. Thus, the state of stress affects the relative ductility of wide and narrow plates in bending, reducing the ductility of the wide plate according to the foregoing mechanism. Similar effects are demonstrated for members subjected to tension and torsion.

The Notch Effect The V-notch requirements stipulated in materials specifications (see also AASHTO Article 10.3.3) articulate the cause of brittle failure in the presence of a notch. The relevance of the state of stress at the base of a notch in a bar under tension explains its diminished ductility. Figure 12-21 represents a notched bar in tension and applies to a notch where behavior is examined in the Charpy test.

The values of the three principal stresses and the maximum shear stress are plotted across a notched tensile test bar from the base of the notch to the center. The principal stresses σ_l, σ_t, and σ_r are labeled in this manner to indicate longitudinal, tangential, and radial effects, respectively. On the surface at the base of the notch, $\sigma_l > \sigma_t > \sigma_r$, so that $\sigma_l = \sigma_1$ and $\sigma_r = \sigma_3$. However, just below the surface σ_t may be less than σ_r so that $\sigma_t = \sigma_3$. The maximum shear stress is always one-half the difference between σ_2 and σ_3.

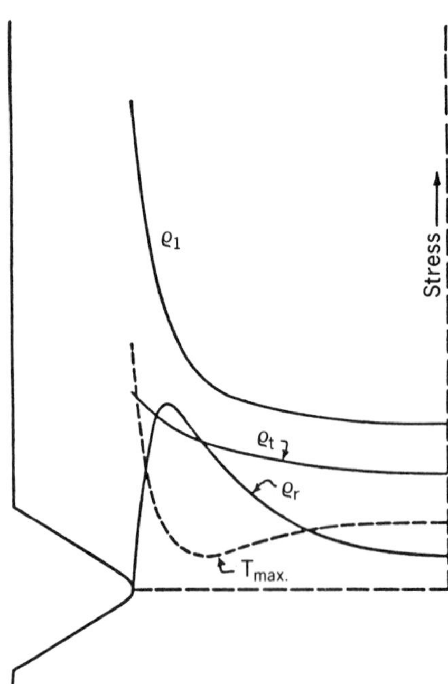

FIGURE 12-21 Variation of state of stress across a notched cylinder in tension, from the base of the notch to the centerline.

The sharper and the deeper the notch is, the higher the values of all three principal stresses just below the surface.

This state of stress is explained noting that at equilibrium there can be no normal stress acting in the direction perpendicular to a free surface. With a sharp notch, the layer of metal along the notch surface will take none of the tensile load imposed, and the entire load must be taken by the area left below the base of the notch. There is therefore a mass of metal, determined by the depth and sharpness of the notch, that is relatively free from applied tensile load. The unloaded mass serves as a ring attached to the load-carrying mass and tries to keep it from contracting. The load-carrying mass responds by developing stresses in both the radial and tangential directions.

A notch is sensitive to metal toughness, ductility, the presence of a fatigue crack, and depth and sharpness, and responds according to a process that tends to change its shape, causing deformations that tend to be less localized and more widely spread.

Resistance to Deformation The overall deformation occurring prior to fracture articulates the ductility of the member. Whether a cohesive failure (crack propagation) will occur, upon reaching a critical value of the greatest principal stress in tension, is a convenient assumption that is not exactly right. Theories of failure, also referred to as yield conditions, are commonly based on the concept of yield strength. This gives the uniaxial yield stress in terms of the three principal stresses, and the yield criterion (also referred to as the Huber–von Mises formula) is stated as

$$\sigma_y^2 = \tfrac{1}{2}\left[(\sigma_1 - \sigma_2)^2 + (\sigma_2 - \sigma_3)^2 + (\sigma_3 - \sigma_1)^2\right] \tag{12-37}$$

where σ_y is the yield stress that may be compared with the uniaxial value F_y. For the usual problems, one of the principal stresses is either zero or small enough to be neglected. For plane stresses, (12-37) is reduced in this case to

$$\sigma_y = \sqrt{\sigma_1^2 + \sigma_2^2 - \sigma_1\sigma_2} \tag{12-38}$$

Flexural stresses on beams and girders assume zero principal stresses normal to the plane of bending. In addition, most structural shapes are comprised of thin plate elements, all subject to (12-38).

Factors affecting resistance to deformation include temperature, amount, and speed of deformation. In general, the higher the temperature the lower the resistance to deformation although exceptions can be noted. At high temperatures flow takes place at very low stress, and resistance to deformation refers to a particular rate of straining. The effect of temperature is better articulated if the other factors are fixed.

The amount of deformation affects resistance in a manner clearly articulated in a stress–strain test. By dividing the applied load by the least actual area at the time the load was read, we can plot the so-called "true stress–strain curve," and therefore understand how resistance to deformation varies over a wide range of deformation.

The speed effect is demonstrated quantitatively and indicates a semilogarithmic relationship. A considerable difference in rate is therefore necessary to produce only a small difference in resistance. Conversely, a small change in stress produces a marked change in speed of deformation. Harder metals are less sensitive to speed changes than softer ones.

Resistance to Fracture Fracture can occur in two ways: the member may be separated along a plane of maximum normal stress (cohesive failure or normal fracture), or one part may slide over the neighboring part until the two parts are separated (shear fracture). Both types are sufficiently different in mechanism and appearance to warrant separate consideration.

In the tension test, a round bar stretches until a crack forms normal to the stress direction. For a ductile member, this occurs after the metal has "necked-down," and the crack starts at the center spreading out. When the crack propagates to the edge of the specimen, the fracture may change from normal to shear type, but this does not affect the load intensity at which breaking begins.

Nadai (1950) summarizes results of tests on yielding and fracture under combined stresses and articulates the criteria and conditions of failure, usually expressed by (12-37). Among these, the best evidence supporting the validity of the maximum normal stress theory comes from tests on thin-walled tubes subjected simultaneously to internal pressure and longitudinal tension or compression. Loading the tube so that the ratio between the tensile stress in the longitudinal direction and the tensile stress in the circumferential direction remained constant, the breaking stress was calculated by dividing the axial load by the actual cross-sectional area.

The amount of deformation in the direction of stretching that precedes fracture depends markedly on the state of stress. Deformation prior to fracture is also affected by anisotropy. If a metal has, for example, two different ductilities in the longitudinal and transverse directions, the deformation preceding fracture cannot be predicted with certainty.

Assuming that fracture will occur when the normal stress reaches a critical value, other factors influencing this value relate to the previous loading history, such as fatigue, the amount of plasticity (plastic deformation), and the cohesive strength. Fracture will depend to a lesser extent on the rate of application of load and changes in temperature. Fracture toughness is discussed in subsequent sections.

Expected Fracture Locations in Bridges (Fatigue Cracking)

Procedures and guidelines are usually independent of potential fracture locations in bridge members. Thus, this review relates primarily to the decision to provide an alternate load path for a proposed or existing bridge consisting of steel girders and beams, and refers mainly to fatigue cracking revealed by inspection. Fatigue cracks are likely to appear at the following locations (Fisher, Hausamann, and Sullivan, 1979; Fisher, 1974, 1984, 1981; PADOT, 1988).

Groove Welds (a) Flange groove welds in relatively older structures with welds made prior to adequate nondestructive inspection techniques; (b) web groove welds as in (a); (c) groove welds in longitudinal stiffeners, treated as structural welds, particularly in older bridges where these components were not inspected; (d) groove welds between longitudinal stiffeners and intersecting members.

Ends of Welded Cover Plates on Tension Flanges (a) At the toe weld or in the weld throat at midwidth of the flange on cover plates with end welds; (b) at the end of longitudinal welds on cover plates without end welds.

Ends of Attachment Plates Welded to Flange or Web (a) Welded splices between adjacent parts, such as gusset plates; (b) repaired flanges or webs with doubler plates; (c) repaired webs using fish plates; (d) attachments for signs, railings, utility supports, and so forth, with the attachment plate parallel to the girders; (e) welded attachment plates perpendicular to the girder, with higher fatigue strength than in (a), (c), and (d).

Diaphragm Connections (a) Ends of welded diaphragm connection plates on girder webs where the plate is not connected to the flange (cracks may occur at the top or bottom of the connection plate when positive attachments are not made to the flange); (b) ends of riveted connection plates on girder webs when the angles are not connected to the flange (web cracks are most likely when connection angles do not overlap the flange angles).

End Connections of Floor Beams (a) Copes and blocked flanges at ends of floor beams (cracks may occur at the reentrant angle of the cope or the blocked flange); (b) connection plates and angles as in diaphragm connections.

Floor Beam Brackets (a) Bracket connections to girder webs, as in diaphragm connections; (b) tie plates connected between top flanges of outside brackets and the floor beams.

Top and Bottom Lateral Bracing Connections (a) Gusset plate bracing connections to girders (web or flange) (when the gusset plate is attached to the web but not to the diaphragm connection plate, cracks may occur in the web gap at the toe of the weld); (b) gusset plate to diaphragm connection plate welds (displacement-induced forces develop in the diaphragms because of differential girder deflections).

Transverse Stiffeners (a) Where these are not connection plates, they normally have adequate fatigue strength; (b) at the end of cut-short stiffeners because of handling, shipping, or web plate vibration (where stiffeners are fitted, cracks may be revealed by cracking of the paint film).

Tack Welds (a) At tack welds used for attaching bridge components during erection (fatigue cracking often originates at these locations); (b) tack welds often occur between gussets and main members, bearing plates and beam flanges, and bolted connection angles and webs.

Plug Welds At any plug weld (made to correct a misplaced drilled hole, or in the field during repairs); where plug welds are painted, fatigue cracking is revealed by cracking of the paint.

Fracture-Critical Bridge Members

AASHTO (1978) provides guidelines for fracture-critical nonredundant steel bridge members, intended to generate a better understanding of the behavior and design of nonredundant bridges (see also Sections 4-2 and 5-17). These guidelines address steel toughness, welding requirements, and welding procedures, consistent with the fact that structural reliability is governed by material properties, design, fabrication, inspection, erection, and usage.

Experience has demonstrated that contributing factors in bridge failures include unqualified inspection techniques and nondestructive testing; design details resulting in notches and difficult joints to weld and inspect; hydrogen-induced cracks; improper fabrication, welding, and weld repair; and lack of base metal and weld metal toughness.

Fracture-critical members are tension components of a bridge whose failure would be expected to result in bridge collapse. The identification of such components is part of the design process, but in order to be critical the component must be in tension. In addition, a fracture-critical member may be either a complete bridge member or it may be a part of a bridge member. Examples are tension ties in arch bridges and tension chords in truss bridges where failure of the tie or chord could cause the bridge to collapse. Some existing complex truss and arch bridges do not depend on any single member (in tension or compression) for structural integrity. Critical tension components usually occur in flexural configurations. One flange of the flexural member must be in tension and is therefore a critical component if failure of the flexural member can cause collapse. Likewise, the web of a flexural member is partially in tension and may thus become the critical component.

Members and components whose failure will not cause bridge collapse are not considered fracture critical. Compression members and components may be such examples, particularly if they fail by yielding or buckling rather than by crack formation. Likewise, longitudinal stiffeners and other accessories welded to the compression area of webs or flanges are not fracture critical.

Design Considerations Essentially, fracture control is implemented through efficient design and detailing. This must also address fatigue requirements that should by necessity be more conservative for fracture-critical members. Although fatigue categories and stress ranges are covered in the specifications, it still remains a design requirement to study each bridge detail to ensure that the detailing will allow effective joining techniques and nondestructive testing of all welded joints.

If fracture-critical members are present, they must be identified and so designated. In addition, the toughness of steel must satisfy the requirements established jointly by AASHTO, the Federal Highway Administration, the American Iron and Steel Institute, and the American Institute of Steel Construction. These requirements include (a) constraint and temperature effects, (b) effect of the rate of loading, (c) correlation between impact-fracture roughness and impact Charpy V-notch energy absorption, and (d) verification of AASHTO toughness requirements.

12-7 DEVELOPMENT OF DATA FOR FATIGUE DESIGN

Fatigue Loadings

The standard AASHTO specifications divide roads into two types, according to Table 10.3.2A. The number of load (stress) cycles is therefore prescribed and serves as a basis for selecting the allowable fatigue stress range given in AASHTO Table 10.3.1A. More specifically, traffic characteristics used to develop fatigue loading data (see also Section 5-14) include volume, composition, time variation, directional distribution, and lane distribution. This record is then used to extrapolate data on fatigue loading caused mainly by trucks, because other vehicles have inconsequential effects.

Traffic Characteristics The usual term used to express traffic volume is *annual average daily traffic* (AADT), which is the total traffic in both directions over a one-year period. The composition, stated in terms of specific categories of vehicles, varies widely with the highway location and type. The FHWA has, however, typical composition data for various highway types, derived from nationwide surveys and updated annually. For example, in a given year, trucks may comprise about 25 percent (interstate), 15 percent (rural primary), and 10 percent (urban primary) of the total highway traffic. If light trucks are added, these percentages may increase from 50 to 100 percent. Approximately two-thirds of trucks are loaded.

Fatigue Loading Data The AASHTO truck volume refers to the average daily truck traffic (ADTT) in one direction, and this is the design truck traffic because it determines the number of stress cycles to be used in selecting allowable fatigue stresses.

Truck Weights Federal amendments have increased the allowable weights as follows: 20 kips for a single axle, 34 kips for a tandem axle, and 80 kips for the gross weight. These limits are for vehicles operating on the interstate system. Figure 12-22 shows probability–density curves of gross weight distribution for three truck types. These curves are normalized by expressing the gross weights as a percentage of the average gross weight for the given type of truck. The area beneath a curve between any normalized weights gives the fraction of trucks within that range. Empty trucks are included and comprise about one-third of the truck traffic.

Fatigue Design Truck This unit represents a given composition of truck traffic and constitutes a gross weight selected so that a given number of unit passages would have the same fatigue effects as the same number of trucks from the given composition. Its gross weight is computed as

$$W = \left[\sum a_i W_i^3\right]^{1/3} \tag{12-39}$$

where a_i is the fraction of trucks within a given weight interval and W_i is the midpoint of that interval. Gross weights for fatigue design trucks range from about 44 kips to 52 kips for rural interstate highways. Similar design trucks have been developed for metropolitan, urban, and rural highways.

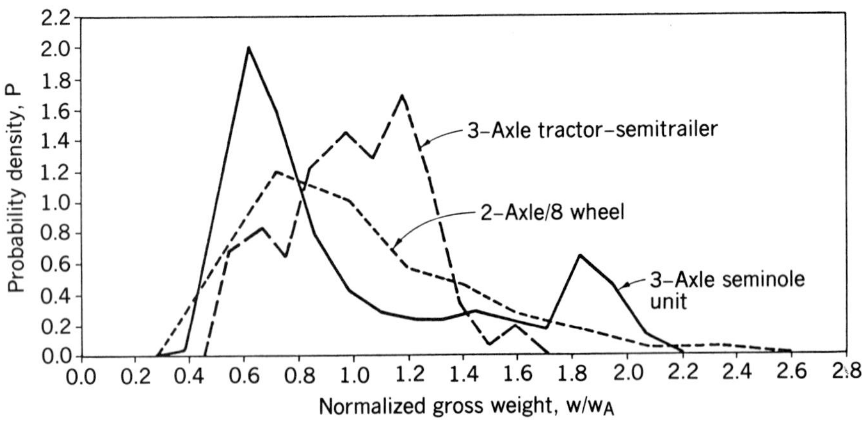

FIGURE 12-22 Gross-weight distribution curves for three truck types: two-axle–eight-wheel, three-axle tractor–semitrailer, and three-axle single unit.

Truck Spacing For light traffic volumes, the spacing of vehicles is predicted by a Poisson distribution (Institute for Transportation Engineers, 1976) giving the fraction of spacings less than a specified value, or

$$F = 1 - \exp[-(qd/v)] \tag{12-40}$$

where F is the fraction of vehicles at a spacing (front-to-front) less than d, v is the average speed of traffic, and q is the one-directional traffic volume, usually for a period of one hour. This formula may be applied to all traffic in one direction regardless of the number of lanes.

Truck spacing is more relevant to fatigue considerations than total vehicular spacing. The distribution of truck spacing can be calculated from (12-40) by using the truck volume q (Goble, Moses, and Pavia, 1974). The average traffic speed v is a function of the total traffic volume (Institute for Traffic Engineers, 1976). The truck spacing distribution is therefore the function of total traffic volume as well as a percentage of trucks in that traffic.

The percentage of closely spaced trucks is quite small for average AADT volumes. On four-lane rural highways with an AADT of 10,000, only about 1.6 percent of truck spacings are less than 100 ft.

Lane Loading Closely spaced traffic can be represented by a uniform lane loading calculated from the average vehicle spacing and weight. The closest possible spacing occurs when traffic stops, but because this condition is rarely encountered in bridges, uniform loading should not be used for fatigue design. An indication of the possible magnitude and frequency of lane loading is provided by considering a highway carrying 2000 vehicles per hour per lane. Such a volume may indicate maximum capacity, however, and may not provide explicit design criteria.

Traffic volume for the 30th highest hour/year is about 15 percent of the AADT. A highway with an AADT of 13,000 vehicles per lane (about 52,000 vehicles for four lanes) should be expected to have a traffic volume exceeding 2000 vehicles per hour per lane only 30 times a year.

Fatigue Reliability Models

In Section 5-14 a brief reference was made to fatigue reliability as it relates to design. At present, most design specifications base fatigue provisions on modified Goodman diagrams, and in many instances the criteria consider a constant range. Invariably, the randomness of fatigue lives and variations in the applied loads are factors to be taken into account.

The simple format shown in Figure 12-23 is suggested by Yao (1974) for fatigue design. The distribution for repeated loads may be obtained from dynamic analysis or field measurements. The acceptable probability of failure, p, is also selected, whereas the stress–life–probability relationships are available from single-stress-level laboratory tests. Referring to the initial

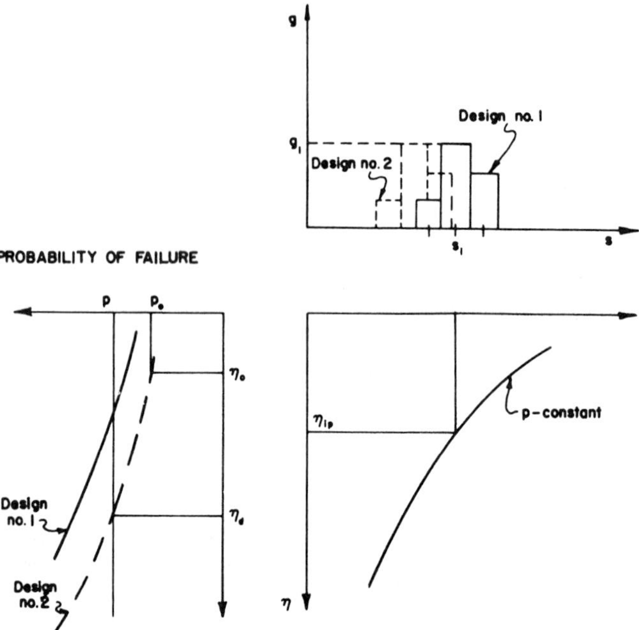

FIGURE 12-23 Graphical illustration of fatigue design procedure. (From Yao, 1974.)

configuration of the structure, the distribution of repeated stress is then computed. The design life can be computed from

$$\eta_D = \eta_0(1 + \beta V_f) \tag{12-41}$$

where η_D is the expected life of the structure, β is a safety index, and V_f is a coefficient of variation for random variables expressing the uncertainty and reflecting the method of analysis, cumulative-damage theory, and so forth. If, compared with the design life, the computed life is unsatisfactory, an iterative procedure is used to obtain acceptable results as shown in Figure 12-23.

Fatigue Life Distribution Among the various mathematical models proposed for the statistical description of the fatigue life, the most widely acceptable and used model is as suggested by ASTM (1963). According to this document, the fatigue life N, or some function of N, is assumed normally distributed, and use is made of the log-normal distribution function. Other models of distribution functions have been suggested by Gumbel (1958, 1963), Freudenthal (1968), Rascon Chavez (1967), Eugene (1965), and Weibull (1939). Among these, the Weibull distribution is most widely used because (a) it is based on experimentally valid assumptions and is relatively easy to use, and (b) it provides better developed statistics than other mathematical models.

Cornell (1969) has suggested the use of a weighted average of two distributions, the Weibull and the gamma distribution. Following results of tests, Korbacher (1971) has found that (a) fatigue life follows a log-normal distribution at high stress levels, but an extremal distribution at lower levels, and (b) fatigue life follows some bimodal distribution at intermediate stress levels.

Random Fatigue Fatigue behavior is better understood through tests and analyses of realistic stress histories. The state of the art in random load fatigue is summarized by Swanson (1968). The linear cumulative damage theory was found to overestimate the fatigue life of specimens subjected to random loads (Hillberry, 1970; Bussa, Sheth, and Swanson, 1972). Sweet and Kozing (1968) have presented a cumulative damage theory that takes into account the randomness of the time to failure and the fact that the failure process can be said to possess memory. Most of these theories are essentially modifications of the Palmgren–Miner linear cumulative damage theory, considering the nonlinearity effect, the order in which large and small stresses are applied, or both. However, none of these theories can be used to describe all known fatigue events. Because it is difficult to include nonlinearity effects or maintain record of the order in which large and small stresses are applied, the Palmgren–Miner criterion appears to be the most widely used reference.

12-8 FATIGUE-INDUCED STRESSES AND FRACTURES

Physical Process of Fatigue

The physical process of fatigue involves two basic phases: (a) crack initiation and (b) crack propagation or subcritical crack growth. Either phase may be relevant in assessing structural performance, depending on the structure type and applied loads. Crack initiation usually refers to the formation of cracks that are easily detectable. Thus, the crack initiation period may extend for a substantial portion of the usable fatigue life in high-cycle fatigue problems where stress fluctuations are low at fatigue-critical locations. On the other hand, where stress fluctuations are high, or in the presence of cracks, notches, and so forth, fatigue cracks will initiate quite early. The two phases are roughly equivalent in terms of order of magnitude and in low-cycle fatigue (less than 100,000 cycles). When welds and other structural details are present, or where defects are unavoidable in the fabrication process, crack propagation can almost begin with the first load application (Hertzberg, 1976; Rolfe and Barsom, 1977; Tiffany and Masters, 1965).

Fracture Approach to Fatigue Flaws are present in many bridge components because of the fabrication process. A common example is the structural weld where defects may develop due to porosity and lack of penetration or fusion (see also Section 12-6).

In fracture mechanics analysis, the basic parameter is the stress intensity factor K, given by

$$K = F(a)\sqrt{\pi a}\,(s) \tag{12-42}$$

where s is the applied stress, $F(a)$ is the finite-geometry correction factor which may depend on a, and a is the crack depth for a surface flaw or half-width of a penetration flaw. The factor K for some common flaws is given by Hertzberg (1976), Rolfe and Barsom (1977), and Tiffany and Masters (1965). From experimental data, the crack growth rate da/dN is related to the stress intensity factor range ΔK by (Paris and Erdogan, 1970)

$$\frac{da}{dN} = C(\Delta K)^n \tag{12-43}$$

where ΔK is obtained from (12-42) by replacing s with the applied stress range S, and C and n are experimental constants depending on the mean cycling stress, the test environment, and the cycling frequency. A graphical presentation of this crack growth relationship is shown in Figure 12-24.

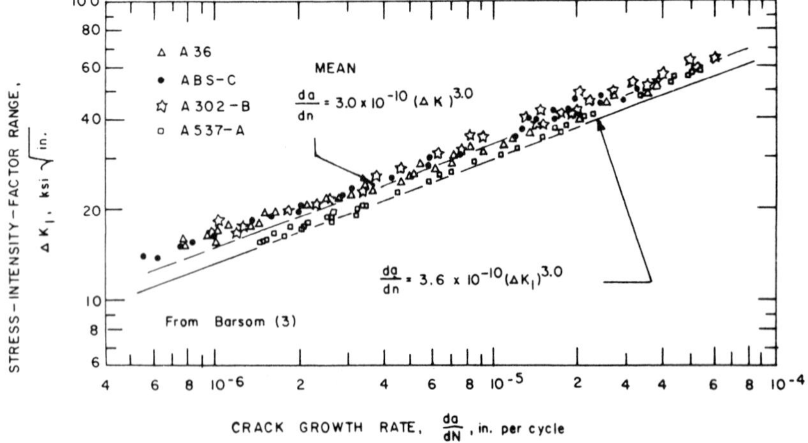

FIGURE 12-24 Fatigue crack growth data for ferrite–pearlite steels. (From Barsom, 1971.)

Rearranging (12-43) and integrating, we obtain

$$NS^m = \frac{1}{C}\int_{a_i}^{a_f} \frac{da}{\left[F(a)\sqrt{\pi a}\right]^n} \tag{12-44}$$

where N is the number of cycles to failure, S is the stress range, and a_i and a_f are the initial and final flow sizes, respectively.

Statistical Distribution of Crack Growth Data The existence of variability in $da/dN - \Delta K$ data is caused by statistical scatter inherent in the material and variation in experimental procedures. Using the range as a basis for determining the standard deviation of da/dN and assuming a log-normal distribution, the coefficient of variation (COV) due to the experimental technique may approach 50 percent.

Scatter inherent in the material has been found to be represented by a COV in da/dN of 0.15 to 0.25.

Mean Time to First Failure The mean number of cycles to first failure may be obtained by considering the order statistics of samples of identical size drawn from a parent population. Let N be a random variable denoting the time to failure of a single element. Also, $F_N(t)$ and $f_N(t)$ are the distribution and density functions of N, respectively. The number of such elements is denoted by k, and let N_1 denote the smallest of N of size k. It can be shown that

$$E(N_1) = E(N)\left(\frac{1}{k}\right)^{1/a_N} \tag{12-45}$$

where $E(N_1)$ is the mean of N_1, and $E(N)$ is the mean of N, which must correspond to the particular loading history.

This concept may be applied to analyze failure in large systems with a large number of identical stress components, assuming that the times to failure of each component are identically distributed and statistically independent random variables. Larger systems will undergo failure sooner than smaller systems if all other factors are the same. In this case N_i are statistically independent as long as the failures in the system are localized and sufficiently separated. This means that the statistical analysis may be more appropriate for estimating the times at which fatigue cracks will appear rather than for studying fatigue failures. In the latter case cracking is likely to be extensive with individual cracks interacting.

Example Let us consider a welded component of a bridge subjected to a predictable constant-amplitude load of 30 kips. Using AWS data, we can write

$$NS^{3.413} = 4.78 \times 10^{10} \qquad s^2 = (0.5106)^2$$

Assuming that N is modeled by a Weilbull distribution, the parameters α (statistical parameter) and $\ln k$ are estimated as

$$\alpha = \frac{\pi}{\sqrt{6}s} = 2.512$$

and

$$\ln k = \frac{0.5772}{\alpha} + \ln\left[4.78 \times 10^{10}(30)^{-3.413}\right] = 13.2$$

and $k = 5.47 \times 10^5$ cycles.

From Weilbull distribution tables, $\mu_N = 4.85 \times 10^5$ cycles (μ_N is the mean value of random variable N). If the bridge contains 10 of these components with nominally identical stress histories, the average period obtained from (12-45) is

$$E(N_1) = 4.85 \times 10^5\left(\tfrac{1}{10}\right)^{1/2.512} = 1.94 \times 10^5 \text{ cycles}$$

This index will indicate the time when first inspection should be scheduled, probably at 190,000 cycles. The probability that fatigue failure may initiate prior to first inspection (190,000 cycles) is

$$P(N < 1.90 \times 10^5) = 0.068$$

If this probability is unacceptable, inspection should be performed earlier. At 100,000 cycles, the probability of fatigue failure is 0.014.

Fatigue Stresses

Observed Stress Spectra A truck crossing a bridge may produce one or more major stress cycles according to the influence lines and may have many smaller cycles superimposed. In every case these stress traces are added to the dead load stress to determine the total stress. In addition, dynamic characteristics result in large vibration stresses (impact), which may persist immediately after the truck leaves the bridge (Oehler, 1957). Typical stress traces are shown in Figure 12-25.

The major stress parameter affecting fatigue is the stress range, that is, the difference between the peak and valley of a stress trace as shown in Figure 12-25. Maximum stress ranges (excluding local stress raisers) as recorded

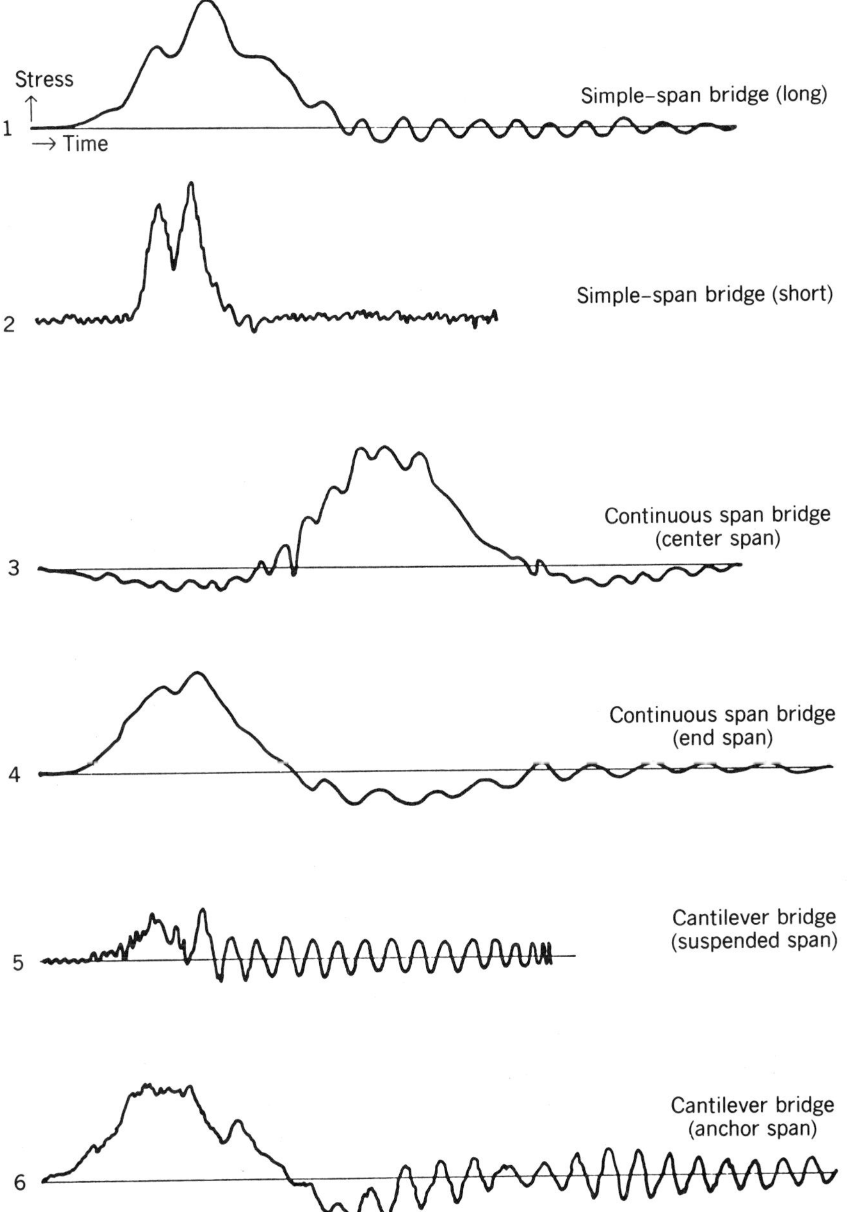

FIGURE 12-25 Typical stress traces for the passage of one single truck across different types of bridges.

TABLE 12-3 Measured Stress Ranges Caused by Traffic

State	Number of Bridges	Number of Histograms	Maximum Stress Range (ksi) Avg.	Maximum Stress Range (ksi) Max.	Effective Stress Range (ksi) Avg.	Effective Stress Range (ksi) Max.
Alabama	3	12	6.5	7.5	2.2	2.8
Connecticut	2	20	3.9	6.0	1.5	2.1
Illinois[a]	2	17	5.3	9.0	2.4	4.9
Louisiana	6	62	3.7	6.0	2.1	2.9
Maryland	6	13	3.6	6.0	1.4	2.4
Michigan	7	14	5.5	7.2	2.0	3.0
Minnesota	1	10	4.3	5.8	1.6	2.0
Ohio	10	49	4.2	6.5	1.3	2.3
Pennsylvania[a]	1[b]	6	6.1	10.5	3.2	4.4
Tennessee	2	4	3.8	4.0	1.6	1.7
Virginia	1	8	3.2	4.5	1.4	2.1
	Total = 41	Total = 215	Avg. = 4.3[c]		Avg. = 1.8[c]	

[a]Traffic includes heavy trucks only.
[b]Bridge was a riveted girder bridge opened to traffic in 1953.
[c]Average weighted according to number of histograms.

from various studies are given in Table 12-3. In this summary individual maximum values range from 0.8 to 10.5 ksi, and average 4.3 ksi with only 8 percent exceeding 7 ksi.

A variable-amplitude stress spectrum can be represented by an effective constant-amplitude stress range that induces about the same damage. This effective stress range is calculated for each histogram, and both average and maximum values are shown in Table 12-3. The individual values range from 0.5 to 4.9 ksi, and average 1.8 ksi.

Stress Spectra Calculated from Loadings A given composition of truck traffic can be represented by a fatigue design truck of average weight, and the resulting stress cycle represents therefore the stress spectrum for the particular composition. The stress range is obtained by positioning the average truck to give maximum positive and negative moments at the selected location. The design requirements are the same as in the design of a member such as a beam, girder, floor beam, stringer, and so on.

Almost 92 to 98 percent of damage is caused by four- and five-axle trucks. This configuration is represented by the AASHTO HS truck if the variable axle spacing at the rear is set at 30 ft. Although, statistically, a reasonable gross weight for the fatigue design truck is 50 kips, the specifications do not make a specific distinction, and the full HS 20 is used in checking allowable fatigue stress ranges. The wheel loads are distributed according to the lateral distribution factors stipulated in the applicable specifications.

A suggested lateral distribution method applicable to bridges with more than two longitudinal beams is shown in the chart of Figure 12-26. The lateral distribution factors F_e and F_i, for exterior and interior beams, respectively, are a function of the lane position factor $P = d/S$, where d is the distance from the exterior beam to the centerline of the outer lane and S is the beam spacing. The distance to the lane centerline is suggested as a parameter for fatigue design because it is the most likely truck position (United States Steel, 1982).

The solid lines represent the upper limit and can be used for an initial check. A second check is made using the dashed line, taking into account the beam span L and the moment of inertia I (both corresponding to the span where the truck is positioned). When P exceeds 0.5, the horizontal line applies to both interior and exterior beams, and this is the case for most bridges. When P is less than 0.5, exterior beams are governed by the sloping solid line. Because the lateral distribution resulting from this procedure gives wheel load fractions less than the AASHTO distribution, the use of the diagram is suggested under the associated explicit implications.

Impact factors for fatigue conditions may be based on typical or average conditions rather than extreme situations. If the impact factor is applied only to the peak live load, it must be large enough to account for the total effect on the stress range. For simple-span bridges, the theoretical increase in stress range is twice the vibration amplitudes because they increase the minimum stress while increasing the maximum stress. For continuous spans, the peak tensile and compressive stresses occur with the truck at different positions. If each peak is increased by an appropriate amount, the total increase in stress is correctly expressed. This suggests that the impact factors stipulated in the standard specifications are appropriate for fatigue design.

Stress Cycles per Truck Passage

Simple Spans The variation of midspan moment as a truck moves slowly across a simple span is shown in Figure 12-27. The stress variation under the same conditions is similar but the curve is more rounded with small vibration stresses superimposed because of dynamic effects. For spans exceeding 60 ft, a single cycle occurs. For spans less than 30 ft (the spacing of the main axles), two major stress cycles are noted. A secondary cycle corresponding to the front axle is superimposed on one of the major cycles. For spans between 30 and 60 ft, the cycle is more complex with two major peaks, and the depth of the valley between these increases as the span increases.

A complex cycle with one or more valleys will induce more fatigue damage than a single simple cycle of the same amplitude and is therefore equivalent to more than one cycle (see also subsequent sections). Vibration stresses have

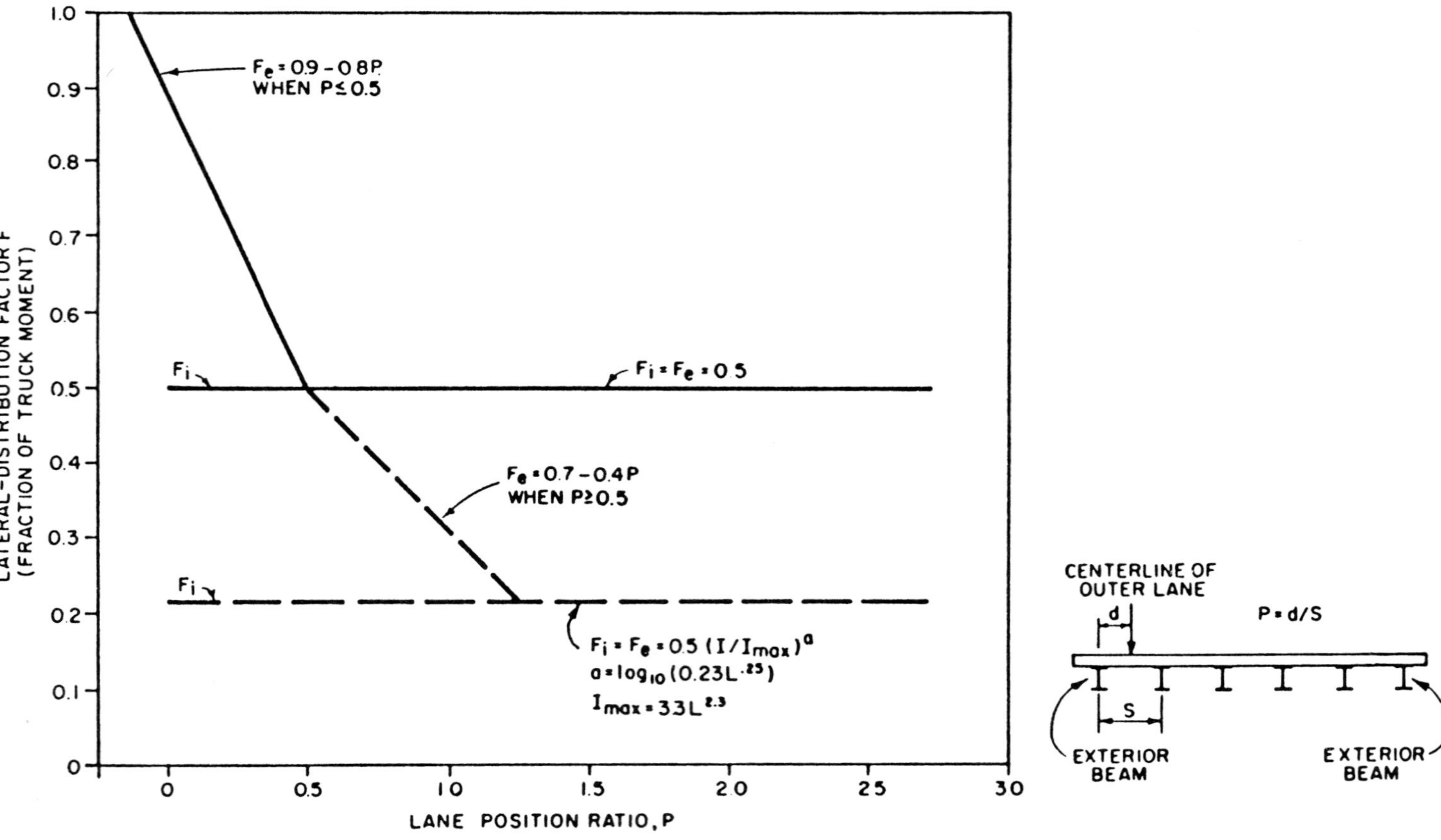

FIGURE 12-26 Lateral distribution diagram (fraction of truck moment) for fatigue design. (From United States Steel, 1982.)

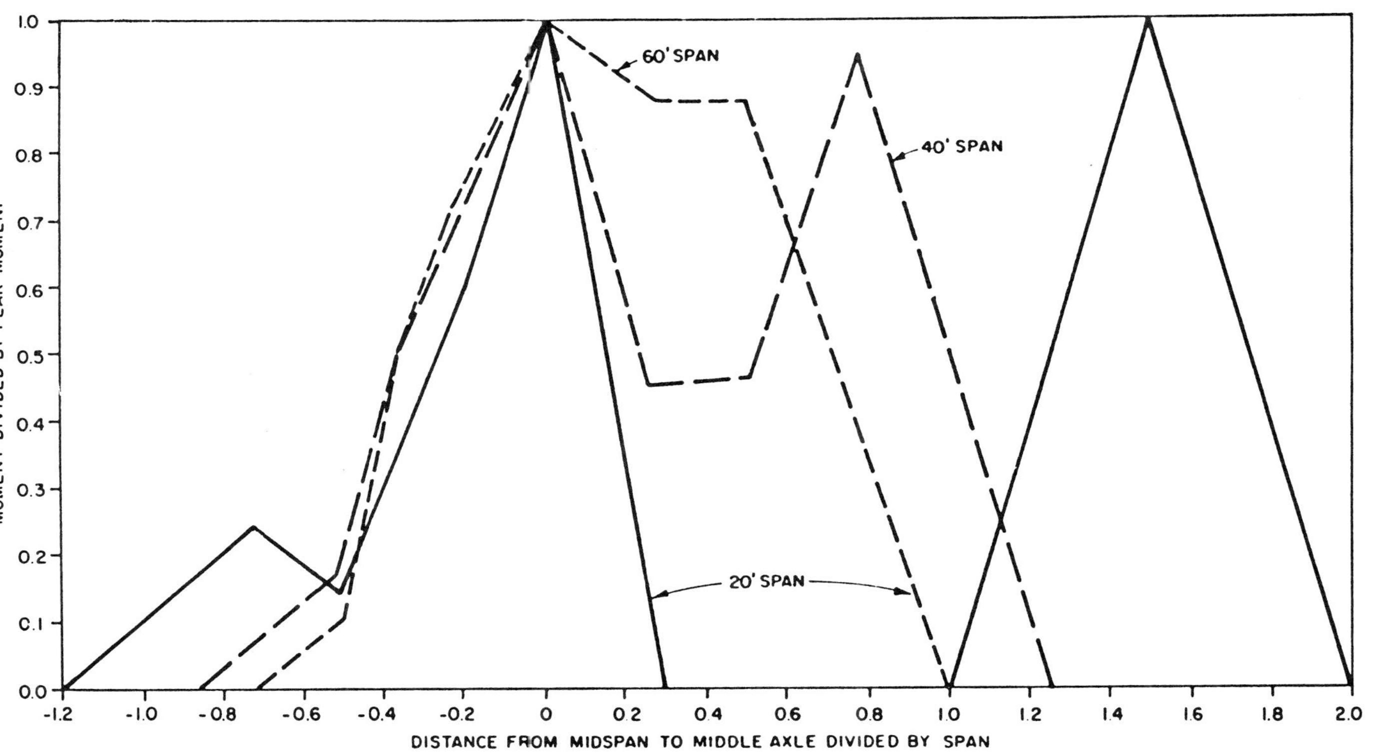

FIGURE 12-27 Midspan moment for simple spans expressed as a function of the truck position (spacing of rear axles 30 ft).

two effects: (a) they increase the maximum stress and the stress range of the complex stress cycle, and (b) they add wiggles to this cycle. The first effect is considered as impact, and the second effect is normally neglected.

Continuous Spans In this case the moment variation is more complex and often results in a complex cycle with several peaks of different heights, usually determined through the use of influence lines. Curves for the equivalent cycles of midspan moments are similar to simple-span bridges and are represented by two equivalent cycles for spans less than 40 ft and by one cycle for longer spans. Interior support moments are also represented by two equivalent cycles for spans less than 40 ft. For spans greater than 60 ft, the equivalent cycle curves increase toward a theoretical maximum value. This limit value is 2.0 for two equal spans, but less for unequal spans or for more than two spans. Normally, 1.5 equivalent cycles may be assumed above 40 ft.

These results are shown graphically in Figure 12-28 for one, two, and three continuous spans and various span lengths. For continuous bridges with more than three spans, or with unequal spans, the curves are essentially similar.

Transverse Members For floor beams, the moment variation for the fatigue truck is determined from the influence lines of the reaction on the transverse member. With longitudinal spans exceeding 30 ft, the equivalent cycles or transverse members are close to 1.0, for both simple and continuous longitudinal members, but increase above 1.0 as the span decreases below 30 ft. For a 20-ft span, the equivalent cycles are 1.11 for two simple spans and 1.03 for two to four continuous spans.

Effect of Closely Spaced Trucks Two closely spaced trucks crossing a bridge may cause a complex stress cycle. For simple spans, this interaction is shown in Figure 12-29 as a function of the truck spacing and the span length. Two fatigue design trucks are used to generate these diagrams, with the second truck placed in the same lane when the front-to-front truck spacing exceeds 50 ft, but adjacent to this lane if the truck spacing is less than 50 ft. Another assumption is that when the second truck is on the adjacent lane, it distributes only 0.3 of its weight to the other lane.

Calculations have been made for two cases that have almost the same distribution of truck spacing (United States Steel, 1982): (a) four-lane rural highways with an annual average daily traffic of 40,000, and (b) four-lane urban highways with an AADT of 60,000. For this traffic volume, the number of trucks spaced closely enough to interact varies from 9 to 31 percent as the span varies from 30 to 240 ft. For lighter traffic corresponding to the mean AADT values, these percentages should be only one-fourth and one-third as great.

It appears that by including the effect of closely spaced trucks, there is only a slight increase in fatigue life when compared with the life for the same

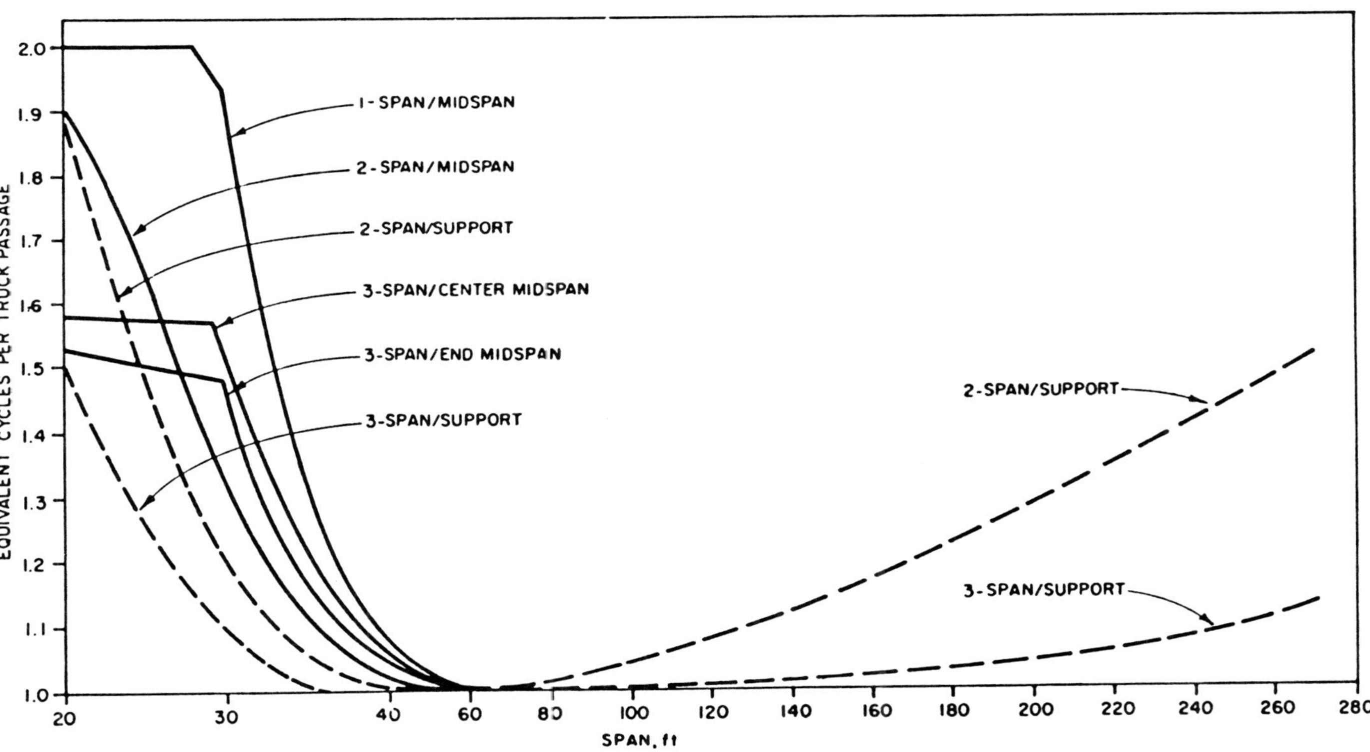

FIGURE 12-28 Equivalent cycles per truck passage shown graphically as a function of the number of spans and span length.

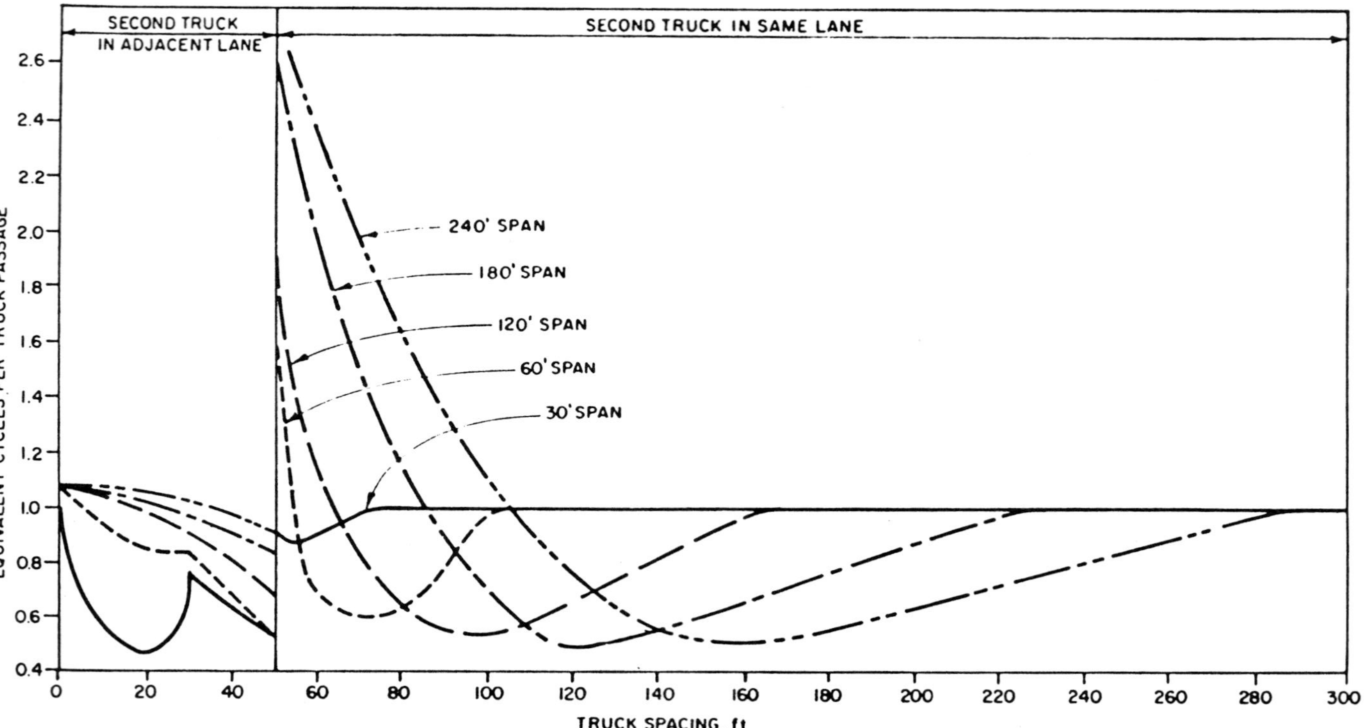

FIGURE 12-29 Interaction of two trucks closely spaced on simple spans as a function of truck spacing and span length.

truck traffic but with all trucks acting independently. This effect is very small because some of the close spacings are beneficial whereas others are detrimental, as shown in Figure 12-29. These conclusions are for simple spans, but may also be applied to continuous bridges.

12-9 FACTORS INFLUENCING FATIGUE

Representation of Fatigue Behavior

In Section 5-14 we presented the concept of the S–N curve, commonly used to express fatigue behavior, where the number of cycles to failure N is plotted versus a cyclic stress parameter S such as the stress amplitude or stress range. A typical S–N diagram is shown in Figure 5-62. Alternatively, it is customary in bridge applications to express the S–N relationship by two straight lines on a log–log plot as shown in Figure 12-30*a*. The horizontal line corresponds to the fatigue limit. A smooth curve is sometimes used instead of the sloping straight line to represent the finite-life portion of the S–N function, although this is seldom needed in bridge applications.

The mathematical characteristics of the curve are established by referring to Figure 12-30. On the log plot shown in Figure 12-30*a*, the sloping solid line is defined by

$$\log N = \log A - B \log S \tag{12-46}$$

where B is the reciprocal of the slope and $\log A$ is the x intercept. The y intercept is $(\log A)/B$. Equation (12-46) is also expressed as

$$N = A/(S^B) \tag{12-47}$$

and the corresponding natural plot is shown in Figure 12-30*b*.

Scatter We have already noted that scatter is adequately represented by a statistical log-normal distribution (probability density) curve (Fuchs and Stephens, 1980; Little and Jebe, 1975; Reemsnyder, 1969). This means that the logarithms of the lives are distributed normally as are the logarithms of the deviations of the individual $\log N$ values from their mean value, as shown in Figure 12-31. The area under the distribution curve between any two values of $\log N - (\log N)_{\text{mean}}$ corresponds to the percentage of occurrence within the same interval.

If several fatigue tests are performed on the same type of specimen but under different cyclic stresses, the resulting fatigue lives will be scattered about a mean, or best fit, line similar to the heavy S–N curve of Figure 12-31. For details such as scatter bands, standard deviation, percentage confidence limits, and so on, see Schilling, Klippstein, Barsom, and Blake (1978) and Arkin and Colton (1970).

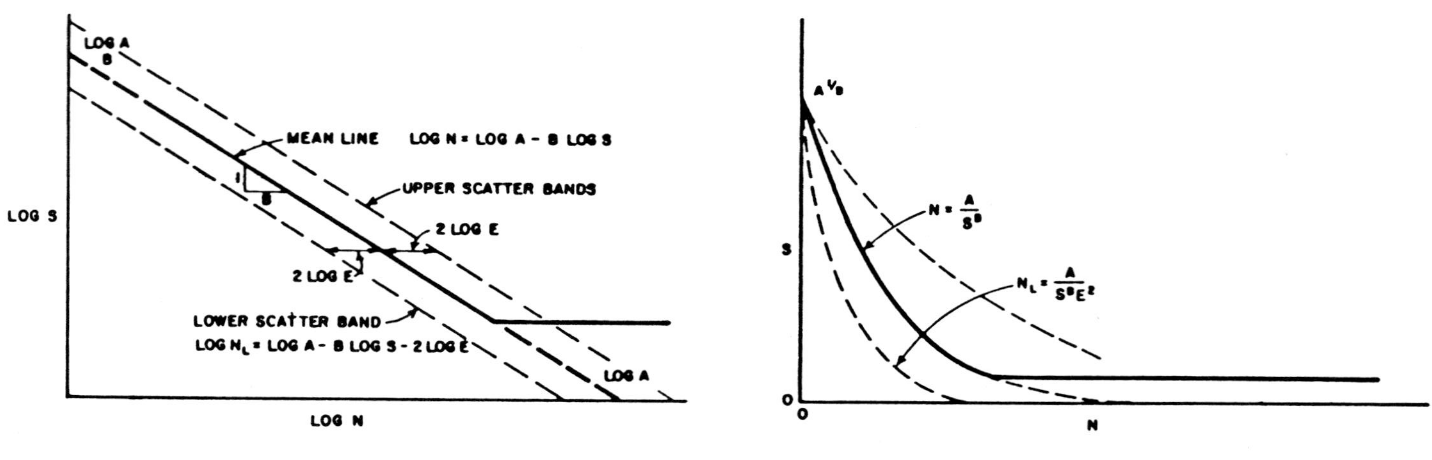

FIGURE 12-30 S–N curves typically used in bridge applications: (*a*) log plot; (*b*) natural plot.

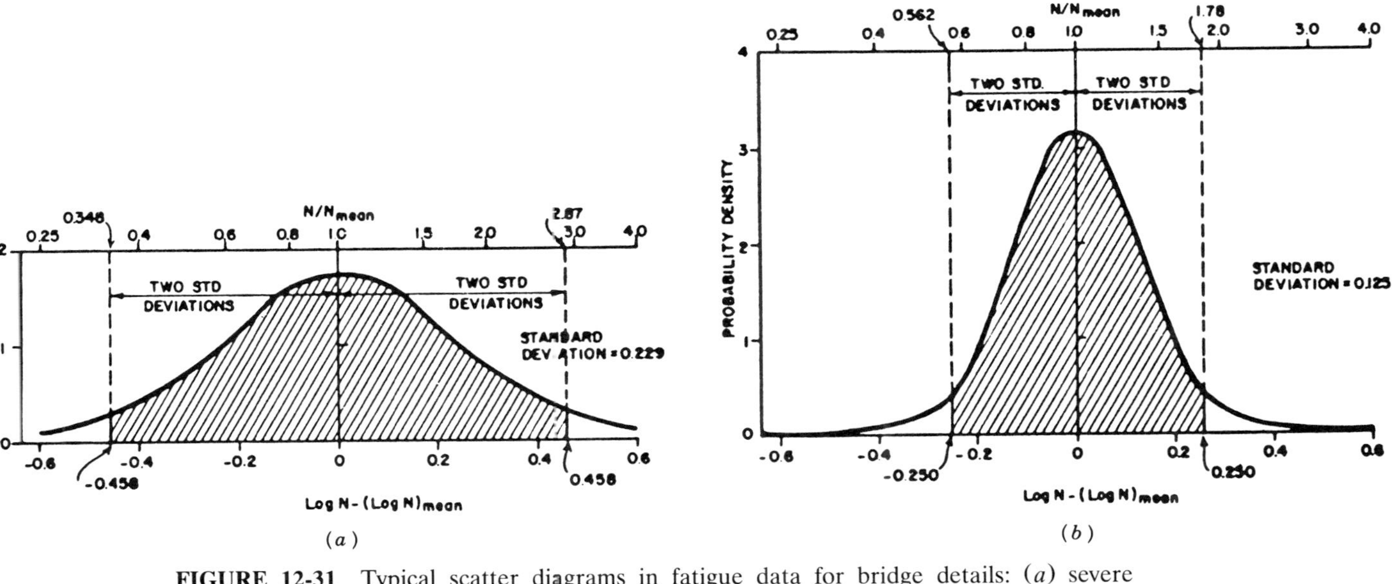

FIGURE 12-31 Typical scatter diagrams in fatigue data for bridge details: (*a*) severe scatter; (*b*) typical scatter.

Variable-Amplitude Loading

Variable-amplitude loading occurs widely in bridges and consists of a series of cycles with different magnitudes, usually applied in a random sequence. A comprehensive review of models used to predict fatigue under variable amplitude and random loading is given by ASCE (1982), and reference to this document is highly recommended.

Articulated forms of variable-amplitude loading are shown in Figure 12-32. One of the least complicated models is shown in Figure 12-32*a* and consists of simple individual cycles superimposed on a constant base. These cycles are either interrupted by pauses or continuously connected. This form of loading has been used in tests of simulated bridge members. A more complex form is shown in Figure 12-32*b* and consists of repeated complex cycles that have secondary reversals between peaks and valleys. This form represents truck loadings on bridges where each truck passage causes one complex cycle. The most complicated form is shown in Figure 12-32*c* and is a continuous complex variation that is not repeated.

Effective Constant-Amplitude Range This was defined in the preceding section, and values calculated from histograms are shown in Table 12-3. The following relationship is used:

$$S_{re} = \left(\sum \alpha_i S_{ri}^B\right)^{1/B} \tag{12-48}$$

where S_{ri} is the midwidth of ith interval, α_i is the frequency of occurrence for that interval, and B is the reciprocal of the slope of the log SN curve as shown in Figure 12-30. For most structural details, B has a value of 3.0. Equation (12-48) is conservative because it neglects any fatigue limit effects on the effective stress range.

Equivalent Number of Cycles A complex cycle can be represented by an equivalent number of simple cycles having the same stress range as the complex cycle. The secondary reversals included in a complex cycle (such as shown in Figure 12-32) have the same fatigue effects as individual simple cycles of the same size. The complex cycle is therefore decomposed into several individual cycles of different sizes as shown in Figure 12-33, and these are represented by an equivalent number of cycles of the same size.

The complex cycle is normally decomposed into a primary cycle and one or more cycles of a higher order. Within each stage, the primary cycle follows a path without reversal, as shown in Figure 12-33. When the path of the complex cycle reverses within a stage, the primary cycle moves horizontally to intersect this path at a later point. This horizontal line is the base stress for the next higher-order cycle. Any complex stress rotation of finite length can be decomposed and represented by an equivalent number of cycles. For most practical cases, the time to initiation is very small compared to the total life

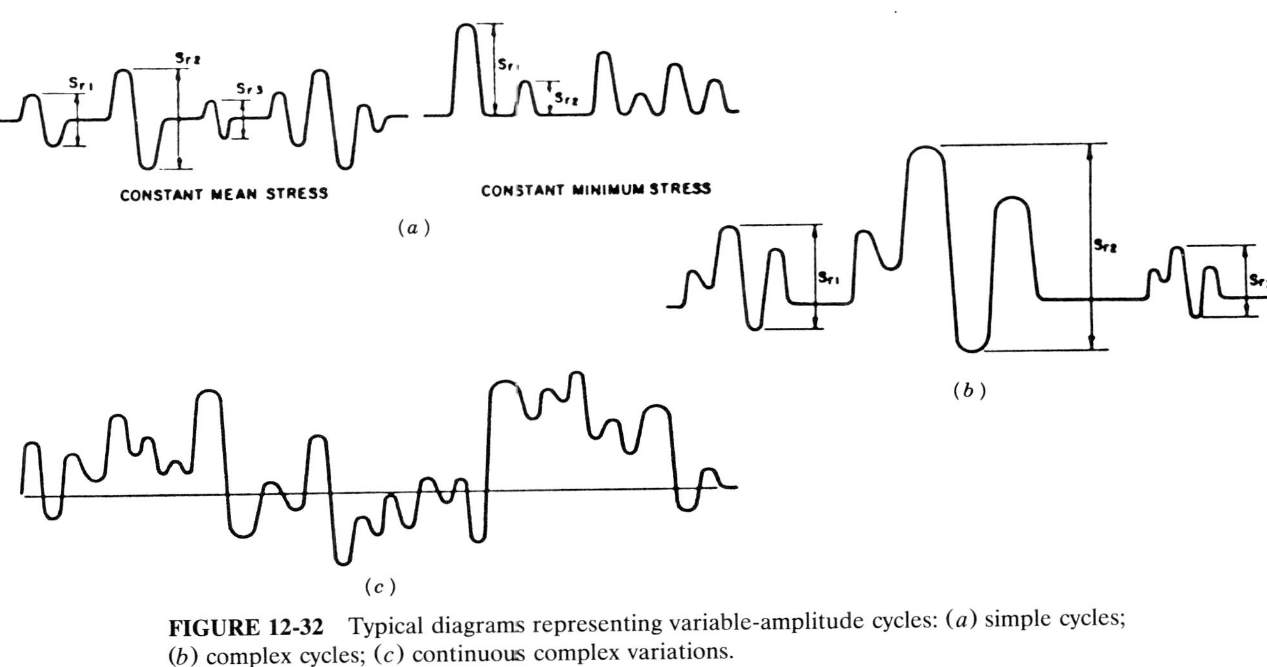

FIGURE 12-32 Typical diagrams representing variable-amplitude cycles: (*a*) simple cycles; (*b*) complex cycles; (*c*) continuous complex variations.

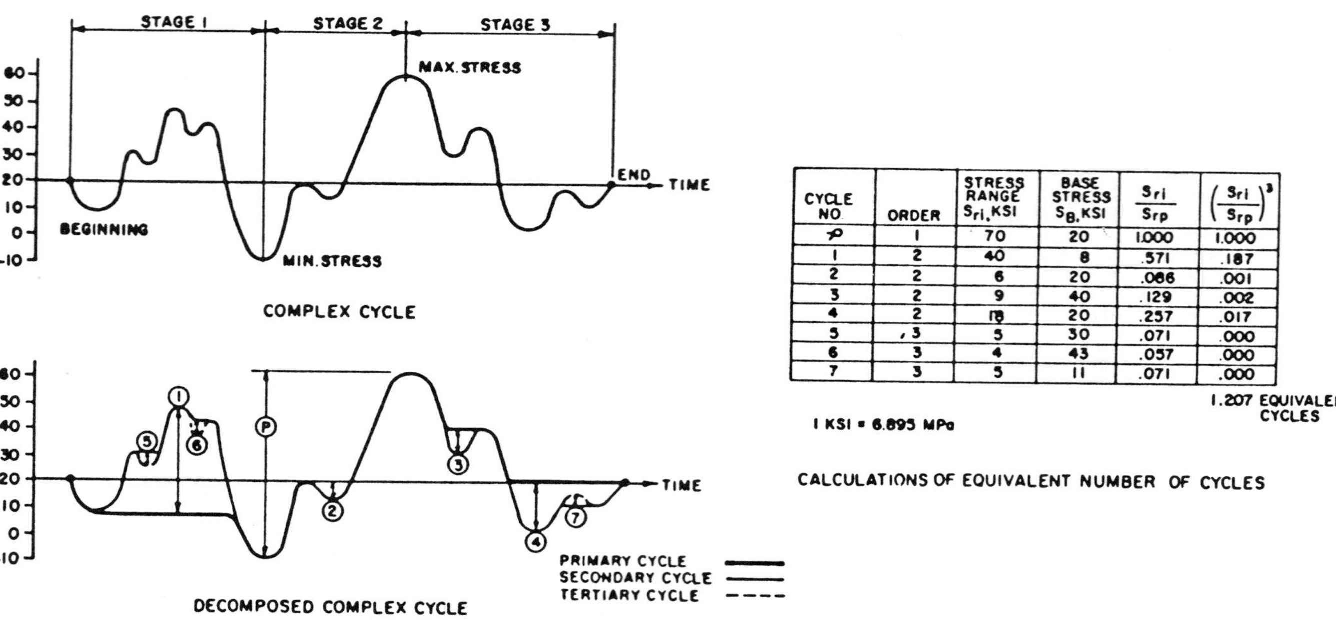

CYCLE NO.	ORDER	STRESS RANGE S_{ri}, KSI	BASE STRESS S_B, KSI	$\frac{S_{ri}}{S_{rp}}$	$\left(\frac{S_{ri}}{S_{rp}}\right)^3$
P	1	70	20	1.000	1.000
1	2	40	8	.571	.187
2	2	6	20	.086	.001
3	2	9	40	.129	.002
4	2	18	20	.257	.017
5	3	5	30	.071	.000
6	3	4	43	.057	.000
7	3	5	11	.071	.000

FIGURE 12-33 Diagram showing the derivation of equivalent number of cycles.

of the bridge, and much of this time involves subcritical crack growth. In this context bridge components can be designed so that cracks do not reach a critical length during the design life. The applicable theoretical criterion is expressed by (12-42) and involves the stress intensity factor range ΔK and the stress range S resulting from the applied load. The crack growth rate is given by (12-43).

In variable-amplitude loading, the ith stress range S_i is a random variable, and the stress range history is a sequence of S_i, where $i = 1, 2, \ldots, k$. An incremental form of the crack growth is given by

$$a_{i+1} = a_i + \Delta a_i = a_i + C\left(F[a_i]S_i\sqrt{\pi a_i}\right)^n \tag{12-49}$$

Computer programs are usually used for the step-by-step evaluation of the crack growth process. The crack extension in a given load cycle (Δa) will depend on (a) the crack geometry including crack length, orientation, shape of crack tip, and plastic deformation; (b) the condition of the crack tip (strain hardening, residual stress, and strain); and (c) the condition of the present load cycle (magnitude, frequency, etc.). Interaction effects in variable-amplitude loading can be accounted for by modifying (12-43) as suggested by Wheeler (1972) and Willenborg et al. (1971).

Factors Relevant to Fatigue

Stress Spectrum Because fatigue is a cyclic phenomenon, the magnitude of the cyclic variation is the most important stress parameter. Minimum, maximum, and mean stresses have a much smaller effect on fatigue life, and for fatigue design of bridges these parameters are usually neglected.

Fatigue failures do not occur unless a part of the stress cycle involves tension. Thus, compressive mean stresses have a marked beneficial effect if they are sufficiently large to cause the entire cycle to be compressive. In practice, however, caution is necessary before dismissing the possibility of fatigue failure in regions of nominal compressive stresses, because unexpected tensile stresses may occur due to the presence of residual stresses or local triaxial stress conditions.

Residual stresses are important in low-stress or long-life exposures. Irrespective of their origin, they remain constant after they are formed unless subsequent higher service loading causes further local yielding. Their effect may be beneficial if they shift the cyclic stresses so that most of the cycle is in the compression region. Conversely, this effect is detrimental if they shift the cyclic stresses upward within the tension region.

Multiaxial stress conditions, such as biaxial and triaxial stress variations, may be represented by an equivalent uniaxial cyclic stress, and ordinarily biaxial or triaxial stress combinations need not be considered. The equivalent uniaxial stress is expressed by (12-37) and (12-38). The maximum principal stress theory gives results similar to those computed using the foregoing criterion, except for principal stresses of opposite signs for which this theory gives higher equivalent stresses. For details involving attachments or similar stress raisers, the tensile stress component perpendicular to the attachment is the most appropriate combined stress parameter.

Impact The adequacy of the AASHTO impact factors suggested in the foregoing sections is confirmed by Schilling (1982b) in theoretical and experimental studies. Schilling, however, recommends that impact factors used for fatigue design should be based on typical or average conditions rather than extreme criteria, which is a view supported by this author. However, caution is necessary for cantilever (suspended-span) bridges, because large vibration stresses can be induced especially after the truck has left the bridge.

Detail Geometry The geometry and configuration of a detail has considerable influence on fatigue strength and behavior. Details may disturb the stress flow and induce local areas of high stresses that can reduce fatigue life.

The stress concentrations caused by notches, holes, and protrusions reduce fatigue strength by a factor K_f which is analogous to the theoretical stress concentration factor K_t (Heywood, 1962). Attachments such as welded, bolted, or riveted plates attached to a longer main plate cause stress concentrations when the main plate is loaded. The detrimental effect of longitudinal attachments may be controlled by using a concave transition radius at the ends and properly grinding the welds at these locations (Fisher et al., 1974).

Surface conditions such as roughness result in stress concentrations affecting the fatigue strength of plates or shapes with milled, weathered, or flame-cut surfaces. The fatigue strength of such plates is less than in plates with machined and polished surfaces. The scatter in fatigue results also tends to be more extensive for such plates and shapes than for welded details. Surface treatments that improve resistance to fatigue include (a) machining, grinding, and polishing; (b) electroplating; (c) shot peening; (d) cold-rolling; (e) surface hardening by flame or induction heating; and (f) carburizing or nitriding. The first two methods reduce surface roughness, and the last four methods induce compressive surface stresses.

Materials Relevant material characteristics affecting fatigue are mainly the tensile strength, the yield strength, the fracture roughness, and the notch sensitivity. The effect of tensile strength decreases with increasing severity of the details. Fatigue tests of welded simulated bridge details within a practical stress range show little effect of the type (strength) of steel. Accordingly, current specifications give the same allowable fatigue stress for all steels.

The basic fatigue strength of steel does not appear to be directly related to its yield strength. There is, however, an interdependence whereby yield strength can have an influence on the effects of stress raisers, residual stresses, and mean stress. The possibility that a reduction in fatigue life may be caused by low fracture toughness exists but is small.

The notch sensitivity index q is often used to articulate the effect of stress concentration on fatigue in terms of the material. This index is given by

$$q = (K_f - 1)/(K_t - 1) \tag{12-50}$$

where notched and unnotched fatigue specimens are tested to obtain K_f. In these tests the value of q varies with the number of cycles at which K_f is defined. Nonetheless, the notch sensitivity index gives a useful and practical indication of fatigue behavior of various steels when stress raisers are present and if the same specimen type is used to evaluate all steels.

Environmental Effects These may cause corrosion and fretting fatigue. Corrosion fatigue is the combined action of corrosion and cyclic loading, and the associated effects are far more serious than either process acting alone. Because of the large number of variables involved in this behavior, it is difficult to provide guidelines for predicting corrosion fatigue effects.

Fretting fatigue can occur when two metal surfaces in contact are subjected to small repetitive sliding movements. These may lead to the initiation of surface cracks that will subsequently reduce fatigue life. Only limited quantitative data are available on fretting to permit reasonable fatigue design calculations. As the contact pressure increases, the fretting fatigue life decreases to a minimum value and remains almost unchanged as the pressure becomes high enough to prevent sliding movements. If these conditions exist in a corrosive environment, the fretting life tends to be reduced. High-hardness, higher-strength steels appear to be more susceptible to fretting than lower-strength structural steels.

12-10 DESIGN APPROACH TO FATIGUE

In addition to the data and fatigue evaluation processes discussed in the foregoing sections, a review of criteria for design based on fatigue reliability is given by ASCE (1982). This review concludes that probabilistic limit state design can consider explicitly the uncertainties affecting performance and include their effects in practical design criteria. Design procedures, however, need not result in additional complexities at the design level, and the format of the provisions may be standardized.

Detail Categories for Load-Induced Fatigue

Currently, four design approaches to fatigue are considered: (a) the fracture mechanics approach, used in a variety of applications; (b) the strain life approach, commonly used in automotive applications; (c) the stress life reduction factor approach, widely used in machine design; and (d) the stress life detail category approach, generally used for structural applications such as bridges and buildings.

Stress Life Detail Category Approach According to this design methodology, typical structural details are grouped according to severity, and allowable stresses corresponding to design lives are assigned to each group. These stresses are determined in fatigue tests on specimens simulating each detail in the group. Stress concentrations, residual stresses, initial flaws, and material variability are directly included in the tests. Scatter is considered in the design approach by using the lower limit of the scatter band.

The fatigue provisions of the present AASHTO specifications are based on this approach. The allowable stresses are based primarily on extensive programs of constant-amplitude fatigue tests (Fisher et al., 1970, 1974, 1979, 1980, 1983; Fisher and Yen, 1977; Fisher, 1977), and are supported by variable-amplitude tests (Schilling, 1978) and by earlier constant-amplitude tests (Munse and Stallmeyer, 1962; Munse, 1964; Haaijer, 1966). The eight detail categories are listed in AASHTO Table 10.3.1A and are described in AASHTO Table 10.3.1B and Figure 10.3.1C. They will not be repeated here except for certain relevant comments.

Category A provides the highest fatigue stress range attainable for bridge members. For Grade 36 and 50 steels, the allowable S–N curve for this category approximates the lower 95 percent confidence limit from tests on rolled shapes. In these tests the fatigue strength was controlled by normal surface imperfections and roughness. Because of the variability in the severity of imperfections and roughness, the scatter in fatigue results for rolled plates and shapes is greater than the scatter for fabricated details. Tests on welded shapes with good-quality flame-cut edges showed that failure was initiated in the flange–web fillet welds rather than in the flame-cut edges. Therefore, flame-cut edges with an ANSI smoothness of 1000 or less are included in category A.

Category B includes base metal and weld metal such as fillet or groove welds, full-penetration transverse groove welds, tapered splices, 24-in. radius curved transitions for flange- or groove-welded attachments, and bolted joints (in this case the weld reinforcement is removed). Category B′ is essentially the same but with backing bars not removed, and in this case a lower stress range is specified. The inclusion of high-strength bolted joints in this category is based on data for symmetric butt splices of two types: (a) slip resistance (friction) joints and (b) bearing joints. In the former fatigue behavior is controlled by gross-section stresses, whereas in the latter fatigue cracks are usually initiated at holes in the net section so that fatigue behavior is controlled by net-section stresses.

Category B is not intended to apply to (a) single-lap joints, (b) joints subjected to prying action, and (c) joints subjected to direct tension perpendicular to the plate surfaces (Fisher and Struik, 1974). Significant bending stresses develop in the plates of such joints, but only membrane stresses occur in symmetric butt splices.

Category C includes transverse stiffeners or attachments, full-penetration groove weld splices with the reinforcement not removed, 6-in.-radius curved transitions for groove-welded attachments, and stud shear connectors. Bending stress at the crack initiation location is the stress parameter for the experimental data on stiffeners and for the corresponding allowable S–N curve.

Category D includes attachments by full- or partial-penetration groove welds where the detail length in the direction of stress is between 2 and 12 times the plate thickness but less than 4 in, curved transitions for groove and fillet welds with a radius between 2 and 6 in, and riveted joints. Although riveting is no longer used to fabricate new bridges, many riveted bridges are still in service. Available fatigue data on riveted symmetric butt splices show that there is an exceptionally large amount of scatter, and that the lower bound S–N curve has an unusually flat slope (Hansen, 1959; Parola et al., 1964; Mindlin, 1968). Among the different types of riveted details in existing bridges are built-up girders, cover plate ends, transverse stiffeners, truss gusset plates, and transverse attachments.

Category E includes the ends of cover plates fillet-welded to flanges less than 0.8 in. thick, attachments longer than 4 in., intermittent longitudinal fillet welds, and fillet-welded lap joints. Variations in the end geometry of cover plates, including tapered and rounded ends, were shown to have little effect on fatigue strength (Fisher, 1977; Munse and Stallmeyer, 1962).

Category E′ includes the ends of cover plates fillet-welded to flanges more than 0.8 in. thick and girder flanges greater than 1 in. thick penetrating through the web of another girder and fillet-welded to each side of that web.

Category F includes only shear stresses on the throat of fillet welds. It applies to continuous or intermittent longitudinal or transverse fillet welds. Shear stress as defined in this category does not control the fatigue strength of fillet-welded details. In such cases the fatigue strength is usually controlled by axial or bending stresses at the toes of transverse welds or at the ends of longitudinal welds as defined in other categories.

Design Methods: Load-Induced Fatigue

Redundancy does not appear to control or otherwise affect fatigue resistance. However, for fracture-critical members the calculated fatigue strength should be decreased to provide a greater safety factor considering the consequences of failure of nonredundant members.

In the present specifications, provisions explicitly relating to fatigue deal only with load-induced fatigue, caused by in-plane stresses for which components and details are properly designed. Distortion-induced fatigue (discussed in subsequent sections) is caused by secondary compatibility stresses not quantified in the typical analysis and design of a bridge. These are stresses resulting from the interaction between longitudinal and transverse members not modified in the line girder analysis.

Present AASHTO Methods For each of the detail categories discussed in the preceding section, the specifications stipulate allowable stress ranges corresponding to the design life groups (stress cycles). The calculated stress range for any detail must be less than the allowable for that detail. This procedure is demonstrated in the design examples of previous sections.

Proposed LRFD Approach The load-induced fatigue provisions are applied only if the detail under consideration is expected to experience a net applied tensile stress. Where compressive dead load stresses occur, fatigue is disregarded only if the compressive dead load stress at the detail is at least two times the maximum tensile live load stress computed using the Fatigue I load combination specified in Table 2-12.

The applied stress range is the basic stress parameter for fatigue design of a steel bridge detail. Residual stresses need not be considered in determining the applied stress at the detail. The articulation of the general condition and detail category shown in AASHTO Table 10.3.1B is the same.

The LRFD specifications introduce the concept of *infinite fatigue life* if the ADTT exceeds 2000 trucks per day. In this case the Fatigue II load combination applies and the fatigue resistance is

$$F_n = (\Delta F)\text{threshold} \tag{12-51}$$

where F_n is the nominal infinite fatigue life resistance and (ΔF)threshold is the constant-amplitude fatigue limit, which is the same as the allowable fatigue stress range in AASHTO Table 10.3.1A for cycles greater than 2,000,000.

If the single-lane ADTT is less than 2000 trucks per day, the bridge is analyzed for *finite fatigue life*, consistent with the projected traffic volume. In this case the Fatigue II load combination applies, and the fatigue resistance is

$$F_n = \Delta F = (A/N)^{1/3} \tag{12-52}$$

where ΔF = nominal finite fatigue life resistance
N = (365)(75)(n)(ADTT)
A = constant taken from Table 12-4
n = number of stress range cycles per truck passage taken from Table 12-5

The fatigue life, in cycles, above the constant-amplitude fatigue threshold is inversely proportional to the cube of the stress range, that is, if the stress

TABLE 12-4 Detail Category Constant, *A*

Detail Category	Constant, A
A	2.5×10^{10}
B	1.2×10^{10}
B′	6.1×10^{9}
C	4.4×10^{9}
C′	4.4×10^{9}
D	2.2×10^{9}
E	1.1×10^{9}
E′	3.9×10^{8}

TABLE 12-5 Cycles per Truck Passage, *n*

	Span Length	
Longitudinal Members	> 40 ft (12 m)	< 40 ft (12 m)
Simple-span girders	1.0	2.0
Continuous girders		
Near interior support	1.5	2.0
Elsewhere	1.0	2.0
Cantilever girders	10.0	
Trusses	1.0	
	Spacing	
Transverse Members	> 20 ft (6 m)	< 20 ft (6 m)
All	1.0	2.0

range is reduced by a factor of 2, the fatigue life increases by a factor 2^3. The design life is considered to be 75 years in the development of these criteria.

Design Methods: Distortion-Induced Fatigue

Field Measurements From a review of field inspection results and bridge details, Fisher, Yen, and Wagner (1987) have identified many existing bridges as susceptible to out-of-plane, distortion-induced fatigue cracking. These out-of-plane deflections and rotations can generate high-magnitude secondary stresses in plate bending. When these are directly superimposed on the primary stresses, fatigue cracking occurs very early.

Floor Beam Connection Plates Extensive cracking has been documented in girder web plates at the ends of floor beam connection plates. Displacements resulting from out-of-plane movement, although very small, concentrate in

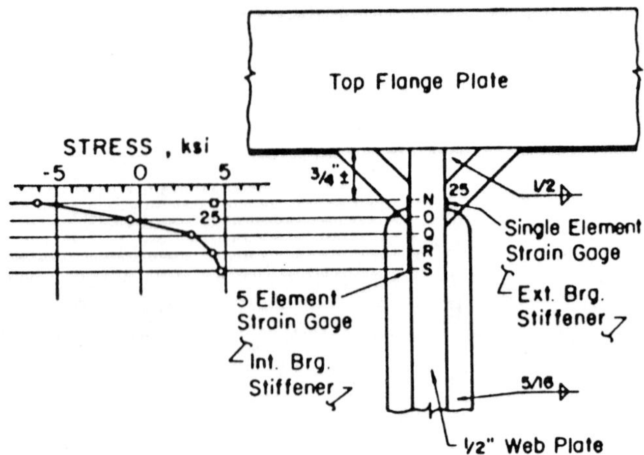

FIGURE 12-34 Typical stress gradient of web at bearing stiffener gap. (From Fisher, Yen, and Wagner, 1987.)

gaps between the floor beam connection plate and the girder tension flange. As is common practice, the connection plate is cut short of the tension flange. An example where this cracking has been found is the Wilson Bridge in Washington D.C. (Fisher, 1983).

Similar cracking has been documented in main girder webs at the floor beam connections of the Poplar Street Bridge (Koob et al., 1985). At piers, floor beams are connected directly to bearing stiffeners, which are tight-fit to both the top and bottom flanges and welded only to the girder webs. Cracks developed in the web at these locations.

Representative stresses at the top flange of one connection are plotted in Figure 12-34. The back-to-back gauges "N" and "25" showed stresses of −6.0 and 4.2 ksi, respectively, indicating that the web was undergoing out-of-plane bending. The maximum measured out-of-plane displacement was 0.001 in. within this region. The projected stress was −14.0 ksi at the longitudinal weld toe. Cracking was also developed in the web at the connection plates of floor beams. In this case typical web behavior is represented in Figure 12-35 for the negative moment region of the girder. The connection plates are tight-fit to, or cut short of, the top (tension) flange. Floor beam end rotations introduce out-of-plane displacement in the web of the girder and cause cracking of the web plate.

Multiple Girder Diaphragm Connection Plates The same cracking potential exists in the gap at the junction of longitudinal girder flanges and the transverse connection plates of diaphragms or cross bracing. As the girders deflect unevenly, the diaphragms transfer lateral forces causing the connection plate to displace and rotate. This distortion causes a similar out-of-plane

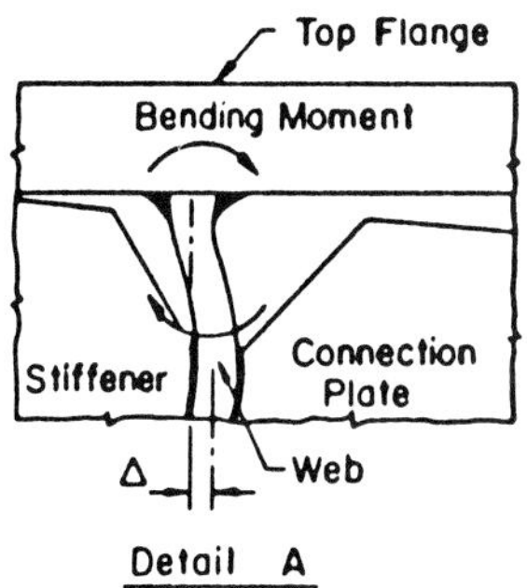

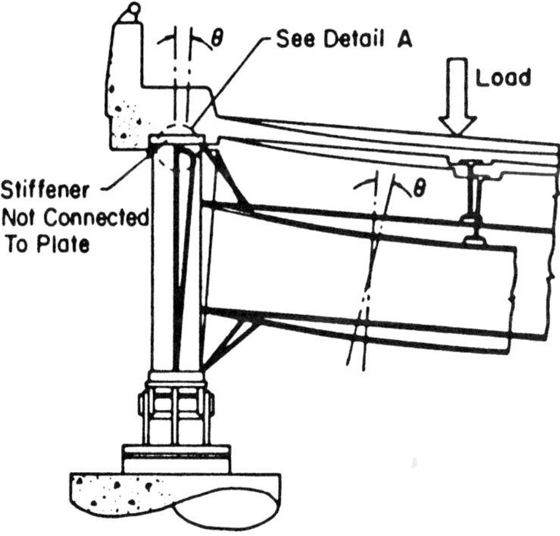

FIGURE 12-35 Representation of floor beam end rotation and web behavior. (From Fisher, Yen, and Wagner, 1987.)

distortion of the web and the development of fatigue cracks in the gap. Cracks of this nature have been documented in the negative moment region of the continuous Beaver Creek Bridge (Lee and Castiglioni, 1986).

Box Member Diaphragms Likewise, cracks have been detected in the web plates of box girders at the gaps of internal diaphragms opposite floor beams attached to the box girder. With the diaphragm plate not connected to the top or bottom flange of the box, web portions or gaps are unstiffened. A combination of out-of-plane displacement and rotation within the gap added to existing relatively large flaws can lead to crack initiation in the gap.

An example is the cracking developed in the box girder of the tie arch structure at Neville Island (Fisher and Pense, 1984), where cracks were found in the welded connections between the diaphragm plates and the outside web of the box. A stress range histogram from the diaphragm is shown in Figure 12-36. The maximum stress range is 5.6 ksi, and the equivalent constant-

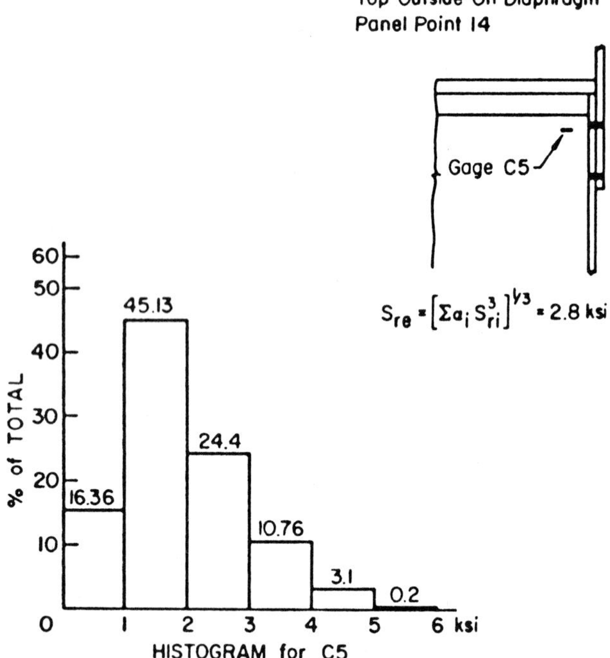

FIGURE 12-36 Histogram showing stress range of box diaphragm.

amplitude stress range is 2.8 ksi. The stress range histogram from the web is shown in Figure 12-37, and articulates a maximum stress range within such gaps between 8 and 11 ksi, with effective stress ranges of 3 to 4 ksi. These stresses caused cracking in the web on both the inner and outer surfaces of the web plate.

The stress gradient generated within the gap as a result of out-of-plane displacement is shown in Figure 12-38. The difference between the maximum

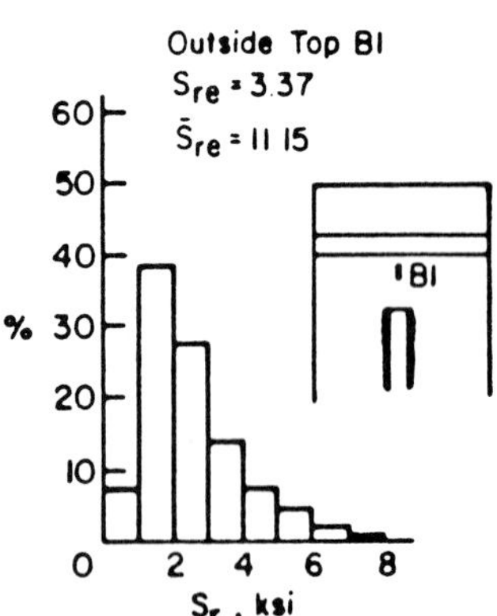

FIGURE 12-37 Histogram showing stress range of box girder web.

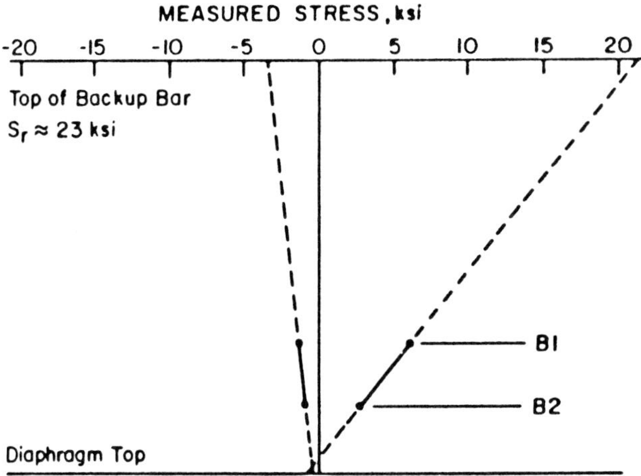

FIGURE 12-38 Diagram showing stress gradients at top outside gap of box girder.

and minimum is the stress range for this cycle of instantaneous stress excursion. Extrapolation (dashed lines) to the backup bar roof shows that likely stress ranges are of the order of 23 ksi, sufficient to cause cracking at the root of the backup bar.

Gusset Plates in Lateral Bracing With attachments through gusset plates connected to web plates, any force or displacement of the bracing members causes out-of-plane movements and associated web stresses. The gusset plate may be welded, bolted, or out free of the transverse stiffener in addition to being connected to the web. An example where this cracking has been observed is the I-79 bridge over Big Sandy Creek (Fisher and Pense, 1984).

LRFD Specifications The approach to distortion-induced fatigue is to preclude these effects through appropriate detailing. Rigid load patterns are stipulated to inhibit the development of secondary stresses that may induce crack growth in either the longitudinal or the transverse member.

Transverse connection plates or transverse stiffeners functioning as connection plates should be welded or bolted to both the tension and compression flange of the cross connection when connecting diaphragms, internal or external diaphragms, and floor beams are attached to these connection plates. The welded or bolted connection should be designed to resist a lateral load of 20 kips.

Lateral connection plates should be kept at least 4 in. from the web and any transverse stiffener, and preferably centered on the stiffener. When they are on the same side of the web as the stiffener, they should be attached to the stiffener. The transverse stiffener at this location should be continuous from compression to tension flange and attached to both flanges. Lateral connection plates attached to unstiffened webs should be located at least 6 in. above or below the flange, but not less than one-half the width of the flange.

Fracture (LRFD Specifications)

Main member components carrying load and subjected to tensile stress will require supplemented impact properties according to the AASHTO material specifications. Thus, all members and connections subjected to tensile stresses under Strength Load Combination I and all components requiring mandatory Charpy V-notch fracture toughness should be identified on the design plans. The appropriate temperature zone should be indicated in the contract documents.

The Charpy V-notch impact requirements vary with the type of steel, type of construction, whether welded or mechanically applied, and the applicable minimum service temperature.

Fracture-critical members or components should be tested according to ASTM A673, frequency P. Members and components not considered fracture critical should be tested according to ASTM A673, frequency H.

Any attachment with length in the direction of the tensile stress more than 4 in. welded to a tension area of a component of a fracture-critical member is considered part of the tension component and therefore fracture critical. In this context tension components refer to tension members and those portions of a flexural member subjected to tensile stresses.

12-11 FATIGUE EFFECTS UNDER VARIABLE-AMPLITUDE LOADING: EXPERIMENTAL INVESTIGATIONS

Welded Steel Beams

Results of laboratory investigations of fatigue effects in welded steel beams subjected to variable-amplitude loadings have been reported by Schilling et al. (1978). These tests involved both small specimens and relatively large beams of various steels, with structural details similar to those tested in NCHRP Reports 102 and 147 (Fisher et al., 1970; Fisher et al., 1974). The main objectives were to develop fatigue data on welded bridge members and to suggest analytical methods for predicting fatigue behavior under variable-amplitude stress spectra but from constant-amplitude fatigue data. This program is shown as NCHRP Project 12-12.

The available results of field movements of stresses in bridges under traffic included 51 data sets covering 37,000 truck passages from six sources. These results showed that the passage of a truck over a bridge produces a single major stress cycle with superimposed vibration stresses, usually small enough to be neglected. The major stress cycles are added to a constant stress induced by the dead load. The difference between the maximum stress and the minimum stress thus obtained has been defined as the stress range S_r.

Specimens and Beams Approximately 156 beams with partial-length cover plates were tested to obtain the lower bound for the variable-amplitude fatigue strength, and 63 welded beams without cover plates were used to obtain the approximate upper bound. The specimens represented A36 and A514 structural steels. Details and fabrication techniques are described by Schilling et al. (1978) and generally followed normal bridge practice.

Welding procedures conformed to the AWS bridge specifications (AWS, 1969, 1971), and quality assurance was provided accordingly. Cover plates were submerged-arc-welded to the main plate along both longitudinal edges, but no welds were placed across the ends of the cover plates. Welded and cover plate beams are shown in Figures 12-39*a* and *b*. For the beam detail shown in Figure 12-39*a*, the flange and web plates are oxygen-cut from larger plates, assembled, tack-welded, and then joined by submerged-arc fillet welds. The cover plate beams were fabricated according to the sketch shown in Figure 12-39*b*. For A514 steel, the fabrication process consisted of welding the cover plate to the flange plate first, and then welding this assembly to the web (those beams are referred to as A beams). Some of the A beams were modified by placing manual fillet welds across the cover plate ends, and some welded beams had cover plates with welds across the ends. The cover plates were attached by submerged-arc fillet welds along both edges but not across the cover plate ends. These beams differ from the A beams only in the assembly sequence. Both types of modified beams are referred to as C beams. The welded beams with cover plates without welds at the ends are referred to as B beams.

Beam Tests The beams were tested as shown in Figures 12-40 and 12-41. For the cover plate beams, the bending stress at the end of the cover plate is the main test parameter. At the left end H, the stress of the cover plate is 50 percent higher than the stress at the right end L. For the welded beams, the nominal bending stress on the outer fibers in the central region of constant moment is the main test parameter. The bending stress in the flange–web fillet weld is about 5 percent less. The testing procedure and sequence for the beams, the cover plate specimens, and the wedge-opening-load (WOL) specimens are detailed by Schilling et al. (1978).

Test Results An extensive summary of the test results is provided by Schilling et al. (1978) for the cover plate specimens and beams. The distribution of failure locations for the welded beams of A514 and A36 steel is shown in Figures 12-42 and 12-43, respectively. The type of cracks in the beams are classified into five categories: (a) edge cracks in the flanges of welded or cover plate beams, (b) intersection cracks spreading outward from the flange–web junction in the welded beams, (c) crescent-shaped cracks initiating at the ends of the welds in the cover plate A and B beams, (d) long cracks occurring along the cross weld in the cover plate C beams, and (e) peeling cracks propagating longitudinally along the throat of the fillet weld connecting the cover plate to the flange.

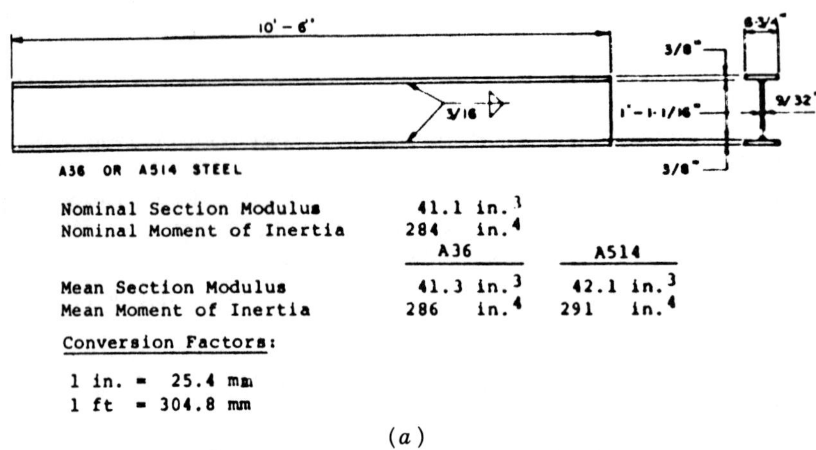

(*a*)

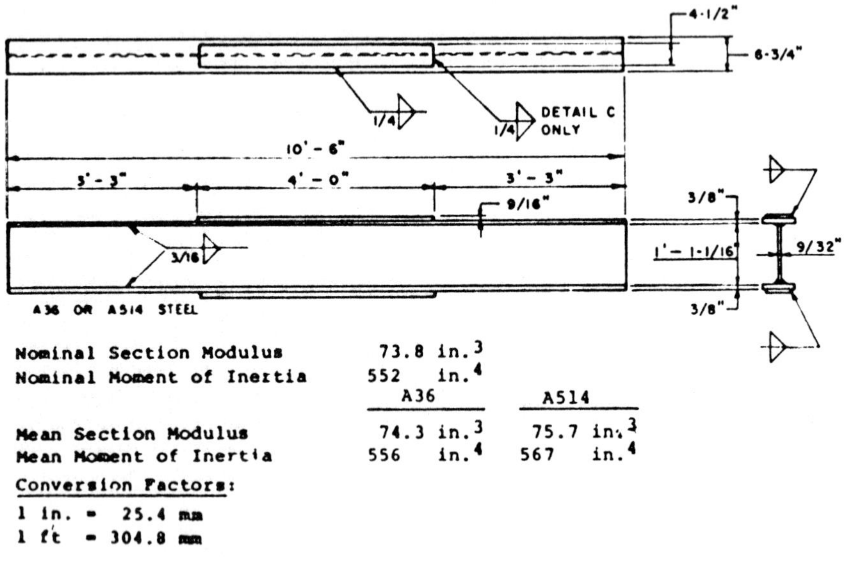

(*b*)

FIGURE 12-39 (*a*) Welded beams; (*b*) cover plate beams.

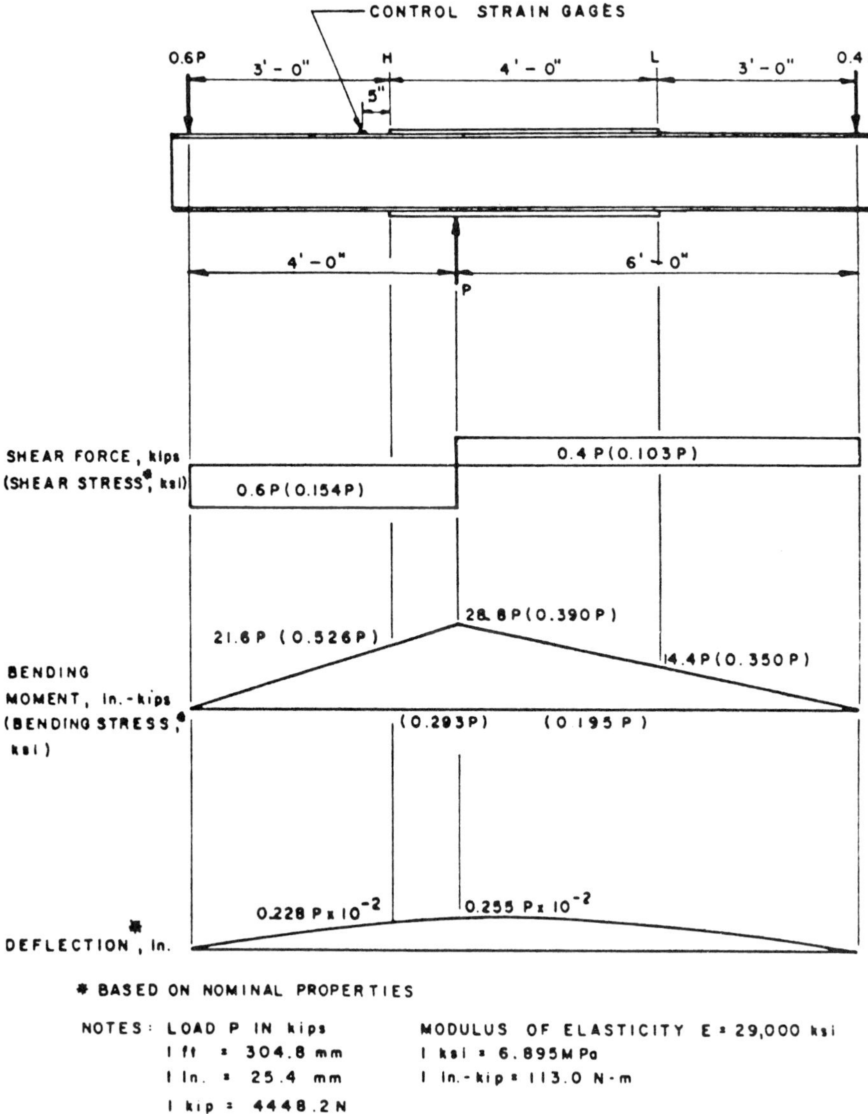

FIGURE 12-40 Loading diagrams for cover plate beams.

Fretting failures occurred at the load points in the first set of welded beams. Therefore, the test setup was modified by adding paper shims and repeating the test under the same conditions. The duplicate set of beams did not show any signs of fretting at the load points but had about the same life as the set with fretting failures.

Relationship Between Constant- and Variable-Amplitude Results Using the effective stress range concept as shown in Figures 12-44 and 12-45,

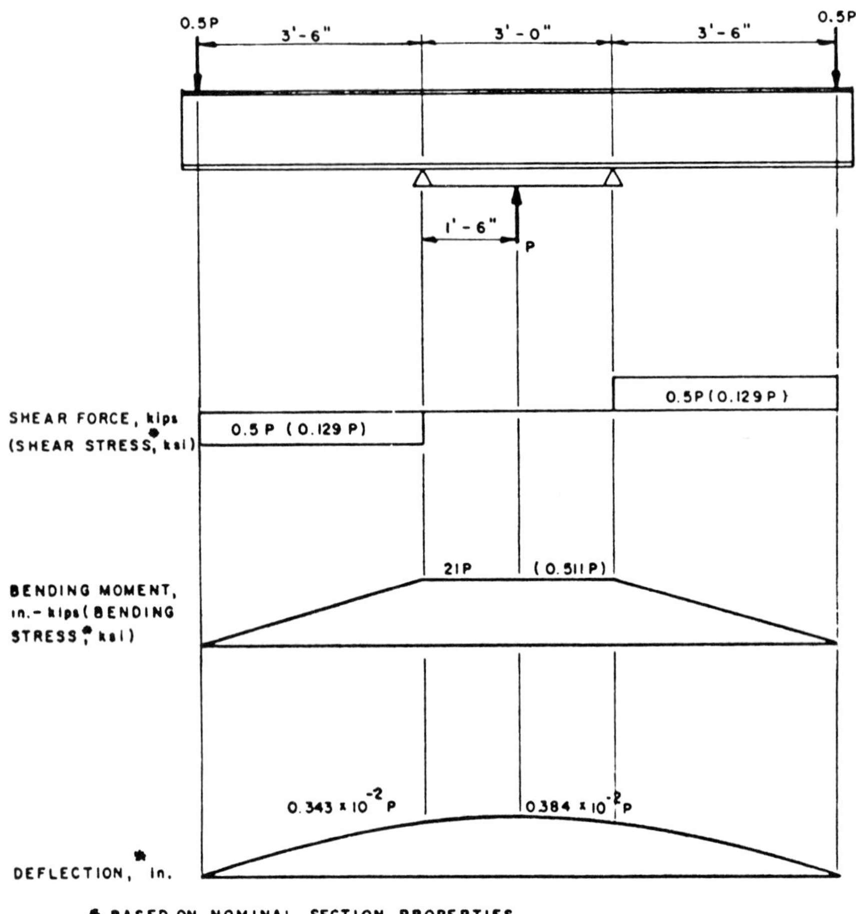

FIGURE 12-41 Loading diagrams for welded beams.

the four separate lines are approximated by a single line relating S_{re} to N. Different methods for calculating S_{re} are as follows.

Rayleigh Distribution In this approach the effective stress range is

$$S_{re} = S_{rm} + CS_{rd} = S_{rm}(1 + CS_{rd}/S_{rm}) \tag{12-53}$$

where S_{rm} = model stress range

S_{rd} = measure of the width of the curve (parameter)

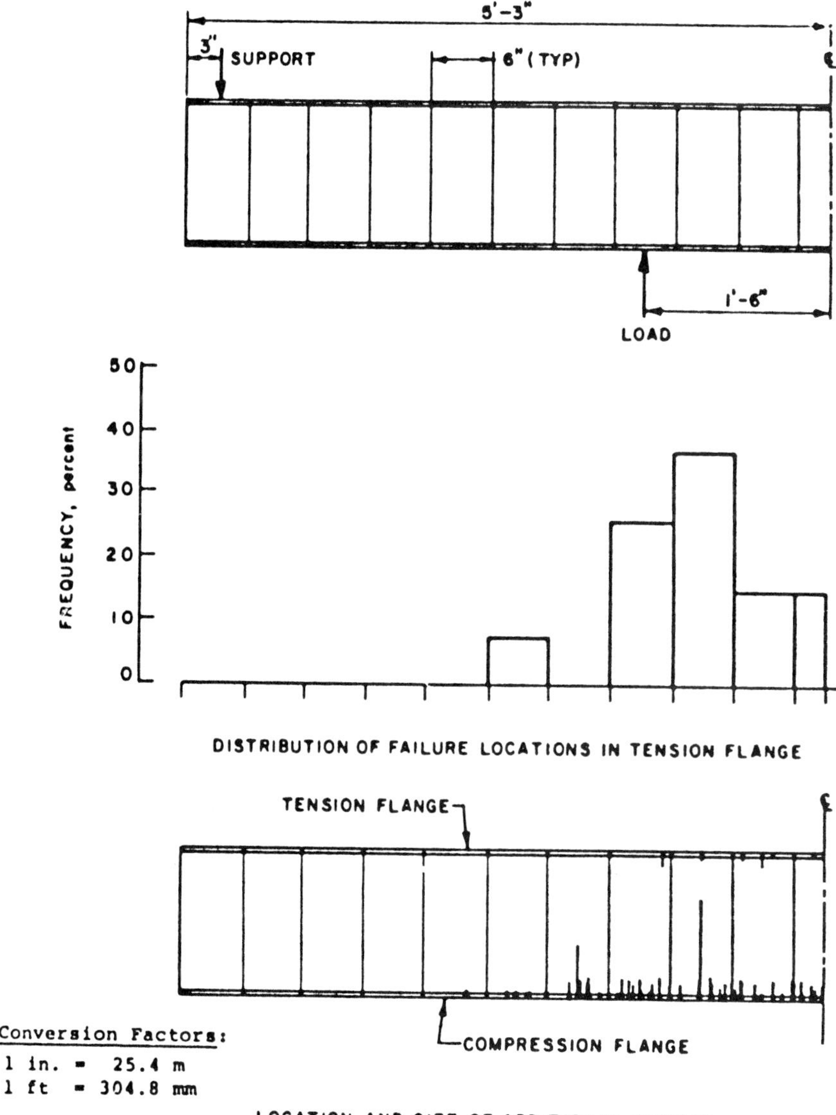

FIGURE 12-42 Distribution of cracks in welded A514 steel beams. (From Schilling et al., 1978.)

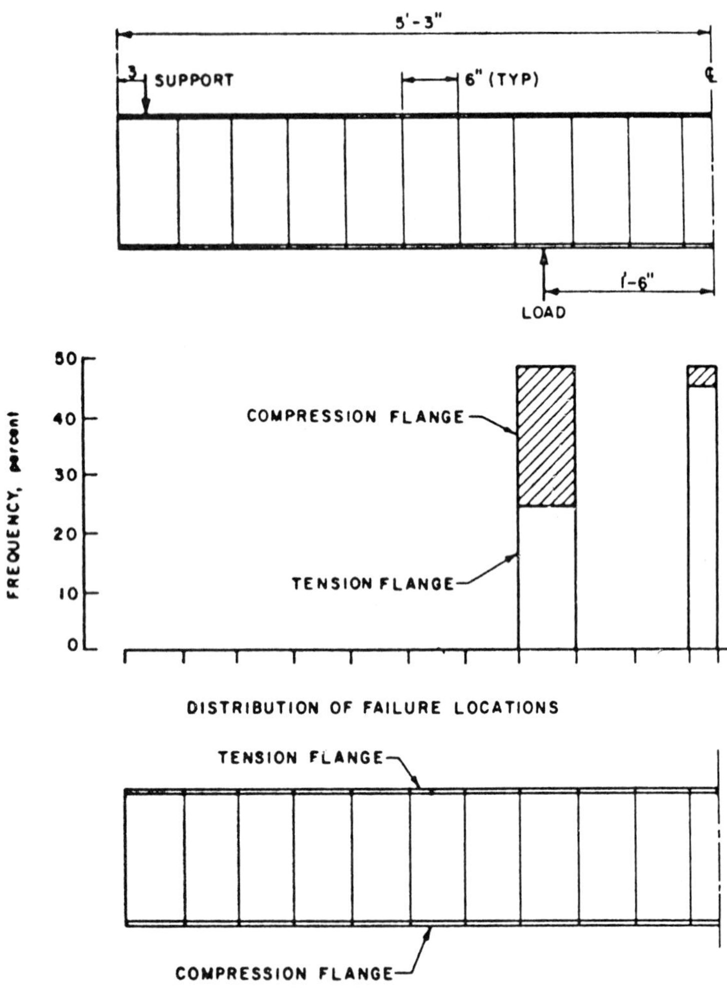

Conversion Factors: LOCATION AND SIZE OF ADDITIONAL CRACKS

1 in. = 25.4 m
1 ft = 304.8 mm

FIGURE 12-43 Distribution of cracks in welded A36 steel beams. (From Schilling et al., 1978.)

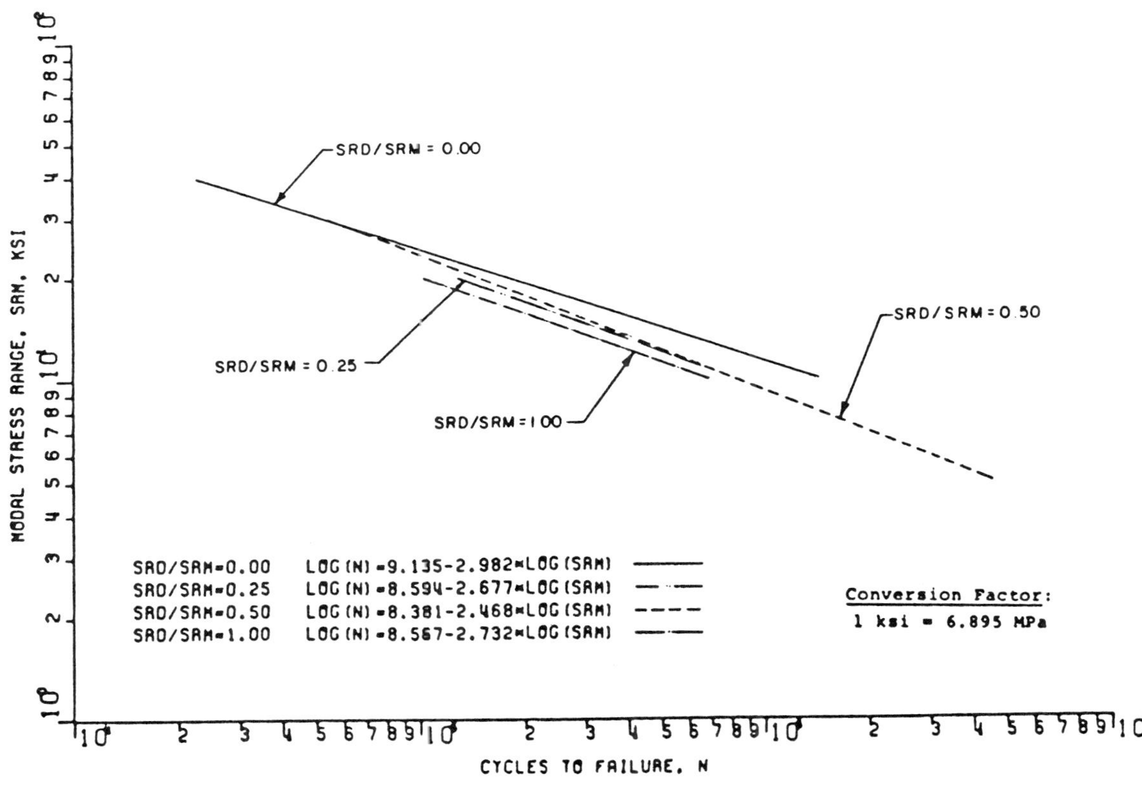

FIGURE 12-44 Modal stress range versus fatigue life for cover plate B and C beams. (From Schilling et al., 1978.)

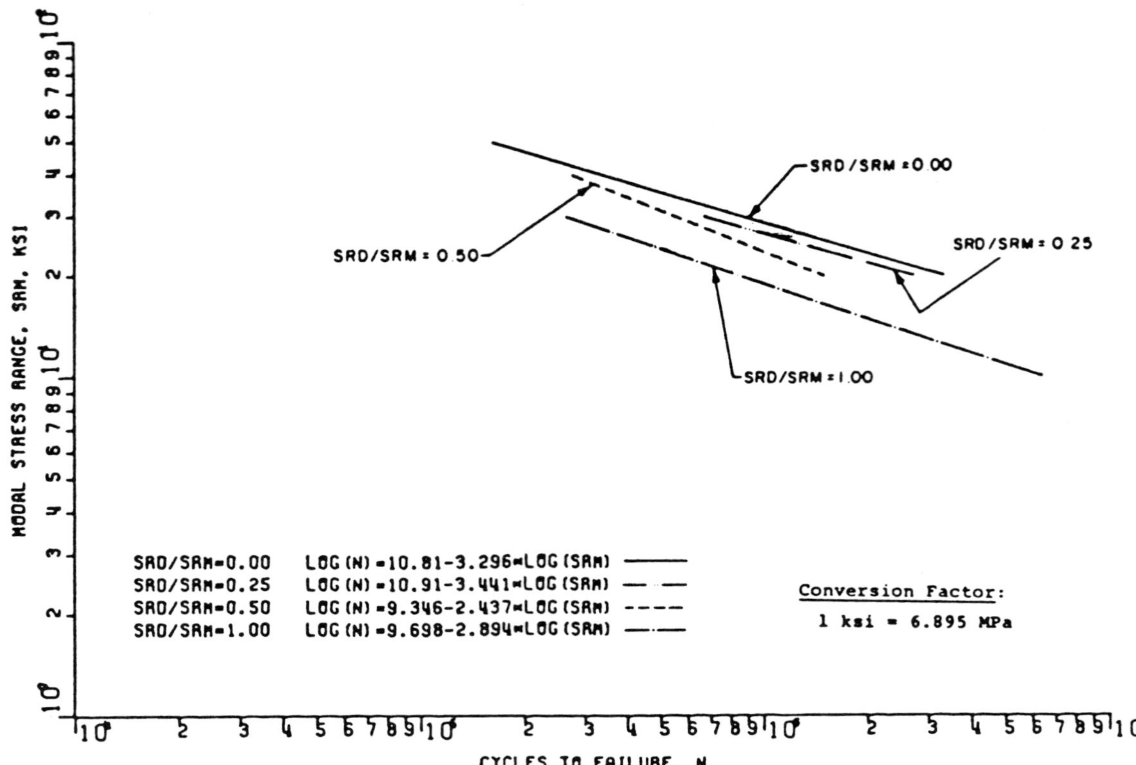

FIGURE 12-45 Modal stress range versus fatigue life for welded beams. (From Schilling et al., 1978.)

C = best-fit value of the correlation factor

Thus, C defines a single stress range that has the same effect on fatigue behavior as the complete spectrum. If $C = 0.378$, S_{re} is the root mean square of the stress ranges in the spectrum. If $C = 0.230$, S_{re} is the mean of the stress ranges.

Effective Stress Range by RMS A version of this approach is expressed by (12-48). Alternatively, the root mean square (RMS) gives

$$S_{re(\text{RMS})} = \left(\sum \alpha_i S_{ri}^2\right)^{1/2} \tag{12-54}$$

where S_{ri} is the ith stress range and α_i is the fraction of stress ranges of that magnitude. The variable-amplitude data points plotted from the tests generally fall within a scatter band bounded by the 95 percent confidence limits for the constant-amplitude data. Thus, the $S_{re(\text{RMS})}$ relates constant- and variable-amplitude data satisfactorily.

Miner's Law This is specifically expressed by (12-48) and has been widely used for many years to show the cumulative effect on fatigue life of stress cycles of different magnitudes.

A comparison of the standard errors of the estimate of the data transformed by the two methods shows that the RMS approach makes the transformed variable-amplitude data fit the constant-amplitude S–N curve somewhat better; hence, the method is recommended in this case. Schilling et al. (1978) also conclude that these details and the corresponding AASHTO Category E (cover plate ends) compare satisfactorily, except that the AASHTO allowable stress line provides an approximate lower limit for the variable- and constant-amplitude test results plotted on the basis of the RMS effective stress range.

Long-Life Tests These tests can show whether curves for bridge members extend to very low stress range levels such as occur in actual bridges, or, conversely, there is a fatigue limit or break in the curves. Long-life test results are shown in Figure 12-46. The best-fit line for the tests at S_{rm} = 10 ksi or above is shown as a solid line within the range of test data. Approximate 95 percent confidence limits for a single test are shown as dashed lines within the range of the test data. Extensions of the best-fit line and confidence limits are indicated as dash–dot lines.

It appears that these data do not indicate a fatigue limit, or significant break in the S–N curve, but suggest that the original S–N curve should be adjusted slightly to better fit the wider range of data. This line is shown as a dash–dot–dot solid line in Figure 12-46. Furthermore, one of the three long-life beams sustained about 104 million cycles without visible cracks at the low-stress end of the cover plate where S_{rm} = 2 ksi and $S_{r(\text{RMS})}$ = 2.8 ksi. The other two beams failed at the low-stress ends at about 60 and 104 million

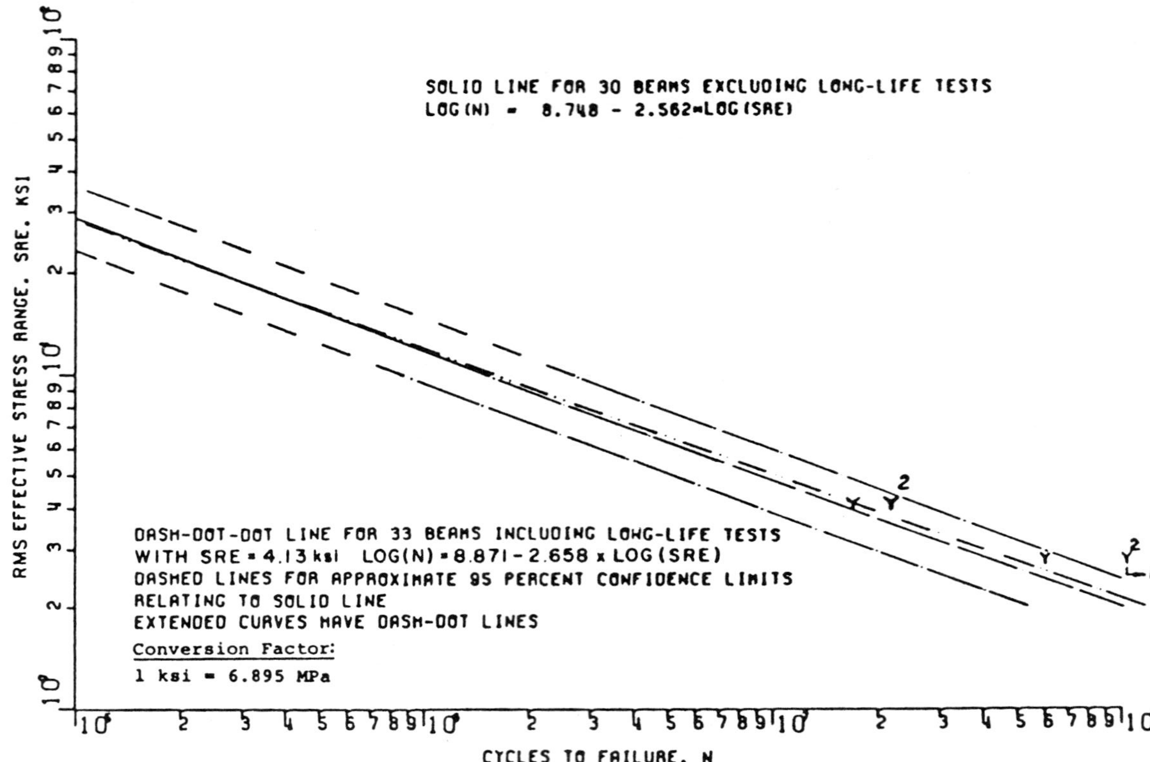

FIGURE 12-46 Long-life test results. (From Schilling et al., 1978.)

cycles. The conclusion is therefore that for this detail a fatigue limit, or break in the S–N curve, occurs at $S_{r(\text{RMS})} = 3$ ksi (approximately).

Crack Growth Tests The crack growth rate da/dN is expressed by (12-43). We recall that a is the crack depth and N is the number of cycles. In these tests crack growth data were sought from cover plate beam specimens. Data, however, from beam tests could not be correlated with crack growth data from the WOL specimens.

The beam tests give information on the initiation and propagation phase of fatigue life for such beams. More specifically, data were obtained showing the crack initiation and propagation phases of the fatigue life for A514 steel cover plate C beams. The ratio of crack length to crack depth l/a for semielliptical cracks is a relevant parameter in crack propagation studies. For the cover plate A and B beams, the results show that the ratio l/a is approximately 4.

Prediction of Beam Fatigue Life Specific guidelines and approaches for predicting crack initiation and propagation or growth are suggested by Schilling et al. (1978) and Barsom (1971, 1974), and illustrate the uncertainties and assumptions involved.

A significant number of cracks occurred in the compression flange of the welded beams and the cover plate C beams, and caused failure in several instances, particularly where a stress reversal occurred (some tensile stress was applied to the compression flange). Similar results have been observed in other tests and do not appear to be inconsistent, although the compression flange failures are not explicitly explained.

Application to Bridge Design Data from these studies may be used to correlate variable-amplitude random-sequence fatigue loadings, which actually occur on bridges, with constant-amplitude fatigue data and allowable stress ranges typically used in routine bridge design. These conclusions are also relevant to estimating the remaining life of existing bridges and to new bridges subjected to unusual loading conditions.

The cyclic stress range corresponding to the passage of a single truck is taken as the main parameter in the fatigue analysis, and stress cycles caused by cars are neglected. The analysis considers the type of histogram shown in Figure 12-47, where the height of each bar represents the percentage of stress ranges within the interval defined by the width of the bar (Cudney, 1968).

Design Example In order to predict the remaining life of a bridge, we must estimate the past and future traffic volumes including expected traffic growth. According to the criteria, the type of bridge must be considered, and in this example the number of stress (daily) cycles is estimated at 1000 and assumed constant throughout the life of the bridge. The stress spectrum summary is

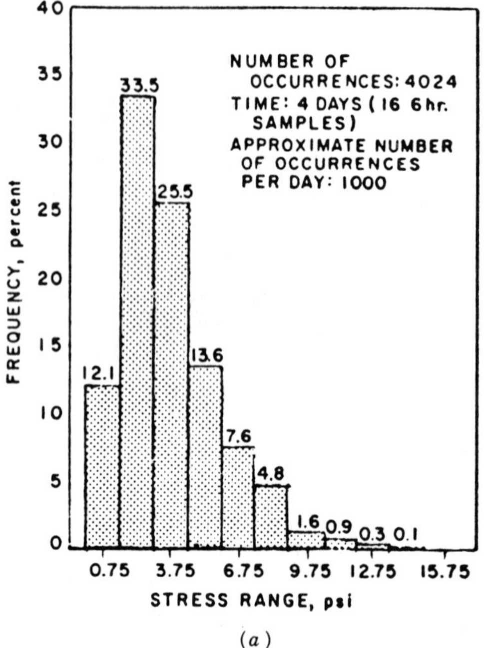

(*a*)

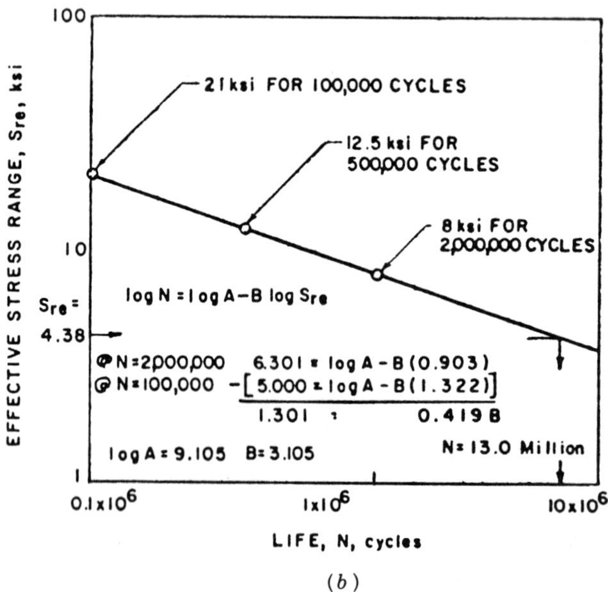

(*b*)

FIGURE 12-47 Example of the estimation of the minimum remaining life: (*a*) histogram for hanger end detail; (*b*) allowable effective stress range for hanger end detail (AASHTO Category E).

TABLE 12-6 Stress Spectrum (ksi) for Design Example

i	α_i	S_{ri}	$\alpha_i S_{ri}^2$	$(\alpha_i S_{ri}^2)^a$
1	0.121	0.75	0.07	—
2	0.335	2.25	1.70	—
3	0.255	3.75	3.59	—
4	0.136	5.25	3.75	3.75
5	0.076	6.75	3.46	3.46
6	0.048	8.25	3.27	3.27
7	0.016	9.75	1.52	1.52
8	0.009	11.25	1.14	1.14
9	0.003	12.75	0.49	0.49
10	0.001	14.25	0.20	0.20
	1.000		19.19	13.83

[a]Considering the fatigue limit.

given in Table 12-6 for both cases, that is, without considering the fatigue limit and considering the fatigue limit.

The effective (constant-amplitude) stress range for the stress specimen defined by a histogram will be taken as the root mean square of the stress ranges in the histogram, and is calculated from (12-54). Thus, the calculations are as follows.

Without considering the fatigue limit,

$$S_{\text{re}} = \sqrt{\sum \alpha_i S_{ri}^2} = \sqrt{19.19} = 4.38 \text{ ksi}$$

$$\log N = 9.105 - 3.105 \log(4.38) \qquad N = 13.0 \times 10^6 \text{ cycles}$$

Then the life in years is

$$Y = \frac{13.0 \times 10^6}{1000 \times 365} = 35.6$$

and the remaining life is 35.6 − 10.0 = 25.6 years

Considering the fatigue limit,

$$S_{re} = \sqrt{13.83} = 3.72 \text{ ksi}$$

$$\log N = 9.105 - 3.105 \log(3.72) \qquad N = 21.6 \times 10^6 \text{ cycles}$$

Then the life in years is

$$Y = \frac{21.6 \times 10^6}{1000 \times 365} = 59.0$$

and the remaining life is 59.0 − 10.0 = 49.0 years.

To obtain the best (most probable) estimate of the remaining life, a curve approximating the mean values of the experimental data should be used. The value of N corresponding to the calculated effective stress range for the spectrum (4.38 ksi) is a conservative estimate of the total number of cycles to failure (13×10^6). The percentage of these cycles falling within any particular interval of stress ranges is given by the histogram. The corresponding expected minimum life in years can be calculated by dividing the total number of cycles (13×10^6) by the estimated number of cycles per year (365×1000) as shown in the calculations. The term "remaining life" indicates that the estimated life is based on the lower limits of the scatter of the fatigue data.

Relevance to Bridge Specifications A suggested approach on how these studies and the associated results may be incorporated into bridge specifications focuses on loading positions, fatigue design truck, design stress range, and design life.

Lane Loading The magnitude and position of the loads for bridge design are intended to represent the most severe condition, and this does not occur frequently enough to influence fatigue considerations. Both the magnitude and position of the live load may be modified to correspond to relevant fatigue events as has been suggested in previous sections.

Fatigue Design Truck A fatigue design truck is analogous to the effective stress range for variable-amplitude spectra and may be calculated by either the RMS or the Miner method. Using the Miner method with $B = 3$ yields (12-39) presented in the form

$$W_F = \left[\sum \alpha_i W_i^3\right]^{1/3} = W_D\left[\sum \alpha_i \phi_i^3\right]^{1/3} \tag{12-55}$$

where α_i is the fraction of trucks with weight W_i, W_D is the weight of an HS 20 truck, and $\phi_i = W_i/W_D$. From various loadometer surveys, $\Sigma \alpha_i \phi_i^3$ varies from 0.3 to 0.5 (Fisher, 1974). The fatigue design truck W_F may be taken as about $0.7W_D$ or 50 kips, which is the value suggested in previous sections.

Eliminating the lane loading from the fatigue design requirements would essentially eliminate fatigue considerations for main members of relatively long span bridges.

Design Stress Range Suggestions emerging from this study with reference to the design stress range are based on field measurements showing that actual stresses in longitudinal beams and stringers are much smaller than calculated by the specification methods (Heins and Sartwell, 1969; Goble, Moses, and Pavia, 1974). Thus, other parameters, such as the total number of beams and the position of the beam under consideration with respect to the traffic lane, may be more relevant than S (beam spacing) in defining the lateral distribution (see also the foregoing sections).

Design S – N Curve and Effect of Fatigue Limit For all details, except AASHTO Category F, the finite-life portion of the S–N curves is defined as

$$N = \frac{A}{F_{\mathrm{sr}}^3} \tag{12-56}$$

where N is the fatigue life, F_{sr} is the stress range, and A is a constant relevant to each detail. Values of A are given by Schilling et al. (1978). The method used to estimate the fatigue design truck and the corresponding design stress range is theoretically applicable when all stress ranges caused by the traffic are above the fatigue limit stress range. In the variable-amplitude spectrum, stress ranges below the fatigue limit do not cause fatigue damage. The $\alpha_i W_i^3$ terms in (12-55) corresponding to these stress ranges are taken as zero. The weight of the fatigue design truck calculated in this manner is denoted as W_F', and the corresponding stress range as F_{sr}'. The correct fatigue life N' is defined by an equation similar to (12-56), even when F_{sr}' is below the fatigue limit (stresses in this case are assumed to cause no fatigue damage) so that

$$N' = \frac{A}{(F_{\mathrm{sr}}')^3} \tag{12-57}$$

When any of the stress ranges in the spectrum is below the fatigue limit, W_F' is less than W_F. The ratio W_F'/W_F depends on the percentage of the stress range distribution curve that is below the constant-amplitude fatigue limit and on the shape of the stress range distribution curve.

Because fatigue is inversely proportional to the stress range cubed, we can write

$$\frac{N'}{N} = \left(\frac{W_F}{W_F'}\right)^3 \tag{12-58}$$

where N' is the life calculated assuming that stress ranges below the fatigue limit cause no fatigue damage, and N is the fatigue life calculated assuming that all stress ranges in the spectrum cause fatigue damage.

The effect of the fatigue limit may be considered in the design in two ways: (a) by using W_F' in conjunction with constant-amplitude S–N curves, and (b) by using W_F' in conjunction with constant-amplitude curves modified according to (12-58).

Design Life According to the suggested method, if the design stress range exceeds the limiting value $F_{sr(L)}$, the estimated minimum life of the detail in years is

$$L = N'/(365TP) \tag{12-59}$$

where T is the average daily truck traffic and P is the average number of loading cycles per truck passage.

Fatigue Limit Effect on Variable-Amplitude Fatigue of Stiffeners

Yamada and Albrecht (1976) have reported the results of constant- and variable-amplitude fatigue tests on 13 butt-welded beams with a flange thickness transition, and 13 cover-plated beams with end welds ground to a 1:3 taper. Variable-amplitude fatigue tests of weldments have been performed by Albrecht and Friedland (1979) to clarify the cyclic behavior of stiffeners near the fatigue limit.

Specimens and Test Data A sketch of the tensile specimens with the attached transverse stiffeners is shown in Figure 12-48. The material is A588 steel with a yield strength F_y = 50 ksi. In addition, the chemical composition conforms with the requirements of the standard specifications for this steel grade.

The variable-amplitude fatigue tests made use of a basic stress range histogram developed by Yamada and Albrecht (1976). This represents the mean of 106 individual stress range histograms for truck traffic across 29 bridges in eight states, as derived by several investigators (Yamada and Albrecht, 1975).

The specimens tested under variable-amplitude fatigue were arranged in a two-way factorial experiment with five equivalent stress ranges, two weld

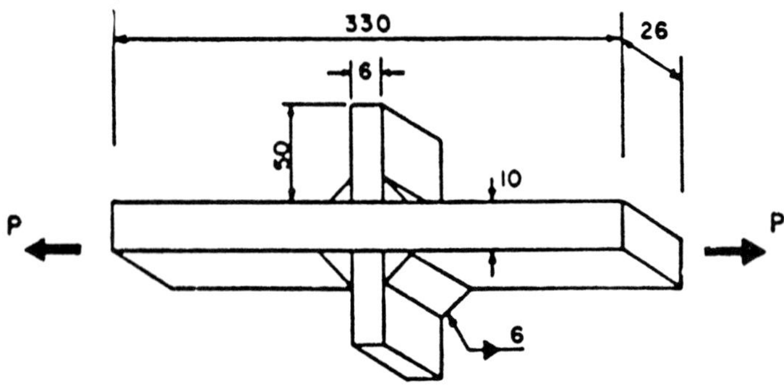

FIGURE 12-48 Tensile specimen with transverse stiffeners (mm). (From Albrecht and Friedland, 1979.)

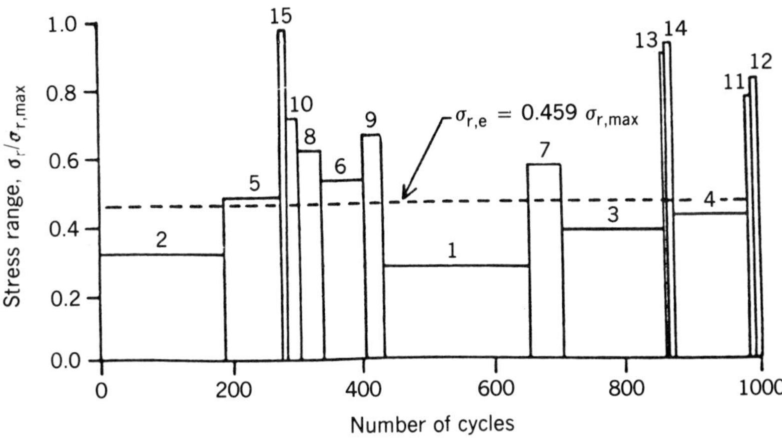

FIGURE 12-49 Basic 15-block loading spectrum used in variable-amplitude fatigue tests. (From Albrecht and Friedland, 1979.)

types, and up to four replicates per cell. The stress ranges were obtained by multiplying the normalized block-loading spectrum of Figure 12-49 by the values of $\sigma_{r(\max)} = 16$, 22, 30, 44, and 56 ksi. The RMC (root mean cube) stress range was computed from $\sigma_{r(\text{RMC})} = 0.459\sigma_{r(\max)}$. Two welding processes were used: submerged-arc and manual.

Constant-amplitude data from fatigue tests on specimens of the same geometry, material, and fabrication are available from Albrecht, Abtahi, and Irwin (1975). Likewise, data for 25 manually welded control specimens from the same batch as their variable-amplitude counterparts are available from Friedland, Albrecht, and Irwin (1979).

Crack Initiation and Propagation Cracks first appeared at one or more points along the weld toe line and then propagated through the thickness of the main plate in a direction perpendicular to the applied force. The specimens fractured when the net ligament underwent ductile rupture at average net-section stresses approaching the ultimate tensile strength. One particular specimen failed after 16,912,000 cycles at a stress range of 7.5 to 30 ksi. A quarter-elliptical crack outline was marked at 16,551,000 cycles and continued from that size to failure after an additional 361,000 cycles.

Test Results The fatigue lives of 10 automatically welded specimens under variable-amplitude loading are plotted in Figure 12-50 versus the RMC stress range of the full stress range histogram. Also shown are the confidence limits for the constant-amplitude test data and the estimated fatigue limit. The stress range is quoted in MPa (1 ksi = 6.9 MPa). The data at the two higher stress range levels fall within the band, whereas at the lowest level tested the

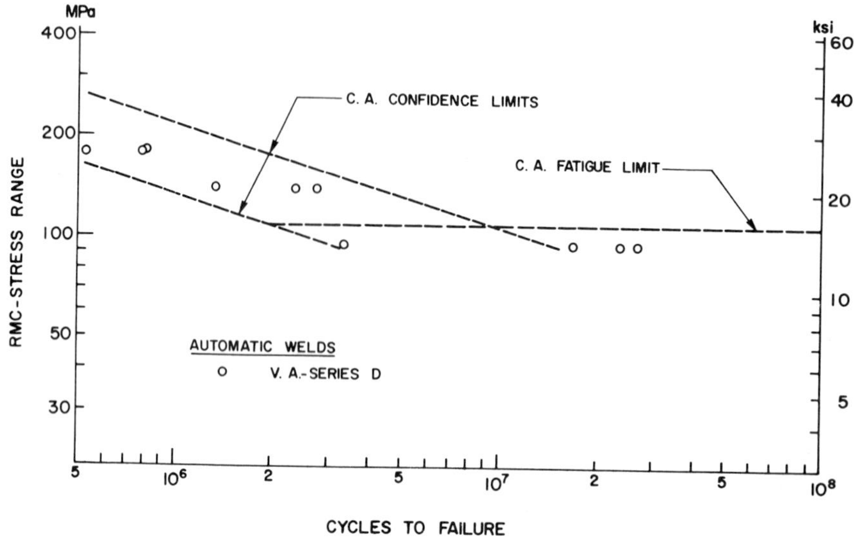

FIGURE 12-50 Variable-amplitude fatigue test data for automatically welded specimens. (From Albrecht and Friedland, 1979.)

data are widely scattered with one point falling near the lower limit and three above the upper limit.

The percentage of the stress range cycles that fall below the fatigue limit for each histogram η is estimated by considering the ratio of the fatigue limit to the maximum stress range, or $\sigma_{r(\max)} = 110/207 = 0.531$, 0.363, and 0.285. These, compared with the normalized stress range ϕ_i (not shown), yield values of $n = 84.4$, 41.3, and 22.6 percent, respectively. We should note that the runout trend for the data in Figure 12-50 becomes stronger with an increasing percentage of cycles below the fatigue limit.

A data summary for all automatically welded specimens subjected to constant- and variable-amplitude fatigue is given in Figure 12-51. The mean of the constant-amplitude test data is extended as a straight line below the range for which data are available until it intersects the fatigue limit. The symbol notation is as follows: (a) triangular symbols represent the log-mean lives of the specimens tested under constant-amplitude cycling, and (b) round and hexagonal symbols represent the variable-amplitude log-mean lives of the specimens tested as series D and C. The vertical-line segments through these points span from the lowest to the highest stress range of the applied variable-amplitude histogram. The values at the lower end indicate the cycles n that fall below the fatigue limit.

Albrecht and Friedland (1979) also report similar comparisons for manually welded specimens tested under constant- and variable-amplitude stress cycling. The same observations with regard to the adequacy and applicability

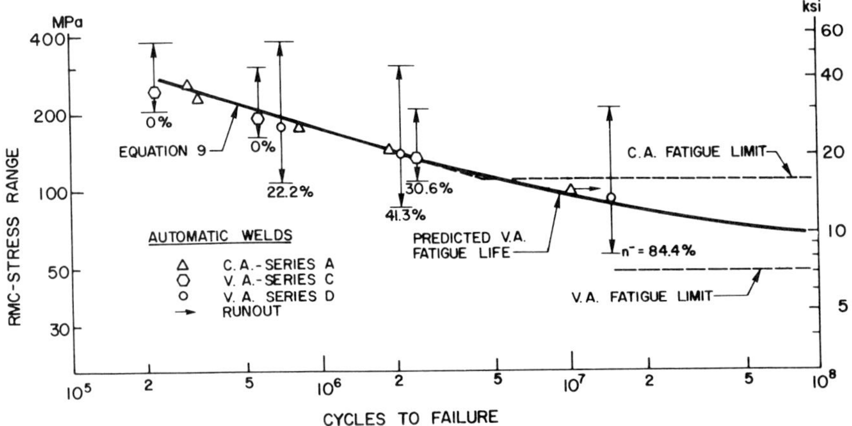

FIGURE 12-51 Comparison of all test data for automatically welded specimens with prediced fatigue lives. (From Albrecht and Friedland, 1979.)

of the RMC stress range in relating variable- and constant-amplitude conditions also apply in this case.

It appears that the RMC stress range of the full histogram provides a good transfer function between constant- and variable-amplitude fatigue provided it is equal to or greater than the fatigue limit. This requirement can be also expressed in terms of a corresponding number of cycles that may fall below the fatigue limit. When the RMC range is less than the fatigue limit, the variable-amplitude data diverge from the straight-line extension of the constant-amplitude mean. These conclusions apply to axially loaded tensile specimens with transverse stiffeners welded either automatically or manually.

Fatigue of Welded Attachments

The fatigue response of steel bridge members under variable-amplitude long-life fatigue loading has also been studied by Fisher, Mertz, and Zhong (1983) as NCHRP Project 12-15(4), with particular emphasis on welded attachments. These tests supplement the relatively shorter-life studies presented in the preceding section by testing nonload-carrying fillet-welded cruciform-type specimens under simple bending using a random variable-amplitude block loading. The tests also provide long-life data.

Beam Tests The study involved rolled-beam sections with as-welded details, focusing on two parameters for the random variable stress spectrum: the frequency of occurrence of stress cycles above the constant-amplitude fatigue limit, and the magnitude of the peak stress range in the stress spectrum. Details of the test beams are shown in Figure 12-52. The beams are W18 × 50 rolled sections, A588 steel.

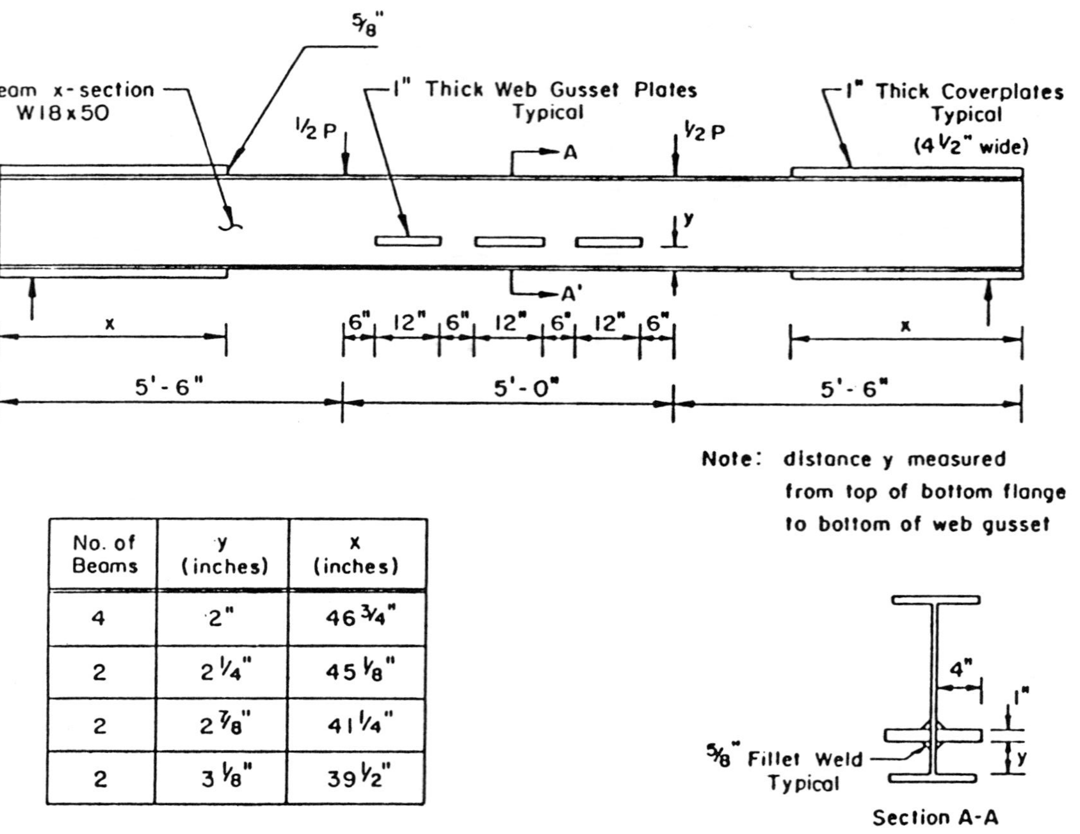

No. of Beams	y (inches)	x (inches)
4	2"	46 3/4"
2	2 1/4"	45 1/8"
2	2 7/8"	41 1/4"
2	3 1/8"	39 1/2"

FIGURE 12-52 Details of test beams.

All beams were tested on the dynamic test bed in the Fritz Engineering Laboratory at Lehigh University. The variable-amplitude cyclical load was applied with an MTS system consisting of two hydraulic jacks each with a capacity of 195 kips (Fisher, Mertz, and Zhong, 1983).

Fatigue Crack Growth Most fatigue crack propagation tests were performed under constant-amplitude cyclic-load fluctuation. Incremental crack lengths and elapsed load cycles provided the crack growth rate da/dN and the corresponding fluctuation of the stress intensity factor ΔK. The RMC stress intensity factor was used to define the effective stress intensity factor corresponding to the crack growth rate increment.

Random Block Loading of Cruciform Specimens The types of joints tested are essentially similar to the cruciform specimens shown in Figure 12-49. Both constant load cycle and variable load cycle tests were carried out. The test specimens were fabricated from A514 steel, and for the variable load cycle tests they were tested using the same variable-amplitude random block-loading sequence involved in the crack growth studies.

Test Results and Conclusions Based on the test results, Fisher, Mertz, and Zhong (1983) summarized their conclusions as follows.

Fatigue Behavior of Web Attachments When the effective stress range for the applied stress spectrum is below the constant-amplitude fatigue limit, both types of attachments either equal or exceed the fatigue resistance provided by an extension of Category E′. If a portion of the applied stress spectrum is above the constant-amplitude fatigue limit, fatigue cracking will occur. Because the scatter in the test data is not much greater than with constant-cycle tests, it is likely that all stress cycles contribute to fatigue damage.

At details where the welds wrap completely around the gusset plate (see also Figure 12-52), the toe of the transverse fillet weld joining the plate to the web provides a potential fatigue crack initiation location. If the web attachment is without such transverse welds, the termination point of the longitudinal welds is the initiation site.

When the effective stress range is above the constant-amplitude fatigue limit, both types of web attachments either equal or exceed the fatigue resistance provided by category E′ when plotted as a function of the effective stress range $S_{r(\mathrm{RMC})}$ versus life.

Cover Plates For an effective stress range below the constant-amplitude fatigue limit, similar behavior regarding straight-line extensions was observed for the S–N lines of Categories E and E′. Hence, as long as a portion of the stress range spectrum exceeds the constant-amplitude fatigue limit, fatigue cracking is likely to occur. Two examples were documented involving cover plates where no stress range cycles exceeded the estimated constant-amplitude fatigue limit for Category E. In this case the cover plate details developed very small semielliptical cracks after 100 million cycles when these details were destructively broken open in an attempt to reveal cracking.

Cover plates having the welds completely wrapped around them provided a fatigue crack initiation site at the toe of the transverse fillet weld joining the plate to the flange. Without transverse welds, the termination of the longitudinal welds was the initiation location.

With the effective stress range above the constant-amplitude limit, both types of cover plate details either equal or exceed the fatigue resistance for Category E′ when first-observed cracking is used.

Crack Growth Under Random Variable Loading Using the effective stress intensity factor derived from the equation $S_{r(\mathrm{RMC})} = (\Sigma \alpha_i S_{ri}^3)^{1/3}$, also referred to as Miner's rule, provides a reasonable correlation with the crack growth relationship applied to constant-cycle tests when extrapolated into the low levels of crack growth rate. On the other hand, the RMC effective stress intensity factor and crack growth behavior appear to be comparable to the behavior of the welded details.

Nonload-Carrying Welds Under Random Variable Block Loading From a limited constant-cycle amplitude test of the cruciform specimens subject to bending, the fatigue resistance was found to be near the upper bound of other tests on similar types of specimens (Goerg, 1963; Mueller and Yen, 1968). The results are compatible with tests reported by these investigators.

The variable-amplitude tests are found to plot near the upper bound of the constant-cycle tests, and the results suggest a constant-cycle fatigue limit close to 22 ksi. Furthermore, smaller simulated specimens tested to date tend to provide longer fatigue lives and higher fatigue resistance when subjected to random variable loading than do full-scale beam specimens.

NCHRP Project 12-15(5)

Keating and Fisher (1987) present the results of research on the behavior of welded bridge details in the high-cycle, long-life regime. In this program large-scale plate girders with cover plates, web attachments, and web stiffeners are subjected to fatigue loading that simulates actual truck traffic. The test specimens selected also allow a study on the effects of distortion-induced fatigue cracking at a connector plate web gap detail. A Rayleigh-type stress spectrum is assumed with the inclusion of occasional overload exceeding the constant-amplitude fatigue limit. This review is an extension of the studies presented in the foregoing sections, and more particularly NCHRP Project 12-15(4).

For comparison, the test results from Project 12-15(4), discussed in the preceding section, are presented in Figure 12-53, recalling that this study

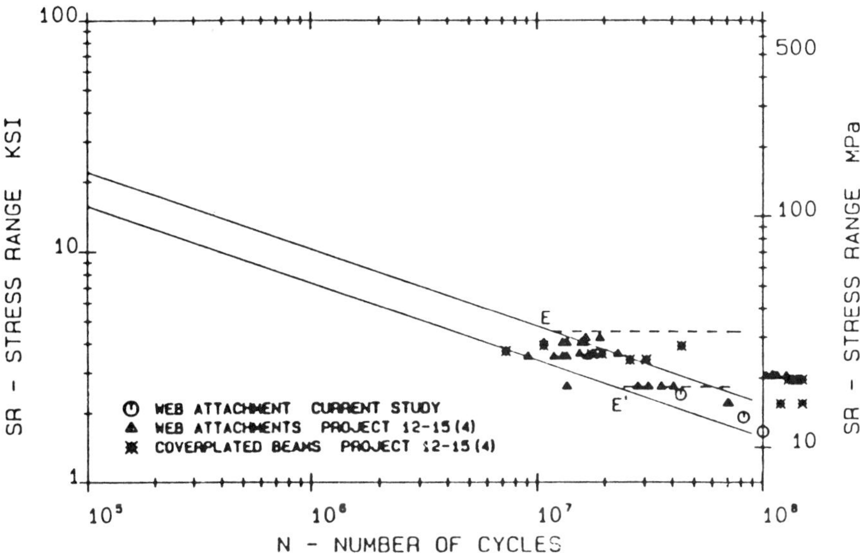

FIGURE 12-53 Fatigue test data, NCHRP Projects 12-15(4) and 12-15(5). (From Keating and Fisher, 1987.)

concluded that the existence of a fatigue limit below which no fatigue crack propagation occurs is assured only if none of the stress cycles exceed this constant-amplitude fatigue limit.

High-Cycle Fatigue Tests The plate girder configuration and welded details are shown in Figure 12-54. The web, flange plates, and attachments are A36 steel. All girders are 26 ft long and are tested in pairs on a 25-ft span under a four-point loading. The cover plates are attached either with or without the transverse end weld. This detail is assigned a Category E′ classification because the flange thickness exceeds 0.8 in. The transverse stiffeners (Category C) are cut short of the tension flange with a 1.5-in. typical gap.

A normalized stress spectrum is used, resulting in an effective stress range of 58 percent of the detail fatigue limit. Overloads are included at rates 0,1 0.05, and 0.01 percent. These result in stress ranges at each detail between 1.2 and 1.5, the assumed constant-amplitude fatigue limit. At the time this review was prepared, the fatigue tests yielded three failure points located at the weld end of three web attachments. One detail, with an effective stress range of 2.4 ksi, failed at 43.6×10^6 cycles. Considering the constant-amplitude limit for Category E′ (2.6 ksi), the position of this particular detail gave an assumed fatigue limit of 4.2 ksi for a 0.1 percent exceedance rate, so that using the actual fatigue limit for this category results in an exceedance rate of 14.6 percent. Fatigue crack growth was observed at another web attachment

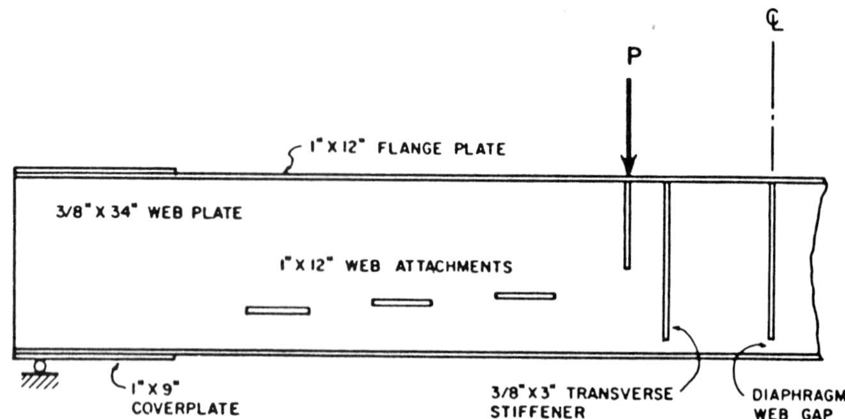

FIGURE 12-54 Plate girder configuration with attachments and welded details.

detail at 81.7×10^6 cycles. At this location the load spectrum gives an effective stress range of 2.2 ksi with a peak value of 4.3 ksi. A third crack was detected at 100.7×10^6 cycles and an effective stress range of 1.4 ksi with a peak value of 3.5 ksi. These data are plotted in Figure 12-53 as open-circle symbols.

For this detail further fatigue crack growth was prevented by placing holes through the web plate at the crack tips (Keating and Fisher, 1987).

Distortion-Induced Fatigue Cracking As shown in Figure 12-54, the test girders have a web gap detail at midspan for a diaphragm connection, where the plate is cut short of the tension flange by 1.5 or 4 in. The diaphram is a W14 × 22 rolled section, and the connection detail has the configuration shown in Figure 12-55. The bottom flange of the test girder is restrained against rotation by bearing on a strut. The in-plane vertical deflection of the girder under the applied load forces the connection plate out of plane by the resisting moment developed at the diaphragm connection. This arrangement simulates the differential displacement of bridge girders and the resulting distortion at diaphragm locations in negative moment regions and at supports.

At the static load that caused an in-plane stress at the connection plate end equivalent to the constant-amplitude fatigue limit (12 ksi), the measured strain corresponded to 43 ksi. Thus, the distortion-induced stress is 20 percent higher than the nominal yield strength of the steel (36 ksi). The measured web gap stress was essentially independent of the vertical bolted position of the diaphragm.

With the introduction of variable-amplitude loading, fatigue cracks were developed quickly in the web gap, with initial cracking detected at 50,000 cycles and extensive cracking occurring at 100,00 cycles. The configuration of distortion-induced fatigue cracks is shown schematically in Figure 12-56.

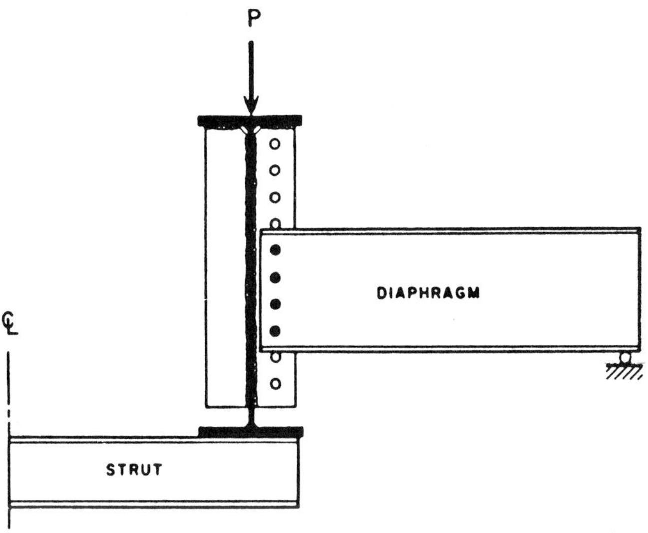

FIGURE 12-55 Test setup for distortion-induced web gap fatigue cracking. (From Keating and Fisher, 1987.)

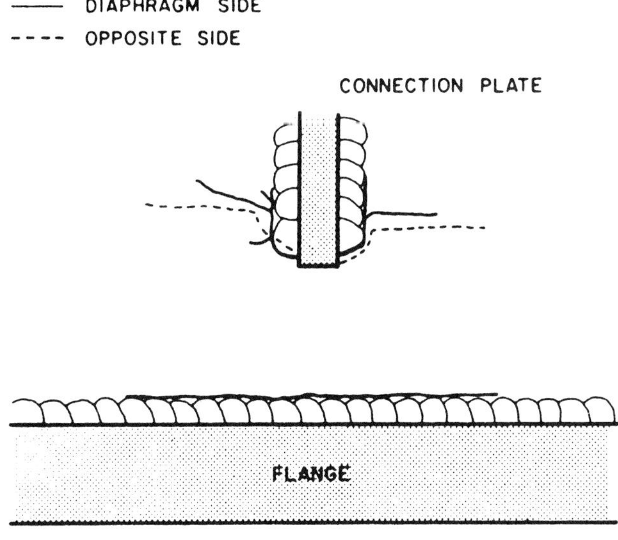

FIGURE 12-56 Configuration and map of distortion-induced cracks in web gap. (From Keating and Fisher, 1987.)

At about 1 million random variable load cycles, the web gap cracking was retrofitted by means of drilled holes at the crack tips. After 500,000 more cycles, the cracks reappeared at the toe of both the connection plate weld and the web–flange weld. The crack tips were again arrested by drilling, and testing continued. Cracks occurred again at 2.1×10^6 and at 2.9×10^6 cycles and included a crack at the drilled hole along the stiffener fillet weld toe. At this stage, the diaphragms were removed from the test girders.

Strain measurements at the top of the uppermost hole revealed a 10 percent increase in the in-plane web stresses due to the presence of the web cracks. Cracks reappeared at the perimeter of the drilled hole at 9.4×10^6 cycles, and when the load reached 20 million cycles, a toe crack reinitiated at the top hole. At 47×10^6 cycles, fatigue cracks initiated at the perimeter of the uppermost hole and at one of the web–flange holes. It appears therefore that when the diaphragms were removed and the in-plane loading continued, the detail had already sustained considerable damage, so that cracks continued to develop and propagate.

12-12 FULL-SCALE WELDED BRIDGE ATTACHMENTS: FATIGUE BEHAVIOR IN EXPERIMENTAL INVESTIGATIONS

This section summarizes the results of NCHRP Project 12-15(3), repeated by Fisher, Barthelemy, Mertz, and Edinger (1980). The study was intended to compile a more comprehensive data base and also to examine the influence of lateral bracing members on the out-of-plane distortion of the lateral plates. In particular, points of interest are (a) the detection of cracks at weld toes; (b) the fatigue behavior of web gusset plates, flange gussets, gusset plates attached to the flange surface, webs with girder flanges framing into the web or piercing through the web, and longitudinal stiffener welds containing lack-of-fusion regions; (c) retrofitting procedures; and (d) applications.

These tests involved 18 full-sized beams consisting of W27 × 145, W27 × 114, and W36 × 160 rolled sections of A36 steel. The experiment design and details are summarized in Figure 12-57.

Cracks at the Weld Toes Each beam was cycled until a crack was detected at the detail. Visual inspection with 10 × magnification and dye penetrant revealed cracks as small as 0.25 in. (6 mm) long. An empirical relationship between the crack width and depth gave an indication of the crack depth. Having defined the fatigue life of the detail, an appropriate retrofitting procedure was used to extend the life of the detail. Although it was possible to retrofit shallow-depth cracks after initial detection, field detection of weld toe cracks for retrofitting purposes was more difficult.

Web Gusset Plates The toe of the transverse fillet weld joining the web gusset plates to the web is a suitable fatigue crack initiation site. Fatigue cracking at this location results from stress concentration due to geometric

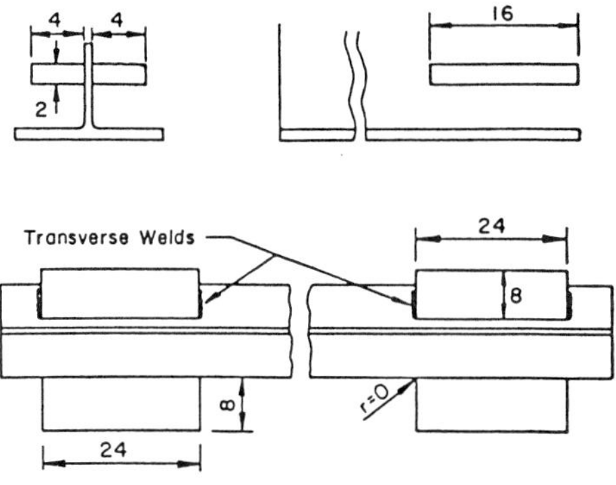

Detail (*a*)

Cross Section	No.	Web Gusset Types			Flange Gusset R(mm, In)	Fillet Welded	Groove Welded	Insert	Flange Plate
	5	X			0				
	2	X			0	X			X
	3		X		5 (2)	X			
W27x145	4		X		5 (2)	X			
	1			X	15 (6)				
	9			X	15 (6)		X	X	
	14	X			0				X
	13	X			0			X	X
	16		X		5 (2)		X	X	
W27x114	18		X		5 (2)	X		X	
	10			X	15 (6)		X		
	11			X	15 (6)			X	X
	15	X			0			X	
	8	X			0		X	X	
	7		X		5 (2)		X		
W36x160	6		X		5 (2)	X			
	12			X	15 (6)			X	X
	17			X	15 (6)			X	

FIGURE 12-57 Experiment design and details, NCHRP Project 12-15(3), (From Fisher, Barthelemy, Mertz, and Edinger, 1980.)

conditions and also because of possible microscopic discontinuities at the fillet weld toe. All types of gusset plates tested either equaled or exceeded the fatigue limit specified for Category E. However, small differences in fatigue resistance were detected, and these were amplified in terms of the applied stress level.

Cracks were developed at the lateral gusset plate ends, and none were detected in the girder web gap adjacent to the transverse stiffener. Variation of the flexural rigidity of the lateral bracing members did not affect the fatigue resistance of the web gusset plate details. During testing, no cracks were detected in the transverse welded connections between the gusset plate and the transverse stiffener, but on completion of the tests destructive inspection showed cracks in two details.

Flange Gussets Detectable fatigue cracking occurred only at the ground "0" radius details (see also Figure 12-57). Considering 24 possible crack sites, three locations were visually detected as cracked, and four more were found using ultrasonic procedures. Subsurface defects were shown to be the origin of crack growth.

All of the "0" radius details with ground ends exceeded the lower confidence level provided by Category E, and no detectable crack growth occurred at the 2-in. (50-mm) and 6-in. (150-mm) radius details. The use of fillet welds instead of groove welds to attach gusset plates to the flange enhanced the probability of crack incidents in the form of weld root flaws. Tests on 4-in. (100-mm) radius transition details attached to the girder web by fillet welds gave fatigue resistance according to Category D. None of the 2-in. and 6-in. radius transition details attached to the girder flange by groove welds developed fatigue cracks within the same cyclic loading.

Gusset Plates Attached to Flange Surface In this arrangement, shown in detail in Figure 12-57*a*, cracking developed from the weld root, propagating through the entire width of the transverse weld. Because the maximum weld size is the plate thickness, the problem cannot be avoided by increasing the weld size. In most cases the result was complete severing of the gusset plate, providing a relatively low fatigue resistance according to Category E′ or less.

Cracking in one test also developed in the beam flange from the transverse weldment, giving a fatigue resistance between Categories E and E′. Although the testing was limited, it suggests that with greater flange thickness, lower fatigue resistance results.

The use of transverse welds only in attaching gusset plates to the flange surface does not improve fatigue resistance. This detail was examined to determine whether longer details may be improved by eliminating the longitudinal welds.

Webs with Girder Flanges Framing or Piercing Web Beams with thin webs (0.25 in.) yielded approximately the same fatigue resistance with the plates either framed into the web or pierced through the web. For the smaller beams, the results conformed to Category E, but for larger-sized girders the results tend to fall at or below the confidence limit provided for Category E. The fatigue resistance of thick flanges welded to the surface of girder webs is better represented by Category E′.

Semicircular end copes at the tips of the thicker (2-in.) flange plates that pierced the web provided a fatigue resistance level less than Category E′. Furthermore, welding of details that penetrate the web on both web sides reduced the unfused region of the detail and improved fatigue resistance to Category E′.

Longitudinal Stiffener Welds Containing Lack-of-Fusion Regions In these details the fatigue crack propagation that developed at the lack-of-fusion regions showed that the crack propagated through the web in a semicircular shape with the crack radius near the tip of the longitudinal stiffener. Because of the difficulty of quantifying the variable lack of fusion and also because of the small number of tests, direct comparison of fatigue resistance with other details was not made.

Retrofitting Procedures Fisher, Barthelemy, Mertz, and Edinger (1980) provide several guidelines on retrofitting methodology in connection with the foregoing details. Among these, the placement of holes at crack tips is essential to the success of retrofitting by drilling holes. If the actual crack tip, however, is difficult to locate and lies beyond the hole, crack propagation will be accelerated.

A threshold level of the ratio of the stress intensity factor fluctuation to the square root of the notch tip radius ($\Delta K/\sqrt{p}$) less than four times the square root of the yield strength ($4\sqrt{\sigma_y}$) was required to ensure that cracks did not start at the welded details.

12-13 SUMMARY AND EVALUATION OF FATIGUE TESTS

Current Provisions

Because the fatigue resistance provisions of the AASHTO specifications were formulated based on data provided in NCHRP Reports 102 and 147 in 1972, several major fatigue studies have been carried out, some of which were presented in summary form in the foregoing sections. The initial fatigue resistance curves were based on approximately 800 fatigue test failure results. Since 1972, more than 1500 additional test results have been added to that data base. New types of details have been added, including longitudinal groove welds in both flat-plate specimens and in box members, internal diaphragms for box-type members, large-scale cover plates and web attachment details, and a wider range of flange attachment details with varying geometries and weld conditions.

The current AASHTO fatigue design curves are shown in Figure 12-58. Because the review of the fatigue data demonstrated that Category B overestimated the fatigue strength of certain longitudinal groove welds, Category B′ was added. These curves were developed using the stress range

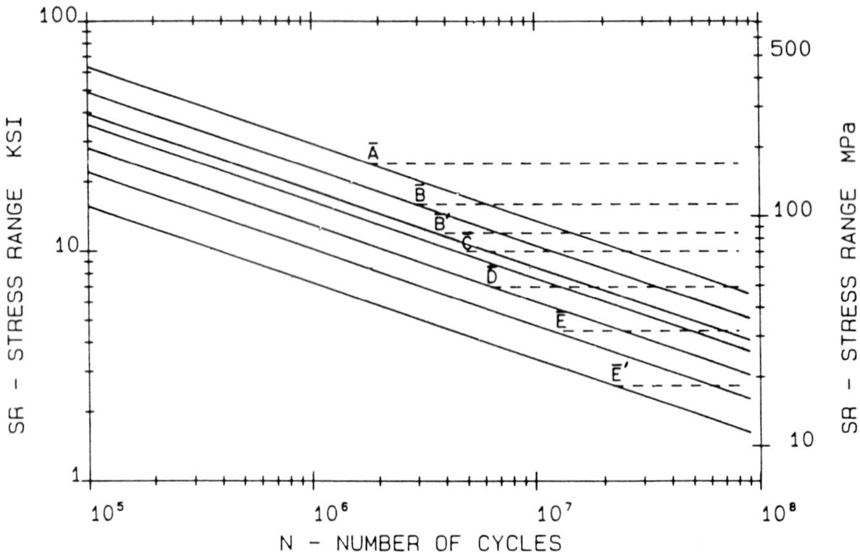

FIGURE 12-58 Current AASHTO fatigue design curves, redundant load path members.

intercept values at 2×10^6 cycles. The constant-amplitude limits for each curve, with the exception of Category E, correspond to their older values. High-cycle fatigue test results of cover-plated beams indicated a fatigue limit of 4.5 ksi (see also the foregoing sections). For Category B′, a constant-amplitude fatigue limit of 12 ksi has been specified. The stress range intercept values for 1×10^5, 5×10^5, and 2×10^6 as well as the constant-amplitude fatigue limits for each curve are as shown in AASHTO Table 10.3.1A.

The major difference between these curves and the older ones is their slope. The majority of the old curves had a slope slightly greater than −3.0. Because of this slope difference, the current curves give a slightly higher fatigue resistance in the low-cycle regime. The new provisions were based on Miner's rule and the assumption that the slope of the fatigue resistance curve was −3.0; hence, the damage estimate was compatible with the damage that will result from the relationship shown in Figure 12-58.

AASHTO (1989) Guide Specifications for Fatigue Design

These specifications contain provisions for a simplified six-step design procedure as follows.

1. Use a fatigue design truck with axle weights of 6, 24, and 24 kips (total gross truck weight 54 kips), and axle spacing of 14 and 30 ft. The gross weight may be changed to reflect data given from the site.

2. Increase the gross weight of the truck by 15 percent to account for impact.
3. Place the truck on the bridge to obtain maximum force and moment effects.
4. If a deck is supported by two members, distribute the moments assuming the deck acts as a simple beam with a single truck placed at the center of the outer traffic lane. If there are more than two members, use a distribution factor s/D, where s is the girder spacing (ft) and D is a parameter that relates to the span.
5. Increase the calculated section properties by 15 percent. If shear connectors are provided, use a full composite section for positive moments and a section including longitudinal reinforcement for negative moments.
6. Compute the design stress range S_r at the critical section as the sum of stresses caused by maximum positive and negative moments. This range must satisfy the criterion $R_s S_r \leq S_{rp}$, where R_s is 1.1 for redundant and 2.0 for nonredundant members, and S_{rp} is the permissible stress range for various detail categories. Values of S_{rp} are given in the appropriate tables in Appendix B of the AASHTO (1989) Guide Specifications for Fatigue Design.

Foreign Data on Fatigue

Japanese Data Extensive fatigue tests have been performed in Japan in conjunction with the Honshu–Shikoku Bridge Authority program to link the major islands with lond-span suspension, cable-stayed, and truss bridges.

In the first group of tests, the specimens were selected to simulate members of welded box girders and other welded joints common to bridge structures. Attention was focused on size effects and the influence of defects on the assessment of fatigue strength. The specimen size was selected to provide data on full residual stresses in the welds. Because most defects are inherent to welds, tests were conducted on specimens with varying defects in order to relate this variation to the fatigue strength. All tests were carried out under constant-amplitude loading. Interestingly, each individual variable tended to provide S–N curves with a slope close to -3.0 (Miki et al., 1982a, 1982b).

The second group of tests involved longitudinally welded box members (Miki et al., 1980, 1983). Two types of longitudinal groove welds were tested: a single-bevel-groove weld and a single-J-groove weld. Several tests were also carried out with fillet welds in the box corners.

Nonload carrying cruciform, fillet-welded joints represent the third basic group of specimens corresponding to the Category C detail. In these tests the best-fit slope is very close to the AASHTO slope with approximately the same variance. The higher fatigue strength established in these specimens resulted from improved electrodes that provided a more favorable weld profile (Shimokawa, Takena, Ho, and Miki, 1984).

ORE Data Test programs carried out by ORE (Office of Research and Development) of the International Union of Railways focused on the bending of structures consisting of two beams welded at right angles. These tests yielded unusually low fatigue results (IUR, 1971). The conclusion was that the fatigue strength was consistently overestimated by most railway specifications and standards, in some cases by a factor of 2.

Subsequent studies attempted to articulate these results and detect a possible bias in interpreting the complexity of the test procedure (ORE, 1974–1979). These studies provided a relatively limited number of test data. As a result, for many details a regression analysis yielded little useful information. In addition, some failures occurred at distinct notches or at load points, thus complicating the evaluation process. However, this program considered a broad variety of detail types tested under a variety of stress ranges and therefore supplements current design curves.

British Fatigue Data A fatigue test program is summarized by the Transport and Road Research Laboratory (TRRL, 1979) and Maddox (1982). This program involved a fatigue study of improved fillet welds. By shot-peening the fillet welds of attachments, the fatigue strength was found to increase. This improvement was more effective on transverse welds than around the ends of longitudinal attachments. As the stress ratio increased, however, the effectiveness of the shot peening was reduced.

The conclusions reached by the report were extrapolated for use in the data base. Thus, results indicated that cracking initiated at the toe of an end section of an intermittent weld, but significant variation in the fatigue behavior of different weld gap patterns was not established. The use of intermittent welds introduces many weld ends, and these can have a wide range of weld termination conditions, resulting in undercutting and other weld defects. The report suggests that the use of this type of weld should not be encouraged, as would be the case if a higher resistance category were assigned to it.

ICOM Fatigue Data Tests have been conducted by ICOM (Swiss Federal Institute of Technology) dealing with the fatigue life of structural details associated with high stress concentrations (Hirt and Crisinel, 1975; Smith, Bremen, and Hirt, 1984; Yamada and Hirt, 1982). The tests established procedures to study and monitor fatigue crack propagation so that analytical methods can be developed and verified. They also considered fatigue strength improvement techniques on structural details fabricated with high-strength steels.

German Fatigue Data Fatigue tests were conducted to study the load capacity of different welded structural components for classification purposes and according to fatigue resistance. The same tests also dealt with the effect of the stress ratio on fatigue strength, taking into account cyclic tension and reversal conditions.

A separate research program focused on the fatigue strength of welded high-strength steels in the as-welded and TIG-dressed condition (Minner and Seeger, 1982). Pilot tests with small specimens showed a significant increase in fatigue strength with TIG or plasma dressing. One objective of the study was to determine the applicability of results from small-scale specimens to full-sized beams and members.

Constant-amplitude fatigue tests on rolled and welded high-strength-steel beams consistently yielded results significantly below the predicted strength for each detail type tested. An explanation for the reduced fatigue strength relates to welding deficiencies such as hydrogen-induced cold cracking and weld undercutting.

Canadian Fatigue Data Two separate studies disseminate the findings from fatigue tests (Comeau and Kulak, 1979; Baker and Kulak, 1984). One dealt with full-scale web attachments, whereas the other examined the effects of backing bars on the fatigue strength of transverse groove welds.

Long web plates attachments with lengths exceeding 4 in. and thicknesses less than 1 in. would normally correspond to Category F, and the test results are in good agreement with this resistance curve. No significant difference in fatigue strength was observed between the through web detail or the discontinuous plate detail.

Other Data A review of fatigue tests on welded details is given by Keating and Fisher (1986) and covers both U.S. and foreign practice. Among the many useful references, the studies by Albrecht and Naemi (1984) and Yamada and Kikuchi (1984) focus on the fatigue characteristics of weathering steels.

Both programs involved plate specimens fabricated with automatic submerged-arc welds. The specimens were tested under constant-amplitude loading, and both programs examined unweathered and weathered conditions. The weathered specimens were subjected to varying levels of atmospheric exposure prior to testing. The Albrecht and Naemi study involved the fatigue testing of 176 specimens, simulating a transverse stiffener detail or an attachment plate. All test results exceeded the Category C resistance curve for stiffeners. The attachment specimens would normally correspond to the Category D detail type. The specimens were fatigue-tested as fabricated (unweathered) after two years of exposure, and after they were weathered for four years. The results compare with the Category D fatigue resistance curve. Details of the specimens are shown in the upper portion of Figure 12-59.

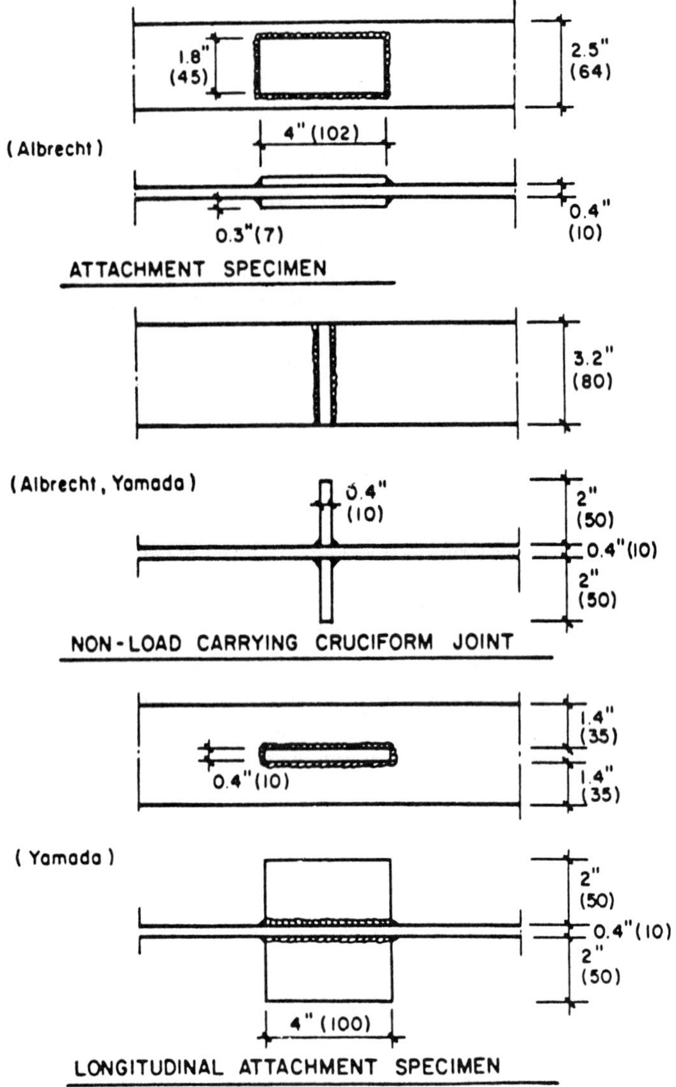

FIGURE 12-59 Test specimens for weathering steel fatigue test programs.

The Yamada and Kikuchi study considered two types of plate specimens: nonload-carrying cruciform joints and a gusset plate as shown in the lower portion of Figure 12-59. Both weathering and standard steel were used. The specimens were tested as fabricated as in the Albrecht and Naemi study. In addition, stiffener-type details were cut out of the web of an actually weathered steel bridge that had completed 5.5 years of service.

All failures plot beyond the Category C curve. In particular, the gusset plate specimen results are compared with the Category D resistance curve, and all data fall significantly above that curve.

Application of Results: Project 12-15(3) Among the conclusions emerging from this study, we summarize the following.

1. Existing structures, with gussets attached to the transverse stiffeners and having intersecting welds, are susceptible to cracks developing from the transverse fillet welds and web intersection.
2. Small ground radii (0.2–0.4 in.) at the ends of rectangular gusset plates provide fatigue resistance equal to Category D.
3. Gusset plates should not be welded to the flange surface with transverse fillet welds alone. At least one longitudinal weldment is necessary to prevent the gusset plate from cracking.
4. Flanges framing into girder webs (whether fillet- or groove-welded to the web surface) should be designed for Category E′ attachments when

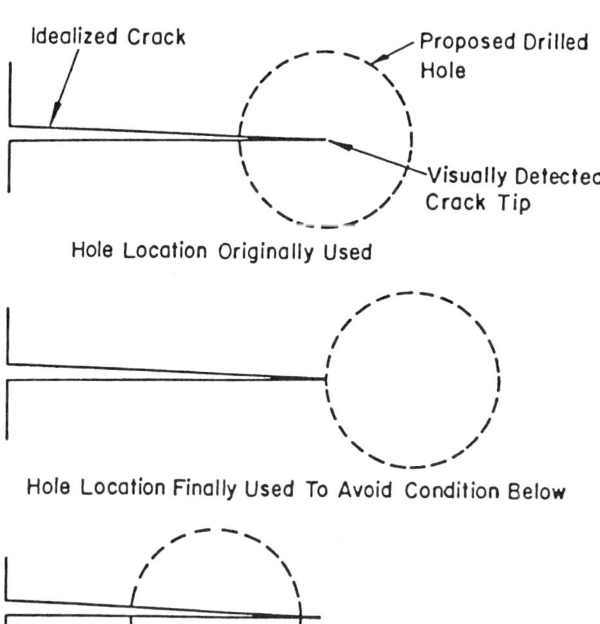

FIGURE 12-60 Crack tip and hole perimeter locations.

the flange thickness exceeds 1 in. Flanges less than 1 in. thick can be designed as Category E details.

5. When flanges pierce through girder webs and are connected to each web surface by fillet welds, they should be designed for Category E′.
6. Holes drilled at the ends of fatigue cracks should have the perimeter of the hole placed at the apparent crack tip, as shown in Figure 12-60.

12-14 OVERLOAD EFFECTS AND FATIGUE OF CONCRETE BRIDGES

Overload Effects

Progressive overload-induced damage to concrete decks has been documented in the field. Interaction between physical damage directly attributable to wheel loads and other damage mechanisms such as corrosion has been confirmed through long-term monitoring and test programs. These phenomena relate directly to the concept of fracture mechanics of concrete and its application to concrete bridge decks.

Deck damage may appear in several forms, but the most important damage mechanism is associated with transverse and longitudinal cracking. Wheel-load-related cracking is more severe in structures with lower ratios of dead to total load. Reinforced concrete decks on steel I beams are more susceptible to this damage than decks on prestressed girders because of the inherent greater flexibility of the steel beams. Overweight vehicle effects on bridges, particularly deck cracking, are manifested in conjunction with other ongoing damage mechanisms but in a complex manner.

Corrosion of reinforcement is promoted by the presence of cracks, and spalling resulting from steel bar corrosion is certainly accelerated by traffic. Cady and Weyers (1984) present a method for predicting the deterioration rate of concrete bridge decks due to corrosion attack, although the method does not include parameters related to traffic density or vehicular overload.

Field Studies of Progressive Damage James, Zimmerman, and McCreary (1987) refer to a Texas–SDHPT-sponsored study (Brown et al., 1978) involving a survey of 25 structures in which the effects of differential truck traffic could be observed by comparison to control bridges. Preliminary results indicate that an observable correlation of some forms of progressive damage induced by heavy trucks could be articulated on some of the bridges surveyed. In particular, higher levels of deck cracking were observed in concrete decks on steel stringers and subjected to heavier traffic.

These observations provided the basis for further studies involving two candidate bridges along with two control structures. All structures are simple spans with concrete decks on steel I beams. The candidate structures experience a much higher level of heavy truck traffic than the control bridges, and this differential traffic is approximately 180 vehicles/hr. It is estimated that the differential traffic has existed for about 26 years. Although both bridges carry loads, for reference the bridge carrying the heavier traffic loads is identified as the loaded structure, and the bridge carrying the light traffic is identified as the unloaded structure.

Total Crack Densities The variation in cracking was observed by dividing the study area of the bridge into transverse and longitudinal regions. Cracks oriented within 30° of a line normal to the direction of traffic are denoted as transverse cracks, whereas all other cracks are diagonal. The crack density is the sum of the length of observed cracks divided by the area of the particular region. The designations left, center, and right refer to the regions between two of the steel beams in relation to the traffic flow.

Invariably, the loaded bridge exhibited more cracking than the unloaded structure for the corresponding regions. For both the loaded and the unloaded bridge, the center regions always exhibited more cracking than the left regions. The center regions usually exhibited more cracking than the right regions in both bridges. Furthermore, the right regions always exhibited more cracking than the left regions in both bridges, although the difference was not the same. The difference in crack density between the loaded and unloaded bridges is much greater in the left regions than in the center or right regions. The more heavily cracked center and right regions have smaller differences between the loaded and unloaded spans (James, Zimmerman, and McCreary, 1987).

Transverse Cracking The loaded bridge exhibited more transverse cracking than the unloaded structure for the same regions, quantified as follows: left regions, 71 percent more; center regions, 8 percent more; and right regions, 11 percent more.

For both the loaded and unloaded bridges, the center regions always exhibited more transverse cracking than the left regions, about 22 and 92 percent more, respectively. The center regions usually exhibited more transverse cracking than the right regions in both bridges, averaging 24 and 20 percent more for the unloaded and the loaded bridge, respectively. For the unloaded bridge, the right regions exhibited 55 percent more cracking than the left regions, but for the loaded bridge there was little difference between right and left regions.

Longitudinal Cracking The same trend is confirmed in longitudinal cracking. Thus, the loaded bridge exhibited more longitudinal cracking than the unloaded structure for both the left and the right regions, or 68 and 23 percent more, respectively. The center regions usually exhibited more cracking in the loaded bridge, by an average of 11 percent.

For both bridges, the center regions always exhibited more longitudinal cracking than the left regions, or 223 and 114 percent more for the unloaded and the loaded bridge, respectively. Likewise, the right regions exhibited much more cracking than the left regions, or 116 and 236 percent more for the loaded and unloaded bridge, respectively.

Diagonal Cracking More diagonal cracking occurred in the loaded bridge than in the unloaded structure in the left regions, averaging 47 percent more. In the more heavily cracked center and right regions, however, this trend was reversed. The center regions were usually cracked less in the loaded bridge by about 17 percent, and the right regions always exhibited less diagonal cracking for the loaded bridge by an average of 55 percent.

Damage Mechanisms Related to Overloads Wegmuller (1977) has developed a finite-element model of a steel–concrete composite bridge, where the concrete, reinforcing bars, and I beams are treated as nonlinear materials. Model tests can be compared to experimental data to demonstrate the modeling of such structures in the postelastic regime. The model is limited to short-term, monotonically increasing loads. Tensile cracking in the concrete is treated merely as complete loss of stiffness after a cracking stress is reached. Strain softening is not included. This work shows that successful nonlinear analysis of bridges subjected to overloads is possible, and up to about 20 percent overload, the structural response may be represented by modeling the slab as an elastic, perfectly plastic material.

An early indication of overload damage is cracking in the slab. Batchelor, Hewitt, and Csagoly (1978) present test results showing that the fatigue endurance limit punching loads for isotropically reinforced slabs with reinforcement ratios of 0.2 percent and for conventionally reinforced slabs are 40 and 50 percent of the static punching failure loads, respectively. The endurance limit for the isotropic slab represents stresses caused by wheel loads almost four times the design wheel load. Observed fatigue failure modes are predominantly punching shear, whereas flexural failure has occurred in some slabs with low reinforcement ratios (see also the foregoing sections). During the first few cycles of repeated loading, cracking was initiated. These cracks widened and spread following repeated loading, and a relatively stable state was reached characterized by little crack growth. Best-fit equations have been developed to represent these data. For conventional orthotropic reinforcement, the ratio of the fatigue strength P_f to the punching resistance P_s' is expressed in terms of the number of cycles to failure N as

$$P_f/P_s' = 1.0 - 0.102 \log N + 0.006(\log N)^2 \qquad (12\text{-}60)$$

The assumption that fatigue failure of decks is unlikely ignores any interaction of fatigue cracking, observed at much lower loads, and corrosion or other environmental effects. Cracking of concrete decks should be expected at normal traffic levels, and corrosion of reinforcing steel can limit the serviceability of the deck. Maeda et al. (1981) suggest that transverse and longitudinal cracking in slabs subjected to normal service loading would not occur closer than 1.3 ft, with crack widths up to 0.008 in. and crack densities up to 170 in./yd^2. More severe cracking, scaling, and spalling of concrete should mean unusual and severe distress.

Tests by Shanafelt and Horn (1985) on prestressed concrete girders have provided useful information on progressive damage mechanisms and helped quantify damage to girders resulting from overweight vehicles. It appears from these tests that prestressed concrete girders and composite reinforced concrete decks are among the most durable and overload-resistant schemes, although the potential for overload-induced progressive damage still exists depending on the level of the live load.

Fracture Mechanisms of Concrete Fracture mechanics theories are applied to plain, reinforced, and prestressed concrete structures in certain fields. Two distinct modeling techniques are emerging. One approach is to model the crack in a continuum using finite elements with free surfaces along element boundaries where the crack is open. The second method, more widely accepted, was proposed by Rashid (1986) and models the crack by a continuous smeared crack band. This is considered to be a better physical model of fracture in concrete than the discrete crack model because there is a zone of some finite thickness surrounding each crack where physical damage reduces the stiffness of the continuum.

Conclusions James, Zimmerman, and McCreary (1987) conclude that cracking may occur at tensile stresses below the expected value of $6\sqrt{f'_c}$ or $7.5\sqrt{f'_c}$. This is based on test results, although it is not known whether residual stresses due to creep or shrinkage are the explanation. Thus, Shanafelt and Horn (1985) have reported cracking at stresses close to $2.7\sqrt{f'_c}$ to $4.5\sqrt{f'_c}$, whereas Bonilla et al. (1986) have observed significant cracking in a one-third-scale reinforced concrete deck subjected to negative moments at stresses close to $3\sqrt{f'_c}$.

Concrete bridge decks appear to respond to overloads by increased densities of longitudinal and transverse cracking, although cracking may also result from normal traffic levels. The increased cracking accelerates corrosion attack and spalling or scaling. Mechanisms are available to articulate the interaction between mechanical and physical effects in progressive overload-induced damage, but specific procedures for studying the rate of deterioration are yet to be developed.

Progressive damage of cracked prestressed concrete girders can be detected and assessed as long as the girder is subjected to moments large enough to reopen the initial cracks. Test data show that the moment necessary to reopen the cracks is lower than the moment necessary to initiate cracking, usually 75 percent of that moment.

Fatigue Strength of Reinforced Concrete Bridges

Fatigue Stress Limits AASHTO Article 8.16.8.3 (load factor design) stipulates the range between a maximum tensile stress and a minimum stress in straight reinforcement caused by live load plus impact at service load. In this regard,

$$f_f = 21 - 0.33f_{\min} + 8(r/h) \tag{12-61}$$

where f_f = stress range (ksi)

$f_{\min}$ = algebraic minimum stress level, tension positive, compression negative (ksi)

r/h = ratio of base radius to height of rolled-on transverse deformations (when the actual value is not known, use 0.3)

The specifications also recommend avoiding bends in the primary reinforcement in regions of high stress range. Fatigue stress limits need not be considered for concrete deck slabs with primary reinforcement perpendicular to traffic and designed according to Article 3.24.3, Case A.

The proposed LRFD specifications articulate fatigue as a limit state. Interestingly, these specifications identify the fatigue design truck as the HS 20 but with a constant rear axle spacing of 30 ft. If a concrete section is in compression under dead load plus 1.5 times the fatigue live load, resulting effects need not be investigated. If the section is in tension under this loading combination, then (a) fatigue should be investigated for a Fatigue I load combination; (b) the section properties should be based on either cracked or uncracked sections, assuming a reduced modulus of rupture equal to $3\sqrt{f_c'}$; and (c) the stress range should be calculated using Fatigue I load combination, provided that this combination produces tension. The stress range is as given by (12-61). Likewise, fatigue need not be investigated in concrete slabs designed according to AASHTO Article 3.24.3, Case A.

The LRFD specifications extend the fatigue stipulations to prestressing tendons and state the assumptions for flexure and axial load design applicable to reinforced and prestressed concrete members for service and fatigue limit states.

Fatigue Strength Fatigue failures of the reinforcement in bridge decks can only occur if one or both extreme stresses in the stress cycle are tensile. Typical S–N curves are shown in Figure 12-61. As in other steel members, there is a fatigue threshold or endurance limit below which fatigue failures should not occur. For straight ASTM A615 bars, this limit is about 24 ksi, essentially the same for Grade 40 and Grade 60 bars.

The fatigue strength of deformed bars decreases as the stress range increases, as the level of the lower stress in the cycle is reduced, and as the ratio of the radius of the fillet at the base of the deformation lugs to the

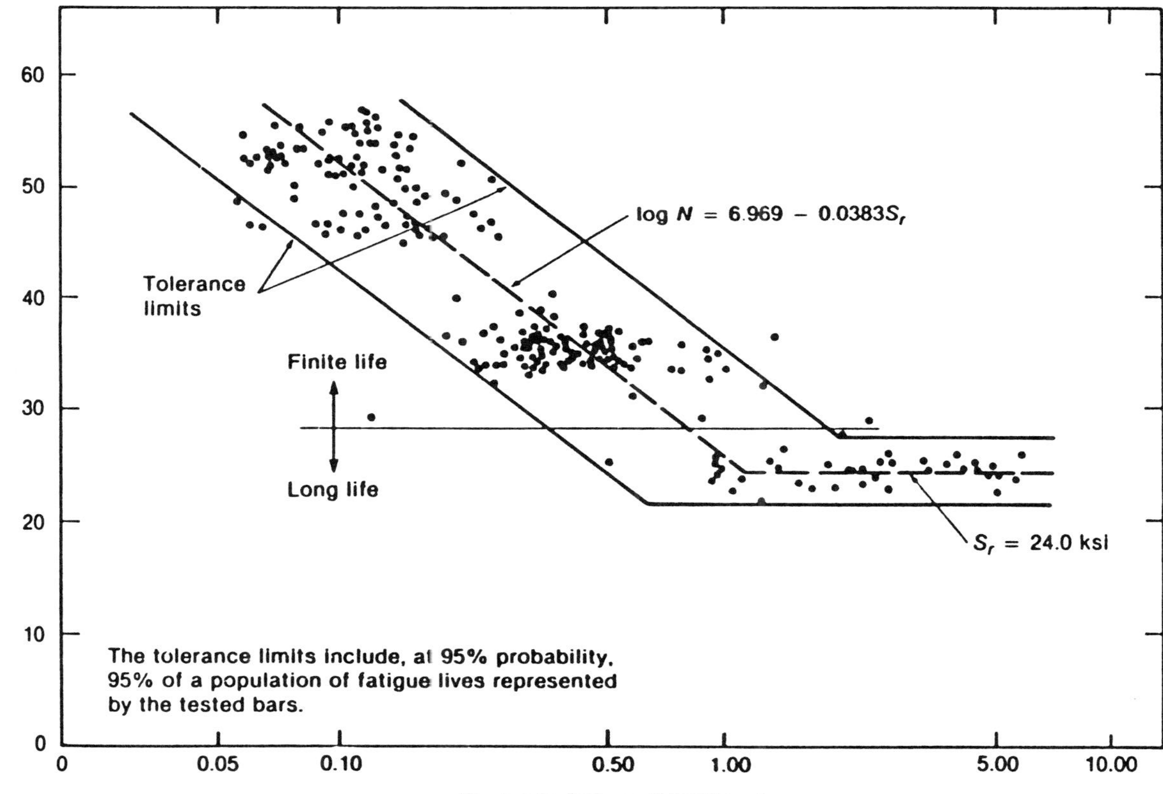

FIGURE 12-61 Typical data on fatigue of deformed bars, showing S–N diagrams.

height of the lugs is decreased. The fatigue strength is basically independent of the yield strength, but is markedly reduced by bends or tack welds in the region of maximum stress. These may reduce fatigue strength by 50 percent.

Fatigue Tests Among early programs are the supplemental tests carried out by AASHTO (1962) following completion of the road test. These resulted in some fractures in fatigue after repeated application of heavy live loads. Following the introduction of high-yield stress reinforcing bars, load factor design methods, and overloads, more tests were initiated to relate fatigue fracture to reinforcing bars. An authoritative report cane from ACI Committee 215 (1974) that dealt with the state of the art on fatigue of plain and reinforced concrete, and provided the basis for the stress range included in the AASHTO 1974 interim specifications.

Another major problem is reported by Helgason et al. (1976) disseminated as NCHRP Report 164. This included a review of existing literature, tests on 353 deformed bars embedded within a concrete beam, and a statistical analysis of the results. Details of these tests are summarized by Corley, Hanson, and Helgason (1978).

It was concluded that the stress range is the predominant factor affecting the fatigue life of reinforcing bars. Thus, a limiting stress range exists above which the bar will have a finite fatigue life and is certain to fracture. At stress ranges below the limit, the bar should be expected to have a long fatigue life and sustain a practically unlimited number of stress cycles. A mean fatigue limit of 5,000,000 cycles was determined for #8 Grade 60 bars.

Results also showed that the magnitude of the mean fatigue limit depends on the geometry of the transverse lugs. This finding was retained in developing design criteria and is articulated by the r/h ratio in (12-61). For #8 Grade 60 bars, the mean fatigue limit at 5,000,000 cycles ranged from 23.0 to 28.5 ksi when the minimum stress level was 6 ksi tension. For #11 Grade 60 bars, fatigue fracture occurred at 1,250,000 cycles under a stress range of 21.3 ksi and a minimum stress of 17.5 ksi tension.

Design Criteria The basic design criteria expressed by (12-61) were developed from a statistical analysis of observed tests results in NCHRP Report 164. The effect of minimum stress $f_{\min}$ and lug geometry r/h on the limiting stress range is shown in Figure 12-62. In these diagrams, lug geometry is assumed to be covered for r/h values between 0.3 and 1.0. The dimensions r and h are clearly identified in Figure 12-62b. Normally, an r/h value of 0.3 would be used because in most cases this ratio is not known. We should note, however, that the maximum allowable stress range, for $r/h = 1.0$ and subjected to tensile stresses only, would be at least 29 ksi (corresponding to $f_{\min} = 0$), which is near the service stress limit for Grade 60 bars under load factor design. This may suggest that improving the geometry and surface characteristics of deformed bars will place fatigue out of the design considerations.

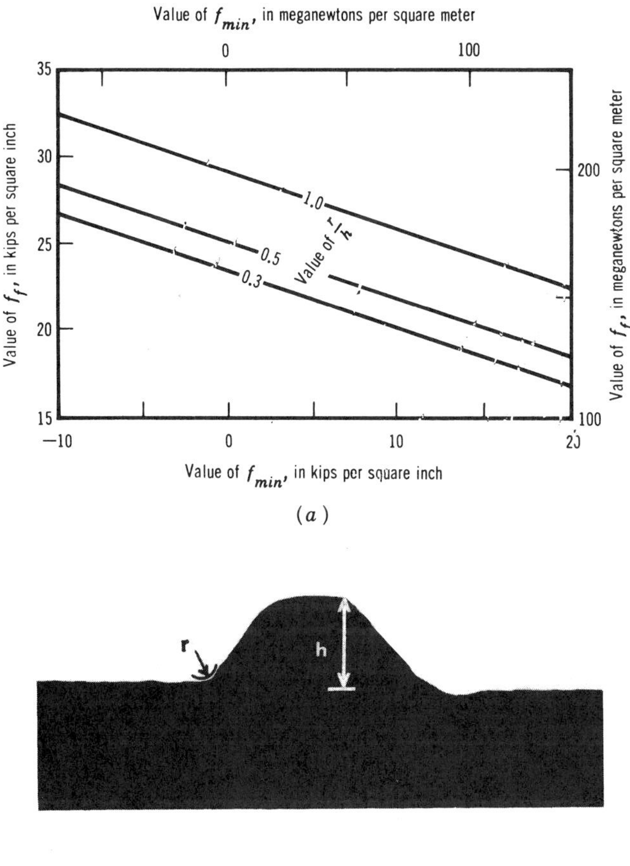

FIGURE 12-62 (*a*) Variation in fatigue stress range f_f with $f_{\min}$ and ratios r/h; (*b*) lug profile showing geometry and parameters r and h.

Design Example

The bridge shown in Figure 12-63 is a two-lane, three-span reinforced concrete slab with spans of 33.5, 40, and 33.5 ft. The superstructure is designed for HS 20 loading plus impact. The objective is to check the design for positive moment reinforcment and cutoff points. Relevant material properties are $f_c' = 3000$ psi and $f_y = 60{,}000$ psi.

The moment envelope curves for service loads are shown in Figure 12-64 and have been obtained from a previous design. These data are tabulated in Table 12-7 and indicate that the controlling section is at point 0.4 of the span.

The reinforcement required for strength in positive moment in the end span is calculated as $A_s = 1.33$ in^2/ft, based on $d = 15.5$ in. The corre-

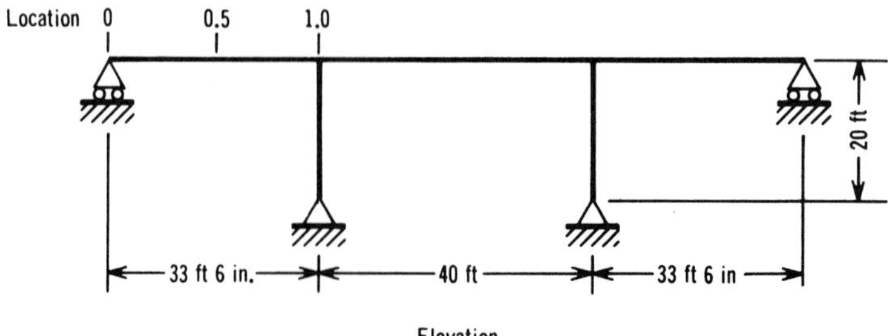

FIGURE 12-63 Elevation and geometry for bridge of design example. (From PCI, 1974.)

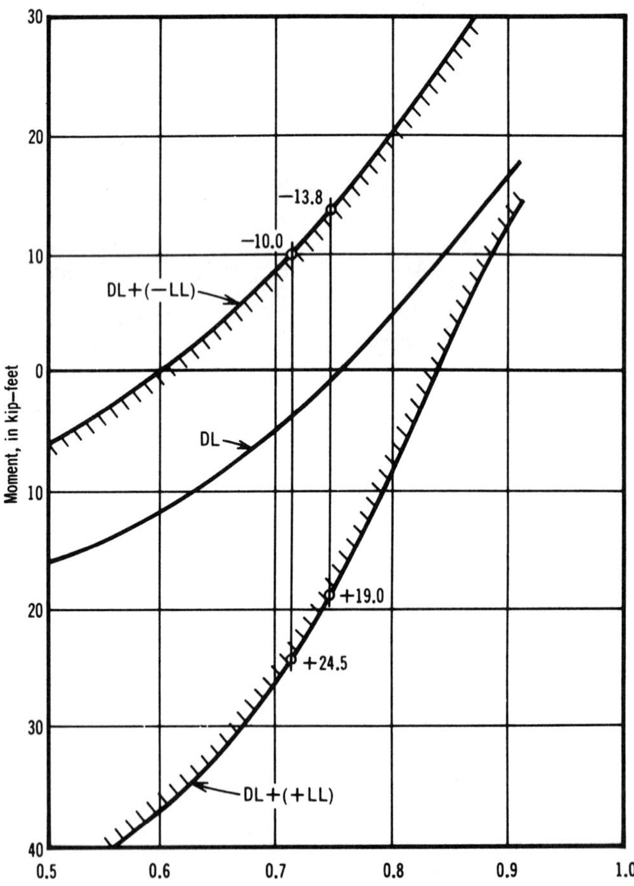

FIGURE 12-64 Service load moment envelopes at interior support of end span.

TABLE 12-7 Extreme Service Load Moments (ft-kips)

Span Location (1)	$D + (L + I)$ (2)	$D + [-(L + I)]$ (3)	Live Load Moment Range (4)
0.3	45.2	11.4	33.8
0.4	46.5	10.2	36.3
0.5	43.4	6.4	37.0
0.6	37.7	0	37.7
0.7	26.2	−8.8	35.0
0.8	8.8	−20.1	28.9
0.9	−11.2	−33.9	−22.7
1.0	−25.9	−54.4	−28.5

sponding maximum range for service load stresses in this span is therefore

$$f_s = M/(A_s jd) = 12M/(1.33 \times 0.9 \times 15.5) = 0.65M$$

The allowable stress range computed from (12-61) is 21.2 ksi. Using this value and the moments tabulated in Table 12-7, we compute the area of reinforcement required to satisfy the fatigue stress requirements. The relevant data are shown in Table 12-8. For example, the reinforcement required to satisfy the fatigue requirements at point 0.4 is

$$A_s = M/(f_f jd) = (36.3 \times 12)/(21.2 \times 0.9 \times 15.5) = 1.47 \text{ in.}^2/\text{ft}$$

Cutoff Points Preliminary bar cutoff locations for the positive reinforcement (#8 bars at 6 in.) have been determined by design. At the theoretical cutoff point for one-half the bars (0.72 points), the extreme service load moments are extrapolated from Figure 12-64 as $+M = 24.5$ ft-kips and $-M =$ 10.0 ft-kips. These moments are also obtained by interpolation from Table 12-7. One-half of the total bars (#8 at 12 in., $A_s = 0.79$ in.^{2}) gives a

TABLE 12-8 Area of Reinforcement Required for Fatigue Criteria

Span Location (1)	f_f (kips/in.^{2}) (2)	f_{min} (kips/in.^{2}) (3)	f_f (Allowable) (kips/in.^{2}) (4)	A_s (in.^{2}/ft) (5)
0.3	21.9	7.4	21.0	1.38
0.4	23.5	6.6	21.2	1.47
0.5	23.9	4.1	22.1	1.44
0.6	24.4	0	23.4	1.39
0.7	17.9	1.0	23.7	
0.8	8.1	2.4	24.2	
NA				
NA				

corresponding stress range in the positive reinforcement of

$$f_f = \frac{+M}{A_s jd} + \frac{-M(k - d'/d)}{A_s jd(1 - k)}$$

$$= \frac{24.5 \times 12}{0.79 \times 0.9 \times 15.5} + \frac{10 \times 12[0.3 - (2.5/15.5)]}{0.72 \times 0.9 \times 15.5(1 - 0.3)} = 29.1 \text{ ksi}$$

The allowable stress range according to (12-61) is calculated as

$$f_f = 21.0 + 0.8 + 2.4 = 24.2 \text{ ksi} \qquad \text{NOT OK}$$

Therefore, the bar cutoff must be extended farther along the span. For a second trial, we compute moments at 0.74 of the span, as +19.0 ft-kips and −13.8 ft-kips. Likewise, we compute the corresponding stress range as

$$f_f = \frac{19.0 \times 12}{0.79 \times 0.9 \times 15.5} + \frac{13.8 \times 12[0.3 - (2.5/15.5)]}{0.72 \times 0.9 \times 15.5(1 - 0.3)} = 24.0 \text{ ksi}$$

From (12-61) the allowable stress range is

$$f_f = 21.1 + 1.1 + 2.4 = 24.5 \text{ ksi} > 24.0 \text{ ksi} \qquad \text{OK}$$

The configuration of the reinforcement bars is shown in Figure 12-65. The compressive stress in the bottom bars used as compressive reinforcement is considered equal to the steel modulus times the strain at the level of the reinforcement ($n = E_s/E_c$). The cutoff for the remainder of the positive reinforcement bars is not affected by fatigue requirements.

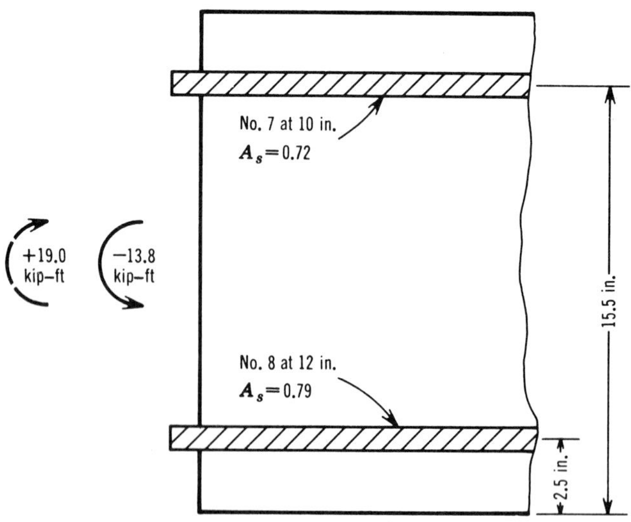

FIGURE 12-65 Bar configuration and slab dimensions near bar cutoff.

REFERENCES

AASHO, 1962. *Standard Specifications for Highway Bridges*, 1962 Edition.

AASHTO, 1978. *Guide Specifications for Fracture Critical Non-Redundant Steel Bridge Members*," *AASHTO, Washington, D.C.*

ACI, 1974. "Considerations for Design of Concrete Structures Subjected to Fatigue Loading," ACI Committee 215, *ACI J., Proc.*, Vol. 71, No. 3, March, pp. 97–121.

Albrecht, P., A. Abtahi, and G. R. Irwin, 1975. "Fatigue Strength of Overloaded Bridge Components," Report No. FHWA-MD-R-76-7, Dept. Civ. Eng., Univ. Maryland, College Park, Oct.

Albrecht, P. and I. M. Friedland, 1979. "Fatigue-Limit Effect on Variable-Amplitude Fatigue of Stiffeners," *J. Struct. Div. ASCE*, Vol. 105, No. ST12, Dec. pp. 2657–2675.

Albrecht, P., and A. H. Naeemi, 1984. "Performance of Weathering Steel in Bridges," NCHRP Report 272.

Anderson, W. J., 1985. "Finite Elements in Mechanical and Structural Design," *Video Lecture Ser. 1002: Dynamic and Nonlinear Analysis*, Automated Analysis Corp., Ann Arbor, Mich., May.

Ang, A. H., 1973. "Structural Risk Analysis and Reliability Based Design," *J. Struct. Div. ASCE*, Vol. 99, No. ST9, Sept., pp. 1891–1910.

Ang, A. H., 1975. "Risk and Reliability Analysis in Engineering Design," *Symp. Struct. and Geotech. Mech.*, Univ. Illinois, Urbana, Oct. 2–3.

Ang, A. H. and C. A. Cornell, 1974. "Reliability Basis of Structural Safety and Design," *J. Struct. Div. ASCE*, Vol. 100, No. ST9, Sept., pp. 1755–1769.

Ang, A. H. and W. H. Tang, 1975. *Probability Concepts in Engineering Planning and Design*, Wiley, New York.

Arkin, H. and R. R. Colton, 1970. *Statistical Methods*, College Outline Series, Barnes and Noble, New York.

ASCE, 1982. "Fatigue and Fracture Reliability," Committee on Fatigue and Fracture Reliability, *J. Struct. Div. ASCE*, Vol. 108, No. ST1, Jan., pp. 3–88.

ASTM, 1963. "A Guide for Fatigue Testing and the Statistical Analysis of Fatigue Data," Committee E-9 on Fatigue, Spec. Publ., No. 91-A, 2nd Ed., Philadelphia.

ASTM, 1981. "Proposed ASTM Test Methods for Measurement of Fatigue Crack Growth Rates," ASTM STP 738, p. 340.

AWS, 1969, 1971. *Welding Specifications for Bridges*, American Welding Society.

Baker, K. A. and G. L. Kulak, 1984. "Fatigue Strength of a Groove Weld on Steel Backing," *Canadian J. Civ. Eng.*, Vol. 11, No. 4, pp. 692–700.

Bakht, B. and P. F. Csagoly, 1980. "Diagnostic Testing of a Bridge," *J. Struct. Div. ASCE*, Vol. 106, No. ST7, July, pp. 1515–1529.

Baldwin, J. W., H. J. Salame, and R. C. Duffield, 1978. "Fatigue Test of a Three-Span Composite Highway Bridge," Report 73-1, Dept. of Civ. Eng., Univ. Missouri, Columbia, June.

Barber, E. S., 1957. "Calculation of Maximum Pavement Temperature From Weather Reports," *Bull.* 168, Hwy. Research Board, pp. 1–8.

Barker, R. M. et al., 1991. "Manuals for the Design of Bridge Foundations," NCHRP 343, TRB, National Research Council, Washington, D.C. Dec.

Barsom, J. M., 1971. "Fatigue-Crack Propagation in Steels of Various Yield Strengths," *J. Eng. Industry ASME*, Nov., pp. 1190–1196.

Batchelor, B. de V., B. E. Hewitt, and P. Csagoly, 1978. "An Investigation of the Fatigue Strength of Deck Slabs of Composite Steel/Concrete Bridges," *Transp. Research Record* 664, TRB, National Research Council, Washington, D.C., pp. 153–161.

Benjamin, J. R. and C. A. Cornell, 1970. *Probability, Statistics, and Decision for Civil Engineers*, McGraw-Hill, New York.

Berger, P., 1982. "Fatigue Testing of Welded Beams," *Proc. IABSE Coll.*, Lausanne, Vol. 37, pp. 323–330.

Berwanger, C., 1970. "Thermal Stresses in Composite Bridges," *Proc. ASCE Specialty Conf. on Steel Structures*, Ser. No. 15, Univ. Missouri, Columbia, June, pp. 27–36.

Bishara, A. G. and S.-Z. Jang, 1980. "Settlement Induced Forces in Concrete Bridges," *J. Struct. Div. ASCE*, Vol. 106, No. ST7, July, pp. 1423–1436.

Black, W., D. S. Moss, and M. Emerson, 1976. "Bridge Temperatures Derived from Measurements of Movement," TRRL Lab. Report 748, Dept. Environment, Berkshire, England.

Bonilla, F. E., et al., 1986. "Composite Action of Precast Panel Bridge Decks in Negative Moment Regions," Research Report 324-2, Texas Transp. Inst., Texas A & M Univ., College Station.

Branson, D. E., et al., 1970. "The Prediction of Creep and Shrinkage Properties of Concrete," Final Report, No. 70-5, Dept. Civ. Eng., Univ. Iowa, Iowa City, Aug.

Brown, J. L., et al., 1978. "Effects of Heavy Trucks on Texas Highways, Interim Report 231 Texas Dept. Highways and Public Transp., Austin.

Bryson, J. O. and R. G. Mathey, 1962. "Surface Condition Effects on Bond Strength of Steel Beams Embedded in Concrete," *ACI J.*, Vol. 59, No. 3, March.

Burdette, E. G. and D. W. Goodpasture, 1971. "Full Scale Bridge Testing: An Evaluation of Bridge Design Criteria," Univ. Tennessee, Knoxville, Dec.

Bussa, S. L., N. J. Sheth, and S. R. Swanton, 1972. "Development of a Random Load Life Prediction Model," *Materials Research and Standards ASTM*, Vol. 12, No. 3, March, pp. 31–43.

Cady, P. D., and R. E. Weyers, 1984. "Deterioration Rates of Concrete Bridge Decks," *J. Transp. Eng. ASCE*, Vol. 110, No. 1, pp. 34–44.

Cicci, F. and P. F., Csagoly, 1974. "Assessment of the Fatigue Life of a Steel Girder Bridge," *Transp. Research Record* 507, TRB, National Research Council, Washington, D.C.

Comeau, M. P., and G. L. Kulak, 1979. "Fatigue Strength of Welded Steel Elements," Tech. Report 79, Univ. Alberta, Canada, Oct.

Corley, W. G., J. M. Hanson, and T. Helgason, 1978. "Design of Reinforced Concrete for Fatigue," *J. Struct. Div. ASCE*, Vol. 104, No. ST6, June, pp. 921–932.

Cornell, C. A., 1969. "Bayesian Statistical Decision Theory and Reliability-Based Design," *Intern. Conf. Str. Safety and Reliab. Eng. Struct.*, Pergamon Press, New York, April, pp. 47–68.

Cudney, G. R., 1968. "Stress Histories of Highway Bridges," *J. Struct. Div. ASCE*, Vol. 94, No. ST12, Dec., pp. 2725–2737.

Douglas, T. R., 1971. "Fatigue of Bridges under Repeated Highway Loadings," Civ. Eng. Dept. Report 54, Univ. Alabama, April.

Duncan, J. M. and C. K. Tan, 1991. "Engineering Manual for Estimating Tolerable Movements of Bridges," NCHRP 343, Transp. Research Board, National Research Council, Washington, D.C.

Ekberg, M., and J. H. Emanuel, 1967. "Thermal Movement of Bridges: A Survey," University of Missouri-Rolla, Oct.

Emerson, M., 1968. "Bridge Temperatures and Movements in the British Isles," Bridge Section-Road Research Lab., RRL Report LR 228, Ministry of Transp.

Emerson, M., 1977. "Temperature Differences in Bridges: Basis of Design Requirements," TRRL Lab. Report 765, Dept. of Transport, Berkshire, England.

Emerson, M., 1979. "Bridge Temperatures for Settling Bearings and Expansion Joints," TRRL Suppl. Report 479, Dept. of Transport, Berkshire, England.

Eugene, J., 1965. "A Statistical Theory of Fatigue Crack Propagation," *Current Aeronautical Fatigue Problems*, Pergamon Press, Inc., New York, p. 215.

European Convention, 1985. "Recommendations for the Fatigue Design of Structures," Committee TC6 Fatigue, European Convention for Construction Steelwork.

Fisher, J. W., 1974. "Guide to 1974 AASHTO Fatigue Specifications," AISC.

Fisher, J. W., 1977. "Bridge Fatigue Guide-Design and Details," American Institute of Steel Construction.

Fisher, J. W., 1981. "Inspecting Steel Bridges for Fatigue Damage," Fritz Eng. Lab. Report 386-15(81), Lehigh Univ., Bethlehem, Pa.

Fisher, J. W., 1983. "Report on Strain Measurements on Woodrow Wilson Bridge," FHWA, U.S. Dept. Transp., Washington, D.C.

Fisher, J. W., et al., 1970. "Effect of Weldments on the Fatigue Strength of Steel Beams," NCHRP Report 102.

Fisher, J. W., et al., 1974. "Fatigue Strength of Steel Beams with Welded Stiffeners and Attachments," NCHRP Report 147.

Fisher, J. W., B. M. Barthelemy, D. R. Mertz, and J. A. Edinger, 1980. "Fatigue Behavior of Full-Scale Welded Bridge Attachments," NCHRP 227, TRB, National Research Council, Washington, D.C., Nov.

Fisher, J. W., J. R. Bellenoit, and J. H. Daniels, 1982. "High Cycle Fatigue Behavior of Steel Bridge Details—Final Report," Fritz Eng. Lab. Report No. 386-13(82), Lehigh Univ., Bethlehem, Pa., Dec. 136 pp.

Fisher, J. W., H. Hausamann, and M. D. Sullivan, 1979. "Detection and Repairs of Fatigue Damage in Welded Highway Bridges," NCHRP Report 206.

Fisher, J. W., D. R. Mertz, and A. Zhong, 1983. "Steel Bridge Members Under Variable Amplitude Long Life Fatigue Loading," NCHRP 267, TRB, National Research Council, Washington, D.C., Dec.

Fisher, J. W., and A. W. Pense, 1984. "Final Report on I-79 Tied Arch Cracking—Nevill Island Bridge," Fritz Eng. Lab. Report 494-1(84), Lehigh Univ., Bethlehem, Pa., Dec.

Fisher, J. W., and J. H. A. Struik, 1974. *Guide to Design Criteria for Bolted and Riveted Joints*, Wiley, New York.

Fisher, J. W., and B. T. Yen, 1977. "Fatigue Strength of Steel Members with Welded Details," *AISC Eng. J.*, vol. 14, 4th Quarter.

Fisher, J. W., B. T. Yen, and D. C. Wagner, 1978: "Review of Field Measurements for Distortion Induced Fatigue Cracking in Steel Bridges," Transp. Research Record 1118, TRB, National Research Council, Washington, D.C., pp. 49–55.

Flügge, W., 1975. *Viscoelasticity*, Blaisdel, Waltham, Mass.

Freudenthal, A. M., 1968. "Prediction of Fatigue Failure," *J. Appl. Phys.*, Vol. 31, No. 12, Dec., pp. 2196–2198.

Friedland, I. M., P. Albrecht, and G. R. Irwin, 1979. "Fatigue Behavior of 2-yr Weathered A588 Specimens with Stiffeners and Attachments," Report No. FHWA-MD-R-79-5, Dept. Civ. Eng., Univ. Maryland, College Park.

Fuchs, H. O. and R. I. Stephens, 1980. *Metal Fatigue in Engineering*, Wiley, New York.

Goble, G. G., F. Moses, and A. Pavia, 1974. "Field Measurements and Laboratory Testing of Bridge Components," Report No. Ohio-DOT-08-74, Dept. Solid Mech., Structures and Mechanical Design, Case Western Reserve Univ., Cleveland, Jan.

Goerg, P., 1963. "Über die Aussagefahigkeit von Dauerversuchen mit Prufkörpern aus Baustahl ST37 und ST52," *Stahlbau*, Vol. 32, No. 2.

"A Guide for Fatigue Testing and the Statistical Analysis of Fatigue Data," ASTM Special Technical Publication 91-A, American Society for Testing and Materials, 1963.

Gumbel, E. J., 1958. *Statistics of Extremes*, Columbia Univ., New York.

Gumbel, E. J., 1963. "Parameters in Distribution of Fatigue Life," *J. Eng. Mech. Div.*, ASCE, Vol. 89, No. EM5, Proc. Paper 3667, Oct., pp. 45–63.

Gumbel, E. J., 1966. *Statistics of Extremes*, Columbia Univ., New York.

Hansen, N. G., 1959. "Fatigue Tests of Joints of High Strength Steels," *J. Struct. Div. ASCE*, Vol. 85, No. ST3, Part 1, March.

Heins, C. P. and A. D. Sartwell, 1969. "Tabulation of Dynamic Strain Data on Three-Span Continuous Bridge Structures," Civ. Eng. Report No. 33, Univ. Maryland, College Park, Nov.

Heins, C. P., and C. H. Yoo, 1970. "Grid Analysis of Orthotropic Bridges," *J. IABSE*, Vol. 30, No. I, pp. 73–91.

Heins, C. P., and J. T. C. Kuo, 1975. "Ultimate Live Load Distribution Factor for Bridges," *J. Struct. Div.*, ASCE, Vol. 101, ST7, July, pp. 1481–1495.

Helgason, T., et al., 1976. "Fatigue Strength of High-Yield Reinforcing Bars," NCHRP Report 164, TRB, Washington, D.C.

Hertzberg, R. W., 1976. *Deformation and Fracture Mechanics of Engineering Materials*, Wiley, New York.

Heywood, R. B., 1962. *Designing Against Fatigue in Metals*, Reinhold, New York.

Highway Research Board, 1962. Special Report 61D: The AASHO Road Test-Bridge Research, HRB, National Research Council, Washington, D.C.

Hillberry, B. M., 1970. "Fatigue Life of 2024-T3 Aluminum Alloy Due to Under Narrow and Random Loading," *ASTM Special Technical Publication* 462, pp. 167–183.

Hirt, M. A. and M. Crisinel, 1975. "La résistance à la fatigue des poutres an âme pleine composées-soudées: Effect des plaguettes et groussels soudes a l'aile," ICOM 017, Swiss Federal Inst., Lausanne.

Hourigan, E. V. and R. C. Holt, 1987. "Design of a Rolled Beam Bridge by New AASHTO Guide Specification for Compact Braced Sections," *AISC Eng. J.*, First Quarter.

Hunt, B. and N. Cooke, 1975. "Thermal Calculations for Bridge Design," *J. Struct. Div. ASCE*, Vol. 101, No. ST9, Sept., pp. 1763–1781.

Imbsen, R. A., D. E. Vandershaf, R. A. Schamber, and R. V. Nutt, 1985. "Thermal Effects in Concrete Bridge Superstructures," NCHRP Report 276, TRB, Washington, D.C., Sept.

Institute for Transportation Engineers, 1976. *Transportation and Traffic Engineering Handbook*, Prentice-Hall, Englewood Cliffs, N.J.

IUR, 1971. "Bending Tests of Structures Consisting of Two Beams Welded at Right Angles," ORE Report D86, Office of Research and Experiments, Intern. Union of Railways.

James, R. W., R. A. Zimmerman, and C. R. McCreary, 1987. "Effects of Overloads on Deterioration of Concrete Bridges," *Transp. Research Record* 1118, TRB, National Research Council, Washington, D.C., pp. 65–72.

Johnson, R. P. and R. J. Buckby, 1986. "Compositive Structures of Steel and Concrete, Volume 2—Bridges," 2nd ed., William Collins Sons, London.

Keating, P. B. and J. W. Fisher, 1986. "Evaluation of Fatigue Tests and Design Criteria on Welded Details," NCHRP 286, TRB, National Research Council, Washington, D.C., Sept.

Keating, P. B. and J. W. Fisher, 1987. "Fatigue Behavior of Variable Loaded Bridge Details Near the Fatigue Limit," *Transp. Research Record* 1118, TRB, National Research Council, Washington, D.C., pp. 56–64.

Keene, P., 1978. "Tolerable Movements of Bridge Foundations," *Transp. Research Record* 678, TRB, National Research Council, Washington, D.C.

Kissane, R. J., 1985. "Lateral Restraint on Noncomposite Beams," Research Report 123, Eng. Research and Development Bureau, New York Dept. Transp., Albany, Aug.

Koob, M. J., P. D. Frey, and J. M. Hanson, 1985. "Evaluation of Web Cracking at Floor Beam to Stiffener Connections of the Poplar Street Bridge Approaches," Illinois Dept. Transp., Springfield, Sept.

Korbacher, G. K., 1971. "On the Modality of Fatigue-Endurance Distributions," *Experimental Mech.*, Dec., pp. 540–547.

Lanigan, A. G., 1973. "The Temperature Response of Concrete Box Girder Bridges," Ph.D. Dissertation, Univ. Auckland, New Zealand.

Lee, J. J., and C. Castiglioni, 1986. "Displacement Induced Stresses in Multigirder Steel Bridges," Fritz Eng. Lab. Report 500-1(86), Lehigh Univ., Bethlehem, Pa., Feb.

Little, R. E. and E. H. Jebe, 1975. *Statistical Designs of Fatigue Experiments*, Wiley, New York.

Liu, Y. N. and W. Zuk, 1963. "Thermoelastic Effects in Prestressed Flexural Members," *PCI J.*, Vol. 8, No. 3, June, pp. 64–85.

Maddox, S. J., 1982. "Improving the Fatigue Lives of Fillet Welds by Shot Peening," *Proc. IABSE Colloquium, Lausanne, Fatigue of Steel and Concrete Structures*, pp. 377–384.

Maeda, Y., et al., 1981. "Deterioration of Highway Bridges in Japan," Technology Reports, Osaka University, Vol. 31, No. 1599, Osaka, Japan, March, pp. 135–144.

Maher, D. R. H., 1970. "The Effects of Differential Temperature on Continuous Prestressed Concrete Bridges," *Civ. Eng. Trans.*, Instit. Engineers, Vol. CE12, No. 1, Paper 2793, pp. 29–32, Australia.

Maier, 1935. *Einfluss des Spannungzustandes auf das Formanderungsvermögen der metallischen Werkstoffe*, Berlin.

Miki, C., T. Nishimura, J. Tajima, and A. Okukawa, 1980. "Fatigue Strength of Steel Members Having Longitudinal Single-Bevel-Groove Welds," *Trans. Japan Welding Soc.* Vol. 11, No. 1, April, pp. 43–56.

Miki, C., F. Nishino, Y. Harabayashi, and H. Ohga, 1982. "Fatigue Strength of Longitudinal Welded Joints Containing Blowholes," *Proc. Japan Soc. Civ. Eng.*, No. 325, Sept., pp. 155–165.

Miki, C., F. Nishino, T. Sasaki, and T. Mori, 1983. "Influence of Root Irregularity of Fatigue Strength of Partially-Penetrated Longitudinal Welds," *Proc. Japan Soc. Civ. Eng.*, No. 337, Sept., pp. 223–226.

Miki, C., J. Tajima, K. Asahi, and H. Takenouchi, 1982. "Fatigue of Large-Size Longitudinal Butt Welds with Partial Penetration," *Proc. Japan Soc. Civ. Eng.*, No. 322, June, pp. 143–156.

Mindlin, H., 1968. "Influence of Details on Fatigue Behavior of Structures," *J. of Struct. Div. ASCE*, Vol. 94, No. ST12, Dec.

Minner, H. H., and T. Seeger, 1982. "Improvement of Fatigue Life of Welded Beams by TIG-Dressing," *Proc. IABSE Colloquium–Lausanne*, pp. 385–392.

Moulton, L. K., H. V. S. GangaRao, and G. T. Halvorsen, 1985. "Tolerable Movement Criteria for Highway Bridges," Report No. FHWA/RD-85/107, FHWA, Washington, D.C., 118 pp.

Munse, W. H. 1964. "Fatigue of Welded Steel Strucutres," Welding Research Council.

Munse W. H., and A. Stallmoyer, 1962. "Effect of End Geometry of Cover Plates on Fatigue," Welding Research Council.

Nadai, A., 1931. *Plasticity*, McGraw-Hill, New York.

Nadai, A., 1937. "Plastic Behavior of Metals in the Strain Hardening Range," *J. Appl. Phys.* pp. 205–213.

Nadai, A., 1950. *Theory of Flow and Fracture of Solids*, McGraw-Hill, New York.

Mueller, J. A. and B. T. Yen, 1968. "Girder Web Boundary Stresses and Fatigue," *Welding Research Council Bull.* 127.

Munse, W. H., 1964. "Fatigue of Welded Steel Structures," Welding Research Council, New York.

Oehler, L. T., 1957. "Vibration Susceptibilities of Various Highway Bridge Types," *J. Struct. Div. ASCE*, Vol. 83, No. ST4. July.

Paris, P. and F. Erdogan, 1970. "A Critical Analysis of Crack Propagation Laws," *J. Basic Eng. ASME*, Dec.

Parola, J. F., Chesson, Jr., E., and W. H. Munse, 1964. "Effect of Bearing Pressure on Fatigue Strength of Riveted Connections," *Civ. Eng. Studies*, Structural Research Series No. 286, Univ. of Illinois, Dec.

Patel, N. J., 1984. "Testing and Evaluation of Irvine Creek Bridge," SRR-84-11, Research and Development Branch, Ontario Ministry Transp. Communications, Canada.

Pauw, A., 1971. "Time Dependent Deflections of a Box Girder Bridge," ACI SP27, pp. 141–158.

PCA, 1974. "Concrete Bridges under Repetitive Loading," Chicago.

Pennsylvania Department of Transportation, 1988. "Guidelines for Fatigue and Fracture Safety Inspection of Bridges," Bridge Management Systems Div., Pennsylvania Dept. Transp.

Precast Segmental Box Girder Bridge Manual. Post Tensioning Institute and Prestressed Concrete Institute (1978) 116 pp.

Priestley, M. J. N., 1972. "Effects of Transverse Temperature Gradients on Bridges," Report No. 394, Ministry of Works, New Zealand, Sept.

Priestley, M. J. N., 1972. "Thermal Gradients in Bridges—Some Design Considerations," *New Zealand Eng.*, Vol. 27, No. 7, July, pp. 228–233.

Priestley, M. J. N., 1976. "Design Thermal Gradients for Concrete Bridges," *New Zealand Eng.* Vol. 31, Part 9, Sept., pp. 213–219.

Priestley, M. J. N., 1976. "Linear Heat Flow and Thermal Stress Analysis of Concrete Bridge Decks," Research Report No. 76/3, Univ. Canterburgy Christchurch, New Zealand, Feb.

Priestley, M. J. N. and I. G. Buckle, 1979. "Ambient Thermal Response of Concrete Bridges," *Road Research Unit Bull.* 42, *Bridge Seminar*, Vol. 2, Road Research Unit, New Zealand.

Rahman, F. and K. P. George, 1979. "Thermal Stress Analysis in Continuous Skew Bridge," *J. Struct. Div. ASCE*, Vol. 105, No. ST7, July, pp. 1525–1542.

Rahman, F. and K. P. George, 1980. "Thermal Stresses in Skew Bridges by Model Tests," *J. Struct. Div. ASCE*, Vol. 106, No. ST1, Jan., pp. 39–58.

Rascon Chavez, O. A., 1967. "Stochastic Model to Fatigue," *J. Eng. Mech. Div. ASCE*, Vol. 93, No. EM3, Proc. Paper 5293, June, pp. 147–156.

Rashid, Y. R., 1986. "Ultimate Strength Analysis of Prestressed Concrete Pressure Vessels," *Nuclear Eng. and Design*, Vol. 7, pp. 334–344.

Reemsnyder, H. S., 1969. "Procurement and Analysis of Structural Fatigue Data," *J. Struct. Div. ASCE*, Vol. 95, No. ST7, July.

Reynolds, J. C. and J. H. Emanuel, 1974. "Thermal Stresses and Movements in Bridges," *J. Struct. Div. ASCE*, Vol. 100, No. ST1, Jan., pp. 63–78.

Roeder, C. W., and L. Eltvik, 1985. "An Experimental Evaluation of Autostress Design," *Transp. Research Record* 1044, TRB.

Rolfe, S. T. and J. M. Barsom, 1977. *Fracture and Fatigue Control of Structures*, Prentice-Hall, Englewood Cliffs, N.J.

Schilling, C. G., 1982. "Impact Factors for Fatigue Design," *J. Struct. Div. ASCE*, Vol. 108, No. ST9, Sept., pp. 2034–2044.

Schilling, C. G., 1982. "Lateral-Distribution Factors For Fatigue Design," *J. Struct. Div. ASCE*, Vol. 108, No. ST9, Sept., pp. 2015–2032.

Schilling, C. G. and K. H. Klippstein, 1972. "NCHRP Project 12-12: Stress Spectrums," U.S. Steel Corp., Res. Lab. Report 76.019-001(3), Sept.

Schilling, C. G. K. H. Klippstein, J. M. Barsom, and G. T. Blake, 1978. "Fatigue of Welded Steel Bridge Members Under Variable-Amplitude Loadings," NCHRP Report 188, TRB, National Research Council, Washington, D.C.

Shanafelt, G. O. and W. B. Horn, 1985. "Guidelines for Evaluation and Repair of Prestressed Concrete Bridge Members," NCHRP Report 280, TRB, National Research Council, Washington, D.C.

Shimokawa, H., K. Takena, F. Ho, and C. Miki, 1984. "Effects of Stress Ratio on the Fatigue Strength of Cruciform Fillet Welded Joints," *Proc. Japan Soc. Civ. Eng.*, No. 344, April, pp. 121–128.

Smith, I. F. C., U. Bremen, and M. A. Hirt, 1984. "Fatigue Thresholds and Improvement of Welded Connections," ICOM 125, Swiss Federal Inst., Lausanne, March.

Suetoh, S., E. G. Burdette, D. W. Goodpasture, and H. Deatherage, 1990. "Unintended Composite Action in Highway Bridges," *Transp. Research Record* 1275, TRB, National Research Council, Washington, D.C., 89–94

Swanson, S. R., 1968. "Random Load Fatigue Testing: A State of the Art Survey," *Materials Research and Standards ASTM*, April, pp. 10–44.

Sweet, A. L. and F. Kozing, 1968. "Investigation of a Random Cumulative Damage Theory," *J. Materials ASTM*, Vol. 3, No. 4, Dec., pp. 802–823.

Tajima, J., A. Okukawa, M. Sugizaki, and H. Takenouchi, 1977. "Fatigue Tests of Panel Point Structures of Truss Made of 80 kg/mn High Strength Steel," IIW DOC. XIII-831-77, July, 37 pp.

Tantani, H. M., 1986. "Ultimate Strength of Highway Girder Bridges," Ph.D. Thesis, Dept. Civ. Eng., Univ. Michigan, Ann Arbor.

Tiffany, C. and J. Masters, 1965. "Applied Fracture Mechanics," *Fracture Toughness Testing*, STP 381, ASTM.

TRRL, 1979. "Fatigue Behavior of Intermittent Fillet Welds," Tech. Report LF860, Transport and Road Research Lab.

United States Steel Corporation, 1982. "Fatigue," *USS Highway Structurés Design Handbook*, Vol. 1, USS, Pittsburgh.

United States Steel Corporation, 1983. "Determining the Need for Lateral Wind Bracing in Plate Girder Bridges," *U.S. Steel. Hwy. Structures Design Handbook*, Vol. II, U.S. Steel, Pittsburgh.

Viest, I. M., 1960. "Review of Research on Composite Steel-Concrete Beams," *J. Struct. Div. ASCE*, Vol. 86, No. ST6, June, pp. 1–21.

Viest, I. M., Siess, C. P., Appleton, J. H., and N. M. Newmark, 1952. "Full-Scale Tests of Channel Shear Connectors and Composite T-Beams," *Bull. 405*, Univ. of Illinois Engineering Experiment Station, Urbana.

Wah, T. and R. E. Kirksey, 1969. "Thermal Characteristics of Highway Bridges," Report No. HR 12-4, Southwest Research Inst., San Antonio, Texas, July.

Walkinshaw, J. L., 1978. "Survey of Bridge Movements in Western United States," *Transp. Research Record* 678, TRB, National Research Council, Washington, D.C.

Wegmuller, A. W., 1977. "Overload Behavior of Composite Steel-Concrete Bridges," *J. Struct. Div. ASCE*, Vol. 103, No. ST9, pp. 1799–1819.

Weibull, W., 1939. "A Statistical Theory of the Strength of Materials," *Proc. Royal Swedish Inst. Eng.* Research, No. 151, Stockholm.

Weibull, W., 1961. *Fatigue Testing and Analysis of Results*, Pergamon Press, Oxford, England.

Weibull, W., 1966. "Analysis of Fatigue Test Results," *First Seminar on Fatigue and Fatigue Design*, Columbia University, New York, pp. 68–69.

Wheeler, O. F., 1972. *J. Basic Eng. ASME*, March, pp. 181–186.

Will, K. M., C. P. Johnson, and H. Matlock, 1977. "Analytical and Experimental Investigation of the Thermal Response of Highway Bridges," Research Report 23-2, Texas Dept. Highways and Public Transp., Univ. Texas, Austin, Feb., 148 pp.

Willenborg, et al., 1971. "A Crack Growth Retardation Model Using an Effective Stress Concept," AFFDL-TM-FBR-71-1, Wright-Patterson AFB, Ohio.

Wirshing, P. H. and J. T. P. Yao, 1970. "Statistical Methods in Structural Fatigue," *J. Struct. Div. ASCE*, Vol. 96, No. ST6, June, pp. 1201–1219.

Yao, J. T. P., 1974. "Fatigue Reliability and Design," *J. Struct. Div. ASCE*, Vol. 100, No. ST9, Sept., pp. 1827–1836.

Yao, J. T. P., 1980. "Reliability Considerations for Fatigue Analysis and Design of Structures," School Civ. Eng., Purdue Univ., June.

Yamada, K., 1975. "Fatigue Behavior of Structural Components Subjected to Variable Amplitude Loading," Ph.D. Thesis, Univ. Maryland, College Park.

Yamada, K. and P. Albrecht, 1975. "A Collection of Live Load Stress Histograms of U.S. Highway Bridges," C. E. Report, Univ. Maryland, College Park.

Yamada, K. and P. Albrecht, 1976. "Fatigue Design of Welded Bridge Details for Service Stresses," *Transp. Research Record* 607, TRB, National Research Council, Washington, D.C.

Yamada, K. and P. Albrecht, 1977. "Fatigue Behavior of Two Flange Details," *J. Struct. Div. ASCE*, Vol. 103, No. ST4, April, pp. 781–791.

Yamada, K. and M. A. Hirt, 1982. "Fatigue Life Estimation Using Fracture Mechanics," *Proc. IABSE Colloquium, Lausanne, Fatigue of Steel and Concrete Structures*, pp. 361–368.

Yamada, K. and Y. Kikuchi, 1984. "Fatigue Tests of Weathered Welded Joints," *J. Struct. Div. ASCE*, Vol. 110, No. 9, Sept., pp. 2164–2177.

Yokel, F. Y., 1990. "Proposed Design Criteria for Shallow Bridge Foundations," Report, U.S. Dept. Commerce, National Inst. Standards and Tech., Gaithersburg, Md.

Zhou, J. and A. S. Nowak, 1987. "Nonlinear Analysis of Highway Bridges," *Transp. Research Record* 1118, TRB, National Research Council, Washington, D.C., pp. 25–29.

Zuk, W., 1961. "Thermal and Shrinkage Stresses in Composite Beams," *ACI J.*, Vol. 58, No. 3, Sept., pp. 327–340.

Zuk, W., 1965. "Simplified Design Check of Thermal Stresses in Composite Highway Bridges," *Hwy. Research Record*, No. 103, pp. 10–13.

Zuk, W., 1965. "Thermal Behavior of Composite Bridges—Insulated and Uninsulated," *Hwy. Research Record*, No. 76, pp. 231–253.

CHAPTER 13

BRIDGE DETAILS

13-1 BASIC TYPES OF BEARINGS

Bridge bearings are mechanical devices designed to transmit the loads to substructure elements and to provide for expansion of the superstructure. They are articulated into four basic types: (a) fixed bearings; (b) hinged bearings; (c) sliding or expansion bearings; and (d) hinged, linked, and roller-jointed devices.

A fixed-end bearing is capable of supplying a vertical and a horizontal reaction plus a restraining moment. Considering the expense of fixing a heavy steel member at the ends, such a bearing is not usually selected for bridges. A hinged bearing will permit rotation of the ends of the member, and this is usually accomplished by a pin. Hinges carrying heavy vertical loads are normally provided with lubrication systems to reduce friction and ensure free rotation without excessive wearing.

Expansion bearings permit rotation as well as movement of the superstructure resulting from temperature change, shrinkage, and deflection. These articulations are usually provided in two forms: the sliding joint and the roller joint. With the introduction of new materials and fabrication methods, sliding joints can accommodate longer spans and heavier loads. Devices classified as expansion generally transmit only friction or longitudinal shear force from the movement of the bearing during longitudinal expansion and contraction, but occasionally must also transmit lateral thrust in the transverse direction between the superstructure and the substructure.

When it is necessary to provide a bearing that offers only normal reaction with no rotational or lateral restraint, a combination of hinge and roller bearing is used. The hinge permits rotation and the roller allows expansion.

For articulations within a truss or a heavy girder, hinges are provided by a pin or a link with two pins. A pin will permit rotational movement transmitting also horizontal and vertical reactions. A link will transmit force only along the length of the link and thus control the direction of the reaction. Although it is possible to design an ordinary hinge or roller joint to accommodate the loads of a truss or a heavy girder, the link connection is usually preferred especially in view of the larger movement afforded by this arrangement.

The span beyond which the simple bearing plate is unsatisfactory is largely a matter of judgment and experience. According to most specifications, bridges with spans less than 50 ft will not require provisions for deflection, and may be arranged to slide upon metal plates with smooth surfaces. For larger spans, the bearings must have any of the configurations discussed in the foregoing paragraphs and must be detailed accordingly. These devices should also be designed to provide sufficient restraint to maintain the stability of the superstructure. Failure of bearings to adequately tie the superstructure to the substructure when subjected to earthquake forces contributed to the collapse of several bridges during the 1971 seismic event in southern California.

Bearing devices are discussed in detail in the following sections, supplemented by design examples.

13-2 SPECIFICATIONS RELEVANT TO BEARINGS

Standard AASHTO Specifications Fixed and expansion bearings are discussed in Article 10.29 in terms of general design requirements. For service load design, the structural aspects of pins, rollers, and expansion rockers are defined in Article 10.32.4. Bearing per linear inch on expansion rockers and rollers should not exceed the values determined from the following.

Diameters up to 25 in.:

$$p = \frac{F_y - 13{,}000}{20{,}000} 600d \tag{13-1a}$$

Diameters 25 to 125 in.:

$$p = \frac{F_y - 13{,}000}{20{,}000} 3000\sqrt{d} \tag{13-1b}$$

where p is the allowable bearing (lb/linear inch), d is the diameter of the rocker or roller (in.), and F_y is the minimum steel yield strength in tension in the roller or bearing plate, whichever is the smaller. The allowable unit bearing stress on concrete masonry or pads is discussed in Article 8.15.2.1.3.

Pot Bearings These may be supplied as fixed bearings, guided expansion bearings, or nonguided expansion bearings. They should provide for the thermal expansion and contraction, rotation, camber changes, and creep and shrinkage of structural members where applicable. The design requirements are discussed in Section 19 (Design Division), and the materials, fabrication and installation in Section 18 (Construction Division).

Disk Bearings These devices consist of a polyether urethane structural element (disk) confined by upper and lower steel bearing plates. They are equipped with a shear restriction mechanism to prevent movement of the disk. Likewise, these bearings must accommodate thermal expansion and contraction, rotation, camber changes, and creep and shrinkage where applicable. AASHTO Section 20 (Design Division) defines the design requirements, and Section 18 (Construction Division) discusses the materials, fabrication, and installation procedures.

Elastomeric Bearings These devices (discussed in detail in the following sections) are made partially or wholly from elastomer. Their main function is to transmit loads and accommodate movement between a bridge and its supporting substructure elements. Section 18 (Construction Division) of the standard AASHTO specifications covers the general requirements of plain and reinforced pads as well as materials, fabrication, and installation procedures.

Proposed LRFD Specifications In general, bearings are required to resist all factored loads and accommodate the design translations and rotations of the structure. The design rotation θ_u is taken as the sum of the service limit state rotations caused by all applicable loads, plus the maximum rotation caused by fabrication and installation tolerances, plus 0.01°.

Characteristics Because bearing systems differ in their ability to support these loads and movements, the suitability of a bearing chosen for a particular application may be inferred from Table 13-1. The ratings in this table reflect general judgment and observation and should be used as a guide in choosing different bearing devices. Certain common bearing types corresponding to the notation of Table 13-1 are shown in Figure 13-1.

Restraint of Movement Movement or restraint of movement at the bearing will induce horizontal forces and moments. Horizontal forces are manifested by sliding friction, rolling friction, or deformation of a flexible element in the bearing.

The sliding friction force is computed from

$$H_u = \mu P_u \tag{13-2}$$

TABLE 13-1 Suitability of Bearing Systems (From LRFD Specifications)

	Movement		Rotation			Load Capacity		
Type of Bearing	Long.	Trans.	Trans.	Long.	Vert.	Vert.	Long.	Trans.
Plain elastomeric pad	S	S	S	S	L	L	L	L
Fiberglass reinforced pad	S	S	S	S	L	L	L	L
Cotton duck reinforced pad	U	U	U	U	U	S	L	L
Steel reinforced elastomeric bearing	S	S	S	S	L	S	L	L
Plane sliding bearing	S	S	U	U	S	S	R	R
Curved sliding spherical bearing	R	R	S	S	S	S	R	R
Curved sliding cylindrical bearing	R	R	S	U	U	S	R	R
Disk bearing	R	R	S	S	L	S	S	R
Double cylindrical bearing	R	R	S	S	U	S	R	R
Pot bearing	R	R	S	S	L	S	S	S
Rocker bearing	S	U	S	U	U	S	R	R
Knuckle bearing	U	U	S	U	U	S	S	R
Single roller bearing	S	U	S	U	U	S	U	R
Multiple roller bearing	S	U	U	U	U	S	U	U

S = suitable
U = unsuitable
L = suitable for limited applications
R = may be suitable, but requires special considerations or additional elements such as sliders or guideways
Long. = longitudinal axis
Trans. = Transverse axis
Vert. = vertical axis

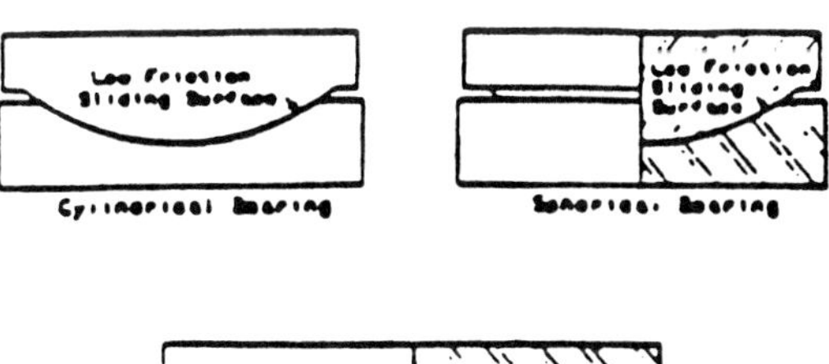

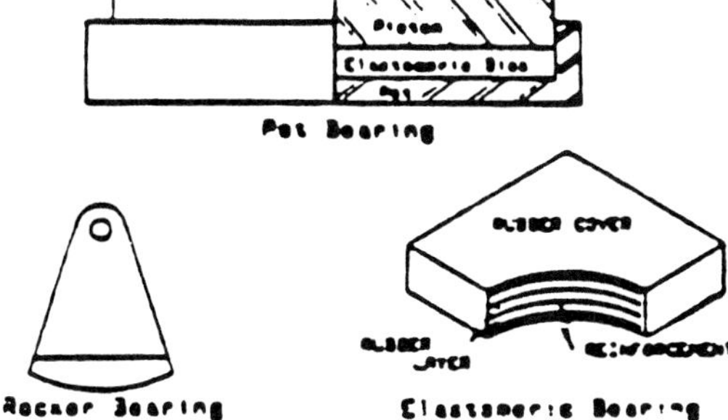

FIGURE 13-1 Common bearing types.

where H_u is the factored horizontal load on the bearing or restraint and P_u is the factored compressive on the bearing at the strength limit state. The rolling friction forces should be determined by tests. The force required to deform an elastomeric element is computed as

$$H_u = GA\frac{\Delta_u}{h_{rt}} \tag{13-3}$$

where G is the shear modulus of the elastomer, A is the plan area of elastomeric bearing, Δ_u is the factored shear deformation of the elastomer under factored load conditions, and h_{rt} is the total elastomer thickness in the elastomeric bearing.

Bridge members affected by the bearing function should be proportioned for a factored moment M_u transferred by the bearing. For curved sliding bearings, M_u is estimated by

$$M_u = \mu P_u R \tag{13-4}$$

where R is the radius of the curved bearing.

For unconfined elastomeric bearings and pads bent about an axis parallel to their long sides, M_u is calculated from

$$M_u = (0.5E_c)\frac{\theta_{rL,x}}{h_{rt}} \tag{13-5}$$

where E_c is the effective modulus in compression and $\theta_{rL,x}$ is the relative rotation of the top and bottom surfaces of the bearing (radians) about the transverse axis under total load.

13-3 SLIDING BEARINGS

Special Characteristics

Until recently, short-span beam bridges were designed to expand and contract by the sliding of one steel plate on another; the rotation of the end of the beam was either neglected or accommodated by a curved bearing surface on one of the plates. Quite often this arrangement resulted in considerable longitudinal forces due to friction between the steel surfaces, which increased markedly as the steel plates rusted.

Since the introduction of the chemical composition Teflon (TFE) or tetrafluorethylene, resistance to sliding is no longer a problem because this material has the lowest coefficient of friction of any solid material. Standard bridge bearings are available with a sliding surface of Teflon combined with other materials to provide appropriate strength. The overall requirement of the design is to produce moderate compressive strength, chemical inertness, and low friction. In the usual forms, this material is bonded to special backing plates of carbon steel, stainless steel, and neoprene. When Teflon sliding surfaces are combined with a neoprene pad, both sliding and rocking are possible.

The interacting sliding surfaces may consist of Teflon on Teflon or one surface of Teflon in contact with a surface of steel (preferably stainless). When stainless-steel mating surfaces are used, they should be polished or rolled as necessary to meet the friction requirements. Where the danger of rusting on the sliding stainless-steel plate exists, both contact surfaces may be Teflon coated. Both arrangements are shown in Figure 13-2.

The usual thickness of a Teflon coating bonded to the steel backing plate is 3/32 in., and this may result in an average coefficient of friction of 0.06 when calculating the longitudinal forces on the substructure (see also the following section). A typical design range of compression on the Teflon layer is 1 to 2 ksi. If a neoprene pad is used under the Teflon, the dimensions of the latter are controlled by the length and width of the neoprene. In order to restrict the tendency of the bridge to move in the transverse direction, anchor bolts can be inserted through the top plate and set in concrete masonry.

TFE bearing surfaces are fully accredited by the AASHTO specifications and covered in Section 18. Testing and acceptance criteria may be specified at the discretion of the designer and should comply with AASHTO Article 18.8.3.

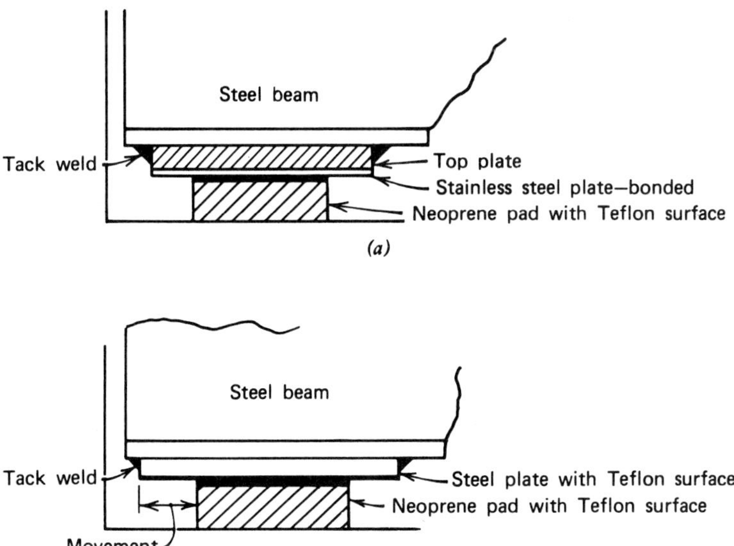

FIGURE 13-2 Expansion bearing arrangements: (*a*) with one thickness of Teflon; (*b*) with two thicknesses of Teflon.

Friction in TFE Sliding Surfaces

A state-of-the-art report (Campbell and Kong, 1987) has identified 14 parameters that affect the coefficient of friction of TFE sliding on a metallic plate. These are lubrication, contact pressure, speed of movement, eccentricity of loading, temperature, creep, roughness of the mating surface, type of TFE, attachment of TFE to the backing plate, surface conditions, length of the travel path, load and travel history, specimen size, and wear. However, these effects are not clearly documented, particularly with regard to the influence of lubricated surfaces.

Campbell, Kong, and Manning (1990) report results of laboratory tests carried out to study the effect of certain parameters considered to be the most influential on the coefficient of friction of TFE. A suggested expression for this coefficient is

$$\mu = QP^{n-1} \tag{13-6}$$

where Q and n are relevant parameters and P is the contact pressure on the TFE. The factor Q is essentially a function of temperature and speed of travel, whereas n is mainly a function of the filler content of the TFE, surface lubrication, surface finish of the metallic sliding plate, and loading history (Taylor, 1972; Price, 1982). The value of n is less than unity.

Test Program The ranges of parameters are given in Table 13-2 and reflect the field conditions to which TFE sliding bearings are likely to be subjected in practice. The maximum pressures of 30 and 45 MPa are specified by the Ontario code for dead load and total load, respectively. The two selected roughness values, measured perpendicular to the direction of polishing, correspond to commercially available #8 and #4 finish stainless-steel plates (ASTM A480-820, 1982).

The specimens consisted of dimpled, lubricated TFE resin with a diameter of 75 mm and a thickness of 4.5 mm, and were recessed to a depth of 2.5 mm in a rigid steel backing plate. A dwell of load period of 12 h was used before testing according to the requirements of the AASHTO specifications. A description of the test procedure is given by Campbell, Kong, and Manning (1990).

TABLE 13-2 Ranges of Parameters: TFE Tests

Parameter	Range
Pressure, P (MPa)	10, 15, 25, 30, 45
Temperature, T (°C)	−25, 20
Speed of travel, V (mm/s)	0.08, 1, 20
Roughness of metallic plate, R (μm)	0.03, 0.34

Test Results The coefficient of friction at the TFE–stainless-steel interface is determined from the ratio of horizontal to vertical load. A typical variation of this coefficient over a complete cycle of movement is shown in Figure 13-3. The static and dynamic coefficients of friction (expressing the forces necessary to start and maintain movement, respectively) are articulated. The static coefficient is the maximum value, and the dynamic coefficient is the minimum value.

Considerable wear of the TFE occurred in a test that involved a smooth stainless-steel plate at a contact pressure of 45 MPa, a sliding speed of 20 mm/s, and a temperature of −25°C. This wear was manifested as a deposit of flakes of TFE on the stainless-steel plate at each end of the stroke, and amounted to a 4 percent loss of weight after 8000 cycles. By contrast, in a corresponding test using rough stainless steel, the loss in weight was only 0.5 percent.

In general, the highest coefficient of friction, both static and dynamic, was recorded during the initial movement. A rapid drop in its magnitude occurred after one cycle of movement and accelerated with the speed of movement. Both coefficients appeared to stabilize and remain fairly constant up to 50 cycles. Data obtained beyond 50 cycles indicate an increase in both coefficients up to about 1000 cycles, after which the values stabilized again and remained constant up to 18,000 cycles.

It appears from these observations that the values of the coefficients of friction after 50 cycles of movement are representative, and this trend is followed by both the static and dynamic values. Data at 50 cycles are shown in Figures 13-4 through 13-6, where the coefficient of friction is plotted versus the contact pressure. Also shown is the best fit to the data of the

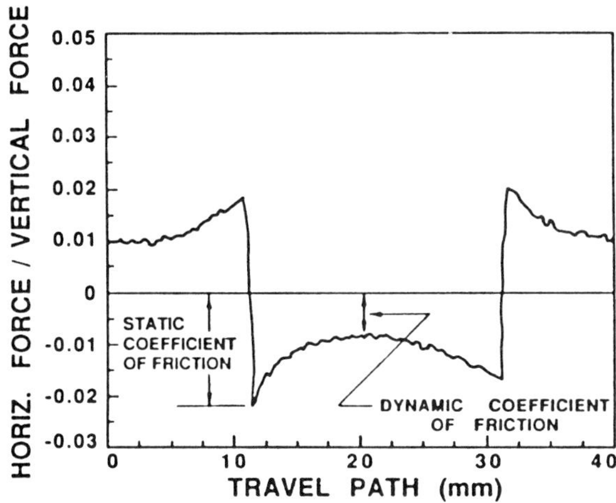

FIGURE 13-3 Variation of the ratio of horizontal to vertical forces over a cycle. (From Campbell, Kong, and Manning, 1990.)

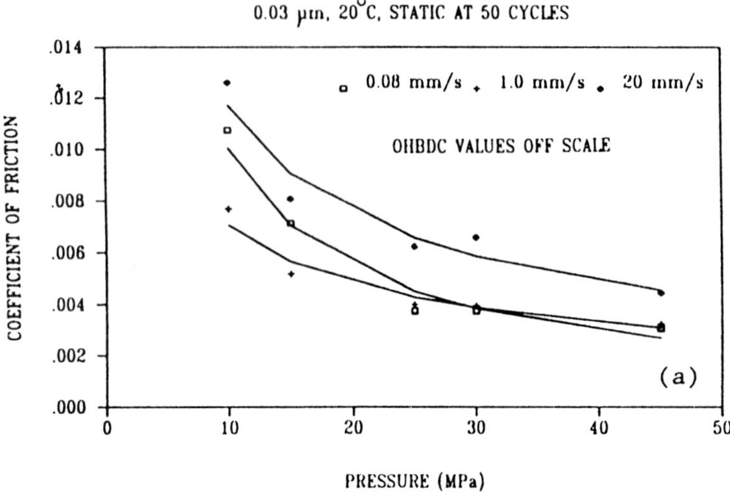

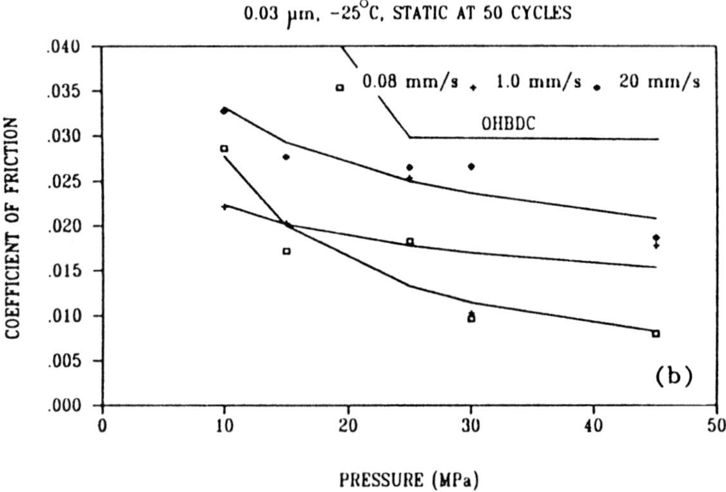

FIGURE 13-4 Influence of pressure, speed of movement, and temperature on the coefficient of friction. (From Campbell, Kong, and Manning, 1990.)

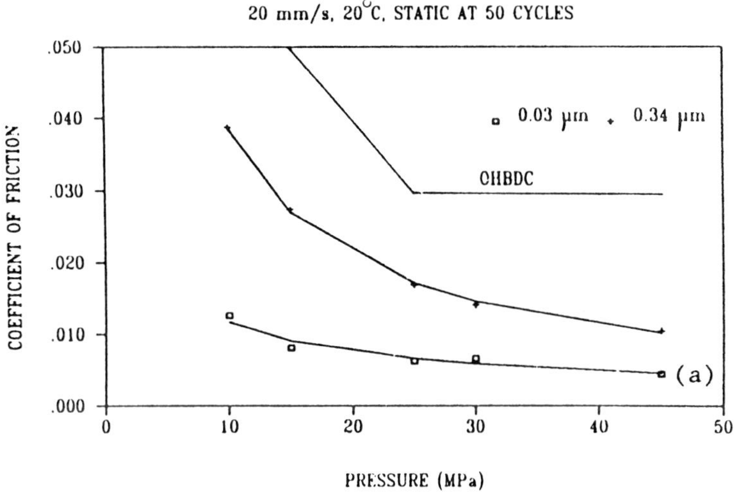

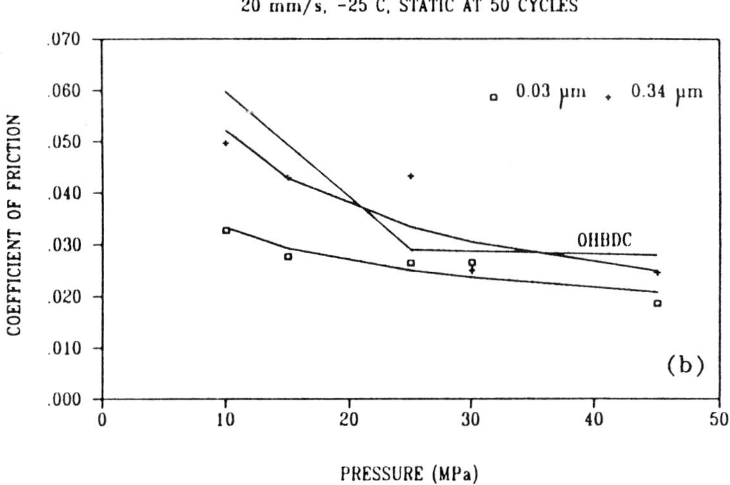

FIGURE 13-5 Influence of pressure, surface roughness, and temperature on the coefficient of friction. (From Campbell, Kong, and Manning, 1990.)

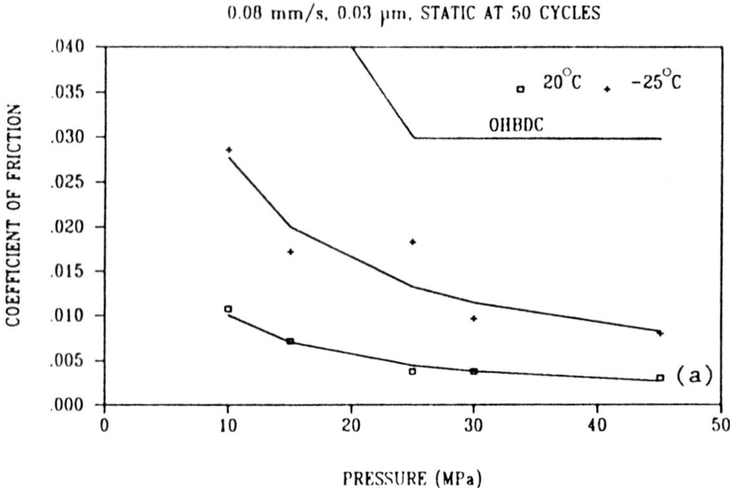

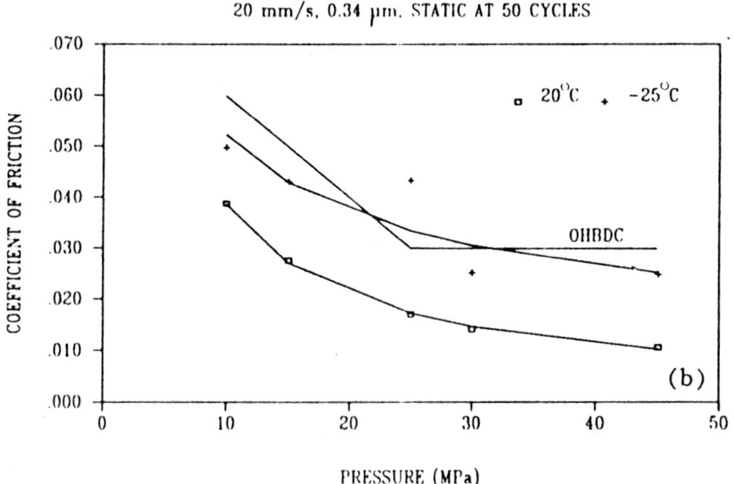

FIGURE 13-6 Influence of pressure, temperature, and surface roughness on the coefficient of friction. (From Campbell, Kong, and Manning, 1990.)

relationship given by (13-6). Reasonable fits are obtained, particularly from the tests at 20°C, but more scatter is apparent from the tests at −25°C. Values of the coefficient of friction specified in the Ontario code (OHBDC) for unfilled lubricated TFE sliding surfaces are also included in the plots.

The effect of speed of movement for a given temperature and surface roughness of the stainless steel is shown in Figure 13-4. The influence of surface roughness for a particular temperature and speed of movement is shown in Figure 13-5. The effect of temperature for a particular speed of movement and surface roughness is indicated in Figure 13-6. All plots confirm the trend that the coefficient of friction of TFE decreases with increasing pressure under all conditions.

Conclusions Campbell, Kong, and Manning (1990) draw the following conclusions.

1. The coefficient of friction decreases with increasing pressure over the range 10 to 45 MPa (1.4–6.4 ksi). A general trend is an increase in this coefficient with sliding speed in the range 0.08 to 20 mm/s.
2. The coefficient of friction increases with roughness of the stainless-steel plate over the range 0.03 to 0.34 μm, and also with decreasing temperature in the range 20 to 25°C.
3. The initial coefficient of friction can be as high as five times the value after 50 cycles of movement.
4. The dynamic coefficient is lower than the static coefficient, but follows the same trend with increasing number of cycles.

Proposed LRFD Specifications

The mating surface specified for use in conjunction with PTFE is stainless steel. Curved mating surfaces should be stainless steel or anodized aluminum. Friction testing is required for the PTFE and its mating surface. Minimum thickness requirements are specified to account for the anticipated wear that usually causes higher friction and reduces the thickness of the remaining PTFE. The thickness of the stainless-steel mating surface should be at least 1/16 in. when the maximum dimension of the surface is less than 12 in., and 1/8 in. when the maximum dimension exceeds 12 in.

Contact Pressure The contact stress between the PTFE and the mating surface is computed at the strength limit state using the nominal area, and is limited to prevent excessive creep or plastic flow of the PTFE. The constant stress at the edge is computed by considering the maximum moment induced by the bearing assuming a linear stress distribution. This stress should not exceed the values given in Table 13-3.

TABLE 13-3 Maximum Strength Limit State Contact Stress for PTFE (psi): LRFD Specifications

Material	Average Contact Stress		Edge Contact Stress	
	Dead Load	All Loads	Dead Load	All Loads
Unconfined PTFE				
Unfilled sheets	2000	3000	2750	3750
Filled sheets[a]	4000	6000	5000	7500
Confined sheet PTFE	4000	6000	5000	7500
Woven PTFE fiber over a metallic substrate	4000	6000	5000	7500

[a]These figures are for maximum filler content. Permissible stresses for intermediate filler contents shall be obtained by linear interpolation.

Coefficient of Friction The design coefficient of friction of the PTFE sliding surface should be determined from Table 13-4. Values somewhat smaller than those given in the table are possible under quality control, but testing is needed to confirm the selected value. Certification testing is thus essential to ensure that the friction actually achieved is appropriate for the bearing and substructure design.

Attachment Recessing is stipulated as the most effective way to prevent creep in unfilled PTFE, and bending helps to retain the PTFE in the recess but makes replacement of the disk more difficult.

The mating surface for flat sliding surfaces should be attached to a backing plate by welding so as to remain flat and in full contact during its service life. The welds used for the attachment should be clear of the contact and sliding area of the PTFE surface.

TABLE 13-4 Design Coefficients of Friction: LRFD Specifications

Material	Average Bearing Stress			
	500 psi	1000 psi	2000 psi	> 3000 psi
Unfilled PTFE				
Unlubricated flat sheet	0.16	0.14	0.12	0.08
Lubricated sheet PTFE	0.10	0.09	0.08	0.06
Lubricated dimpled sheet	0.08	0.07	0.06	0.04
Filled PTFE (sheet or woven)	0.20	0.18	0.15	0.10
Woven fabric from PTFE resin	0.10	0.09	0.08	0.06
Woven fabric from PTFE fiber and metallic substrate	0.08	0.07	0.05	0.04

13-4 ROLLING EXPANSION BEARINGS

General

For considerable loads and when the longitudinal movement cannot be accommodated by the relatively simple bearings, devices consisting of steel rollers are required. For fixed bearings, where longitudinal movement must be prevented, a large steel bearing pin is usually selected. With very large reactions, a set of segmental rollers will satisfy the design requirements. Rolling-type bearings are shown in Figures 13-7*a* and *b*. The segmental rollers shown in Figure 13-7*a* are preferred to round rollers because they occupy less space than the round ones for the same diameter. Because the motion of the rollers is quite small relative to their radius, the arcs that are necessary are also small and the segmental roller is thus efficient. Segmental rollers may be connected with two tie bars as shown to prevent them from moving independently.

Another rolling-type expansion bearing is shown in Figure 13-8*a* and is suitable for larger bridges. Several variations are shown in the view. The sole plate may be bolted to the girder as shown at the left of the figure, or welded as shown at the right. Resistance to uplift is ensured with the use of a hinge plate shown at the left. If such resistance is not needed, lateral movement is prevented by a plate attached as shown at the right. The corresponding fixed bearing is shown in Figure 13-8*b*.

For articulations within a truss or a girder, hinges are provided either as a pin or a link with two pins. A pin permits rotation while also transmitting horizontal and vertical reactions. A link transmits forces only along its length and thus gives the means of controlling the directions of the reactions. In many cases the link connection appears to be the most reliable considering the larger movement afforded in this arrangement.

Usual Problems A common problem in steel bearings is moisture causing rusting. Preventive measures include (a) coating steel surfaces with oil, (b) using steels resistant to rusting, and (c) enclosing the rollers and contact surfaces in a flexible rubber housing. The rusting problem is more severe in regions where the use of salt is mandatory for deicing purposes. Rusting can impede free movement and induce considerable horizontal forces.

In bearings consisting of only one roller, the round surface accommodates the rocking action as well as the longitudinal movement. With two or more rollers, an independent pin must be provided to allow end rotation of the bridge due to deflection. The minimum size of the rollers is stipulated by the applicable specifications.

Design Example: Segmental Roller Bearing

The design will consider four independent bearing parts. The masonry plate rests on the concrete substructure and requires an area of concrete ground

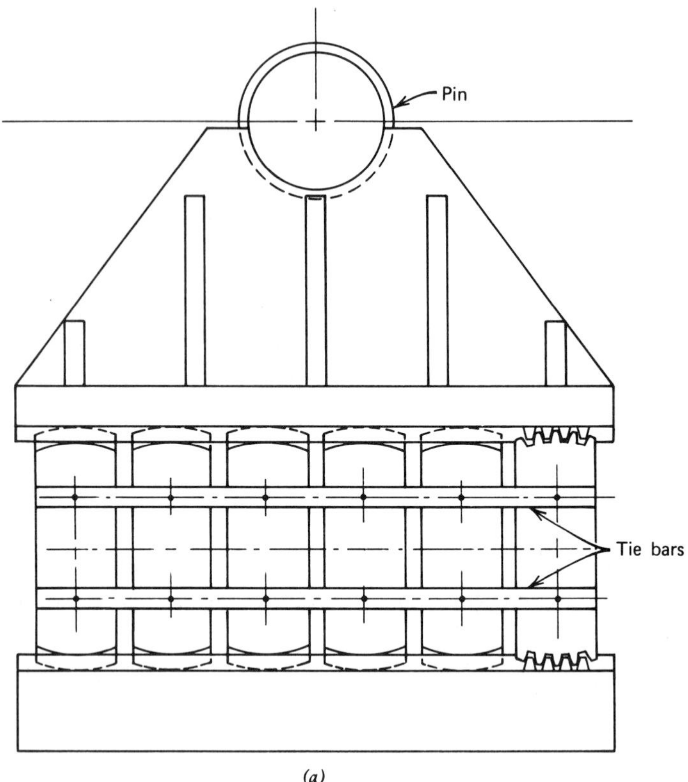

(a)

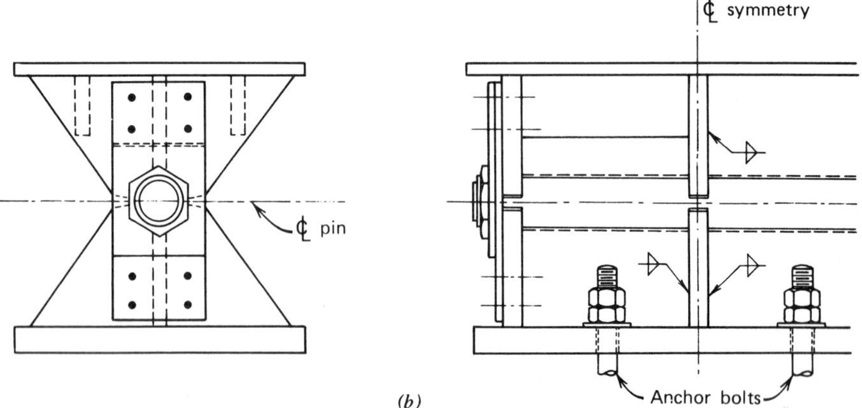

(b)

FIGURE 13-7 Examples of bearings: (*a*) segmental rollers used as expansion bearing; (*b*) pin bearing at fixed locations.

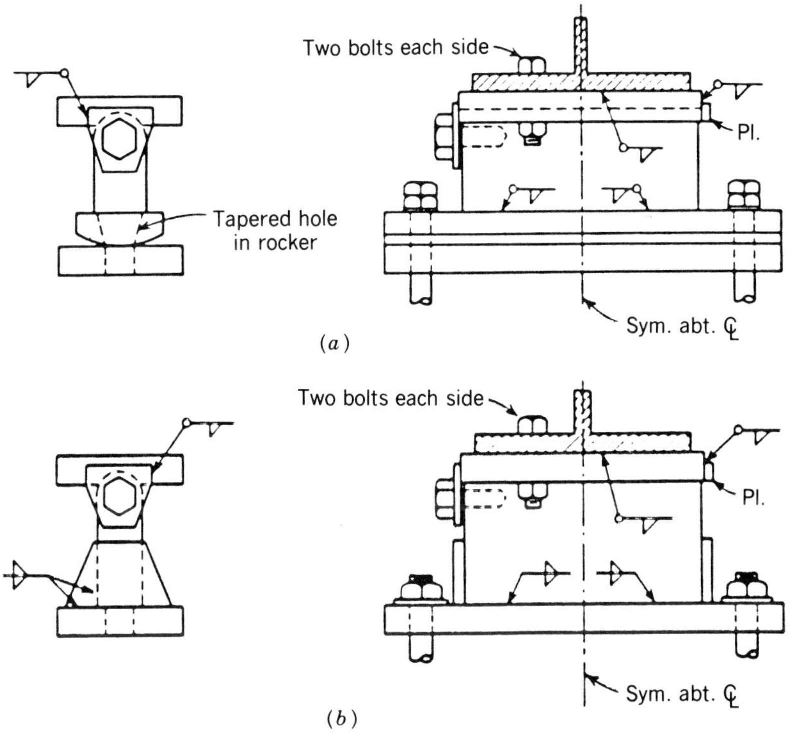

FIGURE 13-8 (*a*) Rolling-type expansion beams; (*b*) fixed bearing.

smooth and level for full contact and uniform bearing. The segmental rollers transfer the vertical load and also permit the horizontal translation of the superstructure with reference to the substructure. The bottom bearing shoe rests on top of the rollers and accepts the reaction from the top bearing shoe through the pin. The forces acting on the assembly are the total reactions from the superstructure.

For this example, the following are given:

Total reaction: 400 kips

Total bridge expansion length: 300 ft

Temperature range: -20 to 100°F; rollers set vertically at 50°F

Movement: $6.5 \times 10^{-6}(70)(300)(12) = 1.64$ in. say 2 in. (conservative)

Bridge inclination (rotation) at the bearing: Assume $1/2°$

Specifications: standard AASHTO

Step 1. *Material and Size of Rollers.* The number, length, and diameter of the rollers are selected to resist the reaction divided by the allowable contact bearing stress.

Using M 102 class G material and $F_y = 50{,}000$ psi, the allowable bearing per inch is obtained from (13-1a) as

$$p = \frac{50 - 13}{20}(600d) = 1110(d)\ \text{lb/in.}$$

Selecting three rollers, each 15 in. long,

$$d = \frac{400{,}000}{1110 \times 15 \times 3} = 8\ \text{in.} \qquad \text{Use 9 in.}$$

For a total movement of 2 in. in either direction, the required circumference length is 4 in., giving an angle of rotation of $2/4.5 = 0.44$ radian $= 25°$, and a roller width of $2 \times 4.5 \times \sin 25° = 3.87$ in.; use 4 in.

Using $f'_c = 3000$ psi for the substructure concrete, the allowable bearing stress f_b is $0.3 \times 3 = 0.9$ ksi, and the required bearing area of the masonry plate is $400/0.9 = 445$ in.^{2}. Note that the minimum size necessary to accommodate the three rollers is 22×20 in. $= 440$ in.^{2} with concrete extending at least 6 in. beyond the edge of the masonry plate. Therefore, the allowable bearing stress may be increased by $\sqrt{A_2/A_1} = 1.57 < 2.0$, or the allowable $f_b = 0.9 \times 1.57 = 1.4$ ksi. The masonry plate dimensions, are therefore 22×20 in.

The thickness of the masonry plate is based on $F_b = 40$ ksi and is computed as 1.05 in.; use 1.25 in. thick. Four 1.5-in.-diameter anchor bolts are provided to fix the plate to the masonry. The general details of the bearing assembly are shown in Figure 13-9a through d.

Step 2. *Selection of Pin.* Because the top and bottom ribs are in the same vertical line, there are no bending stresses in the pin, but only bearing pressure. Equilibrium is achieved when the pin diameter times the total width of the ribs times the allowable bearing stress on the pins subject to rotation is greater than the total reaction.

The allowable bearing on the pins is $0.4F_y = 0.4 \times 50{,}000 = 20{,}000$ psi. The bearing area required is $400/20 = 20$ in.^{2}. For three ribs 1.5 in. thick (each), the required diameter of the pin is $20/(3 \times 1.5) = 4.45$ in.; say 5 in.

The bottom bearing shoe must have sufficient length and width to maintain full contact with the rollers. The thickness of the plate and depth of the bearing shoe are determined by the bending stress at a section through the centerline of the pin. The height of the shoe must be such that a 45° line from the intersection point of the centerline of the outside roller and the bottom plate of the shoe intersects the horizontal diameter of the pin at the edge of the pin. This implies that the load on the rollers is divided equally. Based on this criterion, the configuration of the bottom shoe is as shown in Figure 13-10. Referring to the same figure, the 1-in.

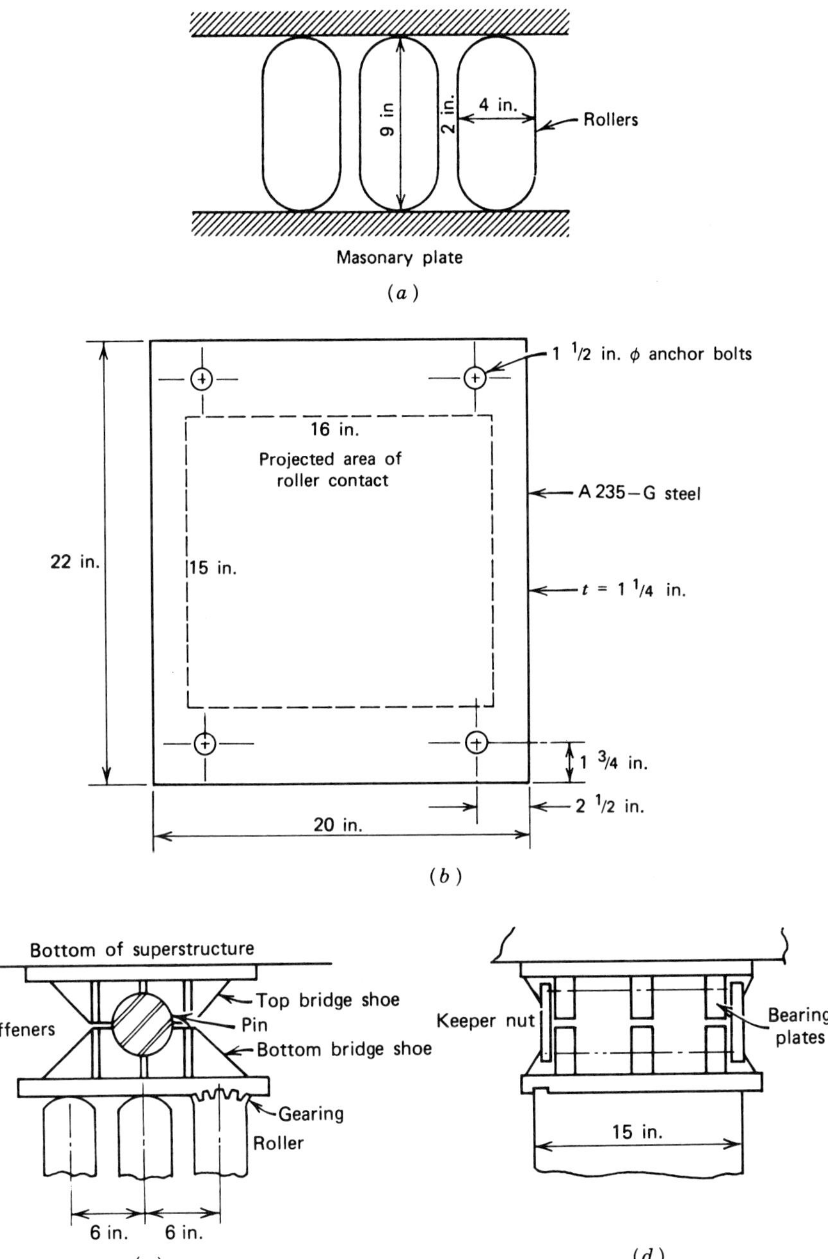

FIGURE 13-9 Bearing details for the design example: (*a*) roller arrangement; (*b*) detail of masonry plate and anchor bolts; (*c*) side view of roller bearing; (*d*) end view of roller bearing.

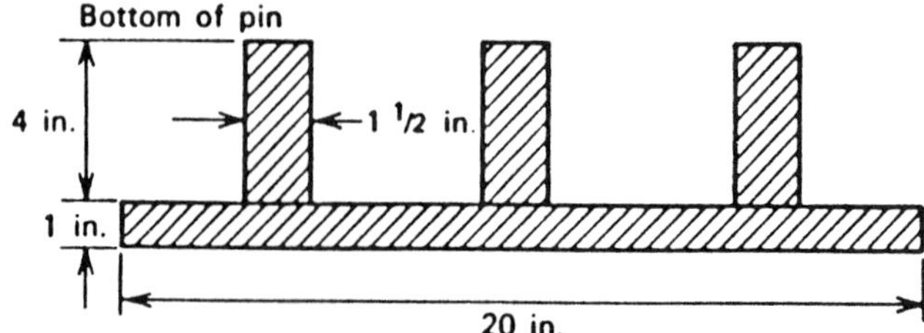

FIGURE 13-10 Bottom shoe; cross section along centerline of pin.

thickness of the bottom plate is checked by computing the maximum bending at the centerline of the pin, which is $400/3 \times 6 = 800$ in.-kips.

Also, $I = 84.9$ in.^{4}, $C_{max} = 3.32$ in., and $f_b = 800 \times 3.32/84.9 = 31.3$ ksi $< 0.8F_y = 40$ ksi.

The top bearing shoe details and dimensions should be similar to the bottom bearing shoe.

13-5 ELASTOMERIC BEARINGS

Special Features of Design

An elastomeric bearing is made partially or wholly of elastomer (synthetic rubber) that has adequate strength to support bridge loads. Two basic types are available: plain pads (consisting of elastomer only) and reinforced bearings (consisting of layers of elastomer restrained at their interfaces by integrally bonded steel or fabric reinforcement).

A parameter relevant to design is the shape factor (at one layer of a bearing) defined as

$$S = \frac{\text{Loaded area}}{\text{Effective area free to bulge}} = \frac{LW}{2t(L + W)} \tag{13-7}$$

where L is the gross length of the rectangular bearing parallel to the longitudinal axis of the bridge (in.), W is the gross width of the rectangular bearing perpendicular to the longitudinal axis of the bridge, and t is the thickness of the individual layer.

Compressive Stress AASHTO stipulates that, unless shear deformation is prevented, the average compressive stress σ_c in any layer should not exceed the value GS/β, nor should it exceed 1000 psi for steel reinforced bearings or 800 psi for fabric reinforced bearings and fabric pads. In this expression G is the shear modulus of the elastomer (psi) at 73°F, β is a modifying factor (1.0 for internal layers or reinforced bearings, 1.4 for cover layers, and 1.8 for plain pads), and S is the shape factor for the thickest layer. The allowable compressive stress may be increased by 10 percent when shear translation is prevented.

Compressive Deflection The elastomeric bearing compresses under application of load. The compressive deflection should be limited to ensure the serviceability of the structure. The effects of creep of the elastomer should be added to the deflection when considering long-term effects.

Other Requirements The shear deformation should include creep, shrinkage, posttensioning, and thermal effects. The bearings are designed so that

$$T \geq 2\Delta_s \tag{13-8}$$

where Δ_s is the shear deformation (in.) and T is the total elastomer thickness of the bearing. Therefore, the bearing pad acts as an expansion bearing up to a total movement of one-half the effective rubber thickness. This expansion produces a longitudinal force as a result of the shearing resistance of the pad. This shear force is approximated by

$$F_s = G\frac{A}{T}\Delta_s \tag{13-9}$$

where F_s is the shear force in the bearing, A is the plan area of the bearing, and the parameters G, T, and Δ_s are as before. Variations of G with temperature should be considered. Equation (13-9) is similar to (13-3).

To ensure stability, the total thickness T of the bearing should not exceed the smallest of

$$L/5,\ W/5,\ \text{or}\ D/6 \qquad \text{for plain pads}$$

and

$$L/3,\ W/3,\ \text{or}\ D/4 \qquad \text{for reinforced bearing}$$

where D is the gross diameter of circular bearings, and other parameters are as given previously.

Steel girders seated on elastomeric bearings must have flanges locally stiff to inhibit damage to the pad. This can be accomplished by a sole plate or by vertical stiffeners. Single-webbed girders symmetric about the minor (vertical) axis and placed symmetrically about the bearing need no additional stiffening if

$$\frac{b_f}{2t_f} \leq \sqrt{\frac{F_{yg}}{3.4\sigma_c}} \tag{13-10}$$

where b_f is the total flange width, t_f is the flange thickness, and F_{yg} is the yield stress of the girder steel.

Proposed LRFD Specifications

Steel Reinforced Bearings Steel reinforced elastomeric bearings are recommended because of their greater strength and superior performance (Roeder, Stanton, and Taylor, 1987; Roeder and Stanton 1991). Holes are strongly discouraged, but if they are necessary their effect should be considered in calculating the shape factor because they reduce the effective area and enhance the tendency to bulge.

The shear modulus G is the most relevant material property for design. Hardness has been found to correlate only loosely with shear modulus, and the ranges given in Table 13-5 represent the variation commonly encountered in practice. The shear modulus of the elastomer at 73°F should be used as the basis for design. If the bearing is specified explicitly by this parameter, that value should govern design, and other properties should be obtained from Table 13-5.

Compressive Stresses The new provisions limit the shear stress and strain in the elastomer. The relationship between the shear stress and the applied compressive load depends on the shape factor, and higher shape factors give higher capacities. Thus, the average compressive stress must satisfy the following.

For bearings subject to shear deformation,

$$\sigma_{\mathrm{CTL}} \leq 1600 \text{ psi} \qquad \sigma_{\mathrm{CTL}} \leq 1.66GS \qquad \sigma_{\mathrm{CLL}} \leq 0.66GS \tag{13-11}$$

For bearings fixed against shear deformation,

$$\sigma_{\mathrm{CTL}} \leq 1750 \text{ psi} \qquad \sigma_{\mathrm{CTL}} \leq 2.00GS \qquad \sigma_{\mathrm{CLL}} \leq 1.00GS \tag{13-12}$$

where σ_c is the average compressive stress, equal to P/A, at the service limit state, TL is the total load, and LL is the live load.

TABLE 13-5 Shear Modulus *G* for Steel Reinforced Elastomeric Bearing: LRFD Specifications

Hardness (Shore A)	50	60	70
Shear modulus (psi)	95–130	130–200	200–300
at 73°F (MPa)	0.68–0.93	0.93–1.43	1.43–2.14
Creep deflection at 25 years			
instantaneous deflection	25%	35%	45%
k	0.75	0.6	0.55

Shear Strain Likewise, the shear strain is limited to $0.5h_t$ in order to avoid rollover at the edges and delamination due to fatigue. Fatigue tests were conducted to 20,000 cycles, representing one expansion–contraction cycle per day for 55 years. This provision is therefore unconservative if the shear deformation is caused by high-cycle loading due to braking forces or vibration. In this case the maximum shear deformation should be limited to $0.10h_t$.

Stability The average compressive stress is limited to half the predicted buckling stress, calculated by the buckling theory developed by Gent (1964) and modified by Stanton et al. (1990a) to account for changes in geometry during compression, and calibrated from experimental results.

Reinforcement The applicable provisions intend to ensure that the reinforcement is adequate to sustain the tensile stresses induced by compression of the bearing. Holes in the reinforcement result in stress concentrations with associated detrimental effects and should be avoided.

Elastomeric Pads Plain elastomeric pads are less strong and more flexible because they are restrained from bulging by friction alone. Inevitably, slip occurs, especially under dynamic load, and causes larger compressive deflections and higher strains. If fiberglass layers are present, they tend to inhibit the deformations of plain pads. The bonding, however, is less strong, and the fiberglass pad cannot carry the same loads as the steel reinforced bearing (Stanton and Roeder, 1982; Crozier, Stoker, Martin, and Nordlin, 1979).

Compressive Stress The average compressive stress σ_c must satisfy the following.

For plain elastomeric pads (PEP),

$$\sigma_{\mathrm{CTL}} \leq 800 \text{ psi} \qquad s_{\mathrm{CTL}} \leq 0.55GS \tag{13-13}$$

For fiberglass pads (FGP),

$$\sigma_{\mathrm{CTL}} \leq 800 \text{ psi} \qquad s_{\mathrm{CTL}} \leq 1.00GS \tag{13-14}$$

For cotton duck pads (CDP),

$$\sigma_{\mathrm{CTL}} \leq 1500 \text{ psi} \tag{13-15}$$

In cotton duck reinforced pads, the shape factor is essentially infinite; hence, there is only one stress limit. In this case 1500 psi is approximately 15 percent of the failure load, and this is consistent with the safety factors applied to steel reinforced elastomeric bearings.

Observed Defects

In the past, where elastomeric bearing pads have been allowed to project beyond the edge of the supported member, horizontal slips developed in the projecting portion of the pad near midthickness after a few years in service. These cracks typically extended inward to the compressed region. These incidents are documented by Byers (1979).

Design Example: AASHTO Criteria

A suggested design methodology includes the following steps.

1. Establish the applicable temperature range from AASHTO Article 3.16. This determines the total horizontal movement of the beam or girder.
2. Calculate the required area of the bearing pad based on the allowable compressive stress. With precast concrete beams make the bearing width the same as the beam width.
3. Compute the shape factor.
4. Determine the compressive stress and strain and adjust the pad size and thickness if necessary until the actual compressive stress and strain are within the limits specified in Articles 14.4.1.1 and 14.4.1.2, respectively.
5. Check for nonparallel load surfaces.
6. Compute the shear forces applied to the substructure. These forces are due to the shear resistance of the pad while it is under strain because of thermal expansion and contraction, and are computed from (13-9).
7. In regions of seismic activity, anchor bolts should be used to attach the beam or girder to the substructure concrete. The load plate in this case can be made wider as shown in Figure 13-11, and the holes are slotted

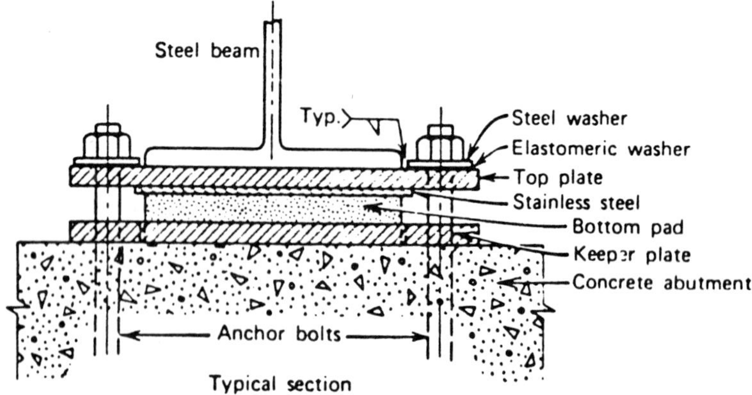

FIGURE 13-11 Anchor bolt detail for fixed attachment of beam to substructure.

to allow for longitudinal adjustment. Where the holes are necessary in the elastomeric pad for anchor bolts, the shape factor should be computed accordingly.

Design Data The two-span continuous prestressed concrete beam bridge shown in Figure 13-12 has expansion bearings at the abutments, points A and C, and a fixed bearing at the center pier, point B. The reactions from the superstructure are as follows.

At support A,

$$\text{Dead load} = 125 \text{ kips} \qquad \text{Live load} = 96 \text{ kips} \qquad (\text{no impact})$$

At support B,

$$\text{Dead load} = 250 \text{ kips} \qquad \text{Live load} = 108 \text{ kips} \qquad (\text{no impact})$$

The bridge is constructed in a moderate climate so that the temperature range is $30 + 40 = 70°\text{F}$.

Elastomeric Pad at Abutments The shear deformation Δ_s due to temperature is $\Delta_s = 0.000006 \times 100 \times 70 = 0.042$ ft $= 0.51$ in. From (13-8),

$$T_{\min} = 2 \times 0.51 = 1.02 \text{ in.}$$

We will, however, use $T = 1.5$ in. to account for creep, shrinkage, and posttensioning, used in 1/2-in. laminations ($t = 1/2$ in.).

The bearing area required A is $221/0.800 = 277$ in.^{2}. Because the bottom flange width of the girder is 28 in., we select a pad 10×28 in., $A = 280$ in.^{2}. From these dimensions the shape factor is computed as

$$S = \frac{10(28)}{2(0.5)(28 + 10)} = 7.4$$

Assuming a hardness of 60, the compressive strain corresponding to $S = 7.4$ is found from AASHTO Figure 14.4.1.2B as 4 percent (instantaneous deflection). Creep deflection at 25 years is taken from AASHTO Table 14.3.1 as 35

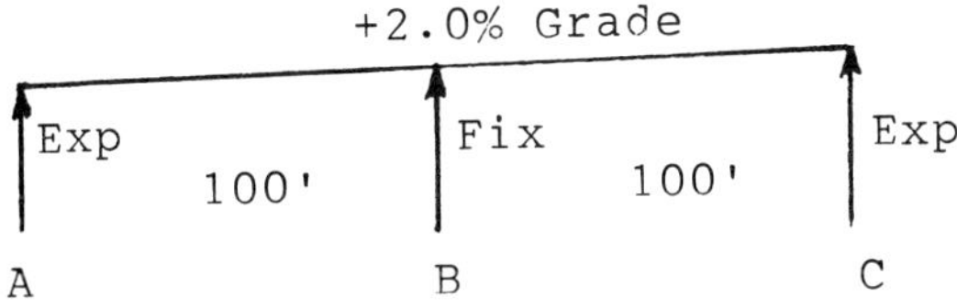

FIGURE 13-12 Two-span continuous prestressed concrete beam bridge.

percent of the instantaneous deflection, or $4 \times 0.35 = 1.4$ percent, giving a total compressive strain of $4 + 1.4 = 5.4$ percent, which is below the maximum value necessary to ensure serviceability.

Using $G = 150$ psi and $\beta = 1.4$, the allowable compressive stress is $150 \times 7.4/1.4 = 793$ psi. The actual stress is $221/280 = 789$ psi, OK.

The nonparallel surfaces and the rotation at A are checked next.

From prestress	$\theta_A = +0.0074$	
From dead load	$\theta_A = -0.0050$	(Computed from beam properties)
Bridge grade	$= +0.0200$	
Net	$= +0.0224$	

The offset at the bearing pad is $0.0224 \times 10 = 0.224$ in. and the allowable offset $2A_c$ is $2 \times 0.04 \times 1.5 = 0.120$ in. Because the relative rotation between the top and bottom surfaces exceeds the allowable rotation, a beveled sole plate will be provided.

Elastomeric Pad at Pier Because the bearing is fixed, we select two pads, each one 1/2 in. thick. The bearing area required A is $358/0.800 = 448$ in.^{2}. Two bearing pads 9×28 give $A = 2 \times 9 \times 28 = 504$ in.^{2}. The shape factor S is $(9 \times 28)/(2 \times 0.5 \times 37) = 6.8$, and the compressive stress is $358/0.504 = 710$ psi. Assuming a hardness of 60 and referring to AASHTO Figure 14.4.1.2B, the compressive strain is 3.2 percent. Checking the nonparallel load surfaces, we likewise determine that the actual rotation exceeds the allowable rotation, and a beveled sole plate will be provided.

The allowable compressive stress is $150 \times 6.8/1.4 = 729$ psi > 710 psi, OK.

Shear Force at Abutment The shear force is computed from (13-9). Expansion is taken from 0 to 40°F, and for this range we use a shear modulus value of 180 for grade 60.

$$\begin{aligned}
&\text{Expansion due to temperature} = 0.00006 \times 40 \times 100 = 0.024 \text{ ft} \\
&\text{Creep and shrinkage} \qquad\qquad\qquad\qquad\quad = 0.028 \text{ ft} \\
&\qquad \text{Total } \Delta_s \qquad\qquad\qquad\qquad\qquad\quad = 0.052 \text{ ft} = 0.63 \text{ in.}
\end{aligned}$$

$$\text{Shear force} \qquad F_s = \frac{180 \times 280 \times 0.63}{1.5} = 21.2 \text{ kips}$$

Typical bearing details for the example are shown in Figure 13-13. The 1992 AASHTO Specifications recommend two design methods (A or B). These should be identified on the plans together with the design load. AASHTO (1992 Commentary) presents design examples solved with the aid of a

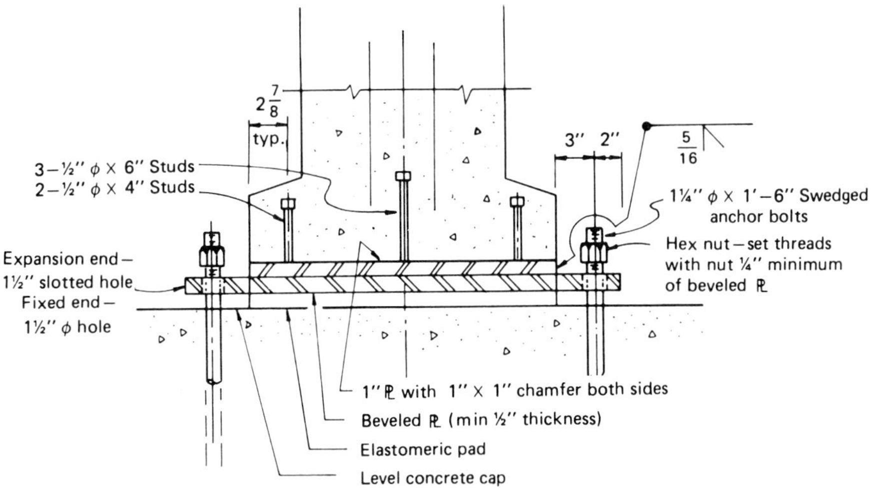

FIGURE 13-13 Typical details of elastomeric bearings for the design example.

microcomputer, using method B, and subsequently checked using method A. The design load is required because it is needed in conjunction with the test procedures.

At present, AASHTO permits only natural rubber (polyisoprene) and neoprene (polychloprene) because these have had an extensive history of satisfactory use. In addition, more field experience is available with these two materials than with any other, and invariably this experience is satisfactory.

13-6 SPECIAL BEARINGS

Pot Bearings

In principle, elastomeric pot bearings have become popular because of their increased load-carrying capacity. A plain rectangle or disk of elastomer with unconfined edges will deform vertically and horizontally under load. If the same material is contained in a pot without freedom to bulge or distort horizontally, its behavior approaches a viscous fluid with an associated higher load capacity. The rotation can be accommodated provided it does not exceed the rotation producing a maximum vertical strain of 10 percent.

The pots may be made of either steel or aluminum. The most commonly used types are (a) those permitting rotations in all directions, but fixed against translation in any direction; (b) those permitting rotations in all directions, guided for translation in one direction, and fixed against translation at right angles; and (c) those permitting rotation and translation in all directions (nonguided expansion bearings). In pot bearings permitting trans-

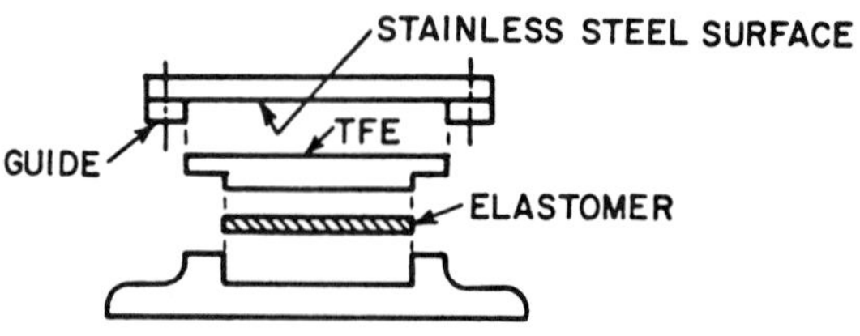

FIGURE 13-14 Elastomeric pot bearing.

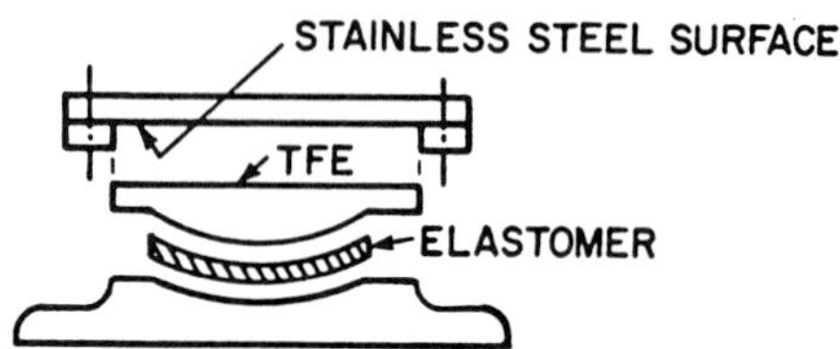

FIGURE 13-15 Elastomeric spherical or cylindrical bearings.

lation, the sliding surfaces are commonly stainless steel and a Teflon (TFE) product, as shown in Figure 13-14. If the anticipated rotation is high, special spherical or cylindrical elastomeric bearings are available as fixed, guided, or floating arrangements, as shown in Figure 13-15.

AASHTO Provisions Design parameters for pot bearings are articulated in AASHTO Article 19.1 for fixed, guided, and nonguided expansion devices. Pot bearings are not considered stiff against bending in their plane. A sole plate, beveled if necessary, on top and a masonry plate at the bottom should be provided.

LRFD Specifications Special emphasis is placed on the rotational function of pot bearings. Because softer elastomers permit rotation more readily, they should be preferred. Thus, the specified nominal hardness is between 50 and 60 on the Shore A scale. Rotational capacity can be increased by using a deeper pot, a thicker elastomeric pad, and a larger vertical clearance between the pot wall and the piston or slider.

Other provisions cover in detail the geometric requirements, allowable average stresses on the elastomer at the service limit state, the use of a seal between the pot and the piston, and the design requirements of the pot and the piston.

Bronze or Copper Alloy Bearings

Very often, in lieu of rockers for expansion bearings, a sliding arrangement using self-lubricating bronze plates is selected in conjunction with the steel components. These plates are lubricated by special thick-film, solid dry lubricants, extruded and compressed into recesses under high pressure. A quoted coefficient of friction is approximately 0.1 between the steel and a lubricating bronze plate.

Bronze or copper alloy sliding surfaces have been used over a long period with relatively satisfactory performance. However, there has been little research to confirm the properties and engineering data. A common problem has been the lack of distinction between the performance of the different materials, especially in view of evidence suggesting that there may be considerable differences between the different types of bearings.

The LRFD specifications identify materials in terms of ASTM classifications and specify a mating surface that has a Rockwell hardness value at least 100 points greater than that of bronze. Unless the coefficient of friction is available from tests, for design purposes its value should be taken as 0.1 for self-lubricating bronze components, and 0.4 for other types.

The nominal bearing stress due to combined dead and live loads at the service limit state should not exceed 2 ksi for C90500 alloy, 2.5 ksi for C91100 alloy, and 8 ksi for C86300 alloy.

Disk Bearings

These function by deformations of the polyether urethane disk that must be stiff enough to resist vertical loads without excessive deformation, and yet flexible enough to accommodate the imposed rotations without lift-off or excessive stress on other components. Disk bearings may be supplied as fixed bearings, guided expansion bearings, and nonguided expansion bearings.

Both the standard AASHTO and the LRFD specifications contain provisions regarding materials and design parameters. The design of the urethane disk may be based on the assumption that it behaves as a linear elastic material, laterally unrestrained at its top and bottom surfaces. Resisting moments computed in this manner have conservative values because they ignore creep that reduces these moments. However, the compressive direction due to creep must be taken into account. The assumption of linear elastic behavior in the urethane disk leads to feasible designs limited to the designated region of Figure 13-16. These data are based on the stress limits due to edge loading effects of the PTFE and the stress limits on the disk. Other constraints are introduced by the fact that no lift-off of components can be tolerated.

The average compressive stress on the disk should not exceed 5000 psi. If the outer surface of the disk is not vertical, the stress should be computed using the smallest plan area of the disk.

Curved Sliding Surfaces (Spherical Bearings)

Bearings with curved sliding surfaces consist of two metal parts with matching curved surfaces and a low-friction sliding interface. The curved surfaces

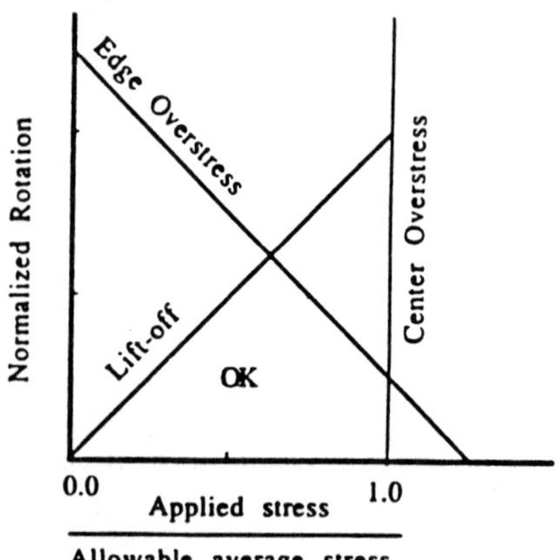

FIGURE 13-16 Feasible parameter combinations of disk bearings.

may be either cylindrical or spherical. The two surfaces of a sliding interface should have the same radius.

For cylindrical bearings, the stress may be calculated from

$$p_u = \frac{P_u}{(DW)} \tag{13-16}$$

For spherical bearings, the stress is calculated from

$$p_u = \frac{4P_u}{\pi D^2} \tag{13-17}$$

where P_u = dead and live load at the strength limit state
p_u = average pressure at the strength limit state
D = diameter of the loaded surface of the bearing
W = width of the bearing in the transverse direction

In bearings that are required to resist horizontal loads, an external restraint system must be provided; otherwise the radius should be limited to

$$R = \frac{D}{2\sin(\beta + \theta_u)} \tag{13-18}$$

where

$$\beta = \tan^{-1}\left(\frac{H_u}{P_{sd}}\right)$$

In the foregoing expressions, θ_u is the relative rotation of the top and bottom surfaces of the bearing at the strength limit state, H_u is the factored horizontal load on the bearing or restraint, and P_{sd} is the dead load at the service limit state. The intent of this restriction is to recognize that the geometry of a curved bearing combined with gravity loads can provide considerable resistance to lateral load. An external restraint may often be a more reliable way to resist large lateral loads.

Example of Spherical Bearing Figures 13-17*a*, *b*, and *c* show the plan, lower casting, and upper weldment details, respectively, for the spherical bearings used in the Daniel Boone Expressway and in the I-75–I-275 four-level interchange in Kentucky. The latter bridge is discussed in the example of Section 5-13 (Xanthakos, 1971). These bearings were used for the cross box girders or transverse bents at the piers. The spherical surfaces, both fixed and expansion, are coated with a resin-bonded molybdenum disulphide and graphite solid-film lubricant. The processing requirements for this coating were covered in the special provisions and apply to materials, allowable pressure, and method of application.

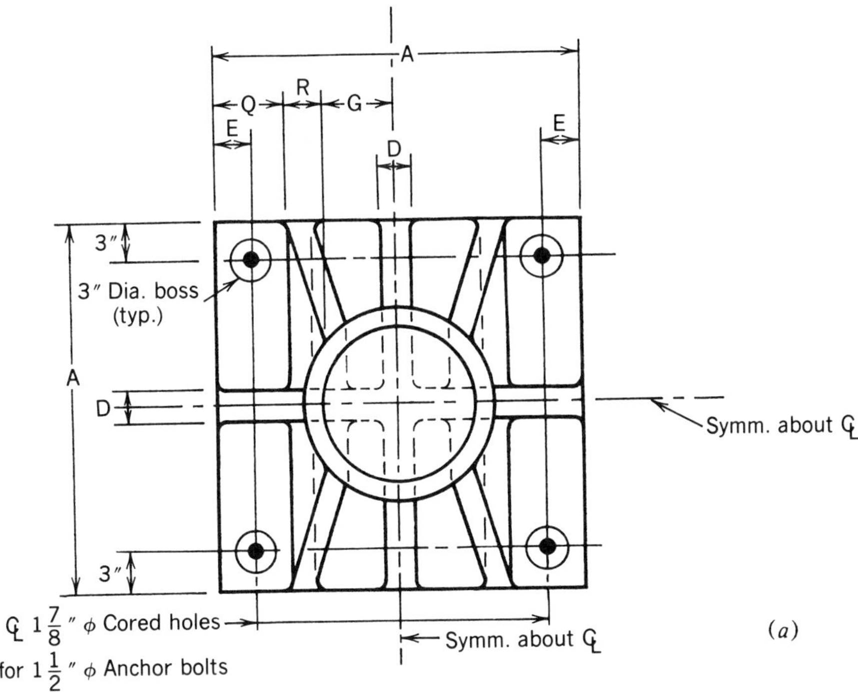

FIGURE 13-17 Spherical fixed bearing: (*a*) general plan; (*b*) lower casting detail; (*c*) upper weldment detail.

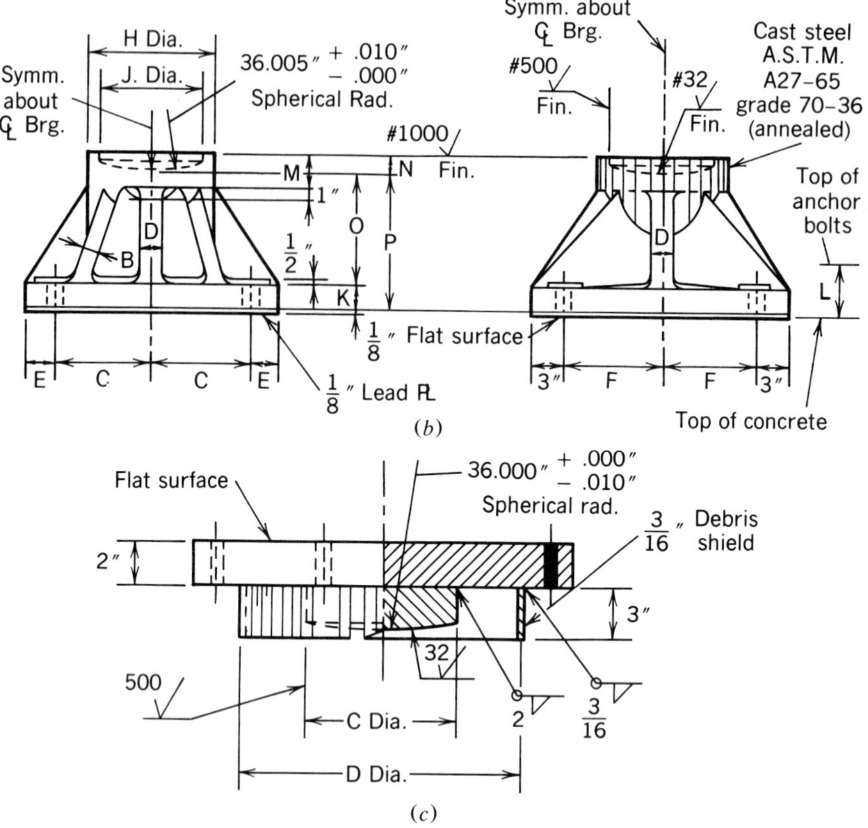

FIGURE 13-17 *(Continued)*

The design of these bearings included the following steps.

1. Select a spherical diameter necessary to accommodate the maximum anticipated rotation of the transverse box girder either in its direction or along the direction of the bridge.
2. Determine the contact area between the upper weldment and the spherical surface of the bearing, so that the actual pressure does not exceed the allowable pressure based on the strength of the film coating.
3. Determine the size of the vertical plates that receive the load from the cylinder and distribute this load to the base plate, and select ribs and diaphragms to reinforce and stiffen the vertical plates and the base plate.
4. Select a bearing height to accommodate an angle of inclination of the vertical plates less than 45°.

5. Select anchor bolts.
6. Proportion the base plate according to the allowable bearing on concrete masonry.
7. For expansion bearings provide a rocker to allow movement in the direction of the bridge. Because in this case movement in the direction of the box girder is restrained, pier columns should be designed for thermal expansion and contraction of the transverse box girders.

For a total reaction of 600 kips and an allowable compressive stress of 6 ksi on the spherical surface, we select dimension C (diameter in upper weldment) to be 12 in. The spherical area is $3.14 \times 6^2 = 113$ in.^{2}, and the load-carrying capacity is $113 \times 6 = 678$ kips, OK.

For the lower casting, we select $J = 13$ in. and $H = 15$ in. Next, we specify $C = 11.25$ in., $E = 2.25$ in., and $F = 10.5$ in., giving $A = 27$ in. The total bearing area is $27 \times 27 = 729$ in.^{2}, so that the bearing on masonry is $600/729 = 823$ psi, OK (for $f'_c = 3000$ psi, the allowable bearing stress is $0.3 \times 3000 = 900$ psi). Other critical dimensions are $M = 3.5$ in., $O = 10.5$ in., $K = 2.5$ in., $N = 1.25$ in., and $P = 13$ in. Note that four 1.5-in.-diameter anchor bolts are provided as shown, and dimension L is 5.25 in.

13-7 BEARING PIN CONNECTION: DESIGN EXAMPLE

A built-up compression member of an arch bridge supplies a reaction of 90 kips. A pin connection will be designed to transfer this load to the foundation concrete masonry. A suggested detail is shown in Figure 13-18. One of the design requirements is to provide a smooth appearing surface on the outside or fascia side of the member.

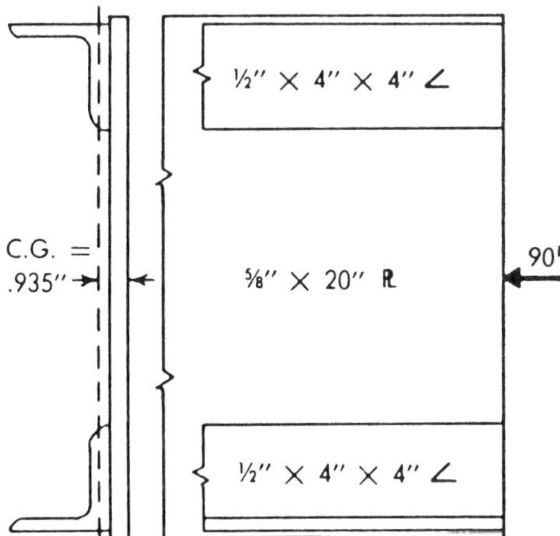

FIGURE 13-18 Detail for pin connection.

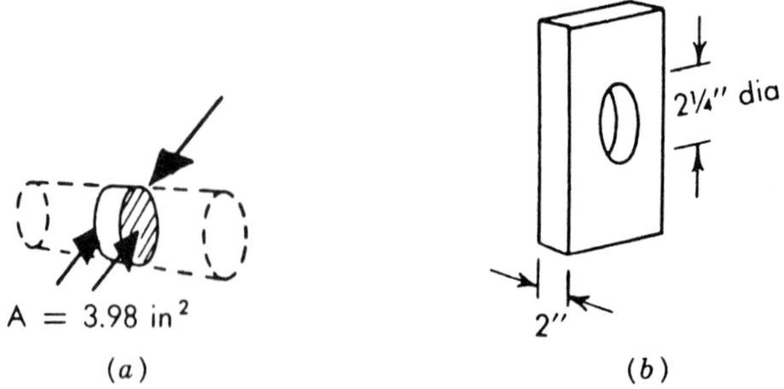

FIGURE 13-19 (*a*) Double shear in bearing pin; (*b*) single bearing in connection plate.

Note that in the cross section of the built-up compression member, the center of gravity is 0.935 in. from the outside face. Thus, by selecting and attaching a plate of sufficient thickness, we can have the center of gravity in line with the center of gravity of the compression member, so that the load can be transferred without any eccentricity.

The bearing pin is subjected to a double shear load, or 45 kips on each side, as shown in Figure 13-19*a*. For A36 steel, the allowable shear stress is $0.40F_y = 0.40 \times 36$ or 14 ksi, giving a required pin area of $45/14 = 3.22$ in.^{2}. Use a 2.25-in.-diameter pin, $A = 3.98$ in.^{2}.

The thickness of the connecting plate is based on the minimum bearing area necessary to resist the applied load. Assuming that the pin is not subject to rotation, the allowable bearing is $0.80F_y$, or 29 ksi. The bearing (minimum) area is therefore $90/29 = 3.10$ in.^{2}. By selecting a 2-in.-thick plate, the center of gravity will line up with the center of gravity of the compression member, and the bearing area supplied is $2 \times 2.25 = 4.50$ in.$^2 > 3.10$.

The next step involves determination of the required depth d of the connecting plate, as shown in Figure 13-20. The connecting plate may be considered a beam supported at its center and withstanding the applied load of 90 kips (assumed equally distributed) as two cantilevers. The angle and the plate are assumed to carry a portion of the 90-kip load according to the ratio of their individual areas. Thus, the compression load carried by each angle is $P_{\text{angle}} = 90 \times 3.75/20 = 16.9$ kips, and the compression load carried by the plate is $P_{\text{plate}} = 90 \times 12.5/20 = 56$ kips.

According to this analysis, there is a uniform load with an intensity of $56/20 = 2.8$ kips/in. along the entire width of the plate, and a concentrated load of 16.9 kips acting at the center of gravity of the angles or at a distance of 8.75 in. from the center of the pin. The resulting cantilever moment is

$$M = 16.9 \times 8.75 + 2.8 \times 10^2/2 = 288 \text{ in.-kips}$$

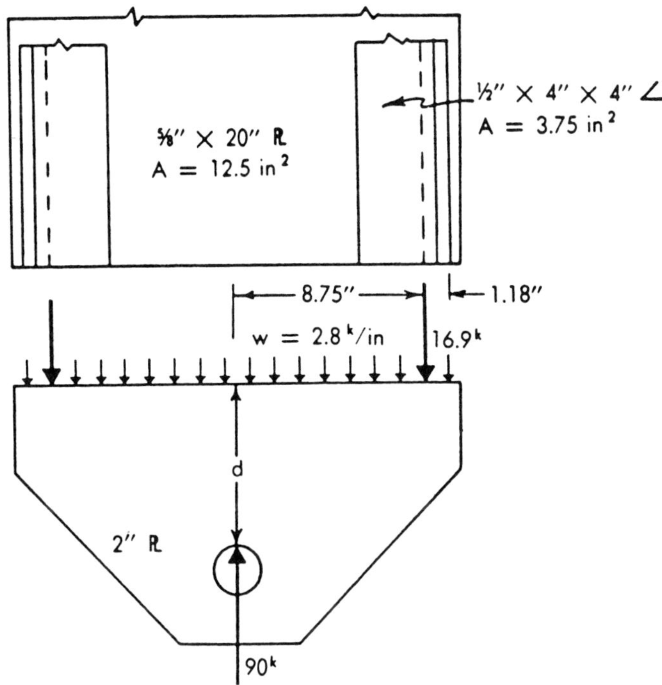

FIGURE 3-20 Connection detail showing the distribution of load along the connecting plate.

The required section modulus is

$$S = 288{,}000/20{,}000 = 14.4 = td^2/6$$

Using $t = 2$ in., we compute $d = 6.6$ in.; use 7 in.

13-8 LINKS AND HANGERS

Links and hangers are commonly used to support suspended spans in multispan bridge units. Pins should be of sufficient length to secure full bearing of all parts connected upon the turned body of the pin. They should be located with respect to the gravity axis of the members so as to reduce to a minimum the stress due to bending. Connected members should be restrained against any lateral movement on the pins and also against lateral distortion due to the skew of the bridge.

Design Example: Expansion Link A typical detail of an expansion link is shown in Figures 13-21*a* and *b*. The left side is the end of the cantilever span, and the right side is the end of the suspended span. The total reaction

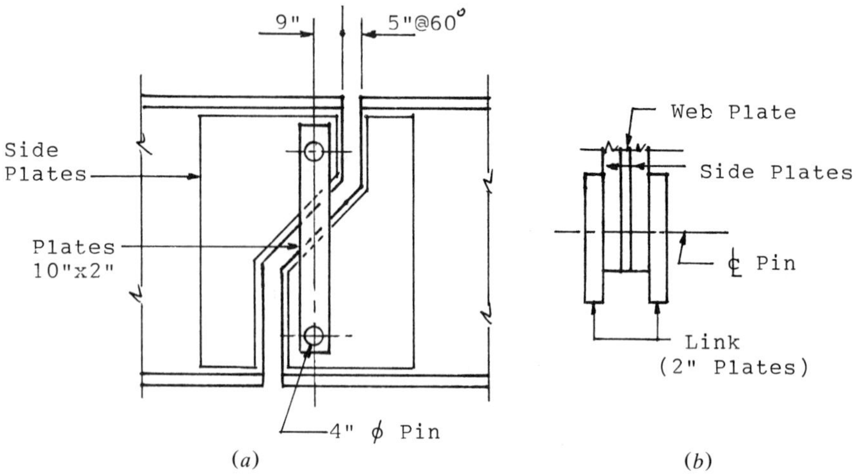

FIGURE 3-21 Link detail: (*a*) elevation; (*b*) section through pin.

to be transferred at the link support is 205 kips. The web of the plate girder is 7/16 in. thick, and the steel is A36 grade.

As in the previous example, the pin is subjected to double shear, so that the required pin area is $205/(2 \times 14) = 7.32$ in.^{2}. Considering the sensitivity of these members to rusting and probable bending stresses, we will choose a pin with a 4-in.-diameter $A = 3.14 \times 2^2 = 12.6$ in.^{2}. For side plates, we select 1 3/4-in.-thick plates. The total thickness of the reinforced web is $1\ 3/4 \times 2 + 7/16 = 3\ 15/16$ in. Because the pin may be subjected to rotation, the allowable bearing is $0.40F_y$, or 14 ksi, and the total bearing to be resisted by the web is $3.93 \times 4 \times 14 = 220$ kips, OK. Likewise, we select plate link hangers 2 in. thick. The allowable (total) bearing in each link (double bearing) is $2 \times 14 \times 4 = 112$ kips > 103 kips.

Design Example: Fixed Hinge The same reaction of 205 kips must be transferred at a fixed hinge, as shown in the detail of Figure 13-22. The web is a 72-in.-by-7/16-in. plate. As in the previous example, we select side plates 1 3/4 in. thick, hanger plates 2 in. thick, and a pin diameter of 4 in. Note that the left side of the detail is the end of the cantilever span, and the right side is the end of the suspended span.

For A36 steel, the allowable basic stress in the weld is $0.27 \times 58 = 15.6$ ksi, assuming that the tensile strength of the electrode classification is equal to the tensile strength of the connected part. We will, however, use $F_v = 12.4$ ksi. Accordingly, the allowable load per inch of fillet weld is as follows: 3/16 in. = 1640 psi; 1/4 in. = 2200 psi; 5/16 in. = 2750 psi; 3/8 in. = 3300 psi; and 1/2 in. = 4375 psi. Note also that the minimum size of the fillet weld for a 1 3/4-in. plate is 3/8 in. The allowable weld stress is limited, however, by

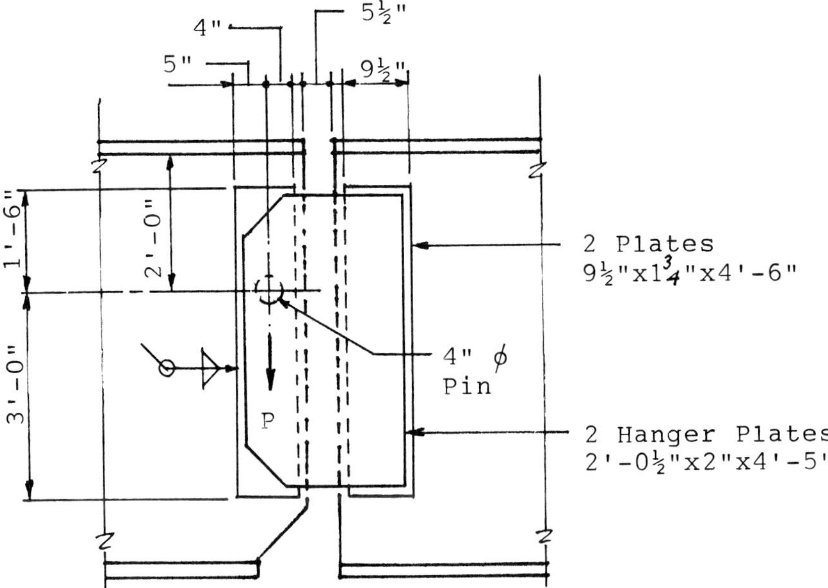

FIGURE 3-22 Fixed hinge detail, pin, hanger, and connection plates.

the web plate. Thus, based on the allowable shear on the 7/16-in. web, the allowable weld stress is 2625 psi.

Considering the plate on one side of the web and assuming the three sides welded to the side plate, the weld system is subjected to an eccentric load $V = 103$ kips as shown in Figure 13-23. The center of gravity of the three weld sides is found by taking moments about the vertical side, or

$$2 \times 9.5 \times 4.5 = 73\bar{x} \qquad \text{or} \qquad x = 90.25/73 = 1.25 \text{ in.}$$

$$I = \frac{1 \times 54^3}{12} + 54 \times 1.25^2 + 2$$

$$\times \frac{1.25^3}{3} + 2 \times \frac{8.25^3}{3} + 2 \times 9.5 \times 27^2 = 27{,}430$$

The moment of V about the center of gravity is $103{,}000 \times 18.25 = 1{,}879{,}750$ in.-lb = 156.6 ft-kips. The weld stress due to moment is

$$\frac{Md}{I} = \frac{1{,}879{,}750 \times 27.5}{27{,}430} = 1884 \text{ psi}$$

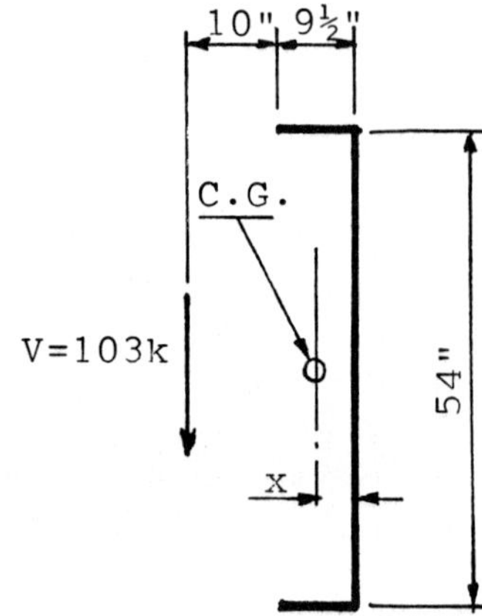

FIGURE 3-23 Weld configuration and dimensions.

The direct stress due to shear is

$$V = \frac{103{,}000}{73} = 1410 \text{ psi}$$

The resultant stress is found graphically as 2650 psi ≈ 2625 psi, OK.

13-9 STRESSES IN PIN CONNECTIONS

In the foregoing examples, bending in the pins was not considered because a much larger diameter or cross-sectional area is provided (12.6/7.32 = 1.7). In many instances, however, and where the load to be transferred is considerable, the pin is relatively long. In these cases the computation of bending stresses in the pin using conventional beam theory is nearly correct and should be considered. Referring to Figure 13-24, the bending moment may be computed assuming that the load in each plate is concentrated at the center as shown in Figure 13-24*b*. When the main web is reinforced for bearing or when several plates are packed together, this assumption may be extremely conservative, and the moment may instead be computed assuming the uniform distribution shown in Figure 13-24*c*.

For a given bending moment and assuming stresses within the elastic range, the maximum bending stress in the pin is

$$f = \frac{Mc}{I} = \frac{M(d/2)}{\pi d^4/64} = \frac{10.2M}{d^3} \tag{13-19}$$

which is rewritten as

$$M = fd^3/10.2 \tag{13-20}$$

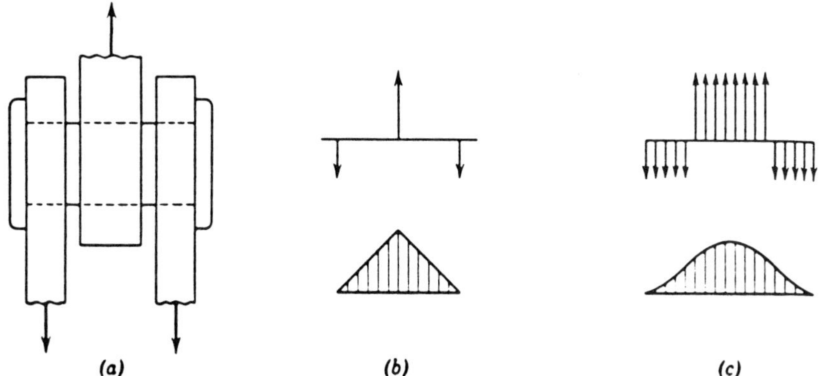

FIGURE 3-24 Moments in pins: (*a*) pin connection; (*b*) moments in pin-assuming concentrated loads; and (*c*) moments in pin-assuming uniform loads.

For the example of Figure 13-21 and using $f = 20$ ksi and $d = 4$ in., we obtain $M = 20 \times 4^3/10.2 = 125.5$ in.-kips. Using the stress distribution shown in Figure 13-24*c*, the maximum moment occurs at the face of the side plate, and is $103 \times 1 = 103$ in-kips < 125.5.

Shear stresses computed by simple beam theory will be in considerable error because of the very small span–depth ratio. For this reason, computing the nominal shear stress based on a uniform stress distribution over the section is acceptable, in which case $f_v = V/A$.

The actual bearing stresses between plates and pins will depend on the accuracy of the fit that can be provided between them. If the two parts fit properly, the nominal bearing stress is simply $f_b = P/td$, where P is the load, t is the total thickness of the plates, and d is the pin diameter.

Combined Stresses Pins are not subjected to tension; hence, it is not necessary to combine shear and tensile stresses. Consideration must be given to the interaction of bending stresses with bearing stresses, because the maximum compressive bending stress invariably occurs in the region of high bearing stress. This interaction, however, is not critical because in this region the pin is subjected to triaxial compression. Because failure in this case would occur when the difference in the principal stresses is large, the combined triaxial compression stresses will not control provided the stress in each direction is kept within its allowable limits.

13-10 REQUIREMENTS OF SHOP AND FIELD SPLICES

General

Shop Splices The location of web shop splices is usually dictated by the available plate lengths, fabrication needs, and costs. The actual location is often left to the fabricator. Web sections are generally connected by full-penetration groove welds, and therefore the resulting girder is verified by appropriate inspection procedures.

The splices of abutting flange plates of different thickness are sloped to avoid abrupt changes in cross section. The transition may be provided by chamfering the thicker plate, by sloping the weld itself, or by a combination thereof. These procedures are governed by applicable AASHTO and AWS specifications.

Field Splices Field splices are normally provided in continuous spans and in single spans that are too long for shipment and erection. Field splices in beams, plate girders, and box girders are usually bolted because of the difficulty sometimes encountered in field welding. Occasionally, the flanges are groove-welded and the web bolted, but flanges that must be welded from both sides present a problem on the inner face because of the difficulty of providing a sound weld and the juncture of the flange and the web.

The strength of members connected by high-strength bolts should be based on the gross section for compression members and for members primarily in bending. In the latter case, if more than 15 percent of the flange area is removed, the amount in excess of that should be deducted from the gross area. The tensile stress on the new section should not exceed $0.50F_u$ under service load conditions or $1.0F_u$ when using the strength design method, where F_u is the minimum tensile strength of the steel.

Splices are designed (service load) for a capacity based on the average of the calculated design stress at the point of splice and the allowable stress of the member at the same point, but not less than 75 percent of the allowable stress. For strength design, splices should be proportioned for not less than the average of the required strength and the strength of the member at the point of splice, but not less than 75 percent of the strength of the member. When a section changes in a splice, the smaller section should be used.

Field splices are usually located at points of dead load contraflexure where the bending moment requirements are the least severe. Certain state standards require specific splice locations, in some instance at the supports, although these are regions of maximum moments. Interestingly, dead load moments and shears due to the weight of a beam or girder should be computed based on the actual erection procedure and the structural configuration of the member before the field splice is made, rather than on continuous beam behavior, unless procedures are introduced to ensure the absence of stresses until the field splice is in place (e.g., temporary beam shoring).

LFRD Specifications

Bolted connections designed to resist shear may be either the slip-critical (friction) or the bearing type. Connections subjected to stress reversal, heavy impact loads, severe vibration, and detrimental effects from joint slippage are

TABLE 13-6 Values of Bolt Tensile Strength: LRFD Specifications

Diameter (in.)	Tensile Strength, F_{ub} (ksi)	
	M164 (A325)	M253 (A490)
5/8–1	120	150
$1\frac{1}{8}$–$1\frac{1}{2}$	105	150

designated as slip critical. Connections of tension members should be designed as slip critical.

Shear Strength The nominal shear strength of fasteners at the strength limit state in joints where the length between extreme fasteners is less than 50 in. is taken as

$$P_n = 0.50A_bF_{ub}$$

when threads are excluded from the shear plane and

$$P_n = 0.40A_bF_{ub}$$

when threads are included in the shear plane, where A_b is the cross-sectional area of the fastener and F_{ub} is the tensile strength of the fastener. Values of F_{ub} are given in Table 13-6.

Slip-Critical Connections The nominal shear strength at the strength limit state must satisfy the foregoing fastener shear strength requirements, and also the strength

$$P_n = 1.13K_nK_sN_BN_sP_b \qquad (13\text{-}21)$$

TABLE 13-7 Required Fastener Tension, P_t

Diameter (in.)	Required Tension, P_t (kips)	
	M164 (A325)	M253 (A490)
5/8	19	24
3/4	28	35
7/8	39	49
1	51	64
1-1/8	56	80
1-1/4	71	102
1-3/8	85	121
1-1/2	103	148

K_n is the hole size factor given in Table 13-8.

TABLE 13-8 Values of Hole Size Factor, K_n

K_n	1	0.85	0.70	0.60
	For normal sized holes	For oversized and short-slotted holes	For long-slotted holes with the slot perpendicular to the direction of the force	For long-slotted holes with the slot parallel of the direction of the force

K_s is the surface condition factor given in Table 13-9.

TABLE 13-9 Values of Surface Conditions Factor, K_s

	0.33	0.50	0.40
K_s	For class A surface conditions	For class B surface conditions	For class C surface conditions

where N_B = number of bolts in joint
N_s = number of slip planes
P_t = required tension in the fasteners (Table 13-7)
K_n = hole side factor (Table 13-8)
K_s = surface condition factor (Table 13-9)

13-11 DESIGN EXAMPLE: BOLTED FIELD SPLICE FOR PLATE GIRDER (SERVICE LOAD)

A bolted field splice, with 7/8-in.-diameter high-strength bolts, is located 25 ft from the interior support of a continuous plate girder. At this location the maximum shear and maximum moments are as follows:

Maximum Shear		Maximum Moments	
DL	$V = 40.3$ kips	Positive	$M = 650$ ft-kips
LL + I	$V = 49.4$ kips	Negative	$M = -460$ ft-kips
Total	$V = 89.7$ kips		

Fatigue need not be considered when computing bolt stresses, because no reduction in these stresses is required. Fatigue should be considered, however, in the design of the splice plates. A fatigue design moment ensures that splice plate stresses will not exceed the allowable fatigue stress for base metal adjacent to friction-type (slip-resistant) connections.

The fatigue design moment M_f is taken as $M_f = M_a(F_s/F_f)$, where M_a is the actual maximum moment, F_s is the allowable tensile stress, and F_f is the allowable fatigue stress. Likewise, the design shear V_f is taken as $V_f = V_a(M_f/M_a)$, where V_a is the actual maximum shear. The parameters M_f and V_f will control the design if they are greater than (a) 75 percent of the capacity of the section or (b) the average of the actual moment or shear and the capacity of the section. In this example the fatigue design moment governs the splice material, and (b) controls the bolt arrangement.

Properties of Net Section (12 × 7/8-in. flanges, 5/16-in. web.) For 7/8 = in.-diameter high-strength bolts, the flange hole are is 1 × 0.875 = 0.875 in.^{2}, and the web hole area is 1 × 5/16 = 0.313 in.^{2}.

$$\sum d^2 \text{ for holes} = 5^2 + 10^2 + 15^2 + 20^2 = 750 \text{ in.}^2$$

$$\sum Ad^2 \text{ web holes} = 20 \times 0.313 \times 750 = 470$$

$$\sum Ad^2 \text{ flange holes} = 2 \times 2 \times 0.875 \times 24.44^2 = \underline{2901}$$

$$I_{\text{holes}} = 2561 \text{ in.}^4$$

The moment of inertia of the gross section (two flange plates 12 × 7/8 in., and web plate 48 × 5/16 in.) is I_y = 15,420 in.^{4}, and the moment of inertia of the net section is therefore I_{bet} = 15,420 − 2561 = 12,859 in.^{4}.

Required Splice Capacity The ratio of the minimum to the maximum stress in the flanges is $R = -460/650 = -0.708$. The allowable fatigue stresses are

$$\text{Tension} \quad F_r = \frac{20{,}500}{1 - 0.55R} = \frac{20.5}{1 - 0.55(-0.708)} = 14.75 \text{ ksi}$$

$$\text{Compression} \quad F_r = \frac{0.55F_y}{1 - \left(\dfrac{0.55F_y}{13.3} - 1\right)R} = \frac{19.8}{1 - \left(\dfrac{19.8}{13.3} - 1\right)(-0.708)} = 14.72 \text{ ksi}$$

For the splice plates, the fatigue design moment is

$$M_f = 650 \times 20.0/14.71 = 884 \text{ ft-kips}$$

The moment capacity of the section (calculated, net section) is

$$M_{net} = (20 \times 12{,}859)/(12 \times 24.88) = 861 \text{ ft-kips} \qquad \text{(Maximum)}$$

or $0.75 \times 861 = 646$ ft-kips. The average moment capacity is now

$$M_{av} = (861 + 646)/2 = 753 \text{ ft-kips} < 884 \text{ ft-kips}$$

The fatigue design moment controls the splice plate design.

The shear capacity corresponding to the average moment is

$$V = 753 \times 89.7/650 = 104 \text{ kips}$$

Web Splice The moment transferred to the web is proportional to the moment of inertia ratio, or

$$M_w = 753 \times 2410/12{,}859 = 142 \text{ ft-kips}$$

$$M_v = 104 \times 3.25/12 \qquad = \underline{29}$$

$$\text{Total } M \qquad = 170 \text{ ft-kips}$$

The moment of inertia of the bolts is

$$I_{x-x} = 2 \times 2 \times 750 = 3000 \qquad I_{y-y} = 18 \times 1.5^2 = 41$$

Total $I = 3041$ in.^{4}. The shear on the bolts is calculated as follows:

$$P_s = 104/18 = 5.78 \qquad \text{(Direct shear)}$$

The load on the uppermost bolt due to moment

$$P_m = 170 \times 12 \times 20.06/3041 = 13.46 \text{ kips}$$

giving a vertical component

$$P_V = 13.46 \times 1.5/20.06 = 1.01 \text{ kips}$$

and a horizontal component

$$P_H = 13.46 \times 20.0/20.06 = 13.42 \text{ kips}$$

The total load on the uppermost bolt is the resultant

$$P = \sqrt{(5.78 + 1.01)^2 + (13.42)^2} = 15.04 \text{ kips}$$

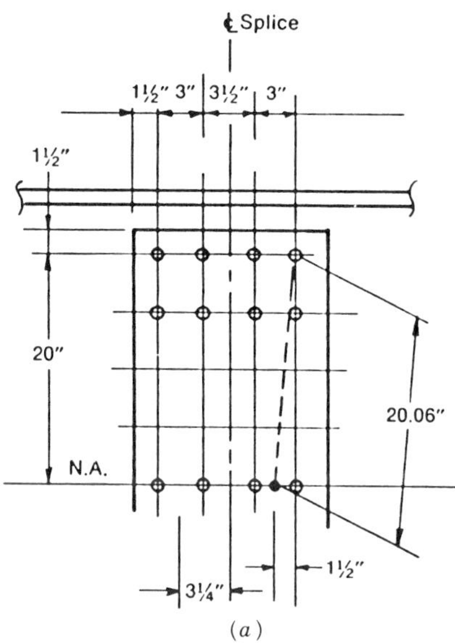

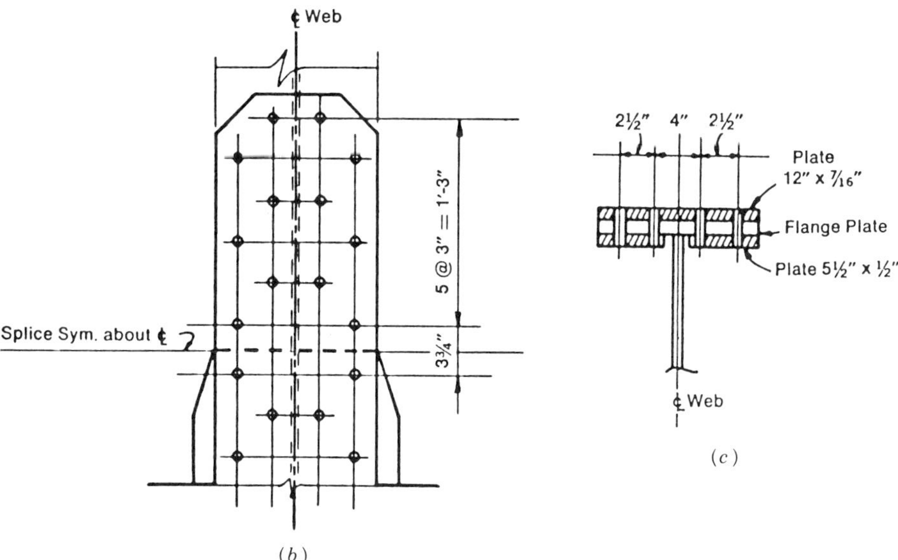

FIGURE 13-25 Field splice detail for example problem: (*a*) web splice; (*b*) plan of flange splice; (*c*) section through flange splice. (From AISC, 1990.)

For 7/8-in.-diameter A325 bolts, excluding threads in the shear plane, the allowable shear is $0.60 \times 19 \times 1.4 = 16.0$ kips, OK.

For the web splice, we select two plates 12 1/2 × 5/16 in., 3 ft 7 in. long,

$$I = 2\left[(0.313 \times 43^3)/12 - 470\right] = 3208 \text{ in.}^4$$

The portion of the fatigue design moment carried by the web plates is $844 \times 2410/12{,}859 = 166$ ft-kips. The maximum stress in these plates is therefore

$$f = (166 \times 12 \times 21.5)/1208 = 13.35 \text{ ksi} < 14.71 \qquad \text{OK}$$

Flange Splice The plan of the flange splice plates is shown in Figure 13-25*b*, and the section in Figure 13-25*c*. The splice plates transmit the axial tension and compression to the girder flanges in double shear by the 7/8-in.-diameter high-strength bolts. The required average moment capacity of the flange splice is $M_f = 753 - 142 = 611$ ft-kips. The axial force in the flanges that forms a couple to supply this moment capacity is $P_f = 611 \times 12/48.88 = 151$ kips. The required number of bolts is $151/16 = 9.4$ bolts; use 12 bolts as shown.

The fatigue design moment capacity required for the flanges is $M_f = 884 - 166 = 718$ ft-kips, or $P_f = 718 \times 12/48.88 = 176$ kips, resulting in a required area $A = 176/20 = 8.80$ in.^{2}. For the flange splice plates shown in Figure 13-25*c*, the cross-sectional area is $(12 - 2)7/16 + (11 - 2)/2 = 8.88$ in.^{2}, OK.

13-12 DESIGN EXAMPLE: BOLTED FIELD SPLICE OF BOX GIRDER (LOAD FACTOR DESIGN)

For load factor design, slip-critical joints must be designed to prevent slip at the overload $D + (5/3)(L + I)$. In addition, the force caused by this overload (H or HS truck only) should not exceed the design slip resistance $\phi F_s = \phi T_b \mu$ (design slip resistance per unit of bolt area given in AASHTO Table 10.57A). For class A (slip coefficient of 0.33), the value of $\phi T_b \mu$ is 21 ksi (A325 bolts), set so that the allowable load for a 7/8-in.-diameter bolt in double shear is $P = 2 \times 0.60 \times 21 = 25.3$ kips/bolt.

The design moment is chosen as the greater of (a) the average of the calculated moment and maximum capacity of the section, and (b) 75 percent of the maximum capacity of the section. The calculated moment is equal to $1.30[D + 1.67(L + I)]$, where the associated moments are at the section under consideration. For this example the section capacity is based on the gross section minus the loss of flange area due to holes in excess of 15 percent.

TABLE 13-10 Table of Bending Moments, 37 ft from Support (ft-kips)

	For Service Loads	Factor	For Overload	Factor	Maximum Design Loads
DL_1	−100	1	−100	1.30	−130
DL_2	50	1	50	1.30	65
$+(L + I)$	1690	$\frac{5}{3}$	2817	1.30	3662
$-(L + I)$	−1050				
Maximum	1640		2767		3597
Minimum	−1100				

The box girder is a two-span continuous unit, each span 120 ft, composite in both the positive and the negative moment regions. The splice is located 37 ft from the interior support. The bending moments and shears are tabulated in Tables 13-10 and 13-11, respectively. The notation DL_1 indicates dead load acting on the steel section, and DL_2 denotes dead load acting on the composite section. Because the positive moment governs, the splice is designed for that moment, and, for simplicity, the composite slab is neglected. A typical section at the splice is shown in Figure 13-26.

Net Section at Top Flange Splice The deduction of flange width at a section through two holes is $2 \times 1 = 2.00$ in. The deduction through a chain of four holes is $4 \times 1 - 2 \times 0.947 = 2.11$ (controls).

The gross flange area is $(9/16)12 = 6.75$ in.^{2}. The area deducted for holes is $(9/16)2.11 = 1.18$ in.^{2}. Likewise, 15 percent of the gross area is $6.75 \times 0.15 = 1.01$ in.^{2}. Therefore, the net deduction for two flanges is $2 \times (1.18 - 1.01) = 0.34$ in.^{2}.

Net Section at Bottom Flange Splice The center of a longitudinal stiffener ST7.5 × 25 (not shown in detail) coincides with the center of gravity of the bolt holes. The following areas are deducted: 16 holes in the bottom flange

TABLE 13-11 Table of Shears, 37 ft from Support (kips)

	For Service Loads	Factor	For Overload	Factor	Maximum Design Loads
DL_1	58.0	1	58.0	1.30	75.4
DL_2	13.8	1	13.8	1.30	17.9
$L + I$	47.8	$\frac{5}{3}$	79.7	1.30	103.6
Total					196.9

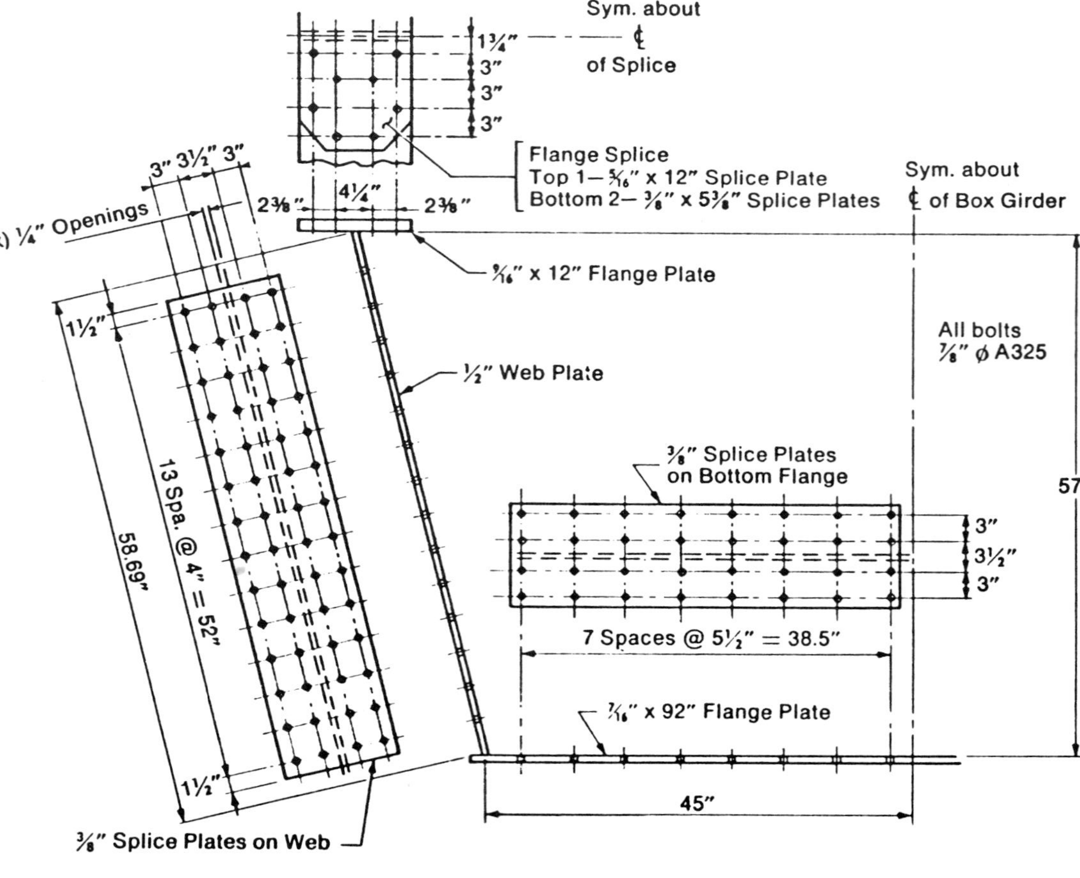

FIGURE 13-26 Section at bolted field splice. (From AISC, 1990.)

TABLE 13-12 Properties of Net Section at the Splice, Steel Only

Material	A	d	Ad	Ad^2	I_o	I
Gross section	119.79		−938			64,287
Top flange bolt holes	−0.34	28.78	−10	−282		−282
Bottom flange bolt holes	−2.20		63	−1,815		−1,815
	117.25 in.^{2}		−885 in.^{3}			62,190

$$d_s = \frac{-885}{117.25} = -7.55 \text{ in.}$$

$$-7.55 \times 855 = \underline{-6{,}682}$$

$$I_{NA} = 55{,}508 \text{ in.}^4$$

$$d_{\text{top of steel}} = 29.06 + 7.55 = 36.61 \qquad d_{\text{bottom of steel}} = 28.94 - 7.55 = 21.39 \text{ in.}$$

$$S_{\text{top of steel}} = \frac{55{,}508}{36.61} = 1{,}516 \text{ in.}^3 \qquad S_{\text{bottom of steel}} = \frac{55{,}508}{21.39} = 2{,}595 \text{ in.}^3$$

plate, two holes in the flange of the stiffener, and two holes in the stiffener web. The deductions in the flange area are computed as follows:

Gross area of bottom flange and stiffener

$$= (7/16)92 + 7.35 = 47.60 \text{ in.}^2$$

$$\text{Area deducted for holes} = (7/16)16 + 2 \times 0.622 + 2 \times 0.55 = 9.34$$

$$\text{15 percent of gross area} = \underline{7.14}$$

$$\text{Net deduction for bottom flange} = 2.20 \text{ in.}^2$$

The properties of the net section at the splice are tabulated in Table 13-12.

Design Moments and Shears The capacity of the net section is governed by the section modulus at the top of the steel. This capacity is (A36 steel)

$$M_{\text{net}} = 36 \times 1516/12 = 4548 \text{ ft-kips}$$

or

$$0.75M_{\text{net}} = 0.75 \times 4548 = 3411 \text{ ft-kips}$$

The average of the calculated moment for the design loads and the net capacity of the section is

$$M_{\text{av}} = (3597 + 4548)/2 = 4073 \text{ ft-kips} \qquad \text{(Controls)}$$

The design shear is calculated as

$$V_d = 196.9 \times 4073/3597 = 223 \text{ kips} \qquad \text{(Two webs)}$$

or, in the plane of each web,

$$V_d = 0.5 \times 223 \times 58.69/57 = 115 \text{ kips}$$

Web Splice The forces acting on the web splice are the design shear, a moment M_v due to the eccentricity of this shear, and a portion M_w of the design moment. We calculate

$$I_w = 15{,}891 + 58.69(7.55)^2 = 19{,}236 \text{ in.}^4$$

Next, we compute

$$\begin{aligned} M_v &= 223 \times 3.25/12 = 60 \\ M_w &= 4073 \times 19{,}236/55{,}508 = 1411 \\ \text{Total } M &= 1471 \text{ ft-kips, or } 736 \text{ ft-kips/web} \end{aligned}$$

We select two web splice plates 3/8 × 55 in., with two rows of bolts as shown in Figure 13-26. Because the area of one hole is 0.375 in.^{2}, the percentage of plate removed from the gross section is

$$\frac{14 \times 0.375}{0.375 \times 55} \times 100 = 25.5 \text{ percent}$$

and the fraction of the hole area that must be deducted is (25.5 − 15.0)/22.5 = 0.41. For the given bolt spacing, we compute $\Sigma Ad^2 = 1119$ in.^{4}, along the slope of the web, or, with respect to the horizontal axis, $\Sigma Ad^2 = 1119 \times (57/58.69)^2 = 1055$ in.^{4}. The properties of the web splice section are computed next and tabulated in Table 13-13.

Actual Stresses in Web Splice The maximum stress due to bending is

$$f_b = 736 \times 12/310 = 28.5 \text{ ksi} < 36 \text{ ksi}$$

The allowable shear stress is $0.55F_y = 19.8$ ksi, and the actual shear stress is $f_v = 115/41.25 = 2.8$ ksi, OK.

The plates are next checked for fatigue under service loads. The range of moment carried by the web is

$$M_w = (1640 + 1100) \times 19{,}236/55{,}508 = 950 \text{ ft-kips, or } 475 \text{ ft-kips/web}$$

TABLE 13-13 Properties of the Web Splice Section

Material	A	d	Ad^2	I_o	I
Two splice plates $\frac{3}{8} \times 55$	41.25	7.55	2,351	9,807	12,158
Area of holes	−4.31	7.55	−246	−1,055	−1,301
					10,857 in.^{4}

$d_{\text{top of splice}} = 27.50 + 7.55 = 35.05$ in. $\quad d_{\text{bottom of splice}} = 27.50 - 7.55 = 19.95$ in.

$S_{\text{top of splice}} = \dfrac{10{,}857}{35.05} = 310$ in.^{3} $\quad S_{\text{bottom of splice}} = \dfrac{10{,}857}{19.95} = 544$ in.^{3}

This gives a maximum bending stress range in the web splice plate of

$$f_{br} = 475 \times 12 \times 35.05/12{,}158 = 16.4 \text{ ksi}$$

The base metal adjacent to friction-type fasteners is classified as category B. For 500,000 cycles, the allowable stress range is 29 ksi, OK.

Web Bolts The bolts must resist the vertical shear, the moment due to the eccentricity of this shear, and the portion of the girder moment transferred to the web. The polar moment of inertia of the bolt group is

$$I = 2 \times 2 \times 1820 + 28 \times 7.55 \times (58.69/57)^2 + 28 \times 1.5^2 = 9035$$

The web moments for overload are calculated next

$$\begin{aligned} M_v &= 151.5 \times 3.25/12 &&= 41 \\ M_w &= 2767 \times 19{,}236/55{,}508 &&= \underline{959} \\ &\text{Total} &&= 1000, \text{ or } 500 \text{ ft-kips/web} \end{aligned}$$

The loads on each bolt are as follows:
Due to shear,

$$P_s = [(151.5)/(2 \times 28)](58.69/57) = 2.8 \text{ kips}$$

Due to moment (outermost bolt),

$$\text{Vertical component} = \frac{500 \times 12 \times 1.5}{5035} \times \frac{58.69}{57} = 1.03 \text{ kips}$$

$$\text{Horizontal component} = \frac{500(58.69/57) \times 12 \times (26 + 7.55 \times 58.69/57)}{9035}$$

$$= 23.09 \text{ kips}$$

The resultant stress on the outermost bolt is therefore

$$P = \sqrt{(2.79 + 1.03)^2 + 23.09^2} = 23.4 \text{ kips } < 25.3 \text{ kips} \qquad \text{OK}$$

Flange Splice The average stress in the top flange under the maximum design load is

$$f_{b(\text{top})} = [4073 \times 12(28.78 + 7.55)]/55{,}508 = 32.0 \text{ ksi}$$

and the total force on the net flange area is

$$P_{\text{top}} = 32 \times (13.50 - 0.34)/2 = 211 \text{ kips}$$

From the total force, we calculate the net area required in the top flange splice plate as

$$A_{\text{top}} = 211/36 = 5.86 \text{ in.}^2$$

We also compute 75 percent of the net area of the top flange, or

$$0.75(13.50 - 0.34)/2 = 4.94 \text{ in.}^2$$

(does not control). Select two 3/8 × 5-3/8-in. inner plates, and one 5/16 × 12-in. outer plate. These give a total net area of 7.50 in.^{2} > 5.86.

The average stress in the bottom flange under the maximum design load is

$$f_{b(\text{bot})} = [4073 \times 12(2872 - 7.55)]/55{,}508 = 18.6 \text{ ksi}$$

and gives a total force on the net flange area of

$$P_{\text{bot}} = 18.6[40.25 - 16 \times (7/16) + 0.15 \times 40.25] = 731 \text{ kips}$$

Likewise, the required net area of the bottom flange splice is

$$A_{\text{bot}} = 731/36 = 20.3 \text{ in.}^2$$

which is less than 75 percent of the net area of the bottom flange. The bottom splice plate area is therefore

$$A_{\text{bot}} = 0.75 \times 39.39 = 29.5 \text{ in.}^2$$

We select two 3/8 × 41-1/2-in. outer plates, and two 3/8 × 41-1/2-in. inner plates. After reduction of the bolt holes in excess of 15 percent, the net area is 59.6 in.^{2}, OK.

The flange splice plates must also be checked for fatigue under service loads. The range of live load moment at the splice is

$$M_{Lr} = 1640 + 1100 = 2740 \text{ ft-kips}$$

However, the corresponding range of stress in both the top and bottom flange splice plate is considerably less than 29 ksi.

Flange Bolts The number of bolts required in the flange splice is determined by the overload criteria. From Table 13-10, the overload moment is 2767 ft-kips. This gives an average stress in the top flange of

$$f_b = 2767 \times 12(28.78 + 7.55)/55{,}508 = 21.7 \text{ ksi}$$

resulting in a total flange force of

$$P_{\text{top}} = 21.7(13.50 - 0.34)/2 = 143 \text{ kips}$$

For this force, the required number of bolts is $143/25.3 = 5.7$ bolts; use 8 bolts.

Likewise, the average stress in the bottom flange is

$$f_b = 2767 \times 12(28.72 - 7.55)/55{,}508 = 12.7 \text{ ksi}$$

resulting in a total flange force of

$$P_{\text{bot}} = 12.7 \times 39.29 = 499 \text{ kips}$$

For this force, the required number of bolts is $499/25.3 = 19.7$ bolts. However, we will use 64 bolts, as shown in the detail of Figure 13-26.

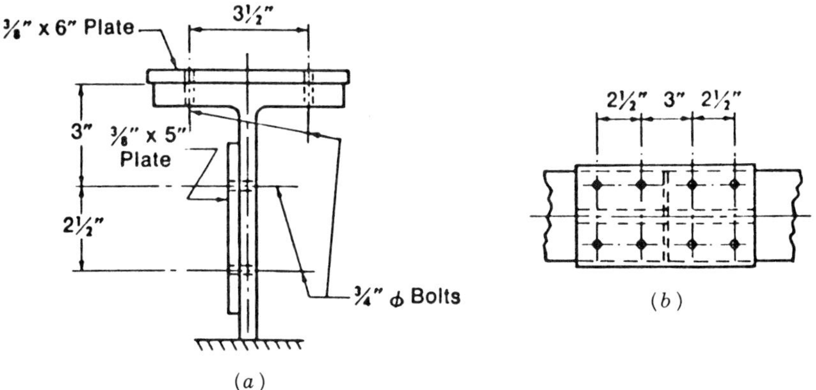

FIGURE 13-27 Splice detail for horizontal bottom flange stiffener: (*a*) cross section; (*b*) plan of stiffener flange splice. (From AISC, 1990.)

Stiffener Splice In a similar procedure, a splice is designed for the ST7.5 × 25 longitudinal bottom flange stiffener. A splice in the stiffener must ensure that the interruption of the stiffener at this point does not become a node for buckling. The splice is designed for the axial capacity of ST7.5 × 25, which is obtained as the product of the critical buckling stress for the bottom flange and the area of the stiffener. The splice detail is shown in Figure 13-27.

13-13 WELDED FIELD SPLICES

AASHTO specifications allow tension and compression members to be spliced by means of full-penetration butt welds, preferably without the use of splice plates, subject to the following requirements.

1. Welded field splices should be arranged to minimize overhead welding.
2. Fillers 1/4 in. or more thick should extend beyond the edges of the splice plate and should be welded to the part on which they are fitted.
3. Welds joining the splice plate to the filler should be sufficient to transmit the splice plate stress.
4. Material of different widths spliced by butt welds should have transitions conforming to AASHTO standards.

The proposed LRFD specifications limit the maximum size of a fillet weld used along edges of connected parts as follows.

- Along edges of material less than 1/4 in. thick, the maximum size should be equal to the thickness of the material.
- Along edges of material 1/4 in. or more in thickness, the maximum size should be 1/16 in. less than the thickness of the material, unless the weld is specifically designated on the plan to be built out to obtain full throat thickness.

The minimum size of fillet welds is determined as follows.

- For base metal thickness $T \leq 3/4$ in., the minimum size is 1/4 in.
- For base metal thickness $T > 3/4$ in., the minimum size is 5/16 in.

In any case the weld size need not exceed the thickness of the thinner part joined, and sufficient preheat is necessary to ensure weld soundness.

Field Welding If welded field splices are considered, they should normally be located on a single plane. Staggering the butt welds of flanges and webs will not improve the performance of the girder. According to welding

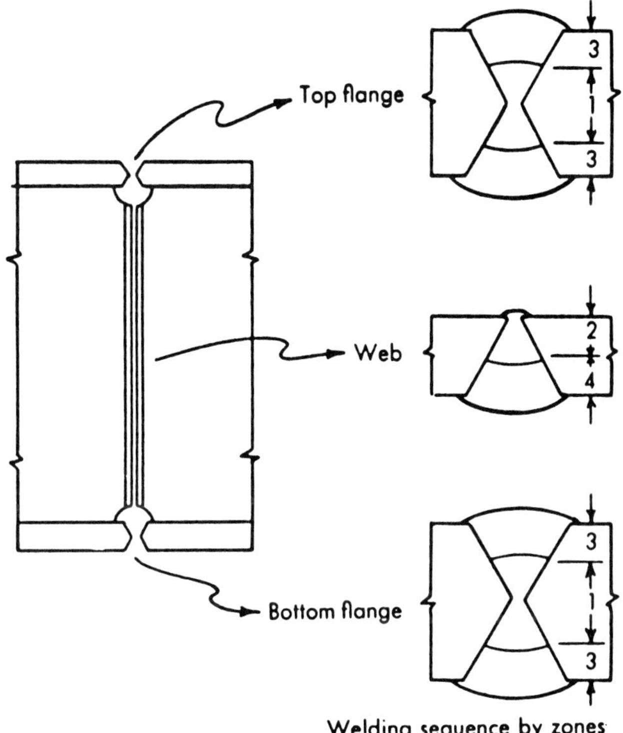

FIGURE 13-28 Usual welding sequence for welded field splice.

procedures, it is easier to prepare the joints and maintain proper fit by flame cutting and leveling when all splices are located in the same plane. An inherent advantage is to extend the fillet welds of flanges to the web to the very end of the girder, because this provides better support when the flanges are clamped together for temporary support during erection.

A usual welding sequence for field splices of beams and girders is based on the following general outline, whereby both the flanges and web are alternately welded to a portion of their depth after they are secured with tack welds, as shown in Figure 13-28.

1. Weld a portion of the thickness of both flanges (1/3–1/2) full width.
2. Weld a portion of the thickness of the weld (about 1/2) full width.
3. Complete the welding of the flanges.
4. Complete the welding of the web.

For deep webs, the vertical welding is sometimes divided into two or more sections following a backstep sequence resulting in a more uniform transverse shrinkage of the joint.

Most butt joints in field splices of webs are the single V type. For thicker webs (usually above 1/2 in.), a double V joint should be specified to reduce the amount of welding and to balance the welding about both sides in order to eliminate angular distortion.

Most flange butt joints to be field-welded depend predominantly on flange thickness increases. A double V joint with half of the welding on both the top and bottom of the joint is suitable to inhibit distortion, but will require considerable overhead welding. For this reason, AWS prequalified joints allow the double V joint to be prepared so that a maximum weld of 3/4 of the flange thickness is on the top and the remaining 1/4 on the bottom.

More data on welded field splices will be found in applicable AASHTO and AWS specifications ISBN (1990).

Design Example A two-span continuous plate girder bridge with span lengths of 100 ft (each span) is composite for positive and negative moments. A field splice (welded) is located 22.5 ft from the center pier and is placed between a diaphragm and a web stiffener close to the dead load inflection point.

The web plate is 3/8 × 42 in. On the positive moment side of the splice, the section has a 3/4 × 10-in. top flange, and a 1 × 10-in. bottom flange. On the negative moment side, the top flange is a 3/4 × 14 in. plate, and the bottom flange is a 1 × 14 in. plate. For a welded field splice, this arrangement offers the advantage that transition slopes are not required.

Section properties are computed for the steel section alone (positive moment, noncomposite), for the steel section with slab longitudinal reinforcement (positive moment of splice), and for the composite section. These properties are shown in Tables 13-14, 13-15, and 13-16, respectively.

TABLE 13-14 Properties of Steel Section on Positive Moment Side of Splice

Material	A	d	Ad	Ad^2	I_o	I
Top flange $\frac{3}{4} \times 10$	7.50	21.37	160	3,425		3,425
Web $\frac{3}{8} \times 42$	15.75				2,315	2,315
Bottom flange 1×10	10.00	−21.50	−215	4,622		4,622
	33.25 in.^{2}		−55 in.^{3}			10,362

$$d_s = \frac{-55}{33.25} = -1.65 \text{ in.} \qquad -1.65(55) = -91$$

$$I_{NA} = 10,271$$

$$d_{\text{top of steel}} = 21.75 + 1.65 = 23.40 \text{ in.} \qquad d_{\text{bottom of steel}} = 22.00 - 1.65 = 20.35 \text{ in.}$$

$$S_{\text{top of steel}} = \frac{10,271}{23.40} = 439 \text{ in.}^3 \qquad S_{\text{bottom of steel}} = \frac{10,271}{20.35} = 505 \text{ in.}^3$$

TABLE 13-15 Properties of Steel Section with Slab Reinforcing, Positive Moment Side of Splice

Material	A	d	Ad	Ad^2	I_o	I
Steel section	33.25		−55			10,362
Reinf. 14-#6	6.16	26.55	164	4,342		4,342
	39.41 in.^{2}		109 in.^{3}			14,704

$$d_s = \frac{109}{39.41} = 2.77 \text{ in.} \qquad -2.77(109) = -\ \ 302$$

$$I_{NA} = 14{,}402 \text{ in.}^4$$

$$d_{\text{top of steel}} = 21.75 - 2.77 = 18.98 \text{ in.} \qquad d_{\text{bottom of steel}} = 22.00 + 2.77 = 24.77 \text{ in.}$$

$$S_{\text{top of steel}} = \frac{14{,}402}{18.98} = 759 \text{ in.}^3 \qquad S_{\text{bottom of steel}} = \frac{14{,}402}{24.77} = 581 \text{ in.}^3$$

$$d_{\text{reinf.}} = 22.25 - 2.77 + 1.00 + 3.30 = 23.78 \text{ in.}$$

$$S_{\text{reinf.}} = \frac{14{,}402}{23.78} = 606 \text{ in.}^3$$

Maximum Moments at Field Splice These moments are summarized as follows:

With positive live load moment =	DL_1	$M = -95$ ft-kips
	DL_2	$M = -20$
	$LL + I$	$M = 575$
With negative live load moment =	DL_1	$M = -95$
	DL_2	$M = -20$
	$LL + I$	$M = -480$

It appears from the foregoing that the location of the field splice near the dead load inflection point renders bending strength noncritical. Fatigue checks, however, should be made for (a) the base metal adjacent to full-penetration groove weld splices (AASHTO category B) and (b) the base metal adjacent to stud-type shear connectors (AASHTO category C). Use 500,000 cycles.

The maximum range of stress in the bottom flange of the girder near the splice is

$$f_{sr} = \frac{480 \times 12}{581} + \frac{575 \times 12}{728} = 19.4 < 29 \text{ ksi}$$

TABLE 13-16 Properties of Composite Section, $3n = 24$, $n = 8$

Composite Section, $3n = 24$						
Material	A	d	Ad	Ad^2	I_o	I
Steel section	33.25		−55			10,362
Conc. $84 \times 7/24$	24.50	26.75	655	17,531	100	17,631
	57.75 in.^{2}		600 in.^{3}			27,993

$$d_{24} = \frac{600}{57.75} = 10.39 \text{ in.} \qquad -10.39 \times 600 = -\ 6{,}234$$

$$I_{NA} = 21{,}759 \text{ in.}^4$$

$$d_{\text{top of steel}} = 21.75 - 10.39 = 11.36 \text{ in.} \qquad d_{\text{bottom of steel}} = 22.00 + 10.39 = 32.39 \text{ in.}$$

$$S_{\text{top of steel}} = \frac{21{,}759}{11.36} = 1{,}915 \text{ in.}^3 \qquad S_{\text{bottom of steel}} = \frac{21{,}759}{32.39} = 672 \text{ in.}^3$$

Composite Section, $n = 8$						
Material	A	d	Ad	Ad^2	I_o	I
Steel section	33.25		−55			10,362
Conc. $84 \times \frac{7}{8}$	73.50	26.75	1,966	52,594	300	52,894
	106.75 in.^{2}		1,911 in.^{3}			63,256

$$d_8 = \frac{1{,}911}{106.75} = 17.90 \text{ in.} \qquad -17.90 \times 1{,}911 = -34{,}210$$

$$I_{NA} = 29{,}046 \text{ in.}^4$$

$$d_{\text{top of steel}} = 21.75 - 17.90 = 3.85 \text{ in.} \qquad d_{\text{bottom of steel}} = 22.00 + 17.90 = 39.90 \text{ in.}$$

$$S_{\text{top of steel}} = \frac{29{,}046}{3.85} = 7{,}544 \text{ in.}^3 \qquad S_{\text{bottom of steel}} = \frac{29{,}046}{39.90} = 728 \text{ in.}^3$$

Assuming that shear connectors are welded to the top flange near the splice, the actual stress range is

$$f_{sr} = \frac{480 \times 12}{759} + \frac{575 \times 12}{7544} = 8.50 < 21 \text{ ksi} \qquad \text{OK}$$

13-14 BEAM–GIRDER CONTINUOUS CONNECTIONS

Examples of these configurations include floor beams piercing through parallel girder systems and longitudinal beams framing into transverse box girders at interior bents. The beams can be made continuous through their girder supports by any of the methods shown in Figure 13-29. The arrangement

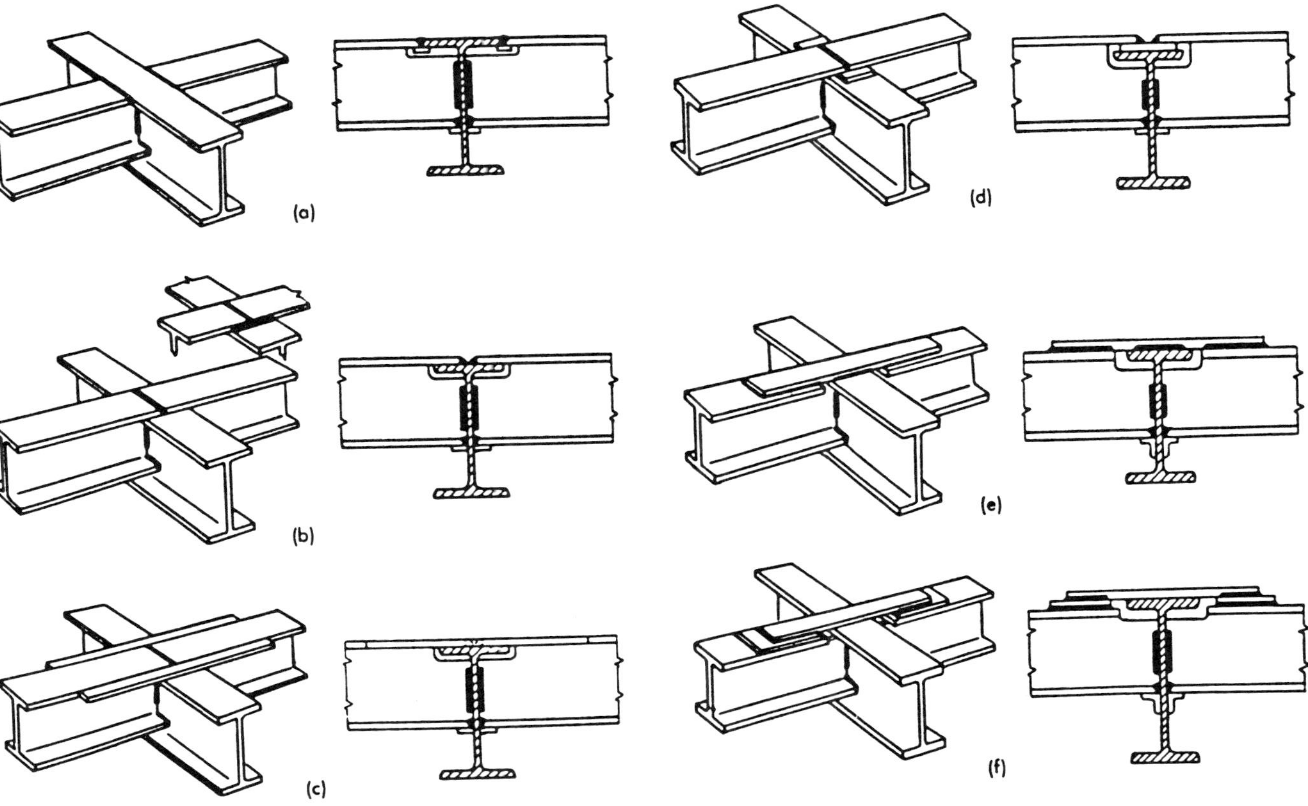

FIGURE 13-29 Various types of beam–girder continuous connections.

shown in Figure 13-29*a* has the beam flange and part of the web below cut back, so that the flange can be butt-welded directly to the edge of the girder flange, which requires that the top surfaces of both members should be on the same level.

In Figure 13-29*b* through *d*, the beam web is cut back just below the top girder flange to allow full contact between the two flanges welded as shown. Likewise, the bridge geometry must accommodate the detail. Additional plates may be used as shown in Figure 13-29*c* along the top after the top flanges are welded to the girder, if this solution is more economical for negative moment regions. The small seat shown in Figures 13-29*c* and *f* facilitates erection and also serves as a backing strip for the groove weld on the lower beam flange. The top connecting plates shown in Figures 13-29*e* and *f* also serve as cover plates to increase the stiffness of the member.

Methods of Analysis Where the supporting girder provides little restraint, if any, for the intersecting beam, the latter may be designed as simply supported even though the flanges are welded to the girder. However, two beam sections framing on opposite sides and loaded (which is the case with bridge framing systems) are restrained, and the resulting end moments must be considered.

The effect of combined stresses is articulated by the example of Figure 13-30*a*. The span length is 20 ft, and the negative moment is taken as $M = W_2 L/12 = 20 \times 240/12 = -400$ in.-kips. The resulting flange forces and stresses are presented diagrammatically in Figure 13-30*b*. The total flange force F is equal to M/d, where $d = 9.65$ in., or $F = 400/9.65 = 41.5$ kips. This gives a normal stress σ_y (tensile), computed as

$$\sigma_y = 41.5/(5.762 \times 0.43) = 16.73 \text{ ksi} \qquad \text{(Tension)}$$

The supporting girder at this location is subjected to a positive moment in its direction, and the resulting stress in the top flange is $\sigma_x = -23.30$ ksi (compression). The two uniaxial stresses σ_x and σ_y are shown acting on an isolated element in the detail of Figure 13-30*b*.

The effect of these stresses on the yielding of the plate can be assessed by the Huber–von Mises formula (also discussed in previous sections). If, for a given combination of normal stresses σ_x and σ_y and shear stress τ_{xy}, the resulting critical stress σ_{cr} equals the yield strength of the steel when tested in uniaxial tension, this combination of stresses is assumed to just produce yielding in the steel. We rewrite the Huber–von Mises formula as

$$\sigma_{cr} = \sqrt{\sigma_x^2 - \sigma_x \sigma_y + \sigma_y^2 + 3\tau_{xy}}$$

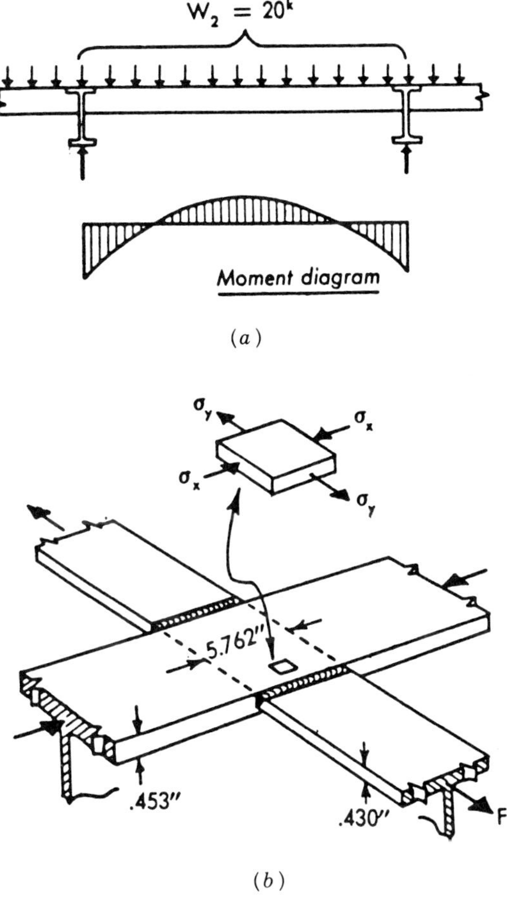

FIGURE 13-30 (*a*) Loading and moment in a beam with continuous connection to supporting girder; (*b*) diagram of forces and stresses acting on top flange.

or

$$\sigma_{cr} = \sqrt{23.3^2 - (-23.3)(16.73) + 16.73^2} = 34.8 \text{ ksi}$$

This result would indicate that the top flange is on the verge of yielding; hence, the tensile flange of the beam should be isolated from the biaxial compressive stress.

In order to demonstrate the sensitivity of the yielding criterion to the actual value of the stresses σ_x and σ_y, we consider a companion beam but in an area where the compressive stress σ_x in the girder flange is considerably less than -23.3 ksi, say -13.0 ksi. Again, the critical stress is computed as

$$\sigma_{cr} = \sqrt{13.0^2 - (13.0)(16.73) + 16.73^2} = 25.8 \text{ ksi}$$

The apparent factor of yielding is $r = \sigma_y/\sigma_{cr} = 36/25.8 = 1.44 \approx 1.50$. This may be acceptable in certain conditions, and it may not be necessary to isolate the two intersecting flanges.

13-15 FIELD SPLICE: BOX GIRDER TO STEEL PIER (LOAD FACTOR METHOD)

Figure 13-31 shows details of a transverse steel box girder at a pier bent framed into a steel column transferring the loads to the foundation. As can be seen the cross girder is integrally joined to the column and extends to the exterior webs of the longitudinal box girders of the bridge.

The steel column and the portion of the cross girder directly above it are fabricated in the shop and erected as one unit. When the longitudinal box girders are fabricated, the portion of the cross girder within them is included in the same unit. The two distinct assemblies (pier column with a portion of the cross girder and longitudinal box girders with the other portion of the cross girder) are connected with field splices.

We will assume that all bending moment is resisted by the flange splice, and all the shear is taken by the web at the connection of the box girder and cross-girder webs.

For the top flange splice, a splice plate connecting the pier column and box girder segment passes over the top flange of the box girder but is not attached to it structurally. The flange of the box girder passes through this region uninterrupted. At the bottom flange, a positive connection is not necessary because this flange is always in compression. The plates can therefore be simply butted together, and the compressive stress transferred by direct bearing.

The splice is designed to carry the larger of (a) 75 percent of the member capacity or (b) the average of the member capacity and the maximum moment. The fasteners are designed for overload with a maximum shear stress of 21 ksi, as in the example of Section 13-12. Likewise, the splice material will be checked for fatigue effects for the base metal adjacent to friction-type fasteners. Splice details are shown in Figure 13-32.

The applied shears and bending moments are tabulated in Table 13-17. The moments are computed for a section through the middle of the inner top flange of the box girder. The box girder reactions are assumed to act at the middepth of the box girder webs.

Maximum Strength of Cross Girder At the face of the column, the cross girder has a section modulus $S = 3159$ in.^{3}, giving a moment capacity $M_u = 36 \times 3159/12 = 9477$ ft-kips, or $0.75M_u = 7108$ ft-kips. The average of M_u and the maximum design moment is $M_{av} = (9477 + 6364)/2 = 7921$ ft-kips > 7108. Therefore, M_{av} controls the design.

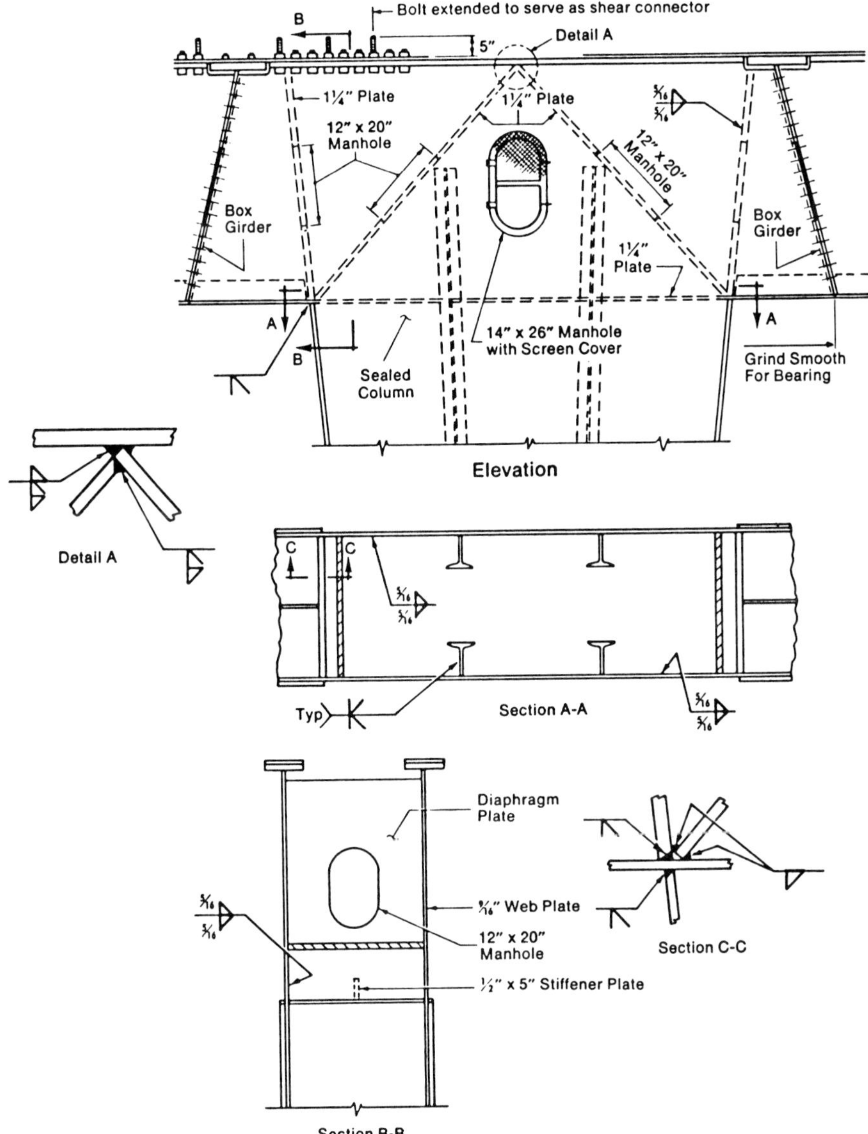

FIGURE 13-31 Cross-girder sections and details at column. (From AISC, 1990.)

Bolted Top Flange Splice. The two top flange splice plates are made of A572, grade 50 steel. Accordingly, the required splice area is

$$A = M_{av}/F_y d = (7921 \times 12)/(50 \times 59.75) = 15.9 \text{ in.}^2$$

We select one 1×17 in. plate in each flange, with a gross area of 17 in.^{2}.

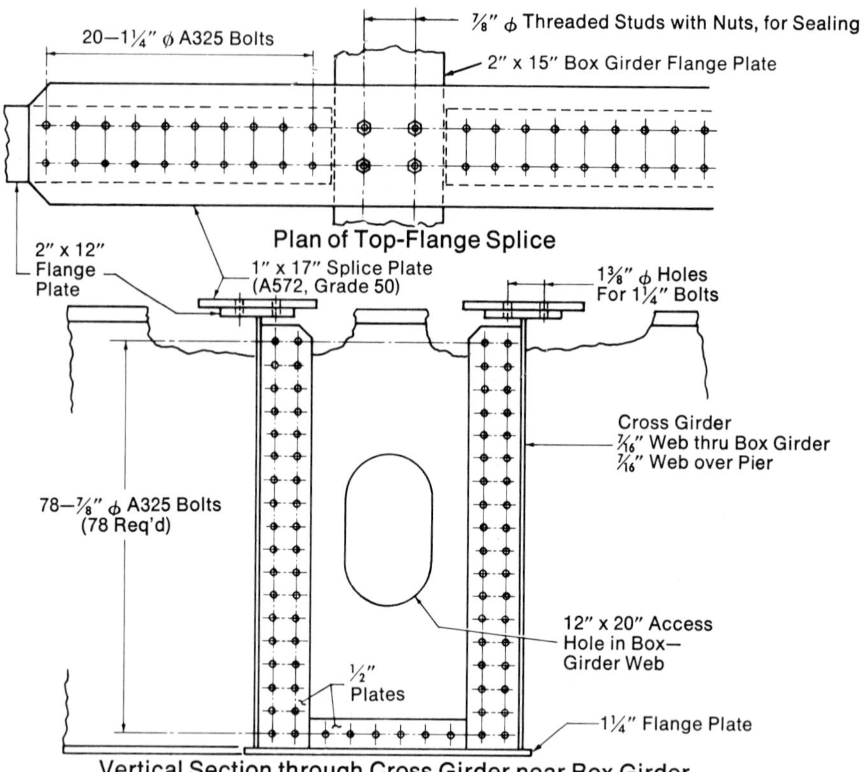

FIGURE 13-32 Cross-girder detail at bolted field splice.

TABLE 13-17 Service and Factored Loads for the Design Example

Service Loads	
Shear (kips)	Moment (kip-ft)
DL_1: 218 + 218 = 436	218 × 9.25 + 218 × 0.58 = 2143
DL_2: 53 + 53 = 106	53 × 9.25 + 53 × 0.58 = 521
$L + I$: 138 + 108 = 246	138 × 9.25 + 108 × 0.58 = 1339
788	4003
Maximum Design Loads: $1.30[D + 5/3(L + I)]$	
Shear (kips)	Moment (ft-kips)
DL_1: 436 × 1.30 = 567	2143 × 1.30 = 2786
DL_2: 106 × 1.30 = 138	521 × 1.30 = 677
$L + I$: 246 × 1.30 × $\frac{5}{3}$ = 533	1339 × 1.30 × $\frac{5}{3}$ = 2901
1238	6364

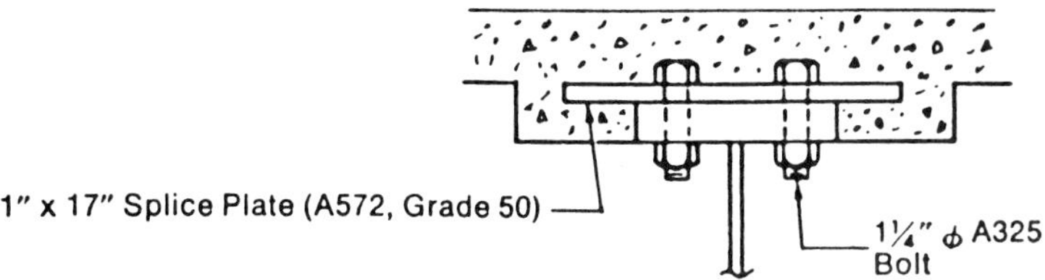

FIGURE 13-33 Section through top flange splice.

The net area is

$$A_{\text{net}} = 17 - (2 \times 1\,3/8 \times 1 - 0.15 \times 17) = 16.8 \text{ in.}^2 > 15.9 \qquad \text{OK}$$

A section through the top flange splice is shown in Figure 13-33.

Note that we select 1.25-in.-diameter A325 bolts. The overload moment is

$$D + (5/3)(L + I) = 2143 + 521 + (5/3) \times 1339 = 4896 \text{ ft-kips}$$

or 2448 ft-kips per flange. The resulting force in the flange is

$$F = M/d = 2448 \times 12/58.25 = 504 \text{ kips}$$

For an allowable stress of 21 ksi under overload and a bolt area of 1.23 in.^{2}, the required number of bolts is $N = 504/(21 \times 1.23) = 20$ bolts.

Web Splice The web splice connection is assumed to carry the entire shear but no moment. As shown in Figure 13-32, 1/2-in. plates are welded to the webs and bottom flange of the cross girder. These plates will be field-bolted to the box girder interior web at the field splices. This arrangement is shown in Figure 13-34.

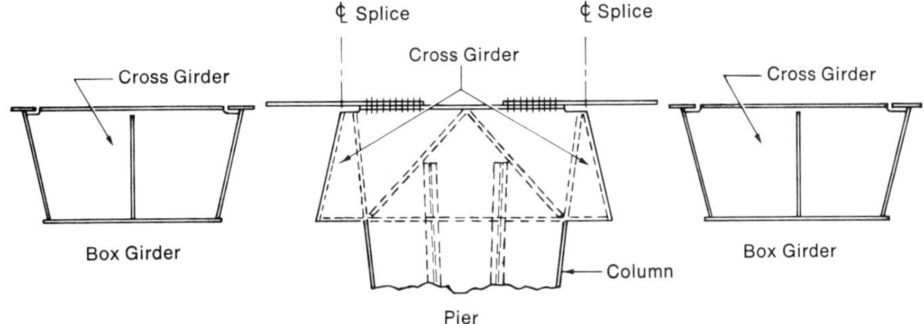

FIGURE 13-34 Field splice arrangement, box girder and cross girder to pier.

The overload shear is obtained from Table 13-17 as

$$436 + 106 + (5/3)246 = 952 \text{ kips}$$

giving a shear on the sloped box girder of $952 \times 58.69/57 = 980$ kips. For this connection, we use 7/8-in.-diameter A325 bolts, as shown in Figure 13-32, for a bolt area of 0.60 in.^{2}. The required number of bolts is $980/(21 \times 0.60) = 77$. Use 78 bolts arranged in two rows of 18 bolts each along each web, and six bolts along the bottom flange.

Fatigue in Bolted Top Flange Splice Fatigue under service loads is checked in the base metal adjacent to friction-type fasteners. The range of the live load moment is 1339 ft-kips, giving a range of force in one flange plate of $1/2 \times (1339 \times 12)/59.75 = 135$ kips. The actual stress range in the gross section of the splice plate is $135/17 = 7.9$ ksi; hence, fatigue considerations do not govern.

Bottom Flange Joint Because this flange is always in compression, the associated compressive force is assumed to be transferred by bearing. For this purpose, abutting plate edges are ground smooth. In order to inhibit buckling due to possible nonuniform bearing stress in the abutting flange plate, a stiffener plate $1/2 \times 5$ in. is provided at midwidth of the cross-girder bottom flange between the inner web of the box girder and the face of the column, as shown in Figure 13-35.

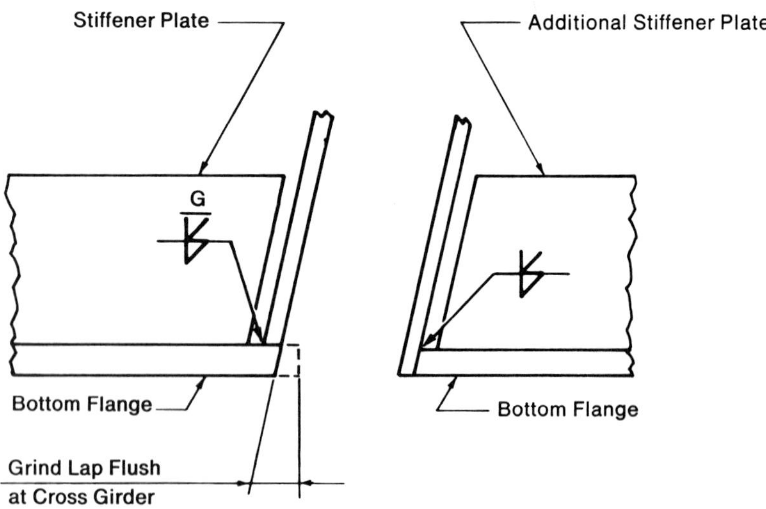

FIGURE 13-35 Detail of bottom flange at welded field splice.

13-16 EXPANSION JOINTS

Requirements and Limitations

Thermal changes cause corresponding changes proportional to the length of the bridge. The structure contracts with the cold and expands with heat, and a typical bridge may be approximately 1 in. longer per 100 feet of length in the summer than in the winter. The bridge will also have daily and short-term changes of a lesser degree according to the temperature changes, and will undergo additional movement associated with elastic deflections.

Changes in length can be compensated for by corresponding deformations within the structure itself, for example, in integral bridges discussed in other sections. At present, however, the usual practice is to provide expansion joints, particularly with large spans and where the forces required to deform the structure are considerable. In this context, bridges are normally designed with provisions for expansion joints at suitable intervals to accommodate movements resulting from expansion and contraction and to relieve the associated forces. Types of joints developed for this purpose vary from open joints, simple planes of weakness, and elastic joints to long interlocking fingered castings and sliding bar joints.

The type of expansion device to be selected for a bridge deck thus depends on the length that must be accommodated by the expansion opening, the skew angle of the bridge, and on whether the joint is to be sealed.

Basic Types of Joints Open deck joints should be located where drainage can be directed to bypass the bearings. They should also permit the free flow of water through the joint. Open deck joints should be avoided where deicing chemicals are applied, and accumulation of water and debris should be prevented.

When joints are located directly above structural members and bearings that may be adversely affected by debris accumulation, closed deck joints should be selected. With deicing chemicals, sealed or waterproof joints are indicated. Sealed deck joints should seal the deck surface completely and should also prevent the accumulation of water and debris that may restrict the operational characteristics of the bridge.

It appears from the foregoing remarks that deck expansion joints may be divided into the following three categories.

1. Modular expansion joint systems including finger and sliding plates, with up to 2 to 16 in. of movement.
2. Metal reinforced expansion joint systems, generally with 2 to 13 in. of movement.
3. Strip seals and armored expansion joint systems, including preformed neoprene seals, with 1/2 to 4 in. of movement.

Alternatively, joints are classified as unsealed (open joint, sliding plate, and finger plate); unsealed joints over bridge seats provided with bridge seat sealer; and sealed joints (preformed joint sealer and neoprene expansion joints).

Applicable Specifications and Standards

AASHTO stipulates that thermal movement must be provided for at the rate of 1.25 in./100 ft. In all bridges proper means should be provided for movement caused by temperature changes. In spans more than 300 ft long, provisions will be made for expansion and contraction in the floor, and this expansion should be secured against lateral movement.

LRFD Specifications According to applicable provisions, deck joints must satisfy the requirements of all specified strength, fatigue and fracture, and service limit states. Joints and their supports must withstand factored force effects over the factored of movements.

Superstructure movements result from the placement of bridge decks, volumetric changes such as shrinkage, temperature, moisture and creep, vehicular passage, wind effects, and the action of earthquakes.

When horizontal movement at the ends of a superstructure is caused by volumetric changes, consideration should be given to the balance of the forces generated within the structure to resist these changes. This provides the basis for determining the neutral point taking into account the relative resistance of bearings and substructure to movement, and allows estimation of the length contributing to movement at a particular joint.

The number of movable deck joints should be kept to a minimum. Preference should thus be given to continuous decks and, where appropriate, integral bridges.

Design Criteria The width of a deck joint, measured normally, should not exceed the following:

- For a single gap, $2.5 + 1.5(1 - 2\sin^2\theta)$
- For modular multiple gaps, $2.0 + 1.0(1 - 2\sin^2\theta)$ where θ is the skew angle of the deck at the joint.

The open width of a transverse deck joint should not be less than 1.0 in. at factored extreme movement.

The opening between fingers of finger plate joints at factored movement should not exceed: (a) 2.0 in. for longitudinal openings greater than 8 in. and (b) 3.0 in. for longitudinal openings 8 in. or less. The finger overlap at factored extreme movement should not be less than 1.5 in.

Joint Components Completely effective joint seals are not available for all situations, particularly with severe skews and where the joints are subjected to considerable movement (see also the following sections). There are instances therefore where an open or closed joint should be considered. In addition, certain abutment designs may have to be changed to improve the performance of deck joints sealed with sheet- or strip-type seals.

The LRFD specifications contain detailed provisions for waterproof joints, joint seals, poured seals, compression and cellular seals, sheet and strip seals, plank seals, and modular seals.

13-17 PERFORMANCE OF BRIDGE DECK EXPANSION JOINTS

A study by Dahir and Mellott (1987) focuses on the characteristics and field performance of modular expansion joints, metal reinforced elastomeric expansion dams, and gland-type bridge expansion dams. The joint rating is according to FHWA standards and covers the general appearance, condition of anchorages, debris accumulation, watertightness, surface damage, noise under traffic, and ease of maintenance.

Typical joint systems considered in this investigation include (a) open joints protected by armored neoprene or preformed neoprene compression seals, shown in Figure 13-36; (b) metal plates with neoprene strips known as "strip seals" anchored to the bridge deck, shown in Figure 13-37; and

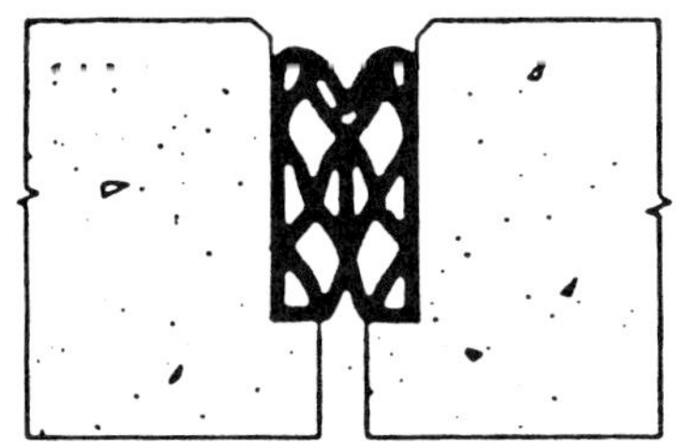

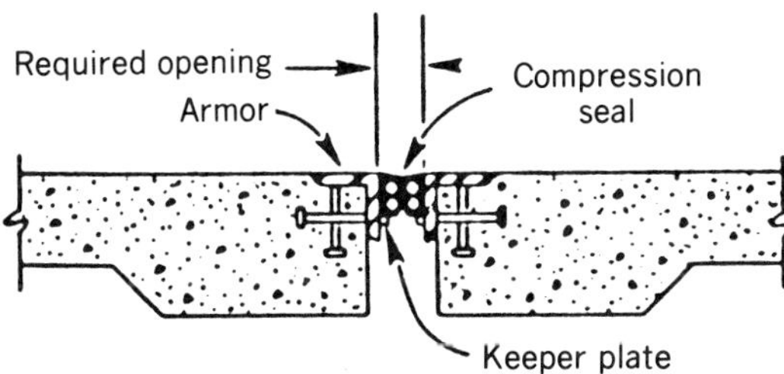

FIGURE 13-36 Typical open joint with conventional seal systems.

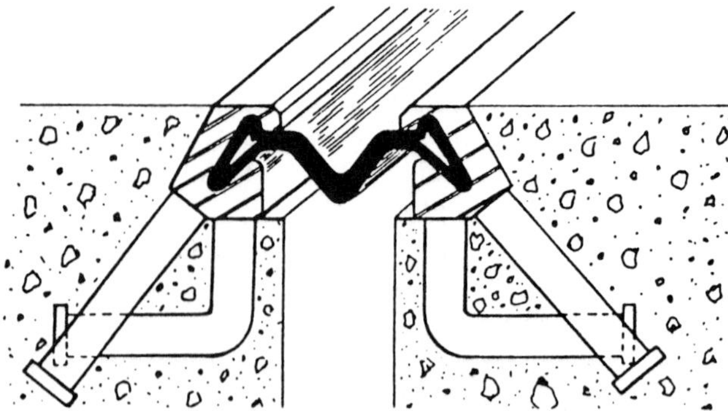

FIGURE 13-37 Strip seal joint system.

(c) toothed bearing supports (finger dams) and modular dam systems, shown in Figure 13-38. A fourth less typical joint system is the metal reinforced and continuous-belt dam shown in Figure 13-39.

The armored neoprene and standard unarmored neoprene compression seals are commonly used with prestressed beams for movement up to 2 in. The strip seals are used for movement up to 4 in. The tooth or finger dams are used for movement exceeding 4 in.

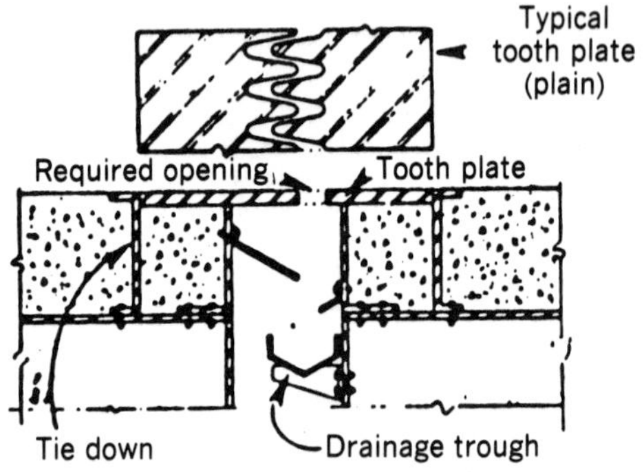

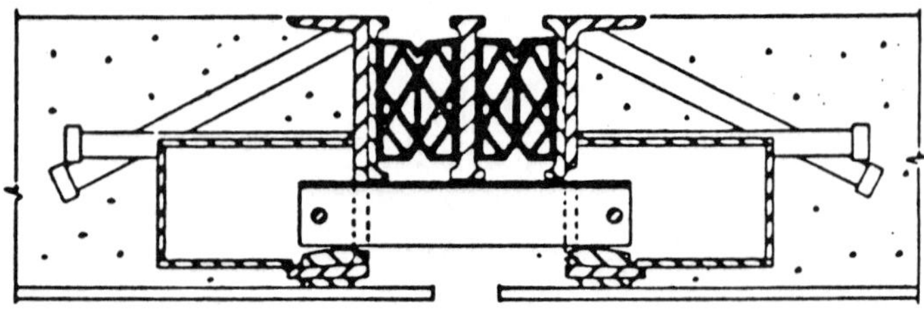

FIGURE 13-38 Typical finger dam and modular dam joints.

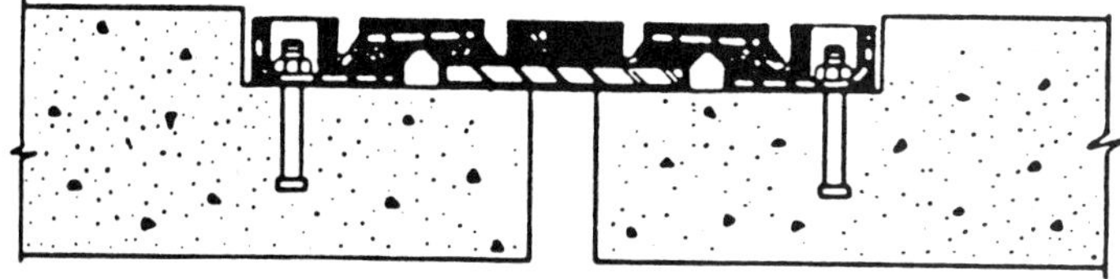

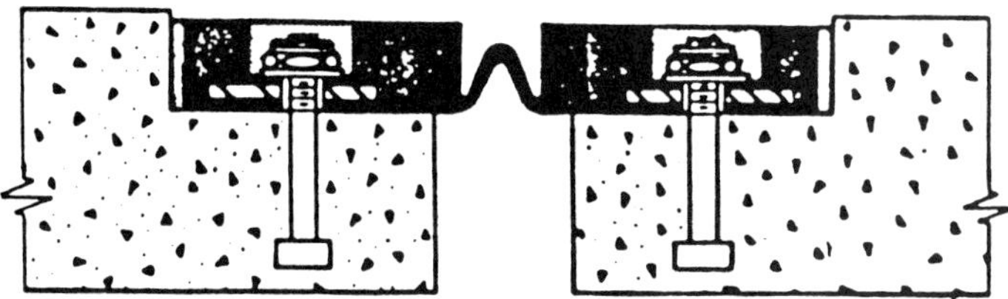

FIGURE 13-39 Metal reinforced and continuous-belt dam system.

Armored Neoprene and Preformed Neoprene Compression Seals These are commonly used in Pennsylvania as well as in other states. Problems may arise if the concrete around the armored channels is not fully consolidated or if the neoprene becomes twisted during construction. These problems can be prevented by appropriate construction controls. Problems relating to rusting are avoided by proper maintenance and periodic painting.

Finger Dams Misalignment during construction may be the cause of early problems. It can be horizontal causing the fingers to jam upon expansion, or vertical causing rough riding, noise, and sometimes fracture. A second possible problem is blockage by debris accumulation. This problem is more serious if contaminated water does not drain to flush loose debris. Debris accumulation further inhibits water drainage and accelerates rebar corrosion.

Some of the cantilever fingers may be bent or even broken and detached from the device under heavy truck traffic and impact unless they are properly designed under fatigue considerations. Proper deck maintenance is essential during service.

Strip Seal or Gland-Type Systems The evaluation by Dahir and Mellot (1987) rates the performance of these joints from fair to quite good. Some of the problems encountered include debris accumulation, leakage, and noise under traffic.

Metal Reinforced Elastomeric and Continuous-Belt Dams These devices, shown in Figure 13-39, are intended to replace the finger dam joint, allowing movement of 4 to 13 in., with nominal need for maintenance. Their performance has varied from poor to fair to barely satisfactory.

Problems associated with their use range from poor anchorage to wear and tear by traffic impact and snow plows. Once the neoprene material is torn and the anchorage becomes loose, the joint will leak and allow debris accumulation.

Modular Expansion Dam System These devices can provide movement up to 26 in. or more through a prefabricated assembly or module consisting of one or more preformed neoprene elastomeric elements secured between contoured transverse load distribution members that bear on specially designed beams, as shown in Figure 13-38. The module is installed between end dams anchored into the concrete deck.

Experience with these systems (FHWA, 1977; TRB, 1983) shows that although some of these joints have performed fairly well, most have had problems as serious as those they were intended to eliminate. Snow plow damage and debris accumulation are two typical examples.

Because it is beyond the scope of this book to provide system ratings and data analysis in terms of the proper installation during construction and periodic maintenance (two aspects critical to the performance of any joint device), we choose to confine this discussion to the functional characteristics of bridge joints. Thus, a problem common with all joint types relates to the skew angle. Many skewed joints with acute angles of 30° to 70° showed buckling or folding of the neoprene and suffered damage from snow plows and traffic. The smaller the skew angle, the more severe was the damage or distortion, especially with heavy traffic. Joints most susceptible to damage include unprotected neoprene compression seal joints (unarmored) and the elastomeric joint types. Edge anchoring of these joints appears to be difficult to maintain.

A final observation associated with the results of this study is the effect of heavy truck traffic on the joint systems. It appears that heavier traffic accelerates and increases joint damage, especially on exposed neoprene. In addition, even some finger dam joints had some of the steel fingers bent under heavy traffic. Sufficient specific data are not available, however, for correlating truck traffic with the extent of damage.

13-18 EXAMPLE OF EXPANSION JOINT

The welded expansion joint shown in Figure 13-40 is made entirely from rolled structural shapes. As detailed, the joint provides for 16-in. movement calculated at the rate of 1.25 in./100 ft for a bridge 1200 ft long.

The joint is made in two halves, each half being symmetrical with respect to the other half by rotating 180°, and is fabricated to fit the roadway crown. The interlocking fingers forming the top surface are flame-cut from the same 28 in.× 1 in.× 24 ft plate as shown in the layout of Figure 13-40*a*, and they

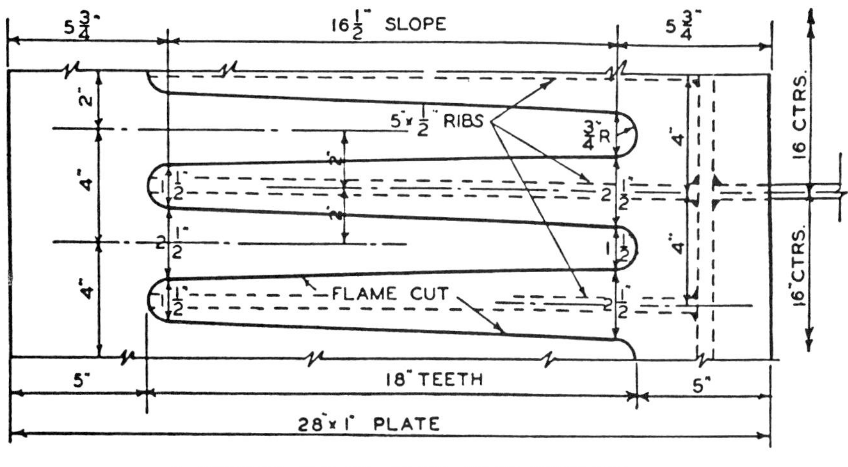

(*a*)

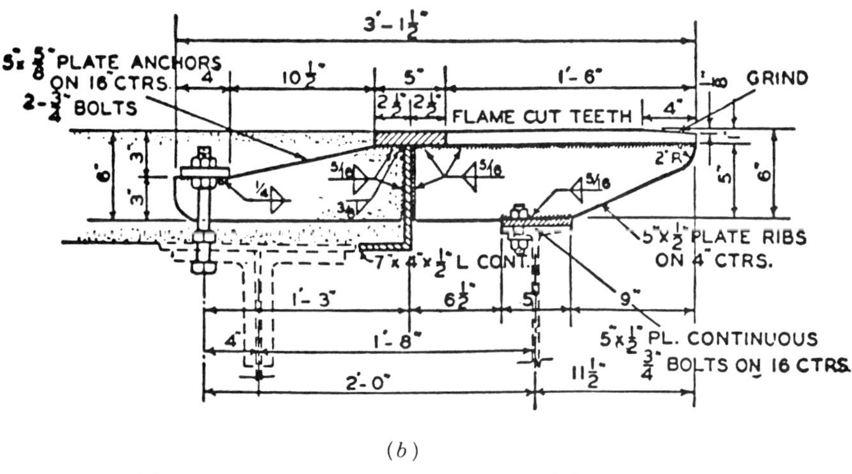

(*b*)

FIGURE 13-40 Finger-type expansion joint, all welded: (*a*) partial plan; (*b*) section along direction of movement.

are spaced at 4-in. centers. The small spacing is selected to allow the distribution of loads over as many fingers as possible. The joint is designed to support a 16-kip wheel load with 100 percent impact. This load is distributed equally to each of five adjacent fingers and is assumed to be applied on a contact area 3 in. long centered 1.5 in. from the end of the fingers. At this position, the fingers on one side of the joint support the entire load. This criterion is the basis for proportioning the size and dimension of the plate, web, and other parts of the joint.

The unusually long cantilever projection of the fingers is balanced by the auxiliary cross beam, supported in turn on the end floor beam as shown in Figure 13-40*b*. The strength of the fingers is derived by continuous groove or fillet welding of the vertical web plate to the underside of the fingers. The near (left) ends of these ribs are anchored for uplift by groove welding to the back of the slab closure angle. This angle is continuously welded to the 1-in. joint plate and also serves as a lateral distribution beam between the plate anchors.

Plate anchors consisting of 5 × 5/8 × 17-in. web plates are welded to the rear of the joint opposite the web of every fourth finger. These plates are therefore spaced at 16-in. centers and engage two 3/4-in. jacking bolts to the flange of the floor beam. These bolts serve as erection bolts for setting the joint to elevation and grade as well as anchor bolts. The entire design should be checked for fatigue under applicable categories and stress ranges.

13-19 DIAPHRAGMS AND CROSS FRAMES: DESIGN REQUIREMENTS

The use of diaphragms and cross frames in bridge design has gained general acceptance. Usually, a series of diahpragms act together with the longitudinal beams or girders to form a system that behaves as a unit.

For concrete T girders and box girders, the diaphragm requirements are specified in AAHSTO Article 8.12. For steel superstructures, diaphragms and cross frames are discussed in AASHTO Article 10.20.

Diaphragms and cross frames, placed at the ends of a span, across interior supports, and intermittently along the span, serve any or all of the following functions: (a) transferring lateral wind loads from the bottom of the girder to the deck and then to the bearing, (b) stabilizing the bottom flange under compressive loads, (c) stabilizing the top flange until the concrete deck is in place, (d) distributing vertical loads, and (e) facilitating the erection process.

Diaphragms and cross frames may be included in the structural model selected to analyze load effects, in particular, the load distribution. In this case they should be designed for the applicable limit state. End cross frames or diaphragms are normally proportioned to transmit the lateral wind to the bearings (see also the examples in previous sections). In addition, support of the end portion of the deck including concentrated live loads should be reflected in the design of end diaphragms. Connections between longitudinal girders and diaphragms should be designed for end moments in the diaphragms. In skewed bridges the longitudinal component of lateral forces should be considered.

If the horizontal wind pressure is not to be resisted by transverse (horizontal) flange bending, the effect of horizontal wind pressure on diaphragms and cross frames should be considered. The LRFD specifications require the wind loads to be carried to the bearings by (a) use of the deck as a horizontal shear diaphragm, (b) use of a top lateral bracing system, or (c) use of cross frames to carry the wind to a bottom lateral bracing.

The wind force on the top flange is assumed to be transmitted directly to the deck slab, and the wind force is transmitted through cross frames also to the slab unless bottom bracing is provided. These assumptions are consistent with past practice and imply that the concrete haunches enveloping the top flange are stiff enough to accommodate the transfer of load.

If lateral wind bracing is provided in one plane (top or bottom flange), the wind forces are assumed to be carried to the supports by the lateral bracing. It is also assumed that the wind force on the other flange will be carried through the cross frames to the plane of lateral bracing. If lateral bracing is absent in the plane of either flange, the wind force on each flange is assumed to be carried to the bridge supports by lateral bending of the flange. If diaphragms are present, the wind loads may be assumed to be resisted equally by all flanges in the same plane.

Straight Box Sections The review of steel box girder bridges in previous sections suggests that diaphragms or cross frames are required within box sections at supports to resist transverse rotation, displacement, and distortion during erection. According to LRFD, however, permanent internal intermediate diaphragms and cross frames are not required for straight steel multiple-box sections properly designed. For single-box sections, internal intermediate diaphragms or cross frames should be spaced at intervals not to exceed 25 ft.

Interior end diaphragms at abutments and piers must be designed to transmit torsional loads from the box to the bearings. These diaphragms serve to retain the box shape and also to transfer vertical, longitudinal, and transverse loads to the bearings. Access holes should be made as small as possible, and their effect on the stress pattern should be investigated.

Half of the wind load on the girder should be applied in the plane of the bottom flange. An interbox lateral bracing should be provided if the combined force caused by the factored horizontal load and the factored dead load exceed the factored resistance.

In general, no lateral bracing system is required between straight box girders. In box girders with sloping webs, the horizontal component of web shear acts as a lateral horizontal force on the flange, and internal lateral bracing or struts may be necessary to resist this force. The need for temporary cross frame to control distortion stresses during erection should be investigated.

13-20 DIAPHRAGMS AND CROSS FRAMES: DESIGN EXAMPLES (LOAD FACTOR METHOD)

Intermediate Cross Frame

Cross-Frame Diagonal Figure 13-41 shows typical cross sections of a three-span continuous welded plate girder bridge. Intermediate cross frames with diagonals forming an inverted V area are spaced at 25-ft intervals. They support the interior stringers and brace the bottom flanges of the main girders transversely. The depth of the cross frames is assumed to be 10 ft. Based on this dimension, the length of the cross-frame diagonal is

$$L_c = \sqrt{10^2 + 9.25^2} = 13.6 \text{ ft}$$

The maximum reaction of a stringer for a cross-frame spacing of 25 ft is 33.4 kips dead load, and 61.1 kips live load plus impact. This reaction is resisted by two compression diagonals, each assumed to carry half the load. For the maximum design load on the stringer, each diagonal must resist a vertical load

$$P_V = 0.5 \times 1.30[33.4 + (5/3)(61.1)] = 88.0 \text{ kips}$$

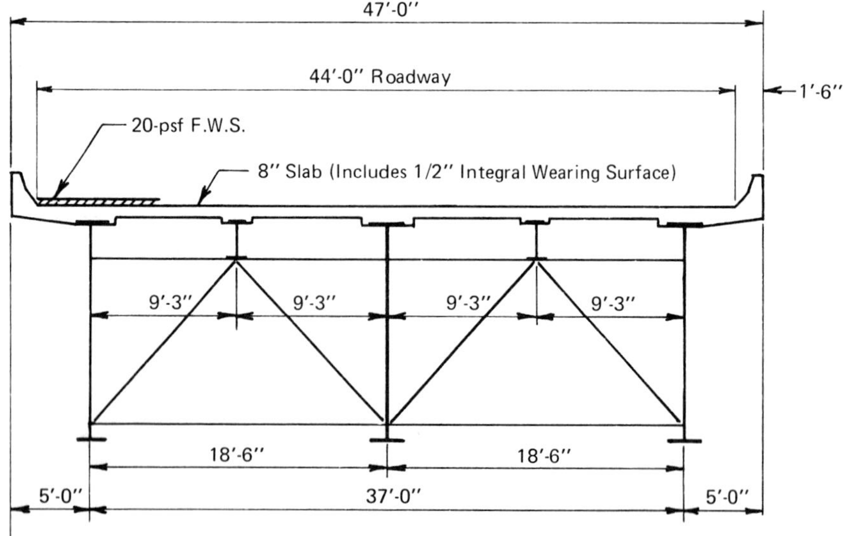

FIGURE 13-41 Typical cross section of bridge in example problem. (From AISC, 1990.)

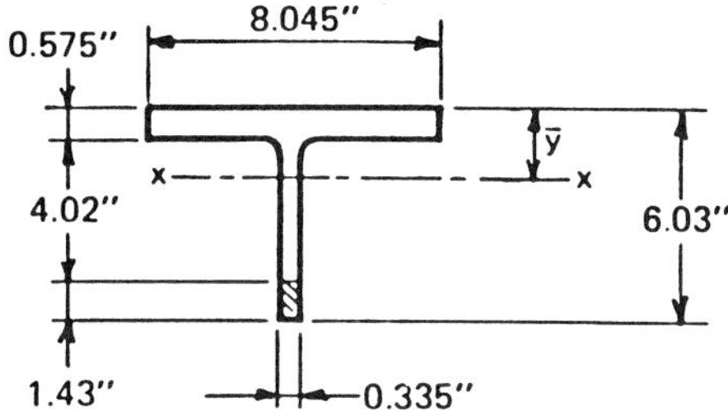

FIGURE 13-42 Cross section of diagonal member of cross frame.

giving an axial load on each diagonal

$$P = 88.0 \times 13.62/10 = 120.0 \text{ kips}$$

The diagonal is connected in the cross frame with 7/8-in.-diameter A325 bolts. For overload, the bolts carry $120/1.3 = 92.1$ kips. We compute the shear capacity, assuming threads in the shear plane, using $\phi F = 28$ ksi, giving a design strength $\phi R = 0.6 \times 28 = 16.8$ kips/bolt.

The required number of bolts is $92.1/16.8 = 5.5$; use 8 bolts. Note that for slip-critical joints, the force caused by $D + (5/3)(L + I)$ must also satisfy the requirements of Article 10.57.3. Using $\phi F_s = 21$ ksi, $A_b = 0.60$ in.^{2}, $N_b = 8$, and $N_s = 1$, the design force ϕR_s is $21 \times 0.6 \times 8 \times 1 = 100.8$ kips > 92.1, OK.

For the diagonal, we select WT6 $\times$ 22.5. For bolted ends $K = 0.75$. Next, referring to Figure 13-42, we calculate $A = 6.13$ in.^{2}, $I = 7.6$ in.^{4}, and the radius of gyration $r = \sqrt{I/A} = \sqrt{7.6/6.13} = 1.11$ in. The slenderness ratio KL_c/r is $0.75 \times 13.62 \times 12/1.11 = 110$. We also compute

$$\sqrt{\frac{2\pi^2 E}{F_y}} = \sqrt{\frac{2\pi^2 \times 29{,}000}{50}} = 107 < 110$$

using $F_y = 50$ ksi. Hence, the critical buckling stress is (AASHTO Article 10.54.1)

$$F_{cr} = \frac{\pi^2 E}{(KL_c/r)^2} = 3.14^2 \times 29{,}000/110^2 = 23.7 \text{ ksi}$$

The maximum capacity of the diagonal is obtained from

$$P_u = 0.85 A_s F_{cr} = 0.85 \times 6.13 \times 23.7 = 123.5 \text{ kips} < 120.0 \qquad \text{OK}$$

As a final check, the effective depth of the stem will now be computed. From Figure 13-42, the ratio d/t_w is assumed to be 12 (d is the effective depth, and t_w is the web thickness), giving an ineffective stem of 1.43×0.335

in. The effective depth must satisfy the relationship $d/t_w = 1625/\sqrt{f_a}$, where f_a, the compressive stress in the section, is $120/6.13 = 19.6$ ksi. From this, we compute $d/t_w = 1625/\sqrt{19{,}600} = 11.6 \approx 12$, OK.

Bottom Strut This member is in tension under the load transmitted by the diagonals. The maximum load on the struts is $P = 120 \times 9.25/13.62 = 81.5$ kips. For WT7 × 19, the slenderness ratio is $L/r = 18 \times 12/1.55 = 139 < 200$. The tensile strength of the struts ($A_s = 5.58$ in.^{2}) is

$$P_u = 5.58 \times 50 = 279 > 81.5 \text{ kips} \qquad \text{OK}$$

Top Strut The top strut of the cross frame is a secondary compression member and must have a slenderness ratio not exceeding 140. The minimum radius of gyration for an unbraced length of 9 ft is $r = 9 \times 12/140 = 0.771$ in. We select a 4 × 4 × 5/16-in. angle, $r_2 = 0.791$, OK.

Cross Frames at Interior Supports

For the same example, bottom bracing is provided as shown in Figure 13-43. Essentially, this bracing consists of a Warren truss of A588 steel in each bay just above the bottom flanges. Using a lateral wind of 0.86 kip/ft of deck length, the shears left and right of the interior support are 68.8 and 69.7 kips, respectively (data obtained from a wind analysis of the lateral bracing). The average panel shear adjacent to the cross frame is $0.5(69.7 + 68.8) = 69.3$ kips (tension). The total wind reaction in the bottom strut is computed as 149.2 kips (compression), giving a net compression in the bottom strut of $149.2 - 69.3 = 79.9$ kips. According to Section 13-19, this wind force must be resisted by the bottom strut in compression, and because the loading is eccentrically applied, the bottom strut acts as a beam–column.

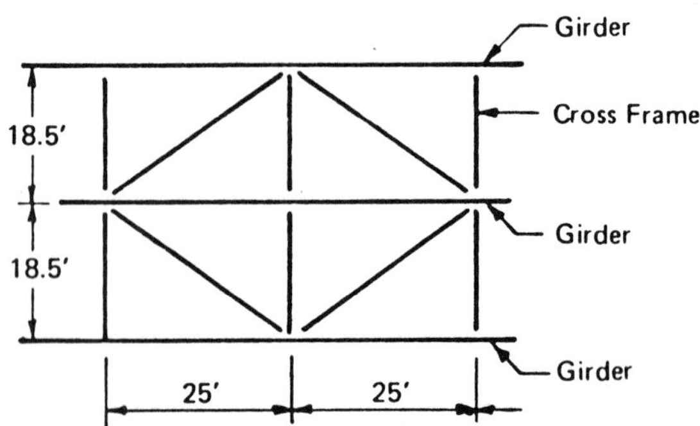

FIGURE 13-43 Bottom braced panel for the bridge of Figure 13-41. (From AISC, 1990.)

The design of the strut under the load strength method must satisfy the requirements of AASHTO Article 10.54.2. The following interaction equations apply:

$$\frac{P}{0.85A_sF_{cr}} + \frac{MC}{M_u(1 - P/A_sF_e)} \le 1.0$$
$$\frac{P}{0.85A_sF_y} + \frac{M}{M_p} \le 1 \qquad (13\text{-}22)$$

where F_{cr} = buckling stress, defined previously
M_u = maximum strength
$F_e = \pi^2E/(KL_c/r)^2$ = Euler buckling stress in plane of bending
C = equivalent moment factor, calculated according to AASHTO Article 10.54.2.2
$M_p = F_yZ$ = full plastic moment of section
Z = plastic section modulus

We select a WT9 × 38, $F_y = 50$ ksi, and $A_s = 11.2$ in.^{2}. The corresponding slenderness ratio in the Y–Y axis is $0.75 \times 18 \times 12/2.61 = 62.1$. Next, we compute $\sqrt{2\pi^2E/F_y} = 107 > 79.4$. Hence, the critical buckling stress for the Y–Y axis is

$$F_{cr} = F_y\left[1 - \frac{F_y}{4\pi^2E}\left(\frac{KL_c}{F_y}\right)^2\right]$$
$$= 50\left[1 - \frac{50}{4 \times 3.14^2 \times 29{,}000}(62.1)^2\right] = 41.58 \text{ ksi}$$

The capacity (maximum strength) of the WT as an unbraced beam is

$$M_u = F_yS\left[1 - \frac{3F_y}{4\pi^2E}\left(\frac{L_b}{0.9b'}\right)^2\right] = 388 \text{ in.-kips}$$

The slenderness ratio in the X–X axis is $KL_c/r_x = 0.75 \times 18 \times 12/2.54 = 63.8$. The Euler buckling stress for this axis is computed as $3.14^2 \times 29{,}000/63.8^2 = 70.32$ ksi.

A section of the WT member is shown in Figure 13-44. The plastic section modulus for this section is computed as $Z = 17.29$ in.^{3}. The full plastic moment of the section as a compact beam is therefore $M_p = F_yZ = 50 \times 17.29 = 865$ in.-kips.

The factored axial wind load is $P = 1.30 \times 79.9 = 103.9$ kips, applied with an eccentricity of 1.80 in. The maximum bending moment in the strut is

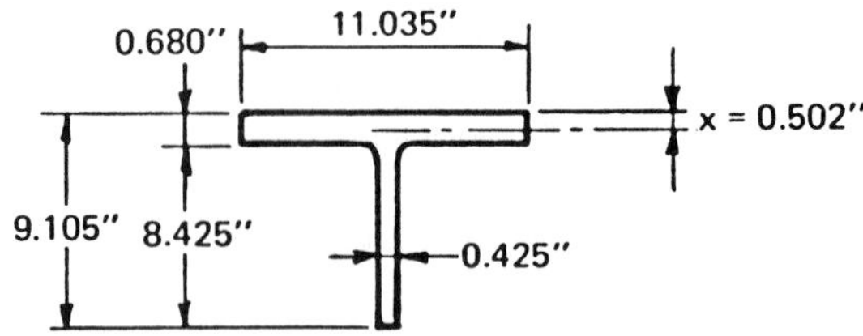

FIGURE 13-44 Typical section of bottom strut WT9 × 38.

therefore

$$M_{\text{ecc}} = 103.9 \times 1.80 = 187.0 \text{ in.-kips}$$

$$M_{\text{DL}} = 1.3 \times 0.038 \times 18^2/8 = -24$$

or total $M = 163.0$ in.-kips.

These results are substituted into (13-22) to check the interaction requirements, giving

$$\frac{103.9}{0.85 \times 11.2 \times 11.58} + \frac{163}{388(1 - 103.9/11.2 \times 70.32)} = 0.746 < 1.0$$

$$\frac{103.9}{0.85 \times 11.2 \times 50} + \frac{187}{865} = 0.434 < 1.0 \qquad \text{OK}$$

Transverse Diaphragm at End Supports

A typical elevation and cross section of the end diaphragm are shown in Figure 13-45. The edge of the slab is haunched down to the beam and transfers the live load directly.

The beam (W12 × 22) directly supports a dead load of 0.43 kip/ft and a wheel load plus impact of 16 × 1.30 = 20.8 kips. These produce the following moments:

$$M_{\text{DL}} = 0.125 \times 0.43 \times 8.5^2 = 3.9 \text{ ft-kips}$$

$$M_{\text{LL}+I} = 20.8 \times 8.5/4 = 44.2 \text{ ft-kips}$$

The required strength is the factored moment

$$M = 1.30(3.9 + 1.67 \times 44.2) = 100.8 \text{ ft-kips}$$

The beam has a plastic section modulus $Z = 29.3$ in.^{3}, an elastic section modulus $S = 25.4$ in.^{3}, and qualifies as a compact section. The strength of

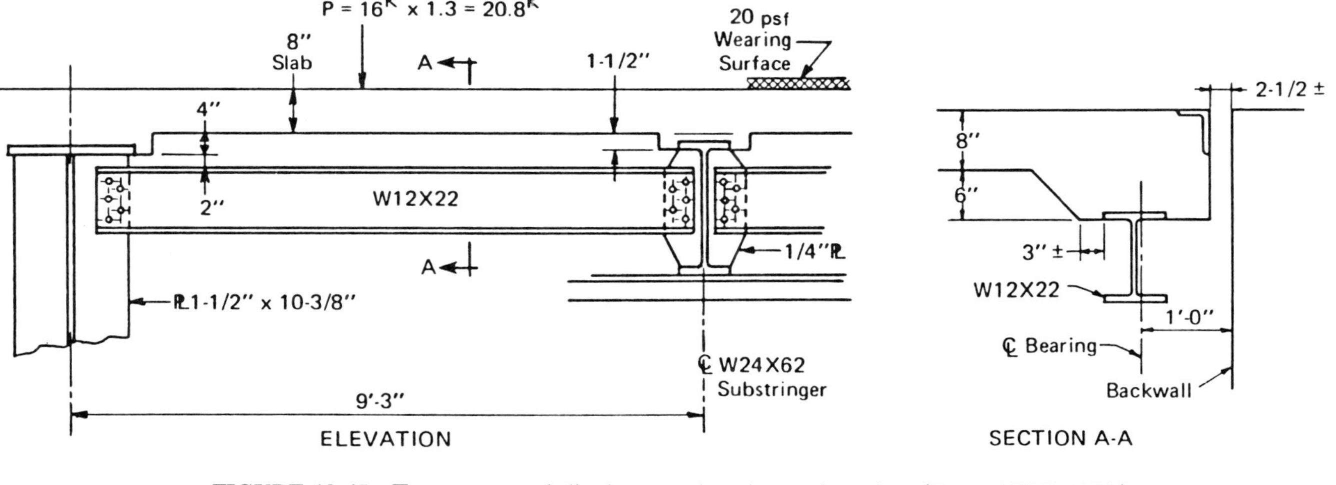

FIGURE 13-45 Transverse end diaphragm, elevation and section. (From AISC, 1990.)

the beam in bending is

$$M_u = F_y Z = 50 \times 29.3/12 = 122.1 \text{ ft-kip} > 100.8 \qquad \text{OK}$$

Likewise, we compute the shears as follows:

$$V_{\text{DL}} = 0.5 \times 0.43 \times 8.5 = 1.8 \text{ kips}$$

$$V_{\text{LL}+I} = 20.8 + 20.8(2.5/8.5) = 26.9 \text{ kips}$$

The factored shear is therefore

$$V = 1.3(1.8 + 1.67 \times 26.9) = 60.7 \text{ kips}$$

The shear capacity of the section is

$$V_u = 0.55 F_y\, dt_w = 0.55 \times 50 \times 12.31 \times 0.26 = 88.0 \text{ kips} > 60.7 \qquad \text{OK}$$

The design of the diaphragm is completed with the selection of the bolted connection as shown in Figure 13-45.

13-21 DIAPHRAGMS AND CROSS FRAMES FOR CURVED GIRDERS

In the foregoing sections we have discussed the optimum diaphragm or cross-frame spacing for curved girder bridges, and it appears that the 12- to 18-ft range is commonly used with curved girders. Lateral bending moments are essentially proportional to the square of the cross-frame spacing. Selection of the cross-frame spacing therefore becomes a matter of comparison and balance between the main girder and cross-frame material.

Because cross frames in curved girders carry calculated stresses, they must be designed as main structural members, with each line of a cross frame extending in a single plane across the width of the bridge. For cross frames, suitable members are provided by tee sections because they are symmetrical about the Y axis and thus carry the loads concentrically. Common cross-frame shapes include X and K configurations selected to best fit the geometry of the bridge. Preferably, they should be full-depth members and should transfer the horizontal and vertical forces to the flanges and web. Cross-frame connections should be attached to both the web and flanges.

In general, the main forces that must be resisted by the cross frames result from dead load, live load plus impact, and from curvature effects. Because these are variable loads, cross frames must also be checked for fatigue.

For expedience in developing a suitable design methodology, the following assumptions are made.

1. The cross frames must resist the effects associated with curvature and resulting from the dead load, live load plus impact, and vertical component of the centrifugal forces.
2. The slab carries wind load on the upper half of the deck, wind on live load, and the horizontal component of the centrifugal forces, and transmits these loads back to the support cross frames and finally to the bearings.
3. According to the wind distribution discussed in the foregoing sections, the support cross frames transmit all wind loads down to the bridge bearings.

TABLE 13-18 Group Loadings for Various Cross-Frame Types

Cross-frame Type	Effects to Be Considered	Group Loadings to Be Checked
End support	Wind, centrifugal force	Group I: $1.3[DR + \frac{5}{3}(TR) + CF_H]$
	Dead load and truck wheel loading	Group II: $1.3[DR + W_T + W_B]$
		Group III: $1.3[DR + TR + CF_H = 0.3(W_T + W_B) + W_L]$
Intermediate	Curvature	Group I: $1.3[D + \frac{5}{3}(L + I) + CF_V]$
		Group II: $1.3[D + W_{Bp}]$
	Wind	Group III: $1.3[D + (L + I) + CF_V + 0.3W_{Bp}]$
Interior support	Curvature	Group I: $1.3[D + \frac{5}{3}(L + I) + CF_V + CF_H]$
	Wind, centrifugal force	Group II: $1.3[D + W_T + W_B]$
		Group III: $1.3[D + (L + I) + CF_V + CF_H + 0.3(W_T + W_B) + W_L]$

D = dead load curvature
$L + I$ = live load plus impact, curvature
W_T = wind on upper half of structure
W_B = wind on lower half of structure
W_L = wind on live load
CF_V = vertical centrifugal force
CF_H = horizontal centrifugal force
DR = dead load on top strut
TR = truck wheel loading on top strut
W_{Bp} = panel wind load, lower half of structure

The result of these assumptions is the summary of group loadings to be checked for cross frames in various parts of the bridge, tabulated in Table 13-18. The loads in each group are factored and correspond to the standard AASHTO specifications.

Method of Analysis In this section the cross-frame forces resulting from curvature will be analyzed according to the theory discussed in Section 6-8. For a four-girder system, the distribution of V loads and torque loads to the girder is as shown in Figure 13-46*a*. The resulting distribution of shears is as

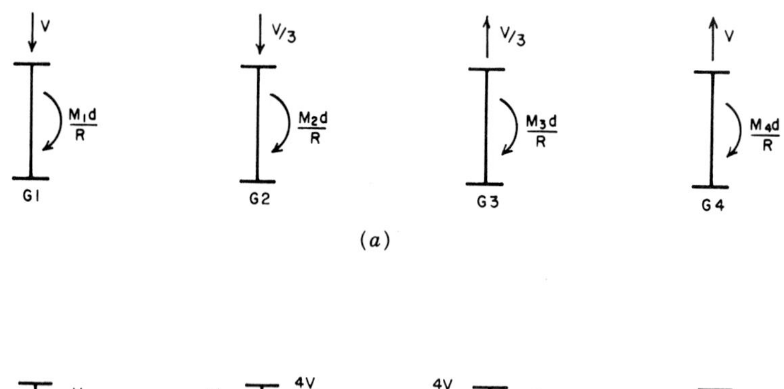

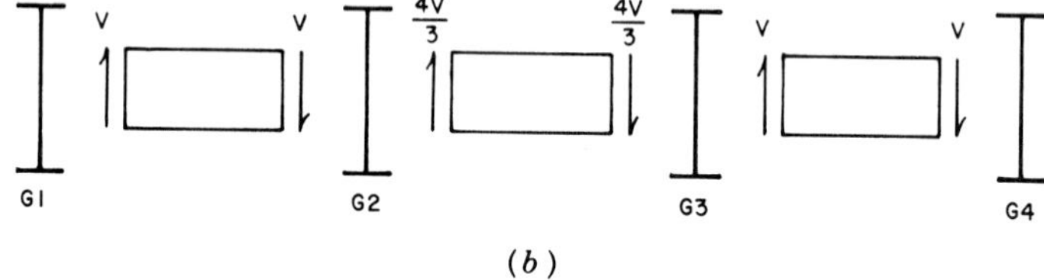

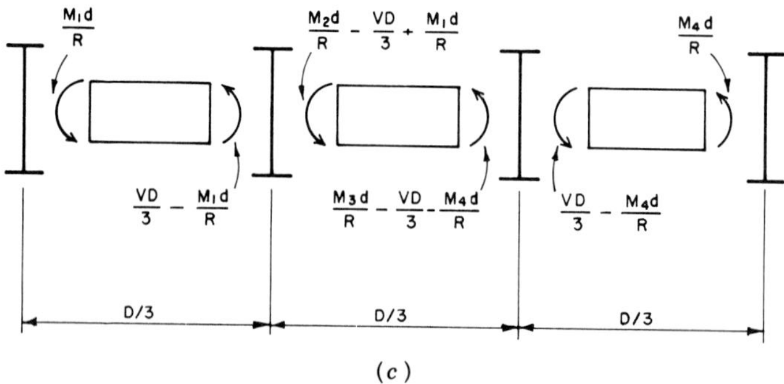

FIGURE 13-46 Four-girder curved bridge system: (*a*) distribution of V loads and torques to girders; (*b*) distribution of end shears on cross frames; (*c*) equilibrium of moments.

shown in Figure 13-46*b*. The end moments on the cross frames must satisfy the equilibrium conditions indicated in Figure 13-46*c*.

The notation used in Figure 13-46 is as follows:

$$V = \frac{M_{1P} + M_{2P} + M_{3P} + M_{4P}}{C(RD/d)}$$

$M_{1P}, M_{2P}, M_{3P}, M_{4P}$ = primary moments in G_1, G_2, G_3, G_4 at the cross frame considered in the analysis

M_1, M_2, M_3, M_4 = total moments in G_1, G_2, G_3, G_4 at the cross frame considered in the analysis

D = width, G_1 to G_4

d = cross-frame spacing along G_1

R = radius of G_1

C = a constant parameter (10/9 for a four-girder system)

The resulting forces from curvature, wind, direct wheel loading, and centrifugal effects are considered according to the group loading.

13-22 DESIGN EXAMPLES OF CROSS FRAME: CURVED BRIDGE (LOAD FACTOR APPROACH)

An intermediate cross frame at point 0.429 of span 1 will be designed for the curved bridge shown in Figure 13-47 (see also Chapter 6). The design criteria are according to Table 13-18, and the analysis is considered in two sections: the curvature effect and the wind effect. The relevant moments at this location and other data are tabulated in Table 13-19.

Group I From the data of Table 13-19, we calculate the following:

$V = \sum M/CR(D/d) = 35.7$ kips $\qquad V/3 = 11.9$ kips

$M_1 = 6260$ ft-kips $\qquad M_1 d/R = 327.8$ ft-kips

$M_2 = 5656$ ft-kips $\qquad M_2 d/R = 296.2$ ft-kips

$M_3 = 4569$ ft-kips $\qquad M_3 d/R = 239.3$ ft-kips

$M_4 = 3598$ ft-kips $\qquad M_4 d/r = 188.4$ ft-kips

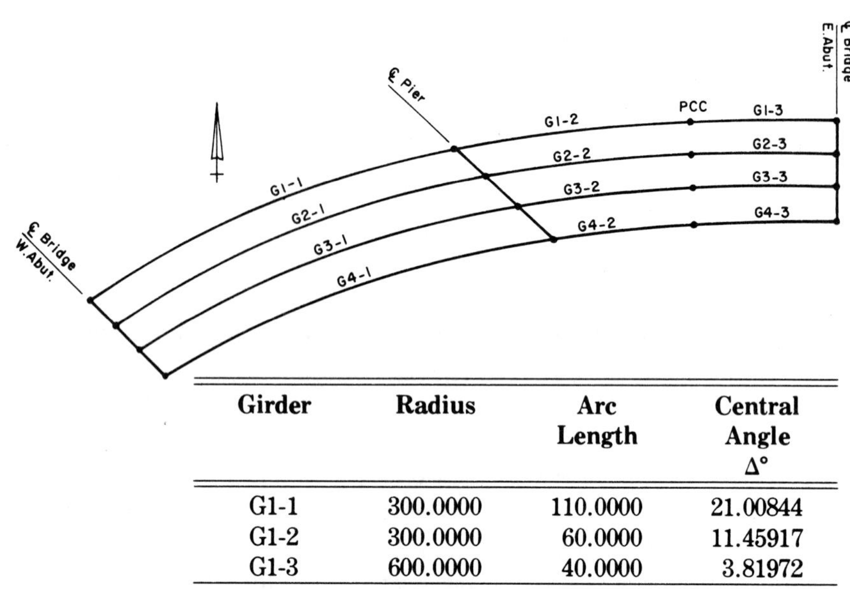

Girder	Radius	Arc Length	Central Angle Δ°
G1-1	300.0000	110.0000	21.00844
G1-2	300.0000	60.0000	11.45917
G1-3	600.0000	40.0000	3.81972
G2-1	291.1667	110.4751	21.73931
G2-2	291.1667	53.1651	10.46183
G2-3	591.1667	39.4111	3.81972
G3-1	282.3333	111.0056	22.52711
G3-2	282.3333	46.2739	9.39067
G3-3	582.3333	38.8222	3.81972
G4-1	273.5000	111.6008	23.37936
G4-2	273.5000	39.3167	8.23650
G4-3	573.5000	38.2333	3.81972

FIGURE 13-47 Plan and data of curved bridge of design example. (From AISC, 1990.)

Group I loading is $1.3[D + (5/3)(L + I) + CF_V)$. The forces on the girders and on the cross frames are shown in Figures 13-48*a* and *b*, respectively. Assuming a cross-frame depth of 3 ft 7 in. (center-to-center of gravity), we compute the following:

$$327.8/3.58 = 91.6 \text{ kips} \qquad 12.8/3.58 = 3.6 \text{ kips}$$

$$309.0/3.58 = 86.3 \text{ kips} \qquad 112.5/3.58 = 31.4 \text{ kips}$$

$$126.8/3.58 = 35.4 \text{ kips} \qquad 188.4/3.58 = 52.6 \text{ kips}$$

These are the internal forces in the cross frames and are indicated in Figure 13-48*c*.

TABLE 13-19 Design Data and Moments for Point 0.429, Span 1

Girder/Description	Moments	Girder/Description	Moments
G1		G1	
$M_{DL_1+DL_2}$	1,481 + 561	$1.3[D + \frac{5}{3}(L + I) + CF_V]$	6,260
M_{L+I}	1,616	$1.3[D]$	2,655
M_{CF_V}	80	$1.3[D + (L + I) + CF_V]$	3,738
G2		G2	
$M_{DL_1+DL_2}$	1,288 + 473	$1.3[K + \frac{5}{3}(L + I) + CF_V]$	5,656
M_{L+I}	1,537	$1.3[D]$	2,289
M_{CF_V}	28	$1.3[D + (L + I) + CF_V]$	3,326
G3		G3	
$M_{DL_1+DL_2}$	1,028 + 368	$1.3[D + \frac{5}{3}(L + I)]$	4,569
M_{L+I}	1,271	$1.3[D]$	1,815
M_{CF_V}	—	$1.3[D + (L + I)]$	3,467
G4		G4	
$M_{DL_1+DL_2}$	755 + 266	$1.3[D + \frac{5}{3}(L + I)]$	3,598
M_{L+I}	1,048	$1.3[D]$	1,327
M_{CF_V}	—	$1.3[D + (L + I)]$	2,690
$\Sigma 1.3[D + \frac{5}{3}(L + I) + CF_V]$	20,082		
$\Sigma 1.3[D]$	8,086		
$\Sigma 1.3[D + (L + I) + CF_V]$	15,340		

$d = 15.71$ ft, $R = 300$ ft, $D = 26.5$ ft, $C = \frac{10}{9}$.

Group II This group involves the loading 1.3(D). Likewise, from the data of Table 13-19, we calculate the following:

$$V = 14.4 \text{ kips} \qquad V/3 = 4.8 \text{ kips}$$

$$M_1 = 2655 \text{ ft-kips} \qquad M_1 d/R = 139.0 \text{ ft-kips}$$

$$M_2 = 2289 \text{ ft-kips} \qquad M_2 d/R = 119.9 \text{ ft-kips}$$

$$M_3 = 1815 \text{ ft-kips} \qquad M_3 d/R = 95.0 \text{ ft-kips}$$

$$M_4 = 1327 \text{ ft-kips} \qquad M_4 d/R = 69.5 \text{ ft-kips}$$

The forces on the girders and on the cross frames are shown in Figures 13-49a and b, respectively. As for group I loading, we compute the following:

$$139.0/3.58 = 38.8 \text{ kips} \qquad 11.8/3.58 = 3.3 \text{ kips}$$

$$131.7/3.58 = 36.8 \text{ kips} \qquad 37.3/3.58 = 10.4 \text{ kips}$$

$$57.7/3.58 = 16.1 \text{ kips} \qquad 69.5/3.58 = 19.4 \text{ kips}$$

These are the internal forces in the cross frames, as shown in Figure 13-49c.

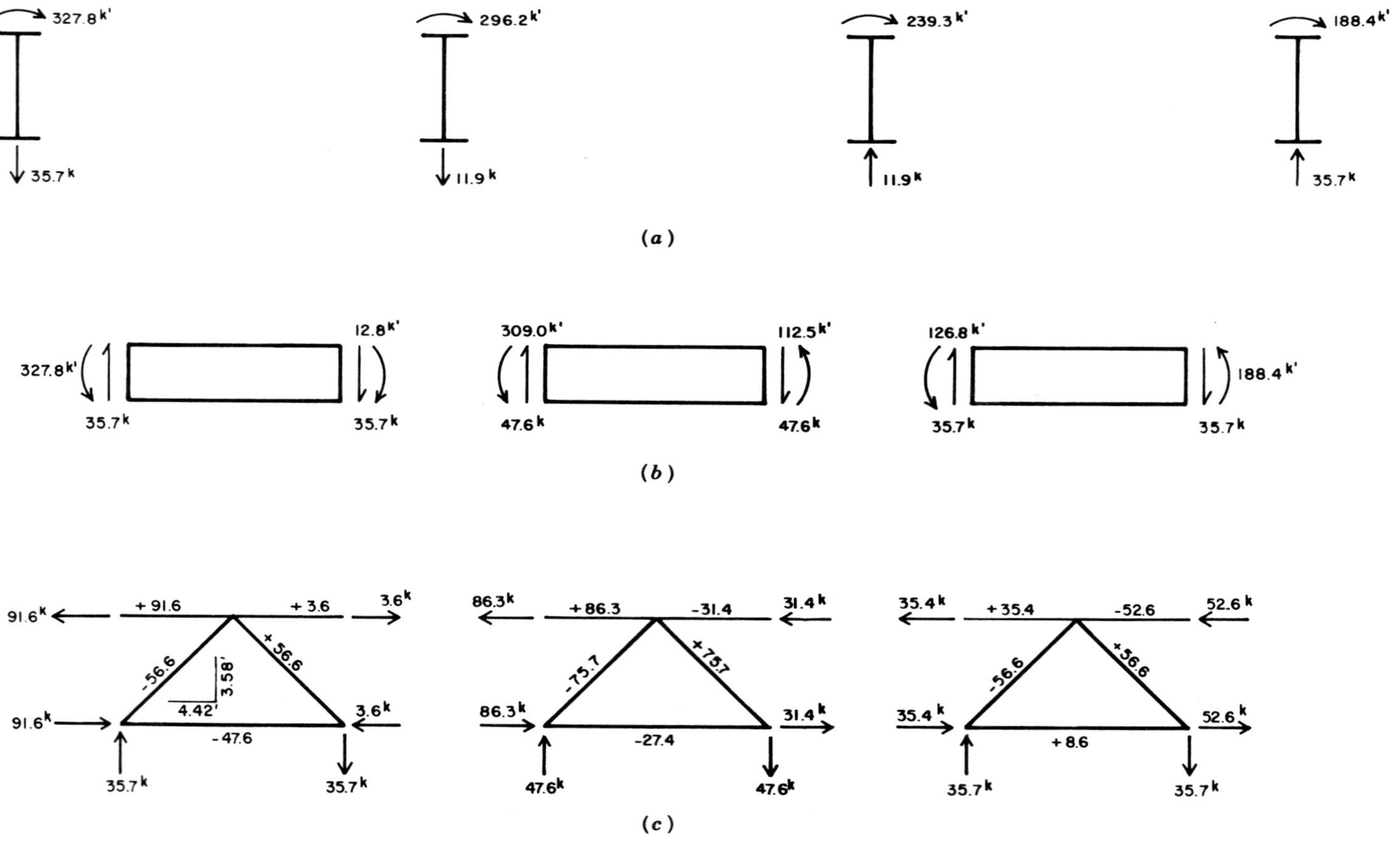

FIGURE 13-48 (*a*) Forces in girders; (*b*) forces in cross frames; (*c*) internal forces in cross frames; group I loading.

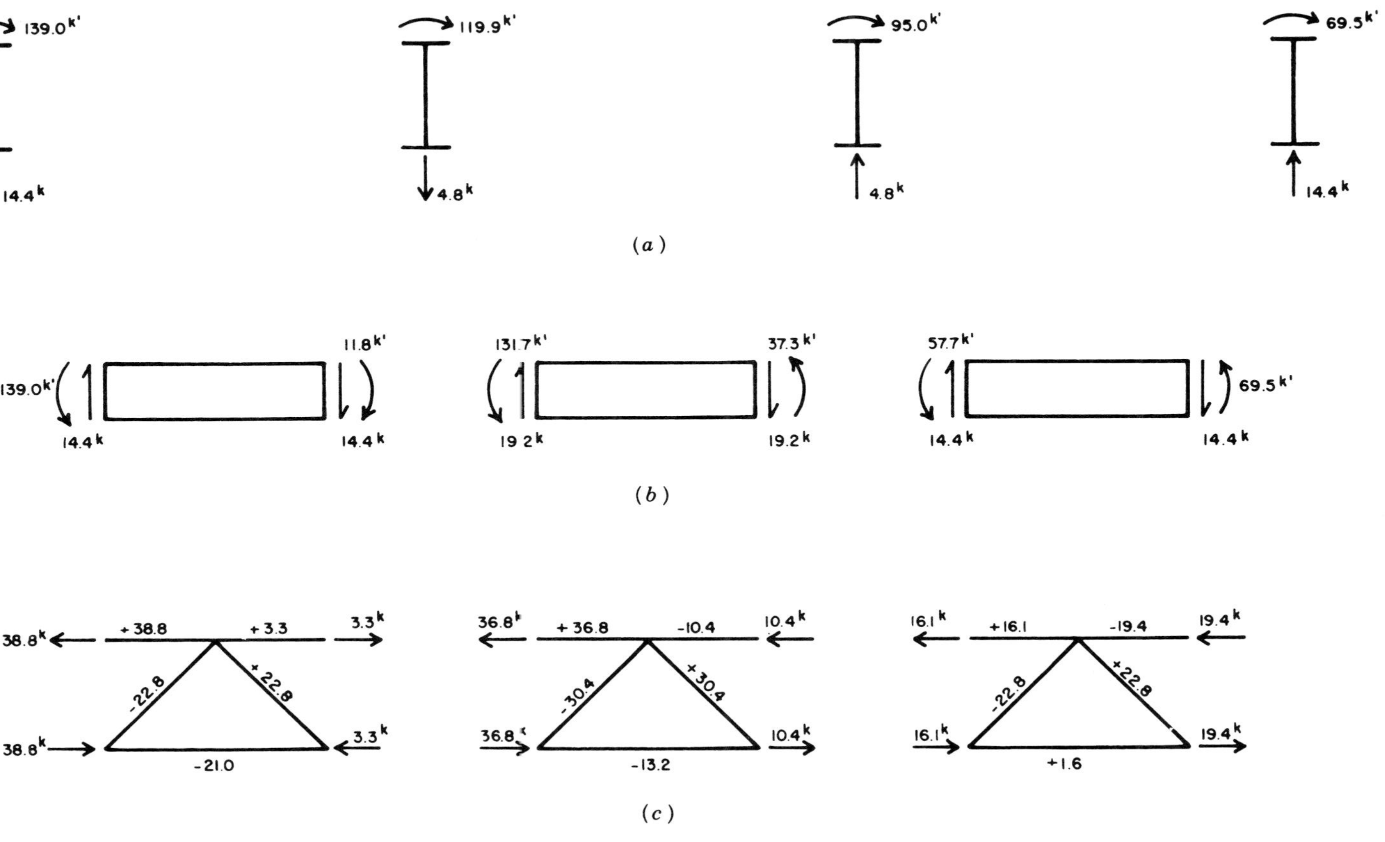

FIGURE 13-49 (*a*) Forces in girders; (*b*) forces in cross frames; (*c*) internal forces in cross frames; group II loading.

Group III Loading This group loading is $1.3[D + (L + I) + CF_V]$. As in the previous groups, we refer to the data of Table 13-19 to compute the following:

$$V = 27.3 \text{ kips} \qquad V/3 = 9.1 \text{ kips}$$
$$M_1 = 4859 \text{ ft-kips} \qquad M_1 d/R = 254.4 \text{ ft-kips}$$
$$M_2 = 4234 \text{ ft-kips} \qquad M_2 d/R = 226.4 \text{ ft-kips}$$
$$M_3 = 3467 \text{ ft-kips} \qquad M_3 d/R = 181.6 \text{ ft-kips}$$
$$M_4 = 2690 \text{ ft-kips} \qquad M_4 d/R = 140.9 \text{ ft-kips}$$

For this group the forces on the girders and on the cross frames are shown in Figures 13-50*a* and *b*, respectively. The resulting internal forces in the cross frames are computed as follows:

$$254.4/3.58 = 71.1 \text{ kips} \qquad 13.3/3.58 = 3.7 \text{ kips}$$
$$239.7/3.58 = 67.0 \text{ kips} \qquad 81.4/3.58 = 22.7 \text{ kips}$$
$$100.2/3.58 = 28.0 \text{ kips} \qquad 140.7/3.58 = 39.4 \text{ kips}$$

These forces are shown in Figure 13-50*c*.

In addition, the intermediate cross frames must transfer the panel wind from the lower half of the structure up to the slab. This loading, designated as W_{Bp}, is shown in Figure 13-51*a*, and the resulting internal forces in the cross frames are shown in Figure 13-51*b*.

Top Strut The forces in the intermediate cross frame are summarized in Table 13-20 for the top strut, the diagonals, and the bottom strut. Group I loading dictates the design. The top strut must resist a compressive force of 52.6 kips and a tensile force of 91.6 kips. The capacity of the member must be computed as follows: $P_u = 0.85 A_s F_{cr}$, for compression, where A_s is the gross effective area and F_{cr} is as defined in AASHTO Article 10.54.1, or $P_u = 0.85 A_s F_y$, for tension, where A_s is the net effective area. In order to prevent local buckling, the width–thickness ratio of the relatively slender stem must satisfy the following relationship:

$$\frac{b'}{t} \leq \frac{2200}{\sqrt{F_y}} \sqrt{\frac{P_u}{P}}$$

We select a WT5 × 15, Grade 50, and we initially check it as a compression member, with $P = 52.6$ kips. The gross and net areas are computed (the net area is the area of the section deducting half the flange coped at the end

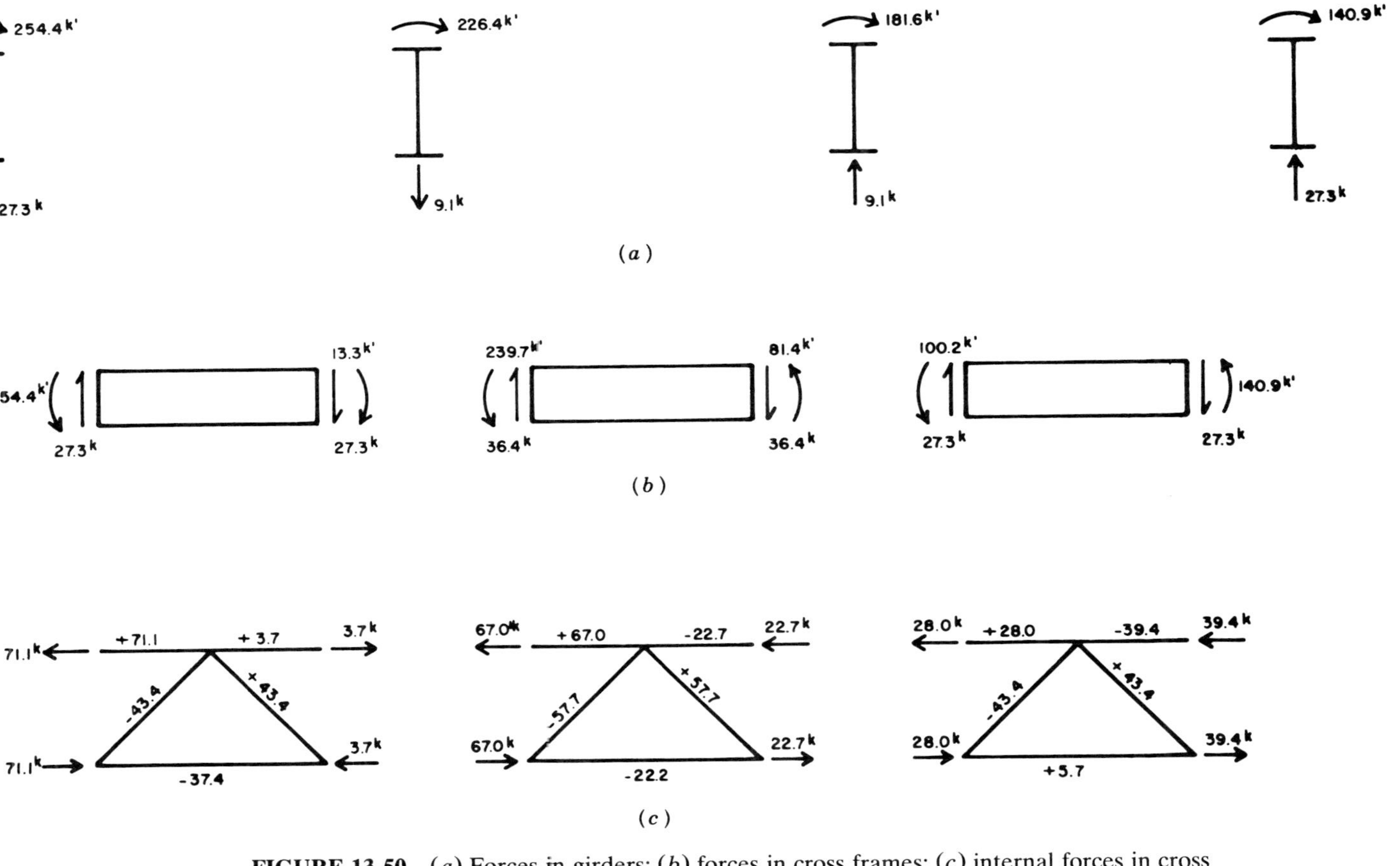

FIGURE 13-50 (*a*) Forces in girders; (*b*) forces in cross frames; (*c*) internal forces in cross frames; group III loading.

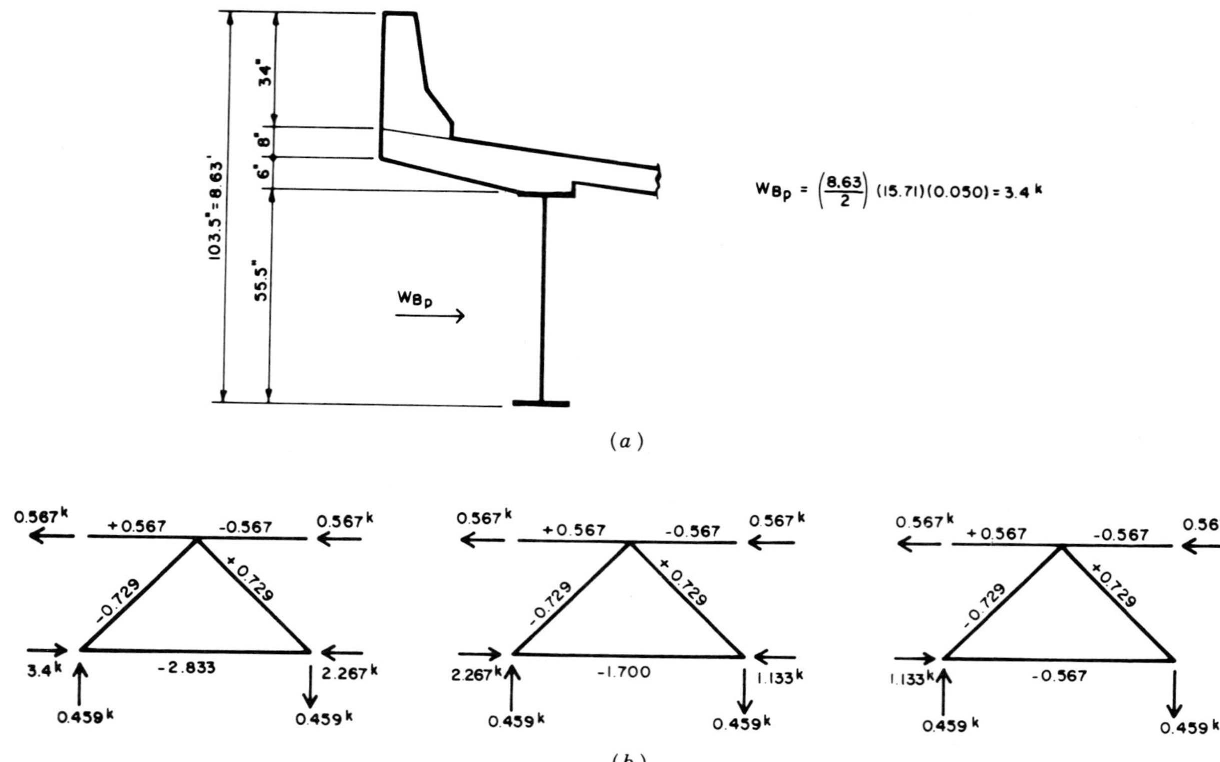

FIGURE 13-51 (*a*) Panel wind loading, W_{Bp}, lower half of structure; (*b*) internal forces in cross frames resulting from W_{Bp}.

TABLE 13-20 Summary of Maximum Design Load for Intermediate Cross Frame of Design Example

Member		Group I $1.3[D + \frac{5}{3}(L + I) + CF_V]$	Group II $1.3[D + W_{Bp}]$	Group III $1.3[D + (L + I) + CF_V + 0.3W_{Bp}]$
Top strut	T	91.6 kips	$38.8 + (1.3)(0.567) = 39.5$ kips	$71.1 + (1.3)(0.3)(0.567) = 71.3$ kips
	C	−52.6 kips	$-19.4 + (1.3)(-0.567) = -20.1$ kips	$-39.4 + (1.3)(0.3)(-0.567) = -39.6$ kips
Diagonal	T	75.7 kips	$30.4 + (1.3)(0.729) = 31.3$ kips	$57.7 + (1.3)(0.3)(0.729) = 58.0$ kips
	C	−75.7 kips	$-30.4 + (1.3)(-0.729) = -31.3$ kips	$-57.7 + (1.3)(0.3)(-0.729) = -58.0$ kips
Bottom strut	T	8.6 kips	$1.6 + (1.3)(-0.567) = 0.9$ kip	$57.7 + (1.3)(0.3)(0.567) = 5.5$ kips
	C	−47.6 kips	$-21.0 + (1.3)(-2.833) = -24.7$ kips	$-37.4 + (1.3)(0.3)(-2.833) = -38.5$ kips

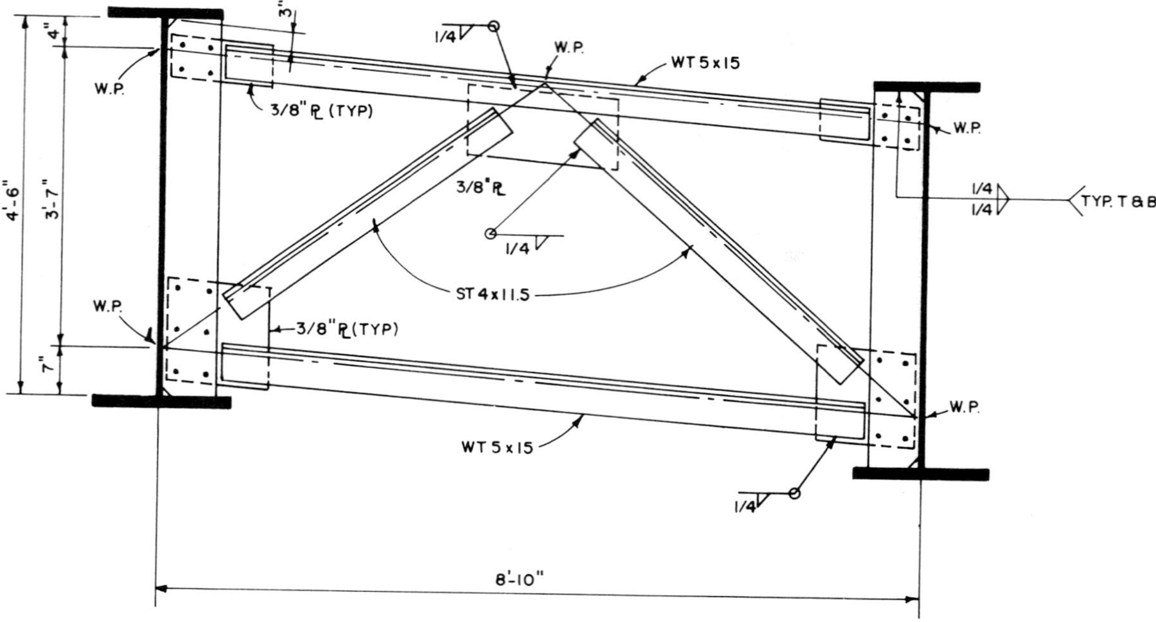

FIGURE 13-52 Detail of intermediate cross frame. (From AISC, 1990.)

connection). Thus, $A_{s(\text{gross})} = 3.42$ in.^{2} and $A_{s(\text{net})} = 3.01$ in.^{2}. Next, we compute KL/r using $K = 0.75$, or $KL/r = 0.75 \times 8.83 \times 12/1.37 = 58 < 120$. Also, $\sqrt{2\pi^2 E/r_y} = 107 > 58$. Therefore, the value of F_{cr} is calculated from AASHTO (10-151), or $F_{cr} = 37.0$ ksi. For compression, the capacity of the member is

$$P_u = 0.85 \times 4.42 \times 37 = 139.0 \text{ kips} > 52.6 \text{ kips} \qquad \text{OK}$$

As a tension member, the strut has a capacity $P_u = A_{s(\text{net})}F_y$, where $A_{s(\text{net})}$ is the net area with half the flange coped and half of the remaining flange discounted. This gives $A_{s(\text{net})} = 2.31$ in.^{2}, or a capacity $P_u = 2.31 \times 50 = 115.5$ kips > 91.6 kips, OK.

The ratio b'/t is $(5.325 - 0.51)/0.300 = 15.8$.

$$\text{Allowable } b'/t = \frac{2200}{\sqrt{50{,}000}}\sqrt{\frac{139}{52.6}} = 16.0 > 15.8 \qquad \text{OK}$$

Diagonal Members and Bottom Strut The design is completed in a similar manner based on the data of Table 13-20. The sections selected are checked for tension and compression.

End Connections Welds, bolts, and connection materials are proportioned for a design load P_0, which is the larger of the following: $0.75P_u$ or $(P + P_u)/2$, where P_u is the capacity of the section and P is the maximum applied load in the member. The design calculations are not repeated. The intermediate cross frame is detailed as shown in Figure 13-52.

13-23 DIAPHRAGMS FOR CURVED STEEL BOX GIRDERS

Curved box girders require internal diaphragms at the supports to resist transverse rotation, displacement, and distortion, and also to transmit the torque to the substructure through the bearing. In addition, as mentioned in Section 6-5, individual box girders not connected by a common flange plate should be provided with intermediate diaphragms or cross frames to distribute the torsional forces unless a rational analysis indicates that they are not needed.

Diaphragms or cross frames serve to restrict the normal and transverse bending stresses associated with distortion and the lateral bending stresses in the narrow top flanges during the wet concrete period. Guidelines for the spacing of intermediate cross frames are based on limiting the distortion stresses to 10 percent of the stress caused by ordinary bending. Accordingly, the cross-frame spacing should not exceed

$$S = L\sqrt{\frac{R}{2000L - 7500}} \leq 25 \text{ ft} \qquad (13\text{-}23)$$

and the cross-sectional area (in.^{2}) of the diagonal should at least be

$$A_b = 75\frac{S_b}{d^2}\frac{t^3}{d + b} \qquad (13\text{-}24)$$

where S_b is the diaphragm spacing (in.), d is the depth of the box (in.), b is the width of the box (in.), t is the thickness of the thickest component of the box girder cross section (in.), R is the radius of the girder (in.), and L is the girder span (in.).

Diaphragms or cross frames should be full-depth members designed as main structural elements. Cross frames, the top lateral bracing members, and the flanges of the box girders should be framed in such a way to transfer the horizontal and vertical forces to the flanges and webs as necessary. Connection plates should be attached to the box girder web and flanges in a manner to prevent distortion of the web at each end of each connection plate.

13-24 DISTORTIONAL RESPONSE OF CURVED CONCRETE BOX GIRDERS

A potential problem in curved concrete box girders is large normal stresses induced by cross-sectional distortions. This problem is avoided if intermediate diaphragms are properly provided (Dabrowski, 1968; Oleinik and Heins, 1974), and in this case conventional curved girder theory may be applied (Vlasov, 1961; Yoo, Evick, and Heins, 1974). Based on load–deformation theories, the induced normal stresses due to distortional effects of the cross section have been found to depend on the second derivative of angular distortion computed in difference form and on the cross-sectional properties. If intermediate diaphragms are present, they inhibit distortional behavior, and the angular deformation at these locations can be assumed to be zero.

Using finite-difference programs, Oleinik and Heins (1974) have studied the distortional response of curved single-cell box sections subjected to dead and live loads. The results were used to formulate empirical relationships that can be used directly in evaluating distortional effects. Thus, the ratio of the maximum distortional stress σ_f to the corresponding bending stress σ_b can be presented in the form

$$\sigma_f/\sigma_b = f(L, R, S, w, WA^*) \qquad (13\text{-}25)$$

where L is the span length, R is the radius of curvature, S is the diaphragm spacing, WA^* is the section property parameter, and w is the applied load.

According to the results of the study, the ratio σ_f/σ_b in the bottom flange is typically greater than the same ratio at other points of the same section. Stress ratio evaluation also requires reference to the longitudinal section. Thus, for a single span the maximum distortional and bending stresses, as well as the maximum stress ratio, should occur at midspan for a uniformly loaded curved box girder.

Likewise, results from the live load study show that bottom flange stresses govern the design. Considering the parameters influencing the ratio σ_f/σ_b in (13-25), Oleinik and Heins (1974) conclude the following.

1. Keeping all other parameters the same and varying the uniform dead load w does not change the stress ratio σ_f/σ_b.
2. The stress ratio varies linearly with the square of the diaphragm spacing, expressed as a decimal fraction of the span length.
3. The cross sections of a curved concrete box girder will deform under load, and unless these deformations are limited by intermediate transverse diaphragms, they can cause distortional stresses that may approach and exceed the normal bending stresses.
4. The addition of diaphragms does not affect the normal bending stress pattern.

13-25 EXAMPLE OF CROSS-FRAME DESIGN: CURVED STEEL BOX GIRDER (ALLOWABLE STRESS METHOD)

A two-span curved box girder bridge (two steel boxes in section) has span lengths of 120 and 120 ft and is composite in both positive and negative moment regions. Three different types of cross frames of A36 steel are selected. Two are used in regions of the box girders where a longitudinal stiffener is provided in the bottom flange, and the third type is used in sections without this stiffener. A cross frame in sections without longitudinal stiffeners is shown in Figure 13-53.

Top Strut For this member we select an angle of $5 \times 5 \times 3/8$ in. with an area of 3.61 in.^{2}. This is investigated (a) for overall buckling, (b) for local buckling, and (c) as a compression member.

For overall buckling and an unbraced length of 118 in., the slenderness ratio L/r is $118/0.99 = 119 < 120$. For local buckling, $b/t = (5 - 0.38)/0.38 = 12.2$, or very close to the limiting value 12.

The force acting on the strut may be computed from $R = 1.2wL$, where w is the load from the wet concrete and L is the cross-frame spacing, or 12.3 ft. From previous designs w due to dead load is computed as 0.60 kip/ft, and

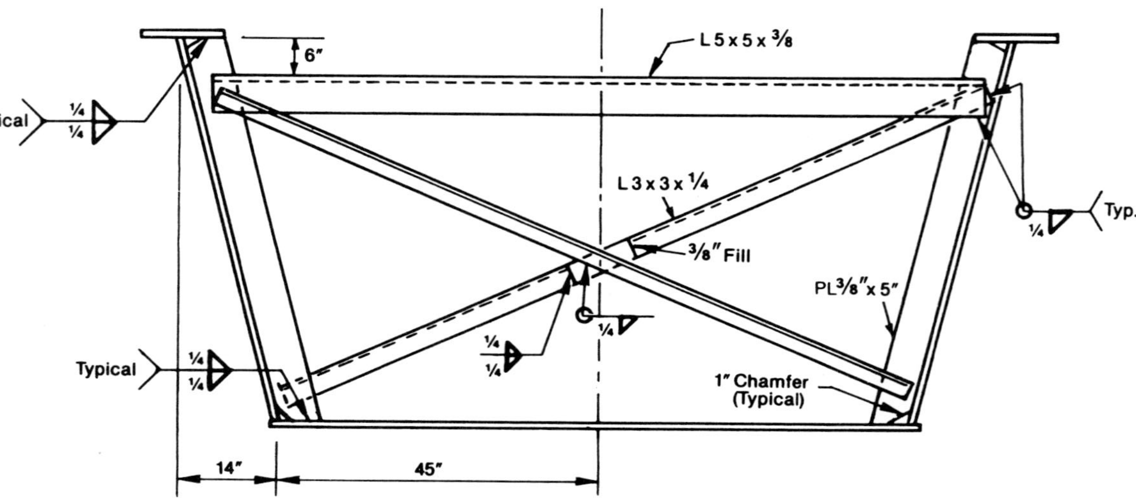

FIGURE 13-53 Cross frame in section without bottom flange longitudinal stiffeners; design example.

for lateral bending w is 0.18 kip/ft. The force in the strut is therefore $R = 1.1(0.18 + 0.60)(12.31) = 10.6$ kips. The allowable stress in the strut is computed for concentrically loaded compression members as

$$F_c = 16{,}980 - 0.53(KL/r)^2 = 16{,}980 - 0.53(119)^2 = 9.47 \text{ ksi}$$

and the allowable force in the strut is

$$R = 9.47 \times 3.61 = 34.2 \text{ kips} > 10.6 \qquad \text{OK}$$

Cross-Frame Diagonals For these members we select an angle of $3 \times 3 \times 1/4$ in., to be checked for overall and local buckling. The diagonal has an area A of 1.44 in.^{2}, and its angle with the horizontal is about 26°. The minimum required area is obtained from (13-24) and is

$$A_b = 75 \times \frac{12.31 \times 12 \times 90}{26.12^2} \times \frac{0.5}{57 + 90} = 0.017 \text{ in.}^2 < 1.44 \qquad \text{OK}$$

Local buckling is checked as follows:

$$b/t = (3 - 0.25)/0.25 = 11 \qquad \text{OK}$$

Bottom Strut The bottom strut may be placed either above the longitudinal stiffener or at the same level as the bottom flange and the stiffener. In the latter case the bottom strut also serves as a bottom transverse stiffener and is attached to the bottom flange by welding. This member is not attached to the longitudinal stiffener. A bottom strut at the flange level is required at the points of maximum flange stress and dead load contraflexure.

13-26 HINGE DETAILS AT BASE OF BRIDGE COLUMNS

Foundations in regions of seismic activity are normally designed to resist the plastic hinge moments developing at the base of the bridge columns. If the columns must be oversized for architectural or other reasons, the design may result in unusually large foundations. In these cases the base of a column may be detailed to respond essentially as a hinge, an arrangement that reduces the plastic moment transferred to the foundation.

The concept involved in the modified hinge detail is to provide a reduced moment capacity in the plastic hinging region at the base of the column, merely by placing a layer of easily compressed material at this location thus introducing partial structural discontinuity between the column and the footing. This discontinuity means a smaller effective cross section at the column base and therefore a reduced hinge capacity. The design of modified hinges, largely left to engineering judgment, may be supplemented with the results of tests of column specimens subjected to axial load and cycled inelastic lateral displacements carried out by Lim, McLean, and Henley (1990). The principal parameters are hinge details affecting moment reduction, column aspect ratio, low axial load, and low-cycle fatigue characteristics.

Relevant Hinge Details The test specimens in this program were arranged in pairs. One specimen incorporated a hinge detail having only horizontal discontinuity (denoted as the CA detail), and the other provided a hinge detail with both horizontal and vertical discontinuities (denoted as the WA detail). These details are shown in Figures 13-54*a* and *b*, where the discontinuity between the column and the footing consists of a layer of compressible material as shown. For the CA detail, only horizontal discontinuity is available; hence, the plastic hinging action in the column is largely concentrated along a horizontal plane at the column–footing interface. Unlike the CA detail, both horizontal and vertical discontinuity mechanisms are provided in the WA detail, where the objective is to distribute plastic stresses over a greater vertical length.

A reference or control column has the detail shown in Figure 13-54*c*. This detail, denoted as CON, has the same dimensions and reinforcement as the columns shown in Figures 13-54*a* and *b*, but lacks the compressible material.

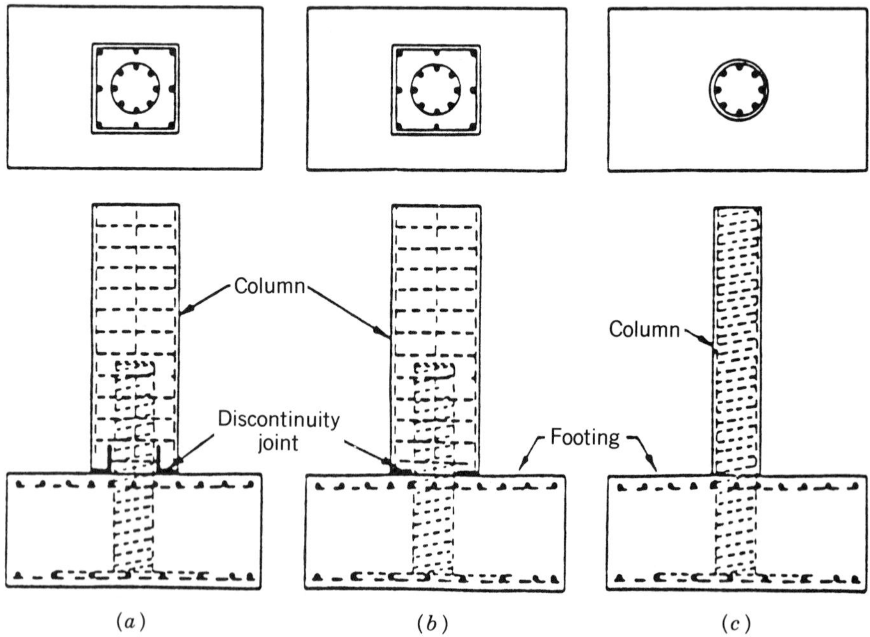

FIGURE 13-54 Modified plastic hinge details: (*a*) WA detail; (*b*) CA detail; (*c*) CON detail. (From Lim, McLean, and Henley, 1990.)

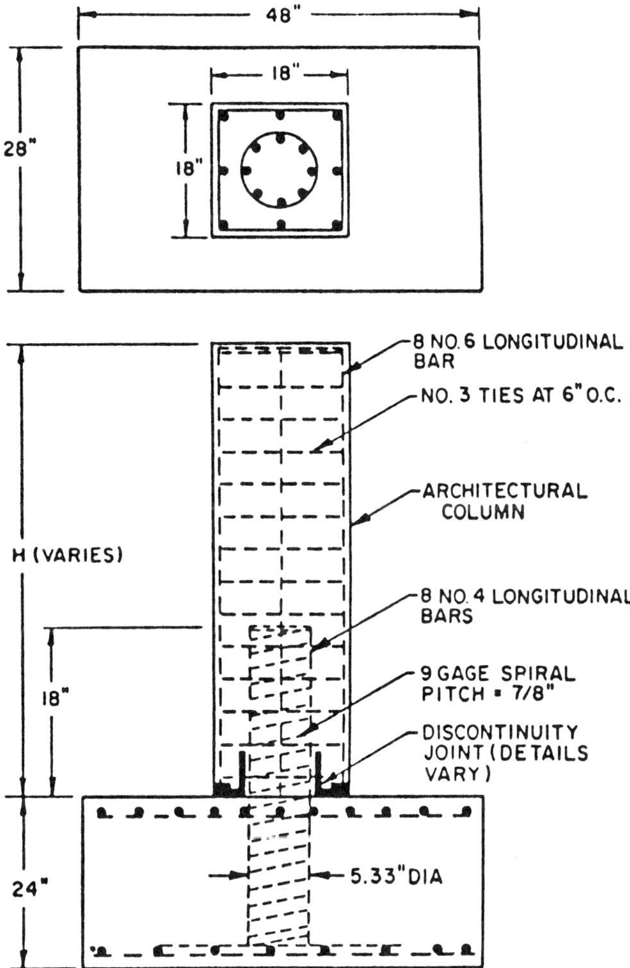

FIGURE 13-55 Dimensions and reinforcements of a typical column. (From Lim, McLean, and Henley, 1990.)

The CON detail can therefore transfer full moments to the footing. Dimensions and reinforcements for a typical column are shown in Figure 13-55.

Height–width ratios for the typical column specimen are 1.25 and 2.50, whereas the column width D is kept constant. Two levels of axial compression are studied, $P/(f_c'A_c) = 0.24$ and 0.35, where A_c is the cross-sectional area of the hinge connection measured out-to-out of spiral. A hinge longitudinal reinforcing steel ratio of 7.2 percent and a hinge spiral reinforcing steel ratio referenced to the area of the hinge connection are included in all specimens.

TABLE 13-21 Summary of Testing Program and Specimen Details (From Lim, McLean, and Henley, 1990)

Specimen Number	Variable Studied	Aspect Ratio (H/D)	Axial Load $(P/f'_c A_c)$	Yield Displacement (in.)	Measured Yield Moment (in.-kips)	Measured Peak Moment (in.-kips)	Maximum Applied Shear Load (kips)
CA1	Aspect ratio	2.50	0.24	0.30	181	270	6.0
WA1	Aspect ratio	2.50	0.24	0.30	133	212	4.7
CON2	Hinge detail	[a]	0.24	0.15	93	209	9.3
CA2	Hinge detail	1.25	0.24	0.15	162	279	12.4
WA2	Hinge detail	1.25	0.24	0.15	142	250	11.1
CA3	Axial load	1.25	0.35	0.15	166	277	12.3
WA3	Axial load	1.25	0.35	0.15	135	240	10.7
CA4	Low-cycle fatigue	1.25	0.24	0.15	138	271	12.0
WA4	Low-cycle fatigue	1.25	0.24	0.15	130	236	10.5

[a]Circular control column with the same height as units CA2 and WA2.

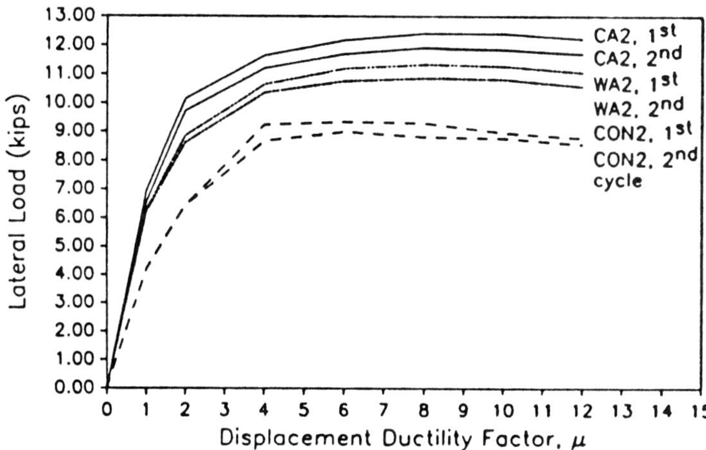

FIGURE 13-56 Shear strength envelope curve for units CA2, WA2, and CON2. (From Lim, McLean, and Henley, 1990.)

The thickness of the discontinuity joint in the CA detail is 1/2 in., and for the WA detail the vertical discontinuity joint is 6 in. high. The horizontal joint thickness in both details is increased to 1 in. at the outer edges of the column to allow hinge rotation without contact between the footing and column. Small-scale tests show that hinge performance is independent of the shape of the column if this contact is prevented. A summary of the testing program and specimen details is given in Table 13-21.

Table 13-21 also gives test results relating to column performance with respect to moment capacity and displacement ductility attained, overall hysteresis behavior, and degradation and energy dissipation characteristics.

Overall Behavior Load–displacement hysteresis curves were produced from the results of tests on all three specimens, WA2, CA2, and CON2, subjected to an axial load of $0.24f_c'A_c$. These curves reflect the true lateral loads including P–Δ effects as well as secondary effects from the axial load, and indicate stable conditions for all three specimens. Evidence of a sudden drop in the load-carrying capacity was not detected, and the plastic hinges continued to absorb energy throughout the tests.

The shear strength envelope curves for these columns are plotted in Figure 13-56, obtained by plotting the maximum shear attained at each peak displacement with respect to the same displacement. For units WA2 and CA2, the degradation in strength is minimal, but for unit CON2, degradation begins after μ = 4. Unit CA2 exhibits the greatest stiffeners, and unit CON2 the least.

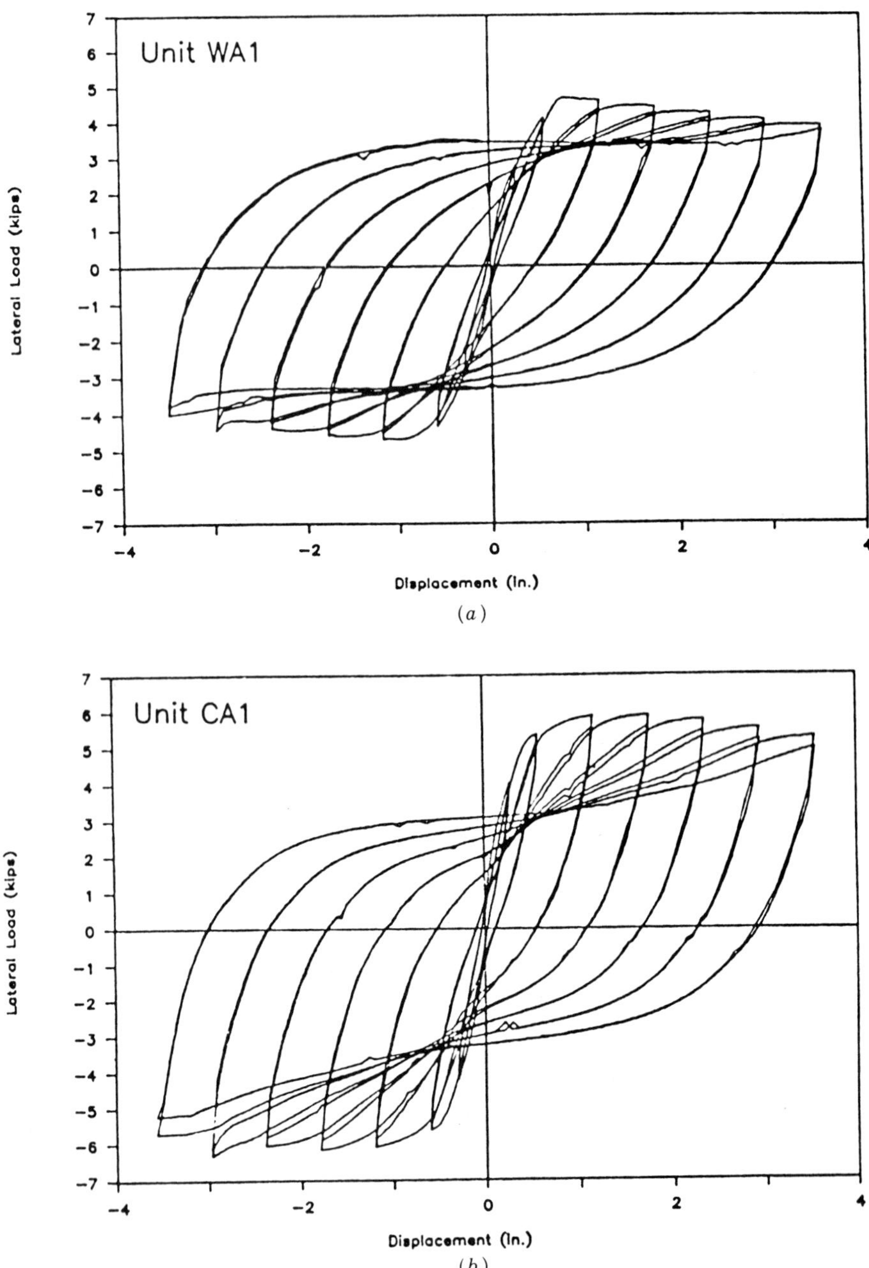

FIGURE 13-57 Lateral load–displacement hysteresis curves: (*a*) unit WA1; (*b*) unit CA1. (From Lim, McLean, and Henley, 1990.)

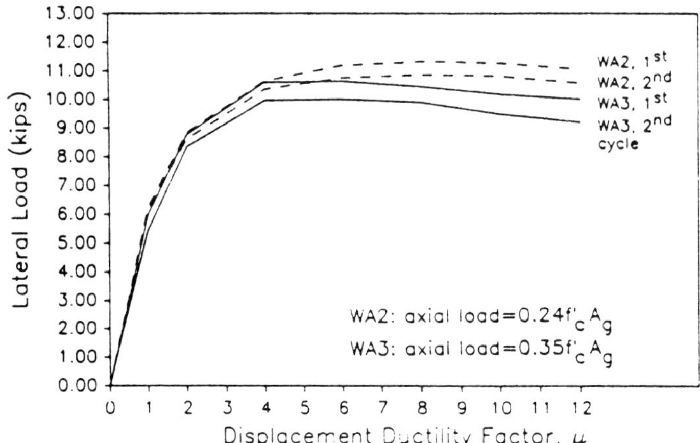

FIGURE 13-58 Shear strength envelope curves for units WA2 and WA3. (From Lim, McLean, and Henley, 1990.)

Effect of Column Aspect Ratio This effect was evaluated for aspect ratios of 2.50 (units WA1 and CA1) and 1.25 (units WA2 and CA2). Hysteresis curves for units WA1 and CA1 are shown in Figures 13.57*a* and *b*, respectively. It appears that flexure dominates column behavior. In addition, the curves for unit CA1 are somewhat narrower than those for unit WA1, indicating decreased energy dissipation characteristics in the column with the CA detail.

Shear strength envelope curves for units WA1 and WA2 and units CA1 and CA2 show that greater degradation in strength occurs in columns with the higher aspect ratio, particularly for the column with the WA detail.

Effect of Axial Load Level Units WA2 and CA2 and units WA3 and CA3 were tested with axial load levels of $0.24f'_cA_c$ and $0.35f'_cA_c$, respectively. Comparison of the hysteresis curves shows that there is only a small difference between specimens with the WA detail, and there is practically no difference for the specimens with the CA detail.

The shear strength envelope curves for these units are shown in Figures 13-58 and 13-59. For the columns with the WA detail, the larger axial load results in a greater drop in strength. For the columns with the CA detail, axial load appears to have little effect on strength. A probable explanation is the confining effect around the hinge region provided by the architectural column, particularly with the CA detail.

Effect of Low-Cycle Fatigue Tests on units WA4 and CA4 carried out to evaluate low-cycle characteristics of moment-reducing hinge details involved multiple cycles at a displacement level of $\mu = 10$. For both specimens very

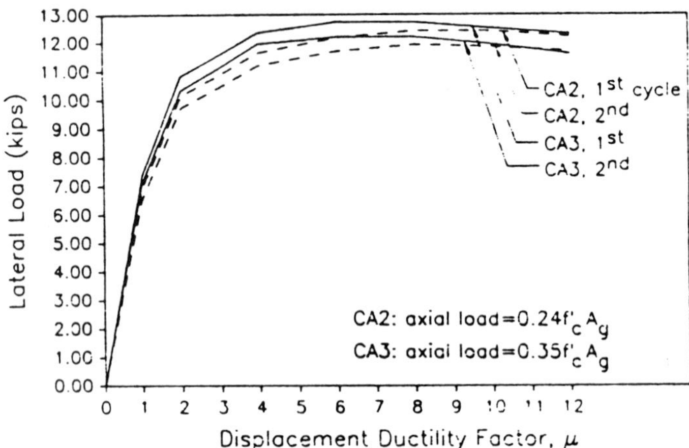

FIGURE 13-59 Shear strength envelope curves for units CA2 and CA3. (From Lim, McLean, and Henley, 1990.)

little degradation occurred after the completion of the second cycle. The hinges continued to exhibit stable plastic behavior even after 16 cycles at this displacement level.

Repeatability of Results For units constructed identically, the repeatability of results is confirmed by the shear strength envelope curves showing essential similarity.

Energy Dissipation Characteristics The energy dissipated by a column during a load cycle is represented by the area confined by the load–displacement hysteresis curve. The energy dissipated by a perfectly elastoplastic system during a complete displacement cycle is shown in Figure 13-60; it is the area of the parallelogram *BCDE*. For a particular displacement ductility factor μ, the ideal plastic energy displacement E_p is expressed as

$$E_p = 4(\mu - 1)V_p \, \Delta y \tag{13-26}$$

where V_p is the maximum shear force attained at that displacement level.

The measured energy dissipation is divided by the E_p value of the column for the same displacement ductility factor. This ratio is referred to as the relative energy dissipation index. Plots of E/E_p values versus the displacement ductility factor μ are shown in Figure 13-61 for units WA2, CA2, and CON2. Evidently, the energy dissipation effectiveness is highest for unit CON2 and is lowest for unit CA2. The reduced effectiveness is attributed to the confinement of the plastic action whereby a reduction in length, over which the plastic hinge is developing, occurs at the base of the column particularly for the CA detail.

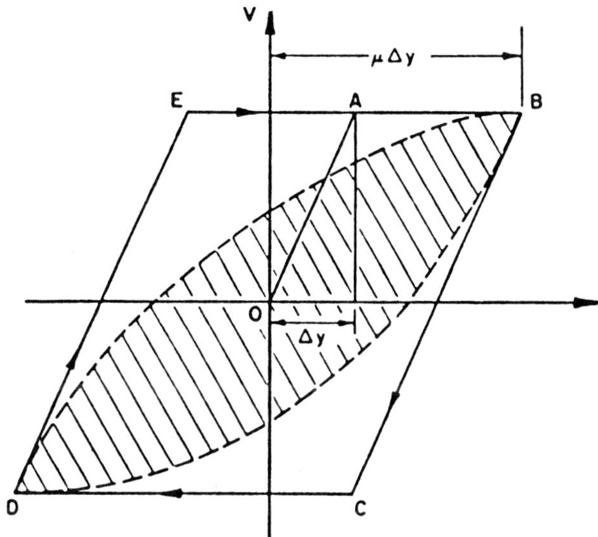

FIGURE 13-60 Actual and idealized perfectly elastoplastic hysteresis curves. (From Lim, McLean, and Henley, 1990.)

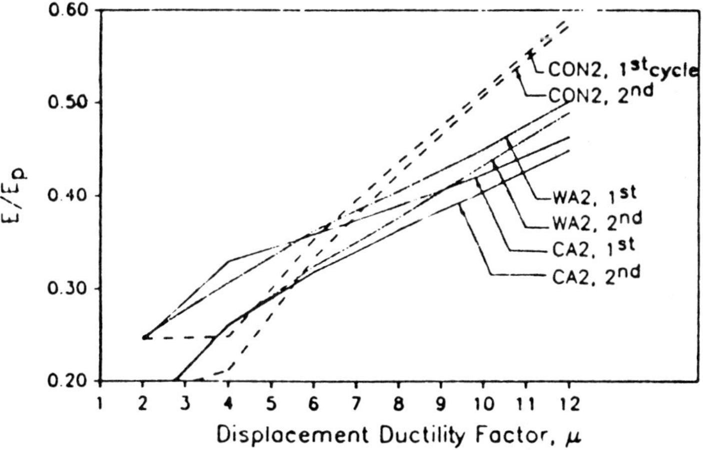

FIGURE 13-61 Relative energy dissipation index curves for units CON2, WA2, and CA2. (From Lim, McLean, and Henley, 1990.)

Conclusion Lim, McLean, and Henley (1990) make the following observations and conclusions.

1. Columns with moment-reducing plastic hinge details, when subjected to cycled inelastic displacement under constant axial load, exhibit a hinging behavior similar to that of an unmodified column of the same dimensions and reinforcement. Stable static load–deflection hysteresis curves are observed even at levels of $\mu = 12$, and the columns continue to absorb energy.

2. Columns with the CA detail develop larger strain values in the longitudinal bars for the same level of displacement than columns with the WA detail or the unmodified detail. Furthermore, columns with the CA detail have lower energy dissipation effectiveness.

3. Flexure dominates the behavior of all the columns investigated in the study. However, a lower aspect ratio results in a greater drop in strength between the first and second cycle of loading at a particular displacement level in columns with the WA detail.

4. Increasing the axial load by 50 percent resulted in a somewhat greater degradation of shear strength in columns with the WA detail, whereas columns with the CA detail remained unaffected by this increase.

5. In the low-cycle fatigue tests, no evidence of distress was noted in columns with WA or CA details when units were cycled up to 16 times at $\mu = 10$.

13-27 CONSTRUCTION ASPECTS OF BRIDGE DECKS

Pouring Sequence in Concrete Decks

For simple spans, concrete is usually deposited beginning at the center of the span and working from the center toward the ends. Concrete in girders should be deposited uniformly for the full depth of the girder and brought up evenly in horizontal layers.

In continuous spans, design considerations are likely to dictate the pouring sequence and time schedule and should be clearly shown on the plans. In continuous spans with composite design only in the positive moment areas, concrete is initially poured in the negative (noncomposite) moment regions, that is, over the interior piers, and usually between the points of contraflexure. In this manner the dead load due to the weight of the concrete slab is resisted only by the steel beams of girders according to the intent of the design. Where the composite design extends throughout the bridge length, the pouring sequence may be chosen to fit the optimum construction schedule provided dead load moments are calculated accordingly and taking into account the actual composite or noncomposite section properties.

Concrete in slab spans should be placed in one continuous operation for each span unless otherwise indicated. All concrete in the same pour should be placed before any concrete in the pour has taken its initial set.

Construction Joints in Deck Slabs

Transverse construction joints are necessary to accommodate the pouring sequence and where the entire deck cannot be placed in a single operation. Most state standards specify that if a deck pour is greater than 200 yd^3 between the longitudinal joints or 300 yd^3 when longitudinal joints are omitted, an optional transverse construction joint must be provided. Its location should be near the point of contraflexure with the pour terminating at the end of the positive moment area in the center span or spans. Shear keys or inclined reinforcement should be provided where necessary to transmit shear or to bond the two sections together. Alternatively, the concrete may be roughened as directed.

Longitudinal bonded joints may be permitted at certain locations. Normally, they are placed in the middle half of the outside framing panels, and, when possible, they should be in line with the edges of the approach pavements.

REFERENCES

AASHTO, 1991. Commentary on the ANSI–AASHTO–AWS Bridge Welding Code D1.5-88.

AISC, 1990. "Highway Structures Design Handbook," *AISCE Marketing*, Pittsburgh.

Ang, B. G., M. J. N. Priestley, and T. Pauley, 1985. "Seismic Shear Strength of Circular Bridge Piers," Research Report 85-5, Dept. Civ. Eng., Univ. Canterbury, Christchurch, New Zealand, July.

ANSI–AASHTO–AWS, 1988. "Bridge Welding Code D1.5-88."

ASTM, 1982.

Bashore, F. J., A. W. Price, and D. E. Branch, 1980. "Determination of Allowable Ratings for Various Proprietary Bridge Deck Expansion Joint Devices at Various Skew Angles," *Michigan Transp. Commi.*, Lancing, May.

Byers, W. G., 1979. "Structural Details and Bridge Performance," *J. Struct. Div. ASCE*, Vol. 105, No. ST7, July, pp. 1393–1404.

Campbell, T. I. and W. L. Kong, 1987. "TFE Sliding Surfaces in Bridge Bearings," Report ME-87-06, Ministry Transp. and Communications, Ontario, Canada, July.

Campbell, T. I., W. L. Kong, and D. G. Manning, 1990. "Laboratory Investigation of the Coefficient of Friction in the Tetrafluorethylene Slide Surface of a Bridge Bearing," Transp. Research Record 1275, TRB, National Research Council, Washington, D.C., pp. 45–52.

Crozier, W. F., J. R. Stoker, V. C. Martin, and E. F. Nordlin, 1979. "A Laboratory Evaluation of Full-Size Elastomeric Bridge Bearing Pads," Research Rep. Calif. Dept. Transp., TL-6574-1-74-26, Hwy. Research Report, June.

Dabrowski, R., 1968. *Curved Thin-Walled Girders—Theory and Analysis*, Springer-Verlag, Berlin.

Dahir, S. H. and D. B. Mellott, 1987. "Bridge Deck Expansion Joints," *Transp. Research Record* 1118, TRB, National Research Council, Washington, D.C., pp. 16–24.

FHWA, 1977. "Watertight Bridge Deck Joint Seals," National Experimental and Evaluation Program, Final Report, Project 11, U.S. Dept. Transp., Washington, D.C., July.

Gent, A. N., 1964. "Elastic Stability of Rubber Compression Springs," *ASME J. Mech. Eng. Sci.*, Vol. 6, No. 4.

ISBN, 1990. "Acceptance Criteria of Steel Bridge Welds," Report No. 0-309-04858-3, TRB, National Research Council, Washington, D.C.

Lim, K. Y., D. I. McLean, and E. H. Henley, 1989. "Plastic Hinge Details for the Bases of Bridge Columns," Small-Scale Test Model, Interim Report, Washington Dept. Transp., Olympia, April.

Lim, K. Y., D. I. McLean, and E. H. Henley, 1990. "Moment-Reducing Hinge Details for the Bases of Bridge Columns," *Transp. Research Record* 1275, TRB, National Research Council, Washington, D.C., pp. 1–11.

Mellott, D. B., 1978. "Status of Bridge Deck Expansion Dam Systems," Research Projects 73-15, 73-16, and 74-20, Pennsylvania *Dept. Transp.*, Nov.

Oleinik, J. O. and C. P. Heins, 1974. "Report No. 58," Dept. Civ. Eng., Univ. Maryland, College Park, Aug.

Park, R. and R. W. G. Blakely, 1979. "Seismic Design of Bridges, Bridge Seminar 1979," Summary Volume 3, Structures Committee, Road Research Unit, National Research Board, New Zealand.

Price, W. I. J., 1982. "Transmission of Horizontal Forces and Movements by Bridge Bearings, Joint Sealing and Bearing Systems for Concrete Structures," *ACI Publication* SP-70, Vol. 2, ACI, pp. 761–784.

Priestley, M. J. N. and R. Park, 1987. "Strength and Ductility of Concrete Bridge Columns Under Seismic Loading," *ACI Struct. J.*, Jan.–Feb.

Roeder, C. W. and J. F. Stanton, 1983. "Elastomeric Bearings: A State of the Art," *J. Struct. Div. ASCE*, Vol. 109, No. ST12, Dec.

Roeder, C. W. and J. F. Stanton, 1991. "State of the Art Elastomeric Bridge Bearing Design," *ACI J.*

Roeder, C. W., J. F. Stanton, and A. W. Taylor, 1987. "Performance of Elastomeric Bearings," NCHRP Report 298, TRB, National Research Council, Washington, D.C., Oct.

Stanton, J. F. et al., 1990. "Elastomeric Bearings: A State of the Art," *J. Struct. Div. ASCE*.

Stanton, J. F. et al., 1990. "Stability of Laminated Elastomeric Bearings," *J. Eng. Mech. ASCE*.

Stanton, J. F. and C. W. Roeder, 1982. "Elastomeric Bearing Design, Construction and Materials," NCHRP Report 248, TRB, National Research Council, Washington, D.C., Aug.

Taylor, M. E., 1972. "PTFE in Highway Bridge Bearings," Report TRRL-LR-491, Transp. and Road Research Lab., Crowthorne, Berkshire, England, 62 pp.

TRB, 1983. "Bridge Inspection and Rehabilitation," Transp. Research Board Report 899, TRB, National Research Council, Washington, D.C.

Vlasov, V. Z., 1961. *Thin-Walled Elastic Beams*, National Science Foundation, Washington, D.C.

Xanthakos, P. P., 1971. "I-75, I-275 4-Level Interchange, Kentucky, Bearing Design," In-House Study.

Yoo, C. H., D. R. Evick, and C. P. Heins, 1974. "Non-Prismatic Curved Girder Analysis," *J. Computers and Structures* Pergamon Press, London.

CHAPTER 14

SUBSTRUCTURES

14-1 FOUNDATION INVESTIGATIONS

The various, and often complex, techniques for conducting subsurface investigations for substructure elements and foundations are outlined in the AASHTO Manual on Subsurface Investigations (1988).

In general, this program includes the following: (a) site reconnaissance and field survey, (b) main field investigation; and (c) a soils and foundation report.

Site Reconnaissance and Field Survey This stage is not intended to provide solutions to foundation problems, but rather emphasizes the need to assess the general site conditions and to articulate the requirements of further investigations. The program addresses the site topography, the local geology of soils, the groundwater conditions, and relevant site characteristics and history.

Main Field Investigation Initially, a boring plan is prepared and detailed according to ground uniformity, expected loads, and type of foundation. With erratic subsurface conditions, the boring plan is more extensive. Where the size of the project and location warrant, the conventional boring program is supplemented by a geophysical survey consisting of cross-hole surveys, seismic exploration, electromagnetic subsurface profiling, and acoustic interpretation.

Foundation exploration usually implements the boring plan and provides pertinent samples. The wide divergence of geological and soil conditions encountered in the United States does not warrant a practical and standard methodology in foundation exploration. The objective, however, is the same: to obtain sufficient data regarding subsurface characteristics to expedite a safe and economical foundation design.

Laboratory testing is part of this phase, its main purpose being to verify soil classification and obtain engineering properties. The tests should also reflect the correlation between field work and foundation planning and should be adequate in simulating the most severe design criteria. Laboratory tests for foundation analysis mainly include classification, strength, and compressibility tests. Dynamic testing of loose sands is indicated if liquefaction or excessive settlement is anticipated in areas of seismic activity and cyclic mobility.

Soils and Foundation Report This report focuses on the scope of the program and addresses the field exploration, laboratory investigations, existing structures and site conditions, analysis of data, and foundation studies. Anticipated construction problems should be identified, and a firm basis should be established for estimating construction costs. The policy of releasing all available information is encouraged, but reliance upon it should essentially remain a matter of judgment. Because the associated legal ramifications are obvious, all information contained in the report should be as accurate as possible.

14-2 ABUTMENT TYPES

Abutments are the end supports of a bridge, and their function is therefore twofold: (a) to transmit the reactions of the superstructure to the foundation and (b) to retain the embankments and provide the end of a bridge where it connects with the approach roadway. Thus, the main loads on the abutment come from the superstructure or are induced as earth pressures.

Typical abutment configurations are shown in Figure 14-1. The gravity abutment shown in Figure 14-1*a* has a bridge seat, back wall and wing walls, and a common footing, resisting all loads as a gravity structure. The U-type abutment shown in Figure 14-1*b* has the wing walls perpendicular to the main structure, and this improves stability against overturning. The spill-through or closed abutment shown in Figure 14-1*c* has the beam seat supported on counterforts resting on a common footing. The pile bent abutment shown in Figure 14-1*d* consists of a pile cap with stub or inclined wings supported on piles. The latter are driven through precored holes if the distance from the bottom of the abutment to the natural ground exceeds about 10 ft.

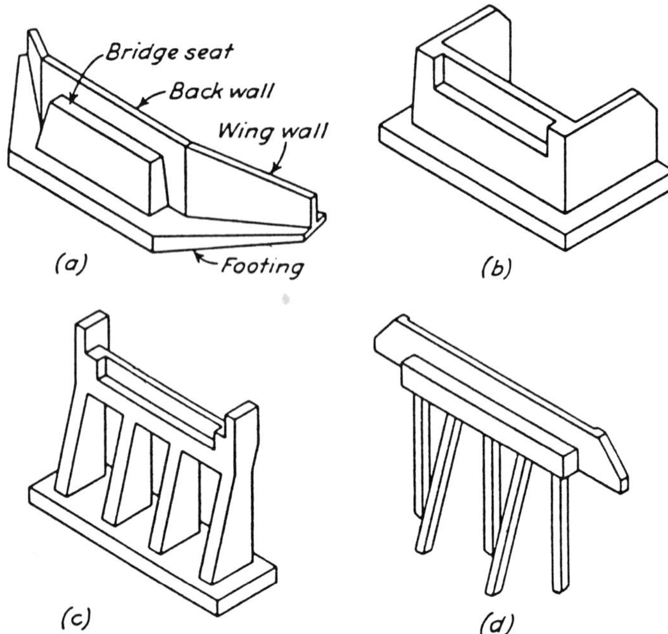

FIGURE 14-1 Various types of abutments: (*a*) typical gravity abutment with wing walls; (*b*) U abutment; (*c*) spill-through abutment; (*d*) pile bent abutment with stub wings.

A modified version is the vaulted abutment shown in Figure 14-2. This utilizes precast prestressed beams to support the abutment span, usually selected when the span at right angles exceeds about 21 ft. Access to the inside of the vault is mandatory. The detail shown in Figure 14-2 is a sand-filled vaulted abutment using a reinforced concrete slab as the abutment slab. This solution is generally indicated when the right-angle design span is less than 21 ft.

A proposed revision to the standard AASHTO specifications (D'Appolonia, Inc., 1989) articulates abutments into four types: stub abutments, partial-depth abutments, full-depth abutments, and integral abutments (see also Section 11-11).

Deflections on abutments may be critical if joints can close causing unwarranted forces to be induced in the superstructure. Because of this, abutments are built with a high degree of stiffness. Closed abutments more than 20 ft high should have counterforts for adequate rigidity. An alternative for the control of deflections is the vaulted type shown in Figure 14-2.

In abutments with wing walls, the length of the wings depends on the arrangement of the embankment slopes. Ordinarily, a slope of 2 : 1 or 1.5 : 1

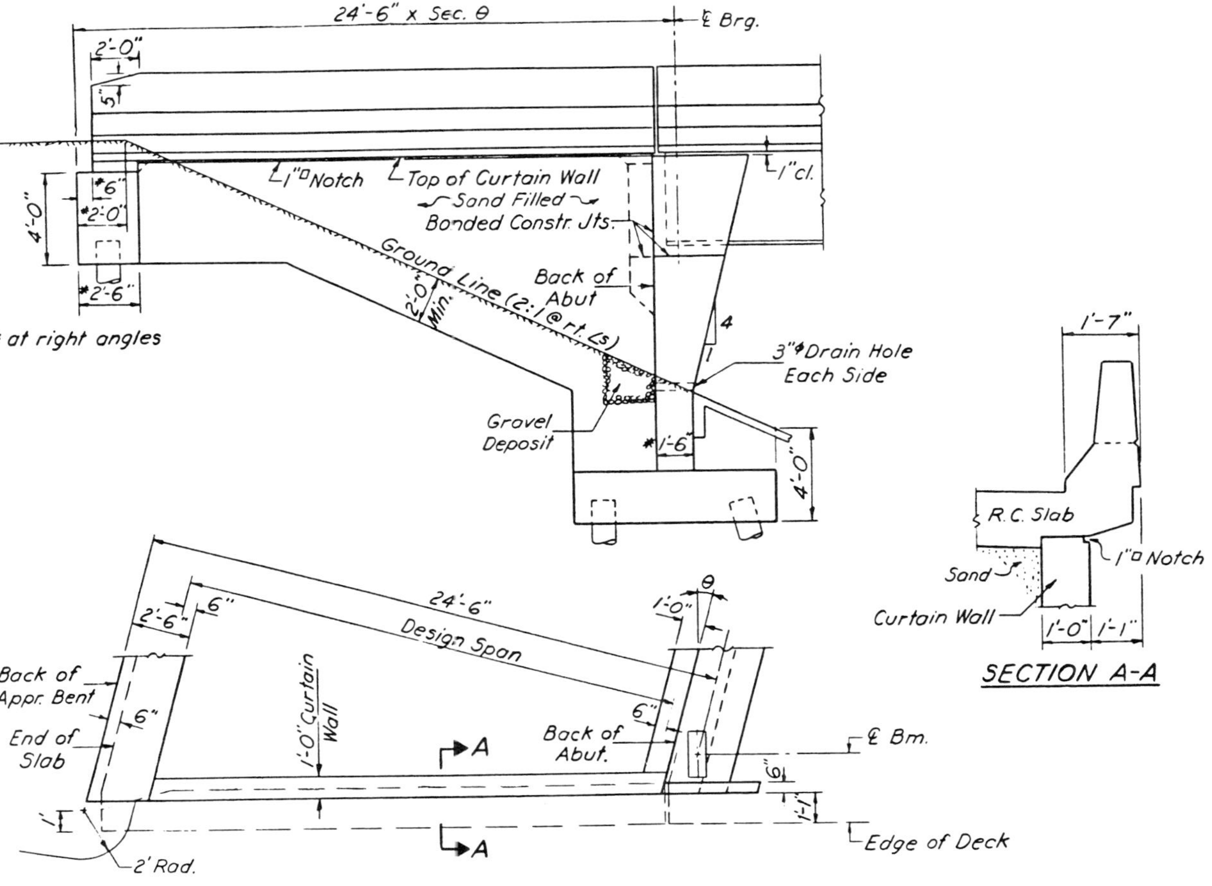

FIGURE 14-2 Filled vaulted abutment; plan and details.

is used resulting in corresponding wing lengths. Abutments should be designed to withstand the following loads and forces.

1. Dead load from the weight of the abutment and bridge superstructure.
2. Live load on the superstructure or on the approach slab.
3. Transverse wind forces and longitudinal forces when the bearings are fixed and longitudinal forces due to friction or shear resistance at expansion bearings.
4. Centrifugal forces for bridges on curves.
5. Earth pressures.

The design should result in a structure that is safe against overturning about the toe of the footing, against sliding along the footing base, or against failure of the foundation soil or overloading of the piles. This safety should be ensured for any combination of loads and forces that can produce the most severe condition of loading.

In calculating stresses, the weight of the fill material directly over an inclined or stepped rear face or over the base of a reinforced concrete spread footing may be considered as part of the effective weight of the abutment. In spread footings, the rear projection is designed as a cantilever fixed at the abutment stem and loaded with the full weight of the superimposed fill material.

For conventional steel or concrete girder superstructures, the abutment type should be independent of the superstructure and should be dictated mainly by the loads and forces, slope arrangement, and economic considerations.

In abutments with expansion bearings, the effect of temperature changes need not be considered beyond the friction forces ordinarily induced at the bearing. Likewise, for design purposes the abutment may be considered as a vertical cantilever free at the top and fixed at the bottom (i.e., friction is assumed to be zero). In integral bridges with both ends having sufficient horizontal stability, the main abutment wall resisting lateral earth pressures may be assumed to be supported at the top and bottom. The top is freely supported by the superstructure, and the bottom may be either fixed or free.

14-3 PIER TYPES

General

Piers are intermediate vertical supports having the following functions: (a) to transfer all vertical loads to the foundation and (b) to resist all horizontal and transverse forces acting on the bridge. Superstructures are normally placed on bearings that sit on pier caps and are capable of transmitting these forces.

In some instances the pier and superstructure may have a continuous connection with adequate reinforcement specially arranged to make framed corners.

Various pier types are shown in Figure 14-3. Most bridge designs have reinforced concrete piers that can accommodate the practical site conditions. The solid or gravity piers shown in Figures 14-3*a* and *b* are usually selected in river crossings. The single-hammerhead type is economical for high river heights, whereas the wall pier is practical for low river heights. In either type,

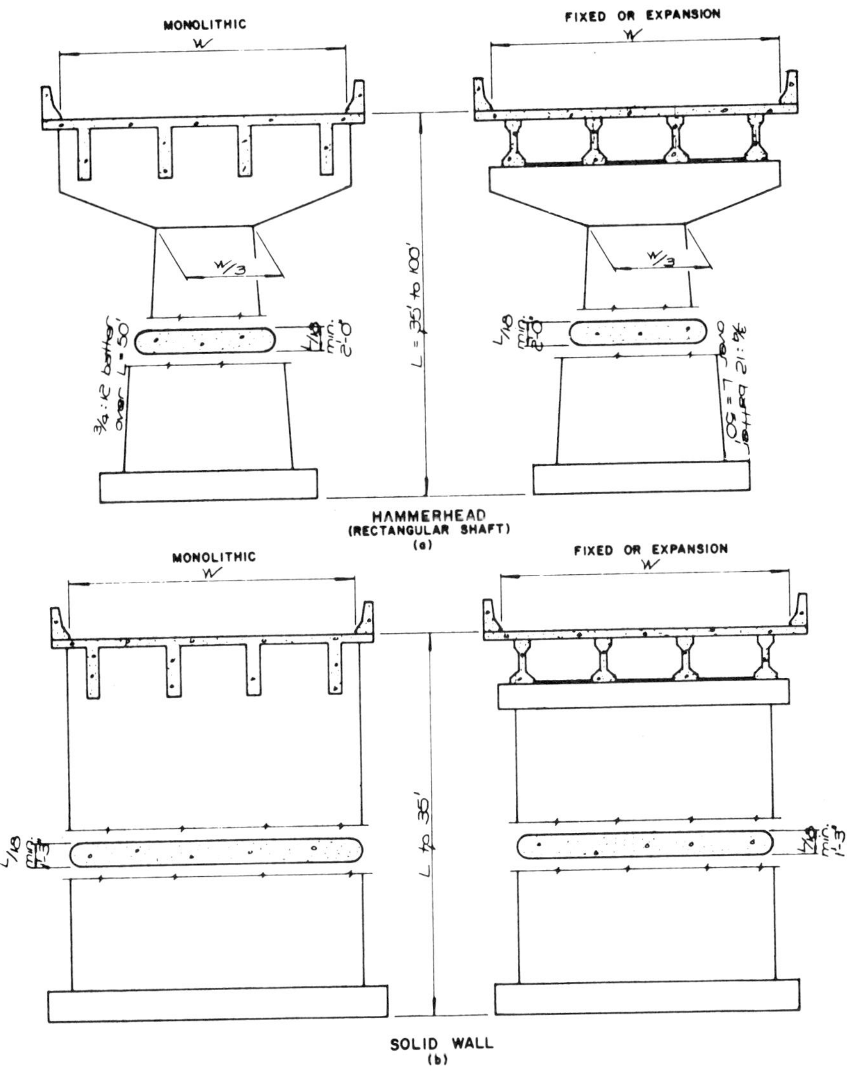

FIGURE 14-3 Various pier types and configurations.

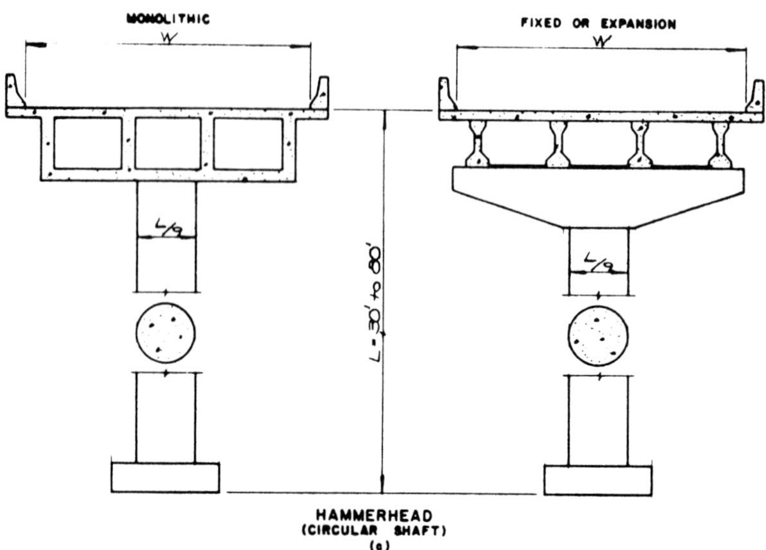

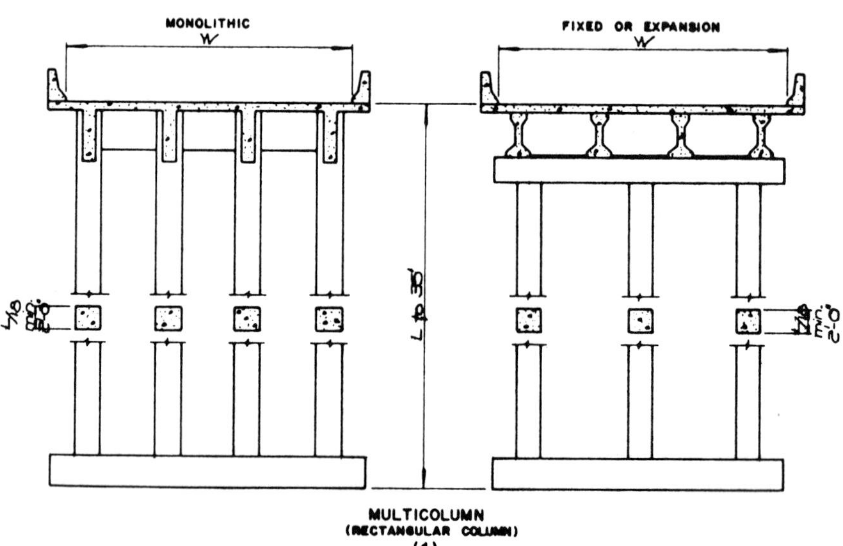

FIGURE 14-3 *(Continued)*

the pier may have a rectangular horizontal section or its ends may be rounded as shown. The dimensions at the bottom of the pier required for stability are usually larger than the top dimensions, resulting in pier sides that are appropriately battered. Normally, the batter varies from 1/4 to 1 in./ft.

The solid pier resists all forces acting on it by its massive configuration. Its design and construction are relatively simple, and the minimum required steel reinforcement is governed by applicable specifications. The main reinforcement consists of vertical bars placed along the perimeter to resist cantilever moments induced by forces acting at the top and perpendicular to the pier. Horizontal steel must be provided to satisfy temperature and shrinkage requirements and also to tie the vertical steel. The pier can sit on spread footings or on a pile cap if piles must be the foundation element. The formwork is simple and economical and results in rapid construction. Although a solid wall pier is practical for low heights, up to about 25 ft, it has been used for higher elevations. Where these economic considerations are no longer justified, the single hammerhead may be selected and will result in less total concrete and thus less weight on the foundation. The hammerhead type can be used with superstructures consisting of multiple-girder systems or of two girders and two trusses, but the cantilever moments and shears on the head are much higher.

The dimensions of the footing or pile cap for a hammerhead pier must be selected to ensure stability in both directions. The main reinforcement in the pier is horizontal steel in the top of the head to resist cantilever moments, and stirrups to resist the shear. Vertical steel around the periphery of the column is likewise provided and is usually determined considering the pier as a tied column subjected to axial loads and bending moments possibly in both directions.

The hammerhead with a circular shaft shown in Figure 14-3*c* may be selected where space use precludes other pier types and where the single-column substructure can ensure stability in both directions. In monolithic construction pier caps are required to transfer moments to the column by torsional action, and on wide caps where more than one girder is framed into them, this transfer may result in special design requirements.

The multiple-column pier bent shown in Figure 14-3*d* is practical on wider structures and may be used in river crossings if debris collection is not a problem. The analysis of this pier involves essentially a rigid frame. The frame is acted upon by loads and forces from the superstructure inducing bending moments and axial loads in the columns and bending moments and horizontal struts in the cap. The analysis is more complicated considering the various load combinations that must be applied, and forming is more difficult than in other types. The common footing improves the base rigidity, particularly against transverse forces.

Where the ground and foundation conditions are favorable or where the distance between columns is appreciable, the columns may rest on individual footings. Individual column footings may be designed for the actual load and moment effects induced in the columns, but the resulting design is more efficient if the ground offers sufficient stiffness for fixed conditions at the base of the columns.

For high piers consideration should be given to a hammerhead type or a stepped shaft. Piers with variable moments of inertia present certain problems in the analysis because of the difficulty of computing critical buckling loads, but they are practical and economical. Another option for high piers (height > 100 ft) is a multistory frame that can provide stability with essentially moderate dimensions. This type lends itself to superstructures of two plate girders or trusses, where columns are centered under them.

In certain pile foundations, the piles can be left extended above ground and interconnected with a reinforced concrete cap (pile bent pier). These piers are economical for moderate heights and nominal transverse forces. Some of the piles may be battered to resist moment effects in either direction. Appearance and strength are improved if the piles are encased in concrete.

Design Requirements

In general, piers should be located to satisfy navigational clearance and to inhibit interference with flood flow. This is accomplished if the piers are placed parallel with the direction of the stream current. Special precautions are warranted against scour, and when this problem is anticipated special investigations are essential to determine the probable depth of scour or flotation of material. The general requirements regarding the depth of pier foundations are summarized in Section 4 of the standard AASHTO specifications.

Piers must be designed to withstand the following loads and forces.

1. Dead load from the weight of the superstructure and the pier.
2. Live load on the superstructure.
3. Wind loads acting on the pier and superstructure.
4. Longitudinal forces at fixed bearings and friction of expansion bearings.
5. Forces due to stream current, floating ice, and drift.
6. Vessel collision forces where applicable.

The base of the foundation must be located below the frost line, and, as mentioned previously, it must also be deep enough to be out of danger from underscouring, which is the most frequent cause of pier failure. Protection against this occurrence is usually provided with riprap or by a cofferdam left in place. The cofferdam should be driven at least 3 ft below the bottom of the foundation and should extend some distance above the top of the footing course.

Where possible, particularly for statically indeterminate structures, piers should be carried to rock, and this is practical when the rock is not more than 20 ft below the river bed. Satisfactory foundations may be obtained on hardpan, gravel, and other firm ground by choosing a proper allowable pressure.

Design Steps Usually, the first step is to establish the pier dimensions from the known layout and geometry of the superstructure. The total pier height is determined from the deck elevation and the bottom footing level. The overall pier height should be kept to a minimum, but the footing must rest on firm material. The minimum width and thickness of the pier head or cap depend on the layout and dimensions at the bridge bearings. For hammerhead-type piers (single shaft), the head is the first item to be designed once the position and magnitude of loads from the superstructure are known. When columns are rigidly connected to the superstructure to form rigid frames, the analysis follows an appropriate procedure. When the vertical supports are separated from the superstructure by bearings, they must be stable without assistance from the superstructure. When the structure is supported on isolated columns, the columns must be strong enough to resist loads and moments assuming the bottom to be fixed and the top free. The footings for individual columns must be designed to resist vertical reactions and bending moments transferred to the bottom of the column.

Multiple-column piers may have each column supporting one beam or girder directly, or when the girders are closely spaced the number of columns may be reduced. In the latter case, a stronger cap is required.

With the dimensions of the head or pier cap selected, a column size may be specified on a trial basis and checked by the design. Usually, transverse column dimensions (width or diameter) are 4 to 6 in. less than the width of the head or cap. Reinforced concrete rectangular shafts or columns may be more efficient than circular elements in resisting moments and are compatible with monolithic rigid frames. Circular shafts or columns used in rigid frames tend to absorb moments and must have adequate capacity to carry them. Their inherent advantages are the ease in forming and the extra confining strength resulting from spirals incorporated in the section. A close spiral spacing also provides increased buckling strength, a characteristic appreciated under seismic forces.

Footing size, dimensions, and shape are selected to satisfy stability against overturning and also to ensure that the soil pressure (with spread footing) is within the allowable limit. The latter is computed as the sum of $P/A + M_x y/I_x \pm M_y x/I_y$. Because there can be no tensile stresses between the footing surface and the ground, only the area of the footing subjected to pressure can be included in calculating the I values. Most designers size the footing so that uplift will not occur. If piles are used in the foundation, pile loads are determined accordingly. Tension in the piles may be accepted if the piles are designed accordingly.

Piers are first checked for Group I loading; in this case the only lateral loads are stream flow and centrifugal forces when the bridge is curved. Because the specifications do not provide factors of safety for stability, a common factor of safety for this condition is 2. The substructure elements are then checked for Group II and III loadings which include wind and longitudinal forces. Under these loadings, allowable stresses and soil pressures may be increased accordingly.

Piers for Special Bridges

Segmental Bridges The segmental concrete bridges reviewed in Chapter 8 may require specific types of piers and foundations, particularly for spans erected by the balanced cantilever method. The substructure may consist of single shafts, double shafts, and moment-resistant piers. The resistance and elastic stability of piers during construction warrant special analyses, and temporary piers or temporary strengthening of permanent piers are commonly used singly or combined to satisfy erection requirements.

The use of full continuity in the superstructure requires corresponding design features to allow for volume changes (shrinkage, creep, and thermal expansion) at the supports. For example, segmental bridges on piers with flexible legs can accommodate full deck continuity and provide frame action between the deck and piers without inhibiting free expansion. Vertical parallel legs may be used on multispan structures because their additional flexibility accommodates larger horizontal displacements. For longer segmental bridges, bearings may be used to provide flexibility.

If single slender piers in the finished structure are designed solely for final bridge loads, they may not be adequate to resist unsymmetrical moments caused by cantilever construction or unbalanced segments and erection equipment. Thus, complicated temporary shoring may be needed, often at considerable cost. In some instances the stability of the cantilever has been ensured by the gantry used for placing the segments.

With double piers, two flexible legs make up the substructure elements, usually supported on a single footing. During construction, stability is maintained with relatively little additional measures and some extra bracing between the slender walls to resist elastic instability.

Moment-resistant piers must withstand unbalanced moments by means of a temporary vertical prestress between the deck and the pier cap resulting in a rigid connection. Where the ratio of span length to pier height is favorable, the rigid connection and associated frame action may be introduced permanently.

An example of a pier for segmental bridges is shown in Figure 14-4, and it appears that piers need not be massive solid structures. The tubular section may in certain cases be replaced by an I section, but caution is warranted because of the reduced torsional resistance of the section.

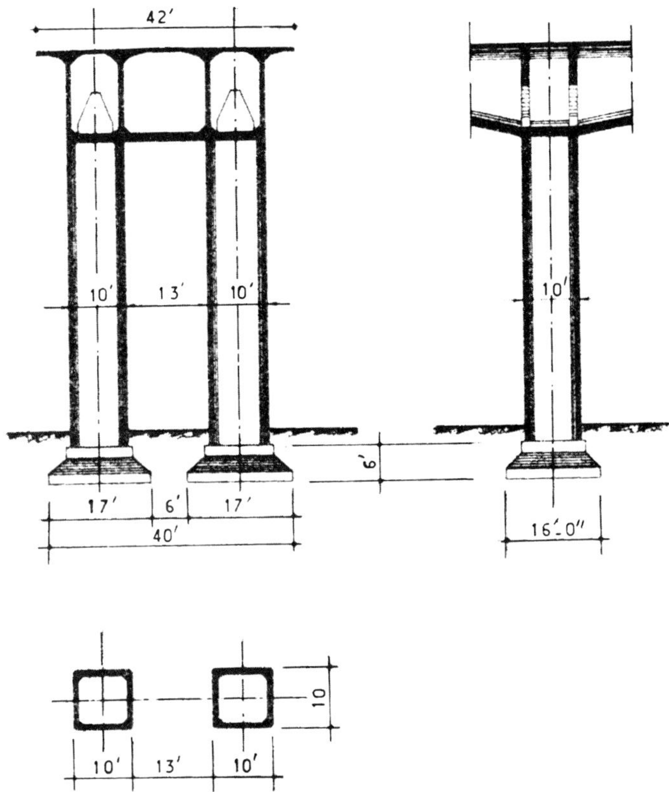

FIGURE 14-4 Pier example for segmental concrete bridges. (From Podolny and Muller, 1982.)

For continuous structures on short stiff piers, length changes of the deck are combined with the effects of longitudinal prestressing to produce considerable bending forces that must be transmitted to the foundation, and the benefits of a rigid connection between the superstructure and the supports are thus lost.

A complete description of pier types and methods of analysis is given by Podolny and Muller (1982).

Trestles Trestles are essentially bridge decks supported on pile bents. They are used extensively for highway crossings over swamps, stagnant waters, and shallow waterways and lakes. Very often they are of considerable length. Superstructure–substructure interaction and cost optimization usually result in relatively short spans. The piles are precast elements and may be circular, octagonal, or rectangular with chamfered edges. The pile dimensions depend not only on the load to be carried but also on the unsupported length.

For bridges with roadways up to 30 ft wide, four piles are used per bent. The piles may be vertical, and, where dictated by stability requirements, the outside piles may be battered.

The bents in a pile trestle should be spaced longitudinally so as to make the cost of the bridge per unit length a minimum. Obviously, where long piles are needed and the installation is difficult, the economical spacing is greater than for short piles. The bent spacing is also related to the type and design of superstructure. If simple slabs are used, the bent spacing may not exceed 20 ft. If precast deck units are to be used, the span length may depend on the available facilities and the maximum practical weight to be handled at the site.

14-4 RELEVANT SPECIFICATIONS

Standard AASHTO Specifications

Service Load Design AASHTO Article 8.15.4 stipulates that the combined flexural and axial load capacity of compression members such as columns must be taken as 35 percent of that computed according to Article 8.16.4 (strength design method). Slenderness effects are considered according to Article 8.16.5. In using these provisions ϕ should be taken as 1.0.

Strength Design Method AASHTO Article 8.16.4 stipulates that the factored axial load P_u at a given eccentricity should not exceed $\phi P_{n(\max)}$, the design axial load strength. Two cases are distinguished. For spiral reinforcement, a coefficient of 0.85 is applied to P_n, and the strength factor ϕ is taken as 0.75. For a column with ties, a coefficient of 0.80 is applied to P_n, and the strength reduction factor ϕ is taken as 0.70.

A provision is also included as a guide to articulate the range of the load–moment interaction relationship for members subjected to combined axial load and flexure. Pure compression involves an axial load at zero eccentricity. For design, pure compression and the associated strength is a hypothetical condition because AASHTO limits the axial load strength of compression members to between 85 and 80 percent of the axial load at zero eccentricity as mentioned previously. Pure flexure assumptions may be used to compute the design moment strength ϕM_n for pure flexure conditions.

It appears that service load procedures are either too conservative or somewhat unrealistic, and this prompts many designers to consider only the load factor approach in the analysis of reinforced concrete pier columns.

LRFD Specifications

As in the standard AASHTO specifications, the LRFD document stipulates the same material resistance factors. Likewise, the value of ϕ may be increased linearly from the value for compression members to the value for flexure as the design axial load strength ϕP_n decreases from $0.10F'_c A_g$ or ϕP_b, whichever is smaller, to zero.

The rectangular compressive stress distribution defined in AASHTO Article 8.16.2.7 may be used in lieu of the more exact concrete stress distribution. Although this pattern does not represent the actual stress distribution in the compression zone at ultimate conditions, it provides essentially the same results obtained in tests. Equations expressing strength are based on the rectangular stress block.

The factor β_1 is clearly defined for rectangular sections. For other configurations, such as flanged sections, β_1 has been shown to be an adequate approximation.

Limits of Reinforcement The amount of reinforcement used for computing moment strength should satisfy the following:

$$c/d_e \leq 0.42 \tag{14-1}$$

where c = depth from extreme compression fiber to the neutral axis at nominal bending resistance of the section

d_g = corresponding effective depth from extreme compression fiber to the centroid of the tensile force in the tensile reinforcement

If (14-1) is not satisfied, the section is considered overreinforced. Overreinforced sections are not allowed in prestressed and partially prestressed sections unless analysis shows that sufficient ductility can be achieved. The definition of d_e can be extended to multilayered systems such as columns having different layers of prestressing reinforcement and reinforcing bars.

For a rectangular section or rectangular section behavior, the nominal flexural resistance for an overreinforced member can be computed as

$$M_n = \left(0.36\beta_1 f'_c b d_e^2\right) \tag{14-2}$$

Compression Members The maximum areas of prestressed and non-prestressed longitudinal reinforcement for noncomposite compression members should satisfy the following expressions:

$$\frac{A_s}{A_g} + \frac{A_{ps}}{A_g}\frac{f_{pu}}{f_y} \leq 0.08 \tag{14-3}$$

and

$$(A_{ps} f_{ps})/(A_g f'_c) \leq 0.3 \tag{14-4}$$

where A_s = area of non-prestressed tension reinforcement (in.^{2})
A_{ps} = area of prestressing steel (in.^{2})
A_g = gross area of section (in.^{2})
f_{pu} = yield strength of prestressing steel (psi)
f_y = specified minimum yield strength of reinforcing bars (psi)
f_{ps} = average stress in prestressing steel at nominal resistance (psi)

The minimum areas of prestressed and non-prestressed longitudinal reinforcement for noncomposite compression members should satisfy the following:

$$\frac{A_s f_y}{A_g f'_c} + \frac{A_{ps} f_{pu}}{A_g f'_c} \geq 0.12 \tag{14-5}$$

According to current AASHTO specifications, the area of longitudinal reinforcement for non-prestressed noncomposite compression members should not be less than 0.01. In most pier columns the dimensions are controlled primarily by bending, and in this case the minimum area limitation does not reflect the influence of the concrete compressive strength. For this strength to be accounted for, the minimum reinforcement in flexural members must be proportional to f'_c/f_y, and this approach is reflected in (14-5).

The minimum number of bars is six in a circular arrangement, and four in a rectangular configuration. The minimum bar size is #5. For structures in Seismic Category A, members are proportioned by considerations other than applied loads. In this case the minimum area of longitudinal reinforcement is the amount required for a member with a reduced effective area of concrete capable of resisting the factored loads. However, the ductility requirements are not changed.

Slenderness Effects For an approximate evaluation of slenderness effects, the criteria and procedures are essentially the same as in AASHTO Article 8.16.5. In structures not braced against sidesway, the flexural members framing into the compression member should be designed for the total magnitude of end moments of the compression member at the joint.

In structures undergoing appreciable lateral deflections resulting from combinations of vertical load or combinations of vertical and lateral load, the moments M_1 and M_2 should be calculated from a second-order analysis. In this case M_1 is the value of the smaller end moment at the strength limit state due to factored load acting on a compression member (positive if the member is bent in single curvature and negative if the member is bent in double curvature). Likewise, M_2 is the value of the larger end moments at the strength limit state due to factored load acting on a compression member, always positive.

Factored Axial Resistance The factored axial resistance is given by

$$P_r = \phi P_n \le \phi P_0 \tag{14-6}$$

where P_r = factored axial resistance, with or without bending
P_n = nominal axial resistance, with or without bending
ϕ = resistance factor
P_0 = nominal axial load strength at zero eccentricity

Spirals and Ties For bridges built within Seismic Categories B, C, or D, the area of steel for spirals and ties is governed by appropriate seismic requirements. When the area of spiral and tie reinforcement is not controlled by seismic considerations or by shear and torsion, it will be designed so that its ratio p_s to the total volume of concrete core measured out-to-out of the spirals should not be less than

$$p_s = 0.45\left(\frac{A_g}{A_c} - 1\right)\frac{f'_c}{f_{yh}} \tag{14-7}$$

where A_c = area of core measured to outside diameter of the spirals (in.^{2})
f_{yh} = specified yield strength of transverse (spiral) reinforcement

Footings The factored shear resistance V_r is taken as

$$V_r = \phi V_n \tag{14-8}$$

where V_n is the nominal shear resistance and ϕ is the resistance factor, in this case $\phi = 0.90$.

In determining the shear resistance of footings in the vicinity of concentrated loads or reactions, the more severe of the following conditions will govern: (a) beam action, with a critical section in a plane across the entire width and located at a distance d from the face of the concentrated load or reaction area; (b) two-way action, with a critical section perpendicular to the plane of the footing so that the perimeter b_0 is a minimum, but not closer than $0.5d$ to the perimeter of the concentrated load or reaction area; and (c) when the footing thickness is not constant, critical sections located at a distance no closer than $0.5d$ from the face of any change in thickness, and located such that the perimeter b_0 is a minimum, should also be investigated.

If a footing is supported on piles, shear on the critical section for column or wall loads should be as follows: (a) the entire reaction from any pile whose center is located at a distance $d_p/2$ or more outside the critical section should be considered as producing shear on that section; (b) the reaction from any pile whose center is located at $d_p/2$ or more inside the critical section may be neglected in computing shear on that section; and (c) for intermediate positions of the pile center, the portion of the pile reaction to be considered as producing shear on the critical section should be based on a linear interpolation between the full value of $d_p/2$ outside the section and the zero value of $d_p/2$ inside the section. In this analysis d_p is the diameter of a round pile or the depth of an H pile at the footing base.

For footings supporting columns with metal base plates, the face or perimeter of the concentrated load should be taken as halfway between the face of the column and the edge of the base plate.

The shear resistance of footings should also be investigated for beam action and two-way action. The current expression for punching shear is retained. If shear perimeters for individual loads overlap or project beyond the edge of the member, the critical perimeter must be taken as that portion of the smallest envelope of an individual shear perimeter actually resisting the critical shear for the group under consideration. An example of this condition is shown in Figure 14-5.

Circular or regular polygon-shaped concrete columns or piers may be treated as square members with the same area, for the location of critical sections for moment, shear, and development of reinforcement. In sloped or stepped footings, the angle of the slope or the depth and location of steps should be detailed to satisfy the design requirements.

Loads and Reactions When a single isolated footing supports a column, pier, or wall, the footing may be assumed to act as a cantilever. When footings support several columns, piers, or walls, the design should consider the actual conditions of continuity and restraint.

The as-built location of pile foundations may differ from the design to the extent of permitted construction tolerances. For the design of a footing, it

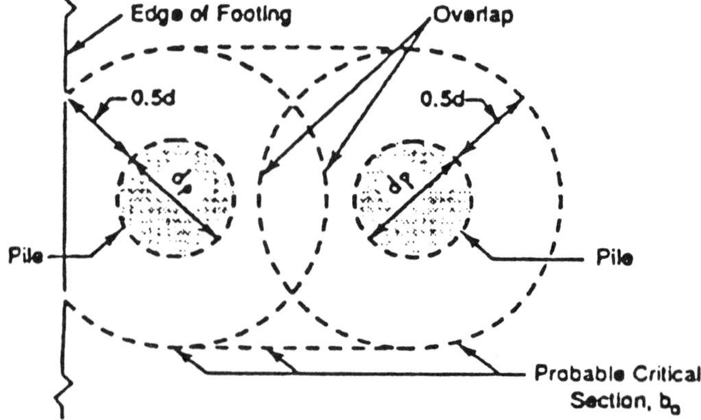

FIGURE 14-5 Modified critical section for shear, with overlapping critical perimeters.

should be assumed therefore that individual driven piles may be out of planned position by 6 in. or 1/4 of the pile diameter, and that the center of gravity of a pile group may be off by as much as 3 in.

Bending Moments External moments are usually computed at the critical section along a vertical plane through the footing. The analysis considers all the forces and stresses acting over the entire area of the footing on one side of the vertical plane. The critical section for bending is taken at the face of the column, pier, or wall. If the columns are not square or rectangular, the critical section is at the side of the concentric square of equivalent area. For footings under masonry walls, the critical section is halfway between the middle and the edge of the wall. If the columns have metallic bases, the critical section is halfway between the column face and the edge of the plate.

The reinforcement should be distributed uniformly across the entire width of the footing in one-way footings and two-way square footings. In two-way rectangular footings, the reinforcement should be distributed as follows: (a) in the long direction, the reinforcement should be placed uniformly across the entire width; (b) in the short direction, a portion of the total reinforcement should be distributed uniformly over a band width equal to the length of the short side of the footing and centered on the centerline of the column or pier. This portion is given by

$$\frac{A_s - (\text{Band width})}{A_s - (\text{Total})} = \frac{2}{\beta + 1} \qquad (14\text{-}9)$$

where β is the ratio of the long side to the short side of the footing. The remainder of reinforcement required in the short direction should be distributed uniformly outside the center band width of the footing.

Transfer of Force at Base of Column Forces and moments acting at the base of a column must be transferred to the top of the footing by bearing on concrete and by dowel reinforcement. Bearing on concrete should not exceed the factored bearing resistance P_r given by

$$P_r = \phi P_n \qquad \text{where} \quad P_n = 0.85 f'_c A_1 m \qquad (14\text{-}10)$$

where P_n = nominal bearing resistance
A_1 = bearing area
ϕ_m = resistance factor taken as 0.70
m = modification factor

Lateral forces should be transferred to supporting footing according to shear transfer provisions. Reinforcement should be provided across the interface between the column and footing or the wall and footing preferably in the form of dowels. This reinforcement should be sufficient to satisfy the following requirements.

1. All forces exceeding the concrete bearing strength should be transferred to the footing by reinforcement.
2. If the loading groups include uplift, the total tensile force must be resisted by reinforcement.
3. The minimum reinforcement should not be less than 0.005 times the gross area of the supported member, with four bars minimum. Where the cross section is determined by considerations other than the structural requirements, an effective area may be used.

If dowels are used, they should have a diameter that does not exceed the diameter of the longitudinal reinforcement by more than 0.15 in.

14-5 LOAD AND MOMENT: ANALYSIS CRITERIA

In general, first-order loads and moments are calculated using elastic analysis, whereas stiffness criteria and assumptions are expected to be reasonable and consistent. The commentary on the Ontario Bridge Code (1983) suggests an uncracked cross section neglecting all reinforcement, or a transformed cracked section throughout. Although the correct representation of the ultimate state would require moment–curvature relationships, reasonable results have been obtained if EI values are assumed to be equal to $E_c I_g(0.2 + 1.2 p_t E_s/E_c)$ for piers and $0.5 E_c I_g$ for superstructure stiffness, where I_g is the gross moment of inertia, $p_t = A_s/A_g$, E_s = 29,000,000 psi, and $E_c = 57{,}000\sqrt{f'_c}$.

The proposed LRFD specifications stipulate that elastic material properties must be consistent with the expected behavior, and changes in these values due to maturity of concrete as well as environmental effects should be included in the analysis. Extreme-event limit states may be accommodated in both the elastic and inelastic ranges. If inelastic analysis is considered, it should be explicitly applied to components containing materials that are innately ductile or can be made to behave in a ductile manner.

A usual assumption in computing loads and moments pertains to the degree of fixity and restraint induced by foundation conditions. Considering the limited data available for determining end fixity, most designers refer to guidelines such as Table 14-1 in conjunction with the following criteria.

1. Piers on multiple rows of piles and drilled shafts may be considered 100 percent fixed at the connections to the piles or shafts.
2. Piers on a single row of piles may be assumed to have pinned conditions at the pile connections with the footings.

TABLE 14-1 Foundation Fixity Parameters (From FHWA, 1969)

G_B	
1.5	Footing on rock anchored
3.0	Footing on rock not anchored
5.0	Footing on soil
1.0	Footing on piles [add 10 ft (3.05 m) to the effective length]

$$G_B = \frac{EI/L \text{ columns}}{EI/L \text{ members resiting column bending at } B \text{ end of column}}$$

(see also Figure 14-6).

3. Piers with spread footings and an allowable foundation pressure at a service load greater than 3 to 6 tons/ft^2 may be assigned 30 percent fixity at the base of the footing.
4. Piers on spread footings with an allowable foundation pressure greater than 6 to 9 tons/ft^2 may be assigned 40 percent fixity at the base of the footing.
5. Piers with spread footings and an allowable foundation pressure exceeding 9 tons/ft^2 may be considered 100 percent fixed at the base of the footing.

Current design philosophies suggest that if the deformation of a structure does not result in a significant change of force effects, due, for example, to an increase in the eccentricity of compressive or tensile forces, such secondary force effects may be ignored. Small deflection theory is thus adequate for the analysis of bridge structures that are insensitive to deformations. Columns and piers, however, where the moment arm of a tensile or compressive force is increased or decreased directly by a deflection, tend to be sensitive to deflection events.

Many bridges are designed to eliminate deck joints (integral bridges). Abutment backwalls may be eliminated by compacting the embankment against the superstructure ends. In this case the restraint against sidesway is minimum, and sidesway and second-order effects should be included in the pier analysis. Large-deflection theory is thus appropriate where the deformation of the structure results in significant changes in force effects. A properly formulated large-deflection model provides all the force effects necessary for design, because these are included directly when equilibrium equations are formulated from the displaced shape. If inelastic behavior is expected at the strength limit state, it should be reflected in the analysis in conjunction with the use of factored loads.

In lieu of a complete second-order analysis, the degree of sensitivity can in many cases be assessed and evaluated by a single-step approximate method, such as the moment magnification factor method discussed in the following section.

Moment Magnification Factor Method

A moment magnifier equation can be derived by referring to Figure 14-6. This analysis is intended to magnify only pier moments resulting from lateral loads. The primary deflection Δ_1 caused by end moment M_1 is

$$\Delta_1 = \frac{M_1 L^2}{8EI}$$

The secondary deflection Δ_2 caused by the axial load influence is

$$\Delta_2 \simeq \frac{0.10P(\Delta_1 + \Delta_2)L^2}{EI}$$

Next, we consider the critical Euler load

$$P_{cr} = \frac{\pi^2 EI}{L^2}$$

written as

$$\frac{L^2}{EI} = \frac{\pi^2}{P_{cr}}$$

and substituting this into the equations for Δ_1 and Δ_2 gives

$$\Delta_1 = \frac{M_1}{8}\left(\frac{\pi^2}{P_{cr}}\right) \quad \text{and} \quad \Delta_2 \simeq 0.10P(\Delta_1 + \Delta_2)\left(\frac{\pi^2}{P_{cr}}\right) \tag{14-11}$$

Summing Δ_1 and Δ_2 gives

$$\Delta_1 + \Delta_2 = \left[\frac{M_1}{8} + P(\Delta_1 + \Delta_2)(0.10)\right]\left(\frac{\pi^2}{P_{cr}}\right) \tag{14-12}$$

The following expression is true:

$$M_c = M_1 + P(\Delta_1 + \Delta_2) \tag{14-13}$$

Combining (14-12) and (14-13), we obtain

$$M_c = M_1 + P\left(\frac{\pi^2}{P_{cr}}\right)\left[\frac{M_1}{8} + P(\Delta_1 + \Delta_2)(0.10)\right]$$

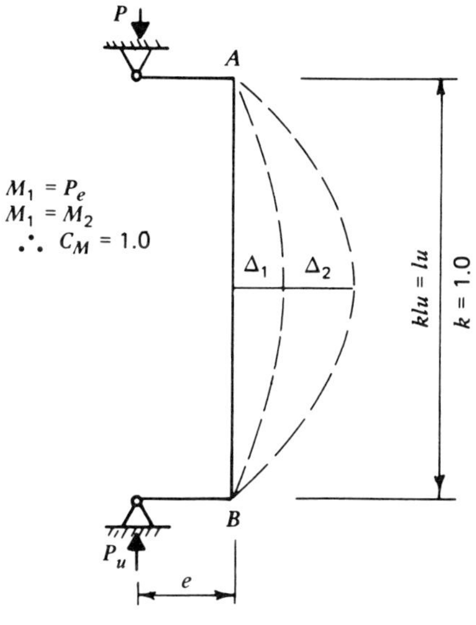

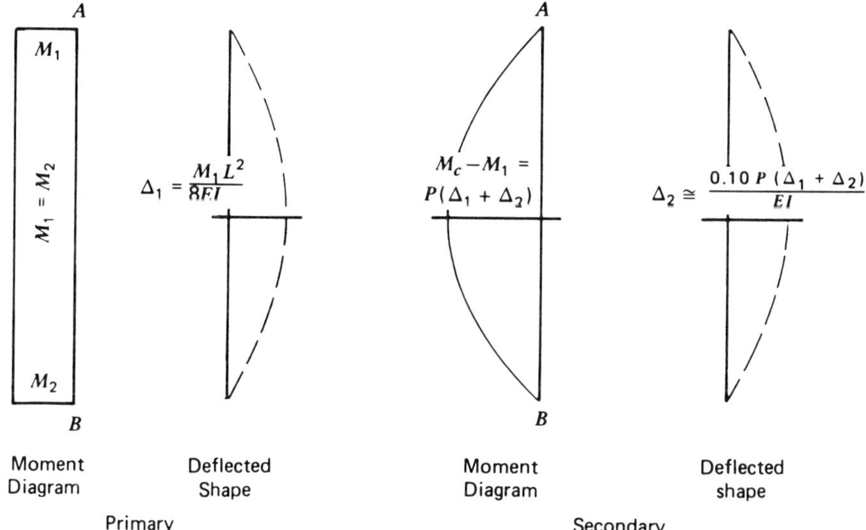

FIGURE 14-6 Deflected shapes of a pier considered in moment magnifier equation. (From Heins and Lawrie, 1984.)

which is reduced to

$$M_c = M_1 + \left(\frac{P}{P_{cr}}\right)(M_1 + M_c - M_1)$$

or

$$M_c = \frac{M_1}{1 - P/P_{cr}} \tag{14-14}$$

If δ is the moment magnifier factor, then $M_c = \delta M$ and

$$\delta = \frac{1}{1 - (P/P_{cr})} \tag{14-15}$$

or, by introducing a load factor ϕ,

$$\delta = \frac{1}{1 - (P/\phi P_{cr})} \tag{14-16}$$

Effective Length Factor K The moment magnifier analysis that led to (14-16) considers columns with pinned ends, single curvature, equal end moments, and no sidesway. The design must therefore be modified by the effective length factor K, included in the equations for allowable compressive stress and applied to the column length KL, according to the degree of fixity at the ends of the column. The resulting column dimensions necessary for stability give a critical buckling load $P_{cr} = (\pi^2 EI)/(KL)^2$. The associated variation in the effective length factors can be large as the end conditions change and as the degree of bracing against sidesway varies.

The effective length factors may be taken from Figure 14-7. The parameter K is the ratio of the effective length of an idealized pinned-end column to the actual length of a column with various other end conditions. KL represents the length between inflection points of a buckled column. Restraint against rotation and translation of the column ends affects the position of the inflection points in the column. The theoretical values of K shown in Figure 14-7 refer to the idealized column end conditions indicated.

Column end conditions seldom comply fully with the idealized restraint against rotation and translation assumed in Figure 14-7; hence, the recommended values are higher than the idealized values. In continuous frames unbraced by sufficient attachment to shear walls, diagonal bracing, or adjacent structures, columns depend on the bending stiffness of the rigidly connected beams for lateral stability. In this case the effective length factor K depends on the amount of stiffness provided by the beams or pier caps at the column ends. If the actual amount of stiffness supplied in this manner is small, the value of K may exceed 2.0.

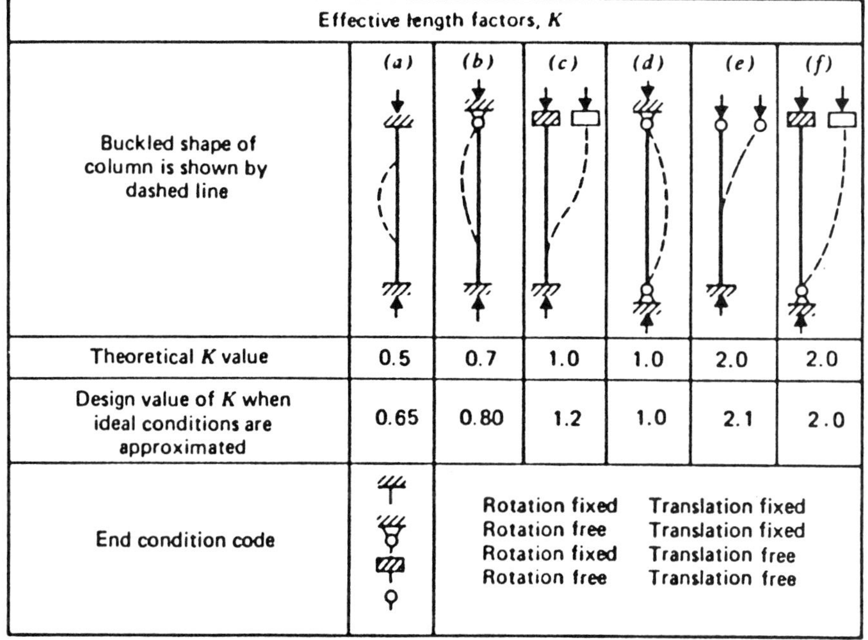

Effective length factors, K						
	(a)	(b)	(c)	(d)	(e)	(f)
Buckled shape of column is shown by dashed line						
Theoretical K value	0.5	0.7	1.0	1.0	2.0	2.0
Design value of K when ideal conditions are approximated	0.65	0.80	1.2	1.0	2.1	2.0
End condition code		Rotation fixed Translation fixed Rotation free Translation fixed Rotation fixed Translation free Rotation free Translation free				

FIGURE 14-7 Effective length factor K.

Assuming that elastic action can occur and that all columns in a frame buckle simultaneously, it can be shown that

$$\frac{G_a G_b (\pi/K)^2 - 36}{6(G_a + G_b)} = \frac{\pi/K}{\tan(\pi/K)} \tag{14-17}$$

where the subscripts a and b refer to the two ends of the column, and

$$G = \sum (I_c/L_c) \Big/ \sum (I_g/L_g) \tag{14-18}$$

where Σ = summation of all members rigidly connected to an end of the column in the plane of bending
I_c = moment of inertia of the column
L_c = unbraced length of the column
I_g = moment of inertia of the beam or other restraining member
L_g = unsupported length of the beam or other restraining member
K = effective length factor

The relationship between K, G_a, and G_b is shown graphically in Figure 14-8,

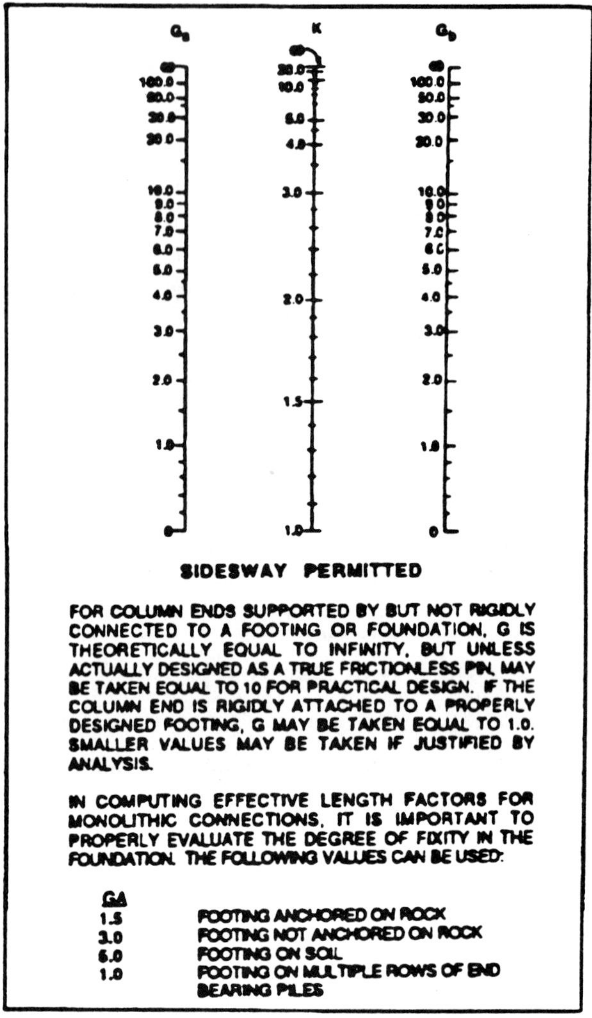

FIGURE 14-8 Relationship between K, G_a, and G_b.

when sidesway is permitted. The chart can be used to obtain values of K readily. In frames with columns in the inelastic buckling range, K may often be reduced. The inelastic buckling range is defined by

$$KL/r < C_c = \sqrt{\frac{2\pi^2 E}{F_y}} \tag{14-19}$$

A procedure for reducing K is given by Yura (1974). The degree of fixity of the foundation, G_B, is essentially the same as in Table 14-1.

Effect of Sidesway In the foregoing analysis, K is determined for unbraced conditions, and the data of Figure 14-8 are only relevant when sidesway is permitted. A braced condition against sidesway could be considered only for a row of columns in a bent strutted by a large drift wall with considerable stiffness against lateral movement in the transverse direction.

In determining the critical buckling load, the formulas for EI at the beginning of this section were developed using the data from Figure 14-9. These curves show that using $EI = E_c I_g/2.5$ is grossly inaccurate for high

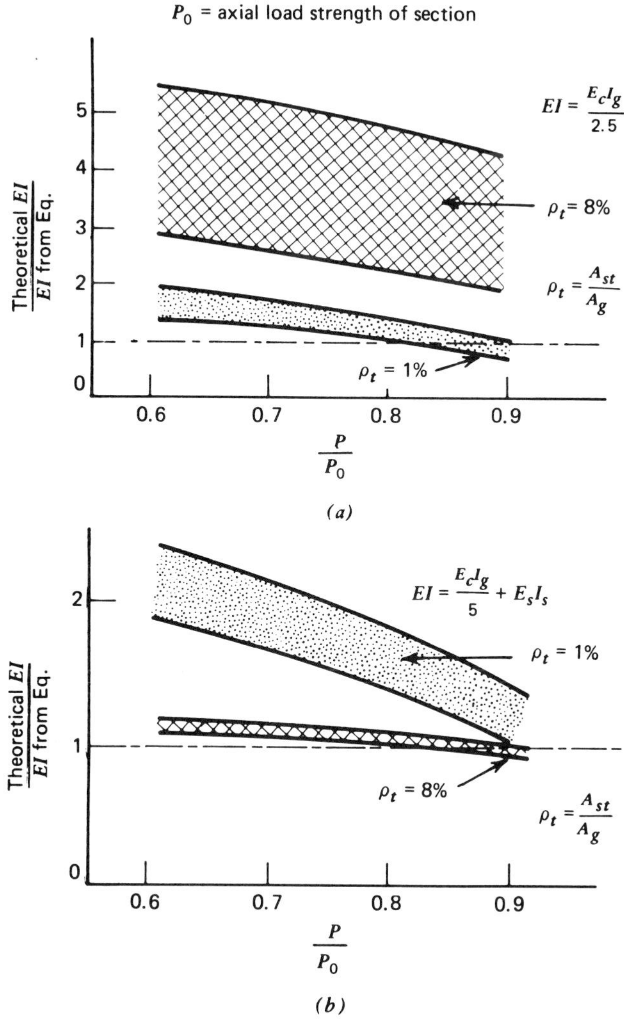

FIGURE 14-9 Comparison of equations for flexural stiffness with theoretical values from moment–curvature diagrams.

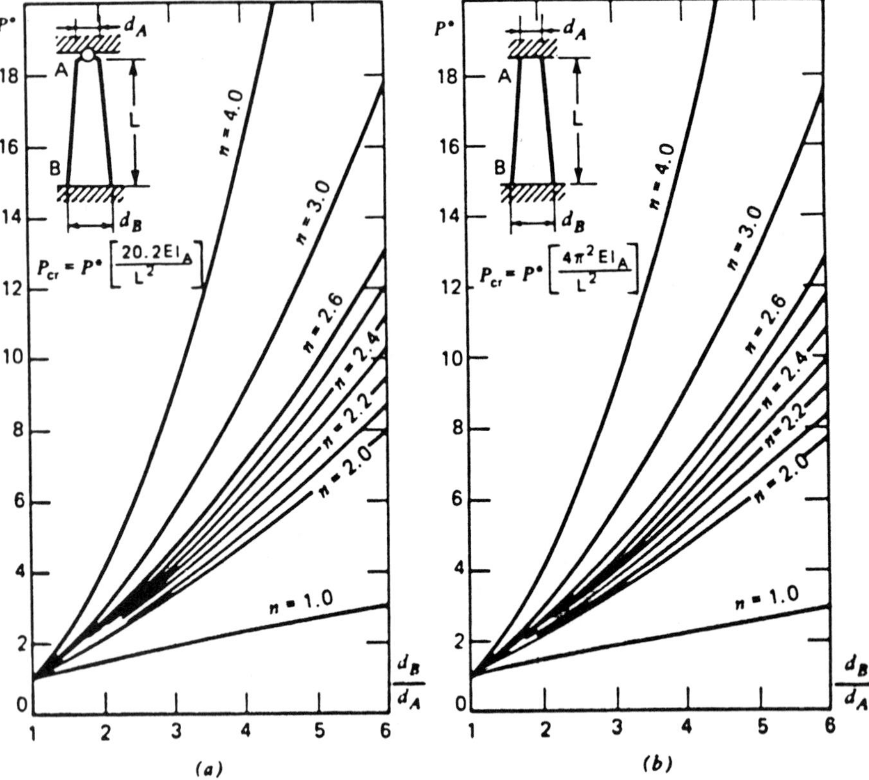

FIGURE 14-10 Critical buckling loads for tapered shafts: (*a*) case 3 column, fixed pinned ends; (*b*) case 4 column, fixed ends.

percentages of steel reinforcement, and using $EI = E_c I_g/5 + E_s I_s$ is somewhat inaccurate for low percentages of steel reinforcement.

It appears therefore that using $EI = E_c I_g/2.5$ is appropriate for pier design with 2 percent reinforcement or less, and $EI = E_c I_g/5 + E_s I_s$ is appropriate for pier design with reinforcement ratio greater than 2 percent. These guidelines should be applied to the computation of critical buckling loads, for columns with constant moment of inertia. If the columns are tapered, the critical buckling loads may be predicted by referring to the charts of Figure 14-10 and Table 14-2. A numerical procedure has also been developed by Newmark (1942) for computing buckling loads for columns and piers with variable moment of inertia.

In a series of piers with variable stiffness attached to a continuous superstructure, the buckling strength should be evaluated considering the entire continuous unit. This procedure is illustrated by Yura (1974) and involves the transfer of lateral thrust through the superstructure to supply additional lateral strength to the more slender piers.

TABLE 14-2 Shape Factors (From Gere and Carter, 1962)

Section	Description	n
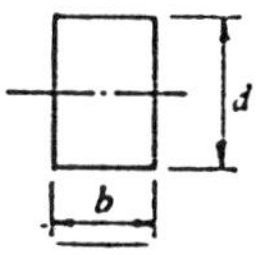 b, d	Solid, rectangular section Constant width b Varying depth d Buckling about horizontal axis	$n = 3$
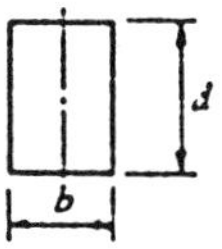 b, d	Solid, rectangular section Constant width b Varying depth d Buckling about vertical axis	$n = 1$
 d	Solid, circular section Varying diameter d	$n = 4$
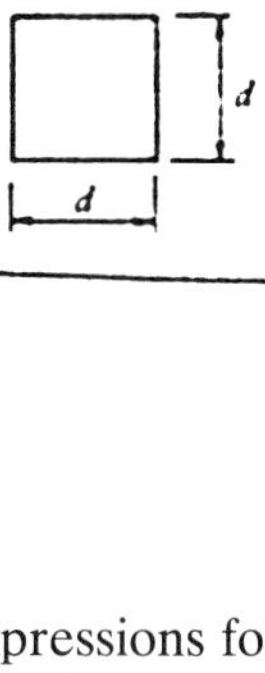 d, d	Solid, square section Varying dimension d	$n = 4$

The expressions for EI in the ACI code (1989) contain the term $(1 + \beta_D)$, accounting for creep effect due to sustained loads. The value of $(1 + \beta_D)$ decreases EI and increases the moment magnification. The factor β_D is the ratio of the maximum factored axial dead load to the maximum design total axial load (always positive), so that β_D is the ratio of dead load to the sum of all axial loads (neglecting signs).

14-6 PIER SECTION DESIGN

Given the most critical conditions of loads and moments, the design of a pier section includes the computation of stresses for serviceability and the prediction of ultimate structural capacity for maximum loads. Very often, piers have variable shape and dimensions, and the analysis must be made for irregular cross sections. The selection of steel reinforcement includes vertical bars as well as lateral hoops, ties, and spirals.

Invariably, pier columns are subjected to axial loads as well as moments from eccentric loading, end rotations, or because of unbraced conditions. It may appear that slender columns are synonymous with columns where deflections have an important effect on member strength. A 1970 survey of the ACI–ASCE Column Committee estimated that 90 percent of all braced columns and 40 percent of all unbraced columns may be designed as short columns. For pier column design, the distinction between braced and unbraced conditions is adequate, and it appears that most pier columns are unbraced. In addition, bridge pier sections are essentially governed by moment rather than by load. Therefore, the most critical load combination is the one producing the largest moment about the weakest pier or column axis. If two loads produce similar moments with variations in load, the load with the maximum eccentricity will normally control.

Length Limitations for Short Columns In the context of design, many engineers believe that slender column effects should be diagnosed before calculation procedures are introduced. Three cases must be distinguished in establishing proper limits on short column length. The first two are for braced frames, with bracing implying negligible lateral movement at the end joints. The third is the general case of unbraced frames that can deflect laterally.

The first case of the two braced conditions, involving single curvature, is the more flexible and includes end moments $M_1 < M_2$ as well as $M_1 = M_2$. The use of $K = 1$, in conjunction with a radius of gyration $r = 0.30h$ for rectangular columns and $0.25h$ for circular columns, is stipulated by ACI Code 318-89 and also allowed by AASHTO. The limiting short column length is defined for braced columns as

$$KL/r = 34 - 12M_1/M_2 \qquad (14\text{-}20)$$

Accordingly, AASHTO allows the effects of slenderness to be disregarded for compression members braced against sidesway if KL/r is less than the second term of (14-20).

The second case involves a reversed curve with an inflection point near the middle. Because the reversed-curvature case induces less column deflection, larger slenderness ratios should normally be permitted. Equation (14-20) may also be used in the case with M_1 negative and M_2 positive, thus increasing KL/r above 34.

The third case applies to all columns in an unbraced frame deflecting laterally such that the top of the column is displaced with respect to its bottom. This sets up a reversed-curvature case with joints at the top and bottom rotating to increase the lateral deflection and the moment resulting from increased eccentricity. The effective length factor is invariably greater than 1 and could only be reduced to 1 with the use of very stiff pier caps.

Criteria for Slenderness Effects In a frame analysis approach, the slender column effect is essentially a problem of adjusting or correcting the results from an ordinary frame study. All specifications, however, encourage the designer to extend the scope of the analysis by considering the influence of axial loads and variable moments of inertia on member stiffness and fixed end moments, as well as the effect of deflections on moments and forces, and finally the effects of duration of loads. The resulting second-order analysis must include the effects of sway deflections on axial loads and moments and must be based on realistic moment–curvature or moment–end rotation relationships. The stiffness (EI) values mentioned at the beginning of Section 14-5 are recommended by ACI Code 318-89 for column–beam frames such as in ordinary bridge piers. In addition, the same code states that it is not necessary to consider the effect of axial loads on the stiffness and carry-over factors for very slender columns ($L/r > 45$).

Such analyses are basically computerized procedures and beyond the scope of this discussion. However, the use of such programs in pier design is gradually increasing because it helps avoid the many detailed requirements of the moment magnifier method. Some engineers have developed in-house procedures, some of them sophisticated and others crude in terms of reflecting the complex variation of stiffness and moments associated with slender columns. Experience and caution would indicate that any second-order analysis procedure developed in conjunction with AASHTO and ACI specifications should be checked against the types of tests and computer solutions used to verify the more approximate moment magnification method. This check is particularly important for problems dealing with unbraced frames where the action of restraining flexural members is known to be critical (Ferguson, Breen, and Jirsa, 1988).

The moment magnification factor δ derived in Section 14-5 and expressed by (14-14) and (14-15) is a simplification of the single-curvature case with equal end moments. Although the axial load is determined from an ordinary analysis, the design moment is based on a magnified or amplified moment δM, where M is the larger end moment on the column calculated by a conventional elastic frame analysis. Note that the value of δ changes as P increases, and increases rapidly as P approaches P_{cr}. When $P = P_{cr}$, the value of δ becomes infinity, signifying the occurrence of stability failure. Values of $P > P_{cr}$, implying negative values of δ, are a physical impossibility and simply mean that failure has occurred. If δ is less than 1, moment reduction is not permitted. The introduction of the factor ϕ in (14-16) reflects the dependence of the Euler load P_{cr} on the column cross-sectional stiffness, tolerances, material variations, and other uncertainties.

If $M_1 \neq M_2$, the maximum moment is not initially at midheight. The largest increase is near midheight, resulting in a smaller total at the new maximum point. The ACI code considers this by including the factor C_m which takes care of different-shaped moment diagrams and shifts in points of maximum moment. Likewise, $M_{max} = \delta M$, where δ is the moment magnifier or amplifier given by

$$\delta = \frac{C_m}{1 - P/\phi P_{cr}} \geq 1.0 \tag{14-21}$$

For members braced against sidesway and without transverse loads between supports, C_m may be taken (ACI code) as

$$C_m = 0.6 + 0.4M_1/M_2 \geq 0.4 \tag{14-22}$$

In unbraced frames, C_m is invariably taken as 1.0. This procedure is included in AASHTO Article 8.16.5.2.7 and approximates the slenderness effects.

With the procedure applied to slender columns, the same methodology may be applied to "short columns" using $\delta = 1$, or to columns where deflections have decisive effects on member strength. However, the moment magnification procedure should not be used for columns with slenderness ratios exceeding 100. Extremely slender columns will require a more detailed analysis. A further consideration is the addition of the term β_D discussed in Section 14-5 accounting for creep.

At $l_u/h = 10$ (l_u is the unsupported length of the compression members), creep effects tend to offset each other. In this process the tendency toward an increase of column curvature from creep is offset by the decreasing distribution factor to the column because of its reducing stiffness. As l_u/h approaches 16 to 20, the creep problem becomes more important, except for columns with reversed moments in braced frames. Reducing the EI value by the $(1 + \beta_D)$ factor in the denominator provides for creep effects. In many cases the β_D correction is a crude approximation and may be grossly conservative because lateral loads seldom manifest creep-related effects. However, with the exception of unusual slender pier columns, the effect of β_D is almost negligible.

Comments on Moment Magnifier The moment magnifier approach is the analysis of an elastic curve with increasing amplitude but without changes in shape as axial load is applied at the column ends or at the joints of the frame. As a result the deflected shape of the concrete will change, but this does not involve serious error. In a general frame, five relevant problems will arise (Ferguson, Breen, and Jirsa, 1988), although the last three are not limited to reinforced concrete. These are as follows.

1. The effective EI of reinforced concrete depends on the magnitude and type of loading as well as on the material properties, and can vary along the column length.
2. Creep in concrete occurs under sustained load and modifies the effective EI needed in column analysis.

3. The effective column length for the calculation of P_{cr} may be either more or less than the unbraced length.
4. The magnified moments must distinguish between sway-producing and nonsway-producing loadings. This distinction articulates single-story and multistory pier bents.
5. The maximum moment does not necessarily occur at midheight.

The analysis is further complicated in columns with variable shape and in tapered shafts. Thus, given the many basic variables, pier design by any approximate methods does not give an exact result. As an example, the amount of pier cap steel or the degree of pier cap cracking can cause a scatter of 10 percent in column strength.

Design Criteria The serviceability design requirements of the standard specifications do not contain provisions for special crack width control at the reinforcement for service level loadings. The proposed LRFD specifications include, however, provisions applied to concrete members subjected to flexure, or flexure and axial loads, or axial tension. For columns with no tension in the cross section, these provisions do not apply.

Crack control is indicated for piers exposed to moisture and freezing and thawing. An expression giving the expected maximum crack width at the tension face of a flexural member is (Gergely and Lutz, 1968)

$$w = 0.076\beta f_s \sqrt[3]{d_c A} \tag{14-23}$$

where w = expected maximum crack width in 0.001-in. units
β = ratio of distances to the neutral axis from the extreme tension fiber and from the centroid of A_s
d_c = cover of outermost bar of A_s measured to the center of the bar
A = tension area per bar measured as specified by ACI code

The ACI code uses (14-23) with $\beta = 1.2$ as the basis for computing limiting values of $f_s \sqrt[3]{d_c A}$, which are set as 175 ksi for interior exposure and 145 ksi for exterior exposure. These values give $w = 0.016$ and 0.013 in., respectively.

The LRFD specifications address the control of cracking in the context of structures subjected to normal exposure. Control of surface cracking is particularly important when using reinforcement with a yield strength greater than 40,000 psi. There appears to be little or no connection between surface crack width and steel corrosion, and thicker or additional cover for reinforcement will result in greater surface crack widths. Structures subject to very aggressive exposure are beyond the scope of the provisions. For example, for bridge piers positioned in water or buried in the ground, a limiting maximum crack width of 0.006 in. is recommended.

Serviceability requirements also stipulate that all piers and columns should be analyzed for stresses. The investigation of stresses at service loads should be based on a straight-line theory of stress and strain in flexure in conjunction with the assumptions given in Article 8.15.3. Ordinarily, stresses due to direct load and bending can be evaluated using procedures provided in the ACI code. Other procedures for computing stresses in piers are formulated by Goodall (1948) and Hu (1955). These methods do not account for double compression in the reinforcing steel, and the results will therefore be conservative in terms of the stress. A method for stress analysis of circular columns, which does include double compression in the reinforcement, has been developed by Toprac (1959).

Axial Load Capacity Pier columns are normally subjected to the simultaneous effects of moments about both major axes. The capacity of the column is computed on the basis of a uniform strain between the extreme concrete fiber and the reinforcing steel location at which the maximum allowable concrete strain is 0.003, which is probably too small according to many investigators. Computation of the biaxial load capacity requires essentially the evaluation of the two uniaxial capacities and the rotation of a curve between the relevant points.

Both the standard AASHTO and the proposed LRFD specifications require that biaxial bending and direct axial load should be subject to a general section analysis based on stress and strain compatibility. In lieu of this approach, the design strength of noncircular members may be computed by the following approximate expressions:

$$\frac{1}{P_{nxy}} = \frac{1}{P_{nx}} + \frac{1}{P_{ny}} - \frac{1}{P_0} \tag{14-24}$$

when the factored axial load $P_u \geq 0.10 f'_c A_g$, and

$$\frac{M_{ux}}{\phi M_{nx}} + \frac{M_{uy}}{\phi M_{ny}} \leq 1 \tag{14-25}$$

where $P_u < 0.10 f'_c A_g$,

where P_{nxy} = nominal axial load strength with biaxial bending

P_{nx} = nominal load strength corresponding to M_{nx} with bending considered in the direction of the x axis only

P_{ny} = nominal axial load strength corresponding to M_{ny} with bending considered in the direction of the y axis only

P_0 = nominal axial load strength of section at zero eccentricity

P_u = factored axial load at given eccentricity

In the ultimate strength analysis of bridge piers, P_u is generally less than $0.10f'_c A_g$. Occasionally, however, the service load (taken as 0.35 times the ultimate load) may cause P_u to be greater than $0.10f'_c A_g$, which would require the use of (14-24).

In the context of the LRFD analysis, the notation is as follows:

P_{rxy} = factored axial resistance with biaxial loading
P_{rx} = factored axial resistance corresponding to M_{rx}
P_{ry} = factored axial resistance corresponding to M_{ry}

Using a PCA computer program, it is possible to simulate the stress and strain compatibility analysis. This program makes use of the equation

$$\left(\frac{M_{ux}}{\phi M_{nx}}\right)^{\log 0.5/\log \beta} + \left(\frac{M_{uy}}{\phi M_{ny}}\right)^{\log 0.5/\log \beta} = 1 \qquad (14\text{-}26a)$$

derived by Parme, Nieves, and Gouwens (1966). Alternatively, it is convenient to use an elliptic equation proposed by Furlong (1979), which gives reasonable results and is expressed as

$$\sqrt{\left(\frac{M_{ux}}{\phi M_{nx}}\right)^2 + \left(\frac{M_{uy}}{\phi M_{ny}}\right)^2} \leq 1 \qquad (14\text{-}26b)$$

A simulation of the interaction plane produced by a stress and strain compatibility analysis is shown in Figure 14-11.

The method relating P_u to the other three loads in (14-24) has been published by Bresler (1960). With charts (or tables) for P_u or P_0 plus the uniaxial moment about the x axis, and similar ones for P_u plus the uniaxial moment about the y axis, values of P_{nx}, P_{ny}, and the theoretical P_0 are easily established. Substitution into the equation gives the desired P_{nxy} or the desired biaxial P_u. As an approximate method, it gives satisfactory values if the necessary charts and tables are available and provided the resulting P_u or ϕP_n is above $0.10f'_c A_g$. We should note that two inaccuracies may result from the biaxial bending and direct load analysis: (a) any approximation of the true shape of the compression block with a rectangular block over a portion of the compressive area tends to underestimate the capacity; and (b) data show that in some cases a strain of 0.4 percent is reasonable as a limiting value before the concrete begins to deteriorate. Because these deviations are on the conservative side, they are generally accepted in the design.

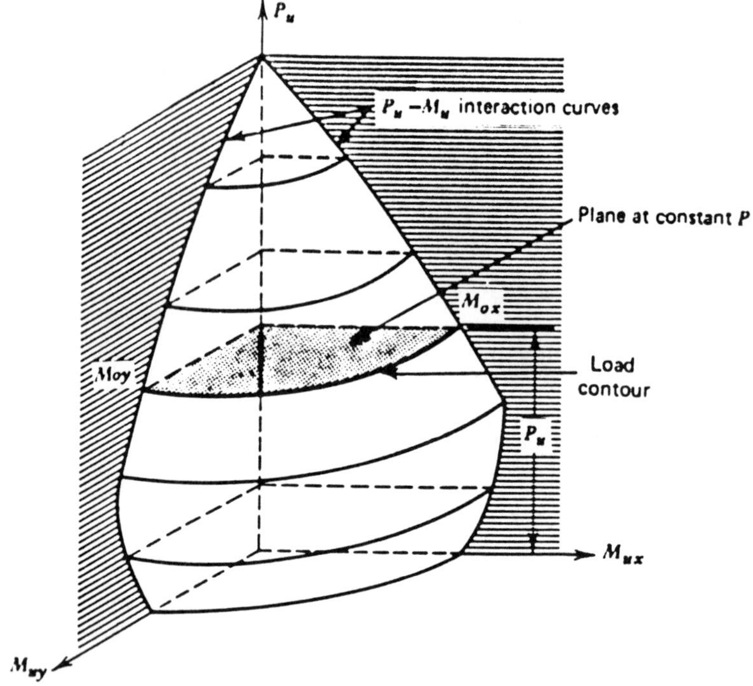

FIGURE 14-11 Load contours for constant P_u on failure surface.

Irregular and Unsymmetrical Shapes The load factor method for irregular or unsymmetrical sections involves considerations similar to biaxial bending analysis. The research conducted in pier design is limited and lacks sufficient experimental documentation. Mathematical modeling has been carried out by Marin (1979) for L-shaped sections, but further tests are warranted to verify this work. One of the problems that merits more studies relates to the shear and lateral reinforcement across thin-walled portions.

Hoops, Ties, and Spirals Until initial concrete deterioration occurs, lateral reinforcement contributes very little to the structural performance of columns, but thereafter its action is significant. Experience has shown that proper reinforcement details often prevent detrimental forms of failure, and in this respect spirals are documented to perform better than hoops and ties. This improvement is probably due to the closer spacing of spirals and the confining conditions afforded by the circular hoops. Studies by Potangaroa et al. (1979) and Gill, Park, and Priestley (1979) also suggest an improved column ductility through the use of closely spaced hoops, ties, and spirals in potential plastic hinge areas. Ductility in such areas can serve the process of energy dissipation under severe dynamic effects and earthquake loadings.

Ductility is measured by the factor μ, defined as the ratio of the maximum displacement under the design earthquake to the theoretical yield displacement. A conservative value of μ for design purposes is 6.

ASSHTO specifications for seismic design recommend the forces resulting from plastic hinging for determining design forces for column piers and pile bents. Alternate conservative design forces are specified if forces resulting from plastic hinging are not calculated.

The application of ductility requirements is indicated in areas of seismic activity. Relevant criteria are included in the New Zealand Concrete Design Code and refer specifically to potential plastic hinge regions. At these locations, the maximum center-to-center spacing of transverse reinforcement should not be less than the larger of one-fifth of the smaller column dimensions (column diameter for circular columns), or six longitudinal bar diameters, or 8 in. (200 mm). In circular columns, p_s should not be less than the smaller of either

$$p_s = 0.45\left(\frac{A_g}{A_c} - 1\right)\frac{f_c'}{f_{yh}}\left(0.5 + 1.25\frac{P_e}{\phi f_c' A_g}\right) \tag{14-27}$$

or

$$p_s = 0.12\frac{f_c'}{f_{yh}}\left(0.5 + 1.25\frac{P_e}{\phi f_c' A_g}\right) \tag{14-28}$$

where P_e should be less than either $\phi 0.7 f_c' A_g$ or $\phi 0.7 P_0$. The value of p_s given by (14-27) is the same as in AASHTO equation (8-63) multiplied by the factor given in the second set of parentheses. Similar criteria are given for rectangular columns and relate the total area of hoops and supplementary cross ties to each of the principal directions of the cross section within a given spacing.

The term P_e in (14-27) and (14-28) is the maximum design compressive load due to gravity and seismic effects acting on a column during an earthquake. Because AASHTO does not provide for vertical seismic loading, the seismic effect may be ignored although its inclusion may provide useful parametric results.

14-7 DESIGN EXAMPLE 1: INTEGRAL ABUTMENT

Figure 14-12 shows a full abutment wall in an integral bridge. For this condition the superstructure is anchored as shown. By placing the top dowel bars near the inside face, the end restraint is essentially removed and the superstructure is free to rotate at the ends. The abutment wall is therefore assumed to have a pinned top. The base of the abutment wall may be

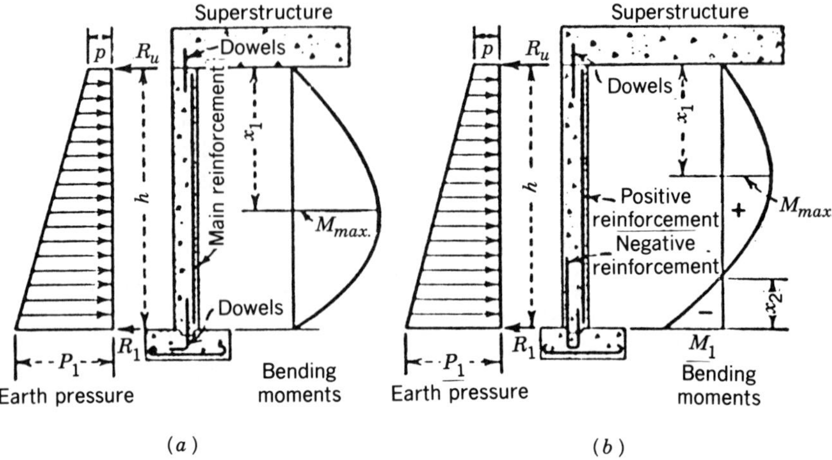

FIGURE 14-12 Abutment wall supported at top: (*a*) wall freely supported at bottom; (*b*) wall fixed at bottom.

designed as pinned, partially fixed, or fully fixed, according to the amount of restraint offered by the ground and as discussed in the foregoing sections. The two most usual conditions are shown in Figures 14-12*a* and *b* for free and fixed bottoms, respectively. The main wall is subjected to lateral earth pressures and reactions from the superstructure. The lateral earth pressure is assumed to be represented by a trapezoidal diagram as shown.

If p and p_1 are the unit earth pressures at the top and bottom of the wall, respectively, and h is the wall height as shown, bending moment equations may be derived for the free and the fixed conditions.

Both Ends Free From Figure 14-12*a*, the reactions are easily obtained as

$$R_u = \tfrac{1}{6}(p_1 + 2p)h \qquad \text{and} \qquad R_l = \tfrac{1}{6}(2p_1 + p)h \tag{14-29}$$

The point of zero shear is also the point of maximum bending moment and is located at distance x_1 from the top so that

$$x_1 = \frac{p}{p_1 - p}\left[-1 + \sqrt{1 + \frac{1}{3}\left(\frac{p_1}{p} + 2\right)\left(\frac{p_1}{p} - 1\right)}\,\right]h \tag{14-30}$$

The maximum bending point at x_1 from the top is

$$M_{\max} = \frac{1}{18}\,\frac{\dfrac{x_1}{h}\left(3\dfrac{p}{p_1 - p} + 2\dfrac{x_1}{h}\right)}{2\dfrac{p}{p_1 - p} + \dfrac{x_1}{h}}(p_1 + 2p)h^2 \tag{14-31}$$

Bottom End Fixed Referring to Figure 14-12*b*, the reactions are computed as

$$R_u = \tfrac{1}{40}(4p_1 + 11p)h \qquad \text{and} \qquad R_l = \tfrac{1}{40}(16p_1 + 9p)h \quad (14\text{-}32)$$

Likewise, the point of zero shear is the point of maximum moment (positive) and is located at distance x_1 from the top given by

$$x_1 = \frac{p}{p_1 - p}\left[-1 + \sqrt{1 + \frac{1}{20}\left(11 + 4\frac{p_1}{p}\right)\left(\frac{p_1}{p} - 1\right)}\,\right]h \quad (14\text{-}33)$$

The maximum positive moment at distance x_1 from the top is

$$M_{max} = \frac{1}{120}\,\frac{\dfrac{x_1}{h}\left(3\dfrac{p}{p_1 - p} + 2\dfrac{x_1}{h}\right)}{2\dfrac{p}{p_1 - p} + \dfrac{x_1}{h}}(11p + 4p_1)h^2 \quad (14\text{-}34)$$

The negative bending moment at the fixed end may be computed from

$$M_1 = -\tfrac{1}{120}(7p + 8p_1)h^2 \quad (14\text{-}35)$$

and the point of contraflexure is located at distance x_2 from the bottom given by

$$x_2 = \frac{p}{p_1 - p}\left[\left(5 + \frac{p_1}{p}\right) - 6\sqrt{1 + \frac{1}{240}\left(11 + 4\frac{p_1}{p}\right)\left(\frac{p_1}{p} - 1\right)}\,\right]h \quad (14\text{-}36)$$

Wall Footing For the freely supported wall at the bottom, the footing is subjected to vertical reactions from the superstructure, the weight of the wall and the footing, and the weight of the fill placed upon the interior part of the footing. The horizontal reaction due to the lateral earth pressure at the bottom is opposed by resistance to sliding and also by a certain passive resistance if the wall has sufficient embedment below the finish grade.

When the wall is fixed at the bottom, in addition to the loads and forces mentioned previously, the footing is subjected to a moment M_1 transferred to it by the vertical wall. Because overturning is not a design condition for either Figure 14-12*a* or *b*, the footing dimensions and projection beyond the wall face are determined to optimize the resulting soil pressure.

Earth Pressure For the abutment configuration shown in Figure 14-12, both the top and the bottom are restrained against movement either into or away from the backfill mass. Therefore, the earth pressure at rest is the concept associated with this condition, and for a well-compacted backfill of granular material, it is reasonable to use $K_0 = 0.5$ and unit weight $\gamma = 120$ pcf. The pressure p at the top of the wall is the result of a live load surcharge (usually 2 ft) with the live load placed in the approach embankment. Friction and cohesion behind the wall is disregarded for simplicity.

An adequate drainage system is assumed behind the abutment wall, and its function is to prevent water pressures from developing behind the wall.

14-8 DESIGN EXAMPLE 2: SPILL-THROUGH ABUTMENT

Spill-through abutments are common types of substructures, selected where shallow foundations are feasible and economical (see also Section 14-2). A

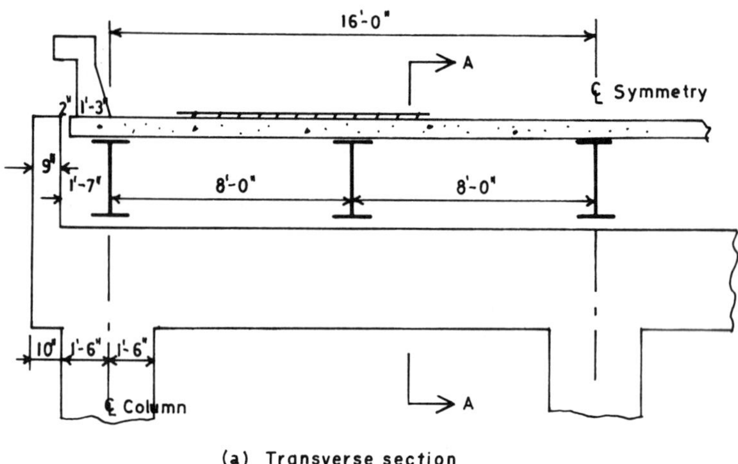

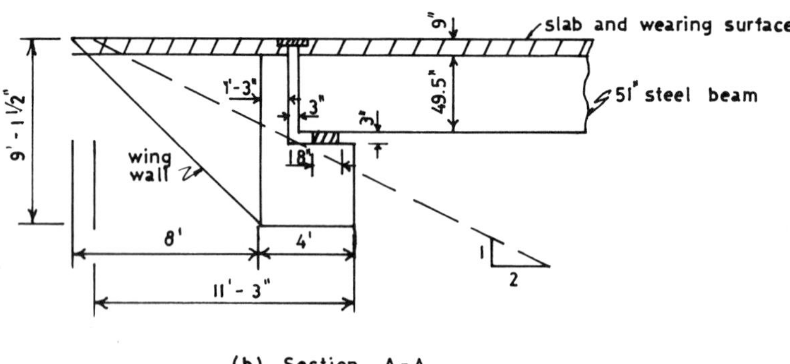

FIGURE 14-13 Spill-through abutment: (a) partial longitudinal elevation; (b) abutment cross section A–A. (From Heins and Firmage, 1979.)

spill-through abutment is presented in Figure 14-13, showing a partial longitudinal elevation (transverse section) and a section through the abutment seat. The shape and dimensions of the abutment accommodate a two-lane bridge superstructure consisting of steel beams and a concrete deck. The bridge has two simple equal spans, each 85 ft long. Three columns are used to support the beam seat as shown in Figure 14-13*a*. Three of the superstructure beams are set directly on the columns, and the other two between columns producing bending moment in the bridge seat.

Dead Loads from Superstructure For the interior beams, the unit dead load from the concrete slab, wearing surface, steel beam, and so forth is computed as 1.10 kips/ft. The unit load for the outside beams is 1.18 kips/ft. This unit load produces dead load reactions at the abutment as follows:

$$\text{Interior beams} \qquad R_{DL} = 1.10 \times 85/2 = 46.8 \text{ kips}$$

$$\text{Outside beams} \qquad R_{DL} = 1.18 \times 85/2 = 50.2 \text{ kips}$$

Live Loads from Superstructure For maximum bending in the beam seat, the live load (HS 20) is placed as shown in Figure 14-14, centered over the interior beams. However, for the superstructure width, the trucks can be placed in different positions across the roadway to produce maximum effects in the elements of the abutment. For maximum live load reactions, one rear axle is placed directly over the bearing line, giving a reaction $P = 16(1 + 71/85) + 4(57/85) = 32.1$ kips. This reaction is conservative because the wheel load position in the deck along the span will be transversely distributed and will not produce concentrated reactions along the abutment beam seat.

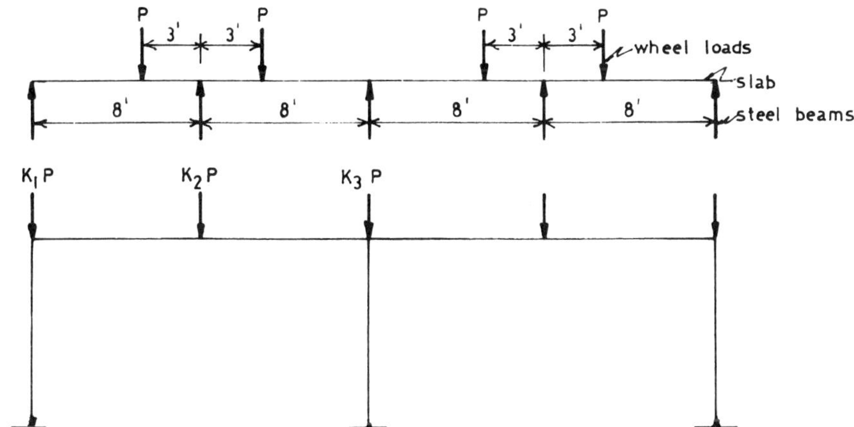

FIGURE 14-14 Live load arrangement for the abutment of Figure 14-13.

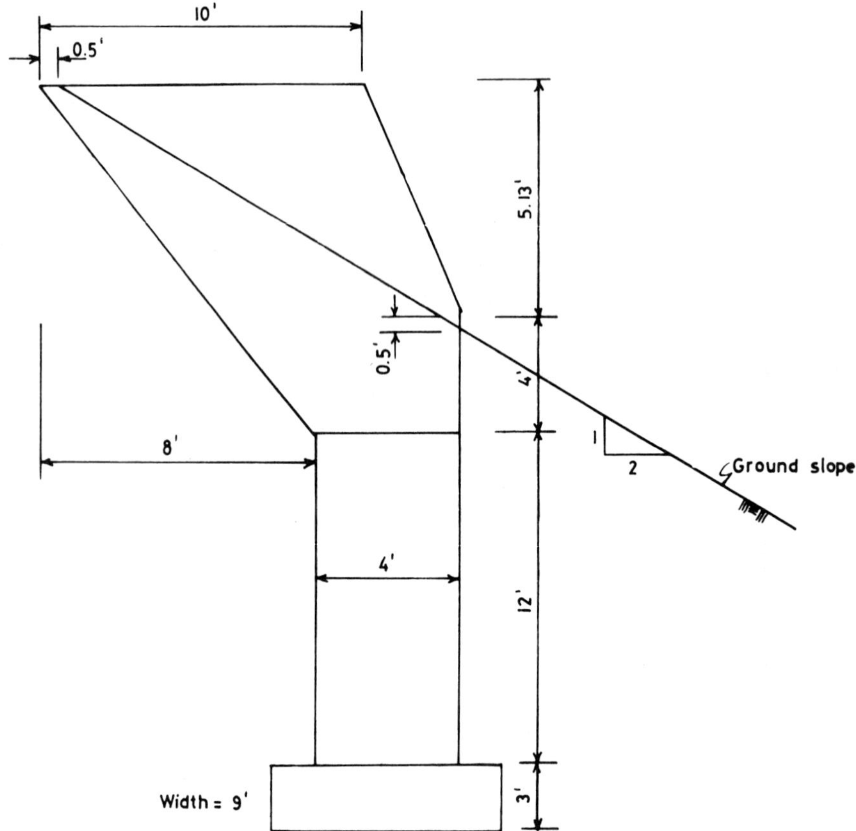

FIGURE 14-15 End view of abutment of Figure 14-13.

Dead Load of Abutment An end view of the abutment is shown in Figure 14-15. From this configuration and also from Figure 14-13, the abutment dead load is computed as follows:

$$\text{Beam seat} = 4 \times 4 \times 0.15 = 2.4 \text{ kips/ft}$$

$$\text{Back wall} = 1.25 \times 4.38 \times 0.15 = 0.82 \text{ kip/ft}$$

$$\text{Wing wall (each)} = [12 \times 9.13 - (2 \times 5.13)/2 - (8 \times 9.13)/2]0.15 = 10.2 \text{ kips}$$

$$\text{Column (each)} = 4 \times 3 \times 12 \times 0.15 = 21.6 \text{ kips}$$

$$\text{Footing (each)} = 3 \times 10 \times 9 \times 0.15 = 40.5 \text{ kips}$$

Longitudinal Force According to the standard AASHTO specifications, this will be taken as 5 percent of two lanes of live load. This force is

$$F_L = 2(0.64 \times 85 + 18)0.05 = 7.3 \text{ kips} \qquad \text{Use 8 kips}$$

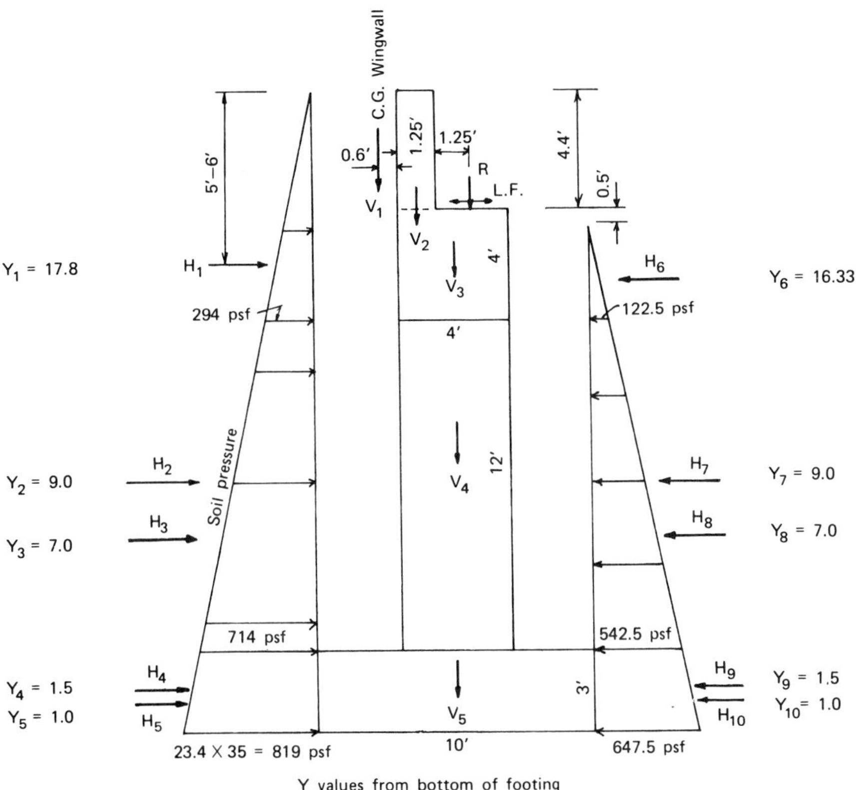

FIGURE 14-16 Loads on forces acting on abutment of Figure 14-13.

Friction This is developed because of expansion and contraction due to temperature changes and is a function of the dead load reaction. The abutment is provided with a Teflon sliding bearing with a coefficient of friction of 0.06. Because the longitudinal forces due to friction oppose each other at the piers, they are transmitted through the beams to the anchor bolts at the abutment bearings. The total friction is 14.5 kips.

Earth Pressures For a spill-through abutment, the earth pressure acting on the beam seat and columns behind the structure is partially offset by the earth pressure in front of the abutment. For simplicity, the earth pressure is taken as 35 lb/ft^3 equivalent fluid pressure, and because there is an approach slab a live load surcharge is not considered. A detailed lateral earth pressure diagram is shown in Figure 14-16.

Base Stability The abutment must be stable against sliding and overturning. The total dead load (superstructure plus abutment) plus earth pressure

TABLE 14-3 Vertical Loads, Lateral Forces, and Moments about Toe of Footing, Design Example of Figure 14-13

Type of Force	Magnitude	(kips)	Arm (ft)	Moment (ft-kip)
H_1: back soil pressure	$\frac{0.294}{2}(36.67)(8.4)$	45.3	17.8	+806
H_2: back soil pressure	$0.294(3 \times 3)(12)$	31.8	9.0	+286
H_3: back soil pressure	$(0.714 - 0.294)(\frac{9}{2})(12)$	22.7	7.0	+159
H_4: back soil pressure	$0.714(3)(3 \times 9)$	57.8	1.5	+87
H_5: back soil pressure	$(0.819 - 0.714)(\frac{3}{2})(27)$	4.3	1.0	+4
H_6: front soil pressure	$0.1225\left(\frac{3.5}{2}\right)(36.67)$	−7.9	16.33	−128
H_7: front soil pressure	$0.1225(3 \times 3)(12)$	−13.2	9.0	−119
H_8: front soil pressure	$(0.5425 - 0.1225)(\frac{9}{2})(12)$	−22.7	7.0	−159
H_9: front soil pressure	$0.5425(3)(3 \times 9)$	−43.9	1.5	−66
H_{10}: front soil pressure	$(0.6475 - 0.5425)(\frac{3}{2})(27)$	−4.2	1.0	−4
Longitudinal force		+14.9	19.0	+283
Total longitudinal force		84.8		
Total overturning moment				1149
V_1: weight of wing walls	2(10.2)	−20.4	7.6	−155
V_2: weight of back wall	0.82(35.17)	−28.8	6.4	−184
V_3: weight of beam seat	2.4(35.17)	−84.4	5.0	−422
V_4: weight of columns	21.6×3	−64.8	5.0	−324
V_5: weight of footings	40.5×3	−121.5	5.0	−607
R: dead load reaction	$3 \times 46.8 + 2 \times 50.2$	−240.8	4.5	−1084
Total vertical force		−560.7		
Total righting moment				−2776

forces acting on the abutment are tabulated in Table 14-3, together with the longitudinal and transverse forces.

The longitudinal force of 84.8 kips must be resisted by friction at the interface of the footing and the supporting soil. The coefficient of friction for concrete on soil varies widely ranging from 0.25 to 0.55. For this example we will use a value of 0.35. Then the factor of safety against sliding is

$$\text{FS} = \frac{0.35 \times 560.7}{84.8} = 2.3$$

which is considered satisfactory because friction at the concrete–soil interface will actually be higher than assumed. The margin of safety against sliding is further improved because the footing is buried and subject to additional restraining effects caused by the soil.

The factor of safety against overturning is

$$FS = 2776/1149 = 2.4, \text{ or close to } 2.5$$

For dead load only, a factor of safety greater than or equal to 2 is recommended. With live load on the bridge, the factor of safety against overturning is improved as follows:

$$\text{Additional balancing moment} = 32.1 \times 4 \times 4.5 = 578 \text{ ft-kips}$$

and

$$FS = (2776 + 578)/1149 = 2.9 \qquad \text{OK}$$

Footings The longitudinal overturning moment about the three footings is the total overturning moment given in Table 14-3 adjusted by the moment caused by the weight of the wing wall and back wall and the beam reactions. The resulting moment is

$$M_L = -1149 - 20.4(2.6) - 28.8(1.37) + 241 \times 0.5 = 1175 \text{ ft-kips}$$

For one footing the overturning moment is

$$M_L = 1175/3 = 392 \text{ ft-kips}$$

Referring to Figure 14-14, the reactions at each column are calculated for $P = 32.1$ kips treating the beam seat as a continuous beam on rigid supports. The coefficients k_1, k_2, and k_3 are determined from conventional structural analysis and are as follows:

$$k_1 = 0.24 \qquad k_2 = 1.50 \qquad k_3 = 0.52$$

or

$$k_1P = 7.8 \text{ kips} \qquad k_2P = 48.8 \text{ kips} \qquad k_3P = 16.9 \text{ kips}$$

The vertical loads and transverse moments on footings caused by frame action are obtained from conventional frame analysis and are as follows:

Outside columns	$P = 200.3$ kips	$M_T = 40$ ft-kips
Inside columns	$P = 290.0$ kips	$M_T = 0$

The soil pressure is calculated from the expression

$$P = \frac{P}{A} \pm \frac{M_L}{S_L} \pm \frac{M_T}{S_T}$$

where the subscripts denote longitudinal and transverse directions. The section properties are calculated as follows:

$$A = 9 \times 10 = 90 \text{ ft}^2 \qquad S_L = 9 \times 10^2/6 = 150 \text{ ft}^3$$
$$S_T = 10 \times 9^2/6 = 135 \text{ ft}^3$$

For outside columns,

$$p = \frac{200.3}{90} \pm \frac{392}{150} \pm \frac{40}{135} = 2.22 \pm 2.61 \pm 0.30 = 5.13 \text{ or } -0.96 \text{ ksf}$$

Likewise, for the inside columns,

$$p = \frac{290}{90} \pm \frac{392}{150} = 3.22 \pm 2.61 = 5.83 \text{ or } 0.61 \text{ ksf}$$

Because there can be no uplift pressure underneath the footing, the outside footings may be reanalyzed ignoring uplift, or some weight of soil above the footing may be included in the calculations. The maximum soil pressure of 5.83 ksf computed for the inside footing is normally within the allowable limit for most soils except for very soft or very loose soils, in which case spread footings would not be used.

Whether the longitudinal moment should be resisted equally by the three columns and their footings depends mainly on the stiffness of the superstructure and the rigidity of the abutment. For this example this assumption is valid. Other possible load conditions to be investigated include live load in one lane only. This would increase the lateral bending moment at the bottom of the outside column by approximately 50 percent, and would induce a lateral moment at the bottom of the inside column of the order of 20 ft-kips. The total vertical load on the outside column would be slightly increased, but the load on the inside column would be slightly decreased. The resulting soil pressure would change slightly, but not enough to warrant redesign of the footing.

Other Members The design of the abutment is completed by calculating shears and moments in the beam seat, back wall, columns, and footings. The final step is to estimate the amount of reinforcement necessary to resist moments and shears in each member.

14-9 DESIGN EXAMPLE 3: LOAD FACTOR METHOD FOR FOUNDATIONS

The application of the load factor method to bridge foundations is detailed and articulated in the proposed LRFD specifications. The related provisions deal essentially with the design of spread footings, driven piles, and drilled-shaft foundations.

In principle, these specifications are based on a probabilistic approach and involve procedures producing an interrelated combination of load, load factor, resistance, resistance factor, and statistical reliability. The resistance procedures may, however, be used in conjunction with locally recognized methods, appropriate for regional conditions, particularly if the statistical nature of these factors is considered through reliability theory and meet regional criteria. Although foundation analysis and design are treated extensively in a companion book and in the context of the load factor approach, this section is included in the present volume to demonstrate the key points of the associated philosophy.

Limit States and Resistance Factors Essentially, the limit states are as specified in other sections, with the following clarifications.

Service Limit States In the context of foundation design, service limit states include (a) settlements and (b) lateral displacements. In the context of serviceability, settlement is investigated as a functional condition and also for economy.

Strength Limit States Strength limit states for foundation design must consider (a) bearing resistance failure, (b) excessive loss of contact, (c) sliding of the base or footing, (d) loss of overall stability, and (e) structural capacity. Foundations must be proportioned so that the factored resistance is not less than the effects of the factored loads.

Resistance Factors In selecting proper resistance factors for a foundation, consideration must be given to the following.

1. The method used to obtain soil properties and design parameters.
2. The different behaviors of sands and clays.
3. The procedure used to predict settlement and to calculate capacity.
4. The failure mode under investigation.
5. The type of foundation.

Resistance factors for different types of foundations at both strength and service limit states are given by the LRFD specifications and include shallow foundations, axially loaded piles, and axially loaded drilled shafts. Statistical data were used where available, combined with reliability theory and engineering judgment in cases where the information was insufficient for calibration.

Example: Pier Foundation for a Two-Span Bridge The bridge in this example (Barker, 1991) consists of two equal 60-ft spans with a continuous superstructure carrying two lanes of traffic in a rural setting. The roadway

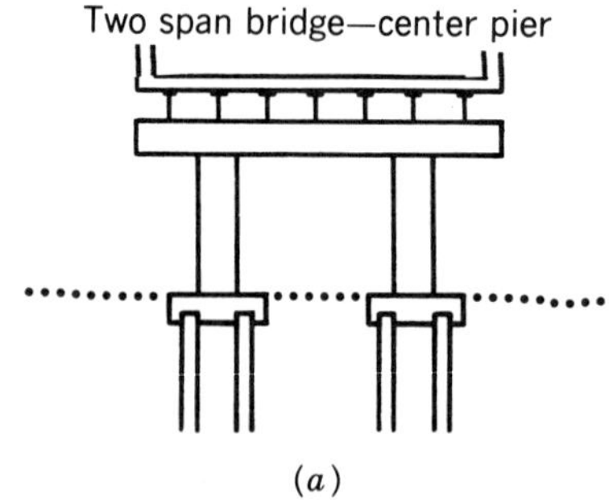

(a)

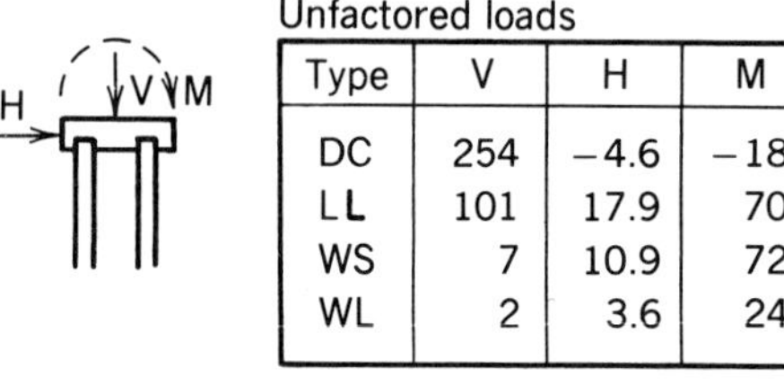

Type	V	H	M
DC	254	−4.6	−18
LL	101	17.9	70
WS	7	10.9	72
WL	2	3.6	24

(b)

FIGURE 14-17 (*a*) Pier elevation and foundation details; (*b*) forces acting on pile cap. (From Barker, 1991.)

width is 26 ft between rails. An 8-in. thick concrete slab provides the deck and is supported by seven 30-in.-deep wide flange beams. The abutments are set on piles.

Figure 14-17*a* shows a pier elevation. The structure consists of a pier cap supported on two columns with individual pile foundations. The pier cap depth is 3.5 ft and matches the 3-ft square columns. The pile cap has a square shape with 8-ft sides and is 2.4 ft thick. The overall height of the pier is 16 ft from the top of the pier cap to the bottom of the pile cap. The unfactored service loads are tabulated in Figure 14-17*b* and act on the pile cap as shown. The live load is as per the LRFD model. The wind load is 50 psf on the superstructure, 40 psf on the substructure, and 100 lb/ft on the live load. There is no impact, and uplift due to wind load is neglected. The vertical dead load includes the weight of the column and the portion of the pile cap.

The load factors and load combinations are shown in Table 14-4 and conform to Tables 2-12 and 2-13. The last group S is the serviceability load combination. The factored forces on the pile cap are tabulated in Table 14-5 and are obtained by matrix multiplication of the table in Figure 14-17*b* and Table 14-4. The forces V_u and H_u are given in kips, and the moment M_u is quoted in ft-kips.

TABLE 14-4 Load Combinations and Load Factors (From Barker, 1991)

Group	DC	LE	WS	WL
I	1.25	1.7	0.0	0.0
II	1.3	0.0	1.25	0.0
III	1.25	1.7	0.4	0.4
S	1.0	1.0	0.4	0.4

TABLE 14-5 Group Combinations and Factored Forces on Pile Cap (From Barker, 1991)

Group	V_u	H_u	M_u
I	489	24.7	97
II	339	7.6	67
III	493	30.5	135
S	359	19.1	90

A group of four piles is selected to support each column arranged as shown in Figure 14-18. The piles are set on a 5-ft square grid as shown. For a number of piles ($n = 4$ and $x = 2.5$ ft), we calculate $\Sigma x^2 = 25$. The axial load per pile is therefore $P_u = V_u/n + M_u x/\Sigma x^2$. The axial and lateral loads per pile are calculated as follows (kips):

Group	max P_u	H_u/n
I	132	6.2
II	92	1.9
III	137	7.6
S	99	4.8

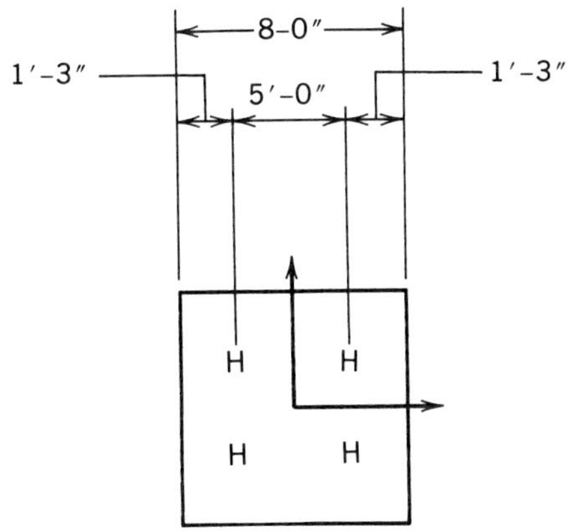

FIGURE 14-18 Footing pier and pile layout. (From Barker, 1991.)

Next, we select HP12 × 53 steel piles. The nominal axial strength of an HP12 × 53 pile is obtained as the product of its gross sectional area times the yield strength of the material. This nominal strength is multiplied by a resistance factor $\phi_a = 0.85$ to obtain the factored axial load strength of the pile. We can therefore write

$$\phi_a P_n = \phi_a(AF_y) = 0.85(15.6 \times 36) = 477 \text{ kips}$$

Referring to Table 14-5, a factored moment M_u is introduced at the pile cap. The bending moment produced in a fixed-head pile by the lateral load is related to the nonlinear response of the soil, as represented by a curve, and to the flexural rigidity EI of the pile, as shown in Figure 14-19. The curve shown is for driven piles (Ooi et al., 1991) and gives the relationship between the lateral load and the maximum bending moment produced in an HP12 × 53 steel pile. For a lateral load of 7.6 kips, the moment produced is 400 in.-kips.

The nominal bending strength of an HP12 × 53 pile is obtained as the product of its plastic section modulus and the yield strength of the material, or $M_n = Z_p F_y$. The nominal strength is then multiplied by a resistance factor ϕ_m (in this case 0.95) to obtain the factored bending strength of the pile. Thus,

$$\phi_m M_n = 0.95(74 \times 36) = 2530 \text{ in.-kips}$$

For the case of combined compression and bending, the structural capacity of the pile is checked using a simple straight-line interaction relationship expressed as

$$(P_u/\phi_a P_n) + (M_u/\phi_m M_n) \leq 1.0$$

stating that if the sum of the two ratios of factored forces to factored resistance is less than 1, the capacity of the member is satisfactory. Using the foregoing data, we can write $(137/477) + (400/2530) = 0.45 < 1.0$, OK.

The last step in the analysis is to check the transfer of load to the surrounding soil and to calculate the soil capacity for a single pile. The soil profile used in determining the soil strength is shown in Figure 14-20. In this simplified presentation the ground consists of an upper 35-ft layer of loose sand overlying a stronger layer of dense sand. The pile penetrates into the dense sand as shown. The nominal soil capacity is the sum of the skin friction and point bearing resistance.

For the conditions shown in Figure 14-20, the nominal soil strength for a single pile is $P_n = Q_{s1} + Q_{s2} + Q_p$ where Q_{s1} is the frictional resistance in loose sand, Q_{s2} is the frictional resistance in dense sand, and Q_p is the direct point bearing. The soil resistance is based on skin friction and end bearing using the SPT method (standard penetration test). According to this method,

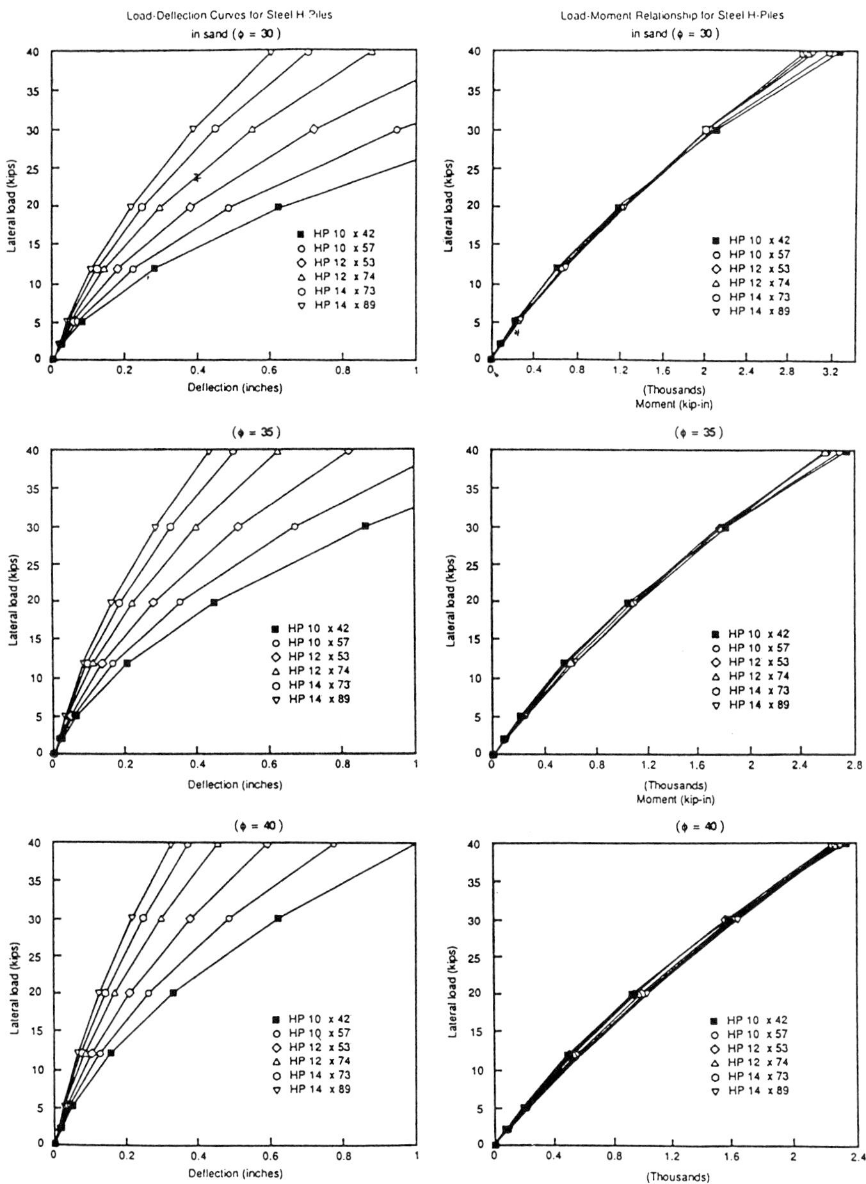

FIGURE 14-19 Load versus moment relationship for steel H piles in sand (ϕ = 30°). (From Ooi et al., 1991.)

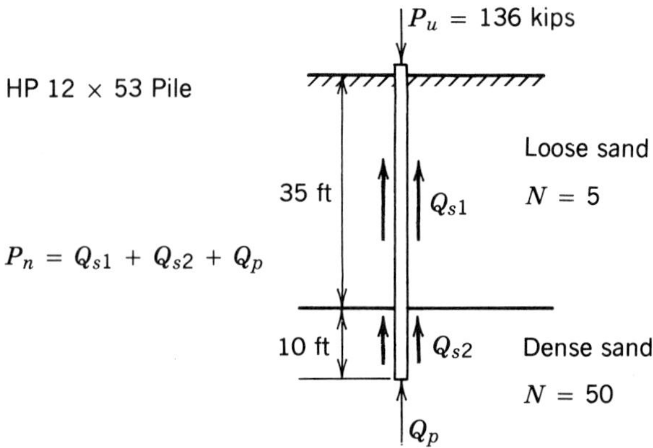

FIGURE 14-20 Soil profile and capacity for a single pile. (From Barker, 1991.)

the resistance is calculated as follows:

$$Q_s = (N/80)(\text{Perimeter})(\text{Length}) \qquad \text{in tons}$$

$$Q_p = 4N_1(\text{Tip area}) \qquad \text{in tons}$$

where N_1 is the penetration resistance factor corrected for overburden pressure, in this case 34.

The soil capacity for a single pile is now calculated as follows:

$$Q_{s1} = (5/80)(4.0)(35) = 8.9 \text{ tons} = 17.5 \text{ kips}$$

$$Q_{s2} = (50/80)(4.0)(10) = 25 \text{ tons} = 50 \text{ kips}$$

$$Q_p = (4)(34)(1.0) = 136 \text{ tons} = 272 \text{ kips}$$

The factored soil capacity is therefore $\phi_g P_n = \phi_g(Q_{S1} + Q_{s2} + Q_p)$, where $\phi_g = 0.45$ and is taken from the LRFD tables for the SPT method, or

$$\phi_g P_n = 0.45(17 + 50 + 272) = 153 \text{ kips} > 137 \text{ kips} \qquad \text{OK}$$

The design must also check the pile group capacity. In sand, this is merely the capacity of a single pile multiplied by the number of piles. For the four-pile group, the strength is $153 \times 4 = 612$ kips > 489 kips, OK.

Pile group settlement is investigated as a service limit state and represents functional requirements. In estimating this settlement, Meyerhof's method is used. This method relates the settlement of a pile group, P, to the SPT blow count of the soil as follows:

$$p = 2qIX^{0.5}/N_{\text{corr}}$$

where q = net foundation pressure (including any negative skin friction) in tons/ft^2 applied at $2D_b/3$, where D_b is the pile length

X = width of the pile group (ft)

I = influence factor of the effective group embedment = $1 - D'/8X$ where $D' = 2D_b/3$

N_{corr} = average corrected penetration resistance

The value of p is therefore $(359)(36)(0.86)(6.0)^{0.5}/34 = 0.6$ in.

The tolerable settlement can be obtained from a consideration of the tolerable angular distortion. For a continuous bridge superstructure, this angular distortion is 0.004. The tolerable differential settlement is therefore 0.004 × span length, or 60 × 12 × 0.004 = 2.9 in. > 0.6 in., OK.

14-10 DESIGN EXAMPLE 4: SINGLE-HAMMERHEAD PIER (LOAD FACTOR DESIGN)

The pier shown in Figure 14-21*a* will be analyzed using load factor procedures according to standard AASHTO criteria. The pier is the center support of a single-lane superstructure bridge supported at the left support and cantilevered to the right of the pier, as shown in Figure 14-21*b* with the pier designated as *BD*. The load is HS 20, with no overload provisions.

Loads and moments are tabulated in Tables 14-6 and 14-7 for the top and the bottom of the pier, respectively. M_L and M_T denote longitudinal and transverse moments, respectively. The various loads are grouped and factored according to Table 2-5. The factor β_D is applied as specified by AASHTO, that is, $\beta_D = 0.75$ to reduce the reaction and produce maximum eccentricity, and $\beta_D = 1.0$ for moments.

Section Capacity Because biaxial bending is the most critical condition, the procedure requires computation of the bending capacity about each axis, and then rotation of a curve between the two axes. We assume that the requirements for capacity are below the balance point on the interaction curve. In this case the capacity can be computed about each axis for a compressive concrete strain of 0.3 percent and a compressive steel strain of 0.00207 at the balance point. Next, we compute the direct load and moment capacity for a concrete compressive strain of 0.003 and a steel compressive strain of 0.005 which are close to a level where the direct load is very small or close to zero. The capacity interaction curve is approximated by a straight line between these two points. In Figure 14-22 the reinforcement bars are located and numbered on the cross section. The strain diagrams are drawn next in a position that is a direct projection of the column section (Heins and Lawrie, 1984).

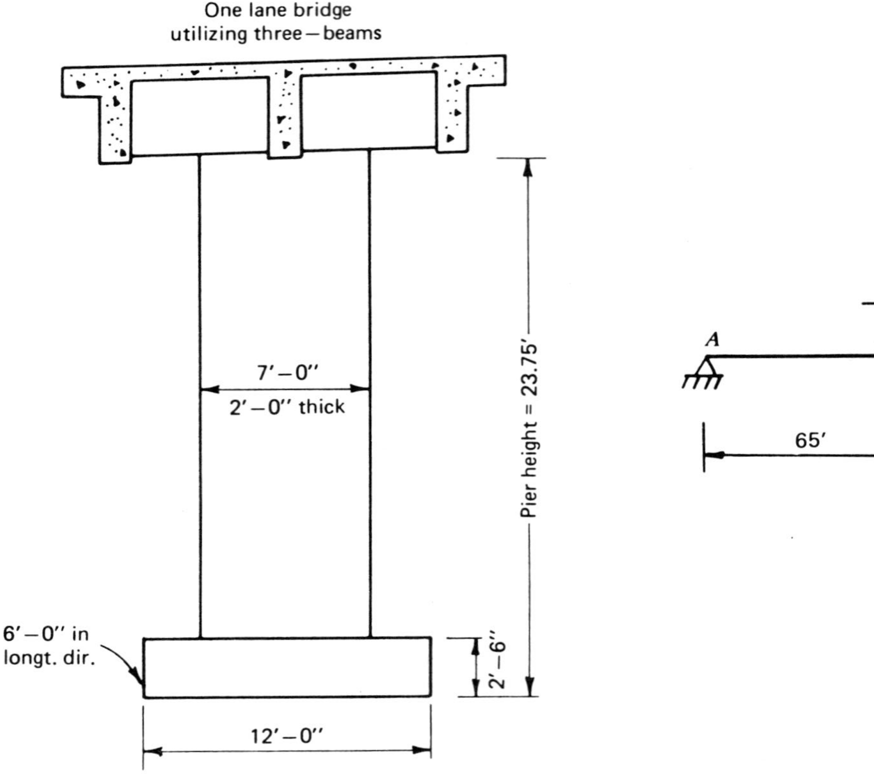

FIGURE 14-21 (*a*) Pier elevation of design example; (*b*) rigid frame of bridge elevation.

TABLE 14-6 Loads and Moments at *BD* Top Pier (HS 20)

LL Case		Case IB			Case II		
Group	Loading	P	M_L	M_T	P	M_L	M_T
I	$\beta_D D$	226.4	+430		226.4	+322	
	$1.67(L + I)$	97.5	+578	391	150.3	−364	601
	Σ	323.9	+1008	391	376.7	−42	601
	$1.3 \times \Sigma$	421.1	+1310	508	489.7	−55	781
II	$\beta_D D$	226.4	+430		+226.4	+322	
	$W + OT$			94			94
	Σ	226.4	+430	94	+226.4	+322	94
	$1.3 \times \Sigma$	294.3	+559	122	294.3	+419	122
III	$\beta_D D$	226.4	+430		+226.4	+322	
	$L + I$	58.4	+346	234	90.0	−218	360
	$0.3W$						
	$0.30T$			28			28
	WL						
	Σ	284.8	+776	262	+316.4	+104	388
	$1.3 \times \Sigma$	370.2	+1009	341	+411.3	+135	504
IV	$\beta_D D$	226.4	+430		+226.4	+322	
	$L + I$	58.4	+346	234	90.0	−218	360
	$S + T$		+227			−556	
	Σ	284.8	+1003	234	316.4	−452	360
	$1.3 \times \Sigma$	370.2	+1304	304	411.3	−588	468
V	Σ Group II	226.4	+430	94	+226.4	+322	94
	$S + T$		+227			−556	
	Σ	226.4	+657	94	+226.4	−234	94
	$1.25 \times \Sigma$	283.0	+821	118	+283.0	−292	118
VI	Σ Group III	284.8	+776	262	+316.4	+104	388
	$S + T$		+227			−556	
	Σ	284.8	+1003	262	+316.4	−452	388
	$1.25 \times \Sigma$	356.0	+1254	328	+395.5	−565	485
VII	$\beta_D D$	226.4	+430		226.4	+322	
	EQ			0			
	Σ			Neglect			
	$1.3 \times \Sigma$						

The strain diagrams of Figure 14-22 are used with a maximum tensile strain of 0.00207 to compute the capacity in the direct load and bending moment about the Y–Y axis. The location of the reinforcement bars is with reference to the centroid of the section. The strains at each bar location are taken from the strain diagram and are given with a + sign for compression and a − sign for tension. Stresses in the bars are computed using E_s = 29,000 ksi and f_y = 60 ksi as maximum stress. The resulting force in the bar

TABLE 14-7 Loads and Moments at *DB* Bottom Pier (HS 20)

LL Case		Case IB			Case II		
Group	Loading	P	M_L	M_T	P	M_L	M_T
I	$\beta_D D$	254.9	+149		254.9	+112	
	$1.67\,L + I$	97.5	+194	391	150.0	−125	360
	Σ	352.4	+343	391	404.9	−13	360
	$1.3 \times \Sigma$	458.1	+446	508	526.4	−17	468
II	$\beta_D D$	254.9	+149				
	W			317			
	OT			94			
	Σ	254.9	+149	411			
	$1.3 \times \Sigma$	331.4	+194	534			
III	$\beta_D D$	254.9	+149		254.9	+112	
	$L + I$	58.4	+116	234	90.0	−75	360
	βW			95			95
	$0.3OT$			28			28
	WL			170			170
	Σ	313.3	+265	527	344.9	+37	653
	$1.3 \times \Sigma$	407.3	+344	685	448.4	+48	849
IV	$\beta_D D$	254.9	+149		254.9	+112	
	$L + I$	58.4	+116	234	90.0	−75	360
	$S + T$		+229			−560	
	Σ	313.3	+494	234	344.9	−223	360
	$1.3 \times \Sigma$	407.3	+642	304	448.4	−290	468
V	Group II	254.9	+149	411			
	$S + T$		+229				
	Σ	254.9	+378	411			
	$1.25 \times \Sigma$	318.6	+472	514			
VI	Group III	313.3	+265	527	344.9	+37	653
	$S + T$		+229			−560	
	Σ	313.3	+494	527	344.9	−523	653
	$1.25 \times \Sigma$	391.6	+617	659	431.1	−654	816
VII	$\beta_D D$	254.9	+149				
	EQ			1161			
	Σ	254.9	+149	1161			
	$1.3 \times \Sigma$	331.4	+194	1509			

is the stress multiplied by the area A_s. The moments induced by these forces are then taken around the centroid. The rectangular concrete compression block is computed from (14-10), and the moment of the compression block is taken around the centroid. These forces and moments are added to obtain the balanced load and moment P_b and M_b. The entire procedure is shown in tabulated form in Table 14-8 for the relevant capacity reductions ϕ. The

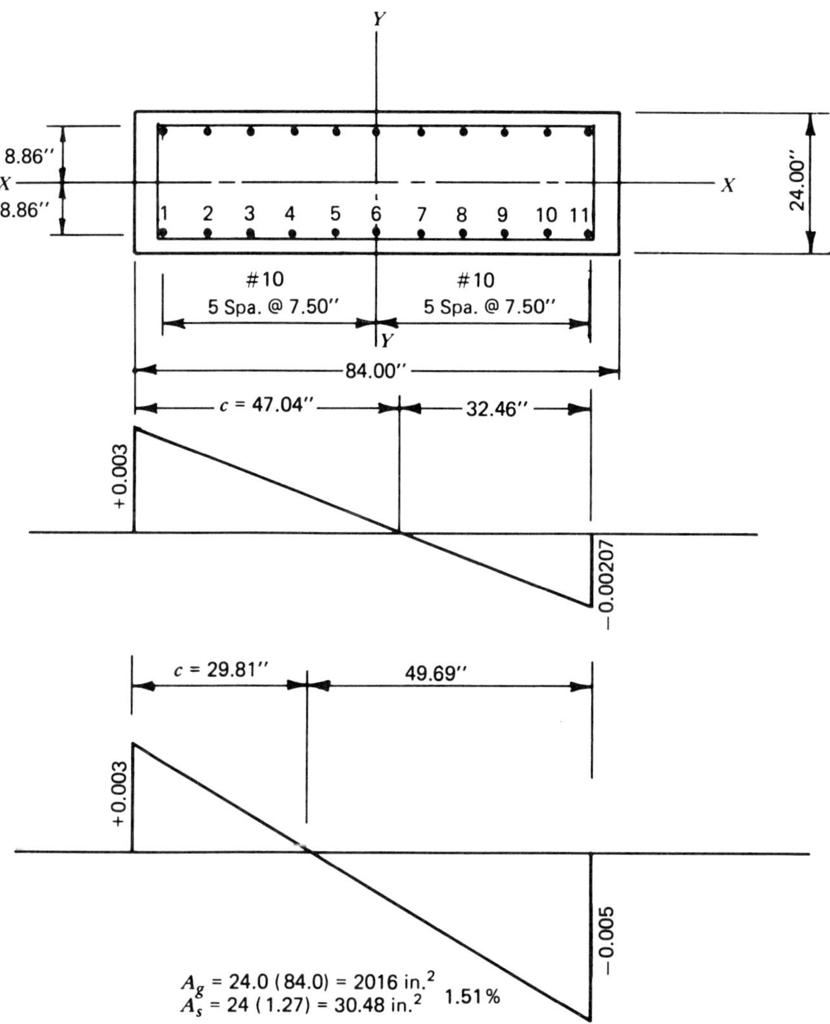

FIGURE 14-22 Reinforcement arrangement and strain diagrams in the transverse direction. (From Heins and Lawrie, 1984.)

procedure shown in Table 14-8 is repeated for an intermediate point with 0.005 as the maximum tensile strain in the reinforcement, and 0.003 strain in the concrete, and the results are tabulated in Table 14-9.

The values computed in Tables 14-8 and 14-9 are plotted in Figure 14-23, for the most critical cases obtained from the table of ultimate loads and moments. The values shown are for bending about the Y–Y axis.

Using the same procedure, the section capacity is computed about the X–X axis (longitudinal direction), and the related data are shown in Tables

TABLE 14-8 Pier Capacity Analysis $\epsilon_s = 0.00207$ (Axis Y–Y)

Reinforcing Location	Distance R/F to Centroid	Reinforcing Area A_s	Strain at R/F		Stress in R/F		Force in R/F		Moment (kip-ft)
1	37.50	3.81	+0.00271		+60.00		+228.6		714
2	30.00	2.54	+0.00223		+60.00		+152.4		381
3	22.50	2.54	+0.00176		+51.04		+129.6		243
4	15.00	2.54	+0.00128		+37.12		+94.3		118
5	7.50	2.54	+0.00080		+23.20		+58.9		37
6	0	2.54	+0.00032		+9.28		+23.6		0
7	7.50	2.54		−0.00016		−4.64		−11.8	7
8	15.00	2.54		−0.00064		−18.56		−47.1	59
9	22.50	2.54		−0.00111		−32.19		−81.8	153
10	30.00	2.54		−0.00159		−46.11		−117.1	293
11	37.50	3.81		−0.00207		−60.00		−228.6	714
							+687.4	−486.4	2719

$$c = 47.04 \qquad \beta_1 = 0.85$$

$$\beta_1 c = 39.98$$

$$A_c = 960$$

$$C = 0.85(3250)(960) = 2652 \text{ kips}$$

$$M_c = 2652[42.0 - (39.98)/2](\tfrac{1}{12}) = 4864 \text{ kip-ft}$$

$$\left.\begin{aligned} \phi P_b &= 2652 + 687 - 486 = 2853 \text{ kips} \\ \phi M_b &= 2719 + 4864 = 7583 \text{ kip-ft} \end{aligned}\right\} \quad \phi = 1.0$$

$$\left.\begin{aligned} \phi P_b &= 2568 \text{ kips} \\ \phi M_b &= 6825 \text{ kip-ft} \end{aligned}\right\} \quad \phi = 0.90$$

$$\left.\begin{aligned} \phi P_b &= 1997 \text{ kips} \\ \phi M_b &= 5308 \text{ kip-ft} \end{aligned}\right\} \quad \phi = 0.70$$

14-10 and 14-11 and in Figure 14-24. These values are likewise plotted, in addition to the ϕP_0 term, which is the direct load capacity, and are shown in Figure 14-25.

The critical loading is selected and compared with a simulated interaction plane, using two interaction equations presented in the foregoing sections. For this example the most critical condition is the Group I loading at the top of the pier.

Section Capacity Analysis at Pier Top The case that controls is Group I, case IB, with biaxial bending. According to this case,

$$P_u = 421 \text{ kips} \qquad M_{u(x-x)} = 1310 \text{ ft-kips} \qquad M_{u(y-y)} = 508 \text{ ft-kips}$$

Moment magnification is considered in the transverse direction only, because the pier column is braced longitudinally against sidesway by the frame action.

TABLE 14-9 Pier Capacity Analysis $\epsilon_s = 0.005$ (Axis Y–Y)

Reinforcing Location	Distance R/F to Centroid	Reinforcing Area A_s	Strain at R/F	Stress in R/F	Force in R/F	Moment (kip-ft)
1	37.50	3.81	+0.00255	+60.00	+228.6	714
2	30.00	2.54	+0.00179	+51.91	+131.9	330
3	22.50	2.54	+0.00104	+30.16	+76.6	144
4	15.00	2.54	+0.00028	+8.12	+20.6	26
5	7.50	2.54	−0.00047	−13.63	−34.6	−22
6	0	2.54	−0.00123	−35.67	−90.6	0
7	7.50	2.54	−0.00198	−57.42	−145.8	91
8	15.00	2.54	−0.00274	−60.00	−152.4	190
9	22.50	2.54	−0.00349	−60.00	−152.4	286
10	30.00	2.54	−0.00425	−60.00	−152.4	381
11	37.50	3.81	−0.00500	−60.00	−228.6	714
					+ 457.7 −956.8	2854

$$c = 29.81$$

$$\beta_1 c = 0.85(29.81) = 25.34 \text{ in.}$$

$$A_c = 25.34(24.0) = 608 \text{ in.}^2$$

$$C = 0.85(3250)(608) = 1680 \text{ kips}$$

$$M_c = 1680[42.0 - (25.34)/2]\left(\tfrac{1}{12}\right) = 4106 \text{ kip-ft}$$

$$\left.\begin{aligned} \phi P_n &= +457.7 - 956.8 + 1680 = 1181 \text{ kips} \\ \phi M_n &= 2854 + 4106 = 6960 \text{ kip-ft} \end{aligned}\right\} \quad \phi = 1.0$$

$$\left.\begin{aligned} \phi P_n &= 1063 \text{ kips} \\ \phi M_n &= 6264 \text{ kip-ft} \end{aligned}\right\} \quad \phi = 0.90$$

$$\left.\begin{aligned} \phi P_n &= 827 \text{ kips} \\ \phi M_n &= 4872 \text{ kip-ft} \end{aligned}\right\} \quad \phi = 0.70$$

For this sidesway, the minimum value of e is 1/10 the pier depth in the transverse direction, or $e = 7 \times 12/10 = 8.4$ in. In this case $M_{u(y-y)} = 421 \times 8.4/12 = 295$ ft-kips (minimum eccentricity does not control).

The *EI* stiffness value is calculated next for the critical buckling loads, and moment magnification is considered only in the transverse direction. The section properties are computed as follows:

$$E_c = 3{,}250{,}000 \text{ psi} \qquad \text{For creep } \beta_D = 0 \qquad E_s = 29{,}000{,}000 \text{ psi}$$

$$I_g = 24 \times 84^3/12 = 1{,}185{,}410 \text{ in.}^4$$

$$I_s = 4(1.27)(7.5^2 + 15.0^2 + 22.5^2 + 30.0^2) = 8572$$

$$= 6(1.27)(3.75^2) = 10{,}716$$

Total $\quad I_s = 19{,}288 \text{ in.}^4$

$$EI = \frac{E_c I_g/5 + E_s I_s}{1 + \beta_D} = 1.33 \times 10^{12} \text{ psi}$$

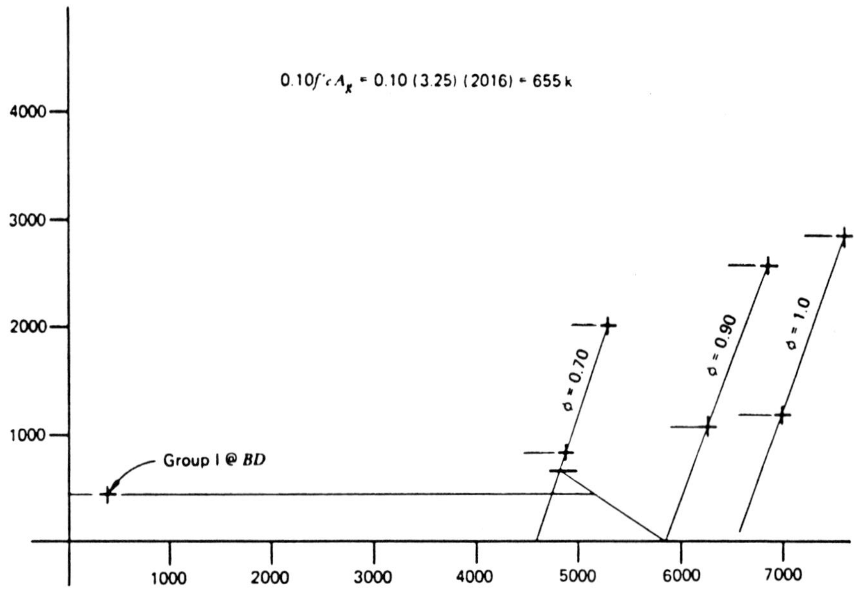

FIGURE 14-23 Procedure for pier section analysis; axis Y–Y (transverse direction). (From Heins and Lawrie, 1984.)

TABLE 14-10 Pier Capacity Analysis ϵ_s = 0.005 (Axis X–X)

Reinforcing Location	Distance R/F to Centroid	Reinforcing Area A_s	Strain at R/F		Stress in R/F		Force in R/F		Moment (kip-ft)
1	8.86	13.97	+0.00180		+52.20		+729.2		538
2	0	2.54		−0.00160		−46.40		−117.9	0
3	8.86	13.97		−0.00500		−60.00		−838.2	619
							+729.2	−956.1	1157

$$c = 7.82$$

$$\beta_1 c = 0.85(7.82) = 6.65$$

$$A_c = 6.65(84.0) = 558$$

$$C = 558(3.25)(0.85) = 1542 \text{ kips}$$

$$M_c = 1542[12.0 - (6.65)/2]\left(\tfrac{1}{12}\right) = 1115 \text{ kip-ft}$$

$$\left.\begin{aligned} \phi P_n &= 1542 + 729.2 - 956.1 = 1315 \text{ kips} \\ \phi M_n &= 1115 + 1157 = 2272 \text{ kip-ft} \end{aligned}\right\} \quad \phi = 1.0$$

$$\left.\begin{aligned} \phi P_n &= 1184 \text{ kips} \\ \phi M_n &= 2045 \text{ kip-ft} \end{aligned}\right\} \quad \phi = 0.90$$

$$\left.\begin{aligned} \phi P_n &= 920 \text{ kips} \\ \phi M_n &= 1590 \text{ kip-ft} \end{aligned}\right\} \quad \phi = 0.70$$

TABLE 14-11 Pier Capacity Analysis $\epsilon_s = 0.00207$ (Axis X–X)

Reinforcing Location	Distance R/F to Centroid	Reinforcing Area A_s	Strain at R/F		Stress in R/F		Force in R/F		Moment (kip-ft)
1	8.86	13.97	+0.00224		+60.00		+838.2		619
2	0	2.54	+0.00008		+2.32		+5.9		0
3	8.86	13.97		−0.00207		−60.00		−838.2	619
							+ 844.1	−838.2	1238

$$c = 12.34 \text{ in.}$$

$$\beta_1 c = 0.85(12.34) = 10.49$$

$$A_c = 84.0(10.49) = 881$$

$$C = 881(3.25)(0.85) = 2434 \text{ kips}$$

$$M_c = 2434[12.0 - (10.40)/2](\tfrac{1}{12}) = 1370 \text{ kip-ft}$$

$$\left.\begin{aligned} \phi P_b &= 2434 + 844.1 - 838.2 = 2440 \text{ kips} \\ \phi M_b &= 1370 + 1238 = 2608 \text{ kip} \end{aligned}\right. \quad \phi = 1.0$$

$$\left.\begin{aligned} \phi P_b &= 2196 \text{ kips} \\ \phi M_b &= 2347 \text{ kip-ft} \end{aligned}\right\} \quad \phi = 0.90$$

$$\left.\begin{aligned} \phi P_b &= 1708 \text{ kips} \\ \phi M_b &= 1826 \text{ kip-ft} \end{aligned}\right\} \quad \phi = 0.70$$

For the critical buckling load, we use $K = 2$. Then

$$P_{cr} = \frac{\pi^2 EI}{(KL)^2} = \frac{\pi^2(1.33 \times 10^{12})}{(2 \times 23.75 \times 12)^2} = 38{,}400 \text{ kips}$$

The magnification factor δ is computed using $\phi = 0.77$ (established by capacity charts) and $C_m = 1.0$ (because the column is in unbraced condition)

From (14-21),

$$\delta = \frac{1.0}{1 - 421/(0.77 \times 38{,}400)} = 1.014$$

Therefore, the load and moment conditions are

$$P_u = 421 \text{ kips} \qquad M_{u(x-x)} = 1310 \text{ ft-kips}$$

$$M_{u(y-y)} = 508 \times 1.014 = 515 \text{ ft-kips}$$

Scaling from the capacity curves is used to obtain the following:

$$\frac{M_{ux}}{\phi M_{nx}} = 0.88 \qquad \text{and} \qquad \frac{M_{uy}}{\phi M_{ny}} = 0.10 \qquad \text{(From plots)}$$

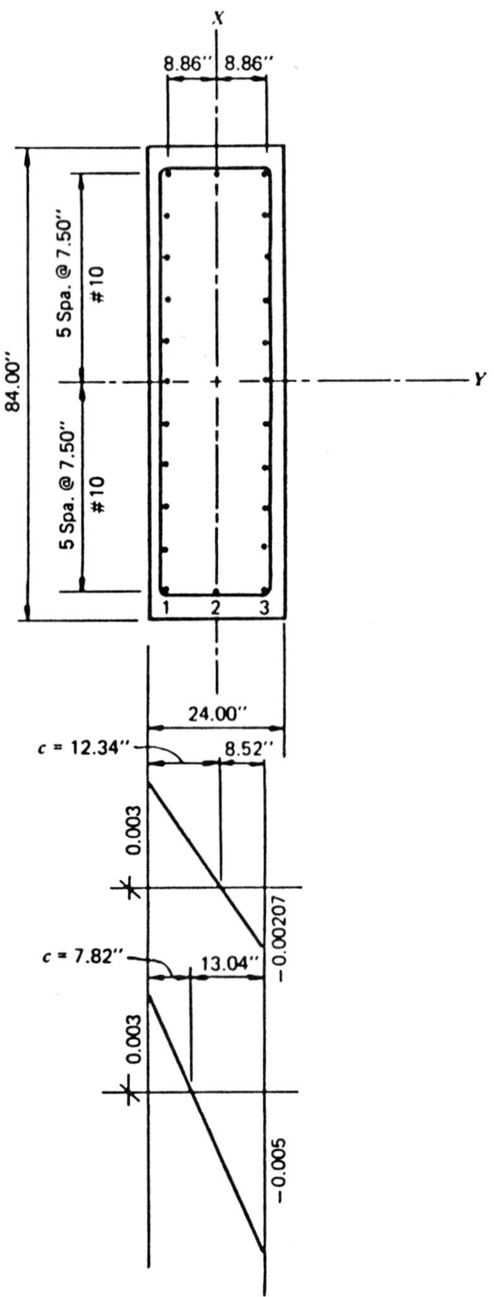

FIGURE 14-24 Reinforcement arrangement and strain diagrams, longitudinal direction. (From Heins and Lawrie, 1984.)

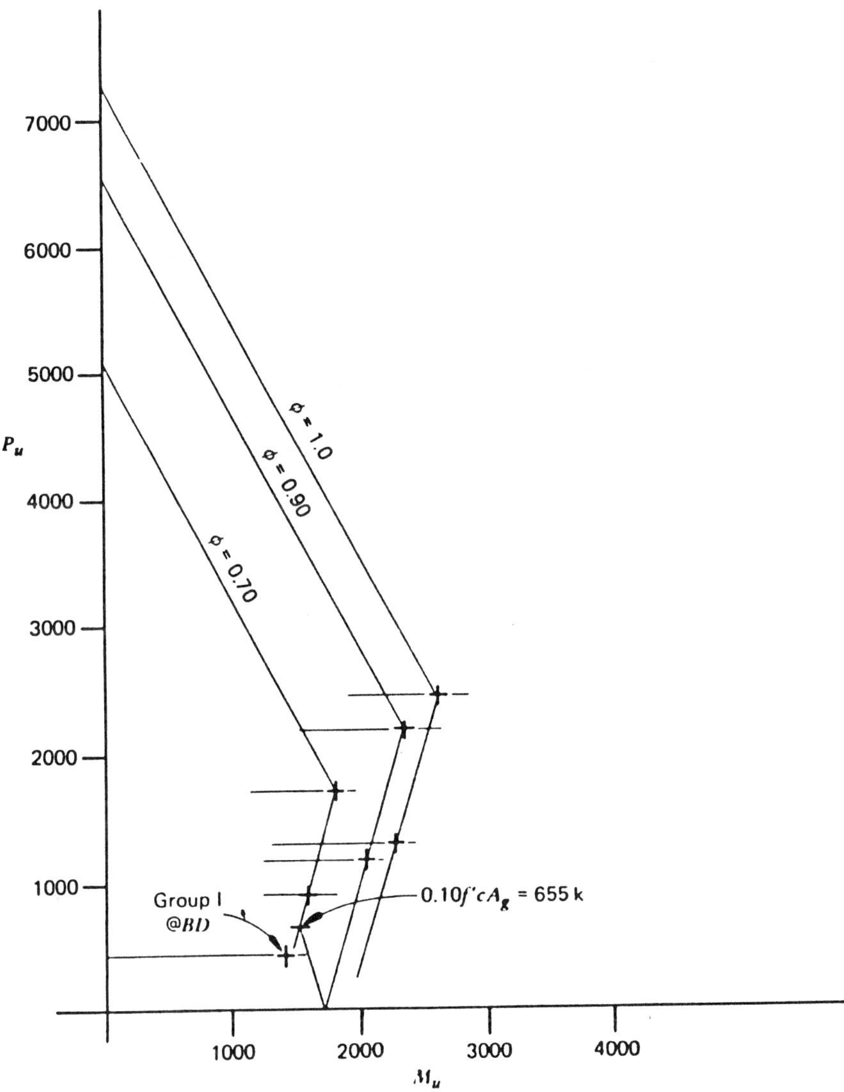

Pier section capacity: axis X-X (longitudinal moment.

$$\phi P_0 = [(0.85)(f'_c)(A_g - A_{st}) + A_{st}f_y]\ \phi;$$

$$\phi P_0 = [(0.85)(3.25)(2016 - 30.48) + 30.48(60)]$$

$$\phi = 7314 \text{ k} \qquad \phi = 1.0,$$

$$= 6582 \text{ k} \qquad \phi = 0.90,$$

$$= 5120 \text{ k} \qquad \phi = 0.70.$$

FIGURE 14-25 Procedure for pier section analysis; axis X–X (longitudinal direction). (From Heins and Lawrie, 1984.)

The biaxial capacity is checked according to two criteria.

1. According to AASHTO [see (14-25)],

$$\frac{M_{ux}}{\phi M_{nx}} + \frac{M_{uy}}{\phi M_{ny}} < 1 \qquad \text{or} \qquad 0.88 + 0.10 = 0.98 < 1.0 \qquad \text{OK}$$

2. According to Furlong [see (14-26)],

$$\sqrt{(0.88)^2 + (0.10)^2} = 0.88 < 1.0 \qquad \text{OK}$$

Analysis by the Hu Method The procedure for computing stresses at a given column section involves the following steps.

1. Compute the section constants. These are

A_c = gross area of cross section
A_s = area of steel
I_x^s = moment of inertia of steel about x–x axis (nontransformed section)
I_y^s = moment of inertia of steel about y–y axis (nontransformed section)
I_x^c = moment of inertia of gross section about x–x axis
I_y^c = moment of inertia of gross section about y–y axis

$$np = \frac{nA_s}{A_c} \qquad nQ_x = \frac{nI_x^s}{I_x^c} \qquad nQ_y = \frac{nI_y^s}{I_y^c}$$

2. Compute the load constants. These are as follows:

$$e_x = \frac{M_x}{Nd} \qquad e_y = \frac{M_y}{Nb} \qquad m = \frac{e_x}{e_y}$$

where M_x = bending moment about x–x axis
M_y = bending moment about y–y axis
N = total direct load
d = length of section normal to x–x axis
b = length of section normal to y–y axis

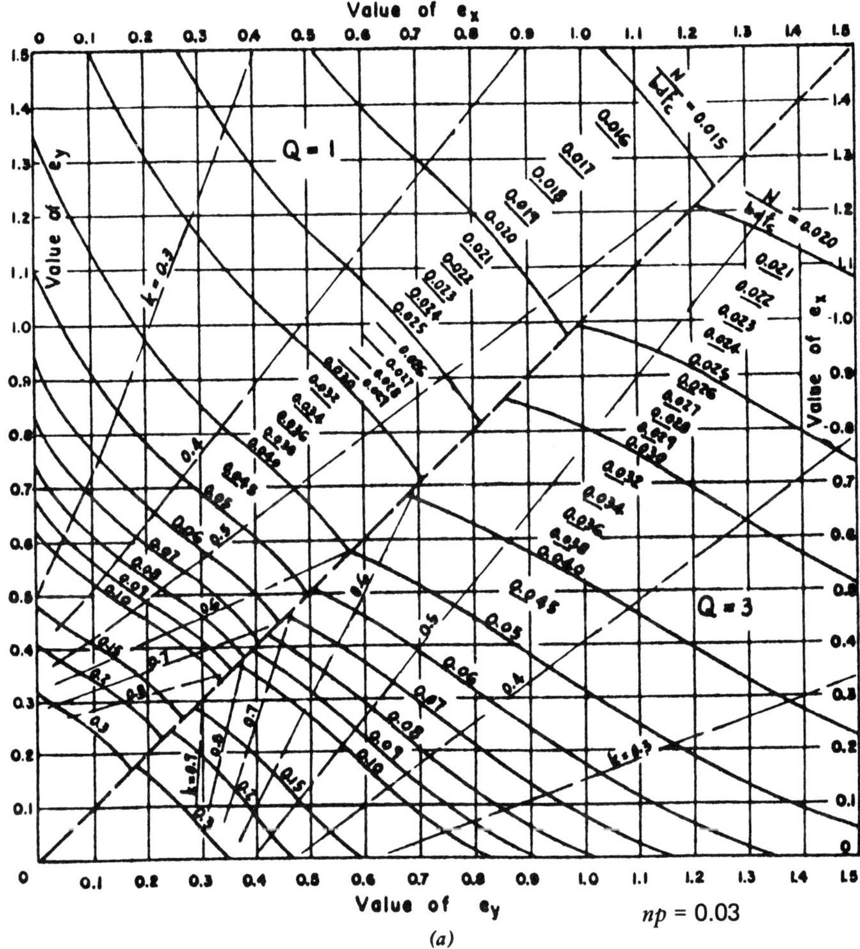

FIGURE 14-26 Charts and bending curves for section analysis, biaxial bending. (From Hu, 1955.)

3. Compute the value of Q from sectional constants and load constants. This value is

$$Q = \left(nq_y + m^2 nq_x\right)/np$$

4. Obtain the values of N/bdf_c and k by reference to Figure 14-26. It is necessary to interpolate between the charts shown in Figure 14-26 for both np and Q as computed from step 3. The e_x and e_y axes are not

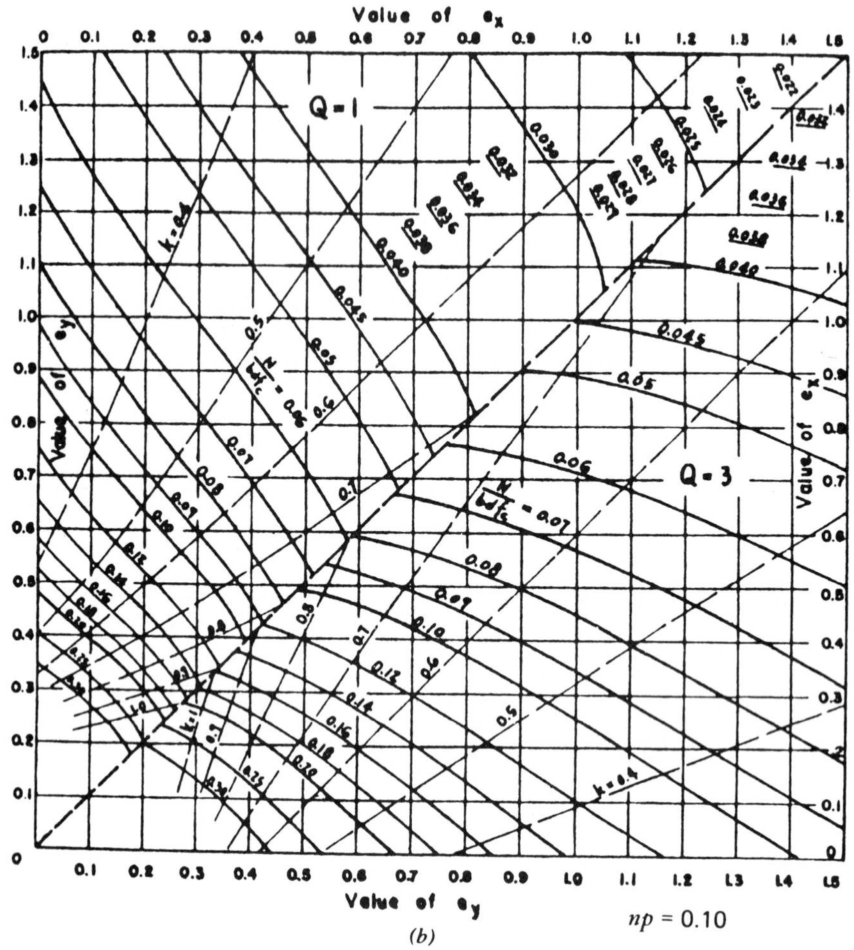

(b)

FIGURE 14-26 *(Continued)*

consistent on these charts. The value of Q may be kept within the chart range by proper selection of the x and y axes. The values of np in the chart are within most practical ranges.

5. Obtain the concrete stress by making use of N/bdf_c. The stress in the steel is computed as

$$f_s = nf_c\left(1 - \frac{1 - a_x}{h} - \frac{1 - a_y}{k}\right)$$

where $h = k/m$ and a_x and a_y are the clearances to the center lines of the bars in the x and y directions, respectively, expressed as a fraction of the section dimension in the same direction.

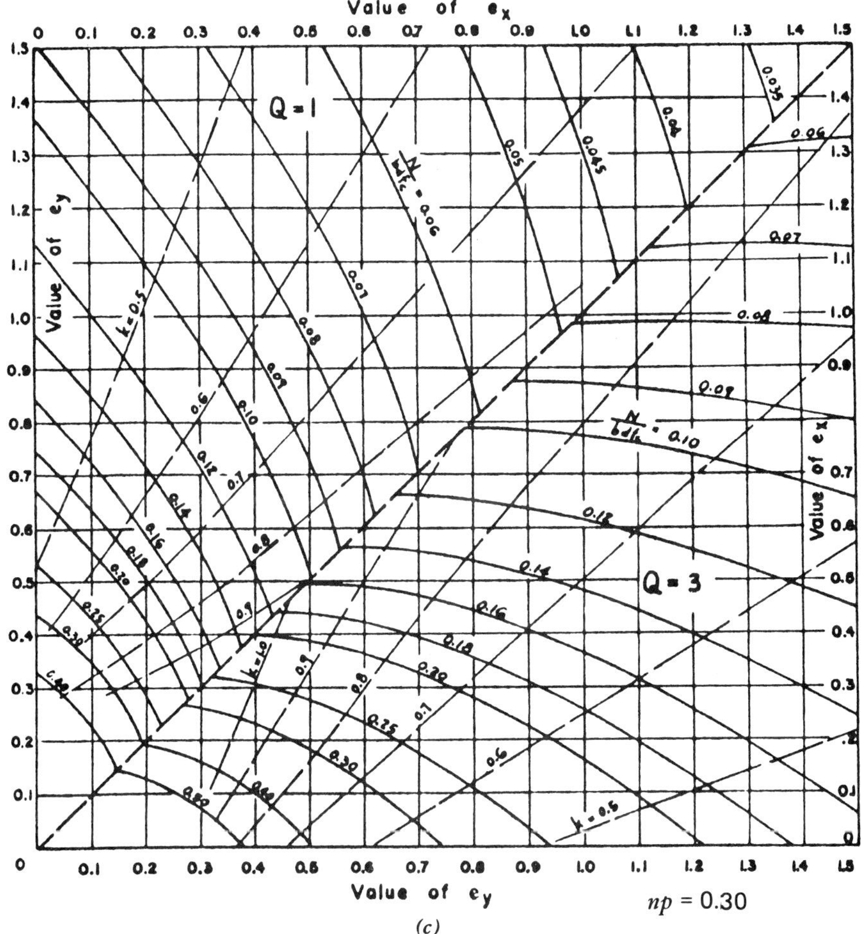

(c)

FIGURE 14-26 *(Continued)*

For the column top section and reinforcement bars arranged as shown in Figure 14-22, we will now check the stresses under service loads. These loads are

$$P = 360 \text{ kips} \qquad M_y = 776 \text{ ft-kips} \qquad M_x = 234 \text{ ft-kips}$$

Also,

$$A_s = 30.48 \text{ in.}^2 \qquad \text{and} \qquad A_c = 2016 \text{ in.}^2$$

The following parameters are computed:

$$I_x^s = 4(1.27)(7.50)^2 + 4(1.27)(22.50)^2 + 4(1.27)(30.00)^2$$
$$+4(1.27)(15.00)^2 + 6(1.27)(37.50)^2 = 19{,}298 \text{ in.}^4$$

$$I_x^c = 24.0(84.0)^3/12 = 1{,}185{,}400 \text{ in.}^4$$

$$I_y^s = 2(11)(1.27)(8.86)^2 = 2193 \text{ in.}^4$$

$$I_y^c = 84(24)^3/12 = 96{,}768 \text{ in.}^4$$

$$np = 9(30.48)/2016 = 0.136$$

$$nq_x = 9(19{,}289)/1{,}185{,}400 = 0.1464$$

$$nq_y = 9(2193)/96{,}768 = 0.204$$

$$e_x = (234)/(360.2 \times 7.00) = 0.0928$$

$$e_y = (776)/(360.2 \times 2.00) = 1.077$$

$$m = 0.0928/1.077 = 0.086$$

$$Q = \frac{0.2040 + (0.086)^2(0.1464)}{0.1361} = 1.51$$

The values of N/bdf_c' and k are computed by interpolation as shown in the following tabulation:

	$np = 0.10$		$np = 0.30$	
	N/bdf_c'	k	N/bdf_c'	k
$Q = 1$	0.072	0.35	0.115	0.45
$Q = 3$	0.115	0.38	0.205	0.52
$Q = 1.51$	0.083	0.36	0.138	0.47

from which the following are obtained

$$np = 0.1361 \qquad N/bdf_c' = 0.093 \qquad k = 0.38$$

Also, the factors h, a_x, and a_y are computed from the foregoing expressions as

$$h = 0.38/0.086 = 4.42$$

$$a_x = 4.5/34.0 = 0.0536$$

$$a_y = 3.14/24.0 = 0.1308$$

The stress in the concrete and the steel can now be computed as follows:

$$f_c = \frac{N}{bd(0.093)} = \frac{360{,}000}{(24.0)(84.0)(0.093)} = 1920 \text{ psi}$$

$$f_s = 9(1920)\left(1 - \frac{1 - 0.0536}{442} - \frac{1 - 0.1308}{0.38}\right) = 26.0 \text{ ksi}$$

Serviceability requirements also stipulate a check on the crack width. According to Kaar and Mattock (1963), it is computed as follows. For the top of the column, $A = 2(3.14)(7.5) = 47.1$ and $w = 0.115\sqrt[4]{47.1}\,(26.0) \times 10^{-3} =$ 0.0078 in., which, using the criteria discussed in the foregoing sections, should be acceptable. [Applying (14-23), w is computed as 0.013 in.].

14-11 ALTERNATE LIMIT STATE APPROACH: SOIL–STRUCTURE SYSTEMS

General Principles

In the basic LRFD methodology, components and members of structures must satisfy the following equation:

$$\sum \gamma_i Q \leq \eta \phi P_n = \eta P_r$$

where γ_i = load factor (a statistically based mulitplier applied to force effects)

ϕ = resistance factor (a statistically based multiplier applied to nominal resistance)

η = resistance modifier (relating to ductility, redundancy, and operational importance)

Q = force effect (a deformation stress or stress resultant, such as thrust, shear, torque, or moment, caused by applied loads, imposed deformations, or volumetric changes)

The four limit states are summarized as follows.

1. Service limit state associated with restrictions on stress, deformation, and crack width under regular service conditions.
2. Strength limit state intended to ensure strength and stability and provide resistance to the statistically significant load combinations.
3. Fatigue and fracture limit state associated with restrictions on the stress range under regular service conditions reflecting the expected stress events.
4. Extreme-event limit state intended to ensure structural survival during major earthquakes or in the event of collision and ice flow.

Alternatively, and in the context of the soil–structure system, a limit state is said to have been reached when the system can no longer satisfy the requirements for which it was designed. From this general definition, limit states are further classified as ultimate limit states which take into account the worst credible values that the associated variables could take, or as serviceability limit states in which the most probable values are used. Although, in principle, all limit states should be examined explicitly, in practice, only one may be more critical than the others.

Soil–Structure System This approach is applied to the combined system of soil and structure, and therefore the interaction between them must be considered and the soil–structure system must be defined quantitatively.

Ultimate Limit States These are manifested by (a) failure in the soil without failure in the structure, involving loss of stability and causing considerable rigid movement of the structure; (b) failure in the structure without failure in the soil, resulting in loss of equilibrium in part of the structure; and (c) failure in the structure and soil together.

Serviceability Limit States These are assumed to have been reached if the following occur: (a) excessive deformation and ground settlement, heave, lateral displacement, and so forth, transferred to the structure or involving adjacent structures; (b) excessive structural deformation; and (c) excessive cracking in the structure.

Loads and Load Effects A load is an external action applied to the soil–structure system; examples include dead, imposed, and wind loads acting on the structure, or gravitational forces and surcharges acting on the soil mass under consideration.

Load effects are necessarily internal to the soil–structure system. In this context internal forces and moments in the structure are considered load effects. In considering the overall stability, the total disturbing moment or force is a load effect.

Characteristic Values These are not intended to have direct design significance. The term is merely a convenient reference item that may have a specified minimum value, a statutory value, or an average value based on testing.

Characteristic Loads For dead, imposed, and wind loads, the characteristic values should be taken as defined in the specifications.

Characteristic Material Properties These involve strength, stiffness, unit weight, and so on. For concrete and steel, they are defined in the specifications. For soils, the characteristic value of any material property is the best estimate of the in situ value. The assessment of test results relates to their reliability and number of tests in relation to the variability of in situ conditions. If sufficient reliable results are available, the characteristic value may be taken as the mean of the results. This is justified because failure in the soil requires the development of limiting states of stress to cover a significantly large area over which the assumption of average values is reasonable.

Characteristic in Situ Values The first requirement is to establish the initial pore water pressure. This would be expected to be the best estimate of pressures considering the available information as well as seasonal or other variations. Characteristic initial vertical stresses may be derived from the characteristic unit weight and pore water pressures. They may also be necessary to establish the characteristic initial horizontal stresses, particularly if the expected deformations are not sufficient for the limiting states of stress to be attained.

We should note that soil pressures acting on a structure are not classified as loads, although they may be considered load effects.

Most Probable Values These are defined as follows.

1. For variables that are constant with time, for example, dead loads and certain material properties, the most probable values are the best estimates of the in situ values.
2. For variables that change with time, for example, live load and wind load and soil strength and stiffness, the most probable values are the best estimates of the extreme values that will occur during the anticipated service time of the structure.

The most probable values of the different variables are obtained by applying partial factors to the characteristic values. In many instances these factors will be close to unity. Exceptions include structural material strength where the characteristic values are defined in terms of a specified minimum strength rather than mean values.

Worst Credible Values For loads and material properties, these have an accepted very small probability of being encountered. As a criterion, the probability should notionally be set at 0.1 percent. The worst credible value of any variable will be either the maximum or the minimum value depending on whether the effect is beneficial or adverse.

The worst credible load effects will be the worst credible combinations of the effects of different variables. Allowance can be made for the reduced probability of the worst credible values of the individual variables occurring together.

For variables such as dead, live, and wind loads and structural material strengths, the worst credible values may be obtained by applying partial safety factors to the characteristic values. For soil strengths, this approach is not always appropriate. For variables such as pore water pressure, material stiffness, and initial stresses, the partial factor approach is impractical. When partial factors are not to be applied, worst credible values should be evaluated directly considering the information available.

Structural Factors and Limit State Requirements

Partial Safety Factors These are introduced in the calculations to reflect the effect of various uncertainties inherent in the design and construction.

Load Factors γ_{f1} *and* γ_{f2} The load variation factor γ_{f1} takes into account the possibility of unfavorable deviations of loads from their characteristic values. The load combination factor γ_{f2} takes into account the reduced probability of loads, which are stochastically independent, occurring at the same time.

Structural Performance Factor γ_p This factor reflects the following effects: (a) inaccurate assessment of loading effects and unforeseen stress redistribution within a system; (b) variations in construction accuracy; (c) the importance of the limit state under consideration; and (d) some systems that provide a warning of approaching a limit state as opposed to others that may reach it suddenly.

Partial Materials Factor γ_m This factor reflects the effect of the following: (a) materials in the system may have a strength lower than indicated by samples, and (b) the structure may be weakened from construction imperfections.

For ultimate limit states, the factors γ_{f1} and γ_{f2} should be applied to the characteristic values of the loads, and the resulting values taken as the worst credible loads or load combinations. The worst credible material strengths can be obtained in the same manner by dividing the characteristic strengths by γ_m.

Ultimate Limit State Requirements These requirements are satisfied if

$$\frac{\text{Worst credible resistance } R}{\text{Worst credible load effect } S} \geq \gamma_p \tag{14-37}$$

where R and S are calculated using the worst credible values and combinations of loads and material strengths. The term γ_p is the structural performance factor.

Serviceability Limit State Requirements These requirements are satisfied if it can be demonstrated that movement, distortion, and cracking of the structure (including surrounding structures) are acceptable. The most probable values would normally be assumed, except where the consequences can be serious and the associated effects particularly sensitive to variations, in which case the most conservative values may be assumed. Where structural elements are considered, the limit state for service conditions will often be deemed to be satisfied once ultimate limit state checks have been made.

For movements in the soil, two approaches may be followed, and in case of uncertainty the assessment of the design may have its basis on the results from both analyses. Thus, it is possible to (a) calculate the most probable settlement, heave, or other movement directly and (b) demonstrate that equilibrium and stability can be maintained with stresses in the soil known from experience of similar problems or on the basis of a theoretical investigation consistent with acceptable criteria. Occasionally, it may be necessary to carry out separate checks for the ultimate limit state.

Numerical Example: Strip Footing

Limit State Design The strip footing shown in Figure 14-27 represents a substructure element for an underpass. The pier is long and continuous. The dead loads and live loads are designated by G_k and Q_k, respectively, shown with the corresponding load factors γ_f. Also shown is the worst credible and most probable water level.

For the shallow foundation conditions involved in this example, we will make use of the bearing capacity theory expressed by the Terzaghi bearing

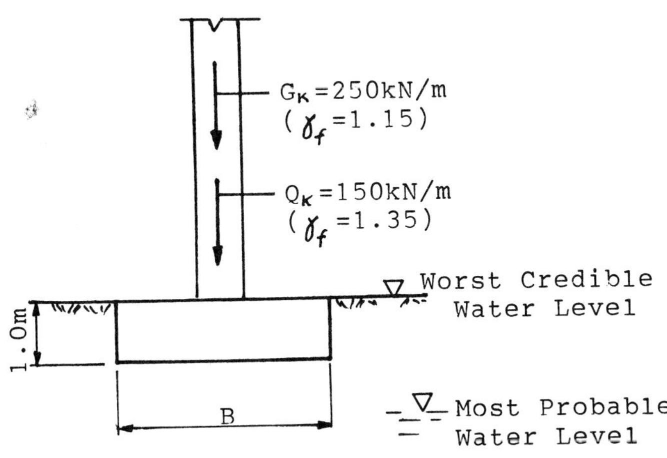

FIGURE 14-27 Strip footing of design example.

capacity equation given as

$$q_{\text{ult}} = cN_c + \gamma' z N_q + 0.5\gamma' B N_\gamma \qquad (14\text{-}38)$$

where q_{ult} = ultimate soil bearing pressure
c = cohesion of soil
z = depth to bottom of footing
γ' = unit weight of soil (submerged weight for soil below water table)
B = width of footing
N_c, N_q, N_γ = bearing capacity factors

The material properties, most probable values, worst credible values, and units are tabulated as follows:

Material Property	Most Probable Value	Worst Credible Value	Unit
Concrete unit weight	24	$\gamma_f = 1.15$	kN/m^3
Concrete strength	30	$\gamma_m = 1.5$	N/mm^2
Steel strength	425	$\gamma_m = 1.15$	N/mm^2
Soil unit weight	20	17	kN/m^3
c'	10	5	kPa
ϕ'	25	21	°
N_c	19	16	°
N_q	12	7	°
N_γ	8	3	°
Water unit weight	10	10	kN/m^3

where γ_f and γ_m are the load and material factors, respectively (in this case specified by relevant design criteria).

Bearing Capacity (Ultimate Limit State) From the loads and the worst credible values, we calculate the worst credible load effect S as follows:

$$S = \left[\frac{(250 \times 1.15)}{B} + \frac{(150 \times 1.35)}{B}\right] + [(24 \times 1.15 \times 1.0) - (10 \times 1.0)]$$

$$= \frac{490}{B} + 17.6$$

Likewise, the design resistance is the worst credible resistance R, computed as

$$R = q_{ult}[(5 \times 16) + (17 - 10) \times 1.0 \times 7 + \{0.5 \times (17 - 10) \times B \times 3\}]$$
$$= 129 + 10.5B$$

For a structural performance factor $\gamma_p = 1.2$ (established from design criteria), the ultimate limit state is satisfied if $R \geq \gamma_p S$, or

$$129 + 10.5B = 1.2\left(\frac{490}{B} + 16.1\right)$$

which solved gives $B = 3.9$ m.

Allowable Stress Design Likewise, the dead load is 250 kN/m, and the live load is 150 kN/m. The water level is taken at the ground line as previously. Material properties are as follows: concrete unit weight, 24 kN/m^3; soil unit weight, 20 kN/m^3; and water unit weight, 10 kN/m^3. For design purposes, the soil parameters are $c' = 10$ kPa, $\phi' = 25°$, giving $N_c = 19$, $N_q = 12$, and $N_\gamma = 8$.

The actual applied soil pressure is

$$q_p = \left[\left(\frac{250 + 150}{B}\right) + (24 - 10) \times 1.0\right] = \frac{400}{B} + 14$$

whereas the ultimate bearing resistance is divided by the factor of safety

$$q_r = \frac{1}{F}[(10 \times 19) + (20 - 10) \times 1.0 \times 12 + \{0.5(20 - 10) \times B \times 8\}]$$
$$= \frac{1}{F}(310 + 40B)$$

Using $F = 3$ and $q_p = q_r$

$$\frac{310 + 40B}{3} = \frac{400}{B} + 14$$

which gives $B = 3.1$ m.

REFERENCES

AASHTO, 1978. "Foundation Investigation Manual."

AASHTO, 1988. "Manual of Subsurface Investigations."

ACI, 1966. "Suggested Design Procedures for Combined Footings and Mats," ACI Comm. 436, *ACI J.*, Vol. 63, No. 10, Oct., pp. 1041–1058.

ACI, 1977. "Analysis and Design of Reinforced Concrete Bridge Structures," ACI Comm. 343, Publ. 343-R77.

ACI, 1980. "Control of Cracking in Concrete Structures," ACI Comm. 224, *Concrete Inter.: Design and Construction*, Vol. 2, No. 10, Oct., pp. 35–76.

ACI, 1984. *Design Handbook*, Vol. I, *Beams, One-Way Slabs—Brackets, Footings, and Pile Caps in Accordance with the Strength Design Method*, ACI 318-83, Publ. SP-17.

ACI, 1989. "Building Code Requirements (318-89)," American Concrete Institute.

Barker, R. M., 1991. "Substructure and Foundation Design," *Proc. Third Bridge Eng. Conf. TRB-FHWA*, March 10–13, Denver.

Barker, R. M., et al., 1991. "Manuals for the Design of Bridge Foundations (Abutments)," NCHRP 343, TRB, National Research Council, Washington, D.C.

Bresler, B., 1960. "Design Criteria for Reinforced Concrete Columns Under Axial Load and Biaxial Bending," *ACI J.* 57, p. 481.

Cornell, C. A., 1969. "A Probability Based Structural Code," *ACI J.*, Vol. 66, No. 12, Dec., pp. 974–985.

Cranston, W. B., 1972. "Analysis and Design of Reinforced Concrete Columns," Research Report No. 20, Paper 41.020, Cement and Concrete Assoc., London, 54 pp.

D'Appolonia, Inc., 1989. "Recommended Specifications for the Design of Foundations, Retaining Walls and Substructures," NCHRP 12-35, Transp. Research Board, National Research Council, Washington, D.C.

Duncan, J. M. and C. K. Tan, 1991. "Engineering Manual for Estimating Tolerable Movements of Bridges," NCHRP 24-4, VPI & SU, May.

Elms, D. G. and G. R. Martin, 1979. "Factors Involved in the Seismic Design of Bridge Abutments," *Appl. Tech. Council Workshop on Earthquake Resistance of Hwy. Bridges*, ATC-6-1, Jan., pp. 230–252.

Ferguson, P. M. and J. E. Breen, 1975. "Investigation of the Long Concrete Column in a Frame Subject to Lateral Loads," *Reinforced Concrete Columns*, SP-50, ACI, Detroit, pp. 75–114.

Ferguson, P. M., J. E. Breen, and J. O. Jirsa, 1988. *Reinforced Concrete Fundamentals*, Wiley, New York.

Ford, J. S., D. C. Chang, and J. E. Breen, 1981. "Behavior of Concrete Columns Under Controlled Lateral Deformation," *ACI J.* 78, No. 1, Jan.–Feb., pp. 3–20.

Ford, J. S., D. C. Chang, and J. E. Breen, 1981. "Design Indications from Tests of Unbraced Multipanel Concrete Frames," *Concrete International: Design and Construction*, Vol. 3, No. 3, March, pp. 37–47.

Furlong, R. W., 1961. "Ultimate Strength of Square Columns Under Biaxially Eccentric Loads," *ACI J.* Vol. 32, No. 9, March, *Proc.* 57, p. 1129.

Furlong, R. W., 1979. "Concrete Columns under Biaxially Eccentric Thrust," *ACI J.*, Oct.

Gergely, P. and L. A. Lutz, 1968. "Maximum Crack Width in Reinforced Concrete Flexural Member," *Causes, Mechanism and Control of Cracking in Concrete*, SP-20, ACI, Detroit, p. 87.

Gill, W. D., R. Park, and M. J. N. Priestley, 1979. "Ductility of Rectangular Reinforced Concrete Columns with Axial Load," Research Report 79-1, Dept. Civ. Eng., Univ. Canterbury, New Zealand, Feb.

Goodall, D. M., 1948. "Analysis of Rectangular Reinforced Concrete Sections Subjected to Direct Stress and Bending in Two Directions," *Public Roads Magazine*, June.

Gouwens, A. J. DATE. "Biaxial Bending Simplified," *Reinforced Concrete Columns*, SP-50, ACI, Detroit.

Granston, W. B., 1972. "Analysis and Design of Reinforced Concrete Columns," Research Report 20, Paper 41.020, Cement and Concrete Assoc., London.

Heins, C. P. and D. A. Firmage, 1979. *Design of Modern Steel Highway Bridges*, Wiley, New York.

Heins, C. P. and R. A. Lawrie, 1984. *Design of Modern Highway Bridges*, Wiley, New York.

Hu, L. S., 1955. "Eccentric Bending in Two Directions of Rectangular Concrete Columns," *ACI J.*, May.

Kaar, P. H. and A. H. Mattock, 1963. "High Strength Bars as Concrete Reinforcement, Part 4, Control of Cracking," *PCA Development Depart. Bull.* D59, Jan.

MacGregor, J. G., 1975. A Reexamination of the EI Value for Slender Columns," *Reinforced Concrete Columns*, SP-50, ACI, Detroit.

MacGregor, J. G., 1983. "Load and Resistance Factors for Concrete Design," *ACI J.*, Vol. 80, No. 4, July–Aug., pp. 279–287.

MacGregor, J. G., J. E. Breen, and E. O. Pfrang, 1970. "Design of Slender Columns," *ACI J.* Vol. 67, No. 1, Jan., p. 6.

Marin, J., 1979. "Design Aids for L-Shaped Reinforced Concrete Shapes," *ACI J.*, Nov.

Newmark, N. M., 1942. "Numerical Procedure for Computing Deflections, Moments, and Buckling Loads," *ASCE Proc.*, May.

Ooi, P. S. K., J. M. Duncan, K. B. Rojiani, and R. M. Barker, 1991. "Engineering Manual for Driven Piles," NCHRP 343, TRB, Washington, D.C.

Pannell, F. N., 1963. "Failure Surface for Members in Compression and Biaxial Bending," *ACI J.*, No. 1, Jan., p. 129.

Park, R. and T. Paklay, 1975. *Reinforced Concrete Structures*, Wiley, New York.

Parme, A. L., J. M. Nieves, and A. Gouwens, 1966. "Capacity of Reinforced Rectangular Columns Subject to Biaxial Bending," *ACI J.*, Sept.

Podolny, W. and J. M. Muller, 1982. *Construction and Design of Prestressed Concrete Segmental Bridges*, Wiley, New York.

Podolny, W. and J. B. Scalzi, 1986. *Construction and Design of Cable-Stayed Bridges*, Wiley, New York.

Portland Cement Association, 1972. Notes on Load Factor Design for Reinforced Concrete Bridge Structures with Design Applications, Chicago.

Potangaroa, R. T., M. J. N. Priestley, and R. Park, 1979. "Ductility of Spirally Reinforced Concrete Columns Under Seismic Loading," *Research Report* 79-8, Dept. of Civ. Eng., Univ. Canterbury, Feb.

Priestley, M. J. N., R. Park, and R. T. Potangaroa, 1981. "Ductility of Spirally Confined Concrete Columns," *Trans. ASCE Struct. Div.*, Vol. 107, Jan., pp. 181–202.

Skogman, B. C., M. K. Tadros, and R. Grasmick, 1988. "Ductility of Reinforced and Prestressed Concrete Flexural Members," *PCI J.*, Vol. 33, No. 6, Nov.–Dec., pp. 94–107.

Snyder, R. and F. Moses, 1978. "Load Factor Design for Substructures and Retaining Walls," FHWA Report No. 79-S0862, Nov.

Toprac, A. A., 1959. "Circular Reinforced Concrete Columns Subjected to Direct Stress and Bending," *Public Roads Magazine*, Dec.

Xanthakos, P. P., 1979. *Slurry Walls*, McGraw-Hill, New York.

Xanthakos, P. P., 1991. *Ground Anchors and Anchored Structures*, Wiley, New York.

Yura, J. A., 1974. "The Effective Length of Columns in Unbraced Frames," *AISC Eng. J.*, April.

ABOUT THE AUTHOR

Petros P. Xanthakos is a consulting engineer practicing in the Washington, D.C., area. He holds B.S., M.S., and Ph.D. degrees in Civil and Structural Engineering and is registered as a structural and professional engineer. In 1975, he was awarded the Octave Chanute Medal by the Western Society of Engineers for the best publication in all fields of engineering.

In 1979 he authored *Slurry Walls*, selected as the outstanding book in civil engineering by the Association of College and Research Libraries. In 1991, he authored *Ground Anchors and Anchored Structures*, a Wiley publication. Dr. Xanthakos has also authored *Slurry Walls as Structural Systems* to be published by McGraw-Hill in 1993, and is also senior author for *Ground Control and Improvement Technologies*, to be published by John Wiley & Sons, Inc. in 1993.

INDEX

Entries having an asterisk (*) have subentries *not* arranged alphabetically but in the order they appear in the pages.